Calculus

SINGLE & MULTIVARIABLE

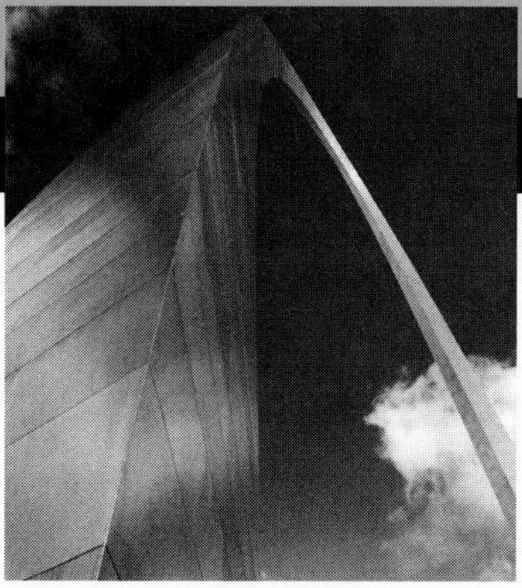

SECOND EDITION

Calculus

SINGLE & MULTIVARIABLE

BRIAN E. BLANK
STEVEN G. KRANTZ
Washington University in St. Louis

WILEY *John Wiley & Sons, Inc.*

Vice President & Executive Publisher	Laurie Rosatone
Senior Acquisitions Editor	David Dietz
Senior Editorial Assistant	Pamela Lashbrook
Development Editor	Anne Scanlan-Rohrer
Marketing Manager	Debi Doyle
Project Manager	Laura Abrams
Production Manager	Dorothy Sinclair
Production Editor	Sandra Dumas
Design Director	Harry Nolan
Senior Designer	Maddy Lesure
Senior Media Editor	Melissa Edwards
Media Specialist	Lisa Sabatini
Production Management Services	MPS Limited, a Macmillan Company
Cover and Chapter Opening Photo	© Walter Bibikow/JAI/Corbis
Text and Cover Designer	Madelyn Lesure

This book was set in 10/12 TimesTen-Roman at MPS Limited, a Macmillan Company, and printed and bound by R. R. Donnelley (Jefferson City). The cover was printed by R. R. Donnelley (Jefferson City).

This book is printed on acid free paper. ♾

ISBN 13 978-0470-45360-5 (Single & Multivariable)
ISBN 13 978-0470-60198-3 (Single Variable)
ISBN 13 978-0470-45359-9 (Multivariable)

Printed in the United States of America

10 9 8 7 6 5 4 3 2

A pebble for Louis. BEB

This book is for Hypatia, the love of my life. SGK

CONTENTS

PREFACE

Calculus is one of the milestones of human thought. In addition to its longstanding role as the gateway to science and engineering, calculus is now found in a diverse array of applications in business, economics, medicine, biology, and the social sciences. In today's technological world, in which more and more ideas are quantified, knowledge of calculus has become essential to an increasingly broad cross-section of the population.

Today's students, more than ever, comprise a highly heterogeneous group. Calculus students come from a wide variety of disciplines and backgrounds. Some study the subject because it is required, and others do so because it will widen their career options. Mathematics majors are going into law, medicine, genome research, the technology sector, government agencies, and many other professions. As the teaching and learning of calculus is rethought, we must keep our students' backgrounds and futures in mind.

In our text, we seek to offer the best in current calculus teaching. Starting in the 1980s, a vigorous discussion began about the approaches to and the methods of teaching calculus. Although we have not abandoned the basic framework of calculus instruction that has resulted from decades of experience, we have incorporated a number of the newer ideas that have been developed in recent years. We have worked hard to address the needs of today's students, bringing together time-tested as well as innovative pedagogy and exposition. Our goal is to enhance the critical thinking skills of the students who use our text so that they may proceed successfully in whatever major or discipline that they ultimately choose to study.

Many resources are available to instructors and students today, from Web sites to interactive tutorials. A calculus textbook must be a tool that the instructor can use to augment and bolster his or her lectures, classroom activities, and resources. It must speak compellingly to students and enhance their classroom experience. It must be carefully written in the accepted language of mathematics but at a level that is appropriate for students who are still learning that language. It must be lively and inviting. It must have useful and fascinating applications. It will acquaint students with the history of calculus and with a sense of what mathematics is all about. It will teach its readers technique but also teach them concepts. It will show students how to discover and build their own ideas and viewpoints in a scientific subject. Particularly important in today's world is that it will illustrate ideas using computer modeling and calculation. We have made every effort to insure that ours is such a calculus book.

To attain this goal, we have focused on offering our readers the following:

- A writing style that is lucid and readable
- Motivation for important topics that is crisp and clean
- Examples that showcase all key ideas
- Seamless links between theory and applications
- Applications from diverse disciplines, including biology, economics, physics, and engineering

- Graphical interpretations that reinforce concepts
- Numerical investigations that make abstract ideas more concrete

Content

In this second edition of *Calculus*, we present topics in a sequence that is fairly close to what has become a standard order for an "early transcendentals" approach to calculus. In the outline that follows, we have highlighted the few topics that arise outside the commonplace order.

In Chapter 1, we review the real number system, the Cartesian plane, functions, inverse functions, graphing, and trigonometry. In this chapter, we introduce infinite sequences (Section 1.4) and parametric curves (Section 1.5).

We begin Chapter 2 with an informal discussion of limits before we proceed to a more precise presentation. After introducing the notion of continuity, we turn to limits of sequences, which we use to define exponential functions. Logarithmic functions are then defined using the discussion of inverse functions found in Chapter 1.

Chapter 3 introduces the derivative and develops the standard rules of differentiation. Numerical differentiation is covered as well. In the section on implicit differentiation, we also consider slopes and tangent lines for parametric curves. The chapter concludes with the differential calculus of inverse trigonometric, hyperbolic, and inverse hyperbolic functions. The remainder of the text has been written so that instructors who wish to omit hyperbolic functions may do so.

After a section on related rates, Chapter 4 presents the standard applications of the differential calculus to the analysis of functions and their graphs. A section devoted to the Newton-Raphson Method of root approximation is optional.

Chapter 5 features a rapid approach to the evaluation of Riemann integrals. The other direction of the Fundamental Theorem of Calculus, the differentiation of an integral with respect to a limit of integration, is presented later in the chapter. In an optional section that follows, an alternative approach to logarithmic and exponential functions is developed. The section on evaluating integrals by substitution includes helpful hints for using the extensive table of integrals that is provided at the back of the book. Chapter 5 concludes with a section on numerical integration.

Chapter 6 treats techniques of integration, including integration by parts and trigonometric substitution. The method of partial fractions is divided into two sections: one for linear factors and one for irreducible quadratic factors. In a streamlined calculus course, the latter can be omitted. Improper integrals are also split into two sections: one for unbounded integrands and one for infinite intervals of integration.

Geometric applications of the integral, including the volume of a solid of revolution, the arc length of the graph of a function, the arc length of a parametric curve, and the surface area of a surface of revolution, are presented in Chapter 7. Other applications of the integral include the average value of a continuous function, the mean of a random variable, the calculation of work, and the determination of the center of mass of a planar region. The final two sections of Chapter 7 are devoted to separable and linear first order differential equations.

After a brief review of infinite sequences, Chapter 8, the last chapter of single variable calculus, discusses infinite series, power series, Taylor polynomials, and Taylor series.

Chapter 9, the first chapter of multivariable calculus, begins with a section that introduces vectors in the plane. It is followed by a section that covers the same material in space. After sections on the dot and cross product, Chapter 9 concludes with a comprehensive account of lines and planes in space.

Chapter 10 is devoted to the differential calculus and geometry of space curves. The final two sections are concerned with curvature and associated concepts, including applications to motion and derivations of Kepler's Laws. Because these two sections are not used later in the text, they may be omitted by instructors who need additional time for other topics.

The differential calculus of functions of two and three variables is taken up in Chapter 11. All topics in the standard curriculum are discussed. One less conventional topic that we treat is the development of order 2 and order 3 Taylor polynomials for functions of two variables. We use these ideas in the discriminant test for saddle points and local extrema but nowhere else. It is therefore feasible to omit the discussion of multivariable Taylor polynomials.

Chapter 12 is devoted to multiple integrals and their applications. In a section that precedes cylindrical and spherical coordinates, we develop polar coordinates *ab initio*. This section may be omitted, of course, if polar coordinates have been introduced earlier in the calculus curriculum.

The concluding chapter on vector calculus, Chapter 13, covers vector fields, line and surface integrals, divergence, curl, flux, Green's Theorem, Stokes's Theorem, and the Divergence Theorem.

Structural Elements

We start each chapter with a preview of the topics that will be covered. This short initial discussion gives an overview and provides motivation for the chapter. Each section of the chapter concludes with three or four *Quick Quiz* questions located before the exercises. Some of these questions are true/false tests of the theory. Most are quick checks of the basic computations of the section. The final section of every chapter is followed by a summary of the important formulas, theorems, definitions, and concepts that have been learned. This end-of-chapter summary is, in turn, followed by a large collection of review exercises, which are similar to the worked examples found in the chapter. Each chapter ends with a section called *Genesis & Development*, in which we discuss the history and evolution of the material of the chapter. We hope that students and instructors will find these supplementary discussions to be enlightening.

Occasionally within the prose, we remind students of concepts that have been learned earlier in the text. Sometimes we offer previews of material still to come. These discussions are tagged *A Look Back* or *A Look Forward* (and sometimes both).

Calculus instructors frequently offer their insights at the blackboard. We have included discussions of this nature in our text and have tagged them *Insights*.

Proofs

During the reviewing of our text, and after the first edition, we received every possible opinion concerning the issue of proofs—from the passionate *Every proof must be included* to the equally fervent *No proof should be presented*, to the seemingly cynical *It does not matter because students will skip over them anyway*. In fact, mathematicians reading research articles often *do* skip over proofs, returning later to those that are necessary for a deeper understanding of the material. Such an approach is often a good idea for the calculus student: Read the statement of a theorem, proceed immediately to the examples that illustrate how the theorem is used, and only then, when you know what the theorem is really saying, turn back to the proof (or sketch of a proof, because, in some cases, we have chosen to omit details that seem more likely to confuse than to enlighten, preferring instead to concentrate on a key, illuminating idea).

Exercises

There is a mantra among mathematicians that calculus is learned by *doing*, not by watching. Exercises therefore constitute a crucial component of a calculus book. We have divided our end-of-section exercises into three types: *Problems for Practice*, *Further Theory and Practice*, and *Calculator/Computer Exercises*. In general, exercises of the first type follow the worked examples of the text fairly closely. We have provided an ample supply, often organized into groups that are linked to particular examples. Instructors may easily choose from these for creating assignments. Students will find plenty of unassigned problems for additional practice, if needed.

The *Further Theory and Practice* exercises are intended as a supplement that the instructor may use as desired. Many of these are thought problems or open-ended problems. In our own courses, we often have used them sparingly and sometimes not at all. These exercises are a mixed group. Computational exercises that have been placed in this subsection are not necessarily more difficult than those in the *Problems for Practice* exercises. They may have been excluded from that group because they do not closely follow a worked example. Or their solutions may involve techniques from earlier sections. Or, on occasion, they may indeed be challenging.

The *Calculator/Computer* exercises give the students (and the instructor) an opportunity to see how technology can help us to see, and to perceive. These are problems for exploration, but they are problems with a point. Each one teaches a lesson.

Notation

Throughout our text, we use notation that is consistent with the requirements of technology. Because square brackets, brace brackets, and parentheses mean different things to computer algebra systems, we do not use them interchangeably. For example, we use only parentheses to group terms.

Without exception, we enclose all functional arguments in parentheses. Thus, we write $\sin(x)$ and not $\sin x$. With this convention, an expression such as $\cos(x)^2$ is unambiguously defined: It must mean the square of $\cos(x)$; it cannot mean the cosine of $(x)^2$ because such an interpretation would understand $(x)^2$ to be the argument of the cosine, which is impossible given that $(x)^2$ is not found inside parentheses. Our experience is that students quickly adjust to this notation because it is logical and adheres to strict, exceptionless rules.

Occasionally, we use $\exp(x)$ to denote the exponential function. That is, we sometimes write $\exp(x)$ instead of e^x. Although this practice is common in the mathematical literature, it is infrequently found in calculus books. However, because students will need to code the exponential function as exp(x) in Matlab and in Maple, and as Exp[x] in Mathematica, we believe we should introduce them to what has become an important alternative notation. Our classroom experience has been that once we make students aware that $\exp(x)$ and e^x mean the same thing, there is no confusion.

The Second Edition

For this second edition, we have moved our coverage of inverse functions to Chapter 1. Logarithmic functions now closely follow the introduction of exponential functions in Chapter 2. The remaining standard transcendental functions of the calculus curriculum now appear in Chapter 3. The topic of related rates has been moved to the chapter on applications of the derivative. We deleted a section entirely devoted to applications of the derivative to economics, but some of the material survives in the exercises.

We did not discuss the use of tables of integrals in the first edition, but we now give examples and exercises in Chapter 5, The Integral. Accordingly, we now provide an extensive Table of Integrals. In the chapter on techniques of integration, the two sections devoted to partial fractions were separated in the first edition but are now contiguous. In Chapter 7, Applications of the Integral, the discussion of density functions has been expanded. The section devoted to separable differential equations has been moved to this chapter, and a new section on linear differential equations has been added. In that chapter, we also have expanded our discussion of density functions. Our treatment of Taylor series in the first edition occupied an entire chapter and was much longer than the norm. We shortened this material so that it now makes up two sections of the chapter on infinite series. Nevertheless, with discussions of Newton's binomial series and applications to differential equations, our treatment remains thorough.

The second edition of this text retains all of the dynamic features of the first edition, but it builds in new features to make it timely and lively. It is an up-to-the-moment book that incorporates the best features of traditional and present-day methodologies. It is a calculus book for today's students and today's calculus classroom. It will teach students calculus and imbue students with a respect for mathematical ideas and an appreciation for mathematical thought. It will show students that mathematics is an essential part of our lives and make them want to learn more.

Supplementary Resources

WileyPLUS is an innovative, research-based, online environment for effective teaching and learning. (To learn more about WileyPLUS, visit www.wileyplus.com.)

What do students receive with WileyPLUS?

- **A research-based design.** WileyPLUS provides an online environment that integrates relevant resources, including the entire digital textbook, in an easy-to-navigate framework that helps students study more effectively.
  - WileyPLUS adds structure by organizing textbook content into smaller, more manageable "chunks."
  - Related media, examples, and sample practice items reinforce the learning objectives.
  - Innovative features such as calendars, visual progress tracking, and self-evaluation tools improve time management and strengthen areas of weakness.
- **One-on-one engagement.** With WileyPLUS for Blank/Krantz, *Calculus* 2nd edition, students receive 24/7 access to resources that promote positive learning outcomes. Students engage with related examples (in various media) and sample practice items, including the following:
  - Animations based on key illustrations in each chapter
  - Algorithmically generated exercises in which students can click on a "help" button for hints, as well as a large cache of extra exercises with final answers
- **Measurable outcomes.** Throughout each study session, students can assess their progress and gain immediate feedback. WileyPLUS provides precise reporting of strengths and weaknesses, as well as individualized quizzes, so that students are confident they are spending their time on the right things. With WileyPLUS, students always know the exact outcome of their efforts.

What do instructors receive with WileyPLUS?

- **Reliable resources.** WileyPLUS provides reliable, customizable resources that reinforce course goals inside and outside of the classroom as well as visibility into individual student progress. Precreated materials and activities help instructors optimize their time:
- **Customizable course plan.** WileyPLUS comes with a precreated course plan designed by a subject matter expert uniquely for this course. Simple drag-and-drop tools make it easy to assign the course plan as-is or modify it to reflect your course syllabus.
- **Course materials and assessment content.** The following content is provided.
  - Lecture Notes PowerPoint Slides
  - Classroom Response System (Clicker) Questions
  - Instructor's Solutions Manual
  - Gradable Reading Assignment Questions (embedded with online text)
  - Question Assignments (end-of-chapter problems coded algorithmically with hints, links to text, whiteboard/show work feature, and instructor-controlled problem solving help)
- **Gradebook.** WileyPLUS provides instant access to reports on trends in class performance, student use of course materials, and progress towards learning objectives, helping inform decisions and drive classroom discussions.

Powered by proven technology and built on a foundation of cognitive research, WileyPLUS has enriched the education of millions of students in more than 20 countries around the world.

For further information about available resources to accompany Blank/Krantz, *Calculus* 2nd edition, see www.wiley.com/college/blank.

Acknowledgments

Over the years of the development of this text, we have profited from the comments of our colleagues around the country. We would particularly like to thank Chi Keung Chung, Dennis DeTurck, and Steve Desjardins for the insights and suggestions they have shared with us. To Chi Keung Chung, Roger Lipsett, and Don Hartig, we owe sincere thanks for preventing many erroneous answers from finding their way to the back of the book.

We would also like to take this opportunity to express our appreciation to all our reviewers for the contributions and advice that they offered. Their duties did not include rooting out all of our mistakes, but they found several. We are grateful to them for every error averted. We alone are responsible for those that remain.

Debut Edition Reviewers

David Calvis, Baldwin-Wallace College
Gunnar Carlsson, Stanford University
Chi Keung Cheung, Boston College
Dennis DeTurck, University of Pennsylvania
Bruce Edwards, University of Florida
Saber Elyadi, Trinity University
David Ellis, San Francisco State University
Salvatrice Keating, Eastern Connecticut State University
Jerold Marsden, California Institute of Technology
Jack Mealy, Austin College
Harold Parks, Oregon State University
Ronald Taylor, Berry College

Second Edition Reviewers

Anthony Aidoo, Eastern Connecticut State University
Alvin Bayless, Northwestern University
Maegan Bos, St. Lawrence University
Chi Keung Cheung, Boston College
Shai Cohen, University of Toronto
Randall Crist, Creighton University
Joyati Debnath, Winona State University
Steve Desjardins, University of Ottawa
Bob Devaney, Boston University
Esther Vergara Diaz, Université de Bretagne Occidentale
Allen Donsig, University of Nebraska, Lincoln
Hans Engler, Georgetown University
Mary Erb, Georgetown University
Aurelian Gheondea, Bikent University

Klara Grodzinsky, Georgia Institute of Technology
Caixing Gu, California Polytechnic State University, San Luis Obispo
Mowaffaq Hajja, Yarmouk University
Mohammad Hallat, University of South Carolina Aiken
Don Hartig, California Polytechnic State University, San Luis Obispo
Ayse Gul Isikyer, Gebze Institute of Technology
Robert Keller, Loras College
M. Paul Latiolais, Portland State University
Tim Lucas, Pepperdine University
Jie Miao, Arkansas State University
Aidan Naughton, University of St. Andrews
Zbigniew Nitecki, Tufts University
Harold Parks, Oregon State University
Elena Pavelescu, Rice University
Laura Taalman, James Madison University
Daina Taimina, Cornell University
Jamal Tartir, Youngstown State University
Alex Smith, University of Wisconsin Eau Claire
Gerard Watts, Kings College, London

ABOUT THE AUTHORS

Brian E. Blank and Steve G. Krantz have a combined experience of more than 60 years teaching experience. They are both award-winning teachers and highly respected writers. Their extensive experience in consulting for a variety of professions enables them to bring to this project diverse and motivational applications as well as realistic and practical uses of the computer.

Brian E. Blank

Brian E. Blank was a calculus student of William O. J. Moser and Kohur Gowrisankaran. He received his B.Sc. degree from McGill University in 1975 and Ph.D. from Cornell University in 1980. He has taught calculus at the University of Texas, the University of Maryland, and Washington University in St. Louis.

Steven G. Krantz

Steven G. Krantz was born in San Francisco, California in 1951. He earned his Bachelor's degree from the University of California at Santa Cruz in 1971 and his Ph.D. from Princeton University in 1974. Krantz has taught at the University of California Los Angeles, Princeton, Penn State, and Washington University in St. Louis. He has served as chair of the latter department. Krantz serves on the editorial board of six journals. He has directed 17 Ph.D. students and 9 Masters students. Krantz has been awarded the UCLA Alumni Foundation Distinguished Teaching Award, the Chauvenet Prize of the Mathematical Association of America, and the Beckenbach Book Award of the Mathematical Association of America. He is the author of 165 scholarly papers and 55 books.

CHAPTER 1

Basics

PREVIEW

Many of the variables encountered in everyday life are subject to continuous change. When traveling along a straight line from one point to another, we must pass through all points in between. If a tire deflates from 28 psi (pounds per square inch) to 10 psi, then at some time in between, it was $19\frac{1}{3}$ psi. Everyone reading this text was, at some time, π years old.

Continuous variables often depend on other continuous variables through precise mathematical laws. The area A of a circle is related to its radius r by the formula $A = \pi r^2$. When a body that is initially at rest moves under constant acceleration g, the distance s that it travels in time t is given by $s = gt^2/2$. Because r determines A and t determines s, we say that A is a *function* of r and s is a *function* of t. Calculus provides tools for understanding how continuous variables change and for investigating the functional relationships between continuous variables.

Before beginning a study of calculus, it is important to understand clearly what is meant by a "function" and what is meant by a "variable." We must also have a way to interpret a functional relationship graphically. These topics are reviewed in the present chapter. Although some of the topics will be familiar to you from previous studies, it is likely that you will learn some new ideas as well.

1.1 Number Systems

The simplest number system is the set of *natural numbers* 0, 1, 2, 3, ..., denoted by the letter $\mathbb{N}$. Multiplication and addition are operations in $\mathbb{N}$, meaning that the sum or product of two natural numbers is a natural number. Subtraction, however, may not make sense if we have *only* the natural numbers at our disposal. For example, $3-5$ has no meaning in $\mathbb{N}$. Therefore we must consider the larger number system $\mathbb{Z}$, the *integers*, consisting of the numbers $\ldots, -3, -2, -1, 0, 1, 2, 3, \ldots$ The set of *positive* integers, $\{1, 2, 3, \ldots\}$, is denoted by $\mathbb{Z}^+$.

Although addition, multiplication, and subtraction make sense in the integers, division may not. For instance, the expression 3/5 does not represent an integer. So we pass to the larger number system $\mathbb{Q}$, which consists of all expressions of the form a/b where a and b are integers, and b is not zero. This is the system of *rational numbers.*

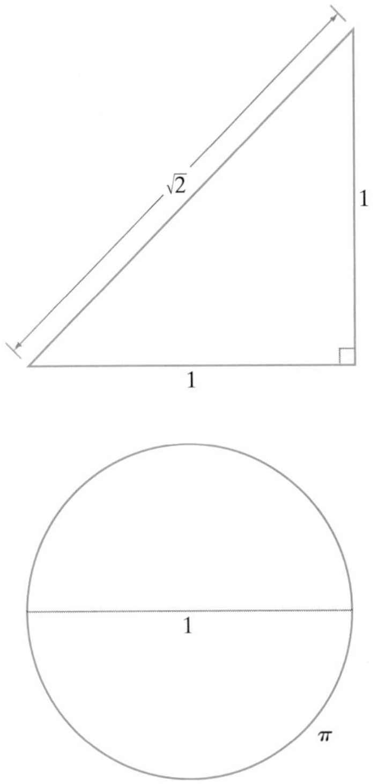

▲ Figure 1 Two numbers that are *not* rational: the hypotenuse of a right triangle with base and height 1 and the circumference of the circle of diameter 1

In general, the numbers that we encounter in everyday life—prices, speed limits, weights, temperatures, interest rates, and so on—are rational numbers. However, there are also numbers that are not rational. As a simple example, the length of a diagonal of a square of side length 1 is $\sqrt{2}$, which is *not* a rational number (see Figure 1 and Exercise 57). As another example, the number π, which is defined as the circumference of a circle of diameter 1, is not a rational number (see Figure 1).

When studying calculus, it is important to remember that the rational numbers are not appropriate for the mathematical process of *taking limits.* For instance, although the sequence of rational numbers

$$3, \frac{31}{10}, \frac{314}{100}, \frac{3141}{1000}, \frac{31415}{10000}, \ldots$$

seems to tend to, or approach, the decimal representation of π, namely $3.14159265\ldots$, π is not a rational number—it cannot be represented as a quotient of integers.

INSIGHT At present, when we use expressions like "tend to" and "approach a limit" or "limits," we are speaking intuitively. Later, when we deal with infinite decimal expansions, we will also do so intuitively. Chapter 2 makes the notion of limit more precise. Chapter 8 discusses in detail the ideas connected with infinite decimal expansions.

Because calculus involves the systematic use of various kinds of limiting processes, we work not with rational numbers but with the larger *real number system* $\mathbb{R}$, which consists of all three types of infinite decimal expansions:

1. *Decimal expansions that, after a finite number of digits, contain only zeros* This type of decimal expansion is called *terminating*. The number 2.657000 ... is an example. We would ordinarily write this number as 2.657 and would recognize it as the rational number 2657/1000. Any other real number with a decimal expansion that is terminating can be displayed as a rational number in the same way.
2. *Nonterminating expansions that, after a finite number of terms, repeat a single block of terms endlessly* An example is the number $x = 8.347626262\ldots$. Notice

that, after the digit 7, the block 62 is repeated endlessly. To understand this number, we write

$$100000x = 834762.626262\ldots$$
$$1000x = 8347.626262\ldots$$

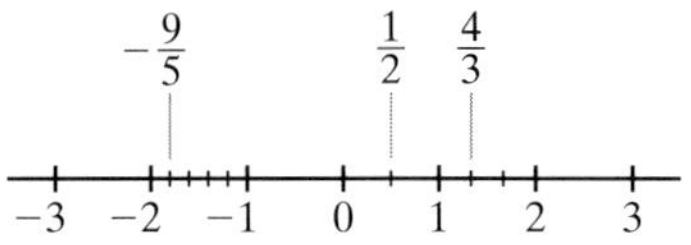

▲ Figure 2

Subtracting the second equation from the first gives $99000x = 826415$, or $x = 826415/99000$. We can adapt this procedure to demonstrate that any number of this type is rational. Conversely, we can use long division to convert every rational number to a decimal expansion that either terminates or repeats.

3. *Nonterminating, nonrepeating decimal expansions* Numbers of this type cannot be rational and are therefore called *irrational*. As was mentioned earlier, the numbers $\sqrt{2}$ and π are irrational: The first several digits of their decimal expansions are 1.41421356... and 3.14159265..., respectively.

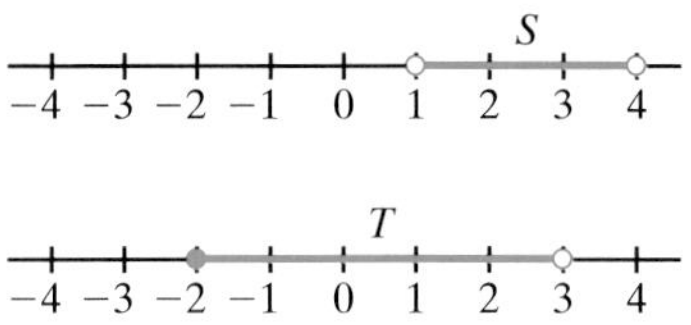

▲ Figure 3

Because any decimal expansion must be one of these three types, every real number is either a rational number or an irrational number. To aid in thinking and to display answers, it helps to exhibit real numbers in a diagram, such as a *number line*. Begin with an infinite line with equally spaced points representing the integers. Subdivide the line to represent fractions (see Figure 2). With sufficiently accurate tools, it is possible to determine which point corresponds to any *terminating* decimal expansion. This still leaves many unlabeled points, which are limits of sequences of the points that have already been labeled. In other words, the unlabeled points correspond to the nonterminating decimal expansions. In this way, we can think of each point of the line as corresponding to a real number and vice versa.

Sets of Real Numbers

The most common way to describe a set of real numbers is with the notation $\{x : P(x)\}$, which is read "the set of all x that satisfy the property $P(x)$." If x belongs to a set S, we say that x is an element, or a member, of S, and we write $x \in S$.

▶ **EXAMPLE 1** Sketch the sets $S = \{s : 1 < s < 4\}$ and $T = \{t : -2 \leq t < 3\}$.

Solution The sets are sketched in Figure 3. A solid dot indicates that -2 is included in T, whereas hollow dots show that 1 and 4 are missing from S and that 3 is missing from T. ◀

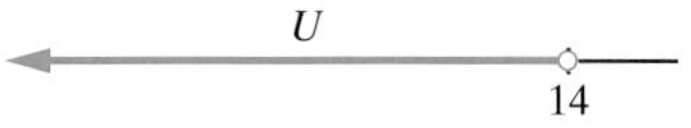

▲ Figure 4

Frequently, the expression $P(x)$ is complicated and requires a calculation before we can determine the set being described.

▶ **EXAMPLE 2** Sketch the set $U = \{u : 2u + 5 > 3u - 9\}$.

Solution We must first simplify the inequality $2u + 5 > 3u - 9$. Add 9 and subtract $2u$ from both sides to obtain $14 > u$. Therefore $U = \{u : u < 14\}$. The sketch is in Figure 4. ◀

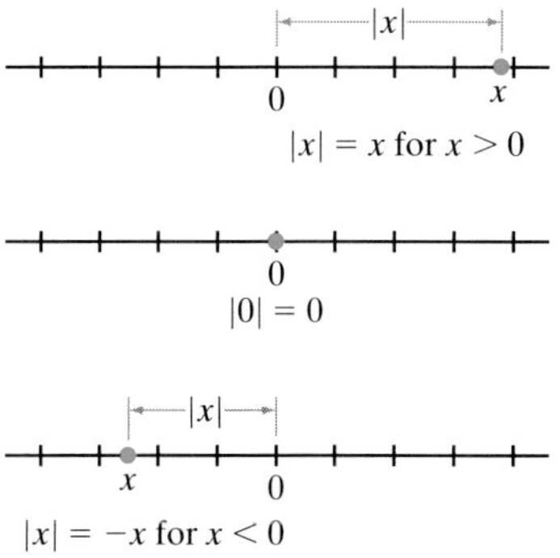

▲ Figure 5

DEFINITION If x is a real number, then the expression $|x|$, called the absolute value of x, represents the linear distance from x to 0 on the number line. If $x \geq 0$, then $|x| = x$. If $x < 0$, then $|x| = -x$. Figure 5 illustrates these cases.

The absolute value arises frequently. For example, both x and $-x$ are square roots of x^2. Because the expression $\sqrt{x^2}$ denotes the *nonnegative* square root of x^2, we may write

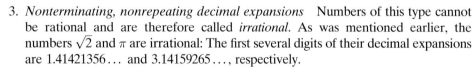

$$\sqrt{x^2} = |x|$$

without having to worry about the sign of x. The following example illustrates another convenient use of absolute value.

▶ **EXAMPLE 3** Sketch the set $V = \{x : |x-4| \le 3\}$.

Solution We must solve the inequality $|x-4| \le 3$.

Case 1. If $x-4>0$, then $|x-4| \le 3$ becomes $(x-4) \le 3$ or $x \le 7$. The conditions $x-4>0$ and $x \le 7$ taken together yield $4<x\le 7$.

Case 2. If $x-4<0$, then $|x-4| \le 3$ becomes $-(x-4) \le 3$ or $-3 \le (x-4)$. This is equivalent to $1 \le x$. The conditions $x-4<0$ and $1 \le x$ taken together yield $1 \le x < 4$.

Case 3. If $x-4=0$, then $x=4$ and $|x-4| \le 3$ becomes $0 \le 3$ which is certainly true. To summarize, we have learned that

$$V = \{x : 4<x \le 7 \text{ or } 1 \le x < 4 \text{ or } x = 4\}.$$

This can be written more concisely as $V = \{x : 1 \le x \le 7\}$. A sketch of V is given in Figure 6. ◀

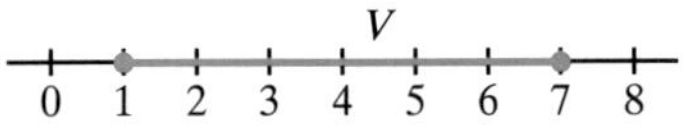

▲ Figure 6

INSIGHT If we redo the analysis of Example 3 using any constant c instead of the number 4 and any nonnegative constant r instead of the number 3, then we see that the inequality $|x-c| \le r$ is equivalent to the two inequalities $-r \le x-c$ and $x-c \le r$. These two inequalities can be expressed concisely as $c-r \le x \le c+r$. That is,

$$\{x : |x-c| \le r\} = \{x : c-r \le x \le c+r\}. \tag{1.1.1}$$

In words, the set we are describing is simply those points with distance from c less than or equal to r.

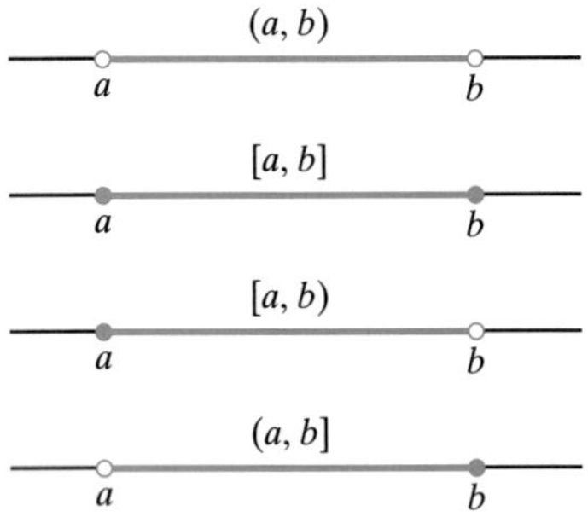

▲ Figure 7

DEFINITION The distance between two real numbers x and y is either $x-y$ or $y-x$, whichever is nonnegative. A convenient way to say this is that the distance between x and y is $|x-y|$.

The definition of distance helps describe the set $V = \{x : |x-4| \le 3\}$ from Example 3 as the set of numbers x with a distance from 4 that is less than or equal to 3. This interpretation reinforces the finding that V is the set of x such that $1 \le x \le 7$.

Intervals

The most frequently encountered sets of real numbers are *intervals*. The sets S, T, U, and V from Examples 1, 2, and 3 are examples of intervals. If a and b are real numbers with $a \le b$, then we can produce the following four types of *bounded intervals* (see also Figure 7).

$$(a,b) = \{x : a<x<b\}$$

$$[a,b] = \{x : a \le x \le b\}$$

$$[a, b) = \{x : a \le x < b\}$$
$$(a, b] = \{x : a < x \le b\}$$

Intervals of the form (a, b) are called *open*, whereas those of the form $[a, b]$ are called *closed*. The other two types are called either *half open* or *half closed* (these two terms are interchangeable). If I is any one of these four types of bounded intervals, then we say that a is the *left endpoint* of I, and b is the *right endpoint* of I. The interval $[a, a]$ is simply the set $\{a\}$, which means it has only the number a in it. Notice that an interval such as (a, a) does not contain any point. When a set has no elements, it is called the *empty set* and is denoted by the symbol $\emptyset$.

If a and b are the left and right endpoints, respectively, of an interval I, then the *midpoint* (or *center*) c of I is given by

$$c = \frac{a+b}{2}.$$

It is easy to verify that c is equally distant from each of the endpoints:

$$|c-a| = \left|\left(\frac{a+b}{2}\right) - a\right| = \frac{b-a}{2} = \left|b - \left(\frac{a+b}{2}\right)\right| = |b-c|.$$

▶ **EXAMPLE 4** Write the interval (3, 11) in the form $\{x : |x-c| < r\}$.

Solution Because $I = \{x : |x-c| < r\}$ is the set of points with a distance from c that is less than r, we can deduce that c is the midpoint of I. The midpoint of the interval $(3, 11)$ is $(3+11)/2$, or 7. Therefore $c = 7$. The distance r between c and the endpoints 3 and 11 is 4. We can therefore write the interval $(3, 11)$ as $\{x : |x-7| < 4\}$. ◀

When working with inequalities, we often want to multiply or divide each side by a nonzero number c. Inequalities are *preserved* under multiplication or division by positive numbers. For example, if $x \le y$ and $0 < c$, then $cx \le cy$. Inequalities are *reversed* under multiplication or division by negative numbers. Thus if $x \le y$ and $c < 0$, then $cx \ge cy$.

▶ **EXAMPLE 5** Write the set $W = \{x : |3x-18| \le 6\}$ as a closed interval $[a, b]$.

Solution We first multiply each side of $|3x-18| \le 6$ by 1/3 to obtain

$$\frac{1}{3}|3x-18| \le \frac{1}{3} \cdot 6, \text{ or}$$
$$\left|\frac{1}{3}(3x-18)\right| \le 2, \text{ or}$$
$$|x-6| \le 2.$$

The inequality $|x-6| \le 2$ is shorthand for the two inequalities $-2 \le x-6 \le 2$. By adding 6 to each side, we obtain $4 \le x \le 8$. Thus W is the closed interval $[4, 8]$. ◀

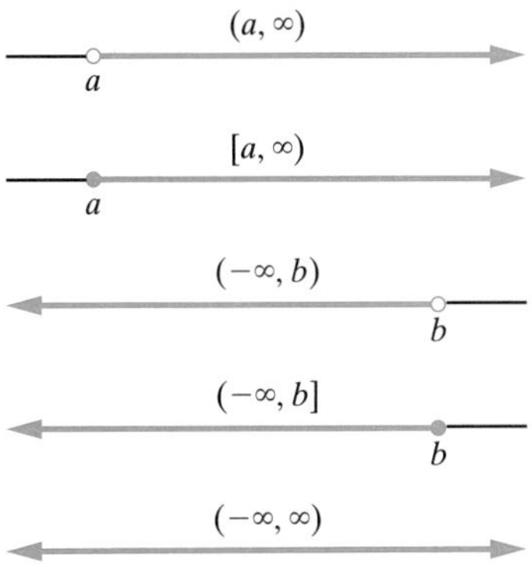

▲ Figure 8

We can use the symbols ∞ and $-\infty$ together with interval notation to describe some intervals with only one endpoint or with no endpoints (see Figure 8). These intervals are

$$(a, \infty) = \{x : a < x\}$$
$$[a, \infty) = \{x : a \leq x\}$$
$$(-\infty, b) = \{x : x < b\}$$
$$(-\infty, b] = \{x : x \leq b\}$$
$$(-\infty, \infty) = \{x : -\infty < x < \infty\}$$

Each of these intervals is an *unbounded interval*, as opposed to the bounded intervals described earlier.

INSIGHT The symbols ∞ and $-\infty$ do not represent real numbers. They are simply a convenient notational device. We never write $[-\infty, a)$ or any other expression suggesting that $+\infty$ or $-\infty$ is an element of a set of real numbers.

▶ **EXAMPLE 6** Solve the inequality

$$\frac{3x-2}{|x+1|} > 1.$$

Solution Observe that x cannot equal -1. For other values of x, we clear the denominator by multiplying both sides of the equation by $|x+1|$. This quantity is never negative, so the multiplication preserves the inequality. The result is $3x-2 > |x+1|$. We divide our analysis of this last inequality into two cases, according to the sign of the expression inside the absolute value.

If $x > -1$, then $x+1 > 0$, and $|x+1| = x+1$. The inequality $3x-2 > |x+1|$ becomes $3x-2 > (x+1)$, which simplifies to $x > 3/2$. The points that satisfy both $x > -1$ and $x > 3/2$ make up the interval $(3/2, \infty)$

In the second case, when $x < -1$, the expression $x+1$ is negative, and, as a consequence, $|x+1| = -(x+1)$. The inequality $3x-2 > |x+1|$ becomes $3x-2 > -(x+1)$, which simplifies to $x > 1/4$. There are, however, no points that satisfy both $x < -1$ and $x > 1/4$.

In summary, we see that the set of points satisfying the given inequality is simply the interval $(3/2, \infty)$ that we found in the first case. ◀

If A is a set of real numbers, and x is a number in A, then we write $x \in A$. Given two sets A and B of real numbers, we can form the *union* $A \cup B = \{x : x \in A \text{ or } x \in B\}$ and the *intersection* $A \cap B = \{x : x \in A \text{ and } x \in B\}$. In words, the union $A \cup B$ consists of every real number that is in at least one of the two sets A and B. The intersection $A \cap B$ consists of every real number that is in both sets A and B.

▶ **EXAMPLE 7** Let $A = \{x : |x| \leq 5\}$ and $B = \{x : |x| > 3\}$. Describe the intersection $A \cap B$ as a union of intervals.

Solution Using set equation (1.1.1) with $r = 5$ and $c = 0$, we have $A = \{x : -5 \leq x \leq 5\} = [-5, 5]$. To determine if a real number x is in the set B, we consider the cases $x > 0$, $x = 0$, and $x < 0$ separately. If $x > 0$, then $|x| = x$. Therefore a positive number x is in B if and only if $x > 3$. If $x = 0$, then $|x| = 0$, and x is not in B. If $x < 0$, then $-x = |x|$. Therefore a negative number x is in B if and only if $-x > 3$, or $x < -3$. Thus $B = (-\infty, -3) \cup (3, \infty)$. Figure 9 shows that $A \cap B = [-5, -3) \cup (3, 5]$. ◀

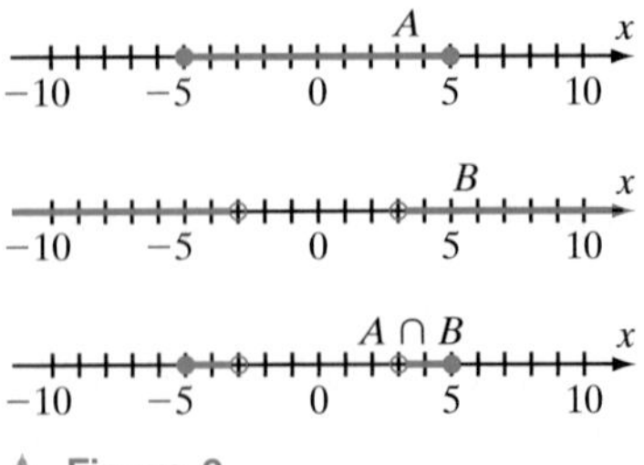

▲ Figure 9

The Triangle Inequality

It is not difficult to see that $|a+b| = |a| + |b|$ when a and b have the same sign (see Exercise 53). If a and b have opposite signs, however, then cancellation occurs in the sum $a+b$ but not in the sum $|a| + |b|$. In this case, $|a+b| < |a| + |b|$. In summary, all real numbers a and b satisfy

$$|a+b| \le |a| + |b|.$$

This relationship, called the *Triangle Inequality*, is a useful tool for estimating the magnitude of a sum in terms of the magnitudes of its summands.

▶ **EXAMPLE 8** Suppose the distance of a to b is less than 3, and the distance of b to 2 is less than 6. Estimate the distance of a to 2.

Solution We have $|a-2| = |(a-b)+(b-2)|$. By the Triangle Inequality,

$$|a-2| \le |a-b| + |b-2| < 3 + 6 = 9.$$

Therefore the distance of a to the number 2 is not greater than 9. ◀

The Triangle Inequality can be extended to three or more summands. For example,

$$|a+b+c| \le |a| + |b| + |c|.$$

Approximation

Calculus deals with many "infinite processes." One infinite process that we have already considered is the decimal expansion of an irrational number. Although we can think abstractly of irrational numbers such as π and $\sqrt{2}$, we cannot write all the digits of their decimal expansions. If a calculator returns the value 3.14159265359 for π, then it is really displaying an *approximation* of π. A small *error* has resulted. In general, the type of error that results when an infinite process is terminated is known as a *truncation error*.

In numerical work, we always have the equation

$$\textit{True value} = \textit{approximate value} + \textit{error}.$$

If a is an approximation of the number x, then $|x-a|$ is called the *absolute error*, and $|x-a|/|x|$ is called the *relative error*. Relative error is often of greater concern than absolute error. For example, an absolute error of $1000 in the national budget is not serious because it is a very small relative error. The same absolute error in your personal budget, however, would be far more significant. When we calculate an approximation, we must ensure that the error term is acceptably small. Often the size of an allowable error is stated in terms of decimal places.

DEFINITION If the absolute error $|x-a|$ is less than or equal to $5 \times 10^{-(q+1)}$, then a approximates (or agrees with) x to q decimal places.

▶ **EXAMPLE 9** Does 22/7 approximate π to as many decimal places as does 3.14? Does 1.418 approximate $\sqrt{2}$ to two decimal places?

Solution Because $|\pi - 22/7| = 1.26\ldots \times 10^{-3}$, we have

$$5\times 10^{-(3+1)} = 5\times 10^{-4} < \left|\pi - \frac{22}{7}\right| < 5\times 10^{-3} = 5\times 10^{-(2+1)}.$$

Therefore 22/7 approximates π to two, but not three, decimal places. Similarly, $\pi - 3.14 = 1.59\ldots \times 10^{-3}$ and so

$$5\times 10^{-(3+1)} = 5\times 10^{-4} < |\pi - 3.14| < 5\times 10^{-3} = 5\times 10^{-(2+1)}.$$

Thus the number 3.14, like 22/7, approximates π to two, but not three, decimal places.

Because $|\sqrt{2} - 1.418| = 3.78\ldots \times 10^{-3} < 5\times 10^{-(2+1)}$, we conclude that 1.418 does approximate $\sqrt{2}$ to two decimal places. ◄

INSIGHT Even if two numbers a and b agree to q decimal places, we cannot be certain that the results of rounding a and b to q decimal places will be the same. In Example 9, we see that $\sqrt{2}$ and 1.418 agree to two decimal places. Rounding these numbers to two decimal places, however, results in 1.41 and 1.42, respectively.

When working with a calculator or computer software, there are several sources of error. The errors that we now consider are caused by the way real numbers are stored in calculators or computers.

The *floating point decimal representation* of a nonzero real number x is

$$x = \pm 0.a_1a_2a_3a_4\ldots \times 10^p$$

where p is an integer, and $a_1, a_2, \ldots$ are natural numbers from 0 to 9 with $a_1 \neq 0$. The decimal point is thought of as "floating" to the left of the first nonzero digit. A number with a floating point decimal representation of $\pm 0.\, a_1a_2\ldots a_k \times 10^p$ is said to have k *significant digits.*

Because the memory of a calculator or computer is finite, it must limit the number of significant digits with which it works. This can lead to a type of error known as *roundoff error*. It is important to understand that a computation can reduce the number of significant digits. Such a *loss of significant digits*, or, more simply, a *loss of significance*, can occur when two very nearly equal numbers are subtracted. For example, 0.124 and 0.123 have three significant digits, whereas their difference, 0.1×10^{-2}, has only one significant digit. The next example illustrates why loss of significance is sometimes called *catastrophic cancellation.*

► **EXAMPLE 10** Suppose that $x = 0.412\times 0.300 - 0.617\times 0.200$. A calculator that displays three significant digits is used to evaluate the products before subtraction. What relative error results?

Solution When rounded to three significant digits, the products $0.412\times 0.300 = 0.1236$ and $0.617\times 0.200 = 0.1234$ become 0.124 and 0.123. The difference of these numbers is $x = 0.124 - 0.123 = 0.001 = 0.1\times 10^{-2}$. Of course, x really equals $0.1236 - 0.1234$, or 0.2×10^{-3}. The relative error is therefore

$$\frac{|x-a|}{|x|} = \frac{|0.2\times 10^{-3} - 0.1\times 10^{-2}|}{|0.2\times 10^{-3}|} = \frac{0.8\times 10^{-3}}{0.2\times 10^{-3}} = 4,$$

or 400%. ◄

Computer Algebra Systems From time to time, this book refers to software packages known as *computer algebra systems.* A computer algebra system not only performs arithmetic operations with specific numbers but also carries out those operations symbolically. When the "factor" command of a computer algebra system is applied to x^6-y^6, for example, the result is

$$(x-y)(x+y)(x^2+xy+y^2)(x^2-xy+y^2).$$

Several calculators now offer some of the capabilities of computer algebra systems.

QUICK QUIZ

1. What interval does the set $\{x : |x-2| < 9\}$ represent?
2. Write the interval $(-5, 1)$ in the form $\{x : |x-c| < r\}$.
3. Write $\{x : |x+8| > 7\}$ as the union of two intervals.
4. For which numbers b is $|-5+b|$ less than $5+|b|$?
5. What is the smallest number that approximates 0.997 to two decimal places? The largest?

Answers

1. $(-7, 11)$ 2. $\{x : |x-(-2)| < 3\}$ 3. $(-\infty, -15) \cup (-1, \infty)$
4. $b > 0$ 5. 0.992, 1.002

EXERCISES

Problems for Practice

In Exercises 1–6, convert the decimal to a rational fraction. (Ellipses are included in some exercises to indicate repetition.)

1. 2.13
2. 0.00034
3. 0.232323...
4. 2.222...
5. 5.001001001...
6. 15.7231231231...

In Exercises 7–12, use long division to convert the rational fraction to a (possibly nonterminating) decimal with a repeating block. Identify the repeating block.

7. 1/40
8. 25/8
9. 5/3
10. 2/7
11. 18/25
12. 31/14

In Exercises 13–18, write the set using interval notation. Use the symbol $\cup$ where appropriate.

13. $\{x : 1 \le x \le 3\}$
14. $\{x : |x-2| < 5\}$
15. $\{t : t > 1\}$
16. $\{u : |u-4| \ge 6\}$
17. $\{y : |y+4| \le 10\}$
18. $\{s : |s-2| > 8\}$

In Exercises 19–24, sketch the set on a real number line.

19. $\{x : 2x-5 < x+4\}$
20. $\{s : |s-2| < 1\}$
21. $\{t : (t-5)^2 < 9/4\}$
22. $\{y : 7y+4 \ge 2y+1\}$
23. $\{x : |3x+9| \le 15\}$
24. $\{w : |2w-12| \ge 1\}$

In Exercises 25–28, write the interval in the form $\{x : |x-c| < r\}$ or $\{x : |x-c| \le r\}$.

25. $[-1, 3]$
26. $[3, 4\sqrt{2}]$
27. $(-\pi, \pi+2)$
28. $(\pi-\sqrt{2}, \pi)$

Further Theory and Practice

29. Explain why the sum and product of two rational numbers are always rational.
30. Give an example of two irrational numbers with a rational product; give an example of two irrational numbers with a rational sum.
31. Two commonly used approximations of π are 22/7 and 3.14. How can you tell at a glance that these approximations cannot be exact?

32. Find the smallest interval of the form $\{x : |x - \pi| \leq r\}$ that contains both 22/7 and 3.14. Is r rational or irrational?

33. A chemistry experiment, if followed exactly, results in the production of 34.5 g of a compound. To allow for a small amount of experimental error, the lab instructor will accept, without penalty, any reported measurement within 1% of the correct mass. In what interval must a measurement lie if a penalty is to be avoided? Describe this interval as a set of the form $\{x : |x - c| \leq \varepsilon\}$.

34. A door designed to be 885 mm wide and 1475 mm high is to fit in a 895 mm × 1485 mm rectangular frame. The clearance between the door and the frame can be no greater than 7 mm on any side for an acceptable seal to result. Under anticipated temperature ranges, the door will expand by at most 0.2% in height and width. The frame does not expand significantly under these temperature ranges. At all times, even when the door has expanded, there must be a 1 mm clearance between the door and the frame on each of the four sides. The door does not have to be manufactured precisely to specifications, but there are limitations. In what interval must the door width lie? The height?

In Exercises 35–43, sketch the set on a real number line.

35. $\{x : x > -2 \text{ and } x^2 < 9\}$

36. $\{s : |s + 3| < |2s + 7|\}$

37. $\{y : y - \sqrt{7} < 3y + 4 \leq 4y + \sqrt{2}\}$

38. $\{t : |t^2 + 6t| \leq 10\}$

39. $\{x : |x^2 - 5| \geq 4\}$

40. $\{s : |s + 5| < 4 \text{ and } |s - 2| \leq 8\}$

41. $\{x : x + 1 \geq 2x + 5 > 3x + 8\}$

42. $\{t : (t - 4)^2 < (t - 2)^2 \text{ or } |t + 1| \leq 4\}$

43. $\{x : |x^2 + x| > x^2 - x\}$

Describe each set in Exercises 44–47 using interval notation and the notation $\{x : P(x)\}$. Use the symbol $\cup$ where appropriate.

44. The set of points with a distance from 2 that does not exceed 4

45. The set of numbers that are equidistant from 3 and −9

46. The set of numbers with a square that lies strictly between 2 and 10

47. The set of all numbers with a distance less than 2 from 3 and with a square that does not exceed 8

In Exercises 48–51, if the set is given with absolute value signs, then write it without absolute value signs. If it is given without absolute value signs, then write it using absolute value signs.

48. $\{x : x + 5 < |x + 1|\}$

49. $\{s : |s - 4| > |2s + 9|\}$

50. $\{t : t^2 - 3t < 2t^2 - 5t\}$

51. $\{w : w/(w + 1) < 0\}$

52. Suppose $p(x) = x^2 + Bx + C$ has two real roots, r_1 and r_2 with $r_1 < r_2$. Write the set $\{x : p(x) \leq 0\}$ as an interval.

53. Show that equality holds in the Triangle Inequality when both summands are nonpositive.

54. Use the Triangle Inequality to prove that

$$|a| - |b| \leq |a + b|$$

for all $a, b \in \mathbb{R}$.

55. Prove that there is no smallest positive real number.

56. At first glance, it may appear that the number $y = 0.\overline{9} = 0.999...$ is just a little smaller than 1. In fact, $y = 1$. Let $x = 1 - y$. Explain why it is true that $x < 10^{-n}$ for every positive n. Deduce that $0 \leq x < b$ for every positive b. Explain why this implies that $y = 1$.

57. About 2400 years ago, the followers of Pythagoras discovered that if x is a positive number such that $x^2 = 2$, then x is irrational. Complete the following outline to obtain a proof of this fact.

a. Suppose that $x = a/b$ where a and b are integers with no common factors.

b. Conclude that $2 = x^2 = a^2/b^2$.

c. Conclude that $2b^2 = a^2$.

d. Conclude that 2 divides a evenly, with no remainder. Therefore $a = 2\alpha$ for some integer α.

e. Conclude that $b^2 = 2\alpha^2$.

f. Conclude that 2 divides b evenly, with no remainder.

g. Notice that parts d and f contradict part a.

58. Explain the error in the following reasoning: Let $x = (\pi + 3)/2$. Then $2x = \pi + 3$, and $2x(\pi - 3) = \pi^2 - 9$. It follows that $x^2 + 2\pi x - 6x = x^2 + \pi^2 - 9$, and $x^2 - 6x + 9 = x^2 - 2\pi x + \pi^2$. Each side is a perfect square: $(x - 3)^2 = (x - \pi)^2$. Therefore $x - 3 = x - \pi$, and $3 = \pi$.

Calculator/Computer Exercises

In Exercises 59–62, determine the interval that y must lie in to agree with x to q decimal places.

59. $x = 0.449$, $q = 3$

60. $x = 24$, $q = 2$

61. $x = 0.999 \times 10^{-3}$, $q = 3$

62. $x = 0.213462 \times 10^{-1}$, $q = 5$

63. Write a number with four significant digits that agrees with $x = 3.996$ to two decimal places but that differs from x in each digit.

64. Let $x = 0.4449$. Round this number *directly* to three, to two, and to one significant digit(s). Now *successively* round x to three, to two, and to one significant digit(s). The mathematical phrasing of this phenomenon is that *rounding is not transitive*. Describe this idea in your own words.

65.

$$\alpha = 1728148040 - 140634693\sqrt{151}$$

and

$$\beta = 1728148040 + 140634693\sqrt{151}.$$

A direct computation (with integers) shows that $\alpha \cdot \beta = 1$. Calculate decimal representations of α and β. Use them to compute the product $\alpha \cdot \beta$. Note: To twenty significant digits, $140634693\sqrt{151}$ equals

$$1728148039.9999999997.$$

Calculators that do not record enough significant digits will give 0 for the value of α.

66. Find two numbers $0.a_1a_2a_3a_4a_5$ and $0.b_1b_2b_3b_4b_5$ with $a_1 = b_1$ that agree when rounded to four significant digits but that do not agree when rounded directly to one significant digit.

The term a_nx^n is said to be the *leading term* of the polynomial

$$a_nx^n + a_{n-1}x^{n-1} + \cdots + a_1x + a_0,$$

where $a_n \neq 0$. The other terms are referred to as *lower-order terms*. In calculus, for large values of $|x|$, neglecting all the lower-order terms of a polynomial introduces only a small relative error; this relative error tends to zero as $|x|$ tends to infinity. In Exercises 67–70, present two tables, one for $x > 0$ and one for $x < 0$, to support this assertion. For example, for $p(x) = 3x^4 + 100x^3 - 1000$, the table below supports the assertion that the relative error $|p(x) - 3x^4| / |p(x)|$ tends to zero as x tends to positive infinity.

x	**Relative Error**
10^3	3.2258×10^{-2}
10^4	3.3223×10^{-3}
10^5	3.3322×10^{-4}
10^6	3.3332×10^{-5}
10^8	3.3333×10^{-7}

67. $x + 10^{10}$
68. $x^2 + 12345x$
69. $-3x^4 + 10^5x^3$
70. $|x|^5 + 1000x^4$

1.2 Planar Coordinates and Graphing in the Plane

A set such as $\{a, b\}$ does not have an implied ordering of its elements. We do not, for example, interpret a as the first element of $\{a, b\}$ and b as the second. When an ordering of a and b is required, we use the *ordered pair* notation (a, b). We say that a is the *first coordinate* and b the *second coordinate* of the ordered pair (a, b). Although $\{a, b\}$ and $\{b, a\}$ are exactly the same set, (a, b) and (b, a) are different ordered pairs (if $a \neq b$). The set of all ordered pairs of real numbers is called the *Cartesian plane* and is denoted by $\mathbb{R}^2$:

$$\mathbb{R}^2 = \{(a, b) : a, b \in \mathbb{R}\}.$$

Just as we can describe a point on a line with a real number, so too can we describe a point in a plane with an ordered pair of real numbers. Begin with intersecting horizontal and vertical lines. Put scales of real numbers on each line so that the zero of each scale is at the point of intersection. The point of intersection of the two lines is called the *origin* and is represented by the symbol 0. The vertical and horizontal lines are called *axes*. It is convenient to name each axis by using a variable. Commonly, the horizontal axis is named the x-axis and the vertical axis is named the y-axis. The Cartesian plane is then called the xy-*plane*. The two axes divide the plane into four regions that are known as *quadrants*, which are ordered as in Figure 1.

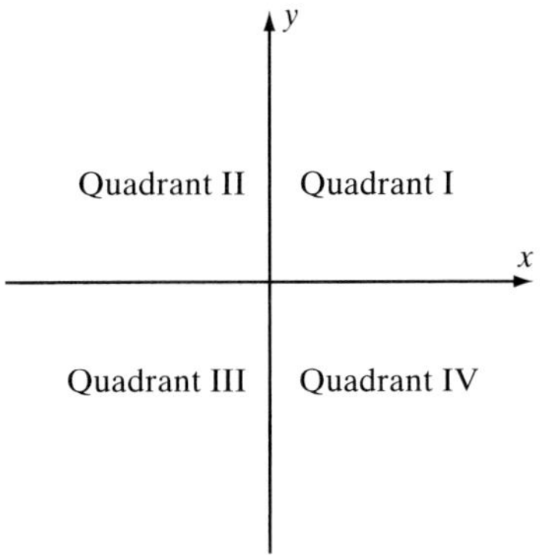

▲ Figure 1 The xy-plane

Each point P is determined by its displacement from the axes. Here the word "displacement" is interpreted in the *signed* sense: Positive displacement is to the right or upward, and negative displacement is to the left or downward. The displacement a of the point P from the y-axis is said to be the x-*coordinate*, or *abscissa*, of P. The displacement b of the point P from the x-axis is said to be the y-*coordinate*, or *ordinate*, of P. We represent the point P by the *ordered pair* (a, b),

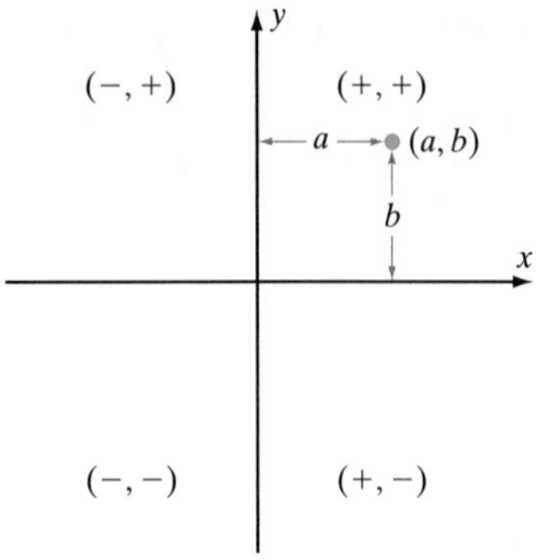

▲ Figure 2

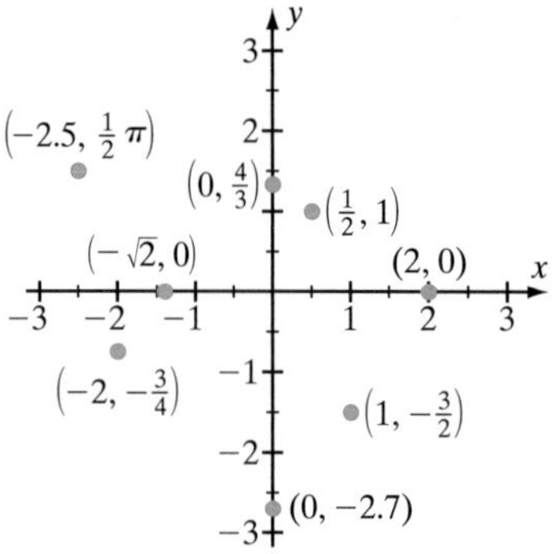

▲ Figure 3

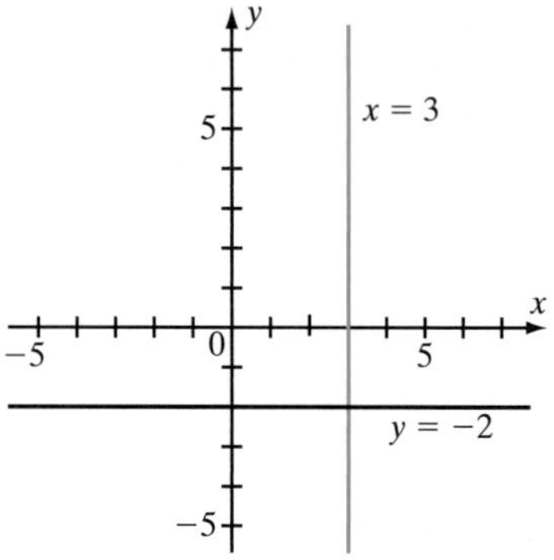

▲ Figure 4

as shown in Figure 2. The signs of the x- and y-coordinates in each of the four quadrants are also illustrated in Figure 2. Figure 3 exhibits several points and their coordinates.

It is easy to plot single points in the *xy-plane*. Often our job, however, is to plot a *set* of points with coordinates that satisfy some given equation in the variables x and y. This set is often referred to as the *locus* of the equation, or the *curve* defined by the equation. The equation is called the *Cartesian equation* of the locus of points. For example, the graph of the Cartesian equation $x = 3$ is the vertical line consisting of every point $(3, y)$. Here y may be any real number because the Cartesian equation $x = 3$ imposes no restrictions on y. The plot of $x = 3$ is shown in Figure 4. In a similar way, we see that the graph of $y = -2$ is the horizontal line that consists of every point $(x, -2)$. This line is also shown in Figure 4. In the next example we consider a Cartesian equation that involves both x and y.

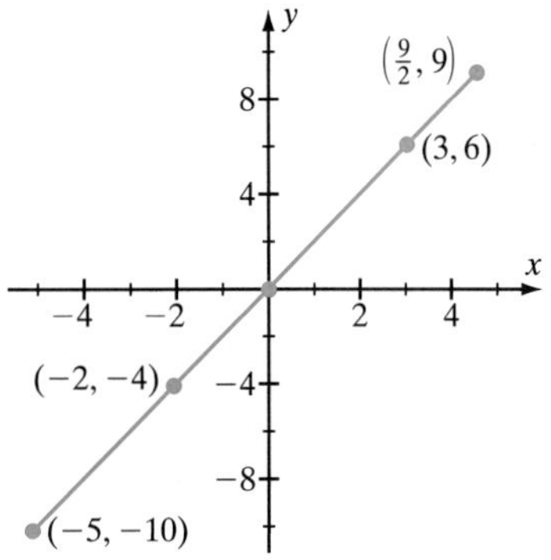

▲ Figure 5

▶ **EXAMPLE 1** Sketch or graph all points (x, y) in the plane that satisfy the equation $y = 2x$.

Solution For now, we perform this task by plotting some points and connecting them. Notice that the points $(0, 0)$, $(-5, -10)$, $(3, 6)$, $(-2, -4)$, and $(9/2, 9)$ all satisfy the equation $y = 2x$. Plotting these points suggests that the collection of all points satisfying $y = 2x$ is a straight line (see Figure 5). ◀

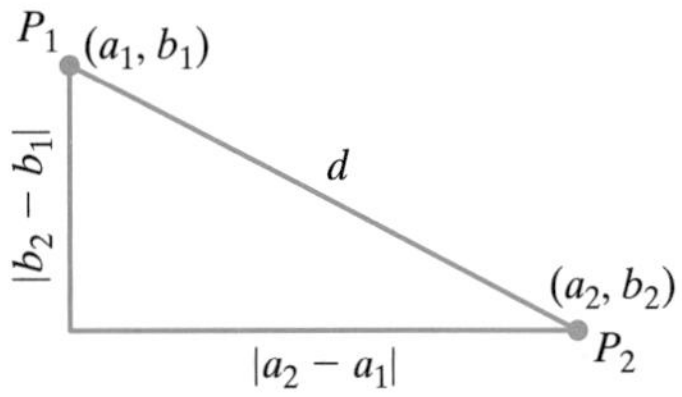

▲ Figure 6

$d = \sqrt{(a_2 - a_1)^2 + (b_2 - b_1)^2}$

INSIGHT When we plot a few points and connect them with a smooth curve, we are assuming that the curve is well behaved between the points we have actually plotted. Later, we will learn how calculus provides ways to know when this assumption is justified.

The Distance Formula and Circles

We can learn more from Cartesian coordinates if we use some geometric insight. First we review the *distance formula*. Let $\overline{P_1P_2}$ denote the line segment between points $P_1 = (a_1, b_1)$ and $P_2 = (a_2, b_2)$ in the plane. The shortest distance between P_1 and P_2 is the length of line segment $\overline{P_1P_2}$, which we denote by $|\overline{P_1P_2}|$. We can calculate this distance by realizing line segment $\overline{P_1P_2}$ as the hypotenuse of a right triangle (see Figure 6). By the Pythagorean Theorem, the length d of the hypotenuse is given by

$$d^2 = |a_2 - a_1|^2 + |b_2 - b_1|^2,$$

or

$$|\overline{P_1P_2}| = \sqrt{(a_2 - a_1)^2 + (b_2 - b_1)^2}.$$

Because the quantities under the square root sign are squared, we may omit the absolute value signs.

▶ **EXAMPLE 2** Calculate the distance between $(2, 6)$ and $(-4, 8)$.

Solution The requested distance is

$$d = \sqrt{(2-(-4))^2 + (6-8)^2} = \sqrt{6^2 + (-2)^2} = \sqrt{40} = 2\sqrt{10}. \quad ◀$$

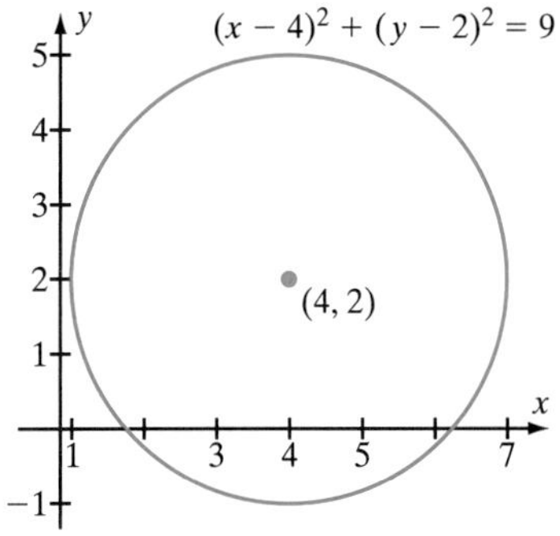

▲ Figure 7

▶ **EXAMPLE 3** Graph the set of points satisfying $(x-4)^2 + (y-2)^2 = 9$.

Solution We rewrite the equation as

$$\sqrt{(x-4)^2 + (y-2)^2} = 3.$$

According to this equation, 3 is the distance between the point $(4, 2)$ and any point (x, y) that satisfies the equation. The collection of such points is a circle of radius 3 and center $(4, 2)$. The locus is exhibited in Figure 7. ◀

The Equation of a Circle

Example 3 illustrates a general principle: An equation of the form

$$(x-h)^2 + (y-k)^2 = r^2 \tag{1.2.1}$$

with $r > 0$ has a graph that is a circle of radius r and center (h, k). Conversely, every circle is described by an equation of this form. We call this equation the *standard form* for a circle. The circle centered at the origin with radius 1 is called the *unit circle*. It is the locus of the equation

$$x^2 + y^2 = 1.$$

▶ **EXAMPLE 4** Write the equation of the circle with center $(8, -3/5)$ and radius 6.

Solution We apply formula (1.2.1) to obtain $(x-8)^2 + \left(y - (-3/5)\right)^2 = 6^2$, or $(x-8)^2 + (y+3/5)^2 = 36$. ◀

The Method of Completing the Square

To work efficiently with expressions of the form $Ax^2 + Bx + C$, we must use an algebraic procedure known as the *method of completing the square*. It is a good idea to review this procedure because it will be needed from time to time. (Although you should not memorize the following formulas, you should certainly remember the steps.)

Completing the Square

Step 1: Starting from $Ax^2 + Bx + C$, group together the terms that involve x, and from this group factor out the coefficient A of x^2:

$$Ax^2 + Bx + C = A\left(x^2 + \frac{B}{A}x\right) + C.$$

Step 2: To the group of terms that involve x, add the square of half the coefficient of x. From the constant term subtract an equal amount. In doing so, do not forget to account for the coefficient A:

$$Ax^2 + Bx + C = A\left(x^2 + \frac{B}{A}x + \left(\frac{B}{2A}\right)^2\right) + C - A\left(\frac{B}{2A}\right)^2.$$

Step 3: Identify the square:

$$\begin{aligned} Ax^2 + Bx + C &= A\left(x + \frac{B}{2A}\right)^2 + C - A\left(\frac{B}{2A}\right)^2 \\ &= A\left(x + \frac{B}{2A}\right)^2 + \frac{4AC - B^2}{4A}. \end{aligned}$$

▶ **EXAMPLE 5** Apply the method of completing the square to $3x^2 + 18x + 16$ and to $5 - 4x - 4x^2$.

Solution We have

$$\begin{aligned} 3x^2 + 18x + 16 &= 3(x^2 + 6x) + 16 && \text{(Step 1)} \\ &= 3\left(x^2 + 6x + \left(\frac{6}{2}\right)^2\right) + 16 - 3 \cdot \left(\frac{6}{2}\right)^2 && \text{(Step 2)} \\ &= 3\left(x + \frac{6}{2}\right)^2 - 11 && \text{(Step 3)} \end{aligned}$$

and

$$\begin{aligned} 5 - 4x - 4x^2 &= -4(x^2 + x) + 5 && \text{(Step 1)} \\ &= -4\left(x^2 + x + \left(\frac{1}{2}\right)^2\right) + 5 + 4\left(\frac{1}{2}\right)^2 && \text{(Step 2)} \\ &= -4\left(x + \frac{1}{2}\right)^2 + 6. && \text{(Step 3)} \end{aligned}$$

◀

▶ **EXAMPLE 6** Suppose that $A \neq 0$. Use the method of completing the square to show that the roots of the quadratic equation $Ax^2 + Bx + C = 0$ satisfy the Quadratic Formula

$$x = \frac{-B - \sqrt{B^2 - 4AC}}{2A} \quad \text{or} \quad x = \frac{-B + \sqrt{B^2 - 4AC}}{2A}.$$

Solution From Step 3, we see that a root x of $Ax^2 + Bx + C = 0$ satisfies

$$A\left(x+\frac{B}{2A}\right)^2+\frac{4AC-B^2}{4A}=0$$

or

$$\left(x+\frac{B}{2A}\right)^2=\frac{B^2-4AC}{4A^2}$$

or

$$\left|x+\frac{B}{2A}\right|=\frac{\sqrt{B^2-4AC}}{2|A|}.$$

It follows that

$$x+\frac{B}{2A}=-\frac{\sqrt{B^2-4AC}}{2A} \quad \text{or} \quad x+\frac{B}{2A}=\frac{\sqrt{B^2-4AC}}{2A},$$

from which the Quadratic Formula results. ◀

▶ **EXAMPLE 7** The locus of the equation $2x^2+2y^2+8x-4y+6=0$ is a circle. Convert this equation to the standard form for a circle, and determine the center and radius.

Solution We complete the square as follows:

$$2\left(x^2+\frac{8}{2}x\right)+2\left(y^2+\frac{-4}{2}y\right)+6=0. \qquad \text{(Step 1)}$$

We may divide by 2 to obtain

$$(x^2+4x)+(y^2-2y)+3=0.$$

Next we have

$$\left(x^2+4x+\left(\frac{4}{2}\right)^2\right)+\left(y^2-2y+\left(\frac{-2}{2}\right)^2\right)+3-\left(\frac{4}{2}\right)^2-\left(\frac{-2}{2}\right)^2=0. \qquad \text{(Step 2)}$$

Thus our equation becomes

$$(x+2)^2+(y-1)^2+3-4-1=0$$

or

$$(x+2)^2+(y-1)^2=2.$$

Finally, we write this as

$$(x-(-2))^2+(y-1)^2=(\sqrt{2})^2.$$

Therefore our equation describes the circle with center $(-2,1)$ and radius $\sqrt{2}$. ◀

INSIGHT It is wrong to suppose that every quadratic equation that we encounter can be put in the form of a circle. Notice that, for the method in Example 7 to work, it is necessary that both x^2 and y^2 appear and that both of these terms have the same coefficient. Also, no xy term ever appears in the equation of a circle.

Parabolas, Ellipses, and Hyperbolas

Parabolas, ellipses, and hyperbolas are curves that, like circles, can be realized by intersecting a cone with a plane (in three-dimensional space). Such curves are known as *conic sections*. We will study the geometric properties of these curves in greater detail later. For now, it is important simply to recognize the Cartesian equations of these curves when they arise. When discussing a conic section, we often refer to an axis of symmetry. For example, we say that a line L is an *axis of symmetry* for a curve C if the reflection of C through L is the same set of points as C itself (see Figure 8).

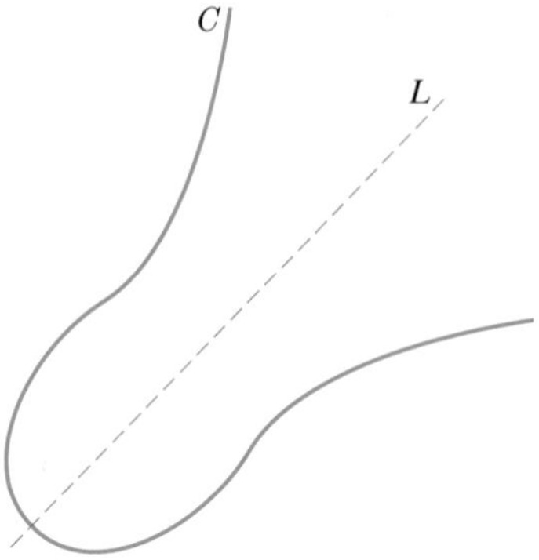

▲ Figure 8 L is an axis of symmetry for C.

DEFINITION Let a and b be positive numbers. An *ellipse* with vertical and horizontal axes of symmetry (see Figure 9) is the locus of an equation of the form

$$\frac{(x-h)^2}{a^2} + \frac{(y-k)^2}{b^2} = 1.$$

A *hyperbola* with vertical and horizontal axes of symmetry (see Figures 10 and 11) is the locus of an equation of the form

$$\frac{(x-h)^2}{a^2} - \frac{(y-k)^2}{b^2} = 1$$

or

$$\frac{(y-k)^2}{b^2} - \frac{(x-h)^2}{a^2} = 1.$$

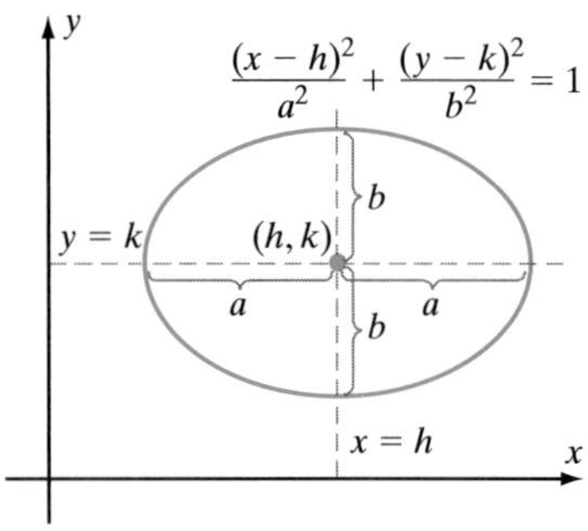

▲ Figure 9

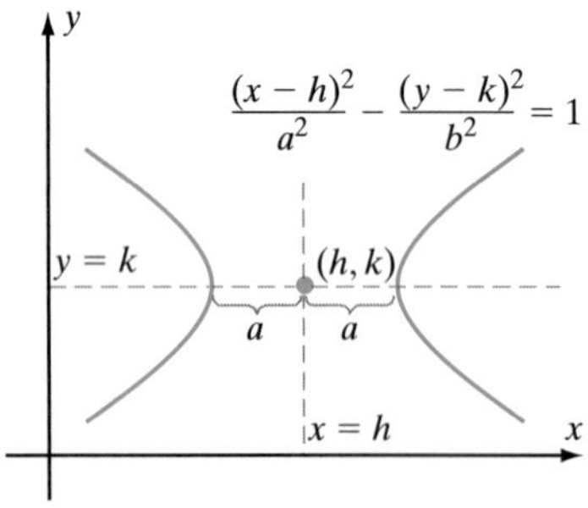

▲ Figure 10

Notice that the Cartesian equations of ellipses and hyperbolas, like those of circles, involve the squares of both variables x and y. The Cartesian equation of an ellipse with vertical and horizontal axes is similar to that of a circle, except that the coefficients of $(x-h)^2$ and $(y-k)^2$ are unequal. Likewise, the equation of a hyperbola is similar except that the coefficients of $(x-h)^2$ and $(y-k)^2$ have different signs. The method of completing the square may be used to obtain the *standard form* of an ellipse or a hyperbola if its Cartesian equation is presented in a different way.

▶ **EXAMPLE 8** The graph of $4x^2 + 8x - 9y^2 = 21$ is a hyperbola. Find the standard form for its Cartesian equation, and use it to determine the axes of symmetry of the hyperbola.

Solution Applying the method of completing the square, we have

$$\begin{aligned} 4x^2 + 8x - 9y^2 &= 4(x^2 + 2x) - 9y^2 \\ &= 4(x^2 + 2x + 1) - 9y^2 - 4 \cdot 1 \\ &= 4(x+1)^2 - 9y^2 - 4. \end{aligned}$$

Thus the equation $4x^2 + 8x - 9y^2 = 21$ is equivalent to $4(x+1)^2 - 9y^2 - 4 = 21$, or $4(x+1)^2 - 9y^2 = 25$, or $4(x+1)^2/25 - 9y^2/25 = 1$, or $(x+1)^2/(25/4) - y^2/(25/9) = 1$. The standard form is therefore

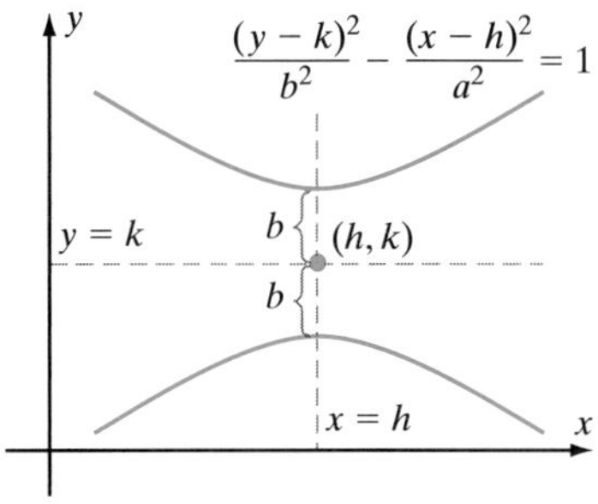

▲ Figure 11

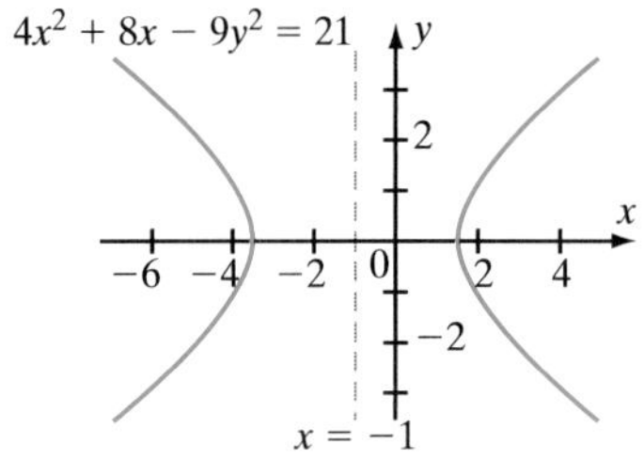

▲ Figure 12

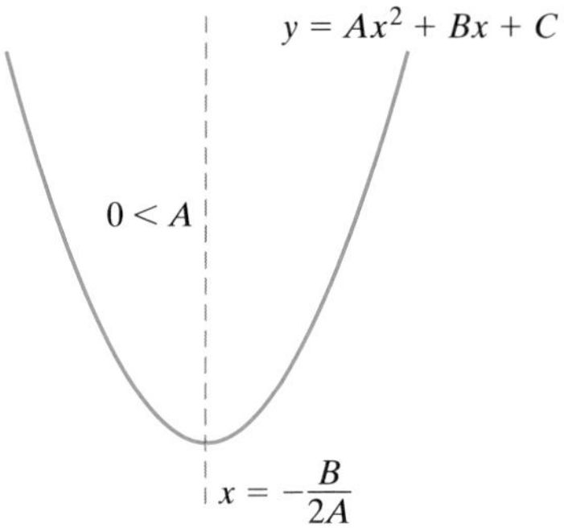

▲ Figure 13a

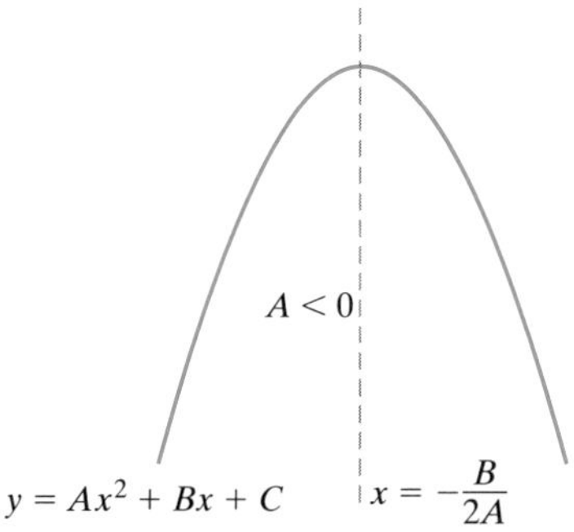

▲ Figure 13b

$$\frac{(x-(-1))^2}{(5/2)^2} - \frac{(y-0)^2}{(5/3)^2} = 1.$$

The axes of symmetry are $x = -1$ and $y = 0$ (see Figure 12). ◀

DEFINITION A *parabola* (with a vertical axis of symmetry) is the locus of an equation of the form $y = Ax^2 + Bx + C$ with $A \neq 0$.

Notice that the equation of a parabola involves the square of *only one* variable. To understand the locus of the equation $y = Ax^2 + Bx + C$, we must again use the method of completing the square.

THEOREM 1 The vertical line $x = -B/(2A)$ is an axis of symmetry of the parabola $y = Ax^2 + Bx + C$ $(A \neq 0)$.

Proof. Let $s = -B/(2A)$. If the horizontal line $y = y_0$ intersects the parabola at two points $P_0 = (x_0, y_0)$, and $P_1 = (x_1, y_0)$ with $x_0 < x_1$, then we can calculate x_0 and x_1 by using the Quadratic Formula. The explicit formulas that we obtain for x_0 and x_1 allow us to verify that

$$|x_0 - s| = |x_1 - s|. \tag{1.2.2}$$

The details of this calculation are outlined in Exercise 76. From equation (1.2.2) we deduce that P_0 and P_1 are an equal distance from the vertical line $x = s$. It follows that $x = s$ is an axis of symmetry of the parabola. ■

DEFINITION The intersection of the parabola with its axis of symmetry is called the *vertex* of the parabola.

If A is positive, then the parabola $y = Ax^2 + Bx + C$ opens upward. In this case, the *minimum* value of y is the ordinate of the vertex. It occurs when $x = -B/(2A)$ (see Figure 13a). If A is negative, then the parabola opens downward. In this case, the *maximum* value of y is the ordinate of the vertex. It occurs when $x = -B/(2A)$ (see Figure 13b).

▶ **EXAMPLE 9** What are the axis of symmetry and vertex of the parabola C that is the graph of the equation $y = 3x^2 + 18x + 16$?

Solution From Example 5, we have

$$\begin{aligned} y &= 3x^2 + 18x + 16 \\ &= 3\left(x + \frac{6}{2}\right)^2 - 11 \\ &= 3(x-(-3))^2 - 11. \end{aligned}$$

We conclude that $x = -3$ is the axis of symmetry of C. If we set $x = -3$ in the equation that defines C, then we find $y = -11$. It follows that the point $(-3, -11)$ is the vertex of C. The graph of C is given in Figure 14. ◀

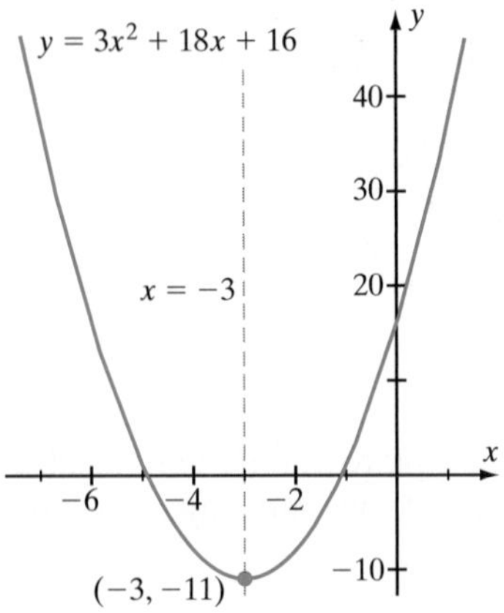

▲ Figure 14

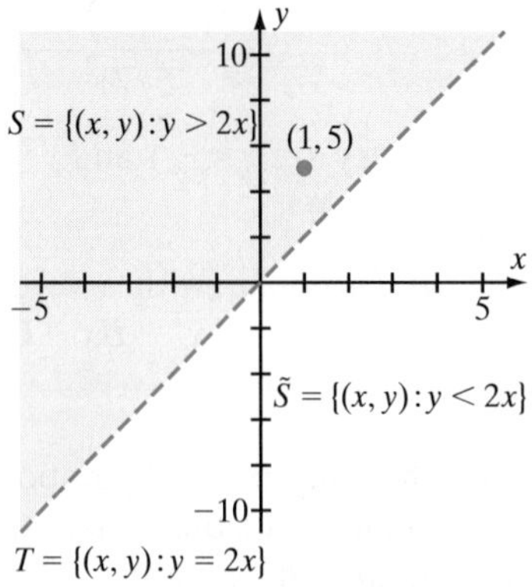

▲ Figure 15

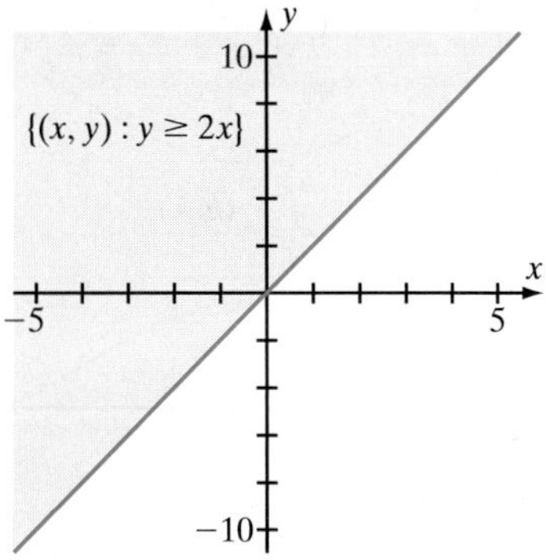

▲ Figure 16

Regions in the Plane

It is sometimes useful to graph *regions* in the plane. These regions are usually described by inequalities rather than equalities.

▶ **EXAMPLE 10** Sketch $S = \{(x, y) : y > 2x\}$.

Solution We first use a dotted line to sketch $T = \{(x, y) : y = 2x\}$ to mark the boundary of our region. The set T divides the plane into two regions: $\tilde{S} = \{(x, y) : y < 2x\}$ and $S = \{(x, y) : y > 2x\}$. We pick an arbitrary test point $(1, 5)$ from the upper region. Because $(1, 5)$ is in S and in the upper region in Figure 14, the upper region must be S. ◀

INSIGHT In Figure 15, we use a dotted line for the boundary of region S. Doing so shows that the boundary is not included in the region. When the boundary is included, we use a solid line. The region $\{(x, y) : y \geq 2x\}$, for example, includes its boundary (see Figure 16).

▶ **EXAMPLE 11** Sketch $G = \{(x, y) : |y| > 2 \text{ and } |x| \leq 5\}$.

Solution The sets $\{(x, y) : |y| > 2\}$ and $\{(x, y) : |x| \leq 5\}$ are illustrated in Figures 17a and 17b. The set G consists of the points common to these two sets (see Figure 17c). ◀

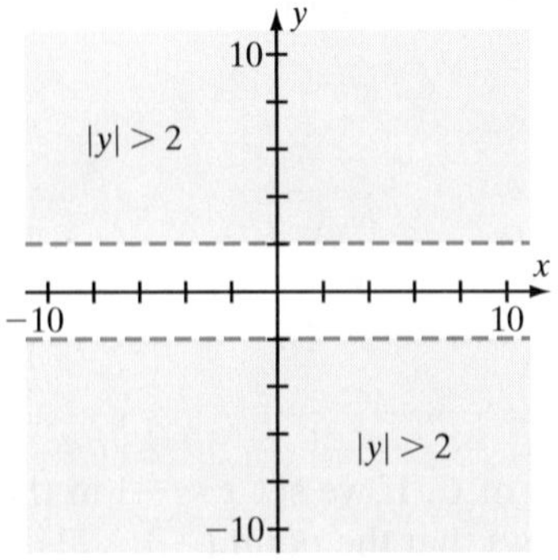

▲ Figure 17a $\{(x, y) : |y| > 2\}$

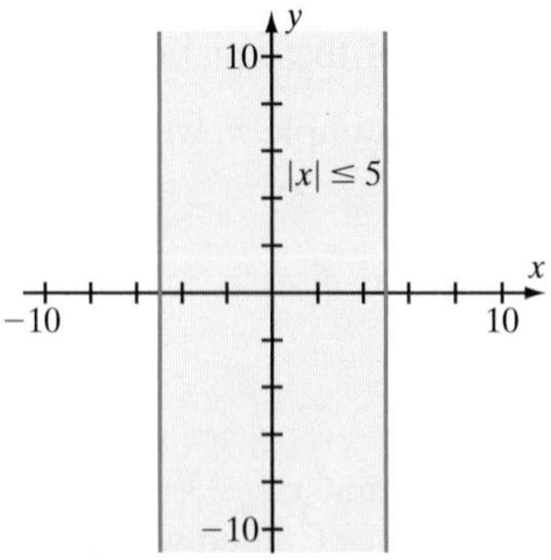

▲ Figure 17b $\{(x, y) : |x| \leq 5\}$

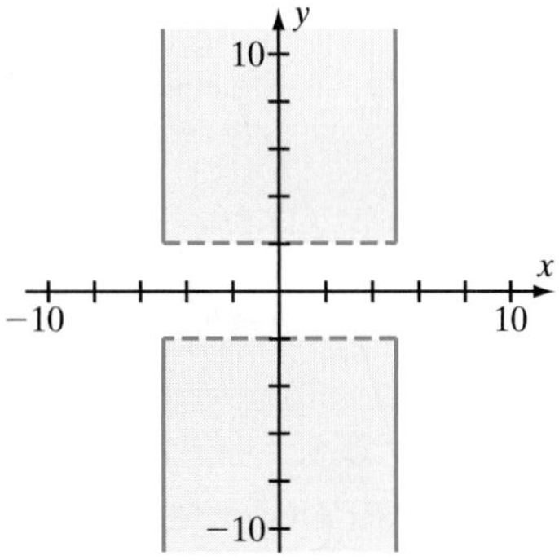

▲ Figure 17c $|y| > 2$

A rectangular region that includes its sides is said to be *closed*. A closed rectangle with sides that are parallel to the coordinate axes has the form

$$\{(x,y) : a \le x \le b \quad \text{and} \quad c \le y \le d\}$$

for some constants a, b, c, and d (see Figure 18). We use the notation $[a,b]\times[c,d]$ for this rectangle.

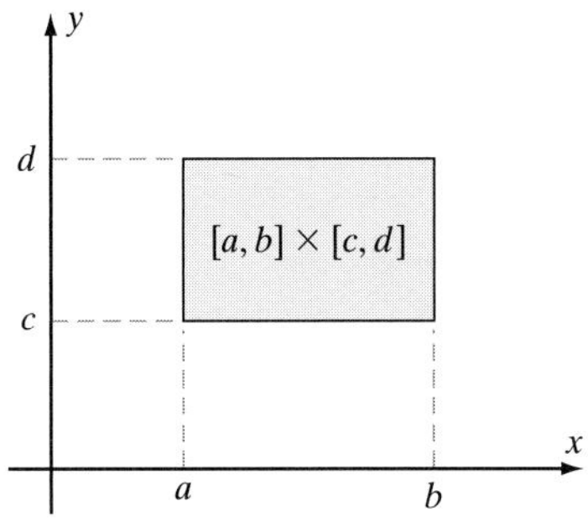

▲ Figure 18 The rectangle $[a,b]\times[c,d] = \{(x,y) : a \le x \le b$ and $c \le y \le d\}$.

Every time we create a plot (whether by hand, graphing calculator, or computer software), we select a rectangle $[a,b]\times[c,d]$ and graph only those portions of the curve that lie within it. The rectangle we choose is called the *viewing window* or the *viewing rectangle*. Sometimes, as in the case of circles and ellipses, we can find a viewing window that captures the entire curve. For curves such as parabolas and hyperbolas, however, the viewing window we choose will show only part of the curve (because such a curve is unbounded). Finding appropriate viewing windows for complicated equations can be tricky. Chapter 4 discusses how to use the ideas of calculus to help in this task.

QUICK QUIZ

1. What is the distance between the points $(1,-2)$ and $(-1,-3)$?
2. Write the equation of a circle passing through the origin and with center 3 units above the origin.
3. Use the method of completing the square to express $x^2 - 12x$ as the difference of squares.
4. Match the descriptions *point*, *circle*, *parabola*, *hyperbola*, and *ellipse* to the Cartesian equations $2x^2 - 2x - 4y^2 - 3y = 25/2$, $2x^2 - 2x + 4y^2 + 3y - 1/2 = 0$, $3x^2 - 3x + 3y^2 + 4y + 2 = 0$, $3x^2 - 3x + 3y^2 + 4y + 25/12 = 0$, and $x^2 + 4x + y = 3$.
5. What is the Cartesian equation of the parabola that opens downward and has its vertex at $(-2,-5)$?

Answers

1. $\sqrt{5}$ 2. $x^2 + (y-3)^2 = 3^2$ 3. $(x-6)^2 - 6^2$ 4. hyperbola, ellipse, circle, point, parabola 5. $y = -5 - (x+2)^2$

EXERCISES

Problems for Practice

1. Plot the points $(2,-4)$, $(7,3)$, $(\sqrt{3},\sqrt{6})$, $(\pi+3, \pi-3)$, $(\pi^2, \sqrt{\pi})$, $(-8/3, -2/\pi)$ on a set of coordinate axes.

2. Graph the set of points that satisfies $y + x = 3$.

3. Graph the set of points that satisfies $y - x = 2$.

4. Which of the points $(2,1)$, $(3,0)$, $(4,-1)$, or $(1/2, 9/2)$ is farthest from the origin? Which is nearest to the origin? Which is farthest from $(-5,6)$? Which is nearest to $(10,7)$?

5. Let $A = (2,3)$, $B = (-4,7)$, and $C = (-5,-6)$. Calculate the distance of each of these points to each of the others.

▶ In each of Exercises 6–15, the Cartesian equation of a circle is given. Sketch the circle and specify its center and radius. ◀

6. $(x+8)^2 + (y-1)^2 = 16$

7. $(x-1)^2 + (y-3)^2 = 9$

8. $x^2 + (y+7)^2 = 1$

9. $x^2 + (y+5)^2 = 2$

10. $(x-3)^2 + y^2 + y = 1$

11. $x^2 + y^2 - y = 0$

12. $x^2 + y^2 - 6x + 8y - 11 = 0$

13. $3x^2 - 6y + 12x + 3y^2 = 2$

14. $-x^2 - 6y = y^2 + x + 7$
15. $4x^2 + 4y^2 + 8y - 16x = 0$

In each of Exercises 16–20, a circle is described in words. Give its Cartesian equation.

16. The circle with center at the origin and radius 2
17. The circle with center $(-3, 5)$ and radius 6
18. The circle with center $(3, 0)$ and diameter 8
19. The circle with radius 5 and center $(-4, \pi)$
20. The circle with center $(0, -1/4)$ and radius $1/4$

In each of Exercises 21–26, the Cartesian equation of a parabola is given. Determine its vertex and axis of symmetry.

21. $y = x^2 - 3$
22. $y = 2(3 - x)^2 + 4$
23. $y = 2x - x^2$
24. $y = 3x^2 - 6x + 1$
25. $2x^2 + 12x + 2y + 9 = 0$
26. $x^2 - x - 3y = 1$

The *center* of an ellipse or hyperbola is the point of intersection of its axes of symmetry. In each of Exercises 27–34, state whether the graph of the given Cartesian equation is an ellipse or hyperbola. Determine its standard form and center.

27. $4x^2 + y^2 = 1$
28. $-(x + 5)^2 + y^2 = 9$
29. $x^2 + x + 9y^2 = 15/4$
30. $9x^2 - y^2 - y = 1/2$
31. $x^2 + 2x + 4y^2 + 24y = 12$
32. $x^2 - 10x - y^2 - 8y = 0$
33. $2x^2 - 3y^2 + 6y = 103$
34. $x^2/4 + x + y^2 + 6y = 6$

In each of Exercises 35–45, sketch the given region.

35. $\{(x, y) : |x| < 3\}$
36. $\{(x, y) : x^2 < y\}$
37. $\{(x, y) : |x| < 7,\ |y + 4| > 1\}$
38. $\{(x, y) : |x| \leq 5,\ |y| > 2\}$
39. $\{(x, y) : x^2 + y^2 > 16\}$
40. $\{(x, y) : x < 2y,\ x \geq y - 3\}$
41. $\{(x, y) : (x - 2)^2 + y^2 \geq 4\}$
42. $\{(x, y) : x + 5y \geq 4,\ x \leq 2,\ y \geq -8\}$
43. $\{(x, y) : |x - y| < 1,\ x \geq 4\}$
44. $\{(x, y) : x^2 + y^2 = 4,\ y > 0,\ x > 1\}$
45. $\{(x, y) : x^2 + y^2 < 9,\ x < 0,\ y > -1\}$

Further Theory and Practice

46. Find a point that is equidistant from the two points $(2, 3)$ and $(8, 10)$.
47. Find a point that is equidistant from the three points $(3, 4)$, $(6, 3)$, and $(-1, -4)$.
48. Show that there can be no point equidistant from $(1, 2)$, $(3, 4)$, $(8, 15)$, and $(6, -3)$.
49. Describe {all points} that are equidistant from $(1, 0)$, $(0, 1)$, and $(0, -1)$.

Which of the equations in Exercises 50–53 are circles? Which are not? Give precise reasons for your answers.

50. $5x^2 - 5y^2 + x + 2y = 5$
51. $5x^2 + 5y^2 = 2x - 3y + 6$
52. $x^2 - y^2 + 6x = -2y^2 - 7$
53. $x^2 + 2y^2 + x - 5y = 7 + 3x^2$

In each of Exercises 54–57, determine a value of k for which the graph of the given Cartesian equation is a point.

54. $x^2 + y^2 - 16x + k = 0$
55. $2x^2 + 18x + 2y^2 + k = 0$
56. $x^2 - 6x + y^2 + 2y + k = 1$
57. $k - x^2 - x - y^2 + y = 1$

58. For what values of k is the graph of $x^2 - 8x + y^2 + 2y = k$ empty?
59. An open vertical strip is a region of the form $\{(x, y): a < x < b\}$ where $a < b$. What values of a and b produce the widest open vertical strip that contains no point of the hyperbola $x^2 + 6x - 4y^2 = 7$?

In Exercises 60–69, sketch the set.

60. $\{(x, y) : |x| < 1,\ |y| < 1\}$
61. $\{(x, y) : |x| \cdot |y| < 1\}$
62. $\{(x, y) : |x| < 1 \text{ or } |y| < 1\}$
63. $\{(x, y) : |x + y| \leq 1\}$
64. $\{(x, y) : 0 < x \leq y < 2x \leq 1\}$
65. $\{(x, y) : 3x - 1 < 0 \leq y + 2x + 5\}$
66. $\{(x, y) : x^2 < y^2\}$
67. $\{(x, y) : x < y^2 < x^2\}$
68. $\{(x, y) : x < y < \sqrt{x}\}$
69. $\{(x, y) : y > 2x,\ x - 3y = 6\}$

70. Find the centers of the two circles of radius $\sqrt{65}$ that pass through the points $(0, -6)$ and $(3, -5)$.
71. From Example 1, we know that the graph of $y = 2x$ is a line. This line intersects each planar circle in zero, one, or two points. Give an algebraic reason that explains this geometric fact.
72. Find the equation of the circle that passes through the points $(-3, 4)$, $(1, 6)$, and $(9, 0)$.
73. Find the equation of the circle that passes through the points $(2, 4)$, $(4, 0)$, and $(-5, -3)$.
74. Suppose that $\alpha \neq 0$. The equation $x = \alpha y^2 + \beta y + \gamma$ is a parabola with horizontal axis of symmetry. What is the Cartesian equation of the axis of symmetry? What are the coordinates of the vertex?
75. Plot the locus of points P with an ordinate that is equal to the distance of the point P to the point $(0, 1)$.

76. Suppose that $A \neq 0$. Let $s = -\frac{B}{2A}$. Let $\mathcal{P}$ be the parabola whose equation is $y = Ax^2 + Bx + C$. Suppose that the horizontal line $y = y_0$ intersects $\mathcal{P}$ at two points $P_0 = (x_0, y_0)$ and $P_1 = (x_1, y_0)$ with $x_0 < x_1$. Show that

$$|x_0 - s| = |\gamma| = |x_1 - s|$$

where

$$\gamma = \frac{1}{2A}\sqrt{B^2 - 4A(C - y_0)}.$$

Calculator/Computer Exercises

In Exercises 77 and 78, plot the given pair of conic sections. Zoom in to approximate the points of intersection to three decimal places.

77. $y = x^2 - 2.2x$, $y = 2.3 - 1.1x - 1.7x^2$

78. $x^2 - \sqrt{5}x + \sqrt{3}y^2 - y = 1/\pi$, $x^2 - \sqrt{2}x - 2y^2 - \pi y = 0$

In Exercises 79 and 80, plot the given region.

79. $\{(x, y): 1 \le 1.3x^2 + 1.7x + y^2 - 0.9y \text{ and } x^2 + 0.5x + 2.1y^2 \le 1\}$

80. $\{(x, y): 1 \le x^2 - 2x - \sqrt{3}y^2 + 2\sqrt{3}y \text{ and } y \le \sqrt{5} + \sqrt{3}x - x^2\}$

81. Plot the parabola with Cartesian equation $y = 2 + x + x^2$ in the viewing window $[0, 2] \times [0, 8]$. Add the plots of $y = 2x$ and $y = 4x$ to the window. Intuition suggests that there is a value of m between 2 and 4 such that the graph of $y = mx$ has exactly one point of intersection with the parabola. Use a graphical method to approximate m to one decimal place. (In Chapter 3, calculus will enable you to determine m exactly.)

1.3 Lines and Their Slopes

As we will see in Chapter 3, calculus answers the geometric question, *What is the line that is tangent to a smooth curve at a given point?* In this section, we review the Cartesian equations that describe lines in the plane. A good understanding of these equations will be an important resource in the applications ahead.

Slopes

When a highway sign reads "Caution: 6% Grade," it tells us that, for every mile of horizontal motion, the road drops 0.06 mile. This information is important for truckers who are carrying heavy loads. In mathematical language, the road has *slope* -0.06. Now look at the line in Figure 1. We want to compute its slope, or the ratio of vertical motion to horizontal motion (*rise over run*). To do this, we choose two points on the line. The *rise* is the difference of y-coordinates, or $y_2 - y_1$. The *run* is the difference of x-coordinates, or $x_2 - x_1$. The *slope* is the ratio of these two quantities:

$$\text{Slope} = \frac{\text{rise}}{\text{run}} = \frac{\Delta y}{\Delta x} = \frac{y_2 - y_1}{x_2 - x_1}.$$

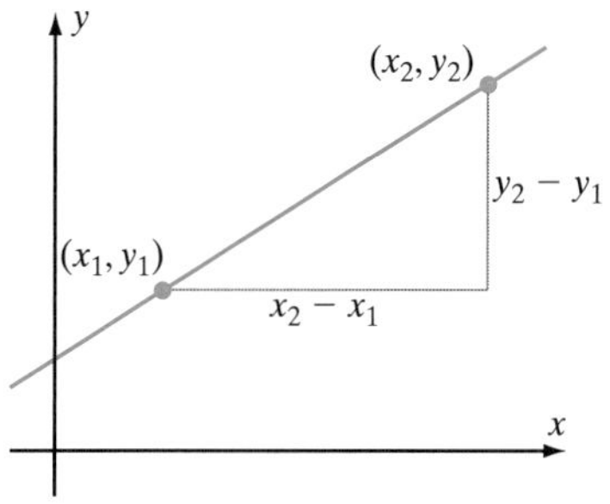

▲ Figure 1
$\text{Slope} = \frac{\text{rise}}{\text{run}} = \frac{\Delta y}{\Delta x} = \frac{y_2 - y_1}{x_2 - x_1}.$

INSIGHT We subtract the x-coordinates in the same order as the y-coordinates: $x_2 - x_1$ and $y_2 - y_1$. However, we *can* reverse the order in *both* the numerator and the denominator because doing so introduces two minus signs that cancel each other.

If we compute the slope of the line in Figure 1 using two other points $(\tilde{x}_1, \tilde{y}_1)$ and $(\tilde{x}_2, \tilde{y}_2)$, do we obtain the same slope? The two triangles ΔABC and $\Delta\tilde{A}\tilde{B}\tilde{C}$ are similar (see Figure 2), so the ratios of the corresponding sides are the same. In other words, rise over run is the same, and we *do* obtain the same slope if we compute using two other points.

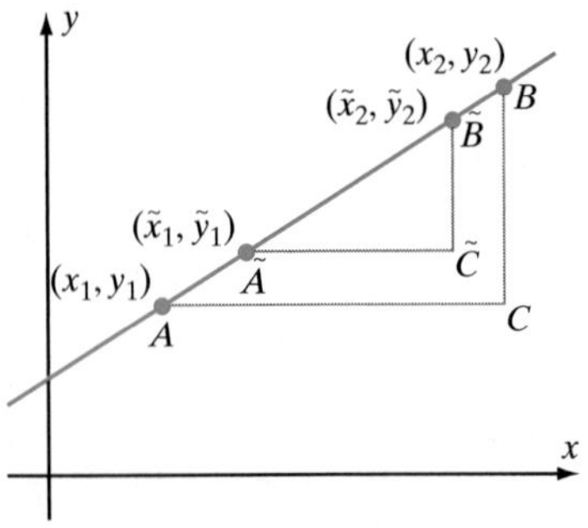

▲ Figure 2 ΔABC and $\Delta\tilde{A}\tilde{B}\tilde{C}$ are similar.

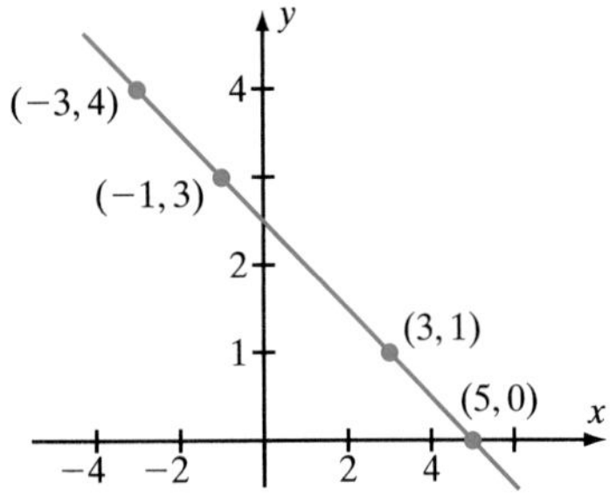

▲ Figure 3

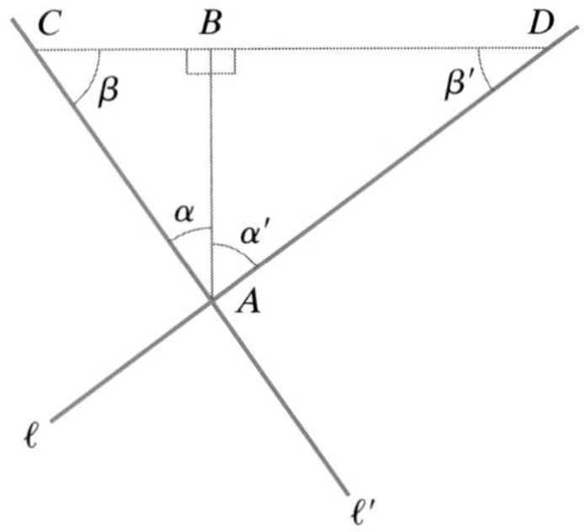

▲ Figure 4

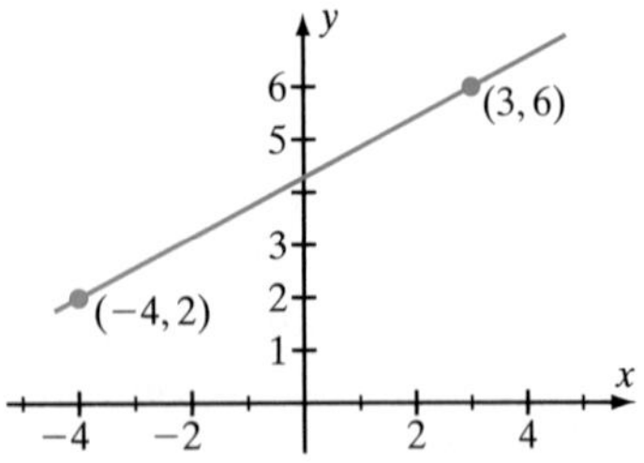

▲ Figure 5

▶ **EXAMPLE 1** Find the slope of the line in Figure 3.

Solution If we use the two points $(3,1)$ and $(-3,4)$, then

$$\text{Slope} = \frac{\text{rise}}{\text{run}} = \frac{4-1}{-3-3} = -\frac{1}{2}.$$

Had we used another pair of points, such as $(5,0)$ and $(-1,3)$, we would have obtained the same slope:

$$\text{Slope} = \frac{\text{rise}}{\text{run}} = \frac{3-0}{-1-5} = -\frac{1}{2}.$$

Even if we had used these points in reverse order, that is, $(-1,3)$ and $(5,0)$, we would have obtained the same slope:

$$\text{Slope} = \frac{\text{rise}}{\text{run}} = \frac{0-3}{5-(-1)} = -\frac{1}{2}. \quad ◀$$

A positive slope indicates that the line *rises* relative to left-to-right motion. A negative slope, as in Example 1, indicates that the line *falls* relative to left-to-right motion. The equation of a horizontal line is of the form $y = b$ for some constant b; the slope of such a line is zero because $\Delta y = 0$. A vertical line has an equation of the form $x = a$ for some constant a; the slope of such a line is *undefined* because $\Delta x = 0$ and division by 0 is not defined.

Slopes of Perpendicular Lines

How are the slopes of two perpendicular lines related? Look at the perpendicular lines ℓ and ℓ', shown in Figure 4. Neither line is vertical or horizontal, so they both have nonzero slope. From point C to point A, the run on line ℓ' is equal to $|\overline{BC}|$, and the rise is the negative of $|\overline{BA}|$. Thus $-\text{slope}\ \ell' = |\overline{BA}|/|\overline{BC}|$. Similarly, we see that slope $\ell = |\overline{BA}|/|\overline{BD}|$. Notice that $\alpha + \alpha' = 90°$ and $\beta' + \alpha' = 90°$. Subtracting these two equations gives $\alpha = \beta'$. The same reasoning also shows that $\beta = \alpha'$. Therefore ΔABC and ΔDBA are similar. Thus

$$\frac{|\overline{BA}|}{|\overline{BC}|} = \frac{|\overline{BD}|}{|\overline{BA}|} = \frac{1}{(|\overline{BA}|/|\overline{BD}|)},$$

or

$$-\text{slope}\ \ell' = \frac{1}{\text{slope}\ \ell}. \tag{1.3.1}$$

It is also true that if the slopes of two lines ℓ and ℓ' satisfy equation (1.3.1), then ℓ and ℓ' are mutually perpendicular. In conclusion:

> **THEOREM 2** Two lines, neither of which is horizontal or vertical, are mutually perpendicular if and only if their slopes are negative reciprocals.

▶ **EXAMPLE 2** Find the slope of any line perpendicular to the line shown in Figure 5.

Solution We start by finding the slope of the line in the figure: $(6-2)/(3-(-4)) = 4/7$. The slope of any perpendicular line is the negative reciprocal of this number, or $-7/4$. ◀

Slopes of Parallel Lines

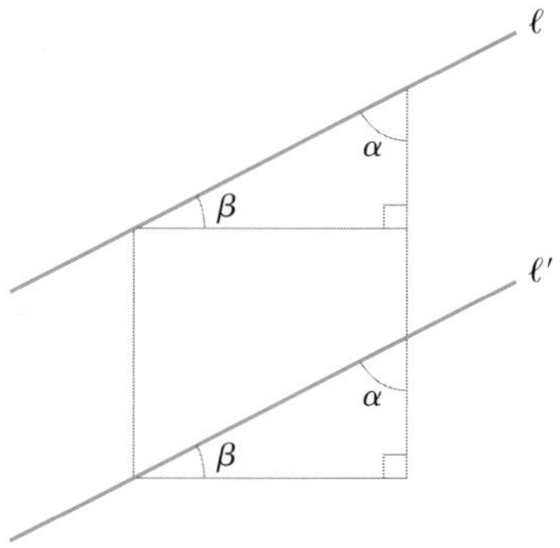

▲ Figure 6

If lines ℓ and ℓ' are parallel, then the two right triangles shown in Figure 6 are congruent. It follows that slope $\ell =$ slope ℓ'. The converse is also easily deduced:

THEOREM 3 Two lines are parallel if and only if they have the same slope.

▶ **EXAMPLE 3** Verify that the line through $(3, -7)$ and $(4, 6)$ is parallel to the line through $(2, 5)$ and $(4, 31)$.

Solution We compute that the slope of the first line is $(6 - (-7))/(4 - 3) = 13$, and the slope of the second line is $(31 - 5)/(4 - 2) = 13$. Because the slopes are equal, the lines are parallel. ◀

Equations of Lines

We need to understand how to write the equations that represent lines. We also want to recognize when a given equation represents a line. There are many geometrical ways to describe a line. For instance, a line is determined by two points through which it passes (see Figure 7a). A line is also determined by a point and a slope (see Figure 7b). No matter how the information for a line ℓ is presented, the basic rule for obtaining the Cartesian equation of ℓ is: *Express the slope of the line in two different ways and equate.*

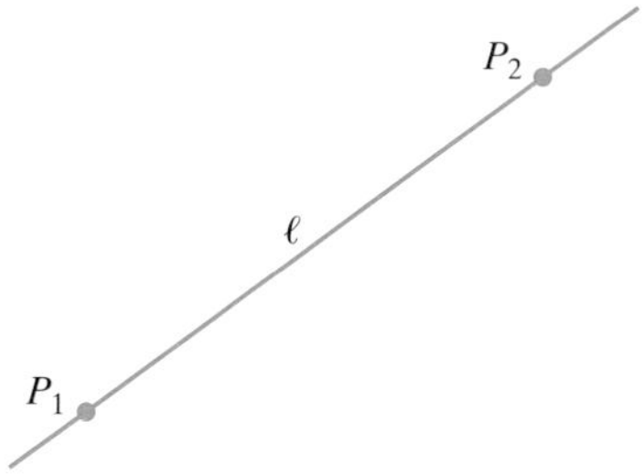

▲ Figure 7a Line ℓ is determined by P_1 and P_2.

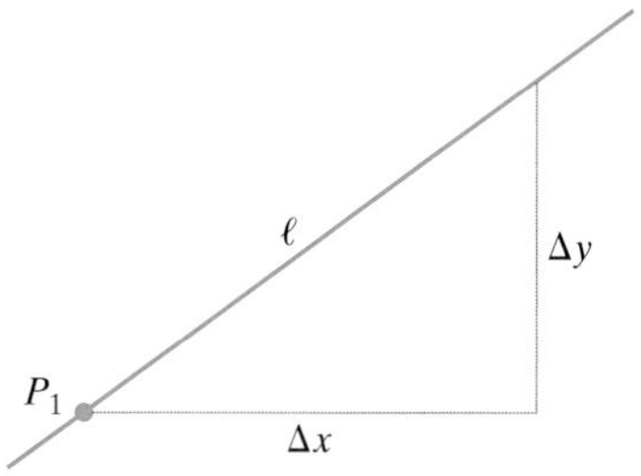

▲ Figure 7b Line ℓ is determined by P_1 and $m = \Delta y/\Delta x$.

The Point-Slope Form of a Line

Suppose line ℓ has given slope m and passes through a given point (x_0, y_0). Let (x, y) be a variable point on line ℓ. One expression for the slope of ℓ is

$$\text{Slope} = \frac{y - y_0}{x - x_0}.$$

The slope is also given to be m:

$$\text{Slope} = m.$$

Equating these two expressions for slope gives

$$\frac{y - y_0}{x - x_0} = m,$$

or $y - y_0 = m(x - x_0)$. By isolating y, we obtain the *point-slope form* for the line that has slope m and passes through point (x_0, y_0):

$$y = m(x - x_0) + y_0 \qquad (1.3.2)$$

▶ **EXAMPLE 4** Find the equation of the line ℓ passing through the point $(2, -4)$ and having slope 3.

Solution Let (x, y) be a variable point on the line. We can compute the slope using $(2, -4)$ and (x, y):

$$\text{Slope} = \frac{y-(-4)}{x-2} = \frac{y+4}{x-2}.$$

We also know that the slope is 3. Equating the two expressions for slope, we find that

$$\frac{y+4}{x-2} = 3,$$

or $y + 4 = 3(x - 2)$, or $y = 3(x - 2) - 4$. ◀

INSIGHT Once equation (1.3.2) has been mastered, there is no need to repeat the steps of Example 4 each time. Substituting the given data into equation (1.3.2) yields the answer immediately. Thus we obtain the solution to Example 4 by substituting $m = 3$, $x_0 = 2$, and $y_0 = -4$ into equation (1.3.2). Because equation (1.3.2) is used so frequently in Chapter 3, it is a good idea to commit it to memory.

The Two-Point Form of a Line

Suppose line ℓ passes through the two points (x_0, y_0) and (x_1, y_1). Let (x, y) be a variable point on line ℓ. One expression for slope is

$$\text{Slope} = \frac{y-y_0}{x-x_0}.$$

We may also use the two given points to calculate slope:

$$\text{Slope} = \frac{y_1-y_0}{x_1-x_0}.$$

Equating these two expressions for slope gives

$$\frac{y-y_0}{x-x_0} = \frac{y_1-y_0}{x_1-x_0},$$

or

$$y - y_0 = \left(\frac{y_1-y_0}{x_1-x_0}\right)(x - x_0).$$

By isolating y, we obtain the *two-point form* for the line that passes through the points (x_0, y_0) and (x_1, y_1):

$$y = \left(\frac{y_1-y_0}{x_1-x_0}\right)(x - x_0) + y_0 \qquad (1.3.3)$$

INSIGHT The two-point and point-slope forms of a line are quite similar. Both have the form $y = \text{slope} \cdot (x - x_0) + y_0$. The difference between them is that, in the two-point form, the given data must be used to calculate the slope. By contrast, the slope m appears explicitly as the coefficient of $x - x_0$ in the point-slope form.

▶ **EXAMPLE 5** Write the equation of the line passing through points $(2,1)$ and $(-4,3)$.

Solution Using $(x_0, y_0) = (2,1)$ and $(x_1, y_1) = (-4,3)$ in formula (1.3.3), we find that the equation for the line is

$$y = \left(\frac{3-1}{-4-2}\right)(x-2) + 1$$
$$= -\frac{1}{3}(x-2) + 1.$$

In a slightly different approach, we might have used the points $(2,1)$ and $(-4,3)$ to compute the slope m and then substuituted into equation (1.3.2). Thus

$$m = \frac{3-1}{-4-2} = \frac{2}{-6} = -\frac{1}{3}$$

and, by (1.3.2),

$$y = -\frac{1}{3}(x-2) + 1.$$

We may also use the other point, $(-4,3)$, as (x_0, y_0) in equation (1.3.3). If we do so, then we obtain

$$y = -\frac{1}{3}\bigl(x - (-4)\bigr) + 3.$$

This last equation does not look exactly like our first answer, but with a little manipulation we see that it is equivalent:

$$y = -\frac{1}{3}\bigl(x - (-4)\bigr) + 3 = -\frac{1}{3}(x - 2 + 6) + 3$$
$$= -\frac{1}{3}(x-2) - \frac{1}{3} \cdot 6 + 3 = -\frac{1}{3}(x-2) + 1. \quad ◀$$

The Slope-Intercept Form of a Line

A point at which the graph of a Cartesian equation intersects a coordinate axis often has special significance. We introduce some terms associated with such a point.

DEFINITION The x-intercept of a line is the x-coordinate of the point where the line intersects the x-axis (provided such a point exists). In general, we say that a is an x-intercept of a curve if the point $(a,0)$ is on the curve. In other words, a is an x-intercept of a curve if the curve intersects the x-axis at $(a,0)$. The y-intercept of a line is the y-coordinate of the point where the line intersects the y-axis (provided such a point exists). In general, we say that b is a y intercept of a curve if the point $(0,b)$ is on the curve. In other words, b is a y-intercept of a curve if the curve intersects the y-axis at $(0,b)$.

The line with slope m and y-intercept b has equation

$$y = mx + b, \tag{1.3.4}$$

as we see by setting $x_0 = 0$ and $y_0 = b$ in equation (1.3.2). Equation (1.3.4) is the *slope-intercept* form of the line.

INSIGHT Every nonvertical line has both a slope m and a unique y-intercept b. Therefore every nonvertical line has a slope-intercept form. By contrast, the slope of a vertical line is not defined. Therefore a vertical line does *not* have a slope-intercept form.

▶ **EXAMPLE 6** What is the y-intercept of the line that passes through the points $(-1,-4)$ and $(4,6)$?

Solution The slope is $(6-(-4))/(4-(-1))$, or 2. Therefore a point-slope form of the line is $y = 2(x-4)+6$. Expanding the right side yields the slope-intercept form: $y = 2x - 8 + 6$, or $y = 2x + (-2)$. The y-intercept is -2. ◀

The General Equation of a Line

If a line is not vertical, then it has a slope-intercept equation $y = mx + b$. We may rewrite this equation in the form $-mx + y = b$, or $Ax + By = D$ where $A = -m$, $B = 1$, and $D = b$. The vertical line $x = a$ may also be written in the form $Ax + By = D$ by setting $A = 1$, $B = 0$, and $D = a$. Thus *every* line can be described by a Cartesian equation of the form

$$Ax + By = D \tag{1.3.5}$$

where A and B are not both zero. Conversely, if A and B are not both zero, then equation (1.3.5) describes a line. We say that equation (1.3.5) is the *general equation* of a line (or the *general linear equation*).

▶ **EXAMPLE 7** What is the slope of the line $4x - 2y = 7$?

Solution We rewrite the general linear equation in slope-intercept form:

$$-2y = -4x + 7, \quad \text{or}$$
$$y = 2x - \frac{7}{2}.$$

This equation tells us that the line has slope 2. As additional information, we see that the y-intercept is $-7/2$. ◀

▶ **EXAMPLE 8** Suppose that A and B are not both zero. Let D_1 and D_2 be any two constants. Show that the lines defined by $Ax + By = D_1$ and $Ax + By = D_2$ are parallel.

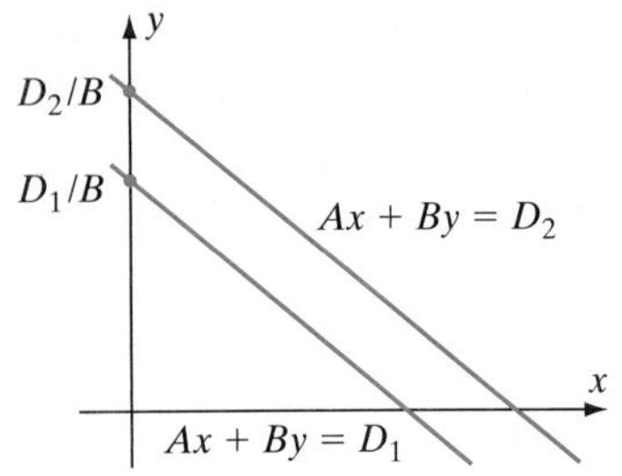

▲ Figure 8

Solution If $B = 0$, then both lines are vertical, hence parallel. Otherwise, the given equations may be rewritten as

$$y = -\frac{A}{B}x + \frac{D_1}{B} \quad \text{and} \quad y = -\frac{A}{B}x + \frac{D_2}{B}.$$

From these equations, which are in slope-intercept form, we can see that the two lines have common slope $-A/B$. They are therefore parallel (see Figure 8). ◀

The Intercept Form of a Line

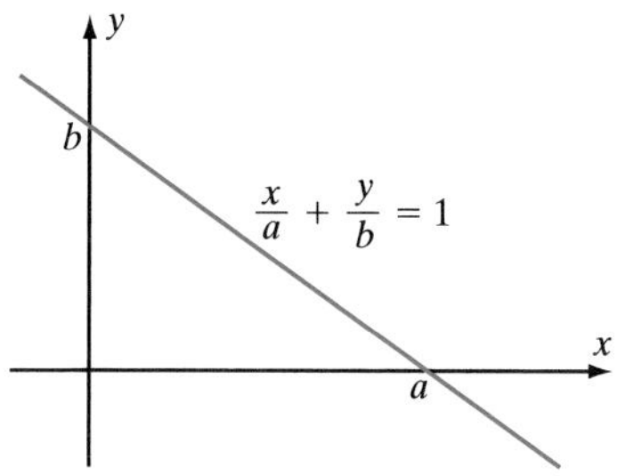

▲ Figure 9

Notice that if a and b are nonzero, then the equation

$$\frac{x}{a} + \frac{y}{b} = 1 \tag{1.3.6}$$

is a particular form of the general equation of a line. Therefore the graph of equation (1.3.6) is a line. Furthermore, the two points $(a,0)$ and $(0,b)$ satisfy equation (1.3.6). It follows that equation (1.3.6) describes the line ℓ with nonzero x- and y-intercepts a and b, respectively. It is the *intercept* form of ℓ (see Figure 9).

▶ **EXAMPLE 9** Find the intercept equation of the line with x-intercept 3 and y-intercept -5. State the slope-intercept form as well.

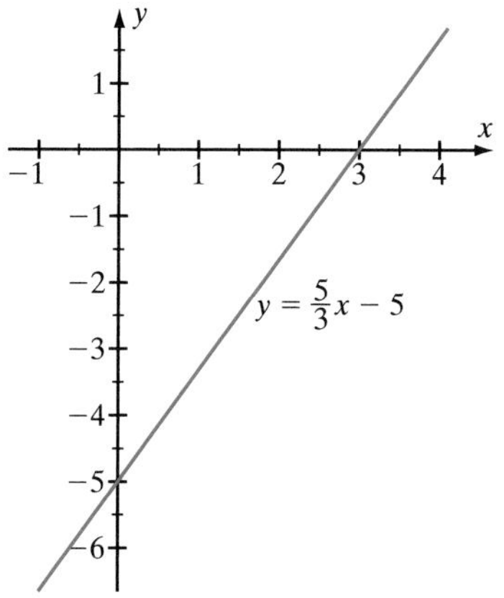

▲ Figure 10

Solution By (1.3.6), the intercept equation is

$$\frac{x}{3} + \frac{y}{-5} = 1.$$

This equation is equivalent to

$$\frac{y}{-5} = -\frac{x}{3} + 1,$$

or

$$y = \frac{5}{3}x - 5,$$

which is the slope-intercept form (see Figure 10). ◀

▶ **EXAMPLE 10** What is the intercept form of the line defined by the equation $2x-3y+6=0$?

Solution First, we rewrite the equation as $2x-3y=-6$. Dividing both sides of the equation by -6 yields the intercept form equation:

$$\frac{x}{-3} + \frac{y}{2} = 1. \quad ◀$$

From this equation, we see at a glance that the line intersects the x-axis at $(-3,0)$ and the y-axis at $(0,2)$. An alternative method is to substitute $y=0$ into the given equation to obtain the x-intercept $a=-3$. Substituting $x=0$ into the given equation leads to the y-intercept $b=2$. We obtain the intercept form of the equation, namely $x/(-3)+y/2=1$, by substituting these values into equation (1.3.6).

Normal Equation of a Line

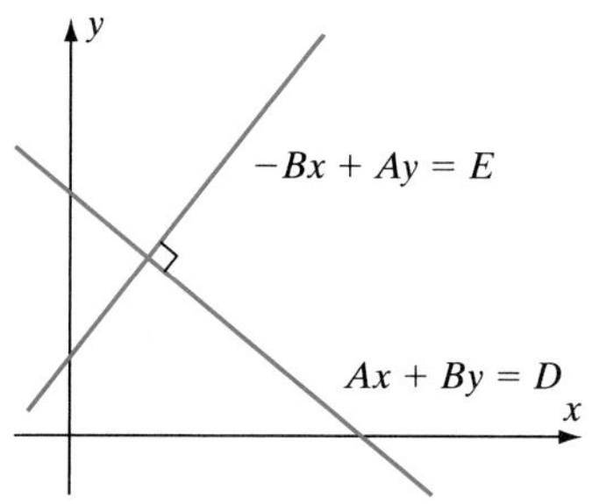

▲ Figure 11

Suppose that A and B are not both zero. By Example 8, the lines described by $Ax+By=D$ for different values of D are parallel. It may be shown (Exercise 65) that the line ℓ' described by

$$-Bx + Ay = E \tag{1.3.7}$$

is, for any constant E, perpendicular (or *normal*) to the line ℓ defined by $Ax+By=D$ (see Figure 11).

▶ **EXAMPLE 11** Write the slope-intercept equation of the line ℓ' that is perpendicular to the line $2x+3y+4=0$ and that passes through point $P=(2,6)$. At what point Q does ℓ' intersect the given line?

Solution A line perpendicular to $2x+3y=-4$ has the form $-3x+2y=E$ for some constant E. Because the point $(2,6)$ is on ℓ', we have $-3(2)+2(6)=E$, or $E=6$. Therefore the equation of ℓ' can be written as $-3x+2y=6$. We isolate y to obtain the slope-intercept equation: $y=(3/2)x+3$.

To find the point of intersection of the two lines, we solve their Cartesian equations simultaneously. By substituting $y=(3/2)x+3$ into the equation of the given line, we obtain $2x+3((3/2)x+3)+4=0$, or $13x/2+13=0$, or $x=-2$. If we substitute this value of x into $2x+3y=-4$, then we obtain $y=0$. Thus $Q=(-2,0)$ is the point of intersection (see Figure 12). ◀

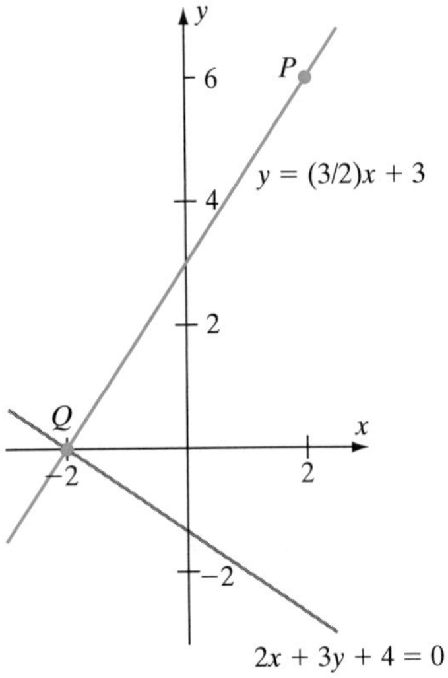

▲ Figure 12

Least Squares Lines

Table 1 displays total annual U.S. income and consumption in billions of dollars for the years 2002–2008.

	2002	**2003**	**2004**	**2005**	**2006**	**2007**	**2008**
Income (x)	35528	36654	38809	41079	43976	46653	48396
Consumption (y)	7351	7704	8196	8694	9207	9710	10059

Table 1

Figure 13 shows a plot of the seven data points (x,y). (Economists refer to the data of Table 1 as *time-series data.* Figure 13 is called a *time-series graph.*) It is apparent that the plotted points lie close to a straight line. The question arises: What equation $y=mx+b$ should we use to describe the approximate relationship between the observed values of x and y? This problem, finding a line that best fits observed data, occurs in many fields. The procedure for a solution that is used most often is called the *method of least squares.*

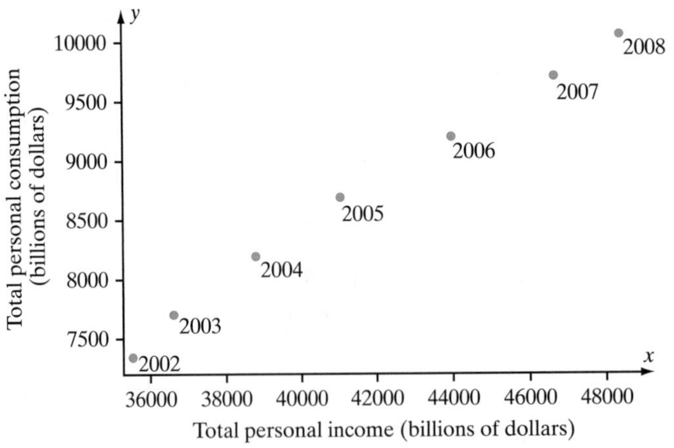

▲ Figure 13

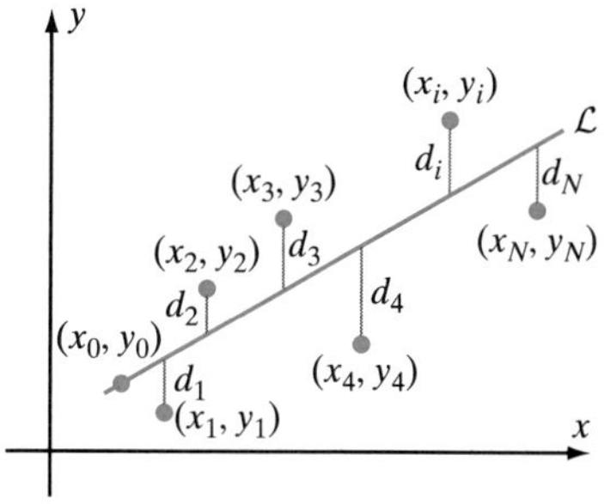

▲ Figure 14

Suppose $N+1$ data points $(x_0, y_0), (x_1, y_1), \ldots, (x_N, y_N)$ lie in the vicinity of a line $\mathcal{L}$. Let $d_i \geq 0$ denote the vertical distance of the data point (x_i, y_i) to $\mathcal{L}$ (as in Figure 14). If (x_i, y_i) is not on the line, then d_i may be regarded as an error. The sum $d_0^2 + d_1^2 + d_2^2 + \cdots + d_N^2$ of these squared errors is an overall measure of how well $\mathcal{L}$ approximates the observed data. The *least squares line* (or the *regression line*) is the line for which the sum $d_1^2 + d_2^2 + \cdots + d_N^2$ is minimized. In Chapter 13, we will learn the Method of Least Squares in full generality. For now, we assume that the line we seek passes through a specified point.

THEOREM 4 Given the $N+1$ data points $(x_0, y_0), (x_1, y_1), \ldots, (x_N, y_N)$, the least squares line through (x_0, y_0) is given by $y = m(x - x_0) + y_0$ where

$$m = \frac{(x_1 - x_0) \cdot (y_1 - y_0) + (x_2 - x_0) \cdot (y_2 - y_0) + \cdots + (x_N - x_0) \cdot (y_N - y_0)}{(x_1 - x_0)^2 + (x_2 - x_0)^2 + \cdots + (x_N - x_0)^2}. \tag{1.3.8}$$

Proof. For any real number m, the equation of line $\mathcal{L}$ through point (x_0, y_0) with slope m is $y = m(x - x_0) + y_0$. The vertical distance of the observation (x_i, y_i) to $\mathcal{L}$ is

$$d_i = |y_i - (m(x_i - x_0) + y_0)| = |(y_i - y_0) - m(x_i - x_0)|.$$

We therefore seek the value m for which the sum

$$S = \big((y_1 - y_0) - m(x_1 - x_0)\big)^2 + \cdots + \big((y_N - y_0) - m(x_N - x_0)\big)^2$$

is minimized. By expanding each square and collecting like powers of m, we find that

$$S = Am^2 + Bm + C$$

where

$$A = (x_1 - x_0)^2 + (x_2 - x_0)^2 + \cdots + (x_N - x_0)^2$$

and

$$B = -2\big((x_1 - x_0)(y_1 - y_0) + (x_2 - x_0)(y_2 - y_0) + \cdots + (x_N - x_0)(y_N - y_0)\big).$$

We do not calculate C because its value is not needed. Because A is positive, we may use Theorem 1 from Section 1.2 to conclude that the quantity $S = Am^2 + Bm + C$ has a minimum value when $m = -B/2A$. Formula (1.3.8) may now be obtained by substituting the expressions for A and B into $-B/2A$ and simplifying. ■

INSIGHT Because d_i measures the amount by which the approximating line misses the ith observation, the expression $d_1 + \cdots + d_n$ represents the total absolute error. This sum may seem to be a simpler measure of error than the sum of the squares of the errors, but, by using $d_1^2 + \cdots + d_n^2$, we can reduce the problem of minimizing the error to a problem that we have already solved—finding the vertex of a parabola.

▶ **EXAMPLE 12** Refer to Table 1 and Figure 13. Use the income-consumption data for 2002–2008 to pass a least squares line through the point corresponding to 2002. In 1995, total personal income equaled \$24609 billion. What consumption does the regression line predict for that year? Given that consumption in 1995 was \$4976 billion, what is the relative error of the regression line prediction?

Solution We set $x_0 = 35528$ (income for 2002) and $y_0 = 7351$ (consumption for 2002). Let $(x_1, y_1) = (36654, 7704)$ represent the income and consumption for 2003. Continue by letting $(x_2, y_2) = (38809, 8196)$ represent the data for 2004, and so on.

Table 2 lists all computations needed for the slope of the required least squares line.

	i	x_i	y_i	$x_i - x_0$	$y_i - y_0$	$(x_i - x_0)^2$	$(x_i - x_0)\cdot(y_i - y_0)$
	1	36654	7704	1126	353	1267876	397478
	2	38809	8196	3281	845	10764961	2772 445
	3	41079	8694	5551	1343	30813601	7454 993
	4	43976	9207	8448	1856	71368704	15679488
	5	46653	9710	11125	2359	123765625	26243875
	6	48396	10059	12868	2708	165585424	34846544
Total						403566191	87394823

Table 2

If we set $m = 87394823/403566191 \approx 0.21656$, then the equation for the least squares line is

$$y = 0.21656(x - 35528) + 7351.$$

Figure 15 illustrates this regression line. The predicted value of y corresponding to $x = 24609$ is

$$0.21656(24609 - 4986) + 7351 = 4986.$$

The relative error is $(100 \times |4976 - 4986|/4976)\%$, or about 0.2%. ◀

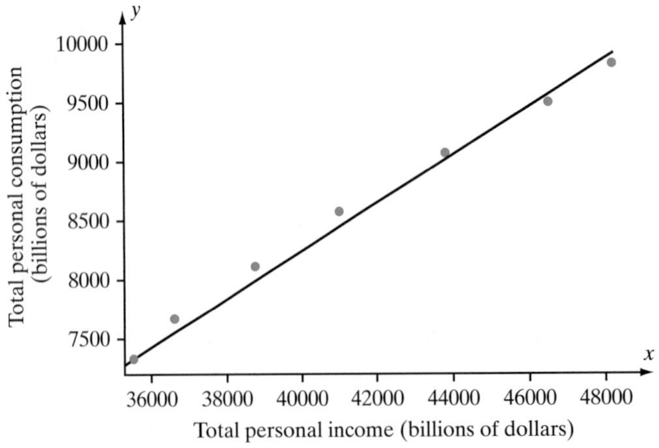

▲ Figure 15

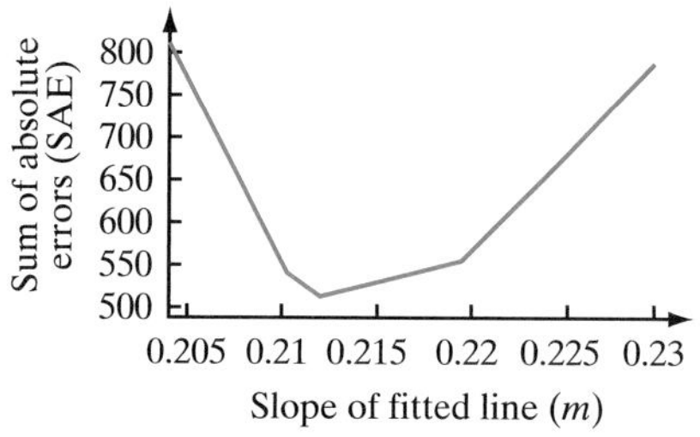

Figure 16a

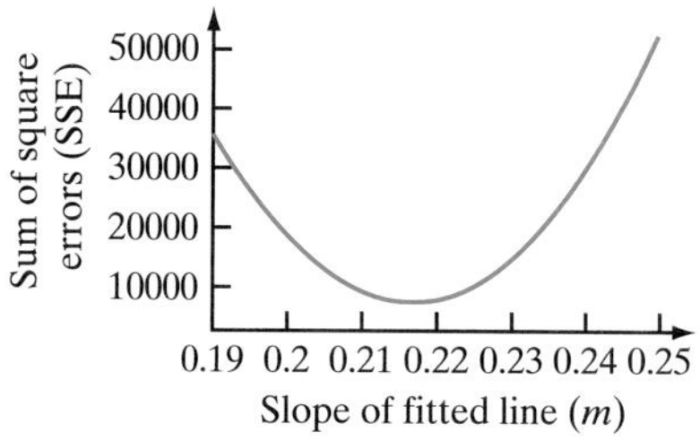

Figure 16b

INSIGHT For each m, we can calculate the sum SAE of the absolute errors: $SAE = d_1 + d_2 + \cdots + d_6$. The graph of the points (m, SAE) appears in Figure 16a. Observe that when we use the sum of the absolute errors, we obtain a jagged curve. By contrast, the sum of the squared errors, $SSE = d_1^2 + d_2^2 + \cdots + d_6^2$, results in a smooth graph, which is a parabola (as we learned in the proof of Theorem 3 and as is shown in Figure 16b). In Chapter 4, we will learn how calculus provides a powerful tool for locating the lowest point on the graph of *any* smooth curve.

QUICK QUIZ

1. What is the slope of the line through the points $(-1,-2)$ and $(3,8)$?
2. What is the slope of the line described by the Cartesian equation $3x + 4y = 7$?
3. What is the slope of the line that is perpendicular to the line described by the Cartesian equation $5x - 2y = 7$?
4. What are the intercepts of the line described by the equation $5x - y/2 = 1$?
5. What is the equation of the least squares line through the origin for the points $(1,2)$, $(2,5)$, and $(3,10)$?

Answers

1. 5/2 2. −3/4 3. −2/5 4. x-intercept: 1/5, y-intercept: −2 5. $y = 3x$

EXERCISES

Problems for Practice

1. Find the slopes of each line in Figure 17.

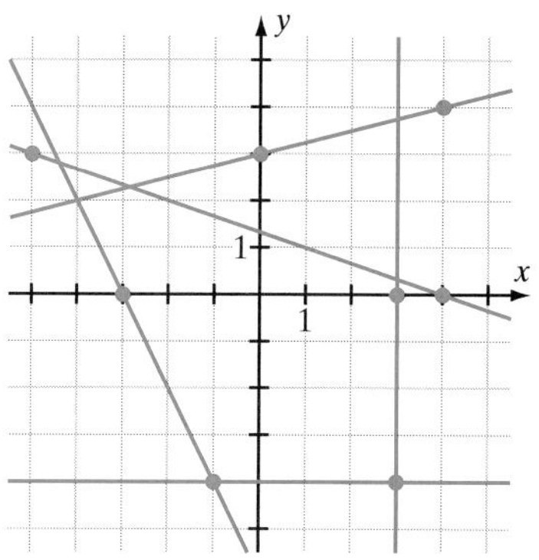

Figure 17

2. Sketch, on the same set of axes, lines passing through point $(1,3)$ and having slopes −3, −2, −1, 0, 1, 2, and 3.

3. Sketch, on the same set of axes, lines having slope −3 and passing through points $(-2,-7)$, $(-1,0)$, $(3,0)$, and $(5,1)$.

4. Sketch the line that passes through point $(-2,5)$ and that rises 7 units for every 2 units of left-to-right motion.

In Exercises 5–8, write the point−slope equation of the line determined by the given data.

5. Slope 5, point $(-3,7)$

6. Slope −2, point $(4.1, 8.2)$

7. Slope −1, point $(-\sqrt{2}, 0)$

8. Slope 0, point $(-\pi, \pi)$

In Exercises 9–12, write the point-slope equation of the line determined by the two given points.

9. $(2,7)$, $(6,-4)$

10. $(12,1)$, $(-4,-4)$

11. $(-7,-4)$, $(9,0)$
12. $(1,-5)$, $(-5,1)$

In Exercises 13–20 write the slope-intercept equation of the line determined by the given data.

13. Slope -4, y-intercept 9
14. Slope π, y-intercept π^2
15. Slope $\sqrt{2}$, y-intercept $-\sqrt{3}$
16. Slope 0, y-intercept 3
17. Slope 3, x-intercept -4
18. Slope -5, x-intercept 5
19. x-intercept -2, y-intercept 6
20. x-intercept -5, y-intercept -12

In Exercises 21–24, write the slope-intercept equation of the line that passes through the two given points.

21. $(2,7)$, $(3,10)$
22. $(1/2,1)$, $(2,7)$
23. $(-1,3)$, $(2,-9)$
24. $(2\pi, -\pi)$, $(0,0)$

In Exercises 25–28, write the intercept form of the equation of the line determined by the given data.

25. x-intercept -2, y-intercept 6
26. x-intercept 5, y-intercept 1/3
27. Slope 3, x-intercept -1
28. Slope -2, y-intercept 5

In Exercises 29–32, write the slope-intercept equation of the line that passes through the given point and that is parallel to the given line.

29. $(1,-2)$, $y=-x/2$
30. $(3,4)$, $x/3+y/2=1$
31. $(2,1)$, $6x-3y-7=0$
32. $(4,0)$, $x+2y-8=0$

In Exercises 33–36, write the slope-intercept equation of the line that passes through the given point and that is perpendicular to the given line.

33. $(1,-2)$, $4x+8y=5$
34. $(0,-5)$, $y=3x+5$
35. $(3,4)$, $x/3+y/2=1$
36. $(-3,0)$, $x=1-6y$

In Exercises 37–40, a general linear equation of a line is given. Find the x-intercept, the y-intercept, and the slope of the line.

37. $3x+4y=2$
38. $x/2+2y=4$
39. $x-3y=3$
40. $5x-2y=10$

In Exercises 41–48, sketch the line whose Cartesian equation is given.

41. $x/2=1$
42. $3y=-7$
43. $y=3(x+1)-2$
44. $y=(x-2)/2+4$
45. $y-2x=4$
46. $x/3+y/9=1$
47. $2x+3y=6$
48. $-x/2-2y=1$

49. Is the line through points $(1,2)$ and $(4,7)$ parallel to the line through points $(-4,-3)$ and $(-7,6)$? Is it perpendicular to the line through points $(7,9)$ and $(5,-7)$?
50. Is the line through points $(-3,8)$ and $(2,23)$ parallel to the line through points $(-4,6)$ and $(3,28)$? Is it perpendicular to the line through points $(-2,4)$ and $(7,1)$?
51. Write the equation of the line perpendicular to the line in Figure 18 and passing through point $(7,-8)$.

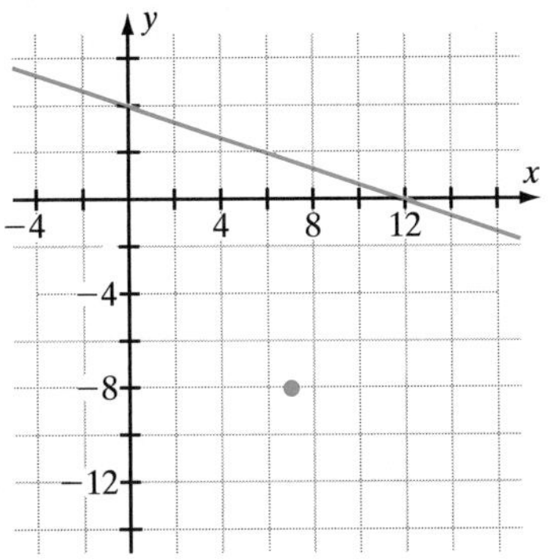

▲ Figure 18

52. What are the slopes of the lines in Figure 19?

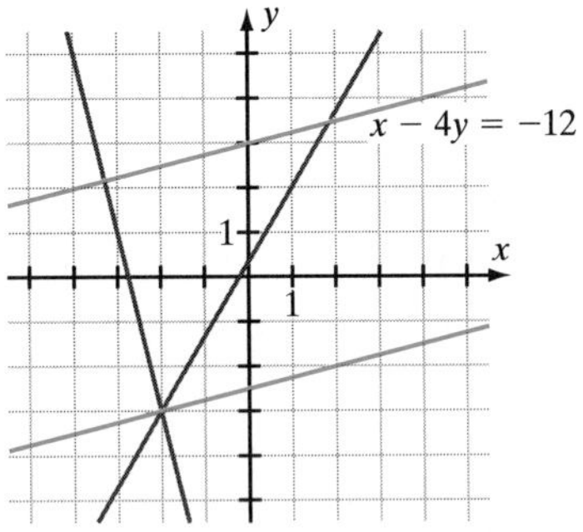

▲ Figure 19

Further Theory and Practice

53. Find a point (a,b) so that the line through (a,b) and $(5,-7)$ has slope 2.
54. Find a point (a,b) so that the line through (a,b) and $(4,2)$ is parallel to the line through $(3,6)$ and $(2,9)$.

55. Find a point (a, b) so that the line through (a, b) and $(-2, 7)$ is perpendicular to the line through $(-2, 7)$ and $(4, 9)$.
56. Find a point (a, b) with distance 1 unit from $(4, 6)$ so that the line through (a, b) and $(4, 6)$ is perpendicular to the line through $(4, 6)$ and $(-8, 4)$.
57. Find a point (a, b) with distance 1 unit from the origin and a point (c, d) with distance 5 units from the origin so that the line through (a, b) and (c, d) has slope 2.
58. Find a point (a, b) on the line through $(-2, 7)$ and $(9, -4)$ so that the line through (a, b) and $(2, 1)$ has slope 8.

In Exercises 59–62, find the point at which the lines determined by the two given equations intersect.

59. $2x + y = 4$, $x + 3y = 7$
60. $3x + 5y = 14$, $x - y = 2$
61. $-2x + 5y = 38$, $5x + 2y = -8$
62. $y = 3x + 1/2$, $y = 2x + 1$

63. Find the point on the line $3x - 8y = 1$ that is nearest to point $(0, 9)$.
64. Where does the line $x - 7y = -15$ intersect the circle $(x-3)^2 + (y+1)^2 = 25$?
65. Suppose that A and B are constants that are not both zero and that D and E are any two constants. Prove that the lines $Ax + By = D$ and $-Bx + Ay = E$ are perpendicular.
66. Physical therapists recommend that ramps for people who use wheelchairs rise not more than 1 in. for each foot of forward motion. Formulate this recommendation in terms of slope. If the entrance to a certain public building is 30 ft from the sidewalk and the front door is 8 ft off the ground (up a steep flight of stairs), how can a suitable wheelchair ramp be built?
67. For what values of y_0 is the distance between the points of intersection of $y = y_0$ with $y = 2x + 1$ and $y = x + 2$ equal to $1{,}000{,}000$?
68. Determine the regression line through $P_0 = (x_0, y_0)$ that is based on the observations $P_1 = (x_0 - h, u)$ and $P_2 = (x_0 + h, v)$. How is the slope of the regression line related to the slopes of the segments P_1P_0 and P_0P_2?
69. Suppose that X and Y are items (or goods) that can be purchased at prices p_X and p_Y, respectively. Suppose that x represents the number of units of good X, and y represents the number of units of good Y that a consumer might purchase. The first quadrant of the xy-plane is known as *commodity space* in economics. If the consumer has a fixed amount C that he may allot to the purchase of goods X and Y, then the locus of all points (x, y) that represent purchasable combinations of these two goods is known as the consumer's *budget line*. (It is actually a line segment.) Determine a Cartesian equation for the budget line. What are its intercepts? What is its slope? If the consumer's circumstances change so that he has a different amount C' that he can use toward the purchase of goods X and Y, then what is the relationship of the new budget line to the old one?
70. The graph of $y = x^2$ is said to be *concave up*. One way to define this property is as follows: If $P = (a, a^2)$ and $Q = (b, b^2)$ are any two points on the graph of $y = x^2$, then the line segment $\overline{PQ}$ lies above the graph of $y = x^2$. What is the equation of the line that passes through P and Q? Algebraically verify that the midpoint of $\overline{PQ}$ lies above the parabola.
71. The distance of a point P to a line ℓ is the distance of P to Q where Q is the point on ℓ such that the line segment $\overline{PQ}$ is perpendicular to ℓ. By elementary geometry $|\overline{PQ}| < |\overline{PQ}'|$ if Q' is any other point on ℓ. Prove that if $Ax + By + C = 0$ is a general linear equation for ℓ and $P = (x_0, y_0)$, then

$$\frac{|Ax_0 + By_0 + C|}{\sqrt{A^2 + B^2}}$$

is the distance from P to ℓ.
72. A *lattice point* in the plane is a point with integer coordinates. Suppose that P and Q are lattice points. What relationship must the coordinates of P and Q satisfy for the midpoint of the line segment $\overline{PQ}$ to be a lattice point? Suppose that $P_1, \ldots, P_5$ are five distinct lattice points. Show that at least one pair determines a line segment whose midpoint is also a lattice point.
73. Let $P = (s, t)$ be a point in the xy-plane. Let $P' = (t, s)$. Calculate the slope of the line ℓ' that passes through P and P'. Deduce that ℓ' is perpendicular to the line ℓ whose equation is $y = x$. Let Q be the point of intersection of ℓ and ℓ'. Show that P and ℓ' are equidistant from Q. (As a result, each of P and P' is the reflection of the other through the line $y = x$.)

C. Calculator/Computer Exercises

In later chapters, we will use calculus to assign a meaning to the "slope" of a planar curve at a point on it. Here we explore this concept graphically. The idea is illustrated by Figure 20. The plot shown appears to be a line segment, but it is actually an arc of the parabola $y = x^2$ in a small viewing

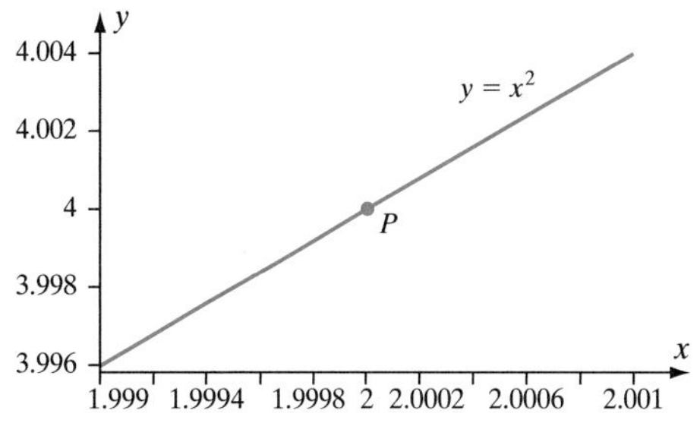

▲ Figure 20

window centered at the point $P=(2,4)$. In each of Exercises 74–79, plot the given curve in a viewing window containing the given point P. Zoom in on the point P until the graph of the curve appears to be a straight line segment. Compute the slope of the line segment: It is an approximation to the slope of the curve at P.

74. $y=x^2$; $P=(2,4)$
75. $y=x^2$; $P=(1,1)$
76. $y=2x/(x^2+1)$; $P=(1/2,4/5)$
77. $y=2x/(x^2+1)$; $P=(0,0)$
78. $y=2x/(x^2+1)$; $P=(1,1)$
79. $y=\sqrt{x}$; $P=(1,1)$

80. The following table records several paired values of automobile mileage (x) measured in thousands of miles and hydrocarbon emissions per mile (y) measured in grams:

x	5.013	10.124	15.060	24.899	44.862
y	0.270	0.277	0.282	0.310	0.345

Plot these points. Determine the regression line through $(44.862, 0.345)$. If this pattern continues for higher mileage cars, about how many grams of hydrocarbons per mile would a car with 100,000 miles emit?

81. The femur is the single large bone that extends from the hip socket to the kneecap. When skeletal remains are incomplete, anthropologists and forensic scientists can estimate height (y) from femur length (x). Plot the four observations (measured in cm):

x	38.6	44.1	48.7	53.4
y	154.3	168.4	178.3	187.5

Use this data to find the regression line that passes through $(38.6, 154.3)$.

82. Let x denote the perimeter of a rectangle, and let y denote the area. Calculate x and y for the five rectangles with (width, height) $=(7,42)$, $(8,23)$, $(11,13)$, $(18,9)$, and $(26,8)$. Plot the five ordered pairs(x,y) that you have calculated. The plot is nearly linear. Find the least squares line through the origin. For a square of side length 2, the perimeter x is 8. What is the regression line approximation of the area y? What is the relative error? What went wrong?

83. Graph the parabola $y=\sqrt{x}$. Add the upper half of the circle centered at $P=(5/4,0)$ with radius 1 to your graph. Notice that the parabola and the circle appear to be tangent at the point of intersection Q. What is the equation of the line through P and Q? What is the line ℓ through Q that is perpendicular to PQ? Add the plot of ℓ to your graph.

84. Graph $y=1/(1+x^2)$. Add the lower half of the circle centered at $P=(9/4,3)$ with radius $5\sqrt{5}/4$ to your graph. Notice that curve and the circle appear to be tangent at the point of intersection Q. What is the equation of the line through P and Q? What is the line ℓ through Q that is perpendicular to PQ? Add the plot of ℓ to your graph.

85. Let $P_0=(x_0,y_0)=(4,16)$, $(x_1,y_1)=(1,2)$, and $(x_2,y_2)=(2,6)$. Obtain an expression for the sum of the squares of the errors $d_1^2+d_2^2$ associated with a line through P_0 that has slope m. With m measured along the horizontal axis and $d_1^2+d_2^2$ measured along the vertical axis, graph $d_1^2+d_2^2$ in the viewing window $[4.5,5]\times[0,1.2]$. Use your plot to find the slope of the regression line $\mathcal{L}$ through P_0. Now plot the sum of the absolute errors d_1+d_2 in the viewing window $[3.5,6]\times[0,7]$. Use your plot to find the value of m that minimizes d_1+d_2. Finally, on the same coordinate axes, plot the three data points, the line through P_0 that minimizes d_1+d_2, and $\mathcal{L}$.

1.4 Functions and Their Graphs

Suppose that S and T are sets (collections of objects). A *function* f on the set S with values in the set T is a *rule* that assigns to each element of S a *unique* element of T. We write $f: S \to T$ and say "f maps S into T." The set S is called the *domain* of f. We will refer to the set T as the *range* of f. Notice that there are three distinct parts to a function: a domain, a range, and a rule that assigns one, and only one, value in the range to each value in the domain. For an element x in S, we write $f(x)$ (read "f at x") for the element in T that is assigned to x by the function f. We say that x is the *argument* of $f(x)$. The *image* of f is the set $\{f(x): x \in S\}$ of all values in T that the function f assumes.

It is useful to think of a function as an "input-output machine": The value x is the input, and the value $f(x)$ is the output. The domain of the function is the set of input values, and the image is the set of output values. The range is a set that contains every value in the image but may also contain some values that the function does not actually assume. We can think of the range as the set of *possible* output values of the function. For example, if a teacher who has 60 students in a class enters exam grades in a column of a spreadsheet, then the grade recorded in the ith row is a function f of i. If the grades are integers between 0 and 100, then it is convenient to take $\{0, 1, 2, 3, \ldots, 100\}$ to be the range of f and write $f\colon \{1, 2, 3, \ldots, 60\} \to \{0, 1, 2, 3, \ldots, 100\}$. The image of f consists of the grades that were actually attained by students. Because there are no more than 60 such grades, the chosen range is certainly larger than the image. Yet we easily indicated the range, whereas we would have found it tiresome to write out all the values of the image.

Examples of Functions of a Real Variable

The domains and ranges of the functions encountered in calculus are usually sets of real numbers; in fact, these sets will often consist of one or more intervals in $\mathbb{R}$. A function with a domain that is a set of real numbers is a *function of a real variable*. A function with a range that is a set of real numbers is a *real-valued function*. Often, we use *arrow notation*, $x \mapsto f(x)$ (the symbol $\mapsto$ is read as "is mapped to"). If we refer to a function f by giving a formula for $f(x)$ without specifying the domain, then the domain of f is understood to be the largest set of real values x for which $f(x)$ makes sense *as a real number*. If the range of a real-valued function f is not explicitly specified, then we understand the range to be the entire real line $\mathbb{R}$. The theorems of calculus often enable us to determine the image of f.

▶ **EXAMPLE 1** Let $F(x) = \sqrt{9 - x^2}$. What is the domain of F?

Solution For $F(x)$ to make sense as a real number, the inequality

$$9 - x^2 \geq 0$$

must be satisfied. Therefore $x^2 \leq 9$, or $-3 \leq x \leq 3$. In summary, the domain of F is the interval $[-3, 3]$. ◀

Example 1 illustrates one of the most common considerations in determining the domain of a function: The argument of an even root such as $\sqrt{\ }$ or $\sqrt[4]{\ }$ must be nonnegative. The example that follows illustrates another situation that must often be considered: The denominator of a fraction may not be zero.

▶ **EXAMPLE 2** Let $G(x) = x^2 + 2x + 4$ and $H(x) = (x^3 - 8)/(x - 2)$. Show that $G(x) = H(x)$ for every $x \neq 2$. Are G and H the same function?

Solution The identity

$$x^3 - 8 = (x^2 + 2x + 4)(x - 2) \tag{1.4.1}$$

is true for every x, as we can verify by expanding the right side and combining terms. However, because division by 0 is not defined, we can divide each side of (1.4.1) by $x - 2$ only when $x \neq 2$. Thus we have

$$H(x) = \frac{x^3 - 8}{x - 2} = \frac{(x^2 + 2x + 4)(x - 2)}{(x - 2)} = x^2 + 2x + 4 = G(x) \quad \text{for} \quad x \neq 2.$$

The domain $\mathcal{D}_G$ of G is the largest set of real numbers for which the expression $x^2 + 2x + 4$ makes sense. Therefore $\mathcal{D}_G = \mathbb{R}$. The domain $\mathcal{D}_H$ of H is the largest set of real numbers for which the expression $(x^3-8)/(x-2)$ makes sense. We must exclude $x=2$ because division by 0 is undefined. Therefore $\mathcal{D}_H = \{x \in \mathbb{R} : x \neq 2\}$. We conclude that G and H are *not* the same function: they have different domains. ◀

INSIGHT Suppose S' is a set that consists of some but not all points of a set S. Given a function G defined on S, we can create a function on the subset S' by assigning the output value $G(x)$ to the input value $x \in S'$. This function is called the *restriction* of G to S'. The notation $G|_{S'}$ is sometimes used to denote the restriction of G to S'. In Example 2, the domain S of G equals $\mathbb{R}$, the subset S' of S is $\{x \in \mathbb{R} : x \neq 2\}$, and H is the restriction of G to S'.

Piecewise-Defined Functions

A function is often described by different rules over different subintervals of its domain. Such a function is said to be a *piecewise-defined* or *multicase function.*

The *absolute value function* $x \mapsto |x|$ is a familiar multicase function that will often be used in this text. It can be written as

$$x \mapsto \begin{cases} x & \text{if} \quad x \geq 0 \\ -x & \text{if} \quad x < 0 \end{cases}$$

Another example of a piecewise-defined function that occurs frequently is the *signum*, or *sign*, function:

$$\text{signum}(x) = \begin{cases} 1 & \text{if} \quad x > 0 \\ 0 & \text{if} \quad x = 0\,. \\ -1 & \text{if} \quad x < 0 \end{cases}$$

The next example illustrates that multicase functions are often encountered in everyday life.

▶ **EXAMPLE 3** Schedule X from the 2008 U.S. income tax Form 1040 is reproduced here. This form helps determine the income tax $T(x)$ of a single filer with taxable income x.

Over	But not over	The tax is	Of the amount over
\$0	\$8025	10%	\$0
\$8025	\$32550	\$802.50 + 15%	\$8025
\$32550	\$78850	\$4481.25 + 25%	\$32550
\$78850	\$164550	\$16056.25 + 28%	\$78850
\$164550	\$357700	\$40052.25 + 33%	\$164550
\$357700		\$103791.75 + 35%	\$357700

Write T using mathematical notation.

Solution The function T is defined for all positive values of x, but there are six different cases. We use the following notation:

$$T(x) = \begin{cases} 0.10x & \text{if } 0 < x \le 8025 \\ 802.50 + 0.15(x - 8025) & \text{if } 8025 < x \le 32550 \\ 4481.25 + 0.25(x - 32550) & \text{if } 32550 < x \le 78850 \\ 16056.25 + 0.28(x - 78850) & \text{if } 78850 < x \le 164550 \\ 40052.25 + 0.33(x - 164550) & \text{if } 164550 < x \le 357700 \\ 103791.75 + 0.35(x - 357700) & \text{if } 357700 < x \end{cases}$$

For a given value of x, there are two steps used to determine $T(x)$. First, we must determine the interval in which the value x lies. Second, we must apply the formula that corresponds to that interval. For example, a single filer with taxable income of $\$30,000$ would observe that $x = 30000$ lies in $(8025, 32550]$ and would compute his tax to be

$$T(30000) = 802.50 + 0.15(30000 - 8025) = \$4098.80. \quad ◀$$

Graphs of Functions

In calculus, it is often useful to represent functions visually. Pictures, or *graphs*, can help us think about functions. For now, we will only graph functions with domains and ranges that are subsets of the real numbers. These real-valued functions of a real variable are graphed in the xy-plane. The elements of a function's domain are thought of as points of the x-axis. A function's values are measured on the y-axis.

DEFINITION The graph of f is the set of all points (x, y) in the xy-plane for which x is in the domain of f and $y = f(x)$. Thus if f is a real-valued function that is defined on a set S of real numbers, then the graph of f is the curve $\{(x, y) : x \in S, y = f(x)\}$.

The graphs of some basic functions are shown in Figure 1. If a number x_0 is not in the domain of a function f, then there is no point of the form (x_0, y) that lies on the graph of f. In other words, the vertical line $x = x_0$ does not intersect the graph of f when x_0 is outside the domain of f. If x_0 is in the domain of f, then there is one and only one point of the form (x_0, y) that lies on the graph of f—namely, the unique point (x_0, y_0) with $y_0 = f(x_0)$.

INSIGHT These considerations lead to the Vertical Line Test: If every vertical line drawn through a curve intersects that curve only once, then the curve is the graph of a function. If there is a vertical line that intersects the curve in more than one point, then the curve is not the graph of a function.

▶ **EXAMPLE 4** Is the graph of the equation $x^2 + y^2 = 13$ the graph of a function?

Solution The graph of $x^2 + y^2 = 13$ is the circle shown in Figure 2. Observe that there are many vertical lines that intersect the graph in more than one point. It takes only one such line to demonstrate that the curve is not the graph of a

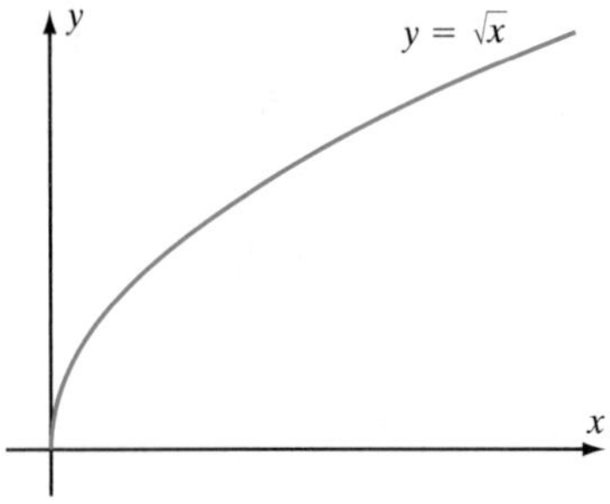

▲ Figure 1a

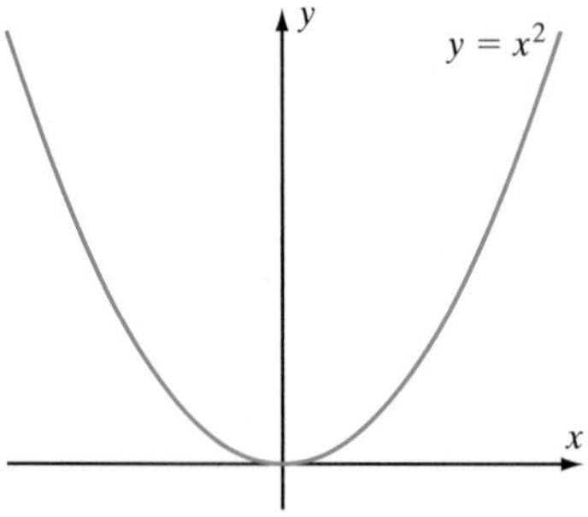

▲ Figure 1b

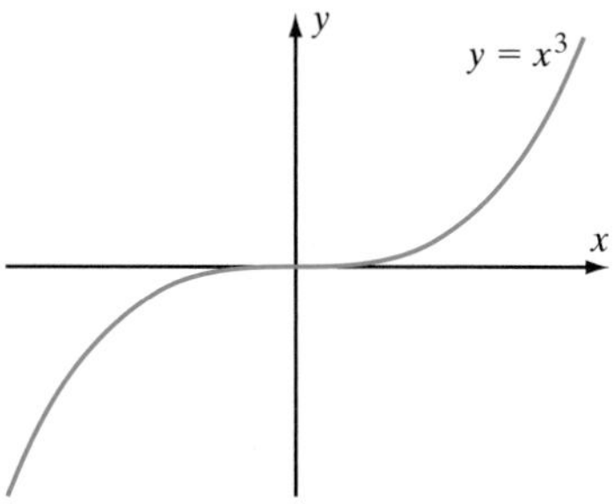

▲ Figure 1c

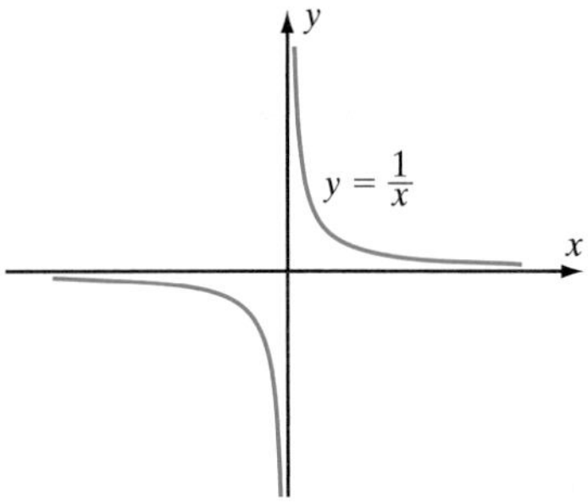

▲ Figure 1d

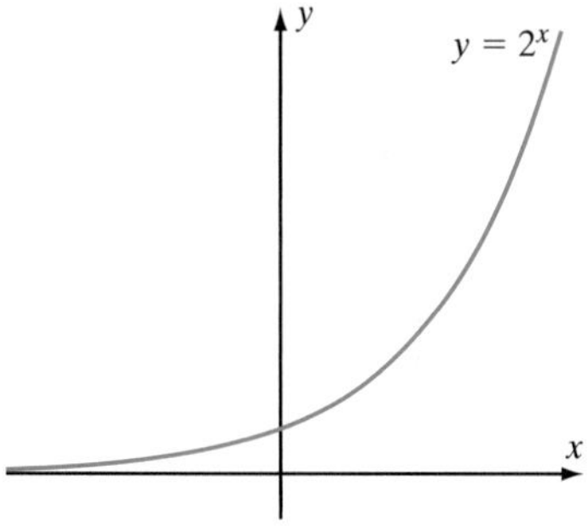

▲ Figure 1e

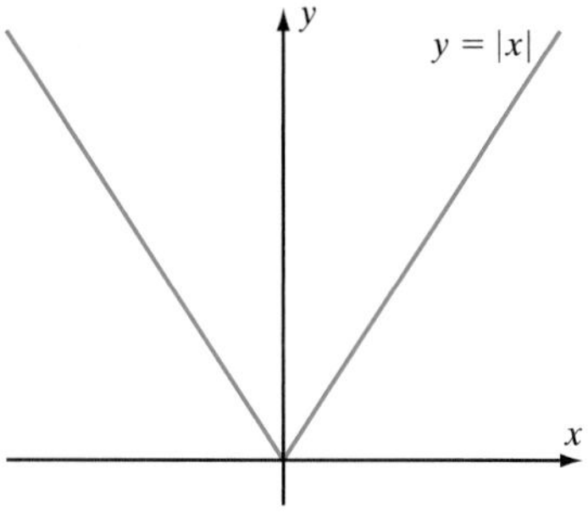

▲ Figure 1f

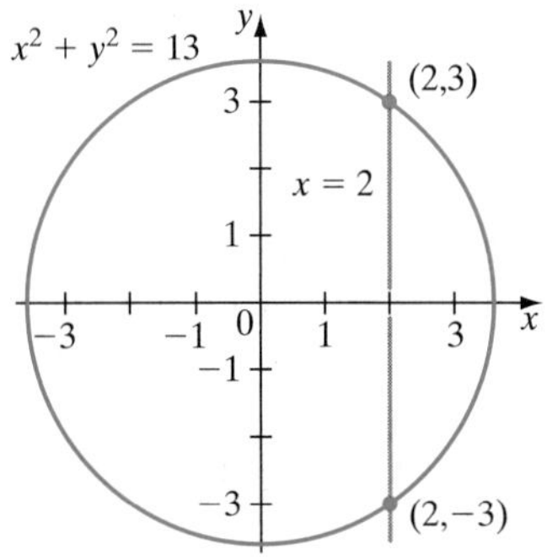

▲ Figure 2

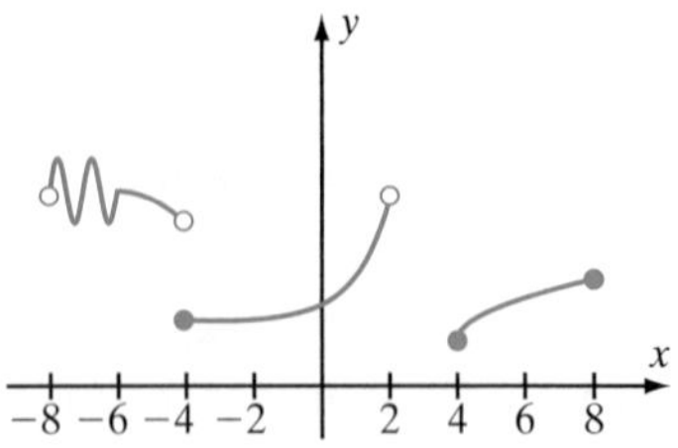

▲ Figure 3

function. Figure 2 illustrates this idea with the vertical line $x = 2$, for which there are two y values, $y = 3$ and $y = -3$, such that (x, y) is on the given circle. ◀

▶ **EXAMPLE 5** Is the curve in Figure 3 the graph of a function?

Solution For each x that lies in either the open interval $(-8, 2)$ or the closed interval $[4, 8]$, there is only one corresponding y value. Thus the curve passes the Vertical Line Test and is the graph of a function $x \mapsto f(x)$ (even though we do not know a formula for $f(x)$). The domain of f is $(-8, 2) \cup [4, 8]$. ◀

Chapter 4 explains some techniques for graphing functions. For now, it is a good idea to learn to recognize and sketch the graphs of basic functions (such as those plotted in Figure 1) and to use a graphing calculator or computer for more complicated functions. When we sketch a graph by hand, we can easily highlight important features (such as isolated points missing from the domain) and noteworthy details (such as intercepts). For instance, the function H from Example 2 is graphed in Figure 4. An open dot indicates that the number 2 is not in the domain of H.

▶ **EXAMPLE 6** Graph the signum function, which is defined in the preceding subsection.

Solution The part of the graph of signum to the left of $x = 0$ lies on the horizontal line $y = -1$. The part to the right of $x = 0$ lies on the horizontal line $y = 1$. A closed

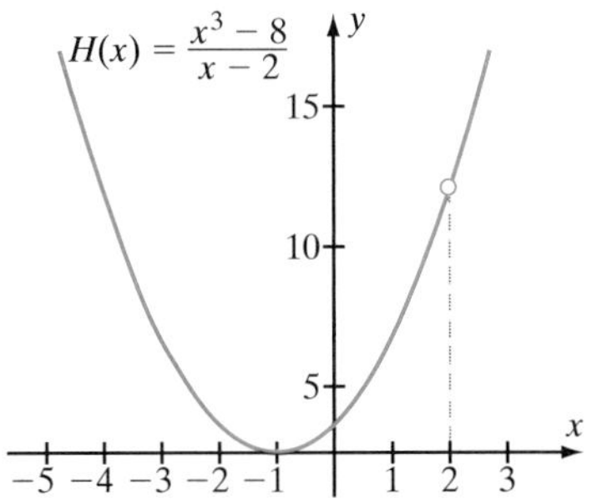

▲ Figure 4

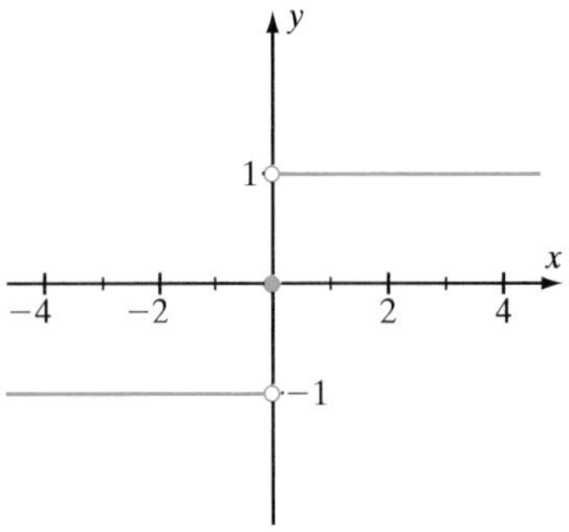

▲ Figure 5 $y = \text{signum}(x)$

dot indicates that $(0, 0)$ is on the graph of signum. Because the points $(0, -1)$ and $(0, 1)$ are not on the graph of signum, we use open dots (see Figure 5). ◀

▶ **EXAMPLE 7** One popular graphing calculator uses the rectangle $[-10, 10] \times [-10, 10]$ as its default viewing window. Explain why no parts of the graphs of $f(x) = \sqrt{x^2 - x - 132}$ and $g(x) = (2x^4 + 100)/(x^2 + 2x + 2)$ appear in this viewing window.

Solution The function f may be written as

$$f(x) = \sqrt{x^2 - x - 132} = \sqrt{(x + 11)(x - 12)}.$$

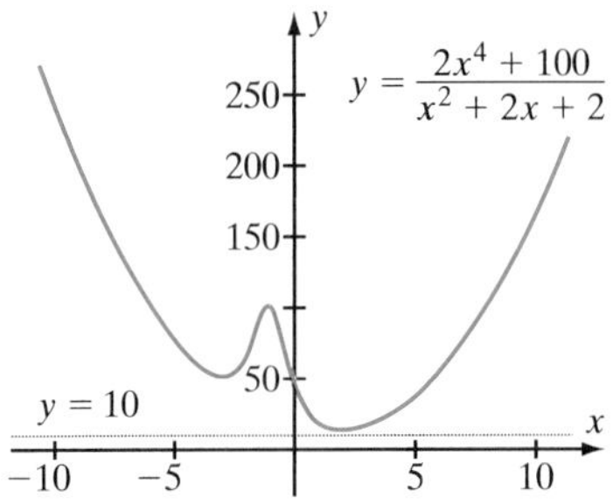

▲ Figure 6

For x to be in the domain of f, the product $(x + 11)(x - 12)$ must be nonnegative. Therefore the factors $x + 11$ and $x - 12$ must have the same sign. This happens only for $x \leq -11$ (both factors nonpositive) and for $x \geq 12$ (both factors nonnegative). No part of the graph of f will appear in any viewing window of the form $[-10, 10] \times [y_{\min}, y_{\max}]$ because $[-10, 10]$ is not in the domain of f.

The problem with the graph of g is different. By completing the square in the denominator, we see that $g(x) = (2x^4 + 100)/((x + 1)^2 + 1)$. The denominator is greater than or equal to 1, hence it is never 0. The domain of g is, therefore, the set of all real numbers. In particular, $[-10, 10]$ is in the domain of g, and we can be sure that *some* viewing rectangle of the form $[-10, 10] \times [y_{\min}, y_{\max}]$ will contain part of the graph of g. Indeed, Figure 6 shows the graph of g in the viewing window $[-10, 10] \times [0, 250]$. We see that $g(x) > 10$ for every value of x in $[-10, 10]$. In other words, the rectangle $[-10, 10] \times [-10, 10]$ lies entirely below the graph of g. ◀

Sequences

A function f with a domain that is the set $\mathbb{Z}^+$ of positive integers is called an *infinite sequence*. In general, a function f is a *sequence* if its domain is a finite or infinite set of consecutive integers. We can write a sequence f by listing its values:

$$f(1), f(2), f(3), \ldots .$$

Thus a sequence f can be thought of as a list, with $f(1)$ being the first element of the list, $f(2)$ the second element, and so on. It is common to abandon the function notation altogether and write the sequence as

$$f_1, f_2, f_3, \ldots ,$$

or sometimes as $\{f_n\}_{n=1}^{\infty}$, or even as $\{f_n\}$. We refer to f_n as the nth *term* of the sequence f. Occasionally, it is useful to begin a sequence with an index different from 1. An example is $\{3n - 5\}_{n=4}^{\infty}$, which denotes the sequence $7, 10, 13, 16, \ldots$.

▶ **EXAMPLE 8** For $n = 1, 2, 3, \ldots$ let f_n be the product of the positive integers less than or equal to n. List the first six terms of $\{f_n\}_{n=1}^{\infty}$.

Solution We calculate $f_1 = 1$, $f_2 = 2 \cdot 1 = 2$, $f_3 = 3 \cdot 2 \cdot 1 = 6$, $f_4 = 4 \cdot 3 \cdot 2 \cdot 1 = 24$, $f_5 = 5 \cdot 4 \cdot 3 \cdot 2 \cdot 1 = 120$, and $f_6 = 6 \cdot 5 \cdot 4 \cdot 3 \cdot 2 \cdot 1 = 720$. ◀

INSIGHT The numbers of Example 8 are known as factorials. They are important in many fields of mathematics and they arise in calculus as well. It is common to write $n!$ (read n factorial) for f_n. Thus $n! = n \cdot (n-1) \cdots 3 \cdot 2 \cdot 1$ for $n = 1, 2, 3, \ldots$. We also define $0!$ to be 1. That is, $0! = 1$.

▶ **EXAMPLE 9** Use the decimal expansion of π to define a sequence of rational numbers $\{q_n\}$ that become progressively closer to π.

Solution We can define such a sequence $\{q_n\}$ by $q_1 = 3.1$, $q_2 = 3.14$, $q_3 = 3.141$, $q_4 = 3.1415$, $q_5 = 3.14159$, and so on. Thus

$$q_n = \text{the first } (n+1) \text{ digits of the decimal expansion of } \pi.$$

This is not a constructive definition because no method of computing the decimal expansion of π has been given. Constructive sequences are given in the Exercises 57 and 58. ◀

A sequence $\{s_n\}$ is *inductively* (or *recursively*) defined if s_{n+1} is defined in terms of some or all of its predecessors $s_1, \ldots, s_n$. In this case, we must specify (or *initialize*) some of the first few values of the sequence $\{s_n\}$.

▶ **EXAMPLE 10** The Fibonacci sequence f_n is defined by $f_{n+2} = f_{n+1} + f_n$ for $n \geq 1$. The values f_1 and f_2 are both initialized to be 1. What is f_7?

Solution We calculate:

$$\begin{aligned}
f_3 &= f_2 + f_1 = 1 + 1 = 2\\
f_4 &= f_3 + f_2 = 2 + 1 = 3\\
f_5 &= f_4 + f_3 = 3 + 2 = 5\\
f_6 &= f_5 + f_4 = 5 + 3 = 8\\
f_7 &= f_6 + f_5 = 8 + 5 = 13
\end{aligned}$$ ◀

Functions from Data

In practice, many functional relationships are not initially discovered in the form of mathematical expressions. For example, suppose that several observations or measurements of two variables x and y are made and that y depends on x by means of an undetermined function f. One way to find an expression for $f(x)$ is to plot a sequence $\{(x_n, y_n)\}$ of the observed values. The result is known as a *scatter plot*, or a *scatter diagram*. Comparison of the scatter plot with known graphs can suggest a suitable formula for f. This formula is then said to be a *mathematical model* of the observations.

▶ **EXAMPLE 11** In 1929, the American astronomer Edwin Hubble presented evidence that indicates the expansion of the universe. Hubble measured the distances and recession velocities of several galaxies relative to a fixed galaxy.

(Recession velocity is the speed at which two galaxies move away from each other.) Some of Hubble's data is tabulated here.

	Virgo	Pegasus	Perseus	Ursa Major 1	Leo	Gemini	Bootes	Hydra	Coma Berenices
d	22	68	108	255	315	405	685	1100	137
v	0.75	2.4	3.2	9.3	12.0	14.4	24.5	38.0	?

In the table, d represents distance in millions of light years (here a light year is the distance that light will travel in one year), and v represents velocity in thousands of miles per second. Find a model for v as a function of d. What is the recession velocity of Coma Berenices?

Solution Figure 7 shows a plot of the tabulated data. Because d is the independent variable and v the dependent variable, we refer to the horizontal axis as the d-axis, the vertical axis as the v-axis, and the plane they determine as the dv-plane. Because the plotted points lie near a line that passes through the origin of the dv-plane, the model we seek has the equation $v = H \cdot d$ where H denotes the slope of the line. The calculated values needed for the slope of the regression line through the origin are recorded in the following table:

	d	v	$d \cdot v$	d^2
	22	0.75	16.5	484
	68	2.4	163.2	4624
	108	3.2	345.6	11664
	255	9.3	2371.5	65025
	315	12.0	3780.0	99225
	405	14.4	5832.0	164025
	685	24.5	16782.5	469225
	1100	38	41800	1210000
Sum			71091.3	2024272

According to formula (1.3.8), we have

$$\begin{aligned} H &= \frac{d_1 v_1 + d_2 v_2 + \cdots + d_8 v_8}{d_1^2 + d_2^2 + \cdots + d_8^2} \\ &= \frac{(22)(0.75) + (68)(2.4) + \cdots + (1100)(38.0)}{22^2 + 68^2 + \cdots + 1100^2} \\ &= \frac{71091.3}{2024272} \\ &= 0.03512. \end{aligned}$$

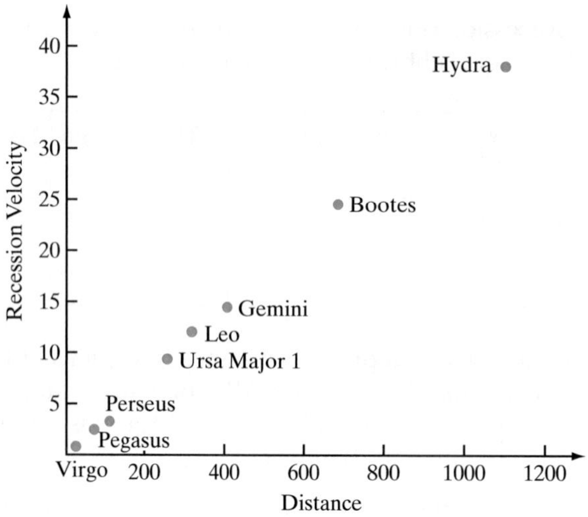

▲ Figure 7

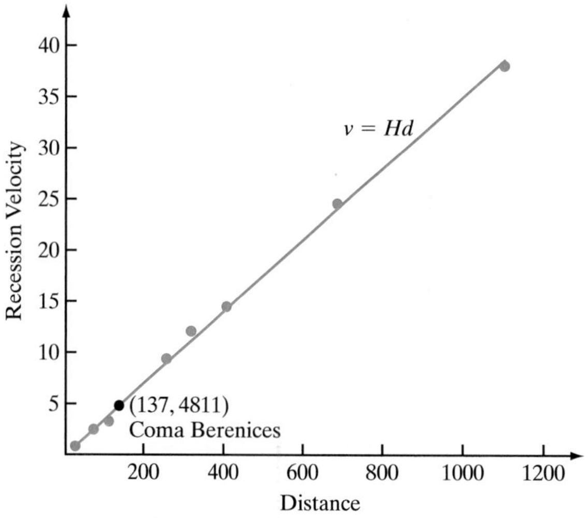

▲ Figure 8

Thus $\upsilon = H \cdot d = 0.03512 \cdot d$. The graph of this equation is given in Figure 8, with Hubble's observations superimposed.

Because $d = 137$ for Coma Berenices, we calculate that the corresponding recession velocity is $\upsilon = 0.035\,12 \cdot 137$, or $\upsilon = 4.811$. Because the units for υ are thousands of miles per second, our estimate amounts to 4811 mi/s. In fact, Hubble measured the recession velocity of Coma Berenices to be 4700 mi/s. ◀

INSIGHT The constant H is known as *Hubble's constant.* A dimension analysis shows that 1/H represents a time because the equation $\upsilon = H \cdot d$ implies

$$\frac{1}{H} = \frac{d}{\upsilon} = \frac{\text{distance}}{\text{distance/time}} = \text{time}.$$

In the Big Bang theory of the universe, $1/H$ represents the age of the universe. Refinements in astronomical measurements have led to corrections to Hubble's data and to revised estimates of H. Data obtained from the Wilkinson Microwave Anisotropy Probe (WMAP) released in February 2008 suggest that the universe is approximately 13.73 billion years old.

By "transforming" data, the method of least squares can often be applied in instances when a scatter plot is not linear. For example, suppose that $z = A \cdot k^x$, where A is a known constant, and k is a fixed base that is to be determined. We apply $\log_{10}$ to each side of the equation $z = A \cdot k^x$. Using the logarithm laws $\log_{10}(u \cdot v) = \log_{10}(u) + \log_{10}(v)$ and $\log_{10}(u^p) = p \cdot \log_{10}(u)$, we obtain

$$\begin{aligned} \log_{10}(z) &= \log_{10}(A \cdot k^x) \\ &= \log_{10}(A) + \log_{10}(k^x) \\ &= \log_{10}(A) + x \cdot \log_{10}(k). \end{aligned}$$

If we set $y = \log_{10}(z)$, $m = \log_{10}(k)$, and $b = \log_{10}(A)$, then the transformed equation becomes the linear equation

$$y = m \cdot x + b$$

in the variables x and y.

▶ **EXAMPLE 12** A radioactive isotope of gold is used to diagnose arthritis. Let $f(x)$ denote the fraction of blood serum gold remaining in the patient at the time x days after the initial dose. Measured values of $f(x)$ for $x = 1, 2, \ldots, 8$ are listed in the table.

Days x	**0**	**1**	**2**	**3**	**4**	**5**	**6**	**7**	**8**
Serum gold (z)	1.0	0.91	0.77	0.66	0.56	0.49	0.43	0.38	0.34

Find a positive constant k such that $z = k^x$ is a model for $f(x)$. Use the model to predict the fraction of the initial dose that remains in the blood serum after ten days.

Solution Figure 9a shows the scatter plot of this data. Following the discussion that precedes this example, we apply $\log_{10}$ to each side of $z = k^x$ to obtain the linear equation

$$y = \log_{10}(z) = \log_{10}(k^x) = x \cdot \log_{10}(k) = m \cdot x$$

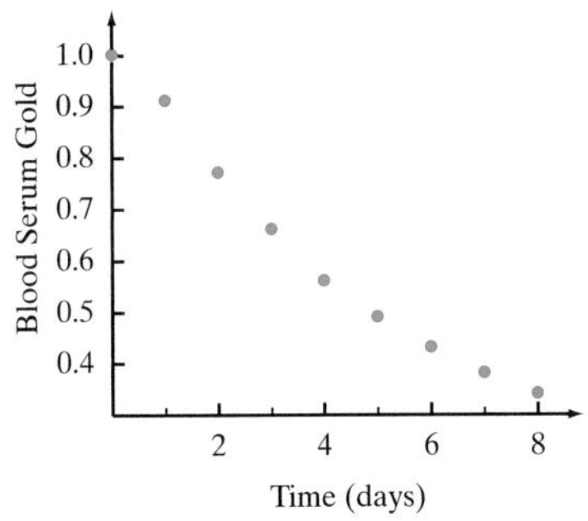

▲ Figure 9a

where $m = \log_{10}(k)$. Because the line $y = m \cdot x$ passes through the origin, we pass the regression line through the data point $(0, 0)$. With $x_0 = 0, y_0 = 0$, and $N = 8$, equation (1.3.8) becomes

$$m = \frac{x_1 y_1 + x_2 y_2 + \cdot \cdot \cdot + x_8 y_8}{x_1^2 + x_2^2 + \cdot \cdot \cdot + x_8^2}.$$

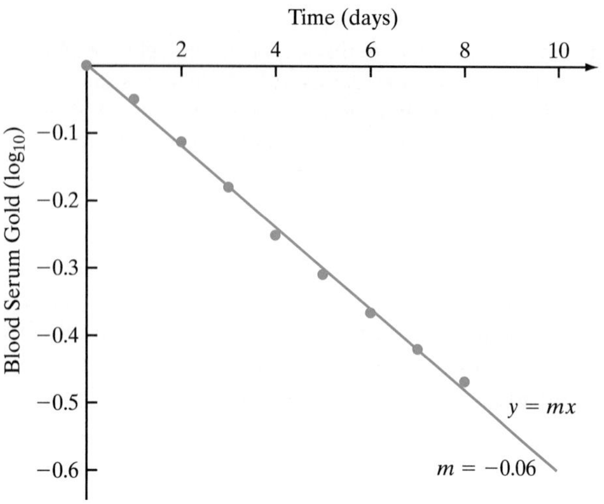

▲ Figure 9b

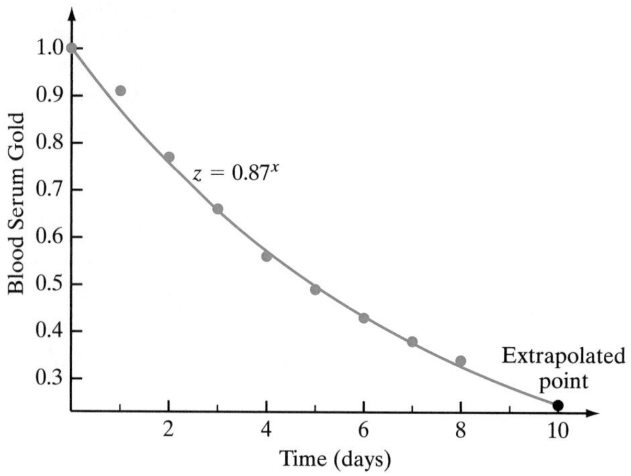

▲ Figure 9c

The values that we need for the calculation of m can be found in the following table.

	x	z	$y = \log_{10}(y)$	$x \cdot y$	x^2
	1	0.91	−0.0409	−0.0409	1
	2	0.77	−0.1135	−0.2270	4
	3	0.66	−0.1804	−0.5414	9
	4	0.56	−0.2518	−1.0073	16
	5	0.49	−0.3098	−1.5490	25
	6	0.43	−0.3665	−2.1992	36
	7	0.38	−0.4202	−2.9415	49
	8	0.34	−0.4685	−3.7482	64
Sum				−12.2545	204

Our estimate of m is $-12.2545/204$, or $m = -0.06$.

Figure 9b shows the transformed data points, along with the least squares line $y = -0.06x$. Our estimate for m provides an estimate for the base k, namely $k = 10^{-0.06} \approx 0.87$. In Figure 9c, the graph of $y = 0.87^x$ is superimposed on the original scatter plot. The predicted value of the serum gold concentration ten days after the initial dose is $f(10) = 0.87^{10} \approx 0.25$. ◀

Functions of Several Variables

Many variables depend on several other variables. For example, the volume V of a right circular cone of height h and base radius r depends on both h and r. We say that V is a function of h and r, and we write $V(h, r)$. In this case, we can describe V by the rule

$$V(h, r) = \frac{1}{3}\pi r^2 h.$$

Similar notation is used for functions of three or more variables. Thus the volume V of a rectangular box is a function of its height h, width w, and depth d:

$$V(h, w, d) = hwd.$$

Although we will occasionally encounter multivariable functions in the next several chapters, we will not study the calculus of multivariable functions until Chapter 13.

QUICK QUIZ

1. True or false: a is in the domain of a real-valued function f of a real variable if and only if the vertical line $x = a$ intersects the graph of f exactly once.
2. True or false: b is in the range of a real-valued function f of a real variable if and only if the horizontal line $y = b$ intersects the graph of f at least once.
3. True or false: b is in the image of a real-valued function f of a real variable if and only if the horizontal line $y = b$ intersects the graph of f exactly once.
4. What is the domain of the function defined by the expression $\sqrt{\dfrac{9 - x^2}{x^2 - 4}}$?
5. Suppose that $\{f_n\}_{n=0}^{\infty}$ is recursively defined by $f_0 = 1$ and $f_n = n \cdot f_{n-1}$ for $n > 0$. What is f_6?

Answers

1. True 2. False 3. False 4. $[-3, -2) \cup (2, 3]$ 5. 720

EXERCISES

Problems for Practice

In Exercises 1–8, state the domain of the function defined by the given expression.

1. $x/(1 + x)$
2. $\sqrt{x^2 + 2}$
3. $\sqrt{x^2 - 2}$
4. $\sqrt{2 - x^2}$
5. $1/(x^2 - 1)$
6. $\sqrt{x}/(x^2 + x - 6)$
7. $\sqrt{x^2 - 4x + 5}$
8. $1/\sqrt{(x^2 - 4)(x - 1)}$

In Exercises 9–14, sketch the graph of the function defined by the given expression.

9. $x^2 + 1$
10. $x^2 - 1$
11. $1 - x^2$
12. $x^2/3 + 3$
13. $3 - x^2/2$
14. $x^2 + 6x + 10$

In Exercises 15–24, plot several points, and sketch the graph of the function defined by the given expression.

15. x^{-2}
16. $\sqrt{x^2}$

17. $\sqrt{2x+4}$

18. $\sqrt{x-2}$

19. $(x-4)^{-1/2}$

20. $(x+1)^{-3}$

21. $1/\sqrt{x+1}$

22. $\text{signum}(|x^2-x|)$

23. $\begin{cases} x^2 & \text{if } x<1 \\ 2-x^2 & \text{if } x \geq 1 \end{cases}$

24. $\begin{cases} x^2-4 & \text{if } x<2 \\ x+2 & \text{if } -2 \leq x < 2 \\ x^2 & \text{if } 2 \leq x \end{cases}$

Further Theory and Practice

25. We say that y is *proportional* to x if $y = kx$ for some constant k (known as the *proportionality constant*). Use the concept of proportionality to determine the arc length $s(\alpha)$ for the arc of a circle of radius r that subtends a central angle measuring α degrees. (Deduce the proportionality constant by using the value of s that corresponds to $\alpha = 360°$.)

26. Let $y = f(x)$ where $f(x) = mx + b$ for constants $m \neq 0$ and b. Show that a change in the value of x from x_0 to $x_0 + \Delta x$ results in a change Δy in the value of $f(x)$ that does not depend on the initial value x_0. In other words, the increment $\Delta y = f(x_0 + \Delta x) - f(x_0)$ depends on the increment Δx but not on the value of x_0.

27. A variable $u = f(x, y)$ is said to be *jointly proportional* to x and y if $f(x, y) = kxy$ for some constant k. The area of a sector of a circle is jointly proportional to its central angle and to the square of the radius r of the circle. What is the area $A(r, \alpha)$ if the degree measure of the central angle of the sector is α? (Deduce the proportionality constant by using the value of A that corresponds to $\alpha = 360°$.)

28. Calculate the area $A(\ell, \alpha)$ of an isosceles triangle having two sides of length ℓ enclosing an angle α.

29. Let r and s be the roots of $x^2 + Ax + B$. Express the coefficients A and B as functions of r and s.

30. Sketch the graph of the tax function T for $0 < x \leq 50000$ (see Example 3).

31. Figure 10 shows the graph of a function f. Give a formula for f.

32. The domain of the function f that is graphed in Figure 10 is $[0, 5]$. For every $x \in [0, 5]$, let $m(x)$ be the slope of the graph of f at $(x, f(x))$, provided the slope exists. Otherwise, $m(x)$ is not defined. What is the domain of m? Sketch the graph of the slope function m. For $x \in [0, 5]$, let $A(x)$ be the area under the graph of f and over the interval $[0, x]$. Sketch the graph of the area function A.

33. Let T denote the tax function described in Example 3. In this exercise, we will restrict the domain of T to $0 < x < 50000$. For x in this interval, excluding the points 8025 and 32550, let $m(x)$ denote the slope of the graph of T at the point $(x, T(x))$. Give a multicase formula for m. Let $A(x)$ be the area under the graph of m and over the interval $(0, x)$. Give a multicase formula for A. What is the relationship of A to T?

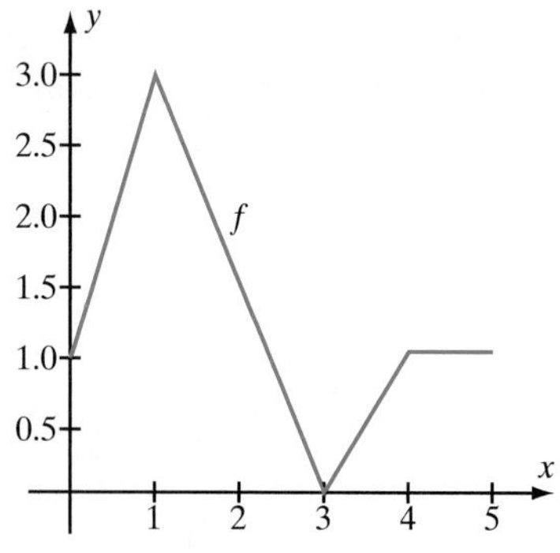

▲ Figure 10

In Exercises 34–40, give a recursive definition of the sequence.

34. $f_n = 2n$, $n = 1, 2, 3, \ldots$

35. $f_n = 2^n$, $n = 1, 2, 3, \ldots$

36. $f_n = 1 + 2 + \cdots + n$, $n = 1, 2, 3, \ldots$

37. $f_n = n!$, $n = 1, 2, 3, \ldots$

38. $f_n = 2^{((-1)^n)}$, $n = 1, 2, 3, \ldots$

39. $f_n = n(n+1)/2$, $n = 1, 2, 3, \ldots$

40. $\{1, 5, 17, 53, 161, \ldots\}$

41. For any real number x, the *greatest integer in* x is denoted by $\lfloor x \rfloor$ and defined to be the unique integer satisfying $\lfloor x \rfloor \leq x < \lfloor x \rfloor + 1$. For example, $\lfloor 3.2 \rfloor = 3$ and $\lfloor -3.2 \rfloor = -4$. Notice that $\lfloor x \rfloor = x$ if and only if x is an integer. The function $x \mapsto \lfloor x \rfloor$ is called the *greatest integer function*. (The expression $\lfloor x \rfloor$ is sometimes read as "the floor of x.") The *integer part* of x, denoted by $\text{Int}(x)$, is that part of the decimal expansion of x to the left of the decimal point. Express $\lfloor x \rfloor$ as a multicase function of $\text{Int}(x)$. Graph $x \mapsto \lfloor x \rfloor$ and $x \mapsto \text{Int}(x)$ for $-3 \leq x \leq 3$.

42. A kilowatt-hour is the amount of energy consumed in 1 hour at the constant rate of 1000 watts. At time $t = 0$ hour, a three-way lamp is turned on at the 50-watt setting. An hour later, the lamp is turned up to 100 watts. Forty minutes after that, the lamp is turned up to 150 watts. Ninety minutes later, it is turned off. Let $E(t)$ be the (cumulative) energy consumption in kilowatt-hours expressed as a function of time measured in hours. Graph $E(t)$ for $0 \leq t \leq 4$.

43. A loan of P dollars is paid back by means of monthly installments of m dollars for n full years. Find a formula for the function $I(P, m, n)$ that gives the total amount of interest paid over the life of the loan.

44. Initially, a solution of salt water has total mass 500 kg and is 99% water. Over time, the water evaporates. Just for fun, estimate to within 25 kg the mass of the solution at the time when water makes up 98% of the solution. Find a formula for the total mass $m(p)$ of the solution as a function of p, the percentage contribution of water to the total mass. Exactly what is $m(98)$? Graph $m(p)$.

45. Recursively define two sequences $\{c_n\}_{n=1}^{\infty}$ and $\{C_n\}_{n=1}^{\infty}$ by initializing $c_0 = C_0 = 1$ and, for each nonnegative integer n, setting

$$c_{n+1} = c_0 c_n + c_1 c_{n-1} + \cdots + c_{n-1} c_1 + c_n c_0$$

and

$$C_{n+1} = \frac{4n+2}{n+2} C_n.$$

Calculate c_n and C_n for $0 \le n \le 8$. (These integer sequences can be shown to be the same for all nonnegative integers n. The members of these sequences are called *Catalan numbers*.)

If $P(x)$ is an assertion about x that is either true or false depending on the value of x, then we can make $P(x)$ into a numerical function by using 1 for true and 0 for false. Thus the function $x \mapsto (x^2 < x)$ has the value 1 for $-1 < x < 1$ and 0 for all other values of x. Some calculators and computer programs use this assignment of truth values as a convenient tool for expressing multicase functions. For instance, we can describe the function

$$F(x) = \begin{cases} -x+3 & \text{if } x<0 \\ 3 & \text{if } 0<x<2 \\ x+1 & \text{if } x \ge 2 \end{cases}$$

with one formula as follows:

$$F(x) = -x \cdot (x<0) + 3(0<x)(x<2) + (x+1) \cdot (x \ge 2).$$

In each of Exercises 46–49, use this method to express the given multicase function with one formula. Plot the function.

46. $H(x) = \begin{cases} 1 & \text{if } x \ge 0 \\ 0 & \text{if } x<0 \end{cases}$ (H is known as the *Heaviside function*).

47. $R(x) = \begin{cases} 1 & \text{if } x>1 \\ x & \text{if } 0<x \le 1 \\ 0 & \text{if } x \le 0 \end{cases}$ (R is the *ramp function*.)

48. $r(x) = \begin{cases} 0 & \text{if } x>1 \\ 1 & \text{if } 0<x \le 1 \\ 0 & \text{if } x \le 0 \end{cases}$ (r is the rectangular bump function.)

49. $b(x) = \begin{cases} 0 & \text{if } x \le 1 \\ x & \text{if } 0<x \le 1 \\ 2-x & \text{if } 1<x \le 2 \\ 0 & \text{if } 2<x \end{cases}$ (b is the *triangular bump function*.)

Calculator/Computer Exercises

50. In the table, x represents cigarette consumption per adult (in hundreds) for 1962 for eight countries. The variable y represents mortality per 100,000 due to heart disease in 1962.

	x	y
Australia	32.2	238.1
Belgium	17.0	118.1
Canada	33.5	211.6
West Germany	18.9	150.3
Ireland	27.7	187.3
Netherlands	18.1	124.7
Great Britain	27.9	194.1
United States	39	256.9

Plot the eight points (x, y). Find a function $f(x) = mx + 12.5$ that could be used to model the relationship between cigarette consumption and mortality due to heart disease.

51. Let $x \in [0, 1]$ be the incidence of a certain disease in the general population. For example, if 132 persons out of 100,000 have the disease, then $x = 0.00132$. Suppose that a screening procedure for the disease results in false positives 1% of the time and in false negatives 2% of the time. Then, according to *Bayes's law*,

$$P(x) = \frac{0.98x}{0.97x + 0.01}$$

is the probability that a person who tests positive for the disease actually has the disease. Graph P in an appropriate window. What does this graph tell you about the value of routinely using this screening procedure if the disease is rare?

52. The *astronomical unit* (AU) is the mean distance of Earth to the sun. The table gives the mean distances (D) in astronomical units between the sun and the six planets that were known at the time of Johannes Kepler.

The table also provides the planets' periods of orbit about the sun (T) in years. Plot the six points (D, T). In the window $[0,10]\times[0,30]$, graph the function $f_q(x)=x^q$ for each $q=m/n$ with $1\leq m$ and $n\leq 3$. For what value of q does the graph of f_q fit the data well? Notice that $T<D$ for $D<1$, and $T>D$ for $D>1$. Explain why this indicates that $q>1$. Which points show that $q<2$?

	T	D
Mercury	0.24	0.387
Venus	0.615	0.723
Earth	1.00	1.00
Mars	1.88	1.524
Jupiter	11.86	5.203
Saturn	29.547	9.539

53. When the argument of a function f changes by an increment h from $x-h$ to x, the value of the function changes by the increment $F(x,h)=f(x)-f(x-h)$. If h is small, then it is to be expected that $F(x,h)$ is also small. In 1615, Kepler noticed that when $f(x_0)$ is a maximum value of f, the increment $F(x_0,h)$ is even smaller than one might expect. In this exercise, we use the function $f(x)=5-24x^2+20x^3-3x^4$ to investigate Kepler's observation.

a. Graph f in the window $[2,5]\times[20,135]$. Using the graph, determine the value x_0 of x for which $f(x)$ is maximized.

b. Calculate $F(x_0,h)$ and $F(3.99,h)$ for $h=10^{-3}$, 10^{-4}, and 10^{-5}. You will notice that for each of these values of h, $F(x_0,h)$ is much smaller than $F(3.99,h)$.

c. Plot $F(3.99,h)$ for $10^{-5}\leq h\leq 10^{-3}$. You will see that $F(3.99,h)\approx mh$ for some constant m.

d. Now plot $F(x_0,h)$ for $10^{-5}\leq h\leq 10^{-3}$. You will see that $F(x_0,h)\approx Ah^2$ for some constant A.The reason for the behavior Kepler observed is that h^2 is negligible compared to h when h is small.

54. In this exercise, you will compute the first several terms of a sequence $\{m_n\}$ of rational numbers that approach $\sqrt{2}$. Let I_1 be the interval $[1,2]$. It contains $\sqrt{2}$. Let $m_1=3/2$ be the midpoint of I_1 so that m_1 divides I_1 into two subintervals. Let I_2 be the subinterval of I_1 that contains $\sqrt{2}$. That is, $I_2=[1,m_1]$ if $2<m_1^2$, and $I_2=[m_1,2]$ otherwise. Let m_2 be the midpoint of I_2. Then m_2 divides I_2 into two subintervals. Let I_3 be the subinterval of I_2 that contains $\sqrt{2}$. Continue this process. Calculate I_n and m_n for $1\leq n\leq 8$. What is the length of the interval I_8? Without calculating $\sqrt{2}$, what is the maximum error that can result if m_8 is used to approximate $\sqrt{2}$? (This procedure for approximating a root is called the *bisection method.*)

55. This exercise is based on an ancient Babylonian method for approximating the square root of a positive number c. Assume that c lies in the interval $(1,4)$ for simplicity. Set $x_0=3/2$. For $n\geq 0$, define

$$x_{n+1}=\frac{1}{2}\left(x_n+\frac{c}{x_n}\right).$$

a. Use the definition of x_{n+1} to obtain the identity $|x_{n+1}-\sqrt{c}|=(x_n-\sqrt{c})^2/2x_n$.

b. Use part (a) to deduce that if x_n approximates $\sqrt{c}$ to k decimal places, then x_{n+1} approximates $\sqrt{c}$ to $2k$ decimal places.

c. Use this algorithm to compute $\sqrt{3.75}$ to 8 decimal places.

56. This exercise is based on an ancient Greek algorithm for approximating $\sqrt{2}$. Let $s_0=d_0=1$. Let $s_n=s_{n-1}+d_{n-1}$ and $d_n=2s_{n-1}+d_{n-1}$ for $n\geq 1$. Let $r_n=d_n/s_n$.

a. Calculate s_n, d_n, and r_n for $1\leq n\leq 10$.

b. Verify the equation

$$d_n^2-2s_n^2=(-1)^{n-1}$$

for $0\leq n\leq 10$. (It can be proved true for all positive integers n.)

c. Compare r_{10} with $\sqrt{2}$. Use the equation of part (b) to explain why r_n is a good rational approximation to $\sqrt{2}$ for large n.

57. This exercise is based on a formula for π published by François Viète in 1593. Let $q_1=1/\sqrt{2}$. For $n\geq 2$, let $q_n=q_1(1+q_{n-1})^{1/2}$. Let $Q_n=q_1q_2\ldots q_n$, and $p_n=2/Q_n$. Calculate p_n for $1\leq n\leq 8$.

58. This exercise describes a very fast procedure for calculating the digits of π (supposing that the digits of $\sqrt{2}$ are already known). Set $a_0=\sqrt{2}$, $b_0=0$, and $p_0=2+\sqrt{2}$. For $n\geq 0$, define

$$a_{n+1}=\frac{1}{2}\left(\sqrt{a_n}+\frac{1}{\sqrt{a_n}}\right)$$

$$b_{n+1}=\sqrt{a_n}\cdot\frac{(1+b_n)}{(a_n+b_n)}$$

$$p_{n+1}=p_n\cdot b_{n+1}\cdot\frac{(a_{n+1}+1)}{(1+b_{n+1})}.$$

Calculate p_n for $n=1,2,$ and 3.

1.5 Combining Functions

Most of the functions that we encounter are built up from simpler functions. This building process is accomplished using arithmetic and other operations. Although these two statements may seem perfectly obvious, it is worth noting explicitly how some of these operations work. The functions that we will be considering from now until Chapter 10 will have both domain and range that are subsets of the real number system. We will not always state this fact explicitly.

Arithmetic Operations

Let c be a constant. Suppose that f and g are functions with the same domain S. For $s \in S$,

$$\begin{aligned} &(f+g)(s) &&\text{means} && f(s)+g(s);\\ &(f-g)(s) &&\text{means} && f(s)-g(s);\\ &(f\cdot g)(s) &&\text{means} && f(s)\cdot g(s);\\ &(cf)(s) &&\text{means} && c\cdot f(s);\ \text{and}\\ &(f/g)(s) &&\text{means} && f(s)/g(s),\ \text{as long as } g(s)\neq 0. \end{aligned}$$

INSIGHT If f and g are functions, then $f+g$, $f-g$, $f\cdot g$, and f/g are new functions. The rules defining the new functions are expressed in terms of the original functions f and g. The next two examples show how these simple ideas are put into practice.

▶ **EXAMPLE 1** Let $f(x) = 2x$, and $g(x) = x^3$. Calculate $f+g$, $f-g$, $f\cdot g$, and f/g.

Solution We calculate that

$$\begin{aligned} (f+g)(x) &= f(x)+g(x) = 2x + x^3;\\ (f-g)(x) &= f(x)-g(x) = 2x - x^3;\\ (f\cdot g)(x) &= f(x)\cdot g(x) = 2x\cdot x^3 = 2x^4;\ \text{and}\\ \left(\frac{f}{g}\right)(x) &= \frac{f(x)}{g(x)} = \frac{2x}{x^3} = \frac{2}{x^2},\ \text{as long as } x\neq 0. \end{aligned}$$ ◀

Given a function f, we let f^2 denote the function $f\cdot f$. That is, $f^2(x) = f(x)^2$. We similarly define f^k for other positive integers k. For instance, if $f(x) = 2x-5$, then $f^4(x) = (2x-5)^4$.

▶ **EXAMPLE 2** Let $f(x) = x^2$, and $g(x) = x+2$. Compute

$$\left(\frac{f+(2g)\cdot f}{g^2}\right)(3).$$

Solution To perform this calculation, we apply the rules to write

$$\left(\frac{f+(2g)\cdot f}{g^2}\right)(3) = \frac{f(3)+2\cdot g(3)\cdot f(3)}{g(3)^2} = \frac{9+2\cdot 5\cdot 9}{5^2} = \frac{99}{25}.$$ ◀

Polynomial Functions

A polynomial function p is a function of the form

$$p(x) = a_N x^N + a_{N-1} x^{N-1} + \cdots + a_1 x + a_0,$$

where N is a nonnegative integer, and a_N does not equal zero. We say that N is the *degree* of the polynomial $p(x)$ and that a_N is the *leading coefficient*. According to this definition, if c is a nonzero constant, then the constant function $q(x) = c$ is a degree 0 polynomial. A degree 1 polynomial p has the form $p(x) = a_1 x + a_0$ and is sometimes called a *linear polynomial* because its graph is a line.

If $p(r) = 0$, then we say that p *vanishes* at r and that r is a *root*, or a *zero*, of p. If r is a root of a polynomial p, then there is a polynomial q of degree one less than that of p such that $p(x) = (x - r) \cdot q(x)$. The polynomial q may be found by division. The Fundamental Theorem of Algebra implies that every polynomial with real coefficients can be factored into a product of polynomials with real-valued coefficients that have degree one or two. A degree 2 polynomial $ax^2 + bx + c$ that cannot be factored as a product of two degree 1 polynomials with real coefficients is said to be *irreducible*. These are the quadratic polynomials with roots

$$\frac{-b + \sqrt{b^2 - 4ac}}{2a} \quad \text{and} \quad \frac{-b - \sqrt{b^2 - 4ac}}{2a}$$

that have nonzero imaginary parts. Such nonreal roots occur if and only if $b^2 < 4ac$.

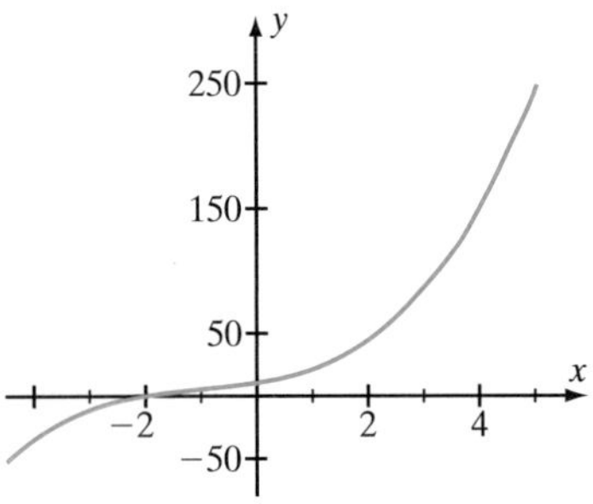

▲ Figure 1
$y = x^3 + 3x^2 + 7x + 10$

▶ **EXAMPLE 3** Factor the polynomial $x^3 + 3x^2 + 7x + 10$ as a product of polynomials with real coefficients.

Solution Because the degree of the given polynomial is greater than two, it can be factored. A plot of the equation $y = x^3 + 3x^2 + 7x + 10$ reveals that -2 is a root, as shown in Figure 1. It follows that $x - (-2) = x + 2$ is a factor. Division of $x + 2$ into $x^3 + 3x^2 + 7x + 10$ yields another factor, $x^2 + x + 5$:

$$\begin{array}{r|l} & x^2 + x + 5 \\ \hline x + 2 & x^3 + 3x^2 + 7x + 10 \\ & x^3 + 2x^2 \\ \hline & \quad\; x^2 + 7x \\ & \quad\; x^2 + 2x \\ \hline & \qquad\; 5x + 10 \\ & \qquad\; 5x + 10 \\ \hline & \qquad\qquad 0 \end{array}$$

According to the Quadratic Formula, the roots of $x^2 + x + 5$ are $(-1 \pm \sqrt{1^2 - 4 \cdot 1 \cdot 5})/(2 \cdot 1)$ or $-1/2 \pm i\sqrt{19}/2$. Because these roots are not real numbers, the polynomial $x^2 + x + 5$ cannot be factored into degree one polynomials with real-valued coefficients. Thus

$$x^3 + 3x^2 + 7x + 10 = (x + 2)(x^2 + x + 5)$$

is the complete factorization over the real numbers. ◀

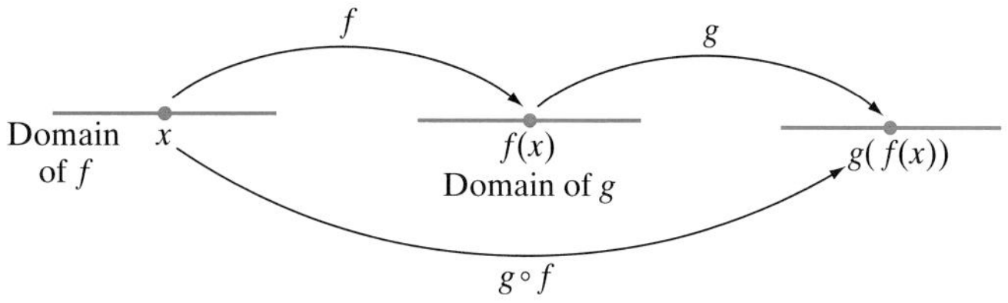

▲ Figure 2

Composition of Functions

Another way of combining functions is by *functional composition*. Suppose that f and g are functions and that the domain of g contains the range of f. This means that if x is in the domain of f, then g may be applied to $f(x)$ (see Figure 2). The result of these two operations, one following the other, is called *g composed with f*, or the *composition* of g with f. We write $g \circ f$ for the composition and define

$$(g \circ f)(x) = g(f(x)).$$

INSIGHT The expression $(g \circ f)(x)$ may look like multiplication of functions, but it is not. For example, if $g(x) = x^2$, and $f(x) = 7x^3$, then

$$(g \cdot f)(x) = g(x) \cdot f(x) = x^2 \cdot 7x^3 = 7x^5,$$

whereas

$$(g \circ f)(x) = g(f(x)) = g(7x^3) = (7x^3)^2 = 49x^6.$$

A good rule when dealing with compositions of functions is to use parentheses to organize formulas. Also, simplify parenthetical expressions by starting from the inside and working out.

▶ **EXAMPLE 4** Let $f(x) = x^2 + 1$, and $g(x) = 3x + 5$. Calculate $g \circ f$ and $f \circ g$.

Solution We calculate that

$$(g \circ f)(x) = g(f(x)) = g(x^2 + 1). \tag{1.5.1}$$

We have started to work inside the parentheses: The first step is to substitute the definition of f, namely $x^2 + 1$, into the equation. The definition of g now says that the evaluation of g at any argument is obtained by multiplying that argument by 3 and then adding 5. This is the rule no matter what the argument of g is. In the present case, we are applying g to $x^2 + 1$. Therefore the right side of equation (1.5.1) equals $3(x^2 + 1) + 5$. This simplifies to $3x^2 + 8$. In conclusion, $(g \circ f)(x) = 3x^2 + 8$. We compute $f \circ g$ in a similar manner:

$$(f \circ g)(x) = f(g(x)) = (g(x))^2 + 1 = (3x + 5)^2 + 1 = 9x^2 + 30x + 26. \quad ◀$$

INSIGHT Example 4 shows that $f \circ g$ can be very different from $g \circ f$. In general, composition of functions is not commutative: The order of the operands g and f matters.

▶ **EXAMPLE 5** How can we write the function $r(x) = (2x + 7)^3$ as the composition of two functions? If $u(t) = 3/(t^2 + 4)$, how can we find two functions v and w for which $u = v \circ w$?

Solution We can think of the function r as two operations applied in sequence. Read the function aloud: First we double and add 7, and then we cube. Thus define $f(x) = 2x + 7$ and $g(x) = x^3$ to get $r(x) = (g \circ f)(x)$, or $r = g \circ f$.

Similarly, read aloud the way we evaluate u at w: First we square t and add 4, and then we divide 3 by the quantity just obtained. As a result, we define $w(t) = t^2 + 4$ and $v(t) = 3/t$. It follows that $u(t) = (v \circ w)(t)$, or $u = v \circ w$. ◀

We can compose three or more functions in a similar way. Thus

$$(H \circ G \circ F)(x) = H(G(F(x))).$$

▶ **EXAMPLE 6** Write the function r from Example 5 as the composition of three functions instead of two.

Solution Read the function aloud: *First* we double, *then* we add 7, and then we cube. We set $F(x) = 2x$, $G(x) = x + 7$, and $H(x) = x^3$. We get $r(x) = (H \circ G \circ F)(x)$. ◀

Inverse Functions

When we work with a function f whose domain is S and range is T, we generally use a rule to calculate the value $f(s)$ in the range that corresponds to a given value s in the domain. Stated in the language of equations, we compute $t = f(s)$ in T for a given value s in S (see Figure 3). There are times when we must "invert" this process. That is, for a given value t in the range of f, we seek a value s for which $f(s) = t$. Stated in geometric terms, we seek the abscissa s of a point (s, t) of intersection of the graph of $y = f(x)$ with the horizontal line $y = t$ (see Figure 4). The following definition describes the functions for which such points of intersection always exist.

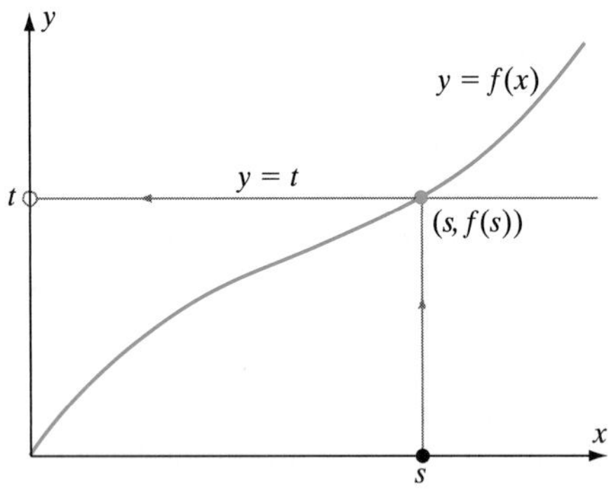

▲ Figure 3

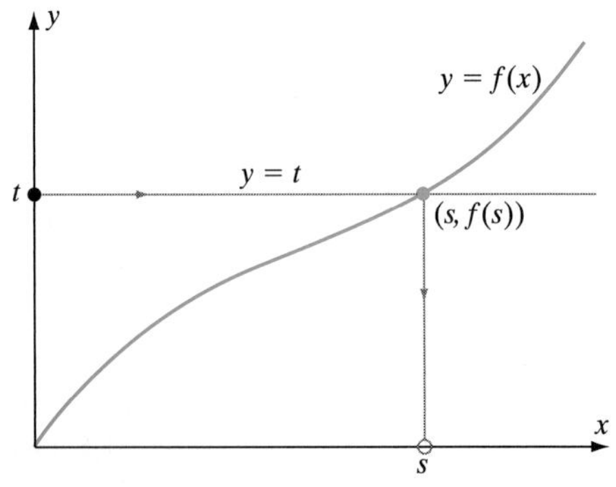

▲ Figure 4

> **DEFINITION** A function $f: S \to T$ is onto if, for every t in T, there is at least one s in S for which $f(s) = t$.

INSIGHT In geometric terms, if f is an onto function and t is any value in the range of f, then the horizontal line $y = t$ intersects the graph of f in at least one point.

▶ **EXAMPLE 7** Let $S = [0, 3]$ and $T = [1, 10]$. Let f and g be functions with domain S and range T defined by $f(x) = 3x + 1$ and $g(x) = 2x^2 - 4x + 4$. Is f onto? What about g?

Solution Figure 5 shows the plots of $y = f(x)$ and $y = g(x)$ for $0 \leq x \leq 3$. We see that every horizontal line $y = t$ with $1 \leq t \leq 10$ intersects the graph of f. Analytically, for every t in the range of f, we can explicitly find a value s in the domain for which $f(s) = t$. We simply set $3s + 1 = t$ and solve $s = (t - 1)/3$. Notice that $0 \leq s$ for $1 \leq t$ and $s \leq 3$ for $t \leq 10$. Thus s is in the domain of f. We conclude that f is onto. On the other hand, we see from Figure 5 that there are values t in T for which the horizontal lines $y = t$ do not intersect the graph of g. We can understand the reason for this as:

$$g(x) = 2(x^2 - 2x) + 4 = 2(x^2 - 2x + 1) + 4 - 2 = 2(x-1)^2 + 2 \geq 2,$$

which shows that the equation $g(s) = t$ has no solution s for any $t < 2$. Figure 5 illustrates this idea with $t = 3/2$. We conclude that g is not onto. ◀

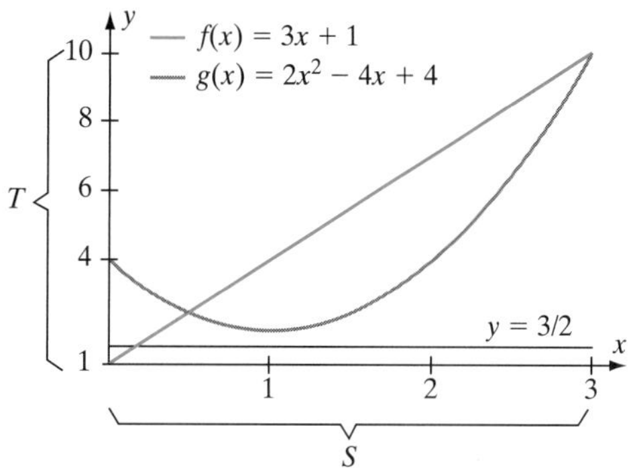

▲ Figure 5

Another difficulty that can arise when we try to invert a function f is that the horizontal line $y = t$ may intersect the graph of $y = f(x)$ in more than one point. The functions for which this complication does not arise are described by the next definition.

DEFINITION A function $f: S \to T$ is *one-to-one* if it takes different elements of S to different elements of T.

INSIGHT In geometric terms, no horizontal line intersects the graph of a one-to-one function in more than one point. Equivalently, a function f is one-to-one if, for any given t, the equation $f(s) = t$ has at most one solution s.

▶ **EXAMPLE 8** Figure 6 shows the graphs of four functions with domain $[0,2]$ and range $[0,4]$. Determine the functions that are onto. Determine the functions that are one-to-one.

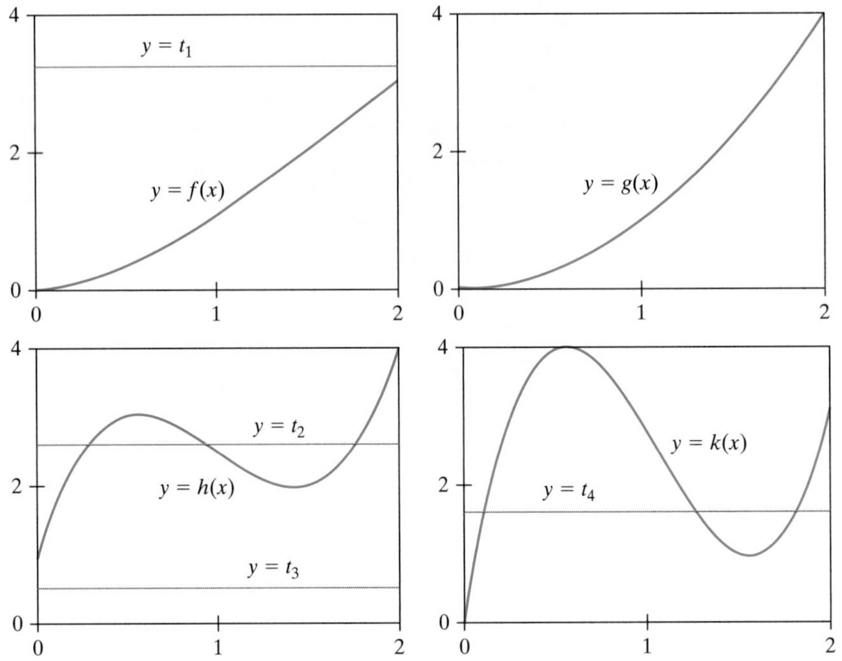

▲ Figure 6

Solution It is evident from the graphs that g and k attain every value in the range. These functions are onto. Because the horizontal line $y = t_1$ does not intersect the graph of f, and the horizontal line $y = t_3$ does not intersect the graph of h, these functions are not onto.

It is evident that the graphs of f and g do not have more than one point of intersection with any horizontal line. Therefore f and g are one-to-one. Because the horizontal line $y = t_2$ intersects the graph of h in more than one point and because the horizontal line $y = t_4$ intersects the graph of k in more than one point, these functions are not one-to-one. In summary, f is one-to-one but not onto, g is one-to-one and onto, h is neither one-to-one nor onto, and k is onto but not one-to-one. ◀

Suppose that a function $f: S \to T$ is both one-to-one and onto. Then, for every t in the range of f, there is exactly one s in the domain of f such that $f(s) = t$. If we denote by $f^{-1}(t)$ the unique value of s such that $f(s) = t$, then we have $f(f^{-1}(t)) = t$ for every t in the range of f. Notice that f^{-1} is a function with domain T and range S. We say that the function f is *invertible*, and $f^{-1}: T \to S$ is the *inverse function* of f.

If $f: S \to T$ is invertible, then the equations $f(s) = t$ and $s = f^{-1}(t)$ yield the equation $f^{-1}(f(s)) = s$ for every s in S. This equation tells us that f is the inverse of its inverse function f^{-1}. In symbols, we have $(f^{-1})^{-1} = f$. A convenient way to think about inverse functions is that f^{-1} "undoes" or "reverses" what f does. Symmetrically, f undoes what f^{-1} does. We summarize our findings:

THEOREM 5 Suppose that a function $f:S \to T$ is both one-to-one and onto. Then there is a unique function $f^{-1}:T \to S$ such that

$$f(f^{-1}(t)) = t \text{ for all } t \text{ in } T \quad \text{and} \quad f^{-1}\big(f(s)\big) = s \text{ for all } s \text{ in } S. \qquad (1.5.2)$$

INSIGHT If $f:S \to T$ is invertible, then $f \circ f^{-1}$ is the identity function on T and $f^{-1} \circ f$ is the identity function on S. That is, the functions $f \circ f^{-1}: T \to T$ and $f^{-1} \circ f: S \to S$ satisfy

$$(f \circ f^{-1})(t) = t \text{ for all } t \text{ in } T \quad \text{and} \quad (f^{-1} \circ f)(s) = s \quad \text{for all } s \text{ in } S.$$

The next two examples illustrate techniques for finding the inverse of an invertible function.

▶ **EXAMPLE 9** The function $f:\mathbb{R} \to \mathbb{R}$ defined by $f(s) = 2s + 3$ is one-to-one and onto. Find its inverse, f^{-1}.

Solution By substituting $f^{-1}(t)$ for s in the equation $f(s) = 2s + 3$, we obtain $f(f^{-1}(t)) = 2f^{-1}(t) + 3$. Because $f(f^{-1}(t)) = t$ by (1.5.2), we see that $2f^{-1}(t) + 3 = t$. Therefore

$$f^{-1}(t) = \frac{t-3}{2}.$$

We may check this solution as follows:

$$f(f^{-1}(t)) = 2 \cdot f^{-1}(t) + 3 = 2 \cdot \frac{(t-3)}{2} + 3 = t$$

and

$$f^{-1}(f(s)) = \frac{(f(s) - 3)}{2} = \frac{(2s + 3 - 3)}{2} = s.$$

INSIGHT It is a common error to suppose that f^{-1} means the reciprocal $1/f$ of f. This is not so. For the function $f(s) = 2s + 3$, $s \in \mathbb{R}$ of Example 9,

$$\left(\frac{1}{f}\right)(t) = \frac{1}{f(t)} = \frac{1}{2t+3} \neq \frac{t-3}{2} = f^{-1}(t).$$

To summarize:

$f(t)^{-1}$ means $1/f(t)$, the reciprocal of the number $f(t)$.

$f^{-1}(t)$ means the value of the function f^{-1} at t.

In the next example, we find an inverse function using essentially the same method as that of the preceding example. The difference is that we use simpler notation.

▶ **EXAMPLE 10** Let $f:[0,2] \to [1,2]$ be defined by $f(x) = (x^3/8) + 1$. Find a formula for the function f^{-1}.

Solution We set $y = f(x) = x^3/8 + 1$ and solve the equation $x^3/8 + 1 = y$ for x in terms of y. We find that $x^3 = 8(y-1)$, or $x = 2(y-1)^{1/3}$. Because $x = f^{-1}(y)$, we conclude that $f^{-1}(y) = 2(y-1)^{1/3}$. ◀

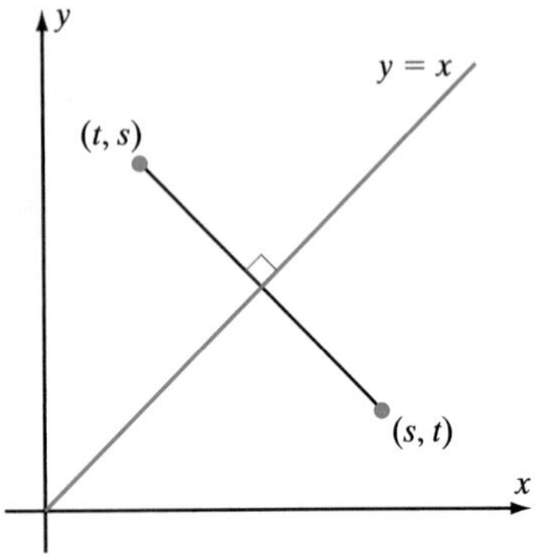

▲ Figure 7

INSIGHT When we graph $y = f(x)$, we measure the function's argument x along the horizontal axis, and we measure the functional value y along the vertical axis. That is, we label the horizontal axis with the independent variable x and the vertical axis with the dependent variable y. When we find an inverse function, we reverse the roles that the variables originally had. Thus in Example 10, the independent variable of the function $f^{-1}(y) = 2(y-1)^{1/3}$ is y. We may plot this function in the usual way and label the horizontal axis by y. Some instructors prefer to consistently use x for the horizontal axis and y for the vertical axis. To follow that convention, simply replace y with x on both sides of the equation $f^{-1}(y) = 2(y-1)^{1/3}$ and plot $f^{-1}(x) = 2(x-1)^{1/3}$. Remember, in a formula for a function, the variable used as the argument serves as a placeholder. If we rename the placeholder, then the function does not change. The functions $f^{-1}(y) = 2(y-1)^{1/3}$ and $f^{-1}(x) = 2(x-1)^{1/3}$ are the same.

The Graph of the Inverse Function

It is not always possible to find an explicit formula for the inverse function f^{-1} of an invertible function $f: S \to T$. Nevertheless, we can often infer useful information from the graph of f^{-1}, which can *always* be obtained from the graph of f. Suppose that (s,t) is a point on the graph of f. This means $t = f(s)$, and, therefore, $s = f^{-1}(t)$. In other words, the point (t,s) is on the graph of f^{-1}. Figure 7 shows that the points (s,t) and (t,s) are reflections of each other through the line $y = x$. (Exercise 60 provides a rigorous justification.) If we apply this observation to all points on the graph of f, then we see that the graph of f^{-1} is the reflection through the line $y = x$ of the graph of f (see Figure 8). We state our findings as a theorem:

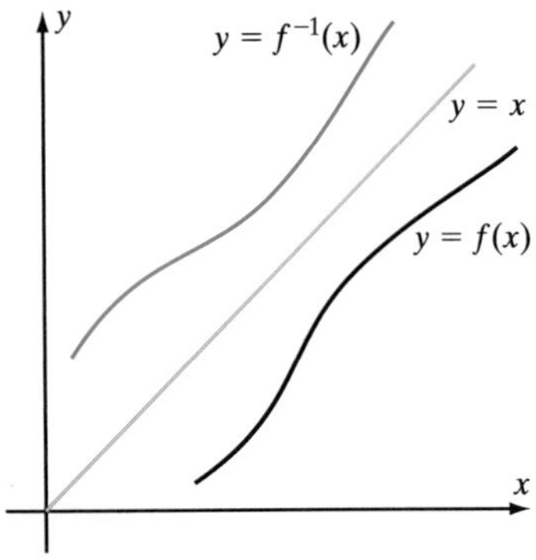

▲ Figure 8

THEOREM 6 If $f: S \to T$ is an invertible function, then the graph of $f^{-1}: T \to S$ is obtained by reflecting the graph of f through the line $y = x$.

▶ **EXAMPLE 11** Use the functions f and f^{-1} of Example 10 to illustrate Theorem 6.

Solution The graphs of $f(x) = x^3/8 + 1$ and $f^{-1}(x) = 2(x-1)^{1/3}$ are shown in Figure 9. Notice that the domain of f^{-1} is the range $[1,2]$ of f, and the range of f^{-1} is the domain $[0,2]$ of f. It is evident that each graph is the reflection of the other through the line $y = x$. ◀

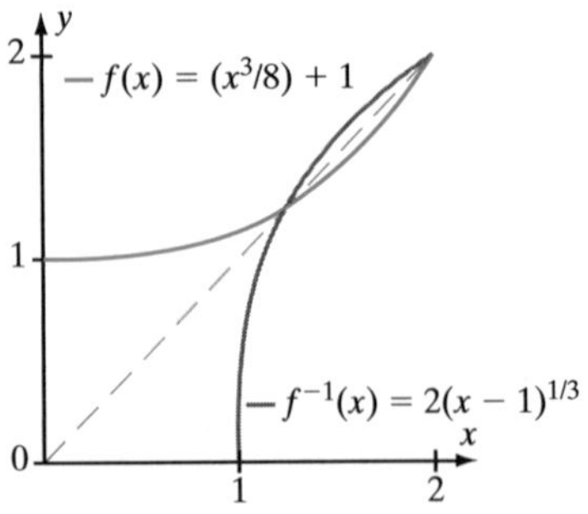

▲ Figure 9

INSIGHT To verify Theorem 6 on a calculator or computer, make sure that each axis is equally scaled. The reflection property will not appear correctly unless the length of 1 unit on the vertical axis is the length of 1 unit on the horizontal axis.

Vertical and Horizontal Translations

It is useful to obtain the graphs of the functions $x \mapsto f(x+h)$ and $x \mapsto f(x)+k$ from the graph of the function f. A point on the graph of $x \mapsto f(x+h)$ is of the form $(x, f(x+h))$. A horizontal shift, or *translation*, by h of this point results in $(x+h, f(x+h))$, which is a point on the graph of f. Vertical translations are treated similarly. We record these observations as a theorem.

THEOREM 7 The graph of $x \mapsto f(x+h)$ is obtained by shifting the graph of f horizontally by an amount h. The shift is to the left if $h>0$ and to the right if $h<0$ (see Figure 10). The graph of $x \mapsto f(x)+k$ is obtained by shifting the graph of f vertically by an amount k. The shift is up if $k>0$ and down if $k<0$ (see Figure 11).

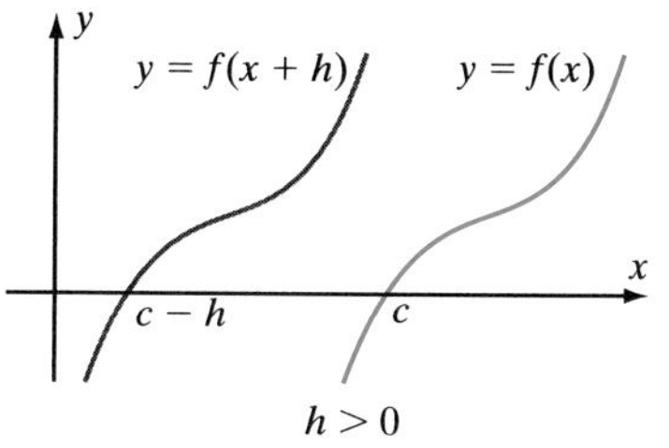

▲ Figure 10a

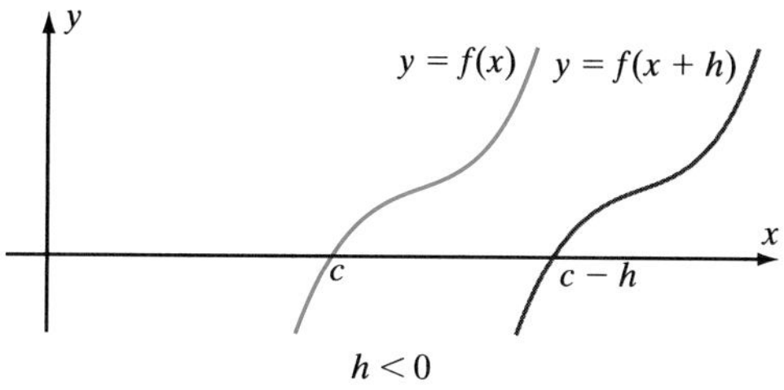

▲ Figure 10b

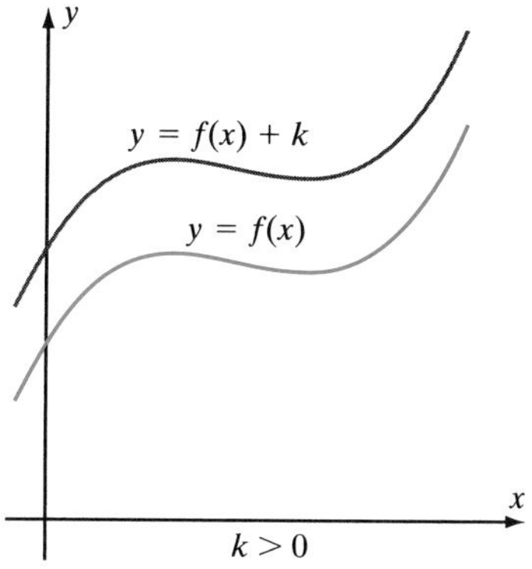

▲ Figure 11a

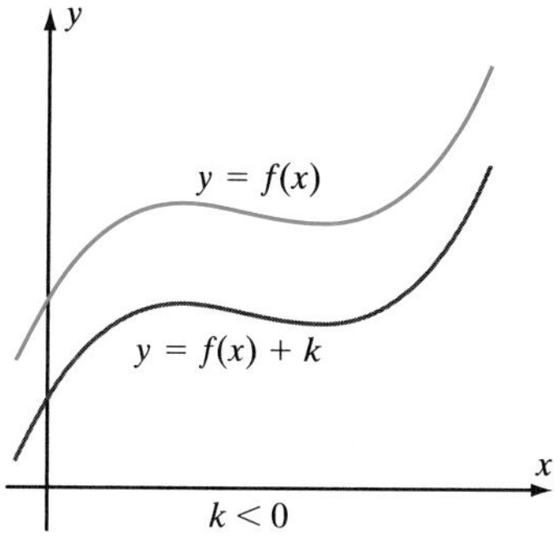

▲ Figure 11b

▶ **EXAMPLE 12** Describe the relationship between the graph of $f(x)=x^2+6x+13$ and the parabola $y=x^2$.

Solution We use the method of completing the square to write

$$\begin{aligned} f(x) &= x^2+6x+13 \\ &= \left(x^2+6x+\left(\frac{6}{2}\right)^2\right)+13-\left(\frac{6}{2}\right)^2 \\ &= (x+3)^2+4. \end{aligned}$$

Thus we obtain the graph of f by shifting the parabola $y=x^2$ to the left 3 units and up 4 units. The graph of f is therefore a parabola with vertex at $(-3,4)$ (see Figure 12). ◀

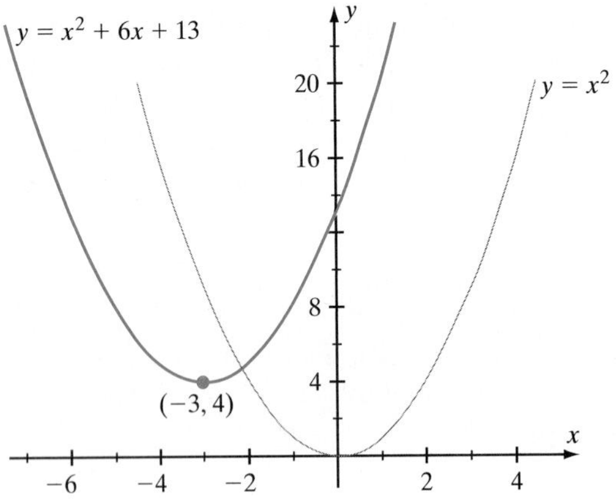

▲ Figure 12

Even and Odd Functions

If the graph of a function has the property of symmetry, then that property can be used as an aid in sketching the graph. It is therefore useful to identify algebraic criteria for symmetry. We discuss two basic symmetries here and explore several others in the exercises.

A function is *even* if $f(-x) = f(x)$ for every x in its domain. Observe that, if n is an even integer, then $f(x) = x^n$ is an even function—that is the reason for the terminology. A function is *odd* if $f(-x) = -f(x)$ for every x in its domain. If n is an odd integer, then $f(x) = x^n$ is an odd function. The graph of an even function is symmetric about the $y-$axis: If the function is plotted only for $x \geq 0$, then its complete graph may be obtained by reflecting across the y-axis. The graph of an odd function is symmetric about the origin: If the function is graphed for $x \geq 0$, then the complete graph may be obtained by rotating 180 degrees, which is a reflection through the origin.

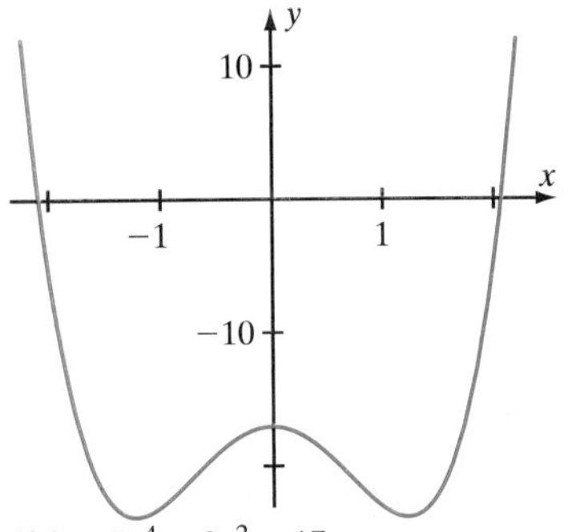

$f(x) = 3x^4 - 9x^2 - 17$

▲ Figure 13

▶ **EXAMPLE 13** What symmetries do the graphs of $f(x) = 3x^4 - 9x^2 - 17$ and $g(x) = (4x^3 - 10x)/(2 + 6x^2)$ possess?

Solution Because $f(-x) = 3(-x)^4 - 9(-x)^2 - 17 = 3x^4 - 9x^2 - 17 = f(x)$, we conclude that f is an even function, and its graph is symmetric about the y-axis. Similarly we calculate that

$$\begin{aligned} g(-x) &= \frac{4(-x)^3 - 10(-x)}{2 + 6(-x)^2} \\ &= \frac{-4x^3 + 10x}{2 + 6x^2} \\ &= -\left(\frac{4x^3 - 10x}{2 + 6x^2}\right) \\ &= -g(x). \end{aligned}$$

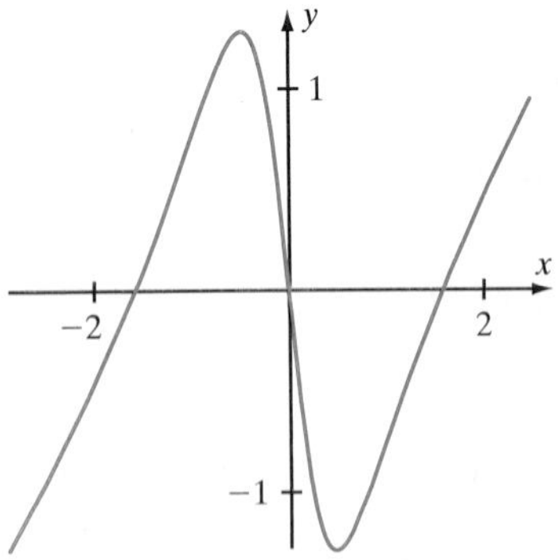

$g(x) = (4x^3 - 10x)/(2 + 6x^2)$

▲ Figure 14

Thus g is an odd function. Its graph is symmetric about the origin (see Figures 13 and 14). ◀

Pairing Functions–Parametric Curves

Given two functions φ_1 and φ_2 with a common domain I, we can plot the points $(\varphi_1(t), \varphi_2(t))$ as t varies through I. The resulting graph is called a *parametric curve*. This notion is made precise by the following definition.

DEFINITION Suppose C is a curve in the plane, and I is an interval of real numbers. If φ_1 and φ_2 are functions with domain I, and if the plot of the points $(\varphi_1(t), \varphi_2(t))$ for t in I coincides with C, then C is said to be *parameterized* by the equations $x = \varphi_1(t)$ and $y = \varphi_2(t)$. These equations are called *parametric equations* for C. The variable t is said to be a *parameter* for the curve, and I is the *domain* of the parameterization of C. We say that C is a *parametric curve*.

Notice that, when a curve is parameterized by the equations $x = \varphi_1(t)$ and $y = \varphi_2(t)$, y is not given as a function of x. Instead, each of the variables x and y is given as a function of the parameter t.

▶ **EXAMPLE 14** Purchasing a stock amounts to purchasing a share of a company. The value of such an investment rises or falls with the value of the company. Purchasing a bond amounts to making a loan; interest is paid to the bondholder until the loan is repaid. In general, bonds are not as rewarding as stocks, but they are less risky. The risk $\varphi_1(t)$ and reward (i.e., investment return) $\varphi_2(t)$ of a combined investment portfolio of stocks and bonds depend on the percentage t of the portfolio that is made up of stocks. The following two tables present risk and reward data for several values of t.

t	0	20	35	50	65	80	100
$\varphi_1(t)$ **(Risk)**	0.1179	0.1121	0.1186	0.1325	0.1513	0.1735	0.2085

t	0	20	35	50	65	80	100
$\varphi_2(t)$ **(Return)**	8.333	9.583	10.92	12.10	13.125	14.25	15.63

Sketch the risk-reward curve.

Solution The common domain of $\varphi_1(t)$ and $\varphi_2(t)$ is the interval $I = [0, 100]$. Therefore the domain of parameterization of the risk-reward curve is $I = [0, 100]$. Figure 15, which illustrates the risk-reward relationship as a parametric curve, was produced by plotting the points $(\varphi_1(t), \varphi_2(t))$ for $t = 0, 20, 35, 50, 65, 80, 100$ and then joining the points by a continuous curve. (Filling in points on a curve between plotted data points is called interpolation.) Figure 15 enables us to understand the risk-reward relationship at a glance. The portion of the curve that corresponds to values of t greater than 20 slopes upward. In other words, as risk increases, the investor's return also increases. (Investors must be compensated for knowingly assuming greater risk.) For $t < 20$, the risk-reward curve slopes downward; return on investment decreases as risk increases. (As the percentage of stocks becomes too small, the portfolio suffers from a lack of diversification, and overall risk

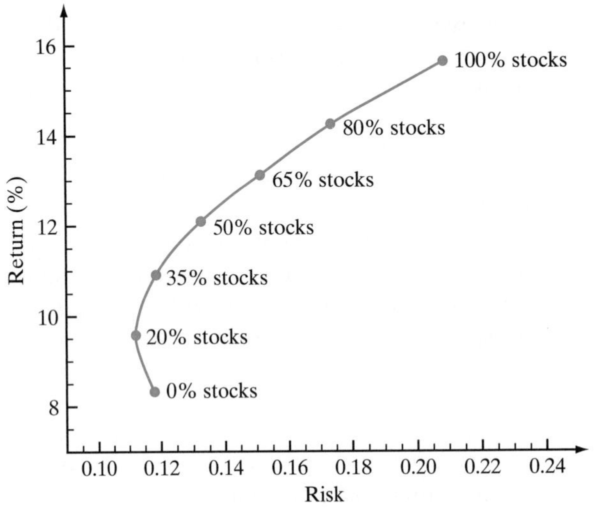

▲ Figure 15 In a mixed portfolio of stocks and bonds, the expected return of the portfolio and the risk of the portfolio are functions of the percentage of stocks.

increases, even though the risks of the components of the portfolio decrease. Investors are often insensitive to, or unaware of, this sort of risk.) ◀

If a curve $\mathcal{C}$ is defined by parametric equations $x = \varphi_1(t)$ and $y = \varphi_2(t)$, then it is often useful to imagine that the parameter t denotes time. We may then think of a particle tracing out the curve $\mathcal{C}$ and occupying the point $\big(\varphi_1(t), \varphi_2(t)\big)$ at time t. In this context, we may refer to $\mathcal{C}$ as a *path* or a *trajectory*.

▶ **EXAMPLE 15** A particle moves in the xy-plane with coordinates given by

$$x = 2 - t^2 \quad \text{and} \quad y = 1 + 3t^2, \qquad -1 \le t \le 2.$$

Describe the particle's path $\mathcal{C}$.

Solution We can rewrite the equation for x as $t^2 = 2 - x$. Substituting this equation into the parametric equation for y results in $y = 1 + 3(2 - x)$, or $y = 7 - 3x$. However, the curve $\mathcal{C}$ is not the entire straight line with Cartesian equation $y = 7 - 3x$. To see this, we analyze the motion of the particle. It starts at the point $(1, 4)$ when $t = -1$. As t increases from -1 to 0, the x-coordinate increases to 2, and the y-coordinate decreases to 1. Thereforethe line segment from $(1, 4)$ to $(2, 1)$ ($\overline{AB}$ in Figure 16) is traced for $-1 \le t \le 0$. Then, as t increases from 0 to 2, x decreases to -2, and y increases to 13 ($\overline{BC}$ of Figure 16). We conclude that $\mathcal{C}$ is the line segment $\overline{ABC}$ with endpoints $(-2, 13)$ and $(2, 1)$. Segment $\overline{AB}$ of $\mathcal{C}$ is traced once in each direction. ◀

INSIGHT In Example 15, we eliminated the parameter t to deduce that the parametric curve $\mathcal{C}$ is the graph of a Cartesian equation. Although this method is often useful, it is not always possible to eliminate the parameter. When we do eliminate the parameter as an aid in sketching the curve, we should not neglect the information that the parameterization carries.

Parameterized Curves and Graphs of Functions

Example 14 shows that a parameterized curve is not necessarily the graph of a function. However, we can always think of the graph of a function $f: S \to T$ as a parameterized curve—we simply use the parametric equations $x = s$, $y = f(s)$ with S as the domain of the parameterization. This idea allows us to use a graphing calculator or computer algebra system to graph the inverse of an invertible function f that has domain S. To do so, we plot the parametric equations $x = f(s)$ and $y = s$ for s in S. These equations do the job because the coordinates of any point (x, y) on the parametric curve satisfy $y = s = f^{-1}(f(s)) = f^{-1}(x)$. Notice that it is *not* necessary to find an explicit formula for f^{-1} to use this technique.

▶ **EXAMPLE 16** Let $f: [0,1] \to [1,3]$ be defined by $f(s) = s^5 + s + 1$, $0 \le s \le 1$. Sketch the graph of the inverse function f^{-1}.

Solution The function f^{-1} maps the interval $[1,3]$ to $[0,1]$ (see Figure 17). The command in the caption for Figure 17 shows how to plot an inverse function using the computer algebra system Maple. ◀

QUICK QUIZ

1. If $g(x) = (2x+1)/(3x+4)$ for $x > 0$ and $f(x) = 1/x$ for $x > 0$, what is $(g \circ f)(x)$?
2. The function $f: [1, \infty) \to (0, 1/2]$ defined by $f(x) = x/(1+x^2)$ is invertible. What is $f^{-1}(x)$?
3. How do the graphs of $f(x) = x^3 + 2$ and $g(x) = (x-1)^3 + 4$ compare?
4. Does either of $f(x) = (x^2+1)/(3x^4+5)$ or $g(x) = x^3/(3x^4+5)$ have a graph that is symmetric with respect to the y-axis? With respect to the origin?
5. Describe the parametric curve $x = |t| + t$, $y = 2t$, $-1 \le t \le 1$.

Answers

1. $(x+2)/(4x+3)$ 2. $(1+\sqrt{1-4y^2})/(2y)$ 3. The graph of g is obtained by shifting the graph of f 1 unit to the right and 2 units up. 4. f is symmetric with respect to the y-axis; g is symmetric with respect to the origin. 5. The vertical line segment from $(0,-1)$ to $(0,0)$ followed by the line segment from $(0,0)$ to $(2,2)$.

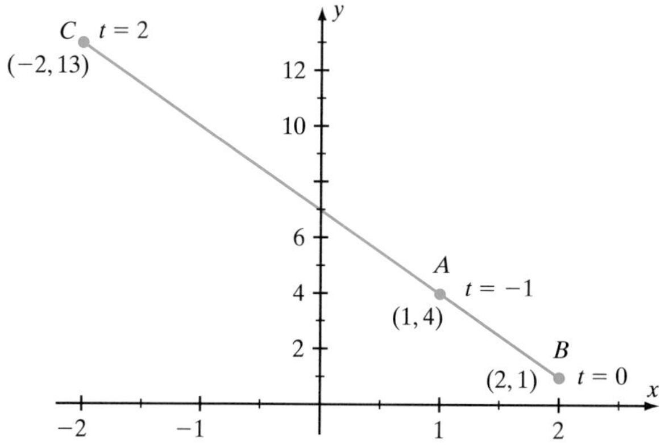

▲ Figure 16

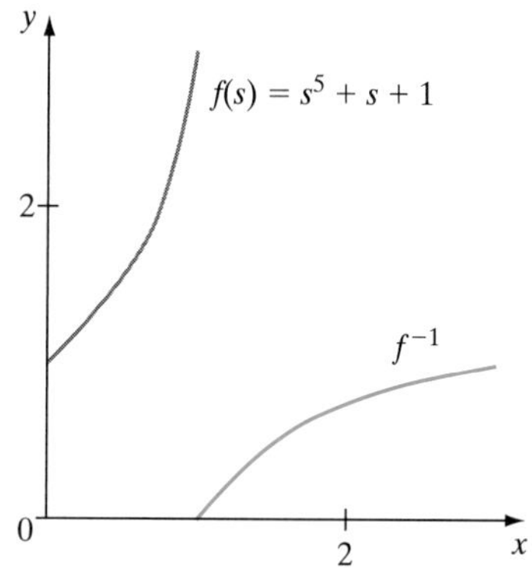

plot ([s, f(s), s = 0..1], view = [0..3, 0..3], scaling=constrained, color = GRAY);

plot([f(s), s, s = 0..1], view=[0..3, 0..3], scaling = constrained, color = BLUE);

▲ Figure 17

EXERCISES

Problems for Practice

Let $F(x)=x^2+5$, $G(x)=(x+1)/(x-1)$, and $H(x)=2x-5$. In Exercises 1–10, calculate the value of the given function at x.

1. $F+G$
2. $F-3H$
3. $G\circ H$
4. $H\circ G$
5. $H\cdot F-H\circ F$
6. G/F
7. $F\circ G\circ H$
8. $H\circ F-F\circ H$
9. $G\circ(1/G)$
10. $H\circ H\circ H-H\circ H$

In Exercises 11–14, write the function h as the composition $h=g\circ f$ of two functions. (There is more than one correct way to do this.)

11. $h(x)=(x-2)^2$
12. $h(x)=2x+7$
13. $h(x)=(x^3+3x)^4$
14. $h(x)=3/\sqrt{x}$

In Exercises 15–18, find a function g such that $h=g\circ f$.

15. $h(x)=3x^2+6x+4$, $f(x)=x+1$
16. $h(x)=x^2+4$, $f(x)=x-1$
17. $h(x)=(x^2+1)/(x^4+2x^2+3)$, $f(x)=x^2+1$
18. $h(x)=2x^2+x-\sqrt[6]{x}+1$, $f(x)=\sqrt{x}$

Let $f(x)=\sqrt{2x+5}$, and $g(x)=x^{-1/3}$. In Exercises 19–22, calculate the given expression.

19. $(f\circ g)(1/8)$
20. $(g\circ f)(2)$
21. $f^2(11)\cdot g^3(54)$
22. $(g\circ g)(512)$

In each of Exercises 23–26, write the given polynomial as a product of irreducible polynomials of degree one or two.

23. x^2+4x-5
24. x^3+x^2-4x-4
25. $x^4+2x^3-2x^2-8x-8$
26. x^4+3x^2+2

In each of Exercises 27–38, a function $f:S\to T$ is specified. Determine if f is invertible. If it is, state the formula for $f^{-1}(t)$. Otherwise, state whether f fails to be one-to-one, onto, or both.

27. $S=[0,\infty)$, $T=[1,\infty)$, $f(s)=s^2+1$
28. $S=[0,2]$, $T=[-1,1]$, $f(s)=(s-1)^2$
29. $S=[0,1]$, $T=[0,2]$, $f(s)=s^2+s$
30. $S=(0,\infty)$, $T=(1,\infty)$, $f(s)=s^4+1$
31. $S=[-2,5)$, $T=[-35,98)$, $f(s)=s^3-27$
32. $S=\mathbb{R}$, $T=\mathbb{R}$, $f(s)=s^4-2s$
33. $S=(4,\infty)$, $T=(1,16/15)$, $f(s)=s^2/(s^2-1)$
34. $S=[0,1]$, $T=[0,1/2]$, $f(s)=s/(s+1)$
35. $S=(1,6)$, $T=(2,3)$, $f(s)=\sqrt{s+3}$
36. $S=(1,4)$, $T=(4,5)$, $f(s)=\sqrt{s}+3$
37. $S=[1,\infty)$, $T=(0,1]$, $f(s)=1/(s^2+1)$
38. $S=(0,\infty)$, $T=(0,1]$, $f(s)=1/(1+s^4)$

39. Examine the graphs in Figure 18 and determine which represent invertible functions. If the graph represents an invertible function, draw a graph of the inverse.

In Exercises 40–43, two functions f and g are given. Find a constant h such that $g(x)=f(x+h)$. What horizontal translation of the graph of f results in the graph of g?

40. $f(x)=2x+1$, $g(x)=2x+5$
41. $f(x)=1-3x$, $g(x)=7-3x$
42. $f(x)=x^2+4$, $g(x)=x^2-6x+13$
43. $f(x)=\sqrt{1-x^2}$, $g(x)=\sqrt{2x-x^2}$

In Exercises 44–47, two functions f and g are given. Find constants h and k such that $g(x)=f(x+h)+k$. Describe the relationship between the plots of f and g.

44. $f(x)=x^2$, $g(x)=x^2+2x+5$
45. $f(x)=3x^2$, $g(x)=3x^2+12x$
46. $f(x)=\sqrt{1-x^2}$, $g(x)=1+\sqrt{2x-x^2}$
47. $f(x)=(x+3)/x$, $g(x)=-x/(x+3)$

In Exercises 48–52, describe the curve that is the graph of the given parametric equations.

48. $x=7$, $y=t^2+1$, $-1\le t\le 2$
49. $x=2t+1$, $y=6t-4$
50. $x=1/t$, $y=3, 0<t\le 1$
51. $x=12t^2+1$, $y=2t$
52. $x=1/(1+t^2)$, $y=1+t^2$

Further Theory and Practice

53. If p and q are nonzero polynomials, express the degree deg $(p\cdot q)$ of $p\cdot q$ in terms of the degrees deg (p) and deg (q) of p and q. Do the same for deg $(p\circ q)$. Is deg$(p\circ q)$ always equal to deg$(q\circ p)$? What is the relationship of deg$(p\pm q)$ to deg(p) and deg(q)?
54. Find all polynomials p such that $(p\circ p)(x)=x$. *Hint:* What degree must p have?
55. Suppose $p(x)$ is a polynomial of degree $n>1$. Let $I(x)=x$, and define $f=p\circ(I+p)$. What is deg (f)? How are the roots of p related to the roots of f? Show that there is a polynomial q such that $f=p\circ q$. Illustrate with $p(x)=x^2-3x-4$.

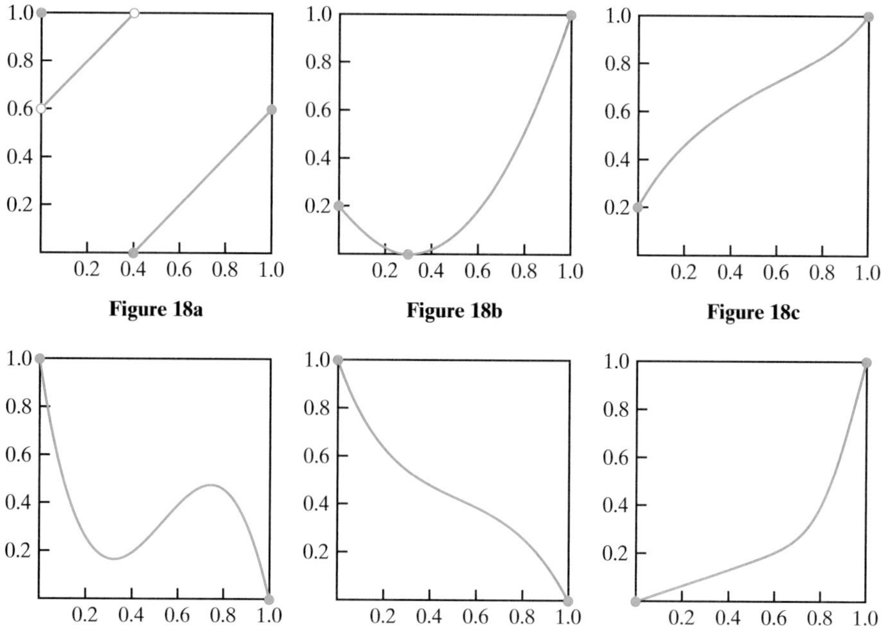

Figure 18a **Figure 18b** **Figure 18c**

Figure 18d **Figure 18e** **Figure 18f**

▲ Figure 18

56. A function that is a quotient of polynomials is called a *rational function*. Every rational function can be written in the form $f(x) + g(x)/h(x)$, where f, g, and h are polynomial functions, and the degree of $g(x)$ is *strictly less than* the degree of $h(x)$. Write $(3x^5 + 2x^4 - x^2 + 6)/(x^3 - x + 3)$ in this form.

57. An *affine* function $f : \mathbb{R} \to \mathbb{R}$ has the form $f(x) = ax + b$, where a and b are constants. Prove that the composition of two affine functions is affine and that the inverse of an invertible affine function is affine.

58. If $f: S \to T$ is a real-valued function, and c is a constant, let $f + c$ be defined by $(f + c)(x) = f(x) + c$. State the domain and range of $f + c$. Show that if f is invertible, then $f + c$ is invertible. What is the relationship between $(f + c)^{-1}$ and f^{-1}?

59. Prove that if $f: (a, b) \to (c, d)$ and $g: (c, d) \to (\alpha, \beta)$ are invertible, then $g \circ f$ is invertible.

60. Let $P = (s, t)$ be a point in the xy-plane. Let $P' = (t, s)$ Calculate the slope of the line ℓ' that passes through P and P' Deduce that ℓ' is perpendicular to the line ℓ whose equation is $y = x$. Let Q be the point of intersection of ℓ and ℓ' Show that P and P' are equidistant from Q. (As a result, P and P' are reflections of each other through the line $y = x$.

In Exercises 61–64, find a function g such that $g \circ f = h$.

61. $f(x) = x + 1$, $h(x) = x^2 + 2x + 3$

62. $f(x) = 2x + 3$, $h(x) = (x + 5)/(x - 5)$

63. $f(x) = x^2 - 9$, $h(x) = 2x^2$

64. $f(x) = (x - 1)/x^2$, $h(x) = x^2/(x - 1)$

In Exercises 65–68, find a function f such that $g \circ f = h$.

65. $g(x) = x^2 + 2$, $h(x) = x^2 - 8x + 18$

66. $g(x) = (x - 1)/(x + 1)$, $h(x) = x^2/(x^2 + 2)$

67. $g(x) = x^3 + 1$, $h(x) = x^2$

68. $g(x) = 4x + 5$, $h(x) = 8x$

In each of Exercises 69–72, find a function f whose graph is the given curve $\mathcal{C}$.

69. $\mathcal{C}$ is obtained by translating the graph of $y = x^2$ to the right by 3 units.

70. $\mathcal{C}$ is obtained by translating the graph of $y = x^3 + 2x$ down 3 units and 2 units to the right.

71. $\mathcal{C}$ is obtained by reflecting the graph of $y = (x^3 + 1)/(x^2 + 1)$ about the y-axis.

72. $\mathcal{C}$ is obtained by reflecting the graph of $y = (x + 1)/(x^4 + 1)$ about the origin.

73. Let $f(x) = x^p$ for some fixed power p. Is it ever true that $f \circ f = f \cdot f$?

74. What composition property of the function $f(x) = (1 - x)/(1 + x)$ is illustrated by Figure 19?

75. Let $f(x) = x^5$, $I(x) = x$, and $p(x) = \pi$ for all x. Find functions φ, ψ, μ and λ such that $\varphi \circ f = p$, $f \circ \psi = p$, $\mu \circ f = I + p$ and $f \circ \lambda = I + p$.

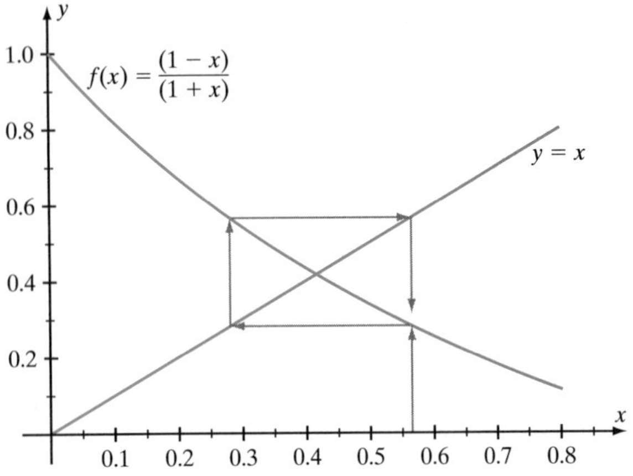

▲ Figure 19

76. Let $\mathcal{T}$ be the collection of functions of the form

$$x \mapsto \frac{ax+b}{cx+d}$$

with $ad-bc=1$. (These functions are called *linear fractional transformations.*) Show that the composition of two functions in $\mathcal{T}$ is also in $\mathcal{T}$. By setting $a=d=1$ and $b=c=0$, we see that the identity function $I(x)=x$ is in $\mathcal{T}$. Show that every function in $\mathcal{T}$ has an inverse that is also in $\mathcal{T}$ and that is obtained by swapping a and d, and negating b and c.

77. Let

$$f(x) = \begin{cases} 2x & \text{if } x \in (0,3] \\ 12-2x & \text{if } x \in (3,6] \\ 0 & \text{if } x \in (0,6] \end{cases}.$$

Graph f. Notice that f is a one-tooth function. Show that $f \circ f$ is a two-tooth function. Graph it.

78. If h is a fixed positive number, then the function r defined

$$r(x) = \frac{f(x+h/2) - f(x-h/2)}{h}$$

measures the *average rate of change* of f over the interval $[x-h/2,\ x-h/2]$. If $f(x)=mx+b$, then what is $r(x)$? If $f(x)=ax^2+mx+b$, then what is $r(x)$? Show that if f is a polynomial of degree n, then r is a polynomial of degree $n-1$.

79. Let p be a positive constant. Find functions F, G, and H that allow you to express the power law for logarithms,

$$\log_{10}(x^p) = p \cdot \log_{10}(x),$$

and the addition law for logarithms,

$$\log_{10}(p \cdot x) = \log_{10}(p) + \log_{10}(x),$$

in the form $\log_{10} \circ G = F \circ \log_{10}$ and $\log_{10} \circ F = H \circ \log_{10}$, respectively.

▶ In Exercises 80–83, describe and sketch the curve that has the given parametric equations. ◀

80. $x = 2t^2 + t - 1,\ y = t + 1$

81. $x = t^2,\ y = t^3$

82. $x = 3\log_{10}(t),\ y = 5\log_{10}(t)$

83. $x = \log_{10}(t)/\log_{10}(2),\ y = 3t$

84. Suppose $A = (x_1, y_1)$ and $B = (x_2, y_2)$ are any two distinct points in the plane. Let

$$\begin{aligned} \varphi_1(t) &= (1-t)\cdot x_1 + t\cdot x_2 \\ \varphi_2(t) &= (1-t)\cdot y_1 + t\cdot y_2 \end{aligned}$$

Show that $x = \varphi_1(t)$, $y = \varphi_2(t)$, $0 \le t \le 1$, is a parameterization of line segment $\overline{AB}$.

85. Let $S(x) = -x$. Express the property that f is an even function by means of the composition of S and f. Express the property that g is an odd function by means of the composition of S and g.

86. Let p be any polynomial.

a. Show that

i. $q(x) = (p(x) + p(-x))/2$ is an even polynomial, and

ii. $r(x) = (p(x) - p(-x))/2$ is an odd polynomial.

Because $p = q + r$, this shows that every polynomial can be written as the sum of an even polynomial and an odd polynomial.

b. Show that q contains only even powers of x, and r contains only odd powers.

c. If p is even, deduce that the coefficient of each odd power of x in $p(x)$ is zero. If p is odd, deduce that the coefficient of each even power of x in $p(x)$ is zero.

d. If p is even, show that there is a polynomial $s(x)$ such that $p(x) = s(x^2)$.

e. If p is odd, show that $p(0) = 0$. Deduce that there is an even polynomial $t(x)$ such that $p(x) = x \cdot t(x)$.

Calculator/Computer Exercises

▶ In Exercises 87–90, graph the curves $\mathcal{C}$ and $\mathcal{C}'$ in the same viewing window. ◀

87. $\mathcal{C} = \{(x,y) : y = \sqrt{x^2+2x+2}\}$; $\mathcal{C}'$ is obtained by translating $\mathcal{C}$ to the right by 2 units and up by 1 unit.

88. $\mathcal{C} = \{(x,y) : y = x^3 + 2x\}$; $\mathcal{C}'$ is obtained by translating $\mathcal{C}$ down 3 units and right 2 units.

89. $\mathcal{C} = \{(x,y) : y = (x^3+1)/(x^2+1)\}$; $\mathcal{C}'$ is obtained by reflecting $\mathcal{C}$ about the y-axis.

90. $\mathcal{C} = \{(x,y) : y = (x+1)/(x^4+1)\}$; $\mathcal{C}'$ is obtained by reflecting $\mathcal{C}$ about the origin.

▶ In each of Exercises 91 and 92, two functions f and g with common domain I are given. Plot the parametric equations $x = f(t)$, $y = g(t)$ for t in I. Follow the accompanying directions. ◀

91. $f(t) = t^4 + t + 1$, $g(t) = t^3 - t$, $I = [-1, 0]$. Find the points for which $y = 18x/25$ and the values of the parameter that correspond to these points.

92. $f(t)=t^4+t+1$, $g(t)=t^3-2t$, $I=[-7/4, 3/2]$. A point P is a *double point* of a parametric curve if there are two values of t in I such that $P=(f(t), g(t))$. Find the double point of the given curve and the two values that parameterize that point.

In Exercises 93 and 94, graph the inverse function of the given function f. Do not attempt to find a formula for f^{-1}.

93. $f(x)=x^5-3x^3+x^2+6x-8,\ 0\le x\le 2$

94. $f(x)=x^3-3x^2+3x, 0\le x\le 1$

1.6 Trigonometry

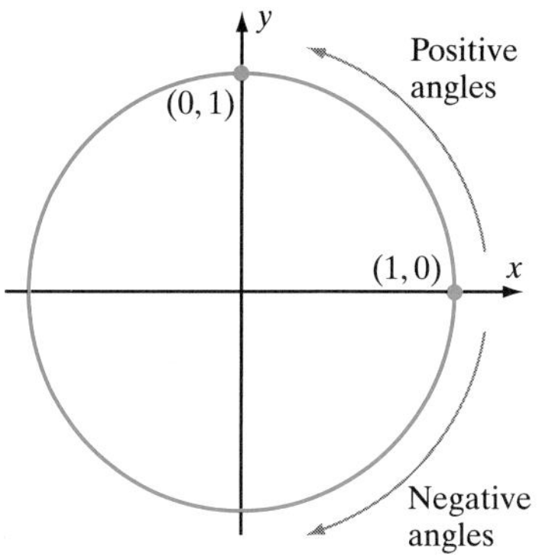

Figure 1 The unit circle

When first learning trigonometry, we study triangles and measure angles in degrees. In calculus, however, it is convenient to study trigonometry in a more general setting and to measure angles differently.

In calculus, we measure angles by rotation along the unit circle in the plane, beginning at the positive x-axis. Counterclockwise rotation corresponds to positive angles, and clockwise rotation corresponds to negative angles (see Figure 1). The *radian measure* of an angle is defined as the length of the arc of the unit circle that it subtends with the positive x-axis (with an appropriate + or − sign).

In degree measure, one full rotation about the unit circle is 360°; in radian measure, one full rotation about the circle is the circumference of the unit circle, namely 2π. Let us use the symbol θ to denote an angle. The principle of proportionality tells us that

$$\frac{\text{degree measure of } \theta}{360^\circ}=\frac{\text{radian measure of } \theta}{2\pi}.$$

In other words,

$$\text{radian measure of } \theta=\frac{\pi}{180}\cdot(\text{degree measure of } \theta)$$

and

$$\text{degree measure of } \theta=\frac{180}{\pi}\cdot(\text{radian measure of } \theta).$$

Figure 2 shows several angles that include both the radian and degree measures. This book always refers to radian measure for angles unless degrees are explicitly specified. Doing so makes the calculus formulas simpler. Thus for example, if we refer to "the angle $2\pi/3$," then it should be understood that this is in radian measure. Likewise, if we refer to "the angle 3," it is also understood to be radian measure. We sketch this last angle (see Figure 3) by noting that a full rotation is $2\pi=6.28\ldots$, and 3 is a little less than half of that.

Sine and Cosine Functions

DEFINITION Let $A=(1,0)$ denote the point of intersection of the unit circle with the positive x-axis. Let θ be any real number. A unique point $P=(x,y)$ on the unit circle is associated with θ by rotating $\overline{OA}$ by θ radians. (Remember that the rotation is counterclockwise for $\theta>0$ and clockwise for $\theta<0$.) (See Figure 4.) The resulting radius $\overline{OA}$ is called the *terminal radius* of θ, and P is called the *terminal point* corresponding to θ. The number y is called the *sine* of θ and is written $\sin(\theta)$. The number x is called the *cosine* of θ and is written $\cos(\theta)$.

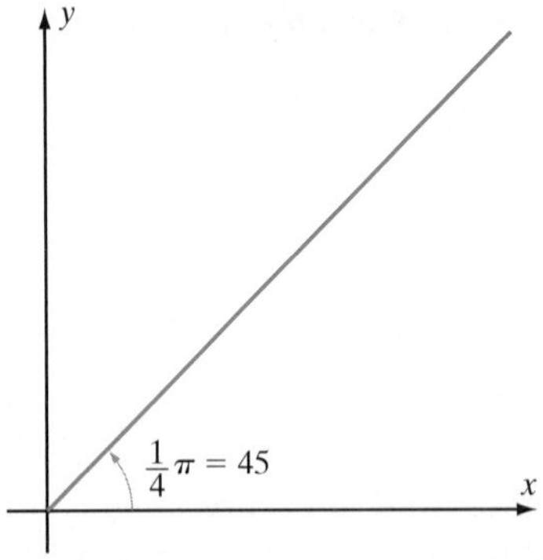

▲ Figure 2a

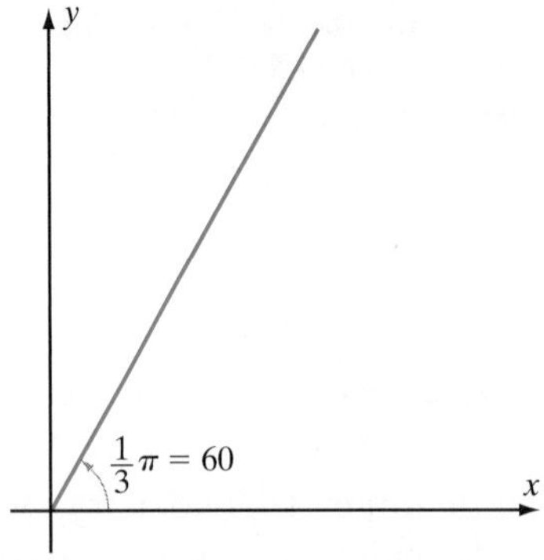

▲ Figure 2b

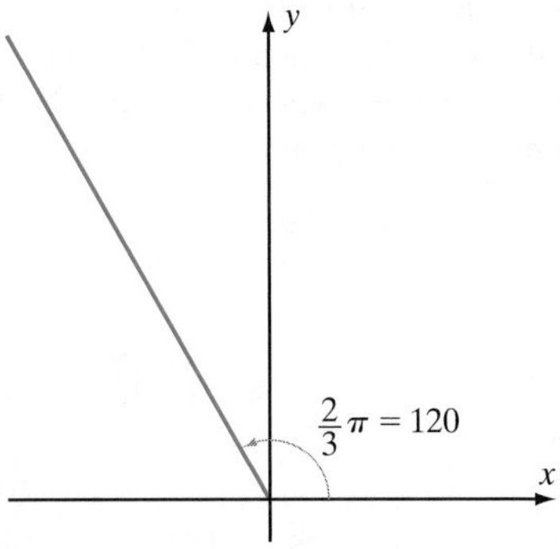

▲ Figure 2c

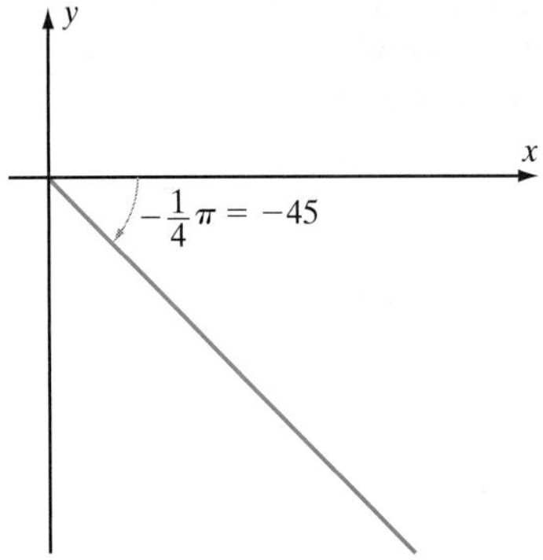

▲ Figure 2d

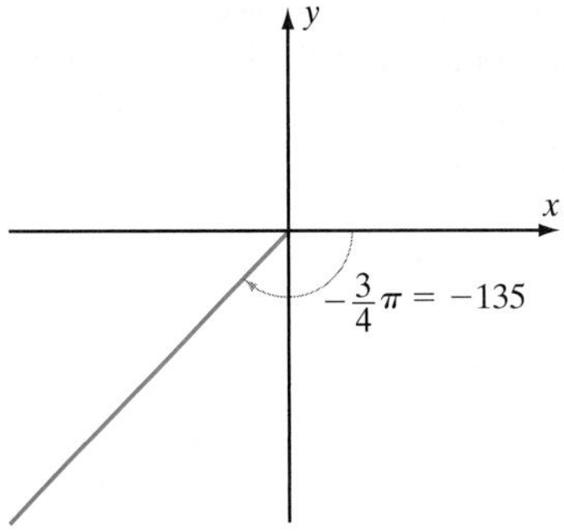

▲ Figure 2e

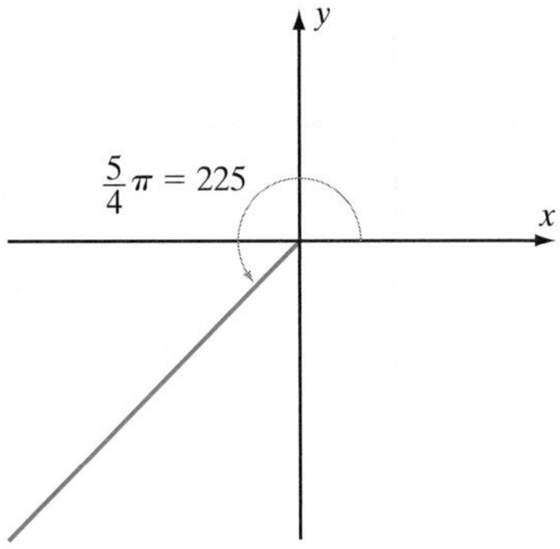

▲ Figure 2f

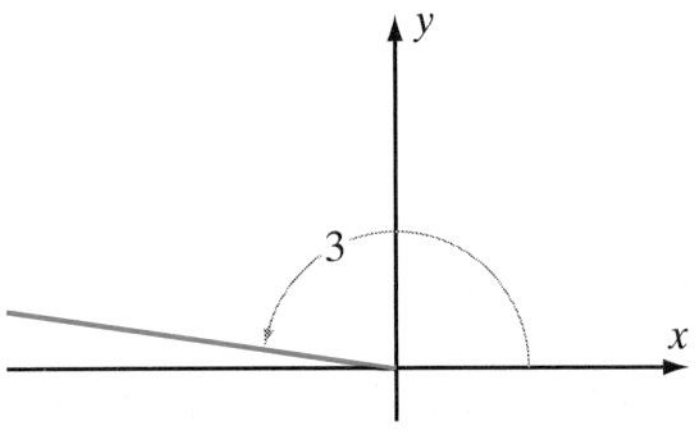

▲ Figure 3

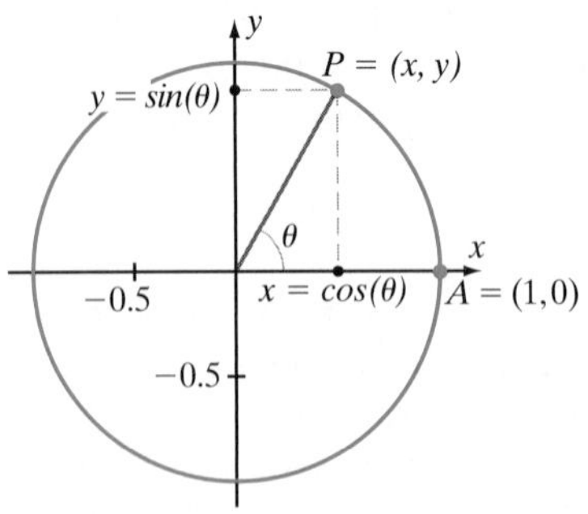

▲ Figure 4

Because the point $(\cos(\theta), \sin(\theta))$ is on the unit circle, the following two fundamental properties are immediate:

$$(\sin(\theta))^2 + (\cos(\theta))^2 = 1 \qquad \text{for any } \theta \text{ in } \mathbb{R}, \qquad (1.6.1)$$

and

$$-1 \le \cos(\theta) \le 1 \text{ and } -1 \le \sin(\theta) \le 1 \qquad \text{for any } \theta \text{ in } \mathbb{R}.$$

It is common to write $\sin^2(\theta)$ to mean $(\sin(\theta))^2$ and $\cos^2(\theta)$ to mean $(\cos(\theta))^2$. This is consistent with the notation for functions described in Section 1.5. Indeed, sine and cosine are both functions; each has domain equal to $\mathbb{R}$ and image equal to $[-1, 1]$.

▶ **EXAMPLE 1** Compute the sine and cosine of $\pi/3$

Solution We sketch the terminal radius and associated triangle (see Figure 5). This is a $\pi/6 - \pi/3 - \pi/2$ $(30° - 60° - 90°)$ triangle with sides of ratio $1{:}\sqrt{3}{:}2$. Thus the side-length x of Figure 5 satisfies $1/x = 2$, or $x = 1/2$. Likewise, the side-length y of Figure 5 satisfies $y/x = \sqrt{3}$, or $y = \sqrt{3}x = \sqrt{3}/2$. It follows that

$$\sin\left(\frac{\pi}{3}\right) = \frac{\sqrt{3}}{2} \quad \text{and} \quad \cos\left(\frac{\pi}{3}\right) = \frac{1}{2}. \quad ◀$$

If θ is any number, then θ and $\theta + 2\pi$ have the same terminal radius and the same terminal point (adding 2π simply adds one more trip around the circle—see Figure 6). As a result,

$$\sin(\theta) = \sin(\theta + 2\pi)$$

and

$$\cos(\theta) = \cos(\theta + 2\pi).$$

We say that the sine and cosine functions have *period* 2π. In other words, the functions repeat themselves every 2π units.

In practice, when we calculate the trigonometric functions of an angle θ, we reduce it by multiples of 2π so that we can consider an equivalent angle θ', called the *associated principal angle*, satisfying $0 \le \theta' < 2\pi$. For instance, $15\pi/2$ has associated principal angle $3\pi/2$ (because $3\pi/2 = 15\pi/2 - 3 \cdot 2\pi$) and $-10\pi/3$ has associated principal angle $2\pi/3$ because $2\pi/3 = -10\pi/3 + 2 \cdot 2\pi$.

How does the concept of angle and sine and cosine presented here relate to the classical notion using triangles? Any angle θ such that $0 \le \theta < \pi/2$ has a right triangle in the first quadrant associated to it, with a vertex on the unit circle such that the base is the segment connecting $(0,0)$ to $(x,0)$, and the height is the segment connecting $(x,0)$ to (x,y) (see Figure 7). Therefore

$$\sin(\theta) = y = \frac{y}{1} = \frac{\text{opposite side}}{\text{hypotenuse}}$$

and

$$\cos(\theta) = x = \frac{x}{1} = \frac{\text{adjacent side}}{\text{hypotenuse}}.$$

Thus for angles θ between 0 and $\pi/2$, the new definition of sine and cosine using the unit circle is equivalent to the classical definition using adjacent and opposite sides and the hypotenuse. For other angles θ, the classical approach is to reduce to this special case by subtracting multiples of $\pi/2$. The approach using the unit circle is considerably clearer because it makes the signs of sine and cosine obvious.

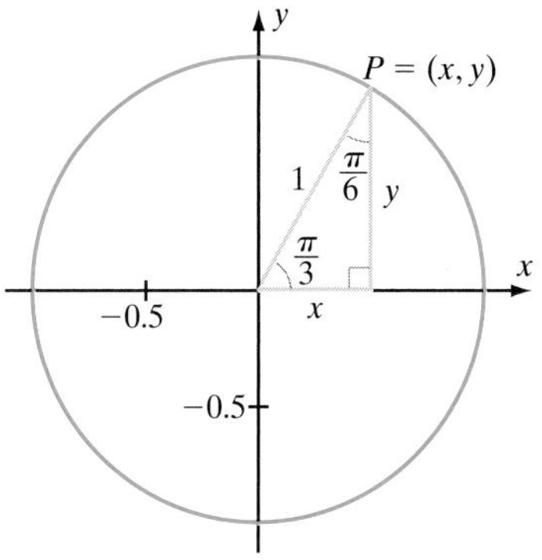

▲ Figure 5

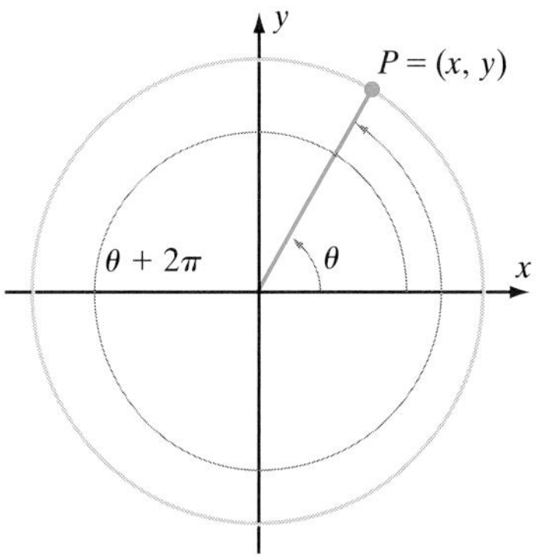

▲ Figure 6

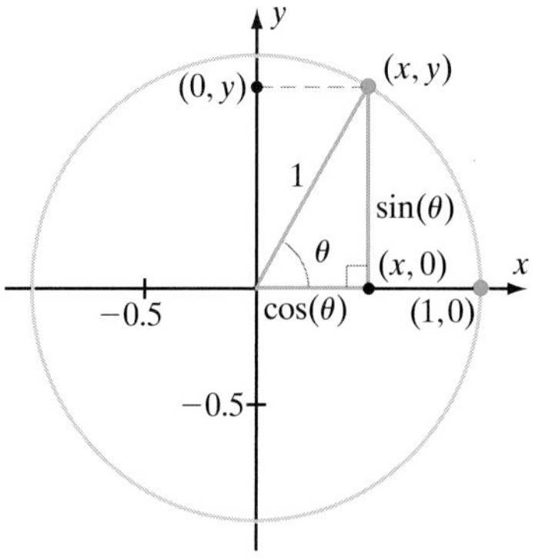

▲ Figure 7

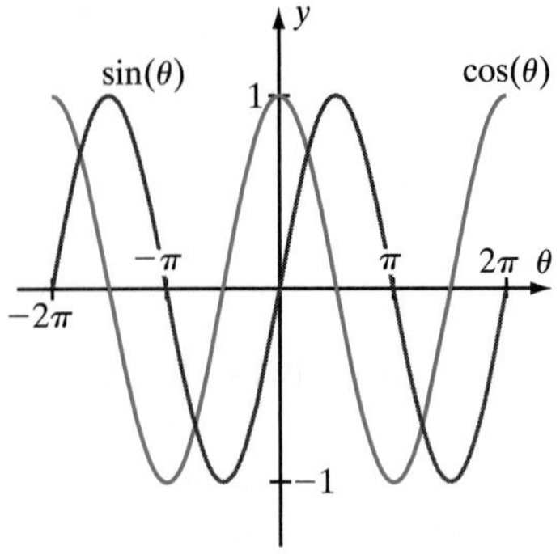

▲ Figure 8a

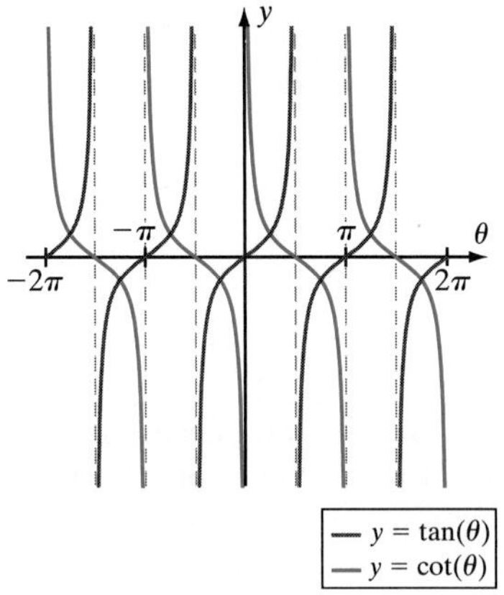

▲ Figure 8b

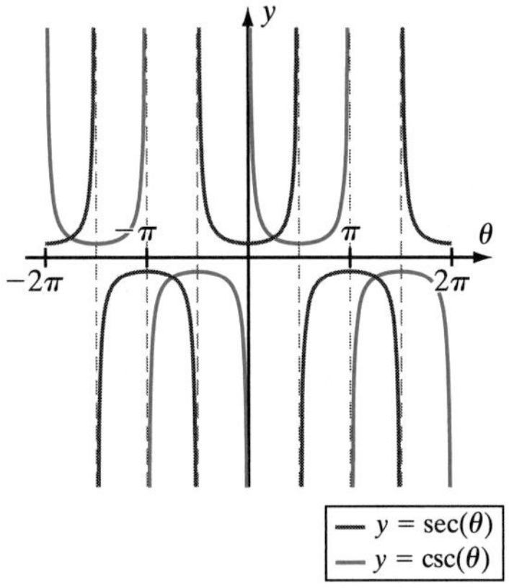

▲ Figure 8c

Other Trigonometric Functions

In addition to sine and cosine, there are four other trigonometric functions—tangent, cotangent, secant, and cosecant—defined as follows:

$$\tan(\theta) = \frac{y}{x} = \frac{\sin(\theta)}{\cos(\theta)}$$

$$\cot(\theta) = \frac{x}{y} = \frac{\cos(\theta)}{\sin(\theta)}$$

$$\sec(\theta) = \frac{1}{x} = \frac{1}{\cos(\theta)}$$

$$\csc(\theta) = \frac{1}{y} = \frac{1}{\sin(\theta)}.$$

Whereas sine and cosine have the entire real line for their domain, $\tan(\theta)$ and $\sec(\theta)$ are undefined at odd multiples of $\pi/2$ (because cosine is zero there). Also, the functions $\cot(\theta)$ and $\csc(\theta)$ are undefined at even multiples of $\pi/2$ (because sine is zero there).

Figure 8 shows the graphs of the six trigonometric functions. Compare the graph of the cosine and sine functions. We obtain the graph of $y = \cos(\theta)$ by shifting the graph of $y = \sin(\theta)$ by $\pi/2$ to the left. According to Theorem 1 from Section 1.5, this means that

$$\sin\left(x + \frac{\pi}{2}\right) = \cos(x).$$

This identity is a special case of the Addition Formula for the sine, which is given in equation (1.6.4) in the next section.

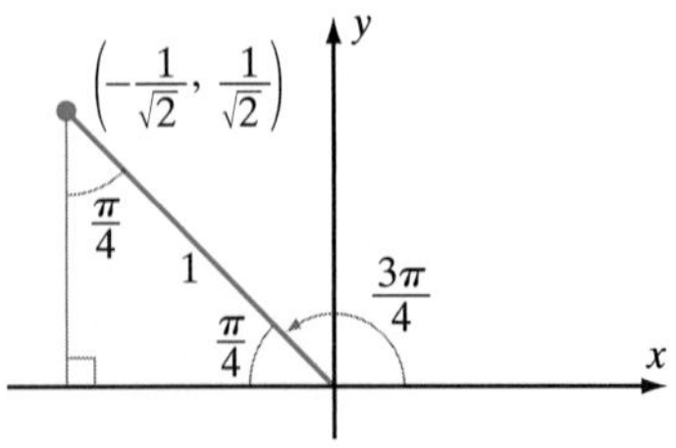

▲ Figure 9

▶ **EXAMPLE 2** Compute all trigonometric functions for the angle $\theta = 11\pi/4$.

Solution Because the principal associated angle is $3\pi/4$, we deal with that angle. Figure 9 shows that the triangle associated to this angle is an isosceles right triangle with hypotenuse 1. Therefore $x = -1/\sqrt{2}$ and $y = 1/\sqrt{2}$. It follows that

$$\sin(\theta) = y = \frac{1}{\sqrt{2}}$$
$$\cos(\theta) = x = -\frac{1}{\sqrt{2}}$$
$$\tan(\theta) = \frac{y}{x} = -1$$
$$\cot(\theta) = \frac{x}{y} = -1$$
$$\sec(\theta) = \frac{1}{x} = -\sqrt{2}$$
$$\csc(\theta) = \frac{1}{y} = \sqrt{2}. \quad ◀$$

Similar calculations allow us to complete Table 1 for the values of the trigonometric functions at the principal angles that are multiples of $\pi/6$ or $\pi/4$.

Angle	**Sine**	**Cosine**	**Tangent**	**Cotangent**	**Secant**	**Cosecant**
0	0	1	0	undef	1	undef
$\pi/6$	$1/2$	$\sqrt{3}/2$	$1/\sqrt{3}$	$\sqrt{3}$	$2/\sqrt{3}$	2
$\pi/4$	$\sqrt{2}/2$	$\sqrt{2}/2$	1	1	$\sqrt{2}$	$\sqrt{2}$
$\pi/3$	$\sqrt{3}/2$	$1/2$	$\sqrt{3}$	$1/\sqrt{3}$	2	$2/\sqrt{3}$
$\pi/2$	1	0	undef	0	undef	1
$2\pi/3$	$\sqrt{3}/2$	$-1/2$	$-\sqrt{3}$	$-1/\sqrt{3}$	-2	$2/\sqrt{3}$
$3\pi/4$	$\sqrt{2}/2$	$-\sqrt{2}/2$	-1	-1	$-\sqrt{2}$	$\sqrt{2}$
$5\pi/6$	$1/2$	$-\sqrt{3}/2$	$-1/\sqrt{3}$	$-\sqrt{3}$	$-2/\sqrt{3}$	2
π	0	-1	0	undef	-1	undef
$7\pi/6$	$-1/2$	$-\sqrt{3}/2$	$1/\sqrt{3}$	$\sqrt{3}$	$-2/\sqrt{3}$	-2
$5\pi/4$	$-\sqrt{2}/2$	$-\sqrt{2}/2$	1	1	$-\sqrt{2}$	$-\sqrt{2}$
$4\pi/3$	$-\sqrt{3}/2$	$-1/2$	$\sqrt{3}$	$1/\sqrt{3}$	-2	$-2/\sqrt{3}$
$3\pi/2$	-1	0	undef	0	undef	-1
$5\pi/3$	$-\sqrt{3}/2$	$1/2$	$-\sqrt{3}$	$-1/\sqrt{3}$	2	$-2/\sqrt{3}$
$7\pi/4$	$-\sqrt{2}/2$	$\sqrt{2}/2$	-1	-1	$\sqrt{2}$	$-\sqrt{2}$
$11\pi/6$	$-1/2$	$\sqrt{3}/2$	$-1/\sqrt{3}$	$-\sqrt{3}$	$2/\sqrt{3}$	-2

Table 1

Trigonometric Identities The trigonometric functions satisfy many useful identities. In addition to equation (1.6.1) already stated, we will need the following important formulas:

$$1 + \tan^2(\theta) = \sec^2(\theta) \tag{1.6.2}$$

$$1 + \cot^2(\theta) = \csc^2(\theta) \tag{1.6.3}$$

$$\sin(\theta + \phi) = \sin(\theta)\cos(\phi) + \cos(\theta)\sin(\phi) \quad \text{(Addition Formula)} \tag{1.6.4}$$

$$\cos(\theta + \phi) = \cos(\theta)\cos(\phi) - \sin(\theta)\sin(\phi) \quad \text{(Addition Formula)} \tag{1.6.5}$$

$$\sin(\theta - \phi) = \sin(\theta)\cos(\phi) - \cos(\theta)\sin(\phi) \quad \text{(Addition Formula)} \tag{1.6.6}$$

$$\cos(\theta - \phi) = \cos(\theta)\cos(\phi) + \sin(\theta)\sin(\phi) \quad \text{(Addition Formula)} \tag{1.6.7}$$

$$\sin(2\theta) = 2\sin(\theta)\cos(\theta) \quad \text{(Double Angle Formula)} \tag{1.6.8}$$

$$\cos(2\theta) = \cos^2(\theta) - \sin^2(\theta) \quad \text{(Double Angle Formula)} \tag{1.6.9}$$

$$\sin(-\theta) = -\sin(\theta) \quad \text{(Sine is an odd function)} \tag{1.6.10}$$

$$\cos(-\theta) = \cos(\theta) \quad \text{(Cosine is an even function)} \tag{1.6.11}$$

$$\sin^2\left(\frac{\theta}{2}\right) = \frac{1 - \cos(\theta)}{2} \quad \text{(Half Angle Formula)} \tag{1.6.12}$$

$$\cos^2\left(\frac{\theta}{2}\right) = \frac{1 + \cos(\theta)}{2} \quad \text{(Half Angle Formula)} \tag{1.6.13}$$

▶ **EXAMPLE 3** Prove identity (1.6.2): $1 + \tan^2(\theta) = \sec^2(\theta)$.

Solution Using identity (1.6.1), we have

$$\begin{aligned} \tan^2(\theta) + 1 &= \frac{\sin^2(\theta)}{\cos^2(\theta)} + 1 \\ &= \frac{\sin^2(\theta)}{\cos^2(\theta)} + \frac{\cos^2(\theta)}{\cos^2(\theta)} \\ &= \frac{\sin^2(\theta) + \cos^2(\theta)}{\cos^2(\theta)} \\ &= \frac{1}{\cos^2(\theta)} \\ &= \sec^2(\theta). \end{aligned}$$ ◀

Identities (1.6.1)–(1.6.13) were known to the astronomer Hipparchus (ca. 180–125 BCE). There are many other trigonometric identities of a more specialized nature.

Modeling with Trigonometric Functions

Although the subject of trigonometry was invented by the ancient Greek astronomers for physical measurement, the trigonometric functions have proved to be useful in many contexts. For example, because the sine function is cyclic with period 2π (i.e., the sine function repeats itself after 2π), it is often used to model cyclic phenomena.

▶ **EXAMPLE 4** The table shows the normal monthly mean temperatures of St. Louis, Missouri, in degrees Fahrenheit.

Jan.	**Feb.**	**March**	**April**	**May**	**June**	**July**	**Aug.**	**Sept.**	**Oct.**	**Nov.**	**Dec.**
29	34	43	56	66	75	79	77	70	58	45	34

Model the mean temperature (T) as a function of time (t) by using the equation

$$T(t) = b + A\sin(\omega \cdot t + \phi) \tag{1.6.14}$$

for suitable constants b, A, ω, and ϕ. Use one month as the unit of time. Locate the origin of the time axis so that $t = 0$ corresponds to January 1, and $t = 12$ corresponds to December 31. Assume that, for each month, the actual temperature in the middle of the month is equal to the mean temperature for that month. That is, construct your model so that $T(1/2) = 29$, $T(3/2) = 34$, $T(5/2) = 43$, and so on.

Solution First observe that

$$\sin\left(\omega\left(t_0 + \frac{2\pi}{\omega}\right) + \phi\right) = \sin(\omega t_0 + 2\pi + \phi) = \sin(\omega t_0 + \phi).$$

On setting $t = t_0 + \frac{2\pi}{\omega}$ in formula (1.6.14), we see that, for every value of t_0,

$$T\left(t_0 + \frac{2\pi}{\omega}\right) = b + A\sin\left(\omega \cdot \left(t_0 + \frac{2\pi}{\omega}\right) + \phi\right) = b + A\sin(\omega \cdot t_0 + \phi) = T(t_0).$$

In other words, the function T repeats itself after $2\pi/\omega$. This duration is called the period of T. Because the period of weather is 12 when measured in months, we set $2\pi/\omega = 12$, or $\omega = \pi/6$. Thus $T(t) = b + A\sin(\pi t/6 + \phi)$. To identify the values of A and b, we note that $b - A \le T(t) \le b + A$ (because $-1 \le \sin(\pi t/6 + \phi) \le 1$). Because the least mean temperature is 29, and the greatest mean temperature is 79, we set $b - A = 29$, and $b + A = 79$. Solving these two equations simultaneously, we obtain $b = 54$, and $A = 25$. Thus

$$T(t) = 54 + 25\sin\left(\frac{\pi}{6}t + \phi\right).$$

To find the value of ϕ, notice that the maximum value of $\sin(\pi t/6 + \phi)$ occurs when the sine takes on its maximum value 1. That is, the expression $\sin(\pi t/6 + \phi)$ is maximized when $\pi t/6 + \phi = \pi/2$. Therefore the function $T(t)$ is maximized when $\pi t/6 + \phi = \pi/2$. Looking at the table, we make the reasonable assumption that this maximum is 79 and occurs in the middle of July, when $t = 13/2$. Thus we set $\pi \cdot (13/2)/6 + \phi = \pi/2$. This gives $\phi = -7\pi/12$. Our model is

$$T(t) = 54 + 25\sin\left(\frac{\pi}{6}t - \frac{7\pi}{12}\right).$$

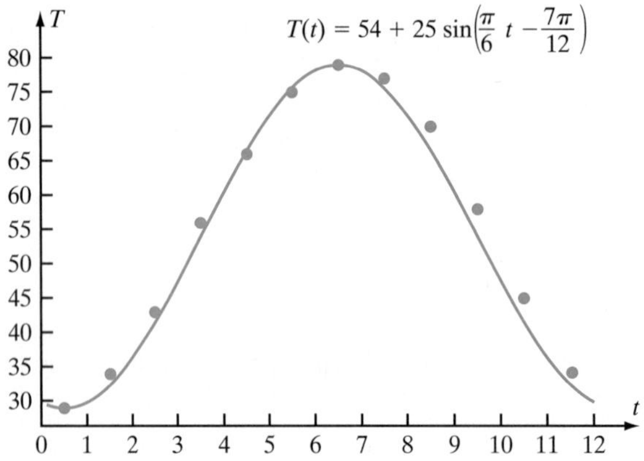

▲ Figure 10

Figure 10 shows how well this model fits the measured data. ◀

QUICK QUIZ

1. What is the domain of the sine function?
2. What is the image of the cosine function?
3. Which trigonometric functions are even? Odd?
4. Simplify $\sin(2\theta)/\sin(\theta)$.

Answers

1. $\mathbb{R}$ 2. $[-1,1]$ 3. even: cos and sec; odd: sin, tan, cot, csc, 4. $2\cos(\theta)$

EXERCISES

Problems for Practice

▶ In Exercises 1–4, calculate each of the six trigonometric functions at angle θ without using a calculator. ◀

1. $\theta = \pi/6$
2. $\theta = \pi/4$
3. $\theta = 2\pi/3$
4. $\theta = 4\pi/3$

▶ In Exercises 5–14, calculate the given expression without using a calculator. ◀

5. $\sin(\pi/3)\sin(\pi/6)$
6. $\cos(0) - \cos(\pi)$
7. $\cos(\pi/6) + \cos(\pi/3)$
8. $\sin(\pi/4)\cos(\pi/4)$
9. $\tan(\pi/3)/\tan(\pi/6)$
10. $\cos(2\pi/3)\csc(2\pi/3)$
11. $\sin(\pi \cdot \sin(\pi/6))$
12. $\sec(-\pi/3)^{\csc(-\pi/2)}$
13. $\sin(19\pi/2)^{\cos(33\pi)}$
14. $4\tan(\pi/4) - \sin(17\pi/2)$

▶ In Exercises 15–20, θ is a number between 0 and $\pi/2$. Calculate the unevaluated trigonometric function from the given information. ◀

15. $\cos(\theta)$; $\sin(\theta) = 1/3$
16. $\tan(\theta)$; $\cos(\theta) = 3/5$
17. $\sin(2\theta)$; $\cos(\theta) = 4/5$
18. $\sin(\theta/2)$; $\cos(\theta) = 3/7$
19. $\cos(\theta/2)$; $\sin(\theta) = 5/13$
20. $\cos(\theta + \pi)$; $\cos(\theta) = 0.1$

▶ In Exercises 21–24, state which of the six trigonometric functions are positive when evaluated at θ in the indicated interval. ◀

21. $\theta \in (0, \pi/2)$
22. $\theta \in (\pi/2, \pi)$
23. $\theta \in (\pi, 3\pi/2)$
24. $\theta \in (3\pi/2, 2\pi)$

▶ In Exercises 25–28, graph the function. ◀

25. $f(t) = \sin(2t)$, $-2\pi \le t \le 2\pi$
26. $f(t) = \cos(t/2)$, $-2\pi \le t \le 2\pi$

27. $f(t) = \sin(t - \pi/6),\ 0 \le t \le 2\pi$

28. $f(t) = \cos(2t + \pi/3),\ 0 \le t \le 2\pi$

In Exercises 29–31, use the cosine and sine functions to give a parameterization of the unit circle such that the domain of parameterization is $[0, 2\pi)$ and the additional requirements are satisfied.

29. The initial point is $(1, 0)$, and the circle is traced counterclockwise as the parameter value increases.

30. The initial point is $(0, 1)$, and the circle is traced counterclockwise as the parameter value increases.

31. The initial point is $(0, 1)$, and the circle is traced clockwise as the parameter value increases.

Further Theory and Practice

32. For each θ in $\{0, \pi/6, \pi/4, \pi/3, \pi/2\}$, find an integer value of n such that $\sin(\theta) = \sqrt{n}/2$ (The pattern found in this exercise is sometimes used as a memory aid.)

33. The length $s(r, \theta)$ of an arc of a circle that subtends a central angle of radian measure θ is proportional to the radius r of the circle and to θ. What is $s(r, \theta)$?

34. For a circle of radius r, the area $A(r, \theta)$ of a sector with central angle measuring θ radians is jointly proportional to θ and to r^2. What is $A(r, \theta)$?

35. The equations $\sin(2\theta) = \sin(\theta)\cos(\theta)$ and $\sin(2\theta) = 2\sin(\theta)\cos(\theta)$ are fundamentally different. Explain why.

36. Find constants A and B such that

$$\sin(3\theta) = A\sin(\theta) - B\sin^3(\theta)$$

for every θ.

In Exercises 37–44, use one or more of the basic trigonometric identities to derive the given identity.

37. $\tan(\theta + \phi) = \dfrac{\tan(\theta) + \tan(\phi)}{1 - \tan(\theta)\tan(\phi)}$

38. $\sin(\theta)\sin(\phi) = \dfrac{\cos(\theta - \phi) - \cos(\theta + \phi)}{2}$

39. $\cos(\theta)\cos(\phi) = \dfrac{\cos(\theta - \phi) + \cos(\theta + \phi)}{2}$

40. $\sin(\theta)\cos(\phi) = \dfrac{\sin(\theta + \phi) + \sin(\theta - \phi)}{2}$

41. $\sin(\theta) = \cos\left(\dfrac{\pi}{2} - \theta\right)$

42. $\sin(\theta + \pi) = -\sin(\theta)$

43. $\cos(\theta + \pi) = -\cos(\theta)$

44. $\tan(\theta + \pi) = \tan(\theta)$

45. Use the identity $\pi/3 + \pi/4 = 7\pi/12$ to calculate the six trigonometric functions at $7\pi/12$.

46. Use the Half Angle Formulas to evaluate the six trigonometric functions at $\pi/8$.

In each of Exercises 47–50, derive the given identity by using the identities in Exercises 38–40.

47. $\sin(\theta) + \sin(\phi) = 2\sin(\frac{\theta + \phi}{2})\cos(\frac{\theta - \phi}{2})$

48. $\sin(\theta) - \sin(\phi) = 2\cos(\frac{\theta + \phi}{2})\sin(\frac{\theta - \phi}{2})$

49. $\cos(\theta) + \cos(\phi) = 2\cos(\frac{\theta + \phi}{2})\cos(\frac{\theta - \phi}{2})$

50. $\cos(\theta) - \cos(\phi) = -2\sin(\frac{\theta + \phi}{2})\sin(\frac{\theta - \phi}{2})$

51. Suppose that A and B are constants. Use the Addition Formula for the sine to find an amplitude C and a phase shift ϕ (both in terms of A and B) such that

$$A\cos(\theta) + B\sin(\theta) = C\sin(\theta + \phi)$$

for every θ.

52. Let $y = mx + b$ be the equation of a nonhorizontal, nonvertical line. Such a line intersects the x-axis making an angle ϕ with the positive x-axis that has radian measure between 0 and π. What is the relationship between m and ϕ? Can this relationship, suitably interpreted, be extended to vertical or horizontal lines? Why is ϕ independent of b?

53. The *angle of elevation* of a point above the earth is the acute angle between the horizontal and the line of sight. Suppose two points on the earth's surface have distance ℓ between them. Line ℓ is small enough for us to assume that the earth is flat between the two points. Suppose that θ and ϕ are the angles of elevation of a mountain peak relative to these two points (see Figure 11). Show that the height h of the mountain is given by

$$h = \frac{\ell}{|\cot(\theta) - \cot(\phi)|}.$$

Suppose that a and b are positive constants. In Exercises 54–57, identify the curve that is parameterized by the equations. In each case, use a trigonometric identity to eliminate the parameter. (It may help to consider the special case $a = b = 1$ first.)

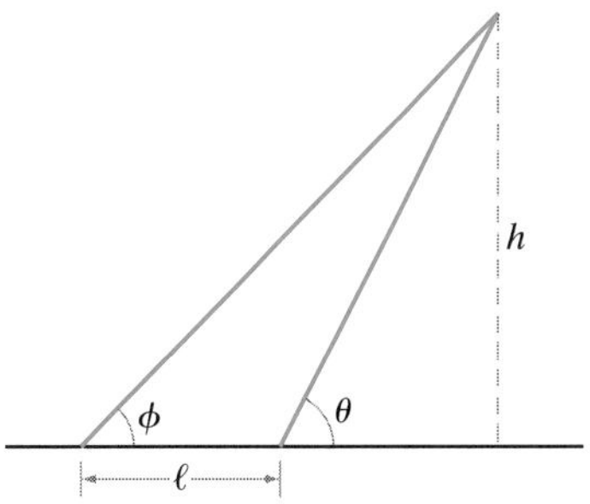

Figure 11

54. $x = a\cos(\theta),\ y = b\sin(\theta),\ \theta \in [0, 2\pi]$

55. $x = a\cos^2(\theta),\ y = b\sin^2(\theta),\ \theta \in [0, \pi]$

56. $x = a\sec(\theta),\ y = b\tan(\theta),\ \theta \in [0, \pi/2)$

57. $x = a\tan(\theta),\ y = b\cot(\theta),\ \theta \in (0, \pi/2)$

58. Suppose that h and k are constants and that a and b are positive constants. Parameterize the ellipse with center (h, k) and semi-axes a and b parallel to the x-axis and y-axis.

A function f is said to have *period* p if there is a smallest positive number p such that $f(x+p) = f(x)$ for all x in the domain of f. In Exercises 59–63, find the period of the function defined by the given expression.

59. $\sin(x+\sqrt{3})$
60. $\cos(2\pi x)$
61. $\tan(x)$
62. $\sin(x) + \tan(x)$
63. $\tan(2x) + \sin(3x)$

64. The lowest monthly normal temperature of Philadelphia is 31°F and occurs in January. The highest monthly normal temperature of Philadelphia is 77°F and occurs in July. Find a model of temperature T as a function of time t that has the form $T(t) = b + A\sin(\omega t + \phi)$.

65. The lowest monthly normal temperature of Nome is 4°F and occurs at the end of December ($t = 12$). The highest monthly normal temperature of Nome is 51°F and occurs at the beginning of July ($t = 6$). Find a model of temperature T as a function of time t that has the form $T(t) = b + A\sin(\omega t + \phi)$.

66. The number of visitors to a tourist destination is seasonal. However, the long-term trend at this particular destination is down. A model for the number of tourists y as a function of time t measured in years is

$$y(t) = 5 \cdot 10^4 \frac{2+\cos(2\pi t)}{1+t^{1/4}}.$$

Here $t = 0$ corresponds to a time at which there were 150,000 tourists. Is y a periodic function of t? (Refer to the instructions for Exercise 59 for the definition of "period.") Sketch the graph of y as a function of t for $0 \le t \le 4$. From the sketch, y has *local maxima* at $t = 1, 2, 3$, and so on. Define in your own words what the statement "a local maximum of y occurs at t_0" means.

67. A polygon is *regular* if all sides have equal length. For example, an equilateral triangle is a regular 3-gon (triangle) and a square is a regular 4-gon (quadrilateral). A polygon is said to be inscribed in a circle if all of its vertices lie on the circle.

a. Show that the perimeter $p(n, r)$ of a regular n-gon inscribed in a circle of radius r is

$$p(n, r) = 2rn\sin\left(\frac{\pi}{n}\right).$$

b. Show that the area $A(n,r)$ of a regular n-gon inscribed in a circle of radius r is

$$A(n, r) = \frac{1}{2}r^2 n\sin\left(\frac{2\pi}{n}\right).$$

Calculator/Computer Exercises

68. Graph $y = 1-\cos(2x)$ and $y = \sin(x)$ in $[0.2, 1]\times[0, 1.5]$. Successively zoom in to the point of intersection to find a value $x_0 \in (0.2, 1)$ that satisfies $1-\cos(2x_0) = \sin(x_0)$. Give x_0 to three decimal places. What is $\sin(x_0)$ to three decimal places? Use a trigonometric identity to find the exact value of $\sin(x_0)$.

69. Calculate the values of the sequence $a_n = n \cdot \sin(1/n)$ for $1 \le n \le 15$. Now calculate a_n for larger values of n by setting $n = 10^k$ for $1 \le k \le 6$. There is a number ℓ, which is called the *limit* of a_n, such that $\ell - a_n$ decreases to 0 as n increases to infinity. What is ℓ Now plot $f(x) = \sin(x)/x$ for $0 \le x \le 1$. Explain how the limiting behavior of a_n can be determined from this graph.

70. Calculate the values of the sequence $a_n = n^2 \cdot (1-\cos(1/n))$ for $1 \le n \le 15$. Now calculate a_n for larger values of n by setting $n = 10^k$ for $1 \le k \le 4$. There is a number ℓ, which is called the *limit* of a_n, such that $\ell - a_n$ decreases to 0 as n increases to infinity. What is ℓ? Now plot $f(x) = (1-\cos(x))/x^2$ for $0 < x \le 1$. Explain how the limiting behavior of a_n can be determined from this graph.

71. Graph $f(x) = 90x + \cos(x)$ in the viewing windows $[-10, 10]\times[-1000, 1000]$ and $[-0.0001, 0.0001]\times[0.5, 1.5]$. Explain why the graph of f appears to be a straight line in each of these windows. Which straight lines do these graphs appear to coincide with? Sketch the graph of the function f in a way that better displays its behavior.

72. In later chapters, we will investigate the approximation of complicated functions by simpler ones. For instance, the function $x \mapsto x$ is a good approximation to the function $x \mapsto \sin(x)\cos(x^2)$ for $|x|$ sufficiently small. Use a graphing utility to find an interval on which x approximates $\sin(x)\cos(x^2)$ with an absolute error of at most 0.01.

73. The monthly normal temperatures in degrees Fahrenheit for Raleigh, North Carolina, are 40, 42, 49, 59, 67, 74, 78, 77, 71, 60, 50, and 42 (starting with January). Plot these temperatures. Find and plot a continuous model.

Cosine waves are functions of the form $x \mapsto A\cos(\nu x + \phi)$ where A is the *amplitude* of the wave, ν the *frequency*, and ϕ the *phase shift*. The physical superposition of two waves is obtained mathematically by adding the two wave functions. Exercises 74–76 concern the superposition of cosine waves. The identities from Exercises 49 and 50 are particularly useful in this context.

74. For $\phi = 0$, $\pi/3$, $2\pi/3$, and $3\pi/2$, graph the superposition $x \mapsto \cos(2x) + \cos(2x + \phi)$, $0 \le x \le 2\pi$. Each of these four curves is the graph of a wave of the form $x \mapsto A \cdot \cos(2x + \theta)$ for some amplitude A and phase shift θ, each depending on ϕ. Use your graph to determine A and θ when $\phi = 2\pi/3$. Use the identity from Exercise 39 to obtain a formula for A and θ. Calculate the magnitude of the superposed wave at $x = 2$ for the four given values of ϕ. Each of these signals is less than 1.5. For what value of ϕ is the magnitude of the superposed wave at $x = 2$ as large as it can be?

75. Suppose that ν_1 and ν_2 are both large compared with their absolute difference $|\nu_1 - \nu_2|$. The superposition $t \mapsto \cos(\nu_1 \cdot t) + \cos(\nu_2 \cdot t)$ might be described as a high-frequency cosine wave in which the beats have low-frequency *amplitude modulation*. Use Exercise 49 to express the superposition in the form $A(t) \cdot \cos(\omega \cdot t)$ where the frequency ω is about the same size as the individual frequencies ν_1 and ν_2 and where the amplitude $A(t)$ of the beats varies in time according to a low-frequency cosine wave. Illustrate with the graph of the superposed wave for $\nu_1 = 8$ and $\nu_2 = 6$. What is ω in this case? Add the graphs of $\pm 2\cos(t)$ to your figure. What is the frequency of the modulated amplitude?

76. An *amplitude modulation* (AM) radio transmitter broadcasts an audio tone (or *baseband signal*) $\cos(\nu \cdot t)$ by modulating the amplitude of a very high-frequency *carrier signal* $A \cdot \cos(\omega \cdot t)$ by a positive factor $(1 + m \cdot \cos(\nu \cdot t))$, where the *modulation index* m is less than 1. Thus the emitted signal has the form

$$S(t) = A \cdot (1 + m \cdot \cos(\nu \cdot t))\cos(\omega \cdot t).$$

a. Use the identity from Exercise 49 to express $S(t)$ as a superposition of three cosine waves. The largest and smallest frequencies of these three waves are called *sidebands*.

b. Set $A = 1, m = 1/2, \nu = 2$, and $\omega = 8$. Graph the signal $S(t)$ and envelopes $\pm(1 + m \cdot \cos(\nu \cdot t))$ for $0 \le t \le 2\pi$. Explain how the sidebands can be determined visually from your graph.

c. Suppose that $\nu_{max} < \omega$ is the highest baseband frequency. The largest and smallest frequencies broadcast are called the *upper* and *lower sidebands*. What are they in terms of ω and ν_{max}? Their difference is called the *bandwidth*.

d. Carrier frequencies for AM radio range from 550 kHz (kilohertz), or 550×10^3 cycles per second, to 1610 kHz. Typically, ν_{max} can be as high as 15 kHz. An AM receiver works by recovering the audio source $\cos(\nu \cdot t)$ from the received signal $S(t)$, a process known as *demodulation*. Demodulation requires that the bandwidths of different radio stations that broadcast to the same location not overlap. What must the minimum frequency separation between the carrier signals of two such stations be?

77. Snell's inequality, discovered in 1621, states that

$$t \le \tan\left(\frac{t}{3}\right) + 2\sin\left(\frac{t}{3}\right),$$

for $0 < t < 3\pi 2$. An even older inequality,

$$\frac{3\sin(t)}{2 + \cos(t)} < t,$$

for all $t > 0$, was obtained in 1458 by Nicholas Cusa (1401–1464). Graphically illustrate each inequality for $0 < t < 0.01$.

Summary of Key Topics in Chapter 1

Number Systems (Section 1.1)

In calculus, we deal with the three types of real numbers:

1. Terminating decimal expansions
2. Nonterminating decimal expansions that, after a finite number of terms, repeat a single block of terms endlessly
3. Nonterminating, nonrepeating decimal expansions

The first two types are rational numbers, and the third type consists of irrational numbers.

Sets of Real Numbers (Section 1.1)

Sets of numbers are usually described with the notation $\{x : P(x)\}$ and are graphed on a number line. Often a set is described by inequalities that must first be solved before the set can be easily sketched. The most frequently encountered sets are intervals:

$$(a, b) = \{x : a < x < b\}$$
$$[a, b] = \{x : a \le x \le b\}$$
$$[a, b) = \{x : a \le x < b\}$$

$$(a,b] = \{x : a < x \le b\}$$
$$(a,\infty) = \{x : a < x\}$$
$$[a,\infty) = \{x : a \le x\}$$
$$(-\infty,a) = \{x : x < a\}$$
$$(-\infty,a] = \{x : x \le a\}$$
$$(-\infty,\infty) = \{x : -\infty < x < \infty\}$$

Points and Loci in the Plane (Section 1.2)

In the plane, points are located using ordered pairs of real numbers, (x, y). The distance between two points (a_1, b_1) and (a_2, b_2) is given by the formula

$$d = \sqrt{(a_2 - a_1)^2 + (b_2 - b_1)^2}.$$

In general, we are interested in graphing sets of points that are described by equations. Such a graph is called the locus of the equation. For a given equation, a point (x, y) is on the graph if and only if the numbers x and y satisfy the equation simultaneously. An important instance is given by equations of the form

$$(x-h)^2 + (y-k)^2 = r^2$$

with $r > 0$. Such an equation describes a circle with center (h, k) and radius r.

Lines and Their Slopes (Section 1.3)

If ℓ is a line in the plane, and if (x_1, y_1) and (x_2, y_2) are points on ℓ, then the slope of ℓ is given by the expression

$$\frac{y_2 - y_1}{x_2 - x_1}.$$

Parallel lines have the same slopes; perpendicular lines have slopes that are negative reciprocals. It takes two pieces of data to describe a line in the plane: a point through which the line passes and the slope, or two points through which the line passes, or two intercepts, or the slope and an intercept.

The equation for a line (except a vertical one) can always be obtained by finding two expressions for the slope and equating them. The following are four particularly useful forms of equations for lines.

1. *The Point-Slope Form* A line with slope m and passing through the point (x_0, y_0) has equation

$$y = m(x - x_0) + y_0.$$

2. *The Two-point Form* A line passing through the points (x_0, y_0) and (x_1, y_1) has equation

$$y = \left(\frac{y_1 - y_0}{x_1 - x_0}\right)(x - x_0) + y_0.$$

3. *The Slope-Intercept Form* A line with slope m and y-intercept $(0, b)$ has equation

$$y = mx + b.$$

4. *The Intercept Form* A line with intercepts $(a, 0)$ and $(0, b)$ where a and b are both nonzero has equation

$$\frac{x}{a} + \frac{y}{b} = 1.$$

5. *The general linear equation* Every line, vertical lines included, has equation $Ax + By = D$.

Functions and Their Graphs (Section 1.4)

If S and T are sets, then a function is a rule that assigns to each element of S a unique element of T. We call S the domain of the function and T the range of the function. In calculus, S and T are usually sets of real numbers, and functions are usually described by formulas. Frequently, the formula is given, and it is up to the reader to infer the domain and range of the function. The graph of a function consists of all points (x, y) in the plane such that x is in the domain of f, and $y = f(x)$.

Combining Functions (Section 1.5)

Most functions that we encounter are built up from simple functions by using operations, such as addition, subtraction, multiplication, division, and composition, on functions. Composition is the most subtle. If f and g are functions, and if the domain of g contains the range of f, then we define $(g \circ f)(x) = g(f(x))$. Care should be taken not to confuse composition with multiplication.

A function $f : S \to T$ is *onto* if for every t in T, there is an s in S such that $f(s) = t$. A function $f : S \to T$ is *one-to-one* if for every t in T, there is no more than one s in S such that $f(s) = t$. If f is both one-to-one and onto, then there is a unique function f^{-1} with domain T and range S such that $f(f^{-1}(t)) = t$ for every t in T and $f^{-1}(f(s)) = s$ for every s in S. In this case, we say that f is *invertible*. The function f^{-1} is the inverse of f (and, symmetrically, f is the inverse of f^{-1}).

Trigonometric Functions (Section 1.6)

Radian measure of an angle is related to the degree measure of an angle by the formula

$$x \text{ radians} = \frac{180x}{\pi} \text{ degrees}.$$

If θ is a real number, let $P(\theta)$ be the point of the unit circle cut off by rotating the positive x-axis by θ radians. The rotation is counterclockwise for $\theta > 0$ and clockwise for $\theta < 0$. We define $\cos(x)$ and $\sin(x)$ to be the abscissa and ordinate, respectively, of $P(\theta)$. The other trigonometric functions are defined as quotients: $\tan(x) = \sin(x)/\cos(x)$, $\cot(x) = \cos(x)/\sin(x)$, $\csc(x) = 1/\sin(x)$, and $\sec(x) = 1/\cos(x)$. There are many identities among the trigonometric functions, and most can be derived from

$$\cos^2(\theta) + \sin^2(\theta) = 1,$$
$$\sin(\theta + \phi) = \sin(\theta)\cos(\phi) + \cos(\theta)\sin(\phi),$$

and

$$\cos(\theta + \phi) = \cos(\theta)\cos(\phi) - \sin(\theta)\sin(\phi).$$

REVIEW EXERCISES for Chapter 1

In each of Exercises 1–8, express the given set as an interval or union of intervals (possibly unbounded).

1. $\{x : |x+5| < 6\}$
2. $\{x : |x-3| \geq 7\}$
3. $\{x : x^2 - 3x \leq 0\}$
4. $\{x : |x+3|/x > 5\}$
5. $\{x : (2x+1)/|x| < 4\}$
6. $\{x : x^2 < 6 + x\}$
7. $\{x : x - 1 \leq 6/x\}$
8. $\{x : |2x+1| > |x-3|\}$

9. What is the distance between the points $(-3, -1)$ and $(9, -6)$?
10. What is the distance between the points of intersection of the line $4x - 3y = 12$ and the coordinate axes?
11. What is the midpoint of the line segment between the points $(-3, -1)$ and $(9, -6)$?
12. What is the distance between the points of intersection of the line $y = 3x - 5$ and the parabola $y = x^2 - 9$?

In each of Exercises 13–16, identify the type of conic section whose Cartesian equation is given. Determine its vertical and horizontal axes of symmetry and, if it is a circle or parabola, its radius or vertex.

13. $x^2 + y^2 + 2y = 2x$
14. $7/4 + 3x - x^2 - y = 0$
15. $9x^2 + 18x + 4y^2 - 16y = 11$
16. $y^2 + 4y = 4x^2$

In each of Exercises 17–20, find the slope-intercept form of the line ℓ determined by the given data.

17. ℓ has slope 3 and x-intercept 2.
18. ℓ is parallel to the line $2x - 4y = 9$ and passes through $(-1, -3)$.
19. ℓ is perpendicular to $2x - 2y = 1$ and passes through $(-1, -3)$.
20. ℓ passes through the points $(1, -1)$ and $(3, -5)$.

In each of Exercises 21–24, use the given point P to write the point-slope form of the line ℓ determined by the given data.

21. ℓ is parallel to the line $x + y/6 = 1$ and passes through $P = (-2, 1)$.
22. ℓ is perpendicular to the line $2x + y = 1$ and passes through $P = (-2, 1)$.
23. ℓ passes through $P = (-2, 1)$ and has y-intercept 5.
24. ℓ passes through $P = (1, 6)$ and has x-intercept 2.

In each of Exercises 25–28, find the intercept form of the equation of the line ℓ determined by the given data.

25. ℓ passes through $(0, -3)$ and $(4, 0)$.
26. ℓ is parallel to the line $y = 2 - x$ and passes through $P = (2, 3)$.
27. ℓ is perpendicular to the line $2y = x$ and passes through $P = (1, 1)$.
28. ℓ passes through $(1, 5)$ and $(3, 1)$.

In each of Exercises 29–32, find the point or points of intersection of the curves whose equations are given.

29. $2x - y = 1$ and $x + y = 5$
30. $x + 2y = 7$ and $3x + y = 1$
31. $y = 3x - 5$ and $y = x^2 - 9$
32. $y = x^2 - 3$ and $y = 1 + 2x - x^2$

In each of Exercises 33–40, find the domain of the function defined by the given expression.

33. $(x^2 - 2x - 2)/(x^2 - 2x + 2)$
34. $x/(1 + 1/x)$
35. $(x^2 - 25)/(x - 5)$
36. $(4 - x^2)^{-1/2}$
37. $(8 + x^3)^{-1/3}$
38. $\sqrt{x^2 + x - 2}$
39. $1/(4 - \sqrt{25 - x^2})$
40. $\sqrt{9 - x^2}/(1 + \sqrt{x^2 - 1})$

In each of Exercises 41–46, calculate f_n for $1 \leq n \leq 5$.

41. $f_n = n/(n+1)$
42. $f_n = 1 + (-1/2)^n$
43. $f_n = 2^n + 2n$
44. $f_n = \dfrac{2 + 1/n}{2 - 1/n}$
45. $f_n = 1 + 2 + 4 + \cdots + 2^{n+1}$
46. $f_n = 1 \cdot 2 \cdot 3 \cdots (n+2)$

In each of Exercises 47–50, a sequence $\{f_n\}$ is defined recursively. Calculate f_5.

47. $f_1 = 1,\ f_n = 2f_{n-1}$
48. $f_1 = 1,\ f_n = n^2 - f_{n-1}$
49. $f_1 = 0,\ f_n = 2f_{n-1} + n$
50. $f_1 = 1,\ f_n = 1 + 2/(1 + 1/f_{n-1})$

In Exercises 51–60, calculate the given expression that is constructed from $f(x) = 2x + 3$, $g(x) = 1 + x^2$ and $h(x) = (1 - x)/(1 + x)$.

51. $f^2(\sqrt{x})$
52. $(f^2 - 4g)(x)$
53. $(h \circ g)(x)$
54. $(g \circ f)(x)$
55. $(f \circ g^2)(x)$

56. $(h \circ (2f))(x)$
57. $(f \circ g^{1/2})\sqrt{x}$
58. $(h \circ h)(x)$
59. $\left(f^{-1} \circ \frac{1}{f}\right)(x)$
60. $(f \circ f \circ f)(x)$

In each of Exercises 61–64, f is an invertible function. The domain S of f is given, as is the formula for $f(s)$ for s in S. State the domain T of f^{-1} and a formula for $f^{-1}(t)$ for t in T.

61. $S = [1,4]$, $f(s) = 1/(1+s)$
62. $S = [1,\infty)$, $f(s) = 3s^2 - 1$
63. $S = [0,1]$, $f(s) = s^2 + 2s + 2$
64. $S = (0,1)$, $f(s) = (1+s)/(2+s)$

In each of Exercises 65–68, a noninvertible function f is given. Determine whether f is onto but not one-to-one, one-to-one but not onto, or neither one-to-one nor onto.

65. $f:[-1,\infty) \to (0,1]$, $f(s) = 1/(1+s^2)$
66. $f:[1,100] \to [1,110]$, $f(s) = s + \log_{10}(s)$
67. $f:[0,\infty) \to [0,1]$, $f(s) = s/(s+1)$
68. $f:[0,2] \to [0,2]$, $f(s) = s^2 - 2s + 2$

In each of Exercises 69–72, state how the graph of g may be obtained from the graph of f by a vertical translation, a horizontal translation, or both.

69. $f(x) = x^2$, $g(x) = (x+2)^2 - 1$
70. $f(x) = (x+1)/(x+2)$, $g(x) = (x-1)/x$
71. $f(x) = x^2$, $g(x) = x^2 - 6x$
72. $f(x) = x^3/(x^4+2)$, $g(x) = f(x+3) + 3$

In each of Exercises 73–76, a parametric curve with domain of parameterization I is given. Describe and sketch it.

73. $x = 2t+1$, $y = 3t+2$, $I = [-2/3, -1/2]$
74. $x = 3\sin(t)$, $y = 3\cos(t)$, $I = [-\pi/2, \pi/2]$
75. $x = \cos^2(t)$, $y = \sin^2(t)$, $I = [0, \pi/2]$
76. $x = |t| + t$, $y = |t| - t$, $I = [-1, 1]$

Each of Exercises 77–80 concerns a right triangle with a hypotenuse of length c and legs of lengths a and b. Let A be the angle opposite to the side of length a. From the given information, calculate the six trigonometric functions of A.

77. $a = 3$, $b = 2$
78. $a = 12$, $b = 5$
79. $b = 3$, $c = \sqrt{10}$
80. $b = 3$, $c = 5$

81. Let α be the radian measure of angle A of Exercise 78. Evaluate the six trigonometric functions at $\alpha + \pi/2$.
82. Let α be the radian measure of angle A of Exercise 80. Evaluate the six trigonometric functions at $\alpha + \pi$.

In each of Exercises 83–86, simplify the given expression.

83. $\tan(\pi/4) - \cot(3\pi/4)$
84. $\sin(\pi/6) + \cos(5\pi/3)$
85. $\csc(\pi/6)\sec(3\pi/4)$
86. $\cot(\pi/3)\cot(4\pi/3)$

In each of Exercises 87–90, use the given information about the acute angle α to calculate the value of r.

87. $\sin(\alpha) = 3/4$, $r = \sin(2\alpha)$
88. $\tan(\alpha) = 1/2$, $r = \cos(2\alpha)$
89. $\cos(\alpha) = 3/5$, $r = \tan(2\alpha)$
90. $\sec(\alpha) = 3/2$, $r = \cos(2\alpha + \pi/2)$

In this first chapter, we reviewed three fundamental developments that underlie calculus: the real number line, the mathematical notion of a function, and the principles of analytic geometry.

Irrational Numbers and Successive Approximation

The history of the real number line begins in the Golden Age of ancient Greece. The first important step came with the realization that not every number is rational. Some time before 410 BCE, the followers of Pythagoras (ca. 580–500 BCE) discovered that $\sqrt{2}$ is irrational. The elegant argument outlined in Exercise 57, Section 1.1, is the earliest proof that has been preserved. It was recorded by Aristotle (384–322 BCE) but is certainly older.

The mathematicians of ancient Greece devised a number of methods for successively approximating $\sqrt{2}$ and other irrational numbers. One of these algorithms is based on finding integer solutions to the equation $x^2-2y^2=1$ (see Exercise 56, Section 1.4). Archimedes (ca. 287–212 BCE) used the same idea to approximate $\sqrt{3}$ to six decimal places. He obtained his estimate, 1351/780, from the solution $x=1351$, $y=780$ of the equation $x^2-3y^2=1$.

Equations of the form $x^2-D\cdot y^2=1$, with D a nonsquare integer, are called *Pell equations*, and Archimedes seems to have had some expertise in their solution. In 1773, a copy of a previously unknown message from Archimedes to the mathematicians of Alexandria was discovered in a German library. In his letter, Archimedes posed a riddle concerning the numbers of eight types of cattle. The key to the *cattle problem*, as it has come to be known, is the solution in integers of the Pell equation $x^2-4729494\cdot y^2=1$. In 1880, A. Amthor obtained the least positive integer solution to this Pell equation:

$$x_0 = 109931986732829734979866232821433543901088049,$$
$$y_0 = 50549485234315033074477819735540408986340,$$

which enabled him to show that the solution to the cattle problem has 206,545 decimal digits. In 1981, a Cray I supercomputer at the Lawrence Livermore National Laboratory calculated this large number, using about 10 minutes of computing time. The life span of the computer that solved this ancient problem was only about a dozen years. Having been purchased in the late 1970s for \$19,000,000, it was auctioned off in 1993 for \$10,000.

The Number π

The irrational number π also featured prominently in the mathematics of the ancient Greeks. Archimedes proved that, if A and C are the area and circumference of a circle of radius r, then the ratios A/r^2 and $C/2r$ are the *same* constant π. (It is not known who discovered this fact or even when the discovery was made.) Archimedes also approximated π to two decimal places by deriving the inequalities $3+10/71<\pi<3+1/7$. The computation of the digits of π has attracted the interest of mathematicians ever since. Sir Isaac Newton (1642–1727), for example, calculated π to 15 decimal places. By 1761, 113 digits of π were known. In that year, the Swiss mathematician Johann Heinrich Lambert (1728–1777) published a proof that π is irrational. In 1806, Adrien-Marie Legendre (1752–1833) proved the stronger result that π^2 is irrational, and he conjectured even more.

A real number is said to be *algebraic* if it is a root of some polynomial with rational coefficients; otherwise, it is said to be *transcendental*. Every rational number is algebraic, but an irrational number may be algebraic or transcendental. For example, the irrational number $\sqrt{2}$ is algebraic because it is a root of x^2-2. At the time Legendre ventured his conjecture, mathematicians did not even know if transcendental numbers existed. It was not until 1844 that Joseph Liouville (1809–1882) established the existence of transcendental numbers. Thirty years later, Georg Cantor's (1845–1918) novel ideas about infinite sets put the matter in a totally new light.

Using simple but original ideas, Cantor proved that algebraic numbers can be put into one-to-one correspondence with natural numbers and that transcendental numbers cannot. We say that the set of algebraic numbers is *countable*, or *denumerable*, because of this property and that the set of transcendental numbers is *uncountable*, or *nondenumerable*. In a very real sense, Cantor proved that there are more transcendental numbers than algebraic numbers. He did not, however, show that any one particular number is transcendental.

Indeed, it is usually very difficult to show that a given number is transcendental. In 1882, Carl Louis Ferdinand von Lindemann (1852–1939) finally proved Legendre's conjecture that π is transcendental. Meanwhile, the digit hunters have continued their work. In 2002, a supercomputer at the University of Tokyo was used to calculate more than 1.2 trillion digits of π.

Zeno's Paradoxes

The fundamental idea of calculus, that which distinguishes it from algebra and classical geometry, is the infinite process. We have mentioned the algorithms by which Greek mathematicians created *infinite* sequences, which tend to $\sqrt{2}$ and $\sqrt{3}$ in the limit. We have discussed the manner in which Archimedes inscribed an *infinite* sequence of regular polygons in a circle in order to *successively approximate* its area and circumference. A number of other infinite processes have been outlined in the exercises of Chapter 1. The idea is not to actually perform an infinite number of operations but to see what results in the limit. That is the essence of calculus. Although the mathematicians of ancient Greece who originated these ideas were well aware of the power of the infinite process, they were also concerned with the logical difficulties that can ensue.

Zeno of Elea (ca. 496–435 BCE) lived in the age of Pericles. Little is known of Zeno's life. As for his death, it is written that Zeno was tortured and executed for his part in a conspiracy against the state. Although the original writings of Zeno have not survived, numerous references to them have. Zeno is best known today for four paradoxes—the *Dichotomy,* the *Achilles,* the *Arrow,* and the *Stade*—that Aristotle retold.

According to the *Dichotomy,* "There is no motion because that which is moved must arrive at the middle before it arrives at the end, and it must traverse the half of the half before it reaches the middle, and so on ad infinitum." Thus if a line is infinitely divisible, motion cannot begin. The *Achilles* asserts that, given a head start, a tortoise will not be overtaken by the runner Achilles because Achilles must first reach the point P_1 at which the tortoise started. In the time required for Achilles to arrive at P_1, the tortoise will have advanced to a point P_2, which Achilles must reach before overtaking the tortoise. By the time Achilles has run from P_1 to P_2, the tortoise will have moved ahead to a point P_3, and so on, ad infinitum. The four paradoxes are "immeasurably subtle and profound," in the words of Bertrand Russell. Aristotle called them fallacies but was not able to refute them. Indeed, it would be a long time before mathematicians created a framework in which they could undertake the infinite processes of calculus free from such logical difficulties.

Analytic Geometry

After the decline of Greek civilization, mathematical progress came slowly. The next significant contribution to the foundation of calculus came in the 14th century when philosophers in England and France attempted to analyze accelerated motion. In doing so, the Norman scholar Nicole Oresme (1323–1382) identified the real numbers with a geometric continuum. After another long hiatus, François Viète (1540–1603) further demonstrated the power of geometrically interpreting the real numbers. His work set the stage for the development of analytic geometry by Pierre de Fermat (1601–1665) and René Descartes (1596–1650).

Analytic geometry, as developed in the 17th century by Fermat and Descartes, was the final breakthrough that permitted the discoveries of calculus. The crucial idea that the solution set of an equation such as $F(x, y) = 0$ can be identified with a curve in the plane appeared in the *Géométrie*, which Descartes published in 1637. In the words of Descartes, "I believed that I could borrow all that was best both in geometrical analysis and in algebra, and correct all the defects of the one by help of the other."

Descartes led something of a fugitive existence. He left France for Holland, where he lived for most of his adult life. Changing residence 24 times in 21 years, he adopted the motto "He who hides well lives well." In 1649, Descartes accepted an invitation to tutor Sweden's Queen Christina in philosophy, but he died of pneumonia soon after his arrival. The mortal remains of Descartes had a macabre afterlife that mirrored his transient lifestyle. Issues of religious affiliation led to an initial burial in a Stockholm cemetery intended for unbaptized children. Sixteen years later, a French delegation was dispatched to repatriate the remains of the country's most distinguished philosopher. As the journey was long, the coffin was short. The bones of Descartes, minus the skull, returned home to a sequence of burials in and disinterments from a museum garden, the courtyard of the Louvre, and four churches. In a chapel of the Church of St. Germain-des-Pres, Paris, a marble slab bearing an 1819 inscription marks the final resting spot. The skull of Descartes was returned to France at a later date and is now in the collection of the Musée de l'Homme in Paris.

The Scientific Revolution

Prior to the 16th century, most scientific teaching in Europe consisted of scholars passing a fixed collection of theories and ideas to their students. These theories were not to be questioned or analyzed—merely learned. The method of experimentation and proof had not yet become a universally accepted part of scientific activity. Ancient authority ruled the sciences.

During the Age of Reason, a revolution took place. The so-called rationalist movement in philosophy demanded that new ideas be demonstrated or proved. Descartes became one of the most influential exponents of the movement. Today, the rationalist methodology is universally accepted. When a calculus instructor insists on *proving* assertions in class, it is because of a deeply held conviction that students should not simply learn facts by rote but should truly understand the concepts being taught.

If the 16th century had brought a revolution in scientific thinking, then the 17th century ushered in the era of *quantitative* relationships. In *The Assayer,* Galileo (1564–1642) published the manifesto of the new science:

> Philosophy is written in this grand book, the universe which stands continually open to our gaze. But the book cannot be understood unless one first learns to comprehend the language and read the letters in which it is composed. It is written in the language of mathematics and its characters are triangles, circles, and other geometric figures.

Scientists now looked for *functional* relationships between the variables of nature. Galileo determined the distance an object falls under gravity as a function of time. Johannes Kepler (1571–1630) discovered that the time in which a planet orbits the sun is a function of its mean distance to the sun (Kepler's Third Law). Christiaan Huygens (1629–1695) determined centrifugal force as a function of radius and velocity, providing a link between the laws of Kepler and Galileo. Before the end of the 17th century, Newton formulated the law of universal gravitation, demonstrating that Kepler's laws and Galileo's law are but different aspects of the same general phenomenon.

The realization that natural phenomena could be expressed through functional relationships was not at all limited to motion. The Dutch scientist Willebrord Snell (1580–1626) found that the angle of refraction of a ray of light is a function of the angle of incidence (Snell's law). Evangelista Torricelli (1608–1647) demonstrated that the height of mercury in a vertical tube is a function of atmospheric pressure. Robert Boyle (1627–1691) postulated the functional relationship among the pressure, volume, and temperature of a gas. Robert Hooke (1635–1703) expressed the force exerted by a spring as a function of its compression or extension. The age of functions had arrived. Before the end of the 17th century, calculus would be invented as a tool to study these functions.

The Completeness Property of the Real Numbers

Although scientific observation validated the phenomena predicted by calculus, the new discipline quickly came under serious criticism concerning the foundation of the entire subject. Just as the ancient Greeks had opened the door to Zeno's paradoxes when they admitted the infinite process, so had the mathematicians of the 17th century set themselves up for a new set of paradoxes by their lack of rigor in working with the real number line. By the 1800s, mathematicians realized that a deeper understanding of the real number system was essential. Their focus turned to what we now call the *completeness property* of the real numbers, which guarantees that there are no holes or gaps in the real number line. Here is one way to describe the completeness of the real numbers: If $\{I_n\}$ is a sequence of closed bounded nonempty real intervals with each interval a subinterval of its predecessor as in the diagram,

$$[\ \ [\ \ \underbrace{[\ \ \underbrace{[\ldots\ldots]}_{I_{n+1}}\ \ ]}_{I_n}\ \ ]\ \ ]$$

then there is at least one point in $\mathbb{R}$ that is in every I_n. The real number system satisfies this property and is complete; the rational number system is not. An understanding of the role of completeness came about in 1872 thanks to several mathematicians, including Georg Cantor, Julius Richard Dedekind (1831–1916), and Karl Weierstrass (1815–1897). Although many further discoveries would be made, by 1872 everything was in place for a rigorous development of calculus, a subject that was then already 200 years old.

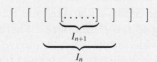

CHAPTER 2

Limits

PREVIEW

The concept of *limit* is the basis for a solid understanding of calculus. We will therefore spend an entire chapter studying limits before proceeding to calculus. The notion of limit originated in Greece approximately 2500 years ago. It arose then for many of the same reasons that we study limits in a modern calculus course: to understand continuous motion and to measure lengths, areas, volumes, and other physical quantities.

Consider a circle $\mathcal{C}$ of radius r. We all learn in geometry that its area A is πr^2, and its circumference C is $2\pi r$, but what exactly do we mean by the area and circumference of a circle? Precise answers can be understood only in terms of limits.

One approach is to approximate the area and circumference of $\mathcal{C}$ by a region with an area and perimeter that we do know. Figure 1 shows an equilateral triangle $\mathcal{P}_3$ inscribed in $\mathcal{C}$. We know the area and perimeter of the triangle $\mathcal{P}_3$, but it is not a very good approximation to the circle. We add sides, one at a time, to the inscribed figure to create inscribed polygons with more and more sides of equal length. It seems clear that the inscribed polygon is becoming closer and closer to the circle. Figure 2 shows that the area inside the inscribed 6-gon is fairly close to the area of the circle, and the area inside the inscribed 12-gon is *extremely* close to that of the circle. When the polygon has one billion sides, we could not possibly draw a figure illustrating the difference between the inscribed polygon and the circle, but there *is*

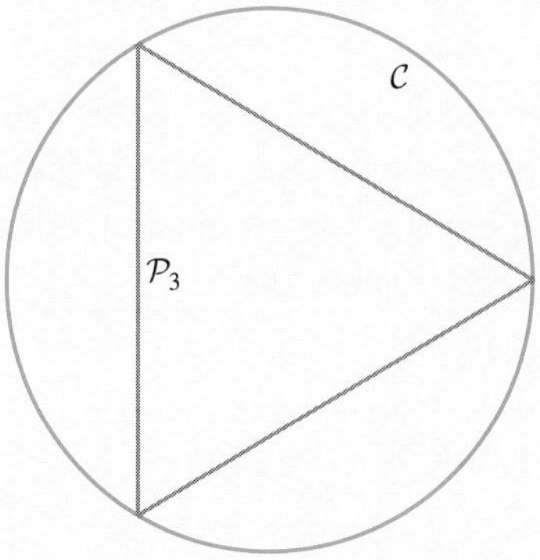

▲ Figure 1 An equilateral triangle inscribed in circle $\mathcal{C}$

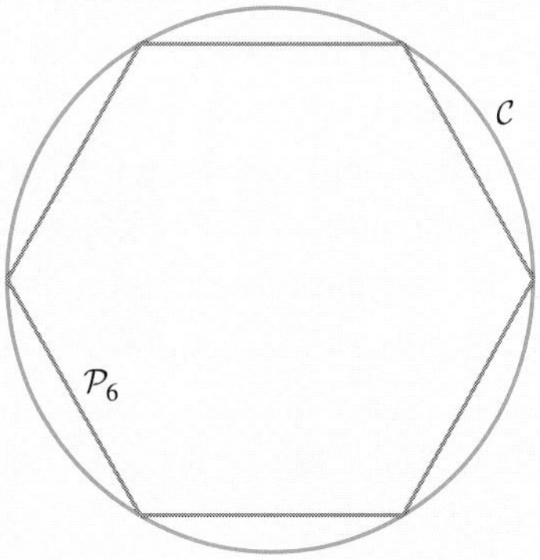

▲ Figure 2a A 6-gon inscribed in circle $\mathcal{C}$

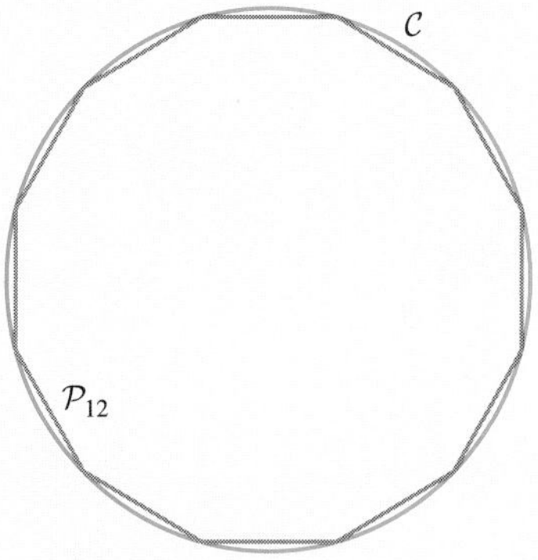

▲ Figure 2b A 12-gon inscribed in circle $\mathcal{C}$

a difference. No matter how small the sides of the polygon become, the polygon will have many small flat sides of equal length, even though the circle is round. It is correct to say that, when the side length of the polygon is close to zero (that is, the number of sides of the polygon is large), the area inside the polygon is close to the area inside the circle. It is *incorrect* to say that the area of any approximating polygon is equal to that of the circle.

The correct mathematical terminology is that, as the number of sides n tends to infinity, the *limit* of the area A_n enclosed by the n-sided polygon $\mathcal{P}_n$ equals the area A inside the circle, and the *limit* of the perimeter $\mathcal{C}_n$ of $\mathcal{P}_n$ equals the circumference C of the circle. We write these statements as

$$A = \lim_{n \to \infty} A_n \quad \text{and} \quad C = \lim_{n \to \infty} \mathcal{C}_n. \tag{2.0.1}$$

For another application of the *limit* concept, suppose that we view a training film of a runner. The film shows elapsed time and distance markers that allow us to measure the distance $s(t)$ the athlete has run in any given time t. It is easy to compute the *average* speed of the runner between any two points: Simply divide the total distance run by the total elapsed time. However, the speed of the runner will vary from one instant of time to another. How can the runner's speed $v(c)$ at a given instant of time c be calculated? For any value $t > c$, the average speed of the runner in the small time interval $[c, t]$ is

$$\frac{s(t) - s(c)}{t - c}.$$

For t close to c, this will give a good approximation to $v(c)$ because the runner's speed does not change much in the small time interval $[c, t]$. If we press this point further, then we intuitively arrive at the concept of *instantaneous velocity* at c:

$$v(c) = \lim_{t \to c} \frac{s(t) - s(c)}{t - c}. \tag{2.0.2}$$

As we will see, equations (2.0.1) and (2.0.2) actually serve to *define* the concepts of area, circumference, and instantaneous velocity. These examples convey an important idea: The concept of *limit* involves a variable that approaches arbitrarily close to a number but does not actually reach it. In equation (2.0.1), the natural number n can get arbitrarily large, but it is never actually equal to infinity. In equation (2.0.2), the number t cannot actually be equal to c because that would result in the meaningless fraction 0/0. The key is first to understand exactly what is meant by the limits in these equations and then to learn methods for computing them.

The purpose of this chapter is to give a precise meaning to the concept of *limit*. That having been done, much of the rest of this textbook concerns ways to calculate limits. Indeed, the notion of limit is used in a fundamental way in each of the remaining chapters.

2.1 The Concept of Limit

This section includes a number of examples that informally explore the idea of limit. The understanding of limits gained by reading this section is intuitive, but it is adequate for most purposes in calculus. Section 2.2 presents a precise formulation of the *limit* concept.

Informal Definition of Limit

Let f be a real-valued function that is defined in an open interval to the left of a real number c and in an open interval to the right of c. We say, "$f(x)$ has the limit ℓ as x tends to c," and we write

$$\lim_{x \to c} f(x) = \ell,$$

if the values $f(h)$ are close to ℓ when x is close to, but not equal to, c. As an alternative, we sometimes write

$$f(x) \to \ell \quad \text{as} \quad x \to c$$

and say "$f(x)$ tends to ℓ as x tends to c" (see Figure 1).

INSIGHT The value $f(c)$ plays no role in the definition of the limit of $f(x)$ as x tends to c. We do *not* even assume that f is defined at c. Indeed, a definition of "limit" that required f to be defined at c would not be relevant to many important applications. Even when $f(c)$ is defined, the value $f(c)$ does not have to be related to $\lim_{x\to c} f(x)$ in any way. In a sense, $\lim_{x\to c} f(x)$ is what we anticipate that $f(x)$ will equal at $x = c$, not necessarily what $f(x)$ actually equals when $x = c$.

▶ **EXAMPLE 1** Let f, g, and h be the three functions defined by

$$f(t) = t^2 \text{ for all } t, \qquad g(t) = t^2 \text{ for all } t \neq 0, \qquad \text{and} \qquad h(t) = \begin{cases} t^2 & \text{if } t \neq 0 \\ 1 & \text{if } t = 0 \end{cases}$$

Discuss the limits of $f(t)$, $g(t)$, and $h(t)$ as t approaches 0.

Solution When $|t| < 1$, we have $t^2 = |t| \cdot |t| < 1 \cdot |t| = |t|$. Thus, when $|t|$ is less than 1 (as is ultimately the case when t approaches 0), the number $f(t) = t^2$ is even closer to 0 than $|t|$ is (see Figure 2). We conclude that, as t approaches 0, so does $f(t)$. We write this as $\lim_{t\to 0} f(t) = 0$.

Next, we consider g. The fact that 0 is not in the domain of g does not affect whether $g(t)$ has a limit at $t = 0$. For $t \neq 0$, we have $g(t) = t^2$, and our calculation of the limit of $f(t)$ at $t = 0$ allows us to conclude that $\lim_{t\to 0} g(t) = \lim_{t\to 0} t^2 = 0$.

Finally, we turn to the function h. The definition of "limit" tells us that the equation $h(0) = 1$ has nothing to do with the existence or value of $\lim_{t\to 0} h(t)$. When

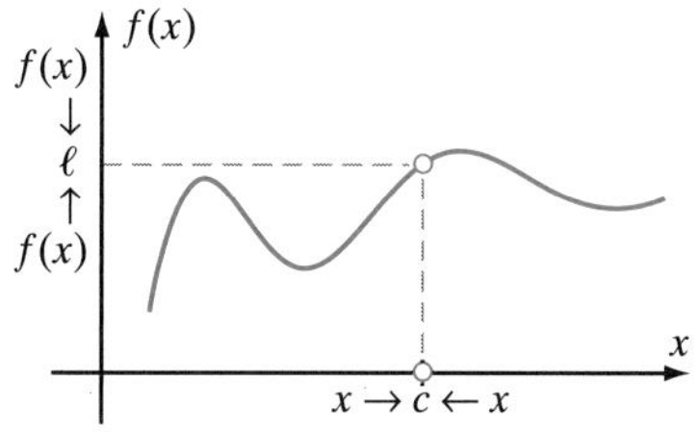

▲ Figure 1 $f(x)$ has the limit ℓ as x tends to c

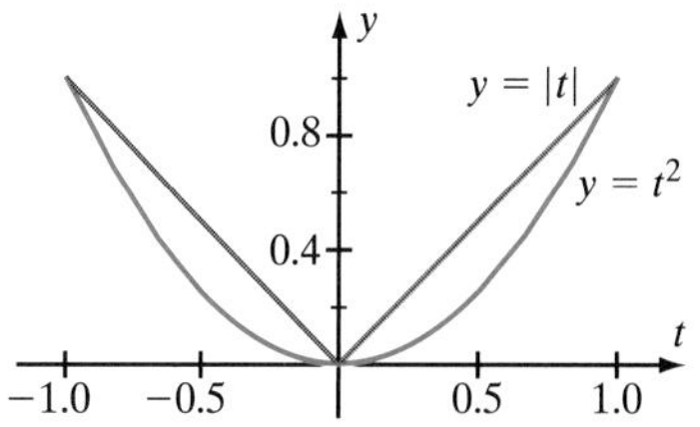

▲ Figure 2

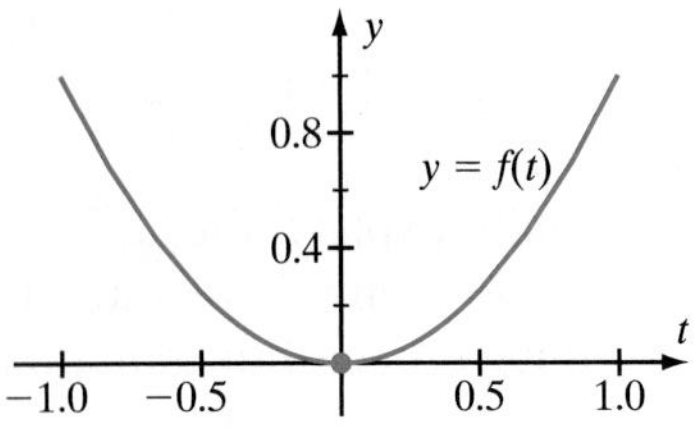

▲ Figure 3a $\lim_{t \to 0} f(t) = 0$

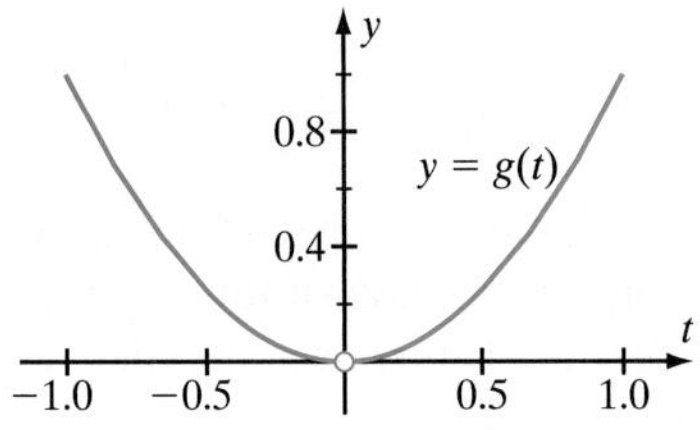

▲ Figure 3b $\lim_{t \to 0} g(t) = 0$

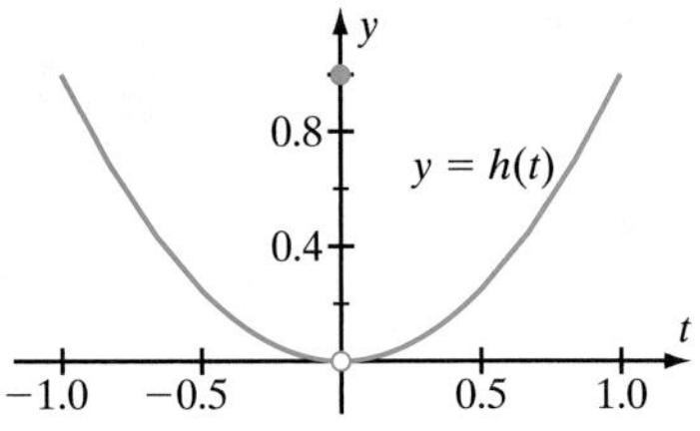

▲ Figure 3c $\lim_{t \to 0} h(t) = 0$

t is near 0 but unequal to 0, we have $h(t) = t^2$. Therefore $\lim_{t \to 0} h(t) = \lim_{t \to 0} t^2 = 0$. Figure 3 illustrates the three limits that we have calculated. ◀

INSIGHT In Example 2, we used the plausible idea that the limit of a sum or difference is the sum or difference of the limits, the limit of a product is the product of the limits, and the limit of a quotient is the quotient of the limits (as long as we do not divide by zero). These limit properties are treated in more detail in Section 2.2.

▶ **EXAMPLE 2** Let $f(x) = (x^2 - 6x + 6)/x$. Evaluate $\lim_{x \to 5} f(x)$.

Solution We look at the components of f. As x tends to 5, the expression x^2 tends to 25, and the expression $6x$ tends to 30. We conclude that

$$\lim_{x \to 5} f(x) = \frac{25 - 30 + 6}{5} = \frac{1}{5}.$$

A numerical investigation, as shown in the table, supports this conclusion.

Approach from the left side of 5 **Approach from the right side of 5**

x	4.99	4.999	4.9999	4.99999	→	5	←	5.00001	5.0001	5.001	5.01
$f(x)$	0.19240	0.19924	0.19992	0.19999	...		...	0.20000	0.20008	0.20076	0.20760

The actual value of $f(x)$ for $x = 5$ has been intentionally omitted from the table. It is not used in the investigation of the limiting value of $f(x)$ as x approaches 5. ◀

One-Sided Limits

Implicit in the limit definition is the fact that a function should have just one limit at a point c. Suppose that $f(x)$ is close to a number ℓ_L when x is close to c and to the left of c, and that $f(x)$ is close to a different number ℓ_R when x is close to c and to the right of c. We say that f has a *left limit* ℓ_L at c and a *right limit* ℓ_R at c, but we do *not* say that f has a limit at c (see Figure 4). These *one-sided limits* are written as $\lim_{x \to c^-} f(x) = \ell_L$ (or $f(x) \to \ell_L$ as $x \to c^-$) for the left limit and $\lim_{x \to c^+} f(x) = \ell_R$ (or $f(x) \to \ell_R$ as $x \to c^+$) for the right limit. From these informal definitions, it is clear that

$$\lim_{x \to c} f(x) = \ell \quad \text{is equivalent to} \quad \left(\lim_{x \to c^-} f(x) = \ell_L, \quad \lim_{x \to c^+} f(x) = \ell_R, \quad \text{and} \quad \ell_L = \ell_R \right).$$

In other words, if both left and right limits exist and are the same, then the limit *does* exist.

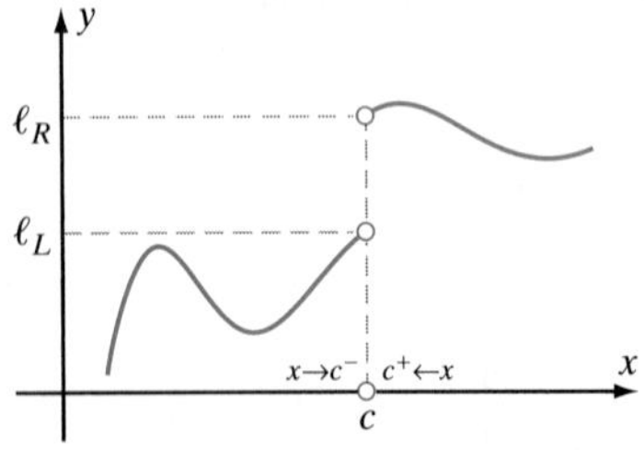

▲ Figure 4 $\lim_{x \to c^+} f(x) = \ell_R$, $\lim_{x \to c^-} f(x) = \ell_L$

INSIGHT The notation for one-sided limits arises from the $c = 0$ case. As x approaches 0 from the left, x assumes negative values; hence, $x \to 0^-$. As x approaches 0 from the right, x assumes positive values; therefore, $x \to 0^+$. Keep this in mind to help remember the notation.

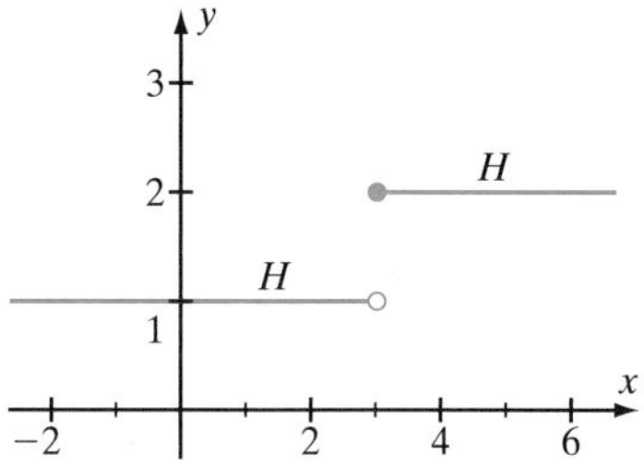

▲ Figure 5 $H(x) = \begin{cases} 1 & \text{if } x < 3 \\ 2 & \text{if } x \geq 3 \end{cases}$

▶ **EXAMPLE 3** Let H and G be defined by

$$H(x) = \begin{cases} 1 & \text{if } x < 3 \\ 2 & \text{if } x \geq 3 \end{cases} \quad \text{and} \quad G(s) = \begin{cases} s^2 + 1 & \text{if } s < 4 \\ 5s - 3 & \text{if } s > 4 \end{cases}.$$

Does $H(x)$ have a limit as x tends to 3? Does $G(s)$ have a limit as s tends to 4?

Solution Refer to Figure 5 for the graph of H. As x approaches (but does not equal) 3 from the left, $H(x)$ tends to 1. On the other hand, as x approaches (but does not equal) 3 from the right, $H(x)$ tends to 2. Thus $\lim_{x \to 3^-} H(x) = 1$, and $\lim_{x \to 3^+} H(x) = 2$. However, $H(x)$ does not have a limit as x tends to 3 because there is *no single value* that $H(x)$ approaches as x tends to 3.

When s is near 4 but to the left of 4, s^2 is near 16, and $s^2 + 1$ is near 17. Thus $\lim_{s \to 4^-} G(s) = 17$. On the other hand, when s is near 4 but to the right of 4, $5s$ is near 20, and $5s - 3$ is near 17. Hence $\lim_{s \to 4^+} G(s) = 17$. Because $G(s)$ tends to 17 both from the left and the right, we can say that $\lim_{s \to 4} G(s) = 17$. This limit can also be seen from the graph of G (see Figure 6). ◀

Limits Using Algebraic Manipulations

In many situations, some algebraic manipulations are required in order to evaluate a limit.

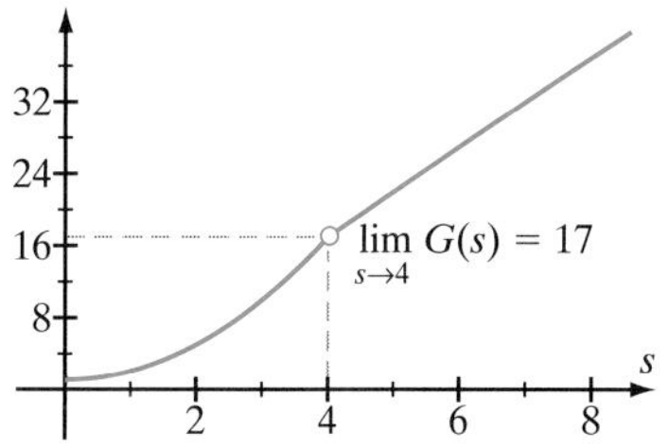

▲ Figure 6

$G(s) = \begin{cases} s^2 + 1 & \text{if } s < 4 \\ 5s - 3 & \text{if } s > 4 \end{cases}$

▶ **EXAMPLE 4** Does

$$f(x) = \frac{x^2 - 4}{x + 2}$$

have a limit as x approaches -2?

Solution This type of limit often occurs in calculus. The function f is *not* defined at the point where the limit is being calculated. We therefore cannot attempt a shortcut by simply substituting $x = -2$. Instead, we attack this problem by noticing that

$$f(x) = \frac{(x + 2)(x - 2)}{(x + 2)}.$$

As long as $x \neq -2$, then $x + 2 \neq 0$. We can divide this common nonzero factor from the numerator and denominator. As a result, $f(x) = x - 2$ when $x \neq -2$. Remember that, in calculating the limit of f at -2, we consider the values of $f(x)$ for x close to, but not equal to, -2. Therefore, as $x \to -2$, $f(x)$ tends to the same limit as $x - 2$ does. In summary,

$$\lim_{h \to -2} \frac{x^2 - 4}{x + 2} = \lim_{x \to -2} (x - 2) = -4.$$

Refer to Figure 7. ◀

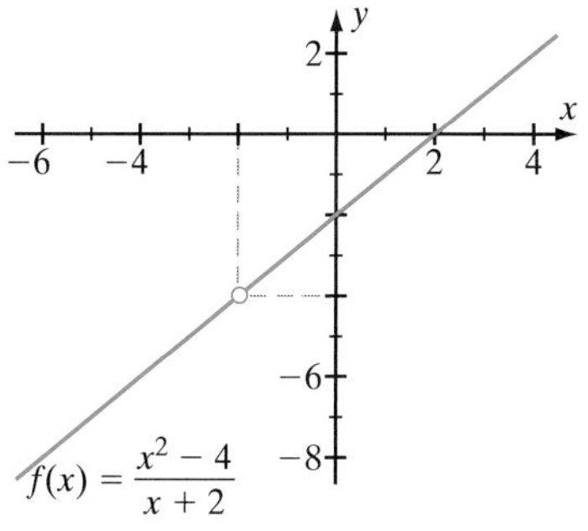

▲ Figure 7

Specified Degrees of Accuracy

We turn now to a practical, computation-oriented approach to the limit concept. By $\lim_{x \to c} f(x) = \ell$, we mean that $f(x)$ tends to ℓ when x tends to c. Now we ask: How close does x need to be to c to force $f(x)$ to stay within 0.1 of ℓ? How close does x need to be to c to force $f(x)$ to stay within 0.01 of ℓ? and so on.

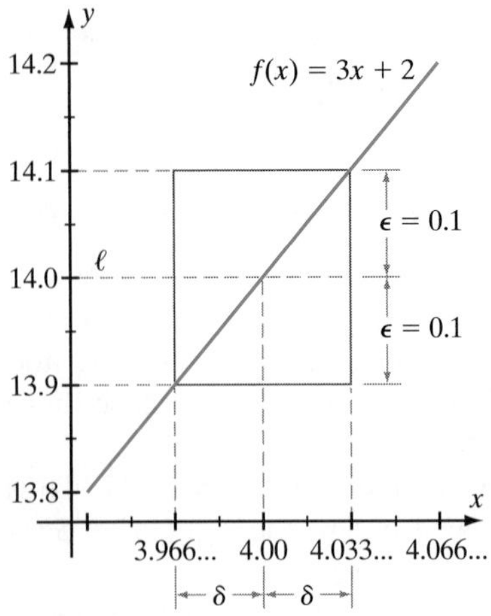

▲ Figure 8
$\varepsilon = 0.1,\ \delta = \frac{0.1}{3} = 0.0333\ldots$

This method of thinking about limits is important for applications and also leads to a deeper understanding. In the examples that follow, the bound on the desired distance of $f(x)$ to ℓ is called ε. The required distance of x to c is called δ.

▶ **EXAMPLE 5** If x is close to 4, then $f(x) = 3x + 2$ is close to 14. We therefore say that $\lim_{x\to 4} f(x) = 14$. How close must x be to 4 to force $f(x) = 3x + 2$ to be within 0.1 of 14? Within 0.01 of 14?

Solution We will denote the limit, 14, by ℓ. To force $f(x) = 3x + 2$ to within $\varepsilon = 0.1$ of ℓ, we solve the inequality $|f(x) - \ell| < \varepsilon$, or $|(3x + 2) - 14| < 0.1$, or $|3x - 12| < 0.1$. The solution is obtained by multiplying both sides of the last inequality by 1/3, which gives us $|x - 4| < \frac{0.1}{3} = 0.0333\ldots$. Therefore provided x stays within $\delta = \frac{0.1}{3} = 0.0333\ldots$ of $c = 4$, $f(x)$ will stay within $\varepsilon = 0.1$ of $\ell = 14$ (see Figure 8). Similarly, to make $f(x)$ stay within $\varepsilon = 0.01$ of $\ell = 14$, we solve the inequality $|(3x + 2) - 14| < 0.01$. The solution is $|x - 4| < \frac{0.01}{3} = 0.00333\ldots$. Therefore $f(x) = 3x + 2$ stays within $\varepsilon = 0.01$ of $\ell = 14$, provided that x stays within $\delta = \frac{0.01}{3} = 0.00333\ldots$ of $c = 4$. There is clearly a pattern here. If we want to make $f(x) = 3x + 2$ stay within a certain distance ε of $\ell = 14$, then we restrict x to lie within $\delta = \varepsilon/3$ of $c = 4$. We can achieve any desired degree of accuracy by taking ε as small as we like. Our calculation offers compelling evidence of a quantitative nature that justifies the intuitive feeling that $\lim_{x\to 4}(3x + 2) = 14$. ◀

▶ **EXAMPLE 6** Let $f(x) = x^2 + 4$. How close to 0 does x have to be to force $f(x)$ to be within $\varepsilon = 0.1$ of $\ell = 4$?

Solution Although it is intuitively clear that $\lim_{x\to 0} f(x) = 4$, let us examine the matter from the point of view of approximation, as in Example 5. To force $f(x) = x^2 + 4$ to be within $\varepsilon = 0.1$ of $\ell = 4$, we solve the inequality

$$|f(x) - 4| < 0.1.$$

The solution is $|(x^2 + 4) - 4| < 0.1$, or $|x^2| < 0.1 = 10/100$, or $|x| < \sqrt{10}/10$. Because we are interested in the distance of x from 0, we write

$$|x - 0| < \frac{\sqrt{10}}{10}.$$

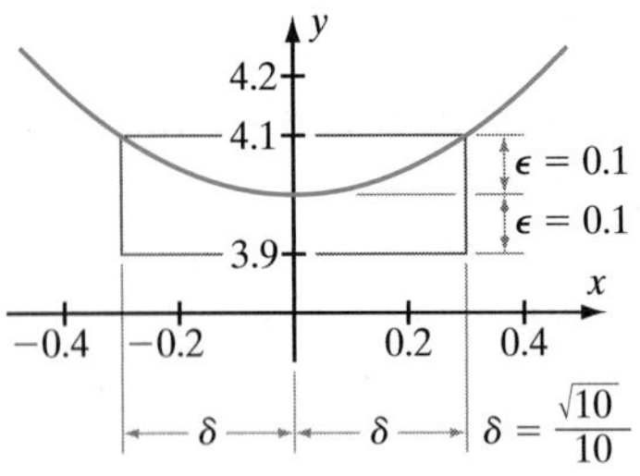

▲ Figure 9

Thus $f(x) = x^2 + 4$ is within $\varepsilon = 0.1$ of $\ell = 4$ when x is within $\delta = \sqrt{\varepsilon} = \sqrt{10}/10$ of $c = 0$ (see Figure 9). We arrive at our conclusion by working backward from the desired estimate on the function x^2. Let us now check our work: If $|x - 0| < \sqrt{10}/10$, then $|x^2| < 10/100$, or $|(x^2 + 4) - 4| < 0.1$. This means that $|f(x) - 4| < 0.1$, which is the desired result. Once we become accustomed to this process, we can work backward with confidence and not bother to check the work by repeating the calculation in the forward direction. ◀

INSIGHT The degree of closeness for x to c in Example 6 (namely, $\delta = \sqrt{\varepsilon}$) is different from that in Example 5 (namely, $\delta = \varepsilon/2$). Different functions approach their limits at different rates. This will be one of the main points in the limit problems that we study.

Graphical Methods

Examples 5 and 6 provide algebraic solutions to the question, "For a given small number ε, how can we take x close enough to c to force $f(x)$ to be within ε of ℓ?" We may also answer this question by graphical methods. Indeed, the condition that $f(x)$ stays within ε of ℓ when x stays within δ of c is easily visible when we plot f in the viewing rectangle: $[c-\delta, c+\delta] \times [\ell-\varepsilon, \ell+\varepsilon]$: The graph exits the window only through the lateral sides—not through the top or bottom of the viewing window.

▶ **EXAMPLE 7** Let $f(x) = (x^2 - 6x + 6)/x$. In Example 2, we observed that $\lim_{x\to 5} f(x) = 1/5$. Find a positive number δ such that $f(x)$ stays within 0.01 of 1/5 when x stays within δ of 5.

Solution In this example, we set $c = 5$, $\ell = 1/5$, and $\varepsilon = 0.01$. We begin by making an arbitrary choice of δ: $\delta = 0.05$. Figure 10 shows the graph of f in the viewing window $[5-0.05, 5+0.05] \times [1/5-0.01, 1/5+0.01]$. Because the graph of f exits this window through its top and bottom, we must use a smaller value for δ. To determine such a value, we observe that, on the horizontal axis in Figure 10, the distance between tick marks is 0.004. From the plot, we can see that $f(x)$ stays within $\varepsilon = 0.01$ of $\ell = 1/5$, provided that x stays within 3×0.004 of $c = 5$. We therefore set $\delta = 0.012$. The resulting graph of f in the viewing rectangle $[5-0.012, 5+0.012] \times [1/5-0.01, 1/5+0.01]$ shows that our choice $\delta = 0.012$ suffices (see Figure 11). ◀

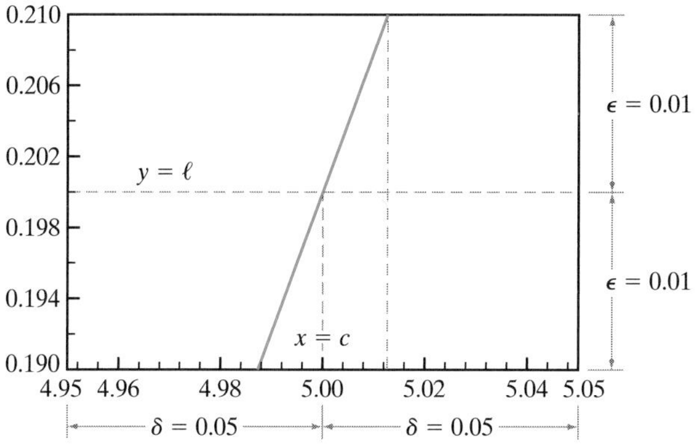

▲ Figure 10 Viewing rectangle $[5-\delta, 5+\delta] \times [1/5-\varepsilon, 1/5+\varepsilon]$

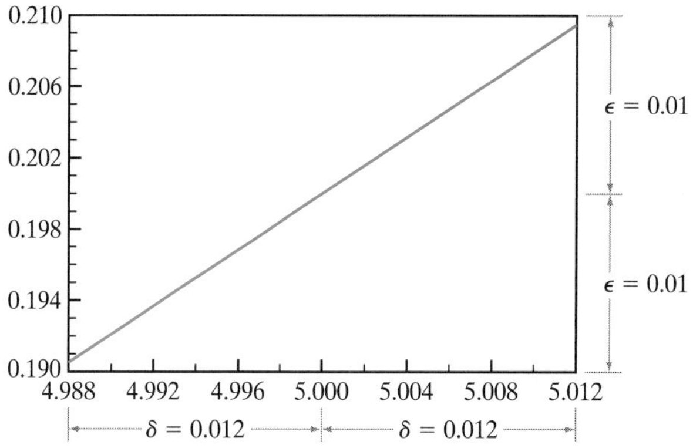

▲ Figure 11 Viewing rectangle $[5-\delta, 5+\delta] \times [1/5-\varepsilon, 1/5+\varepsilon]$ for a smaller δ.

Some Applications of Limits

Numerical calculations of the type we have been examining come up frequently in applications. Following are some examples.

▶ **EXAMPLE 8** An unmanned spacecraft is to be sent to a distant planet. The ground control personnel compute that an error of size d degrees in the spacecraft's initial angle of trajectory will result in the spacecraft missing its target by $E(d) = 30 \cdot (10^d - 1)$ miles. The spacecraft must land within 0.1 mile of its target for the mission to be considered successful. What error in the measurement d of the trajectory angle is allowable?

Solution To answer this question, we require $30 \cdot (10^d - 1) < 0.1$. This inequality is equivalent to $10^d - 1 < 1/300$, or $10^d < 1 + 1/300$. The solution set is $d < \log_{10}(1 + 1/300) \approx 0.00144524073$. Thus $d = 0.00144$ will suffice. Trajectory angle

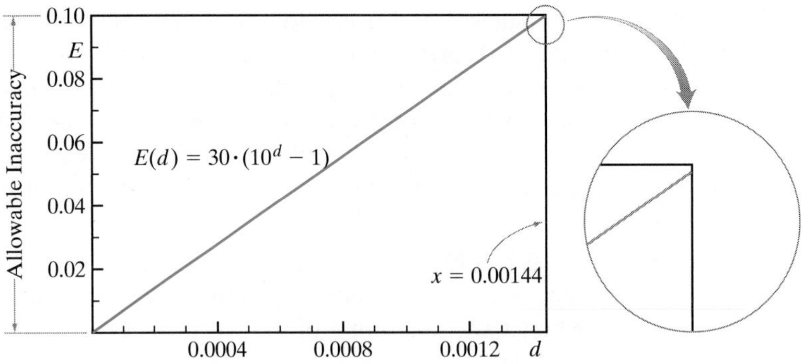

▲ Figure 12

measurements on this flight can be measured to within an accuracy of 0.00144 degrees without affecting the success of the mission (see Figure 12). ◀

▶ **EXAMPLE 9** In a controlled reaction of hydrogen with bromine, the rate R at which hydrogen bromide is produced (in moles per liter per second) is given by $R(c) = 0.08\sqrt{c}$ where c is the concentration of bromine (in moles per liter). How close to 0.16 must the bromine concentration be maintained to ensure that the rate of production of hydrogen bromide is within 0.001 of 0.032? Answer in the form of an interval $(0.16 - \delta, 0.16 + \delta)$.

Solution We require that $|0.08\sqrt{c} - 0.032| < 0.001$. From the Insight following Example 3, Section 1.1, we see that this inequality is equivalent to $-0.001 < 0.08\sqrt{c} - 0.032 < 0.001$, or, after adding 0.032 to each side, $0.031 < 0.08\sqrt{c} < 0.033$. Dividing by the positive number 0.08 results in $0.3875 < \sqrt{c} < 0.4125$. Finally, by squaring each side, we obtain $0.15015625 < c < 0.17015625$. Let $\delta_1 = 0.16 - 0.15015625 = 0.00984375$, and $\delta_2 = 0.17015625 - 0.16 = 0.01015625$. We have found that $R(c)$ is within 0.001 of 0.032 for c in the interval $I = (0.16 - \delta_1, 0.16 + \delta_2)$. The interval $(0.16 - \delta, 0.16 + \delta)$ we seek must be contained in I and centered at 0.16. We therefore take δ to be the smaller of δ_1 and δ_2: $\delta = 0.00984375$. In summary, the rate of production of hydrogen bromide is within 0.001 of 0.032 when the concentration of bromine is within 0.00984375 of 0.16. A plot of $R(c)$ in the viewing window $[0.16 - 0.00984375, 0.16 + 0.00984375] \times [0.032 - 0.001, 0.032 + 0.001]$ confirms our analysis (see Figure 13). ◀

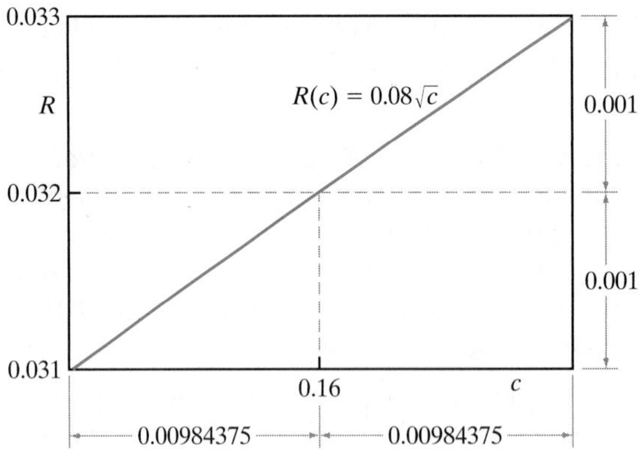

▲ Figure 13

QUICK QUIZ

1. True or false: $\lim_{x\to 0} f(x)$ can exist even if f does not have the point 0 in its domain.
2. Calculate $\lim_{t\to 5}(t^2+25)/(t+5)$.
3. Calculate $\lim_{t\to 5}(t^2-25)/(t-5)$.
4. How close must a positive number x be to 2 to ensure that x^2 is within 1/100 of 4?

Answers

1. True 2. 5 3. 10 4. $\sqrt{401}/10-2$

EXERCISES

Problems for Practice

In Exercises 1–8, decide whether the indicated limit exists. If the limit does exist, compute it.

1. $\lim_{x\to 2}(x+3)$

2. $\lim_{s\to 9}(s^2-6s+10)$

3. $\lim_{h\to 4}(3h^2+2h+1)$

4. $\lim_{s\to 0}(2+s)/s$

5. $\lim_{h\to 1}\dfrac{h-3}{h+1}$

6. $\lim_{x\to -3}\dfrac{x-3}{x^2-9}$

7. $\lim_{x\to 2} g(x)$ for

$$g(x)=\begin{cases} 4 & \text{if } x<0\\ -4 & \text{if } x\geq 0\end{cases}$$

8. $\lim_{x\to -1} f(x)$ for

$$f(x)=\begin{cases} 6 & \text{if } x\leq -1\\ 10 & \text{if } x>-1\end{cases}$$

In Exercises 9–14, some algebraic manipulation is necessary to determine whether the indicated limit exists. If the limit does exist, compute it and supply reasons for each step of your answer. If the limit does not exist, explain why.

9. $\lim_{x\to 5}\dfrac{(x^2-25)}{(x-5)}$

10. $\lim_{t\to 7}\dfrac{(t+7)}{(t^2-49)}$

11. $\lim_{t\to -7}\dfrac{(t+7)}{(t^2-49)}$

12. $\lim_{h\to 0}\dfrac{(h^6+h^8)}{(h^8-3h^3)}$

13. $\lim_{x\to -4}\dfrac{(x^2+6x-8)}{(x^2-2x-24)}$

14. $\lim_{x\to -3}\dfrac{(x^2-9)^2}{(x+3)^2}$

In Exercises 15–18, determine if $\lim_{x\to c} f(x)$ exists. Consider separately the values f takes when x is to the left of c and the values f takes when x is to the right of c. If the limit exists, compute it.

15. $f(x)=\begin{cases} x^2-3x & \text{if } x<2\\ -\dfrac{x}{x-1} & \text{if } x\geq 2\end{cases}$ $\quad c=2$

16. $f(x)=\begin{cases} \dfrac{x^2+6x-7}{x^2-1} & \text{if } x<1\\ (x+2)^2 & \text{if } x\geq 1\end{cases}$ $\quad c=1$

17. $f(x)=\begin{cases} \dfrac{x^2-4}{3} & \text{if } x\leq 5\\ \dfrac{x^2-3x-10}{x^2-9x+20} & \text{if } x>5\end{cases}$ $\quad c=5$

18. $f(x)=\begin{cases} \dfrac{x^2-9}{x^2+9} & \text{if } x\leq 3\\ \dfrac{(x-3)^2}{x^2-9} & \text{if } x>3\end{cases}$ $\quad c=3$

In Exercises 19–26, calculate how close x needs to be to c to force $f(x)$ to be within 0.01 of ℓ.

19. $f(x)=x+1, c=2, \ell=3$

20. $f(x)=2x-3, c=-2, \ell=-7$

21. $f(x)=5x+2, c=3, \ell=17$

22. $f(x)=x^2-1, c=0, \ell=-1$

23. $f(x)=5-x^2/4, c=0, \ell=5$

24. $f(x)=1+\sqrt{|x|}, c=0, \ell=4$

25. $f(x)=\begin{cases} x & \text{if } x<1\\ 10x-9 & \text{if } x\leq 1\end{cases}$ $\quad c=0.999, \ell=0.999$

26. $f(x)=\begin{cases} 4x-3 & \text{if } x\leq 1\\ x & \text{if } x>1\end{cases}$ $\quad c=1, \ell=1$

In Exercises 27 and 28, demonstrate that the limit does not exist by considering one-sided limits.

27. $\lim_{x\to 2} f(x)$ where

$$f(x)=\begin{cases} \dfrac{x^2-4}{x+2} & \text{if } x\leq 2\\ \dfrac{x^2-4}{x-2} & \text{if } x>2\end{cases}$$

28. $\lim_{x\to -5} f(x)$ where

$$f(x) = \begin{cases} 4x & \text{if } x \le -5 \\ 3x-4 & \text{if } x > -5 \end{cases}$$

Further Theory and Practice

In Exercises 29–32, you are given four functions f, g, h, and k, and a point c.

a. State the domain of each function.

b. Identify two of these functions that are the same. Explain why. In what way do the other functions differ from these two functions?

c. For each function, decide if it has a limit at c. Identify the limit if it exists.

29. $g(x) = \dfrac{x^2-1}{x-1}, h(x) = \begin{cases} \dfrac{x^2-1}{x-1} & \text{if } x \ne 1 \\ 2 & \text{if } x = 1 \end{cases}$

$f(x) = x+1, k(x) = \begin{cases} \dfrac{x^2-1}{x-1} & \text{if } x \ne 1 \\ 1 & \text{if } x = 1 \end{cases} \quad c = 1$

30. $f(x) = \dfrac{x^2+x-6}{x-2}, h(x) = \begin{cases} \dfrac{x^2+x-6}{x-2} & \text{if } x \ne 2 \\ 0 & \text{if } x = 2 \end{cases}$

$g(x) = x+3, k(x) = \begin{cases} \dfrac{x^2+x-6}{x-2} & \text{if } x \ne 2 \\ 5 & \text{if } x = 2 \end{cases} \quad c = 2$

31. $f(x) = \begin{cases} \dfrac{|x|}{\text{signum}(x)} & \text{if } x \ne 0 \\ 0 & \text{if } x = 0 \end{cases}, \quad h(x) = \dfrac{|x|}{\text{signum}(x)},$

$f(x) = \begin{cases} \dfrac{|x|}{\text{signum}(x)} & \text{if } x \ne 0 \\ 1 & \text{if } x = 0 \end{cases}, \quad k(x) = x, c = 0$

32. $f(x) = \begin{cases} \dfrac{x^3-x^2}{x^2} & \text{if } x \ne 0 \\ -1 & \text{if } x = 0 \end{cases}, \quad h(x) = \dfrac{x^3-x^2}{x^2},$

$f(x) = \begin{cases} \dfrac{x^3-x^2}{x^2} & \text{if } x \ne 0 \\ 0 & \text{if } x = 0 \end{cases}, \quad k(x) = x-1, \; c = 0$

33. The *Heaviside function* $H(x)$ is defined by

$$H = (x)\begin{cases} 1 & \text{if } x > 0 \\ 0 & \text{if } x < 0. \end{cases}$$

Compute $\lim_{x\to 0^+} H(x)$ and $\lim_{x\to 0^-} H(x)$. Does $\lim_{x\to 0} H(x)$ exist?

34. Suppose a sample of gas is held at constant pressure. Let V_0 denote the volume when the temperature is 0 degrees centigrade. The *Law of Charles and Gay-Lussac* relates the volume V and temperature T (measured in degrees centigrade) of the given gas sample by the equation $V = V_0 + V_0 \cdot T/273$. Use this law to show that T belongs to an interval of the form $T > \tau$. The number τ is known as *absolute zero*. Discuss the one-sided limit of V as T tends to absolute zero.

35. One foot is 0.3048 meter. How precisely should a length be measured in meters to guarantee accuracy of 0.001 foot?

36. One ounce is 28.350 grams. How precisely should weight be measured in ounces to guarantee accuracy of 0.001gram?

37. Let $\lfloor x \rfloor$ be the greatest integer that is less than or equal to x. For instance, $\lfloor 3/2 \rfloor = 1, \lfloor 0 \rfloor = 0, \lfloor -1/2 \rfloor = -1$. Discuss each limit.

a. $\lim_{x\to 0} \lfloor x \rfloor$

b. $\lim_{x\to 1/2} \lfloor x \rfloor$

c. $\lim_{x\to 1} 1/\lfloor x \rfloor$

d. $\lim_{x\to -1/2} 1/\lfloor x \rfloor$

38. Let n be an integer. For the greatest integer function $\lfloor x \rfloor$ defined in Exercise 37, discuss each limit.

a. $\lim_{x\to n^+} \lfloor x \rfloor$

b. $\lim_{x\to n^-} \lfloor x \rfloor$

c. $\lim_{x\to n} \lfloor x \rfloor$

39. Graph the greatest integer function $\lfloor x \rfloor$, defined in Exercise 37. What is the graphical indication that $\lim_{x\to n} \lfloor x \rfloor$ does not exist for an integer n?

40. The graph of a function f that is defined on $(1,4) \cup (4,5]$ appears in Figure 14. Determine all values of c at which the limit $\lim_{x\to c} f(x)$ exists. Determine all values of c at which the left limit $\lim_{x\to c^-} f(x)$ exists. Determine all values of c at which the right limit $\lim_{x\to c^+} f(x)$ exists.

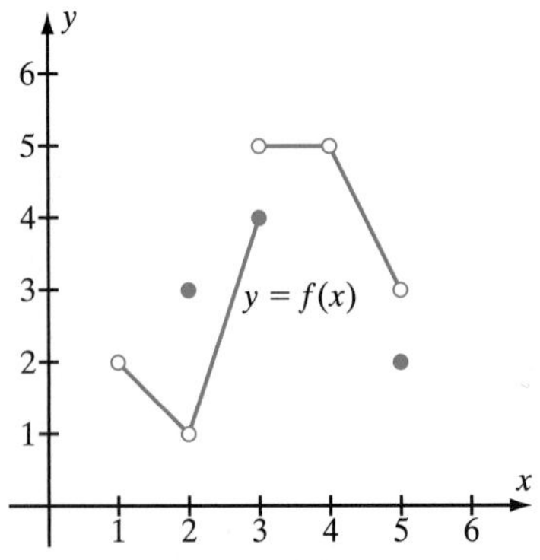

▲ Figure 14

In each of Exercises 41 and 42, a particle is moving in the xy-plane in such a way that it is at the specified point at time t.

a. Compute the average velocity of the particle from $t=1$ to $t=2$.
b. Compute the average velocity from $t=1$ to $t=1.5$.
c. Compute the average velocity from $t=1$ to $t=1.1$.
d. Compute the average velocity from $t=1$ to $t=1+h$ where h is a small nonzero number.
e. What is the limit of these average velocities as $h\to 0$?

41. $(t^2, 0)$
42. $(0, 4t^2-3)$

In each of Exercises 43–46, a particle moves on an axis. Its position $p(t)$ at time t is given. For the given value t_0 and a positive h, the average velocity over the time interval $[t_0, t_0+h]$ is $\overline{v}(h) = \dfrac{p(t_0+h)-p(t_0)}{h}$.

a. Calculate $\overline{v}(h)$ explicitly, and use the expression you have found to calculate $v_0 = \lim_{h\to 0^+}\overline{v}(h)$.
b. How small does h need to be for $\overline{v}(h)$ to be between v_0 and $v_0+0.1$?
c. How small does h need to be for $\overline{v}(h)$ to be between v_0 and $v_0+0.01$?
d. Let $\varepsilon>0$ be a small positive number. How small does h need to be to guarantee that $\overline{v}(h)$ is between v_0 and $v_0+\varepsilon$?

43. $p(t)=t^2+2;\quad t_0=2$
44. $p(t)=3t^2;\quad t_0=1$
45. $p(t)=t^2+2t;\quad t_0=3$
46. $p(t)=1+5t+t^2;\quad t_0=0$

In Exercises 47–50, a function f and a point c are given.

a. What is the slope of the line passing through $(c, f(c))$ and $(c+h, f(c+h))$?
b. What is the limit of these slopes as $h\to 0$?

47. $f(x)=x^2;\quad c=3$
48. $f(x)=x^2;\quad c=2$
49. $f(x)=x^2+2x;\quad c=1$
50. $f(x)=x-x^2;\quad c=-1$
51. Discuss $\lim_{x\to 0}\dfrac{x^{5/3}}{|x|}$.
52. Discuss $\lim_{x\to 1}\dfrac{\sqrt{|x-1|}}{\sqrt{|x^2-1|}}$.

Calculator/Computer Exercises

In each of Exercises 53–58, a function f, a point c, and a positive ε are given. Graphically determine a δ for which $f(x)$ is within ε of $f(c)$ for every x that is within δ of c.

53. $f(x)=4/x^2, c=2, \varepsilon=0.01$
54. $f(x)=\sqrt{x}, c=1, \varepsilon=0.1$
55. $f(x)=x^2, c=10, \varepsilon=0.01$
56. $f(x)=1/x, c=0.2, \varepsilon=0.01$
57. $f(x)=1/\sqrt{x}, c=9, \varepsilon=0.01$
58. $f(x)=\sqrt{x^2+16}, c=3, \varepsilon=0.01$

59. A machinist must do all of his work to within a tolerance of 10^{-3} mm. The calibrations on his machine are such that if the machinist's settings are accurate to within m mm, then the dimensions of the product have a tolerance within $1.2|m|+100m^2$ mm. What accuracy of the settings is required (i.e., how small must the machinist make m) to produce work of the desired tolerance?

In Exercises 60–64, a function f and a point c *not* in the domain of f are given. Analyze $\lim_{x\to c} f(x)$ as follows.

a. Evaluate $f(c-1/10^n)$ and $f(c+1/10^n)$ for $n=2,3,4$.
b. Formulate a guess for the value $\ell=\lim_{x\to c} f(x)$.
c. Find a value δ such that $f(x)$ is within 0.01 of ℓ for every x that is within δ of c.
d. Graph $y=f(x)$ for x in $[c-1/10^4, c)\cup(c, c+1/10^4]$ to verify visually that the limit of f at c exists.

60. $f(x)=\sin(x)/x,\ c=0$
61. $f(x)=\dfrac{\cos(x)}{x-\pi/2},\ c=\pi/2$
62. $f(x)=\dfrac{\sin(x)}{x-\pi},\ c=\pi$
63. $f(x)=\dfrac{x-\sin(x)}{x^3},\ c=0$
64. $f(x)=\dfrac{\cos(x)-1}{x^2},\ c=0$

In each of Exercises 65–68, a particle moves on an axis. Its position $p(t)$ at time t is given. For a positive h, the average velocity over the time interval $[2, 2+h]$ is $\overline{v}(h)=\dfrac{p(2+h)-p(2)}{h}$.

a. Numerically determine $v_0=\lim_{h\to 0+}\overline{v}(h)$.
b. How small does h need to be for $\overline{v}(h)$ to be between v_0 and $v_0+0.1$?
c. How small does h need to be for $\overline{v}(h)$ to be between v_0 and $v_0+0.01$?

65. $p(t)=t^3-8t$
66. $p(t)=(2t)^{3/2}-2t$
67. $p(t)=2t-\dfrac{24}{\pi}\cos\left(\dfrac{\pi t}{12}\right)$
68. $p(t)=9t^2/(t+1)$

69. A spring moves so that the amount of extension x is given by $x(t)=\sin(t)$. (A negative value of x corresponds to compression.) In this exercise, you will approximate the instantaneous velocity $v(t)$ of the spring at time t by its average velocity $\overline{v}(t)$ over the time interval $[t-0.0001, t+0.0001]$.

a. Write $\overline{v}(t)$ explicitly, and graph $\overline{v}(t)$ for $0\le t\le 2\pi$.
b. When, approximately, is $\overline{v}(t)$ the greatest? The most negative?

c. When, approximately, is $\overline{v}(t)$ zero?

d. Graph x and $\overline{v}$ in the viewing window $[0, 2\pi] \times [-1.1, 1.1]$. For what subinterval I of $[0, 2\pi]$ is $\overline{v} < 0$? What is the behavior of x as t increases through this subinterval I? What does $\overline{v}(t) < 0$ mean?

e. Based on the graph of $\overline{v}$, what function does $\overline{v}$ appear to approximate?

2.2 Limit Theorems

In Section 2.1, we discussed limits intuitively and numerically. Now we learn a precise definition of the statement

$$\lim_{x \to c} f(x) = \ell.$$

It will be written with mathematical symbols, but it will mean "$f(x)$ can be forced arbitrarily close to ℓ by making x sufficiently close to c."

We measure the distance of $f(x)$ to ℓ by the absolute difference $|f(x) - \ell|$. When we say that $f(x)$ can be forced arbitrarily close to ℓ, we mean that we can make $|f(x) - \ell|$ smaller than ε, no matter how small the positive number ε is. Thus "$f(x)$ can be forced arbitrarily close to ℓ" translates as "for any given positive ε, we can make $|f(x) - \ell| < \varepsilon$."

"By making x sufficiently close to c" translates to "we are considering only those x that lie within a specified distance δ of c." In other words, we refer to those x for which $0 < |x - c| < \delta$. Now we are ready for the complete definition.

DEFINITION Let f be a real-valued function that is defined in an open interval just to the left of a real number c and in an open interval just to the right of c. We say, "The limit of $f(x)$ is equal to ℓ as x approaches c," and we write $\lim_{x \to c} f(x) = \ell$ if, for any $\varepsilon > \delta$, there is a $\delta > 0$ such that

$$|f(x) - \ell| < \varepsilon$$

for all values of x such that

$$0 < |x - c| < \delta.$$

The limit can also be written as $f(x) \to \ell$ as $x \to c$ (see Figure 1).

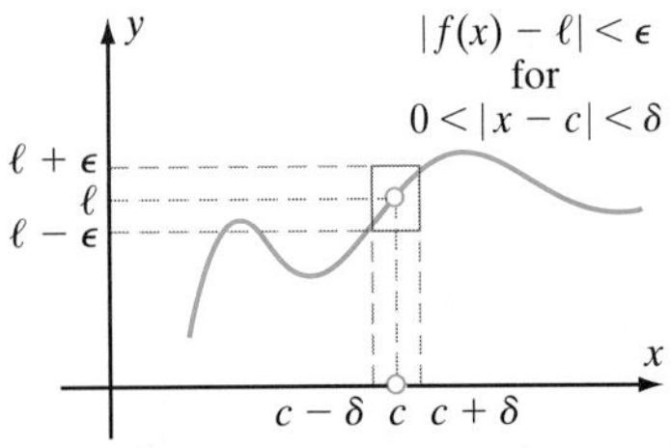

▲ Figure 1 $f(x)$ has the limit ℓ as x tends to c

This definition is often referred to as the ε-δ definition of *limit*. Notice that the inequality $0 < |x - c| < \delta$ means we are considering values of x that are near to c but *not* equal to c. An ε-δ verification of a limit starts with the assumption that an arbitrary positive ε is given. Then a δ corresponding to that ε must be found. This is exactly what we did in the quantitative examples from Section 2.1 in which specific values of ε were given. As you read the next example, notice that we find δ in terms of ε for a general value of ε.

▶ **EXAMPLE 1** Let $f(x) = 3x + 2$. Using the rigorous definition of limit, verify that $\lim_{x \to 4} f(x) = 14$.

Solution Let $\varepsilon > 0$. We must find a $\delta > 0$ such that $0 < |x-4| < \delta$ implies $|f(x) - 14| < \varepsilon$. As is usually the case, we work backward to find a δ that corresponds to the given ε. First, notice that there is a direct relationship between $|x-4|$ and $|f(x) - 14|$:

$$|f(x) - 14| = |3x + 2 - 14| = |3x - 12| = 3|x - 4|.$$

From this, we see that $|f(x) - 14| < \varepsilon$ if $|x - 4| < \varepsilon/3$. Therefore $\delta = \varepsilon/3$ is the value for δ that we want. That is, if $0 < |x - 4| < \delta = \varepsilon/3$, then $|f(x) - 14| < \varepsilon$. ◀

Compare Example 1 with Example 5 of Section 2.1. Both examples involve the same function f, the same point c, and the same limit ℓ. In the preceding section, we specified a particular numerical value for ε, and then determined a corresponding numerical value for δ. In the present discussion, we start with a general value of ε. Here is how we ue the information from Example 1 in a numerical way. Suppose we want to force $f(x) = 3x + 2$ to be within 0.06 of 14. We set $\varepsilon = 0.06$. The calculation in the example tells us that $\delta = \varepsilon/3 = 0.02$. In short, if x is within $\delta = 0.02$ of $c = 4$, then $f(x) = 3x + 2$ is forced to within $\varepsilon = 0.06$ of $\ell = 14$ (see Figure 2). Now suppose instead that we want to force $f(x)$ to be within 0.003 of 14. To do this, we simply set $\varepsilon = 0.003$. Our computation tells us that when x is within of $\delta = \varepsilon/3 = 0.001$ of $c = 4$, then $f(x) = 3x + 2$ is forced to within $\varepsilon = 0.003$ of $\ell = 14$. When refining our degree of accuracy, we do not have to keep repeating the calculation. We have done the calculation once and for all but with symbols ε and δ instead of specific numbers. We can substitute numbers later, depending on the degree of accuracy desired.

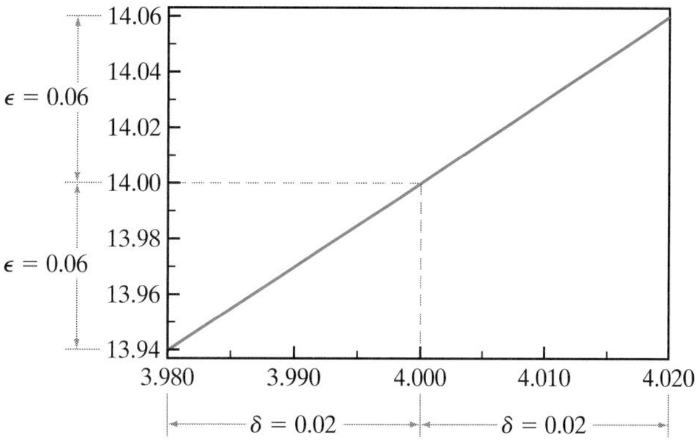

▲ Figure 2

INSIGHT In Example 1, our intuition allows us to easily determine the required limit. Why, then, has the ε-δ definition been introduced? The answer is that there are many limits that are not obvious. (See, for example, the trigonometric limits at the end of this section.) Having a rigorous definition allows us to verify limits in such situations.

One-Sided Limits

Many useful functions that do not have limits at certain points *do* have one-sided limits at those points. Furthermore, many functions are defined on an interval of the form (a, b), $[a, b]$, $(a, b]$, $[a, b)$ or on a half line of the form (a, ∞), $[a, \infty)$, $(-\infty, b)$, $(-\infty, b]$. The concept of one-sided limit is necessary to investigate the behavior of such a function at an endpoint of its domain of definition.

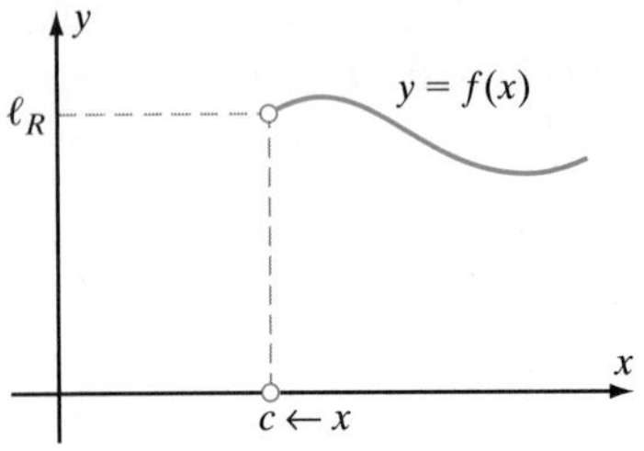

▲ Figure 3 $\lim_{x\to c^+} f(x) = \ell_R$

DEFINITION Suppose that f is a function defined in an interval just to the right of c. If, for every $\varepsilon > 0$, there is a $\delta > 0$ such that $|f(x) - \ell_R| < \varepsilon$ for each x that satisfies $c < x < c + \delta$, then we say that f has ℓ_R as a limit from the right, or f has right limit ℓ_R at c, and we write

$$\lim_{x\to c^+} f(x) = \ell_R.$$

Figure 3 shows a function f with a right limit ℓ_R at c.

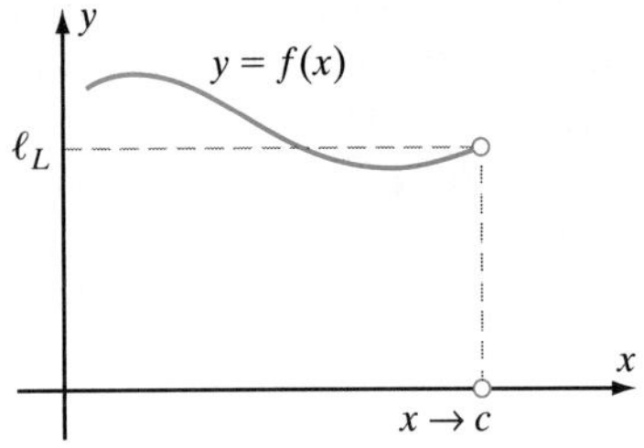

▲ Figure 4 $\lim_{x\to c^-} f(x) = \ell_L$

DEFINITION Let f be a function defined in an interval just to the left of c. If, for every $\varepsilon > 0$, there is a $\delta > 0$ such that $|f(x) - \ell_L| < \varepsilon$ for each x that satisfies $c - \delta < x < c$, then we say that f has ℓ_L as a limit from the left, or f has left limit ℓ_L at c, and we write

$$\lim_{x\to c^-} f(x) = \ell_L.$$

Figure 4 illustrates a function f with left limit ℓ_L at c.

By comparing the definitions of one-sided and two-sided limits, we see that

$$\lim_{x\to c} f(x) = \ell \text{ is equivalent to } \left(\lim_{x\to c^-} f(x) = \ell \quad \text{and} \quad \lim_{x\to c^+} f(x) = \ell \right).$$

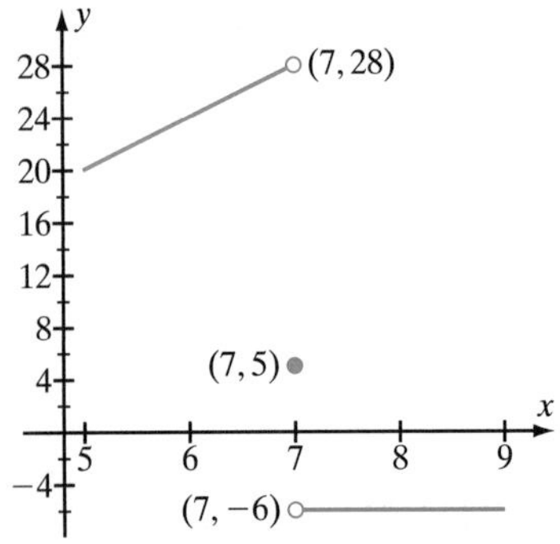

▲ Figure 5

$$f(x) = \begin{cases} 4x & \text{if } x < 7 \\ 5 & \text{if } x = 7 \\ -6 & \text{if } x > 7 \end{cases}$$

▶ **EXAMPLE 2** Determine the one-sided limits of $f(x)$ at 7 if f is defined by

$$f(x) = \begin{cases} 4x & \text{if } x < 7 \\ 0 & \text{if } x = 7. \\ -6 & \text{if } x > 7 \end{cases}$$

Solution As with all limit problems, we first look at this one intuitively. To the right of $c = 7$, $f(x)$ is equal to -6 for every x, so it appears that $f(x)$ has right limit $\ell_R = -6$ at $c = 7$. To the left of $c = 7$, $f(x)$ is equal to $4x$, so it appears that $f(x)$ has left limit $\ell_L = 4 \cdot 7$, or $\ell_L = 28$ at $c = 7$. Now let us carefully verify these assertions by using the definitions (see Figure 5).

First we consider the right limit. Let $\varepsilon > 0$. We must find a positive δ such that, if x is to the right of 7 by an amount less than δ, then $f(x)$ is within ε of $\ell_R = -6$. Because $f(x) = -6$ for every value of x greater than 7, we have $|f(x) - (-6)| = |-6 - (-6)| = 0$ for $x > 7$. Therefore no matter what positive value we choose for δ, the inequality $|f(x) - (-6)| < \varepsilon$ holds for $7 < x < 7 + \delta$. We may select $\delta = 1$, for example, but any other choice of positive number for δ would do just as well. By finding an appropriate δ for a given positive ε, we have shown that $\lim_{x\to 7^+} f(x) = -6$.

We now consider the left limit. Let $\varepsilon > 0$. We must find a positive δ such that, if x is to the left of 7 by an amount less than δ, then $f(x)$ is within ε of $\ell_L = 28$, or $|f(x) - 28| < \varepsilon$. Because $f(x) = 4x$ when x is to the left of 7, the desired inequality is $|4x - 28| < \varepsilon$, or, on multiplying each side of the inequality by 1/4, $|x - 7| < \varepsilon/4$. We therefore choose $\delta = \varepsilon/4$ and retrace our steps. Thus if $|x - 7| < \delta$, then we have $4|x - 7| < 4\delta$, or $|4x - 28| < \varepsilon$, or $|f(x) - \ell_L| < \varepsilon$. By definition, then, $\lim_{x\to 7^-} f(x) = 28$. Notice in this example that the ordinary (two-sided) limit, $\lim_{x\to 7} f(x)$, does not exist. ◀

Basic Limit Theorems

There are a few simple theorems that save a great deal of work in calculating limits. The first tells us that a function can have at most one limit at a point c.

THEOREM 1 Suppose a function f is defined on an open interval (a, c) to the left of c and on an open interval (c, b) to the right of c. Then f cannot have two distinct limits at c.

Theorem 1 will come up often, at least implicitly, as we study calculus. It assures us that if we compute a limit by two different methods, then both methods will give the same answer. A proof of Theorem 1 is outlined in Exercise 77.

Now we turn our attention to various arithmetic properties of limits. For example, suppose that $\lim_{x\to c} f(x) = \ell$. and $\lim_{x\to c} g(x) = m$. Given $\varepsilon > 0$, we can take x close enough to c to ensure that $f(x)$ is within $\varepsilon/2$ of ℓ and $g(x)$ is within $\varepsilon/2$ of m. Using the Triangle Inequality, we see that $(f+g)(x)$ is within ε of $\ell + m$ for these values of x:

$$\begin{aligned} \left|\big(f(x)+g(x)\big)-(\ell+m)\right| &= \left|\big(f(x)-\ell\big)+\big(g(x)-m\big)\right| \\ &\le |f(x)-\ell| + |g(x)-m| && \text{Triangle Inequality} \\ &< \frac{\varepsilon}{2}+\frac{\varepsilon}{2} \\ &= \varepsilon. \end{aligned}$$

This tells us that the limit of the sum $f(x)+g(x)$ as x approaches c is the sum of the limits of $f(x)$ and $g(x)$ at c. We state this result, as well as some similar assertions, in the following theorem, which tells us that limits of algebraic combinations of functions behave in the expected way.

THEOREM 2 If f and g are two functions, c is a real number, and $\lim_{x\to c} f(x)$ and $\lim_{x\to c} g(x)$ exist, then the following equations hold

a. $\lim_{x\to c}(f+g)(x) = \lim_{x\to c} f(x) + \lim_{x\to c} g(x)$
   $\lim_{x\to c}(f-g)(x) = \lim_{x\to c} f(x) - \lim_{x\to c} g(x)$;
b. $\lim_{x\to c}(f\cdot g)(x) = \big(\lim_{x\to c} f(x)\big)\cdot\big(\lim_{x\to c} g(x)\big)$;
c. $\lim_{x\to c}\left(\frac{f}{g}\right)(x) = \dfrac{\lim_{x\to c} f(x)}{\lim_{x\to c} g(x)}$, provided that $\lim\limits_{x\to c} g(x) \neq 0$; and
d. $\lim_{x\to c}\big(\alpha\cdot f(x)\big) = \alpha\cdot\big(\lim_{x\to c} f(x)\big)$ for any constant α.

By successively applying Theorem 2a as many times as necessary, we can handle the sum of any finite number of functions. A similar remark applies to Theorem 2b and any finite product of functions. For example, because we know that $\lim_{x\to c} x = c$, we can deduce that $\lim_{x\to c} x^n = c^n$ for any positive integer n. From these observations and the other parts of Theorem 2, we obtain the next theorem.

THEOREM 3 If $p(x)$ and $q(x)$ are polynomials with $q(c) \neq 0$, then

$$\lim_{x \to c} p(x) = p(c)$$

and

$$\lim_{x \to c} \frac{p(x)}{q(x)} = \frac{p(c)}{q(c)}.$$

INSIGHT Functions of the form $x \mapsto p(x)/q(x)$ where $p(x)$ and $q(x)$ are polynomials are called *rational functions*. Theorem 3 gives a large class of functions (the rational functions) with limits that can be obtained by evaluation. Later in this section, we will see other important functions that have this convenient property.

▶ **EXAMPLE 3** Use Theorem 3 to compute $\lim_{x \to -3} \dfrac{2x^2 + 1}{6x - 5}$.

Solution Let $p(x) = 2x^2 + 1$ and $q(x) = 6x - 5$. Notice that $q(-3) = 6(-3) - 5 = -23$. Because this quantity is nonzero, it follows from Theorem 3 that

$$\lim_{x \to -3} \frac{2x^2 + 1}{6x - 5} = \lim_{x \to -3} \frac{p(x)}{q(x)} = \frac{p(-3)}{q(-3)} = -\frac{19}{23}. \quad ◀$$

Notice that Theorems 2 and 3 allow us to calculate limits without reference to the ε-δ definition. Our next theorem can be used in the same way.

THEOREM 4 Let n be a positive integer and c a real number (that is positive if n is even). Then

$$\lim_{x \to c} \sqrt[n]{x} = \sqrt[n]{c}.$$

▶ **EXAMPLE 4** Calculate

$$\lim_{x \to 9} \frac{\sqrt{x} - 3}{x - 9}.$$

Solution Because the limit of the denominator is 0, Theorem 2c cannot be applied to calculate the limit of the quotient. We must first prepare the quotient by means of an algebraic manipulation that is frequently helpful:

$$\frac{\sqrt{x} - 3}{x - 9} = \frac{\sqrt{x} + 3}{\sqrt{x} + 3} \cdot \frac{\sqrt{x} - 3}{x - 9} = \frac{(\sqrt{x})^2 - (3)^2}{(\sqrt{x} + 3)(x - 9)} = \frac{x - 9}{(\sqrt{x} + 3)(x - 9)} = \frac{1}{\sqrt{x} + 3}.$$

Therefore using Theorems 2c, 2a, and 4 in succession, we have

$$\lim_{x \to 9} \frac{\sqrt{x} - 3}{x - 9} = \lim_{x \to 9} \frac{1}{\sqrt{x} + 3} = \frac{1}{\lim_{x \to 9}(\sqrt{x} + 3)} = \frac{1}{(\lim_{x \to 9} \sqrt{x}) + 3} = \frac{1}{6}. \quad ◀$$

A Rule That Tells When a Limit Does Not Exist

Sometimes it is useful to know a rule that shows that a certain limit does not exist.

THEOREM 5 If $\lim_{x \to c} g(x) = 0$, and if $f(x)$ does not have limit 0 as x approaches c, then $\lim_{x \to c} \dfrac{f(x)}{g(x)}$ does not exist.

Proof. Let $h(x) = f(x)/g(x)$. To show that $h(x)$ does not have a limit as x approaches c, we write $f(x) = g(x) \cdot h(x)$ and argue by contradiction. If $h(x)$ did have a limit ℓ as x approached c, then Theorem 2b would give

$$\lim_{x \to c} f(x) = \lim_{x \to c} h(x) \cdot \lim_{x \to c} g(x) = \ell \cdot 0 = 0,$$

contradicting the hypothesis about f. ■

INSIGHT Here is an equivalent formulation of Theorem 5 that is frequently useful: If $\lim_{x \to c} \dfrac{f(x)}{g(x)}$ exists, and if $\lim_{x \to c} g(x) = 0$, then $\lim_{x \to c} f(x) = 0$.

▶ **EXAMPLE 5** Does $\lim_{x \to -1} \dfrac{2+x}{(1+x)^2}$ exist?

Solution We notice that

$$\lim_{x \to -1} (2 + x) = 2 + (-1) = 1 \neq 0 \text{ and } \lim_{x \to -1} (x+1)^2 = \left((-1)+1\right)^2 = 0^2 = 0.$$

According to Theorem 5, $\lim_{x \to -1} \dfrac{(2+x)}{(1+x)^2}$ does not exist. ◀

The Pinching Theorem

THEOREM 6 (**The Pinching Theorem**). Suppose that the functions f, g, and h are defined on an open interval (a, c) to the left of c and on an open interval (c, b) to the right of c. Let S denote the union of these two intervals. Assume further that $g(x) \leq f(x) \leq h(x)$ for all x in S (see Figure 6). If $\lim_{x \to c} g(x) = \ell$, and $\lim_{x \to c} h(x) = \ell$, then $\lim_{x \to c} f(x) = \ell$.

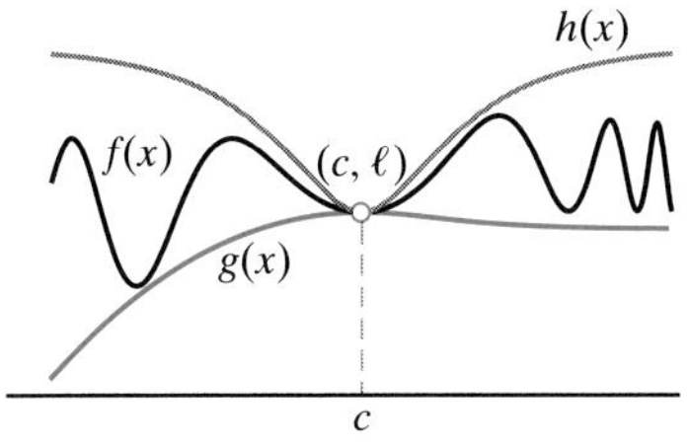

▲ Figure 6 $\lim_{x \to c} f(x) = \ell$

The truth of the Pinching Theorem is geometrically clear (refer to Figure 6). We may apply the Pinching Theorem to one-sided limits as well. For example, if $g(x) \leq f(x) \leq h(x)$ for all x in (c, b), if $\lim_{x \to c^+} g(x) = \ell$, and if $\lim_{x \to c^+} h(x) = \ell$, then we may conclude that $\lim_{x \to c^+} f(x) = \ell$.

▶ **EXAMPLE 6** Let $f(x) = x \cdot \sin(1/x)$ when $x \neq 0$. Use the Pinching Theorem to compute $\lim_{x \to 0} f(x)$.

Solution Notice that, because $|\sin(1/x)| \leq 1$ for every nonzero x, we have

$$f(x) \leq \left| x \cdot \sin\left(\frac{1}{x}\right) \right| = |x| \cdot \left| \sin\left(\frac{1}{x}\right) \right| \leq |x|$$

and

$$-|x| \leq -|x| \cdot \left| \sin\left(\frac{1}{x}\right) \right| = -\left| x \cdot \sin\left(\frac{1}{x}\right) \right| \leq f(x).$$

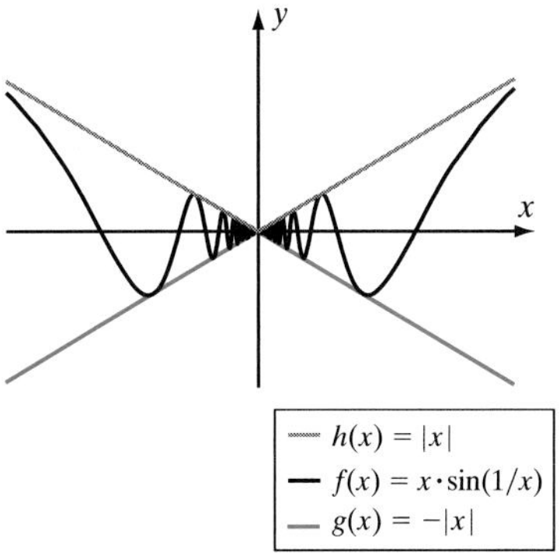

▲ Figure 7

In short,

$$-|x| \leq f(x) \leq |x|. \tag{2.2.1}$$

We have "pinched" $f(x)$ between $h(x) = |x|$ and $g(x) = -|x|$ (see Figure 7). Now it is easy to see that

$$\lim_{x \to 0} |x| = 0 \quad \text{and} \quad \lim_{x \to 0} (-|x|) = 0. \tag{2.2.2}$$

Equations (2.2.1) and (2.2.2), together with the Pinching Theorem, allow us to see that $\lim_{x \to 0} x \cdot \sin(1/x) = 0$. ◀

Some Important Trigonometric Limits

THEOREM 7 For every c, we have $\lim_{t \to c} \sin(t) = \sin(c)$ and $\lim_{t \to c} \cos(t) = \cos(c)$.

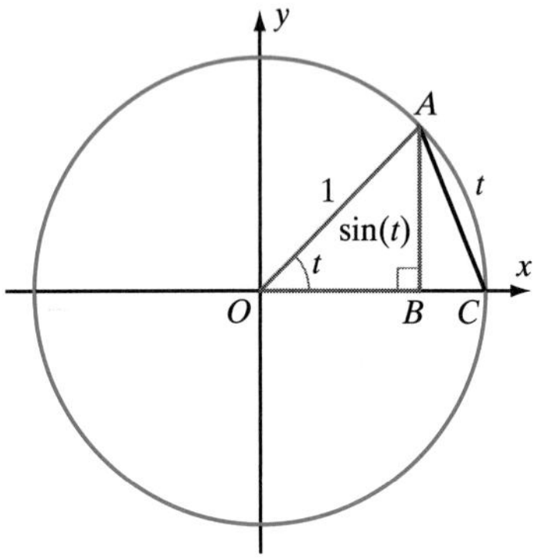

▲ Figure 8 $0 \le \sin(t) = |\overline{AB}| < |\overline{AC}| < t$

Proof. Let t be an angle, measured in radians, between 0 and $\pi/2$. Figure 8 exhibits the unit circle. Because t represents a small *positive* angle measured in radians, the length of the arc that the angle subtends is t. We see that $0 \le \sin(t) = |\overline{AB}| < |\overline{AC}| < t$. A similar picture shows that $t < \sin(t) \le 0$ when t is negative. Therefore $-|t| < \sin(t) < |t|$ for all t in $(-\pi/2, \pi/2)$. By the Pinching Theorem,

$$\lim_{t \to 0} \sin(t) = 0. \qquad (2.2.3)$$

Replacing θ with t in Half Angle Formula (1.6.12) and solving for $\cos(t)$, we obtain

$$\cos(t) = 1 - 2\sin^2\left(\frac{t}{2}\right).$$

It follows that

$$\lim_{t \to 0} \cos(t) = \lim_{t \to 0}\left(1 - 2\sin^2\left(\frac{t}{2}\right)\right) = 1 - 2\lim_{t \to 0}\left(\sin^2\left(\frac{t}{2}\right)\right) = 1 - 2 \cdot 0^2 = 1. \qquad (2.2.4)$$

Now let c be arbitrary. For every t, let $h = t - c$. Notice that $h \to 0$ as $t \to c$. We have

$$\begin{aligned}
\lim_{t \to c} \sin(t) &= \lim_{t \to c} \sin(c + (t - c)) \\
&= \lim_{h \to 0} \sin(c + h) && \text{Definition of } h \\
&= \lim_{h \to 0} (\sin(c)\cos(h) + \cos(c)\sin(h)) && \text{Addition Formula (1.8)} \\
&= \lim_{h \to 0} (\sin(c)\cos(h)) + \lim_{h \to 0} (\cos(c)\sin(h)) && \text{Theorem 2a} \\
&= \sin(c) \cdot \lim_{h \to 0} \cos(h) + \cos(c) \cdot \lim_{h \to 0} \sin(h) && \text{Theorem 2d} \\
&= \sin(c) \cdot 1 + \cos(c) \cdot 0 && \text{Equations (2.2.3) and (2.2.4)} \\
&= \sin(c)
\end{aligned}$$

A similar calculation using Addition Formula (1.9) shows that $\lim_{t \to c} \cos(t) = \cos(c)$ for every real c. ■

INSIGHT Because $\csc(x) = 1/\sin(x)$, $\sec(x) = 1/\cos(x)$, $\tan(x) = \sin(x)/\cos(x)$, and $\cot(x) = \cos(x)/\sin(x)$, we may use Theorem 2c to conclude that, if f is any trigonometric function, then

$$\lim_{x \to c} f(x) = f(c)$$

at each point c in the domain of f.

We now have a large class of functions with limits that can be computed simply by evaluation: polynomials, rational functions, n^{th} roots, trigonometric functions, and all the combinations thereof that can be formed with the arithmetic operations described in Theorem 2. However, many limits in calculus are not obtained so easily. The next theorem gives two important examples.

THEOREM 8 If t is measured in radians, then

$$\lim_{t\to 0}\frac{\sin(t)}{t}=1 \quad \text{and} \quad \lim_{t\to 0}\frac{1-\cos(t)}{t^2}=\frac{1}{2}. \tag{2.2.5}$$

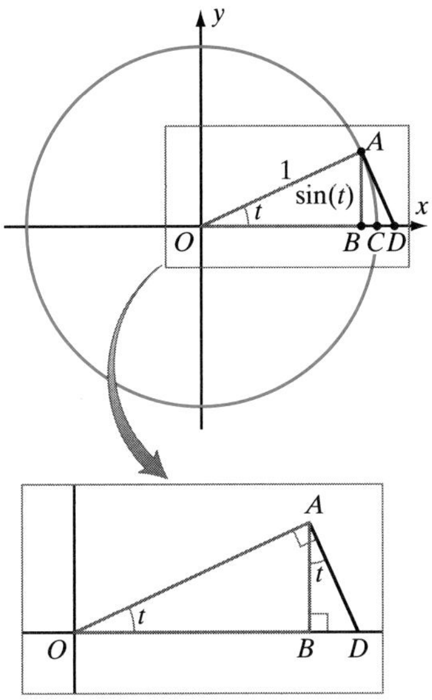

▲ Figure 9 $t = \angle AOD = \angle BAD = \frac{\pi}{2} - \angle ADO$

Proof. Let $f(t) = \sin(t)/t$, $t \neq 0$. On a calculator we compute that

t	0.1	0.01	0.001	0.0001
$f(t)$	0.9983341664	0.9999833334	0.9999998333	0.9999999983

to ten decimal places. On the basis of this numerical evidence, it is reasonable to guess that $\lim_{t\to 0} f(t) = 1$. Let us investigate why.

Refer again to Figure 8. As we observed in the course of proving Theorem 7, Figure 8 shows that $\sin(t) < t$ or, equivalently,

$$f(t) < 1. \tag{2.2.6}$$

This pinches $f(t)$ from above by the constant function $h(t) = 1$. Now we will pinch $f(t)$ from below. See Figure 9, which is similar to the preceding figure except chord AC of Figure 8 has been replaced with tangent line segment $\overline{AD}$. Because $\angle OAD$ is a right angle, we see that ΔABD is similar to ΔOBA. It follows that

$$\frac{|\overline{BD}|}{|\overline{AD}|} = \frac{\sin(t)}{1}.$$

Therefore $|\overline{BC}| < |\overline{BD}| = |\overline{AD}| \cdot \sin(t)$. However, we see from ΔOAC that $|\overline{AD}| = \tan(t)$. It follows that

$$|\overline{BC}| < \tan(t) \cdot \sin(t).$$

Therefore

$$|\overline{BC}| + \sin(t) < \tan(t) \cdot \sin(t) + \sin(t) = \sin(t)(1 + \tan(t)). \tag{2.2.7}$$

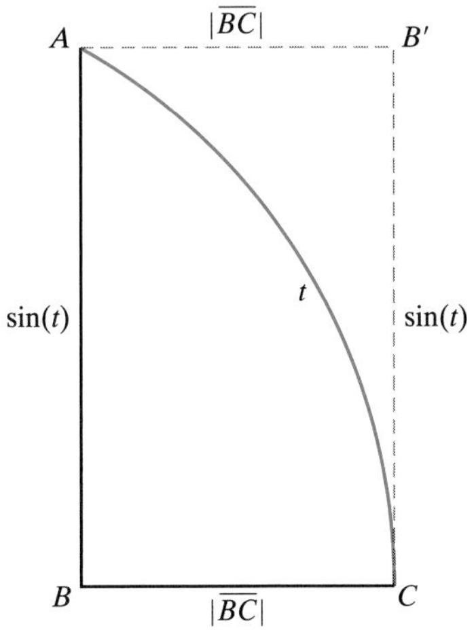

▲ Figure 10

Now observe that $t < |\overline{BC}| + \sin(t)$ (see Figure 10). This inequality, together with inequality (2.2.7), shows that

$$t < \sin(t)(1 + \tan(t)).$$

Divide each side of this inequality by the positive quantity $t(1 + \tan(t))$ to obtain

$$\frac{1}{1+\tan(t)} < \frac{\sin(t)}{t} = f(t).$$

Thus for small *positive* t, we have pinched $f(t)$ from below by $g(t) = 1/(1 + \tan(t))$. Because $\lim_{t\to 0} 1/(1 + \tan(t)) = 1/(1 + \tan(0)) = 1$, the Pinching Theorem tells us that we have $\lim_{t\to 0^+} f(t) = 1$. But $f(-t) = f(t)$. It follows that $\lim_{t\to 0^-} f(t) = \lim_{t\to 0^+} f(t) = 1$. We conclude that the two-sided limit of $f(t)$ at $t = 0$ exists and equals 1, which is the first limit formula of Theorem 8.

For the second limit formula, we use the Half Angle Formula $1 - \cos(t) = 2\sin^2(t/2)$ to obtain

$$\frac{1-\cos(t)}{t^2} = \frac{2\sin^2(t/2)}{t^2} = \frac{1}{2}\left(\frac{\sin(t/2)}{t/2}\right)^2. \tag{2.28}$$

Let $x = t/2$ and observe that $x \to 0$ as $t \to 0$. Using equation (2.2.8) and the limit formula for $\sin(x)/x$ that we established in the first part of this proof, we conclude that

$$\begin{aligned}\lim_{t\to 0}\frac{1-\cos(t)}{t^2} &= \lim_{t\to 0}\frac{1}{2}\left(\frac{\sin(t/2)}{t/2}\right)^2 = \lim_{x\to 0}\frac{1}{2}\left(\frac{\sin(x)}{x}\right)^2 \\ &= \frac{1}{2}\lim_{x\to 0}\left(\frac{\sin(x)}{x}\right)^2 = \frac{1}{2}\left(\lim_{x\to 0}\frac{\sin(x)}{x}\right)^2 = \frac{1}{2}\cdot 1^2 = \frac{1}{2}. \quad \blacksquare\end{aligned}$$

▶ **EXAMPLE 7** Show that

$$\lim_{t\to 0}\frac{1-\cos(t)}{t} = 0.$$

Solution By Theorem 8 and Theorem 2b, we have

$$\lim_{t\to 0}\frac{1-\cos(t)}{t} = \lim_{t\to 0}\frac{1-\cos(t)}{t^2}\cdot t = \lim_{t\to 0}\frac{1-\cos(t)}{t^2}\cdot \lim_{t\to 0} t = \frac{1}{2}\cdot 0 = 0. \quad ◀$$

QUICK QUIZ

1. True or false: The limit $\lim_{x\to c} f(x)$ may exist even if c is *not* in the domain of f.
2. True or false: If $\lim_{x\to c} f(x) = \ell$ *and* c is in the domain of f, then $\ell = f(c)$.
3. True or false: If f has both a left limit and a right limit at c, then $\lim_{x\to c} f(x)$ exists.
4. Evaluate $\lim_{x\to 0}\dfrac{x}{\sin(x)}$.

Answers

1. True 2. False 3. False 4. 1

EXERCISES

Problems for Practice

▶ In Exercises 1–8, evaluate the given limit. ◀

1. $\lim_{x\to 3} 2x$
2. $\lim_{x\to 1}(6x - 1)$
3. $\lim_{x\to 6}(x/3 + 2)$
4. $\lim_{x\to 12}(3 - x/4)$
5. $\lim_{x\to 0}(\sqrt{2} - \pi x)$
6. $\lim_{x\to 3}\dfrac{x^2 - 4x + 3}{x - 3}$
7. $\lim_{x\to -2}\dfrac{x^2 - 4}{x - 2}$
8. $\lim_{x\to 2}\dfrac{x^2 - 4}{x - 2}$

In Exercises 9–12, the indicated limit $\lim_{x\to c} f(x)$ does not exist. Prove it by evaluating the one-sided limits $\lim_{x\to c^-} f(x)$ and $\lim_{x\to c^+} f(x)$.

9. $\lim_{x\to -5} f(x)$ where

$$f(x) = \begin{cases} 3x & \text{if } x<5 \\ 4 & \text{if } x>5 \end{cases}$$

10. $\lim_{x\to -3} f(x)$ where

$$f(x) = \begin{cases} x^2 & \text{if } x<-3 \\ 4x & \text{if } x>-3 \end{cases}$$

11. $\lim_{x\to 1} f(x)$ where

$$f(x) = \begin{cases} 0 & \text{if } x<1 \\ -4 & \text{if } x>1 \end{cases}$$

12. $\lim_{x\to 0} f(x)$ where

$$f(x) = \begin{cases} x-1 & \text{if } x<0 \\ x^3 & \text{if } x\geq 0 \end{cases}$$

In Exercises 13–16, show that $\lim_{x\to 4} f(x)$ exists by calculating the one-sided limits $\lim_{x\to 4^-} f(x)$ and $\lim_{x\to 4^+} f(x)$.

13. $f(x) = \begin{cases} x^2+1 & \text{if } x<4 \\ 5x-3 & \text{if } x>4 \end{cases}$

14. $f(x) = \begin{cases} 2x+3 & \text{if } x<4 \\ x+7 & \text{if } x>4 \end{cases}$

15. $f(x) = \begin{cases} x^3 & \text{if } x<4 \\ -64 & \text{if } x=4 \\ 4x^2 & \text{if } x>4 \end{cases}$

16. $f(x) = \begin{cases} 8 & \text{if } x<4 \\ 5 & \text{if } x=4 \\ 2x & \text{if } x>4 \end{cases}$

In Exercises 17–24, use Theorem 2 to evaluate the limit.

17. $\lim_{x\to 4}(2x+6)$

18. $\lim_{x\to -3}(-4x+5)$

19. $\lim_{x\to 1}(x^2-6)$

20. $\lim_{x\to 2}(x+1)/x$

21. $\lim_{x\to 0}(x^2+2)/(x+1)$

22. $\lim_{x\to 7}(x+3)(x-7)$

23. $\lim_{x\to 4}\dfrac{(x-5)x}{x+1}$

24. $\lim_{x\to -2}(x^3+1)^3$

In each of Exercises 25–32, either determine that the limit does not exist or state its value. Justify your answer using Theorems 3, 4, and 5.

25. $\lim_{x\to 1}\dfrac{(x-2)^2}{x+1}$

26. $\lim_{x\to 4}\dfrac{x+4}{x-4}$

27. $\lim_{x\to 3}\dfrac{x^2-9}{x-3}$

28. $\lim_{x\to -2}\dfrac{x^2-4}{x-2}$

29. $\lim_{x\to -1}\dfrac{x^2-1}{x+1}$

30. $\lim_{x\to 16}\dfrac{1/\sqrt{x}}{1/(x^{1/4}-10)}$

31. $\lim_{x\to \pi}\dfrac{x-\sqrt[3]{x}}{\sqrt{x}-3}$

32. $\lim_{x\to \pi}\dfrac{x^2-9}{(x-\pi)^2}$

Use the Pinching Theorem to evaluate the limits in Exercises 33–38.

33. $\lim_{x\to 0} x^3\cos(1/x)$

34. $\lim_{x\to 1}(x-1)\cdot\sin(1/(x-1))$

35. $\lim_{x\to 5}\left(1+(x-5)^2\sin(\csc(\pi x))\right)$

36. $\lim_{x\to -9}\left(4+(x+9)\cos(1/(9+x))\right)$

37. $\lim_{x\to 2}\left((x+1)+(x-2)^2\sin(\sec(\pi/x))\right)$

38. $\lim_{x\to 0}\left(\cos(1+x)+x^{1/3}\cos(1/x)\right)$

In Exercises 39–42, decide whether the limit can be determined from the given information. If the answer is yes, then find the limit.

39. $2-|x-2|^3 \leq f(x) \leq 2+|x-2|^2$; $\lim_{x\to 2} f(x)$

40. $2-|x-2|^3 \leq f(x) \leq 2+|x-2|^2$; $\lim_{x\to 4} f(x)$

41. $-3|x|+5|x-5| \leq f(x) \leq 6|x|+(x-5)^2$; $\lim_{x\to 0} f(x)$

42. $20-|x-2| \leq f(x) \leq x^2-4x+24$; $\lim_{x\to 2} f(x)$

In Exercises 43–46, investigate the one-sided limits of the function at the endpoints of its domain of definition.

43. $f(x) = \sqrt{(x-1)(2-x)}$

44. $f(x) = \dfrac{x-1}{x^2-1}, \quad -1<x<1$

45. $f(x) = \dfrac{x^2-4}{x^2+2x}, \quad -2<x<2, \quad x\neq 0$

46. $f(x) = \sqrt{x}+\dfrac{1}{\sqrt{1-x}}, \quad 0<x<1$

Further Theory and Practice

In Exercises 47–56, use your intuition to decide whether the limit exists. Justify your answer by using the rigorous definition of limit.

47. $\lim_{x\to -4} |x|/x$

48. $\lim_{x\to 0} |x|/x$

49. $\lim_{x\to -7}(|x|-3x)$

50. $\lim_{x\to 1}\dfrac{x^2-1}{|x-1|}$

51. $\lim_{x\to -3}(x-|x|)^2$

52. $\lim_{x\to -3} f(x)$ where

$$f(x)=\begin{cases} 6 & \text{if } x\le -3 \\ -2x & \text{if } x>-3 \end{cases}$$

53. $\lim_{x\to 3} f(x)$ where

$$f(x)=\begin{cases} x^2 & \text{if } x\le 0 \\ 1-x^2 & \text{if } x>0 \end{cases}$$

54. $\lim_{x\to 4} f(x)$ where

$$f(x)=\begin{cases} x^2-1 & \text{if } x<4 \\ 3x+3 & \text{if } x\ge 4 \end{cases}$$

55. $\lim_{x\to 5} f(x)$ where

$$f(x)=\begin{cases} x+1 & \text{if } x<5 \\ x-1 & \text{if } x\ge 5 \end{cases}$$

56. $\lim_{x\to 2} f(x)$ where

$$f(x)=\begin{cases} \dfrac{x^2+x-6}{x-2} & \text{if } x<2 \\ \dfrac{x^3-2x^2+x-2}{x-2} & \text{if } x>2 \end{cases}$$

▸ In Exercises 57–66, use the basic limits of Theorem 8 to evaluate the limit. Note: x° means "x degrees." ◂

57. $\lim_{x\to 0}\dfrac{\sin(3x)}{x}$

58. $\lim_{x\to 0}\dfrac{\tan(2x)}{x}$

59. $\lim_{x\to 0}\dfrac{\sin(2x)}{\sin(x)}$

60. $\lim_{x\to 0}\dfrac{\sin^2(x)}{1-\cos(x)}$

61. $\lim_{x\to 0}\dfrac{1-\cos(x)}{\tan(x)}$

62. $\lim_{x\to 0}\dfrac{\sin(x)-\tan(x)}{x^3}$

63. $\lim_{x\to 0}\dfrac{\sin(x^{\circ})}{x}$

64. $\lim_{x\to 0}\dfrac{1-\cos(x^{\circ})}{x^2}$

65. $\lim_{x\to 0}\dfrac{\sin(2x)\tan(3x)}{x^2}$

66. $\lim_{x\to 0}\dfrac{\sin^2(3x^2)}{x^4}$

▸ In Exercises 67–73, use algebraic manipulation (as in Example 5) to evaluate the limit. ◂

67. $\lim_{x\to 1}\left(\dfrac{x^2-1}{\sqrt{x}-1}\right)$

68. $\lim_{x\to 4}\left(\dfrac{x-4}{\sqrt{x}-2}\right)$

69. $\lim_{h\to 0}\dfrac{h}{\sqrt{1+2h}-1}$

70. $\lim_{t\to 0}\left(\dfrac{\sqrt{3+t}-\sqrt{3}}{t}\right)$

71. $\lim_{h\to 0}\left(\dfrac{2}{h\sqrt{4+h}}-\dfrac{1}{h}\right)$

72. $\lim_{x\to 4}\left(\dfrac{\sqrt{8-x}-2}{\sqrt{5-x}-1}\right)$

73. $\lim_{x\to 4}\dfrac{\sqrt{x}-2}{\sqrt{x+5}-3}$

▸ In Exercises 74–80, evaluate the one-sided limits. ◂

74. $\lim_{x\to 3^+}\dfrac{x-3}{|x-3|}$

75. $\lim_{x\to 3^-}\dfrac{x-3}{|x-3|}$

76. $\lim_{x\to 3^-}\left(\dfrac{6+x^2}{5}+\sqrt{13-2x}\right)$

77. $\lim_{x\to 4^-}\dfrac{x+\sqrt{16-3x}}{x+\sqrt{16-x^2}}$

78. $\lim_{x\to 0^+}\dfrac{\sqrt{\sin(x)}}{\sqrt{x}}$

79. $\lim_{x\to 0^+}\dfrac{\sqrt{\sin(x)}}{\sqrt{x}}$

80. $\lim_{x\to 1^-}\dfrac{(\sqrt{x}-1)\sqrt{x^2-3x+2}}{(1-x)^{3/2}}$

81. Suppose that ℓ_1 and ℓ_2 are two unequal numbers. $\varepsilon_0=|\ell_1-\ell_2|/2$. Notice that $\varepsilon_0>0$. Suppose that $\lim_{x\to c} f(x)=\ell_1$. Then there is a $\delta_1>0$ such that $|f(x)-\ell_1|<\varepsilon_0$ for all values of x satisfying $0<|x-c|<\delta_1$. Show that there is no $\delta_2>0$ such that $|f(x)-\ell_2|<\varepsilon_0$ for all values of x satisfying $0<|x-c|<\delta_2$. Conclude that $\lim_{x\to c} f(x)=\ell_2$ is impossible.

82. Prove the following variant of Theorem 5: If $\lim_{x\to c} f(x)/g(x)$ exists, and $\lim_{x\to c} g(x)=0$, then $\lim_{x\to c} f(x)=0$.

▸ In Exercises 83–86, use the Pinching Theorem to establish the required limit. ◂

83. If $0\le f(x)\le x^2$ for all x, then $\lim_{x\to 0} f(x)/x=0$.

84. If $|g(x)-f(x)|\le\sqrt{|x|}$, and if $\lim_{x\to 0} f(x)=\ell$, then $\lim_{x\to 0} g(x)=\ell$.

85. If $\lim_{x\to c}|f(x)|=0$, then $\lim_{x\to c} f(x)=0$.

86. If $\lim_{x\to c} f(x)=0$, then $\lim_{x\to c} f^3(x)=0$.

Computer/Calculator Exercises

87. Using Theorem 3, we obtain

$$\lim_{x\to 2^+} \frac{x^3}{x^2-2.9x+2} = 40.$$

As $x\to 2^+$ about how close to 2 must x be to ensure that

$$\left|\frac{x^3}{x^2-2.9x+2} - 40\right| < 0.1?$$

Illustrate this by a graph of $f(x) = x^3/(x^2-2.9x+2)$ in an appropriate viewing rectangle.

88. It can be shown that

$$\lim_{x\to\pi^+} \frac{\sqrt{x-\pi}}{\sqrt{\sin(2x)}} = \frac{1}{\sqrt{2}}.$$

Find an interval I with center π such that

$$\left|\frac{\sqrt{x-\pi}}{\sqrt{\sin(2x)}} - \frac{1}{\sqrt{2}}\right| < 0.01$$

for x in I. Illustrate this by a graph of $f(x) = \sqrt{x-\pi}/\sqrt{\sin(2x)}$ in an appropriate viewing rectangle.

89. Calculate $\ell = \lim_{x\to 2}(x^3-3x^2+2x+1)$. Find an interval I with center 2 such that

$$|(x^3-3x^2+2x+1)-\ell| < 0.1$$

for x in I. Illustrate this by a graph of $f(x) = x^3 - 3x^2 + 2x + 1$ in an appropriate viewing rectangle.

90. Verify that $\lim_{x\to 1}(2x^3/(x+1)) = 1$. For $\varepsilon = 1/100$, use a computer algebra system to determine exactly what δ (in the definition of limit) should be.

91. Graph $g(x) = \sin(x^\circ)/x$ for $-45 \le x \le 45$ where x° means "x degrees." Use your graph to determine the value of $\lim_{x\to 0} g(x)$.

92. Graph $g(x) = (1-\cos(x^\circ))/x^2$ for $-45 \le x \le 45$, where x° means "x degrees." Use your graph to determine the value of $\lim_{x\to 0} g(x)$.

In each of Exercises 93–96, plot the given functions g, f, and h in a common viewing rectangle that illustrates f being pinched at the point c. Determine $\lim_{x\to c} f(x)$.

93. $g(x) = 2$, $f(x) = 2 + (x-1)^2\cos^2(x/(x-1))$, $h(x) = 2 + |x-1|/4 \quad c = 1$

94. $g(x) = (x-2)^2$, $f(x) = (x-2)^2\left(2 + \sin\left(\tan(\pi/x)\right)\right)\cdot(x-2)^2$, $h(x) = 3(x-2)^2 \quad c = 2$

95. $g(x) = -2|x|$, $f(x) = \cos(x-1/x) - \cos(x+1/x)$, $h(x) = 2|x| \quad c = 0$

96. $g(x) = 2 - 4|x|$, $f(x) = (2+x^2)/(1+x^2) + \sin(x+1/x) + \sin(x-1/x)$, $h(x) = 2(1+|x|) \quad c = 0$

2.3 Continuity

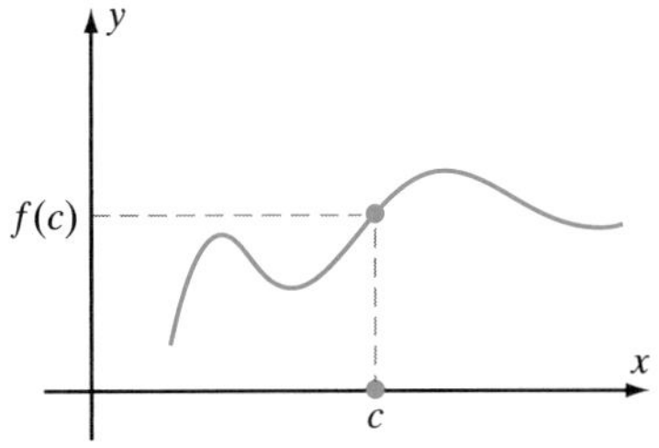

▲ Figure 1 $f(x)$ has the limit $f(c)$ as x tends to c.

When we drive along a high-speed road, we frequently cannot see very far ahead. We learn to drive as though the portion of the road immediately beyond our vision is the obvious extension of the portion of the road that we *can* see. What we are doing is computing the limit of what we see. Usually, the actual state of the road coincides with what we have anticipated it will be (i.e., the road is continuous). If the road has been damaged, or if a bridge is out, then the actual state of the road does *not* coincide with what we have anticipated (the road is discontinuous). In this section, we develop mathematical analogues of these ideas.

The Definition of Continuity at a Point

Here is a precise definition of continuity of a function at a point.

DEFINITION Suppose a function f is defined on an open interval that contains the point c. We say that f is *continuous* at c, provided that

$$\lim_{x\to c} f(x) = f(c).$$

See Figure 1.

> **INSIGHT** When we say that f is continuous at c, we are asserting *three* things:
> 1. f is defined at c;
> 2. there is a certain value $\lim_{x\to c} f(x)$ that we expect f to take at c (i.e., $\lim_{x\to c} f(x)$ exists); and
> 3. the expected value $\lim_{x\to c} f(x)$ agrees with the value $f(c)$ which value f actually takes at c ($\lim_{x\to c} f(x) = f(c)$).

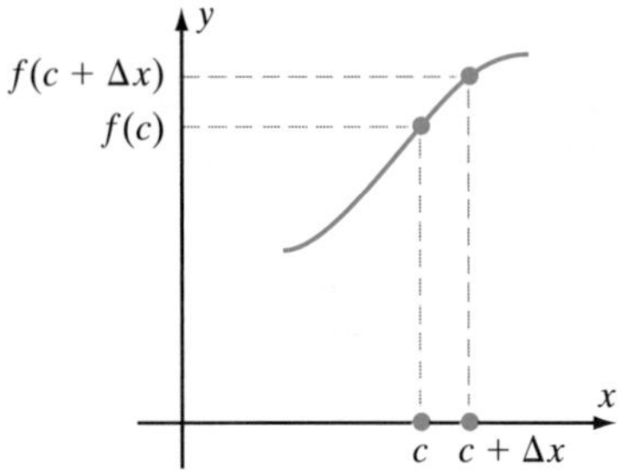

▲ Figure 2 A small input error near a point of continuity results in a small output error.

We consider continuity of a function *only* at points of its domain. For instance, we would not say that the function $f(x) = \sin(x)/x$ is continuous at 0 because 0 is not in the domain of f. If a point c is in the domain of f, and f is not continuous at c, then we say that f is *discontinuous* at c. Such a point is called a *point of discontinuity* of f. If c is a point of discontinuity of f, then either $f(x)$ does not have a limit as x approaches c, or the limit exists but is not equal to $f(c)$.

When a function f is continuous at *every* point of its domain, then we say that f is a *continuous function*. The theorems about limits that were developed in Section 2.2 show that polynomials, rational functions, n^{th} roots, and trigonometric functions are continuous functions.

An Equivalent Formulation of Continuity

The continuity of f at c can be expressed in another form. Let Δx denote an increment in x. Write $x = c + \Delta x$, and observe that $x \to c$ is equivalent to $\Delta x \to 0$. With this notation, the continuity of f at c is equivalent to

$$\lim_{\Delta x\to 0} f(c+\Delta x) = f(c).$$

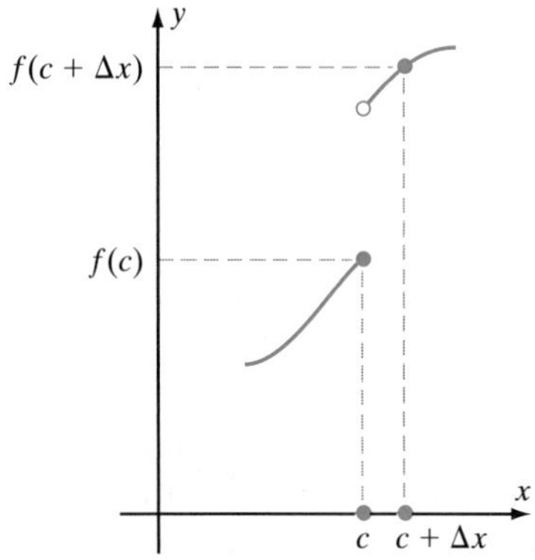

▲ Figure 3 A small input error near a point of discontinuity can result in a large output error.

Imagine an experiment to determine the mass $y = f(x)$ of a compound that is produced by a chemical reaction with a substance that has mass x. Suppose the value of x in the experiment is taken to be c. Inevitably, there will be a small measurement error Δx. With care, the *actual* value of x, $c + \Delta x$, will be close to the desired value c, but it will usually not be *exactly* c. Although the result of the experiment is intended to be $f(c)$, it is really $f(c + \Delta x)$. However, the result of the experiment is not necessarily unreliable. If c is a value at which f is continuous, then the difference between $f(c + \Delta x)$ and $f(c)$ can be made small by controlling the size of the measurement error Δx. In other words, sufficiently small errors in the independent variable x result in correspondingly small errors in the dependent variable y (see Figure 2). By contrast, if c is a point at which f is discontinuous, then a very small nonzero value of Δx can result in a substantial error, $f(c + \Delta x) - f(c)$, in the outcome of the experiment (see Figure 3).

▶ **EXAMPLE 1** Let

$$F(x) = \begin{cases} -4 & \text{if } x < 1 \\ -x-3 & \text{if } x \geq 1 \end{cases}.$$

Verify that F is continuous at every c in $\mathbb{R}$.

Solution If $c < 1$, then

$$\lim_{x\to c} F(x) = \lim_{x\to c} (-4) = -4 = F(c).$$

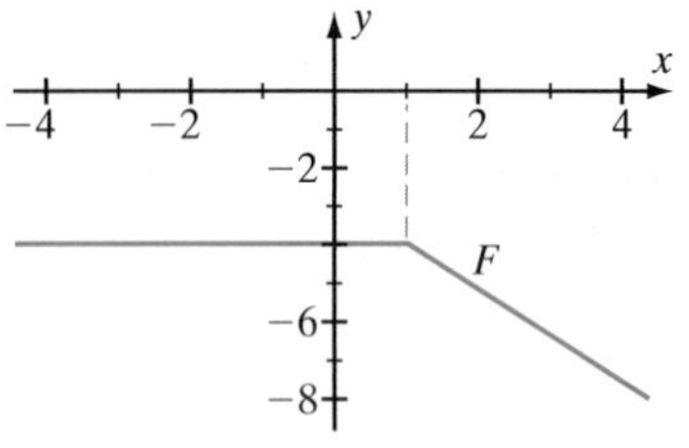

▲ Figure 4

If $c > 1$, then

$$\lim_{x \to c} F(x) = \lim_{x \to c} (-x - 3) = -c - 3 = F(c).$$

If $c = 1$, then we see, by checking both on the left and on the right, that

$$\lim_{x \to 1^-} F(x) = \lim_{x \to 1^-} (-4) = -4 = F(1) \quad \text{and} \quad \lim_{x \to 1^+} F(x) = \lim_{x \to 1^+} (-x - 3) = -4 = F(1).$$

Therefore F is continuous at every c in $\mathbb{R}$ (see Figure 4).

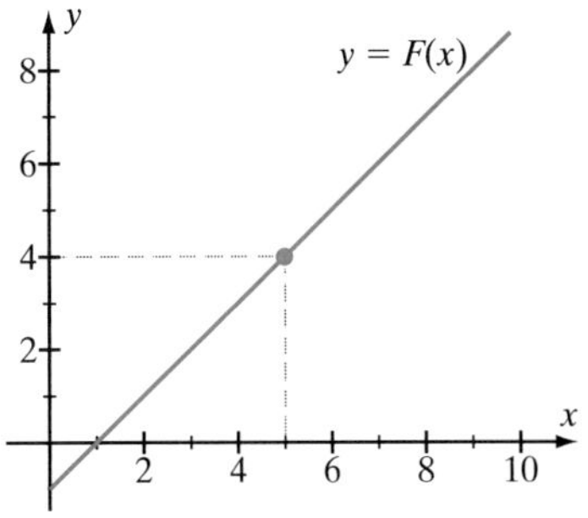

▲ Figure 5

$$F(x) = \begin{cases} \dfrac{x^2 - 6x + 5}{x - 5} & \text{if } x \neq 5 \\ 4 & \text{if } x = 5 \end{cases}$$

▶ **EXAMPLE 2** Let

$$F(x) = \begin{cases} \dfrac{x^2 - 6x + 5}{x - 5} & \text{if } x \neq 5 \\ 4 & \text{if } x = 5 \end{cases}.$$

Is F a continuous function?

Solution We must investigate F at each point c of its domain $\mathbb{R}$. First suppose that $c \neq 5$. Because the denominator of $F(x)$ is never 0 when x is near such a value of c, we may apply Theorem 2c from Section 2.2:

$$\lim_{x \to c} F(x) = \frac{c^2 - 6c + 5}{c - 5} = F(c).$$

Next we consider $c = 5$. Remember that, when we compute $\lim_{x \to 5} F(x)$, we use values of x *near* 5 but *not* equal to 5. Consequently, the quantity $x - 5$ is nonzero. Observe that this expression appears not only in the denominator of $F(x)$, but also in the numerator $x^2 - 6x + 5$, which factors as $(x - 5)(x - 1)$. Of course, any nonzero expression that appears in both the numerator and denominator of a fraction may be cancelled. Therefore

$$\lim_{x \to 5} F(x) = \lim_{x \to 5} \frac{x^2 - 6x + 5}{x - 5} = \lim_{x \to 5} \frac{(x - 5)(x - 1)}{x - 5} = \lim_{x \to 5} (x - 1) = 4 = F(5),$$

which shows that F is continuous at $c = 5$. In summary, we have demonstrated that F is continuous at every point of its domain. By definition, it follows that F is a continuous function. Figure 5 shows the graph of F in the viewing window $[0, 10] \times [-1, 9]$.

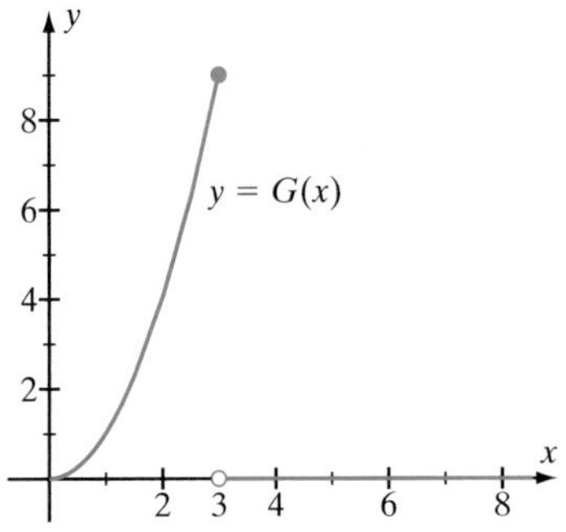

▲ Figure 6

$$G(x) = \begin{cases} x^2 & \text{if } x < 3 \\ 9 & \text{if } x = 3 \\ 0 & \text{if } x > 3 \end{cases}$$

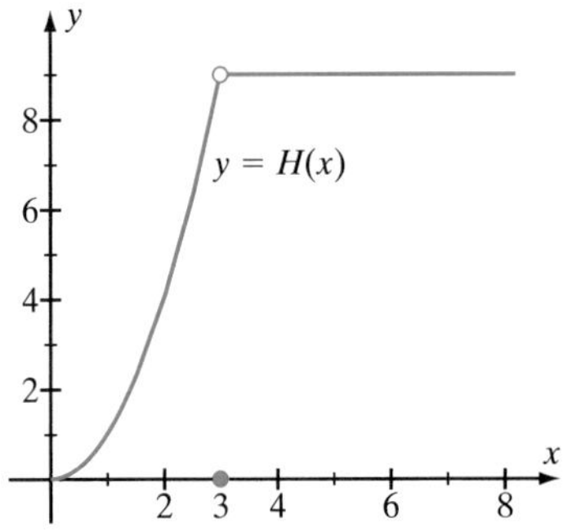

▲ Figure 7

$$H(x) = \begin{cases} x^2 & \text{if } x < 3 \\ 0 & \text{if } x = 3 \\ 9 & \text{if } x > 3 \end{cases}.$$

▶ **EXAMPLE 3** Let

$$G(x) = \begin{cases} x^2 & \text{if } x < 3 \\ 9 & \text{if } x = 3 \\ 0 & \text{if } x > 3 \end{cases} \quad \text{and} \quad H(x) = \begin{cases} x^2 & \text{if } x < 3 \\ 0 & \text{if } x = 3 \\ 9 & \text{if } x > 3 \end{cases}.$$

See Figures 6 and 7. Discuss the continuity of G and H at $c = 3$.

Solution Observe that $\lim_{x \to 3} G(x)$ does not exist because $\lim_{x \to 3^+} G(x) = 0$ and $\lim_{x \to 3^-} G(x) = 9$. So it is certainly not the case that $\lim_{x \to 3} G(x) = G(3)$. Thus G is discontinuous at $c = 3$. Although $\lim_{x \to 3} H(x)$ exists, $\lim_{x \to 3} H(x) = 9 \neq H(3)$. Therefore H is also discontinuous at $c = 3$. ◀

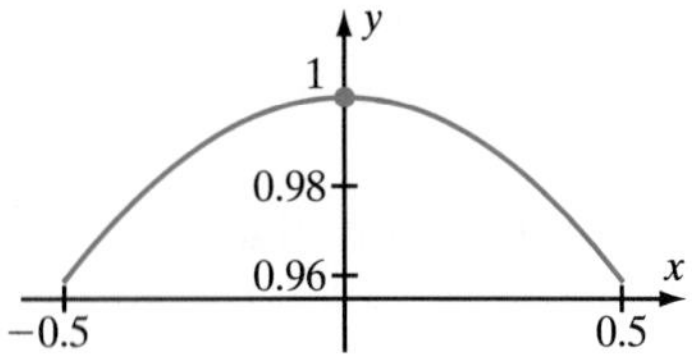

▲ Figure 8

$$F(x) = \begin{cases} \dfrac{\sin(x)}{x} & \text{if } x \neq 0 \\ 1 & \text{if } x = 0 \end{cases}$$

▶ **EXAMPLE 4** Show that the function

$$F(x) = \begin{cases} \dfrac{\sin(x)}{x} & \text{if } x \neq 0 \\ 1 & \text{if } x = 0 \end{cases}$$

is continuous at $c = 0$.

Solution Theorem 8 from Section 2.2 tells us that $\lim_{x\to 0} F(x) = 1 = F(0)$. Thus F is continuous at 0 (see Figure 8). ◀

Continuous Extensions

Consider a continuous function f that is defined on an interval (a, c) to the left of c and on an interval (c, b) to the right of c. Can we assign a value to f at c to create a continuous function on (a, b)? More precisely, is there a continuous function F that is defined on the entire interval (a, b) such that $F(x) = f(x)$ for $x \neq c$? The answer is that F exists precisely when $\lim_{x\to c} f(x)$ exists. If we let $\ell = \lim_{x\to c} f(x)$, then the function F defined by

$$F(x) = \begin{cases} f(x) & \text{if } x \neq c \\ \ell & \text{if } x = c \end{cases}$$

is continuous. We refer to F as the *continuous extension* of f. This situation is illustrated in Figure 9. Because the domain of F is larger than the domain of f, we see that F is not the same function as f. Refer back to Example 2, and you will see that the function F of that example is a continuous extension of $x \mapsto (x^2 - 6x + 5)/\ (x - 5)$. Similarly, the function F of Example 4 is a continuous extension of $x \mapsto \sin(x)/x$.

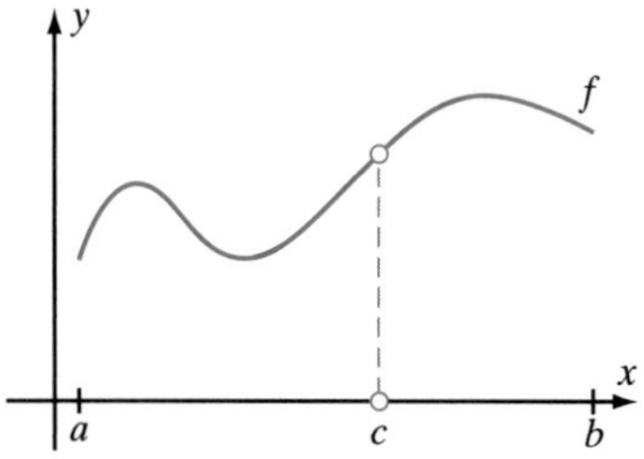

▲ Figure 9a

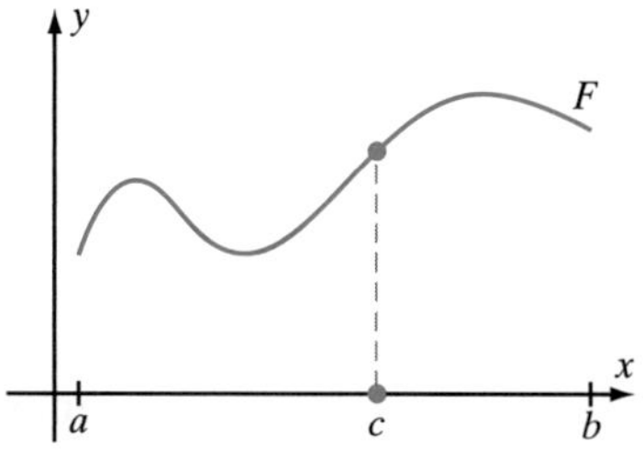

▲ Figure 9b The continuous extension of f

▶ **EXAMPLE 5** Let

$$f(x) = \begin{cases} x^2 - 14 & \text{if } x < -4 \\ x + 7 & \text{if } x > -4 \end{cases} \quad \text{and} \quad g(x) = \begin{cases} x^2 - 3 & \text{if } x < -4 \\ 9 - x & \text{if } x > -4 \end{cases}.$$

Is it possible to define a continuous extension of f? Of g?

Solution Because, the left limit, $\lim_{x\to(-4)^-} f(x) = (-4)^2 - 14 = 2$, and the right limit, $\lim_{x\to(-4)^+} f(x) = (-4) + 7 = 3$, are unequal, we see that $\lim_{x\to -4} f(x)$ does not exist. Therefore no continuous extension exists. On the other hand, as x approaches -4 from the left, the value $g(x) = x^2 - 3$ approaches 13. In addition,

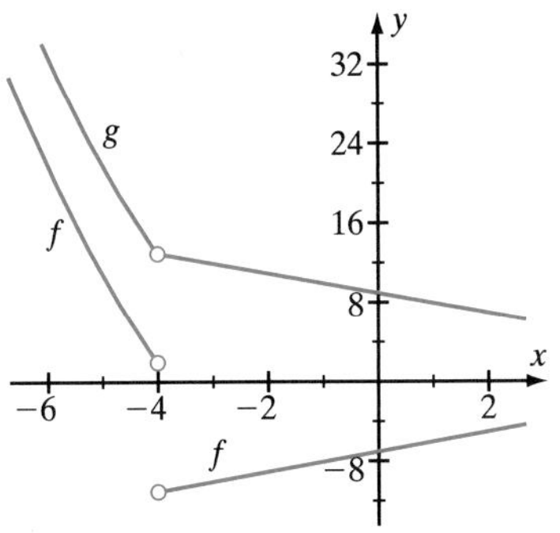

▲ Figure 10
$f(x) = \begin{cases} x^2 - 14 & \text{if } x < -4 \\ x + 7 & \text{if } x > -4 \end{cases}$
$g(x) = \begin{cases} x^2 - 3 & \text{if } x < -4 \\ 9 - x & \text{if } x > -4 \end{cases}$

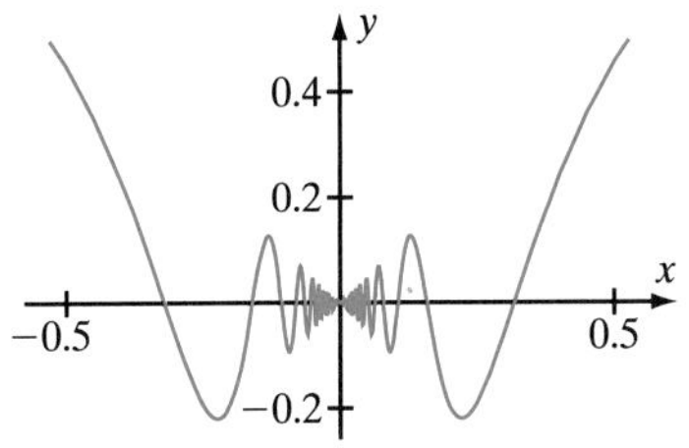

▲ Figure 11
$F(x) = \begin{cases} x \cdot \sin\left(\frac{1}{x}\right) & \text{if } x \neq 0 \\ 0 & \text{if } x = 0 \end{cases}$

as x approaches -4 from the right, the value $g(x) = 9 - x$ also approaches 13. Thus $\lim_{x \to -4} g(x) = 13$, and the function

$$G(x) = \begin{cases} g(x) & \text{if } x \neq -4 \\ 13 & \text{if } x = -4 \end{cases}$$

is a continuous extension of g. The graphs of f and g (see Figure 10) clearly illustrate the different behaviors of the functions at $x = -4$. ◀

We sometimes say that a function is continuous if its graph has no "jumps" or if the graph can be drawn without lifting the pencil from the paper. These intuitive concepts are helpful in treating some of the preceding examples. However, the continuous function

$$F(x) = \begin{cases} x \cdot \sin\left(\frac{1}{x}\right) & \text{if } x \neq 0 \\ 0 & \text{if } x = 0 \end{cases}$$

that we considered in Example 6 from Section 2.2 is much more difficult to understand in this informal manner. The graph oscillates rapidly near the origin, as shown in Figure 11, and the truth is that we cannot really draw the graph *with or without* lifting the pencil from the paper. It is precisely because of functions like these that we have gone to the trouble of learning careful definitions of "limit" and "continuous." These careful definitions enable us to say exactly at which points a function is continuous and at which points it is not.

One-Sided Continuity

In calculus, we frequently study functions that are defined on intervals that have an endpoint. In such cases, we can approach the endpoint only from one side. There are other situations in which we can approach a point in the domain from each side, but the behavior of the function is quite different on the two sides. We require a concept of continuity that is appropriate for such cases.

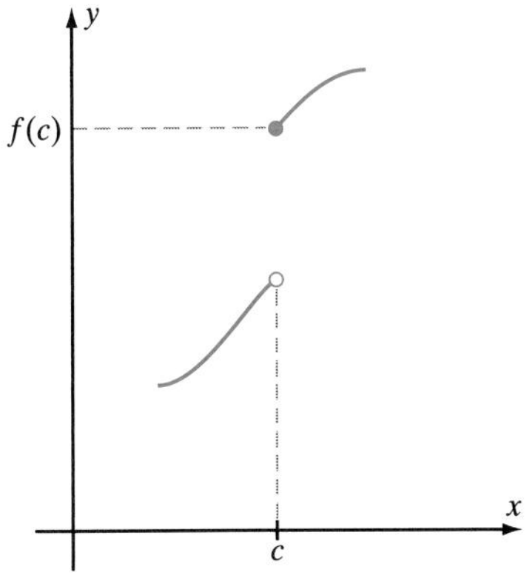

▲ Figure 12 f is right-continuous at c.

DEFINITION Suppose f is a function with a domain that contains $[c, b)$. If

$$\lim_{x \to c^+} f(x) = f(c),$$

then we say that f is continuous from the right at c, or f is right-continuous at c. Figure 12 exhibits a function that is right-continuous at c.

DEFINITION Suppose f is a function with a domain that contains $(a, c]$. If

$$\lim_{x \to c^-} f(x) = f(c),$$

then we say that f is continuous from the left at c, or f is left-continuous at c. Figure 13 shows a function that is left-continuous at c.

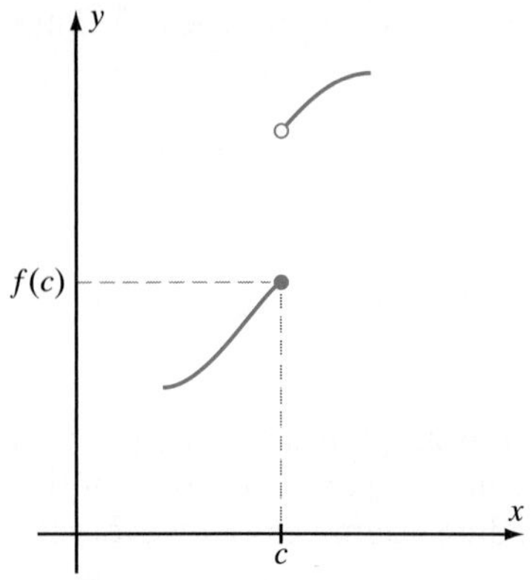

▲ **Figure 13** f is left-continuous at c.

Just as in the definition of continuity, the notions of left and right continuity at c entail three requirements:

1. The point c is in the domain of f.
2. The one-sided limit at c exists.
3. The one-sided limit must equal $f(c)$.

Also notice that "f is continuous at c" is equivalent to "f is left continuous at c and right-continuous at c." If $\lim_{x\to c^+} f(x)$ and $\lim_{x\to c^-} f(x)$ both exist but are different numbers, then f is said to have a *jump discontinuity* at c (see Figure 14).

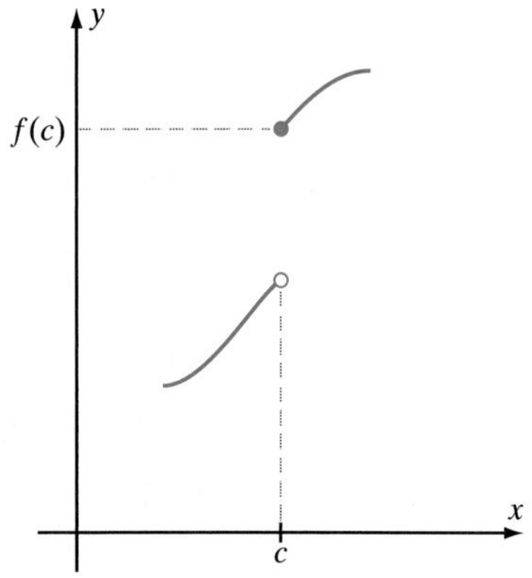

▲ **Figure 14a** Examples of functions with jump discontinuities

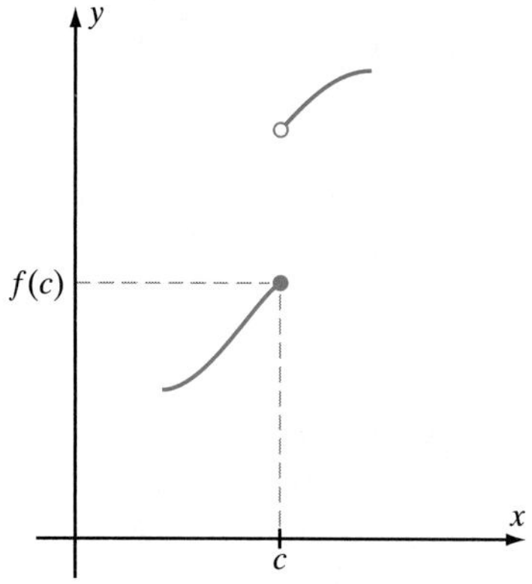

▲ **Figure 14b**

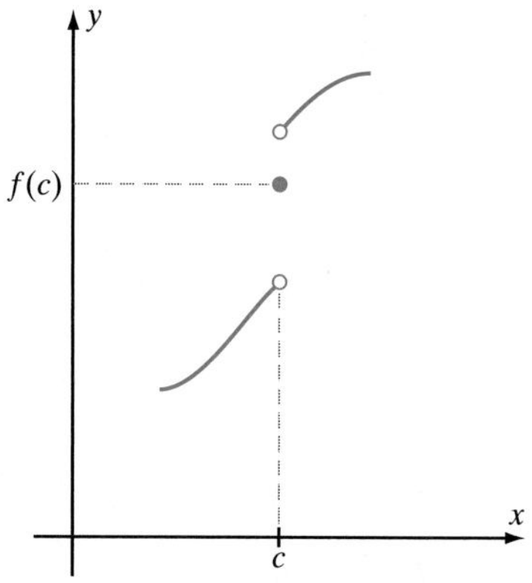

▲ **Figure 14c**

▶ **EXAMPLE 6** Is the function

$$g(x) = \begin{cases} 2x & \text{if } x < 4 \\ x^2 & \text{if } x \geq 4 \end{cases}$$

continuous from the left at 4? Is it continuous from the right at 4?

Solution The function g satisfies

$$\lim_{x\to 4^-} g(x) = \lim_{x\to 4^-} 2x = 8 \neq 16 = g(4).$$

Thus g is not continuous from the left at 4. However,

$$\lim_{x\to 4^+} g(x) = \lim_{x\to 4^+} x^2 = 16 = g(4).$$

Therefore g is continuous from the right at 4. ◀

Some Theorems about Continuity–Arithmetic Operations and Composition

Our work with the concept of continuity would be tedious if we had to check, from the definition, whether each function that we encounter is continuous. Fortunately, the limit theorems from Section 2.2 give rise to analogous theorems for continuity.

THEOREM 1 Let α be a constant.

a. If f and g are functions that are continuous at $x = c$, then so are $f + g, f - g$, $\alpha \cdot f$, and $f \cdot g$. If $g(c) \neq 0$, then f/g is also continuous at $x = c$.
b. If f and g are both right-continuous at $x = c$, then so are $f + g, f - g, \alpha \cdot f$, and $f \cdot g$. If $g(c) \neq 0$, then f/g is also right-continuous at $x = c$. The same is true if right continuity is replaced with left continuity.

Theorem 1 follows from the corresponding results about limits in Theorem 2 from Section 2.2.

▶ **EXAMPLE 7** Let $F(x) = \sin(x)$, $G(x) = x^2 - 6x + 5$, and $H(x) = F(x)/G(x)$. For which values of x is H continuous?

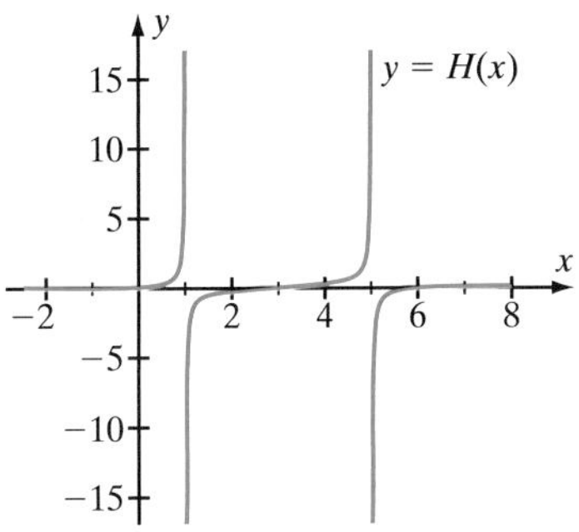

▲ Figure 15
$H(x) = \sin(x)/(x^2 - 6x + 5)$

Solution As was mentioned earlier in this section, the limit theorems from Section 2.2 tell us that trigonometric functions and polynomials are continuous. In particular, the functions F and G are continuous at every point of the real line. Therefore the function H is continuous at every value of x for which $G(x) = x^2 - 6x + 5 = (x - 1)(x - 5)$ is nonzero. Thus H is continuous at each point different from 1 and 5. The points 1 and 5 are not in the domain of H, so we do not speak about the continuity of H at these points. The plot of H in the viewing window $[-2, 8] \times [-15, 15]$ (see Figure 15) indicates that H does not have a continuous extension at the exceptional points. ◀

If the image of a function f is contained in the domain of a function g, then we may consider the composition $(g \circ f)(x) = g(f(x))$. This equation defines an operation that consists of first applying f and then applying g. In general, using the limits of f and g to calculate the limit of the composite function $g \circ f$ can be tricky. Exercises 55–57 at the end of this section explore the possibilities. The following theorem assures us that nothing unexpected happens when both f and g are continuous: Composition preserves continuity.

THEOREM 2 Suppose the image of a function f is contained in the domain of a function g. If f is continuous at c, and g is continuous at $f(c)$, then the composed function $g \circ f$ is continuous at c.

▶ **EXAMPLE 8** Is $x \mapsto \cos(x^2 + 1)$ continuous at every x?

Solution The functions $f(x) = x^2 + 1$ and $g(x) = \cos(x)$ have the entire real line as their domain, and they are continuous at each point. Therefore $(g \circ f)(x) = \cos(x^2 + 1)$ is defined for every value of x, and, by Theorem 2, it is continuous at every x. ◀

Advanced Properties of Continuous Functions

The next two theorems about continuity depend on a deeper property of the real numbers—the *completeness* property. When we imagine the real numbers, we are accustomed to picturing a line (the real number line) that has no gaps in it. The analytic formulation of this idea is the completeness property of the real numbers. There are a number of equivalent ways to express the property precisely. One formulation is the *nested interval property*, which is discussed in Genesis & Development, Chapter 1. Another is the *least upper bound property* of the real numbers—see Genesis & Development, Chapter 2. A third formulation of the completeness property is discussed in Section 2.6.

Like many theorems in calculus, the two theorems that follow require that we use the notion of a function f being continuous on a *closed* interval $[a, b]$. This means that f is continuous at every point in the open interval (a, b), $f(x)$ is right-continuous at a, and $f(x)$ is left-continuous at b. In particular, if f is continuous on $[a, b]$, then the functional values $f(a)$ and $f(b)$ at the endpoints are the same as the one-sided limits of f at the endpoints.

THEOREM 3 **(The Intermediate Value Theorem).** Suppose that f is a continuous function on the closed interval $[a, b]$. Suppose further that $f(a) \neq f(b)$. Then, for any number γ between $f(a)$ and $f(b)$, there is a number c between a and b such that $f(c) = \gamma$.

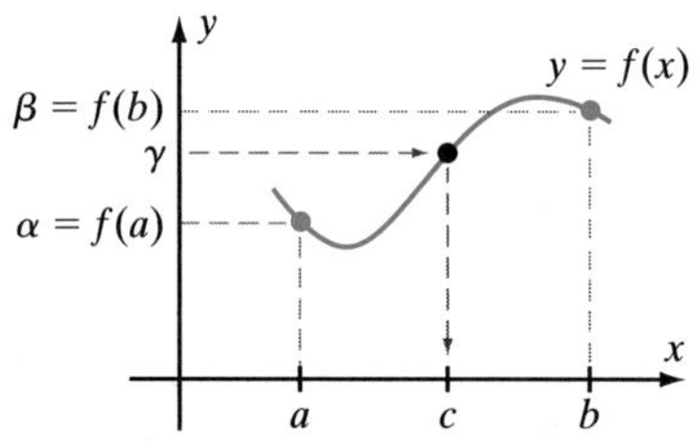

▲ Figure 16

INSIGHT In simple language, Theorem 3 states that continuous functions do not skip values. If f is continuous and takes the values α and β, then f takes all the values between α and β (see Figure 16). In even plainer language, the graph of f does not have jumps—a jump would correspond to skipped values. This explains why we sometimes say that a continuous function can be graphed without lifting the pencil from the paper. Although we now understand this statement to be purely informal, we can use it to guide our intuition.

▶ **EXAMPLE 9** Is there a point in the interval $(0, 2)$ at which the function $f(x) = (x^3+1)^2$ takes the value 10?

Solution Because it is a polynomial, the function $f(x) = (x^3 + 1)^2$ is continuous. We can easily find values a and b such that $\gamma = 10$ is between $f(a)$ and $f(b)$. For example, if $a = 0$, and $b = 2$, then $f(a) = 1 < 10 < 81 = f(b)$. According to the Intermediate Value Theorem, we can be sure that there is a number c between 0 and 2 such that $f(c) = 10$. ◀

INSIGHT The Intermediate Value Theorem gives us conditions under which we can be certain that the equation $f(c) = \gamma$ has a solution c. In Example 9, we can actually find a formula for the value c for which $f(c) = \gamma$. With the aid of a little algebra, we obtain $c = \sqrt[3]{\sqrt{10} - 1} \simeq 1.2931\ldots$. When we use the Intermediate Value Theorem, we usually cannot find an explicit formula for the value of a solution c of the equation $f(c) = \gamma$. In Chapter 4, we will learn how to find good approximations to the solutions of such equations.

▶ **EXAMPLE 10** Mr. Woodman weighs 150 pounds at noon on January 1 $(t = 1)$, 130 pounds at noon on January 15 $(t = 15)$, and 140 pounds at noon on January 30 $(t = 30)$. His weight is a continuous function of time t. How often during the month of January can we be sure that he will weigh 132 pounds?

Solution Let $w(t)$ be Mr. Woodman's weight in pounds at time t where t is measured in days. Our hypotheses may be written as

$$w(1) = 150, \qquad w(15) = 130, \qquad w(30) = 140.$$

Because 132 is between 150 and 130, there is a time t_1, $1 < t_1 < 15$, such that $w(t_1) = 132$. Likewise, 132 is between 130 and 140. Thus there is a time t_2, $15 < t_2 < 30$, such that $w(t_2) = 132$. In other words, we have found that between noon on January 1 and noon on January 30, Mr. Woodman will weigh 132 pounds at least two times. As Figure 17 illustrates, Mr. Woodman may weigh 132 pounds at other times between January 1 and January 30. Without further information, however, we can only say for certain that there are at least two times. ◄

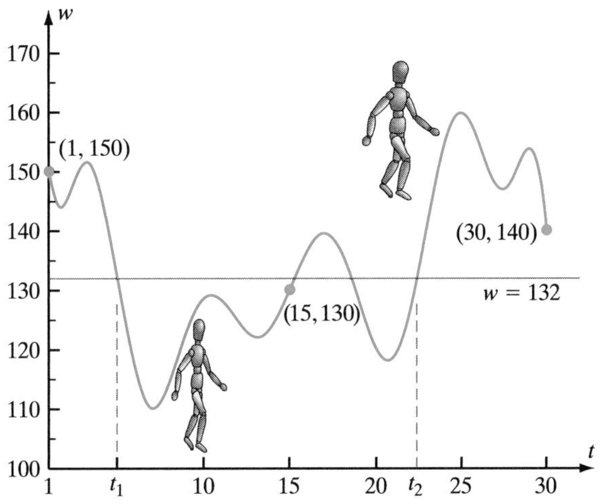

▲ Figure 17

INSIGHT As Example 10 illustrates, the Intermediate Value Theorem says nothing about uniqueness. It gives us conditions under which there is sure to be a solution c of the equation $f(c) = \gamma$, but it does not tell whether there is only one solution or several solutions.

We have already mentioned that, if n is a positive integer, then $x \mapsto \sqrt[n]{x}$ is a continuous function (Theorem 4 from Section 2.2). Until now, we have implicitly assumed the existence of nth roots. How can we be sure, however, that there is a number c such that $c^{17} = \pi$? The Intermediate Value Theorem can be used to prove the existence of such roots, as shown in the next example.

► **EXAMPLE 11** Let γ be a positive number and n a positive integer. Show that there is a number c such that $c^n = \gamma$.

Solution Let $f(x)$ be defined on $[0, \infty)$ by $f(x) = x^n$. We know that f is a continuous function by Theorem 3 from Section 2.2. Let $a = 0$, $b = \gamma + 1$, $\alpha = f(a)$, and $\beta = f(b)$. Then,

$$\alpha = f(0) = 0 < \gamma < \gamma + 1 < (\gamma+1)^n = f(b) = \beta.$$

The Intermediate Value Theorem tells us that there is a c in (a, b) such that $f(c) = \gamma$. Thus $c^n = \gamma$. ◄

The Intermediate Value Theorem is said to be an *existence theorem* because it asserts the existence of a value without specifying it. We will encounter several theorems of this type in our study of calculus. Indeed, the theorem that follows asserts the existence of a highest and a lowest point on the graph of a function that is continuous on a closed, bounded interval.

DEFINITION Suppose that f is a function with domain S. If there is a point α in S such that $f(\alpha) \leq f(x)$ for all x in S, then the point α is called a minimum for the function f, and $m = f(\alpha)$ is called the minimum value of f. If there is a point β in S such that $f(\beta) \geq f(x)$ for all x in S, then the point β is called a maximum for the function f, and $M = f(\beta)$ is called its maximum value.

THEOREM 4 **(The Extreme Value Theorem)**. If f is a continuous function on the closed interval $[a, b]$, then f has a minimum α in $[a, b]$ and a maximum β in $[a, b]$.

INSIGHT A continuous function on a closed interval $[a, b]$ has a greatest value $f(\beta)$ and a least value $f(\alpha)$. These values are also known as extreme values, or extrema, of f.

▶ **EXAMPLE 12** Let $f(x) = x^3 - 7x^2 + 3x - 1$ on $[0, 4]$. Does f take a maximum value on this interval? Does f take a minimum value?

Solution Because f is continuous (it is a polynomial), it takes a greatest value at some point β in $[0, 4]$ and a least value at some point α in $[0, 4]$. We can be sure that this is true by Theorem 4, even though we do not yet know how to find α and β. ◀

INSIGHT One important application of calculus is to actually find α and β. This, in turn, will enable us to solve many interesting problems. (Chapter 4 covers this idea in more detail.)

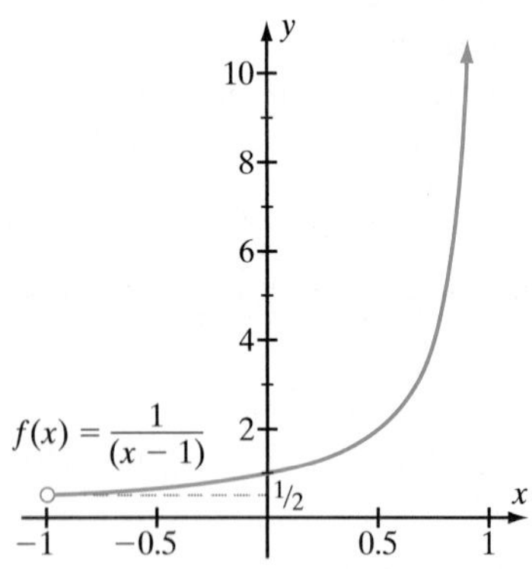

▲ Figure 18

▶ **EXAMPLE 13** Let $f(x) = 1/(1 - x)$ (see Figure 18). Are there maximum and minimum values for f on the interval $(-1, 1)$?

Solution Here the function f is continuous, and the interval $(-1, 1)$ is bounded. However, this interval is not closed because it does not contain its endpoints. Because the Extreme Value Theorem may only be applied to closed intervals, we may not use it here. Instead, we must investigate the behavior of f. We observe that f has no greatest value on the interval $(-1, 1)$ because $f(x)$ becomes large without bound as x approaches 1 from the left (e.g., $f(0.9) = 10$, $f(0.9999) = 10000$, etc.). Also, f has no least value on $(-1, 1)$, although this is so for a more subtle reason—if x is near -1 and to the right of -1, then $f(x)$ is nearly 1/2. Indeed, the values of f become arbitrarily close to 1/2 when x is sufficiently close to -1 on the right, but f never actually assumes the value 1/2 at any point of the interval $(-1, 1)$. So, f does not assume a minimum value on $(-1, 1)$. ◀

INSIGHT Although the function f of Example 13 does not have a maximum or minimum value on $I = (-1, 1)$, the Extreme Value Theorem is not contradicted because I is not closed. It is incorrect to assume, however, that there cannot be a maximum or minimum value when the Extreme Value Theorem does not apply. The function $x \mapsto x^2$ has a minimum value 0 on I, the function $x \mapsto 1 - x^2$ has a maximum value 1 on I, and the function $x \mapsto \cos(2\pi x)$ has a maximum value 1 and a minimum value -1 on I.

QUICK QUIZ

1. True or false: If c is in the domain of f, and $\lim_{x\to c} f(x)$ exists, then f is continuous at c.
2. At what points is $f(x) = \sin(x^2)/(x^4 - 16)$ continuous? Is f a continuous function?
3. True or false: If $f : [0, 3] \to (0, 1]$, then f has a minimum value.
4. True or false: If $f : [0, 3] \to (0, 1]$ is continuous, then f has a minimum value.
5. True or false: If $f : (0, 1] \to [0, 3]$ is continuous, then f has a maximum value.

Answers

1. False 2. At every point of its domain, $\{x \in \mathbb{R} : x \neq \pm 2\}$; Yes 3. False 4. True 5. False

EXERCISES

Problems for Practice

In each of Exercises 1–4, determine at which of the points -1, 1, and 2 the given function f is continuous.

1. $f(x) = \begin{cases} x^2 - 1 & \text{if } x < 2 \\ 3 & \text{if } x = 2 \\ 3\sin(\pi/x) & \text{if } 2 < x \end{cases}$
2. $f(x) = \begin{cases} 1 - x & \text{if } x \leq -1 \\ \sqrt{5 - x^2} & \text{if } -1 < x < 1 \\ 2 - \cos(\pi x) & \text{if } 1 \leq x \end{cases}$
3. $f(x) = \begin{cases} (x^2 - 1)/(x^2 + 1) & \text{if } x \leq -1 \\ (3x+2)^{1/3} & \text{if } -1 < x < 2 \\ 4 - x & \text{if } 2 \leq x \end{cases}$
4. $f(x) = \begin{cases} (x^3 - x)/(x^2 - 1) & \text{if } x < -1 \\ 2x + 1 & \text{if } -1 \leq x \leq 2 \\ (x^2 - 4)/(x - 2) & \text{if } 2 < x \end{cases}$

In Exercises 5–24, determine the values at which the given function f is continuous. Remember that if c is not in the domain of f, then f cannot be continuous at c. Also remember that the domain of a function that is defined by an expression consists of all real numbers at which the expression can be evaluated.

5. $f(x) = x^2 + 4$
6. $f(x) = 2x - 9$
7. $f(x) = 1/(x + 1)$
8. $f(x) = x/(x^2 - x - 5)$
9. $f(x) = 1/(x^2 + 1)$
10. $f(x) = 1/(1 - x^2)$
11. $f(x) = \dfrac{x^2 - 5x + 6}{x^2 + x - 12}$
12. $f(x) = \begin{cases} x^2 + 1 & \text{if } x < 0 \\ 1/(2x + 1) & \text{if } x \geq 0 \end{cases}$
13. $f(x) = \begin{cases} x + 1 & \text{if } x \leq 3 \\ 6 & \text{if } x > 3 \end{cases}$
14. $f(x) = \begin{cases} x^2 - 2 & \text{if } x \leq 3 \\ 8 & \text{if } x > 3 \end{cases}$
15. $f(x) = \begin{cases} (x+1)^4 & \text{if } x < -4 \\ -20x + 1 & \text{if } x \geq -4 \end{cases}$
16. $f(x) = \begin{cases} (x^2/4) - 7 & \text{if } x < 6 \\ 2 & \text{if } x = 6 \\ 9 - x & \text{if } x > 6 \end{cases}$
17. $f(x) = \begin{cases} 2x - 19 & \text{if } x < 7 \\ -5 & \text{if } x = 7 \\ 2 - x & \text{if } x > 7 \end{cases}$
18. $f(x) = \begin{cases} 2x + 1 & \text{if } x < 0 \\ 4 & \text{if } x = 0 \\ x^2 - 2 & \text{if } x > 0 \end{cases}$
19. $f(x) = \begin{cases} (x^2 - 1)/(x + 1) & \text{if } x \neq -1 \\ 2 & \text{if } x = -1 \end{cases}$
20. $f(x) = \begin{cases} (x^2 - 1)/(x + 1) & \text{if } x \neq -1 \\ \cos(\pi x) - 1 & \text{if } x = -1 \end{cases}$
21. $f(x) = \begin{cases} \sin(x)/x & \text{if } x \neq 0 \\ 0 & \text{if } x = 0 \end{cases}$
22. $f(x) = \begin{cases} \tan(x)/x & \text{if } x \neq 0 \\ 1 & \text{if } x = 0 \end{cases}$

23. $f(x) = \begin{cases} x^2 - 3 & \text{if } x \le 3 \\ (x^2 - 9)/(x - 3) & \text{if } x > 3 \end{cases}$

24. $f(x) = \sqrt{x} \cdot \tan(x) + \sqrt{\pi - x} \cdot \csc(x)$

In each of Exercises 25–28, define $F(2)$ to obtain a continuous extension F of f.

25. $f(x) = \begin{cases} x^2 + 7 & \text{if } x < 2 \\ x^3 + 3 & \text{if } x > 2 \end{cases}$

26. $f(x) = \begin{cases} 5/(x^2 + 6) & \text{if } x < 2 \\ 1/x & \text{if } x > 2 \end{cases}$

27. $f(x) = \begin{cases} x^6/(18 - x) & \text{if } x < 2 \\ x^4/(x + 2) & \text{if } x > 2 \end{cases}$

28. $f(x) = \begin{cases} x^3/(x^3 + 1) & \text{if } x < 2 \\ x^4/(x^4 + 2) & \text{if } x > 2 \end{cases}$

For Exercises 29–32, determine the points at which the function is left-continuous, the points at which the function is right-continuous, and the points at which the function is continuous. Give reasons for your answers.

29. $f(x) = \begin{cases} 2 & \text{if } x \le 5 \\ 3 & \text{if } x > 5 \end{cases}$

30. $g(x) = \begin{cases} x^2 - 15 & \text{if } x < -4 \\ 1 & \text{if } x \ge -4 \end{cases}$

31. $f(x) = \begin{cases} x + 1 & \text{if } x < 3 \\ 2 & \text{if } x \ge 3 \end{cases}$

32. $g(x) = \begin{cases} x^2 & \text{if } x < 0 \\ 2x & \text{if } x \ge 0 \end{cases}$

In Exercises 33–38, a continuous function f is defined on a closed, bounded interval I. Determine the extreme values of the function f. Sketch the graph of f and label the points at which f assumes its extreme values.

33. $f(x) = x^2 + 2x + 3, I = [-3, -2]$

34. $f(x) = x^2 + 2x + 3, I = [-2, 1]$

35. $f(x) = x^2 + 2x + 3, I = [1, 2]$

36. $f(x) = 11 + 6x - x^2, I = [1, 4]$

37. $f(x) = 2 - \cos(x), I = [0, 2\pi]$

38. $f(x) = x + \sin(x), I = [-\pi, \pi]$

In each of Exercises 39–42, a continuous function f is defined on a closed, bounded interval I. A consequence of the Intermediate Value Theorem is that there is an interval $[\alpha, \beta]$ with the property that, for every γ between α and β, the equation $f(x) = \gamma$ has a solution $x = c$ with c in I. What are α and β?

39. $f(x) = \sin(x), \quad I = [-\pi/6, \pi/4]$

40. $f(x) = x + 24/x, \quad I = [2, 6]$

41. $f(x) = (7x - 3x^2)/8, \quad I = [1, 2]$

42. $f(x) = \dfrac{x^2 + 1}{x^2 + x + 2}, \quad I = [-3, 5]$

Further Theory and Practice

In each of Exercises 43–46, a function, f with domain $(-\infty, 1) \cup (3, \infty)$ is given. Define $F(x) = ax + b$ for $1 \le x \le 3$ and $F(x) = f(x)$ for x not in $[1, 3]$. Determine a and b so that F is continuous.

43. $f(x) = \begin{cases} 1 & \text{if } x < 1 \\ -6 & \text{if } x > 3 \end{cases}$

44. $f(x) = \begin{cases} x^2 + 16 & \text{if } x < 1 \\ x - 4 & \text{if } x > 3 \end{cases}$

45. $f(x) = \begin{cases} (x + 1)/(x^2 + 2) & \text{if } x < 1 \\ x^2 & \text{if } x > 3 \end{cases}$

46. $f(x) = \begin{cases} \cos(\pi x) & \text{if } x < 1 \\ \sin(\pi x) & \text{if } x > 3 \end{cases}$

In each of Exercises 47–50, determine at which points the given function is discontinuous.

47. $f(x) = \begin{cases} \sin(\pi x/2) & \text{if } x \text{ is not an integer} \\ 0 & \text{if } x \text{ is an integer} \end{cases}$

48. $f(x) = \begin{cases} \cos(\pi/2) & \text{if } x \text{ is not an integer} \\ 1 & \text{if } x \text{ is an integer} \end{cases}$

49. $f(x) = \begin{cases} 2\cos(\pi x/3) & \text{if } x \text{ is not an integer} \\ 1 & \text{if } x \text{ is an integer} \end{cases}$

50. $f(x) = \begin{cases} \cos(\pi x/4) & \text{if } x \text{ is not an integer} \\ \sin(\pi x/4) & \text{if } x \text{ is an integer} \end{cases}$

51. For each of the four functions,

$$f(x) = \frac{x^2 - 1}{x - 1}, \quad g(x) = \begin{cases} x + 1 & \text{if } x \ne 1 \\ 2 & \text{if } x = 1 \end{cases},$$

$$h(x) = x + 1, \text{ and } k(x) = \begin{cases} x + 1 & \text{if } x \ne 1 \\ 1 & \text{if } x = 1 \end{cases},$$

compute the left and right limits at $c = 1$, determine if it has a limit at $c = 1$, and determine if it is continuous at $c = 1$. (Pay particular attention to the domains of the functions.)

52. Show that the single filer tax function T of Example 3, Section 1.4 is a continuous function.

53. Consider the single filer tax function T that is given in Example 3 from Section 1.4. There are six tax brackets that correspond to six marginal tax rates: 10%, 15%, 25%, 28%, 33%, and 35%. Suppose the tax brackets are not altered, but the marginal tax rates are reduced to 8%, 12%, 16%, 20%, 25%, and 30%. What is the new tax function, assuming it is continuous?

54. Prove that a function f is continuous at c if and only if f is both left-continuous at c and right-continuous at c.

55. For any real number x, let $\lfloor x \rfloor$ denote the greatest integer which does not exceed x.

a. What is $\lim_{x \to 3^-} \lfloor x \rfloor$? Prove it.

b. What is $\lim_{x \to 2^+} \lfloor 4 - 2x \rfloor$? Prove it.

c. What is $\lim_{x \to 0^-} \lfloor 1/\lfloor x \rfloor \rfloor$? Prove it.

d. What is $\lim_{x \to 0^-} \lfloor \lfloor -x \rfloor \rfloor / x$? Prove it.

56. The function $f(x) = 4 - (x-2)^2$ assumes a maximum value on the interval $I = (0,4)$. What is that value? By determining the image of $f(x)$ for x in I, show that f assumes no minimum value on I. Why is the Extreme Value Theorem not contradicted?

57. Suppose a train pulls out of a station and comes to a stop at the next station 80 km away exactly 1 h later. Assuming the velocity of the train is a continuous function, prove that, at some moment, the train must have been traveling at a velocity of precisely 60 km/h.

58. Let p be a polynomial of odd degree and leading coefficient 1. It can be shown that $\lim_{x\to\infty} p(x) = \infty$ and $\lim_{x\to-\infty} p(x) = -\infty$. Use these facts and the Intermediate Value Theorem to prove that p has at, least one real root.

59. Show that $10x^3 - 7x^2 + 20x - 14 = 0$ has a root in $(0,1)$. Deduce that $10\sin^3(x) - 7\sin^2(x) + 20\sin(x) - 14 = 0$ has a solution.

60. Use the Intermediate Value Theorem to prove that $p(x) = 3x^5 - 7x^2 + 1$ has three real roots and that they are located in the intervals $(-1,0)$, $(0,1)$, and $(1,2)$.

61. Because $p(x) = x^3 + x + 1$ has odd degree, we can be sure (by Exercise 58) that it has at least one real root. Determine a finite interval that contains a root of p. Explain why.

62. Use the Intermediate Value Theorem to show that $x/2 = \sin(x)$ has a positive solution. (Consider the function $f(x) = x/2 - \sin(x)$ and the points $a = \pi/6$ and $b = 2.1$.)

63. Use the Intermediate Value Theorem to show that $p(x) = 10x^4 + 46x^2 - 39x^3 - 39x + 36$ has at least two roots between 1 and 3.

In each of Exercises 64–67, an assertion is made about a function p that is continuous on the closed interval $I = [-1,3]$ and for which $p(-1) = 4$ and $p(3) = -2$. If the statement is true, explain why. Otherwise, sketch a function p that shows it is false.

64. p has at least one root in I.
65. p has at most one root in I.
66. p has exactly one root in I.
67. p has an old number of roots in I.

In each of Exercises 68–73, an assertion is made about a function f that is defined on a closed, bounded interval. If the statement is true, explain why. Otherwise, sketch a function f that shows it is false. (Note: $|f|$ is defined by $|f|(x) = |f(x)|$.)

68. If f is continuous, then $|f|$ is continuous.
69. If f is continuous, then f is continuous.
70. If f is continuous, then f^2 is continuous.
71. If f^2 is continuous, then f is continuous.
72. If f is continuous, then its image is a closed, bounded interval.
73. If f has no maximum, then f is discontinuous at some point of interval.

74. Suppose that α is a constant in the interval $(0,1)$. By applying the Intermediate Value Theorem to $f(x) = \alpha - \sin(x)/x$ and using $\lim_{x\to 0}\sin(x)/x = 1$, show that the equation $\alpha = \sin(x)$ has a positive solution in x.

Exercises 75–77 illustrate the following theorem: If the range of f is contained in the domain of g, if $\lim_{x\to c} f(x) = \ell$, and if $\lim_{y\to\ell} g(y) = L$, then $\lim_{x\to c} g(f(x)) = g(\ell)$ or $\lim_{x\to c} g(f(x)) = L$ or $\lim_{x\to c} g(f(x))$ does not exist.

75. Let $g(x) = \begin{cases} 0 & \text{if } x \neq 0 \\ 1 & \text{if } x = 0 \end{cases}$ and $f(x) = 0$ for all x. Let $c = 0$. Evaluate ℓ and L. Show that $\lim_{x\to 0} g(f(x)) = g(\ell)$, and compare this result with Theorem 2.

76. Let g be defined as in Exercise 75. Let $f(x) = x$. Let $c = 0$. Evaluate ℓ and L. Show that $\lim_{x\to 0} g(f(x)) = L$, and compare this result with Theorem 2.

77. Let g be defined as in Exercise 75. Let

$$f(x) = \begin{cases} x & \text{if } x \text{ is irrational} \\ 0 & \text{if } x \text{ is rational} \end{cases}.$$

Show that $\lim_{x\to 0} g(f(x))$ does not exist, and compare this result with Theorem 2.

78. Let $f : [0,1] \to [0,1]$ be a continuous function. Prove that there is a number c, $0 \le c \le 1$, such that $f(c) = c$. Such a value is said to be a *fixed point* of f. (*Hint:* Think about the function $g(x) = f(x) - x$.)

79. A hiker walks up a mountain path. He starts at the bottom of the path at 8:00 AM and reaches the top at 6:00 PM. The next morning, he starts down the same path at 8:00 AM and reaches the bottom at 6:00 PM. Show that there is at least one time of the day such that on each day the hiker's elevation was the same. Illustrate this by giving a representative sketch of the ascent and descent elevation functions in the rectangle $[0\,\text{h}, 10\,\text{h}] \times [0, L]$ where L is the elevation of the peak.

80. The polynomial $p(x) = x^{51} + x - 10^{51}$ has exactly one real root c. Find an integer k such that c is in the interval $(k, k+1)$.

81. Prove that, if f is a continuous function on $[0,1]$, and $f(0) = f(1)$ then there is a value c in $(0,1)$ such that $f(c) = f(c + 1/2)$. This is a special case of the *Horizontal Chord Theorem.* (*Hint:* Apply the Intermediate Value Theorem to the function g defined on $[0, 1/2]$ by $g(x) = f(x + 1/2) - f(x)$.)

Calculator/Computer Exercises

82. Locate, to four decimal places of accuracy, the maximum and minimum values of the function $h(x) = x^4 - 5x^3 + 7x + 9$ on the interval $[0,4]$.

83. Use Exercise 58 to explain why $p(\tan(x)) = c$ has a real solution x for any real value of c and any cubic real polynomial p. Approximate a solution to $\tan^3(x) + 3\tan(x) = 19$.

84. Graph $f(x) = (x^3 - 3x + 1)/(x^3 - 3x - 1)$ in the viewing rectangle $[-3, 3] \times [-15, 15]$. Locate the points, to four decimal places, at which f does not have a continuous extension.

In Exercises 85–88, graph the given function f over an interval centered about the given point c, and determine if f has a continuous extension at c.

85. $f(x) = x/|x|, c = 0$

86. $f(x) = (x^4 - 6x^3 + 7x^2 + 4x - 4)/(x-2)^2, c = 2$

87. $f(x) = (x^4 - 6x^3 + 7x^2 + 4x - 4)/(x - 2), c = 2$

88. $f(x) = x^2 \sin(x)/(\sqrt{1 + x^2} - 1), c = 0$

89. Use the factorization

$$x^4 - 10x^3 + 38x^2 - 64x + 40 = (x^2 - 6x + 10)(x-2)^2$$

to show that

$$p(x) = x^4 - 10x^3 + 38x^2 - 64x + 50 \geq 10.$$

For what values of γ does $p(x) = \gamma$ have a real solution? Show that $p(x) = 20$ has a solution in each of the intervals $(0, 1)$ and $(4, 5)$. Find these solutions to four decimal places.

90. Let $f(x) = (3x - \sin(2x))/x$ for $-2 \leq x \leq 4$, $x \neq 0$. Locate the extreme values of f to four decimal places.

91. Let $f(x) = \cos^3(x - \pi/3)\sin(x)/x$ for $-2 \leq x \leq 2$, $x \neq 0$. Locate the extreme values of the continuous extension of f to four decimal places.

2.4 Infinite Limits and Asymptotes

Consider the graph of the equation $y = 1/x$ in Figure 1. Observe that y becomes arbitrarily large and positive as $x \to 0^+$. Pictorially, the graph approaches the positive y-axis. Also, as $x \to 0^-$, the ordinate y becomes negative without bound. Pictorially, the graph approaches the negative y-axis. Also, as $x \to \pm\infty$, the graph approaches the horizontal line $y = 0$. In this section, we develop a precise language for describing this type of behavior of graphs.

Infinite-Valued Limits

It is convenient to *extend* the notion of limit. Up to now, we have agreed that the limits $\lim_{x\to 0^+} 1/x$ and $\lim_{x\to 0^-} 1/x$ do not exist. For the limit $\lim_{x\to 0^+} 1/x$ to exist, there would have to be a real number ℓ to which the values of $1/x$ would tend as x approached 0 from the right. As the graph of $y = 1/x$ in Figure 1 shows, no such real number can exist. The graph is, in fact, unbounded. For similar reasons, $\lim_{x\to 0^-} 1/x$ cannot equal a real number. The definition of "limit" that we now give extends our existing definition so that ℓ is allowed to be infinite.

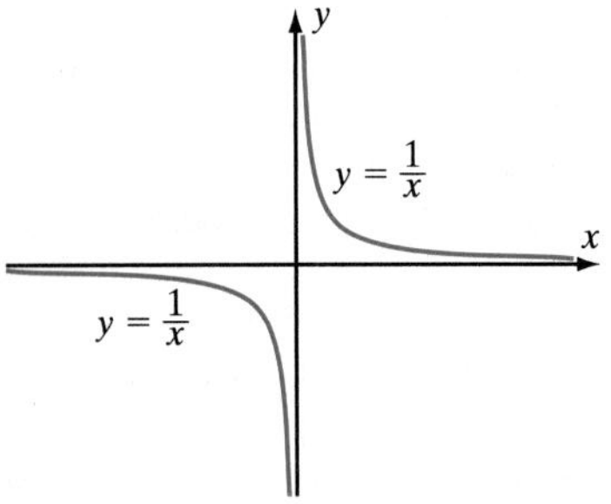

▲ Figure 1

DEFINITION Let f be a function that is defined on an open interval just to the left of c and also on an interval just to the right of c. We write

$$\lim_{x\to c} f(x) = +\infty$$

if $f(x)$ becomes arbitrarily large and positive as $x \to c$. We write

$$\lim_{x\to c} f(x) = -\infty$$

if $f(x)$ becomes arbitrarily large and negative without bound as $x \to c$.

More rigorously, $\lim_{x\to c} f(x) = +\infty$ if, for any $N > 0$, there is a $\delta > 0$ such that $f(x) > N$ whenever $0 < |x - c| < \delta$. Also, $\lim_{x\to c} f(x) = -\infty$ if for any $M < 0$ there is a $\delta > 0$ such that $f(x) < M$ whenever $0 < |x - c| < \delta$.

The one-sided limits $\lim_{x\to c^+} f(x) = \pm\infty$ and $\lim_{x\to c^-} f(x) = \pm\infty$ are defined similarly.

INSIGHT The symbol ∞ is used many times in this section. It does not represent a real number. It is merely a convenient piece of notation for certain unbounded behavior.

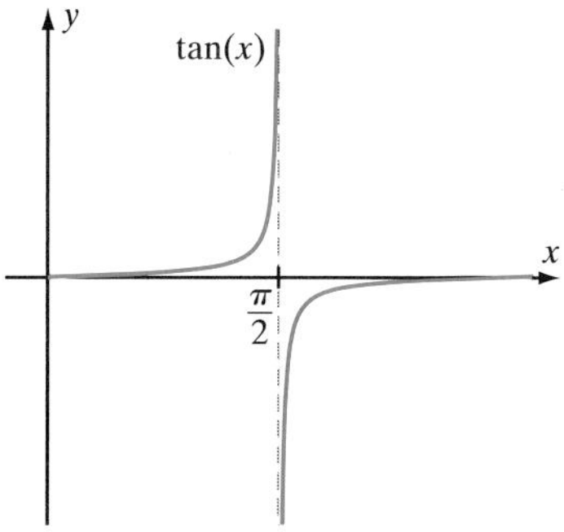

▲ Figure 2
$\lim_{x\to\pi/2^+} \tan(x) = -\infty$

The definitions of $\lim_{x\to c} f(x) = \pm\infty$ provide an analytic basis with which to investigate infinite limits. In practice, the graph of a function with an infinite limit will clearly indicate this behavior. Often we determine where a denominator of $f(x)$ vanishes in order to locate possible points c at which $\lim_{x\to c} f(x)$ might be $\pm\infty$. For example, the denominator $\cos(x)$ in $\sin(x)/\cos(x)$ vanishes at $x = \pi/2$ and

$$\lim_{x\to\pi/2^+} \tan(x) = \lim_{x\to\pi/2^+} \frac{\sin(x)}{\cos(x)} = -\infty,$$

as shown in Figure 2.

▶ **EXAMPLE 1** Analyze the following limits:

a. $\lim_{x\to -1} \dfrac{1}{(x+1)^2}$
b. $\lim_{x\to\pi/2^-} \sec(x)$
c. $\lim_{x\to\pi/2^+} \sec(x)$
d. $\lim_{x\to 2} -\dfrac{1}{(x-2)^4}$

Solution By inspection, we find:

a. $\lim_{x\to -1} \dfrac{1}{(x+1)^2} = +\infty$ (see Figure 3),
b. $\lim_{x\to\pi/2^-} \sec(x) = +\infty$ (see Figure 4),
c. $\lim_{x\to\pi/2^+} \sec(x) = -\infty$ (see Figure 5),
d. $\lim_{x\to 2} -\dfrac{1}{(x-2)^4} = -\infty$ (see Figure 6). ◀

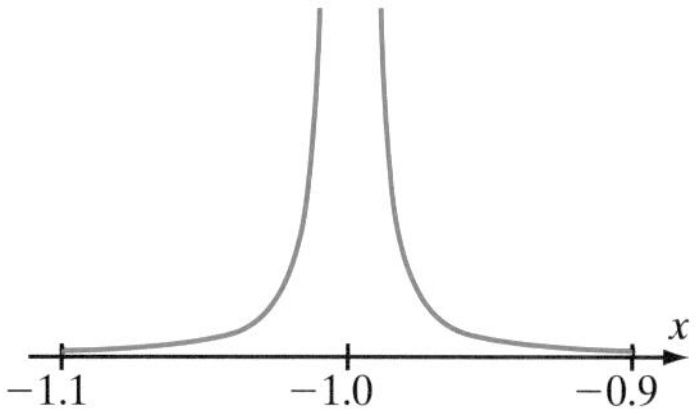

▲ Figure 3 The plot of $1/(x+1)^2$ in the viewing rectangle $[-1.1, -0.9] \times [0, 1000]$

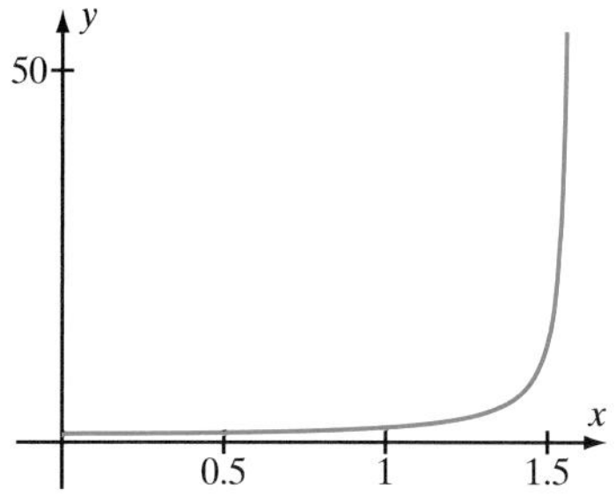

▲ Figure 4 The plot of $\sec(x)$ in the viewing rectangle $[0, \pi/2] \times [0, 50]$

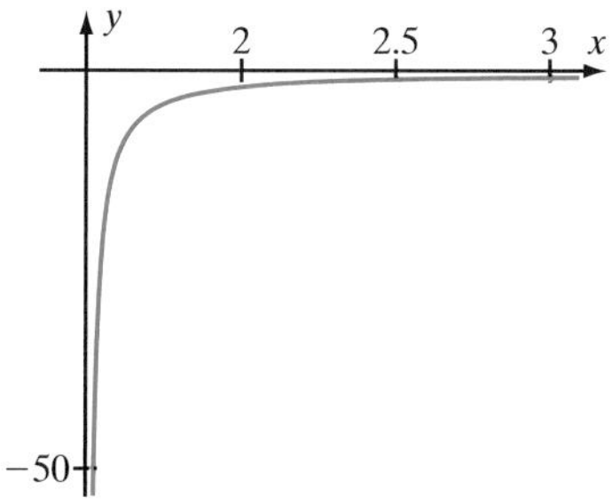

▲ Figure 5 The plot of $\sec(x)$ in the viewing rectangle $[\pi/2, \pi] \times [-50, 0]$

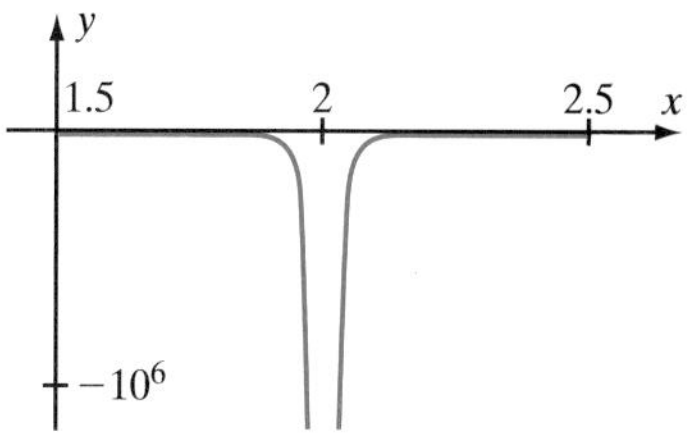

▲ Figure 6 The plot of $-1/(x-2)^4$ in the viewing rectangle $[1.5, 2.5] \times [-10^6, 0]$

Vertical Asymptotes The close connection between the figures and our perception of these limits suggests the following definition.

> **DEFINITION** If a function f has a one-sided or two-sided infinite limit as $x \to c$, then the line $x = c$ is said to be a vertical asymptote of f.

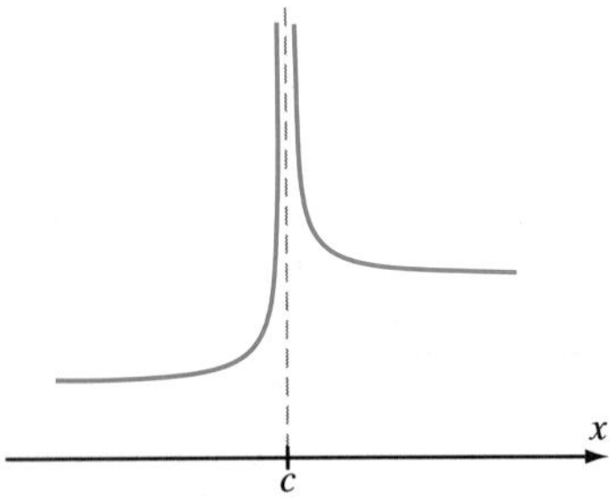

▲ Figure 7 $x = c$ is a vertical asymptote.

In this circumstance, the graph of f becomes arbitrarily close to the vertical line $x = c$, and the sketch of the graph of f should exhibit this line. In Figure 7, the vertical asymptote is rendered as a dotted line. When $\lim_{x \to c} f(x) = +\infty$, we sometimes say the asymptote is *upward vertical*. When $\lim_{x \to c} f(x) = -\infty$, we say the asymptote is *downward vertical*. In the case of one-sided limits, we have four possible behaviors: *upward vertical from the left, upward vertical from the right, downward vertical from the left,* and *downward vertical from the right.*

▶ **EXAMPLE 2** Discuss vertical asymptotes for the functions analyzed in Example 1.

Solution The graph of $y = 1/(x+1)^2$ has $x = -1$ as an upward vertical asymptote because the denominator becomes arbitrarily small and positive when x is close to, and on either side of, -1 (refer to Figure 3).

The function $y = \sec(x)$ has $x = \pi/2$ as an upward vertical asymptote from the left and a downward vertical asymptote from the right (refer to Figures 4 and 5). When x is just to the left of $\pi/2$, the denominator of $1/\cos(x)$ is small and positive; when x is just to the right of $\pi/2$, however, the denominator is small in absolute value and negative.

Finally, $y = -1/(x-2)^4$ has $x = 2$ as a two-sided downward vertical asymptote (refer to Figure 6). The denominator is small and positive when x is close to 2. ◀

▶ **EXAMPLE 3** Examine the function $g(x) = x^2/(x^2 - 3x - 4)$ for vertical asymptotes.

Solution The denominator factors as $(x+1)\cdot(x-4)$; hence it vanishes only at $x = -1$ and $x = 4$. At all other values of x, the function g will be continuous; in particular, it will have a finite limit and will certainly not have a vertical asymptote. The only candidates for vertical asymptotes are $x = -1$ and $x = 4$.

Write the function as

$$g(x) = \frac{x^2}{(x+1)(x-4)}.$$

When x is just to the left of -1, x^2 is positive (approximately 1), $(x+1)$ is negative and small in absolute value, and $(x-4)$ is negative. Therefore

$$\lim_{x \to 1^-} g(x) = \lim \frac{+}{(\text{small} -)\cdot(-)} = +\infty.$$

Similar reasoning shows that

$$\lim_{x \to 1^+} g(x) = \lim \frac{+}{(\text{small} +)\cdot(-)} = -\infty$$

(the only difference this time is that $(x+1) > 0$).

When x is just to the left of 4, x^2 is positive, $(x+1)$ is positive, and $(x-4)$ is negative and small in absolute value. Therefore

$$\lim_{x\to 4^-} g(x) = \lim \frac{+}{(+)\cdot(\text{small } -)} = -\infty.$$

Similar reasoning shows that

$$\lim_{x\to 4^+} g(x) = \lim \frac{+}{(+)\cdot(\text{small } +)} = +\infty.$$

We conclude that g has $x=-1$ and $x=4$ as vertical asymptotes. It should be noted that the asymptotes are one-sided. We obtain Figure 8 by using the preceding information and plotting some points. ◀

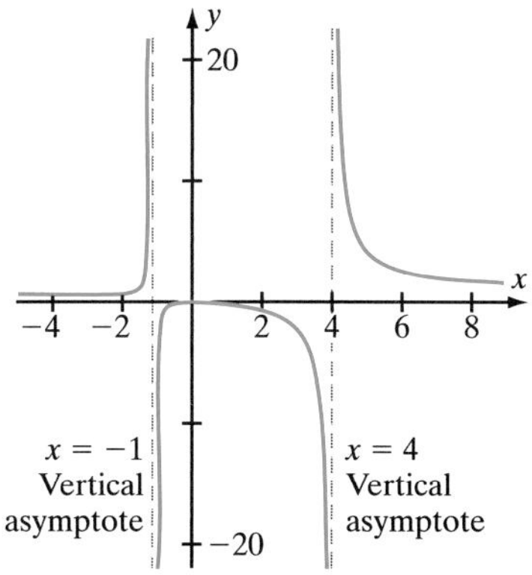

▲ Figure 8 Graph of $g(x) = x^2/(x^2-3x-4)$

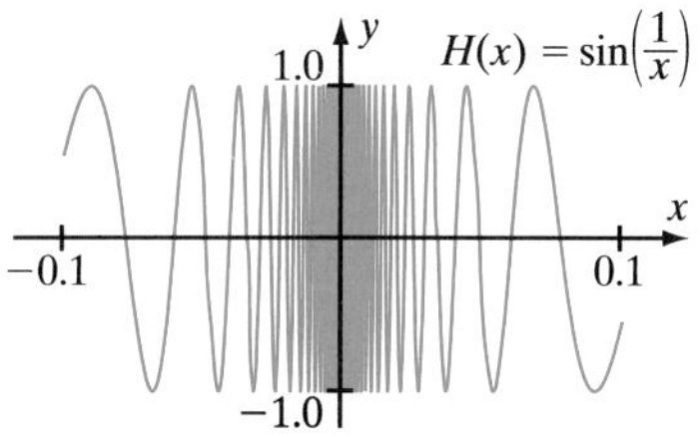

▲ Figure 9 Infinitely many oscillations and no asymptotes

INSIGHT The function g in Example 3 has a vertical asymptote at each point at which it is undefined. The same is true of the functions in Example 1. However, it is incorrect to think that a function must have a vertical asymptote at a point at which it is not defined. Consider the function $H(x) = \sin(1/x)$, which is undefined at $x=0$. Because $|H(x)|$ is never greater than 1, H certainly cannot have $+\infty$ or $-\infty$ as a limit. So $H(x)$ has no vertical asymptote at $x=0$. In fact, H has no limit of any kind at 0, as shown in Figure 9.

Limits at Infinity

Now we turn to the problem of identifying horizontal asymptotes. This requires another notion of limit.

DEFINITION Assume that the domain of f contains an interval of the form (a, ∞) and let α be a finite real number. We say that

$$\lim_{x\to +\infty} f(x) = \alpha$$

if $f(x)$ approaches α when x becomes arbitrarily large and positive.

Now assume that the domain of g contains an interval of the form $(-\infty, b)$. Let β be a finite real number. We say that

$$\lim_{x\to -\infty} g(x) = \beta$$

if $g(x)$ approaches β when x is negative and becomes arbitrarily large in absolute value.

More rigorously, $\lim_{x\to+\infty} f(x) = \alpha$ if, for any $\varepsilon > 0$, there is an N such that $|f(x)-\alpha| < \varepsilon$ whenever $N < x < \infty$. Also, $\lim_{x\to-\infty} g(x) = \beta$ if, for any $\varepsilon > 0$, there is an M such that $|g(x)-\beta| < \varepsilon$ whenever $-\infty < x < M$.

▶ **EXAMPLE 4** Examine the limits at infinity for the function

$$f(x) = \frac{3x^2-6x+8}{x^2+4x+6}.$$

Solution A good rule of thumb when studying the limits at infinity of a *rational function* (the quotient of polynomials) is to divide the numerator and denominator by the highest power of x that appears in the denominator. We therefore divide the numerator and denominator of f by x^2 (the highest power that appears in the denominator) to obtain

$$f(x) = \frac{3 - 6/x + 8/x^2}{1 + 4/x + 6/x^2}.$$

Now it is clear that as $x \to +\infty$ or $x \to -\infty$, all terms vanish except the 3 in the numerator and the 1 in the denominator. Thus we see that

$$\lim_{x \to +\infty} f(x) = \frac{3+0+0}{1+0+0} = 3 \quad \text{and} \quad \lim_{x \to -\infty} f(x) = \frac{3+0+0}{1+0+0} = 3.$$

This limiting behavior is visible in the plot of f in Figure 10, in which $\alpha = 3$. ◀

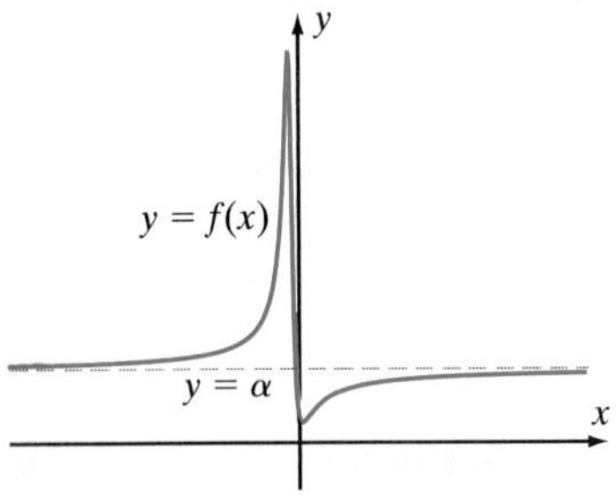

▲ Figure 10 $\lim_{x \to -\infty} f(x) = \alpha$ and $\lim_{x \to +\infty} f(x) = \alpha$

Horizontal Asymptotes

DEFINITION If

$$\lim_{x \to +\infty} f(x) = \alpha,$$

then we say that the line $y = \alpha$ is a horizontal asymptote of f. Similarly, if

$$\lim_{x \to -\infty} f(x) = \beta,$$

then we say that the line $y = \beta$ is a horizontal asymptote of f.

If the line L defined by $y = \alpha$ is a horizontal asymptote of f, then the graph of f becomes arbitrarily close to L. When we sketch the graph of f, we include the horizontal asymptote L as a dotted line (refer to Figure 10). It may happen that both $\lim_{x \to +\infty} f(x) = \alpha$ and $\lim_{x \to -\infty} f(x) = \alpha$. In this case, the line $y = \alpha$ is certainly a horizontal asymptote. However, the line would still be called an asymptote if only one of these limits existed. If $\lim_{x \to +\infty} f(x) = \alpha$, and $\lim_{x \to -\infty} f(x) = \beta$ with $\alpha \neq \beta$, then f has two distinct horizontal asymptotes.

In Figure 10, we would say that $y = \alpha$ is a right upper asymptote and a left lower asymptote. Of course, another graph could have a right lower asymptote or a left upper asymptote. The next example shows that a horizontal asymptote need not be either upper or lower.

▶ **EXAMPLE 5** Analyze the graph of $g(x) = \sin(2x)/(1 + x^2)$ for horizontal asymptotes.

Solution We divide the numerator and denominator by the highest power of x that appears in the denominator (namely, x^2):

$$g(x) = \frac{\sin(2x)/x^2}{(1/x^2) + 1}.$$

Because $|\sin(2x)| \leq 1$, it follows that $\sin(2x)/x^2 \to 0$ as $x \to \pm\infty$. Similarly, $1/x^2 \to 0$ as $x \to \pm\infty$. Using limit rules that are parallel to those from Section 2.2 for finite limits, we obtain

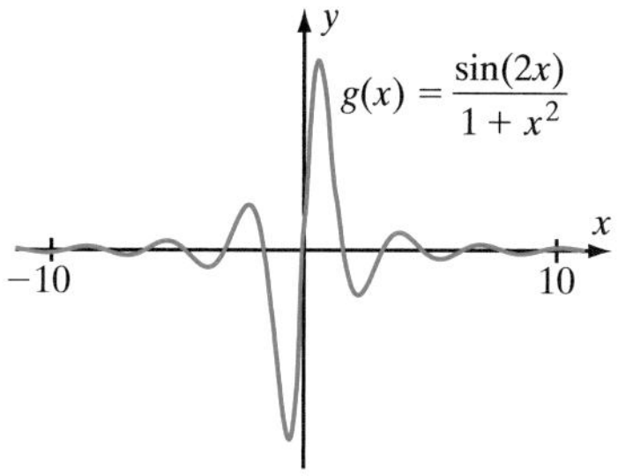

▲ Figure 11

$$\lim_{x \to +\infty} g(x) = \frac{0}{0+1} = 0 \quad \text{and} \quad \lim_{x \to -\infty} g(x) = \frac{0}{0+1} = 0.$$

These limits show that the graph of g has the line $y = 0$ as a left and a right horizontal asymptote. Because the graph of $g(x)$ oscillates about its asymptote, we do not speak of $y = 0$ as either an upper or a lower asymptote (see Figure 11).

▶ **EXAMPLE 6** Determine all horizontal and vertical asymptotes for the function

$$f(x) = \frac{7x^3 - 1}{2x^3 + 12x^2 + 18x}.$$

Solution First, if we divide the numerator and denominator by x^3, we obtain

$$f(x) = \frac{7 - 1/x^3}{2 + 12/x + 18/x^2}.$$

It is then clear that

$$\lim_{x \to +\infty} f(x) = \frac{7}{2} \quad \text{and} \quad \lim_{x \to -\infty} f(x) = \frac{7}{2}.$$

Therefore f has the line $y = 7/2$ as a horizontal asymptote.

For vertical asymptotes, we factor the denominator of $f(x)$:

$$f(x) = \frac{7x^3 - 1}{2x(x+3)^2}.$$

We notice that f is continuous except at 0 and -3. When x is close to and *on either side of* -3, $7x^3 - 1$ is negative, $2x$ is negative, and $(x+3)^2$ is positive and arbitrarily small. It follows that

$$\lim_{x \to -3} f(x) = \lim \frac{(-)}{(-) \cdot (\text{small} +)} = +\infty.$$

Therefore the line $x = -3$ is an upward vertical asymptote.

Also, when x is just to the left of 0, $7x^3 - 1$ is negative, $2x$ is negative and arbitrarily small in absolute value, and $(x+3)^2$ is positive. Therefore

$$\lim_{x \to 0^-} f(x) = \lim \frac{(-)}{(\text{small} -) \cdot (+)} = +\infty.$$

Similarly, when x is just to the right of 0, $7x^3 - 1$ is negative, $2x$ is positive and arbitrarily small, and $(x+3)^2$ is positive. Thus

$$\lim_{x \to 0^+} f(x) = \lim \frac{(-)}{(\text{small} +) \cdot (+)} = -\infty.$$

We conclude that the line $x = 0$ is a vertical asymptote for f. Notice that the asymptotic behavior on the left of 0 is different from that on the right of 0: On the left, the graph goes up; on the right, it goes down. The graph of f in Figure 12 exhibits the vertical and horizontal asymptotes that we have found. ◀

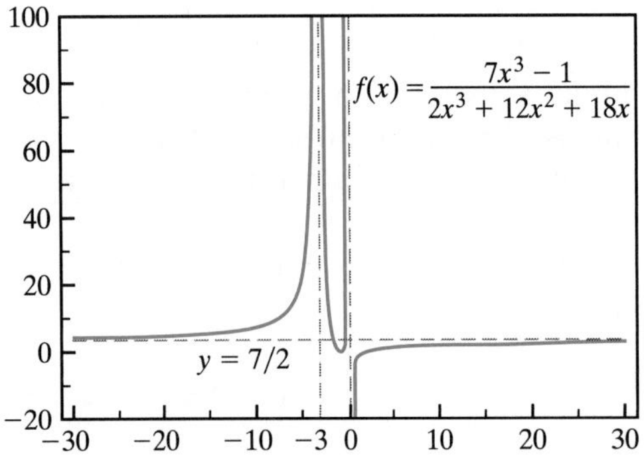

▲ Figure 12

INSIGHT When we seek horizontal asymptotes of a rational function

$$\frac{a_n x^n + a_{n-1}x^{n-1} + \cdots + a_1 x + a_0}{b_m x^m + b_{m-1}x^{m-1} + \cdots + b_1 x + b_0} \quad (a_n, b_m \neq 0),$$

it is useful to divide each term by x^m. The resulting expression can be easily analyzed for asymptotes. The result is shown in Table 1.

Comparison of Degrees	Horizontal Asymptotes
$n > m$	No asymptotes
$n = m$	$y = a_n/b_m$ as $x \to \pm\infty$
$n < m$	$y = 0$ as $x \to \pm\infty$

Table 1

QUICK QUIZ

1. True or false: If f and g are continuous functions, and $\lim_{x\to c} g(x) = 0$, then $y = c$ is a vertical asymptote for f/g.
2. True or false: If n and m are positive integers, and $(x^n + 1)/(x^m + 2)$ does not have a horizontal asymptote, then $n > m$.
3. Does the function $f(x) = \sin(x)$ have either horizontal or vertical asymptotes?
4. What are the asymptotes for $g(x) = (x^2 - 5)/(2x^2 - 8)$?

Answers

1. False 2. True 3. Neither 4. $x = -2$, $x = 2$, $y = 1/2$

EXERCISES

Problems for Practice

► In Exercises 1–20, determine whether the given limit exists. If it does exist, then compute it. ◄

1. $\lim_{x\to 6} \dfrac{3}{(x-6)^2}$

2. $\lim_{x\to -\infty} \dfrac{4x}{4x-7}$

3. $\lim_{x\to +\infty} \dfrac{x^4 + 3x - 92}{x^2 - 7x^2 + 44}$

4. $\lim_{x\to -1^+} (x+1)^{-1}$

5. $\lim_{x\to +\infty} \dfrac{x + \sqrt{x}}{x^2 - \sqrt{x}}$

6. $\lim_{x\to \pi} \dfrac{3}{\sin(x)}$

7. $\lim_{x\to\infty} \dfrac{x+\cos(x)}{x-\sin(x)}$
8. $\lim_{x\to+\infty} \dfrac{\sqrt{x}}{x+1}$
9. $\lim_{x\to 1^+} \dfrac{1}{\sqrt{x}-1}$
10. $\lim_{x\to 0+} \csc(x)$
11. $\lim_{x\to+\infty} \dfrac{x^2-4x+9}{3x^2-8x+18}$
12. $\lim_{x\to 3^-} \dfrac{\cos(x)}{x-3}$
13. $\lim_{x\to 0^+} \csc^2(x)$
14. $\lim_{x\to+\infty} \dfrac{x^2+5x-7}{3x^2+4}$
15. $\lim_{x\to 0} \dfrac{3}{\sqrt{|x|}}$
16. $\lim_{x\to+\infty} \dfrac{\sqrt{x}-5}{\sqrt{x}+4}$
17. $\lim_{x\to 0^-} \cot(x)$
18. $\lim_{x\to-\infty} \dfrac{x^{4/3}}{x^2+\sin(x)}$
19. $\lim_{x\to 2^-} \tan(\pi/x)$
20. $\lim_{x\to+\infty} \dfrac{\sin(x)}{x}$

In each of Exercises 21–34, a function f is given. Find all horizontal and vertical asymptotes of the graph of $y=f(x)$. Plot several points and sketch the graph.

21. $f(x) = \dfrac{x}{x-7}$
22. $f(x) = \dfrac{x^2-1}{x^2-4}$
23. $f(x) = \dfrac{\sqrt{|x|}}{x}$
24. $f(x) = \dfrac{\sqrt{|x|}}{x+3}$
25. $f(x) = \dfrac{(x-1)^{2/3}}{(x^2+8)^{1/3}}$
26. $f(x) = \dfrac{x}{x^2+1}$
27. $f(x) = \dfrac{1}{x^2+1}$
28. $f(x) = \dfrac{x^2}{x^2-7x+12}$
29. $f(x) = \dfrac{x^{1/3}}{|x|}$
30. $f(x) = \dfrac{x}{(x+1)(x+4)}$
31. $f(x) = \dfrac{(x+4)^{-1/3}}{x\cdot(x+1)}$
32. $f(x) = \dfrac{x^3+1}{x^3-1}$
33. $f(x) = \cot(x),\ -\dfrac{\pi}{2} < x < \dfrac{\pi}{2}$
34. $f(x) = \sec(x),\ -\dfrac{\pi}{2} < x < \dfrac{\pi}{2}$

Further Theory and Practice

35. Formulate and prove a version of Theorem 1 (the Limit Uniqueness Theorem) from Section 2.2 for $\lim_{x\to+\infty} f(x)$ and $\lim_{x\to-\infty} f(x)$.
36. Formulate and prove a version of Theorem 2 (the sum, difference, product, quotient, and scalar multiplication rules for limits) from Section 2.2 for the limits $\lim_{x\to+\infty} f(x)$ and $\lim_{x\to-\infty} f(x)$.
37. Formulate and prove a version of the Pinching Theorem from Section 2.2 for the limits $\lim_{x\to+\infty} f(x)$ and $\lim_{x\to-\infty} f(x)$.
38. We do not attempt to formulate sum, difference, product, or quotient rules for infinite valued limits (as in the first definition of "limit" in this section) because there are serious ambiguities in doing arithmetic with the symbols $+\infty$ and $-\infty$. Give examples to illustrate these ambiguities.
39. Prove that, if $\lim_{x\to c} F(x) = \ell \in \mathbb{R}$, and if $\lim_{x\to c} G(x) = +\infty$, then $\lim_{x\to c}\big(F(x)+G(x)\big) = +\infty$. Similarly, if $\lim_{x\to c} H(x) = -\infty$, then prove that $\lim_{x\to c}\big(F(x)+H(x)\big) = -\infty$.

In each of Exercises 40–45, a function f is given. Find all horizontal and vertical asymptotes of the graph of $y=f(x)$. Plot several points, and sketch the graph.

40. $f(x) = \sin(x)/x$
41. $f(x) = x/\sin(x)$
42. $f(x) = |x|^{1/2}/\sin(x)$
43. $f(x) = \cos^2(x)/x$
44. $f(x) = |x|/\sin(x)$
45. $f(x) = 1/|\sin(x)|$

In Exercises 46–48, graph the function f. Refer to the definition of "vertical asymptote" to decide if $x=0$ is a vertical asymptote of the graph of f.

46. $f(x) = \sin(1/x)$
47. $f(x) = |\sin(1/x)/x|$
48. $f(x) = \big(1.1+\sin(1/x)\big)/x$

49. Let $f(x) = x/10^{100}$ and $g(x) = 10^{100}/x$. What are $f(10^{80})$ and $g(10^{80})$? Now compute $\lim_{x\to\infty} f(x)$ and $\lim_{x\to\infty} g(x)$. Is it significant that $f(x)$ is much smaller than $g(x)$ for all $0<x<10^{80}$? (The number 10^{80} was chosen simply because it is large. It is on the order of the number of protons in the universe.)

Exercises 50–52 concern physical laws that entail natural one-sided limits and vertical asymptotes. In each exercise, sketch the graph and indicate the vertical asymptote.

50. $m = m_0/\sqrt{1-(v/c)^2}$ (Einstein's Relativistic Mass Law, which relates the mass m of an object at velocity v to its rest mass m_0 and to the speed of light c.)
51. $P = RT/(V-b) - a/V^2$ (This is Van der Waals's equation relating the pressure P and volume V of 1 mole of a gas at constant temperature T. The constant R is

the universal gas constant, and the constant a depends on the particular gas under consideration. The constant b is the volume occupied by the gas molecules themselves.)

52. $H = 2gR^2/(2gR - v_0^2) - R$ (This is the equation for the height H a missile rises above Earth's surface as a function of its initial velocity v_0; g is the acceleration due to Earth's gravity, and R is Earth's radius.)

53. If f is a continuous function on a closed interval $[a, b]$?, can the graph of f have a vertical asymptote $x = c$ where c is a value in $[a, b]$? Explain your reasoning.

54. In an enzyme reaction, the reaction rate V is given in terms of substrate amount S by

$$V = \frac{V_* S}{K + S}.$$

Identify V_* as a limit as $S \to \infty$. Although V does not have a maximum value, V_* is called the *maximum velocity* (where *velocity* refers to reaction rate). Explain why.

In Exercises 55–58, you are asked to investigate the notion of a *skew asymptote*, which is also called a *slant asymptote* and an *oblique asymptote*.

55. We say that the graph of a function f has the line $y = mx + b$ as its skew asymptote if either

$$\lim_{x \to +\infty} \big(f(x) - (mx + b)\big) = 0$$

or

$$\lim_{x \to -\infty} \big(f(x) - (mx + b)\big) = 0$$

a. If $m = 0$, prove that this notion of skew asymptote coincides with the notion of horizontal asymptote.

b. If $f(x) = p(x)/q(x)$ is a quotient of polynomials (a rational function), then prove that a *necessary condition* for the graph of f to have $mx + b$ as a skew asymptote, with $m \neq 0$, is that the degree of p is precisely one greater than the degree of q.

c. If the degree of p exceeds the degree of q by precisely two, then we could say that the graph of f is asymptotic to a parabola. Explain this idea in more detail, and give a precise definition.

56. Find all horizontal, vertical, and skew asymptotes of the graphs of the following functions. Then plot some points and give a sketch.

a. $f(x) = \dfrac{x^2}{x - 5}$

b. $g(x) = \dfrac{x^{3/2}}{x^{1/2} + 7}$

c. $k(x) = \dfrac{(x^2 + 4x + 3)^{4/3}}{(x - 9)^{5/3}}$

d. $m(x) = (x + 5) \cdot \dfrac{\sin(1/x)}{(1/x)}$

57. Find all parabolic asymptotes for the functions $F(x) = x^3/(x + 3)$, $G(x) = x^4/(x^2 + 5x)$, and $H(x) = (x^{7/3} + x^2 + x^{1/3})/(x^{1/3} + 1)$. (Refer to Exercise 55, part c, for terminology.) Sketch the graphs of H and its parabolic asymptote.

58. Consider the following three attempts at determining the skew asymptote of the function

$$f(x) = \frac{x^2 + 3x + 2}{x - 1}.$$

Determine which argument is correct and explain what is wrong with the others.

a. Because

$$\begin{aligned} f(x) &= \frac{x^2 + 3x + 2}{x - 1} \\ &= \frac{x^2}{x}\frac{(1 + 3/x + 2/x^2)}{(1 - 1/x)} \\ &= x\frac{(1 + 3/x + 2/x^2)}{(1 - 1/x)} \end{aligned}$$

and because the fraction tends to 1 as $x \to \infty$, the skew asymptote must be $x \cdot 1$.

b. Because

$$\begin{aligned} f(x) &= \frac{x^2 + 3x + 2}{x - 1} = \frac{x}{x}\frac{(x + 3 + 2/x)}{(1 - 1/x)} \\ &= \frac{(x + 3 + 2/x)}{(1 - 1/x)} \end{aligned}$$

and because the numerator tends to $x + 3$ and the denominator to 1, the skew asymptote must be $x + 3$.

c. Because

$$f(x) = \frac{x^2 + 3x + 2}{x - 1} = x + 4 + \frac{6}{x - 1},$$

the skew asymptote must be $x + 4$.

Calculator/Computer Exercises

Sometimes the asymptotic behavior of a graph takes place very slowly. In Exercises 59 and 60, you are asked to perform a numerical calculation to determine how slowly the limit is being achieved.

59. We know that

$$\lim_{x \to +\infty} \frac{\sqrt{x}}{x + 1} = 0.$$

How large must x be to guarantee that the function is smallert than 0.001?

60. We know that

$$\lim_{x \to 0-} \frac{|x|^{1/4} + 6}{x^{1/3}} = -\infty.$$

How close must x be to 0 to guarantee that $(|x|^{1/4} + 6)/x^{1/3} < -10^8$?

▶ In Exercises 61–63, the function f has one or more horizontal asymptotes. Plot f and its horizontal asymptote(s). Specify another window in which f and its right horizontal asymptote appear to nearly coincide. Repeat for the left horizontal asymptote if it is different. ◀

61. $f(x) = \dfrac{3x^2 + x\cos(x)}{x^2+1}$

62. $f(x) = \dfrac{2x^2 + \sin(x)}{|x|^3+1}$

63. $f(x) = \dfrac{\sqrt{x^4 + x\sin(x)}}{x^2+2x+2}$

▶ In Exercises 64–66, the function f has one or more vertical asymptotes. Plot f and its vertical asymptote(s). ◀

64. $f(x) = \dfrac{x^2+x+3}{x^3-x-3}$

65. $f(x) = \dfrac{4+\cos(x)}{\sin(x)+\cos(2x)}, \ 0 \le x \le 3$

66. $f(x) = \dfrac{x^4+1}{x^3-3x+1}$

2.5 Limits of Sequences

This section addresses limits of infinite sequences of real numbers. Recall that an infinite sequence f of real numbers is a function with domain $\mathbb{Z}^+$and range $\mathbb{R}$ (as discussed in Section 1.4). Following the common convention for sequences, we will write f_n instead of $f(n)$ and refer to the sequence as $\{f_n\}_{n=1}^{\infty}$ or just $\{f_n\}$ for short. Our interest here is whether the numbers f_n tend to some limit as n tends to infinity. Before giving a precise definition of "limit," let's apply our intuition to some examples.

▶ **EXAMPLE 1** We suppose that a quantity of radioactive material is decaying. At the beginning of each week, there is half as much material as there was at the beginning of the previous week. The initial quantity is 5 g. Use sequence notation to express the amount of material at the beginning of the nth week. Does this sequence have a limit?

Solution The amount a_1 of material at the start of the first week is 5 g. The amount a_2 of material at the start of the second week is 5/2 g. The amount a_3 of material at the start of the third week is 5/4 g. Similarly, the amounts at the beginnings of the fourth and fifth weeks are $a_4 = 5/8$ g and $a_5 = 5/16$ g. In general, the amount a_n of material at the start of the nth week is given by $a_n = 5\cdot(1/2)^{n-1}$. The magnitude of each term is half of its predecessor. Our intuitive conclusion is that the amount of radioactive material tends to 0 as the number of weeks tends to infinity. ◀

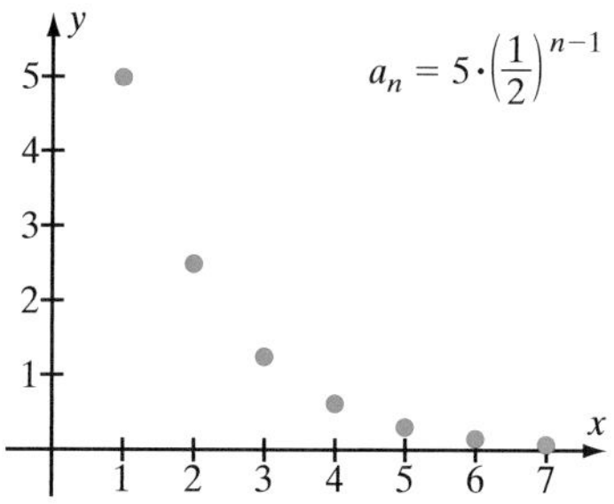

▲ Figure 1

INSIGHT A sequence $\{a_n\}_{n=1}^{\infty}$ can be graphed in the same way as any other real-valued function. We simply plot the points $(1, a_1)$, $(2, a_2)$, $(3, a_3)$, and so on. Figure 1 shows the points $(1, 5)$, $(2, 5/2)$, $(3, 5/4)$, $(4, 5/8)$, $(5, 5/16)$, $(6, 5/32)$, and $(7, 5/64)$ of the sequence $\{a_n\}_{n=1}^{\infty}$ from Example 1. Each point is half the distance to the x-axis of the preceding point. The graph confirms our intuitive conclusion that a_n tends to 0 as n tends to infinity.

▶ **EXAMPLE 2** Set $a_j = (-1)^j$. Does the sequence $\{a_j\}_{j=1}^{\infty}$ tend to a limit as j tends to infinity?.

Solution By calculating $a_j = (-1)^j$ for $j = 1, 2, 3, 4, \ldots$, we may write out this sequence as $-1, 1, -1, 1, \ldots$. The sequence does not seem to tend to any limit: Half of the time the value is 1, and the other half, the value is -1. The sequence does not become and remain close to a single value. Therefore we say that it has no limit (see Figure 2). ◀

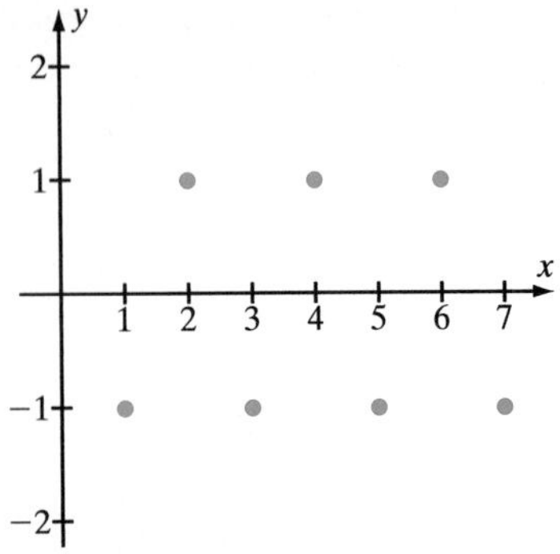

▲ Figure 2

INSIGHT In Example 1, we used n as the index variable of a sequence, whereas in Example 2, we used j. There is no significance to the letter that is used because the index variable merely serves as a placeholder for the integers 1, 2, 3, ... The letters i, j, k, m, and n are all commonly used as index variables of sequences.

A Precise Discussion of Convergence and Divergence

DEFINITION Suppose $\{a_n\}_{n=1}^{\infty}$ is a sequence, and ℓ is a real number. We say the sequence has limit ℓ (or converges to ℓ) if, for each $\varepsilon > 0$, there is an integer N such that $|a_n - \ell| < \varepsilon$ for all $n \geq N$ (see Figure 3).

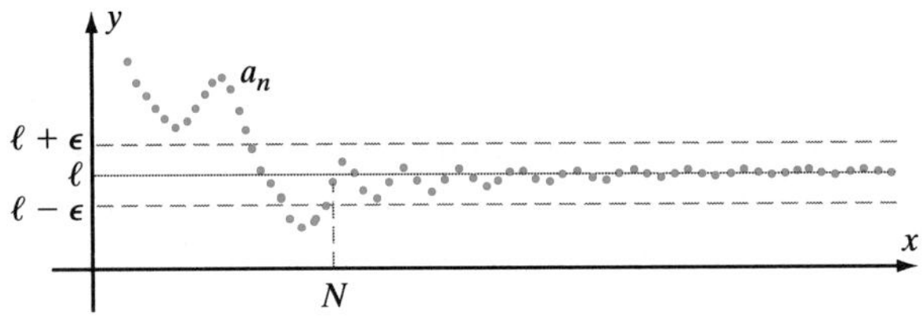

▲ Figure 3 $|a_n - \ell| < \varepsilon$ for $n \geq N$

When the sequence $\{a_n\}_{n=1}^{\infty}$ has limit ℓ, we write

$$\lim_{n \to \infty} a_n = \ell,$$

or

$$a_n \to \ell \quad \text{as} \quad n \to \infty.$$

A sequence that has a limit is said to be convergent. If a sequence does not converge, then we say that it diverges, and we call it a divergent sequence.

INSIGHT In Section 2.4, we discussed the limit at infinity, $\lim_{x \to \infty} f(x)$, for a function f of a continuous real variable x. The definition of $\lim_{n \to \infty} a_n$ is very similar. The difference is that a_n is defined for only positive integer values of n.

▶ **EXAMPLE 3** Discuss convergence for the sequence 1, 1/2, 1/3, 1/4, ...

Solution Let $a_n = 1/n$ for all n. The terms a_n get progressively closer to 0. Let us now verify that $\lim_{n \to \infty} a_n = 0$. We set $\ell = 0$ and suppose that ε is a given positive number. We reason backward to select N. To guarantee that $|a_n - \ell| < \varepsilon$, we need. $|1/n - 0| < \varepsilon$. This simply means that $1/n < \varepsilon$. This last inequality is satisfied if $n > 1/\varepsilon$. Therefore we select N to be the first integer that is greater than $1/\varepsilon$. If $n \geq N$, then $n > 1/\varepsilon$ is true. Therefore $1/n < \varepsilon$ and $|a_n - \ell| < \varepsilon$. In conclusion, $\lim_{n\to\infty} a_n = 0$. ◀

The Tail of a Sequence

Intuitively, we say that $\{a_n\}_{n=1}^{\infty}$ converges to ℓ if a_n is as close as we please to ℓ when n is large enough. A crucial feature of this idea is that convergence depends only on what $\{a_n\}_{n=1}^{\infty}$ does when n is large. The values of the first ten thousand (or first million, or first billion, etc.) terms are irrelevant. If we fix any N, then we refer to the terms

$$a_{N+1}, a_{N+2}, a_{N+3}, \ldots$$

as the *tail end* (or simply the *tail*) of the sequence $\{a_n\}$. Of course, since N is an arbitrary positive integer, there are infinitely many tail ends of a sequence.

▶ **EXAMPLE 4** Find the limit of the sequence defined by

$$a_n = \begin{cases} 0 & \text{if } 1 \leq n \leq 100000 \\ 10 & \text{if } n \geq 100001 \end{cases}.$$

Solution The sequence $\{a_n\}_{n=1}^{\infty}$ converges to 10. To see this, pick $\varepsilon > 0$. Let $N = 100001$. If $n \geq N$, then $a_n = 10$ and $|a_n - 10| = |10 - 10| = 0 < \varepsilon$, as desired. ◀

Some Special Sequences

A number of special sequences occur repeatedly in our work. We now collect several of these sequences for easy reference. Some of these sequences diverge because their terms grow without bound. It is convenient to extend the limit notation to include these types of divergent sequences.

DEFINITION If, for any $M > 0$, there is an N such that $a_n > M$ for all indices $n \geq N$, then we write

$$\lim_{n\to\infty} a_n = \infty$$

and say that the sequence $\{a_n\}$ tends to infinity. If for any $M < 0$, there is an N such that $a_n < M$ for all indices $n \geq N$, then we write

$$\lim_{n\to\infty} a_n = -\infty$$

and say that the sequence $\{a_n\}$ tends to minus infinity.

▶ **EXAMPLE 5** Suppose that $\{a_n\}$ is a sequence of nonzero numbers such that $\lim_{n\to\infty} a_n = \infty$. Show that $\lim_{n\to\infty} \dfrac{1}{a_n} = 0$.

Solution It is evident that, as the terms of a sequence get larger, their reciprocals get smaller. It is not hard to verify this deduction using the precise definitions that have been given. Given $\varepsilon > 0$, let M be an integer that satisfies $M > 1/\varepsilon$.

By hypothesis, there is an integer N such that $a_n > M$ for all $n \geq N$. For these values of n, we have

$$\left|\frac{1}{a_n} - 0\right| = \frac{1}{a_n} < \frac{1}{M} < \varepsilon,$$

which establishes the asserted limit. ◄

THEOREM 1 Let α and r be any real numbers. Assume that a and p are positive. Then:

a. $\lim_{n\to\infty} \alpha = \alpha$.
b. $\lim_{n\to\infty} n^p = \infty$.
c. $\lim_{n\to\infty} 1/n^p = 0$.
d. *If* $|r| > 1$, then the sequence $\{r^n\}_{n=1}^{\infty}$ diverges, and $\lim_{n\to\infty} |r|^n = \infty$.
e. *If* $|r| < 1$, then $\lim_{n\to\infty} r^n = 0$.
f. $\lim_{n\to\infty} a^{1/n} = 1$.

Each conclusion of Theorem 1 is intuitively clear. For example, part a tells us that a constant sequence converges to that constant. Part b states that, as n grows without bound, so does any positive power of n. Part d states that when a number greater than 1 is raised to successively higher exponents, then the resulting powers grow without bound. Parts c and e follow from parts b and d, using Example 5. The next example illustrates the use of Theorem 1.

► **EXAMPLE 6** Discuss the convergence of $\{1/\sqrt{n}\}, \{\sqrt[100]{n}\}, \{(-99/100)^n\}$, and $\{(101/100)^n\}$.

Solution Theorem 1 can be applied to each of these sequences. The sequence $\{1/\sqrt{n}\}$ converges to 0 by Theorem 1c with $p = 1/2$. We write $\sqrt[100]{n} = n^{1/100}$ and conclude that $\{\sqrt[100]{n}\}$ diverges to infinity by using Theorem 1b with $p = 1/100$. The sequence $\{(-99/100)^n\}$ can be written as $\{r^n\}$ with $r = -99/100$. Because $|r| = 99/100 < 1$, we conclude that the sequence $\{(-99/100)^n\}$ converges to 0 by Theorem 1e. Finally, we write $\{(101/100)^n\}$ as $\{r^n\}$ with $r = 101/100 > 1$. This sequence tends to infinity by Theorem 1d. ◄

INSIGHT A numerical investigation of the convergence of a sequence requires some caution. The evaluations $(101/100)^{50} = 1.6446\ldots$ and $(101/100)^{150} = 4.4484\ldots$ hardly suggest that the sequence $\{(101/100)^n\}$ tends to infinity. As the table indicates, it is necessary to calculate $(101/100)^n$ for rather large values of n to obtain convincing numerical evidence.

n	200	500	1000	10000
$(101/100)^n$	7.3160	1000	20959	1.6358×10^{43}

Limit Theorems

There are a few rules that enable us to easily evaluate the limits of many common sequences. The limit rules for the sequences that follow are similar to the limit rules for the functions in Section 2.2.

THEOREM 2 Suppose that $\{a_n\}_{n=1}^{\infty}$ and $\{b_n\}_{n=1}^{\infty}$ are convergent sequences. Then

a. $\lim_{n\to\infty}(a_n + b_n) = \lim_{n\to\infty} a_n + \lim_{n\to\infty} b_n$.
b. $\lim_{n\to\infty}(a_n - b_n) = \lim_{n\to\infty} a_n - \lim_{n\to\infty} b_n$.
c. $\lim_{n\to\infty}(a_n \cdot b_n) = (\lim_{n\to\infty} a_n) \cdot (\lim_{n\to\infty} b_n)$.
d. $\lim_{n\to\infty} \dfrac{a_n}{b_n} = \dfrac{\lim_{n\to\infty} a_n}{\lim_{n\to\infty} b_n}$, provided that $\lim_{n\to\infty} b_n \neq 0$.
e. $\lim_{n\to\infty} \alpha \cdot a_n = \alpha \cdot \lim_{n\to\infty} a_n$ for any real number α.
f. $\lim_{n\to\infty} a_n$ is unique.

From now on, we will usually calculate the limit of a sequence by using Theorem 2, together with the known limits of several basic sequences, such as those given in Theorem 1. Although we will not explicitly mention Theorem 2f, we will use it frequently, because it shows that if we calculate a limit by any particular method, the answer will be the same as the answer obtained by a different method.

▶ **EXAMPLE 7** Use Theorem 2 to analyze the limit

$$\lim_{n\to\infty} \frac{n^3 + 4n - 6}{3n^3 + 2n}.$$

Solution The numerator and the denominator of this limit become ever larger with n. Therefore $\{n^3 + 4n - 6\}$ and $\{3n^3 + 2n\}$ are not convergent sequences, and Theorem 2d cannot be applied directly. Before calling on Theorem 2, we divide the numerator and denominator by the largest power of n that appears (namely, n^3):

$$\begin{aligned}
\lim_{n\to\infty} \frac{n^3 + 4n - 6}{3n^3 + 2n} &= \lim_{n\to\infty} \frac{1 + 4n^{-2} - 6n^{-3}}{3 + 2n^{-2}} \\
&= \frac{\lim_{n\to\infty}(1 + 4n^{-2} - 6n^{-3})}{\lim_{n\to\infty}(3 + 2n^{-2})} && \text{Theorem 2d} \\
&= \frac{\lim_{n\to\infty} 1 + 4 \cdot \lim_{n\to\infty} n^{-2} - 6 \cdot \lim_{n\to\infty} n^{-3}}{\lim_{n\to\infty} 3 + 2 \lim_{n\to\infty} n^{-2}} && \text{Theorem 2a, 2b and 2e} \\
&= \frac{1 - 4 \cdot 0 + 6.0}{3 + 2 \cdot 0} && \text{Theorem 1a} \\
&= \frac{1}{3}. && \text{Simplify}
\end{aligned}$$

◀

INSIGHT If the graph of a function f has a horizontal asymptote $y = \ell$ as $x \to \infty$, then the sequence $\{a_n\}$ defined by $a_n = f(n)$ satisfies

$$\lim_{n\to\infty} a_n = \lim_{n\to\infty} f(n) = \lim_{x\to\infty} f(x) = \ell.$$

In other words, $\{a_n\}$ has limit ℓ. We know from the Insight following Example 6 of Section 2.4 that $f(x) = \dfrac{x^3 + 4x - 6}{3x^3 + 2x}$ has $y = 1/3$ as a horizontal asymptote. Therefore $\lim_{n\to\infty} \dfrac{n^3 + 4n - 6}{3n^3 + 2n} = \dfrac{1}{3}$ (see Figure 4).

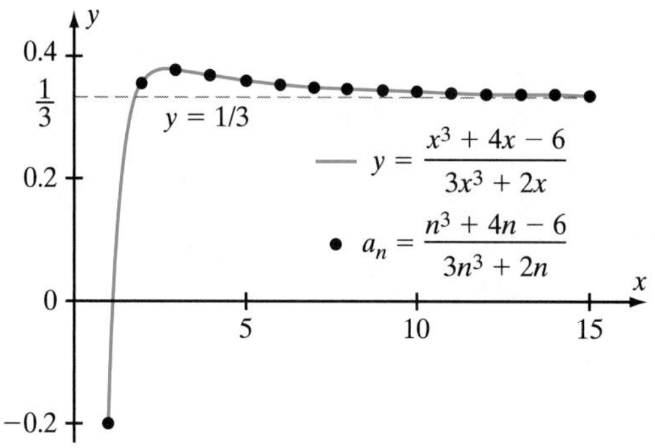

▲ Figure 4

▶ **EXAMPLE 8** Calculate $\lim_{n\to\infty}(2^n + 3^n)/4^n$.

Solution We begin by dividing the numerator and denominator by the largest term, 4^n. Using Theorem 2a, we obtain

$$\lim_{n\to\infty}\frac{2^n + 3^n}{4^n} = \lim_{n\to\infty}\frac{(2/4)^n + (3/4)^n}{(4/4)^n}$$
$$= \lim_{n\to\infty}\left(\left(\frac{1}{2}\right)^n + \left(\frac{3}{4}\right)^n\right) = \lim_{n\to\infty}\left(\frac{1}{2}\right)^n + \lim_{n\to\infty}\left(\frac{3}{4}\right)^n.$$

Each of these limits is 0 by Theorem 1e. The requested limit is therefore 0 as well. ◀

Our next theorem is the analogue for sequences of Theorem 6 from Section 2.2.

THEOREM 3 **(Pinching Theorem for Sequences).** If $\lim_{n\to\infty} a_n = \lim_{n\to\infty} b_n = \ell$ and $a_n \le c_n \le b_n$ for all n, then $\lim_{n\to\infty} c_n = \ell$.

▶ **EXAMPLE 9** Evaluate $\lim_{n\to\infty}\sin(n)/n^2$.

Solution Let $a_n = -1/n^2$, $b_n = 1/n^2$, and $c_n = \sin(n)/n^2$. Theorem 1c with $p = 2$ tells us that $\lim_{n\to\infty} b_n = 0$. Theorem 2e with $\alpha = -1$ then tells us that $\lim_{n\to\infty} a_n = -\lim_{n\to\infty} b_n = 0$. Because $-1 \le \sin(n) \le 1$, we see that $a_n \le c_n \le b_n$ (see Figure 5). It follows from the Pinching Theorem that $c_n = \sin(n)/n^2$ has the same limit, 0, when n tends to infinity. ◀

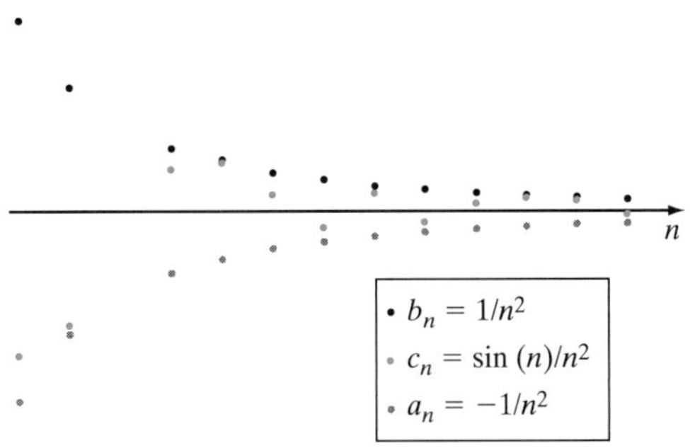

▲ Figure 5

Geometric Series

Suppose that an arrow is shot at a target 10 meters away. (That distance is 1 dkm.) Before hitting its target, the arrow must first travel 1/2 the distance, followed by 1/2 the remaining distance (or 1/4 dkm), and then 1/2 the remaining distance (or 1/8 dkm), and so on (see Figure 6). By this reasoning, we conclude that the distance the arrow travels is $1/2 + 1/4 + 1/8 + 1/16 + \cdots$ (without end) dkm. On the other hand, when it hits the target, the distance the arrow has traveled is 1 dkm. In other words, it appears that

$$\frac{1}{2} + \frac{1}{4} + \frac{1}{8} + \frac{1}{16} + \cdots = 1.$$

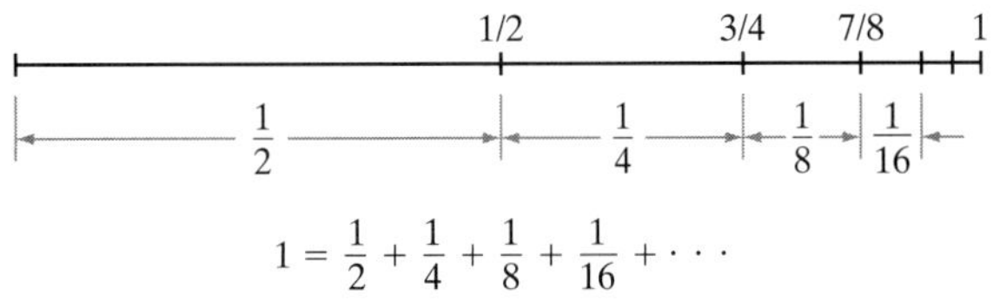

▲ Figure 6

Before we can make sense of such an equation, we must give the left side a precise meaning. We certainly do *not* assert that infinitely many numbers are added up in the usual fashion on the left. After all, how could such an infinite sum be accomplished? What we can do is add the N numbers

$$\frac{1}{2} + \frac{1}{4} + \frac{1}{8} + \frac{1}{16} + \cdots + \frac{1}{2^N}$$

for any positive integer N, no matter how large. In other words, we can form the infinite sequence $\{s_N\}_{N=1}^{\infty}$ where $s_N = 1/2 + 1/4 + 1/8 + 1/16 + \cdots + 1/2^N$. If we can show that $\lim_{N\to\infty} s_N$ exists, then we can define $1/2 + 1/4 + 1/8 + 1/16 + \cdots$ to be this limit:

$$\frac{1}{2} + \frac{1}{4} + \frac{1}{8} + \frac{1}{16} + \cdots = \lim_{N\to\infty} s_N = \lim_{N\to\infty}\left(\frac{1}{2} + \frac{1}{4} + \frac{1}{8} + \frac{1}{16} + \cdots + \frac{1}{2^N}\right). \qquad (2.5.1)$$

Our discussion suggests that this limit exists and is equal to 1. The next theorem will allow us to verify our conclusion (in Example 10).

THEOREM 4 Suppose that $r \neq 1$.

a. For each positive integer N

$$1 + r + r^2 + \cdots + r^{N-1} = \frac{r^N - 1}{r - 1}. \tag{2.5.2}$$

b. If $|r| < 1$, then

$$\lim_{N \to \infty} (1 + r + r^2 + \cdots + r^{N-1}) = \frac{1}{1 - r}. \tag{2.5.3}$$

Proof. We calculate

$$\begin{aligned}(r-1)(1 + r + r^2 + \cdots + r^{N-1}) &= r \cdot (1 + r + r^2 + \cdots + r^{N-1}) \\ &\quad - 1 \cdot (1 + r + r^2 + \cdots + r^{N-1}) \\ &= r + r^2 + r^3 + \cdots + r^{N-1} + r^N \\ &\quad - (1 + r + r^2 + \cdots + r^{N-1}).\end{aligned}$$

We see that each of the terms $r, r^2, \ldots, r^{N-1}$ appears twice on the right side: once with a $+$ sign and once with a $-$ sign. After cancellation, only the terms r^N and -1 remain on the right side. Thus we are left with

$$(r-1)(1 + r + r^2 + \cdots + r^{N-1}) = r^N - 1.$$

Dividing each side of this identity by $r - 1$, which is nonzero, we obtain (2.5.2).

For the second assertion, we have

$$\begin{aligned}\lim_{N \to \infty} (1 + r + r^2 + \cdots + r^{N-1}) &= \lim_{N \to \infty} \frac{r^N - 1}{r - 1} && \text{Part a} \\ &= \frac{1}{r-1} \cdot \lim_{N \to \infty} r^N - \frac{1}{r-1} && \text{Theorem 2 a, e, f} \\ &= \frac{1}{r-1} \cdot 0 - \frac{1}{r-1} && \text{Theorem 1e} \\ &= \frac{1}{1-r} && \text{Simplify.}\end{aligned}$$ ■

The left sides of formulas (2.5.2) and (2.5.3) are called *geometric series*. Notice that geometric series arise by adding the terms of a geometric progression $1, r, r^2, r^3 \ldots$.

▶ **EXAMPLE 10** Verify that

$$\frac{1}{2} + \frac{1}{4} + \frac{1}{8} + \frac{1}{16} + \cdots = 1.$$

Solution Let $c_N = 1/2 + 1/4 + 1/8 + \cdots + 1/2^N$. According to definition (2.5.1), we must show that $\lim_{N \to \infty} c_N = 1$. Observe that for $r = 1/2$,

$$\begin{aligned}c_N &= \frac{1}{2} + \frac{1}{4} + \frac{1}{8} + \cdots + \frac{1}{2^N} = r + r^2 + r^3 + \cdots + r^N \\ &= (1 + r^2 + r^3 + \cdots + r^{N-1}) + (r^N - 1) = a_N + b_N,\end{aligned}$$

where $a_N = 1 + r^2 + r^3 + \cdots + r^{N-1}$, and $b_N = r^N - 1$. Using Theorem 4b with $r = 1/2$, we see that $\lim_{N\to\infty} a_N = 1/(1-1/2) = 2$. Using Theorem 1e with $r = 1/2$, Theorem 1a with $\alpha = 1$, and Theorem 2b, we see that $\lim_{N\to\infty} b_N = 0 - 1 = -1$. Now, using Theorem 2a, we have

$$\frac{1}{2} + \frac{1}{4} + \frac{1}{8} + \frac{1}{16} + \cdots = \lim_{N\to\infty} c_N = \lim_{N\to\infty} a_N + \lim_{N\to\infty} b_N = 2 + (-1) = 1. \blacktriangleleft$$

INSIGHT A discussion of Zeno's paradoxes (to which Example 10 is related) may be found in Genesis and Development in Chapter 1.

▶ **EXAMPLE 11** What rational number does the repeating decimal expansion 0.999 ... represent?

Solution The number 0.999 ... represents

$$\begin{aligned}
&\lim_{N\to\infty}\left(\frac{9}{10} + \frac{9}{100} + \frac{9}{1000} + \cdots + \frac{9}{10^N}\right) \\
&= \frac{9}{10}\lim_{N\to\infty}\left(1 + \frac{1}{10} + \frac{1}{100} + \cdots + \frac{1}{10^{N-1}}\right) \\
&= \frac{9}{10}\cdot\frac{1}{1-1/10} \qquad \text{Theorem 4b} \\
&= 1. \blacktriangleleft
\end{aligned}$$

◀ **A LOOK BACK AND A LOOK FORWARD** ▶ When we refer to an "infinite sum" $a_1 + a_2 + a_3 + \cdots$, we actually mean the limit $\lim_{N\to\infty}(a_1 + a_2 + \cdots + a_N)$ of a finite sum with an increasing number of terms. Chapter 8 is devoted to a detailed study of such sums.

Using Continuous Functions to Calculate Limits

Consider the sequence $\{\cos(\frac{1}{n})\}_{n=1}^{\infty}$. Because the cosine is continuous, we know that $\lim_{x\to 0}\cos(x) = \cos(\lim_{x\to 0} x) = \cos(0) = 1$. We also know that $\lim_{n\to\infty}(1/n) = 0$. We therefore expect that $\lim_{n\to\infty}\cos(1/n) = \cos\big(\lim_{n\to\infty}(1/n)\big) = \cos(0) = 1$. The next theorem justifies this calculation.

THEOREM 5 Let $\{b_n\}_{n=1}^{\infty}$ be a sequence that converges to a limit ℓ. Suppose that $\{b_n\}$ and ℓ are contained in the domain of a function f that is continuous at ℓ. Then $\lim_{n\to\infty} f(b_n) = f(\lim_{n\to\infty} b_n) = f(\ell)$.

Proof. Let ε be a positive number. Let $a_n = f(b_n)$. We are to show that $\lim_{n\to\infty} a_n = f(\ell)$. Because f is continuous at ℓ, there is a positive number δ such that $|f(x) - f(\ell)| < \varepsilon$ for all x in the interval $I = [\ell - \delta, \ell + \delta]$. Because $\lim_{n\to\infty} b_n = \ell$, there is an integer N such that b_n is in I for each $n \geq N$. It follows that

$$|a_n - f(\ell)| = |f(b_n) - f(\ell)| < \varepsilon$$

for all $n \geq N$, which proves the asserted limit. ■

▶ **EXAMPLE 12** Evaluate $\lim_{n\to\infty}\sqrt{4+3/n}$.

Solution We set $b_n = 4 + 3/n$ and $f(t) = \sqrt{t}$. Then we see from Theorems 1 and 2 that $\ell = \lim_{n\to\infty} b_n = 4$. Because f is continuous on the interval $(0, \infty)$, an interval that contains $\{b_n\}$ and ℓ, we may apply Theorem 5. We conclude that

$$\lim_{n\to\infty}\sqrt{4+\frac{3}{n}} = \lim_{n\to\infty} f(b_n) = f(\lim_{n\to\infty} b_n) = f(4) = \sqrt{4} = 2.$$

Our next theorem uses the function $t \mapsto 1/t$, which transforms the sequence of positive integers to a sequence that converges to 0 from the right. ◀

THEOREM 6 If $\lim_{x\to 0^+} f(x) = \ell$, and $a_n = f(1/n)$, then $\lim_{n\to\infty} a_n = \ell$.

▶ **EXAMPLE 13** Calculate $\lim_{n\to\infty} n\sin(1/n)$.

Solution We let

$$a_n = n\sin\left(\frac{1}{n}\right) = \frac{\sin(1/n)}{1/n} = f\left(\frac{1}{n}\right)$$

for $f(x) = \sin(x)/x$. The first of the two limit formulas in line (2.2.5) states that $\lim_{x\to 0} f(x) = 1$. Theorem 6 therefore tells us that $\lim_{n\to\infty} a_n = \lim_{x\to 0} f(x) = 1$. ◀

QUICK QUIZ

1. True or false: If $\lim_{n\to\infty} a_n = 0$ then there is a tail of the sequence for which the absolute value of every term is less than $1/10^{100}$.
2. Evaluate $\lim_{n\to\infty} 10^{100/n}$.
3. Express 0.777 ... as a rational number.
4. True or false: If $\lim_{n\to\infty} a_n = \ell$, then a_{n+1} is closer to ℓ than a_n for all n.

Answers

1. True 2. 1 3. 7/9 4. False

EXERCISES

Problems for Practice

▶ In Exercises 1–15, determine whether the sequence $\{a_n\}$ converges. If it does, state the limit. ◀

1. $a_n = n/(n^2+1)$
2. $a_n = n+5$
3. $a_n = 3n/(2n+1)$
4. $a_n = 3-(-1)^n$
5. $a_n = 1/(n+2)$
6. $a_n = 1-1/n$
7. $a_n = \cos(n\pi)$
8. $a_n = \sin(n\pi)$
9. $a_n = 3^{-n}+2^{-n}$
10. $a_n = 1/n - 4^{1/n}$
11. $a_n = (3n^3-5)/(4n^2+5)$
12. $a_n = (n^2-n)/(n^2+n)$
13. $a_n = 4 + \dfrac{(-n)^3}{n^3+1}$
14. $a_n = n^{1/100}$
15. $a_n = (1/100)^{1/n}$

▶ In Exercises 16–38, calculate $\lim_{n\to\infty} a_n$. ◀

16. $a_n = 3+4/n$
17. $a_n = 2^{-n}-6$
18. $a_n = (n-2n^2)/(n+n^2)$
19. $a_n = (4n^2+6n+3)/(8n^2+3)$
20. $a_n = (7n^5+6n^4+n^2)/(3n^5+11)$
21. $a_n = \left(3+\dfrac{1}{n}\right)\cdot\left(2-\dfrac{5}{n^2}\right)$
22. $a_n = (10^{-n}+10^n)/10^n$
23. $a_n = 2^{-n}\cdot n/(n+4)$
24. $a_n = \sqrt{4n+1/\sqrt{n}}$

25. $a_n = (2-(1/2)^n)/(4+(1/3)^n)$
26. $a_n = 8^{1/n} + 1/8^n$
27. $a_n = (2^n - 3^n)/(3^n + 4^n)$
28. $a_n = \dfrac{\cos(1/n)}{2+\csc(1/n)}$
29. $a_n = 2\sin(1/n) + 3\cos(1/n)$
30. $a_n = 5\tan(\pi^{1/n}) - 3\tan(\pi/n)$
31. $a_n = \tan(\pi n/(3n+1))$
32. $a_n = \cot(n^2\pi/(4n^2-3))$
33. $a_n = n^8/(2n^4+n^2+1)^2$
34. $a_n = 4n/\sqrt{n^2+5n+2}$
35. $a_n = \tan(\pi \cdot 2^{1/n})$
36. $a_n = \cos(\pi\sec(n/(n^2+1)))$
37. $a_n = \sin\left(\pi\sin\left(\dfrac{\pi n}{6n+2}\right)\right)$
38. $a_n = (12\cdot 7^n - 5^n)/(4\cdot 7^n + 9\cdot 2^n)$

In each of Exercises 39–44, use the Pinching Theorem to evaluate $\lim_{n\to\infty} a_n$.

39. $a_n = \cos(n)/n$
40. $a_n = \sin(2n)/2^n$
41. $a_n = 2^n/(3\cdot 2^n + \cos(2^n))$
42. $a_n = \dfrac{(\cos(3^n) + 2\cdot 3^n)}{3^n}$
43. $a_n = (-1)^n n/(5+6(-1)^n n)$
44. $a_n = (2n^2 + (-1)^n n + 1)/n^2$

In Exercises 45–48, evaluate the geometric series.

45. $1 + 1/3 + 1/9 + 1/27 + \cdots$
46. $1 - 1/3 + 1/9 - 1/27 + \cdots$
47. $1 + 1/\sqrt{3} + 1/3 + 1/(3\sqrt{3}) + \cdots$
48. $1/5 + 1/25 + 1/125 + 1/625 + \cdots$

49. Express 1.11111111... as a rational number.
50. Express 1.01010101... as a rational number.

Further Theory and Practice

In Exercises 51–54, determine the value of the given limit. Then verify your answer using the precise definition of limit.

51. $\lim_{n\to\infty} \dfrac{1}{n+7}$
52. $\lim_{n\to\infty} \dfrac{n}{n^2+1}$
53. $\lim_{n\to\infty} \dfrac{2n+3}{n+5}$
54. $\lim_{n\to\infty} \dfrac{n^2}{3n^2+2}$

In Exercises 55–60, evaluate the given sum.

55. $1/4 + 1/8 + 1/16 + 1/32 + \cdots$
56. $1/\sqrt{2} + 1/2 + 1/(2\sqrt{2}) + 1/4 + \cdots$
57. $16/3 + 32/9 + 64/27 + 128/81 + \cdots$
58. 3.512121212...
59. 123.01232323...
60. $\alpha^2/(1+\alpha^2)^3 + \alpha^3/(1+\alpha^2)^5 + \alpha^4/(1+\alpha^2)^7 + \cdots$

61. Find a divergent sequence $\{a_n\}$ such that $\{a_n^2\}$ is convergent.
62. Prove that, if $\{a_n\}_{n=1}^{\infty}$ converges to a real number ℓ, then $\lim_{n\to\infty}(a_n - \ell) = 0$.
63. Prove that, if $\{a_n\}_{n=1}^{\infty}$ converges to a real number ℓ, then $\lim_{n\to\infty}(a_n - a_{n+1}) = 0$.
64. Suppose that $\{a_n\}_{n=1}^{\infty}$ and $\{b_n\}_{n=1}^{\infty}$ are sequences satisfying $|a_n - b_n| < 1/n$ for every n. Prove that either both sequences converge to the same limit or both sequences diverge.
65. Suppose that $a_n \to \ell \neq 0$. Show that $\lim_{n\to\infty}(a_n/a_{n+1})$ converges.
66. Let α be a positive real number. Set $a_1 = 1$ and $a_n = (\alpha + a_{n-1})/2$ for $n > 1$. Find a formula for $|\alpha - a_n|$. Prove that $\{a_n\}$ converges.
67. Suppose $p(x)$ is a polynomial of degree m and $q(x)$ is a polynomial of degree n, and prove the following assertions.

a. If $m < n$, then $\lim_{j\to\infty} \dfrac{p(j)}{q(j)} = 0$.

b. If $m = n$, then $\lim_{j\to\infty} \dfrac{p(j)}{q(j)} = \dfrac{p_m}{q_n}$ where p_m is the leading coefficient of p, and q_n is the leading coefficient of q.

c. If $m > n$, then $\lim_{j\to\infty} \dfrac{p(j)}{q(j)} = \text{signum}\left(\dfrac{p_m}{q_n}\right)\infty$.

68. If p is a polynomial, show that $\lim_{j\to\infty} \dfrac{p(j+1)}{q(j)} = 1$.
69. Many doubly indexed sequences $\{a_{n,m}\}$ arise in mathematics. Often we need to know if

$$\lim_{n\to\infty}\lim_{m\to\infty} a_{n,m} = \lim_{m\to\infty}\lim_{n\to\infty} a_{n,m}.$$

Show that this equality can be false by considering $a_{n,m} = n/(n+m)$.
70. Suppose that s is a fixed positive number. Calculate

$$\lim_{n\to\infty}\left(\frac{s+2s+3s+\cdots+ns}{n^2}\right).$$

Calculator/Computer Exercises

There are many sequences with convergence properties that are *not* obvious. The sequences in Exercises 71–75 converge, but it is quite tricky to determine what the limit is and to prove the answer. Approximate the limit to four decimal places. State what value of n you are using to approximate the limit.

71. $\lim_{n\to\infty} n.\sin(1/n)$
72. $\lim_{n\to\infty} n^2.(1-\cos(1/n))$
73. $\lim_{n\to\infty} n^{1/n}$
74. $\lim_{x\to\infty} (1-1/n)^n$
75. $\lim_{n\to\infty} (\sqrt{n}/\sqrt{n+1})^{\sqrt{n}}$

76. Graph the function $f(x) = (1+x)^{1/x}$ for $0 < x < 1$. Focus on the behavior of $f(x)$ for small positive values of x by replotting f in the window [0, 0.01, 2.705, 2.719]. What does the limiting behavior of $f(x)$ as $x \to 0^+$ appear to be? What does this behavior tell you about the limit of the sequence $\{(1+1/n)^n\}_{n=1}^{\infty}$?

77. Graph the function $f(x) = x^x$ for $x > 0$. What does the limiting behavior of $f(x)$ as $x \to 0^+$ appear to be? What does this behavior tell you about the limit of the sequence $\{n^{1/n}\}_{n=1}^{\infty}$?

78. Here is an interesting way to use a sequence to solve a quadratic equation: Consider the equation $x^2 - 6x + 5 = 0$. Write $x^2 = 6x - 5$, or $x = 6 - 5/x$. For the x on the right, substitute $6 - 5/x$. The result is

$$x = 6 - \frac{5}{6 - 5/x}.$$

Again, for the x on the right, substitute $x = 6 - 5/x$. The result is

$$x = 6 - \frac{5}{6 - \dfrac{5}{6 - 5/x}}.$$

This process, which we may continue indefinitely, suggests that x is the limit of the sequence

$$6,\ 6 - \frac{5}{6},\ 6 - \frac{5}{6 - 5/6},\ 6 - \frac{5}{6 - \dfrac{5}{6 - 5/6}},\ \ldots$$

This is called a *continued fraction expansion*. Compute the first 20 terms of this sequence. Do they seem to be converging to something? Is the limit a root of the original quadratic equation? Can you discover the other root in the same way?

2.6 Exponential Functions and Logarithms

We begin this section by extending our study of sequential limits. In the preceding section, our method for demonstrating the convergence of a sequence was to explicitly produce its limit. Now we will learn a theorem that establishes the convergence of certain sequences even when their limits have not been identified. With this theorem we are able to define the expression a^x for any positive number a and any real number x. By fixing $a \neq 1$ and regarding x as a variable, we obtain the invertible function $x \mapsto a^x$. This function, which is called the *exponential function* with *base a*, and its inverse, the *logarithm* with *base a*, are the main focus of this section. In some applications, particular choices of the base, such as $a = 2$ or $a = 10$, are convenient or traditional. In calculus, there is a particular base denoted by e that is especially useful because it leads to formulas that are simpler than their analogues with other bases.

The Monotone Convergence Property of the Real Numbers

Many sequences have the property that their terms get steadily larger. Other sequences have terms that continually get smaller. Our next definition introduces the terminology that describes such sequences.

DEFINITION A sequence $\{a_n\}$ is said to be:

- *increasing* if $a_n < a_{n+1}$ for every n;
- *nondecreasing* if $a_n \leq a_{n+1}$ for every n;
- *decreasing* if $a_n > a_{n+1}$ for every n;
- *nonincreasing* if $a_n \geq a_{n+1}$ for every n.

These four types of sequences are said to be *monotonic* or *monotone*. Figure 1 shows an increasing sequence $\{a_n\}$ and a decreasing sequence $\{b_n\}$.

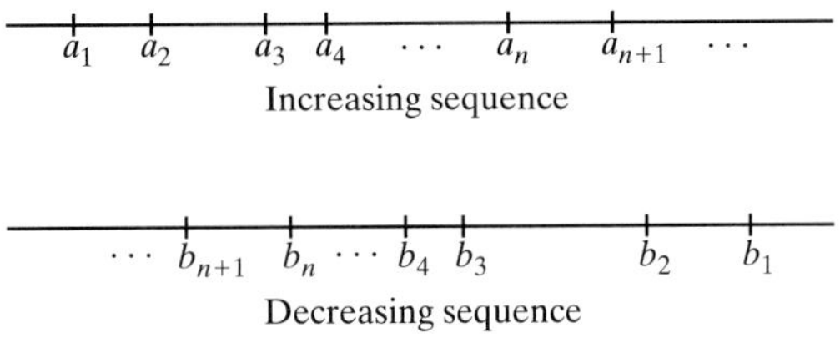

▲ Figure 1

A real number r with a nonterminating decimal expansion gives rise to a nondecreasing sequence in a natural way. Consider $r = 1/3 = 0.3333\ldots$, for example. If we set $a_1 = 0.3$, $a_2 = 0.33$, $a_3 = 0.333$, and so on, then we obtain a sequence with each term greater than the preceding one. Notice that, even though this sequence is increasing, its terms do *not* grow arbitrarily large. In fact, every term in the sequence is less than 1/3. Our next definition describes sequences that do not have arbitrarily large terms.

DEFINITION A sequence $\{a_n\}$ is bounded if there is a real number U such that $|a_n| \leq U$ for all n. We refer to U as a bound for the sequence and say that $\{a_n\}$ is bounded by U.

Notice that a sequence $\{a_n\}_{n=1}^{\infty}$ that is bounded by U is contained in the closed interval $[-U, U]$. The converse is also true: If $\{a_n\}_{n=1}^{\infty}$ is contained in the interval $[\alpha, \beta]$, then $\{a_n\}_{n=1}^{\infty}$ is bounded by the maximum value of $|\alpha|$ and $|\beta|$. Our next theorem guarantees that a monotone, bounded sequence converges to a real number.

THEOREM 1 **(Monotone Convergence Property).** Suppose that $\{a_n\}_{n=1}^{\infty}$ is monotone and bounded. Then $\{a_n\}_{n=1}^{\infty}$ converges to some real number ℓ. If I is a closed interval containing $\{a_n\}_{n=1}^{\infty}$, then ℓ belongs to I.

The Monotone Convergence Property is one way of expressing the completeness of the real number system. Because completeness is a vital requirement for many of the important theorems about continuous functions, such as the Intermediate and Extreme Value Theorems of Section 2.3, it is convenient to have several formulations of completeness. Different, but equivalent, versions of the completeness property are discussed in the *Genesis & Development sections* for Chapters 1 and 2. Now we will study further applications of completeness (in the form of the Monotone Convergence Property). Our first example shows that every decimal expansion corresponds to a real number.

▶ **EXAMPLE 1** Suppose that m is a nonnegative integer and that $\{d_n\}_{n=1}^{\infty}$ is a sequence of decimal digits. That is, each d_n is an integer such that $0 \leq d_n \leq 9$. Interpret the decimal expansion $m.d_1d_2d_3\ldots$ as the limit of a monotone, bounded sequence.

Solution For every positive integer n, we truncate the decimal expansion after the nth decimal place. In this way, we obtain a sequence $\{a_n\}_{n=1}^{\infty}$ with $a_n = m.d_1d_2d_3\ldots d_n$. Our sequence is nondecreasing because $a_{n+1} = a_n + d_{n+1}/10^{n+1} \geq a_n$. Also, $\{a_n\}_{n=1}^{\infty}$ is bounded by $m+1$. The Monotone Convergence Property tells us that $\{a_n\}_{n=1}^{\infty}$ converges to some real number ℓ. It is this number ℓ that the decimal expansion $m.d_1d_2d_3\ldots$ represents. ◀

Irrational Exponents Suppose that a is a positive number. If x is equal to an integer, then

$$a^x = \underbrace{a \cdot a \cdots a}_{x \text{ factors}} \quad \text{if } x > 0, \quad a^x = 1 \quad \text{if } x = 0, \quad \text{and} \quad a^x = \frac{1}{\underbrace{a \cdot a \cdots a}_{|a| \text{ factors}}} \quad \text{if } x < 0.$$

If a and b are positive bases, then elementary algebra shows that the laws of exponents,

$$a^0 = 1, \tag{2.6.1}$$
$$a^1 = a, \tag{2.6.2}$$
$$a^{x+y} = a^x a^y, \tag{2.6.3}$$
$$a^{x-y} = \frac{a^y}{a^y}, \tag{2.6.4}$$
$$(a^x)^y = a^{xy}, \tag{2.6.5}$$
$$(a \cdot b)^x = a^x \cdot b^x \tag{2.6.6}$$

hold for all integers x and y. These laws guide us in extending the definition of a^x to any rational number $x = p/q$ where p and q are integers with $q > 0$. The first step is to define $a^{1/q}$ to be the qth root $\sqrt[q]{a}$ of a (so that equation (2.6.5) holds with $x = 1/q$ and $y = q$). As a result, we have

$$a^{p/q} = (a^p)^{1/q} = \sqrt[q]{a^p}.$$

It again requires only elementary algebra to show that the exponent laws (2.6.1) through (2.6.6) remain valid for rational values of x and y. Additionally, if $x \neq 0$, then we have

$$a^x = b \qquad \text{if and only if} \qquad b^{1/x} = a. \tag{2.6.7}$$

The terminology that we introduce now to describe the order properties of exponentials will be used frequently in later chapters.

DEFINITION Suppose that the domain of a real-valued function f is a set S of real numbers. We say that f is increasing if $f(s) < f(t)$ for all s and t in S with $s < t$. We say that f is decreasing if $f(s) > f(t)$ for all s and t in S with $s < t$.

▶ **EXAMPLE 2** Let a be a positive number not equal to 1. Use the laws of exponents to show that the exponential function $x \mapsto a^x$ with base a is an increasing function on the set of rational numbers if $a > 1$ and a decreasing function if $a < 1$.

Solution For now, the domain S of the exponential function with base a is the set of all rational numbers. Suppose that $1 < a$. If q is a positive integer, then

$1 < a = (a^{1/q})^q$. If $a^{1/q} \leq 1$ were true, then a, the product of q numbers less than or equal to 1, would also be less than or equal to 1. We conclude that $1 < a^{1/q}$. If p is a positive integer, then $a^{p/q}$, which is equal to $(a^{1/q})^p$, is the product of p numbers greater than 1. It follows that $1 < (a^{1/q})^p$. That is, $1 < a^r$ for any positive rational number $r = p/q$. As a result, for rational numbers s and t with $s < t$, we have $1 < a^{t-s}$. On multiplying each side of this inequality by the positive quantity a^s, we obtain the inequality $a^s < a^s \cdot a^{t-s} = a^{s+(t-s)} = a^t$. This inequality shows that an exponential function with base greater than 1 is increasing. Now suppose that $a < 1$, and set $b = 1/a$. Because $b > 1$, the exponential function with base b is increasing (by the first part of this example). Thus $b^s < b^t$ holds for rational numbers s and t with $s <$ t. It follows that $a^s = 1/b^s > 1/b^t = a^t$. ◀

We must still address the problem of defining a^x for irrational values of x. What, for example, does the expression $3^{\sqrt{2}}$ mean?

We answer this question by approximating $3^{\sqrt{2}}$ by numbers of the form 3^x where x is a rational approximation of $\sqrt{2}$. The better x approximates $\sqrt{2}$, the better 3^x approximates $3^{\sqrt{2}}$. Let us see how this idea can be put into practice. Because $\sqrt{2} = 1.41421356\ldots$ is between the rational numbers $1.4 = 14/10$ and $1.5 = 15/10$, we see that $3^{\sqrt{2}}$ must be between $3^{14/10} = 4.655536\ldots$ and $3^{15/10} = 5.196152\ldots$. We narrow the range in which $3^{\sqrt{2}}$ must lie by using better rational approximations of $\sqrt{2}$. Thus because $\sqrt{2}$ is between $1.41 = 141/100$ and $1.42 = 142/100$, we see that $3^{\sqrt{2}}$ must be between $3^{141/100} = 4.706965\ldots$ and $3^{142/100} = 4.758961\ldots$. From these estimates, we infer that $3^{\sqrt{2}} = 4.7$ to one decimal place. Continuing this process, we let x_n be the rational number obtained by terminating the decimal expansion of $\sqrt{2}$ after the nth decimal place. We also let $y_n = x_n + 1/10^n$. Then $3^{x_n} < 3^{\sqrt{2}} < 3^{y_n}$ for every positive integer n. In the following table, we record our calculations for $n \leq 8$.

n	x_n	y_n	3^{x_n}	3^{y_n}
1	1.4	1.5	4.655536722	5.196152423
2	1.41	1.42	4.706965002	4.758961394
3	1.414	1.415	4.727695035	4.732891793
4	1.4142	1.4143	4.728733930	4.729253463
5	1.41421	1.41422	4.728785881	4.728837832
6	1.414214	1.414215	4.728806661	4.728811856
7	1.4142136	1.4142137	4.728804583	4.728805103
8	1.41421356	1.41421357	4.728804376	4.728804427

Because the number $3^{\sqrt{2}}$ lies between the last two numbers of each line of the table, we see that $3^{\sqrt{2}}$ must equal 4.728804 to six decimal places. Furthermore, we deduce that we can define $3^{\sqrt{2}}$ precisely by passing to the limit.

In general, if a is any positive number, and x is any irrational number, then we define a^x by

$$a^x = \lim_{n \to \infty} a^{x_n} \tag{2.6.8}$$

where $\{x_n\}$ is an increasing sequence of rational numbers such that $\lim_{n\to\infty} x_n = x$. Example 2 tells us that $\{a^{x_n}\}$ is a monotonic sequence. The Monotone Convergence Property therefore assures us that the limit in formula (2.6.8) exists. Moreover, it is not difficult to show that this limit does *not* depend on the particular sequence $\{x_n\}$ that approaches x. If it did, then the right side of formula (2.6.8) would not define a^x in an unambiguous way, and, as a result, formula (2.6.8) would not be an acceptable definition.

With a^x extended to all real values of x by formula (2.6.8), the laws of exponents recorded in lines (2.6.1) through (2.6.7) become valid for all real values of x and y. In Chapter 5, we will use calculus to provide an alternative development of irrational exponents. At that time we will prove the exponent laws.

▶ **EXAMPLE 3** Simplify

$$\frac{3^{\pi+1}\cdot\pi^{3-\sqrt{2}}}{3^{\pi-1}\cdot\pi^{1-\sqrt{2}}} \qquad \text{and} \qquad (\sqrt[3]{\pi}\cdot 2^{1/6})^3.$$

Solution We calculate

$$\frac{3^{\pi+1}\cdot\pi^{3-\sqrt{2}}}{3^{\pi-1}\cdot\pi^{1-\sqrt{2}}} = 3^{(\pi+1)-(\pi-1)}\cdot\pi^{(3-\sqrt{2})-(1-\sqrt{2})} = 3^2\cdot\pi^2 = 9\pi^2$$

and

$$\left(\sqrt[3]{\pi}\cdot 2^{1/6}\right)^3 = (\pi^{1/3})^3\cdot(2^{1/6})^3 = \pi\sqrt{2}.$$

Exponential and Logarithmic Functions

We continue to let a denote a positive constant not equal to 1. Our present goal is to describe the basic properties of the exponential functions with base a. For now, we will be content to list the properties that are exhibited by their graphs.

Figure 2 shows the graph of $x\mapsto a^x$ for $a>1$. Notice that

$$\lim_{x\to-\infty} a^x = 0 \quad \text{and} \quad \lim_{x\to\infty} a^x = \infty \quad \text{for } 1<a.$$

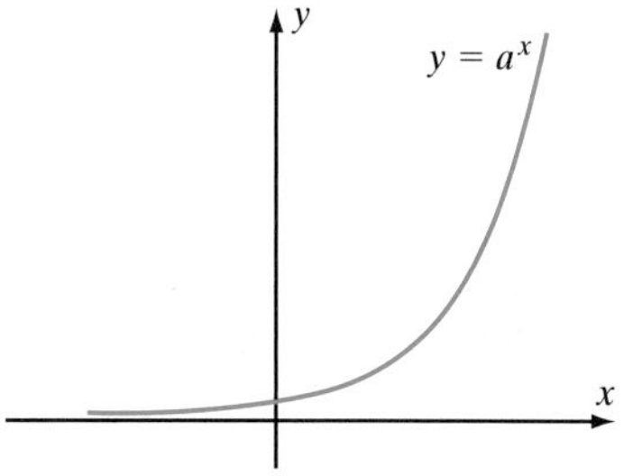

▲ Figure 2 The graph of $y=a^x$ for $a>1$

Figure 3 shows the graph of $x\mapsto a^x$ for $0<a<1$. Here we see that

$$\lim_{x\to-\infty} a^x = \infty \qquad \text{and} \qquad \lim_{x\to\infty} a^x = 0 \qquad \text{for } 0<a<1.$$

As the figures indicate, the exponential function $x\mapsto a^x$ is continuous: For each c, we have

$$\lim_{x\to c} a^x = a^c. \tag{2.6.9}$$

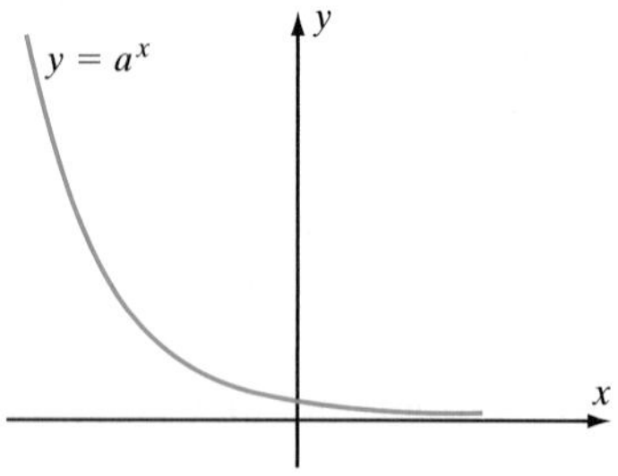

▲ Figure 3 The graph of $y=a^x$ for $0<a<1$

These observations, together with the Intermediate Value Theorem, allow us to conclude that, for any positive base (not equal to 1), the image of an exponential function is the set $\mathbb{R}^+ = (0, \infty)$ of positive real numbers.

The graphs of the exponential functions also exhibit another important property: In Figures 2 and 3 we see that, for $a>1$, the graph of $y=a^x$ is everywhere rising, and, for $0<a<1$, the graph of $y=a^x$ is everywhere falling. These properties are the graphical representations of our observation in Example 2 that an exponential function with base greater than 1 is increasing, and an exponential function with base less than 1 is decreasing. (When we solved Example 2, the domain of $x\mapsto a^x$ was limited to rational numbers. By continuity, the order relationships of the exponential functions remain true when their domains are extended to all real numbers.)

A function that is either increasing on its entire domain or decreasing on its entire domain is necessarily one-to-one. Consequently, the exponential function with base a is an invertible function with domain $\mathbb{R}$ and image $\mathbb{R}^+$. Its inverse is called the *logarithm with base a* and is denoted by $y \mapsto \log_a(y)$. The domain of a logarithm function is $\mathbb{R}^+$, and the image is $\mathbb{R}$. The inverse relationship between the exponential function with base a and the logarithm with base a is expressed by the following two identities:

$$a^{\log_a(y)} = y \quad \text{for any positive number } y \quad \text{and} \quad \log_a(a^x) = x \text{ for any real number } x.$$

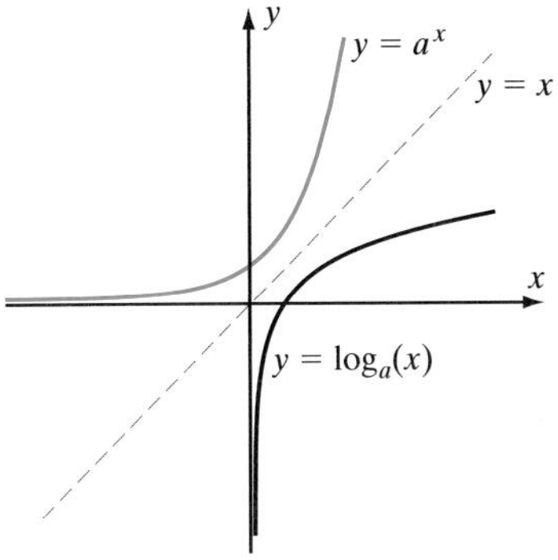

▲ Figure 4

As with any pair of inverse functions, the graphs of $y = a^x$ and $y = \log_a(x)$ are reflections of each other through the line $y = x$ (see Figures 4 and 5).

▶ **EXAMPLE 4** Calculate $\log_3(81)$.

Solution In general, $\log_3(y)$ is the power to which we must raise 3 in order to obtain y. For $y = 81$, we have $3^4 = 81$ and therefore $\log_3(81) = 4$. ◀

Now we introduce some useful properties of logarithm functions.

THEOREM 2 Let a and b be positive numbers not equal to *1*. If $x > 0$, and $y > 0$, then

a. $\log_a(1) = 0$,
b. $\log_a(a) = 1$,
c. $\log_a(x \cdot y) = \log_a(x) + \log_a(y)$, and
d. $\log_a(x/y) = \log_a(x) - \log_a(y)$.
e. For any exponent p, $\log_a(x^p) = p \cdot \log_a(x)$,
f. $\log_a(x) = \dfrac{\log_b(x)}{\log_b(a)}$, and
g. $\log_a(b) = \dfrac{1}{\log_b(a)}$.

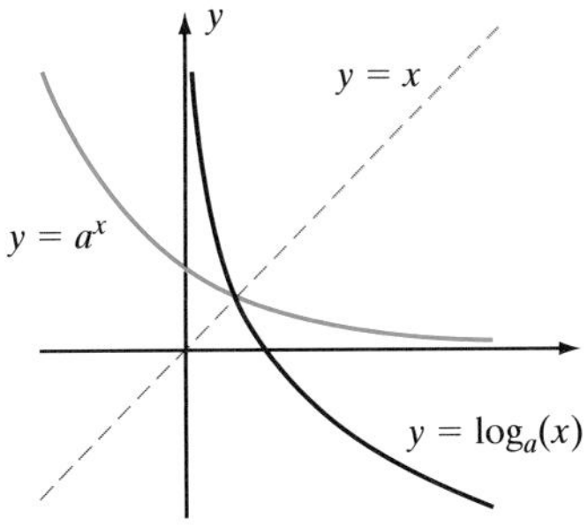

The graph of $y = \log_a(x), 0 < a < 1$

▲ Figure 5

Proof. Assertions a and b are just new ways of stating that $a^0 = 1$ and $a^1 = a$. Properties c, d, and e can be derived from a corresponding property of the exponential function with base a. As a representative example, we will show how to deduce assertion c from equation (2.6.3). We let $u = \log_a(x \cdot y)$, $s = \log_a(x)$, and $t = \log_a(y)$. It follows that $a^u = x \cdot y = a^s a^t = a^{s+t}$. Because the exponential function is one-to-one, we conclude that $u = s + t$, or $\log_a(x \cdot y) = \log_a(x) + \log_a(y)$. To prove part f, we use part e with $p = \log_a(x)$ to see that

$$b^{\log_b(a)\log_a(x)} = b^{\log_b(a^{\log_a(x)})} = a^{\log_a(x)} = x,$$

which means that $\log_b(a)\log_a(x) = \log_b(x)$. Assertion g is the special case of assertion f that results when x is taken to be b. ■

Equation (2.6.3) may be expressed in words as:

The exponential of a sum is the product of the exponentials.

Similarly, Part c of Theorem 2 states the following:

The logarithm of a product is the sum of the logarithms.

We can see that these two statements really say the same thing if we keep in mind the fact that logarithm and exponential are inverse operations.

Logarithms were invented because there was a practical application for a function with property c of Theorem 2. In the late 16th century, multiplication was a tedious process known only to scholars. John Napier (1550–1617) invented logarithms to simplify multiplication. With tables that convert a positive number u to $\log_{10}(u)$ and back, property c can be used to reduce the operation of multiplication to one of addition. The next example illustrates this idea.

▶ **EXAMPLE 5** A table of logarithms base 10 contains the entries:

x	$\log_{10}(x)$
⋮	⋮
6975	3.8435442119
⋮	⋮
7891	3.8971320434
⋮	⋮
55039725	7.7406762553
⋮	⋮

Use these entries to find the product of 6975 and 7891.

Solution The table tells us that 3.8435442119 and 3.8971320434 are the logarithms base 10 of 6975 and 7891, respectively. The sum of these two logarithms is 7.7406762553. This is the logarithm of the number, 55039725, to its left. Thus

$$\begin{aligned}\log_{10}(55039725) &= 7.7406762553 = 3.8435442119 + 3.8971320434 \\ &= \log_{10}(6975) + \log_{10}(7891) = \log_{10}(6975 \cdot 7891).\end{aligned}$$

The number 55039725 is therefore the product of 6975 and 7891. (Logarithms were widely used in this way from their first publication in 1614 until the advent of hand-held scientific calculators in the early 1970s.) The slide rule was a mechanical implementation of this idea. ◀

▶ **EXAMPLE 6** Simplify the expression

$$\log_3(27) - 4\log_2(4) - 3\log_5(5^2).$$

Solution The expression equals

$$\begin{aligned}\log_3(3^3) - 4\log_2(2^2) - 3\log_5(5^2) &= 3\log_3(3) - 4\cdot\big(2\log_2(2)\big) - 3\big(2\log_5(5)\big) \\ &= 3\cdot 1 - 4\cdot 2\cdot 1 - 3\cdot 2\cdot 1 \\ &= -11.\end{aligned}$$

◀

▶ **EXAMPLE 7** Solve the equation $2^x \cdot 3^{2x} = 5/4^x$ for the unknown x.

Solution We take the logarithm base 2 of both sides to obtain $\log_2(2^x \cdot 3^{2x}) = \log_2(5/4^x)$. Applying the rules for logarithms from Theorem 2 results in

$$\log_2(2^x) + \log_2(3^{2x}) = \log_2(5) - \log_2(4^x),$$

or

$$x \cdot \log_2(2) + x \cdot \log_2(3^2) = \log_2(5) - x \cdot \log_2(4).$$

Gathering together all the terms involving x yields

$$x \cdot (\log_2(2) + \log_2(9) + \log_2(4)) = \log_2(5),$$

or

$$x \cdot \log_2(72) = \log_2(5).$$

Dividing by $\log_2(72)$ and using rule f of Theorem 2, we obtain

$$x = \frac{\log_2(5)}{\log_2(72)} = \log_{72}(5) = 0.37633\ldots\ .$$

To verify this numerical calculation with a standard scientific calculator, use Theorem 2f with $x = 5$, $a = 72$, and $b = 10$ to express $\log_{72}(5)$ as $\log_{10}(5)/\log_{10}(72)$. ◀

INSIGHT Applying a logarithm, as in Example 7, is a standard technique for moving a variable from an exponent position. The choice of base 2 in Example 7 was arbitrary. Any other base would have led to the same result. It is a good idea to trace through the solution of Example 7 using the application of the logarithm base a as the first step. The value obtained for x turns out to be $\log_a(5)/\log_a(72)$, but this expression does not depend on a since $\log_a(5)/\log_a(72) = \log_{72}(5)$ by Theorem 2f. This is precisely the answer we obtained with our particular choice of base 2.

The Number e

The graphs of $y = (1+1/x)^x$ and $y = (1-1/x)^{-x}$ in Figure 6 suggest that $(1+1/x)^x$ is an increasing function of x, that $(1-1/x)^{-x}$ is a decreasing function of x, and that there is a number e such that $y = e$ is a common horizontal asymptote:

$$e = \lim_{x\to\infty}\left(1+\frac{1}{x}\right)^x \tag{2.6.9}$$

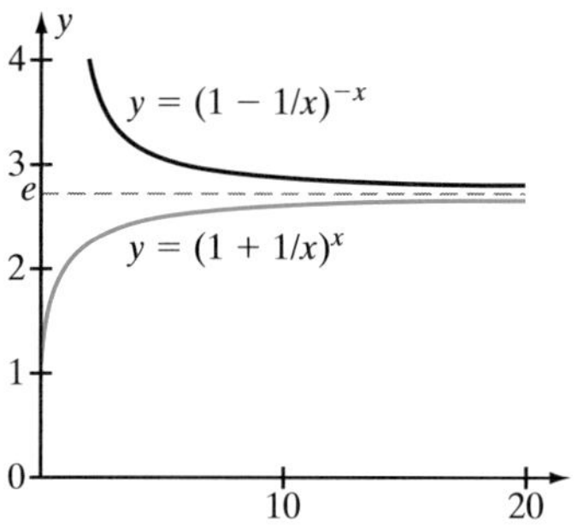

▲ Figure 6

and

$$e = \lim_{x\to\infty}\left(1-\frac{1}{x}\right)^{-x}. \tag{2.6.10}$$

These inferences are correct and can be verified by elementary (albeit technical) arguments. Calculus will provide us with an easier method for establishing these facts in Chapter 5.

In practice, we often use equations (2.6.9) and (2.6.10) with a variable that tends to infinity through integer values. Replacing x by n, a more conventional symbol for a discrete variable, and letting n tend to infinity through integer values, we obtain

$$e = \lim_{n \to \infty} \left(1 + \frac{1}{n}\right)^n \tag{2.6.11}$$

and

$$e = \lim_{n \to \infty} \left(1 - \frac{1}{n}\right)^{-n}. \tag{2.6.12}$$

The existence of the limits in formulas (2.6.11) and (2.6.12) can also be proved by showing that $\{(1+1/n)^n\}_{n=1}^{\infty}$ is a bounded, increasing sequence and that $\left\{(1-1/n)^{-n}\right\}_{n=1}^{\infty}$ is a bounded, decreasing sequence. Formulas (2.6.11) and (2.6.12) then follow from the Monotone Convergence Property. To approximate the number e, we can calculate $(1+1/n)^n$ for several values of n (see the following table).

n	**(1 + 1/n)n**
7	2.546499...
8	2.565784...
9	2.581174...
10	2.593742...
10^6	2.718280...
10^7	2.7182816...

Further calculation yields the decimal expansion

$$e = 2.71828182845904523536\ldots$$

to 20 places. (It is known that e is an irrational number.) The function $x \mapsto e^x$ is called the *natural exponential function* or simply the *exponential function*. It is often convenient to use exp to denote the exponential function. Thus

$$\exp(x) = e^x.$$

This alternative notation is particularly helpful in typesetting complicated expressions. Compare, for example,

$$e^{\frac{(x^3+1)^2}{3(x-1)}} \quad \text{and} \quad \exp\left(\frac{(x^3+1)^2}{3(x-1)}\right).$$

▶ **EXAMPLE 8** Show that

$$e^u = \lim_{x \to \infty} \left(1 + \frac{u}{x}\right)^x \quad \text{and} \quad e^u = \lim_{x \to \infty} \left(1 - \frac{u}{x}\right)^{-x} \tag{2.6.13}$$

for every number u.

Solution We first observe that equation (2.6.13) is obvious for $u = 0$. Suppose that $u > 0$.

Because power functions are continuous, we have

$$e^u \overset{(2.6.11)}{=} \left(\lim_{t \to \infty} \left(1 + \frac{1}{t}\right)^t \right)^u = \lim_{t \to \infty} \left(\left(1 + \frac{1}{t}\right)^t \right)^u \overset{(2.6.5)}{=} \lim_{t \to \infty} \left(1 + \frac{1}{t}\right)^{ut}.$$

and

$$e^u \overset{(2.6.12)}{=} \left(\lim_{t \to \infty} \left(1 - \frac{1}{t}\right)^{-t} \right)^u = \lim_{t \to \infty} \left(\left(1 - \frac{1}{t}\right)^{-t} \right)^u \overset{(2.6.5)}{=} \lim_{t \to \infty} \left(1 - \frac{1}{t}\right)^{-ut}.$$

Set $x = ut$, and observe that $1/t = u/x$. Additionally, notice that the variable x tends to infinity as t does. It follows that

$$e^u = \lim_{t \to \infty} \left(1 + \frac{1}{t}\right)^{ut} = \lim_{x \to \infty} \left(1 + \frac{u}{x}\right)^x,$$

and

$$e^u = \lim_{t \to \infty} \left(1 - \frac{1}{t}\right)^{-ut} = \lim_{x \to \infty} \left(1 - \frac{u}{x}\right)^{-x},$$

which is equation (2.6.13) for $u > 0$. The $u < 0$ case is proved in a similar way. ◄

INSIGHT If $u > 0$, then $x \mapsto f(x)^u$ is increasing if f is increasing, and $x \mapsto g(x)^u$ is decreasing if g is decreasing. If we apply this observation to $f(x) = (1 + 1/x)^x$ and $g(x) = (1 - 1/x)^{-x}$, then the solution of Example 8 shows that $(1 + \frac{u}{x})^x$ increases to e^u when $u > 0$ and $(1 - \frac{u}{x})^{-x}$ decreases to e^u when $u > 0$. These Monotone Convergence Properties will be important in Section 3.4 in Chapter 3 when we study the rate of change of the exponential function.

An Application: Compound Interest

If P_0 dollars are deposited in a bank account that pays r percent *simple interest* per year then, after one year, the account grows to

$$P_0 + \frac{r}{100} P_0 = P_0 \cdot \left(1 + \frac{r}{100}\right) \text{ dollars.}$$

This represents the return of the original amount P_0 (known as the *principal*) plus the interest $(r/100) \cdot P_0$. For an account in which the interest is *compounded n* times yearly, the year is divided into n equal pieces, and, at the end of each time interval of length $1/n$, an interest payment of r/n percent is added to the account. Each time this fraction of the interest is added to the account, the money in the account is multiplied by

$$1 + \frac{r/n}{100}.$$

Because this is done n times during the year, the account grows to

$$P(1) = P_0 \cdot \left(1 + \frac{r}{100n}\right)^n \tag{2.6.14}$$

dollars at the end of one year.

For example, under semiannual (twice yearly) compounding, the account is worth $P \cdot (1 + r/200)$ after six months. This becomes the principal on which interest is paid for the second half of the year. Therefore at the end of the year, the account is worth

$$P_0 \cdot \left(1 + \frac{r}{200}\right)\left(1 + \frac{r}{200}\right) = P \cdot \left(1+\frac{r}{200}\right)^2.$$

The same reasoning shows that if r percent annual interest is compounded n times a year, then at the end of t years, the initial principal P_0 grows to

$$P(t) = P_0 \cdot \left(1+\frac{r}{100n}\right)^{nt}.$$

Compounding can be performed daily, which corresponds to $n = 365$ in formula (2.6.14). By setting $n = 365 \times 24$, we would obtain hourly compounding and so on. *Continuous compounding* is the limiting result of letting n tend to infinity in formula (2.6.14). Thus under continuous compounding, the account is worth

$$P(1) = \lim_{n\to\infty} P_0 \cdot \left(1+\frac{r}{100n}\right)^n$$

dollars at the end of one year and

$$P(t) = \lim_{n\to\infty} P_0 \cdot \left(1+\frac{r}{100n}\right)^{nt}$$

dollars after t years. These limits can be evaluated using equation (2.6.13) The next theorem summarizes our findings.

THEOREM 3 If the interest on principal P_0 accrues at an annual rate of r percent compounded n times per year, then the value $P(t)$ of the deposit after t years is given by

$$P(t) = P_0 \cdot \left(1+\frac{r}{100n}\right)^{nt}. \qquad (2.6.15)$$

If the interest accrues at an annual rate of r percent compounded continuously, then the value $P(t)$ of the deposit after t years is given by

$$P(t) = P_0 \cdot e^{rt/100}. \qquad (2.6.16)$$

► **EXAMPLE 9** The sum of \$5000 is placed in a savings account with 6% annual interest. If simple interest is paid, then how large is the account after 4.5 years? What if the interest is compounded continuously?

Solution Use (2.6.15) with $P_0 = 5000$, $r = 6$, $n = 1$, and $t = 4.5$. We find that the size of the account is given by

$$P(4.5) = 5000 \cdot \left(1+\frac{6}{100}\right)^{4.5} \approx \$6499.00.$$

When the compounding is continuous, formula (2.6.16) with $P_0 = 5000$, $r = 6$, and $t = 4.5$ tells us that the size of the account is

$$P(4.5) = 5000 \cdot e^{6\cdot(4.5)/100} \approx \$6549.82. \quad ◄$$

▶ **EXAMPLE 10** A woman wishes to set up an endowment to pay her nephew \$50,000 in cash on the day of his 21st birthday. The endowment is set up on the day of his birth and is locked in at 9% annual interest, compounded continuously. How much principal should be put into the account to yield the desired sum?

Solution Let P_0 be the initial principal deposited in the account on the day of the nephew's birth. After twenty-one years at 9% annual interest, compounded continuously, the principal is to have grown to \$50,000:

$$50000 = P_0 \cdot e^{(0.09)\cdot 21}.$$

Solving for P_0 gives

$$P_0 = 50000 \cdot e^{-0.09\cdot 21} = 50000 \cdot e^{-1.89} \approx 7553.59.$$

If the account were drawing only simple annual interest of 9%, then the compounding would be done only at the end of every year. The initial principal would be determined by the equation $50000 = P_0 \cdot (1+.09)^{21}$. Solving this equation for P_0, we find that a deposit of \$8184.90 would be required, which is 8.35% more than the sum needed under continuous compounding. ◀

The Natural Logarithm

The inverse of the exponential function is called the *natural logarithm* and is generally denoted by $x \mapsto \ln(x)$ instead of $x \mapsto \log_e(x)$. The facts that we obtained for logarithms with base greater than 1 hold for the natural logarithm. Thus the domain of the natural logarithm is $\mathbb{R}^+ = (0, \infty)$, and its image is $\mathbb{R} = (-\infty, \infty)$. For every real s and positive t, we have

$$\ln(e^s) = s \qquad \text{and} \qquad e^{\ln(t)} = t.$$

If x and y are positive and p is any real number, then

$$\ln(1) = 0, \tag{2.6.17}$$

$$\ln(e) = 1, \tag{2.6.18}$$

$$\ln(x \cdot y) = \ln(x) + \ln(y), \tag{2.6.19}$$

$$\ln\left(\frac{x}{y}\right) = \ln(x) - \ln(y), \tag{2.6.20}$$

$$\ln(x^p) = p \cdot \ln(x), \tag{2.6.21}$$

$$\log_a(x) = \frac{\ln(x)}{\ln(a)}, \text{ and} \tag{2.6.22}$$

$$\ln(b) = \frac{1}{\log_b(e)}. \tag{2.6.23}$$

▶ **EXAMPLE 11** Express

$$Q = \frac{4 \cdot \log_3(5) - (1/2) \cdot \log_3(25)}{2 \cdot \log_3(6) + (1/3) \cdot \log_3(64)}$$

in the form $\ln(u)/\ln(v)$.

Solution The numerator of Q equals

$$\log_3(5^4) - \log_3(25^{1/2}) = \log_3(625) - \log_3(5) = \log_3(625/5) = \log_3(125).$$

Similarly, the denominator can be rewritten as

$$\log_3(6^2) + \log_3(64^{1/3}) = \log_3(36) + \log_3(4) = \log_3(36 \cdot 4) = \log_3(144).$$

Putting these two results together and using Theorem 2 (vi) and (2.6.21), we see that

$$Q = \frac{\log_3(125)}{\log_3(144)} = \log_{144}(125) = \frac{\ln(125)}{\ln(144)}. \quad ◀$$

Exponential Growth

A quantity $P(t)$ is said to *grow exponentially if* $P(t) = Ae^{kt}$ for positive constants A and k. Notice that $A = P(0)$. If we let

$$T = \frac{\ln(2)}{k}, \tag{2.6.24}$$

then

$$P(t+T) = Ae^{k(t+T)} = Ae^{kt}e^{kT} = P(t)e^{kT} = P(t)e^{k\cdot\ln(2)/k} = P(t)e^{\ln(2)} = 2 \cdot P(t)$$

for every t. We call k the *growth constant* and T the *doubling time* of $P(t)$. When we use the doubling time instead of the growth constant, it is sometimes convenient to rewrite $P(t)$ in terms of the exponential function with base 2. We do that as follows:

$$P(t) = Ae^{kt} = A\exp\left(\frac{\ln(2)}{T}t\right) = A \cdot \left(e^{\ln(2)}\right)^{t/T} = A \cdot 2^{t/T}. \tag{2.6.25}$$

▶ **EXAMPLE 12** In a yeast nutrient broth, the population of a colony of the bacterium Escherichia Coli (E.Coli) grows exponentially and doubles every twenty minutes. If there are 6000 bacteria at 10:00 A.M., then how many will there be at noon?

Solution To answer this question, let $P(t)$ be the number of bacteria at time t (in hours) where 10:00 A.M. corresponds to $t = 0$, and noon corresponds to $t = 2$. We have $A = P(0) = 6000$. We have been told that the doubling time T is twenty minutes. We convert this to the units with which we are working: $T = 1/3$ hour. Substituting this value of T and $A = 6000$ into equation (2.6.25) results in

$$P(t) = 6000 \cdot 2^{3t}.$$

We conclude that $P(2) = 6000 \cdot 2^6 = 384{,}000$. ◀

INSIGHT Do not make the mistake of thinking that population problems can be done using just arithmetic. The reasoning "if the population doubles in twenty minutes, then it will quadruple in forty minutes" is generally wrong because populations usually do not grow linearly.

Exponential Decay

A quantity $m(t)$ is said to *decay exponentially* if $m(t) = A\exp(-\lambda t)$ for positive constants A and λ. Notice that $A = m_0$ where $m(0) = m_0$. If we let

$$\tau = \frac{\ln(2)}{\lambda}, \tag{2.6.26}$$

then

$$m(t+\tau) = Ae^{-\lambda(t+\tau)} = Ae^{-\lambda t}e^{-\lambda\tau} = m(t)e^{-\lambda\tau} = m(t)e^{-\lambda \ln(2)/\lambda} = m(t)e^{-\ln(2)} = \frac{1}{2}\cdot m(t)$$

for every t. We call λ the *decay constant* and τ the *half-life* of $m(t)$. When we use the half-life instead of the decay constant, it is sometimes convenient to rewrite $m(t)$ in terms of the exponential function with base 2. We do that as follows:

$$m(t) = A\exp(-\lambda t) = A\exp\left(-\frac{\ln(2)}{\tau}t\right) = A\left(e^{-\ln(2)}\right)^{t/\tau} = \frac{A}{2^{t/\tau}}. \qquad (2.6.27)$$

▶ **EXAMPLE 13** Eight grams of a certain radioactive isotope decay exponentially to 6 grams in 100 years. After how many more years will only 4 grams remain?

Solution First note that the answer is not "we lose 2 grams every hundred years so after 100 more years, the isotope will have decayed from 6 grams to 4 grams." Instead, we let $m(t) = m_0\exp(-\lambda t) = 8\exp(-\lambda t)$ denote the amount of radioactive material at time t. Because $m(100) = 6$, we have

$$6 = m(100) = 8e^{-100\lambda} = 8(e^{-\lambda})^{100}.$$

We conclude that

$$e^{-\lambda} = \left(\frac{6}{8}\right)^{1/100} = \left(\frac{3}{4}\right)^{1/100}.$$

Thus the formula for the amount of isotope present at time t is

$$m(t) = 8e^{-\lambda t} = 8(e^{-\lambda})^t = 8\left(\frac{3}{4}\right)^{t/100}.$$

Now we have complete information about the function m, and we can answer the original question. There will be 4 grams of material present when

$$4 = m(t) = 8\cdot\left(\frac{3}{4}\right)^{t/100} \qquad \text{or} \qquad \frac{1}{2} = \left(\frac{3}{4}\right)^{t/100}.$$

We solve for t by taking the natural logarithm of both sides:

$$\ln(1/2) = \ln\left(\left(\frac{3}{4}\right)^{t/100}\right) = \frac{t}{100}\cdot\ln(3/4),$$

or

$$t = 100\cdot\frac{\ln(1/2)}{\ln(3/4)} \approx 240.942.$$

Thus at $t = 240.942$, which is to say after 140.942 more years, there will be 4 grams of the isotope remaining. ◀

QUICK QUIZ

1. True or false: Every monotone sequence converges.
2. True or false: Every bounded sequence converges.
3. Evaluate $\lim_{n\to\infty}(1+1/n)^n$.
4. Solve $2\exp(x) = 3^x$.

Answers

1. False 2. False 3. e 4. $\ln(2)/(\ln(3)-1)$

EXERCISES

Problems for Practice

In Exercises 1–8, simplify the given expression.

1. $\sqrt{2}^{\sqrt{3}} \cdot \sqrt{2}^{\sqrt{3}}$
2. $4^{\pi} \cdot 4^{e}$
3. $(1/8)^{-\pi/3}$
4. $(8^{\sqrt{3}} \cdot 4^{\sqrt{7}})/2^{\pi}$
5. $(\sqrt{11}^{\sqrt{2}})^{\sqrt{2}}$
6. $(\sqrt{e}^{\sqrt{2}})^{2}$
7. $(3^4 \cdot 9^3/27^2)^{1/2}$
8. $\left((5^{3/4})^3/\sqrt{\sqrt{5}}\right)^{1/2}$

In Exercises 9–22, rewrite the given expression without using any exponentials or logarithms.

9. $\log_5(1/125)$
10. $\log_7(\sqrt{7})$
11. $e^{\ln(3)}$
12. $\exp(-3\ln(2))$
13. $\log_{27}(9/3^x)$
14. $\log_8(64 \cdot 4^{2x} \cdot 2^{-6})$
15. $\log_2(\log_2(4))$
16. $\log_{1/4}(8^{2x} \cdot 2^{-4x})$
17. $\log_{4/9}(4^{x/2} \cdot 3^{4x} \cdot 2^{-5x})$
18. $\log_4(16^x) - \log_3(27) + 4^{\log_4(5)}$
19. $2^{\log_8(27x^3)}$
20. $\left(e^{\log_2(7)}\right)^{\ln(2)}$
21. $\exp\left(\sqrt{\ln(3^{\ln(81)})}\right)$
22. $(e^3)^{\ln(4)} - \left(3^{\ln(59)}\right)^{\log_3(e)}$

In Exercises 23–26, solve for x.

23. $1/2^x = 8 \cdot 2^x$
24. $5^x = 125^{(1-2x)}$
25. $4^x \cdot 3^{2x} = 8 \cdot 6^{-x}$
26. $3^{-x} \cdot 4^{2x} = 4/9$

In Exercises 27–32, sketch the graph of the given equation. Label salient points.

27. $y = 3^x + 1$
28. $y = 2^{-x}$
29. $y = 3 - e^x$
30. $y = \exp(1 - x)$
31. $y = 3 + \log_2(2x)$
32. $y = \ln(x + e)$

In Exercises 33–42, find the limit.

33. $\lim_{n\to\infty}(1 + 1/n)^n$
34. $\lim_{n\to\infty}(1 + 2/n)^n$
35. $\lim_{n\to\infty}(1 + e/n)^n$
36. $\lim_{n\to\infty}(1 + 1/(2n))^n$
37. $\lim_{n\to\infty}(1 + 3/(4n))^n$
38. $\lim_{n\to\infty}(1 - 1/n)^n$
39. $\lim_{n\to\infty}(1 - \pi/n)^n$
40. $\lim_{n\to\infty}(1 - 1/(2n))^n$
41. $\lim_{n\to\infty}(1 - \pi/(4n))^n$
42. $\lim_{n\to\infty}(1 - \ln(3)/n)^n$

In Exercises 43–52, find the limit.

43. $\lim_{x\to e} 2^{\ln(x^3)}$
44. $\lim_{x\to e^-} (\pi)^{1/(x-e)}$
45. $\lim_{x\to e^+} (e/\pi)^{1/(x-e)}$
46. $\lim_{x\to 0} 3^{\sin(x)/x}$
47. $\lim_{x\to 1} 5^{\log_3(x)}$
48. $\lim_{x\to\infty}(12 + 7e^{-3x})$
49. $\lim_{x\to-\infty}(1 + 3/x)^x$
50. $\lim_{x\to-\infty}(1 - 1/(\pi x))^x$
51. $\lim_{x\to\infty} \dfrac{e^{2x} - e^{-2x}}{e^{2x} + e^{-2x}}$
52. $\lim_{x\to\infty} \dfrac{10 \cdot 3^{2x} - 2^x}{2 \cdot 9^x + 5^{-x}}$

In Exercises 53–56, find all asymptotes of the graph of the given equation.

53. $y = (e^{7x} + e^{-7x})/(e^{7x} - e^{-7x})$
54. $y = e^{-x^2}$
55. $y = 32/(8 - 2^x)$
56. $y = (3^x + 2^x)/(3^x - 3)$

In each of Exercises 57–60, an initial investment P_0, an annual interest rate, and a number m are given. Determine the value of the investment after m months under the following conditions:

a. Simple interest
b. Semiannual compounding
c. Quarterly compounding
d. Daily compounding
e. Continuous compounding

57. $P_0 = 1000$, 6%, $m = 12$
58. $P_0 = 1000$, 7%, $m = 60$
59. $P_0 = 5000$, 5.25%, $m = 36$
60. $P_0 = 10000$, 7.5%, $m = 24$

In each of Exercises 61–64, you are given certain information about a population of bacteria. Assuming exponential growth, find the number of bacteria at 8:00 P.M.

61. 5000 bacteria at noon;
8000 bacteria at 3:00 P.M.

62. 4000 bacteria at 5:00 P.M.;
population doubles every two hours

63. 6500 bacteria at 7:00 P.M.;
8000 bacteria at 9:00 P.M.

64. 7000 bacteria at 4:00 P.M.;
population triples every 8 hours

In each of Exercises 65–68 you are given a certain amount (in grams) of a radioactive isotope, and some information about its rate of decay. Assuming exponential decay, determine how much of the substance is present in the year 2010.

65. 5 grams in 1955;
4 grams in 1996

66. 8 grams in 1994;
half-life is 30 years

67. 12 grams in 2028;
10 grams in 2040

68. 15 grams in 2000;
10 grams in 2100

In each of Exercises 69–72, determine the half-life from the given information about the rate of decay of a radioactive substance.

69. 12 grams in 1960;
8 grams in 2000

70. 10 grams in 1960;
7 grams in 2010

71. 14 grams in 1880;
10 grams in 1980

72. 20 grams in 1900;
15 grams in 1950

Further Theory and Practice

In Exercises 73–76, sketch the graph of the given equation.

73. $y = |e^x - 1|$

74. $y = e^{|x|}$

75. $y = \ln(|x - 2|)$

76. $y = |\ln(|e - x|)|$

In Exercises 77–80, evaluate the given limit.

77. $\lim_{n\to\infty} \left(\dfrac{n+1}{n}\right)^n$

78. $\lim_{n\to\infty} \left(\dfrac{n}{n+1}\right)^n$

79. $\lim_{n\to\infty} \left(1+\dfrac{3}{n}\right)^{2n}$

80. $\lim_{n\to\infty} \left(1+\dfrac{1}{n^2}\right)^{n^2}$

81. Suppose that $a_n \to 1$ as $n \to \infty$. Because any power of 1 is also 1, it might seem as if $a_n^n \to 1$. To the contrary, for any $v > 0$, it is possible to find a sequence $\{a_n\}$ such that $a_n \to 1$, but $a_n^n \to v$. Use formula (2.6.11) to demonstrate this fact.

If an amount A is to be received at time T in the future, then the *present value* of that payment is the amount P_0 that, if deposited immediately with the current interest rate locked in, will grow to A by time T under continuous compounding. The present value of an income stream is the sum of the present values of each future payment. In each of Exercise 82–85, calculate the present value of the specified income stream.

82. Mr. Woodman wants to give \$100,000 to his son Chip when Chip turns 25 years old. Given that current interest rates are 5%, and Chip has just turned 18, what is the present value of the gift?

83. Mrs. Woodman wants to present a \$144,000 college fund to her newborn daughter Sylvia when Sylvia turns 18 years old. Given that current interest rates are 6.5%, what is the present value of the gift?

84. A winning lottery ticket pays \$300,000 immediately and four more payments of the same amount at yearly intervals. If the current interest rate is 4.5%, what is the present value of the winnings?

85. Mr. Woodman pledges three equal payments of \$1000 at yearly intervals to a forrest conservation organization. If the first installment is to be paid in two years, and if the current interest rate is 4%, what is the present value of the donation?

86. Every exponential function can be expressed by means of the natural exponential function. To be specific, for any fixed $a > 0$, there is a constant k such that $a^x = e^{kx}$ for every x. Express k in terms of a.

87. Investment advisors refer to the following rule of thumb: If an investment is left to compound continuously at a constant annual rate of r%, then the time required for the investment to double in value is $72/r$ years. What is the exact rule that this "rule of 72" approximates? What integer gives a better approximation than 72? (The number 72 is used because 72 is divisible by r for $r = 3, 4, 6, 8, 9$, and 12.)

88. If at time $t = 0$, an object is placed in an environment that is maintained at a constant temperature T_∞, then according to Newton's Law of Heating and Cooling, the temperature $T(t)$ of the object at time t is

$$T(t) = T_\infty + (T(0) - T_\infty)e^{-kt}$$

for some constant $k > 0$. What is the limiting temperature of the object as $t \to \infty$? What asymptote does the graph of T have? Sketch the graph of T assuming that $T(0) > T_\infty$. Sketch the graph of T assuming that $T(0) < T_\infty$.

89. Suppose M and P_0 are constants such that $M > P_0 > 0$. The size of a population P is given as a function of time t by

$$P(t) = \frac{P_0 M}{P_0 + (M - P_0)e^{-kt}}$$

for some positive constant k. This equation for $P(t)$ is known as the *logistic growth formula*. In a population model, time is generally restricted to an interval with a left endpoint, but assume that $-\infty < t < \infty$ here. By calculating $\lim_{t\to\infty} P(t)$ and $\lim_{t\to-\infty} P(t)$, determine the horizontal asymptotes of the graph of P.

90. If the air resistance to a falling object is proportional to the speed of that object with positive proportionality constant k, then the speed of that object when dropped from a height is

$$v(t) = kg(1 - e^{-t/k}).$$

What is the limit of $v(t)$ as $t \to \infty$? (This limiting value is known as the *terminal velocity* of the falling object.) Sketch the graph of v. What horizontal asymptote does it have?

91. When a drug is intravenously introduced into a patient's bloodstream at a constant rate, the concentration C of the drug in the patient's body is typically given by

$$C(t) = \frac{\alpha}{\beta}(1 - e^{-\beta t})$$

where α and β are positive constants. What is the limiting concentration $\lim_{t\to\infty} C(t)$? Sketch the graph of C. What horizontal asymptote does it have?

92. The diffusion of a solute through a cell membrane is described by Fick's Law:

$$c(t) = (c(0) - C)\exp(-kt) + C,$$

where $c(t)$ is the concentration of the solute in the cell, C is the concentration of the solute outside the cell, and k is a positive constant. What is the limit of the concentration of the solute in the cell? Sketch the graph of c when $c(0) > C$ and when $c(0) < C$. What horizontal asymptote does each graph have?

93. Suppose that $\alpha > 1$. Let $x_0 = 1$. For $n > 1$, let $x_n = (\alpha + x_{n-1})/2$. Show that $\{x_n\}$ is a bounded increasing sequence. To what number does $\{x_n\}$ converge?

94. Let $s_n = 1 + 1/2! + 1/3! + \cdots + 1/n!$ for every integer n. It is evident that $\{s_n\}$ is an increasing sequence. To prove that it is also bounded, show that $k! \geq 2^{k-1}$ for every positive integer k. It then follows that

$$s_n < 1 + 1/2 + 1/4 + 1/8 + \cdots.$$

Use this inequality and formula (2.5.3) to obtain an upper bound U for $\{s_n\}$. The Monotone Convergence Property tells us that $\{s_n\}$ converges to a number not greater than U. In Chapter 10, we will see that the limit is $e - 1$.

▶ **Radiocarbon Dating** Two isotopes of carbon, ^{12}C, which is stable, and ^{14}C, which decays exponentially with a 5700-year half-life, are found in a known fixed ratio in living matter. After death, carbon is no longer metabolized, and the amount $m(t)$ of ^{14}C decreases due to radioactive decay. In the analysis of a sample performed T years after death, the mass of ^{12}C, unchanged since death, can be used to determine the mass m_0 of ^{14}C that the sample had at the moment of death. The time T since death can then be calculated from the law of exponential decay and the measurement of $m(T)$. Use this information for solving Exercises 95–98. ◀

95. Until relatively recently, mammoths were thought to have become extinct 10,000 years ago. In 1991, fossils of dwarfed woolly mammoths found on Wrangel Island, which is located in the Arctic Ocean off the coast of Siberia, were analyzed. From this radiocarbon dating, it was determined that mammoths survived on Wrangel Island until at least 1700 BCE. What percentage of m_0 was the measured amount of ^{14}C in these fossils?

96. In 1994, a parka-clad mummified body of a girl was found in a subterranean meat cellar near Barrow, Alaska. Radiocarbon analysis showed that the girl died around CE 1200. What percentage of m_0 was the amount of ^{14}C in the mummy?

97. Prehistoric cave art has recently been found in a number of new locations. At Chauvet, France, the amount of ^{14}C in some charcoal samples was determined to be $0.0879 \cdot m_0$. About how old are the cave drawings?

98. The skeletal remains of a human ancestor, Lucy, are reported to be 3,180,000 years old. Explain why the radiocarbon dating of matter of this age would be futile.

Calculator/Computer Exercises

99. Plot $y = \exp(x)$ for $0 \leq x \leq 2$. Let $P(c)$ denote the point $(c, \exp(c))$ on the graph. The purpose of this exercise is to graphically explore the relationship between $\exp(c)$ and the slope of the tangent line at $P(c)$. For $c = 1/2$, 1, and 3/2, calculate the slope $m(c)$ of the secant line that passes through the pair of points $P(c - 0.001)$ and $P(c + 0.001)$. For each c, calculate $|\exp(c) - m(c)|$ to see that $m(c)$ is a good approximation of $\exp(c)$. Add the three secant lines to your viewing window. For each of $c = 1/2$, 1, and 3/2, add to the viewing window the line through $P(c)$ with slope $\exp(c)$. As we will see in Chapter 3, these *are* the tangent lines at $P(1/2)$, $P(1)$, and $P(3/2)$. It is likely that they cannot be distinguished from the secant lines in your plot.

100. Plot $y = \ln(x)$ for $1/2 \leq x \leq 3$. Let $P(c)$ denote the point $(c, \ln(c))$ on the graph. The purpose of this exercise is to graphically explore the relationship between 1/c and the slope of the tangent line at $P(c)$. For $c = 1$, 3/2, and 2, calculate the slope $m(c)$ of the secant line that passes through the pair of points $P(c-0.001)$ and $P(c+0.001)$. For each c, calculate $|1/c - m(c)|$ to see that $m(c)$ is a good approximation of 1/c. Add the three secant lines to your viewing window. For each of $c = 1$, 3/2, and 2, add to the viewing window the line through $P(c)$ with slope 1/c. As we will see in Chapter 3, these *are* the tanget lines at $P(1)$, $P(3/2)$, and $P(2)$. It is likely that they cannot be distinguished from the secant lines in your plot.

Summary of Key Topics in Chapter 2

Limits *(Sections 2.1–2.2)*

The expression $\lim_{x\to c} f(x) = \ell$ means that $f(x)$ approaches ℓ as x approaches c. We say that f has limit ℓ at c. More precisely, $\lim_{x\to c} f(x) = \ell$ means that given, $\varepsilon > 0$, there is a $\delta > 0$ such that if $0 < |x - c| < \delta$, then $|f(x) - \ell| < \varepsilon$.

One-Sided Limits *(Section 2.2)*

We say that f has right limit ℓ at c if the domain of f contains the interval (c, b), and given $\varepsilon > 0$, there is a $\delta > 0$ such that if $c < x < c + \delta$, then $|f(x) - \ell| < \varepsilon$. The definition of left limit is similar. These limits are written $\lim_{x\to c^+} f(x)$ and $\lim_{x\to c^-} f(x)$, respectively. Left and right limits have properties similar to those of two-sided limits. A function has limit ℓ at c if and only if it has left limit ℓ at c and right limit ℓ at c.

Uniqueness of Limits *(Section 2.2)*

A function has at most one limit at a point: If $\lim_{x\to c} f(x) = \ell$ and $\lim_{x\to c} f(x) = m$, then $\ell = m$.

Algebraic Properties of Limits *(Section 2.2)*

a. $\lim_{x\to c}(f(x) + g(x)) = \lim_{x\to c} f(x) + \lim_{x\to c} g(x)$

b. $\lim_{x\to c}(f(x) \cdot g(x)) = (\lim_{x\to c} f(x)) \cdot (\lim_{x\to c} g(x))$

c. $\lim_{x\to c} f(x)/g(x) = \lim_{x\to c} f(x)/\lim_{x\to c} g(x)$, provided that $\lim_{x\to c} g(x) \neq 0$

d. $\lim_{x\to c}(\alpha \cdot f(x)) = \alpha \cdot \lim_{x\to c} f(x)$ for any constant $\alpha \in \mathbb{R}$

Nonexistence of Limits *(Section 2.2)*

If $\lim_{x\to c} f(x) \neq 0$ and $\lim_{x\to c} g(x) = 0$, then $\lim_{x\to c} f(x)/g(x)$ does not exist.

The Pinching Theorem *(Section 2.2)*

If $g(x) \leq f(x) \leq h(x)$ and $\lim_{x\to c} g(x) = \lim_{x\to c} h(x) = \ell$, then $\lim_{x\to c} f(x) = \ell$.

Two Important Trigonometric Limits *(Section 2.2)*

$$\lim_{x\to 0} \frac{\sin(x)}{x} = 1 \quad \text{and} \quad \lim_{x\to 0} \frac{1-\cos(x)}{x^2} = \frac{1}{2}$$

Continuity *(Section 2.3)*

A function f is continuous at c if $\lim_{x\to c} f(x) = f(c)$.

One-Sided Continuity (Section 2.3)

A function f with a domain that contains $[c, b)$ is said to be right-continuous at c if $\lim_{x\to c^+} f(x) = f(c)$. The definition of left-continuous is similar. Left- and right-continuous functions have properties similar to those for continuous functions. A function is continuous at c if and only if it is both left- and right-continuous at c.

Properties of Continuous Functions (Section 2.3)

If the functions f and g are continuous at c, then so are $f+g$, $f-g$, and $f \cdot g$. Furthermore, f/g is continuous at c provided that $g(c) \neq 0$.

The Intermediate Value Theorem (Section 2.3)

If f is a continuous function with a domain that contains $[a, b]$, if $f(a) = \alpha$ and $f(b) = \beta$, and if γ is a real number between α and β, then there is a number c between a and b such that $f(c) = \gamma$.

The Extreme Value Theorem (Section 2.3)

If f is a continuous function with a domain that contains $[a, b]$, then there are numbers α and β in $[a, b]$ such that $f(\alpha) \leq f(x) \leq f(\beta)$ for all x in $[a, b]$.

Infinite Limits and Limits at Infinity (Section 2.4)

The expression $\lim_{x\to c} f(x) = \infty$ means that $f(x)$ gets arbitrarily large as x approaches c. More precisely, given $K > 0$, there is a $\delta > 0$ such that if $0 < |x - c| < \delta$, then $f(x) > K$. The definitions of the one-sided infinite limits $\lim_{x\to c^+} f(x) = \infty$ and $\lim_{x\to c^-} f(x) = \infty$ are similar. Analogous definitions exist for $\lim_{x\to c} f(x) = -\infty$, $\lim_{x\to c^+} f(x) = -\infty$ and $\lim_{x\to c^-} f(x) = -\infty$. In any of these cases, we say that $x = c$ is a vertical asymptote of the graph of f.

The expression $\lim_{x\to\infty} f(x) = \ell$ means that $f(x)$ gets arbitrarily close to ℓ as x gets arbitrarily large through positive values. More precisely, given $\varepsilon > 0$, there is $K > 0$ such that if $x > K$, then $|f(x) - \ell| < \varepsilon$. The definition of $\lim_{x\to -\infty} f(x) = \ell$ is analogous. In either case, we say that $y = \ell$ is a horizontal asymptote of the graph of f.

Limits of Sequences (Section 2.5)

The expression

$$\lim_{j\to\infty} a_j = \ell$$

means that a_j gets arbitrarily close to the real number ℓ as the positive integer variable j gets arbitrarily large. More precisely, given $\varepsilon > 0$, there is a $J > 0$ such that if $j > J$, then $|a_j - \ell| < \varepsilon$. We say that the sequence $\{a_j\}$ is convergent.

The algebraic properties of limits discussed in Section 2.2 are also valid for limits of sequences, as is the Pinching Theorem.

The Monotone Convergence Property (Section 2.6)

A sequence $\{a_j\}$ is said to be monotone if it is decreasing ($a_{j+1} \leq a_j$ for every j) or if it is increasing ($a_{j+1} \geq a_j$ for every j). A sequence $\{a_j\}$ is said to be bounded if there is a K such that $|a_j| < K$ for every j. The Monotone Convergence Property of the real number system states that every monotone bounded sequence is convergent.

Exponential Functions
(Section 2.6)

If $a > 0$ and x is irrational, then we may define a^x by

$$a^x = \lim_{x\to\infty} a^{x_n}$$

where $\{x_n\}$ is a sequence of rational numbers that converges to x. All the usual rules of exponents remain valid:

$$a^{x+y} = a^x a^y, \quad a^{-x} = \frac{1}{a^x}, \quad \text{and} \quad (a^x)^y = a^{xy}.$$

The Number e
(Section 2.6)

The number e is defined by

$$e = \lim_{n\to\infty}\left(1+\frac{1}{n}\right)^n.$$

It is an irrational number with a decimal representation that begins with 2.7182818284590452. In general, for any real number u,

$$e^u = \lim_{x\to\infty}\left(1+\frac{u}{x}\right)^x.$$

Logarithms
(Section 2.6)

If a is a positive number not equal to 1, then the exponential function with base a, namely $x \mapsto a^x$, is invertible. The inverse function is written as $y \mapsto \log_a(y)$ if $a \neq e$. When e is used as a base, $x \mapsto e^x$ is called the exponential function, and its inverse, which is called the natural logarithm function, is denoted by $y \mapsto \ln(y)$. The natural logarithm satisfies the two identities

$$\ln(xy) = \ln(x) + \ln(y)$$

and

$$\ln(x^p) = p\ln(x).$$

Exponential Growth and Decay
(Section 2.6)

A quantity $P(t)$ grows exponentially if $P(t) = Ae^{kt}$ for positive constants A and k. The doubling time of P is given by

$$T = \frac{\ln(2)}{k}$$

and has the property that, for every t,

$$P(t+T) = 2\cdot P(t).$$

A quantity $m(t)$ decays exponentially if $m(t) = A\exp(-\lambda t)$ for positive constants A and λ. The half-life

$$\tau = \frac{\ln(2)}{\lambda}$$

has the property that, for every t,

$$m(t+\tau) = \frac{1}{2}\cdot m(t).$$

Review Exercises for Chapter 2

In each of Exercises 1–30, a function $f(x)$ and a value c are given. Determine whether or not $\lim_{x\to c} f(x)$ exists. If the limit does exist, compute it.

1. $f(x) = 1/\sin(2x)$, $c = \pi$
2. $f(x) = x^2 - 6x + 8$, $c = 5$
3. $f(x) = (x^2 + 7x + 10)/(x+5)$, $c = -4$
4. $f(x) = (x - 18)/(x+2)$, $c = 3$
5. $f(x) = (x^2 - 36)/(x+6)$, $c = -6$
6. $f(x) = (x-1)^2/(x+1)$, $c = -1$
7. $f(x) = (x+5)/(x^2-25) =$, $c = -5$
8. $f(x) = \begin{cases} \dfrac{x^2-5x+6}{x-2} & \text{if } x<2 \\ \sec(\pi x/2) & \text{if } x \ge 2 \end{cases}$ $\quad c = 2$
9. $f(x) = \begin{cases} \dfrac{x^2-7}{3} & \text{if } x \le 5 \\ \dfrac{x^2+2x-35}{x^2-8x+15} & \text{if } x>5 \end{cases}$ $\quad c = 5$
10. $f(x) = |x| - x + |-x| + \sqrt{x^2}$, $c = -3$
11. $f(x) = 3x + \tan(\pi x)$, $c = 1/2$
12. $f(x) = 3x + \tan(\pi x)$, $c = 2$
13. $f(x) = x/|x|$, $c = -3$
14. $f(x) = x/|x|$, $c = 0$
15. $f(x) = x/\sin(x)$, $c = 0$
16. $f(x) = \sin(x)/x$, $c = \pi/6$
17. $f(x) = x^2/(1-\cos(x))$, $c = 0$
18. $f(x) = |x|\csc(x)$, $c = 0$
19. $f(x) = (x^{1/3} + 28)/\sqrt{x}$, $c = 64$
20. $f(x) = (x^2-9)/(x-3)^2$, $c = 3$
21. $f(x) = (1-\cos(x))/x$, $c = 0$
22. $f(x) = \sin(2x)/x^2$, $c = 0$
23. $f(x) = \sin(7x)/\sin(3x)$, $c = 0$
24. $f(x) = x\sin(x)/(\cos(x)-1)$, $c = 0$
25. $f(x) = (1-\cos(x))/\sin(x)$, $c = 0$
26. $f(x) = \tan(x)/x$, $c = 0$
27. $f(x) = \tan(3x)/x$, $c = 0$
28. $f(x) = \sin(x)/x$, $c = \pi/2$
29. $f(x) = (\sin(x) + \cos(x))/x$
30. $f(x) = x/(1-\cos(x))$, $c = 0$

In Exercises 31–34, use algebraic manipulation to evaluate the given limit.

31. $\lim_{x\to 4}\left(\dfrac{x^2-16}{\sqrt{x}-2}\right)$
32. $\lim_{x\to 9}\left(\dfrac{x-3}{\sqrt{x}-9}\right)$
33. $\lim_{x\to 0}\dfrac{12x}{\sqrt{1+6x}-1}$
34. $\lim_{x\to 0}\left(\dfrac{\sqrt{5+x}-\sqrt{5}}{x}\right)$

In Exercises 35–40, evaluate the given one-sided limits.

35. $\lim_{x\to 5^+}\dfrac{x-5}{|x-5|}$
36. $\lim_{x\to 5^-}\dfrac{x-5}{|x-5|}$
37. $\lim_{x\to 4^-}\left(\dfrac{\sqrt{16-4x}}{\sqrt{16-x^2}}\right)$
38. $\lim_{x\to 0^+}\dfrac{\sqrt{x}}{\sqrt{\sin(x)}}$
39. $\lim_{x\to 0^-}\sqrt{x^2}/x$
40. $\lim_{x\to(\pi/2)^-}(1/\sqrt{2})^{\tan(x)}$

In Exercises 41–54, determine whether or not the given limit exists. If the limit does exist, compute it. Where appropriate, answer with ∞ or $-\infty$.

41. $\lim_{x\to-\infty}(9 - e^{2x})$
42. $\lim_{x\to-\infty}(\pi - e^{-3x})$
43. $\lim_{x\to\infty}(\pi + 2^x)$
44. $\lim_{x\to-\infty}(7 - 2^x)$
45. $\lim_{x\to\infty}(\cos(x) + 2e^{-x})$
46. $\lim_{x\to\infty}\dfrac{5x^3+2x+1}{3x^3+x^2+x+2}$
47. $\lim_{x\to\infty}\dfrac{7x^2+2x+1}{x^3+x+2}$
48. $\lim_{x\to\infty}\dfrac{4x^3+x+3}{x^2+x+2}$
49. $\lim_{x\to\infty}\dfrac{3x+2^x}{3^x+2}$
50. $\lim_{x\to\infty}\dfrac{10+\ln(x)}{5+\ln(x)}$
51. $\lim_{x\to\infty}(1-1/x)^x$
52. $\lim\limits_{x\to-\infty}\left(1+\dfrac{\ln(2)}{x}\right)^x$
53. $\lim_{x\to 0} x\cos^2(1/x)$
54. $\lim_{x\to 0} x\csc(x)\cos(1/x)$

In each of Exercises 55–58, calculate how close x needs to be to c to force $f(x)$ to be within 0.01 of ℓ.

55. $f(x) = x+1 \quad c=-1 \quad \ell = 0$
56. $f(x) = 2x+13 \quad c=-7 \quad \ell=-1$
57. $f(x) = 2x^2 \quad c=-5 \quad \ell=50$
58. $f(x) = \begin{cases} 4x-3 & \text{if } x \le 1 \\ x^2 & \text{if } x>1 \end{cases} \quad c=1 \quad \ell=1$

In each of Exercises 59–62, determine whether the given function f has a continuous extension F that is defined on the entire real line. If it does, answer with the value of F at the point that is not in the domain of f.

59. $f(x) = \begin{cases} -x/\sqrt{x^2} & \text{if } x<0 \\ \sin(x)/x & \text{if } x>0 \end{cases}$

60. $f(x) = \begin{cases} (x^2-3x+2)/(x-2) & \text{if } x<2 \\ x+1 & \text{if } x>2 \end{cases}$

61. $f(x) = \begin{cases} (x^2-1)/(x+1) & \text{if } x<-1 \\ x^2+x-2 & \text{if } x>-1 \end{cases}$

62. $f(x) = \lim_{x\to 0} \sin^2(4x^2)/x^4$

In each of Exercises 63–68, find all horizontal and vertical asymptotes for the graph of the function that is defined by the given expression.

63. $(x+4)/(x+3)$
64. $(x^2+9)/(x^2-9)$
65. $(2x+3)/(x^2+1)$
66. $3x^2/(x^2-7x+6)$
67. $(3x^2-9x+6)/(x^2+x-2)$
68. $\dfrac{\exp(5x)+\exp(-5x)}{\exp(5x)-\exp(-5x)}$

In Exercises 69–72, simplify the given expression.

69. $(1/27)^{-2/3}$
70. $\left(\sqrt{e}^{\sqrt{3}}\right)2\sqrt{3}$
71. $(2^4\cdot 4^3/8^2)^{3/2}$
72. $\left(\sqrt{8}^{\sqrt{2}}\right)^{4\sqrt{2}/3}$

In Exercises 73–76, rewrite the given expression without using any exponentials or logarithms.

73. $\log_4(8\cdot 4^{2x}\cdot 2^{1-2x})$
74. $\exp(\log_3(6)/\log_3(e))$
75. $\log_{1/\sqrt{3}}(\log_4(64))$
76. $(\sqrt{e})^{\ln(4x^4)}$
77. Solve for x: $(3^x)^2 = 18\cdot 2^x$.
78. Solve for x: $4^x = 2\cdot 8^{-x}$.
79. Describe the function $f(x) = \lim_{y\to x^+}\lfloor y\rfloor - \lim_{y\to x^-}\lfloor y\rfloor$.

In each of Exercises 80–85, calculate $\lim_{n\to\infty} a_n$ for the given sequence $\{a_n\}$.

80. $\left\{\dfrac{3n+2}{n+6}\right\}$
81. $\{100^{1/n}\}$
82. $\left\{\left(\dfrac{n+2}{n}\right)^{1/n}\right\}$
83. $\left\{5^{-n}+\dfrac{2}{7^n}\right\}$
84. $\left\{\dfrac{3^n-4^n}{2^{2n}+4^n}\right\}$
85. $\left\{\dfrac{3n^2-5}{4n^3+5}\right\}$

86. The volume $V(t)$ of harvestable timber in a young forest grows exponentially; that is, $V(t) = Ae^{kt}$ for positive constants A and k. If the yearly rate of increase is equal to 3.5%, then what percentage increase is expected in 10 years?

87. Learning curves are sometimes exponential in form. Suppose that after t weeks in a classroom, the class average (out of 100) on a standard aptitude test is

$$A(t) = 100(1-e^{Kt}).$$

A particular class has an average score of 75 after 6 weeks. In how many more weeks will the average score exceed 90?

88. Moore's law (after Gordon Moore, cofounder of Intel) asserts that the number of transistors that can be pressed onto a fixed area doubles every 18 months. Assuming that this observation remains valid, in how many months will the number of transistors that can be pressed onto a fixed area increase by a factor of 100?

89. The amount of ^{226}Ra decays exponentially. The half-life of ^{226}Ra is 1620 years. An area is contaminated with a level that is five times greater than the maximum safe level. Without a cleanup, how long will this area remain unsafe?

In Exercises 90–96, which concern radiocarbon dating, m_0 refers to the mass of ^{14}C that the analyzed sample had at the moment of death. (Refer to the instructions for Exercises 95–98 of Section 2.6 for further details.) The half-life of ^{14}C is 5700 years.

90. The *Shroud of Turin* is a cloth believed by some to be the burial cloth of Jesus. In 1988, strands of the cloth were given to four different institutions for dating. Each institution declared the *Shroud of Turin* to be a medieval forgery from around 1325 CE. Approximately what percentage of m_0 was measured?

91. In 1995, the Dead Sea Scroll text known as 4Q258 was radiocarbon dated. The result suggested a date of about 180 CE. Paleographic evidence had placed the date at about 100 BCE.
  a. What fraction of m_0 would have validated the paleographic evidence?
  b. What fraction was actually found?

92. The oldest North American mummy, the Spirit Cave Man, was discovered near Fallon, Nevada in 1940. In the days before radiocarbon dating, Nevada State Museum anthropologists estimated the mummy to be about 2000 years old. In 1996, radiocarbon dating of hair and textile strands that were buried with the mummy placed the age at 9400 years. What fraction of m_0 was found?

93. On September 19, 1991, a mummified human body was discovered in an Austrian glacier near the Tisenjoch pass. Radiocarbon analysis has shown that the Iceman, as he has come to be called, died around 3325 BCE. What percentage of m_0 was measured?

94. After World War II, to avoid imprisonment for selling art work to the Nazis, Dutch artist Han van Meegeren (1889–1947) confessed to having forged the paintings he sold as Vermeers. Given that Vermeer died in 1675, what is the maximum percentage of m_0 that could have been found in the pigments of one of his paintings in 1945? What percentage of m_0 would fakes from 1930 have had in 1945?

95. When samples, all having equal ^{12}C content, were extracted from pigments coming from different areas of a cave drawing in Chauvet, one sample had ^{14}C content that was 1.837 times greater than the ^{14}C content found in the other samples. About how many years after the original work was the "touch-up" work done?

96. Cro-Magnon Man migrated to Europe about 40,000 years ago. For a long time scientists believed that the Neanderthals vanished soon afterwards. However, recent radiocarbon dating of a European Neanderthal fossil resulted in a ^{14}C value of $0.026\,m_0$. For about how long did the European Neanderthals and Cro-Magnons coexist (at the least)?

Mathematical standards did not always require concepts to be clearly defined and results to be correctly demonstrated. Even though the *limit* is basic to calculus, a rigorous treatment of limits was not developed until the 19th century. By that time, most of the topics that are studied in a modern calculus course had long been known. For 200 years, mathematicians relied on an intuitive understanding of limits, but it eventually became clear that a more precise viewpoint was necessary.

The Controversy over Infinitesimals

The fundamental concept of differential calculus is the limit

$$\ell = \lim_{\Delta x \to 0} \frac{f(c + \Delta x) - f(c)}{\Delta x}.$$

We have already used a limit of this form in our discussion of instantaneous velocity (see Section 2.1). We will consider such limits in much greater detail in Chapter 3. For a long time, the most common approach to such a limit was through the mechanism of *infinitesimals* (quantities that are infinitely small). The limit ℓ was envisioned as a quotient in which the numerator was the infinitesimal increment dy in the variable $y = f(x)$ caused by incrementing the value c of the variable x by the infinitesimal amount dx. For example, if $y = x^2$, then

$$dy = (c+dx)^2 - c^2 = 2c \cdot dx + (dx)^2,$$

and

$$\ell = \frac{dy}{dx} = \frac{2c \cdot dx + (dx)^2}{dx} = 2c + dx.$$

Under the rules of infinitesimal algebra, a non-infinitesimal quantity remains unchanged when an infinitesimal is added to it. Thus the limit ℓ is $2c$.

Although the infinitesimal method yields the correct answer, a logical dilemma is present: If $2c + dx = 2c$, then dx is really 0, in which case, the quotient $\frac{dy}{dx}$ is not algebraically permitted. On the other hand, if the quotient $\frac{dy}{dx}$ is permitted because $dx \neq 0$, then $2c + dx \neq 2c$.

In 1734, George Berkeley (1685–1753), Bishop of Cloyne, Ireland, published *The Analyst,* a tract addressed to an "infidel mathematician," generally presumed to be the astronomer Edmond Halley. In it, Berkeley attacked the hypocrisy of those who scorned religious belief but who put their faith in infinitesimals, which were not then precisely conceived or employed in a consistent manner:

> The Imagination, which faculty derives from sense, is very much strained and puzzled to frame clear ideas of the least particles of time ... The further the mind analyseth and pursueth these fugitive ideas, the more it is lost and bewildered; the objects at first fleeting and minute, soon vanishing out of sight ... I have no controversy about your conclusions, but only about your logic and method: how you demonstrate? what objects you are conversant with, and whether you conceive them clearly? what principles you proceed upon; how sound they may be; and how you apply them?

The criticisms of Berkeley initiated a flurry of interest in the foundations of calculus. There was also a second stimulus that pointed to the need for more precision. In 1747, Jean le Rond d'Alembert (1717–1783) published both the law governing the motion of a vibrating string and its general solution. Although d'Alembert's equation did not come into question, his solution to it did. In 1748, Leonhard Euler and, in 1753, Daniel Bernoulli (1700–1782) proposed different solutions. Mathematics was in the embarrassing situation of having three of its greatest proponents debate mathematical truths as they would points of metaphysics.

Bolzano and the Intermediate Value Theorem

By the end of the 18th century, mathematicians had started to look for ways to formulate the concepts and theorems of calculus without reference to *infinitesimals*. The approaches taken by Bernard Bolzano (1781–1848) and Augustin-Louis Cauchy (1789–1857) stand out. Bolzano gave the first "modern" definition of "continuity" in his 1817 paper on the Intermediate Value Theorem. In Bolzano's definition, f is continuous at x "if the difference $f(x + \omega) - f(x)$ can be made smaller than any given magnitude by taking ω as small as is wished." In the title of his paper, Bolzano stressed that he was giving an *analytic* proof of the Intermediate Value Theorem. The theorem was known previously, but it was always considered self-evident on geometric grounds. Bolzano realized that the theorem must depend on both the properties of the functions under consideration and on the completeness property of the real numbers. Here, for the first time, we find explicit awareness that a completeness property of the real number system requires proof.

One of Bolzano's approaches to completeness was the *least upper bound property* of the real number system. Let S be a bounded set of real numbers. A number U is said to be the *least upper bound* of S if U is the least number such that $x \leq U$ for each x in S. For example, if $S = \{x \in \mathbb{R} : x^2 < 2\}$, then every element x of S satisfies $|x| < \sqrt{2}$, so $\sqrt{2}$ is an upper bound of S. It is not hard to show that no smaller number is an upper bound of S.

The least upper bound property states that every nonempty set of real numbers that is bounded above has a least upper bound. The rational number system $\mathbb{Q}$ does not have this property. For example, if $S' = \{x \in \mathbb{R} : x^2 < 2\}$, then every rational number larger than $\sqrt{2}$ is an upper bound but not a least upper bound. The Least Upper Bound Property, the Monotone Convergence Property, and the nested interval property, which was discussed in the Genesis & Development for Chapter 1, are equivalent properties of the real number system.

Bolzano's position was Professor of Theology at the University of Prague. However, after 1819 when he was dismissed for heresy, he was subjected to police supervision, and his writing was curtailed. In addition to his careful derivation of the Intermediate Value Theorem from the least upper bound property of the real numbers, he made several original discoveries concerning the foundations of mathematical analysis. Nevertheless, because he worked and published in relative obscurity, his ideas did not receive their due until long after his death.

Cauchy and Weierstrass

Credit, if not priority, for the movement toward rigor in calculus must be given to Augustin-Louis Cauchy, the most prominent mathematician in the greatest center of mathematics of his day. His *Cours d'Analyse* (1821) and subsequent Ecole Polytechnique lecture notes alerted a generation of mathematicians to the possibility of attaining a rigor in calculus that was comparable to that of classical geometry. Although the verbal definition of "limit" that Cauchy gave in his *Cours* was scarcely better than that given by his predecessors, and certainly worse than that given by Bolzano, it is clear from his translation of his verbal definition into mathematical inequalities that he meant exactly the definition of limit as given in Section 2.2. Indeed, it was Cauchy who introduced the symbols and ε and δ (with ε standing for *"erreur"* and δ for "*différence*"). Many of the theorems of calculus received their first modern formulations in Cauchy's work.

A completely correct presentation of the foundations of calculus was finally given by Karl Weierstrass. We have already mentioned Weierstrass in the Genesis & Development for Chapter 1 in connection with his construction of the real number system. Weierstrass had an unusual career as a research mathematician. Born in the village of Ostenfelde, Weierstrass entered the University of Bonn in 1834 as a law student. He did not begina serious study of mathematics until 1839, when he was 24. From 1840 until 1856, Weierstrass taught science, mathematics, and physical fitness at a high school. He became a professor at the University of Berlin in 1864, at the relatively advanced age of 49. It was in his lectures at the University of Berlin that Weierstrass clarified the fundamental structures of mathematical analysis. Of primary importance in this regard was Weierstrass's careful delineation of different types of convergence. In the words of David Hilbert (1862–1943),

> It is essentially a merit of the scientific activity of Weierstrass that mathematicians are now in complete agreement and certainty concerning types of analytic reasoning based on the concepts of irrational number and, more generally, limit.

At one time, mathematical theory required successive revision as mathematicians delved deeper into existing theory. The formulation of a precise definition of limit has given a permanent character to the foundation of calculus.

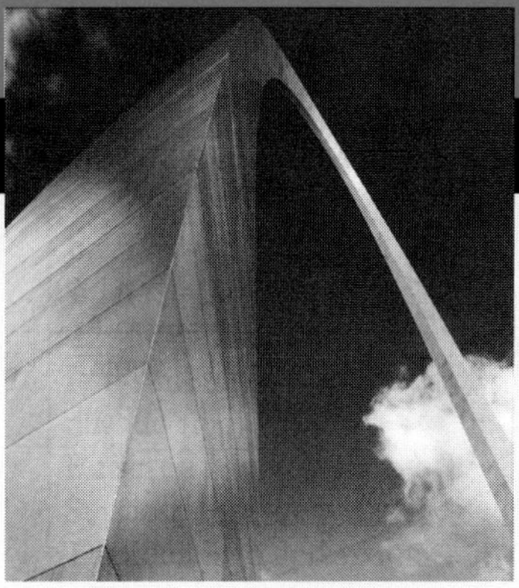

CHAPTER 3

The Derivative

PREVIEW

Suppose that c is a point in the domain of a function f. As the independent variable x of $f(x)$ changes from c to $c+\Delta x$, the value of $f(x)$ changes from $f(c)$ to $f(c+\Delta x)$. The *average* rate of change of $f(x)$ is obtained by dividing the change in $f(x)$ by the change in x:

$$\frac{f(c+\Delta x)-f(c)}{\Delta x} \tag{3.0.1}$$

This average rate of change over the interval $[c, c+\Delta x]$ can be useful, but there are times when the rate of change *at* the particular value c has greater importance. Because Δx appears in the denominator of formula (3.0.1), we cannot simply set $\Delta x = 0$ to obtain the rate of change of f at c. Instead, we let Δx *tend* to 0 and use the limit

$$\lim_{\Delta x \to 0} \frac{f(c+\Delta x)-f(c)}{\Delta x} \tag{3.0.2}$$

to define the *instantaneous rate of change* of $f(x)$ at $x = c$. We call limit (3.0.2) the *derivative of f* at c and denote it by $f'(c)$. As we will see, $f'(c)$ also represents the slope of the tangent line to the graph of the equation $y = f(x)$ at the point $(c, f(c))$.

The branch of mathematics devoted to the study of derivatives is called *differential calculus*. In this chapter, you will learn methods for evaluating the derivatives of common functions. Because the direct calculation of limits such as (3.0.2) can be difficult, we will develop rules that allow us to compute derivatives without actually referring to limits.

3.1 Rates of Change and Tangent Lines

Imagine car A driving down a road, as in Figure 1. We are standing at the side of the road, and we wish to determine the velocity of car A at any given moment. Here is how we proceed. We record the position $p(t)$ of car A as time elapses. Time $t = 0$ corresponds to the time that our observations begin (see Table 1).

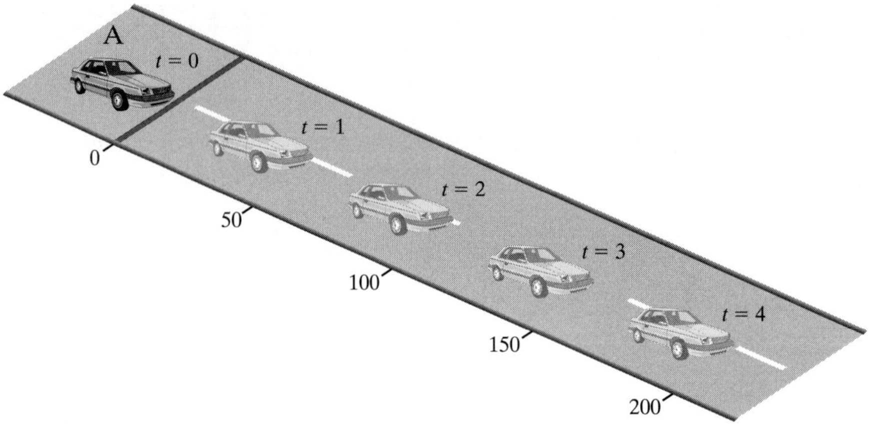

▲ Figure 1

Time (seconds)	0	1	2	3	4
Distance (feet)	0	50	100	150	200

Table 3.1

Analyzing the speed of car A is simple because it is traveling at a constant rate: 50 feet in 1 second, 100 feet in 2 seconds, 150 feet in 3 seconds, 200 feet in 4 seconds, and so on. At all observed times,

$$\text{rate} = \frac{\text{distance traveled}}{\text{time elapsed}} = 50 \text{ feet/second.}$$

Suppose that a second car B travels down the road, as in Figure 2, with position $p(t) = 8t^2$ feet if t is measured in seconds. Some values of $p(t)$ are recorded in Table 2.

With each passing second, car B covers more ground than in the previous second. For example, car B moves $8 - 0 = 8$ feet in the first second, $32 - 8 = 24$ feet in the second second, $72 - 32 = 40$ feet in the third second, and $128 - 72 = 56$ feet in the fourth second. In other words, car B is *accelerating*. Because the velocity of car B is ever changing, how can we calculate the velocity of car B at a given instant of time, say $t = 1$?

To begin to answer this question, we can calculate the *average velocity* of car B between times $t = 1$ and $t = 2$. This is

$$\text{rate} = \frac{\text{distance traveled}}{\text{time elapsed}} = \frac{(32-8)\text{ft}}{(2-1)\,s} = 24\frac{\text{ft}}{\text{s}}.$$

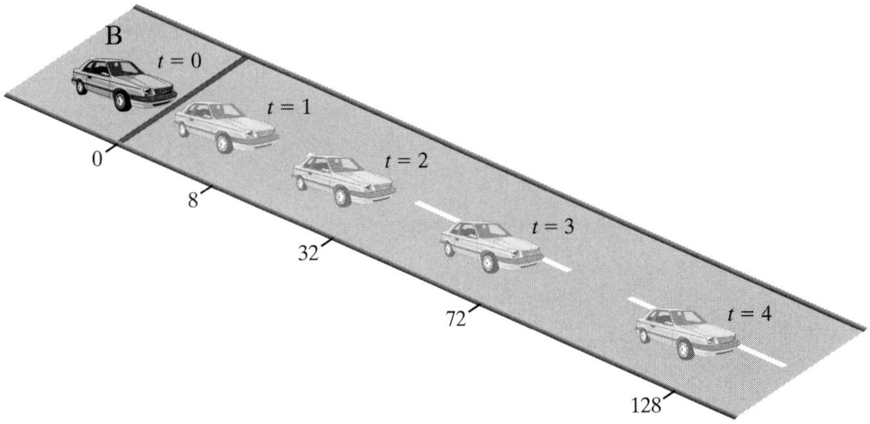

▲ Figure 2

Time (seconds)	0	1	2	3	4
Distance (feet)	0	8	32	72	128

Table 3.2

This calculation gives a rough approximation to the velocity at time $t = 1$ but does not tell us the *exact* velocity at the instant $t = 1$. However, it suggests an idea: instead of calculating average velocity over a long time interval (of length one second), let us calculate average velocity over a short time interval. This should give a better approximation to the velocity at the instant $t = 1$. We will use the symbol Δt to denote the elapsed time. Over the time interval from $t = 1$ to $t = 1 + \Delta t$, car B travels from $p(1)$ to $p(1 + \Delta t)$. The average velocity in ft/s is then

$$\begin{aligned} \text{rate} &= \frac{\text{distance traveled}}{\text{elapsed time}} \\ &= \frac{p(1 + \Delta t) - p(1)}{\Delta t} \\ &= \frac{8(1+\Delta t)^2 - 8 \cdot 1^2}{\Delta t} \\ &= \frac{8 + 16 \cdot \Delta t + 8(\Delta t)^2 - 8}{\Delta t} \\ &= \frac{16 \cdot \Delta t + 8(\Delta t)^2}{\Delta t} \\ &= 16 + 8(\Delta t). \end{aligned}$$

We have found that the average velocity over the time interval from $t = 1$ to $t = 1+\Delta t$ is equal to $16 + 8(\Delta t)$ feet per second. Notice that if we let Δt tend to 0, then the average velocity, $16 + 8(\Delta t)$ ft/s, has a limit: 16 ft/s. It is therefore reasonable to say that the velocity of car B at the instant $t = 1$ is the limit of the average velocity over the time interval from $t = 1$ to $t = 1 + \Delta t$ as $\Delta t \to 0$. With this definition, the velocity of car B at $t = 1$ is $\lim_{\Delta t \to 0}(16 + 8(\Delta t))$ ft/s, or 16 ft/s.

The Definition of Instantaneous Velocity

We now summarize our investigation of instantaneous velocity with a definition.

DEFINITION Suppose that the position of a body moving along an axis is described by a function p of time t. The *instantaneous velocity* of the body at time c is

$$\lim_{\Delta t \to 0} \frac{p(c+\Delta t) - p(c)}{\Delta t},$$

provided that this limit exists and is finite.

For ease of reference, we denote the instantaneous velocity at time c by $p'(c)$. Thus

$$p'(c) = \lim_{\Delta t \to 0} \frac{p(c+\Delta t) - p(c)}{\Delta t}. \tag{3.1.1}$$

Our first two examples show that limit formula (3.1.1) agrees with our intuitive understanding of velocity in some simple situations.

▶ **EXAMPLE 1** Let α be a constant. Suppose that the position of car C is given by $p(t) = \alpha$ ft for all values of t. Because $p(t)$ is constant, car C is not moving. Common sense tells us that its velocity is 0 at every instant c. Verify that formula (3.1.1) leads to the same conclusion.

Solution Let c denote any instant of time. We calculate:

$$p'(c) = \lim_{\Delta t \to 0} \frac{p(c+\Delta t) - p(t)}{\Delta t} = \lim_{\Delta t \to 0} \frac{\alpha - \alpha}{\Delta t} = \lim_{\Delta t \to 0} \frac{0}{\Delta t} = \lim_{\Delta t \to 0} 0 = 0 \quad ◀$$

▶ **EXAMPLE 2** Suppose that the position of car A is given in feet by $50t$ when t is measured in seconds (as in Table 1). Verify that the instantaneous velocity of car A is 50 feet per second for every value of t.

Solution Because $p(t) = 50t$ feet, the instantaneous velocity in feet per second at time $t = c$ is

$$\begin{aligned} p'(c) &= \lim_{\Delta t \to 0} \frac{p(c+\Delta t) - p(c)}{\Delta t} \\ &= \lim_{\Delta t \to 0} \frac{50(c+\Delta t) - 50c}{\Delta t} \\ &= \lim_{\Delta t \to 0} \frac{50\Delta t}{\Delta t} \\ &= \lim_{\Delta t \to 0} 50 \\ &= 50. \quad ◀ \end{aligned}$$

INSIGHT In Example 2 it is useful to notice that we can replace $p(t) = 50t$ with $p(t) = \alpha t$ where α is an arbitrary constant. By doing so we see that

$$\text{If} \quad p(t) = \alpha t, \quad \text{with } \alpha \text{ constant, then } p'(c) = \alpha. \tag{3.1.2}$$

Our next two examples call for velocity calculations that require the precise definition of instantaneous velocity, as given by formula (3.1.1).

▶ **EXAMPLE 3** As in the discussion at the beginning of this section, the position of car B is given by $p(t) = 8t^2$. What is its velocity when $t = 1$? When $t = 2$?

Solution By formula (3.1.1), the instantaneous velocity of car B at time $t = c$ is

$$\begin{aligned} p'(c) &= \lim_{\Delta t \to 0} \frac{p(c+\Delta t) - p(c)}{\Delta t} \\ &= \lim_{\Delta t \to 0} \frac{8(c+\Delta t)^2 - 8c^2}{\Delta t} \\ &= 8 \cdot \lim_{\Delta t \to 0} \frac{\left(c^2 + 2c\Delta t + (\Delta t)^2\right) - c^2}{\Delta t} \\ &= 8 \cdot \lim_{\Delta t \to 0} (2c + \Delta t) \\ &= 8 \cdot 2c. \end{aligned}$$

If we set $c = 1$ in this formula for $p'(c)$, then we arrive at $p'(1) = 16$, which agrees with the value we obtained in the discussion at the beginning of this section. In this example, however, we have done much more. By calculating $p'(c)$ for a general value c, we can state the velocity of car B at any other time with no extra work. Thus to calculate the velocity of car B when $t = 2$, we set $c = 2$ in our formula for $p'(c)$ and obtain $8 \cdot 2 \cdot 2$, or 32. ◀

INSIGHT In Example 3, it is useful to notice that we can replace $p(t) = 8t^2$ with $p(t) = \alpha t^2$ where α is an arbitrary constant. By repeating the calculation with the arbitrary constant α instead of the specific constant 8, we see that

$$\text{If } p(t) = \alpha t^2, \quad \text{with } \alpha \text{ constant, then} \quad p'(c) = \alpha \cdot 2c. \tag{3.1.3}$$

▶ **EXAMPLE 4** Relative to a specified horizontal coordinate system, the position of a point Q on a piston is given by

$$p(t) = 18 + 4t - 2t^2,$$

in inches, for $0 \le t \le 4$ seconds. See Figure 3. What is the instantaneous velocity of the piston at time $t = 3$?

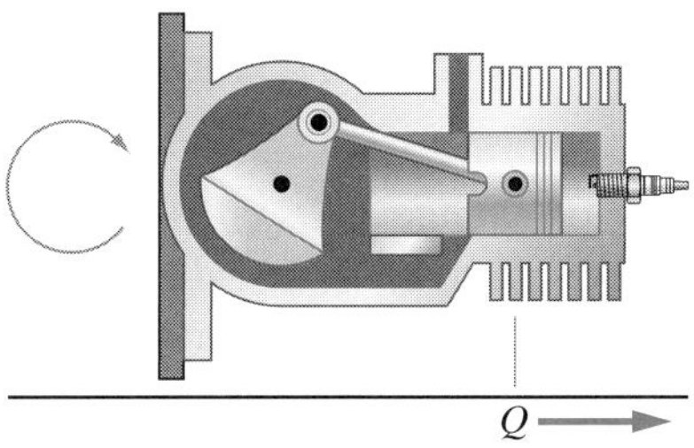

▲ **Figure 3a** Positive velocity

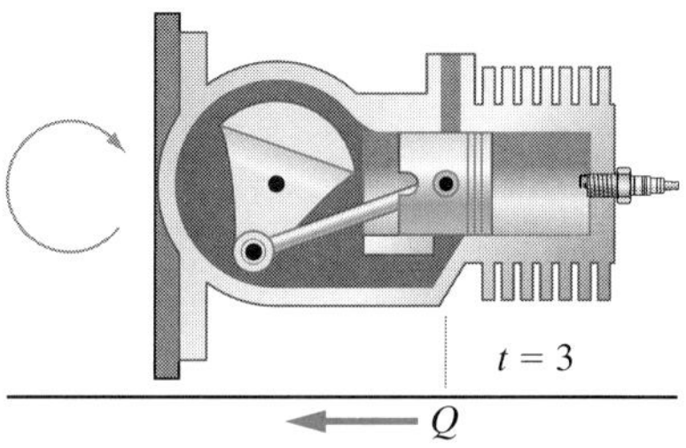

▲ **Figure 3b** Negative velocity

Solution Because $p(t)$ is measured in inches and t in seconds, we see from formula (3.1.1) that the units of $p'(3)$ will be inches per second. We calculate $p(3) = 18 + 4 \cdot 3 - 2 \cdot 3^2 = 12$ and

$$\begin{aligned} p(3 + \Delta t) &= 18 + 4(3 + \Delta t) - 2(3+\Delta t)^2 \\ &= 18 + 12 + 4\Delta t - 2\left(9 + 6\Delta t + (\Delta t)^2\right) \\ &= 12 - 8\Delta t - 2(\Delta t)^2. \end{aligned}$$

Therefore

$$\begin{aligned} p'(3) &= \lim_{\Delta t \to 0} \frac{p(3 + \Delta t) - p(3)}{\Delta t} \\ &= \lim_{\Delta t \to 0} \frac{\left(12 - 8\Delta t - 2(\Delta t)^2\right) - 12}{\Delta t} \\ &= \lim_{\Delta t \to 0} \frac{-8\Delta t - 2(\Delta t)^2}{\Delta t} \\ &= \lim_{\Delta t \to 0} (-8 - 2\Delta t) \\ &= -8. \end{aligned}$$

Thus $p'(3) = -8$ in/s.

INSIGHT The meaning of negative velocity of an object can be understood by looking at the definition of velocity. If $\Delta t > 0$, then the average velocity

$$\frac{p(c + \Delta t) - p(c)}{\Delta t}$$

at time $t = c$ will be positive if $\left(p(c + \Delta t) - p(c)\right) > 0$ (the object is moving forward) and negative if $\left(p(c + \Delta t) - p(c)\right) < 0$ (the object is moving backward). Passing to the limit, we see that positive instantaneous velocity corresponds to forward motion whereas negative instantaneous velocity corresponds to backward motion. In Example 4, the piston is moving to the left (in the negative direction in the coordinate system) at the instant $t = 3$ (see Figure 3b). In everyday speech, we do not distinguish between speed and velocity. In mathematics and physics, however, *speed* refers to the absolute value of velocity. Thus speed is always nonnegative.

Instantaneous Rate of Change

We have just defined instantaneous velocity as a limit of average velocities. It is useful to extend this construction to other functions f. In the present discussion, we do not assume that the variable x represents time. Additionally, the expression $f(x)$ is not assumed to be a distance. Whatever quantities x and $f(x)$ do measure, it is still true that the quotient

$$\frac{f(c + \Delta x) - f(c)}{\Delta x}$$

is the average rate of change of $f(x)$ as x changes from c to $c + \Delta x$. By letting $\Delta x \to 0$, we obtain the *instantaneous rate of change* $f'(c)$ of $f(x)$ at $x = c$:

$$f'(c) = \lim_{\Delta x \to 0} \frac{f(c + \Delta x) - f(c)}{\Delta x}, \tag{3.1.4}$$

provided this limit exists and is finite.

Notice that the limits in formulas (3.1.1) and (3.1.4) are mathematically the same. That is because instantaneous velocity is a particular example of the instantaneous rate of change of a function. Other instantaneous rates of change often are of interest. In economics, the function $f(x)$ of formula (3.1.4) might be the cost of producing x units of a commodity. In that case, the instantaneous rate of change $f'(c)$ is called the *marginal cost* of producing the cth unit. In pharmacology, $f(x)$ might be the measurable effect of a drug on a patient at time x. The instantaneous rate of change $f'(c)$ would then be called the *sensitivity* of the patient to the drug at time c.

▶ **EXAMPLE 5** The volume of a sphere of radius r is $4\pi r^3/3$. What is the instantaneous rate of change of the volume with respect to r when $r = 3$?

Solution The volume $f(r)$ of a sphere of radius r is given by $f(r) = 4\pi r^3/3$. Let us denote the constant $4\pi/3$ by α for short, so that $f(r) = \alpha r^3$. We are required to calculate $f'(3)$. We will carry out the calculation for a general value c of r and answer the question by substituting $c = 3$. To do so, we refer to formula (3.1.4). It is more natural to continue using the variable r for the radius rather than switching to x. Letting Δr denote the increment in the radius, we have

$$f'(c) = \lim_{\Delta r \to 0} \frac{f(c+\Delta r) - f(c)}{\Delta r} = \lim_{\Delta r \to 0} \frac{\alpha(c+\Delta r)^3 - \alpha c^3}{\Delta r}$$
$$= \alpha \lim_{\Delta r \to 0} \frac{\left(c^3 + 3c^2 \cdot \Delta r + 3c(\Delta r)^2 + (\Delta r)^3\right) - c^3}{\Delta r}.$$

Because the numerator simplifies to $3c^2 \cdot \Delta r + 3c(\Delta r)^2 + (\Delta r)^3$, or $(\Delta r)\left(3c^2 + 3c(\Delta r) + (\Delta r)^2\right)$, it follows that

$$f'(c) = \alpha \lim_{\Delta r \to 0} \frac{(\Delta r)\left(3c^2 + 3c(\Delta r) + (\Delta r)^2\right)}{\Delta r} = \alpha \lim_{\Delta r \to 0} \left(3c^2 + 3c(\Delta r) + (\Delta r)^2\right)$$
$$= \alpha \cdot 3c^2.$$

In summary:

$$\text{If } f(x) = \alpha x^3 \text{ with } \alpha \text{ constant, then } f'(c) = \alpha \cdot 3c^2. \tag{3.1.5}$$

Because $\alpha = 4\pi/3$, we see that

$$f'(c) = \frac{4\pi}{3} \cdot 3c^2 = 4\pi c^2.$$

Thus $f'(3) = 4\pi \cdot 3^2 = 36\pi$ is the instantaneous rate of change of volume with respect to r when $r = 3$. ◀

◀ **A LOOK BACK** Examples 1, 2, 3, and 5 are concerned with the instantaneous rates of change of functions of the form $f(x) = \alpha x^n$ for $n = 0, 1, 2,$ and 3. A careful examination reveals that there is a pattern to the formulas for $f'(c)$. The key is to restate line (3.1.5) as follows: If α is a constant, if $n = 3$, and if $f(x) = \alpha x^n$, then $f'(c) = \alpha \cdot nc^{n-1}$. Lines (3.1.3) and (3.1.2) show that this relationship between f and f' also holds for $n = 2$ and $n = 1$, respectively. Furthermore, if we set $n = 0$, then $f(x) = \alpha x^n = \alpha x^0 = \alpha$ is the constant function α, and $f'(c) = \alpha \cdot nc^{n-1} = \alpha \cdot 0 \cdot c^{n-1} = 0$ is the function that is identically 0. In Example 1, we verified that the instantaneous rate of change of a constant function is 0. Thus the rule we have found for f' holds when $n = 0$. In summary:

$$\text{If } f(x) = \alpha \cdot x^n \quad \text{for } n = 0, 1, 2, \text{ or } 3, \quad \text{then } f'(c) = \alpha \cdot nc^{n-1}. \tag{3.1.6}$$

Sums of Functions

Our next theorem tells us that, if α is a constant, then the instantaneous rate of change of $(\alpha f)(x)$ is α times the instantaneous rate of change of $f(x)$. It also tells us that we can calculate the instantaneous rate of change of a sum by adding the instantaneous rates of change of its summands.

THEOREM 1 Suppose that $f'(c)$ and $g'(c)$ exist, and that α and β are constants. Then

$$(\alpha f)'(c) = \alpha \cdot f'(c), \tag{3.1.7}$$

$$(f + g)'(c) = f'(c) + g'(c), \tag{3.1.8}$$

and, more generally,

$$(\alpha \cdot f + \beta \cdot g)'(c) = \alpha \cdot f'(c) + \beta \cdot g'(c). \tag{3.1.9}$$

Proof. When we calculate an instantaneous rate of change using definition (3.1.4), we are evaluating a limit. We therefore make use of appropriate limit laws. Thus to calculate $(\alpha \cdot f + \beta \cdot g)'(c)$ we use parts (a) and (d) of Theorem 2, Section 2.2:

$$\begin{aligned}
(\alpha \cdot f + \beta \cdot g)'(c) &= \lim_{\Delta x \to 0} \frac{(\alpha \cdot f + \beta \cdot g)(c + \Delta x) - (\alpha \cdot f + \beta \cdot g)(c)}{\Delta x} \\
&= \lim_{\Delta x \to 0} \frac{\alpha\big(f(c + \Delta x) - f(c)\big) + \beta\big(g(c + \Delta x) - g(c)\big)}{\Delta x} && \text{rearranging terms} \\
&= \lim_{\Delta x \to 0} \frac{\alpha\big(f(c + \Delta x) - f(c)\big)}{\Delta x} + \lim_{\Delta x \to 0} \frac{\beta\big(g(c + \Delta x) - g(c)\big)}{\Delta x} && \text{using Theorem 2a, Section 2.2} \\
&= \alpha \cdot \lim_{\Delta x \to 0} \frac{f(c + \Delta x) - f(c)}{\Delta x} + \beta \cdot \lim_{\Delta x \to 0} \frac{g(c + \Delta x) - g(c)}{\Delta x} && \text{using Theorem 2d, Section 2.2} \\
&= \alpha \cdot f'(c) + \beta \cdot g'(c) && \text{by definition (3.1.4),}
\end{aligned}$$

which proves (3.1.9). Equation (3.1.8) is the special case of (3.1.9) that results from setting $\alpha = \beta = 1$. Equation (3.1.7) is the special case of (3.1.9) that results from setting $\beta = 0$. ■

▶ **EXAMPLE 6** What is the instantaneous rate of change of $F(x) = 7x^3 - 5x^2$ at $x = -1$?

Solution We are asked for $F'(-1)$. Equation (3.1.4) defines this quantity as a limit, but we now have several rules that allow us to calculate instantaneous rate of change without referring to the limit process. Let $f(x) = 7x^3$ and $g(x) = -5x^2$. Then $F(x) = f(x) + g(x)$. By equation (3.1.6) with $\alpha = 7$ and $n = 3$, we have $f'(-1) = 7 \cdot 3(-1)^{3-1} = 21$. By equation (3.1.6) with $\alpha = -5$ and $n = 2$, we have $g'(-1) = (-5) \cdot 2(-1)^{2-1} = 10$. Thus by applying equation (3.1.8), we obtain $F'(-1) = 21 + 10 = 31$. ◀

We can easily extend formulas (3.1.8) and (3.1.9) to three or more summands. For example, by applying (3.1.8) twice, we can handle the sum of three terms as follows:

$$(f + g + h)'(c) = \big((f + g) + h\big)'(c) = (f + g)'(c) + h'(c) = f'(c) + g'(c) + h'(c). \quad (3.1.10)$$

▶ **EXAMPLE 7** Let $p(t) = 18 + 4t - 2t^2$. In Example 4, we obtained $p'(3) = -8$ by evaluating a limit. Derive this result by using equations (3.1.6) and (3.1.10) instead.

Solution Let $f(t) = 18 = 18t^0$, $g(t) = 4t = 4t^1$, and $h(t) = -2t^2$. According to equation (3.1.6), we have $f'(3) = 18 \cdot 0 \cdot (3)^{0-1} = 0$, $g'(3) = 4 \cdot 1 \cdot (3)^{1-1} = 4$, and $h'(3) = (-2) \cdot 2 \cdot (3)^{2-1} = -12$. Therefore by equation (3.1.10), we have $p'(3) = 0 + 4 + (-12) = -8$. ◀

The Concept of Tangent Line

It is easy to understand the idea of the tangent line to a circle. Here are three good ways to think about the tangent line to the circle C at the point P (illustrated in Figure 4).

1. The tangent line is perpendicular to the radius at P and passes through P.
2. The tangent line passes through P and intersects C at just that one point.
3. The tangent line passes through P, and the circle lies on one side of the tangent line.

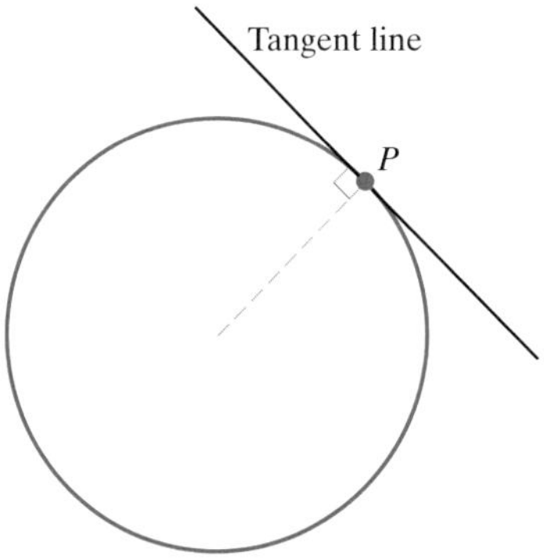

▲ Figure 4

All of these descriptions are correct, and all of them uniquely determine the tangent line to the circle C at the point P. But each of them is useless in determining the tangent line to the curve in Figure 5 at the point P. Intuition tells us that the dotted line in the figure is the tangent line. But this line in no sense satisfies statements 1, 2, or 3; we need a new mathematical description of tangent line.

We use a limiting process to determine the tangent line to the graph of $y = f(x)$ at a point $P = \big(c, f(c)\big)$. Consider the graph in Figure 5. The dotted line indicates what our intuition tells us that the tangent line is—the dotted line through P that "most nearly approximates" the curve. The trouble is that it takes *two* pieces of information to determine a line. We only know one: the point P through which the line passes. We get a second piece of information, namely the slope m, by considering nearby *secant lines*, such as the line passing through points P and $\tilde{P}$ in Figure 6.

A secant line is a line passing through two points of the curve—in this case $P = \big(c, f(c)\big)$ and $\tilde{P} = \big(c + \Delta x, f(c + \Delta x)\big)$. The slope of this secant line is

$$\frac{f(c + \Delta x) - f(c)}{(c + \Delta x) - c}, \quad \text{or} \quad \frac{f(c + \Delta x) - f(c)}{\Delta x}.$$

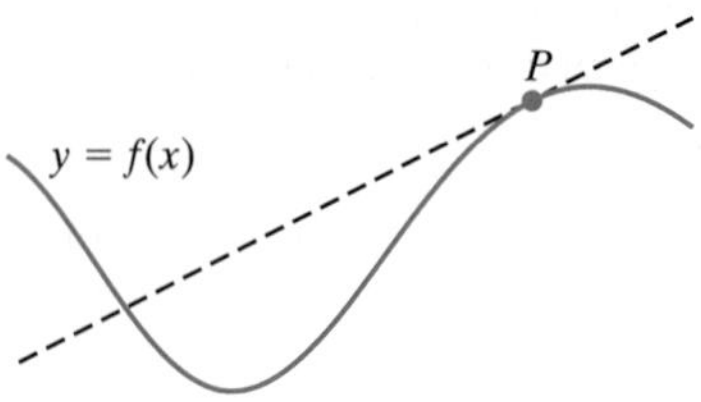

▲ Figure 5

What we see geometrically is that, as $\Delta x \to 0$ (that is, as $\tilde{P} \to P$), the slope of the secant line tends to the slope of the tangent line at P. In other words, we are saying that the slope m of the tangent line to $y = f(x)$ at $P = (c, f(c))$ is given by

$$m = \lim_{\Delta x \to 0} \frac{f(c + \Delta x) - f(c)}{\Delta x}. \tag{3.1.11}$$

Refer to Figure 7.

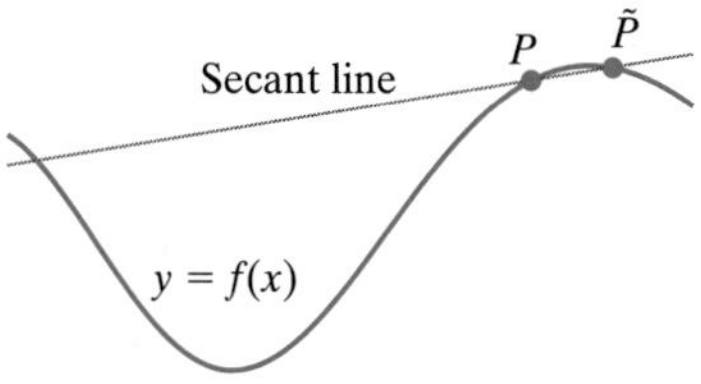

▲ Figure 6

INSIGHT Observe that the right side of equation (3.1.11) is identical to the right side of equation (3.1.4). That is, the slope of the tangent line to the graph of f at $(c, f(c))$ is the *same* as the instantaneous rate of change of $f(x)$ with respect to x at $x = c$. This may seem surprising at first—especially when $f(x)$ has some other physical meaning (such as volume, as in Example 5). But when we graph the equation $y = f(x)$, the value $f(x)$, in addition to any other meaning it may have, represents the distance of the point $(x, f(x))$ from the x-axis. The increment $f(c + \Delta x) - f(c)$ is therefore a vertical increment Δy, and the quotient $\dfrac{f(c + \Delta x) - f(c)}{\Delta x}$ is a slope $\dfrac{\Delta y}{\Delta x}$. Thus $f'(c)$ is both an instantaneous rate of change of f *and* a slope.

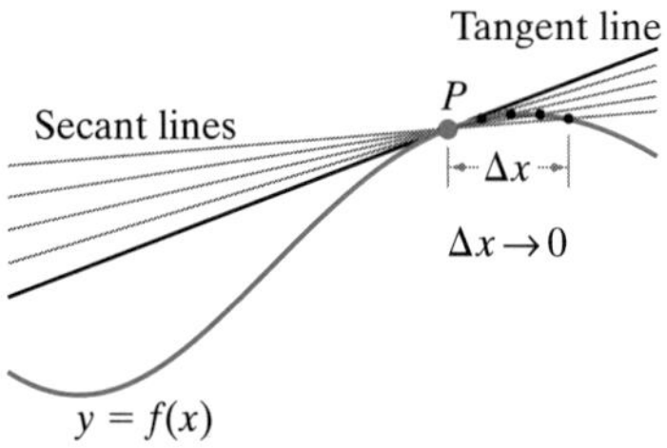

▲ Figure 7

Using the point-slope equation of a line (Section 1.3) and formula (3.1.11), we may now define the tangent line to the graph of a function at a point.

DEFINITION Let f be defined on an open interval containing the point c. Suppose that the limit

$$f'(c) = \lim_{\Delta x \to 0} \frac{f(c + \Delta x) - f(c)}{\Delta x}$$

exists and is finite. Then the *tangent line* to the graph of f at the point $(c, f(c))$ is the line with Cartesian equation

$$y = f'(c) \cdot (x - c) + f(c). \tag{3.1.12}$$

In particular,

$f'(c) =$ the slope of the tangent line to the graph of f at the point $(c, f(c))$.

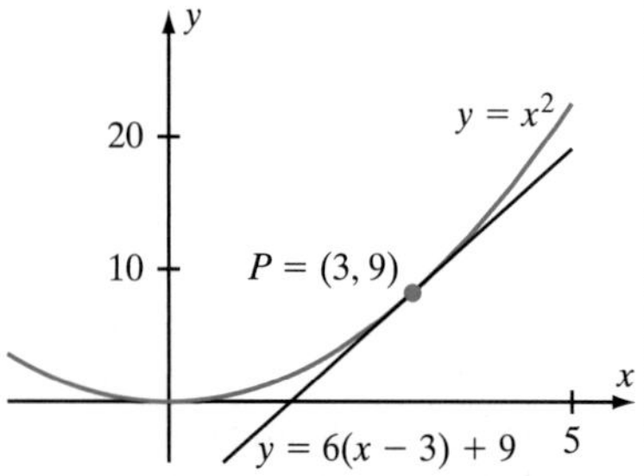

▲ Figure 8 From the tickmarks, notice that the axes have different scales. That is why the slope of the tangent line, which is actually 6, appears to be close to 1.

▶ **EXAMPLE 8** Find the equation of the tangent line to the graph of $y = x^2$ at the point $P = (3, 9)$.

Solution Let $f(x) = x^2$. The slope of the tangent line at $P = (3, 9)$ is

$$f'(3) = \lim_{\Delta x \to 0} \frac{f(3 + \Delta x) - f(3)}{\Delta x}.$$

To evaluate this limit, we can use line (3.1.6) with $\alpha = 1$, $n = 2$, and $c = 3$. We obtain $f'(3) = 1 \cdot 2 \cdot (3^1) = 6$. According to formula (3.1.12), the equation of the tangent line is $y = 6(x - 3) + 9$. The graph of f and its tangent line are shown in Figure 8. ◀

▶ **EXAMPLE 9** Find the tangent line to the curve $y = 1/x$ at the point $P = (-3, -1/3)$.

Solution Let $f(x) = 1/x$. Let $c \neq 0$ be a point in the domain of f. We calculate

$$\begin{aligned} f'(c) &= \lim_{\Delta x \to 0} \frac{f(c+\Delta x) - f(c)}{\Delta x} \\ &= \lim_{\Delta x \to 0} \frac{1}{\Delta x}\left(\frac{1}{c+\Delta x} - \frac{1}{c}\right) \\ &= \lim_{\Delta x \to 0} \frac{1}{\Delta x}\left(\frac{-\Delta x}{c(c+\Delta x)}\right) \\ &= -\lim_{\Delta x \to 0} \frac{1}{c(c+\Delta x)} \\ &- \frac{1}{c^2}. \end{aligned}$$

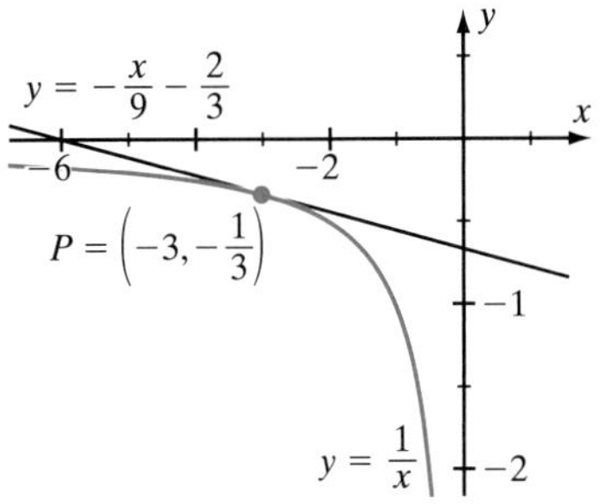

▲ Figure 9

Setting $c = -3$, we see that the slope of the tangent line at P is $f'(-3) = -1/(-3)^2 = -1/9$. According to definition (3.1.12), the equation of the tangent line is $y = (-1/9)\,(x - (-3)) + (-1/3)$, or $y = -x/9 - 2/3$. Figure 9 illustrates the graph of $y = 1/x$ and its tangent line at P. ◀

INSIGHT From the solution to Example 9, we see that $f'(c) = -1/c^2$ if $f(x) = 1/x$. Using equation (3.1.7), we deduce that, if α is a constant and $f(x) = \alpha/x$, then $f'(c) = -\alpha/c^2$. Letting $n = -1$, we may restate this result as follows: if $f(x) = \alpha \cdot x^n$, then $f'(c) = \alpha \cdot n\,c^{n-1}$. In other words, formula (3.1.6), which previously we have shown to hold for $n = 0, 1, 2, 3$, is also true for $n = -1$:

$$\text{If } f(x) = \alpha \cdot x^n \quad \text{for } n = -1, 0, 1, 2, \text{ or } 3, \quad \text{then } f'(c) = \alpha \cdot n\,c^{n-1}. \tag{3.1.13}$$

Normal Lines to Curves

A line L is said to be *perpendicular* or *normal* to a curve $y = f(x)$ at a point $P = (c, f(c))$ if L is perpendicular to the tangent line to the curve at the point P. If the tangent line is the horizontal line $y = f(c)$, then the normal line is the vertical line $x = c$. If the slope $f'(c)$ of the tangent line is nonzero, then its negative reciprocal $-1/f'(c)$ is the slope of the normal line (by Theorem 1 of Section 1.3). Thus, if $f'(c) \neq 0$, then

$$y = -\frac{1}{f'(c)}(x - c) + f(c) \tag{3.1.14}$$

is the normal line to the graph of f at the point $(c, f(c))$.

▶ **EXAMPLE 10** Find the line perpendicular to the curve $y = x^3/2$ at the point $P = (-2, -4)$.

Solution Let $f(x) = x^3/2$. Using formula (3.1.6) with $\alpha = 1/2$ and $n = 3$, we have

$$f'(c) = \frac{1}{2} \cdot 3c^2 = \frac{3}{2}c^2.$$

Therefore $f'(-2) = 3 \cdot (-2)^2/2 = 6$ is the slope of the tangent line to the graph at P. The slope of the normal line is the negative reciprocal of this number, or $-1/6$. The equation of the normal line to the graph of $y = x^3/2$ at P is, by (3.1.14), $y = (-1/6)\,(x - (-2)) + (-4)$, or $y = -x/6 - 13/3$. A sketch is shown in Figure 10. ◀

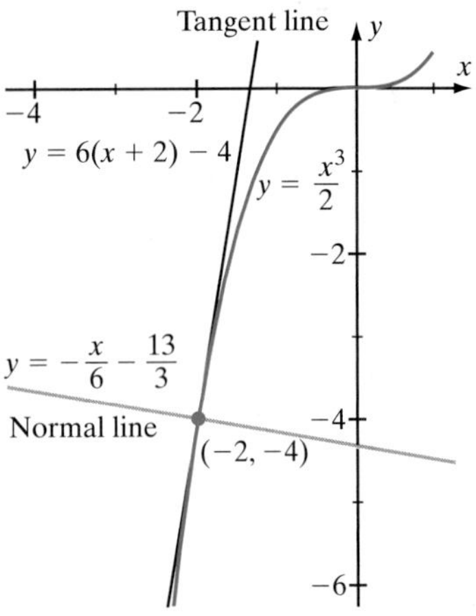

▲ Figure 10

Corners and Vertical Tangent Lines

The definition of the tangent line of f at $(c, f(c))$ requires the two-sided limit (3.1.11) for the slope. This limit fails to exist if the one-sided limits

$$\ell_L = \lim_{\Delta x \to 0^-} \frac{f(c + \Delta x) - f(c)}{\Delta x} \qquad \text{and} \qquad \ell_R = \lim_{\Delta x \to 0^+} \frac{f(c + \Delta x) - f(c)}{\Delta x}$$

exist and are finite but unequal. In this case, the line $y - f(c) = \ell_L(x - c)$ is "tangent" to that part of the graph of f to the left of c, and the line $y - f(c) = \ell_R(x - c)$ is "tangent" to that part of the graph of f to the right of c. But there is no line that is tangent to *both* sides of the graph of f. Therefore f does not have a tangent line at $(c, f(c))$. We say that f has a *corner* at $(c, f(c))$. A familiar example is the vee-shaped graph of $y = |x|$, which has a corner at (0, 0): see Figure 11.

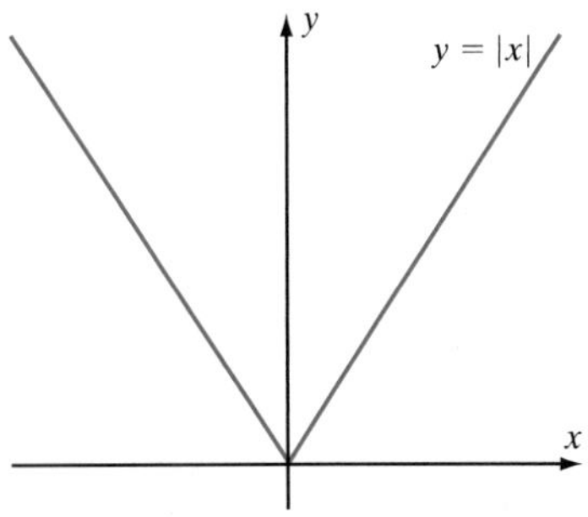

▲ Figure 11

▶ **EXAMPLE 11** Does the graph of the function f defined by

$$f(x) = \begin{cases} x^2 & \text{if } \ x < 2 \\ 5 - x^2/4 & \text{if } \ x \geq 2 \end{cases}$$

have a tangent line at the point $P = (2, 4)$?

Solution Notice that the left and right limits of $f(x)$ at $x = 2$ exist and equal 4. Because these limits exist and agree, f is continuous at 2. For $c = 2$, we have

$$\begin{aligned} \lim_{\Delta x \to 0^-} \frac{f(c + \Delta x) - f(c)}{\Delta x} &= \lim_{\Delta x \to 0^-} \frac{((2 + \Delta x)^2 - 4)}{\Delta x} \\ &= \lim_{\Delta x \to 0^-} \frac{(\Delta x)^2 + 4\Delta x}{\Delta x} \\ &= \lim_{\Delta x \to 0^-} (4 + \Delta x) \\ &= 4. \end{aligned}$$

On the other hand,

$$\begin{aligned} \lim_{\Delta x \to 0^+} \frac{f(c + \Delta x) - f(c)}{\Delta x} &= \lim_{\Delta x \to 0^+} \frac{\left(5 - \frac{1}{4}(2 + \Delta x)^2\right) - \left(5 - \frac{1}{4} \cdot 2^2\right)}{\Delta x} \\ &= \lim_{\Delta x \to 0^-} \frac{-\Delta x - \frac{1}{4}(\Delta x)^2}{\Delta x} \\ &= \lim_{\Delta x \to 0^+} \left(-1 - \frac{1}{4}\Delta x\right) \\ &= -1. \end{aligned}$$

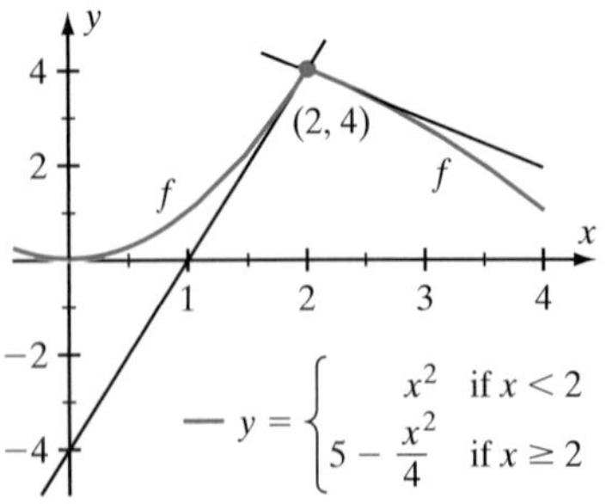

▲ Figure 12

Because the one-sided limits exist but are unequal, the graph of f has a corner at P. The graph of f is shown in Figure 12. The lines that are "one-sided" tangents are included in the plot. The graph does *not* have a tangent line at P. ◀

Our final example concerns a function for which formula (3.1.11) does not evaluate to a finite limit.

▶ **EXAMPLE 12** Discuss the tangent line of $f(x) = 3 + (x-2)^{1/3}$ at the point $P = (2, 3)$.

Solution Because

$$\frac{f(2+\Delta x) - f(2)}{\Delta x} = \frac{\left(3 + (\Delta x)^{1/3}\right) - 3}{\Delta x} = \frac{1}{(\Delta x)^{2/3}},$$

we may use Theorem 5 of Section 2.2 to conclude that

$$\lim_{\Delta x \to 0} \frac{f(2+\Delta x) - f(2)}{\Delta x} = +\infty.$$

Our definition of tangent line therefore does not apply. Nevertheless, the graph of f (Figure 13) indicates that f has a vertical tangent line, $x = 2$, at P. It is possible to develop the concept of the tangent line to include vertical tangent lines, but because we do not need the greater generality, we will not do so. ◀

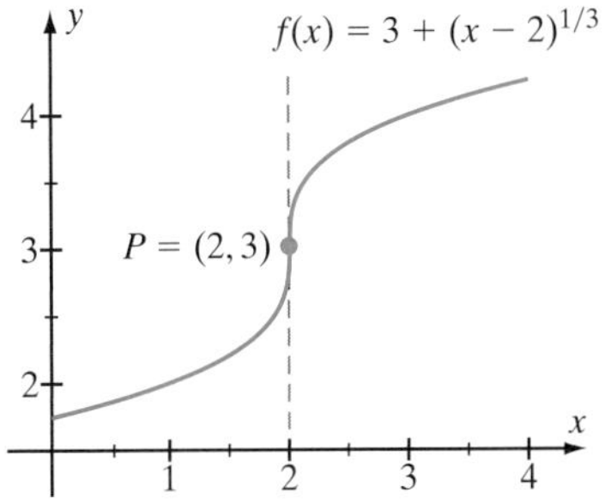

▲ Figure 13

QUICK QUIZ

1. A particle moves along an axis with position $p(t) = 4t^2$ at time t. What is its velocity when $t = 3$?
2. The position of a particle moving along the x-axis is $24t - 2t^3$ at time t. For what values of t is the particle moving to the left?
3. What is the instantaneous rate of change of the area of a circle with respect to its radius when its circumference is 10?
4. What is the tangent line to the graph of $y = 2x^3 + 3$ at the point $(-1, 1)$?

Answers

1. 24 2. $(-\infty, -2) \cup (2, \infty)$ 3. 10 4. $y = 6(x+1) + 1$

EXERCISES

Problems for Practice

▶ In Exercises 1–12, $p(t)$ describes the position of an object at time t. Calculate the instantaneous velocity at time c. ◀

1. $p(t) = 5$ $\quad c = 2$
2. $p(t) = -3t$ $\quad c = 5$
3. $p(t) = -7t^2$ $\quad c = 3$
4. $p(t) = -4t^3$ $\quad c = 2$
5. $p(t) = 6t + 8$ $\quad c = 3$
6. $p(t) = -5t^2 + 2$ $\quad c = 1$
7. $p(t) = 2t^3 - 17t$ $\quad c = 2$
8. $p(t) = 2t^3 - 3t^2$ $\quad c = -1$
9. $p(t) = t^2 - 6t + 10$ $\quad c = 2$
10. $p(t) = t^3 + 2t^2 + 3t + 4$ $\quad c = 2$
11. $p(t) = 1/t$ $\quad c = 1/2$
12. $p(t) = 2t - 3/t$ $\quad c = 3$

▶ In Exercises 13–16, $p(t)$ describes the position of a moving body at time t. Determine whether, at time $t = 4$, the body is moving forward, backward, or neither. ◀

13. $p(t) = 6t + 3$
14. $p(t) = t^2 - 8t$
15. $p(t) = -t^3 + 5t^2$
16. $p(t) = 1/t$

▶ In each of Exercises 17–20, a function f and a point c are given. Calculate $f'(c)$. ◀

17. $f(x) = 5x^2 - 21x$ $\quad c = 3$
18. $f(x) = 2x^3 + 3x^2$ $\quad c = -3$
19. $f(x) = 3x^2 + 2/x$ $\quad c = -2$
20. $f(x) = 5x^3 + 4x^2 + 3x + 2$ $\quad c = -2$

In Exercises 21–24, find the slope of the tangent line to the graph of the given function at the given point P.

21. $f(x) = x^2$ $P = (3, 9)$
22. $f(x) = x^3 - 2x$ $P = (1, -1)$
23. $f(x) = 3x^2 + 6$ $P = (-1, 9)$
24. $f(x) = -4x^2 + x + 1$ $P = (2, -13)$

In each of Exercises 25–28, a function f and a point P are given. Find the point-slope form of the equation of the tangent line to the graph of f at P.

25. $f(x) = 2x^2$ $P = (5, 50)$
26. $f(x) = x^3/6$ $P = (2, 4/3)$
27. $f(x) = -3x^2 + 5$ $P = (-2, -7)$
28. $f(x) = 1/x$ $P = (-1, -1)$

In each of Exercises 29–32, a function f and a point P are given. Find the slope-intercept form of the equation of the tangent line to the graph of f at P.

29. $f(x) = 5x^2$ $P = (2, 20)$
30. $f(x) = x^3/3$ $P = (-3, -9)$
31. $f(x) = 3x^2 + 2x$ $P = (1, 5)$
32. $f(x) = 2x - 2/x$ $P = (-1/2, 3)$

In each of Exercises 33–36, a function f and a point P are given. Find the point-slope form of the equation of the normal line to the graph of f at P.

33. $f(x) = 2x^2$ $P = (5, 50)$
34. $f(x) = x^3/6$ $P = (2, 4/3)$
35. $f(x) = -3x^2 + 5$ $P = (-2, -7)$
36. $f(x) = 1/x$ $P = (-1, -1)$

In each of Exercises 37–40, a function f and a point P are given. Find the slope-intercept form of the equation of the normal line to the graph of f at P.

37. $f(x) = 3x^2$ $P = (-1, 3)$
38. $f(x) = x^3/2$ $P = (2, 4)$
39. $f(x) = 3x^3 - 2x^2 - 10$ $P = (2, 6)$
40. $f(x) = x^2 - 3/x$ $P = (-3, 10)$

Further Theory and Practice

41. The instantaneous rate of change of velocity is acceleration. For the position function $p(t) = t^3$, what is the acceleration at time $t = 1$?
42. The population of a colony of bacteria after t hours is $B(t) = 5000 + 6t^3$. At what rate is the population changing after 2 hours?
43. If $C(x)$ is the cost of producing x units of an item, then the *marginal cost* of the xth item is defined to be $C'(x)$. Suppose that the cost in cents of producing x pencils is $C(x) = 5 + 0.1x - 0.001x^2$ for $x \leq 50$. What is the marginal cost when $x = 25$?
44. A large herd of reindeer is dying out. The number of reindeer in the herd at time t (measured in months), $0 \leq t \leq 15$, is

$$r(t) = 25000 - 800t - 40t^2 - t^3.$$

At what rate are the reindeer dying out after 11 months?

In Exercises 45–50, $p(t)$ describes the position of an object at time t. Calculate the instantaneous velocity at time c.

45. $p(t) = t(t+1)$ $c = 2$
46. $p(t) = t^2(2t - 3)$ $c = 3$
47. $p(t) = t^2(3 - 2/t)$ $c = 3$
48. $p(t) = (t+2)(t+3)$ $c = -1$
49. $p(t) = (t^2 + 9)/t$ $c = 2$
50. $p(t) = t(t+1)(t+2)$ $c = 2$

In Exercises 51–54, find a line that is tangent to the graph of the given function f and that is parallel to the line $y = 12x$.

51. $f(x) = 3x^2 + 1$
52. $f(x) = x^2 - 4x + 2$
53. $f(x) = x^3 - 15x + 20$
54. $f(x) = 11x - 4/x$

55. Six points are labeled on the graph of the function f in Figure 14. The instantaneous rates of change of f with respect to x at the six points are -3, -1, 0, 3, 10, and 20. Match each point to the corresponding rate of change.

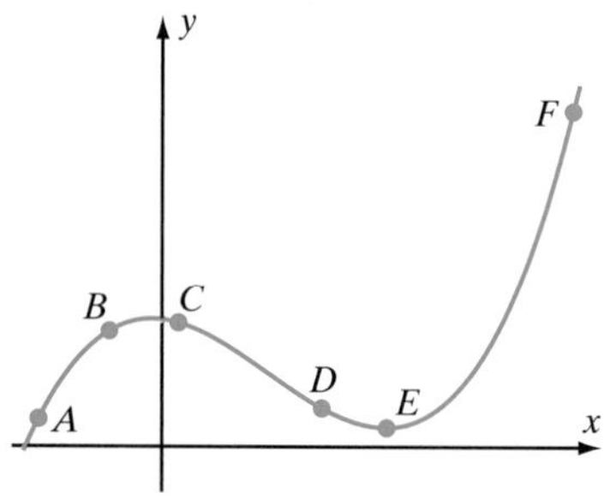

▲ Figure 14

56. What is the rate of change of the area of a square with respect to its side length when the side length is 8 centimeters?
57. What is the rate of growth of the surface area of a sphere with respect to the radius when the radius is 8 inches? (The surface area of a sphere of radius r is $4\pi r^2$.)
58. What is the rate of change of the area of an equilateral triangle with respect to its side length when that side length is 8 inches?
59. Let $p(t) = t^2 + t$ denote the position of a moving body. Determine for which values of t the velocity of the body is positive and for which values of t the velocity is negative.

60. Let $p(t) = t^2 - 6t^3$ denote the position of a moving body. Determine for which values of t the velocity of the body is positive and for which values of t the velocity is negative.

61. Find the equation of the line that is tangent to the graph of $f(x) = 3x^3 + 12$ and that passes through the origin.

62. Find the equations of the tangent lines to the graph of $f(x) = 3x^2$ that pass through the point $(1, -9)$.

63. Let $f(x) = x^2$. For what value(s) of c does the tangent line to the graph of f at $(c, f(c))$ pass through the point $(3, 5)$?

64. For what values of A and C does the graph of $y = Ax^2 + C$ pass through the point $P = (1, 1)$ and have the same tangent line at P as the graph of $y = x^3$?

65. Find the points on the graph of the function $f(x) = x^3 - 2x^2 - 8x + 3$ at which the tangent line is parallel to the graph of $y = 4 - 9x$.

66. Find all values of c for which the tangent lines to the graphs of $f(x) = x^2 - 7x + 9$ and $g(x) = 9/x$ at $(c, f(c))$ and $(c, g(c))$ are parallel.

67. Find all values of c for which the tangent lines to the graphs of $f(x) = x^3 - 8x + 3$ and $g(x) = 4/x$ at $(c, f(c))$ and $(c, g(c))$ are parallel.

68. Let $f(x) = x^2$, and suppose that a and b are different constants. Find a formula for the point of intersection of the tangent line to the graph of f at $x = a$ and the tangent line to the graph of f at $x = b$.

69. Suppose that c is a positive constant. Let L_c be the line that is tangent to the graph of $f(x) = 1/x$ at $P = (c, 1/c)$. Show that the area of the triangle formed by L_c and the positive axes is independent of c. Compute that area.

70. At Wrigley field in Chicago, Cubs fans throw the ball back onto the field when a visiting team hits a home run. Suppose that the height H above field level of a ball thrown back by a Cubs fan is given in feet by

$$H(t) = 18 + 13.8t - 16t^2$$

when t is measured in seconds.

a. How high above field level is the fan sitting?
b. What is the rate at which the ball rises as a function of time?
c. At what time does the ball reach its maximum height? What is this maximum height?
d. What is the average rate of change of H during the ball's upward trajectory?
e. How long is the ball in the air?
f. What is the rate of change of H at the moment the ball hits the ground?
g. What is the average rate of change of H over the entire trajectory of the ball?

71. Suppose $a \neq 0$. What relationship between a and b is a necessary and sufficient condition for the graph of $f(x) = ax^2 + b$ to have a tangent line that passes through the origin?

72. Suppose that f is a function whose graph has a tangent line at each point. If $g(x) = f(x) + \alpha$ for some constant α, show that the graph of g has a tangent line at each point and that the slope of the tangent line to the graph of g at $(c, g(c))$ is the same as the slope of the tangent line to the graph of f at $(c, f(c))$. Explain this geometrically.

73. Let $f(x)$ be a quadratic polynomial. Show that the tangent line to the graph of f at any point P can only intersect the graph of f at P.

74. **Finding a function from the tangent lines to its graph.** If f is a continuous function with $f(2) = 5$, and if the slope of the tangent line to the graph of f at $(c, f(c))$ is -2 for $-\infty < c < 1$, 1 for $1 < c < 3$, and -1 for $3 < c < \infty$, find f.

75. Suppose that f is defined on an open interval centered at c. Suppose also that

$$\ell_R = \lim_{h \to 0^+} \frac{f(c+h) - f(c)}{h}$$

and

$$\ell_L = \lim_{h \to 0^-} \frac{f(c+h) - f(c)}{h}$$

exist. Let

$$T_R(x) = f(c) + \ell_R(x - c) \quad \text{for } x \geq c$$
$$T_L(x) = f(c) + \ell_L(x - c) \quad \text{for } x \leq c.$$

Define $\alpha_f(c)$ to be the radian measure of the angle through which T_R must be rotated counterclockwise about $(c, f(c))$ to coincide with T_L. We may think of $\alpha_f(c)$ as the angle of the corner at $P = (c, f(c))$.

a. For what values of $\alpha_f(c)$ is there actually a corner at P? Explain.
b. For what value of $\alpha_f(c)$ is there a tangent line at P. Explain.
c. If the graph of f has a vertical tangent at P, is $\alpha_f(c)$ defined? Explain.

76. Suppose that the graph of f has a nonvertical tangent T_P at each point P on it. Thus for each P there are two numbers $m(P)$ and $b(P)$ such that $T_P(x) = m(P)\,x + b(P)$.

a. Define the "tangents to the graph of f at infinity and minus infinity," $\lim_{P \to \infty} T_P$ and $\lim_{P \to -\infty} T_P$, in an appropriate way.
b. Show that if $f(x) = 1/x$, then $\lim_{P \to \infty} T_P$ and $\lim_{P \to -\infty} T_P$ are horizontal asymptotes of the graph of f.
c. Is the converse to (b) true? Think of $f(x) = \sin(x)/x$. Explain your answer.

Calculator/Computer Exercises

77. The trajectory of a fly ball is such that the height in feet above ground is $H(t) = 4 + 72t - 16t^2$ when t is measured in seconds.

a. Compute the average velocity in the following time intervals:
i. $[2, 3]$ **iii.** $[2, 2.01]$
ii. $[2, 2.1]$ **iv.** $[2, 2.001]$
b. Compute the instantaneous velocity at $t = 2$.

78. For the trajectory in the preceding problem,

a. Compute the average velocity in the following time intervals:

i. $[3, 3.1]$ **iii.** $[2.9, 3]$
ii. $[3, 3.01]$ **iv.** $[2.99, 3]$

b. Compute the instantaneous velocity at $t = 3$.

79. The position of an oscillating body is given by $p(t) = \sin(2t + \pi/6)$. Calculate the average velocity of the body over a time interval of the form $[0, \Delta t]$ for $\Delta t = 10^{-n}$, $n = 0, 1, 2, 3$ and 4. Display your results in the form of a table. Formulate a guess v for the instantaneous velocity of the body at time $t = 0$. Plot p. In the same coordinate plane, add the graph of the straight line that passes through $(0, 1/2)$ and that has slope v. Does the resulting figure support your conjectured value? Explain.

80. The position of a moving body is given by $p(t) = (2.718281828)^t$. Calculate the average velocity of the body over a time interval of the form $[1, 1 + \Delta t]$ for a sequence of small values of Δt. Display your results in the form of a table. Formulate a guess m for the instantaneous velocity of the body at time $t = 1$. Plot p. In the same coordinate plane, add the graph of the straight line that passes through $(1, 2.718281828)$ and that has slope m. Does the resulting figure support your conjectured value? Explain.

81. Repeat the preceding exercise with $t = 2$ and $t = 3$. What is the apparent relationship between the position of the body $p(t)$ and the instantaneous velocity at t?

82. Graph the function $f(x) = x/(x + 1)$, and zoom in on the point $(-2, 2)$. Formulate a guess for the slope of the tangent line L at that point. Now examine the quotient $(f(-2 + \Delta x) - f(-2))/\Delta x$ for various small values of Δx. Display your results in a table. Use this data to estimate the slope of L. How do your visual and numerical estimates compare?

83. Repeat the preceding exercise for the function $f(x) = \tan(x)$ at the point $(\pi/4, 1)$.

84. Suppose that $a > 0$. Tangent lines to the exponential functions $f(x) = a^x$ are investigated in this exercise.

a. Use the formula for the slope of a tangent line to show that the slope of the tangent line to f at $(c, f(c))$ is

$$f'(c) = a^c \lim_{h \to 0} \frac{a^h - 1}{h}.$$

b. Show that the slope of the tangent line to f at $(0, 1)$ is

$$\lim_{h \to 0} \frac{a^h - 1}{h}.$$

c. This limit cannot be computed by substituting $h = 0$ in the expression because that results in the meaningless expression 0/0. We will learn how to compute this limit in Section 3.6. For now, we will investigate it numerically. To be specific, let $a = 2$. To identify the limit, graph the function $h \mapsto (2^h - 1)/h$ in a window whose horizontal range is a small interval centered at 0. From the graph (zooming in if necessary), identify

$$\lim_{h \to 0} \frac{2^h - 1}{h}$$

to four decimal places.

d. Graph $f(x) = 2^x$ tangent lines to the graph of f at $(0, 1)$ and $(2, 4)$ in the same viewing window.

e. Repeat parts (c) and (d) with $a = 3$. Use $(-1, 1/3)$ and $(1, 3)$ as the points of tangency.

3.2 The Derivative

In Section 3.1, we saw that the limit

$$f'(c) = \lim_{\Delta x \to 0} \frac{f(c + \Delta x) - f(c)}{\Delta x}$$

represents the instantaneous rate of change of the function f at the point c. It is also the slope of the tangent line to the graph of f at $(c, f(c))$. Later in this chapter, and especially in Chapter 4 and beyond, we will see many other applications of this limit. Because of its importance, we give it a name and study ways to evaluate it.

DEFINITION Let f be a function that is defined in an open interval that contains a point c. If the limit

$$\lim_{\Delta x \to 0} \frac{f(c+\Delta x) - f(c)}{\Delta x}$$

exists and is finite, then we say that f is *differentiable* at c. We call this limit the *derivative* of the function f at the point c, and we denote it by the symbol $f'(c)$, as in Section 3.1:

$$f'(c) = \lim_{\Delta x \to 0} \frac{f(c+\Delta x) - f(c)}{\Delta x} \tag{3.2.1}$$

The process of calculating $f'(c)$ is called *differentiation.*

INSIGHT The property of differentiability can also be stated in geometric terms: the function f is differentiable at c if and only if the graph of f has a *nonvertical* tangent line at the point $(c, f(c))$.

It is often convenient to express $f'(c)$ by means of formulas that differ somewhat from (3.2.1). For example, the letter h is often used instead of Δx to indicate an increment:

$$f'(c) = \lim_{h \to 0} \frac{f(c+h) - f(c)}{h}. \tag{3.2.2}$$

Also, if we set $x = c + \Delta x$, then $\Delta x = x - c$ and

$$\frac{f(c+\Delta x) - f(c)}{\Delta x} = \frac{f(x) - f(c)}{x - c}.$$

The statement $\Delta x \to 0$ is equivalent to $x \to c$. Thus the definition of the derivative of f at c can be stated as

$$f'(c) = \lim_{x \to c} \frac{f(x) - f(c)}{x - c}. \tag{3.2.3}$$

In Section 3.1 we calculated $f'(c)$ for several functions. For now, if f is not one of the functions we have studied, we must compute $f'(c)$ by referring to definition (3.2.1). Our first example is an illustration of such a calculation.

► **EXAMPLE 1** Calculate $f'(5)$ for $f(x) = \sqrt{4x - 11}$.

Solution Here $c = 5$. Starting from formula (3.2.1), we have

$$\begin{aligned} f'(5) &= \lim_{\Delta x \to 0} \frac{f(5+\Delta x) - f(5)}{\Delta x} \\ &= \lim_{\Delta x \to 0} \frac{\sqrt{4(5+\Delta x) - 11} - \sqrt{4(5) - 11}}{\Delta x} \\ &= \lim_{\Delta x \to 0} \frac{\sqrt{9 + 4\Delta x} - 3}{\Delta x} \end{aligned}$$

$$= \lim_{\Delta x \to 0} \left(\frac{(\sqrt{9+4\Delta x}+3)}{\sqrt{9+4\Delta x}+3} \cdot \frac{\sqrt{9+4\Delta x}-3}{\Delta x} \right)$$

$$= \lim_{\Delta x \to 0} \frac{(\sqrt{9+4\Delta x})^2 - 3^2}{(\sqrt{9+4\Delta x}+3)\Delta x}$$

$$= \lim_{\Delta x \to 0} \frac{9+4\Delta x - 9}{(\sqrt{9+4\Delta x}+3)\Delta x}$$

$$= 4 \lim_{\Delta x \to 0} \frac{1}{\sqrt{9+4\Delta x}+3}$$

$$= 4 \cdot \frac{1}{3+3},$$

or $f'(5) = 2/3$. ◄

INSIGHT As Example 1 demonstrates, calculating a derivative by working directly from definition (3.2.1) may require several algebraic steps and some specialized manipulations. One of the goals of this chapter is to develop formulas that enable us to calculate derivatives easily, without referring to the limit definition.

Other Notations for the Derivative

There are several alternative notations for the derivative $f'(c)$ of f at c. For example, $f'(c)$ is often denoted by $D(f)(c)$. *Leibniz notation*, which is named after Gottfried Wilhelm von Leibniz (1646–1716), one of the coinventors of calculus, is also widely used. It can take the following forms

$$\frac{df}{dx}(c) \quad \text{or} \quad \left.\frac{df}{dx}\right|_{x=c}.$$

(The vertical bar with subscript $x = c$ to denote evaluation at c was introduced in the 19th century.) One advantage of Leibniz notation over $f'(c)$ is that Leibniz notation tells us explicitly what the variable of differentiation is (in this case it is x). This reminder can be helpful when several variables are under simultaneous consideration. It is also a suggestive notation. After all, if we let Δf denote the increment $f(c+\Delta x) - f(c)$, then

$$\frac{df}{dx}(c) = \lim_{\Delta x \to 0} \frac{f(c+\Delta x) - f(c)}{\Delta x} = \lim_{\Delta x \to 0} \frac{\Delta f}{\Delta x}. \tag{3.2.4}$$

To Leibniz, the symbol df/dx in (3.2.4) actually represented a ratio of infinitesimal quantities df and dx. These quantities are known as *differentials*, but we will not give any meaning to them as separate infinitesimal entities. Later we will find that there are circumstances in which manipulation of these differentials will be useful.

A curve C in the xy-plane is not necessarily the graph of a function. But such a curve can still have a tangent line. When we think of the variables x and y as being related by virtue of the ordered pair (x, y) being on the curve C, we write

$$\frac{dy}{dx}(x_0) \quad \text{or} \quad \left.\frac{dy}{dx}\right|_{(x_0, y_0)} \tag{3.2.5}$$

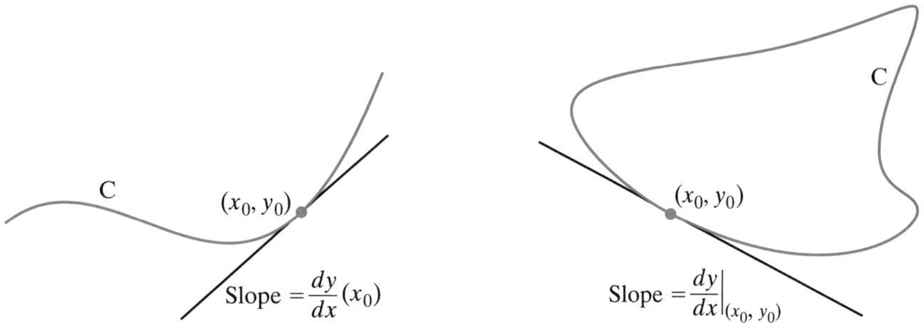

▲ Figure 1

for the slope of a tangent line to the curve at the point $(x_0, y_0) \in C$. If there is more than one value of y such that (x_0, y) is on C, then the second expression in (3.2.5) should be used—see Figure 1.

In the case that g is a function of the time variable t, then we sometimes adopt the *Newton notation* and write the derivative as $\dot{g}(t)$. This notational device is named after Sir Isaac Newton (1642–1727), the other coinventor of calculus. Newton's notation is commonly used in physics today.

The Derived Function

If f has a derivative at every point of S, then we say that f is *differentiable on* S. In the most common situation that is encountered in this text, S is an open interval. If f is differentiable on S, then we can form a function f' on S that is defined by

$$f'(x) = \lim_{\Delta x \to 0} \frac{f(x + \Delta x) - f(x)}{\Delta x} \qquad \text{for } x \in \in S.$$

This function f' is said to be the *derived function* of f or the *derivative* of f. Other notations for this function are

$$\frac{df}{dx} \quad \frac{d}{dx}f \quad \text{and} \quad D(f).$$

INSIGHT The two expressions

$$\frac{df}{dx}(x) \qquad \text{and} \qquad \frac{d}{dx}f(x)$$

mean the same thing. However, some care must be taken when using Leibniz notation to signify a derivative at a *fixed* point c in the domain of f'. The expression $\frac{d}{dx}f(c)$ can cause confusion about the order in which the operations of differentiation and evaluation are performed. Does the expression $\frac{d}{dx}f(c)$ mean that x should be set equal to c in $\frac{d}{dx}f(x)$? Or does $\frac{d}{dx}f(c)$ signify that the constant $f(c)$ should be differentiated with respect to x? These different interpretations generally lead to different results. The ambiguity can be avoided by using one of the following notations:

$$\frac{df}{dx}(c) \qquad \text{or} \qquad \left.\frac{d}{dx}f(x)\right|_{x=c} \qquad \text{or} \qquad \left.\frac{df}{dx}\right|_{x=c}.$$

The vertical bar notation is especially clear in signifying that evaluation is the final operation.

Differentiability and Continuity

Continuity and differentiability are properties of a function that are reflected in its graph. Suppose that f is defined on an open interval containing c, and let $P = (c, f(c))$. If f is continuous at c, then we know that if we follow the points $(x, f(x))$ as $x \to c$, those points approach P. Continuity may be interpreted as *predictability*.

We also know that, when a function f is differentiable at c, then its graph has a well-defined tangent line at P. Differentiability may be interpreted as a degree of *smoothness*. This is a stronger property than predictability. If f is discontinuous at c, then Figure 2 suggests that there is no line that is tangent to the graph of f at P. In that figure, line L through P is the only candidate to be a tangent line. However, line L cannot be regarded as a tangent line because it does not have one of the properties we expect from a tangent line: for x just to the left of c, the point $(x, f(x))$ on the graph of f is *not* near to L. We state our conclusions as a theorem.

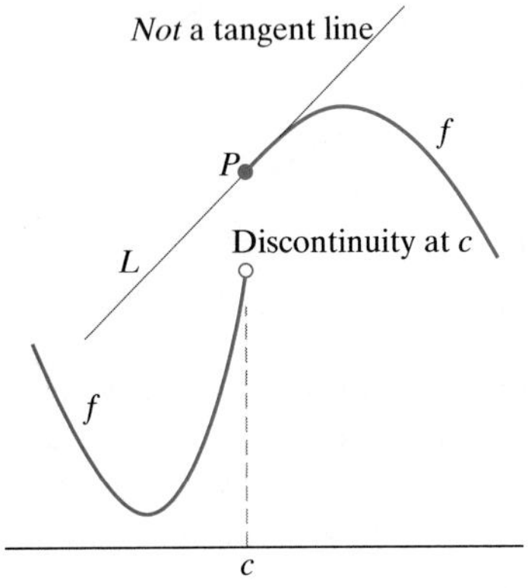

▲ Figure 2

THEOREM 1 If f is not continuous at a point c in its domain, then f is not differentiable at c. In other words, if f is differentiable at a point c in its domain, then f is continuous at c.

Proof. The two statements of the theorem are logically equivalent. We prove the second. To that end, suppose that f is differentiable at a point c in its domain. Using formula (3.2.3) for $f'(c)$, we see that $\lim_{x\to c}(f(x) - f(c))/(x - c)$ exists. Because $\lim_{x\to c}(x - c) = 0$, we conclude that $\lim_{x\to c}(f(x) - f(c)) = 0$ by Theorem 5 of Section 2.2. Therefore $\lim_{x\to c} f(x) = f(c)$. ■

Theorem 1 guarantees that, if f is differentiable on an interval (a, b), then f is also continuous on (a, b). Continuous functions, however, are not necessarily differentiable. For example, the graph of $f(x) = 3 + (x-2)^{1/3}$ has a *vertical* tangent at $P = (2, 3)$ (Example 12, Section 3.1). This implies that f is *not* differentiable at $x = 2$, even though f is continuous there. Corners are also points of continuity but not differentiability. Here is a simple example.

▶ **EXAMPLE 2** Let $f(x) = |x|$. Certainly f is continuous at all points. Show that $f'(0)$ does not exist.

Solution The relevant limit as $\Delta x \to 0$ through *positive* values, is

$$\begin{aligned}\lim_{\Delta x\to 0^+} \frac{f(0+\Delta x) - f(0)}{\Delta x} &= \lim_{\Delta x\to 0^+} \frac{|\Delta x| - 0}{\Delta x}\\ &= \lim_{\Delta x\to 0^+} \frac{\Delta x}{\Delta x}\\ &= \lim_{\Delta x\to 0^+} 1\\ &= 1.\end{aligned}$$

But, as $\Delta x \to 0$ through *negative* values, the limit is

$$\lim_{\Delta x \to 0^-} \frac{f(0+\Delta x) - f(0)}{\Delta x} = \lim_{\Delta x \to 0^-} \frac{|\Delta x| - 0}{\Delta x}$$
$$= \lim_{\Delta x \to 0^-} \frac{-\Delta x}{\Delta x}$$
$$= \lim_{\Delta x \to 0^-} -1$$
$$= -1.$$

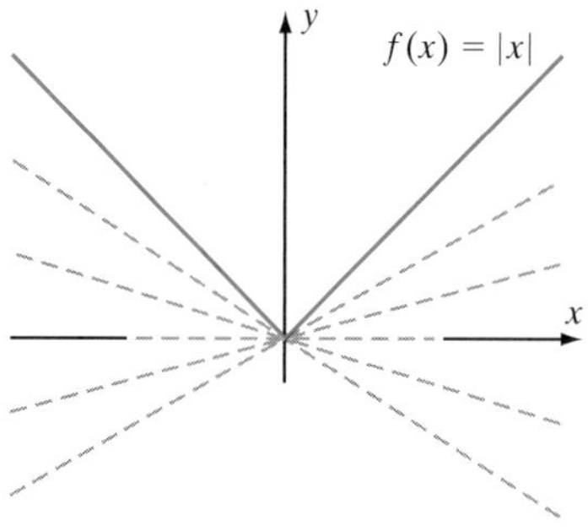

▲ Figure 3

Because the left and right limits do not agree, the limit does not exist. Therefore the derivative does not exist at 0. The fact that the function $f(x) = |x|$ does not have a derivative at $x = 0$ has a simple geometrical interpretation: the dotted lines in Figure 3 show that there is no geometrically satisfactory way of defining a tangent line to the graph of f at (0, 0). This signifies that $f'(0)$ does not exist. ◀

The domain of a derived function f' is the subset of the domain of f that is obtained by removing the points where f is not differentiable. The domain of f' can be smaller than the domain of f.

▶ **EXAMPLE 3** Let $f(x) = |x|$. Completely describe the function f'.

Solution Example 2 shows that $f'(0)$ does not exist. Therefore 0 is not in the domain of f'. If $x > 0$, then the graph of f near the point $(x, f(x))$ is just a straight line segment of slope 1. If $x < 0$, then the graph of f near the point $(x, f(x))$ is just a straight line segment of slope -1. These considerations suggest that $f'(x) = 1$ for $x > 0$, and $f'(x) = -1$ for $x < 0$, as can be verified using the limit definition of $f'(x)$. We conclude that the domain of the derived function f' is the union $(-\infty, 0) \cup (0, \infty)$ of the two open half-lines to the left and right of 0 and

$$f'(x) = \begin{cases} -1 & \text{if } x < 0 \\ 1 & \text{if } x > 0 \end{cases}.$$

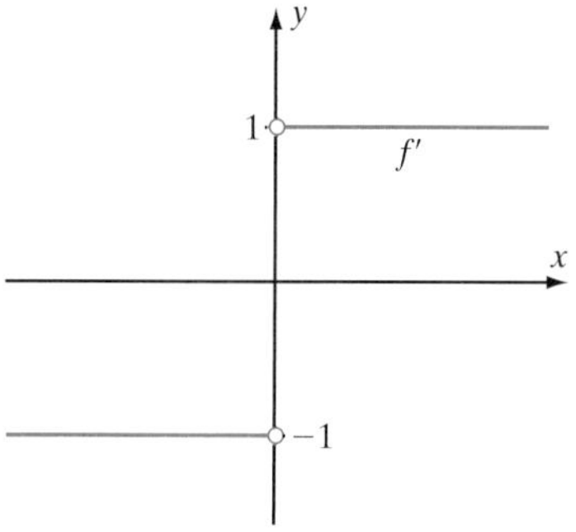

▲ Figure 4 The graph of f' where $f(x) = |x|$

The graph of f' appears in Figure 4. ◀

◀ **A LOOK BACK** Differentiability is a stronger property than continuity (Theorem 1). Continuous functions are appealing for their predictability. Differentiable functions have this predictability *and* a powerful geometric property: a well-defined tangent.

Investigating Differentiability Graphically

To use graphing techniques to examine the differentiability of a continuous function f at a point c, we define a new function ϕ by

$$\phi(x) = \frac{f(x) - f(c)}{x - c} \quad \text{for } x \neq c.$$

Then ϕ is a continuous function, but the point c is *not* in its domain. In Section 2.3, we learned that ϕ can be continuously extended at c if and only if

$$\lim_{x \to c} \phi(x) = \lim_{x \to c} \frac{f(x) - f(c)}{x - c}$$

exists. In view of formula (3.2.3) for $f'(c)$, we can restate our conclusion as follows:

> The function f is differentiable at c if and only if ϕ can be continuously extended at c.

We can often use a graphing calculator or software to decide whether or not ϕ can be continuously extended and, therefore, whether or not f is differentiable at c. The next example illustrates these ideas.

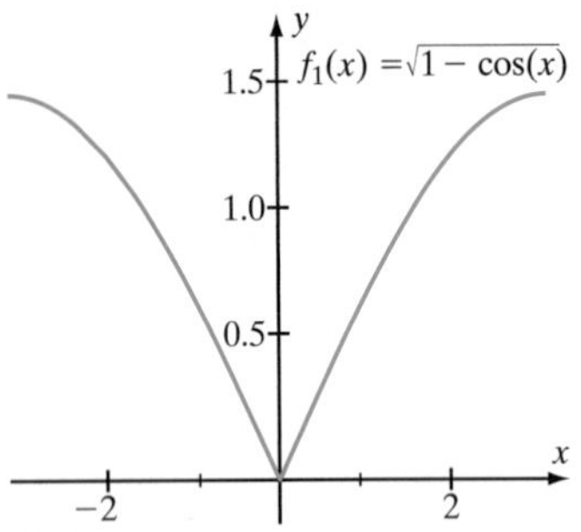

▲ Figure 5

▶ **EXAMPLE 4** Determine whether or not the functions

$$f_1(x) = \sqrt{1-\cos(x)} \quad \text{and} \quad f_2(x) = \begin{cases} x\sin(1/x) & \text{if } x \neq 0 \\ 0 & \text{if } x = 0 \end{cases}$$

are differentiable at $x = 0$.

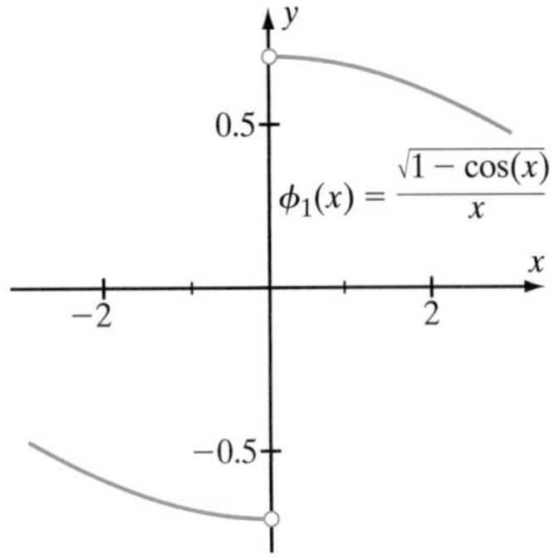

▲ Figure 6

Solution The graph of f_1, exhibited in Figure 5, does not seem to be smooth at x = 0. This suggests that f_1 is not differentiable at $x = 0$. Let

$$\phi_1(x) = \frac{f_1(x) - f_1(0)}{x - 0} = \frac{\sqrt{1-\cos(x)} - 0}{x - 0} = \frac{\sqrt{1-\cos(x)}}{x} \quad \text{for } x \neq 0.$$

The graph of ϕ_1, seen in Figure 6, shows that $\phi_1(x)$ cannot be continuously extended to $x = 0$. This confirms that f_1 is not differentiable at $x = 0$.

The graph of f_2 in Figure 7 indicates that f is continuous at $x = 0$, but differentiability is far from transparent. Let

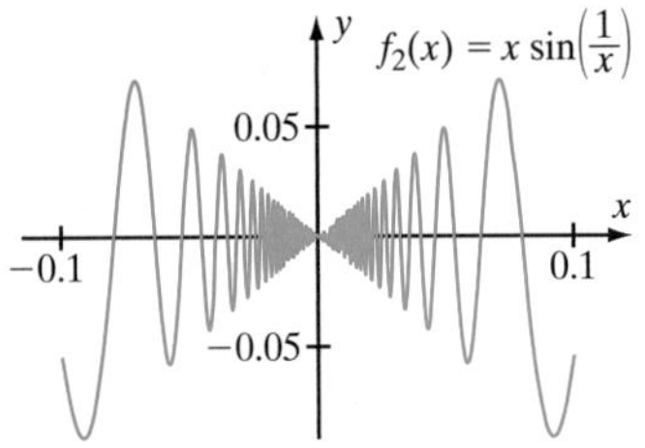

▲ Figure 7

$$\phi_2(x) = \frac{x\sin(1/x) - 0}{x - 0} = \sin\left(\frac{1}{x}\right) \quad \text{for } x \neq 0.$$

It is evident from the graph of ϕ_2 (Figure 8) that ϕ_2 cannot be continuously extended at $x = 0$. This shows that f_2 is not differentiable at $x = 0$. ◀

Derivatives of Sine and Cosine

In Section 2.2 (Theorem 8 and Example 8), we went to some trouble to obtain the limit formulas

$$\lim_{\Delta x \to 0} \frac{\sin(\Delta x)}{\Delta x} = 1 \quad \text{and} \quad \lim_{\Delta x \to 0} \left(\frac{1-\cos(\Delta x)}{\Delta x}\right) = 0. \tag{3.2.6}$$

These limit formulas can now be used to evaluate the derivatives of sine and cosine.

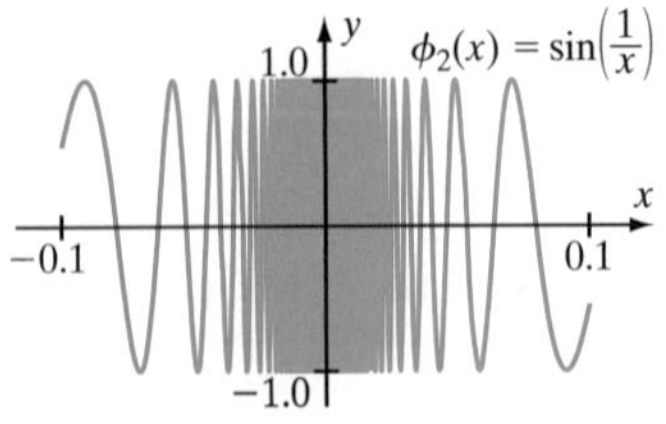

▲ Figure 8

THEOREM 2 For each x,

$$\frac{d}{dx}\sin(x) = \cos(x) \tag{3.2.7}$$

and

$$\frac{d}{dx}\cos(x) = -\sin(x). \tag{3.2.8}$$

Proof. To prove equation (3.2.7), we start with the definition of the derivative and calculate as follows:

$$
\begin{aligned}
&\frac{d}{dx}\sin(x)\\
&= \lim_{\Delta x\to 0}\frac{\sin(x+\Delta x)-\sin(x)}{\Delta x}\\
&= \lim_{\Delta x\to 0}\frac{\Big(\sin(x)\cos(\Delta x)+\cos(x)\sin(\Delta x)\Big)-\sin(x)}{\Delta x} && \text{Addition Formula (1.6.4)}\\
&= \lim_{\Delta x\to 0}\frac{\sin(x)(\cos(\Delta x)-1)}{\Delta x}+\lim_{\Delta x\to 0}\frac{\cos(x)\sin(\Delta x)}{\Delta x}\\
&= -\sin(x)\lim_{\Delta x\to 0}\frac{(1-\cos(\Delta x))}{\Delta x}+\cos(x)\lim_{\Delta x\to 0}\frac{\sin(\Delta x)}{\Delta x}\\
&= -\sin(x)\cdot 0+\cos(x)\cdot 1 && \text{Equations (3.2.6)}\\
&= \cos(x).
\end{aligned}
$$

In a similar manner, we derive formula (3.2.8) by using Addition Formula (1.6.4) for the cosine, written in the form $\cos(x+\Delta x)=\cos(x)\cos(\Delta x)-\sin(x)\sin(\Delta x)$, together with equations (3.2.6):

$$
\begin{aligned}
\frac{d}{dx}\cos(x) &= \lim_{\Delta x\to 0}\frac{\cos(x+\Delta x)-\cos(x)}{\Delta x}\\
&= \lim_{\Delta x\to 0}\frac{\Big(\cos(x)\cos(\Delta x)-\sin(x)\sin(\Delta x)\Big)-\cos(x)}{\Delta x}\\
&= \lim_{\Delta x\to 0}\frac{\cos(x)\cos(\Delta x)-\cos(x)}{\Delta x}+\lim_{\Delta x\to 0}\frac{-\sin(x)\sin(\Delta x)}{\Delta x}\\
&= -\cos(x)\lim_{\Delta x\to 0}\frac{1-\cos(\Delta x)}{\Delta x}-\sin(x)\lim_{\Delta x\to 0}\frac{\sin(\Delta x)}{\Delta x}\\
&= -\cos(x)\cdot 0-\sin(x)\cdot 1\\
&= -\sin(x).
\end{aligned}
$$

■

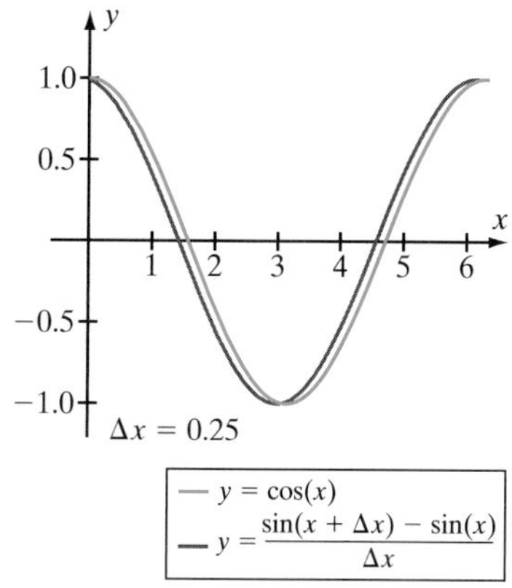

▲ Figure 9a

▶ **EXAMPLE 5** The derivative formula

$$\frac{d}{dx}\sin(x)=\cos(x)$$

was obtained by algebraic manipulations involving trigonometric identities. Find visual evidence for its correctness.

Solution By definition,

$$\frac{d}{dx}\sin(x)=\lim_{\Delta x\to 0}\frac{\sin(x+\Delta x)-\sin(x)}{\Delta x}.$$

We can *approximate* this limit by selecting a small value for Δx. Figure 9a illustrates the approximation when $\Delta x = 0.25$. For this choice of Δx, which is not

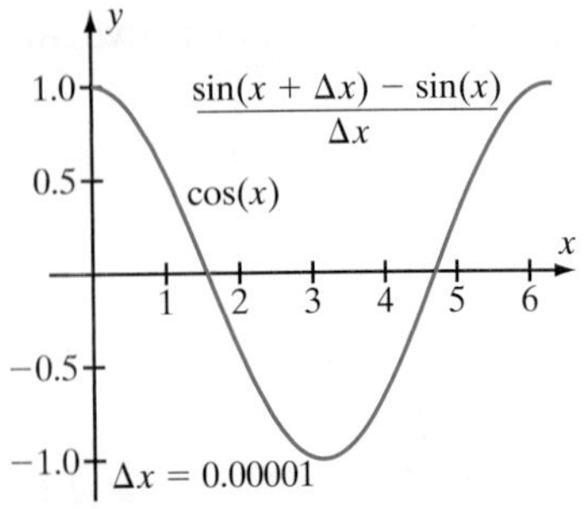

▲ Figure 9b

particularly small, the approximation is not extremely accurate. Nevertheless, it does give the idea. In Figure 9b, we set $\Delta x = 10^{-5}$ and graphed *both* the approximate derivative

$$\frac{\sin(x+10^{-5})-\sin(x)}{10^{-5}}$$

and the exact derivative $\cos(x)$. Only one curve appears in the figure because the amount by which the approximate derivative deviates from $\cos(x)$ is too minute to be displayed. ◀

Derivatives of Expressions

In calculus, we often use expressions as a shorthand for functions. Thus the expression $x^3 + \cos(\sqrt{x+1})$ is frequently interpreted to be the function f defined by $f(x) = x^3 + \cos(\sqrt{x+1})$ for all values of x for which the expression makes sense. With such an interpretation, we will write

$$\frac{d}{dx}\left(x^3 + \cos(\sqrt{x+1})\right)$$

as an equivalent form of $\frac{df}{dx}(x)$. This has the advantage of making the relationship between the expressions defining f and f' more clear. With this notation, some of the limits that we learned in Section 3.1 can be expressed by formulas for the derivatives of expressions. Thus (3.1.6) becomes

$$\frac{d}{dx}(\alpha \cdot x^n) = \alpha \cdot n x^{n-1} \qquad \text{for} \qquad n = -1, 0, 1, 2, 3. \tag{3.2.9}$$

Also, now that we are thinking of the derivative as a function, formulas in Section 3.1 that we stated for a fixed argument c can be restated in terms of a variable argument x. Thus if α and β are constants, then equations (3.1.7), (3.1.8), and (3.1.9) become

$$(\alpha f)'(x) = \alpha \cdot f'(x), \tag{3.2.10}$$

$$(f+g)'(x) = f'(x) + g'(x), \tag{3.2.11}$$

and, more generally,

$$(\alpha \cdot f + \beta \cdot g)'(x) = \alpha \cdot f'(x) + \beta \cdot g'(x). \tag{3.2.12}$$

Summary of Differentiation Formulas

The differentiation formulas that appear in the following table are used frequently and should be memorized. Several additional formulas will be added to the table in later sections.

$$\frac{d}{dx}(x^n) = n x^{n-1} \qquad \text{for} \qquad n = -1, 0, 1, 2, 3$$

$$\frac{d}{dx}\sin(x) = \cos(x)$$

$$\frac{d}{dx}\cos(x) = -\sin(x)$$

QUICK QUIZ

1. Calculate $\dfrac{d}{dx}(3\sin(x))$.
2. Calculate $\dfrac{d}{dx}\left(7x^3 - \dfrac{5}{x}\right)$.
3. If $f(x) = \cos(x)$, then what is $f'(\pi/3)$?
4. What is the equation of the tangent line to the graph of $y = 8\sin(x)$ at $(\pi/3, 4\sqrt{3})$?

Answers

1. $3\cos(x)$ 2. $21x^2 + 5/x^2$ 3. $-\sqrt{3}/2$ 4. $y = 4(x - \pi/3) + 4\sqrt{3}$

EXERCISES

Problems for Practice

In each of Exercises 1–10, compute the indicated derivative for the given function by using the formulas and rules that are summarized at the end of this section.

1. $f'(-3)$, $f(x) = \sin(x) - \cos(x)$
2. $\dot{g}(\pi/2)$, $g(t) = t^3/3 + \cos(t)$
3. $\dfrac{dF}{du}(\frac{\pi}{6})$, $F(u) = 5u + 6\cos(u)$
4. $D(f)(0)$, $f(x) = -3\sin(x)$
5. $\dot{g}(5)$, $g(t) = 8t^3 - 8t$
6. $D(H)(1)$, $H(x) = 10/x$
7. $f'(c)$, $f(x) = 3\sin(x) + 4\cos(x)$
8. $\dfrac{d\varphi}{dw}(\frac{\pi}{4})$, $\varphi(\omega) = 2\sin(\omega) + 5$
9. $\dfrac{dg}{ds}|_{s=-2}$, $g(s) = 3s^3 - \cos(s)$
10. $\dfrac{df}{dx}|_{x=0}$, $f(x) = 10x - 3\sin(x)$

In Exercises 11–14, calculate $g'(x)$ by using the formulas and rules that are summarized at the end of this section.

11. $g(x) = -\pi x + 1/(\pi x)$
12. $g(x) = x^2 - 4x$
13. $g(x) = 4x^3 + 6x^2 + 1$
14. $g(x) = 2x + \sin(x)$

In each of Exercises 15–18, find a point x where $f'(x) = 6$.

15. $f(x) = x^2 + 5$
16. $f(x) = 12\sin(x)$
17. $f(x) = x^3 + 1$
18. $f(x) = 3x^2 - 6x + 50$

In each of Exercises 19–22, calculate $f'(x)$, and sketch the graph of the equation $y = f'(x)$.

19. $f(x) = 3 - 2x^3$
20. $f(x) = x^2 + 3x$
21. $f(x) = x^2 + |x|$
22. $f(x) = 3x - |x|$

In each of Exercises 23–26, a function f and a value c are given. Find an equation of the tangent line to the graph of f at $(c, f(c))$.

23. $f(x) = \sin(x), c = \pi/3$
24. $f(x) = \cos(x), c = \pi/6$
25. $f(x) = 2\cos(x) - 4\sin(x)$, $c = 3\pi/4$
26. $f(x) = 3x - \cos(x)$, $c = \pi/2$

In each of Exercises 27–30, use the method of Example 1 to calculate $f'(x)$ for the given function f.

27. $f(x) = \sqrt{x}$
28. $f(x) = \sqrt{x+3}$
29. $f(x) = \sqrt{3x+7}$
30. $f(x) = \sqrt{9-x}$

Further Theory and Practice

31. Figure 10 illustrates the graphs of four functions: F, G, H, and K. Three of these functions are derivatives of the others. Identify these relationships with equations of the form $g = f'$ where g and f are chosen from among F, G, H, and K.

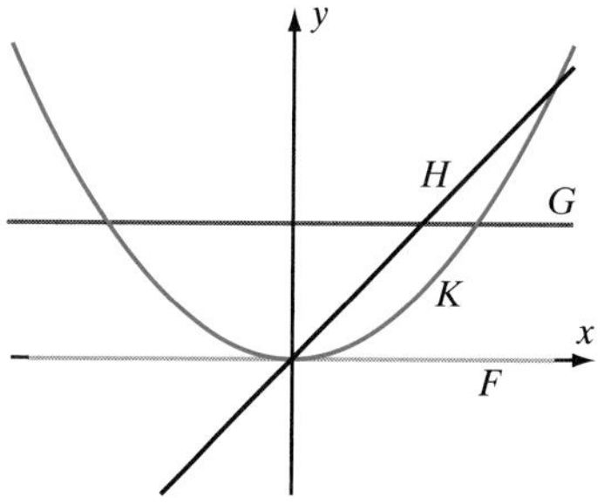

Figure 10

In Exercises 32–35, calculate the derivative of the given function f at the given point c.

32. $f(x) = |x|^{3/2}$, $c = 0$

33. $f(x)=x\cdot|x|,\ c=0$
34. $f(x)=x^{3/2},\ c=1$
35. $f(x)=1/\sqrt{x},\ c=1/16$

36. Suppose that $f(x)=|x-2|+|x-3|$. What is the domain of f'? Sketch the graph of f'.
37. Find a polynomial function $p(x)$ of degree 1 such that $p(2)=6$ and $p'(4)=-5$.
38. Find a polynomial function of degree 2 such that $p(3)=4$, $p(5)=-10$, and $p'(-3)=7$.

In each of Exercises 39–42 a function f is given. Calculate $f'(x)$.

39. $f(x)=1/(1+x)$
40. $f(x)=1/(1+x^2)$
41. $f(x)=1/(1+\sqrt{x})$
42. $f(x)=\sqrt{1+x^2}$

In each of Exercises 43–46, a multicase function f is defined. Is f differentiable at $x=0$? Give a reason for your answer.

43. $f(x)=\begin{cases} x^2 & \text{if } x\le 0 \\ x & \text{if } x>0 \end{cases}$

44. $g(x)=\begin{cases} x^3 & \text{if } x\le 0 \\ x^2 & \text{if } x>0 \end{cases}$

45. $f(x)=\begin{cases} \sin(x) & \text{if } x\le 0 \\ x^2+x & \text{if } x>0 \end{cases}$

46. $f(x)=\begin{cases} \sin(|x|) & \text{if } x\le 0 \\ 1-x & \text{if } x>0 \end{cases}$

In Exercises 47–50, specify a function f and a value c for which the given limit equals $f'(c)$. (You need not evaluate the limit.)

47. $\lim_{h\to 0}\dfrac{\sqrt{4+h}-2}{h}$

48. $\lim_{h\to 0}\dfrac{10^h-1}{h}$

49. $\lim_{h\to 0}\dfrac{1/(5+h)-1/5}{h}$

50. $\lim_{h\to 0}\dfrac{\sin(h)}{h}$

51. An increasing, differentiable function may have a decreasing derivative. That is, it may happen that $f(a)<f(b)$ but $f'(a)f'(b)$ for all a and b in the domain of f with $a<b$. Show that $f(x)=\sqrt{x}$ has these properties.
52. If $f(-x)=f(x)$ for every x, then f is said to be an *even* function. If $f(-x)=-f(x)$, for every x, then f is said to be an *odd* function. Assume that f is differentiable. Show that if f is even, then f' is odd. Show that if f is odd, then f' is even.
53. Evaluate $f'(0)$ if $f(x)=x^\alpha$ where $\alpha\ge 1$. Show that $f(x)=x^\alpha$ is not differentiable at 0 if $\alpha<1$.
54. Find a continuous function f on $\mathbb{R}$ that is differentiable on $(-\infty,0)\cup(0,\infty)$ such that $f(0)=0$ and $f'(x)=H(x)$ for $x\neq 0$. Here H is the Heaviside function:

$$H(x)=\begin{cases} 0 & \text{if } x<0 \\ 1 & \text{if } x>0 \end{cases}.$$

55. Let $f(x)=x/(1+x^2)$. Use the identity

$$\frac{x/(1+x^2)-c/(1+c^2)}{x-c}=\frac{1-xc}{(1+x^2)(1+c^2)}$$

to compute $f'(c)$.
56. Let $f(x)=x/(1+x^2)^2$. Use the identity

$$\frac{x/(1+x^2)^2-c/(1+c^2)^2}{x-c}=\frac{1-cx^3-c^2x^2-2xc-xc^3}{(1+x^2)^2(1+c^2)^2}$$

to compute $f'(c)$.
57. Suppose that f is differentiable at c. Let $g(x)=f(x+k)$ where K is a constant. Show that g is differentiable at $c-k$, and evaluate $g'(c-k)$ in terms of $f'(c)$.
58. Suppose that f is a differentiable function and that K is a constant. Define $g(x)=f(kx)$. Show that g is differentiable, and find a formula for $g'(x)$ in terms of $f'(kx)$.
59. In economics, if $q(p)$ is the *demand* for a product at price p; that is, the number of units of the product that are sold at price P, then

$$E(p)=-\lim_{\Delta p\to 0}\frac{(q(p+\Delta p)-q(p))/q(p)}{\Delta p/p}$$

is defined to be the *elasticity of demand*. Compute $E(p)$ in terms of the derivative of the demand function q.
60. Let $C(x)$ denote the cost of producing x units of a commodity. Although x usually refers to a nonnegative integer, it is common in economics to treat x as a continuous variable. The marginal cost $M(x)$ refers to the cost of producing an $x+1^{\text{st}}$ unit after x units are produced.
a. Describe M in terms of C.
b. Write down the limit definition of $C'(x)$. What is the relationship between this quantity and $M(x)$?
c. Under what circumstances can we plausibly use $C'(x)$ as an expression for the marginal cost?
61. Let K and M be positive constants. Suppose that as population $P(t)$ increases from $M/10$ to M, it satisfies the equation

$$P'(t)=k\cdot P(t)\cdot(M-P(t)).$$

Show that the instantaneous rate of change of the population is greatest when the population size is $M/2$.
62. Let the function f be specified by

$$f(x)=\begin{cases} x^2+2 & \text{if } x\le 1 \\ ax+b & \text{if } x>1. \end{cases}$$

Select a and b so that $f'(1)$ exists.
63. Prove that the function

$$f(x)=\begin{cases} x\cdot\sin(1/x) & \text{if } x\neq 0 \\ 0 & \text{if } x=0 \end{cases}$$

is continuous at $x=0$ but not differentiable at $x=0$.

64. Prove that the function

$$f(x) = \begin{cases} x^2 \cdot \sin(1/x) & \text{if} \quad x \neq 0 \\ 0 & \text{if} \quad x = 0 \end{cases}$$

is differentiable at $x = 0$, *and* compute $f'(0)$.

65. Let f be differentiable at c. Let $y = ax + b$ be the equation of the tangent line to the graph of f at $(c, f(c))$. Prove that

$$\lim_{x \to c} \frac{f(x) - (ax + b)}{x - c} = 0.$$

Calculator/Computer Exercises

In each of Exercises 66–69, a function f, a viewing rectangle R, and a point c are specified. Graph both f and the tangent to the graph of f at $(c, f(c)$ in R.

66. $f(x) = \sqrt{x}, R = [0, 4] \times [0, 2], c = 2$
67. $f(x) = x^2 - 2x, R = [0, 1] \times [-1, 0], c = 1/2$
68. $f(x) = \sin(x), R = [0, \pi/2] \times [0, 1], c = \pi/4$
69. $f(x) = 1/x, R = [1/2, 2] \times [0, 2], c = 1$

In each of Exercises 70–73 a function f and four values c, x_1, x_2 and x_3 are given. Using an appropriate viewing rectangle centered about the point $P = (c, f(c))$, graph f and the three secant lines passing through P that are determined by $(x_1, f(x_1))$, $(x_2, f(x_2))$, and $(x_3, f(x_3))$. Use the secant line through $(x_3, f(x_3))$ to estimate $f'(c)$.

70. $f(x) = \sqrt{x + 1}, \ c = 0, \ x_1 = 0.5, \ x_2 = 0.3, \ x_3 = 0.1$
71. $f(x) = x/(x + 1), \ c = 1, \ x_1 = 1.8, \ x_2 = 1.5, \ x_3 = 1.2$
72. $f(x) = 1 + \tan(x), \ c = 0, \ x_1 = -0.5, \ x_2 = -0.3, \ x_3 = -0.1$
73. $f(x) = \sin(x), \ c = \pi/4, \ x_1 = 1.2, \ x_2 = 1, \ x_3 = 0.8$

In Exercises 74–77, approximate $f'(c)$ for the given f and c in the following way: find a small viewing window with $P = (c, f(c))$ near the center. The window should be small enough so that the graph of f appears to be a straight line. Let Q and R be the endpoints of the graph of f as it exits this window. Use the slope of QR as the approximation to $f'(c)$.

74. $f(x) = \tan\left(\frac{\pi}{4} \sin(\pi x/2)\right), \ c = 1$
75. $f(x) = (x^2 + x + 1)/(x^2 + 2), \ c = 1$
76. $f(x) = x - 1/x, \ c = 1$
77. $f(x) = \csc(\pi x), \ c = 0.999$

In each of Exercises 78–83, a function f and a point c are given. Graph the function

$$\phi(x) = \frac{f(x) - f(c)}{x - c}$$

in an appropriate viewing window centered about the line $x = c$. Use the graph of ϕ to decide whether or not $f'(c)$ exists. Explain the reason for your answer, and, if your answer is that $f'(c)$ exists, use the graph of ϕ to approximate the value of $f'(c)$.

78. $f(x) = |x|, \ c = 0$
79. $f(x) = \sqrt{x^3 - 3x + 2}, \ c = 1$
80. $f(x) = \sqrt{x^4 - 2x^3 + 2x - 1}, \ c = 1$
81. $f(x) = \sqrt[3]{\cos(x)}, \ c = \pi/2$
82. $f(x) = \begin{cases} \sin^2(x)/x & \text{if} \quad x \neq 0 \\ 0 & \text{if} \quad x = 0 \end{cases}, \quad c = 0$
83. $f(x) = \begin{cases} \sin(x)/x & \text{if} \quad x \neq 0 \\ 1 & \text{if} \quad x = 0 \end{cases}, \quad c = 0$

In Exercises 84–87, graph the given function f in the suggested viewing rectangle R. From this graph, you will be able to detect at least one point at which f may not be differentiable. By zooming in, if necessary, identify each point c for which $f'(c)$ does not exist. Sketch or print your final graph, and explain what feature of the graph indicates that f is not differentiable at c.

84. $f(x) = x\sqrt{|\cos(x)|}, \ R = [-3, 3] \times [-3, 3]$
85. $f(x) = \cos^{4/5}(x), \ R = [0, 3] \times [0, 1]$
86. $f(x) = x \sin^{1/3}(x), \ R = [1, 5] \times [-5, 2.5]$
87. $f(x) = \sqrt{|x|}\cos^{1/3}(x^2), \ R = [-1, 2] \times [-1.4, 1]$

88. Graph $f(x) = -\sin(x)$ and $g(x) = 10^4(\cos(x + .0001) - \cos(x))$ in the viewing rectangle $[0, 2\pi] \times [-1, 1]$. What differentiation formula does the graph illustrate?

89. Let $f(x) = \sin(2x)$. Graph

$$g(x) = 10^4\left(f(x + 10^{-4}) - f(x)\right)$$

for $0 \leq x \leq \pi$. Use your graph to identify $f'(x)$.

3.3 Rules for Differentiation

In Sections 1 and 2, we learned how to differentiate some specific functions. For example, we proved that if α is any real number, then

$$\frac{d}{dx}(\alpha \cdot x^n) = \alpha \cdot n x^{n-1} \qquad \text{for} \qquad n = -1, 0, 1, 2, 3.$$

The particular case in which $n = 0$ is a rule that we use frequently: If $f(x) = \alpha$ for all x, then $f'(x) = 0$ for all x. In other words,

The derivative of a constant is zero.

We also derived the formulas

$$\frac{d}{dx}\sin(x) = \cos(x) \quad \text{and} \quad \frac{d}{dx}\cos(x) = -\sin(x).$$

However, if we want to know the derivative of other functions f, then, in general, we must calculate

$$f'(x) = \lim_{\Delta x \to 0} \frac{f(x + \Delta x) - f(x)}{\Delta x}.$$

This is the definition of the derivative, and all information about the derivative follows from it. Working directly with this limit, however, can be difficult or tedious. It is therefore important to know some shortcuts, or *rules*, for calculating derivatives. In this section, we will learn rules that tell us how to differentiate algebraic combinations of basic functions.

Addition, Subtraction, and Multiplication by a Constant

In Section 3.1, we learned that $(f+g)'(x) = f'(x) + g'(x)$ and $(f-g)'(x) = f'(x) - g'(x)$. We can state these differentiation rules as follows:

The derivative of a sum (or difference) is the sum (or difference) of the derivatives.

Moreover, if α is any constant, then $(\alpha \cdot f)'(x) = \alpha \cdot f'(x)$. That is:

The derivative of a constant times a function is the constant times the derivative of the function.

Of course, we can also express these rules using Leibniz notation:

$$\frac{d}{dx}(f+g)(x) = \frac{d}{dx}f(x) + \frac{d}{dx}g(x), \quad \frac{d}{dx}(f-g)(x) = \frac{d}{dx}f(x) - \frac{d}{dx}g(x), \quad \text{and}$$

$$\frac{d}{dx}(\alpha \cdot f)(x) = \alpha \cdot \frac{d}{dx}f(x).$$

As we noted with formula (3.2.12), the last three equations can be combined into one general form: If both α and β are constants, then $(\alpha \cdot f + \beta \cdot g)'(x) = \alpha \cdot f'(x) + \beta \cdot g'(x)$. A function of the form $\alpha \cdot f + \beta \cdot g$ is said to be a *linear combination* of f and g. Using this terminology, the last derivative law may be stated as:

The derivative of a linear combination of functions is that linear combination of the derivatives of the functions.

As we observed in line (3.1.10), the rule for differentiating a linear combination can be extended to three or more summands.

▶ **EXAMPLE 1** Calculate

$$\frac{d}{dx}\left(6x^3 - 7x^2 + 2x - 5 - \frac{3}{x}\right).$$

Solution We reduce our differentiation problem to simpler differentiations that we already know. Thus

$$\begin{aligned}\frac{d}{dx}\left(6x^3 - 7x^2 + 2x - 5 - \frac{3}{x}\right) &= 6\frac{d}{dx}x^3 - 7\frac{d}{dx}x^2 + 2\frac{d}{dx}x - \frac{d}{dx}5 - 3\frac{d}{dx}x^{-1} \\ &= 6\cdot(3x^2) - 7\cdot(2x) + 2\cdot(1) - 0 - 3\cdot\left(\frac{-1}{x^2}\right) \\ &= 18x^2 - 14x + 2 + \frac{3}{x^2}.\end{aligned}$$ ◀

Products and Quotients

Now we turn to the differentiation of products and quotients. Because the derivative of a sum and a difference both turned out to be easy and rather obvious, you might assume that the derivative of a product or a quotient is equally simple. *But this is not so.* For example, let $f(x) = x^2$, and $g(x) = x$. Then we have

Derivative of Product	**Product of Derivatives**
$\frac{d}{dx}(x^2 \cdot x) = \frac{d}{dx}(x^3) = 3x^2$	$\frac{d}{dx}(x^2) \cdot \frac{d}{dx}(x) = 2x \cdot 1 = 2x.$

Plainly the derivative of a product is *not* the product of the derivatives. Also

Dervative of Quotient	**Quotient of Derivatives**
$\frac{d}{dx}\left(\frac{x^2}{x}\right) = \frac{d}{dx}(x) = 1$	$\frac{\frac{d}{dx}(x^2)}{\frac{d}{dx}(x)} = \frac{2x}{1} = 2x.$

So the derivative of a quotient is *not* the quotient of the derivatives.

The correct formulas are a bit surprising but not difficult to use.

THEOREM 1 If $f'(c)$ and $g'(c)$ exist, then $(f\cdot g)'(c)$ also exists, and

$$(f \cdot g)'(c) = f'(c)\cdot g(c) + f'(c)\cdot g'(c).$$

Using Leibniz notation, the Product Rule is

$$\frac{d}{dx}(f \cdot g) = \frac{df}{dx} \cdot g + f \cdot \frac{dg}{dx}.$$

In words, the Product Rule states:

We differentiate a product by adding the second function times the derivative of the first to the first function times the derivative of the second.

A proof of the Product Rule may be found in the *Genesis and Development* section at the end of this chapter. In the meantime, let us see how this rule is used.

▶ **EXAMPLE 2** Differentiate $H(x) = x\cos(x)$.

Solution We think of $H(x)$ as the product $f(x)\cdot g(x)$, where $f(x) = x$ and $g(x) = \cos(x)$. Therefore

$$\begin{aligned} H'(x) &= f'(x)\cdot g(x) + f(x)\cdot g'(x) \\ &= \left(\frac{d}{dx}x\right)\cdot \cos(x) + x\cdot \frac{d}{dx}\cos(x) \\ &= 1\cdot \cos(x) + x\cdot(-\sin(x)) \\ &= \cos(x) - x\sin(x). \quad ◀ \end{aligned}$$

Sometimes it is necessary to differentiate a product of three (or more) functions. To treat such a problem, we apply the Product Rule successively. Here is an example.

▶ **EXAMPLE 3** Differentiate $H(x) = x^2\sin(x)\cos(x)$ with respect to x.

Solution We think of $H(x)$ as the product $f(x)\cdot g(x)$, where $f(x) = x^2\sin(x)$ and $g(x) = \cos(x)$. A first application of the Product Rule gives us

$$\begin{aligned} \frac{d}{dx}H(x) &= \frac{d}{dx}(f\cdot g)(x) \\ &= \frac{d}{dx}f(x)\cdot g(x) + f(x)\cdot \frac{d}{dx}g(x) \\ &= \left(\frac{d}{dx}(x^2\sin(x))\right)\cdot \cos(x) + (x^2\sin(x))\cdot \frac{d}{dx}\cos(x) \\ &= \left(\frac{d}{dx}(x^2\sin(x))\right)\cdot \cos(x) + (x^2\sin(x))\cdot(-\sin(x)) \\ &= \left(\frac{d}{dx}(x^2\sin(x))\right)\cdot \cos(x) - x^2\sin^2(x). \end{aligned}$$

To complete the differentiation, we apply the Product Rule one more time:

$$\begin{aligned} \frac{d}{dx}H(x) &= \left(\left(\frac{d}{dx}x^2\right)\sin(x) + x^2\frac{d}{dx}\sin(x)\right)\cdot \cos(x) - x^2\sin^2(x) \\ &= (2x\sin(x) + x^2\cos(x))\cdot \cos(x) - x^2\sin^2(x) \\ &= 2x\sin(x)\cos(x) + x^2\cos^2(x) - x^2\sin^2(x) \\ &= 2x\sin(x)\cos(x) + x^2\left(\cos^2(x) - \sin^2(x)\right). \quad ◀ \end{aligned}$$

Now we will learn to differentiate the quotient f/g of two functions. First we will investigate the special case that is obtained by taking $f(x) = 1$. It is useful, so we record it separately:

THEOREM 2 **The Reciprocal Rule** If $g'(c)$ exists, and if $g(c) \neq 0$, then $(1/g)'(c)$ exists. Moreover,

$$\left(\frac{1}{g}\right)'(c) = -\frac{g'(c)}{g^2(c)}.$$

Proof. We start with the definition of $(1/g)'(c)$:

$$\begin{aligned}
\left(\frac{1}{g}\right)'(c) &= \lim_{\Delta x \to 0} \frac{1/g(c+\Delta x) - 1/g(c)}{\Delta x} \\
&= -\lim_{\Delta x \to 0} \frac{(g(c+\Delta x) - g(c))}{\Delta x \cdot g(c) \cdot g(c+\Delta x)} \\
&= -\lim_{\Delta x \to 0} \frac{(g(c+\Delta x) - g(c))}{\Delta x} \cdot \lim_{\Delta x \to 0} \frac{1}{g(c) \cdot g(c+\Delta x)} \\
&= -g'(c) \cdot \frac{1}{g(c)^2}.
\end{aligned}$$

■

▶ **EXAMPLE 4** Show that

$$\frac{d}{dx}\sec(x) = \sec(x)\tan(x).$$

Solution Because $\sec(x) = 1/\cos(x)$, the Reciprocal Rule applies:

$$\begin{aligned}
\frac{d}{dx}\sec(x) &= -\frac{\frac{d}{dx}\cos(x)}{\cos^2(x)} \\
&= -\frac{-\sin x}{\cos^2(x)} \\
&= \frac{\sin x}{\cos^2(x)} \\
&= \frac{1}{\cos(x)} \cdot \frac{\sin(x)}{\cos(x)} \\
&= \sec(x) \cdot \tan(x).
\end{aligned}$$

This formula will be added to the list of basic derivatives that we are compiling. ◀

THEOREM 3 The Quotient Rule: If $f'(c)$ and $g'(c)$ exist, and if $g(c) \neq 0$, then $(f/g)'(c)$ exists. Moreover,

$$\left(\frac{f}{g}\right)'(c) = \frac{g(c) \cdot f'(c) - f(c) \cdot g'(c)}{g^2(c)}.$$

Proof. Notice that the quotient f/g may be written as the product $f \cdot (1/g)$. The Reciprocal Rule tells us that $(1/g)'(c)$ exists. The Product Rule now tells us that $(f \cdot (1/g))'(c)$ exists and

$$\begin{aligned}\left(\frac{f}{g}\right)'(c) &= \left(f\cdot\left(\frac{1}{g}\right)\right)'(c)\\ &= f'(c)\cdot\left(\frac{1}{g}\right)(c)+f(c)\cdot\left(\frac{1}{g}\right)'(c)\\ &= f'(c)\cdot\left(\frac{1}{g(c)}\right)+f(c)\cdot\left(\frac{-g'(c)}{g^2(c)}\right)\\ &= f'(c)\cdot\left(\frac{g(c)}{g^2(c)}\right)+f(c)\cdot\left(\frac{-g'(c)}{g^2(c)}\right)\\ &= \frac{g(c)\cdot f'(c)-f(c)\cdot g'(c)}{g^2(c)}\end{aligned}$$

(combining terms, using common denominator $g^2(c)$). ■

▶ **EXAMPLE 5** Calculate the derivative of

$$H(x)=\frac{x^2+1}{x^3}.$$

Solution Observe that $H(x)=f(x)/g(x)$ where $f(x)=x^2+1$ and $g(x)=x^3$. Then

$$\begin{aligned}H'(x) &= \frac{g(x)\cdot f'(x)-f(x)\cdot g'(x)}{(g(x))^2}\\ &= \frac{(x^3)\cdot(2x)-(x^2+1)\cdot(3x^2)}{(x^3)^2}.\\ &= \frac{-x^4-3x^2}{x^6}=-\frac{(x^2+3)}{x^4}.\end{aligned}$$

Notice that this calculation makes sense only if $g(x)\neq 0$; that is, when $x\neq 0$. ◀

Numeric Differentiation

The rules that we have learned enable us to differentiate extremely complicated functions. The application of these rules, however, can be laborious. Moreover, even when we use a differentiation rule to obtain an exact evaluation of a derivative, we may have to approximate constants such as $\sqrt{2}$ and π that appear in our answer. It is therefore convenient to have a method for approximating the numerical value $f'(c)$ right from the outset. Such a procedure is known as *numeric differentiation.*

Suppose that f is a function that is defined on the interval (a,b) and differentiable at a point c in the interval. Because

$$f'(c)=\lim_{h\to 0}\frac{f(c+h)-f(c)}{h},$$

we can approximate $f'(c)$ by the difference quotient $(f(c+h)-f(c))/h$ for a small value of h. When $h>0$ and c is fixed, the quotient

$$D_{+}f(c,h)=\frac{f(c+h)-f(c)}{h}$$

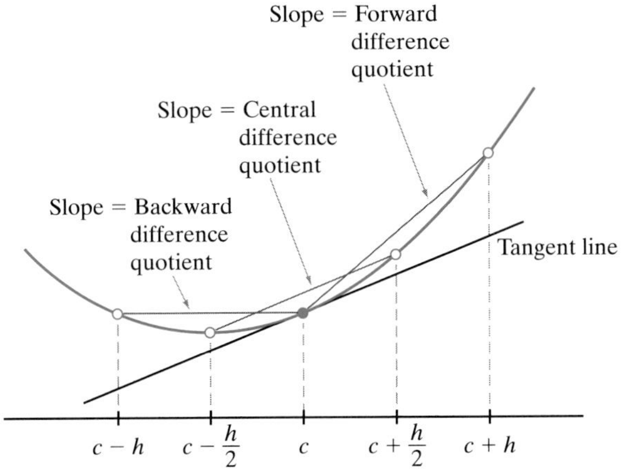

Figure 1

is known as a *forward difference quotient*. The *backward difference quotient* is defined by

$$D_{-}f(c,h) = \frac{f(c) - f(c-h)}{h}$$

where $h > 0$. The *central difference quotient* is defined by

$$D_0 f(c,h) = \frac{f(c+h/2) - f(c-h/2)}{h}. \tag{3.3.1}$$

Each of these three difference quotients can be used to approximate $f'(c)$. However, the situation illustrated by Figure 1 is typical: For a given value of h, the central difference quotient generally gives the best approximation to the derivative.

When we use the approximation

$$D_0 f(c,h) \approx f'(c),$$

the question of *how* to choose h in an optimal way is not easy to answer. Consider $f(x) = x^4$ and $c = 1/2$. The exact value of $f'(c)$ is 1/2. Now consider the following table of difference quotients (which were computed using ten significant digits, of which six are exhibited):

h	$D_{+}f(c,h)$	$D_{-}f(c,h)$	$D_0 f(c,h)$
10^{-1}	0.671	0.369	0.505
10^{-2}	0.505201	0.485199	0.50005
10^{-3}	0.501502	0.498502	0.500001
10^{-4}	0.50015	0.49985	0.5
10^{-6}	0.500002	0.499998	0.5
10^{-7}	0.499999	0.499999	0.500002
10^{-9}	0.500222	0.499995	0.500222
10^{-11}	0.500222	0.500222	0.477485
10^{-12}	0.454747	0.500222	0

Table 1

As you read down each column, you will notice that, with decreasing h, each difference quotient seems to approach the limit 0.5 at first. Observe that the central difference quotient converges quickest. In each column, however, accuracy is degraded when h becomes *too small*. This is the result of *loss of significance*, a problem that is discussed in Section 1.1.

It is difficult to predict the value of h at which accuracy begins to deteriorate. There are two types of error that result when we use $D_0 f(c,h)$ to approximate $f'(c)$.

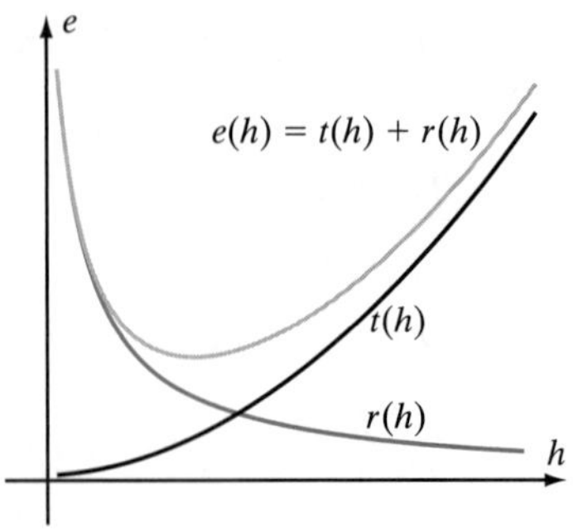

▲ Figure 2

As we have seen, there is a *roundoff error* $r(h)$. Terminating the limit process $f'(c) = \lim_{h\to 0} D_0 f(c,h)$ by stopping at a particular value of h also introduces a *truncation error* $t(h)$. The total error $e(h)$ is given by $e(h) = t(h) + r(h)$. Typical graphs of these error functions can be seen in Figure 2. Unfortunately, there is no way to ensure that $t(h)$ and $r(h)$ are *both* very small. One strategy is to let h become smaller and smaller until either the desired accuracy is attained or the total error begins to increase.

▶ **EXAMPLE 6** Let $V(x) = \sqrt{2+\sin(x)}$. Approximate $V'(0.4174)$ to four significant digits.

Solution Table 2 presents the results of our central difference quotient calculations. (Ten digits have been carried in the computations, six digits are tabulated, but no more than four are significant.)

h	0.1	0.01	0.001	0.0001	0.00001	0.000001
$\boldsymbol{D_0V(0.4174, h)}$	0.294630	0.294708	0.294709	0.294710	0.294700	0.295000

Table 2

Because the leftmost four digits of $D_0V(0.4174, h)$ agree when $h = 10^{-2}$ and 10^{-3}, we can be reasonably certain that $V'(0.4174) = 0.2947$ to four significant digits (and that the last column represents a decline in accuracy caused by using too small a value for h). In Section 3.5, we will be able to show that $V'(x) = \cos(x)/\left(2\sqrt{2+\sin(x)}\right)$. With this formula for $V'(x)$, we can verify that our numerical differentiation is correct to four significant digits. ◀

Summary of Differentiation Rules

Sum Rule $(f+g)'(x) = f'(x) + g'(x)$ or $\dfrac{d}{dx}(f+g)(x) = \dfrac{df}{dx}(x) + \dfrac{dg}{dx}(x)$

Difference Rule $(f-g)'(x) = f'(x) - g'(x)$ or $\dfrac{d}{dx}(f-g)(x) = \dfrac{df}{dx}(x) - \dfrac{dg}{dx}(x)$

Constant Multiple Rule $(\alpha\cdot f)'(x) = \alpha\,(f'(x))$ or $\dfrac{d}{dx}(\alpha\cdot f)(x) = \alpha\,\dfrac{df}{dx}(x)$

Constant Rule $(\alpha)' = 0$ or $\dfrac{d}{dx}\alpha = 0$ if α is any constant

Product Rule $(f\cdot g)'(x) = f'(x)\,g(x) + f(x)\,g'(x)$ or $\dfrac{d}{dx}(f\cdot g)(x) = \left(\dfrac{df}{dx}(x)\right) g(x) + f(x)\,\dfrac{dg}{dx}(x)$

Quotient Rule $\left(\dfrac{f}{g}\right)'(x) = \dfrac{g(x)f'(x) - f(x)g'(x)}{g(x)^2}$ or $\dfrac{d}{dx}\left(\dfrac{f}{g}\right)(x) = \dfrac{g(x)\dfrac{df}{dx}(x) - f(x)\dfrac{dg}{dx}(x)}{g(x)^2}$

Reciprocal Rule $\left(\frac{1}{g}\right)'(x) = -\frac{g'(x)}{g(x)^2}$ or $\frac{d}{dx}\left(\frac{1}{g}\right)(x) = -\frac{\frac{dg}{dx}(x)}{g(x)^2}$

Approximate Derivative Rule $f'(c) \approx \frac{f(c+h/2) - f(c-h/2)}{h}$

Derivatives of Basic Functions

$$\frac{d}{dx}(x^n) = nx^{n-1} \quad \text{for} \quad n = -1, 0, 1, 2, 3$$

$$\frac{d}{dx}\sin(x) = \cos(x)$$

$$\frac{d}{dx}\cos(x) = -\sin(x)$$

$$\frac{d}{dx}\sec(x) = \sec(x)\tan(x)$$

QUICK QUIZ

1. What is the derivative of $x\sin(x)$?
2. What is the derivative of $\cos(x)/x$?
3. What is the derivative of $\sec(x)$?
4. Use the central difference quotient with $h = 0.1$ to approximate the derivative of $\tan(x)$ at $x = 0$.

Answers

1. $\sin(x) + x\cos(x)$ 2. $-(x\sin(x) + \cos(x))/x^2$ 3. $\sec(x)\tan(x)$ 4. 1.00083

EXERCISES

Problems for Practice

In Exercises 1–6, use the rules for differentiating sums and differences, as in Example 1, to compute the derivative of the given expression with respect to x.

1. $5x^3 - 3x^2 + 9$
2. $6 - 4/x + 4x - x^2/2 - 2x^3$
3. $x^3/\pi + \pi\cos(x) + \sqrt{\pi}$
4. $3x^3 - 2x^2 + \pi\sin(x) + 1/\pi$
5. $\frac{1}{5}(4\sin(x) - 3\cos(x)) + 5x$
6. $\sqrt{2}(\sin(x) - 2\cos(x)) + 2x^2 - 3/x$

In Exercises 7–18, use the Product Rule to compute the derivative of the given expression with respect to x. (In each of Exercises 7, 8, 14, 16, and 18, do not avoid using the Product Rule by first expanding the product.)

7. $x^2(x+3)$
8. $4x(x^2+5)$
9. $x^3\cos(x)$
10. $5\sin(x) - 6x\cos(x)$
11. $\sin^2(x)$
12. $\cos^2(x)$
13. $\sin(x)\cos(x)$
14. $(x^2-5)(3x+2)$
15. $7(x^2+x)\sin(x)$
16. $3x^3(x^3+7)$
17. $(1/x)\cdot\sin(x)$
18. $(1/x)\cdot(5x^3+4/x)$

In Exercises 19–26, use the Reciprocal Rule to compute the derivative of the given expression with respect to x.

19. $1/(x+1)$
20. $1/(1+x^2)$
21. $2/(3x^2+4)$
22. $3/(2+\sin(x))$
23. $1/(x+3\cos(x))$
24. $5/(\cos(x)+\sin(x))$
25. $9/(3\cos(x)+x^3)$
26. $1/(x^3+x^2+1)$

In Exercises 27–38, use the Quotient Rule to compute the derivative of the given expression with respect to x.

27. $x/(x+1)$
28. $3x/(x^2+1)$
29. $(x-1)/(x+1)$
30. $(x^2+1)/(x^2+2)$
31. $(5x-2/x)/x$
32. $x/\sin(x)$
33. $\sin(x)/x$
34. $\cos(x)/(x-5)$
35. $(x^2+7)/\cos(x)$
36. $2x^3/(1+\cos(x))$
37. $(1-\cos(x))/(1+\cos(x))$
38. $(x+\sin(x))/(x+\cos(x))$

Calculate the derivative of each of the expressions in Exercises 39–44 by applying both the Product and Quotient Rules.

39. $x\sin(x)/(x+1)$
40. $x\cos(x)/(x^2+3)$
41. $x(x^2+1)(x^3+2)$
42. $\sin^2(x)/x$
43. $\sin^2(x)\cdot\sec(x)$
44. $x\sin(x)/(x+\sin(x))$

In Exercises 45–50, find the tangent line to the graph of the given function at the given point.

45. $f(x)=(x+3)/(x+1)$, $P=(1,2)$
46. $f(x)=x\cos(x)$, $P=(\pi,-\pi)$
47. $f(x)=6/(1+x^2)$, $P=(1,3)$
48. $f(x)=4\sin(x)\cos(x)$, $P=(\pi/4,2)$
49. $f(x)=\sin(x)/(x+1)$, $P=(0,0)$
50. $f(x)=(3x^2-5)/(x+4)$, $P=(1,-2/5)$

In Exercises 51–54, use the result of Example 4 to compute the derivative of the given expression with respect to x.

51. $x\sec(x)$
52. $\sec(x)/x$
53. $1/(1+2\sec(x))$
54. $(\sec(x)-\cos(x))/\sin(x)$

In Exercises 55–58, use the given information to estimate $f'(c)$ at the given point c.

55. $f(4)=5.7$ and $f(4.1)=6.2$, $c=4$
56. $f(4)=5.7$ and $f(4.1)=6.2$, $c=4.1$
57. $f(\pi+0.01)=f(\pi)+0.2$, $c=\pi$
58. $f(3.47)=2.61$ and $f(3.49)=2.67$, $c=3.48$

Further Theory and Practice

59. Given $f(2)=3$, $f'(2)=5$, $g(2)=8$, and $g'(2)=-6$, find $(f\cdot g)'(2)$ and $(f/g)'(2)$.

60. Given $f(3)=-4$ and $f'(3)=6$, find $(1/f)'(3)$ and $(f^2)'(3)$.

61. Given $f(-1)=5$, $g(-1)=4$, $f'(-1)=2$, and $g'(-1)=-8$, find $(f\cdot g)'(-1)$, $(f^2)'(-1)$, and $(g^2)'(-1)$.

62. Given $f(4)=2$, $g(4)=1$, $f'(4)=-5$, and $g'(4)=-9$, find $(f\cdot g^2)'(4)$, $(g\cdot f^2)'(4)$, $(1/f^2)'(4)$, and $(g/f)'(4)$.

63. Use the Quotient Rule to differentiate

$$\frac{x+1/x}{1/x}.$$

Verify your answer by simplifying the expression and then differentiating.

64. Derive a formula for $\frac{d}{dx}x^4$ by applying the Product Rule to $x\cdot x^3$. Verify your answer by applying the Product Rule to $x^2\cdot x^2$.

In Exercises 65–70, calculate the derivative of the given expression.

65. x^{-6}
66. $(x^2+1)(x^3+2)\sin(x)$
67. $x\sin(x)\cos(x)$
68. $\sin^3(x)$
69. $(x^2+1)^3$
70. $\sin(2x)$

71. The graph of $f(x)=2x/(x+1)$ and the point $P=(2,4)$, which lies above the graph, are shown in Figure 3. The purpose of this exercise is to find the equation of every tangent line to the graph of f that passes through P. Copy Figure 3 on your paper. To your sketch, add the tangent lines that pass through P. How many do there appear to be? You will use this predicted number N as a "reality check" in part c.

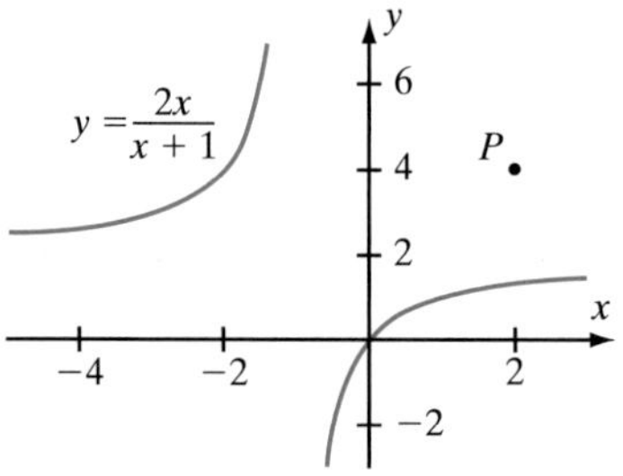

▲ Figure 3 The graph of $f(x)=\frac{2x}{x+1}$

a. What is the slope of the tangent line to the graph of f at $(c, f(c))$? (By inspection, the tangents to the graph of f all slope upward. Before continuing, make sure that your algebraic formula for the slope is consistent with this property.)

b. What is the equation of the tangent line to the graph of f at $(c, f(c))$.

c. What equation must c satisfy if the tangent line at $(c, f(c))$ passes through P? (How many solutions does this equation have? Before continuing, make sure that the equation for c has the same number of solutions as your predicted number N.)

d. Find the equations of all tangent lines to the graph of f that pass through P.

72. If f is a differentiable function, find a formula for $(f^2)'(x)$. Next find a formula for $(f^3)'(x)$. What do you expect the formula for $(f^n)'(x)$ is? (Predict by analogy, but do not prove this last formula. These matters will be taken up later in the chapter.)

73. Find constants A and B such that $f(x) = A\cos(x) + B\sin(x)$ satisfies

$$f'(x) + f(x) = 3\cos(x) - 5\sin(x).$$

74. Suppose that $f(x) = (x - r)^3 p(x)$ where $p(x)$ is a differentiable function with $p(r) \neq 0$. Show that there is a differentiable function q with $q(r) \neq 0$ such that $f'(x) = (x - r)^2 q(x)$.

75. Show that if $f(x) = |x|$, then $(f^2)'(0)$ exists even though $f'(0)$ does not.

76. We know that

$$\frac{d}{dx}x = 1, \quad \frac{d}{dx}1 = 0, \text{ and}$$

$$\frac{d}{dx}\{f_1(x) + \cdots + f_n(x)\} = \frac{d}{dx}f_1(x) + \cdots + \frac{d}{dx}f_n(x).$$

Explain what is wrong with the following reasoning:

$$\frac{d}{dx}x = \frac{d}{dx}\left\{\underbrace{1 + \cdots + 1}_{x \text{ summands}}\right\} = \left\{\underbrace{0 + \cdots + 0}_{z}\right\} = 0$$

77. Assume that g is differentiable and nonvanishing and that $1/g$ is differentiable. Derive the Reciprocal Rule for g from the Product Rule applied to g and $1/g$.

78. Show that if $f(x) = x^2$, then the central difference quotient $D_0 f(c, h)$ for approximating $f'(c)$ is exactly equal to $f'(c)$.

79. Show that the central difference quotient $D_0 f(c, h)$ is an average of a forward and a backward difference quotient.

80. Suppose that a consumer's satisfaction with purchasing x units of good X and y units of good Y can be quantified. The locus of points (x, y) for which the consumer's satisfaction is a fixed value is called an *indifference curve*: The consumer is equally satisfied at all points on the curve. A typical indifference curve is illustrated in Figure 4. Most consumers have a fixed budget for the purchase of goods X and Y. The locus of points that represent the combination of goods X and Y that the consumer can actually purchase on his budget is a straight line called the *budget line*. The point of tangency between the budget line and an indifference curve is called the *consumer equilibrium*. See Figure 4. If $(2, 1/5)$ is the equilibrium point for a consumer whose indifference curve is

$$3y(x^2 + 1) + 5x(y + 1) = 15,$$

find the consumer's budget line in the form

$$Ax + By = C.$$

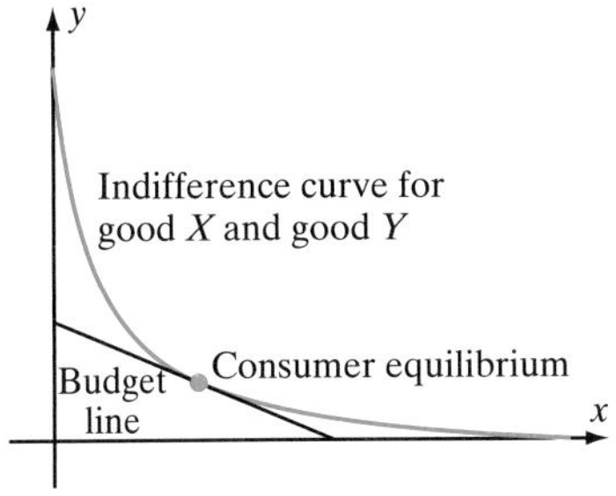

▲ Figure 4

81. If $q(p)$ is the demand for a product at price p, that is, the number of units of the product that are sold at price p, then

$$E(p) = -q'(p) \cdot \frac{p}{q(p)}$$

is defined to be the *elasticity of demand*.

a. Let Δp denote a small change in p, and let $\Delta q = q(p + \Delta p) - q(p)$ be the corresponding change in q. Show that, for Δp sufficiently small, we have the approximation

$$E(p) \approx -\frac{\Delta q / q(p)}{\Delta p / p}.$$

b. Elasticity of demand for potatoes at standard supermarket prices has been estimated to be 0.31. Approximately what percentage change in potato consumption will result from a 2% increase in the price of potatoes?

c. Over a fixed period, a television repair service recorded the number of repairs that it made at three different prices:

Price	Contracts Obtained
\$90	96
\$80	200
\$70	300

Estimate the elasticity of demand at $p = \$70$ and at $p = \$90$. Use a central difference quotient to estimate elasticity of demand at $p = \$80$. (Elasticity of demand for television repair was estimated to be 3.84 in the 1970s.)

82. Let $f_n(x) = x^n$ for each nonnegative integer n. Notice that $f_{n+1}(x) = x f_n(x)$. Use the Product Rule to show that

$$f'_{n+1}(x) = f_n(x) + f'_n(x).$$

Use this equation to show that $f'_{n+1}(x) = (n+1)x^n$ if $f'_n(x) = nx^{n-1}$. In other words, if the formula

$$\frac{d}{dx}x^n = nx^{n-1}$$

is valid for a nonnegative integer n, then it is valid for the next integer $n+1$. In fact, we know that this differentiation formula is true for $n = 0, 1, 2,$ and 3. It follows that the formula is true for the next integer 4. Because the formula is true for 4, it is true for the next integer 5, and so on. We deduce that

$$\frac{d}{dx}x^n = nx^{n-1} \qquad \text{for all nonnegative integers } n.$$

Calculator/Computer Exercises

▶ In Exercises 83–86, a function f and a point c are given. Prepare a table of the forward, backward, and central difference quotients, $D_+f(c,h)$, $D_-f(c,h)$, $D_0f(c,h)$ respectively, for $h = 10^{-n}$, $1 \le n \le 5$. ◀

83. $f(x) = \cos(\pi x)$, $c = 1$

84. $f(x) = \sin(\pi x)$, $c = 1/4$

85. $f(x) = \sqrt{x + 1/x}$, $c = 3$

86. $f(x) = (x^3 + 3x^2 + 2)/(x^2 - 2)$, $c = 2$

▶ In Exercises 87–90, find $f'(c)$ (exactly) for the given function f and the given value c of x. Then approximate $f'(c)$ to 5 decimal places by (1) finding a floating point evaluation of the exact answer and (2) using a central difference quotient $D_0f(c, h)$. Record the value of h used. ◀

87. $f(x) = \sin(x)$, $c = \pi/6$

88. $f(x) = \sqrt{2x}$, $c = 1$

89. $f(x) = 1/x$, $c = 1/\sqrt[4]{2}$

90. $f(x) = \cos(x)$, $c = \pi/4$

91. Let $f(x) = \cos(x)$. Graph $x \mapsto$ to $D_0f(x, 10^{-5})$ for $0 \le x \le 2\pi$. What is $f'(x)$? Add the graph of $f'(x)$ to your viewing window. (The graphs should nearly coincide.)

92. Let $f(x) = x^4/4$. Graph $x \mapsto D_0f(x, 10^{-5})$ for $-1.5 \le x \le 1.5$. What is $f'(x)$? Add the graph of $f'(x)$ to your viewing window. (The graphs should nearly coincide.)

93. The purpose of this exercise is to locate a value a_0 such that

$$\left.\frac{d}{dx}a_0^x\right|_{x=0} = 1.$$

Let $f_a(x) = a^x$. Set $L(a) = D_0f_a(0, 10^{-5})$. Compute $L(2.70)$ and $L(2.73)$. Graph L for $2.70 \le a \le 2.73$ to see that there is a value a_0 in $[2.70, 2.73]$ such that $L(a_0) = 1$. Narrow down the interval in which a_0 must lie until you have found a_0 to three decimal places.

94. Let $f(x) = \sin^2(x)$. Graph $x \mapsto D_0f(x, 10^{-5})$ for $0 \le x \le 2\pi$. Now compute f', and add the graph to the viewing window. (The graphs should nearly coincide.)

3.4 Differentiation of Some Basic Functions

In the preceding section, we learned how to differentiate functions that are formed arithmetically from simpler functions, provided that we know how to differentiate the simpler ones. The purpose of this section is to build up the library of basic functions that we can differentiate. Your work in calculus will be easier if you commit these derivatives to memory.

Powers of x By now we have learned to differentiate x^p for the integers $p = -1, 0, 1, 2,$ and 3. For these values of p, we know that

$$\frac{d}{dx}x^p = p\,x^{p-1}. \tag{3.4.1}$$

In our first example, we extend equation (3.4.1) to include *all* positive integer values of p.

▶ **EXAMPLE 1** Suppose that p is a positive integer. Let $f(x) = x^p$. Show that $f'(c) = pc^{p-1}$.

Solution Because we already know that the asserted formula is true for $p = 1$, we assume that $1 < p$. Because $p-1$ is positive, we have $\lim_{x\to 0} x^{p-1} = 0$. Thus formula (3.2.3) with $c = 0$ simplifies as follows:

$$f'(0) = \lim_{x\to 0} \frac{f(x) - f(0)}{x - 0} = \lim_{x\to 0} \frac{x^p - 0}{x - 0} = \lim_{x\to 0} x^{p-1} = 0 = p \cdot 0^{p-1}.$$

Having established the desired formula for $c = 0$, we now assume that $c \neq 0$ and set $r = x/c$. Notice that $r \to 1$ as $x \to c$. Using formula (3.2.3), we have

$$f'(c) = \lim_{x\to c} \frac{f(x) - f(c)}{x - c} = \lim_{x\to c} \frac{x^p - c^p}{x - c} = \lim_{x\to c} \frac{c^p}{c} \frac{(x/c)^p - 1}{x/c - 1} = c^{p-1} \lim_{r\to 1} \frac{r^p - 1}{r - 1}.$$

We now appeal to equation (2.5.2) with $N = p$ to obtain

$$f'(c) = c^{p-1} \lim_{r\to 1} \left(\underbrace{1 + r + r^2 + \cdots + r^{p-1}}_{p \text{ summands}} \right)$$

$$= c^{p-1} \cdot \left(\underbrace{1 + 1 + 1^2 + \cdots + 1^{p-1}}_{p \text{ summands}} \right) = p\, c^{p-1}. \quad ◀$$

INSIGHT Equation (3.4.1) refers to a variable x. By contrast, the derivative formula of Example 1 was obtained by treating c as a fixed point. This choice was made to facilitate the notation used in the solution. We should now observe that Example 1 establishes $f'(c) = pc^{p-1}$ for *every* value of c. In other words, we may regard c as a variable in the formula. With that understanding, we may replace c with the more conventional choice of variable x. We then obtain $f'(x) = px^{p-1}$, which is precisely equation (3.4.1) with a different choice of notation for the derivative.

We now know that equation (3.4.1) holds for all integer values of p with $-1 \le p$. In our next example, we extend this result to *all* integer values of p.

▶ **EXAMPLE 2** Use the Reciprocal Rule to show that formula (3.4.1) holds for all integer values of p with $p \le -2$.

Solution Suppose that $p \le -2$. Write $p = -|p|$, and set $g(x) = x^{|p|}$. From Example 1, we know that $g'(x) = |p|\, x^{|p|-1}$. Therefore using the Reciprocal Rule, we have

$$\frac{d}{dx} x^p = \frac{d}{dx} x^{-|p|} = \frac{d}{dx} \left(\frac{1}{x^{|p|}} \right) = \frac{d}{dx} \left(\frac{1}{g(x)} \right)$$

$$= -\frac{g'(x)}{g^2(x)} = -\frac{|p|\, x^{|p|-1}}{x^{2|p|}} = px^{-|p|-1} = px^{p-1}.$$

The next example shows that formula (3.4.1) is also valid for $p = 1/2$. ◀

▶ **EXAMPLE 3** For $p = 1/2$, formula (3.4.1) can be written as

$$\frac{d}{dx} \sqrt{x} = \frac{1}{2\sqrt{x}}. \tag{3.4.2}$$

Verify that equation (3.4.2) holds for all $x > 0$.

Solution We verify this formula by referring to the definition of the derivative. Since

$$\frac{\sqrt{x+\Delta x}-\sqrt{x}}{\Delta x}=\frac{(\sqrt{x+\Delta x}-\sqrt{x})(\sqrt{x+\Delta x}+\sqrt{x})}{\Delta x(\sqrt{x+\Delta x}+\sqrt{x})}=\frac{(\sqrt{x+\Delta x})^2-(\sqrt{x})^2}{\Delta x(\sqrt{x+\Delta x}+\sqrt{x})}$$

$$=\frac{(x+\Delta x)-x}{\Delta x(\sqrt{x+\Delta x}+\sqrt{x})}=\frac{\Delta x}{\Delta x(\sqrt{x+\Delta x}+\sqrt{x})}=\frac{1}{\sqrt{x+\Delta x}+\sqrt{x}},$$

we conclude that

$$\frac{d}{dx}\sqrt{x}=\lim_{\Delta x\to 0}\frac{\sqrt{x+\Delta x}-\sqrt{x}}{\Delta x}=\lim_{\Delta x\to 0}\frac{1}{\sqrt{x+\Delta x}+\sqrt{x}}=\frac{1}{\sqrt{x+0}+\sqrt{x}}=\frac{1}{2\sqrt{x}}.$$ ◀

Now that we have verified formula (3.4.1) for one value of p that is not an integer, it is reasonable to wonder if (3.4.1) is valid for other values of p. Indeed, it is. We record this fact as a theorem for now, but we defer the proof until Section 5.6. In the meantime, you should use this differentiation formula whenever it is needed.

THEOREM 1 **(The Power Rule)** If p is any real number, then

$$\frac{d}{dx}x^p=p\,x^{p-1}.$$

▶ **EXAMPLE 4** Differentiate

$$\frac{1}{4x^{12}}+\frac{6x^{7/3}-2}{x^2}$$

with respect to x.

Solution After rewriting the given expression as a sum of three monomials, we use Theorem 1 with $p=-12$, $p=1/3$, and $p=-2$:

$$\frac{d}{dx}\left(\frac{1}{4x^{12}}+\frac{6x^{7/3}-2}{x^2}\right)=\frac{d}{dx}\left(\frac{1}{4}x^{-12}+6x^{1/3}-2x^{-2}\right)$$

$$=\frac{1}{4}(-12)x^{-13}+6\left(\frac{1}{3}\right)x^{-2/3}-2(-2)x^{-3}$$

$$=-\frac{3}{x^{13}}+\frac{2}{x^{2/3}}+\frac{4}{x^3}.$$ ◀

▶ **EXAMPLE 5** In an adiabatic (no heat in or out) argon system, the third power of pressure (P) is inversely proportional to the fifth power of volume (V). At a pressure of 1000 atmospheres (atm), the volume of the argon sample is 34 cm^3. What is the rate of change of pressure with respect to volume when the sample is 50 cm^3?

Solution According to the given information, there is a proportionality constant k such that $P^3=k/V^5$, or $P=cV^{-5/3}$ where $c=k^{1/3}$. Our first step is to find the value of c by substituting $P=1000$ atm and $V=34$ cm^3 into the equation $P=cV^{-5/3}$. We obtain

$$1000\,\text{atm} = c \cdot (34\,\text{cm}^3)^{-5/3},$$

or $c = 1000 \cdot 34^{5/3}$ atm $\times$ cm^5. Therefore using Theorem 1 with $p = -5/3$, we calculate

$$\begin{aligned}\frac{d}{dV}P &= \frac{d}{dV}(cV^{-5/3})\\ &= -\frac{5c}{3}V^{-8/3}\\ &= -\frac{5000\cdot 34^{5/3}}{3}V^{-8/3}\text{atm cm}^5\end{aligned}$$

and

$$\left.\frac{dP}{dV}\right|_{V=50\,\text{cm}^3} = -\frac{5000\cdot 34^{5/3}}{3} \times (50\,\text{cm}^3)^{-8/3}\ \text{atm} \times \text{cm}^5 \approx -17.5\,\frac{\text{atm}}{\text{cm}^3}. \quad ◀$$

Before we turn our attention to other types of functions, we prove one simple consequence of the Power Rule.

THEOREM 2 The derivative p' of a polynomial p is a polynomial. If p is not a constant, then the degree of p' is one less than the degree of p. If p is a constant, then p' is identically 0.

Proof. If $p(x) = a_N x^N + a_{N-1}x^{N-1} + \cdots + a_2x^2 + a_1x + a_0$ with $a_N \neq 0$, then, on applying the Power Rule to each term of $p(x)$, we obtain

$$p'(x) = N\cdot a_N x^{N-1} + (N-1)\cdot a_{N-1}x^{N-2} + \cdots + 2a_2x + a_1,$$

from which the asserted results follow. ■

Trigonometric Functions

We continue to build our library of differentiation formulas by finding the derivatives of the remaining trigonometric functions. We have already seen how to differentiate sine and cosine (Section 3.2) and secant (Example 4 of Section 3.3). The other trigonometric functions can now be handled easily. Before doing so, recall that we often write $\sin^2(x)$ and $\tan^2(x)$ rather than $(\sin(x))^2$ and $(\tan(x))^2$.

THEOREM 3 The derivatives of $\tan(x)$, $\cot(x)$, and $\csc(x)$ are

$$\frac{d}{dx}\tan(x) = \sec^2(x), \tag{3.4.3}$$

$$\frac{d}{dx}\cot(x) = -\csc^2(x), \tag{3.4.4}$$

and

$$\frac{d}{dx}\csc(x) = -\csc(x)\cot(x). \tag{3.4.5}$$

Proof. We use the Quotient Rule to prove formula (3.4.3):

$$\begin{aligned}
\frac{d}{dx}\tan(x) &= \frac{d}{dx}\left(\frac{\sin(x)}{\cos(x)}\right) \\
&= \frac{\cos(x)\cdot\frac{d}{dx}(\sin(x)) - \sin(x)\cdot\frac{d}{dx}(\cos(x))}{\cos^2(x)} \\
&= \frac{\cos^2(x)+\sin^2(x)}{\cos^2(x)} \\
&= \frac{1}{\cos^2(x)} \\
&= \sec^2(x).
\end{aligned}$$

The proof of formula (3.4.4) is similar and is left as Exercise 66. The proof of equation (3.4.5) is analogous to the derivation of formula (3.3.1) and is left as Exercise 67. ■

▶ **EXAMPLE 6** Find the tangent line to the graph of $y = x\cot(x)$ at the point $(\pi/6, \pi\sqrt{3}/6)$.

Solution We first find the slope. Using the Product Rule, we calculate:

$$\frac{dy}{dx} = \frac{d}{dx}(x\cot(x)) = 1\cdot\cot(x) + x\cdot\left(-\csc^2(x)\right) = \cot(x) - x\cdot\csc^2(x).$$

Evaluating at $x = \pi/6$, we find the slope to be $\sqrt{3} - (\pi/6)\cdot 2^2 = \sqrt{3} - 2\pi/3$. According to formula (3.1.12), the point-slope equation of the tangent line is

$$y = \left(\sqrt{3} - \frac{2\pi}{3}\right)\cdot\left(x - \frac{\pi}{6}\right) + \frac{\pi\sqrt{3}}{6}. \quad ◀$$

The Derivative of the Natural Exponential Function

In Section 2.6, we introduced the number e, and, in Example 8 from that section, derived the basic formulas

$$e^x = \lim_{n\to\infty}\left(1+\frac{x}{n}\right)^n \quad \text{and} \quad e^x = \lim_{n\to\infty}\left(1-\frac{x}{n}\right)^{-n}. \tag{3.4.6}$$

We will now see that the exponential function $x \mapsto e^x$, or $\exp(x)$, has the remarkable property that it is its own derivative.

THEOREM 4 For every x, we have

$$\frac{d}{dx}e^x = e^x \tag{3.4.7}$$

or, equivalently,

$$\frac{d}{dx}\exp(x) = \exp(x). \tag{3.4.8}$$

Proof. Let $f(x) = e^x$. We will first show that (3.4.7) holds for $x = 0$. Because $f(0) = e^0 = 1$ and

$$f'(0) = \lim_{h\to 0} \frac{e^h - e^0}{h} = \lim_{h\to 0} \frac{e^h - 1}{h},$$

we must show that

$$\lim_{h\to 0} \frac{e^h - 1}{h} = 1. \tag{3.4.9}$$

As a first step, we will prove the one-sided limit

$$\lim_{h\to 0^+} \frac{e^h - 1}{h} = 1. \tag{3.4.10}$$

The numerical evidence supporting this equation is certainly compelling:

h	10^{-3}	10^{-4}	10^{-5}	10^{-6}	10^{-7}	10^{-8}
$\frac{e^h-1}{h}$	1.00050017	1.00005000	1.00000500	1.00000050	1.00000005	1.00000000

So is the graphical evidence. In Figure 1, we see that

$$1 + h < e^h < \frac{1}{1-h} \qquad (0 < h < 1) \tag{3.4.11}$$

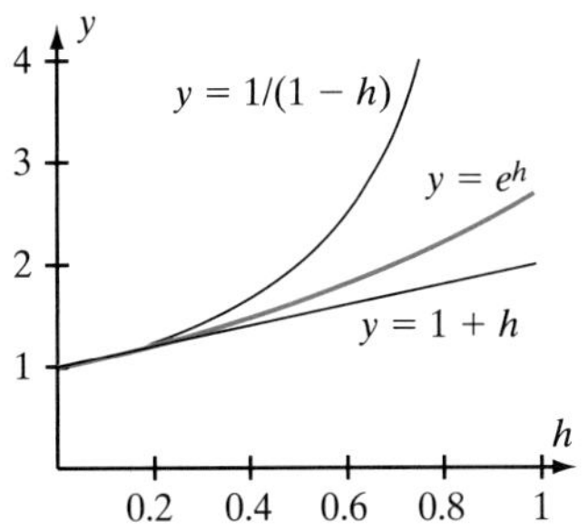

▲ Figure 1

In fact, we know that $(1+h/x)^x$ increases to e^h as x tends to infinity—refer to the *Insight* following Example 8 of Section 2.6. Taking $x = 1$, we see that $1 + h < e^h$, which is the first inequality of line (3.4.11). From Section 2.6 we also know that $(1 - h/x)^{-x}$ decreases to e^h as x tends to infinity. Taking $x = 1$, we see that $e^h < (1-h)^{-1}$, which is the second inequality of line (3.4.11). Subtracting 1 from each part of line (3.4.11) and simplifying $1/(1-h) - 1$ to $h/(1-h)$, we have

$$h < e^h - 1 < \frac{h}{1-h}.$$

Dividing by h (which is positive), we obtain

$$1 < \frac{e^h - 1}{h} < \frac{1}{1-h}.$$

Because $\lim_{h\to 0^+}(1/(1-h)) = 1$, the Pinching Theorem of Section 2.2 establishes limit formula (3.4.10).

Next, we complete the proof of two-sided limit (3.4.9). We note that if $u = -h$, then $u \to 0^-$ as $h \to 0^+$. It follows that

$$\lim_{u\to 0^-} \frac{e^u - 1}{u} = \lim_{u\to 0^-} \frac{1 - e^u}{-u} = \lim_{u\to 0^-} e^u \left(\frac{e^{-u} - 1}{-u}\right)$$

$$= \lim_{u\to 0^-} e^u \lim_{u\to 0^-} \frac{e^{-u} - 1}{-u} = 1 \cdot \lim_{h\to 0^+} \frac{e^h - 1}{h} \overset{(3.4.10)}{=} 1.$$

Together, this limit formula and (3.4.10) imply the two-sided limit (3.4.9).

Finally, we obtain (3.4.7) for a general value of x by reducing to the $x = 0$ case as follows:

$$\begin{aligned} f'(x) &= \lim_{h\to 0} \frac{f(x+h) - f(x)}{h} \\ &= \lim_{h\to 0} \frac{e^{x+h} - e^x}{h} \\ &= \lim_{h\to 0} \frac{e^x e^h - e^x}{h} \\ &= e^x \lim_{h\to 0} \frac{e^h - 1}{h} \\ &= e^x \end{aligned}$$

by (3.4.9). ■

▶ **EXAMPLE 7** Suppose that an object vibrates about its equilibrium position according to the formula $x(t) = e^{-t} \sin(t)$ mm. (Resistance to movement gives rise to the "damping" factor e^{-t}.) At what speed is the object moving when $t = 1$?

Solution We compute $x'(t)$ by means of the Quotient Rule:

$$\begin{aligned} x'(t) &= \frac{d}{dt}\left(\frac{\sin(t)}{e^t}\right) \\ &= \frac{e^t \dfrac{d}{dt}\sin(t) - \sin(t)\dfrac{d}{dt}e^t}{(e^t)^2} \\ &= \frac{e^t\cos(t) - e^t\sin(t)}{e^{2t}} \\ &= \frac{\cos(t) - \sin(t)}{e^t}. \end{aligned}$$

Therefore

$$x'(1) = \frac{\cos(1) - \sin(1)}{e} \approx -0.1108. \quad ◀$$

Summary of Derivatives of Basic Functions

$$\frac{d}{dx}(x^p) = px^{p-1} \qquad -\infty < p < \infty$$

$$\frac{d}{dx}\sin(x) = \cos(x)$$

$$\frac{d}{dx}\cos(x) = -\sin(x)$$

$$\frac{d}{dx}\tan(x) = \sec^2(x)$$

$$\frac{d}{dx}\cot(x) = -\csc^2(x)$$

$$\frac{d}{dx}\sec(x) = \sec(x)\tan(x)$$

$$\frac{d}{dx}\csc(x) = -\csc(x)\cot(x)$$

$$\frac{d}{dx}e^x = e^x$$

QUICK QUIZ

1. Differentiate $x^{3/2}$ with respect to x.
2. Differentiate $\csc(x)\cot(x)$ with respect to x.
3. What is $\lim_{x\to 0}(e^x - 1)/x$?
4. Differentiate xe^x with respect to x.

Answers

1. $(3/2)\,x^{1/2}$ 2. $-\csc(x)\cot^2(x) - \csc^3(x)$ 3. 1 4. $(x+1)e^x$

EXERCISES

Problems for Practice

In Exercises 1–28 differentiate the given expression with respect to x.

1. $8x^{10} - 6x^{-5}$
2. $2x^{3/2} - 1/x^{3/2}$
3. $6x^{5/3} - 25x^{3/5}$
4. $\sqrt{x^5} + 3/\sqrt{x}$
5. $2x + e^x$
6. xe^x
7. $e^x/x^{3/2}$
8. $\sqrt{x}\csc(x)$
9. $x^{-9}\cot(x)$
10. $x/\tan(x)$
11. $\sec(x)/\sqrt{x}$
12. $\sec(x) - \tan(x)$
13. $\csc(x) + \cot(x)$
14. $x\cot(x) - \csc(x)$
15. $\csc(x)\cot(x)$
16. $x^2\tan(x)$
17. $\tan(x)\sec(x)$
18. $x^{-5}e^x$
19. $e^x\sin(x)$
20. $2/(\cot(x) - \csc(x))$
21. $3/(5 + 7e^x)$
22. $x/(2 + \tan(x))$
23. $(1 + \sec(x))/\tan(x)$
24. $(1 - \sqrt{x})/(1 + \sqrt{x})$
25. $x^2/(x + e^x)$
26. $e^{-x}\cos(x)$
27. $x^{1/3}/(1 + x^{2/3})$
28. $(e^x - 1)/(e^x + 1)$

In Exercises 29–36, find the tangent line to the graph of $y = f(x)$ at P.

29. $f(x) = 3x^{2/3}$, $P = (8, 12)$
30. $f(x) = 6/(1 + \sqrt{x})$, $P = (4, 2)$
31. $f(x) = x^{-3}\sin(x)$, $P = (\pi, 0)$
32. $f(x) = x^7 + \tan(x)$, $P = (0, 0)$
33. $f(x) = \tan(x)\sec(x)$, $P = (\pi/3, 2\sqrt{3})$
34. $f(x) = e^x$, $P = (0, 1)$
35. $f(x) = 8x\tan(x)$, $P = (\pi/4, 2\pi)$
36. $f(x) = \csc(x)/x$, $P = (\pi/6, 12/\pi)$

In each of Exercises 37–40, use the Product Rule to differentiate the given expression with respect to x.

37. e^{2x}
38. $\tan^2(x)$
39. $\csc^2(x)$
40. $\tan(x) + \tan^2(x)$

Further Theory and Practice

In Exercises 41–44, find a polynomial whose derivative is the given polynomial.

41. $7x^6 - 4x + 6$
42. $x^9 - 2x^3 - 1$
43. $x^8 + 6x^5 - 3x$
44. $10x^6 + x^2 + 4x - 3$

In Exercises 45–48, find a function whose derivative is the given function.

45. $1/\sqrt{x}$
46. $3/x^3$
47. $-8\csc^2(x)$
48. $\cos(x)\,(5\csc^2(x) - 1)$

49. Suppose that p is a polynomial. We say that α is a root of p of *multiplicity* m if there is a polynomial q with $q(\alpha) \neq 0$ and $p(x) = (x-\alpha)^m q(x)$. Prove that if α is a root of p of multiplicity m, then α is a root of p' of multiplicity $m-1$.
50. What is the equation of the line that is tangent to the graph of $y = e^x$ at the point (t, e^t)? What are the x- and y-intercepts of this line?
51. Find a function whose derivative is $\tan^2(x)$.
52. Let p be a fixed number. Let C be any number. Show that $y(x) = Cx^p$ is a solution of the differential equation

$$x\frac{dy}{dx} = p\,y.$$

In Exercises 53–56, compute $f'(c)$ for the given f and c.

53. $f(x) = x^5(2x + 1)$, $c = -1$
54. $f(x) = x^3(2x - 1)^2$, $c = 2$
55. $f(x) = (x^5 + 2x^2 - 3)(2x^3 + 7x - 2)$, $c = 1$
56. $f(x) = (x^5 + x - 1)(2x^2 - 3)^2$, $c = -1$

57. Let $f(x) = \sec^2(x) - \tan^2(x)$. Use the Product Rule to differentiate each summand of f. Simplify to show that $f'(x) = 0$ for each x in the domain of f. How can you obtain this result more easily?
58. Let $f_n(x) = x^n e^x$ for every positive integer n. Find f_n' in terms of f_n and f_{n-1}.
59. Relate the limit

$$\lim_{h\to 0}\frac{e^{h/2} - e^{-h/2}}{h}$$

to a derivative and evaluate it.
60. Let $f(x) = \tan(x)$. For every $t \in (-\pi/2, \pi/2)$, write down the equation of the tangent line to the graph of f at $(t, f(t))$ in the form

$$x = M(t)(y - \tan(t)) + t.$$

What function is $M(t)$? What line results when you compute

$$x = \lim_{t\to(\pi/2)^-}(M(t)(y - \tan(t)) + t)?$$

Is it an asymptote of f?
61. Show that

$$\frac{d}{dx}|x|^p = p\frac{|x|^p}{x} \qquad (x \neq 0).$$

62. Suppose that p is a degree n polynomial, and q is a degree m polynomial. What is the degree of $(p \cdot q)'$? Of $(p \circ q)'$? Of the numerator and denominator of $(p/q)'$ (before performing any cancellations)?
63. Find all values of c such that the intercepts of the tangent line to $y = e^x$ at (c, e^c) are equidistant from the origin.
64. Let $g(x) = e^{kx}$ where k is a constant. Using the definition of the derivative, show that

$$g'(x) = g'(0)g(x).$$

65. Suppose that $0 < p < 2$. For $x > 0$, let

$$f(x) = \frac{x^p}{x^2 + 1}.$$

Find the point at which the graph of f has a horizontal tangent line.
66. Write $\cot(x) = \cos(x)/\sin(x)$, and use the Quotient Rule, as in the derivation of equation (3.4.3), to verify the formula

$$\frac{d}{dx}\cot(x) = -\csc^2(x).$$

67. Write $\sec(x) = 1/\cos(x)$, and use the Reciprocal Rule to verify the formula

$$\frac{d}{dx}\csc(x) = -\csc(x)\cot(x).$$

If x and $f(x)$ measure physical quantities that bear units (such as meters, grams, seconds, etc.), then the difference quotients that are used in definition (3.2.1) also bear units, which are imparted to $f'(x)$. Exercises 68–70 pertain to *dimensional analysis*, which requires that the dimensions of both sides of an equation are the same.

68. Suppose that $y = x^p$. Use dimensional analysis to predict the correct value of α in the equation $dy/dx = px^{\alpha}$.

69. In the Product Rule for $(f \cdot g)'(t)$, there is a term $f'(t) \cdot g(t)$ and a term $f(t) \cdot g'(t)$ but no term involving $f'(t) \cdot g'(t)$. Use dimensional analysis to explain why no such term should be present.

70. Show that the Quotient Rule for $(f/g)'(t)$ is dimensionally correct. That is, the Quotient Rule leads to the correct units for the derivative of a quotient.

71. If $q(p)$ is the demand for a product at price p—that is, the number of units of the product that are sold at price p—then $E(p) = -q'(p) \cdot p/q(p)$ is called the *elasticity of demand* for the product at price p. Suppose that a and b are positive constants.

a. What is $E(p)$ if $q(p) = a - bp$?

b. What is $E(p)$ if $q(p) = a/p^b$?

72. The *hyperbolic sine* and *hyperbolic cosine* functions are defined by

$$\cosh(x) = \frac{e^x + e^{-x}}{2}$$

and

$$\sinh(x) = \frac{e^x - e^{-x}}{2}.$$

Express $\cosh'(x)$ in terms of $\sinh(x)$ and $\sinh'(x)$ in terms of $\cosh(x)$.

73. The *hyperbolic tangent* (tanh) and *hyperbolic secant* (sech) are defined by

$$\tanh(x) = \frac{\sinh(x)}{\cosh(x)} = \frac{e^x - e^{-x}}{e^x + e^{-x}}$$

and

$$\operatorname{sech}(x) = \frac{1}{\cosh(x)} = \frac{2}{e^x + e^{-x}}.$$

Express $\tanh'(x)$ and $\operatorname{sech}'(x)$ in terms of $\tanh(x)$ and $\operatorname{sech}(x)$.

Calculator/Computer Exercises

74. Verify

$$\left.\frac{d}{dx}e^{-x}\right|_{x=0} = -1$$

numerically by computing central difference quotients.

For each of the given functions in Exercises 75–78, graph f and f' in the given viewing rectangle R. Fill in the following table.

Interval where f increases	Interval where f decreases	Point at which f has a horizontal tangent
Interval where $f' > 0$	Interval where $f' < 0$	Point(s) at which $f' = 0$

Use the table to draw inferences that relate the sign of f' to the behavior of f. (These relationships will be studied in Chapter 4.)

75. $f(x) = 2x + 6 - 5/(x^4 + 1)$, $R = [-1.3, 0.5] \times [-3.5, 4.2]$

76. $f(x) = 1 + (x^3 - 1)/(x^4 + 1)$, $R = [-3, 1.6] \times [-0.6, 1.9]$

77. $f(x) = x^{4/3}/(2 + \sin(x))$, $R = [0, 7] \times [-2.5, 8.25]0$

78. $f(x) = x^2 \sin(x)/(1 + x + x^4)$, $R = [-3, 1.2] \times [-0.9, 2.1]$

79. Let $g_k(x) = e^{kx}$. Plot

$$k \mapsto D_0 g_k(0, 10^{-4})$$

for $-2 \le k \le 2$. Use your plot to determine $g_k'(0)$.

80. For what values of $x > 1$, if any, does $x \mapsto e^x$ grow faster than $x \mapsto x^{20}$? When, if ever, does e^x catch up in size to x^{20}?

81. Let p_0, τ_0, μ, and Δ be fixed constants. The set $\mathcal{H}$ of points (p, τ) in the first quadrant of the $p\tau$-plane that satisfy

$$(\tau - \mu^2\tau_0)p - (\tau_0 - \mu^2\tau)p_0 + 2\mu^2\Delta = 0$$

is called the *Hugoniot curve*. It arises in the study of gas dynamics produced by combustion.

a. Find $d\tau/dp$.

b. Let $Q = (p_1, \tau_1)$ be a point on the Hugoniot curve $\mathcal{H}$. What is the equation of the line that is tangent to $\mathcal{H}$ at Q?

c. Assume that $\Delta < 0$. (This is the mathematical condition for an *exothermic* reaction.) Show that the point (p_0, τ_0) lies below the Hugoniot curve.

d. Continue to suppose that $\Delta < 0$. Solve for those points Q on the Hugoniot curve $\mathcal{H}$ such that the line tangent to $\mathcal{H}$ at Q passes through the point $P = (p_0, \tau_0)$. The points that you find are known as the *Chapman-Jouguet points*. The tangent lines that you find are known as the *Rayleigh lines*.

e. Illustrate the Hugoniot curve and its Rayleigh lines for the specifications $p_0 = 2.1$, $\tau_0 = 1.7$, $\mu = 1/2$, and $\Delta = -2$. Label the Chapman-Jouguet points.

82. Graph $f(x) = (x + 1)^{4/3}(x - 1)^{2/3}$ for $-1.5 \le x \le 1.5$. Does the graph of f have a tangent line at the point $(-1, 0)$? At the point $(1, 0)$? Use the Power Rule to explain the behavior of the graph of f at these two points.

3.5 The Chain Rule

In principle, we know how to differentiate $H(x) = (x^3 + x)^{100}$. All we have to do is expand the expression and then differentiate. But this would be a great deal of work, and it turns out that we can avoid it.

Instead, we think of H as the composition of two functions: $H = g \circ f$ where $g(u) = u^{100}$ and $f(x) = x^3 + x$. (Recall, that in Section 1.5, we learned how to recognize compositions of functions.) We know how to differentiate both f and g. Let's see how we can use this information to differentiate their composition H.

A Rule for Differentiating the Composition of Two Functions

Imagine car A driving down the road with position given by $a(t)$. Imagine a second car B driving down the road with position given by $b(t)$. At a given instant, if B is traveling at a rate of 8 mi/hr ($db/dt = 8$), and if A is traveling twice as fast as B, then it is clear that A is traveling at the rate of 16 mi/hr. In other words, if we know the rate of change of a with respect to b (in this case, 2) and we know the rate of change of b with respect to t (in this case, 8), then we know the rate of change of a with respect to t (namely, $2 \cdot 8 = 16$). Notice that the third quantity is the *product* of the first two. Schematically we have

$$\begin{pmatrix} \text{rate of change of} \\ a \\ \text{with respect to} \\ t \end{pmatrix} = \begin{pmatrix} \text{rate of change of} \\ a \\ \text{with respect to} \\ b \end{pmatrix} \times \begin{pmatrix} \text{rate of change of} \\ b \\ \text{with respect to} \\ t \end{pmatrix}$$

or

$$\frac{da}{dt} = \frac{da}{db} \cdot \frac{db}{dt}.$$

This example suggests that

The derivative of a composition is the product of the derivatives of the component functions.

This formula is known as the *Chain Rule*. It gives us a way to differentiate the composition of two differentiable functions. The formulation of the Chain Rule in Leibniz notation is particularly suggestive. It looks as though we are canceling fractions. As a matter of fact we are *not*. Derivatives expressed in Leibniz notation are *not* fractions. However the "difference quotients" which converge to the derivative *are* fractions, and we *can* cancel those. That observation, if executed with some care, can be used to prove the Chain Rule. A somewhat different proof may be found in the *Genesis and Development* for this chapter.

THEOREM 1 **Chain Rule** Suppose that the range of the function f is contained in the domain of the function g. If f is differentiable at c and if g is differentiable at $f(c)$, then $g \circ f$ is differentiable at c. Moreover

$$\frac{d}{dx}(g\circ f)\bigg|_{x=c} = \left(\frac{dg}{du}\bigg|_{u=f(c)}\right)\cdot\left(\frac{df}{dx}\bigg|_{x=c}\right). \tag{3.5.1}$$

Recall from Section 3.2 that the vertical bar notation indicates the point at which the derivative is evaluated. It is important to notice that the two derivatives on the right side of (3.5.1) are evaluated at *different* points. When we calculate $(g\circ f)(x) = g(f(x))$, we evaluate g at $f(x)$, and we evaluate f at x. This evaluation scheme is preserved in the application of the Chain Rule: In computing the derivative of $g\circ f$ at c, the derivative of g is evaluated at $f(c)$ and the derivative of f is evaluated at c. Thus while the basic idea of the Chain Rule is that the derivative of a composition is the product of the derivatives, the statement of the Chain Rule is a bit complicated because we have to be careful about where we evaluate the derivatives!

With primes denoting derivatives, the Chain Rule becomes

$$(g\circ f)'(c) = g'(f(c))\cdot f'(c).$$

INSIGHT It is sometimes useful to think of a composition as having an "outside function" and an "inside function." The outside function is the *last* function applied: In the composition $g\circ f$, it would be g. In applying the Chain Rule to such a composition, we differentiate the outside function first, holding the inside function fixed. *Then* we differentiate the inside function (which would be f in the composition $g\circ f$). Keep this point of view in mind when reading the examples that follow.

Some Examples

▶ **EXAMPLE 1** Differentiate $H(x) = \sin(3x^2+1)$.

Solution We write $H = g\circ f$ where $g(u) = \sin(u)$ is the "outside function" and $f(x) = 3x^2+1$ is the "inside function." Then

$$\frac{dg}{du} = \cos(u) \qquad \text{and} \qquad \frac{df}{dx} = 6x.$$

As a result,

$$\begin{aligned}\frac{dH}{dx} &= \frac{dg}{du}\bigg|_{u=f(x)}\cdot\frac{df}{dx}\\ &= \cos(f(x))\cdot(6x)\\ &= (\cos(3x^2+1))\cdot 6x.\end{aligned}$$

Using the idea in the preceding *Insight*, we have the following solution scheme:

$$\underbrace{H'(x)}_{\substack{\nearrow \\ \text{derivative} \\ \text{of} \\ \text{composed} \\ \text{function}}} = \underbrace{\cos}_{\substack{\uparrow \\ \text{derivative} \\ \text{of} \\ \text{outside} \\ \text{function}}}\Big(\underbrace{3x^2+1}_{\substack{\uparrow \\ \text{inside} \\ \text{function}}}\Big) \cdot \underbrace{6x}_{\substack{\nwarrow \\ \text{derivative} \\ \text{of} \\ \text{inside} \\ \text{function}}}.$$

◀

▶ **EXAMPLE 2** Differentiate $H(x) = (x^3 + x)^{100}$.

Solution We write $H = g \circ f$ where $g(u) = u^{100}$ and $f(x) = x^3 + x$. According to the Chain Rule,

$$\frac{dH}{dx} = \left(\frac{dg}{du}\bigg|_{u=f(x)}\right) \cdot \left(\frac{df}{dx}\right).$$

Now

$$\frac{dg}{du} = 100u^{99} \qquad \text{and} \qquad \frac{df}{dx} = 3x^2 + 1.$$

Therefore

$$\begin{aligned} \frac{dH}{dx} &= (100 \cdot u^{99})\Big|_{u=f(x)} \cdot (3x^2+1) \\ &= \left(100 \cdot f(x)^{99}\right) \cdot (3x^2+1) \\ &= \left(100(x^3+x)^{99}\right) \cdot (3x^2+1) \cdot \end{aligned}$$

◀

INSIGHT The merit of using functional notation when applying the Chain Rule is that evaluation issues can be incorporated directly into the notation. There is an alternative notation that employs variables rather than explicit functions. It is encountered frequently because it is more streamlined and easier to remember. The idea is that if y depends on u, and u depends on x, then y depends on x. The Chain Rule expresses the relationship between the derivatives of these variables as follows:

$$\frac{dy}{dx} = \frac{dy}{du}\frac{du}{dx}. \tag{3.5.2}$$

In using formula (3.5.2), it must be understood that the variable y on the left side is a function of x. Its rate of change must therefore be expressed in terms of x alone. Any u that remains on the right side of (3.5.2) must be written in terms of x in order to complete the calculation. Thus in Example 2, we set $y = u^{100}$, $u = x^3 + x$, and, after applying (3.5.2), we replace u with $x^3 + x$:

$$\begin{aligned} \frac{dy}{dx} = \frac{dy}{du}\frac{du}{dx} &= \left(\frac{d}{du}u^{100}\right)\left(\frac{du}{dx}(x^3+x)\right) \\ &= 100u^{99}(3x^2+1) = 100(x^3+x)^{99}(3x^2+1). \end{aligned}$$

▶ **EXAMPLE 3** Find the tangent line to the graph of $H(x) = \tan^7(x)$ at the point $(\pi/4, 1)$.

Solution Writing $\tan^7(x)$ as $(\tan(x))^7$ makes it clearer that $u \mapsto u^7$ is the "outside" function. Thus $H = g \circ f$ where $g(u) = u^7$, and $f(x) = \tan(x)$. Now $g'(u) = 7u^6$, and $f'(x) = \sec^2(x)$. Notice that $g'(f(x)) = 7f(x)^6 = 7f^6(x)$. We can now compute $H'(x) = (g \circ f)'(x)$:

$$\begin{aligned} H'(x) &= g'(f(x)) \cdot f'(x) \\ &= 7 \cdot f^6(x) \cdot \sec^2(x) \\ &= 7 \cdot \tan^6(x) \cdot \sec^2(x). \end{aligned}$$

It follows that the slope of the tangent line to the graph of H at $(\pi/4, 1)$ is

$$H'\left(\frac{\pi}{4}\right) = 7\tan^6\left(\frac{\pi}{4}\right)\sec^2\left(\frac{\pi}{4}\right) = 7 \cdot 1^6 \cdot (\sqrt{2})^2 = 14.$$

Thus the tangent line's equation in point-slope form is $y = 14 \cdot (x - \pi/4) + 1$. ◀

INSIGHT Make sure that you understand the difference between $g'(f(x))$ and $(g \circ f)'(x)$. Because f plays a role in the formation of the function $g \circ f$, its rate of change $f'(x)$ is a component of the rate of change $(g \circ f)'(x)$. On the other hand, $g'(f(x))$ means the rate of change of g at the point $f(x)$: the rate of change of f plays no role in this last quantity.

Examples 2 and 3 involve an application of the Chain Rule that arises frequently. In each example, we considered a function f that has been raised to a fixed power p. Because this type of composition occurs so often, it is worthwhile stating this special case of the Chain Rule:

The Chain Rule—Power Rule

$$\frac{d}{dx} f(x)^p = p\, f(x)^{p-1} \frac{d}{dx} f(x) \tag{3.5.3}$$

Alternatively, if we write $u = f(x)$, then formula (3.5.3) becomes

$$\frac{d}{dx} u^p = p\, u^{p-1} \frac{du}{dx}. \tag{3.5.4}$$

▶ **EXAMPLE 4** Use formula (3.5.3) to derive the Reciprocal Rule of Section 3.3.

Solution If we take $p = -1$ in the Chain Rule–Power Rule, then we have

$$\frac{d}{dx}\left(\frac{1}{f(x)}\right) = \frac{d}{dx}(f(x))^{-1} = (-1)\, f(x)^{-2} \frac{d}{dx} f(x) = -\frac{f'(x)}{f(x)^2},$$

as required. ◀

Multiple Compositions

Of course, some functions are the composition of more than two simpler functions. To differentiate such a function, we must apply the Chain Rule several times.

▶ **EXAMPLE 5** Differentiate $Q(x) = \sin^3(5x)$.

Solution We can write Q as $Q = H \circ (G \circ F)$ where $H(s) = s^3$, $G(u) = \sin(u)$, and $F(x) = 5x$. With this grouping, Q is the composition of two functions, H and $G \circ F$. By the Chain Rule,

$$\frac{dQ}{dx} = \left(\left.\frac{dH}{ds}\right|_{s=G \circ F(x)}\right) \cdot \left(\frac{d(G \circ F)}{dx}\right). \tag{3.5.5}$$

Now we apply the Chain Rule a second time to the composition $G \circ F$ to obtain

$$\frac{d(G \circ F)}{dx} = \left(\left.\frac{dG}{du}\right|_{u=F(x)}\right) \cdot \left(\frac{dF}{dx}\right).$$

Substituting this last equation into equation (3.5.5) now yields

$$\frac{dQ}{dx} = \left(\left.\frac{dH}{ds}\right|_{s=G \circ F(x)}\right) \cdot \left(\left.\frac{dG}{du}\right|_{u=F(x)}\right) \cdot \left(\frac{dF}{dx}\right).$$

Finally, we calculate that

$$\frac{dH}{ds}(s) = 3s^2, \quad \frac{dG}{du}(u) = \cos(u), \quad \text{and} \quad \frac{dF}{dx}(x) = 5.$$

Therefore

$$\begin{aligned} \frac{dQ}{dx} &= 3\big(G(F(x))\big)^2 \cdot \cos\big(F(x)\big) \cdot 5 \\ &= 15 \sin^2(5x) \cdot \cos(5x). \end{aligned}$$ ◀

INSIGHT It is worth noticing that the philosophy of differentiating the outside function first and working toward the inside applies even in the case of multiple compositions. You should review Example 5 with this observation in mind.

An Application

The Chain Rule allows us to investigate new kinds of rate-of-change problems. We conclude with an example.

▶ **EXAMPLE 6** An oil slick spreads so that it forms a disk centered around the point of contamination. As the fixed volume of spilled oil disperses, and the thickness of the slick decreases, the radius of the slick increases at the rate of 2 ft per minute. How fast is the area of polluted surface water growing when the radius is 500 ft?

Solution The area of polluted surface water is given by $A = \pi r^2$. The radius r of the slick is a function of time t (with $dr/dt = 2$ ft/min)}. It follows that A depends on t as well. We use Leibniz notation and differentiate both sides of the equation $A = \pi r^2$ with respect to t to obtain

$$\frac{dA}{dt} = \frac{dA}{dr} \cdot \frac{dr}{dt} = \frac{d}{dr}(\pi r^2) \cdot \frac{dr}{dt} = (\pi \cdot 2r) \cdot \frac{dr}{dt}.$$

Now, substituting $r = 500$ ft and $dr/dt = 2$ ft/min into this last equation, gives

$$\frac{dA}{dt} = \pi \cdot 1000\,\text{ft} \cdot 2\,\frac{\text{ft}}{\text{min}} = 2000\pi\,\frac{\text{ft}^2}{\text{min}}.$$

Therefore the surface area of the oil slick is increasing at the rate of 2000π square feet per minute. ◄

INSIGHT Example 6 poses a question about the rate of change of A with respect to t. The answer therefore requires the derivative of A with respect to t. Because the formula for A, namely πr^2, involves r, not t, the Chain Rule must be applied to effect the correct differentiation. Leibniz notation aids in keeping track of the terms.

A LOOK FORWARD ► When two variables are related by an equation, and one variable changes, then the other variable must change in order to satisfy the equation. The task of determining the unknown rate of change of one variable from the known rate of change of the other, as in Example 6, is called a *related rates* problem. In Section 4.1 we will examine this application of differentiation in greater detail.

Derivatives of Exponential Functions

In Section 3.4 we learned the derivative of the natural exponential function. Now we use that formula together with the Chain Rule to calculate the derivative of an exponential function with an arbitrary base.

THEOREM 2 Suppose that a is a positive constant. Then

$$\frac{d}{dx}a^x = a^x \cdot \ln(a). \tag{3.5.6}$$

Proof. Because the range of the natural exponential is the entire real line, there is, for every x, a number u such that

$$e^u = a^x.$$

If we apply the natural logarithm to both sides, then we obtain $u = \ln(e^u) = \ln(a^x) = x \cdot \ln(a)$. In other words,

$$a^x = e^u = e^{x \cdot \ln(a)}.$$

We differentiate each side of this equation with respect to x and apply the Chain Rule to obtain

$$\frac{d}{dx}(a^x) = \frac{d}{dx}e^{x\cdot\ln(a)} = \frac{d}{dx}e^u = \frac{d}{du}e^u \cdot \frac{du}{dx} = e^u \cdot \frac{du}{dx}$$

$$= e^{x\cdot\ln(a)} \cdot \frac{d}{dx}\left(x \cdot \ln(a)\right) = e^{x\cdot\ln(a)} \cdot \ln(a) = a^x \cdot \ln(a).$$ ■

INSIGHT Notice that Theorem 2 generalizes our earlier formula for the derivative of the natural exponential function: when a is taken to be e, the factor $\ln(a)$ becomes $\ln(e)$, which is 1, and formula (3.5.6) reduces to (3.4.7). This gain in simplicity is one reason that e is the preferred base for the exponential function in calculus.

The proof of Theorem 2 employs a formula that is important in its own right. The equation

$$a^x = e^{x\ln(a)} \tag{3.5.7}$$

tells us that we can always rewrite an exponential function so that e becomes the base.

▶ **EXAMPLE 7** Calculate

$$\frac{d}{dx}3^x \quad \text{and} \quad \frac{d}{dx}5^{\sin(x)}.$$

Solution For the first problem, we use formula (3.5.6) with $a = 3$:

$$\frac{d}{dx}3^x = 3^x \cdot \ln(3).$$

For the second problem, we set $u = \sin(x)$ and use the Chain Rule before we apply formula (3.5.6) with $a = 5$:

$$\frac{d}{dx}5^{\sin(x)} = \frac{d}{dx}5^u = \frac{d}{du}5^u \cdot \frac{d}{dx}u = \ln(5)\cdot 5^u \cdot \frac{d}{dx}\sin(x) = \ln(5)\cdot 5^{\sin(x)} \cdot \cos(x).$$ ◀

Summary of Derivatives of Basic Functions

$$\frac{d}{dx}(x^p) = px^{p-1} \qquad -\infty < p < \infty$$

$$\frac{d}{dx}\sin(x) = \cos(x)$$

$$\frac{d}{dx}\cos(x) = -\sin(x)$$

$$\frac{d}{dx}\tan(x) = \sec^2(x)$$

$$\frac{d}{dx}\cot(x) = -\csc^2(x)$$

$$\frac{d}{dx}\sec(x) = \sec(x)\tan(x)$$

$$\frac{d}{dx}\csc(x) = -\csc(x)\cot(x)$$

$$\frac{d}{dx}e^x = e^x$$

$$\frac{d}{dx}a^x = a^x \cdot \ln(a)$$

QUICK QUIZ

1. If $y = u^4$, and $u = x^2 + 3x + 1$, then what is dy/dx?
2. If $g(u) = \sin(u)$, and $f(x) = 3x^5$, then what is $(g \circ f)'(x)$?
3. Evaluate $\frac{d}{dx} 2^{3x}$.
4. If $f(x) = \sqrt{x^2 + 7}$, then what is $f'(3)$?

Answers

1. $4(x^2 + 3x + 1)^3(2x + 3)$ 2. $15x^4 \cos(3x^5)$ 3. $3 \cdot 2^{3x} \ln(2)$ 4. 3/4

EXERCISES

Problems for Practice

In Exercises 1–18, calculate the derivative of the given expression with respect to x.

1. $\sin(3x)$
2. $\cos(2x + \pi/3)$
3. e^{5x}
4. $\sec(x/4)$
5. $\cos(1/x)$
6. $\sin(\sqrt{x})$
7. $\sin(x^2 + 3x)$
8. $\tan(\sin(x))$
9. $\sec(\cos(x))$
10. $\cot(5-x^5)$
11. $\tan(1 - 7x)$
12. $\tan(x^3)$
13. $x/\sqrt{1+x}$
14. $x^2/\sqrt{1-3x}$
15. $\cot(x^2 + 4) + \csc(x^2 + 4)$
16. $\exp(x^2 + 1)$
17. $\exp(-1/x^2)$
18. $x^4 \sin(1/x^2)$

In Exercises 19–26, use the Chain Rule–Power Rule to differentiate the given expression with respect to x.

19. $(2x + \sin(x))^3$
20. $(x^2 + 3x)^{10}$
21. $\sqrt{4 + x^2}$
22. $1/\tan^5(x)$
23. $\sin^3(x)$
24. $(x - 1/x)^{5/2}$
25. $12\sqrt{1 + e^x}$
26. $\left(\frac{x}{x+1}\right)^4$

In each of Exercises 27–30, compute $(f \circ g)'$ and $(g \circ f)'$.

27. $f(x) = x^4 + 7x^2$, $g(x) = \sqrt{x}$
28. $f(x) = 3x^5 - 2x^2$, $g(x) = \sin(x)$
29. $f(x) = x/(x + 1)$, $g(x) = x^3$
30. $f(x) = \sin(x)$, $g(x) = 3x^5$

In Exercises 31–40, calculate the derivative of the given expression with respect to x.

31. 2^x
32. $(1/3)^x$
33. 3^{2x}
34. 5^{2-x}
35. $8^x - x^8$
36. $3^{(x^2)}$
37. $(1 + e)^{1/x}$
38. $\sin(\pi^x)$
39. $(3x + 2)/5^x$
40. $10^{2x} \cdot 2^{10x}$

In each of Exercises 41–44, use the specified value of c and the given information about f and g to compute $(g \circ f)'(c)$.

41. $g(3) = 2$, $g'(2) = 3$, $g'(3) = 4$, $f(2) = 3$, $f'(2) = 8$, $f'(3) = 5$, $c = 2$
42. $g(9) = -2$, $g'(-2) = 7$, $g'(9) = -3$, $f(-2) = 9$, $f'(-2) = 4$, $f'(9) = 5$, $c = -2$
43. $g(1/2) = 6$, $g'(1/2) = 1/3$, $g'(6) = -3$, $f(6) = 1/2$, $f'(1/2) = 9$, $f'(6) = 12$, $c = 6$
44. $g(-5) = -6$, $g'(-6) = 12$, $g'(-5) = -7$, $f(-6) = -5$, $f'(-5) = 1/3$, $f'(-6) = -9$, $c = -6$

In each of Exercises 45–50 use the Chain Rule repeatedly to determine the derivative with respect to x of the given expression.

45. $\sqrt{1 + (1+2x)^5}$
46. $\sin(\cos(3x))$
47. $\cos^3(2x)$
48. $(1 + \cos^2(x))^{3/2}$
49. $2\tan(\sqrt{3x} + 2)$
50. $(x+\sqrt{1-x})^4$

Further Theory and Practice

In Exercises 51–54, let $c = 4$, $F(x) = 1 + 3x$, $G(x) = \sqrt{x}$, and $H(x) = x/(x + 5)$. Calculate the requested derivative.

51. $(F \circ G \circ H)'(c)$

52. $\left(\frac{d}{dx}F\circ\left(\frac{H}{G}\right)\right)(c)$

53. $(G\circ F\circ F)'(c)$

54. $\left(\frac{d}{dx}F\circ\left(\frac{1}{H\circ G}\right)\right)(c)$

The basic "dimensions" of mechanics are length, mass, and time. Trigonometric functions are unitless because they represent the ratio of two lengths—the units cancel. Arguments of trigonometric functions are angles and so unitless. Likewise, exponential functions are unitless and take unitless arguments. Exercises 55–58 are concerned with dimensional analysis.

55. An oscillation about equilibrium is given by $x(t) = A\sin(\omega t)$ when t is measured in seconds and x in meters. What units do the constants A and ω bear? Use the Chain Rule to explain why $x'(t)$ is dimensionally correct.

56. Suppose that t represents time, y distance. If we change the units of either t or $y=f(t)$ or both, then we change the numerical value of dy/dt. For example, 60 miles per hour is the same speed as 88 feet per second. Explain how the Chain Rule accounts for the numerical change in dy/dt when we change the units of y and t.

57. An object oscillates about its equilibrium position according to the formula

$$x(t) = Ae^{-kt}\sin(wt)$$

when t is measured in seconds and x in centimeters. Describe the units of the positive constants K, A, and ω. Use the Chain Rule to explain why $x'(t)$ is dimensionally correct.

58. According to the Theory of Relativity, the mass m of an object at speed v is given by

$$m(v) = \frac{m_0}{\sqrt{1-v^2/c^2}}$$

where c is the speed of light, and m_0 is its rest mass. There is a function φ such that

$$m'(v) = m_0\,\varphi(v)\left(1-\frac{v^2}{c^2}\right)^{-3/2}.$$

What are the dimensions of $m'(v)$, $\varphi(v)$, and $(1-v^2/c^2)^{-3/2}$? Calculate $m'(v)$, identifying each application of the Chain Rule. What is $\varphi(v)$? Verify that $\varphi(v)$ has the expected dimensions.

59. At time $t>0$ when t is measured in years, the mass of a 100 g sample of the isotope ^{14}C is $m(t) = 100\cdot e^{-0.001213\cdot t}$ g. Calculate $m'(t)/m(t)$, $\lim_{t\to\infty} m'(t)$, and $\lim_{t\to\infty} m'(t)$.

60. The diameter of a certain spherical balloon is decreasing at the rate of 2 inches per second. How fast is the volume of the balloon decreasing when the diameter is 8 feet?

61. The leg of an isosceles right triangle increases at the rate of 2 inches per minute. At the moment when the hypotenuse is 8 inches, how fast is the area changing?

62. If f is a differentiable function that never takes the value 0, define $g(x) = 1/f(1/x)$ for $x\neq 0$. Compute $g'(x)$.

63. Suppose that f is a differentiable function. Let $g(t) = f(t^2)$. Simplify $f'(t^2) - g'(t)$ and $f'(t) - g'(\sqrt{t})$.

64. The formulas for $(f\circ g)'(c)$ and $(g\circ f)'(c)$ both involve a product of the derivatives of f and g. In what way do the formulas differ?

In Exercises 65–74, find $f'(x)$ for the given function f.

65. $f(x) = 4\sec^5(\sqrt{7x})$

66. $f(x) = \tan(\sin(e^{3x}))$

67. $f(x) = 2\tan(\sqrt{x}) + 6\sqrt{\tan(x)}$

68. $f(x) = 1/\sqrt{\exp(\cos(2x))}$

69. $f(x) = (5x^3+1)^2/\sqrt{x^2+1}$

70. $f(x) = \sqrt{\sin^2(3x)+\cos^2(3x)}$

71. $f(x) = \sin^2(\pi\sin(\pi x^2))$

72. $f(x) = \cos^2(\sqrt{2x^2+3})$

73. $f(x) = 6\sqrt{\sqrt{2x-1}+\sqrt{2x+1}}$

74. $f(x) = (2x^2+1)^{1/3}(3x^3+1)^{4/3}$

75. Suppose that $f(x) = \begin{cases} (cx+1)^3 & \text{if } x\le 0 \\ x+1 & \text{if } x>0 \end{cases}$. Can we choose c so that f is a differentiable function? If yes, then specify c. Otherwise, explain why not.

76. Suppose that $f(x) = \begin{cases} e^{ax} & \text{if } x\le 0 \\ 1+\sin(bx) & \text{if } x>0 \end{cases}$. How must we choose a and b so that f is a differentiable function?

77. Suppose that the size $P(t)$ of a population at time t is

$$P(t) = \frac{P_0 M}{P_0 + (M-P_0)e^{-kMt}}$$

for some positive constants K, P_0, and $M>P_0$. (This growth model is known as *logistic growth.*

a. What is the initial population size? What is the limiting size of the population as $t\to\infty$?

b. Verify that $P'(t) = k\,P(t)\,(M-P(t))$.

78. If a unit mass is dropped from a height, and if the air resistance is proportional to the velocity $v(t)$ of that object, then

$$v(t) = kg(1-e^{-t/k})$$

in the downward direction.

a. What is the limiting (or terminal) velocity v_∞ as $t\to\infty$?

b. Calculate the acceleration $v'(t)$ of the object.

c. What is the limit of the acceleration (rate of change of velocity) as $t\to\infty$?

79. If a unit mass is dropped from a height, and if the air resistance is proportional to the square of the velocity $v(t)$ of that object, then

$$v(t) = \sqrt{\frac{g}{\kappa}}\left(\frac{1-\exp(-2t\sqrt{g\kappa})}{1+\exp(-2t\sqrt{g\kappa})}\right)$$

in the downward direction. Here κ is a positive constant that depends on the aerodynamic properties of the object as well as the density of the air.

a. What is the limiting (or terminal) velocity v_∞ as $t \to \infty$?
b. Calculate the acceleration $v'(t)$ of the object.
c. What is the limit of the acceleration as $t \to \infty$?
d. Verify that v satisfies the differential equation

$$v'(t) = g - kv^2(t).$$

80. A drug is intravenously introduced into a patient's bloodstream at a constant rate beginning at $t = 0$. The concentration C of the drug in the patient's body is given by the formula

$$C(t) = \frac{\alpha}{\beta}(1 - e^{-\beta t})$$

where α and β are positive constants.

a. Verify that $C(t)$ is a solution of the differential equation

$$C'(t) = \alpha - \beta \cdot C(t).$$

b. Use the differential equation to deduce the rate at which the drug is administered. How is the rate at which the body eliminates the drug from the bloodstream related to the concentration of the drug?
c. In the long run, what is the (approximate) concentration of the drug in the patient's bloodstream?

81. The hyperbolic sine function (sinh) and hyperbolic cosine function (cosh) are defined in Exercise 72, Section 3.4. Let a be a constant. Express $\frac{d}{dx}\cosh(ax)$ and $\frac{d}{dx}\sinh(ax)$ in terms of $\sinh(ax)$ and $\cosh(ax)$.

82. The hyperbolic tangent function (tanh) and hyperbolic cosine function (sech) are defined in Exercise 73, Section 3.4. Let a be a constant. Express $\dfrac{d}{dx}\tanh(ax)$ and $\dfrac{d}{dx}\operatorname{sech}(ax)$ in terms of $\tanh(ax)$ and $\operatorname{sech}(ax)$.

83. The hyperbolic tangent function (tanh) is defined in Exercise 73, Section 3.4. Verify that $v(t) = \sqrt{\frac{g}{\kappa}}\tanh(t\sqrt{g\kappa})$ satisfies the differential equation, $v'(t) = g - \kappa v^2(t)$. (See Exercise 79 for the significance of this differential equation.)

Computer/Calculator Exercises

84. Let $f(x) = \tan(x^3 + x + 1)$. To five decimal places, find the unique value of x_0 in $(0, 0.45)$ for which $f'(x_0) = 20$.

85. Suppose that $g(x) = (x-1)/(1+x^2)$ and $f(x) = (x-2)/(1+x^2)$ for x in the interval $I = [0, 2]$. Calculate $g'(x)$, $f'(x)$, and $(g \circ f)'(x)$. Where are g, f, and $g \circ f$ increasing most rapidly? (The maximum rate of increase of $g \circ f$ does not, in general, occur where either g or f attains its maximum rate of increase.)

86. The average temperature in degrees Fahrenheit of the town of Rainy Lake is modeled by

$$F(t) = 47 - 43\cos\left(\frac{2\pi}{365}(t - 33)\right).$$

In this expression, t is measured in days with $t = 0$ corresponding to the start of the new year. On what day is the average temperature the greatest? On what days is the average temperature increasing the fastest? Decreasing the fastest? Stable? What are the net temperature changes on the days during which the temperature is stable?

87. One gram of the isotope ^{238}U decreases according to the exponential decay law $m(t) = e^{-kt}$g. After 4.51 billion years, 0.5 g remains. Approximately what is the rate of change of this ^{238}U sample when its mass is 1/8 g?

88. Refer to Exercise 59 for the mass of a sample of the isotope ^{14}C that is initially 100 g.

a. When does the mass decrease at the rate of 1 gram per millennium? (A millennium is 1000 years).
b. In which 1000-year interval, expressed in the form $[t_0, t_0 + 1000]$, is the mass reduced by exactly 1 gram?

89. A casserole dish at temperature $T_0 = 350°$F is plunged into a large basin of hot water whose temperature is $T_\infty = 115°$F. The temperature of the dish satisfies Newton's Law of Cooling:

$$T(t) = T_\infty + (T_0 - T_\infty)e^{-kt}.$$

In 25 seconds, the temperature of the dish is 175°F.

a. Graph the temperature T.
b. Graph the rate T' at which the temperature decreases.
c. What is the rate of change of T when T is equal to 175°F? 155°F? 135°F? 125°F? 115.5°F?

90. Two ceramic objects are removed from a $T_0 = 450°$F oven into a room at $T_\infty = 72°$F. They are to be glued together, but the glue is not reliable when applied to objects whose temperatures are greater than 85°F. Five minutes after being taken from the oven, the temperature T of the ceramic objects is 300°F. Refer to the preceding exercise for Newton's Law of Cooling. Graph the temperature T. Graph the rate T' at which the temperature decreases. What is the rate of change of T when the objects are cool enough to be glued? If the thermometer used is accurate to three significant digits, when are the objects at 75°F?

91. After jumping from an airplane, a skydiver has velocity given by

$$v(t) = -256(1 - e^{-t/8})\ \text{ft/s}.$$

Eight seconds into her descent, she opens her parachute. For $t \geq 8$, her velocity is given by

$$v(t) = -16 - (240e - 256)e^{-2t+15}\ \text{ft/s}.$$

Graph $v(t)$. Is it everywhere continuous? Differentiable? Graph $v'(t)$ (the acceleration). Is it everywhere continuous? Differentiable?

3.6 Derivatives of Inverse Functions

The Sum Rule, Product Rule, Quotient Rule, and Chain Rule are essential tools in the theory of differentiation. The key idea behind them is that the differentiation of a complicated function can be reduced to calculating the derivatives of its basic building blocks. Thus we have learned how to obtain the derivatives $(f+g)'$, $(f\cdot g)'$, $(f/g)'$, and $(g\circ f)'$ from the derivatives f' and g'. In this section, we continue this program by developing a formula that expresses the derivative of an inverse function f^{-1} in terms of the derivative of f. As an important application of this theory, we consider the function $f(x)=e^x$ and its inverse function $f^{-1}(x)=\ln(x)$ in detail and obtain information about the derivative of the logarithm function.

Continuity and Differentiability of Inverse Functions

Suppose that S and T are open intervals and that $f: S \to T$ is an invertible function. This means that, for every t in T, there is exactly one s in S such that $f(s)=t$. Equivalently, there is a function $f^{-1}: T \to S$ such that $f(f^{-1}(t))=t$ for all t in T, and $f^{-1}(f(s))=s$ for all s in S. Our immediate goal is to find a relationship between $(f^{-1})'$ and f'. A simple example illustrates what we have in mind.

▶ **EXAMPLE 1** Let $S=(0,\infty)$, $T=(9,\infty)$, and let $f: S\to T$ be defined by $f(s)=s^2+9$. Calculate $(f^{-1})'(25)$.

Solution Setting $t=s^2+9$, we find

$$s=\sqrt{t-9}, \tag{3.6.1}$$

which tells us that $f^{-1}(t)=\sqrt{t-9}$. We calculate $(f^{-1})'(t)$ by using the Chain Rule–Power Rule:

$$\frac{d}{dt}f^{-1}(t)=\frac{1}{2}(t-9)^{-1/2}\frac{d}{dt}(t-9)=\frac{1}{2}(t-9)^{-1/2}=\frac{1}{2\sqrt{t-9}}. \tag{3.6.2}$$

Therefore

$$(f^{-1})'(25)=\frac{1}{2\sqrt{25-9}}=\frac{1}{8}. \quad ◀$$

INSIGHT The solution of Example 1 relies on an explicit formula for $f^{-1}(t)$. This method has a serious drawback: It is not always possible to find a formula for an inverse function. Notice, however, that if equation (3.6.1) is used to simplify equation (3.6.2), then we obtain

$$\frac{d}{dt}f^{-1}(t)=\frac{1}{2\sqrt{t-9}}=\frac{1}{2s}.$$

Looking for a relationship between $(f^{-1})'$ and f', we calculate

$$f'(s)=\frac{d}{ds}(s^2+9)=2s$$

and notice that this expression is the denominator of our formula for $f^{-1}(t)$. That is,

$$\frac{d}{dt}f^{-1}(t)=\frac{1}{\frac{d}{ds}f(s)}.$$

Observe that, with this equation, we do not need a formula for $f^{-1}(t)$ because the right-hand side does not involve $f^{-1}(t)$. It is precisely this relationship between $(f^{-1})'(t)$ and $f'(s)$ that we will now examine in greater generality.

Before seeking a formula for $(f^{-1})'(t)$, it is appropriate that we first investigate when the limit that defines the derivative exists. Our next theorem states a condition that guarantees the differentiability of f^{-1}.

THEOREM 1 Suppose that S and T are open real intervals and that $f: S \to T$ is invertible.

(a) If f is continuous on S, then f^{-1} is continuous on T.
(b) If f is differentiable on S and f' is never zero on S, then f^{-1} is differentiable on T.

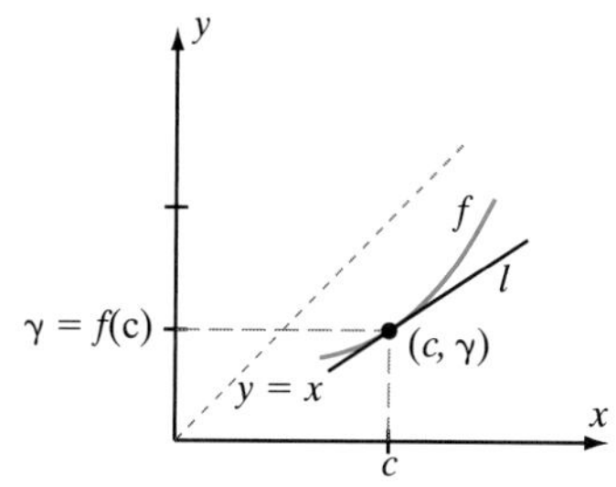

▲ Figure 1

Proof. We will sketch a proof of part (b), which is our primary interest in this section. Let c be a point in S, and set $\gamma = f(c)$. Supposing that $f'(c)$ exists, and $f'(c) \neq 0$, we infer that the graph of f has a nonhorizontal tangent line ℓ at (c, γ). See Figure 1. In Section 1.5 we learned that the graph of $f^{-1}: T \to S$ is the reflection of the graph of f through the line $y = x$. In this reflection through $y = x$, the nonhorizontal line ℓ is transformed to a nonvertical tangent line L to the graph of f^{-1} at (γ, c). See Figure 2. This is precisely the geometric criterion for the existence of the derivative of f^{-1} at γ. ■

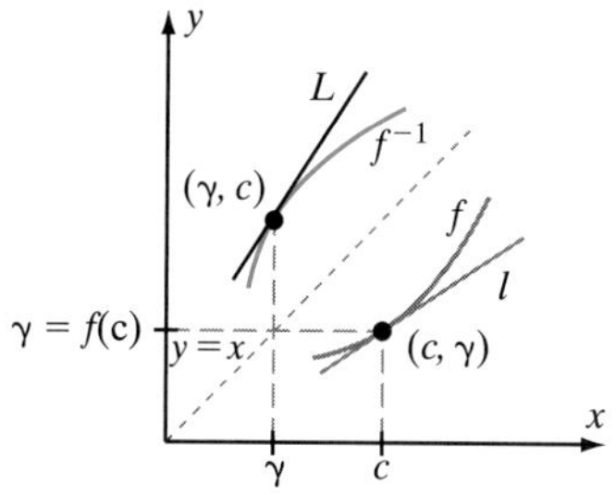

▲ Figure 2

The relationship between the graphs of f and f^{-1} also allows us to deduce the value of $(f^{-1})'(\gamma)$. Reflection through the line $y = x$ has the effect of interchanging x and y coordinates. Therefore if $\Delta y/\Delta x$ is the slope of line ℓ, then $\Delta x/\Delta y$ is the slope of the reflected line L. See Figure 3. In other words, the slope of L is the reciprocal of the slope of ℓ. Because $f'(c)$ is the slope of ℓ, it follows that L must have slope $1/f'(c)$. That is,

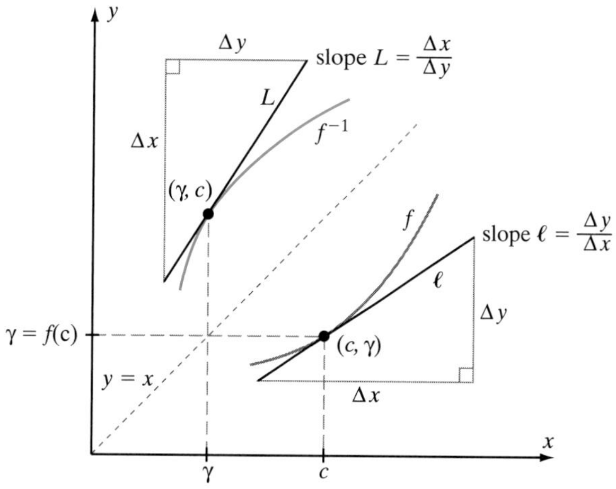

▲ Figure 3

$$(f^{-1})'(\gamma) = \frac{1}{f'(c)} \tag{3.6.3}$$

where c and γ are related by the equations $\gamma = f(c)$ and $c = f^{-1}(\gamma)$.

Let us use calculus to verify our deduction. To begin, we differentiate both sides of the basic equation $(f \circ f^{-1})(t) = t$ with respect to t. Notice that we will have to use the Chain Rule on the left side. We get $f'(f^{-1}(t)) \cdot (f^{-1})'(t) = 1$ or

$$(f^{-1})'(t) = \frac{1}{f'\left(f^{-1}(t)\right)}. \tag{3.6.4}$$

Notice that equation (3.6.4) validates our geometric argument: when we set $t = \gamma$ in equation (3.6.4) and remember that $c = f^{-1}(\gamma)$, we obtain equation (3.6.3). If we let $s = f^{-1}(t)$, then we can write formula (3.6.4) in Leibniz notation as

$$\frac{d(f^{-1})}{dt}(t) = \frac{1}{\frac{df}{ds}(s)}.$$

We now have an equation for the derivative of the inverse of a function. We summarize it in the following theorem:

THEOREM 2 **The Inverse Function Derivative Rule** Let f be an invertible function that is defined on an open interval containing the point s. Let $t = f(s)$. If f is differentiable at $s = f^{-1}(t)$ and if df/ds is nonzero at this point, then the derivative of f^{-1} at t is given by

$$\left.\frac{d(f^{-1})}{dt}\right|_t = \frac{1}{\left.\frac{df}{ds}\right|_{s=f^{-1}(t)}}. \tag{3.6.5}$$

INSIGHT This new formula brings to mind one feature of the Chain Rule: We must keep careful track of where we should be evaluating $\frac{d}{dt}f^{-1}$. We should evaluate this derivative at the same place where we evaluate f^{-1}, which would be at an element t in T. But then $s = f^{-1}(t)$ is the corresponding element of S, and that is where we should evaluate f or $\frac{df}{ds}$. That is why $\frac{df}{ds}$ appears in formula (3.6.5) with the argument $f^{-1}(t)$.

▶ **EXAMPLE 2** Let $f(s) = s^3$. Use the Inverse Function Derivative Rule to calculate $(f^{-1})'(t)$. Verify your work by finding a formula for $f^{-1}(t)$ and then differentiating $f^{-1}(t)$ directly.

Solution When we use the Inverse Function Derivative Rule to calculate $(f^{-1})'(t)$, we first calculate $f'(s)$. In this example, we have $f'(s) = 3s^2$. The Inverse Function Derivative Rule tells us that if $\left.\frac{df}{ds}\right|_{s=f^{-1}(t)} \neq 0$, then

$$\left.\frac{d(f-1)}{dt}\right|_t = \frac{1}{\left.\frac{df}{ds}\right|_{s=f^{-1}(t)}} = \frac{1}{3s^2|_{s=f^{-1}(t)}}.$$

Now, if $s = f^{-1}(t)$, then $t = f(s) = s^3$. Solving this equation for s, we obtain $s = t^{1/3}$, or $f^{-1}(t) = t^{1/3}$. Thus

$$\left.\frac{d(f^{-1})}{dt}\right|_t = \frac{1}{3(t^{1/3})^2} = \frac{1}{3}t^{-2/3} \text{ for } t \neq 0.$$

To verify this formula for $(f^{-1})'(t)$, we can apply the Power Rule directly to $f^{-1}(t)$, which we have already found. We obtain

$$(f^{-1})'(t) = \frac{d}{dt}t^{1/3} = \frac{1}{3}t^{1/3-1} = \frac{1}{3}t^{-2/3} \text{ for } t \neq 0,$$

which agrees with the formula we derived from the Inverse Function Derivative Rule. ◀

▶ **EXAMPLE 3** The function $f(x) = (x^5 + x + 2)^{5/2}$ is increasing and therefore invertible. Calculate $(f^{-1})'(32)$.

Solution In this example, we cannot compute an explicit formula for f^{-1}. However, the Inverse Function Derivative Rule—in the form of equation (3.6.3)—tells us that $(f^{-1})'(32) = 1/f'(c)$ where c is the unique number such that $f(c) = 32$, or $(c^5 + c + 2)^{5/2} = 32$. Raising each side to the $2/5^{\text{th}}$ power, we obtain $c^5 + c + 2 = 4$. In this case, we can spot that $c = 1$. Using the Chain Rule–Power Rule, we calculate

$$f'(x) = \frac{5}{2}\cdot(x^5+x+2)^{3/2}\cdot(5x^4+1) \tag{3.6.6}$$

and therefore

$$f'(c) = f'(1) = \frac{5}{2}\cdot(1^5+1+2)^{3/2}\cdot(5\cdot 1^4+1) = 120.$$

We conclude that $(f^{-1})'(32) = 1/f'(c) = 1/120$. ◀

INSIGHT In Example 3, we were unable to evaluate $s = f^{-1}(t)$ for a general value of t. That inability prevented us from calculating $(f^{-1})'(t)$ directly—we did not have an expression to differentiate. Instead, we applied the Inverse Function Derivative Rule. To do so, we had to find the value $c = f^{-1}(\gamma)$ for $\gamma = 32$. In other words, we had to solve the equation $s = f^{-1}(t)$ for a *specific* value γ of t. That value, 32, may have seemed artificial. Indeed, it was chosen so that we could solve $f(c) = \gamma$ by inspection. In practice, however, other values of γ would have posed no greater difficulty. That is because we can turn to the "solve" command of a calculator or computer if, for a given γ, we cannot find an exact solution c of the equation $f(c) = \gamma$. For example, suppose that we require $(f^{-1})'(35)$. The command `fsolve((x^5 + x + 2)^(5/2) = 35, x)` in the computer algebra system *Maple* tells us that $x = 1.0234$ solves the equation $f(x) = 35$ (to four decimal places). Substituting this value of x in formula (3.6.6) gives us $f'(x) = 136.86$. It follows from the Inverse Function Derivative Rule that $(f^{-1})'(35) \approx 1/136.86$.

Derivatives of Logarithms

The Inverse Function Derivative Rule can be used to calculate the derivative of the logarithm functions that we studied in Section 2.6. We begin with the natural logarithm, which is the logarithm with base e.

THEOREM 3 For every $t>0$, we have

$$\frac{d}{dt}\ln(t) = \frac{1}{t}. \tag{3.6.7}$$

Proof. Let $f(s) = e^s$. Then $f^{-1}(t) = \ln(t)$. Applying the Inverse Function Derivative Rule, we have

$$\left.\frac{d(f^{-1})}{dt}\right|_t = \frac{1}{\left.\frac{df}{ds}\right|_{s=f^{-1}(t)}} = \frac{1}{e^s|_{s=f^{-1}(t)}} = \frac{1}{e^{f^{-1}(t)}} = \frac{1}{e^{\ln(t)}} = \frac{1}{t}.$$ ■

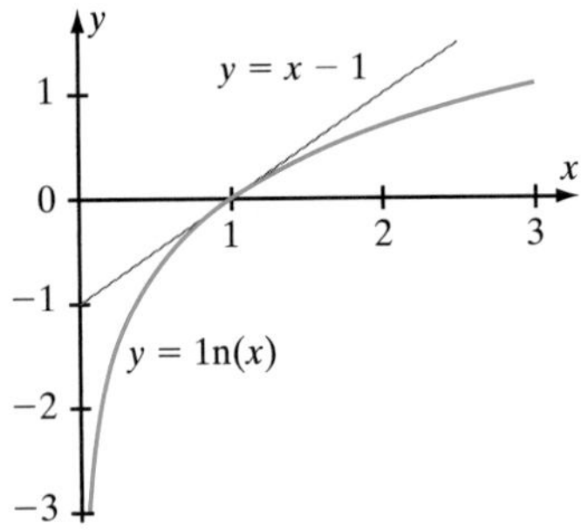

▲ Figure 4

▶ **EXAMPLE 4** Find the equation of the tangent line to the graph of $y = \ln(x)$ at its x-intercept.

Solution Earlier in this section, we observed that $\ln(1) = 0$. Because the natural logarithm is an increasing function, the point $(1, 0)$ is the only x-intercept of the curve $y = \ln(x)$. Because

$$\left.\frac{d}{dx}\ln(x)\right|_{x=1} = \left.\frac{1}{x}\right|_{x=1} = 1,$$

the Point-Slope Equation for the required tangent line is $y = 1\cdot(x-1)+0$ or $y = x - 1$. See Figure 4. ◀

▶ **EXAMPLE 5** Calculate

$$\frac{d}{dx}\ln\left(\sqrt{x}\sec(x)\right).$$

Solution A glance at the argument of the logarithm may suggest that the Product Rule is needed for the calculation. In fact, that is not the case. By applying properties (2.6.17) and (2.6.19) of the natural logarithm prior to differentiation, we can simplify the differentiation:

$$\begin{aligned}
\frac{d}{dx}\ln\left(\sqrt{x}\sec(x)\right) &= \frac{d}{dx}\left(\ln(\sqrt{x}) + \ln\left(\sec(x)\right)\right) \\
&= \frac{d}{dx}\left(\ln(x^{1/2}) + \ln\left(\sec(x)\right)\right) \\
&= \frac{1}{2}\frac{d}{dx}\ln(x) + \frac{d}{dx}\ln(u) \qquad (u = \sec(x)) \\
&= \frac{1}{2x} + \frac{d}{du}\ln(u)\frac{du}{dx} \\
&= \frac{1}{2x} + \frac{1}{u}\sec(x)\tan(x) \\
&= \frac{1}{2x} + \tan(x).
\end{aligned}$$ ◀

THEOREM 4 Let a be a positive constant not equal to 1. For every $t > 0$, we have

$$\frac{d}{dt}\log_a(t) = \frac{1}{t \cdot \ln(a)}. \tag{3.6.8}$$

Proof. Let $f(s) = a^s$. Then $f^{-1}(t) = \log_a(t)$. Using equation (3.5.6) for $f'(s)$ and applying the Inverse Function Derivative Rule, we have

$$\left.\frac{d(f^{-1})}{dt}\right|_t = \frac{1}{\left.\frac{df}{ds}\right|_{s=f^{-1}(t)}} = \frac{1}{a^s \cdot \ln(a)|_{s=f^{-1}(t)}} = \frac{1}{a^{f^{-1}(t)} \cdot \ln(a)} = \frac{1}{a^{\log_a(t)} \cdot \ln(a)} = \frac{1}{t \cdot \ln(a)}.$$

■

INSIGHT Notice that Theorem 4 generalizes formula (3.6.7) for the derivative of the natural logarithm function: when a is taken to be e, the factor $\ln(a)$ becomes $\ln(e)$, which is 1, and formula (3.6.8) reduces to (3.6.7). This gain in simplicity is one reason that e is the preferred base for the logarithm function in calculus.

▶ **EXAMPLE 6** Calculate

$$\frac{d}{dx}\log_2(\sqrt{1+2^x}).$$

Solution Before differentiating, we use Theorem 2(*e*) of Section 2.6 to simplify the logarithm. Then we set $u = 1 + 2^x$ and use formula (3.6.8) together with the Chain Rule:

$$\begin{aligned}\frac{d}{dx}\log_2(\sqrt{1+2^x}) &= \frac{1}{2}\frac{d}{dx}\log_2(1+2^x) = \frac{1}{2}\frac{d}{du}\log_2(u)\cdot\frac{du}{dx}\\ &= \frac{1}{2}\frac{1}{u\cdot\ln(2)}\cdot 2^x\ln(2) = \frac{2^{x-1}}{1+2^x}.\end{aligned}$$

◀

Logarithmic Differentiation

We next show how to use the logarithm as an aid to differentiation. The key idea is that if f is a function taking positive values, then we can exploit the formula

$$\frac{d}{dx}\ln(f(x)) = \frac{f'(x)}{f(x)}. \tag{3.6.9}$$

DEFINITION The derivative of the logarithm of f is called the *logarithmic derivative* of f. The process of differentiating $\ln(f(x))$ is called *logarithmic differentiation.*

As the right side of equation (3.6.9) indicates, the logarithmic derivative of f measures the *relative* rate of change of f. This quantity often conveys more useful information than f' itself. It is used frequently in biology, pharmacology, medicine, and economics. For example, consider an item whose price is increasing at the rate of \$100 per month. Is that a significant rate of change? The answer depends on the price of the item. For a \$150,000 house, the relative rate of change is only 1/1500 per month, or 0.80% per year. If the item is a \$1500 computer system, then the relative rate of change is 1/15 per month, or 80% per year.

Logarithmic differentiation is also useful in calculus, where we use the algebraic properties of the logarithm to simplify complicated products and quotients *before* differentiating. Logarithmic differentiation can also be an effective technique for finding the derivative of an expression that involves an exponent that is a function. The next two examples will illustrate these ideas.

▶ **EXAMPLE 7** Calculate the derivative of the function $f(x) = x^x$.

Solution First we take the natural logarithm of both sides:

$$\begin{aligned}\ln(f(x)) &= \ln(x^x)\\ &= x \cdot \ln(x).\end{aligned}$$

Then we differentiate each side, using the Product Rule to calculate the derivative on the right side:

$$\frac{d}{dx}\ln(f(x)) = 1 \cdot \ln(x) + x \cdot \frac{1}{x} = \ln(x) + 1.$$

Next we use formula (3.6.9) to simplify the left side:

$$\frac{f'(x)}{f(x)} = \ln(x) + 1.$$

Finally we solve for $f'(x)$:

$$f'(x) = f(x) \cdot (\ln(x) + 1) = x^x \cdot (\ln(x) + 1).$$ ◀

INSIGHT Notice that we cannot solve Example 7 by using the Power Rule (Theorem 1 from Section 3.4):

$$\frac{d}{dx}x^x \neq x \cdot x^{x-1} = x^x.$$

This incorrect computation treats the exponent x as if it were constant. *The Power Rule can be used only when the exponent is constant.* Nor can we use formula (3.5.6) with $a = x$:

$$\frac{d}{dx}x^x \neq \ln(x) \cdot x^x.$$

This incorrect computation treats the base x as if it were constant. *Formula (3.5.6) can be used only when the base is constant.*

The difficulty of Example 7 is that *both* the base and the exponent involve the variable of differentiation. Applying the logarithm succeeds by removing the variable from the exponent position. Another approach is to use formula (3.5.7) with $a = x$ to remove the variable from the base position. We obtain

$$x^x = e^{x\ln(x)},$$

from which we obtain the result of Example 7 as follows:

$$\frac{d}{dx}x^x = \frac{d}{dx}e^{x\ln(x)} = e^{x\ln(x)}\frac{d}{dx}(x\ln(x)) = e^{x\ln(x)} \cdot (\ln(x) + 1) = x^x \cdot (\ln(x) + 1).$$

▶ **EXAMPLE 8** Calculate the derivative of

$$f(x) = \frac{x^3 \cdot e^x}{1 + x^2}.$$

Solution This calculation can be done in a straightforward but tedious fashion by means of the Quotient Rule and the Product Rule. Use of Logarithmic Differentiation reduces the algebra that is necessary. To start, we apply the natural logarithm to each side:

$$\ln(f(x)) = \ln(x^3) + \ln(e^x) - \ln(1+x^2) = 3\ln(x) + x - \ln(1+x^2).$$

Next, we differentiate each side of the last equation with respect to x. We obtain

$$\frac{f'(x)}{f(x)} = \frac{3}{x} + 1 - \frac{2x}{1+x^2}.$$

Finally,

$$\begin{aligned} f'(x) &= f(x) \cdot \left(\frac{3}{x} + 1 - \frac{2x}{1+x^2}\right) \\ &= \frac{x^3 \cdot e^x}{1+x^2} \cdot \left(\frac{3}{x} + 1 - \frac{2x}{1+x^2}\right) \end{aligned}$$ ◀

Summary of Derivatives of Basic Functions

$$\frac{d}{dx}(x^p) = p\,x^{p-1} \qquad -\infty < p < \infty$$

$$\frac{d}{dx}\sin(x) = \cos(x)$$

$$\frac{d}{dx}\cos(x) = -\sin(x)$$

$$\frac{d}{dx}\tan(x) = \sec^2(x)$$

$$\frac{d}{dx}\cot(x) = -\csc^2(x)$$

$$\frac{d}{dx}\sec(x) = \sec(x)\tan(x)$$

$$\frac{d}{dx}\csc(x) = -\csc(x)\cot(x)$$

$$\frac{d}{dx}e^x = e^x$$

$$\frac{d}{dx}a^x = a^x \cdot \ln(a)$$

$$\frac{d}{dx}\ln(x) = \frac{1}{x}$$

$$\frac{d}{dx}\log_a(x) = \frac{1}{x \cdot \ln(a)}$$

QUICK QUIZ

1. If $f(x) = x + \exp(x)$, then what is $(f^{-1})'(1)$?
2. Calculate $f'(x)$ for $f(x) = \ln(x) + 1/x$.
3. Calculate $f'(x)$ for $f(x) = \log_3(x^2)$.
4. Express π^7 as a power of e.

Answers

1. 1/2 2. $(x-1)/x^2$ 3. $2/(x\ln(3))$ 4. $e^{7\ln(\pi)}$

EXERCISES

Problems for Practice

In Exercises 1–8, assume that $f:\mathbb{R} \to \mathbb{R}$ is invertible and differentiable. Compute $(f^{-1})'(4)$ from the given information.

1. $f^{-1}(4) = 1, f'(1) = 2$
2. $f^{-1}(4) = 3, f'(3) = 1/6$
3. $f^{-1}(4) = -1, f'(-1) = 3$
4. $f^{-1}(4) = 7, f'(7) = -3/8$
5. $f^{-1}(4) = 1, f'(s) = 5s^4 + 1$
6. $f^{-1}(4) = 1, f'(s) = 4 + \pi\cos(\pi s)$
7. $f(\sqrt{3}) = 4, f'(s) = (2 + s^2)/(1 + s^2)$
8. $f(0) = 4, f'(s) = 2e^{2s} + e^{-s}$

In Exercises 9–26, use the Inverse Function Derivative Rule to calculate $(f^{-1})'(t)$.

9. $f:(-\infty,\infty) \to (-\infty,\infty), f(s) = 3s - 5$
10. $f:(0,\infty) \to (0,\infty), f(s) = \sqrt{s}$
11. $f:(1,\infty) \to (0, 1/2), f(s) = 1/\sqrt{s+3}$
12. $f:(0,\infty) \to (4,\infty), f(s) = s^2 + 4$
13. $f:(0,3) \to (2, 245), f(s) = s^5 + 2$
14. $f:(-\infty,\infty) \to (0,\infty), f(s) = \exp(1 - s)$
15. $f:(1,3) \to (3, 15), f(s) = s^2 + 2s$
16. $f:(1,8) \to (1/64, 1), f(s) = 1/s^2$
17. $f:(0,4) \to (3, 5), f(s) = \sqrt{s^2 + 9}$
18. $f:(1/2,5) \to (-1/3, 2/3), f(s) = (s-1)/(s+1)$
19. $f:(1,2) \to (2, 5/2), f(s) = (s^2 + 1)/s$
20. $f:(1,\infty) \to (-\infty,\infty), f(s) = \ln(s^2 - 1)$
21. $f:(0,1) \to (1, e), f(s) = \exp(\sqrt{s})$
22. $f:(0,\infty) \to (1,\infty), f(s) = \exp(s^2)$
23. $f:(0,\infty) \to (0,\infty), f(s) = \log_2(1 + s)$
24. $f:(1,\infty) \to (2,\infty), f(s) = \log_3(6 + 3^s)$
25. $f:(-\infty,\infty) \to (0,\infty), f(s) = 2^{s^3}$
26. $f:(-\pi/2,\pi/2) \to (-\infty,\infty), f(s) = \tan(s)$

In each of Exercises 27–49, differentiate the given expression with respect to x.

27. $\ln(2x)$
28. $\ln(x^2+e)$
29. $x\ln(x) - x$
30. $\ln(1 + \exp(x))$
31. $\ln(x)/x$
32. $x/\ln(x)$
33. $6\sqrt{\ln(x)}$
34. $\ln^3(x)$
35. $\ln(\sec(x))$
36. $\cos(\ln(x))$
37. $\ln(\ln(x))$
38. $\ln\left(\dfrac{x-1}{x+1}\right)$
39. $\ln(x + \exp(x))$
40. $2^{\ln(x)}$
41. $3^{\log_2(x)}$
42. $\log_2(5x)$
43. $\log_{10}(5x + 3)$
44. $1/(5 + \log_{10}(x))$
45. $2^x\log_2(1 + 4x)$
46. $x\log_3(x)$
47. $x^3\log_{10}(6-x)$
48. $\log_3(\sqrt{1 + 9^x})$
49. $\log_5(\log_2(x))$

In each of Exercises 50–55, a function f is given. Use logarithmic differentiation to calculate $f'(x)$.

50. $f(x) = (2x + 1)^3(x^2 + 2)^3(x^3 + 2x + 1)^2$
51. $f(x) = (x^2+x)^4(x^3+x^2 - 1)^3/(x^2 + 2)^2$
52. $f(x) = (x^2 + 1)^3\sin^2(x)\cos^3(x)$
53. $f(x) = x^{3x}$
54. $f(x) = x^{\sin(x)}$
55. $f(x) = x^{(x^2)}$

Further Theory and Practice

In Exercises 56–59, evaluate the derivative f' of the given function f in two ways. First, apply the Chain Rule to $f(x)$, without simplifying $f(x)$ in advance. Second, simplify $f(x)$, and then differentiate the simplified expression. Verify that the two expressions are equal.

56. $f(x) = \ln(3x)$
57. $f(x) = \ln(x^2)$

58. $f(x) = \ln(xe^x)$
59. $f(x) = \ln(1/\sqrt{x})$

▶ In each of Exercises 60–63, the given function f is invertible on an open interval containing the given point c. Write the equation of the tangent line to the graph of f^{-1} at the point $(f(c), c)$. ◀

60. $f(s) = \sqrt{s}$, $c = 9$
61. $f(s) = 1/\sqrt{s+3}$, $c = 6$
62. $f(s) = \sqrt{s^2+9}$, $c = 4$
63. $f(s) = s^5 + 2$, $c = 1$

▶ In Exercises 64–71, find $f^{-1}(\gamma)$ for the given f and γ (but do not try to calculate $f^{-1}(t)$ for a general value of t). Then calculate $(f^{-1})'(\gamma)$. ◀

64. $f(s) = s^5 + 2s^3 + 2s + 3$, $\gamma = 3$
65. $f(s) = s^3 + 2s - 7$, $\gamma = 5$
66. $f(s) = \pi s - \cos(\pi s/2)$, $\gamma = \pi$
67. $f(s) = (s^3+s^2+1)/(s^2+1)$, $\gamma = 3/2$
68. $f(s) = s+e^s$, $\gamma = 1$
69. $f(s) = s\ln(s)$, $\gamma = 0$
70. $f(s) = s + \log_2(s)$, $\gamma = 11$
71. $f(s) = \log_2(s) + \log_4(s)$, $\gamma = 9$

72. Suppose that f is differentiable and invertible, that $P = (2, 3)$ is a point on the graph of f, and that the slope of the tangent to the graph of f at P is 1/7. Use this information to find the equation of a line that is tangent to the graph of $y = f^{-1}(x)$.

73. If $f(x) = 4e^{x-1} + 3\ln(x)$ for $x > 0$, then what is $(f^{-1})'(4)$?

▶ In each of Exercises 74–79, use logarithmic differentiation to calculate the derivative of the given function. ◀

74. $\sin^x(x)$
75. $x^{\ln(x)}$
76. $(x^2+1)^{(x^3+1)}$
77. $\log_2^x(x)$
78. $\sin^{\cos(x)}(x)$
79. $\ln(x)^{\ln(x)}$

80. Show that there is a value c such that the tangent lines to the graphs of $y = e^x$ and $y = \ln(x)$ at (c, e^c) and $(c, \ln(c))$ are parallel.

81. Suppose that $f\colon (a,b) \to (c,d)$ and $g\colon (c,d) \to (\alpha, \beta)$ are invertible. Prove that $g \circ f$ is invertible. Derive a formula for $((g \circ f)^{-1})'$ in terms of f' and g'.

82. Suppose that u and υ are differentiable functions with $u(x) > 0$ for all x. Let $f(x) = u(x)^{\upsilon(x)}$. Find a formula for $f'(x)$.

83. The invertible function $f\colon [1,\infty) \to [0,\infty)$ defined by $f(s) = s\ln(s) - s + 1$ is used in statistical physics for estimating entropy. Find c and γ with $f(c) = \gamma$ so that $s = (t-\gamma) + c$ is a tangent line to the graph of $s = f^{-1}(t)$ in the ts-plane.

84. At time t, the bloodstream concentration $C(t)$ of an intravenously administered drug is

$$C(t) = \frac{\alpha}{\beta}(1 - e^{-\beta t}) \qquad (0 \le t)$$

for positive constants α and β.

a. Given that the dimension of C is mass/volume, and an exponential is unitless with unitless argument, what are the dimensions of the constants β and α?

b. By comparing $C(s)$ with $C(u)$ for $s < u$, show that the concentration of the drug is increasing.

c. What is the horizontal asymptote of the graph of $y = C(t)$ in the ty-plane?

d. What are the domain and image of C?

e. Suppose that the range of C is the image of C. Then C is invertible. Describe C^{-1} completely by stating its domain, stating its range, and giving an explicit formula for $C^{-1}(s)$. (Note: The roles of s and t in this application are reversed from the discussion of inverses in this section. That is because it is suggestive to denote *time* by t, and *time* is the variable of the function being inverted.)

f. Calculate $C'(t)$ and $(C^{-1})'(s)$.

85. According to Fick's Law, the diffusion of a solute through a cell membrane is described by

$$c(t) = C + (c(0) - C)\exp(-kAt/V) \qquad 0 \le t < \infty.$$

Here $c(t)$ is the concentration of the solute, C is the concentration outside the cell, A is the area of the cell membrane, V is the volume of the cell, and k is a positive constant. Suppose that $C < c(0)$.

a. Given that the dimensions of C and $c(t)$ are both mass/volume, and an exponential is unitless with unitless argument, what is the dimension of the constant k?

b. Show that $c(t)$ is increasing on $[0, \infty)$ if $c(0) < C$. Show that $c(t)$ is decreasing on $[0, \infty)$ if $c(0) > C$.

c. What is the horizontal asymptote of the graph of $y = c(t)$ in the ty-plane?

d. What are the domain and image of c?

e. Suppose that the range of c is the image of c. Then c is invertible. Describe c^{-1} completely by stating its domain, stating its range, and giving an explicit formula for $c^{-1}(s)$. (Note: The role reversal of s and t is discussed in Exercise 84e.)

f. Calculate $c'(t)$ and $(c^{-1})'(s)$.

86. An object is dropped from a height h. At time t, its height is $y(t)$, and its velocity is $\upsilon(t) = y'(t)$. If air resistance is $\kappa\,\upsilon(t)^2$ for a positive constant κ, then

$$y(t) = h + t\sqrt{\frac{g}{\kappa}} - \frac{1}{\kappa}\ln\left(\frac{1 + e^{2t\sqrt{g\kappa}}}{2}\right).$$

a. Show that

$$\upsilon(t) = -\sqrt{\frac{g}{\kappa}}\left(\frac{1 - \exp(-2t\sqrt{g\kappa})}{1 + \exp(-2t\sqrt{g\kappa})}\right).$$

b. Calculate the terminal velocity, $\lim_{t\to 0} \upsilon(t)$.
c. Verify that $\upsilon'(t) = -g + \kappa\upsilon(t)^2$.
d. Show that $\upsilon(t) = -\sqrt{g/\kappa}\, \tanh(t\sqrt{g\kappa})$. (See Exercise 73, Section 3.4.)

87. If the number q of units of an ordinary product that are sold at price p is denoted by $f(p)$, then f is a decreasing function of p. This relationship between p and q is known as the *Law of Demand*. From it, we may infer that price is a function of demand: $p = f^{-1}(q)$. The graph of $p = f^{-1}(q)$ in the qp-plane is said to be the *demand curve*. The *elasticity of demand* at price p is defined to be $E(p) = -p \cdot f'(p)/f(p)$. Express the equation of the tangent line to the demand curve at the point (q_0, p_0) in terms of q_0, p_0, and $E(p_0)$.

Calculator/Computer Exercises

In each of Exercise 88–91, a point γ is given, as well as a function f that is invertible on an open interval containing $c = f^{-1}(\gamma)$. Approximate $(f^{-1})'(\gamma)$.

88. $f(s) = 2s + \sqrt{s^3+1}$, $\gamma = 4$
89. $f(s) = s^3 + 2s + \sin(s)$, $\gamma = 1$
90. $f(s) = s^7 + s^3 - 2$, $\gamma = 1$
91. $f(s) = s^5 + 4s - 2$, $\gamma = 15$

In Exercises 92–95, a function f and its domain are given. Graph $x = f^{-1}(y)$ in the yx-plane. (Do not attempt to find a formula for f^{-1}. Recall that a procedure for graphing an inverse function has been described in Section 1.5.) Find the equation of the tangent line to the graph of $x = f^{-1}(y)$ at $(2, f^{-1}(2))$. Add the graph of the tangent line to your plot.

92. $f(x) = 2x^5 - x^2 + 3x$, $0 \le x \le 1$
93. $f(x) = 1 + x\exp(x)/2$, $-1 \le x \le 1.25$
94. $f(x) = \sqrt{x} + 3\ln(x)$, $1/2 \le x \le 2$
95. $f(x) = \sin^2(x) - \cos(x) + 2.6$, $0 \le x \le 2$

96. Approximate the value c such that the tangent lines to the graphs of $y = e^x$ and $y = \ln(x)$ at (c, e^c) and $(c, \ln(c))$ are parallel. Illustrate with a graph of the two functions and their tangent lines.

97. Graph $y = 10^5(\ln(x+10^{-5}) - \ln(x-10^{-5}))/2$ for $0.1 \le x \le 10$. Add the graph of $y = 1/x$ to the same viewing window. Which differentiation rule is illustrated?

3.7 Higher Derivatives

The derivative of a function is itself a function. For example, the derivative of the function

$$p(t) = t^3 - 5t^2 + 7$$

is

$$p'(t) = 3t^2 - 10t.$$

The function p' is differentiable, so we may consider its derivative $(p')'(t) = 6t - 10$. We simplify the notation by writing p'' for $(p')'$. Of course the second derivative p'' is also a function, and we may differentiate it to obtain the *third derivative*

$$p'''(t) = 6.$$

This procedure may be carried on indefinitely: The derivative of the $(k-1)^{\text{th}}$ derivative is called the kth derivative. However, writing more than two or three primes becomes tedious, and it is even more tedious to read. So we have other ways to write the higher derivatives.

Notation for Higher Derivatives

DEFINITION The derivative of f' is called the *second derivative* of f. In general, the derivative of the $(k-1)^{th}$ derivative of f, if it exists, is called the kth derivative. We write the kth derivative of f as

$$f''^{\cdots} \ (k \text{ primes}) \quad \text{or} \quad f^{(k)} \quad \text{or} \quad \frac{d^kf}{dx^k} \quad \text{or} \quad \frac{d^k}{dx^k}f \quad \text{or} \quad D^k(f).$$

Sometimes you will see roman numerals used. For example, $f^{(iv)}$ might be used instead of $f^{(4)}$. The number k is called the *order* of the derivative $f^{(k)}$. To distinguish f' from higher order derivatives, we often call f' the *first derivative* of f. The function f itself is sometimes called the *zeroeth derivative* of f. When the notation $f^{(k)}$ is used, it is important to include the parentheses, for f^k has already been defined to be the kth power of f.

▶ **EXAMPLE 1** Define $p(x) = 5x^4 - 8x^2 - 7x - 11$. Find $p^{(3)}(4)$.

Solution We have, in order,

$$p^{(1)}(x) = 20x^3 - 16x - 7, \qquad p^{(2)}(x) = 60x^2 - 16, \qquad p^{(3)}(x) = 120x.$$

Therefore $p^{(3)}(4) = 480$. ◀

INSIGHT Notice that, when we calculate a higher derivative, we do not substitute in a value for x until we are finished differentiating. If you substitute in the value of x too soon, then you end up with a constant function, and the next derivative will be 0.

When f is a function and $f^{(1)}(c), f^{(2)}(c), \ldots, f^{(k)}(c)$ exist, then we say that f is k times differentiable at c. Suppose that f is two times differentiable at c. This means that $f'(c)$ and $f''(c)$ exist. In particular, the function f' is differentiable at c. According to the definition of the derivative, this means that the function f' is defined on an open interval centered at c. Thus hidden in this definition is the requirement that f is not only differentiable at c but also on an open interval centered at c. Similarly, if f is k times differentiable at c, then $f, f', \ldots, f^{(k-1)}$ must be differentiable in an open interval centered at c.

Velocity and Acceleration

If p is position, and t is time, then $v(t) = p'(t)$ is the rate of change of position with respect to time, or *velocity* (as in Section 3.1). The rate of change of velocity with respect to time is *acceleration* $a(t)$. Thus $a(t) = v'(t) = p''(t)$: acceleration is the first derivative of velocity with respect to time and the second derivative of position with respect to time.

▶ **EXAMPLE 2** If position (measured in feet) at time t (measured in seconds) is given by $p(t) = \tan(t)$ for small values of t, then find the acceleration at time $t = 0.01$ seconds.

Solution We compute

$$\begin{aligned}\frac{d^2p}{dt^2} &= \frac{d}{dt}\left(\frac{dp}{dt}\right) = \frac{d}{dt}\sec^2(t) \overset{(3.5.3)}{=} 2\sec(t)\cdot\left(\frac{d}{dt}\sec(t)\right)\\ &= 2\sec(t)\cdot(\sec(t)\tan(t)) = 2\sec^2(t)\tan(t).\end{aligned}$$

At time $t = 0.01$, we have

$$a(0.01) = \left.\frac{d^2p}{dt^2}\right|_{0.01} = 2\sec^2(0.01)\tan(0.01) \approx 0.0200027 \text{ ft/s}^2. \quad ◀$$

Newton's Notation Yet another notation for higher derivatives is due to Newton. If $p(t)$ is a function of the time variable t, then the first and second derivatives are represented by $\dot{p}(t)$ and $\ddot{p}(t)$, respectively. This notation is often used in physics and engineering.

▶ **EXAMPLE 3** An arrow is shot straight up. Its height at time t seconds is given in feet by the formula $H(t) = -16t^2 + 160t + 5$. Compute its velocity and its acceleration. Discuss the physical significance of these quantities.

Solution The velocity is $v(t) = \dot{H}(t) = -32t + 160$ ft/s. The acceleration is $a(t) = \ddot{H}(t) = -32$ ft/s^2. Some interesting physical data may be read from this information. For instance, when the arrow reaches its maximum height, then, at that instant, it has neither upward nor downward motion: Its velocity is zero. We solve $0 = \dot{H}(t) = -32t + 160$ to obtain that the maximum height occurs when $t = 5$. At this time, the height is $H(5) = 405$ feet. Notice that at that time, the *acceleration* is -32 ft/s^2—the acceleration is constantly equal to this value—but the velocity is equal to zero. ◀

INSIGHT Although one can compute second, third, fourth, and higher derivatives, the physical significance of these quantities is increasingly difficult to understand. It turns out that, in some areas of engineering, the third derivative is useful: It is called *jerk* (and the terms *jolt*, *surge*, and *lurch* have also been used). When you are in an elevator that is not working properly, and your stomach feels as if it were left behind on the previous floor, then you are experiencing jerk. In the past, engineers have built "jerkmeters" in order to subject space vehicles to vibration tests. No term for the rate of change of jerk, the fourth derivative of position, is in widespread usage, although *jounce* has been suggested.

Approximation of Second Derivatives Higher derivatives can also be approximated by difference quotients. We will illustrate the procedure by approximating the second derivative. We start by writing $f''(c)$ as $(f')'(c)$. We can approximate $(f')'(c)$ by the central difference quotient $D_0(f')\,(c, h)$ of f' (as discussed in Section 3.3). Thus

$$f''(c) \approx \frac{f'(c + h/2) - f'(c - h/2)}{h}.$$

Next, approximate $f'(c + h/2)$ and $f'(c - h/2)$ by the central difference quotients $D_0 f(c + h/2, h)$ and $D_0 f(c - h/2, h)$:

$$f''(c) \approx \frac{\frac{f(c+h) - f(c)}{h} - \frac{f(c) - f(c-h)}{h}}{h},$$

or

$$f''(c) \approx \frac{(f(c+h) - f(c)) - (f(c) - f(c-h))}{h^2}. \tag{3.7.1}$$

This is the *second central difference quotient* approximation to $f''(c)$. The algebraic simplification

$$f''(c) \approx \frac{f(c+h) - 2f(c) + f(c-h)}{h^2}$$

may also be used, but, because the number $f(c+h)+f(c-h)$ is very close to $2f(c)$ when h is small, the subtraction can result in a loss of significant digits. Therefore the second central difference quotient approximation must be implemented with care.

▶ **EXAMPLE 4** The height of an object propelled upwards from ground level is approximately

$$f(t) = 33.33 \ln(54.25 \cos(0.5424t - 1.552))\,\text{m} \quad \text{for} \quad 0 \leq t \leq 2.862\,s.$$

Estimate the acceleration of the object at time $t = 2$.

Solution We are asked to approximate $f''(c)$ for $c = 2$. (Because the formula for $f(t)$ is based on an approximation, using the differentiation rules of calculus to obtain an "exact" value of $f''(2)$ would suggest an accuracy that cannot be obtained.) Figure 1 is a screen capture of a *Maple* session. In the first line, the second central difference quotient has been defined and named $D2$. In the next line, f has been defined. The third line sets the number of digits used in the computations to 15 to help guard against loss of significance (Section 1.1). The fourth line evaluates the second central difference quotients that correspond to $h = 0.01, 0.001, \ldots, 0.00000001$. The first four values of h indicate that $f''(2) \approx -12.3$. The two smallest values of h, 10^{-7} and 10^{-8}, result in wildly inaccurate approximations due to loss of significance. We conclude that $f''(2) \approx -12.3$. (The negative sign indicates that the object is decelerating, as we would expect). The last line of the *Maple* session evaluates $f''(2)$ using the derivative rules of calculus. The result confirms our numerical approximation. ◀

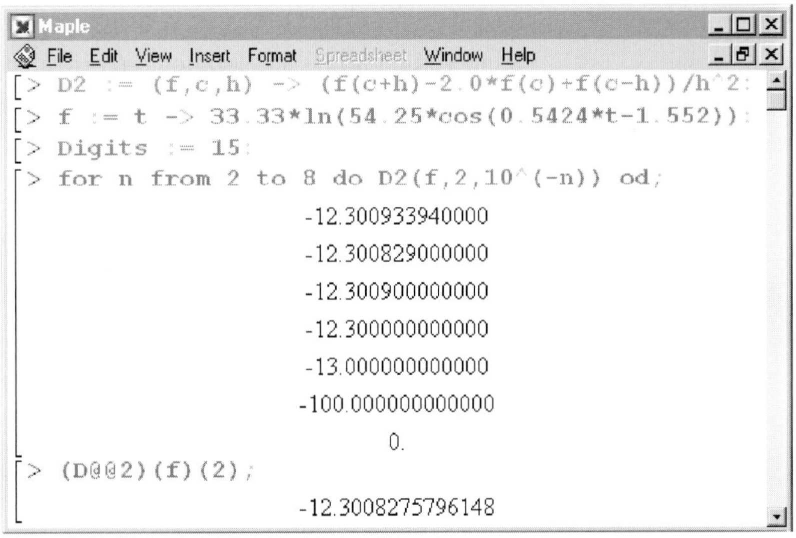
Maple
File Edit View Insert Format Spreadsheet Window Help
> D2 := (f,c,h) -> (f(c+h)-2.0*f(c)+f(c-h))/h^2:
> f := t -> 33.33*ln(54.25*cos(0.5424*t-1.552)):
> Digits := 15:
> for n from 2 to 8 do D2(f,2,10^(-n)) od;
-12.300933940000
-12.300829000000
-12.300900000000
-12.300000000000
-13.000000000000
-100.000000000000
0.
> (D@@2)(f)(2);
-12.3008275796148

▲ Figure 1

Leibniz's Rule

The Product Rule tells us how to compute the *first* derivative of a product $f \cdot g$. In 1695, Leibniz derived a rule for computing *higher* derivatives of a product. Leibniz's Rule tells us how to calculate $(f \cdot g)^{(n)}(c)$ in terms of $f'(c), f''(c), \ldots, f^{(n)}(c)$, and $g'(c), g''(c), \ldots, g^{(n)}(c)$. For now, we will state Leibniz's Rule only for second derivatives. Exercise 62 explains the general rule and asks you to verify it for $n = 2$.

THEOREM 1 Leibniz's Rule for Second Derivatives. If f and g are 2 times differentiable at c, then so is $f \cdot g$ and

$$(f \cdot g)''(c) = f''(c)g(c) + 2f'(c)g'(c) + f(c)g''(c). \tag{3.7.2}$$

▶ **EXAMPLE 5** A mass on a spring oscillates about its equilibrium position according to the formula

$$x(t) = e^{-t/2}\cos(t\sqrt{2}) \text{ cm}$$

when t is measured in seconds. What is the acceleration of the mass at time t?

Solution Let $f(t) = e^{-t/2}$, and $g(t) = \cos(t\sqrt{2})$. Then $f^{(1)}(t) = -e^{-t/2}/2, f^{(2)}(t) = e^{-t/2}/4$, $g^{(1)}(t) = -\sqrt{2}\sin(t\sqrt{2})$, and $g^{(2)}(t) = -2\cos(t\sqrt{2})$. Thus

$$\begin{aligned}\frac{d^2}{dt^2}\left(e^{-t/2}\cos(t\sqrt{2})\right) &= g(t)f^{(2)}(t) + 2f^{(1)}(t)g^{(1)}(t) + f(t)g^{(2)}(t)\\ &= \frac{1}{4}e^{-t/2}\cos(t\sqrt{2}) + 2\left(-\frac{1}{2}e^{-t/2}\right)\left(-\sqrt{2}\sin(t\sqrt{2})\right) - 2e^{-t/2}\cos(t\sqrt{2})\\ &= \sqrt{2}e^{-t/2}\sin(t\sqrt{2}) - \frac{7}{4}e^{-t/2}\cos(t\sqrt{2})\text{cm/s}^2.\end{aligned}$$ ◀

QUICK QUIZ

1. Calculate $\dfrac{d^2}{dx^2}\exp(x^2)$.
2. Suppose that $a(t)$ is acceleration, $v(t)$ is velocity, and $p(t)$ is position. Correct the following three equations by inserting appropriate differentiation symbols: $p(t) = v(t)$, $v(t) = a(t)$, $p(t) = a(t)$.
3. What is $\dfrac{d^{17}}{dx^{17}}(3x^{16} + 11x^{15} - 17x^{13})$?
4. Use Leibniz's Rule to calculate $\dfrac{d^2}{dx^2}\left(x^2\cos(x)\right)$.

Answers

1. $2(1 + 2x^2)\exp(x^2)$ 2. $p'(t) = v(t)$, $v'(t) = a(t)$, $p''(t) = a(t)$ 3. 0
4. $2\cos(x) - 4x\sin(x) - x^2\cos(x)$

EXERCISES

Problems for Practice

▶ In each of Exercises 1–26, an expression for $f(x)$ is given. Compute the first, second, and third derivatives of $f(x)$ with respect to x. ◀

1. $5x^2 - 6x + 7$
2. $4x^3 - 7x^{-5}$
3. $1/x$
4. $x^{7/3}$
5. $\ln(x)$
6. $x\ln(x)$
7. $\sqrt{6x+5}$
8. 2^{3x}
9. $\cos(3x)$
10. $\exp(-x)$
11. $\exp(2x)$

12. $\exp(x^2)$
13. $\ln(x+1)$
14. $(x^2-1)(x+5)$
15. $(x+1)/(x-1)$
16. $x/(2-x)$
17. $27(x^2+1)^{5/3}$
18. $x\sin(x)$
19. $x\cos(2x)$
20. $\tan(x)$
21. 2^x
22. $x/3^x$
23. $e^x\sin(x)$
24. $2^x\cos(3x)$
25. $\log_3(x)$
26. $x\log_2(x)$

In Exercises 27–36, calculate the requested derivative.

27. $f^{(5)}(x)$ where $f(x)=\sin(x)$
28. $\dfrac{d^2f}{dx^2}$ where $f(x)=\dfrac{x^2}{x-1}$
29. $\ddot{g}(1)$ where $g(t)=(3t^2-2)^{-4}$
30. $f'''(\pi/6)$ where $f(t)=\sin(t)+\cos(2t)$
31. $H^{(4)}(x)$ where $H(x)=-x^6+7x^5-9x^2+11$
32. $\dfrac{d^2f}{dx^2}$ where $f(x)=\sec(5x)$
33. $f'''(x)$ where $f(x)=\cos(4x+3)$
34. $\dfrac{d^3}{dt^3}(t^{-5/3}+4t^4-8t^{7/2})$
35. $g''(\pi/6)$ where $g(x)=\cos(\pi\sin(x))$
36. $H^{(2)}(\pi/4)$ where $H(x)=x\tan(x)$

In Exercises 37–40, a velocity $\upsilon(t)$ is given. Calculate the acceleration (in m/s^2).

37. $\upsilon(t)=24$ m/s
38. $\upsilon(t)=t^2-5t$ m/s
39. $\upsilon(t)=-16t+5$ m/s
40. $\upsilon(t)=8e^{-t/2}$ m/s

In Exercises 41–44, a position $p(t)$ is given. Calculate the acceleration (in m/s^2).

41. $p(t)=15+24t$ m
42. $p(t)=3\sin(2t)$ m
43. $p(t)=3t^2+16t$ m
44. $p(t)=10-3e^{-2t}$ m

45. The position of a bicycle at time t (measured in minutes) is $p(t)=2t^3+t^2+6t$ meters for $0<t<10$. What is the acceleration of the bicycle at time $t=3$s? At time $t=6$s?
46. The position of a moving body is described by $p(t)=at^2+bt+c$. If $p(0)=3$, then what is c? If $p'(0)=6$, then what is b? If $p''(0)=-5$, then what is a? Where is the body at time $t=6$?

Exercises 47–50 concern an object that is propelled straight up. Its height at time t seconds is given in feet by $H(t)=-16t^2+128t+68$.

47. Compute the velocity and acceleration of the object.
48. For how many seconds does the object rise?
49. What is the maximum height of the object?
50. Supposing that the trajectory is such that the object is able to fall to the ground (height 0 feet), for how many seconds does the object fall?

51. An object is thrown straight down. Its height at time t seconds is given in feet by $H(t)=-16t^2-160t+84$. With what velocity does it impact the earth (at height 0).

Exercises 52–54 are concerned with a car's advance $p(t)$ during its period of deceleration. Suppose that during the first t seconds of braking, the car moved forward $p(t)=48t-3t^2-t^3/2$ ft. Suppose also that this formula was in effect until the car came to a stop.

52. What was the car's velocity at the moment brakes were applied?
53. What was its acceleration $a(t)$ at time t? (The negative sign of $a(t)$ means that the car was decelerating.)
54. How long was the period of deceleration?

Further Theory and Practice

55. Let

$$f(x)=\begin{cases} x^3 & \text{if } x>0 \\ -x^3 & \text{if } x\le 0.\end{cases}$$

a. Prove that $f'(0)$ exists and equals 0.
b. Prove that $f''(0)$ exists and equals 0.
c. Prove that $f'''(0)$ does not exist.

56. Suppose that f and g are twice differentiable. Calculate a formula for $(f/g)''$.
57. Suppose that $p(x)$ is a nonzero polynomial. Prove that $p(x)$ has degree precisely k if and only if
a. $p^{(k+1)}(x)=0$ for all x; and
b. $p^{(k)}(x_0)\neq 0$ for some x_0.
58. Find an explicit formula for the polynomial $p(x)$ of degree 2 such that $p(0)=2$, $p'(0)=4$, and $p''(0)=-7$.
59. Find an explicit formula for the polynomial $p(x)$ of degree 3 such that $p(1)=1$, $p'(2)=-2$, $p''(-1)=-14$, and $p'''(5)=6$.
60. Suppose that $p(x)$ is a polynomial, and $|p''(x)|\le 1$ for all x. Prove that p is of degree at most 2, and the coefficient of its degree 2 term is less than or equal to 1/2 in absolute value.

In each of Exercises 61–66, use equation (3.7.2) to compute the requested derivative.

61. $D^2(f\cdot g)(x)$ where $f(x)=x^2$ and $g(x)=\cos(x)$
62. $\dfrac{d^2}{dx^2}(x\sqrt{x+1})$
63. $(f\cdot g)''(x)$ where $f(x)=1/x$ and $g(x)=\ln(x)$
64. $D^2(f\cdot g)\,(x)$ where $f(x)=x^3$and $g(x)=\cos(x)$

65. $(f\cdot g)^{(2)}(x)$ where $f(x)=\sqrt{x}$ and $g(x)=\exp(x)$

66. $\dfrac{d^2}{dx^2}\left((x+3)^2(2x+1)^3\right)$

67. Suppose that f is twice differentiable at c. Show that the graphs of $y=f(x)$ and $y=f(c)+f'(c)(x-c)+f''(c)(x-c)^2/2$ both pass through the point $(c, f(c))$ and have the same tangent line there.

Exercises 68–73 involve the factorial numbers $n!$, which were introduced in Section 1.4. They can be defined by $0! = 1$ and, for a positive integer n, $n! = n(n-1)\ \dots\ 3\dots\ 2\cdot\ 1$.

68. Let n be a positive integer. If k is a positive integer no greater than n, then the expression

$$\binom{n}{k}=\frac{n!}{k!(n-k)!}=\frac{n(n-1)(n-2)\cdots(n-k+1)}{k\cdot(k-1)\cdots 3\cdot 2\cdot 1}$$

is called a *binomial coefficient*. For example, if $n=7$, and $k=3$, then $n-k+1=5$, and

$$\binom{7}{3}=\frac{7\cdot 6\cdot 5}{3\cdot 2\cdot 1}=35.$$

By definition $\binom{n}{0}=1$. Leibniz's Rule states that $(f\cdot g)^{(n)}(c)$ is the sum of the $n+1$ expressions that are obtained by substituting $k=0,1,2,\dots,n$ in the formula $\binom{n}{k}\cdot f^{(n-k)}(c)\cdot g^{(k)}(c)$. Explicitly write out Leibniz's Rule for $n=1, 2,$ and 3. Observe that the $n=1$ case is the Product Rule, the $n=2$ case is equation (3.7.2). Verify Leibniz's Rule for $n=2$.

69. The Legendre polynomials P_n, introduced in 1782 by A.M. Legendre, can be defined by the following formula, which was discovered by Benjamin Olinde Rodrigues (1794–1850 or 1851):

$$P_n(x)=\frac{1}{2^n n!}\frac{d^n}{dx^n}(x^2-1)^n \qquad n=0,1,2,3,\dots .$$

Calculate $P_0(x)$, $P_1(x)$, $P_2(x)$, $P_3(x)$, and $P_4(x)$.

70. In general, the Legendre polynomial $P_n(x)$ (as defined in the preceding exercise) satisfies the differential equation

$$(1-x^2)P_n''(x)-2xP_n'(x)+n(n+1)P_n(x)=0.$$

Verify this equation for $P_5(x)=(63x^5-70x^3+15x)/8$.

71. For each nonnegative integer n, the Hermite polynomial H_n is defined by

$$H_n(x)=(-1)^n e^{x^2}\frac{d^n}{dx^n}e^{-x^2}.$$

Calculate $H_0(x)$, $H_1(x)$, $H_2(x)$, $H_3(x)$, and $H_4(x)$.

72. In general, the Hermite polynomial $H_n(x)$ (as defined in the preceding exercise) satisfies the differential equation

$$H_n''(x)-2xH_n'(x)+2nH_n(x)=0.$$

Verify this equation for $H_5(x)=32x^5-160x^3+120x$.

73. Prove that any polynomial $p(x)$ of degree k satisfies

$$p(x)=p(0)+p'(0)\cdot x+\frac{p''(0)}{2!}\cdot x^2+\dots+\frac{p^{(k)}(0)}{k!}\cdot x^k.$$

74. Show that there is a polynomial $p(x)$ of degree $n+1$ such that

$$\frac{d^n}{dx^n}\tan(x)=p(\tan(x)).$$

75. Let $p(x)$ be a polynomial. We say that p has a root of order k at a if there is a polynomial q such that $p(x)=(x-a)^k\ q(x)$ and $q(a)\neq 0$. Prove that p has a root of order 2 at $x=a$ if and only if $p(a)=0$, $p'(a)=0$, and $p''(a)\neq 0$. (In general, p has a root of order k at a if and only if $p(a)=p'(a)=p^{(k-1)}(a)=0$ and $p^{(k)}(a)\neq 0$.)

76. Let $f(x)=\left(\left(x^{150}+x^{100}\right)^6+x^{93}+x^{90}+16\right)^{42}$. Compute $f^{(83)}(0)$. (Hint: Simple observations go farther than brute force.)

77. Suppose that f is invertible and twice differentiable, that $f(c)=\gamma$, and that $f'(c)\neq 0$. Show that

$$(f^{-1})''(\gamma)=-\frac{f''(c)}{(f'(c))^3}.$$

78. Verify that

$$y(t)=h+t\sqrt{\frac{g}{\kappa}}-\frac{1}{\kappa}\ln\left(\frac{1+\exp(2t\sqrt{g\kappa})}{2}\right)$$

satisfies the differential equation

$$y''(t)=-g+k(y'(t))^2.$$

Calculator/Computer Exercises

In Exercises 79–82, you are given a function f, a value c, and a viewing rectangle R containing the point $P=(c,f(c))$. In R, graph the four functions f, g, h, and k, where $g(x)=f(c)+f'(c)(x-c)$, $h(x)=g(x)+½f''(c)\ (x-c)^2$, and $k(x)=h(x)+⅙f'''(c)\ (x-c)^3$. The graphs of g, h, and k are, respectively, linear, parabolic, and cubic approximations of the graph of f near c. The method of constructing these approximating functions, which are called *Taylor polynomials of f with base point c*, is studied in Chapter 8.

79. $f(x)=\sqrt{x}$, $c=0.5$, $R=[0,1]\times[0,1.1]$

80. $f(x)=x^{3/2}$, $c=1$, $R=[0, 4]\times[-1/2, 9]$

81. $f(x)=\sin^2(x)$, $c=\pi/3$, $R=[0, 2]\times[-0.2, 1.2]$

82. $f(x)=xe^x$, $c=1$, $R=[0, 2]\times[-2.7, 14.8]$

In Exercises 83–86, use second central difference quotients to approximate $f''(c)$ to four decimal places.

83. $f(x)=\sqrt{x+1/x}$, $c=e$

85. $f(x)=\ln(\sqrt{x}+1/\sqrt{x})$, $c=3$

86. $f(x)=\tan\left(\sin^2(x)\right)$, $c=\pi/4$

3.8 Implicit Differentiation

Examine the locus of the circle $x^2+y^2=4$ in Figure 1. We know that, at each point of the graph, there is a tangent line to the curve. However, we cannot calculate the equation of the tangent line using the techniques of calculus that we have learned so far because the circle is not the graph of a function. One way to remedy this problem is to solve for y in terms of x. We obtain two curves: the upper half of the circle, which is the graph of $y=+\sqrt{4-x^2}$, and the lower half of the circle, which is the graph of $y=-\sqrt{4-x^2}$. Thus we have realized the circle as the union of the graphs of two functions. Refer to Figure 2.

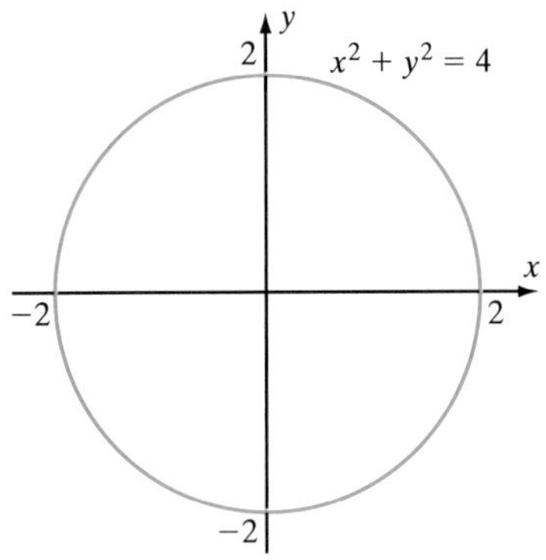

▲ Figure 1

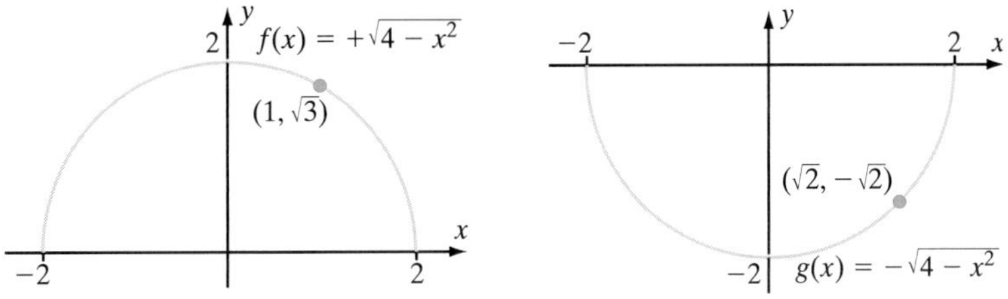

▲ Figure 2

Finding the slope of the tangent line to the circle at the point $P=(1,\sqrt{3})$ is just the same as finding the slope of the tangent line to the graph of $f(x)=+\sqrt{4-x^2}$ at P. Similarly, finding the tangent line to the graph of $x^2+y^2=4$ at $Q=(\sqrt{2},\ -\sqrt{2})$ is just the same as finding the slope of the tangent line to the graph of the function $g(x)=-\sqrt{4-x^2}$. Differentiating using the Chain Rule–Power Rule, we find that

$$f'(x) = \frac{d}{dx}(4-x^2)^{1/2} = \frac{1}{2}\cdot(4-x^2)^{-1/2}\cdot(-2x) = \frac{-x}{\sqrt{4-x^2}}$$

and

$$g'(x) = -f'(x) = \frac{x}{\sqrt{4-x^2}}.$$

Because $f'(1)=-1/\sqrt{3}$ and $g'(\sqrt{2})=1$, the slopes of the tangent lines to the circle at $(1,\sqrt{3})$ and $(\sqrt{2},\ -\sqrt{2})$ are $-1/\sqrt{3}$ and 1, respectively.

It turns out that there is a simpler way to handle the graph of an equation that does not express y explicitly as a function of x. Called the method of implicit differentiation, this technique eliminates the need to solve for y in terms of x.

The Main Idea of Implicit Differentiation

The example $x^2+y^2=4$ that we have been considering suggests a framework for the general problem we wish to investigate. Because the expression x^2+y^2 depends on the variables x and y, we may write it as $F(x,y)$, using the notation for functions of two variables (Section 1.4). The right side of $x^2+y^2=4$ is a constant, which we may represent by C. Thus $x^2+y^2=4$ is an example of an equation that has the form $F(x,y)=C$.

It should be understood that, when we write $F(x,y)$, the two variables x and y are independent variables. Assigning a value to one of them has no bearing on the value of the other. Thus if we set $x = 1$ in $F(x, y) = x^2 + y^2$, the equation $F(1, y) = 1 + y^2$ imposes no restriction on the value of y, which clearly is *not* a function of x. The matter is quite different when the equation $F(x, y) = C$ holds. For example, setting $x = 1$ in the equation $x^2 + y^2 = 4$ forces y to be $\sqrt{3}$ if (x, y) lies on the upper semicircle and $-\sqrt{3}$ if (x, y) lies on the lower semicircle. This suggests that, on each of the two semicircles, y is a function of x. Indeed, as we have already noticed, we can solve for y and obtain *explicit* formulas, $y = f(x) = \sqrt{4 - x^2}$ and $y = g(x) = -\sqrt{4 - x^2}$, for y in terms of x.

In general, we cannot obtain an *explicit* expression for y from an equation of the form $F(x, y) = C$. Instead, we understand the equation $F(x, y) = C$ to *implicitly* define y as a function of x. It may seem impossible to obtain an expression for dy/dx if we do not have a formula for y, but we *can* find dy/dx, if it exists, by differentiating each side of the equation $F(x, y) = C$ with respect to x. When we do this, we treat y as though it were a differentiable function of x on an open interval centered at x_0. This is the method of implicit differentiation.

Some Examples

In illustrating the method of implicit differentiation, we will use Leibniz notation because it contains a built-in reminder of the variable with respect to which we differentiate.

► **EXAMPLE 1** Use the method of implicit differentiation to find the slopes of the tangent lines to the curve $x^2 + y^2 = 4$ at the points $(1, \sqrt{3})$ and $(\sqrt{2}, -\sqrt{2})$.

Solution We apply d/dx to both sides of the equation. Remembering to treat y as though it were a function of x, we have

$$\frac{d}{dx}(x^2 + y^2) = \frac{d}{dx}(4)$$

or

$$2x \cdot \frac{dx}{dx} + 2y \cdot \frac{dy}{dx} = 0\cdot$$

Take particular notice that we have applied the Chain Rule–Power Rule to the first term, even though this was not strictly necessary. There was no need to write $\frac{dx}{dx}$, but in doing so, we have made our treatment of the term x^2 just the same as our treatment of the term y^2. As we work our first examples, this will lead to less confusion. Because $dx/dx = 1$, our equation becomes

$$2x \cdot + 2y \cdot \frac{dy}{dx} = 0$$

or

$$(2y) \cdot \frac{dy}{dx} = -2x.$$

We may solve for dy/dx by dividing both sides by $2y$ if $y \neq 0$. The result is

$$\frac{dy}{dx} = -\frac{x}{y} \text{ if } y \neq 0.$$

Now it is a simple matter to find the slope of the tangent line to the curve at $(1,\sqrt{3})$, for at this point we have

$$\frac{dy}{dx} = -\frac{x}{y} = -\frac{1}{\sqrt{3}}.$$

This is the same answer that we obtained in the introduction, but the technique is easier.

Similarly, we find the slope of the tangent line to the curve at $(\sqrt{2}, -\sqrt{2})$ to be

$$\frac{dy}{dx} = -\frac{x}{y} = -\frac{\sqrt{2}}{-\sqrt{2}} = 1. \quad \blacktriangleleft$$

INSIGHT Compare Example 1 with the discussion at the beginning of this section. Notice that the method of implicit differentiation does not require separate consideration of the two functions $f(x) = \sqrt{4-x^2}$ and $g(x) = -\sqrt{4-x^2}$ whose graphs make up the curve $x^2 + y^2 = 4$.

Knowing how to use the method of implicit differentiation includes knowing when the method cannot be applied. There are a number of reasons why

$$\left.\frac{dy}{dx}\right|_{P_0}$$

may fail to exist at a point $P_0 = (x_0, y_0)$ on the graph of $F(x, y) = C$. In using the method of implicit differentiation, we assume that y is a function of x on an open interval centered at x_0. This may not actually be the case. In the case of the circle $x^2 + y^2 = 4$, when $P_0 = (-2, 0)$ or $P_0 = (2, 0)$, neither $x_0 = -2$ nor $x_0 = 2$ is the center of an x-interval on which y is defined. Even if y is a function of x on an interval centered at x_0, it may happen that y is not differentiable with respect to x at P_0. This occurs if there is no tangent line at P_0 or if the tangent line exists but is vertical.

Implicit differentiation always expresses the derivative dy/dx as a fraction involving x or y or both. The points *on the curve* where the numerator of this fraction is nonzero but where the denominator is zero correspond to places where the tangent line is vertical or where it fails to exist. (No simple rule is available for points at which both the numerator and denominator are zero.)

► **EXAMPLE 2** At what points on the curve $(y-1)^3 + y(x-2) =$ does dy/dx not exist?

Solution If we differentiate the equation $(y-1)^3 + y(x-2) = 0$ implicitly with respect to x, using the Chain Rule–Power Rule for the first summand and the Product Rule for the second, then we obtain

$$3(y-1)^2 \cdot \frac{dy}{dx} + \frac{dy}{dx} \cdot (x-2) + y \cdot (1-0) = 0,$$

or

$$\left(3(y-1)^2 + (x-2)\right)\frac{dy}{dx} = -y.$$

Thus if $3(y-1)^2 + (x-2) \neq 0$, the derivative dy/dx exists and

$$\frac{dy}{dx} = -\frac{y}{3(y-1)^2 + (x-2)}. \tag{3.8.1}$$

We must investigate the points (x, y) that satisfy the equations

$$3(y-1)^2 + (x-2) = 0 \tag{3.8.2}$$

(because equation (3.8.1) is not valid at such points) and

$$(y-1)^3 + y(x-2) = 0 \tag{3.8.3}$$

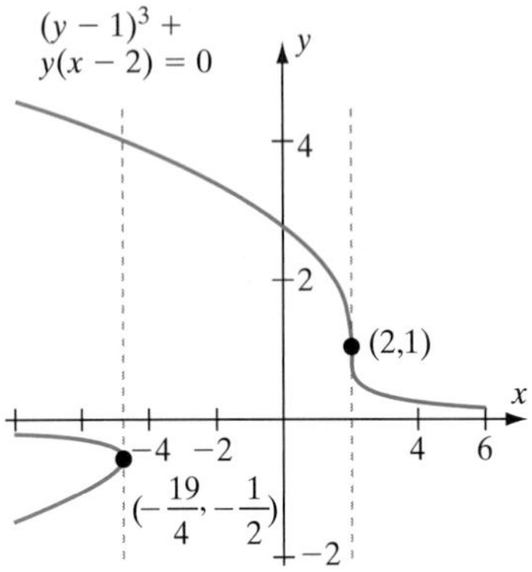

Figure 3

(because we consider only points on the given curve). Equation (3.8.2) gives us

$$x - 2 = -3\,(y-1)^2. \tag{3.8.4}$$

Substituting this formula for $x-2$ into equation (3.8.3) results in $(y-1)^3 - 3y \cdot (y-1)^2 = 0$. Factoring, we obtain $(y-1)^2\,(y-1-3y) = 0$, or $(y-1)^2(2y+1) = 0$. The solutions to this last equation are $y = 1$, and $y = -1/2$. Substituting these values back into (3.8.4) allows us to determine that $x = 2$ when $y = 1$, and $x = -19/4$ when $y = -1/2$. We have graphed portions of the curve $(y-1)^3 + y(x-2) = 0$ in Figure 3. We see that, although y is a function of x near the point (2, 1), dy/dx fails to exist because the tangent line is vertical. On the other hand, we cannot even define y as a continuous function of x on an open interval containing $-19/4$ with the graph of y passing through the point $(-19/4, -1/2)$. Therefore dy/dx has no meaning at this point. ◀

▶ **EXAMPLE 3** Find the slope of the tangent line to the curve $8xy^{9/2} = 9 - x^{-7/5}$ at the point $(1, 1)$.

Solution Using implicit differentiation, we have

$$\frac{d}{dx}(8xy^{9/2}) = \frac{d}{dx}(9 - x^{-7/5}),$$

or, by using the Product Rule and the Chain Rule–Power Rule on the left side,

$$8y^{9/2} + 8x \cdot \frac{9}{2} \cdot y^{7/2} \cdot \frac{dy}{dx} = 0 + \frac{7}{5} \cdot x^{-12/5}.$$

To make the algebra simpler, we will substitute in the point (1,1) into this equation *before* we solve for dy/dx (but notice that we *have* finished differentiating). We obtain $8 + 36 \cdot dy/dx = 7/5$, or $dy/dx = -11/60$. ◀

INSIGHT You might have noticed that the equation considered in Example 3 is not of the form $F(x, y) = C$, as specified in the method of implicit differentiation. Although it would have been easy to rearrange the equation as $8xy^{9/2} + x^{-7/5} = 9$, which does have the standard form $F(x, y) = C$, it is not necessary to bring all variables to one side of the equation in order to use the method of implicit differentiation. We may differentiate the equation as it is presented.

▶ **EXAMPLE 4** Find the equation of the tangent line to the curve

$$xy^4 + x = 17 \tag{3.8.5}$$

at each point where $x = 1$.

Solution If we substitute $x = 1$ into the equation, then we obtain $y^4 + 1 = 17$. This simplifies to $y^4 = 16$ or $y = \pm 2$. Thus there are two points, (1, 2) and (1, −2), with $x = 1$. Now we differentiate equation (3.8.5) to obtain

$$\frac{d}{dx}(xy^4 + x) = \frac{d}{dx}(17).$$

Performing the differentiations, we have

$$y^4 + x \cdot 4y^3 \cdot \frac{dy}{dx} + 1 = 0,$$

or

$$\frac{dy}{dx} = \frac{-1 - y^4}{4xy^3}.$$

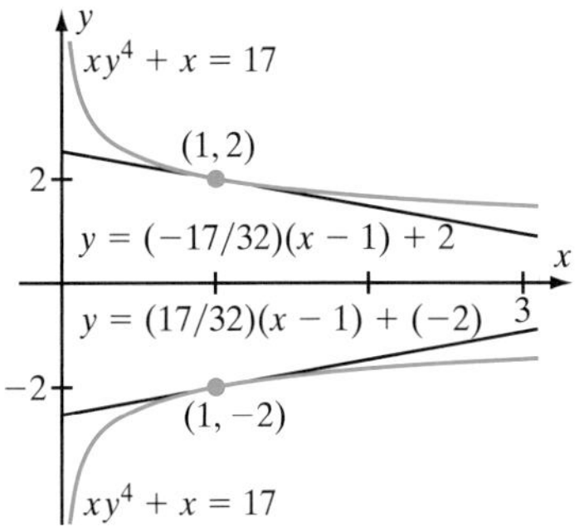

▲ Figure 4

Thus at the point (1, 2), the tangent line has slope −17/32; whereas, at the point (1,−2), the tangent line has slope 17/32. The equations of these tangent lines are $y = (-17/32) \cdot (x - 1) + 2$ and $y = (17/32)\ (x - 1) + (-2)$, respectively. The curve and its tangents are illustrated in Figure 4. ◀

INSIGHT When we apply the method of implicit differentiation to find the slope of the tangent line to $F(x, y) = C$ at (x_0, y_0), the expression for

$$\left.\frac{dy}{dx}\right|_{(x_0, y_0)}$$

will, in general, involve y_0 as well as x_0. As we see in Example 4, there may be more than one y for a given x_0 such that (x_0, y) is on the curve. The value of dy/dx will depend on which value of y is used. Geometrically, this corresponds to the vertical line $x = x_0$ intersecting the curve $F(x, y) = C$ in more than one point. The tangents to the curve at these points need not be parallel, which is why the value of y sometimes must be specified when we calculate dy/dx by the method of implicit differentiation. This is unlike the situation for the explicitly defined functions that we considered prior to this section. When y is defined as an explicit function of x with x_0 in its domain, the vertical line $x = x_0$ intersects the curve in only one point, and dy/dx is determined by x_0 alone.

Calculating Higher Derivatives

We can use the method of implicit differentiation to calculate higher derivatives.

▶ **EXAMPLE 5** Consider the equation $\sin(y) + x = 1/\sqrt{2}$. Calculate d^2y/dx^2 at the point $(0, \pi/4)$.

Solution Differentiating both sides with respect to x yields

$$\cos(y)\frac{dy}{dx} + 1 = 0. \tag{3.8.6}$$

We do not yet substitute in the coordinates of the point we wish to study. We must differentiate another time. Using the Product Rule to differentiate the first summand on the left side of equation (3.8.6), we obtain

$$\left(-\sin(y).\frac{dy}{dx}\right)\left(\frac{dy}{dx}\right) + \cos(y)\left(\frac{d^2y}{dx^2}\right) + 0 = 0. \tag{3.8.7}$$

Now that we have finished differentiating, it is correct to put in our numerical values: Setting $y = \pi/4$ in equation (3.8.6) yields

$$\frac{1}{\sqrt{2}} \frac{dy}{dx}\bigg|_{(0,\ \pi/4)} + 1 = 0,$$

or

$$\frac{dy}{dx}\bigg|_{(0,\ \pi/4)} = -\sqrt{2}.$$

Now we substitute this value for $\frac{dy}{dx}$, together with $x = 0$ and $y = \pi/4$, into equation (3.8.7) to obtain

$$\left(-\frac{1}{\sqrt{2}}\right)(-\sqrt{2})^2 + \left(\frac{1}{\sqrt{2}}\right)\left(\frac{d^2y}{dx^2}\bigg|_{(0,\ \pi/4)}\right) = 0,$$

or

$$\frac{d^2y}{dx^2}\bigg|_{(0,\ \pi/4)} = 2. \quad ◀$$

INSIGHT There is a subtlety in going from line (3.8.6) to line (3.8.7). When we differentiated the term $\cos(y) \cdot dy/dx$, we had to apply the Product Rule to this term. The differentiation of the factor $\cos(y)$ then led to another factor of dy/dx while the differentiation of the factor dy/dx led to a factor of d^2y/dx^2. In this step, we must be careful to keep track of both types of derivative that arise.

Parametric Curves Our focus in this section has been the method of implicit differentiation. Placed in a broader context, our discussion has concerned the calculation of tangent lines to curves that are not presented in the form $y = f(x)$. It is therefore appropriate to extend that discussion by studying the tangent line problem for parametric curves. Recall from Section 1.5 that these are curves that have the form $x = \varphi_1(t)$, $y = \varphi_2(t)$.

THEOREM 1 Suppose that φ_1 and φ_2 are differentiable functions on an interval $I = (a, b)$ and that $\mathcal{C} = \{(\varphi_1(t), \varphi_2(t)) : a < t < b\}$. Let $P_0 = (x_0, y_0) = (\varphi_1(t_0), \varphi_2(t_0))$ for some value t_0 in I. If the parametric equations $x = \varphi_1(t)$, $y = \varphi_2(t)$ determine y as a differentiable function f of x on an arc of $\mathcal{C}$ that extends beyond P_0 on each side and if $\varphi_1'(t_0) \neq 0$, then

$$\frac{dy}{dx}\bigg|_{(x_0,\ y_0)} = \frac{\varphi_2'(t_0)}{\varphi_1'(t_0)}. \tag{3.8.8}$$

Proof. By hypothesis, $\varphi_2(t) = y = f(x) = f(\varphi_1(t))$ for some differentiable function f defined on an open interval containing t_0. Differentiating each side of the equation $\varphi_2(t) = f(\varphi_1(t))$ with respect to t, we obtain $\varphi_2'(t) = f'(\varphi_1(t)) \cdot \varphi_1'(t)$. Evaluating at $t = t_0$, we have $\varphi_1'(t_0) \cdot f'(\varphi_1(t_0)) = \varphi_2'(t_0)$, or $\varphi_1'(t_0) \cdot f'(x_0) = \varphi_2'(t_0)$. If $\varphi_1'(t_0) \neq 0$, then

$$\frac{dy}{dx}\bigg|_{(x_0,\ y_0)} = f'(x_0) = \frac{\varphi_2'(t_0)}{\varphi_1'(t_0)}. \quad ■$$

▶ **EXAMPLE 6** At the beginning of this section, we determined that the slope of the tangent line to the graph of $x^2 + y^2 = 4$ at $P = (1,\sqrt{3})$ is $-1/\sqrt{3}$. Derive this result using the parametric equations $x = 2\cos(t)$, $y = 2\sin(t)$.

Solution We may verify that $\varphi_1(t) = 2\cos(t)$ and $\varphi_2(t) = 2\sin(t)$, then $\varphi_1{}^2(t) + \varphi_2{}^2(t) = 4(\cos^2(t) + \sin^2(t)) = 4$. Because $\varphi_1(\pi/3) = 2\cos(\pi/3) = 1$ and $\varphi_2(\pi/3) = 2\sin(\pi/3) = \sqrt{3}$, the point $P = (1, \sqrt{3})$ corresponds to $t_0 = \pi/3$. Formula (3.8.8) tells us that the required slope is

$$\left.\frac{dy}{dx}\right|_{(1,\sqrt{3})} = \frac{\varphi_2'(t_0)}{\varphi_1'(t_0)} = \left.\frac{2\cos(t)}{-2\sin(t)}\right|_{t=t_0} = -\left.\frac{\cos(t)}{\sin(t)}\right|_{t=\pi/3} = -\frac{\cos(\pi/3)}{\sin(\pi/3)} = -\frac{1}{\sqrt{3}}. \quad ◀$$

INSIGHT In practice, we often do not give names such as φ_1 and φ_2 to the functions that define x and y parametrically. When we forgo the extra notation, formula (3.8.8) simplifies to the more intuitive (and memorable) equation

$$\frac{dy}{dx} = \frac{dy/dt}{dx/dt}. \qquad (3.8.9)$$

Thus in Example 6, the parametric equations $x = 2\cos(t)$ and $y = 2\sin(t)$ lead to $dx/dt = -2\sin(t)$, $dy/dt = 2\cos(t)$, and therefore, by (3.8.9),

$$\frac{dy}{dx} = \frac{2\cos(t)}{-2\sin(t)} = -\frac{\cos(t)}{\sin(t)}.$$

The result is, of course, the same.

QUICK QUIZ

1. The equation $\ln(y) + 4y - x = 1$ implicitly defines $y = f(x)$. What is $f(3)$?
2. Calculate dy/dx at (2, 1) if $\exp(y - 1) + y = x$.
3. Suppose that $y^3 + y = 10x$. Calculate dy/dx and d^2y/dx^2 when $x = 1$.
4. If $x = t^2 + 1$, and $y = t^3$, calculate dy/dx at (5, −8).

Answers

1. 1 2. 1/2 3. $dy/dx = 10/13$ and $d^2y/dx^2 = -120/13^3$ 4. −3

EXERCISES

Problems for Practice

▶ In each of Exercises 1–6, y is a function of x. Calculate the derivative of the given expression with respect to x. (Your answer *should* contain the term dy/dx.) ◀

1. $4y^{3/2}$
2. $\cos(y)$
3. $2x/\sqrt{y}$
4. $(y^3 - 1)/y$
5. xe^y
6. $\ln(y)/x$

▶ In each of Exercises 7–14, use the method of implicit differentiation to calculate dy/dx at the point P_0 ◀

7. $xy^2 + yx^2 = 6$ $\quad P_0 = (1, 2)$
8. $\sin(\pi xy) - xy^2 + 2y = 1$ $\quad P_0 = (1, 1)$
9. $x^{3/5} + 4y^{3/5} = 12$ $\quad P_0 = (32, 1)$
10. $x^4 - y^4 = -15$ $\quad P_0 = (1, 2)$
11. $xy^3 + 3y^2 + x^2 = 21$ $\quad P_0 = (1, 2)$
12. $x^2y + \ln(y) = 1$ $\quad P_0 = (-1, 1)$
13. $xe^{y-1} + y^2 = 4$ $\quad P_0 = (3, 1)$
14. $5x^3\dfrac{x + 2y}{x - y} = 0$ $\quad P_0 = (1, 2)$

▶ In Exercises 15–24, use implicit differentiation to find the tangent line to the given curve at the given point P_0. ◀

15. $xy - 1 = \sqrt{x} + \sqrt{y}$ $\quad P_0 = (4, 1)$
16. $x^3 - y^3 + 4y = 5$ $\quad P_0 = (2, -1)$

17. $\sin^2(x) + \cos^2(y) = 5/4$ $P_0 = (\pi/3, \pi/4)$
18. $7x^{1/2} - xy^{1/2} = 3$ $P_0 = (9, 4)$
19. $x + 3/y + \ln(y) = 5$ $P_0 = (2, 1)$
20. $x + y/x - \exp(y^2) = 2$ $P_0 = (3, 0)$
21. $e^{xy} = 2y^2 - 1$ $P_0 = (0, -1)$
22. $2^{x-y} = xy^3$ $P_0 = (2, 1)$
23. $\ln(xy - 1) + y^2 = 4$ $P_0 = (1, 2)$
24. $xe^{x-1} - \ln(xy) = 1$ $P_0 = (1, 1)$

In Exercises 25–28, use implicit differentiation to find the normal line to the given curve at the given point P_0.

25. $xy^4 - x^3y = 16$ $P_0 = (2, 2)$
26. $x^4 - 8y + y^6 = 9$ $P_0 = (2, 1)$
27. $x - y + \sin(2y) = 1$ $P_0 = (\pi/4, \pi/4)$
28. $x^2 - 3xy^2 + 1 = 1/y$ $P_0 = (3, 1)$

In Exercises 29–34, find dy/dx and d^2y/dx^2 at the point P_0 by implicit differentiation.

29. $y^3 + y + x = 4$ $P_0 = (2, 1)$
30. $y^3 - 2xy = 20$ $P_0 = (-3, 2)$
31. $xy - 6/y = 4$ $P_0 = (2, 3)$
32. $y - \sin(2y) = x\ln(x)$ $P_0 = (1, 0)$
33. $2e^y = 3 - x + y$ $P_0 = (1, 0)$
34. $2x + y^{1/3} + y = 0$ $P_0 = (1, -1)$

In Exercises 35–46, find the tangent line to the parametric curve $x = \varphi_1(t)$, $y = \varphi_2(t)$ at the point corresponding to the given value t_0 of the parameter.

35. $\varphi_1(t) = t^2 + 1,\ \varphi_2(t) = t^3 + 1$ $t_0 = 1$
36. $\varphi_1(t) = t^3 + t,\ \varphi_2(t) = t^2 + 1$ $t_0 = 1$
37. $\varphi_1(t) = t^3 + t,\ \varphi_2(t) = t^3 - t$ $t_0 = 2/3$
38. $\varphi_1(t) = \cos(2t),\ \varphi_2(t) = \sin(3t)$ $t_0 = \pi/4$
39. $\varphi_1(t) = t + \cos(t),\ \varphi_2(t) = t + \sin(t)$ $t_0 = \pi/4$
40. $\varphi_1(t) = t/(1 + t^2),\ \varphi_2(t) = t^2/(1 + t^2)$ $t_0 = 1/2$
41. $\varphi_1(t) = t/(1 + t^2),\ \varphi_2(t) = t^3/(1 + t^2)$ $t_0 = 2/3$
42. $\varphi_1(t) = te^t,\ \varphi_2(t) = t + e^t$ $t_0 = 0$
43. $\varphi_1(t) = \exp(\cos(t)),\ \varphi_2(t) = \exp(\sin(t))$ $t_0 = \pi/6$
44. $\varphi_1(t) = 1/(1 + t^2),\ \varphi_2(t) = t(1 + \ln(t))$ $t_0 = 1$
45. $\varphi_1(t) = 2^t + 3,\ \varphi_2(t) = 3^t + 2$ $t_0 = 1$
46. $\varphi_1(t) = \log_2(t),\ \varphi_2(t) = t\log_3(t)$. $t_0 = 3$

Further Theory and Practice

47. Show that the tangent lines to the curve $x^2 - 4xy + y^2 = 9$ at the points where the curve crosses the x-axis are parallel.
48. Find all points on the curve $xy - 2x^2 + y^2 = 8$ where the tangent line has slope $-2, -1, 0, 1$, or 2. Show that no tangent line to the curve is vertical.
49. The curve $x^2 - xy + y^2 = 4$ is an ellipse. Notice that $P = (a, b)$ is on the ellipse if and only if $Q = (-a, -b)$ is on the ellipse. Assuming that P and Q are both points on the ellipse, show that the tangent lines at P and Q are parallel.
50. Suppose that (x_0, y_0) is a point on the ellipse $x^2/a^2 + y^2/b^2 = 1$. Show that the tangent line to the ellipse at (x_0, y_0) is given by

$$\frac{x_0x}{a^2} + \frac{y_0y}{b^2} = 1.$$

51. For each positive value of α, the equation

$$x^3 - 3\alpha xy + y^3 = 0$$

defines a curve known as a Folium of Descartes (see Figure 4 of this chapter's *Genesis and Development* for a plot). Suppose that the point $P = (a, b)$ is on the Folium of Descartes. Because the equation is symmetric in x and y, it follows that $Q = (b, a)$ is also on the curve. What is the product of the slopes of the tangent lines at P and Q?
52. Let S and T be open intervals in $\mathbb{R}$. Let $f: S \to T$ be a function that is differentiable and invertible. Show how to derive the formula for $\frac{d}{dt}f^{-1}$ that we learned in Theorem 2 of Section 3.6 by applying implicit differentiation to the equation $t = f(s)$.
53. The graph of the curve $x^y = y^x$ $(x, y > 0)$ consists of the ray $x = y > 0$ and a curve C. Find dy/dx at the point P on C whose abscissa is 2. (Hint: Take the natural logarithm of each side first.)
54. Use the method of implicit differentiation to find the slope of the curve $x^{2/3} + y^{2/3} = 25$ at the points $(27, 64)$ and at $(27, -64)$. Verify your answers by explicitly solving for y and differentiating the explicit expression. Why must you use two explicit formulas for y?
55. Let a be a positive constant. Let T be any tangent line to

$$\sqrt{x} + \sqrt{y} = a.$$

Compute the sum of the intercepts of T on the coordinate axes. (The answer does not depend on the choice of T.)
56. Suppose that $A \neq 0$. The locus of

$$x^{2/3} + y^{2/3} = a^{2/3}$$

is called an astroid $\mathcal{A}_a$. Let T be a tangent line to the astroid $\mathcal{A}_a$. Find a formula for the distance between the two intercepts of T. How does this distance depend on T?
57. Saha's equation

$$\frac{1 - y}{y^2} = \frac{A\exp(b/x)}{x^{3/2}}$$

describes the degree of ionization within stellar interiors. In this equation, A and b are constants, y represents the fraction of ionized atoms in the star, and x represents stellar temperature in degrees Kelvin. Find dy/dx.
58. Let p and q be nonzero integers. Apply implicit differentiation to the equation $x^p - y^q = 0$ to obtain a proof of the Power Rule for rational exponents:

$$\frac{d}{dx}\left(x^{p/q}\right) = \frac{p}{q} \cdot x^{(p/q)-1}.$$

59. By the *demand curve* for a given commodity, we mean the set of all points (p, q) in the pq−plane where q is the number of units of the product that can be sold at price p. The elasticity of demand for the product at price p is defined to be $E(p) = -q'(p) \cdot p/q(p)$.

a. Suppose that a demand curve for a commodity is given by

$$p + q + 2p^2q + 3pq^3 = 1000$$

when p is measured in dollars, and the quantity q of items sold is measured by the 1000. For example, the point $(p, q) = (6, 3.454)$ is on the curve. That means that 3454 items are sold at \$6. What is the slope of the demand curve at the point (6, 3.454)?

b. What is elasticity of demand for the product of part (a) at $p = \$6$.

60. Let a be a positive constant. The *Kulp Quartic* is the graph of the equation $x^2y^2 = a^2(a^2 - y^2)$. Show that this curve is also described by the parametric equations $x = a \tan(t)$, $y = a\cos(t)$. Show that the Kulp Quartic is, furthermore, the union of the graphs of two (explicit) functions. Set $a = 2$. Find the slope of the tangent line at $(2, \sqrt{2})$ by three different methods.

Computer/Calculator Exercises

In each of Exercises 61–64, a curve is given as well as an abscissa x_0. Find the ordinate y_0 such that (x_0, y_0) is on the curve, and determine the slope of the tangent line to the curve at (x_0, y_0).

61. $x^3 - 3xy + y^3 = 0$ $\quad x_0 = 10.000$

62. $x^y - y^x = 0,\ x \neq y$ $\quad x_0 = 3.0000$

63. $x^3 - x^2y^2 + y^3 = 0$ $\quad x_0 = -1.4649$

64. $\ln(xy) + y = 5$ $\quad x_0 = 2.124$

In each of Exercises 65–68, a curve is given as well as an abscissa x_0. Find the ordinate y_0 such that (x_0, y_0) is on the curve, and determine the equation of the tangent line to the curve at (x_0, y_0).

65. $x^3 - 2xy + y^3 = 0$ $\quad x_0 = 10.000$

66. $x^y - y^x = 0,\ x \neq y$ $\quad x_0 = 3.1416$

67. $x^3 - x^2y^2 + y^3 = 0$ $\quad x_0 = -2.0125$

68. $e^{xy} + y = 10$ $\quad x_0 = 1.5727$

In Exercises 69–72, plot the given curve in a viewing rectangle that contains the given point P_0. Then add a plot of the tangent line to the curve at P_0.

69. $x^3 - 2xy + y^3 = 0$ $\quad P_0 = (0.5, 0.9304\ \ldots)$

70. $x^y - y^x = 0,\ x \neq y$ $\quad P_0 = (3.7500, 2.0860)$

71. $x^3 - x^2y^2 + y^3 = 0$ $\quad P_0 = (2.1125, 1.9289)$

72. $xe^{xy} + y = 2$ $\quad P_0 = (1.0000, 0.44285)$

Implicit Differentiation

In Exercises 73–77, plot the given parametric curve $x = \varphi_1(t)$, $y = \varphi_2(t)$ in a viewing rectangle that contains the given point P_0. Find the equation of the tangent line at P_0. Add the tangent line to your plot.

73. $\varphi_1(t) = 8\cos(t)$, $\varphi_2(t) = 8\sin(t)(\sin(t/2))^2$, $P_0 = (4, \sqrt{(3)}$ (Tear Drop Curve)

74. $\varphi_1(t) = 100t/(1 + t^2)^2$, $\varphi_2(t) = 100t^2/(1 + t^2)^2$, $P_0 = (32, 16)$ $(t = 1/2)$ (Regular Bifolium)

75. $\varphi_1(t) = (1 + t^2)/(1 + 3t^2)$, $\varphi_2(t) = 4t/(1 + 3t^2$, $P_0 = (1/2, 1)$ $(t = 1)$

76. $\varphi_1(t) = (3 - t^2)/(1 + t^2)$, $\varphi_2(t) = 2t(3 - t^2)/(1 + t^2)$, $P_0 = (1, 2)$ $(t = 1/2)$ (Maclaurin's Trisectrix)

77. $\varphi_1(t) = 1 - 3t^2$, $\varphi_2(t) = t - 3t^3$, $P_0 = (1/4, -1/8)$ $(t = -1/2)$ (Cubic of Tschirnhaus)

78. Graphically locate the points on the curve $x^4 + x + 5xy^3 = 1$ where the tangent line is vertical. Confirm that the method of implicit differentiation yields the same points.

79. Graphically locate the points on the curve $x^3 - 6xy^2 = 4$ where the tangent line is vertical. Confirm that the method of implicit differentiation yields the same points.

80. Plot the *ampersand curve*

$$(x^2 - y^2)(y - 1)(2y - 3) = 2(x^2 + y^2 - 2y)^2$$

in the viewing window $[-3/2, 3/2] \times [-1/4, 3/2]$. Notice that $P_0 = (1, 1)$ lies on the ampersand. However, the general formula for dy/dx at P_0 results in the meaningless 0/0. This complication is not surprising: we expect some unusual behavior at P_0 because the ampersand has two tangent lines at P_0, as is evident from the plot. Find points $P_1 = (x_1, 0.9999)$ and $P_2 = (x_2, 0.9999)$ on the ampersand with $0 < x_1 < 1$ and $1 < x_2$. Calculate the slopes m_1 and m_2 of the tangent lines at P_1 and P_2. As approximations to the two tangent lines at P_0, add to your plot the two lines passing through P_0 with slopes m_1 and m_2.

81. Let $q(p)$ be the demand function of Exercise 59a. Use the method of implicit differentiation to find $q'(4)$. Find $q(4.2)$ and $q(3.8)$. Use these values to approximate $q'(4)$.

82. Saha's equation describes the degree of ionization within stellar interiors. For our own sun,

$$\frac{1 - y}{y^2} = \frac{0.787 \times 10^9 \exp(158000/x)}{x^{3/2}}$$

where y represents the fraction of ionized atoms in the sun, and x represents solar temperature in degrees Kelvin. Find dy/dx when $x = 6 \times 10^6$.

3.9 Differentials and Approximation of Functions

Loosely interpreted, the equation

$$f'(c) = \lim_{\Delta x \to 0} \frac{f(c+\Delta x) - f(c)}{\Delta x}$$

says that

$$f'(c) \approx \frac{f(c+\Delta x) - f(c)}{\Delta x} \tag{3.9.1}$$

when Δx is small and nonzero. A choice of a fairly small value of Δx often gives a good approximation. With a little algebraic manipulation, we can convert approximation (3.9.1) into an approximation of $f(c+\Delta x)$:

$$f(c+\Delta x) \approx f(c) + f'(c) \cdot \Delta x. \tag{3.9.2}$$

We may interpret equation (3.9.2) as follows: If we know the values of $f(c)$ and $f'(c)$, then we can estimate the value of $f(x)$ at a nearby point $x = c + \Delta x$. In this context, the point c is sometimes called a *base point*. When Δx can take any value in an interval $(-h, h)$, so that $x = c + \Delta x$ can take any value in $(c-h, c+h)$, we refer to c as the *center* of approximation (3.9.2). Sometimes we abbreviate $f(x+\Delta x) - f(x)$ as $\Delta f(x)$. Notice that $\Delta f(x)$ is the increment in the functional value when the independent variable x is incremented by Δx. Using this notation and replacing c with x, approximation (3.9.2) becomes

$$\Delta f(x) \approx f'(x)\Delta x, \tag{3.9.3}$$

which is illustrated in Figure 1. We call this approximation scheme the *method of increments*.

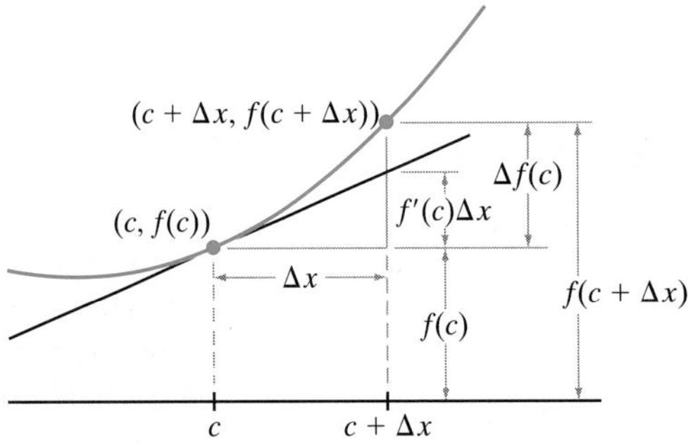

▲ **Figure 1** $f(c+\Delta x) - f(c) = \Delta f(c) \approx f'(c)\Delta x$

▶ **EXAMPLE 1** Use approximation (3.9.2) to estimate the value of $\sqrt{4.1}$.

Solution To make the left side, $f(c+\Delta x)$, of approximation (3.9.2) equal to the quantity $\sqrt{4.1}$ that we are to approximate, we set $f(x) = x^{1/2}$ and $c + \Delta x = 4.1$. We calculate $f'(x) = (1/2)\,x^{-1/2}$. We must now select two numbers c and Δx that sum to 4.1. If we choose $c = 4$ and $\Delta x = 0.1$, then $f(c) = 4^{1/2} = 2$, $f'(c) = (1/2) \cdot 4^{-1/2} = 1/4$, and approximation (3.9.2) becomes

$$\sqrt{4.1} = f(c+\Delta x) \approx f(c) + f'(c)\Delta x = 2 + \frac{1}{4}(0.1) = 2.025.$$

A calculator shows that our simple approximation is accurate to within 0.001. ◀

INSIGHT The selection of $c = 4$ in Example 1 is no accident. It is easy for us to evaluate f and f' at this particular c. Moreover, $c = 4$ is near to 4.1, which means that Δx is small. Making Δx small tends to increase accuracy. If, for example, we apply the method of increments to estimate $\sqrt{10}$ using the same choice of c as in Example 1, namely $c = 4$, then $\Delta x = 6$, and we obtain $\sqrt{10} \approx 4^{1/2} + (1/2)4^{-1/2}(6) = 3.5$. The true value of $\sqrt{10}$ is 3.162. . . . Our approximation is therefore off by more than 0.33. A much better estimate is obtained by taking $c = 9$ and $\Delta x = 1$. The resulting approximation is

$$\sqrt{10} \approx 9^{1/2} + \left(\frac{1}{2}\right)9^{-1/2}(1) = 3.166. . . ,$$

which is correct to two decimal places. The accuracy of the method of increments strongly depends on the size of the increment Δx. In general, the smaller the increment, the more accurate the approximation.

▶ **EXAMPLE 2** A machine shop builds triangular jigs of isosceles shape. The bases are precut and have length exactly 10 inches. The heights vary. During a rush period, the measurement of the two sides gets sloppy, and they are off by as much as 0.1 inch. By about how much will the area of the resulting jig deviate from the desired area?

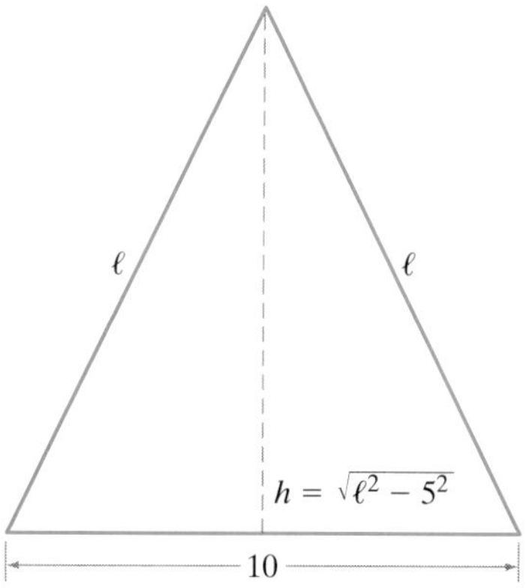

▲ Figure 2

Solution The ideal length of the two equal sides is denoted by ℓ in Figure 2. The height of the triangle is $h = \sqrt{\ell^2 - 5^2}$, and the area is therefore

$$A(\ell) = \frac{1}{2} \cdot 10 \cdot h = 5 \cdot \sqrt{\ell^2 - 25}.$$

Notice that $A'(\ell) = 5\ell(\ell^2 - 25)^{-1/2}$. If the length measurement is off by $\Delta\ell$, then the area is

$$A(\ell + \Delta\ell) \approx A(\ell) + A'(\ell) \cdot \Delta\ell = A(\ell) + 5\ell \cdot (\ell^2 - 25)^{-1/2} \cdot \Delta\ell.$$

The increment in area is therefore

$$\Delta A \equiv A(\ell + \Delta\ell) - A(\ell) \approx 5\ell \cdot (\ell^2 - 25)^{-1/2} \cdot \Delta\ell.$$

Because $|\Delta\ell| \le 0.1$, the approximate deviation ΔA of the area can be as much as $5\ell \cdot (\ell^2 - 25)^{-1/2} \cdot 0.1$, or

$$\frac{\ell}{2\sqrt{\ell^2 - 25}}.$$

For example, if the desired length ℓ is 20 inches, then we substitute $\ell = 20$ into this formula and find that the error in area is approximately 0.5164 in^2. ◀

INSIGHT In Example 2, the error in area can be found from the area equation $A(\ell) = 5\sqrt{\ell^2 - 25}$. Thus,

$$\text{error in area} = A(\ell + 0.1) - A(\ell) = 5\sqrt{(\ell + 0.1)^2 - 25} - 5\sqrt{\ell^2 - 25}.$$

However, this formula for the error does not allow us to easily understand the dependence on ℓ. The important feature of the method of increments in Example 2 is that it yields a simple formula that works for *all* triangles under consideration with a minimum of calculation.

Increments in Economics

If we take $\Delta x = 1$ in (3.9.2), then we obtain $f(c+1) \approx f(c) + f'(c)$ or

$$f(c+1) - f(c) \approx f'(c). \tag{3.9.4}$$

This approximation is often used in economics. For example, if $f(x)$ represents the quantity of goods produced as a function of the amount x of labor, then f is called the *product of labor*. The increment in production that is achieved by adding one more unit of labor is known as the *marginal product of labor (MPL)*. Thus

$$MPL(c) = f(c+1) - f(c).$$

If we compare this definition of MPL with approximation (3.9.4), then we obtain

$$MPL(c) \approx f'(c).$$

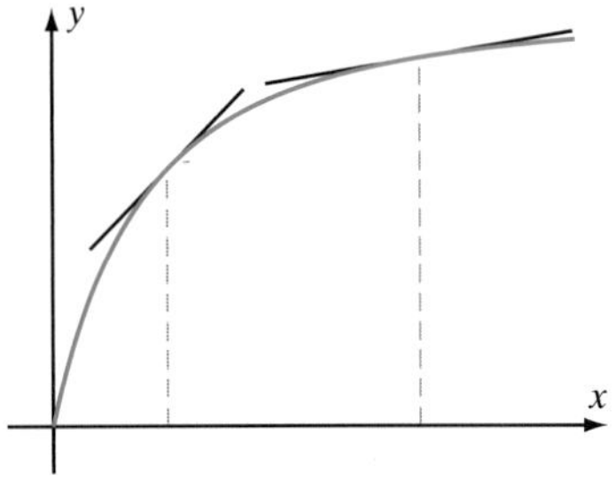

▲ Figure 3 Level of output as a function of the quantity of labor input. The slope of the tangent line equals the marginal product of labor.

Many economics texts define $MPL(c)$ to be $f'(c)$ straight away. Figure 3 is a graphic that is typically found in textbooks about macroeconomics.

Another example of this construction is *marginal cost.* If $C(x)$ is the cost of producing x units of an item, then the *marginal cost* is the cost $C(x+1) - C(x)$ of producing one more item. In practice, the approximation $C'(x)$ is generally used instead of $C(x+1) - C(x)$. Similarly, if $R(x)$ and $P(x)$ are the revenue and profit, respectively, produced by selling x items, then $R'(x)$ and $P'(x)$ are the *marginal revenue* and *marginal profit*, respectively.

Linearization

Let's look at the approximation (3.9.2) again. If we let $x = c + \Delta x$, then we get

$$f(x) = f(c + \Delta x) \approx f(c) + f'(c) \cdot \Delta x = f(c) + f'(c) \cdot (x - c)$$

or

$$f(x) \approx f(c) + f'(c) \cdot (x - c) \tag{3.9.5}$$

for x close to c. The right hand side of this approximation is a function of x which we will denote by L:

$$L(x) = f(c) + f'(c) \cdot (x - c). \tag{3.9.6}$$

Notice that the graph of L is the tangent line to the graph of f at $(c, f(c))$.

DEFINITION The function L defined by (3.9.6) is said to be the *linearization* of f at c. The approximation $L(x) \approx f(x)$ for x close to c is called the *tangent line approximation* to f at c. Sometimes the term *best linear approximation* is used.

In the leftmost viewing window of Figure 4 we see the graph of a function f together with its linearization at c. We have zoomed in around the point $(c, f(c))$ in the middle view. Notice how good an approximation L is of f in this viewing

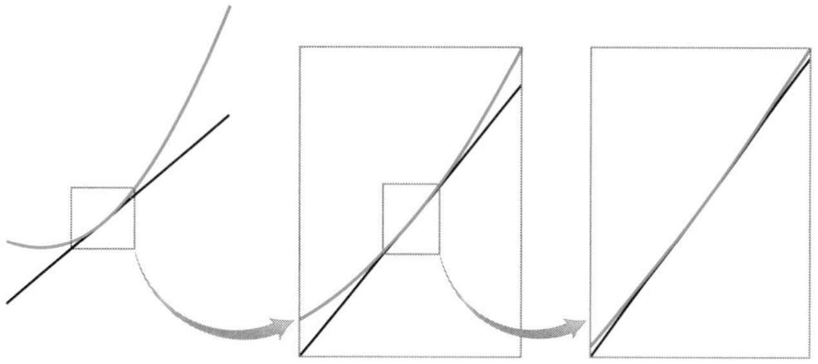

▲ Figure 4

window. In the right view, we have zoomed in still further. Distinguish between L and f. There is, to be sure, a difference between these two functions—it is just too minute to show up on screen.

Important Linearizations

It is useful to know certain linearizations because of the frequency with which they occur in calculus and other subjects.

▶ **EXAMPLE 3** Show that

$$(1+x)^p \approx 1 + p \cdot x$$

for values of x near 0.

Solution Let $f(x) = (1+x)^p$ and $c = 0$. Using the Chain Rule–Power Rule, we have

$$f'(x) = \frac{d}{dx}(1+x)^p = p(1+x)^{p-1}\frac{d}{dx}(1+x) = p(1+x)^{p-1}.$$

Therefore $f'(c) = f'(0) = p(1+0)^{p-1} = p$. Approximation (3.9.5) takes the form

$$(1+x)^p = f(x) \approx f(c) + f'(c)\cdot(x-c) = 1 + p\cdot(x-0) = 1 + p\cdot x,$$

as required. Figure 5 illustrates the approximation when $p = 1/2$. ◀

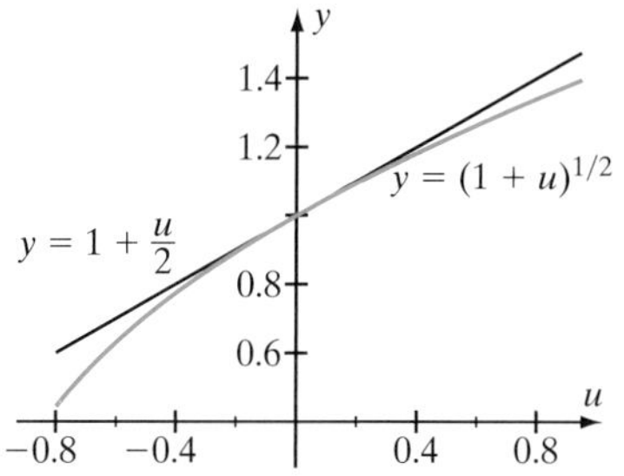

▲ Figure 5

▶ **EXAMPLE 4** Show that the linearizations of $\sin(x)$, $\cos(x)$, and $\tan(x)$ at $x = 0$ are

$$\sin(x) \approx x, \quad \cos(x) \approx 1, \quad \text{and} \quad \tan(x) \approx x$$

Solution Let $c = 0$. For $f(x) = \sin(x)$, approximation (3.9.5) becomes

$$\sin(x) = f(x) \approx f(c) + f'(c)\cdot(x-c) = \sin(0) + \cos(0)\cdot(x-0) = x.$$

Similarly, for $f(x) = \cos(x)$, approximation (3.9.5) becomes

$$\cos(x) = f(x) \approx f(c) + f'(c)\cdot(x-c) = \cos(0) + \sin(0)\cdot(x-0) = 1$$

Finally, for $f(x) = \tan(x)$, we have

$$\tan(x) = f(x) \approx f(c) + f'(c)\cdot(x-c) = \tan(0) + \sec^2(0)\cdot(x-0) = x. \quad ◀$$

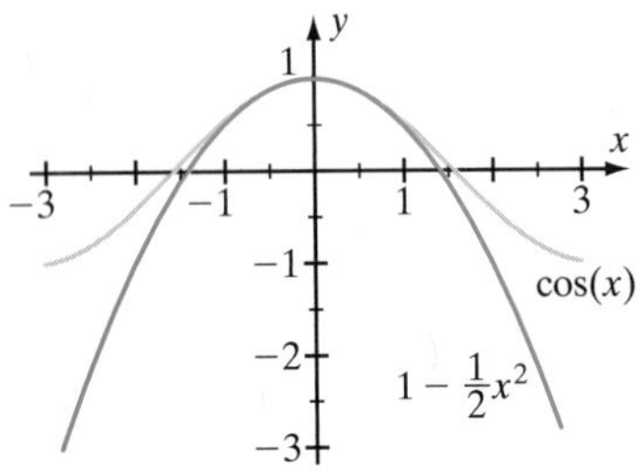

▲ Figure 6

INSIGHT It should be noticed that we have already obtained an estimate that is more precise than the linearization $\cos(x) \approx 1$. Recall the basic limit formula:

$$\lim_{x\to 0} \frac{1-\cos(x)}{x^2} = \frac{1}{2}$$

that we derived in Section 2.3. This limit says that

$$\cos(x) \approx 1 - \frac{1}{2}x^2$$

for small x. The two functions $f(x) = \cos(x)$ and $g(x) = 1 - \frac{1}{2}x^2$ have been graphed in Figure 6. Notice how well g approximates f for small values of x—but *only* for small values of x. Clearly this approximation of $\cos(x)$ is better than the linearization $L(x) = 1$. In Chapter 10, we will learn a general approach, called the Taylor expansion, for obtaining estimates that are more precise than the tangent line approximation.

Differentials

In the early days of calculus, the ideas were not made precise. There was no concept of "limit." Instead, the derivative was obtained as a quotient of infinitesimally small quantities:

$$\frac{f(x+dx)-f(x)}{dx}.$$

Here dx represents a quantity that is positive but smaller than every positive real number. What does this mean? Nobody really knew. Nevertheless, calculations of derivatives of specific functions were done with these mysterious quantities. Many times these calculations involved strange steps that could not be rigorously justified.

It took nearly 200 years for the puzzling theory of differentials (which is what we call expressions like "dx") to be replaced by the more precise notions of *increment* and *limit*. More recently, in 1960, it was determined that the notion of differential *could* be made precise. However, the modern theory of infinitesimals is a subject for an advanced course in mathematical logic, and we cannot discuss it here. What you need to know is that the *notation* of differentials is still used today, purely as an intuitive device. It is reassuring to know that differentials can be treated rigorously, but for us, they will be just an economical way to express the method of increments. Here is how we do it.

As Δx becomes smaller in (3.9.3), the estimate becomes more accurate. When Δx becomes "infinitesimal," estimate (3.9.3) becomes an equality. We represent the infinitesimal increment in x as dx and the resulting infinitesimal change in f as df. Thus (3.9.3) becomes

$$df(x) = f'(x)dx. \tag{3.9.7}$$

It is tempting to divide this equation through by dx to yield the identity

$$\frac{df(x)}{dx} = f'(x).$$

This explains the origin of the Leibniz notation for the derivative. But, for our purposes, we think of equation (3.9.7) as nothing more than another way to write approximation (3.9.3). Indeed, formula (3.9.7) is sometimes referred to as a *differential approximation.*

A LOOK BACK It may help to remember that there is only one key idea behind all the terminology that has been introduced. *Linearization, best linear approximation, tangent line approximation*, and *differential approximation* all refer to the same thing: If f is differentiable at c, if the point (x, y) is on the tangent line to the graph of f at $(c, f(c))$, and if x is close to c, then y is close to $f(c)$.

We conclude by stating the rules of differentiation in differential form. In the following table, c represents a constant.

$$\begin{aligned} d(c) &= 0 & d(\sin(u)) &= \cos(u)du \\ d(cu) &= cd(u) & d(\cos(u)) &= -\sin(u)du \\ d(u+v) &= du+dv & d(\tan(u)) &= \sec^2(u)du \\ d(uv) &= udv+vdu & d(\sec(u)) &= \sec(u)\tan(u)du \\ d(u/v) &= (vdu-udv)/v^2 & d(\cot(u)) &= -\csc^2(u)du \\ d(1/v) &= -dv/v^2 & d(\csc(u)) &= -\csc(u)\cot(u)du \\ d(u^p) &= pu^{p-1}du & d(\exp(u)) &= \exp(u)du \\ d(\ln(u)) &= (1/u)du & d(a^u) &= a^u\ln(a)du \end{aligned}$$

QUICK QUIZ

1. Use differentials to approximate $7.9^{1/3}$.
2. What is the linearization of $\cos(x)+\sin(x)$ at 0?
3. What is the best linear approximation of $f(x)=\ln(1+x)$ at $x=0$?
4. What is the differential approximation of $f(x)=e^x$?

Answers

1. $2+(1/3)\,8^{-2/3}(-0.1)$ 2. $1+x$ 3. x 4. $e^x dx$

EXERCISES

Problems for Practice

In Exercises 1–24, use the method of increments to estimate the value of $f(x)$ at the given value of x using the known value $f(c)$.

1. $f(x)=\sqrt{x}, c=4, x=3.9$
2. $f(x)=\sqrt{x}, c=9, x=8.95$
3. $f(x)=1/x^{1/3}$, $c=8$, $x=8.07$
4. $f(x)=x^{-3/2}$, $c=4$, $x=4.21$
5. $f(x)=x^{2/3}$, $c=8$, $x=8.15$
6. $f(x)=(1+x)^{-1/4}$, $c=15$, $x=16$
7. $f(x)=(1+7x^{2})^{1/3}$, $c=3$, $x=2.8$
8. $f(x)=\sin(x)$, $c=0$, $x=0.02$
9. $f(x)=\cos(x)$, $c=\pi/3$, $x=1.06$
10. $f(x)=\sin(x)-\cos(x)$, $c=\pi/4$, $x=0.75$
11. $f(x)=\tan(x)$, $c=\pi/4$, $x=0.8$
12. $f(x)=\cot(x)$, $c=\pi/3$, $x=1$
13. $f(x)=\sec(x)$, $c=\pi/6$, $x=0.5$
14. $f(x)=\csc(\pi x/4)$, $c=1$, $x=0.94$
15. $f(x)=\ln(x)$, $c=e^3$, $x=20$
16. $f(x)=x\ln(x)$, $c=1$, $x=0.92$
17. $f(x)=e^x$, $c=0$, $x=-0.17$
18. $f(x)=xe^x$, $c=0$, $x=0.12$
19. $f(x)=16^x$, $c=0$, $x=0.13$
20. $f(x)=x^2 2^x$, $c=0$, $x=0.15$
21. $f(x)=\sqrt{2}\cos(\pi/(x+3))$, $c=1$, $x=0.81$
22. $f(x)=\sin(\sqrt{\pi x})$, $c=\pi/16$, $x=0.2$
23. $f(x)=\sin^2(x)$, $c=\pi/4$, $x=0.75$
24. $f(x)=\sqrt{x}/(1+\sqrt{x})$, $c=9$, $x=8.6$

In Exercises 25–30, choose an appropriate function f and point c, and use the differential approximation of f in order to estimate the given number. Compute the absolute error.

25. $\sqrt{24}$
26. $\sqrt[3]{-7.5}$
27. $\sqrt{1+\sqrt{9.7}}$

28. $\sin(0.48\pi)$
29. $\cos(55°)$
30. $\tan(0.23\pi)$

In Exercises 31–36, calculate the linearization $L(x) = f(0)+f'(0)\cdot x$ for the given function f at $c = 0$.

31. $f(x) = 1/(3+2x)$
32. $f(x) = x\sqrt{1+x}$
33. $f(x) = x/(4+x)^{3/2}$
34. $f(x) = x/e(x)$
35. $f(x) = 1/(x+\cos(x))$
36. $f(x) = \cos(x)/(1+\sin(2x))$

In Exercises 37–42, calculate the linearization $L(x) = f(c)+f'(c),\cdot(x-c)$ for the given function f at the given value c.

37. $f(x) = \cos(x)$, $c = \pi/3$
38. $f(x) = \tan(x)$, $c = \pi/4$
39. $f(x) = (25/9)^x$, $c = 1/2$
40. $f(x) = x\ln(x)$, $c = e$
41. $f(x) = e(x-1)/x$, $c = 1$
42. $f(x) = \cos^{2}(x)$, $c = \pi/4$

Further Theory and Practice

43. Suppose that a, b, and p are constants. What is the linearization of $f(x) = \left(\dfrac{1+ax}{1+bx}\right)^p$ at 0?

44. Suppose that f, g, and h are differentiable functions and that $f = g\cdot h$. Suppose that $g(4) = 6$, $dg(4) = 7, dx$, $h(4) = 5$, and $dh(4) = -2dx$. What is $df(4)$?

45. Explain how the linearizations of the differentiable functions f and g at c may be used to discover the product rule for $(f\cdot g)'(c)$.

46. Explain how the linearizations of the differentiable functions f and g at c may be used to discover the quotient rule for $(f/g)'(c)$ at a point c for which $g(c)\neq 0$.

Suppose that $(u, v)\mapsto F(u, v)$ is a given function of two variables. Suppose that y is an unknown function of x that satisfies the differential equation $y'(x) = F(x, y(x))$ and initial condition $y(x_0) = y_0$. (Together these two equations constitute an *initial value problem*. We will study such problems in greater detail in Chapter 7.) The method of increments can be used to approximate $y(x_1)$ where $x_1 = x_0+\Delta x$:

$$y(x_1) \approx y(x_0) + y'(x_0)\Delta x = \underbrace{y_0 + F(x_0, y_0)\Delta x}_{y_1.}$$

We call this *Euler's Method* of approximating the unknown function y. In each of Exercises 47–50, an initial value problem is given. Calculate the Euler's Method approximation y_1 of $y(x_1)$.

47. $dy/dx = x+y$, $y(1) = 2$, $x_1 = 1.2$
48. $dy/dx = x-y$, $y(-2) = -1$, $x_1 = -2.15$
49. $dy/dx = x^2-2y$, $y(0) = 3$, $x_1 = 1/4$
50. $dy/dx = 1+y/x$, $y(2) = 1/2$, $x_1 = 3/2$

By the *demand curve* for a given commodity, we mean the set of all points (p, q) in the pq plane where q is the number of units of the product that can be sold at price p. In Exercises 51–54 use the differential approximation to estimate the demand $q(p)$ for a commodity at a given price p.

51. Suppose that a demand curve for a commodity is given by

$$p+q+2p^2q+3pq^3 = 1000$$

when p is measured in dollars and the quantity q of items sold is measured by the 1000. The point $(p_0, q_0) = (6.75, 3.248)$ is on the curve. That means that 3248 items are sold at \$6.75. What is the slope of the demand curve at the point $(6.75, 3.248)$? Approximately how many units will be sold if the price is increased to \$6.80? Decreased to \$6.60?

52. Suppose that a demand curve for a commodity is given by

$$2p^2q+p\sqrt{q}/100 = 500005$$

when p is measured in dollars. This tells us that about 10000 units are sold at \$5. Approximately how many units of the commodity can be sold at \$5.25? At \$4.75?

53. The demand curve for a commodity is given by

$$\frac{pq^2}{190950}+q\sqrt{p} = 8019900$$

when p is measured in dollars. Use the differential approximation to estimate the number of units that can be sold at \$9.75.

54. The demand curve for a commodity is given by

$$\frac{p^2q}{10}+5p\sqrt{q} = 39000$$

when p is measured in dollars. Approximately how many units of the commodity can be sold at \$1.80?

Calculator/Computer Exercises

In each of Exercises 55–58 a function f, a point c, an increment Δx, and a positive integer n are given. Use the method of increments to estimate $f(c+\Delta x)$. Then let $h = \Delta x/N$. Use the method of increments to obtain an estimate y_1 of $f(c+h)$. Now, with $c+h$ as the base point and y_1 as the value of $f(c+h)$, use the method of increments to obtain an estimate y_2 of $f(c+2h)$. Continue this process until you obtain an estimate y_N of $f(c+N\cdot h) = f(c+\Delta x)$. We say that we have taken *N steps* to obtain the approximation. The number h is said to be the *step size*. Use a calculator or computer to evaluate $f(c+\Delta x)$ directly. Compare the accuracy of the one-step and N-step approximations.

55. $f(x) = x^{1/3}$, $c = 27$, $\Delta x = 0.9$, $N = 3$
56. $f(x) = \sqrt{x}$, $c = 4$, $\Delta x = 0.5$, $N = 5$

57. $f(x) = \ln(x)$, $c = e$, $\Delta x = 3 - e$, $N = 2$

58. $f(x)\, 1/\sqrt[3]{x}$, $c = -8$, $\Delta x = 1$, $N = 4$

In each of Exercises 59–62, an initial value problem is given, along with its exact solution. (Read the instructions for Exercises 47–50 for terminology.) Verify that the given solution is correct by substituting it into the given differential equation and the initial value condition. Calculate the Euler's Method approximation $y_1 = y_0 + F(x_0, y_0)\Delta x$ of $y(x_1)$ where $\Delta x = x_1 - x_0$. Let $m_1 = (F(x_0, y_0) + F(x_1, y_1))/2$ and $z_1 = y_0 + m_1 \Delta x$. This is the *Improved Euler Method* approximation of $y(x_1)$. Calculate z_1. By evaluating $y(x_1)$, determine which of the two approximations, y_1 or z_1, is more accurate.

59. $dy/dx = x + y$, $y(1) = 2$, $x_1 = 1.2$; Exact solution: $y(x) = 4\exp(x-1) - x - 1$

60. $dy/dx = x - y$, $y(-2) = -1$, $x_1 = -2.15$; Exact solution: $y(x) = 2\exp(-2-x) + x - 1$

61. $dy/dx = x^2 - 2y$, $y(0) = 3$, $x_1 = 1/4$; Exact solution: $y(x) = x^2/2 - x/2 + 1/4 + 11/4\exp(-2x)$

62. $dy/dx = 1 + y/x$, $y(2) = 1/2$, $x_1 = 3/2$; Exact solution: $y(x) = x\ln(x) + x(1/4 - \ln(2))$

In each of Exercises 63–65, a demand curve is given. Use the method of implicit differentiation to find dq/dp. For the given price p_0, solve the demand equation to find the corresponding demand q_0. Then use the differential approximation with base point p_0 to estimate the demand at price p_1. Find the exact demand at price p_1. What is the relative error that results from the differential approximation?

63. $p^2 q + 2p\,q^{1/4} = 250100$, $p_0 = 5.10$, $p_1 = 5$

64. $2p\,q + p\,q^{1/3} =$, 160200, $p_0 = 9.75$, $p_1 = 10$

65. $p^2 q/10 + 5pq^{1/5} = 1280200$, $p_0 = 1.80$, $p_1 = 2$

3.10 Other Transcendental Functions

Exponential functions, logarithm functions, and trigonometric functions belong to a large class of functions whose members are known as *transcendental functions* in the mathematical literature. The functions we introduce in this section—inverse trigonometric functions and hyperbolic functions—also belong to that class. With these additions, we complete the list of transcendental functions that traditionally receive detailed coverage in a first-year calculus course. Students who pursue further courses in quantitative subjects will encounter many other transcendental functions that arise in advanced classes or specialized studies.

Inverse Trigonometric Functions

Figure 1 shows the graphs of the six trigonometric functions. Notice that each graph has the property that some horizontal line intersects the graph at least twice. Therefore no trigonometric function is invertible. Another way of seeing this point is that each trigonometric function f satisfies the equation $f(x + 2\pi) = f(x)$ and is therefore *not* one-to-one.

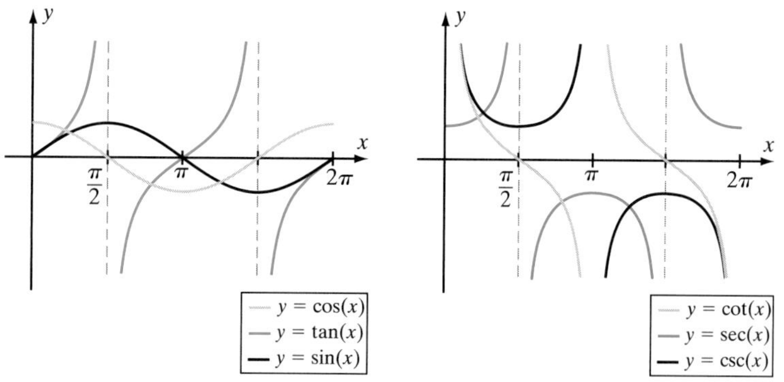

▲ Figure 1

We must restrict the domains of the trigonometric functions if we want to discuss inverses for them. In this section, we learn the standard methods for performing this restriction operation. As you read this section, you may find it useful to refer back to Section 1.6 to review the properties of the trigonometric functions.

Inverse Sine and Cosine

Consider the sine function with domain restricted to the interval $[-\pi/2, \pi/2]$. We will denote this restricted function by $x \mapsto \mathrm{Sin}(x)$ (capitalized to emphasize that its domain differs from that of the ordinary sine function). The graph of $y = \mathrm{Sin}(x)$ in Figure 2 shows that $x \mapsto \mathrm{Sin}(x)$ is a one-to-one function from $[-\pi/2, \pi/2]$ onto $[-1, 1]$. Therefore it is invertible. Its inverse function, $\mathrm{Sin}^{-1}:[-1, 1] \to [-\pi/2, \pi/2]$, is increasing, one-to-one, and onto. The expression $\mathrm{Sin}^{-1}(x)$ is read as 'the inverse sine of x" or "the arcsine of x" and is often denoted by $\arcsin(x)$. We can obtain the graph of $y = \arcsin(x)$ by reflecting the graph of $y = \sin(x)$ in the line $y = x$ (Figure 3).

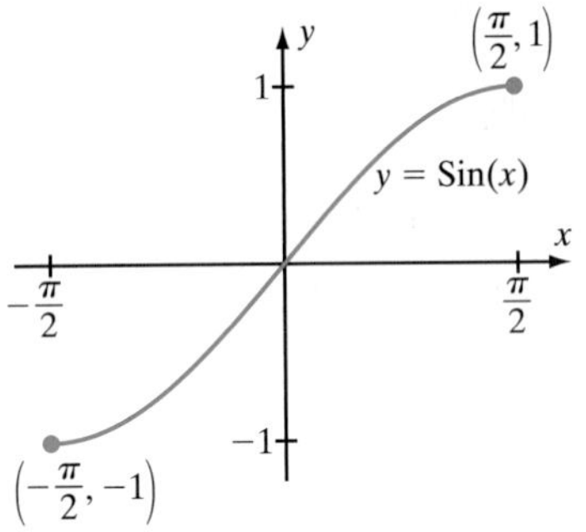

▲ Figure 2

▶ **EXAMPLE 1** Calculate $\arcsin(\sqrt{2}/2)$, $\arcsin(-\sqrt{3}/2)$, and $\arcsin(\sin(3\pi/2))$.

Solution To calculate $\arcsin(t)$, we look for the $s \in [-\pi/2, \pi/2]$ for which $t = \sin(s)$. Because $\sin(\pi/4) = \sqrt{2}/2$ and $\pi/4$ is in $[-\pi/2, \pi/2]$, we have $\arcsin(\sqrt{2}/2) = \pi/4$. Notice that, even though $\sin(x)$ takes the value $\sqrt{2}/2$ at many different values of the variable x, the function $\mathrm{Sin}(x)$ equals $\sqrt{2}/2$ only at $x = \pi/4$. See Figure 4.

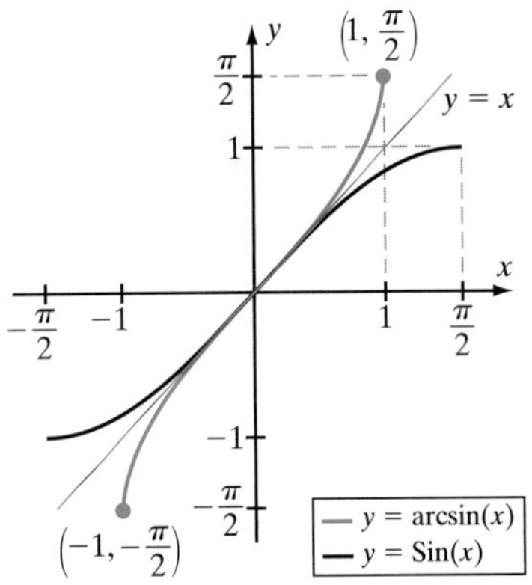

▲ Figure 3

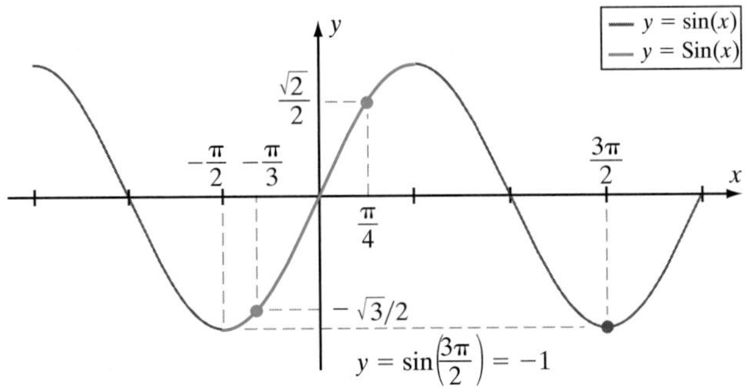

▲ Figure 4

In a similar way, we calculate $\arcsin(-\sqrt{3}/2) = -\pi/3$. The last evaluation requires particular care:

$$\arcsin\left(\sin\left(\frac{3\pi}{2}\right)\right) = \arcsin(-1) = -\frac{\pi}{2}. \quad ◀$$

INSIGHT The function $x \mapsto \arcsin(x)$ is the inverse function of $x \mapsto \mathrm{Sin}(x)$, not of $x \mapsto \sin(x)$. Otherwise, we would have $\arcsin(\sin(3\pi/2)) = 3\pi/2$ instead of the correct value $-\pi/2$ obtained in Example 1. It is always true that $\sin(\arcsin(t)) = t$, but $\arcsin(\sin(s)) = s$ is true only when s is in the interval $[-\pi/2, \pi/2]$.

The study of the inverse of cosine involves similar considerations, but we must select a different domain for our function. We define $\mathrm{Cos}(x)$ to be the cosine

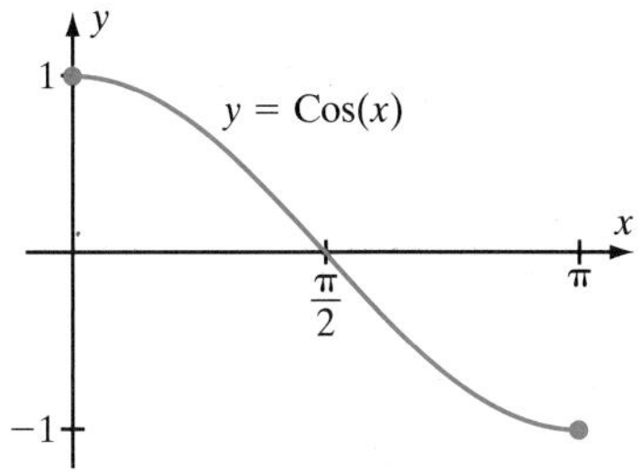

▲ Figure 5

function restricted to the interval $[0, \pi]$. Then, as Figure 5 shows, $x \mapsto \mathrm{Cos}(x)$ is a one-to-one function. It takes on all the values in the interval $[-1, 1]$. Thus $\mathrm{Cos}: [0, \pi] \to [-1, 1]$ is both one-to-one and onto, hence invertible. Its inverse function, $\mathrm{Cos}^{-1}: [-1, 1] \to [0, \pi]$, is decreasing, one-to-one, and onto. The inverse cosine $x \mapsto \mathrm{Cos}^{-1}(x)$ is also called the *arccosine* and denoted by $\arccos(x)$. We reflect the graph of $y = \mathrm{Cos}(x)$ in the line $y = x$ to obtain the graph of $y = \mathrm{Cos}^{-1}(x)$. The result is shown in Figure 6.

▶ **EXAMPLE 2** Calculate $\arccos(-\sqrt{2}/2)$, $\arccos(0)$, and $\arccos(\sqrt{3}/2)$.

Solution We calculate as in Example 1 except that here we are looking for values in the interval $[0, \pi]$. We have $\arccos(-\sqrt{2}/2) = 3\pi/4$, $\arccos(0) = \pi/2$, and $\arccos(\sqrt{3}/2) = \pi/6$. ◀

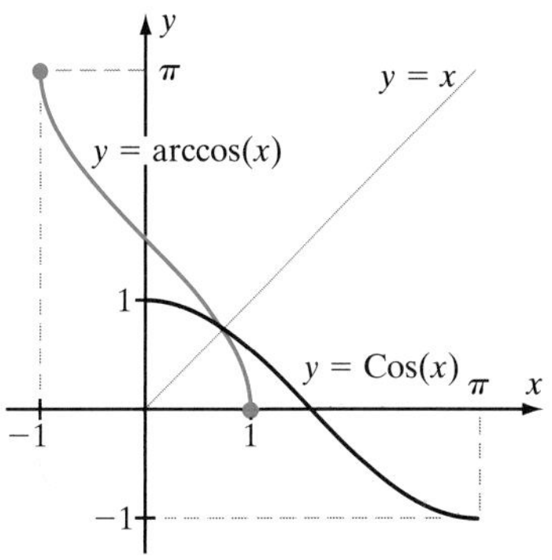

▲ Figure 6

THEOREM 1 The functions $t \mapsto \arcsin(t)$ and $t \mapsto \arccos(t)$ are differentiable on the open interval $(-1, 1)$ and

$$\frac{d}{dt}\arcsin(t) = \frac{1}{\sqrt{1-t^2}} \tag{3.10.1}$$

and

$$\frac{d}{dt}\arccos(t) = -\frac{1}{\sqrt{1-t^2}}. \tag{3.10.2}$$

Proof. For t in the open interval $(-1, 1)$, let $s = \arcsin(t)$. Then s is in the open interval $(-\pi/2, \pi/2)$ and $\cos(s)$ is positive. Using this observation together with the identity $\cos^2(s) = 1 - \sin^2(s)$, we calculate that

$$\cos(s) = \sqrt{\cos^2(s)} = \sqrt{1 - \sin^2(s)} = \sqrt{1 - \Big(\sin\big(\arcsin(t)\big)\Big)^2} = \sqrt{1-t^2}.$$

We may now obtain the formula for the derivative of $\arcsin(t)$ by using the Inverse Function Derivative Rule that we learned in Section 3.6:

$$\frac{d}{dt}\arcsin(t) = \frac{1}{\frac{d}{ds}\sin(s)\big|_{s=\arcsin(t)}} = \frac{1}{\cos(s)\big|_{s=\arcsin(t)}} = \frac{1}{\sqrt{1-t^2}}.$$

The derivative of $\arccos(t)$ is calculated in the same way. ■

From formula (3.10.1), we see that $\arcsin'(t) > 0$. In Chapter 4, we will learn that this inequality is the analytic characterization of a graphical property that we have already observed: The inverse sine is an increasing function. Similarly, formula (3.10.2) shows that $\arccos'(t) < 0$ for all $t \in (-1, 1)$. In Chapter 4, we will learn that this inequality implies that the inverse cosine is a decreasing function, which we have already observed from its graph.

▶ **EXAMPLE 3** Calculate

$$\frac{d}{dx}\arcsin(x)\Big|_{x=0} \quad \text{and} \quad \frac{d}{dx}\arccos(x)\Big|_{x=1/2}.$$

Solution From (3.10.1) and (3.10.2), we obtain

$$\frac{d}{dx}\arcsin(x)\Big|_{x=0} = \frac{1}{\sqrt{1-x^2}}\Big|_{x=0} = 1$$

and

$$\frac{d}{dx}\arccos(x)|_{x=1/2} = -\frac{1}{\sqrt{1-x^2}}\Big|_{x=1/2} = -\frac{1}{\sqrt{1-(1/2)^2}} = -\frac{2}{\sqrt{3}}. \blacktriangleleft$$

In Chain Rule form, the derivative formulas of Theorem 1 become

$$\frac{d}{dx}\arcsin(u) = \frac{1}{\sqrt{1-u^2}}\cdot\frac{du}{dx} \qquad (3.10.3)$$

and

$$\frac{d}{dx}\arccos(u) = -\frac{1}{\sqrt{1-u^2}}\cdot\frac{du}{dx}. \qquad (3.10.4)$$

▶ **EXAMPLE 4** Calculate the following derivatives:

$$\frac{d}{dx}\arcsin\left(\frac{1}{x}\right)\Big|_{x=-2/\sqrt{3}} \quad \text{and} \quad \frac{d}{dx}\arccos(\sqrt{x})\Big|_{x=1/4}.$$

Solution Using (3.10.3) with $u = 1/x$ and $du/dx = -1/x^2$, we obtain

$$\frac{d}{dx}\arcsin\left(\frac{1}{x}\right)\Big|_{x=-2/\sqrt{3}} = \frac{1}{\sqrt{1-(1/x)^2}}\cdot\left(-\frac{1}{x^2}\right)\Big|_{x=-2/\sqrt{3}}$$
$$= \frac{1}{\sqrt{1-3/4}}\cdot\left(-\frac{1}{4/3}\right) = -\frac{1}{1/2}\cdot\frac{3}{4} = -\frac{3}{2}.$$

Using (3.10.4) with $u = \sqrt{x}$ and $du/dx = (1/2)x^{-1/2}$, we have

$$\frac{d}{dx}\arccos(\sqrt{x})\Big|_{x=1/4} = -\frac{1}{\sqrt{1-(\sqrt{x})^2}}\cdot\left(\frac{1}{2}x^{-1/2}\right)\Big|_{x=1/4}$$
$$= -\frac{1}{\sqrt{1-(1/2)^2}}\cdot\left(\frac{1}{2\sqrt{1/4}}\right) = -\frac{2}{\sqrt{3}}. \blacktriangleleft$$

▶ **EXAMPLE 5** Let a be a positive constant. Show that

$$\frac{d}{dx}\arcsin\left(\frac{x}{a}\right) = \frac{1}{\sqrt{a^2-x^2}}. \qquad (3.10.5)$$

Solution Setting $u = x/a$ in formula (3.10.3), we calculate

$$\frac{d}{dx}\arcsin\left(\frac{x}{a}\right) = \frac{d}{dx}\arcsin(u) = \frac{1}{\sqrt{1-u^2}}\cdot\frac{du}{dx} = \frac{1}{\sqrt{1-(x/a)^2}}\cdot\frac{1}{a} = \frac{1}{\sqrt{a^2-x^2}}. \blacktriangleleft$$

The Inverse Tangent Function

Define the function $x \mapsto \mathrm{Tan}(x)$ to be the restriction of $x \mapsto \tan(x)$ to the interval $(-\pi/2, \pi/2)$. See Figure 7 for the graph of $y = \mathrm{Tan}(x)$. Because the function Tan is a one-to-one, onto function with domain $(-\pi/2, \pi/2)$ and range $(-\infty,\infty)$, the inverse function Tan^{-1} exists. We read $\mathrm{Tan}^{-1}(x)$ as "the inverse tangent of x" or "the arctangent of x." The notation $\arctan(x)$ is often used instead of $\mathrm{Tan}^{-1}(x)$. The domain of $\mathrm{Tan}^{-1}(x)$ is $(-\infty, \infty)$, and its image is $(-\pi/2, \pi/2)$. The graph of the arctangent is shown in Figure 8.

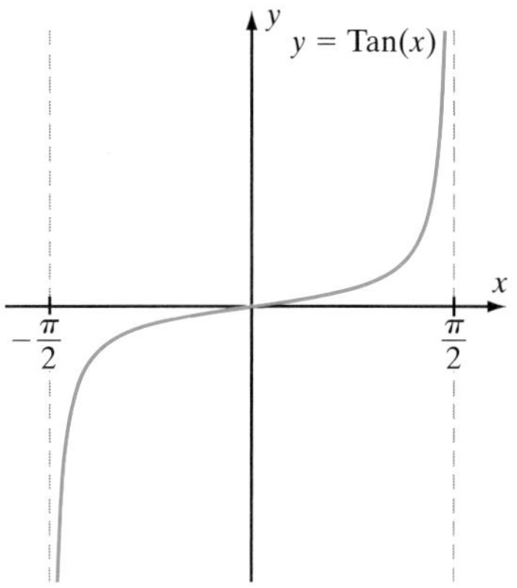

▲ Figure 7

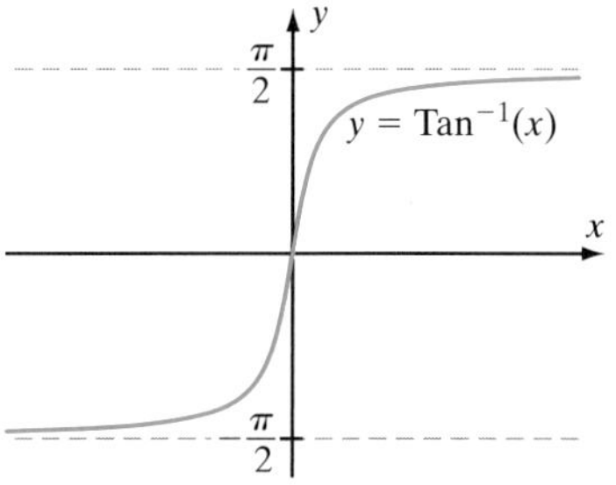

▲ Figure 8

▶ **EXAMPLE 6** Calculate $\arctan(1)$, $\arctan(\sqrt{3})$, and $\arctan(-1/\sqrt{3})$.

Solution We calculate $\arctan(t)$ by finding the unique s between $-\pi/2$ and $\pi/2$ for which $\tan(s) = t$. Thus $\arctan(1) = \pi/4$, $\arctan(\sqrt{3}) = \pi/3$, and $\arctan(-1/\sqrt{3}) = -\pi/6$. ◀

THEOREM 2 The function $t \mapsto \arctan(t)$ is differentiable for every t and

$$\frac{d}{dt}\arctan(t) = \frac{1}{1+t^2}. \tag{3.10.6}$$

Proof. As usual, we calculate the derivative of a new inverse function by using the method of Section 3.6:

$$\begin{aligned}\frac{d}{dt}\arctan(t) &= \frac{1}{\frac{d}{ds}\mathrm{Tan}(s)\big|_{s=\arctan(t)}} = \frac{1}{\sec^2(s)\big|_{s=\arctan(t)}} = \frac{1}{1+\tan^2(s)\big|_{s=\arctan(t)}} \\ &= \frac{1}{1+(\tan(s))^2\big|_{s=\arctan(t)}} = \frac{1}{1+(\tan(\arctan(t)))^2} = \frac{1}{1+t^2}.\end{aligned}$$

■

The Chain Rule form of the inverse tangent derivative formula is

$$\frac{d}{dx}\arctan(u) = \frac{1}{1+u^2}\cdot\frac{du}{dx}. \tag{3.10.7}$$

▶ **EXAMPLE 7** Let a be a positive constant. Show that

$$\frac{d}{dx}\left(\frac{1}{a}\arctan\left(\frac{x}{a}\right)\right) = \frac{1}{a^2+x^2}. \tag{3.10.8}$$

Solution Using equation (3.10.7) with $u = x/a$, we calculate

$$\begin{aligned}\frac{d}{dx}\left(\frac{1}{a}\arctan\left(\frac{x}{a}\right)\right) &= \frac{1}{a}\cdot\frac{d}{dx}\arctan(u) = \frac{1}{a}\cdot\frac{1}{1+u^2}\cdot\frac{du}{dx} \\ &= \frac{1}{a}\cdot\frac{1}{1+(x/a)^2}\cdot\frac{1}{a} = \frac{1}{a^2+x^2}.\end{aligned}$$

◀

Other Inverse Trigonometric Functions

We will say just a few words about the inverse cotangent, the inverse secant, and the inverse cosecant (because they do not arise frequently). We use $x \mapsto \mathrm{Cot}(x)$ to denote the restriction of the cotangent function to the interval $(0, \pi)$. From the graph of $y = \mathrm{Cot}(x)$ in Figure 9, we see that $x \mapsto \mathrm{Cot}(x)$ is a decreasing function that takes on all real values. Consequently, the inverse function $\mathrm{Cot}^{-1} : (-\infty, \infty) \to (0, \pi)$ exists. Refer to Figure 10 for its graph. We sometimes write $\mathrm{Cot}^{-1}(x)$ as $\operatorname{arccot}(x)$. A computation similar to that of Theorem 2 yields the derivative of the inverse cotangent:

$$\frac{d}{dt}\operatorname{arccot}(t) = -\frac{1}{1+t^2}. \tag{3.10.9}$$

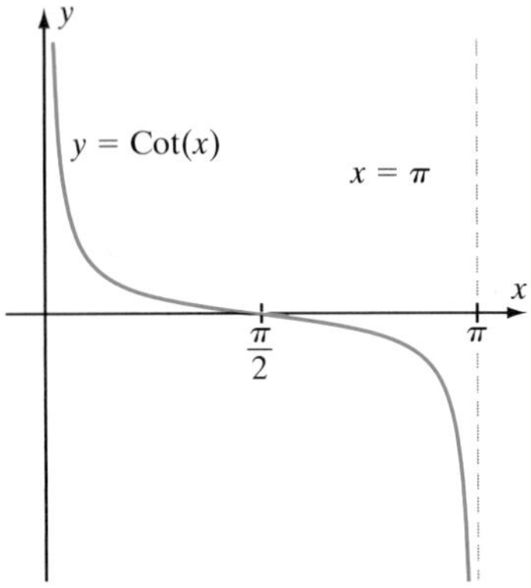

▲ Figure 9

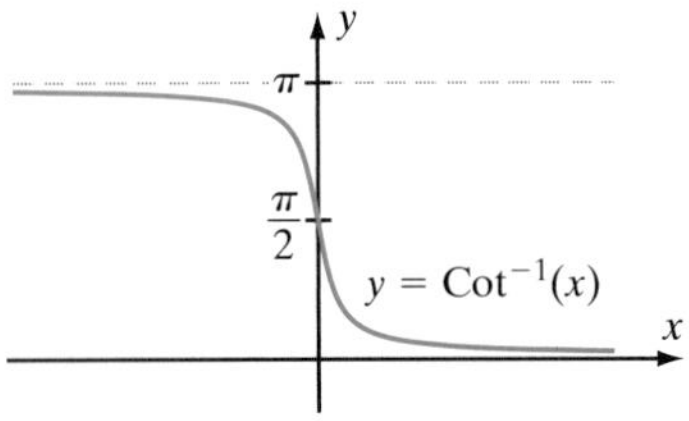

▲ Figure 10

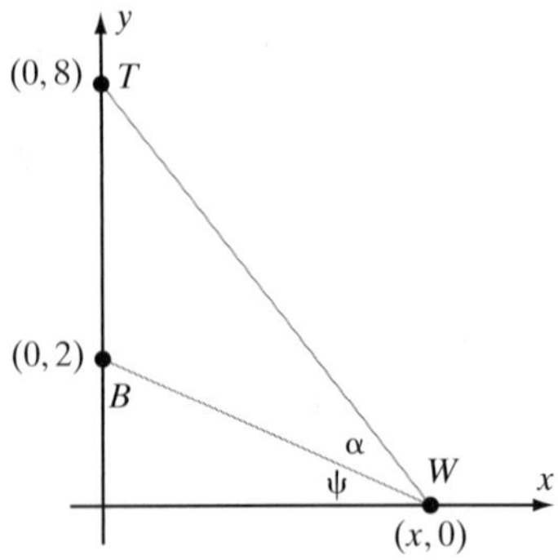

▲ Figure 11

▶ **EXAMPLE 8** Figure 11 shows two fixed points $T = (0, 8)$ and $B = (0, 2)$ on the y-axis and a variable point $W = (x, 0)$ on the x-axis. The radian measure α of $\angle TWB$ is a function of x. Calculate $d\alpha/dx$.

Solution From Figure 11, we see that $\cot(\alpha + \psi) = x/8$, and $\cot(\psi) = x/2$. It follows that

$$\alpha = (\alpha + \psi) - \psi = \operatorname{arccot}\left(\frac{x}{8}\right) - \operatorname{arccot}\left(\frac{x}{2}\right).$$

Both inverse cotangents in this equation have the form $\operatorname{arccot}(x/a)$. Letting $u = x/a$, we use equation (3.10.9) and the Chain Rule to calculate

$$\begin{aligned} \frac{d}{dx}\operatorname{arccot}\left(\frac{x}{a}\right) &= \frac{d}{dx}\operatorname{arccot}(u) = \frac{d}{du}\operatorname{arccot}(u)\cdot\frac{du}{dx} = -\frac{1}{1+u^2}\cdot\frac{1}{a} \\ &= -\frac{1}{1+(x/a)^2}\cdot\frac{1}{a} = -\frac{a}{a^2+x^2}. \end{aligned} \tag{3.10.10}$$

We may replace a with 2 and 8 in (3.10.10). If we subtract the resulting formulas, then we obtain

$$\frac{d\alpha}{dx} = \frac{d}{dx}\left(\operatorname{arccot}\left(\frac{x}{8}\right) - \operatorname{arccot}\left(\frac{x}{2}\right)\right) = \frac{2}{2^2+x^2} - \frac{8}{8^2+x^2}.$$

We complete the calculation by expressing the right side of this last equation in terms of a common denominator. We have

$$\begin{aligned} \frac{d\alpha}{dx} &= \frac{2(64+x^2) - 8(4+x^2)}{(4+x^2)(64+x^2)} = \frac{96-6x^2}{(4+x^2)(64+x^2)} \\ &= -6\cdot\frac{(x^2-16)}{(4+x^2)(64+x^2)}. \end{aligned} \tag{3.10.11}$$ ◀

For the inverse secant, we define $x \mapsto \mathrm{Sec}(x)$ to be the restriction of the secant to the set $[0, \pi/2) \cup (\pi/2, \pi]$. From Figure 12, we see that $x \mapsto \mathrm{Sec}(x)$ is one-to-one with image consisting of all numbers greater than or equal to 1 and all numbers less than or equal to -1. The inverse function

$$\mathrm{Sec}^{-1} : (-\infty, -1] \cup [1, \infty) \to [0, \pi/2) \cup (\pi/2, \pi]$$

is plotted in Figure 13. We sometimes write $\operatorname{arcsec}(x)$ for $\mathrm{Sec}^{-1}(x)$ and read "arcsecant" instead of "inverse secant ." The derivative of the arcsecant is given by

$$\frac{d}{dt}\operatorname{arcsec}(t) = \frac{1}{|t| \cdot \sqrt{t^2 - 1}}\ , \ |t| > 1. \tag{3.10.12}$$

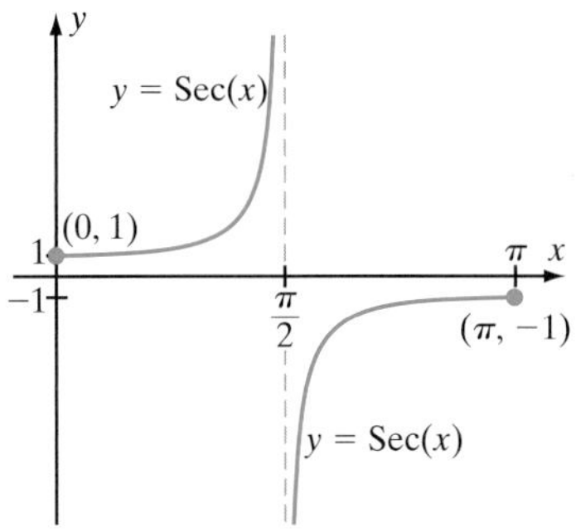

▲ Figure 12

▶ **EXAMPLE 9** Derive formula (3.10.12).

Solution The derivation is quite similar to the ones we have already carried out, but the absolute value in formula (3.10.12) alerts us to the need for paying careful attention to the signs of the terms that arise. Applying the Inverse Function Derivative Rule, we have

$$\begin{aligned}\frac{d}{dt}\operatorname{arcsec}(t) &= \frac{1}{\frac{d}{ds}\mathrm{Sec}(s)\Big|_{s=\operatorname{arcsec}(t)}} = \frac{1}{\sec(s)\tan(s)|_{s=\operatorname{arcsec}(t)}} \\ &= \frac{1}{\sec\left(\operatorname{arcsec}(t)\right)\tan\left(\operatorname{arcsec}(t)\right)} = \frac{1}{t \cdot \tan\left(\operatorname{arcsec}(t)\right)}.\end{aligned} \tag{3.10.13}$$

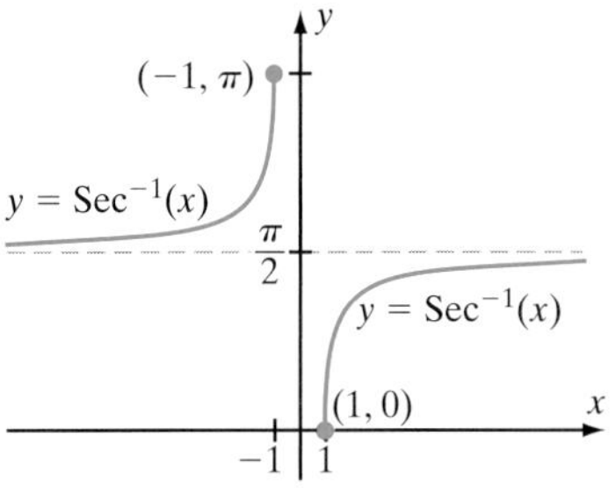

▲ Figure 13

Next, trigonometric identity (1.6.2) with $\theta = \operatorname{arcsec}(t)$ gives us $\tan^2(\operatorname{arcsec}(t)) = \sec^2(\operatorname{arcsec}(t)) - 1 = t^2 - 1$, or $\tan(\operatorname{arcsec}(t)) = \pm\sqrt{t^2 - 1}$. If $t > 1$, then $0 < \operatorname{arcsec}(t) < \pi/2$ and $\tan(\operatorname{arcsec}(t)) > 0$. Therefore $\tan\left(\operatorname{arcsec}(t)\right) = +\sqrt{t^2 - 1}$ for $t > 1$. In this case, $t = |t|$ and (3.10.13) becomes

$$\frac{d}{dt}\operatorname{arcsec}(t) = \frac{1}{t \cdot \tan\left(\operatorname{arcsec}(t)\right)} = \frac{1}{|t| \cdot (+\sqrt{t^2 - 1})} = \frac{1}{|t| \cdot \sqrt{t^2 - 1}}.$$

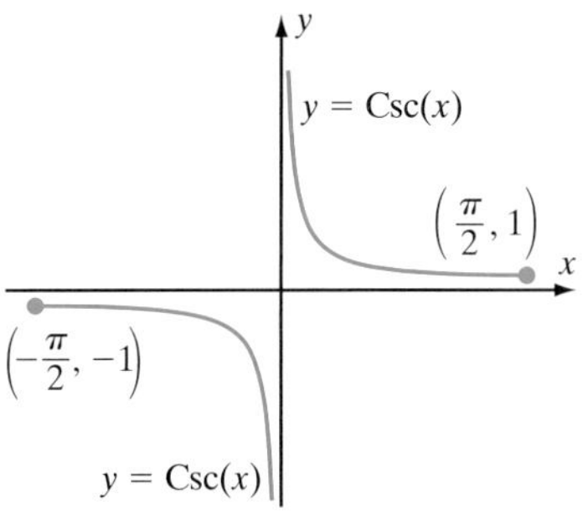

▲ Figure 14

If $t < -1$, then $\pi/2 < \operatorname{arcsec}(t) < \pi$ and $\tan(\operatorname{arcsec}(t)) < 0$. Therefore $\tan(\operatorname{arcsec}(t)) = -\sqrt{t^2 - 1}$ for $t < -1$. In this case, $t = -|t|$ and (3.10.13) becomes

$$\begin{aligned}\frac{d}{dt}\operatorname{arcsec}(t) &= \frac{1}{t \cdot \tan\left(\operatorname{arcsec}(t)\right)} = \frac{1}{t \cdot (-\sqrt{t^2 - 1})} \\ &= \frac{1}{(-t) \cdot \sqrt{t^2 - 1}} = \frac{1}{|t| \cdot \sqrt{t^2 - 1}}.\end{aligned}$$ ◀

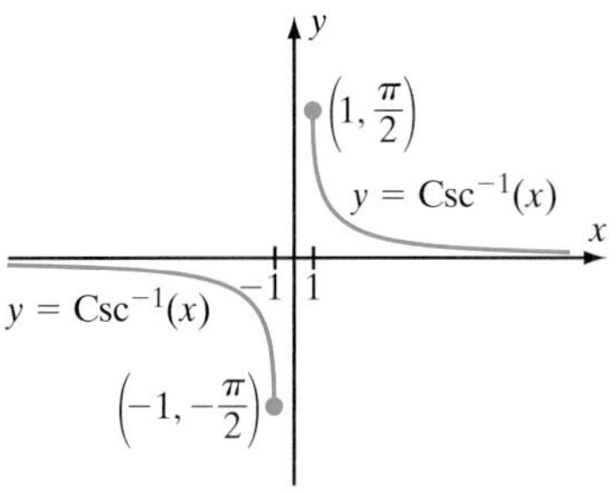

▲ Figure 15

We obtain the final inverse trigonometric function by defining $x \mapsto \mathrm{Csc}(x)$ to be the restriction of the cosecant to the set $[-\pi/2, 0) \cup (0, \pi/2]$. The graph is exhibited in Figure 14. We see that $x \mapsto \mathrm{Csc}(x)$ is one-to-one and therefore has an inverse function Csc^{-1} that maps the image of Csc to the domain of Csc. As x ranges over the values in the domain of Csc, $\sin(x)$ takes on all values in the interval $[-1, 1]$ except for 0. Therefore the cosecant takes on all values greater than or equal to 1 and all values less than or equal to -1. Thus Csc^{-1} has domain $(-\infty, -1] \cup [1, \infty)$ and range $[-\pi/2, 0) \cup (0, \pi/2]$. Its graph is shown in Figure 15. The derivative of the inverse cosecant is given by

$$\frac{d}{dx}\mathrm{Csc}^{-1}(x) = -\frac{1}{|x|\cdot\sqrt{x^2-1}}, \qquad |x| > 1. \tag{3.10.14}$$

Notation

In this section we have carefully distinguished between the trigonometric functions and their restrictions to domains on which they are invertible. This has led us to use the notations Sin^{-1}, Cos^{-1}, Tan^{-1}, and Sec^{-1}. It is standard practice among mathematicians and scientists to simply write $\sin^{-1}$, $\cos^{-1}$, $\tan^{-1}$, and $\sec^{-1}$. When you see an expression such as $\tan^{-1}(x)$, remember that it refers to the inverse of a restriction of the tangent. It is especially important not to confuse an inverse trigonometric function (such as $\mathrm{Tan}^{-1}(x)$) with the reciprocal of the corresponding trigonometric function (such as $(\mathrm{Tan}(x))^{-1} = 1/\mathrm{Tan}(x)$). The "arc" notation eliminates this danger.

The Hyperbolic Functions

In engineering, physics, and other applications, certain special functions called the hyperbolic functions arise frequently. Here we shall define and briefly discuss them.

DEFINITION The *hyperbolic cosine* of t is defined to be

$$\cosh(t) = \frac{e^t + e^{-t}}{2}$$

for all real numbers t. The *hyperbolic sine* of t is defined to be

$$\sinh(t) = \frac{e^t - e^{-t}}{2}.$$

The graphs of the hyperbolic sine and cosine are given in Figures 16 and 17. Notice that the hyperbolic cosine, like the ordinary cosine, is an even function. That is, $\cosh(-t) = \cosh(t)$. The hyperbolic sine, like the ordinary sine, is an odd function: $\sinh(-t) = -\sinh(t)$.

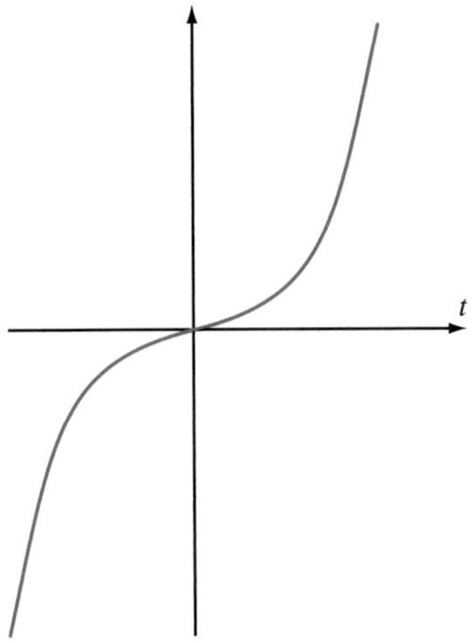

▲ Figure 16 $y = \sinh(t)$

INSIGHT Why is the term "hyperbolic" used? Recall that, for classical trigonometry based on the unit circle, the numbers

$$x = \cos(t) \quad \text{and} \quad y = \sin(t)$$

satisfy $x^2+y^2 = 1$. That is, the point $(\cos(t), \sin(t))$ lies on the unit circle. By contrast, the numbers

$$x = \cosh(t) \quad \text{and} \quad y = \sinh(t)$$

satisfy

$$x^2 - y^2 = \left(\frac{e^t+e^{-t}}{2}\right)^2 - \left(\frac{e^t-e^{-t}}{2}\right)^2 = \frac{e^{2t}+2+e^{-2t}}{4} - \frac{e^{2t}-2+e^{-2t}}{4} = \frac{4}{4} = 1.$$

The hyperbolic cosine and hyperbolic sine therefore satisfy the identity

$$\cosh^2(t) - \sinh^2(t) = 1. \tag{3.10.15}$$

In other words, the point (cosh (t), sinh (t)) lies on the "unit hyperbola" whose equation is $x^2 - y^2 = 1$.

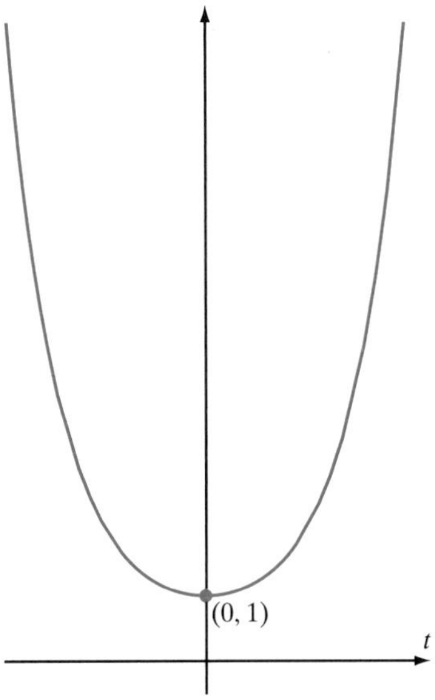

▲ Figure 17 $y = \cosh(t)$

▶ **EXAMPLE 10** Show that

$$\cosh(2t) = \cosh^2(t) + \sinh^2(t) \quad \text{and} \quad \sinh(2t) = 2\sinh(t)\cosh(t).$$

Solution The basic method of establishing identities among the hyperbolic trigonometric functions is to convert them to exponential functions. Thus

$$\begin{aligned}\cosh^2(t) + \sinh^2(t) &= \left(\frac{e^t + e^{-t}}{2}\right)^2 + \left(\frac{e^t - e^{-t}}{2}\right)^2 \\ &= \left(\frac{1}{4}e^{2t} + \frac{1}{2} + \frac{1}{4e^{2t}}\right) + \left(\frac{1}{4}e^{2t} - \frac{1}{2} + \frac{1}{4e^{2t}}\right).\end{aligned}$$

On simplification, we obtain

$$\cosh^2(t) + \sinh^2(t) = \frac{1}{2}e^{2t} + \frac{1}{2}e^{-2t} = \cosh(2t).$$

Similarly

$$2\sinh(t)\cosh(t) = 2\left(\frac{e^t - e^{-t}}{2}\right)\left(\frac{e^t + e^{-t}}{2}\right) = \frac{1}{2}(e^{2t} - e^{-2t}) = \sinh(2t). \quad ◀$$

INSIGHT There are many analogies between the standard trigonometric functions and the hyperbolic trigonometric functions. Compare the identities in Example 10 with the circular trigonometric identities

$$\cos(2t) = \cos^2(t) - \sin^2(t) \quad \text{and} \quad \sin(2t) = 2\sin(t)\cos(t).$$

As a rule of thumb, there is a hyperbolic trigonometric formula for every circular trigonometric formula. But you must exercise some caution because there are some differences in the occurrence of minus signs.

Just as the tangent, the cotangent, the secant, and the cosecant can be defined as quotients involving the sine and cosine, so can the hyperbolic tangent (tanh), hyperbolic cotangent (coth), hyperbolic secant (sech), and hyperbolic cosecant (csch):

$$\begin{aligned}\tanh(t) &= \frac{\sinh(t)}{\cosh(t)}, && \text{all real } t;\\ \coth(t) &= \frac{\cosh(t)}{\sinh(t)}, && t \neq 0;\\ \operatorname{sech}(\mathrm{t}) &= \frac{1}{\cosh(t)}, && \text{all real } t;\\ \operatorname{csch}(\mathrm{t}) &= \frac{1}{\sinh(t)}, && t \neq 0.\end{aligned}$$

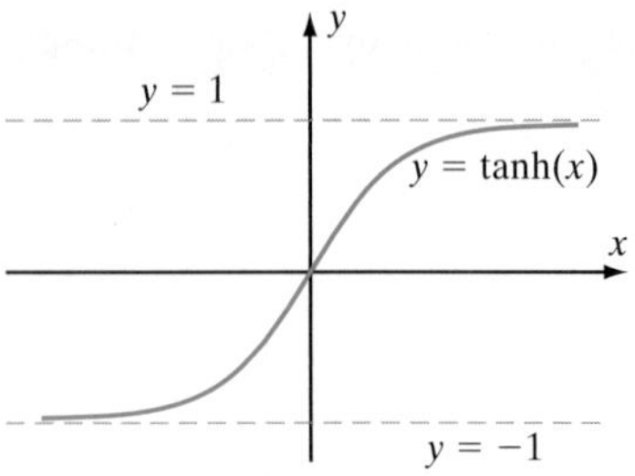

▲ Figure 18

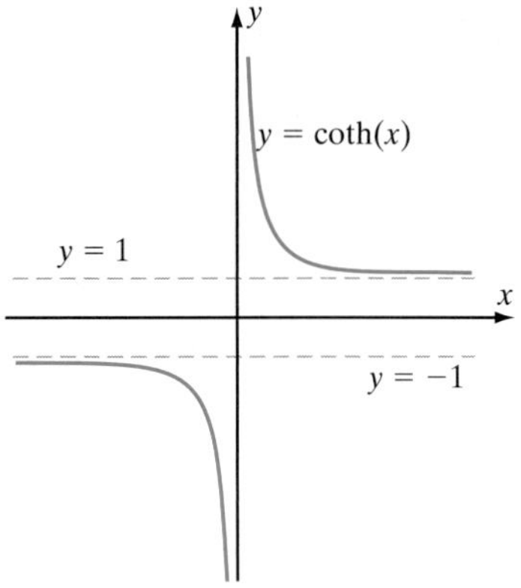

▲ Figure 19

The graphs of these hyperbolic functions are given in Figures 18, 19, 20, and 21. Notice that the graphs of $y = \tanh(x)$ and $y = \coth(x)$ have the lines $y = 1$ and $y = -1$ as horizontal asymptotes:

$$\lim_{x\to\infty} \tanh(x) = \lim_{x\to\infty} \coth(x) = 1 \qquad \text{and} \qquad \lim_{x\to-\infty} \tanh(x) = \lim_{x\to-\infty} \coth(x) = -1.$$

The graphs of sech and csch have the line $y = 0$ as a horizontal asymptote:

$$\lim_{x\to\infty} \operatorname{sech}(x) = \lim_{x\to-\infty} \operatorname{sech}(x) = 0 \qquad \text{and} \qquad \lim_{x\to\infty} \operatorname{csch}(x) = \lim_{x\to-\infty} \operatorname{csch}(x) = 0.$$

The graphs of coth and csch have the line $x = 0$ as a vertical asymptote:

$$\lim_{x\to0^+} \operatorname{csch}(x) = \lim_{x\to0^+} \coth(x) = \infty \qquad \text{and} \qquad \lim_{x\to0^-} \operatorname{csch}(x) = \lim_{x\to0^-} \coth = -\infty.$$

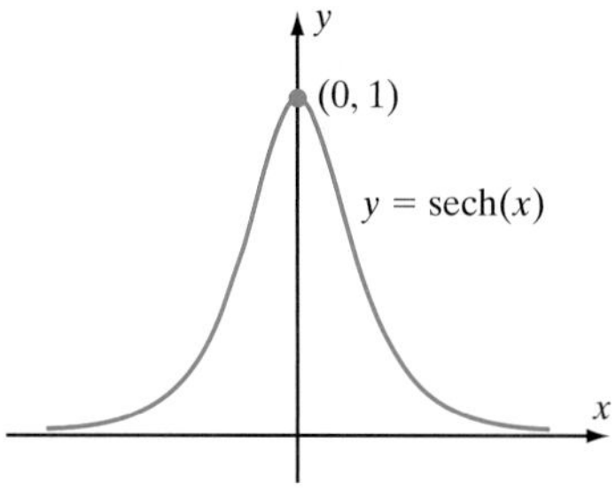

▲ Figure 20

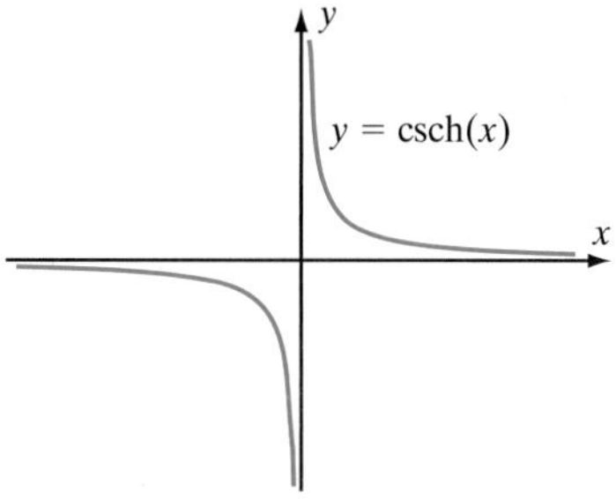

▲ Figure 21

Derivatives of the Hyperbolic Functions

It is easy to calculate that

$$\frac{d}{dt}\cosh(t) = \frac{d}{dt}\left(\frac{e^t + e^{-t}}{2}\right) = \frac{e^t - e^{-t}}{2} = \sinh(t).$$

Similarly,

$$\frac{d}{dt}\sinh(t) = \frac{d}{dt}\left(\frac{e^t - e^{-t}}{2}\right) = \frac{e^t + e^{-t}}{2} = \cosh(t).$$

The calculation of the derivatives of the remaining four hyperbolic functions follow from these formulas and the Quotient Rule. For example,

$$\begin{aligned}\frac{d}{dt}\tanh &= \frac{d}{dt}\left(\frac{\sinh(t)}{\cosh(t)}\right)\\ &= \frac{\cosh(t)\sinh'(t) - \sinh(t)\cosh'(t)}{\cosh^2(t)}\\ &= \frac{\cosh^2(t) - \sinh^2(t)}{\cosh^2(t)}\\ &= \frac{1}{\cosh^2(t)}\\ &= \operatorname{sech}^2(t).\end{aligned}$$

The six differentiation formulas for the hyperbolic trigonometric functions are listed in the following table.

$$\frac{d}{dt}\sinh(t) = \cosh(t) \qquad \frac{d}{dt}\coth(t) = -\operatorname{csch}^2(t)$$

$$\frac{d}{dt}\cosh(t) = \sinh(t) \qquad \frac{d}{dt}\operatorname{sech}(t) = -\operatorname{sech}(t)\cdot\tanh(t)$$

$$\frac{d}{dt}\tanh(t) = \operatorname{sech}^2(t) \qquad \frac{d}{dt}\operatorname{csch}(t) = -\operatorname{csch}(t)\cdot\coth(t)$$

The Inverse Hyperbolic Functions

Because the hyperbolic sine is increasing on the entire real line, it is one-to-one. We may therefore discuss its inverse, $\sinh^{-1}(x)$ (or $\operatorname{arcsinh}(x)$). The other hyperbolic functions are either not one-to one, or not onto $\mathbb{R}$. But if their domains and ranges are suitably restricted, then their inverses may be defined. These inverses may be calculated explicitly in terms of functions that we already understand.

▶ **EXAMPLE 11** Show that

$$\sinh^{-1}(x) = \ln\left(x + \sqrt{x^2+1}\right).$$

Solution Let $y = \sinh^{-1}(x)$. Then $\sinh(y) = x$, or $e^y - e^{-y} = 2x$. If we let $u = e^y$, then our last equation becomes $u - u^{-1} = 2x$, or $u^2 - 2x\cdot u - 1 = 0$. We may solve this quadratic equation in u by using the Quadratic Formula:

$$u = \frac{2x \pm \sqrt{(-2x)^2 - 4(1)(-1)}}{2} = x \pm \sqrt{x^2+1}.$$

Because $u = e^y$ cannot be negative, the positive sign is used. Thus $e^y = u = x + \sqrt{x^2+1}$. From this we conclude that $y = \ln(x + \sqrt{x^2+1})$. ◀

For reference, the following list provides the formulas for the inverses of the hyperbolic functions.

$$\sinh^{-1}(x) = \ln(x + \sqrt{x^2+1}), \quad \text{all real } x;$$

$$\cosh^{-1}(x) = \ln(x + \sqrt{x^2-1}), \quad x \geq 1;$$

$$\tanh^{-1}(x) = \frac{1}{2}\ln\left(\frac{1+x}{1-x}\right), \quad -1 < x < 1;$$

$$\coth^{-1}(x) = \frac{1}{2}\ln\left(\frac{x+1}{x-1}\right), \quad |x| > 1;$$

$$\operatorname{sech}^{-1}(x) = \ln\left(\frac{1+\sqrt{1-x^2}}{x}\right), \quad 0 < x \leq 1;$$

$$\operatorname{csch}^{-1}(x) = \ln\left(\frac{1}{x} + \frac{\sqrt{1+x^2}}{|x|}\right), \quad x \neq 0.$$

The following list shows the basic derivative rules for the inverse hyperbolic functions,

$$\frac{d}{dx}\sinh^{-1}(x) = \frac{1}{\sqrt{1+x^2}} \qquad \text{all real } x;$$

$$\frac{d}{dx}\cosh^{-1}(x) = -\frac{1}{\sqrt{x^2-1}} \qquad |x|>1;$$

$$\frac{d}{dx}\tanh^{-1}(x) = \frac{1}{1-x^2} \qquad -1<x<1;$$

$$\frac{d}{dx}\operatorname{sech}^{-1}(x) = -\frac{1}{x\sqrt{1-x^2}} \qquad 0<x<1$$

The next example will derive these equations in a somewhat more general form.

▶ **EXAMPLE 12** Let a be a positive constant. Show that

$$\frac{d}{dx}\sinh^{-1}\left(\frac{x}{a}\right) = \frac{1}{\sqrt{a^2+x^2}}, \tag{3.10.16}$$

$$\frac{d}{dx}\left(\frac{1}{a}\tanh^{-1}\left(\frac{x}{a}\right)\right) = \frac{1}{a^2-x^2}, \tag{3.10.17}$$

and

$$\frac{d}{dx}\left(\frac{1}{a}\operatorname{sech}^{-1}\left(\frac{x}{a}\right)\right) = -\frac{1}{x\sqrt{a^2-x^2}}. \tag{3.10.18}$$

Solution We can use the explicit formulas for the inverse hyperbolic functions. For example, because $\sinh^{-1}(x) = \ln(x+\sqrt{x^2+1})$, we have

$$\sinh^{-1}\left(\frac{x}{a}\right) = \ln\left(\frac{x}{a} + \sqrt{\frac{x^2}{a^2}+1}\right) = \ln\left(\frac{x}{a} + \frac{\sqrt{x^2+a^2}}{a}\right)$$

$$= \ln\left(\frac{x+\sqrt{x^2+a^2}}{a}\right) = \ln\left(x+\sqrt{x^2+a^2}\right) - \ln(a).$$

Therefore for $u = x+\sqrt{x^2+a^2}$, we have

$$\frac{d}{dx}\sinh^{-1}\left(\frac{x}{a}\right) = \frac{d}{dx}\left(\ln(x+\sqrt{x^2+a^2}) - \ln(a)\right) = \frac{d}{dx}\ln(u) - \frac{d}{dx}\ln(a) = \frac{d}{dx}\ln(u).$$

We apply the Chain Rule to obtain

$$\frac{d}{dx}\sinh^{-1}\left(\frac{x}{a}\right) = \frac{d}{du}\ln(u)\cdot\frac{du}{dx} = \frac{1}{u}\cdot\frac{du}{dx} = \frac{1}{x+\sqrt{x^2+a^2}}\cdot\frac{d}{dx}\left(x+\sqrt{x^2+a^2}\right)$$

$$= \frac{1}{x+\sqrt{x^2+a^2}}\cdot\left(1+\frac{x}{\sqrt{x^2+a^2}}\right).$$

The rest of the calculation consists of algebraic simplification:

$$\frac{d}{dx}\sinh^{-1}(x) = \frac{1}{x+\sqrt{x^2+a^2}}\cdot\left(1+\frac{x}{\sqrt{x^2+a^2}}\right)$$

$$= \frac{1}{x+\sqrt{x^2+a^2}}\cdot\left(\frac{x+\sqrt{x^2+a^2}}{\sqrt{x^2+a^2}}\right) = \frac{1}{\sqrt{x^2+a^2}}.$$

Similarly, from

$$\tanh^{-1}\left(\frac{x}{a}\right) = \tanh^{-1}(v)\big|_{v=x/a} = \frac{1}{2}\ln\left(\frac{1+v}{1-v}\right)\bigg|_{v=x/a}$$

$$= \frac{1}{2}\ln\left(\frac{1+x/a}{1-x/a}\right) = \frac{1}{2}\ln\left(\frac{a+x}{a-x}\right) = \frac{1}{2}\ln(a+x) - \frac{1}{2}\ln(a-x),$$

we obtain

$$\frac{d}{dx}\tanh^{-1}(x) = \frac{1}{2}\frac{d}{dx}\left(\ln(a+x) - \frac{1}{2}\ln(a-x)\right) = \frac{1}{2}\left(\frac{1}{a+x} + \frac{1}{a-x}\right) = \frac{a}{a^2-x^2}$$

and (3.10.17) follows. These derivatives can also be calculated using the Inverse Function Derivative Rule. Let us use that rule for the remaining calculation. We will need the identity

$$1 - \tanh^2(t) = \operatorname{sech}^2(t), \tag{3.10.19}$$

which is obtained by dividing each term of identity (3.10.15) by $\cosh^2(t)$. Let x be positive (so that it is in the domain of the hyperbolic secant), and set $y = \operatorname{sech}^{-1}(x)$. Notice that y is positive and therefore $\tanh(y)$ is also positive. From this observation and identity (3.10.19), it follows that $\tanh(y) = \sqrt{1-\operatorname{sech}^2(y)}$. Because $y = \operatorname{sech}^{-1}(x)$, we have $x = \operatorname{sech}(y)$ and

$$\frac{d}{dx}\operatorname{sech}^{-1}(x) = \frac{1}{\frac{d}{dy}\operatorname{sech}(y)} = -\frac{1}{\operatorname{sech}(y)\tanh(y)}$$

$$= -\frac{1}{\operatorname{sech}(y)\sqrt{1-\operatorname{sech}^2(y)}} = -\frac{1}{x\sqrt{1-x^2}}.$$

Finally, applying the Chain Rule form of this derivative rule with $u = x/a$, we have

$$\frac{d}{dx}\operatorname{sech}^{-1}\left(\frac{x}{a}\right) = \frac{d}{dx}\operatorname{sech}^{-1}(u)\cdot\frac{du}{dx} = -\frac{1}{u\sqrt{1-u^2}}\cdot\frac{1}{a}$$

$$= -\frac{1}{(x/a)\sqrt{1-(x/a)^2}}\cdot\frac{1}{a} = -\frac{a}{x\sqrt{a^2-x^2}}$$

and formula (3.10.18) results.

A LOOK FORWARD ▶ It is certainly possible to do without the hyperbolic trigonometric functions because they are merely arithmetic combinations of functions that we already know and understand. However, knowledge of these new functions enables us to recognize certain patterns and expressions that arise frequently in applications. They also simplify many formulas, as we will see in Chapters 4 and 6.

Summary of Derivatives of Basic Functions

$$\frac{d}{dx}(x^p) = px^{p-1} \qquad -\infty < p < \infty$$

$$\frac{d}{dx}\sin(x) = \cos(x)$$

$$\frac{d}{dx}\cos(x) = -\sin(x)$$

$$\frac{d}{dx}\tan(x) = \sec^2(x)$$

$$\frac{d}{dx}\cot(x) = -\csc^2(x)$$

$$\frac{d}{dx}\sec(x) = \sec(x)\tan(x)$$

$$\frac{d}{dx}\csc(x) = -\csc(x)\cot(x)$$

$$\frac{d}{dx}e^x = e^x$$

$$\frac{d}{dx}a^x = a^x \cdot \ln(a)$$

$$\frac{d}{dx}\ln(x) = \frac{1}{x}$$

$$\frac{d}{dx}\log_a(x) = \frac{1}{x \cdot \ln(a)}$$

$$\frac{d}{dx}\arcsin(x) = \frac{1}{\sqrt{1-x^2}}$$

$$\frac{d}{dx}\arccos(x) = -\frac{1}{\sqrt{1-x^2}}$$

$$\frac{d}{dx}\arctan(x) = \frac{1}{1+x^2}$$

$$\frac{d}{dx}\text{arcsec}(x) = \frac{1}{|x| \cdot \sqrt{x^2-1}}$$

QUICK QUIZ

1. True or false: $\sin(\arcsin(t)) = t$ for every t in $[-1, 1]$.
2. True or false: $\arccos(\cos(s)) = s$ for every real number s.
3. True or false: $\text{Tan}^{-1}(x) = \cot(x)$.
4. Simplify: $\cosh(\ln(x))$.

Answers

1. True 2. False 3. False 4. $(x^2+1)/(2x)$

EXERCISES

Problems for Practice

▶ In Exercises 1–18, calculate the value of the given inverse trigonometric function at the given point. ◀

1. $\arcsin(1)$
2. $\arcsin(\sqrt{3}/2)$
3. $\arcsin(-1)$
4. $\arccos(-1)$
5. $\arccos(1/2)$
6. $\arccos(-\sqrt{3}/2)$
7. $\arccos(\sqrt{2}/2)$
8. $\arctan(-1)$
9. $\arctan(-\sqrt{3})$
10. $\operatorname{arcsec}(2)$
11. $\operatorname{arcsec}(\sqrt{2})$
12. $\operatorname{arccsc}(1)$
13. $\operatorname{arccsc}(-\sqrt{2})$
14. $\operatorname{arccot}(-\sqrt{3})$
15. $\arcsin(\sin(5\pi/4))$
16. $\arctan(\tan(-3\pi/4))$
17. $\operatorname{arcsec}(\sec(-5\pi/6))$
18. $\arccos(\cos(4\pi/3))$

▶ In Exercises 19–34, differentiate the given expression with respect to x. ◀

19. $\arcsin(1-x)$
20. $\arcsin(\sqrt{x})$
21. $x\arcsin(x)$
22. $\arccos(x^2)$
23. $\log_2(\arccos(x^2))$
24. $\cos(\arctan(x))$
25. $\arctan(e^x)$
26. $\arctan(2/x)$
27. $x\arctan(x)$
28. $\arctan(x)/x$
29. $x^2\operatorname{arcsec}(x)$
30. $\operatorname{arcsec}(1/x)$
31. $\operatorname{arccsc}(x^2)$
32. $\operatorname{arccsc}(\sin(x))$
33. $\operatorname{arccot}(\sqrt{x})$
34. $\operatorname{arccot}(1/x^2)$

▶ In Exercises 35–54, differentiate the given expression with respect to x. ◀

35. $\sinh(3x)$
36. $\sinh(\ln(x))$
37. $\ln(\sinh(x))$
38. $\cosh(x-2)$
39. $\cosh(x)/x$
40. $\cosh^2(x)$
41. $\ln(\tanh(x))$
42. $\tanh(\ln(x+2))$
43. $\tanh(\tan(x))$
44. $x\operatorname{sech}(x)$
45. $\operatorname{sech}\sqrt{x}$
46. $\coth(x^2)$
47. $\sinh^{-1}(5x)$
48. $x\operatorname{sech}^{-1}(3x)$
49. $\arctan(\sinh(x))$
50. $\arcsin(\tanh(x))$
51. $\tanh^{-1}(\cos(x))$
52. $\tanh^{-1}(\coth(x))$
53. $\tanh^{-1}(\cosh(x))$
54. $\sinh^{-1}(\sqrt{x^2-1})$

Further Theory and Practice

▶ In Exercises 55–58, find the linearization of the given function $f(x)$ at the given point c. ◀

55. $f(x) = \arcsin(x)$, $c = 1/2$
56. $f(x) = \arcsin(x)$, $c = -1/\sqrt{2}$
57. $f(x) = \arctan(x)$, $c = \sqrt{3}$
58. $f(x) = \arctan(x)$, $c = -1$

59. Show that $\arcsin(x) = \pi/2 - \arccos(x)$ for all x in $[-1, 1]$.
60. Show that $\arctan(x) = \operatorname{arccot}(1/x)$ for all $x > 0$ and $\arctan(x) = \operatorname{arccot}(1/x) - \pi$ for all $x < 0$.
61. Show that $\cos^2(\arcsin(x))$ is the restriction to the interval $[-1, 1]$ of a polynomial in x.
62. Show that $\cos(2\arccos(x))$ is the restriction to the interval $[-1, 1]$ of a polynomial in x.
63. Verify the identity $\cosh(x+y) = \cosh(x)\cosh(y) + \sinh(x)\sinh(y)$.
64. Verify the identity $\sinh(x+y) = \sinh(x)\cosh(y) + \cosh(x)\sinh(y)$.
65. Verify the identity $\cosh(2x) = 2\cosh^2(x) - 1$.
66. Verify the identity $\cosh(2x) = 2\sinh^2(x) + 1$.
67. Verify the identity
$$\tanh(x+y) = \frac{\tanh(x) + \tanh(y)}{1 + \tanh(x)\tanh(y)}.$$
68. Show that
$$\left(\cosh(x) + \sinh(x)\right)^n = \cosh(nx) + \sinh(nx)$$
for any n, integer or not.
69. Show that $y = \sinh(\omega x)$ and $y = \cosh(\omega x)$ are solutions of the differential equation $y''(x) - \omega^2 y(x) = 0$.

Let a be a positive constant, and set
$$T(x) = a \cdot \operatorname{sech}^{-1}\left(\frac{x}{a}\right) - \sqrt{a^2 - x^2}.$$

▶ Exercises 70 and 71 are concerned with the graph of $y = T(x)$. Such a curve is called a *tractrix*. ◀

70. Show that $T'(x) = -\sqrt{a^2 - x^2}/x$.
71. The tip of a boat is at the point $(a, 0)$ on the x-axis. A tether of length a is attached to the tip of the boat and extends to

the origin. The boat is pulled by the tether as in Figure 22. Show that the tip of the boat moves along a tractrix.

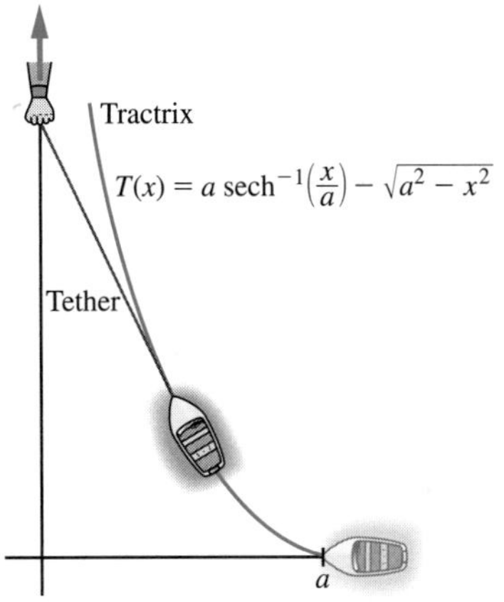

▲ Figure 22

72. The velocity of a water wave of wavelength λ as a function of the depth d of the water is given by

$$v = \sqrt{\frac{g}{2\pi}\lambda \tanh\left(\frac{2\pi d}{\lambda}\right)}.$$

Use the linear approximation of $\tanh(x)$ at $x = 0$ to show that $v(d) \approx \sqrt{gd}$ for small d. By considering the horizontal asymptote of the hyperbolic tangent, show that $v \approx \sqrt{\frac{g}{2\pi}\lambda}$ for large d.

The *gudermanian function gd* defined by

$$gd(u) = \arctan(\sinh(u))$$

links the usual trigonometric functions and the hyperbolic functions. Exercises 73–75 concern the gudermanian.

73. Show that

$$\begin{aligned}\cosh(u) &= \sec(gd(u))\\ \tanh(u) &= \sin(gd(u))\\ \coth(u) &= \csc(gd(u))\\ \mathrm{sech}(u) &= \cos(gd(u))\\ \mathrm{csch}(u) &= \cot(gd(u)).\end{aligned}$$

74. Show that $gd'(u) = \mathrm{sech}(u)$.

75. Use the Inverse Function Derivative Rule to show that

$$\frac{d}{d\theta} gd^{-1}(\theta) = \sec(\theta).$$

▶ A real-valued function f of a real variable x is said to be *algebraic* if there is a polynomial $p(u, v)$ with integer coefficients such that $p(x, f(x)) = 0$ for all x. For example, $f(x) = 2x + \sqrt{x^2 + 1}$ is algebraic because

$$\begin{aligned}p(x, f(x)) &= 3x^2 + \left(2x + \sqrt{x^2+1}\right)^2\\ &\quad - 4x\left(2x + \sqrt{x^2+1}\right) - 1\\ &\equiv 0\end{aligned}$$

for $p(u, v) = 3u^2 + v^2 - 4uv - 1$. A function that is not algebraic is said to be *transcendental*. In Exercises 76–79, find a polynomial that shows that the given expression is algebraic. ◀

76. $x - \sqrt{2}$

77. $\sqrt{x} + \sqrt{3}$

78. $2 + \sqrt{1 + x/x}$

79. $\sqrt[3]{x + \sqrt{x}}$

Calculator/Computer Exercises

▶ In Exercises 80–85, use a central difference quotient to approximate $f'(c)$ for the given f and c. Plot the function and the tangent line at $(c, f(c))$. ◀

80. $f(x) = \arcsin\left(\dfrac{x^2}{x^2+1}\right), \quad c = 1.3$

81. $f(x) = x\,\mathrm{arcsec}\,(1 + x^4), \quad c = 2.1$

82. $f(x) = \arcsin(2\arctan(x)/\pi), \quad c = 1.7$

83. $f(x) = \sinh\left(\dfrac{x}{1+\sqrt{x}}\right), \quad c = 4.5$

84. $f(x) = \arccos(\tanh(x)), \quad c = 0.7$

85. $f(x) = \sinh^{-1}(\log_2(x)), \quad c = 2.5$

86. Graph the six hyperbolic functions in the rectangle $[0, 1.5] \times [0, 2]$. Identify the seven points of intersection in the first quadrant.

87. The *Gudermannian function gd* is defined on the real line by $gd(u) = \arctan(\sinh(u))$. Its inverse gd^{-1} is defined on $(-\pi/2, \pi/2)$.

a. Graph gd.
b. Graph gd^{-1} by means of the parametric equations $x = gd(t)$ and $y = t$.

Summary of Key Topics in Chapter 3

Instantaneous Velocity (Section 3.1) If the position of a moving body at time t is given by a function $p(t)$, then the quantity

$$\lim_{\Delta t \to 0} \frac{p(t + \Delta t) - p(t)}{\Delta t},$$

if it exists, is called the instantaneous velocity of the moving body.

Tangent and Normal Lines
(Section 3.1)

If f is a function whose domain contains (a, b) and if $c \in (a, b)$, then the slope of the tangent line to the graph of f at c is equal to $f'(c)$, provided that the derivative exists. Thus the tangent line has equation

$$y = f'(c) \cdot (x - c) + f(c).$$

The normal line to the curve at the point c is the line through $(c, f(c))$, which is perpendicular to the tangent line at that point.

Yet another interpretation of the derivative involves rates of change. If f is a function of a parameter x, then $f'(c)$ represents the instantaneous rate of change of f with respect to x at the point c, provided that the derivative exists.

The Derivative
(Section 3.2)

In general, if f is a function whose domain contains (a, b), if $c \in (a, b)$, and if

$$\lim_{\Delta x \to 0} \frac{f(c + \Delta x) - f(c)}{\Delta x}$$

exists, then this quantity is called the derivative of f at the point c. It is denoted by any of the notations

$$f'(c) \text{ or } D(f)(c) \text{ or } \left.\frac{df}{dx}\right|_c \text{ or } \dot{f}(c).$$

The process of calculating the derivative is called differentiation. A function that possesses the derivative is called differentiable.

Derivatives of Special Functions
(Sections 3.2, 3.4, 3.5, 3.6)

By direct calculation, we can determine the derivatives of a number of particular functions. These are:

$$\frac{d}{dx} x^p = p \cdot x^{p-1} \qquad \text{for any real number } p.$$

Also

$$\frac{d}{dx} \sin(x) = \cos(x)$$

$$\frac{d}{dx} \cos(x) = -\sin(x)$$

$$\frac{d}{dx} \tan(x) = \sec^2(x)$$

$$\frac{d}{dx} \cot(x) = -\csc^2(x)$$

$$\frac{d}{dx} \sec(x) = \sec(x) \cdot \tan(x)$$

$$\frac{d}{dx} \csc(x) = -\csc(x) \cdot \cot(x)$$

$$\frac{d}{dx} e^x = e^x$$

$$\frac{d}{dx} a^x = a^x \cdot \ln(a)$$

$$\frac{d}{dx}\ln(x) = \frac{1}{x}$$
$$\frac{d}{dx}\log_a(x) = \frac{1}{x \cdot \ln(a).}$$

Rules for Taking Derivatives (Section 3.3)

We have

$$(f+g)' = f' + g'$$
$$(f-g)' = f' - g'$$
$$(\alpha \cdot f)' = \alpha \cdot (f'), \qquad \alpha \text{ constant}$$
$$(f \cdot g)' = f' \cdot g + f \cdot g'$$
$$\left(\frac{f}{g}\right)' = \frac{g \cdot f' - f \cdot g'}{g^2}.$$

The Chain Rule (Section 3.5)

If $g(u)$ and $f(x)$ are differentiable functions, if the range of f is in the domain of g, and if c is in the domain of f, then $g \circ f$ is differentiable at c and

$$(g \circ f)'(x) = g'(f(x)) \cdot f'(x).$$

In Leibniz notation, the Chain Rule is

$$\left.\frac{d(g \circ f)}{dx}\right|_{x=c} = \left.\frac{dg}{du}\right|_{u=f(c)} \cdot \left.\frac{df}{dx}\right|_{x=c}.$$

If we use variables $u = f(x)$ and $y = g(u)$, then we may write the Chain Rule as

$$\frac{dy}{dx} = \frac{dy}{du} \cdot \frac{du}{dx}.$$

The Chain Rule—Power Rule is the case of the Chain Rule in which $g(u) = u^p$:

$$\frac{d}{dx}u^p = pu^{p-1} \cdot \frac{du}{dx}.$$

The Inverse of a Function and Its Derivative (Section 3.6)

If S, T are open intervals, if $f: S \to T$ is invertible and differentiable, and if f' never vanishes on S, then f^{-1} is differentiable on T. The derivative is given by the formula

$$\frac{d}{dt}f^{-1}(t) = \frac{1}{\left.\frac{df}{ds}\right|_{s=f^{-1}(t)}}$$

provided that the denominator on the right is nonzero. Using prime notation, if $f(c) = \gamma$ and $f'(c) \neq 0$, then

$$(f^{-1})'(\gamma) = \frac{1}{f'(c)}.$$

The exponential function $\exp(x) = e^x$ has an inverse function with domain $(0, \infty)$. It is called the natural logarithm and is denoted by ln. (Its derivative has been recorded earlier.)

Higher Derivatives
(Section 3.7)

The derivative of a given function $f(x)$ is a new function $f'(c)$. This new function can in turn be differentiated, provided the derivative exists. The resulting function is called the second derivative. It is denoted by

$$f''(x) \quad \text{or} \quad f^{(2)}(x) \quad \text{or} \quad \frac{d^2f}{dx^2} \quad \text{or} \quad \ddot{f}(x) \quad \text{or} \quad D^2f(x).$$

The second derivative function can in turn be differentiated, and so on. The derivative of the $(k-1)^{\text{th}}$ derivative function is called the kth derivative function. It is denoted by the symbols

$$f''^{\cdots\prime}\ (k \text{ primes}) \quad \text{or} \quad f^{(k)} \quad \text{or} \quad \frac{d^kf}{dx^k} \quad \text{or} \quad D^kf.$$

Leibniz's rule for the second derivative of a product asserts that

$$(f\cdot g)^{(2)} = f^{(2)}g^{(0)} + 2f^{(1)}g^{(1)} + f^{(0)}g^{(2)}.$$

Implicit Differentiation
(Section 3.8)

If $F(x, y) = C$ is an equation which expresses y implicitly as a function of x, then we can differentiate it without first solving for y as a function of x. This is the method of implicit differentiation.

If x and y are expressed in terms of a parameter t by the equations $x = \varphi_1(t)$ and $y = \varphi_2(t)$, and if y is a differentiable function of x, then

$$\frac{dy}{dx} = \frac{\varphi_2'(t)}{\varphi_1'(t)} = \frac{dy/dt}{dx/dt}$$

Differential Approximation
(Section 3.9)

If f is differentiable at c, then f may be approximated at a nearby point $c + \Delta x$ by

$$f(c+\Delta x) \approx f(c) + f'(c)\cdot\Delta x,$$

or

$$f(x) \approx f(c) + f'(c)\cdot(x-c).$$

Writing $\Delta f(x)$ for $f(x+h) - f(x)$, we have

$$\Delta f(x) \approx f'(x)\cdot\Delta x$$

which is written in terms of differentials as

$$df(x) = f'(x)dx.$$

Inverse Trigonometric Functions
(Section 3.10)

We obtain an inverse of each trigonometric function by restricting its domain. The name of the inverse function is obtained by prefixing "arc" to the name of the trigonometric function. The derivative formulas are

$$\frac{d}{dx}\arcsin(x) = \frac{1}{\sqrt{1-x^2}}$$

$$\frac{d}{dx}\arccos(x) = -\frac{1}{\sqrt{1-x^2}}$$

$$\frac{d}{dx}\arctan(x) = \frac{1}{1+x^2}$$

$$\frac{d}{dx}\operatorname{arccot}(x) = -\frac{1}{1+x^2}$$
$$\frac{d}{dx}\operatorname{arcsec}(x) = \frac{1}{|x|\sqrt{x^2-1}}$$
$$\frac{d}{dx}\operatorname{arccsc}(x) = -\frac{1}{|x|\sqrt{x^2-1}}.$$

Hyperbolic Trigonometric Functions (3.10)

The basic hyperbolic trigonometric functions are defined by

$$\cosh(x) = \frac{e^x + e^{-x}}{2} \quad \text{and} \quad \sinh(x) = \frac{e^x - e^{-x}}{2}.$$

The four other hyperbolic trigonometric functions are defined as ratios that are analogous to those of standard trigonometry: $\tanh(x) = \sinh(x)/\cosh(x)$, $\coth(x) = \cosh(x)/\sinh(x)$, $\operatorname{sech}(x) = 1/\cosh(x)$, and $\operatorname{csch}(x) = 1/\sinh(x)$. The derivative formulas are

$$\frac{d}{dx}\sinh(x) = \cosh(x) \qquad \frac{d}{dx}\coth(x) = -\operatorname{csch}^2(x)$$
$$\frac{d}{dx}\cosh(x) = \sinh(x) \qquad \frac{d}{dx}\operatorname{sech}(x) = -\operatorname{sech}(x)\cdot\tanh(x)$$
$$\frac{d}{dx}\tanh(x) = \operatorname{sech}^2(x) \qquad \frac{d}{dx}\operatorname{csch}(x) = -\operatorname{csch}(x)\cdot\coth(x).$$

If the domains and ranges of the hyperbolic trigonometric functions are suitably restricted, then their inverses may be defined. The basic derivative rules for these inverse hyperbolic functions are

$$\frac{d}{dx}\sinh^{-1}(x) = \frac{1}{\sqrt{1+x^2}} \qquad \text{all real } x$$
$$\frac{d}{dx}\cosh^{-1}(x) = -\frac{1}{\sqrt{x^2-1}} \qquad |x|>1$$
$$\frac{d}{dx}\tanh^{-1}(x) = \frac{1}{1-x^2} \qquad -1<x<1$$
$$\frac{d}{dx}\operatorname{sech}^{-1}(x) = -\frac{1}{x\sqrt{1-x^2}} \qquad 0<x<1.$$

Review Exercises for Chapter 3

In Exercises 1–12, calculate the slope-intercept form of the tangent line to the graph of f at $(c, f(c))$

1. $f(x) = x^3$, $c = -2$
2. $f(x) = 3x^{-4/3}$, $c = 27$
3. $f(x) = 12/\sqrt{x}$, $c = 16$
4. $f(x) = \cos(2x)$, $c = \pi/3$
5. $f(x) = 12x\sin(\pi x)$, $c = 1/6$
6. $f(x) = e^x$, $c = \ln(2)$
7. $f(x) = (2x^2 - 3)/x$, $c = 3$
8. $f(x) = \sqrt{6x^2+1}$, $c = 2$
9. $f(x) = 2\ln(x)/e - 1/x$, $c = e$
10. $f(x) = e^{3x}/x$, $c = 1/3$
11. $f(x) = 3/2^x$, $c = 0$
12. $f(x) = 5(1+x^2)/(3+x)$, $c = 2$

▶ In Exercises 13–62, differentiate the given expression with respect to x. ◀

13. $\sqrt{x}$
14. $1/\sqrt{2x+1}$
15. $(3+x)/x^2$
16. $\sqrt{x}(3x^{3/2}+5x^{5/2})$
17. x^3+1/x^3
18. $1/\sqrt{16-x^4}$
19. $(x-1)(x+2)/x$
20. $x/(1+x^3)$
21. $(2+x^3)^{-3}$
22. $(1+x^3)/(2+x^2)$
23. $\sin(\pi - x)$
24. $\cos(3x)$
25. $\cos(2x+\pi/4)$
26. $\tan(\pi x)$
27. $\sec(2x)$
28. $\csc(\pi x/2)$
29. $\cot(1/x)$
30. $\sin^3(x)$
31. $\sin^3(5x^2)$
32. $x^2\cos(2x)$
33. $x^3\sin(x^2)$
34. $\sin(x)/x^2$
35. $(1+\sin(x))/(1+\cos(x))$
36. $\tan(x)/(1+\tan(x))$
37. $\sqrt{1+\cos(3x^2)}$
38. $\exp(-2x)$
39. xe^{3x}
40. x^2e^{-x}
41. $xe^x/(1+x)$
42. $\exp(x+\ln(x))$
43. 2^{-x}
44. 10^x-10^{-x}
45. $1/x+\ln(x)$
46. $\ln(3x+2)$
47. $x^3\ln(x)$
48. $\ln(3/x)+3/\ln(x)$
49. $\ln(2\ln(3x))$
50. $\log_3(1+3^x)$
51. $x^2\log_2(x+2)$
52. $(2+x)^x$
53. $2^x \cdot 3^x \cdot 5^x$
54. $\arcsin(x/2)$
55. $\arctan(2/x)$
56. $\operatorname{arcsec}(x^2)$
57. $\arcsin(\cos(x))$
58. $\cosh(2x)-\sinh(3x)$
59. $\cosh(x^2)$
60. $\tanh(\ln(x))$
61. $\sinh^{-1}(e^x)$
62. $\tanh^{-1}(1/x)$

▶ In Exercises 63–68, a particle's position along the x-axis is given at time t. Calculate the particle's velocity and acceleration. ◀

63. $55+60t-16t^2$
64. $18\sin(3t+\pi/2)$
65. $20/(1+t^2)$
66. $\exp(2t)/(1+\exp(2t))$
67. $25-e^{-5t}$
68. $\ln(1+t^4/4)$

▶ In Exercises 69–74, find the point c such that $\gamma = f(c)$, and use it to calculate $(f^{-1})'(\gamma)$. ◀

69. $f(x)=x^3+3x-9,\ \gamma=5$
70. $f(x)=4/(1+e^x),\ \gamma=2$
71. $f(x)=x^3/(4+x^2),\ \gamma=1$
72. $f(x)=x^2\ln(x),\ \gamma=0$
73. $f(x)=xe^x,\ \gamma=e$
74. $f(x)=2^{10-x^3},\ \gamma=4$

▶ In Exercises 75–80, use the method of implicit differentiation to find dy/dx. ◀

75. $\exp(y)-x+y=1$
76. $y\ln(y)=1+x^2$
77. $x^2-y+y^2=16$
78. $x^2+xy+y^2=12$
79. $y+1/y=1+\ln(x)$
80. $1/x+1/y=4/(1+y)$

▶ In Exercises 81–84, use the method of implicit differentiation to find dy/dx and d^2y/dx^2. ◀

81. $y^3+y-x=15$
82. $y^2+xy-2x=6$
83. $\sin(y)+2\cos(x)=1$
84. $e^y-xy=1$

▶ In Exercises 85–90, find the slope-intercept form of the tangent line to the given parametric curve at the given value t_0 of the parameter. ◀

85. $x=t^2-t,\ y=t^3-6,\ t_0=2$
86. $x=t^3+2,\ y=t^3-t,\ t_0=-1$
87. $x=t-\sin(t),\ y=1-\cos(t),\ t_0=\pi/3$
88. $x=1+te^t,\ y=2t+e^t,\ t_0=0$
89. $x=t+\ln(t),\ y=t+2/t,\ t_0=1$
90. $x=2t/(1+t^2),\ y=(1+t)/(1+t^2),\ t_0=1$

▶ In Exercises 91–94, use the differential approximation with base point c to estimate $f(c+\Delta x)$. ◀

91. $f(x)=\sqrt{1+7x},\ c=5,\ \Delta x=-0.3$
92. $f(x)=(5+x^3)^{1/5},\ c=3,\ \Delta x=0.3$
93. $f(x)=8x/\sqrt{9-x},\ c=5,\ \Delta x=0.2$
94. $f(x)=(15+4x)^{1/3}/x,\ c=3,\ \Delta x=-0.4$

▶ In each of Exercises 95–98, a function f, a point c, and an increment h are given. Calculate $f(c-h/2)$, $f(c)$, and $f(c+h/2)$, and use these values to approximate $f'(c)$. Then calculate $f(c-h)$ and $f(c+h)$. Use these values to estimate $f''(c)$. ◀

95. $f(x) = x^x$, $c = 1.418$, $h = 0.002$

96. $f(x) = \sin^x(x)$, $c = 0.5236$, $h = 0.0004$

97. $f(x) = \left(x\exp(x) + 1\right)/\left(\ln(x) + \sqrt{x}\right)$, $c = 3.14$, $h = 0.0002$

98. $f(x) = \sqrt{x^2 + 4\sqrt{x^2 + \sqrt{4 + 5x^2}}}$, $c = 3.0$, $h = 0.002$

The Tangent Problem in Antiquity

Apollonius (ca. 260 B.C., Perga - ?) was the greatest geometer of antiquity. His most important work was *Conics*, written in eight books. The first four of these books have been preserved in Greek, the next three in ninth-century Arabic translations. Book VIII has been lost. Before Apollonius, the geometers of ancient Greece did not find an acceptable definition of the tangent line to a curve. Apollonius, however, found a sophisticated way to approach the difficulty. His insight was to start with the normal line N. Given a curve C and a point Q not on C, Apollonius constructed N by solving a minimization problem: He found the point P on C with minimum distance to Q (as in Figure 1). After Apollonius obtained the normal line $N = PQ$, he took the tangent to be the line through P perpendicular to N.

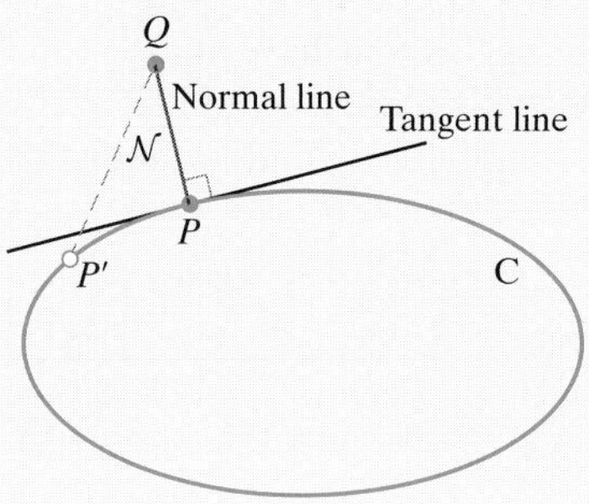

▲ Figure 1

The Solutions of Fermat and Descartes to the Tangent Problem

The introduction of analytic geometry revived the tangent problem in the 1630s. Descartes described the problem as "... the most useful and general that I have ever wanted to know in geometry." To find the tangent at a point P of a curve C, Descartes would find a circle that just *touched* C at P without crossing C. Let O be the center of this circle. Descartes took the line through P perpendicular to OP to be the tangent of C. In effect, Descartes had "reduced" the problem of finding a tangent line to the seemingly more difficult problem of finding a tangent circle. This method appeared in *La Géométrie* (1637), accompanied by illustrative examples that did not reveal how cumbersome a method it really was.

We will illustrate the method of Descartes by finding the tangent line to $y = \sqrt{x}$ at the point $P = (c, \sqrt{c})$. We seek the circle with center $(a, 0)$ and radius r that just touches the given parabola at P. After writing a as $c + h$, the equation of the circle becomes

$$\left(x-(c+h)\right)^2 + y^2 = r^2.$$

The parameters h and r are to be determined in terms of the given value c. At the point P of intersection of the circle and the parabola $y = \sqrt{x}$, the ordinate satisfies $y^2 = x$. Therefore the abscissa c of P satisfies the second degree equation

$$Q(x) = (x-(c+h))^2 + x - r^2 = 0.$$

This means that $x - c$ is a factor of $Q(x)$. Descartes reasoned that because the circle *touches* but does not *cross* the parabola, the root c must have *even* multiplicity m. Because the degree of Q is 2, the only possibility is that $m = 2$ and $Q(x) = (x - c)^2$. Thus $(x - c)^2 = x^2 + (1 - 2(c + h))x + (c + h)^2 - r^2$. Expanding the left side, we obtain

$$x^2 - 2cx + c^2 = x^2 + \left(1 - 2(c+h)\right)x + \left((c+h)^2 - r^2\right).$$

Each power of x must appear with the same coefficient on both sides of this identity. This leads to the simultaneous equations: $-2c = 1-2(c+h)$ and $c^2 = (c+h)^2 - r^2$. Solving these equations, we obtain $h = 1/2$. The line N determined by the two points $P = (c, \sqrt{c})$ and $(c + h, 0)$, therefore, has slope

$$\frac{\sqrt{c} - 0}{c - (c+h)} = \frac{\sqrt{c} - 0}{c - (c + 1/2)} = -2\sqrt{c}.$$

The slope of the line perpendicular to N, that is, the slope of the tangent line, is therefore the negative reciprocal of this number, or $1/(2\sqrt{c})$. See Figure 2. Of course, nowadays we have a much easier time of it: The Power Rule tells us that if $f(x) = \sqrt{x} = x^{1/2}$, then $f'(c) = (1/2)\, c^{-1/2} = 1/(2\sqrt{c})$.

The method of Descartes becomes entirely unwieldy when tried on even slightly more complicated examples. One contemporary protested that the method gives rise to computations that are a "... labyrinth from which it is extraordinarily difficult to emerge." When Fermat learned of the work of Descartes on tangents in 1637, he published his own

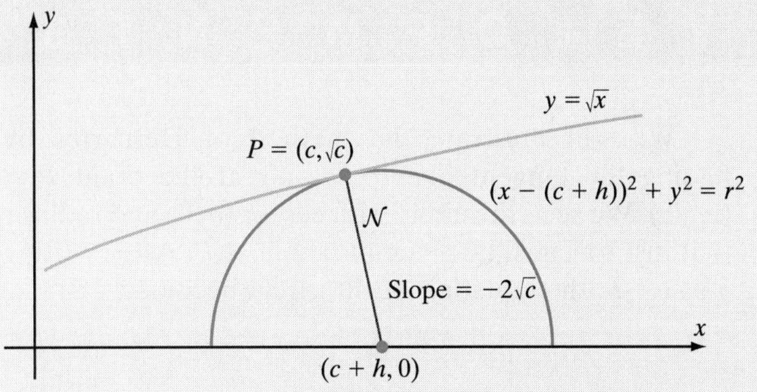

▲ Figure 2

method for finding the tangent line to the graph of a function f at c, a method that he had discovered some time earlier. Fermat did not use the slope m of the tangent line as a parameter. Instead, he used the length s of the *subtangent*. See Figure 3. Because m and s are related by the equation $m = f(c)/s$, the difference is not significant. Using the notation of Figure 3, we have $s/h = f(c)/(H - f(c))$ by similar triangles. This allows us to solve

$$m = \frac{f(c)}{s} = \frac{H - f(c)}{h}.$$

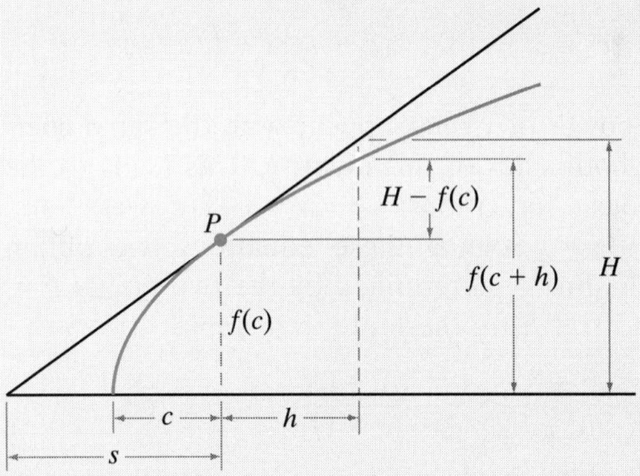

▲ Figure 3

Fermat argued that, when h is very small, $H \approx f(c+h)$. He called this approximation an *adequation*. Fermat made the substitution $H \approx f(c+h)$, obtaining

$$m \approx \frac{f(c+h) - f(c)}{h}.$$

For f a polynomial, Fermat expanded $f(c+h)$ to obtain a polynomial in h with constant term $f(c)$:

$$f(c+h) = f(c) + f_1(c)h + f_2(c)h^2 + \ldots .$$

Thus

$$m \approx \frac{f_1(c)h + f_2(c)h^2 + \ldots}{h}.$$

He then eliminated h from the denominator by cancellation

$$m \approx f_1(c) + f_2(c)h + \ldots$$

At this stage, Fermat either set $h = 0$ or simply discarded the terms with h as a factor—it is not clear from his writing. In any event, Fermat changed the "adequation" to the equality $m = f_1(c)$. Now, when f is a polynomial, it is easy to calculate each of the terms $f_1(c)$, $f_2(c)$, ... And of course it turns out that $f_1(c) = f'(c)$.

Fermat and Descartes became embroiled in a dispute over the tangent problem. It began with a letter of January 18, 1638 in which Descartes attributed the success of Fermat's method to luck. When other French mathematicians sided with Fermat, Descartes wrote

> "I admire that [Fermat's method] ... has found defenders; when M.Descartes will have understood the method, then he will cease to admire *only* that the method has found its defenders and also admire the method itself."

Descartes chose Gérard Desargues (1593 − 1662, Lyon) to referee the dispute. In April of 1638, Desargues concluded that, although Fermat's method was correct, his explanation was faulty and therefore "M. des Cartes is right and M. de Fermat is not wrong."

Descartes was still not persuaded: "His supposed rule is not so general as he makes it seem, and it can be applied only to the easiest problems, not to any question which is the least bit difficult." With that Descartes posed a challenge problem for Fermat: to find the tangent lines of $x^3 + y^3 = nxy$, a curve that is now known as the *Folium of Descartes*. Descartes must have expected that Fermat's method would be as futile as his own when applied to this curve. In fact, Fermat had not the slightest difficulty (Figure 4). He quickly sent Descartes a solution together with a revised explanation of his method. Descartes conceded, not too gracefully: "... had you so explained [your method] in the first place, I would not have objected at all." In fact, Descartes continued to object in his correspondence with others. "Is it not a great marvel," he wrote, "that in six months Fermat has found a new slant to justify his method." In a subsequent letter, Descartes described Fermat's tangent construction as "the most

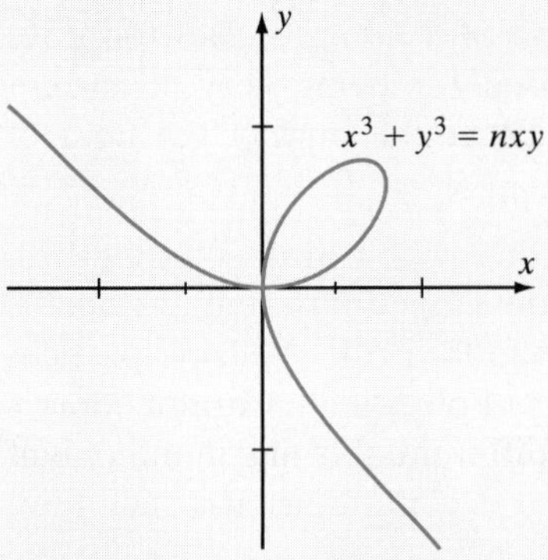

▲ Figure 4

ridiculous gibberish." By 1641, Descartes finally realized the futility of attacking the mathematics of Fermat and switched tactics: "I believe that he [Fermat] does know mathematics, but I still maintain that in philosophy he reasons badly."

Newton's Method of Differentiation

So far, we have described the tangent problem only from a geometric point of view. But the kinematic viewpoint is also important. Among the early contributors to the differential calculus, Roberval and Torricelli are noteworthy for being the first to link the tangent problem with the notion of instantaneous velocity. To Sir Isaac Newton, that connection would be fundamental.

The mathematical language of Newton has not, for the most part, survived. Newton called the variables x and y *fluents*, imagining them to trace a curve by their movement. The velocities of the *fluents* were called *fluxions* and denoted by $\dot{x}$ and $\dot{y}$. Newton's infinitesimals were called *moments of fluxions* and represented by $\dot{x}o$ and $\dot{y}o$ where o is not 0 but an "infinitely small quantity." To illustrate, let us use

$$F(x, y) = x^3 - ax^2 + axy - y^3 = 0,$$

the example given by Newton in his *Methodus Fluxionum* written in 1670. Newton substituted $x + \dot{x}o$ for x and $y + \dot{y}o$ for y in the equation $F(x, y) = 0$. On expanding the left side of the equation $F(x + \dot{x}o, y + \dot{y}o) = 0$, remembering that $F(x, y) = 0$, we arrive at

$$(3x^2 - 2ax + ay)\dot{x}o + (ax - 3y^2)\dot{y}o + a \cdot \dot{x}o \cdot \dot{y}o + (\dot{x}o)^2(3x - a + \dot{x}o) - (\dot{y}o)^2(3y + \dot{y}o) = 0.$$

Newton divided by o, obtaining

$$(3x^2 - 2ax + ay)\dot{x} + (ax - 3y^2)\dot{y} + o\left(a\dot{x}\dot{y} + \dot{x}^2(3x - a + \dot{x}o) - \dot{y}^2(3y + \dot{y}o)\right) = 0.$$

He then *discarded* all terms with o as a factor. In his words,

> "But whereas o is supposed to be infinitely little . . . , the terms which are multiplied by it will be nothing in respect to the rest. Therefore I reject them and there remains: $(3x^2 - 2ax + ay)\dot{x} + (ax - 3y^2)\dot{y} = 0$."

The result is that

$$\frac{\dot{y}}{\dot{x}} = -\frac{3x^2 - 2ax + ay}{ax - 3y^2},$$

which is the slope of the tangent line.

Newton, like Fermat before him and Leibniz after him, employed the suspect procedure of dividing by a quantity that he would later take to be 0. Newton's attempt to explain this undesirable situation is not very convincing. From his *Principia* of 1686, we find:

> "Those ultimate ratios with which quantities vanish are not truly the ratios of ultimate quantities, but limits toward which the ratios of quantities, decreasing without limit, do always converge, and to which they approach nearer than by any given difference, but never go beyond, nor in effect attain to, until the quantities have diminished in infinitum. Quantities, and the ratio of quantities, which in any finite time converge continually to equality, and before the end of that time approach neare the one to the other than by any given difference, become ultimately equal."

Bishop Berkeley, the critic quoted in *Genesis and Development 2*, had this to say in response:

> "The great author of the method of fluxions felt this difficulty and therefore he gave in to those nice abstractions and geometrical metaphysics without which he saw nothing could be done on the received principles . . . It must, indeed, be acknowledged that he used fluxions like the scaffold of a building, as things to be laid aside or got rid of . . . And what are these fluxions? . . . They are neither finite quantities nor quantities infinitely small, nor yet nothing. May we not call them the ghosts of departed quantities . . . ?"

As we have described in *Genesis and Development 2*, this controversy eventually led to the formalization of the limit concept.

The Product Rule and the Chain Rule

The rules of differential calculus that have been presented in this chapter were developed by Newton and, independently, at a later date, by Leibniz. The Product Rule, for example, appeared in Leibniz's work as $d(xv) = xdv + vdx$. Leibniz obtained this equation by treating $d(xv)$ as the infinitesimal increment in xv that arises by incrementing x by dx and v by dv. He then *discarded* the *second* order infinitesimal $dx\ dv$ that arises in the calculation. Thus

$$d(xv) = (x+dx)(v+dv) - xv = xv + xdv + vdx + (dxdv) - xv = xdv + vdx.$$

Leibniz introduced the notation dx in a manuscript of 11 November 1675. In that manuscript, Leibniz struggled to obtain the rules of differential calculus such as the Product Rule. It is not easy to determine Leibniz's progress on that particular day. His initial guess was that $d(xv) = dx \cdot dv$. By the end of the manuscript, he seems to have been aware of the incorrectness of his conjecture. His initial guess for the Quotient Rule was also wrong. But, by the end of November 1675, Leibniz had discovered a correct calculus of differentials.

Newton evidently felt constrained to account for the $dx\ dv$ term that Leibniz simply discarded. He offered the following demonstration (which we have converted into Leibniz's infinitesimal notation):

$$\begin{aligned} d(xv) &= \left(x+\frac{1}{2}dx\right)\left(v+\frac{1}{2}dv\right) - \left(x-\frac{1}{2}dx\right)\left(v-\frac{1}{2}dv\right) \\ &= xv + \frac{1}{2}x\,dv + \frac{1}{2}v\,dx + \frac{1}{4}dx\,dv - xv + \frac{1}{2}x\,dv \\ &\quad + \frac{1}{2}v\,dx - \frac{1}{4}dx\,dv \\ &= x\,dv + v\,dx. \end{aligned}$$

Berkeley specifically cited this derivation as an example of Newton's mathematical sleight of hand. In 1862, Sir William Rowan Hamilton (1805 – 1865, Dublin), himself a major figure in the development of calculus and physics, was also troubled by this computation:

> "It is very difficult to understand the *logic* by which Newton proposes to prove [the Product Rule]. His mode of getting rid of [$dxdv$] appeared to me … to involve so much artifice, as to deserve to be called sophistical … But by what right or by what reason other than to give an unreal air of simplicity to the calculation does he prepare the products thus? … it quite masks the notion of a limit … Newton does not seem to have cared for being very consistent in his philosophy, if he could anyway get hold of truth … "

In the early 1900s, Constantin Carathéodory (1873, Berlin – 1950, Munich) developed differential calculus in such a way that the standard rules became particularly easy to prove. Following Carathéodory's approach, f is said to be differentiable at c if and only if there is a function Φ_f that is continuous at c such that

$$f(x) = f(c) + (x-c)\cdot\Phi_f(x).$$

The derivative of f at c is then defined by $f'(c) = \Phi_f(c)$. Let us use this interpretation of the derivative to derive two theorems whose proofs we deferred.

Proof of the Product Rule

Suppose that f and g are differentiable at c. Thus there are continuous functions Φ_f and Φ_g such that

$$f(x) = f(c) + (x-c)\cdot\Phi_f(x) \quad \text{and} \quad g(x) = g(c) + (x-c)\cdot\Phi_g(x).$$

Therefore

$$\begin{aligned} (f\cdot g)(x) &= f(x)g(x) \\ &= \big(f(c) + (x-c)\cdot\Phi_f(x)\big)\big(g(c) + (x-c)\cdot\Phi_g(x)\big) \\ &= f(c)g(c) + (x-c)\cdot f(c)\cdot\Phi_g(x) \\ &\quad + (x-c)\cdot g(c)\cdot\Phi_f(x) + (x-c)^2\cdot\Phi_f(x)\cdot\Phi_g(x) \\ &= (f\cdot g)(c) + (x-c)\cdot \\ &\quad \big(f(c)\cdot\Phi_g(x) + g(c)\cdot\Phi_f(x) + (x-c)\cdot\Phi_f(x)\cdot\Phi_g(x)\big). \end{aligned}$$

Because the function

$$\Phi_{f\cdot g}(x) = f(c)\cdot\Phi_g(x) + g(c)\cdot\Phi_f(x) + (x-c)\cdot\Phi_f(x)\cdot\Phi_g(x)$$

is continuous, it follows that $f\cdot g$ is differentiable and

$$\begin{aligned} (f\cdot g)'(c) &= \Phi_{f\cdot g}(c) \\ &= f(c)\cdot\Phi_g(c) + g(c)\cdot\Phi_f(c) \\ &\quad + (c-c)\cdot\Phi_f(c)\cdot\Phi_g(c) \\ &= f(c)\cdot g'(c) + g(c)\cdot f'(c). \end{aligned}$$

Proof of the Chain Rule

Recall that the Chain Rule asserts that if f is differentiable at c and if g is differentiable at $f(c)$, then $g\circ f$ is differentiable at c and $(g\circ f)'(c) = g'(f(c))\cdot f'(c)$.

Assuming the stated differentiability hypotheses for f and g, there are continuous functions Φ_f and Φ_g such that

$$f(x) = f(c) + (x - c) \cdot \Phi_f(x), \quad f'(c) = \Phi_f(c)$$

and

$$g(x) = g(f(c)) + (x - f(c)) \cdot \Phi_g(x), \quad g'(f(c)) = \Phi_g(f(c)).$$

Therefore

$$\begin{aligned} g(f(x)) &= g(f(c)) + (f(x) - f(c)) \cdot \Phi_g(f(x)) \\ &= g(f(c)) + ((x - c) \cdot \Phi_f(x)) \cdot \Phi_g(f(x)) \\ &= g(f(c)) + (x - c) \cdot (\Phi_f(x) \cdot (\Phi_g \circ f)(x)). \end{aligned}$$

Because $x \mapsto \Phi_f(x) \cdot (\Phi_g \circ f)(x)$ is continuous at c, it follows that $g \circ f$ is differentiable at c and that

$$\begin{aligned} (g \circ f)'(c) &= \Phi_f(c) \cdot (\Phi_g \circ f)(c) \\ &= \Phi_g(f(c)) \cdot \Phi_f(c) = g'(f(c)) \cdot f'(c). \end{aligned}$$

CHAPTER 4

Applications of the Derivative

PREVIEW

The need to maximize and minimize functions often arises in science, engineering, and commerce. For example, a company developing a new product might want to maximize its profit. An ecologist might want the company to minimize its consumption of resources. In short, procedures for optimizing functions are among the most vital applications of mathematics. In this chapter, we will see that the derivative provides a powerful tool for analyzing the extreme values of functions.

We will learn that, to maximize or minimize a differentiable function f, we must first find the roots of the equation $f'(x) = 0$. Solving such an equation can present formidable difficulties, but, here too, the derivative can be an essential aid. In this chapter, we will learn that differential calculus provides a method for approximating the roots of equations.

One of the main considerations of the present chapter will be the use of the derivative as an aid in sketching the graph of a function. We will learn to combine several applications of the derivative to sketch the graphs of functions. Graphing calculators and software are certainly useful tools, but they are not a substitute for this knowledge.

4.1 Related Rates

In a related rates problem, two variables are related by means of an equation. When one variable changes, so must the other to satisfy the equation. The question we study in this section is, *How can we use the rate of change of one variable to calculate the rate of change of the other variable*?

The Role of the Chain Rule in Related Rates Problems

In the most basic related rates problem, two variables x and y are related by a function: $y = g(x)$. If x depends on a third variable t, then so does y through the equation $y = g(x)$. The Chain Rule tells us the relationship between the rates of change of x and y with respect to t:

$$\frac{dy}{dt} = \frac{dg}{dx}\frac{dx}{dt}.$$

▶ **EXAMPLE 1** If a train that is approaching a platform with speed s (measured in m/s) sounds a horn with frequency 500 Hz, then, according to the Doppler effect, the frequency ν heard by a stationary observer on the platform is

$$\nu = 500 \cdot \left(\frac{345}{345 - s}\right) \text{ Hz.}$$

If the train is decelerating at a constant rate of 4 m/s^2, what is the rate of change of ν when the train's speed is 25 m/s?

Solution

1. The quantities involved are the variables ν and s. These are functions of time t.
2. The relationship between the variables is expressed by the equation $\nu = 500 \cdot 345/(345 - s) = 172500/(345 - s)$. In this example, one variable, ν, has been given explicitly in terms of the other.
3. We implicitly differentiate with respect to t. Using the Chain Rule and the Reciprocal Rule, we calculate

$$\frac{d\nu}{dt} = 172500\frac{d}{dt}\left(\frac{1}{345 - s}\right) = 172500\frac{d}{ds}\left(\frac{1}{345 - s}\right)\frac{ds}{dt}$$

$$= -172500\,\frac{\frac{d}{ds}(345 - s)}{(345 - s)^2} \cdot \frac{ds}{dt} = \frac{172500}{(345 - s)^2} \cdot \frac{ds}{dt}.$$

4. We substitute the given data into the formula for $d\nu/dt$ that we obtained in Step 3. We find that, when $s = 25$ and $ds/dt = -4$, the rate of change $d\nu/dt$ of the observed frequency, in Hz/s, is

$$\frac{d\nu}{dt} = \frac{172500}{(345 - 25)^2}(-4) = -\frac{1725}{256} \approx -6.74.$$ ◀

The Role of Implicit Differentiation

The Method of Implicit Differentiation is used to solve related rates problems in which two variables x and y are related by an equation

$$F(x, y) = C, \tag{4.1.1}$$

where C is a constant. Each side of equation (4.1.1) is differentiated with respect to t. After differentiation, the right side of the equation is 0, and the left side is an expression involving x, y, dx/dt, and dy/dt. We may write this last equation as

$$G\left(x, y, \frac{dx}{dt}, \frac{dy}{dt}\right) = 0. \tag{4.1.2}$$

Typically, one of the derivatives, dx/dt or dy/dt, is specified, and the other must be determined from equation (4.1.2). To do so, it may be necessary to use equation (4.1.1) if the value of only one of the variables, x or y, has been specified. The next example will illustrate these ideas.

▶ **EXAMPLE 2** Suppose that $y^3 + xy - 4x = 0$. If $x = 4$ and $dx/dt = 3$ when $t = 5$, what is dy/dt when $t = 5$?

Solution

1. The quantities involved are the variables x and y. These are functions of time t.
2. The relationship between the variables is expressed by the equation $y^3 + xy - 4x = 0$. We can use the form that has been given without explicitly solving for one variable in terms of the other.
3. We implicitly differentiate with respect to t. Using the Chain Rule-Power Rule for the term y^3 and the Product Rule and Chain Rule for the term xy, we obtain

$$3y^2\frac{dy}{dt} + \left(\frac{dx}{dt}\cdot y + x\frac{dy}{dt}\right) - 4\frac{dx}{dt} = 0. \tag{4.1.3}$$

   (The left side of equation (4.1.3) is the expression $G(x, y, dx/dt, dy/dt)$ to which we referred in the general discussion.)
4. We know that, at $t = 5$, we have $dx/dt = 3$ and $x = 4$. To use the equation obtained in Step 3, we also need to know the value of y at $t = 5$. To determine this value, we turn to the given equation $y^3 + xy - 4x = 0$ and substitute $x = 4$. We obtain $y^3 + 4y - 16 = 0$, or $(y - 2)(y^2 + 2y + 8) = 0$, or $y = 2$. We now substitute all known values into equation (4.1.3) of Step 3:

$$3(2)^2\frac{dy}{dt} + \left((3)\cdot(2) + (4)\frac{dy}{dt}\right) - 4(3) = 0.$$

   Solving for dy/dt, we obtain $dy/dt = 3/8$. ◀

INSIGHT Figure 1 shows the plot of the graph $\mathcal{C}$ of the equation $y^3 + xy - 4x = 0$ from Example 2 in the viewing rectangle $[1/4, 8] \times [3/4, 2.5]$. We may think of a particle moving upward and to the right along $\mathcal{C}$ with position (x, y) at time t. From this point of view, the variable t is a parameter for $\mathcal{C}$. When t is equal to 5, the particle passes through the point $P_0 = (4, 2)$. Because explicit parametric equations for x and y are not given, the value 5 of t is used only as a way of referring to P_0. Had the particle passed through P_0 at a different time such as 7, or even at an unspecified instant t_0, the calculations would have been exactly the same. Formula (3.8.9) in Chapter 3 tells us that the slope of the tangent line to $\mathcal{C}$ at P_0 is

$$\left.\frac{dy}{dx}\right|_{P_0} = \frac{dy/dt|_{t=5}}{dx/dt|_{t=5}} = \frac{3/8}{3} = \frac{1}{8}.$$

The equation of the tangent line to $\mathcal{C}$ at P_0 is therefore $y = (1/8)(x - 4) + 2$, and the line described by this equation is also plotted in Figure 1.

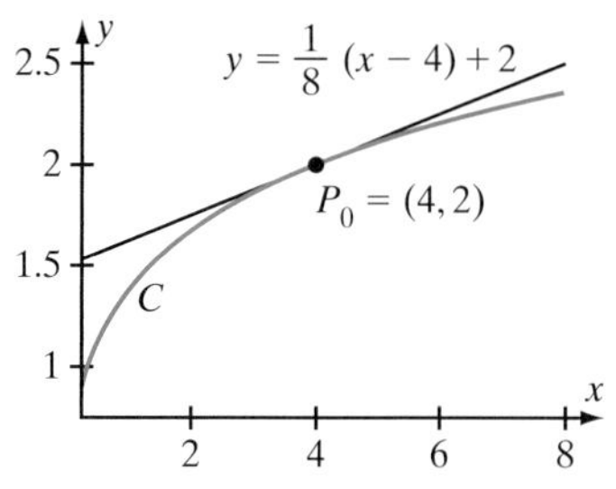

▲ Figure 1

▶ **EXAMPLE 3** The relationship between pressure P (measured in atmospheres) and volume V (measured in liters) of 1 mole of carbon dioxide at 0°c is approximated by van der Waal's equation

$$\left(P + \frac{a}{V^2}\right)(V - b) = 22.40,$$

where $a = 3.592$, and $b = 0.04267$. At a certain time t_0, the volume V of a carbon dioxide sample is 4 liters, and the pressure is increasing at the rate of 0.1 atmosphere per second. At what rate is V decreasing when $t = t_0$?

Solution

1. The quantities involved are the pressure P and the volume V. These are functions of time t.
2. The relationship between the variables is expressed by van der Waal's equation. We can use the form that has been given without explicitly solving for one variable in terms of the other.
3. We implicitly differentiate with respect to t. Using the Product Rule, we obtain

$$\left(\frac{dP}{dt} - 2\frac{a}{V^3}\frac{dV}{dt}\right)(V - b) + \left(P + \frac{a}{V^2}\right)\frac{dV}{dt} = 0.$$

4. We know that at t_0, $dP/dt = 0.1$ and $V = 4$. The constants a and b are given. To use the equation obtained in Step 3, we also need to know the value of P at t_0. We use van der Waal's equation to determine the value of P:

$$\left(P + \frac{3.592}{4^2}\right)(4 - 0.04267) = 22.40$$

or $P = 5.436$. We now substitute all known values into the equation of Step 3:

$$\left(0.1 - 2\frac{3.592}{4^3}\frac{dV}{dt}\right)(4 - 0.04267) + \left(5.436 + \frac{3.592}{4^2}\right)\frac{dV}{dt} = 0.$$

We solve for dV/dt, obtaining $dV/dt = -7.586 \times 10^{-2}$ liters per second. Thus the volume of the carbon dioxide sample is decreasing at the rate of 7.586×10^{-2} liters per second. ◀

Basic Steps for Solving a Related Rates Problem

The analysis that enabled us to solve Examples 1–3 can be systematized as follows:

1. Identify the quantities that are varying, and identify the variable (this is often "time") with respect to which the change in these quantities is taking place.
2. Establish a relationship (an equation) between the quantities isolated in Step 1.
3. Differentiate the equation from Step 2 with respect to the variable identified in Step 1. Be sure to apply the Chain Rule carefully.
4. Substitute the numerical data into the equation from Step 3, and solve for the unknown rate of change.

We will consider two more examples that illustrate this approach.

▶ **EXAMPLE 4** A sample of a new polymer is in the shape of a cube. It is subjected to heat so that its expansion properties may be studied. The surface area of the cube is increasing at the rate of 6 square inches per minute. How fast is the volume increasing at the moment when the surface area is 150 square inches?

Solution The primary interest of this example is that surface area and volume are related indirectly. We use the Chain Rule, but we do not have to differentiate implicitly.

1. The functions are surface area A and volume V. It is useful to introduce the function s for side length because A and V are related by means of s. The variable is time t.
2. We have $V = s^3$ *and* $A = 6s^2$. (The last equation arises because a cube has six faces, each with area s^2.)
3. We differentiate both the equations from Step 2 with respect to the variable t. Thus

$$\frac{dV}{dt} = 3s^2 \cdot \frac{ds}{dt} \qquad \text{and} \qquad \frac{dA}{dt} = 6 \cdot 2s \cdot \frac{ds}{dt}.$$

We use the second of these equations to solve for ds/dt:

$$\frac{ds}{dt} = \frac{1}{12s} \cdot \left(\frac{dA}{dt}\right).$$

We substitute this expression for ds/dt into the equation that we obtained for dV/dt:

$$\frac{dV}{dt} = 3s^2 \cdot \left(\frac{1}{12s} \cdot \left(\frac{dA}{dt}\right)\right) = \frac{s}{4} \cdot \frac{dA}{dt}.$$

4. We notice that, at the instant with which the problem is concerned, $A = 150$, hence $s = 5$. Also, $dA/dt = 6$. Substituting this information into the last equation in Step 3 gives

$$\frac{dV}{dt} = \frac{5}{4} \cdot 6 = \frac{15}{2} \text{ (cu. in.)/min.}$$

The volume of the cube of polymer is expanding at the rate of 15/2 cubic inches per minute at the moment in question. ◄

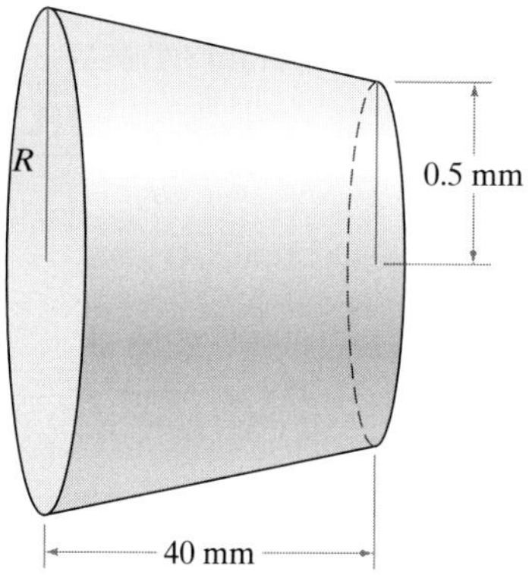

▲ Figure 2

► **EXAMPLE 5** (**An Application to Biology**) A blood vessel is slightly tapered in shape, like a truncated cone, as illustrated in Figure 2. The effect of a certain pain-relieving drug is to increase the larger radius at the rate of 0.01 millimeter per minute. During this expansion process, the length of the blood vessel remains fixed at 40 millimeters, and the smaller radius remains fixed at 0.5 millimeters. At the moment when the larger radius is 2 millimeters, how is the volume of the blood vessel changing?

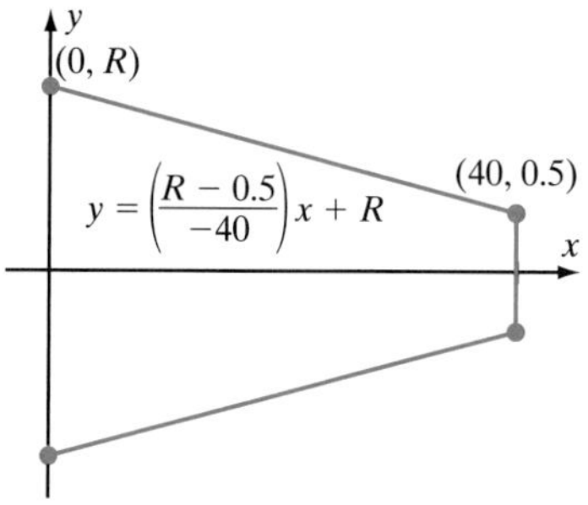

▲ Figure 3

Solution

1. The functions involved in this problem are the larger radius R and the volume of the blood vessel V. The variable is t.
2. To calculate the volume of the blood vessel for a given R, we use the cross-sectional diagram in Figure 3. The edge of the blood vessel is given by a line of slope $(R - 0.5)/(0 - 40)$ and y-intercept R. From this information, we see that the slope-intercept equation of the line is

$$y = \left(\frac{R - 0.5}{-40}\right)x + R.$$

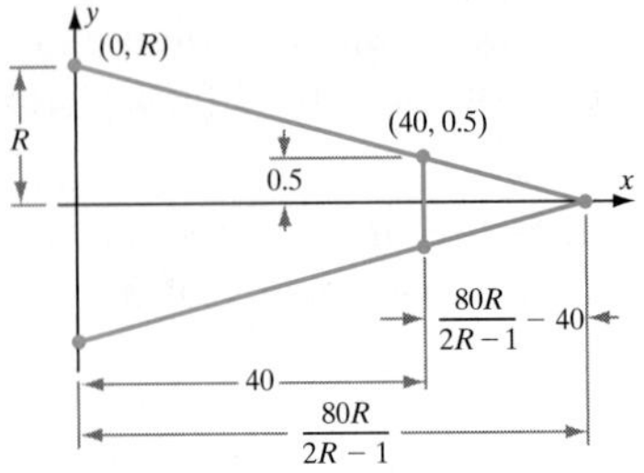

Figure 4

Substituting $y = 0$ in this equation and solving for x, we find that the line has x-intercept $40R/(R - 0.5)$, or $80R/(2R - 1)$. From Figure 4, we see that

$$V = \begin{pmatrix}\text{volume of cone with} \\ \text{radius} = R, \\ \text{height} = \dfrac{80R}{2R-1}\end{pmatrix} - \begin{pmatrix}\text{volume of cone with} \\ \text{radius} = 0.5, \\ \text{height} = \dfrac{80R}{2R-1} - 40\end{pmatrix}$$

$$= \left(\frac{1}{3}\pi R^2\left(\frac{80R}{2R-1}\right)\right) - \left(\frac{1}{3}\pi(0.5)^2\left(\frac{80R}{2R-1} - 40\right)\right)$$

$$= \frac{1}{3}\pi\left(R^2 \cdot \frac{80R}{2R-1} - 0.25 \cdot \frac{40}{2R-1}\right)$$

$$= \frac{10}{3}\pi\left(R^2 \cdot \frac{8R}{2R-1} - \frac{1}{2R-1}\right)$$

$$= \frac{10}{3}\pi\frac{8R^3 - 1}{2R-1}$$

$$= \frac{10}{3}\pi(4R^2 + 2R + 1) \quad \text{(by long division of polynomials).}$$

3. We may now differentiate V with respect to t. Doing so, we obtain

$$\frac{dV}{dt} = \frac{d}{dt}\left(\frac{10\pi}{3}(4R^2 + 2R + 1)\right) = \frac{10\pi}{3}\left(\frac{d}{dR}(4R^2 + 2R + 1)\right)\frac{dR}{dt}$$

$$= \frac{10\pi}{3}(8R + 2)\frac{dR}{dt} = \frac{20\pi}{3}(4R + 1)\frac{dR}{dt}.$$

4. We substitute the values $R = 2$ and $dR/dt = 0.01$ to obtain

$$\frac{dV}{dt} = \frac{20\pi}{3}(4 \cdot 2 + 1)(0.01) = \frac{20\pi}{3}(9)(0.01) = 60\pi(0.01) = 0.6\pi.$$

Thus the volume of the blood vessel is increasing at the rate of 0.6π cubic millimeters per minute. ◀

QUICK QUIZ

1. Suppose that $y = x^3$, $x = 2$, and $dy/dt = 60$ when $t = 10$. What is dx/dt when $t = 10$?
2. Suppose that x and y are positive variables, that $2x + y^3 = 14$, that x is a function of t, and that $dx/dt = 4$ for all t. What is dy/dt when $x = 3$?

Answers

1. 5 2. −2/3

EXERCISES

Problems for Practice

▶ In each of Exercises 1–6, the variable y is given as a function of x, which depends on t. The values x_0 and v_0 of, respectively, x and dx/dt are given at a value t_0 of t. Use this data to find dy/dt at t_0. ◀

1. $y = x^3$, $x_0 = 2$, $v_0 = 5$
2. $y = x^2 - 2x$, $x_0 = 2$, $v_0 = -3$
3. $y = \cos(x)$, $x_0 = \pi/6$, $v_0 = -2$
4. $y = \sqrt{2x - 1}$, $x_0 = 5$, $v_0 = 7$
5. $y = x \ln(2x)$, $x_0 = 1/2$, $v_0 = 3$
6. $y = x^{3/2} - \exp(-2x)$, $x_0 = 1$, $v_0 = 4$

▶ In each of Exercises 7–10, the variable y is given as a function of x, which depends on t. The values y_0 and s_0 of, respectively, y and dy/dt are given at a value t_0 of t. Use this data to find dx/dt at t_0. ◀

7. $y = 14 - 11x$, $y_0 = 12$, $s_0 = 22$
8. $y = x^3$, $y_0 = -8$, $s_0 = 5$
9. $y = 6\exp(-3x)$, $y_0 = 6$, $s_0 = 54$
10. $y = 4x + \sin^2(x)$, $y_0 = \pi + 1/2$, $s_0 = 30$

▶ In each of Exercises 11–16, variables x and y, which depend on t, are related by a given equation. A point P_0 on the graph of that equation is also given, as is *one* of the following two values:

$$v_0 = \left.\frac{dx}{dt}\right|_{P_0} \quad \text{or} \quad s_0 = \left.\frac{dy}{dt}\right|_{P_0}.$$

Find the other. ◀

11. $x^2 + y^3 = 3$, $P_0 = (2, -1)$, $v_0 = 3$
12. $3x^3 + y^3 = 3$, $P_0 = (-2, 3)$, $v_0 = 2$
13. $x^2 + xy = 2$, $P_0 = (-2, 1)$, $s_0 = 3$
14. $2\sqrt{x} - \sqrt{3y} = 1$, $P_0 = (4, 3)$, $s_0 = 5$
15. $2x^3 - x^2y + 3y = 13$, $P_0 = (2, 3)$, $v_0 = -2$
16. $x^2 + x\exp(y) + y = 6$, $P_0 = (2, 0)$, $s_0 = -1$

▶ In each of Exercises 17–22, variables x and y are functions of a parameter t and are related by a given equation. A point P_0 on the graph of that equation is also given, as is *one* of the following two values:

$$v_0 = \left.\frac{dx}{dt}\right|_{P_0} \quad \text{or} \quad s_0 = \left.\frac{dy}{dt}\right|_{P_0}.$$

Find the other value. Also, referring to the *Insight* that follows Example 2, find the tangent line to the graph of the given equation at P_0. ◀

17. $x^2 + y^3 = 9$, $P_0 = (-1, 2)$, $v_0 = 3$
18. $2x^3 + y^3 = 8$, $P_0 = (2, -2)$, $v_0 = -1$
19. $3x^2 - xy = 1$, $P_0 = (1, 2)$, $s_0 = -2$
20. $2x - \sqrt{x} - \sqrt{y} = 3$, $P_0 = (4, 9)$, $s_0 = 2$
21. $2x^2 - xy^2 + \ln(y/2) = 6$, $P_0 = (3, 2)$, $v_0 = -2$
22. $x^2 - 3x\exp(y) + y = 10$, $P_0 = (-2, 0)$, $s_0 = -1$

23. The diagonal of a square is increasing at a rate of 2 inches per minute. At the moment when the diagonal measures 5 inches, how fast is the area of the square increasing?
24. The length of each leg of an isosceles right triangle is increasing at the rate of 2 mm/s. How fast is the hypotenuse increasing at the moment when each leg is 10 mm?
25. The side length of an equilateral triangle is decreasing at the rate of 3 cm/s. How fast is the area decreasing at the moment when the area is $27\sqrt{3}\text{cm}^2$?
26. The volume of a cube is increasing at the rate of 60 mm^3/s. How fast is the surface area of the cube increasing when each edge is 20 mm?
27. Let P denote the position of particle that, from its initial point $Q = (1, 0)$, moves counterclockwise along the unit circle. If θ denotes the angle subtended at the origin O by the circular arc QP, then the radial velocity $d\theta/dt$ of the particle is constantly equal to 6. How fast is the area of the sector QPO swept out by the particle changing?
28. If an airplane climbs at 550 mph making a $\pi/6$ angle with the horizontal, how fast is it gaining altitude?
29. A container is pulled up a 10 ft long ramp from the ground to a loading dock 4 ft above the ground. At the moment it is halfway along the ramp, it is moving at the rate of 1.2 ft/s. How fast is it rising at that instant?
30. A rocket that is rising vertically is being tracked by a ground-level camera located 3 mi from the point of blast-off. When the rocket is 2 mi high, its speed is 400 mph. At what rate is the (acute) angle between the horizontal and the camera's line of sight changing?
31. A spherical balloon is losing air at the rate of 6 cm^3/min. At what rate is its radius r decreasing when $r = 24$ cm?
32. A spherical balloon is inflated so that its volume increases at the rate of 6 cm^3/min. How fast is the surface area of the balloon increasing when $r = 24$ cm?
33. Suppose that grain pouring from a chute forms a conical heap in such a way that the height is always 2/3 the radius of the base. At the moment when the conical heap is 3 m high, its height is rising at the rate of 1/2 m/min. At what rate (in m^3/min) is the grain pouring from the chute?
34. Particle A moves down the x-axis in the positive direction at a rate of 5 units per second. Particle B walks up the y-axis in the positive direction at a rate of 8 units per second. At the moment when A is at $(4, 0)$ and B is at $(0, 9)$, how rapidly is the distance between A and B changing?
35. A policeman stands 60 feet from the edge of a straight highway while a car speeds down the highway. Using a radar gun, the policeman ascertains that, at a particular instant, the car is 100 feet away from him and the distance between himself and the car is changing at a rate of 92

feet per second. At that moment, how fast is the car traveling down the highway?

36. A 5 foot, 10 inch tall woman is walking away from a wall at the rate of 4 ft/s. A light is attached to the wall at a height of 10 feet. How fast is the length of the woman's shadow changing at the moment when she is 12 feet from the wall?

37. A spherical raindrop is evaporating. Suppose that units are chosen so that the rate at which the volume decreases is numerically equal to the surface area. At what rate does the radius decrease?

38. A lighthouse is 100 feet tall. It keeps its beam focused on a boat that is sailing away from the lighthouse at the rate of 300 feet per minute. If θ denotes the acute angle between the beam of light and the surface of the water, then how fast is θ changing at the moment the boat is 1000 feet from the lighthouse?

39. A particle moves around the curve $x^2 + 2y^3 = 25$ in such a way that $dy/dt = y$. What is the rate of change of the particle's x-coordinate at the moment when $x = 3$?

Further Theory and Practice

40. A spherical balloon is being inflated. At a certain instant its radius is 12 cm and its area is increasing at the rate of $24\,\text{cm}^2/\text{min}$. At that moment, how fast is its volume increasing?

41. A conical tank filled with water is 5 m high. The radius of its circular top is 3 m. The tank leaks water at the rate of $120\ \text{cm}^3/\text{min}$. When the surface of the remaining water has area equal to $\pi\ \text{m}^2$, how fast is the depth of the water decreasing?

42. A particle moves along the curve $y = \sqrt{x}$ in an xy-plane in which each horizontal unit and each vertical unit represents 1 cm. At the moment the particle passes through the point $(4, 2)$, its x-coordinate increases at the rate 7 cm/s. At what rate does the distance between the particle and the point $(0, 5)$ change?

43. If R_1 and R_2 are parallel variable resistances, then the resulting resistance R satisfies $1/R = 1/R_1 + 1/R_2$. If R_1 increases at the rate of $0.6\,\Omega/\text{s}$ when $R_1 = 40\,\Omega$, and $R_2 = 20\,\Omega$, then at what rate must R_2 decrease if R remains constant? (The unit by which resistance in an electric circuit is usually measured, the *ohm*, is denoted by Ω.)

44. At 1 PM, a car traveling east at a constant speed of 30 mph passes through an intersection. At 2 PM, a car traveling south at a constant speed of 40 mph passes through the same intersection. How fast is the distance between the two cars changing at 3 PM?

45. A particle moves along the curve $y = \sqrt{x}$. At what point are the rates of change of the particle's coordinates equal?

46. In a hemispherical tank of radius 20 feet, the volume of water is $\pi h^2(60 - h)/3$ cubic feet when the depth is h feet at the deepest point. If the water is draining at the rate of 5 cubic feet per minute, how fast is the area of the water on the surface decreasing when the water is 10 feet deep?

47. When heated, the height of a certain solid cylindrical rod increases at twice the rate at which the radius increases. If the volume increases at the constant rate of $60\pi\ \text{cm}^3/\text{min}$, at what rate is the radius increasing when $r = 3\,\text{cm}$ and $h = 17\,\text{cm}$?

48. The *demand curve* for a given commodity is the set of all points (p, q) in the pq-plane where q is the number of units of the product that can be sold at price p. The *elasticity of demand* for the product at price p is defined to be $E(p) = -q'(p) \cdot p/q(p)$.

a. Suppose that a demand curve for a commodity is given by

$$p + q + 2p^2q + 3pq^3 = 1000$$

when p is measured in dollars and the quantity q of items sold is measured by the 1000. For example, the point $(p, q) = (6, 3.454)$ is on the curve. That means that 3454 items are sold at \$6. What is the slope of the demand curve at the point $(6, 3.454)$?

b. What is elasticity of demand for the product of part at $p = \$6$?

49. If a volume V_0 of oil spills from a tanker at sea, then there are positive constants t_0 and h_1 such that, for $t \geq t_0$, the spilled oil spreads as a cylindrical disk of height $h(t) = h_1/\sqrt{t}$. Show that, for $t > t_0$, the radius $r(t)$ of this disk satisfies

$$r'(t) = \frac{1}{4t^{3/4}}\sqrt{\frac{V_0}{\pi h_1}}.$$

50. The heat index is the apparent temperature T (°F) that a human feels. A rough aproximation of T in terms of actual temperature x (°F) and relative humidity h, where h is a number in the interval $[0, 100]$, is $T = 2.7 + 0.885x - 0.787h + 0.012xh$. Suppose that the actual temperature is increasing at the rate of 1.7°F/hr at the moment $x = 88$ and $h = 60$. How fast must humidity decrease for the heat index to be unchanged?

51. The radius of a certain cone is increasing at a rate of 6 centimeters per minute while the height is decreasing at a rate of 4 centimeters per minute. At the instant when the radius is 9 centimeters and the height is 12 centimeters, how is the volume changing?

52. Under the cover of darkness, a bug has made its way to the center of a round table of radius 2/3 m. Suddenly, a light 4/3 m directly over the bug is switched on. The surprised bug scampers in a direct line toward the edge of the table at the rate of 12 cm/s. When the distance between the bug and the light bulb is z meters, the intensity I of illumination is given by $I = 120\cos(\theta)/z^2$ (see Figure 5). At what rate is I decreasing when the bug reaches the edge of the table?

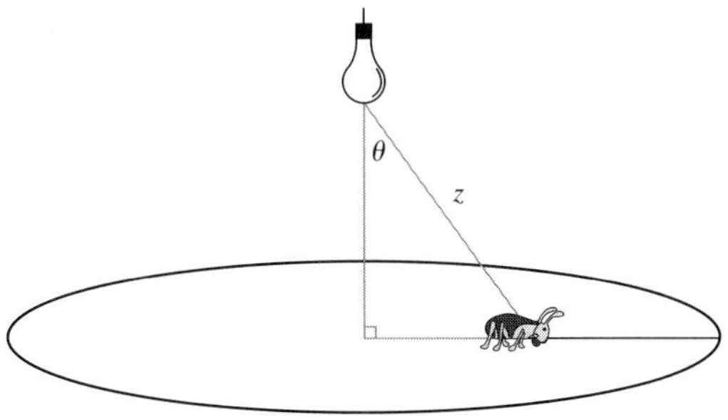

▲ Figure 5

Calculator/Computer Exercises

53. The sum of the volumes of a cube and a ball is, and remains, 24000 cm^3. The radius of the ball increases at a rate of 2 cm/s. At the moment when the diameter is 26 cm, how fast is the side length of the cube decreasing?

54. An approximation of the heat index T in terms of actual temperature x (°F) and relative humidity h, where h is a number in the interval $[0, 100]$, is

$$\begin{aligned} T = -42.4 + 2.05x + 10.1h - 0.225\,xh - 6.84 \cdot 10^{-3}x^2 \\ - 5.48 \cdot 10^{-2} \cdot h^2 + 1.23 \cdot 10^{-3} \cdot x^2h \\ + 8.53 \cdot 10^{-4}xh^2 - 1.99 \cdot 10^{-6}x^2h^2. \end{aligned}$$

(See Exercise 50 for a simpler approximation.) Suppose that the actual temperature is increasing at the rate of 1.7°F/hr at the moment $x = 88$ and $h = 60$. How fast must humidity decrease for the heat index to be unchanged?

4.2 The Mean Value Theorem

In this section, we begin to investigate the properties of the graph of a function f that can be deduced from the derivative f'. Our first example is simple enough to be analyzed without calculus. It will, however, suggest ways in which powerful methods of calculus can be used to understand more complicated functions.

▶ **EXAMPLE 1** Population biologist Raymond Pearl observed that the mass x of a yeast culture grows at the rate $R(x) = k \cdot x \cdot (2c - x)$, where c and k are positive constants. For what value of the mass x is its rate of change $R(x)$ greatest?

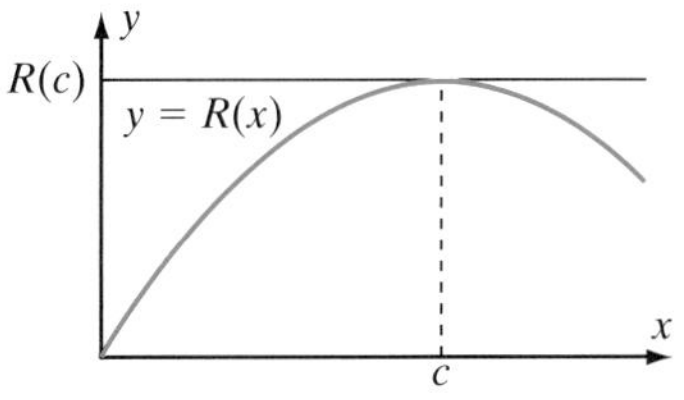

▲ Figure 1

Solution Figure 1 shows the graph of $y = R(x)$. Because $R(x)$ is a quadratic expression in x, we may complete the square as follows:

$$\begin{aligned} R(x) &= k \cdot x \cdot (2c - x) \\ &= kc^2 - k(c^2 - 2cx + x^2) \\ &= kc^2 - k(x - c)^2. \end{aligned}$$

Because $k \cdot (x - c)^2 > 0$ for $x \neq c$, we see that

$$R(x) = kc^2 - k(x-c)^2 < kc^2 - 0 = kc^2 = R(c) \qquad \text{for } x \neq c.$$

Thus the maximum value of $R(x)$ is attained when $x = c$. ◀

INSIGHT In Example 1, the tangent line to the graph of R at $(c, R(c))$ appears to be horizontal, or, equivalently, has zero slope (see Figure 1). Indeed, we calculate that

$$\frac{d}{dx}R(x) = \frac{d}{dx}\left(kc^2 - k(x-c)^2\right) = \frac{d}{dx}(kc^2) - k\frac{d}{dx}(x-c)^2 = 0 - k \cdot 2(x-c)^1\frac{d}{dx}(x-c) = -2k(x-c),$$

which is zero for $x = c$. We see that the maximum value of $R(x)$ occurs at *precisely* the value of x at which $R'(x)$ is zero. What we have just observed is not a coincidence: In this section, we will learn why the maximum value of a differentiable function is located at a point where its derivative vanishes.

Maxima and Minima

To use the derivative for identifying maximum and minimum values of a function, we must first develop an appropriate notion of maximum and minimum. Remember that $f'(c)$ is defined as the limit of $(f(x)-f(c))/(x-c)$ as x approaches c. We cannot expect $f'(c)$ to contain information about f far away from c. The information that $f'(c)$ contains must be *local* in nature.

DEFINITION Let f be a function with domain S. We say that f has a *local maximum* at the point c in S if there is a $\delta > 0$ such that $f(x) \leq f(c)$ for all x in S such that $|x-c| < \delta$. We call $f(c)$ a *local maximum value* for f. The term *relative maximum*, which has the same meaning as *local maximum*, is also used. If $f(x) \leq f(c)$ for all x in S, then we say that f has an *absolute maximum* at c and that $f(c)$ is the *absolute maximum value* for f. Another term for *absolute maximum* is *global maximum* (refer to Figure 2).

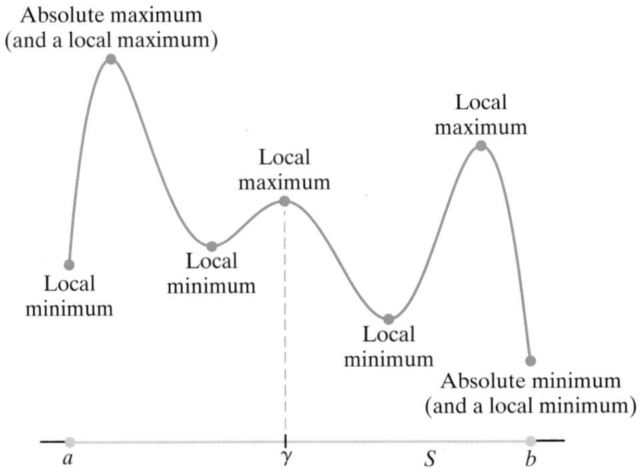

▲ Figure 2

DEFINITION Let f be a function with domain S. We say that f has a *local minimum* (or *relative minimum*) at the point c in S if there is a $\delta > 0$ such that $f(x) \geq f(c)$ for all x, in S such that $|x-c| < \delta$. We call $f(c)$ a *local minimum value* for f. If $f(x) \geq f(c)$ for *all* x in S, then we say that f has an *absolute minimum* (or *global minimum*) at c and that $f(c)$ is the *absolute minimum value* for f (refer to Figure 2).

The term *extremum* refers to either a local maximum or a local minimum. The plural forms of "extremum," "maximum," and "minimum," are "extrema," "maxima," and "minima," respectively.

If f has a local maximum at c, then f takes its greatest value at c only when compared with nearby points. If f has a local minimum at c, then f takes its least value at c when compared with values of $f(x)$ for x near c. By contrast, an *absolute* maximum or minimum at c is determined by comparing $f(c)$ with the values of f at *all* points of the function's domain. In Figure 2, $f(a)$ is *not* the least value that the function f takes, but it is the least when compared to nearby values in the domain of f. Therefore f has a local minimum at a but not an absolute minimum. Similarly, $f(\gamma)$ is not the greatest value that the function f takes, so f does not have an absolute maximum at γ. However, $f(\gamma)$ is the greatest value when compared to nearby values. Therefore f has a local maximum at γ.

If $f(c)$ is the greatest value of f when compared to all points of the domain, then it is certainly the greatest value when compared to all nearby points. In other words, an absolute maximum value is a local maximum value. The converse is not necessarily true, as Figure 2 shows. Similarly, if an absolute minimum value occurs at a point c, then a local minimum value also occurs at c, although the converse is not true.

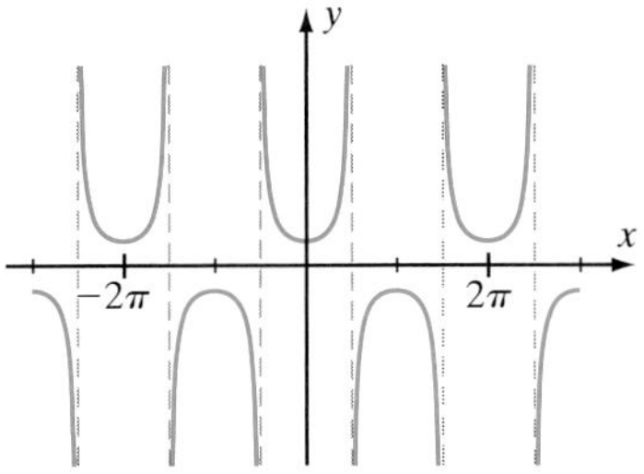

▲ Figure 3 $f(x) = \sec(x)$

▶ **EXAMPLE 2** Discuss local and absolute extreme values for the function $f(x) = \sec(x), \quad -3\pi \le x \le 3\pi$.

Solution Examine the graph of f in Figure 3. The function f has many local maxima, for instance, at -3π, $-\pi$, π, and 3π. It also has many local minima, for instance, at -2π, 0, and 2π. None of these local maxima and minima are absolute maxima and minima, for $\sec(x)$ takes both positive and negative values that are large in absolute value, without bound. ◀

Locating Maxima and Minima

We want to know how to find the local maxima and minima of a given function. The next theorem identifies some of the points we should examine.

THEOREM 1 (**Fermat**) Let f be defined on an open interval that contains the point c. Suppose that f is differentiable at c. If f has a local extreme value at c, then $f'(c) = 0$.

Proof. Suppose that f has a local maximum value at c. By definition, there is a $\delta > 0$ such that $f(x) \le f(c)$ for all x within δ of c. If Δx is smaller than δ, then $x = c + \Delta x$ is within δ of c and so $f(c+\Delta x) - f(c) \le 0$. When we write $f'(c) = \lim_{\Delta x \to 0} \left(f(c+\Delta x) - f(c)\right)/\Delta x$ in terms of its two one-sided limits, we obtain

$$f'(c) = \lim_{\Delta x \to 0^+} \frac{f(c+\Delta x) - f(c)}{\Delta x} = \lim_{\Delta x \to 0^+} \frac{\text{negative numerator}}{\text{positive denominator}} \le 0$$

and

$$f'(c) = \lim_{\Delta x \to 0^-} \frac{f(c+\Delta x) - f(c)}{\Delta x} = \lim_{\Delta x \to 0^-} \frac{\text{negative numerator}}{\text{negative denominator}} \ge 0.$$

Taken together, these two inequalities imply that $f'(c) = 0$. Similar reasoning leads to the same conclusion when f has a local minimum value at c. ■

INSIGHT Fermat's Theorem has a geometric interpretation: The graph of a differentiable function has a horizontal tangent at a local extremum (as we observed in Figure 1). Fermat's Theorem, it should be noted, does *not* apply to endpoints, such as a and b in Figure 2.

We must be careful when applying Theorem 1. Notice the direction of the implication: *If a local extreme value occurs at c, then* $f'(c) = 0$. Suppose we differentiate a function f and find the places where $f'(x) = 0$. Fermat's Theorem says that these are the points we should be considering to identify the local extrema of f. However, Fermat's Theorem does *not* tell us that these candidates must actually be extrema. Figure 4 shows us some of the possibilities that can occur. Notice in particular that $f'(\gamma) = 0$, but f does not have a local extremum at γ. In summary, Fermat's Theorem *cannot* be used for concluding that a local extremum occurs at a point; it can only be used to locate *candidates* for a local extremum.

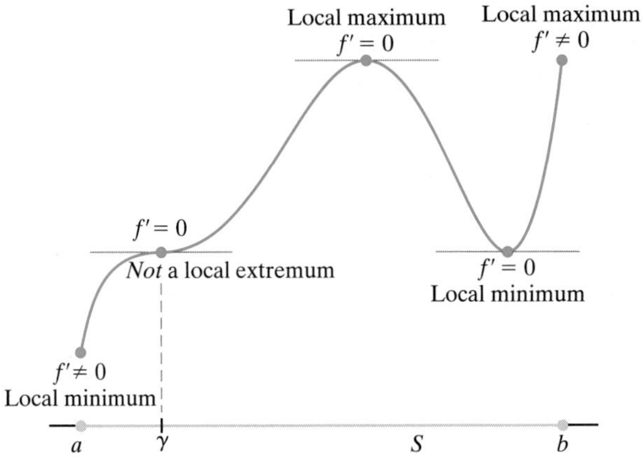

▲ Figure 4

▶ **EXAMPLE 3** Use the first derivative to locate local and absolute extrema for the function $f(x) = \cos(x)$.

Solution First note that f is differentiable on the entire real line, so Fermat's Theorem applies. We calculate that $f'(x) = -\sin(x)$; therefore, $f'(x) = 0$ when $x = \ldots, -3\pi, -2\pi, -\pi, 0, \pi, 2\pi, \ldots$. Are there local maxima or local minima at these points, or neither? First, there are local minima (indeed absolute minima) at the points $\ldots, -3\pi, -\pi, \pi, \ldots$. This is true because the value of f is -1 at these points, which is certainly the absolute minimum value for cosine because $-1 \le \cos(x) \le 1$ for all x. Second, there are local maxima (indeed absolute maxima) at the points $\ldots, -2\pi, 0, 2\pi, \ldots$. This is true because the value of f is $+1$ at these points, which is certainly an absolute maximum value for cosine. Are there any other local maxima and minima for f? If there were, then Fermat's Theorem guarantees that f' would vanish there. But we have already checked all the points where f' vanishes, so we have found all local maxima and minima. ◀

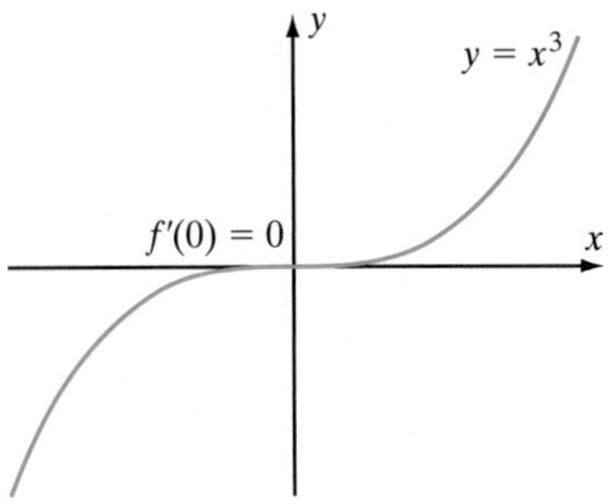

▲ Figure 5

▶ **EXAMPLE 4** Use Fermat's Theorem to determine whether the function $f(x) = x^3$ has any local or absolute extrema.

Solution Observe that f is differentiable at every point. Therefore we may use Fermat's Theorem. We calculate that $f'(x) = 3x^2$. Thus the only zero of f' is the point $x = 0$. If f has a local extremum, then it must occur at $x = 0$. A glance at the graph of f in Figure 5 shows that there is neither a local maximum nor a local minimum at 0. Does this contradict Fermat's Theorem? The theorem says that *if* f has a local maximum or minimum at $x = c$, then $f'(c) = 0$. It does *not* say that if $f'(c) = 0$, then there must be a local maximum or a minimum at c. Thus there is no contradiction. ◀

INSIGHT It is important to understand that we really did use Fermat's Theorem in Example 4. We used it to conclude that no value of x, except *possibly* $x = 0$, could result in a local extremum of f. After using Fermat's Theorem to rule out most points, we must then investigate the points that remain. Fermat's Theorem says nothing about the behavior of f at these remaining points.

Rolle's Theorem and the Mean Value Theorem

In Section 4.3, we will resume our investigation of extreme values. For now we continue to study what the derivative says about the graph of a function f.

THEOREM 2 **(Rolle's Theorem)** Let f be a function that is continuous on $[a,b]$ and differentiable on (a,b). If $f(a) = f(b)$, then there is a number c in the open interval (a,b) such that $f'(c) = 0$.

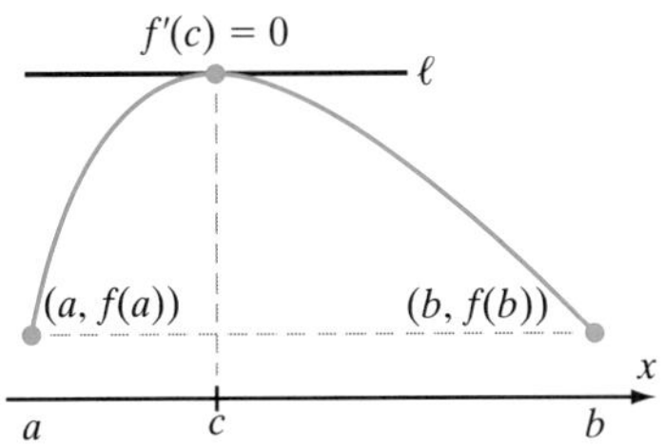

▲ Figure 6 Rolle's Theorem

Proof. Because f is continuous on $[a,b]$, the Extreme Value Theorem (Section 2.3 of Chapter 2) guarantees that f has an absolute minimum and an absolute maximum on the interval $[a,b]$. If one of these extreme values is assumed at a point c inside the open interval (a,b), then Fermat's Theorem tells us that $f'(c) = 0$ (see Figure 6). Otherwise, one of these extreme values occurs at a and the other at b. In this case, the maximum and minimum values of f are the same because $f(a) = f(b)$. We deduce that f is constant and $f'(c) = 0$ for any choice of c in the interval (a,b). Either way, there is a point in the interior of the interval at which f' is zero. ■

▶ **EXAMPLE 5** A ball is thrown straight up from ground level. It returns under the force of gravity to ground level at some time t_0. Assume only that the motion is described by a differentiable function. Is the velocity of the ball ever 0?

Solution If $h(t)$ is height, then $h(0) = h(t_0) = 0$. By Rolle's Theorem, there is a time c between 0 and t_0 when $h'(c) = 0$. In other words, the velocity of the ball is zero at time c. Our physical intuition reassures us on this point: The time c is the time when the ball reaches its highest point. At that instant the ball is traveling neither up nor down. ◀

When interpreted geometrically, Rolle's Theorem says that, if f is a differentiable function with $f(a) = f(b)$, then there is a point c between a and b at which the graph of f has a horizontal tangent line ℓ (refer to Figure 6). Notice that ℓ is parallel to the line segment that joins the endpoints $(a, f(a))$ and $(b, f(b))$ of the graph. The next theorem states that there is an analogous result when $f(a) \neq f(b)$.

THEOREM 3 **(The Mean Value Theorem)** If f is a function that is continuous on $[a,b]$ and differentiable on (a,b), then there is a number c in the open interval (a,b) such that

$$f'(c) = \frac{f(b) - f(a)}{b - a}.$$

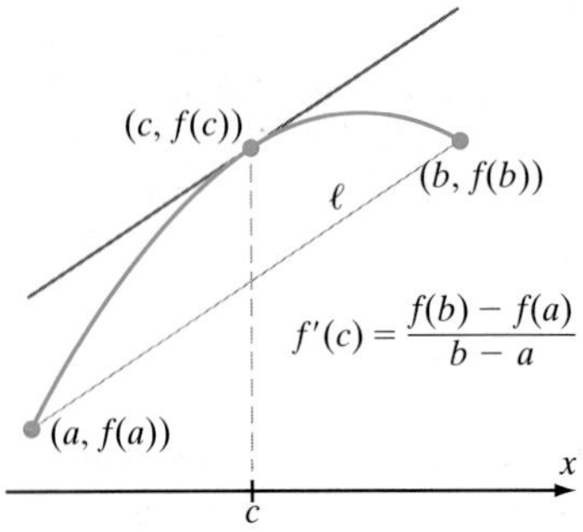

▲ Figure 7

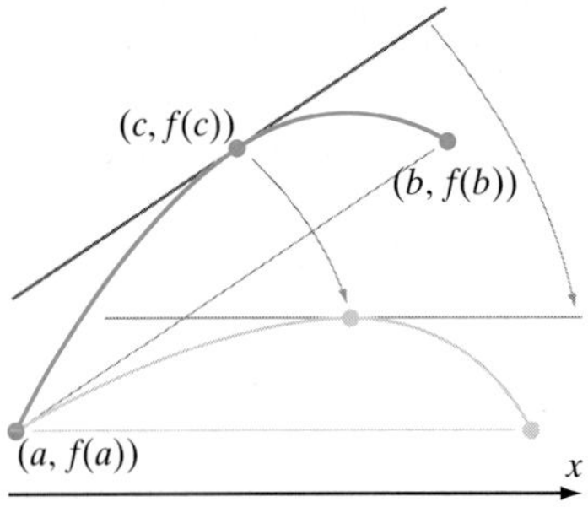

▲ Figure 8

INSIGHT Look at Figure 7. The Mean Value Theorem says that there is a number c between a and b for which the tangent line at $(c, f(c))$ is parallel to the line ℓ passing through the points $(a, f(a))$ and $(b, f(b))$. This assertion is really Rolle's Theorem in disguise. To see this, simply rotate Figure 7 so that the line ℓ is horizontal (see Figure 8)—the picture is essentially the same as that for Rolle's Theorem. Informally, the Mean Value Theorem is a "rotated" version of Rolle's Theorem. See Exercise 82 for an analytical proof.

▶ **EXAMPLE 6** An automobile travels 120 miles in three hours. Assuming that the position function p is continuous on the closed interval $[0, 3]$ and differentiable on the open interval $(0, 3)$, can we conclude that at some moment in time, the car is going precisely 40 miles per hour?

Solution Common sense tells us that the answer must be "yes." If the car were always traveling at less than 40 mph, then in 3 hours it would travel less than 120 miles. Similarly, if the car were always traveling faster than 40 mph, then in 3 hours it would go more than 120 miles. Therefore the car either travels at a constant speed of 40 mph, in which case the answer to the question is obviously "yes," or there is a time when the car travels at a speed less than 40 mph and another time when it travels at more than 40 mph. Because we expect the speed to be a continuous function, we expect it to assume the intermediate value 40.

Let us use the Mean Value Theorem to give a short, more direct answer. We are given that $p(3) = p(0) + 120$. By the Mean Value Theorem, there is a point c in $(0, 3)$ such that

$$p'(c) = \frac{p(3) - p(0)}{3 - 0} = \frac{(120 + p(0)) - p(0)}{3} = \frac{120}{3} = 40,$$

as required. ◀

An Application of the Mean Value Theorem

We know that the derivative of a constant function is identically zero. The Mean Value Theorem enables us to see that the converse is also true:

THEOREM 4 Let f be a differentiable function on an interval (α, β). If $f'(x) = 0$ for each x in (α, β), then f is a constant function.

Proof. Let a and b be any two points in (α, β). By the Mean Value Theorem, there is a c between a and b such that

$$\frac{f(b) - f(a)}{b - a} = f'(c).$$

But, by hypothesis, $f'(c) = 0$. Thus $f(b) - f(a) = 0$, or $f(b) = f(a)$. Therefore f takes the same value at any two points a, b in the interval (α, β). This is just another way of saying that f is constant. ■

▶ **EXAMPLE 7** In Section 2.6, we used properties of the exponential function to show that

$$\ln(xy) = \ln(x) + \ln(y)$$

for all positive x and y. Use Theorem 4 to derive this identity in a different way.

Solution Fix y and consider the function $f(x) = \ln(xy) - \ln(x)$ for x in $(0, \infty)$. Letting $u = xy$, we calculate

$$\begin{aligned} f'(x) &= \frac{d}{dx}\ln(xy) - \frac{d}{dx}\ln(x) \\ &= \frac{d}{dx}\ln(u) - \frac{1}{x} \\ &= \left(\frac{d}{du}\ln(u)\right)\frac{du}{dx} - \frac{1}{x} \\ &= \frac{1}{u}\frac{du}{dx} - \frac{1}{x} \\ &= \frac{1}{xy}\frac{d}{dx}(xy) - \frac{1}{x} \\ &= \frac{1}{xy}\cdot y - \frac{1}{x} \\ &= 0. \end{aligned}$$

It follows from Theorem 4 that $f(x)$ is constant. To evaluate that constant, we may choose any convenient value in the domain of f. Selecting the value 1, we calculate

$$f(1) = \left(\ln(xy) - \ln(x)\right)\Big|_{x=1} = \ln(y) - \ln(1) = \ln(y) - 0 = \ln(y).$$

Because $f(x)$ is constant, we have $f(x) = f(1)$, or $\ln(xy) - \ln(x) = \ln(y)$, for every positive x. Adding $\ln(x)$ to each side of this equation yields the required identity. ◀

We prove our next theorem now because it follows quickly from Theorem 4. We will not use it until Section 4.9 but, beginning with that section, it will play an important role in our studies.

THEOREM 5 If F and G are differentiable functions such that $F'(x) = G'(x)$ for every x in (α, β), then there is a constant C such that $G(x) = F(x) + C$ for every x in the interval (α, β).

Proof. Theorem 4 applied to the function $f(x) = G(x) - F(x)$ tells us that there is a constant C such that $f(x) = C$ for each x in (α, β). Thus $G(x) - F(x) = C$, or $G(x) = F(x) + C$. ■

▶ **EXAMPLE 8** Suppose that $F'(x) = 2x$ for all x and that $F(1) = 10$. Find $F(4)$.

Solution If we set $G(x) = x^2$, then $G'(x) = 2x$. Thus $F'(x) = G'(x)$ for every x. According to Theorem 5, the function F that we seek must have the form $x^2 = F(x) + C$ where C is a constant. Now we use the information $F(1) = 10$ to determine the specific value of C. Substituting $x = 1$ in $x^2 = F(x) + C$, we obtain $1^2 = F(1) + C$, or $F(1) = 1 - C$. Equating our two values for $F(1)$, we have $10 = 1 - C$, or $C = -9$. Therefore $x^2 = F(x) - 9$, or $F(x) = x^2 + 9$. It follows that $F(4) = 16 + 9 = 25$. ◀

QUICK QUIZ

1. Suppose that f is differentiable on an open interval containing the point c.
   a. True or false: if $f'(c) = 0$, then a local extremum of $f(x)$ occurs at $x = c$.
   b. True or false: if a local extremum of $f(x)$ occurs at $x = c$, then $f'(c) = 0$.
2. The function $f(x) = xe^{-x}$ has a local maximum. Where does it occur?
3. What upper bound on the number of local extrema of $f(x) = x^4 + ax^3 + bx^2 + cx + d$ can be deduced from Fermat's Theorem?
4. If f is continuous on $[7, 13]$, differentiable on $(7, 13)$, and if $f(7) = 11$ and $f(13) = 41$, then what value must $f'(x)$ assume for some x in $(7, 13)$?
5. If $F'(x) = 2\cos(x)$ and $F(\pi) = 3$, then what is $F(\pi/6)$?

Answers

1. a. false; b. true 2. 1 3. 3 4. 5 5. 4

EXERCISES

Problems for Practice

In each of Exercises 1–6, a function f is given. Locate each point c for which $f(c)$ is a local extremum for f. (Calculus is not needed for these exercises.)

1. $f(x) = (x - 2)^2 + 3$
2. $f(x) = 2 - 4(x - 6)^2$
3. $f(x) = 1/(x^2 + 1)$
4. $f(x) = 1/((x - 2)^2 + 4)$
5. $f(x) = 3\sin(4x)$
6. $f(x) = \exp(-|x|)$

In each of Exercises 7–22, use Fermat's Theorem to locate each c for which $f(c)$ is a candidate extreme value of the given function f.

7. $f(x) = 2x^2 - 24x + 36$
8. $f(x) = 12x^2 + 48x$
9. $f(x) = x^4 - 2x^2 + 1$
10. $f(x) = \cot(x)$
11. $f(x) = (x - 3)(x + 5)^3$
12. $f(x) = x/(x^2 + 1)$
13. $f(x) = x - \ln(x)$
14. $f(x) = x + 1/x$
15. $f(x) = e^x - x$
16. $f(x) = xe^x$
17. $f(x) = e^x/x$
18. $f(x) = x\ln(x)$
19. $f(x) = \ln(x)/x$
20. $f(x) = \sqrt{x} + 2/\sqrt{x}$
21. $f(x) = x^{3/2} - 9x^{1/2}$
22. $f(x) = x/(1 + x + x^2)$

In each of Exercises 23–28, verify that the hypotheses of the Rolle's Theorem hold for the given function f and interval I. The theorem asserts that $f'(c) = 0$ for some c in I. Find such a c.

23. $x + 1/x$, $I = [1/3, 3]$
24. $\exp(x) + \exp(-x)$, $I = [-1, 1]$
25. $x - x^{3/2}$, $I = [0, 1]$
26. $\cos(2x) + \sin(x)$, $I = [\pi/6, 5\pi/6]$
27. $f(x) = x^3 + 4x^2 + x - 6$, $I = [-2, 1]$
28. $\ln(x^2 - 4x + 7)$, $I = [1, 3]$

In each of Exercises 29–34, verify that the hypotheses of the Mean Value Theorem hold for the given function f and interval I. The theorem asserts that, for some c in I, the derivative $f'(c)$ assumes what value?

29. $f(x) = x^4 + 7x^3 - 9x^2 + x$, $I = [0, 1]$
30. $f(x) = x^2 \cdot \sin(x)$, $I = [0, \pi/2]$
31. $f(x) = x^{1/5}$, $I = [1, 32]$
32. $f(x) = -(x-2)^2 + 4$, $I = [-2, 4]$
33. $f(x) = -(x-1)^4 + 1$, $I = [-1, 2]$
34. $f(x) = x^2 + 2x + 3$, $I = [-1, 4]$

In each of Exercises 35–40, an expression $f(x)$ is given. Find all functions F such that $F'(x) = f(x)$.

35. 5
36. $3x$
37. $x^2 + \pi$
38. $4x^{1/2} + 3$
39. $\cos(x)$
40. $3\sin(x) - 4x$

In each of Exercises 41–48, use the given information to find $F(c)$.

41. $F'(x) = 3x^2$, $F(2) = -4$, $c = 3$
42. $F'(x) = 4\sin(x)$, $F(\pi/3) = 3$, $c = \pi$
43. $F'(x) = 4/x$, $F(e^2) = 7$, $c = e^3$
44. $F'(x) = \sec(x)^2$, $F(\pi/4) = 0$, $c = \pi/3$
45. $F'(x) = e^x$, $F(2) = 2 + e^2$, $c = 3$
46. $F'(x) = 2/\sqrt{x}$, $F(4) = 6$, $c = 9$

47. $F'(x) = 2/x^2$, $F(-1) = 4$, $c = 4$
48. $F'(x) = 2^x \cdot \ln^2(2)$, $F(1) = 1 + 2\ln(2)$, $c = 3$

Further Theory and Practice

In Exercises 49–52 find a value c the existence of which is guaranteed by Rolle's Theorem applied to the given function f on the given interval I.

49. $f(x) = e^x \sin(x)$, $I = [0, \pi]$
50. $f(x) = x^3 - x$, $I = [0, 1]$
51. $f(x) = \sin(x) + \cos(x)$, $I = [-\pi/4, 3\pi/4]$
52. $f(x) = (x^2 + x)/(x^2 + 1)$, $I = [-1, 0]$

In Exercises 53–56, find a value c whose existence is guaranteed by the Mean Value Theorem applied to the given function f on the interval $I = [a, b]$.

53. $f(x) = x/(x-1)$, $I = [2, 4]$
54. $f(x) = Ax^2 + B \quad (A \neq 0)$, $I = [a, b]$
55. $f(x) = x^3 + 3x - 1$, $I = [1, 7]$
56. $f(x) = x + 1/x$, $I = [1, 2]$

In each of Exercises 57–60, a continuous function f is given on an interval $I = [a, b]$. Sketch the graph of $y = f(x)$ for x in I. In each case, explain why there can be no c in (a, b) for which equation (4.2.1) holds. Explain why the Mean Value Theorem is not contradicted.

57. $f(x) = |x - 5|$, $I = [4, 7]$
58. $f(x) = (x^2 - 2x + 1)^{1/4}$, $I = [0, 2]$
59. $f(x) = \exp(-|x|)$, $I = [-1, 1]$
60. $f(x) = \begin{cases} x - 5 & \text{if } -3 \leq x \leq -1 \\ -x - 7 & \text{if } -1 < x \leq 2 \end{cases}$, $I = [-3, 2]$

61. Use Theorem 4 with $f(x) = \cos^2(x) + \sin^2(x)$ to prove the identity $\cos^2(x) + \sin^2(x) = 1$.
62. Use Theorem 5 with $F(x) = \sec^2(x)$ and $G(x) = \tan^2(x)$ to prove the identity $1 + \tan^2(x) = \sec^2(x)$.
63. Show that $3x^4 - 4x^3 + 6x^2 - 12x + 5 = 0$ has at most two real valued solutions.
64. Show that $x^3 - 3x^2 + 4x - 1 = 0$ has exactly one real root.
65. Show that $x^9 + 3x^3 + 2x + 1 = 0$ has exactly one real root.
66. Use Rolle's Theorem to show that $p(x) = x^3 + ax^2 + b$ cannot have three negative roots.
67. Use Rolle's Theorem to prove that $p(x) = x^3 + ax^2 + b$ cannot have two distinct negative roots if $a < 0$.
68. Show that $p(x) = x^3 + ax + b$ cannot have three negative roots.
69. Use Rolle's Theorem to prove that $p(x) = x^{3+} ax + b$ can have only one real root if $a > 0$.
70. Suppose that $f(x)$, $g(x)$, and $h(x)$ are continuous for $x \geq a$ and differentiable for $x > a$. Use the Mean Value Theorem to show that if $h(a) = 0$ and $0 < h'(x)$ for $a < x$, then $0 < h(x)$ for $a < x$. Deduce that if $g(a) \leq f(a)$ and $g'(x) < f'(x)$ for $a < x$, then $g(x) < f(x)$ for $a < x$.
71. Bernoulli's inequality states that $1 + px < (1 + x)^p$ for $0 < x$ and $1 < p$. Use Exercise 70 to prove this.
72. Use the Mean Value Theorem to show that $(1 + x)^p < 1 + px$ for $0 < x$ and $p < 1$.
73. Use the Mean Value Theorem to show that $\sin(x) < x$ for $0 < x$.
74. Suppose that f satisfies the conditions of the Mean Value Theorem on $[a, b]$. Show that if $0 < h < b - a$, then there exists a $\theta \in (0, 1)$ such that $f(a + h) = f(a) + h \cdot f'(a + \theta h)$.
75. Suppose that p is a differentiable function such that $p(a) = p(b) = 0$. Let k be *any* real number. Use the function $x \mapsto e^{-kx} p(x)$ to show that there is a c between a and b such that $p'(c) = kp(c)$.
76. Define $f(x) = \sin(1/x)$ for x in the interval $(0, 1)$. Use the Mean Value Theorem to show that f' takes on all real values.
77. Use the Mean Value Theorem to prove that

$$\lim_{x \to +\infty} (\sqrt{x+1} - \sqrt{x}) = 0.$$

78. Consider $f(x) = \sqrt{1 + x}$ on the interval $[h/2, h]$. Use the Mean Value Theorem to prove that

$$\lim_{x \to 0} \left(\frac{\sqrt{1+h} - \sqrt{1 + h/2}}{h} \right) = \frac{1}{4}.$$

79. Suppose that $f: \mathbb{R} \to \mathbb{R}$ is a differentiable function and that $|f'(x)| \leq C_1$ for all x, where C_1 is a numerical constant. Prove that

$$|f(s) - f(t)| C_1 \cdot |s - t|$$

for all $s, t \in \mathbb{R}$.
80. Use Theorem 4 to show that

$$\arctan(x) + \arctan(1/x) = \frac{\pi}{2}, \quad x > 0.$$

81. Use Theorem 5 to show that

$$2 \cdot \arctan(\sqrt{x}) = \arcsin\left(\frac{x-1}{x+1}\right) + \frac{\pi}{2}, \quad x \geq 0.$$

82. Analytic Proof of the Mean Value Theorem Suppose that f satisfies the hypotheses of the Mean Value Theorem. Define g by

$$g(x) = f(x) - \left(f(a) + \frac{f(b) - f(a)}{b - a} \cdot (x - a) \right)$$

Apply Rolle's Theorem to g. Deduce the assertion of the Mean Value Theorem.
83. Suppose that f satisfies the hypotheses of Rolle's Theorem on $[a, b]$. In this exercise we will construct a sequence $\{[a_n, b_n]\}_{n=0}^{\infty}$ of intervals with the properties
(1) $[a_n, b_n] \subset [a_{n-1}, b_{n-1}]$ for $n \geq 1$;
(2) $b_n - a_n = (b_{n-1} - a_{n-1})/2$ for $n \geq 1$;
(3) $f(a_n) = f(b_n)$ for $n \geq 0$; and

(4) There exists a point $c \in (a,b)$ for which $c = \lim_{n\to\infty} a_n = \lim_{n\to\infty} b_n$ and

$$f'(c) = \lim_{n\to\infty} \frac{f(b_n) - f(a_n)}{b_n - a_n} = 0.$$

If $f((a+b/2) = f(a)$, then set $a_0 = (a+b)/2$ and $b_0 = b$. Otherwise, set $a_0 = a$ and $b_0 = b$. Let $g(x) = f(x + (b_0 - a_0)/2) - f(x)$. Use the Intermediate Value Theorem to find an $a_1 \in [a_0, (a_0 + b_0)/2)$ such that $g(a_1) = 0$. Let $b_1 = a_1 + (b-a)/2$. Show that $f(a_1) = f(b_1)$. Replacing a_0 and b_0 with a_1 and b_1 in the definition of g, find a_2 and b_2 for which properties (1), (2), and (3) hold for $n = 2$. Repeat this construction indefinitely. Prove property (4), thereby establishing Rolle's Theorem.

▶ **Discrete Dynamical Systems.** Suppose that $\Phi : \mathbb{R} \to \mathbb{R}$ is a continuous function. The collection of sequences $\{x_n\}_{n\geq 0}$ that satisfies $x_{n+1} = \Phi(x_n)$ for $n \geq 0$ is called the *discrete dynamical system associated with* Φ. Notice that each element $\{x_n\}$ of a discrete dynamical system is determined by its first element x_0. A number x_* such that $\Phi(x_*) = x_*$ is called an *equilibrium point* of the dynamical system. We say that an equilibrium point x_* of Φ is *stable* if there is a $\delta > 0$ such that each element $\{x_n\}$ of the dynamical system associated to Φ converges to x_* provided that $|x_* - x_0| < \delta$. Exercises 84–89 are concerned with these ideas. ◀

84. Suppose that x_* is an equilibrium point of the dynamical system. Show that if $x_N = x_*$, then $x_n = x_*$ for all $n \geq N$.

If $x_n = x_*$, then $x_{N+1} = \Phi(x_N) = \Phi(x_*) = x_*$. Continuing, $x_{N+2} = \Phi(x_{N+1}) = \Phi(x_*) = x_*$, and so on. (In words, once a member of the sequence $\{x_n\}$ equals x_*, its successor equals x_* as well.) It follows that once a member of the sequence $\{x_n\}$ equals x_*, all subsequent members equal x_*

85. Suppose that $\{x_n\}_{n\geq 0}$ is an element of the discrete dynamical system associated with Φ. Suppose that $\xi = \lim_{n\to\infty} x_n$. Show that ξ is an equilibrium point of the discrete dynamical system associated with Φ.

86. Suppose that x_* is an equilibrium point of the dynamical system Φ. Suppose that Φ' exists and is continuous on an open interval centered at x_*. Suppose also that $|\Phi'(x_*)| < 1$. Show that there are positive numbers δ and K with $K < 1$ such that

$$|\Phi'(x)| < K \quad \text{for every } x \text{ in } (x_* - \delta, x_* + \delta).$$

Illustrate this with a sketch.

87. *Continuation:* Suppose that x_0 is in the interval $(x_* - \delta, x_* + \delta)$. Deduce that $x_1 = \Phi(x_0)$ is also in this interval by observing that

$$|x_1 - x_*| = |\Phi(x_0) - \Phi(x_*)| \leq K|x_0 - x_*|.$$

Explain the reasoning behind this. Illustrate the position of x_*, x_0, and x_1 with a sketch.

88. *Continuation:* Follow the steps of the preceding exercise to deduce that

$$|x_n - x_*| \leq K^n |x_0 - x_*|.$$

89. *Continuation:* Prove the following theorem: If x_* is an equilibrium point of the dynamical system Φ such that (1) Φ' exists and is continuous on an open interval centered at x_* and (2) $|\Phi'(x_*)| < 1$, then x_* is a stable equilibrium of Φ.

90. Let a be a positive constant and set

$$T(x) = a \cdot \operatorname{sech}^{-1}\left(\frac{x}{a}\right) - \sqrt{a^2 - x^2}.$$

The curve $y = T(x)$ is known as a *tractrix*. Use Theorem 5 to show that if y satisfies the two equations

$$\frac{d}{dx} y(x) = -\frac{\sqrt{a^2 - x^2}}{x}, \quad y(a) = 0,$$

then $y(x) = T(x)$.

Calculator/Computer Exercises

▶ In each of Exercises 91–94, calculate and plot the derivative f' of the given function f. Use this plot to identify candidates for the local extrema of f. Add the plot of f to the window containing the graph of f'. From this second plot, determine the behavior of f at each candidate for a local extremum. ◀

91. $f(x) = x^3 - 2x + \cos(x)$

92. $f(x) = x^2 - 2x\ln(1 + x^2) + x - 4$

93. $f(x) = x - 2\exp(-x^2)$

94. $f(x) = \sin^2(x) - x^3 + 5x + 20$

▶ In Exercises 95–98, approximate the value c guaranteed by the application of the Mean Value Theorem to the given function f on the given interval $I = [a, b]$. Graph the function, the tangent line at $(c, f(c))$, and the line segment between $(a, f(a))$ and $(b, f(b))$. ◀

95. $f(x) = x^5 + x^3 + 1$, $I = [0, 2]$

96. $f(x) = xe^x$, $I = [-3/2, 1]$

97. $f(x) = x^4 - 2x^3 + x^2 - 2x + 13$, $I = [-1, 3]$

98. $f(x) = x\sin(1/x)$, $I = [1/(4\pi), \pi/24]$

4.3 Maxima and Minima of Functions

In Section 4.2, we learned that the derivative is a useful tool for *finding* candidates for local extrema. However, we did not learn much about *identifying* these extrema—the subject of this section. We begin with a consideration of what the derivative tells us about the graph of a function. Recall the following definition from Section 2.6 in Chapter 2:

DEFINITION A function f is *increasing* on an interval I if $f(s) < f(t)$ whenever s and t are points in I with $s < t$ (see Figure 1). We say that f is *decreasing* if $f(s) > f(t)$ whenever $s < t$ (see Figure 2).

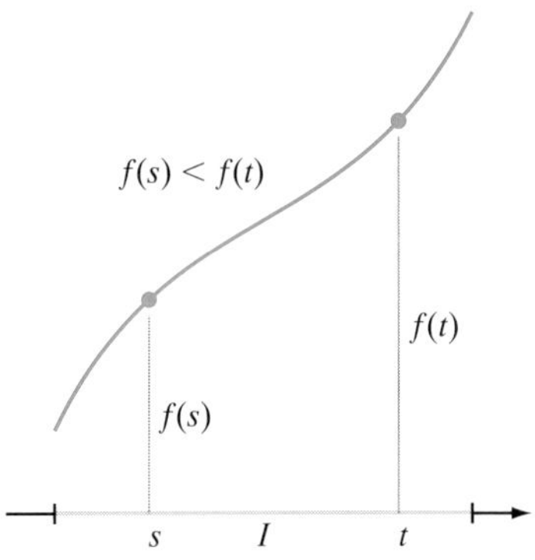

Figure 1 f is increasing on I.

▶ **EXAMPLE 1** Where is the function $f(x) = \sin(x)$ increasing or decreasing on the interval $(0, 2\pi)$?

Solution As Figure 3 shows, the function f is increasing on the intervals $(0, \pi/2)$ and $(3\pi/2, 2\pi)$. It is decreasing on the interval $(\pi/2, 3\pi/2)$. ◀

The statements in Example 1 about the sine function are obvious because it is a familiar function. What we need is a technique for telling where *any* differentiable function is increasing or decreasing.

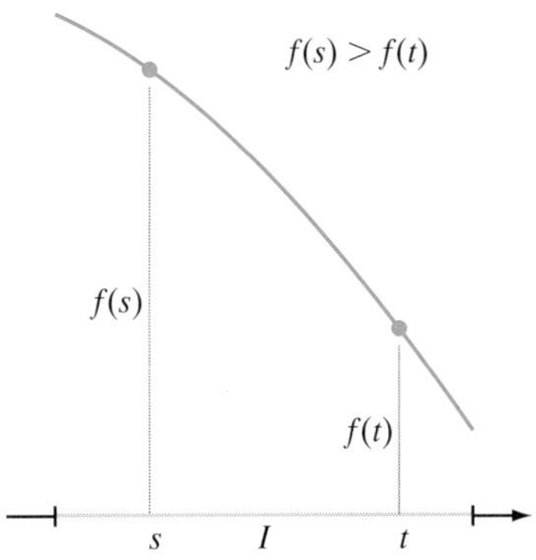

Figure 2 f is decreasing on I.

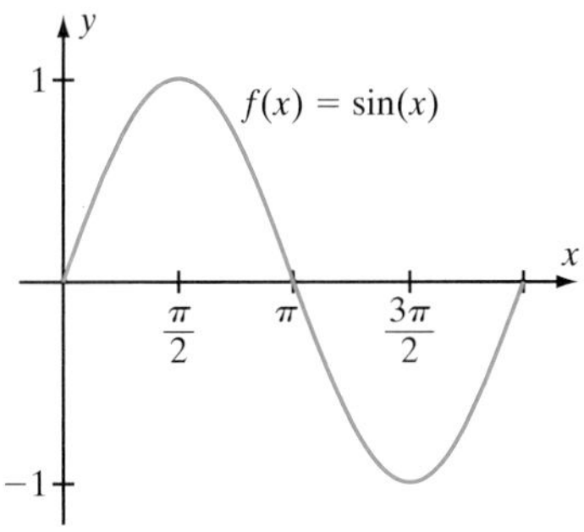

Figure 3

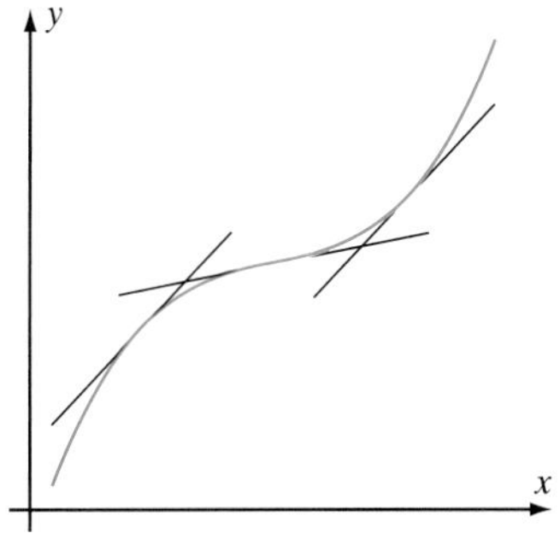

Figure 4

Using the Derivative to Tell When a Function Is Increasing or Decreasing

Geometric intuition suggests that rising tangent lines to the graph of $y = f(x)$ occur only where f is increasing (see Figure 4). Because the tangent line at $(c, f(c))$ rises if and only if its slope is positive, which is to say $f'(c) > 0$, we can use the derivative f' as a tool for locating intervals on which f increases. Similarly, falling tangent lines occur only where f is decreasing. The condition for a tangent line at $(c, f(c))$ to fall is that its slope $f'(c)$ is negative (Figure 5). The next theorem makes the relationship between the sign of f' and the behavior of f precise.

THEOREM 1 If $f'(x) > 0$ for each x in an interval I, then f is increasing on I. If $f'(x) < 0$ for each x in I, then f is decreasing on I.

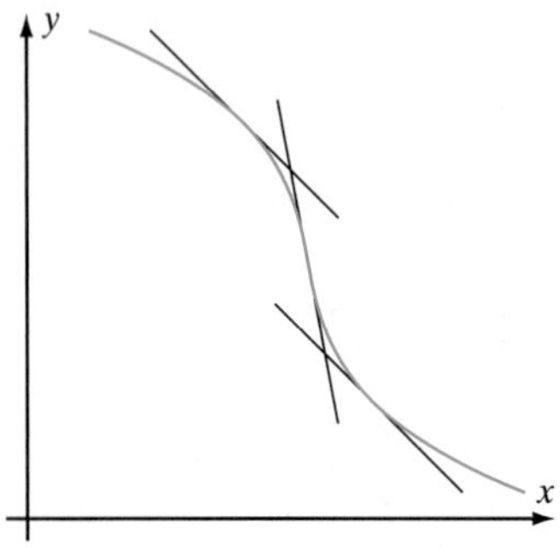

▲ Figure 5

Proof. Let $s < t$ be points in I with $s < t$. By the Mean Value Theorem applied to f on the interval $[s, t]$, there is a number c between s and t such that

$$\frac{f(t) - f(s)}{t - s} = f'(c).$$

But, by hypothesis, $f'(c) > 0$ and $t - s > 0$. It follows that

$$f(t) - f(s) = \underbrace{f'(c)}_{\text{positive}} \cdot \underbrace{(t - s)}_{\text{positive}} > 0$$

or $f(t) > f(s)$. Therefore f is increasing on I. The other statement may be proved similarly. ■

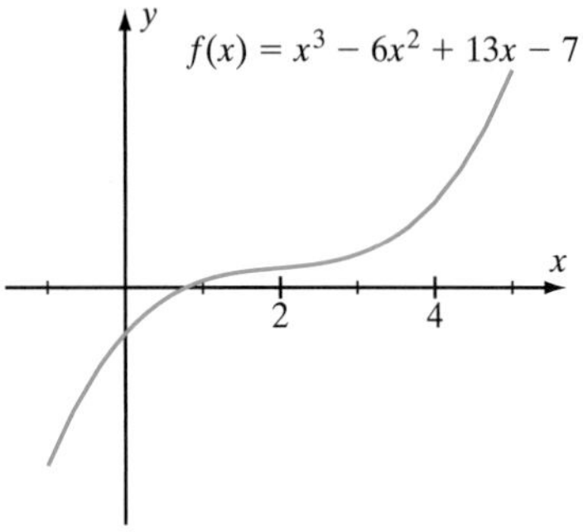

▲ Figure 6

▶ **EXAMPLE 2** Examine the function $f(x) = x^3 - 6x^2 + 13x - 7$ to determine intervals on which it is increasing or decreasing.

Solution The graph of f in Figure 6 suggests that f is increasing on the interval $[-1, 5]$ over which it is plotted. However, the plot of f over a finite interval cannot exhibit the behavior of f over the entire real line. Instead, we appeal to Theorem 1. By completing the square, we see that the derivative of f is given by

$$\begin{aligned} f'(x) &= \frac{d}{dx}(x^3 - 6x^2 + 13x - 7) \\ &= 3x^2 - 12x + 13 = 3(x^2 - 4x) + 13 \\ &= 3(x^2 - 4x + 4) - 3 \cdot (4) + 13 = 3(x-2)^2 + 1. \end{aligned}$$

Because $(x-2)^2 \geq 0$, we see that $f'(x) > 0$ for all x in $\mathbb{R}$. By Theorem 1, we conclude that f is increasing on all of $\mathbb{R}$. ◀

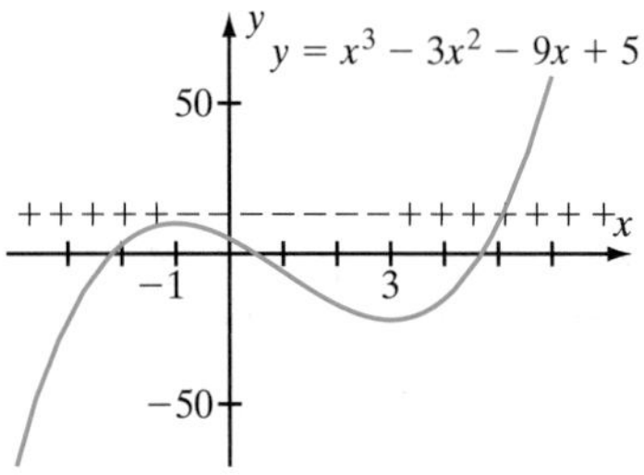

▲ Figure 7

It is sometimes helpful to notice that if $a < b$, then the two factors of $(x - a)(x - b)$ have the same sign if and only if $x < a$, in which case the factors $(x - a)$ and $(x - b)$ are both negative, or $x > b$, in which case the factors are both positive. We therefore see that

$$(x-a)(x-b) \quad \begin{cases} > \ 0 & \text{if } x < a \\ < \ 0 & \text{if } a < x < b \\ > \ 0 & \text{if } b < x. \end{cases}$$

▶ **EXAMPLE 3** On which intervals is the function $f(x) = x^3 - 3x^2 - 9x + 5$ increasing? On which intervals is it decreasing?

Solution Begin by calculating

$$f'(x) = 3x^2 - 6x - 9 = 3(x+1)(x-3) = 3\big(x - (-1)\big)(x-3).$$

Using the reasoning that precedes this example, we conclude that f' is positive on $(-\infty, -1)$ and on $(3, \infty)$, and f' is negative on $(-1, 3)$. Therefore f is increasing on the intervals $(-\infty, -1)$ and $(3, \infty)$ and decreasing on the interval $(-1, 3)$. Figure 7 exhibits the graph, which includes + signs to indicate where f' is positive and − signs to indicate where f' is negative. ◀

Critical Points and the First Derivative Test for Local Extrema

Now that we have learned to identify where a differentiable function f is increasing or decreasing, we can apply our knowledge to determine where f has a local maximum or minimum. We can even apply Theorem 1 to a continuous function that is not differentiable everywhere. This observation is important because many of the functions that we encounter in practice are not differentiable at every point of their domains. For example, the function $f(x) = |x|$ is not differentiable at $x = 0$, but, as we see in Figure 8, it has an absolute minimum there. Our tests for local extrema must therefore take into account isolated points of nondifferentiability. We begin with a definition.

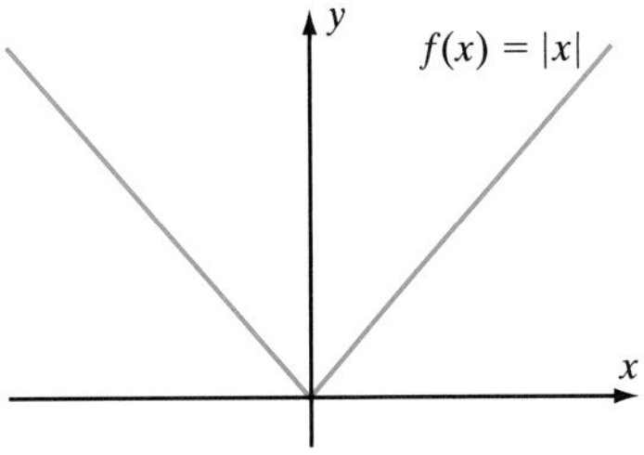

▲ Figure 8

DEFINITION Let c be a point in an open interval on which f is continuous. We call c a *critical point* for f if one of the following two conditions holds:

a. f is not differentiable at c; or
b. f is differentiable at c and $f'(c) = 0$.

Fermat's Theorem implies that, if the domain of a function f is an open interval, then the critical points for f are the only places that have to be examined in the search for the local extrema of f. The next theorem provides criteria for determining the behavior of a function at a critical point.

THEOREM 2 **(The First Derivative Test)** Suppose that $I = (\alpha, \beta)$ is an open interval contained in the domain of a continuous function f, that a point c in I is a critical point for f, and that f is differentiable at every point of I other than c.

a. If $f'(x) < 0$ for $\alpha < x < c$ and $f'(x) > 0$ for $c < x < \beta$, then f has a local minimum at c.
b. If $f'(x) > 0$ for $\alpha < x < c$ and $f'(x) < 0$ for $c < x < \beta$, then f has a local maximum at c.
c. If $f'(x)$ has the same sign on both sides of c, then f has *neither* a local minimum *nor* a local maximum (even if $f'(c) = 0$).

The rationale for the test is obvious. Look at Figure 9. The hypotheses of part a of the First Derivative Test imply that f is decreasing just to the left of c and

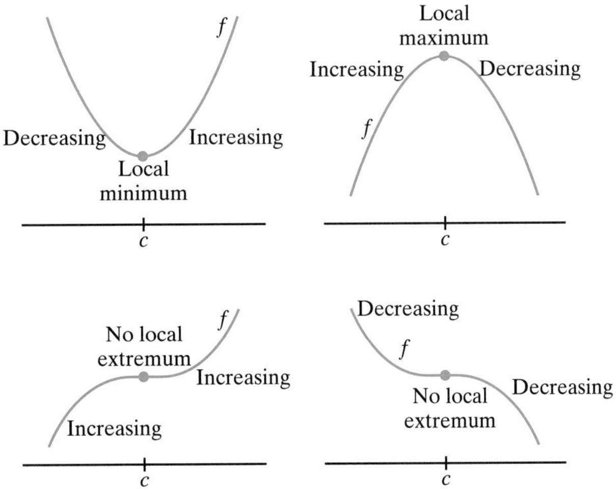

▲ Figure 9

increasing just to the right of c. It is evident that a continuous function must have a local minimum at such a point. The hypotheses of part b of the First Derivative Test imply that f is increasing just to the left of c and decreasing just to the right of c. A continuous function must have a local maximum at such a point. Finally, the hypothesis of part c implies that either f is increasing both to the left and to the right of c, or f is decreasing both to the left and to the right of c. A continuous function cannot have a local extremum at such a point.

INSIGHT When the First Derivative Test is applied, the interval I will usually be only a part of the domain of the function, as shown in Figure 10. The length of I is unimportant, but it must extend to each side of the critical point. As x approaches the critical point c_1 from the left in Figure 10, the values $f(x)$ decrease and so $f'(x)<0$. As x passes through and moves to the right from c_1, the values $f(x)$ increase and so $f'(x)>0$. A similar analysis at the critical point c_2 in Figure 10 leads us to these restatements, in plain English, of parts (a) and (b) of the First Derivative Text:

If $f'(x)$ changes from negative to positive as x passes through a critical point from left to right, then f has a local minimum at the critical point.

If $f'(x)$ changes from positive to negative as x passes through a critical point from left to right, then f has a local maximum at the critical point.

Notice that $f'(x)<0$ on both sides of critical point c_3. Because the sign of the $f'(x)$ does not change as x passes through this critical point, part (c) of the First Derivative Test tells us that there is no local extremum at this critical point.

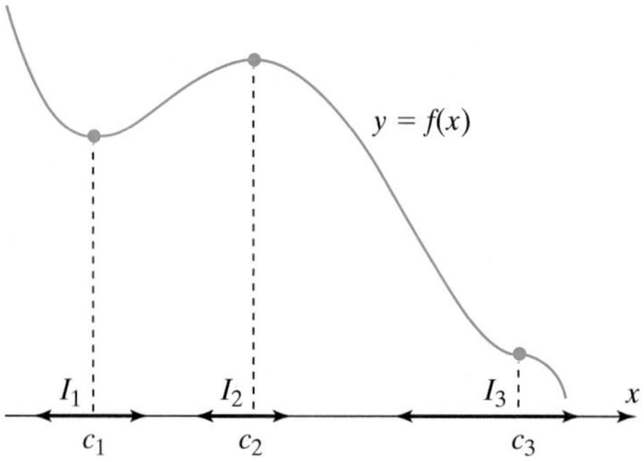

▲ Figure 10

▶ **EXAMPLE 4** Suppose that a river's current has speed u and that a fish is swimming upstream with speed v relative to the water. This implies that $v>u$. (If v were equal to u, then the fish would be at a standstill relative to land. If v were less than u, then the fish would be carried downstream by the current.) The energy E expended in such a migration is

$$E(v)=\alpha\frac{v^3}{v-u}$$

where α is a positive constant. For what value of v is E minimized?

Solution The domain of E is the open interval (u,∞). Also, E is differentiable at all points in its domain. By Fermat's Theorem, any local extremum will occur at a zero of E'. Using the Quotient Rule, we calculate

$$E'(v) = \alpha \frac{(v-u)\cdot 3v^2 - v^3(1-0)}{(v-u)^2} = \alpha \frac{2v^3 - 3uv^2}{(v-u)^2} = 2\alpha v^2 \frac{v - 3u/2}{(v-u)^2}.$$

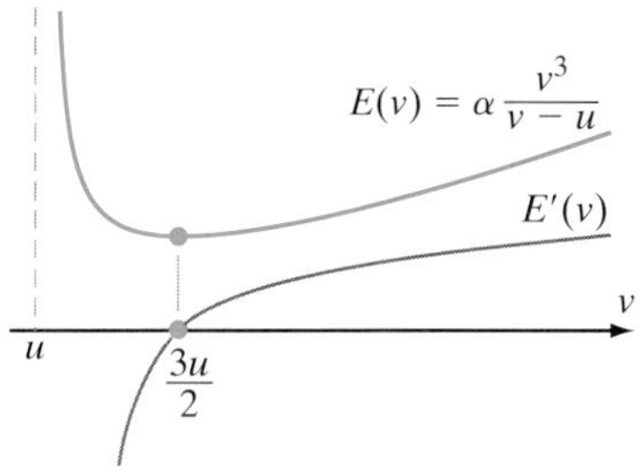

▲ Figure 11
$E'(v) < 0$ for $v < 3u/2$ and
$E'(v) > 0$ for $v > 3u/2$

Setting $E'(v) = 0$ yields $v = 3u/2$. Notice that the sign of $E'(v)$ is the same as the sign of $v - 3u/2$. Thus $E'(v) < 0$ for $v < 3u/2$ and $E'(v) > 0$ for $v > 3u/2$. According to the First Derivative Test, $E(v)$ has a local minimum at $v = 3u/2$. The graphs of E and E' in Figure 11 bear this out. ◀

▶ **EXAMPLE 5** Find and analyze the critical points for the function $f(x) = x - \sin(x)$.

Solution Because $f(x)$ is differentiable at every x, the critical points of f are the zeros of f'. Because $f'(x) = 1 - \cos(x)$, we see that the zeros of f' are the numbers x for which $\cos(x) = 1$. Thus the critical points are the numbers $\ldots, -4\pi, -2\pi, 0, 2\pi, 4\pi, \ldots$. At every other point, we have $\cos(x) < 1$ and $f'(x) = 1 - \cos(x) > 0$. Therefore $f'(x)$ does *not* change sign at any critical point. Part (c) of the First Derivative Test tells us that none of the critical points is a local extremum. The graph of f, shown in Figure 12, confirms this conclusion. Because f' is never negative, the graph of f increases from left to right. ◀

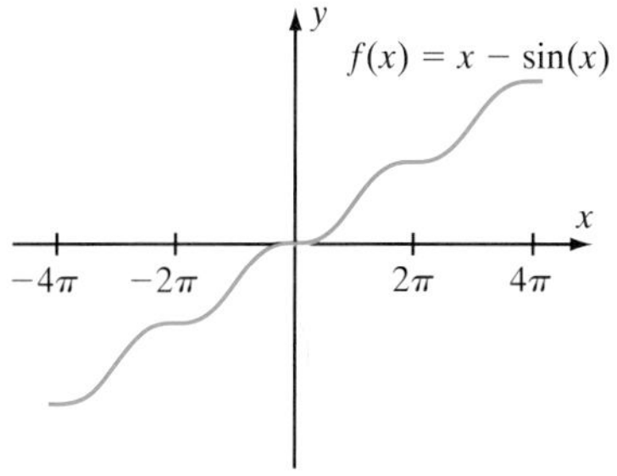

▲ Figure 12

▶ **EXAMPLE 6** Find and analyze the critical points for the function $f(x) = x \cdot (x-1)^{1/3}$.

Solution As long as $x \neq 1$, we have

$$\begin{aligned} f'(x) &= 1 \cdot (x-1)^{1/3} + x \cdot \left(\frac{1}{3} \cdot (x-1)^{-2/3}\right) \\ &= \frac{(x-1) + x/3}{(x-1)^{2/3}} \\ &= \frac{(4/3)x - 1}{(x-1)^{2/3}}. \end{aligned}$$

Thus f' can vanish only when the numerator $4x/3 - 1$ of this fraction vanishes, or when $x = 3/4$. The function f therefore has two critical points: $c = 3/4$, where $f' = 0$, and $c = 1$, where f is not differentiable. Notice that the denominator of the expression for f' is always positive when $x \neq 1$ (because the exponent has an even numerator). Refer to Figure 13 as you work through the analysis.

Begin with the critical point 3/4. When x is slightly to the left of 3/4, the numerator of $f'(x)$ is negative so $f'(x) < 0$. When x is slightly to the right of 3/4, the numerator of $f'(x)$ is positive, so $f'(x) > 0$. By the First Derivative Test, there is a local minimum at $c = 3/4$.

Next, consider the critical point $c = 1$. When x is slightly to the left of 1, the numerator of $f'(x)$ is positive. The same is true when x is slightly to the right of 1. By the First Derivative Test, $c = 1$ is not a local extremum. ◀

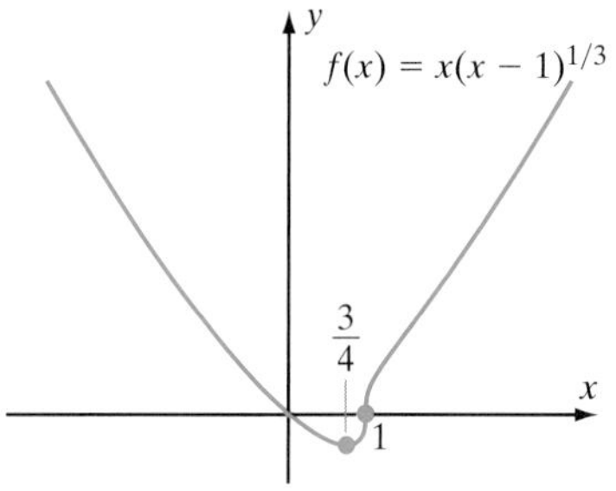

▲ Figure 13

In later sections, we will learn how to use the second derivative to identify local maxima and minima. We will also learn how to locate absolute maxima and minima.

QUICK QUIZ

1. True or false: If $f(x)$ is differentiable for every x in $\mathbb{R}$ and if $f'(c)=0$, then f has either a local minimum or a local maximum when $x=c$.
2. True or false: If $f(x)$ is differentiable for every x in $\mathbb{R}$, if $f'(x)>0$ for all $x<c$, and if $f'(x)>0$ for all $x>c$, then $f(x)$ has a local minimum at $x=c$.
3. True or false: If $f(x)$ is defined for every x in $\mathbb{R}$, if $f'(x)>0$ for all $x<c$, and if $f'(x)<0$ for all $x>c$, then $f(x)$ has a local maximum at $x=c$.
4. Find and analyze the local extrema for the function $f(x)=x^3/3+x^2-15x+5$.

Answers

1. False 2. False 3. True 4. Local maximum at $x=-5$; local minimum at $x=3$.

EXERCISES

Problems for Practice

In each of Exercises 1–28, use the first derivative to determine the intervals on which the given function f is increasing and on which f is decreasing. At each point c with $f'(c)=0$, use the First Derivative Test to determine whether $f(c)$ is a local maximum value, a local minimum value, or neither.

1. $f(x)=x^2+x$
2. $f(x)=3x^3+x^2+4$
3. $f(x)=x^3+3x^2-45x+2$
4. $f(x)=4x^3-6x^2+8$
5. $f(x)=3x^4+20x^3+36x^2$
6. $f(x)=x^5-5x^4$
7. $f(x)=-x/(x^2+25)$
8. $f(x)=x/(x+1)$
9. $f(x)=(x+1)/(x-1)$
10. $f(x)=x(x+2)^2$
11. $f(x)=(x+1)(x+2)^2$
12. $f(x)=(x+1)^2(x+2)^2$
13. $f(x)=(x^2-3)/(x+2)$
14. $f(x)=1/(x^4+6)$
15. $f(x)=4x+9/x$
16. $f(x)=x-\sqrt{x}$
17. $f(x)=3x+2\sin(x)$
18. $f(x)=2\cos^2(x)+3x$
19. $f(x)=e^x-x$
20. $f(x)=xe^{-x}$
21. $f(x)=x^2e^{-x}$
22. $f(x)=\exp(x)/x^2$
23. $f(x)=2^x-x$
24. $f(x)=2^x-4^x$
25. $f(x)=x\log_2(x)$
26. $f(x)=x^2\ln(x)$
27. $f(x)=x/\ln(x)$
28. $f(x)=\ln(x)/x$

In Exercises 29–40, find each critical point c of the given function f. Then use the First Derivative Test to determine whether $f(c)$ is a local maximum value, a local minimum value, or neither.

29. $f(x)=x(x+1)^{3/2},\ -1<x$
30. $f(x)=x^{1/3}(x+2)^{3/2},\ -2<x$
31. $f(x)=x^{1/5}(x-12)$
32. $f(x)=x^{2/3}(x+10)$
33. $f(x)=3x-x^{1/3}$
34. $f(x)=x^{3/2}-6x^{1/2}, 0<x$
35. $f(x)=x^{2/3}-4x^{1/3}$
36. $f(x)=3x^{1/3}-5x^{1/5}$
37. $f(x)=(x-1)^{1/3}(x+5)^{2/3}$
38. $f(x)=(x+1)^{2/3}/x$
39. $f(x)=(x-4)^{1/5}(3x+1)^{2/3}$
40. $f(x)=|x+1|/(x^2+15)$

41. A function f' has been plotted in Figure 14. Determine on what intervals f is increasing and on what intervals f is decreasing.

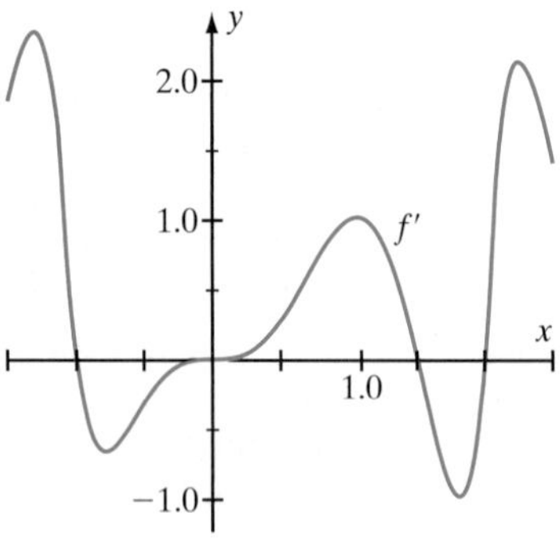

▲ Figure 14

42. A derivative f' has been plotted in Figure 15. Determine the points where $f(x)$ changes (as x moves to the right) from being increasing to being decreasing. Determine the

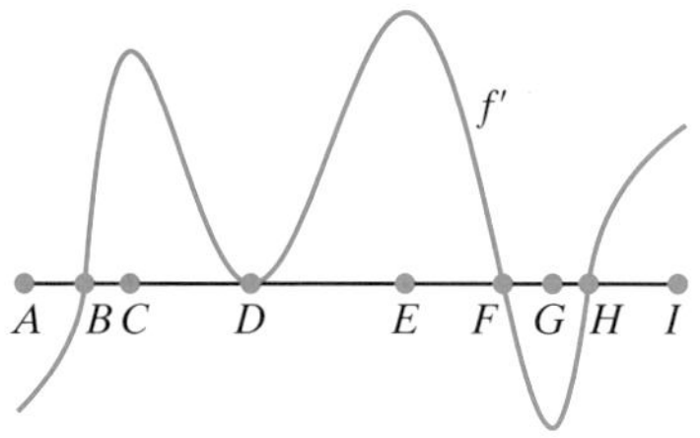

Figure 15

points where f changes from being decreasing to being increasing.

Further Theory and Practice

Find and test the critical points of the functions in Exercises 43–47:

43. $f(x) = 2x^2 - 2x^2$
44. $f(x) = |x-7| + x^2$
45. $f(x) = 5x - |8-3x|$
46. $f(x) = \sqrt{|2x+9|}$
47. $f(x) = x^{1/3} - |x|$

In each of Exercises 48–53, a function f is given. In each case, the domain of f and the value of $f(x)$ depend on a positive constant b. Determine each point at which f has a local extremum. Use the First Derivative Test to classify the extremum.

48. $f(x) = x + 4b^2/x,\ b < x$
49. $f(x) = x^2\sqrt{b^2 - x^2},\ 0 < x < b$
50. $f(x) = x^2/(3x-b),\ b/3 < x$
51. $f(x) = (x^2 + 3b^2)/(b-x),\ x < b$
52. $f(x) = (x+b)/(x^2 + 15b^2),\ -b < x$
53. $f(x) = (x^4 + 8b^4)/(x^2 + b^2),\ b < x$

In each of Exercises 54–60, determine each point c, where the given function f satisfies $f'(c) = 0$. At each such point, use the First Derivative Test to determine whether f has a local maximum, a local minimum, or neither.

54. $f(x) = \arctan(x) - x/5$
55. $f(x) = \arcsin(x) - \sqrt{x}$
56. $f(x) = 2\arctan(x) - \arcsin(x)$
57. $f(x) = \sinh(x) + 2\cosh(x)$
58. $f(x) = \tanh(x) - \exp(2x)$
59. $f(x) = 2\sinh^{-1}(x) - \ln(x)$
60. $f(x) = \tanh^{-1}(\sqrt{x}) - 2\sqrt{x}$

61. Let f be differentiable on, $\mathbb{R}$. Suppose that $f'(2) > 0$. Is $f(2.000001) > f(2)$? Explain your answer.

62. *True or false*: If $f\colon \mathbb{R} \to \mathbb{R}$ is a polynomial and if f has two local minima, then it has a local maximum. Explain your answer.

63. Prove that if $f\colon \mathbb{R} \to \mathbb{R}$ is differentiable, and $f' > 0$ everywhere, then f is one-to-one. Prove that if $f' < 0$ everywhere, then f is one-to-one.

64. If f is increasing on an interval I, does it follow that f^2 is increasing? What if the range of f is $(0, \infty)$?

65. Prove that if a differentiable function f is increasing on an interval I, then f^3 is increasing.

66. Suppose that a real-valued function f is defined and increasing on an interval (a, b). Let J be the image of f. Without using calculus, prove that $f^{-1}\colon J \to (a, b)$ exists and is increasing. Now assume that f' exists and is positive on the interval (a, b). Give a calculus proof that f^{-1} is increasing.

67. Without using calculus, prove that the composition of two increasing functions is increasing. Now assume that f and g are differentiable functions with positive derivatives and that $g \circ f$ is defined. Use calculus to show that $g \circ f$ is increasing.

68. Assume that f and g are decreasing, differentiable functions and that $g \circ f$ is defined. Use calculus to show that $g \circ f$ is increasing.

69. Let $f(x) = ax^2 + bx + c$ with $a \neq 0$. By completing the square, determine the interval(s) on which f is increasing and the interval(s) on which f is decreasing. Use the derivative of f to verify your conclusions.

70. The reaction rate V of a common enzyme reaction is given in terms of substrate level s and the Michaelis constant K by,

$$V = \frac{V_* S}{K + S} \quad (S \geq 0).$$

Show that V is an increasing function of S. It follows that V has no absolute maximum value. Is V bounded? What is $\lim_{S\to\infty} V$?

71. Suppose that $f(x) = (x-a)\ (x-b)\ q(x)$ where q is a polynomial that is positive on an interval (α, β) containing $[a, b]$. Show that $f'(a)\, f'(b) < 0$.

72. The function f' has been plotted in Figure 16. It is known that $f'(A) = 0$, $f'(B) = 0$, $f'(c) = 0$, and $f'(D) = 0$. Determine whether f has any local extreme values at the points A, B, C, and D. Explain your reasoning.

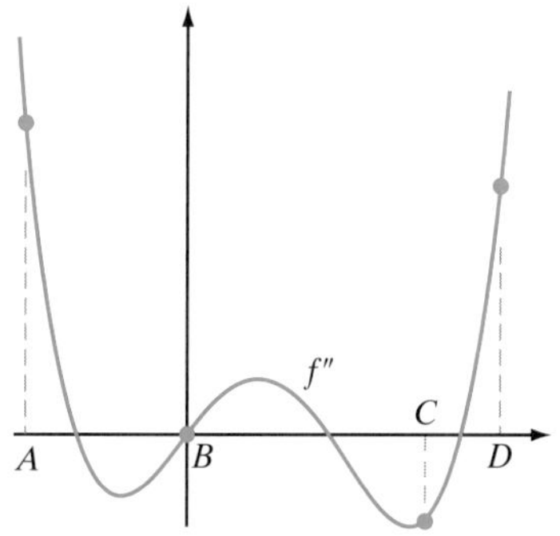

Figure 16

73. In economic theory, the *product of labor* $f(n)$ measures commodity output as a function of the level n of labor input. The *marginal product of labor (MPL)* represents the extra output that results from one additional unit of work. In general, the function *MPL* has a certain property that is illustrated in three different ways by the graphs in Figure 17. What is the property? Explain how the graphs illustrate it.

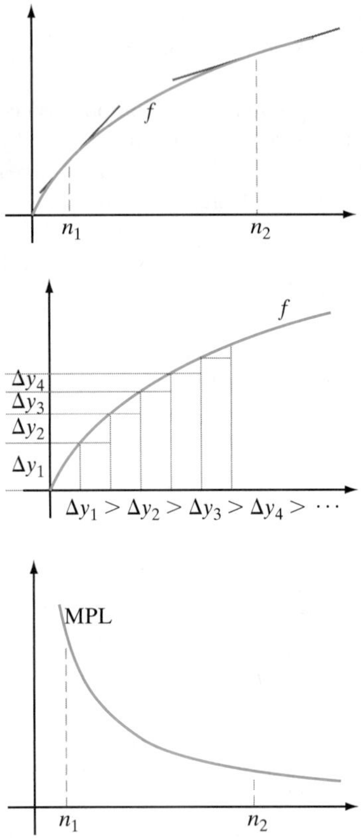

▲ Figure 17

74. Give necessary and sufficient algebraic conditions on the coefficients of the polynomial

$$p(x) = ax^3 + bx^2 + cx + d$$

that guarantee that p is increasing on the entire real line.

75. Let $f\colon \mathbb{R} \to (0, \infty)$ be a differentiable function. Compare the local extrema of f with those for $1/f$. The local maxima for f become what for $1/f$? The local minima for f become what for $1/f$?

76. Suppose that $f(c)$ and $g(c)$ are positive. The Product and Quotient Rules state

$$(f \cdot g)'(c) = \frac{+f'(c) \cdot g(c) + f(c) \cdot g'(c)}{1}$$

and

$$\left(\frac{f}{g}\right)'(c) = \frac{+g(c) \cdot f'(c) - g'(c) \cdot f(c),}{g(c)^2}$$

respectively. The unusual presentation here serves to highlight some similarities between the two rules. Notice that both $(f \cdot g)'(c)$ and $(f/g)'(c)$ have a positive denominator. Also notice that each of the terms $f'(c) \cdot g(c)$ and $f(c) \cdot g'(c)$ appears in both formulas. Using what you learned about the derivative in this section, explain why each of the four signs makes sense.

77. If T is the temperature in degrees Kelvin of a white dwarf star, then the rate at which T changes as a function of time t is given by $dT/dt = -\alpha T^{7/2}$ where α is a positive constant. Do white dwarfs cool or heat up? Is the rate at which a white dwarf's temperature changes increasing or decreasing? Explain your answers.

78. According to Saha's equation, the fraction $x \in (0, 1)$ of ionized hydrogen atoms in a stellar mass depends on temperature T according to the equation

$$\frac{1-x}{x^2} = A \cdot T^{-3/2} \cdot \exp\left(\frac{k}{T}\right)$$

for positive constants A and K.

a. Without differentiating, explain why the right side of this equation is a decreasing function of T.

b. Without differentiating, explain why the left side of this equation is a decreasing function of x.

c. Without differentiating, explain why x is an increasing function of T.

d. Use part c to explain why Saha's equation determines x as a function of T.

e. By differentiating implicitly, verify that x is an increasing function of T.

79. Show that the function $y \mapsto y \exp(y)$ is increasing for $y > 0$. Deduce that for every positive x, there is a unique y such that $y \exp(y) = x$. This relationship implicitly determines a function that is often denoted by W and is called Lambert's W function:

$$W(x)\exp\big(W(x)\big) = x \quad (x > 0).$$

Use implicit differentiation to show that W is an increasing function. Show that

$$W'(x) = \frac{W(x)}{x\big(1 + W(x)\big)}.$$

80. In this exercise, you will show that

$$\lim_{n \to \infty} n^{1/n} = 1$$

following a proof of Alan Beardon. Suppose that $0 < p < 1$. Define $f(x) = px - x^p$ for $\leq x < \infty$. By calculating $f'(x)$, show that $f(1) \leq f(x)$. Use this inequality to deduce that $x^p \leq px + 1 - p$. In this inequality, set $x = \sqrt{n}$ and $p = 1/n$, and use the Pinching Theorem to show that $\lim_{n \to \infty} n^{1/n} = 1$.

81. Darboux's Theorem (after Jean Darboux, 1842–1917) states that if f' exists at every point of an open interval containing $[a,b]$, and if γ is between $f'(a)$ and $f'(b)$, then there is a c in the interval (a,b) such that $f'(c)=\gamma$. The existence of such a c follows from the Intermediate Value Theorem if f' is assumed continuous. Darboux's Theorem, which you will prove in this exercise, tells us that the assumption of continuity is not needed for functions that are derivatives. Suppose for definiteness that $f'(a)<\gamma<f'(b)$. Define g on $[a,b]$ by $g(x)=f(x)-\gamma x$.

a. Explain why g has a minimum value that occurs at some point c in (a,b).

b. Show that $f'(c)=\gamma$.

c. Use Darboux's Theorem to show that if f increases on the interval (l,c) and decreases on the interval (c,r) (or vice versa), then $f'(c)=0$.

Calculator/Computer Exercises

In each of Exercises 82–89, use the first derivative to determine the intervals on which the function is increasing and on which the function is decreasing.

82. $f(x)=x^4-x^2-7.1x+3.2$

83. $f(x)=x^4-4x^3+3.94x^2-3.87x$

84. $f(x)=2x^5-1.2x^4-13.7x+5.2$

85. $f(x)=\dfrac{x+e^x}{x^2+e^x}$

86. $f(x)=(x^2+x+1)/(x^4+1)$

87. $f(x)=x^3-x^2+e^{-x}$

88. $f(x)=(4x^3-3)\exp(-x^2)$

89. $f(x)=x^{1/3}-\ln(1+x^2)$

In each of Exercises 90–93, a function f is given. Calculate f' and plot $y=f'(x)$ in a suitable viewing window. Use this plot to identify the points at which f has local extrema.

90. $f(x)=x^3-x+\sin^2(\pi x/2),\quad -1<x<1$

91. $f(x)=2+x^3/3-\sin^2(x+1),\quad -3<x<3$

92. $f(x)=\ln(x)\ln(1-x^2)$

93. $f(x)=\exp(-x^2)+\exp\left(-(x-1)^2\right)+\exp\left(-(x-3)^2\right),$ $-2<x<5$

94. For each given function f, graph the function $S(x)=\text{signum}(f'(x))$ for x in the given interval I. Use the graph of S to determine and classify the local extrema of f.

a. $f(x)=x^5-12x^4+55x^3-120x^2+124x-48,$ $I=[0.8,4.1]$

b. $f(x)=4x^4-24x^3+51x^2-44x+12,\quad I=[0.3,2.6]$

c. $f(x)=x^4-x^3-7x^2+x+6,\quad I=[-2.4,3.2]$

4.4 Applied Maximum-Minimum Problems

In this section, we learn how to apply the techniques of calculus to study a variety of practical problems, all of which involve finding the maximum or minimum of some quantity. Therefore we will certainly make use of the ideas that were introduced in Section 4.3. In this section, however, we concentrate on finding absolute extrema. Our first example illustrates the important role that endpoints play in such a search.

▶ **EXAMPLE 1** The total mechanical power P (in watts) that an individual requires for walking a fixed distance at constant speed v (in kilometers per hour) with step length s (in meters) is

$$P(s)=\alpha\frac{v^3}{s}+\beta vs\quad\left(0<s\leq L\right),$$

where L is the individual's maximum step size. In this example, we use $L=1\ m$. The positive constants α and β depend on the distance walked and the individual's gait. In this example, we use $\alpha=1.4$ and $\beta=45$. For what s is $P(s)$ minimized if $v=5$? If $v=6$?

Solution We calculate that

$$P'(s)=\frac{d}{ds}\left(\alpha\frac{v^3}{s}+\beta vs\right)=-\alpha\frac{v^3}{s^2}+\beta v=\frac{\beta v}{s^2}\left(s^2-\frac{\alpha}{\beta}v^2\right)=\frac{\beta v}{s^2}(s+s_0)(s-s_0),$$

where $s_0 = v\sqrt{\alpha/\beta}$. From this formula, we see that $s = s_0$ is the unique positive solution of the equation $P'(s) = 0$. For $v = 5$, we have $s_0 = v\sqrt{\alpha/\beta} = 5\sqrt{1.4/45} \approx 0.88$, which is less than L and therefore in the domain of P. For $0 < s < s_0$, we have

$$P'(s) = \underbrace{\frac{\beta v}{s^2}(s + s_0)}_{\text{positive}} \underbrace{(s - s_0)}_{\text{negative}} < 0,$$

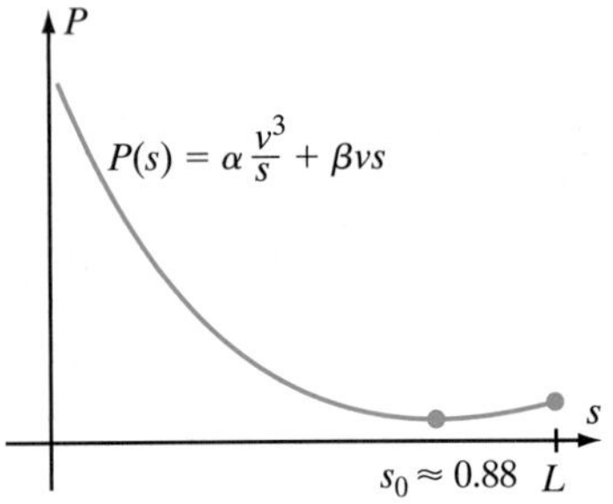

▲ Figure 1

and for $s_0 < s \leq L$, we have

$$P'(s) = \underbrace{\frac{\beta v}{s^2}(s + s_0)}_{\text{positive}} \underbrace{(s - s_0)}_{\text{positive}} > 0.$$

We conclude that P decreases on the interval $(0, s_0)$ and increases on the interval $(s_0, L]$. Therefore P has an absolute minimum at $s = s_0$. The graph of P confirms our analysis (see Figure 1).

For $v = 6$, the value of $s_0 = v\sqrt{\alpha/\beta} = 6\sqrt{1.4/45}$ is about 1.06, which is greater than L and, therefore *outside* the domain of P. Of course, P cannot have a minimum at a point that is not in its domain. To determine the point at which $P(s)$ is minimized, observe that, for all $0 < s \leq L$,

$$P'(s) = \underbrace{\frac{\beta v}{s^2}}_{\text{positive}}(s + s_0)(s - s_0) < \underbrace{\frac{\beta v}{s^2}(s + s_0)}_{\text{positive}} \underbrace{(L - s_0)}_{\text{negative}} < 0.$$

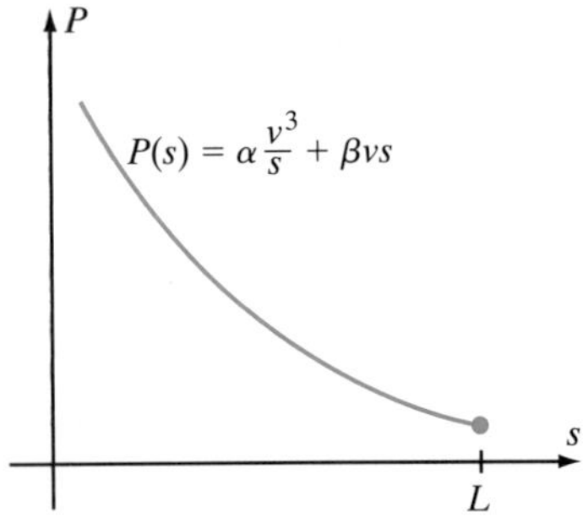

▲ Figure 2

We conclude that P is decreasing on its entire domain $(0, L]$. Therefore $P(s)$ is minimized when $s = L$ (see Figure 2). ◄

INSIGHT Example 1 shows us that when we determine the absolute extrema of a function, we must consider its domain. If a function is defined at an endpoint of its domain, then an extreme value of the function may occur there.

Closed Intervals

Let us consider the absolute extrema of a continuous function f on a *closed* interval $[a, b]$. The Extreme Value Theorem (Section 2.3 of Chapter 2) guarantees that f has both an absolute maximum and an absolute minimum in $[a, b]$. Also, the techniques of Section 4.3 guarantee that if these extrema are inside the open interval (a, b), then they will be found among the critical points. It is possible, however, that the absolute maximum or absolute minimum of f could be at a or at b. We therefore rely on the following procedure.

Basic Rule for Finding Extrema of a Continuous Function on a Closed Bounded Interval

To find the extrema of a continuous function f on a closed interval $[a, b]$, we test

a. The points in (a, b) where f is not differentiable;
b. The points in (a, b) where f' exists and equals 0;
c. The endpoints a and b.

In brief, we should test the *critical points* and the *endpoints*.

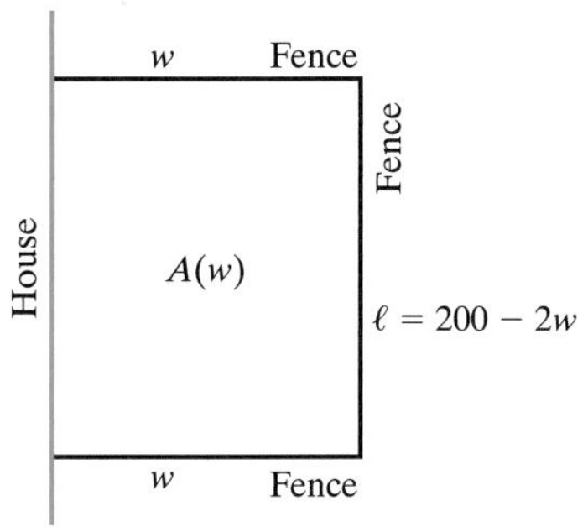

▲ Figure 3

▶ **EXAMPLE 2** A man builds a rectangular garden along the side of his house. He will put fencing on three sides (see Figure 3). If he has 200 feet of fencing available, then what dimensions will yield the garden of greatest area?

Solution The function that we must maximize is A, the area inside the fence. If ℓ denotes length, and w denotes width, as in Figure 3, then $A = \ell \cdot w$. The trouble is that A seems to be a function of two variables, and we have no techniques for maximizing such a function. We must use the additional piece of information that $\ell + 2w = 200$, or $\ell = 200 - 2w$. We substitute this expression for ℓ into the formula for A. Thus our job is to maximize

$$A(w) = (200 - 2w) \cdot w = 200w - 2w^2, \quad 0 \le w \le 100.$$

Notice that we are careful to write down the range of values of w: This is necessary to apply the Basic Rule. (Because ℓ cannot be negative, the equation $\ell + 2w = 200$ tells us that the greatest possible value for w is 100.)

Now A is a polynomial, so it is differentiable at all values of w. Also $dA/dw = 200 - 4w$. Clearly, dA/dw vanishes at $w = 50$. Thus $w = 50$ is a critical point.

By the Basic Rule, the maximum occurs at either the critical point $w = 50$ or at one of the endpoints $w = 0$ or $w = 100$. To determine which, we calculate the value of $A(w)$ at each of these three points: $A(0) = 0$, $A(50) = 5000$, and $A(100) = 0$. Clearly, the maximum occurs when $w = 50$. For this value of w, we calculate $\ell = 200 - 2 \cdot 50 = 100$. ◀

INSIGHT Even this straightforward example contains all the major ingredients of the examples that we will see in this section. Let's outline them:

1. Determine the function f to be maximized or minimized.
2. Identify the relevant variable(s).
3. If f depends on more than one variable, find relationships among the variables that will allow substitution into the expression for f. The methods developed in this chapter require you to obtain a function of one variable to be maximized or minimized.
4. Determine the set of allowable values for the variable (this is usually an interval).
5. Find all places where the function is not differentiable, all places where the derivative is zero, and the endpoints. These are the points to test.
6. Use the First Derivative Test, or simply substitute in values as in Example 2, to determine which of the points found in Step 5 solves the maximum-minimum problem being studied.

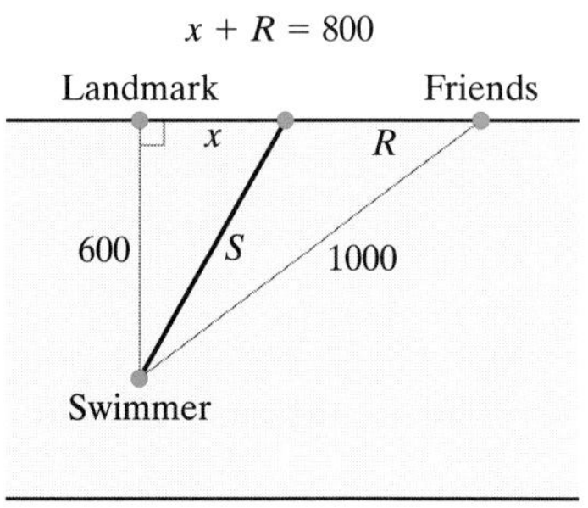

▲ Figure 4

▶ **EXAMPLE 3** A swimmer is 600 m straight out from a landmark on shore, as shown in Figure 4. She wants to meet some friends who are 800 m down the beach from the landmark. The swimmer can swim at a rate of 100 m/min and run on the beach at the rate of 200 m/min. Toward what point on shore should she swim to minimize the time it takes for her to join her friends?

Solution

Step 1. The function to be minimized is the time T that it takes the swimmer to swim and run to her friends.

Step 2. The relevant variables are S, the distance swum, and R, the distance run. Then the time spent swimming is $S/100$, and the time spent running is $R/200$. Therefore

$$T = \frac{S}{100} + \frac{R}{200}. \tag{4.4.1}$$

Step 3. Let x be the distance between the point where the swimmer hits the beach and the landmark. Then, by the Pythagorean Theorem,

$$x = (S^2 - 600^2)^{1/2}.$$

Also, $x + R = 800$, or

$$R = 800 - x = 800 - (S^2 - 600^2)^{1/2}.$$

Substituting this expression for R into equation (4.4.1) gives

$$T(S) = \frac{S}{100} + \frac{800 - (S^2 - 600^2)^{1/2}}{200}. \tag{4.4.2}$$

Step 4. From Figure 4, the swimmer cannot swim any less than her 600 m distance from the shore, nor more than the hypotenuse length, $\sqrt{600^2 + 800^2}$, or 1000, of the larger right triangle in Figure 4. Therefore the range of values for S is $600 \leq S \leq 1000$.

Step 5. The function T is differentiable at all points of $(600, 1000)$, and we compute

$$\begin{aligned}\frac{dT}{dS} &= \frac{1}{100} - \frac{(1/2)\cdot(S^2 - 600^2)^{-1/2}\cdot 2S}{200} \\ &= \frac{2}{200} - \frac{S}{200\cdot(S^2 - 600^2)^{1/2}} \\ &= \frac{\left(2(S^2 - 600^2)^{1/2} - S\right)}{200\cdot(S^2 - 600^2)^{1/2}}.\end{aligned}$$

This expression can vanish only if

$$2(S^2 - 600^2)^{1/2} = S$$

or

$$4(S^2 - 600^2) = S^2$$

or

$$3S^2 = 4\cdot 600^2.$$

Thus the only root of dT/dS in the interval $(600, 1000)$ is $S = 400\sqrt{3}$. This critical point and the endpoints $S = 600$ and $S = 1000$ are the candidates for the minimum we seek.

Step 6. Using equation (4.4.2), we calculate that $T(600) = 10$, $T(1000) = 10$, and

$$T(400\sqrt{3}) = 4 + 3\sqrt{3} \approx 9.196 < 10.$$

Thus $S = 400\sqrt{3}$ gives the minimum. The value of x that we seek is therefore $x = \sqrt{(400\sqrt{3})^2 - 600^2} = 200\sqrt{3}$. ◀

INSIGHT The graph of T from Example 3 appears in Figure 5. By repeatedly zooming in on the low point of the graph, we can approximate the coordinates $(692.8, 9.196)$. A numerical algorithm for approximating extreme values is described in the instructions for Exercises 61 and 62.

Examples with the Solution at an Endpoint

▶ **EXAMPLE 4** A 10-inch piece of wire can be bent into a circle or a square, or it can be cut into two pieces. In this case, a circle will be formed from the first piece and a square from the other. What is the maximal area that can result?

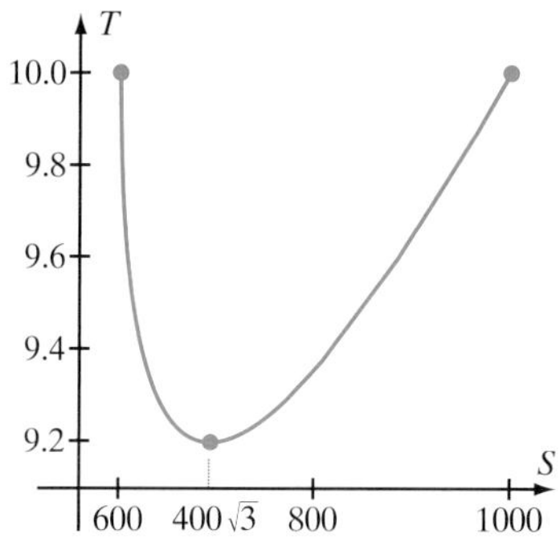

▲ Figure 5
$T(S) = \frac{S}{100} + \frac{800 - (S^2 - 600^2)^{1/2}}{200}$

Solution

Step 1. The function to be maximized is the sum of the two areas (one of which will be 0 if the wire is not cut). Call this function A.

Step 2. The relevant variables are the lengths c (for the part of the wire that will form the circle) and s (for the part of the wire that will form the square). If the circle is made from the piece of length c, then the circumference of the circle is c so that the radius of the circle is $c/(2\pi)$. Therefore

$$\text{Area of the circle} = \pi \cdot \text{radius}^2 = \pi \cdot \left(\frac{c}{2\pi}\right)^2 = \frac{c^2}{4\pi}.$$

The square is then manufactured from the piece of wire of length s. The perimeter of the square is then s, and the side length is $s/4$. The area of the square is $(s/4)^2$, or $s^2/16$. In conclusion, the function that we need to maximize is

$$A = \frac{c^2}{4\pi} + \frac{s^2}{16}.$$

Step 3. The variables c and s are related by the equation $c + s = 10$, or $s = c - 10$. Substituting this into the formula for A gives

$$A = \frac{c^2}{4\pi} + \frac{(c-10)^2}{16}. \tag{4.4.3}$$

Step 4. Clearly $0 \le c \le 10$ because the wire is 10 inches long.

Step 5. The function A is a polynomial hence is differentiable everywhere. The only candidates for the maximum we seek are the endpoints 0 and 10 and the zeros of A'. Now

$$A'(c) = \frac{2c}{4\pi} + \frac{2(c-10)}{16}$$
$$= \frac{(4+\pi)c - 10\pi}{8\pi}.$$

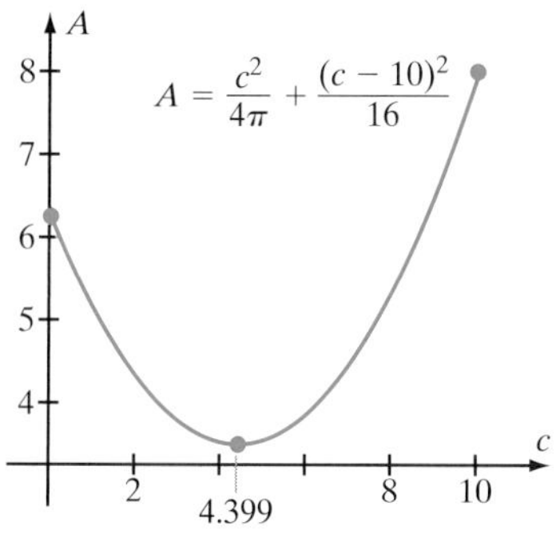

▲ Figure 6

The only root of $A'(c) = 0$ occurs when $(4+\pi)c - 10\pi = 0$, or equals zero, or $c = 10\pi/(4+\pi) \approx 4.399$. This number is certainly in $[0, 10]$; hence, it is a critical point.

Step 6. Referring to equation (4.4.3), we see that $A(0) = 25/4 = 6.25$, $A(10) = 25/\pi \simeq 7.958$, and $A(10\pi/(4+\pi)) = 25/(4+\pi) \simeq 3.501$. Clearly the greatest area is achieved when $c = 10$, or when we use the entire wire to make a circle but no square. The plot of equation (4.4.3), shown in Figure 6, confirms our analysis. ◀

Our next example appears to be quite similar to Example 3. However, there is a surprising and instructive difference.

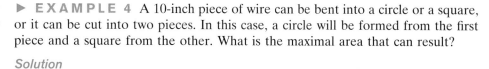

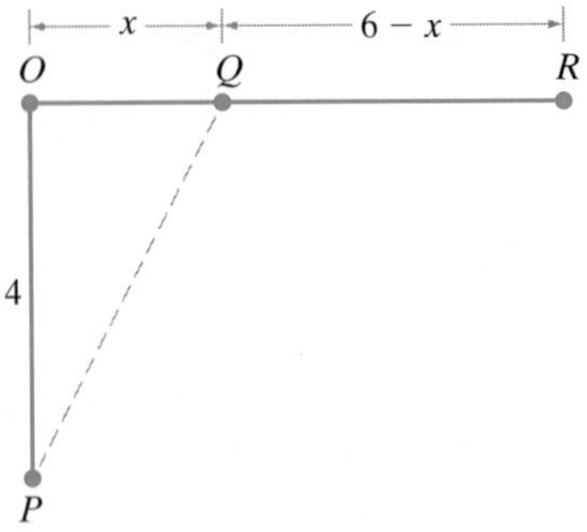

▲ Figure 7

▶ **EXAMPLE 5** A jogger is at point P, 4 miles due south of point O. His destination, point R, is 6 miles due east of O. He can run along roads from P to O and from O to R at the rate of 5 miles per hour. Or he can hike across the field from P to a point Q at the rate of 3 miles per hour and then follow the road to point R at the rate of 5 miles per hour (see Figure 7). How quickly can the jogger get from P to R?

Solution

Step 1. The function to be minimized is the sum of the times required to run from P to Q and from Q to R. Let us call this total time T.

Step 2. Let x be the distance of Q from O.

Step 3. When $x = 0$, the runner jogs 4 miles north and then 6 miles east, all at 5 mph. His total time is 2 hours along this route. Thus $T(0) = 2$. Now suppose that $x > 0$. The distance of Q to R is $6 - x$, and the time required to jog this distance is $(6 - x)/5$. By the Pythagorean Theorem, the distance of P to Q is $\sqrt{4^2 + x^2}$, and the time required to hike this distance is $(\sqrt{4^2 + x^2})/3$. Thus the time T required for the journey is

$$T(x) = \frac{\sqrt{4^2 + x^2}}{3} + \frac{6 - x}{5}$$

when $0 < x \leq 6$. (All values of $T(x)$ are stated in hours.)

Step 4. Clearly $0 \leq x \leq 6$.

Step 5. We calculate

$$T'(x) = \frac{x}{3\sqrt{16 + x^2}} - \frac{1}{5} = \frac{5x - 3\sqrt{16 + x^2}}{15\sqrt{16 + x^2}}$$

for $0 < x < 6$. In order that $T'(x) = 0$, we must have $5x = 3\sqrt{16 + x^2}$, or $25x^2 = 9(16 + x^2)$, or $16x^2 = 9 \cdot 16$. The positive solution is $x = 3$. The candidates for the location of an extremum are therefore 3 and the endpoints 0 and 6.

Step 6. Because $T'(x) < 0$ for x in the interval $(0, 3)$, and $T'(x) > 0$ in the interval $(3, 6)$, we deduce that T decreases on $(0, 3)$ and increases on $(3, 6)$. If T has a minimum, then it must occur at $x = 3$ or $x = 0$ because T is discontinuous at $x = 0$. In fact, $T(0) = 2$ hr and

$$T(3) = \frac{\sqrt{4^2 + 3^2}}{3} + \frac{6 - 3}{5} = \frac{34}{15} \approx 2.2667 \text{ hr}.$$

We conclude that $T(0) = 2$ is the minimum time in which to get from P to R. ◀

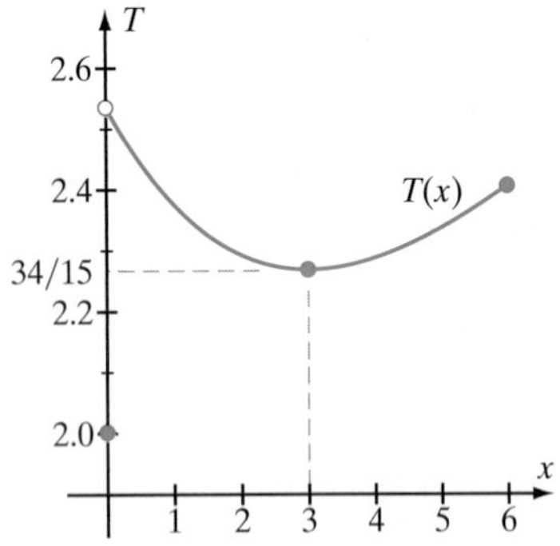

▲ Figure 8

INSIGHT The graph of T from Example 5 appears in Figure 8. Notice that T is discontinuous. The Extreme Value Theorem therefore does not apply to T, and we cannot be certain, prior to our analysis, that T has any extreme value on $[0, 6]$. In fact, T does not assume a maximum value, but, as we have found, T does assumes a minimum value on $[0, 6]$.

Examples 1, 4, and 5 demonstrate that checking the endpoints is not simply an academic exercise. As we have seen, the solution to a perfectly reasonable problem can occur at an endpoint.

Profit Maximization

Let p be the price of a product, and let x denote the number of units of that product that are sold. For an ordinary item, if the price p increases, then the sales level x decreases. In other words, x is a decreasing function $\mathcal{D}$ of p:

$$x = \mathcal{D}(p). \tag{4.4.4}$$

Equation (4.4.4) is called the *demand equation* and the function $\mathcal{D}$ is known as the *demand function*. Because $\mathcal{D}$ is a decreasing function, it has an inverse:

$$p = \mathcal{D}^{-1}(x). \tag{4.4.5}$$

The plot of equation (4.4.5) with price measured along the vertical axis is called a *demand curve* (see Figure 9).

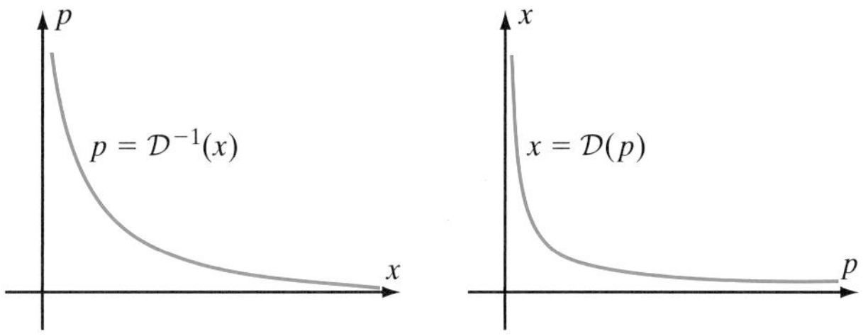

Figure 9 Equivalent representations of the demand equation

The cost $C(x)$ of producing x units of a product includes *fixed costs*, which are independent of the number of units produced. For example, if a new automobile motor is produced, then there will be design and testing costs that are the same regardless of how many engines are manufactured. There will also be *incremental costs* that rise with each additional unit produced. For example, each new motor requires a certain amount of raw materials, labor, and plant costs. The derivative $C'(x)$ is called the *marginal cost* (as discussed in Section 3.9 of Chapter 3). In a simple model, each additional item produced will require an incremental cost equal to m. In this model, the cost function is given by

$$C(x) = \underbrace{C_0}_{\text{fixed costs}} + \underbrace{m \cdot x}_{\text{(marginal cost)}\cdot\text{(quantity)}}.$$

Notice that, in this model, the marginal cost is the constant m. Marginal cost, however, is often nonconstant.

The total revenue R is the number of units sold times the price per unit. Thus $R = x \cdot p$. Equations (4.4.4) and (4.4.5) allow us to express revenue in terms of price or quantity:

$$R = x \cdot \mathcal{D}^{-1}(x) \quad \text{or} \quad R = \mathcal{D}(p) \cdot p. \tag{4.4.6}$$

The profit P is equal to revenue less cost. That is, $P = R - C$. By using equations (4.4.4), (4.4.5), and (4.4.6), we can write P in terms of either x or p:

$$P = x \cdot \mathcal{D}^{-1}(x) - C(x) \quad \text{or} \quad P = \mathcal{D}(p) \cdot p - C\big(\mathcal{D}(p)\big). \tag{4.4.7}$$

Determining the value p that maximizes profit is clearly an important business problem.

▶ **EXAMPLE 6** Suppose that the total dollar cost for producing x copies of a particular book is $C(x) = 12000 + 3x$. Past experience indicates that the number x of books sold is related to price p by the demand equation $x = 6000\,(1 - p/30)$. How should the book be priced to optimize profit?

Solution

Step 1. We are to maximize the profit function $P = \mathcal{D}(p) \cdot p - C(\mathcal{D}(p))$ given in line (4.4.7).

Step 2. The variables are the book price p and the number x of books sold.

Step 3. The demand equation $x = \mathcal{D}(p) = 6000\,(1 - p/30)$ expresses x in terms of p. Substituting this formula for $\mathcal{D}(p)$ in the equation for the profit function results in

$$P = \mathcal{D}(p) \cdot p - C(\mathcal{D}(p)) = 6000\left(1 - \frac{p}{30}\right) \cdot p - \left(12000 + 3 \cdot 6000\left(1 - \frac{p}{30}\right)\right).$$

After expanding and combining terms, we find that

$$P(p) = 6600p - 200p^2 - 30000$$

is the function to be maximized.

Step 4. The domain for the function P is the set of all positive p.

Step 5. The domain of P, namely the open interval $(0, \infty)$, has no endpoint. Because P is differentiable at all points in its domain, the only critical points are the zeros of $dP/dp = 6600 - 400p$. Thus $p = 6600/400 = 33/2$ is the unique critical point of P.

Step 6. Because $dP/dp > 0$ for $p < 33/2$ and $dP/dp < 0$ for $p > 33/2$, the First Derivative Test tells us that profit is maximized when $p = 33/2$, or \$16.50. ◀

An Example Involving a Transcendental Function

▶ **EXAMPLE 7** Mr. Woodman is viewing a 6-foot-long tapestry that is hung lengthwise on a wall of a 20 ft by 20 ft room. The bottom end of the tapestry is 2 feet above his eye level (see Figure 10). At what distance from the tapestry should Mr. Woodman stand to obtain the most favorable view? That is, for what value of x is angle α maximized?

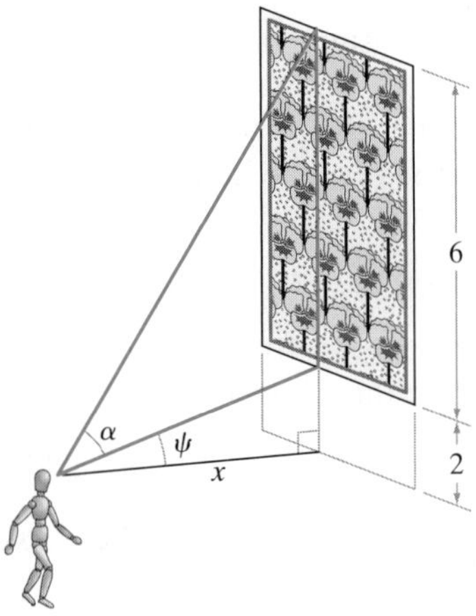

▲ Figure 10

Solution

Step 1. From Figure 10, we see that $\cot(\alpha + \psi) = x/8$, or $\alpha = \operatorname{arccot}(x/8) - \psi$.

Step 2. The variables that appear in our formula for α are x and ψ.

Step 3. The variable ψ is related to x by the equation $\cot(\psi) = x/2$, or $\psi = \operatorname{arccot}(x/2)$. Substituting this value for ψ in the formula for α obtained in Step 1, we obtain $\alpha = \operatorname{arccot}(x/8) - \operatorname{arccot}(x/2)$. This is the function of x that we are to maximize.

Step 4. The variable x is in the range $(0, 20)$. (In practice, Mr. Woodman might not be able to approach the wall with the tapestry or the opposite wall so closely. If the remainder of our analysis were to turn up a value of x close to 0 or 20, then we would need to be more accurate about the domain.)

Step 5. The domain of the function to be maximized does not contain an endpoint. Because α is differentiable at all points of its domain, the only critical points are the zeros of $d\alpha/dx$. Using formula (3.10.11), we have,

$$\frac{d\alpha}{dx} = -6\frac{(x^2 - 16)}{(4 + x^2)(64 + x^2)} = -6\frac{(x + 4)}{(4 + x^2)(64 + x^2)} \cdot (x - 4).$$

Thus $x = 4$ is the unique critical point of P.

Step 6. From the formula for $d\alpha/dx$, we see that

$$\frac{d\alpha}{dx} = -6\frac{(+)}{(+)(+)} \cdot (-) > 0 \text{ for } x < 4 \text{ and}$$

$$\frac{d\alpha}{dx} = -6\frac{(+)}{(+)(+)} \cdot (+) < 0 \text{ for } x > 4.$$

Thus α is an increasing function of x on $(0,4)$ and a decreasing function of x on $(4,20)$. It follows that $x = 4$ ft is the optimal distance at which Mr. Woodman should stand. ◀

QUICK QUIZ

1. True or false: If f is continuous on $[a,b]$, then its maximum and minimum must occur at critical points of the function.
2. True or false: If f is defined and continuous on $[1,5]$, if f is differentiable on the intervals $(1,2)$ and $(2,5)$, and if $f'(3) = 0$, then the maximum value of $f(x)$ must be among the numbers $f(1)$, $f(2)$, $f(3)$, and $f(5)$.
3. True or false: If f is defined and continuous on $[2,4]$, if $f(2) = 0$ and $f(4) = 3$, and if there is no x in the interval $(2,4)$ for which $f'(x) = 0$, then the maximum value of $f(x)$ must be 3.
4. Calculate the minimum values of $f(x) = x + 100/x$ on the intervals $[1,5]$, $[5,12]$, and $[12,20]$.

Answers

1. False 2. True 3. False 4. 25, 20, 61/3

EXERCISES

Problems for Practice

▶ Solve each of the maximum-minimum problems in Exercises 1–20. Some may not have a solution, whereas others may have their solution at the endpoint of the interval of definition. ◀

1. A rectangle is to have perimeter 100 m. What dimensions will give it the greatest area?

2. What positive number plus its reciprocal gives the least sum?

3. What first quadrant point on the curve $xy^2 = 1$ is closest to the origin?

4. Find the longest vertical chord between the curves $y = x^3/3 - 3x^2/2 + 2x - 3/4$ and $y = 3x^3/4 - 15x^2 + 5x$, $0 \le x \le 3$.

5. A box (with no lid) is to be constructed from a sheet of cardboard by cutting squares from the corners and folding up the sides (Figure 11). If the original sheet of cardboard measures 20 inches by 20 inches, then what should be the size of the squares removed to maximize the volume of the resulting box?

6. Of all right triangles with hypotenuse 100 cm, which has greatest area?

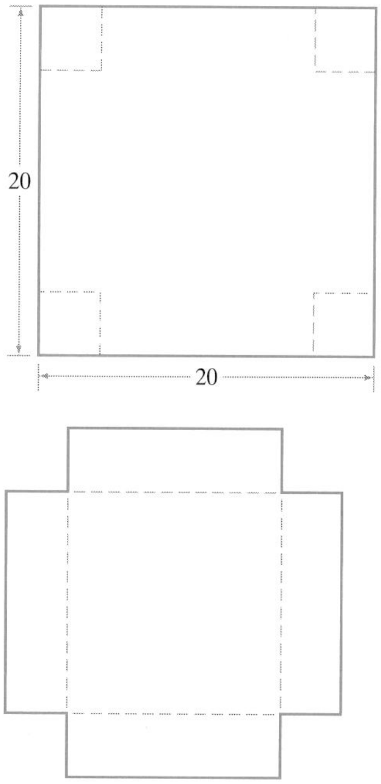

▲ Figure 11

7. A line with negative slope passes through the point $(1, 2)$ and has x-intercept a and y-intercept b. For what slope is the product ab minimized?
8. A cylindrical can is constructed with tin sides and an aluminum top and bottom. If aluminum costs 0.2 cents per square inch, if tin costs 0.1 cents per square inch, and if the can is to hold 20 cubic inches, what dimensions will minimize the cost of the can?
9. Which nonnegative number exceeds its cube by the greatest amount?
10. A right triangle is to contain an area of 50 cm^2. What dimensions will minimize the sum of the lengths of its two legs?
11. A printed page is to have 1 inch margins on all sides. The page should contain 8 square inches of type. What dimensions of the page will minimize the *area* of the page while still meeting these other requirements?
12. A church window will be in the shape of a rectangle surmounted by a semi-circle (Figure 12). The area of the window is to be 15 square meters. What should be the dimensions of the window to minimize the perimeter?

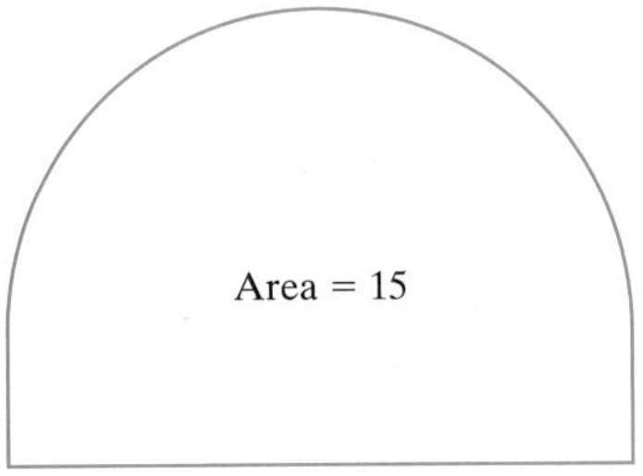

▲ Figure 12

13. A rectangle has its base on the x-axis, and its upper corners are on the graph of $y = 4 - x^2$. What dimensions for the rectangle will give it maximal area?
14. A box with square base and rectangular sides is being designed. The material for the sides costs 10 cents per square inch and that for the top and bottom costs 4 cents per square inch. If the box is to hold 100 cubic inches, then what dimensions will minimize the cost of materials for the box?
15. An open-topped rectangular planter is to hold 3 m^3. Its concrete base is square. Each brick side is rectangular. If the unit cost of brick is twelve times that of concrete, what dimensions result in the cheapest material cost?
16. What is the right triangle of least area that can be inscribed in a circle of radius 3 meters?
17. A pentagonal shape consisting of a rectangle adjoining an equilateral triangle (as in Figure 13) is to enclose 9801 m^2. What base length will minimize the perimeter?
18. What is the rectangle (with sides parallel to the axes) of greatest area that can be inscribed in the ellipse $x^2 + 4y^2 = 16$?

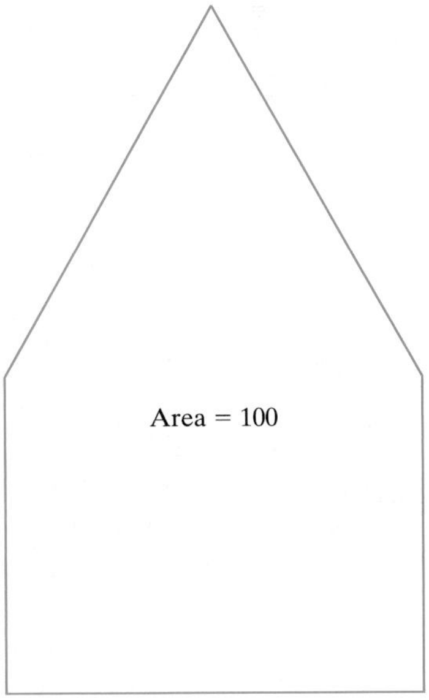

▲ Figure 13

19. A cosmetic container is to be in the shape of a right circular cylinder surmounted by a hemisphere (see Figure 14). If the surface area is 5π square inches, what is the greatest volume that this container can hold?

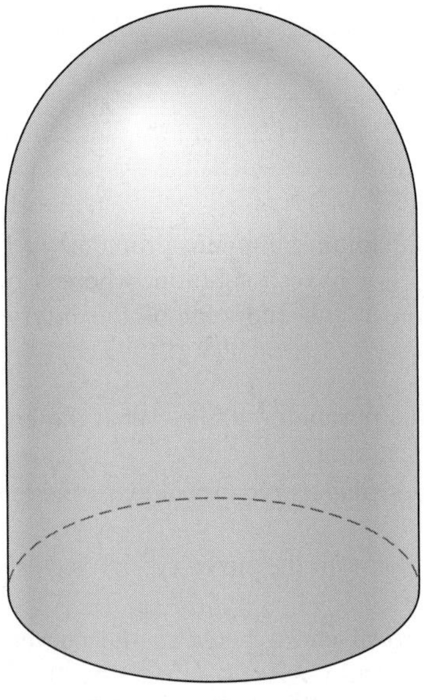

▲ Figure 14 Cylinder surmounted by hemisphere

20. Assume that if the price of a certain book is p dollars, then it will sell x copies where $x = 7000 \cdot (1 - p/35)$. Suppose that the dollar cost of producing those x copies is $15000 + 2.5x$. Finally, assume that the company will not

sell this book for more than \$35. Determine the price for the book that will maximize profit.

In each of Exercises 21–24, a cost function c and a form of the demand equation are given. Calculate the sales level x that maximizes profits.

21. $C(x) = 1000 + 2x$, $D^{-1}(x) = 1602 - 8x$
22. $C(x) = 800 + 3x$, $D^{-1}(x) = 26403 - 12x$
23. $C(x) = 1200 + 8x$, $D(p) = 51 - p/8$
24. $C(x) = 7000 + x$, $D(p) = 2402 - 2p$

In each of Exercises 25–34, find the absolute minimum value and absolute maximum value of the given function on the given interval.

25. $f(x) = x^2 + x$; $[-1, 1]$
26. $f(x) = x^3 + 3x^2 - 45x + 2$; $[-6, 4]$
27. $f(x) = x^3 - 3x^2/2 + 2$; $[-1, 3]$
28. $f(x) = x^5 - 5x^4$; $[-2, 3]$
29. $f(x) = x(x + 2)^2$; $[-3, 1]$
30. $f(x) = (x^2 - 3)/(x + 2)$; $[-3/2, 1]$
31. $f(x) = 4x + 9/x$; $[1, 2]$
32. $f(x) = x - 4\sqrt{x}$; $[1, 3]$
33. $f(x) = e^x - x$; $[-1, 1]$
34. $f(x) = x^2 e^{-x}$; $[-2, 3]$

Further Theory and Practice

35. The relationship between the speed s of cars making their way through a tunnel at a constant rate and the number x of cars per km is expressed by the equation $s = \alpha - \beta x$, where α is a positive constant that carries the units of km/hr, and β is a positive constant that carries the units of km^2/hr. What value of x results in the greatest number of cars passing through the tunnel per hour?

36. A rectangle is to have one corner on the positive x-axis, one corner on the positive y-axis, one corner at the origin, and one corner on the line $y = -5x + 4$. Which such rectangle has the greatest area?

37. Prove the *Maximum Profit Principle*: Marginal revenue $R'(x_0)$ equals marginal cost $C'(x_0)$ at the production level x_0 that maximizes profit.

38. The fixed cost of producing a certain compact disc is \$12000. The marginal cost per disc is \$1.20. Market research and previous sales suggest that the demand function for this particular disc will be $x = 19000 - 600p$. At what production level x_0 is the marginal cost equal to the marginal revenue? Verify that profit is maximized when $x = x_0$.

39. If the cost of producing x units of a commodity is $C(x)$, then the average cost $\overline{C}(x)$ of producing those x units is $\overline{C}(x) = C(x)/x$. Prove the *Minimum Average Cost Principle*: When minimized, the average cost equals the marginal cost.

40. A right circular cone of height h and base radius r has volume $\pi r^2 h/3$ and lateral surface area $\pi r\sqrt{r^2 + h^2}$. What is the greatest volume that such a cone can have if its surface area, including the base, is 2π?

41. In a model for optimizing the angle of release θ of a basketball shot, suppose that a and b are positive constants. Let θ_0 be the value of θ in the interval $(\arctan(b/a), \pi/2)$ for which

$$f(\theta) = \left(a\sin(\theta)\cos(\theta) - b\cos^2(\theta)\right)^{-1}$$

is minimized. What is $\tan(\theta_0)$?

42. A piece of paper is circular in shape, with radius 6 inches. A sector is cut from the circle and the two straight edges taped together to form a cone (see Figure 15). What angle for the sector will maximize the volume inside the cone?

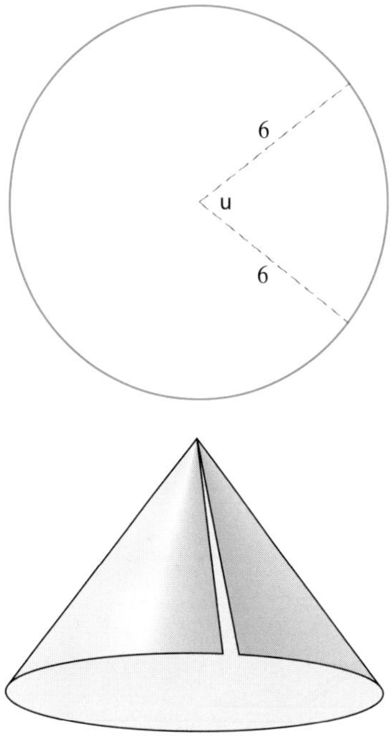

▲ Figure 15

43. A wire of length L can be shaped into a circle or a square, or it can be cut into two pieces, one of which is formed into a circle and the other a square. How is the minimum total area obtained?

44. A wire of length L can be shaped into a (closed) semicircle or a square, or it can be divided into two pieces. With one, a (closed) semicircle is formed and with the other a square. Find the largest and smallest areas that are possible.

45. A wire of length L can be shaped into a (closed) semicircle or a circle. The wire can also be divided into two pieces, with one piece forming a (closed) semicircle, and the other, a circle. Find the largest and smallest areas that are possible.

46. A battery that produces V volts with internal resistance r ohms is connected in series to an external device with resistance R ohms. The power output of the battery is then $V^2R/(r+R)^2$. What value of R results in maximum power?

47. The drag on an airplane at a given altitude is given by $(a + b/v^4)\,v^2$ where a and b are positive constants, and v is the velocity of the plane. At what speed is drag minimized?

48. In a certain process, the optimal time t for removing an object from a heat source x units away is obtained by maximizing

$$H = \frac{A}{\sqrt{t}}\exp\left(-\frac{cx^2}{t}\right).$$

Here A and c are positive constants. What value of t maximizes H?

49. The potential energy of a diatomic molecule is $U = A\,(b/r^{12} - 1/r^6)$, where A and b are positive constants, and r is the inter-atomic distance. What value of r minimizes U?

50. Carpenters will need to carry rods of various lengths around a hall corner as shown in Figure 16. One hall is 6 feet wide, and the other is 8 feet wide. Ignoring the thickness of the rod, and assuming that the rod is carried parallel to the floor, determine the longest rod that the carpenters will be able to carry around the corner.

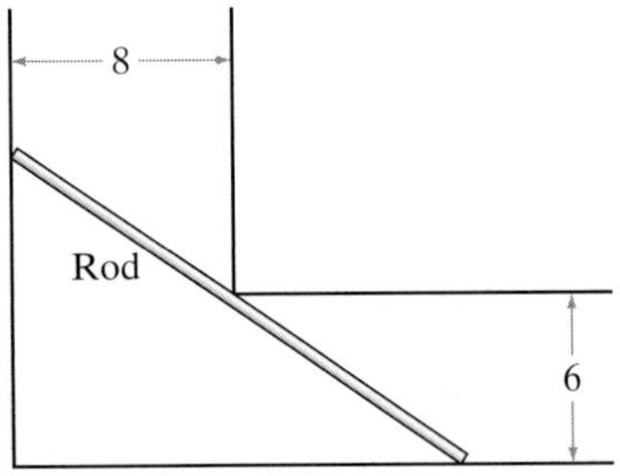

▲ Figure 16

51. A garden is to be bounded on one side by part of a river that is shaped in a circular arc of radius 80 feet. The other three sides will be straight fencing at right angles. The configuration is shown in Figure 17. Suppose that 100 feet of fence is available. Carry out the steps for determining the largest possible garden, but do not solve for the critical points.

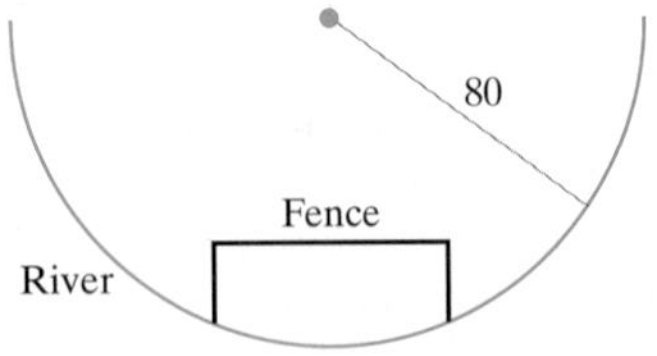

▲ Figure 17

52. When a person coughs, air pressure in the lungs increases, and the radius r of the windpipe decreases from the normal value ρ. Units can be chosen so that the flow F through the windpipe resulting from a cough is then given by

$$F(r) = (\rho - r)r^4$$

for $0 < r \le \rho$. What value of r results in the most productive cough?

53. In Figure 18, B is a vascular branch point on a main artery from the heart H. The branch angle $\theta \in (0, \pi/2)$ determines the resistance Ω to blood flow to the organ O along the path HBO according to

$$\Omega(\theta) = \frac{a - b\cot(\theta)}{R^4} + \frac{b}{r^4 \sin(\theta)}.$$

For what value of $\cos(\theta)$ is resistance minimized?

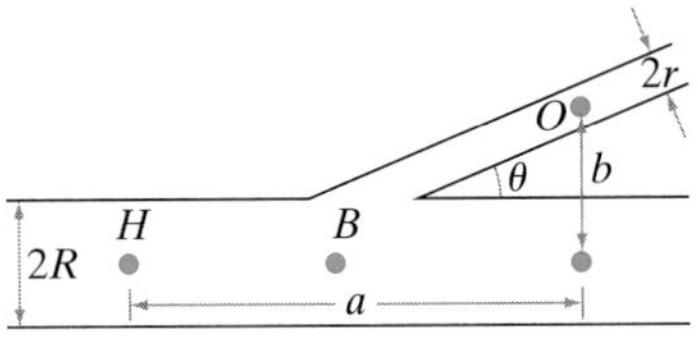

▲ Figure 18

54. (E. V. Huntington's problem) A circular pool has radius r and center at point O. Suppose that Earl is at point P at the edge of the pool. He wishes to get to the diametrically opposite point R in the least amount of time. He can swim from P to Q and then run to R. Suppose that Earl's maximum swimming speed is u, and his maximum running speed is v. Let θ be the angle $\angle OPQ$.

- **a.** Find an expression $f(\theta)$, $0 \le \theta \le \pi/2$ for the time required for Earl to swim from P to Q in a straight line and then to run from Q to R along the edge of the pool.
- **b.** Show that if $u < v$, then f has one critical point in the interval $(0, \pi/2)$, and this critical point corresponds to a maximum time.
- **c.** Show that if $u \ge v$, then f has no critical point in the interval $(0, \pi/2)$.
- **d.** For which values of the parameters u and v does $\theta = 0$ give the minimum time? For which values of the parameters u and v does $\theta = \pi/2$ give the maximum time?
- **e.** How quickly can Earl get from P to R?

55. (V. L. Klee's problem) A rectangle is to be formed by 300 m of fencing of which a 100 m straight length is already in place. The additional 200 m can be used to form the three other sides or some of it can be used to extend the existing length. How large can the enclosed region be?

▶ Exercises 56–58 concern the most economical dimensions of a cylindrical tin can that holds a predetermined volume V. ◀

56. What are the dimensions if the amount of metal used in the can is to be minimized? Express your answer as a height h to radius r ratio.

57. (P. L. Roe's problem) When the circular top and bottom is cut from a large sheet of metal, there will be wastage. Thus the amount of metal necessary to make a can is actually larger then the amount of metal used in the can itself. Suppose that the top and bottom circles are each cut from squares whose side length is the diameter of the circle. What is h/r when the amount of metal (including wastage) is minimized?

58. (Continuation of P. L. Roe's problem) Suppose that the metal necessary to make the cylindrical can is $4\sqrt{3}r^2 + 2\pi rh$, including wastage. In forming the can, a side join of length h and two circular joins of length $2\pi r$ each are necessary. If a length k join costs the same as a unit area of the metal, then the total cost C of production is

$$C = 4\sqrt{3}r^2 + 2\pi rh + (4\pi r + h)/k.$$

Write C as a function of r, and find an equation for the critical point of C. Show that there are constants a, b, c, and d such that

$$\frac{h}{r} = \frac{ark + b}{crk + d}$$

when r is the critical point. In the case of very inexpensive side joins ($k \to \infty$), what is the approximate ratio h/r? In the case of very expensive side joins ($k \to 0$), what is the approximate ratio h/r?

59. Suppose that a uniform rod of length ℓ and mass m can rotate freely about one end. If a point mass $6m$ is attached to the rod a distance x from the pivot, then the period of small oscillations is equal to

$$2\pi\sqrt{\frac{2}{3g}} \cdot \sqrt{\frac{\ell^2 + 18x^2}{12x + \ell}}.$$

For what value of x is the period least?

60. A volume V_0 of gas is held at pressure p_0 in a reservoir. The gas is discharged through a nozzle of opening area A into a region at lower pressure p. Then the rate of discharge (in units of weight/time) is given by

$$A\sqrt{\frac{2g\gamma}{\gamma - 1}\frac{p_0}{V_0}\left(\left(\frac{p}{p_0}\right)^{2/\gamma} - \left(\frac{p}{p_0}\right)^{(\gamma+1)/\gamma}\right)}$$

where γ is the adiabatic constant of the gas. What is p/p_0 when the rate of discharge is greatest?

61. When a boat travels through open water, it creates a trailing wedge of interior angle $2\beta_0$ in which there is constructive interference between waves. This is known as the Kelvin wedge. The value of $\tan(\beta_0)$ is the maximum value of

$$\frac{\tan(\alpha)}{2 + \tan^2(\alpha)}.$$

Show that $\sin(\beta_0) = 1/3$. (The Kelvin wedge therefore encloses an angle of about 39°, independent of the speed of the boat.)

62. A road is to be built from $P = (0, b)$ to $Q = (a, 0)$. The cost of new road construction is c_1 per unit distance. There is an existing straight road from the origin $(0, 0)$ to Q. To take advantage of this existing route, highway engineers are considering building a road from P to some point $R = (c, 0)$ on the existing road. In that case, the existing road between R and Q must be upgraded to handle the increased traffic flow. The cost of this upgrade is c_2 per unit distance where $c_2 < c_1$. Determine the minimum cost route.

Calculator/Computer Exercises

63. Find the absolute minimum value and absolute maximum value of

$$f(x) = \frac{x^2 + \sin(x)}{\sqrt{x^4 + 2x + 2}}, \qquad -4 \le x \le 4.$$

64. Suppose that the cost function for producing a certain product is $C(x) = 100000 + 1000\sqrt{x} + x^2/40$. Find the production level x_0 at which average cost $\overline{C}(x) = C(x)/x$ is minimized. What is $\overline{C}(x_0)$? What is the marginal price at this production level?

65. Two heat sources separated by a distance 10 cm are located on the x-axis. The heat received by any point on the x-axis from each of these sources is inversely proportional to the square of its distance from the source. Suppose that the heat received a distance 1cm from one source is twice that received 1 cm from the other source. What is the location of the coolest point between the two sources?

66. Suppose that, for a beam of length 11 meters, the deflection from the horizontal of a point a distance x from one end is proportional to

$$2x^4 - 33x^3 + 1331x.$$

Determine the point at which the deflection is greatest.

67. Supposes that a shotputter releases the shot at height h, with angle of inclination α, and initial speed v. Then the horizontal distance R that the shot travels is given by

$$R = \frac{v^2\sin(2\alpha) + v\sqrt{v^2\sin^2(2\alpha) + 8gh\cos^2(\alpha)}}{2g}.$$

Use a computer algebra system to find the value of α that maximizes R.

68. Let α_0 denote the solution of $\tan(\alpha_0) = 1/3$ in $(0, \pi/2)$. Plot

$$E(\alpha) = \frac{\tan(\alpha)\big(1 - 3\tan(\alpha)\big)}{3 + \tan(\alpha)}$$

for $0 < \alpha \le \alpha_0$. Use the plot of E to find the maximum value of $E(\alpha)$ to three decimal places. Calculate $E'(\alpha)$. Show that E(alpha) is maximized when α satisfies

$$\tan(\alpha)^2 + 6\tan(\alpha) - 1 = 0.$$

Solve for $\tan(\alpha)$ and find the *exact* maximum value of $E(\alpha)$. (This example, with E representing the efficiency of a screw jack, μ the coefficient of friction, and α the pitch, arises in mechanical engineering.)

69. A tetramer is a protein with four subunits. In the study of tetramer binding, the equation

$$Y = \frac{x + 3x^2 + 3x^3 + 10x^4}{1 + 4x + x^2 + 4x^3 + 10x^4}$$

expresses a typical relationship between saturation Y and ligand concentration x $(x \geq 0)$. Ordinarily, the variable Y/x is plotted as a function of x. Explain why Y/x has an absolute maximum value. Find it to three decimal places.

70. An underground pipeline is to be built between two points $P = (1,4)$ and $Q = (2,0)$. The subsurface rock formation under P is separated from what lies under Q by a curve $\mathcal{C}$ of the form $y = x^2$. The cost per unit distance of laying pipeline from P to the graph of $\mathcal{C}$ is \$5000. The unit cost from the graph of $\mathcal{C}$ to Q is \$3000. Assuming that the pipeline will consist of two straight line segments, analyze the minimum cost route.

▶ Exercises 71 and 72 utilize a minimization algorithm called the *Golden Ratio Search*. Let ρ denote the Golden Ratio, $(\sqrt{5} - 1)/2$. Let $r = \rho/(\rho + 1) \approx 0.381966$. Suppose that f is continuous on $I = [a_1, b_1]$, has a minimum at c in (a_1, b_1), is decreasing on (a_1, c), and is increasing on (c, b_1). (Such a function is said to be *unimodal*.) Let $\Delta x_1 = b_1 - a_1$, $\alpha_1 = a_1 + r \cdot \Delta x_1$, and $\beta_1 = b_1 - r \cdot \Delta x_1$. Calculate $f(\alpha_1)$ and $f(\beta_1)$. If $f(\alpha_1) < f(\beta_1)$, then let $a_2 = a_1$ and $b_2 = \beta_1$. Otherwise, let $a_2 = \alpha_1$ and $b_2 = b_1$. Repeat the procedure, letting $\Delta x_2 = b_2 - a_2 = \rho\Delta x_1$ $\alpha_2 = a_2 + r \cdot \Delta x_2$, and $\beta_2 = b_2 - r \cdot \Delta x_2$. Calculate $f(\alpha_2)$ and $f(\beta_2)$. (From the point of view of efficiency, it is useful to observe that one of these two values has been computed in the last step: If $a_2 = a_1$, then $\beta_2 = \alpha_1$; if $a_2 = \alpha_1$, then $\alpha_2 = \beta_1$.) If $f(\alpha_2) > f(\beta_2)$, then let $a_3 = a_2$ and $b_3 = \beta_2$. Otherwise, let $a_3 = \alpha_2$ and $b_3 = b_2$. Repeat the procedure. After every step, the point c is in the interval $I_n = [a_n, b_n]$. Because $b_n - a_n = \rho^{n-1} \cdot (b - a)$, the midpoint) $c_n = (a_n + b_n)/2$ of I_n satisfies $|c - c_n| \leq \rho^{n-1} \cdot (b - a)/2$. Thus n can be chosen large enough so that c_n approximates c with any specified accuracy. For the given function f on [0, 1]), locate c to two decimal places. ◀

71. $f(x) = 3 - 2\sqrt{x} + x^3$

72. $f(x) = 2 + x^2 - \sin(x)$

4.5 Concavity

We have learned that we can use the sign of the first derivative f' of a differentiable function f to determine where the graph of f rises and where it falls. However, much more can be said about the shape of a graph. In this section, we study the properties of the graph of a function that are revealed by the second derivative.

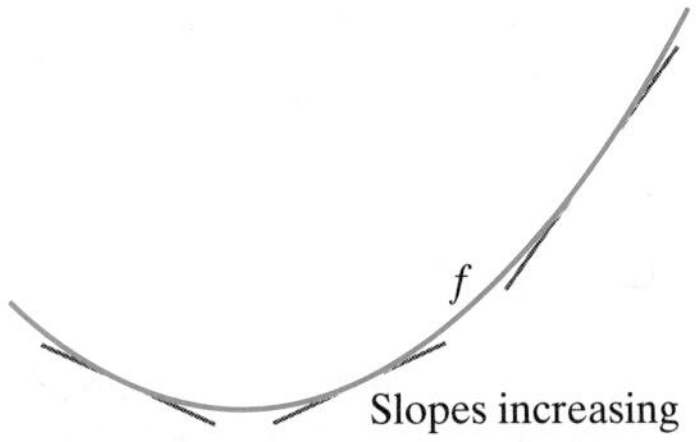

▲ Figure 1 The graph of f is concave up.

DEFINITION Let the domain of a differentiable function f contain an open interval I. If f'(the slope of the tangent line to the graph) increases as x moves from left to right in I, then the graph of f is said to be *concave up* on I (see Figure 1). If f' decreases as x moves from left to right in I, then the graph of f is said to be *concave down* on I (see Figure 2).

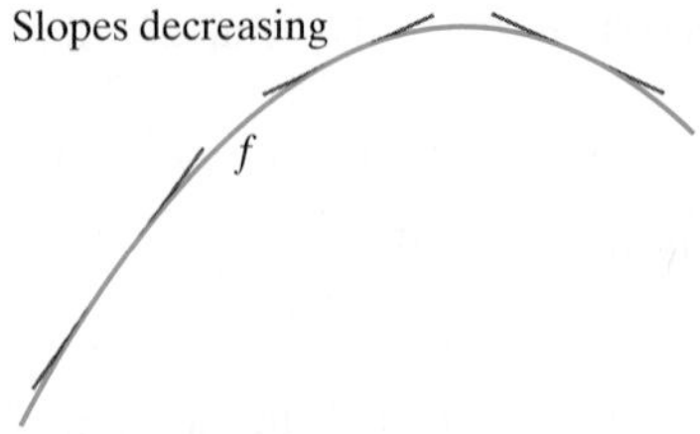

▲ Figure 2 The graph of f is concave down.

There are many ways to think about the concept of concavity. If P is a point on the graph of f, then another way to define "f is concave up at P" is to require that the tangent line to the graph at P lie below the graph just to the left and to the right of P (see Figure 1). Similarly, "f is concave down at P" could be defined to mean that the tangent line to the graph at P lies above the graph just to the left and to the right of P (as shown in Figure 2). Alternatively, we could define the graph of f to be concave up (respectively down) if the graph of f lies below (respectively above) every chord joining two points of the graph of f (see Figure 3). These other definitions represent alternative points of view and will not be discussed further.

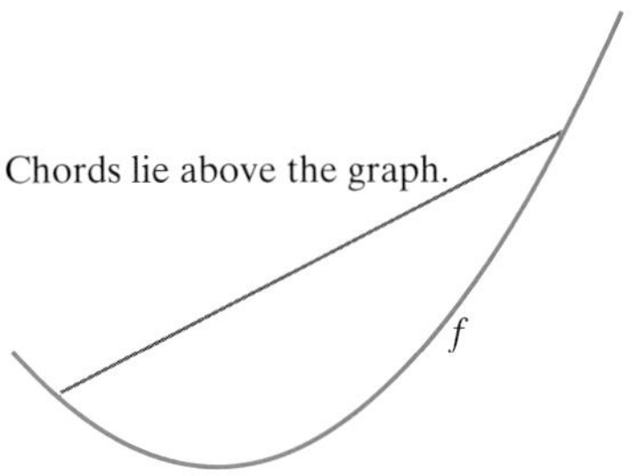

▲ Figure 3a The graph of f is concave up.

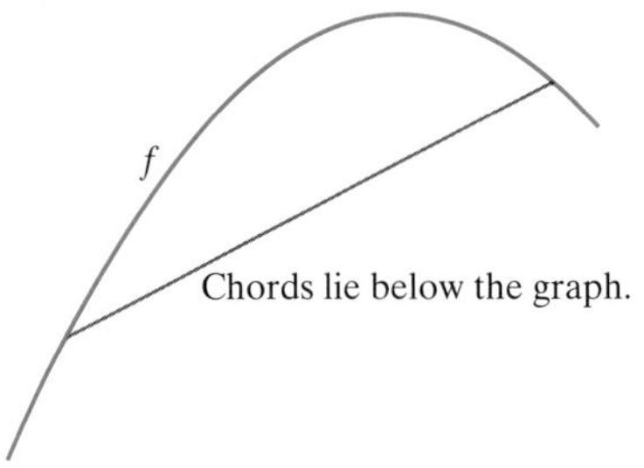

▲ Figure 3b The graph of f is concave down.

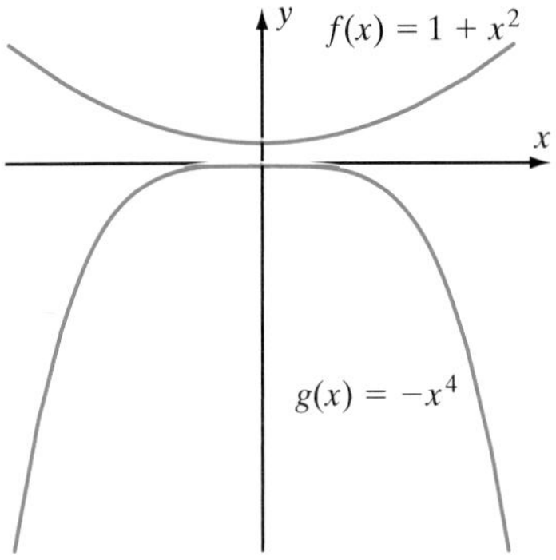

▲ Figure 4

▶ **EXAMPLE 1** Discuss concavity for the graphs of $f(x) = 1 + x^2$ and $g(x) = -x^4$.

Solution We calculate that $f'(x) = 2x$, which is an increasing function on the entire real line. Therefore the graph of f is concave up on the interval $(-\infty, \infty)$ (see Figure 4). Similarly, the function $g(x) = -x^4$ satisfies $g'(x) = -4x^3$, which is a decreasing function on $(-\infty, \infty)$. Therefore the graph of g is concave down on $(-\infty, \infty)$, as can also be seen in Figure 4. ◀

INSIGHT At first, it is easy to confuse "increase" and "decrease" of a given function with the "concave up" and "concave down" properties of that function. The first two properties are independent of the second two. This means that a function can be increasing and concave down, as the function g from Example 1 illustrates on the interval $(-\infty, 0)$, or decreasing and concave down (g on the interval $(0, \infty)$). Similarly, a function can be increasing and concave up (the function f from Example 1 on the interval $(0, \infty)$) or decreasing and concave up (f on the interval $(-\infty, 0)$).

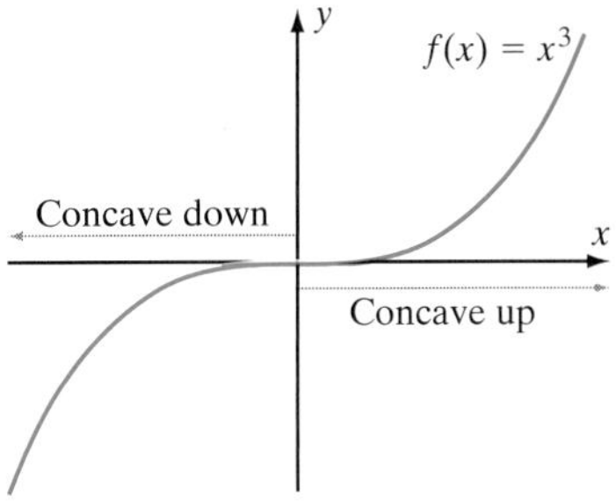

▲ Figure 5

▶ **EXAMPLE 2** Discuss concavity for the graph of the function $f(x) = x^3$.

Solution First of all, $f'(x) = 3x^2$. On the interval $(0, \infty)$, the function $f'(x) = 3x^2$ is increasing. Therefore the graph of f is concave up on $(0, \infty)$. On the interval $(-\infty, 0)$, the function $f'(x) = 3x^2$ is decreasing (because it becomes less and less positive when x is negative and increasing). Therefore the graph of f is concave down on $(-\infty, 0)$ (see Figure 5). ◀

INSIGHT Example 2 shows that the graph of a function can be concave up on one interval in its domain and concave down on another.

Using the Second Derivative to Test for Concavity

An important use of the second derivative is that it enables us to recognize the two types of concavity.

THEOREM 1 **(The Second Derivative Test for Concavity)** Suppose that the function f is twice differentiable on an open interval I.

a. If $f''(x) > 0$ for every x in I, then the graph of f is concave up on I.
b. If $f''(x) < 0$ for every x in I, then the graph of f is concave down on I.

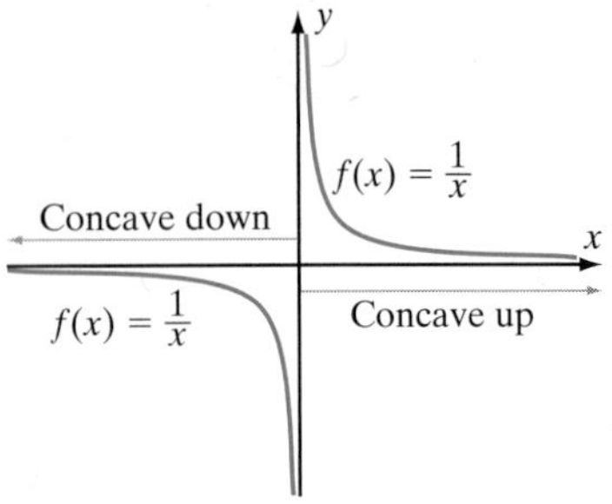

▲ Figure 6

Proof. Notice that $f'' = (f')'$. Thus the hypothesis that $f''(x) > 0$ on I implies that $f'(x)$ is increasing on I, or, equivalently, that f is concave up. Similarly the hypothesis that $f'' < 0$ means that $f'(x)$ is decreasing, or that f is concave down.

▶ **EXAMPLE 3** Apply the Second Derivative Test for Concavity to the function $f(x) = 1/x$.

Solution Provided that $x \neq 0$, we calculate that $f'(x) = -1/x^2$ and $f''(x) = 2/x^3$. Notice that $f''(x) > 0$ on the interval $(0, \infty)$. Hence, the graph of f is concave up on that interval. Also $f''(x) < 0$ on the interval $(-\infty, 0)$. Therefore the graph of f is concave down on that interval. A glance at the graph of f in Figure 6 confirms these calculations. ◀

Points of Inflection

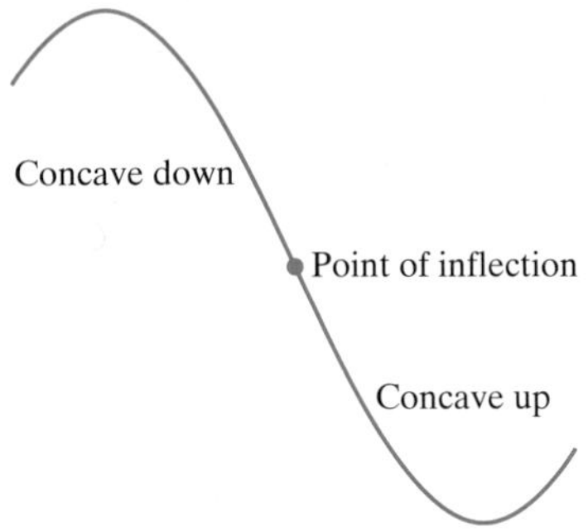

▲ Figure 7

DEFINITION Let f be a continuous function on an open interval I. If the graph of $y = f(x)$ changes concavity (from positive to negative or from negative to positive) as x passes from one side to the other of a point c in I, then the point $(c, f(c))$ on the graph of f is called a *point of inflection* (or *an inflection point*). We also say that f has a point of inflection at c (see Figure 7).

Notice that the function $f(x) = 1 + x^2$ from Example 1 (graphed in Figure 4) is concave up everywhere; hence, it has no points of inflection. Similarly, the function $g(x) = -x^4$ of Example 1 (also in Figure 4) has no points of inflection because it is concave down everywhere. On the other hand, the function $f(x) = x^3$ from Example 2 (graphed in Figure 5) changes concavity at $c = 0$, so $(0, 0)$ is a point of inflection. Finally, the graph of $f(x) = 1/x$ in Example 3 (graphed in Figure 6) is concave down to the left of 0 and concave up to the right of 0; however, 0 is not in the domain of f, so f does *not* have a point of inflection at 0.

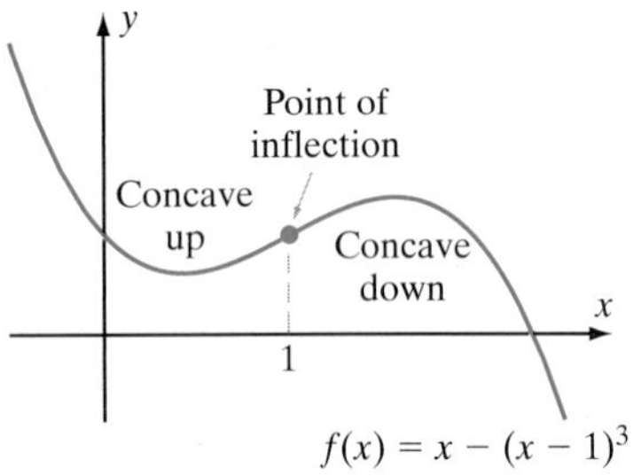

▲ Figure 8

▶ **EXAMPLE 4** Examine the graph of $f(x) = x - (x-1)^3$ for concavity and points of inflection.

Solution We calculate that $f'(x) = 1 - 3(x-1)^2$ and $f''(x) = -6(x-1)$. Therefore $f'' > 0$ on the interval $(-\infty, 1)$ and $f'' < 0$ on the interval $(1, \infty)$. We conclude that the graph of f is concave up when $x < 1$ and concave down when $x > 1$. It follows that f has a point of inflection at 1, as shown in Figure 8. ◀

INSIGHT The critical points of the function f from Example 4 are the solutions of the equation $f'(x) = 1 - 3(x-1)^2 = 0$. Using the Quadratic Formula, we see that the critical points of f are located at $x = 1 \pm 1/\sqrt{3}$. However, the inflection point is at $x = 1$. Inflection points are *not* the same as critical points.

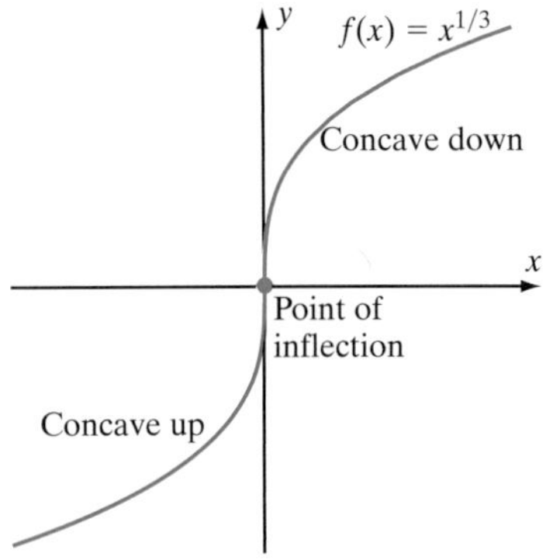

▲ Figure 9

▶ **EXAMPLE 5** Examine the function $f(x) = x^{1/3}$ for concavity and points of inflection.

Solution For $x \neq 0$, we have $f'(x) = (1/3) \cdot x^{-2/3}$ and $f''(x) = (-2/9) \cdot x^{-5/3}$. Therefore $f'' > 0$ on the interval $(-\infty, 0)$, and $f'' < 0$ on the interval $(0, \infty)$. Thus the graph of f is concave up on $(-\infty, 0)$ and concave down on $(0, \infty)$. Therefore f has a point of inflection at 0. Notice that f is not differentiable at 0, but 0 is still an inflection point because f is continuous and the concavity changes there (see Figure 9). ◀

Strategy for Determining Points of Inflection

Suppose that f has a point of inflection at c and that f'' exists on both sides of c. Because the graph of f is concave up on one side of c and concave down on the other, we conclude that, as x moves from one side of c to the other, the value of $f''(x)$ changes sign. If $f''(c)$ exists, then we expect it to be equal to 0, which is true. (See Exercise 81 of Section 4.3.) For instance, in Example 4, the inflection point of the function $f(x) = x - (x-1)^3$ occurs at $x = 1$, which is a root of the equation $f''(x) = 0$. We have also seen (in Example 5) that an inflection point can occur at a point where $f''(x)$ fails to exist. One of these two possibilities must occur at a point of inflection. These considerations lead to the following strategy for determining points of inflection of a continuous function f on an open interval I:

1. Locate all points of I at which $f'' = 0$ or f'' is undefined.
2. At each of these points, check to see if f'' changes sign.

INSIGHT It is a common error to think that $f''(c) = 0$ implies that $(c, f(c))$ is a point of inflection for f. The function $g(x) = -x^4$ of Example 1 (Figure 4) shows why this thinking is wrong: The graph of g is concave down everywhere, so there are no points of inflection. In particular, g does not have a point of inflection at 0 even though $g''(0) = 0$.

It may help to compare the tasks of locating points of inflection and local extrema. Recall that local extrema occur at points where f' changes sign. To find such points, we must first locate the points at which either $f' = 0$ or f' does not exist; these are the points at which f' *might* change sign. Having located these points, we must then check each one of them to see if f' really does change sign. The search for points of inflection is completely analogous. Step 1 of the strategy—locating all points at which $f'' = 0$ or f'' is undefined—produces the *candidates* that might be points of inflection. Step 2 of the strategy—checking to see if f'' changes sign—determines whether or not a candidate is an actual point of inflection.

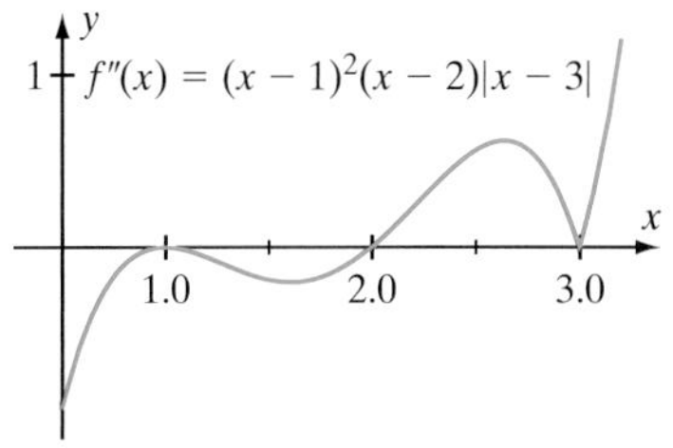

Figure 10

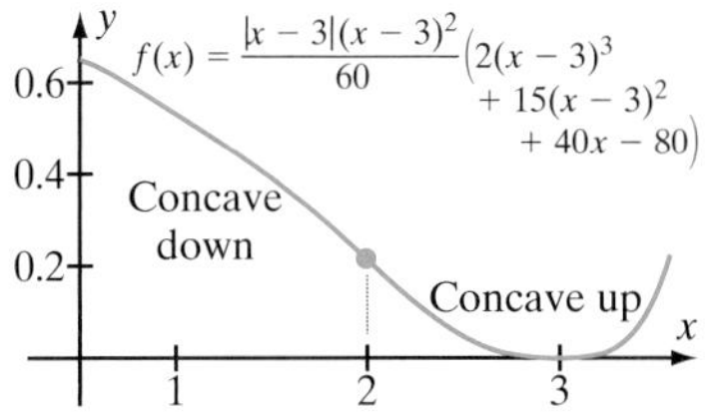

Figure 11 Point of inflection

EXAMPLE 6 A function f is twice differentiable with $f''(x) = (x-1)^2(x-2) \cdot |x-3|$ for every x. Determine each x at which f has a point of inflection.

Solution By hypothesis, $f''(x)$ exists for every x. Therefore a point of inflection c must satisfy $f''(c) = 0$. Clearly the points 1, 2, and 3 are the only candidates. Because the factors $(x-1)^2$ and $|x-3|$ are nonnegative, we see that the sign of $f''(x)$ is the same as the sign of $(x-2)$. This tells us that $f''(x)$ changes sign only at $x = 2$ (as can be seen from the plot of f'' in Figure 10). Therefore f has a point of inflection only at $x = 2$. Figure 11 contains the plot of one function f that has the specified second derivative. It supports our conclusions. ◀

The Second Derivative Test at a Critical Point

We now apply our knowledge of concavity to the classification of extrema.

THEOREM 2 **(The Second Derivative Test for Extrema)** Let f be twice differentiable (both f' and f'' exist) on an open interval containing a point c at which $f'(c) = 0$.

a. If $f''(c) > 0$, then c is a local minimum.
b. If $f''(c) < 0$, then c is a local maximum.
c. If $f''(c) = 0$, then no conclusion is possible from *this* test.

Proof. The inequality $f''(c) > 0$ means $(f')'(c) > 0$ so f' is increasing at c. Because $f'(c) = 0$, it follows that f' must be going from negative to positive as x passes from left to right through c. By the First Derivative Test, there is a local minimum at c. Similarly, $f''(c) < 0$ means $(f')'(c) < 0$ so f' is decreasing at c. Because $f'(c) = 0$, f' must be going from positive to negative as x passes from left to right through c. By the First Derivative Test, there is a local maximum at c.

To understand the last statement of the Second Derivative Test, notice that $f'(0) = 0$ and $f''(0) = 0$ for each of the three functions $f(x) = x^4$, $f(x) = -x^4$, and $f(x) = x^3$. The first function obviously has a minimum at 0, the second has a maximum, and the third has neither (see Figure 12). By themselves, the two equations $f'(c) = 0$ and $f''(c) = 0$ do not distinguish between any of the three possible behaviors. ■

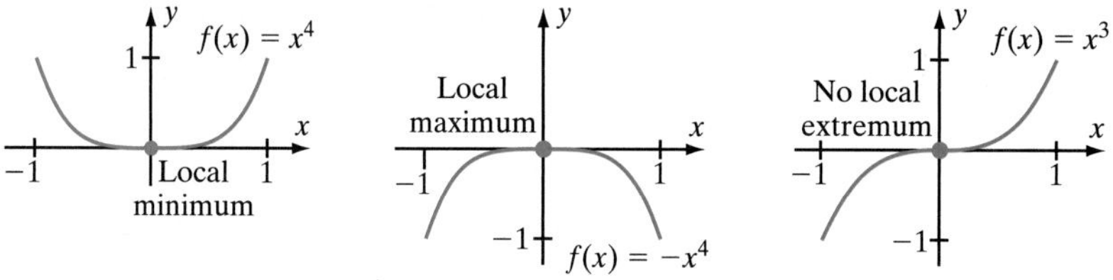

▲ Figure 12 $f'(0) = 0$ and $f''(0) = 0$ for all three functions.

INSIGHT At first, it is surprising that $f''(c) > 0$ corresponds to a minimum while $f''(c) < 0$ corresponds to a maximum. Psychologically, this seems backwards. However, the inequality $f''(c) > 0$ means that the graph is concave up at c, which indicates a minimum rather than a maximum. Similarly $f''(c) < 0$ means that the graph is concave down at c, which indicates a maximum rather than a minimum.

Do not interpret the third part of the Second Derivative Test to mean that no conclusion is possible at all. We are able to analyze the behavior of each of the three functions $f(x) = x^4$, $f(x) = -x^4$, and $f(x) = x^3$ at its critical point 0 even though the Second Derivative Test is inconclusive for each of them.

▶ **EXAMPLE 7** Use Fermat's Theorem and the Second Derivative Test for Extrema to determine the local extrema of $f(x) = 2x^3 + 15x^2 + 24x + 23$.

Solution This function is a polynomial; it is defined on the entire real line and is differentiable everywhere. Therefore the only critical points are the zeros of f'. We calculate

$$f'(x) = 6x^2 + 30x + 24 = 6(x + 1)(x + 4),$$

from which we see that the zeros of f' are -1 and -4. By Fermat's Theorem, f can have a local extremum only at these two points. To use the Second Derivative Test for Extrema, we calculate $f'(x) = 12x + 30$. Because $f''(-1) = 18 > 0$, the Second Derivative Test for Extrema says that there is a local minimum at $c = -1$. Because $f''(-4) = -18 < 0$, the Second Derivative Test for Extrema says that there is a local maximum at $c = -4$ (see Figure 13). ◀

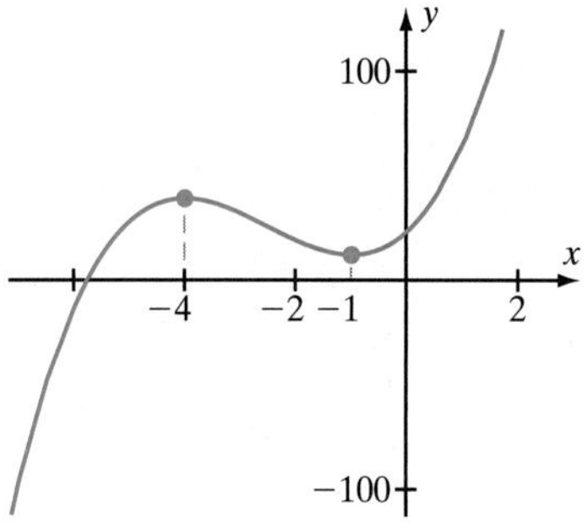

▲ Figure 13

A LOOK BACK Although the Second Derivative Test for Extrema is convenient, it does not supersede the First Derivative Test (Theorem 2 of Section 4.3). That is because the First Derivative Test (part a) requires less computation than the Second Derivative Test, (part b) can be applied when the second derivative does not exist, and (part c) can be applied when the second derivative exists but is 0.

QUICK QUIZ

1. True or false: $f''(4) = 0$ tells us that $(4, f(4))$ is a point of inflection for $y = f(x)$.
2. True or false: $f''(x) < 0$ on $(1,5)$ and $f''(x) > 0$ on $(5, 5.1)$ tells us that $(5, f(5))$ is a point of inflection for $y = f(x)$.
3. Suppose that $f''(x) = (x+2)^4(x-3)$. Does $f(x)$ have a point of inflection at $x = -2$? At $x = 3$?
4. True or false: $f''(4) < 0$ tells us that $f(x)$ has a maximum at $x = 4$.
5. True or false: $f'(4) = 0$ and $f''(4) > 0$ tell us that $f(x)$ has a maximum at $x = 4$.

Answers

1. False 2. True 3. No at $x = -2$, yes at $x = 3$ 4. False 5. False

EXERCISES

Problems for Practice

In each of Exercises 1–36, determine the intervals on which the given function f is concave up, the intervals on which f is concave down, and the points of inflection of f. Find all critical points. Use the Second Derivative Test to identify the points x at which $f(x)$ is a local minimum value and the points at which $f(x)$ is a local maximum value.

1. $f(x) = x - 4\sqrt{x}$
2. $f(x) = 1/x - 1/x^2$
3. $f(x) = \sqrt{x} + 3/\sqrt{x}$
4. $f(x) = x\sqrt{x-2}$
5. $f(x) = x^2 + 54/x$
6. $f(x) = 1/(x^2+1)$
7. $f(x) = x^3 + 9x^2 - 21x + 15$
8. $f(x) = -2x^3 + 12x^2 + 7$
9. $f(x) = 2x^3 - 3x^2 - 12x + 1$
10. $f(x) = x/(x-4)$
11. $f(x) = x^4 - 2x^3 - 9x^2$
12. $f(x) = x^4 - 8x^3 + 9$
13. $f(x) = 3x^5 - 10x^3 + 15x - 8$
14. $f(x) = 2x^5 + 15x^4 + 30x^3 - 42$
15. $f(x) = x/(x^2+3)$
16. $f(x) = x/\sqrt{x^2+1}$
17. $f(x) = (x^2-1)/(x^2+1)$
18. $f(x) = (x+1)^2/x$
19. $f(x) = \cos(x) + x,\ 0 < x < 2\pi$
20. $f(x) = x/2 - \sin(x),\ 0 < x < 2\pi$
21. $f(x) = \tan(x/2),\ -\pi < x < \pi$
22. $f(x) = \tan(x) - x,\ -\pi < x < \pi$
23. $f(x) = \sin^2(x),\ 0 < x < \pi$
24. $f(x) = x - e^x$
25. $f(x) = e^x + e^{-x}$
26. $f(x) = xe^{-x}$
27. $f(x) = x - \ln(x)$
28. $f(x) = 1/x + \ln(x)$
29. $f(x) = x^2 - 8\ln(x)$
30. $f(x) = x - \ln(x) + 2/x$
31. $f(x) = x^{1/3}(x-5)$
32. $f(x) = x^{2/3}(x-4)^{1/3}$
33. $f(x) = x + 1/2^x$
34. $f(x) = \arctan(1 + x^2)$
35. $f(x) = \arcsin(x) - x^2,\ -1 < x < 1$
36. $f(x) = \cosh(x) - 2\exp(x)$

Further Theory and Practice

In Exercises 37–42, the derivative f' of a function f is given. Determine and classify all local extrema of f.

37. $f'(x) = x(x-1)$
38. $f'(x) = x^2(x-1)$
39. $f'(x) = x^2 - 1$
40. $f'(x) = (x^2-1)^2$
41. $f'(x) = (x^2-4)^5$
42. $f'(x) = (x-1)(x-2)^2(x-3)^3(x-4)^4$

In Exercises 43–48, the second derivative f'' of a function f is given. Determine every x at which f has a point of inflection.

43. $f''(x) = x(x+1)$
44. $f''(x) = x^2(x+2)$
45. $f''(x) = x^2 + 4$
46. $f''(x) = (x^2-1)^2$
47. $f''(x) = (x^2-25)^3$
48. $f''(x) = x(x-1)^2(x-2)^3(x-3)^4$

49. For a positive constant a, the *Witch of Agnesi* is the curve whose equation is $y = 8a^3/(4a^2+x^2)$. Find the points of inflection of this curve. On what intervals is this curve concave up? Concave down?

50. Suppose that a function f has a point of inflection at c. Can f have a local extremum at c?

51. Suppose that $f(x) > 0$ and $f''(x) > 0$ for all x. Describe the concavity of f^2.

52. Suppose that f and g are twice differentiable functions that are concave up. Is $f+g$ concave up? Is $f \cdot g$ concave up?

53. Let k and c be positive constants. Suppose that a yeast population has mass m that satisfies the inequalities $0 < m(t) < 2c$ and the differential equation

$$m'(t) = k \cdot m(t) \cdot (2c - m(t))$$

for all values of t. By completing the square on the right side, determine the ordinate of the point of inflection of the graph of $y = m(t)$. Discuss the concavity of this graph.

54. For every positive x, the number $W(x)$ is defined to be the unique positive number such that

$$W(x)\exp(W(x)) = x.$$

The function $x \mapsto W(x)$ is known as Lambert's W function. Show that

$$W'(x) = \frac{1}{x + \exp(W(x))}$$

Deduce that W is an increasing function, and show that the graph of W is concave down.

55. Suppose that x and y are twice differentiable functions of a parameter t. Show that

$$\frac{d^2y}{dx^2} = \frac{d}{dx}\left(\frac{\dot{y}}{\dot{x}}\right)$$

where Newton's notation indicates differentiation with respect to t.

56. Suppose that x and y are twice differentiable functions of a parameter t. Use Exercise 55 to show that

$$\frac{d^2y}{dx^2} = \frac{\dot{x}\ddot{y} - \dot{y}\ddot{x}}{(\dot{x})^3}$$

where Newton's notation indicates differentiation with respect to the parameter t.

57. Let a be a positive constant. One arch of the cycloid is defined by the equations $x = a(t - \sin(t))$, $y = a(1 - \cos(t))$ for $0 < t < 2\pi$. Use the formula of Exercise 56 to analyze the concavity of the graph. Identify any local extrema.

58. Let C be the parametric curve defined by $x = t^2 - 1$, $y = t^3 - t$ for $-\infty < t < \infty$. Notice that the points of C corresponding to $t = -1$, $t = 0$, and $t = 1$ lie on the x-axis. Use the formula of Exercise 56 to show that C is concave up for $t > 0$ and concave down for $t < 0$. Use these observations to sketch C.

59. Let C be the parametric curve defined by $x = \tan(t)$, $y = \cos^2(t)$ for $-\pi/2 < t < \pi/2$. Use the formula of Exercise 56 to determine the interval on which C is concave down and the intervals on which C is concave up. Notice that $0 < y \le 1$ with the maximum value occuring at the point $(0, 1)$. Calculate the horizontal asymptote of C. Use all this information to sketch C.

▶ The formula

$$\frac{|f''(x)|}{\left(1+(f'(x))^2\right)^{3/2}}$$

measures the *curvature* of the graph of $y = f(x)$ at the point $(x, f(x))$. In Exercises 60–63, compute the curvature of the given function. ◀

60. $f(x) = x^3 + x + 2$
61. $f(x) = \cos(x)$
62. $f(x) = \sqrt{x}$
63. $f(x) = 1/x$

▶ In each of Exercises 64–67, find the point at which f has maximum curvature. (See the instructions to Exercises 60–63.) ◀

64. $f(x) = \ln(x)$
65. $f(x) = x^3/3,\ 0 < x$
66. $f(x) = \sin(x),\ 0 < x < \pi$
67. $f(x) = e^x$

68. Show that the curvature of the circle $x^2 + y^2 = r^2$ is $1/r$ at all points (x, y) with $y \neq 0$.

69. Assuming that the range of f is contained in the domain of g, find a formula for $(g \circ f)''(c)$. Suppose that f and g are concave up. What property of g ensures that $g \circ f$ is also concave up?

70. The *product of labor* $f(n)$ and the *marginal product of labor* (MPL) have been defined in Section 4.3, Exercise 60. Economists generally assume that the graph of the function f has a certain property. The three graphs that appear in Figure 17 of Section 4.3 illustrate this property in different ways. What is the property?

71. Let $R(q)$ denote the revenue realized by selling q units of an item. Let $c(q)$ denote the total cost incurred in selling those q units. Suppose that the profit $R-C$ is maximized at q_0. The second derivatives R'' and C'' are plotted in Figure 14. Identify which curve is R'' and which curve is C'' and explain your reasoning.

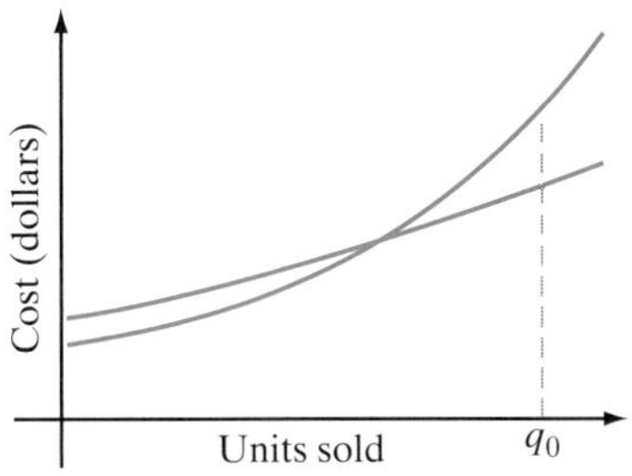

▲ Figure 14 Plots of R'' and C''

Calculator/Computer Exercises

In Exercises 72–79, approximate the critical points and inflection points of the given function f. Determine the behavior of f at each critical point.

72. $f(x) = \arcsin((x+1)/(x^2+2))$

73. $f(x) = x/(1+e^x)$

74. $f(x) = x^4 + 3x^2 - 2x + 4$

75. $f(x) = x^4 - 2e^x + 3x$

76. $f(x) = 6x^5 - 45x^4 + 80x^3 + 30x^2 - 30x + 2$

77. $f(x) = (x^4+1)/(x^4+x+2)$

78. $f(x) = 2\arctan(x) - \arctan(x+1)$

79. $f(x) = \tanh(x) - x^2/(x^2+1)$

80. For a function f that is twice differentiable, define $C(x) = a \cdot \text{signum}(f''(x))$, where a is a positive constant. Explain how the graph of C may be used to determine the concavity and the points of inflection of the graph of f. Illustrate by plotting

$$f(x) = 6x^6 + x^5 - 60x^4 - 35x^3 + 120x^2 + 52x + 160$$

and $c(x) = 300\ \text{signum}(f''(x))$ in the viewing window $[-2.3, 2.8] \times [-310, 310]$. (The choice of 300 for a scales the graph of C so that we can see it in the same viewing window as f.)

4.6 Graphing Functions

Curve sketching is one of the most important techniques of analytical thinking. The skills of creating and understanding graphs are used in all of the physical sciences and many of the social sciences as well. In previous sections, we have learned that certain aspects of the graph of a function f can be determined from the first and second derivatives of f. We have seen that graphs of functions can have vertical asymptotes or horizontal asymptotes or both. In this section, we will combine all these concepts and use them to sketch the graphs of functions. Even if you have a graphing calculator or software with graphing capabilities, the knowledge that you will gain in this section will be useful. The best way to learn how to interpret a graph is to learn how it is created.

Basic Strategy of Curve Sketching

The following steps can be used to sketch the graphs of an extensive range of functions.

Basic Steps in Sketching a Graph

1. Determine the domain and (if possible) the range of the function.
2. Find all horizontal and vertical asymptotes.
3. Calculate the first derivative, and find the critical points for the function.
4. Find the intervals on which the function is increasing or decreasing.
5. Calculate the second derivative, and find the intervals on which the function is concave up or concave down.
6. Identify all local maxima, local minima, and points of inflection.
7. Plot these points, as well as the y-intercept (if applicable) and any x-intercepts (if they can be computed). Sketch the asymptotes.
8. Connect the points plotted in Step 7, keeping in mind concavity, local extrema, and asymptotes.

If you follow this systematic procedure in sketching your graphs, you will find that you can handle a wide variety of interesting examples.

▶ **EXAMPLE 1** Graph the function $f(x) = 5x/(x-2)^2$.

Solution We follow the outline just given:

1. The domain of f consists of all real numbers except 2. When x is near the point 2, the function $f(x)$ takes arbitrarily large positive values.
2. We notice that $\lim_{x\to+\infty} f(x) = \lim_{x\to-\infty} f(x) = 0$. Therefore the line $y = 0$ is a horizontal asymptote for the graph. Also $\lim_{x\to 2} f(x) = +\infty$. Therefore the line $x = 2$ is an upward vertical asymptote for f.
3. We calculate that

$$f'(x) = \frac{(x-2)^2 \cdot 5 - 5x \cdot 2(x-2)}{(x-2)^4} = \frac{-5(x+2)}{(x-2)^3}.$$

   The first derivative is defined at all points of the domain of f. Because f' vanishes at -2, this is the only critical point. In Step 6, we will determine whether or not f has a local extremum at $x = -2$.
4. The first derivative can change sign only at -2 (the critical point) and $+2$ (the point where f is undefined). Because $f'(-3) = -1/25$, we conclude that $f' < 0$ on the interval $(-\infty, -2)$; hence f is decreasing there. Because $f'(0) = 5/4 > 0$, we conclude that $f' > 0$ on the interval $(-2, 2)$; hence f is increasing there. Finally, because $f'(3) = -25 < 0$, we see that $f' < 0$ on the interval $(2,\infty)$; hence f is decreasing there.
5. The second derivative is

$$f''(x) = \frac{(x-2)^3 \cdot (-5) - (-5(x+2)) \cdot \left(3(x-2)^2\right)}{(x-2)^6}. = \frac{10 \cdot (x+4)}{(x-2)^4}$$

   Notice that the denominator of f'' is always positive on the domain of f. Clearly $f'' < 0$ on the interval $(-\infty, -4)$ because the numerator is; hence f is concave down on that interval. Also $f'(x) > 0$ when $x > -4$ (except at $x = 2$, the point where f, f', f'' are undefined); hence f is concave up on each of the intervals $(-4, 2)$ and $(2, \infty)$.
6. Because $f''(-2) = 5/64 > 0$, there is a local minimum at the critical point $x = -2$. The point $(-2, f(-2)) = (-2, -5/8)$ will be labeled on our graph. From Step 5, the concavity changes at $x = -4$. Therefore f has a point of inflection at $x = -4$. The point $(-4, f(-4)) = (-4, -5/9)$ will be labeled on our graph. The concavity does not change at $x = 2$.
7. The y-intercept is $(0, f(0)) = (0, 0)$. Because $x = 0$ is the only solution to $f(x) = 0$, the point $(0, 0)$ is also the only x-intercept. The points $(0, 0)$, $(-2, -5/8)$, and $(-4, -5/9)$ have been plotted in Figure 1.
8. The final sketch of the graph of f is in Figure 1. All of the pertinent data are labeled. Notice that we can deduce from all the information gathered that f has an absolute minimum at $x = -2$. There is no absolute maximum. ◀

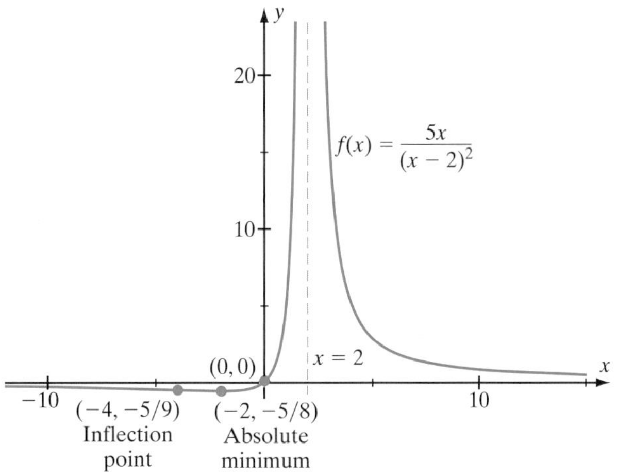

▲ Figure 1

▶ **EXAMPLE 2** Sketch the graph of $g(x) = 4x^3 + x^4$.

Solution

1. Because g is a polynomial, the domain of g consists of all real numbers. When x is large, the term x^4 forces g to take arbitrarily large positive values.
2. By the comment in Step 1, g has no finite limit as $x \to \pm\infty$. Thus there are no horizontal asymptotes. Also, g is continuous at every point because it is a polynomial. Consequently, there are no vertical asymptotes.
3. We have $g'(x) = 12x^2 + 4x^3 = 4x^2(3 + x)$. Because $g'(x)$ is defined for all x, the only critical points are the zeros of g'. These are 0 and -3. In Step 6, we will determine whether g has a local extremum at either of these points.
4. From the formula for $g'(x)$ in Step 3, we see that the first derivative can change sign only at -3 and at 0. Because $g'(-4) = -64 < 0$, we conclude that $g' < 0$ on the interval $(-\infty, -3)$. Therefore g is decreasing on $(-\infty, -3)$. Because $g'(-1) = 8 > 0$, we conclude that $g' > 0$ on the interval $(-3, 0)$. Therefore g is increasing on $(-3, 0)$. Because $g'(1) = 16 > 0$, we conclude that $g' > 0$ on $(0, \infty)$. Therefore g is increasing on $(0, \infty)$.
5. The second derivative is $g''(x) = 24x + 12x^2 = 12x(2 + x)$. From this formula, we see that g'' can change sign only at the points 0 and -2 where it vanishes. Because $g''(-3) = 36 > 0$, we conclude that $g'' > 0$ on the interval $(-\infty, -2)$; hence, g is concave up on $(-\infty, -2)$. Because $g''(-1) = -12 < 0$, we conclude that $g'' < 0$ on the interval $(-2, 0)$; hence, g is concave down on $(-2, 0)$. Because $g''(1) = 36 > 0$, we conclude that $g'' > 0$ on the interval $(0, \infty)$; hence, g is concave up on $(0, \infty)$.
6. By Step 5, we know that the concavity changes at -2 and at 0. These are the inflection points of g. We will label the points $(-2, g(-2)) = (-2, -16)$ and $(0, g(0)) = (0, 0)$ on our graph. In Step 4, we saw that $g' < 0$ to the left of -3 and $g' > 0$ to the right of -3. Therefore a local minimum occurs at the critical point -3. We will label the point $(-3, g(-3)) = (-3, -27)$ on our graph. We have already noted that the other critical point 0 is an inflection point.
7. The intercepts are $(0, 0)$ and $(-4, 0)$. They will also be labeled in our graph.
8. Figure 2 shows the sketch of the graph of g. All pertinent data have been indicated. We can deduce that there is an absolute minimum at -3. There is no absolute maximum. ◀

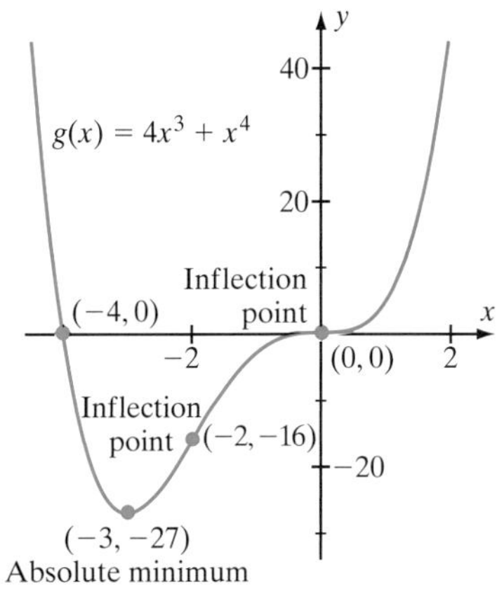

▲ Figure 2

INSIGHT In each of Examples 1 and 2, we stated that the information gathered allowed us to determine that a particular point is an absolute minimum of the function under consideration. The graphs of each function certainly support our conclusion about these points. Note, however, that because a graph can only be given in a finite viewing rectangle, the graphs cannot by themselves prove assertions concerning absolute extrema. If you have not already done so, it will be instructive for you to explain why $(-2, -5/8)$ must be an absolute minimum of the function f from Example 1 and why $(-3, -27)$ must be an absolute minimum of the function g from Example 2.

▶ **EXAMPLE 3** Sketch the graph of $f(x) = (4x+1)/\sqrt{x^2-3x+2}$.

Solution

1. Because $x^2 - 3x + 2 = (x-1)(x-2)$ is negative between its two roots, the interval $(1, 2)$ is not in the domain of f. The endpoints of this interval are also excluded because the denominator is 0 at these values. The domain of f consists of the union of the open intervals $(-\infty, 1)$ and $(2, \infty)$. The function f is continuous on these intervals.
2. By calculating $\lim_{x\to 1^-} f(x) = \infty$ and $\lim_{x\to 2^+} f(x) = \infty$, we see that the vertical lines $x = 1$ and $x = 2$ are upward vertical asymptotes. (It makes no sense to consider $\lim_{x\to 1^+} f(x)$ and $\lim_{x\to 2^-} f(x)$ because such limits involve x in the interval $(1, 2)$, which is not in the domain of f.) By writing
$$f(x) = \frac{x(4+1/x)}{\sqrt{x^2(1-3/x+2/x^2)}} = \frac{x}{|x|}\cdot\frac{4+1/x}{\sqrt{1-3/x+2/x^2}}$$
and noting that $x/|x| = 1$ for $0 < x$ and $x/|x| = -1$ for $x < 0$, we deduce that $\lim_{x\to\infty} f(x) = 4$ and $\lim_{x\to-\infty} f(x) = -4$. The lines $y = 4$ and $y = -4$ are therefore horizontal asymptotes.
3. After some simplification, we find that
$$f'(x) = -\frac{1}{2}\,\frac{14x-19}{(x^2-3x+2)^{3/2}}.$$
Because f' is defined for all x in the domain of f, the only critical points are the zeros of f'. The expression
$$-\frac{1}{2}\,\frac{14x-19}{(x^2-3x+2)^{3/2}}$$
has one zero, $x = 19/14$, but this zero is not in the domain of f' since it is between 1 and 2 and therefore not in the domain of f. We conclude that f has no critical points.
4. The denominator of $f'(x)$ is always positive. The sign of $f'(x)$ is therefore the sign of its numerator $14x - 19$. Remembering that the domain of f is $(-\infty, 1) \cup (2, \infty)$, we deduce that $f'(x) > 0$ for $x < 1$ and $f'(x) < 0$ for $x > 2$. Therefore f is increasing on $(-\infty, 1)$ and decreasing on $(2, \infty)$.
5. After simplification, we find that
$$f''(x) = \frac{1}{4}\left(\frac{56x^2-156x+115}{(x^2-3x+2)^{5/2}}\right).$$
Note that f'' exists at every point in the domain of f. Because $156^2 - 4\cdot 56\cdot 115 = -1424 < 0$, the Quadratic Formula tells us that f'' has no real zeros.

6. By Step 5, we know that there are no points of inflection. Because $f''(0) > 0$, we deduce that the graph of f is concave up everywhere.
7. The intercepts $(-1/4, 0)$ and $(0, \sqrt{2}/2)$ and the vertical and horizontal asymptotes are plotted in Figure 3.

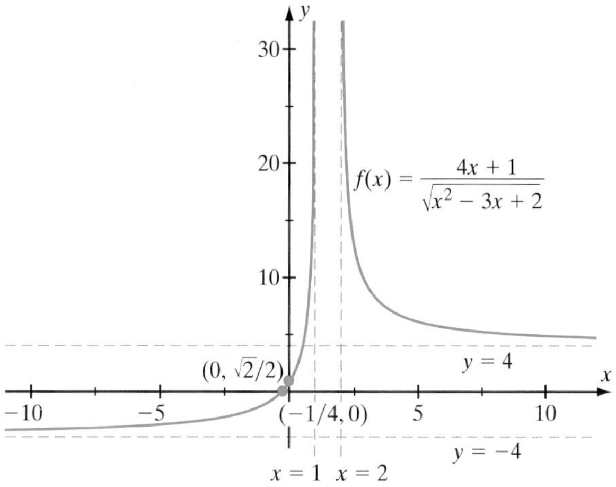

▲ Figure 3

8. The final sketch of the graph of f appears in Figure 3. There are no local extrema. Notice that, although $f(x) > -4$ for all x in the domain of f, the number -4 is not a value of $f(x)$ for any x. Therefore -4 is not the minimum value of f. ◀

Periodic Functions

A function f is said to be *periodic* if there is a positive number p such that

$$f(x+p) = f(x), \qquad x \in \mathbb{R}. \tag{4.6.1}$$

The smallest nonnegative number p for which equation (4.6.1) is true is called the *period* of f. The sine, cosine, secant, and cosecant functions all have period 2π. However, we observe that

$$\begin{aligned}\tan(x+\pi) &= \frac{\sin(x+\pi)}{\cos(x+\pi)} = \frac{\sin(x)\cos(\pi)+\cos(x)\sin(\pi)}{\cos(x)\cos(\pi)-\sin(x)\sin(\pi)} \\ &= \frac{\sin(x)(-1)+0}{\cos(x)(-1)-0} = \frac{\sin(x)}{\cos(x)} = \tan(x),\end{aligned}$$

which shows that the the period of the tangent is either π, or possibly a smaller number. However, because the derivative of $\tan(x)$ is $\sec^2(x)$, which is positive on $(-\pi/2, \pi/2)$, we know that $\tan(x)$ increases on an interval of length π. Therefore $\tan(x+p) = \tan(x)$ is impossible for $0 \le p < \pi$. It follows that $\tan(x)$ has period π, as does its reciprocal $\cot(x)$.

To graph a function f whose period is p, we plot f over the interval $[0, p]$ and then translate the plot of this restriction by p units (Figure 4). If f is continuous on $[0, p]$ then it attains a maximum value and a minimum value by the Extreme Value Theorem. The next example illustrates these points.

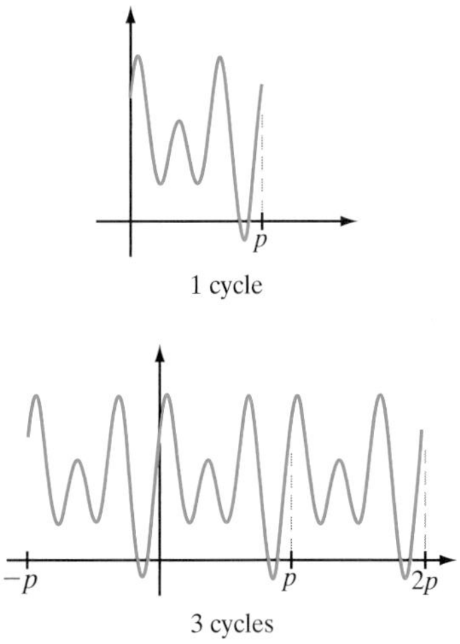

▲ Figure 4

▶ **EXAMPLE 4** Sketch the graph of the function $f(x) = \sin(x)/(2 + \cos(x))$.

Solution

1. The function $f(x)$ is defined for all real x because its denominator is never zero. Observe that $f(x + 2\pi) = f(x)$ for all x. This tells us that f is periodic, and the entire graph of f is determined by the part that lies over $[0, 2\pi]$.
2. Because f is continuous at every x, there are no vertical asymptotes. Because f is periodic and nonconstant, we conclude that f has no horizontal asymptotes.
3. We calculate that
$$f'(x) = \frac{1 + 2\cos(x)}{(2 + \cos(x))^2}.$$
The zeros of f' occur when $\cos(x) = -1/2$. The solutions of this equation in the interval $[0, 2\pi]$ are $x = 2\pi/3$ and $x = 4\pi/3$.
4. Observe that the sign of $f'(x)$ is the same as the sign of $1 + 2\cos(x)$. This expression is positive for x in the intervals $[0, 2\pi/3)$ and $(4\pi/3, 2\pi]$; it is negative for x in the interval $(2\pi/3, 4\pi/3)$. Therefore f is increasing on $[0, 2\pi/3)$ and $(4\pi/3, 2\pi]$ and f is decreasing on $(2\pi/3, 4\pi/3)$.
5. After differentiation and algebraic simplification, we find that
$$f''(x) = -2\sin(x)\frac{(1 - \cos(x))}{(2 + \cos(x))^3}.$$
Because the expressions $1 - \cos(x)$ and $2 + \cos(x)$ are nonnegative, we see that the signs of $f''(x)$ and $\sin(x)$ are opposite. Therefore $f''(x) < 0$ for $0 < x < \pi$ and $f''(x) > 0$ for $\pi < x < 2\pi$. We conclude that the graph of f is concave down on the interval $(0, \pi)$ and concave up on the interval $(\pi, 2\pi)$.
6. Because the critical point $2\pi/3$ is in the interval on which $f'' < 0$, the Second Derivative Test tells us that $f(x)$ has a local maximum at $x = 2\pi/3$. Similarly, we see that f has a local minimum at $4\pi/3$. From the formula for $f'(x)$ that we obtained in Step 5, we notice that the sign of $f''(x)$ changes at each point where $\sin(x)$ vanishes. Therefore the points 0, π, and 2π are the points of inflection that f has in the interval $[0, 2\pi]$.
7. The y-intercept is the origin. The x-intercepts are the points of the form $(k\pi, 0)$ where k is an integer. Because $f(0) = 0$, $f(2\pi/3) = \sqrt{3}/3$, $f(4\pi/3) = -\sqrt{3}/3$, and $f(2\pi) = 0$, we deduce that f has an absolute maximum value of $\sqrt{3}/3$ and an absolute minimum value of $-\sqrt{3}/3$.
8. Figure 5 contains a sketch of that part of the graph that lies over $[0, 2\pi]$. This portion of the graph is repeated periodically in Figure 6, which shows four cycles of f. ◀

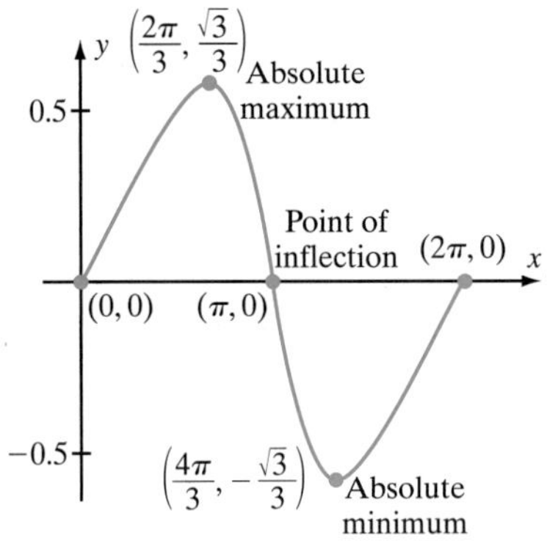

▲ Figure 5

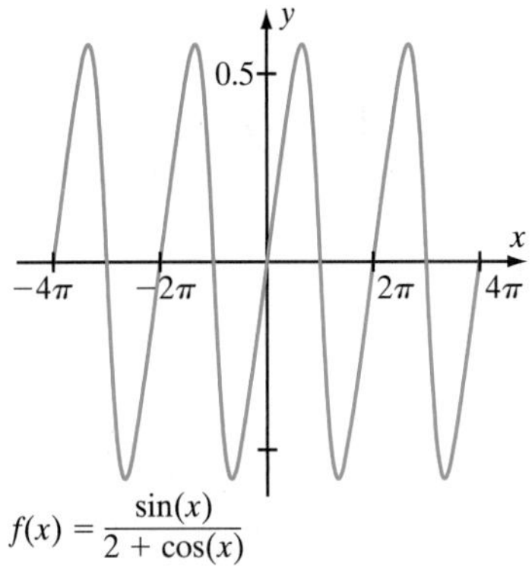

▲ Figure 6

Skew-Asymptotes

The line $y = 2x + 3$ and the graph of $f(x) = 2x + 3 + 1/x$ are shown in Figure 7. Notice that the graph of f becomes closer to the straight line $y = 2x + 3$ as $x \to \infty$ and $x \to -\infty$. This leads to a definition.

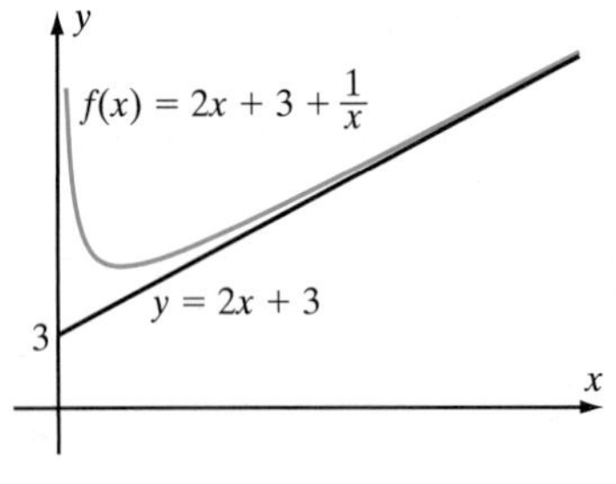

▲ Figure 7

DEFINITION The line $y = mx + b$ is a *skew-asymptote* of f if
$$\lim_{x \to \infty} (mx + b - f(x)) = 0$$
or if
$$\lim_{x \to -\infty} (mx + b - f(x)) = 0.$$
Other terms for this line are *slant-asymptote* and *oblique-asymptote*.

With the introduction of this concept, Step 2 of the eight basic steps of graph-sketching should be expanded to include *all* asymptotes. The most common situation in which a skew-asymptote occurs is in graphing a rational function with a numerator of degree one more than its denominator. In this case, division allows us to determine the skew-asymptote.

▶ **EXAMPLE 5** Sketch the graph of $f(x) = x^3/(x^2 - 4)$.

Solution We follow the outline given above:

1. The domain of f consists of all real numbers except $x = +2$ and $x = -2$.
2. The lines $x = 2$ and $x = -2$ are vertical asymptotes. Because division gives us
$$\frac{x^3}{x^2-4} = x + \frac{4x}{x^2-4},$$
we see that $y = x$ is a skew-asymptote:
$$\lim_{x\to\pm\infty}(x - f(x)) = \lim_{x\to\pm\infty}\left(x - \left(x + \frac{4x}{x^2-4}\right)\right) = -\lim_{x\to\pm\infty}\frac{4x}{x^2-4}$$
$$= -\lim_{x\to\pm\infty}\frac{4x/x^2}{1-4/x^2} = -4\lim_{x\to\pm\infty}\frac{1/x}{1-4/x^2} = -4\cdot\frac{0}{1-0} = 0.$$
3. We calculate that
$$\frac{d}{dx}\left(\frac{x^3}{x^2-4}\right) = \frac{x^2(x^2-12)}{(x^2-4)^2}.$$
The first derivative is defined at all points of the domain of f. Because f' vanishes at 0 and $\pm\sqrt{12}$, these are the only critical points.
4. The first derivative changes sign at $-\sqrt{12}$ and $\sqrt{12}$ but not at 0. We see that $f' > 0$ on the intervals $(-\infty, -\sqrt{12})$ and $(\sqrt{12}, \infty)$. We see that $f' < 0$ on the intervals $(-\sqrt{12}, -2), (-2, 2)$, and $(2, \sqrt{12})$.
5. Differentiating the expression in Step 3, we find that the second derivative of f is
$$f''(x) = 8x\,\frac{x^2+12}{(x^2-4)^3}.$$
The numerator $x^2 + 12$ is always positive and doesn't affect the sign of f''. The sign of f'' can change only at a zero of f'' or at a point where f'' doesn't exist. There are three such points: $x = 0$ is a zero of $f''(x)$ and $f''(x)$ doesn't exist at $x = -2$ and $x = 2$. To determine the sign of f'' on the intervals $(-\infty, -2)$, $(-2, 0)$, $(0, 2)$, and $(2, \infty)$, we need only evaluate f'' at one point in each interval. Because $f''(-3) = -504/125$, $f''(-1) = 104/27$, $f''(1) = -104/27$, and $f''(3) = 504/125$, we deduce that f is concave up over the intervals $(-2, 0)$ and $(2, \infty)$ and concave down over the intervals $(-\infty, -2)$ and $(0, 2)$.
6. From Step 5, we know that the concavity changes at $-2, 0$, and 2. Because -2 and 2 are not in the domain of f, we see that $(0, 0)$ is the only point of inflection. From Step 4 we see that a local maximum occurs at $-\sqrt{12}$ and a local minimum at $\sqrt{12}$.
7. The graph of f crosses the axes only at $(0, 0)$. In Figure 8, we have plotted this point together with the point of inflection and the local extrema. We have also drawn in the vertical asymptotes and the skew-asymptote.
8. The final sketch of the graph of f is in Figure 8. All of the pertinent data have been indicated. ◀

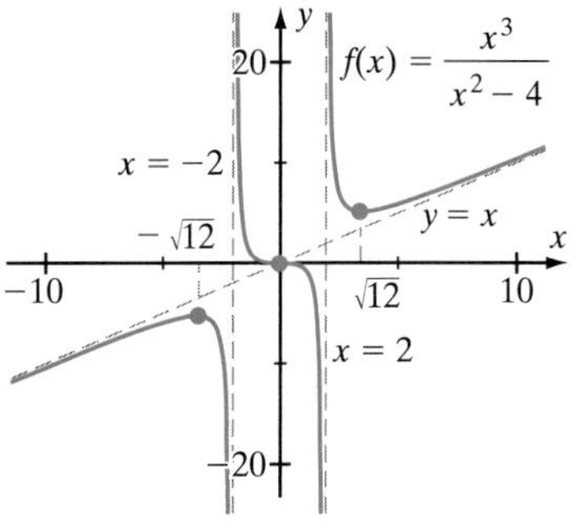

▲ Figure 8

Graphing Calculators/Software

Graphing calculators and plotting software are convenient tools. However, we must understand their limitations and, when necessary, be able to correct the plots that they render. Figure 9 is a screen capture in which a computer algebra system has been used to plot

$$f(x) = \frac{x(\tan(x) - \sin(x))^2}{\tan(\sin(x)) - \sin(\tan(x))}$$

over the interval $[-0.01, 0.01]$. Even though 20 digits have been carried in the computation, a *loss of significance* (Section 1.1) has resulted in a wildly inaccurate

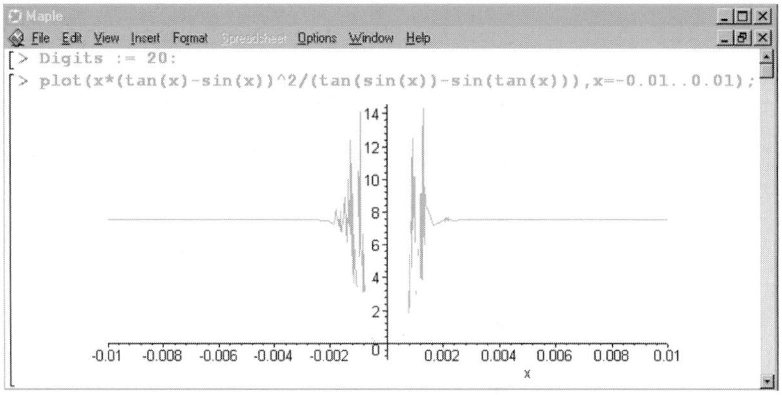

▲ Figure 9

plot. Although a computer algebra system allows us to increase the number of digits until a reliable plot is attained, this resource may not be an option with a graphing calculator. In this case, an important approximation technique of calculus (to be learned in Chapter 8) shows that

$$f(x) = \frac{15}{2} - \frac{205}{42}x^2 + \frac{85403}{42336}x^4 + E(x),$$

where the summand $E(x)$ is negligible over the interval $[-0.1, 0.1]$. By discarding the term $E(x)$, we obtain a good polynomial approximation that is easily and accurately plotted. Although Figure 10 shows only one curve, we are actually viewing both the plot of f, rendered using 40 digits, and the plot of $y = 15/2 - 205x^2/42 + 85403x^4/42336$. The differences between the values of $f(x)$ and the approximating polynomial cannot be detected in the plot.

There are many situations in which calculator and computer plots can be deceptive. Often the solution is to find a better viewing window. In Figure 11 we have plotted the function $f(x) = 100x$ and the function $g(x) = 100x + \sin(x^4)$ over the interval $[0, 2]$. The graphs of f and g over this interval appear to be the same line segment. Yet when we zoom in to the subinterval $[1.1, 1.2]$, as is also shown in Figure 11, the graphs of f and g appear to be two different line segments. Of course, no part of the graph of g is actually a line segment. Over the interval [10, 12] the plots of f and g once again appear virtually indistinguishable (Figure 12). When we zoom in to the subinterval $[10.1, 10.11]$, we can see the expected sinusoidal oscillations that the plot of g has about the line $y = 100x$.

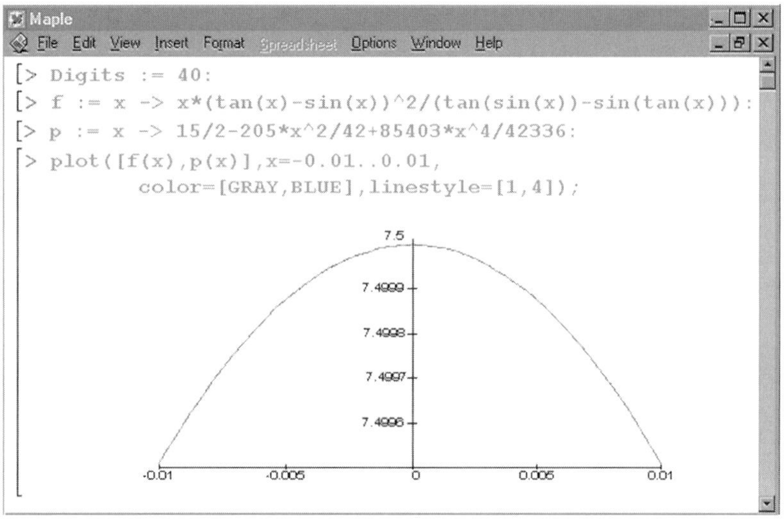

▲ Figure 10

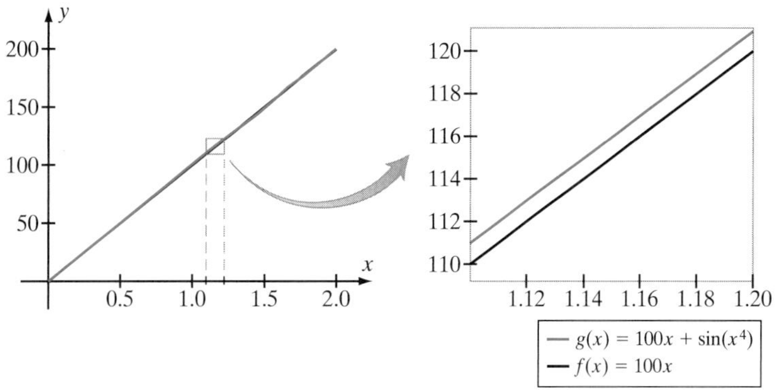

▲ Figure 11

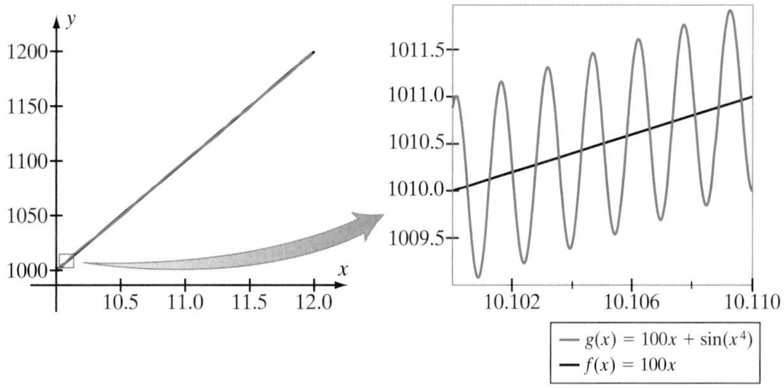

▲ Figure 12

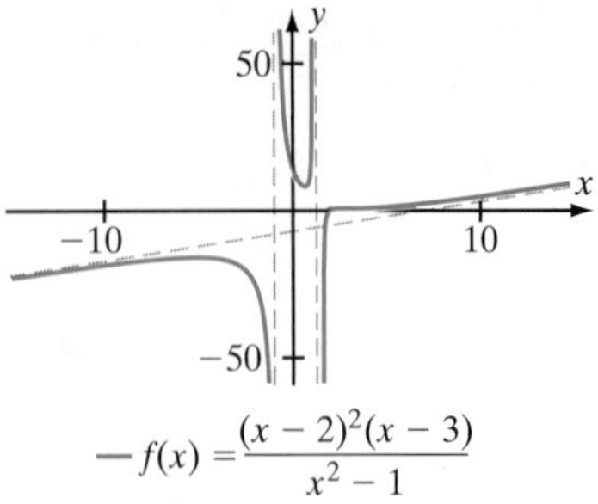

▲ Figure 13

Sometimes a sketch is more useful than an accurate graph. In Figure 13 the function

$$f(x) = \frac{(x-2)^2(x-3)}{(x^2-1)}$$

has been plotted in the viewing rectangle $[-15,15]\times[-50,50]$. This plot reveals many features of f: a local maximum near $x=-5$, a local minimum near $x=1/2$, vertical asymptotes $x=-1$ and $x=1$, and a skew-asymptote $y=x-7$. However, this plot hides important information and, as a result, makes $f(x)$ appear to be increasing for $x>1$. When we plot f over the interval $[1.8, 3.2]$, as in Figure 14, we discover two local extrema, a point of inflection, and a second x-intercept, none of which is visible in Figure 13. We may incorporate all the information about f into a *sketch* as in Figure 15. But Figure 15 has been obtained by distorting the distance between the two x-intercepts. There is simply no scale available that allows *all* features of the graph of f to be clearly visible in one viewing window.

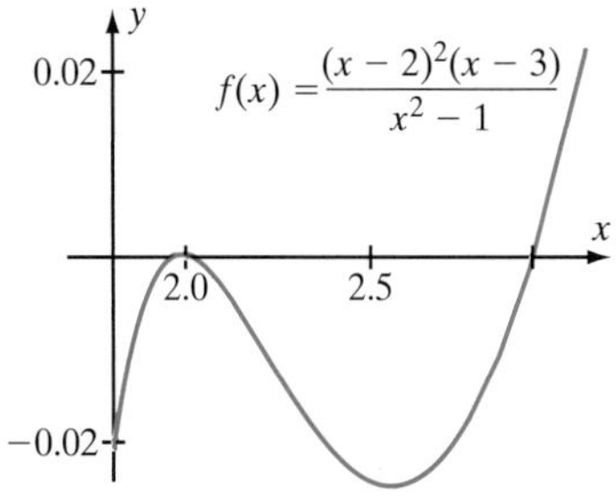

▲ Figure 14

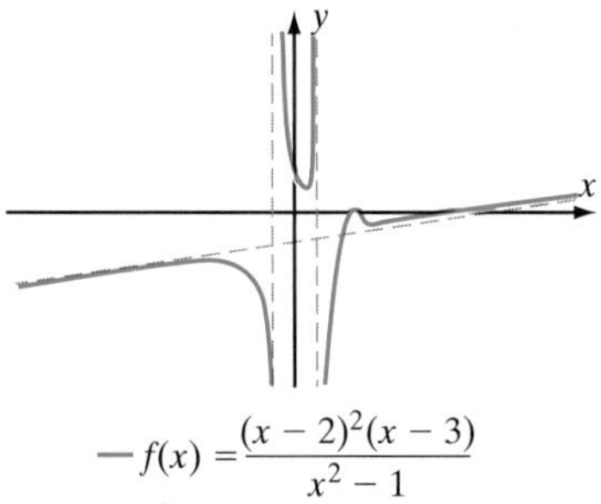

▲ Figure 15

QUICK QUIZ

1. What are the periods of $\sin(x/2)$, $\sec(x)$, and $\cot(x)$?
2. What is the skew-asymptote of $f(x) = (3x^2 - 2x + 5\cos(x))/x$?
3. If $f(x) = 2x^3/(x-1)^2$, then $f'(x) = 2x^2(x-3)/(x-1)^3$ and $f''(x) = 12x/(x-1)^4$. Determine (i) the interval or intervals on which f increases, (ii) the interval or intervals on which f decreases, (iii) all local extrema, (iv) the interval or intervals on which the graph of f is concave up, (v) the interval or intervals on which the graph of f is concave down, (vi) all points of inflection, (vii) all horizontal and vertical asymptotes, and (viii) all skew-asymptotes.

Answers

1. 4π, 2π, π 2. $y = 3x - 2$ 3. (i) $(-\infty, 1)$ and $(3, \infty)$, (ii) $(1,3)$, (iii) local minimum at $x=3$, (iv) $(0,1)$ and $(1, \infty)$, (v) $(-\infty, 0)$, (vi) 0, (vii) $x=1$ (vertical asymptote), (viii) $y = 2x + 4$

EXERCISES

Problems for Practice

▶ Follow the outline given in this section to give a careful sketch of the graph of each of the functions in Exercises 1–24. Your sketch should exhibit, and have labeled, all of the following: a) local and global extrema, b) inflection points, c) intervals on which function is increasing or decreasing, d) intervals on which function is concave up or concave down, e) horizontal and vertical asymptotes. ◀

1. $f(x) = x^3 - 3x^2 - 9x + 7$
2. $f(x) = x^3(x - 8)$
3. $f(x) = (x + 1)(x - 2)^2$
4. $f(x) = x^5 - 6x^4 + 8x^3$
5. $f(x) = x/(x^2 + 4)$
6. $f(x) = (x + 1)/(x - 1)$
7. $f(x) = (x^2 + 4)/(x^2 - 4)$
8. $f(x) = (x + 1)/(x + 2)^2$
9. $f(x) = x^2 - x^{1/2}$
10. $f(x) = x/(x - 4)^2$
11. $f(x) = \sin(x) - x$
12. $f(x) = \cos(x) - \sin(x)$
13. $f(x) = (x^{1/3} + 1)/x$
14. $f(x) = x^{-4/3}(x + 2)$
15. $f(x) = x^{1/3}/(x + 3)$
16. $f(x) = (x^3 - 12x)/(x^2 + 4)$
17. $f(x) = x(x + 1)^{1/5}$
18. $f(x) = |x|/(x + 1)$
19. $f(x) = x^{1/3}(x - 2)$
20. $f(x) = \sqrt{x}/(\sqrt{x} - 1)$
21. $f(x) = \sqrt{x}/(x + 1)$
22. $f(x) = x + 54/\sqrt{x}$
23. $f(x) = x^2 + 16/x$
24. $f(x) = 1/x - 1/x^2$

Further Theory and Practice

Sketch the graph of each of the functions in Exercises 25–40, exhibiting and labeling: a) all local and global extrema; b) inflection points; c) intervals on which the function is increasing or decreasing; d) intervals on which the function is concave up or concave down; e) all horizontal and vertical asymptotes.

25. $f(x) = x^{1/3}/(x - 4)$
26. $f(x) = \frac{1}{2}x^{2/3} - x^{1/3}$
27. $f(x) = x^2(1 - x^2)^{1/2}$
28. $f(x) = x^2/(x^{1/3} + 8)$
29. $f(x) = \sqrt{x^2 + 8x + 18/x}$
30. $f(x) = |x^2 - 9|$
31. $f(x) = (x - 5)\sqrt{|x + 2|}$
32. $f(x) = \cos^4(x),\ 0 < x < 2\pi$
33. $f(x) = (x^3 + x + 1)/x$
34. $f(x) = \sin(x) + \tan(x),\ -\pi < x < \pi$
35. $f(x) = \tan(x) + \sec(x)$
36. $f(x) = |x^{1/3} - 4|$
37. $f(x) = |x| \cdot (x + 1)^2$
38. $f(x) = (x + 1)/\sqrt{(x-1)^2}$
39. $f(x) = (13x + 14)/(x^2 - 1)$
40. $f(x) = (3x + 2)/|x^2 - 1|$

In Exercises 41 and 42, follow the instructions for Exercises 25–40. Additionally, determine the skew-asymptote of the given function.

41. $f(x) = (x^2 - 4)/(2x)$
42. $f(x) = (x^3 + x^2 + 2)/x^2$

43. Suppose that $f: \mathbb{R} \to \mathbb{R}$ is twice differentiable and $f''(x) \geq 1$ for all x. Prove that the graph of f cannot have any horizontal asymptotes.

44. A graph is called *symmetric with respect to the* y-axis if $(-x, y)$ lies on the graph whenever (x, y) lies on the graph. We test for symmetry with respect to the y axis by replacing x with $-x$ in the equation for the graph. If the equation remains unchanged, the symmetry property holds. Explain the reasoning behind this test. A graph is called *symmetric with respect to the origin* if $(-x, -y)$ lies on the graph whenever (x, y) lies on the graph. We test for symmetry with respect to the origin by replacing x with $-x$ and y with $-y$ in the equation for the graph. If the equation remains unchanged, the symmetry property holds. Explain the reasoning behind this test. Test each of the following for symmetry in the y-axis and symmetry in the origin. Give a sketch of each graph.
a. $9y^2 - x^2 = 36$
b. $x^3 + y^2 = 4$
c. $x^2 - 4y^2 = 16$
d. $x + \sin(y) = 3y$

Calculator/Computer Exercises

Follow the outline given in this section to give a careful sketch of the graph of each of the functions in Exercises 45–54. Your sketch should exhibit, and have labeled, all of the following: (i) all local and global extrema, (ii) inflection points, (iii) intervals on which function is increasing or decreasing, (iv) intervals on which function is concave up or concave down, (v) all horizontal, vertical, and skew-asymptotes.

45. $f(x) = x^4 + 6x^3 + x^2 + x - 4$
46. $f(x) = x^5 + x^3 + 2x^2$
47. $f(x) = x^5 + x^4 - 5x^3 - 5x^2 + 6x + 6$
48. $f(x) = \dfrac{(x - 1)(x - 4)}{(x - 3)(x - 2)}$
49. $f(x) = x^2 + x\sqrt{1 - x^2}$
50. $f(x) = \dfrac{(x - 3)^2(x - 4)}{(x - 1)(x - 2)}$
51. $f(x) = \dfrac{\sqrt{4 - x^2}}{x^2 + 1}$
52. $f(x) = \dfrac{x^3 + x + 1}{x^2 + 4x + 5}$
53. $f(x) = \dfrac{x^3 + 20x + 15}{\sqrt{16 + x^2}}$
54. $f(x) = \dfrac{x^2 - x + 2}{\sqrt{x^4 + 2x^2 + 2}}$

55. Plot

$$f(x) = \frac{x^2 - 1}{x^2 + 1} \text{ and } g(x) = \frac{x^4 - 102x^2 + 100}{x^4 - 99x^2 - 100}$$

over the intervals [0, 8.5] and [8.5, 9.9]. Explain why the graphs over the first interval are so similar and why the graphs over the second interval become very different.

56. Plot the function

$$f(x) = \frac{3x^5 + 10000x^4}{x^5 + 10000x^4 + x + 10000}$$

over the interval $[-15, 15]$. What appears to be the asymptote for f? Analyze f for asymptotes and plot f in a window that better illustrates its asymptote.

57. Plot $f(x) = 12x^5 - 2565x^4 + 146200x^3 + 1$ for $x \in I = [0, 300]$. Plot f' and f'' for $x \in I$. Is f increasing on I? Is $f' > 0$ on I? Is the graph of f concave up on I? Is $f'' > 0$ on I?

4.7 l'Hôpital's Rule

Consider the limit

$$\lim_{x \to c} \frac{f(x)}{g(x)}. \tag{4.7.1}$$

If $\lim_{x\to c} f(x)$ exists and if $\lim_{x\to c} g(x)$ exists and is not zero, then the limit (4.7.1) is straightforward to evaluate (Section 2.2). However, the situation is more complicated when $\lim_{x\to c} g(x)=0$. For example, if $f(x) = \sin(x), g(x) = x$, and $c = 0$, then we know that limit (4.7.1) exists and equals 1 (Section 2.2). However, if $f(x) = x$ and $g(x) = x^2$, then the limit of the quotient as $x \to 0$ does not exist.

In this section we learn a rule for evaluating limits of the type (4.7.1) when either $\lim_{x\to c} f(x) = \lim_{x\to c} g(x) = 0$ or $\lim_{x\to c} f(x) = \lim_{x\to c} g(x) = \pm\infty$. In these cases, when the limits of the numerator and denominator of (4.7.1) are evaluated independently, the quotient takes the form $\frac{0}{0}$ or $\frac{\infty}{\infty}$. Such forms of the limit are called *indeterminate forms* because the symbols $\frac{0}{0}$ and $\frac{\infty}{\infty}$ have no meaning. The limit might actually exist and be finite or might not exist. One cannot analyze such a form simply by evaluating the limits of the numerator and denominator and forming their quotient.

l'Hôpital's Rule for the Indeterminate Form $\frac{0}{0}$

Suppose that f' and g' are continuous at c and $f(c) = g(c) = 0$. Provided there are no divisions by zero, we have

$$\lim_{x\to c} \frac{f(x)}{g(x)} = \lim_{x\to c} \frac{f(x) - f(c)}{g(x) - g(c)} = \lim_{x\to c} \frac{\dfrac{f(x) - f(c)}{x - c}}{\dfrac{g(x) - g(c)}{x - c}} = \frac{\lim_{x\to c} \dfrac{f(x) - f(c)}{x - c}}{\lim_{x\to c} \dfrac{g(x) - g(c)}{x - c}} = \frac{f'(c)}{g'(c)}$$

and

$$\lim_{x\to c} \frac{f'(x)}{g'(x)} = \frac{\lim_{x\to c} f'(x)}{\lim_{x\to c} g'(x)} = \frac{f'(c)}{g'(c)}.$$

Therefore

$$\lim_{x\to c} \frac{f(x)}{g(x)} = \lim_{x\to c} \frac{f'(x)}{g'(x)}.$$

This observation suggests the following procedure for calculating the limit (4.7.1):

THEOREM 1 **(l'Hôpital's Rule)** Let f and g be differentiable functions on an interval (a, c) to the left of c and on an interval (c, b) to the right of c. If

$$\lim_{x\to c} f(x) = \lim_{x\to c} g(x) = 0$$

then

$$\lim_{x\to c} \frac{f(x)}{g(x)} = \lim_{x\to c} \frac{f'(x)}{g'(x)}$$

provided that the limit on the right exists as a finite or infinite limit.

A proof of l'Hôpital's Rule is outlined in Exercises 86–88. Meanwhile, let's see some examples in which it is used.

▶ **EXAMPLE 1** Evaluate

$$\lim_{x\to 1} \frac{\ln(x)}{x^2 - 1}.$$

Solution We first notice that both the numerator and denominator have limit 0 as $x \to 1$. Thus the quotient is indeterminate at 1 and of the form $\frac{0}{0}$. L'Hôpital's Rule applies and the limit equals

$$\lim_{x\to 1} \frac{\ln(x)}{x^2 - 1} = \lim_{x\to 1} \frac{\frac{d}{dx}\ln(x)}{\frac{d}{dx}(x^2 - 1)},$$

provided this last limit exists. We calculate

$$\lim_{x\to 1} \frac{\frac{d}{dx}\ln(x)}{\frac{d}{dx}(x^2 - 1)} = \lim_{x\to 1} \frac{1/x}{2x} = \frac{1}{2}.$$

The requested limit has value 1/2. ◀

INSIGHT When we apply l'Hôpital's Rule to a limit of type (4.7.1) as in Example 1, we calculate the derivative of the numerator f and the derivative of the denominator g. We do *not* calculate the derivative of the quotient f/g.

▶ **EXAMPLE 2** Evaluate the limit

$$\lim_{x\to 0} \frac{x}{x - \sin(x)}.$$

Solution As $x \to 0$ both numerator and denominator tend to zero, so the quotient is indeterminate at 0 of the form 0/0. Thus l'Hôpital's Rule applies. Our limit equals

$$\lim_{x\to 0} \frac{\frac{d}{dx}x}{\frac{d}{dx}\left(x - \sin(x)\right)},$$

provided that this last limit exists. It equals

$$\lim_{x\to 0} \frac{1}{1-\cos(x)} = +\infty .$$

We conclude that the original limit equals $+\infty$. ◀

▶ **EXAMPLE 3** Calculate

$$\lim_{x\to 0} \frac{\sin(3x)}{\sin(2x)} \qquad \text{and} \qquad \lim_{x\to \pi/6} \frac{\sin(3x)}{\sin(2x)} .$$

Solution With the first limit we may proceed as in Examples 1 and 2. We summarize the calculations:

$$\lim_{x\to 0} \frac{\sin(3x)}{\sin(2x)} = \lim_{x\to 0} \frac{\frac{d}{dx}\sin(3x)}{\frac{d}{dx}\sin(2x)} = \lim_{x\to 0} \frac{3\cos(3x)}{2\cos(2x)} = \frac{3}{2}.$$

For the second limit, we calculate $\lim_{x\to\pi/6} \sin(3x) = \sin(\pi/2) = 1$ and $\lim_{x\to\pi/6} \sin(2x) = \sin(\pi/3) = \sqrt{3}/2$. Therefore l'Hôpital's Rule does not apply. However, we do not need it because

$$\lim_{x\to \pi/6} \frac{\sin(3x)}{\sin(2x)} = \frac{\lim_{x\to\pi/6} \sin(3x)}{\lim_{x\to\pi/6} \sin(2x)} = \frac{1}{\sqrt{3}/2} = \frac{2}{\sqrt{3}}. \quad ◀$$

INSIGHT It is tempting to apply l'Hôpital's Rule to any limit that is written as a quotient. However, l'Hôpital's Rule is valid *only* when its hypotheses are satisfied. If we had applied l'Hôpital's Rule to the second limit in Example 3, then we would have obtained

$$\lim_{x\to \pi/6} \frac{\sin(3x)}{\sin(2x)} = \lim_{x\to \pi/6} \frac{\frac{d}{dx}\sin(3x)}{\frac{d}{dx}\sin(2x)} = \lim_{x\to \pi/6} \frac{3\cos(3x)}{2\cos(2x)} = \frac{0}{1} = 0,$$

which is wrong. The error here is that we did not check that the numerator and denominator both have limit zero; in other words, we did not check that the quotient is indeterminate.

When we apply l'Hôpital's Rule we may end up with another indeterminate form. Sometimes we need to apply l'Hôpital's Rule two or more times to evaluate a limit:

▶ **EXAMPLE 4** Evaluate the limit

$$\lim_{x\to \pi} \frac{1+\cos(x)}{(x-\pi)^2} .$$

Solution Notice that both the numerator and denominator tend to zero as $x \to \pi$. Therefore the quotient is indeterminate at π of the form $\frac{0}{0}$. So we may apply l'Hôpital's Rule to evaluate the limit. It equals

$$\lim_{x \to \pi} \frac{-\sin(x)}{2(x-\pi)},$$

provided that this limit exists. However both the numerator and denominator of this last expression tend to zero as $x \to \pi$, so we apply l'Hôpital's Rule again. The last limit equals

$$\lim_{x \to \pi} \frac{-\cos(x)}{2},$$

provided that this latest limit exists. But it clearly does exist, and it equals 1/2. We conclude that

$$\lim_{x \to \pi} \frac{1+\cos(x)}{(x-\pi)^2} = \frac{1}{2}. \quad ◀$$

l'Hôpital's Rule for the Indeterminate Form $\frac{\infty}{\infty}$

There is another version of l'Hôpital's Rule, which can be used for the indeterminate form $\frac{\infty}{\infty}$ (and the related forms $\frac{\infty}{-\infty}$ and $\frac{-\infty}{\infty}$).

THEOREM 2 Let $f(x)$ and $g(x)$ be differentiable functions on an interval (a, c) to the left of c and on an interval (c, b) to the right of c. If $\lim_{x \to c} f(x)$ and $\lim_{x \to c} g(x)$ both exist and equal $+\infty$ or $-\infty$ (they may have the same sign or different signs) then

$$\lim_{x \to c} \frac{f(x)}{g(x)} = \lim_{x \to c} \frac{f'(x)}{g'(x)},$$

provided this last limit exists either as a finite or infinite limit.

▶ **EXAMPLE 5** Evaluate

$$\lim_{x \to 0} \frac{4+1/x^2}{\sqrt{2+9/x^4}}.$$

Solution In this example, we set $f(x) = 4 + 1/x^2$ and $g(x) = \sqrt{2+9/x^4}$. Because $\lim_{x \to 0} f(x) = \infty$ and $\lim_{x \to 0} g(x) = \infty$, we may apply Theorem 2. We obtain

$$\begin{aligned}
\lim_{x \to 0} \frac{4+1/x^2}{\sqrt{2+9/x^4}} &= \lim_{x \to 0} \frac{\frac{d}{dx}(4+1/x^2)}{\frac{d}{dx}\sqrt{2+9/x^4}} \\
&= \lim_{x \to 0} \frac{-2/x^3}{-18/(x^3\sqrt{2x^4+9})} \\
&= \frac{-2}{-18} \lim_{x \to 0} \sqrt{2x^4+9} \\
&= \frac{1}{9}\sqrt{9} \\
&= \frac{1}{3}. \quad ◀
\end{aligned}$$

The Indeterminate Form $0 \cdot \infty$

L'Hôpital's Rule applies only to quotients. Nevertheless, we may be able to rewrite a non-quotient as a quotient so that l'Hôpital's Rule may be applied. The indeterminate form $0.\infty$ is often handled in this way. The next example illustrates the technique.

▶ **EXAMPLE 6** Evaluate the limit $\lim_{x\to 0} x^2 \cdot \ln(x^2)$.

Solution We may rewrite the orginal limit as

$$\lim_{x\to 0} \frac{\ln(x^2)}{1/x^2}.$$

Notice that the numerator tends to $-\infty$ and the denominator tends to $+\infty$ as $x \to 0$. Thus the quotient is indeterminate at 0 of the form $\frac{-\infty}{\infty}$. Theorem 2 therefore applies. Accordingly, we have

$$\lim_{x\to 0} \frac{\ln(x^2)}{1/x^2} = \lim_{x\to 0} \frac{\frac{d}{dx}\ln(x^2)}{\frac{d}{dx}(1/x^2)} = \lim_{x\to 0} \frac{(1/x^2)(2x)}{-2x^{-3}} = -\lim_{x\to 0} x^2 = 0. \quad ◀$$

INSIGHT In Example 6, a formal limit computation results in $\lim_{x\to 0} x^2 \cdot \ln(x^2) = 0 \cdot (-\infty)$. Although the expressions $0 \cdot \infty$ and $0 \cdot (-\infty)$ are also called indeterminate forms, l'Hôpital's Rule does not directly address them. An algebraic manipulation is required to rewrite the limit in a way that is appropriate for l'Hôpital's Rule.

l'Hôpital's Rule for Limits at ∞

THEOREM 3 Let f and g be differentiable functions. If $\lim_{x\to+\infty} f(x) = \lim_{x\to+\infty} g(x) = 0$, or if $\lim_{x\to+\infty} f(x) = \pm\infty$ and $\lim_{x\to+\infty} g(x) = \pm\infty$, then

$$\lim_{x\to+\infty} \frac{f(x)}{g(x)} = \lim_{x\to+\infty} \frac{f'(x)}{g'(x)}$$

provided that this last limit exists either as a finite or infinite limit. The same result holds for the limit as $x \to -\infty$.

▶ **EXAMPLE 7** Evaluate $\lim_{x\to+\infty} x^2/e^x$.

Solution We first notice that both the numerator and the denominator tend to $+\infty$ as $x \to +\infty$. Thus the quotient is indeterminate at $+\infty$ of the form $+\infty/+\infty$. Therefore Theorem 3 applies and our limit equals $\lim_{x\to+\infty} 2x/e^x$. Again the numerator and denominator tend to $+\infty$ as $x \to +\infty$ so we again apply Theorem 3. The limit equals $\lim_{x\to+\infty} 2/e^x = 0$. We conclude that $\lim_{x\to+\infty} x^2/e^x = 0$. ◀

The Indeterminate Form 0^0

If x is any positive number, then $x^0 = 1$. It follows that $\lim_{x\to 0^+} x^0 = 1$, which may suggest the (false) conclusion that the form 0^0 is 1. We can see that the situation is not so straightforward by observing that $\lim_{x\to 0^+} 0^x = \lim_{x\to 0^+} 0 = 0$, which may suggest the (equally false) conclusion that the form 0^0 is 0. In fact, 0^0 is indeterminate and must be analyzed carefully. The usual technique is to first apply the logarithm, which reduces an expression involving exponentials to one involving a product or a quotient. A further algebraic manipulation may be required to rewrite the limit in a way that is appropriate for l'Hôpital's Rule.

▶ **EXAMPLE 8** Evaluate $\lim_{x\to 0^+} x^x$.

Solution We study the limit of $f(x)=x^x$ by considering $\ln(f(x))=x\cdot\ln(x)$. We rewrite this as

$$\lim_{x\to 0^+}\ln\big(f(x)\big)=\lim_{x\to 0^+}\frac{\ln(x)}{1/x}.$$

The numerator tends to $-\infty$ and the denominator tends to $+\infty$, so the quotient is indeterminate of the form $\frac{-\infty}{+\infty}$. L'Hôpital's Rule therefore applies. We calculate

$$\lim_{x\to 0^+}\frac{\ln(x)}{1/x}=\lim_{x\to 0^+}\frac{\frac{d}{dx}\ln(x)}{\frac{d}{dx}(1/x)}=\lim_{x\to 0^+}\frac{1/x}{-1/x^2}=\lim_{x\to 0^+}(-x)=0.$$

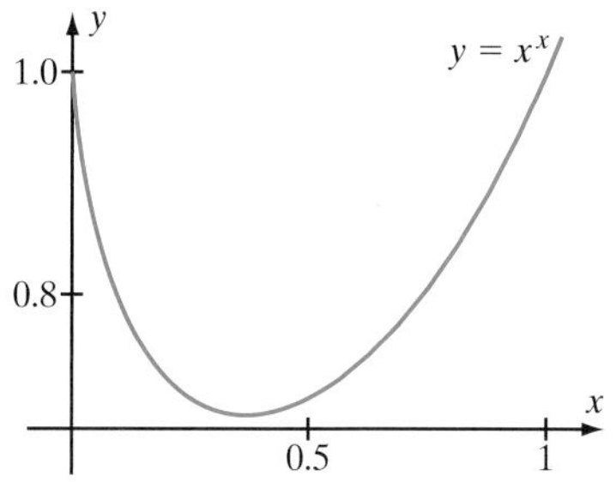

▲ Figure 1

Now the only way that $\ln(f(x))$ can tend to zero is if $f(x)=x^x$ tends to 1. We conclude that $\lim_{x\to 0^+} x^x=1$. ◀

INSIGHT The limit in Example 8 is not obvious. A plot of x^x for $0<x>1$ (Figure1) provides us with visual evidence that our calculation is correct. Moreover, numerical evidence also supports our conclusion:

x	0.1	0.01	0.001	0.0001	0.00001
x^x	0.7943	0.9550	0.9931	0.9991	0.9999

The Indeterminate Form 1^∞

Because $1^x=1$ for any real number x, it might seem that the indeterminate form 1^∞ should equal 1. The next example will prove otherwise. As with the form 0^0, the technique for handling 1^∞ is to first apply the logarithm.

▶ **EXAMPLE 9** Use l'Hôpital's Rule to evaluate $\lim_{x\to\infty}(1+3/x)^x$.

Solution Let $f(x)=(1+3/x)^x$ and consider $\ln(f(x))=\ln((1+3/x)^x)=x\cdot\ln(1+3/x)$. This expression is indeterminate of the form $\infty\cdot 0$.

We rewrite it as

$$\lim_{x\to\infty}\frac{\ln(1+3/x)}{1/x},$$

so that both the numerator and denominator tend to 0. In this form l'Hôpital's Rule applies and we have

$$\lim_{x\to\infty}\ln(f(x))=\lim_{x\to\infty}\frac{\frac{d}{dx}\ln\left(1+\frac{3}{x}\right)}{\frac{d}{dx}\left(\frac{1}{x}\right)}=\lim_{x\to\infty}\frac{\frac{1}{1+3/x}\cdot\left(-\frac{3}{x^2}\right)}{-\frac{1}{x^2}}=3\lim_{x\to\infty}\frac{1}{1+3/x}=3.$$

Because the only way $\ln(f(x))$ can tend to 3 is if $f(x)$ tends to e^3, we conclude that $\lim_{x\to\infty}(1+3/x)^x=e^3$, which agrees with formula (2.6.13). ◀

The Indeterminate Form ∞^0

As with the forms 0^0 and 1^∞, the technique for handling ∞^0 is to first apply the logarithm.

▶ **EXAMPLE 10** Calculate $\lim_{x\to\infty} x^{1/x}$.

Solution We study the limit of $f(x) = x^{1/x}$ by considering $\ln(f(x)) = \ln(x)/x$, which gives us an indeterminate form that is suitable for l'Hôpital's Rule. We have

$$\lim_{x\to\infty} \ln(f(x)) = \lim_{x\to\infty} \frac{\ln(x)}{x} = \lim_{x\to\infty} \frac{\frac{d}{dx}\ln(x)}{\frac{d}{dx}x} = \lim_{x\to\infty} \frac{1/x}{1} = 0.$$

Now the only way that $\ln(f(x))$ can tend to zero as x tends to infinity is if $f(x) = x^{1/x}$ tends to 1. We conclude that $\lim_{x\to\infty} x^{1/x} = 1$. ◀

Putting Terms over a Common Denominator

Many times, a simple algebraic manipulation will put a limit into a form that can be studied using l'Hôpital's Rule. The next example illustrates this idea.

▶ **EXAMPLE 11** Evaluate the limit

$$\lim_{x\to 0}\left(\frac{1}{\sin(x)} - \frac{1}{x}\right).$$

Solution We put the fractions over a common denominator to rewrite our limit as

$$\lim_{x\to 0}\left(\frac{x-\sin(x)}{x\cdot\sin(x)}\right).$$

Both numerator and denominator vanish as $x\to 0$. Thus the quotient has indeterminate form 0/0. By l'Hôpital's rule, the limit is therefore equal to

$$\lim_{x\to 0}\left(\frac{x-\sin(x)}{x\cdot\sin(x)}\right) = \lim_{x\to 0}\frac{\frac{d}{dx}\big(x-\sin(x)\big)}{\frac{d}{dx}\big(x\sin(x)\big)} = \lim_{x\to 0}\frac{1-\cos(x)}{\sin(x)+x\cos(x)}.$$

This quotient is still indeterminate; we apply l'Hôpital's rule again to obtain

$$\lim_{x\to 0}\left(\frac{x-\sin(x)}{x\cdot\sin(x)}\right) = \lim_{x\to 0}\frac{\frac{d}{dx}\big(1-\cos(x)\big)}{\frac{d}{dx}\big(\sin(x)+x\cos(x)\big)} = \lim_{x\to 0}\frac{\sin(x)}{2\cos(x)-x\sin(x)} = \frac{0}{2-0} = 0.$$ ◀

▶ **EXAMPLE 12** Calculate $\lim_{x\to+\infty}(\sqrt{x^2+6x}-x)$.

Solution The algebraic technique with which we treat this difference is known as *conjugation*. We have previously used this method in Example 4 of Section 2.2. In this technique, we multiply and divide $\sqrt{x^2+6x}-x$ by its conjugate, $\sqrt{x^2+6x}+x$. Thus

$$\lim_{x\to+\infty}(\sqrt{x^2+6x}-x) = \lim_{x\to+\infty}\frac{(\sqrt{x^2+6x}-x)(\sqrt{x^2+6x}+x)}{\sqrt{x^2+6x}+x} = \lim_{x\to+\infty}\frac{6x}{\sqrt{x^2+6x}+x}.$$

This limit has indeterminate form ∞/∞. However, repeated applications of l'Hôpital's Rule do not lead to a simpler limit, as you may wish to verify. Instead, we calculate as follows:

$$\lim_{x\to+\infty}\frac{6x}{\sqrt{x^2+6x}+x} = \lim_{x\to+\infty}\frac{6x}{\sqrt{x^2(1+6/x)}+x} = \lim_{x\to+\infty}\frac{6x}{|x|\sqrt{1+6/x}+x}$$
$$= \lim_{x\to+\infty}\frac{6x}{x\sqrt{1+6/x}+x} = \lim_{x\to+\infty}\frac{6}{\sqrt{1+6/x}+1} = \frac{6}{\sqrt{1+0}+1} = 3.$$

(In this calculation, notice that the expression $|x|$ simplifies to x because x assumes positive values as $x \to +\infty$.) ◀

INSIGHT In each of Examples 11 and 12, the original difference leads to the indeterminate form $\infty - \infty$. However, even though the two terms of the difference individually have infinite limits, their difference results in a subtle cancellation that yields a finite limit.

◀ **A LOOK BACK** We have encountered seven types of indeterminate forms. The two basic indeterminate forms $\frac{0}{0}$ and $\frac{\infty}{\infty}$ can be handled directly with l'Hôpital's Rule. The indeterminate form $0 \cdot \infty$ is generally converted to one of the basic forms by placing the reciprocal of one of the factors in the denominator. That is, $0 \cdot \infty$ is treated as

$$\frac{\infty}{1/0} \quad \text{or} \quad \frac{0}{1/\infty}.$$

Some algebraic manipulation is necessary before l'Hôpital's Rule can be applied to the indeterminate form $\infty - \infty$. The three indeterminate forms that involve exponents, namely 0^0, 1^∞, and ∞^0, are converted to the indeterminate form $0 \cdot \infty$ by applying the logarithm.

QUICK QUIZ

1. If $\lim_{x\to c} f(x) = \lim_{x\to c} g(x) = \alpha$, then what must α be to apply l'Hôpital's Rule to the limit $\lim_{x\to c}(f(x)/g(x))$?
2. Evaluate $\lim_{x\to 0} \tan(3x)/\tan(2x)$.
3. For which indeterminate forms is it helpful to first apply the logarithm?
4. Evaluate $\lim_{x\to +\infty} (\sqrt{x^2 + 12x} - x)$.

Answers

1. $\alpha = 0$, $\alpha = -\infty$, or $\alpha = +\infty$ 2. 3/2 3. 0^0, 1^∞, and ∞^0 4. 6

EXERCISES

Problems for Practice

▶ In each of Exercises 1–16, use l'Hôpital's Rule to find the limit, if it exists. ◀

1. $\lim_{x\to 0}(1 - e^x)/x$

2. $\lim_{x\to +\infty} \ln(x)/\sqrt{x}$

3. $\lim_{x\to 5} \dfrac{\ln(x/5)}{x - 5}$

4. $\lim_{x\to 0} \dfrac{x + \sin(5x)}{x - 3\sin(4x)}$

5. $\lim_{x\to \pi/2} \dfrac{\ln(\sin(x))}{(\pi - 2x)^2}$

6. $\lim_{x\to 0} \dfrac{\cos(x) - 1}{e^x - 1}$

7. $\lim_{x\to -1} \dfrac{\cos(x+1) - 1}{x^3 + x^2 - x - 1}$

8. $\lim_{x\to 0} \dfrac{e^x - e^{-x}}{x}$

9. $\lim_{x\to 0} \dfrac{e^x - e^{-x}}{x^2}$

10. $\lim_{x\to -1} \dfrac{x + x^2}{\ln(2 + x)}$

11. $\lim_{x\to 1} \dfrac{\ln(x)}{x - \sqrt{x}}$

12. $\lim_{x\to 0} \dfrac{\sin^2(3x)}{1 - \cos(4x)}$

13. $\lim_{x\to +\infty} \dfrac{\sin(3/x)}{\sin(9/x)}$

14. $\lim_{x\to+\infty} \dfrac{\sin^2(2/x)}{3/x}$

15. $\lim_{x\to-\infty} \dfrac{\ln(1+1/x)}{\sin(1/x)}$

16. $\lim_{x\to\pi/2} \tan(2x)/\cot(x)$

In each of Exercises 17–24, apply l'Hôpital's Rule repeatedly (when needed) to evaluate the given limit, if it exists.

17. $\lim_{x\to\infty} x^2/e^{3x}$

18. $\lim_{x\to\infty} x^3/e^{2x}$

19. $\lim_{x\to0} \dfrac{\sin^2(x)}{x^2}$

20. $\lim_{x\to1} \dfrac{(x-1)^3}{\ln^2(x)}$

21. $\lim_{x\to1} \dfrac{(x-1)^2}{\cos^2(\pi x/2)}$

22. $\lim_{x\to0} \dfrac{1-\cos(4x)}{\sin^2(x)}$

23. $\lim_{x\to0} \dfrac{\sin(2x)-2x}{x^3}$

24. $\lim_{x\to0} \dfrac{e^x-e^{-x}-2x}{\sin(x^3)}$

In each Exercises 25–32, use an algebraic manipulation to put the limit in a form which can be treated using l'Hôpital's Rule; then evaluate the limit.

25. $\lim_{x\to+\infty} x\cdot e^{-2x}$

26. $\lim_{x\to0} (e^x-e^{-x})\cdot x^{-1}$

27. $\lim_{x\to\infty} e^{-x}\ln(x)$

28. $\lim_{x\to+\infty} x^{-2}\ln(x)$

29. $\lim_{x\to1^+} (x-1)^{-1}\ln(x)$

30. $\lim_{x\to0} \sin(x)\cot(3x)$

31. $\lim_{x\to-2} (x+2)\tan(\pi x/4)$

32. $\lim_{x\to\pi} \tan(x/2)\sin(3x)$

In each of Exercises 33–40, use the logarithm to reduce the given limit to one that can be handled with l'Hôpital's Rule.

33. $\lim_{x\to0^+} x^{\sqrt{x}}$

34. $\lim_{x\to0^+} \ln^x(x)$

35. $\lim_{x\to\infty} \left(\dfrac{x-5}{x}\right)^x$

36. $\lim_{x\to+\infty} \left(x\cdot\ln(x)\right)^{-1/x}$

37. $\lim_{x\to+\infty} x^{1/x}$

38. $\lim_{x\to-1^+} (2+x)^{\ln(x+1)}$

39. $\lim_{x\to+\infty} x^{1/\sqrt{x}}$

40. $\lim_{x\to-\infty} (x^2)^{(e^x)}$

In each of Exercises 41–46, put the fractions over a common denominator and use l'Hôpital's Rule to evaluate the limit, if it exists.

41. $\lim_{x\to\infty} \left(\dfrac{x^3}{\sqrt{x^2+5}} - \dfrac{x^3}{\sqrt{x^2+3}}\right)$

42. $\lim_{x\to0} \left(\dfrac{1}{x} - \dfrac{1}{\ln(x+1)}\right)$

43. $\lim_{x\to0} \left(\dfrac{x}{1-\cos(x)} - \dfrac{2}{x}\right)$

44. $\lim_{x\to0} \left(\dfrac{1}{3x} - \dfrac{1}{3\sin(x)}\right)$

45. $\lim_{x\to0} \left(\dfrac{1}{\sin(x)} - \dfrac{1}{\sin(2x)}\right)$

46. $\lim_{x\to0} \left(\dfrac{1}{e^x+e^{-x}-2} - \dfrac{1}{x^2}\right)$

In each of Exercises 47–52, use an algebraic manipulation to reduce the limit to one that can be treated with l'Hôpital's Rule.

47. $\lim_{x\to+\infty} (\sqrt{4x-5}-2\sqrt{x})$

48. $\lim_{x\to+\infty} (\sqrt{x^2+6x}-x)$

49. $\lim_{x\to-\infty} (x+1)\ln\left(\dfrac{x}{x+1}\right)$

50. $\lim_{x\to+\infty} (\sqrt{x^2+x}-\sqrt{x^2+1})$

51. $\lim_{x\to+\infty} (\sqrt{e^{2x}-e^x}-e^x)$

52. $\lim_{x\to+\infty} \left(\sqrt{x^6+4x}-(x^3+1)\right)$

In Exercises 53–60, calculate the given limit.

53. $\lim_{x\to0} \dfrac{\arctan(x)}{x}$

54. $\lim_{x\to0} \dfrac{\arcsin(3x)}{\arctan(2x)}$

55. $\lim_{x\to0} \dfrac{\sinh(x)}{\exp(x)-1}$

56. $\lim_{x\to0} \dfrac{\cosh(x)-1}{\sinh(x)}$

57. $\lim_{x\to0} \dfrac{\cosh(x)-1}{\sinh^2(x)}$

58. $\lim_{x\to\infty} \dfrac{\tanh(x)}{\arctan(x)}$

59. $\lim_{x\to0} \dfrac{\arcsin(x)-x}{\arctan(x)-x}$

60. $\lim_{x\to0} \dfrac{\operatorname{arcsinh}(x)-\arcsin(x)}{x^3}$

In Exercises 61–64, use the logarithm to reduce the indeterminate form 1^∞ to one that can be handled with l'Hôpital's Rule.

61. $\lim_{x\to0} (\cos(2x))^{1/x^2}$

62. $\lim_{x\to\infty} (\sqrt{1-1/x})^x$

63. $\lim_{x\to\infty}\left(\dfrac{5+x}{2+x}\right)^x$

64. $\lim_{x\to 0}\left(1+\sin(x)\right)^{\csc(x)}$

Further Theory and Practice

65. Use l'Hôpital's Rule to check that $\lim_{h\to 0^+}(1+h)^{1/h}=e$. Now calculate $\lim_{h\to 0^+}(1+hx)^{1/h}$ for any fixed x.

66. Use l'Hôpital's Rule to calculate

$$\lim_{h\to 0}\frac{f(x+h)-2f(x)+f(x-h)}{h^2}$$

for a function f that is twice continuously differentiable at x.

67. Let a and b be constants with $b\neq 1$. Use l'Hôpital's Rule to calculate

$$\lim_{x\to 0}\frac{a\sin(x)-\sin(ax)}{\tan(bx)-b\tan(x)}.$$

68. For any real number k and any positive number a prove that

$$\lim_{x\to+\infty}\frac{x^k}{e^{ax}}=0.$$

In each of Exercises 69–76, use l'Hôpital's Rule to evaluate the one-sided limit.

69. $\lim_{x\to(\pi/2)^-}(x-\pi/2)\cot(x)$

70. $\lim_{x\to\pi^+}(x-\pi)\tan(x/2)$

71. $\lim_{x\to 0^+}\dfrac{\ln(1+x)}{\ln(1+3x)}$

72. $\lim_{x\to 0^+}\dfrac{\sin(x)}{\ln(1+x)}$

73. $\lim_{x\to 0^+}\dfrac{2\ln(x)}{\ln(2x)}$

74. $\lim_{x\to 0^+}\ln(1+x)\ln(x)$

75. $\lim_{x\to 0^+}x^{-1/\ln(x)}$

76. $\lim_{x\to 0^+}x^{5/\ln(2x)}$

In each of Exercises 77–82, a function f with domain either $I=(-\infty,\infty)$ or $I=(0,\infty)$ is given. Sketch the graph of f. (The set C of critical points of f and the set I of inflection points of f are provided in some cases.) Use l'Hôpital's Rule to determine the horizontal asymptote of the graph. If $I=(0,\infty)$, use l'Hôpital's Rule to determine $\lim_{x\to 0^+}f(x)$.

77. $f(x)=x/e^x$; $I=(-\infty,\infty)$

78. $f(x)=x^2/e^x$; $I=(-\infty,\infty)$

79. $f(x)=e^x/(e^x+x)$; $I=(-\infty,\infty)$; $I=\{2.269\ldots\}$

80. $f(x)=\ln(x)/x$; $I=(0,\infty)$

81. $f(x)=x\ln(x)/(1+x^2)$; $I=(0,\infty)$; $C=\{0.301\ldots,3.319\ldots\}$; $I=\{0.720\ldots,5.486\ldots\}$

82. $f(x)=\ln(x)/(1/x+e^{x-1})$; $I=(0,\infty)$; $C=\{0.730\ldots\}$; $I=\{1.646\ldots\}$

83. Suppose that r is a constant greater than 1. Let α be any constant. Show that

$$\lim_{x\to\infty}\frac{x^\alpha}{r^x}=0.$$

Deduce that

$$\lim_{x\to\infty}\frac{a\cdot r^x+p(x)}{b\cdot r^x+q(x)}=\frac{a}{b}.$$

for any polynomials $p(x)$ and $q(x)$ and any constants a and b with $b\neq 0$.

84. Evaluate

$$\lim_{x\to\infty}\frac{x^{(x+1)/x}}{\sqrt{1+x^2}}.$$

85. For any constants α and β with $\beta>0$, show that

$$\lim_{x\to\infty}\frac{\ln(x)^\alpha}{x^\beta}=0.$$

86. Cauchy's Mean Value Theorem states the following: If $f(x)$ and $g(x)$ are functions that are continuous on the closed interval $[a,b]$ and differentiable on the open interval (a,b), then there is a point $\xi\in(a,b)$ such that $(g(b)-g(a))\,f'(\xi)=(f(b)-f(a))\,g'(\xi)$. Notice that, if both $g(a)\neq g(b)$ and $g'(\xi)\neq 0$, then

$$\frac{f'(\xi)}{g'(\xi)}=\frac{f(b)-f(a)}{g(b)-g(a)}.$$

Complete the following outline to obtain a proof of Cauchy's Mean Value Theorem

a. Let $r(x)=$ $\big(f(b)-f(a)\big)\big(g(x)-g(a)\big)-\big(f(x)-f(a)\big)\big(g(b)-f(a)\big)$. Check that $r(a)=r(b)$.

b. Apply Rolle's Theorem to the function r on the interval $[a,b]$ to conclude that there is a point $\xi\in(a,b)$ such that $r'(\xi)=0$.

c. Rewrite the conclusion of (b) to obtain the conclusion of Cauchy's Mean Value Theorem.

87. Write the equation of the line through the points $(g(a), f(a))$ and $(g(b), f(b))$. If $P=(x,y)$ is a point in the plane then calculate the vertical distance from P to this line. Now give a geometric interpretation of the function r in Exercise 86(a). Explain the motivation of the proof of Cauchy's Mean Value Theorem.

88. Use Cauchy's Mean Value Theorem to prove the first form of l'Hôpital's Rule (Theorem 1).

Calculator/Computer Exercises

In each of Exercises 89–92, investigate the given limit numerically and graphically.

89. $\lim_{x\to 0}(1-\cos(x))/\sin(x)^2$

90. $\lim_{x\to 0}\tan(2\sin(x))/\sin(2\tan(x))$

91. $\lim_{x\to 0^+} x^{(1-\sin(x)/x)}$

92. $\lim_{x\to 0}(1+\tan(2x))^{1/x}$

In each of Exercises 93–95, illustrate the assertion of l'Hôpital's Rule by plotting f/g and f'/g' in an interval centered at c.

93. $f(x)=(1-\cos(x))$, $g(x)=x\tan(x)$, $c=0$

94. $f(x)=(1+x-e^x)$, $g(x)=x\sin(x)$, $c=0$

95. $f(x)=(1+\cos(x)), g(x)=(x-\pi)\sin(x), c=\pi$

96. Example 10 shows that $\lim_{x\to 0^+} x^x = 1$. Plot $y=x^x$ for values of x that have the form $-p/q$ where p and q are positive integers with q odd. Let $\{x_n\}$ be a sequence of such negative rational numbers with $\lim_{n\to\infty} x_n = 0$. What can be said of $\lim_{n\to\infty} x_n^{x_n}$?

4.8 The Newton-Raphson Method

All quadratic equations $ax^2+bx+c=0$ can be solved explicitly by means of the Quadratic Formula:

$$x=\frac{-b\pm\sqrt{b^2-4ac}}{2a}.$$

There are formulas for solving cubic and fourth degree equations as well. However, there are no elementary formulas for solving general polynomial equations of degree five and higher. If we turn to *transcendental* equations, such as

$$\frac{\pi}{3}=x-\frac{1}{2}\sin(x),$$

then matters are even more obscure. Yet there is often a need to solve such equations, as we will see in several applied problems.

If we settle for an approximate solution of an equation (with, say, a specified number of decimal places of accuracy), then calculus enables us to find sufficiently accurate solutions for many equations that come up in practice. The technique that we will study in this section is called the *Newton-Raphson Method* (or sometimes just *Newton's Method*). It is one of the methods used by calculators and computer algebra systems to solve equations.

The Geometry of the Newton-Raphson Method

See Figure 1 for the fundamental idea behind the Newton-Raphson Method: the graph of a differentiable function f is approximated well, near a point $(x_1, f(x_1))$, by the tangent line ℓ at the point $(x_1, f(x_1))$. The picture suggests that, if x_1 is near a point c where the graph of f crosses the x-axis, then the point x_2 at which ℓ intersects the axis will be even closer. This simple observation is the basic idea of the Newton-Raphson Method.

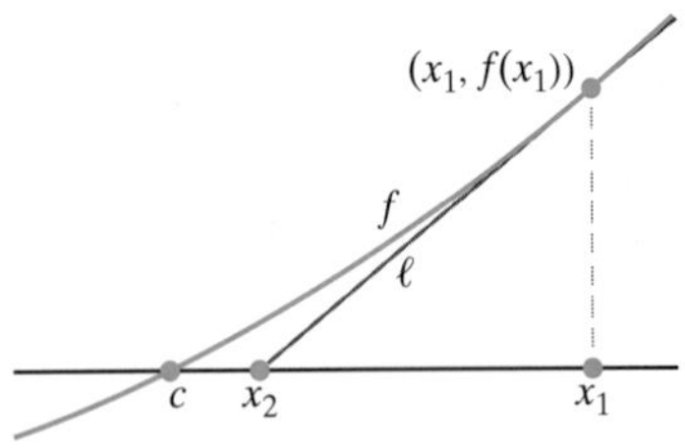

▲ Figure 1

▶ **EXAMPLE 1** Let P denote the point at which the graphs of $y=\cos(x)$ and $y=x$ cross. A very rough estimate of the x-coordinate of P is $\pi/3$. Apply the fundamental idea behind the Newton-Raphson method to find a better approximation.

Solution The idea of the Newton-Raphson Method is to use a tangent line approximation to improve an initial estimate of a root of an equation $f(x)=0$. We start by casting the current problem into this form. The two curves intersect at the value of x for which $x=\cos(x)$, or $x-\cos(x)=0$. If we define $f(x)=x-\cos(x)$, then the value c of x that we are seeking is the solution of $f(x)=0$. Let us begin with the given estimate $x_1=\pi/3$ of c. The tangent line at $(x_1, f(x_1))$ that is, at $(\pi/3, \pi/3-1/2)$, has slope

$$f'\left(\frac{\pi}{3}\right) = \frac{d}{dx}(x - \cos(x))\bigg|_{x=\pi/3} = (1 + \sin(x))\big|_{x=\pi/3} = 1 + \frac{\sqrt{3}}{2}.$$

The point-slope equation of the tangent line is therefore

$$y = \left(1 + \frac{\sqrt{3}}{2}\right)\left(x - \frac{\pi}{3}\right) + \left(\frac{\pi}{3} - \frac{1}{2}\right).$$

The line intersects the x-axis when $y = 0$—that is, at

$$x_2 = \frac{\pi}{3} + \frac{1/2 - \pi/3}{1 + \sqrt{3}/2} \approx 0.75395527.$$

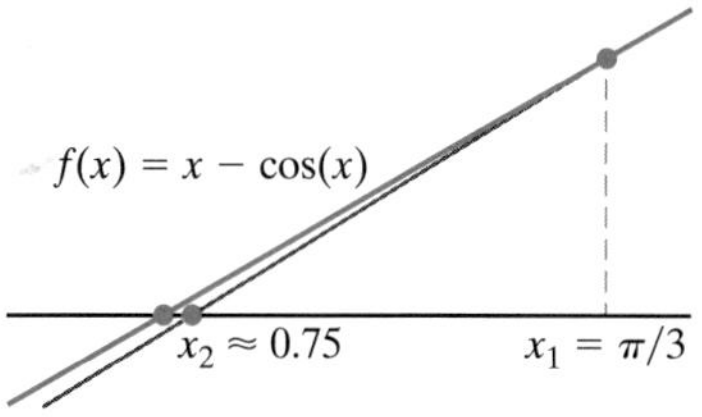

▲ Figure 2

See Figure 2. Now examine the table:

	x	$\cos(x)$
x_1	1.0471976	0.5
x_2	0.75395527	0.72898709

We see that x_1 was a rather poor estimate: the values of x and $\cos(x)$ are quite far apart for that value of x. However, applying the idea of the Newton-Raphson Method results in a remarkable improvement: the values of x and $\cos(x)$ differ by less than 0.025 at x_2. ◀

What if even greater accuracy is required? In that case, we can replace x_1 by x_2 and repeat the procedure to get a better approximation x_3. We keep repeating until the desired degree of accuracy is obtained. That procedure is the Newton-Raphson Method.

Calculating with the Newton-Raphson Method

To put the Newton-Raphson Method to use, we need to develop an *algorithm*. If we are given a differentiable function f and an initial estimate x_1 of a root c of f, then the tangent line at $(x_1, f(x_1))$ has point-slope equation

$$y = f'(x_1)(x - x_1) + f(x_1).$$

Let x_2 be the x-intercept of this tangent line. By substituting $y = 0$ and $x = x_2$ in the equation of the tangent line, we find that

$$x_2 = x_1 - \frac{f(x_1)}{f'(x_1)}, \tag{4.8.1}$$

provided that $f'(x_1) \neq 0$. If we replace x_1 with x_2 on the right side of (4.8.1), then the resulting expression is used to define the next approximation x_3, and so on. Let us summarize what we have learned.

The Newton-Raphson Method

If f is a differentiable function, then the $(n+1)^{\text{st}}$ estimate x_{n+1} for a zero of f is obtained from the nth estimate x_n by the formula

$$x_{n+1} = x_n - \frac{f(x_n)}{f'(x_n)},$$

provided that $f'(x_n) \neq 0$.

We can formulate the Newton-Raphson Method in a way that highlights its iterative nature. Let

$$\Phi(x) = x - \frac{f(x)}{f'(x)}. \tag{4.8.2}$$

Then, starting from a first estimate x_1 of a root c of $f(x) = 0$, we generate subsequent approximations, $x_2, x_3, \ldots$ by the formula

$$x_{j+1} = \Phi(x_j). \tag{4.8.3}$$

▶ **EXAMPLE 2** Use the Newton-Raphson Method to determine $\sqrt{3}$ to within an accuracy of 10^{-7}.

Solution The problem is equivalent to finding the positive value where the function $f(x) = x^2 - 3$ vanishes. We first calculate $f'(x) = 2x$ and, following equation (4.8.2), define

$$\Phi(x) = x - \left(\frac{x^2 - 3}{2x}\right) = x - \frac{x}{2} + \frac{3}{2x} = \frac{1}{2}\left(x + \frac{3}{x}\right).$$

For a first estimate, we take $x_1 = 1.7$. Then, according to equation (4.8.3), we get

$$x_2 = \Phi(x_1) = \frac{1}{2}\left(1.7 + \frac{3}{1.7}\right) = 1.73235294.$$

A check with a calculator reveals that x_2 already agrees with $\sqrt{3}$ to three decimal places of accuracy. Applying the Newton-Raphson Method a second time, we obtain our third estimate

$$x_3 = \Phi(x_2) = \frac{1}{2}\left(1.73235294 + \frac{3}{1.73235294}\right) = 1.7320508.$$

This value agrees with $\sqrt{3}$ to seven decimal places of accuracy, as may be verified with a calculator. ◀

Accuracy

If we do not already know the answer in advance, or if we do not have a calculator handy, how do we determine the accuracy of the approximation? If the Newton-Raphson Method is to be useful in practice, then this is clearly a question that must be answered.

Suppose that our sequence $\{x_n\}$ of Newton-Raphson estimates lies in an interval on which the values of $|f'(x)|$ are bounded by a constant C_1. Suppose also that x_n and x_{n+1} agree to k decimal places of accuracy. Then

$$x_{n+1} = x_n + \varepsilon$$

where $|\varepsilon| < 5 \times 10^{-(k+1)}$. Thus

$$x_n - \frac{f(x_n)}{f'(x_n)} = x_n + \varepsilon,$$

or

$$f(x_n) = -f'(x_n) \cdot \varepsilon.$$

It follows that

$$\begin{aligned}|f(x_n)| &= |f'(x_n)\cdot\varepsilon| \\ &= |f'(x_n)|\cdot|\varepsilon| \\ &\le C_1\cdot\left(5\times 10^{-(k+1)}\right) \\ &= 5\cdot C_1\cdot 10^{-(k+1)}.\end{aligned}$$

This inequality allows us to estimate how close $f(x_n)$ is to zero. In practice we may just continue the iterations until two or three successive approximations are within $5\times 10^{-(k+1)}$ of each other. This is usually the signal that any further refinement will take place beyond the kth decimal place.

If f has a continuous second derivative, then we can say more about the rate at which the Newton-Raphson iterates $\{x_n\}$ approach a root c of f. Let us denote the absolute difference $|c - x_n|$ by ε_n. If $f'(c)$ and $f''(c)$ are nonzero, and if the Newton-Raphson iterations $\{x_n\}$ converge to c, then it may be shown that

$$\lim_{j\to\infty}\frac{\varepsilon_{n+1}}{\varepsilon_n^2} = \left|\frac{f''(c)}{2f'(c)}\right|. \tag{4.8.4}$$

We will use this result only to explain why the Newton-Raphson Method often yields a very accurate approximation after only a few iterations. To that end, notice that the right side of equation (4.8.4) is just a positive number—let us call it C. Equation (4.8.4) says that $\varepsilon_{n+1}\approx C\cdot\varepsilon_n^2$ for large n. Thus when the Newton-Raphson Method is successful, the $(n+1)^{\text{st}}$ error ε_{n+1} is approximately proportional to the square ε_n^2 of the nth error. For example, if ε_n is on the order of 10^{-3}, then ε_{n+1} will be on the order of 10^{-6} and ε_{n+2} will be on the order of 10^{-12}. A loose rule of thumb is that, in a successful application of the Newton-Raphson Method, the number of decimal places of accuracy doubles with each iteration. The technical description is that the Newton-Raphson Method is a *second order process.*

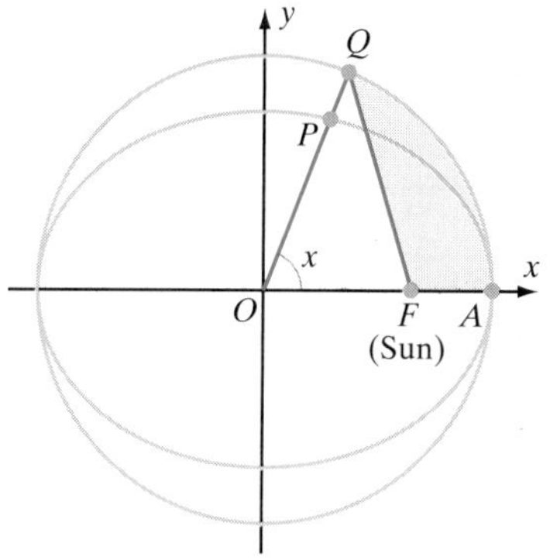

Figure 3
When a planet is at position P, its mean anomaly M is defined by $M = 2(\text{Area}(AFQ)/\overline{OA}^2)$.

▶ **EXAMPLE 3** Kepler's equation

$$M = x - e\cdot\sin(x)$$

relates the mean anomaly M and the eccentric anomaly x of a planet travelling in an elliptic path of eccentricity e. See Figure 3. It is important in predictive astronomy to be able to solve for x if M is given. Kepler expressed his opinion that an exact solution for x is impossible: "... if anyone would point out my mistake and show me the way, then he would be a great Apollonius in my opinion." Instead of finding an exact solution, find x to four decimal places when $M = 1.0472$ and $e = 0.2056$ (the eccentricity of Mercury's orbit).

Solution Let $f(x) = 1.0472 - x + 0.2056\cdot\sin(x)$. Because $f(0) = 1.0472 > 0$ and $f(\pi/2) = -0.3180 < 0$, the Intermediate Value Theorem tells us that f has a root c in the interval $(0, \pi/2)$. We calculate $f'(x) = -1 + 0.2056\cdot\cos(x)$. Because $f'(x) < -1 + 0.2056\cdot 1 < 0$, we can deduce that f is decreasing and the equation $f(x) = 0$ has *only* one solution. The graph of f in Figure 4 confirms our analysis. From the graph, it appears that $x_1 = 1.2$ is a good first estimate. We let

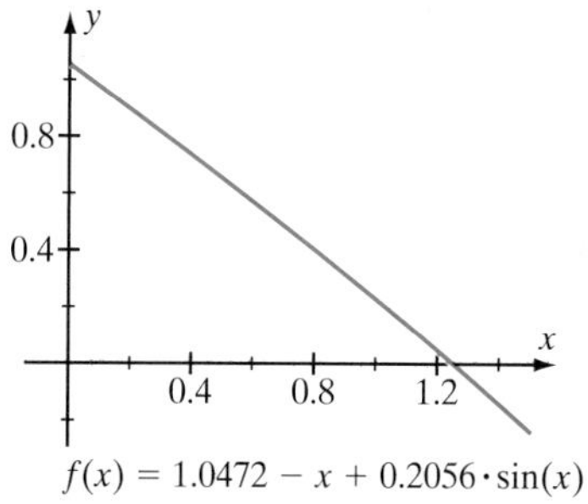

Figure 4

$$\Phi(x) = x - \frac{f(x)}{f'(x)} = x - \frac{1.0472 - x + 0.2056\cdot\sin(x)}{-1 + 0.2056\cdot\cos(x)}$$

and calculate

$$\begin{aligned} x_2 &= \Phi(x_1) = 1.2419527 \\ x_3 &= \Phi(x_2) = 1.2417712 \\ x_4 &= \Phi(x_3) = 1.2417712. \end{aligned}$$

Because $|x_4 - x_3| < 5 \times 10^{-5}$, we can be reasonably sure that $c = 1.24177$ is correct to four decimal places. ◀

Pitfalls of the Newton-Raphson Method

Obviously the Newton-Raphson Method will go badly wrong if $f'(x_j) = 0$ for some index j. Look at Figure 5. We see that the tangent line at x_j is horizontal, so of course it never intersects the x-axis, and the method breaks down. In fact, there can be trouble if $f'(x_j)$ is small, even if it is not zero, because $f(x_j)/f'(x_j)$ can be large. If this is the case, then x_{j+1} will be far from x_j even though the root may be close to x_j, as is illustrated in Figure 6. In these circumstances, we say that the Newton-Raphson Method *diverges*. The *overshoot* that causes this divergence can result in convergence to a root other than the one that is sought (Figure 7). If an inflection point is located close to a root, then the phenomenon of *cycling* may occur (see Figure 8). The best thing to do when presented with one of these problems is to try a different initial estimate.

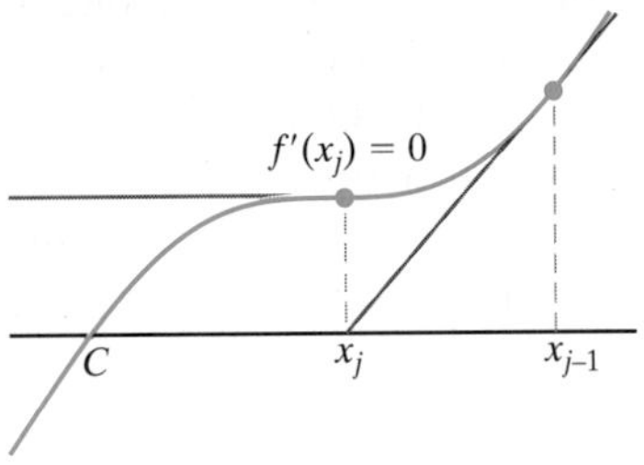

▲ Figure 5

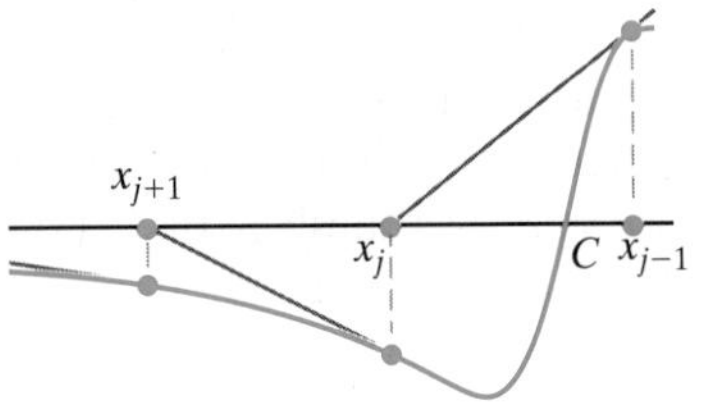

▲ Figure 6

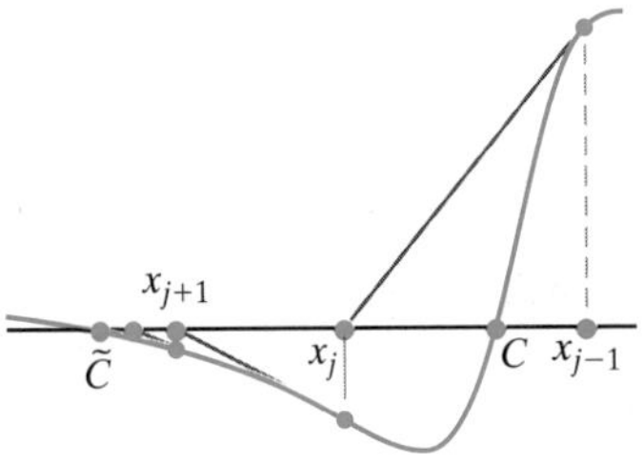

▲ Figure 7

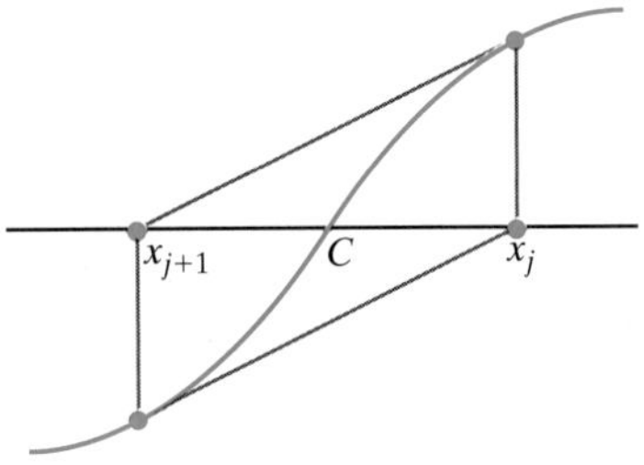

▲ Figure 8

A Computer Implementation

The Newton-Raphson Method can be easily implemented on a calculator or computer. The next example shows how to put the method into action using the computer algebra system *Maple*.

▶ **EXAMPLE 4** The index of refraction r of a triangular glass prism satisfies

$$r \sin\left(\alpha - \frac{\theta_1}{r}\right) = \sin(\theta_4),$$

where the angles α, θ_1, and θ_4 are as in Figure 9. If α, θ_1, and θ_4 are measured to be 45°, 46°, and 54° respectively, then what is r to eight decimal places?

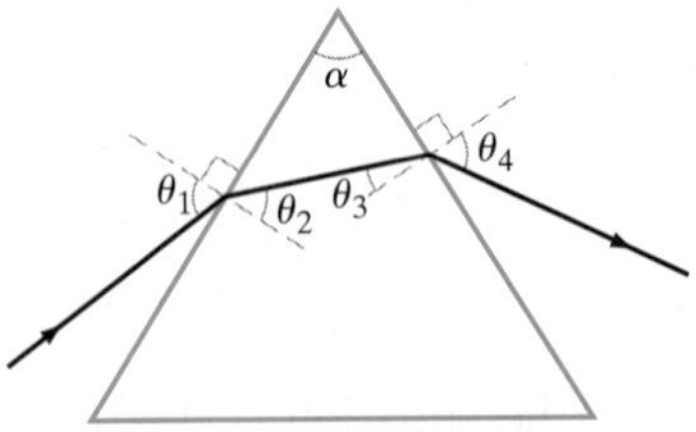

▲ Figure 9 Refraction of light by a triangular prism

Solution Let

$$f(r) = r \sin\left(\frac{\pi}{4} - \frac{46\,\pi}{180\,r}\right) - \sin\left(\frac{54\pi}{180}\right).$$

As in most computations that involve the derivatives of trigonometric functions, we convert to radian measurement of angles. Figure 10 then shows a solution using *Maple*. Because $|x_5 - x_4| < 5 \times 10^{-9}$, we can be reasonably certain that $r = 2.08020645$ is correct to 8 decimal places. (However, if α, θ_1, and θ_4 have not been measured to more than 2 significant digits, then the extra digits in our answer have no reliability.) ◀

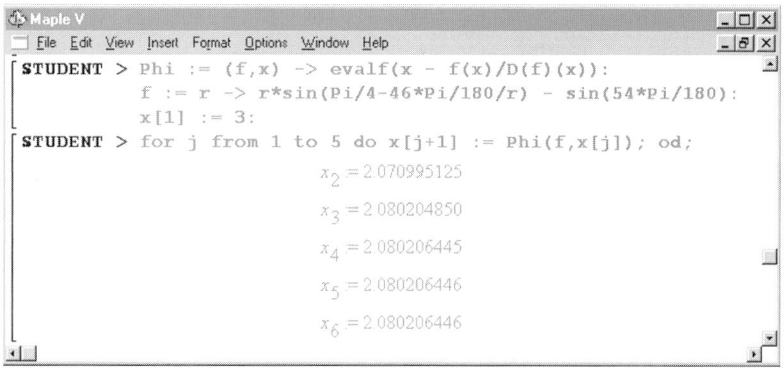
Maple V
File Edit View Insert Format Options Window Help
STUDENT > Phi := (f,x) -> evalf(x - f(x)/D(f)(x)):
f := r -> r*sin(Pi/4-46*Pi/180/r) - sin(54*Pi/180):
x[1] := 3:
STUDENT > for j from 1 to 5 do x[j+1] := Phi(f,x[j]); od;
x2 = 2.070995125
x3 = 2.080204850
x4 = 2.080206445
x5 = 2.080206446
x6 = 2.080206446

▲ Figure 10

An Application in Economics: Bond Valuation

When a corporation or government issues a bond and sells it to an investor, the issuer agrees to pay back (or *redeem*) the *face value* P to the investor in a specified number n of years (at which time the bond is said to *mature*). If the *coupon rate* of the bond is $100r\%$ then, until the bond matures the issuer makes semi-annual interest payments amounting to $rP/2$ each. For example, a bond with a \$5000 face value that pays \$125 in interest every six months is said to have a 5% coupon rate because $r \cdot 5000/2 = 125$ implies that $r = 0.05$. After a bond has been issued, the redemption value P and the coupon rate r do not change.

Because bondholders often want to sell bonds before their maturity, determining the market value of a bond is an important problem. The market price V of a bond is rarely its face value P. As interest rates vary, the market price V of a bond must adjust accordingly. For example, if interest rates have gone up because B was issued, new bonds will have a higher coupon rate than B. The only way to sell bond B is to offer it at a price V that is below its face value P.

An investor who redeems a bond will enjoy a *capital gain* (or suffer a *capital loss*) if he has purchased the bond at a market price V that is less than (or greater than) its redemption price P. This capital gain or loss must be considered in the determination of the *effective yield* x of the bond. The effective yield x is expressed by the formula

$$V = P \cdot \left((1+x)^{-n} + \frac{r}{2} \cdot \frac{1-(1+x)^{-n}}{\sqrt{1+x}-1} \right).$$

In practice, the investor will know the price V at which a bond is offered as well as its coupon rate r, its face value P, and its maturity n. To make an informed decision, the investor must determine the effective yield x. Because it is not feasible to solve exactly for x, a root approximation method such as the Newton-Raphson Method is necessary.

▶ **EXAMPLE 5** A bond with a face value \$10000, coupon rate $6\frac{3}{4}\%$, and maturity of 20 years sells for \$9125. What is the effective yield of the bond? (Bond yields are customarily stated with two decimal places of accuracy.)

Solution Here, $P = 10000$, $r = 0.0675$, $n = 20$, and $V = 9125$. Let

$$f(x) = 10\,000\left((1+x)^{-20} + \frac{.0675}{2}\cdot\frac{1-(1+x)^{-20}}{\sqrt{1+x-1}}\right) - 9125$$

and

$$\Phi(x) = x - \frac{f(x)}{f'(x)}.$$

Because there is a capital gain, the effective yield is greater than r. It therefore makes sense to start our approximation process with a number x_1 greater than r. If we choose $x_1 = 0.07$, then we calculate that

$$\begin{aligned} x_2 &= \Phi(0.07) = 0.07711 \\ x_3 &= \Phi(0.07711) = 0.07753 \\ x_4 &= \Phi(0.07753) = 0.07753. \end{aligned}$$

(Clearly, f is quite complicated and Φ is even more so. The recorded values of $\Phi(x_n)$ have been obtained by using a computer algebra system.) The agreement between x_3 and x_4 to two decimal places indicates that it is time to stop the algorithm. The effective yield x is 7.75.% ◀

QUICK QUIZ

1. If we wish to apply the Newton-Raphson Method to calculate $4^{1/5}$ by setting $x_1 = 1$ and letting $x_{n+1} = \Phi(x_n)$ for $n \geq 1$, then what function Φ can we use?
2. If we use the Newton-Raphson Method to find a root of $f(x) = x^5 + 2x - 40$ and $x = 2$ is our first estimate of the root, then what is our second?

Answers

1. $x - (x^5 - 4)/(5x^4)$ 2. 84/41

EXERCISES

Problems for Practice

▶ In each of Exercises 1–6, use formulas (4.8.2) and (4.8.3) together with the given initial estimate x_1 to find the next two estimates, x_2 and x_3, of a root of the given function f. In each case, x_3 will be accurate to at least two decimal places. ◀

1. $f(x) = x^2 - 6$, $x_1 = 2$
2. $f(x) = x^3 + 2$, $x_1 = -1$
3. $f(x) = x^2 - 5x + 2$, $x_1 = 4$
4. $f(x) = x^3 + x + 1$, $x_1 = -1$
5. $f(x) = x^5 - 24$, $x_1 = 2$
6. $f(x) = x^4 - 70$, $x_1 = 3$

▶ In each of Exercises 7–12, a number is specified in the form a^b. Use formulas (4.8.2) and (4.8.3), the given initial estimate x_1, and a function $f(x) = x^p - a^q$ with p and q positive integers to find the next two estimates, x_2 and x_3, of the given number. ◀

7. $4^{1/3}$, $x_1 = 2$
8. $(-5)^{1/3}$, $x_1 = -2$
9. $6^{1/2}$, $x_1 = 2.5$
10. $(0.5)^{3/2}$, $x_1 = 0.5$
11. $\sqrt{10}$, $x_1 = 2$
12. $(-7)^{3/5}$, $x_1 = -3$

Further Theory and Practice

13. If f is a nonconstant linear function, how many iterations of the Newton-Raphson Method are required before the root of f is found? Does your answer depend on the initial estimate?

14. Suppose you are applying the Newton-Raphson Method to a function f and that after five iterations you land precisely on a zero of f. What value will the sixth and subsequent iterations have?

15. Suppose that a given function f satisfies $f(c) = 0$ and that, in addition, there is an interval I about c such that

$$\left|\frac{f(x)f''(x)}{f'(x)^2}\right| < 1.$$

Then it is known that if the initial estimate is a point chosen from the interval I then the Newton-Raphson Method will certainly converge to c. Test each of the following functions for this condition on the given interval:

a. $f(x) = \sin x$, $c = 0$, $I = (-\pi/4, \pi/4)$
b. $f(x) = x^2 - x$, $c = 0$, $I = (1/2, 2)$
c. $f(x) = x^3 + 8$, $c = -2$, $I = (-3, -1)$
d. $f(x) = 3x^2 - 12$, $c = 2$, $I = (1.5, 2.5)$

16. It is interesting to observe that division can be accomplished numerically by an implementation of the Newton-Raphson Method that does not involve any division at all. Suppose that we wish to calculate α/β where β is nonzero. The idea is to approximate $1/\beta$ and then multiply by α. For the approximation of $1/\beta$, apply the Newton-Raphson method to $f(x) = \beta - 1/x$. What is the function $\Phi(x)$ of formula (4.8.2)? (The formula involves multiplication and subtraction, but not division.) If $\beta = 3$ and $x_1 = 0.4$, what value does formula (4.8.3) yield for x_3?

17. A Babylonian approximation to $\sqrt{c}$ from the second millennium B.C. consists of starting with a first estimate x_1 and then computing the subsequent approximations that are defined iteratively by

$$x_{j+1} = \frac{1}{2}\left(x_j + \frac{c}{x_j}\right).$$

Show that this ancient algorithm is the Newton-Raphson Method applied to $f(x) = x^2 - c$. Of course there was no notion of derivative when this algorithm was discovered. Considering that x_{j+1} is the average of x_j and c/x_j, what might Babylonian mathematicians have had in mind when they devised this algorithm?

18. Define

$$g(x) = \begin{cases} (x-4)^{1/2} & \text{if x} \geq 4 \\ -(4-x)^{1/2} & \text{if x} < 4 \end{cases}$$

Prove that, for any initial guess other than $x = 4$, the Newton-Raphson Method will not converge.

19. If f is a function which is always concave up or always concave down, if f' has no root, and if f has a root, then the Newton-Raphson Method always converges to a root no matter what the first estimate. Explain why.

In doing Exercises 20–23, refer to Exercises 84–89 of Section 4.2 for background on discrete dynamical systems. Assume that f'' exists and is continuous.

20. Let $\Phi(x) = x - f(x)/f'(x)$. Suppose that $f'(x_*) \neq 0$. Show that x_* is an equilibrium point of the dynamical system associated to Φ if and only if x_* is a root of f.

21. Suppose that x_* is a root of f and that $|\Phi'(x_*)| < 1$ and that Φ' is continuous on an open interval centered at x_*. Show that the sequence $\{x_n\}$ of Newton-Raphson iterates will converge to x_* provided that x_1 is sufficiently close to x_*.

22. Show that $|\Phi'(x_*)| < 1$ is equivalent to

$$\left|\frac{f(x_*)f''(x_*)}{f(x_*)^2}\right| < 1.$$

23. Deduce that under the hypotheses of the headnote, the Newton-Raphson iterates will converge to the root x_* of f provided that x_1 is chosen sufficiently close to x_*.

Calculator/Computer Exercises

24. Plot

$$f(x) = x^5 + 4x^4 + 6x^3 + 4x^2 + 1$$

in the viewing window $[-2.3, 0.5] \times [-3.25, 3]$. You will see that f has a root near -2.1. Apply the Newton-Raphson Method using the initial estimate $x_1 = 0$. What happens? Now try the initial estimate $x_1 = -1.5$ and calculate x_2 through x_{10}. Use the graph of f to explain why -1.5, although not too distant from the root, is a poor choice for x_1. (In fact, when $x_1 = -1.5$ initializes the Newton-Raphson process, x_{54} is the first approximation that is closer to the root than the initial estimate.) Now use the initial estimate $x_1 = -6$. What approximation does x_{10} give? (Notice that, because of the geometry of the graph of f, the number -6, though not as close to the root as -1.5, is a more efficient choice for the initial estimate.)

In each of Exercises 25–28, the given function has one real root. Approximate it by making an initial estimate x_1 and applying the Newton-Raphson Method until an integer n is found such that $|x_n - x_{n-1}| < 5 \times 10^{-7}$. State $x_1, x_2, x_3, \ldots, x_n$.

25. $f(x) = x^5 + x^3 + 2x - 5$
26. $f(x) = 2x + \cos(x)$
27. $f(x) = 5x^5 - 9x^4 + 15x^3 - 27x^2 + 10x - 18$
28. $f(x) = x^7 + 3x^4 - x^2 + 3x - 2$

In Exercises 29 and 30 find the largest real root of the given function $f(x)$.

29. $f(x) = x + \sin(4.6x)/x$
30. $f(x) = x + \tan(x)$, $x \in (-\pi, \pi)$.

31. Use the Newton-Raphson Method to approximate the positive solution of $\sqrt{x^3+1} = \sqrt[3]{x^2+1}$ to 5 decimal places.

32. The polynomial $f(x) = x^{11} + x^7 - 1$ has one root γ in the interval $[0, 1]$. Find this root to five decimal places using the Newton-Raphson Method.

If $(x-c)^2$ is a factor of a polynomial $p(x)$ but $(x-c)^3$ is not, then c is a root of $p(x)$ of *multiplicity* 2. The graph of $y = p(x)$ *touches* the x-axis at a root of multiplicity 2 but does not *cross* the x-axis there. In each of Exercises 33–36, plot the given polynomial $p(x)$ in the specified viewing rectangle. Identify a rational number c that is a root of p with multiplicity 2. Use the Newton-Raphson Method with initial estimate $x_1 = c + 1/2$ to obtain iterates $x_2, x_3, \ldots, x_n$. Terminate the process at the smallest value of n for which $|x_N - c| > 5 \times 10^{-4}$. What is N? You will notice that the convergence is slow. Record the value of N so that it can be used for comparison in Exercise 37.

33. $p(x) = x^4 - 2x^3 + 3x^2 - 4x + 2$, $[-2, 3] \times [-5, 50]$
34. $p(x) = x^4 + x^3 - 6x^2 - 4x + 8$, $[-3, 0] \times [-2, 20]$
35. $p(x) = 4x^4 - 4x^3 - 35x^2 + 36x - 9$, $[-3.2, 3.2] \times [-125, 70]$
36. $p(x) = x^4 - 8x^3 + 23x^2 - 28x + 12$, $[0.5, 3.5] \times [-0.25, 2.8]$

37. A variant of Newton-Raphson for roots of higher multiplicity. Each polynomial in Exercises 33–36 has one root c of multiplicity $m = 2$. Approximate each such root using the initial estimate $x_1 = c + 1/2$ and the formula

$$x_{j+1} = x_j - m\frac{p(x_j)}{p'(x_j)}.$$

In each case, terminate the process at the smallest value of M for which $|x_M - c| < 5 \times 10^{-4}$. What is M? Compare the number M with the corresponding number N that results when the unmodified Newton-Raphson Method is used.

38. The *atomic packing factor* (APF) of a crystal is the volume of a unit cube that is occupied by atoms. The APF of NaCl is known to be 2/3. From the geometry of the salt crystal, it can also be shown that the APF of NaCl is given by

$$\frac{2\pi(1 + r^3)}{3(1 + r)^3},$$

where r is the ion size ratio of Na^+ and Cl^-. Use the Newton-Raphson Method to approximate r to 5 decimal places.

39. The period T of a pendulum of length ℓ and maximum deviation φ (measured in radians) from the vertical is given approximately by

$$T = 2\pi\sqrt{\frac{\ell}{g}}\left(1 + \frac{1}{16}\varphi^2 + \frac{11}{3072}\varphi^4\right).$$

If $\ell = 0.8555$ m. and $T = 2.000$ seconds, what is φ. Use $g = 9.807$ m./sec^2. and compute φ to four significant digits.

40. A boat starts for the opposite shore of a 2 km wide river so that (a) the boat always heads toward the point that is directly across from the original launch point, and (b) the speed of the boat is three times faster than the current of the river. If we set up on an xy-coordinate system so that the boat's initial position is $(2, 0)$, the landing site is at $(0, 0)$, and the current is in the positive y-direction, then the path of the boat is given by

$$y = \left(\frac{x}{2}\right)^{2/3} - \left(\frac{x}{2}\right)^{4/3}.$$

What are the x coordinates at the two points at which the boat is 0.137 km downstream.

41. In each of the following parts, use the Newton-Raphson Method to determine to 2 decimal places the effective yield of the bond that is described.

a. A bond with a face value \$10000, coupon rate $6\frac{3}{4}\%$, and maturity of 20 years that sells for \$9125.
b. A bond with a face value of \$10000, coupon rate 6.5%, and maturity 10 years that sells for \$10575.

In each of Exercises 42–45, two \$10000 bonds with the same maturity are offered. Determine which is the better investment by calculating the effective yield of each.

42. Price = \$10531, coupon rate = 7% or Price = \$9146, coupon rate = 5%, $n = 10$
43. Price = \$8681, coupon rate = 5% or Price = \$10674, coupon rate = 7%, $n = 20$
44. Price = \$9518, coupon rate = $6\frac{1}{2}\%$ or Price = \$8705, coupon rate = 6%, $n = 30$
45. Price = \$10679, coupon rate = 8% or Price = \$11052, coupon rate = 9%, $n = 8$

46. The polynomial

$$x^4 + 3.6995x^3 - 4.04035x^2 - 16.13693x + 6.83886$$

has two negative roots that are quite close to each other and two positive roots. Use the Newton-Raphson method to find all four roots to five decimal places. Use a plot to obtain suitable initial estimates.

47. Figure 8 illustrates the phenomenon of cycling with the function $f(x) = x + \sin(x)$ and an initial estimate $x_1 \in (0, 2)$ that leads, after two iterations, to $x_3 = x_1$. What equation does x_1 satisfy? Use the Newton-Raphson Method to approximate x_1.

48. For every positive x, the number $W(x)$ is defined to be the unique positive number such that

$$W(x)\exp(W(x)) = x.$$

The function $x \mapsto W(x)$ is known as Lambert's W function. Compute $W(1)$ and $W(e)$.

49. In biological kinetics, the equation

$$Y = \frac{x + 3x^2 + 3x^3 + 10x^4}{1 + 4x + x^2 + 4x^3 + 10x^4}$$

represents a typical relationship between saturation Y and dimensionless ligand concentration x ($x \geq 0$). Use the Newton-Raphson Method to find the absolute maximum value of Y/x.

4.9 Antidifferentiation and Applications

It is useful in calculus to be able to reverse the differentiation process. That is, given a function f, we wish to find a function F such that $F' = f$. In this section, we discuss this technique and find some applications for it.

Antidifferentiation

Let f be defined on an open interval I. If F is a differentiable function such that $F'(x) = f(x)$ for all x in I, then F is said to be an *antiderivative* for F on I. An antiderivative is certainly not unique. That is because, if C is any constant, then

$$\frac{d}{dx}(F + C)(x) = \frac{d}{dx}\left(F(x) + C\right) = \frac{d}{dx}F(x) + \frac{d}{dx}C = f(x) + 0 = f(x).$$

In other words, if F is an antiderivative for f on an interval I, then so is $F + C$ for any constant C. *Every* antiderivative G for f on I can, in fact, be accounted for in this way: if $F'(x) = f(x)$ and $G'(x) = f(x)$ for all x in I, then $F'(x) = G'(x)$ and Theorem 5 from Section 4.2 tells us that $G(x) = F(x) + C$.

DEFINITION The collection of all antiderivatives of a function f is denoted by

$$\int f(x)\,dx. \tag{4.9.1}$$

This expression is called the *indefinite integral* of f. We refer to $f(x)$ as the *integrand* of expression (4.9.1). If F is a particular antiderivative of f, then, in accordance with the preceding discussion, we write

$$\int f(x)\,dx = F(x) + C, \tag{4.9.2}$$

where C represents an arbitrary constant. We refer to C as a *constant of integration*. The right side of (4.9.2) is interpreted to be the collection that consists of each function of the form $F(x) + C$ for some constant C.

▶ **EXAMPLE 1** Calculate $\int x^3\,dx$.

Solution Because differentiation of a power of x reduces the exponent by one, we reverse our reasoning and guess that x^4, which is one degree greater than x^3, will be an antiderivative. However,

$$\frac{d}{dx}x^4 = 4x^3,$$

which is off by a factor of 4. We therefore adjust our guess to

$$F(x) = \frac{1}{4}x^4.$$

Now we have

$$F'(x) = \frac{1}{4}\cdot 4x^3 = x^3$$

as desired. Because all other antiderivatives for x^3 differ from this one by a constant, we write our solution as

$$\int x^3 dx = \frac{1}{4}x^4 + C. \quad ◄$$

INSIGHT You may be surprised that the antidifferentiation process looks a bit indirect. In fact, it is, but we will learn to antidifferentiate in an organized way. Chapter 6 is devoted to a wide variety of techniques for obtaining the antiderivatives of functions.

Antidifferentiating Powers of x

Example 1 is an instance of a general principle that is worth isolating:

Indefinite Integrals of Powers of x

The antiderivative of x^m is

$$\int x^m dx = \frac{1}{m+1}x^{m+1} + C \quad \text{for any } m \neq -1. \tag{4.9.3}$$

Indeed, for $m \neq -1$, we have

$$\frac{d}{dx}\left(\frac{1}{m+1} \cdot x^{m+1}\right) = \frac{1}{m+1} \cdot (m+1)x^m = x^m.$$

Observe that we must exclude $m = -1$ because the factor $m+1$ in the denominator is 0 for $m = -1$. The next example fills in this gap.

► **EXAMPLE 2** Show that

$$\int \frac{1}{x}\, dx = \ln(|x|) + C. \tag{4.9.4}$$

Solution Let $f(x) = 1/x$ for $x \neq 0$. In Section 3.6 we learned that

$$\frac{d}{dx}\ln(x) = \frac{1}{x}$$

for $x > 0$. This formula tells us that $\ln(x)$ is an antiderivative of $f(x)$ on the interval $(0, \infty)$. Note, however, that $1/x$ is also defined on $(-\infty, 0)$, but $\ln(x)$ is not. We can repair this mismatch of domains by letting $F(x) = \ln(|x|)$. Because $|x| = x$ on $(0, \infty)$, we see that nothing is changed on that interval: $F(x) = \ln(|x|) = \ln(x)$ is an antiderivative of $1/x$ for $x > 0$. On the other hand, for $x < 0$ and $u = |x|$, we have $u = -x$ and

$$\frac{d}{dx}F(x) = \frac{d}{dx}\ln(u) = \frac{d}{du}\ln(u) \cdot \frac{du}{dx} = \frac{1}{u}\frac{d}{dx}(-x) = \frac{1}{-x}(-1) = \frac{1}{x} = f(x).$$

In summary,

$$\frac{d}{dx}\ln(|x|) = \frac{1}{x}$$

for every $x \neq 0$, which implies formula (4.9.4). ◄

INSIGHT Together, formulas (4.9.3) and (4.9.4) tell us

$$\int x^m\,dx = \begin{cases} \dfrac{1}{m+1}x^{m+1} + C & \text{if } m \neq -1 \\ \ln(|x|) + C & \text{if } m = -1. \end{cases}$$

We can restate some of the basic rules of differentiation in a form that is helpful for antidifferentiaton:

THEOREM 1 Let f and g be functions whose domains contain an open interval I. Let F be an antiderivative for f and G an antiderivative for g. Then

a. $\displaystyle\int \big(f(x) + g(x)\big)\,dx = F(x) + G(x) + C$;

b. $\int \alpha \cdot f(x)\,dx = \alpha \cdot F(x) + C$ for any constant α.

INSIGHT You may wonder why part a does not have two constants (because there are two indefinite integrals). The answer is that we can represent the sum of two arbitrary constants by just another (single) arbitrary constant. Similarly, because a multiple of an arbitrary constant is another arbitrary constant, there is no need to write the constant in part b as $\alpha \cdot C$.

▶ **EXAMPLE 3** Calculate $\int (3x^5 - 7x^{3/2} - 4)\,dx$.

Solution According to Theorem 1a, we can solve this problem by antidifferentiating in pieces. We can use formula (4.9.3) for each:

a. An antiderivative of x^5 is $x^6/6$. Therefore an antiderivative of $3x^5$ is $3x^6/6$, or $x^6/2$.

b. An antiderivative for $7x^{3/2}$ is $7 \cdot \frac{2}{5}x^{5/2}$, or $\frac{14}{5}x^{5/2}$.

c. An antiderivative for 4 is $4x$. (We can see that this antiderivative is correct by differentiating it, but we can also derive it. If we write 4 as $4x^0$, then formula (4.9.3) with $m = 0$ tells us that $\int x^0\,dx = x + C$. Applying Theorem 1b with $\alpha = 4$ and $f(x) = x^0$, we obtain $\int 4x^0\,dx = 4x + C$.)

Putting calculations a, b, and c together yields the antiderivative

$$F(x) = \frac{1}{2}x^6 - \frac{14}{5}x^{5/2} - 4x.$$

We check our work by differentiating:

$$\frac{d}{dx}\left(\frac{1}{2}x^6 - \frac{14}{5}x^{5/2} - 4x\right) = \frac{1}{2}\cdot 6x^5 - \frac{14}{5}\cdot\frac{5}{2}\cdot x^{3/2} - 4 = 3x^5 - 7x^{3/2} - 4.$$

Because all other antiderivatives differ by a constant from the one we have found, we write the solution as

$$\int (3x^5 - 7x^{3/2} - 4)\,dx = \frac{1}{2}x^6 - \frac{14}{5}x^{5/2} - 4x + C. \quad ◀$$

▶ **EXAMPLE 4** Calculate the indefinite integral

$$\int \frac{\sqrt{t} - t^2}{t^4}\,dt.$$

Solution Algebraic manipulations can often simplify an integration problem. After rewriting the indefinite integral, we can use formula (4.9.2) and Theorem 1a as follows:

$$\int \frac{\sqrt{t}-t^2}{t^4}\,dt = \int\left(\frac{t^{1/2}}{t^4} - \frac{t^2}{t^4}\right)dt = \int (t^{-7/2} - t^{-2})\,dt$$

$$= \frac{t^{-5/2}}{(-5/2)} - \frac{t^{-1}}{(-1)} + C = -\frac{2}{5}t^{-5/2} + t^{-1} + C. \quad ◀$$

Antidifferentiation of other Functions

We may also antidifferentiate trigonometric functions and exponential functions. For each differentiation formula from Chapter 3, there is a corresponding antidifferentiation formula.

Derivative Formula	***Antiderivative Formula***
$\frac{d}{dx}\sin(x) = \cos(x)$	$\int \cos(x)\,dx = \sin(x) + C$
$\frac{d}{dx}\cos(x) = -\sin(x)$	$\int \sin(x)\,dx = -\cos(x) + C$
$\frac{d}{dx}\tan(x) = \sec^2(x)$	$\int \sec^2(x)\,dx = \tan(x) + C$
$\frac{d}{dx}\cot(x) = -\csc^2(x)$	$\int \csc^2(x)\,dx = -\cot(x) + C$
$\frac{d}{dx}\sec(x) = \sec(x)\tan(x)$	$\int \sec(x)\tan(x)\,dx = \sec(x) + C$
$\frac{d}{dx}\csc(x) = -\csc(x)\cot(x)$	$\int \csc(x)\cot(x)\,dx = -\cos(x) + C$
$\frac{d}{dx}e^x = e^x$	$\int e^x\,dx = e^x + C$
$\frac{d}{dx}a^x = \ln(a)\cdot a^x$	$\int a^x dx = \frac{1}{\ln(a)}a^x + C$
$\frac{d}{dx}\arcsin\left(\frac{x}{a}\right) = \frac{1}{\sqrt{a^2-x^2}}$	$\int \frac{1}{\sqrt{a^2-x^2}}\,dx = \arcsin\left(\frac{x}{a}\right) + C$
$\frac{d}{dx}\frac{1}{a}\arctan\left(\frac{x}{a}\right) = \frac{1}{a^2+x^2}$	$\int \frac{1}{a^2+x^2}\,dx = \frac{1}{a}\arctan\left(\frac{x}{a}\right) + C$
$\frac{d}{dx}\frac{1}{a}\operatorname{arcsec}\left(\frac{x}{a}\right) = \frac{1}{\lvert x\rvert\sqrt{x^2-a^2}}$	$\int \frac{1}{\lvert x\rvert\sqrt{x^2-a^2}}\,dx = \frac{1}{a}\operatorname{arcsec}\left(\frac{x}{a}\right) + C$
$\frac{d}{dx}\frac{1}{a}\operatorname{arctanh}\left(\frac{x}{a}\right) = \frac{1}{a^2-x^2}$	$\int \frac{1}{a^2-x^2}\,dx = \frac{1}{a}\operatorname{arctanh}\left(\frac{x}{a}\right) + C$
$\frac{d}{dx}\operatorname{arcsinh}\left(\frac{x}{a}\right) = \frac{1}{\sqrt{a^2+x^2}}$	$\int \frac{1}{\sqrt{a^2+x^2}}\,dx = \operatorname{arcsinh}\left(\frac{x}{a}\right) + C$

▶ **EXAMPLE 5** Calculate $\int \sin(5x)\,dx$.

Solution We recall that the derivative of $\cos(5x)$ is $-5\sin(5x)$. We need to adjust our guess to eliminate the factor of -5. Therefore the antiderivative of $\sin(5x)$ should be $-\frac{1}{5}\cos(5x)$. We check that

$$\frac{d}{dx}\left(-\frac{1}{5}\cos(5x)\right) = -\frac{1}{5}\left(-5\sin(5x)\right) = \sin(5x),$$

as desired. We write our solution as

$$\int \sin(5x)\,dx = -\frac{1}{5}\cos(5x) + C. \quad ◄$$

Velocity and Acceleration

If a body moves in a linear path with position $p(t)$ at time t, then the instantaneous velocity of the body at time t is the instantaneous rate of change of position, or the derivative of p:

$$v(t) = \frac{d}{dt}p(t).$$

Similarly the acceleration of the body at time t is the instantaneous rate of change of velocity, or the derivative of v:

$$a(t) = \frac{d}{dt}v(t) = \frac{d}{dt}\left(\frac{d}{dt}p(t)\right) = \frac{d^2}{dt^2}p(t).$$

In the language of antiderivatives, we have

Velocity is an antiderivative of acceleration.
Position is an antiderivative of velocity.

INSIGHT If we measure distance p in feet and time t in seconds, then the average rate of change of p with respect to t is expressed as

$$\frac{\Delta p \text{ feet}}{\Delta t \text{ seconds}}.$$

This explains why the unit of measure for velocity is "feet per second" or ft/s.

Likewise, if velocity v is measured in ft/s, then acceleration is measured by considering

$$\frac{\Delta v \text{ ft/s}}{\Delta t \text{ s}}.$$

This explains why the unit of measure for acceleration is "feet per second per second" or ft/s^2.

► **EXAMPLE 6** A police car accelerates from rest at a rate of 3 mi/min^2. How far will it have traveled at the moment that it reaches a velocity of 65 mi/hr?

Solution To solve this problem, we let $p(t)$ be the position of the police car at time t (measured in minutes), $v(t)$ its speed, and $a(t)$ its acceleration. We begin with the information

$$\frac{d}{dt}v(t) = a(t) = 3.$$

Antidifferentiating, we find that $v(t) = 3t + C$, where C is the constant of integration. We recall that the car begins at rest, which means that the initial velocity is 0. Thus $0 = v(0) = 3 \cdot 0 + C$, or $C = 0$. We have learned that $v(t) = 3t$, or

$$\frac{d}{dt}p(t) = 3t.$$

Antidifferentiating again, we find that $p(t) = 3t^2/2 + D$, where D is a constant of integration. If we substitute $t = 0$, we find that $p(0) = 3 \cdot 0^2/2 + \mathrm{D}$, or $p(0) = D$. Thus

$$p(t) = \frac{3}{2}t^2 + p(0),$$

where $p(0)$ is the initial position of the car. To answer the original question, we convert the question to common units of minutes: How far has the police car traveled when it is going 65/60 mi./min.? The time at which the car has reached this velocity is obtained by solving $v(t) = 3t = 65/60$. We find that $t = 65/180$ min. The car reaches the desired speed in 65/180 minutes (or 21.67 seconds). The distance the car has traveled in this time is

$$p(65/180) - p(0) = \frac{3}{2}(65/180)^2 = \frac{12675}{64800} \approx 0.1956 \text{ miles.} \quad ◀$$

▶ **EXAMPLE 7** An object is dropped from a window on a calm day. It strikes the ground precisely 4 seconds later. From what height was the object dropped?

Solution To answer this question, we need the fact (obtained from experimentation) that the acceleration due to gravity of a falling body near the earth's surface is about 32.16 ft/s^2. If $h(t)$ is the height of the body at time t seconds then we have

$$\frac{d^2h}{dt^2} = a(t) = -32.16.$$

Notice the minus sign, which indicates that the body is falling in the direction of decreasing height. We antidifferentiate to find that

$$\frac{dh}{dt} = v(t) = -32.16t + C.$$

However, the initial velocity of the falling body is zero, so we find that $0 = v(0) = (-32.16) \cdot 0 + C$; hence $C = 0$ and

$$\frac{dh}{dt} = -32.16t.$$

Antidifferentiating again yields $h(t) = -16.08\mathrm{t}^2 + D$, where D is the constant of integration. We do not know the initial height (this is what we seek), but we know that after 4 seconds the height is zero. Therefore we may solve

$$0 = h(4) = -16.08 \cdot 4^2 + D$$

yielding $D = 257.28$. Thus the height of the falling object at any time t is given by $h(t) = -16.08t^2 + 257.28$. The answer to the original question is that the initial height is $h(0) = 257.28$ feet. ◀

▶ **EXAMPLE 8** It is known that the acceleration due to gravity near the surface of Mars is about 3.72 m/s^2. If an object is propelled downward from a height of 12 m with a speed of 72 km/hr, in how many seconds will it hit the surface of the

Red Planet? What is the velocity at impact? (The data given describe the last 12 meters of the journey of Mars Pathfinder, which blasted into space on December 4 1996 and entered the Martian atmosphere on July 4 1997.)

Solution The solution is parallel to that of Example 7, but the acceleration will be different because the events are taking place on Mars.

Let $h(t)$ denote the height of the body at time t. We begin with the equation $h''(t) = -3.72 \text{ m/s}^2$. We antidifferentiate to find that $h'(t) = v(t) = (-3.72t + C)$ m/s. We are told that the initial velocity of the falling body is -72 km/hr. We convert this to meters per second:

$$v(0) = -72\frac{\text{km}}{\text{hr}} = -72 \cdot \frac{1000\,\text{m}}{3600\,\text{s}} = -20\frac{\text{m}}{\text{s}}.$$

On the other hand, $v(0) = (-3.72 \cdot 0 + C)$ m/s, or $v(0) = C$ m/s. It follows that $C = -20$. Substituting this value for C in the equation for h' gives us $h'(t) = (-3.72 \cdot t - 20)$ m/s. Antidifferentiating this equation results in $h(t) = (-1.86 \cdot t^2 - 20t + D)$ m, where D is the constant of integration. We are given that the initial height is 12 meters. Therefore $12\,\text{m} = h(0) = (-1.86 \cdot 0^2 - 20 \cdot 0 + D)$ m, or $D = 12$ m. Thus the height of the falling object at time t is $h(t) = (-1.86 \cdot t^2 - 20t + 12)$ m. From this equation we see that the body strikes the surface of Mars when $0 = h(t) = (-1.86 \cdot t^2 - 20t + 12)$. We solve this equation using the quadratic formula, to find that $t = -11.3$ or $t = 0.570$ (to three significant digits). Only the second of these solutions is physically meaningful. We conclude that it takes 0.570 seconds for the body to strike the surface of Mars. Finally, in the course of our calculation we determined that $v(t) = h'(\text{t}) = (-3.72 \cdot t - 20)$ m/s. The velocity at impact is therefore $v(0.570) = (-3.72 \cdot 0.570 - 20)$ m/s $= -22.1$ m/sec. ◀

QUICK QUIZ

1. Suppose that $F(x)$ is an antiderivative of $\cos(x)$ such that $F(\pi/4) = \sqrt{2}$. What is $F(\pi/3)$
2. True or false: There is exactly one antiderivative $F(x)$ of x^{-2} on $(0, \infty)$ such that $F(1) = 100$.
3. True or false: There is no antiderivative $F(x)$ of $1/x$ on $\{x : x \neq 0\}$ such that $F(-1) = 200$ and $F(1) = 100$.
4. True or false: $\int \ln(|x|)dx = 1/x + C$.

Answers

1. $(\sqrt{2} + \sqrt{3})/2$ 2. True 3. False 4. False

EXERCISES

Problems for Practice

▶ In each of Exercises 1–20, calculate the indefinite integral. ◀

1. $\int (x^2 - 5x)dx$
2. $\int (3\sin(x) - 5\cos(x) + 1)dx$
3. $\int e^x dx$
4. $\int \sec(x)\tan(x)dx$
5. $\int \sqrt{x+2}\,dx$
6. $\int (x^2 - x^{-2} + x^{1/2} - x^{-1/2})\,dx$
7. $\int \left(\frac{x^2 + x^{-3}}{x^4}\right)dx$
8. $\int x^{3/2}(x^{-3} - 4x^{-2} + 2x^{-1})\,dx$
9. $\int (x+1)^2 dx$
10. $\int x(x+1)(x+2)\,dx$
11. $\int \exp(e \cdot x)\,dx$
12. $\int x^{1/2}(x+1)\,dx$
13. $\int (x^{-7/3} - 4x^{-2/3})\,dx$
14. $\int \csc^2(x)\,dx$

15. $\int (3\cos(4x) + 2x)\,dx$
16. $\int (3\sin(7x) + 7\sin(3x))\,dx$
17. $\int \sec^2(8x)\,dx$
18. $\int \csc(2x)\cot(2x)\,dx$
19. $\int (3x-2)^3 dx$
20. $\int x^2(\sqrt{x}+1)^2 dx$

In each of Exercises 21–25, compute $F(c)$ from the given information.

21. $F'(x) = 6x^2,\ F(1) = 3,\ c = 0$
22. $F'(x) = 2x + 3,\ F(1) = 2,\ c = -1$
23. $F'(x) = \cos(x),\ F(\pi/2) = -1,\ c = \pi/6$
24. $F'(x) = 1/x,\ F(1) = 3,\ c = -e^2$
25. $F'(x) = 6e^{2x},\ F(0) = -1,\ c = 1/2$

26. A car accelerates from rest at the constant rate of 2 mi/min^2. After the car has traveled one eighth of a mile, how fast will it be traveling?
27. A sprinter accelerates during the first 20 m of a race at the rate of 4 m/s^2. Of course, she begins at rest. How fast is she running at the moment she hits the 20 m mark?
28. An object is dropped from a window 100 ft above the ground. At what speed is the object traveling at the moment of impact with the ground?
29. A baseball is thrown straight up with initial velocity 100 ft/s. How high will the ball go before it begins its fall back to Earth? What will be the velocity of the ball, on the way down, when it is at height 25 ft?
30. An automobile is cruising at a constant speed of 55 mi/hr. To pass another vehicle, the car accelerates at a constant rate. In the course of one minute the car covers 1.3 miles. What is the rate at which the car is accelerating? What is the speed of the car at the end of this minute?
31. A car traveling at a speed of 50 mi/hr must come to a halt in 1200 ft. If the vehicle will decelerate at a constant rate, what should that rate be?
32. A baseball is dropped from the top of a building. When it strikes the ground its instantaneous velocity is – 75 ft./sec. How tall is the building?
33. A baseball is dropped from the top of a building which is 361 ft tall. How long will it take the ball to strike the ground?
34. The acceleration due to gravity near the surface of the moon is about -5.3064 ft/s^2. If an object is dropped from the top of a cliff on the moon, and takes 5 seconds to strike the surface of the moon, then how high is the cliff?

Further Theory and Practice

Use trigonometric identities to compute the indefinite integrals in Exercises 35–41.

35. $\int \sin(x)\cos(x)\,dx$
36. $\int (\cos^2(x) - \sin^2(x))\,dx$
37. $\int \tan^2(x)\,dx$
38. $\int \sqrt{1+\cos(2x)}\,dx$
39. $\int \cos^2(x)\,dx$
40. $\int \cot^2(x)\,dx$
41. $\int \dfrac{\sin(x)}{\cos^2(x)}\,dx$

42. Evaluate $\int \frac{1}{2^x}\,dx$.
43. Evaluate $\int 2^{x\ln(2)}\,dx$.
44. Evaluate $\int \dfrac{10^x}{3^{2x}}\,dx$
45. Suppose that f is continuous on $\mathbb{R}$, that f is positive on $(-\infty, -3)$, $(2, 4)$, and $(4, \infty)$, and that f is negative on $(-3, 2)$. If F is an antiderivative of f on $\mathbb{R}$, classify the local extrema of F.
46. If G is the antiderivative of g is $G \circ f$ the antiderivative of $g \circ f$? Why or why not?
47. Suppose that $\int g(x)\,dx = G(x) + C$. What is $\int g(f'(x))f(x)dx$?

In each of Exercises 48–52, use the Chain Rule to calculate the given indefinite integral.

48. $\int x(x^2+1)^{100}\,dx$
49. $\int 2x\cos(x^2+7)\,dx$
50. $\int \sin^2(x)\cos(x)\,dx$
51. $\int x\exp(x^2)\,dx$
52. $\int \cos(\sin(x))\cos(x)\,dx$

53. Suppose that f and g are antiderivatives of f on an open interval I and that $F(x_0) = G(x_0)$ for some point x_0 in I. Show that $F(x) = G(x)$ for every x in I. Show that this conclusion is false if I is the union of two nonoverlapping open intervals.
54. Find the constant A such that
$$\int e^{kt}(1+e^{kt})^n dt = A\,(1+e^{kt})^{n+1} + C$$
for $n \neq -1$ and $k \neq 0$.
55. The error function, denoted $\operatorname{erf}(x)$, is defined to be the antiderivative of $2\exp(-x^2)/\sqrt{\pi}$ whose value at 0 is 0. What is the derivative of $x \mapsto erf(\sqrt{x})$?
56. Refer to the preceding exercise for the definition of the error function erf. Where is erf increasing? On what interval(s) is the graph of erf concave up? Concave down?
57. An arrow is shot at a 60° angle to the horizontal with initial velocity 190 ft/s. How high will the arrow travel? What will be the horizontal component of its velocity at height 50 feet (going up)?
58. In a light rail system, the Transit Authority schedules a 10 mile trip between two stops to be made in 20 minutes. The train accelerates and decelerates at the rate of 1/3 mi/min^2.

If the engineer spends the same amount of time accelerating and decelerating, what is the top speed during the trip?

59. The engineer of a freight train needs to stop in 1700 feet in order not to strike a barrier. The train is traveling at a speed of 30 mi/hr. The engineer applies one set of brakes, which causes him to decelerate at the rate of 1/5 mi/min^2. Fifteen seconds later, realizing that he is not going to make it, he applies a second set of brakes, which, together with the first set, causes the train to decelerate at a rate of 3/10 mi/min^2. Will the train strike the barrier? If not, with how many feet to spare? If so, at what speed is the impact?

60. A speeding car passes a policeman who is equipped with a radar gun. The policeman determines that the car is doing 85 mi/hr. By the time the policeman is ready to give chase, the car has a 15 second lead. The policeman has been trained to catch vehicles within 2 minutes of the beginning of pursuit. At what constant rate does the police car need to be able to accelerate to catch the speeding car?

61. A woman decides to determine the depth of an empty well by dropping a stone into the well and measuring how long it takes the stone to hit bottom. She drops the stone into the well and hears it hit bottom 6 seconds later. However sound travels at the approximate speed of 660 mi/hr. So there is a delay from the time that the stone hits bottom to the time that she hears it. Taking this delay into account, determine the depth of the well.

62. Visitors to the top of the Empire State Building in New York City are cautioned not to throw objects off the roof. How much damage could a one ounce weight dropped from the roof do? The height of the building is 1250 feet. Using the quantity mass $\times$ velocity, which is *momentum*, as a measure of the impact of the weight when it strikes, compare your answer with the impact of a ten ounce hammer being swung at a speed of ten feet per second.

63. A model rocket is launched straight up with an initial velocity of 100 ft/s. The fuel on board the rocket lasts for 8 seconds and maintains the rocket at this upward velocity (countering the negative acceleration due to gravity). After the fuel is spent then gravity takes over. What is the greatest height that the rocket reaches? With what velocity does the rocket strike the ground?

64. The initial temperature $T(0)$ of an object is 50°C. If the object is cooling at the rate

$$T'(t) = -0.2e^{-0.02t}$$

when measured in degrees centigrade per second, then what is the limit of its temperature as t tends to infinity?

65. At a given moment in time, two race cars are abreast and traveling at the speed of 90 mi/hr. Car A then begins to accelerate at a constant rate of 6 mi/min^2 and Car B begins to accelerate at a constant rate of 9 mi/min^2. The cars are driving on a track of circumference 2 miles. How long will it take Car B to lap Car A three times?

66. If a car that is initially moving at 100 km/hr decelerates to 0 at a constant rate r in 60 m, what is r?

67. If the air resistance to a falling object is proportional to the velocity of that object, then the velocity of that object when dropped from a height H is

$$v(t) = -kg(1 - e^{-t/k}).$$

Here k is a positive constant and g is the constant acceleration due to gravity. Calculate the height y of the object as a function of t.

68. If the air resistance to a falling object is proportional to the square of the velocity of that object, then the acceleration of that object is given by

$$a(t) = -4ge^{2\sqrt{g\kappa}t}(e^{2\sqrt{g\kappa}t} + 1)^{-2}$$

where g is the acceleration due to gravity and κ is a positive constant. Find the velocity v of the object as a function of t assuming that $v(0) = 0$. Hint: First do Exercise 54.

Calculator/Computer Exercises

69. Suppose that the rate of change of the mass m of a sample of the isotope ^{14}C satisfies

$$m'(t) = -0.1213 \cdot e^{-0.0001213t} \frac{\text{g}}{\text{yr}}$$

when t is measured in years. If $m(0) = 1000$ g, then for what value of t is $m(t)$ equal to 800 g?

70. The amount y of an isotope satisfies

$$\frac{dy}{dt} = -k \cdot y(0) \cdot e^{-kt} \frac{\text{g}}{\text{yr}}$$

when t is measured in years and $y(0)$ is measured in grams. If, after one year, the amount y is one-third the initial amount $y(0)$, then what is the value of k?

71. The initial temperature $T(0)$ of an object is 50°C. If the object is cooling at the rate

$$T'(t) = -0.2e^{-0.02t}$$

when measured in degrees centigrade per second, what is the limit T_∞ of its temperature as t tends to infinity? How much time τ_1 does it take for the object to cool 4°C (from 50°C to 46°C)? How much additional time τ_2 elapses before the object cools an additional 2°C? How much additional time τ_3 elapses before the object cools an additional 1°C (from 44°C to 43°C)?

72. Suppose the velocity of an object when dropped from a height 100 m is

$$v(t) = -19.6(1 - e^{-t/2})m/s.$$

(This equation can arise when air resistance is proportional to velocity.) Calculate the height y of the object as a function of t. Use the Newton-Raphson Method to compute the time t of descent. Plot $y(t)$ for $0 \le t \le T$.

73. Suppose the acceleration of an object when dropped from a height 100 m is

$$a(t) = -39.2 \cdot e^{4.427t}(1+e^{4.427t})^{-2}.$$

(This equation can arise when air resistance is proportional to the square of the velocity.) Calculate the downward velocity v if $v(0) = 0$. (Look at Exercise 54 to do this part.) Plot $v(t)$ as a function of t. What is the "terminal velocity"

$$v_{\infty} = \lim_{t \to \infty} v(t)?$$

Use the Newton-Raphson Method to determine the value of t for which $v(t) = 0.999\, v_{\infty}$.

74. After jumping from an airplane, a skydiver has downward velocity given by

$$v(t) = -192(1 - e^{-t/6})\ \text{ft/s}.$$

Twelve seconds into his descent, the skydiver opens his parachute. For $t > 12$ his velocity is given by

$$v(t) = (16 - 192(1 - e^{-2}))e^{24-2t} - 16\ \text{ft/s}.$$

a. Graph the skydiver's velocity. Is it everywhere continuous? Differentiable?

b. Graph the skydiver's acceleration. Is it everywhere continuous? Differentiable?

c. Determine the skydiver's "terminal velocity"

$$v_{\infty} = \lim_{t \to \infty} v(t).$$

d. If the skydiver jumped from 3000 feet, calculate his height function.

e. How long does the first half of his jump, that is the first 1500 feet, take? How long does the second half take?

75. The sine integral Si is defined to be the antiderivative of $\sin(x)/x$ such that $\text{Si}(0) = 0$. Analyze the graph of $\text{Si}(x)$ over $-4\pi \le x \le 4\pi$ for intervals of increase and decrease and for upward and downward concavity. Explain your analysis. Then use a computer algebra system to graph $\text{Si}(x)$ over this interval.

Summary of Key Topics in Chapter 4

Related Rates *(Section 4.1)*	If two variables x and y are related by an equation $F(x, y) = C$, if x depends on a third variable t, and if dx/dt is known, then the Chain Rule together with the method of implicit differentiation (when needed) can be used to find dy/dt.
Local and Absolute Maxima and Minima *(Section 4.2)*	Let f be a function with domain S. We say that f has a local maximum at a point $c \in S$ if there is a $\delta > 0$ such that $f(c) \ge f(x)$ for all $x \in S$ such that $\lvert x - c \rvert < \delta$. We say that f has a local minimum at c if there is a $\delta > 0$ such that $f(c) \le f(x)$ for all $x \in S$ such that $\lvert x - c \rvert < \delta$. We call c an absolute minimum or maximum in case the above inequalities hold for all x in S. The locations of local/global maxima and minima are called local/global extrema. If f is differentiable on an open interval I and if $c \in I$ is an extremum then $f'(c) = 0$.
Rolle's Theorem and the Mean Value Theorem *(Section 4.2)*	Let f be continuous on $[a, b]$ and differentiable on (a, b) and suppose that $f(a) = f(b)$. Then there is a number $c \in (a, b)$ such that $f'(c) = 0$. More generally, let f be continuous on $[a, b]$ and differentiable on (a, b). Then there is a $c \in (a, b)$ such that $$f'(c) = \frac{f(b) - f(a)}{b - a}.$$

A consequence of the mean value theorem is that if f is differentiable on an open interval I and $f' = 0$ on I then f is a constant function. As a consequence, if two functions f and g have the same derivative function then they differ by a constant.

Critical Points (Section 4.3)

A point where either $f' = 0$ or f' is undefined is called a critical point.

Increasing and Decreasing Functions (Section 4.3)

If $f' > 0$ on an open interval I then f is increasing on I.
If $f' < 0$ on an open interval I then f is decreasing on I.

The First Derivative Test for Extrema (Section 4.3)

Let c be a critical point for a differentiable function f.

a. If there is a $\delta > 0$ such that $f' < 0$ on $(c - \delta, c)$ and $f' > 0$ on $(c, c + \delta)$, then c is a local minimum.
b. If there is a $\delta > 0$ such that $f' > 0$ on $(c - \delta, c)$ and $f' < 0$ on $(c, c + \delta)$, then c is a local maximum.

Applied Maximum-Minimum Problems (Section 4.4)

The ability to locate extrema of functions enables us to solve a variety of applied problems. In these applied problems, we often must find the extrema of a continuous function on a closed and bounded interval. We seek the extrema at the endpoints and the critical points.

Concavity, Points of Inflection, Second Derivative Test (Section 4.5)

If f' increases on an open interval I then the graph of f is said to be concave up on I. If f' decreases on an open interval I then the graph of f is said to be concave down on I.

In case $f'' > 0$ on I then f is concave up on I; if $f'' < 0$ on I then f is concave down on I. A point of inflection for a function is a point where the concavity changes from down to up or up to down.

If f is differentiable on an open interval containing c and if c is a critical point then we have:

a. If $f''(c)$ exists and is positive then c is a local minimum.
b. If $f''(c)$ exists and is negative then c is a local maximum.
c. If $f''(c)$ exists and equals zero then no conclusion is possible.

Graphing (Sections 4.2, 4.3, 4.5, 4.6)

We note that differentiable functions are continuous, making graphing simpler. The concepts of increasing, decreasing, critical point, point of inflection, concave up, and concave down enable us to draw very sophisticated sketches of curves.

l'Hôpital's Rule (Section 4.7)

If c is either a real number or $\pm\infty$ and if $\lim_{x\to c} f(x) = \lim_{x\to c} g(x) = 0$ or $\lim_{x\to c} f(x) = \lim_{x\to c} g(x) = \infty$ then

$$\lim_{x\to c} \frac{f(x)}{g(x)} = \lim_{x\to c} \frac{f'(x)}{g(x)}.$$

With algebraic manipulation, l'Hôpital's Rule can also be used to treat the indeterminate forms $0 \cdot \infty$ and $\infty - \infty$. After an application of the logarithm, l'Hôpital's Rule can be used to treat the indeterminate forms 0^0, 1^∞, and ∞^0.

The Newton-Raphson Method (Section 4.8)

If x_1 is a good initial estimate to a root c of a differentiable function f, then the sequence of iterates $\{x_n\}$ obtained by

$$x_{n+1} = x_n - \frac{f(x_n)}{f'(x_n)}$$

is often used to approximate c.

Antidifferentiation (Section 4.9)

Given a function f, it is useful to be able to determine functions F such that $F' = f$. We call f an antiderivative for f and write $\int f(x)dx = F(x) + C$. The expression $\int f(x)\,dx$ is called an indefinite integral. Indefinite integrals can be used to study problems about velocity and acceleration of moving bodies.

Review Exercises for Chapter 4

In each of Exercises 1–4, variables x and y are related by a given equation. A point P_0 on the graph of that equation is also given. Use the given value of $v_0 = \left.\frac{dx}{dt}\right|_{P_0}$ to calculate $\left.\frac{dy}{dt}\right|_{P_0}$.

1. $x + y^2 = 1 + x^3$, $P_0 = (3, 5)$, $v_0 = 2$
2. $xy = y^3 + 2$, $P_0 = (5, 2)$, $v_0 = -3$
3. $y/x + 2/y = 6$, $P_0 = (1/4, 1/2)$, $v_0 = -3$
4. $y + \exp xy = 3$, $P_0 = (0, 2)$, $v_0 = 2$

5. A snowball is melting so that its surface area decreases at the rate of $3\,\text{cm}^2$/min. At what rate is the radius r decreasing when $r = 4$?
6. As the sun sets, rays passing just over the top of a 20 m building make an angle θ with Earth that changes at the rate of $-2\pi/(24 \cdot 60)$ radians per minute. How fast is the shadow of the building lengthening when $\theta = \pi/6$?
7. A heap of sand in the shape of a right circular cone of height h and base radius r has volume $\pi r^2 h/3$. The volume of the heap is constantly 8π\m^3, but, as surface grains slide down the side, the radius r increases at the rate of 3/160 m/min. At what rate is the height decreasing when $r = 3/2$ m?
8. The lateral surface area of a right circular cone of height h and base radius r is $\pi r\sqrt{r^2 + h^2}$. The cone increases in size in such a way that $h = 4r/3$ and $dA/dt = 20\pi\,\text{cm}^2$/s. How fast is the radius increasing when the height is 24 cm?
9. A particle moves along the curve $y = 1/x$. When $t = 7$, the x-coordinate of the particle's position is 1/6 and $dx/dt = 3$. At $t = 7$ how fast is the sum of the particle's coordinates changing?
10. A 13 foot ladder leans against a wall. The foot of the ladder is being pushed toward the wall at the rate of 1/2 ft/s. When the foot of the ladder is 5 ft from the wall, how fast is the top of the ladder rising?
11. A rope extends in a straight line from the deck of a boat to a pulley at the edge of the dock. The level of the pulley is 6 ft above the boat's deck. The rope is being reeled in at the rate of 2/5 ft per second. How fast is the boat approaching the dock when it is 8 ft away?
12. A photographer stands at point Q that is 6 ft away from point P on a straight track (with $\overline{PQ}$ being perpendicular to the track). The photographer focuses a camera on a runner whose rate is 18 ft/s. When the runner is 8 ft from P, how fast is the camera rotating?

In each of Exercises 13–16, calculate the values c in I and $f'(c)$ that arise when the Mean Value Theorem is applied to the given function f on the given interval I.

13. $f(x) = x^3$, $I = [1, 4]$
14. $f(x) = 1/\sqrt{x}$, $I = [1/16, 1/4]$
15. $f(x) = 1/x$, $I = [-27, -3]$
16. $f(x) = \ln(|x|)$, $I = [-e^3, -1]$

In each of Exercises 17–20, calculate the value c in I that arises when Rolle's Theorem is applied to the given function f on the given interval I.

17. $f(x) = x + 1/x, \quad I = [-3, -1/3]$
18. $f(x) = x(1 - x^2), \quad I = [0, 1]$
19. $f(x) = \dfrac{x - e}{1 - e} + \ln(x), \quad I = [1, e]$
20. $f(x) = x - \sqrt{x}, \quad I = [1/16, 9/16]$

In each of Exercises 21–24, use the given information to find $F(c)$.

21. $F'(x) = -1/x^2 + 1/x, \quad F(1) = 7, \quad c = 1/2$
22. $F'(x) = 4\sin(x), \quad F(\pi/3) = 3, \quad c = \pi$
23. $F'(x) = 4/x, \quad F(e^2) = 7, \quad c = e^3$
24. $F'(x) = \sec(x)^2, \quad F(\pi/4) = 0, \quad c = \pi/3$

In each of Exercises 25–38, find all critical points of the function f defined by the given expression. Then use the First Derivative Test to determine whether each critical point is a local maximum for f, a local minimum, or neither.

25. $x^3 - 12x + 1$
26. $4x^3 - x^4$
27. $x^3 - 48\ln(x)$
28. $3x^4 - 4x^3 - 30x^2 - 36x + 7$
29. $x^3 + 48/x$
30. $\ln(x)/x^2$
31. $\ln^2(x)/x$
32. $x - 12x^{1/3}$
33. $\sqrt{x}/(x + 5)$
34. $(x^3 - 3x^2 - 4x - 8)/(x + 1)$
35. $x^2/2 - 4x + 6\ln(x + 1)$
36. $(6x^2 - x^3)^{1/3}$
37. $x^6 - 3x^5 + 3x^4 - x^3 + 2$
38. $(x^2 + 6x - 6)/e^x$

In each of Exercises 39–42, calculate the minimum and maximum values of the given function f on the given interval I.

39. $f(x) = x^3 - 12x + 2, \quad I = [-3, 5]$
40. $f(x) = 2x^3 - 3x^2, \quad I = [-1, 2]$
41. $f(x) = xe^{-x}, \quad I = [-1, 2]$
42. $f(x) = x(x-1)^{2/3}, \quad I = [1/2, 3/2]$

43. A Norman window has the shape of a rectangle surmounted by a semicircle. If the perimeter is 5 m, what base length maximizes the area?

44. When a person coughs, air pressure in the lungs increases and the radius r of the trachea decreases from the normal value ρ. Units can be chosen so that the speed v of air in the trachea is given by $v = (\rho - r)r^2$. For what value of r/ρ is v maximized?

45. The surface area A of a cell in a beehive depends on two side lengths, a and b, and an acute angle θ that determines the shape of the opening to the cell. The formula is

$$A(\theta) = 6ab + \frac{3}{2}b^2\left(\sqrt{3}\csc(\theta) - \cot(\theta)\right).$$

Show that for any positive constants a and b, the surface area is minimized when $\sec(\theta) = \sqrt{3}$.

46. A 12 inch piece of wire can be bent into a circle or a square. Or, it can be cut into two pieces. In this case a circle will be formed from the first piece and a square from the other. What is the *minimal* area that can result? (Example 4 from Section 4.4 is similar but examines the maximal area instead.)

47. Consider the two curves $y = \sqrt{x}$ and $y = x^2$ for $0 \le x \le 1$. What is the longest horizontal line segment between the two curves?

48. A rectangular box is twice as wide as it is deep. If the volume of the box is 144 in^3, then what is the smallest total surface area of its six faces?

49. A rectangular box is twice as wide as it is deep. If the total surface area of its six faces is 144 in^2, then what is the smallest volume?

In Exercises 50–59, find (a) the interval(s) on which the graph of the given function f is concave up, (b) the interval(s) on which the graph of f is concave down, (c) and the inflection points of f, if any.

50. $f(x) = x^3 + 9x^2 - x + 1$
51. $f(x) = x^4 - 6x^2 + 7x + 3$
52. $f(x) = x/(x^2 + 27)$
53. $f(x) = e^x/\sqrt{2e^6 + e^{2x}}$
54. $f(x) = x^2e^{-x}$
55. $f(x) = 3/x + \ln(x)$
56. $f(x) = x^3 + 48\ln(x)$
57. $f(x) = \ln(x)/x^2$
58. $f(x) = x^{1/3} + x^2/9$
59. $f(x) = x^{4/5}(x - 27)$

In Exercises 60–73, evaluate the given limit. (For some of these limits, l'Hôpital's Rule is *not* the most effective method.)

60. $\lim_{x\to\infty} \dfrac{\sqrt{9x^2 + 1}}{2x}$
61. $\lim_{x\to\infty} \dfrac{2e^x + x}{e^x + 2x}$
62. $\lim_{x\to 0} \dfrac{\sin(5x)}{\sin(3x)}$
63. $\lim_{x\to\pi/8} \dfrac{\cos(12x)}{\cos(4x)}$
64. $\lim_{x\to 0} \dfrac{\tan(20x)}{\tan(4x)}$
65. $\lim_{x\to\pi/2} \dfrac{(2x - \pi)}{\cot(x)}$
66. $\lim_{x\to 0} \dfrac{x\sin(x)}{\sin(x^2)}$

67. $\lim_{x\to\infty} x^3 e^{-x/2}$

68. $\lim_{x\to 0} \dfrac{e^{3x}+e^{-2x}-2-x}{x^2}$

69. $\lim_{x\to+\infty} (\sqrt{x^2+8x}-x)$

70. $\lim_{x\to+\infty} (\sqrt{x^2+6x}-\sqrt{x^2+9})$

71. $\lim_{x\to\infty} \left(\dfrac{x-2}{x+1}\right)^x$

72. $\lim_{x\to 0^+} x^{2/\ln(x)}$

73. $\lim_{x\to 0} \left(\dfrac{1}{x}-\dfrac{1}{e^x-1}\right)$

In Exercises 74–77, find the next two approximations to the root of $f(x)=0$ using the Newton-Raphson Method.

74. $f(x)=x-\exp(-x)$, $x_1=1.0$

75. $f(x)=2+\ln(x)-x$, $x_1=4.0$

76. $f(x)=xe^x-3$, $x_1=1.0$

77. $f(x)=\sqrt{x}+1/\sqrt{x}-3$, $x_1=8.0$

In each of Exercises 78–90, calculate the indefinite integral.

78. $\int e^{-x}\,dx$

79. $\int \sec(x)\tan(x)\,dx$

80. $\int \sqrt{x+2}\,dx$

81. $\int (3x+1)^2\,dx$

82. $\displaystyle\int \frac{x^2+3x^{-2}}{x^4}\,dx$

83. $\int x(x-1)(x+1)\,dx$

84. $\int 2^{x+3}\,dx$

85. $\int (x^{4/3}-4x^{-1/3})\,dx$

86. $\int (3\sin(x)+\cos(3x))\,dx$

87. $\int \sec^2(3x)\,dx$

88. $\int \csc(x)\cot(x)\,dx$

89. $\int x^3(\sqrt{x}+2)\,dx$

90. $\int (-x^{-1})\,dx$

Fermat's Life

In the center of Toulouse, France, lies a beautiful civic building called the Capitole; in this building is a large statue of Pierre de Fermat, seated with a sign declaring him to be the father of differential calculus. A muse is nearby, showing her abundant appreciation for his mental powers. Whereas mathematicians today either work for universities, government, or industry, in Fermat's day many mathematicians were amateurs. Fermat was a judge by profession, and rather a wealthy man. He was a cultured man and a supporter of the arts. He wrote his native French elegantly and was fluent in Italian, Spanish, Latin, and classical Greek. Although he developed a reputation as a classicist and a poet, his special passion was for mathematics.

Fermat was born in the small town of Beaumont-de-Lomagne in the year 1601. Sometime in the 1620s, Fermat studied at the University of Bordeaux with followers of François Viète, the leading French mathematician of an earlier generation. Fermat went on to study law at the university at Orléans, attaining his degree in 1631. He served as magistrate in a special court in Castres that had been created to settle disputes among Protestants and Catholics. It was in Castres, in 1665, that Fermat died. Ten years later, his remains were transferred to the Church of the Augustines in Toulouse.

Fermat's Investigation of Extrema

In 1629 Fermat joined a project that aimed to use the surviving commentaries of Pappus to reconstruct the lost books of Apollonius. While working on this project, sometime before 1636, Fermat hit upon the idea of analytic geometry. He wrote up his discovery in 1636 and circulated it in Paris in 1637, prior to the publication of *La Géometrie* of Descartes. In 1637 Fermat also published his method of maxima and minima, which we now call Fermat's Theorem.

After demonstrating his method on a simple classical problem, Fermat proposed and answered the following problem: divide a line segment of length b into two pieces such that the product of the length of one by the square of the length of the other is a maximum. In other words, maximize

$$f(x) = (b - x) \cdot x^2 = bx^2 - x^3, \qquad 0 \le x \le b.$$

Supposing that this maximum occurs at $x = a$, Fermat set $f(a) = f(a + h)$ and worked with the equation

$$\frac{f(a+h) - f(a)}{h} = 0$$

or

$$\frac{b(a+h)^2 - (a+h)^3 - (ba^2 - a^3)}{h} = 0.$$

This simplifies to $(2ab - 3a^2) + (b - 3a)h - h^2 = 0$. Either by discarding the remaining terms that involve h or by setting $h = 0$, Fermat concluded that $2ab - 3a^2 = 0$ and $a = 2b/3$. Fermat did not use language that suggests that he considered h to be infinitesimal or even small: his approach was an algebraic one that he developed before his discovery of analytic geometry. The derivative had not yet been invented and it was this very investigation that led Fermat to introduce it, staking his claim to be "father of the differential calculus.'

Fermat's Investigation of Points of Inflection

The notion of point of inflection arose very early in the study of the tangent problem. In letters written to Fermat in 1636, Roberval suggested that the curve known as the *conchoid* has two points at which there are no tangent lines. What Roberval had discovered and was referring to is that the conchoid has two points of inflection. The graph of the conchoid (together with one of its tangent lines at an inflection point) appears in Figure 1. Roberval called attention to the change in concavity at B by describing the conchoid as "convex on the outside" from A to B and "convex on the inside" from B through c to infinity.

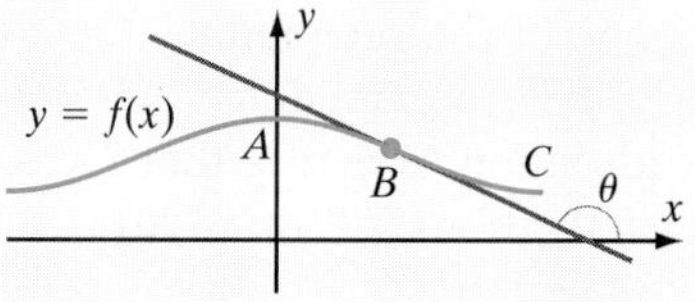

▲ Figure 1 A conchoid and a tangent at a point of inflection

Fermat addressed this issue in a letter to Mersenne dated 20 April 1638. Fermat's discussion indicates that he continued to think of his Tangent Method as a by-product of his Method of Maxima/Minima. He wrote to Mersenne: "...in order to properly graph the [conchoid], it is appropriate that we investigate the points of inflection, where the curvature becomes concave or convex, or vice versa. This question is elegantly resolved by the Method of Maxima/Minima by virtue of the following general lemma." Fermat continued by asserting that among all tangents to the conchoid, the tangent at the inflection point is distinguished by minimizing the angle θ made with the positive x-axis. This is the same thing as minimizing $\tan(\theta) = f'(x)$. Because Fermat's Theorem gives $(f')'(x) = 0$ as a necessary condition for this minimum, Fermat had, in effect, discovered a necessary condition for $(x, f(x))$ to be a point of inflection, namely $f''(x) = 0$.

Fermat's Principle of Least Time

Although Fermat had little interest in physics, he did turn his consideration to one particular question in optics. Therein lies another bitter dispute with Descartes. This particular quarrel began with an early correspondent of Fermat, Jean de Beaugrand. Beaugrand's major work, *Géostatique* (1636), had received severe criticism from Descartes, who wrote

> "Although I have seen many squarings of the circle, perpetual motions, [etc.] ... I can say nonetheless that I have never seen so many errors combined in one postulate. I can therefore conclude that the contents of *Géostatique* are so irrelevant, ridiculous, and mistaken, that I wonder if any honest man has ever troubled himself to read it."

Beaugrand retaliated by circulating three pamphlets in which he charged Descartes with plagiarism. Also, using his position as *Secrétaire du Roi,* Beaugrand obtained a printer's copy of Descartes's *La Dioptrique* prior to its publication. Beaugrand distributed copies in the hope that embarrassing prepublication criticism would ensue. Fermat was one of the recipients.

In an April or May 1637 commentary on *La Dioptrique*, Fermat drew attention to errors that Descartes had made in his derivations of the Laws of Reflection and Refraction. Descartes was not pleased. The situation between the two grew more tense when, after reading *La Géometrie*, Fermat sent Descartes documents that established his priority in the matters of analytic geometry and the tangent problem.

Eight years after the death of Descartes, in 1658, the editor of the *Letters of Descartes,* Claude Clerselier, found two letters from Fermat among the papers of Descartes. He asked Fermat for copies of any other letters that he might have sent to Descartes. Because Fermat had only written the two that had been found, he interpreted the request to be a reopening of the optics controversy.

Fermat argued that there were three fundamental flaws that invalidated Descartes's treatment of refraction: Descartes did not understand what a valid demonstration should be, Descartes had not argued consistently even within his own incorrect logical framework, and Descartes had based his derivation on the patently false notion that the denser is the medium, the easier is the passage of light through it. Experts warned Fermat to be more cautious: experimentation confirmed the Law of Refraction that Descartes had published.

At about this time, Fermat received the *Traité de la Lumié* of Marin Cureau de la Chambre (1594, Mans - 1669, Paris). In it was a derivation of the Law of Reflection based on the method of minimization. Suppose that a ray of light emitted from A is reflected by the straight line ECF and arrives at B, as in Figure 2.

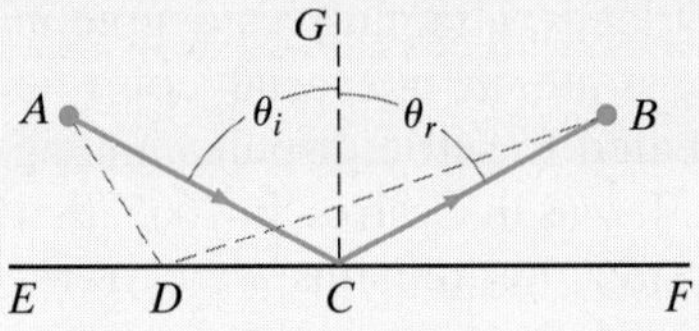

▲ Figure 2

The law of reflection states that the angle of incidence ACG equals the angle of reflection GCB. Cureau derived this from the postulate, that light travels from A to EF to B in the *least distance*. That is, for any other point D on EF, $AC + CB < AD + DB$. This inequality follows easily from classical geometry.

Fermat saw that, if he reformulated Cureau's principle as a *least time* postulate, then he would be able to apply calculus to the more complicated phenomenon of refraction. With refraction we suppose that the points A and B are in different media, separated by boundary EF as in Figure 3. Light travels with speed v_A in the medium containing point A and with speed v_B in the medium containing B. The Law of Refraction asserts that

$$\frac{\sin(\alpha)}{\sin(\beta)} = \frac{v_A}{v_B}$$

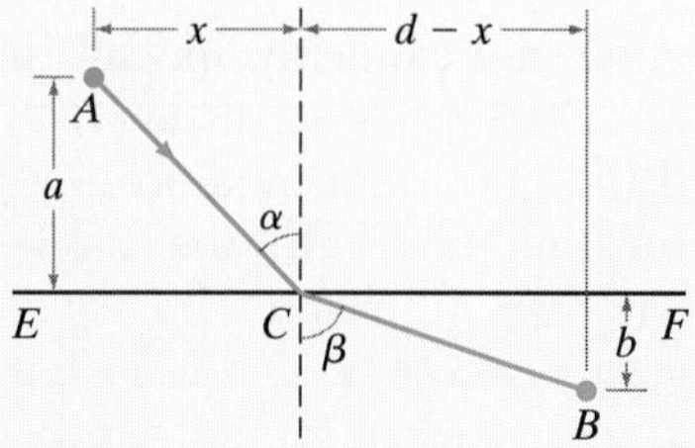

▲ Figure 3

where α is the angle of incidence and β is the angle of refraction (as in Figure 3). Fermat supposed that the point C at which the ray of light impinged on EF was such that the light ray travelled from A to C to B in *least time*. This general principle is now known as *Fermat's Principle of Least Time*. From this principle, Fermat derived the Law of Refraction. We see from Figure 3 that the total time required for the light ray to reach B from A by impinging on EF at a point whose abscissa is x beyond A is

$$f(x) = \frac{\sqrt{a^2 + x^2}}{v_A} + \frac{\sqrt{b^2 + (d-x)^2}}{v_B}.$$

We therefore set

$$f'(x) = \frac{x}{v_A\sqrt{a^2 + x^2}} - \frac{d-x}{v_B\sqrt{b^2 + (d-x)^2}} = 0$$

or

$$\frac{\sin(\alpha)}{v_A} - \frac{\sin(\beta)}{v_B} = 0,$$

which is the Law of Refraction.

The Law of Refraction, published by Fermat in February 1662, is known as *Snell's Law* after Willebrord van Snel van Roijen (1580–1626, Leiden), who had discovered a graphical procedure for determining the direction of a refracted ray. (Snel published his papers using the Latin version, Snellius, of his name and that accounts for the common spelling). Given the blunders in Descartes's "derivation" of Snell's Law, Fermat was astounded to find himself reaching the same conclusion. As he wrote to Clerselier

> "The reward of my work has been the most extraordinary, unexpected, and fortunate that I have ever obtained. Because, after running through all the equations, multiplications, antitheses, and other operations of my method, and having finally brought this problem to a conclusion [as enclosed], I found that my principle gave precisely the same ratio of refractions that M. Descartes had obtained. I was so surprised by this unexpected event that I scarcely recovered from my astonishment."

Fermat's Principle of Least Time was not accepted by his contemporaries. Clerselier wrote to Fermat on 6 May 1662:

> "The principle which you take as the basis of your demonstration ... is nothing but a moral principle and not a physical one; it is not, and cannot be, the cause of any effect of Nature."

In Fermat's last scientific letter, it is not known to whom, he wrote "[the Cartesians] prefer to leave the matter undecided ... Let posterity be the judge."

Rolle's Theorem and The Mean Value Theorem

Michel Rolle (1652, Ambert - 1719, Paris) published his *Traité d'Algèbre* in 1690. In a follow-up one year later, Rolle demonstrated that if $P(x)$ is a polynomial, then $P'(x)$ has a real root between any two real roots of $P(x)$. He did not, however, state his result, which we now call *Rolle's Theorem*, in the language of derivatives. Instead, he used his "Method of Cascades," a formal *algebraic* operation on a polynomial $P(x)$. Indeed, Rolle was a vigorous opponent of infinitesimal methods, which he did not believe to have a solid foundation.

The Mean Value Theorem first appeared (in geometric form) in the *Geometria Indivisibilibus Continuorum* of 1635 of Bonaventura Cavalieri (1598 (?), Milan–1647, Bologna). Its analytic formulation is sometimes attributed to Lagrange, sometimes to the physicist André-Marie Ampère (1775, Lyon–1836, Marseille), who used it in a paper written in 1806.

The Newton-Raphson Method

Newton's method of approximating real roots of polynomials appeared in a paper of 1669 but the first published account appeared in Wallis's *Algebra* of 1685. Wallis demonstrated the technique with Newton's example: $y^3 - 2y - 5 = 0$. Newton chose 2 as the first approximation to the root and substituted $2 + p$ for y, obtaining the equation $p^3 + 6p^2 + 10p - 1 = 0$ for the error p. By neglecting the terms involving p^2 and p^3, Newton solved $p = 0.1 + q$ (where q is an error term). He substituted the right side into the cubic polynomial for p, obtaining $q^3 + 6.3q^2 + 11.23q + 0.061 = 0$. He carried out the next step in the same way, writing $q = -0.061/11.23 + r$ or $q = -0.0054 + r$. Newton continued with one more iteration, obtaining a term $s = -0.00004854$. The resulting approximation to the root is $2 + p = 2 + 0.1 + q = 2 + 0.1 - 0.0054 + r = 2 + 0.1 - 0.0054 - 0.00004854 = 2.09455146$, which agrees with the exact root to 8 decimal places.

Notice that, in Newton's original algorithm, the function under consideration changes at every step. Moreover, the derivative does not explicitly appear. In a booklet that appeared in 1690, Joseph Raphson (1648 (?)–1715 (?)) introduced the important modifications that transformed Newton's method into the easily implemented form we now use. Raphson's improvement is that the function $\Phi(x) = x - f(x)/f'(x)$ that is used remains the same at every step of the iteration.

CHAPTER 5

The Integral

PREVIEW

Up to now, our discussion of calculus has been limited to the derivative and its applications. In this chapter, we learn about one of the most powerful and pervasive tools in all of mathematical science: the definite integral of a function f. The definite integral represents a limit of the form

$$\lim_{\Delta x \to 0} \Big(f(x_1)\,\Delta x + f(x_2)\,\Delta x + \cdots + f(x_N)\,\Delta x \Big),$$

where Δx is an increment of x, as usual. Like the derivative, the definite integral involves the key concept of calculus: the *limit*. However, addition and multiplication are the algebraic operations of the definite integral. This contrasts with the definition of the derivative,

$$f'(x) = \lim_{\Delta x \to 0} \frac{f(x + \Delta x) - f(x)}{\Delta x},$$

which involves subtraction and division.

In Section 5.1, we motivate the definition of the integral by considering the problem of calculating area. We learn that, if f is a positive function defined on the interval $[a, b]$, then we can think of the definite integral as the area of the planar region that lies under the graph of $y = f(x)$, above the x-axis, and between the vertical lines $x = a$ and $x = b$. However, keep in mind that the integral is useful in analyzing many different mathematical and physical problems. We begin with area simply because it provides us with an accessible and geometrical place to start, but area is only the first of many interpretations of the definite integral that we will learn.

The contrasting algebraic operations involved in the derivative and the integral suggest that the processes of differentiation and integration are in some way *inverse* to one another. The Fundamental Theorem of Calculus links the two main branches of calculus by making the inverse relationship between derivative and integral precise. In doing so, the theorem provides a rule that allows us to calculate many definite integrals without referring to the limit definition. For definite integrals that lie beyond the power of the Fundamental Theorem of Calculus, it is important to have efficient numerical approximation methods. In this chapter, we study several of these numerical techniques as well.

5.1 Introduction to Integration–The Area Problem

We calculate the area of a rectangle by multiplying length by width. We find other areas, such as the area of a trapezoid or of a triangle, by performing geometric constructions to reduce the problem to the calculation of the area of a rectangle (see Figure 1). If we wish to calculate the area of the region under the graph of a function $y = f(x)$, then there is no hope of dividing the region and reassembling it into *one* rectangle (see Figure 2). Instead we attempt to divide the region into *many* rectangles.

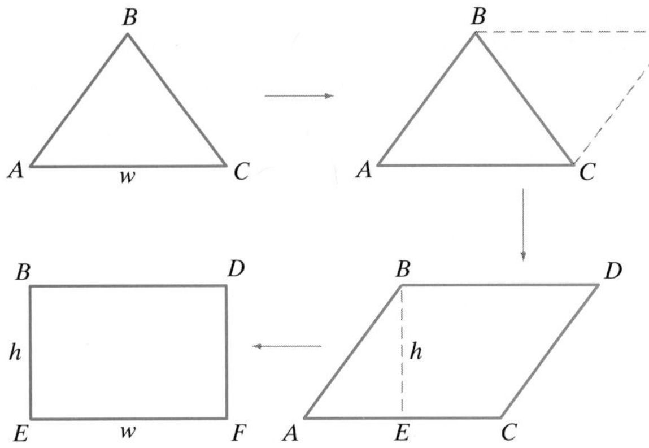

▲ Figure 1 Area $(\Delta ABC) = \frac{1}{2}$ area $(EBDF) = \frac{1}{2}hw$

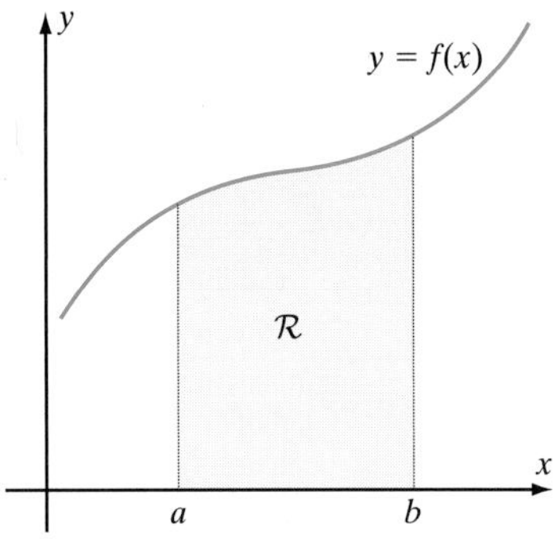

▲ Figure 2

Figure 3 suggests the procedure that we will follow: The area of the shaded region is approximated by the sum of the areas of the rectangles. Of course, there is some inaccuracy: Notice that the rectangles do not quite fit. Parts of some of the rectangles protrude from the region while other parts fall short. If we take a great many rectangles, as in Figure 4, it appears that the sum total of all these errors is quite small. We might hope that, as the number of rectangles becomes ever larger, the total error tends to zero. If so, this procedure would tell us the area of region $\mathcal{R}$ in Figure 2.

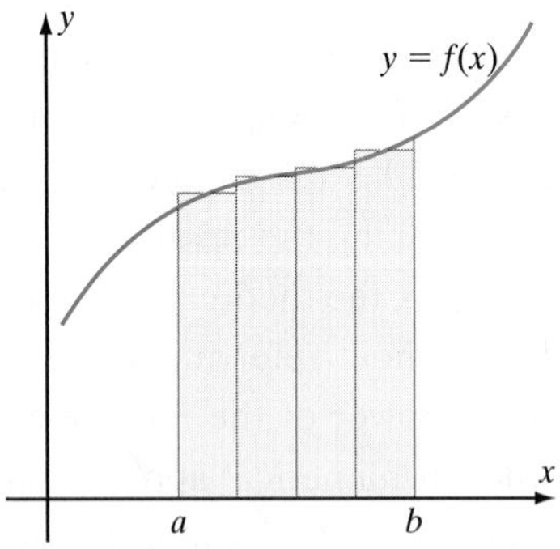

▲ Figure 3

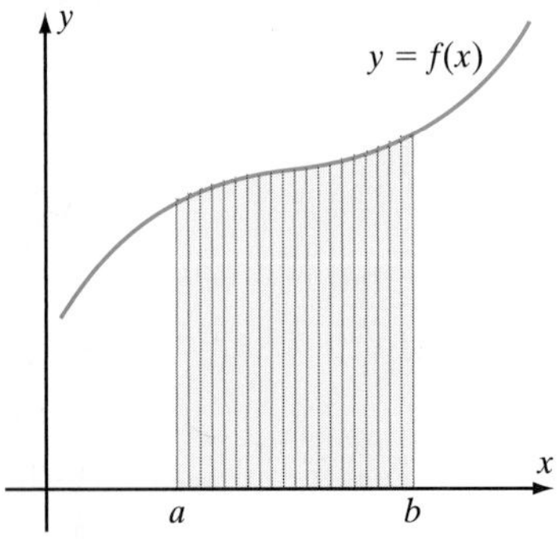

▲ Figure 4

Approximating an area by subdividing into simple regions (rectangles or triangles) is an ancient program. It is the method by which Archimedes obtained the formula for the area inside a circle. This chapter presents the method precisely. Before we begin, however, it will be convenient to learn a notation for the sum of a large or indefinite number of terms.

Summation Notation

If $a_1, a_2, a_3, \ldots, a_N$ is a (finite) sequence of numbers, then we denote the sum, $a_1 + a_2 + a_3 + \cdots + a_N$, of these numbers by $\sum_{i=1}^{N} a_i$. For example, if you turn in eight assignments in a course and receive a_1 points for the first assignment, a_2 points for the second assignment, and so on, then your total homework score for the course, $a_1 + a_2 + a_3 + a_4 + a_5 + a_6 + a_7 + a_8$, can be expressed concisely as $\sum_{i=1}^{8} a_i$. In general, if M and N are integers with $M \leq N$ and if $a_M, a_{M+1}, a_{M+2}, \ldots, a_N$ are numbers (or algebraic expressions), then we read the notation $\sum_{i=M}^{N} a_i$ as "the sum of the numbers (or algebraic expressions) a_i for i equal to M, $M+1$, $M+2$, $\ldots$, up to i equal to N." Thus

$$\sum_{i=M}^{N} a_i = a_M + a_{M+1} + a_{M+2} + \cdots + a_N.$$

▶ **EXAMPLE 1** Evaluate the sums $\sum_{i=1}^{5} i^2$ and $\sum_{i=5}^{8} (3i + 4)$.

Solution The sum $\sum_{i=1}^{5} i^2$ is a shorthand notation for $1^2 + 2^2 + 3^2 + 4^2 + 5^2$, which equals 55. The expression $\sum_{i=5}^{8} (3i + 4)$ means $(3 \cdot 5 + 4) + (3 \cdot 6 + 4) + (3 \cdot 7 + 4) + (3 \cdot 8 + 4)$, which equals 94. ◀

The symbol Σ, which is the capital Greek letter "sigma," stands for "summation." A certain amount of flexibility is inherent in the sigma notation. The sum can begin at any integer M and can end at any integer N with $N \geq M$. The letter i in the sum $\sum_{i=M}^{N} a_i$ is called the *index of summation*. Because the summation index i is simply a placeholder for the integers from M to N, another letter could be used with no change in meaning. Thus

$$\sum_{i=M}^{N} a_i = \sum_{j=M}^{N} a_j.$$

Also, the distributive law and the associativity of addition can be used to manipulate sums. Thus if $a_M, a_{M+1}, \ldots, a_N, b_M, b_{M+1}, \ldots, b_N$, and are real numbers, then

$$\sum_{i=M}^{N} (a_i + b_i) = \sum_{i=M}^{N} a_i + \sum_{i=M}^{N} b_i \qquad (5.1.1)$$

and

for any constant α

$$\sum_{i=M}^{N} \alpha \cdot a_i = \alpha \cdot \sum_{i=M}^{N} a_i \qquad (5.1.2)$$

▶ **EXAMPLE 2** Evaluate the sum $\sum_{j=1}^{4} (3j^2 - 5j)$.

Solution Using (5.1.1) and (5.1.2), we calculate

$$\begin{aligned}\sum_{j=1}^{4} (3j^2 - 5j) &= 3 \cdot \sum_{j=1}^{4} j^2 - 5 \cdot \sum_{j=1}^{4} j \\ &= 3(1^2 + 2^2 + 3^2 + 4^2) - 5(1 + 2 + 3 + 4) = 3 \cdot 30 - 5 \cdot 10 = 40.\end{aligned}$$ ◀

Some Special Sums

There are certain sums that we use frequently. For example, we have already considered the sum $1+r+r^2+\cdots+r^{N-1}$, which arises when we add the terms of the geometric sequence $\{r^j\}_{j=0}^{N-1}$ with $r \neq 1$ (see Section 2.5 in Chapter 2). Using sigma notation, we can express formula (2.5.2) for the sum of a geometric series as

$$\sum_{j=0}^{N-1} r^j = \frac{r^N - 1}{r-1} \qquad (r \neq 1). \tag{5.1.3}$$

The following are two more sums that will be useful in this section. (Proofs of these formulas are outlined in Exercises 53–55.)

$$\sum_{j=1}^{N} j = \frac{N(N+1)}{2} \tag{5.1.4}$$

$$\sum_{j=1}^{N} j^2 = \frac{N(N+1)(2N+1)}{6} \tag{5.1.5}$$

▶ **EXAMPLE 3** Use formulas (5.1.4) and (5.1.5) to calculate $1+2+3+\cdots+100$ and $100+121+144+169+\cdots+400$.

Solution By formula (5.1.4),

$$1+2+3+\cdots+100 = \sum_{j=1}^{100} j = \frac{100\cdot(100+1)}{2} = 5050.$$

The second sum is equal to $10^2+11^2+12^2+13^2+\cdots+20^2$. To apply formula (5.1.5), which tells us how to evaluate the sum of consecutive squares when the first term is 1^2, we must first rewrite the given sum:

$$\begin{aligned} 10^2+11^2+12^2+13^2+\cdots+20^2 &= \sum_{j=10}^{20} j^2 \\ &= \left(\sum_{j=1}^{20} j^2\right) - \sum_{j=1}^{9} j^2 \\ &= \frac{20\cdot 21\cdot 41}{6} - \frac{9\cdot 10\cdot 19}{6} \\ &= 2585. \end{aligned}$$ ◀

The sum in the next example is called a *collapsing sum* or *telescoping sum*.

▶ **EXAMPLE 4** Suppose that $x_0, x_1, x_2, \ldots, x_N$ are points in the domain of a function F. Show that

$$\sum_{j=1}^{N} \Big(F(x_j) - F(x_{j-1})\Big) = F(x_N) - F(x_0). \tag{5.1.6}$$

Solution We can see the pattern from the cases $N=2$ and $N=3$:

$$\sum_{j=1}^{2} \Big(F(x_j) - F(x_{j-1})\Big) = \Big(F(x_1) - F(x_0)\Big) + \Big(F(x_2) - F(x_1)\Big) = F(x_2) - F(x_0)$$

and

$$\sum_{j=1}^{3}\Big(F(x_j)-F(x_{j-1})\Big)=\Big(F(x_1)-F(x_0)\Big)+\Big(F(x_2)-F(x_1)\Big)+\Big(F(x_3)-F(x_2)\Big)$$
$$=F(x_3)-F(x_0). \qquad (5.1.7)$$

Notice that, in every pair of consecutive summands, the second term of the second summand cancels with the first term of the first summand. For example, look at the second summand, $(F(x_2)-F(x_1))$, on the right side of equation (5.1.7). Observe that the term $F(x_2)$ cancels with the term $-F(x_2)$ of the succeeding summand, and the term $-F(x_1)$ cancels with the term $F(x_1)$ of the preceding summand. When all the cancellations are performed, the only terms that remain are $-F(x_0)$, which does not have a predecessor with which to cancel, and $F(x_N)$, which does not have a successor with which to cancel. ◀

Approximation of Area

Let f be a positive, continuous function defined on an interval $[a,b]$. Let $\mathcal{R}$ be the region lying above the x-axis, below the graph of the function $y=f(x)$, and between the vertical lines $x=a$ and $x=b$, as shown earlier in Figure 2. Our goal is to determine the area of $\mathcal{R}$. As a first step, we will approximate $\mathcal{R}$ by rectangles. The terminology and notation we now introduce will describe and denote the bases of the approximating rectangles.

DEFINITION Let N be a positive integer. The uniform partition $\mathcal{P}$ of order N of the interval $[a,b]$ is the set $\{x_0, x_1, \ldots, x_N\}$ of $N+1$ equally spaced points such that

$$a=x_0<x_1<x_2<\cdots<x_{N-1}<x_N=b.$$

See Figure 5.

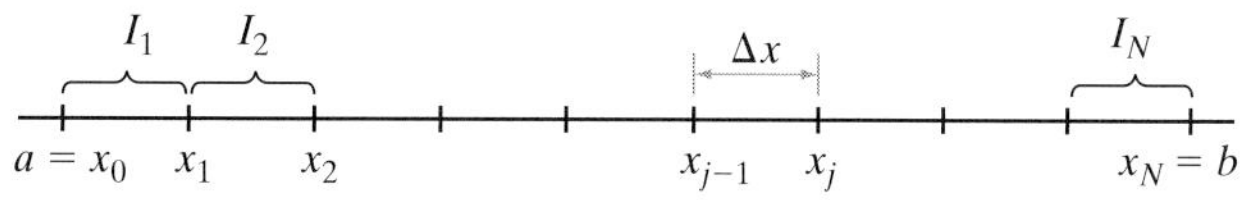

▲ **Figure 5** In the uniform partition of order N of the interval $[a,b]$, each subinterval has width $\Delta x=(b-a)/N$.

If $\mathcal{P}$ is a uniform partition of order N of $[a,b]$, then the points of $\mathcal{P}$ divide $[a,b]$ into N subintervals,

$$I_1=[x_0,x_1],\quad I_2=[x_1,x_2],\quad \ldots,\quad I_N=[x_{N-1},x_N],$$

of equal length Δx, where

$$\Delta x=\frac{b-a}{N}. \qquad (5.1.8)$$

Refer again to Figure 5. Notice that each point x_j of $\mathcal{P}$ with $j\geq 1$ can be obtained from the preceding point x_{j-1} by adding Δx:

$$x_j=x_{j-1}+\Delta x \qquad (1\leq j\leq N). \qquad (5.1.9)$$

Formula (5.1.9) is convenient for successively calculating the points of $\mathcal{P}$. Alternatively, each point x_j of $\mathcal{P}$ can be expressed in terms of its distance $j \cdot \Delta x$ from a:

$$x_j = a + j \cdot \Delta x \qquad (0 \le j \le N). \tag{5.1.10}$$

To approximate the area of region $\mathcal{R}$ of Figure 2, we choose a positive integer N and erect a rectangle over each of the subintervals $I_1, I_2, \ldots, I_N$ that arise from the uniform partition of order N. How tall should each rectangle be? We want each rectangle to be tall enough to touch the graph. One convenient way to achieve this is to take the height of the jth rectangle—that is, the rectangle that has subinterval $I_j = [x_{j-1}, x_j]$ as its base—to be $f(x_j)$, as shown in Figure 6. Thus the jth rectangle (shaded most darkly in Figure 6) has height equal to $f(x_j)$, base equal to Δx, and area equal to the product $f(x_j) \cdot \Delta x$ of the height and the base. The sum of all the rectangular areas,

$$\sum_{j=1}^{N} f(x_j) \cdot \Delta x, \tag{5.1.11}$$

is said to be a *right endpoint approximation* of the area of the region bounded above by the curve $y = f(x)$, below by the x-axis, and on the left and right by the vertical lines $x = a$ and $x = b$.

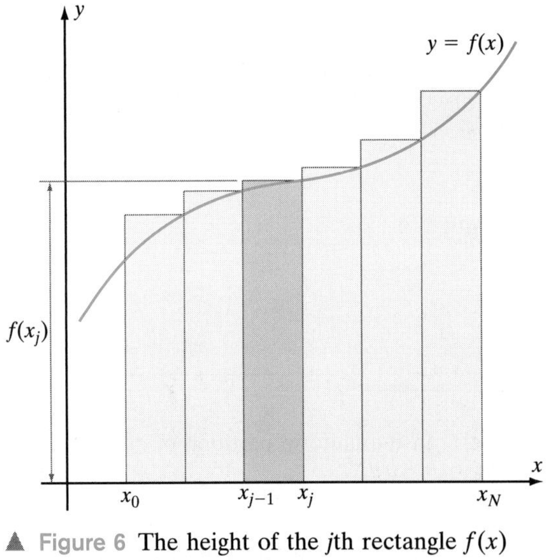

▲ Figure 6 The height of the jth rectangle $f(x)$

▶ **EXAMPLE 5** Using the uniform partition of order 6 of the interval $[1, 4]$, find the right endpoint approximation to the area A of the region bounded above by the graph of $f(x) = x^2 + x$, below by the x-axis, on the left by the vertical line $x = 1$, and on the right by the vertical line $x = 4$.

Solution Making the correspondence between the given data and the general notation used in formula (5.1.11), we have $a = 1, b = 4, N = 6$, and $\Delta x = (b - a)/N$

$= (4-1)/6 = 1/2$. The first point of the specified partition is $x_0 = a = 1$, and then we calculate successively

$$x_1 = x_0 + \Delta x = 1 + \frac{1}{2} = \frac{3}{2}, \qquad x_2 = x_1 + \Delta x = \frac{3}{2} + \frac{1}{2} = 2, \qquad x_3 = x_2 + \Delta x = 2 + \frac{1}{2} = \frac{5}{2},$$

$$x_4 = x_3 + \Delta x = \frac{5}{2} + \frac{1}{2} = 3, \qquad x_5 = x_4 + \Delta x = 3 + \frac{1}{2} = \frac{7}{2}, \qquad x_6 = x_5 + \Delta x = \frac{7}{2} + \frac{1}{2} = 4.$$

These points divide the interval $[1, 4]$ into these six subintervals:

$$I_1 = \left[1, \frac{3}{2}\right], \qquad I_2 = \left[\frac{3}{2}, 2\right], \qquad I_3 = \left[2, \frac{5}{2}\right],$$

$$I_4 = \left[\frac{5}{2}, 3\right], \qquad I_5 = \left[3, \frac{7}{2}\right], \qquad I_6 = \left[\frac{7}{2}, 4\right].$$

Each subinterval is the base of a rectangle that is used in the approximation. Formula (5.1.11) gives us the right endpoint approximation:

$$\begin{aligned}\sum_{j=1}^{6} f(x_j)\cdot\Delta x &= f(x_1)\cdot\Delta x + f(x_2)\cdot\Delta x + f(x_3)\cdot\Delta x + f(x_4)\cdot\Delta x + f(x_5)\cdot\Delta x + f(x_6)\cdot\Delta x \\ &= \left(\left(\frac{3}{2}\right)^2 + \frac{3}{2}\right)\cdot\left(\frac{1}{2}\right) + \left(2^2+2\right)\cdot\left(\frac{1}{2}\right) + \left(\left(\frac{5}{2}\right)^2 + \frac{5}{2}\right)\cdot\left(\frac{1}{2}\right) \\ &\quad + \left(3^2+3\right)\cdot\left(\frac{1}{2}\right) + \left(\left(\frac{7}{2}\right)^2 + \frac{7}{2}\right)\cdot\left(\frac{1}{2}\right) + \left(4^2+4\right)\cdot\left(\frac{1}{2}\right) \\ &= \frac{265}{8} \\ &= 33.125.\end{aligned}$$

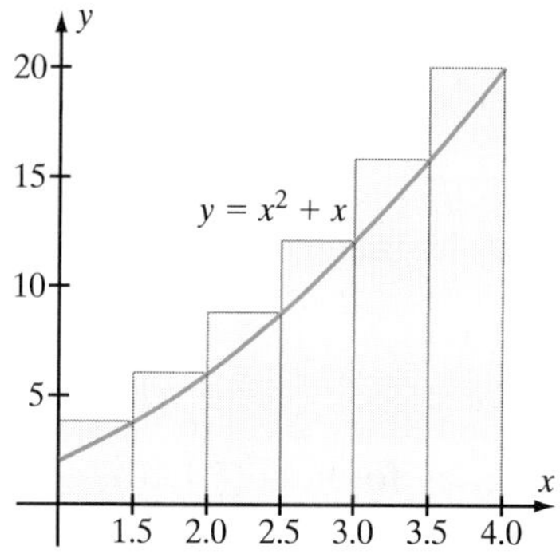

▲ Figure 7

This quantity, the sum of the areas of the shaded rectangles shown in Figure 7, is the requested approximation of A. Figure 7 shows that this estimate is not very accurate (as is to be expected when Δx is not very small). Exercise 46 asks you to show that $A = 57/2$ (exactly) using techniques that are developed in Examples 6 and 7 of this section. ◀

A Precise Definition of Area

Figures 3 and 4, shown earlier, suggest that, as N gets larger, the sums in formula (5.1.11) tend to approximate the area of the region under the graph of f more and more accurately. Actually, that area is not yet a defined concept. Our primitive notion of area allows us to compare the areas of different regions—to say that one region has greater area than another or that two regions have equal areas or nearly equal areas. However, our intuition does not help us assign numerical values to the areas of curved regions. To do that, we approximate the region by rectangles, sum the area of those rectangles, as in formula (5.1.11), and take the limit of these sums as the common length of the bases of the approximating rectangles decreases to 0.

DEFINITION The area A of the region that is bounded above by the graph of f, below by the x-axis, and laterally by the vertical lines $x = a$ and $x = b$, is defined as the limit

$$A = \lim_{N\to\infty} \sum_{j=1}^{N} f(x_j) \cdot \Delta x. \qquad (5.1.12)$$

As defined in Section 2.5, this means that, for every $\varepsilon > 0$, we can force the area of the rectangular approximation, namely $\sum_{j=1}^{N} f(x_j) \cdot \Delta x$, to be within ε of the quantity A by taking a sufficiently large number N of rectangles.

INSIGHT Suppose that we did not define the area to be this exotic-looking limit. What would the area of a region bounded by curves *mean*? The answer is that it would not mean anything—there would be no prior definition for this concept. If we wish to consider area, we must first define the concept. Having done so, we must show that our intuitive ideas about area are consistent with the definition we have adopted.

Granted that *some* definition is needed, we may still question the particular definition adopted. Does limit (5.1.12) exist? *Yes*, for continuous functions, it does exist. Would the limit be different if we had determined the height of the rectangle with base $I_j = [x_{j-1}, x_j]$ by using some point in I_j other than the right endpoint x_j? *No*, the limit would be the same. The left endpoint, the midpoint, or any other point of I_j would serve just as well in limit (5.1.12). We discuss these issues in greater detail in Section 5.2.

We now use this definition of the area concept to calculate some areas.

▶ **EXAMPLE 6** Calculate the area A of the region that is bounded above by the graph of $y = 2x$, below by the x-axis, and on the sides by the vertical lines $x = 0$ and $x = 6$.

Solution In this example, we take $a = 0$, $b = 6$, and $f(x) = 2x$. When we divide the interval $[0, 6]$ into N equal subintervals, the common length is $\Delta x = (6-0)/N = 6/N$. The points of the uniform partition of order N are, by formula (5.1.10),

$$x_j = a + j \cdot \Delta x = 0 + j \cdot \frac{6}{N} = \frac{6j}{N} \qquad (0 \le j \le N).$$

The next step is to simplify the sum that appears on the right side of definition (5.1.12). We calculate

$$\begin{aligned}\sum_{j=1}^{N} f(x_j) \cdot \Delta x &= \sum_{j=1}^{N} f\left(\frac{6j}{N}\right) \cdot \frac{6}{N} \\ &= \sum_{j=1}^{N} 2 \cdot \frac{6j}{N} \cdot \frac{6}{N} \\ &= \frac{72}{N^2} \cdot \sum_{j=1}^{N} j \\ &\overset{(5.1.4)}{=} \frac{72}{N^2} \cdot \frac{N(N+1)}{2}\end{aligned}$$

$$= 36\frac{N+1}{N}$$

$$= 36\left(1 + \frac{1}{N}\right).$$

The values of this expression are tabulated for seven choices of N:

N	10	20	100	1000	10 000	100 000	1 000 000
36 (1 + 1/N)	39.6	37.8	36.36	36.036	36.0036	36.00036	36.000036

Figure 8 shows the approximation for $N = 20$. The area of the region inside the 20 approximating rectangles is, according to the table, 37.8. Moving to the right along the second row of the table, we see that the approximations become closer to 36 as N increases. We calculate the *exact* area using definition (5.1.12):

$$A = \lim_{N\to\infty} \sum_{j=1}^{N} f(x_j) \cdot \Delta x = \lim_{N\to\infty} 36\left(1 + \frac{1}{N}\right) = 36(1 + 0) = 36. \quad ◀$$

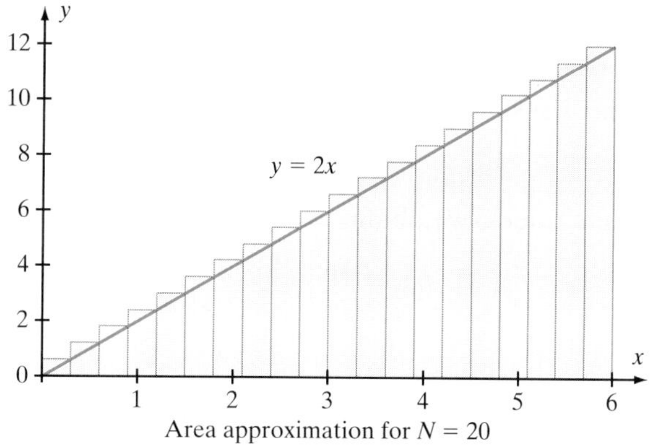

▲ Figure 8

INSIGHT The test of any new concept is twofold: Is it consistent with simpler concepts when those simpler concepts are applicable? Does it succeed in any cases for which the simpler concepts fail? In Example 6, the region for which we calculated area is a right triangle with base 6 and altitude 12. The familiar formula for the area of a triangle, half the base times the height, tells us that, for the triangular region of Example 6, $A = (1/2)(6)(12)$, or $A = 36$. Thus our new definition of area agrees with the standard definition in this example. In the next example, we use limit (5.1.12) to evaluate an area that *requires* the limit concept.

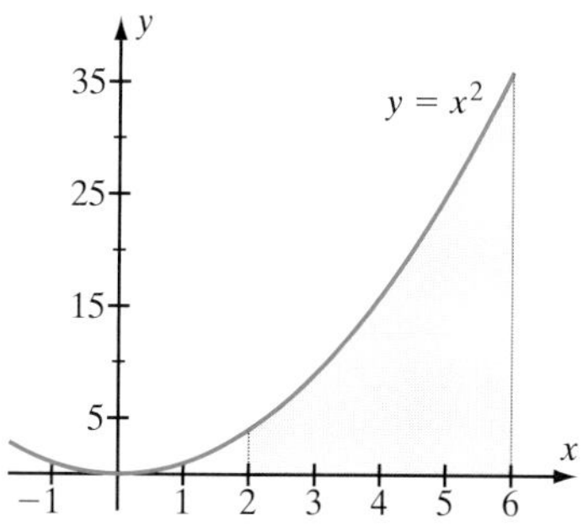

▲ Figure 9

▶ **EXAMPLE 7** Calculate the area A of the region (shaded in Figure 9) that is under the graph of $f(x) = x^2$, above the x-axis, and between the vertical lines $x = 2$ and $x = 6$.

Solution Here $a = 2$, and $b = 6$. We divide the interval $[2, 6]$ into N subintervals of equal length $\Delta x = (b - a)/N = (6 - 2)/N = 4/N$ by means of the uniform partition with points

$$x_j = a + j \cdot \Delta x = 2 + j \cdot \frac{4}{N} = 2 + \frac{4j}{N} \qquad (0 \leq j \leq N).$$

Using definition (5.1.12) for the area, we calculate

$$\begin{aligned}
A &= \lim_{N\to\infty} \sum_{j=1}^{N} f(x_j) \cdot \Delta x \\
&= \lim_{N\to\infty} \sum_{j=1}^{N} f\left(2 + \frac{4j}{N}\right) \cdot \frac{4}{N} \\
&= \lim_{N\to\infty} \sum_{j=1}^{N} \left(2 + \frac{4j}{N}\right)^2 \cdot \frac{4}{N} \\
&= \lim_{N\to\infty} \sum_{j=1}^{N} \left(4 + \frac{16j}{N} + \frac{16j^2}{N^2}\right) \cdot \frac{4}{N} \\
&= \lim_{N\to\infty} \sum_{j=1}^{N} \left(\frac{16}{N} + \frac{64}{N^2} j + \frac{64}{N^3} j^2\right) \\
&= \lim_{N\to\infty} \left(\frac{16}{N} \cdot \sum_{j=1}^{N} 1 + \frac{64}{N^2} \cdot \sum_{j=1}^{N} j + \frac{64}{N^3} \cdot \sum_{j=1}^{N} j^2\right).
\end{aligned}$$

The first sum on the right side is $\underbrace{1 + 1 + \cdots + 1}_{N \text{terms}}$, or N. Formulas (5.1.4) and (5.1.5) give us the values of the other two sums. When we replace these three sums with their values, we obtain

$$\begin{aligned}
A &= \lim_{N\to\infty} \left(\frac{16}{N} \cdot N + \frac{64}{N^2} \cdot \frac{N(N+1)}{2} + \frac{64}{N^3} \cdot \frac{N(N+1)(2N+1)}{6}\right) \\
&= \lim_{N\to\infty} \left(16 + 32 \cdot \frac{(N+1)}{N} + \frac{32}{3} \cdot \frac{(N+1)}{N} \cdot \frac{(2N+1)}{N}\right) \\
&= \lim_{N\to\infty} \left(16 + 32 \cdot \left(1 + \frac{1}{N}\right) + \frac{32}{3} \cdot \left(1 + \frac{1}{N}\right) \cdot \left(2 + \frac{1}{N}\right)\right) \\
&= \lim_{N\to\infty} \left(16 + 32 \cdot 1 + \frac{32}{3} \cdot 1 \cdot 2\right) \\
&= \frac{208}{3}. \quad ◀
\end{aligned}$$

Concluding Remarks

The method that we have used to define area can be used to attack many different types of problems in addition to the calculation of area. For example, later in this chapter, we will apply the ideas of this section to analyze income distribution and compute cardiac output. In Chapter 7, we will see that the right side of equation (5.1.12) can be used to calculate other geometric quantities such as volume, arc length, and surface area, as well as physical quantities such as work and center of mass. In the rest of the present chapter, we develop a general tool, called the *Riemann integral*, that enables us to study all of these types of problems. We also investigate a method of evaluating Riemann integrals that does not require the elaborate and tedious algebra used to solve Example 7.

QUICK QUIZ

1. Calculate $\sum_{j=0}^{3}\cos(j\pi/4)$.
2. Calculate $\sum_{j=0}^{3}(2j+j^2+2^j)$.
3. What is the fifth subinterval of $[3,9]$ when the uniform partition of order 7 is used?
4. What is the right endpoint approximation of the area of the region under the graph of $y=1/x$ and over the interval $[1,2]$ when the uniform partition of order 3 is used?

Answers

1. 1 2. 41 3. [45/7, 51/7] 4. 37/60

EXERCISES

Problems for Practice

In Exercises 1–10, write out the sum, and perform the addition.

1. $\sum_{j=1}^{4}3j$
2. $\sum_{j=0}^{6}(2j-1)$
3. $\sum_{j=2}^{5}(-2j^2)$
4. $\sum_{\ell=4}^{6}\ell/(\ell-3)$
5. $\sum_{n=2}^{5}2n/(n-1)$
6. $\sum_{k=2}^{4}(k^3-6k)$
7. $\sum_{m=3}^{6}(2m^2-3m)$
8. $\sum_{j=1}^{5}j\cdot\sin(j\pi/2)$
9. $\sum_{j=1}^{3}j\cdot\sin^2(j\pi/6)$
10. $\sum_{j=0}^{4}1/(2j-3)$

In Exercises 11–16, use summation notation to express the sum.

11. $2+3+4+5+6$
12. $3+6+9+12+15$
13. $9+13+17+21+25+29$
14. $9+16+25+36$
15. $1/4+1/5+1/6+1/7+1/8$
16. $2/5+3/7+4/9+5/11$

Use formulas (5.1.4) and (5.1.5) to calculate the sums in Exercises 17–22.

17. $\sum_{j=1}^{12}j^2$
18. $\sum_{j=1}^{12}(j^2-4)$
19. $\sum_{j=1}^{12}(j^2+2j)$
20. $\sum_{j=1}^{10}(3j^2+2j+1)$
21. $\sum_{j=12}^{24}(3j-2)$
22. $\sum_{j=7}^{16}(2j^2-21j)$

In Exercises 23–26, use an identity to simplify the sum.

23. $\sum_{j=0}^{19}\exp(j)$
24. $\sum_{j=0}^{5}\frac{2^{2j-1}}{3^{j+1}}$
25. $\sum_{j=2}^{6}\ln(j)$
26. $\sum_{j=7}^{27}\ln\left(\frac{j+1}{j}\right)$

In each of Exercises 27–38, calculate the right endpoint approximation of the area of the region that lies below the graph of the given function f and above the given interval I of the x-axis. Use the uniform partition of given order N.

27. $f(x)=x^2-2x$ $I=[3,5]$, $N=2$
28. $f(x)=1/x$ $I=[2,3]$, $N=2$
29. $f(x)=-\log_2(x)$ $I=[1/2,1]$, $N=2$
30. $f(x)=2x^2+1$ $I=[0,9/2]$, $N=3$
31. $f(x)=2+\sin(2x)$ $I=[\pi/2,2\pi]$, $N=3$
32. $f(x)=4-3\cos(x)$ $I=[0,4\pi]$, $N=3$
33. $f(x)=x/(x+1)$ $I=[-4,-2]$, $N=4$
34. $f(x)=2x/(2x-1)$ $I=[-3,-1]$, $N=4$
35. $f(x)=x\sin(x)$ $I=[-\pi,\pi]$, $N=4$
36. $f(x)=\sec(x)$ $I=[-\pi/3,\pi/3]$, $N=4$
37. $f(x)=x^3-6x+6$ $I=[-1,5/2]$, $N=7$
38. $f(x)=x+\cos(2x)$ $I=[0,3\pi]$, $N=6$

Further Theory and Practice

Suppose that $f(x)\geq 0$ for x in $I=[a,b]$. If for each subinterval $[x_{j-1},x_j]$ that arises from the uniform partition $\mathcal{P}=\{x_0,x_1,\ldots,x_N\}$ of I, we use the left endpoint x_{j-1} instead of the right endpoint in formula (5.1.9), then we obtain the *left endpoint approximation*

$$\sum_{j=1}^{N} f(x_{j-1}) \cdot \Delta x$$

of the area A of the region below the graph of f and above I (see Figure 10). As can be seen from Figure 10, if f increases on I, then the right endpoint approximation overestimates A, and the left endpoint approximation underestimates A. In each of Exercises 39–44, calculate the average of the left and right endpoint approximations. (For purposes of comparison, the exact value of A is given. Notice that your answer is more accurate than both the left and right endpoint approximations.) ◀

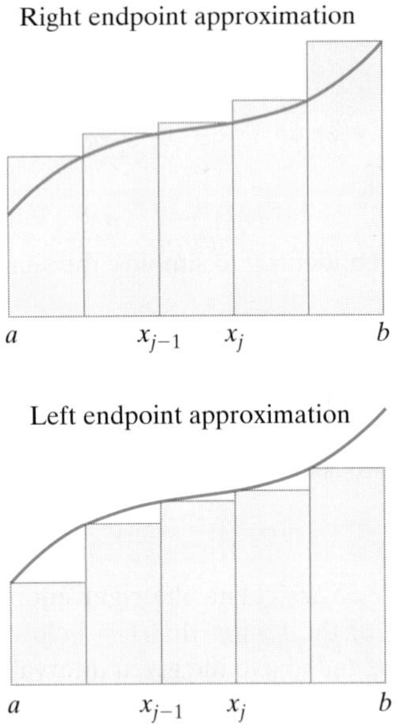

▲ Figure 10

39. $f(x) = \sin(x)$ $\quad I = [0, \pi/3]$, $N = 2$, $A = 1/2$
40. $f(x) = 2 + \tan(x)$ $\quad I = [-\pi/4, \pi/4]$, $N = 2$, $A = \pi$
41. $f(x) = (2x+1)/(x+1)$ $\quad I = [0, 2]$, $N = 2$, $A = 4 - \ln(3)$
42. $f(x) = 1 + x^3$ $\quad I = [-1, 3]$, $N = 3$, $A = 24$
43. $f(x) = 1 - 1/x^2$ $\quad I = [1, 2]$, $N = 3$, $A = 1/2$
44. $f(x) = 16^x$ $\quad I = [0, 1]$, $N = 4$, $A = 15/\ln(16)$

45. Following the technique from Example 7, calculate the area of the region that lies below the graph of $f(x) = 4 - x^2$ and above the interval $[0, 2]$ of the x-axis.

46. Following the technique from Examples 6 and 7, show that the area of the region that lies below the graph of $f(x) = x^2 + x$ and above the interval $[1, 4]$ of the x-axis is $57/2$.

47. Find p such that $e^p = e \cdot e^2 \cdot e^3 \ldots e^{100}$. Find q so that $\ln(q) = \sum_{n=1}^{100} \ln(n)$.

48. Calculate the sum $S = \sum_{j=1}^{N} (2j-1)$ of the first N odd positive integers.

49. Calculate the sum $S = \sum_{j=1}^{N} (2j)^2$ of the first N even positive square integers. Subtract your value of S from the value of $\sum_{j=1}^{2N} j^2$ to calculate the sum $\sum_{j=1}^{N} (2j-1)^2$ of the first N odd positive square integers.

50. Which is larger, $\sum_{j=1}^{N} j^2$ or $\sum_{j=1}^{N^2} j$? Explain why.

51. Suppose that $\{a_j\}_{j=0}^{\infty}$ is a sequence of real numbers. Suppose that M and N are integers such that $0 < M < N$. Show that

$$\sum_{j=M}^{N} (a_j - a_{j-1}) = a_N - a_{M-1}.$$

Hint: Look at the cases $N = M+1$ and $N = M+2$ to determine the pattern.

52. Evaluate $1/2 + 1/6 + 1/12 + \cdots + 1/(1000 \cdot 1001)$ by writing the jth term as

$$\frac{1}{j \cdot (j+1)} = \frac{1}{j} - \frac{1}{j+1}$$

and using Exercise 51.

53. Write

$$S = 1 + 2 + 3 + \cdots + N$$

forward and backward:

$$\begin{array}{ccccccc} S & = & 1 & + & 2 & + \cdots + & N \\ S & = & N & + & N-1 & + \cdots + & 1. \\ \hline 2S & = & N+1 & + & N+1 & + \cdots + & N+1 \end{array}$$

Add vertically. Derive formula (5.1.4) by solving for S.

54. Derive formula (5.1.4) by noticing that

$$\sum_{j=1}^{N} (2j-1) = \sum_{j=1}^{N} (j^2 - (j-1)^2).$$

Write the sum on the left in terms of $S = 1 + 2 + 3 + \cdots + N$. Evaluate the sum on the right using Exercise 51. Solve for S.

55. Derive formula (5.1.5) by noticing that

$$\sum_{j=1}^{N} (3j^2 - 3j + 1) = \sum_{j=1}^{N} (j^3 - (j-1)^3).$$

Write the expression on the left as a combination of three sums, one of which is $3S = 3(1^2 + 2^2 + 3^2 + \cdots + N^2)$. Use formula (5.1.4) to evaluate $\sum_{j=1}^{N} 3j$. Evaluate the sum on the right using Exercise 51. Solve for S.

56. Suppose that $b > 0$. Following the technique from Example 6, show that the area of the region that lies below the graph of $f(x) = x^2$ and above the interval $[0, b]$ of the x-axis is $b^3/3$.

57. Let $f(x) = x^2$. Suppose that $b > 0$ and that $F(x)$ is an antiderivative of $f(x)$. In this exercise, we will approximate the area A of the region that lies below the graph of f and above the interval $I = [0, b]$ of the x-axis. We will use one rectangle with base I. For the height, we use $f(c)$ where c is the point in $(0, b)$ obtained by applying the Mean Value Theorem to F on I—see Figure 11. Use Exercise 56 to show that this approximation is, in fact, exact.

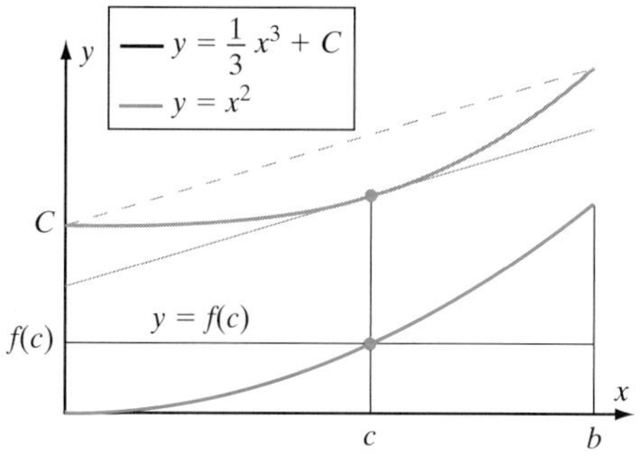

Figure 11

58. Show that

$$\sum_{j=1}^{N}(4j^3 - 6j^2 + 4j - 1) = \sum_{j=1}^{N}(j^4 - (j-1)^4).$$

Evaluate the sum on the right using Exercise 51. Write the expression on the left as a combination of four sums, one of which is $4S = 3(1^3 + 2^3 + 3^3 + \cdots + N^3)$. Use formula (5.1.4) to evaluate $\sum_{j=1}^{N} 4j$. Use formula (5.1.5) to evaluate $\sum_{j=1}^{N} 6j^2$. Solve for S to obtain the identity

$$\sum_{j=1}^{N} j^3 = \left(\frac{N(N+1)}{2}\right)^2.$$

59. Suppose $0 \le a < b$. Following the technique of Example 7, use the formula from Exercise 58 to show that the area under the curve $y = x\, y = x^3$ from $x = 0$ to $x = b$ is $b^4/4$. Deduce that the area under the curve $y = x^3$ from $x = a$ to $x = b$ is $(b^4 - a^4)/4$.

60. Suppose m and k are positive constants. For $f(x) = mx + k$, let $A(b)$ denote the area under the graph of f, above the x-axis, and between $x = 0$ and $x = b$. Calculate $A(b)$, and show that $A'(b) = f(b)$.

61. For $f(x) = x^2$, let $A(b)$ denote the area under the graph of f, above the x-axis, and between $x = 0$ and $x = b$. Show that $A'(b) = f(b)$.

62. Notice that, in limit (5.4), $\lim_{N\to\infty} \Delta x = 0$. It may seem that $\lim_{N\to\infty} \sum_{j=1}^{N} f(x_j)\Delta x = 0$ in view of the following calculation:

$$\begin{aligned}\lim_{N\to\infty} \sum_{j=1}^{N} f(x_j)\Delta x &= \lim_{N\to\infty} \sum_{j=1}^{N} f(x_j) \lim_{N\to\infty} \Delta x \\ &= \left(\lim_{N\to\infty} \sum_{j=1}^{N} f(x_j)\right) \cdot 0 \\ &= 0.\end{aligned}$$

Explain what is wrong with this reasoning.

63. For any positive integer k, let

$$S_N(k) = 1^k + 2^k + 3^k + \cdots + N^k.$$

It is known that

$$S_N(k) = \frac{1}{k+1} N^{k+1} + P_k(N)$$

where P_k is a polynomial of degree k. Accept this fact without proof (but notice that for $k = 1$, 2, and 3, this assertion follows from formula (5.1.4), formula (5.1.5), and Exercise 58, respectively).

a. Show that $\lim_{N\to\infty} S_N(k)/N^{k+1} = 1/(k+1)$.

b. Suppose $0 \le a < b$. Show that the area under the curve $y = x^k$ from $x = 0$ to $x = b$ is $b^{k+1}/(k+1)$.

c. Deduce that the area under the curve $y = x^k$ from $x = a$ to $x = b$ is $(b^{k+1} - a^{k+1})/(k+1)$.

64. Calculate $\lim_{N\to\infty} (1/N)\sum_{j=1}^{N} \sqrt{1 - (j/N)^2}$ by identifying this number as the limit of right endpoint approximations of the area of a region with known area.

65. Let $f\colon [0, a] \to [0, b]$ be a continuous, increasing, onto function. Let A denote the area in the xy-plane that lies under the graph of $y = f(x)$ and above the interval [0, a] of the x-axis. What is the area in the yx-plane that lies under the graph of $x = f^{-1}(y)$ and above the interval $[0, b]$ of the y-axis?

66. Use Exercise 51 and the product formula

$$\sin(A)\cos(B) = \frac{1}{2}\Big(\sin(A+B) + \sin(A-B)\Big)$$

with $A = t/2$ and $B = kt$ to show that

$$\sum_{k=1}^{N} \sin\left(\frac{t}{2}\right)\cos(kt) = \frac{1}{2}\left(\sin\left(\left(N + \frac{1}{2}\right)t\right) - \sin\left(\frac{1}{2}t\right)\right).$$

Use this summation formula, together with the product formula, with $A + B = (N + 1/2)t$ and $A - B = -t/2$ to show that

$$\sum_{k=1}^{N} \cos(kt) = \frac{\sin(Nt/2)\cos((N+1)t/2)}{\sin(t/2)}.$$

67. Suppose $0 < b \le \pi/2$. Use the last formula from Exercise 66 with $t = b/N$ to show that the area of the region under the graph of $y = \cos(x)$ and over the interval $[0, b]$ is $\sin(b)$.

Calculator/Computer Exercises

In a computer algebra system, a right endpoint approximation can be implemented by means of a one-line command. For example, if the real numbers a and b, the positive integer N, and the function f, have been defined, then the command

```
evalf((b-a)/N*add(f(a+j*(b-a)/N),j=1..N));
```

calculates formula (5.1.9) in Maple. In Exercises 68–71, an interval $[a, b]$, and a function f are given. The function is positive

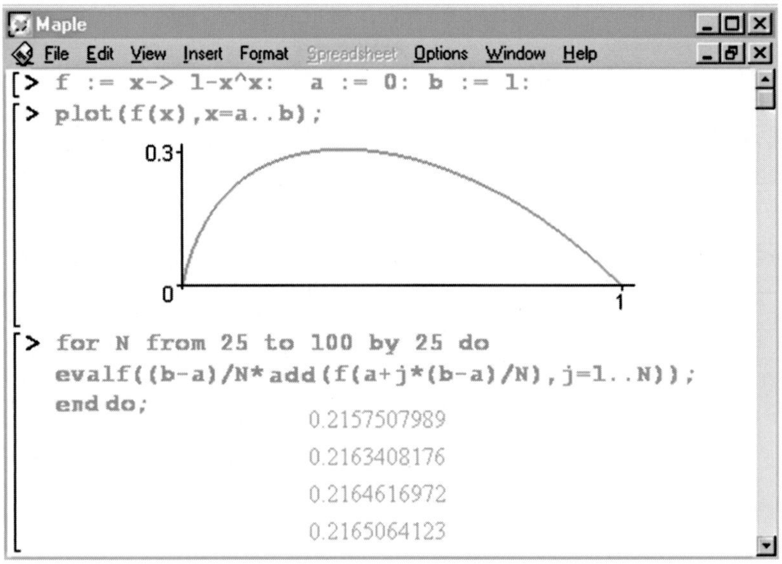

▲ Figure 12

on (a,b) and it is 0 at the endpoints. Approximate the area under $y=f(x)$ and over the interval $[a, b]$ by using the right endpoint approximation, starting with $N=25$. Increment N by 25 until the first two decimal places of the sum remain the same for three consecutive calculations. Figure 12 shows a Maple implementation for the function $f(x)=1-x^x, 0\le x\le 1$. (This procedure does not guarantee two decimal places of accuracy. Section 5.7 presents several methods that can be used to achieve a prescribed accuracy.) ◀

68. $f(x)=\sqrt{16-x^4},\ -2\le x\le 2$

69. $f(x)=\sqrt{x}-x, 0\le x\le 1$

70. $f(x)=(2-x^2)/(2+x^2),\ -\sqrt{2}\le x\le\sqrt{2}$

71. $f(x)=\sin^2(x), 0\le x\le\pi$

▶ In each of Exercises 72–75, locate the left x-intercept a and the right x-intercept b of the graph of the given function f. Then follow the instructions for Exercises 68–71. ◀

72. $f(x)=1+2x-x^4$

73. $f(x)=2+x-\exp(x)$

74. $f(x)=10-x^2-1/x^3$

75. $f(x)=x\exp(-x^2)-0.1$

5.2 The Riemann Integral

We now define a mathematical operation on functions that captures the essence of the method from Section 5.1, but that can be applied to several other situations as well.

Riemann Sums

Let f be *any* function defined on the interval $[a,b]$. (We no longer assume that f is a nonnegative function.) Let N be a positive integer. As in Section 5, the uniform partition of order N of the interval $[a,b]$ is the set of $N+1$ equally spaced points

$$x_j=a+j\cdot\frac{b-a}{N}\qquad(0\le j\le N)$$

that divides $[a,b]$ into the N subintervals

$$I_1=[x_0,x_1],\quad I_2=[x_1,x_2],\quad \ldots,\quad I_N=[x_{N-1},x_N]$$

of equal length

$$\Delta x = \frac{b-a}{N}.$$

A *choice of points* $\mathcal{S}_N$ associated with the uniform partition of order N is a sequence $s_1, s_2, \ldots, s_N$ of points with s_j in I_j for each j. The point s_j can be chosen arbitrarily from the subinterval I_j. Thus s_j can be a right endpoint or a left endpoint or *any* interior point. Figure 1 illustrates one possible choice of points. If $\mathcal{S}_N$ is a choice of points, then the expression

$$\mathcal{R}(f, \mathcal{S}_N) = \sum_{j=1}^{N} f(s_j) \cdot \Delta x \tag{5.2.1}$$

is called a *Riemann sum* of f. The notation $\mathcal{R}(f, \mathcal{S}_N)$ indicates that a Riemann sum depends on the function f and the choice of points $\mathcal{S}_N$.

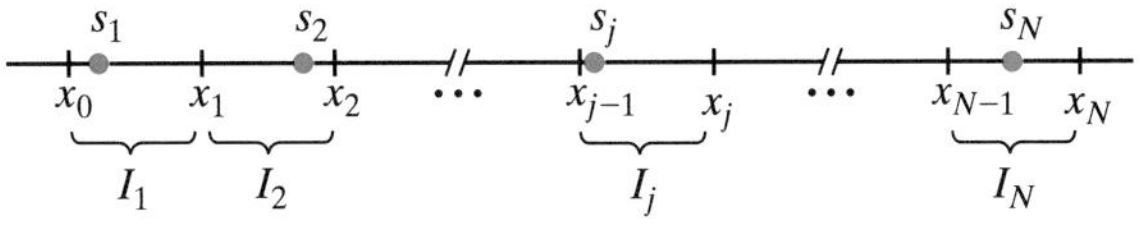

▲ Figure 1

▶ **EXAMPLE 1** Write a Riemann sum for the function $f(x) = x^2 - 4$ and the interval $[a, b] = [-5, 3]$ using the partition $\{-5, -3, -1, 1, 3\}$.

Solution Observe that

$$I_1 = [-5, -3], \quad I_2 = [-3, -1], \quad I_3 = [-1, 1], \quad I_4 = [1, 3],$$

and $\Delta x = 2$. To form a Riemann sum, we must first make a selection of points $s_j (1 \leq j \leq 4)$ with s_j in I_j. According to the definition, we may select *any* point s_1 in I_1, *any* point s_2 in I_2, and so on. Here is our *arbitrary* choice:

$$s_1 = -3 \in I_1, \quad s_2 = -\frac{3}{2} \in I_2, \quad s_3 = 0 \in I_3, \quad s_4 = 1 \in I_4.$$

The Riemann sum for this collection of points $\mathcal{S}_4$ is

$$\begin{aligned}
\mathcal{R}(f, \mathcal{S}) &= \sum_{j=1}^{4} f(s_j) \cdot \Delta x \\
&= f(-3) \cdot 2 + f\left(-\frac{3}{2}\right) \cdot 2 + f(0) \cdot 2 + f(1) \cdot 2 \\
&= (9-4) \cdot 2 + \left(\frac{9}{4} - 4\right) \cdot 2 + (0-4) \cdot 2 + (1-4) \cdot 2 \\
&= 10 - \frac{7}{2} - 8 - 6 \\
&= -\frac{15}{2}.
\end{aligned}$$

Of course, a different choice of points s_1, s_2, s_3, s_4 would lead to a different Riemann sum. ◀

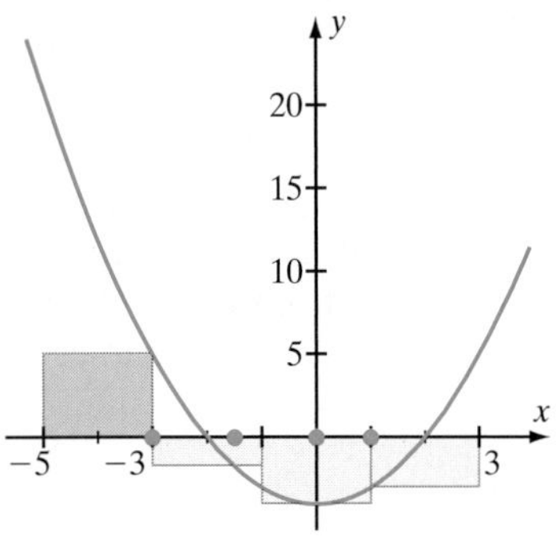

▲ Figure 2

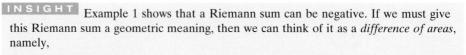

INSIGHT Example 1 shows that a Riemann sum can be negative. If we must give this Riemann sum a geometric meaning, then we can think of it as a *difference of areas*, namely,

$$\mathcal{R}(f,\mathcal{S}_4) = \underbrace{10}_{\text{Area of rectangle above } x\text{-axis}} - \underbrace{\left(\frac{7}{2}+8+6\right)}_{\text{Area of rectangles below } x\text{-axis}}.$$

See Figure 2. In later chapters, we will use Riemann sums to represent physical quantities other than area. In those cases, *it is usually not appropriate* to associate Riemann sums with the area concept.

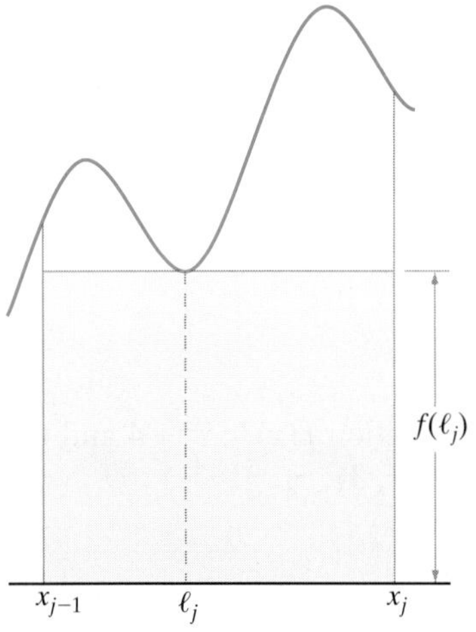

▲ Figure 3a

Suppose that f is continuous. Let ℓ_j be a point in I_j at which f attains its minimum value on I_j (see Figure 3a), and let u_j be a point in I_j at which f achieves its maximum value on I_j (see Figure 3b). The Extreme Value Theorem (Section 2.3, in Chapter 3) guarantees that such points ℓ_j and u_j exist. We denote the resulting choices of points associated to the uniform partition by $\mathcal{L}_N = \{\ell_1, \ell_2, \ldots, \ell_N\}$ and $\mathcal{U}_N = \{u_1, u_2, \ldots, u_N\}$. The Riemann sums

$$\mathcal{R}(f,\mathcal{L}_N) = \sum_{j=1}^{N} f(\ell_j)\Delta x$$

and

$$\mathcal{R}(f,\mathcal{U}_N) = \sum_{j=1}^{N} f(u_j)\Delta x$$

are respectively called the *lower Riemann sum* and *upper Riemann sum* of order N. They are the least and greatest Riemann sums of order N, as we will prove in Theorem 1. Figures 4 and 5 illustrate typical lower and upper Riemann sums.

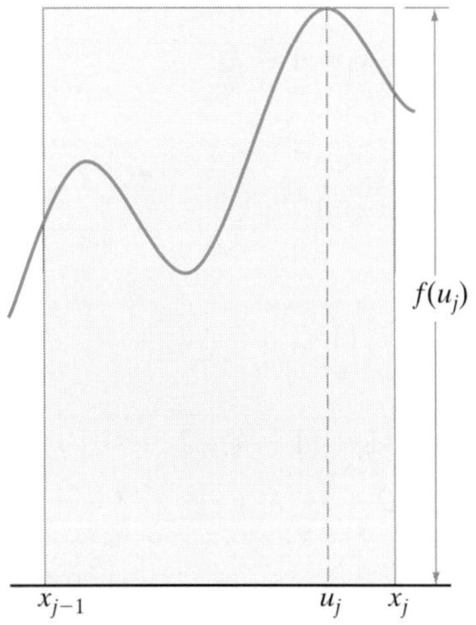

▲ Figure 3b

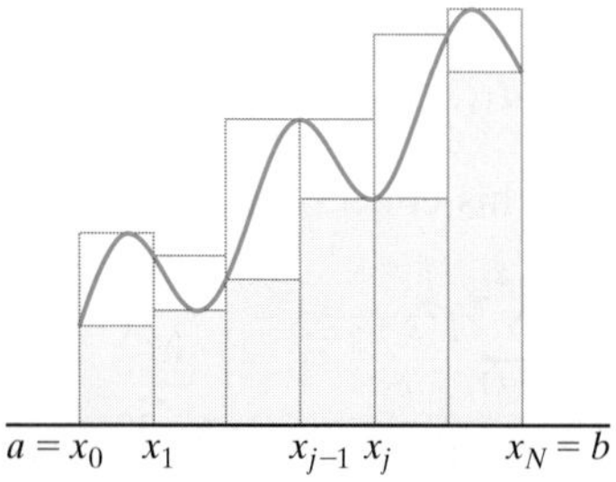

▲ Figure 4 The area of the shaded rectangles is a lower Riemann sum.

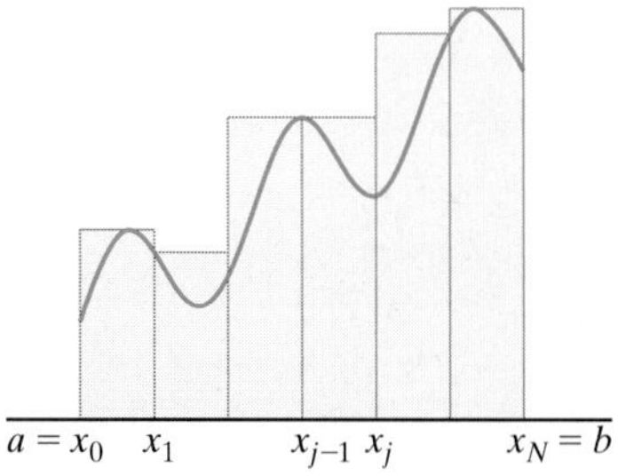

▲ Figure 5 The area of the shaded rectangles is an upper Riemann sum.

▶ **EXAMPLE 2** Consider the continuous function $f(x) = x^2 - 4$ on the interval $[-5,3]$ (as in Example 1). Find the upper and lower Riemann sums of order 4.

Solution The four equal-length subintervals of $[-5,3]$ are $I_1 = [-5,-3]$, $I_2 = [-3,-1]$, $I_3 = [-1,1]$, and $I_4 = [1,3]$. The graph of f is shown in Figure 6, in which the four subintervals have been separated. By inspection, f attains its maximum value on I_1 at $u_1 = -5$ and its minimum value at $\ell_1 = -3$. Continuing to the other three subintervals, we find that $u_2 = -3$, $\ell_2 = -1$, $u_3 = -1$, $\ell_3 = 0$, $u_4 = 3$, and $\ell_4 = 1$. (Because $f(x)$ is maximized on I_3 at both $x = -1$ and $x = 1$, we could equally well have set $u_3 = 1$.) Substituting these values into the formulas for the upper and lower Riemann sums, we obtain

$$\begin{aligned}\mathcal{R}(f,\mathcal{U}_4) &= \sum_{j=1}^{4} f(u_j)\Delta x \\ &= f(-5)\cdot 2 + f(-3)\cdot 2 + f(-1)\cdot 2 + f(3)\cdot 2 \\ &= 2(21 + 5 - 3 + 5) \\ &= 56\end{aligned}$$

and

$$\begin{aligned}\mathcal{R}(f,\mathcal{L}_4) &= \sum_{j=1}^{4} f(\ell_j)\Delta x \\ &= f(-3)\cdot 2 + f(-1)\cdot 2 + f(0)\cdot 2 + f(1)\cdot 2 \\ &= 2(5 - 3 - 4 - 3) \\ &= -10.\end{aligned}$$

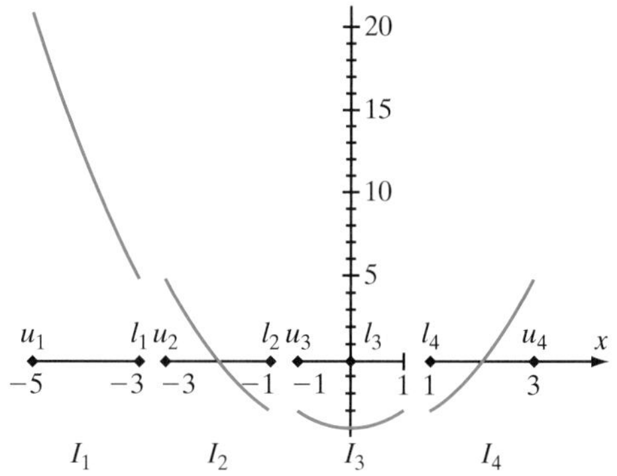

▲ Figure 6

INSIGHT Notice that the Riemann sum $\mathcal{R}(f, \mathcal{S}_4) = -15/2$ calculated in Example 1 lies between the upper and lower Riemann sums $\mathcal{R}(f, \mathcal{U}_4) = 56$ and $\mathcal{R}(f, \mathcal{L}_4) = -10$, which are calculated in Example 2. The following theorem generalizes this observation and states another key fact about upper and lower Riemann sums.

THEOREM 1 Suppose that f is continuous on an interval $[a, b]$.

a. If $\mathcal{S}_N = \{s_1, \cdots, s_N\}$ is any choice of points associated with the uniform partition of order N, then

$$\mathcal{R}(f,\mathcal{L}_N) \le \mathcal{R}(f,\mathcal{S}_N) \le \mathcal{R}(f,U_N). \tag{5.2.2}$$

b. The numbers $\mathcal{R}(f,\mathcal{L}_N)$ and $\mathcal{R}(f,\mathcal{U}_N)$ become arbitrarily close to each other for N sufficiently large. That is,

$$\lim_{N\to\infty} \big(\mathcal{R}(f,\mathcal{U}_N) - \mathcal{R}(f,\mathcal{L}_N)\big) = 0. \tag{5.2.3}$$

Proof. It follows from the definitions of ℓ_j and u_j that

$$f(\ell_j) \le f(s_j) \le f(u_j)$$

for every s_j in I_j. If we multiply each term by the positive quantity Δx and sum over $j = 1, 2, \ldots, N$, then we obtain

$$\sum_{j=1}^{N} f(\ell_j)\Delta x \le \sum_{j=1}^{N} f(s_j)\Delta x \le \sum_{j=1}^{N} f(u_j)\Delta x,$$

which is the assertion of Theorem 1a.

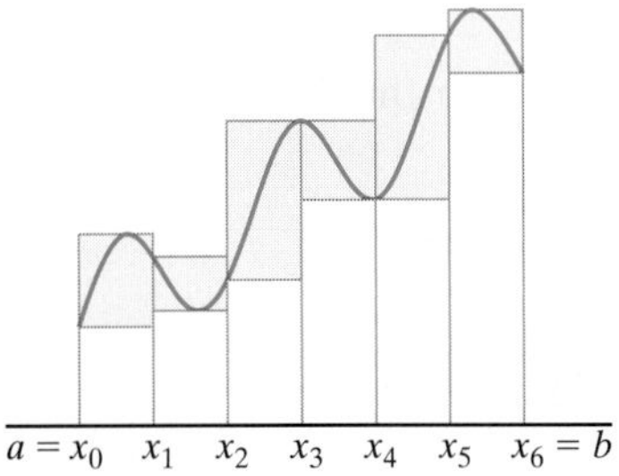

▲ Figure 7 The total area enclosed by the shaded rectangles is $\mathcal{R}(f, \mathcal{U}_6) - \mathcal{R}(f, \mathcal{L}_6)$.

Theorem 1b appears to be plausible when we look at examples. Refer again to Figures 4 and 5, which illustrate typical upper and lower Riemann sums. The area of the shaded rectangles in Figure 7 represents the difference $\mathcal{R}(f, \mathcal{U}_N) - \mathcal{R}(f, \mathcal{L}_N)$ for $N = 6$. In Figure 8, we increase the number of subintervals to 30. Observe that the difference between the upper and lower Riemann sums, again represented by the area of the shaded rectangles, is *significantly smaller.* Theorem 1b follows from an important property of continuous functions on closed, bounded intervals. According to that property, which is proved in texts on advanced calculus, given a positive ε, there is a positive δ such that $f(s)$ is within ε of $f(t)$, provided that s and t are points in $[a, b]$ within δ of each other. If we take N large enough so that $\Delta x = (b - a)/N < \delta$, then u_j and ℓ_j are within δ of each other for each j. Therefore

$$\begin{aligned} 0 \le \mathcal{R}(f, U_N) - \mathcal{R}(f, \mathcal{L}_N) &= \sum_{j=1}^{N} (f(u_j) - f(\ell_j)) \cdot \Delta x < \sum_{j=1}^{N} \varepsilon \cdot \Delta x \\ &= \varepsilon \sum_{j=1}^{N} \Delta x = \varepsilon \cdot (b - a). \end{aligned}$$

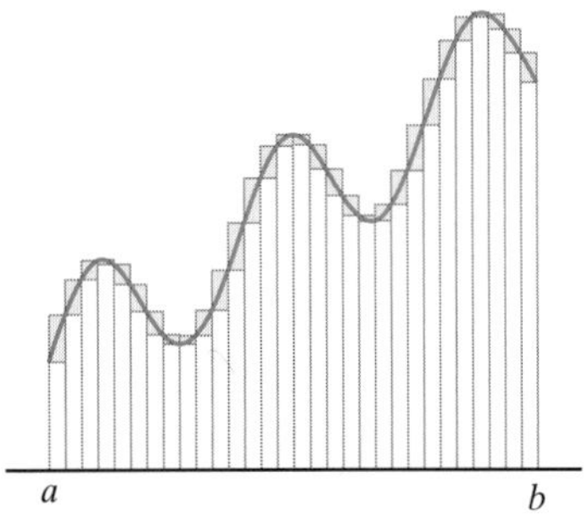

▲ Figure 8 The area of the shaded region decreases as the number N of subdivisions increases.

Because ε is arbitrary, we have shown that $\mathcal{R}(f, \mathcal{U}_N) - \mathcal{R}(f, \mathcal{L}_N)$ can be made arbitrarily small by taking N sufficiently large, which is Theorem 1b. ■

The Riemann Integral

In Section 5.1, we defined the area under the graph of a positive function to be the *limit* as N tends to infinity of right endpoint approximations. This definition will be the basis of our limit definition for general Riemann sums.

DEFINITION Suppose that f is a function defined on the interval $[a, b]$. We say that the Riemann sums $\mathcal{R}(f, \mathcal{S}_N)$ *tend to* the real number ℓ (or that ℓ is the *limit* of the Riemann sums $\mathcal{R}(f, \mathcal{S}_N)$) as N tends to infinity, if, for any $\varepsilon > 0$, there is a positive integer M such that

$$|\mathcal{R}(f, \mathcal{S}_N) - \ell| < \varepsilon$$

for all N greater than or equal to M and any choice of $\mathcal{S}_N$. If this is the case, we say that f is *integrable* on $[a, b]$, and we denote the limit ℓ by the symbol

$$\int_a^b f(x)\, dx.$$

This numerical quantity is called the *Riemann integral* of f on the interval $[a, b]$. The operation of going from the function f to the number $\int_a^b f(x)\, dx$ is called *integration.*

Let us examine the components of the notation for the Riemann integral. The elongated "S" symbol is called the *integral sign* and is used to remind us that the integral is a limit of *sums*. The left and right endpoints of the interval $[a, b]$ are called the *limits of integration*. We refer to a as the *lower limit of integration* and b as the *upper limit of integration*. The expression $f(x)$ is called the *integrand*. For now, the expression dx serves only to remind us of the variable of integration. We may think of dx as the infinitesimal limit of Δx as Δx tends to zero. In later work, dx will be a useful device for helping us transform integrals into new integrals.

Only the presence of limits of integration serves to distinguish the Riemann integral $\int_a^b f(x)dx$ from the *indefinite integral* $\int f(x)dx$ from Section 4.9 (in Chapter 4). To emphasize the distinction, we sometimes refer to the Riemann integral $\int_a^b f(x)dx$ as a *definite integral*. As the nearly identical notations for $\int f(x)dx$ and $\int_a^b f(x)dx$ suggest, there is an important relationship between the two types of integrals. Later in this section, we learn that indefinite integrals may be used to calculate definite integrals. First, we state a theorem that assures the existence of the definite integral $\int_a^b f(x)dx$ for most of the functions that we encounter in calculus.

THEOREM 2 If f is continuous on the interval $[a, b]$, then f is integrable on $[a, b]$. That is, the Riemann integral $\int_a^b f(x)dx$ exists.

Proof. The proof of this assertion involves technical details that are best left to a text on advanced calculus. The following simple observations, however, give a good idea of what is involved. Suppose f is increasing and positive on $[a, b]$. The areas of the shaded regions in Figures 9a, 9b, and 9c represent $\mathcal{R}(f, \mathcal{L}_N)$ for $N = 1, 2, 4$. The amount by which $\mathcal{R}(f, \mathcal{L}_2)$ exceeds $\mathcal{R}(f, \mathcal{L}_1)$ is represented by the area of the rectangle that lies above the dashed line in Figure 9b. Similarly, the amount by which $\mathcal{R}(f, \mathcal{L}_4)$ exceeds $\mathcal{R}(f, \mathcal{L}_2)$ is represented by the area of the rectangles that lie above the dashed lines in Figure 9c. If we continue to bisect each subinterval, we obtain an *increasing* sequence of lower Riemann sums that is bounded above by

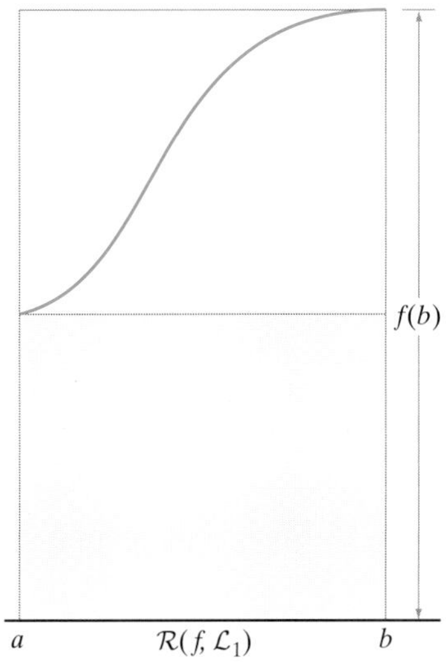

▲ Figure 9a

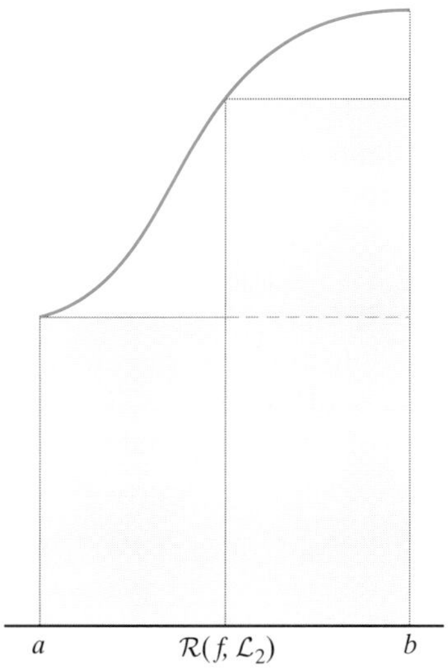

▲ Figure 9b

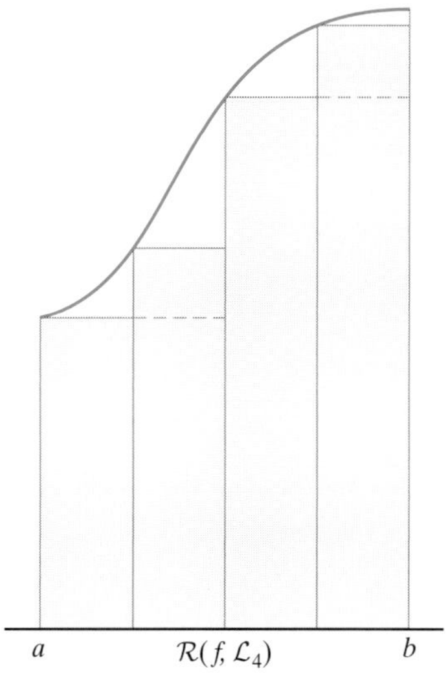

▲ Figure 9c

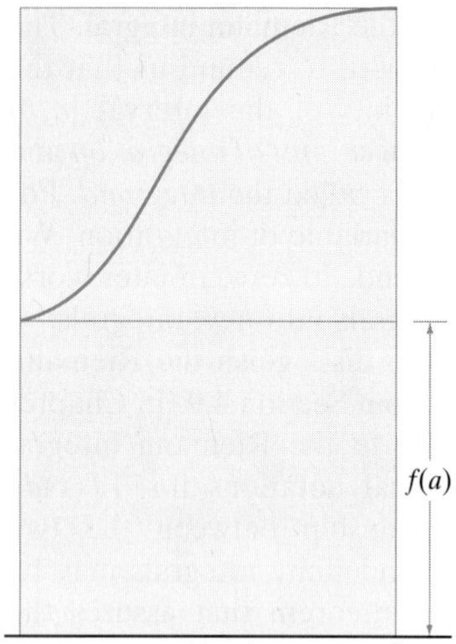

▲ Figure 10a

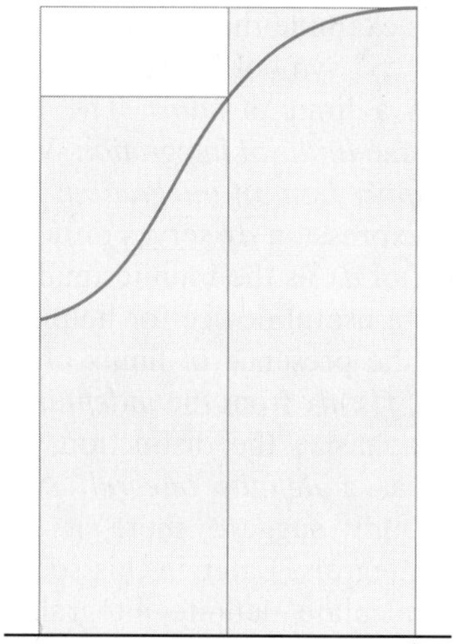

▲ Figure 10b

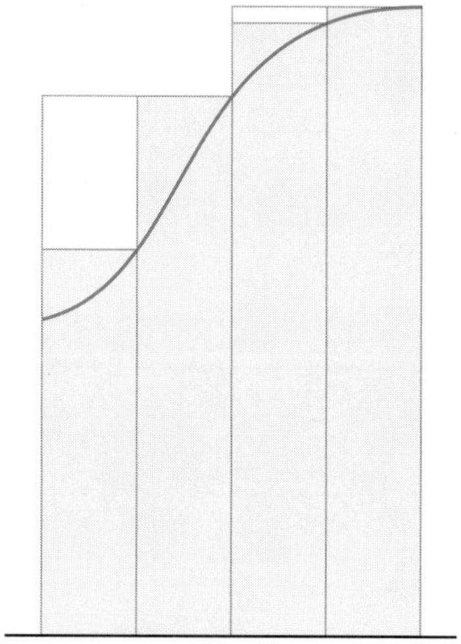

▲ Figure 10c

$f(b)\cdot(b-a)$ (the area of the circumscribed rectangle shown in Figure 9a). We also obtain a *decreasing* sequence of upper Riemann sums that is bounded below by $f(a)\cdot(b-a)$, as shown in Figure 10. According to the Monotone Convergence Property of the real numbers (Section 2.6 in Chapter 2), both sequences of upper and lower Riemann sums converge. We infer from limit (5.2.3) that the sequences must converge to the same number. Call this common limit ℓ. Let ε be an arbitrary positive number. Because the sequences $\{\mathcal{R}(f, \mathcal{L}_N)\}$ and $\{\mathcal{R}(f, \mathcal{U}_N)\}$ converge to ℓ, there is an integer M such that $0 \le \ell - \mathcal{R}(f, \mathcal{L}_N) < \varepsilon/2$ and $0 \le \mathcal{R}(f, \mathcal{U}_N) - \ell < \varepsilon/2$ for all N with $M \le N$. Using inequality (5.2.2), we see that, for every N with $M \le N$ and any choice of points $\mathcal{S}_N$, we have

$$\begin{aligned} |\mathcal{R}(f, \mathcal{S}_N) - \ell| &\le \mathcal{R}(f, \mathcal{U}_N) - \mathcal{R}(f, \mathcal{L}_N) \\ &= \mathcal{R}(f, \mathcal{U}_N) - \ell + \ell - \mathcal{R}(f, \mathcal{L}_N) \\ &= \varepsilon/2 + \varepsilon/2 \\ &= \varepsilon, \end{aligned}$$

which shows that f has a Riemann integral on the interval $[a, b]$. ■

Calculating Riemann Integrals

Calculating definite integrals by the method of Riemann sums is tedious and usually difficult. Fortunately, our next theorem shows how indefinite integrals, or antiderivatives, can be used to evaluate definite integrals without considering Riemann sums. Recall from Section 4.9 in Chapter 4 that F is an antiderivative of f on an open interval (a, b) if $F'(x) = f(x)$ for all x in (a, b). If we say that F is an antiderivative of f on the *closed* interval $[a, b]$, then we mean that F is continuous on $[a, b]$ and $F' = f$ on (a, b). As you read the proof of Theorem 3, notice how crucial it is to have flexibility in the choice of points $\{s_j\}$ that we use to form the Riemann sums.

THEOREM 3 Suppose that F is an antiderivative of a continuous function f on $[a, b]$. Then,

$$\int_a^b f(x)\,dx = F(b) - F(a). \tag{5.2.4}$$

Proof. Let $\{a = x_0, x_1, \ldots, x_N = b\}$ be the uniform partition of order N. By applying the Mean Value Theorem to F on each subinterval I_j, we obtain a point s_j in I_j such that

$$\frac{F(x_j) - F(x_{j-1})}{x_j - x_{j-1}} = F'(s_j).$$

Therefore

$$F(x_j) - F(x_{j-1}) = F'(s_j) \cdot (x_j - x_{j-1}) = f(s_j)\Delta x.$$

The Riemann sum corresponding to the choice of points $\mathcal{S}_N = \{s_1, s_2, \ldots, s_N\}$ is given by

$$\begin{aligned}\mathcal{R}(f, \mathcal{S}_N) &= \sum_{j=1}^{N} f(s_j)\Delta x \\ &= \sum_{j=1}^{N} \Big(F(x_j) - F(x_{j-1})\Big) \\ &= \Big(F(x_1) - F(x_0)\Big) + \Big(F(x_2) - F(x_1)\Big) + \Big(F(x_3) - F(x_2)\Big) \\ &\quad + \cdots + \Big(F(x_{N-1}) - F(x_{N-2})\Big) + \Big(F(x_N) - F(x_{N-1})\Big).\end{aligned}$$

This sort of telescoping sum has been considered in Example 4 from Section 5.1. Observe that the terms $-F(x_0)$ and $F(x_N)$ appear once in this sum. The other terms, $F(x_j)$ with $0 < j < N$, all appear twice: once with a positive sign and once with a negative sign. Therefore after the ensuing cancellations, only the terms $-F(x_0)$ and $F(x_N)$ remain. We obtain

$$\mathcal{R}(f, \mathcal{S}_N) = F(x_N) - F(x_0) = F(b) - F(a).$$

From this equation and Theorem 2, we deduce that

$$\int_a^b f(x)dx = \lim_{N\to\infty} \mathcal{R}(f, \mathcal{S}_N) = \lim_{N\to\infty} \underbrace{(F(b) - F(a))}_{\text{independent of } N} = F(b) - F(a). \quad \blacksquare$$

INSIGHT Formula (5.2.4) holds for every antiderivative of f. Thus if F and G are both antiderivatives of f, then $F(b) - F(a)$ must equal $G(b) - G(a)$ because, according to Theorem 3, each difference is equal to $\int_a^b f(t)dt$. We can also deduce this fact from our work in Section 4.2: Because $F' = G'(= f)$, Theorem 5 from Section 4.2 tells us that there is a constant C such that $G = F + C$. It follows that

$$G(b) - G(a) = (F(b) + C) - (F(a) + C) = F(b) - F(a).$$

The significance for us is that we do not have to choose a particular antiderivative when we apply Theorem 3.

Theorem 3 provides us with a very powerful tool for calculating definite integrals. The right side of equation (5.2.4) occurs so frequently in calculus that a special notation has been introduced: $F\big|_a^b$ and $F(x)\big|_{x=a}^{x=b}$ are both used to symbolize the difference $F(b) - F(a)$. That is,

$$F\big|_a^b = F(b) - F(a) \quad \text{and} \quad F(x)\big|_{x=a}^{x=b} = F(b) - F(a).$$

Theorem 3 states the first of two important relationships between the two main branches, differentiation and integration, of calculus. The second of these relationships will be discussed in Section 5.4. Together, they make up the *Fundamental Theorem of Calculus.*

▶ **EXAMPLE 3** Evaluate $\int_1^{27} \frac{1}{\sqrt[3]{x}}\,dx$.

Solution We calculate the antiderivative $F(x) = \frac{3}{2}x^{2/3}$ of the integrand $f(x) = x^{-1/3}$. From Theorem 3, it follows that

$$\int_1^{27} \frac{1}{\sqrt[3]{x}}\,dx = F(27) - F(1) = \frac{3}{2}\,(27^{2/3} - 1^{2/3})$$
$$= \frac{3}{2}\,(9-1) = 12. \quad ◀$$

▶ **EXAMPLE 4** Evaluate $\int_0^{\pi} \sin(x)dx$.

Solution Because $F(x) = -\cos(x)$ is an antiderivative of $f(x) = \sin(x)$, it follows that

$$\int_0^{\pi} \sin(x)dx = F(\pi) - F(0) = -\cos(\pi) - (-\cos(0)) = -(-1) - (-1) = 2. \quad ◀$$

Using the Fundamental Theorem of Calculus to Compute Areas

Let f be a positive continuous function defined on an interval $[a, b]$. In Section 5.1, we defined the area A of the region under the graph of f to be a limit of Riemann sums (as we now call them). Theorem 2 tells us that this limit exists. If we can find an antiderivative F of f, then we can use Theorem 3 to calculate $A = \int_a^b f(x)dx = F(b) - F(a)$.

▶ **EXAMPLE 5** Calculate the area A of the region that lies under the graph of $f(x) = x^2$, above the x-axis, and between the vertical lines $x = 2$ and $x = 6$.

Solution In Example 7 from Section 5.1, we found that $A = 208/3$ by means of a somewhat tedious computation. We can now use Theorem 3 to calculate A much more easily. Because $F(x) = x^3/3$ is an antiderivative of $f(x) = x^2$, as we know from formula (4.13), we have

$$A = \int_2^6 x^2 dx = F(6) - F(2) = \frac{6^3}{3} - \frac{2^3}{3} = \frac{208}{3}. \quad ◀$$

▶ **EXAMPLE 6** Calculate the area A of the shaded region in Figure 11.

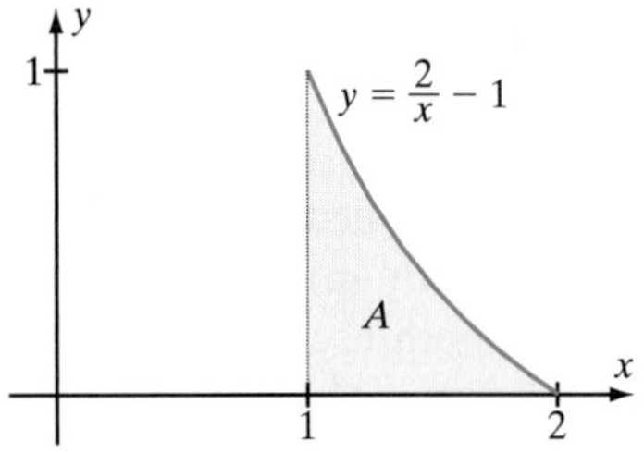

▲ Figure 11

Solution Because $2\ln(x) - x$ is an antiderivative of $2/x - 1$, we have

$$A = \int_1^2 \left(\frac{2}{x} - 1\right) dx = (2\ln(x) - x)\Big|_{x=1}^{x=2} = (2\ln(2) - 2) - \underbrace{(2\ln(1) - 1)}_{\text{Note parentheses}}.$$

Therefore

$$A = (2\ln(2) - 2) - (0 - 1) = 2\ln(2) - 2 + 1 = 2\ln(2) - 1. \quad ◀$$

INSIGHT The minus sign in the expression $F(x)\Big|_{x=a}^{x=b} = F(b) - F(a)$ is the source of many algebraic errors. It is a good idea to enclose $F(a)$ in parentheses when it is negative or when it involves a sum of terms. Doing so will help you distribute the minus sign correctly.

QUICK QUIZ

1. If $\{s_1, s_2, s_3, s_4\}$ is a choice of points associated with the order 4 uniform partition of the interval $[-1,2]$, what is the smallest possible value that can be chosen for s_3? The largest?
2. What are the lower and upper Riemann sums for $f(x) = x^2$ when the order 4 uniform partition of $[-1,2]$ is used?
3. Evaluate $\int_{-1}^{2} (3x^2 - 2x)dx$.
4. True or false: $\int_a^b f(x)dx = \int_a^b f(u)du$.
5. True or false: $\int_a^b f(u)dx = \int_a^b f(x)du$.

Answers

1. Smallest: 1/2; Largest: 5/4 2. Lower: 45/32; Upper: 327/64 3. 6
4. True 5. False

EXERCISES

Problems for Practice

In Exercises 1–4, for the given function f, interval I, and uniform partition of order N:

a. Evaluate the Riemann sum $\mathcal{R}(f, \mathcal{S})$ using the choice of points $\mathcal{S}$ that consists of the midpoints of the subintervals of I.

b. Evaluate the definite integral that $\mathcal{R}(f, \mathcal{S})$ approximates.

1. $f(x) = 1/x$ $\quad \mathcal{S} = \{4, 8, 12\}, I = [2, 14], N = 3$

2. $f(x) = 3 - x^2/2$ $\quad \mathcal{S} = \{-4, -2, 0\}, I = [-5, 1], N = 3$

3. $f(x) = x^2$ $\quad \mathcal{S} = \{-1, 1, 3, 5\}, I = [-2, 6], N = 4$

4. $f(x) = x^2 + x$ $\quad \mathcal{S} = \{-3, 1, 5, 9\}, I = [-5, 11], N = 4$

In Exercises 5–8, for the given function f, interval I, and uniform partition of order N:

a. Evaluate the Riemann sum $\mathcal{R}(f, \mathcal{S})$ using the choice of points $\mathcal{S}$ that consists of the left endpoints of the subintervals of I.

b. Evaluate the definite integral that $\mathcal{R}(f, \mathcal{S})$ approximates.

5. $f(x) = x^3 + 4x$ $\quad \mathcal{S} = \{1, 3, 5\}, I = [1, 7], N = 3$

6. $f(x) = \sin(x)$ $\quad \mathcal{S} = \{0, \pi/3, \pi 2/3, \pi\}, N = 4$

7. $f(x) = |x|$ $\quad \mathcal{S} = \{-7, -4, -1, 2\}, I = [-7, 5], N = 4$

8. $f(x) = 1/x$ $\quad \mathcal{S} = \{1, 5/4. 3/2, 7/4, 2\}, I = [1, 9/4], N = 5$

In Exercises 9–12, for the given function f, interval I, and uniform partition of order N:

a. Evaluate the Riemann sum $\mathcal{R}(f, \mathcal{S})$ using the choice of points $\mathcal{S}$ that consists of the right endpoints of the subintervals of I.

b. Evaluate the definite integral that $\mathcal{R}(f, \mathcal{S})$ approximates.

9. $f(x) = e^x$ $\quad \mathcal{S} = \{-2, 0, 2\}, I = [-4, 2], N = 3$

10. $f(x) = 1/x$ $\quad \mathcal{S} = \{1, 5/4, 3/2, 7/4, 2\}, I = [3/4, 2], N = 5$

11. $f(x) = \cos(x)$ $\quad \mathcal{S} = \{-\pi/3, \pi/3, \pi, 5\pi/3\}, I = [-\pi, 5\pi/3], N = 4$

12. $f(x) = x^2 - x$ $\quad \mathcal{S} = \{-2, -1, 0, 1\}, I = [-3, 1], N = 4$

In Exercises 13–16, calculate the lower and upper Riemann sums for the given function f, interval I, and uniform partition of order 2.

13. $f(x) = x^2$ $\quad I = [-1/2, 2]$

14. $f(x) = x^2 - 2x + 2$ $\quad I = [0, 3]$

15. $f(x) = \cos(x)$ $\quad I = [0, 4\pi/3]$

16. $f(x) = x + 1/x$ $\quad I = [1/2, 2]$

In each of Exercises 17–23, sketch the integrand of the given definite integral over the interval of integration. Evaluate the integral by calculating the area it represents.

17. $\int_1^3 \sqrt{2}\, dx$

18. $\int_1^5 (\cos^2(x) + \sin^2(x))dx$

19. $\int_{-1}^3 |x| dx$

20. $\int_{-2}^3 (2 + \text{signum}(x))dx$

21. $\int_{-1}^2 (1 + x)dx$

22. $\int_{-1}^1 \sqrt{1 - x^2} dx$

23. $\int_{-2}^3 |x - 1| dx$

In each of Exercises 24–44, evaluate the given definite integral by finding an antiderivative of the integrand and applying Theorem 3.

24. $\int_{\pi}^{3\pi} \pi\, dx$
25. $\int_1^2 (6x^2 - 2x)\, dx$
26. $\int_{-1}^{0} x^{99} dx$
27. $\int_1^4 \sqrt{x}\, dx$
28. $\int_1^9 (x - 1/\sqrt{x})\, dx$
29. $\int_0^{\pi/4} \sec^2(x)\, dx$
30. $\int_{\pi/3}^{3\pi} \sin(x)\, dx$
31. $\int_0^{\pi/3} \sec(x)\tan(x)\, dx$
32. $\int_1^{16} x^{-3/4}\, dx$
33. $\int_8^{27} x^{1/3}\, dx$
34. $\int_2^3 (6/x^2)\, dx$
35. $\int_0^1 e^x\, dx$
36. $\int_1^4 (1/x)\, dx$
37. $\int_{-e}^{-1} (1/x)\, dx$
38. $\int_{-1}^{0} e^{1+x}\, dx$
39. $\int_0^3 2^x\, dx$
40. $\int_1^2 (1/10^x)\, dx$
41. $\int_{\pi/4}^{\pi/3} \csc^2(x)\, dx$
42. $\int_{\pi/6}^{\pi/4} \csc(x)\cot(x)\, dx$
43. $\displaystyle\int_2^{-1} \frac{x^2+5x+6}{x+2}\, dx$
44. $\displaystyle\int_1^2 \frac{x^2+2x+1}{x}\, dx$

Further Theory and Practice

In each of Exercises 45–50, a function f, an interval $[a, b]$, and a positive integer N are given. Calculate $\mathcal{R}(f, \mathcal{L}_N)$, the Riemann sum $\mathcal{R}(f, \mathcal{S})$ using the midpoint of each subinterval for the choice of points, and $\mathcal{R}(f, \mathcal{U}_N)$. (You will notice that the inequalities of line (5.2.2) hold.)

45. $f(x) = e^x$ $\quad I = [0, 1],\ N = 1$
46. $f(x) = \log_{10}(x)$ $\quad I = [10, 100],\ N = 1$
47. $f(x) = 1/(1 + x)$ $\quad I = [1, 3],\ N = 2$
48. $f(x) = \sin(x)$ $\quad I = [0, \pi/2],\ N = 2$
49. $f(x) = \ln(1 + x)$ $\quad I = [0, 3],\ N = 3$
50. $f(x) = 1 + \sin(x)$ $\quad I = [-\pi/2, \pi/2],\ N = 4$

In each of Exercises 51–56, evaluate the given integral.

51. $\int_0^{\pi/3} \sin(3x)\, dx$
52. $\int_{\pi/6}^{\pi/4} \cos(2x)\, dx$
53. $\int_0^{\pi/8} \sec^2(2x)\, dx$
54. $\int_{1/4}^{1/2} \csc(\pi x)\cot(\pi x)\, dx$
55. $\int_0^1 e^{-x}\, dx$
56. $\int_{-1}^{1} (1/2^x)dx$

57. Using $\{0, 1, 2, 3\}$ for the partition, compute the upper and lower Riemann sums for

$$f(x) = 2x^3 - 9x^2 + 12x + 1$$

and the interval $[0, 3]$. Verify that $\int_0^3 f(x)\, dx$ is between these two Riemann sums.

In each of Exercises 58–61, a function f and an interval $[a, b]$ are given. Let $\Delta x = (b - a)/2$. Use the Mean Value Theorem to find points s_1 in $[a, (a + b)/2]$ and s_2 in $[(a + b)/2, b]$ such that the Riemann sum

$$\mathcal{R}(f, \{s_1, s_2\}) = f(s_1) \cdot \Delta x + f(s_2) \cdot \Delta x$$

is equal to the Riemann integral of f over $[a, b]$. (*Hint*: The method for finding the required points can be found in the proof of Theorem 3.)

58. $f(x) = x^2$ $\quad [0, 2]$
59. $f(x) = 4x^3$ $\quad [1, 5]$
60. $f(x) = 4x + 3x^2$ $\quad [0, 2]$
61. $f(x) = 1/x$ $\quad [1, 2]$

62. Suppose that $a < b$. Obtain a formula for $\int_a^b x^k\, dx$ that is valid for every $k > 0$. Now suppose that $k < 0$. What assumptions on a and b must you make in this case to apply Theorem 3? With these assumptions, evaluate $\int_a^b x^k\, dx$ for $k < 0$, making sure to treat $k = -1$ as a special case.

63. Suppose $0 < b < a$. With respect to the ellipse and circumscribed circle plotted in Figure 12, show that

$$y_E = \frac{b}{a} \cdot y_C.$$

By comparing Riemann sums for the ellipse with Riemann sums for the circle, deduce that the area enclosed by the ellipse

$$\frac{x^2}{a^2} + \frac{y^2}{b^2} = 1$$

is πab.

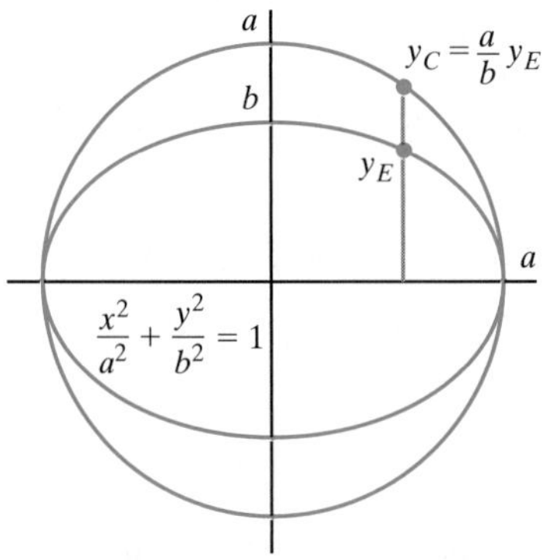

▲ Figure 12

64. Let $x_j = 1 + j/N$ for $0 \le j \le N$. Let $\mathcal{S}_N = \{s_j = \sqrt{x_{j-1} \cdot x_j} : 1 \le j \le N\}$.
   a. Verify that $\mathcal{S}_N$ is a valid choice of points for the order N uniform partition of interval $[1, 2]$.

b. Let $f(x) = 1/x^2$. Show that

$$\mathcal{R}(f, \mathcal{S}_N) = N \sum_{j=1}^{N} \left(\frac{1}{N+j-1} - \frac{1}{N+j} \right).$$

c. For the same f, prove that $\mathcal{R}(f, \mathcal{S}_N) = 1/2$ for every N. (Use part b and your result from Exercise 37 from Section 5.1.)

d. Use part c to evaluate $\int_1^2 1/x^2 dx$.

65. Let c be the midpoint of $[a, b]$. Let m and h be any constants. Show that

$$\int_a^b (m(x-c)+h)\, dx = h(b-a).$$

66. Suppose that f is a continuous, positive function on $[a, b]$. Let c be the midpoint of $[a, b]$. Let L denote the line that is tangent to $y = f(x)$ at the point $(c, f(c))$. Suppose that the graph of f lies above L. Use the preceding exercise to deduce that

$$f\left(\frac{a+b}{2}\right) \cdot (b-a) \le \int_a^b f(x)\, dx.$$

67. Suppose that f is a positive function whose graph over the interval $[a, b]$ lies under the chord joining $(a, f(a))$ to $(b, f(b))$. By calculating the area of the trapezoid with vertices $(a, 0)$, $(b, 0)$, $(b, f(b))$, and $(a, f(a))$, deduce that

$$\int_a^b f(x)\, dx \le (b-a) \cdot \left(\frac{f(a)+f(b)}{2} \right).$$

68. Suppose f is continuous and increasing on $[a, b]$. Show that

$$\mathcal{R}(f, \mathcal{U}_N) - \mathcal{R}(f, \mathcal{L}_N) = (f(b) - f(a)) \cdot \frac{b-a}{N}.$$

69. Suppose that f is continuous on $[a, b]$, differentiable on (a, b), and that $|f'(x)| \le M_1$ for x in (a, b). Let $\{\ell_1, \ldots, \ell_N\}$ and $\{u_1, \ldots, u_N\}$ be the points that give rise to the lower and upper Riemann sums of order N. For each $j = 1 \ldots N$, apply the Mean Value Theorem to f on the interval with endpoints ℓ_j and u_j. Deduce that $f(u_j) - f(\ell_j) \le M_1 \cdot \Delta x$. Use this inequality to prove that

$$\mathcal{R}(f, \mathcal{U}_N) - \mathcal{R}(f, \mathcal{L}_N) \le M_1 \frac{(b-a)^2}{N}.$$

Calculator/Computer Exercises

In each of Exercises 70–73, a Riemann sum $\mathcal{R}(f, \{s_1, \ldots, s_N\})$ is to be calculated as an approximation to the given definite integral. Use the given value of N and, for each j between 1 and N, choose s_j to be the midpoint of the jth subinterval of the uniform partition of order N. Figure 13 shows how Maple can be used to do this for $\int_0^{\pi/2} \ln(1+\sin(x))\, dx$ and $N = 100$.

70. $\int_0^{\sqrt{3}} \sqrt{1+x^2}\, dx$ $\quad N = 80$

71. $\int_0^1 \cos^4(\pi x)\, dx$ $\quad N = 100$

72. $\int_0^1 \exp(-x^2)\, dx$ $\quad N = 40$

73. $\int_0^{\pi} x \sin(x)\, dx$ $\quad N = 60$

In Exercises 74–77, plot the function f over the interval $[a, b]$. Calculate the lower and upper Riemann sums $\mathcal{R}(f, \mathcal{L}_{50})$ and $\mathcal{R}(f, \mathcal{U}_{50})$. (The exact value of $\int_a^b f(x)\, dx$ lies between these two estimates.)

74. $f(x) = \sqrt{1+\sqrt{x}}$ $\quad I = [0, 9]$

75. $f(x) = \left(2+\sin(x)\right)^{4/3}$ $\quad I = [-\pi/2, \pi/2]$

76. $f(x) = 1/(3 - x^{1/3})$ $\quad I = [1, 8]$

77. $f(x) = \sec(x^2/4)$ $\quad I = [0, \sqrt{\pi}]$

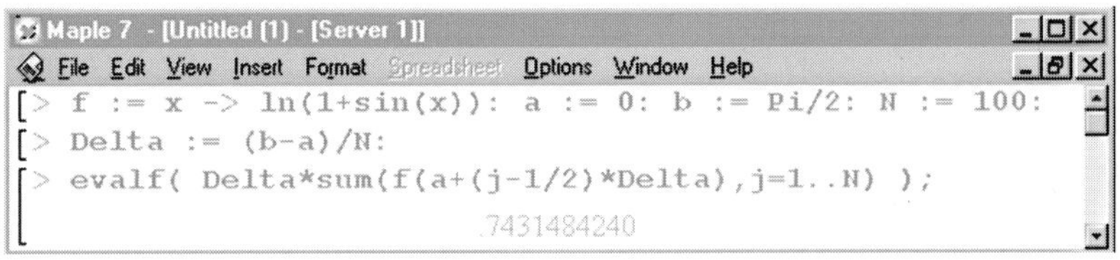

Figure 13

5.3 Rules for Integration

The calculation of integrals can be simplified if we take note of certain rules.

THEOREM 1 If f and g are integrable functions on the interval $[a, b]$ and if α is a constant, then $f+g$, $f-g$, and $\alpha \cdot f$ are integrable, and

a. $\int_a^b (f(x)+g(x))dx = \int_a^b f(x)dx + \int_a^b g(x)dx$ *and*
$\int_a^b (f(x)-g(x))dx = \int_a^b f(x)dx - \int_a^b g(x)dx$;
b. $\int_a^b \alpha \cdot f(x)dx = \alpha \cdot \int_a^b f(x)dx$;
c. $\int_a^b \alpha dx = \alpha \cdot (b-a)$;
d. $\int_a^a f(x)dx = 0$; *and*
e. *If* $a \le c \le b$, *then* $\int_a^c f(x)dx + \int_c^b f(x)dx = \int_a^b f(x)dx$.

Notice that Theorem 1a says that $f+g$ and $f-g$ are integrable, provided that f and g can be integrated separately. It also tells how to calculate the integrals of $f+g$ and $f-g$. Likewise, Theorem 1b says that αf is integrable, provided that f is. It tells how to calculate the integral of αf. Rules a and b are consequences of analogous limit rules that we learned in Section 2.2 of Chapter 2.

Theorem 1c states that integrating a constant α over an interval I results in α times the length of I. In a sense, this rule simply states that the area of a rectangle is the length of the base times the height. This may be proved either by referring to the definition of the integral as a limit or by applying Theorem 3 from Section 5.2 after noting that αx is an antiderivative of α.

Theorem 1d says that the integral over an interval of zero length is zero, whereas rule 1e says that contiguous intervals of integration may be added. If we think of the integral as representing area, for instance, then rules 1d and 1e are geometrically plausible (see Figure 1).

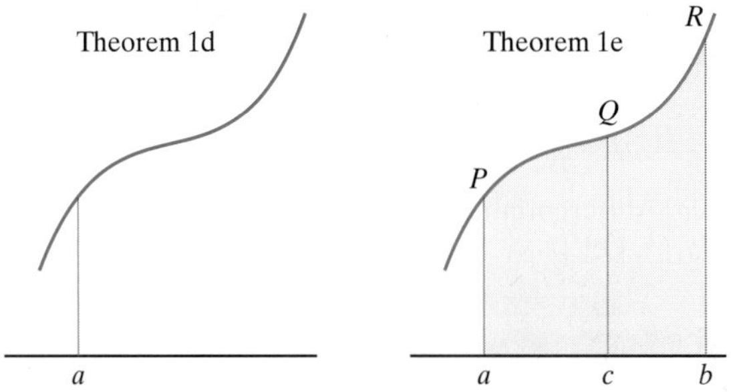

▲ **Figure 1** Theorem 1d: Area = base × height = $0 \times f(a) = 0$.
Theorem 1e: Area ($abRP$) = area ($acQP$) + area ($cbRQ$)

▶ **EXAMPLE 1** Calculate the integral

$$\int_1^2 \frac{3+5t^3}{t^2}\,dt.$$

Solution We use Theorem 1a and 1b to rewrite the integral as

$$\int_1^2 \frac{3+5t^3}{t^2}\,dt = \int_1^2 \frac{3}{t^2}\,dt + \int_1^2 5\frac{t^3}{t^2}\,dt = 3\cdot\int_1^2 t^{-2}dt + 5\cdot\int_1^2 t\,dt.$$

Because $-t^{-1}$ is an antiderivative of t^{-2} and $t^2/2$ is an antiderivative of t, we can evaluate each integral using the Fundamental Theorem of Calculus (Theorem 3, Section 5.2):

$$\int_1^2 \frac{3+5t^3}{t^2}\,dt = 3\cdot\int_1^2 t^{-2}dt + 5\cdot\int_1^2 t\,dt = 3\cdot\left(-t^{-1}\Big|_{t=1}^{t=2}\right) + 5\left(\frac{t^2}{2}\Big|_{t=1}^{t=2}\right)$$
$$= 3\left(-\frac{1}{2}-(-1)\right)+5\left(\frac{2^2}{2}-\frac{1}{2}\right) = 9. \quad ◀$$

▶ **EXAMPLE 2** Can we use the equations $\int_0^1 x\,dx = 1/2$ and $\int_0^1 x^2 dx = 1/3$ to evaluate the integral $\int_0^1 x^3 dx$ by observing that its integrand x^3 is the product of x and x^2?

Solution No. It is tempting, but wrong, to think that we can somehow factor this integral into two simpler integrals that we already know. Using the Fundamental Theorem of Calculus, we obtain

$$\int_0^1 x^3 dx = \frac{x^4}{4}\Big|_{x=0}^{x=1} = \frac{1}{4} - 0 = \frac{1}{4},$$

which is certainly not $(1/2) \times (1/3)$. Thus

$$\int_a^b f(x)g(x)\,dx \neq \int_a^b f(x)\,dx \cdot \int_a^b g(x)\,dx$$

in general. In other words, the integral of a product is *not* the product of the integrals. In Chapter 7, we will learn techniques for handling integrals of products. ◀

▶ **EXAMPLE 3** Let f be a continuous function on the interval $[-2, 5]$ such that

$$\int_{-2}^1 f(x)\,dx = 6 \quad\text{and}\quad \int_1^5 f(x)\,dx = -2.$$

Use this information to evaluate $\int_{-2}^5 f(x)\,dx$.

Solution Using Theorem 1e, we have

$$\int_{-2}^5 f(x)\,dx = \int_{-2}^1 f(x)\,dx + \int_1^5 f(x)\,dx = 6 + (-2) = 4. \quad ◀$$

▶ **EXAMPLE 4** Suppose that g is a continuous function on the interval $[4, 9]$ that satisfies

$$\int_4^9 g(x)\,dx = 8 \quad\text{and}\quad \int_4^7 g(x)\,dx = -7.$$

Calculate $\int_7^9 g(x)\,dx$.

Solution We use Theorem 1e to see that

$$8 = \int_4^9 g(x)\,dx = \int_4^7 g(x)\,dx + \int_7^9 g(x)\,dx = -7 + \int_7^9 g(x)\,dx.$$

Therefore $\int_7^9 g(x)\,dx = 15$. ◀

▶ **EXAMPLE 5** Evaluate $\int_{\pi/2}^{\pi/2} t^{5/2}\sin(t^2)\,dt$.

Solution Theorem 1d tells us that $\int_a^a f(x)\,dx = 0$, no matter what the integrand is. Therefore

$$\int_{\pi/2}^{\pi/2} t^{5/2}\sin(t^2)\,dt = 0. \quad ◀$$

Reversing the Direction of Integration

In the definition of the integral, the upper limit of integration is taken to be greater than or equal to the lower limit. However, it is sometimes convenient to allow the upper limit to be less than the lower limit. If a is *greater* than b, then we *define* $\int_a^b f(x)\,dx$ for a function f integrable on $[b, a]$ by

$$\int_a^b f(x)\,dx = -\int_b^a f(x)\,dx. \tag{5.3.1}$$

▶ **EXAMPLE 6** Evaluate $\int_1^{-4} t^2\,dt$.

Solution Notice that $F(t) = t^3/3$ is an antiderivative of t^2. Thus

$$\int_1^{-4} t^2\,dt = -\int_{-4}^{1} t^2\,dt = -(F(1) - F(-4)) = -\left(\frac{1}{3} - \left(\frac{-64}{3}\right)\right) = -\frac{65}{3}. \quad ◀$$

INSIGHT Remember that if f is nonnegative and $a < b$, then the integral of f over $[a, b]$ represents area and is positive. In Example 6, $f(t) = t^2$ is nonnegative, but $\int_1^{-4} t^2 dt$ is negative. This is because the lower limit of integration is greater than the upper limit.

Let us verify that the Fundamental Theorem of Calculus (Theorem 3 from Section 5.2) remains true regardless of the order of the limits of integration.

THEOREM 2 If f is continuous on an interval that contains the points a and b (in any order) and if F is an antiderivative of f, then

$$\int_a^b f(x)\,dx = F(b) - F(a). \tag{5.3.2}$$

Proof. If $a < b$, then equation (5.3.2) is precisely formula (5.2.4). If $a = b$, then the right side of equation (5.3.2) is 0 and so, by Theorem 1d, is the left side. In the remaining case, $a > b$, we have

$$\int_a^b f(x)\,dx \overset{(5.3.1)}{=} -\int_b^a f(x)\,dx \overset{(5.2.4)}{=} -\big(F(a) - F(b)\big) = F(b) - F(a). \quad ■$$

Now that we have defined the definite integral even when the lower limit of integration is greater than the upper limit of integration, we may ask if Theorem 1e continues to hold no matter what the order of a, b, and c may be. The answer is "yes," and we state the result as a theorem.

THEOREM 3 If f is integrable on an interval that contains the three points a, b, and c (in any order), then

$$\int_a^b f(x)\,dx + \int_b^c f(x)\,dx = \int_a^c f(x)\,dx. \qquad (5.3.3)$$

Rather than give a general proof of Theorem 3, we will work out one example.

▶ **EXAMPLE 7** Suppose that f is integrable on $[1,5]$. Show that equation (5.3.3) is valid for $a=5$, $b=3$, and $c=1$.

Solution We have

$$\begin{aligned}
\int_5^3 f(x)\,dx + \int_3^1 f(x)\,dx &= -\int_3^5 f(x)\,dx - \int_1^3 f(x)\,dx && \text{By formula (5.3.1)}\\
&= -\left(\int_1^3 f(x)\,dx + \int_3^5 f(x)\,dx\right) && \text{Reorder the terms}\\
&= -\int_1^5 f(x)\,dx && \text{By Theorem 1e}\\
&= \int_5^1 f(x)\,dx. && \text{By formula (5.3.1)}
\end{aligned}$$
◀

Order Properties of Integrals

In this subsection, we collect a number of facts concerning integrals and order relations.

THEOREM 4 If f, g, and h are integrable on $[a,b]$ and

$$g(x) \le f(x) \le h(x) \quad \textit{for} \quad x \textit{ in } [a,b]$$

then

$$\int_a^b g(x)\,dx \le \int_a^b f(x)\,dx \le \int_a^b h(x)\,dx. \qquad (5.3.4)$$

In particular, if f is integrable on $[a,b]$ and if m and M are constants such that

$$m \le f(x) \le M$$

for all $x \in [a,b]$, then

$$m \cdot (b-a) \le \int_a^b f(x)\,dx \le M \cdot (b-a). \qquad (5.3.5)$$

Proof. If $\mathcal{S}_N = \{s_j\}$ is a choice of points associated with the order N uniform partition of $[a,b]$, then

$$\sum_{j=1}^N g(s_j)\Delta x \le \sum_{j=1}^N f(s_j)\Delta x \le \sum_{j=1}^N h(s_j)\Delta x,$$

or $\mathcal{R}(g,\mathcal{S}_N) \le \mathcal{R}(f,\mathcal{S}_N) \le \mathcal{R}(h,\mathcal{S}_N)$. Line (5.3.4) results by letting N tend to infinity.

In the particular case when $g(x)=m$ and $h(x)=M$ for all x, line (5.3.4) simplifies to become line (5.3.5). ■

▶ **EXAMPLE 8** Use the inequalities of line (5.3.5) to estimate $\int_0^2 \sqrt{1+x^3}\,dx$.

Solution Because $\sqrt{1+x^3}$ is increasing on $[0,2]$, we have

$$1=\sqrt{1+0^3}\le\sqrt{1+x^3}\le\sqrt{1+2^3}=3 \text{ for } 0\le x\le 2.$$

We use the estimates of line (5.3.5) with $a=0$, $b=2$, $m=1$, and $M=3$ to obtain

$$2=1\cdot(2-0)\le\int_0^2\sqrt{1+x^3}\,dx\le 3\cdot(2-0)=6.$$

The value of the given integral is between 2 and 6. ◀

INSIGHT The integrand of Example 8 does not have an elementary antiderivative. Therefore we cannot use the Fundamental Theorem of Calculus to evaluate the integral exactly. In such cases, we must rely on an approximation. As the estimates obtained in Example 8 suggest, Theorem 4 usually gives us only an order of magnitude estimate. In Section 5.8, we study techniques for achieving accurate approximations of definite integrals.

The following theorem, which is used often, is obtained by taking $m=0$ in the first inequality of line (5.3.5).

THEOREM 5 If f is integrable on $[a,b]$ and if $f(x)\ge 0$, then $\int_a^b f(x)\,dx\ge 0$.

The next theorem may be deduced from line (5.3.4) by taking $g(x)=-|f(x)|$ and $h(x)=|f(x)|$.

THEOREM 6 If f is integrable on $[a,b]$, then

$$\left|\int_a^b f(x)\,dx\right|\le\int_a^b |f(x)|\,dx. \tag{5.3.6}$$

INSIGHT Thinking of the definite integral as a limit of Riemann sums, we may regard inequality (5.3.6) as a continuous analogue of the (discrete) triangle inequality $|A_1+\cdots+A_N|\le|A_1|+\cdots+|A_N|$. In fact, if we set $A_j=f(s_j)\cdot\Delta x$ for $1\le j\le N$ and let $N\to\infty$, then we obtain an alternative proof of inequality (5.3.6).

▶ **EXAMPLE 9** Verify inequality (5.3.6) for $f(x)=x^2+x$ and the interval $[-1, 2]$.

Solution Because $x^3/3+x^2/2$ is an antiderivative of $f(x)$, Theorem 3 from Section 5.2 gives us

$$\int_{-1}^2 f(x)\,dx=\left(\frac{x^3}{3}+\frac{x^2}{2}\right)\Bigg|_{-1}^2=\left(\frac{8}{3}+\frac{4}{2}\right)-\left(\frac{-1}{3}+\frac{1}{2}\right)=\frac{9}{2}.$$

Because $x^2 + x = (x+1)x$, we see that $f(x)$ is the product of one positive term and one negative term when $-1 < x < 0$. Thus $f(x) < 0$ for $-1 < x < 0$. Also, $f(x)$ is clearly nonnegative for $x \geq 0$. From these observations, it follows that

$$|f(x)| = \begin{cases} -(x^2 + x) & \text{if } -1 < x < 0 \\ x^2 + x & \text{if } 0 \leq x \end{cases}$$

Therefore

$$\int_{-1}^{2} |f(x)|\, dx = \int_{-1}^{0} -(x^2 + x)\, dx + \int_{0}^{2} (x^2 + x)\, dx = \frac{29}{6}.$$

As a result, we have

$$\left| \int_{-1}^{2} f(x)\, dx \right| = \left| \frac{9}{2} \right| = \frac{9}{2} = \frac{27}{6} < \frac{29}{6} = \int_{-1}^{2} |f(x)|\, dx,$$

which agrees with inequality (5.3.6). ◄

The Mean Value Theorem for Integrals

In Chapter 4, we learned a Mean Value Theorem for derivatives. We will now demonstrate an analogous theorem for integrals.

THEOREM 7 **(Mean Value Theorem for Integrals)**. Let f be continuous on $[a, b]$. There is a value c in the open interval (a, b) such that

$$f(c) = \frac{1}{b-a} \int_a^b f(x)\, dx. \tag{5.3.7}$$

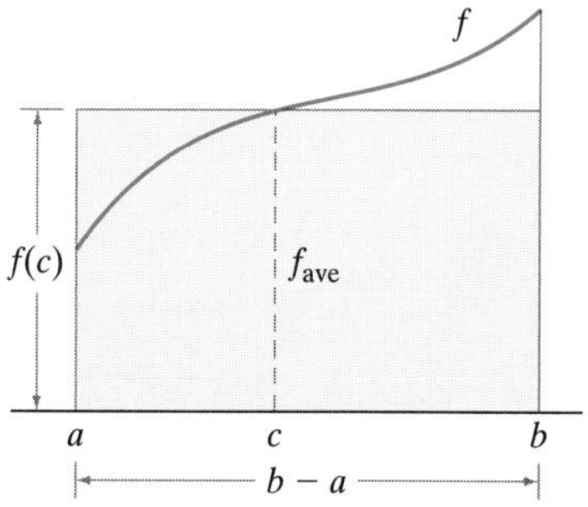

▲ **Figure 2** $f_{\text{avg}} \cdot (b-a) =$ area of shaded rectangle $= \int_a^b f(x)\, dx$

Proof. Let m and M be the minimum and maximum values of f on $[a, b]$. These values exist by the Extreme Value Theorem (Section 2.3 of Chapter 2). The function $g(x) = (b-a) \cdot f(x)$ is continuous on $[a, b]$. The minimum and maximum values of g on $[a, b]$ are $\alpha = m \cdot (b-a)$ and $\beta = M \cdot (b-a)$, respectively. According to line (5.3.5), the number $\int_a^b f(x)\, dx$ lies between α and β. The Intermediate Value Theorem assures us that such an intermediate number can be realized as a value of g. In other words, there is a point c in the interval (a, b) such that

$$g(c) = \int_a^b f(x)\, dx.$$

Because $g(c) = (b-a) f(c)$, equation (5.3.7) follows. ■

The *average value* f_{avg} of f on $[a, b]$ is defined to be

$$f_{\text{avg}} = \frac{1}{b-a} \int_a^b f(x)\, dx.$$

The Mean Value Theorem for Integrals tells us that a *continuous* function assumes its average value. There is no such theorem for discrete functions. For example, if you take two exams and score 80 and 90, then your average is 85. However, you did not score 85 on either exam—the average value 85 is not assumed. In Section 7.3 in Chapter 7, we motivate the definition of *average value*. For now, we are content with the geometric interpretation of average value: If f is positive on $[a, b]$, then the area under the graph of f over $[a, b]$ equals the area of the rectangle with base $[a, b]$ and height f_{avg}:

$$\int_a^b f(x)\,dx = f_{\text{avg}} \cdot (b - a).$$

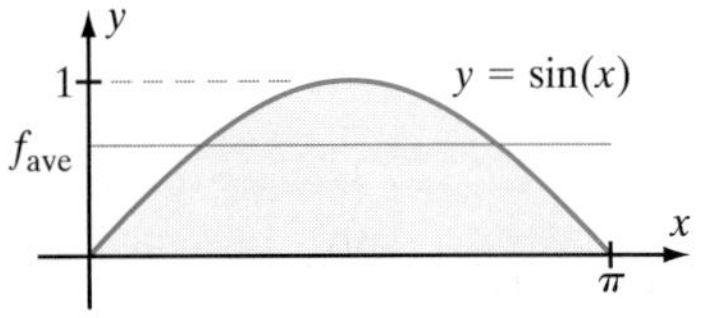

This is illustrated in Figure 2.

▶ **EXAMPLE 10** Compute the average value of $f(x) = \sin(x)$ for $0 \le x \le \pi$.

Solution Because $-\cos(x)$ is an antiderivative of $\sin(x)$, we use Theorem 3 from Section 5.2 to calculate the average value of f on $[0, \pi]$ as follows:

$$f_{\text{avg}} = \frac{1}{\pi - 0}\int_0^{\pi} \sin(x)\,dx = \frac{1}{\pi} \cdot \left(-\cos(x)\Big|_0^{\pi}\right) = \frac{1}{\pi}(-(-1) - (-1)) = \frac{2}{\pi}.$$

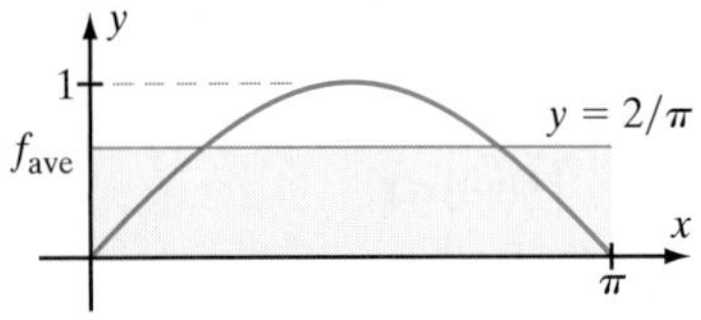

▲ Figure 3 The shaded regions have equal areas.

See Figure 3. Notice that the average value is *not* the average of the values of f at the endpoints (which is 0 in this case). It is *not* the average of the maximum and minimum values of f on the interval (which is 1/2 in this case). Also, it is *not* the value of f at the midpoint of the interval (which is 1 in this case). ◀

QUICK QUIZ

1. What is $\int_5^5 \tan^3(x)\,dx$
2. Calculate $\int_3^{-3} x^2 dx$.
3. If $\int_0^2 f(x)\,dx = 3$ and $\int_0^7 f(x)\,dx = -5$, then what is $\int_7^2 3f(x)\,dx$?
4. What is the average value of $f(t) = t^2$ for $1 \le t \le 4$?

Answers

1. 0 2. −18 3. 24 4. 7

EXERCISES

Problems for Practice

1. Suppose that $\int_1^3 f(x)\,dx = -8$ and $\int_3^7 f(x)\,dx = 12$. Evaluate $\int_1^7 f(x)\,dx$.

2. Suppose that $\int_2^{12} g(x)\,dx = -6$ and $\int_2^6 g(x)\,dx = -12$. Evaluate $\int_6^{12} g(x)\,dx$.

3. Suppose that $\int_{-7}^3 f(x)\,dx = -7$ and $\int_{-7}^3 g(x)\,dx = -4$. Evaluate $\int_{-7}^3 (4f(x) - 9g(x))\,dx$.

4. Suppose that $\int_4^8 f(x)\,dx = 6$. Evaluate $\int_8^4 f(x)\,dx$.

5. Suppose that $\int_7^{-2} f(x)\,dx = 6$ and $\int_7^9 f(x)\,dx = -4$. Evaluate $\int_9^{-2} f(x)\,dx$.

6. Suppose that $\int_{-3}^{-7} g(x)\,dx = 5$ and $\int_{-3}^{-5} g(x)\,dx = 12$. Evaluate $\int_{-7}^{-5} g(x)\,dx$.

7. Suppose that $\int_9^4 f(x)\,dx = 5$ and $\int_9^4 g(x)\,dx = 15$. Evaluate $\int_4^9 (6f(x) - 7g(x))\,dx$.

8. Suppose that $\int_3^5 f(x)\,dx = 2$. Evaluate $\int_5^3 -4f(x)\,dx$.

9. Suppose that $\int_5^{-3} (-3f(x)/4)\,dx = 7$. Evaluate $\int_5^{-3} (6f(x) + 1)\,dx$.

10. Suppose that $\int_2^{-9} f(x)\,dx = 5$. Evaluate $\int_{-9}^2 (3f(x) - 5x)\,dx$.

11. Suppose that $\int_1^4 f(x)/3dx = 2$. Evaluate $\int_4^1 3f(x)\,dx$.

12. Suppose that $\int_0^4 (2f(x) - x^2)\,dx = 6$. Evaluate $\int_0^4 f(x)\,dx$.

13. Suppose that $\int_2^1 f(x)\,dx = 0$ and $\int_2^1 g(x)\,dx = 0$. Evaluate $\int_2^1 (f(x) - 3g(x) + 5)\,dx$.

14. Suppose that $\int_6^8 (3f(x) - x)\,dx = 6$ and $\int_8^6 (2x + 4g(x))\,dx = -8$. Evaluate $\int_8^6 (f(x) - 5g(x))\,dx$.

▶ In Exercises 15–20, compute the average value f_{avg} of f over $[a, b]$ and find a point c in (a, b) for which $f(c) = f_{\text{avg}}$. Illustrate the geometric significance of c with a sketch accompanied by a description. ◀

15. $f(x) = 1 + x$ $\quad a = 1, b = 5$

16. $f(x) = x^3$ $a = 0,\ b = 2$
17. $f(x) = \sqrt{x}$ $a = 1, b = 4$
18. $f(x) = 1/x^2$ $a = 1/2,\ b = 2$
19. $f(x) = 12x^2 + 5$ $a = 0,\ b = 3$
20. $f(x) = 2x + 2/x$ $a = 1,\ b = e$

In Exercises 21–28, compute the average value of f over $[a, b]$

21. $f(x) = 8 - x^3$ $a = -4,\ b = 2$
22. $f(x) = x^2 + 1/x^2$ $a = 1,\ b = 3$
23. $f(x) = \sqrt{x} - 1/\sqrt{x}$ $a = 1, b = 4$
24. $f(x) = e^x$ $a = 0,\ b = 2$
25. $f(x) = \cos(x)$ $a = 0,\ b = \pi/2$
26. $f(x) = \sin(x)$ $a = -\pi/2,\ b = \pi/3$
27. $f(x) = \sec^2(x)$ $a = 0,\ b = \pi/4$
28. $f(x) = \sec(x)\tan(x)$ $a = \pi/4,\ b = \pi/3$

In each of Exercises 29–32, an integral is given. Do not attempt to calculate its value V. Instead, use the inequalities of line (5.3.5) (as in Example 8) to find numbers A and B such that $A \le V \le B$.

29. $\int_0^3 \sqrt{9 + x^3}\, dx$
30. $\int_1^6 x\sqrt{3 + x}\, dx$
31. $\int_0^4 (25 - x^2)^{-1/2}\, dx$
32. $\int_{\pi/4}^{\pi/3} \sin^2(x)\, dx$

Further Theory and Practice

In Exercises 33–36, use the information to determine $\int_a^b f(x)\, dx$ and $\int_a^b g(x)\, dx$.

33. $\int_a^b (f(x) - 3g(x))\, dx = 3,\ \int_a^b (-6g(x) + 9f(x))\, dx = 6$
34. $\int_a^b (f(x) + 4g(x))\, dx = 5,\ \int_a^b 3g(x)\, dx = -2$
35. $\int_a^b (f(x) + 2g(x))\, dx = -7,\ \int_a^b (g(x) - f(x))\, dx = 4$
36. $\int_a^b (f(x) - 7g(x))\, dx = -9,\ \int_a^b (g(x) + f(x))\, dx = 0$

In Exercises 37–44, use Theorem 3 from Section 5.2 with one or more parts of Theorem 1 from this section to calculate the given definite integral.

37. $\displaystyle\int_1^3 \frac{3x - 5}{x}\, dx$
38. $\int_1^3 x(4x^2 - 2)\, dx$
39. $\int_{-1}^0 (6x^2 + e^x)\, dx$
40. $\int_1^2 (x - 2)(x + 3)\, dx$
41. $\int_0^1 \sqrt{x}(1 - x)\, dx$
42. $\displaystyle\int_0^{1/4} \frac{x - 1}{\sqrt{x} - 1}\, dx$
43. $\int_0^1 3e^x(2 - e^{-x})\, dx$
44. $\displaystyle\int_{\pi/4}^{\pi/2} \frac{x\sin(x) + 1}{3x}\, dx$

45. Suppose that p and q are constants with $p > 0$ and $q > 0$. Is it ever true that

$$\int_0^1 x^p \cdot x^q\, dx = \int_0^1 x^p\, dx \cdot \int_0^1 x^q\, dx?$$

46. When is it true that

$$\int_a^b f(x) \cdot 1\, dx = \int_a^b f(x)\, dx \cdot \int_a^b 1\, dx?$$

47. Find an example of a function f, a function g, and values of a and b such that

$$\int_a^b \frac{f(x)}{g(x)}\, dx \neq \frac{\int_a^b f(x)\, dx}{\int_a^b g(x)\, dx}.$$

In each of Exercises 48–51, a definite integral is given. Do not attempt to calculate its value V. Instead, find the extreme values of the integrand on the interval of integration, and use these extreme values together with the inequalities of line (5.3.5) to obtain numbers A and B such that $A \le V \le B$.

48. $\displaystyle\int_{-1}^2 \frac{x^2 + 5}{x + 2}\, dx$
49. $\displaystyle\int_1^4 \frac{1}{x^2 - 4x + 5}\, dx$
50. $\displaystyle\int_0^3 \frac{(x + 1)^2}{x^2 + 1}\, dx$
51. $\int_1^3 xe^{-x/2} dx$

52. Use trigonometric identities to evaluate $\int_0^{\pi/2} \cos^2(x/2)\, dx$ and $\int_0^{\pi} \sin^2(x)\, dx$.

In each of Exercises 53–58, $F(x)$ is a function of a variable x that appears in a limit (or in the limits) of integration of a given definite integral. Express $F(x)$ explicitly by calculating the integral.

53. $F(x) = \int_1^x (3t^2 + 1)dt$
54. $F(x) = \int_{\pi/6}^x 2\cos(t)dt$
55. $F(x) = \int_x^1 (2t + 1/t^2)dt$
56. $F(x) = \int_x^{2x} (2t + 3t^2)dt$
57. $F(x) = \displaystyle\int_1^{x^2} \frac{1}{\sqrt{t}}\, dt$
58. $F(x) = \int_x^{x^2} (2 + \sqrt{t})dt$

In each of Exercises 59–62, verify that inequality (5.3.6) holds for the given integrand f and interval I of integration.

59. $f(x) = \pi x - 3$ $I = [0,\ 1]$
60. $f(x) = \sin(x) - \cos(x)$ $I = [0,\ \pi/3]$
61. $f(x) = 1 - 9x^2$ $I = [-2,\ 1]$
62. $f(x) = e^x - 7$ $I = [0,\ 2]$

63. Let $P_2(x) = (3x^2 - 1)/2$. This polynomial is called the degree 2 Legendre polynomial. Show that if $Q(x) = Bx + C$ for constants B and C, then

$$\int_{-1}^{1} P_2(x) \cdot Q(x)\, dx = 0.$$

64. Let $P_3(x) = (5x^3 - 3x)/2$. This polynomial is called the degree 3 Legendre polynomial. Show that if $Q(x) = Ax^2 + Bx + C$ for constants A, B, and C, then

$$\int_{-1}^{1} P_3(x) \cdot Q(x)\, dx = 0.$$

Calculator/Computer Exercises

In each of Exercises 65–68, a definite integral is given. Do not attempt to calculate its value V. Instead, find the extreme values of the integrand on the interval of integration, and use these extreme values together with the inequalities of line (5.3.5) to obtain numbers A and B such that $A \leq V \leq B$.

65. $\int_0^2 \frac{1+x}{1+x^4}\, dx$

66. $\int_0^{3/2} (\sqrt{x} - \sin(x))\, dx$

67. $\int_0^{\pi/2} (3\sin(x^2) + \cos(x))\, dx$

68. $\int_{1/2}^{1} e^{-x} \ln(1+x)\, dx$

In Exercises 69–72, compute the average value f_{avg} of f over $[a,b]$, and find a value of c in (a,b) at which f attains this average value. Illustrate the geometric meaning of the Mean Value Theorem for Integrals with a graph.

69. $f(x) = x + \sin(x)$ $a = 0,\ b = \pi/2$

70. $f(x) = x^2 + \cos(x)$ $a = 0,\ b = \pi$

71. $f(x) = x^2 + 4/x$ $a = 2$ $b = 6$

72. $f(x) = (x^3 - 4x + 6)/3$ $a = 0,\ b = 2$

5.4 The Fundamental Theorem of Calculus

Theorem 3 from Section 5.2 tells us that, if f is continuous on $[a,b]$ and *if* f has an antiderivative F on $[a,b]$, then

$$\int_a^b f(t)dt = F(b) - F(a). \tag{5.4.1}$$

This equation is the first part of the Fundamental Theorem of Calculus. In this section, we learn that *every* continuous function f has an antiderivative. This fact constitutes the second part of the Fundamental Theorem of Calculus.

In the analytic proof of the second part of the Fundamental Theorem of Calculus, we assume only that f is a continuous function. However, we will motivate this theorem geometrically by considering areas. Therefore for the purposes of this discussion, we suppose that f is positive in addition to being continuous. To this function f we associate an *area function* F that is defined by

$$F(x) = \int_a^x f(t)dt. \tag{5.4.2}$$

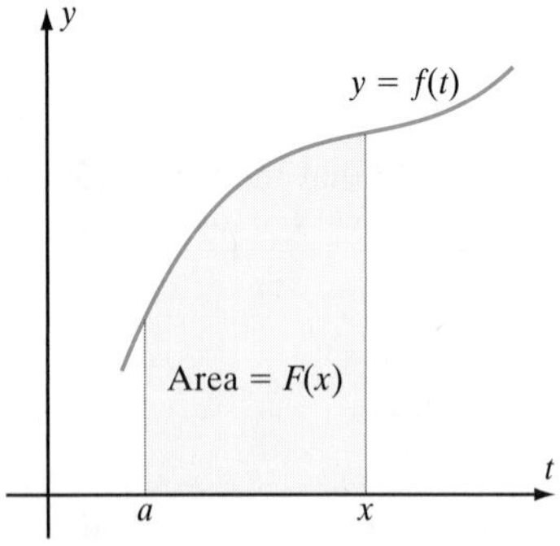

Figure 1

Figure 1 shows the graph of $y = f(t)$ in the ty-plane. As Figure 1 indicates, the value of $F(x)$ is the area of the region that lies below the graph of f, above the t-axis, and between the vertical lines $t = a$ and $t = x$. Notice that F is a function of the upper limit x of integration, *not* of the symbol t that appears within the integral. Also, we do *not* assume that an explicit formula for $F(x)$ can be found. It is known, for example, that the area function $F(x) = \int_a^x (2 + \cos(t^2))dt$ cannot be expressed in terms of finitely many elementary functions. It is precisely such area functions that we wish to understand.

The main result of this section is that F is an antiderivative of f. In other words, F is a differentiable function and $F' = f$. To prove these assertions, we refer to the definition of the derivative and look at the difference quotient:

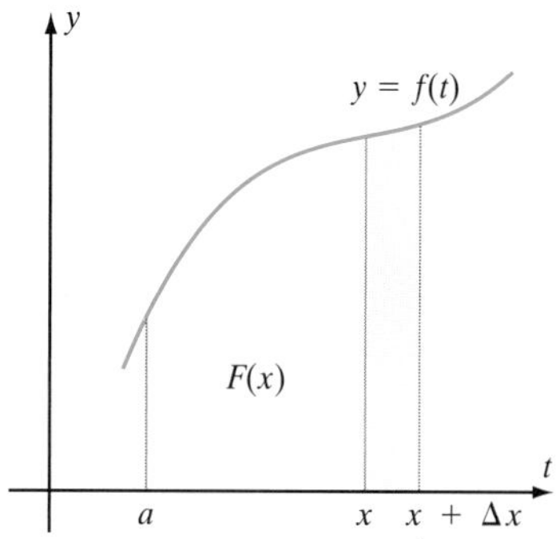

▲ Figure 2 Area $= \int_x^{x+\Delta} f(t)\,dt$

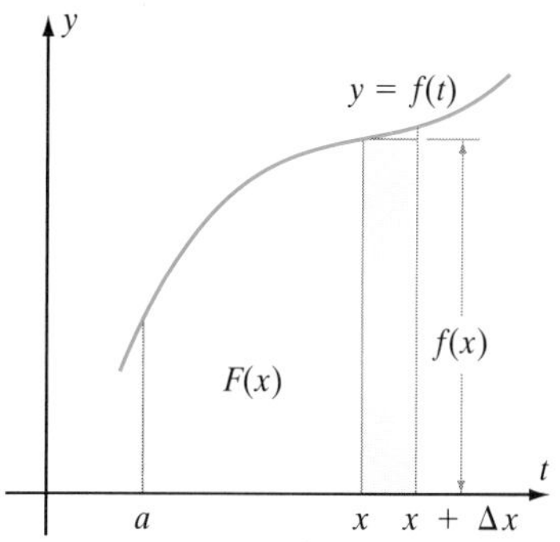

▲ Figure 3 $\int_x^{x+\Delta} f(t)\,dt$ is approximately the area $f(x)\,\Delta x$ of the shaded rectangle.

$$\frac{F(x+\Delta x)-F(x)}{\Delta x}=\frac{\int_a^{x+\Delta x} f(t)dt-\int_a^x f(t)dt}{\Delta x}.$$

We now use Theorem 3 from Section 5.3 to rewrite the numerator of the right side of this equation. We obtain

$$\frac{F(x+\Delta x)-F(x)}{\Delta x}=\frac{\int_x^{x+\Delta x} f(t)dt}{\Delta x}. \tag{5.4.3}$$

The numerator $\int_x^{x+\Delta x} f(t)dt$ represents the area of the shaded region in Figure 2. When Δx is small, the integral $\int_x^{x+\Delta x} f(t)dt$ is nearly equal to the area of a rectangle of base Δx and height $f(x)$ (see Figure 3). Therefore

$$\frac{F(x+\Delta x)-F(x)}{\Delta x}\approx\frac{f(x)\cdot\Delta x}{\Delta x}$$

as $\Delta x \to 0$. Thus we have intuitively deduced that

$$\lim_{\Delta x\to 0}\frac{F(x+\Delta x)-F(x)}{\Delta x}=f(x).$$

We can make this argument rigorous by appealing to the Mean Value Theorem for Integrals, which tells us that there exists a c between x and $x+\Delta x$ such that

$$\frac{\int_x^{x+\Delta x} f(t)dt}{\Delta x}=\frac{1}{(x+\Delta x)-x}\int_x^{x+\Delta x} f(t)\,dt=f(c). \tag{5.4.4}$$

By combining equations (5.4.3) and (5.4.4), we find that for any Δx, there is a c between x and $x+\Delta x$ such that

$$\frac{F(x+\Delta x)-F(x)}{\Delta x}\overset{(5.4.3)}{=}\frac{\int_x^{x+\Delta x} f(t)dt}{\Delta x}\overset{(5.4.4)}{=}f(c). \tag{5.4.5}$$

Because c lies between x and $x+\Delta x$, it follows that $c \to x$ as $\Delta x \to 0$. From these observations, we obtain

$$\lim_{\Delta x\to 0}\frac{F(x+\Delta x)-F(x)}{\Delta x}=\lim_{\Delta x\to 0} f(c)=\lim_{c\to x} f(c)=f(x),$$

where the last equality results from the continuity of f. This equation means that the derivative of F at x exists and equals $f(x)$. In short, F is an antiderivative for f. The theorem that follows combines the two parts of the Fundamental Theorem of Calculus.

THEOREM 1 **(Fundamental Theorem of Calculus).** Let f be a continuous function on the interval $[a,b]$.

a. If F is any antiderivative of f on $[a,b]$, then

$$\int_a^b f(t)dt = F(b) - F(a).$$

b. The function F defined by

$$F(x) = \int_a^x f(t)dt \quad (a \leq x \leq b)$$

is an antiderivative of f. This function F is continuous on $[a, b]$ and

$$F'(x) = f(x) \quad (a < x < b).$$

In Leibniz notation,

$$\frac{d}{dx}\int_a^x f(t)\,dt = f(x). \tag{5.4.6}$$

Together, the two parts of the Fundamental Theorem inform us that differentiation and integration are *inverse operations*. Let us apply the first part of the Fundamental Theorem to a continuous derivative F'. Because F is an antiderivative of F', we obtain

First differentiate F.

$$\int_a^x F'(t)\,dt \quad = \quad F(x) - F(a).$$

Then integrate F'. The result is F (plus a constant).

Thus if we differentiate F and integrate the derivative, then we recover the function F with which we started (plus the constant $-F(a)$). The second part of the Fundamental Theorem, equation (5.4.6), tells us that if we integrate the function f and then differentiate with respect to the upper limit of integration, then we recover the function f with which we started:

First integrate f.

$$\frac{d}{dx}\int_a^x f(t)\,dt \quad = \quad f(x).$$

Then differentiate. The result is f.

We therefore think of integration and differentiation as processes that are *inverse* to one another.

Examples Illustrating the First Part of the Fundamental Theorem

We have been using the first part of the Fundamental Theorem of Calculus since Section 5.2. The integrands in the examples we now consider require greater care than those found in previous sections.

▶ **EXAMPLE 1** Compute $\int_0^1 e^{2t}dt$.

Solution To use the first part of the Fundamental Theorem, we need to find an antiderivative for the function $f(t) = e^{2t}$. Notice that

$$\frac{d}{dt}e^{2t} = 2e^{2t}, \quad \text{or} \quad \frac{d}{dt}\left(\frac{1}{2}e^{2t}\right) = e^{2t}.$$

Therefore $F(t) = \frac{1}{2}e^{2t}$ is a suitable antiderivative, and

$$\int_0^1 e^{2t}\,dt = F(t)\Big|_{t=0}^{t=1} = \frac{1}{2}e^2 - \frac{1}{2}e^0 = \frac{1}{2}(e^2 - 1). \quad ◀$$

INSIGHT Remember that a *definite* integral represents a definite number or expression. Including an arbitrary constant of integration in the final answer (as we would for an *indefinite* integral) is wrong. In Example 1, we used the Fundamental Theorem of Calculus to compute the *definite* integral $\int_0^1 f(t)dt$. Accordingly, we first computed the *indefinite* integral (or antiderivative) $\int f(t)dt$. At this stage, had we added a constant C to the *indefinite* integral $F(t) = e^{2t}/2$, it would not have mattered:

$$\int_0^1 e^{2t}dt = \left(\frac{1}{2}e^{2t} + C\right)\Big|_{t=0}^{t=1} = \left(\frac{1}{2}e^2 + C\right) - \left(\frac{1}{2}e^0 + C\right)$$

$$= \left(\frac{1}{2}e^2 + C\right) - \left(\frac{1}{2} - C\right) = \frac{1}{2}(e^2 - 1).$$

The constant C cancels out and does not appear in the final answer. The definite integral $\int_0^1 e^{2t}\,dt$ represents a unique number. The *only* correct answer is $(e^2 - 1)/2$. In particular, the answer "$(e^2 - 1)/2 + C$ where C is an arbitrary constant" is incorrect.

▶ **EXAMPLE 2** Calculate $\int_0^\pi \sin^2(t)dt$.

Solution Because it is not immediately clear what an antiderivative for $\sin^2(t)$ might be, we first rewrite the integrand using the Half-Angle Formula (equation (1.14) from Section 1.6 in Chapter 1),

$$\sin^2\left(\frac{\theta}{2}\right) = \frac{1 - \cos(\theta)}{2}.$$

Setting $\theta = 2t$, we obtain $\sin^2(t) = (1 - \cos(2t))/2$. We substitute this identity into our original integral and find the required antiderivative:

$$\int_0^\pi \sin^2(t)\,dt = \int_0^\pi \frac{1}{2}(1 - \cos(2t))\,dt = \frac{1}{2}\left(t - \frac{1}{2}\sin(2t)\right)\Big|_{t=0}^{t=\pi} = \frac{\pi}{2}. \quad ◀$$

INSIGHT The Solution to Example 2 begins with the remark that finding an antiderivative of $\sin^2(t)$ is not obvious. A common error is to think that because $t^3/3$ is an antiderivative of t^2, it must follow that $\sin^3(t)/3$ is an antiderivative of $\sin^2(t)$. *This reasoning is incorrect.* We must use the Chain Rule when we differentiate $\sin^3(t)/3$. We obtain

$$\frac{d}{dt}\left(\frac{1}{3}\cdot\sin^3(t)\right) = \frac{1}{3}\cdot 3\sin^2(t)\cdot\frac{d}{dt}\sin(t) = \sin^2(t)\cos(t),$$

which shows that $\sin^3(t)/3$ is an antiderivative of $\sin^2(t)\cos(t)$, *not* of the given integrand $\sin^2(t)$.

Examples of the Second Part of the Fundamental Theorem

In the second part of the Fundamental Theorem, we define a function in terms of an integral. It is important that you understand this part because it will be the key to many applications of the integral and also the key to our rigorous approach to the logarithm function in the next section.

▶ **EXAMPLE 3** The hydrostatic pressure exerted against the side of a certain swimming pool at depth x is

$$P(x) = \int_0^x t \cdot (\sin(2t) + 1)dt.$$

What is the rate of change of P with respect to depth when the depth is 4?

Solution We have not yet learned how to calculate an antiderivative of $f(t) = t \cdot (\sin(2t) + 1)$. However, the second part of the Fundamental Theorem tells us that we do not need to evaluate the integral that defines $P(x)$ in order to compute $P'(x)$. Because $P(x) = \int_0^x f(t)\,dt$, Theorem 1b tells us that

$$P'(x) = f(x) = x \cdot (\sin(2x) + 1).$$

In particular, $P'(4) = 4 \cdot (\sin(8) + 1) \approx 7.957$. ◀

▶ **EXAMPLE 4** Calculate the derivative of $F(x) = \int_\pi^{x^2} \sin(\sqrt{t})\,dt$ with respect to x.

Solution The Fundamental Theorem tells us how to differentiate an integral with respect to its upper limit of integration. This example involves an upper limit that is a function of the variable of differentiation. We must use the Chain Rule. If we let $u = x^2$, then

$$\begin{aligned}\frac{d}{dx}F(x) &= \frac{d}{dx}\int_\pi^u \sin(\sqrt{t})\,dt = \frac{d}{du}\left(\int_\pi^u \sin(\sqrt{t})\,dt\right) \cdot \frac{du}{dx} = \sin(\sqrt{u}) \cdot 2x \\ &= \sin(\sqrt{x^2}) \cdot 2x = 2x\sin(|x|).\end{aligned}$$ ◀

Example 4 illustrates an extension of formula (5.4.6). If u is a function of x, then

$$\frac{d}{dx}\int_a^{u(x)} f(t)\,dt = f(u(x)) \cdot \frac{d}{dx}u(x). \qquad (5.4.7)$$

▶ **EXAMPLE 5** Calculate the derivative of $F(x) = \int_x^0 \exp(-t^2)\,dt$ with respect to x.

Solution The Fundamental Theorem tells us how to differentiate an integral with respect to its *upper* limit of integration, but in this example, the variable of differentiation is the lower limit of integration. There is a simple remedy: We reverse the order of integration. Thus

$$\frac{d}{dx}\int_x^0 \exp(-t^2)\,dt = -\frac{d}{dx}\int_0^x \exp(-t^2)\,dt = -\exp(-x^2).$$ ◀

Example 5 illustrates another extension of formula (5.4.6). We can combine the ideas of Examples 4 and 5 to state the extension as follows:

$$\frac{d}{dx}\int_{u(x)}^b f(t)\,dt = -f(u(x)) \cdot \frac{d}{dx}u(x). \qquad (5.4.8)$$

▶ **EXAMPLE 6** Let F be defined by $F(x) = \int_{\cos(x)}^{\sin(x)} \sqrt{1-t^2}\,dt$ for $0, < x < \pi/2$. Differentiate $F(x)$ with respect to x.

Solution This example involves *both* upper and lower limits of integration that depend on x. None of the formulas (5.4.6), (5.4.7), and (5.4.8) is immediately applicable. However, if we choose any fixed number a and, using (5.3.1), write

$$\begin{aligned} F(x) &= \int_{\cos(x)}^{\sin(x)} \sqrt{1-t^2}\,dt = \int_{\cos(x)}^{a} \sqrt{1-t^2}\,dt + \int_{a}^{\sin(x)} \sqrt{1-t^2}\,dt \\ &= \int_{a}^{\sin(x)} \sqrt{1-t^2}\,dt - \int_{a}^{\cos(x)} \sqrt{1-t^2}\,dt, \end{aligned}$$

then we can differentiate each of the two summands using formula (5.4.7):

$$\begin{aligned} \frac{d}{dx}F(x) &= \sqrt{1-\sin^2(x)}\,\frac{d}{dx}\sin(x) - \sqrt{1-\cos^2(x)}\,\frac{d}{dx}\cos(x) \\ &= \sqrt{\cos^2(x)}\cos(x) - \sqrt{\sin^2(x)}(-\sin(x)) \\ &= |\cos(x)|\cos(x) + |\sin(x)|\sin(x) \\ &= \cos^2(x) + \sin^2(x) \qquad \text{because } 0 < x < \pi/2 \\ &= 1. \end{aligned}$$ ◀

INSIGHT In Chapter 6, we learn how to directly calculate the integral that defines the function F from Example 6. But, as with Examples 3, 4, and 5, it is not necessary to compute F to obtain F'. It is therefore interesting to observe that our calculation $F'(x) \equiv 1$ actually gives us an indirect way to calculate $F(x)$. Because $G(x) = x$ is also an antiderivative of the constant function 1, Theorem 5 from Section 4.1 of Chapter 4 tells us that $F(x) = G(x) + C = x + C$ for some constant C. Substituting $x = 0$ into this last equation, we obtain

$$C = F(0) = \int_{\cos(0)}^{\sin(0)} \sqrt{1-t^2}\,dt = \underbrace{-\int_0^1 \sqrt{1-t^2}\,dt}_{\substack{\text{Area of 1/4-unit circle} \\ \text{(see Figure 4)}}} = -\frac{\pi}{4}.$$

Therefore $F(x) = x - \pi/4$.

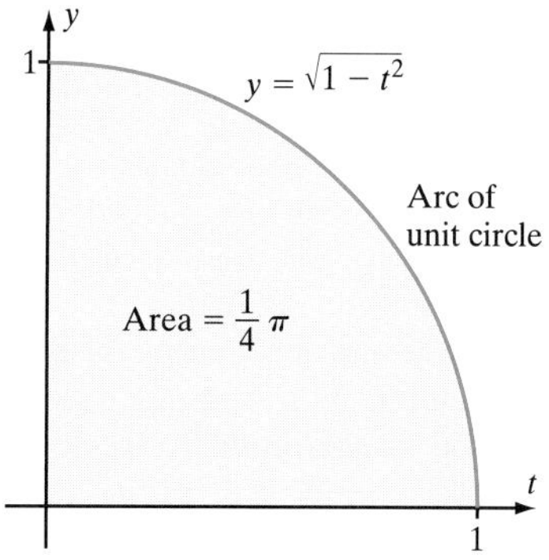

▲ Figure 4

QUICK QUIZ

1. Suppose that $F(x) = \int_\pi^x \sin(t^2)\,dt$. What is $F'(2)$?
2. Suppose that $F(x) = \int_x^0 (1+t^2)^{-1}\,dt$. What is $F'(2)$?
3. Suppose that $F(x) = \displaystyle\int_x^{2x} \frac{2+t}{1+t}\,dt$. What is $F'(1)$?
4. Evaluate $\displaystyle\int_0^1 \frac{d}{dx}(xe^x)\,dx$.

Answers

1. sin(4) 2. −1/5 3. 7/6 4. e

EXERCISES

Problems for Practice

▶ Use the first part of the Fundamental Theorem of Calculus together with the ideas of Examples 1 and 2 to evaluate the definite integrals in Exercises 1–10. ◀

1. $\int_0^2 e^{3x}\,dx$
2. $\int_{-1}^0 e^{-x}\,dx$
3. $\int_{\pi/6}^{\pi/4} 12\cos(2x)\,dx$
4. $\int_0^{\pi} \sin(x/2)\,dx$
5. $\int_{-\pi/3}^{\pi/2} 4\sin(2x)\,dx$
6. $\int_{-\pi/4}^{\pi/2} \cos^2(x)\,dx$
7. $\int_{\pi/2}^{\pi} \sec(x/3)\tan(x/3)\,dx$
8. $\int_{-\pi/8}^{\pi/8} \sec^2(2x)\,dx$
9. $\int_0^{\pi/4} \sin^2(2x)\,dx$
10. $\int_{-\pi/4}^{\pi/4} \cos^2(4x)\,dx$

▶ In Exercises 11–20, calculate $F(x) = \int_a^x f(t)\,dt$. ◀

11. $f(t) = 3t^2 + 1$	$a = -1$	
12. $f(t) = \sin(t)$	$a = \pi/4$	
13. $f(t) = 4t^{1/3}$	$a = 8$	
14. $f(t) = e^{-t}$	$a = -1$	
15. $f(t) = (e^t + e^{-t})/2$	$a = 0$	
16. $f(t) = \sec^2(t)$	$a = 0$	$(-\pi/2 < x < \pi/2)$
17. $f(t) = \sec(t)\tan(t)$	$a = -\pi/4$	$(-\pi/2 < x < \pi/2)$
18. $f(t) = 1/t$	$a = e$	$(0 < x)$
19. $f(t) = 15\sqrt{t}(1+t)$	$a = 1$	$(0 < x)$
20. $f(t) = (3t + 1/2)/\sqrt{t}$	$a = 4$	$(0 < x)$

▶ In each of Exercises 21–28, calculate the derivative of $F(x)$ with respect to x. ◀

21. $F(x) = \int_{-2}^x (2t+1)(t+2)\,dt$
22. $F(x) = \int_1^x t^{1/3}(\sqrt{t}+3)\,dt$
23. $F(x) = \displaystyle\int_{-1}^x \frac{t^3 + 2t^2 + 3t + 6}{t+2}\,dt$
24. $F(x) = \int_0^x \tan^2(t)\,dt$
25. $F(x) = \int_0^x \tan(t^2)\,dt$
26. $F(x) = \int_{-2}^x \sin(t^3)\,dt$
27. $F(x) = \int_{-1}^x t\sqrt{1+t}\,dt$
28. $F(x) = \int_1^x e^t \ln(t)\,dt$

▶ In each of Exercises 29–34, calculate the derivative of $F(x)$ with respect to x. ◀

29. $F(x) = \int_x^{\pi/4} \cos(4t)\,dt$
30. $F(x) = \int_x^{\pi/4} \sqrt{1+\cos(t)}\,dt$
31. $F(x) = \int_x^{\pi/4} \cot(t)\,dt$
32. $F(x) = \int_x^5 \sqrt{2t^2 - 1}\,dt$
33. $F(x) = \int_x^5 \sqrt{2 - \sin^2(t)}\,dt$
34. $F(x) = \int_x^{\pi} \cos(\sqrt{t})\,dt$

▶ In each of Exercises 35–42, follow the method of Example 6 to calculate $F'(x)$. ◀

35. $F(x) = \int_x^{2x} \sin(2t)\,dt$
36. $F(x) = \int_{-x}^x \sqrt{1+t}\,dt$
37. $F(x) = \int_x^{x^2} \cot(t)\,dt$
38. $F(x) = \int_x^{\sqrt{x}} (1+t^2)^{-1/2}\,dt$
39. $F(x) = \int_{\exp(-x)}^{\exp(x)} \ln(t)\,dt$
40. $F(x) = \int_{2x}^{3x} \cos(t^2)\,dt$
41. $F(x) = \int_{2x}^{\sqrt{x}} \sqrt{2t^2 - 1}\,dt$
42. $F(x) = \int_{x^2}^{x^3} \sqrt{1+t^2}\,dt$

Further Theory and Practice

▶ In Exercises 43–46, compute F' and F''. Determine the intervals on which F is increasing, decreasing, concave up, and concave down. ◀

43. $F(x) = \int_0^x t(t-1)\,dt$
44. $F(x) = \int_0^x t\exp(-t)\,dt$
45. $F(x) = \int_1^x t\ln(t)\,dt \quad 0 < x$
46. $F(x) = \int_0^x (\sqrt{1+t^2} - t)\,dt$
47. Suppose that f is continuous on $(-a, a)$ and that f is an odd function: $f(-t) = -f(t)$. For x in $(-a, a)$, let $F(x) = \int_{-x}^x f(t)\,dt$. Show that $F' = 0$, and use this result to evaluate F.
48. Suppose that f is continuous on $(-a, a)$ and that f is an even function: $f(-t) = f(t)$. Differentiate $F(x) = \int_0^x f(t)\,dt - \int_{-x}^0 f(t)\,dt$ with respect to x, and use your result to show that $\int_{-x}^0 f(t)\,dt = \int_0^x f(t)\,dt$.
49. Suppose a body travels in a line with position $p(t)$ at time t and velocity $v(t)$ at time t. Show that
$$\int_a^b v(t)\,dt = p(b) - p(a).$$

▶ In Exercises 50–53, verify that point P is on the graph of function F, and calculate the tangent line to the graph of F at P. ◀

50. $F(x) = \int_1^x 6\sqrt{t}\,dt$	$P = (4, 28)$
51. $F(x) = \int_1^x 1/t\,dt$	$P = (e, 1)$
52. $F(x) = \int_4^x t^2\,dt$	$P = (7, 93)$
53. $F(x) = \int_0^x \cos(t)\,dt$	$P = (\pi/6, 1/2)$

54. Calculate $F(x) = \int_{-1}^{x} f(t)\, dt$ where f is the function with the graph that appears in Figure 5.

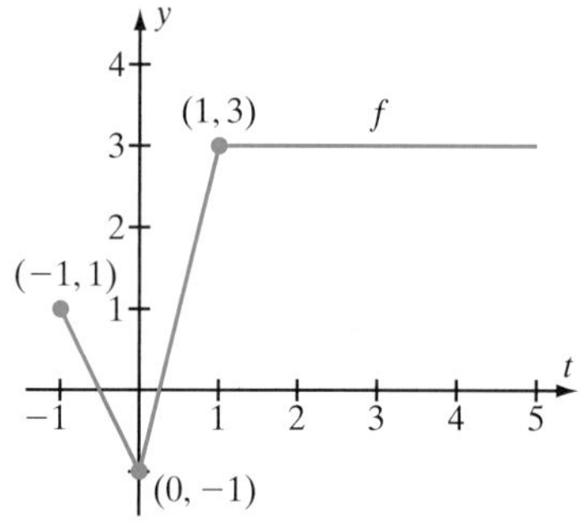

Figure 5

55. Find a function f such that $F(x) = \int_0^x f(t)\, dt$ for the function F with the graph that appears in Figure 6.

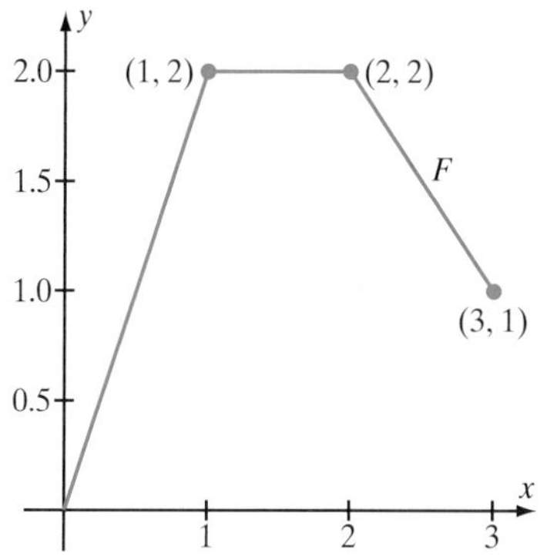

Figure 6

56. For the function f shown in Figure 7, determine on what interval(s) $F(x) = \int_0^x f(t)\, dt$ is increasing, decreasing, concave up, and concave down.

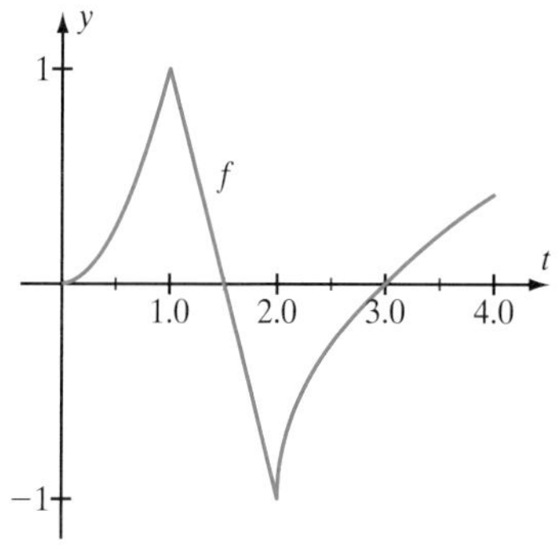

Figure 7

57. Consider the integration

$$\int_{-2}^{2} t^{-4}\, dt = -\frac{1}{3}t^{-3}\Big|_{-2}^{2} = -\frac{1}{12}.$$

Integrating a positive function from left to right should not result in a negative number. What error has led to the incorrect negative answer?

58. The Fundamental Theorem of Calculus guarantees the existence of a function F with a derivative of $|x|$. Write an explicit formula for F. Your formula should contain no integrals.

59. Let g and h be differentiable functions, and let f be a continuous function. Suppose that the range of h is contained in the domain of g. Find a formula for

$$\frac{d}{dx}\int_a^{g(h(x))} f(t)\, dt.$$

60. Let g and h be differentiable functions, and let f be a continuous function. Use the method of Example 6 to find a formula for

$$\frac{d}{dx}\int_{h(x)}^{g(x)} f(t)\, dt.$$

61. Suppose that f is a differentiable function with continuous derivative f'. What is the average rate of change of the function f over the interval $[a, b]$? (Refer to Section 3.1 in Chapter 3, if necessary.) What is the average value of f' over $[a, b]$? (Refer to Section 5.3, if necessary.) Prove that these two quantities are equal.

62.

$$G(a) = \frac{d}{dx}\int_a^x f(t)\, dt\Big|_{x=0}.$$

What is $G'(a)$?

63. Suppose f and g are functions with continuous derivatives on an interval containing $[a, b]$. Prove that if $f(a) \le g(a)$ and if $f'(x) \le g'(x)$ for all x in $[a, b]$, then $f(x) \le g(x)$ for all x in $[a, b]$.

64. In probability and statistics, the error function (erf) is defined for $x \ge 0$ by

$$\mathrm{erf}(x) = \frac{2}{\sqrt{\pi}}\int_0^x \exp(-t^2)\, dt.$$

Show that the graph of erf is concave down over $[0, \infty)$.

65. *Dawson's integral* is the function defined for $x \ge 0$ by

$$F(x) = \exp(-x^2)\int_0^x \exp(t^2)\, dt.$$

Compute $F'(x)$.

66. The *Fresnel sine integral*, defined by

$$\text{Fresnel } S(x) = \int_0^x \sin\left(\frac{\pi}{2}t^2\right) dt,$$

is an important function in the theory of optical diffraction. Determine the intervals on which this function is concave up.

67. The function

$$C(x) = \int_0^x \frac{t^4}{\sqrt{1+t^2}}\,dt \quad (x \geq 0)$$

arises in the computation of pressure within a white dwarf. Show that C is an increasing function with a graph that is concave up.

68. The *sine integral* is the function Si, defined by $\mathrm{Si}(x) = \int_0^x \sin(t)/t\,dt$. Calculate $\lim_{x\to 0} \mathrm{Si}(x)/x$.

69. Let f be continuous on an interval I that contains a and b. For x in I set $F(x) = \int_a^x f(t)\,dt$ and $G(x) = \int_b^x f(t)\,dt$. Use the Fundamental Theorem of Calculus to show that there is a constant C such that $F = G + C$. Using Theorem 3 from Section 5.3 to give a second proof.

Averaging data smooths it out. The Mean Value Theorem for Integrals tells us that for a fixed h, we can think of the integral

$$A_f(x) = \frac{1}{2h}\int_{x-h}^{x+h} f(t)\,dt$$

as an operation that averages the values of f over the interval $[x-h, x+h]$. Figure 8 shows a plot of the price f of one share of Microsoft's stock at the close of ten consecutive trading days in the summer of 2009. Figure 9 shows a plot of the average price A_f that is obtained by using $h = 1/4$. (This choice of h is arbitrary.) In each of Exercises 70–72, calculate and plot A_f over $[-1, 1]$ for the given function f and $h = 1/2$.

70. $f(x) = |x|$

71. $f(x) = \begin{cases} 0 & \text{if } x < 0 \\ x & \text{if } 0 \leq x \end{cases}$

72. $f(x) = \mathrm{signum}(x)$

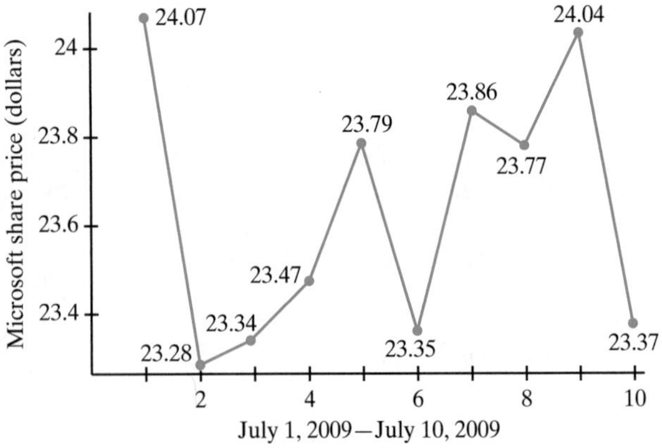

▲ Figure 8

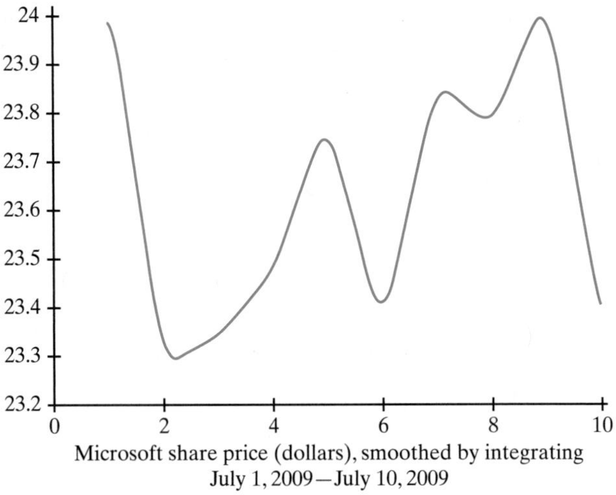

▲ Figure 9

73. Let $f(x) = \lfloor x \rfloor$ be the greatest integer function. Show that

$$F(x) = \int_0^x f(t)\,dt$$

is not differentiable at $x = 1$.

74. Let b be a constant, and set

$$f(t) = \begin{cases} 0 & \text{if } t < 0 \\ x + b & \text{if } 0 \leq t \end{cases}.$$

Show that $F(x) = \int_{-1}^x f(t)\,dt$ is continuous at $x = 0$ for all values of b but differentiable at $x = 0$ only if $b = 0$.

Calculator/Computer Exercises

In Exercises 75–78, plot F' and F''. Determine the intervals on which F is increasing, decreasing, concave up, and concave down.

75. $F(x) = \int_{-3}^x (t^3 - 3t^2 + 3t + 4)\,dt$

76. $f(x) = \int_0^x (t+1)/(t^4+1)\,dt$

77. $F(x) = \int_0^x (t^3 - t)\,e^t\,dt$

78. $F(x) = \int_1^x (t^2 - 3)\ln(t)\,dt \quad x > 0$

79. In a particular regional climate, the temperature varies between $-28°C$ and $46°C$ and averages $13°C$. The number of days $N(T)$ in the year on which the temperature remains below T degrees centigrade is given (approximately) by

$$N(T) = \int_{-28}^T f(x)\,dx \quad (-28 \leq T \leq 46)$$

where

$$f(x) = 12.66\exp\left(-\frac{(x-13)^2}{265.8}\right).$$

Plot $y = N(T)$ for $-28 \leq T \leq 46$. On about how many days does the temperature reach at least 37°C?

80. Find positive numbers a and b for which the sine integral defined in Exercise 68 is concave down on the interval $(0, a)$, concave up on the interval (a, b), and concave down again just to the right of b.

81. The Fresnel sine integral, FresnelS, is defined in Exercise 66. The Fresnel cosine integral, FresnelC, is defined analogously with $\cos\left(\frac{\pi}{2}t^2\right)$ replacing $\sin\left(\frac{\pi}{2}t^2\right)$. Plot the parametric equations $x = \text{FresnelC}(t)$, $y = \text{FresnelS}(t)$ for $0 \leq t \leq 6$. (Your plot is an arc of a curve that is known as the *Cornu spiral*.)

In each of Exercises **82–85**, a function f and a point c are given. Let F be defined by formula (5.4.2) with $a = 1/4$. (This assignment of a is for the sake of being definite. A different value would not change anything in these exercises.) For $h = 0.001$, numerically calculate $\int_{c-h/2}^{c+h/2} f(t)\,dt$, and use this value to obtain a central difference quotient approximation of $F'(c)$. Use the value of $f(c)$ together with equation (5.4.6) to verify your approximation of $F'(c)$.

82. $f(t) = \cos(\pi t^4)$ $\quad c = 1$

83. $f(t) = \sqrt{9 + t^4}$ $\quad c = 2$

84. $f(t) = t^{t-1}$ $\quad c = 3$

85. $f(t) = \sqrt{t}e^{t-4}$ $\quad c = 4$

5.5 A Calculus Approach to the Logarithm and Exponential Functions

In this section, we take a second look at exponential and logarithm functions. In our new approach, we use calculus to define the exponential and logarithm functions and to develop their properties. In doing so, we can fill in some details that so far have been missing. Of course, everything you will learn in this section is consistent with the notions we have already developed.

First, let us summarize our previous approach. In Section 2.6 of Chapter 2, we introduced the special base

$$e = \lim_{n\to\infty}\left(1 + \frac{1}{n}\right)^n \tag{5.5.1}$$

and demonstrated that

$$e^u = \lim_{n\to\infty}\left(1 + \frac{u}{n}\right)^n \tag{5.5.2}$$

for all u. In Section 3.4 of Chapter 3, we used formula (5.5.2) to prove that

$$\frac{d}{dx}e^x = e^x.$$

We then defined the natural logarithm to be the inverse function of the exponential function and derived the formula

$$\frac{d}{dx}\ln(x) = \frac{1}{x}$$

from the Inverse Function Derivative Rule.

The approach that we have summarized is a valid one, but it depends on formula (5.5.1), which we did not prove. It is instructive to see how calculus can be applied to finesse such difficulties. For the purposes of this section we will put aside everything that we have learned about the exponential and logarithm functions and redevelop these functions *from scratch*. Our point of departure is to define the natural logarithm function by the construction that is the subject of the second part of the Fundamental Theorem of Calculus:

$$\ln(x) = \int_1^x \frac{1}{t}\,dt, \quad 0 < x < \infty. \tag{5.5.3}$$

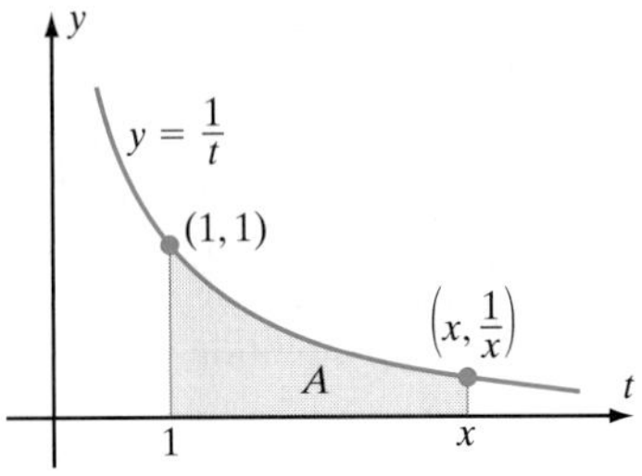

▲ Figure 1 $\ln(x) = A$

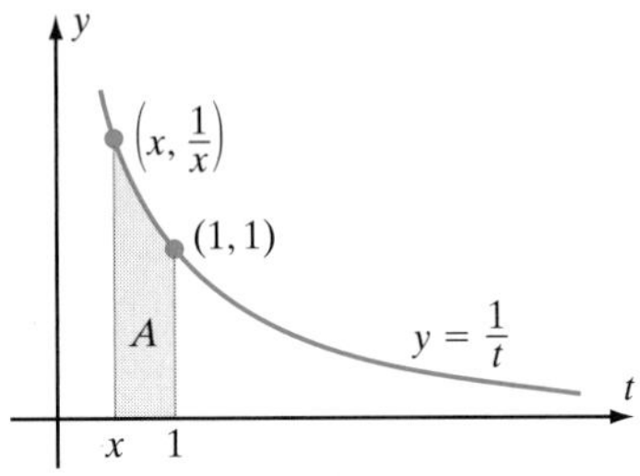

▲ Figure 2 $\ln(x) = -A (0 < x < 1)$

For $x > 1$, it is perfectly all right to think of $\ln(x)$ as the area of the region that lies under the graph of $y = 1/t$ and above the interval $[1, x]$ (Figure 1). For $0 < x < 1$, the value $\ln(x)$ is the *negative* of the area between the graph and the x-axis (because in this case the integration of the positive integrand $1/t$ is from right to left). See Figure 2. As you look at Figures 1 and 2, imagine sliding the point labeled x to the left or to the right on the horizontal axis. The area of the shaded region will certainly change. That is the functional relationship captured by formula (5.5.3). Although this definition of the natural logarithm may not be an intuitive one, it does lead quickly to the properties of the logarithm that we require. Indeed, a number of basic facts about the natural logarithm can be deduced immediately.

THEOREM 1 The natural logarithm has the following properties:

a. If $x > 1$, then $\ln(x) > 0$.
b. If $x = 1$, then $\ln(x) = 0$.
c. If $0 < x < 1$, then $\ln(x) < 0$.
d. The natural logarithm has a continuous derivative:

$$\frac{d}{dx}\ln(x) = \frac{1}{x}, \quad 0 < x.$$

e. The natural logarithm is an increasing function: $\ln(x_1) < \ln(x_2)$ if $0 < x_1 < x_2$.

Proof. Parts a and c may be understood geometrically in terms of areas, as has been discussed. Part b follows from Theorem 1d of Section 5.3. The derivative formula of part d is equation (5.4.6) with $f(t) = 1/t$. Finally, we deduce that $\ln(x)$ is an increasing function of x because its derivative $1/x$ is everywhere positive. ■

As the computation in Example 2 from Section 4.9 of Chapter 4 shows, the derivative formula of Theorem 1d can be generalized to

$$\frac{d}{dx}\ln(|x|) = \frac{1}{x},$$

which is true for all $x \neq 0$. Therefore the equation

$$\int \frac{1}{x}\,dx = \ln(|x|) + C$$

is the preferred antiderivative formulation of Theorem 1d.

Properties of the Natural Logarithm

Using definition (5.5.3), we can derive many useful properties of the natural logarithm.

THEOREM 2 Let x and y be positive. Then

$$\ln(x \cdot y) = \ln(x) + \ln(y), \tag{5.5.4}$$

$$\ln\left(\frac{1}{x}\right) = -\ln(x), \tag{5.5.5}$$

$$\ln\left(\frac{x}{y}\right) = \ln(x) - \ln(y), \tag{5.5.6}$$

and, for any number p,

$$\ln(x^p) = p \cdot \ln(x). \tag{5.5.7}$$

Proof. To obtain identity (5.5.4), we fix y and use the Chain Rule to calculate

$$\frac{d}{dx}\ln(xy) = \frac{1}{xy} \cdot \frac{d}{dx}(xy) = \frac{1}{xy} \cdot y = \frac{1}{x}.$$

In other words, for each fixed y, $\ln(xy)$ is an antiderivative of $1/x$. Because $\ln(xy)$ and $\ln(x)$ are antiderivatives of the same expression, $1/x$, it follows that

$$\ln(xy) = \ln(x) + C \tag{5.5.8}$$

for some constant C. By setting $x = 1$ in equation (5.5.8), we see that $\ln(y) = 0 + C$, or $C = \ln(y)$. Substituting this value for C in equation (5.5.8) yields (5.5.4).

Next, we set $y = 1/x$ in equation (5.5.4), obtaining

$$\ln(x) + \ln\left(\frac{1}{x}\right) = \ln\left(x \cdot \frac{1}{x}\right) = \ln(1) = 0,$$

which is equivalent to equation (5.5.5). To obtain equation (5.5.6), we write x/y as $x \cdot (1/y)$ and use equations (5.5.4) and (5.5.5) as follows:

$$\ln\left(\frac{x}{y}\right) = \ln\left(x \cdot \frac{1}{y}\right) \overset{(5.5.4)}{=} \ln(x) + \ln\left(\frac{1}{y}\right) \overset{(5.5.5)}{=} \ln(x) - \ln(y).$$

Equation (5.5.7) is proved in several stages. As you read the demonstration, notice how earlier steps are used to prove later steps. First, if $p = 0$, then $\ln(x^p) = \ln(1) = 0 = 0 \cdot \ln(x)$. Thus equation (5.5.7) is true for $p = 0$. Next observe that

$$\ln(x^2) = \ln(x \cdot x) = \ln(x) + \ln(x) = 2 \cdot \ln(x).$$

Similarly,

$$\ln(x^3) = \ln(x^2 \cdot x) = \ln(x^2) + \ln(x) = 2 \cdot \ln(x) + \ln(x) = 3\ln(x)$$

This idea can be repeated, showing that (5.5.7) holds for every positive integer. Now suppose that p is a negative integer. Then $p = -|p|$ with $|p|$ a positive integer. By the positive integer case of (5.5.7), which has just been proved, we have

$$\ln\left((x^{-1})^{|p|}\right) = |p|\ln(x^{-1}).$$

As a result,

$$\ln(x^p) = \ln(x^{-|p|}) = \ln\left((x^{-1})^{|p|}\right) = |p|\ln(x^{-1}) = |p|\ln\left(\frac{1}{x}\right) \overset{(5.5.5)}{=} -|p|\ln(x) = p\ln(x).$$

This equation completes the proof that identity (5.5.7) is valid for all integers p. Now suppose that p is a rational number. That is, suppose that there are two integers m and $n \neq 0$ such that $p = m/n$. Then, using the integer case of (5.5.7) twice, we have

$$m \cdot \ln(x) = \ln(x^m) = \ln(x^{p \cdot n}) = \ln\left((x^p)^n\right) = n \cdot \ln(x^p).$$

By dividing the first and last terms in this chain of equalities by n, we obtain $(m/n) \cdot \ln(x) = \ln(x^p)$, or $p \cdot \ln\,(x) = \ln(x^p)$. We have now proved that equation (5.5.7) holds for all rational values of p. Later in this section, we will rigorously define irrational exponents in a way that ensures the validity of property (5.5.7) for irrational values of p as well. ■

Graphing the Natural Logarithm Function

Now we examine the graph of $y = \ln\,(x)$. We have already learned that the natural logarithm is an increasing function with x-intercept 1 (Theorem 1e and 1b). Because

$$\frac{d^2}{dx^2}\ln(x) = \frac{d}{dx}\left(\frac{1}{x}\right) = -\frac{1}{x^2} < 0,$$

we infer that the graph of the natural logarithm is concave down. There are no relative maxima or minima because the derivative $1/x$ is never 0.

Let us now investigate the behavior of $\ln(x)$ as x tends to infinity. Because the natural logarithm is an increasing function, $\ln(x)$ either increases to a finite bound as x tends to infinity or $\ln(x)$ increases without bound. We can rule out the first possibility by considering the sequence $\{\ln(2^n)\}$. Because $2 > 1$, we deduce that $\ln(2) > \ln(1) = 0$. Therefore

$$\ln(2^n) \overset{(5.5.7)}{=} n \cdot \ln(2) \to \infty.$$

We conclude that

$$\lim_{x\to\infty} \ln(x) = \infty.$$

We can complete our sketch of $y = \ln(x)$ for $0 < x < \infty$ by investigating the behavior as $x \to 0^+$. In fact, a change of variable allows us to determine the behavior at 0^+ from the limit at infinity:

$$\lim_{x\to 0^+} \ln(x) \overset{u=1/x}{=} \lim_{u\to+\infty} \ln\left(\frac{1}{u}\right) \overset{(5.5.5)}{=} -\lim_{u\to\infty} \ln(u) = -\infty.$$

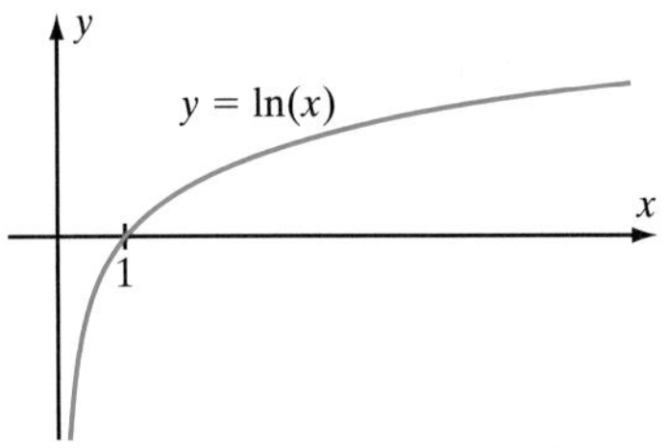

▲ Figure 3

In other words, the y-axis is a vertical asymptote. A sketch of the graph of $y = \ln(x)$ appears in Figure 3.

Because $x \mapsto \ln(x)$ is a continuous function that takes on arbitrarily large positive and negative values, the Intermediate Value Theorem (Section 2.3 of Chapter 2) tells us that the equation $\ln(x) = \gamma$ has a solution for each real number γ. Because the natural logarithm is an increasing function, the solution is unique. We summarize our findings in Theorem 3.

THEOREM 3 The natural logarithm is an increasing function with domain equal to the set of positive real numbers. Its range is the set of all real numbers. The equation $\ln(x) = \gamma$ has a *unique* solution $x \in \mathbb{R}^+$ for every $\gamma \in \mathbb{R}$. The graph of the natural logarithm function is concave down. The y-axis is a vertical asymptote.

The Exponential Function

In the preceding subsection, we deduced that the equation $\ln(x) = \gamma$ has a unique positive solution x for every real number γ. In other words, the natural logarithm is an invertible function with domain $\mathbb{R}^+$ and image $\mathbb{R}$. The inverse of the natural logarithm is called the *exponential function* (or *natural exponential function*) and is written $x \mapsto \exp(x)$. The domain of the exponential function is the entire real line; the image of the exponential function is the set of positive real numbers.

Recall that, when we say that the exponential function is the inverse function of the natural logarithm, we mean that a real number a and a positive number b are related by the equation

$$b = \exp(a)$$

if and only if

$$a = \ln(b)$$

We may also express the inverse relationship between the natural exponential and logarithm functions as follows:

$$\ln(\exp(a)) = a \quad \text{for all } a \quad \text{and} \quad \exp(\ln(b)) = b \quad \text{for all } b > 0.$$

Before continuing, remember that we have started from scratch in this section. In the current approach, we do not yet have the number e. The idea is to use the exponential function to define e. After that is done, we will show that $\exp(x) = e^x$. These developments, however, are still ahead.

Properties of the Exponential Function

The graph of $y = \exp(x)$ appears in Figure 4. It is obtained by reflecting the graph of $y = \ln(x)$ through the line $y = x$. (Recall that this is the general procedure for finding the graph of an inverse function.) Because the graph of $y = \ln(x)$ lies below the line $y = x$ and is concave down and rising, we conclude that the graph of $y = \exp(x)$ lies above the line $y = x$ and is concave up and rising. In particular, the natural exponential is an increasing function. Because $\ln(1) = 0$, it follows that $\exp(0) = 1$. In other words, 1 is the y-intercept of the graph of $y = \exp(x)$. The exponential function only assumes positive values, so there is no x-intercept.

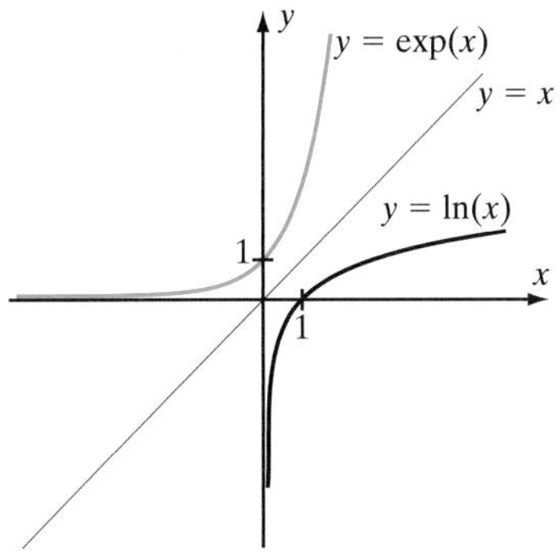

▲ Figure 4

Next, we turn to some of the algebraic properties of the exponential function. The basic exponential law is

$$\exp(s + t) = \exp(s) \cdot \exp(t), \tag{5.5.9}$$

which holds for all real numbers s and t. Identity (5.5.9) is the analogue of equation (5.5.4) for the logarithm, from which it can be derived. Indeed, notice that

$$\ln(\exp(s + t)) = s + t = \ln(\exp(s)) + \ln(\exp(t)) \overset{(5.5.4)}{=} \ln(\exp(s) \cdot \exp(t))$$

Because the natural logarithm is a one-to-one function, it follows that $\exp(s + t)$ must equal $\exp(s) \cdot \exp(t)$, as identity (5.5.9) asserts.

A few other useful identities can be obtained from equation (5.5.9). For instance, when s equals $-t$, equation (5.5.9) becomes $\exp(0) = \exp(-t) \cdot \exp(t)$ or

$$\exp(-t) = \frac{1}{\exp(t)}, \tag{5.5.10}$$

which is the analogue of logarithm identity (5.5.5). Formulas (5.5.9) and (5.5.10) can be combined to produce

$$\exp(s-t)=\frac{\exp(s)}{\exp(t)}, \tag{5.5.11}$$

which is the analogue of logarithm identity (5.5.6). Finally, we obtain the exponential version of logarithm equation (5.5.7) by noting that

$$\ln(\exp(s)^t) \overset{(5.5.8)}{=} t\ln(\exp(s)) = ts = st = \ln(\exp(st)).$$

Because the logarithm is one-to-one, we deduce that

$$(\exp(s))^t = \exp(st) \tag{5.5.12}$$

for all real s and t.

Derivatives and Integrals Involving the Exponential Function

Just as the algebraic properties of the exponential function follow from corresponding properties of the logarithm, so too can the calculus properties of the exponential function be deduced from their logarithm counterparts.

THEOREM 4 The exponential function satisfies

$$\frac{d}{dx}\exp(x) = \exp(x) \quad \text{and} \quad \int \exp(x)\,dx = \exp(x) + C.$$

More generally,

$$\frac{d}{dx}\exp(u) = \exp(u)\frac{du}{dx} \quad \text{and} \quad \int \exp(u)\frac{du}{dx}\,dx = \exp(u) + C.$$

Proof. When we studied inverse functions in Section 3.6 of Chapter 3, we learned that, if g is the inverse function of f, if f is differentiable at $g(c)$, and if $f'(g(c)) \neq 0$, then g is differentiable at c, and

$$g'(c) = \frac{1}{f'(g(c))}.$$

Let $g(x) = \exp(x)$ and $f(x) = \ln(x)$. Then

$$\left.\frac{d}{dx}\exp(x)\right|_{x=c} = \frac{1}{\left.\frac{d}{dx}\ln(x)\right|_{x=\exp(c)}} = \frac{1}{\left.\frac{1}{x}\right|_{x=\exp(c)}} = \frac{1}{1/\exp(c)} = \exp(c).$$

In other words,

$$\frac{d}{dx}\exp(x) = \exp(x),$$

which is the first assertion. The other assertions follow from this one. ■

The Number e

In our present approach, we define the number e by the simple equation

$$e = \exp(1). \tag{5.5.13}$$

See Figure 5a. Applying the natural logarithm to each side of equation (5.5.13) and using the inverse relationship of the natural logarithm and exponential functions, we obtain

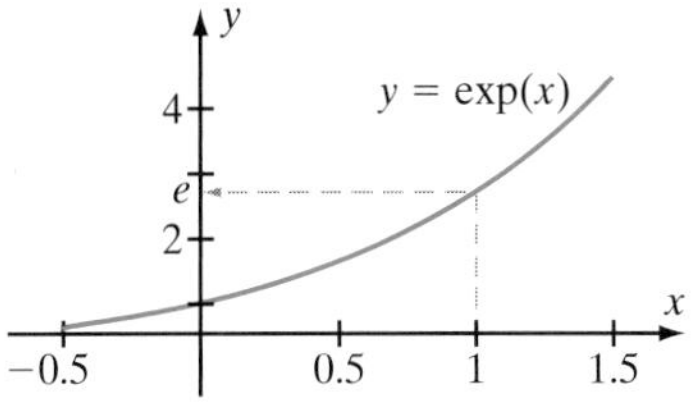

▲ Figure 5a $e = \exp(1)$

$$\ln(e) = 1. \tag{5.5.14}$$

See Figure 5b. By noting that $\ln(e) = \int_1^e 1/x\, dx$, we see that e is the number such that the area under the graph of $y = 1/x$ and over the interval $[1, e]$ is 1. Figure 5c illustrates this third representation of the number e.

One advantage of the present development is that we can derive equation (5.5.1) in a few steps. To do so, we first examine the definition of the derivative of $\ln(x)$ at $x = 1$:

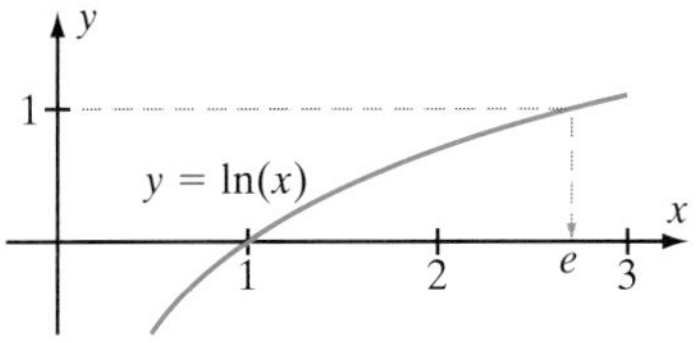

▲ Figure 5b $\ln(e) = 1$

$$\lim_{h\to 0} \frac{\ln(1+h) - \ln(1)}{h} = \frac{d}{dx}\ln(x)\bigg|_{x=1} = \frac{1}{x}\bigg|_{x=1} = 1.$$

In this equation, we will let $h = 1/n$ where n tends to infinity through positive integers. Thus

$$1 = \lim_{n\to\infty} \frac{\ln\Big(1 + (1/n)\Big) - \ln(1)}{1/n} = \lim_{n\to\infty} \frac{\ln(1 + 1/n)}{1/n}$$
$$= \lim_{n\to\infty} n \ln\left(1 + \frac{1}{n}\right) \overset{(5.5.7)}{=} \lim_{n\to\infty} \ln\left(\left(1 + \frac{1}{n}\right)^n\right).$$

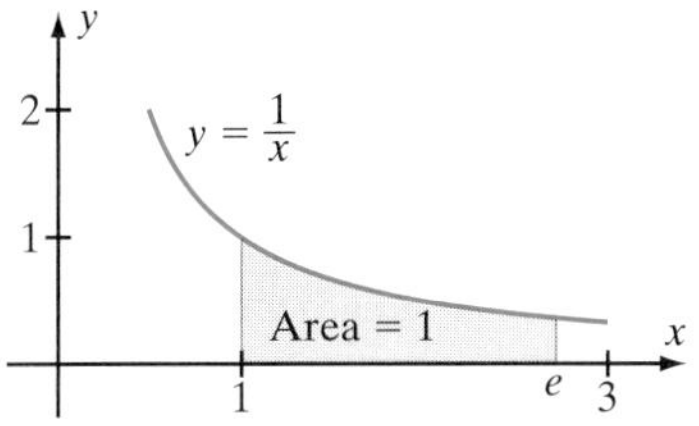

▲ Figure 5c $\int_1^e \frac{1}{x} dx = 1$

Now we exponentiate both sides of this equation and use the continuity of the exponential function to obtain

$$e \overset{(5.5.13)}{=} \exp(1) = \exp\left(\lim_{n\to\infty} \ln\left(\left(1 + \frac{1}{n}\right)^n\right)\right)$$
$$= \lim_{n\to\infty} \exp\left(\ln\left(\left(1 + \frac{1}{n}\right)^n\right)\right) = \lim_{n\to\infty} \left(1 + \frac{1}{n}\right)^n,$$

which is formula (5.5.1).

Logarithms and Powers with Arbitrary Bases

Until now, we have understood a^x to be a limit when the value of x is irrational. In our new treatment, we use the logarithm and exponential functions to define a^x in a more elegant way. To do so, we observe that, for any positive number a and any rational number x,

$$a^x = \exp(\ln(a^x)) \overset{(5.5.7)}{=} \exp(x \cdot \ln(a)). \tag{5.5.15}$$

Notice that the right side of equation (5.5.15) is meaningful even when x is irrational. We therefore define a^x for *any* real number x by

$$a^x = \exp(x \cdot \ln(a)) \qquad (x \in \mathbb{R}) \tag{5.5.16}$$

Bear in mind that equation (5.5.16) extends the definition of exponentiation to irrational values of x and agrees with the elementary, algebraic notion of exponentiation when x is rational.

As an immediate consequence of this extension, we see that

$$\ln(a^x) = \ln(\exp(x \cdot \ln(a))) = x \cdot \ln(a)$$

which shows that formula (5.5.7) is valid even for irrational values of x. As the next example shows, it is easy to use definition (5.5.16) when working with exponents.

▶ **EXAMPLE 1** Use formula (5.5.16) to derive the Power Rule of differentiation:

$$\frac{d}{dx}x^p = p \cdot x^{p-1}.$$

Solution Since Chapter 3, we have used this basic differentiation rule without verification. Now we can prove it as follows:

$$\begin{aligned}
\frac{d}{dx}x^p &= \frac{d}{dx}\Big(\exp\left(p \cdot \ln(x)\right)\Big) && \text{(by (5.5.16))}\\
&= \exp\big(p \cdot \ln(x)\big) \cdot \frac{d}{dx}\big(p \cdot \ln(x)\big) && \text{(Chain Rule)}\\
&= \exp\big(p \cdot \ln(x)\big) \cdot \left(p \cdot \frac{1}{x}\right) && \\
&= x^p \cdot \left(p \cdot \frac{1}{x}\right) && \text{(by (5.5.16))}\\
&= p \cdot x^{p-1}. && \text{(algebraic simplification)}
\end{aligned}$$ ◀

Notice that when a is taken to be e, we have $\ln(a) = \ln(e) = 1$ by equation (5.5.14). As a result, formula (5.5.16) simplifies to

$$e^x = \exp(x).$$

We may use this observation to rewrite equation (5.5.16) as

$$a^x = e^{x\ln(a)}. \tag{5.5.17}$$

The laws of exponents can all be quickly derived from equation (5.5.14) and the corresponding properties of the exponential function. We state these laws in the next theorem and leave their proofs as Exercises 53–56.

THEOREM 5 **(Laws of Exponents)** If $a, b > 0$, and $x, y \in \mathbb{R}$, then

a. $a^0 = 1$
b. $a^1 = a$
c. $a^{x+y} = a^x \cdot a^y$
d. $a^{x-y} = \dfrac{a^x}{a^y}$
e. $(a^x)^y = a^{x \cdot y}$
f. $a^x = b$ if and only if $b^{1/x} = a$ (provided $x \neq 0$)
g. $(a \cdot b)^x = a^x \cdot b^x$

Logarithms with Arbitrary Bases

If a is any positive number other than 1, then we define the logarithm function $\log_a$ with base a by

$$\log_a(x) = \frac{\ln(x)}{\ln(a)} \qquad (x \in \mathbb{R}^+). \tag{5.5.18}$$

Notice that $\ln(a)$ makes sense in (5.5.18) because a is assumed to be positive. Furthermore, because we have made the assumption $a \neq 1$, $\ln(a)$ is nonzero, and the division in equation (5.5.18) is permitted. Observe that $\log_e(x)$ reduces to the natural logarithm $\ln(x)$ because $\ln(a) = 1$ when $a = e$.

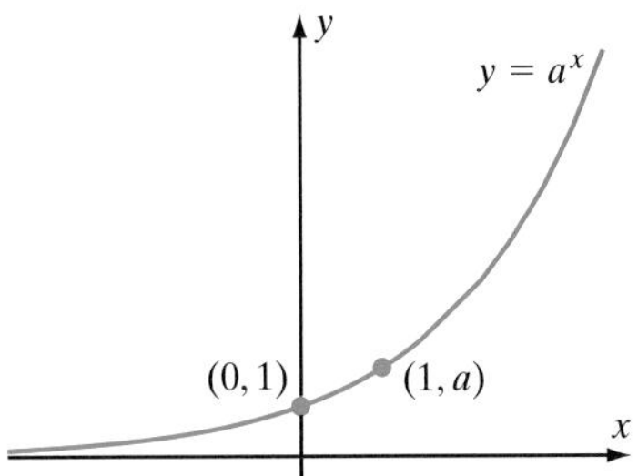

▲ **Figure 6** The graph of $y = a^x$, $a > 1$

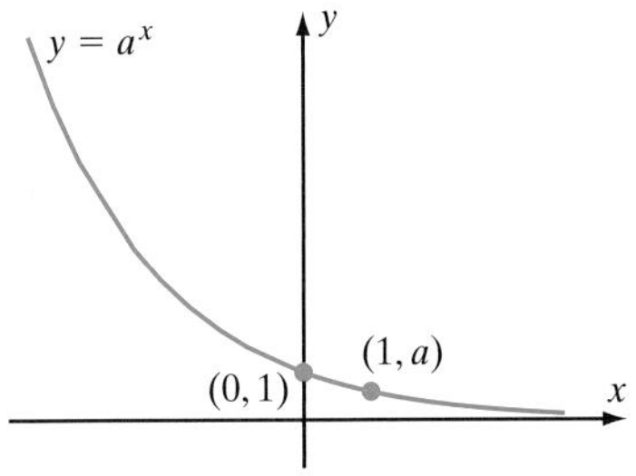

▲ **Figure 7** The graph of $y = a^x$, $0 < a < 1$

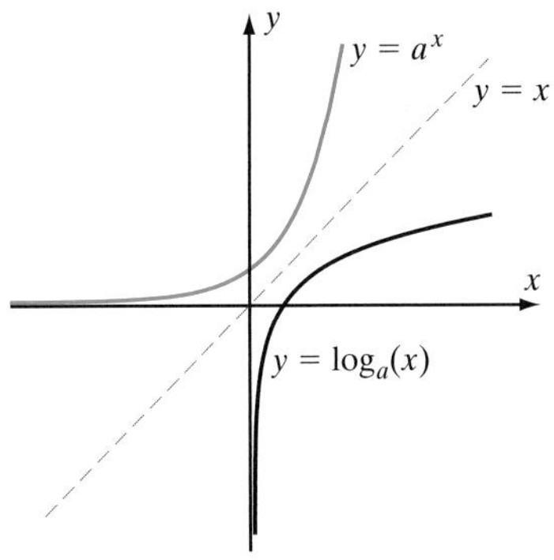

▲ **Figure 8a** $y = \log a(x), 1 < a.$

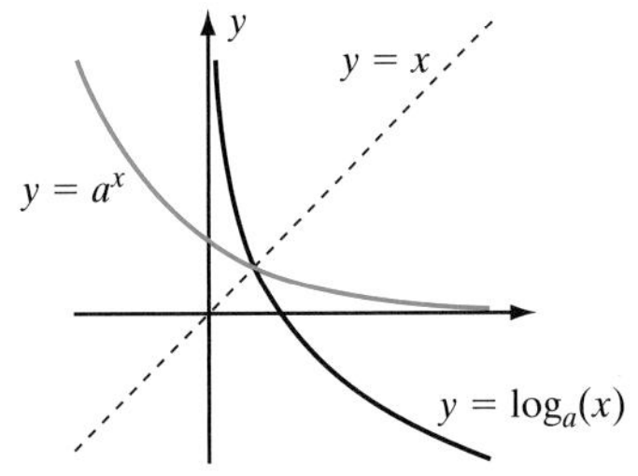

The graph of $y = \log_a(x)$, $0 < a < 1$

▲ **Figure 8b**

THEOREM 6 For any fixed positive $a \neq 1$, the function $x \mapsto a^x$ has domain $\mathbb{R}$ and image $\mathbb{R}^+$. The function $x \mapsto \log_a(x)$ has domain $\mathbb{R}^+$ and image $\mathbb{R}$. The two functions satisfy the identities

$$a^{\log_a(y)} = y \quad (y \in \mathbb{R}^+) \quad \text{and} \quad \log_a(a^x) = x \quad (x \in \mathbb{R}).$$

In words, the functions $x \mapsto a^x$ and $y \mapsto \log_a(y)$ are *inverse* to each other.

Proof. We obtain the inverse relationship between $x \mapsto a^x$ and $y \mapsto \log_a(y)$ from the inverse relationship between the natural logarithm and exponential functions. For $y > 0$, we use equation (5.5.17) with $x = \ln(y)/\ln(a)$ to obtain

$$a^{\log_a(y)} = a^{\ln(y)/\ln(a)} \overset{(5.5.17)}{=} \exp\left(\frac{\ln(y)}{\ln(a)} \cdot \ln(a)\right) = \exp(\ln(y)) = y.$$

In the other direction, we have, for any x,

$$\log_a(a^x) \overset{(5.5.18)}{=} \frac{\ln(a^x)}{\ln(a)} \overset{(5.5.17)}{=} \frac{\ln(\exp(x \cdot \ln(a)))}{\ln(a)} = \frac{x \cdot \ln(a)}{\ln(a)} = x. \qquad \blacksquare$$

From Theorem 6 and the discussion of inverse functions in Section 1.5 of Chapter 1, we see that the graphs of $y = a^x$ and $y = \log_a(x)$ are reflections of each other through the line $y = x$. Figures 6, 7, and 8 illustrate these reflections.

The laws of logarithms with arbitrary bases can all be quickly derived from equation (5.5.18) and the corresponding properties of the natural logarithm function. We state these properties in the next theorem and leave their proofs as Exercises 57–60.

THEOREM 7 If $x, y, a > 0$, and $a \neq 1$, then

a. $\log_a(a) = 1$
b. $\log_a(1) = 0$
c. $\log_a(xy) = \log_a(x) + \log_a(y)$
d. $\log_a\left(\frac{1}{x}\right) = -\log_a(x)$
e. $\log_a\left(\frac{x}{y}\right) = \log_a(x) - \log_a(y)$
f. $\log_a(x^p) = p\log_a(x)$

Differentiation and Integration of a^x and $\log_a(x)$

Differentiation formulas for a^x and $\log_a(x)$ are readily obtained from the derivatives of $\exp(x)$ and $\ln(x)$ and the Chain Rule. For a fixed positive constant $a \neq 1$, we have

$$\frac{d}{dx}a^x = \ln(a) \cdot a^x \tag{5.5.19}$$

and

$$\frac{d}{dx}\log_a(x) = \frac{1}{x \cdot \ln(a)}. \tag{5.5.20}$$

We obtain formula (5.5.19) by setting $u = x \cdot \ln(a)$ and calculating with the Chain Rule as follows:

$$\frac{d}{dx}(a^x) = \frac{d}{dx}e^{x \cdot \ln(a)} = \frac{d}{dx}e^u = \frac{d}{du}e^u \cdot \frac{du}{dx} = e^u \cdot \frac{d}{dx}(x \cdot \ln(a)) = e^{x \cdot \ln(a)} \cdot \ln(a) = a^x \cdot \ln(a).$$

The derivation of (5.5.20) is straightforward:

$$\frac{d}{dx}\log_a(x) = \frac{d}{dx}\left(\frac{\ln(x)}{\ln(a)}\right) = \frac{1}{\ln(a)}\frac{d}{dx}\ln(x) = \frac{1}{\ln(a)} \cdot \frac{1}{x}.$$

Formula (5.5.19) is often combined with the Chain Rule and written in the form

$$\frac{d}{dx}a^u = \ln(a) \cdot a^u \cdot \frac{du}{dx}. \tag{5.5.21}$$

Finally, it is useful to rewrite formula (5.5.19) in antiderivative form as

$$\int a^x \, dx = \frac{a^x}{\ln(a)} + C.$$

QUICK QUIZ

1. Suppose that $\int_1^a 1/t\,dt = 3$, and $\int_1^b 1/t\,dt = 5$. What is $\int_1^{ab} 1/t\,dt$?
2. For what value of a is $\int_1^a 1/t\,dt = 1$?
3. Suppose that a is a positive constant. For what value of u is $a^x = e^u$?
4. Suppose that a is a positive constant not equal to 1. For what value of k is $\log_a(x) = k \cdot \ln(x)$?

Answers

1. 8 2. e 3. $x\ln(a)$ 4. $1/\ln(a)$

EXERCISES

Problems for Practice

In each of Exercises 1–12, the numbers u and v satisfy the equations $\int_1^u (1/t)dt = 4$ and $\int_1^v (1/t)\,dt = 12$. Use these equations, formula (5.5.3), and the properties of logarithms to calculate the integral of $1/t$ over the given interval.

1. $[1, 1/u]$
2. $[1, ue]$
3. $[1, u/e]$
4. $[u, 2u]$
5. $[3, 3u]$
6. $[1, uv]$
7. $[1, v/u]$
8. $[u, v]$
9. $[u, uv]$
10. $[1, v^2]$
11. $[1, 2v]$
12. $[2u, 2v]$

In each of Exercises 13–30, calculate the derivative with respect to x of the given expression.

13. $\ln(4x)$
14. $\ln(-3x)$
15. $\ln(2x+3)$
16. $\ln(\sec(x))$
17. $\ln(\sqrt{1-x})$
18. $\ln(1/x)$
19. $1/(4+\ln(x))$
20. $\ln(\ln(x))$
21. $\ln^2(x)$
22. 5^x
23. $x2^x$
24. $3^x \cdot \exp(x^2)$
25. $2^{\ln(x)}$
26. $\ln(1+3^x)$
27. $\log_{10}(5/x)$
28. $\log_5(5+2x)$
29. $\log_2(3x)$
30. $\log_2(2+3^x)$

In each of Exercises 31–34, calculate the given definite integral.

31. $\int_0^1 2^x\,dx$
32. $\int_0^3 (9/3^x)\,dx$
33. $\displaystyle\int_0^1 \frac{1+10^x}{10^x}\,dx$
34. $\displaystyle\int_0^1 \frac{2^x+3^x}{5^x}\,dx$

Further Theory and Practice

In each of Exercises 35–42, calculate the first and second derivatives of the given expression, and classify its local extrema.

35. $2^x - x$
36. $x2^x$
37. $x10^{-x}$
38. $\log_2(x) - x$
39. $x\log_{10}(x)$
40. $\log_3(x)\,/x$
41. $\ln^2(x)$
42. $x^2 - 2\log_8(x)$

In each of Exercises 43–48, calculate the first and second derivatives of $F(x) = \int_a^{u(x)} f(t)\,dt$ for the given functions u and f.

43. $u(x) = \ln(x)$ $\quad f(t) = 1/t$
44. $u(x) = \log_2(x)$ $\quad f(t) = 1/t$
45. $u(x) = 1/x$ $\quad f(t) = \ln(t)$
46. $u(x) = 2^x$ $\quad f(t) = 2^t$
47. $u(x) = \ln(x)$ $\quad f(t) = \ln(t)$
48. $u(x) = \log_2(x)$ $\quad f(t) = \log_3(t)$

49. Let A and B be constants. Show that $y = A\cos(\ln(x)) + B\sin(\ln(x))$ is a solution of the equation

$$x^2y'' + xy' + y = 0.$$

50. Suppose that $a > 1$. Let m be any positive number. Show that the graph of $y = \log_a(x)$ has a tangent line with slope m.

51. Show that

$$\int \frac{1}{|x|}\,dx = \ln(|x|) + C$$

is not a valid integration formula.

52. It is known that $\int_1^e \ln(x)\,dx = 1$. If a is a positive constant, what is $\int_1^e \ln(ax)\,dx$?

53. Let A and B be constants. Calculate the derivative of $A \cdot x\ln(x) + B \cdot x$ with respect to x. Show that there are values of A and B such that

$$\int \ln(x)\,dx = A \cdot x\ln(x) + B \cdot x + C$$

where C is an arbitrary constant.

54. Fix a positive h, and graph $f(x) = 1/(1+x)$ for $0 \le x \le h$. By comparing the area under the graph of f with an inscribed rectangle and a circumscribed rectangle, show that

$$\frac{h}{1+h} < \ln(1+h) < h \quad (0 < h).$$

55. Let f be a differentiable function that is defined on $\mathbb{R}^+$ and that satisfies identity $f(xy) = f(x) + f(y)$. This exercise outlines a proof that

$$f(x) = f'(1) \cdot \ln(x).$$

a. By differentiating both sides of the given equation with respect to x, treating y as constant, show that $f'(xy) \cdot y = f'(x)$.
b. By setting $x = t$ and $y = 1/t$ in the equation obtained in part a, show that f is an antiderivative of the function $t \mapsto f'(1)/t$.
c. Use the result of part c to deduce that $f(t) = f'(1) \cdot \ln(t) + C$ where C is a constant. By evalutating $f(t)$ at $t = 1$, show that $C = f(1)$.
d. By considering the given property of f with $x = y = 1$, deduce that $f(1) = 0$.

56. Find the minimum value m of $f(x) = x - \ln(x)$. Use the inequality $x - \ln(x) \ge m$ to deduce that $\ln(1+x) \le x$ for $-1 < x$. Prove that there is no real value of x such that $\ln(x) = x$.

57. Use equations (5.5.16) and (5.5.9) to prove Theorem 5c.

58. Use equations (5.5.16) and (5.5.11) to prove Theorem 5d.

59. Use equations (5.5.16) and (5.5.12) to prove Theorem 5e.

60. Use equations (5.5.16), (5.5.4), and (5.5.9) to prove Theorem 5g.

61. Use equations (5.5.18) and (5.5.4) to prove Theorem 7c.

62. Use equations (5.5.18) and (5.5.5) to prove Theorem 7d.

63. Use equations (5.5.18) and (5.5.6) to prove Theorem 7e.

64. Use equations (5.5.18) and (5.5.7) to prove Theorem 7f.

▶ Exercises 65–73 concern a function called the *Lambert W function*, introduced in Exercise 61 and denoted here by W. The Lambert W function is useful for the solution of many equations involving exponentials and logarithms. ◀

65. Show that $V:(-1,\infty) \to (-1/e,\infty)$ defined by $V(x) = x\exp(x)$ is an invertible function. Denote V^{-1} by W. Thus for any $-1/e > y < \infty$, and in particular for any nonnegative y, there is a unique value $W(y)$ of x for which $x\exp(x) = y$.
66. Implicitly differentiate the equation $W(y) \cdot \exp(W(y)) = y$ to show that

$$W'(y) = \frac{1}{y + \exp(W(y))} = \frac{W(y)}{y(1 + W(y))}.$$

67. Use l'Hôpital's Rule to show that $\lim_{y \to 0} W(y)/y = 1$.
68. Suppose that $1 < a$ and $0 < y$. Show that $W(y \cdot \ln(a))/\ln(a)$ is the unique solution of $x \cdot a^x = y$.
69. Show that for any nonnegative number y there is a unique value of x such that $x\ln(x) = y$. Use the Lambert W function to express the solution as a function of y.
70. Show that for any real number u, there is a unique value of x such that $x + \ln(x) = u$. Use the Lambert W function to express the solution as a function of u.
71. Show that for any real number u, the equation $x + e^x = u$ has a unique solution. Verify that $x = u - W(e^u)$ is the solution.
72. Suppose that a and b are constants greater than 1. Determine where the minimum value of $a^x - \log_b(x)$ occurs.
73. In this exercise, you will establish an inequality that is useful in analyzing motion with air resistance.
    a. Show that 2 is the minimum value of $x + 1/x$ for $x > 0$.
    b. Prove that $c^t + c^{-t} > 2$ if $t > 0$ and $c > 1$. Hence, $c^t - 1 > 1 - c^{-t}$ if $t > 0$ and $c > 1$.
    c. Suppose that a and b are positive constants, that $c > 1$, and that

    $$\int_0^a (c^t - 1)\,dt = \int_0^b (1 - c^{-t})\,dt.$$

    Show that $b > a$.
    d. Show that if a and b are positive, and if

    $$e^a - e^{-b} = a + b,$$

    then $b > a$.

Calculator/Computer Exercises

74. Use the Newton-Raphson Method to find an approximation to the solution of the equation $\ln(x) = 3 - x$. How can you be sure that there is *exactly* one real solution?
75. Use the first derivative to minimize $\sqrt{x-1}/\log_3(x)$.
76. Use the first derivative to maximize $\log_2(1 + x^2)/2^x$, $0 \le x$.
77. Find the local extrema of $x + \log_2(1 + x + x^2)$.
78. The integral $\int_a^b (1/\ln(x))\,dx$ approximates the number of primes p in the interval $[a,b]$. For example, the 8000th and 9000th primes are 81799 and 93179. Thus there are exactly 1000 primes in the interval $[81800, 93179]$. The approximation $\int_{81800}^{93179} (1/\ln(x))\,dx = 1000.04\ldots$ is quite good. Using a Riemann sum with $N = 4$ and the midpoints of the subintervals as a choice of points, approximate the number of primes that lie in the interval $[50000, 60000]$. (The exact number of primes in the interval is 924.)

5.6 Integration by Substitution

The method of substitution, which is also called "change of variable," provides a way to simplify or transform an integrand. We begin by illustrating the method of substitution with an example.

Consider the integral

$$\int \cos(x^2 + 1) \cdot 2x\,dx.$$

The function $\phi(x) = x^2 + 1$ and its derivative $\phi'(x) = 2x$ *both* appear in the integrand. Suppose we denote the expression $\phi(x) = x^2 + 1$ by u. We write

$$u = x^2 + 1. \tag{5.6.1}$$

We then have $\dfrac{du}{dx} = 2x$. It is suggestive to write this as

$$\frac{du}{dx}\,dx = 2x\,dx$$

and to abbreviate this last formula as

$$du = 2x\,dx. \tag{5.6.2}$$

Now we may use equations (5.6.1) and (5.6.2) to rewrite $\int \cos(x^2+1)\cdot 2x\,dx$ as $\int \cos(u)\,du$. The transformed integral is considerably simpler. Because an antiderivative for $\cos(u)$ is $\sin(u)$, we have

$$\int \cos(u)\,du = \sin(u) + C.$$

Rewriting this in terms of the original variable x, we have

$$\int \cos(x^2+1)\cdot 2x\,dx = \sin(x^2+1) + C.$$

As with all indefinite integral problems, we can check the answer by differentiation:

$$\frac{d}{dx}(\sin(x^2+1)) = \cos(x^2+1)\cdot\frac{d}{dx}(x^2+1) = \cos(x^2+1)\cdot 2x,$$

as required. To save space, we will not include verifications of our work in later examples. However, you should get into the habit of checking the antiderivatives that you obtain.

Let us summarize the method of substitution, which is sometimes referred to as the *u-substitution* because the variable that effects the substitution is commonly designated by u.

Key Steps for the Method of Substitution

1. To apply the method of substitution to an integral of the form $\int f(x)\,dx$, find an expression $\phi(x)$ in the integrand that has a derivative $\phi'(x)$ that also appears in the integrand.
2. Substitute u for $\phi(x)$ and du for $\phi'(x)dx$.
3. Do not proceed unless the entire integrand is expressed in terms of the new variable u. (*No* xs can remain.)
4. Evaluate the new integral to obtain an answer expressed in terms of u.
5. Resubstitute to obtain an answer in terms of x.

It is an error to omit step 5. In the example that we worked at the start of this section, the first four steps result in

$$\int \cos(x^2+1)\cdot 2x\,dx = \int \cos(u)\,du = \sin(u) + C. \tag{5.6.3}$$

In line (5.6.3), the indefinite integral in the middle is certainly equal to the antiderivative on the right. However, these expressions in u are not literally equal to $\int \cos(x^2+1)\cdot 2x\,dx$, which signifies an antiderivative with respect to x. The antiderivative $\sin(u)+C$ can equal the given integral only after Step 5 has been performed.

Some Examples of Indefinite Integration by Substitution

▶ **EXAMPLE 1** Evaluate the indefinite integral $\int \sin^4(x)\cos(x)\,dx$.

Solution We notice that the expression $\sin(x)$, together with its derivative $\cos(x)$, appears in the integrand. This suggests setting

$$u = \sin(x) \quad \text{and} \quad du = \left(\frac{d}{dx}\sin(x)\right)dx = \cos(x)\,dx.$$

The integral then becomes $\int u^4\,du$, which is readily calculated to be $u^5/5 + C$. Resubstituting to obtain an expression in x, we conclude that

$$\int \sin^4(x)\cos(x)\,dx = \frac{1}{5}\sin^5(x) + C. \quad ◀$$

▶ **EXAMPLE 2** Compute the integral

$$\int \frac{\sqrt{x}}{(2+x^{3/2})^3}\,dx.$$

Solution We notice that the derivative of the expression $(2 + x^{3/2})$ is $(3/2)x^{1/2}$ and that the latter is similar (up to a constant multiple) to the numerator. This motivates the substitution

$$u = 2 + x^{3/2} \quad \text{and} \quad du = \frac{d}{dx}(2 + x^{3/2})\,dx = \frac{3}{2}x^{1/2}\,dx.$$

The formula for du may be rewritten as

$$\sqrt{x}\,dx = \frac{2}{3}\,du.$$

Therefore upon substitution, our integral becomes

$$\int \frac{(2/3)\,du}{u^3} = \frac{2}{3}\int u^{-3}\,du = \frac{2}{3}\left(-\frac{1}{2}u^{-2}\right) + C = -\frac{1}{3}u^{-2} + C.$$

Resubstituting the x-variable gives

$$\int \frac{\sqrt{x}}{(2+x^{3/2})^3}\,dx = -\frac{1}{3}(2+x^{3/2})^{-2} + C. \quad ◀$$

The Method of Substitution for Definite Integrals

The method of substitution also applies well to definite integrals; the only new feature is that when we change variables, we must take the limits of integration into account.

▶ **EXAMPLE 3** Evaluate the integral $\int_3^4 x \cdot \sqrt{25 - x^2}\,dx$.

Solution We observe that the derivative of the expression $25 - x^2$ is $-2x$. This motivates the substitution

$$u = 25 - x^2 \quad \text{and} \quad du = \frac{d}{dx}(25 - x^2)\,dx = -2x\,dx.$$

However, we must not stop here. In the x-integral, the limits of integration are 3 and 4. When $x = 3$, $u = 25 - 3^2 = 16$; when $x = 4$, $u = 25 - 4^2 = 9$. Making our substitutions, we obtain

$$\begin{aligned} \int_3^4 x \cdot \sqrt{25 - x^2}\,dx &= -\frac{1}{2}\int_3^4 \sqrt{25 - x^2}(-2x)\,dx \\ &= -\frac{1}{2}\int_{16}^9 \sqrt{u}\,du \end{aligned}$$

$$= -\left(\frac{1}{2}\cdot\frac{2}{3}u^{3/2}\right)\Bigg|_{u=16}^{u=9}$$
$$= \left(-\frac{9^{3/2}}{3}\right) - \left(-\frac{16^{3/2}}{3}\right)$$
$$= -\frac{27}{3} + \frac{64}{3}$$
$$= \frac{37}{3}.$$

INSIGHT Example 3 demonstrates that a change of variable can convert an integral of the form $\int_a^b f(x)\,dx$ with $a<b$ into an integral of the form $\int_c^d g(u)\,du$ with $c>d$. In other words, a change of variable can reverse the direction of integration. This is perfectly all right. Calculate in the usual way, and do not omit the minus sign that arises in the calculation of du.

When we apply the method of substitution to a definite integral, it is essential that we take into account the effect that the change of variables has on the limits of integration. Example 3 illustrates how to do this *while* the substitution is being performed. An alternative method is to calculate the indefinite integral in the original variable and then use the original limits of integration, as is done in the next example.

▶ **EXAMPLE 4** Calculate $\int_0^2 xe^{x^2}\,dx$.

Solution Let

$$u = x^2 \quad \text{and} \quad du = \frac{d}{dx}(x^2)\,dx = 2x\,dx.$$

When $x = 2$, we have $u = 2^2 = 4$; when $x = 3$, we have $u = 3^2 = 9$. It follows that

$$\int_2^3 xe^{x^2}\,dx = \frac{1}{2}\int_4^9 e^u\,du = \frac{1}{2}e^u\Big|_{u=4}^{u=9} = \frac{1}{2}(e^9 - e^4).$$

Alternatively, we can use the method of substitution to calculate an antiderivative of the given integrand, a procedure that entails resubstitution. Then we use this antiderivative with the original limits of integration to evaluate the given definite integral. Thus with $u = x^2$ and $du = 2x\,dx$, we have

$$\int xe^{x^2}\,dx = \frac{1}{2}\int e^u du = \frac{1}{2}e^u + C = \frac{1}{2}e^{x^2} + C.$$

We may omit the additive constant C when calculating the definite integral. The result is

$$\int_2^3 xe^{x^2}\,dx = \frac{1}{2}e^{x^2}\Big|_{x=2}^{x=3} = \frac{1}{2}(e^9 - e^4),$$

as we obtained with the first method. ◀

INSIGHT When evaluating a definite integral by making a substitution, it is important to use limits of integration that are appropriate for the variable of the integration. If you resubstitute so that the antiderivative is expressed in terms of the original variable of integration, then the original limits of integration should be used. If the antiderivative is expressed in terms of a variable of substitution u, and you do not resubstitute, then limits of integration must be calculated for u. In general, they will not be the same as the limits for the original variable, as Example 4 shows.

In the next example, the method of substitution works in a not-so-obvious way.

▶ **EXAMPLE 5** Evaluate the integral

$$\int \frac{x^2}{\sqrt{x+5}}\,dx.$$

Solution We do not immediately see an expression in the integrand whose derivative also appears. We instead set $u = x + 5$ because this substitution *simplifies* the radical. It follows that $du = dx$ and $x = u - 5$. The integral now has the form

$$\int \frac{(u-5)^2}{\sqrt{u}}\,du.$$

If we expand the numerator, this integral becomes

$$\int \frac{u^2 - 10u + 25}{\sqrt{u}}\,du = \int (u^{3/2} - 10u^{1/2} + 25u^{-1/2})\,du = \frac{2}{5}u^{5/2} - \frac{20}{3}u^{3/2} + 50u^{1/2} + C.$$

Resubstituting the x variable gives a final answer of

$$\int \frac{x^2}{\sqrt{x+5}}\,dx = \frac{2}{5}(x+5)^{5/2} - \frac{20}{3}(x+5)^{3/2} + 50(x+5)^{1/2} + C. \quad ◀$$

The Role of the Chain Rule in the Method of Substitution

We have not said much about *why* the method of substitution works. All that the method really amounts to is an application of the Chain Rule. For instance, suppose that we want to study the integral

$$\int F(\phi(x)) \cdot \phi'(x)\,dx. \tag{5.6.4}$$

If G is an antiderivative for F, then

$$(G \circ \phi)'(x) = G'(\phi(x))\phi'(x) = F(\phi(x)) \cdot \phi'(x).$$

Thus $G(\phi(x))$ is an antiderivative of $F(\phi(x)) \cdot \phi'(x)$ and

$$\int F(\phi(x)) \cdot \phi'(x)\,dx = G(\phi(x))$$

The method of substitution is merely a device for seeing this procedure more clearly. We let

$$u = \phi(x) \quad \text{and} \quad du = \phi'(x)\,dx$$

to transform integral (5.6.4) to

$$\int F(u)\,du = G(u) + C = G(\phi(x)) + C.$$

When an Integration Problem Seems to Have Two Solutions

It is frustrating when your solution to an exercise is different from the solution that a friend has obtained or from the one in the back of the book. The next example illustrates how this can happen.

▶ **EXAMPLE 6** Compute the indefinite integral

$$\int \sin(x)\cos(x)\,dx.$$

Solution We use the substitution $u = \sin(x)$, $du = \cos(x)\,dx$. The integral becomes

$$\int \sin(x)\cos(x)\,dx = \int u du = \frac{u^2}{2} + C = \frac{\sin^2(x)}{2} + C.$$

This problem can be done in several different ways. Another solution can be derived by setting $u = \cos(x)$, $du = -\sin(x)\,dx$. The integral then becomes

$$\int \sin(x)\cos(x)\,dx = -\int u\,du = -\frac{u^2}{2} + C = -\frac{\cos^2(x)}{2} + C.$$

Although both of these solution methods are correct, the answer $\sin^2(x)/2 + C$ appears to be quite different from the answer $-\cos^2(x)/2 + C$. How do we reconcile this apparent contradiction?

We observe that

$$\frac{\sin^2(x)}{2} + C = \frac{1 - \cos^2(x)}{2} + C = -\frac{\cos^2(x)}{2} + \left(\frac{1}{2} + C\right).$$

Now the two solutions are starting to look the same. Remember that C is an arbitrary constant. Hence $D = 1/2 + C$ is also an arbitrary constant. Thus we have discovered that $\sin^2(x)/2 + C$ is precisely the same solution as $-\cos^2(x)/2 + D$.

In summary, although two methods lead to the same solution, it requires a trigonometric identity to see that they are the same. Put another way, any two solutions of our indefinite integral problem will differ by a constant. ◀

Integral Tables

Extensive tables of definite and indefinite integrals have been compiled. For more than 200 years, they have proved to be useful resources for the evaluation of integrals. Typically, tables of integrals are organized by grouping together integrands that share one or more expressions of the same type. For example, the brief table of integrals that is found at the back of this text has a section devoted to integrands that involve $\sin(x)$, another section for integrands that involve $\cos(x)$, and an additional section for integrands that involve both $\sin(x)$ and $\cos(x)$.

An integral formula often can be found in a table of integrals in exactly the form that is needed. Consider the integral, $\int (x^2/\sqrt{x+5})\,dx$, of Example 5. To evaluate this integral using the table in this text, we identify the salient expression, $\sqrt{x+5}$, and browse the formulas that involve the term $\sqrt{a+bx}$. We find equation 76, which tells us that

$$\int \frac{x^2}{\sqrt{a+bx}}\,dx = \frac{2}{b^3}\left(\frac{1}{5}(a+bx)^{5/2} - \frac{2a}{3}(a+bx)^{3/2} + a^2\sqrt{a+bx}\right) + C.$$

If we substitute $a = 5$ and $b = 1$ into this formula, then we obtain

$$\int \frac{x^2}{\sqrt{x+5}}\,dx = \frac{2}{1^3}\left(\frac{1}{5}(x+5)^{5/2} - \frac{2(5)}{3}(x+5)^{3/2} + 5^2\sqrt{x+5}\right) + C,$$

which agrees with our solution of Example 5.

Sometimes an entry from an integral table is only the first step of a solution. The next example illustrates this situation.

▶ **EXAMPLE 7** Use a table of integrals to evaluate

$$\int \frac{\sqrt{5+3t}}{\sqrt{7+4t}}\,dt.$$

Solution Again, we focus on the term $\sqrt{a+bx}$ in the integrand. Equation 86,

$$\int \frac{\sqrt{a+bx}}{\sqrt{1+x}}\,dx = \sqrt{1+x}\sqrt{a+bx} + \frac{(a-b)}{\sqrt{b}}\ln\left(\sqrt{a+bx} + \sqrt{b}\sqrt{1+x}\right) + C,$$

is promising, but we must first do some work to rewrite the given integrand so that it exactly matches the tabulated form. To that end, we write $\sqrt{7+4t} = 7\sqrt{1+4t/7}$, or $\sqrt{7+4t} = 7\sqrt{1+x}$ for $x = 4t/7$. Next, we make the substitution $t = 7x/4$, $dt = (7/4)\,dx$ in the original integral:

$$\int \frac{\sqrt{5+3t}}{\sqrt{7+4t}}\,dt = \int \frac{\sqrt{5+3(7x/4)}}{\sqrt{7+4(7x/4)}}\left(\frac{7}{4}\right)dx$$
$$= \frac{7}{4}\int \frac{\sqrt{5+21x/4}}{\sqrt{7}\sqrt{1+x}}\,dx = \frac{\sqrt{7}}{8}\int \frac{\sqrt{20+21x}}{\sqrt{1+x}}\,dx.$$

We may now evaluate the integral on the right by applying equation 86 with $a = 20$ and $b = 21$. We obtain

$$\frac{\sqrt{7}}{8}\int \frac{\sqrt{20+21x}}{\sqrt{1+x}}\,dx = \frac{\sqrt{7}}{8}\Bigg(\sqrt{1+x}\sqrt{20+21x}$$
$$- \frac{1}{\sqrt{21}}\ln\left(\sqrt{20+21x} + \sqrt{21}\sqrt{1+x}\right)\Bigg) + C.$$

On resubstituting $x = 4t/7$, we find the value of the original integral:

$$\int \frac{\sqrt{5+3t}}{\sqrt{7+4t}}\,dt = \frac{\sqrt{7}}{8}\Bigg(\sqrt{1+4t/7}\sqrt{20+21(4t/7)}$$
$$- \frac{1}{\sqrt{21}}\ln\left(\sqrt{20+21(4t/7)} + \sqrt{21}\sqrt{1+4t/7}\right)\Bigg) + C$$
$$= \frac{\sqrt{7}}{8}\left(\frac{1}{\sqrt{7}}\sqrt{7+4t}\sqrt{20+12t} - \frac{1}{\sqrt{21}}\ln\left(\sqrt{20+12t} + \frac{\sqrt{21}}{\sqrt{7}}\sqrt{7+4t}\right)\right) + C$$
$$= \frac{1}{8}\left(2\sqrt{7+4t}\sqrt{5+3t} - \frac{1}{\sqrt{3}}\ln\left(2\sqrt{5+3t} + \sqrt{3}\sqrt{7+4t}\right)\right) + C. \quad ◀$$

As the next example demonstrates, using an integral table to its full advantage can sometimes require considerable resourcefulness.

▶ **EXAMPLE 8** Use a table of integrals to evaluate

$$\int \frac{e^{3t}}{1-e^{4t}}\,dt.$$

Solution Turning to the table of integrals at the back of this text, we see no suitable formula among those pertaining to integrands with the term e^x. However, if we let $x = e^t$, then the denominator of the given integrand becomes $1-x^4$. Because the table contains a section for integrands involving a^4-x^4, the substitution $x = e^t$, or $t = \ln(x)$, $dt = dx/x$, is a lead worth pursuing. With this substitution, we obtain

$$\int \frac{e^{3t}}{1-e^{4t}}\,dt = \int \frac{x^3}{1-x^4}\frac{1}{x}\,dx = \int \frac{x^2}{1-x^4}\,dt.$$

Equation 61 of the table of integrals tells us that

$$\int \frac{x^2}{a^4-x^4}\,dx = \frac{1}{4a}\ln\left(\left|\frac{a+x}{a-x}\right|\right) - \frac{1}{2a}\arctan\left(\frac{x}{a}\right) + C.$$

With $a = 1$, this equation becomes

$$\int \frac{x^2}{1-x^4}\,dx = \frac{1}{4}\ln\left(\left|\frac{1+x}{1-x}\right|\right) - \frac{1}{2}\arctan(x) + C.$$

By resubstituting $x = e^t$, we obtain

$$\int \frac{e^{3t}}{1-e^{4t}}\,dt = \frac{1}{4}\ln\left(\left|\frac{1+e^t}{1-e^t}\right|\right) - \frac{1}{2}\arctan(e^t) + C.$$ ◀

INSIGHT Tables of integrals have remained useful even after the emergence of powerful computer algebra systems. One reason is that computers evaluate integrals in a way that differs qualitatively from the human approach. Computer algebra systems must use algorithms that are designed to handle a wide array of integrands. These general procedures may not be optimized for the particular integrand at hand. Thus Maple correctly computes the evaluation

$$\int \sec(x)\tan^3(x)\,dx = \frac{1}{3}\frac{\sin^4(x)}{\cos^3(x)} - \frac{1}{3}\frac{\sin^4(x)}{\cos(x)} - \frac{1}{3}\sin^2(x)\cos(x) - \frac{2}{3}\cos(x).$$

However, formula 193 of our table,

$$\int \sec(x)\tan^3(x)\,dx = \frac{1}{3}\sec^3(x) - \sec(x) + C,$$

is certainly preferable. Of course, we do not have to limit ourselves to only one of these integration aids. Computer algebra systems have proved to be valuable in detecting unsuspected errors in tables of integrals. And that is a good reminder—whichever of these tools you use, your last step should be the same as if you calculated the integral by hand: Verify the answer by differentiation.

Integrating Trigonometric Functions

Although we know how to differentiate each trigonometric function, we have up until now only learned to integrate the sine and cosine functions. Using the method of substitution, we can obtain antiderivatives for the remaining trigonometric functions.

▶ **EXAMPLE 9** Show that

$$\int \tan(x)\,dx = \ln(|\sec(x)|) + C. \tag{5.6.5}$$

Solution When we are presented with an integrand that is made up of trigonometric functions, it is sometimes helpful to rewrite the expression entirely in terms of sines and cosines. In this case, we have

$$\int \tan(x)\,dx = \int \frac{\sin(x)}{\cos(x)}\,dx.$$

Now the substitution $u = \cos(x)$, $du = -\sin(x)\,dx$ converts the last integral to

$$\int -\frac{1}{u}\,du = -\ln(|u|) + C.$$

Resubstituting $u = \cos(x)$ yields the solution:

$$\begin{aligned}\textstyle\int \tan(x)\,dx &= -\ln(|\cos(x)|) + C\\ &= \ln(|\cos(x)|^{-1}) + C\\ &= \ln(|\sec(x)|) + C.\end{aligned}$$ ◀

INSIGHT Remember that, as a rule of thumb, the tangent and secant functions occur together in differentiation and integration formulas. For example, we have already learned that

$$\int \sec^2(x)\,dx = \tan(x) + C \quad \text{and} \quad \int \sec(x)\tan(x)\,dx = \sec(x) + C.$$

Equation (5.6.5) is another instance in which tan(x) and sec(x) appear together in an integral formula.

The formula for the antiderivative of the secant, which we derive in the next example, is more complicated.

▶ **EXAMPLE 10** Use a substitution to derive the formula

$$\int \sec(x)\,dx = \ln(|\sec(x) + \tan(x)|) + C. \tag{5.6.6}$$

Solution By examining the right side of formula (5.6.6) and deducing that it arises as the antiderivative of $\int (1/u)\,du$ where $u = \sec(x) + \tan(x)$, we can spot the way to proceed, which otherwise would be far from obvious. We need $\sec(x) + \tan(x)$ in the denominator, so the trick is to multiply and divide the integrand by this expression:

$$\int \sec(x)\,dx = \int \sec(x)\cdot\frac{\sec(x)+\tan(x)}{\sec(x)+\tan(x)}\,dx = \int \frac{\sec(x)\tan(x)+\sec^2(x)}{\sec(x)+\tan(x)}\,dx.$$

Now we make the substitution $u = \sec(x) + \tan(x)$, $du = (\sec(x)\tan(x) + \sec^2(x))\,dx$. The last integral becomes $\int (1/u)\,du = \ln(|u|) + C$. When we resubstitute $u = \sec(x) + \tan(x)$, we obtain formula (5.6.6). A detailed historical discussion of this formula may be found in Genesis and Development 6. ◄

The antiderivative formulas for the cotangent and cosecant may be derived by following the methods for the tangent and secant, respectively. They are included in the table that follows.

Summary of Integration Rules

$$\int \tan(x)\,dx = \ln(|\sec(x)|) + C$$
$$\int \cot(x)\,dx = -\ln(|\csc(x)|) + C$$
$$\int \sec(x)\,dx = \ln(|\sec(x) + \tan(x)|) + C$$
$$\int \csc(x)\,dx = -\ln(|\csc(x) + \cot(x)|) + C$$

QUICK QUIZ

1. The method of substitution is the antiderivative form of what differentiation rule?
2. Evaluate $\int \cos^3(x)\sin(x)\,dx$.
3. Evaluate $\int_3^5 x\sqrt{x^2-9}\,dx$.
4. Evaluate $\int_0^{\pi/4} \tan(x)\,dx$.

Answers

1. Chain Rule 2. $-\cos^4(x)/4$ 3. 64/3 4. $\ln(2)/2$

EXERCISES

Problems for Practice

► In Exercises 1–10, determine a substitution that will simplify the integral. In each problem, record your choice of u and the resulting expression for du. Then evaluate the integral. ◄

1. $\int \sin(3x)\,dx$

2. $\int 24\sec^2(4t)\,dt$

3. $\int 64(x^8+1)^{-5}x^7\,dx$

4. $\int 24t\sqrt{t^2+4}\,dt$

5. $\int 30x^2(x^3-5)^{3/2}\,dx$

6. $\int (\sqrt{t}+4)^6 t^{-1/2}\,dt$

7. $\int \sin(s)\cos^4(s)\,ds$

8. $\displaystyle\int 24\frac{\sqrt{\sqrt{t}+1}}{\sqrt{t}}\,dt$

9. $\displaystyle\int \frac{\sin(\pi/t)}{t^2}\,dt$

10. $\int 24t^2\sec(t^3)\tan(t^3)\,dt$

► Use the method of substitution to calculate the indefinite integrals in Exercises 11–22. ◄

11. $\int 16x(x^2+1)^7\,dx$

12. $\int 24x^2(5-4x^3)^{-2}\,dx$

13. $\displaystyle\int \frac{x}{\sqrt{1+x^2}}\,dx$

14. $\int 24\cos(3x)\sin(3x)\,dx$

15. $\displaystyle\int \frac{\sin(x)}{\cos^2(x)}\,dx$

16. $\displaystyle\int \frac{\ln(x)}{x}\,dx$

17. $\int 24\sin^5(2x)\cos(2x)\,dx$

18. $\int 24x\cos(4x^2-5)\,dx$

19. $\displaystyle\int \frac{24x}{\sqrt{1+x^2}}\,dx$

20. $\displaystyle\int \frac{24\sin(x)}{\sqrt{2+\cos(x)}}\,dx$

21. $\int \frac{\sin(\sqrt{x})}{\sqrt{x}}\,dx$

22. $\int \frac{\sin(t)-\cos(t)}{(\sin(t)+\cos(t))^2}\,dt$

▶ Use the method of substitution to evaluate the definite integrals in Exercises 23–36. ◀

23. $\int_1^2 (t^2-t)^5(2t-1)\,dt$

24. $\int_0^5 \sqrt{1+3x}\,dx$

25. $\int_0^4 x\sqrt{\frac{1}{2}x^2+1}\,dx$

26. $\int_{\pi/3}^{\pi} \cos^3(5x)\sin(5x)\,dx$

27. $\int_{\sqrt{\pi/4}}^{\sqrt{3\pi/8}} 24x\cos(4x^2-\pi)\,dx$

28. $\int_0^3 \frac{\sin(\pi\sqrt{t+1})}{\sqrt{t+1}}\,dt$

29. $\int_{-1}^0 24\frac{x^2}{(x^3-1)^5}\,dx$

30. $\int_0^1 24\frac{\exp(x)}{(1+\exp(x))^2}\,dx$

31. $\int_{\pi^2/4}^{\pi^2} \frac{\cos(\sqrt{x})}{\sqrt{x}}\,dx$

32. $\int_0^2 (8x+5)(4x^2+5x+1)^{-2/3}\,dx$

33. $\int_0^{\pi/3} \sec^3(\theta)\tan(\theta)\,d\theta$

34. $\int_4^9 t^{-3/2}\sin(\pi/\sqrt{t})dt$

35. $\int_0^{\pi/4} \frac{24\tan(x)\sec^2(x)}{(1+2\tan(x))^2}\,dx$

36. $\int_1^e \frac{\sqrt{\ln(x)}}{x}\,dx$

▶ Evaluate the definite integrals in Exercises 37–44. ◀

37. $\int_0^{\pi/8} \tan(2x)\,dx$

38. $\int_{\pi/2}^{\pi} \cot(x/2)\,dx$

39. $\int_0^{\pi} \sec(x/4)\,dx$

40. $\int_{\pi/2}^{\pi} \csc(x/3)\,dx$

41. $\int_0^{1/2} x\tan(\pi x^2)\,dx$

42. $\int_2^4 24\cot(\pi/x)/x^2\,dx$

43. $\int_0^{\pi/2} \cos(x)\sec(\pi\sin(x)/4)\,dx$

44. $\int_1^2 \frac{24}{x^2}\csc\left(\frac{\pi}{3}\frac{2x-1}{x}\right)dx$

▶ In Exercises 45–52, use the indicated formula from the table of integrals at the back of the text to evaluate the given integral. ◀

45. $\int \frac{1}{(4t+3)(t+3)}\,dt$ (Formula 23)

46. $\int \frac{3t+2}{2t+1}\,dt$ (Formula 22)

47. $\int \frac{9+6t+t^2}{25+6t+t^2}\,dt$ (Formula 35)

48. $\int \frac{t}{1-16t^4}\,dt$ (Formula 60)

49. $\int \frac{1}{4+\exp(2t)}\,dt$ (Formula 39)

50. $\int \frac{1}{(1+4t^2)^2}\,dt$ (Formula 32)

51. $\int \frac{1}{2+\sqrt{t}}\,dt$ (Formula 7)

52. $\int \frac{\cot(t)}{3+2\sin(t)}\,dt$ (Formula 19)

Further Theory and Practice

▶ Calculate the integrals in Exercises 53–76. ◀

53. $\int x\cdot(2x+3)^{1/2}\,dx$

54. $\int \frac{x^2}{(x-1)^3}\,dx$

55. $\int \frac{x}{\sqrt{x+3}}\,dx$

56. $\int x^3\cdot(x^2-1)^{1/2}\,dx$

57. $\int x^5\cdot(6+x^2)^{-1/2}\,dx$

58. $\int x^2(x+2)^{1/3}\,dx$

59. $\int (x+2)\sqrt{x-5}\,dx$

60. $\int \frac{24x^3}{\sqrt{1+x^2}}\,dx$

61. $\int \frac{\cos(x)}{\sqrt{4-\sin^2(x)}}\,dx$

62. $\int_0^{\pi/3} \frac{\sin(x)}{1-\sin^2(x)}\,dx$

63. $\int \frac{\ln(x^{11})-\ln(x^7)}{x}\,dx$

64. $\int \left(\frac{\ln(x)}{x}-\frac{1}{x\ln(x)}\right)dx$

65. $\int \left(\frac{2}{x\ln(x^2)}-\frac{2}{x\ln^2(x)}\right)dx$

66. $\int \left(\frac{x}{x^2+1}\right)\left(1-\frac{1}{x^2}\right)dx$

67. $\int \frac{\arctan(x)}{1+x^2}\,dx$

68. $\int \frac{\exp(x)}{1+\exp(x)}\,dx$

69. $\int \frac{\exp(x)}{1+\exp(2x)}\,dx$

70. $\int \frac{\exp(2x)}{1+\exp(x)}\,dx$

71. $\displaystyle\int \frac{\exp(x)}{1+2\exp(x)+\exp(2x)}\,dx$

72. $\displaystyle\int \frac{x+\arcsin(x)}{\sqrt{1-x^2}}\,dx$

73. $\displaystyle\int \frac{6x^3+3x+1}{\sqrt{4-x^2}}\,dx$

74. $\displaystyle\int \frac{\exp(x)}{\sqrt{1-\exp(2x)}}\,dx$

75. $\displaystyle\int \frac{\exp(2x)}{\sqrt{1-\exp(2x)}}\,dx$

76. $\displaystyle\int \frac{24\exp(2x)}{\sqrt{1-\exp(x)}}\,dx$

In each of Exercises 77–82, evaluate the given integral by applying a substitution to a formula from a table of integrals.

77. $\displaystyle\int \frac{\tan(t)}{\sqrt{4+\cos(t)}}\,dt$

78. $\int \sqrt{1+\sqrt{t}}\,dt$

79. $\displaystyle\int \frac{12}{t\sqrt{1+t^4}}\,dt$

80. $\displaystyle\int \frac{1}{(2+\exp(t))^2}\,dt$

81. $\displaystyle\int \frac{\sec^3(t)\tan(t)}{1+\sec^2(t)}\,dt$

82. $\int \cos(\ln(t))\,dt$

83. Evaluate the integral $\int_{-1}^{1} 3t^2\,dt$ directly. Calculate it again by performing the integration separately on the two subintervals $[-1, 0]$ and $[0, 1]$ and making the substitution $u=t^2$ on each.

84. Let $f:[-a,a]\to\mathbb{R}$ be continuous and odd $(f(-x)=-f(x))$. Show that

$$\int_{-a}^{a} f(x)\,dx = 0.$$

85. Let $f:[-a,a]\to\mathbb{R}$ be continuous and even $(f(-x)=f(x))$. Show that

$$\int_{-a}^{a} f(x)\,dx = 2\int_{0}^{a} f(x)\,dx.$$

86. Let $f:[-1, 1]\to\mathbb{R}$ be continuous. Evaluate

$$\int_{-\pi/2}^{\pi/2} x\cdot f(\cos(x))\,dx.$$

(See Exercise 84.)

87. Suppose that f is continuous on $[a,b]$. Use a substitution to show that

$$\int_a^b f(x)\,dx = \int_0^{b-a} f(b-x)\,dx.$$

88. Suppose that f is continuous on $[a,b]$. Use a substitution to show that

$$\int_a^b f(x)\,dx = \int_a^b f(a+b-x)\,dx.$$

89. Suppose that f is continuous on $[a, b]$. Use a substitution to show that

$$\int_a^b f(x)\,dx = (b-a)\int_0^1 f(a+(b-a)x)\,dx.$$

90. Suppose that f is continuous on $[a,b]$. let $\delta=(b-a)/2$ and $\mu=(a+b)/2$. Use a substitution to show that

$$\int_a^b f(u)\,du = \delta\int_{-1}^{1} f(\delta x+\mu)\,dx$$

91. Suppose that α and β are different constants. Use the identity

$$\sin(A)\cos(B) = \frac{\sin(A+B)+\sin(A-B)}{2}$$

to evaluate $\int \sin(\alpha x)\cos(\beta x)\,dx$.

92. Evaluate

$$\int \sec^2(x)\tan(x)\,dx$$

by using the change of variable $u=\tan(x)$. Then evaluate the integral by making the change of variable $v=\sec(x)$. Verify that the two answers that you obtain are equivalent.

93. Verify that

$$\frac{2t^3+5t^2+2t+1}{(t^2+1)(t+1)^2} = \frac{2t}{t^2+1}+\frac{1}{(t+1)^2}.$$

Use this decomposition to evaluate

$$\int \frac{2t^3+5t^2+2t+1}{(t^2+1)(t+1)^2}\,dt.$$

(This example shows how beneficial Theorem 1a from Section 5.3 can be.)

94. Evaluate $\int_0^1 x\sqrt{1-\sqrt{1-x^2}}\,dx$. Start with the substitution $u=\sqrt{1-x^2}$, and finish with a second substitution.

Calculator/Computer Exercises

In Exercises 95–98, determine the value of the upper limit of integration b for which a substitution converts the integral on the left to the integral on the right.

95. $\int_0^b (3t^2+1)\sec^2(t^3+t)\tan^2(t^3+t)\,dt = \int_0^1 u^2\,du$

96. $\int_0^b (4x^3+2x)\sin(x^4+x^2)\,dx = \int_0^\pi \sin(u)\,du$

97. $\int_0^b (1+e^x)\sqrt{x+e^x}\,dx = \int_1^9 \sqrt{u}\,du$

98. $\int_1^b (x+1)\exp(1/x-\ln(x))/x^2\,dx = \int_{1/4}^{1} \exp(u)\,du$

5.7 More on the Calculation of Area

If a function f is continuous and nonnegative on an interval $[a, b]$, then $\int_a^b f(x)\,dx \geq 0$. This is clear because each of the summands in the Riemann $\sum_{j=0}^{N} f(s_j)\Delta x$ is nonnegative (because f is nonnegative). Similarly, if f is continuous and $f \leq 0$ on an interval $[c, d]$, then $\int_c^d f(x)\,dx \leq 0$. Again, this is justified by the fact that each summand in the Riemann Sum $\sum_{j=0}^{N} f(s_j)\Delta x$ is nonpositive (because f is nonpositive).

In the first case, the integral represents area (as discussed in Section 5.1). The same ideas show that when $f \leq 0$, the integral represents the *negative* of the area because the summands $f(s_j)\Delta x$ are negatives of the areas of the corresponding rectangles.

We obtain from these remarks the following guidelines for determining area.

Rules for Calculating Area

1. If $f(x) \geq 0$ for x in $[a, b]$, then $\int_a^b f(x)\,dx$ equals the area of the region that is below the graph of f, above the x-axis, and between $x = a$ and $x = b$.
2. If $f(x) \leq 0$ for x in $[a, b]$, then $\int_a^b f(x)\,dx$ equals the *negative* of the area of the region that is above the graph of f, below the x-axis, and between $x = a$ and $x = b$.

We may use these observations to solve some fairly sophisticated area problems.

▶ **EXAMPLE 1** What is the area of the region bounded by the graph of $f(x) = \sin(x)$ and the x-axis between the limits $x = \pi/3$ and $x = 3\pi/2$?

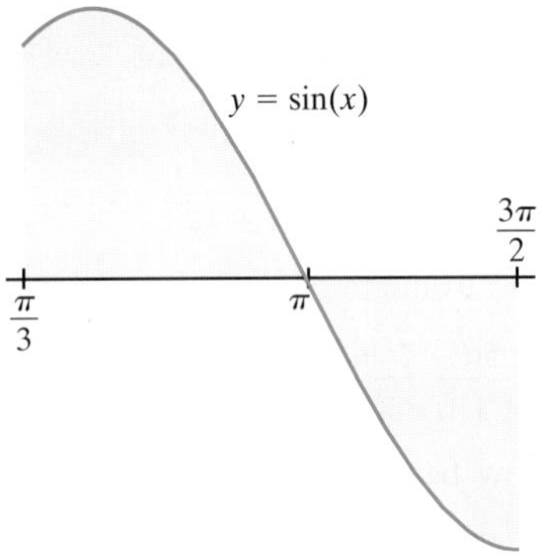

▲ Figure 1

Solution Look at Figure 1. Notice that $f(x) \geq 0$ for $x \in [\pi/3, \pi]$ and $f(x) \leq 0$ for $x \in [\pi, 3\pi/2]$. Thus the (positive) area of the region lying above $[\pi/3, \pi]$ is

$$\int_{\pi/3}^{\pi} \sin(x)\,dx = -\cos(x)\Big|_{\pi/3}^{\pi} = -(-1) - \left(-\frac{1}{2}\right) = \frac{3}{2}.$$

On the other hand, the (positive) area of the region lying *below* $[\pi, 3\pi/2]$ is

$$-\int_{\pi}^{3\pi/2} \sin(x)\,dx = -\left(-\cos(x)\right)\Big|_{\pi}^{3\pi/2} = 0 - (-1) = 1.$$

The total area is the sum of the two component areas, or $3/2 + 1$, or $5/2$. ◀

Theorem 1e from Section 5.3 suggests a way in which we can calculate areas (and integrals) for piecewise-defined functions.

▶ **EXAMPLE 2** Define

$$f(x) = \begin{cases} x + 7 & \text{if } -2 \leq x \leq 1 \\ 9 - x^2 & \text{if } x < -2 \text{ or } x > 1. \end{cases}$$

What is the area of the region below the graph of f and above the x-axis?

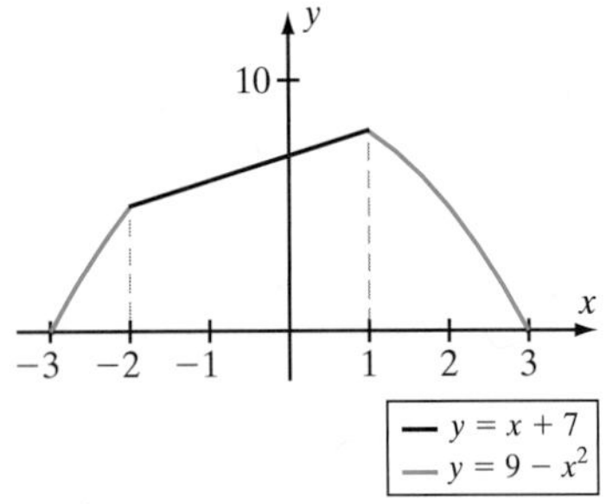

▲ Figure 2

Solution From the definition of f, we see that $f \geq 0$ for $-3 \leq x \leq 3$. (The graph of f over this interval appears in Figure 2.) We see that the problem naturally divides into three pieces:

1. The component of area above the interval $[-3,-2]$
2. The component of area above the interval $[-2, 1]$
3. The component of area above the interval $[1, 3]$

We thus do three separate calculations.

1. When x is in $[-3,-2)$, we have $f(x)=9-x^2$ and

$$\int_{-3}^{-2} f(x)\,dx = \left(9x-\frac{x^3}{3}\right)\Bigg|_{x=-3}^{x=-2} = \left(-18+\frac{8}{3}\right)-\left(-27+\frac{27}{3}\right)=\frac{8}{3}.$$

2. When x is in $[-2, 1]$, we have $f(x)=x+7$ and

$$\int_{-2}^{1} f(x)\,dx = \left(\frac{1}{2}x^2+7x\right)\Bigg|_{x=-2}^{x=1} = \left(\frac{1}{2}+7\right)-\left(\frac{4}{2}-14\right)=\frac{39}{2}.$$

3. When x is in $(1,3]$, we have $f(x)=9-x^2$ and

$$\int_{1}^{3} f(x)\,dx = \left(9x-\frac{x^3}{3}\right)\Bigg|_{x=1}^{x=3} = \left(27-\frac{27}{3}\right)-\left(9-\frac{1}{3}\right)=\frac{28}{3}.$$

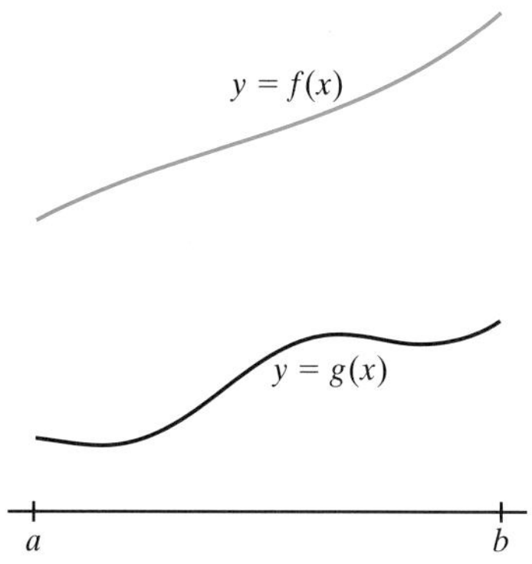

▲ Figure 3

Thus the total area between the graph of f and the x-axis is $8/3+39/2+28/3$, or $63/2$. ◄

The Area between Two Curves

Examples 1 and 2 are special cases of the general problem of finding the area between two curves. Suppose f and g are continuous functions with domains that each contain the interval $[a,b]$. Suppose further that $f(x)\geq g(x)$ on this interval (see Figure 3). To estimate the area below the graph of f and above the graph of g on this interval, we let $\mathcal{P}=\{x_0,\ldots,x_N\}$ be a uniform partition of $[a,b]$. We select points $s_j\in I_j=[x_{j-1},x_j]$ and erect rectangles with base I_j and height $f(s_j)-g(s_j)$, as in Figure 4. The area we wish to estimate is approximately given by the sum of the areas of these rectangles, or

$$\sum_{j=1}^{N}\Big(f(s_j)-g(s_j)\Big)\cdot\Delta x.$$

As Δx tends to 0, we expect that this sum will become a more accurate approximation to the desired area. However, the sum also happens to be a Riemann sum for the integral

$$\int_a^b \Big(f(x)-g(x)\Big)\,dx.$$

We have therefore derived the following result, stated as a theorem.

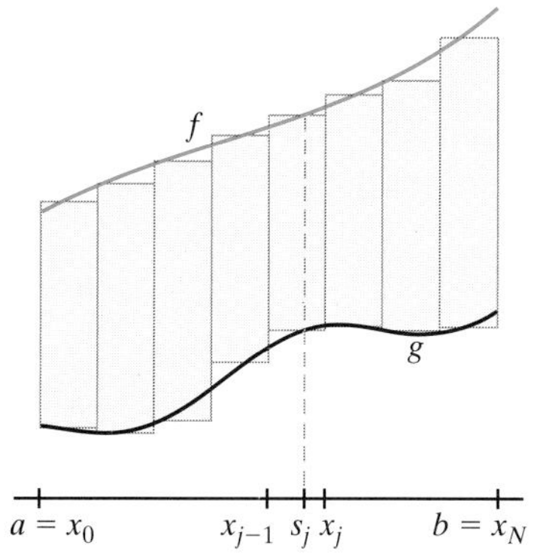

▲ Figure 4

THEOREM 1 Let f and g be continuous functions on the interval $[a,b]$ and suppose that $f(x)\geq g(x)$ for all x in $[a,b]$. The area of the region under the graph of f and above the graph of g on the interval $[a,b]$ is given by

$$\int_a^b \Big(f(x)-g(x)\Big)\,dx.$$

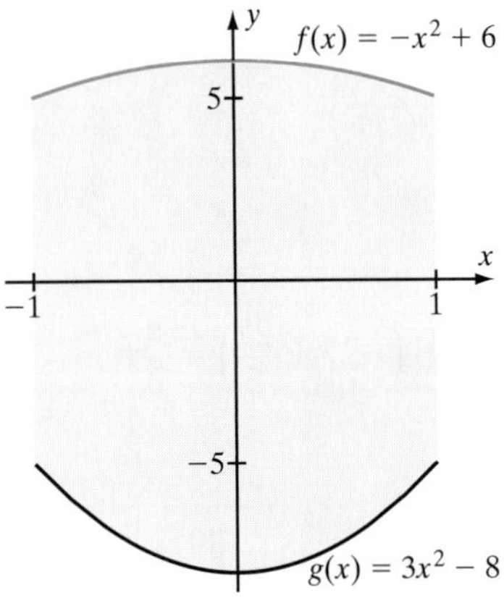

▲ Figure 5

▶ **EXAMPLE 3** Calculate the area A of the region between the curves $f(x) = -x^2 + 6$ and $g(x) = 3x^2 - 8$ for x in the interval $[-1, 1]$.

Solution We see from the sketch in Figure 5 that $f(x) \geq g(x)$ on this interval. According to Theorem 1, the desired area is given by

$$\int_{-1}^{1} (f(x) - g(x))\,dx = \int_{-1}^{1} ((-x^2+6) - (3x^2-8))\,dx = \int_{-1}^{1} (-4x^2+14)\,dx.$$

Therefore

$$A = \left(-\frac{4}{3}x^3 + 14x\right)\Bigg|_{-1}^{1} = \left(\left(-\frac{4}{3}\right)\cdot 1 + 14\cdot 1\right) - \left(\left(-\frac{4}{3}\right)\cdot(-1) + 14\cdot(-1)\right) = \frac{76}{3}. \quad ◀$$

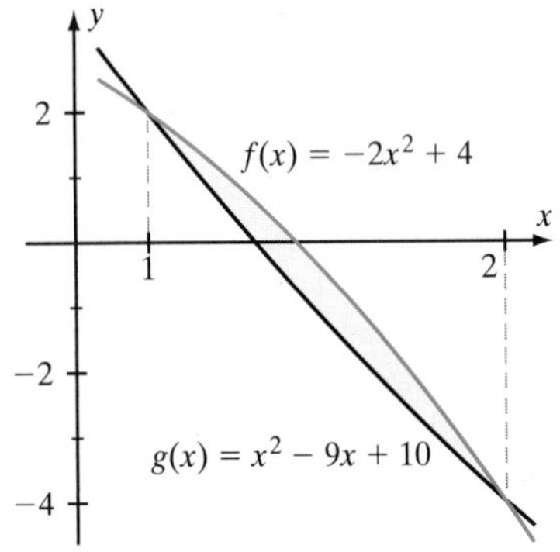

▲ Figure 6

▶ **EXAMPLE 4** Calculate the area between the parabolas $f(x) = -2x^2 + 4$ and $g(x) = x^2 - 9x + 10$.

Solution The region is shaded in Figure 6. We have not specified an interval on which to work, but we can calculate the interval of integration: The parabolas intersect when $-2x^2 + 4 = x^2 - 9x + 10$, or $x^2 - 3x + 2 = 0$. The roots are $x = 1$ and $x = 2$, as can be seen either by factoring or by using the Quadratic Formula. On the interval $[1, 2]$, $f(x) \geq g(x)$. Thus the desired area is given by

$$\begin{aligned}\int_{1}^{2} (f(x) - g(x))\,dx &= \int_{1}^{2} ((-2x^2+4) - (x^2-9x+10))\,dx \\ &= \int_{1}^{2} (-3x^2+9x-6)\,dx \\ &= \left(-x^3 + \frac{9}{2}x^2 - 6x\right)\Bigg|_{x=1}^{x=2} \\ &= (-8+18-12) - \left(-1 + \frac{9}{2} - 6\right) \\ &= \frac{1}{2}. \quad ◀\end{aligned}$$

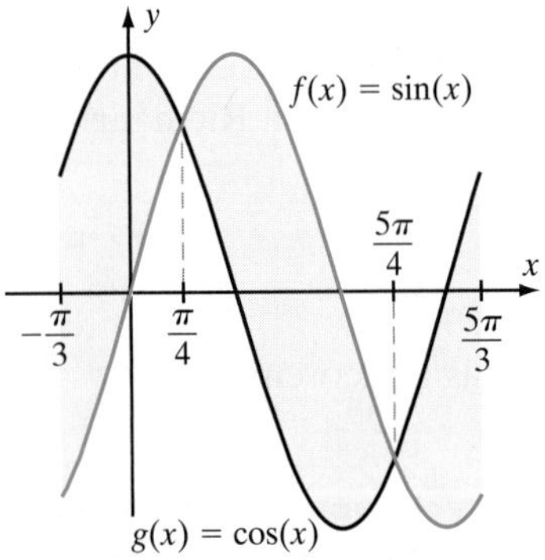

▲ Figure 7

▶ **EXAMPLE 5** Calculate the area between the curves $f(x) = \sin(x)$ and $g(x) = \cos(x)$ on the interval $[-\pi/3, 5\pi/3]$.

Solution The graphs of f and g appear in Figure 7. Notice that $f(x) \geq g(x)$ at some points while $g(x) \geq f(x)$ at other points. We need to break up the interval $[-\pi/3, 5\pi/3]$ into subintervals on which either one or the other of these inequalities is true, but not both. Thus we need to find the points of intersection of the graphs of f and g. Setting $\sin(x) = \cos(x)$, we see that the graphs intersect at the points $x = \pi/4$ and $x = 5\pi/4$ in the interval $[-\pi/3, 5\pi/3]$. We therefore make separate calculations of the area on $[-\pi/3, \pi/4]$, the area on $[\pi/4, 5\pi/4]$, and the area on $[5\pi/4, 5\pi/3]$. On $[-\pi/3, \pi/4]$, we have $\cos(x) \geq \sin(x)$ (as we can see from Figure 7), so the corresponding area is

$$\int_{-\pi/3}^{\pi/4} \big(\cos(x) - \sin(x)\big)\,dx = \big(\sin(x) + \cos(x)\big)\Big|_{-\pi/3}^{\pi/4} = \sqrt{2} + \frac{\sqrt{3}-1}{2}.$$

On $[\pi/4, 5\pi/4]$, we see that $\sin(x) \geq \cos(x)$, and the corresponding area is

$$\int_{\pi/4}^{5\pi/4} \Big(\sin(x) - \cos(x)\Big)\, dx = \Big(-\cos(x) - \sin(x)\Big)\Big|_{\pi/4}^{5\pi/4} = 2\sqrt{2}.$$

On $[5\pi/4, 5\pi/3]$, $\cos(x) \geq \sin(x)$. The corresponding area is, thus,

$$\int_{5\pi/4}^{5\pi/3} \Big(\cos(x) - \sin(x)\Big)\, dx = \Big(\sin(x) + \cos(x)\Big)\Big|_{5\pi/4}^{5\pi/3} = \sqrt{2} + \frac{1-\sqrt{3}}{2}.$$

Adding the three areas that we have computed gives the total area between the curves:

$$\left(\sqrt{2} + \frac{\sqrt{3}-1}{2}\right) + (2\sqrt{2}) + \left(\sqrt{2} + \frac{1-\sqrt{3}}{2}\right) = 4\sqrt{2}. \blacktriangleleft$$

Reversing the Roles of the Axes

To calculate the area between certain pairs of curves, it is convenient to interchange the roles of the x-axis and the y-axis. Consider the following example.

▶ **EXAMPLE 6** Compute the area between the curves $x = y^2 - y - 4$ and $x = -y^2 + 3y + 12$.

Solution The curves in question are parabolas. By setting the expressions for x equal to each other, we obtain

$$y^2 - y - 4 = -y^2 + 3y + 12,$$

or

$$y^2 - 2y - 8 = 0.$$

The roots are $y = -2$ and $y = 4$. The parabolas intersect at $(2, -2)$ and $(8, 4)$. Figure 8 exhibits this information, as well as the region whose area we wish to calculate.

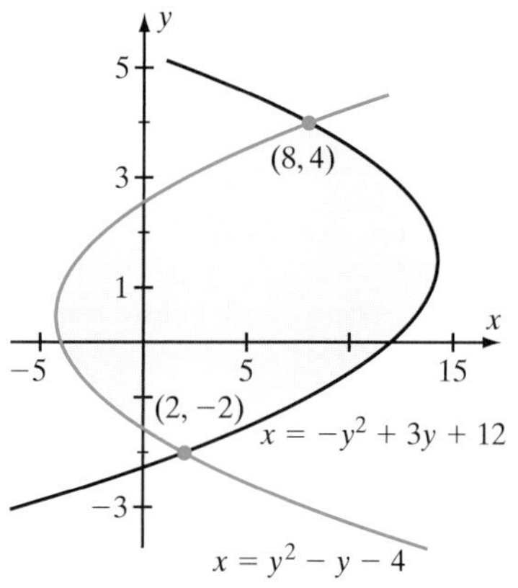

▲ Figure 8

If we were to calculate the area of the region using an x-integral, then we would need to solve for the upper and lower branch of each parabola, which would make the algebra complicated. Instead, we think of x as a function of y and approximate the area by horizontal rectangles (see Figure 9). Reasoning as before, we are led to consider the integral

$$\int_{-2}^{4} ((-y^2 + 3y + 12) - (y^2 - y - 4))dy.$$

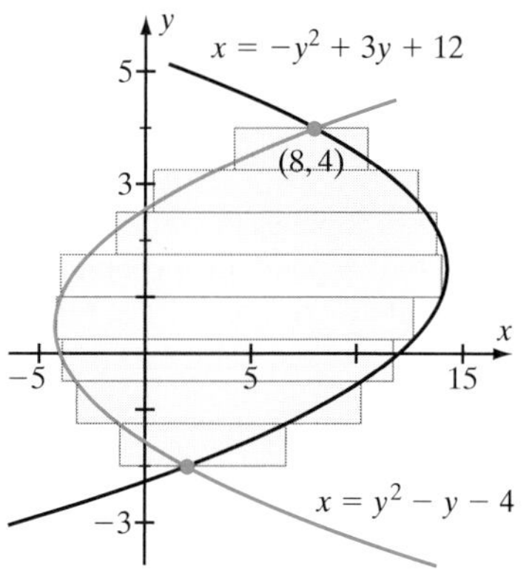

▲ Figure 9

Notice that we subtract the expression describing the parabola on the left from that describing the parabola on the right—this corresponds to subtracting lesser x-coordinates from greater x-coordinates, so that we will be computing the *positive* area. Simplifying the integrand of the last integral and then integrating term by term, we obtain

$$\begin{aligned} \int_{-2}^{4} (-2y^2 + 4y + 16)dy &= \left(-\frac{2}{3}y^3 + 2y^2 + 16y\right)\Big|_{y=-2}^{y=4} \\ &= \left(\frac{-128}{3} + 32 + 64\right) - \left(\frac{16}{3} + 8 - 32\right) = 72 \end{aligned}$$

for the area between the two parabolas. ◀

QUICK QUIZ

1. Suppose that f and g are continuous functions on $[a, b]$ with $g(x) \le f(x)$ for every x in $[a, b]$. The area of the region that is bounded above and below by the graphs of f and g, repectively, is equal to $\int_a^b h(x)\,dx$ for what expression $h(x)$?
2. What is the area between the x-axis and the curve $y = \sin(x)$ for $0 \le x \le 2\pi$?
3. What is the area between the y-axis and the curve $x = 1 - y^2$?
4. The area of a region is obtained as $\int_0^2 (4 - x^2)\,dx$ by integrating with respect to x. What integral represents the area if the integration is performed with respect to y?

Answers

1. $f(x) - g(x)$ 2. 4 3. 4/3 4. $\int_0^4 \sqrt{4 - y}\,dy$

EXERCISES

Problems for Practice

In each of Exercises 1–12, a function f is defined on a specified interval $I = [a, b]$. Calculate the area of the region that lies between the vertical lines $x = a$ and $x = b$ and between the graph of f and the x-axis.

1. $f(x) = 2\cos(x)$ $I = [\pi/4, 2\pi/3]$
2. $f(x) = 3x^2 - 3x - 6$ $I = [-2, 4]$
3. $f(x) = 2x^2 - 8$ $I = [-3, 5]$
4. $f(x) = 4/x - x$ $I = [1, 3]$
5. $f(x) = x^3 + x + 2$ $I = [-2, 1]$
6. $f(x) = x/(x^2 + 1)^2$ $I = [-1, 3]$
7. $f(x) = 8x \cdot (1 - x^2)^2$ $I = [-1/2, 1]$
8. $f(x) = \exp(x) - e$ $I = [-1, 2]$
9. $f(x) = x(4 - x^2)^3$ $I = [-2, 3]$
10. $f(x) = (x^2 - 1)(x - 2)$ $I = [-2, 2]$
11. $f(x) = 12 - 9x - 3x^2$ $I = [-2, 2]$
12. $f(x) = 8x - 5/(1 + x^2)$ $I = [-1, 2]$

In each of Exercises 13–18, a function f is defined piecewise on an interval $I = [a, b]$. Find the area of the region that is between the vertical lines $x = a$ and $x = b$ and between the graph of f and the x-axis.

13. $f(x) = \begin{cases} -x^2 & \text{if } -3 \le x \le 1 \\ 2x - 3 & \text{if } 1 < x \le 4 \end{cases}$ $I = [-3, 4]$

14. $f(x) = \begin{cases} -x + 3 & \text{if } -2 \le x < 0 \\ -x^2 + 3x + 3 & \text{if } 0 \le x \le 3 \end{cases}$ $I = [-2, 3]$

15. $f(x) = \begin{cases} \sin(x) & \text{if } 0 \le x \le \pi/4 \\ \cos(x) & \text{if } \pi/4 < x \le \pi \end{cases}$ $I = [0, \pi]$

16. $f(x) = \begin{cases} x - 2 & \text{if } 0 \le x \le 2 \\ \sin(\pi x) & \text{if } 2 < x \le 4 \end{cases}$ $I = [0, 4]$

17. $f(x) = \begin{cases} 4 - x^2 & \text{if } -2 \le x \le 3 \\ x^2 - 8x + 10 & \text{if } 3 < x \le 6 \end{cases}$ $I = [-2, 6]$

18. $f(x) = \begin{cases} \sec(x) & \text{if } 0 \le x \le \pi/3 \\ 4\cos(x) & \text{if } \pi/3 < x \le \pi \end{cases}$ $I = [0, \pi]$

In each of Exercises 19–26, find the area of the region that is bounded by the graphs of $y = f(x)$ and $y = g(x)$ for x between the abscissas of the two points of intersection.

19. $f(x) = x^2 - 1$ $g(x) = 8$
20. $f(x) = 15/(1 + x^2)$ $g(x) = 3$
21. $f(x) = x^2 + x - 1$ $g(x) = 2x + 1$
22. $f(x) = 2x + 3$ $g(x) = 9 + x - x^2$
23. $f(x) = x^2 + x + 1$ $g(x) = 2x^2 + 3x - 7$
24. $f(x) = x^2$ $g(x) = -2x^2 - 15x - 18$
25. $f(x) = x^2 + 5$ $g(x) = 2x^2 + 1$
26. $f(x) = x^2 + 5x + 9$ $g(x) = -x^2 + 19x - 15$

In each of Exercises 27–30, the graphs of $y = f(x)$ and $y = g(x)$ intersect in more than two points. Find the total area of the regions that are bounded above and below by the graphs of f and g.

27. $f(x) = x^3 - 3x^2 - x + 4$ $g(x) = -3x + 4$
28. $f(x) = x^3 + x^2$ $g(x) = x^2 + x$
29. $f(x) = x^4 - 5x^2$ $g(x) = -4$
30. $f(x) = 2\sin(x)$ $g(x) = 6(\pi - x)/(5\pi)$

In each of Exercises 31–34, determine the area between the two curves over the range of x.

31. $f(x) = -\sqrt{3}\cos(x)$ $g(x) = \sin(x),\ -\pi \le x \le 2\pi/3$
32. $f(x) = \sqrt{3}\sin(x)$ $g(x) = \cos(x),\ -\pi/2 \le x \le 7\pi/6$
33. $f(x) = \sin(2x)$ $g(x) = \cos(x), 0 \le x \le \pi/2$
34. $f(x) = \cos(2x)$ $g(x) = \sin(x),\ -\pi/2 \le x \le \pi/2$

In Exercises 35–40, treat the y variable as the independent variable and the x variable as the dependent variable. By integrating with respect to y, calculate the area of the region that is described.

35. The region between the curves $x = y^2$ and $x = -y^2 + 4$
36. The region between the curves $y = x$ and $x = y^2 - 2$
37. The region between the curves $x = \cos(y)$ and $x = \sin(y)$, $\pi/4 \le y \le 5\pi/4$
38. The region between the curves $x = y^2 + 1$ and $x + y = 3$

39. The region between the curves $y = \ln(x)$ and $y = \dfrac{x-1}{e-1}$

40. The region in the first quadrant between the curves $y = \pi x/2$ and $y = \arcsin(x)$

Further Theory and Practice

In Exercises 41–44, find the area of the region(s) between the two curves over the given range of x.

41. $f(x) = x(1 + x^2)$ $\quad g(x) = x/2,\ 0 \le x \le 1$
42. $f(x) = x \cdot \cos(x^2)$ $\quad g(x) = x\sin(x^2),\ 0 \le x \le \sqrt{5\pi/2}$
43. $f(x) = 2\sin(x)$ $\quad g(x) = \sin(2x),\ 0 \le x \le \pi$
44. $f(x) = (x^3 - 8)/x$ $\quad g(x) = 7(x - 2),\ 1 \le x \le 4$

In Exercises 45–48, calculate the area of the region between the pair of curves.

45. $x = y^2 + 6$ $\quad x = -y^2 + 14$
46. $y = (x - 3)/2$ $\quad x = y^2$
47. $x = y^2$ $\quad x = y^3$
48. $x = y$ $\quad x = y^4$

In each of Exercises 49–52, the given integral $\int_0^b f(x)\,dx$ represents the area of the region in the xy-plane that lies below the graph of f and above the interval $[0, b]$ of the x-axis. Express the area as an integral of the form $\int_c^d g(y)dy$. For example, the integral $\int_0^1 2x\,dx$ represents the area of the triangle with vertices (0, 0), (1, 0), and (1, 2). This area can also be represented as $\int_0^2 (1 - y/2)\,dy$ (see Figure 10).

49. $\int_0^4 \sqrt{x}\,dx$
50. $\int_0^1 (e^x - 1)\,dx$
51. $\int_0^1 \arcsin(x)\,dx$
52. $\int_0^2 (x^2 = 2x)\,dx$

In Exercises 53–56, the integral $\int_a^b (f_1(x) - f_2(x))\,dx$ represents the area of a region in the xy-plane that is bounded by the graphs of f_1 and f_2. Express the area of the region as an integral of the form $\int_c^d (g_1(y) - g_2(y))\,dy$. For example, the integral $\int_0^1 (x - x^2)\,dx$ represents the area of the shaded region in Figure 11. This area can also be represented as $\int_0^1 (\sqrt{y} - y)\,dy$. (*Hint*: A sketch of the graphs of f_1 and f_2 over the interval of integration is helpful. Calculating the given integral is *not* helpful. You need not evaluate the integral with respect to y that you obtain.)

53. $\int_0^4 (\sqrt{x} - x/2)\,dx$
54. $\int_0^1 (2\sqrt{x} - 2x^3)\,dx$
55. $\int_{-3}^3 (3 - |x|)\,dx$
56. $\int_{-1}^1 (e - \exp(|x|))\,dx$

In each of Exercises 57–60, express the area of the given region as a sum of integrals of the form $\int_a^b f(x)\,dx$.

57. The triangle with vertices (1, 0), (3, 0), (2, 1)
58. The triangle with vertices (1, 0), (3, 1), (2, 2)
59. The region enclosed by $y = |x|$ and $y = 2 - x^2$
60. The larger of the two pieces of the disk $x^2 + y^2 \le 1$ that is formed when the disk is cut by the line $y = x + 1$

In Exercises 61–64, express the area of the region as an integral of the form $\int_c^d g(y)\,dy$ or as a sum of such integrals.

61. The region of Exercise 57
62. The region of Exercise 58
63. The region of Exercise 59
64. The region of Exercise 60

In each of Exercises 65–70, a sum of integrals of the form $\int_a^b f(x)\,dx$ is given. Express the sum as a single integral of form $\int_c^d g(y)\,dy$.

65. $\int_0^2 \sqrt{x}\,dx = \int_2^4 \sqrt{4 - x}\,dx$
66. $\int_{-2}^0 \sqrt{x + 2}\,dx + \int_0^2 (\sqrt{x + 2} - x)\,dx$

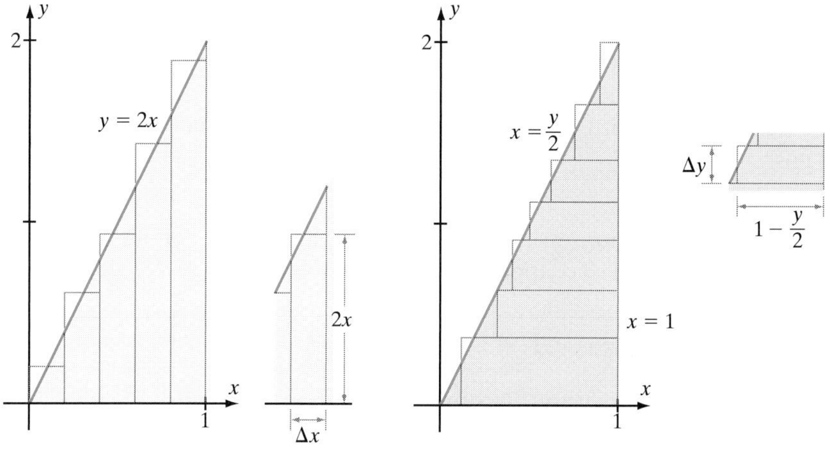

▲ Figure 10

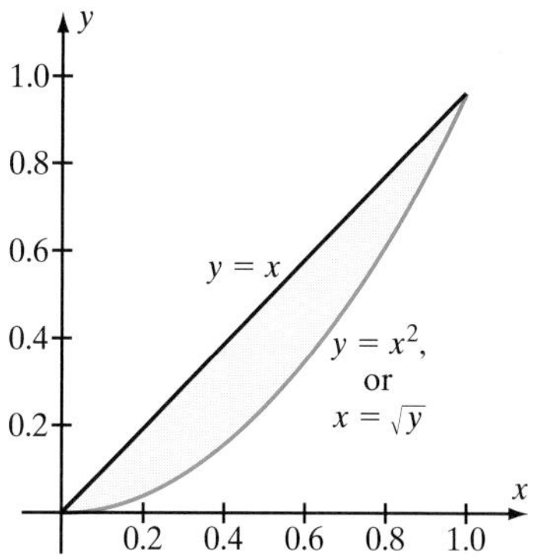

▲ Figure 11

67. $\int_0^{\sqrt{2}/2} \arcsin(x)\,dx + \int_{\sqrt{2}/2}^{1} \arccos(x)\,dx$
68. $\int_1^2 \sqrt{x-1}\,dx = \int_2^3 (3-x)\,dx$
69. $\int_{-1}^0 (2-2\sqrt{-x})\,dx + \int_0^1 (2-2x^2)\,dx$
70. $\int_6^{10} \sqrt{x-6}\,dx + \int_{10}^{14} \sqrt{14-x}\,dx$

71. Find the area between the curves $x-4y=1$ and $x^2+2xy+y^2-2x+3y=0$.
72. The graphs of $y=x+2/x$ and $y=3+(x-2)^2$ intersect at $P=(2,3)$, at one point to the left of P, and at one point to the right of P. Find the total area of the regions that are between the two graphs.

Calculator/Computer Exercises

In Exercises 73–79, plot the graphs as indicated. Then find the abscissas a and b of the two points of intersection. Find the area bounded by the two graphs for $a \le x \le b$.

73. $y=x^3+x,\ y=6-x^4$ $\quad -2.2 \le x \le 1.4$
74. $y=1+2x,\ y=\exp(x)$ $\quad -0.1 \le x \le 1.3$
75. $y=\ln(x)/x,\ y=(x-1)/8$ $\quad 0.9 \le x \le 4$
76. $y=\sin(x),\ y=1-3\sin(x)$ $\quad 0 \le x \le \pi$
77. $y=|x|,\ y=1-x^2-x^3$ $\quad -1 \le x \le 0.7$
78. $y=\sqrt{x},\ y=x+x^2$ $\quad 0 \le x \le 0.5$
79. $y=\sec^2(x),\ y=x+2$ $\quad -1 \le x \le 1$

80. Let b denote the smallest positive solution of $\sin^2(x) = \cos(x^2)$. Approximate the area between $f(x)=\cos(x^2)$ and $g(x)=\sin^2(x)$ over the interval $[0, b]$.

5.8 Numerical Techniques of Integration

Although the Fundamental Theorem of Calculus provides a powerful rule for evaluating Riemann integrals, many important definite integrals cannot be calculated *exactly*. The inability to calculate an integral exactly occurs when it is impossible to express the antiderivative of the integrand in terms of finitely many of the functions that we know. Even integrands that do not appear to be especially complicated can fall into this category. For example, the distance traveled by a satellite in an elliptical trajectory involves an integral of the form

$$\int_0^{\pi/2} \sqrt{1-k^2\sin^2(\theta)}\,d\theta$$

where k is a positive constant less than 1. The value of this integral is required for many applications, yet no elementary antiderivative exists. In fact, many real-world problems involve integrands that have no elementary antiderivatives. Therefore it is important to be able to approximate a definite integral to any specified degree of accuracy. The right endpoint approximation that we learned in Section 5.1 will accomplish that goal but not very efficiently. The three approximation techniques that are introduced in this section are all better choices because they will match the accuracy of the right endpoint approximation but at a substantially smaller computational cost.

The Midpoint Rule

Let f be a continuous function on the interval $[a,b]$, and let N denote a positive integer that is to be chosen. To approximate the integral $\int_a^b f(x)\,dx$, we use the uniform partition

$$a = x_0 < x_1 < x_2 < \cdots < x_N = b,$$

which divides the interval $[a,b]$ into N subintervals of equal length

$$\Delta x = \frac{b-a}{N}.$$

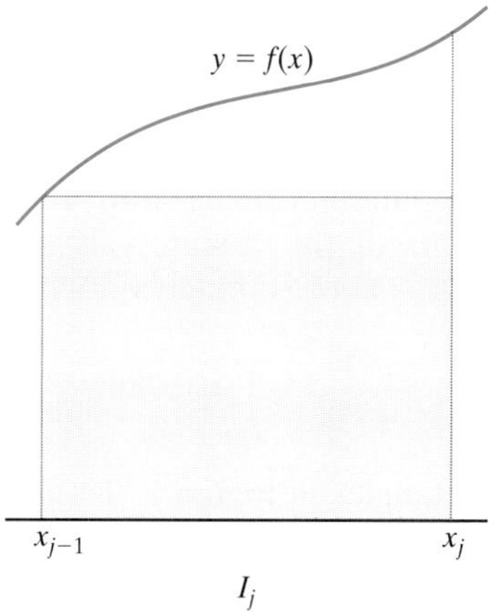

▲ Figure 1a Left endpoint approximation

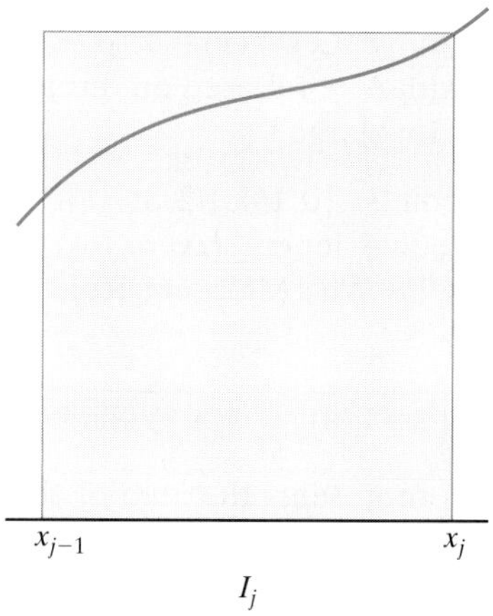

▲ Figure 1b Right endpoint approximation

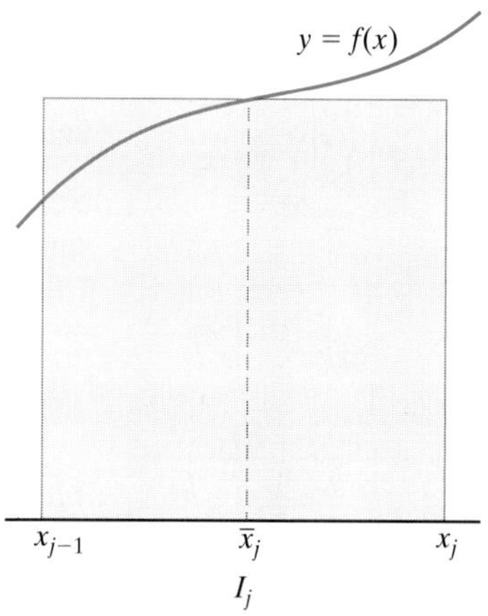

▲ Figure 1c Midpoint approximation

Because the definite integral $\int_a^b f(x)\,dx$ is defined as the limit $\lim_{N\to\infty}\sum_{j=1}^{N} f(s_j)\Delta x$ of Riemann sums, it is natural to choose the points $s_j \in I_j = [x_{j-1}, x_j]$ in a convenient manner and then use the Riemann sum $\sum_{j=1}^{N} f(s_j)\Delta x$ to approximate the integral. The candidates for s_j that readily come to mind are the left endpoint x_{j-1} of I_j, the right endpoint x_j of I_j, and the midpoint

$$\overline{x}_j = \frac{x_{j-1}+x_j}{2} = a + \left(j - \frac{1}{2}\right)\Delta x. \tag{5.8.1}$$

Because a continuous function typically either increases over an entire *small* interval or decreases over the interval, the midpoint is the preferred choice (as indicated by the shaded area of Figure 1). The resulting Riemann sum

$$\mathcal{M}_N = \Delta x \cdot (f(\overline{x}_1) + f(\overline{x}_2) + \cdots + f(\overline{x}_N)) \tag{5.8.2}$$

is called the *midpoint approximation* of order N. In general, the approximation becomes more accurate as N increases. However, we do not want to choose N so large that the calculation of $\mathcal{M}_N$ becomes impractical. We may therefore state the basic approximation problem in this way: How do we determine the smallest value of N that allows us to be sure that $\mathcal{M}_N$ is an acceptable approximation? The following theorem states an error estimate that is the key to solving this problem for the midpoint approximation.

THEOREM 1 **(Midpoint Rule)** Let f be a continuous function on the interval $[a, b]$. Let N be a positive integer. For each integer j with $1 \le j \le N$, let $\overline{x}_j$ be defined by equation (5.8.1). Let $\mathcal{M}_N$ be the midpoint approximation

$$\mathcal{M}_N = \Delta x \cdot (f(\overline{x}_1) + f(\overline{x}_2) + \cdots + f(\overline{x}_N)).$$

If C is a constant such that $|f''(x)| \le C$ for $a \le x \le b$, then

$$\left| \int_a^b f(x)\,dx - \mathcal{M}_N \right| \le \frac{C}{24} \cdot \frac{(b-a)^3}{N^2}. \tag{5.8.3}$$

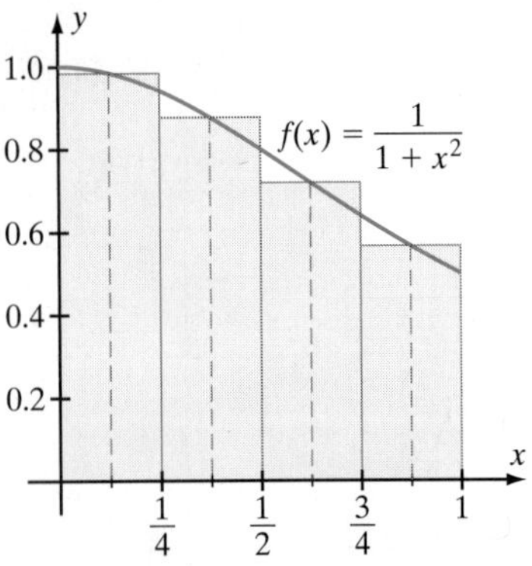

▲ Figure 2

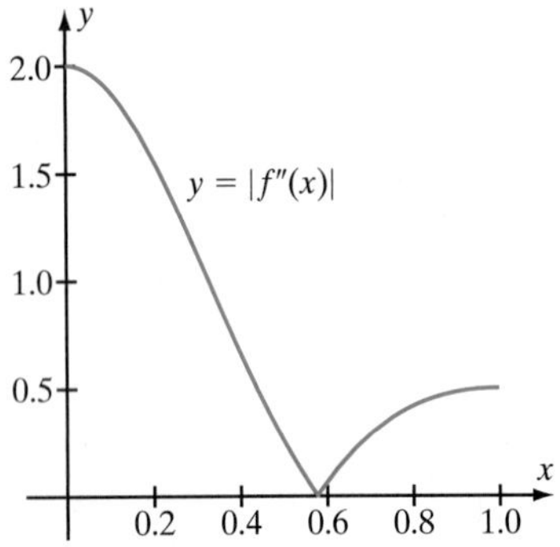

▲ Figure 3 The plot of $y = |f''(x)|$ where $f(x) = 1/(1+x^2)$

▶ **EXAMPLE 1** Let $f(x) = 1/(1+x^2)$. Estimate $A = \int_0^1 f(x)\,dx$ using the Midpoint Rule with $N = 4$. Based on this approximation, how small might the exact value of A be? How large?

Solution The points $\{0, 1/4, 1/2, 3/4, 1\}$ partition the interval $[0,1]$ into four subintervals of equal length ($\Delta x = 1/4$). The midpoints of the subintervals are 1/8, 3/8, 5/8, and 7/8. The Midpoint Rule tells us that A is approximately equal to

$$\mathcal{M}_4 = \frac{1}{4}\cdot\left(\frac{1}{1+(1/8)^2} + \frac{1}{1+(3/8)^2} + \frac{1}{1+(5/8)^2} + \frac{1}{1+(7/8)^2}\right) \approx 0.7867.$$

The number $\mathcal{M}_4$ represents the area of the shaded rectangles in Figure 2. To find out what the exact value of A might be, we calculate that $f'(x) = -2x/(1+x^2)^2$ and

$$f''(x) = \frac{6x^2-2}{(1+x^2)^3}.$$

The graph of $|f''|$ in Figure 3 reveals that the maximum value of $|f''(x)|$ for $0 \le x \le 1$ occurs at $x = 0$ and is equal to 2. We may therefore use $C = 2$ in inequality (5.8.3), which tells us that the approximation $\mathcal{M}_4$ differs from A by no more than

$$\frac{C\cdot(b-a)^3}{24N^2} = \frac{2\cdot 1^3}{24\cdot 4^2} \approx 0.0052.$$

We conclude that

$$\underbrace{0.7867}_{\text{Midpoint approximation}} - \underbrace{0.0052}_{\text{Error estimate}} \le A \le \underbrace{0.7867}_{\text{Midpoint approximation}} + \underbrace{0.0052}_{\text{Error estimate}}. \qquad (5.8.4)$$

By calculating the difference that begins line (5.8.4) and the sum that terminates it, we see that the exact value of A may be as small as 0.7815 or as large as 0.7919. It so happens that in this example, we know that $F(x) = \arctan(x)$ is an antiderivative of the integrand $f(x)$. Therefore the exact value of A is $F(1) - F(0)$, or $\pi/4$. In other words, $A = \pi/4 = 0.785\ldots$, which is indeed between our lower and upper estimates. The merit of Theorem 1 is that we do not need to know the exact value of A to be sure that it lies within the estimates. ◀

INSIGHT An exact value of an integral, such as $A = \pi/4$ in Example 1, must often be decimalized in numerical work. By rounding an infinite decimal expansion to m decimal places, we introduce an error on the order of $5 \times 10^{-(m+1)}$. According to inequality (5.8.1), if we take N large enough so that the inequality

$$\frac{C\cdot(b-a)^3}{24N^2} < 5 \times 10^{-(m+1)}$$

holds, then the midpoint approximation generates an error less than $5 \times 10^{-(m+1)}$. In other words, the midpoint approximation allows us to evaluate an indefinite integral with as much precision as we would be able to obtain from the decimalization of an exact answer.

▶ **EXAMPLE 2** Let $f(x) = \cos(x^2)$. A plot of $y = |f''(x)|$ in the viewing window $[0.2, 1.4] \times [0, 4]$ shows that $|f''(x)| < 3.8442$ for $0.2 \le x \le 1.4$ (see Figure 4). Use a midpoint approximation to calculate $\int_{0.2}^{1.4} f(x)\,dx$ to two decimal places.

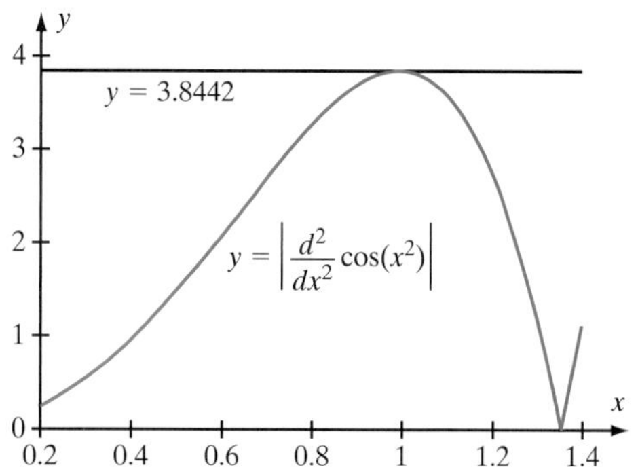

▲ Figure 4

Solution We may assign the value 3.8442 to the constant C that appears in inequality (5.8.3). Then, with $a = 0.2$ and $b = 1.4$, inequality (5.8.3) becomes

$$\left|\int_{0.2}^{1.4} \cos(x^2)\,dx - \mathcal{M}_N\right| \le \frac{C}{24}\frac{(b-a)^3}{N^2} = \frac{3.8442}{24}\frac{(1.4-0.2)^3}{N^2} = \frac{0.2767824}{N^2}.$$

To approximate $\int_{0.2}^{1.4} \cos(x^2)\,dx$ to two decimal places, we must find an N such that

$$\left|\int_{0.2}^{1.4} \cos(x^2)\,dx - \mathcal{M}_N\right| \le 5 \times 10^{-3}.$$

To obtain this inequality, we choose N so that $0.2767824/N^2 \le 5 \times 10^{-3}$, or $276.7824/5 \le N^2$, or $7.44\ldots \le N$. The smallest integer satisfying this inequality is $N = 8$. On substituting this value of N in the inequalities that we have obtained, we see that

$$\left|\int_{0.2}^{1.4} \cos(x^2)\,dx - \mathcal{M}_8\right| \le \frac{0.2767824}{8^2} < 5 \times 10^{-3}.$$

Thus $\mathcal{M}_8$ approximates the given integral to two decimal places. To evaluate $\mathcal{M}_8$, we calculate $\Delta x = (1.4 - 0.2)/8 = 0.15$, $\bar{x}_1 = 0.2 + \frac{1}{2}0.15 = 0.275$, $\bar{x}_2 = 0.275 + 0.15 = 0.425$, $\bar{x}_3 = 0.425 + 0.15 = 0.575$, $\bar{x}_4 = 0.575 + 0.15 = 0.725$, $\bar{x}_5 = 0.725 + 0.15 = 0.875$, $\bar{x}_6 = 0.875 + 0.15 = 1.025$, $\bar{x}_7 = 1.025 + 0.15 = 1.175$, and $\bar{x}_8 = 1.175 + 0.15 = 1.325$. Our approximation is

$$\mathcal{M}_8 = 0.15\left(\cos(0.275^2) + \cos(0.425^2) + \cos(0.575^2) + \cdots + \cos(1.325^2)\right) = 0.7522\ldots.$$

A more accurate approximation shows that $\int_{0.2}^{1.4} \cos(x^2)\,dx = 0.7498\ldots$. From this evaluation, we see that the absolute error resulting from our approximation is about 0.0024, which is indeed less than the specified 0.005. The importance of inequality (5.8.3) is that it allows us to attain a specified accuracy even when the value of the integral is not known. ◀

The Trapezoidal Rule

Suppose that f is positive over the subinterval $I_j = [x_{j-1}, x_j]$. We can approximate the area under the graph of f and over I_j by the area of a trapezoid, as indicated in

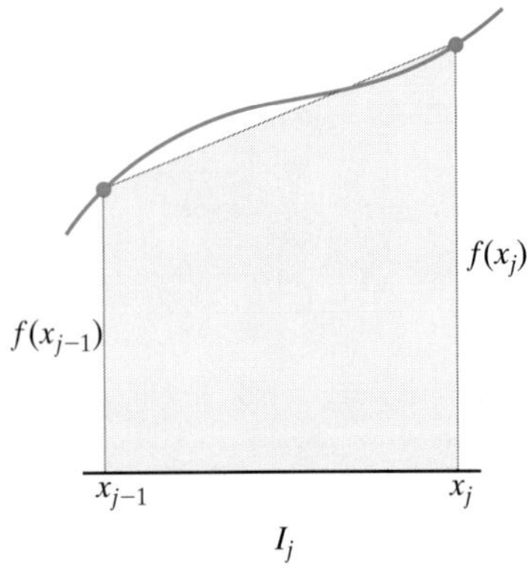

▲ Figure 5

Figure 5. The trapezoidal area is the width Δx of the base times the average height of the trapezoid:

$$A_j = \frac{f(x_{j-1}) + f(x_j)}{2} \cdot \Delta x.$$

The order N *trapezoidal approximation* $\mathcal{T}_N$ is defined by summing these trapezoidal areas:

$$\begin{aligned}\mathcal{T}_N &= A_1 + A_2 + \cdots + A_N \\ &= \frac{1}{2}\cdot\Big(f(x_0)+f(x_1)\Big)\cdot\Delta x + \frac{1}{2}\cdot\Big(f(x_1)+f(x_2)\Big)\cdot\Delta x + \cdots \\ &\quad + \frac{1}{2}\cdot\Big(f(x_{N-1})+f(x_N)\Big)\cdot\Delta x.\end{aligned}$$

After combining terms, we obtain the formula

$$\mathcal{T}_N = \frac{\Delta x}{2}\cdot\Big(f(x_0) + 2f(x_1) + 2f(x_2) + \cdots + 2f(x_{N-1}) + f(x_N)\Big). \qquad (5.8.5)$$

It should be noted that approximation (5.8.5) is equally valid for functions that are not positive everywhere. For example, if $f(x)$ is negative on the subinterval $I_j = [x_{j-1}, x_j]$, then $\int_{x_{j-1}}^{x_j} f(x)\,dx$ represents the negative of the area of the region that lies under I_j and above the graph of f. This negative contribution to the value of $\int_a^b f(x)\,dx$ is reflected by the contribution of the negative values $f(x_{j-1})$ and $f(x_j)$ to $\mathcal{T}_N$. Similar considerations apply if $f(x_{j-1})$ and $f(x_j)$ have opposite signs.

THEOREM 2 **(Trapezoidal Rule)** Let f be a continuous function on the interval $[a,b]$. Let N be a positive integer. Let $\mathcal{T}_N$ be the trapezoidal approximation

$$\mathcal{T}_N = \frac{\Delta x}{2}\cdot\Big(f(x_0) + 2f(x_1) + 2f(x_2) + \cdots + 2f(x_{N-1}) + f(x_N)\Big).$$

If $|f''(x)| \le C$ for all x in the interval $[a,b]$, then the order N trapezoidal approximation $\mathcal{T}_N$ is accurate to within $C\cdot(b-a)^3/(12N^2)$. In other words,

$$\left|\int_a^b f(x)\,dx - \mathcal{T}_N\right| \le \frac{C}{12}\cdot\frac{(b-a)^3}{N^2}. \qquad (5.8.6)$$

It is instructive to compare the error estimates for the midpoint approximation and the trapezoidal approximation:

$$\frac{C\cdot(b-a)^3}{24N^2} \qquad \text{Midpoint Rule}$$

$$\frac{C\cdot(b-a)^3}{12N^2}. \qquad \text{Trapezoidal Rule}$$

As its larger denominator suggests, the Midpoint Rule is usually more accurate than the Trapezoidal Rule. Even so, the Trapezoidal Rule can be more suitable for analyzing discrete data. We illustrate this point with an example from economics.

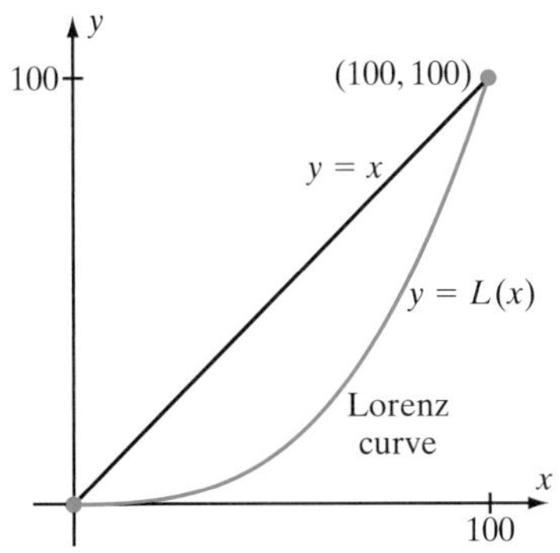

▲ Figure 6

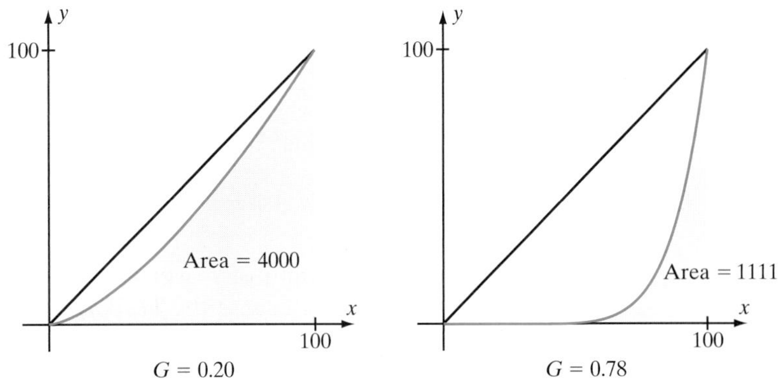

▲ Figure 7

▶ **EXAMPLE 3** For each x between 0 and 100, let $L(x)$ denote the percentage of a nation's total annual income that is earned by the lowest-earning $x\%$ of its inhabitants during a specified year. Economists call the increasing function L: $[0, 100] \to [0, 100]$ a *Lorenz function*. The graph of the equation $y = L(x)$ is called a *Lorenz curve*. Notice that $L(0) = 0$ and $L(100) = 100$. If all personal income were equally distributed, then the *Lorenz curve* would be $y = x$. In actuality, the *Lorenz curve* lies below the line $y = x$ for $0 < x < 100$ (see Figure 6). The area A of the region below the *Lorenz curve* is used to measure the disparity of income distribution. Notice that, because A is between 0 and 5000, the number $G = 1 - A/5000$, known as the *Gini coefficient* or the *coefficient of inequality*, is between 0 and 1. If the *Lorenz curve* is $y = x$ (perfectly equal income distribution), then $A = 5000$ and $G = 0$. As Figure 7 shows, when income distribution becomes increasingly unequal, the *Lorenz curve* approaches the bottom $(y = 0)$ and right side $(x = 100)$ of the triangle, the area A approaches 0, and the coefficient of inequality G approaches 1. Use the tabulated data to estimate the coefficient of inequality for the United States in 2006.

Lowest Fifth	Second Fifth	Third Fifth	Fourth Fifth	Highest Fifth
3.4	8.6	14.6	22.9	50.5

Percent Distribution of Total Income 2006

Solution Income data is generally given by quintile (as in the table) and not in the cumulative form needed for the Lorenz function. Nevertheless, L can be obtained by a little bit of addition. We know that $L(0) = 0$, and the table tells us that $L(20) = 3.4$. To this number, we add the percentage of income earned by the second fifth of workers: $L(40) = 3.4 + 8.6 = 12.0$. Similarly, we obtain $L(60)$ by adding to $L(40)$ the percentage of income earned by the middle fifth. Thus $L(60) = 12.0 + 14.6 = 26.6$. Continuing in this way, we find $L(80) = L(60) + 22.9 = 26.6 + 22.9 = 49.5$, and $L(100) = \mathrm{L}(80) + 50.5 = 49.5 + 50.5 = 100$. Using the partition $\{0, 20, 40, 60, 80, 100\}$ of $[0, 100]$ with $N = 5$ and $\Delta x = 20$, we obtain the following approximation from the Trapezoidal Rule:

$$A = \int_0^{100} L(x)\,dx \approx \frac{20}{2} \cdot (0 + 2 \cdot 3.4 + 2 \cdot 12.0 + 2 \cdot 26.6 + 2 \cdot 49.5 + 100) = 2830.0.$$

Therefore the coefficient of inequality is approximately $1 - 2830/5000 \approx 0.434$. ◀

INSIGHT Example 3 is typical of many discrete data problems in which the Midpoint Rule does not permit effective use of the data. In Example 3, the midpoints of the partitioning subintervals are $\overline{x}_1 = 10, \overline{x}_2 = 30, \ldots, \overline{x}_5 = 90$. Notice that the values $L(\overline{x}_j)$ are not available. Yet these are the very values of L that we would need to apply the Midpoint Rule.

Simpson's Rule

The graph of f over a *small* subinterval will typically be concave up or concave down. Because the Trapezoidal Rule and Midpoint Rule approximations are based on straight line approximations to the graph of f, neither rule is able to reflect concavity. However, if we approximate the graph of f over a small subinterval by a *parabola,* as in Figure 8, then we can take into account the concavity of f. This idea leads to the most accurate of the approximation rules that we will discuss—Simpson's Rule.

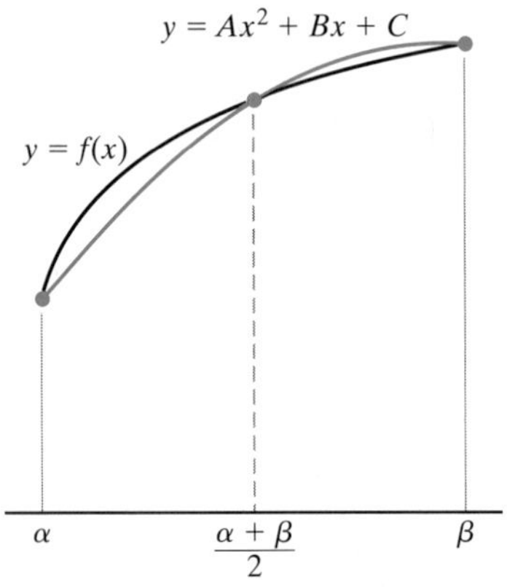

▲ Figure 8 Approximating the arc of a curve with parabolic arc

To derive a formula for Simpson's Rule, we need to know the area under an arc of a parabola. The required formula is stated in the next theorem. It is used only in the derivation of Simpson's Rule and need not be memorized.

THEOREM 3 If $P(x) = Ax^2 + Bx + C$, and if $I = [\alpha, \beta]$ is an interval with midpoint γ, then

$$\int_\alpha^\beta P(x)\,dx = \frac{\beta-\alpha}{6}\cdot(P(\alpha) + 4P(\gamma) + P(\beta)).$$

Proof. We calculate

$$\int_\alpha^\beta P(x)\,dx = \left(\frac{1}{3}A\beta^3 + \frac{1}{2}B\beta^2 + C\beta\right) - \left(\frac{1}{3}A\alpha^3 + \frac{1}{2}B\alpha^2 + C\alpha\right)$$

$$= \frac{1}{6}(\beta-\alpha)(2A\alpha^2 + 3\alpha B + 2\alpha A\beta + 6C + 3\beta B + 2A\beta^2) \quad \text{after factoring.}$$

It is now simply a matter of patient algebraic calculation to show that $P(\alpha) + 4P(\gamma) + P(\beta)$ equals the last factor. ■

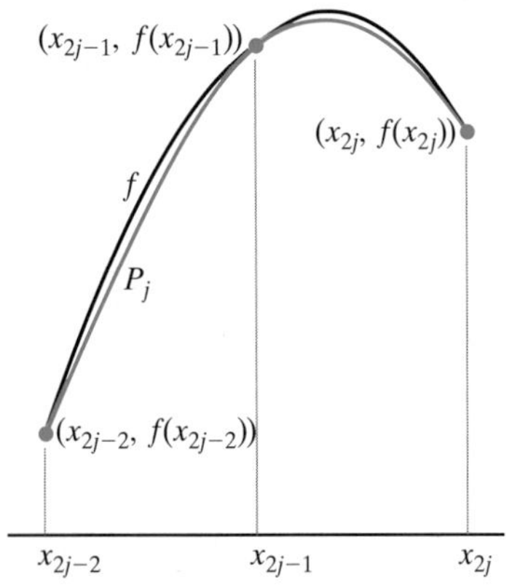

▲ Figure 9

To formulate Simpson's Rule, we select a partition of $[a, b]$ with an *even* number ($N = 2\ell$) of subintervals of equal length Δx. We pair up the first and second intervals, the third and fourth intervals, and so on. Over each pair of intervals, we approximate f by a parabola passing through the points on the graph of f corresponding to the three endpoints of the intervals. (Because the equation of a parabola $y = Ax^2 + Bx + C$ has three coefficients, a parabola is uniquely determined by any three points through which it passes.) The paired intervals are

$$[x_{2j-2}, x_{2j-1}] \quad \text{and} \quad [x_{2j-1}, x_{2j}] \quad (j = 1, \ldots, \ell)$$

For each j from 1 to ℓ, the approximating parabola P_j passes through the three points $(x_{2j-2}, f(x_{2j-2}))$, $(x_{2j-1}, f(x_{2j-1}))$, and $(x_{2j}, f(x_{2j}))$ (as in Figure 9). By applying Theorem 3, with $\alpha = x_{2j-2}, \gamma = x_{2j-1}$, and $\beta = x_{2j}$, we see that the integral of the parabola over the interval $[x_{2j-2}, x_{2j}]$ is

$$\begin{aligned}\frac{x_{2j}-x_{2j-2}}{6}&\left(P_j(x_{2j-2})+4P_j(x_{2j-1})+P_j(x_{2j})\right)\\ &=\frac{2\Delta x}{6}\left(f(x_{2j-2})+4f(x_{2j-1})+f(x_{2j})\right)\\ &=\frac{\Delta x}{3}\left(f(x_{2j-2})+4f(x_{2j-1})+f(x_{2j})\right).\end{aligned}$$

Finally, we add the integrals from $j=1$ to $j=\ell$ to obtain Simpson's approximation of $\int_a^b f(x)\,dx$:

$$\begin{aligned}\mathcal{S}_N = \frac{\Delta x}{3}\cdot\Big(f(x_0)+4f(x_1)+2f(x_2)+4f(x_3)+\cdots+2f(x_{N-2})\\ +4f(x_{N-1})+f(x_N)\Big).\end{aligned} \tag{5.8.7}$$

THEOREM 4 **(Simpson's Rule).** Let f be continuous on the interval $[a,b]$. Let N be a positive *even* integer. If C is any number such that $|f^{(4)}(x)| \le C$ for $a \le x \le b$, then

$$\left|\int_a^b f(x)\,dx - \mathcal{S}_N\right| \le \frac{C}{180}\cdot\frac{(b-a)^5}{N^4}. \tag{5.8.8}$$

▶ **EXAMPLE 4** Let $f(x)=1/(1+x^2)$. Estimate $A=\int_0^1 f(x)\,dx$ using Simpson's Rule with $N=4$. In what interval of real numbers does A lie? (In Example 1, we answered this question using the Midpoint Rule.)

Solution Because $N=4$ is even, we can apply Simpson's Rule. The length of each subinterval in the order 4 uniform partition of the interval of integration, $[0,1]$, is $\Delta x = 1/4$. The points of the partition are $x_0=0$, $x_1=1/4$, $x_2=1/2$, $x_3=3/4$, and $x_4=1$. At these points, we calculate $f(x_0)=1$, $f(x_1)=1/(1+(1/4)^2)=16/17$, $f(x_2)=1/(1+(1/2)^2)=4/5$, $f(x_3)=1/(1+(3/4)^2)=16/25$, and $f(x_4)=1/2$. According to Simpson's Rule:

$$\int_0^1 \frac{1}{1+x^2}\,dx \approx S_4 = \frac{1/4}{3}\left(1+4\cdot\frac{16}{17}+2\cdot\frac{4}{5}+4\cdot\frac{16}{25}+\frac{1}{2}\right)=\frac{1}{12}\cdot\frac{8011}{850}\approx 0.78539. \quad ◀$$

We now estimate how accurate this approximation is. With successive applications of the Quotient Rule, we calculate that

$$f^{(4)}(x) = 24\cdot\frac{5x^4-10x^2+1}{(1+x^2)^5}.$$

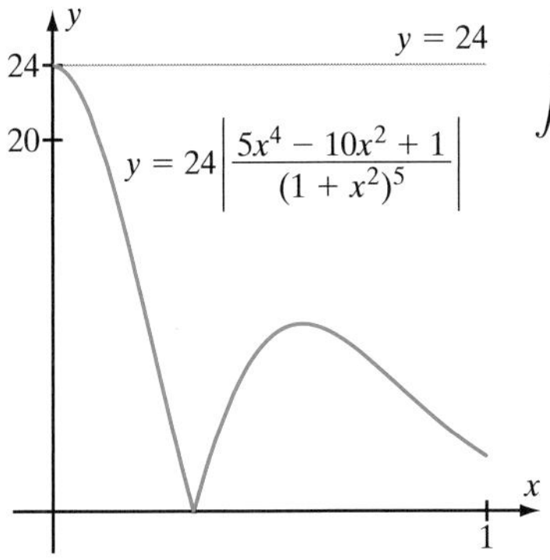

▲ Figure 10

Using the methods from Section 4.3 in Chapter 4, we can determine that, for x in $[0,1]$, $|f^{(4)}(x)|$ attains a maximum value of 24. Plotting $|f^{(4)}|$ also reveals this bound (see Figure 10). Taking $C=24$ and $N=4$ in inequality (5.8.8) shows that Simpson's Rule with $N=4$ is accurate to within

$$\frac{24 \cdot 1^5}{180 \cdot 4^4} \approx 0.00052.$$

The exact value of A is therefore between $S_4 - 0.00052 = 0.78487$ and $S_4 + 0.00052 = 0.78591$. In other words, A lies in the interval $[0.78487, 0.78591]$. As noted in Example 2, the exact value of A is $\pi/4 = 0.78539\ \ldots$, which is indeed between the upper and lower bounds we determined. ◄

INSIGHT The error bounds of inequalities (5.8.3), (5.8.6), and (5.8.8) are *worst-case* estimates. An approximation rule often yields an estimate of an integral that is much better than the error bound suggests. For instance, the actual error in Example 4 is $\pi/4 - 0.78539216 = 6.0034\ldots \times 10^{-6}$, a number that is significantly smaller than 0.00052, the upper bound for the error that is furnished by inequality (5.8.8).

Using Simpson's Rule with Discrete Data

In many practical instances, the value of a definite integral $\int_a^b f(x)\,dx$ is needed even though a formula for $f(x)$ is not known. Instead, the values of f are measured at a discrete set of points. Under these circumstances, the integral *must* be approximated. In Example 2, we considered a discrete data problem that arises in economics. The following example from physiology is also typical.

The rate r at which blood is pumped from the heart is called *cardiac output.* The units of cardiac output are volume/time (such as liters per minute). In the *impulse injection method* of measuring cardiac output, a mass M of dye (or radioactive isotope) is injected into a pulmonary vein. Let $t = 0$ correspond to the time of injection, and let $t = T$ correspond to the time required to pump all the dye from the heart. The concentration $c(t)$ of dye (in units of mass/volume) that is pumped from the heart into the aorta is monitored. The mass of dye pumped from the heart in the time interval $[t, t + \Delta t]$ is approximately

$$\underbrace{\underset{\frac{\text{mass of dye}}{\text{volume}}}{c(t)} \cdot \underbrace{\underset{\frac{\text{volume}}{\text{time}}}{r} \cdot \underset{\text{time}}{\Delta t}}_{\text{Volume}}}_{\text{Mass of dye.}}$$

If we partition the time interval $[0, T]$ into N subintervals of equal length Δt, then the Riemann sum $\sum_{j=1}^{N} c(t_{j-1}) r \Delta t$ approximates the amount of dye pumped from the heart. Letting $\Delta t \to 0$, we obtain

$$\underset{\text{Mass of dye injected}}{M} = \underset{\text{Mass of dye pumped out.}}{\int_0^T r \cdot c(t)dt}$$

In obtaining this formula, we have assumed that M is small enough that recirculation does not occur before time T. (Blood can travel from the heart to the farthest extremity of the circulatory system and back in as little as 12 seconds.) We solve for r in the last equation and find that cardiac output is given by

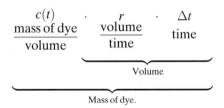

$$r = \frac{M}{\int_0^T c(t)\,dt}.$$

▶ **EXAMPLE 5** A doctor injects 3 mg of dye into a vein near the patient's heart. The aortic concentration measurements have been tabulated, with time t given in seconds and concentration $c(t)$ given in milligrams per liter. Estimate the patient's cardiac output.

t	0	1	2	3	4	5	6	7	8
$c(t)$	0	0.54	2.7	6.6	8.9	6.4	3.1	1.2	0

Solution Because the partition has nine equally spaced points forming an even number ($N = 8$) of equal length subintervals, Simpson's Rule is applicable:

$$\int_0^8 c(t)\,dt \approx \frac{8-0}{3\cdot 8}(1(0) + 4(0.54) + 2(2.7) + 4(6.6) + 2(8.9) + 4(6.4)$$
$$+ 2(3.1) + 4(1.2) + 1(0))$$
$$\approx 29.453.$$

Therefore r is approximately 3/29.453 L/s, or 180/29.453 = 6.11 L/min. ◀

INSIGHT In Example 5, notice that the two data points equal to 0 have been included in the sum just as they would have been had they had nonzero values. It is a common error to omit data points that are 0 because they seem not to contribute to the sum. In fact, each member of the sequence of data points, $f(x_0), f(x_1), f(x_2), f(x_3), \ldots$ must be multiplied by the corresponding member of the sequence of coefficients, 1, 4, 2, 4, … (in the case of Simpson's Rule). Omitting an $f(x_j)$ that is 0 will lead to the subsequent data points being multiplied by the wrong coefficients.

QUICK QUIZ

1. Approximate $\int_1^5 (6/x)\,dx$ by the Midpoint Rule with $N = 2$.
2. Approximate $\int_1^5 (15/x)\,dx$ by the Trapezoidal Rule with $N = 2$.
3. Approximate $\int_1^5 (45/x)\,dx$ by Simpson's Rule with $N = 4$.
4. Approximate $\int_{2.5}^{3.5} f(x)\,dx$ by a) the Midpoint Rule, b) the Trapezoidal Rule, and c) Simpson's Rule, using, in each case, as much of the data, $f(2.5) = 24$, $f(3.0) = 36$, and $f(3.5) = 12$, as possible.

Answers

1. 9 2. 28 3. 73 4. a) 36, b) 27, c) 30

EXERCISES

Problems for Practice

▶ In each of Exercises **1–10**, a function f, an interval I, and an even integer N are given. Approximate the integral of f over I by partitioning I into N equal length subintervals and using the Midpoint Rule, the Trapezoidal Rule, and then Simpson's Rule. ◀

1. $f(x) = x^2$ $\quad I = [1, 3], N = 2$

2. $f(x) = x^{1/2}$ $\quad I = [0, 2], N = 2$

3. $f(x) = \sqrt{2+x+x^2}$ $I = [-1, 1], N = 2$
4. $f(x) = (1+x^3)^{(1/3)}$ $I = [-1, 1], N = 2$
5. $f(x) = \ln(x)$ $I = [1, 3], N = 2$
6. $f(x) = 15x/(1+x)$ $I = [0, 4], N = 4$
7. $f(x) = \cos(x)$ $I = [\pi/4, 5\pi/4], N = 4$
8. $f(x) = \sqrt{1+x}$ $I = [0, 8], N = 4$
9. $f(x) = x^3$ $I = [-1/2, 7/2], N = 4$
10. $f(x) = 1/x$ $I = [2, 6], N = 4$

In each of Exercises 11–14, a function $f(x)$ and an interval $I = [a, b]$ are given. Also given is the approximation $\mathcal{M}_{10}$ of $A = \int_a^b f(x)\,dx$ that is obtained by using the Midpoint Rule with a uniform partition of order 10.

a. Use inequality (5.8.3) to find a lower bound α for A.
b. Use inequality (5.8.3) to find an upper bound β for A.
c. Calculate A and ascertain that it lies in the interval $[\alpha, \beta]$.

11. $f(x) = x + 1/x$ $I = [1/2, 7/2]$, $\mathcal{M}_{10} = 7.93202\ldots$
12. $f(x) = \sqrt{1+x}$ $I = [0, 8]$, $\mathcal{M}_{10} = 17.34205\ldots$
13. $f(x) = (2+x)^{(-1/3)}$ $I = [-1, 3]$, $\mathcal{M}_{10} = 2.88409\ldots$
14. $f(x) = x^2 + \cos(x)$ $I = [0, \pi/2]$, $\mathcal{M}_{10} = 2.28972\ldots$

In each of Exercises 15–18, a function $f(x)$ and an interval $I = [a, b]$ are given. Also given is the approximation $\mathcal{T}_{10}$ of $A = \int_a^b f(x)\,dx$ that is obtained by using the Trapezoidal Rule with a uniform partition of order 10.

a. Use inequality (5.8.6) to find a lower bound α for A.
b. Use inequality (5.8.6) to find an upper bound β for A.
c. Calculate A and ascertain that it lies in the interval $[\alpha, \beta]$.

15. $f(x) = x^3$ $I = [-5, 1]$, $\mathcal{T}_{10} = -158.160\ldots$
16. $f(x) = 1/(10-x)$ $I = [1, 8]$, $\mathcal{T}_{10} = 1.51366\ldots$
17. $f(x) = x\exp(x^2)$ $I = [3/2, 2]$, $\mathcal{T}_{10} = 22.64660\ldots$
18. $f(x) = \sec(x)$ $I = [0, \pi/4]$, $\mathcal{T}_{10} = 0.88209\ldots$

In each of Exercises 19–22, a function $f(x)$ and an interval $I = [a, b]$ are given. Also given is the approximation $\mathcal{S}_{10}$ of $A = \int_a^b f(x)\,dx$ that is obtained by using Simpson's Rule with a uniform partition of order 10.

a. Use inequality (5.8.8) to find a lower bound α for A.
b. Use inequality (5.8.8) to find an upper bound β for A.
c. Calculate A and ascertain that it lies in the interval $[\alpha, \beta]$.

19. $f(x) = x^5$ $I = [1, 4]$, $\mathcal{S}_{10} = 682.54050\ldots$
20. $f(x) = 1/x$ $I = [2, 10]$, $\mathcal{S}_{10} = 1.61008\ldots$
21. $f(x) = \sqrt{(x)}$ $I = [1, 9]$, $\mathcal{S}_{10} = 17.33275\ldots$
22. $f(x) = 10\sin(x)$ $I = [0, 3\pi/2]$, $\mathcal{S}_{10} = 10.00281\ldots$

Income data for three countries are given in the following tables. In each table, x represents a percentage, and $L(x)$ is the corresponding value of the Lorenz function, as described in Example 3. In each of Exercises 23–27, use the specified approximation method to estimate the coefficient of inequality for the indicated country. (The values $L(0) = 0$ and $L(100) = 100$ are not included in the tables, but they should be used.)

x	20	40	60	80
$L(x)$	5	20	30	55

Income Data, Country A

x	25	50	75
$L(x)$	15	25	40

Income Data, Country B

x	10	20	30	40	50	60	70	80	90
$L(x)$	4	8	14	22	32	42	56	70	82

Income Data, Country C

23. Country A, Trapezoidal Rule
24. Country B, Trapezoidal Rule
25. Country C, Trapezoidal Rule
26. Country B, Simpson's Rule
27. Country C, Simpson's Rule

In Exercises 28–30, use Simpson's Rule to estimate cardiac output based on the tabulated readings (with t in seconds and $c(t)$ in mg/L) taken after the injection of 5 mg of dye.

28.

t	0	1.5	3	4.5	6	7.5	9
$c(t)$	0	2.4	6.3	9.7	7.1	2.3	0

29.

t	0	1	2	3	4	5	6	7	8
$c(t)$	0	1.9	5.8	9.4	10.4	9.1	5.9	2.1	0

30.

t	0	1	2	3	4	5	6	7	8	9	10
$c(t)$	0	3.8	6.8	8.6	9.7	10.2	9.4	8.2	6.1	3.1	0

Further Theory and Practice

31. Use inequality (5.8.6) to determine how large N should be to ensure that the Trapezoidal Rule approximates $\int_1^9 \exp(-x/2)\,dx$ to within 10^{-4}? How about for Simpson's Rule?

In each of Exercises 32–35, a function $f(x)$ is given. Use inequality (5.8.3) to determine how large N must be to guarantee that the Midpoint Rule approximation of $\int_{-2}^{2} f(x)\,dx$ is accurate to within 10^{-3}.

32. $f(x) = 1/(3 + x)$
33. $f(x) = \sqrt{6 + x}$
34. $f(x) = e^x$
35. $f(x) = \sin(\pi x)$

In each of Exercises 36–39, a function $f(x)$ is given. Use inequality (5.8.8) to determine how large N must be to guarantee that the Simpson's Rule approximation of $\int_{-2}^{2} f(x)\,dx$ is accurate to within 10^{-3}.

36. $f(x) = 1/(3 + x)$
37. $f(x) = \sqrt{6 + x}$
38. $f(x) = e^x$
39. $f(x) = \sin(\pi x)$

40. Show that the Simpson's Rule approximation is exact when applied to a polynomial of degree 3 or less. (*Hint:* Use inequality (5.8.4) to show that the error is 0.)

41. Show that the average of x^3 over the interval $[\gamma - h, \gamma + h]$ is $\gamma\ (\gamma^2 + h^2)$. Show that this quantity is also equal to $((\gamma - h)^3 + 4\ \gamma^3 + (\gamma + h)^3)/6$. Deduce that Theorem 3, which was stated for polynomials of degree 2, is also valid for polynomials $P(x) = Ax^2 + Bx + C + Dx^3$ of degree 3. As a result, Simpson's Rule is exact when applied to cubic polynomials.

42. Applying Simpson's Rule to the right side of

$$\pi = 4\int_0^1 \sqrt{1 - x^2}\,dx$$

gives an approximation of π. In practice, this method of approximation is not useful because of its computational cost. For example, with $N = 100$, the approximation is $3.1411\ldots$, which is accurate to only three decimal places. Examine the graph of $y = \sqrt{1 - x^2}$ for $0 \le x \le 1$. What geometric property does this graph have that graphs of approximating parabolas do not have? Why is error bound (5.8.4) of no use here?

43. Simpson's Rule is *generally* more accurate than the Midpoint Rule, but it is not *always* more accurate. Calculate $A = \int_{-1}^{1} \sqrt{|x|}\,dx$. With $N = 2$, estimate A using both the Midpoint Rule, $\mathcal{M}_2$, and Simpson's Rule, $\mathcal{S}_2$. What are the absolute errors? Repeat with $N = 4$.

44. Calculate $A = \int_{-1}^{1} |x|\,dx$. Let N be an even positive integer that is not divisible by 4. Show that if a uniform partition of order N is used, then the Midpoint Rule approximation of A is exact, but the Simpson's Rule approximation is not.

45. Suppose f has derivatives of order 4 in an interval containing $[a, b]$. *Chevilliet's form of the error in Simpson's Rule* states that the error in using Simpson's Rule is *exactly*

$$\frac{(b - a)^4}{180\,N^4}\,(f'''(\xi) - f'''(\eta))$$

for two points ξ and η in $[a, b]$. Show that the error estimate as stated in the text (sometimes known as *Stirling's form*) follows from Chevilliet's form of the error.

In Exercises 46 and 47, several values of the Lorenz function L have been tabulated (refer to Example 2). Use trapezoidal approximations to estimate the coefficient of inequality that corresponds to the given data. (Note: The tables represent partitions that are *not* uniform. Also, the data points (0, 0) and (100, 100) have not been included in the tables but should be used in the calculations.)

46.

x	15	25	50	75	90
$L(x)$	3	19	25	42	70

47.

x	16	28	51	75	88	97
$L(x)$	3	8	24	46	69	88

48. Figure 11 shows a map of the province of Manitoba flipped onto its western boundary. Distances are in kilometers. The southern portion of the province, appearing to the right in Figure 11, is essentially a trapezoid. Use Simpson's Rule to estimate the area of the northern portion of the province. Approximately what is the area of Manitoba when estimated in this way? (The actual area is 649,953 km^2.) Repeat with the Trapezoidal Rule. Repeat with the Midpoint Rule using increments of 156 km along the S-axis. (For more precise estimation, the small arc of the eastern boundary that is not the graph of a function must be taken into account.)

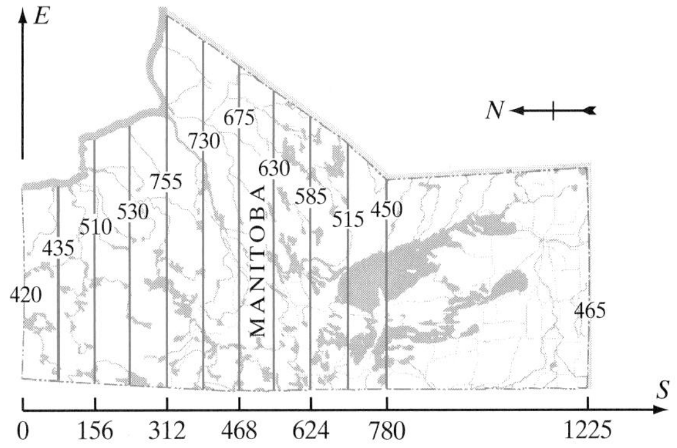

Figure 11

49. A 6 m wide swimming pool is illustrated in Figure 12, on next page. Depths are given every 2 m. Estimate the volume of water in the pool when it is filled. Use the Midpoint, Trapezoidal, and Simpson's Rules.

50. Let N be an even integer. The three approximation methods discussed in this section satisfy

$$\mathcal{S}_N = \frac{1}{3}\mathcal{T}_{N/2} + \frac{2}{3}\mathcal{M}_{N/2}. \qquad (5.8.9)$$

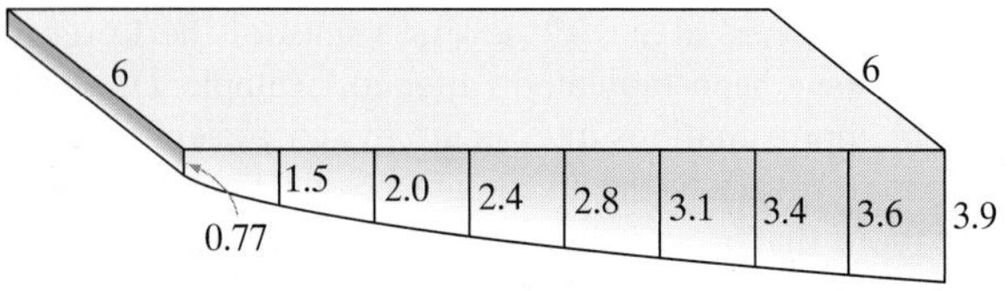

▲ Figure 12

Use equation (5.8.9) to show that

$$(\mathcal{S}_N - \mathcal{T}_{N/2}) = 2\,(\mathcal{M}_{N/2} - \mathcal{S}_N).$$

Let N be an even integer. From equation (5.8.9) we may deduce that one of the estimates $\mathcal{T}_{N/2}$ and $\mathcal{M}_{N/2}$ is greater than or equal to $\mathcal{S}_N$, and the other is less than or equal to $\mathcal{S}_N$. In each of Exercises 51–54, calculate $\mathcal{S}_N$, $\mathcal{M}_{N/2}$, and $\mathcal{T}_{N/2}$ for the given even value of N and verify this deduction.

51. $\int_0^4 x^4\,dx \quad N = 4$
52. $\int_0^4 \sqrt{x}\,dx \quad N = 2$
53. $\int_1^7 1/x\,dx \quad N = 6$
54. $\int_0^\pi \sin(x)\,dx \quad N = 4$
55. During a 6-minute span, at intervals of 1 minute, the speedometer of a car reads

Time (minutes)	0	1	2	3	4	5	6
Speed (km/hr)	90	80	75	80	80	70	60

Use Simpson's Rule to estimate the distance traveled by the car during that 6-minute period.

56. Speed measurements of a runner taken at half-second intervals during the first 5 seconds of a sprint are provided in the following table.

Time (s)	0	0.5	1.0	1.5	2.0	2.5	3.0	3.5	4.0	4.5	5.0
Speed (m/s)	0	5.26	6.67	7.41	8.33	8.33	9.52	9.52	10.64	10.64	10.87

About how many meters did the athlete run during that 5-second interval? Use Simpson's Rule.

Calculator/Computer Exercises

In each of Exercises 57–60, a function f and an interval $[a, b]$ are specified. Plot $y = |f''(x)|$ for $a \le x \le b$. Use your plot and inequality (5.8.3) to determine the number N of subintervals required to guarantee an absolute error no greater than 5×10^{-4} when the Midpoint Rule is used to approximate $\int_a^b f(x)\,dx$. State N and the resulting approximation.

57. $f(x) = \sqrt{1+x} - \sqrt{x} \quad [1.25, 4]$
58. $f(x) = x^3/\sqrt{1+x^2} \quad [1/4, 4]$
59. $f(x) = \sin(x)/x \quad [1, 5]$
60. $f(x) = \sin(x^2) \quad [1, 2]$

In each of Exercises 61–64, a function f and an interval $[a, b]$ are specified. Calculate the Simpson's Rule approximations of $\int_a^b f(x)\,dx$ with $N = 10$ and $N = 20$. If the first five decimal places do not agree, increment N by 10. Continue until the first five decimal places of two consecutive approximations are the same. State your answer rounded to four decimal places.

61. $f(x) = \ln(1+x^2) \quad [0, 5]$
62. $f(x) = \exp(\sqrt{x}) \quad [1, 4]$
63. $f(x) = \sqrt{1+x^3} \quad [0, 4]$
64. $f(x) = \sin(\pi\cos(x)) \quad [0, \pi/3]$

In each of Exercises 65–68, an integral $\int_a^b f(x)\,dx$ and a positive integer N are given. Compute the exact value of the integral, the Simpson's Rule approximation of order N, and the absolute error ε. Then find a value c in the interval (a, b) such that $\varepsilon = (b-a)^5|f^{(4)}(c)|/(180 \cdot N^4)$. (This form of the error, which resembles the Mean Value Theorem, implies inequality (5.8.4).)

65. $\int_1^4 \sqrt{x}\,dx \quad N = 6$
66. $\int_1^e 1/x\,dx \quad N = 4$
67. $\int_1^8 (1+x)^{1/3}\,dx \quad N = 4$
68. $\int_8^{15} 1/\sqrt{1+x}\,dx \quad N = 4$
69. Data is said to be *normally distributed* with *mean* μ and *standard deviation* $\sigma > 0$ if, for every $a < b$, the probability that a randomly selected data point lies in the interval $[a, b]$ is

$$\frac{1}{\sigma\sqrt{2\pi}} \int_a^b \exp\left(-\frac{1}{2}\left(\frac{x-\mu}{\sigma}\right)^2\right) dx.$$

Suppose the scores on an exam are normally distributed with mean 70 and standard deviation 15. Use Simpson's Rule with $N = 6$ to approximate the probability that a randomly selected exam has a grade in the range [55, 85]. (The answer, the probability of being within one standard deviation of the mean, does not depend on the particular values 15 and 70 of the parameters that are used in this exercise.)

70. Suppose the scores on an exam are normally distributed with mean 80 and standard deviation 10. (Refer to Exercise 69.) What is the probability that a randomly selected exam has a grade in the range [60, 100]? Use

Simpson's Rule with $N = 10$. (The answer, the probability of being within two standard deviations of the mean, does not depend on the particular values 10 and 80 of the parameters that are used in this exercise.)

71. The partition $\{0, 15, 25, 50, 75, 90, 100\}$ of $[0, 100]$ given in Exercise 46 precludes an application of Simpson's Rule because it is not uniform. However, because the number of subintervals is even, the *idea* behind Simpson's Rule can be used. Find the degree 2 polynomial p_1 that passes through the three points (0, 0), (15, 3), and (25, 19). The *interpolation* command of a computer algebra system will make short work of finding this polynomial. Calculate $\int_0^{25} p_1(x)\,dx$. Next, find the degree 2 polynomial p_2 that passes through the three points (25, 19), (50, 25), and (75, 42). Calculate $\int_{25}^{75} p_2(x)\,dx$. Repeat this procedure with the last three points, (75, 42), (90, 75), and (100, 100). What approximation of $\int_0^{100} L(x)\,dx$ results?

72. The polynomial $p_1(x)$ of Exercise 71 is negative on the subinterval $(0, 11.42857\ldots)$. As a result, the value of $\int_0^{11.42857\ldots} p_1(x)\,dx$ underestimates $\int_0^{11.42857\ldots} L(x)\,dx$. To compensate, averaging $\int_0^{25} p_1(x)\,dx$ with the trapezoidal approximation of $\int_0^{25} L(x)\,dx$, which is an overestimate, might be considered. When the calculation of Exercise 71 is amended in this way, what approximation of $\int_0^{100} L(x)\,dx$ results?

73. In a particular regional climate, the temperature varies between $-22°$C and $40°$C, averaging $\mu = 11°$C. The number of days $F(T)$ in the year on which the temperature remains below T degrees centigrade is given (approximately) by

$$F(T) = \int_{-22}^{T} f(x)\,dx \qquad (-22 \le T \le 40),$$

where

$$f(x) = 12.72 \exp\left(-\frac{(x-11)^2}{266.4}\right).$$

Notice that $F(T)$ is the sort of *area integral* that we studied in Section 5.4.

a. Use Simpson's Rule with $N = 20$ to approximate $F(40)$. What should the exact value of $F(40)$ be?

b. Heat alerts are issued when the daily high temperature is 36°C or more. On about how many days a year are heat alerts issued?

c. Suppose that global warming raises the average temperature by 1°C, shifting the graph of f by 1 unit to the right. The new model may be obtained by simply replacing μ with 12 and using $[-21, 41]$ as the domain (see Figure 13). What is the percentage increase in heat alerts that will result from this 1°C shift in temperature?

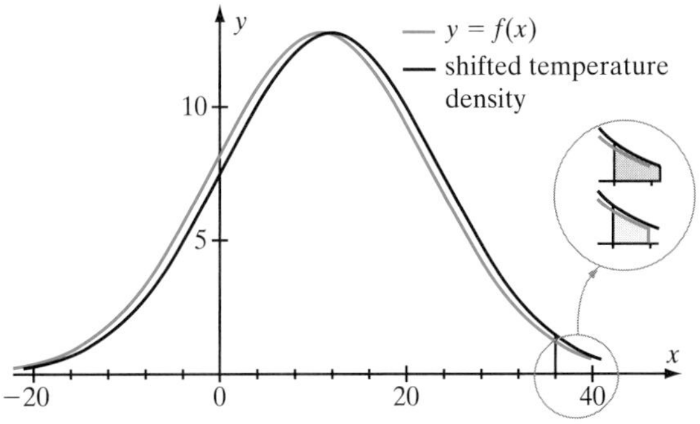

Figure 13

Summary of Key Topics in Chapter 5

Sigma Notation and Partitions
(Sections 5.1, 5.2)

The notation

$$\sum_{j=M}^{N} a_j$$

means $a_M + a_{M+1} + a_{M+2} + \cdots + a_{N-1} + a_N$.

The uniform partition of order N of an interval $[a, b]$ is the set $\{x_0, x_1, \ldots, x_N\}$ of points x_j defined by

$$x_j = a + j \cdot \frac{b-a}{N} \qquad (0 \le j \le N).$$

These points divide $[a, b]$ into N subintervals, $I_j = [x_{j-1}, x_j]$ $(0 \le j \le N)$. Each subinterval has length $\Delta x = (b-a)/N$.

Area
(Section 5.1)

Let f be a continuous positive function on $[a, b]$. The area of the region under the graph of f, above the x-axis, and between the vertical lines $x=a$ and $x=b$ is the limit as N tends to infinity of the expression

$$\sum_{j=1}^{N} f(x_j) \cdot \Delta x$$

where $\Delta x = (b-a)/N$ and $x_1 = a + \Delta x,\ x_2 = a + 2\Delta x,\ \ldots,\ x_N = a + N\Delta x = b.$

The Riemann Integral
(Section 5.2)

If f is any function on an interval $[a, b]$ that has been divided into N equal width subintervals, and if for each j a points s_j in the jth subinterval is chosen, then the expression

$$\mathcal{R}(f, S_N) = \sum_{j=1}^{N} f(s_j) \cdot \Delta x$$

is called a Riemann sum for f. Here we let S_N denote $\{s_1, \ldots, s_N\}$. If the Riemann sums for f have a limit as the number N of subintervals tends to infinity, then that limit is called the *Riemann integral* for f on $[a, b]$, and f is said to be integrable on $[a, b]$. The Riemann integral of f on $[a, b]$ is denoted by

$$\int_a^b f(x)\, dx$$

and is said to be the *definite integral* of f over the interval $[a, b]$. A definite integral is *always* a number. (By contrast, the *indefinite integral* of f, $\int f(x)\, dx$, denotes a collection of functions of the variable of integration.)

Fundamental Theorem of Calculus
(Sections 5.2 and 5.4)

A continuous function f is integrable. The Fundamental Theorem of Calculus consists of two formulas that relate differentiation and integration. Together they imply that differentiation and integration are inverse operations. If f is continuous on an interval I that contains the point a, then the function F defined on I by

$$F(x) = \int_a^x f(t)dt,$$

is an antiderivative of f:

$$F'(x) = f(x).$$

If F is *any* antiderivative of f, then the definite integral of f can be calculated by

$$\int_a^b f(x)\, dx = F(b) - F(a).$$

Properties of the Integral
(Section 5.3)

The integral satisfies the following five properties. If f and g are integrable on $[a, b]$ and if α is a constant, then

1. $\int_a^b (f(x) + g(x))\, dx = \int_a^b f(x)\, dx + \int_a^b g(x)\, dx$ and
   $\int_a^b (f(x) - g(x))\, dx = \int_a^b f(x)\, dx - \int_a^b g(x)\, dx,$
2. $\int_a^b \alpha f(x)\, dx = \alpha \cdot \int_a^b f(x)\, dx,$

3. $\int_a^b \alpha\, dx = \alpha(b-a)$,
4. $\int_a^a f(x)\, dx = 0$, and
5. $\int_a^c f(x)\, dx + \int_c^b f(x)\, dx = \int_a^b f(x)\, dx$.

If $b < a$, then we define $\int_a^b f(x)\, dx$ to be equal to $-\int_b^a f(x)\, dx$. If f is continuous on $[a, b]$, then the Mean Value Theorem for Integrals states that there is a c in (a, b) such that

$$f(c) = \frac{1}{b-a}\int_a^b f(x)\, dx.$$

Logarithmic and Exponential Functions (Section 5.5)

The *natural logarithm* can be *defined* for positive x by

$$\ln(x) = \int_1^x \frac{1}{t}\, dt.$$

All properties of the logarithm can be derived from this formula. The inverse of the natural logarithm is called the *natural exponential function*, or *exponential function* for short. It is denoted by *exp*. The number e is defined by

$$e = \exp(1).$$

If a is any positive number, and x is any real number, then we define

$$a^x = \exp(x \ln(a)).$$

The familiar laws of exponents all follow from this definition.

Substitution (Section 5.6)

If $f(x)$ has the form $F(\varphi(x)) \cdot \varphi'(x)$, then the calculation of the integral of f can be simplified by a substitution of the form

$$u = \varphi(x) \quad \text{and} \quad du = \varphi'(x)\, dx.$$

The Calculation of Area (Section 5.7)

If $f(x) \geq 0$ on $[a, b]$, then

$$\int_a^b f(x)\, dx$$

represents the area under the graph of f, above the x-axis, and between $x = a$ and $x = b$. If $f(x) \leq 0$ on $[a, b]$, then the integral represents the negative of the area above the graph of f, below the x-axis, and between $x = a$ and $x = b$.

If $f(x) \geq g(x)$ on $[a, b]$, then the area between the two graphs and between the vertical lines $x = a$ and $x = b$ is given by

$$\int_a^b (f(x) - g(x))\, dx.$$

To calculate the area between two graphs, it is necessary to break up the problem into integrals over intervals on which one of the two functions is greater than or equal to the other.

For some types of area problems, it is useful to integrate in the y-variable, treating x as the dependent variable.

Numerical Integration (Section 5.8) Methods of approximating definite integrals are particularly useful when an antiderivative of the integrand cannot be found. Let f be a continuous function on $[a, b]$. For a positive integer N, let $\Delta x = (b - a)/N$, $x_j = a + j\Delta x$, and $\bar{x}_j = (x_{j-1} + x_j)/2$. The definite integral $\int_a^b f(x)\,dx$ can be approximated by the Midpoint Rule:

$$\Delta x \cdot (f(\bar{x}_1) + f(\bar{x}_2) + \cdots + f(\bar{x}_N)).$$

The Trapezoidal Rule for approximating $\int_a^b f(x)\,dx$ is

$$\frac{\Delta x}{2}\left(f(x_0) + 2f(x_1) + 2f(x_2) + \cdots + 2f(x_{N-1}) + f(x_N)\right).$$

Of these two methods of approximation, the Midpoint Rule is usually more accurate. Still more accurate is Simpson's Rule, which can be applied when N is even:

$$\int_a^b f(x)\,dx \approx \frac{\Delta x}{3}\left(f(x_0) + 4f(x_1) + 2f(x_2) + 4f(x_3) + \cdots + 2f(x_{N-2}) + 4f(x_{N-1}) + f(x_N)\right).$$

For $k = 2$ and $k = 4$, let C_k be a number such that the kth derivative $f^{(k)}$ satisfies $|f^{(k)}(x)| \leq C_k$ on the interval $[a, b]$ of integration. The errors that result from using the Midpoint Rule, the Trapezoidal Rule, and Simpson's Rule are no larger than

$$\frac{C_2}{24} \cdot \frac{(b-a)^3}{N^2}, \quad \frac{C_2}{12} \cdot \frac{(b-a)^3}{N^2}, \quad \text{and} \quad \frac{C_4}{180} \cdot \frac{(b-a)^5}{N^4},$$

respectively.

REVIEW EXERCISES for Chapter 5

In each of Exercises 1–4, use sigma notation to express the sum.

1. $1 + 4 + 9 + 16 + \cdots + 25^2$
2. $120 + 122 + 124 + 126 + \cdots + 240$
3. $1/2 + 1/3 + 1/4 + 1/5 + 1/6 + \cdots + 1/100$
4. $1 - 2 + 3 - 4 + 5 - 6 + \cdots - 100$

In each of Exercises 5–8, calculate the right endpoint approximation of the area of the region that lies below the graph of the given function f and above the given interval I of the x-axis. Use the uniform partition of given order N.

5. $f(x) = x^3$ $\quad I = [2, 5], N = 3$
6. $f(x) = 1/x$ $\quad I = [1, 2], N = 2$
7. $f(x) = (x - 1)/(x + 1)$ $\quad I = [1, 4], N = 3$
8. $f(x) = \sin(x)$ $\quad I = [0, \pi/2], N = 3$

In each of Exercises 9–12, for the given function f, interval I, and uniform partition of order N, evaluate the Riemann sum $\mathcal{R}(f, S)$ using the choice of points S that consists of the midpoints of the subintervals of I.

9. $f(x) = \sin(x), S = \{\pi/6, \pi/2\}$ $\quad I = [0, 2\pi/3], N = 2$
10. $f(x) = (7 - x)/x, S = \{2, 4, 6\}$ $\quad I = [1, 7], N = 3$
11. $f(x) = 16x^2, S = \{5/4, 7/4, 9/4, 11/4\}$ $\quad I = [1, 3], N = 4$
12. $f(x) = x - x^2, S = \{1/8, 3/8, 5/8, 7/8\}$ $\quad I = [0, 1], N = 4$

In each of Exercises 13–28, evaluate the given definite integral.

13. $\int_1^3 6x^2\,dx$
14. $\int_{-1}^2 (10x^4 + 9x^2)\,dx$
15. $\int_{-2}^{-1} (1/x)\,dx$
16. $\int_4^9 \sqrt{x}\,dx$
17. $\int_1^4 (2x - 1/\sqrt{x})\,dx$
18. $\int_{\pi/6}^{\pi/3} \sec^2(x)\,dx$
19. $\int_{\pi/6}^{\pi/3} (4\sin(x) - \cos(x))\,dx$

20. $\int_0^{\pi/4} \sec(x)\tan(x)\,dx$

21. $\int_{16}^{81} (1/x^{1/4})\,dx$

22. $\int_1^8 x^{2/3}\,dx$

23. $\int_1^2 (24/x^2)\,dx$

24. $\int_0^{\ln(3)} e^x\,dx$

25. $\int_0^2 3^x\,dx$

26. $\int_{\pi/4}^{\pi/2} \csc^2(x)\,dx$

27. $\int_{\pi/4}^{\pi/2} \csc(x)\cot(x)\,dx$

28. $\int_1^2 (x^2+x+2)/x\,dx$

In Exercises 29–32, compute the average value of f over $[a, b]$, and find a point c in (a, b) at which f attains this average value.

29. $f(x) = 1/x^{2/3}$ $\quad a = 1, b = 8$

30. $f(x) = x^2$ $\quad a = 1, b = 5$

31. $f(x) = \sqrt{x}$ $\quad a = 1, b = 4$

32. $f(x) = 1/x^2$ $\quad a = 1/2, b = 3$

In each of Exercises 33–44, a function $F(x)$ and a point c are given. Calculate $F'(c)$.

33. $F(x) = \displaystyle\int_0^x \frac{t+3}{t+1}\,dt$ $\quad c = 3$

34. $F(x) = \int_1^x t^{1/3}\sqrt{\sqrt{t}+1}\,dt$ $\quad c = 64$

35. $F(x) = \int_{-1}^x (t^3+20)/(t+2)\,dt$ $\quad c = 2$

36. $F(x) = \int_x^{\pi/4} \tan^2(t)\,dt$ $\quad c = \pi/3$

37. $F(x) = \int_x^2 \sin^2(\pi/t)\,dt$ $\quad c = 4$

38. $F(x) = \int_x^6 \sqrt{2t^2-1}\,dt$ $\quad c = 5$

39. $F(x) = \int_1^{3x} \ln(1+t)\,dt$ $\quad c = 1$

40. $F(x) = \int_0^{x^2} \sqrt{1+\sqrt{t}}\,dt$ $\quad c = 3$

41. $F(x) = \int_{6x}^{10} (t^2+t+2)^{2/3}\,dt$ $\quad c = 1/3$

42. $F(x) = \int_x^{2x} t\sin(t)\,dt$ $\quad c = \pi/4$

43. $F(x) = \int_{-x}^x \sqrt{5+t^2}\,dt$ $\quad c = 2$

44. $F(x) = \int_{\sqrt{x}}^{x\sqrt{x}} \ln(t)\,dt$ $\quad c = 9$

In each of Exercises 45–49 use the equations $\int_1^\alpha (1/t)\,dt = 3$ and $\int_1^\beta (1/t)\,dt = 6$ to evaluate $\int_a^b (1/t)\,dt$ over the given interval $[a, b]$.

45. $[1, \alpha\beta]$

46. $[1, \beta/\alpha]$

47. $[\alpha, \beta]$

48. $[\alpha, \alpha\beta]$

49. $[1, \alpha^4]$

In each of Exercises 50–70, make a substitution, and evaluate the given integral.

50. $\int 6x\cos(x^2)\,dx$

51. $\displaystyle\int \frac{\cos(x)}{\sin^3(x)}\,dx$

52. $\int 24x\sqrt{1+x^2}\,dx$

53. $\int 24x^3/\sqrt{2+x^4}\,dx$

54. $\displaystyle\int \frac{x+2}{x^2+4x+5}\,dx$

55. $\displaystyle\int \frac{\sin(x)}{3+2\cos(x)}\,dx$

56. $\int 24\sin(3x)\cos^5(3x)\,dx$

57. $\displaystyle\int \frac{(\sqrt{x}+1)^{1/3}}{\sqrt{x}}\,dx$

58. $\displaystyle\int 2\pi\,\frac{\cos(\pi/x^2)}{x^3}\,dx$

59. $\displaystyle\int \frac{2}{x\ln^2(x)}\,dx$

60. $\int (\ln(2x)/x)\,dx$

61. $\int 12x\sqrt{4x+1}\,dx$

62. $\displaystyle\int \frac{x^2}{(x+1)^3}\,dx$

63. $\displaystyle\int \frac{x}{\sqrt{x+2}}\,dx$

64. $\int 60x^3\sqrt{x^2+1}\,dx$

65. $\displaystyle\int \frac{\arctan^2(x)}{1+x^2}\,dx$

66. $\displaystyle\int \frac{\exp(x)}{\sqrt{1+\exp(x)}}\,dx$

67. $\displaystyle\int \frac{\arcsin(x)}{\sqrt{1-x^2}}\,dx$

68. $\displaystyle\int \frac{x+\exp(2x)}{x^2+\exp(2x)}\,dx$

69. $\displaystyle\int \frac{1-2x}{\sqrt{1-x^2}}\,dx$

70. $\displaystyle\int \frac{2+x}{4+x^2}\,dx$

Evaluate the definite integrals in Exercises 71–76.

71. $\int_0^{\pi/9} 24\tan(3x)\,dx$

72. $\int_{\pi/2}^{\pi} 24\cot(x/3)\,dx$

73. $\int_0^{\pi/2} \sec(x/3)\,dx$

74. $\displaystyle\int_{\exp(2)}^{\exp(4)} \frac{1}{x\ln(x)}\,dx$

75. $\int_0^2 x^3\sqrt{9+x^4}\,dx$

76. $\int_0^1 x\sqrt[3]{1-x}\,dx$

In each of Exercises 77–80, find the area of the region that is bounded by the graphs of $y = f(x)$ and $y = g(x)$ for x between the abscissas of the two points of intersection.

77. $f(x) = 4 - x^2$ $\quad g(x) = 3x$

78. $f(x) = x^2 + x + 2 \quad g(x) = 18 + x - x^2$
79. $f(x) = -x^4 + x^2 + 16 \quad g(x) = 2x^4 - 2x^2 - 20$
80. $f(x) = 2^x \quad g(x) = 1 + 2x - x^2$

In each of Exercises 81–84, find the total area of the bounded regions whose upper and lower boundaries are arcs of the graphs of the given functions f and g.

81. $f(x) = x^3 - 2x^2 - 3x \quad g(x) = 0$
82. $f(x) = x^3 - 2x^2 + 3 \quad g(x) = 1 + x$
83. $f(x) = x^3 - x^2 - 9x + 4 \quad g(x) = -5$
84. $f(x) = x^4 - 2x^2 + 1 \quad g(x) = 1 - x^3$

In each of Exercises 85–88, a function f, an interval I, and an even integer N are given. Approximate the integral of f over I by partitioning I into N equal length subintervals and using the Midpoint Rule, the Trapezoidal Rule, and then Simpson's Rule.

85. $f(x) = 8 - x^3 \quad I = [-1, 3], N = 2$
86. $f(x) = 4x^3 + 1 \quad I = [1, 2], N = 2$
87. $f(x) = (1 + x)/(1 + 4x) \quad I = [0, 2], N = 4$
88. $f(x) = 18x/(1 + 8x) \quad I = [0, 1], N = 4$

In each of Exercises 89–92, a function f and an interval $I = [a, b]$ are given. Also given is the approximation $\mathcal{M}_{10}$ of $A = \int_a^b f(x)\,dx$ that is obtained by using the Midpoint Rule with a uniform partition of order 10. Use the error estimate for the Midpoint Rule approximation to find a lower bound α and and upper bound β for A.

89. $f(x) = \sqrt{5 + x} \quad I = [-1, 11], \ \mathcal{M}_{10} = 37.34080\ldots$
90. $f(x) = x\,(1+x)^3 \quad I = [1, 3], \ \mathcal{M}_{10} = 138.08676\ldots$
91. $f(x) = (1+x)^{1/3} \quad I = [0, 7], \ \mathcal{M}_{10} = 11.25500\ldots$
92. $f(x) = \ln(1 + x) \quad I = [1, 6], \ \mathcal{M}_{10} = 7.23877\ldots$

93. From west to east, a city is 5.1 km wide. North-south measurements at 0.85 km intervals are given in Figure 1.

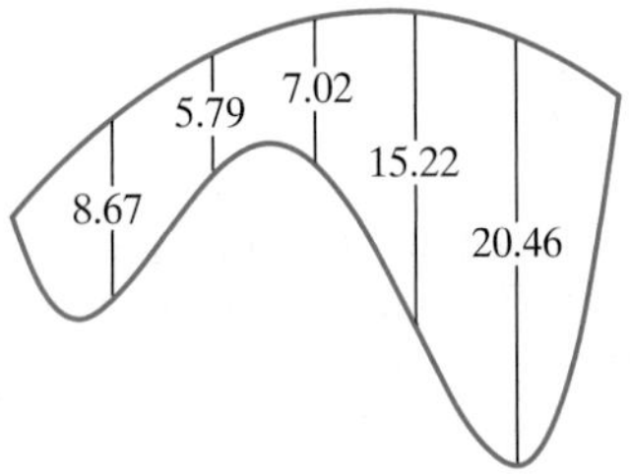

Figure 1

Use the Trapezoidal Rule and Simpson's Rule to estimate the city's area.

94. During a 5-minute span, at intervals of 1 minute, the speedometer of a car reads

Time (minutes)	0	1	2	3	4	5
Speed (km/h)	80	74	60	54	46	0

Use the Trapezoidal Rule to estimate the distance traveled by the car during that 5-minute period.

95. Speed measurements of a runner taken at half-second intervals during the last 5 seconds of a sprint are provided in the following table.

Time (s)	5.0	5.5	6.0	6.5	7.0	7.5	8.0	8.5	9.0	9.5	10.0
Speed (m/s)	10.66	10.69	10.86	10.98	10.96	10.94	10.80	10.71	10.64	10.58	10.43

About how many meters did the athlete run during that 5-second interval? Use Simpson's Rule.

In Exercises 96–99, use a table of integrals to calculate the given integral.

96. $\displaystyle\int \frac{t^5}{(1+t^3)^2}\,dt$

97. $\displaystyle\int \frac{2}{(2x+1)(x+3/2)^2}\,dx$

98. $\displaystyle\int \frac{36\cos(t)}{\left(3+\sin^2(t)\right)^2}\,dt$

99. $\displaystyle\int \frac{\exp(3t)}{4 - \exp(2t)}\,dt$

Archimedes

The Greek geometer Eudoxus (ca. 408–ca. 355 BCE) used the limiting process to rigorously demonstrate area and volume formulas that Democritus (ca. 460–ca. 370 BCE) had discovered earlier. The arguments that Eudoxus introduced came to be known as the *Method of Exhaustion*. In the hands of Archimedes, this method became a powerful technique.

In one of his earliest works, Archimedes derived a formula for the area of a parabolic sector. An arc $\widehat{AVB}$ of a parabola is drawn in Figure 1. The point V is known as the vertex of the segment and is characterized as follows: Among all line segments from the chord $\overline{AB}$ to the parabolic arc $\widehat{AB}$ that are perpendicular to chord $\overline{AB}$, the one terminating at V is longest. Archimedes demonstrated that the area $\mathcal{A}$ enclosed by line segment $\overline{AB}$ and parabolic arc $\widehat{AVB}$ is $2hb/3$, or $(4/3) \times |\Delta AVB|$ (using the notation $|R|$ to denote the area of a planar region R). The introduction of V creates two parabolic subarcs, $\widehat{AV}$ and $\widehat{VB}$. Each has a vertex; these vertices are labelled $V1$ and $V2$ in Figure 2a. Archimedes proved from a few easily derived geometric properties of the parabola that

$$|\Delta AV_1V| + |\Delta VV_2B| = \frac{1}{4}|\Delta AVB|.$$

At the next iteration, illustrated in Figure 2b, Archimedes interposed four new vertices V_3, V_4, V_5, and V_6 and concluded that

$$|\Delta AV_3V_1| + |\Delta V_1V_4V| = |\frac{1}{4}\Delta AV_1V| \quad \text{and}$$

$$|\Delta VV_5V_2| + |\Delta V_2V_6B| = \tfrac{1}{4}|\Delta VV_2B|.$$

Combining these two equalities yields

$$|\Delta AV_3V_1| + |\Delta V_1V_4V| + |\Delta VV_5V_2| + |\Delta V_2V_6B|$$
$$= \frac{1}{4}(|\Delta AV_1V| + |\Delta VV_2B|) = \frac{1}{4^2}|\Delta AVB|.$$

Continuing, Archimedes "exhausted" the parabolic segment by a sequence $\{R_n\}$ of regions composed of triangles: $R_0 = \Delta AVB$, $R_1 = \Delta AV_1V \cup \Delta VV_2B$, $R_2 = \Delta AV_3V_1 \cup \Delta V_1V_4V \cup \Delta VV_5V_2 \cup \Delta V_2V_6B, R_3, \ldots.$ In general, $|R_n| = 4^{-n}|R_0|$. Using the formula for the sum of a geometric progression, Archimedes obtained

$$|R_1 \cup R_2 \cup \cdots \cup R_N| = \left(1 + \frac{1}{4} + \frac{1}{4^2} + \cdots + \frac{1}{4^N}\right) \times |\Delta AVB|$$
$$= \frac{4}{3}\left(1 - \frac{1}{4^{N+1}}\right)|\Delta AVB|.$$

He finished this proof with an airtight argument that $1/4^{N+1} \to 0$ and $\mathcal{A} = 4|\Delta AVB|/3$.

The Method

The translation of Archimedes' works into Latin in the late 1500s introduced European mathematicians to the power of the infinite process. Although these mathematicians found elegant formulas, rigorous proofs, and glimpses of what might be attained, they did not find the methods by which Archimedes *discovered* his results. The 17th-century mathematician John Wallis even protested that Archimedes "covered up the traces of his investigations, as if he has grudged posterity the secret of his method of inquiry." So the matter stood for 200 more years.

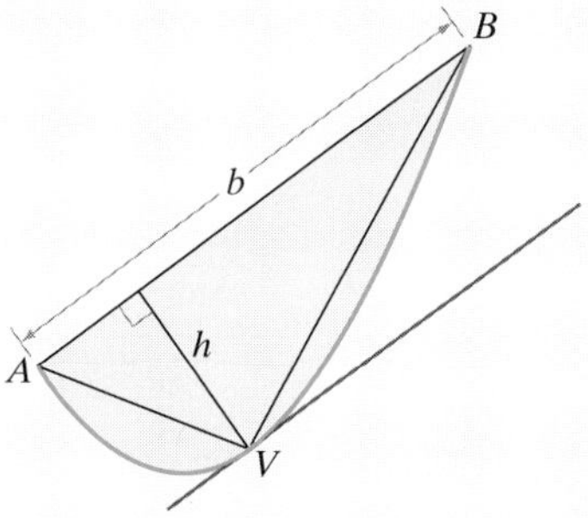

▲ Figure 1 Archimedes proved that the area of the parabolic segment AVB is equal to (2/3) hb

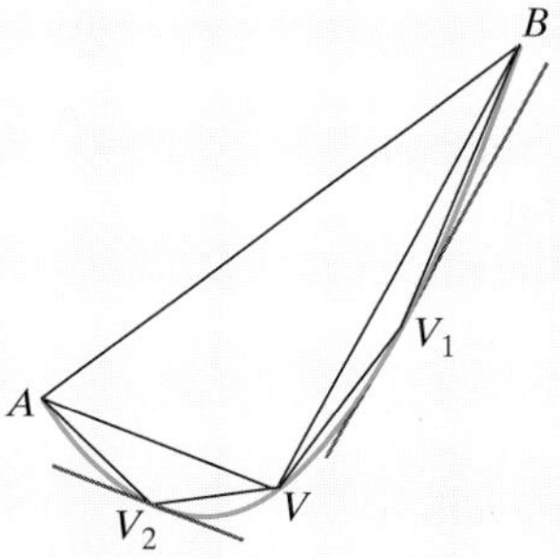

▲ Figure 2a Vertex V creates two parabolic segments with vertices $V1$ and $V2$.

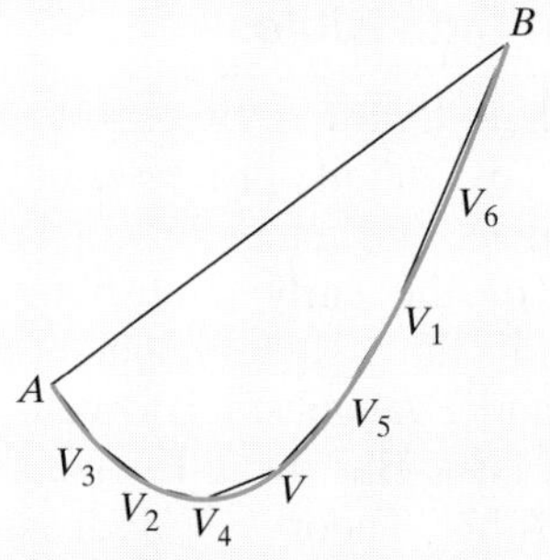

▲ Figure 2b Repeat the process of adding vertices.

In 1906, Danish scholar Johan Ludvig Heiberg learned of a manuscript that had been catalogued in a cloister in Constantinople (now Istanbul). The manuscript contained a religious text of the Eastern Orthodox Church and was a so-called *palimpsest*—a parchment of which the original script had been washed and overwritten. Because it was reported that the original mathematical text had been imperfectly washed, Heiberg voyaged from Copenhagen to Constantinople to investigate. What Heiberg uncovered was a mathematical manuscript containing numerous works of Archimedes. It had been copied in Greek in the 10th century only to be washed and written over in the 13th century. As luck would have it, most of the original text could be restored.

Heiberg's chief find was the *Method Concerning Mechanical Theorems, dedicated to Eratosthenes.* Scholars had known from the references of later Greek mathematicians that such a work had existed until at least the 4th century CE but, until Heiberg's discovery, the text was presumed to be lost. In the *Method,* we find, to use the words of Wallis, Archimedes writing for posterity:

> Archimedes to Eratosthenes, Greeting: ... since I see that you are an excellent scholar... and lover of mathematical research, I have deemed it well to explain to you and put down in this same book, a special method whereby the possibility will be offered to you to investigate any mathematical question by means of mechanics.... [I]f one has previously gotten a conception of the problem by this method, it is easier to produce the proof than to find it without a provisional conception. ... [W]e feel obliged to make the method known partly... in the conviction that there will be instituted thereby a matter of no slight utility in mathematics.

Theory of Indivisibles

The first half of the 17th century saw a number of attempts to use infinite processes in the computation of areas, volumes, and centers of gravity. Among the most important of the early contributors were Pierre de Fermat and Gilles Roberval in France, Bonaventura Cavalieri and Evangelista Torricelli in Italy, and John Wallis and Isaac Barrow in England. In 1635, Cavalieri introduced his Theory of Indivisibles, a relatively effective albeit cumbersome procedure for calculating areas and volumes. Using his method, Cavalieri became the first to show that $\int_0^1 x^n\,dx = 1/(n+1)$ for natural numbers n. Although Cavalieri's student Evangelista Torricelli became the foremost proponent of the Theory of Indivisibles, Torricelli believed that Cavalieri had only rediscovered a method that Archimedes must have known:

> I should not dare confirm that this geometry of indivisibles is actually a new discovery. I should rather believe that the ancient geometers availed themselves of this method in order to discover the more difficult theorems, although in their demonstration they may have preferred another way, either to conceal the secret of their art or to afford no occasion for criticism by invidious detractors.

Fermat and the Integral Calculus

The extension of Cavalieri's formula $\int_0^1 x^n\,dx = 1/(n+1)$ from natural numbers to rational values of n was obtained independently by Fermat and Torricelli. Fermat's clever demonstration runs as follows: Given natural numbers p and $q \neq 0$, let $f(x) = x^{p/q}$. Fix a number $r \in (0, 1)$, and set $x_k = r^{kq}$ for $k = 0, 1, 2, \ldots$. The points $\{x_k\}$ serve to (nonuniformly) partition $[0, 1]$ into infinitely many subintervals. Fermat's idea was to use the sum $\sum_{k=0}^{N} f(x_k)(x_k - x_{k+1})$ to approximate the area under $y = x^{p/q}$ for $0 \leq x \leq 1$. Figure 3a shows the approximation for $r = 0.9$. By taking a larger value of r (with r still less than 1), we obtain a better approximation (see Figure 3b). In the limit, we obtain the exact area:

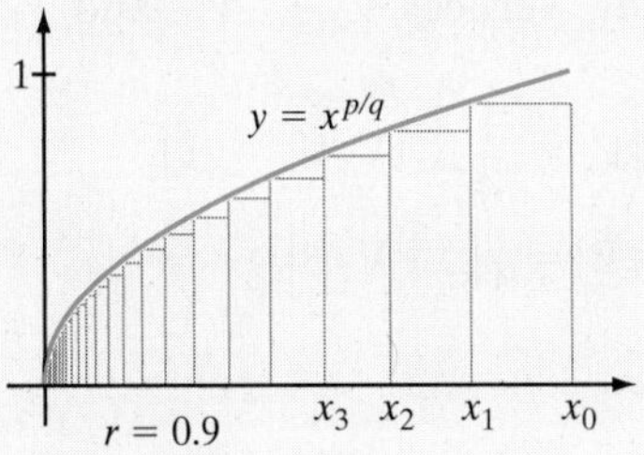

▲ Figure 3a

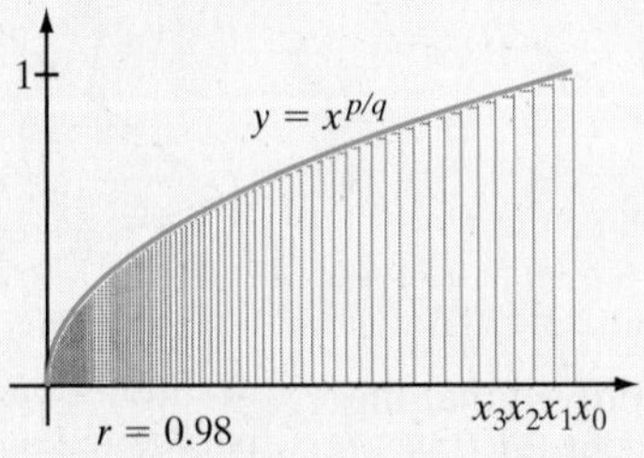

▲ Figure 3b

$$\int_0^1 f(x)\,dx = \lim_{r\to 1^-} \lim_{N\to\infty} \sum_{k=0}^{N} f(x_k)(x_k - x_{k+1}).$$

For $f(x) = x^{p/q}$, we calculate

$$\begin{aligned}\int_0^1 x^{p/q}\,dx &= \lim_{r\to 1^-} \lim_{N\to\infty} \sum_{k=0}^{N} (r^{kq})^{p/q}(r^{kq} - r^{(k+1)q}) \\ &= \lim_{r\to 1^-} \lim_{N\to\infty} \sum_{k=0}^{N} r^{kp} r^{kq}(1 - r^q) \\ &= \lim_{r\to 1^-} (1 - r^q) \lim_{N\to\infty} \sum_{k=0}^{N} r^{(p+q)k}.\end{aligned}$$

Because $0 < r^{p+q} < 1$, we may apply formula (2.5.3) to obtain

$$\int_0^1 x^{p/q}\,dx = \lim_{r\to 1^-} \frac{1 - r^q}{1 - r^{p+q}}.$$

Although *we* know how to compute the indeterminate form on the right by using l'Hôpital's Rule, Fermat had to resort to the following tricky backward application of formula (2.5.2):

$$\begin{aligned}\int_0^1 x^{p/q}\,dx &= \lim_{r\to 1^-} \frac{1 - r^q}{1 - r^{p+q}} = \lim_{r\to 1^-} \frac{(1 - r^q)/(1 - r)}{(1 - r^{p+q})/(1 - r)} \\ &= \frac{\lim_{r\to 1^-} \sum_{k=0}^{q-1} r^k}{\lim_{r\to 1^-} \sum_{k=0}^{p+q-1} r^k} = \frac{q}{p+q} = \frac{1}{p/q + 1}.\end{aligned}$$

Kepler and Roberval

Before the discovery of the Fundamental Theorem of Calculus, many mathematicians succeeded in evaluating particular integrals by ad hoc methods. For example, in his last great astronomical work, the *Epitome Astronomiae Copernicanae* (1618–1621), Kepler calculated what we would now call a limit of Riemann sums for $\int_0^\beta \sin(t)\,dt$. In effect, Kepler discovered the formula

$$\int_0^\beta \sin(t)\,dt = 1 - \cos(\beta).$$

One curve that attracted the attention of prominent mathematicians throughout the 17th century was the *cycloid*, which is the locus of points generated by a fixed point on a circle as the circle rolls along a line at constant speed without spinning or slipping (see Figure 4). Early in the 17th century, Galileo conjectured that the area under one arch of the cycloid is three times the area of the rolling circle. To reach this conjecture, this leading scientist resorted to cut-out paper figures. By 1636, Roberval was able to rigorously verify Galileo's conjecture. Roberval's work is a measure of how much progress had been made in a very short period of time.

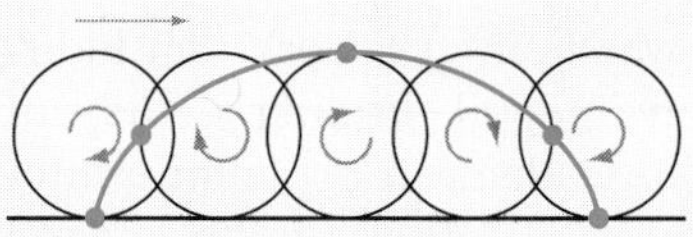
▲ Figure 4 A point on a rolling circle generates a cycloid.

Newton and the Fundamental Theorem of Calculus

Sir Isaac Newton's first contact with higher mathematics came as a college senior in 1664 when he was 21 years old. He learned the tangent problem from the *Géométrie* of Descartes, and he learned Cavalieri's Theory of Indivisibles from Wallis's *Arithmetica Infinitorum*. It was Wallis's work that, in the winter of 1664–1665, led Newton, who was still an undergraduate, to evaluate the area of the region under the graph of $y = (1 - t^2)^\alpha$ and over the t-interval $[0, x]$. In doing so, he had already taken the decisive step of letting the endpoint of integration be a variable, a key idea of the Fundamental Theorem of Calculus.

When plague closed Cambridge University in 1665 and 1666, Newton returned home to reorganize his discoveries. He wrote his discoveries in "The October 1666 Tract on Fluxions," a manuscript that was not published until 1967. It was in this tract that Newton first stated both parts of the Fundamental Theorem of Calculus. This remarkable manuscript also contains several other theorems, including the Chain Rule and Integration by Substitution.

Notation

The notation that we use today for the indefinite integral was introduced by Gottfried Wilhelm Leibniz (1646–1716) in a manuscript of 1675. At that time, Leibniz did not include the differential. For example, he wrote $\int x^2 = x^3/3$. By the time his work appeared in print in 1686, however, Leibniz included the differential in his notation and stressed its importance. The modern notation $\int_a^b f(x)\,dx$ for the definite integral—that is, the positioning of the limits of integration on the integral sign—was introduced in 1822 by Joseph

Fourier (1768–1830). It was Fourier's work that indirectly led to the Riemann integral. Fourier's study of heat transfer required the evaluation of definite integrals of the form $\int_0^{2\pi} f(x)\sin(nx)\,dx$ for quite general functions *f*. However, the definition of integration then in use was far too limited for the desired generality. Georg Friedrich Bernhard Riemann (1826–1866) realized that, to further develop Fourier's work, a new concept of the integral was needed. Toward that end, in 1854, Riemann introduced what we now call the Riemann integral.

Bernhard Riemann

Riemann's short life was marked by nervous breakdowns and serious illness—pleurisy, jaundice, and, ultimately, tuberculosis. Although he began his studies in theology, he was able to convince his father to allow him to pursue what was then an "economically valueless" subject—mathematical physics. In quantity, his mathematical output was relatively modest for a mathematician of his stature. Yet his work has assumed fundamental importance in several fields—he is remembered for the Riemann integral in analysis, the Riemann zeta function in number theory, Riemann surfaces and Cauchy-Riemann equations in the theory of functions of a complex variable, and Riemannian manifolds in geometry. The importance that his work has had in physics is also noteworthy. Indeed, Riemannian geometry proved to be essential for Einstein's theory of relativity, as well as for much of the physics that has been done since.

CHAPTER 6

Techniques of Integration

PREVIEW

The purpose of the present chapter is to develop some facility with calculating integrals. Until now, the integrals that we have calculated have been of two types. Either we have immediately recognized an antiderivative, or we have converted an integral into one of that type by making a direct substitution. In fact, an array of diverse procedures can be brought to bear on the calculation of integrals. In this chapter, we will study the most important of those methods.

Tables of integrals and computer algebra systems are valuable resources for calculating integrals. However, neither renders obsolete the need for a basic working knowledge of integration theory. In this section, you will learn fundamental algebraic and trigonometric manipulations that are the basis for evaluating integrals that arise frequently in mathematical applications. But the methods you will learn in this section have a significance that transcends integration theory. Indeed, these techniques are needed for the development of many important topics in engineering, science, and mathematics.

6.1 Integration by Parts

The integral of the sum of two functions is the sum of their integrals. This simple observation enables us to break up some complicated integration problems into several simple ones. But what if we want to integrate the *product* of two functions? A technique called *integration by parts* enables us to evaluate many integrals that involve a product of two functions.

The Product Rule in Reverse

If u and v are two differentiable functions, then we know that

$$\frac{d}{dx}(u \cdot v) = \frac{du}{dx} \cdot v + \frac{dv}{dx} \cdot u.$$

We rewrite this equation as

$$u \cdot \frac{dv}{dx} = \frac{d}{dx}(u \cdot v) - v \cdot \frac{du}{dx}.$$

Integrating both sides gives

$$\int u \cdot \frac{dv}{dx}\, dx = \int \frac{d}{dx}(u \cdot v)\, dx - \int v \cdot \frac{du}{dx}\, dx$$

Observing that $u \cdot v$ is the antiderivative of $\frac{d}{dx}(u \cdot v)$, we obtain the *integration by parts* formula:

$$\int u \cdot \frac{dv}{dx}\, dx = u \cdot v - \int v \cdot \frac{du}{dx}\, dx. \tag{6.1.1}$$

INSIGHT Often we use "differential notation" and summarize the integration by parts formula as

$$\int u\, dv = uv - \int v\, du. \tag{6.1.2}$$

Here dv is shorthand for $\frac{dv}{dx}dx$, and du is shorthand for $\frac{du}{dx}dx$. This compact form of the integration by parts formula is the easiest to remember. Bear in mind, however, that it conceals the original variable of integration (x in our discussion).

We may use a version of formula (6.1.2) for definite integrals provided that we use appropriate limits of integration:

$$\int_{x=a}^{x=b} u\, dv = u \cdot v\Big|_{x=a}^{x=b} - \int_{x=a}^{x=b} v\, du. \tag{6.1.3}$$

Some Examples

▶ **EXAMPLE 1** Calculate $\int x\cos(x)\,dx$.

Solution We integrate by parts. Our first job is to decide what function will play the role of u and what function will play the role of v. Let us take $u = x$ and $dv = \frac{dv}{dx}dx = \cos(x)\,dx$. Then $\int x \cdot \cos(x)\,dx = \int u\,dv$. Therefore according to equation (6.1.2),

$$\int x\cos(x)\,dx = uv - \int v\,du. \tag{6.1.4}$$

Our task is to work out the right-hand side. We already know what u is. What is v? Because $dv = \cos(x)\,dx$, we see that $v = \int dv = \int \cos(x)\,dx$ is an antiderivative of $\cos(x)$.

Therefore we may take $v = \sin(x)$. Likewise, because $u = x$, it follows that $du = \frac{du}{dx} \cdot dx = 1 \cdot dx$. If we substitute for u, v, and du on the right side of equation (6.1.4), then we obtain

$$\int x\cos(x)\,dx = x\sin(x) - \int \sin(x) \cdot 1\,dx.$$

We have converted our original integration problem into a much easier one. Because $\int \sin(x)\,dx = -\cos(x) + C$, we conclude that

$$\int x\cos(x)\,dx = x\sin(x) - (-\cos(x) + C) = x\sin(x) + \cos(x) - C.$$

If you prefer, you may replace the arbitrary constant C with $-C$. Then, the answer takes on the more familiar form $x\sin(x) + \cos(x) + C$. As usual with indefinite integrals, we can check our work by differentiating our answer and verifying that the result equals the integrand:

$$\frac{d}{dx}\Big(x\sin(x) + \cos(x) - C\Big) = \Big(x\cos(x) + \sin(x)\Big) + \Big(-\sin(x)\Big) - 0 = x\cos(x). \quad ◀$$

INSIGHT The purpose of integration by parts is to convert a difficult integral into a new integral that will be easier. We must bear this notion in mind as we make our choice of u and v. In Example 1, if we had selected $u = \cos(x)$ and $dv = x\,dx$, then we would calculate $du = \frac{du}{dx}dx = -\sin(x)\,dx$, $v = \int dv = \int x\,dx = x^2/2$, and the new integral

$$\int v\,du = -\int \frac{x^2}{2}\sin(x)\,dx$$

would have been *harder,* not easier, than the one we started with.

When a power of x is present in the integrand, it is *often* a good rule to choose u to be the power of x. Then, when integration by parts is applied, the power of x will be decreased. In this way, we guarantee in advance that *one* factor becomes simpler. Of course, we must also consider the role of dv. As the next example shows, that consideration may be more significant.

▶ **EXAMPLE 2** Calculate $\int_1^4 2x\ln(x)\,dx$.

Solution In this example, it would be foolish to take $dv = \ln(x)\,dx$ because we do not know an antiderivative for $\ln(x)$. So we take $dv = 2x\,dx$ and $u = \ln(x)$. Then we have

$$u = \ln(x) \qquad dv = 2x\,dx$$
$$du = (1/x)\,dx \qquad v = x^2.$$

Thus

$$\int_1^4 2x\ln(x)\,dx = \int_1^4 \ln(x)\,2x\,dx = \int_1^4 u\,dv = uv\Big|_1^4 - \int_1^4 v\,du = \ln(x)\cdot x^2\Big|_1^4 - \int_1^4 x^2 \cdot \frac{1}{x}\,dx.$$

Once again, we have converted a difficult integral into a new integral that is much simpler. We continue calculating:

$$\int_1^4 2x\ln(x)\,dx = x^2\ln(x)\Big|_1^4 - \int_1^4 x^2\cdot\frac{1}{x}\,dx = \Big(4^2\ln(4) - 1^2\cdot 0\Big) - \Big(8 - \frac{1}{2}\Big) = 16\ln(4) - \frac{15}{2} = 32\ln(2) - \frac{15}{2}. \quad ◀$$

▶ **EXAMPLE 3** Calculate $\int \ln(x)\,dx$

Solution This is a new type of problem for us. On the one hand, we do not know an antiderivative for $\ln(x)$. On the other hand, the integrand does not appear to be a product because we see only the one function, $\ln(x)$. However, we can rewrite the integral as

$$\int \ln(x) \cdot 1\,dx.$$

We take $u = \ln(x)$ and $dv = 1\ dx$. We have

$$\begin{array}{ll} u = \ln(x) & dv = 1\,dx \\ du = (1/x)\,dx & v = x. \end{array}$$

Thus

$$\int \ln(x) \cdot 1\,dx = \int u\,dv = u \cdot v - \int v\,du = \ln(x) \cdot x - \int x \cdot \frac{1}{x}\,dx$$
$$= x\ln(x) - \int 1\,dx = x\ln(x) - x + C. \quad ◀$$

Advanced Examples Sometimes applying integration by parts just once is not sufficient to solve the problem at hand, and further applications might be needed. The next two examples will illustrate.

▶ **EXAMPLE 4** In atomic measurements, it is sometimes convenient to use the Bohr radius $\alpha_0 = 0.52917725 \times 10^{-10}$m as a unit of distance. Let ρ denote any (unitless) positive number. At a given instant of time, the probability of finding the hydrogen electron at a distance no greater than $\rho \cdot \alpha_0$ from the atomic nucleus is

$$\int_0^{\rho} 4r^2 e^{-2r}\,dr.$$

What is the probability p that the hydrogen electron is less than twice the Bohr radius from the nucleus?

Solution The answer is given by

$$p = \int_0^2 4r^2 e^{-2r}\,dr = 4\int_0^2 r^2 e^{-2r}\,dr = 4J \quad \text{where} \quad J = \int_0^2 r^2 e^{-2r}\,dr.$$

To calculate this integral, we take $u = r^2$ and $dv = e^{-2r}dr$. Then

$$\begin{array}{ll} u = r^2 & dv = e^{-2r}\,dr \\ du = 2r\,dr & v = -e^{-2r}/2. \end{array}$$

We therefore have

$$J = \int_0^2 r^2 e^{-2r}\,dr = \int_{r=0}^{r=2} u\,dv = u \cdot v\Big|_{r=0}^{r=2} - \int_{r=0}^{r=2} v\,du$$
$$= r^2 \cdot \left(-\frac{e^{-2r}}{2}\right)\Bigg|_{r=0}^{r=2} - \int_0^2 \left(-\frac{e^{-2r}}{2}\right) \cdot 2r\,dr = -2e^{-4} + \int_0^2 r\,e^{-2r}\,dr.$$

Thus $J = K - 2e^{-4}$ where $K = \int_0^2 r\,e^{-2r}\,dr$. We see that our new integral K is *simpler* than the one we started with (because the power of r is lower), but it still requires work. Another application of integration by parts is needed. We apply the integration by parts formula to K, taking $u = r$ and $dv = e^{-2r}dr$ We now have

$$u = r \qquad dv = e^{-2r}\,dr$$
$$du = 1\,dr \qquad v = -e^{-2r}/2.$$

Therefore

$$K = \int_0^2 r\,e^{-2r}\,dr = \int_{r=0}^{r=2} u\,dv = u \cdot v\Big|_{r=0}^{r=2} - \int_{r=0}^{r=2} v\,du$$
$$= r\left(\frac{e^{-2r}}{-2}\right)\Big|_{r=0}^{r=2} - \int_0^2 \left(\frac{e^{-2r}}{-2}\right)dr = -e^{-4} + \frac{1}{2}\int_0^2 e^{-2r}dr$$
$$= -e^{-4} + \frac{1}{2}\left(\frac{e^{-2r}}{-2}\right)\Big|_{r=0}^{r=2} = -e^{-4} + \left(\frac{1}{2}\left(\frac{e^{-4}}{-2}\right) - \frac{1}{2}\left(\frac{1}{-2}\right)\right) = \frac{1}{4} - \frac{5}{4}e^{-4}.$$

Finally, the required probability p is given by

$$p = 4J = 4(K - 2e^{-4}) = 4\left(\left(\frac{1}{4} - \frac{5}{4}e^{-4}\right) - 2e^{-4}\right) = 1 - 13e^{-4} \approx 0.7618967. \quad ◀$$

INSIGHT If we organize our work properly, keeping repeated applications of the integration by parts formula separate, then there is no harm in reusing the symbols u and v each time. Also notice that we can use the integration by parts formula with a variable of integration other than x.

In our next example, it will appear that we have evaluated an integral indirectly. After integrating by parts, the original integral arises on the right side of the equation with a coefficient not equal to 1. We treat the original integral as an unknown variable, isolate it on one side of the equation, and divide the equation by its coefficient.

▶ **EXAMPLE 5** Calculate the integral

$$S = \int e^x \cos(x)\,dx.$$

Solution We take $u = e^x$ and $dv = \cos(x)\,dx$. Then

$$u = e^x \qquad dv = \cos(x)\,dx$$
$$du = e^x\,dx \qquad v = \sin(x).$$

Therefore

$$S = \int u\,dv = u \cdot v - \int v\,du = e^x \sin(x) - \int \sin(x)\,e^x\,dx.$$

We have converted the original integral problem S to a new integration problem, but the new one does not look any simpler than the original—indeed, all we have done is to replace $\cos(x)$ with $\sin(x)$. Nevertheless, we integrate by parts again, taking $u = e^x$ and $dv = \sin(x)\,dx$. As a result,

$$\begin{aligned} u &= e^x & dv &= \sin(x)\,dx \\ du &= e^x\,dx & v &= -\cos(x). \end{aligned}$$

We have

$$S = e^x \sin(x) - \int \sin(x)\, e^x\,dx = e^x \sin(x) - \left(-e^x \cos(x) + \int \cos(x) e^x\,dx\right).$$

Using the symbol S, we can rewrite this last equation as

$$S = e^x \sin(x) + e^x \cos(x) - S.$$

We can solve this equation for S, obtaining

$$S = \frac{1}{2}\left(e^x \sin(x) + e^x \cos(x)\right).$$

Finally, let us observe that, because S is an indefinite integral, a constant of integration should be added to obtain the most general solution:

$$S = \frac{1}{2} e^x \left(\sin(x) + \cos(x)\right) + C. \quad ◀$$

INSIGHT The integral that we have just calculated arises naturally in signal processing and may be found in many engineering texts. A remarkable feature of the method of Example 5 is that we never "calculate" the integral in the usual way. We express the integral (which we think of as the unknown S) in terms of itself, and then we obtain the solution by using algebra.

The method of Example 5 can also be applied to integrals of expressions of the form $\sin(ax)\cdot\cos(bx)$, $\sin(ax)\cdot\sin(bx)$, and $\cos(ax)\cdot\cos(bx)$. Because these integrals are sometimes needed in applications, we will record them for reference (*not* for memorization):

$$\int \sin(ax)\cos(bx)\,dx = -\frac{1}{2}\frac{\cos((a-b)x)}{a-b} - \frac{1}{2}\frac{\cos((a+b)x)}{a+b} + C, \quad a \neq \pm b \quad (6.1.5)$$

$$\int \sin(ax)\sin(bx)\,dx = \frac{1}{2}\frac{\sin((a-b)x)}{a-b} - \frac{1}{2}\frac{\sin((a+b)x)}{a+b} + C, \quad a \neq \pm b \quad (6.1.6)$$

$$\int \cos(ax)\cos(bx)\,dx = \frac{1}{2}\frac{\sin((a-b)x)}{a-b} + \frac{1}{2}\frac{\sin((a+b)x)}{a+b} + C, \quad a \neq \pm b. \quad (6.1.7)$$

In our next example, we must appeal to a trigonometric identity before we can apply the method of Example 5.

▶ **EXAMPLE 6** Show that

$$\int \sec^3(x)\,dx = \frac{1}{2}\sec(x)\tan(x) + \frac{1}{2}\ln\left(|\sec(x)+\tan(x)|\right) + C. \qquad (6.1.8)$$

Solution Because $\sec^2(x)$ is the derivative of $\tan(x)$, we write $\sec^3(x)\,dx = \sec(x)\cdot\sec^2(x)\,dx$ and set $u = \sec(x)$ and $dv = \sec^2(x)\,dx$. We then have $du = \sec(x)\tan(x)\,dx$ and $v = \tan(x)$. The integration by parts formula gives us

$$\int \sec^3(x)\,dx = \sec(x)\cdot\tan(x) - \int \tan(x)\cdot\sec(x)\tan(x)\,dx = \sec(x)\cdot\tan(x) - \int \tan^2(x)\cdot\sec(x)\,dx.$$

We now use the identity $1+\tan^2(x) = \sec^2(x)$, or $\tan^2(x) = \sec^2(x) - 1$, and substitute for $\tan^2(x)$ in the integral on the right. With S denoting $\int \sec^3(x)\,dx$, we have

$$\begin{aligned} S &= \sec(x)\cdot\tan(x) - \int \left(\sec^2(x) - 1\right)\cdot\sec(x)\,dx \\ &= \sec(x)\cdot\tan(x) - \int \left(\sec^3(x) - \sec(x)\right)dx \\ &= \sec(x)\cdot\tan(x) - S + \int \sec(x)\,dx \\ &= \sec(x)\cdot\tan(x) - S + \ln\left(|\sec(x)+\tan(x)|\right) + C. \end{aligned}$$

After bringing S to the left side (and changing its coefficient from -1 to $+1$ in doing so), we obtain

$$2S = \sec(x)\cdot\tan(x) + \ln\left(|\sec(x)+\tan(x)|\right) + C,$$

from which formula (6.1.8) follows. ◀

Reduction Formulas

In Example 4, we evaluated the integral $\int r^2 e^{-2r}\,dr$ by applying the integration by parts formula twice. The first application results in an integral of the form $\int re^{-2r}\,dr$, which is just like the original integral except that the power of r has been reduced by one. When we perform this second integration by parts, we reduce it to $\int r^0 e^{-2r}\,dr$, or $\int e^{-2r}\,dr$. This last integral is a standard one that does not require further reduction. In summary, for $n = 2$ and $n = 1$, applying integration by parts to $\int r^n e^{-2r}\,dr$ results in a formula involving the integral $\int r^{n-1}e^{-2r}\,dr$, which is just like the original except for the lower power. Such a formula is known as a *reduction formula.*

▶ **EXAMPLE 7** Let a be a nonzero constant. Derive the reduction formula

$$\int x^n e^{ax}\,dx = \frac{1}{a}x^n e^{ax} - \frac{n}{a}\int x^{n-1}e^{ax}\,dx. \qquad (6.1.9)$$

Solution We let $u = x^n$ and $dv = e^{ax}dx$. Then $du = nx^{n-1}\,dx$ and $v = e^{ax}/a$. The integration by parts formula gives us

$$\int x^n e^{ax}\,dx = x^n\cdot\frac{1}{a}e^{ax} - \int \frac{1}{a}e^{ax}\cdot nx^{n-1}\,dx,$$

which, after a little rearrangement, is equation (6.1.9). ◀

▶ **EXAMPLE 8** Use formula (6.1.9) to evaluate $\int x^3 e^{-x}\,dx$.

Solution Note that $a = -1$. Setting $n = 3$ in equation (6.1.9), we obtain

$$\int x^3 e^{ax}\,dx = \frac{1}{a}x^3 e^{ax} - \frac{3}{a}\int x^2 e^{ax}\,dx. \tag{6.1.10}$$

We now use equation (6.1.9) with $n = 2$ to reduce the integral on the right side of (6.1.10):

$$\int x^3 e^{ax}\,dx = \frac{1}{a}x^3 e^{ax} - \frac{3}{a}\left(\frac{1}{a}x^2 e^{ax} - \frac{2}{a}\int x e^{ax}\,dx\right) = \frac{1}{a}x^3 e^{ax} - \frac{3}{a^2}x^2 e^{ax} + \frac{6}{a^2}\int x e^{ax}\,dx.$$

The integral on the right side of the last equation can be reduced by applying equation (6.1.9) with $n = 1$:

$$\begin{aligned}
\int x^3 e^{ax}\,dx &= \frac{1}{a}x^3 e^{ax} - \frac{3}{a^2}x^2 e^{ax} + \frac{6}{a^2}\left(\frac{1}{a}x e^{ax} - \frac{1}{a}\int x^0 e^{ax}\,dx\right)\\
&= \frac{1}{a}x^3 e^{ax} - \frac{3}{a^2}x^2 e^{ax} + \frac{6}{a^3}x e^{ax} - \frac{6}{a^3}\int e^{ax}\,dx\\
&= \frac{1}{a}x^3 e^{ax} - \frac{3}{a^2}x^2 e^{ax} + \frac{6}{a^3}x e^{ax} - \frac{6}{a^4}e^{ax} + C\\
&= \left(\frac{1}{a}x^3 - \frac{3}{a^2}x^2 + \frac{6}{a^3}x - \frac{6}{a^4}\right)e^{ax} + C.
\end{aligned}$$

We now set $a = -1$ to obtain

$$\int x^3 e^{-x}\,dx = -(x^3 + 3x^2 + 6x + 6)e^{-x} + C. \quad ◀$$

QUICK QUIZ

1. The integration by parts formula is another way of looking at what formula for the derivative?
2. An application of integration by parts leads to an equation of the form

$$\int x^5 e^{-3x}\,dx = Ax^5 e^{-3x} + B\int x^4 e^{-3x}\,dx.$$

What are A and B?

3. An application of integration by parts leads to an equation of the form

$$\int x^5 \ln(x)\,dx = \frac{1}{6}x^6 \ln(x) - \int \psi(x)\,dx.$$

What is $\psi(x)$?

4. Evaluate $\int \ln(x)\,dx$.

Answers

1. Product Rule 2. $A = -1/3$; $B = 5/3$ 3. $x^5/6$ 4. $x\ln(x) - x + C$

EXERCISES

Problems for Practice

In Exercises 1–10, integrate by parts to evaluate the given indefinite integral.

1. $\int xe^x\,dx$
2. $\int xe^{-x}\,dx$
3. $\int (2x+5)e^{x/3}\,dx$
4. $\int x\sin(x)\,dx$
5. $\int (4x+2)\sin(2x)\,dx$
6. $\int 9x\cos(3x)\,dx$
7. $\int x\ln(4x)\,dx$
8. $\int \ln(x)x^2dx$
9. $\int 9x^2\ln(x)\,dx$
10. $\int 16x^3\ln(2x)\,dx$

In Exercises 11–20, integrate by parts to evaluate the given definite integral.

11. $\int_0^1 xe^x\,dx$
12. $\int_0^2 xe^{-x/2}\,dx$
13. $\int_0^\pi x\cos(x)\,dx$
14. $\int_0^{\pi/4} x\cos(2x)\,dx$
15. $\int_0^{\pi/2} 4x\sin(x/3)\,dx$
16. $\int_1^e \ln(x)\,dx$
17. $\int_1^e x\ln(x)\,dx$
18. $\int_{1/3}^1 x\ln(3x)\,dx$
19. $\displaystyle\int_1^4 \frac{\ln(x)}{\sqrt{x}}\,dx$
20. $\int_0^1 x3^x\,dx$

In Exercises 21–30, integrate by parts successively to evaluate the given indefinite integral.

21. $\int x^2e^x\,dx$
22. $\int x^2e^{-x}\,dx$
23. $\int x^2\cos(x)\,dx$
24. $\int x^2\sin(x)\,dx$
25. $\int 4x^2\cos(2x)\,dx$
26. $\int 27x^2\sin(3x)\,dx$
27. $\int x^3e^x\,dx$
28. $\int x^3\sin(x)\,dx$
29. $\int 16x^3\ln^2(x)\,dx$
30. $\int \ln^3(x)\,dx$

In Exercises 31–36, integrate by parts to evaluate the given definite integral.

31. $\int_0^1 2\arctan(x)\,dx$
32. $\int_0^1 \arcsin(x)\,dx$
33. $\int_0^{\sqrt{3}} \arcsin(x/2)\,dx$
34. $\int_0^2 \operatorname{arccot}(x/2)\,dx$
35. $\int_{1/2}^1 2\arccos(x)\,dx$
36. $\int_{\sqrt{2}}^2 2x\operatorname{arcsec}(x)\,dx$

In Exercises 37–40, use reduction formula (6.1.9) to evaluate the given integral.

37. $\int_0^1 x^2e^{x/2}\,dx$
38. $\int_0^1 4x^2e^{-2x}\,dx$
39. $\int_0^1 x^3e^x\,dx$
40. $\int_0^1 8x^3e^{2x}\,dx$

Further Theory and Practice

In each of Exercises 41–44, simplify the integrand before integrating by parts.

41. $\int \ln(\sqrt{x})\,dx$
42. $\int x\ln(x^5)\,dx$
43. $\int x^3\ln(1/x)\,dx$
44. $\int 4x\sin(x)\cos(x)\,dx$

Evaluate each of the integrals in Exercises 45–54.

45. $\int (x/2^x)\,dx$
46. $\displaystyle\int \frac{\ln(x)}{\sqrt{x}}\,dx$
47. $\displaystyle\int \left(\frac{\ln(x)}{x}\right)^2 dx$
48. $\int 9\sqrt{x}\,\ln(x)\,dx$
49. $\int x\sec(x)\tan(x)\,dx$
50. $\int x\sec^2(x)\,dx$
51. $\int x\cos(x+\pi/4)\,dx$
52. $\int 9x\sin(3x+4)\,dx$
53. $\int 2\cos(\ln(x))\,dx$
54. $\displaystyle\int \left(\frac{\ln(x)}{x}+\frac{\ln(x)}{x^2}\right) dx$
55. Evaluate $\int_{-2}^1 x(x+3)^{-1/2}\,dx$. For the first step, integrate by parts with $u=x$.
56. Evaluate $\int_1^2 x(x-1)^{1/2}dx$. For the first step, integrate by parts with $u=x$.

In Exercises 57–60, make an appropriate substitution before integrating by parts.

57. $\int \sin(\sqrt{x})\,dx$
58. $\int \exp(\sqrt{x})\,dx$
59. $\int 2x^3\exp(x^2)\,dx$

60. $\int 3x^5 \sin(x^3)\,dx$

61. Evaluate $\int_0^1 (e^x - x)^2\,dx$.

62. Evaluate $\int_1^2 3(x-2)^2 \ln(x)\,dx$.

63. Evaluate $\int x^p \ln(x)\,dx$.

In each of Exercises 64–69, integrate by parts. This will result in an integrand of the form $P(x)/Q(x)$ where $P(x)$ and $Q(x)$ are polynomials with the degree of $P(x)$ greater than or equal to the degree of $Q(x)$. Such an integrand is handled by performing polynomial division to put $P(x)/Q(x)$ into the form $r(x) + s(x)/Q(x)$ where $r(x)$ and $s(x)$ are polynomials with the degree of $s(x)$ less than the degree of $Q(x)$.

64. $\int_0^1 x \arctan(x)\,dx$

65. $\int_0^1 2x \arctan(x/\sqrt{3})\,dx$

66. $\int_0^1 6x^2 \arctan(x)\,dx$

67. $\int_0^1 \ln(1+x^2)\,dx$

68. $\int_0^1 x\ln(1+x)\,dx$

69. $\int_{-1}^0 9x^2 \ln(2+x)\,dx$

70. Suppose that a is a nonzero constant. Calculate $\int \sec^3(ax)\,dx$.

71. Derive the reduction formula

$$\int \ln^n(x)\,dx = x\ln^n(x) - n\int \ln^{n-1}(x)\,dx.$$

72. Suppose that a is a nonzero constant. Derive the formula

$$\int \ln(ax+b)\,dx = \frac{1}{a}(ax+b)\ln(ax+b) - x + C.$$

73. Use the formula of Exercise 72 to evaluate $\int \ln(x^2-1)\,dx$.

74. Use the formula of Exercise 72 to evaluate $\int \ln(x^2+7x+10)\,dx$.

In each of Exercises 75–78 a region $\mathcal{R}$ is described. Calculate its area.

75. $\mathcal{R}$ is the region that is bounded above by $y = \pi x/2$ and below by $y = \arcsin(x)$ for $0 \le x \le 1$.

76. $\mathcal{R}$ is the region that is bounded above by $y = 2\arctan(x)$ and below by $y = \pi x/2$ for $0 \le x \le 1$.

77. $\mathcal{R}$ is the region that is bounded above by $y = \cos(\pi x/2)$ and below by $y = x\cos(\pi x/2)$ for $-1 \le x \le 1$.

78. $\mathcal{R}$ is the region that is bounded above by $y = \ln(1+x)$ and below by $y = \ln(1+x^2)$ for $0 \le x \le 1$.

79. Let $\alpha = \int_0^1 (1+x^4)^{-1}\,dx$. Express $\int_0^1 \ln(1+x^4)\,dx$ in terms of α. (It is known that $\alpha = \sqrt{2}(\pi + 2\ln(1+\sqrt{2}))/8$, but do not attempt this evaluation.)

80. Integrate by parts repeatedly, each time with u equal to a power of $1-x$, to prove that

$$\int_0^1 \sqrt{x}(1-x)^4\,dx = \frac{8}{3}\cdot\frac{6}{5}\cdot\frac{4}{7}\cdot\frac{2}{9}\cdot\frac{2}{11}.$$

81. Derive the reduction formula

$$\int \frac{1}{(x^2+a^2)^{n+1}}\,dx = \frac{1}{2a^2 n}\,\frac{x}{(x^2+a^2)^n} + \frac{2n-1}{2a^2 n}\int \frac{1}{(x^2+a^2)^n}\,dx.$$

82. Derive formula (6.1.5).

83. Derive formula (6.1.6).

84. Derive formula (6.1.7).

85. Suppose that a and b are constants, not both 0. Derive the formula

$$\int e^{ax}\cos(bx)\,dx = \frac{e^{ax}\big(a\cos(bx) + b\sin(bx)\big)}{a^2+b^2} + C.$$

86. Suppose that a and b are constants, not both 0. Derive the formula

$$\int e^{ax}\sin(bx)\,dx = \frac{e^{ax}\big(a\sin(bx) - b\cos(bx)\big)}{a^2+b^2} + C.$$

87. The Fresnel sine and Fresnel cosine functions, denoted by FresnelS and FresnelC, are important in the theory of optics. They are defined by

$$\text{FresnelS}(x) = \int_0^x \sin\left(\frac{\pi}{2}t^2\right)dt$$

and

$$\text{FresnelC}(x) = \int_0^x \cos\left(\frac{\pi}{2}t^2\right)dt.$$

a. Integrate by parts to calculate $\int \text{FresnelS}(x)\,dx$. (Your answer will involve FresnelS(x) and elementary functions.)

b. Integrate by parts to calculate $\int \text{FresnelC}(x)\,dx$. (Your answer will involve FresnelC(x) and elementary functions.)

c. Integrate by parts to calculate $\int 2x\text{FresnelS}(x)\,dx$. (Your answer will involve both Fresnel functions and elementary functions.)

d. Integrate by parts to calculate $\int 2x\text{FresnelC}(x)\,dx$. (Your answer will involve both Fresnel functions and elementary functions.)

88. The *error function* (denoted by erf) is central to the subjects of probability and statistics. It is defined by

$$\text{erf}(x) = \frac{2}{\sqrt{\pi}}\int_0^x \exp(-t^2)\,dt.$$

a. Integrate by parts to calculate $\int \text{erf}(x)\,dx$. (Your answer will involve erf (x) and elementary functions.)

b. Integrate by parts to calculate $\int x^2\exp(-x^2)\,dx$. (Take $u = x$, $dv = x\exp(-x^2)\,dx$. Your answer will involve erf (x) and elementary functions.)

c. Integrate by parts to calculate $\int x^2\text{erf}(x)\,dx$. (You will need the result of part b. Your answer will involve erf (x) and elementary functions.)

89. The sine integral Si (x) is defined by

$$\mathrm{Si}(x) = \int_0^x \frac{\sin(t)}{t}\,dt.$$

(The integrand has a continuous extension at 0.)

a. Integrate by parts to calculate $\int \mathrm{Si}(x)\,dx$. (Your answer will involve $\mathrm{Si}(x)$ and elementary functions.)

b. Integrate by parts to calculate $\int x\mathrm{Si}(x)\,dx$. (Your answer will involve $\mathrm{Si}(x)$ and elementary functions.)

Calculator/Computer Exercises

90. Refer to Example 4. For approximately what value ρ is the probability of finding the hydrogen electron within a distance $\rho \cdot \alpha_0$ of the nucleus equal to 1/2?

91. Find the area of the region that is bounded above by $y = 1 + 2x$ and below by $y = xe^x$.

92. Find the total area of the regions that are bounded above and below by $y = x\sin(\pi x)$ and $y = x\sin(\pi x^2)$ for $0 \le x \le 1$.

93. Plot $\arctan(x) - 1$ and $\ln(x)/x$ for $0.7 \le x \le 5$. Find the two points of intersection. Calculate the area of the region enclosed by the two curves.

94. Suppose that $f(x)$ is an odd function: $f(-x) = -f(x)$. Let

$$b_n = \frac{1}{\pi}\int_{-\pi}^{\pi} f(x)\sin(nx)\,dx \quad (n \ge 1).$$

The function

$$\sum_{n=1}^{N} b_n \sin(nx)$$

is said to be the order N *Fourier sine-polynomial approximation* of f. Using the viewing window $[-2, 2] \times [-2, 2]$, plot the order N Fourier sine-polynomial approximation of $f(x) = x$ for $N = 3$ and 5.

95. Suppose that $f(x)$ is an even function: $f(-x) = f(x)$. Let

$$a_n = \frac{1}{\pi}\int_{-\pi}^{\pi} f((x)\cos(nx)\,dx \quad (n \ge 0).$$

The function

$$\frac{a_0}{2} + \sum_{n=1}^{N} a_n \cos(nx)$$

is said to be the order N *Fourier cosine-polynomial approximation* of f. Using the viewing window $[-2, 2] \times [-0.2, 4]$, plot the order N Fourier cosine-polynomial approximation of $f(x) = x^2$ for $N = 3$ and 5.

96. Let the Fourier coeffcients $a_n\,(n \ge 0)$ and $b_n\,(n \ge 1)$ of a function f be defined as in Exercises 94 and 95. If f is a differentiable function on the interval $(-\pi, \pi)$, then it can be shown that the degree N Fourier polynomial

$$\frac{a_0}{2} + \sum_{n=1}^{N}\Big(a_n\cos(nx) + b_n\sin(nx)\Big)$$

converges to $f(x)$ as N tends to infinity. Calculate the degrees 2 and 3 Fourier polynomials of $f(x) = \exp(x)$. Plot these Fourier polynomials and the function f in the viewing window $[-2.6, 2.6] \times [-2.25, 14.25]$. (With these small values of N, you can only begin to see the Fourier polynomials taking on the shape of f. In Figure 1, we have plotted f and the approximating Fourier polynomials of degree 5 and 20.)

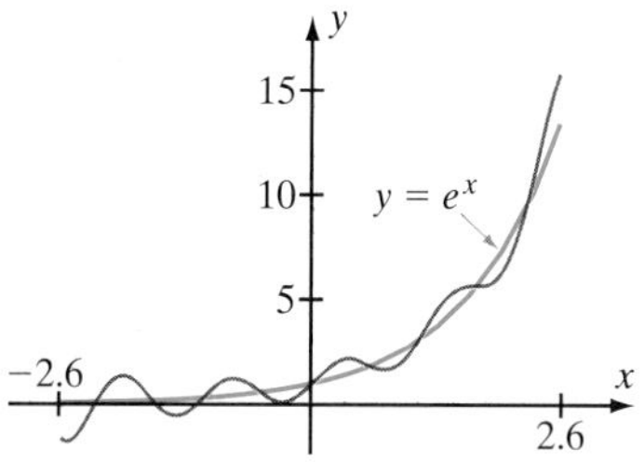

▲ Figure 1a The order 5 Fourier polynomial of the exponential function.

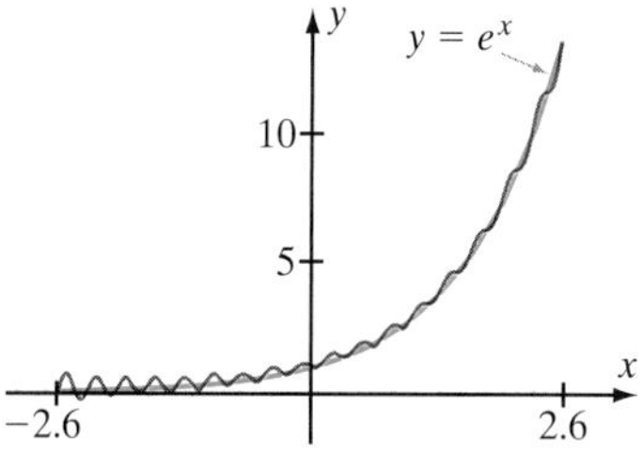

▲ Figure 1b The order 20 Fourier polynomial of the exponential function.

6.2 Powers and Products of Trigonometric Functions

This section introduces several procedures for treating integrals involving powers of sines and cosines. The techniques we use are a combination of trigonometric identities and substitutions. We will group the problems by type. You will want to study the technique for each type of problem. There are no new formulas to memorize in this section. To master this material you must learn how and when to apply each technique.

Squares of Sine, Cosine, Secant, and Tangent

In this subsection, you will learn how to integrate the expressions $\sin^2(\theta)$ and $\cos^2(\theta)$ by using the equations

$$\sin^2(\theta) = \frac{1 - \cos(2\theta)}{2} \tag{6.2.1}$$

and

$$\cos^2(\theta) = \frac{1 + \cos(2\theta)}{2}, \tag{6.2.2}$$

which are obtained by replacing θ with 2θ in Half-Angle Formulas (1.6.10) and (1.6.11). Notice that the right sides of identities (6.2.1) and (6.2.2) differ only in the sign that precedes the term $\cos(2\theta)$. In particular, formula (6.2.1) for $\sin^2(\theta)$ involves $\cos(2\theta)$ and *not* $\sin(2\theta)$.

▶ **EXAMPLE 1** Show that

$$\int \sin^2(x)\,dx = \frac{1}{2}x - \frac{1}{4}\sin(2x) + C \tag{6.2.3}$$

and

$$\int \cos^2(x)\,dx = \frac{1}{2}x + \frac{1}{4}\sin(2x) + C. \tag{6.2.4}$$

Solution Identities (6.2.1) and (6.2.2) are used to derive formulas (6.2.3) and (6.2.4), respectively. Let us illustrate with the second formula:

$$\begin{aligned} \int \cos^2(x)\,dx &= \int \frac{1 + \cos(2x)}{2}\,dx \\ &= \int \frac{1}{2}\,dx + \int \frac{\cos(2x)}{2}\,dx \\ &= \frac{1}{2}x + \frac{1}{4}\sin(2x) + C. \quad ◀ \end{aligned}$$

INSIGHT In the next section, you will find that it is sometimes best to express the antiderivatives of $\sin^2(x)$ and $\cos^2(x)$ in terms of trigonometric functions of the argument x (instead of $2x$). Using the identity $\sin(2x) = 2\sin(x)\cos(x)$, we may rewrite formulas (6.2.3) and (6.2.4) as

$$\int \sin^2(x)\,dx = \frac{1}{2}\Big(x - \sin(x)\cos(x)\Big) + C \tag{6.2.5}$$

and

$$\int \cos^2(x)\,dx = \frac{1}{2}\Big(x + \sin(x)\cos(x)\Big) + C. \tag{6.2.6}$$

We conclude our discussion of $\sin^2(x)$ and $\cos^2(x)$ by using equations (6.2.5) and (6.2.6) to evaluate some definite integrals that arise frequently:

$$\int_0^{\pi/2} \cos^2(x)\,dx = \int_0^{\pi/2} \sin^2(x)\,dx = \frac{\pi}{4}, \tag{6.2.7}$$

$$\int_0^{\pi} \cos^2(x)\,dx = \int_0^{\pi} \sin^2(x)\,dx = \frac{\pi}{2}, \tag{6.2.8}$$

and

$$\int_0^{2\pi} \cos^2(x)\,dx = \int_0^{2\pi} \sin^2(x)\,dx = \pi. \tag{6.2.9}$$

We now turn our attention to $\int \sec^2(x)\,dx$ and $\int \tan^2(x)\,dx$. The first of these is a basic indefinite integral that we already know: $\int \sec^2(x)\,dx = \tan(x) + C$. In the next example, we will see how this formula can be used to obtain an antiderivative of $\tan^2(x)$.

▶ **EXAMPLE 2** Show that

$$\int \tan^2(x)\,dx = \tan(x) - x + C. \tag{6.2.10}$$

Solution As in Example 6 of Section 1, we use the identity $\tan^2(x) = \sec^2(x) - 1$. We have

$$\int \tan^2(x)\,dx = \int (\sec^2(x) - 1)\,dx = \int \sec^2(x)\,dx + \int (-1)\,dx = \tan(x) - x + C.$$

Higher Powers of Sine, Cosine, Secant, and Tangent

We now know how to evaluate $\int f^n(x)\,dx$ where $f(x)$ is one of the trigonometric functions $\sin(x)$, $\cos(x)$, $\sec(x)$, $\tan(x)$, and $n = 0$, 1, or 2. If n is a higher integer power, then we can use a reduction formula, repeatedly if necessary, to reduce the calculation to one we already know. Here, for future reference, are the reduction formulas we need:

$$\int \sin^n(x)\,dx = -\frac{1}{n}\sin^{n-1}(x)\cos(x) + \frac{n-1}{n}\int \sin^{n-2}(x)\,dx \quad (n \neq 0) \tag{6.2.11}$$

$$\int \cos^n(x)\,dx = \frac{1}{n}\sin(x)\cos^{n-1}(x) + \frac{n-1}{n}\int \cos^{n-2}(x)\,dx \quad (n \neq 0) \tag{6.2.12}$$

$$\int \sec^n(x)\,dx = \frac{1}{n-1}\sec^{n-2}(x)\tan(x) + \frac{n-2}{n-1}\int \sec^{n-2}(x)\,dx \quad (n \neq 1) \tag{6.2.13}$$

$$\int \tan^n(x)\,dx = \frac{1}{n-1}\tan^{n-1}(x) - \int \tan^{n-2}(x)\,dx \quad (n \neq 1). \tag{6.2.14}$$

▶ **EXAMPLE 3** Derive reduction formula (6.2.11).

Solution Let $I_n = \int \sin^n(x)\,dx$ denote the left side of formula (6.2.11). We apply the integration by parts formula as follows:

$$I_n = \int \underbrace{\sin^{n-1}(x)}_{u}\,\underbrace{\sin(x)\,dx}_{dv} = \underbrace{\sin^{n-1}(x)}_{u} \cdot \underbrace{\Big(-\cos(x)\Big)}_{v} - \int \underbrace{\Big(-\cos(x)\Big)}_{v}\,\underbrace{(n-1)\sin^{n-2}(x)\cos(x)\,dx}_{du}.$$

This equation simplifies to

$$I_n = -\sin^{(n-1)}(x)\cos(x) + (n-1)\int \sin^{(n-2)}\cos^2(x)\,dx.$$

We replace $\cos^2(x)$ with $1 - \sin^2(x)$ and expand the integral:

$$\begin{aligned} I_n &= -\sin^{(n-1)}(x)\cos(x) + (n-1)\int \sin^{(n-2)}(1-\sin^2(x))\,dx \\ &= -\sin^{(n-1)}(x)\cos(x) + (n-1)\int \sin^{(n-2)}\,dx - (n-1)\int \sin^n(x)\,dx \\ &= -\sin^{(n-1)}(x)\cos(x) + (n-1)\int \sin^{(n-2)}\,dx - (n-1)I_n. \end{aligned}$$

We now bring the unknown I_n that appears on the right side of the equation over to the left side. We obtain

$$I_n + (n-1)I_n = -\sin^{(n-1)}(x)\cos(x) + (n-1)\int \sin^{(n-2)}\,dx.$$

Because the left side of this equation simplifies to $n \cdot I_n$, dividing through by n results in formula (6.2.11). ◄

In practice, reduction formulas (6.2.11) through (6.2.14) are used successively until the power of the trigonometric function has been reduced to 0 or 1.

► **EXAMPLE 4** Evaluate $\int \cos^6(x)\,dx$.

Solution After one application of formula (6.2.12), we obtain

$$\int \cos^6(x)\,dx = \frac{1}{6}\sin(x)\cos^5(x) + \frac{5}{6}\int \cos^4(x)\,dx.$$

When we apply formula (6.2.12) to $\int \cos^4(x)\,dx$, we obtain

$$\int \cos^6(x)\,dx = \frac{1}{6}\sin(x)\cos^5(x) + \frac{5}{6}\left(\frac{1}{4}\sin(x)\cos^3(x) + \frac{3}{4}\int \cos^2(x)\,dx\right)$$

or

$$\int \cos^6(x)\,dx = \frac{1}{6}\sin(x)\cos^5(x) + \frac{5}{24}\sin(x)\cos^3(x) + \frac{5}{8}\int \cos^2(x)\,dx.$$

At this point, we can exploit equation (6.2.6). Alternatively, reduction formula (6.2.12) can be applied to $\int \cos^2(x)\,dx$, resulting in

$$\int \cos^2(x)\,dx = \frac{1}{2}\sin(x)\cos(x) + \frac{1}{2}\int \cos^0(x)\,dx = \frac{1}{2}\sin(x)\cos(x) + \frac{1}{2}x + C.$$

Thus

$$\int \cos^6(x)\,dx = \frac{1}{6}\sin(x)\cos^5(x) + \frac{5}{24}\sin(x)\cos^3(x) + \frac{5}{16}\sin(x)\cos(x) + \frac{5}{16}x + C. \quad ◄$$

Odd Powers of Sine and Cosine

Although reduction formulas (6.2.11) and (6.2.12) can be used to evaluate $\int \sin^n(x)\,dx$ and $\int \cos^n(x)\,dx$ when n is any positive integer, there is a simpler, less tedious method available when n is odd. We will use the identities

$$\cos^2(x) = 1 - \sin^2(x) \text{ and } \sin^2(x) = 1 - \cos^2(x).$$

Suppose that k and ℓ are positive integers. We integrate an odd power $\sin^{2k+1}(x)$ of $\sin(x)$ by writing

$$\sin^{2k+1}(x) = \left(\sin^2(x)\right)^k \sin(x) = \left(1-\cos^2(x)\right)^k \sin(x). \tag{6.2.15}$$

We integrate an odd power $\cos^{2\ell+1}(x)$ of $\cos(x)$ by writing

$$\cos^{2\ell+1}(x) = \left(\cos^2(x)\right)^\ell \cos(x) = \left(1-\sin^2(x)\right)^\ell \cos(x). \tag{6.2.16}$$

Let us see how these formulas help us in practice:

▶ **EXAMPLE 5** Use formula (6.2.16) to calculate $\int \cos^3(x)\,dx$.

Solution Formula (6.2.16) with $\ell = 1$ tells us that

$$\int \cos^3(x)\,dx = \int (\cos^2(x)) \cdot \cos(x)\,dx = \int (1-\sin^2(x)) \cdot \cos(x)\,dx.$$

We make the substitution $u = \sin(x), du = \cos(x)\,dx$, which results in

$$\int (1-u^2)\,du = u - \frac{1}{3}u^3 + C.$$

Resubstituting the x variable, we have

$$\int \cos^3(x)\,dx = \sin(x) - \frac{1}{3}\sin^3(x) + C. \quad ◀$$

Here is a slightly more complicated example:

▶ **EXAMPLE 6** Calculate $\int \sin^5(x)\,dx$.

Solution Using formula (6.2.15) with $k = 2$, we write

$$\int \sin^5(x)\,dx = \int \left(\sin^2(x)\right)^2 \cdot \sin(x)\,dx = \int \left(1-\cos^2(x)\right)^2 \cdot \sin(x)\,dx.$$

The substitution $u = \cos(x)$, $du = -\sin(x)\,dx$ results in

$$-\int \left(1-u^2\right)^2 du = \int \left(-1+2u^2-u^4\right) du = -u + \frac{2}{3}u^3 - \frac{1}{5}u^5 + C.$$

Resubstituting the expressions in x yields the final solution:

$$\int \sin^5(x)\,dx = -\cos(x) + \frac{2}{3}\cos^3(x) - \frac{1}{5}\cos^5(x) + C. \quad ◀$$

Integrals That Involve Both Sine and Cosine

In the preceding subsections, we saw how to integrate positive powers of $\sin(x)$ and $\cos(x)$. Next we consider integrals

$$\int \sin^m(x)\cos^n(x)\,dx$$

that involve positive integer powers of both $\sin(x)$ and $\cos(x)$. There are two possibilities. Either at least one of the numbers m and n is odd, or both m and n are even. Here is what we do in each of these circumstances:

If at least one of m or n is odd, then we apply the odd-power technique that we discussed in the preceding subsection. (If both are odd, then we apply the technique to the *smaller* of the two.)

If both m and n are even, then we use the identity $\cos^2(x) + \sin^2(x) = 1$ to convert the integrand to a sum of even powers of sine or of cosine. Then we apply the appropriate reduction formula.

Our next two examples illustrate these techniques.

▶ **EXAMPLE 7** Evaluate the integral $\int \cos^3(x)\sin^4(x)\,dx$.

Solution Because cosine is raised to an odd power, we apply formula (6.2.16). We have

$$\int \cos^3(x)\sin^4(x)\,dx = \int \sin^4(x)\cos^2(x)\cos(x)\,dx = \int \sin^4(x)\big(1-\sin^2(x)\big)\cos(x)\,dx.$$

The substitution $u = \sin(x)$, $du = \cos(x)\,dx$ converts the last integral to

$$\int u^4(1-u^2)\,du = \int (u^4 - u^6)\,du = \frac{1}{5}u^5 - \frac{1}{7}u^7 + C.$$

Therefore after resubstituting the x variable, we have

$$\int \cos^3(x)\sin^4(x)\,dx = \frac{1}{5}\sin^5(x) - \frac{1}{7}\sin^7(x) + C. \quad ◀$$

▶ **EXAMPLE 8** Evaluate $\int \sin^4(x)\cos^6(x)\,dx$.

Solution Here both powers are even. For the first step, it is more efficient to work with the *smaller* power. We write

$$\begin{aligned}\int \sin^4(x)\cos^6(x)\,dx &= \int \big(\sin^2(x)\big)^2\cos^6(x)\,dx \\ &= \int \big(1-\cos^2(x)\big)^2\cos^6(x)\,dx \\ &= \int \big(1-2\cos^2(x)+\cos^4(x)\big)\cos^6(x)\,dx \\ &= \int \cos^6(x)\,dx - 2\int \cos^8(x)\,dx + \int \cos^{10}(x)\,dx. \qquad (6.2.17)\end{aligned}$$

We can apply reduction formula (6.2.12) to each of the three integrals in the last line, but here it is most efficient to work with the *highest* power of cosine. Thus

$$\int \cos^{10}(x)\,dx = \frac{1}{10}\sin(x)\cos^9(x) + \frac{9}{10}\int \cos^8(x)\,dx \qquad (6.2.18)$$

and, by substituting equation (6.2.18) into the right side of (6.2.17), we obtain

$$\begin{aligned}\int \sin^4(x)\cos^6(x)\,dx &= \int \cos^6(x)\,dx - 2\int \cos^8(x)\,dx + \left(\frac{1}{10}\sin(x)\cos^9(x) + \frac{9}{10}\int \cos^8(x)\,dx\right) \\ &= \frac{1}{10}\sin(x)\cos^9(x) + \int \cos^6(x)\,dx + \left(-2+\frac{9}{10}\right)\int \cos^8(x)\,dx \\ &= \frac{1}{10}\sin(x)\cos^9(x) + \int \cos^6(x)\,dx - \frac{11}{10}\int \cos^8(x)\,dx. \qquad (6.2.19)\end{aligned}$$

We apply reduction formula (6.2.12) again, this time with $n = 8$:

$$\int \cos^8(x)\,dx = \frac{1}{8}\sin(x)\cos^7(x) + \frac{7}{8}\int \cos^6(x)\,dx. \tag{6.2.20}$$

Therefore, by substituting equation (6.2.20) into (6.2.19), we have

$$\begin{aligned}\int \sin^4(x)\cos^6(x)\,dx &= \frac{1}{10}\sin(x)\cos^9(x) + \int \cos^6(x)\,dx - \frac{11}{10}\left(\frac{1}{8}\sin(x)\cos^7(x) + \frac{7}{8}\int \cos^6(x)\,dx\right)\\ &= \frac{1}{10}\sin(x)\cos^9(x) - \frac{11}{80}\sin(x)\cos^7(x) + \left(1 - \frac{77}{80}\right)\int \cos^6(x)\,dx\\ &= \frac{1}{10}\sin(x)\cos^9(x) - \frac{11}{80}\sin(x)\cos^7(x) + \frac{3}{80}\int \cos^6(x)\,dx.\end{aligned}$$

Instead of using reduction formula (6.2.12) three more times to evaluate $\int \cos^6(x)\,dx$, we can now use the result of Example 4:

$$\begin{aligned}\int \sin^4(x)\cos^6(x)\,dx = {}& \frac{1}{10}\sin(x)\cos^9(x) - \frac{11}{80}\sin(x)\cos^7(x)\\ &+ \frac{3}{80}\left(\frac{1}{6}\sin(x)\cos^5(x) + \frac{5}{24}\sin(x)\cos^3(x) + \frac{5}{16}\sin(x)\cos(x) + \frac{5}{16}x\right) + C.\end{aligned}$$ ◀

Converting to Sines and Cosines

Sometimes, when other trigonometric functions are converted to sines and cosines, the methods that have been discussed can be brought to bear. The next example illustrates this idea.

▶ **EXAMPLE 9** Evaluate $\int \tan^5(x)\sec^3(x)\,dx$.

Solution We have

$$\tan^5(x)\sec^3(x) = \frac{\sin^5(x)}{\cos^5(x)}\sec^3(x) = \cos^{-8}(x)\left(\sin^2(x)\right)^2\sin(x) = \cos^{-8}(x)\left(1 - \cos^2(x)\right)^2\sin(x).$$

The substitution $u = \cos(x)$, $du = -\sin(x)\,dx$ therefore converts the given integral to

$$-\int u^{-8}(1-u^2)^2\,du = -\int (u^{-8} - 2u^{-6} + u^{-4})du = \frac{1}{7u^7} - \frac{2}{5u^5} + \frac{1}{3u^3} + C.$$

We resubstitute $u = \cos(x)$ to obtain

$$\int \tan^5(x)\sec^3(x)\,dx = \frac{1}{7\cos^7(x)} - \frac{2}{5\cos^5(x)} + \frac{1}{3\cos^3(x)} + C,$$

or

$$\int \tan^5(x)\sec^3(x)\,dx = \frac{1}{7}\sec^7(x) - \frac{2}{5}\sec^5(x) + \frac{1}{3}\sec^3(x) + C.$$

QUICK QUIZ

1. What trigonometric identity is used to evaluate $\int \sin^2(x)\,dx$ without using a reduction formula?
2. Evaluate $\int_0^{2\pi} \cos^2(x)\,dx$.

3. The equation $\int_0^{\pi/2} \sin^2(x)\cos^3(x)\,dx = \int_0^1 (u^2 - u^4)\,du$ results from what substitution?
4. For what value c is $\int_0^{\pi} \sin^8(x)\,dx = c\int_0^{\pi} \sin^6(x)\,dx$? (Do not integrate!)

Answers

1. $\sin^2(x) = \big(1-\cos(2x)\big)/2$ 2. π 3. $u = \sin(x)$ 4. 7/8

EXERCISES

Problems for Practice

▶ In each of Exercises 1–6, evaluate the given integral. ◀

1. $\int \sin^2(x/2)\,dx$
2. $\int \sin^2(4x)\,dx$
3. $\int \cos^2(2x)\,dx$
4. $\int \cos^2(x+\pi/3)\,dx$
5. $\int \tan^2(x/2)\,dx$
6. $\int \big(\cos^2(x) - \sin^2(x)\big)\,dx$

▶ In each of Exercises 7–16, evaluate the given indefinite integral. ◀

7. $\int 6\sin^3(2x)\,dx$
8. $\int 3\cos^3(x+2)\,dx$
9. $\int \sin^3(x/3)\,dx$
10. $\int 5\cos^5(x/3)\,dx$
11. $\int 3\sin^5(x/5)\,dx$
12. $\int 7\sin^7(x/10)\,dx$
13. $\int 5\sin^2(x/3)\cos^3(x/3)\,dx$
14. $\int 7\sin^3(2x/3)\sqrt{\cos(2x/3)}\,dx$
15. $\displaystyle\int \frac{\cos^3(x)}{\sin(x)}\,dx$
16. $\displaystyle\int \frac{5\cos^3(2x)}{\sqrt{\sin(2x)}}\,dx$

▶ In each of Exercises 17–22, evaluate the given definite integral. ◀

17. $\int_0^{\pi} \sin^3(x)\cos^4(x)\,dx$
18. $\int_0^{\pi/2} \sin^2(x)\cos^3(x)\,dx$
19. $\int_0^{\pi} \sin^3(x)\cos^6(x)\,dx$
20. $\int_0^{\pi/2} \sqrt{\sin(x)}\cos^3(x)\,dx$
21. $\int_0^{\pi/2} 15\sin^2(x)\cos^5(x)\,dx$
22. $\int_0^{3\pi/2} 5\sin^2(x/3)\cos^7(x/3)\,dx$

23. Use a reduction formula to show that

$$\int \cos^4(u)\,du = \frac{1}{4}\cos^3(u)\sin(u) + \frac{3}{8}\cos(u)\sin(u) + \frac{3}{8}u + C.$$

24. Use a reduction formula to show that

$$\int \sin^4(u)\,du = -\frac{1}{4}\sin^3(u)\cos(u) - \frac{3}{8}\cos(u)\sin(u) + \frac{3}{8}u + C.$$

▶ In each of Exercises 25–29, use Exercise 23 or 24 to evaluate the given definite integral. ◀

25. $\int_0^{2\pi} \cos^4(x)\,dx$
26. $\int_0^{1} \cos^4(\pi x)\,dx$
27. $\int_{\pi/2}^{\pi} \sin^4(x)\,dx$
28. $\int_{\pi/2}^{\pi} \sin^4(x/2)\,dx$
29. $\int_0^{\pi/4} \cos^2(t)\sin^2(t)\,dt$

▶ In each of Exercises 30–33, use a reduction formula and Exercise 23 or 24 to evaluate the given definite integral. ◀

30. $\int_0^{\pi/4} 8\sin^6(x)\,dx$
31. $\int_0^{\pi} \cos^6(x/2)\,dx$
32. $\int_0^{\pi/2} \cos^4(t)\sin^2(t)\,dt$
33. $\int_{\pi/6}^{\pi/3} \cos^2(x)\sin^4(x)\,dx$

▶ In each of Exercises 34–37, use reduction formula (6.2.13), formula (6.2.14), or a substitution to evaluate the given indefinite integral. ◀

34. $\int 2\tan^3(x)\,dx$
35. $\int \tan^4(x)\,dx$
36. $\int 3\sec^4(x)\,dx$
37. $\int \sec^3(x)\tan(x)\,dx$

▶ In each of Exercises 38–41, use reduction formula (6.2.13) or formula (6.2.14) to evaluate the given indefinite integral. ◀

38. $\int_0^{\pi/4} 2\tan^3(x)\,dx$
39. $\int_0^{\pi/4} \sec^4(x)\,dx$
40. $\int_0^{\pi/4} 12\tan^4(x)\,dx$
41. $\int_0^{\pi/3} 10\sec^6(x)\,dx$

▶ In each of Exercises 42–45, use the identity $1+\tan^2(x) = \sec^2(x)$ to convert the given integral to one that involves only

$\tan(x)$ or only $\sec(x)$. Then use reduction formula (6.2.13) or formula (6.2.14) to evaluate the given indefinite integral.

42. $\int_0^{\pi/4} \sec^2(x)\tan^2(x)\,dx$

43. $\int_0^{\pi/4} \sec^4(x)\tan^2(x)\,dx$

44. $\int_0^{\pi/4} \sec^2(x)\tan^4(x)\,dx$

45. $\int_0^{\pi/3} 2\sec(x)\tan^2(x)\,dx$

In each of Exercises 46–56, evaluate the given integral by converting the integrand to an expression in sines and cosines.

46. $\int 3\tan(x)\sec^3(x)\,dx$

47. $\int 8\tan(x)\sec^4(x)\,dx$

48. $\int \tan^3(x/2)\,dx$

49. $\int 3\tan^3(x)\sec(x)\,dx$

50. $\int \tan(x)\sin(x)\,dx$

51. $\int \cot(x/3)\csc^3(x/3)\,dx$

52. $\int 4\cot(x/2)\csc^4(x/2)\,dx$

53. $\int \cot^3(x/3)\csc(x/3)\,dx$

54. $\int 2\sin^2(x)\tan(x)\,dx$

55. $\int 6\cos^3(x)\tan^2(x)\,dx$

56. $\int 5\sin^5(x/3)\cot^2(x/3)\,dx$

Further Theory and Practice

In each of Exercises 57–61, the denominator of the given integrand is of the form $a \pm b$. Multiply numerator and denominator by $a \mp b$ to obtain a difference of squares in the denominator. Then use an appropriate trigonometric identity before integrating.

57. $\displaystyle\int \frac{1}{1-\sin(x)}\,dx$

58. $\displaystyle\int \frac{1}{1+\cos(x)}\,dx$

59. $\displaystyle\int \frac{\cos(x)}{1-\cos(x)}\,dx$

60. $\displaystyle\int \frac{1}{\sec(x)-1}\,dx$

61. $\displaystyle\int \frac{\tan(x)}{\sec(x)+1}\,dx$

Identities (6.2.1) and (6.2.2) are the basis of an alternative method for calculating the integrals $\int \sin^{2k}(x)\,dx$ and $\int \cos^{2\ell}(x)\,dx$ for positive integers k and ℓ. The cited identities yield the equations

$$\sin^{2k}(x) = \left(\sin^2(x)\right)^k = \frac{1}{2^k}\left(1-\cos(2x)\right)^k$$

and

$$\cos^{2\ell}(x) = \left(\cos^2(x)\right)^\ell = \frac{1}{2^\ell}\left(1+\cos(2x)\right)^\ell.$$

After the binomials on the right sides have been expanded, the same manipulation is applied to all even powers of $\cos(2x)$. The process is continued until only odd powers of $\cos(2x)$, $\cos(4x)$, ... remain. The method of Example 5 is then used to integrate the odd powers of cosine. Use the procedure just outlined to calculate the integrals in Exercises 62–65.

62. $\int \sin^4(x)\,dx$

63. $\int \cos^6(x)\,dx$

64. $\int \sin^6(x)\,dx$

65. $\int \cos^2(x)\sin^4(x)\,dx$

When $\sin(x)$ and $\cos(x)$ are both raised to the same positive power in an integrand, the identity $\sin(2x) = 2\sin(x)\cos(x)$ may be used to simplify the integral. Use this observation as the basis for calculation of the integrals in Exercises 66–69.

66. $\int \cos(x)\sin(x)\,dx$

67. $\int \cos^2(x)\sin^2(x)\,dx$

68. $\int \cos^3(x)\sin^3(x)\,dx$

69. $\int \cos^4(x)\sin^4(x)\,dx$

70. Sketch the graphs of $y=\sin^2(x)$ and $y=\cos^2(x)$ for $0 \le x \le 2\pi$. Use the identity $\sin^2(x)+\cos^2(x)=1$ to give a geometric proof of the equations in line (6.2.9).

71. By making the substitution $u=\sin(x)$, show that

$$\int \sin(x)\cos(x)\,dx = \frac{1}{2}\sin^2(x) + C.$$

By making the substitution $u=\cos(x)$, show that

$$\int \sin(x)\cos(x)\,dx = -\frac{1}{2}\cos^2(x) + C.$$

By using the formula for $\sin(2x)$, show that

$$\int \sin(x)\cos(x)\,dx = -\frac{1}{4}\cos(2x) + C.$$

Explain why these answers are not contradictory.

72. Calculate $\int_0^{\pi/2} 4x\cos^2(x)\,dx$.

73. Calculate $\int_{-\pi/2}^{\pi/2} \sqrt{2+2\sin(x)}\cos^3(x)\,dx$.

74. Find the area of the region bounded by $y=\sin(x)$ and $y=\sin^2(x)$ for $0 \le x \le \pi/2$.

75. Find the area of the region bounded by $y=\sin^2(x)$ and $y=\sin^3\sin^3(x)$ for $0 \le x \le \pi/2$.

76. Derive reduction formula (6.2.12).

77. Derive reduction formula (6.2.13).

78. Derive reduction formula (6.2.14).

79. Let m and n be integers such that $m \ne -n$ Prove the reduction formula

$$\int \sin^n(x)\cos^m(x)\,dx = -\frac{\sin^{n-1}(x)\cos^{m+1}(x)}{m+n} + \frac{n-1}{m+n}\int \sin^{n-2}(x)\cos^m(x)\,dx.$$

80. Suppose that j and k are nonnegative integers, not both zero. Derive the reduction formula

$$\int \tan^{2k}(x)\sec^{2j+1}(x)\,dx = \frac{1}{2k+2j}\tan^{2k+1}(x)\sec^{2j-1}(x) + \frac{2j-1}{2k+2j}\int \tan^{2k}(x)\sec^{2j-1}(x)\,dx.$$

81. Suppose that A and B are constants, not both of which are 0. Notice that

$$A\cos(x) + B\sin(x) = \sqrt{A^2+B^2}\,(\cos(x)\cdot\xi + \sin(x)\cdot\eta)$$

where $\xi = A/\sqrt{A^2+B^2}$ and $\eta = B/\sqrt{A^2+B^2}$. By observing that (ξ, η) is a point on the unit circle, deduce that $A\cos(x) + B\sin(x) = \sqrt{A^2+B^2}\cos(x+\phi)$ for some ϕ. Evaluate

$$\int \frac{1}{(A\cos(x)+B\sin(x))^k}\,dx$$

for $k = 1$ and $k = 2$.

82. For each natural number n, let

$$W_n = \int_0^{\pi/2} \sin^n(x)\,dx.$$

a. Use formula (6.2.11) to show that

$$W_n = \frac{n-1}{n}W_{n-2} \quad (n \geq 2).$$

b. By calculating W_0 and using part a of this exercise, show that

$$W_{2N} = \frac{1\cdot 3\cdot 5\cdots(2N-1)}{2\cdot 4\cdot 6\cdots(2N)}\cdot\frac{\pi}{2}.$$

c. By calculating W_1 and using part a of this exercise, show that

$$W_{2N+1} = \frac{2\cdot 4\cdot 6\cdots(2N)}{3\cdot 5\cdot 7\cdots(2N+1)}.$$

d. By comparing integrands, deduce that $W_{2N+1} \leq W_{2N} \leq W_{2N-1}$.

e. Use parts b, c, and d to show that

$$\frac{2^2}{1^2}\cdot\frac{4^2}{3^2}\cdot\frac{6^2}{5^2}\cdots\frac{(2N)^2}{(2N-1)^2}\cdot\frac{1}{(2N+1)} \leq \frac{\pi}{2}$$

and

$$\frac{\pi}{2} \leq \frac{2^2}{1^2}\cdot\frac{4^2}{3^2}\cdot\frac{6^2}{5^2}\cdots\frac{(2N)^2}{(2N-1)^2}\cdot\frac{1}{2N}.$$

f. Prove that there is a number θ_N in $[0,1]$ such that

$$\frac{\pi}{2} = \frac{2^2}{1^2}\cdot\frac{4^2}{3^2}\cdot\frac{6^2}{5^2}\cdots\frac{(2N)^2}{(2N-1)^2}\cdot\frac{1}{2N+\theta_N}.$$

g. Deduce the following formula for π that John Wallis published in 1655:

$$\pi = \lim_{N\to\infty}\left(\frac{2^2}{1^2}\cdot\frac{4^2}{3^2}\cdot\frac{6^2}{5^2}\cdots\frac{(2N)^2}{(2N-1)^2}\cdot\frac{1}{N}\right).$$

Note: Wallis's limit formula is beautiful, but because the convergence is so slow, it is useless for numerically approximating π.

Calculator/Computer Exercises

83. Let a be the least positive abscissa of a point of intersection of the curves $y = \cos^3(x)$ and $y = \sin^2(x)$. Find the area of the region between the two curves for $-a \leq x \leq a$.

84. The curves $y = 10\sin^3(x) + 2$ and $y = \sec^3(x)$ have a point of intersection with abscissa a in $[-0.5, 0]$ and a point of intersection with abscissa b in $[1, 1.5]$. Find the area of the region between the two curves for $a \leq x \leq b$.

85. The curves $y = \sin^2(x)\cos^3(x)$ and $y = \sin^5(x)\cos^2(x)$ have a point of intersection with abscissa b in $[0.5, 1]$. Find the area of the region between the two curves for $0 \leq x \leq b$.

86. The line $y = 0.6x$ intersects the curve $y = \sin^4(x)$ at the origin and at two points with positive coordinates. Find the total area of the regions between the two curves in the first quadrant.

87. Let x_0 be the least positive value of x such that the tangent line to $y = \sin^4(x)$ at $(x_0, \sin^4(x_0))$ passes through the origin. Find the area enclosed by this tangent line and $y = \sin^4(x)$ for $0 \leq x \leq x_0$.

6.3 Trigonometric Substitution

In this section, you will learn to treat integrands that involve expressions of the form

$$Ax^2 + Bx + C, \quad A \neq 0.$$

We have already encountered a few isolated instances of integrals that contain quadratic expressions. For example, if $A = -1$ or 1, $B = 0$, and $C = a^2 > 0$, then $Ax^2 + Bx + C$ becomes $a^2 - x^2$ or $a^2 + x^2$. Formulas (3.10.5) and (3.10.8) of Section 3.10 in Chapter 3 tell us that

$$\int \frac{dx}{\sqrt{a^2 - x^2}} = \arcsin\left(\frac{x}{a}\right) + C \tag{6.3.1}$$

and

$$\int \frac{dx}{a^2 + x^2} = \frac{1}{a}\arctan\left(\frac{x}{a}\right) + C. \tag{6.3.2}$$

As these two examples show, inverse trigonometric functions can appear in the antiderivatives of functions that involve the expression $Ax^2 + Bx + C$.

Trigonometric Substitution

Suppose that a is a positive constant. Consider the three quadratic polynomials (a) $a^2 - x^2$ where $-a \le x \le a$, (b) $a^2 + x^2$, and (c) $x^2 - a^2$ where $a \le |x|$. By making an appropriate substitution in each of these expressions, we may exploit a trigonometric identity to reduce these sums to a single square:

a. The expression $a^2 - x^2$ is simplified by substituting $x = a\sin(\theta)$ with $-\pi/2 \le x \le \pi/2$. The simplification is realized as follows:

$$a^2 - x^2 = a^2 - a^2\sin^2(\theta) = a^2\left(1 - \sin^2(\theta)\right) = a^2\cos^2(\theta).$$

Notice that the range of θ implies that $0 \le \cos(\theta)$. Therefore $\sqrt{a^2 - x^2} = \sqrt{a^2\cos^2(\theta)} = |a\cos(\theta)| = a\cos(\theta)$.

b. The expression $a^2 + x^2$ is simplified by substituting $x = a\ \tan(\theta)$ with $-\pi/2 < x < \pi/2$. The simplification is realized as follows:

$$a^2 + x^2 = a^2 + a^2\tan^2(\theta) = a^2\left(1 + \tan^2(\theta)\right) = a^2\sec^2(\theta)$$

Notice that the range of θ implies that $0 < \sec(\theta)$. Therefore $\sqrt{a^2 + x^2} = \sqrt{a^2\sec^2(\theta)} = |a\sec(\theta)| = a\sec(\theta)$.

c. The expression $x^2 - a^2$ is simplified by substituting $x = a\sec(\theta)$. The simplification is realized as follows:

$$x^2 - a^2 = a^2\sec^2(\theta) - a^2 = a^2(\sec^2(\theta) - 1) = a^2\tan^2(\theta).$$

If $a \le |x|$, then we take θ in the interval $[0, \pi/2)$. For these values of θ, we have $\sqrt{x^2 - a^2} = \sqrt{a^2\tan^2(\theta)} = |a\tan(\theta)| = a\tan(\theta)$.

These calculations suggest a new substitution technique. To integrate a function that contains the expression $a^2 - x^2$, we consider the change of variable $x = a\sin(\theta)$, $dx = a\cos(\theta)\,d\theta$. Similarly, the expression $a^2 + x^2$ in an integral suggests the change of variable $x = a\tan(\theta)$, $dx = a\sec^2(\theta)\,d\theta$. Finally, the expression $x^2 - a^2$ points to the substitution $x = a\sec(\theta)$, $dx = a\sec(\theta)\tan(\theta)\,d\theta$. These substitutions are examples of a type of change of variable that is known as the *method of inverse (or indirect) substitution*. Here is a summary of the trigonometric substitutions that we will use:

Expression	**Substitution**	**Result of Substitution**	**After Simplification**
$a^2 - x^2$	$x = a\sin(\theta)$, $dx = a\cos(\theta)d\theta$	$a^2\left(1 - \sin^2(\theta)\right)$	$a^2\cos^2(\theta)$
$a^2 + x^2$	$x = a\tan(\theta)$, $dx = a\sec^2(\theta)d\theta$	$a^2\left(1 + \tan^2(\theta)\right)$	$a^2\sec^2(\theta)$
$x^2 - a^2$	$x = a\sec(\theta)$, $dx = a\sec(\theta)\tan(\theta)d\theta$	$a^2\left(\sec^2(\theta) - 1\right)$	$a^2\tan^2(\theta)$

Table 1

If you already know the fundamental trigonometric identities that are the basis of these three substitutions, then Table 1 requires no memorization. Instead, work enough practice problems so that you recognize when a particular trigonometric substitution is called for.

▶ **EXAMPLE 1** Calculate

$$\int \frac{x^2}{\sqrt{1-x^2}}\,dx.$$

Solution The expression $1-x^2$ suggests the substitution $x = 1\cdot\sin(\theta)$, $dx = \cos(\theta)\,d\theta$ for $-\pi/2 < \theta < \pi/2$. With this change of variable, the given integral becomes

$$\int \frac{\sin^2(\theta)}{\sqrt{1-\sin^2(\theta)}}\cos(\theta)\,d\theta = \int \frac{\sin^2(\theta)}{\sqrt{\cos^2(\theta)}}\cos(\theta)\,d\theta = \int \frac{\sin^2(\theta)}{\cos(\theta)}\cos(\theta)\,d\theta = \int \sin^2(\theta)\,d\theta = \frac{1}{2}\left(\theta - \sin(\theta)\cos(\theta)\right) + C.$$

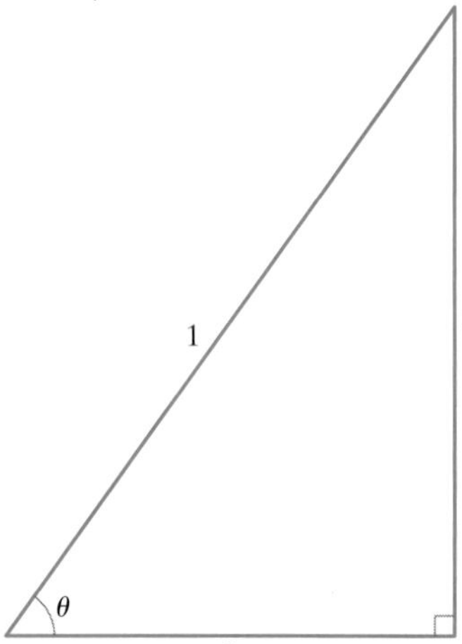

▲ Figure 1 $\frac{x}{1} = \sin(\theta) \Rightarrow \cos(\theta) = \frac{b}{1} = \sqrt{1-x^2}$

We finish the calculation by resubstituting. Because $x/1 = \sin(\theta)$, we draw a right triangle with angle θ, opposite side x, and hypotenuse 1. See Figure 1. By Pythagoras, the adjacent side b equals $\sqrt{1-x^2}$. It follows that $\cos(\theta) = \sqrt{1-x^2}/1$. We therefore resubstitute $\theta = \arcsin(x)$, $\sin(\theta) = x$, and $\cos(\theta) = \sqrt{1-x^2}$ to obtain

$$\int \frac{x^2}{\sqrt{1-x^2}}\,dx = \frac{1}{2}\left(\arcsin(x) - x\sqrt{1-x^2}\right) + C. \quad ◀$$

▶ **EXAMPLE 2** Calculate

$$\int_1^{\sqrt{3}} \sqrt{1+x^2}\,dx.$$

Solution The expression $1+x^2$ suggests the substitution $x = 1\cdot\tan(\theta)$, $dx = \sec^2(\theta)d\theta$. To obtain the limits of integration for θ, we note that $\tan(\theta) = 1$ when $\theta = \pi/4$ and $\tan(\theta) = \sqrt{3}$ when $\theta = \pi/3$. The given integral becomes

$$\int_1^{\sqrt{3}} \sqrt{1+x^2}\,dx = \int_{\pi/4}^{\pi/3} \sqrt{1+\tan^2(\theta)}\sec^2(\theta)\,d\theta = \int_{\pi/4}^{\pi/3} \sqrt{\sec^2(\theta)}\sec^2(\theta)\,d\theta = \int_{\pi/4}^{\pi/3} \sec^3(\theta)\,d\theta.$$

We can now finish the calculation by appealing to formula (6.1.8):

$$\int_1^{\sqrt{3}} \sqrt{1+x^2}\,dx = \frac{1}{2}\left(\sec(\theta)\tan(\theta) + \ln(|\sec(\theta)+\tan(\theta)|)\right)\Big|_{\theta=\pi/4}^{\theta=\pi/3} = \frac{1}{2}\left(2\sqrt{3}-\sqrt{2}\right) + \frac{1}{2}\ln\left(\frac{2+\sqrt{3}}{1+\sqrt{2}}\right). \quad ◀$$

▶ **EXAMPLE 3** Calculate

$$\int_{2\sqrt{2}}^{4} \frac{1}{x\sqrt{x^2-4}}\,dx.$$

Solution The expression x^2-4, when written as x^2-2^2, suggests the substitution $x = 2\sec(\theta)$, $dx = 2\cdot\sec(\theta)\tan(\theta)d\theta$. The limit of integration $x = 2\sqrt{2}$, corresponds to $\sec(\theta) = \sqrt{2}$ or $\theta = \pi/4$. The limit of integration $x = 4$ corresponds to $\sec(\theta) = 2$, or $\theta = \pi/3$. Therefore

$$\int_{2\sqrt{2}}^{4} \frac{1}{x\sqrt{x^2-4}}\,dx = \int_{\pi/4}^{\pi/3} \frac{2\sec(\theta)\tan(\theta)}{2\sec(\theta)\sqrt{4\sec^2(\theta)-4}}\,d\theta = \int_{\pi/4}^{\pi/3} \frac{\tan(\theta)}{2\sqrt{\sec^2(\theta)-1}}\,d\theta$$

$$= \int_{\pi/4}^{\pi/3} \frac{\tan(\theta)}{2\sqrt{\tan^2(\theta)}}\,d\theta = \frac{1}{2}\int_{\pi/4}^{\pi/3} 1\,d\theta$$

$$= \frac{1}{2}\theta\Big|_{\pi/4}^{\pi/3} = \frac{1}{2}\left(\frac{\pi}{3}-\frac{\pi}{4}\right) = \frac{\pi}{24}.$$ ◀

▶ **EXAMPLE 4** Calculate the integrals

$$\int \frac{1}{\sqrt{9+x^2}}\,dx \qquad \text{and} \qquad \int \frac{x}{\sqrt{9+x^2}}\,dx.$$

Solution Both integrands involve the expression $9+x^2$, which we treat as 3^2+x^2. Therefore the substitution $x=3\tan(\theta)$, $dx=3\sec^2(\theta)d\theta$ is indicated. With this substitution, the first integral becomes

$$\int \frac{1}{\sqrt{9+9\tan^2(\theta)}}\,3\sec^2(\theta)\,d\theta = \int \frac{\sec^2(\theta)}{\sqrt{1+\tan^2(\theta)}}\,d\theta = \int \frac{\sec^2(\theta)}{\sqrt{\sec^2(\theta)}}\,d\theta$$

$$= \int \frac{\sec^2(\theta)}{|\sec(\theta)|}\,d\theta = \int \sec(\theta)\,d\theta = \ln\big(|\sec(\theta)+\tan(\theta)|\big) + C.$$

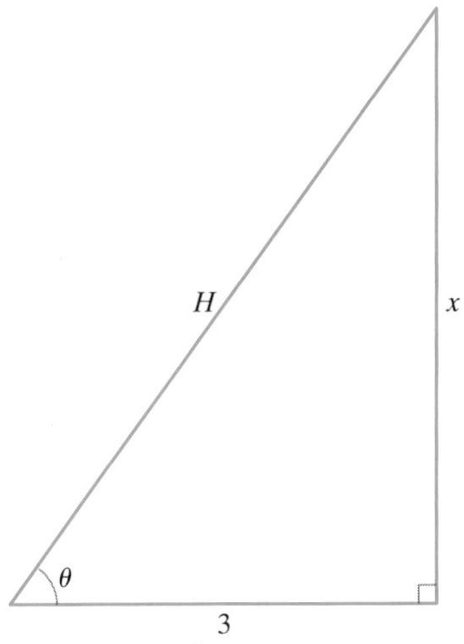

▲ Figure 2 $\frac{x}{3} = \tan(\theta) \Rightarrow \sec(\theta) = \frac{H}{3} = \frac{\sqrt{3^2+x^2}}{3}$

We complete the integration by resubstituting. Because $x/3=\tan(\theta)$, we draw a right triangle with angle θ, opposite side x, adjacent side 3, and hypotenuse H (Figure 2). By Pythagoras, H equals $\sqrt{9+x^2}$. It follows that $\sec(\theta) = \sqrt{9+x^2}/3$. Therefore

$$\int \frac{1}{\sqrt{9+x^2}}\,dx = \ln\left(\left|\frac{\sqrt{9+x^2}}{3} + \frac{x}{3}\right|\right) + C = \ln\left(\left|\frac{x+\sqrt{9+x^2}}{3}\right|\right) + C.$$

Using the logarithm law $\ln(u/v) = \ln(u) - \ln(v)$, we can simplify this a bit with

$$\int \frac{1}{\sqrt{9+x^2}}\,dx = \ln\left(\left|x+\sqrt{9+x^2}\right|\right) - \ln(3) + C = \ln\left(\left|x+\sqrt{9+x^2}\right|\right) + \tilde{C},$$

where we have written the (arbitrary) constant $C-\ln(3)$ as $\tilde{C}$. Finally, we observe that $\sqrt{9+x^2} > \sqrt{x^2} = |x|$. Therefore $x+\sqrt{9+x^2} > 0$, and we may omit the unnecessary absolute values from our answer. Writing the arbitrary constant as C again, we have

$$\int \frac{1}{\sqrt{9+x^2}}\,dx = \ln\left(x+\sqrt{9+x^2}\right) + C.$$

It is possible to calculate the second integral, $\int (x/\sqrt{9+x^2})dx$, by making the same trigonometric substitution, $x=3\tan(\theta)$. However, we notice that a constant multiple of the derivative $2x$ of $9+x^2$ is present in the integral. The direct substitution $u=9+x^2$, $du=2x\,dx$ is therefore more efficient. With this substitution, the integral $\int \frac{x}{\sqrt{9+x^2}}\,dx$ becomes $\frac{1}{2}\int u^{-1/2}\,du$, or $\sqrt{u}+C$. Therefore

$$\int \frac{x}{\sqrt{9+x^2}}\,dx = \sqrt{9+x^2} + C.$$ ◀

◀ **A LOOK BACK** In the method of substitution (Section 5.6 of Chapter 5), we select an expression that is already present in a given integral and replace that expression with a new variable. For example, we treat the integral $\int \frac{x}{\sqrt{9+x^2}}\,dx$ of Example 4 by making the (direct) substitution $u = 9 + x^2$ for an expression that is already in the integral. In the method of inverse (or indirect) substitution, we replace the original variable of integration with a function that is nowhere to be found in the original integral. In the integral $\int \frac{1}{\sqrt{9+x^2}}\,dx$ of Example 4, we make the indirect substitution $x = 3\tan(\theta)$. In this substitution, the term $\tan(\theta)$ does not have any apparent connection to the integrand. In theory, the two methods of substitution are distinct. However, in practice, the two methods are carried out in exactly the same manner.

General Quadratic Expressions That Appear Under a Radical

Because a general quadratic expression $Ax^2 + Bx + C$ can be converted (by completing the square) to a sum or difference of squares, the trigonometric substitutions that we have just learned can be applied to integrals that involve these general quadratic expressions. Let us see how this works in practice.

▶ **EXAMPLE 5** Calculate the integral

$$\int \frac{dt}{\sqrt{t^2 - 8t + 25}}.$$

Solution The first step is to complete the square under the radical sign:

$$t^2 - 8t + 25 = \left(t^2 - 8t + (8/2)^2\right) + 25 - (8/2)^2 = (t-4)^2 + 9.$$

To evaluate the integral

$$\int \frac{dt}{\sqrt{(t-4)^2 + 9}},$$

we first make the direct substitution $x = t - 4$, $dx = dt$. This results in the integral

$$\int \frac{dx}{\sqrt{x^2 + 9}}.$$

Here the substitution $x = 3\tan(\theta)$, $dx = 3\sec^2(\theta)\,d\theta$ is called for. With that substitution, we calculated this integral to be $\ln(x + \sqrt{9+x^2}) + C$ in Example 3. Resubstituting $x = t - 4$, we obtain

$$\int \frac{dt}{\sqrt{t^2 - 8t + 25}} = \ln\left(t - 4 + \sqrt{(t-4)^2 + 9}\right) + C = \ln\left(t - 4 + \sqrt{t^2 - 8t + 25}\right) + C. \quad ◀$$

▶ **EXAMPLE 6** Calculate $\int \sqrt{21 + 8x - 4x^2}\,dx$.

Solution First we complete the square:

$$21 + 8x - 4x^2 = 21 - 4(x^2 - 2x) = 21 + 4 - 4(x^2 - 2x + 1) = 25 - 4(x-1)^2 = 4\left(\left(\frac{5}{2}\right)^2 - (x-1)^2\right).$$

Therefore

$$\int \sqrt{21+8x-4x^2}\,dx = 2\int \sqrt{\left(\frac{5}{2}\right)^2 - (x-1)^2}\,dx.$$

Next, we make the substitution $x-1=(5/2)\sin(\theta)$, $dx=(5/2)\cos(\theta)\,d\theta$. The last integral becomes

$$2\int \sqrt{\left(\frac{5}{2}\right)^2 - \left(\frac{5}{2}\right)^2 \sin^2(\theta)}\cdot \frac{5}{2}\cos(\theta)\,d\theta = 2\cdot\frac{5}{2}\cdot\frac{5}{2}\int \sqrt{1-\sin^2(\theta)}\cos(\theta)\,d\theta = \frac{25}{2}\int \cos^2(\theta)\,d\theta.$$

The last of these integrals evaluates to

$$\frac{25}{4}\left(\theta + \sin(\theta)\cos(\theta)\right) + C \tag{6.3.3}$$

by equation (6.2.6). The value of the original integral is obtained by resubstituting for θ in expression (6.3.3). The change of variable $x-1=(5/2)\sin(\theta)$ gives us

$$\sin(\theta) = \frac{x-1}{5/2} \quad \text{and} \quad \theta = \arcsin\left(\frac{x-1}{5/2}\right). \tag{6.3.4}$$

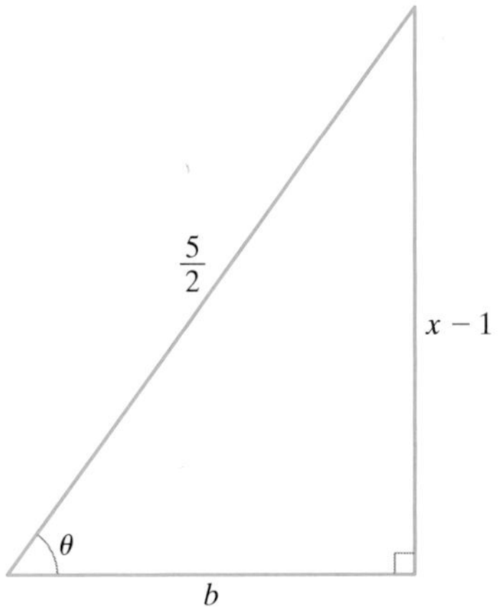

▲ Figure 3
$(x-1)/(5/2)=\sin(\theta) \Rightarrow \cos(\theta) = b/(5/2) = \frac{\sqrt{(5/2)^2-(x-1)^2}}{(5/2)}$

To resubstitute for $\cos(\theta)$ in expression (6.3.3), we refer to the triangle associated with the change of variable, which is shown in Figure 3. By calculating the length of the adjacent side, we find that

$$\cos(\theta) = \frac{\sqrt{(5/2)^2-(x-1)^2}}{5/2}. \tag{6.3.5}$$

Using lines (6.3.4) and (6.3.5) to substitute for θ, $\sin(\theta)$, and $\cos(\theta)$ in expression (6.3.3), we have

$$\int \sqrt{21+8x-4x^2}\,dx = \frac{25}{4}\left(\arcsin\left(\frac{x-1}{5/2}\right) + \frac{(x-1)}{(5/2)}\cdot\frac{\sqrt{(5/2)^2-(x-1)^2}}{(5/2)}\right) + C.$$

After a little bit of algebraic simplification, this equation becomes

$$\int \sqrt{21+8x-4x^2}\,dx = \frac{25}{4}\arcsin\left(\frac{2}{5}(x-1)\right) + \frac{(x-1)}{2}\sqrt{21+8x-4x^2} + C. \quad ◀$$

Quadratic Expressions Not Under a Radical Sign

Suppose that $P(x)$ is a polynomial. To calculate

$$\int \frac{P(x)}{Ax^2+Bx+C}\,dx,$$

it is best not to attempt trigonometric substitutions immediately. Instead, use the following three steps:

1. If the degree of P is less than or equal to 1, then proceed directly to Step 2. Otherwise, *divide* the denominator into the numerator. This division will result in

$$Q(x) + \frac{R(x)}{Ax^2+Bx+C}$$

where $Q(x)$ is a polynomial and $R(x)$ has degree less than or equal to 1. The polynomial $Q(x)$ is easy to integrate. Continue by applying Step 2 to the new fraction $R(x)/(Ax^2 + Bx + C)$.

2. If the degree of $P(x)$ is 0, or if you have divided in Step 1 and the degree of $R(x)$ is 0, then proceed to Step 3. If $P(x)$ or $R(x)$ has degree 1, then separate the numerator into a multiple of $(2Ax+B)$ plus a constant K. Integrate the expression
$$\frac{2Ax + B}{Ax^2 + Bx + C}$$
by making the substitution $u = Ax^2+Bx+C$, $du = (2Ax+B)\,dx$.
3. Integrate the expression
$$\frac{K}{Ax^2 + Bx + C}$$
by completing the square in the denominator.

The only way to understand these steps is to see them used in practice:

▶ **EXAMPLE 7** Integrate
$$\int \frac{2x^3 - 8x^2 + 20x - 5}{x^2 - 4x + 8}\,dx.$$

Solution The degree of the numerator is not of 0 or 1, so we start with Step 1. First we divide:

$$\begin{array}{r} 2x \\ x^2 - 4x + 8 \,\overline{)\,2x^3 - 8x^2 + 20x - 5} \\ -(2x^3 - 8x^2 + 16x) \\ \hline 4x - 5 \end{array}.$$

After division, we obtain
$$\frac{2x^3 - 8x^2 + 20x - 5}{x^2 - 4x + 8} = 2x + \frac{4x - 5}{x^2 - 4x + 8}.$$

Therefore
$$\int \frac{2x^3 - 8x^2 + 20x - 5}{x^2 - 4x + 8}\,dx = x^2 + \int \frac{4x - 5}{x^2 - 4x + 8}\,dx.$$

Note that the derivative of the denominator in the last integrand is $2x - 4$. Therefore we rewrite the integral on the right side as
$$\int \frac{2(2x - 4) + 3}{x^2 - 4x + 8}\,dx = 2 \cdot \int \frac{(2x - 4)}{x^2 - 4x + 8}\,dx + \int \frac{3}{x^2 - 4x + 8}\,dx. \tag{6.3.6}$$

Now we can see the motivation for this step more clearly. With the substitution $u = x^2 - 4x + 8$, $du = 2x - 4\,dx$, the first integral in (6.3.6) is
$$2\int \frac{du}{u} = 2 \cdot \ln(|u|) + C = 2\ln(x^2 - 4x + 8) + C. \tag{6.3.7}$$

In the last logarithm, we discard the redundant absolute values because we observe by completing the square that $x^2 - 4x + 8 = (x - 2)^2 + 4$ is positive.

According to Step 3, we write the second integral on the right side of (6.3.6) as

$$\int \frac{3}{x^2 - 4x + 8}\, dx = \int \frac{3}{(x-2)^2 + 4}\, dx.$$

We next make the direct substitution $u = x - 2$, $du = dx$, obtaining

$$\int \frac{3}{u^2 + 4}\, du = \frac{3}{2}\arctan\left(\frac{u}{2}\right) + C.$$

We resubstitute for x to obtain

$$\int \frac{3}{x^2 - 4x + 8}\, dx = \frac{3}{2}\arctan\left(\frac{x-2}{2}\right) + C. \qquad (6.3.8)$$

Substituting the information from lines (6.3.7) and (6.3.8) into equation (6.3.6), we have

$$\int \frac{4x - 5}{x^2 - 4x + 8}\, dx = 2\ln(x^2 - 4x + 8) + \frac{3}{2}\arctan\left(\frac{x-2}{2}\right) + C.$$

Therefore

$$\int \frac{2x^3 - 8x^2 + 20x - 5}{x^2 - 4x + 8}\, dx = x^2 + 2\ln(x^2 - 4x + 8) + \frac{3}{2}\arctan\left(\frac{x-2}{2}\right) + C. \quad ◀$$

QUICK QUIZ

1. What indirect substitution is appropriate for $\int \frac{1}{\sqrt{1+x^2}}\, dx$?
2. What indirect substitution is appropriate for $\int \frac{1}{\sqrt{1-4x^2}}\, dx$?
3. What indirect substitution is appropriate for $\int \frac{1}{\sqrt{x^2-2}}\, dx$?
4. What indirect substitution is appropriate for $\int \frac{1}{\sqrt{10+2x+x^2}}\, dx$?

Answers

1. $x = \tan(\theta)$ 2. $x = \frac{1}{2}\sin(\theta)$ 3. $x = \sqrt{2}\sec(\theta)$ 4. $x + 1 = 3\tan(\theta)$

EXERCISES

Problems for Practice

▶ In each of Exercises 1–10, evaluate the given integral by making a trigonometric substitution (even if you spot another way to evaluate the integral). ◀

1. $\int \frac{x}{\sqrt{9-x^2}}\, dx$

2. $\int \frac{1}{\sqrt{9-x^2}}\, dx$

3. $\int \frac{4}{\sqrt{4-x^2}}\, dx$

4. $\int \frac{2}{\sqrt{1-4x^2}}\, dx$

5. $\int \frac{x^2}{\sqrt{1-x^2}}\, dx$

6. $\int \sqrt{25-x^2}\, dx$

7. $\int \frac{x^2}{(1-x^2)^{3/2}}\, dx$

8. $\int (4-x^2)^{3/2}\, dx$

9. $\int \sqrt{1-4x^2}\, dx$

10. $\int (1-9x^2)^{3/2}\, dx$

▶ In each of Exercises 11–20, evaluate the given integral by making a trigonometric substitution (even if you spot another way to evaluate the integral). ◀

11. $\int \frac{4x}{x^2+1}\,dx$
12. $\int \frac{12x}{x^2+9}\,dx$
13. $\int \frac{6}{9x^2+1}\,dx$
14. $\int \frac{2}{x^2+4}\,dx$
15. $\int 4\sqrt{1+4x^2}\,dx$
16. $\int \frac{1}{\sqrt{1+2x^2}}\,dx$
17. $\int \frac{1}{(1+x^2)^{3/2}}\,dx$
18. $\int \frac{1}{(5+x^2)^{3/2}}\,dx$
19. $\int \frac{2x^2}{\sqrt{1+x^2}}\,dx$
20. $\int \frac{1}{x\sqrt{1+x^2}}\,dx$

In each of Exercises **21–30**, evaluate the given integral by making a trigonometric substitution (even if you spot another way to evaluate the integral).

21. $\int_{2\sqrt{2}}^{4} \frac{4}{x\sqrt{x^2-4}}\,dx$
22. $\int_{\sqrt{2}}^{2} \frac{1}{\sqrt{x^2-1}}\,dx$
23. $\int_{1/\sqrt{2}}^{\sqrt{5}/2} \frac{2}{\sqrt{4x^2-1}}\,dx$
24. $\int_{\sqrt{5}}^{\sqrt{13}} \frac{x}{\sqrt{x^2-4}}\,dx$
25. $\int_{\sqrt{2}}^{\sqrt{5}} \frac{2x^2}{\sqrt{x^2-1}}\,dx$
26. $\int_{\sqrt{2}}^{\sqrt{10}} \frac{x^3}{\sqrt{x^2-1}}\,dx$
27. $\int_{\sqrt{5}}^{\sqrt{10}} \frac{6}{(x^2-1)^{3/2}}\,dx$
28. $\int_{\sqrt{3}}^{\sqrt{6}} \frac{2}{(x^2-2)^{3/2}}\,dx$
29. $\int_1^{\sqrt{2}} \sqrt{x^2-1}\,dx$
30. $\int_1^5 \sqrt{2x^2-1}\,dx$

In each of Exercises **31–38**, evaluate the given integral.

31. $\int \frac{2x+3}{\sqrt{1-x^2}}\,dx$
32. $\int \frac{x^2+x}{\sqrt{1-x^2/4}}\,dx$
33. $\int \frac{2x+3}{x^2+1}\,dx$
34. $\int \frac{x^2+2}{x^2+1}\,dx$
35. $\int \frac{2x^2+1}{\sqrt{x^2+1}}\,dx$
36. $\int \frac{2x^2+1}{\sqrt{x^2+4}}\,dx$
37. $\int \frac{x^2+1}{(9-x^2)^{3/2}}\,dx$
38. $\int \frac{8x^2+4x+3}{(1-4x^2)^{3/2}}\,dx$

Further Theory and Practice

Each of the integrands in Exercises **39–46** involves an expression of the form $a^2 - b^2x^2$, $a^2 + b^2x^2$, or $b^2x^2 - a^2$. Use an indirect substitution of the form $x = (a/b)\sin(\theta)$, $x = (a/b)\tan(\theta)$, or $x = (a/b)\sec(\theta)$ to calculate the given integral.

39. $\int_0^{1/\sqrt{2}} \frac{1}{16x^2+8}\,dx$
40. $\int_0^1 \sqrt{25-16x^2}\,dx$
41. $\int_0^1 \frac{3}{\sqrt{16+9x^2}}\,dx$
42. $\int_{4/3}^{\sqrt{2}} \sqrt{9x^2-16}\,dx$
43. $\int_0^2 \frac{6}{(9+4x^2)^{3/2}}\,dx$
44. $\int_{4/5}^1 \frac{(25x^2-16)^{1/2}}{x}\,dx$
45. $\int_1^2 \frac{10}{\sqrt{25x^2-16}}\,dx$
46. $\int_0^1 \frac{16x^2}{\sqrt{9-4x^2}}\,dx$

In each of Exercises **47–50**, calculate the given integral by first integrating by parts and then making a trigonometric substitution.

47. $\int_{\sqrt{2}}^{2} \operatorname{arcsec}(x)\,dx$
48. $\int_0^{1/\sqrt{2}} x\arcsin(x)\,dx$
49. $\int_0^{1/2} x^2\arcsin(x)\,dx$
50. $\int_{\sqrt{2}}^{2} x^2\operatorname{arcsec}(x)\,dx$

In Exercises **51–58**, calculate the given integral.

51. $\int \frac{2}{(x^2+1)^2}\,dx$
52. $\int \frac{x^2-1}{(x^2+1)^2}\,dx$
53. $\int \frac{\sqrt{1+x^2}}{x}\,dx$
54. $\int \frac{2x^2}{\sqrt{x^2-1}}\,dx$

55. $\int \frac{\sqrt{1+x^2}}{x^2}\,dx$

56. $\int \frac{2x}{\sqrt{x^4+1}}\,dx$

57. $\int \frac{x^2}{(1+x^2)^{3/2}}\,dx$

58. $\int 8x^2\sqrt{1-x^2}\,dx$

In Exercises 59–74, calculate the given integral.

59. $\int \frac{1}{\sqrt{2x-x^2}}\,dx$

60. $\int 2\sqrt{x^2+2x+2}\,dx$

61. $\int \frac{x}{\sqrt{x^2+2x+2}}\,dx$

62. $\int 8\sqrt{x^2-x-1}\,dx$

63. $\int \frac{4}{(x^2-2x+2)^2}\,dx$

64. $\int \sqrt{x^2-2x-3}\,dx$

65. $\int \frac{6x}{x^2-14x+58}\,dx$

66. $\int \frac{2x^2}{x^2+2x+5}\,dx$

67. $\int \frac{x^2-6x+8}{x^2+4}\,dx$

68. $\int \frac{2x+4}{x^2+4x+13}\,dx$

69. $\int \frac{12x}{2x^2+6x+5}\,dx$

70. $\int \frac{5x^2+20x-3}{x^2+4x+13}\,dx$

71. $\int \frac{8x}{(x^2+2x+2)^3}\,dx$

72. $\int \frac{x}{\sqrt{x^2+4x-5}}\,dx$

73. $\int \frac{x+1}{\sqrt{10x-x^2}}\,dx$

74. $\int \frac{x}{\sqrt{16x-x^2-48}}\,dx$

In each of Exercises 75–78, divide before integrating.

75. $\int \frac{1+x^2}{4+x^2}\,dx$

76. $\int \frac{2x^3+x^2}{9+x^2}\,dx$

77. $\int \frac{1+x^2}{5+4x+x^2}\,dx$

78. $\int \frac{x^3+1}{5-2x+x^2}\,dx$

In each of Exercises 79–84, make a (nontrigonometric) indirect substitution $x=\phi(u)$ to evaluate the given integral.

79. $\int \frac{1}{1+\sqrt{x}}\,dx$

80. $\int \sqrt{1+\sqrt{x}}\,dx$

81. $\int \frac{1-\sqrt{x}}{1+\sqrt{x}}\,dx$

82. $\int \frac{x^{1/4}+1}{x^{1/2}+1}\,dx$

83. $\int \frac{\sqrt{1+x}}{x}\,dx$

84. $\int \frac{1}{1+x^{1/3}}\,dx$

85. Suppose that a is a positive constant. Show that the substitution $x=a\sin(\theta)$ leads to the formula

$$\int \frac{1}{a^2-x^2}\,dx = \frac{1}{a}\ln\left(\frac{x+a}{\sqrt{a^2-x^2}}\right)+C.$$

Show that the substitution $x=a\sec(\theta)$ leads to the formula

$$\int \frac{1}{a^2-x^2}\,dx = \frac{1}{a}\ln\left(\frac{x+a}{\sqrt{x^2-a^2}}\right)+C.$$

Use these formulas to evaluate $\alpha=\int_0^{1/2}\frac{1}{1-x^2}\,dx$ and $\beta=\int_{\sqrt{2}}^{2}\frac{1}{1-x^2}\,dx$. Show that the identity

$$\frac{1}{a^2-x^2} = \frac{1}{2a}\left(\frac{1}{x+a}-\frac{1}{x-a}\right)$$

leads to the formula

$$\int \frac{1}{a^2-x^2}\,dx = \frac{1}{2a}\ln\left(\left|\frac{x+a}{x-a}\right|\right)+C$$

and use it to verify your answers for α and β.

Calculator/Computer Exercises

86. Let b be the abscissa of the point of intersection of the curves $y=2x^2/\sqrt{4-x^2}$ and $y=x$ in the first quadrant. Find the area of the region between the two curves for $0\le x\le b$.

87. Find the area of the region that is bounded above by $y=(1+x^2)^{-3/2}$ and below by $y=x^2/\sqrt{1-x^2}$.

88. Let a be the smaller abscissa, b the larger, of the two first-quadrant points of intersection of the curves $y=3-x/\sqrt{1-x^2}$ and $y=\sqrt{1-x^2}/x$. Find the area of the region between the two curves for $a\le x\le b$.

89. Let b be the abscissa of the point of intersection of the curves $y=2x-x^2$ and $y=x^2/(2+2x+x^2)$ in the first quadrant. Find the area of the region between the two curves for $0\le x\le b$.

90. Let a be the negative x-intercept of $y=\dfrac{x^3-2x+1}{x^2-2x+2}$. Find the total area of the regions between the curve and the x-axis for $a\le x\le 1$.

6.4 Partial Fractions–Linear Factors

In this section, you will begin to learn how to integrate quotients of polynomials; such functions are called *rational functions*. The basic scheme, called *partial fractions*, consists of writing any rational function as a sum of building blocks, which we already know how to integrate. Let us now describe what some of these building blocks are.

Simple Linear Building Blocks

$$\frac{A}{x-a},$$

where A and a are constants. The integral of this building block is

$$A \cdot \ln(|x-a|).$$

Repeated Linear Building Blocks

$$\frac{A}{(x-a)^m},$$

where m is an integer greater than 1. The integral of such a building block is

$$-\frac{A}{m-1} \cdot \frac{1}{(x-a)^{m-1}}.$$

These are all the linear building blocks, and you can see that they are easy to understand. A great many rational functions can be written as a sum of linear building blocks and repeated linear building blocks. Let us see how this is done.

The Method of Partial Fractions for Linear Factors

▶ **EXAMPLE 1** Integrate

$$\int \frac{3}{(x-1)(x+2)}\, dx.$$

Solution The scheme is to rewrite the integrand as

$$\frac{3}{(x-1)(x+2)} = \frac{A}{x-1} + \frac{B}{x+2}.$$

Here A and B are constants that we must determine. We put the terms on the right over a common denominator:

$$\frac{3}{(x-1)(x+2)} = \frac{A(x+2) + B(x-1)}{(x-1)(x+2)}.$$

Both sides of this last equation have the same denominator. The only way that equality can hold is if the numerators are equal:

$$3 = A(x+2) + B(x-1). \qquad (6.4.1)$$

Equation (6.4.1) can be reorganized to more clearly reflect its status as an identity of polynomials in x:

$$0x + 3 = (A + B)x + (2A - B).$$

For two polynomials to be equal, their constant terms must be equal, their linear terms must be equal, and so on. Thus

$$3 = 2A - B \quad \text{and} \quad 0 = A + B.$$

We solve these two linear equations simultaneously. The second equation tells us that $B = -A$; substituting this value for B in the first equation results in $3 = 2A - (-A)$, or $3 = 3A$. Thus $A = 1$ and $B = -1$. We have discovered that

$$\frac{3}{(x-1)(x+2)} = \frac{1}{x-1} + \frac{-1}{x+2}.$$

Therefore

$$\begin{aligned}\int \frac{3}{(x-1)(x+2)}\,dx &= \int \frac{1}{x-1}\,dx - \int \frac{1}{x+2}\,dx \\ &= \ln(|x-1|) - \ln(|x+2|) + C \\ &= \ln\left(\left|\frac{x-1}{x+2}\right|\right) + C. \quad ◀\end{aligned}$$

INSIGHT It is important to understand that we do not solve equation (6.4.1) for x. The procedure is to find values for A and B that convert equation (6.4.1) into an *identity* in x. In other words, we are to find A and B so that equation (6.4.1) is valid for *every* x. Substituting any value for x in equation (6.4.1) results in an equation in the two unknowns A and B. This leads to a very simple alternative method: We solve for each unknown by choosing a value of x that eliminates the other unknown. Thus when $x = 1$, the coefficient of B in (6.4.1) is 0, and the equation reduces to $3 = 3A$ or $A = 1$. Letting x be -2 results in $3 = -3B$ or $B = -1$. Notice that there is no need to solve simultaneous equations with this technique.

The process used in Example 1 relies on a general algebraic technique for handling rational functions, which we now begin to describe. In the first case that we discuss, we assume that the denominator factors as $(x - a_1)\,(x - a_2)\,\ldots\,(x - a_K)$ with no repetition among the roots $a_1, a_2, \ldots, a_K$.

The Method of Partial Fractions for Distinct Linear Factors

To integrate a function of the form

$$\frac{p(x)}{(x-a_1)(x-a_2)\cdots(x-a_K)},$$

where $p(x)$ is a polynomial and the a_j are distinct real numbers, first perform the following two algebraic steps:

1. Be sure that the degree of p is *less* than the degree of the denominator; if it is not, divide the denominator into the numerator. This step must be performed even if the degrees of the denominator and the numerator are equal.
2. Decompose the function into the form

$$\frac{A_1}{x-a_1} + \frac{A_2}{x-a_2} + \cdots + \frac{A_K}{x-a_K},$$

solving for the numerators $A_1, A_2, \ldots, A_K$. The result is called the *partial fraction decomposition* of the original rational function.

▶ **EXAMPLE 2** Calculate the integral

$$\int \frac{2x^3 - 9x^2 - 5x + 9}{x^2 - 4x - 5}\, dx.$$

Solution We first notice that the degree of the numerator is not less than the degree of the denominator, so we must divide. We find that

$$\frac{2x^3 - 9x^2 - 5x + 9}{x^2 - 4x - 5} = 2x - 1 + \frac{x+4}{x^2 - 4x - 5} = 2x - 1 + \frac{x+4}{(x+1)(x-5)}. \qquad (6.4.2)$$

Observe that the rational function that ends line (6.4.2) has a numerator with degree less than the degree of the denominator and a denominator that is the product of distinct linear factors. We may therefore apply the method of partial fractions for distinct linear factors. The next step is to solve for the unknowns A and B in the decomposition

$$\frac{x+4}{(x+1)(x-5)} = \frac{A}{x+1} + \frac{B}{x-5}.$$

Putting the right side over a common denominator

$$\frac{x+4}{(x+1)(x-5)} = \frac{A(x-5) + B(x+1)}{(x+1)(x-5)}$$

and equating numerators leads to

$$x + 4 = A(x-5) + B(x+1). \qquad (6.4.3)$$

Let us continue with the simple method that we described in the preceding *Insight*. One at a time, we substitute $x = -1$ and $x = 5$ into equation (6.4.3). The substitution $x = -1$ results in $-1 + 4 = A(-1-5) + B \cdot 0$, or $3 = -6A$, or $A = -1/2$. The substitution $x = 5$ results in $5 + 4 = A \cdot 0 + B(5+1)$, or $9 = 6B$, or $B = 3/2$. In conclusion,

$$\begin{aligned}\int \frac{2x^3 - 9x^2 - 5x + 9}{x^2 - 4x - 5}\, dx &= \int \left(2x - 1 + \frac{(-1/2)}{x+1} + \frac{3/2}{x-5}\right) dx \\ &= \int (2x-1)\, dx + \int \frac{-1/2}{x+1}\, dx + \int \frac{3/2}{x-5}\, dx \\ &= x^2 - x - \frac{1}{2}\ln(|x+1|) + \frac{3}{2}\ln(|x-5|) + C.\end{aligned}$$ ◀

INSIGHT Notice the way the values of x are chosen in Example 2. We choose $x = -1$ so that the coefficient of B will be 0. The result is an equation that involves only A. Similarly, the choice $x = 5$ eliminates the unknown A, leaving us with an equation in B alone. As in the *Insight* to Example 1, the values substituted for x are the roots of the original denominator.

Heaviside's Method In Examples 1 and 2, we described a shortcut for determining the partial fraction decomposition

$$\frac{p(x)}{(x-a_1)(x-a_2)\cdots(x-a_K)} = \frac{A_1}{x-a_1} + \frac{A_2}{x-a_2} + \cdots + \frac{A_K}{x-a_K} \tag{6.4.4}$$

when $a_1, a_2, \ldots, a_K$ are distinct real numbers. In that method, we evaluate a suitable expression at each of the roots $a_1, a_2, \ldots, a_K$. The idea is that we can pick off the coefficients $A_1, A_2, \ldots, A_K$ without solving multiple equations simultaneously. Of course, we cannot set $x = a_j$ in equation (6.4.4) itself because that would involve division by 0. It turns out that there is an efficient algorithm for determining equations that are suitable for the substitutions $x = a_j$. The procedure is named after the electrical engineer Oliver Heaviside (1850–1925), who used it systematically in his work. Let $q_j(x)$ denote what is left of the denominator $(x-a_1)(x-a_2)\cdots(x-a_K)$ after the factor $(x-a_j)$ is removed. To determine the coefficient A_j, simply multiply both sides of equation (6.4.4) by $x - a_j$. This results in

$$\frac{p(x)}{q_j(x)} = A_j + (x-a_j)\cdot(\text{terms that involve the other coefficients}).$$

Now set $x = a_j$. We see immediately that

$$A_j = \frac{p(a_j)}{q_j(a_j)}. \tag{6.4.5}$$

▶ **EXAMPLE 3** Use Heaviside's method to find the partial fraction decomposition for

$$\frac{3x^2+x-1}{x(x-3)(x+2)}.$$

Solution We note that the numerator has smaller degree than the denominator and that the denominator factors into three distinct linear terms. We may proceed with Heaviside's method. We write

$$\frac{3x^2+x-1}{x(x-3)(x+2)} = \frac{A}{x-0} + \frac{B}{x-3} + \frac{C}{x-(-2)}. \tag{6.4.6}$$

The roots of the denominator polynomial are $a_1 = 0$, $a_2 = 3$, and $a_3 = -2$. We let $p(x)$ denote the numerator $3x^2 + x - 1$. To determine A, multiply both sides by x. The result is

$$\frac{3x^2+x-1}{(x-3)(x+2)} = A + x\cdot\left(\frac{B}{x-3} + \frac{C}{x+2}\right).$$

Now we substitute $x = 0$ into this equation to find that $A = -1/(-6)$. To find B, multiply equation (6.4.6) through by $x - 3$. The result is

$$\frac{3x^2+x-1}{x(x+2)} = B + (x-3)\cdot\left(\frac{A}{x} + \frac{C}{x+2}\right).$$

Now substituting in $x = 3$ yields $B = 29/15$. A similar computation shows that $C = 9/10$. Try it!

INSIGHT After a little practice, you will be able to carry out Heaviside's method with ease. If you had looked at the final decomposition of Example 5 without seeing the calculation, then you might have thought tedious algebra was required to obtain coefficients such as 1/6, 29/15, and 9/10. But, as we have seen, Heaviside's method is as easy to apply when the coefficients are "messy" as it is when the coefficients are simple integers. Remember, however, that the procedure we have described works only when the denominator of the integrand factors into distinct linear factors.

Repeated Linear Factors When one or more of the linear factors in the denominator is repeated, then we must use a slightly different partial fractions scheme.

Partial Fractions When Repeated Linear Factors Are Present

Consider the rational function

$$\frac{p(x)}{(x-a_1)^{m_1}(x-a_2)^{m_2}\cdots(x-a_K)^{m_K}},$$

where $p(x)$ is a polynomial, the a_j are distinct real numbers, and the m_j are positive integers (some of them exceeding 1). To integrate a function of this type, perform the following two steps:

1. Be sure that the degree of p is less than the degree of the denominator; if it is not, *divide the denominator into the numerator.*
2. For each of the factors $(x-a_j)^{m_j}$ in the denominator of the rational function being considered, the partial fraction decomposition must contain terms of the form

$$\frac{A_1}{(x-a_j)^1}+\frac{A_2}{(x-a_j)^2}+\cdots+\frac{A_{m_j}}{(x-a_j)^{m_j}}.$$

▶ **EXAMPLE 4** Evaluate the integral

$$\int \frac{5x^2+18x-1}{(x+4)^2(x-3)}\,dx.$$

Solution We must decompose the integrand into the form

$$\frac{5x^2+18x-1}{(x+4)^2(x-3)}=\frac{A_1}{x+4}+\frac{A_2}{(x+4)^2}+\frac{B}{x-3}.$$

Notice that, because $(x+4)$ is a repeated factor in the denominator of the integrand, our partial fraction decomposition must contain *all powers* of $(x+4)$ up to and including the highest power occurring in the denominator of the integrand.

Now, putting the right-hand side over a common denominator (and equating numerators), leads to

$$5x^2+18x-1=A_1(x+4)(x-3)+A_2(x-3)+B(x+4)^2.$$

Equating the coefficients of like powers of x gives the system

$$\begin{aligned} A_1 \qquad\quad + B &= 5\\ A_1 + A_2 + 8B &= 18\\ -12A_1 - 3A_2 + 16B &= -1. \end{aligned}$$

The first equation allows us to eliminate $B = 5 - A_1$ from the remaining two equations. They become

$$-7A_1 + A_2 = -22 \quad \text{and} \quad -28A_1 - 3A_2 = -81.$$

We solve these equations in standard fashion to obtain $A_1 = 3$ and $A_2 = -1$. Finally $B = 5 - A_1 = 2$. Thus

$$\int \frac{5x^2+18x-1}{(x+4)^2(x-3)}\,dx = \int \frac{3}{x+4}\,dx + \int \frac{(-1)}{(x+4)^2}\,dx + \int \frac{2}{x-3}\,dx = 3\ln(|x+4|) + \frac{1}{(x+4)} + 2\ln(|x-3|) + C.$$

INSIGHT In Example 4, you might be tempted to leave out one of the terms $A_1/(x+4)$ or $A_2/(x+4)^2$ on the right-hand side. Try it (Exercises 65 and 66). You will find that you are unable to solve the resulting algebraic system for the coefficients. The more complicated partial fraction decomposition is really necessary. Algebraic theory tells us that the coefficients of a partial fractions expansion are unique. You cannot succeed when you start out with the wrong form—unless an omitted coefficient just happens to be 0.

Summary of Basic Partial Fraction Forms

We conclude by summarizing the techniques of this section with several quick instances of the different partial fraction forms that can arise:

$$\frac{3x-1}{(x-4)(x+7)} = \frac{A}{x-4} + \frac{B}{x+7}$$

$$\frac{x^2+x+1}{x(x^2-1)} = \frac{A}{x} + \frac{B}{x+1} + \frac{C}{x-1}$$

$$\frac{x+3}{(x-2)^2(x+1)} = \frac{A}{x-2} + \frac{B}{(x-2)^2} + \frac{C}{x+1}$$

$$\frac{x^4-2x-1}{(x+8)^3(x-1)^2} = \frac{A}{x+8} + \frac{B}{(x+8)^2} + \frac{C}{(x+8)^3} + \frac{D}{x-1} + \frac{E}{(x-1)^2}.$$

When you begin to do the exercises, refer to these examples to tell which form of the partial fractions technique you should be using. Also notice this unifying theme: The number of unknown coefficients in any partial fractions problem is equal to the degree of the denominator.

QUICK QUIZ

1. True or false: If $q(x)$ is a polynomial of degree n that factors into linear terms and if $p(x)$ is a polynomial of degree m with $m < n$, then an explicitly calculated partial fraction decomposition of $p(x)/q(x)$ requires the determination of n unknown constants.
2. For what values of A, B, and C is $A(x+1)(x+2) + Bx(x+2) + Cx(x+1) = 4x^2 + 11x + 4$ an identity in x?
3. What is the form of the partial fraction decomposition of $\dfrac{x^2+3}{x(x^2-4)}$?
4. What is the form of the partial fraction decomposition of $\dfrac{2x^2+1}{x^3(x+1)}$?

Answers

1. True 2. $A = 2$, $B = 3$, $C = -1$
3. $\dfrac{A}{x} + \dfrac{B}{x-2} + \dfrac{C}{x+2}$ 4. $\dfrac{A}{x} + \dfrac{B}{x^2} + \dfrac{C}{x^3} + \dfrac{D}{x+1}$

EXERCISES

Problems for Practice

▶ In each of Exercises 1–10, write down the *form* of the partial fraction decomposition of the given rational function. Do not explicitly calculate the coefficients. ◀

1. $\dfrac{x+2}{x(x+1)}$
2. $\dfrac{2x+3}{x^2-1}$
3. $\dfrac{3x^2+1}{x^2-4}$
4. $\dfrac{21}{(x-2)(x+5)}$
5. $\dfrac{x^3}{(2x-5)(x+4)}$
6. $\dfrac{x^2+2x+2}{(x-4)^2(x+2)}$
7. $\dfrac{x^3+7}{x^2(x+1)(x-2)}$
8. $\dfrac{x^5+1}{x^2(x+1)^2}$
9. $\dfrac{x^6-17}{(x-5)^2(x+3)^2}$
10. $\dfrac{2x^4}{(x-7)^3(x+1)}$

▶ In each of Exercises 11–20, decompose the given rational function into partial fractions. Calculate the coefficients. ◀

11. $\dfrac{1}{x(x+1)}$
12. $\dfrac{5-x}{x^2-1}$
13. $\dfrac{x+6}{x^2-4}$
14. $\dfrac{21}{x^2+3x-10}$
15. $\dfrac{11x}{(2x-5)(x+3)}$
16. $\dfrac{36}{(x-4)^2(x+2)}$
17. $\dfrac{2x^3+x^2-5x+2}{x^2(x+1)(x-2)}$
18. $\dfrac{1}{(x^2+x)^2}$
19. $\dfrac{2x^3-4x^2-13x+76}{(x^2-4x+4)(x^2+6x+9)}$
20. $\dfrac{8}{(x-1)^3(x+1)}$

▶ In each of Exercises 21–26, use Heaviside's method to calculate the partial fraction decomposition of the given rational function. ◀

21. $\dfrac{x+1}{(2x+7)(2x+9)}$
22. $\dfrac{2x+1}{(x-2)(x+1)}$
23. $\dfrac{5x^2+3x+1}{(x-2)(x+3)(x+4)}$
24. $\dfrac{2x^2+8x+13}{(x+1)(x+3)(x+5)}$
25. $\dfrac{7x^2+9x+5}{4x^3-x}$
26. $\dfrac{3x^3+5x^2+7x+7}{(x-1)x(x+1)(x+2)}$

▶ In each of Exercises 27–38, use the method of partial fractions to calculate the given integral. ◀

27. $\displaystyle\int \frac{3x+1}{x^2-1}\,dx$
28. $\displaystyle\int \frac{3x+4}{x^2+x-6}\,dx$
29. $\displaystyle\int \frac{9x+18}{(x-3)(x+6)}\,dx$
30. $\displaystyle\int \frac{x^3-2x^2-2x-2}{x^2-x}\,dx$
31. $\displaystyle\int \frac{3x^2}{x^2+2x+1}\,dx$
32. $\displaystyle\int \frac{7x^2+7x-2}{x(x-1)(x+2)}\,dx$
33. $\displaystyle\int \frac{x^2+14x+39}{(x+2)(x+3)(x+5)}\,dx$
34. $\displaystyle\int \frac{7x^2+4x+5}{x^3-x}\,dx$
35. $\displaystyle\int \frac{x^2+1}{x^3+x^2}\,dx$
36. $\displaystyle\int \frac{9x}{(x-2)^2(x+1)}\,dx$
37. $\displaystyle\int \frac{5x-6}{(x-2)^2(x+2)}\,dx$
38. $\displaystyle\int \frac{25}{(x+1)(x-4)^2}\,dx$

▶ Calculate each of the definite integrals in Exercises 39–44. ◀

39. $\displaystyle\int_2^3 \frac{4x-7}{(x-1)(x-4)}\,dx$
40. $\displaystyle\int_1^2 \frac{2x+1}{x(x+1)}\,dx$
41. $\displaystyle\int_{-1}^1 \frac{x+6}{(x-2)(x+2)}\,dx$

42. $\int_0^1 \frac{1}{(x+1)(2x+1)}\,dx$

43. $\int_1^2 \frac{2x^2+5x+4}{x(x+1)(x+2)}\,dx$

44. $\int_1^2 \frac{x^2-2x-2}{x^2(x+1)}\,dx$

Further Theory and Practice

Calculate each of the indefinite integrals in Exercises 45–48.

45. $\int \frac{4x^2+x+2}{x^3(x+2)}\,dx$

46. $\int \frac{3x^2+5x+3}{x^3(x+1)}\,dx$

47. $\int \frac{3x^3-16x^2+26x-14}{(x-1)^2(x-2)^2}\,dx$

48. $\int \frac{5x^3+5x^2-x-2}{(x^2+x)^2}\,dx$

Calculate each of the definite integrals in Exercises 49–52.

49. $\int_1^2 \frac{2x^3+6x^2+9x+2}{x(x+1)^3}\,dx$

50. $\int_2^3 \frac{x^2+10x+1}{(x^2-1)^2}\,dx$

51. $\int_0^1 \frac{3x^2+8x+6}{(x+1)^2(x+2)}\,dx$

52. $\int_4^9 \frac{64x}{(x+1)(x-3)^3}\,dx$

In each of Exercises 53–56, make a substitution before applying the method of partial fractions to calculate the given integral.

53. $\int \frac{\exp(x)}{\exp(2x)-1}\,dx$

54. $\int \frac{2^{x+2}}{4^x-4}\,dx$

55. $\int \frac{\cos(x)}{\sin^2(x)-5\sin(x)+6}\,dx$

56. $\int \frac{2\ln(x)+5}{x\left(\ln^2(x)+\ln(x)\right)}\,dx$

In each of Exercises 57–61, two functions f and g are given. Calculate $\int f(x)\,dx$ by first making the substitution $u=g(x)$ and then applying the method of partial fractions.

57. $f(x)=1/(x^{3/2}-x)$; $g(x)=x^{1/2}$

58. $f(x)=\frac{1}{x^{4/3}-x^{2/3}}$; $g(x)=x^{1/3}$

59. $f(x)=\frac{1}{\sqrt{x+2}(x+1)}$; $g(x)=\sqrt{x+2}$

60. $f(x)=\frac{\sqrt{x}+2}{\sqrt{x}(x-1)}$; $g(x)=\sqrt{x}$

61. $f(x)=\frac{1}{x^{1/2}-x^{1/6}}$; $g(x)=x^{1/6}$

62. What happens if you attempt a partial fraction decomposition of $1/(x-3)^4$ into

$$\frac{A}{x-3}+\frac{B}{(x-3)^2}+\frac{C}{(x-3)^3}+\frac{D}{(x-3)^4}?$$

(A long calculation should be your last resort.)

63. Assume that a is a positive constant and that $|x|<a$. Use the method of partial fractions to derive the formula

$$\int \frac{dx}{a^2-x^2}=\frac{1}{2a}\ln\left(\frac{a+x}{a-x}\right)+C.$$

64. Find the partial fraction decomposition of $\frac{1}{x^3-3x^2+2}$.

The correct partial fraction decomposition of $\frac{5x^2+18x-1}{(x+4)^2(x-3)}$ is of the form $\frac{A_1}{x+4}+\frac{A_2}{(x+4)^2}+\frac{B}{x-3}$. The values of the coefficients A_1, A_2, and B are determined in Example 6. In Exercises 65 and 66, show that the given incorrect form does not admit a solution for the unknown coefficients.

65. $\frac{A_1}{x+4}+\frac{B}{x-3}$

66. $\frac{A_2}{(x+4)^2}+\frac{B}{x-3}$

67. Mr. Woodman set up a partial fraction decomposition

$$\frac{5x^2-4x+2}{x^2(x-1)^2}=\frac{A}{x^2}+\frac{B}{(x-1)^2}$$

and correctly solved $A=2$, $B=3$.

a. What is wrong with his method of solution?

b. Why did the right answer result even though the method of solution was incorrect?

c. Show that if the numerator is $5x^2-4x+1$ instead of $5x^2-4x+2$, then the procedure used will fail. What is the correct partial fraction decomposition in this case?

Calculator/Computer Exercises

68. Calculate $\int_3^5 \frac{x+3}{x^3-3x^2+3}\,dx$ by finding a partial fraction decomposition of the integrand. (Find the roots of the denominator of the integrand numerically. Then use Heaviside's method to determine the coefficients of the partial fraction decomposition.)

69. Let $q(x)=(x+2)(x+3)(x+5)$. Find the area of the region in the first quadrant that is bounded on the left by the y-axis, above by $y=(2x^2+5)/q(x)$, and below $y=(x^3+2)/q(x)$, and below by $y=(x^3+2)/q(x)$.

70. Find the area of the region that is bounded above by the graph of $y = \ln(x)$ and below by the graph of $y = x^3/(x^2 + 3x + 2)$.

71. Let $f(x) = 4 + 2x - x^2$, let $g(x) = \dfrac{2x^3 + x + 2}{x^4 - 5x^3 + 6x - 1}$, and let a be the solution of $f(x) = g(x)$ in [0.25, 0.5]. Use a partial fraction decomposition to find the area of the region bounded above by the graph of $y = f(x)$, $a \le x \le 1$ and below by the graph of $y = g(x)$, $a \le x \le 1$.

▶ Computer algebra systems can calculate the partial fraction decomposition of a rational function when the arithmetic can be done with rational numbers. For example, the Maple command for obtaining the partial fraction decomposition of a rational expression R in the variable x is

```
convert(R, parfrac, x);
```

In Exercises **72–75**, use a computer algebra system to find the partial fraction decomposition of the given rational functions. ◀

72. $\dfrac{x^3 - 9x^2 + 25x - 19}{x^4 - 5x^3 + 6x^2 + 4x - 8}$

73. $\dfrac{x^3 + 10x^2 + 27x + 24}{x^4 + 10x^3 + 35x^2 + 50x + 24}$

74. $\dfrac{x^5 + 2x^4 - 10x^3 - 7x^2 + 9x - 24}{x^4 + 2x^3 - 11x^2 - 12x + 36}$

75. $\dfrac{3x^4 + 12x^3 + 17x^2 + 14x + 2}{x^6 + 2x^5 - x^4 - 4x^3 - x^2 + 2x + 1}$

6.5 Partial Fractions–Irreducible Quadratic Factors

In Section 6.4, you learned how to evaluate integrals of the form

$$\int \frac{p(x)}{(x - a_1)^{m_1}(x - a_2)^{m_2} \cdots (x - a_K)^{m_K}}\, dx, \qquad (6.5.1)$$

where $p(x)$ is a polynomial. To do so, we decomposed the integrand into a sum of simpler expressions using the method of partial fractions. The purpose of this section is to extend that technique to denominators that include irreducible quadratic factors. Because the Fundamental Theorem of Algebra tells us that any polynomial can be split into a product of linear and irreducible quadratic factors, our study of the method of partial fractions will therefore be complete. As a result, we will be able to integrate any rational function $p(x)/q(x)$.

Rational Functions with Quadratic Terms in the Denominator

In evaluating integrals of the form (6.5.1), we have made use of two simple algebraic expressions: the simple linear building block $A/(x - a)$ and the repeated linear building block $A/(x - a)^m$. To find a partial fraction decomposition of the rational function $p(x)/q(x)$ when $q(x)$ has one or more irreducible quadratic factors, we need two new building blocks.

Simple Quadratic Building Blocks

$$\frac{Bx + C}{ax^2 + bx + c}, \qquad (6.5.2)$$

assuming that the denominator $ax^2 + bx + c$ cannot be factored; otherwise, we would be in the situation of linear building blocks that we studied in Section 6.4. Recall that we learned how to integrate expressions of the form (6.5.2) in Section 6.3.

Repeated Quadratic Building Blocks

$$\frac{Bx + C}{(ax^2 + bx + c)^n},$$

where n is an integer greater than 1. Integrands of this form have been treated in Section 6.3. Because $n > 1$ we will have to make certain adjustments in the partial fractions representation.

We can now state our complete partial fractions rule. You will find that this rule just builds upon steps that you are already familiar with: The presence of irreducible quadratic terms in the denominator does not change the way we treat the linear terms.

The Method of Partial Fractions

To integrate a rational function of the form

$$\frac{p(x)}{(x-a_1)^{m_1}\cdots(x-a_K)^{m_K}\cdot(a_1x^2+b_1x+c_1)^{n_1}\cdots(a_Lx^2+b_Lx+c_L)^{n_L}},$$

where the factors $x-a_1$, $x-a_2$, $\ldots, x-a_K$, $a_1x^2+b_1x+c_1$, $a_2x^2+b_2x+c_2$, $\ldots, a_Lx^2+b_Lx+c_L$ are all distinct, perform the following four steps:

1. Be sure that the degree of the numerator p is less than the degree of the denominator; if it is not, then divide the denominator into the numerator.
2. Be sure that the quadratic factors $a_jx^2+b_jx+c_j$ cannot be factored into linear factors with real coefficients. (Do this by verifying that $b_j^2-4a_jc_j<0$; if $b_j^2-4a_jc_j\geq 0$ for any quadratic term $a_jx^2+b_jx+c_j$, then that expression must be factored into a product of two linear terms.)
3. For each of the factors $(x-a_j)^{m_j}$ in the denominator of the rational function being considered, the partial fraction decomposition must contain terms of the form

$$\frac{A_1}{(x-a_j)^1}+\frac{A_2}{(x-a_j)^2}+\cdots+\frac{A_{m_j}}{(x-a_j)^{m_j}}.$$

4. For each of the factors $(x^2+b_jx+c_j)^{n_j}$ in the denominator of the integrand being considered, the partial fraction decomposition must contain terms

$$\frac{B_1x+C_1}{a_jx^2+b_jx+c_j}+\frac{B_2x+C_2}{(a_jx^2+b_jx+c_j)^2}+\cdots+\frac{B_{n_j}x+C_{n_j}}{(a_jx^2+b_jx+c_j)^{n_j}}.$$

The first three examples will give you a more concrete idea of how to set up decompositions into partial fractions (without the distraction of determining the unknown coefficients or performing subsequent integrations).

▶ **EXAMPLE 1** State the form of the partial fraction decomposition of

$$\frac{8x^9}{(x+4)(x+5)^3(x^2+6)(x^2+7)^2}.$$

Solution The degree of the denominator is 10. Because the degree of the numerator is less than 10, division is not needed. The factors of the denominator include a nonrepeated linear term, $x+4$, and a repeated linear term, $(x+5)^3$. In a partial fraction decomposition, the numerators of the summands that correspond to such factors are always constants. Among the factors of the denominator of the given rational function, we also find a nonrepeated irreducible quadratic, x^2+6, and a repeated irreducible quadratic, $(x^2+7)^2$. In a partial fraction decomposition, the numerators of the summands that correspond to such factors are always linear polynomials. Thus

$$\frac{8x^9}{(x+4)(x+5)^3(x^2+6)(x^2+7)^2} = \frac{A}{x+4} + \frac{B}{x+5} + \frac{C}{(x+5)^2} + \frac{D}{(x+5)^3} + \frac{Ex+F}{x^2+6} + \frac{Gx+H}{x^2+7} + \frac{Ix+J}{(x^2+7)^2}.$$

Notice that the number of unknown factors in the numerators is 10, which is the same as the degree of the denominator polynomial in the original rational function. ◀

Checking for Irreducibility It is very important to begin with the right form of a partial fraction decomposition before attempting to solve for the unknown coefficients. Usually, it is impossible to solve for the unknown coefficients when the partial fraction decomposition has not been set up correctly.

▶ **EXAMPLE 2** Find the correct form of the partial fraction decomposition for the rational expression

$$\frac{3x^2+11x+10}{(x+1)(x^2+4x+3)}.$$

Solution Because the degree of the numerator is 2 and the degree of the denominator is 3, a preliminary division is not necessary. Notice that the correct partial fraction decomposition is *not*

$$\frac{3x^2+11x+10}{(x+1)(x^2+4x+3)} = \frac{A}{x+1} + \frac{Bx+C}{x^2+4x+3}. \tag{6.5.3}$$

The reason is that we have not factored the denominator as far as possible. The quadratic polynomial x^2+4x+3 factors as

$$x^2+4x+3 = (x+1)(x+3).$$

Thus the given rational function really falls under the case of distinct linear factors, which we studied in Section 6.4. The correct form of the partial fraction decomposition is

$$\frac{3x^2+11x+10}{(x+1)(x^2+4x+3)} = \frac{3x^2+11x+10}{(x+1)^2(x+3)} = \frac{A}{x+1} + \frac{B}{(x+1)^2} + \frac{C}{x+3}. \tag{6.5.4}$$

You may want to practice the methods of determining the unknown coefficients that we studied in Section 6.4. You will find that the values $A=2$, $B=1$, and $C=1$ turn equation (6.5.4) into an identity in x. By contrast, there are no values of A, B, and C for which equation (6.5.3) is an identity in x. ◀

INSIGHT Remember that the Quadratic Formula tells us whether a quadratic factor ax^2+bx+c is irreducible or not. If the discriminant b^2-4ac is negative, then the polynomial is irreducible; otherwise, it can be factored into linear terms. A quick calculation of the discriminant of x^2+4x+3 in Example 2 would result in the positive number $4^2-4\cdot 1\cdot 3=4$. That would alert us to the reducibility of x^2+4x+3.

▶ **EXAMPLE 3** Find the correct form of the partial fraction decomposition of $3/(x^3+1)$.

Solution This expression may appear to have an irreducible denominator, but it does not. Cubic (and higher-degree) polynomials are *never* irreducible. Notice that -1 is a root of the polynomial x^3+1. Therefore $(x+1)$ is a divisor. After performing the polynomial division, we find that

$$x^3+1=(x+1)(x^2-x+1).$$

The quadratic factor has negative discriminant $(-1)^2-4\cdot 1\cdot 1=-3$, so it is irreducible. Thus we see that the correct partial fraction decomposition for our rational expression is

$$\frac{3}{x^3+1}=\frac{A}{x+1}+\frac{Bx+C}{x^2-x+1}. \qquad (6.5.5)$$ ◀

INSIGHT It is important for you to realize that the Fundamental Theorem of Algebra guarantees that any real polynomial can be factored into linear and quadratic factors. In the theory of partial fractions, we never have to use building blocks that involve irreducible cubic, quartic, or higher order factors—they do not exist.

Calculating the Coefficients of a Partial Fraction Decomposition

Now that we have considered the forms that decompositions into partial fractions can take, let us see how the unknown coefficients can be computed in practice.

▶ **EXAMPLE 4** Find the partial fraction decomposition of the rational function $3/(x^3+1)$.

Solution In equation (6.5.5) of Example 3, we determined the correct form of the partial fraction decomposition of $3/(x^3+1)$. The right side of equation (6.5.5) can be written as

$$\frac{A(x^2-x+1)+(Bx+C)(x+1)}{(x+1)(x^2-x+1)}$$

or, after expanding both the numerator and denominator,

$$\frac{(A+B)x^2+(-A+B+C)x+(A+C)}{x^3+1}.$$

The numerator of this expression must equal the numerator of the left side of (6.5.5):

$$0x^2+0x+3=(A+B)x^2+(-A+B+C)x+(A+C).$$

Equating the coefficients of like powers of x on each side of this equation leads to the simultaneous equations

$$A+B=0, \quad -A+B+C=0, \quad A+C=3.$$

The first equation allows us to replace B with $-A$ in the second equation to obtain $-2A+C=0$. Thus $C=2A$. The third equation then tells us that $A+(2A)=3$ or $A=1$. Consequently, $B=-A=-1$ and $C=2A=2$. In conclusion, the partial fraction decomposition is

$$\frac{3}{x^3+1}=\frac{1}{x+1}+\frac{-x+2}{x^2-x+1}.$$ ◀

▶ **EXAMPLE 5** Calculate

$$\int \frac{x^3-x^2-3x+3}{(x^2+4x+5)(x^2+9)}\,dx.$$

Solution The degree of the denominator is 4, one greater than the degree of the numerator. Division is therefore not necessary. Because the discriminants of $x^2 + 4x + 5$ and $x^2 + 9$ are negative numbers, $4^2 - 4 \cdot 1 \cdot 5 = -4$ and $0^2 - 4 \cdot 1 \cdot 9 = -36$, respectively, we can be sure that further factoring of the denominator is impossible. We may therefore proceed directly to the partial fraction decomposition:

$$\frac{x^3 - x^2 - 3x + 3}{(x^2 + 4x + 5)(x^2 + 9)} = \frac{Ax + B}{x^2 + 4x + 5} + \frac{Cx + D}{x^2 + 9}.$$

We combine the two summands on the right to obtain

$$\begin{aligned}\frac{x^3 - x^2 - 3x + 3}{(x^2 + 4x + 5)(x^2 + 9)} &= \frac{(Ax + B)(x^2 + 9) + (Cx + D)(x^2 + 4x + 5)}{(x^2 + 4x + 5)(x^2 + 9)} \\ &= \frac{(A + C)x^3 + (B + 4C + D)x^2 + (9A + 5C + 4D)x + (9B + 5D)}{(x^2 + 4x + 5)(x^2 + 9)}.\end{aligned}$$

We may now equate the numerators on the left with the numerator on the right:

$$x^3 - x^2 - 3x + 3 = (A + C)x^3 + (B + 4C + D)x^2 + (9A + 5C + 4D)x + (9B + 5D).$$

By equating coefficients of like powers of x, we obtain the four equations

$$A + C = 1, \ B + 4C + D = -1, \ 9A + 5C + 4D = -3, \ 9B + 5D = 3.$$

We may use the first and second of these equations to solve for A and B in terms of C and D. We obtain $A = 1 - C$ and $B = -1 - 4C - D$. By using these expressions in C and D to replace A and B in the third and fourth equations, we obtain two equations in the two unknowns C and D. Thus we have

$$9(1 - C) + 5C + 4D = -3, \quad \text{or} \quad -4C + 4D = -12$$

and

$$9(-1 - 4C - D) + 5D = 3, \quad \text{or} \quad -36C - 4D = 12.$$

We solve these last two equations in C and D, obtaining $C = 0$ and $D = -3$. It follows that $A = 1 - C = 1 - 0 = 1$ and $B = -1 - 4C - D = -1 - 4(0) - (-3) = 2$. Thus

$$\frac{x^3 - x^2 - 3x + 3}{(x^2 + 4x + 5)(x^2 + 9)} = \frac{x + 2}{x^2 + 4x + 5} - \frac{3}{x^2 + 9}$$

and

$$\int \frac{x^3 - x^2 - 3x + 3}{(x^2 + 4x + 5)(x^2 + 9)}\,dx = \int \frac{x + 2}{x^2 + 4x + 5}\,dx - \int \frac{3}{x^2 + 9}\,dx. \qquad (6.5.6)$$

We complete the square $x^2 + 4x + 5 = (x + 2)^2 + 1$ and make the substitution $u = x + 2$, $du = dx$ to convert the first integral on the right side of equation (6.5.6) to

$$\int \frac{u}{u^2 + 1}\,du.$$

By means of the substitution $w = u^2 + 1$, $dw = 2u\,du$, this last integral is converted to

$$\frac{1}{2}\int \frac{1}{w}\,dw = \frac{1}{2}\ln(|w|) + C.$$

We resubstitute, observing that $w = u^2 + 1 = x^2 + 4x + 5$ is positive, to obtain

$$\int \frac{x+2}{x^2+4x+5}\,dx = \frac{1}{2}\ln\,(x^2+4x+5)+C.$$

Finally, we use formula (6.3.2) to see that $\int 3/(x^2+9)\,dx = \arctan(x/3)+C$. After substituting for the two integrals on the right side of equation (6.5.6), we have

$$\int \frac{x^3-x^2-3x+3}{(x^2+4x+5)(x^2+9)}\,dx = \frac{1}{2}\ln(x^2+4x+5)-\arctan\left(\frac{x}{3}\right)+C. \quad ◀$$

▶ **EXAMPLE 6** Calculate the integral

$$\int \frac{x-1}{x(x^2+1)^2}\,dx.$$

Solution The degree of the numerator is less than that of the denominator, so we proceed directly to the partial fraction decomposition:

$$\frac{x-1}{x(x^2+1)^2} = \frac{A}{x}+\frac{Bx+C}{x^2+1}+\frac{Dx+E}{(x^2+1)^2}.$$

Putting the right side over a common denominator leads to

$$\begin{aligned} x-1 &= A(x^2+1)^2+(Bx+C)(x^2+1)x+(Dx+E)x \\ &= A(x^4+2x^2+1)+(Bx+C)(x^3+x)+Dx^2+Ex \\ &= x^4(A+B)+x^3(C)+x^2(2A+B+D)+x(C+E)+A. \end{aligned}$$

Equating like coefficients of x leads to the system

$$\begin{aligned} &A+B=0 \\ &C=0 \\ &2A+B+D=0 \\ &C+E=1 \\ &A=-1. \end{aligned}$$

We conclude that $A=-1$, $B=1$, $C=0$, $D=1$, $E=1$. Therefore

$$\int \frac{x-1}{x(x^2+1)^2}\,dx = \int \frac{-1}{x}\,dx + \int \frac{x}{x^2+1}\,dx + \int \frac{x+1}{(x^2+1)^2}\,dx,$$

or

$$\int \frac{x-1}{x(x^2+1)^2}\,dx = -\ln(|x|)+\frac{1}{2}\int \frac{2x}{x^2+1}\,dx+\frac{1}{2}\int \frac{2x}{(x^2+1)^2}\,dx+\int \frac{1}{(x^2+1)^2}\,dx. \qquad (6.5.7)$$

The first and second integrals on the right side are handled with the substitution $u=x^2+1$, $du=2x\,dx$. The results of these integrations are $\int \frac{2x}{x^2+1}\,dx = \ln(x^2+1)+C$ and $\int \frac{2x}{(x^2+1)^2}\,dx = C-\frac{1}{x^2+1}$. For the last integral on the right side of equation (6.5.7), we make the substitution $x=\tan(\theta)$, $dx=\sec^2(\theta)d\theta$. This substitution converts the integral into

$$\int \frac{\sec^2(\theta)}{(\tan^2(\theta)+1)^2}\,d\theta = \int \frac{\sec^2(\theta)}{\sec^4(\theta)}\,d\theta = \int \frac{1}{\sec^2(\theta)}\,d\theta = \int \cos^2(\theta)\,d\theta \overset{(6.2.6)}{=} \frac{1}{2}\theta+\frac{1}{2}\sin(\theta)\cos(\theta)+C.$$

After resubstituting for x, we have

$$\int \frac{1}{(x^2+1)^2}\,dx = \frac{1}{2}\arctan(x) + \frac{1}{2}\frac{x}{x^2+1} + C.$$

Assembling all the pieces, we conclude that our original integral calculates to

$$\int \frac{x-1}{x(x^2+1)^2}\,dx = -\ln(|x|) + \frac{1}{2}\left(\ln(x^2+1) - \frac{1}{(x^2+1)} + \frac{x}{(x^2+1)} + \arctan(x)\right) + C.$$

QUICK QUIZ

1. True or false: If $q(x)$ is a polynomial of degree n that factors into irreducible quadratic terms and if $p(x)$ is a polynomial of degree m with $m<n$, then an explicitly calculated partial fraction decomposition of $p(x)/q(x)$ requires the determination of n unknown constants.
2. What is the form of the partial fraction decomposition of $\dfrac{x^2}{(x^2-3x-4)(x^2-3x+4)}$?
3. What is the form of the partial fraction decomposition of $\dfrac{2x^2+1}{(x^2+1)(x^2+4)}$?
4. What is the form of the partial fraction decomposition of $\dfrac{2x^2+1}{x^2(x^2+4)^2}$?

Answers

1. True 2. $\dfrac{A}{x+1} + \dfrac{B}{x-4} + \dfrac{Cx+D}{x^2-3x+4}$ 3. $\dfrac{Ax+B}{x^2+1} + \dfrac{Cx+D}{x^2+4}$

4. $\dfrac{A}{x} + \dfrac{B}{x^2} + \dfrac{Cx+D}{x^2+4} + \dfrac{Ex+F}{(x^2+4)^2}$

EXERCISES

Problems for Practice

In each of Exercises 1–10, write down the *form* of the partial fraction decomposition of the given rational function. Do *not* explicitly calculate the coefficients.

1. $\dfrac{2x^3+x+1}{(x^2+1)(x^2+4)}$

2. $\dfrac{2x^4}{(x^2+2x+2)(2x^2+5x+3)}$

3. $\dfrac{2x^3+x+1}{(x^2+1)(x^2+x+1)^2}$

4. $\dfrac{2x^4}{(x^2+1)(5x^2+4x+1)}$

5. $\dfrac{2x+1}{(x^2+x+3)(x-4)}$

6. $\dfrac{3x^3+x+1}{x^2(x+1)^2}$

7. $\dfrac{2x^6}{(x^2+4)^3(x-2)}$

8. $\dfrac{3x^5+x+1}{x^2(x^2-1)^2}$

9. $\dfrac{3x^5+x+1}{x^2(x^2-1)(x^2+1)}$

10. $\dfrac{7x^8}{(x-3)^3(x^2+6x+10)^3}$

In each of Exercises 11–20, explicitly calculate the partial fraction decomposition of the given rational function.

11. $\dfrac{3x^2-5x+4}{(x-1)(x^2+1)}$

12. $\dfrac{x^2+2}{(x^2+1)x^2}$

13. $\dfrac{7x^3-3x^2+9x-6}{(x^2+2)(x^2+1)}$

14. $\dfrac{x^2+2x}{(x^2+1)^2}$

15. $\dfrac{x^3-x}{(x^2+1)^2}$

16. $\dfrac{2x^2}{(x+1)^2(x^2+1)}$

17. $\dfrac{x^3+12x^2-9x+48}{(x-3)(x^2+4)}$

18. $\dfrac{2x^2+4x+2}{(x^2+1)^3}$

19. $\dfrac{3x^3-5x^2+10x-19}{(x^2+4)(x^2+3)}$

20. $\dfrac{2x^4+15x^2+30}{(x^2+4)(x^2+3)^2}$

In each of Exercises 21–30, use the method of partial fractions to decompose the integrand. Then evaluate the given integral.

21. $\displaystyle\int \frac{3x^2-5x+4}{(x-1)(x^2+1)}\,dx$

22. $\displaystyle\int \frac{x^2+2}{(x^2+1)x^2}\,dx$

23. $\displaystyle\int \frac{7x^3+9x-3x^2-6}{(x^2+2)(x^2+1)}\,dx$

24. $\displaystyle\int \frac{x^2+2x}{(x^2+1)^2}\,dx$

25. $\displaystyle\int \frac{x^3-x}{(x^2+1)^2}\,dx$

26. $\displaystyle\int \frac{2x^2}{(x+1)^2(x^2+1)}\,dx$

27. $\displaystyle\int \frac{x^3+12x^2-9x+48}{(x-3)(x^2+4)}\,dx$

28. $\displaystyle\int \frac{2x^2+4x+2}{(x^2+1)^3}\,dx$

29. $\displaystyle\int \frac{3x^3-5x^2+10x-19}{(x^2+4)(x^2+3)}\,dx$

30. $\displaystyle\int \frac{2x^4+15x^2+30}{(x^2+4)(x^2+3)^2}\,dx$

In each of Exercises 31–38, evaluate the given definite integral.

31. $\displaystyle\int_0^1 \frac{x^2+3}{(x^2+1)(x^2+2)}\,dx$

32. $\displaystyle\int_2^3 \frac{2x^3-x^2+2x+1}{(x^2-1)(x^2+1)}\,dx$

33. $\displaystyle\int_0^1 \frac{7x^2+4x+6}{(x+1)(x^2+2)}\,dx$

34. $\displaystyle\int_1^2 \frac{3x^4+8x^2+3}{x(x^2+1)^2}\,dx$

35. $\displaystyle\int_1^2 \frac{2x^2-x+2}{x(x^2+1)}\,dx$

36. $\displaystyle\int_1^2 \frac{7x^2+6}{x^2(x^2+1)}\,dx$

37. $\displaystyle\int_0^1 \frac{x^2+x+1}{(x^2+1)^2}\,dx$

38. $\displaystyle\int_0^1 \frac{x^3+x+1}{(x^2+1)^2}\,dx$

Further Theory and Practice

In Exercises 39–44, calculate the given integral.

39. $\displaystyle\int \frac{3x^2+6x+4}{(x+1)(x^2+2x+2)}\,dx$

40. $\displaystyle\int \frac{4x^2+5x+3}{(x+1)(x^2+x+1)}\,dx$

41. $\displaystyle\int \frac{2x^2+4x+9}{x^3-1}\,dx$

42. $\displaystyle\int \frac{5x^2-2x+2}{x^3+1}\,dx$

43. $\displaystyle\int \frac{8x}{x^5-x^4-x+1}\,dx$

44. $\displaystyle\int \frac{48}{(x^2+1)^4}\,dx$

45. Show that equation (6.5.3) is not an identity in x for any values of A, B, and C.

46. What happens if you attempt a partial fraction decomposition of $1/(x^2+3)^4$ into

$$\frac{A_1x+B_1}{x^2+3}+\frac{A_2x+B_2}{(x^2+3)^2}+\frac{A_3x+B_3}{(x^2+3)^3}+\frac{A_4x+B_4}{(x^2+3)^4}?$$

(Brute force calculation should be a last resort.)

Calculator/Computer Exercises

47. Let b be the abscissa of the point of intersection of the graphs of $y=x^4+2$ and $y=\dfrac{5x^3+3x+2}{x^3+x^2+x+1}$ for $0<x<1$. Calculate the area of the region between the two curves for $0\le x\le b$.

48. Let b be the abscissa of the first-quadrant point of intersection of the graphs of $y=x^2$ and $y=\dfrac{3x^3+4x}{x^4+3x^2+2}$. Calculate the area of the region between the two curves for $0\le x\le b$.

49. Find the area of the region in the first quadrant that is bounded above by $y=1-x^2$ and below by $y=\dfrac{3x^2+4x+5}{x^3+3x^2+7x+5}$.

50. Find the four points of intersection of the curves $y=1-x^2/16$ and $y=\dfrac{x^3+x^2+x+2}{x^4+3x^2+2}$. Let a be the least abscissa of these points and b the greatest. Calculate the total area of the regions that are bounded above by one of the curves and below by the other for $a\le x\le b$.

6.6 Improper Integrals–Unbounded Integrands

The theory of the integral that you learned in Chapter 5 enables you to integrate a continuous function $f(x)$ on a closed, bounded interval $[a, b]$. However, often it is necessary to integrate an unbounded function, or a function that is defined on an unbounded interval. In this section and the next, you learn to do so.

Integrals with Infinite Integrands

Let f be a continuous function on the interval $[a, b)$. Suppose that f is unbounded as x approaches b from the left side (see Figure 1). It can be shown that the Riemann sums of an unbounded function do *not* have a limit as the order of the uniform partition tends to infinity. In other words, f does *not* have a Riemann integral over the interval $[a, b]$. We say that $\int_a^b f(x)\,dx$ is an *improper integral* with infinite integrand at b. Figure 2 illustrates the method by which we define and evaluate this type of integral. The idea is to integrate almost up to the singularity—within ε—and then to let ε tend to 0. Thus let ε be a small positive number. Because f is continuous on the interval $[a, b-\varepsilon]$, we can form the Riemann integral $\int_a^{b-\varepsilon} f(x)\,dx$. Now we let the upper limit of integration, $b-\varepsilon$, approach b, which is the same thing as letting ε tend to 0. Recall that we write this limit as $\varepsilon \to 0^+$ because ε approaches 0 through positive values. If the limit exists, then we say that the improper integral converges. Otherwise we say it diverges.

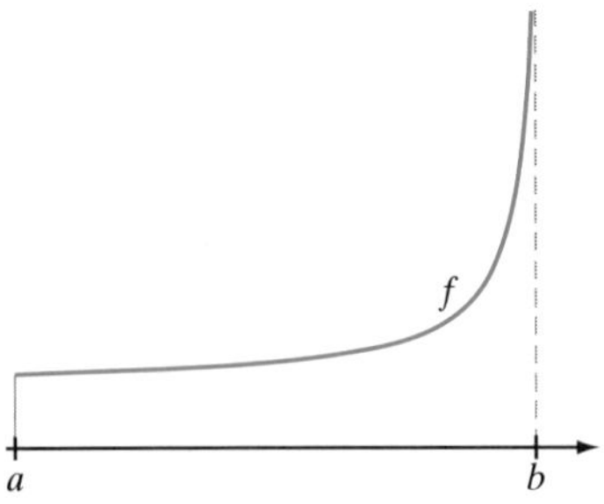

▲ Figure 1 The line $x = b$ is a vertical a asymptote.

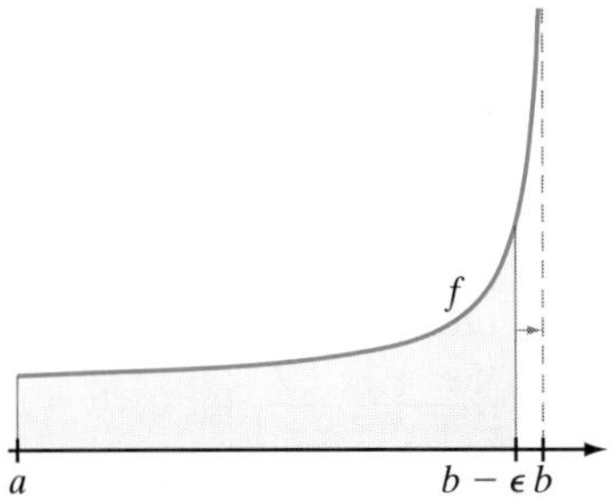

▲ Figure 2 The improper integral $\int_a^b f(x)\,dx$ is defined to be the limit of the area of the shaded region as $b-\varepsilon$ approaches b.

DEFINITION If $f(x)$ is continuous on $[a, b)$, and unbounded as x approaches b from the left, then the value of the improper integral $\int_a^b f(x)\,dx$ is defined to be

$$\int_a^b f(x)\,dx = \lim_{\varepsilon \to 0^+} \int_a^{b-\varepsilon} f(x)\,dx, \qquad (6.6.1)$$

provided that this limit exists and is finite. In this case, we say that the improper integral *converges.* Otherwise, the integral is said to *diverge.*

▶ **EXAMPLE 1** Evaluate the integral $\int_0^8 (8-x)^{-1/3}\,dx$.

Solution The given integral is improper with infinite integrand $f(x) = (8-x)^{-1/3}$ as x approaches 8. Using the antiderivative $F(x) = -\dfrac{(8-x)^{2/3}}{2/3} = -\dfrac{3}{2}(8-x)^{2/3}$ of $f(x)$ on the interval [0, 8), we calculate

$$\int_0^8 (8-x)^{-1/3}\,dx = \lim_{\varepsilon \to 0^+} \int_0^{8-\varepsilon} (8-x)^{-1/3}\,dx = \lim_{\varepsilon \to 0^+} \left(-\frac{3}{2}(8-x)^{2/3} \Big|_0^{8-\varepsilon} \right)$$

$$= -\frac{3}{2} \lim_{\varepsilon \to 0^+} \left(\varepsilon^{2/3} - 8^{2/3}\right) = -\frac{3}{2}(0-4) = 6.$$

We conclude that the given improper integral is convergent and its value is 6. ◀

INSIGHT In Example 1, we calculate the antiderivative $F(x)$ of $f(x) = (8-x)^{-1/3}$ and apply the Fundamental Theorem of Calculus on the closed interval $[0, 8-\varepsilon]$. We are justified in doing so because f is continuous on that interval. However, because f is not continuous on $[0, 8]$, we may not apply the Fundamental Theorem of Calculus directly to $\int_0^8 f(x)\,dx$. Nevertheless, it so happens that in this case, as in many others, the correct answer does result from such an application. That is because, in Example 1, the antiderivative $F(x)$ of $f(x)$ is continuous at the singularity $b = 8$. In general, if a continuous integrand f is infinite at b, and if the antiderivative F of f on $[a, b)$ is continuous at b, then

$$\int_a^b f(x)\,dx = \lim_{\varepsilon\to 0^+} \int_a^{b-\varepsilon} f(x)\,dx = \lim_{\varepsilon\to 0^+} \Big(F(b-\varepsilon) - F(a)\Big) \overset{\text{(continuity)}}{=} F(b) - F(a).$$

This is precisely the answer that would have resulted had we applied the Fundamental Theorem of Calculus *directly* to the improper integral $\int_a^b f(x)\,dx$ without justification. Of course, the continuity of F at an endpoint singularity of f should not be taken for granted.

▶ **EXAMPLE 2** Analyze the integral $\int_1^3 (x-3)^{-2}\,dx$.

Solution This is an improper integral with infinite integrand at 3. We evaluate this integral by considering

$$\begin{aligned}\lim_{\varepsilon\to 0^+} \int_1^{3-\varepsilon} \frac{1}{(x-3)^2}\,dx &= \lim_{\varepsilon\to 0^+} -(x-3)^{-1}\Big|_1^{3-\varepsilon} \\ &= \lim_{\varepsilon\to 0^+} (\varepsilon^{-1} - 2^{-1}) \\ &= +\infty.\end{aligned}$$

We conclude that the given improper integral diverges. ◀

An improper integral with integrand that is infinite at the left endpoint of integration is handled in a manner similar to the right endpoint case.

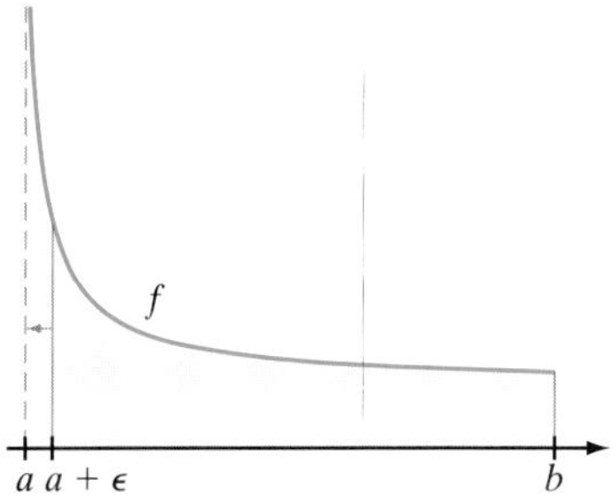

▲ Figure 3 The improper integral $\int_a^b f(x)\,dx$ is defined to be the limit of the area of the shaded region as $a + \varepsilon$ approaches a.

DEFINITION If $f(x)$ is continuous on $(a, b]$ and unbounded as x approaches a from the right, then the value of the improper integral $\int_a^b f(x)\,dx$ is defined by

$$\int_a^b f(x)\,dx = \lim_{\varepsilon\to 0^+} \int_{a+\varepsilon}^b f(x)\,dx, \tag{6.6.2}$$

provided that this limit exists and is finite. In this case, the integral is said to *converge*. Otherwise, the integral is said to *diverge*. Figure 3 illustrates this idea.

▶ **EXAMPLE 3** Evaluate $\int_0^9 x^{-1/2}\,dx$.

Solution This integral is improper with infinite integrand at 0. Following the definition, we calculate

$$\int_0^9 x^{-1/2}\,dx = \lim_{\varepsilon\to 0^+} \int_{0+\varepsilon}^9 x^{-1/2}\,dx = \lim_{\varepsilon\to 0^+} 2x^{1/2}\Big|_\varepsilon^9 = \lim_{\varepsilon\to 0^+} \left(2\sqrt{9} - 2\sqrt{\varepsilon}\right) = 6. \quad ◀$$

Integrands with Interior Singularities

An integrand frequently has a singularity at an interior point of an interval of integration. In these circumstances, we divide the interval of integration into two subintervals, one on each side of the singularity. We then integrate over each subinterval separately. If both of these integrals converge, then the original integral is said to converge. Otherwise, we say that the original integral diverges.

▶ **EXAMPLE 4** Evaluate the improper integral $\int_{-3}^{2} 8(x+1)^{-1/5}\,dx$.

Solution The integrand is unbounded as x tends to -1. Therefore we evaluate separately the two improper integrals $\int_{-3}^{-1} 8(x+1)^{-1/5}\,dx$ and $\int_{-1}^{2} 8(x+1)^{-1/5}\,dx$. For the first of these, we have

$$\begin{aligned}\int_{-3}^{-1} 8(x+1)^{-1/5}\,dx &= \lim_{\varepsilon\to 0^+}\int_{-3}^{-1-\varepsilon} 8(x+1)^{-1/5}\,dx = \lim_{\varepsilon\to 0^+}\left.\frac{8(x+1)^{4/5}}{4/5}\right|_{-3}^{-1-\varepsilon}\\ &= 10\lim_{\varepsilon\to 0^+}\left((-\varepsilon)^{4/5}-(-2)^{4/5}\right) = -10\cdot 2^{4/5}.\end{aligned}$$

The second integral is evaluated in a similar way:

$$\begin{aligned}\int_{-1}^{2} 8(x+1)^{-1/5}\,dx &= \lim_{\varepsilon\to 0^+}\int_{-1+\varepsilon}^{2} 8(x+1)^{-1/5}\,dx = \lim_{\varepsilon\to 0^+}\left(\left.\frac{8(x+1)^{4/5}}{4/5}\right|_{-1+\varepsilon}^{2}\right)\\ &= 10\lim_{\varepsilon\to 0^+}(3^{4/5}-\varepsilon^{4/5}) = 10\cdot 3^{4/5}.\end{aligned}$$

We conclude that the original integral converges and

$$\begin{aligned}\int_{-3}^{2} 8(x+1)^{-1/5}\,dx &= \int_{-3}^{-1} 8(x+1)^{-1/5}\,dx + \int_{-1}^{2} 8(x+1)^{-1/5}\,dx\\ &= -10\cdot 2^{4/5}+10\cdot 3^{4/5} = 10\left(3^{4/5}-2^{4/5}\right).\end{aligned}$$ ◀

It is dangerous to try to save work by not dividing the integral at the singularity. The next example illustrates what can go wrong.

▶ **EXAMPLE 5** Evaluate the improper integral $\int_{-2}^{2} x^{-4}\,dx$.

Solution What we *should* do is divide the given integral into the two integrals $\int_{-2}^{0} x^{-4}\,dx$ and $\int_{0}^{2} x^{-4}\,dx$. Suppose that instead we try to save work and just antidifferentiate:

$$\int_{-2}^{2} x^{-4}\,dx = -\frac{1}{3}x^{-3}\Big|_{-2}^{2} = -\frac{1}{3}\left(\frac{1}{8}-\left(\frac{1}{-8}\right)\right) = -\frac{1}{12}.$$

Clearly something is wrong. The function x^{-4} is positive; hence, its integral, if it exists, should be positive too. What has happened is that an incorrect method has led to an incorrect negative answer. In fact, the calculation

$$\lim_{\varepsilon\to 0^+}\int_{-2}^{0-\varepsilon} x^{-4}\,dx = \lim_{\varepsilon\to 0^+}\left(-\frac{1}{3}x^{-3}\Big|_{-2}^{-\varepsilon}\right) = \lim_{\varepsilon\to 0^+}\left(\frac{1}{3\varepsilon^3}-\frac{1}{24}\right) = +\infty$$

shows that $\int_{-2}^{0} x^{-4}\,dx$ diverges. Therefore by definition, the improper integral $\int_{-2}^{2} x^{-4}\,dx$ also diverges. The moral of this example is always to divide an integral at an interior singularity. ◀

INSIGHT In Example 5, we determined that $\int_{-2}^{2} x^{-4}\,dx$ is divergent by integrating on only one side of the singularity at 0. A calculation similar to that of $\int_{-2}^{0} x^{-4}\,dx$ would have shown that $\int_{0}^{2} x^{-4}\,dx$ also diverges. That calculation turned out to be unnecessary and was therefore omitted. For a conclusion of *divergence*, it is enough to show that the integral diverges on one side of the singularity. By contrast, for a conclusion of *convergence*, we must show that the integrals on *both* sides of an interior singularity converge.

Integrands That Are Unbounded at Both Endpoints

When an integrand f is continuous on an open interval (a, b) and unbounded at both endpoints a and b, then we choose an interior point c—*any* interior point will do—and investigate the two improper integrals $\int_a^c f(x)\,dx$ and $\int_c^b f(x)\,dx$. If *both* converge, then $\int_a^b f(x)\,dx$ is said to converge. In this case, $\int_a^b f(x)\,dx$ is defined to be the sum $\int_a^c f(x)\,dx + \int_c^b f(x)\,dx$. The value of this sum does not depend on the particular interior point c that has been chosen. If one of the two integrals, $\int_a^c f(x)\,dx$ and $\int_c^b f(x)\,dx$ diverges, or if both diverge, then we say that $\int_a^b f(x)\,dx$ also diverges.

▶ **EXAMPLE 6** Determine whether the improper integral $\displaystyle\int_0^1 \frac{1}{x(1-x)^{1/3}}\,dx$ converges or diverges.

Solution The integrand of the given integral is singular at both endpoints 0 and 1. Accordingly, we choose $c = 1/2$ (or any other point between the limits of integration) and begin our analysis with the improper integral $\displaystyle\int_0^{1/2} \frac{1}{x(1-x)^{1/3}}\,dx$. Notice that $0 < (1-x)^{1/3} < 1$ for $0 < x < 1$. Therefore $1 < (1-x)^{-1/3}$ and

$$\frac{1}{x} < \frac{1}{x(1-x)^{1/3}}$$

for $0 < x < 1$. It follows that

$$\int_0^{1/2} \frac{1}{x(1-x)^{1/3}}\,dx = \lim_{\varepsilon\to 0^+} \int_\varepsilon^{1/2} \frac{1}{x(1-x)^{1/3}}\,dx \geq \lim_{\varepsilon\to 0^+} \int_\varepsilon^{1/2} \frac{1}{x}\,dx = \lim_{\varepsilon\to 0^+} \left(\ln\left(\frac{1}{2}\right) - \ln(\varepsilon)\right) = \infty,$$

because $\lim_{\varepsilon\to 0^+} \ln(\varepsilon) = -\infty$. We conclude that $\displaystyle\int_0^{1/2} \frac{1}{x(1-x)^{1/3}}\,dx$ diverges. Because one of the component integrals of $\displaystyle\int_0^1 \frac{1}{x(1-x)^{1/3}}\,dx$ diverges, we conclude that the full integral on $[0, 1]$ diverges. ◀

INSIGHT If f is unbounded at both a and b, and if c is a point between, then it does not matter which of $\int_a^c f(x)\,dx$ or $\int_c^b f(x)\,dx$ is examined first. If the first integral analyzed is divergent (as in Example 6), then $\int_a^b f(x)\,dx$ diverges; an investigation of the behavior on the other side of c is not needed in this case. However, if the first integral analyzed is convergent, then it is the integral on the other side of c that tells you the behavior of $\int_a^b f(x)\,dx$. In Example 6, had we initiated our investigation of $\displaystyle\int_0^1 \frac{1}{x(1-x)^{1/3}}\,dx$ with $\displaystyle\int_{1/2}^1 \frac{1}{x(1-x)^{1/3}}\,dx$, we would have found the integral to the right of 1/2 convergent (Exercise 63). No conclusion would have been possible from that information. We would have had to turn to the integral to the left of 1/2 before coming to a conclusion.

Proving Convergence Without Evaluation

Suppose that f and g are continuous, positive functions on $[a, b)$, unbounded as $x \to b^-$ and that $0 \leq f(x) \leq g(x)$ for all $a < x < b$. If we are able to find an antiderivative of g and use it to show that $\lim_{\varepsilon\to 0^+} \int_a^{b-\varepsilon} g(x)\,dx$ exists and is finite, then we may assert the same conclusions for f. The reason is that $\int_a^{b-\varepsilon} f(x)\,dx$ increases as ε decreases (because f is positive and the area under its graph increases as its right endpoint moves further to the right). Furthermore, $\int_a^b g(x)\,dx$ is an upper bound for these integrals because

$$\int_a^{b-\varepsilon} f(x)\,dx \le \int_a^{b-\varepsilon} g(x)\,dx \le \int_a^b g(x)\,dx < \infty\,.$$

It follows from the completeness property of the real numbers that $\lim_{\varepsilon\to 0^+} \int_a^{b-\varepsilon} f(x)\,dx$ exists and is finite. The next theorem states this observation in a somewhat more general form.

THEOREM 1 **(Comparison Theorem for Unbounded Integrands)** Suppose that f and g are continuous functions on the interval (a, b), that $0 \le f(x) \le g(x)$ for all $a < x < b$, and that $f(x)$ and $g(x)$ are unbounded as $x \to a^+$, or as $x \to b^-$, or as $x \to a^+$ and $x \to b^-$.

a. If $\int_a^b g(x)\,dx$ is convergent, then $\int_a^b f(x)\,dx$ is also convergent and $\int_a^b f(x)\,dx \le \int_a^b g(x)\,dx$.
b. If $\int_a^b f(x)\,dx$ is divergent, then $\int_a^b g(x)\,dx$ is also divergent.

Notice that the Comparison Theorem does *not* say anything about $\int_a^b f(x)\,dx$ if $\int_a^b g(x)\,dx$ is convergent and g is *less* than f. To use the Comparison Theorem to show that an improper integral $\int_a^b f(x)\,dx$ is convergent, we must find a function g that is greater than or equal to f and for which $\int_a^b g(x)\,dx$ is convergent. If A and p are positive constants with $p < 1$, then $g(x) = A/x^p$ is often a useful comparison function for proving convergence on an interval of the form $[0, b]$. See Exercise 55. With the same constraints on A and p, the function $g(x) = A/(b-x)^p$ is often a useful comparison function for proving convergence on the interval $[a, b]$. See Exercise 56.

▶ **EXAMPLE 7** Show that the improper integral $\displaystyle\int_0^9 \frac{1+\cos(x^2)}{\sqrt{x}}\,dx$ is convergent.

Solution The integrand $f(x) = \left(1+\cos(x^2)\right)/\sqrt{x}$ is unbounded at its left endpoint $a = 0$. Consequently, formula (6.6.2) applies. The problem is that we cannot find an antiderivative for $f(x)$. As a result, we cannot calculate $\int_{0+\varepsilon}^9 f(x)\,dx$. We therefore need an indirect way to show that $\int_0^9 f(x)\,dx$ is convergent. The solution is to find a simpler function g to which we can compare f. To that end, we replace the numerator of f with something that is less complicated. Because $-1 \le \cos(x^2) \le 1$, we have

$$0 \le \frac{1+\cos(x^2)}{\sqrt{x}} \le \frac{2}{\sqrt{x}}.$$

We therefore take $g(x) = 2/\sqrt{x}$ as the function for comparison. From Example 3, we see that $\int_0^9 g(x)\,dx = 2\int_0^9 x^{-1/2} = 12$. Therefore by the Comparison Theorem, we conclude that $\displaystyle\int_0^9 \frac{1+\cos(x^2)}{\sqrt{x}}\,dx$ is convergent and $\displaystyle\int_0^9 \frac{1+\cos(x^2)}{\sqrt{x}}\,dx \le 12\,.$ ◀

INSIGHT The upper bound $\int_a^b g(x)\,dx$ stated in part a of the Comparison Theorem is not necessarily a good approximation of $\int_a^b f(x)\,dx$. For example, numerical techniques show that $\displaystyle\int_0^9 \frac{1+\cos(x^2)}{\sqrt{x}}\,dx \approx 7.663$, which is quite a bit smaller than 12, the upper bound obtained in Example 7. Nevertheless, the Comparison Theorem can be used as part of an effective approximation strategy. See Exercises 72–77.

QUICK QUIZ

1. Calculate $\int_0^9 x^{-1/2}\,dx$.
2. Calculate $\int_0^{16} (16-x)^{-1/4}\,dx$.
3. True or false: If f is unbounded at both a and b, and if c is a point in between, then $\int_a^b f(x)\,dx$ is divergent if and only if both $\int_a^c f(x)\,dx$ and $\int_c^b f(x)\,dx$ are divergent.
4. Use the Comparison Theorem to determine which of the following improper integrals converge:

a) $\displaystyle\int_0^{\pi/4} \frac{\sec(x)}{x^2}\,dx$; b) $\displaystyle\int_{-1}^{8} \frac{x^{4/3}}{\sqrt{1+x}}\,dx$; c) $\displaystyle\int_0^{1} \frac{\exp(x)}{\sqrt{x}}\,dx$; d) $\displaystyle\int_0^{1} \frac{\sqrt{1+x}}{x}\,dx$.

Answers

1. 6 2. 32/3 3. False 4. (b) and (c)

EXERCISES

Problems for Practice

In each of Exercises 1–10, determine whether the given improper integral is convergent or divergent. If it converges, then evaluate it.

1. $\int_1^5 (x-5)^{-4/3}\,dx$
2. $\displaystyle\int_{-3}^{-2} \frac{1}{x+2}\,dx$
3. $\int_2^4 (4-x)^{-0.9}\,dx$
4. $\int_1^2 (x-2)^{-1/5}\,dx$
5. $\int_0^{\pi/2} \tan(x)\,dx$
6. $\int_0^{\pi/2} \sec^2(x)\,dx$
7. $\displaystyle\int_0^1 \frac{x}{(1-x^2)^{1/4}}\,dx$
8. $\displaystyle\int_0^1 \frac{1}{\sqrt{1-x^2}}\,dx$
9. $\displaystyle\int_0^2 \frac{1}{4-x^2}\,dx$
10. $\displaystyle\int_{-2}^{-1} \frac{1}{x\sqrt{x^2-1}}\,dx$

In each of Exercises 11–20, determine whether the given improper integral is convergent or divergent. If it converges, then evaluate it.

11. $\int_{-3}^2 (x+3)^{-1.1}\,dx$
12. $\int_{-4}^0 (x+4)^{-0.1}\,dx$
13. $\int_0^8 x^{-1/3}\,dx$
14. $\int_{-1}^3 (x+1)^{-3}\,dx$
15. $\displaystyle\int_0^1 \frac{\ln(x)}{x}\,dx$
16. $\displaystyle\int_4^{13} \frac{1}{\sqrt{x-4}}\,dx$
17. $\int_0^3 x^{-1/2}(1+x)\,dx$
18. $\displaystyle\int_1^e \frac{1}{x\cdot\ln^{1/3}(x)}\,dx$
19. $\int_0^1 \ln(x)\,dx$
20. $\displaystyle\int_1^{\sqrt{2}} \frac{1}{x\sqrt{x^2-1}}\,dx$

In each of Exercises 21–30, determine whether the given improper integral is convergent or divergent. If it converges, then evaluate it.

21. $\displaystyle\int_0^2 \frac{1}{x-1}\,dx$
22. $\int_{-1}^3 6x^{-1/7}\,dx$
23. $\displaystyle\int_{-5}^{-1} \frac{3}{(x+3)^{2/5}}\,dx$

24. $\int_3^6 (x-4)^{-2}\,dx$
25. $\int_{-2}^4 (x+1)^{-2/3}\,dx$
26. $\displaystyle\int_{1/e}^{e} \frac{1}{x\ln^{2/7}(x)}\,dx$
27. $\displaystyle\int_0^3 \frac{x}{x^2-2}\,dx$
28. $\int_{-3}^3 x^{-1/3}(x+1)\,dx$
29. $\displaystyle\int_1^{16} \frac{1}{\sqrt{x}}\left(\frac{1}{2-\sqrt{x}}\right)^{1/3} dx$
30. $\displaystyle\int_0^2 \frac{1}{1-x^{1/3}}\,dx$

In each of Exercises 31–40, determine whether the given improper integral is convergent or divergent. If it converges, then evaluate it.

31. $\displaystyle\int_{-1}^1 \frac{1}{1-x^2}\,dx$
32. $\displaystyle\int_{-2}^2 \frac{1}{\sqrt{4-x^2}}\,dx$
33. $\displaystyle\int_0^1 \frac{1}{\sqrt{x}(1-x)}\,dx$
34. $\displaystyle\int_1^2 \frac{1}{\sqrt{3x-x^2-2}}\,dx$
35. $\displaystyle\int_{-1}^1 \frac{x}{(1-x^2)^{1/4}}\,dx$
36. $\displaystyle\int_0^1 \frac{1}{x\ln(x)}\,dx$
37. $\displaystyle\int_{-1}^4 \frac{1}{\sqrt{1+x}\sqrt{4-x}}\,dx$
38. $\displaystyle\int_{-1}^0 \frac{3}{x^{1/3}(x+1)}\,dx$
39. $\displaystyle\int_0^4 \frac{1}{\sqrt{x}}\left(\frac{1}{2-\sqrt{x}}\right)^{1/3} dx$
40. $\displaystyle\int_0^1 \frac{1}{\sqrt{x}\sqrt{1-x}}\,dx$

Further Theory and Practice

In each of Exercises 41–54, determine whether the given improper integral is convergent or divergent. If it converges, then evaluate it.

41. $\int_{-5}^2 \ln(x+5)\,dx$
42. $\displaystyle\int_0^1 \frac{\exp(-1/x)}{x^2}\,dx$
43. $\displaystyle\int_0^2 \frac{\ln^{1/3}(x)}{x}\,dx$
44. $\int_0^1 \ln^2(x)\,dx$
45. $\int_0^4 x^{-1/2}\ln(x)\,dx$
46. $\int_0^9 x^{-3/2}\ln(x)\,dx$
47. $\displaystyle\int_0^8 \frac{1}{x^{1/3}}\ln\left(\frac{1}{x^{1/3}}\right) dx$
48. $\int_{-1}^3 \ln(|x|)\,dx$
49. $\int_{-1}^1 \ln(1-x^2)\,dx$
50. $\displaystyle\int_0^{\pi/2} \frac{\cos(x)}{\sqrt{\sin(x)}}\,dx$
51. $\displaystyle\int_0^{\pi/2} \frac{\cos^3(x)}{\sin^{1/3}(x)}\,dx$
52. $\displaystyle\int_0^1 \frac{1}{(1+x^2)\sqrt{\arctan(x)}}\,dx$
53. $\displaystyle\int_0^1 \frac{\arcsin(x)}{\sqrt{1-x^2}}\,dx$
54. $\displaystyle\int_0^1 \frac{1}{\sqrt{\sqrt{x+1}-1}\sqrt{x+1}}\,dx$
55. Suppose that p and b are positive numbers. Show that $\displaystyle\int_0^b \frac{1}{x^p}\,dx$ is convergent if and only if $p<1$.
56. Suppose that p is positive and that $0<a<b$. Show that $\displaystyle\int_a^b \frac{1}{(b-x)^p}\,dx$ is convergent if and only if $p<1$.
57. For what positive values of p is the improper integral $\displaystyle\int_1^e \frac{1}{x\ln^p(x)}\,dx$ convergent?

In each of Exercises 58–69 use the Comparison Theorem to determine whether the given improper integral is convergent or divergent. In some cases, you may have to break up the integration before applying the Comparison Theorem.

58. $\displaystyle\int_0^{\pi/2} \frac{\cos(x)}{\sqrt{x}}\,dx$
59. $\displaystyle\int_0^1 \frac{e^x}{\sqrt{x}}\,dx$
60. $\displaystyle\int_2^6 \sqrt{\frac{3+\sin(x)}{x-2}}\,dx$
61. $\displaystyle\int_0^3 \frac{\sqrt{1+x}}{x^{2/3}}\,dx$
62. $\displaystyle\int_0^1 \frac{\sin(x)}{(1-x)^{2/3}}\,dx$
63. $\displaystyle\int_{1/2}^1 \frac{1}{x(1-x)^{1/3}}\,dx$
64. $\displaystyle\int_0^{\pi/2} \frac{1}{\sqrt{x}+\sin(x)}\,dx$
65. $\displaystyle\int_0^{1/3} \frac{\sec(\pi x)}{1-3x}\,dx$

66. $\int_0^1 x^{-1/2}(1-x)^{-3/4}\,dx$

67. $\displaystyle\int_1^2 \frac{1}{\sqrt{x^3+x^2-2}}\,dx$

68. $\displaystyle\int_0^{\pi/2} \frac{\sqrt{x}}{\sin(x)}\,dx$

69. $\displaystyle\int_1^3 \frac{1}{(x-1)^{5/3}\sqrt{3-x}}\,dx$

70. Evaluate the improper integral $\int_5^9 \sqrt{\frac{x+7}{x-5}}\,dx$ by making the direct substitution $u=x-5$ followed by the indirect substitution $u=12\tan^2(\theta)$.

71. Suppose that p is positive and q is nonnegative. Let B denote the *beta function* defined by

$$B(p,q)=\int_0^1 x^{-1+p}(1-x)^{1+q}dx.$$

a. Show that

$$B(p,0)=\frac{1!}{p(p+1)}.$$

b. For $q\geq 1$, show that

$$B(p,q)=\frac{1+q}{p}B(p+1,q-1).$$

c. Use parts a and b to find formulas for $B(p,1)$ and $B(p,2)$.

Calculator/Computer Exercises

In each of Exercises 72–77, the given function $f(x)$ is unbounded as $x\to 0^+$. Determine a function $g(x)=cx^p$ such that (a) $0\leq f(x)\leq g(x)$ for each x in $(0,1]$, and (b) $\int_0^1 g(x)\,dx$ is convergent. This shows that $\int_0^1 f(x)\,dx$ is convergent by the Comparison Theorem. By determining a positive ε such that $\int_0^\varepsilon g(x)\,dx<5\times 10^{-4}$, approximate $\int_0^1 f(x)\,dx$ to three decimal places.

72. $f(x)=x^{-1/2}\cos(x^2)$

73. $f(x)=\sqrt{1+1/x}$

74. $f(x)=(2+\sin(x))/x^{1/3}$

75. $f(x)=\sqrt{x}\csc(x)$

76. $f(x)=\arccos(x)/x^{5/6}$

77. $f(x)=\ln(2+x)/x^{\ln(2)}$

6.7 Improper Integrals—Unbounded Intervals

Suppose that we want to calculate the integral of a continuous function over an unbounded interval of the form $(-\infty,b)$ or $(a,+\infty)$. Like the integrals with unbounded integrand that we studied in the preceding section, these integrals are said to be improper. The theory of the integral that you learned in Chapter 5 does not cover these improper integrals, and some new concepts are needed.

The Integral on an Infinite Interval

The idea that enables us to define improper integrals over unbounded intervals is similar to the one we used in the preceding section for improper integrals with unbounded functions. Instead of trying to integrate over an unbounded interval all at once, we integrate over a bounded interval and then we let an endpoint tend to infinity.

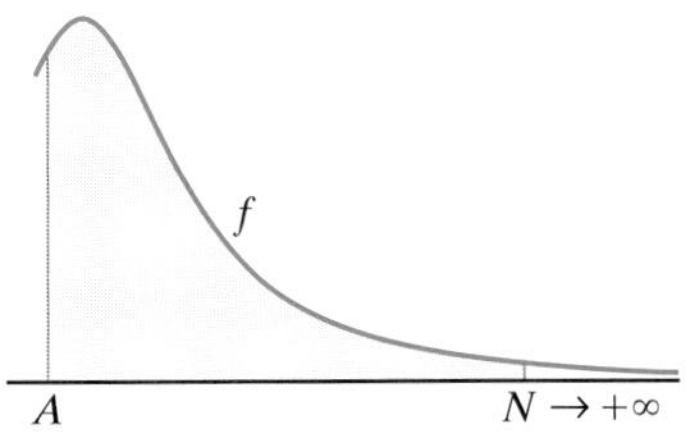

▲ Figure 1a The improper integral $\int_A^\infty f(x)\,dx$ is defined to be $\lim_{N\to+\infty}\int_A^N f(x)\,dx$.

DEFINITION Let f be a continuous function on the interval $[a,\infty)$. The improper integral $\int_a^\infty f(x)\,dx$ is defined by

$$\int_a^\infty f(x)\,dx=\lim_{N\to+\infty}\int_a^N f(x)\,dx$$

provided that the limit exists and is finite. When the limit exists, the integral is said to *converge*. Otherwise, it is said to *diverge* (see Figure 1a). Similarly, if g is a continuous function on the interval $(-\infty,b)$, then the value of the improper integral $\int_{-\infty}^b g(x)\,dx$ is defined to be

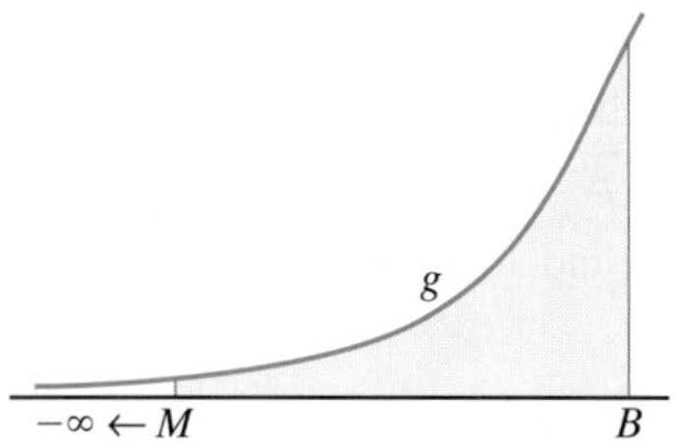

▲ Figure 1b The improper integral $\int_{-\infty}^{B} g(x)\,dx$ is defined to be $\lim_{M\to+\infty} \int_{M}^{B} g(x)\,dx$.

$$\lim_{M\to-\infty} \int_{M}^{b} g(x)\,dx$$

provided that the limit exists and is finite. When the limit exists, the integral is said to *converge*. Otherwise it is said to *diverge* (see Figure 1b).

▶ **EXAMPLE 1** Calculate the improper integral $\int_1^{\infty} x^{-2}\,dx$.

Solution According to the definition, we have

$$\int_1^{\infty} x^{-2}\,dx = \lim_{N\to\infty} \int_1^{N} x^{-2}\,dx = \lim_{N\to\infty} \left(-\frac{1}{x}\Big|_1^N \right) = \lim_{N\to\infty} \left(-\frac{1}{N} - \left(-\frac{1}{1}\right)\right) = 1.$$

We conclude that the integral converges and has value 1. ◀

▶ **EXAMPLE 2** Determine whether the improper integral $\int_{-\infty}^{-8} x^{-1/3}\,dx$ converges or diverges.

Solution We do this by evaluating the limit

$$\lim_{M\to-\infty} \int_{M}^{-8} x^{-1/3}\,dx = \lim_{M\to-\infty} \left(\frac{x^{2/3}}{2/3}\Big|_M^{-8} \right) = \frac{3}{2} \lim_{M\to-\infty} \left((-8)^{2/3} - M^{2/3} \right) = -\infty.$$

We conclude that the integral diverges. ◀

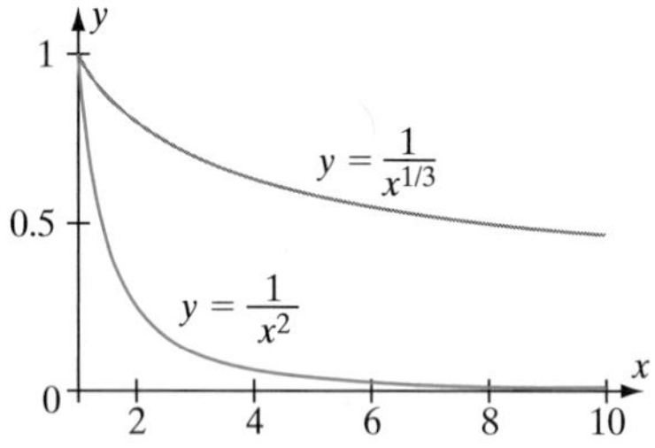

▲ Figure 2

INSIGHT Examples 1 and 2 are concerned with integrands of the form $1/x^p$ (with $p = 2$ and $p = 1/3$, respectively). In general, if $p > 0$, then $f(x) = 1/x^p$ decays to 0 as $x \to \infty$. However, Example 2 shows that the limit property $\lim_{x\to\infty} f(x) = 0$ does not by itself guarantee that $\int_a^{\infty} f(x)\,dx$ converges. In the case of $f(x) = 1/x^{1/3}$, the integrand does not decay to 0 sufficiently rapidly as $x \to \infty$. By contrast, $1/x^2$ tends to 0 much more quickly, as Figure 2 shows. This more rapid decay results in the convergence of $\int_1^{\infty} (1/x^2)\,dx$, as we know from Example 1. The next example analyzes the convergence of $\int_1^{\infty} (1/x^p)\,dx$ in complete generality.

▶ **EXAMPLE 3** Suppose that $a > 0$. Show that the improper integral $\displaystyle\int_a^{\infty} \frac{1}{x^p}\,dx$ diverges for $0 < p \le 1$ and converges for $1 < p$.

Solution For $p = 1$, we have

$$\int_a^{\infty} \frac{1}{x^p}\,dx = \lim_{N\to\infty} \int_a^{N} \frac{1}{x}\,dx = \lim_{N\to\infty} \left(\ln(|x|)\big|_a^N \right) = \lim_{N\to\infty} \left(\ln(N) - \ln(a) \right) = \infty,$$

and the improper integral is divergent. For $p \ne 1$, we have

$$\lim_{N\to\infty} \int_a^{N} \frac{1}{x^p}\,dx = \lim_{N\to\infty} \left(\frac{x^{1-p}}{1-p}\Big|_a^N \right) = \lim_{N\to\infty} \left(\frac{N^{1-p}}{1-p} + \frac{a^{1-p}}{p-1} \right) = \frac{a^{1-p}}{p-1} + \lim_{N\to\infty} \frac{N^{1-p}}{1-p}.$$

If $0<p<1$, then $1-p>0$, and $\lim_{N\to\infty} N^{1-p} = \infty$. Therefore $\int_a^\infty \frac{1}{x^p}\,dx$ is divergent for $0<p<1$. If $1<p$, then $0<p-1$, and $\lim_{N\to\infty} N^{1-p} = \lim_{N\to\infty} \frac{1}{N^{p-1}} = 0$. It follows that $\int_a^\infty \frac{1}{x^p}\,dx$ is convergent for $1<p$. Moreover, $\int_a^\infty \frac{1}{x^p}\,dx = \frac{a^{1-p}}{p-1}$. ◀

An Application to Finance

If an initial investment P_0 is allowed to compound continuously at an annual interest rate of r percent, then formula (2.6.16) tells us it will grow to $P_0 e^{rt/100}$ in t years. In other words, if we let $\rho = r/100$, then the growth factor is $e^{\rho t}$. To attain a sum equal to P in t years, we must invest $Pe^{-\rho t}$ now. We say that $Pe^{-\rho t}$ is the *present value* of an amount P that is to be received t years from now. In this context, the interest rate, r percent, is called the *discount rate*.

Next, instead of a single payment in the future, consider a future *income stream*. We assume that the income will commence T_1 years in the future and that it will terminate at time $T_2 > T_1$. Let $f(t)$ denote the rate at which the income arrives at time t for $0 \le T_1 \le t \le T_2 < \infty$. Let us figure out the present value of this income stream. Divide the interval $[T_1, T_2]$ into small subintervals by means of the uniform partition $T_2 = t_0 < t_1 < \ldots < t_N = T_2$. If we choose a point s_j in $[t_{j-1}, t_j]$ for each j, then we can approximate the money earned over the time interval $[t_{j-1}, t_j]$ by $f(s_j)\cdot\Delta t$. The present value of this amount is about $(f(s_j)\cdot\Delta t)\cdot\exp(-\rho s_j)$. Thus the present value of the entire income stream is about

$$\sum_{j=1}^{N} f(s_j)\exp(-\rho s_j)\,\Delta t.$$

This quantity is a Riemann sum for the integral $\int_{T_1}^{T_2} f(t)e^{-\rho t}\,dt$ that results when $N\to\infty$ and $\Delta t \to 0$. We have found that, if the current annual interest rate is r percent and $\rho = r/100$, then the present value PV of the income stream $f(t)$ for $T_1 \le t \le T_2$ is given by

$$PV = \int_{T_1}^{T_2} f(t)e^{-\rho t}\,dt.$$

When an income stream is to continue in perpetuity, we let T_2 tend to infinity. In other words, the improper integral $\int_{T_1}^{\infty} f(t)e^{-\rho t}\,dt$ is the present value of an income stream that begins T_1 years in the future and continues in perpetuity. (Such perpetuities are rare, but they do exist. One British government perpetuity issued in 1752 still trades.)

▶ **EXAMPLE 4** **(Present Value of a Perpetuity)** Suppose that a trust is established that, beginning immediately, pays $2t + 50$ dollars per year in perpetuity, where t is time measured in years with $t = 0$ corresponding to the present. Assuming a discount rate of 6%, what is the present value of all the money paid out by this trust account?

Solution Here $T_1 = 0$, $r = 6$, and $\rho = r/100 = 0.06$. The present value PV of the perpetuity is given by

$$PV = \int_0^\infty (2t+50)e^{-0.06t}\,dt = \lim_{N\to\infty}\int_0^N \underbrace{(2t+50)}_{u}\,\underbrace{e^{-0.06t}\,dt}_{dv} = \lim_{N\to\infty}\left(\underbrace{(2t+50)}_{u}\underbrace{\frac{e^{-0.06t}}{-0.06}}_{v}\Bigg|_0^N - \int_0^N \underbrace{\frac{e^{-0.06t}}{-0.06}}_{v}\underbrace{2\,dt}_{du}\right).$$

Thus

$$\begin{aligned}
PV &= \lim_{N\to\infty}\left((2t+50)\frac{e^{-0.06t}}{-0.06} - 2\frac{e^{-0.06t}}{(-0.06)^2}\bigg|_0^N\right)\\
&= \lim_{N\to\infty}\left(\left((2N+50)\frac{e^{-0.06N}}{-0.06} - 2\frac{e^{-0.06N}}{(-0.06)^2}\right) - \left((50)\frac{e^{-0}}{-0.06} - 2\frac{e^{-0}}{(-0.06)^2}\right)\right)\\
&= \frac{2}{-0.06}\left(\lim_{N\to\infty} N e^{-0.06N}\right) - (-1388.89).
\end{aligned}$$

We finish our computation by using l'Hôpital's Rule to evaluate the limit, which is the indeterminate form $\infty \cdot 0$. Thus

$$\lim_{N\to\infty} N e^{-0.06N} = \lim_{N\to\infty}\frac{N}{e^{0.06N}} = \lim_{N\to\infty}\frac{\dfrac{d}{dN}N}{\dfrac{d}{dN}e^{0.06N}} = \lim_{N\to\infty}\frac{1}{0.06e^{0.06N}} = 0.$$

It follows that the present value of the annuity is \$1388.89. ◀

Integrals Over the Entire Real Line

To integrate a continuous function f over the entire real line $(-\infty,\infty)$, we choose a point c—*any* point will do—and investigate the two improper integrals $\int_{-\infty}^{c} f(x)\,dx$ and $\int_{c}^{\infty} f(x)\,dx$. If *both* converge, then $\int_{-\infty}^{\infty} f(x)\,dx$ is said to converge. In this case, $\int_{-\infty}^{\infty} f(x)\,dx$ is defined to be the sum $\int_{-\infty}^{c} f(x)\,dx + \int_{c}^{\infty} f(x)\,dx$. The value of this sum does not depend on the particular point c that has been chosen. Thus

$$\int_{-\infty}^{\infty} f(x)\,dx = \lim_{M\to-\infty}\int_{M}^{c} f(x)\,dx + \lim_{N\to+\infty}\int_{c}^{N} f(x)\,dx \qquad (6.7.1)$$

provided that both limits exist and are finite. Otherwise we say that $\int_{-\infty}^{\infty} f(x)\,dx$ diverges.

▶ **EXAMPLE 5** Determine whether the improper integral $\displaystyle\int_{-\infty}^{\infty}\frac{x}{1+x^2}\,dx$ converges or diverges.

Solution We may divide the line at any point. The number 0 makes a natural choice. We begin our analysis with the improper integral $\displaystyle\int_{0}^{\infty}\frac{x}{1+x^2}\,dx$. Using the substitution $u = 1+x^2$, $du = 2x\,dx$, we calculate

$$\lim_{N\to\infty}\int_0^N \frac{x}{1+x^2}\,dx = \frac{1}{2}\lim_{N\to\infty}\int_1^{1+N^2}\frac{1}{u}\,du = \frac{1}{2}\lim_{N\to\infty}\left(\ln(|u|)\Big|_1^{1+N^2}\right) = \frac{1}{2}\lim_{N\to\infty}\ln(1+N^2) = \infty.$$

Because $\displaystyle\int_0^{\infty}\frac{x}{1+x^2}\,dx$ diverges, we conclude that $\displaystyle\int_{-\infty}^{\infty}\frac{x}{1+x^2}\,dx$ also diverges. (A similar computation shows that $\displaystyle\lim_{M\to-\infty}\int_M^0 \frac{x}{1+x^2}\,dx = -\infty$, but this calculation is not necessary. If even one of the limits on the right side of (6.7.1) is infinite, then that is enough to tell us that the integral on the left side of (6.7.1) is divergent.) ◀

INSIGHT It is important that the limits on the right side of equation (6.7.1) be calculated independently. In particular, taking $M = -N$ and calculating one limit is an incorrect shortcut. In the case of Example 5, we have $\int_{-N}^{N} \frac{x}{1+x^2}\,dx = 0$ for every N. It follows that $\lim_{N\to\infty} \int_{-N}^{N} \frac{x}{1+x^2}\,dx = 0$. As Example 5 shows, the improper integral $\int_{-\infty}^{\infty} \frac{x}{1+x^2}\,dx$ is divergent. It does not have 0 or any other number as its value.

▶ **EXAMPLE 6** Evaluate the improper integral $\int_{-\infty}^{\infty} \frac{1}{1+x^2}\,dx$.

Solution To evaluate this integral, we must break up the interval $(-\infty, \infty)$ into two pieces. The subintervals $(-\infty, 0]$ and $[0, \infty)$ are a natural choice, but the selection of 0 as a place to break the interval is not important: Any other point would do in this example. Next, we evaluate separately the improper integrals $\int_{0}^{\infty} \frac{1}{1+x^2}\,dx$ and $\int_{-\infty}^{0} \frac{1}{1+x^2}\,dx$ that result from our choice. For the first of these two improper integrals, we have

$$\int_{0}^{\infty} \frac{1}{1+x^2}\,dx = \lim_{N\to\infty} \int_{0}^{N} \frac{1}{1+x^2}\,dx = \lim_{N\to\infty} \arctan(x)\Big|_{0}^{N} = \lim_{N\to\infty} \left(\arctan(N) - 0\right) = \frac{\pi}{2}.$$

The second integral evaluates to $\pi/2$ in a similar manner. Because each of the integrals on the half line is convergent, we conclude that the original improper integral over the entire real line is convergent and that its value is $\pi/2 + \pi/2$, or π. ◀

Proving Convergence Without Evaluation

In the preceding section, you learned a comparison theorem for unbounded integrands. Similar theorems pertain to integrals over infinite intervals. We state such a theorem for the interval $[a, \infty)$. Analogous results hold for improper integrals over $(-\infty, b]$ and $(-\infty, \infty)$.

THEOREM 1 (Comparison Theorem for Integrals over Unbounded Intervals) Suppose that f and g are continuous functions on the interval $[a, \infty)$ and that $0 \le f(x) \le g(x)$ for all $a < x$.

a. If $\int_a^\infty g(x)\,dx$ is convergent, then $\int_a^\infty f(x)\,dx$ is also convergent and $\int_a^\infty f(x)\,dx \le \int_a^\infty g(x)\,dx$.
b. If $\int_a^\infty f(x)\,dx$ is divergent, then $\int_a^\infty g(x)\,dx$ is also divergent.

Notice that the Comparison Theorem does *not* say anything about $\int_a^\infty f(x)\,dx$ if $\int_a^\infty g(x)\,dx$ is convergent and g is *less* than f. To use the Comparison Theorem to show that an improper integral $\int_a^\infty f(x)\,dx$ is convergent, we must find a function g that is greater than or equal to f and for which $\int_a^b g(x)\,dx$ is convergent. If A and p are positive constants with $p > 1$, then $g(x) = A/x^p$ is often a useful comparison function for proving convergence on an interval of the form $[a, \infty)$. Refer to Example 3.

▶ **EXAMPLE 7** Show that $\int_0^\infty e^{-x^2}\,dx$ is convergent.

Solution We will compare $f(x) = e^{-x^2}$ to $g(x) = e^{-x}$ on the interval $[1, \infty)$. Notice that $x \le x^2$ for $1 \le x$. Therefore $-x^2 \le -x$ and $e^{-x^2} \le e^{-x}$ for $1 \le x$. Next, we calculate

$$\lim_{N\to\infty}\int_1^N g(x)\,dx = \lim_{N\to\infty}\int_1^N e^{-x}\,dx = \lim_{N\to\infty}\left(-e^{-x}\Big|_1^N\right) = \frac{1}{e} - \lim_{N\to\infty}\left(\frac{1}{e^N}\right) = \frac{1}{e}.$$

From this calculation, we conclude that $\int_1^\infty g(x)\,dx$ is convergent. From the Comparison Theorem, we conclude that $\int_1^\infty f(x)\,dx$ is also convergent. In other words, $\lim_{N\to\infty}\int_1^N e^{-x^2}\,dx$ exists and is finite. It follows that

$$\lim_{N\to\infty}\int_0^N e^{-x^2}\,dx = \lim_{N\to\infty}\left(\int_0^1 e^{-x^2}\,dx + \int_1^N e^{-x^2}\,dx\right) = \int_0^1 e^{-x^2}\,dx + \lim_{N\to\infty}\int_1^N e^{-x^2}\,dx$$

exists and is finite, which establishes that $\int_0^\infty e^{-x^2}\,dx$ is convergent.

INSIGHT The Comparison Theorem can be very useful for approximating a convergent improper integral that cannot be evaluated exactly. (See Exercises 81–84.) As it happens, there are several advanced techniques that yield the exact value of the improper integral in Example 7. The resulting formula,

$$\int_0^\infty e^{-x^2}\,dx = \frac{1}{2}\sqrt{\pi}, \tag{6.7.2}$$

has fundamental importance in probability and statistics.

QUICK QUIZ

1. Calculate $\int_0^\infty e^{-x}\,dx$.
2. Calculate $\int_{-\infty}^1 \frac{1}{1+x^2}\,dx$.
3. True or false: If f is continuous on $(0, \infty)$, unbounded at 0, and if $c > 0$, then $\int_0^\infty f(x)\,dx$ is convergent if and only if both improper integrals $\int_0^c f(x)\,dx$ and $\int_c^\infty f(x)\,dx$ are convergent.
4. Use the Comparison Theorem to determine which of the following improper integrals converge:

   a. $\displaystyle\int_1^\infty \frac{\sin(1/x)}{x^2}\,dx$; b. $\displaystyle\int_1^\infty \frac{2+\sin(x)}{x}\,dx$; c. $\displaystyle\int_1^\infty \frac{1+x}{1+x^2}\,dx$;

   d. $\displaystyle\int_0^\infty \frac{\exp(-x^2)}{1+x}\,dx$.

Answers

1. 1 2. $3\pi/4$ 3. True 4. (a) and (d)

EXERCISES

Problems for Practice

▶ In each of Exercises 1–20, determine whether the given improper integral converges or diverges. If it converges, then evaluate it. ◀

1. $\int_3^\infty x^{-3/2}\,dx$

2. $\displaystyle\int_9^\infty \frac{1}{\sqrt{x}}\,dx$

3. $\displaystyle\int_{-1}^\infty \frac{1}{(3+x)^{3/2}}\,dx$

4. $\int_4^\infty \frac{1}{1+x}\,dx$
5. $\int_0^\infty \frac{1}{1+x^2}\,dx$
6. $\int_0^\infty \frac{x}{1+x^2}\,dx$
7. $\int_0^\infty \frac{x}{(1+x^2)^2}\,dx$
8. $\int_2^\infty e^{-x/2}\,dx$
9. $\int_2^\infty \frac{1}{x(x-1)}\,dx$
10. $\int_0^\infty \frac{e^x}{e^{2x}+1}\,dx$
11. $\int_0^\infty \frac{e^x}{(e^x+1)^3}\,dx$
12. $\int_{-1}^\infty \frac{1}{x^2+5x+6}\,dx$
13. $\int_0^\infty \frac{6x^2+8}{(x^2+1)(x^2+2)}\,dx$
14. $\int_0^\infty \left(\frac{x+1}{x^2+1}\right)^2 dx$
15. $\int_1^\infty xe^{-3x^2}\,dx$
16. $\int_0^\infty xe^{-2x}\,dx$
17. $\int_e^\infty \frac{1}{x\ln(x)}\,dx$
18. $\int_e^\infty \frac{1}{x\ln^2(x)}\,dx$
19. $\int_1^\infty \frac{\arctan(x)}{1+x^2}\,dx$
20. $\int_0^\infty (2/3)^x\,dx$

In each of Exercises 21–36, determine whether the given improper integral converges or diverges. If it converges, then evaluate it.

21. $\int_{-\infty}^{-2} x^{-3}\,dx$
22. $\int_{-\infty}^{-2} x^{-1/3}\,dx$
23. $\int_{-\infty}^{-2} \frac{1}{(1+x)^{4/3}}\,dx$
24. $\int_{-\infty}^{2} \frac{1}{\sqrt{3-x}}\,dx$
25. $\int_{-\infty}^{2} \frac{1}{(3-x)^{3/2}}\,dx$
26. $\int_{-\infty}^{1} \frac{x}{(1+x^2)^2}\,dx$
27. $\int_{-\infty}^{0} \frac{1}{(1+x^2)^{3/2}}\,dx$
28. $\int_{-\infty}^{0} \frac{1}{(1+x^2)^2}\,dx$
29. $\int_{-\infty}^{4} e^{x/3}\,dx$
30. $\int_{-\infty}^{0} x\exp(-x^2)\,dx$
31. $\int_{-\infty}^{0} x\exp(x/2)\,dx$
32. $\int_{-\infty}^{0} e^{-x}\,dx$
33. $\int_{-\infty}^{-1} \frac{\sin(\pi/x)}{x^2}\,dx$
34. $\int_{-\infty}^{\ln(3)/2} \frac{e^x}{e^{2x}+1}\,dx$
35. $\int_{-\infty}^{0} \frac{1}{(2-x)\ln^2(2-x)}\,dx$
36. $\int_{-\infty}^{0} \frac{\ln(2-x)}{2-x}\,dx$

In each of Exercises 37–44, determine whether the given improper integral converges or diverges. If it converges, then evaluate it.

37. $\int_{-\infty}^{\infty} xe^{-x^2}\,dx$
38. $\int_{-\infty}^{\infty} e^{-x}\,dx$
39. $\int_{-\infty}^{\infty} \frac{1}{4+x^2}\,dx$
40. $\int_{-\infty}^{\infty} \frac{x}{1+x^2}\,dx$
41. $\int_{-\infty}^{\infty} \frac{x}{(1+x^2)^2}\,dx$
42. $\int_{-\infty}^{\infty} \frac{1}{x^2+2x+10}\,dx$
43. $\int_{-\infty}^{\infty} \frac{1}{x^4+3x^2+2}\,dx$
44. $\int_{-\infty}^{\infty} \frac{x+2}{(x^2+1)^2}\,dx$

In each of Exercises 45–48, an income stream $f(t)$ is given (in dollars per year with $t=0$ corresponding to the present). The income will commence T_1 years in the future and continue in perpetuity. Calculate the present value of the income stream assuming that the discount rate is 5%.

45. $f(t)=1000;\ T_1=0$
46. $f(t)=1000;\ T_1=20$
47. $f(t)=1000+50t;\ T_1=0$
48. $f(t)=1000+50t;\ T_1=20$

Further Theory and Practice

In each of Exercises 49–54, determine whether the given improper integral converges or diverges. If it converges, then evaluate it.

49. $\int_1^\infty 4x^{-2}\arctan(x)\,dx$
50. $\int_1^\infty x^{-2}\ln(x)\,dx$
51. $\int_{-\infty}^{0} x2^x\,dx$
52. $\int_{-\infty}^{0} x^2e^{x+1}\,dx$

53. $\int_0^{\infty} e^{-x}\cos(x)\,dx$

54. $\displaystyle\int_0^{\infty} \frac{1}{e^x+1}\,dx$

Each of the improper integrals in Exercises 55–60 is improper for two reasons. Determine whether the integral converges or diverges. If it converges, then evaluate it.

55. $\displaystyle\int_0^{\infty} \frac{\exp(-1/x)}{x^2}\,dx$

56. $\displaystyle\int_0^{\infty} \frac{\exp(-\sqrt{x})}{\sqrt{x}}\,dx$

57. $\displaystyle\int_1^{\infty} \frac{1}{x\ln^2(x)}\,dx$

58. $\displaystyle\int_1^{\infty} \frac{1}{x\ln^{1/3}(x)}\,dx$

59. $\displaystyle\int_1^{\infty} \frac{2}{x\sqrt{x^2-1}}\,dx$

60. $\displaystyle\int_2^{\infty} \frac{2}{x^2\sqrt{x^2-4}}\,dx$

61. Show that there is no value of p for which $\displaystyle\int_0^{\infty} \frac{1}{x^p}\,dx$ is convergent.

62. For which values of p is $\displaystyle\int_e^{\infty} \frac{1}{x\ln^p(x)}\,dx$ convergent?

In each of Exercises 63–68, use the Comparison Theorem to establish that the given improper integral is convergent.

63. $\displaystyle\int_1^{\infty} \frac{x}{1+x^3}\,dx$

64. $\displaystyle\int_1^{\infty} \frac{2+\sin(x)}{x^2}\,dx$

65. $\displaystyle\int_1^{\infty} \frac{1}{\sqrt{1+x^{5/2}}}\,dx$

66. $\displaystyle\int_1^{\infty} \frac{x}{1+e^{2x}}\,dx$

67. $\displaystyle\int_1^{\infty} \frac{e^{-x}}{\sqrt{x}}\,dx$

68. $\displaystyle\int_1^{\infty} \frac{2^x}{3^x+1}\,dx$

In each of Exercises 69–74, use the Comparison Theorem to establish that the given improper integral is divergent.

69. $\displaystyle\int_1^{\infty} \frac{1}{\sqrt{1+x^2}}\,dx$

70. $\displaystyle\int_e^{\infty} \frac{1}{\ln(x)}\,dx$

71. $\displaystyle\int_1^{\infty} \frac{\sin^2(x)+x}{x^{3/2}}\,dx$

72. $\displaystyle\int_1^{\infty} \frac{3-\sin(x)}{2x-1}\,dx$

73. $\displaystyle\int_1^{\infty} \frac{\exp(x)}{x\exp(x)-1}\,dx$

74. $\displaystyle\int_1^{\infty} \frac{\arctan(x)}{\sqrt{1+x^2}}\,dx$

75. Use formula (6.7.2) to evaluate the (doubly) improper integral $\displaystyle\int_0^{\infty} \frac{1}{\sqrt{x}}e^{-x}\,dx$.

76. Let μ and σ be real constants with $\sigma > 0$. Use formula (6.7.2) to evaluate

$$\int_{-\infty}^{\infty} \exp\left(-\frac{1}{2}\left(\frac{x-\mu}{\sigma}\right)^2\right)dx.$$

77. Let μ and σ be real constants with $\sigma > 0$. Use formula (6.7.2) to evaluate

$$\int_{-\infty}^{\infty} x\exp\left(-\frac{1}{2}\left(\frac{x-\mu}{\sigma}\right)^2\right)dx.$$

78. Show that the improper integral $\Gamma(s) = \int_0^{\infty} x^{s-1}e^{-x}dx$ is convergent for $s > 0$. This function of s is called the *gamma function*.

79. The gamma function Γ is defined in the preceding exercise. Integrate by parts to derive the functional equation $\Gamma(s+1) = s\Gamma(s)$ for $s > 0$. By calculating $\Gamma(1)$ and using the functional equation for Γ, deduce that $\Gamma(n+1) = n!$ for each natural number n.

80. Suppose that f is a continuous nonnegative function such that $\int_0^{\infty} f(x)\,dx$ is convergent. At first one might expect that $\lim_{x\to\infty} f(x) = 0$, but, in fact, this limit does not have to hold. For each $n \in \mathbb{Z}^+$ let $V_n = (n, 1)$, $P_n = (n - 1/2^n, 0)$, and $Q_n = (n + 1/2^n, 0)$. Let O denote the origin. Finally, let f be the nonnegative continuous function defined on $[0, \infty)$ so that its graph consists of the line segments $\overline{OP_1}$, $\overline{P_1V_1}$, $\overline{V_1Q_1}$, $\overline{Q_1P_2}$, $\overline{P_2V_2}$, $\overline{V_2Q_2}$, $\overline{Q_2P_3}$, ... (ad infinitum). Show that $\lim_{x\to\infty} f(x)$ does not exist. Show that $\int_0^{\infty} f(x)\,dx$ is convergent, and calculate its value.

Calculator/Computer Exercises

In each of Exercises 81–84, a continuous function $f(x)$ is given. Determine a function $g(x) = cx^p$ such that (a) $0 \le f(x) \le g(x)$ for each x in $[1, \infty)$, and (b) $\int_1^{\infty} g(x)\,dx$ is convergent. This shows that $\int_1^{\infty} f(x)\,dx$ is convergent by the Comparison Theorem. By determining a positive ε such that $\int_{\varepsilon}^{\infty} g(x)\,dx < 5 \times 10^{-3}$, approximate $\int_1^{\infty} f(x)\,dx$ to two decimal places.

81. $f(x) = 1/\sqrt{1+x^5}$

82. $f(x) = (3+\sin(x))/x^2$

83. $f(x) = 1/\,x^{3x}$

84. $f(x) = x^{-6}\sqrt{1+3x^4}$

Summary of Key Topics in Chapter 6

Integration by Parts
(Section 6.1)

If u and v are continuously differentiable functions, then

$$\int (u \cdot v')\,dx = uv - \int (v \cdot u')\,dx.$$

This formula is often used to treat integrals of products of functions. The integration by parts formula can be written concisely as

$$\int u\,dv = uv - \int v\,du.$$

Powers and Products of Sines and Cosines
(Section 6.2)

It is possible to evaluate integrals of the form

$$\int \sin^m(x)\cos^n(x)\,dx,$$

for m and n nonnegative integers by using one of three procedures. If m is odd, then $m = 2k+1$ for some nonnegative integer k. We rewrite the integral as

$$\int \sin^m(x)\cos^n(x)\,dx = \int \left(\sin^2(x)\right)^k \cos^n(x)\sin(x)\,dx = \int \left(1-\cos^2(x)\right)^k \cos^n(x)\sin(x)\,dx.$$

The substitution $u = \cos(x)$, $du = -\sin(x)\,dx$ can be applied to this last integral. If $n = 2\ell + 1$ is odd, we proceed in a similar manner. We write

$$\int \sin^m(x)\cos^n(x)\,dx = \int \left(\cos^2(x)\right)^\ell \sin^m(x)\cos(x)\,dx = \int \left(1-\sin^2(x)\right)^\ell \sin^m(x)\cos(x)\,dx$$

and make the substitution $u = \sin(x)$, $du = \cos(x)\,dx$. If both powers $m = 2k$ and $n = 2\ell$ are even with $\ell \le k$, we write the integral as

$$\int \sin^m(x)\cos^{2\ell}(x)\,dx = \int \sin^m(x)\left(1-\sin^2(x)\right)^\ell dx,$$

expand the integrand, and integrate the resulting terms using the reduction formula

$$\int \sin^n(x)\,dx = -\frac{1}{n}\sin(x)^{n-1}\cos(x) + \frac{n-1}{n}\int \sin(x)^{n-2}\,dx.$$

If $k < \ell$, we write the integral as

$$\int \sin^m(x)\cos^{2\ell}(x)\,dx = \int \left(1-\cos^2(x)\right)^k \cos^{2\ell}(x)\,dx,$$

expand the integrand, and integrate the resulting terms using the reduction formula

$$\int \cos^n(x)\,dx = \frac{1}{n}\sin(x)\cos^{n-1}(x) + \frac{n-1}{n}\int \cos^{n-2}(x)\,dx.$$

Trigonometric Substitution (Section 6.3)

If a is a positive constant, then we may treat an integrand involving an expression of the form $a^2 - x^2$, $a^2 + x^2$, or $x^2 - a^2$ by making the indirect substitution that is indicated in the following table:

Expression	Substitution	Result of Substitution	After Simplification
$a^2 - x^2$	$x = a\sin(\theta),\ dx = a\cos(\theta)\,d\theta$	$a^2(1 - \sin^2(\theta))$	$a^2\cos^2(\theta)$
$a^2 + x^2$	$x = a\tan(\theta),\ dx = a\sec^2(\theta)\,d\theta$	$a^2(1 + \tan^2(\theta))$	$a^2\sec^2(\theta)$
$x^2 - a^2$	$x = a\sec(\theta),\ dx = a\sec(\theta)\tan(\theta)\,d\theta$	$a^2(\sec^2(\theta) - 1)$	$a^2\tan^2(\theta)$

To treat an integral that involves the expression $\sqrt{\alpha x^2 + \beta x + \gamma}$, we complete the square under the radical. Doing so allows us to write $\alpha x^2 + \beta x + \gamma = \alpha \cdot (a^2 \pm u^2)$ where $u = x + \beta/(2\alpha)$. One of the three trigonometric substitutions listed in the preceding table may then be applied to $a^2 \pm u^2$.

Partial Fractions (Sections 6.4 and 6.5)

To integrate a rational function of the form

$$\frac{p(x)}{(x-a_1)^{m_1}\cdots(x-a_k)^{m_k}\cdot(\alpha_1 x^2+\beta_1 x+\gamma_1)^{n_1}\cdots(\alpha_\ell x^2+\beta_\ell x+\gamma_\ell)^{n_\ell}},$$

perform the following steps:

1. Be sure that the degree of p is less than the degree of the denominator; if it is not, divide the denominator into the numerator.
2. Be sure that the quadratic factors are irreducible. Factor them if not.
3. For each of the factors $(x - a_j)^{m_j}$ in the denominator of the rational function being considered, the partial fraction decomposition must contain terms of the form

$$\frac{A_1}{(x-a_j)^1} + \frac{A_2}{(x-a_j)^2} + \cdots + \frac{A_{m_j}}{(x-a_j)^{m_j}}.$$

4. For each of the irreducible factors $(\alpha_j x^2 + \beta_j x + \gamma_j)^{m_j}$ in the denominator of the integrand being considered, the partial fraction decomposition must contain terms

$$\frac{B_1 x + C_1}{\alpha_j x^2 + \beta_j x + \gamma_j} + \frac{B_2 x + C_2}{(\alpha_j x^2 + \beta_j x + \gamma_j)^2} + \cdots + \frac{B_{n_j} x + C_{n_j}}{(\alpha_j x^2 + \beta_j x + \gamma_j)^{n_j}}.$$

Improper Integrals (Sections 6.6, 6.7)

If $f(x)$ is a continuous function on $[a, b)$ that is unbounded as $x \to b^-$, then the integral $\int_a^b f(x)\,dx$ is defined to be

$$\lim_{\varepsilon \to 0^+} \int_a^{b-\varepsilon} f(x)\,dx,$$

provided that this limit exists. A similar definition is used for functions that are unbounded at the left endpoint of the interval of integration.

If the singularity occurs in the middle of the interval of integration, then the integral should be broken up at the singularity to reduce to the two previous cases. If the integrand is unbounded at both endpoints, then the integral should be broken up at a chosen point in between to reduce to the two basic cases.

If f is a continuous function on the interval $[a, \infty)$, then the integral $\int_a^\infty f(x)\,dx$ is defined to be

$$\lim_{N \to +\infty} \int_a^N f(x)\,dx$$

provided that this limit exists. A similar definition applies to integrals on an interval of the form $(-\infty, b]$. The integral $\int_{-\infty}^{\infty} f(x)\,dx$ is handled by choosing a point c and evaluating the sum of the improper integrals $\int_{-\infty}^{c} f(x)\,dx$ and $\int_c^\infty f(x)\,dx$.

Review Exercises for Chapter 6

In Exercises 1–96, evaluate the given definite integral.

1. $\int_0^1 16xe^{4x}\,dx$
2. $\int_0^2 xe^{-x/2}\,dx$
3. $\int_0^{\pi/2} 4x\sin(x/3)\,dx$
4. $\int_{\pi/2}^{\pi} x\cos(2x)\,dx$
5. $\int_{1/3}^{1} 2x\ln(3x)\,dx$
6. $\int_1^2 3x^2\ln(x/2)\,dx$
7. $\int_{-1}^{1} xe^{-x}\,dx$
8. $\int_0^{1/2} xe^{2x}\,dx$
9. $\int_0^{\pi} x\sin(x)\,dx$
10. $\int_0^{\pi/2} x\cos(x)\,dx$
11. $\int_1^{3e} \ln(x/3)\,dx$
12. $\int_1^{2e} 4\,x\ln(x/2)\,dx$
13. $\int_0^1 x^2e^x\,dx$
14. $\int_0^{\pi/2} x^2\cos(x)\,dx$
15. $\int_0^{\pi} x^2\sin(x)\,dx$
16. $\int_1^{e} \ln^2(x)\,dx$
17. $\int_0^{\pi/6} 12(1+\sec(x))^3\,dx$
18. $\int_0^{\pi/4} 2(1+\tan(x))^3\,dx$
19. $\int_0^{\pi/4} \sin^2(2x)\,dx$
20. $\int_{\pi/2}^{\pi} 2\cos^2(x/2)\,dx$
21. $\int_{\pi/4}^{\pi/2} 12\sin^3(x)\,dx$
22. $\int_0^{\pi} \cos^3(x/2)\,dx$
23. $\int_0^{\pi/2} \sin^2(x)\cos^3(x)\,dx$
24. $\int_0^{\pi} \sin^3(x/2)\cos^2(x/2)\,dx$
25. $\int_0^{\pi} (\cos(x)+\sin(x))^2\,dx$
26. $\int_0^{\pi/2} \sin^3(x)\cos^3(x)\,dx$
27. $\int_0^{\pi} \sin^3(x)\cos^4(x)\,dx$
28. $\int_0^{\pi} \sin^3(x)\cos^6(x)\,dx$
29. $\int_0^{\pi/2} \sin^{3/2}(x)\cos^3(x)\,dx$
30. $\int_0^{\pi/4} 15\sin^2(2x)\cos^5(2x)\,dx$
31. $\int_{\pi/2}^{\pi} \sin^2(x)\cos^7(x)\,dx$
32. $\int_{1/2}^{1} \cos^4(\pi x)\,dx$
33. $\int_0^{2\pi} \sin^4(x)\,dx$
34. $\int_{\pi/4}^{\pi/2} \cos^2(x)\sin^2(x)\,dt$
35. $\int_0^{2\pi} \cos^6(x/4)\,dx$
36. $\int_0^{\pi} \cos^4(x)\sin^2(x)\,dx$
37. $\int_0^{\pi/3} \cos^2(x)\sin^4(x)\,dx$
38. $\int_0^{\pi/2} \tan^3(x/2)\,dx$
39. $\int_0^{\pi/3} \sec^4(x)\,dx$
40. $\int_0^{\pi/3} \tan^4(x)\,dx$
41. $\int_0^{\pi/3} \sec^2(x)\tan^2(x)\,dx$
42. $\int_0^{\pi/3} 10\sec^2(x)\tan^4(x)\,dx$
43. $\int_0^{\pi/4} 2\sec(x)\tan^2(x)\,dx$
44. $\int_0^{\pi/4} 3\tan(x)\sec^3(x)\,dx$
45. $\int_0^{\pi/4} \tan(x)\sec^4(x)\,dx$
46. $\int_0^{\pi/4} 3\tan^3(x)\sec(x)\,dx$
47. $\int_0^{\pi/4} 2\tan(x)\sin(x)\,dx$
48. $\int_{\pi/4}^{\pi/2} 3\cot(x)\csc^3(x)\,dx$
49. $\int_{\pi/4}^{\pi/2} 4\cot(x)\csc^4(x)\,dx$
50. $\int_{\pi/4}^{\pi/2} 3\cot^3(x)\csc(x)\,dx$
51. $\int_0^{\pi/4} 4\sin^2(x)\tan(x)\,dx$
52. $\int_0^{\pi/4} 12\cos^3(x)\tan^2(x)\,dx$
53. $\displaystyle\int_0^{\pi/4} \frac{3\sin(x)+1}{1-\sin^2(x)}\cos(x)\,dx$
54. $\displaystyle\int_0^{\pi/6} \frac{5\sin(x)-7}{\sin^2(x)-3\sin(x)+2}\cos(x)\,dx$
55. $\displaystyle\int_0^1 \frac{1+x^{1/2}}{1+x^{1/4}}\,dx$
56. $\displaystyle\int_0^{1/\sqrt{2}} \frac{8x^2}{\sqrt{1-x^2}}\,dx$

57. $\int_0^{1/\sqrt{2}} \dfrac{2x^2}{1-x^2}\,dx$

58. $\int_4^5 \sqrt{x^2-16}\,dx$

59. $\int_{\sqrt{2}}^{2/\sqrt{3}} \dfrac{1}{x\sqrt{x^2-1}}\,dx$

60. $\int_0^{\sqrt{3}} 6\sqrt{4-x^2}\,dx$

61. $\int_0^3 \dfrac{1}{\sqrt{16+x^2}}\,dx$

62. $\int_5^{12} \dfrac{13}{x\sqrt{169-x^2}}\,dx$

63. $\int_0^{1/2} \sqrt{1-4x^2}\,dx$

64. $\int_{\sqrt{2}}^{\sqrt{5}} 2(x^2-1)^{-3/2}\,dx$

65. $\int_0^{\sqrt{3}} 4(4-x^2)^{3/2}\,dx$

66. $\int_{\sqrt{10}}^5 \dfrac{1}{\sqrt{x^2-9}}\,dx$

67. $\int_{\sqrt{10}}^5 \dfrac{3}{x^2-9}\,dx$

68. $\int_0^1 2\sqrt{1+x^2}\,dx$

69. $\int_0^4 (9+x^2)^{-3/2}\,dx$

70. $\int_{-1}^1 \dfrac{x^2}{\sqrt{1+x^2}}\,dx$

71. $\int_2^3 \dfrac{3x^3}{\sqrt{x^2-1}}\,dx$

72. $\int_0^1 \dfrac{2x^2}{\sqrt{2-x^2}}\,dx$

73. $\int_0^1 \dfrac{x^2}{2-x^2}\,dx$

74. $\int_{-1}^0 \dfrac{3}{2-x-x^2}\,dx$

75. $\int_{-1}^0 \dfrac{1+x}{1-2x+x^2}\,dx$

76. $\int_1^2 \dfrac{2x+3}{x(1+x)}\,dx$

77. $\int_1^2 \dfrac{3x^2-2x+3}{x(1+x^2)}\,dx$

78. $\int_{-1}^1 \dfrac{3x^2+4}{x^4+3x^2+2}\,dx$

79. $\int_{-1}^0 \dfrac{2x+4}{x^2+2x+2}\,dx$

80. $\int_0^1 \dfrac{2x^2+5x}{x^3-3x-2}\,dx$

81. $\int_0^1 \dfrac{2x^3+2x+2}{(1+x^2)^2}\,dx$

82. $\int_0^1 \dfrac{x^4+4x^3+4x^2+4x+4}{(1+x^2)(2+x^2)^2}\,dx$

83. $\int_0^1 \dfrac{2x^3+x^2+2x+2}{(1+x^2)(2+x^2)}\,dx$

84. $\int_0^{16} \dfrac{x^{1/4}}{1+x^{1/2}}\,dx$

85. $\int_1^4 \dfrac{1}{x+x^{1/2}}\,dx$

86. $\int_0^4 \dfrac{1}{x+x^{1/2}}\,dx$

87. $\int_0^4 \dfrac{1}{x^{1/2}}\,dx$

88. $\int_{-5}^4 \dfrac{1}{\sqrt{4-x}}\,dx$

89. $\int_0^{2\sqrt{2}} \dfrac{x}{(8-x^2)^{1/3}}\,dx$

90. $\int_0^{\pi^2} \dfrac{1}{\sqrt{x}}\sin(\sqrt{x})\,dx$

91. $\int_0^\infty x\exp(-x/3)\,dx$

92. $\int_1^e \dfrac{1}{x\ln^{2/3}(x)}\,dx$

93. $\int_0^\infty x^3\exp(-x^4)\,dx$

94. $\int_0^\infty \dfrac{x}{(1+x^2)^2}\,dx$

95. $\int_1^\infty \dfrac{1}{x\sqrt{1+x^2}}\,dx$

96. $\int_1^\infty \dfrac{3x-3}{x^3+1}\,dx$

97. Find the area of the region that is bounded above by $y=1-\sin^3(x)$ and below by $y=\cos^2(x)$ for $0\le x\le \pi/2$.

98. Calculate the total area of the regions between $y=x-x^3$ and $y=x^2\sqrt{1-x^2}$ for $0\le x\le 1$.

99. Find the area of the region that is bounded above by $y=\ln(1+x)$ and below by $y=x\ln(1+x)$ for $0\le x\le 1$.

100. Find the area of the region that is bounded above by $y=\dfrac{30}{(1+x^2)(2+x)}$ and below by $y=\dfrac{30}{(1+x)(2+x)}$ for $0\le x\le 1$.

Each technique of integration that you have learned in Chapter 6 was discovered early in the history of calculus. The integration by parts formula, to cite one example, appeared in Leibniz's first development of calculus (1673). By then, particular instances of integrating by parts had already been published. What distinguishes the contributions of Leibniz and Newton is their search for general techniques that could be applied to a large class of functions. Before their work, the integration formulas that were known tended to be isolated results that were discovered in the course of specific practical and scientific investigations. For example, Kepler discovered the equation

$$\int_0^{\theta} \sin(\phi)\, d\phi = 1 - \cos(\theta)$$

because it arose in his study of the orbit of Mars. Similarly, the formula

$$\int_0^{\theta} \sec(\phi)\, d\phi = \ln\bigl(|\sec(\theta) + \tan(\theta)|\bigr) \qquad (1)$$

became known because of a problem of maritime navigation.

In 1569, Gerardus Mercator (1512–1594) introduced a world map based on a new projection. Prior to Mercator's projection, nautical maps were unreliable at best. The great advantage of the Mercator map is that a line of constant bearing intersects meridians at constant angles. But there is a catch: Navigation by the Mercator map depends on the evaluation of the secant integral on the left side of equation (1). Two English mathematicians who put to sea on separate naval expeditions in the late 1500s are known to have performed that calculation.

When the English Navy needed skilled navigators for a raid against the Spanish fleet in 1591, the Queen granted the mathematician Edward Wright (1561–1615) a sabbatical from Cambridge so that he could serve on board a man-of-war. Although the campaign went badly for the English, Wright survived the fighting. In 1599, he organized his navigational work into a book titled *Certaine Errors of Navigation Corrected.* Wright described one computation as "the perpetual addition of the secantes answerable to the latitudes of each point or parallel into the summe compounded of all former secantes." Translated into the language of calculus, Wright approximated the integral $\int_0^{\theta} \sec(\phi)\, d\phi$ by using Riemann sums in which the increment $\Delta\phi$ was taken to be one 360th of a degree.

A similar numerical calculation of the secant integral was carried out a few years earlier by another English mathematician, Thomas Harriot (1560–1621), who served as navigator, cartographer, and surveyor on Sir Walter Raleigh's expedition to Virginia in 1585. After returning to England, Harriot assumed the intellectual leadership of the circle that assembled around Sir Walter Raleigh. When Shakespeare wrote

Oh paradox! Black is the badge of hell
The hue of dungeons and the school of night

in *Love's Labour's Lost* (IV, iii, 250), he was referring to the evening sessions at Raleigh's country estate during which Harriot presided over discussions of astronomy, geography, chemistry, and philosophy. It was a dangerous circle. One of its members, the poet and playwright Christopher Marlowe, died in knife-play over a tavern bill. Raleigh was sentenced for sedition in 1603, locked up in the Tower of London, and hanged in 1618. Harriot's own patron, the Earl of Northumberland, spent 16 years in prison for his part in a conspiracy to blow up the English houses of parliament.

Harriot's scientific reputation suffered less from his own brief incarceration than from the careless disposition of his last effects. After his death, Harriot's scientific papers were split among two heirs. One parcel was soon misplaced. By the time it turned up in 1784, mixed in among stable accounts, priority for Harriot's discoveries had already been claimed by others. Although some of the papers from the second parcel were published in 1631, most were lost. His calculation of the secant integral, for example, remained unknown until it turned up in the 1960s. Today, Harriot's most obvious legacies are the symbols that we use to express inequalities.

Useful as the calculations of Wright and Harriot were, an analytic proof of equation (1) was still needed. One was finally found by James Gregory in 1668. However, the famous astronomer Edmund Halley described Gregory's derivation as "A long train of Consequences and Complication of Proportions whereby the evidence of the Demonstration is in a great measure lost and the Reader wearied before he attain it." A few years later, Isaac Barrow, Newton's predecessor at Cambridge, published a more satisfactory calculation of the secant integral. Barrow's argument went as follows:

$$\begin{aligned}\int_0^\theta \sec(\phi)\,d\phi &= \int_0^\theta \frac{\cos(\phi)}{\cos^2(\phi)}\,d\phi = \int_0^\theta \frac{\cos(\phi)}{1-\sin^2(\phi)}\,d\phi \\ &= \frac{1}{2}\int_0^\theta \frac{\cos(\phi)}{1-\sin(\phi)}\,d\phi + \frac{1}{2}\int_0^\theta \frac{\cos(\phi)}{1+\sin(\phi)}\,d\phi \\ &= \frac{1}{2}\left(-\ln(1-\sin\theta)+\ln(1+\sin\theta)\right) \\ &= \frac{1}{2}\ln\left(\frac{1+\sin(\theta)}{1-\sin(\theta)}\right),\end{aligned}$$

which is equivalent to formula (1).

Notice that Barrow's proof uses the method of partial fractions, a technique that Leibniz systematically used later in the 17th century. The theoretical basis for the method of partial fractions is the Fundamental Theorem of Algebra. This theorem states that every nonconstant polynomial with complex coefficients has a complex root. Such a result was conjectured by Harriot at the beginning of the 17th century and finally proved in 1797 by Carl Friedrich Gauss (1777–1855), who was then only 20 years old.

"Archimedes, Newton, and Gauss, these three, are in a class by themselves among the great mathematicians, and it is not for ordinary mortals to attempt to range them in order of merit." So wrote, E. T. Bell, one of the most widely read mathematical expositors. Several anecdotes suggest that Gauss's mathematical genius was evident from early childhood. According to one story, Gauss was in elementary school when his class was assigned the task of calculating $1+2+3+\ldots+100$. Gauss arrived at the correct answer nearly instantaneously, having already determined the rule for the sum of an arithmetic progression.

At the age of 14, Gauss discovered that, for any two nonnegative numbers a and b, the two sequences

$$a_n = \begin{cases} a & n=0 \\ \frac{1}{2}(a_{n-1}+b_{n-1}) & n>0 \end{cases} \quad \text{and} \quad b_n = \begin{cases} b & n=0 \\ \sqrt{a_{n-1}b_{n-1}} & n>0 \end{cases}$$

have a common limit $M(a,b)$. We call $M(a,b)$ the *arithmetic-geometric* mean of a and b. Eight years later, Gauss proved that

$$\int_0^\pi \frac{d\phi}{\sqrt{1-k^2\sin^2(\phi)}} = \frac{\pi}{M(1+k,1-k)} \qquad 0\le k<1.$$

The definite integral on the left side of Gauss's equation is important in several branches of mathematics and physics. Although mathematicians had investigated it since the 1660s, Gauss was the first to find an effective method for calculating it.

In 1801, Gauss made a contribution to celestial mechanics that brought him great fame in scientific circles. At the time, seven planets of the solar system were known, but it was suspected that there was an as yet undetected planet between the orbits of Mars and Jupiter. That suspicion arose because of *Bode's law*, which is the approximation $B_m = 0.4 + 0.3\cdot 2^{m-2}$ for the distance in astronomical units (au) between the sun and the mth planet. This formula works quite well for Venus through Mars, as the table shows.

From the table we can also see that Bode's law would work equally well through Uranus if there was a planet between Mars and Jupiter, around 2.8 au from the sun.

On the first night of the 19th century, January 1st, 1801, Giuseppe Piazzi (1746–1826) noticed a faint body that had not been previously catalogued. He

	Mercury	**Venus**	**Earth**	**Mars**	*	**Jupiter**	**Saturn**	**Uranus**
order from sun	1	2	3	4	*	5	6	7
m	1	2	3	4	5	6	7	8
B_m	0.55	0.7	1	1.6	2.8	5.2	10	19.6
distance to sun	0.39	0.72	1	1.52	*	5.20	9.54	19.18

Table 1: Bode's Law and the Seven Known Planets of 1801

continued to watch it until it faded from his sight on February 11, 1801. Piazzi had tracked it for 41 nights over an arc of only 3°. In June of 1801, he published his observations, but they did not help other astronomers relocate the object, which was thought to be the undiscovered planet predicted by Bode's law.

Where astronomers with telescopes failed, Gauss succeeded, using mathematics alone. In particular, Gauss applied a mathematical tool of his own invention—the Method of Least Squares. On December 7, 1801, the first clear viewing night after Piazzi's object emerged from the shadow of the sun, it was spotted exactly where Gauss had predicted it would be. As Gauss remarked, "the fugitive was restored to observation." (Piazzi's object, however, turned out to be only an asteroid, and Bode's law is now known to be invalid.)

Many descendants of Gauss are scattered across the United States. The two youngest of his four sons, Eugen and Wilhelm, emigrated to the United States. Writing from New York City in 1831, Eugen informed his father that "Your name is well known even here in this wilderness." Eugen's army enlistment took him to Fort Snelling, Minnesota, where he served under the command of future president General Zachary Taylor. He settled in St. Charles, Missouri, where the house that he built still stands.

CHAPTER 7

Applications of the Integral

PREVIEW

There is a common philosophical thread that runs through the first five sections of this chapter. Namely, in each of these sections, we introduce certain physical quantities. We approximate these quantities in a plausible way by a mathematical procedure that leads to a Riemann sum. By letting the subinterval lengths that appear in the sum tend to 0, we are in each case led to an integral. We then declare that integral to, in fact, represent the physical quantity that we started out to analyze. What gives us the right to do this?

There is a good practical answer. In each of the instances discussed in this chapter, the mathematical model for the physical quantity being discussed gives the same measurement in practical examples as one or more independent methods. For instance, we develop several mathematical formulas for the volumes of solids of revolution. When applied to calculate the volume inside a sphere, the formula generates the same answer as can be obtained by immersing a sphere in water and measuring the displacement. We also develop a formula for the length of a curve. In practice, this formula gives the same answers as one obtains when measuring the length of the same curves with a tape measure. Similar comments apply to the other physical quantities—work and center of mass—which are discussed here.

When applying mathematics, it is always important to have objective methods for testing the validity of the mathematical models. Scientists are constantly challenging, checking, and revising their methods for modeling. This is a crucial step in the advancement of scientific knowledge.

In the last two sections of this chapter, you learn methods that enable you to determine an unknown function $y(x)$ from information about its rate of change. In these applications, the starting point is a model for the derivative of the function. We deduce or observe an equation involving x, $y(x)$, and $y'(x)$. Such a relationship is called a *differential equation*. The integral is then used to solve the differential equation and find a formula for the function itself. Solving differential equations constitutes one of the most important applications of calculus. For many students, a specialized course in differential equations is the natural continuation of their calculus studies.

7.1 Volumes

In this section, we will think about volume in much the same way that we thought about area in Chapter 5. That is, we will be calculating the volume of a solid by cutting up the solid into elementary pieces.

Volumes by Slicing–The Method of Disks

You may have already learned that

$$V = \frac{1}{3}\pi r^2 h \tag{7.1.1}$$

is the formula for the volume V of a right circular cone of height h and radius r (Figure 1). Let us see for ourselves how we can deduce this formula. We begin by slicing up the cone into N equally thick pieces as in Figure 2. The volume of each slice is approximately equal to the volume of a disk—see Figure 2. If the jth disk has radius r_j and thickness Δx, then its volume is the product $(\pi r_j^2) \cdot \Delta x$ of its cross-sectional area πr_j^2 and its thickness Δx. To obtain an approximation to the volume of the cone, we add up the volumes of the disks:

$$\sum_{j=1}^{N} \pi r_j^2 \, \Delta x. \tag{7.1.2}$$

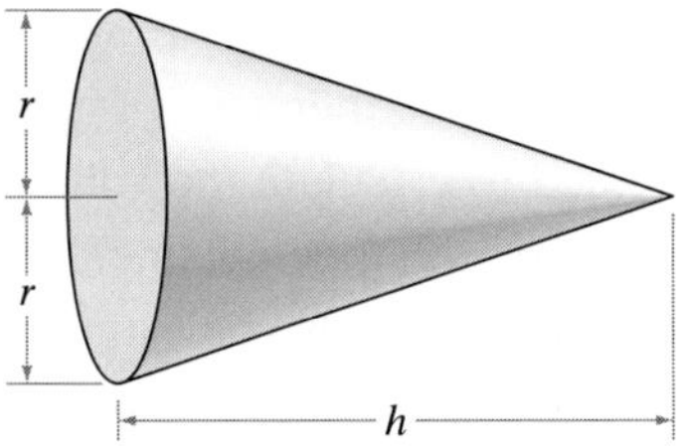

▲ Figure 1

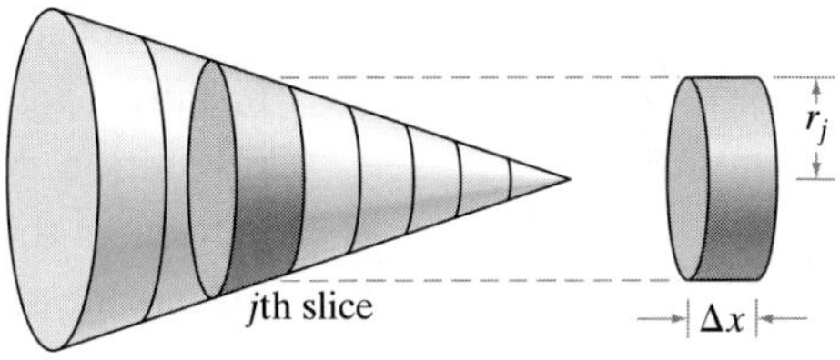

▲ Figure 2 A disk is used to approximate a slice of the cone.

To get a better approximation, we make the disks thinner by increasing the value of N. To get the precise volume V, we let the thickness of the disks tend to 0 by letting N tend to infinity:

$$V = \lim_{N \to \infty} \sum_{j=1}^{N} \pi r_j^2 \, \Delta x. \tag{7.1.3}$$

To compute this limit, we identify expression (7.1.2) as a Riemann sum. Then formula (7.1.3) expresses the volume of the cone as a limit of Riemann sums—that is to say, an integral (as you learned in Chapter 5). Example 1 provides the details of the technique just described.

▶ **EXAMPLE 1** Calculate the volume V of a right circular cone that has height 11 and base of radius 5.

Solution Figure 3 depicts the xy-plane and the cone in question. The cone is positioned so that the x-axis is its axis of symmetry, and the origin is the center of its base. The line segment that is formed by the intersection of the edge of the cone and the xy-plane has slope $-5/11$ and y-intercept $(0, 5)$. Its equation is therefore $y = -5x/11 + 5$.

In Figure 3, we have partitioned the interval $[0, 11]$ into N subintervals of equal length Δx. For each j from 1 to N, a point s_j is chosen in the jth subinterval. (Because the particular value of s_j will not matter, a formula for s_j is not needed.) As Figure 3 indicates, we approximate the jth slice by means of a disk of thickness Δx and radius r_j equal to the vertical distance from the point $(s_j, 0)$ to the edge of the cone. The approximating sum (7.1.2) becomes

$$\sum_{j=1}^{N} \pi \left(\frac{-5s_j}{11} + 5 \right)^2 \cdot \Delta x. \tag{7.1.4}$$

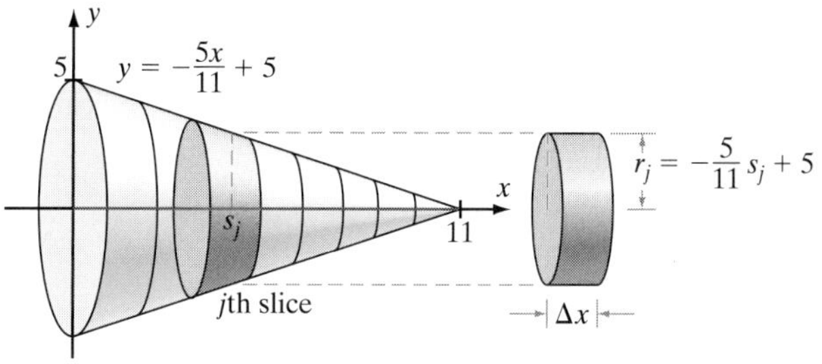

▲ Figure 3

As N increases and Δx becomes smaller, this sum gives a more accurate approximation to the desired volume; however, expression (7.1.4) is also a Riemann sum for the integral

$$\int_0^{11} \pi \left(\frac{-5x}{11} + 5 \right)^2 dx. \tag{7.1.5}$$

Thus both the volume V of the cone and the integral (7.1.5) are equal to the limit of sum (7.1.4) as N tends to infinity. We conclude that V is equal to

$$\int_0^{11} \pi \left(\frac{-5x}{11} + 5 \right)^2 dx = \pi \left(\frac{5}{11} \right)^2 \int_0^{11} (-x + 11)^2 \, dx = \pi \left(\frac{5}{11} \right)^2 \frac{(-x + 11)^3}{-3} \Bigg|_0^{11}$$

$$= \pi \left(\frac{5}{11} \right)^2 \left(\frac{0}{-3} - \frac{11^3}{-3} \right) = \frac{1}{3} \pi \cdot 5^2 \cdot 11. \quad ◀$$

INSIGHT In Example 1, if we replace each occurrence of 5 with r and each 11 with h, then formula (7.1.5) becomes $\int_0^h \pi \left(\frac{-rx}{h} + r \right)^2 dx$, and we find that the volume of the right circular cone with base r and height h is

$$V = \int_0^h \pi \left(\frac{-rx}{h} + r \right)^2 dx = \pi \left(\frac{r}{h} \right)^2 \int_0^h (-x + h)^2 \, dx = -\frac{1}{3} \pi \left(\frac{r}{h} \right)^2 (-x + h)^3 \Bigg|_{x=0}^{x=h} = \frac{1}{3} \pi r^2 h,$$

as given in formula (7.1.1).

Solids of Revolution

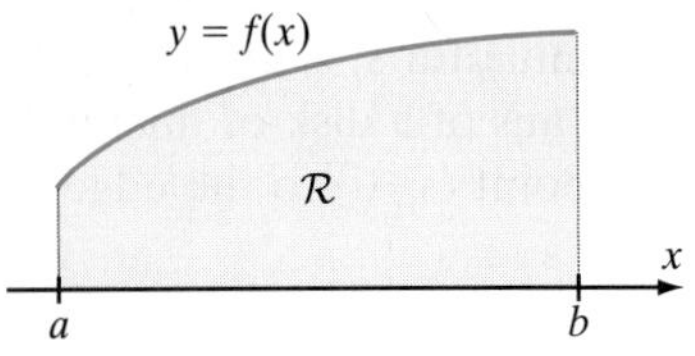

▲ Figure 4

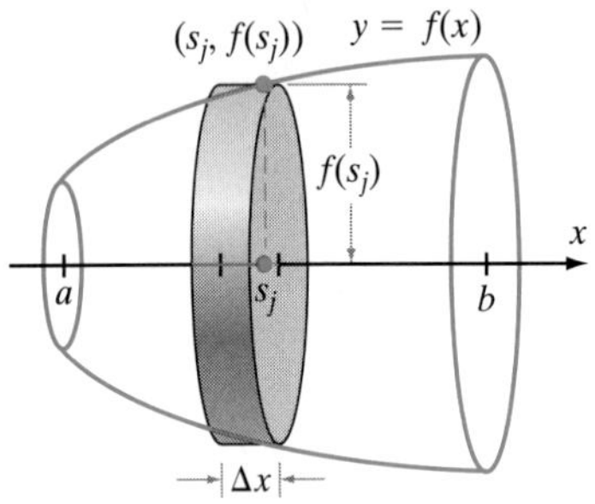

▲ Figure 5

With little modification, the method that we have developed for the cone can be used to find the volumes of other solids. Glance again at Figure 3. The key step is to recognize that, if $\mathcal{R}$ is the region in the xy-plane that lies below the graph of $y = -5x/11 + 5$ and above the interval $[0, 11]$, then the solid cone can be generated by rotating $\mathcal{R}$ about the x-axis. In general, a solid that is obtained by rotating a figure in the xy-plane about a line in the xy-plane is called a *solid of revolution*.

Consider, for example, a nonnegative continuous function f that is defined on an interval $[a, b]$ of the x-axis. The graph of f, the x-axis, and the vertical lines $x = a$ and $x = b$ bound a region $\mathcal{R}$ in the plane, as shown in Figure 4. Let us see how to calculate the volume of the solid of revolution that is generated by rotating the planar region $\mathcal{R}$ about the x-axis.

As before, we partition the interval $[a, b]$ into N subintervals of equal length Δx. For each j from 1 to N, we choose a point s_j in the jth subinterval. Figure 5 indicates that we approximate the jth slice of the solid of revolution with a disk of radius $f(s_j)$ and thickness Δx. The contribution to volume from this disk is the product of its cross-sectional area and thickness, or $(\pi f(s_j)^2) \cdot \Delta x$. The sum of the volumes of the N disks is

$$\sum_{j=1}^{N} \pi f(s_j)^2 \cdot \Delta x.$$

As we let N tend to infinity, the union of the approximating disks becomes closer to the solid of revolution. We therefore *define* the volume V of the solid of revolution to be

$$V = \lim_{N \to \infty} \sum_{j=1}^{N} \pi f(s_j)^2 \cdot \Delta x.$$

Notice that the expression in this limit is a Riemann sum for the integral

$$\int_a^b \pi f(x)^2 \, dx.$$

Because

$$\int_a^b \pi f(x)^2 \, dx = \lim_{N \to \infty} \sum_{j=1}^{N} \pi f(s_j)^2 \cdot \Delta x,$$

we see that the volume of the solid of revolution is represented by a Riemann integral. We may state our conclusion as a theorem.

THEOREM 1 **(Method of Disks: Rotation About the x-Axis)** Suppose that f is a nonnegative, continuous function on the interval $[a, b]$. Let R denote the region of the xy-plane that is bounded above by the graph of f, below by the x-axis, on the left by the vertical line $x = a$, and on the right by the vertical line $x = b$. Then, the volume V of the solid obtained by rotating $\mathcal{R}$ about the x-axis is given by

$$V = \pi \int_a^b f(x)^2 \, dx.$$

▶ **EXAMPLE 2** Calculate the volume of the solid of revolution that is generated by rotating about the x-axis the region of the xy-plane that is bounded by $y = x^2$, $y = 0$, $x = 1$, and $x = 3$.

Solution Figure 6 shows the region to be revolved as well as the solid of revolution. According to Theorem 1, the volume of this solid is

$$V = \pi \int_1^3 (x^2)^2\,dx = \pi \int_1^3 x^4\,dx = \pi\left(\frac{x^5}{5}\Big|_{x=1}^{x=3}\right) = \pi\left(\frac{3^5-1}{5}\right) = \frac{242}{5}\pi. \quad ◀$$

In some problems, it is useful to think of the curve as $x = g(y)$ and to rotate about the y-axis (see Figure 7). Reasoning similar to Theorem 1 gives the following result for the volume of the solid that is obtained when an arc of the curve $x = g(y)$ is rotated about the y-axis.

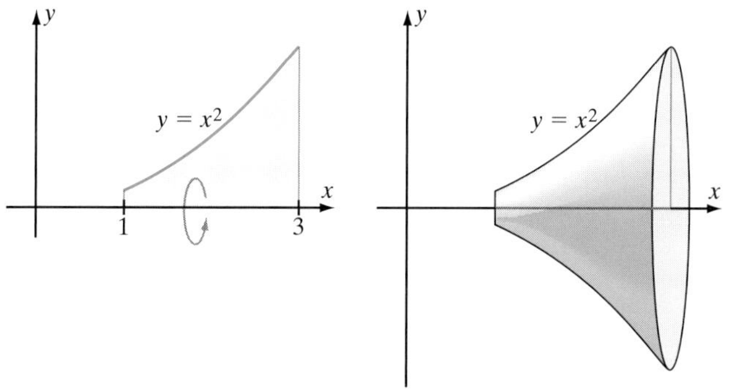

▲ Figure 6

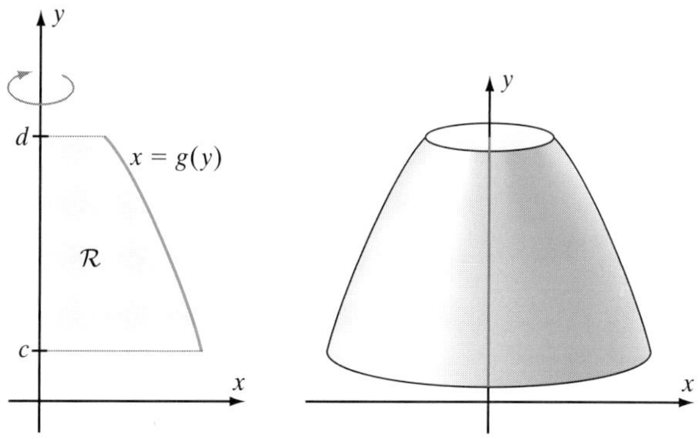

▲ Figure 7 The planar region $\mathcal{R}$ and the solid that $\mathcal{R}$ generates when rotated about the y-axis

THEOREM 2 **(Method of Disks: Rotation About the y-axis)** Suppose that $g(y)$ is a nonnegative continuous function on the interval $c \le y \le d$. Let $\mathcal{R}$ denote the region of the xy-plane that is bounded by the graph of $x = g(y)$, the y-axis, and the horizontal lines $y = c$ and $y = d$. Then the volume V of the solid obtained by rotating $\mathcal{R}$ about the y-axis is given by

$$V = \pi \int_c^d g(y)^2\,dy.$$

▶ **EXAMPLE 3** Calculate the volume enclosed when the graph of $y = x^3$, $2 \le x \le 4$, is rotated about the y-axis.

Solution Refer to Figure 8. Because we are rotating about the y-axis, we want to use Theorem 2. Therefore we express x as a function of y:

$$x = g(y) = y^{1/3}.$$

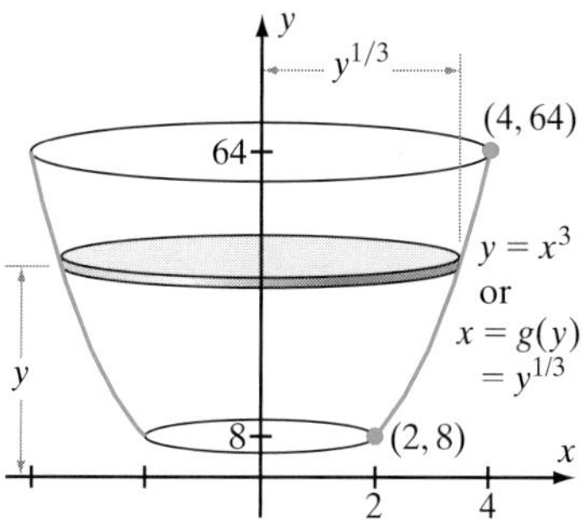

▲ Figure 8

Notice that $x = 2$ corresponds to $y = 8$, and $x = 4$ corresponds to $y = 64$. Then, according to Theorem 2, the desired volume is

$$V = \pi \int_8^{64} (y^{1/3})^2 dy = \frac{3}{5}\pi y^{5/3}\Big|_{y=8}^{y=64} = \frac{3}{5}\pi\left(64^{5/3} - 8^{5/3}\right) = \frac{3}{5}\pi(4^5 - 2^5) = \frac{2976}{5}\pi.$$ ◀

The Method of Washers

Sometimes we want to generate a solid of revolution by rotating the region *between* two graphs. Here is an example.

▶ **EXAMPLE 4** Let $\mathcal{D}$ be the region of the xy-plane that is bounded above by $y = 8\sqrt{x}$ and below by $y = x^2$. Calculate the volume of the solid of revolution that is generated when $\mathcal{D}$ is rotated about the x-axis.

Solution When we decompose this solid of revolution into thin slices, we obtain pieces that are approximately washers, not disks (Figure 9). To approximate the solid, we use washers that correspond to values of x from $x = 0$ to $x = 4$—these values are found by solving the equation $8\sqrt{x} = x^2$. The outer radius of the washer located at x is $8\sqrt{x}$ while the inner radius is x^2 (see Figure 9). The volume contribution of this washer is obtained by multiplying the area between the circles by the thickness of the washer:

$$\left(\pi(8\sqrt{x})^2 - \pi(x^2)^2\right) \cdot \Delta x.$$

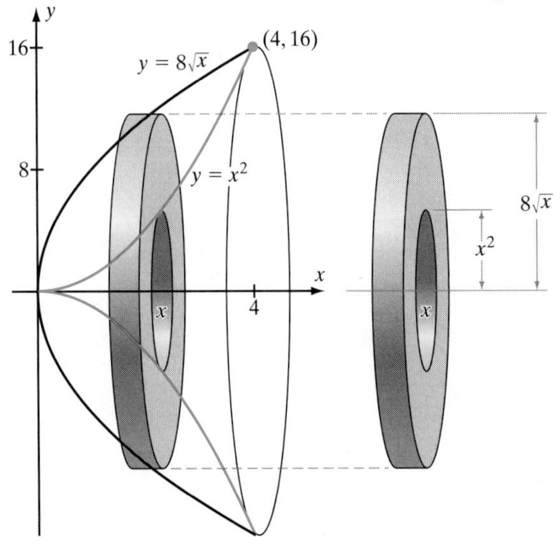

▲ Figure 9

Therefore the volume of the solid is

$$\pi \int_0^4 \left((8\sqrt{x})^2 - (x^2)^2\right) dx = \pi \int_0^4 (64x - x^4)\, dx = \pi\left(32x^2 - \frac{x^5}{5}\right)\Big|_{x=0}^{x=4} = \frac{1536}{5}\pi.$$ ◀

INSIGHT Notice that it would have been an error in Example 4 to use Theorem 1 with the function $f(x) = 8\sqrt{x} - x^2$. This is because we find the area between two circles by subtracting the area inside the smaller circle from the area inside the larger; we do not do it by subtracting the radii and squaring the difference.

The reasoning of Example 4 may be used to derive the following theorem.

THEOREM 3 (**Method of Washers**) Suppose that U and L are nonnegative, continuous functions on the interval $[a, b]$ with $L(x) \leq U(x)$ for each x in this interval. Let $\mathcal{R}$ denote the region of the xy-plane that is bounded above by the graph of U, below by the graph of L, and on the sides by the vertical lines $x = a$ and $x = b$. Then the volume V of the solid obtained by rotating $\mathcal{R}$ about the x-axis is given by

$$V = \pi \int_a^b \left(U(x)^2 - L(x)^2\right) dx. \tag{7.1.6}$$

INSIGHT Equation (7.1.6) may be written as

$$V = \pi \int_a^b U(x)^2\, dx - \pi \int_a^b L(x)^2\, dx.$$

This formula is to be expected because we can obtain the solid of Theorem 3 by first rotating the region under the graph of $y = U(x)$ and then removing the solid that is obtained by rotating the region under the graph of $y = L(x)$.

▶ **EXAMPLE 5** Let $\mathcal{R}$ be the region of the xy-plane that is bounded above by $y = e^x$, $0 \leq x \leq 1$ and below by $y = \sqrt{x}e^{x^2}, 0 \leq x \leq 1$. Calculate the volume of the solid of revolution that is generated when $\mathcal{R}$ is rotated about the x-axis.

Solution The plots of $U(x) = e^x$, $0 \leq x \leq 1$ and $y = L(x) = \sqrt{x}e^{x^2}$ are shown in Figure 10. To the right of these plots are the surfaces they generate when rotated about the x-axis. According to Theorem 3, the volume V of the solid bounded by these surfaces is given by

$$V = \pi \int_0^1 \left(U(x)^2 - L(x)^2\right) dx = \pi \int_0^1 \left(e^{2x} - xe^{2x^2}\right) dx$$

$$= \pi \left(\frac{e^{2x}}{2} - \frac{e^{2x^2}}{4}\right)\Bigg|_{x=0}^{x=1} = \pi \left(\frac{e^2}{2} - \frac{e^2}{4}\right) - \pi \left(\frac{1}{2} - \frac{1}{4}\right) = \pi\, \frac{e^2 - 1}{4}. \quad ◀$$

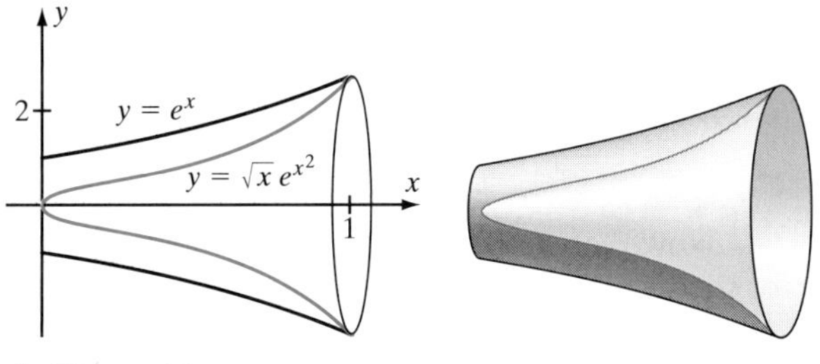

▲ Figure 10

Rotation about a Line that Is Not a Coordinate Axis

With only a small modification, the analysis leading to Theorems 1 and 2 can be applied to calculate the volume of a solid of revolution about a line that is parallel to (but not equal to) one of the coordinate axes. Here is an example.

▶ **EXAMPLE 6** Rotate the parallelogram bounded by $y = 3$, $y = 4$, $y = x$, and $y = x - 1$ about the line $x = 1$, and find the resulting volume V.

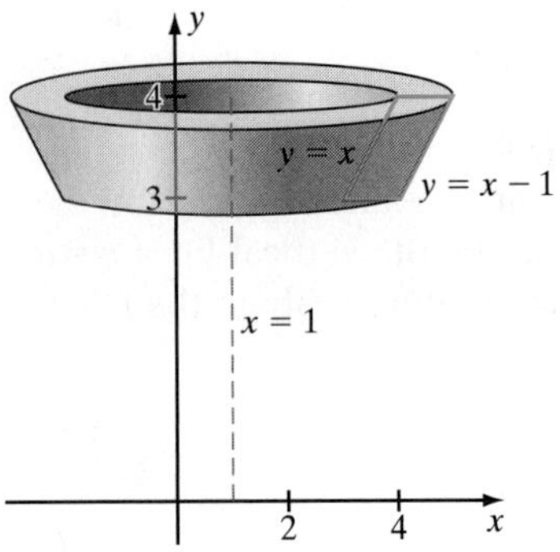

▲ Figure 11

Solution The parallelogram and solid of revolution are illustrated in Figure 11. We approximate the solid by a union of washers corresponding to values of y between 3 and 4. A typical washer at height y is shown in Figure 12. Because y will be the variable of integration, we must express the volume of this washer in terms of y. Refer to Figure 12, noting that the distance of the outer edge of the washer to the y-axis is the x-coordinate of point P. Because point P is on the line $y = x - 1$, that coordinate, expressed in terms of y, is $y + 1$. From this, we must subtract 1 to find the distance of P to the axis of rotation $x = 1$. We conclude that the outer radius of the washer is $(y + 1) - 1$, or y. Similarly, the inner radius of the washer is 1 unit less than the x-coordinate of point Q. Expressed in terms of y, the inner radius is $y - 1$. The volume contribution of this washer is therefore

$$\left(\pi y^2 - \pi(y-1)^2\right) \cdot \Delta y.$$

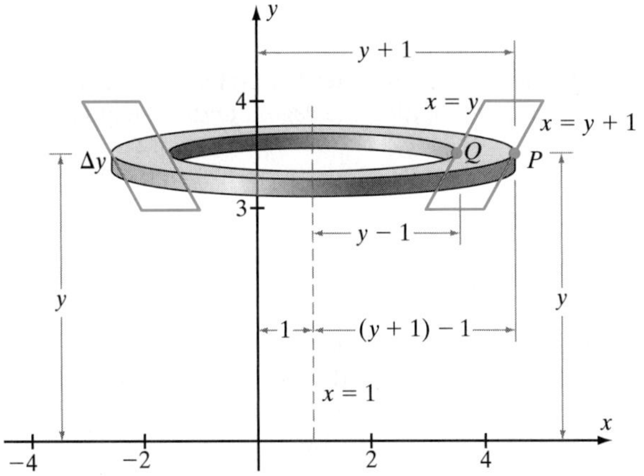

▲ Figure 12 Washer at height y

As before, we sum these volumes and let Δy tend to 0. It follows that

$$\begin{aligned} V &= \int_3^4 \left(\pi y^2 - \pi(y-1)^2\right) dy = \pi \int_3^4 \left(y^2 - (y^2 - 2y + 1)\right) dy \\ &= \pi \int_3^4 (2y - 1)\, dy = \pi(y^2 - y)\Big|_{y=3}^{y=4} = \pi(16-4) - \pi(9-3) = 6\pi. \end{aligned}$$ ◀

Now that we have seen several instances of calculating volume by the method of slicing, it is a good idea to summarize the basic steps that constitute the method.

Steps in Calculating Volume by The Method of Slicing

1. Identify the shape of each slice.
2. Identify the independent variable that gives the position of each slice.
3. Write an expression, in terms of the independent variable, which describes the cross-sectional area of each slice.
4. Identify the interval $[a, b]$ over which the independent variable ranges.
5. With respect to the independent variable of Step 2, integrate the expression for the cross-sectional area from Step 3 over the interval $[a, b]$ from Step 4.

Now let us do another example in which we explicitly follow these five steps. You will notice that, if each step can be completed, then the method of slicing can be applied even if the slices are neither disks nor washers.

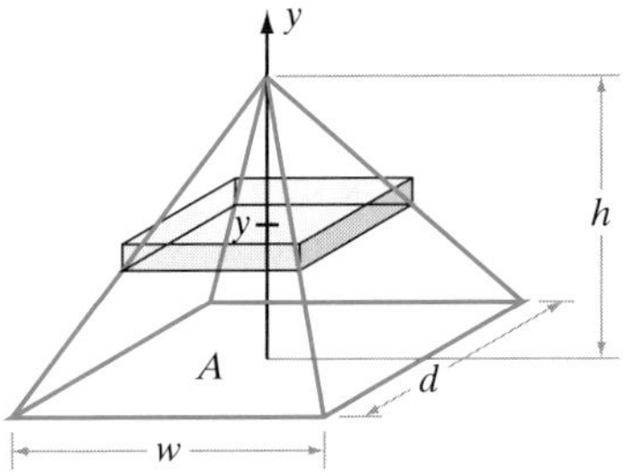

▲ Figure 13

▶ EXAMPLE 7 **(A Solid That Is Not Generated by Rotation)** Let V be the volume of a solid pyramid that has height h and rectangular base of area A. Then,

$$V = \frac{1}{3}Ah.$$

Verify this formula for a solid pyramid of height 5 if the width and depth of the base are 2 and 3, respectively.

Solution

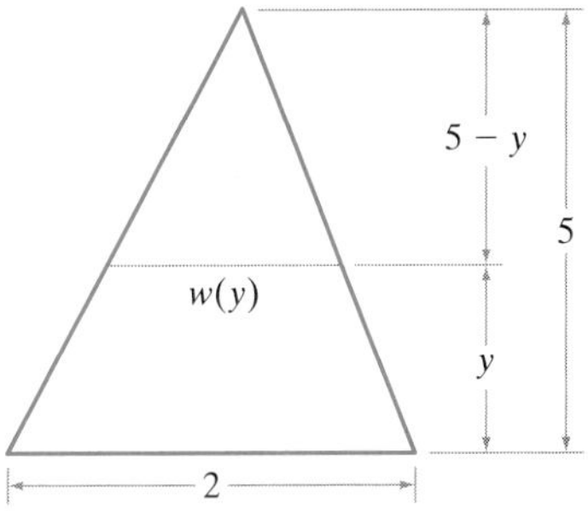

▲ Figure 14

1. The cross-section of each slice will be a rectangle—see Figure 13.
2. Because the slices are perpendicular to the y-axis, each slice is located by its position on the y-axis. So we make y the independent variable.
3. Let $w(y)$ and $d(y)$ denote the width and depth of the rectangular cross-section of the slice at height y. A similar triangular argument (see Figure 14) reveals that $w(y)/2 = (5 - y)/5$ or $w(y) = 2 \cdot (5 - y)/5 = 2(1 - y/5)$. Likewise, $d(y) = 3 \cdot (5 - y)/5 = 3(1 - y/5)$. Therefore the cross-sectional area of the slice is
$$w(y)d(y) = 2 \cdot 3 \cdot \left(1 - \frac{y}{5}\right)^2.$$
4. The range of the independent variable y is given by the extreme values of y: $y \in [0, 5]$.
5. The volume is
$$\int_0^5 6 \cdot \left(1 - \frac{y}{5}\right)^2 dy = -\frac{1}{3} 6 \cdot 5 \cdot \left(1 - \frac{y}{5}\right)^3 \Bigg|_{y=0}^{y=5} = \frac{1}{3} 6 \cdot 5 = \frac{1}{3} A \cdot h. \quad ◀$$

The Method of Cylindrical Shells

For certain solids of revolution, the method of disks or washers is either difficult or not feasible. To handle such cases, we will develop an alternative method for calculating the volumes of solids of revolution. It is called the *method of cylindrical shells* because it uses these pipe-like objects (Figure 15) rather than disks or washers as basic building blocks. This new method rests on a simple observation: The volume of a *thin* cylindrical shell of radius r, height h, and thickness t is approximately $(2\pi r) \cdot h \cdot t$. (Imagine cutting straight down the side of the cylindrical shell and flattening it out. The result is approximately a parallelepiped with side lengths $2\pi r$, h, and t as shown in Figure 15.)

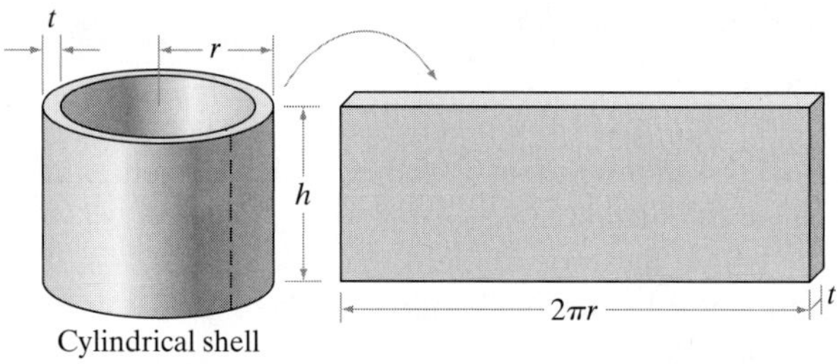

▲ Figure 15

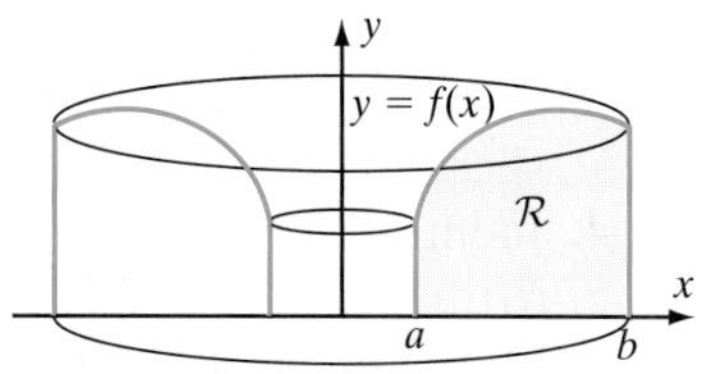

▲ Figure 16

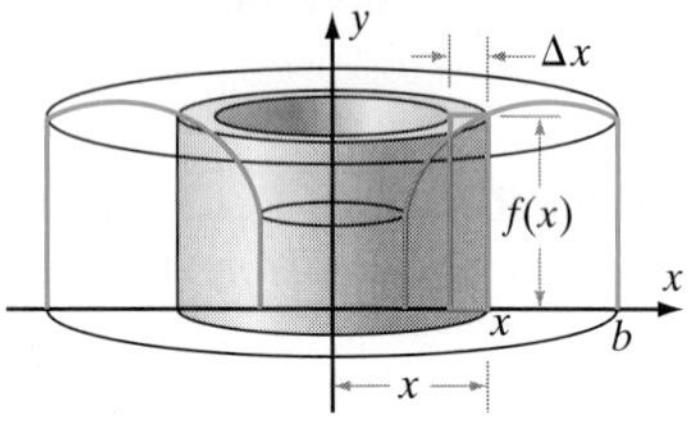

▲ Figure 17

Let f be a nonnegative continuous function on an interval $[a,b]$ of nonnegative numbers. Consider the planar region $\mathcal{R}$ (shaded in Figure 16) that is bounded by the graph of $y=f(x)$, $x=a$, $x=b$, and the x-axis. We seek the volume V of the solid of revolution that is generated by rotating $\mathcal{R}$ about the y-axis as in Figure 16. To approximate V, we partition the interval $[a,b]$ into N subintervals of equal width Δx. Corresponding to each subinterval of the partition, we form a thin rectangle as in Figure 17. When region $\mathcal{R}$ is rotated, the thin rectangle sweeps out a cylindrical shell (Figure 17). The volume contribution of this cylindrical shell is approximately $2\pi x \cdot f(x) \cdot \Delta x$. Summing the contributions from all the cylindrical shells and letting N tend to infinity, we obtain $V = \int_a^b 2\pi x f(x)\,dx$. We summarize our findings in the following theorem.

THEOREM 4 **(The Method of Cylindrical Shells: Rotation About the y-axis)** Let f be a nonnegative continuous function on an interval $[a,b]$ of nonnegative numbers. Let V denote the volume of the solid generated when the region below the graph of f and above the interval $[a,b]$ is rotated about the y-axis. Then,

$$V = 2\pi \int_a^b x f(x)\,dx.$$

▶ **EXAMPLE 8** Calculate the volume generated when the region bounded by $y=x^3+x^2$, the x-interval $[0,1]$, and the vertical line $x=1$ is rotated about the y-axis.

Solution Theorem 4 applies with $f(x)=x^3+x^2$, $a=0$ and $b=1$:

$$V = 2\pi\int_0^1 x\cdot(x^3+x^2)\,dx = 2\pi\int_0^1 (x^4+x^3)\,dx = 2\pi\left(\frac{x^5}{5}+\frac{x^4}{4}\right)\Bigg|_{x=0}^{x=1} = \frac{9}{10}\pi. \quad ◀$$

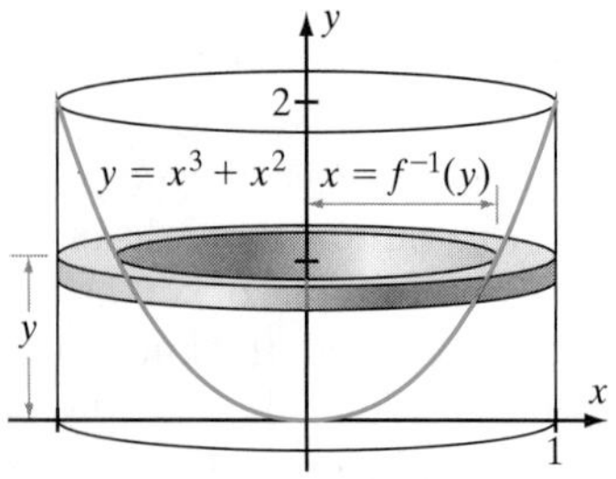

▲ Figure 18

INSIGHT In principle, we could have done this problem by the method of washers (Figure 18). The formula would have been

$$V = \pi\int_0^2 \left(1^2 - f^{-1}(y)^2\right) dy.$$

However, finding a formula for $f^{-1}(y)$ would entail solving the cubic equation $y=x^3+x^2$ for x. Doing so is no easy feat, and the resulting formula is extremely complicated. We have been able to avoid these difficulties by using the method of cylindrical shells.

Similar reasoning applies to the situation in which we rotate a region between $x = g(y)$ and the y-axis about the x-axis. We have the following theorem.

THEOREM 5 **(The Method of Cylindrical Shells: Rotation About the x-Axis)** Suppose that $0 < c < d$. Let g be a nonnegative continuous function on the interval $[c, d]$ of nonnegative numbers. The volume of the solid generated when the region bounded by $x = g(y)$, the y-axis, $y = c$, and $y = d$ is rotated about the x-axis is

$$2\pi \int_c^d y\,g(y)\,dy.$$

▶ **EXAMPLE 9** Let $\mathcal{R}$ be the region that is bounded above by the horizontal line $y = \pi/2$, below by the curve $y = \arcsin(x)$, $0 \le x \le 1$, and on the left by the y-axis. Use the method of cylindrical shells to calculate the volume V of the solid that results from rotating $\mathcal{R}$ about the x-axis.

Solution Region $\mathcal{R}$ is shown in Figure 19a. When we rotate $\mathcal{R}$ about the x-axis we obtain the solid that lies between the cylinder and the "sorcerer's hat" that are shown in Figure 19b. In Figure 19c, we have shifted our perspective a bit and cut away part of the solid's outer boundary to reveal one of the cylindrical shells that we use to calculate the volume. According to Theorem 5 with $c = 0$, $d = \pi/2$, and $g(y) = \sin(y)$, we have $V = 2\pi \int_0^{\pi/2} y \sin(y)\,dy$. On integrating by parts with $u = y$ and $dv = \sin(y)\,dy$, we obtain $du = dy$, $v = -\cos(y)$, and

$$V = 2\pi\left(-y\cos(y)\Big|_0^{\pi/2} + \int_0^{\pi/2} \cos(y)\,dy \right) = 2\pi\big(-y\cos(y) + \sin(y)\big)\Big|_0^{\pi/2} = 2\pi. \quad ◀$$

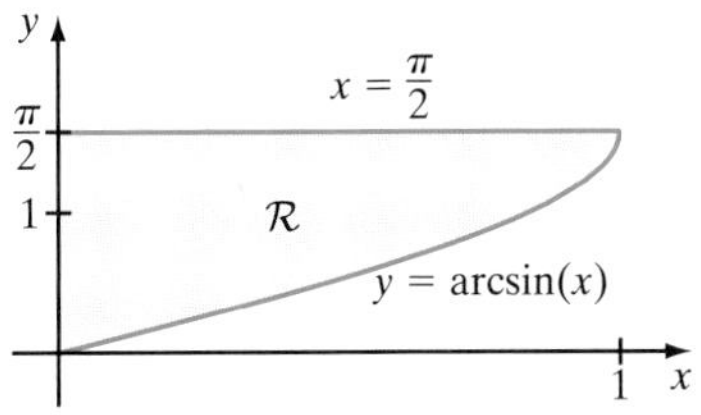

▲ Figure 19a

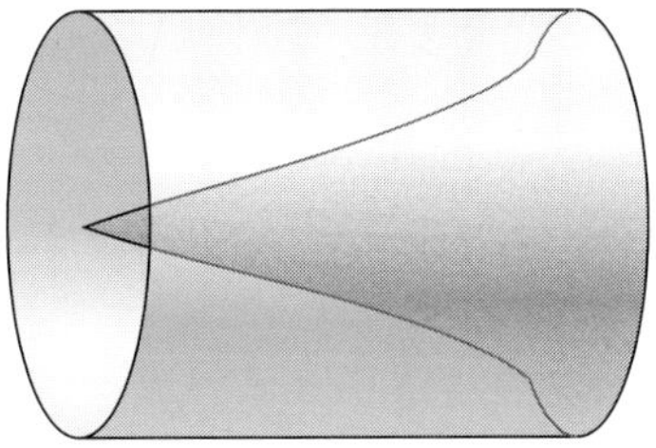
▲ Figure 19b

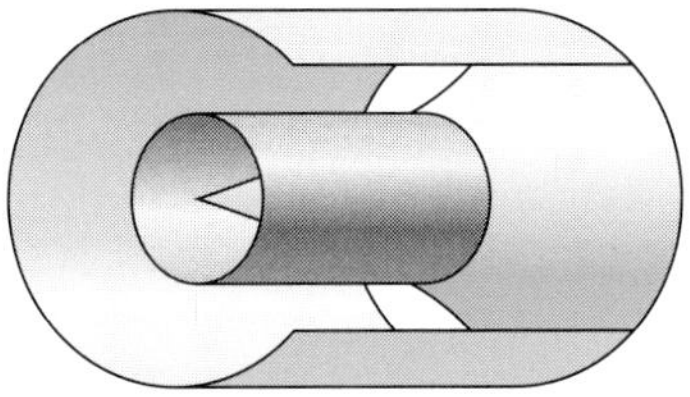
▲ Figure 19c

As we have already seen in the case of washers, the methods of this section apply to rotation about lines other than the coordinate axes. Here is an example in the context of cylindrical shells.

▶ **EXAMPLE 10** Use the method of cylindrical shells to calculate the volume of the solid obtained when the region bounded by $x = y^2$ and $x = y$ is rotated about the line $y = 2$.

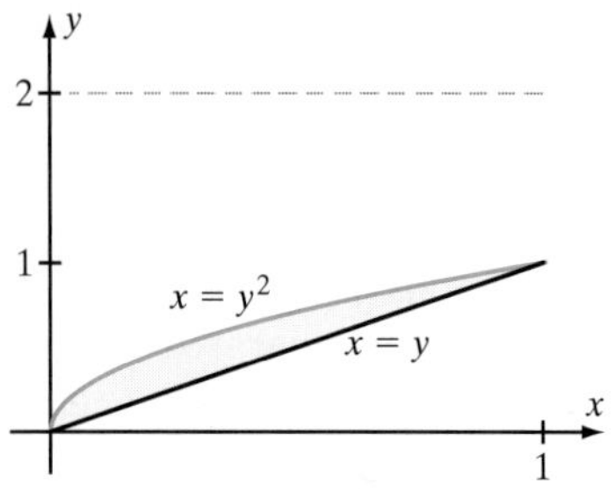

▲ Figure 20

Solution The region that is to be rotated is sketched in Figure 20. The formula of Theorem 5 does *not* apply on two counts: The region that we are rotating is not bounded on the left by the y-axis, and the axis of rotation is not the x-axis. (Either reason alone would be enough to disqualify the use of the theorem.) Nevertheless, we can adapt the method of cylindrical shells to the solid at hand. Figure 21 depicts a cylindrical shell centered about the axis of rotation. Let y denote the height of the bottom edge, and let Δy denote the thickness of the shell. Because we will let Δy tend to 0, the resulting integral will be with respect to the variable y. We notice that the curves intersect when $y = 0$ and $y = 1$ (Figure 20), so these values will be our limits of integration. Everything that we need can be deduced from the coordinates (which we express in terms of the variable of integration) of the points that are labeled P and Q in Figure 21. The height of the cylinder is $y - y^2$ and the *radius* is $2 - y$. The volume V of the solid of revolution is therefore approximately equal to a sum of terms of the form

$$2\pi(2-y)(y-y^2)\Delta y.$$

By letting Δy tend to 0, we obtain

$$V = \int_0^1 2\pi(2-y)(y-y^2)\,dy = 2\pi\int_0^1 (2y - 3y^2 + y^3)\,dy = 2\pi\left(y^2 - y^3 + \frac{y^4}{4}\right)\Bigg|_{y=0}^{y=1} = \frac{1}{2}\pi. \quad ◀$$

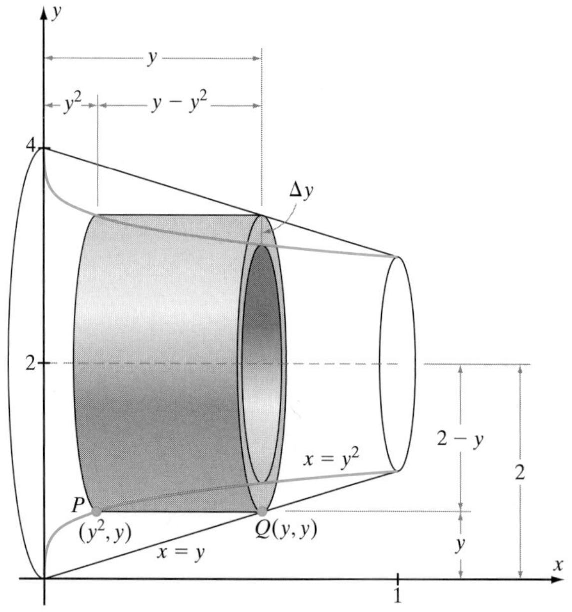

▲ Figure 21

A Final Remark

It is best not to memorize the theorems in this section. It is better to reason about disks, washers, and cylinders each time you calculate a volume. That way, you will be sure that you have taken into account the position of the axis of rotation and the relative positions of the graphs involved.

QUICK QUIZ

1. A region in the xy-plane is rotated about a vertical axis. If the method of disks is used to calculate the volume of the resulting solid of revolution, what is the variable of integration?
2. A region in the xy-plane is rotated about a horizontal axis. If the method of cylindrical shells is used to calculate the volume of the resulting solid of revolution, what is the variable of integration?
3. The region that is bounded above by $y = \sin(x)$ and below by the interval $[0, \pi]$ of the x-axis is rotated about the line $x = -2$. The volume of the resulting solid of revolution is $2\pi \int_0^{\pi} \phi(x) \sin(x)\, dx$ when expressed by the method of cylindrical shells. What is $\phi(x)$?
4. The region that is bounded above by $y = \sqrt{9 + x^2}$ and below by $y = \sqrt{9 - 2x}$ for $0 \le x \le 4$ is rotated about the x-axis. The volume of the resulting solid of revolution is $\pi \int_0^4 \psi(x)\, dx$ when expressed by the method of washers. What is $\psi(x)$?

Answers

1. y 2. y 3. $x + 2$ 4. $x^2 + 2x$

EXERCISES

Problems for Practice

In each of Exercises 1–6, use the method of disks to calculate the volume V of the solid that is obtained by rotating the given planar region $\mathcal{R}$ about the x-axis.

1. $\mathcal{R}$ is the region below the graph of $y = \sqrt{x}$, above the x-axis, and between $x = 1$ and $x = 3$.
2. $\mathcal{R}$ is the region between the x-axis and the parabola $y = 4 - x^2$ for $-2 \le x \le 2$.
3. $\mathcal{R}$ is the region between the x-axis and $y = \sqrt{\sin(x)}$, $0 \le x \le \pi$.
4. $\mathcal{R}$ is the region above the x-axis, below the graph of $y = \sec(x)$, to the right of $x = 0$, and to the left of $x = \pi/4$.
5. $\mathcal{R}$ is the region above the x-axis, below the graph of $y = \exp(x)$, to the right of $x = 0$, and to the left of $x = \ln(2)$.
6. $\mathcal{R}$ is the region below the graph of $y = x^{1/3}$, above the x-axis, and between $x = 1$ and $x = 8$.

In each of Exercises 7–12, use the method of disks to calculate the volume V of the solid that is obtained by rotating the given planar region $\mathcal{R}$ about the y-axis.

7. $\mathcal{R}$ is the region in the first quadrant that is bounded on the left by the y-axis, on the right by the graph of $y = \arcsin(x)$, and above by $y = \pi/2$.
8. $\mathcal{R}$ is the region in the first quadrant that is bounded on the left by the y-axis, on the right by the graph of $x = \cos(y)$, and that is above the line $y = \pi/4$.
9. $\mathcal{R}$ is the region to the right of the y-axis, to the left of the curve $y = \ln(x)$, above the x-axis, and below $y = 1$.
10. $\mathcal{R}$ is the region that is bounded on the left by the y-axis, on the right by the curve $y = x^2$, $1 \le x \le 2$, and that is between the horizontal lines $y = 1$ and $y = 4$.
11. $\mathcal{R}$ is the region bounded above by $y = 2$, below by $y = \sqrt{x}$, and on the left by the y-axis.
12. $\mathcal{R}$ is the region bounded above by $y = 2$, below by $y = 1/2$, and between the y-axis and the graph of $y = 1/x$.

In each of Exercises 13–18, use the method of washers to calculate the volume V of the solid that is obtained by rotating the given planar region $\mathcal{R}$ about the x-axis.

13. $\mathcal{R}$ is the region between the curves $y = x$ and $y = x^2$, $0 \le x \le 1$.
14. $\mathcal{R}$ is the region between the curves $y = x^2$ and $y = x^4$, $0 \le x \le 1$.
15. $\mathcal{R}$ is the region bounded above by the curve $y = 4 - x^2$ and below by the line $y = x + 2$.
16. $\mathcal{R}$ is the region bounded above by the line $y = 2$ and below by the curve $y = x^2 + 1$.
17. $\mathcal{R}$ is the region bounded above by the line $y = x + 1$ and below by the curve $y = 2^x$.
18. $\mathcal{R}$ is the region bounded above by the line $y = 5 - x$ and below by the curve $y = 6/x$.

In each of Exercises 19–24, use the method of washers to calculate the volume V obtained by rotating the given planar region $\mathcal{R}$ about the y-axis.

19. $\mathcal{R}$ is the region between the graphs of $x = 2y$ and $x = y^2$, $0 \le y \le 2$.
20. $\mathcal{R}$ is the region between the curves $x = y^2$ and $x = y^3$, $0 \le y \le 1$.
21. $\mathcal{R}$ is the region that is bounded on the left by the curve $x = y^2 + 2$ and on the right by the curve $x = 4 - y$.
22. $\mathcal{R}$ is the region between the curves $y = x - 1$ and $y = \log_2(x)$, $1 \le x \le 2$.

23. $\mathcal{R}$ is the region in the first quadrant that is bounded on the left by $y = 4x$, on the right by $y = x^2$, and above by $y = 4$.
24. $\mathcal{R}$ is the region between the curves $x = y^{3/2}$ on the left, $x = 2y$ on the right, and $y = 2$ below.

In each of Exercises 25–30, use the method of cylindrical shells to calculate the volume V of the solid that is obtained by rotating the given planar region $\mathcal{R}$ about the y-axis.

25. $\mathcal{R}$ is the region below the graph of $y = \exp(x^2)$, above the x-axis, and between $x = 0$ and $x = 1$.
26. $\mathcal{R}$ is the region bounded above by the curve $y = 2x - x^2$ and below by the x-axis.
27. $\mathcal{R}$ is the region below the graph of $y = x^2 + 1$, above the x-axis, and between $x = 2$ and $x = 4$.
28. $\mathcal{R}$ is the region below the graph of $y = (1 + x^2)^{-2}$, $0 \le x \le 1$ and above the x-axis.
29. $\mathcal{R}$ is the region below the graph of $y = \sin(x)/x$, $0 < x \le \pi$ and above the x-axis.
30. $\mathcal{R}$ is the region below the graph of $y = 2/x$, $1 \le x \le 2$ and above the x-axis.

In each of Exercises 31–36, use the method of cylindrical shells to calculate the volume V of the solid that is obtained by rotating the given planar region $\mathcal{R}$ about the x-axis.

31. $\mathcal{R}$ is the region in the first quadrant that is to the left of the line $y = x - 2$ and below $y = 2$.
32. $\mathcal{R}$ is the region that is to the right of the y-axis, to the left of the curve $y = \sqrt{x}$, and between the lines $y = 1$ and $y = 2$.
33. $\mathcal{R}$ is the region in the first quadrant that is to the left of the curve $y = \sqrt{x - 1}$ and below $y = 1$.
34. $\mathcal{R}$ is the region that is to the right of the y-axis, to the left of the curve $y = 216x^3$, and between the lines $y = 1$ and $y = 8$.
35. $\mathcal{R}$ is the region between the y-axis and the curve $x = y \exp(y^3)$, $0 \le y \le 1$.
36. $\mathcal{R}$ is the region between the y-axis and the curve $x = 2 + y + y^2$, $0 \le y \le 2$.

In each of Exercises 37–42 use the method of cylindrical shells to calculate the volume of the solid that is obtained by rotating the given planar region $\mathcal{R}$ about the y-axis.

37. $\mathcal{R}$ is the region in the first quadrant that is bounded by $y = x$ and $y = x^2$.
38. $\mathcal{R}$ is the region between $y = x$, $y = x^2$, $x = 2$, and $x = 3$.
39. $\mathcal{R}$ is the region below the graph of $y = x^2 + 1$, above the graph of $y = x$, and between $x = 1$ and $x = 3$.
40. $\mathcal{R}$ is the region in the first quadrant below the graph of $y = 1 - x^3$ and above the graph of $y = 1 - x$.
41. $\mathcal{R}$ is the region in the first quadrant below the graph of $y = \sqrt{9 + x^2}$ and above the graph of $y = x + 1$.
42. $\mathcal{R}$ is the region below the graph of $y = 4x - x^2 - 2$ and above the graph of $y = 1$.

In each of Exercises 43–48, use the method of cylindrical shells to calculate the volume V of the solid that is obtained by rotating the given planar region $\mathcal{R}$ about the x-axis.

43. $\mathcal{R}$ is the region that is bounded by the curve $y = x^2$ and the lines $y = 1$, $y = 4$, and $x = 3$.
44. $\mathcal{R}$ is the region that is to the right of $y = 5x - 4$, to the left of $y = 10 - x^2$, and above $y = 1$.
45. $\mathcal{R}$ is the region that is bounded on the left by the curve $y = 5 - x^2$ for $-2 \le x \le -1$, on the right by the curve $y = 5 - x^2$ for $1 \le x \le 2$, above by the line segment $y = 4$ for $-1 \le x \le 1$, and below by the line segment $y = 1$ for $-2 \le x \le 2$.
46. $\mathcal{R}$ is the region that is bounded on the left by the line $y = x + 4$ and on the right by the curve $x = 6y - y^2 - 8$.
47. $\mathcal{R}$ is the region in the first quadrant that is bounded by the y-axis and the curve $x = 4y - y^2 - 3$.
48. $\mathcal{R}$ is the region in the first quadrant that is bounded by the y-axis and the curve $x = 4y^2 - y^3 - 5y + 2$.

In each of Exercises 49–52, calculate the volume V of the solid that is obtained by rotating the planar region between the curves $y = x^2$ and $y = x^{1/2}$ about the given line.

49. $x = 2$
50. $x = -1$
51. $y = 1$
52. $y = -2$

In each of Exercises 53–56, $\mathcal{R}$ is the planar region that is bounded on the left by $y = 4/x^2$, on the right by $x = 2$, and above by $y = 4$. Calculate the volume V of the solid that is obtained by rotating $\mathcal{R}$ about the given line.

53. $x = -1$
54. $x = 3$
55. $y = 4$
56. $y = -1$

Further Theory and Practice

In each of Exercises 57–60, use the method of disks to calculate the volume V of the solid obtained by rotating the given planar region $\mathcal{R}$ about the x-axis.

57. $\mathcal{R}$ is the region below the graph of $y = \sin(x)$, $0 \le x \le \pi$ and above the x-axis.
58. $\mathcal{R}$ is the region below the graph of $y = \cos(x)$, $\pi/6 \le x \le \pi/3$ and above the x-axis.
59. $\mathcal{R}$ is the first quadrant region between the x-axis, the curve $y = \sqrt{x} \cdot \exp(x)$, and the line $x = 1$.
60. $\mathcal{R}$ is the region between the x-axis, the curve $y = x\sqrt{\sin(x)}, 0 \le x \le \pi$.

In each of Exercises 61–64, use the method of disks to calculate the volume obtained by rotating the given planar region $\mathcal{R}$ about the y-axis.

61. $\mathcal{R}$ is the region between the curves $y = \exp(x)$, the y-axis, and the line $y = e$.
62. $\mathcal{R}$ is the first quadrant region between the curve $y = \sin(x^2)$, the y-axis, and the line $y = 1$.

63. $\mathcal{R}$ is the region between the curve $y = x^2/(1-x^2)$, the y-axis, and the line $y = 1/3$.

64. $\mathcal{R}$ is the region in the first quadrant that is bounded by the coordinate axes and the curve $x = (1-y^2)^{3/4}$.

In each of Exercises 65–68, use the method of cylindrical shells to calculate the volume obtained by rotating the given planar region $\mathcal{R}$ about the given line ℓ.

65. $\mathcal{R}$ is the region between the curves $y = x^2 - 4x - 5$ and $y = -x^2 + 4x + 5$; ℓ is the line $x = -3$.

66. $\mathcal{R}$ is the region between the graphs of $y = \exp(x)$, $y = x\exp(x)$, and the y-axis; ℓ is the line $x = 1$.

67. $\mathcal{R}$ is the region between the curves $y = x^2 - x - 5$ and $y = -x^2 + x + 7$; ℓ is the line $x = 5$.

68. $\mathcal{R}$ is the region between the curves $x = y^3$ and $x = -y^2$; ℓ is the line $y = 3$.

In each of Exercises 69–76, calculate the volume of the solid obtained when the region $\mathcal{R}$ is rotated about the given line ℓ.

69. $\mathcal{R}$ is the region between $y = 6 - x^2$ and $y = -x$; ℓ is the line $x = -4$.

70. $\mathcal{R}$ is the region that is bounded above by $y = x^2 + 3$, below by $y = 2x^2 - 1$, and on the left by $x = 1$; ℓ is the line $x = 1$.

71. $\mathcal{R}$ is the region between the curve $y = \cos(x)$ and the x-axis, $\pi/2 \le x \le 3\pi/2$; ℓ is the line $x = 3\pi$.

72. $\mathcal{R}$ is the region in the first quadrant that is bounded on the left by the y-axis, on the right by the curve $x = \tan(y)$, and above by the line $y = \pi/4$; ℓ is the line $x = 1$.

73. $\mathcal{R}$ is the region bounded by the curve $x = \sin(y)$, the y-axis, and the line $y = \pi/2$; ℓ is the line $y = 2$.

74. $\mathcal{R}$ is the region between the curves $y = \sin(x)$ and $y = \cos(x)$, $0 \le x \le \pi/4$; ℓ is the line $y = \sqrt{2}$.

75. $\mathcal{R}$ is the first quadrant region between the curve $y = x^3 + x$, the line $x = 1$, and the x-axis; ℓ is the line $x = -1$.

76. $\mathcal{R}$ is the first quadrant region between the curve $y = 1/(1+x^2)$, the line $y = 1/2$, and the y-axis; ℓ is the line $x = 0$.

77. Suppose $R > r > 0$. Calculate the volume of the solid obtained when the disc $\{(x,y) : x^2 + y^2 \le r^2\}$ is rotated about the line $x = R$. (This solid is called a *torus*)

78. Calculate the volume of the solid obtained when the triangle with vertices $(2,5)$, $(6,1)$, $(4,4)$ is rotated about the line $x = -3$.

79. Calculate the volume obtained when the region outside the square $\{(x,y) : |x| < 1, |y| < 1\}$ and inside the circle $\{(x,y) : x^2 + y^2 \le 4\}$ is rotated about the line $y = -3$.

80. A solid has as its base the ellipse $x^2 + 4y^2 = 16$. The vertical slices *parallel to the line* $y = 2x$ are equilateral triangles. Find the volume.

81. The base of a solid S is the disk $x^2 + y^2 \le 25$. For each $k \in [-5, 5]$, the plane through the line $x = k$ and perpendicular to the xy-plane intersects S in a square. Find the volume of S.

82. A solid has as its base the region bounded by the parabola $x - y^2 = -8$ and the left branch of the hyperbola $x^2 - y^2 - 4 = 0$. The vertical slices perpendicular to the x-axis are squares. Find the volume of the solid.

83. Old Boniface, he took his cheer, Then he drilled a hole in a solid sphere, Clear through the center straight and strong And the hole was just 10 inches long. Now tell us when the end was gained What volume in the sphere remained. Sounds like you've not been told enough. But that's all you need, it's not too tough.

84. An open cylindrical beaker with circular base has height L and radius r. It is partially filled with a volume V of a fluid. Consider the parameters L, r, and V to be constant. The axis of symmetry of the beaker is along the positive y-axis and one diameter of its base is along the x-axis. When the tank is revolved about the y-axis with angular speed ω, the surface of the fluid assumes a shape that is the paraboloid of revolution that results when the curve

$$y = h + \omega^2 x^2/(2g), \quad 0 \le x \le r$$

is revolved about the y-axis. This formula is valid for angular speeds at which the surface of the fluid has not yet touched the base or the mouth of the beaker. The number $h = h(\omega)$ is in the interval $[0, V/(\pi r^2)]$ and depends on ω. (When $\omega = 0$, then $h = V/(\pi r^2)$. As ω increases, h decreases.)

a. Find a formula for $h(\omega)$.

b. At what value ω_S of ω does spilling begin, assuming that $h(\omega) > 0$ for $\omega > \omega_S$?

c. At what value ω_B of ω does the surface touch the bottom of the beaker, assuming that spilling does not occur for $\omega < \omega_B$?

d. As ω increases, does the surface of the fluid touch the bottom of the beaker or the mouth of the beaker first?

Calculator/Computer Exercises

85. The region below the graph of $y = \sqrt{x}$ and above the graph of $y = x\exp(x)$ is rotated about the y-axis. Use Simpson's Rule from Section 5.8 in Chapter 5 to calculate the resulting volume to four decimal places.

86. The region below the graph of $y = \exp(-x^2)$, $-1 \le x \le 1$ is rotated about the x-axis. Use Simpson's Rule to calculate the resulting volume to four decimal places.

87. A flashlight reflector is made of an aluminum alloy whose mass density is 3.74 g/cm^3. The reflector occupies the solid region that is obtained when the region bounded by $y = 2.05\sqrt{x} + 0.496$, $y = 2.05\sqrt{x} + 0.546$, $x = 0$, and $x = 2.80$ cm is rotated about the x-axis. What is the mass of the reflector?

88. The equation of the St. Louis Gateway Arch is

$$y = 693.8597 - 34.38365\,(e^{kx} + e^{-kx})$$

for $k = 0.0100333$ and $-299.2239 < x < 299.2239$, where both x and y are measured in feet. Rotate this curve about its vertical axis of symmetry and compute the resulting volume.

7.2 Arc Length and Surface Area

Just as we have approximated area by a sum of areas of rectangles, so can we approximate the length of a curve by a sum of lengths of line segments. This process leads to an integral that is used to calculate the length of a graph of a function. Analogous ideas can also be applied to calculate the area of a surface of revolution.

The Basic Method for Calculating Arc Length

Suppose that f is a function with continuous derivative on a domain that contains the interval $[a, b]$. Let us calculate the length L of the graph of f over this interval. Fix a positive integer N, and let $a = x_0 < x_1 < x_2 < \ldots < x_{N-1} < x_N = b$ be a uniform partition of the interval $[a, b]$. As Figure 1 suggests, we will use a piecewise linear curve to approximate the graph of f. To be specific, we will use the line segment with endpoints $P_{j-1} = (x_{j-1}, f(x_{j-1}))$ and $P_j = (x_j, f(x_j))$ to approximate the part of the graph of f that lies over the jth subinterval $[x_{j-1}, x_j]$ (see Figure 2). The length ℓ_j of this line segment then approximates the length of the arc of the graph between P_{j-1} and P_j. We sum the lengths ℓ_j to obtain an approximate length for the curve:

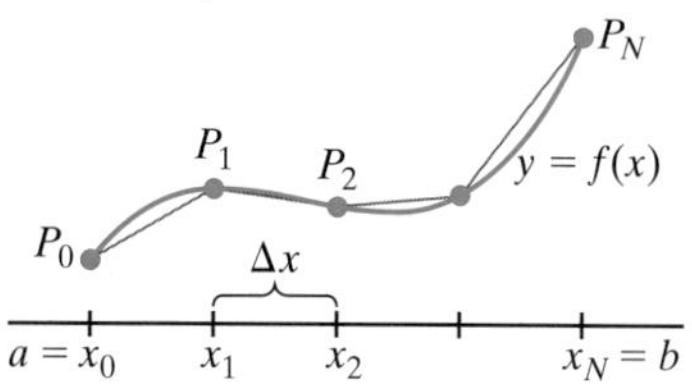

▲ Figure 1

$$L \approx \sum_{j=1}^{N} \ell_j.$$

The accuracy of this approximation is improved by increasing the number N of subintervals (Figure 3). As N increases to infinity and Δx tends to 0, these approximating sums tend to what we think of as the length of the curve:

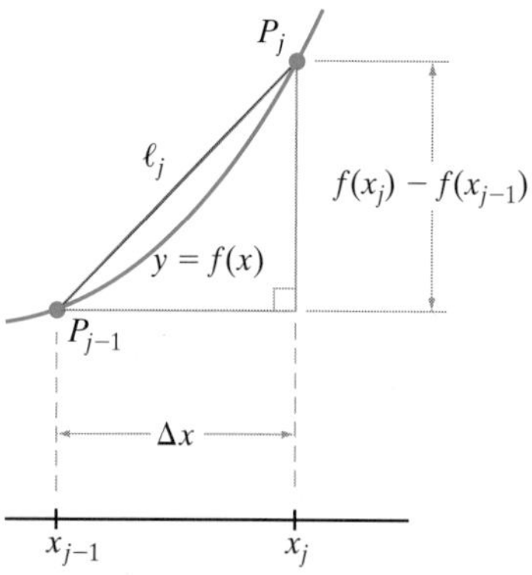

▲ Figure 2

$$L = \lim_{N\to\infty} \sum_{j=1}^{N} \ell_j. \tag{7.2.1}$$

We will use equation (7.2.1) to *define* the length L of the graph of f. Of course, we must find an analytic form of the right side that is suitable for calculation.

Refer again to Figure 2. Notice that the length ℓ_j is given by the usual planar distance formula:

$$\ell_j = \sqrt{(x_j - x_{j-1})^2 + \left(f(x_j) - f(x_{j-1})\right)^2}.$$

We denote the quantity $x_j - x_{j-1}$ by Δx and apply the Mean Value Theorem to the expression $f(x_j) - f(x_{j-1})$ to obtain $f(x_j) - f(x_{j-1}) = f'(s_j)\Delta x$ for some s_j between x_{j-1} and x_j. Now we may rewrite the formula for ℓ_j as

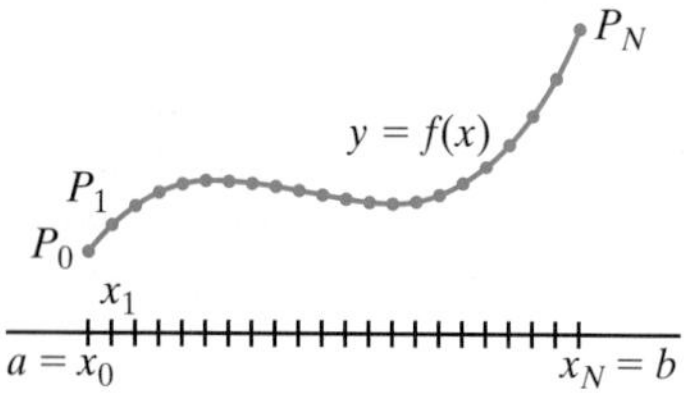

▲ Figure 3

$$\ell_j = \sqrt{(\Delta x)^2 + (f'(s_j)\,\Delta x)^2} = \sqrt{1 + f'(s_j)^2}\,\Delta x.$$

If we use this formula to substitute for ℓ_j in equation (7.2.1), then we have

$$L = \lim_{N\to\infty} \sum_{j=1}^{N} \sqrt{1 + f'(s_j)^2}\,\Delta x. \tag{7.2.2}$$

Because the sums on the right side of equation (7.2.2) are Riemann sums for the integral $\int_a^b \sqrt{1 + f'(x)^2}\,dx$, we conclude that this integral is the value of limit formula (7.2.2) for L. These observations lead to the following definition.

DEFINITION If f has a continuous derivative on an interval containing $[a, b]$, then the *arc length* L of the graph of f over the interval $[a, b]$ is given by

$$L = \int_a^b \sqrt{1 + f'(x)^2}\, dx. \tag{7.2.3}$$

Some Examples of Arc Length

The formula for arc length leads to integrals that are often difficult or impossible to evaluate exactly. In such cases, the techniques of numerical integration that we studied in Section 5.8 in Chapter 5 may be used to obtain accurate approximations. In the present section, we concentrate on simple examples in which the integrals work out neatly.

▶ **EXAMPLE 1** Calculate the arc length L of the graph of $f(x) = 2x^{3/2}$ over the interval $[0, 7]$.

Solution By definition, the length is

$$L = \int_0^7 \sqrt{1 + f'(x)^2}\, dx = \int_0^7 \sqrt{1 + (3x^{1/2})^2}\, dx = \int_0^7 (1 + 9x)^{1/2}\, dx.$$

In this last integral, we make the change of variable $u = 1 + 9x$, $du = 9\, dx$ and note that u varies from 1 to 64 as x varies from 0 to 7. The result is

$$L = \int_1^{64} u^{1/2} \frac{1}{9}\, du = \frac{1}{9} \cdot \frac{2}{3} u^{3/2} \Big|_{u=1}^{u=64} = \frac{2}{27}\left(64^{3/2} - 1\right) = \frac{2}{27}(512 - 1) = \frac{1022}{27}. \quad ◀$$

▶ **EXAMPLE 2** Calculate the length L of the graph of the function $f(x) = (e^x + e^{-x})/2$ over the interval $[1, \ln(8)]$.

Solution We first calculate that $f'(x) = (e^x - e^{-x})/2$ and

$$1 + f'(x)^2 = 1 + \frac{(e^x - e^{-x})^2}{4} = \frac{4 + e^{2x} - 2 + e^{-2x}}{4} = \frac{e^{2x} + 2 + e^{-2x}}{4} = \left(\frac{e^x + e^{-x}}{2}\right)^2.$$

Thus

$$L = \int_1^{\ln(8)} \sqrt{1 + f'(x)^2}\, dx = \int_1^{\ln(8)} \left(\frac{e^x + e^{-x}}{2}\right) dx = \frac{e^x - e^{-x}}{2} \Big|_1^{\ln(8)} = \frac{1}{2}\left(\left(8 - \frac{1}{8}\right) - \left(e - \frac{1}{e}\right)\right). \quad ◀$$

Sometimes an arc length problem is more conveniently solved if we think of the curve as being the graph of $x = g(y)$.

DEFINITION If g has a continuous derivative on an interval containing $[c, d]$, then the arc length L of the graph of $x = g(y)$ for $c \le y \le d$ is given by

$$L = \int_c^d \sqrt{1 + g'(y)^2}\, dy. \tag{7.2.4}$$

This is only a restatement of the definition of arc length and does not require additional analysis.

▶ **EXAMPLE 3** Calculate the length L of that portion of the graph of the curve $9x^2 = 4y^3$ between the points $(0,0)$ and $(2/3, 1)$.

Solution We express the curve as $x = (2/3)y^{3/2}$ and set $g(y) = (2/3)y^{3/2}$. Then $g'(y) = y^{1/2}$. Formula (7.2.4) becomes

$$L = \int_0^1 \sqrt{1 + (y^{1/2})^2}\, dy = \int_0^1 (1+y)^{1/2}\, dy.$$

This integral can be evaluated easily by means of the substitution $u = 1 + y$, $du = dy$. Noting that the limits of integration for u corresponding to $y = 0$ and $y = 1$ are $u = 1$ and $u = 2$, respectively, we have

$$L = \int_1^2 u^{1/2}\, du = \frac{2}{3}u^{3/2}\Big|_{u=1}^{u=2} = \frac{2}{3}\left(2\sqrt{2} - 1\right). \quad ◀$$

INSIGHT In Example 3, we might have solved the equation $9x^2 = 4y^3$ for y and obtained $y = \left(\frac{3}{2}x\right)^{2/3}$. Setting $f(x) = \left(\frac{3}{2}x\right)^{2/3}$, we would obtain $f'(x) = \left(\frac{3}{2}\right)^{2/3} \cdot \frac{2}{3}x^{-1/3} = \left(\frac{3}{2}x\right)^{-1/3}$, and L would be given by

$$L = \int_0^{2/3} \left(1 + \left(\frac{3}{2}x\right)^{-2/3}\right)^{1/2} dx.$$

This integral would have been considerably more difficult to evaluate! (If you wish to try your hand at it, then consider the indirect substitution $x = (2/3)\tan^{-3}(\theta)$.) In general, there is no method for predicting whether it is advantageous to set up an arc length integral in a particular way. The fact is, it is usually impossible to calculate arc length *exactly* whichever variable of integration is chosen.

Parametric Curves

The ideas that we have developed for the arc length of the graph of $y = f(x)$ or $x = g(y)$ can also be applied to a general parameterized curve. Recall from Section 1.5 in Chapter 1 that a parametric curve consists of the points (x, y) arising from parametric equations $x = \varphi_1(t)$ and $y = \varphi_2(t)$. Supposing that $\varphi_1(t)$ and $\varphi_2(t)$ have continuous derivatives on an interval that contains $I = [\alpha, \beta]$, we calculate the length of the parametric curve $\mathcal{C} = \{(\varphi_1(t), \varphi_2(t)) : \alpha \le t \le \beta\}$ as follows.

For a fixed positive integer N, let $\alpha = t_0 < t_1 < t_2 < \ldots < t_{N-1} < t_N = \beta$ be the order N uniform partition of I. We use the line segment with endpoints $P_{j-1} = (\varphi_1(t_{j-1}), \varphi_2(t_{j-1}))$ and $P_j = (\varphi_1(t_j), \varphi_2(t_j))$ to approximate the part of $\mathcal{C}$ corresponding to values of t in the jth subinterval $[t_{j-1}, t_j]$ of I (see Figure 4). The length

$$\ell_j = \sqrt{(\Delta x)^2 + (\Delta y)^2} = \sqrt{\left(\frac{\Delta x}{\Delta t}\right)^2 + \left(\frac{\Delta y}{\Delta t}\right)^2}\, \Delta t$$

of this line segment then approximates the length of the arc of $\mathcal{C}$ between P_{j-1} and P_j. By expressing Δx in terms of φ_1 and Δy in terms of φ_2, we have

$$\ell_j = \sqrt{\left(\frac{\varphi_1(t_j) - \varphi_1(t_{j-1})}{\Delta t}\right)^2 + \left(\frac{\varphi_2(t_j) - \varphi_2(t_{j-1})}{\Delta t}\right)^2}\, \Delta t.$$

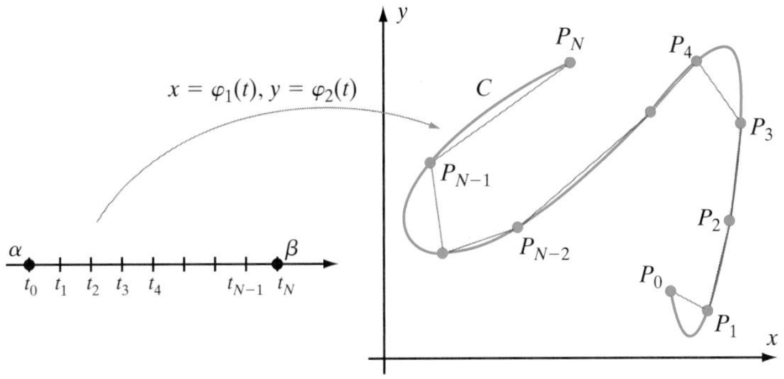

▲ Figure 4

Denoting the midpoint of $[t_{j-1}, t_j]$ by s_j, we observe that $\dfrac{\varphi_1(t_j) - \varphi_1(t_{j-1})}{\Delta t}$ and $\dfrac{\varphi_2(t_j) - \varphi_2(t_{j-1})}{\Delta t}$ are the central difference quotient approximations of $\varphi_1'(s_j)$ and $\varphi_2'(s_j)$, respectively. Thus

$$\ell_j \approx \sqrt{\left(\varphi_1'(s_j)\right)^2 + \left(\varphi_2'(s_j)\right)^2}\,\Delta t.$$

We sum the lengths ℓ_j to obtain an approximate length for the curve:

$$L \approx \sum_{j=1}^{N} \sqrt{\left(\varphi_1'(s_j)\right)^2 + \left(\varphi_2'(s_j)\right)^2}\,\Delta t.$$

As N increases to infinity and Δt tends to 0, these approximating sums tend to our intuitive concept of the arc length L of $\mathcal{C}$:

$$L = \lim_{N\to\infty} \sum_{j=1}^{N} \sqrt{\left(\varphi_1'(s_j)\right)^2 + \left(\varphi_2'(s_j)\right)^2}\,\Delta t.$$

Because the sums on the right side of this formula are Riemann sums for the integral $\int_\alpha^\beta \sqrt{\left(\varphi_1'(t)\right)^2 + \left(\varphi_2'(t)\right)^2}\,dt$, we are led to the following definition.

DEFINITION If φ_1 and φ_2 have continuous derivatives on an interval that contains $I = [\alpha, \beta]$, then the *arc length* L of the parametric curve $\mathcal{C} = \{(\varphi_1(t), \varphi_2(t)) : \alpha \le t \le \beta\}$ is given by

$$L = \int_\alpha^\beta \sqrt{\left(\varphi_1'(t)\right)^2 + \left(\varphi_2'(t)\right)^2}\,dt. \tag{7.2.5}$$

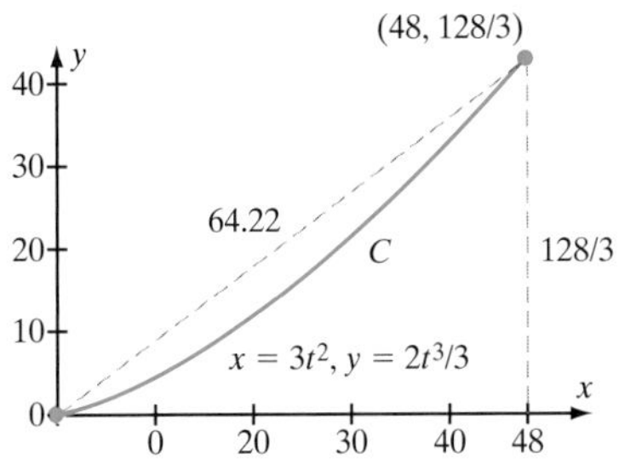

▲ Figure 5

▶ **EXAMPLE 4** Calculate the length L of the parametric curve $\mathcal{C} = \{(\varphi_1(t), \varphi_2(t)) : 0 \le t \le 4\}$ where $\varphi_1(t) = 3t^2$ and $\varphi_2(t) = 2t^3/3$.

Solution The graph of $\mathcal{C}$ in Figure 5 leads us to expect a value for L that does not greatly exceed the straight line distance, $\sqrt{48^2 + (128/3)^2}$, or 64.22 ..., between the endpoints $(0,0)$ and $(48,128/3)$ of $\mathcal{C}$. For an exact evaluation of L using formula (7.2.5), we first calculate $\varphi_1'(t) = 6t$ and $\varphi_2'(t) = 2t^2$. It follows that

$$\sqrt{(\varphi_1'(t))^2 + (\varphi_2'(t))^2} = \sqrt{(6t)^2 + (2t^2)^2} = \sqrt{36t^2 + 4t^4} = \sqrt{4t^2(9+t^2)} = 2t\sqrt{9+t^2}.$$

Making the substitution $u = 9 + t^2$, $du = 2t\,dt$ and noting that $u = 9$ and $u = 25$ are the limits of integration corresponding to $t = 0$ and $t = 4$, we have

$$L = \int_0^4 \sqrt{(\varphi_1'(t))^2 + (\varphi_2'(t))^2}\,dt = \int_0^4 2t\sqrt{9+t^2}\,dt = \int_9^{25} u^{1/2}\,du = \frac{2}{3}u^{3/2}\Big|_9^{25} = \frac{196}{3} = 65.\overline{3}\,. \quad ◀$$

In practice, we often do not give names such as φ_1 and φ_2 to the functions that define x and y parametrically. When we dispense with the extra notation, formula (7.2.5) takes the simpler form

$$L = \int_\alpha^\beta \sqrt{\left(\frac{dx}{dt}\right)^2 + \left(\frac{dy}{dt}\right)^2}\,dt. \tag{7.2.6}$$

▶ **EXAMPLE 5** Calculate the length L of the parametric curve $\mathcal{C}$ defined by the parametric equations $x = 1 + 2t^{3/2}$ and $y = 3 + 2t$ for $0 \le t \le 5$.

Solution We begin by calculating

$$\sqrt{\left(\frac{dx}{dt}\right)^2 + \left(\frac{dy}{dt}\right)^2} = \sqrt{(3t^{1/2})^2 + (2)^2} = (9t+4)^{1/2}.$$

Using equation (7.2.6) with the substitution $u = 9t + 4$, $du = 9dt$ and noting that $u = 4$ and $u = 49$ are the limits of integration corresponding to $t = 0$ and $t = 5$, we have

$$L = \int_0^5 (9t+4)^{1/2}\,dt = \frac{1}{9}\int_4^{49} u^{1/2}\,dt = \frac{2}{27}u^{3/2}\Big|_{u=4}^{u=49} = \frac{2}{27}(343-8) = \frac{670}{27}. \quad ◀$$

Surface Area

Let f be a nonnegative function with continuous derivative on the interval $[a,b]$. Imagine rotating the graph of f about the x-axis. This procedure will generate a *surface of revolution*, as shown in Figure 6. We will develop a procedure for determining the area of such a surface. As with previously developed methods that have led to integral formulas, we first divide the geometric object into a number of pieces that can be approximated by a simple shape. The basic shape that we will use here, a *frustum*, can be seen in Figure 7. It results when a conical surface is cut by two planes that are perpendicular to the axis of symmetry. The formula for the surface area S of a frustum is

$$\text{Surface Area of frustum} = 2\pi\left(\frac{r+R}{2}\right)s \tag{7.2.7}$$

where r and R are the radii of the circular edges and s is the *slant height*, as indicated in Figure 7. Notice that this is the surface area of a cylinder with height equal to the slant height of the frustum and radius equal to the average radius of

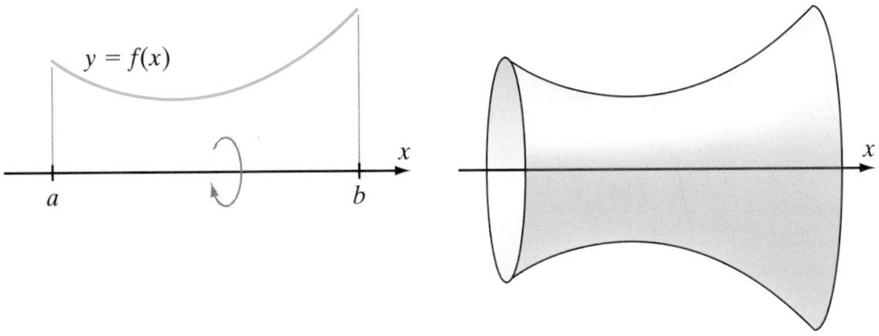

▲ Figure 6 The graph of a function and the surface it generates when rotated about the x-axis.

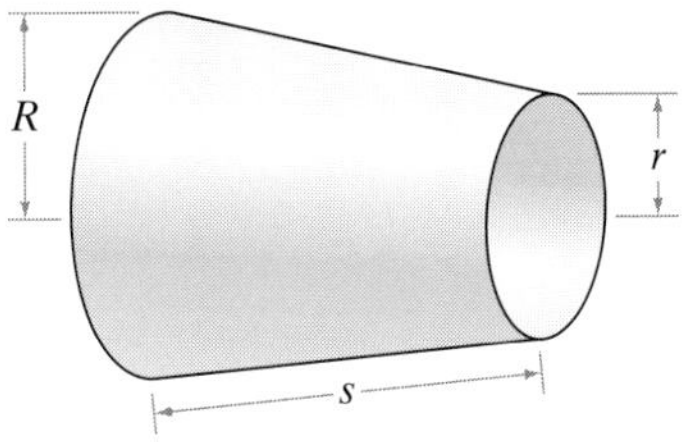

▲ Figure 7 Surface area of frustum = $2\pi\left(\dfrac{r+R}{2}\right)s$

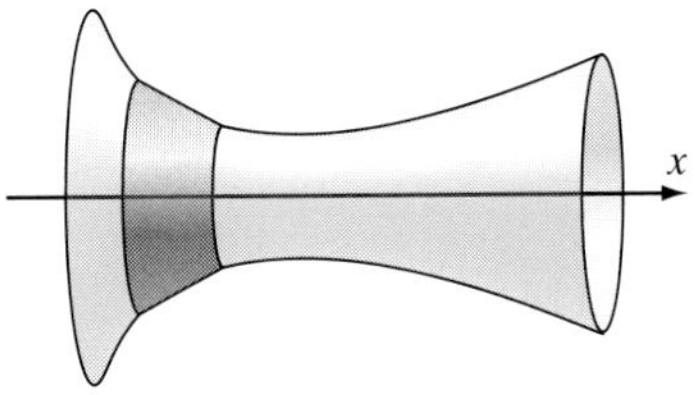

▲ Figure 8 Slice of surface approximated by a frustum of a cone.

the cross-sections of the frustum. Formula (7.2.7) is elementary in the sense that its derivation (as outlined in Exercise 62) does not require calculus.

For each positive integer N, we form a uniform partition

$$a = x_0 < x_1 < x_2 < \ldots < x_{N-1} < x_N = b$$

of the interval $[a, b]$. Let Δx denote the common length of the N subintervals. When the arc of the curve that lies over the jth subinterval is rotated about the x-axis, we obtain a surface that we can approximate by a frustum (as in Figure 8). The slant height of this frustum can be approximated by the arc length ℓ_j of the graph of f from $(x_{j-1}, f(x_{j-1}))$ to $(x_j, f(x_j))$. In the discussion of arc length earlier in this section, we reasoned that there is a point s_j in the jth subinterval such that

$$\ell_j \approx \sqrt{1 + f'(s_j)^2}\,\Delta x.$$

We may use the value $f(s_j)$ at this interior point s_j to approximate the average value $(f(x_{j-1}) + f(x_j))/2$ of f at the endpoints. It follows from formula (7.2.7) that the area contribution of the jth increment of our surface is about

$$2\pi \cdot f(s_j)\sqrt{1 + f'(s_j)^2}\,\Delta x.$$

If we sum up the area contribution from each subinterval of the partition, we find that the area of our surface of revolution is about

$$\sum_{j=1}^{N} 2\pi \cdot f(s_j)\sqrt{1 + f'(s_j)^2}\,\Delta x.$$

The accuracy of this approximation is improved by increasing the number N of subintervals. As N increases to infinity and Δx tends to 0, these approximating sums tend to what we think of as surface area S of the surface of revolution:

$$S = \lim_{N\to\infty} \sum_{j=1}^{N} 2\pi \cdot f(s_j)\sqrt{1 + f'(s_j)^2}\,\Delta x.$$

But these sums are also Riemann sums for the integral

$$2\pi \int_a^b f(x)\sqrt{1 + f'(x)^2}\,dx.$$

These observations lead us to the definition of the surface area of a surface of revolution:

DEFINITION If f is a nonnegative function which has a continuous derivative on an interval containing $[a, b]$, then the *surface area* of the surface of revolution obtained when the graph of f over $[a, b]$ is rotated about the x-axis is given by

$$2\pi \int_a^b f(x)\sqrt{1 + f'(x)^2}\,dx. \tag{7.2.8}$$

▶ **EXAMPLE 6** Let $f(x) = x^3$ for $1 \le x \le 2$. Calculate the area S of the surface that is obtained by rotating the graph of f about the x-axis.

Solution We have

$$S = 2\pi \int_1^2 f(x)\sqrt{1 + f'(x)^2}\,dx = 2\pi \int_1^2 x^3\left(1 + (3x^2)^2\right)^{1/2} dx = \frac{\pi}{27}\int_1^2 \frac{3}{2}(1 + 9x^4)^{1/2}(36x^3)\,dx.$$

We calculate this integral by making the substitution $u = 9x^4$, $du = 36x^3\,dx$. With this substitution, the limits of integration for u become 9 and 144. Therefore

$$S = \frac{\pi}{27}\int_9^{144} \frac{3}{2}(1 + u)^{1/2}\,du = \frac{\pi}{27}(1 + u)^{3/2}\Big|_9^{144} = \frac{\pi}{27}(145^{3/2} - 10^{3/2}). \quad ◀$$

▶ **EXAMPLE 7** Show that $S = 4\pi r^2$ is the surface area of a sphere of radius r.

Solution It is convenient to think of the sphere as the surface of revolution generated by rotating the graph of the function $f(x) = \sqrt{r^2 - x^2}$, $-r \le x \le r$ about the x-axis. Then

$$1 + f'(x)^2 = 1 + \left(\frac{-x}{\sqrt{r^2 - x^2}}\right)^2 = 1 + \left(\frac{x^2}{r^2 - x^2}\right) = \frac{(r^2 - x^2) + x^2}{r^2 - x^2} = \frac{r^2}{r^2 - x^2}.$$

It follows that $\sqrt{1 + f'(x)^2} = r/\sqrt{r^2 - x^2}$ and

$$S = 2\pi \int_{-r}^{r} f(x)\sqrt{1 + f'(x)^2}\,dx = 2\pi \int_{-r}^{r} \sqrt{r^2 - x^2}\,\frac{r}{\sqrt{r^2 - x^2}}\,dx = 2\pi r \int_{-r}^{r} 1\,dx = 4\pi r^2. \quad ◀$$

We may also consider the area of a surface obtained by rotating the graph of a function about the y-axis. We do so by using y as the independent variable. Here is an example:

▶ **EXAMPLE 8** Set up, but do not evaluate, the integral for finding the area of the surface obtained when the graph of $f(x) = x^4$, $1 \le x \le 5$, is rotated about the y-axis.

Solution We write $y = x^4$ and solve for x, obtaining $x = y^{1/4}$. We think of the curve that is to be rotated as the graph of $g(y) = y^{1/4}$, $1 \le y \le 625$. The surface of rotation can be seen in Figure 9. The formula for surface area is

$$2\pi \int_1^{625} g(y)\sqrt{1 + g'(y)^2}\,dy.$$

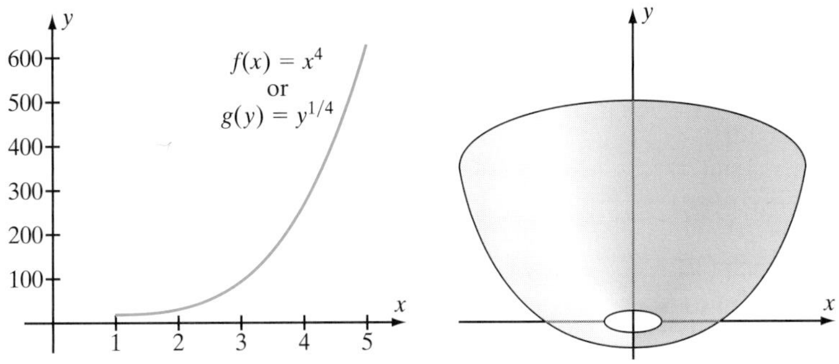

▲ Figure 9 The surface generated by rotating the graph of f about the y-axis. The xy-plane is tilted to reveal the opening at the bottom of the surface

Calculating $g'(y)$ and substituting, we find that the desired surface area is the value of the integral

$$2\pi \int_1^{625} y^{1/4} \sqrt{1 + \left(\frac{1}{4} y^{-3/4}\right)^2}\, dy. \blacktriangleleft$$

QUICK QUIZ

1. The arc length of the graph of $y = x^3$ between $(0,0)$ and $(2,8)$ is equal to $\int_0^2 g(x)\, dx$ for what function $g(x)$?
2. The arc length of the graph of the parametric curve $x = e^t$, $y = t^2$ between $(1,0)$ and $(e,1)$ is equal to $\int_0^1 g(t)\, dt$ for what function $g(t)$?
3. The graph of $y = x^3$ between $(0,0)$ and $(2,8)$ is rotated about the x-axis. The area of the resulting surface of revolution is $2\pi \int_0^2 g(x)\, dx$ for what function $g(x)$?
4. What is the area of the surface of revolution that results from rotating the graph of $y = mx$, $0 \le x \le h$ about the x-axis?

Answers

1. $\sqrt{1 + 9x^4}$ 2. $\sqrt{e^{2t} + 4t^2}$ 3. $x^3\sqrt{1 + 9x^4}$ 4. $\pi m h^2 \sqrt{1 + m^2}$

EXERCISES

Problems for Practice

▶ In each of Exercises 1–4, the graph of the given function f with given domain I is a line segment. Use formula (7.2.3) to calculate the arc length of the graph of f. Verify that this length is the distance between the two endpoints. ◀

1. $f(x) = 3x$ $I = [1, 4]$
2. $f(x) = 2x + 1$ $I = [0, 4]$
3. $f(x) = 3 - x$ $I = [-1, 2]$
4. $f(x) = 10 - 2x$ $I = [0, 5]$

▶ In each of Exercises 5–12, calculate the arc length L of the graph of the given function over the given interval. ◀

5. $f(x) = 2 + x^{3/2}$ $I = [1, 4]$
6. $f(x) = (x - 1)^{3/2}$ $I = [1, 5]$
7. $f(x) = 3 + (2x + 1)^{3/2}$ $I = [0, 4]$
8. $f(x) = (5 - 2x)^{3/2}$ $I = [1/2, 2]$
9. $f(x) = \ln(\cos(x))$ $I = [0, \pi/3]$
10. $f(x) = \ln(\sin(x))$ $I = [\pi/4, \pi/2]$
11. $f(x) = \frac{2}{3}(x^2 + 1)^{3/2}$ $I = [0, 1]$
12. $f(x) = 2(x^2 + 1/3)^{3/2}$ $I = [1, 3]$

▶ In each of Exercises 13–16, calculate the arc length of the graph of the given equation. ◀

13. $x = y^{3/2} - 1$ $0 \le y \le 1$
14. $x = (y - 8)^{3/2} + 2$ $8 \le y \le 12$
15. $x^2/y^3 = 4$ $1/3 \le y \le 7$
16. $\exp(x) = \cos(y)$ $0 \le y \le \pi/4$

In each of Exercises 17–24, calculate the length of the given parametric curve.

17. $x = 3t - 7$ $\quad y = 5 - 4t$ $\quad -1 \le t \le 1$
18. $x = 5t - 7$ $\quad y = 12t$ $\quad 2 \le t \le 7$
19. $x = 4\cos(2t)$ $\quad y = 4\sin(2t)$ $\quad -\pi/6 \le t \le \pi/6$
20. $x = \sin(\pi t^2)$ $\quad y = \cos(\pi t^2)$ $\quad 0 \le t \le 1$
21. $x = 3t^2$ $\quad y = 2t^3$ $\quad 0 \le t \le 1$
22. $x = \exp(2t)$ $\quad y = \exp(3t)$ $\quad -1 \le t \le 1$
23. $x = e^t\cos(t)$ $\quad y = e^t\sin(t)$ $\quad 0 \le t \le 1$
24. $x = \cos^3(t)$ $\quad y = \sin^3(t)$ $\quad 0 \le t \le \pi/2$

In each of Exercises 25–30, calculate the area S of the surface obtained when the graph of the given function is rotated about the x-axis.

25. $f(x) = 3x - 1$ $\quad [1, 4]$
26. $f(x) = x^3/3$ $\quad [0, 1]$
27. $f(x) = 3x^{1/2}$ $\quad [7/4, 4]$
28. $f(x) = x^3/3 + 1/(4x)$ $\quad [1, 2]$
29. $f(x) = e^x/2 + e^{-x}/2$ $\quad [-1, 1]$
30. $f(x) = (x+1)^{1/2}$ $\quad [1, 11]$

In each of Exercises 31–34, calculate the area S of the surface obtained when the given function, over the given interval, is rotated about the y-axis.

31. $f(x) = x/4 - 2$ $\quad [8, 12]$
32. $f(x) = x^2$ $\quad [0, \sqrt{2}]$
33. $f(x) = x^{1/3}$ $\quad [0, 8]$
34. $f(x) = (3x)^{1/3}$ $\quad [0, 1/3]$

Further Theory and Practice

In Exercises 35–38, calculate the arc length L of the graph of the given equation.

35. $y = \dfrac{1}{2}x^2$ $\quad 0 \le x \le 1.$
36. $y = \ln(x)$ $\quad 1 \le x \le e.$
37. $y = 2\sqrt{x}$ $\quad 1/3 \le x \le 1.$
38. $y = \ln(x^2 - 1)$ $\quad \sqrt{2} \le x \le \sqrt{3}.$

In each of Exercises 39–44, calculate the arc length of the graph of the given function over the given interval. (In these exercises, the functions have been contrived to permit a simplification of the radical in the arc length formula.)

39. $f(x) = (1/3)(x^2 + 2)^{3/2}$ $\quad I = [0, 1]$
40. $f(x) = (1/6)x^3 + (1/2)x^{-1}$ $\quad I = [1, 2]$
41. $f(x) = e^x/2 + e^{-x}/2 + 7$ $\quad I = [0, 1]$
42. $f(x) = (2^x + 2^{-x})/(2\ln(2))$ $\quad I = [0, 2]$
43. $f(x) = 2x^4 + 1/(64x^2)$ $\quad I = [1/2, 1]$
44. $f(x) = \left(2x^{1/2} + \dfrac{9}{4}\right)^{3/2}$ $\quad I = [1/4, 1]$

In each of Exercises 45–52, calculate the length L of the given parametric curve.

45. $x = t\cos(t)$ $\quad y = t\sin(t)$ $\quad 0 \le t \le 3\pi$
46. $x = t^2\cos(t)$ $\quad y = t^2\sin(t)$ $\quad 0 \le t \le \sqrt{5}$
47. $x = \dfrac{1}{2}t^2 - t$ $\quad y = \dfrac{1}{2}t^2 + t$ $\quad 0 \le t \le 1$
48. $x = \cos(t) + t\sin(t)$ $\quad y = \sin(t) - t\cos(t)$ $\quad 0 \le t \le 2\pi$
49. $x = t - \sin(t)$ $\quad y = 1 - \cos(t)$ $\quad 0 \le t \le 2\pi$
50. $x = 2\arctan(t)$ $\quad y = \ln(1 + t^2)$ $\quad 0 \le t \le 1$
51. $x = \cos(t) + \ln(\tan(t/2))$ $\quad y = \sin(t)$ $\quad \pi/3 \le t \le \pi/2$
52. $x = 3\exp(t)$ $\quad y = 2\exp\left(\dfrac{3}{2}t\right)$ $\quad 0 \le t \le 1$

53. Find the arc length of the graph of $y^4 - 16x + 8/y^2 = 0$ between $P = (9/8, 2)$ and $Q = (9/16, 1)$.

54. Find the arc length of the graph of $30yx^3 - x^8 = 15$ between $P = (0.1, 500.0\ \ldots)$ and $Q = (1, 8/15)$.

55. Suppose that f is a function with a continuous derivative on an open interval containing $[a, b]$. Let $\mathcal{C}$ be the graph of f over the interval $[a, b]$, and let $\mathcal{C}'$ be the curve with endpoints $(a + h, f(a) + v)$ and $(b + h, f(b) + v)$, is obtained by translating $\mathcal{C}$ horizontally by an amount h and vertically by an amount v. Prove that the lengths of $\mathcal{C}$ and $\mathcal{C}'$ are equal.

56. Suppose that $a > b > 0$. Let $\varepsilon = \sqrt{a^2 - b^2}/a$. Show that, for any $0 < \xi < a$, the length $L(\xi)$ of the arc of the ellipse $x^2/a^2 + y^2/b^2 = 1$ that lies above the interval $[0, \xi]$ of the x-axis integral is given by

$$L(\xi) = \int_0^{\xi} \sqrt{\frac{a^2 - \varepsilon^2 x^2}{a^2 - x^2}}\,dx.$$

Make a change of variable to write $L(\xi)$ in terms of the *elliptic integral*

$$E(\varepsilon, \eta) = \int_0^{\eta} \sqrt{1 - \varepsilon^2 \sin^2(t)}\,dt.$$

Although elliptic integrals cannot be evaluated using elementary functions, they are built into computer algebra systems, and their values have been tabulated.

57. Suppose that $a > b > 0$. Let $\varepsilon = \sqrt{a^2 - b^2}/a$, as in the preceding exercise. Parameterize the ellipse $x^2/a^2 + y^2/b^2 = 1$ by the equations $x = a\sin(t)$, $y = b\cos(t)$, $(0 \le t \le 2\pi)$. Let $\mathcal{L}(\tau)$ denote the arc length of that part of the ellipse that is given by $0 \le t \le \tau \le \pi/2$. Express $\mathcal{L}(\tau)$ in terms of the function $E(\varepsilon, \eta)$ of the preceding exercise.

In each of Exercises 58–61, calculate the area S of the surface obtained when the graph of the given function is rotated about the x-axis.

58. $f(x) = \exp(x)$ $\quad 0 \le x \le 1.$
59. $f(x) = x^4 + \dfrac{1}{32}x^{-2}$ $\quad 1/2 \le x \le 1$
60. $f(x) = \sin(x)$ $\quad 0 \le x \le \pi/2$
61. $f(x) = \dfrac{3x^4 + 1}{6x}$ $\quad 1 \le x \le 2$

62. When the right circular cone of slant height ℓ and base radius R is cut along a lateral edge and flattened, it becomes a sector of a circle of radius ℓ and central angle θ, as in Figure 10. Prove that $\theta = 2\pi R/\ell$. Use this observation to derive formula (7.2.7) for the surface area of a frustum of a cone.

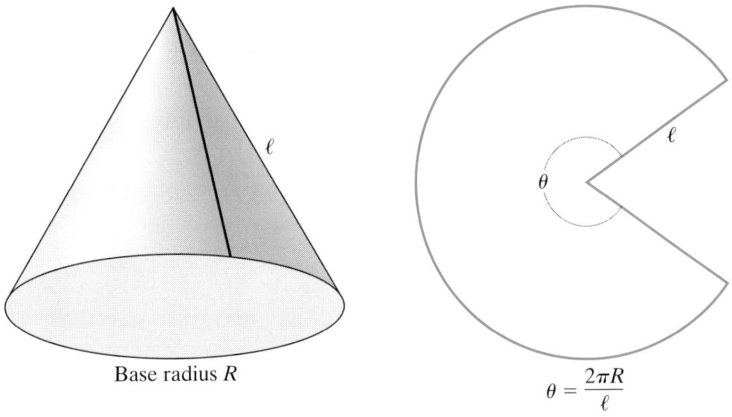

▲ Figure 10

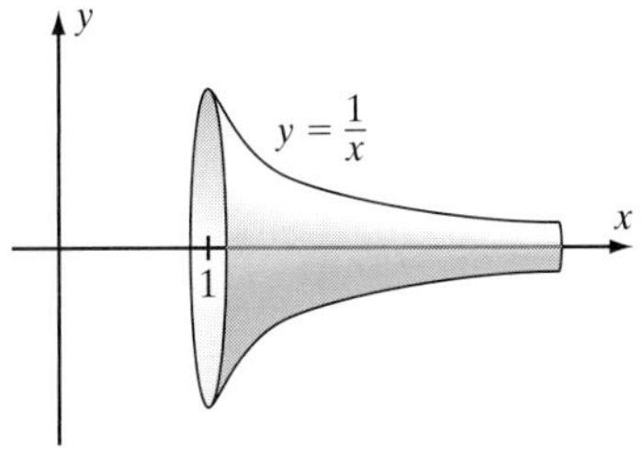

▲ Figure 11 A section of Torricelli's infinitely long solid

63. By means of integration, calculate the surface area of a right circular cone of radius R and height h. (The formula you derive can, of course, be obtained directly from formula (7.2.7) with $r = 0$.)

64. Let $\mathcal{C}$ be the curve parameterized by $x = \varphi_1(t)$, $y = \varphi_2(t)$, $\alpha \le t \le \beta$. Suppose that $0 < \varphi_1{}'(t)$ for all t, and set $a = \varphi_1(\alpha)$ and $b = \varphi_1(\beta)$. Then, $\mathcal{C}$ is the graph of $y = (\varphi_2 \circ \varphi_1^{-1})(x)$, $a \le x \le b$. Show that formula (7.2.5) for the length of $\mathcal{C}$ can be obtained from formula (7.2.3) with $f = \varphi_2 \circ \varphi_1^{-1}$.

65. *Torricelli's infinitely long solid* T is the solid of revolution obtained by rotating about the x-axis the first quadrant region that is bounded above by the graph of $y = 1/x$, below by the x-axis, and on the left by $x = 1$ (see Figure 11). Let V_N and S_N denote the volume and surface area of that part of Torricelli's solid for which $1 \le x \le N$. Calculate the finite number $V = \lim_{N \to \infty} V_N$. Show that $\ln(N) < S_N$, and therefore $\lim_{N \to \infty} S_N = \infty$. Thus Torricelli's solid has finite volume, but its surface area is infinite.

Calculator/Computer Exercises

66. Use Simpson's Rule to calculate the arc length L of the ellipse

$$\frac{x^2}{16} + \frac{y^2}{4} = 1$$

to 2 decimal places of accuracy.

67. Use Simpson's Rule to calculate the arc length of the graph of $y = \sin(x)$, $0 \le x \le 2\pi$, to four decimal places of accuracy.

68. Plot $T(x) = 4x^3 - 3x$ for $-1 \le x \le 1$. Notice that the plot is contained in the square $[-1, 1] \times [-1, 1]$. Of all degree 3 polynomials that have this containment property, T has the longest arc length. Use Simpson's Rule to calculate the arc length of the graph of $y = T(x)$, $-1 \le x \le 1$ to four decimal places of accuracy.

69. The surface of a flashlight reflector is obtained when $y = 2.05\sqrt{x} + 0.496, 0.02 \le x \le 2.80$ cm is rotated about the x-axis. Calculate its surface area to two significant digits.

70. The equation of the St. Louis Gateway Arch is

$$y = 693.8597 - 34.38365\,(e^{kx} + e^{-kx})$$

for $k = 0.0100333$ and $-299.2239 < x < 299.2239$. Find an equation of a parabola with the same vertex and x-intercepts. Plot both curves on the same set of coordinate axes. Compute the length of each.

71. Calculate to four significant digits the surface area that results when the graph of the Gateway Arch equation (given in the preceding exercise) is rotated about the y-axis.

7.3 The Average Value of a Function

If we took five temperature readings during the day that were 50, 58, 68, 70, and 54 (in degrees Fahrenheit), then the *average value* of our readings would be

$$\frac{50 + 58 + 68 + 70 + 54}{5} = \frac{300}{5} = 60.$$

Frequently we need to average infinitely many values. For instance, consider the plot of temperature that is given in Figure 1. To speak of an average temperature during the day, we must average the temperature at infinitely many instants of time. The method for averaging infinitely many values is to use an integral, which is the subject of this section.

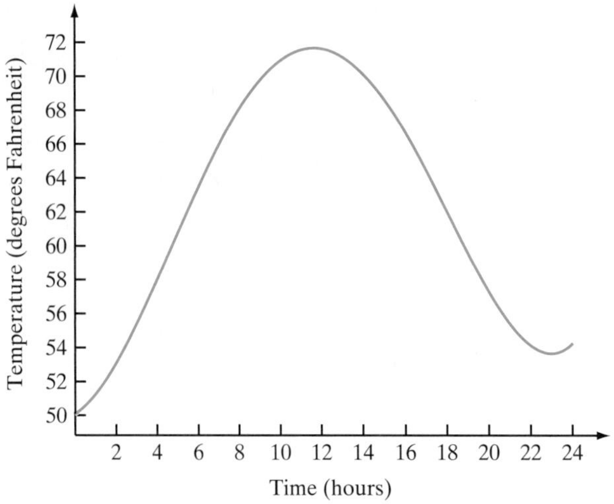

▲ Figure 1

The Basic Technique

Suppose that f is a continuous function whose domain contains the interval $[a, b]$. To estimate the average value of f on this interval, we take a uniform partition of the interval:

$$a = x_0 < x_1 < x_2 < \ldots < x_{N-1} < x_N = b.$$

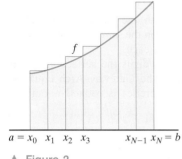

▲ Figure 2

As usual, let $\Delta x = (b - a)/N$ denote the common length of the subintervals that are formed by the partition. If we add up the values of f at the right endpoints of the partition (Figure 2) and divide by the number of points, then the resulting expression is an approximation to the average value f_{avg} of f:

$$f_{\text{avg}} \approx \frac{f(x_1) + f(x_2) + \cdots + f(x_{N-1}) + f(x_N)}{N}.$$

As we let N increase to infinity, our approximation uses more and more points from the interval $[a, b]$ and becomes more accurate. We therefore define the average value f_{avg} of $f(x)$ for x in $[a, b]$ to be

$$f_{\text{avg}} = \lim_{N \to \infty} \frac{f(x_1) + f(x_2) + \cdots + f(x_{N-1}) + f(x_N)}{N}.$$

We may rewrite this expression as

$$\frac{f(x_1) + f(x_2) + \cdots + f(x_{N-1}) + f(x_N)}{N} = \frac{1}{N} \sum_{j=1}^{N} f(x_j) = \frac{1}{N\Delta x} \sum_{j=1}^{N} f(x_j)\,\Delta x = \frac{1}{b-a} \sum_{j=1}^{N} f(x_j)\,\Delta x.$$

Because $\sum_{j=1}^{N} f(x_j)\,\Delta x$ is a Riemann sum for the integral $\int_a^b f(x)\,dx$, we conclude that

$$f_{\text{avg}} = \lim_{N\to\infty} \frac{1}{b-a}\sum_{j=1}^{N} f(x_j)\,\Delta x = \frac{1}{b-a}\int_a^b f(x)\,dx.$$

We summarize the result of our calculation in the following definition.

DEFINITION Suppose that f is a Riemann integrable function on the interval $[a,b]$. The *average value* of f on the interval $[a,b]$ is the number

$$f_{\text{avg}} = \frac{1}{b-a}\int_a^b f(x)\,dx. \tag{7.3.1}$$

▶ **EXAMPLE 1** What is the average value f_{avg} of the function $f(x) = x^2$ on the interval $[3,6]$?

Solution The average is

$$f_{\text{avg}} = \frac{1}{6-3}\int_3^6 x^2\,dx = \frac{1}{3}\frac{x^3}{3}\bigg|_{x=3}^{x=6} = \frac{1}{9}(6^3 - 3^3) = 21. \quad ◀$$

▶ **EXAMPLE 2** A rod of length 9 cm has temperature distribution $T(x) = (2x - 6x^{1/2})^{\circ}$C for $0 \le x \le 9$. This means that, at position x on the rod, the temperature is $(2x - 6x^{1/2})$°C. Calculate the average temperature of the rod.

Solution The computation

$$T_{\text{avg}} = \frac{1}{9-0}\int_0^9 (2x - 6x^{1/2})\,dx = \frac{1}{9}(x^2 - 4x^{3/2})\bigg|_{x=0}^{x=9} = \frac{1}{9}\big((81 - 4\cdot 27) - (0 - 4\cdot 0)\big) = -3$$

shows that the average temperature of the rod is 3°C below zero. ◀

▶ **EXAMPLE 3** Suppose that, at a weather station, the temperature f in degrees Fahrenheit at the time x is recorded to be

$$f(x) = 0.00103265x^4 - 0.04678932x^3 + 0.50831530x^2 + 0.64927850x + 50$$

for $0 \le x \le 24$ hours. What is the average temperature for that day?

Solution The graph of f was given in Figure 1. By scanning the values taken by f, we might estimate the average value to be between 60 and 64. At the beginning of this section, we used the five values $f(0) = 50$, $f(4) = 58$, $f(8) = 68$, $f(14) = 70$, and $f(22) = 54$, to obtain a sample average of 60. Now we can calculate the average over *all* values of f:

$$f_{\text{avg}} = \frac{1}{24}\int_0^{24} f(x)\,dx.$$

Using a computer algebra system to evaluate the integral (or a calculator to perform the arithmetic calculations that arise when the integral is evaluated by means of the Fundamental Theorem of Calculus), we obtain $f_{\text{avg}} = 62.21$ (to two decimal places). ◀

INSIGHT The expected cost of heating a building during the winter is often calculated using the concept of "degree days." A simple definition of *degree day* for a given day is 65 degrees Fahrenheit less the average outdoor temperature for that day. For example, if the cost of heating the weather station of Example 3 is \$0.35 per degree day, then we would expect the heating bill for the day in question to be about $\$0.35 \cdot (65 - 62.205687) \approx \$0.98$.

▶ **EXAMPLE 4** Let f_{avg} denote the average value of f over the interval $[a, b]$. Show that the function f and the constant function $g(x) = f_{\text{avg}}$ have the same integral over $[a, b]$.

Solution We obtain the required equality as follows:

$$\int_a^b g(x)\,dx = \int_a^b f_{\text{avg}}\,dx = (b-a)\cdot f_{\text{avg}} = (b-a)\cdot\left(\frac{1}{(b-a)}\int_a^b f(x)\,dx\right) = \int_a^b f(x)\,dx. \quad ◀$$

INSIGHT This result is the continuous analogue of a well-known fact about discrete averages. Namely, if $\bar{y}$ is the average value of the N numbers $y_1, y_2, \ldots, y_N$, then

$$\sum_{j=1}^{N} \bar{y} = N \cdot \bar{y} = N \cdot \left(\frac{1}{N}\sum_{j=1}^{N} y_j\right) = \sum_{j=1}^{N} y_j.$$

It is often the case that a theorem about finite sums has a continuous analogue in which the sums are replaced by integrals.

◀ **A LOOK BACK** In Section 5.3 of Chapter 5, you learned the Mean Value Theorem for Integrals. That theorem asserted that, if f is continuous on an interval $[a, b]$, then there is a value c in the interval such that

$$f(c) = \frac{1}{b-a}\int_a^b f(x)\,dx. \tag{7.3.2}$$

Because the right side of equation (7.3.2) is f_{avg}, we may interpret the Mean Value Theorem for Integrals in this way: A continuous function on a closed bounded interval *assumes* its average value. This is a property that discrete samples do not necessarily have. The average of the five temperatures of Example 3 is 60, which does not equal any of the five sample temperatures.

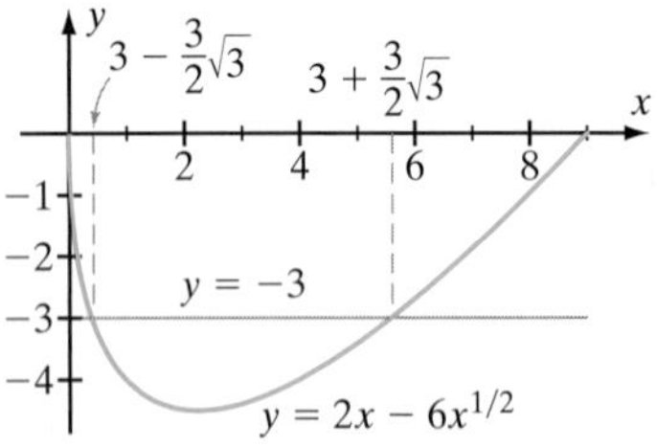

▲ Figure 3

▶ **EXAMPLE 5** At what point (or points) is the temperature of the metal rod of Example 2 equal to the average temperature of the rod?

Solution Because the temperature of the rod, $2x - 6x^{1/2}$, is continuous, we can be sure that there is at least one point at which the temperature is equal to the average temperature $-3°\text{C}$ (a value that was calculated in Example 2). Figure 3 shows that there are actually two points at which the average temperature is attained. To find them, we solve the equation $2x - 6x^{1/2} = -3$. We rewrite this as $2x + 3 = 6x^{1/2}$. If we square each side, then we obtain the quadratic equation $(2x+3)^2 = 36x$ or $4x^2 - 24x + 9 = 0$. According to the Quadratic Formula, the roots are $3 - 3\sqrt{3}/2 \approx 0.40192$ and $3 + 3\sqrt{3}/2 \approx 5.59808$. As expected, both roots are in the interval $[0, 9]$. ◀

Random Variables

Many important variables assume their values in a "random" manner. For example, a company may be able to say quite a bit *in general* about the longevity of the computer hard drives it produces. The manufacturer cannot, however, say anything definite about the lifetime X of the particular hard drive in *your* computer. We say that X is a *random variable*. Similarly, a paleontologist may know what range of values the heights of velociraptors lie in. What cannot be specified for certain is the height of the next specimen to be discovered: It is a random variable.

The Theory of Probability is largely concerned with the analysis of random variables. The field of Statistics uses probability theory to make estimates and inferences about random variables based on observed values. In this subsection, you will learn how calculus plays a role in determining the average value of a random variable.

Suppose that X is a random variable whose values lie in an interval $I = [a, b]$. In other words, I is the range of X. If $[\alpha, \beta]$ is a subinterval of I, then we write $P(\alpha \leq X \leq \beta)$ to denote the probability that the random variable X takes on a value in the subinterval $[\alpha, \beta]$. Just as in everyday informal language, a probability in mathematics is a number between 0 and 1. For example, we have $P(a \leq X \leq b) = 1$ because it is certain, according to our assumptions, that X takes a value in this range.

DEFINITION Suppose that X is a random variable all of whose values lie in an interval I. If there is a nonnegative function f such that

$$P(\alpha \leq X \leq \beta) = \int_{\alpha}^{\beta} f(x)\, dx \tag{7.3.3}$$

for every subinterval $[\alpha, \beta]$ of I, then we say that f is a *probability density function* of X. The abbreviation p.d.f. is commonly used. Sometimes the notation f_X is used to emphasize the association between f and X.

▶ **EXAMPLE 6** In a large class, the grades on a particular exam are all between 38 and 98. Let X denote the score of a randomly selected student in the class. Suppose that the probability density function f for X is given by $f(x) = (136x - 3724 - x^2)/36000$, $38 \leq x \leq 98$. What is the probability that the grade on a randomly selected exam is between 72 and 82?

Solution Using formula (7.3.3), we obtain the probability of a random variable X assuming a value in $[\alpha, \beta]$ by integrating the probability density function of X over $[\alpha, \beta]$. Thus taking $\alpha = 72$ and $\beta = 82$ in formula (7.3.3), we have

$$P(72 \leq X \leq 80) = \int_{72}^{82} \frac{136x - 3724 - x^2}{36000}\, dx = \frac{1}{36000}\left(68x^2 - 3724x - \frac{1}{3}x^3\right)\Bigg|_{72}^{82} = \frac{152}{675} \approx 0.225.$$

That is, the probability that the indicated grade will be between 72 and 82 is about 0.225. ◀

As has been discussed, if f is a probability density function of a random variable X whose values all lie in $I = [a, b]$, then $\int_a^b f(x)\,dx = 1$. The converse to this observation is also known: If f is any continuous, nonnegative function such that $\int_a^b f(x)\,dx = 1$, then there is a random variable X that has range I and probability density function f.

▶ **EXAMPLE 7** Suppose that $g(x) = 1/x^2$. For what number c is $f(x) = c \cdot g(x)$ a probability density function for a random variable whose values all lie in the interval $[1, 3]$?

Solution We calculate

$$\int_1^3 g(x)\,dx = \int_1^3 \frac{1}{x^2}\,dx = \left(-\frac{1}{x}\right)\Big|_1^3 = -\frac{1}{3} - \left(\frac{1}{1}\right) = \frac{2}{3}.$$

Therefore

$$\int_1^3 f(x)\,dx = \int_1^3 c\,g(x)\,dx = c\int_1^3 g(x)\,dx = \frac{2c}{3}.$$

For f to be a probability density function of a random variable with range $[1, 3]$, it is necessary and sufficient that $\int_1^3 f(x)\,dx = 1$. We therefore set $2c/3 = 1$ and find that $c = 3/2$. Thus $f(x) = 3/(2x^2)$ is a probability density function on $[1, 3]$. ◀

In many applications, the range of a random variable is an infinite interval of the form $[a, \infty)$ or $(-\infty, \infty)$. The next example illustrates this idea with an important type of random variable known as a *waiting time*.

▶ **EXAMPLE 8** Suppose that $f(x) = \dfrac{1}{20000} e^{-x/20000}$ is the probability density function of the time to failure X, measured in hours, of an electronic component that has been placed into service. Verify that f is a probability density function on the interval $[0, \infty)$, and calculate the probability that the component will fail within its first 10,000 hours.

Solution Because f is nonnegative and satisfies

$$\begin{aligned}
\int_0^\infty \frac{1}{20000} e^{-x/20000}\,dx &= \frac{1}{20000} \lim_{N\to\infty} \int_0^N e^{-x/20000}\,dx \\
&= \frac{1}{20000} \lim_{N\to\infty} \left(-20000 \exp\left(-\frac{1}{20000}x\right)\Big|_0^N\right) \\
&= \lim_{N\to\infty} \left(1 - \exp\left(-\frac{N}{20000}\right)\right) \\
&= 1,
\end{aligned}$$

we conclude that it is a probability density function. The probability that the component will fail within its first 10,000 hours is

$$P(0 \le X \le 10000) = \int_0^{10000} \frac{1}{20000} e^{-x/20000}\,dx = -e^{-x/20000}\Big|_{x=0}^{x=10000} = 1 - e^{-1/2} \approx 0.3935. \quad ◀$$

Average Values in Probability Theory

Suppose that X is a random variable with range $I = [a, b]$ and probability density function f. Our goal is to determine the average value μ of X. Notice that we cannot simply use our previous formula and integrate X because we do not have a deterministic formula for X. We can calculate the average value f_{avg} of f, but this quantity is *not* the average value μ of X.

To calculate μ, we again rely on a discrete model. If we made a large number N of observations $X_1, \ldots, X_N$ of X, then we would expect

$$\mu \approx \frac{X_1 + X_2 + \ldots + X_N}{N} \tag{7.3.4}$$

to be a good approximation. Form a partition $a = x_0 < x_1 < \ldots < x_M = b$ of the interval I into M subintervals of equal length Δx. For each j from 1 to M, we approximate the contribution of the terms in the numerator of formula (7.3.4) that lie in the jth subinterval $I_j = [x_{j-1}, x_j]$. First we estimate the number of these observations. To do so, we note that the probability that an observation of x will fall in I_j is

$$P(x_{j-1} \le X \le x_j) = \int_{x_{j-1}}^{x_j} f(x)\,dx = f(c_j)\Delta x$$

where $f(c_j)$ is the average value of f in the subinterval I_j. It follows that about $N \cdot (f(c_j)\Delta x)$ of the observations will lie in I_j. Each of these observations will be between x_{j-1} and x_j. Because c_j is in this small interval, we may use c_j as a representative value of the observations in I_j. The contribution of the terms in I_j is therefore about $c_j \cdot (N \cdot f(c_j)\,\Delta x)$. By grouping the summands of $X_1 + X_2 + \ldots + X_N$ according to the subintervals in which they lie, we may rewrite our approximation of μ as

$$\mu \approx \frac{1}{N}\sum_{j=1}^{M} c_j \cdot \Big(N \cdot f(c_j)\Delta x\Big) = \sum_{j=1}^{M} c_j \cdot f(c_j)\Delta x.$$

Because $\sum_{j=1}^{M} c_j \cdot f(c_j)\Delta x$ is a Riemann sum for the integral $\int_a^b x f(x)\,dx$, we are led to the following definition.

DEFINITION If f is the probability density function of a random variable X that takes values in an interval $I = [a, b]$, then the *average* (or *mean*) μ of X is defined to be

$$\mu = \int_a^b x f(x)\,dx. \tag{7.3.5}$$

This value is also said to be the *expectation* of X. Other notations for the expectation of X are $\overline{X}$ and $E(X)$.

▶ **EXAMPLE 9** Let X denote the fraction of total impurities that are filtered out in a particular purification process. Suppose that X has probability density function $f(x) = 20x^3(1-x)$ for $0 \le x \le 1$. What is the average of X?

Solution According to our definition, the average is

$$\overline{X} = \int_0^1 x \cdot 20x^3(1-x)\,dx = \int_0^1 (20x^4 - 20x^5)\,dx = 20\left(\frac{x^5}{5} - \frac{x^6}{6}\right)\Bigg|_{x=0}^{x=1} = \frac{2}{3}. \quad ◀$$

Population Density Functions

The idea of a *density* function arises in many topics outside of probability theory. In the next section, we will encounter mass densities, which are important in physics and engineering. We conclude the present section with a discussion of population densities, which are used for societal planning.

In a simple model for the growth of a city, the municipality's land mass expands radially in all directions from the central point of its initial foundation. We represent the city's center by the origin in the plane and suppose that there are no geographic or economic factors that favor one radial direction over another for settlement. As a result of this assumption, the number of people living in a small square $\mathcal{R}$ of area A is about $f(r)\cdot A$, where $f(r)$ is a proportionality constant depending on the distance r of the center of $\mathcal{R}$ from the origin but not on the particular location of the center on the circle $x^2+y^2=r^2$ (see Figure 4). Let us use this *population density* f to calculate the number $P(R)$ of inhabitants living up to a distance R from the city center.

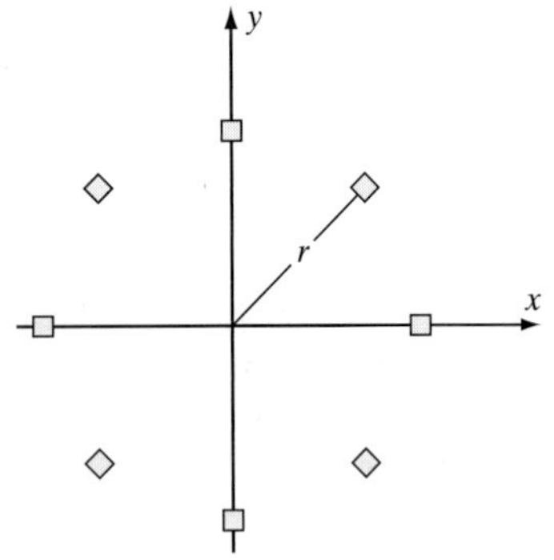

The population in each shaded square of area A is about $f(r)\cdot A$

▲ Figure 4

The partition $x_0=0, x_1=R/N, x_2=2R/N, \ldots, x_{N-1}=(N-1)R/N, x_N=R$ divides the interval $[0, R]$ of the x-axis into N subintervals of equal length $\Delta x = R/N$. We can use this partition of $[0,R]$ to divide the disk of radius R centered at the origin into N concentric rings, as shown in Figure 5. Let s_j be the midpoint of the jth subinterval $[x_{j-1}, x_j]$ of $[0, R]$. Then, for $2 \le j \le N$, the area A_j of the jth ring is equal to $\pi\left(s_j+\frac{1}{2}\Delta x\right)^2 - \pi\left(s_j-\frac{1}{2}\Delta x\right)^2$, or $2\pi s_j\,\Delta x$. We can also use this expression as an approximation to the area A_1 of the first "ring," which is actually a disk. In Figure 6, several squares are drawn inside the jth ring. Each has side length Δx and a center that is a distance s_j from the origin. If we continue to fill in the ring with such squares until the ring is approximately covered, then each square in the covering has area $A=(\Delta x)^2$ and approximately $f(s_j)\cdot A$ inhabitants. If we sum the contributions of these squares to the total population P_j of the jth ring, then, after factoring the common coefficient $f(s_j)$, we obtain $P_j \approx f(s_j)\times A_j$, or $P_j \approx 2\pi s_j f(s_j)\,\Delta x$. Because the total population $P(R)$ in the disc of radius R centered at the origin is equal to $P_1+P_2+\ldots+P_N$, we see that $P(R)\approx \sum_{j=1}^{N} 2\pi s_j f(s_j)\,\Delta x$. Observe that this expression is a Riemann sum for the integral $\int_0^R 2\pi x f(x)\,dx$. Thus on letting N tend to infinity, we obtain

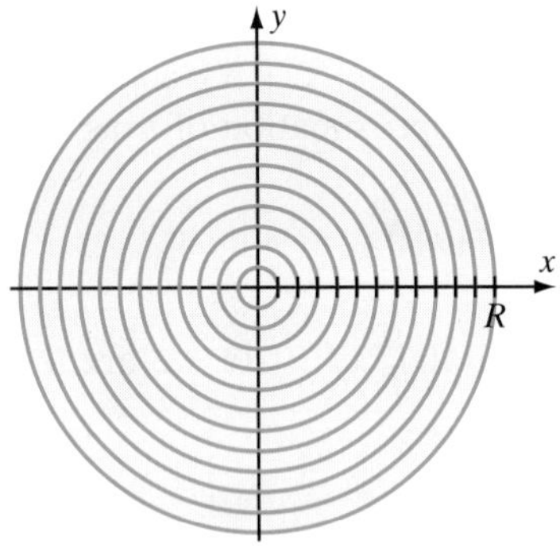

▲ Figure 5

$$P(R) = 2\pi\int_0^R x f(x)\,dx. \tag{7.3.6}$$

INSIGHT Despite the apparent similarity between the integrals of formulas (7.3.5) and (7.3.6), there are significant differences that should be noted. The probability density function of formula (7.3.5) is integrated over the *range* of the random variable X. The population density function of formula (7.3.6) is integrated over a subinterval of the *domain* of the population function P. Equation (7.3.5) is a formula for the *average* value of X. Equation (7.3.6) can be used to calculate *all* values of P.

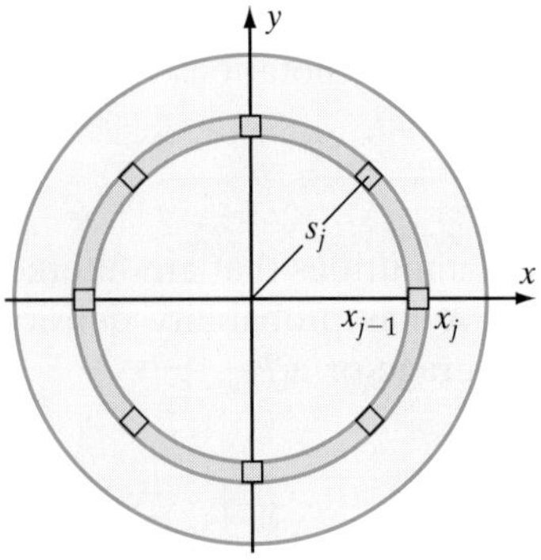

▲ Figure 6

▶ **EXAMPLE 10** In 1950, the population density of Tulsa was given by $f(x)=28000e^{-4x/5}$, where x represents the distance in miles from the central business district. About how many people lived within 20 miles of the city center?

Solution According to formula (7.3.6), $P(20)=2\pi\int_0^{20} 28000\,x\,e^{-4x/5}\,dx$. Using formula (6.1.9) with $n=1$ and $a=-4/5$, we have

$$\begin{aligned} P(20) &= 56000\,\pi\left(-\frac{5}{4}xe^{-4x/5}\Big|_0^{20} + \frac{5}{4}\int_0^{20} e^{-4x/5}\,dx\right) \\ &= 56000\,\pi\left(-25\,e^{-16} + \left(\frac{5}{4}\left(-\frac{5}{4}\,e^{-4x/5}\right)\Big|_0^{20}\right)\right) \\ &= 56000\,\pi\left(-25\,e^{-16} + \left(\frac{25}{16} - \frac{25}{16}e^{-16}\right)\right) \\ &= 56000\cdot\frac{25}{16}\cdot\pi\,(1-17e^{-16}) \\ &\approx 274,889. \end{aligned}$$

QUICK QUIZ

1. What is the average of $\sin(x)$ over the interval $[0, \pi]$?
2. For what c in $I = [0, 3]$ is $f(c)$ the average value of $f(x) = x^2$ on I?
3. Suppose that the probability density of a nonnegative random variable X is $f(x) = \exp(-x)$, $0 \le x < \infty$. What is the probability that $X \le 1$?
4. What is the mean of a random variable that has probability density function $f(x) = x/2$ for $0 \le x \le 2$?

Answers

1. $2/\pi$ 2. $\sqrt{3}$ 3. $1 - 1/e$ 4. $4/3$

EXERCISES

Problems for Practice

In each of Exercises 1–12, calculate the average value of the given function on the given interval.

1. $f(x) = \cos(x)$ $I = [0, \pi/2]$
2. $f(x) = x^2$ $I = [3, 7]$
3. $f(x) = 1/x$ $I = [1, 4]$
4. $f(x) = 3x^2 - 6x + 1$ $I = [3, 7]$
5. $f(x) = \sin(x)$ $I = [\pi/3, \pi]$
6. $f(x) = 1 + \cos(x)$ $I = [0, 2\pi]$
7. $f(x) = (x-1)^{1/2}$ $I = [2, 5]$
8. $f(x) = x^{1/2} - x^{1/3}$ $I = [0, 64]$
9. $f(x) = 60/x^2$ $I = [1, 3]$
10. $f(x) = e^x$ $I = [-1, 1]$
11. $f(x) = \ln(x)$ $I = [1, e]$
12. $f(x) = x\sqrt{169 - x^2}$ $I = [5, 12]$

In each of Exercises 13–16, a function f and an interval $I = [a, b]$ are given. Find a number c in (a, b) for which $f(c)$ is the average value of f on I.

13. $f(x) = 1/x$ $I = [1, 3]$
14. $f(x) = e^x$ $I = [\ln(2), \ln(4)]$
15. $f(x) = x^2 - 10x/3$ $I = [1, 3]$
16. $f(x) = \ln(x)$ $I = [1, e]$

In each of Exercises 17–22, the probability density function f of a random variable X with range $I = [a, b]$ is given. Calculate $P(\alpha \le X \le \beta)$ for the given subinterval $J = [\alpha, \beta]$ of I.

17. $f(x) = 3x^2/8$ $I = [0, 2]$ $J = [1, 2]$
18. $f(x) = 12x^2(1-x)$ $I = [0, 1]$ $J = [0, 1/2]$
19. $f(x) = \sin(x)/2$ $I = [0, \pi]$ $J = [\pi/4, \pi/2]$
20. $f(x) = 3(1+x)/(8\sqrt{x})$ $I = [0, 1]$ $J = [1/4, 1/2]$
21. $f(x) = e^{1-x}/(e-1)$ $I = [0, 1]$ $J = [0, 1/2]$
22. $f(x) = \ln(x)$ $I = [1, e]$ $J = [3/2, 2]$

In each of Exercises 23–28, a function g and an interval I are specified. The function g is nonnegative on I. Find a number c such that $f(x) = cg(x)$ is a probability density function on I.

23. $g(x) = \sec^2(x)$ $I = [0, \pi/3]$
24. $g(x) = 1/x^2$ $I = [1, 2]$
25. $g(x) = 9 - x^2$ $I = [-1, 3]$
26. $g(x) = e^{2x}$ $I = [0, 1]$
27. $g(x) = e^{-2x}$ $I = [0, \infty]$
28. $g(x) = 1/(1+x^2)$ $I = (-\infty, \infty)$

In each of Exercises 29–36, calculate the mean of the random variable whose probability density function is given.

29. $f(x) = 3x^2$ $\quad I = [0, 1]$
30. $f(x) = 2(1 - x)$ $\quad I = [0, 1]$
31. $f(x) = 1/(3\sqrt{x})$ $\quad I = [1/4, 4]$
32. $f(x) = 6x(1 - x)$ $\quad I = [0, 1]$
33. $f(x) = x^2/3$ $\quad I = [-1, 2]$
34. $f(x) = 3/(\pi(1 + x^2)$ $\quad I = [0, \sqrt{3}]$
35. $f(x) = \dfrac{2\ln(x)}{x}$ $\quad I = [1, e]$
36. $f(x) = \dfrac{2}{\pi\sqrt{1 - x^2}}$ $\quad I = [0, 1]$

37. In 1980, the population density for the central core of Houston, a disk with a 6-mile radius, was $f(x) = 5860 \exp(-0.148x)$. What was the population of that part of Houston?

38. In 1960, the population density for the central core of Denver, a 72 mi^2 disk, was $f(x) = 14000 \exp(-x/4)$. What was the population of that part of Denver?

Further Theory and Practice

39. In 1950, the population density in the central core of Akron, a disk with a 4-mile radius, was $f(x) = 13000 \exp(-0.38x)$. In 1970, the population density was $f(x) = 10000 \exp(-0.29x)$. Did the population in that part of Akron increase or decrease? By about how many people?

40. If a city's population density is given by $f(x) = A/(B + Cx)$ for positive constants A, B, and C, then how many people live farther than a distance a from the center but closer than a distance b from the center?

41. A patient with a fever has temperature at time t (measured in hours) given by

$$T(t) = 99.6 - t + 0.8t^2 \text{ degrees,}$$

where temperature is measured in degrees Fahrenheit. Find the patient's average temperature as t ranges from 0 to 3.

42. The temperature distribution on a uniform rod of length 4 meters is $T(s) = 18 + 6s^2 - 2s + \sqrt{s}$, $0 \le s \le 4$. What is the average temperature of the rod?

In each of Exercises 43–52 calculate the average of the given expression over the given interval.

43. $x\exp(x^2)$ $\quad 0 \le x \le 2$
44. $x\sin(x)$ $\quad 0 \le x \le \pi$
45. $3\pi\cos(x)\sqrt{1 + \sin(x)}$ $\quad 0 \le x \le \pi/2$
46. $\sin^3(x)$ $\quad 0 \le x \le \pi$
47. $\cos^2(x)\sin^3(x)$ $\quad 0 \le x \le \pi$
48. $\sin^2(2x)$ $\quad 0 \le x \le \pi/2$
49. $4x\ln(x)$ $\quad 1 \le x \le e$
50. $\ln(x)/x$ $\quad 1 \le x \le e$
51. $\dfrac{5x + 2}{x^2 + x}$ $\quad 1 \le x \le 3$
52. $24x \arcsin(x)$ $\quad 0 \le x \le 1/2$

53. For what value of c is $-1/2$ the average value of $(x - c)\sin(x)$ over the interval $[0, \pi/3]$?

54. Is the average value of $\cos(x)$ for $0 \le x \le \pi/4$ equal to the reciprocal of the average value of $1/\cos(x)$ over the same x-interval?

55. The outdoor temperature in a certain town is given by

$$T(t) = 40 - (t - 45)^2/200$$

for t ranging from 0 to 72 hours. Use the definition of degree days given in Example 3 to calculate the degree days for each of the three successive 24-hour periods. If fuel costs a certain homeowner \$0.30 per degree day, what does it cost her to heat her house during these three days?

56. The temperature of an aluminum rod at point x is given by

$$T(x) = \begin{cases} t^3 - t^2 + 32, & 0 \le t \le 2 \\ 36 - 2t + t^2, & 2 < t \le 8. \end{cases}$$

Find the average temperature of the rod.

57. For what a, $0 \le a \le \pi$, does the function $f(x) = \sin(x) - \cos(x)$ have the greatest average over the interval $[a, a + \pi]$ of length π?

58. Given $c > 0$, for what value b, $0 < b < c$, is $\exp(b)$ equal to the average of $\exp(x)$ for $0 \le x \le c$.

In each of Exercises 59–64, a nonnegative continuous function g and an interval I are specified. The function g is nonnegative on I. Find a positive real number c for which $f(x) = c\, g(x)$ is a probability density function.

59. $g(x) = x\ln(x)$ $\quad [1, e]$
60. $x^3(1 - x)^2$ $\quad [0, 1]$
61. $x^2(1 - x^3)^{-1/2}$ $\quad [0, 1]$
62. $(9 + x^2)^{-1/2}$ $\quad [0, 4]$
63. $\arcsin(x)$ $\quad [0, 1]$
64. $x(1 - x)^{-1/2}$ $\quad [0, 1]$

In each of Exercises 65–74 calculate the expectation of a random variable whose probability density function is given.

65. $\dfrac{1}{2}\sin(x)$ $\quad 0 \le x \le \pi$
66. $\dfrac{1}{2\pi}(1 + \cos(x))$ $\quad 0 \le x \le 2\pi$
67. $\dfrac{4x}{(x^2 + 1)^2}$ $\quad 0 \le x \le 1$
68. $e^{1-x}/(e - 1)$ $\quad 0 \le x \le 1$
69. $\dfrac{3}{16}\sqrt{4 - x}$ $\quad 0 \le x \le 4$
70. $\dfrac{4\arctan(x)}{\pi - 2\ln(2)}$ $\quad 0 \le x \le 1$
71. $2e^{-2x}$ $\quad 0 \le x < \infty$
72. $\dfrac{2}{\sqrt{\pi}}\exp(-x^2)$ $\quad 0 \le x < \infty$

73. $\dfrac{4}{\pi(1+x^2)^2}$ $\quad 0 \le x < \infty$

74. $\dfrac{2x}{(1+x^2)^2}$ $\quad 0 \le x < \infty$

If $P(X \le m) = P(X \ge m) = 1/2$, then we say that m is the *median* of a random variable X. In each of Exercises 75–78, calculate the median of a random variable whose probability density function f is given.

75. $f(x) = \cos(x)$ $\quad 0 \le x \le \pi/2$
76. $f(x) = 3x^2/26$ $\quad 1 \le x \le 3$
77. $f(x) = e^{1-x}/(e-1)$ $\quad 0 \le x \le 1$
78. $f(x) = 4x(1-x^2)$ $\quad 0 \le x \le 1$

79. Let p be a positive constant. Let X be a random variable with range $[0,1]$ and probability density function $f(x) = (p+1)x^p$, $0 \le x \le 1$. Show that $\overline{X} \ne f_{\text{avg}}$. (In general, the average of a random variable is not equal to the average of its probability density function.)

80. Let p be a positive constant. Suppose that X is a random variable with probability density function $f(x) = (p+1)x^p$ for $0 \le x \le 1$.

a. Show that $g(u) = 1 + u - 2^u$ satisfies $0 < g'(u)$ for $0 < u < \log_2(1/\ln(2)) \approx 0.529$, and $g'(u) < 0$ for $\log_2(1/\ln(2)) < u$.

b. Use the values $g(0) = g(1) = 0$ together with the inequalities of part a to deduce that $0 < g(u)$ for $0 < u < 1$.

c. By setting $u = 1/(p+1)$ in the inequality $0 < g(u)$, deduce that

$$2^{1/(p+1)} < \frac{p+2}{p+1}.$$

d. Show that the mean of X is not equal to the median of X. (In general, the mean of a random variable is not equal to its median.)

81. Suppose that f is a continuous positive function on the unbounded interval $[a, \infty)$. Is it appropriate to make the definition

$$f_{\text{avg}} = \lim_{N\to\infty} \frac{1}{N-a}\int_a^N f(x)\,dx?$$

Discuss why or why not.

82. Let λ be a positive constant. Show that $f(x) = \lambda \exp(-\lambda x), 0 \le x < \infty$ is a probability density function. Show that the mean of a random variable with probability density function f is $1/\lambda$.

If μ is any real number, and σ is any positive real number, then

$$f_{\mu,\sigma}(x) = \frac{1}{\sqrt{2\pi}\sigma}\exp\left(-\frac{1}{2}\left(\frac{x-\mu}{\sigma}\right)^2\right), \quad -\infty < x < \infty$$

is a probability density function. Exercises 83 and 84 are concerned with a random variable X that has probability density function $f_{\mu,\sigma}$. Such an X is called a *normal* or *Gaussian* random variable. If $\mu = 0$ and $\sigma = 1$, we use the term *standard normal*. The parameter σ is called the *standard deviation* of X.

83. Show that μ is the mean of a random variable that has probability density function $f_{\mu,\sigma}$.

84. Suppose that X is a random variable that has probability density function $f_{\mu,\sigma}$. Prove that

$$P(\mu - k\sigma \le X \le \mu + k\sigma) = \frac{1}{\sqrt{2\pi}}\int_{-k}^{k} \exp\left(-\frac{1}{2}x^2\right)dx.$$

Calculator/Computer Exercises

In each of Exercises 85–88, a function f is given. Let $A_f(c)$ denote the average value of f over the interval $[c - 1/4, c + 1/4]$. Plot $y = f(x)$ and $y = A_f(x)$ for $-1 \le x \le 1$. The resulting plot will illustrate the gain in smoothness that results from averaging.

85. $f(x) = |x|$

86. $f(x) = \sqrt{|x|}$

87. $f(x) = \begin{cases} 0 & \text{if } x < 0 \\ 1 & \text{if } 0 \le x \end{cases}$

88. $f(x) = \begin{cases} x^2 & \text{if } x < 0 \\ x^2 + x & \text{if } 0 \le x \end{cases}$

In each of Exercises 89–92, a function f and an interval I are given. Calculate the average f_{avg} of f over I, and find a value c in I such that $f(c) = f_{\text{avg}}$. State your answers to three decimal places.

89. $f(x) = \sqrt{x}\exp(-x)$ $\quad I = [0, 4]$
90. $f(x) = \sin(\pi(x^2 - x^3))$ $\quad I = [0, 1]$
91. $f(x) = x/(16 + x^3)$ $\quad I = [2, 4]$
92. $f(x) = \dfrac{\ln(x)}{1+x}$ $\quad I = [1, e]$

93. What is the probability that the value of a normal random variable is within one standard deviation of the mean? Refer to Exercise 84.

94. What is the probability that the value of a normal random variable is within two standard deviations of the mean? Refer to Exercise 84.

7.4 Center of Mass

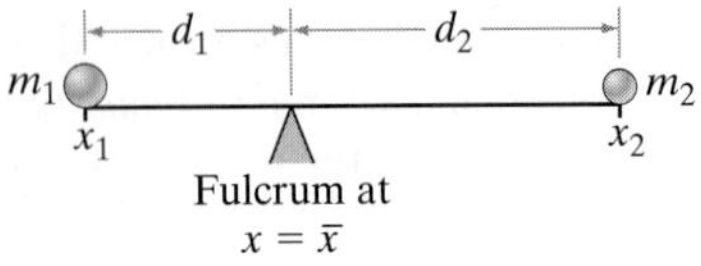

▲ Figure 1

When two children of different masses m_1 and m_2 play on a seesaw, they find that they can balance the seesaw if they position themselves appropriately. Let us set up a simple model of the seesaw. Suppose that two point masses m_1 and m_2 are situated at endpoints of an interval $[x_1, x_2]$ on the x-axis, as in Figure 1. If a fulcrum could be positioned underneath the axis, and if the interval could pivot about the fulcrum, then at what coordinate $\overline{x}$ should we place the fulcrum so as to achieve a balance? According to the lever law of physics, the masses are in balance if and only if the distances d_1 and d_2 to the fulcrum satisfy the equation $m_1 d_1 = m_2 d_2$. In this section, we will use this simple principle, together with calculus, to determine the balancing point of a planar region such as the one shown in Figure 2.

Moments (Two Point Systems)

Before turning our attention to planar figures, it is useful to study the seesaw of Figure 1 in greater detail. The point $\overline{x}$ is said to be the *center of mass* of the system. We may rewrite the equation $m_1 d_1 = m_2 d_2$ in terms of the center of mass as follows: $m_1 \cdot (\overline{x} - x_1) = m_2 \cdot (x_2 - \overline{x})$ or

$$0 = m_1 \cdot (x_1 - \overline{x}) + m_2 \cdot (x_2 - \overline{x}) \qquad (7.4.1)$$

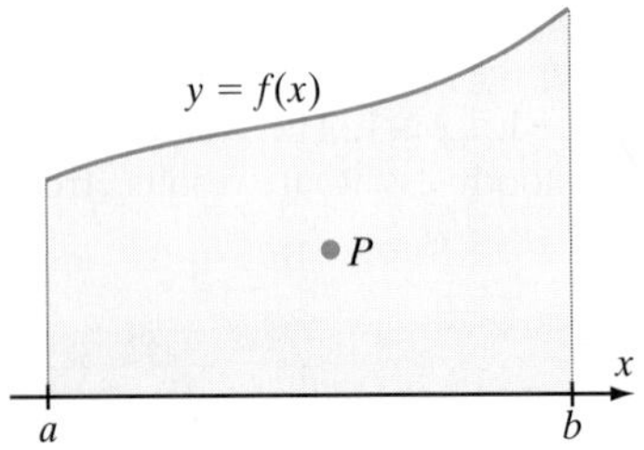

▲ Figure 2 The shaded region balances at point P.

The right side of equation (7.4.1) is said to be the *moment* about the vertical axis $x = \overline{x}$. The lever law as expressed by equation (7.4.1) tells us that our system will balance if and only if the moment about the axis $x = \overline{x}$ is zero. In general, if we position the fulcrum at any point $x = c$, then we define the *moment* $M_{x=c}$ about the axis $x = c$ to be the expression

$$M_{x=c} \;=\; m_1 \cdot (x_1 - c) + m_2 \cdot (x_2 - c). \qquad (7.4.2)$$

The moment about the y-axis,

$$M_{x=0} \;=\; m_1 \cdot x_1 + m_2 \cdot x_2, \qquad (7.4.3)$$

is an important special case. Moments may be positive, negative, or 0. In the first two cases, the axis will swing about the fulcrum as shown in Figure 3.

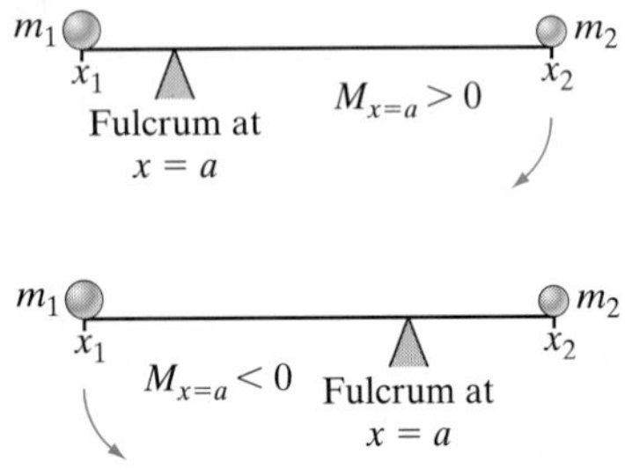

▲ Figure 3

Equations (7.4.1) and (7.4.2) provide us with a method of determining the center of mass $\overline{x}$ of our two point system; we set $M_{x=c} = 0$ in equation (7.4.2) and solve for c. Thus $0 = m_1 \cdot (x_1 - c) + m_2 \cdot (x_2 - c)$, or $m_1 c + m_2 c = m_1 \cdot x_1 + m_2 \cdot x_2$, or $c = (m_1 \cdot x_1 + m_2 \cdot x_2)/(m_1 + m_2)$. Thus

$$\overline{x} = \frac{m_1 \cdot x_1 + m_2 \cdot x_2}{m_1 + m_2}.$$

Notice that the numerator of this expression for $\overline{x}$ is the right side of equation (7.4.3). If we let $M = m_1 + m_2$ denote the total mass of the system, then we can rewrite our formula for $\overline{x}$ in the form

$$\overline{x} = \frac{M_{x=0}}{M}.$$

Moments

Now let's consider the region $\mathcal{R}$ shown in Figure 4. We suppose that the region has uniform mass density δ. By this we mean that the mass of any subset of $\mathcal{R}$ is δ times the area of the subset. Imagine the xy-plane to be horizontal and position a fulcrum underneath the axis $x = c$ (Figure 5). What must c be so that region $\mathcal{R}$ does not swing to one side or the other? To answer this question, we form a partition

$a = x_0 < x_1 < \ldots < x_N = b$ of the interval $[a, b]$. Let Δx denote the common length of the subintervals, and let s_j be the midpoint of the jth subinterval I_j. We divide region $\mathcal{R}$ into the N pieces that lie over the subintervals $I_1, I_2, \ldots, I_N$. The piece that lies over I_j is approximately a rectangle with base Δx and height $f(s_j)$—see Figure 6. The mass of this piece is therefore approximately equal to $\delta f(s_j)\Delta x$. Notice that, if Δx is small, then the points in this j^{th} piece of $\mathcal{R}$ are all approximately a (signed) distance $(s_j - c)$ from the axis $x = c$. The contribution that this piece makes to the tendency of $\mathcal{R}$ to pivot about $x = c$ is therefore about $(s_j - c) \cdot \delta f(s_j)\Delta x$. The total tendency (or moment) $M_{x=c}$ to pivot about the axis $x = c$ is approximately

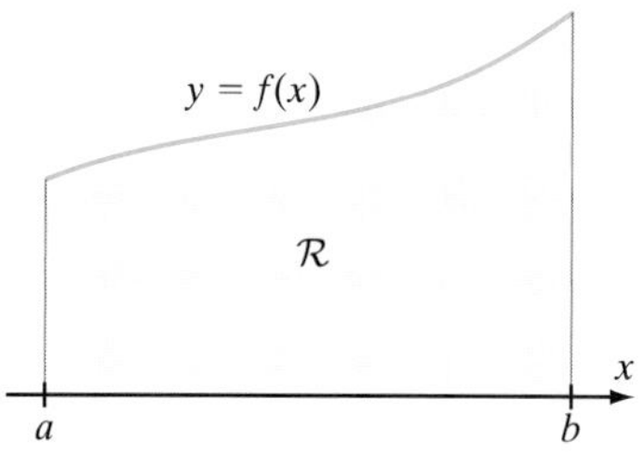

▲ Figure 4

$$\sum_{j=1}^{N} (s_j - c) \cdot \delta f(s_j)\Delta x. \tag{7.4.4}$$

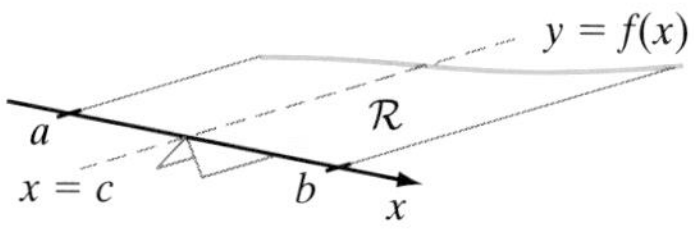

▲ Figure 5

As N increases, Δx decreases, and expression (7.4.4) becomes a better approximation to the tendency of $\mathcal{R}$ to rotate about the line $x = c$. Because expression (7.4.4) is a Riemann sum for the integral $\int_a^b (x - c)\delta f(x)\,dx$, we are led to the following definition.

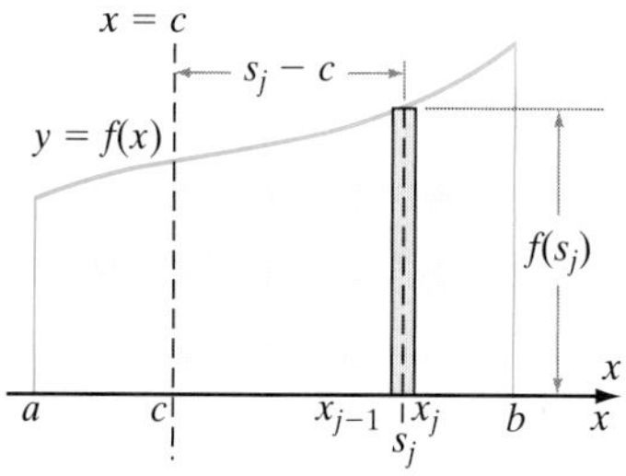

▲ Figure 6

DEFINITION Let c be any real number. Suppose that f is continuous and nonnegative on the interval $[a, b]$. Let $\mathcal{R}$ denote the planar region bounded above by the graph of $y = f(x)$, below by the x-axis, and on the sides by the line segments $x = a$ and $x = b$. If $\mathcal{R}$ has a uniform mass density δ, then the moment $M_{x=c}$ of $\mathcal{R}$ about the axis $x = c$ is defined by

$$M_{x=c} = \int_a^b (x - c)\,\delta f(x)\,dx = \delta \int_a^b (x - c) f(x)\,dx.$$

Notice that the moment $M_{x=c}$ is defined even if the number c is not between a and b.

▶ **EXAMPLE 1** Let $\mathcal{R}$ be the region bounded by $y = x - 1$, $y = 0$, and $x = 6$ (as shown in Figure 7). Suppose that $\mathcal{R}$ has uniform mass density $\delta = 2$. Calculate the moments about the axes $x = 5$ and $x = 0$.

Solution The required moments are

$$M_{x=5} = 2\int_1^6 (x - 5) \cdot (x - 1)\,dx = 2\int_1^6 (x^2 - 6x + 5)\,dx = 2\left(\frac{1}{3}x^3 - 3x^2 + 5x\right)\Bigg|_1^6 = -\frac{50}{3}$$

and

$$M_{x=0} = 2\int_1^6 (x - 0) \cdot (x - 1)\,dx = 2\int_1^6 (x^2 - x)\,dx = 2\left(\frac{1}{3}x^3 - \frac{1}{2}x^2\right)\Bigg|_1^6 = \frac{325}{3}.$$ ◀

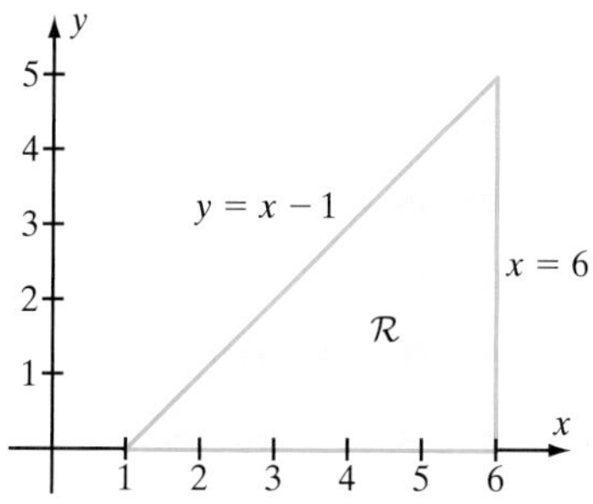

▲ Figure 7

INSIGHT In Example 1, the signs of $M_{x=5}$ and $M_{x=0}$ could have been determined simply by inspecting Figure 7. Imagine a horizontal plate in the shape of $\mathcal{R}$. Our experience tells us that, if the plate were allowed to pivot about the axis $x = 5$, then the left side would fall. From this, we conclude that $M_{x=5} < 0$. (It may be useful to refer back to Figure 3, which illustrates the analogous behavior of a seesaw.) Similar reasoning leads to the conclusion that $M_{x=0}$ must be positive.

Consider again the region $\mathcal{R}$ that is shown in Figure 4. We would like to define the moment $M_{y=d}$ of $\mathcal{R}$ about a horizontal axis $y = d$. Unfortunately, we cannot do so in a way that is analogous to the definition of moments about vertical axes. The reason can be seen in Figure 6. The key idea behind the definition of the moment $M_{x=c}$ is that, if the rectangle is thin, then the point masses within the rectangle are nearly equidistant from the vertical axis $x = c$. Clearly that cannot be true for any horizontal axis.

A thorough solution to the problem of defining $M_{y=d}$ requires multivariable calculus. For now, let us consider only the moment $M_{y=0}$, the moment about the x-axis. Glance at Figure 8. The average distance to the x-axis of the points in the rectangle is $f(s_j)/2$. It therefore seems plausible to use

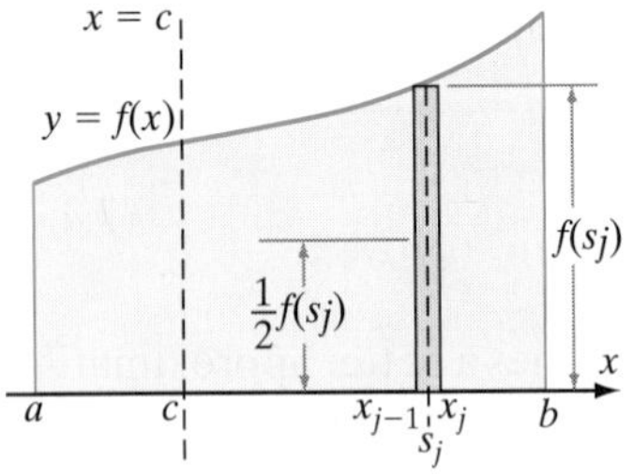

▲ Figure 8

$$\underbrace{\frac{1}{2}f(s_j)}_{\text{Average distance}} \cdot \underbrace{\delta f(s_j)\Delta x}_{\text{Approximate mass}}$$

as an approximation of the moment of the rectangle about the x-axis. Then the resulting approximation to the total moment of $\mathcal{R}$ about the x-axis is

$$\sum_{j=1}^{N} \frac{1}{2} f(s_j) \cdot \delta f(s_j)\Delta x.$$

Because this is a Riemann sum of the integral $\int_a^b \frac{1}{2}\delta f(x)^2\,dx$, we are led to define $M_{y=0}$ by the formula

$$M_{y=0} = \frac{1}{2}\delta \int_a^b f(x)^2\,dx. \tag{7.4.5}$$

After we have studied some multivariable calculus, we will have the tools to rigorously justify this formula.

▶ **EXAMPLE 2** Let $\mathcal{R}$ be the region bounded by $y = x - 1$, $y = 0$, and $x = 6$ (as in Example 1). Suppose that $\mathcal{R}$ has uniform mass density $\delta = 2$. Calculate the moment $M_{y=0}$.

Solution According to formula (7.4.5), the required moment is

$$M_{y=0} = \frac{1}{2}(2)\int_1^6 (x-1)^2\,dx = \left.\frac{(x-1)^3}{3}\right|_1^6 = \frac{125}{3}. \quad ◀$$

Center of Mass

The center of mass $(\bar{x}, \bar{y})$ of a region $\mathcal{R}$ is the point at which the region balances (as in Figure 2). To find the center of mass, we must translate this physical property into mathematical equations.

DEFINITION Let $\mathcal{R}$ be a region as shown in Figure 4. Then the *center of mass* of $\mathcal{R}$ is the point $(\bar{x}, \bar{y})$ whose coordinates satisfy $M_{x=\bar{x}} = 0$ and $M_{y=\bar{y}} = 0$.

THEOREM 1 Let f be a continuous nonnegative function on the interval $[a, b]$. Let $\mathcal{R}$ denote the region bounded above by the graph of $y = f(x)$, below by the x-axis, and on the sides by the line segments $x = a$ and $x = b$. Let M denote the mass of $\mathcal{R}$. If $\mathcal{R}$ has a uniform mass density δ, then the x-coordinate $\overline{x}$ of the center of mass of $\mathcal{R}$ is given by

$$\overline{x} = \frac{M_{x=0}}{M} = \frac{\int_a^b xf(x)\,dx}{\int_a^b f(x)\,dx}, \tag{7.4.6}$$

and the y-coordinate $\overline{y}$ of the center of mass of $\mathcal{R}$ is given by

$$\overline{y} = \frac{M_{y=0}}{M} = \frac{\frac{1}{2}\int_a^b f(x)^2\,dx}{\int_a^b f(x)\,dx}. \tag{7.4.7}$$

Proof. By definition, $M_{x=\overline{x}} = 0$. To determine $\overline{x}$, we solve for $\overline{x}$ in the equation $\delta \int_a^b (x - \overline{x}) f(x)\,dx = 0$. After expanding the integrand and expressing the resulting integral as a sum, we obtain

$$\delta \int_a^b xf(x)\,dx - \overline{x} \cdot \delta \int_a^b f(x)\,dx = 0.$$

Solving for $\overline{x}$, we find that

$$\overline{x} = \frac{\int_a^b xf(x)\,dx}{\int_a^b f(x)\,dx} = \frac{M_{x=0}}{M}.$$

Notice that the uniform mass density δ has been eliminated and does not appear in the final formula for $\overline{x}$.

The equation $\overline{y} = M_{y=0}/M$ is obtained analogously. Into this equation, we substitute the value for $M_{y=0}$ given by formula (7.4.5). The second quotient of line (7.4.7) is the result. ■

▶ **EXAMPLE 3** Let $\mathcal{R}$ be the region bounded by the lines $y = x - 1$, $y = 0$, and $x = 6$ (as in Example 1). Suppose that $\mathcal{R}$ has uniform mass density $\delta = 2$ (as in Example 1). Calculate the center of mass of $\mathcal{R}$.

Solution In Example 1, we calculated $M_{x=0} = 325/3$. The total mass M of $\mathcal{R}$ is computed as follows:

$$M = \delta \int_1^6 (x-1)\,dx = 2\left(\frac{1}{2}x^2 - x\right)\Bigg|_1^6 = 25.$$

Therefore

$$\overline{x} = \frac{325/3}{25} = \frac{13}{3}.$$

In Example 2, we showed that $M_{y=0} = 125/3$. Therefore

$$\overline{y} = \frac{125/3}{25} = \frac{5}{3}.$$

Thus the center of mass is $(13/3, 5/3)$. ◀

The next theorem generalizes formulas (7.4.6) and (7.4.7) by removing the requirement that the lower boundary of $\mathcal{R}$ be the x-axis.

THEOREM 2 Let f and g be continuous functions on the interval $[a, b]$ with $g(x) \le f(x)$ for all x in $[a, b]$. Let $\mathcal{R}$ denote the region bounded above by the graph of $y = f(x)$, below by the graph of $y = g(x)$, and on the sides by the line segments $x = a$ and $x = b$. If $\mathcal{R}$ has uniform mass density, then the x-coordinate $\bar{x}$ of the center of mass of $\mathcal{R}$ is given by

$$\bar{x} = \frac{\int_a^b x\big(f(x) - g(x)\big)\,dx}{\int_a^b \big(f(x) - g(x)\big)\,dx}. \tag{7.4.8}$$

The y-coordinate $\bar{y}$ of the center of mass of $\mathcal{R}$ is given by

$$\bar{y} = \frac{\frac{1}{2}\int_a^b \left(f(x)^2 - g(x)^2\right)dx}{\int_a^b \left(f(x) - g(x)\right)dx}. \tag{7.4.9}$$

Proof. Let Δx denote the common length of the subintervals that result from the uniform partition $a = x_0 < x_1 < \ldots < x_N = b$ of the interval $[a, b]$. Let s_j be the midpoint of the jth subinterval I_j. The height of the thin rectangle shown in Figure 9 is $f(s_j) - g(s_j)$, its area is $(f(s_j) - g(s_j))\Delta x$, its mass is $\delta(f(s_j) - g(s_j))\Delta x$, and its moment about the y-axis is $\delta s_j(f(s_j) - g(s_j))\Delta x$. The total moment resulting from the partition is therefore $\sum_{j=1}^{N} \delta\, s_j\,(f(s_j) - g(s_j))\,\Delta x$. It follows that

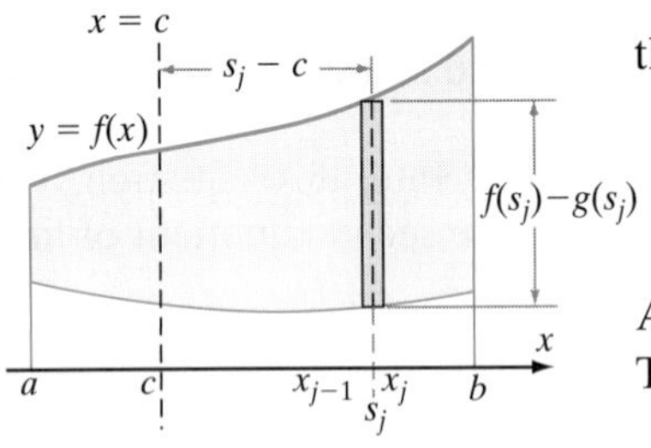

▲ Figure 9

$$M_{x=0} = \lim_{N\to\infty} \sum_{j=1}^{N} \delta\, s_j\,(f(s_j) - g(s_j))\,\Delta x = \delta \int_a^b x\,(f(x) - g(x))\,dx.$$

Also, the mass M of $\mathcal{R}$ is the product of δ and the area of $\mathcal{R}$, or $\delta \int_a^b (f(x) - g(x))\,dx$. Thus

$$\bar{x} = \frac{M_{x=0}}{M} = \frac{\delta \int_a^b x\,(f(x) - g(x))\,dx}{\delta \int_a^b (f(x) - g(x))\,dx} = \frac{\int_a^b x\,(f(x) - g(x))\,dx}{\int_a^b (f(x) - g(x))\,dx}.$$

To calculate the moment of the thin rectangle about the x-axis, we form the product of the mass $\delta\,(f(s_j) - g(s_j))\,\Delta x$ of the rectangle and the average distance, $\dfrac{f(s_j) + g(s_j)}{2}$, of points of the rectangle to the x-axis. See Figure 10. Thus

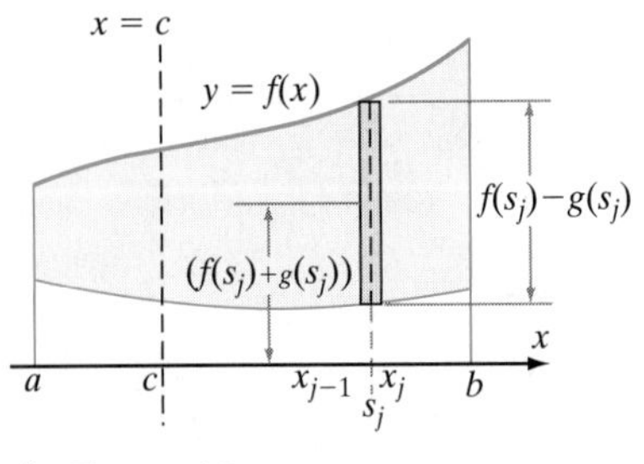

▲ Figure 10

$$\begin{aligned} M_{y=0} &= \lim_{N\to\infty} \sum_{j=1}^{N} \delta \left(\frac{f(s_j) + g(s_j)}{2}\right)(f(s_j) - g(s_j))\,\Delta x \\ &= \delta \int_a^b \left(\frac{f(x) + g(x)}{2}\right)(f(x) - g(x))\,dx \\ &= \frac{\delta}{2} \int_a^b \left(f(x)^2 - g(x)^2\right)dx \end{aligned}$$

and

$$\bar{y} = \frac{M_{y=0}}{M} = \frac{\frac{\delta}{2}\int_a^b \left(f(x)^2 - g(x)^2\right) dx}{\delta \int_a^b \left(f(x) - g(x)\right) dx} = \frac{\frac{1}{2}\int_a^b \left(f(x)^2 - g(x)^2\right) dx}{\int_a^b \left(f(x) - g(x)\right) dx}.$$ ■

▶ **EXAMPLE 4** Let $\mathcal{R}$ be the region bounded above by $y = x + 1$ and below by $y = (x-1)^2$. Suppose that $\mathcal{R}$ has unit mass density. What is the center of mass $(\bar{x}, \bar{y})$ of $\mathcal{R}$?

Solution The two graphs intersect when $x + 1 = (x-1)^2$, or $x + 1 = x^2 - 2x + 1$, or $3x = x^2$. The solutions are $x = 0$ and $x = 3$. Next, we set $f(x) = x + 1$ and $g(x) = (x-1)^2$ and calculate

$$\int_0^3 \left(f(x) - g(x)\right) dx = \int_0^3 \left((x+1) - (x-1)^2\right) dx = \frac{9}{2}.$$

The numerator of formula (7.4.8) is

$$\int_a^b x\left(f(x) - g(x)\right) dx = \int_0^3 x\left((x+1) - (x-1)^2\right) dx = \int_0^3 (3x^2 - x^3)\, dx = \frac{27}{4}.$$

Therefore

$$\bar{x} = \frac{27/4}{9/2} = \frac{3}{2}.$$

The numerator of formula (7.4.9) is

$$\frac{1}{2}\int_a^b \left(f(x)^2 - g(x)^2\right) dx = \frac{1}{2}\int_0^3 \left((x+1)^2 - (x-1)^4\right) dx$$
$$= \frac{1}{2}\int_0^3 (-x^4 + 4x^3 - 5x^2 + 6x)\, dx = \frac{36}{5}.$$

Therefore

$$\bar{y} = \frac{36/5}{9/2} = \frac{8}{5}.$$

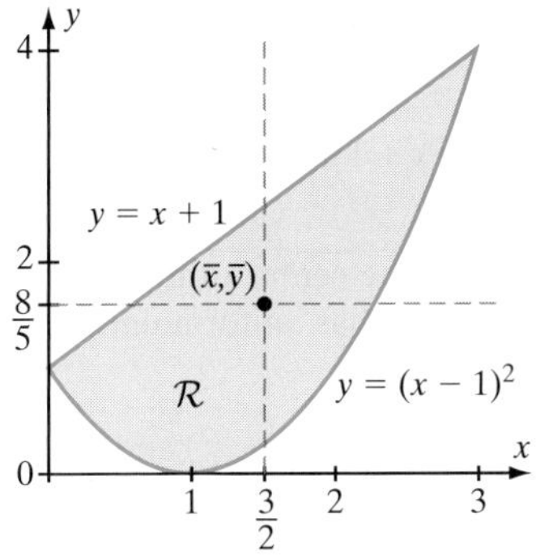

▲ Figure 11

The center of mass of $\mathcal{R}$ is $(3/2, 8/5)$. See Figure 11. ◀

QUICK QUIZ

1. Let $\mathcal{R}$ denote the triangle with vertices $(0,0)$, $(2,0)$, and $(2,6)$. The x-center of mass $\bar{x}$ of $\mathcal{R}$ is given by the equation $\bar{x} = \frac{1}{6}\int_0^2 g(x)\, dx$ for what function $g(x)$?
2. Let $\mathcal{R}$ denote the triangle with vertices $(0,0)$, $(2,0)$, and $(2,6)$. The x-center of mass $\bar{y}$ of $\mathcal{R}$ is given by the equation $\bar{y} = \frac{1}{12}\int_0^2 g(x)\, dx$ for what function $g(x)$?
3. Let $\mathcal{R}$ denote the region that is bounded above by $y = 2/x$, below by $y = 1/x$, on the left by $x = 1$, and on the right by $x = 2$. The area of $\mathcal{R}$ is $\ln(2)$. The x-center of mass $\bar{x}$ of $\mathcal{R}$ is given by the equation $\ln(2)\cdot\bar{x} = \alpha$ for what value of α?
4. Let $\mathcal{R}$ denote the region that is bounded above by $y = 2/x$, below by $y = 1/x$, on the left by $x = 1$, and on the right by $x = 2$. The area of $\mathcal{R}$ is $\ln(2)$. The y-center of mass $\bar{y}$ of $\mathcal{R}$ is given by the equation $2\ln(2)\cdot\bar{y} = \int_1^2 g(x)\, dx$ for what function $g(x)$?

Answers

1. $3x^2$ 2. $9x^2$ 3. 1 4. $3/x^2$

EXERCISES

Problems for Practice

In each of Exercises 1–8, find the moment of the given region $\mathcal{R}$ about the given vertical axis. Assume that $\mathcal{R}$ has uniform unit mass density.

1. $\mathcal{R}$ is the triangular region with vertices $(0,0), (0,2)$, and $(6,0)$; about $x = 3$.

2. $\mathcal{R}$ is the triangular region with vertices $(0,0), (0,2)$, and $(6,0)$; about $x = -1$.

3. $\mathcal{R}$ is the triangular region with vertices $(0,0)$, $(0,2)$, and $(6,0)$; about $x = 2$.

4. $\mathcal{R}$ is the first quadrant region bounded by $y = 4x - x^3$ and the x-axis; about $x = 1$.

5. $\mathcal{R}$ is the first quadrant region bounded by $y = 4x - x^3$ and the x-axis; about $x = 3$.

6. $\mathcal{R}$ is the first quadrant region bounded above by $y = x - x^2$ and the x-axis; about $x = 0$.

7. $\mathcal{R}$ is the region bounded above by $y = 1/x$, below by the x-axis, and on the sides by the vertical lines $x = 1$ and $x = 2$; about $x = -3$.

8. $\mathcal{R}$ is the first quadrant region bounded above by $y = \dfrac{\sin(x)}{x}$, below by the x-axis, and on the sides by $x = \pi/6$ and $x = \pi/2$; about $x = 0$.

In each of Exercises 9–16, find the moment of the given region $\mathcal{R}$ about the x-axis. Assume that $\mathcal{R}$ has uniform unit mass density.

9. $\mathcal{R}$ is the triangular region with vertices $(0,0), (0,2)$, and $(6,0)$.

10. $\mathcal{R}$ is the first quadrant region bounded above by $y = \sqrt{x}$, below by the x-axis, and on the right by $x = 4$.

11. $\mathcal{R}$ is the region bounded above by $y = 1/x$, below by the x-axis, and on the sides by the vertical lines $x = 1$ and $x = 2$.

12. $\mathcal{R}$ is the first quadrant region bounded above by $y = e^x$, below by the x-axis, and on the sides by $x = 1$ and $x = 2$.

13. $\mathcal{R}$ is the region bounded above by $y = \sqrt{4 - x^2}$ and below by the x-axis.

14. $\mathcal{R}$ is the first quadrant region bounded above by $y = 1 - x^3$ and below by the x-axis.

15. $\mathcal{R}$ is the first quadrant region bounded above by $y = \sqrt{x}(1 - x^2)^2$ and below by the x-axis.

16. $\mathcal{R}$ is the first quadrant region bounded above by $y = 6x/(1 + x^3)$, below by the x-axis, and on the right by $x = 1$.

In each of Exercises 17–28, find the center of mass $(\bar{x}, \bar{y})$ of the given region $\mathcal{R}$, assuming that it has uniform mass density.

17. $\mathcal{R}$ is the triangular region with vertices $(0,0), (0,2)$, and $(6,0)$.

18. $\mathcal{R}$ is the region bounded by $y = 4 - x^2$ and the x-axis.

19. $\mathcal{R}$ is the first quadrant region bounded above by $y = \sqrt{4 - x^2}$ and below by the x-axis.

20. $\mathcal{R}$ is the region bounded above by $y = \cos(x)$, $0 \le x \le \pi/2$, and below by the x-axis.

21. $\mathcal{R}$ is the region bounded above by $y = 8 - 2x - x^2$ and below by the x-axis.

22. $\mathcal{R}$ is the region bounded by $y = x - x^3$, $0 \le x \le 1$, and the x-axis.

23. $\mathcal{R}$ is the region bounded by $y = 1/x$, $1 \le x \le 2$, and the x-axis.

24. $\mathcal{R}$ is the region bounded by $y = \sqrt{x}, x = 4$, $x = 9$, and the x-axis.

25. $\mathcal{R}$ is the region bounded above by $y = e^x$, below by the x-axis, on the left by $x = 0$, and on the right by $x = 1$.

26. $\mathcal{R}$ is the region bounded by $y = 1/\sqrt{x}$, $4 \le x \le 9$, and the x-axis.

27. $\mathcal{R}$ is the region bounded above by $y = 1/\sqrt{1 - x^2}$, below by the x-axis, on the left by $x = 0$, and on the right by $x = 1/\sqrt{2}$.

28. $\mathcal{R}$ is the region bounded by $y = x + 1/x$, $x = 1/2$, $x = 2$, and the x-axis.

In each of Exercises 29–36, find the center of mass $(\bar{x}, \bar{y})$ of the given region $\mathcal{R}$, assuming that it has uniform unit mass density.

29. $\mathcal{R}$ is the region bounded above by $y = 4(1 - x^2)$ and below by $y = 1 - x^2$.

30. $\mathcal{R}$ is the region bounded above by $y = 2x$ and below by $y = x^2$.

31. $\mathcal{R}$ is the region bounded above by $y = 2\sqrt{x}$ and below by $y = x$.

32. $\mathcal{R}$ is the region bounded above by $y = x$, below by $y = \sqrt{x}$, on the left by $x = 1$, and on the right by $x = 4$.

33. $\mathcal{R}$ is the region bounded above by $y = x + 5$, below by $y = 4 - x^2$, on the left by $x = -1$, and on the right by $x = 2$.

34. $\mathcal{R}$ is the region bounded above by $y = 5 - x$, below by $y = \sqrt{1 + x}$, and on the left by $x = 0$.

35. $\mathcal{R}$ is the region bounded above by $y = 2$, below by $y = \sqrt{4 - x}$, and on the right by $x = 3$.

36. $\mathcal{R}$ is the region bounded above by $y = 2(1+x^2)$, below by $y = 1 - x^2$, on the left by $x = 0$, and on the right by $x = 1$.

Further Theory and Practice

▶ In each of Exercises 37–44, find the center of mass of the given region $\mathcal{R}$, assuming that it has uniform unit mass density. ◀

37. $\mathcal{R}$ is the region bounded above by $y = \cos(x)$ for $0 \le x \le \pi/2$, below by the x-axis, and on the left by $x = 0$.
38. $\mathcal{R}$ is the region bounded above by $y = \ln(x)$, below by the x-axis, and on the sides by $x = 1$ and $x = e$.
39. $\mathcal{R}$ is the region bounded above by $y = 1/(1+x^2)$, below by the x-axis, and on the sides by $x = 0$ and $x = 1$.
40. $\mathcal{R}$ is the region bounded above by $y = x/\sqrt{1+2x^2}$, below by the x-axis, and on the sides by $x = 0$ and $x = 2$.
41. $\mathcal{R}$ is the region bounded above by $y = (1-x^2)^{3/2}$ and below by the x-axis.
42. $\mathcal{R}$ is the region bounded above by $y = (16+x^2)^{-1/2}$, below by the x-axis, on the left by $x = 0$, and on the right by $x = 3$.
43. $\mathcal{R}$ is the region bounded above by $y = x - 1/x$, below by the x-axis, on the left by $x = 1$, and on the right by $x = 2$.
44. $\mathcal{R}$ is the region in the first quadrant that is bounded above by $y = x\sqrt{1-x^2}$ and below by the x-axis.

▶ In each of Exercises 45–52, find the center of mass of the given region $\mathcal{R}$, assuming that it has uniform unit mass density. ◀

45. $\mathcal{R}$ is the region bounded above by $y = 1$ for $-1 \le x \le 0$ and by $y = 1 - x$ for $0 \le x \le 1$, below by the x-axis, and on the left by $x = -1$.
46. $\mathcal{R}$ is the region bounded above by $y = 1 + |x|$ $(-1 \le x \le 2)$, below by the x-axis, and on the sides by $x = -1$ and $x = 2$.
47. $\mathcal{R}$ is the region bounded above by $y = 4 - x^2$ for $-2 \le x \le 1$, below by $y = 3x$ for $0 \le x$, and below by the x-axis for $x < 0$.
48. $\mathcal{R}$ is the region bounded above by $y = 3x$ for $0 \le x \le 1$, above by $y = 4 - x^2$ for $1 \le x \le 2$ and below by the x-axis.
49. $\mathcal{R}$ is the region bounded above by $y = 2 - x^2$, below by $y = -x$ for $-1 \le x \le 0$, and below by $y = \sqrt{x}$ for $0 \le x \le 1$.
50. $\mathcal{R}$ is the region bounded above by

$$y = \begin{cases} x^2 & \text{if } 0 \le x \le 2 \\ 4(x-3)^2 & \text{if } 2 \le x \le 3 \end{cases}$$

and below by the x-axis.
51. $\mathcal{R}$ is the region bounded above by $y = 2(1+|x|)$ for $-1 \le x \le 4$, below by $y = (x-1)^2$ for $-1 \le x \le 4$, and on the right by $x = 4$.
52. $\mathcal{R}$ is the region bounded above by $y = \sqrt{x+2}$ for $-2 \le x \le 2$, below by the x-axis for $-2 \le x \le 1$, and below by $y = 2\sqrt{x-1}$ for $1 \le x \le 2$.
53. Suppose that f is the probability density function of a random variable X on an interval $[a,b]$. Let $\bar{x}$ be the x-coordinate of the center of mass of the region $\mathcal{R}$ bounded above by the graph of f, below by the x-axis, and on the sides by $x = a$ and $x = b$. Assuming that $\mathcal{R}$ has uniform unit mass density, what is the relationship between $\overline{X}$ and $\bar{x}$?

▶ The moments $M_{x=c} = \int_a^b (x-c)f(x)\,dx$ that have been defined in this section are *first moments*. If f is a continuous function on $[a,b]$, then the *second moment* of f about the vertical axis $x = c$ is defined to be $I_{x=c} = \int_a^b (x-c)^2 f(x)\,dx$. Second moments are used in moment of inertia calculations in engineering and physics. In Exercises 54–57, calculate the second moment of the given function f about the vertical axis $x = c$ for the given c. ◀

54. $f(x) = 1 + \sqrt{x}$	$1 \le x \le 4$	$c = 0$
55. $f(x) = x^2$	$1 \le x \le 2$	$c = -1$
56. $f(x) = x^3 + 1$	$-1 \le x \le 2$	$c = 1$
57. $f(x) = \sin(x)$	$0 \le x \le \pi$	$c = 2\pi$

▶ Second moments, as defined in the instructions to Exercises 54–57, are used throughout probability and statistics. If f is the probability density function of a random variable X with range $[a,b]$ and mean μ_X, then the variance $\mathrm{Var}(X)$ of X is defined to be the second moment of f about the vertical axis $x = \mu_X$:

$$\mathrm{Var}(X) = \int_a^b (x-\mu_X)^2 f(x)\,dx$$

Notice that $0 \le \mathrm{Var}(X)$ because $(x-\mu_X)^2$ and $f(x)$ are both nonnegative. The *standard deviation* σ_X of X is defined by $\sigma_X = \sqrt{\mathrm{Var}(X)}$. (As a result, the variance is also denoted by σ_X^2.) In each of Exercises 58–61, calculate the variance $\mathrm{Var}(X)$ of a random variable X whose probability density function is the given function f. ◀

58. $f(x) = x/4$	$1 \le x \le 3$
59. $f(x) = 3x^2$	$0 \le x \le 1$
60. $f(x) = 8/(3x^3)$	$1 \le x \le 2$
61. $f(x) = 1/(b-a)$	$a \le x \le b$

▶ In Exercises 62–65, refer to the instructions for Exercises 58–61 for the definition of $\mathrm{Var}(X)$. ◀

62. Let p be a positive constant. Suppose that a random variable X has probability function $f(x) = cx^p(1-x)$ for $0 \le x \le 1$. Find formulas for c, μ_X, and $\mathrm{Var}(X)$ in terms of p.
63. If X is a random variable with values in $[a,b]$ and probability density function f, then X^2 is a random variable and $E(X^2) = \int_a^b x^2 f(x)\,dx$. Use this fact to deduce that $\mathrm{Var}(X) = E(X^2) - E(X)^2$.
64. Suppose that $0 < \lambda$ and $f(x) = \lambda\exp(-\lambda x)$, $0 \le x < \infty$ is the probability density function of a random variable X. Calculate μ_X and $\mathrm{Var}(X)$.

65. Suppose that μ is a real number and $0 < \sigma$. It can be shown that

$$f(x) = \frac{1}{\sqrt{2\pi}\sigma} \exp\left(-\frac{1}{2}\left(\frac{x-\mu}{\sigma}\right)^2\right), \quad -\infty \le x \le \infty$$

is the probability density function of a random variable X. Show that $\mu = E(X)$ and $\sigma^2 = \mathrm{Var}(X)$.

Calculator/Computer Exercises

▶ In Exercises 66–70, find the center of mass of the given region $\mathcal{R}$. ◀

66. $\mathcal{R}$ is bounded above by $y = 6x/\exp(x) - 1$ and below by the x-axis.

67. $\mathcal{R}$ is bounded above by $y = 1 + x$ and below by $y = \exp(x^2)$.

68. $\mathcal{R}$ is bounded above by $y = 4 + 2x - x^4$ and below by $y = x - 1$.

69. $\mathcal{R}$ is bounded above by $y = \ln(1 + x + x^2)$ and below by $y = x^2 - 3$.

70. $\mathcal{R}$ is bounded above by $y = \arctan(x^3)$ and below by $y = x^2 + x - 4$.

71. The equation of the Gateway Arch is

$$y = 693.8597 - 34.38365\big(\exp(kx) + \exp(-kx)\big)$$

for $k = 0.0100333$ and $-299.2239 < x < 299.2239$. Find the center of mass for the region that lies above the x-axis and below this curve.

7.5 Work

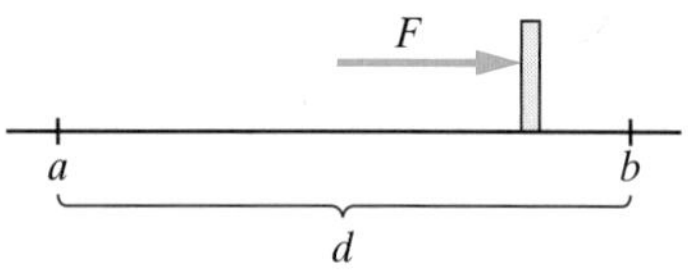

▲ Figure 1

Suppose that a body is moved a distance d along a straight line while being acted on by a force of *constant* magnitude F in the direction of motion (Figure 1). Then, by definition, the *work* performed in this move is force times distance:

$$W = F \cdot d. \tag{7.5.1}$$

Bear in mind that this is a technical definition of work that differs somewhat from everyday usage. For example, a person has to exert a force of 100 pounds to hold up a 100-pound weight. This may be a strain, and the person may tire from the exertion—but if the weight does not move, then no work is done.

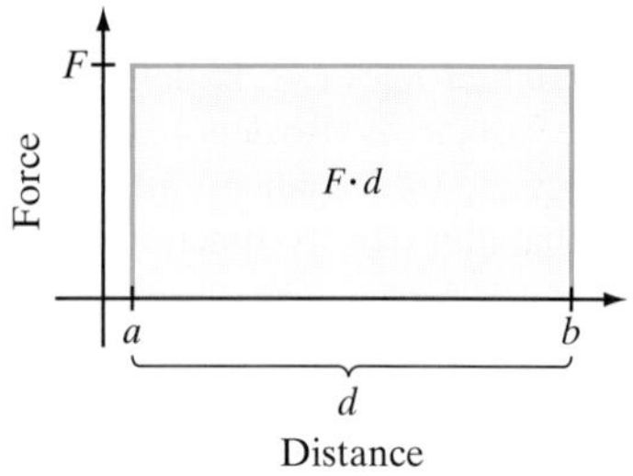

▲ Figure 2

If we graph the constant function F as in Figure 2, then we see that work is related to area. When we think of area under a graph, however, the distance from 0 to 1 on each coordinate axis represents a unit length. The unit used to express area must represent (length)^2. The square-foot and square-meter are examples. In contrast, it is evident from formula (7.5.1) that a unit of work must represent the product of a force and a length. In Figure 2, the unit of measurement along the vertical axis is a unit of force, not length.

Any combination of units that represents force $\times$ distance or, in basic components mass $\times$ (distance/time)^2 can be used to represent work. However, the unit of work can be expressed most simply by consistently using units from one of three systems of measurement. In the United States customary system, the standard unit of force is the *pound* (abbreviated lb). This leads to the *foot-pound* (ft-lb) as the unit of measure for work: it is the amount of work done when a constant force of one pound is applied for one foot. In the *mks system* of measurement, force is measured in newtons (N) and distance is measured in meters (m). The unit of work is the *newton-meter*, which is also called the *joule*(J): 1 joule is the amount of work done when a constant force of 1 newton is applied for 1 meter. In the *cgs system*, 1 *erg* is the amount of work done by a force of one dyne acting over a distance of 1 centimeter. Also, 1 joule is equal to 10^7 ergs and approximately 0.73756215 foot-pounds.

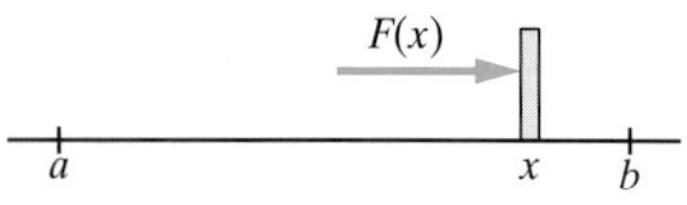

▲ Figure 3

Using Integrals to Calculate Work

Suppose that a body is being moved from $x = a$ to $x = b$ along the x-axis and that the force applied in the direction of motion at the point x is $F(x)$ (see Figure 3).

How much work is done altogether? If F is not a constant function, then formula (7.5.1) cannot be applied. Instead, we examine this problem by partitioning the interval into N subintervals of equal length Δx. If Δx is sufficiently small, then $F(x)$ will not vary greatly over each subinterval of the partition. In other words, if s_j is any point in the jth subinterval, then the force applied throughout the jth subinterval may be approximated by the constant force $F(s_j)$. The work done in the incremental move over the jth subinterval is therefore about $F(s_j) \cdot \Delta x$ (Figure 4). Summing over j, we obtain an approximation to the total work:

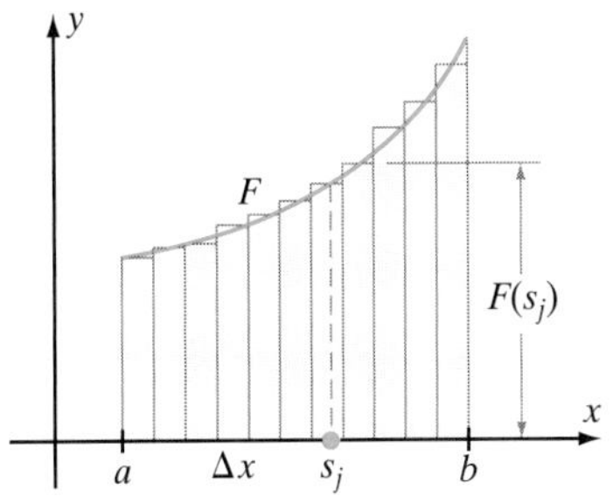

Figure 4

$$W \approx \sum_{j=1}^{N} F(s_j)\,\Delta x. \tag{7.5.2}$$

As N increases and Δx becomes smaller, approximation (7.5.2) becomes more accurate. Notice that these approximating sums are also Riemann sums for the integral $\int_a^b F(x)\,dx$. From Chapter 5, we know that these Riemann sums tend to the integral as the number N of subintervals tends to infinity. We conclude that $\int_a^b F(x)\,dx$ equals the total work. Observe that there is again an analogy between work and area under the force curve.

DEFINITION Suppose that a body is moved linearly from $x = a$ to $x = b$ by a force in the direction of motion. If the magnitude of the force at each point $x \in [a, b]$ is $F(x)$, then the total work W performed is

$$W = \int_a^b F(x)\,dx. \tag{7.5.3}$$

Examples with Weights That Vary

Newton's Law, *force equals mass times acceleration*, or $F = m \cdot a$, tells us that weight and mass are proportional if acceleration remains constant. If for some reason the mass of a load decreases, as in the first example, then the weight of the load decreases.

▶ **EXAMPLE 1** A man carries a leaky 50 pound sack of sand straight up a 100 foot ladder that runs up the side of a building. He climbs at a constant rate of 20 feet per minute. Sand leaks out of the sack at a rate of 4 pounds per minute. Ignoring the man's own weight, how much work does he perform on this trip?

Solution The definition of work that is given previously is applicable whenever a force is applied along a straight line—it does not matter if the line is horizontal or vertical (or any other). The force that the man applies is the weight of the sack, which, because of the leak, is not constant. Let $F(y)$ denote the weight of the sack of sand when the man is y feet above the ground. Because the man is $y = 20t$ ft above the ground at time t when t is measured in minutes, and because the weight of the sack is $50 - 4t$ lb when t is measured in minutes, we have

$$F(y) = 50 - 4\left(\frac{y}{20}\right) = 50 - \frac{y}{5} \text{ lb.}$$

The total work that is performed therefore equals

$$W = \int_0^{100} \left(50 - \frac{y}{5}\right) dy = \left(50y - \frac{y^2}{10}\right)\Bigg|_{y=0}^{y=100} = 4000 \text{ ft-lb.}$$

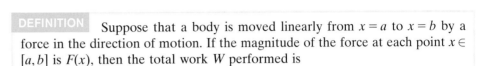

INSIGHT Suppose that we did not ignore the man's weight in Example 1. If he weighs 200 pounds, then he has to lift his own weight when climbing the ladder also. In this case, the work performed is

$$\begin{aligned} W &= \int_0^{100} \left(\left(50 - \frac{y}{5} \right) + 200 \right) dy \\ &= \int_0^{100} \left(50 - \frac{y}{5} \right) dy + \int_0^{100} 200\, dy \\ &= 4000 + 200 \cdot 100 \\ &= 24000 \text{ ft-lb.} \end{aligned}$$

At the surface of Earth, acceleration due to gravity is denoted by g and is equal to 9.80665 m/s^2 in the mks system of measurement. The force needed to lift a mass m is therefore $F = mg$ at Earth's surface. *Near* the surface of Earth, we may continue to use the equation $F = mg$ as an acceptable approximation to the weight-mass relationship. That is why we are able to treat the weight of the man as constant in the *Insight* to Example 1. In the next example, the distance from the surface of Earth becomes so great that the assumption of constant gravitational acceleration is not acceptable.

▶ **EXAMPLE 2** According to Newton's Law of Gravitation, the magnitude of the gravitational force F (measured in newtons) exerted by Earth on a mass m (measured in kilograms) is

$$F = 3.98621 \times 10^{14} \frac{m}{r^2} \tag{7.5.4}$$

when the mass is a distance r (measured in meters) from the center of Earth. Determine how much work is performed in lifting a 5000 kilogram rocket to a height 200 kilometers above the surface of the earth, assuming that the direction of the force, as well as the motion, is straight up. Use 6375.58 km for the radius of Earth.

Solution We convert all units to the mks system. By doing so, we will arrive at a value of work that is measured in joules. Because $r = 6375580$ m at the surface of the earth and $r = 6375580 + 200000$ m, or $r = 6575580$ m, when the rocket is 200 km above the surface of the earth, we have

$$\begin{aligned} W &= \int_{6375580}^{6575580} F(r)\, dr \\ &= \int_{6375580}^{6575580} 3.98621 \times 10^{14} \frac{5000}{r^2}\, dr \\ &= -3.98621 \times 10^{14} \frac{5000}{r} \Bigg|_{r=6375580}^{r=6575580} \\ &= -3.98621 \times 10^{14} \cdot 5000 \left(\frac{1}{6575580} - \frac{1}{6375580} \right) \\ &= 9.50838 \times 10^9 \text{ J.} \end{aligned}$$ ◀

INSIGHT At the surface of Earth, r equals 6375580 meters, and, for this value of r, equation (7.5.4) becomes $F = mg$ with $g = 9.80665 \text{ m/s}^2$. At the top of Mount Everest, r equals 6375580 + 8848, or $r = 6384428$. Because $(6375580/6384428)^2 = 0.997\ldots$, formula (7.5.4) tells us that an object weighs about 0.3% less at the summit of Everest than at the surface of Earth. A similar calculation shows that, when the rocket is 200 kilometers above the surface of Earth, its weight is only 94% of its weight on Earth.

An Example Involving a Spring

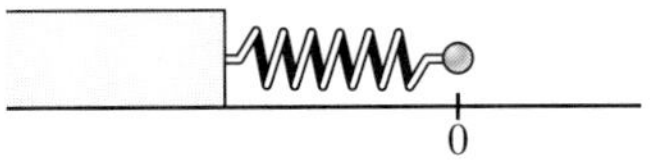

Equilibrium or rest position

▲ Figure 5a Spring at rest

▲ Figure 5b Stretched spring $(x > 0)$

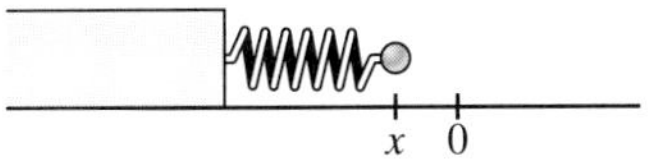

▲ Figure 5c Compressed spring $(x > 0)$

Consider a spring attached to a rigid support with one free end, as in Figure 5a. According to Hooke's Law, the force F that the spring exerts is proportional to the amount that it has been extended or compressed. Although Hooke's Law is not an exact physical law, it is a very good approximation for small extensions and compressions. When working problems with springs, it is usually convenient to set up a coordinate axis with the origin at the free end of the relaxed spring, a point which is sometimes called the *equilibrium position*. Let x denote the coordinate of the free end of the spring. Then x represents the amount of extension when $x > 0$ and the amount of compression when $x < 0$. Hooke's Law may be written as

$$F = -kx \tag{7.5.5}$$

for a positive constant k (known as the *spring constant*). The negative sign in Hooke's Law is required to reflect the direction of the force. It is a restorative force in that it is always directed toward the equilibrium position. Thus when the spring is extended and $x > 0$, the force is directed toward the negative x-axis (Figure 5b). When the spring is compressed and $x < 0$, the force is directed toward the positive x-axis (Figure 5c). Because the left side of equation (7.5.5) carries a unit of force (such as *pound or newton*), so does the right side. It follows that the constant k carries units of force divided by length. The spring constant therefore represents a stiffness density.

To stretch or compress a spring, a force must be exerted that is equal in magnitude to the restorative spring force but opposite in direction. It follows from Theorem 1 that, if $0 < a < b$, then the work done in stretching a spring from a to b is

$$W = \int_a^b kx\,dx.$$

▶ **EXAMPLE 3** If 5 J work is done in extending a spring 0.2 m beyond its equilibrium position, then how much additional work is required to extend it a further 0.2 m?

Solution We first use the given information to solve for the spring constant:

$$5\text{J} = \int_{0\,\text{m}}^{0.2\,\text{m}} kx\,dx = \frac{1}{2}kx^2\Big|_{x=0\,\text{m}}^{x=0.2\,\text{m}} = 0.02k\,\text{m}^2$$

or

$$k = \frac{5}{0.02}\,\text{J}\cdot\text{m}^{-2} = 250\,\text{N}\cdot\text{m}\cdot\text{m}^{-2} = 250\,\text{N}\cdot\text{m}^{-1}.$$

Finally, the required work W is calculated as follows:

$$W = \int_{0.2\,\text{m}}^{0.4\,\text{m}} kx\,dx = \frac{1}{2}kx^2\Big|_{x=0.2\,\text{m}}^{x=0.4\,\text{m}} = \frac{1}{2}k(0.4^2\text{m}^2 - 0.2^2\text{m}^2) = 0.06k\,\text{m}^2.$$

We obtain our final answer by substituting in the value for the spring constant:

$$W = (0.06)(250\,\mathrm{N}\cdot\mathrm{m}^{-1})\,\mathrm{m}^2 = 15\,\mathrm{N}\cdot\mathrm{m} = 15\,\mathrm{J}. \quad ◀$$

Examples that Involve Pumping a Fluid from a Reservoir

The analysis that results in the definition of work as an integral can be used to solve a variety of problems. Until now, we have considered examples that concern a variable force. In our next application, pumping a fluid to the top of a reservoir, it is the distance that varies because the fluid at each horizontal level must be raised a distance equal to its depth. In the two examples that follow, we will need the weight density of water: 9806.65 newtons per cubic meter in the mks system and 62.428 pounds per cubic foot in the U.S. customary system.

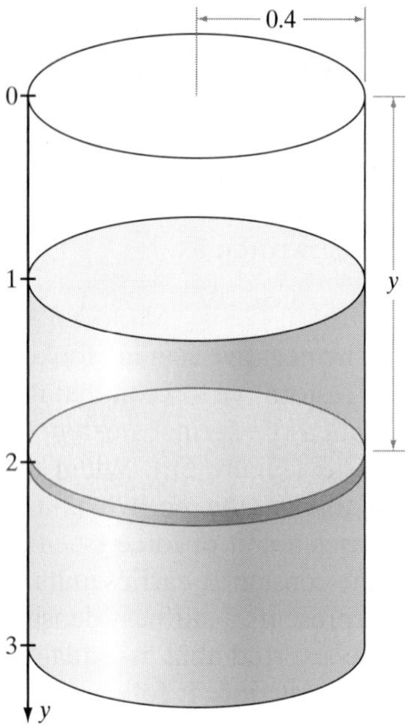

▲ Figure 6

▶ **EXAMPLE 4** A cylindrical sump pit is 3 meters deep; its radius is 0.4 meters. Water has accumulated in the pit; the depth of the water is 2 meters. A sump pump floats on the surface of the water and pumps water to the top of the pit. How much work is done in pumping out half the water?

Solution In this type of problem, it is often convenient to orient the positive y-axis downward, as is indicated in Figure 6. The origin is frequently positioned at the level to which the fluid must be pumped. Figure 6 also illustrates a thin "slice" of water y meters below the top of the pit. If the thickness of the slice is Δy meters, then its volume is $\pi(0.4)^2\Delta y\,\mathrm{m}^3$, and its weight is $\pi(0.4)^2\,\Delta y\,\mathrm{m}^3 \times 9806.65\,\mathrm{N\,m}^{-3} = 1569.06\,\pi\Delta y$ N. The work done to pump this slice y meters to the top is therefore *about* $1569.06\,\pi\,\Delta y$ J by equation (7.5.1). The reason this expression is not exact is that the volume pumped does have some thickness; consequently, not all points are exactly y meters from the top. To approximate the total work done in pumping out half the pit, we form the sum $\sum 1569.06\,\pi\,y\cdot\Delta y$ of the contributions of the slices from $y=1$ to $y=2$. Our approximation improves as Δy tends to 0, and in passing to the limit, we obtain

$$W = \int_1^2 1569.06\,\pi\,y\,dy = 1569.06\,\pi\left.\frac{y^2}{2}\right|_{y=1}^{y=2} = 2353.59\,\pi\,\mathrm{N} \approx 7394\,\mathrm{N}. \quad ◀$$

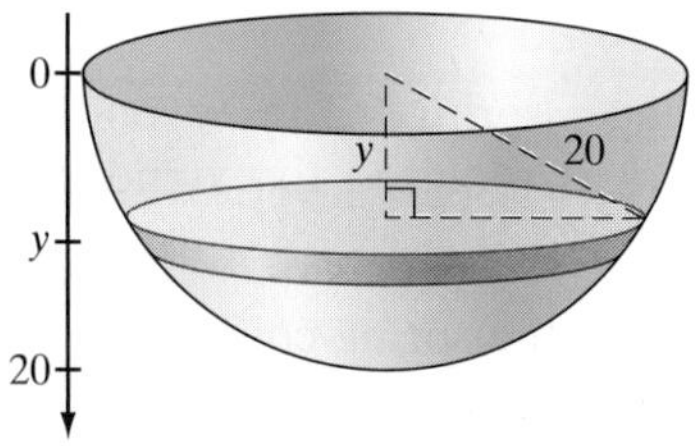

▲ Figure 7

▶ **EXAMPLE 5** A tank full of water is in the shape of a hemisphere of radius 20 feet. A pump floats on the surface of the water and pumps the water from the surface to the top of the tank, where the water just runs off. How much work is done in emptying the tank?

Solution The tank is sketched in Figure 7. As in the preceding example, we have positioned a downward-oriented vertical axis that has the origin at the top of the reservoir. Consider a thin horizontal "slice" at depth y. By solving for the base of the right triangle indicated in Figure 7, we find that the cross-section of the slice is a circle of radius $\sqrt{20^2 - y^2}$. If the thickness of the slice is Δy, then the volume of the slice is $\pi\left(\sqrt{20^2 - y^2}\right)^2\Delta y$ cubic feet. It follows that the weight of the slice is $62.428\cdot\pi(20^2 - y^2)\,\Delta y$ pounds, and the work done in raising this slice to the surface is about $62.428\cdot\pi\,(20^2 - y^2)\,\Delta y\cdot y$ foot-pounds. Passing to an integral as in the preceding example, we see that the total work is

$$\int_0^{20} 62.428\,\pi(20^2 - y^2)y\,dy = 62.428\pi \int_0^{20} (400y - y^3)\,dy$$
$$= 62.428\pi\left(200y^2 - \frac{y^4}{4}\right)\Bigg|_{y=0}^{y=20} = 62.428\pi\left(200 \cdot 20^2 - \frac{20^4}{4}\right),$$

which comes to about 7.84493×10^6 foot-pounds. ◀

QUICK QUIZ

1. To overcome friction, a force of $12 - 3x^2$ N is used to push an object from $x = 0$ to $x = 2$ m. How much work is done?
2. If the amount of work in stretching a spring 0.02 meter beyond its equilibrium position is 8 J, then what force is necessary to maintain the spring at that position?
3. A cube of side length 1 m is filled with a fluid that weighs 1 newton per cubic meter. What work is done in pumping the fluid to the surface?

Answers

1. 16 J 2. 800 N 3. $\frac{1}{2}$ J

EXERCISES

Problems for Practice

1. A steam shovel lifts a 500 pound load of gravel from the ground to a point 80 feet above the ground. The gravel is fine, however, and it leaks from the shovel at the rate of 1 pound per second. If it takes the steam shovel one minute to lift its load at a constant rate, then how much work is performed?

2. On the surface of the earth, a rocket weighs 10,000 newtons. How much work is performed lifting the rocket to a height 100 kilometers above the surface of the earth, assuming that the direction of the force, as well as the motion, is straight up?

3. An object that weighs W_0 at the surface of the earth weighs

$$15697444 \cdot \frac{W_0}{(3962 + y)^2}$$

when it is y miles above the surface of the earth. How much work is done lifting an object from the surface of the earth, where it weighs 100 pounds, straight up to a height 100 miles above Earth?

4. When a mass M measured in slugs is y miles above the surface of the earth, its weight in pounds is

$$502318208 \cdot \frac{M}{(3962 + y)^2}.$$

How much work is done lifting a satellite from the surface of the earth, where it weighs 800 pounds, to an orbit 200 miles above Earth?

5. To lift an object straight up from the surface of the earth to a height 25 km above the surface of the earth requires 58610091 J of work. What is the mass of the object?

6. A 100 kg mass is lifted to a height at which its weight is 98% of its weight at the surface of Earth. How much work is done?

7. A rocket climbs straight up. Its initial total weight, including fuel, is 7000 pounds. Fuel is consumed at the constant rate of 30 pounds per mile. Taking into account the decrease in weight due to fuel consumption but disregarding the decrease in weight due to increasing elevation, approximate the work done in lifting the rocket the first 20 miles into space.

8. A man stands at the top of a tall building and pulls a chain up the side of the building. The chain is 50 feet long and weighs 3 pounds per linear foot. How much work does the man do in pulling the chain to the top?

9. With regard to the preceding exercise, how much work did the man do in pulling up the first 30 feet of the chain to the top of the building?

10. A heavy uniform cable is used to lift a 300 kg load from ground level to the top of a building that is 40 m high. If the cable weighs 220 newtons per linear meter, then how much work is done?

11. With regard to the preceding exercise, how much work is done in lifting the load from ground level to a height 24 meters above ground level?

12. With regard to the cable and load of Exercises 10 and 11, how much work was done in lifting the load from a height

24 meters above ground level to a height of 30 meters above ground level?

13. A heavy uniform cable is used to lift a 300 pound load from ground level to the top of a 100 foot high building. If the cable weighs 20 pounds per linear foot, then how much work is done?

14. With regard to the preceding exercise, how much work is done in lifting the load from ground level to a height 30 feet above ground level?

15. With regard to the cable and load of Exercises 13 and 14, how much work is done in lifting the load from a height 30 feet above ground level to a height 80 feet above ground level?

16. A crane lifts a large boat out of its berth to place it in dry dock. The vessel, including its contents, weighs 80 tons. But as the boat is lifted, it releases water from its holds at the constant rate of 50 cubic feet per minute. The crane raises the boat a distance of 50 feet at the constant rate of 5 feet per minute. How much work is performed in lifting the boat?

17. If a spring with spring constant 8 pounds per inch is stretched 7 inches beyond its equilibrium position, then how much work is done?

18. If a spring with spring constant 20 newtons per centimeter is stretched 12 cm beyond its equilibrium position, then how much work is done?

19. A spring is stretched 10 cm beyond its equilibrium position. If the force required to maintain it in its stretched position is 240 N, then how much work was done?

20. A spring is stretched 2 inches beyond its equilibrium position. If the force required to maintain it in its stretched position is 60 pounds, then how much work was done?

21. A force of 280 pounds compresses a spring 4 inches from its natural length. How much work is done compressing it a further 4 inches?

22. A force of 10^3 newtons compresses a spring 10 cm from its natural length. How much work is done compressing it a further 10 cm?

23. If 40 joules of work are done in stretching a spring 8 cm beyond its natural length, then how much work is done stretching a further 4 cm?

24. If 30 foot-pounds of work are done in stretching a spring 3 inches beyond its natural length, then how much work is done stretching it 1 inch more?

25. The work done in extending a spring at rest 2 inches beyond its equilibrium position is equal to 2/3 ft-lb. How much force is required to maintain the spring in that position?

26. The work done in extending a spring at rest 4 cm beyond its equilibrium position is 1.2 N. What force is required to maintain the spring in that position?

27. A swimming pool full of water has square base of side 15 feet. It is 10 feet deep and is being pumped dry. The pump floats on the surface of the water and pumps the water to the top of the pool, at which point the water runs off. How much work is done in emptying the pool?

28. A tank in the shape of a box has square base of side 5 m. It is 4 m deep and is being pumped dry. The pump floats on the surface of the water and pumps the water to the top of the tank, at which point the water runs off. How much work is done in emptying the tank?

29. A swimming pool has rectangular base with side lengths of 6 m and 10 m. The pool is 3 m deep but not completely filled: the depth of the water it contains is 2 m. A pump floats on the surface of the water and pumps the water to the top of the pool, at which point the water runs off. How much work is done emptying the pool?

30. A swimming pool has rectangular base with side lengths of 16 and 30 feet. The pool is 10 feet deep but not completely filled: the depth of the water it contains is 8 feet. A pump floats on the surface of the water and pumps the water to the top of the pool, at which point the water runs off. How much work does the pump perform in emptying the pool?

31. Referring to the swimming pool of the preceding problem, if power is interrupted when half the water has been pumped, then how much work has been done?

32. A cube-shaped tank has side length 4 m. The depth of water it holds is 3 m. A pump floats on the surface of the water and pumps the water to the top of the tank, at which point the water runs off. How much work is performed removing 32 m^3 of water?

33. A semi-cylindrical tank filled with water is 16 feet long. It has rectangular horizontal cross-sections and vertical cross-sections that are semicircular of radius 6 feet (see Figure 8). A pump floats on the surface of the tank water and pumps water to the top, at which point the water runs off. How much work is done in emptying the tank?

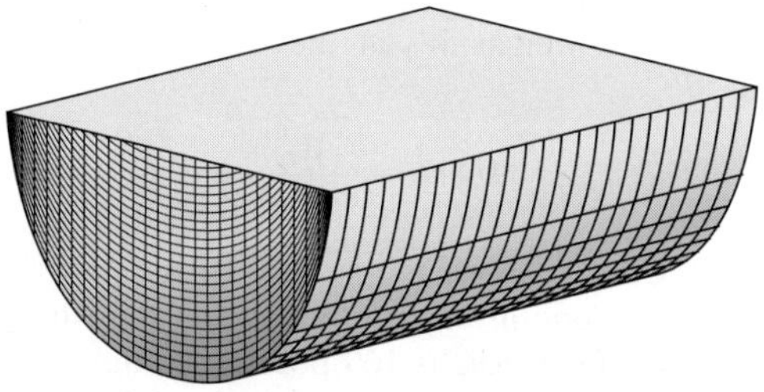

Semi-cylindrical tank

▲ Figure 8

34. A semi-cylindrical tank filled with water is shaped as in Figure 8 but with different dimensions: It is 5 m long, has rectangular horizontal cross-sections, and has vertical cross-sections that are semicircular of radius 2 m. A pump floats on the surface of the water and pumps water to the top, at which point the water runs off. How much work is done in emptying the tank?

35. A filled reservoir is in the shape of a frustum of an inverted cone as shown in Figure 9. The radius of the circular base of the cone is 100 feet. The radius of the circular top of the frustum and the depth at the center are both 50 feet. A pump that floats on the surface of the water pumps water to the top of the reservoir, at which point the water runs off. How much work is done emptying the reservoir?

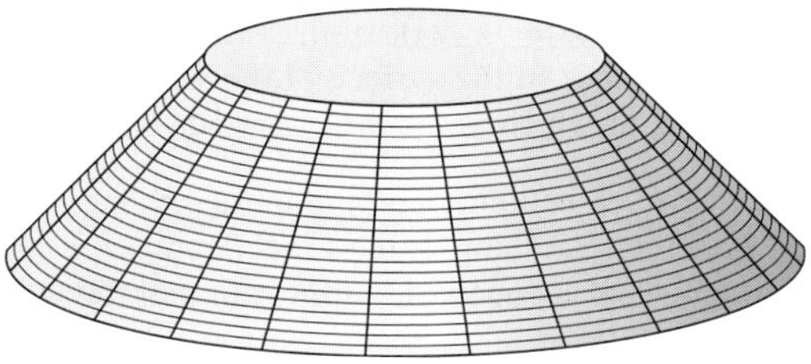

Frustum of an inverted cone

▲ Figure 9

36. A tank filled with water is in the shape of a frustum of pyramid with square base. The base of the pyramid measures 8 m on a side, the side length of the square top is 4 m, and the height of the frustum is 6 m (see Figure 10). A pump floats on the surface of the water and pumps water to the upper edge. If the depth of water in the tank is reduced to 2 m, then what amount of work has been performed?

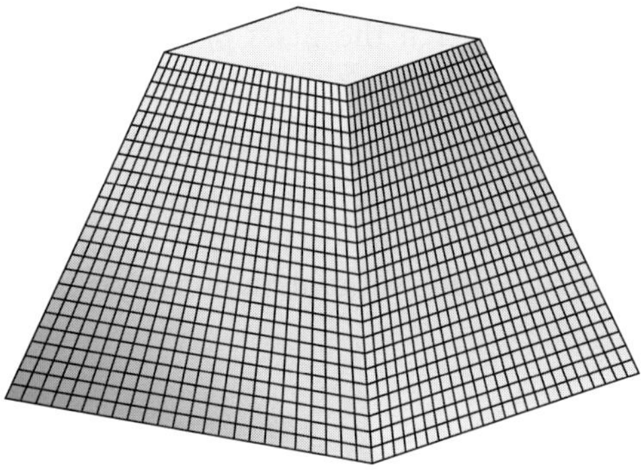

Frustum of a pyramid

▲ Figure 10

Further Theory and Practice

37. When a mass M measured in slugs is y feet above the surface of the earth, its weight in pounds is

$$1.4077 \times 10^{16} \cdot \frac{M}{(2.0917 \times 10^7 + y)^2}.$$

(One pound of force accelerates one slug one foot per second squared.) How much work is done lifting a satellite from the surface of the earth, where it weighs 800 pounds, to an orbit 200 miles above Earth?

38. The work done lifting a 1 kg mass straight up from the surface of the earth is 1202260 J. What height above the surface of the earth does the mass reach?

39. A spring is stretched a certain length beyond its natural length. What percentage of the total amount of work is expended by the first half of the stretch?

40. A stonemason carries a 50-pound sack of mortar 120 feet straight up a ladder at the rate of 20 feet per minute. The sack leaks at the rate of 2 pounds per minute. When the mason is 60 feet off the ground, he shifts the sack and accidentally enlarges the hole. Now the sack leaks at the rate of 3 pounds per minute, but he continues on up at the same rate. If the mason himself weighs 180 pounds, how much work does he perform climbing the ladder with the sack?

41. Suppose a cylinder with cross-sectional area A in^2 has one fixed end (the cylinder head). Suppose that a movable piston closes the other end a variable distance x inches away from the cylinder head. The pressure (force per area) p of gas confined in the cylinder is a continuous function of volume: $p = p(v)$. If $0 < x_1 < x_2$, show that the work done by the piston in compressing the gas from a volume Ax_2 to Ax_1 is

$$W = A \int_{x_1}^{x_2} p(Ax)dx.$$

42. Suppose that a cylinder with cross-sectional area 2 square inches contains 20 cubic inches of a gas under 50 pounds per square inch pressure. If the expression $pv^{1.4}$ is constant as the piston moves, then how much work does a piston do in compressing the gas to 4 cubic inches? (Refer to Exercise 4.1.)

▶ A tank has the shape of a paraboloid of revolution that results from rotating about the y-axis the region that is bounded above by the horizontal line $y = 18$, on the left by the y-axis, and on the right by the graph of $y = 2x^2$ (x and y measured in feet). In Exercises 43–46, a pump floats on the surface of the water and pumps the water to the top of the tank. Calculate the work that is done in performing the task that is described. ◀

43. The tank, initially filled, is pumped until the remaining water is 3 feet deep at the center.

44. The tank, initially filled, is pumped until the remaining water has half the original volume.

45. The tank, initially filled to a depth of 4 feet at the center, is pumped until it is empty.

46. The tank, initially filled to a depth of 4 feet at the center, is pumped until the remaining water is 2 feet deep at the center.

47. A force $F(x) > 0$ causes a mass m, initially at rest at $x = 0$, to move along the x-axis with velocity $v(x)$. Let $W(b)$ be the work done in moving the body from $x = 0$ to $x = b$. Show that $W(b) = mv(b)^2/2$. In other words, the work done is equal to the gain in kinetic energy. *Hint:* Start from Newton's Law, which involves the derivative of velocity with respect to time. Use the Chain Rule to calculate the derivative of v with respect to x.

48. The gas tank of a tractor is in the shape of a cylinder lying on its side. The tank is positioned on the tractor so that its central axis is 5 feet above the ground. The radius of the tank is 1.5 feet, and its length is 5 feet. For winter storage of the tractor, gas is pumped straight up out of the tank, from the surface of the gasoline into a holding tank with opening 15 feet from ground level. If gasoline weighs 35 pounds per cubic foot, then how much work is done in emptying a full tank of gas on the tractor into the holding tank?

49. The magnitude of a force that stretches a spring is given by $F(x) = kx$ when the spring is extended a distance x beyond its equilibrium position. Let $W(x)$ denote the work done in stretching the spring a distance x from its equilibrium position. Describe the curve that results from plotting the parametric equations $F = F(x)$, $W = W(x)$ in the FW-plane.

50. An object sits on a flat surface that presents an irregular frictional force. The object is pushed forward. In pushing it x units from its rest position, the work done is $x + \arctan(x)$. Describe the pushing force as a function of x.

Calculator/Computer Exercises

Exercises 51–54 are concerned with a tank filled to the top with water. Its shape is obtained by rotating about the y-axis the region that is bounded above by the horizontal line $y = 10(1-1/e^2)$, on the left by the y-axis, and on the right by the graph of $y = 10(1 - \exp(-x^2/2))$. Both x and y are measured in feet. A pump floats on the surface of the water and pumps water to the top, at which point the water runs off.

51. How much work is done pumping out a volume of water that leaves the remaining water 4 feet deep at the center?

52. How much work is done in pumping out half the water in the tank?

53. How much work is done in pumping out all the water in the tank?

54. In pumping out all the water in the tank, how deep is the remaining water at the center when half the work is done?

7.6 First Order Differential Equations–Separable Equations

A *differential equation* is an equation that involves one or more derivatives of an unknown function. For example, the rate at which the mass m of a radioactive substance decreases at time t is proportional to the value of m at t. That is, m satisfies the differential equation $\frac{dm}{dt} = -\lambda m$ where λ is a positive constant. This differential equation is a *first order differential equation* because the first derivative is the highest order derivative of the unknown function that appears in the equation. In this section, we will study first order differential equations that have the form

$$\frac{dy}{dx} = F(x, y). \tag{7.6.1}$$

Here $F(x, y)$ is an expression that involves both x and y (in general). For example, $F(x, y)$ could be $2 - xy^3$. It is also possible for the expression $F(x, y)$ to depend on only one of the variables. Thus if $F(x, y) = 1/x$, then equation (7.6.1) becomes $\frac{dy}{dx} = \frac{1}{x}$. Similarly, if $F(x, y) = 2y(10 - y)$, then equation (7.6.1) becomes $\frac{dy}{dx} = 2y(10 - y)$.

Solutions of Differential Equations

Solving an algebraic equation generally consists of finding the finite collection of *numbers* that satisfy the equation. By contrast, to solve a differential equation, we must find the collection of *functions* that satisfy the equation.

> **DEFINITION** We say that a differentiable function y is a *solution* of differential equation (7.6.1) if $y'(x) = F(x, y(x))$ for every x in some open interval. The graph of a solution is said to be a *solution curve* of the differential equation.

Our first example shows that there can be infinitely many different functions that satisfy a given differential equation.

▶ **EXAMPLE 1** Let C denote an arbitrary constant. Verify that the function $y(x) = x + Ce^{-x} - 1$ is a solution of the differential equation $\dfrac{dy}{dx} = x - y$.

Solution We calculate the left side of the differential equation,

$$\frac{dy}{dx} = \frac{d}{dx}(x + Ce^{-x} - 1) = 1 - Ce^{-x},$$

and the right side,

$$x - y = x - (x + Ce^{-x} - 1) = 1 - Ce^{-x}.$$

Because these expressions are equal, it follows that the function $y(x) = x + Ce^{-x} - 1$ satisfies the given differential equation. Notice that this verification does not show how the solution $y(x) = x + Ce^{-x} - 1$ is *found*. The technique for discovering this solution is the subject of Section 7.7. ◀

Slope Fields

We can represent equation (7.6.1) graphically by interpreting dy/dx as the slope of a tangent line to a solution curve. To implement this observation, we select a grid of points in the plane, and, for every point (x, y) in the grid, we draw a short line segment with slope $F(x, y)$ and initial point (x, y). The lengths of the line segments have no significance. They are usually drawn so that they have the same length, which is chosen to be short enough relative to the chosen grid so that different line segments do not intersect. The collection of line segments drawn in this way is called a *slope field*.

▶ **EXAMPLE 2** Sketch a slope field for the differential equation, $\dfrac{dy}{dx} = x - y$, of Example 1.

Solution For this differential equation, the right side of equation (7.6.1) is $F(x, y) = x - y$. For points (x, y) on the x-axis, we have $y = 0$ and $F(x, 0) = x$. This tells us that the line segments with initial points along the x-axis have slopes that increase as x increases, from negative values when $x < 0$ to positive values when $x > 0$. Along the line $y = x$, we have $F(x, x) = 0$ for every x. As a result, the line segments with initial points on the line $y = x$ are all horizontal. In Figure 1, we start our sketch of the slope field by drawing the two families of line segments that we have described. The other line segments in the slope field are determined in a similar manner. For example, the line segments with initial points on the line $y = x - 1$ all have slope 1 because $F(x, x - 1) = x - (x - 1) = 1$. After we fill in the slopes for the rest of the grid, we obtain our final sketch in Figure 2. ◀

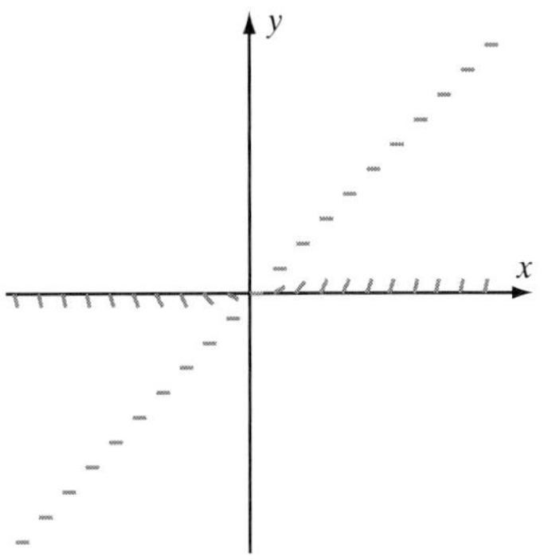

▲ Figure 1

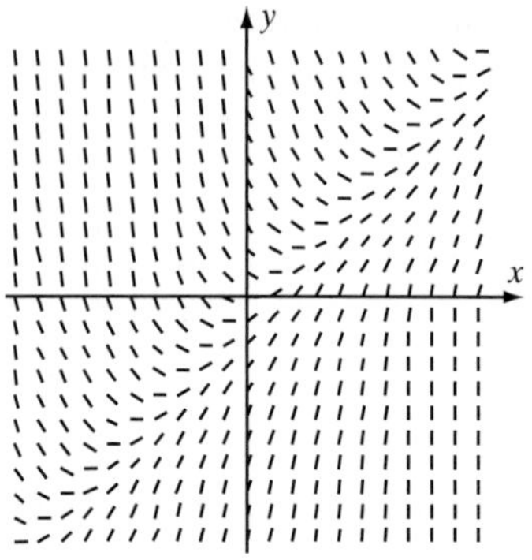

▲ Figure 2

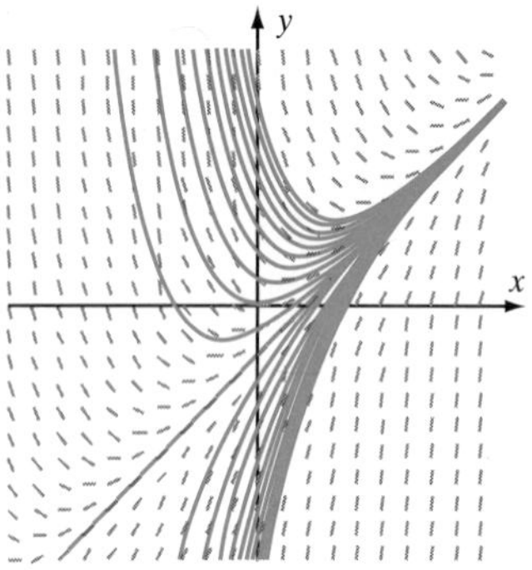

▲ Figure 3

If a solution curve $y(x)$ to equation (7.6.1) is drawn in a slope field and passes through the point (x_0, y_0), then $y(x_0) = y_0$ and $y'(x_0) = F(x_0, y_0)$. In other words, the line segment with initial point (x_0, y_0) is tangent to the solution curve. With a little practice, we can use the slope field to predict the shape of the solution curves to a differential equation without actually finding an analytic solution to the equation.

▶ **EXAMPLE 3** Superimpose the plots of several solution curves of the differential equation, $\frac{dy}{dx} = x - y$, from Example 1 on a slope field plot for this differential equation.

Solution We verified in Example 1 that the equation $y = x + Ce^{-x} - 1$ is a solution curve for any constant C. The graphs of these equations for $C = -5, -9/2, -4, \ldots, 9/2, 5$ have been superimposed on the slope field of Figure 2, resulting in Figure 3. ◀

INSIGHT An experienced eye can often use a slope field to detect the behavior of a solution to a differential equation (even when an explicit formula for the solution has not been found). For example, the "flow" toward $y = x$ of the line segments in Figure 2 suggests that $\lim_{x\to\infty} \frac{y(x)}{x} = 1$ for every solution $y(x)$ of $\frac{dy}{dx} = x - y$. The solution curves in Figure 3 provide supporting evidence for this inference. A rigorous demonstration can be obtained from the explicit formula for the solution curves:

$$\lim_{x\to\infty} \frac{y(x)}{x} = \lim_{x\to\infty} \frac{x + Ce^{-x} - 1}{x} = 1 + \lim_{x\to\infty} \frac{Ce^{-x}}{x} - \lim_{x\to\infty} \frac{1}{x} = 1 + 0 + 0 = 1.$$

Initial Value Problems

When differential equations arise in practice, we often know the value y_0 of the unknown function $y(x)$ at one particular value of x_0. For example, if x represents time, and y represents the depth of water in a tank, then a leak may develop that leads to the differential equation $\frac{dy}{dx} = -\sqrt{y}$. If the tank has height y_0 and is initially filled, and if the leak begins at time $x = x_0$, then we know that $y(x_0) = y_0$. Such an equation is called an *initial condition*.

DEFINITION If x_0 and y_0 are specified values, then the pair of equations

$$\frac{dy}{dx} = F(x, y) \quad \text{(differential equation)} \quad , \quad y(x_0) = y_0 \quad \text{(initial condition)} \quad (7.6.2)$$

is said to be an *initial value problem* (IVP). We say that a differentiable function y is a *solution of initial value problem* (7.6.2) if $y(x_0) = y_0$ and $y'(x) = F(x, y(x))$ for all x in some open interval containing x_0.

It can be shown that, under mild restrictions on F (satisfied by all examples in this book), the initial value problem (7.6.2) has a unique solution.

▶ **EXAMPLE 4** Use the computation of Example 1 to solve the initial value problem $\frac{dy}{dx} = x - y, y(0) = 2$.

Solution In Example 1, we verified that $y(x) = x + Ce^{-x} - 1$ satisfies the given differential equation for every constant C. Notice that $y(0) = 0 + Ce^{-0} - 1 = C - 1$. Therefore to satisfy the initial condition $y(0) = 2$, we set $C - 1 = 2$ and find that $C = 3$. Thus $y(x) = x + 3e^{-x} - 1$ solves the given initial value problem. ◀

Separable Equations

There is no single technique by which equation (7.6.1) can be solved for every $F(x, y)$. Several methods have been developed to handle special cases according to the form of the expression $F(x, y)$. In this section, we will study the case in which $F(x, y) = g(x)\,h(y)$. The expressions

$$F(x,y) = 4\cos(x) = \underbrace{4\cos(x)}_{g(x)} \cdot \underbrace{1}_{h(y)}, \qquad F(x,y) = 7y^3 = \underbrace{7}_{g(x)} \cdot \underbrace{y^3}_{h(y)},$$

$$F(x,y) = \frac{2+x}{1+y^2} = \underbrace{(2+x)}_{g(x)} \cdot \underbrace{\left(\frac{1}{1+y^2}\right)}_{h(y)}$$

are all of this type. When $F(x, y)$ factors as $g(x)\,h(y)$, the differential equation

$$\frac{dy}{dx} = F(x,y) = g(x)\,h(y) \tag{7.6.3}$$

is said to be *separable* because the expressions involving x may be placed on one side of the equation and the terms involving y may be moved to the other side of the equation. To carry out this separation of variables, we rewrite equation (7.6.3) as

$$\frac{1}{h(y)}dy = g(x)dx.$$

This equation is only suggestive, but it does hint at the answer,

$$\int \frac{1}{h(y)}dy = \int g(x)dx.$$

Let us see how we can obtain this result more carefully. Assuming that $h(y) \neq 0$, we rewrite equation (7.6.3) as

$$\frac{1}{h(y)}\frac{dy}{dx} = g(x)$$

and integrate with respect to x to obtain

$$\int \frac{1}{h(y)}\frac{dy}{dx}dx = \int g(x)dx. \tag{7.6.4}$$

Let $H(u)$ be an antiderivative of $1/h(u)$, and let $G(x)$ be an antiderivative of $g(x)$. Using the Chain Rule, we have

$$\frac{d}{dx}H(y) = H'(y)\frac{dy}{dx} = \frac{1}{h(y)}\frac{dy}{dx}.$$

Therefore equation (7.6.4) can be stated as

$$H(y) = G(x) + C. \tag{7.6.5}$$

(There is no need to add a constant of integration to each side; if that were done, then the two constants could be combined into one.) This procedure for solving a differential equation is called the *method of separation of variables*. When we use this method, we will generally simplify the notation and write

$$\int \frac{1}{h(y)}\,dy \quad \text{for} \quad \int \frac{1}{h(y)}\frac{dy}{dx}\,dx.$$

▶ **EXAMPLE 5** Solve the initial value problem

$$\frac{dy}{dx} = \frac{x}{y^2+1}, \quad y(0) = 3.$$

Solution The given differential equation is separable because it may be written as $\frac{dy}{dx} = g(x)\,h(y)$ with $g(x) = x$ and $h(y) = \frac{1}{(y^2+1)}$. Following the method of solution that we have just discussed, we separate variables:

$$\int (y^2+1)\underbrace{\frac{dy}{dx}\,dx}_{\text{treat as } dy} = \int x\,dx,$$

or

$$\frac{1}{3}y^3 + y = \frac{1}{2}x^2 + C.$$

If this equation is to satisfy the given initial condition $y(0) = 3$, then C must satisfy

$$\frac{1}{3}(3)^3 + (3) = \frac{1}{2}(0)^2 + C$$

or $C = 12$. Thus $\frac{1}{3}y^3 + y = \frac{1}{2}x^2 + 12$ is the solution of the given initial value problem. The solution curve is shown in Figure 4. ◀

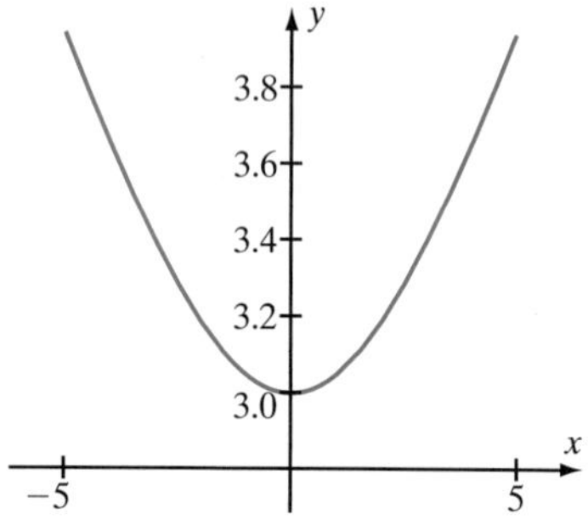

▲ Figure 4 The graph of $\frac{1}{3}y^3 + y = \frac{1}{2}x^2 + 12$

INSIGHT As can be seen in Example 5, the method of separation of variables does not generally yield a solution $y(x)$ of the differential equation $dy/dx = g(x)h(y)$ in *explicit* form. That is, the method of separation of variables does not ordinarily lead to an equation of the form $y(x) =$ (expression in x). Instead, it usually results in an equation of the form

$$(\text{expression in } y) = (\text{expression in } x).$$

Such an equation defines y *implicitly* as a function of x. It is possible to show that the explicit solution to the initial value problem of Example 5 is $y(x) = u(x) - 1/u(x)$ where

$$u(x) = \frac{1}{2}\left(144 + 6x^2 + 2\sqrt{5200 + 432x^2 + 9x^4}\right)^{1/3}.$$

The explicit formula for $y(x)$ is so complicated that it is a chore to even verify that it is correct. By contrast, it is easy to implicitly differentiate the equation $\frac{1}{3}y^3 + y = \frac{1}{2}x^2 + 12$ and verify that the implicitly defined function y does solve the given differential equation.

Equations of the Form $dy/dx = g(x)$

The differential equation $dy/dx = g(x)$ is nothing more than the statement "$y(x)$ is an antiderivative of $g(x)$." Theorem 1 provides the solution of the corresponding initial value problem.

THEOREM 1 If g is a continuous function on an open interval containing a, then the initial value problem

$$\frac{dy}{dx} = g(x), \quad y(a) = b$$

has a unique solution:

$$y(x) = b + \int_a^x g(t)dt. \tag{7.6.6}$$

Proof. According to the Fundamental Theorem of Calculus, $\int_a^x g(t)dt$ is an antiderivative of g. All other antiderivatives of g have the form $\int_a^x g(t)dt + C$. Thus $y(x) = \int_a^x g(t)dt + C$ for some constant C. We use the initial condition to solve for C,

$$b = y(a) = \int_a^a g(t)dt + C = 0 + C.$$

We obtain equation (7.6.6) by substituting $C = b$ in our formula for $y(x)$. ■

Many useful functions in science and engineering are defined by equation (7.6.6) for particular choices of g: the Fresnel functions

$$\text{FresnelC}(x) = \int_0^x \cos\left(\frac{\pi}{2}t^2\right)dt \quad \text{and} \quad \text{FresnelS}(x) = \int_0^x \sin\left(\frac{\pi}{2}t^2\right)dt$$

in optics, the Debye function

$$x \mapsto \int_0^x \left(\frac{t}{e^t - 1}\right)^2 e^t dt$$

in thermodynamics, and so on. Although the function y defined by equation (7.6.6) might seem complicated or abstract at first, remember that the way it is defined allows us to instantly calculate its derivative. By now we have accumulated a good deal of experience using the derivative $y'(x)$ to deduce properties of $y(x)$. It is through those properties that the function y becomes familiar to us. If you review Section 5.5 with Theorem 1 in mind, you will see that all the properties of the logarithm and exponential functions were developed by defining $y(x) = \ln(x)$ to be the solution of the initial value problem $\frac{dy}{dx} = \frac{1}{x}, y(1) = 0.$

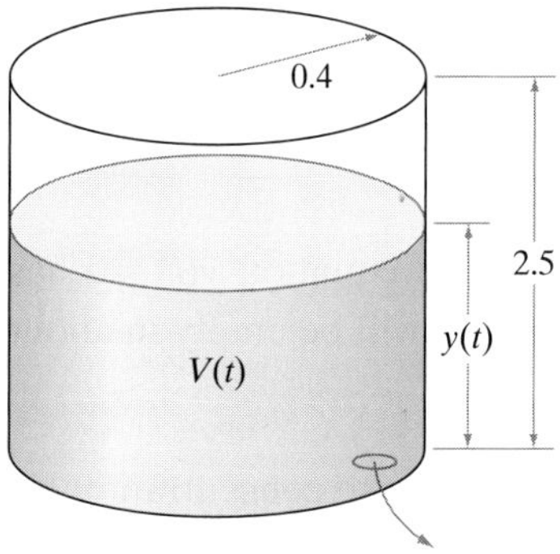

▲ Figure 5 $\frac{dV}{dt} = -2.6 \cdot A \cdot \sqrt{y}$

Examples from the Physical Sciences

In the physical sciences, often the most natural way to describe the relationship between two variables is by means of a differential equation. The differential equation is then solved to determine explicitly the relationship between the variables. The next three examples illustrate these ideas.

▶ **EXAMPLE 6** According to Torricelli's Law, the rate at which water flows out of a tank from a small hole in the bottom is proportional to the area A of the hole and to the square root of the height $y(t)$ of the water in the tank (Figure 5). That is, there is a positive constant k such that

$$\frac{d}{dt}V(t) = -k \cdot A \cdot \sqrt{y(t)},$$

where $V(t)$ is the volume of water in the tank at time t. Consider a cylindrical tank of height 2.5 m and radius 0.4 m that has a hole 2 cm in diameter in its bottom. Suppose that the tank is full at time $t = 0$. If $k = 2.6\ \text{m}^{1/2}/\text{s}$, find the height $y(t)$ of water as a function of time. How long will it take for the tank to empty?

Solution The volume $V(t)$ of water in the tank at time t is related to the height $y(t)$ of the water by the equation $V(t) = \pi(0.4)^2 y(t)$. Because $A = \pi(0.01)^2\ \text{m}^2$, Torricelli's Law tells us that

$$\pi(0.4)^2 \frac{dy}{dt} = -2.6 \cdot \pi(0.01)^2 \cdot \sqrt{y},$$

or

$$\frac{dy}{dt} = -2.6 \frac{\pi(0.01)^2}{\pi(0.4)^2} \sqrt{y} = -0.001625\sqrt{y}.$$

On separating variables, we have

$$\int y^{-1/2}\, dy = \int (-0.001625)\, dt + C,$$

or

$$2y(t)^{1/2} = -0.001625t + C.$$

When we substitute $t = 0$ into this equation, we obtain $C = 2\sqrt{y(0)}$. But $y(0) = 2.5$ because the tank is initially filled to its height of 2.5 meters. Therefore $C = 2\sqrt{2.5}$ and $2y(t)^{1/2} = -0.001625t + 2\sqrt{2.5}$, or

$$y(t) = \left(\sqrt{2.5} - \frac{0.001625}{2} t\right)^2.$$

This is the height of the water in the tank at time t. The equation $y(t) = 0$ has solution $t = 2\sqrt{2.5}/0.001625 \approx 1946$ seconds. Thus the tank will be empty in about $1946/60 \approx 32.4$ minutes. ◀

▶ **EXAMPLE 7** In chemical kinetics, the rate at which the concentration $[S]$ of a substance S changes is often described by means of a differential equation. For example, the dimerization of butadiene

$$C_4H_6 \rightarrow \frac{1}{2} C_8H_{12}$$

has reaction rate

$$\frac{d[C_4H_6]}{dt} = -k[C_4H_6]^2,$$

where k is a positive constant. If $[C_4H_6] = 1/60$ moles/L when $t = 0$ seconds, and $[C_4H_6] = 1/140$ moles/L when $t = 6000$ seconds, then determine $[C_4H_6]$ as a function of time. When is $[C_4H_6] = 1/200$?

Solution Let $y = [C_4H_6]$. We solve as follows

$$\frac{dy}{dt} = -ky^2 \Rightarrow \int \frac{1}{y^2}\, dy = -\int k\, dt \Rightarrow -\frac{1}{y(t)} = -kt + C,$$

where C is the constant of integration. We can determine the two constants C and k from the two given observations: $y(0) = 1/60$ and $y(6000) = 1/140$. The first data point yields

$$-\frac{1}{1/60} = -k \cdot 0 + C$$

or $C = -60$. The second data point yields

$$-\frac{1}{1/140} = -k \cdot 6000 - 60$$

or $k = 1/75$. Thus $-1/y(t) = -t/75 - 60$. Solving for $y(t)$, we obtain $y(t) = 75/(t + 4500)$. Therefore $y(10500) = 75/(10500 + 4500) = 1/200$ moles/L. ◄

► **EXAMPLE 8** Let $y(t)$ and $v(t) = dy/dt$ be the height and velocity of a projectile that has been shot straight up from the surface of Earth with initial velocity v_0. Newton's Law of Gravitation states that

$$\frac{dv}{dt} = -\frac{gR^2}{(R+y)^2},$$

where R is the radius of Earth, and g is the acceleration due to gravity at Earth's surface. Supposing that $v_0 < \sqrt{2gR}$, what is the maximum height attained by the projectile?

Solution At the instant the projectile reaches its maximum height and begins to fall back to Earth, its velocity v is 0. Therefore our strategy is to find v as a function of y and solve for the value of y for which $v = 0$. As a first step, we use the Chain Rule to express dv/dt in terms of dv/dy:

$$\frac{dv}{dt} = \frac{dv}{dy} \cdot \frac{dy}{dt} = \frac{dv}{dy} \cdot v.$$

By equating this expression for dv/dt with the one given by Newton's Law of Gravitation, we obtain

$$v \cdot \frac{dv}{dy} = -\frac{gR^2}{(R+y)^2}.$$

This differential equation is separable. Following the general procedure, we obtain

$$\int v\, dv = \int \left(-\frac{gR^2}{(R+y)^2}\right) dy$$

or

$$\frac{1}{2}v^2 = gR^2(R+y)^{-1} + C.$$

When $y = 0$, we have $v = v_0$. Therefore

$$\frac{1}{2}v_0^2 = gR^2(R+0)^{-1} + C,$$

or $C = \frac{1}{2}v_0^2 - gR$. It follows that

$$\frac{1}{2}v^2 = gR^2(R+y)^{-1} + \left(\frac{1}{2}v_0^2 - gR\right). \tag{7.6.7}$$

Now we set $v=0$ in equation (7.6.7) and solve for y. We find that the maximum height is $v_0^2 R/(2gR - v_0^2)$. ◄

Logistic Growth

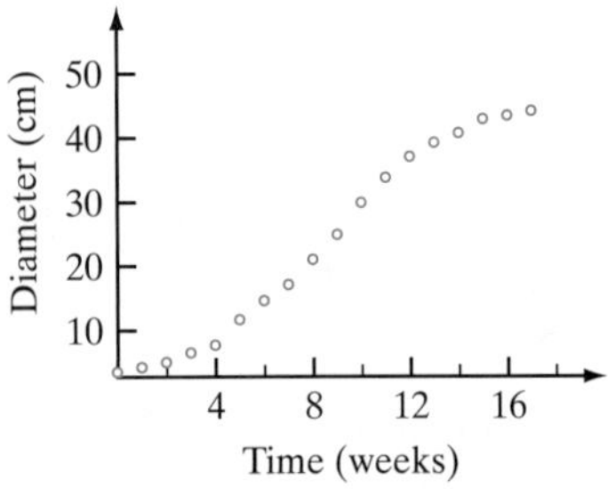

▲ Figure 6 Pumpkin growth

Figure 6 shows the growth of a pumpkin in a controlled agricultural experiment. At first the growth of the pumpkin appears to be exponential. But the growth plot eventually becomes concave down and can no longer be modeled by exponential growth. To obtain a more accurate model, scientists often make the following growth assumptions:

a. There is a size P_∞ that cannot be exceeded. In population models, the constant P_∞ is called the *carrying capacity*;
b. $P'(t) \to 0$ as $P(t) \to P_\infty$. In words, as the function $P(t)$ approaches the carrying capacity P_∞, its rate of growth decreases to 0.

One way to incorporate these assumptions into a growth model is to assume that the rate of population increase is jointly proportional to the population $P(t)$ and to $P_\infty - P(t)$. The resulting differential equation

$$\frac{dP}{dt} = k \cdot P(t) \cdot \big(P_\infty - P(t)\big), \qquad P(0) = P_0 \tag{7.6.8}$$

is called the *logistic growth equation*. In this differential equation k, P_0, and P_∞ represent fixed positive constants. In our discussion, we will assume that $P_0 < P_\infty$. Using the method of separation of variables, we can rewrite equation (7.6.8) as

$$\int \frac{1}{P \cdot (P_\infty - P)} \frac{dP}{dt}\, dt = \int k\, dt,$$

or

$$\int \frac{1}{P \cdot (P_\infty - P)}\, dP = \int k\, dt. \tag{7.6.9}$$

▶ **EXAMPLE 9** Use the method of partial fractions to solve equation (7.6.9).

Solution We apply partial fractions and write

$$\frac{1}{P \cdot (P_\infty - P)} = \frac{A}{P} + \frac{B}{P_\infty - P}.$$

Before we go any further, it is a good idea to make sure that we understand the role of each quantity in this equation. Here P_∞ is a fixed constant that we regard as known, even though we are not given a specific numerical value. The quantity P is a variable. We are to find values for the unknown constants A and B so that the equation is valid for all values of P. Putting the right side over a common denominator and equating numerators leads to the equation

$$1 = A \cdot (P_\infty - P) + B \cdot P.$$

If we let P be 0, then we find that $A = 1/P_\infty$. Letting P be P_∞ results in $B = 1/P_\infty$. With these values, equation (7.6.9) becomes

$$\int \frac{1}{P_\infty} \cdot \left(\frac{1}{P} + \frac{1}{P_\infty - P}\right) dt = \int k\, dt,$$

or

$$\frac{1}{P_\infty}\int \frac{1}{P}\,dt + \frac{1}{P_\infty}\int \frac{1}{P_\infty - P}\,dt = kt + C,$$

or

$$\frac{1}{P_\infty}\cdot\Big(\ln(P) - \ln(P_\infty - P)\Big) = kt + C.$$

We have omitted absolute values in the logarithms because $P(t)$ is in the range $0 < P(t) < P_\infty$. We can combine the logarithms to obtain

$$\ln\left(\frac{P(t)}{P_\infty - P(t)}\right) = P_\infty \cdot (kt + C)$$

or, on exponentiating and letting $A = \exp(P_\infty \cdot C)$,

$$\frac{P(t)}{P_\infty - P(t)} = \exp(P_\infty \cdot kt + P_\infty \cdot C) = \exp(P_\infty \cdot kt)\exp(P_\infty \cdot C) = A \cdot \exp(P_\infty \cdot kt).$$

We may solve for the constant A by substituting $t = 0$:

$$\frac{P_0}{P_\infty - P_0} = \frac{P(0)}{P_\infty - P(0)} = A.$$

Therefore

$$\frac{P(t)}{P_\infty - P(t)} = \frac{P_0}{P_\infty - P_0}\cdot\exp(P_\infty \cdot kt).$$

Isolating $P(t)$ is elementary but somewhat tedious. The result is

$$P(t) = \frac{P_0 \cdot P_\infty}{P_0 + (P_\infty - P_0)\exp(-k \cdot P_\infty \cdot t)}. \qquad (7.6.10) \blacktriangleleft$$

The function P defined by equation (7.6.10) is said to be a *logistic growth function*. The term *sigmoidal growth* is also used in biology. Notice that there are three positive constants that determine logistic growth: the proportionality constant k, the initial value P_0 of P, and the carrying capacity P_∞. The graph of $P(t)$

a. is everywhere rising,

b. is concave up for $P_0 \le P < \frac{1}{2}P_\infty$

c. is concave down for $\frac{1}{2}P_\infty < P < P_\infty$, and

d. has $y = P_\infty$ as a horizontal asymptote.

We can deduce property a by noticing that, because $P_\infty - P_0$, k, and P_∞ are all positive, the expression $(P_\infty - P_0)\exp(-k \cdot P_\infty \cdot t)$ is decreasing. Because this expression appears in the denominator of formula (7.6.10) for $P(t)$, we conclude that $P(t)$ is increasing. Property d can be expressed as

$$\lim_{t\to\infty} P(t) = P_\infty,$$

which is the reason for the subscript ∞. This limiting value for $P(t)$ is an immediate consequence of formula (7.6.10):

$$\lim_{t\to\infty} P(t) = \lim_{t\to\infty} \frac{P_0 \cdot P_\infty}{P_0 + (P_\infty - P_0)\exp(-k \cdot P_\infty \cdot t)} = \lim_{t\to\infty} \frac{P_0 \cdot P_\infty}{P_0 + (P_\infty - P_0)\cdot 0} = P_\infty.$$

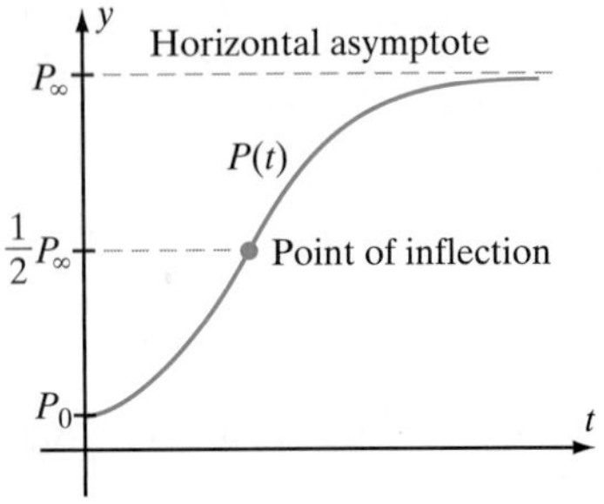

▲ Figure 7 Logistic growth curve

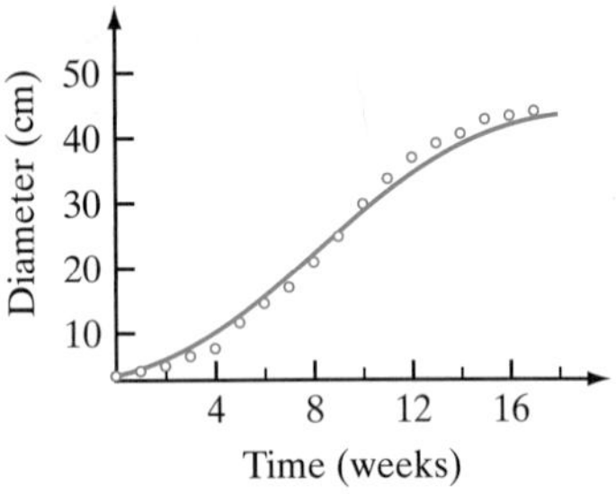

▲ Figure 8 Pumpkin growth

A proof of properties b and c is outlined in Exercise 55. A typical logistic curve is plotted in Figure 7. In Figure 8, we have replotted the pumpkin-growth data, adding the graph of the logistic function determined by $P_0 = 3.6$, $P_\infty = 46$, and $k = 0.0065$.

▶ **EXAMPLE 10** Suppose that 100 children attend a day-care center. Let $P(t)$ be the fraction of the day-care population with a viral infection at time t (measured in days). Suppose that, at 9 AM on Monday ($t = 0$), only 1 child was infected. Suppose that, at 9 AM on Tuesday ($t = 1$), there were 3 infected children (including the original child). Assume that the rate of change of P is jointly proportional to P (because P represents the fraction of the population capable of spreading the infection) and $1 - P$ (because $1 - P$ represents the fraction of the population that can still be infected). By when did the infection spread to half the day-care population?

Solution Because the rate of change of P is jointly proportional to P and $1 - P$, we have

$$\frac{dP}{dt} = k \cdot P(t) \cdot (1 - P(t)),$$

for some proportionality constant k. Because 1 child out of 100 was infected at time $t = 0$, we have $P(0) = 1/100$. Thus the initial value problem for $P(t)$ is

$$\frac{dP}{dt} = k \cdot P(t) \cdot (1 - P(t)) \quad , \quad P(0) = 1/100.$$

Notice that this initial value problem is logistic growth equation (7.6.8) with $P_\infty = 1$ and $P_0 = 1/100$. The solution is given by formula (7.6.10):

$$P(t) = \frac{1/100}{1/100 + (1 - 1/100)\exp(-k \cdot 1 \cdot t)} = \frac{1}{1 + 99e^{-kt}}. \tag{7.6.11}$$

We may now use the observation $P(1) = 3/100$ to solve for k. Because formula (7.6.11) gives us $P(1) = 1/(1 + 99e^{-k})$, we obtain

$$\frac{3}{100} = \frac{1}{1 + 99e^{-k}}, \quad \text{or} \quad 1 + 99e^{-k} = \frac{100}{3}, \quad \text{or} \quad e^{-k} = \frac{1}{99}\left(\frac{100}{3} - 1\right) = \frac{97}{297}.$$

By applying the natural logarithm to each side of the last equation, we see that $-k = \ln(97/297)$. Our final expression for P is

$$P(t) = \frac{1}{1 + 99e^{t\ln(97/297)}} = \frac{1}{1 + 99\exp\left(\ln\left((97/297)^t\right)\right)} = \frac{1}{1 + 99 \cdot (97/297)^t}.$$

The solution to our problem is found by solving the equation

$$\frac{1}{1 + 99 \cdot (97/297)^t} = \frac{1}{2}$$

for t. After some simplification, we obtain $99 = (297/97)^t$ or

$$t = \frac{\ln(99)}{\ln(297/97)} \approx 4.1.$$

According to our model, about 4.1 days after the first child was infected, that is to say, by about 11:30 Friday morning, the fiftieth child became infected. ◀

QUICK QUIZ

1. To which of the following differential equations can we apply the method of separation of variables: a) $dy/dx = \exp(xy)$, b) $dy/dx = \exp(x+y)$, c) $dy/dx = \ln(x^y)$, or d) $dy/dx = \ln(x+y)$?
2. Solve $4\,dy/dx = 3x^2/(y^3+y+5)$.
3. Solve the initial value problem $dy/dx = x/y, y(2) = 3$.
4. If $dy/dx = \sin(x^3)$ and $y(\pi) = 2$, then for what α, β, and C is $y(x) = C + \int_\alpha^\beta \sin(t^3)\,dt$?

Answers

1. b and c 2. $y^4 + 2y^2 + 20y = x^3 + C$ 3. $y^2 = x^2 + 5$
4. $\alpha = \pi$, $\beta = x$, and $C = 2$

EXERCISES

Problems for Practice

In Exercises 1–8, verify that the given function y satisfies the given differential equation. In each expression for $y(x)$, the letter C denotes a constant.

1. $\dfrac{dy}{dx} = xy, y(x) = Ce^{x^2/2}$
2. $\dfrac{dy}{dx} = 2xy^2, y(x) = 1/(C - x^2)$
3. $\dfrac{dy}{dx} = x - 3y, y(x) = x/3 - 1/9 + Ce^{-3x}$
4. $\dfrac{dy}{dx} = e^x + y, y(x) = e^x(x + C)$
5. $\dfrac{dy}{dx} = x + y, y(x) = Ce^x - x - 1$
6. $\dfrac{dy}{dx} = x + xy, y(x) = Ce^{x^2/2} - 1$
7. $\dfrac{dy}{dx} = y + x^2, y(x) = Ce^x - x^2 - 2x - 2$
8. $\dfrac{dy}{dx} = \dfrac{2x - y}{x + y}, y = -x + \sqrt{3x^2 + 2C}$

In each of Exercises 9–18, solve the given differential equation.

9. $\dfrac{dy}{dx} = 6\sqrt{xy}$
10. $(4 + y^2)\dfrac{dy}{dx} = x^2$
11. $(2 + x)\dfrac{dy}{dx} = y^2$
12. $\dfrac{dy}{dx} + 3xy^2 = 0$
13. $\dfrac{dy}{dx} - 3x^2y = 0$
14. $x\dfrac{dy}{dx} - y = 3x^3y$
15. $\exp(2x + 3y)\dfrac{dy}{dx} = 1$
16. $\cot(x)\dfrac{dy}{dx} = \tan(y)$
17. $\dfrac{dy}{dx} = x\sec(y)$
18. $\dfrac{dy}{dx} = y\sec(x)$

In Exercises 19–36, find the solution of the given initial value problem.

19. $y'(x) = 2x$	$y(1) = 3$
20. $y'(x) = 2x + 1$	$y(1) = 5$
21. $y'(x) = \cos(x)$	$y(0) = 2$
22. $y'(x) = \sec^2(x)$	$y(\pi/4) = 3$
23. $y'(x) = x/y(x)$	$y(0) = 1$
24. $y'(x) = xy^2(x)$	$y(1) = 2$
25. $y'(x) = y^2(x)\sin(x)$	$y(0) = 2$
26. $x^3\dfrac{dy}{dx} = \cos^2(y)$	$y(0) = 1$
27. $y\dfrac{dy}{dx} = \dfrac{2x}{\sqrt{1+y^2}}$	$y(0) = 0$
28. $y'(x) = \cos(x)/y(x)$	$y(0) = 2$
29. $\dfrac{dy}{dx} = y + \dfrac{1}{y}$	$y(0) = 3$
30. $(1 + x^2)^4\,y'(x) = 6xy(x)$	$y(0) = 1$
31. $x^2y'(x) = y(x)\sin(1/x)$	$y(2/\pi) = 3$
32. $\dfrac{dy}{dx} = \exp(x + y)$	$y(0) = 0$
33. $y'(x) - xy^2 = xy(4x - y)$	$y(0) = 1$
34. $\dfrac{dy}{dx} = x - 1 + yx - y$	$y(2) = 3$
35. $\dfrac{dy}{dx} = \dfrac{1}{y(1 + x^2)}$	$y(1) = \sqrt{\pi}$
36. $\dfrac{dy}{dx} = 2x\cos(y(x))$	$y(0) = 0$

Further Theory and Practice

37. Match the functions $F_1(x,y) = x$, $F_2(x,y) = y$, $F_3(x,y) = xy$, $F_4(x,y) = x/y$, $F_5(x,y) = x^2 - y^2$, and $F_6(x,y) = y^2$ to the graphs of the slope fields of equation (7.6.1) that are plotted in Figures 9a through 9f.

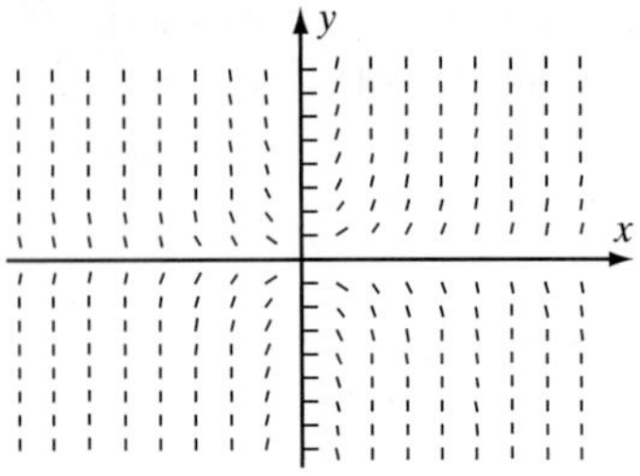

▲ Figure 9a

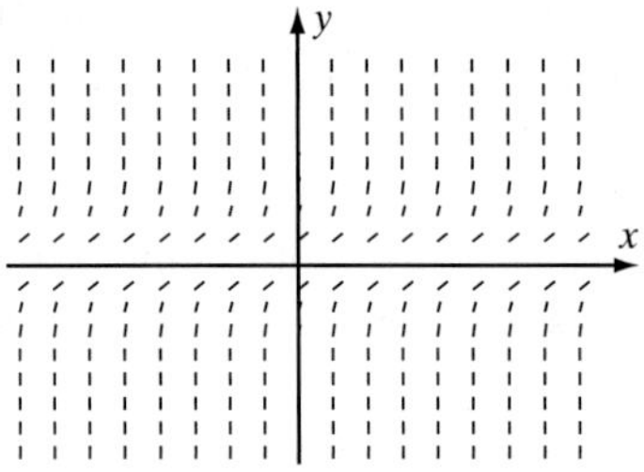

▲ Figure 9b

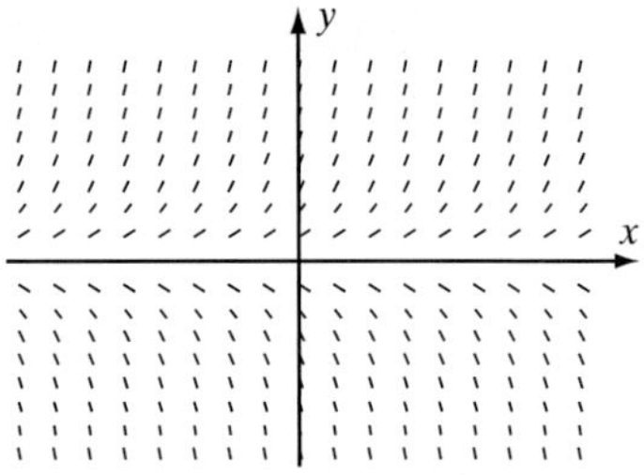

▲ Figure 9c

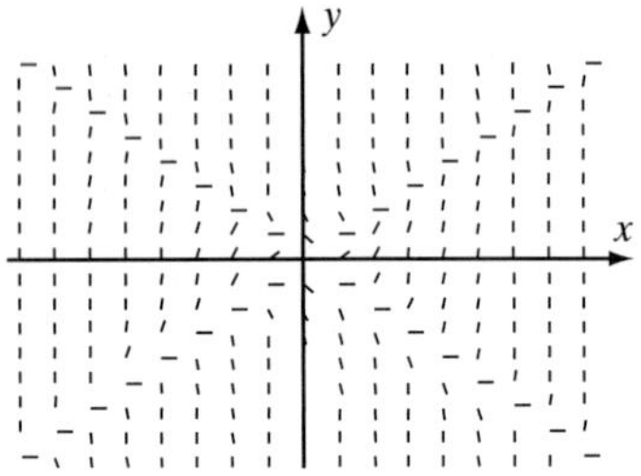

▲ Figure 9d

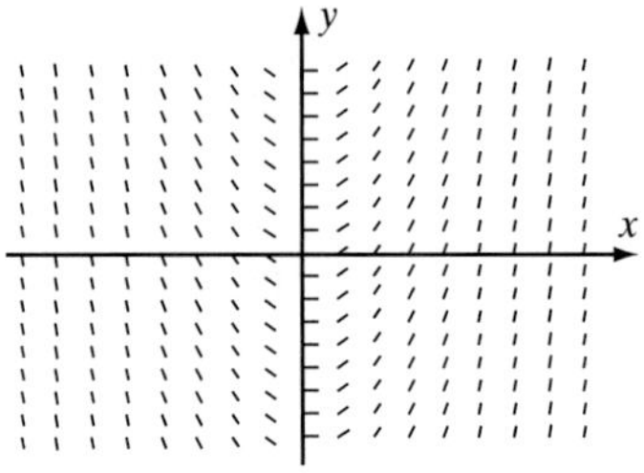

▲ Figure 9e

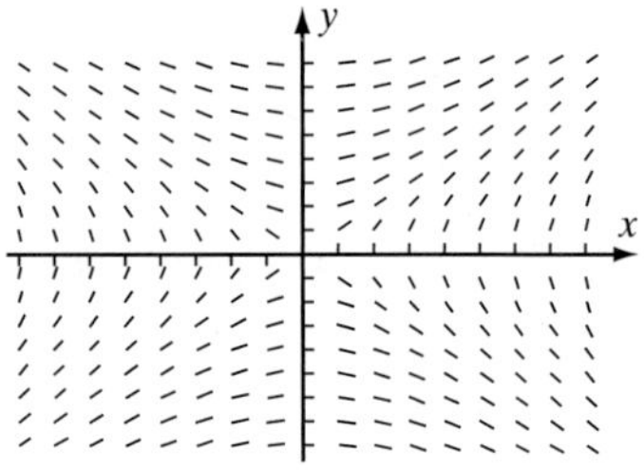

▲ Figure 9f

38. Match the graphs of the solution curves in Figures 10a through 10f to the slope fields of Figures 9a through 9f.

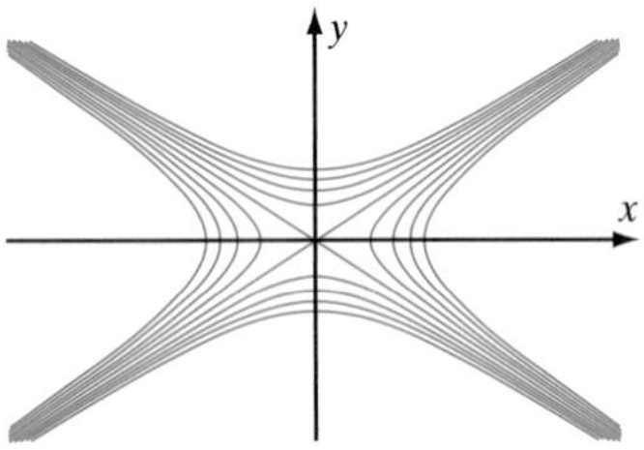

▲ Figure 10a

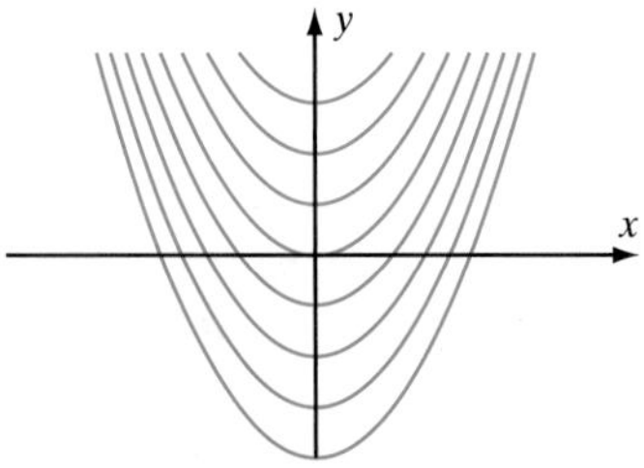

▲ Figure 10b

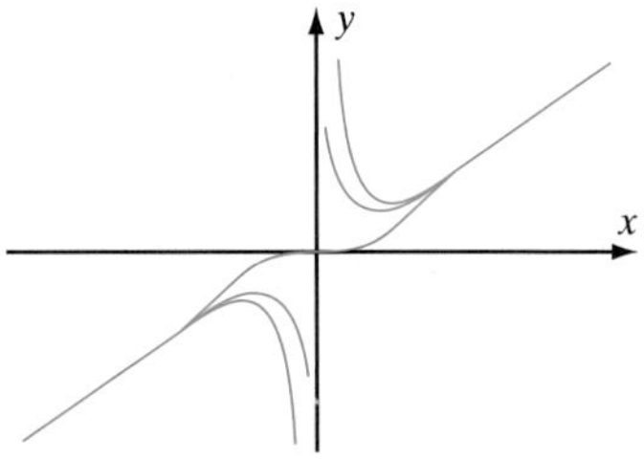

▲ Figure 10c

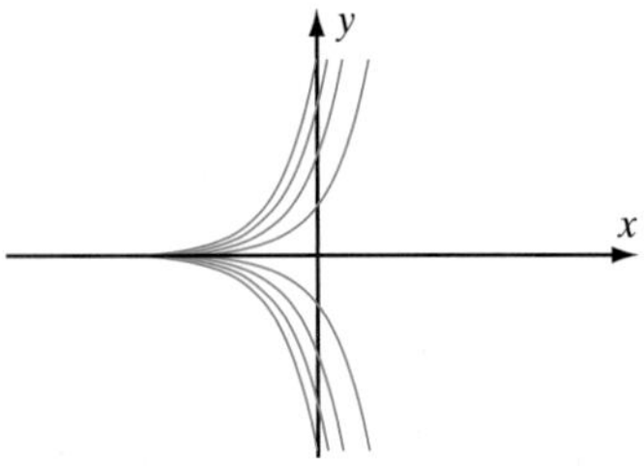

▲ Figure 10d

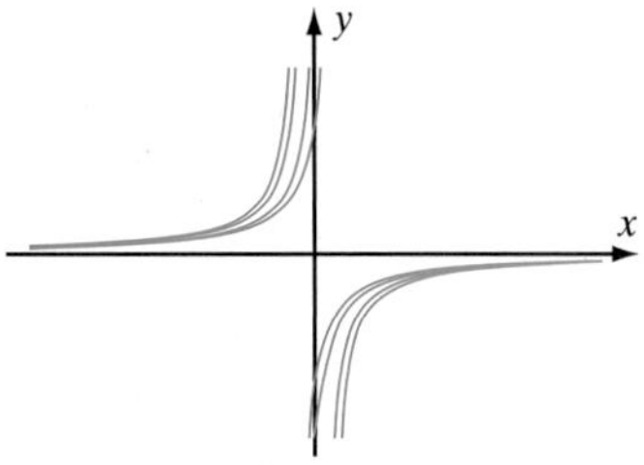

▲ Figure 10e

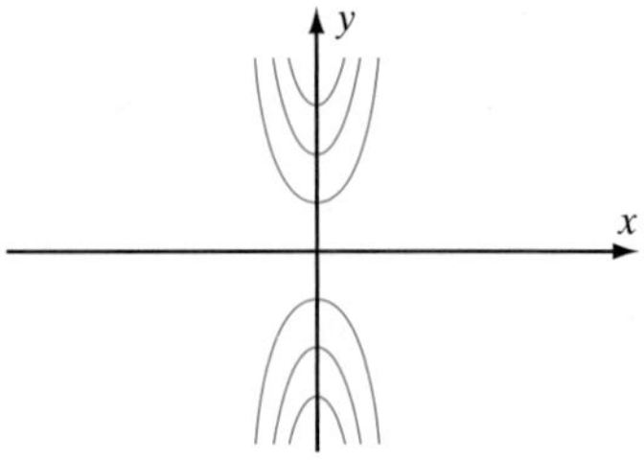

▲ Figure 10f

In each of Exercises 39–44, solve the given differential equation.

39. $\exp(x+y)\dfrac{dy}{dx} = x$

40. $y\dfrac{dy}{dx} = x^3\csc(y)$

41. $\dfrac{dy}{dx} = 3x^2(y^2 - y)$

42. $(4+e^{2x})\dfrac{dy}{dx} = y$

43. $(4+e^{2x})\dfrac{dy}{dx} = ye^x$

44. $(4+e^{2x})\dfrac{dy}{dx} = ye^{2x}$

45. Solve the initial value problem $y'(x) = 4f^3(x)f'(x)$, $f(1) = 2, y(1) = 24$. Verify your formula with $f(x) = 2\sqrt{x}$.

46. Solve the initial value problem $y'(x) = x\,g'(x^2)$, $g(4) = 6$, $y(2) = 12$. Verify your formula with $g(x) = x^2 - 2x - 2$.

47. Solve the initial value problem $y'(x) = f'(x)g'(f(x))$, $f(0) = 3, g(3) = 2, y(0) = 6$. Verify your formula with $f(x) = 3\cos(x)$ and $g(x) = \sqrt{1+x}$.

48. Show that, when applied to the initial value problem equation

$$\frac{dy}{dx} = \frac{3x^2+2}{2y-2},\ y(0) = y_0,$$

the method of separation of variables results in the same equation,

$$y^2(x) - 2y(x) = x^3 + 2x,$$

when $y_0 = 0$ and when $y_0 = 2$. And yet, the solutions to the two initial value problems clearly must differ. Explain how this is possible.

49. According to Torricelli's Law, if there is a hole of area a at the bottom of a tank, then the volume $V(t)$ and height $y(t)$ of water in the tank at time t are related by

$$\frac{dV}{dt} = -a \cdot \sqrt{2gy}$$

where g is equal to gravitational acceleration at Earth's surface. (Notice that the differential equation is dimensionally correct—both sides bear units of L^3/T where L is length and t is time.) Use the Fundamental Theorem of Calculus together with the Chain Rule to show that

$$A(y)\frac{dy}{dt} = -a \cdot \sqrt{2gy}.$$

50. A cubic tank filled with water has side length 1 m. At time $t = 0$, a circular hole of radius 4 cm opens up on the bottom of the tank. If the proportionality constant k in Torricelli's Law is given by $k = 5/\pi\ \mathrm{m}^{1/2}/\mathrm{s}$, how long does it take the tank to drain?

51. A tank in the shape of a conical frustum filled with water is one meter deep at its central axis, has radius 2 m at its top, and one meter at its bottom. At time $t = 0$, a circular hole of radius 1 cm opens up at the bottom. Suppose that the proportionality constant k in Torricelli's Law is given by $k = 32/15\ \mathrm{m}^{1/2}/\mathrm{s}$. Find an equation for the height $y(t)$ of water in the tank at time t. How long does it take the tank to completely drain?

52. The escape velocity is the minimum initial velocity v_e at which a projectile launched straight up will never return to Earth. Use equation (7.6.7) to find a formula for v_e in terms of g and R.

53. A population $P(t)$ satisfies logistic growth equation (7.6.8) with $k = 1/2$ and $P_\infty = 2P_0$. At what time is $P(t) = (3/2)\,P_0$?

54. A population $P(t)$ satisfies the logistic growth equation $P'(t) = \alpha \cdot P(t) - \beta \cdot P^2(t)$ for positive constants α and β. What is the carrying capacity?

55. By completing the square on the right side of logistic growth equation (7.6.8), show that a solution $P(t)$ has exactly one inflection point: $(t_I, P(t_I))$ where $P(t_I) = P_\infty/2$.

56. According to the preceding exercise, the solution $P(t)$ to logistic growth equation (7.6.8) has one inflection point, $(t_I, P(t_I))$, where $P(t_I) = P_\infty/2$. Show that

$$t_I = \frac{1}{kP_\infty}\ln\left(\frac{P_\infty - P_0}{P_0}\right).$$

57. Let P be the logistic growth function defined by equation (7.6.10). Let t_I be the time at which P has an inflection point. (Refer to the two preceding exercises.) Show that $P(t)$ can be expressed in terms of the parameters k, P_∞, and t_I by means of the formula

$$P(t) = \frac{P_\infty}{1 + \exp(-k \cdot P_\infty \cdot (t - t_I))}.$$

58. If a population is to be modeled by the logistic growth function P defined by equation (7.6.10), then the three parameters P_0, k, and P_∞ must be determined. It is easiest to solve for these constants by measuring $P(t)$ at equally spaced times $t_0 - \tau$, t_0, and $t_0 + \tau$. Having done so, we position the origin of the time axis so that $t = 0$ corresponds to the time t_0, then $A = P(-\tau)$, $P_0 = P(0)$, and $B = P(\tau)$ are the measured population values.

a. Use the equation $B = P(\tau)$ to express $\exp(-k \cdot P_\infty \cdot \tau)$ in terms of B, P_0, and P_∞.

b. Use the equation $A = P(-\tau)$ to express $\exp(k \cdot P_\infty \cdot \tau)$ in terms of A, P_0, and P_∞.

c. The product of $\exp(-k \cdot P_\infty \cdot \tau)$ and $\exp(k \cdot P_\infty \cdot \tau)$ is 1. Use this observation together with parts a and b to show that

$$P_\infty = \frac{(A+B)P_0^2 - 2ABP_0}{P_0^2 - AB}.$$

d. Use your formula from part b to show that

$$\exp(-k \cdot P_\infty \cdot t) = \left(\frac{P_0}{A} \cdot \frac{P_\infty - A}{P_\infty - P_0}\right)^{-t/\tau}.$$

e. Deduce that

$$P(t) = \frac{P_0 \cdot P_\infty}{P_0 + (P_\infty - P_0)\left(\frac{P_0}{A} \cdot \frac{P_\infty - A}{P_\infty - P_0}\right)^{-t/\tau}}.$$

59. A town's population (in thousands) was 36 in 1980, 60 in 1990, and 90 in 2000. Assuming that the growth of the population of this town is logistic, what will the population be in the year 2020? (Use the preceding exercise.)

60. In Exercise 58, the logistic growth formula determined in part e is obtained without solving for the growth constant k in terms of the measured data A, P_0, and B.

a. Use the formula in part d of Exercise 58 with $t = \tau$ to obtain a formula for k.

b. Use the logistic growth equation to express k in terms of the data P_0, P_∞, and $P'(0)$.

c. Assuming that τ is small, approximate k using part b and the central difference approximation for the first derivative.

61. The plot of the pumpkin data in Figure 6 has an inflection point near $(9, 25)$. For simplicity, let us use these coordinates for the inflection point. The initial diameter $P(0)$ is 3.6 cm. Using Exercises 55 and 56, find P_∞ and k.

62. A population P satisfies the differential equation

$$P'(t) = 10^{-5} \cdot P(t) \cdot (15000 - P(t)).$$

For what value $P(0)$ of the initial population is the initial growth rate $P'(0)$ greatest?

63. When sucrose is dissolved in water, glucose and fructose are formed. The reaction rate satisfies

$$\frac{d[C_{12}H_{22}O_{11}]}{dt} = -5.7 \times 10^{-5} \cdot [C_{12}H_{22}O_{11}].$$

For what value of t is the concentration of sucrose, $[C_{12}H_{22}O_{11}]$, equal to 1/3 its initial value?

64. The chemical reaction

$$CH_3COCH_3 + I_2 \rightarrow CH_3COCH_2I + HI$$

between acetone and iodine is governed by the differential equation

$$\frac{d[I_2]}{dt} = -0.41 \times 10^{-3} \cdot [I_2]^2$$

(in moles per liter per second). If the initial iodine concentration is c_0, solve for $[I_2]$ as a function of time. Suppose that $[I_2]$ is plotted as a function of time. What is the slope of the graph when $[I_2] = 1/10$ mole/liter? Suppose that $1/[I_2]$ is plotted as a function of time. What is the slope of the graph when $[I_2] = 1/10$ mole/liter?

65. In the decomposition of dinitrogen pentoxide,

$$N_2O_5 \rightarrow 2NO_2 + \frac{1}{2}O_2,$$

the reaction rate satisfies the equation

$$\frac{d[N_2O_5]}{dt} = -k \cdot [N_2O_5]$$

for some positive constant k. If $[N_2O_5]$ is equal to 2.32 moles per liter when $t = 0$ seconds and if $[N_2O_5]$ is equal to 0.37 moles per liter when $t = 3000$ seconds, then what is $[N_2O_5]$ as a function of time?

66. The reaction rate of

$$2NO + H_2 \rightarrow N_2O + H_2O$$

is given by

$$\frac{d[N_2O]}{dt} = 0.44 \times 10^{-6}(h - [N_2O])^3$$

where h is the number of moles per liter of H_2 that are initially present. What is the initial concentration of $[N_2O]$? Solve for $[N_2O]$ as an explicit function of t. In doing so, you will need to extract a square root. Decide which sign, + or − to use; explain your answer.What is $\lim_{t\to\infty}[N_2O]$?

67. Suppose that ϕ is a function of one variable. The differential equation $dy/dx = \phi(y/x)$ is said to be *homogeneous of degree* 0. Let $w(x) = y(x)/x$. Differentiate both sides of the equation $y(x) = x \cdot w(x)$ with respect to x. By equating the resulting expression for dy/dx with $\phi(y/x)$, show that $w(x)$ is the solution of a separable differential equation. Illustrate this theory by solving the differential equation $dy/dx = 2xy/(x^2 + y^2)$. For this example, $\phi(w) = 2w/(1 + w^2)$.

68. Suppose that ψ is a function of one variable and that a, b, and c are constants. If $dy/dx = \psi(a + bx + cy)$, then show that $u(x) = a + bx + cy(x)$ is the solution of a separable equation. Illustrate by solving the equation $dy/dx = 1 + x + y$.

69. The pressure P and temperature T in the outer envelope of a white dwarf (star) are related by the differential equation

$$\frac{dP}{dT} = C\frac{T^{7.5}}{P}, \quad P(0) = 0$$

where C is a constant. Find P as a function of T.

70. The interior temperature T_I (in degrees K) of a cooling white dwarf (star) satisfies the differential equation

$$\frac{dT_I}{dt} = -k\left(\frac{T_I}{7 \times 10^{7\circ}\text{K}}\right)^{7/2}.$$

Here k is a constant with units degrees K per year, and t represents time in years. Solve for T_I. If $k = 6\,^\circ$K/yr and if $T_I(0) = 10^8\,^\circ$K, then in how many years will the star cool to $10^4\,^\circ$K?

71. A spherical star of radius R is in equilibrium if the pressure gradient dP/dr exactly opposes gravity. In the case of Newtonian gravitation, this amounts to

$$\frac{dP}{dr} = -\frac{Gm(r)\rho(r)}{r^2}, \quad P(R) = 0.$$

Here G is the gravitational constant, r represents distance to the center, $m(r)$ represents the mass of that part of the star that is within distance r of the center, and $\rho(r)$ represents the mass density at radius r. Compute $P(r)$ under the assumption that the mass density is a constant ρ. Show that the pressure at the center of the star is $2\pi G\rho^2 R^2/3$ (in terms of the stellar radius R) or $(\pi/6)^{1/3} G M^{2/3} \rho^{4/3}$ (in terms of the stellar mass M).

72. A bullet is fired into a sand pile with an initial speed of 1024 feet per second. The rate at which its velocity is decreased after entering the sand pile is 500 times the $\frac{4}{5}^{\text{th}}$ power of the velocity if units of feet and seconds are used. How long is it before the bullet comes to rest?

73. Refer to the preceding exercise. How far does the bullet travel in the sand before coming to rest?

74. The standard normal probability density f satisfies the differential equation

$$f'(x) = -xf(x), \; f(0) = \frac{1}{\sqrt{2\pi}}.$$

Find $f(x)$.

75. Suppose that, in a certain ecosystem, there is one type of predator and one type of prey. Let $x \cdot 100$ denote the number of predators and $y \cdot 1000$ denote the number of prey. The Austrian mathematician A.J. Lotka (1880–1949) and the Italian mathematician Vito Volterra (1860–1940) proposed the following relationship, the so-called *Lotka-Volterra Equation*, between the two population sizes:

$$\frac{dy}{dx} = \frac{y \cdot (a - bx)}{x \cdot (cy - d)}.$$

Separate variables and solve this equation. What is the predator-prey relationship if the initial prey population is 1500, the initial predator population is 200, and if $a = 6$, $b = 2$, $c = 4$, and $d = 7$? Figure 11 illustrates the solution curve of this Lotka-Volterra equation. Its potato-pancake shape is typical.

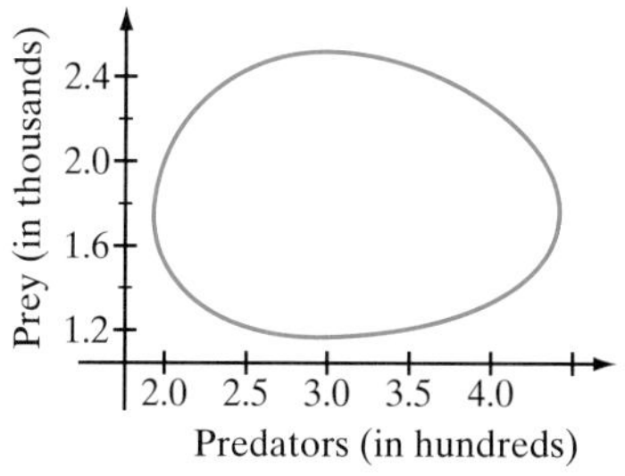

▲ Figure 11 Solution of Lotka-Volterra equation

76. Suppose that λ, A, and p are positive constants. Solve the differential equation

$$\frac{dy}{dt} = \frac{\lambda}{p} \cdot t^{p-1} \cdot (A - y).$$

A solution of this equation is known as a *Janoschek growth function*. Such functions are often used to model animal growth. The parameter A is called "the size of the adult." Why?

77. *The Tractrix*. The tip of a boat is at the point $(L, 0)$ on the positive x-axis. Attached to the tip is a chord of length L. A man starting at $(0, 0)$ pulls the boat by the chord as he walks up the y-axis. The chord is always tangent to the path of the tip of the boat. Show that the position $y(x)$ of the tip of the boat is given by

$$\frac{dy}{dx} = -\frac{\sqrt{L^2 - x^2}}{x}, \quad y(L) = 0.$$

The graph of y is called the *tractrix* (Figure 12).

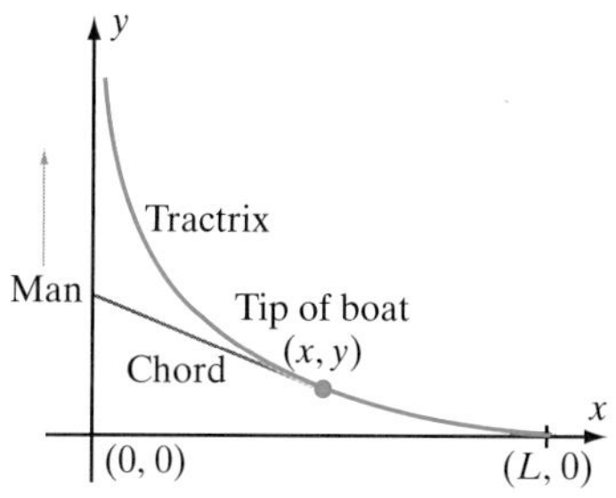

▲ Figure 12

78. Actuaries use mortality tables to show the expected number of survivors of an initial group. Let $L(x)$ denote the number from that group who are living at age x. Then $-L'(x)$ is the instantaneous death rate and the percentage rate, $-L'(x)/L(x)$ is called the *force of mortality*. The first significant formula was derived by B. Gompertz assumed that the force of mortality increases with age according to an expression of the form Bg^x for some constants $B > 0$ and $g > 1$. Solve for $L(x)$ under Gompertz's assumption.

79. M.W. Makeham emended Gompertz's assumption for the force of mortality to take into account accidental deaths—refer to the preceding exercise for background. Under Makeham's assumption, the force of mortality has the form $M + Bg^x$ for some positive constants M, B, and g with $g > 1$. Solve for $L(x)$ under this assumption. The resulting formula is known as *Makeham's Law*.

80. Du Nouy's equation for the decreasing surface tension T of blood serum is

$$\frac{dT}{dt} = -k\frac{T}{\sqrt{t}}$$

for some constant k. Solve this equation if $T = \tau$ at $t = 0$.

81. Let x and y be the measures of two body parts with relative growth rates that are proportional to a common factor $\Phi(t)$:

$$\frac{1}{x}\cdot\frac{dx}{dt} = \alpha\cdot\Phi(t) \quad \text{and} \quad \frac{1}{y}\cdot\frac{dy}{dt} = \beta\cdot\Phi(t).$$

Show that x and y satisfy the Huxley Allometry Equation $y = kx^p$ for suitable constants k and p.

82. The shape of a Bessel horn is determined by the equation

$$\frac{da}{dx} = -\gamma\cdot\frac{a}{x+x_0}$$

where $a(x)$ is the bore radius at distance x from the edge of the bell of the horn, and x_0 and γ are positive constants. Solve for a as a function of x. Determine $a(x)$ for a trumpet with $\gamma = 0.7$, $a(0) = 10$, and $a(30) = 2$. Sketch the graph of $y = a(x)$ for $0 \le x \le 30$.

83. Verify that $y(x) = \exp(-x^2)\int_0^x \exp(t^2)dt$, which is known as *Dawson's Integral*, is the solution of the initial value problem $\frac{dy}{dx} = 1 - 2xy,\ y(0) = 0$.

84. For each $x > 0$, there is a unique $y > 0$ such that $ye^y = x$. Denote this value of y by $W(x)$. Show that the function W (called *Lambert's W function*) is a solution of the differential equation

$$\frac{dy}{dx} = \frac{y}{x(1+y)}.$$

85. Suppose an object of mass m is propelled upwards from the surface of the earth with initial velocity v_0. Suppose that the (downward) force of air resistance $R(v)$ is proportional to the square of the speed: $R(v) = -k\cdot v^2$, where k is a positive constant that carries the units of mass/length. (This is the quadratic drag law.) Solve the initial value problem for motion:

$$m\frac{dv}{dt} = -kv^2 - mg, \quad v(0) = v_0.$$

86. Suppose that the force R of air resistance on a 0.2 kg object thrown straight up with initial velocity 28 m/s is given by the quadratic drag law,

$$R(v) = -kv^2,$$

where $k = 4\times 10^{-6}$ kg/m. Find an equation for the time τ at which the object begins to fall.

87. Let $v(t)$ denote the velocity at time t of an object of mass m that is dropped from a positive height at time $t = 0$. Assume that, throughout its fall, the (upward) force of air resistance on the object can be approximated by the formula $R(v) = +k\cdot v^2$, where k is a positive constant that carries the units of mass/length. (As in Exercises 85 and 86, this equation is the quadratic drag law. Because drag is in the opposite direction of motion, the sign for upward motion is opposite to that for downward motion.) Show that

$$v(t) = \sqrt{\frac{mg}{k}}\tanh\left(-t\sqrt{\frac{kg}{m}}\right)$$

until impact.

88. The rate of elimination of alcohol from the bloodstream is proportional to the amount A that is present. That is,

$$\frac{dA}{dt} = -\frac{1}{k}A,$$

where k is a time constant that depends on the drug and the individual. If k is 1/2 hour for a certain person, how long will it take for his blood alcohol content to reduce from 0.12% to 0.06%?

89. A column of air with cross-sectional area 1 cm^2 extends indefinitely upwards from sea level (i.e. to infinity). The *atmospheric pressure* p at height y above sea level is the weight of that part of this infinite column of air that extends from y to infinity. Assume that the density of the air is proportional to the pressure. (This is an approximation that follows from Boyle's law, assuming constant temperature.) Show that there is a positive constant k such that

$$\frac{dp}{dy} = -kp.$$

Under this model, what is the mathematical expression for atmospheric pressure? What physical dimensions does K carry?

90. Suppose that α and β are positive constants. The differential equation

$$P'(t) = \alpha\cdot e^{-\beta t}\cdot P(t)$$

for a positive function P is known as the *Gompertz growth equation*. (It is named for Benjamin Gompertz (1779–1865), a self-educated scholar of wide-ranging interests.) Find an explicit formula for $P(t)$. Use your explicit solution to show that there is a number P_∞ (known as the *carrying capacity*) such that

$$\lim_{t\to\infty} P(t) = P_\infty.$$

Show that the Gompertz growth equation may be written in the form

$$P'(t) = k\cdot P(t)\cdot\ln\left(\frac{P_\infty}{P(t)}\right).$$

91. In forestry, the function $E(T) = C\cdot e^{\alpha T/(T+\beta)}$ has been used to model water evaporation as a function of temperature T. Here α and β are constants with $\alpha > 2$ and $\beta > 0$. Show that

$$E'(T) = \frac{\alpha\beta}{(T+\beta)^2}E(T)$$

and

$$E''(T) = \frac{2((\alpha/2-1)\beta - T)}{(T+\beta)^4} E(T).$$

Deduce that E is an increasing function of T with a sigmoidal (or S-shaped) graph. What is the point of inflection of the graph of E?

92. Suppose that h_∞, k, and q are positive constants with $q < 1$. Show that the *Chapman-Richards function*, defined by

$$h(t) = h_\infty \cdot \left(1 - \exp(-q\,k\,t)\right)^{1/q},$$

is the solution of the initial value problem

$$h'(t) = k \cdot h(t) \cdot \left(\left(\frac{h_\infty}{h(t)} \right)^q - 1 \right), \quad h(0) = 0,$$

which is used in forest management to model tree growth. Here t represents time, and h measures tree height (or some other growth indicator).

Calculator/Computer Exercises

In each of Exercises 93–96, solve the given initial value problem for $y(x)$. Determine the value of $y(2)$.

93. $dy/dx = (1+x^2)\,/(1+y^4)$ $\quad y(0) = 0$

94. $dy/dx = (x^2)\,/(1+\sin(y))$ $\quad y(0) = 0$

95. $y^2 \cdot dy/dx = (1+y)/(1+2x)$ $\quad y(0) = 0$

96. $dy/dx = x/(1+e^y)$ $\quad y(0) = 0$

97. The weight y of a goose is determined as a function of time by the initial value problem $dy/dt = 0.09 \cdot \sqrt{t} \cdot (5000-y), y(0) = 175$ when t is measured in weeks after birth, and y is measured in grams. In how many weeks does the goose reach half of its mature weight?

98. Suppose that in a certain ecosystem, $x \cdot 100$ denotes the number of predators, and $y \cdot 1000$ denotes the number of prey. Suppose further that the relationship between the two population sizes satisfies the following Lotka-Volterra equation:

$$\frac{dy}{dx} = \frac{y \cdot (6-2x)}{x \cdot (4y-5)}.$$

Separate variables, and solve this equation. If the initial prey population is 1500, and the initial predator population is 200, what are the possible sizes of the population of prey when the predator size is 300?

99. As the result of pollution reduction in the Great Lakes region, certain species of wildlife that had been threatened have made a comeback. Figure 13 contains a plot of the number P of double-crested cormorant nests in the Great Lakes region for the years 1979 through 1993. A logistic growth curve has been superimposed. In this exercise, you will obtain the formula for the logistic curve

$$P(t) = \frac{P_0 \cdot P_\infty}{P_0 + (P_\infty - P_0)\exp(-k \cdot P_\infty \cdot (t-1979))}.$$

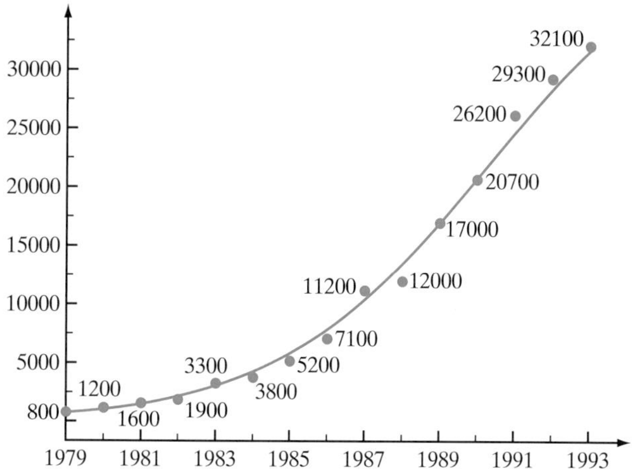

Figure 13 Cormorant nests (1979–1993)

Use $P_\infty = 44000$ for the carrying capacity, and use $P_0 = P(1979) = 800$ for the value of the initial population.

a. For $t = 1984, 1985, \ldots, 1988$, use a central difference quotient to approximate $P'(t)$. For example,

$$P'(1980) \approx \frac{P(1981) - P(1979)}{1981 - 1979} = \frac{1600 - 800}{2}.$$

The logistic differential equation $P'(t) = kP(t)(P_\infty - P(t))$ then permits an estimate for k. Average the five estimates to get a working value of k.

b. Using the value of k obtained in part a, calculate $P(t)$ for $t = 1990, \ldots, 1993$. (Although the observed values for these years appear in Figure 13, they were *not* used to obtain the model. Comparing the predicted values with the observed values is a reality check for the model.)

c. Plot the logistic growth curve for $1979 \le t \le 1993$.

d. According to your logistic model, about when did the number of nests reach 43,000?

7.7 First Order Differential Equations–Linear Equations

In the preceding section, you learned a method for solving the first order differential equation

$$\frac{dy}{dx} = F(x, y)$$

when $F(x, y) = g(x) \cdot h(y)$. There are several other methods that are similarly adapted to other particular forms assumed by $F(x, y)$. In this section, we will study the case in which $F(x, y) = q(x) - p(x)\,y$.

Solving Linear Differential Equations

We say that a first order differential equation of the form

$$\frac{dy}{dx} = q(x) - p(x)\,y$$

is *linear*. We often write a linear equation in the standard form

$$\frac{dy}{dx} + p(x)\,y = q(x). \qquad (7.7.1)$$

This algebraic step does not separate variables, but it does eliminate the y variable from the right side of the equation, and it prepares the left side for multiplication by an "integrating factor" followed by a decisive application of the Product Rule for differentiation.

THEOREM 1 Suppose that $p(x)$ and $q(x)$ are continuous functions. Let $P(x)$ be any antiderivative of $p(x)$. Then the general solution of the linear equation

$$\frac{dy}{dx} + p(x)\,y = q(x)$$

is

$$y(x) = e^{-P(x)} \int e^{P(x)} q(x)\,dx + Ce^{-P(x)}, \qquad (7.7.2)$$

where C is an arbitrary constant.

Proof. The presence of both $y'(x)$ and $y(x)$ in the sum on the left side of equation (7.7.1) leads us to consider the Product Rule for differentiation. If u is a function of x that is never 0, then we have

$$\frac{d}{dx}(uy) = u\frac{dy}{dx} + \left(\frac{du}{dx}\right)y = u\left(\frac{dy}{dx} + \frac{1}{u}\left(\frac{du}{dx}\right)y\right).$$

The idea is to find a function u so that the factor of y on the right side of the preceding equation matches the factor of y in (7.7.1). That is, we want to find a function u such that

$$\frac{1}{u(x)}\left(\frac{du}{dx}\right) = p(x).$$

As we learned in Section 7.6, we solve this separable differential equation by rewriting it as

$$\int \frac{1}{u}\,du = \int p(x)\,dx.$$

Because $P(x)$ is an antiderivative of $p(x)$, the general solution of the preceding equation is $\ln(|u|) = P(x) + C_0$, where C_0 is a constant of integration. Because any particular solution u will serve our purpose, we simplify the algebra by choosing a solution u with $u(x) > 0$ and $C_0 = 0$. With these choices, we have $\ln(u) = P(x)$, or $u(x) = e^{P(x)}$. This is our integrating factor. On multiplying each side of equation (7.7.1) by $e^{P(x)}$, we obtain

$$p(x)e^{P(x)} \cdot y + e^{P(x)}\frac{dy}{dx} = e^{P(x)}q(x).$$

Thus

$$\frac{d}{dx}\left(e^{P(x)} \cdot y\right) = \left(\frac{d}{dx}e^{P(x)}\right) \cdot y + e^{P(x)}\frac{dy}{dx} = P'(x)e^{P(x)} \cdot y + e^{P(x)}\frac{dy}{dx} = p(x)e^{P(x)} \cdot y + e^{P(x)}\frac{dy}{dx} = e^{P(x)}q(x).$$

In other words, $e^{P(x)} \cdot y$ is an antiderivative of $e^{P(x)}q(x)$. That is,

$$e^{P(x)} \cdot y = \int e^{P(x)}q(x)\,dx + C,$$

from which formula (7.7.2) is obtained on dividing each side by $e^{P(x)}$. ■

▶ **EXAMPLE 1** Solve the initial value problem

$$\frac{dy}{dx} + \frac{2y}{x} = 4x, \quad y(2) = 3.$$

Solution The given differential equation is linear: It has the form of equation (7.7.1) with $p(x) = 2/x$ and $q(x) = 4x$. Because

$$\int p(x)\,dx = \int \frac{2}{x}\,dx = 2\ln(|x|) + C,$$

and because we may choose any antiderivative of $p(x)$ for $P(x)$, we set $P(x) = 2\ln(|x|)$. We note from the initial condition that the positive number 2 must belong to the interval on which the required solution $y(x)$ is defined. This interval of definition cannot contain any negative number because, if it did, then 0 would be in the interval. However, we see that the differential equation is not even defined when $x = 0$, so it cannot hold when $x = 0$. Thus the domain of definition of y is limited to positive values. We may therefore dispense with the absolute value function and take $P(x) = 2\ln(x)$. Thus,

$$e^{P(x)} = e^{2\ln(x)} = e^{\ln(x^2)} = x^2,$$

and equation (7.7.2) becomes

$$y(x) = x^{-2}\int x^2 \cdot 4x\,dx + C \cdot x^{-2} = \frac{4}{x^2}\int x^3\,dx + \frac{C}{x^2} = \frac{4}{x^2} \cdot \frac{x^4}{4} + \frac{C}{x^2} = x^2 + \frac{C}{x^2}.$$

Because $y(2) = 2^2 + C/2^2 = 4 + C/4$, the initial condition tells us that $4 + C/4 = 3$. It follows that $C = -4$, and $y(x) = x^2 - 4/x^2$. ◀

▶ **EXAMPLE 2** Solve the initial value problem

$$x\frac{dy}{dx} + y = \sin(x), \quad y(\pi/2) = 2.$$

Solution To put this equation into the form of equation (7.7.1), we divide by x and obtain

$$\frac{dy}{dx} + \frac{1}{x}\cdot y = \frac{\sin(x)}{x}.$$

We set $p(x) = 1/x$, $q(x) = \sin(x)/x$, and take $P(x) = \ln(x)$ for an antiderivative of $p(x)$. (The reasoning for taking $\ln(x)$ instead of $\ln(|x|) + C$ is similar to that of Example 1.) It follows that $e^{P(x)} = e^{\ln(x)} = x$ and equation (7.7.2) become

$$y(x) = x^{-1}\int x\cdot\frac{\sin(x)}{x}\,dx + C\cdot x^{-1} = \frac{1}{x}\int \sin(x)\,dx + \frac{C}{x} = \frac{C - \cos(x)}{x}.$$

Because $y(\pi/2) = 2$, we have

$$\frac{C - \cos(\pi/2)}{\pi/2} = 2,$$

or $C = \pi$. Thus

$$y(x) = \frac{\pi - \cos(x)}{x}.$$ ◀

▶ **EXAMPLE 3** Find the general solution of the linear differential equation

$$\frac{dy}{dx} + \tan(x)\cdot y = \sec(x) \qquad \left(0 < x < \frac{\pi}{2}\right).$$

Do not use equation (7.7.2). Instead, use the technique found in the proof of Theorem 1.

Solution We recall that $\ln(\sec(x))$ is an antiderivative of $\tan(x)$. In the notation of Theorem 1, $q(x) = \sec(x)$, $p(x) = \tan(x)$, and $P(x) = \ln(\sec(x))$. We multiply each side of the given differential equation by $e^{P(x)}$. Observing that $e^{P(x)} = e^{\ln(\sec(x))} = \sec(x)$, we obtain

$$\sec(x)\tan(x)\cdot y + \sec(x)\frac{dy}{dx} = \sec^2(x),$$

or

$$\left(\frac{d}{dx}\sec(x)\right)\cdot y + \sec(x)\frac{dy}{dx} = \sec^2(x),$$

or

$$\frac{d}{dx}(\sec(x)\cdot y) = \sec^2(x).$$

It follows that

$$\sec(x)\cdot y = \int \sec^2(x)\,dx = \tan(x) + C,$$

or

$$y = \frac{\tan(x)}{\sec(x)} + \frac{C}{\sec(x)} = \sin(x) + C\cos(x). \quad ◀$$

An Application: Mixing Problems

Mixing problems occur frequently in biology, chemistry, medicine, and pharmacology. In a basic mixing problem, a container contains a substance in solution. There is an inlet to the container as well as an outlet (see Figure 1). When fluid enters and exits the tank through these channels, the solution in the tank can become enriched or diluted. If m is the mass of the dissolved substance, then the basic equation becomes

$$\frac{dm}{dt} = (\text{rate in}) - (\text{rate out}). \tag{7.7.3}$$

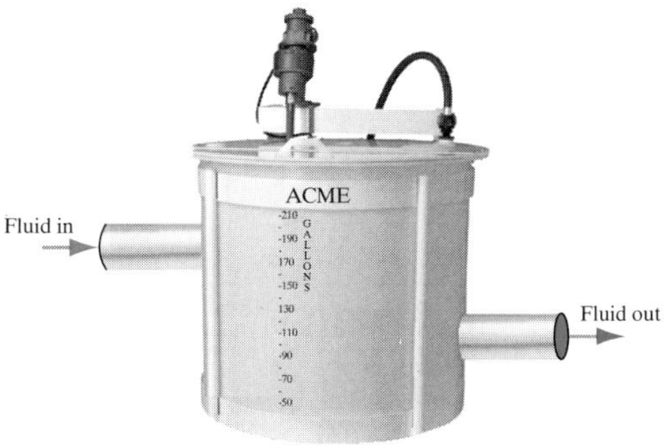

▲ Figure 1

To apply the theory of linear equations, we must assume that the solution is thoroughly mixed at all times so that the concentration of the solution is homogeneous throughout the container. (Nearly all mathematical models involve some approximation. Many involve a compromise between accuracy and simplicity. A completely accurate model that is too difficult to solve may do us little good. By simplifying the model somewhat, we may be able to obtain a useful solution without sacrificing too much accuracy.)

▶ **EXAMPLE 4** A 200 gallon tank is filled with a salt solution initially containing 10 pounds of salt. An inlet pipe brings in a solution of salt at the rate of 10 gallons per minute. The concentration of salt in the incoming solution is $1 - e^{-t/60}$ pounds per gallon when t is measured in minutes. An outlet pipe prevents overflow by allowing 10 gallons per minute to flow out of the tank. How many pounds of salt are in the tank at time t? Long term, about how many pounds of salt will be in the tank?

Solution Let $m(t)$ be the amount of salt in the tank at time t. To calculate the amounts of salt entering the tank through one pipe and exiting through another, we multiply the rate at which the fluid enters and exits by the appropriate concentrations. The rate at which salt enters the tank is

$$10\,\frac{\text{gal}}{\text{min}} \cdot \left(1 - e^{-t/60}\right)\frac{\text{lb}}{\text{gal}},$$

or $10 \cdot (1 - e^{-t/60})$ pounds per minute. The rate at which salt leaves the tank is

$$10 \frac{\text{gal}}{\text{min}} \cdot \frac{m}{200} \frac{\text{lb}}{\text{gal}},$$

or $m/20$ pounds per minute. Equation (7.7.3) becomes

$$\frac{dm}{dt} = 10 \cdot \left(1 - e^{-t/60}\right) - \frac{m}{20},$$

or

$$\frac{dm}{dt} + \frac{1}{20} m = 10 \cdot \left(1 - e^{-t/60}\right),$$

which is a linear differential equation. Of course, the use of t rather than x as an independent variable does not affect the applicability of Theorem 1. We simply use t instead of x and set $p(t) = 1/20$ and $q(t) = 10(1 - e^{-t/60})$. Then $P(t) = t/20$ and equation (7.7.2) gives us

$$m(t) = e^{-t/20} \int e^{t/20} 10\left(1 - e^{-t/60}\right) dt + Ce^{-t/20} = 10\, e^{-t/20} \int \left(e^{t/20} - e^{t/30}\right) dt + Ce^{-t/20},$$

or

$$m(t) = 100\left(2 - 3e^{-t/60}\right) + Ce^{-t/20}.$$

Because $m(0) = 10$, we have $10(20 - 30e^{-0/60}) + Ce^{-0/20} = 10$, or $10(20 - 30) + C = 10$, or $C = 110$. Thus

$$m(t) = 100\left(2 - 3e^{-t/60}\right) + 110e^{-t/20}.$$

Notice that

$$\lim_{t \to \infty} \left(100\left(2 - 3e^{-t/60}\right) + 110e^{-t/20}\right) = \left(100(2 - 3 \cdot 0) + 110 \cdot 0\right) = 200.$$

Thus the amount of salt in the tank approaches 200 pounds. Figure 2 plots the increasing quantity of salt for the first five hours of the process. ◀

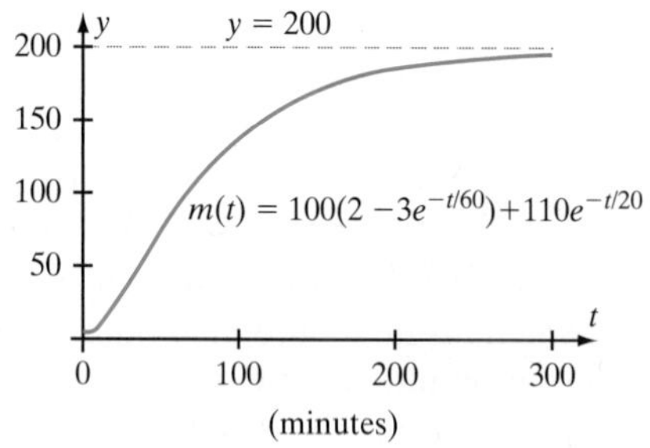

▲ Figure 2

An Application: Electric Circuits

Figure 3 is the schematic diagram of a simple electric circuit involving an electromotive force of $E(t)$ volts (such as produced by a battery or a generator), a resistor with a resistance of R ohms (Ω), and an inductor with inductance of L

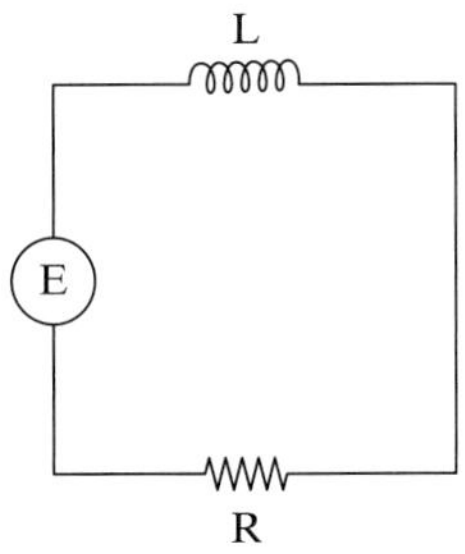

▲ Figure 3

henrys (h). It is a consequence of Kirchhoff's voltage law that the current I (measured in amperes) satisfies

$$L\frac{dI}{dt} + R \cdot I = E(t).$$

▶ **EXAMPLE 5** In the circuit of Figure 3, what is $I(t)$ if $L = 3\,h$, $R = 6\,\Omega$, and $E(t) = 12\cos(2t)V$?

Solution After dividing by L and substituting the given values, we obtain

$$\frac{dI}{dt} + 2 \cdot I = 4\cos(2t).$$

In the notation of Theorem 1 (with I and t taking the roles of y and x, respectively), we have $q(t) = 4\cos(2t)$, $p(t) = 2$, $P(t) = 2t$, and $e^{P(t)} = e^{2t}$. Theorem 1 tells us that the integral

$$\int e^{P(t)} q(t)\,dt = \int e^{2t}\cos(2t)\,dt$$

arises in the solution. To evaluate this integral, we make the substitution $x = 2t$, $dx = 2dt$, which, by Example 5 of Section 6.1 in Chapter 6, results in $\frac{1}{2}\int e^x \cos(x)\,dx$, or $\frac{1}{4}e^x\big(\sin(x) + \cos(x)\big) + C$. On resubstituting, we obtain

$$\int e^{P(t)} q(t)\,dt = \int e^{2t}\cos(2t)\,dt = \frac{1}{4}e^{2t}\big(\sin(2t) + \cos(2t)\big) + C.$$

Equation (7.7.2) of Theorem 1 then gives us

$$I(t) = e^{-P(t)}\int e^{P(t)} q(t)\,dt + Ce^{-P(t)} = 4e^{-2t}\int e^{2t}\cos(2t)\,dt + Ce^{-2t} = \cos(2t) + \sin(2t) + Ce^{-2t}.$$ ◀

Linear Equations with Constant Coefficients

Linear equations with constant coefficients arise frequently in applications. These differential equations have the form $dy/dt + \beta y = \alpha$. By rewriting this equation in the form $dy/dt = \alpha - \beta y$, we see that it is a separable differential equation (in addition to being linear). Although we may solve it by using the techniques we learned in Section 7.6, we will simplify the calculation by appealing to Theorem 1.

THEOREM 2 Suppose that α and β are constants with $\beta \neq 0$. Then the linear differential equation

$$\frac{dy}{dt} + \beta y = \alpha \tag{7.7.4}$$

has general solution

$$y(t) = \frac{\alpha}{\beta} + Ce^{-\beta t}. \tag{7.7.5}$$

The initial value problem

$$\frac{dy}{dt} + \beta y = \alpha, \qquad y(0) = y_0 \tag{7.7.6}$$

has unique solution

$$y(t) = \frac{\alpha}{\beta} + \left(y_0 - \frac{\alpha}{\beta}\right)e^{-\beta t}. \tag{7.7.7}$$

Proof. Equation (7.7.4) is the special case of equation (7.7.1) that results from setting $q(t) = \alpha$ and $p(t) = \beta$. Because $P(t) = \beta t$ is an antiderivative of $p(t)$, equation (7.7.2) tells us that equation (7.7.4) has solution

$$y(t) = e^{-\beta t} \int e^{\beta t} \alpha \, dt + Ce^{-\beta t} = \frac{\alpha}{\beta} + Ce^{-\beta t},$$

which establishes equation (7.7.5). If $y(0) = y_0$, then $\alpha/\beta + Ce^{-\beta \cdot 0} = y_0$, or $C = y_0 - \alpha/\beta$. Substituting this formula for C into equation (7.7.5) results in formula (7.7.7). ■

We will illustrate Theorem 2 with several different applications. In a textbook, it is useful to quote formulas such as (7.7.5) and (7.7.7) to avoid consecutive repetitions of what would be essentially the same calculation. Nevertheless, a student will find it more useful to understand the general technique of solving equation (7.7.4) than to memorize the actual solution. Many equations and formulas appear in calculus textbooks. The study of calculus is simplified by mastering the basic ones and learning the techniques that treat the variants.

Newton's Law for Temperature Change

Suppose that T is the temperature of an object that is surrounded by a body (e.g., water, air, etc.) at *constant* temperature T_∞—the reason for using this subscript will become apparent. Newton's Law states that the rate of change of T is proportional to the difference between T and T_∞. That is,

$$\frac{dT}{dt} = k \cdot (T_\infty - T), \tag{7.7.8}$$

where the constant of proportionality, k, is positive. Note that if $T(t) < T_\infty$, then the right side of equation (7.7.8) is positive and therefore $0 < T'(t)$. As a result, $T(t)$ increases, and we refer to *Newton's Law of Heating*. If, instead, $T_\infty < T(t)$, then $T'(t) < 0$ and T decreases. In this case, we refer to *Newton's Law of Cooling*.

THEOREM 3 Suppose that $T_0 = T(0)$ is the temperature of an object when it is placed in an environment that has constant temperature T_∞. Suppose that the temperature $T(t)$ of the object is governed by Newton's law of temperature change. That is, suppose that $T(t)$ is a solution of equation (7.7.8). Then

$$T(t) = T_\infty + (T_0 - T_\infty)e^{-kt} \qquad (0 \le t < \infty). \tag{7.7.9}$$

Proof. The right side of equation (7.7.8) can be rewritten as $k \cdot T_\infty - k \cdot T$, or $\alpha - \beta T$ with $\alpha = k \cdot T_\infty$ and $\beta = k$. With these values of α and β, equation (7.7.8) becomes

$$\frac{dT}{dt} + \beta T = \alpha.$$

Formula (7.7.7) gives a formula for the unique solution of the initial value problem:

$$T(t) = \frac{\alpha}{\beta} + \left(T_0 - \frac{\alpha}{\beta}\right)e^{-\beta t} = \frac{k \cdot T_\infty}{k} + \left(T_0 - \frac{k \cdot T_\infty}{k}k\right)e^{-kt} = T_\infty + (T_0 - T_\infty)e^{-kt}. \quad ■$$

Notice that $e^{-kt} \to 0$ as $t \to \infty$ and therefore

$$\lim_{t\to\infty} T(t) = \lim_{t\to\infty} \left(T_\infty + (T_0 - T_\infty)e^{-kt}\right) = T_\infty + (T_0 - T_\infty) \lim_{t\to\infty} e^{-kt} = T_\infty + (T_0 - T_\infty)\cdot 0 = T_\infty.$$

Thus as $t \to \infty$, the temperature of the object approaches the ambient temperature T_∞.

▶ **EXAMPLE 6** A thermometer is at room temperature (20.0°C). One minute after being placed in a patient's throat, it reads 38.0°C. One minute later, it reads 38.3°C. Is this second reading an accurate measure (to three significant digits) of the patient's temperature?

Solution Let $T(t)$ denote the temperature of the thermometer at time t with $t = 0$ corresponding to the moment it is placed in the patient's throat. Let T_∞ denote the actual temperature of the patient. Then,

$$T(t) = T_\infty + (T_0 - T_\infty)e^{-kt}$$

with $T_0 = 20$. Although there are two unknowns (k and T_∞) in this problem, we are given two observations with which to determine them:

$$38 = T(1) = T_\infty + (20 - T_\infty)e^{-k}$$

and

$$38.3 = T(2) = T_\infty + (20 - T_\infty)e^{-2k}. \tag{7.7.10}$$

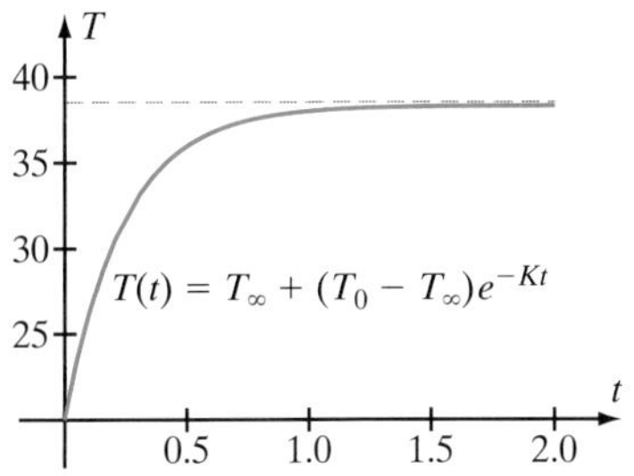

▲ Figure 4a

The first of these equations may be written as

$$\frac{38 - T_\infty}{20 - T_\infty} = e^{-k}. \tag{7.7.11}$$

Because e^{-2k} equals $(e^{-k})^2$, we may replace e^{-2k} in equation (7.7.10) with the square of the left side of (7.7.11). The result is

$$38.3 = T_\infty + (20 - T_\infty)\left(\frac{38 - T_\infty}{20 - T_\infty}\right)^2.$$

This equation simplifies to a linear equation in the unknown T_∞, which we solve to find $T_\infty = 38.305\ \ldots$. The reading at $t = 2$ is therefore an accurate measure of the patient's temperature to three significant digits. ◀

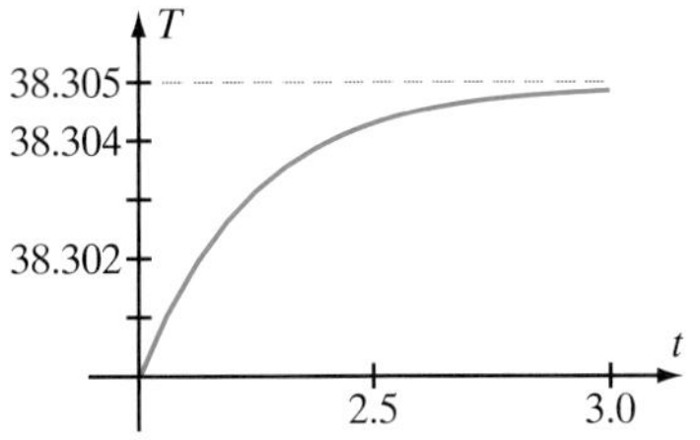

▲ Figure 4b

INSIGHT In Example 6, we may calculate $e^{-k} = 0.016662\ \ldots$ from equation (7.7.11) and the previously calculated $T_\infty = 38.305$. When we substitute the values of T_0, T_∞, and e^{-k} into equation (7.7.9), we find that $T(t) = 38.305 - 18.305 \cdot (0.016662)^t$. The graph of $T(t)$ in the window $[0, 2] \times [30, 40]$ (see Figure 4a) indicates that $T(t)$ is very close to T_∞ after $t = 1.3$. A closer look using the window $[1.5, 2.5] \times [38.266, 38.305]$ reveals that $T(t)$ does continue to increase, but very slowly (Figure 4b). This phenomenon is familiar to anyone who has watched the readout of a digital thermometer while waiting for its beep.

The Linear Drag Law

We consider vertical motion above Earth's surface. Let $y(t)$ denote the height of an object above Earth's surface at time t and $v = dy/dt$ the velocity. The force R of air resistance (or *air drag*) can often be approximated by a *linear velocity law*:

$$R(v) = -k \cdot v. \tag{7.7.12}$$

Here k is a positive constant whose units are mass $\times$ time^{-1}. Notice that the minus sign in formula (7.7.12) ensures that $R(v)$ and v have opposite signs. This means that the direction of air drag is opposite to the direction of motion. (In upward motion, y is increasing, $v = y'$ is positive, and $R(v) = -kv$ is negative, hence a downward force. In downward motion, y is decreasing, $v = y'$ is negative, and $R(v) = -kv$ is positive, hence an upward force.)

▶ **EXAMPLE 7** Let $v(t)$ denote the velocity at time t of an object of mass m that is dropped from a great height. Assuming that air resistance is given by the formula (7.7.12), determine $v(t)$. What is the behavior of $v(t)$ for large t?

Solution Denoting the acceleration due to gravity by the positive constant g, we write the downward force of gravity as $-m \cdot g$. The equation of motion is therefore

$$\underbrace{\underbrace{m \cdot \frac{dv}{dt}}_{\text{mass} \times \text{acceleration}}}_{\text{Total Force}} = \underbrace{-k \cdot v}_{\text{Resistive Force}} + \underbrace{(-m \cdot g)}_{\text{Force of gravity}} .$$

On dividing by m and bringing v to the left side, we obtain the initial value problem

$$\frac{dv}{dt} + \frac{k}{m} \cdot v = -g, \quad v(0) = 0.$$

This initial value problem is of the form (7.7.6) with unknown function v instead of y. To complete the correspondence, we set $\alpha = -g$, $\beta = k/m$, and $y_0 = v(0) = 0$ in (7.7.6). The solution to our initial value problem is then obtained by using these values for the constants and replacing y with v in equation (7.7.7). We obtain

$$v(t) = \underbrace{\frac{\alpha}{\beta} + \left(y_0 - \frac{\alpha}{\beta}\right) e^{-\beta t}}_{\text{right side of (7.7.7)}} = \frac{-g}{k/m} + \left(0 - \frac{(-g)}{k/m}\right) e^{-kt/m},$$

or

$$v(t) = -\frac{mg}{k}\left(1 - e^{-kt/m}\right). \tag{7.7.13}$$

From this formula, we can calculate

$$\lim_{t \to \infty} v(t) = -\frac{mg}{k}.$$

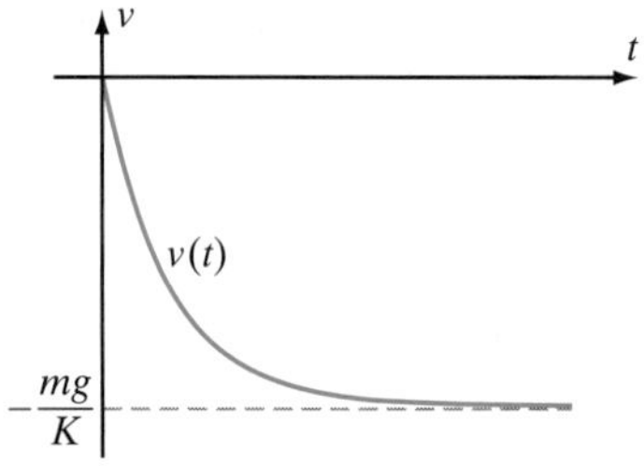

▲ Figure 5 Plot of velocity during a fall under gravity, assuming the linear drag law

The quantity $-mg/k$ is said to be the *terminal velocity* of the falling object. This terminology is somewhat misleading: Terminal velocity is approached arbitrarily closely but is not attained. Figure 5 shows the graph of v and its horizontal asymptote $v = -mg/k$. ◀

INSIGHT One of the reasons that students from so many different backgrounds study calculus is its wide applicability. Mathematicians will often state a theorem in a general form (such as the formulation of Theorem 2) in order to maximize the theorem's scope. We have already seen that the differential equation studied in Theorem 2 applies to problems concerning heat transfer and motion. Here are four more applications of the same differential equation.

1. A drug delivered intravenously to a patient enters his bloodstream at a constant rate. The drug is broken down and eliminated from the patient's body at a rate that

is proportional to the amount of the drug in the bloodstream. If $m(t)$ is the amount of the drug in the bloodstream, then

$$\frac{dm}{dt} = \alpha - \beta \cdot m \tag{7.7.14}$$

for constants α and β.

2. According to Fick's Law, the diffusion of a solute through a cell membrane is described by

$$\frac{dc}{dt} = k\left(c_{\infty} - c(t)\right). \tag{7.7.15}$$

Here $c(t)$ is the concentration of the solute in the cell, c_{∞} is the concentration outside the cell, and k is a positive constant.

3. A pacemaker attached to the heart creates a potassium current I that is determined by the differential equation

$$\frac{dI}{dt} = \alpha(1 - I) - \lambda I \tag{7.7.16}$$

involving positive constants α and λ.

4. In an evolving galaxy, the rate equation for the production and absorption of electromagnetic radiation N has the form

$$\frac{dN}{dt} = \alpha \cdot (1 - k \cdot N) \tag{7.7.17}$$

where α and k are positive constants.

QUICK QUIZ

1. What is the integrating factor for the linear differential equation $\frac{dy}{dx} - \frac{2}{x}y = x^3$?
2. If $P(x) = \int p(x)\,dx$, then what is the antiderivative of $e^{P(x)}\frac{dy}{dx} + e^{P(x)}p(x)y$?
3. What is the general solution of $\frac{dy}{dx} + \frac{1}{x}y = 2$?
4. If $y = 2 + Ce^{-3t}$ is the general solution of $\frac{dy}{dt} + \beta t = \alpha$, then what are α and β?

Answers

1. $\exp(1/x^2)$ 2. $e^{P(x)}y + C$ 3. $y = x + C/x$ 4. $\alpha = 6$; $\beta = 3$

EXERCISES

Problems for Practice

In each of Exercises 1–16, write the linear differential equation in the standard form that is given by equation (7.7.1). Determine the integrating factor $e^{P(x)}$ (in the notation of Theorem 1) and use it, as illustrated by Example 3, to solve the given equation.

1. $\frac{dy}{dx} = 3y$

2. $\frac{dy}{dx} + \frac{2}{3}y = 0$

3. $\frac{dy}{dx} + 4y = 0$

4. $\frac{dy}{dx} + 2y = 1$

5. $\frac{dy}{dx} = 1 + \frac{y}{2}$

6. $\frac{dy}{dx} = 15x^2 - \frac{2xy}{1 + x^2}$

7. $\frac{dy}{dx} = 1 - \frac{y}{x}$

8. $\frac{dy}{dx} + \frac{2}{x}y = 7\sqrt{x}$

9. $\dfrac{dy}{dx} + xy = x$
10. $\dfrac{dy}{dx} + 2xy = \exp(-x^2)$
11. $\dfrac{dy}{dx} - 3y = e^{3x}$
12. $\dfrac{dy}{dx} + \dfrac{1}{2}y = e^{-x/2}$
13. $\dfrac{dy}{dx} + y = 6e^{2x}$
14. $\dfrac{dy}{dx} = \csc(x) + y\cot(x)$
15. $\dfrac{dy}{dx} + 2xy = 8x^3$
16. $\dfrac{dy}{dx} + \dfrac{y}{x\ln(x)} = 1$

In each of Exercises **17–28**, solve the given initial value problem.

17. $\dfrac{dy}{dx} + 2y = 3$ $\quad y(0) = 1/2$
18. $\dfrac{dy}{dx} + 2 = y$ $\quad y(0) = -1$
19. $x\dfrac{dy}{dx} + y = 6x\sqrt{3+x^2}$ $\quad y(1) = 7$
20. $\dfrac{dy}{dx} + \dfrac{y}{x} = 2\cos(\pi x^2)$ $\quad y(2) = -1$
21. $\dfrac{dy}{dx} + \dfrac{2}{x}y = 7\sqrt{x}$ $\quad y(4) = 17$
22. $\dfrac{dy}{dx} + 2xy = \exp(-x^2)$ $\quad y(0) = 5$
23. $\dfrac{dy}{dx} - y = e^x$ $\quad y(0) = -3$
24. $\dfrac{dy}{dx} + y = 3 + e^{-x}$ $\quad y(0) = -1$
25. $\dfrac{dy}{dx} + 2y = 4x$ $\quad y(0) = 3$
26. $\dfrac{dy}{dx} + 2x = y$ $\quad y(0) = 7$
27. $\dfrac{dy}{dx} = y + x^2$ $\quad y(0) = 1$
28. $\dfrac{dy}{dx} + y = (1 + e^{2x})^{-1}$ $\quad y(0) = 0$

29. A mixing tank contains 200 L of water in which 10 kg salt is dissolved. At time $t = 0$, a valve is opened, and a solution with 1 kg salt per 10 L water enters the tank at the rate of 20 L per minute. An outlet pipe maintains the volume of fluid in the tank by allowing 20 L of the thoroughly mixed solution to flow out each minute. What is the mass $m(t)$ of salt in the tank at time t? What is $m_\infty = \lim_{t\to\infty} m(t)$?
30. A mixing tank contains 100 gallons of water in which 60 pounds of salt is dissolved. At time $t = 0$, a valve is opened, and a solution with 1 pound of salt per 5 gallons of water enters the tank at the rate of 2.5 gallons per minute. An outlet pipe maintains the volume of fluid in the tank by allowing 2.5 gallons of the thoroughly mixed solution to flow out each minute. What is the weight $m(t)$ of salt in the tank at time t? What is $m_\infty = \lim_{t\to\infty} m(t)$?
31. A mixing tank contains 100 gallons of water in which 60 pounds of salt is dissolved. At time $t = 0$, a valve is opened, and water enters the tank at the rate of 2.5 gallons per minute. An outlet pipe maintains the volume of fluid in the tank by allowing 2.5 gallons of the thoroughly mixed solution to flow out each minute. What is the weight $m(t)$ of salt in the tank at time t? What is $m_\infty = \lim_{t\to\infty} m(t)$?
32. A mixing tank contains 200 L of water in which 10 kg salt is dissolved. At time $t = 0$, a valve is opened and water enters the tank at the rate of 20 L per minute. An outlet pipe maintains the volume of fluid in the tank by allowing 20 L of the thoroughly mixed solution to flow out each minute. What is the mass $m(t)$ of salt in the tank at time t? What is $m_\infty = \lim_{t\to\infty} m(t)$?
33. A hot bar of iron with temperature 150°F is placed in a room with constant temperature 70°F. After 4 minutes, the temperature of the bar is 125°F. Use Newton's Law of Cooling to determine the temperature of the bar after 10 minutes.
34. A hot rock with initial temperature 160°F is placed in an environment with constant temperature 90°F. The rock cools to 150°F in 20 minutes. Use Newton's Law of Cooling to determine how much longer it take the rock to cool to 100°F.

Further Theory and Practice

35. After falling for 10 seconds, a dropped object hits the ground at 99% of its terminal velocity. If the linear drag coefficient k is 2 kg/s, then what is the mass of the object?
36. If a mass m that is dropped falls with linear drag coefficient k, then at what time does the object reach half its terminal velocity?
37. A lake has a volume of 3×10^9 cubic feet. The concentration of pollutants in the lake is 0.02 % by weight. An inflowing river brings clean water in at the rate of 1 million cubic feet per day. It flows out at the same rate of 1 million cubic feet of water a day. How long will it be before the pollution level is 0.001%? (A cubic foot of water weighs 62.43 pounds.)
38. A casserole dish at temperature $T_0 = 180°C$ is plunged into a large basin of hot water whose temperature is $T_\infty = 48°C$. The temperature of the dish satisfies Newton's Law of Cooling. In 30 seconds, the temperature of the dish is 80°C. How long after the immersion is it before the temperature of the dish cools to 60°C?
39. A thermometer is brought from a 72°F room to the outdoors. After 1 minute, the thermometer reads 50°F, and after a further 30 seconds, the thermometer reads 44°F.

What is the actual temperature outside? (The equation to be solved may appear to be complicated, but it can be reduced to a quadratic equation.)

40. Suppose that $R = 30\,\Omega$, $L = 6\,\text{h}$, and $E(t) = 12\,\text{V}$ in Figure 3. What is $I(t)$?

41. Suppose that $R = 60\,\Omega$, $L = 12\,\text{h}$, and $E(t) = 24e^{-5t}\,\text{V}$ in Figure 3. What is $I(t)$?

42. Suppose that $R = 30\,\Omega$, $L = 15\,\text{h}$, and $E(t) = 5t\,\text{V}$ in Figure 3. If $I(0) = 0$, what is $I(t)$?

43. Suppose that $R = 15\,\Omega$, $L = 5\,\text{h}$, and $E(t) = 20\sin(t)$ V in Figure 3. What is $I(t)$? (Use the integration formula of Exercise 86, Section 6.1 of Chapter 6).

44. A 1000 L tank initially contains a 200 L solution in which 24 kg of salt is dissolved. Beginning at time $t = 0$, an inlet valve allows fresh water to flow into the tank at the constant rate of 12 L/min, and an outlet valve is opened so that 10 L/min of the solution is drained. When the tank has been filled, how much salt does it contain?

45. A 1000 L tank initially contains a 200 L solution in which 24 kg of salt is dissolved. Beginning at time $t = 0$, an inlet valve allows fresh water to flow into the tank at the constant rate of 12 L/min, and an outlet valve is opened so that 16 L/min of the solution is drained. How much salt does the tank contain after 25 minutes?

46. In Hull's learning model, habit strength $H(r)$ depends on the number r of reinforcements according to $H(r) = \lambda(1 - e^{-\lambda r})$. Show that H satisfies differential equation (7.7.4). What are the constants α and β?

47. Several memory experiments have shown that when a pattern is shown to a subject and then withdrawn, the probability that the subject will forget the pattern in the time interval $[0, t]$ is

$$F(t) = p - c \cdot \lambda^t$$

where p, c, and λ are positive constants with $c < p$ and $\lambda < 1$. Show that F satisfies differential equation (7.7.4). What are the constants α and β?

48. Suppose that during wartime, a particular type of weapon is produced at a constant rate μ and destroyed in battle at a constant relative rate δ. If there were N_0 of these weapons at the outbreak of the war ($t = 0$), then how many were there at time t? If the war is a lengthy one, about how many of these weapons will be on hand at time t when t is large?

49. Let $m(t)$ be the amount of a drug in the bloodstream. If the drug is delivered intravenously at a constant rate α and if it is eliminated from the patient's body at a constant relative rate β, then $m(t)$ satisfies equation (7.7.14). What, approximately, is the amount of the drug that is maintained in the bloodstream of a long-term patient?

50. Solve equation (7.7.15), which expresses Fick's Law for the diffusion of a solute through a cell membrane.

51. Solve equation (7.7.16) for the potassium current $I(t)$ induced by a pacemaker.

52. Solve equation (7.7.17) for the electromagnetic radiation in an evolving galaxy.

53. A tank holds a salt water solution. At time $t = 0$, there are L liters in the tank; dissolved in the L liters are m_0 kilograms of salt. Fresh water enters the tank at a rate of ρ liters per hour and is mixed thoroughly with the brine solution. At the same time, the mixed solution is drained at the rate of r liters per hour. Let $m(t)$ denote the mass of the salt in the tank at time t.

a. Assuming perfect mixing prior to draining, show that $m(t)$ satisfies the differential equation

$$m'(t) = -\frac{r}{L + (\rho - r)t}\, m(t).$$

b. What is $m(t)$ in the case that $\rho = r$?

c. Supposing that $\rho \neq r$, solve for $m(t)$.

54. An object of mass m is dropped from height H. Let v denote its velocity at time t. Assume that air drag is linear: $R(v) = -kv$. Show that the height $y(t)$ of the object at time t is

$$y(t) = H - \frac{mg}{k}t + \frac{m^2 g}{k^2}\left(1 - e^{-kt/m}\right).$$

55. Consider the object discussed in Exercise 54. Suppose that at time τ, the value of the air drag coefficient changes from k to κ. (This happens, for example, when a skydiver opens his parachute.) Let $\eta(t)$ denote the object's height above Earth at time t, and let $\sigma(t)$ denote the object's velocity at time t. For $0 \leq t \leq \tau$, $\sigma(t) = v(t)$ where $v(t)$ is given by formula (7.7.13), and $\eta(t) = y(t)$ where $y(t)$ is given by the displayed equation of Exercise 54. At time τ, the velocity of the object is

$$v_\tau = v(\tau) = -\frac{mg}{k}\left(1 - e^{-k\tau/m}\right),$$

and its height above Earth is

$$y_\tau = y(\tau) = H - \frac{mg}{k}\tau + \frac{m^2 g}{k^2}\left(1 - e^{-k\tau/m}\right).$$

The initial value problems that describe the motion after this time τ are

$$m\frac{d\sigma}{dt} = -mg - \kappa\sigma, \quad \sigma(\tau) = v_\tau$$

and

$$\frac{d\eta}{dt} = \sigma, \quad \eta(\tau) = y_\tau.$$

Show that for $t \geq \tau$,

$$\sigma(t) = -\frac{mg}{\kappa} - \left(\frac{mg}{\kappa} + v_\tau\right)\exp\left(-\frac{\kappa}{m}(t - \tau)\right)$$

and

$$\eta(t) = y(\tau) + \int_\tau^t \sigma(u)\, du.$$

For later reference, obtain the explicit form of $\eta(t)$:

$$\eta(t) = H - \frac{mg\tau}{k} + \frac{m^2 g}{k^2}\left(1 - e^{-k\tau/m}\right) - \frac{mg}{\kappa}(t-\tau) - \frac{m}{\kappa^2}(mg + v_\tau \kappa)\left(1 - e^{-(t-\tau)\kappa/m}\right).$$

Let k denote a positive constant. An object is thrown straight up from the ground with initial velocity v_0. Suppose that air resistance gives rise to the force $-kv$. Let T_u denote the time for the upward trajectory, and T_d the time for the downward trajectory. Exercises 56–59 outline a derivation of the somewhat surprising inequality $T_u < T_d$.

56. Show that

$$v(t) = \frac{mg}{k}\left(\left(1 + \frac{kv_0}{mg}\right)e^{-kt/m} - 1\right).$$

What is the value of $v(T_u)$? Show that

$$T_u = \frac{m}{k}\ln\left(1 + \frac{k}{mg}v_0\right).$$

57. Show that the height $y(t)$ of the projectile at time t is

$$y(t) = \frac{mg}{k^2}\left(\left(m + \frac{kv_0}{g}\right)\left(1 - e^{-kt/m}\right) - kt\right)$$

for $0 \le t \le T_u + T_d$. In terms of the parameters of m, g, and K, what is the height H that the projectile reaches before beginning to fall? Use the formula for T_u to express $y(t)$ in the following form:

$$y(t) = \frac{mg}{k^2}\left(me^{kT_u/m}\left(1 - e^{-kt/m}\right) - kt\right).$$

58. Show that the time T_d for the downward trajectory is

$$T_d = -\frac{m}{k}\ln\left(1 + \frac{k}{mg}v_I\right)$$

where v_I is the velocity at the time of impact with the ground.

59. Use the second formula for $y(t)$ found in Exercise 57, together with the equation $y(T_u + T_d) = 0$, to show that

$$\left(e^{kT_u/m} - e^{-kT_d/m}\right) = \frac{kT_u}{m} + \frac{kT_d}{m}.$$

Use Exercise 69 of Section 5.5 in Chapter 5 to deduce that $T_d > T_u$. In words, it takes *longer* to come down than to go up!(It may also be shown that $v_0 > v_I$.)

Since the 1920s, mathematical models have been developed to study running performance. These models rely on physical theory, empirical evidence, and statistical analysis. In 1973, J. B. Keller modified an existing mathematical theory of running to make it more appropriate for sprints. Exercises 60–62 are concerned with the Hill-Keller theory of sprinting. In these exercises, $v(t)$ denotes the speed of a runner at time t, $p(t)$ denotes the horizontal component of the propulsive force per unit mass that is exerted by the runner at time t, and $R(v)$ denotes the total resistive force per unit mass.

60. Assume that $R(v) = v/\tau$ for some positive constant τ (the dimension of which is time). Suppose also that $p(t) = P$ for a positive constant P (the unit of which is distance/time^2). Obviously, this assumption on $p(t)$ is not tenable for races longer than a sprint. Show that

$$\frac{dv}{dt} = P - \frac{v}{\tau}.$$

Find an explicit solution for $v(t)$. (What initial condition must you use?)

61. Use the explicit solution for $v(t)$ found in Exercise 60 to show that

$$v_\infty = \lim_{t\to\infty} v(t)$$

exists. This "terminal velocity" is called *maximum velocity* in the track community.

62. Let $x(t)$ denote the distance a sprinter has run in time t. Show that

$$x(t) = v_\infty\left(t - \tau\left(1 - e^{-t/\tau}\right)\right).$$

Calculator/Computer Exercises

63. A 1 kg object is dropped. After 1 second, its (downward) speed is 9.4 m/s. Assuming that the drag is proportional to the velocity, what is the terminal velocity of the object?

64. A 0.5 kg object is dropped. Its terminal velocity is 49 m/s. Assuming that the drag is proportional to the velocity and that it was dropped from a sufficiently great height, at what time after being dropped is its speed 40 m/s?

65. On December 7, 1941, during the attack on Pearl Harbor, an 800 kg bomb that was dropped from an altitude of 3170, meters struck the battleship *USS Arizona* and set off a series of catastrophic explosions. The ship sank in nine minutes with a death toll of 1177. According to an analysis of the attack that was published in 1997, the flight of the bomb lasted 26 seconds. Assume the Linear Drag Law $R(v) = -kv$.

a. What was the value of k? (Use Exercise 58. Include the units of k in your answer.)

b. What was the theoretical terminal velocity of the bomb.

c. What was the actual velocity of the bomb when it struck the *USS Arizona*?

d. Plot the height $y(t)$ of the bomb for $0 \le t \le 26$.

e. What would the time of fall have been had there been no air drag? The velocity on impact?

66. Suppose that for a skydiver, the value of k in the linear drag law $R(v) = -kv$ is 0.682 lb·s/ft. Let $P(t)$ denote the percentage of terminal velocity of a 120 pound skydiver t seconds into the jump. Plot the function P. How many seconds into the jump is it when the skydiver reaches 95% of terminal velocity? How many feet has the skydiver fallen by this time?

67. A 0.1 kg projectile is thrown up into the air with an initial speed of 20 m/s. Assume that $R(v) = -kv$ where $k = 0.03$ kg/s^{-1}. Approximate the maximum height H above ground that the object reaches, the impact velocity v_I with which the projectile strikes the ground, and the times T_u and T_d of the upward and downward trajectories. Refer to Exercises 56–59.

68. Let $y(t)$ and $v(t)$ denote the height and velocity of the object of Exercise 67. In the same coordinate plane, plot the kinetic energy $T(t) = \mathrm{mv}(t)^2/2$, the potential energy $U(t) = mg \cdot y(t)$, and the total energy $E(t) = T(t) + U(t)$ of the object. Separately, plot $E'(t)$. The behavior of $E(t)$ and the range of values of $E'(t)$ reflect a property of resistive forces that is called *nonconservative*. Explain.

69. At the 1987 World Championships in Rome, split times for the Men's 100 Meters competition show that Ben Johnson (Canada) and Carl Lewis (United States) each had maximum velocity v_∞ equal to 11.8 m/s. (Refer to Exercises 60–62 for the Hill-Keller theory of sprinting). Johnson's winning time was 9.83 s, and Lewis's second place time was 9.93 s.

a. What were the approximate values of P and τ for these men? According to the Hill-Keller model, can Johnson's victory be attributed to greater propulsive force, lesser resistance to motion, or to both factors?

b. In the same coordinate plane, plot $v(t)$ for both runners for $0 \le t \le 9.83$.

c. In meters, what was Johnson's winning margin?

70. As of this writing, former president George Herbert Bush has been the only American president to have jumped from an airplane. During World War II, the future president parachuted from his airplane when it was shot down. On March 25, 1997, at the age of 72, the former president made a recreational skydive/parachute jump. In doing this problem, refer to Exercises 54 and 55 for notation and formulae. Assume that the mass of the president and his jumping gear was 7.03 lb $\cdot\, s^2$/ft. Assume linear drag: $R(v) = -kv$ during the skydive phase of the jump and $R(v) = -kv$ after deployment of the parachute. Use the value $k = 1.4$ lb/s.

a. The president jumped from a height of 12,500 feet and did not open his parachute until he reached 4,500 feet. Compute the duration τ of this skydive. According to the model, what was the president's velocity $v(\tau)$ when he opened his parachute? Plot $v(t)$ for $0 \le t \le \tau$.

b. The total time of the president's dive and jump was about 9 minutes. In the absence of more precise data, use 540 seconds as the elapsed time between jumping from the plane and touching down on the ground. As in Exercise 55, use $\sigma(t)$ and $\eta(t)$ to denote President Bush's velocity and height above ground for $0 \le t \le 540$. From the equation $\eta(540) = 0$, determine k to three significant digits.

c. According to this model, what was the terminal velocity of the president's parachute jump? With what velocity did he touch down?

d. Plot $\sigma(t)$ and $\eta(t)$ for $0 \le t \le 540$.

71. A curling stone released at the front hog line comes to a stop in the button, 93 ft from the line, 10 seconds later. If the deceleration of the stone is

$$\frac{dv}{dt} = -k\,v^{0.05}$$

for some positive constant k, then what was the speed of the stone when it was released?

Summary of Key Topics in Chapter 7

Solids of Revolution: The Method of Disks (Section 7.1)

Let f be a continuous positive function on the interval $[a, b]$. Let $\mathcal{R}$ be the region bounded above by the graph of $y = f(x)$, below by the x-axis, and on the sides by $x = a$ and $x = b$. Then the volume V of the solid generated by rotating $\mathcal{R}$ about the x-axis is

$$V = \pi \int_a^b f(x)^2\,dx.$$

Let g be a continuous positive function on the interval $[c, d]$. Let $\mathcal{R}$ be the region bounded on the right by the graph of $x = g(y)$, on the left by the y-axis, above by $y = d$, and below by $y = c$. Then the volume V of the solid generated by rotating $\mathcal{R}$ about the y-axis is

$$V = \pi \int_c^d g(y)^2\,dy.$$

Solids of Revolution: The Method of Washers (Section 7.1)

Let f and g be continuous functions with $0 \le g(x) \le f(x)$ for x in the interval $[a, b]$. Let $\mathcal{R}$ be the region bounded above by the graph of $y = f(x)$, below by the graph of $y = g(x)$, and on the sides by $x = a$ and $x = b$. Then the volume V of the solid generated by rotating $\mathcal{R}$ about the x-axis is

$$V = \pi \int_a^b \left(f(x)^2 - g(x)^2\right) dx.$$

Solids of Revolution: The Method of Cylindrical Shells (Section 7.1)

Let f and g be continuous functions with $0 \le g(x) \le f(x)$ for x in the interval $[a, b]$. Let $\mathcal{R}$ be the region bounded above by the graph of $y = f(x)$, below by the graph of $y = g(x)$, and on the sides by $x = a$ and $x = b$. If $c \le a$ or $b \le c$, then the volume V of the solid generated by rotating $\mathcal{R}$ about the line $x = c$ is given by

$$V = 2\pi \int_a^b |x - c| \left(f(x) - g(x)\right) dx.$$

Arc Length (Section 7.2)

If f is a continuously differentiable function on the interval $[a, b]$, then the arc length of the graph of f is

$$\int_a^b \sqrt{1 + f'(x)^2}\, dx.$$

Surface Area (Section 7.2)

If f is a nonnegative, continuously differentiable function on the interval $[a, b]$, then the surface area of the surface obtained when the graph of f is rotated about the x-axis is

$$\int_a^b 2\pi f(x) \sqrt{1 + f'(x)^2}\, dx.$$

Average Value of a Function (Section 7.3)

If $f(x)$ is a continuous function on an interval $[a, b]$, then the average value of f on $[a, b]$ is

$$f_{\text{avg}} = \frac{1}{b - a} \int_a^b f(x)\, dx.$$

If X is a random variable that takes values in an interval $[a, b]$, and if f is a nonnegative function on $[a, b]$ such that the probability that X takes a value in any subinterval $[\alpha, \beta]$ is $\int_\alpha^\beta f(x)\, dx$, then the mean μ_X of X (or expectation of X) is

$$\mu_X = \int_a^b x f(x)\, dx.$$

Center of Mass (Section 7.4)

If $g(x) \le f(x)$ for x in $[a, b]$; if c is any real number; if $\mathcal{R}$ is the region bounded above by the graph of $y = f(x)$, below by the graph of $y = g(x)$, and on the sides by the line segments $x = a$ and $x = b$; and if the mass density of $\mathcal{R}$ is constant; then the center of mass $(\bar{x}, \bar{y})$ of $\mathcal{R}$ is given by the equations

$$\bar{x} = \frac{1}{A}\int_a^b x\big(f(x) - g(x)\big)\,dx$$

and

$$\bar{y} = \frac{1}{2A}\int_a^b \big(f(x)^2 - g(x)^2\big)\,dx$$

where A is the area of $\mathcal{R}$.

Work (Section 7.5)

If a body travels a linear path with position given by x (measured in feet), $a \le x \le b$, under the influence of a force in the direction of motion which at position x is $F(x)$, then the work performed in traveling from a to b is $\int_a^b F(x)\,dx$.

First Order Differential Equations (Sections 7.6., 7.7)

A first order differential equation is an equation of the form

$$\frac{dy}{dx} = F(x, y),$$

where $F(x, y)$ is an expression that involves both x and y (in general). A differentiable function y is a solution of this differential equation if $y'(x) = F\big(x, y(x)\big)$ for every x in some open interval. If $F(x, y) = g(x)h(y)$, then the equation is separable, and the (possibly implicit) solution is found by isolating the variables on opposite sides of the equation and integrating:

$$\int \frac{1}{h(y)}\,dy = \int g(x)\,dx.$$

If $F(x, y) = q(x) - p(x)y$, then the equation, which is said to be linear, becomes

$$\frac{dy}{dx} + p(x)y = q(x)$$

and is solved by multiplying through by $e^{P(x)}$ where $P(x)$ is an antiderivative of $p(x)$. The left side is then rewritten as a derivative,

$$\frac{d}{dx}\left(e^{P(x)}y\right) = e^{P(x)}q(x),$$

and the solution is obtained by integrating both sides:

$$e^{P(x)}y = \int e^{P(x)}q(x)\,dx + C.$$

Review Exercises For for Chapter 7

In each of Exercises 1–20, a planar region $\mathcal{R}$ is described and an axis ℓ is given. Calculate the volume V of the solid of revolution that results when $\mathcal{R}$ is rotated about ℓ.

1. $\mathcal{R}$ is the region bounded above by the curve $y = \sqrt{x-1}$, below by the x-axis, and on the right by the vertical line $x = 2$; ℓ is the line $y = 0$.
2. $\mathcal{R}$ is the region that is bounded by the x-axis, the curve $y = \sqrt{x}$, and the line $x = 4$; ℓ is the line $x = 0$.
3. $\mathcal{R}$ is the region in the first quadrant that is bounded above by $y = 1 + x^2$ and on the right by $x = 1$; ℓ is the line $y = 0$.
4. $\mathcal{R}$ is the region in the first quadrant that is bounded above by $y = 1 + x^2$ and on the right by $x = 1$; ℓ is the line $x = 0$.

5. $\mathcal{R}$ is the region in the first quadrant that is bounded above by $y = 1 + x^2$ and on the right by $x = 1$; ℓ is the line $x = 1$.
6. $\mathcal{R}$ is the region in the first quadrant that is bounded above by $y = 1 + x^2$ and on the right by $x = 1$; ℓ is the line $y = 2$.
7. $\mathcal{R}$ is the region in the first quadrant that is bounded above by $y = 5x/(2 + x^3)$, below by the x-axis, and on the right by $x = 2$; ℓ is the line $y = 0$.
8. $\mathcal{R}$ is the region in the first quadrant that is bounded above by $y = 5x/(2 + x^3)$, below by the x-axis, and on the right by $x = 2$; ℓ is the line $x = 0$.
9. $\mathcal{R}$ is the region bounded above by $y = 4x - x^2 - 3$ and below by the x-axis; ℓ is the line $x = 0$.
10. $\mathcal{R}$ is the first quadrant region bounded above by $y = 2$, below by $y = 2^x$, and on the left by $x = 0$; ℓ is the line $y = 2$.
11. $\mathcal{R}$ is the region bounded above by $y = 1$, below by $y = 1/x$, and on the right by $x = 2$; ℓ is the line $y = 0$.
12. $\mathcal{R}$ is the region bounded above by $y = 1$, below by $y = 1/x$, and on the right by $x = 2$; ℓ is the line $y = 1/2$.
13. $\mathcal{R}$ is the region bounded above by $y = 1$, below by $y = 1/x$, and on the right by $x = 2$; ℓ is the line $x = 2$.
14. $\mathcal{R}$ is the region bounded below by the graph of $y = 2x^2 - 8x + 7$ and above by the graph of below the graph of $y = 1$; ℓ is the line $x = 0$.
15. $\mathcal{R}$ is the region bounded above by the curve $y = 9 - x^2$ and below by the line $y = x + 7$; ℓ is the line $y = 0$.
16. $\mathcal{R}$ is the region between the curves $y = x^{1/4}$ and $y = x^{1/2}$; ℓ is the line $y = -1$.
17. $\mathcal{R}$ is the region bounded above by the graph of $y = x\sqrt{\ln(x)}$ for $1 \le x \le e$, below by the x-axis, and on the right by $x = e$; ℓ is the line $y = 0$.
18. $\mathcal{R}$ is the region bounded above by the line $y = 3 - x$, below by the line $y = \sqrt{x - 1}$, and on the left by the line $x = 1$; ℓ is the line $x = 3$.
19. $\mathcal{R}$ is the region in the first quadrant below the graph of $y = 1 + \sin(x)$, above the graph of $y = \cos(x)$, and to the left of $x = \pi/2$; ℓ is the line $y = 0$.
20. $\mathcal{R}$ is the region between the curve $y = 2\sin(x)$, $\pi/6 \le x \le 5\pi/6$ and the line $y = 1$; ℓ is the line $y = -1$.

In Exercises 21–26, calculate the arc length L of the given curve.

21. $y = \dfrac{1}{3}(4x + 1)^{3/2}$ $\quad 0 \le x \le 1/4.$
22. $y = \dfrac{1}{2}\ln(\cos(2x))$ $\quad 0 \le x \le \pi/6.$
23. $y = \left(\dfrac{2}{3}x^{1/2} + \dfrac{1}{36}\right)^{3/2}$ $\quad 1/4 \le x \le 1.$
24. $y = \dfrac{1}{4}x^4 + \dfrac{1}{8}x^{-2}$ $\quad 1/2 \le x \le 1.$
25. $y = \dfrac{2}{5}x^{5/2} + \dfrac{1}{2}x^{-1/2}$ $\quad 1/4 \le x \le 1.$
26. $y = \dfrac{2}{3}x^{3/2} - \dfrac{1}{2}x^{1/2}$ $\quad 1 \le x \le 4.$

In Exercises 27–30, calculate the area S of the surface obtained when the graph of the given function f is rotated about the x-axis.

27. $f(x) = \cos(x)$ $\quad -\pi/2 \le x \le \pi/2$
28. $f(x) = \sqrt{25 - x^2}$ $\quad 2 \le x \le 3$
29. $f(x) = 4\sqrt{x}$ $\quad 0 \le x \le 5$
30. $f(x) = x^3 + \dfrac{1}{12x}$ $\quad 1/2 \le x \le 1$

In each of Exercises 31–34, calculate the length of the given parametric curve.

31. $x = t^2 - 1, y = 2 - 3t^2$ $\quad 0 \le t \le \sqrt{2}$
32. $x = 1 + 2\cos(3 + t), y = 3 - 2\sin(3 + t)$ $\quad 1 \le t \le 3$
33. $x = t^3, y = 3t^2$ $\quad 0 \le t \le 1$
34. $x = \dfrac{1}{3}\exp(3t), y = \dfrac{1}{2}\exp(2t)$ $\quad 0 \le t \le 1/2$

In each of Exercises 35–42, find the average value f_{avg} of the given function f on the given interval I.

35. $f(x) = 4x - x^2 - 3$ $\quad I = [1, 3]$
36. $f(x) = x^2(4 - x^3)^3$ $\quad I = [-1, 2]$
37. $f(x) = 3/(1 + x)$ $\quad I = [1, 7]$
38. $f(x) = 2^x$ $\quad I = [-1, 3]$
39. $f(x) = x/(1 + 2x^2)^2$ $\quad I = [0, 2]$
40. $f(x) = 6/(9 - x^2)$ $\quad I = [-1, 2]$
41. $f(x) = \ln(x)$ $\quad I = [1/2, 2]$
42. $f(x) = xe^x$ $\quad I = [0, \ln(3)]$

In each of Exercises 43–50, calculate the mean of the random variable X that has the given probability density function f on the given interval I.

43. $f(x) = \sin(x)$ $\quad I = [0, \pi/2]$
44. $f(x) = \dfrac{3}{14}\sqrt{x}$ $\quad I = [1, 4]$
45. $f(x) = \dfrac{3}{4}(4x - x^2 - 3)$ $\quad I = [1, 3]$
46. $f(x) = \dfrac{\ln(2)}{7}2^{x+1}$ $\quad I = [-1, 2]$
47. $f(x) = \dfrac{4}{\pi}(1 - x^2)^{-1/2}$ $\quad I = [0, 1/\sqrt{2}]$
48. $f(x) = \dfrac{2}{\pi - 2}\arcsin(x)$ $\quad I = [0, 1]$
49. $f(x) = \ln(x)$ $\quad 1 \le x \le e$
50. $f(x) = \dfrac{4}{\ln(3)(4 - x^2)}$ $\quad I = [0, 1]$

In each of Exercises 51–56, find the center of mass $(\bar{x}, \bar{y})$ of the given region $\mathcal{R}$, assuming that it has uniform unit mass density.

51. $\mathcal{R}$ is bounded above by $y = 1/(1 + x)$, below by the x-axis, on the left by $x = 1$, and on the right by $x = 2$.
52. $\mathcal{R}$ is bounded above by $y = x^{3/2}$, below by the x-axis, on the left by $x = 1$, and on the right by $x = 4$.

53. $\mathcal{R}$ is bounded above by $y = 2\sin(x)$ for $0 \leq x \leq \pi/2$, below by the x-axis, and on the right by $x = \pi/2$.

54. $\mathcal{R}$ is bounded above by $y = 4$, below by $y = \sqrt{x}$, and on the left by the y-axis.

55. $\mathcal{R}$ is bounded above by $y = 4 - x^2$ and below by $y = 3x$.

56. R is bounded above by $y = 6 - x$, below by $y = x^2$, and on the left by the y-axis.

57. An object that weighs W_0 pounds at the surface of the earth weighs $15697444 \cdot W_0 \cdot (3962 + y)^{-2}$ pounds when it is y miles above the surface of the earth. How much work is done lifting an object from the surface of the earth, where it weighs 1600 pounds, straight up to a height 160 miles above Earth?

58. An object of mass m kg weighs $3.98621 \times 10^{14} \cdot m \cdot r^{-2}$ N when it is a distance r m from Earth's center. How much work is done lifting a 200 kg object straight up from the surface of the earth to a height 300 km above the surface? (Use 6375580 m for the radius of Earth.)

59. How high will 3×10^8 J of work lift a 100 kg mass straight up from the surface of the earth? (Refer to the preceding exercise for pertinent information.)

60. A man on the roof of a 30 ft high building uses a chain to pull up a 60 lb container of tar. The chain weighs 2 pounds per linear foot. How much work does the man do in pulling the container from ground level to the roof?

61. A heavy uniform cable is used to lift a 200 kg load from ground level to the top of a building that is 20 m high. If the cable weighs 160 newtons per linear meter, then how much work is done?

62. A heavy uniform cable weighing 20 pounds per linear foot was used to lift a 300 pound load from ground level to the top of a 100 foot high building. How much work was done by the time the load was 60 ft above ground?

63. If a spring with spring constant 10 pounds per inch is stretched 4 inches beyond its equilibrium position, then how much work is done?

64. If a spring with spring constant 24 newtons per centimeter is stretched 6 cm beyond its equilibrium position, then how much work is done?

65. A spring is stretched 12 cm beyond its equilibrium position. If the force required to maintain it in its stretched position is 360 N, then how much work was done?

66. A spring is stretched 2 inches beyond its equilibrium position. If the force required to maintain it in its stretched position is 60 pounds, then how much more work will be done if the spring is stretched 2 more inches?

67. A force of 864 newtons compresses a spring 12 cm from its natural length. How much more work is done compressing it a further 6 cm?

68. If 24 joules of work are done in stretching a spring 6 cm beyond its natural length, then how much more work is done stretching a further 4 cm?

69. The work done in extending a spring 5 cm beyond its equilibrium position is 2 J. What force is required to maintain the spring in that position?

70. A tank in the shape of a box has square base of side 5 m. It is 5 m deep, contains 75 m^3 of water, and is being pumped dry. The pump floats on the surface of the water and pumps the water to the top of the tank, at which point the water runs off. How much work is done in emptying the tank?

71. A cylindrical tank is 2 m high, has base radius 0.3 m, and is initially filled with water. A pump floating on the surface pumps water to the top, at which point the water runs off. How much work is done pumping out half the water?

72. A tank in the shape of a conical frustum is 2 m high. The radius at the top is 1 m, and the radius at the bottom is 0.8 m. The tank initially held water 1.2 m deep at the center of the tank. A surface pump reduced the depth of the water at the center to 0.4 m. How much work was done?

In Exercises 73–80, find the general solution of the given differential equation.

73. $\dfrac{dy}{dx} = xy^2 + y^2$

74. $\dfrac{dy}{dx} = x^2y + 2y$

75. $\dfrac{dy}{dx} = y + 2x$

76. $x\dfrac{dy}{dx} = x - 2y$

77. $\dfrac{dy}{dx} = 2e^{2x-y}$

78. $\dfrac{dy}{dx} = x - y/x$

79. $\dfrac{dy}{dx} = y^2/(1 + x^2)$

80. $\dfrac{dy}{dx} = y\ln(x)$

In Exercises 81–86, solve the given initial value problem.

81. $\dfrac{dy}{dx} = 2y + x \qquad y(0) = 1/2$

82. $\dfrac{dy}{dx} = 2y^2x \qquad y(2) = 1/2$

83. $\dfrac{dy}{dx} = 3 + 2y \qquad y(0) = -1$

84. $\dfrac{dy}{dx} = \dfrac{2xy}{1 + y^2} \qquad y(1/2) = 1$

85. $\dfrac{dy}{dx} + y = 2e^x \qquad y(0) = 5$

86. $\dfrac{dy}{dx} = x + 1 + 2y/x \qquad y(1) = 2$

87. A mixing tank contains 400 L of water in which 25 kg salt is dissolved. At time $t = 0$, a valve is opened, and a solution with 1/10 kg salt per liter of water enters the tank at the rate of 32 L per minute. An outlet pipe maintains the volume of fluid in the tank by allowing 32 L of the thoroughly mixed solution to flow out each minute. What is the mass $m(t)$ of salt in the tank at time t? What is $m_\infty = \lim_{t\to\infty} m(t)$?

88. A mixing tank contains 100 gallons of water in which 40 pounds of salt is dissolved. At time $t = 0$, a valve is opened, and a solution with 1 pound of salt per 8 gallons of water enters the tank at the rate of 4 gallons per minute. An outlet pipe maintains the volume of fluid in the tank by allowing 4 gallons of the thoroughly mixed solution to flow out each minute. What is the weight $m(t)$ of salt in the tank at time t? What is $m_\infty = \lim_{t\to\infty} m(t)$?

89. A mixing tank contains 200 gallons of water in which 100 pounds of salt is dissolved. At time $t = 0$, a valve is opened, and water enters the tank at the rate of 5 gallons per minute. An outlet pipe maintains the volume of fluid in the tank by allowing 5 gallons of the thoroughly mixed solution to flow out each minute. What is the weight $m(t)$ of salt in the tank at time t? What is $m_\infty = \lim_{t\to\infty} m(t)$?

90. A mixing tank contains 100 L of water in which 10 kg salt is dissolved. At time $t = 0$, a valve is opened, and water enters the tank at the rate of 32 L per minute. An outlet pipe maintains the volume of fluid in the tank by allowing 32 L of the thoroughly mixed solution to flow out each minute. What is the mass $m(t)$ of salt in the tank at time t? What is $m_\infty = \lim_{t\to\infty} m(t)$?

Archimedes and the Sphere

Arc lengths, surface areas, and volumes—these concepts were all investigated by the geometers of ancient Greece. Archimedes, for example, demonstrated that the volume enclosed by a sphere is 2/3 the volume of the circumscribing cylinder and that the surface area of a sphere is (also) 2/3 the area of the circumscribing cylinder (including its circular ends). Stated more explicitly, he proved that the volume and surface area of a sphere of radius r are $4\pi r^3/3$ and $4\pi r^2$, respectively.

These results must have pleased Archimedes, for he requested that his tomb be decorated with a sphere and cylinder and that the ratio giving their relationship be inscribed. His wishes were carried out, as we learn from Cicero, who in 75 B.C. rediscovered and repaired the overgrown, forgotten site:

> "I remembered some verses which I had been informed were engraved on his monument, and these set forth that on the top of the tomb there was placed a sphere with a cylinder. ... I observed a small column standing out a little above the briers with the figure of a sphere and a cylinder upon it... When we could get at it. ... I found the inscription, though the latter parts of all the verses were almost half effaced. Thus one of the noblest cities of Greece, and one which at one time likewise had been very celebrated for learning, had known nothing of the monument of its greatest genius... ."

The restored grave of Archimedes eventually vanished again. However, in modern times, workers excavating the site of a hotel in Syracuse reported coming upon a stone tablet inscribed with a sphere and a cylinder.

Kepler and Solids of Revolution

In 1612, Johannes Kepler (1571–1630) took up residence in Linz. As the wine harvest was good that year, the Danube River banks were filled with wine casks. "This being so," in the words of Kepler's biographer, Max Caspar, "as husband and good *pater-familias*, he considered it his duty to provide his home with the necessary drink." Accordingly, Kepler had some wine barrels installed in his house. After seeing a wine merchant estimate the volume of a cask by simply inserting a measuring stick, Kepler set himself the problem of developing a more scientific system of calculating volumes.

Inspired by these wine barrels, Kepler derived formulas for the volumes of several solids of revolution (not all of which were suitable for wine storage). Among other results, he rediscovered Pappus's formula, $2\pi b \cdot \pi a^2$, for the volume of the *torus*, an inner-tube-shaped solid that is generated when a disk of radius a is revolved about an axis a distance $b > a$ from its center (Figure 1). The rigorous deductions of Archimedes were somewhat beyond Kepler, who resorted to the very free use of infinitesimal arguments. In Kepler's own words: *"we could obtain absolute and perfect proofs from the books of Archimedes if we did not shrink from the thorny reading of his writings."* Kepler published the results of his investigations in 1615 at his own expense. Despite Kepler's strenuous efforts to recoup his outlay, sales were limited to one nobleman, two mathematicians, and a library.

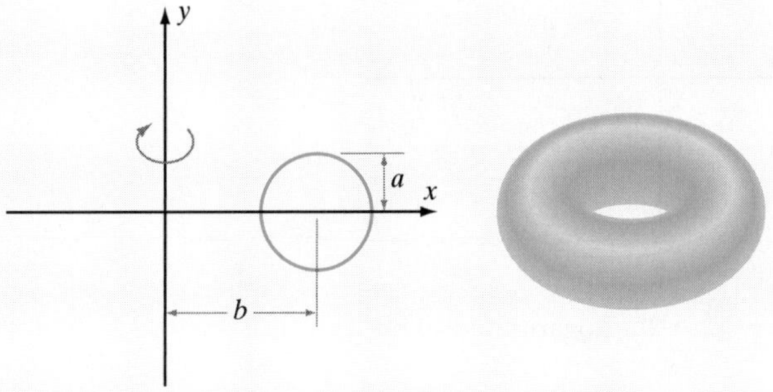

▲ Figure 1

Arc Length

The arc length problem (known then as *rectification*) was also revived by mathematicians in the 1600s. Some leading philosophers believed rectification to be insoluble. Descartes, for one, was emphatic on this point in his *Géométrie:* "The ratios between straight and curved lines are not known, and I believe cannot be discovered by human minds." The idea that it might be possible to compare the arc lengths of two different curves seems to have originated with another philosopher, Thomas Hobbes (1588–1679). Hobbes communicated his idea to Roberval in 1642, and, by the end of that year, Roberval was able to produce equal length arcs of the Archimedean spiral $\sqrt{x^2 + y^2} = \arctan(y/x)$ (Figure 2a) and of the parabola $y = x^2/2$. Three years later,

Evangelista Torricelli calculated the arc length of the logarithmic spiral shown in Figure 2b.

After a twelve-year hiatus, a number of arc length calculations were published in quick succession. In 1657, William Neile (1637–1670) rectified the *semi-cubical parabola* $y^3 = ax^2$. One year later, Christopher Wren (1632–1723), better known as the architect of St. Paul's Cathedral, computed the arc length of the cycloid. In 1659, the general arc length integral, $\int \sqrt{1 + (dy/dx)^2}\, dx$, was obtained more or less simultaneously by Hendrik van Heuraet (1633–?) and Blaise Pascal.

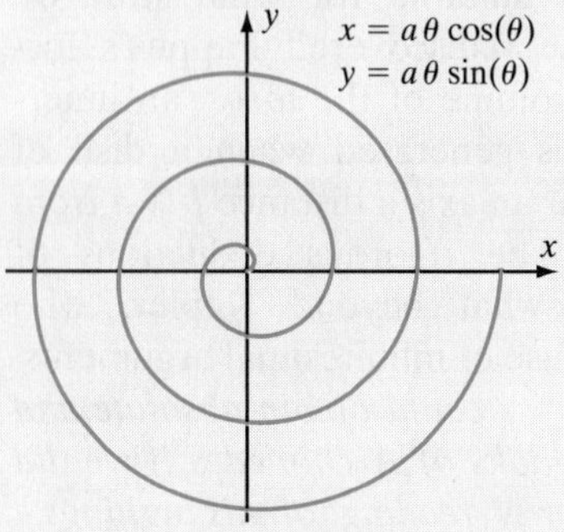

▲ Figure 2a Archimedean spiral

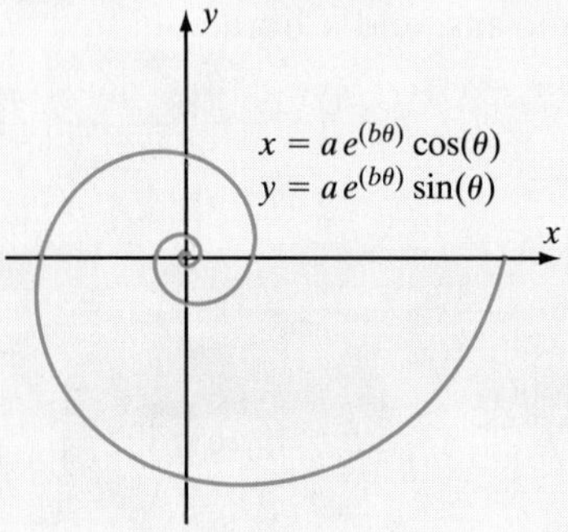

▲ Figure 2b Logarithmic spiral

The *characteristic triangle* or *differential triangle*, the right "triangle" with dx and dy for legs and ds for "hypotenuse" (Figure 3), appeared in the writings of Blaise Pascal, who became interested in rectification after learning of Wren's result. During his study of Pascal's work in 1672, Leibniz realized that the differential triangle could become the basis of a method of quadrature that supplemented its role in rectification. It was this insight about arc length that led Leibniz to the formula for integration by parts and provided him with his first recognition of the relationship between tangents (derivatives) and areas (integrals).

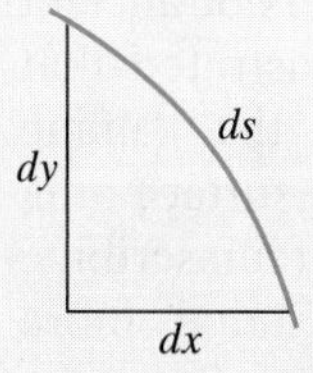

▲ Figure 3 Leibniz's characteristic triangle

Torricelli's Infinitely Long Solid of Revolution

In 1641, Evangelista Torricelli studied the infinitely long solid that is obtained by rotating about the x-axis the region in the first quadrant that is bounded above by the graph of $y = 1/x$, below by the x-axis, and on the left by $x = 1$ (see Figure 4). What is particularly interesting about this solid is that despite its infinite length and infinite surface area, it has *finite* volume. As had been the case in ancient Greece, philosophy was not yet prepared for the mathematical consequences of the infinite. A spirited and personal exchange between the philosopher Thomas Hobbes and the mathematician John Wallis erupted. As Wallis bluntly framed the issue, an understanding of Torricelli's solid *"requires more Geometry and Logic than Mr. Hobs is Master of."* The reply of Hobbes was equally pointed: "*I think Dr. Wallis does wrong ... for, to understand this for sense, it is not required that a man should be a geometrician or a logician but that he should be mad.*"

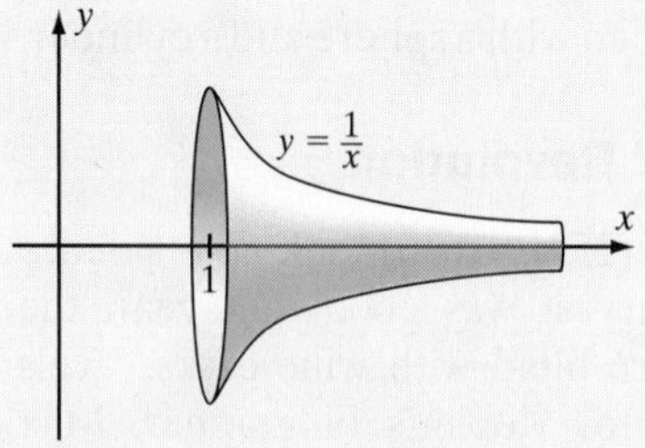

▲ Figure 4 A section of Torricelli's infinitely long solid

The Catenary

The shape of a cable hanging freely under gravity (Figure 5) was first determined in 1690 by Leibniz, who coined the name "catenary," and, independently, by

Jakob Bernoulli. One interesting method of determining the mathematical formula of the catenary relies on a "variational principle" similar to Fermat's Principle of Least Time (*Genesis and Development 4*). Among all curves of fixed length l joining two points P and Q, the catenary arc of length l with endpoints P and Q has lowest center of gravity, or, equivalently, the least potential energy.

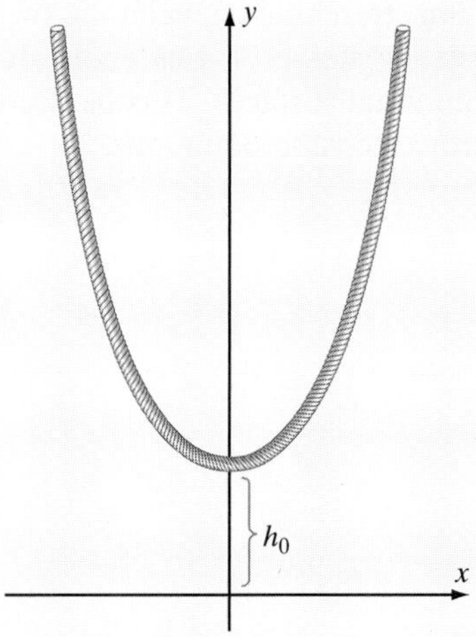

▲ Figure 5

The theory of differential equations developed in this chapter provides an alternative method for deriving the equation of the catenary. Let $y(x)$ be the function whose graph represents the hanging cable. Set $q(x)=y'(x)$. Two physical quantities are relevant: the weight w of the cable per unit of length and the horizontal tension T_1 at the lowest point of the cable (see Figure 6). Because there is no lateral movement of the cable, the horizontal forces must balance. When translated into mathematics, the equation of horizontal forces can be expressed by the separable first order differential equation.

$$\frac{dq}{dx} = \frac{w}{T_1}(1+q^2)^{1/2}.$$

We rewrite this equation as

$$\int \frac{1}{(1+q^2)^{1/2}}\,dq = \int \frac{w}{T_1}\,dx.$$

Using integration formula (3.10.16) with $a = 1$, we obtain

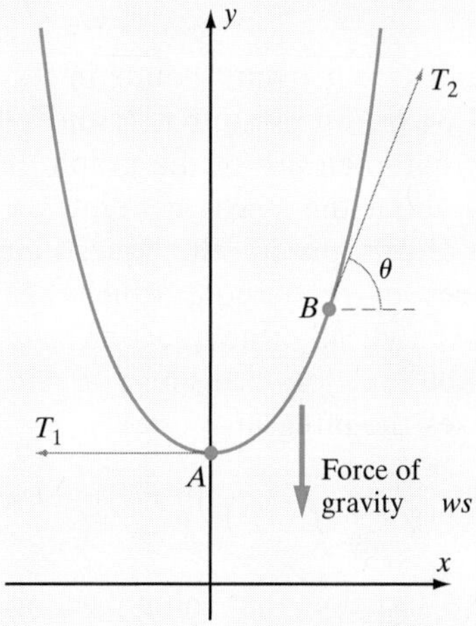

▲ Figure 6

$$\sinh^{-1}(q) = \frac{w}{T_1}x + C.$$

We know that the cable has a horizontal tangent when $x = 0$, the abscissa of point A in Figure 6. Thus we have $q(0) = y'(0) = 0$. From this initial condition, we see that $C = 0$. Thus our solution becomes $\sinh^{-1}(q) = wx/T_1$ or $q(x) = \sinh(wx/T_1)$. But $q = y'$. Therefore

$$\frac{dy}{dx} = \sinh\left(\frac{w}{T_1}x\right).$$

We integrate this last equation to obtain

$$y(x) = \frac{T_1}{w}\cosh\left(\frac{w}{T_1}x\right) + h,$$

where h is a constant of integration. (If h_0 is the height of the point A from the x-axis, then $h = h_0 - T_1/w$.) In general, a curve whose equation is of the form

$$y = h + a \cdot \cosh\left(\frac{x}{a}\right)$$

for positive constants a and h is called a *catenary*. The Gateway Arch in St. Louis, Missouri is very nearly an inverted catenary. Its architectural plans specify the equation

$$y = 693.8597 - 68.7672 \cdot \cosh\left(\frac{3.0022}{299.2239}x\right)$$

for the centroids of its cross-sections (when x and y are measured in feet).

The Catenoid

Suppose that $P=(a,\alpha)$ and $Q=(b,\beta)$ are points in the plane that lie above the x-axis. Suppose that P and Q are joined by a curve $\mathcal{C}$ that is an arc of the graph of a function f. Rotating $\mathcal{C}$ about the x-axis generates a surface of revolution. Euler posed the question, "Among all such surfaces of revolution, which has minimal surface area?" Analytically, the question may be asked in this way: Which function f satisfies $f(a)=\alpha$ and $f(b)=\beta$ and minimizes the integral

$$I(f) = 2\pi \int_a^b f(x)\sqrt{1 + \left(f'(x)^2\right)}\,dx\,?$$

The task of finding a function that minimizes an integral such as $I(f)$ is substantially more complicated than finding an extremum of a numerical function. This kind of problem is studied in a branch of mathematics called the *calculus of variations*. Euler solved his own problem, proving that $I(f)$ is minimized when $\mathcal{C}$ is the arc of a catenary. The surface of revolution is called a *catenoid*. It arises naturally—indeed, you can generate a catenoid with some wire and soapy water. Form wires in the shape of two circles with different radii, bring them together in a soap solution, and draw them apart so that they are circles of revolution about the same axis. The resulting soap film stretching between the two circular wires will assume the shape of a catenoid. In general, soap films give minimal surfaces. This fact can be proved from their surface tension equations.

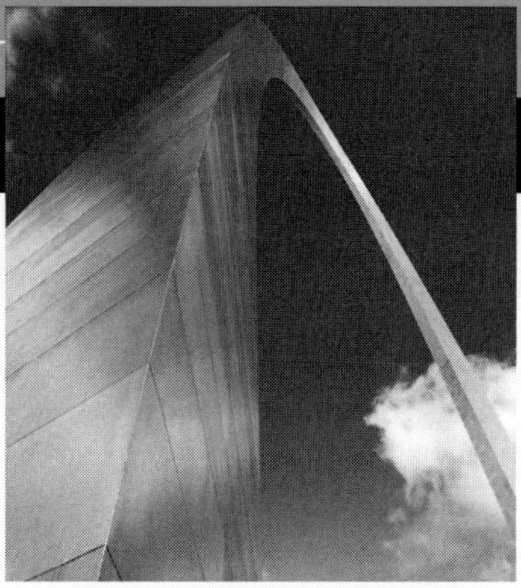

CHAPTER 8

Infinite Series

PREVIEW

When we learn to write fractions in decimal form, we soon become aware that some numbers result in a nonterminating sequence of digits. For example, when we decimalize 1/3, we apply the long division algorithm. No matter how many times we divide, we are left with a remainder. If we terminate the procedure after any division, then we are left with 0.33 ... 33, which is a little too small. If we round up to 0.33 ... 34, then the result is a number that is a little too great. We must therefore accept that the decimalization of 1/3 is 0.333... without end.

Of course, in grade school we do not pause long to reflect on the infinite process that is symbolized by the trailing dots. Intuitively, we understand that the notation must stand for the "infinite sum"

$$\frac{3}{10} + \frac{3}{100} + \frac{3}{1000} + \frac{3}{10000} + \cdots,$$

even if we have only a fuzzy idea of what that might mean.

Now that we have studied some calculus we are prepared to understand exactly what such sums signify. The concept of "limit" that is needed for the infinite processes of calculus is just what is needed to understand the infinite sums that arise in grade school and elsewhere. In fact, our approach is nothing more than a framework for turning our intuition into something precise. If you think of the sequence

$$0.3 = \frac{3}{10}, \quad 0.33 = \frac{3}{10} + \frac{3}{100}, \quad 0.333 = \frac{3}{10} + \frac{3}{100} + \frac{3}{1000}, \ \ldots$$

as a sequence of approximations of 1/3 that become ever more accurate, then you already have the key idea of understanding an "infinite sum" as a limit of its "partial sums."

Infinite sums are important in calculus, but they are not confined to mathematical disciplines. If you study scientific subjects such as biology or thermodynamics, then you are likely to encounter infinite sums in those pursuits. If you learn about the money supply in economics or annuities in finance, then you will see that infinite sums are needed for those concepts. This chapter will provide you with a good basis for your understanding of infinite sums, wherever they arise.

In this chapter, we also study infinite sums $\sum_{n=0}^{\infty} a_n(x-c)^n$ for which x is treated as a variable lying in an interval centered at c. Such *power series*, as they are called, have found many applications in mathematics and science. Several important differential equations from physics and engineering are best solved by power series methods. Functions defined by power series have become fundamental tools in the study of diverse subjects such as civil engineering, aeronautics, structural geology, and celestial mechanics.

Power series are not used only to create new functions. In many cases, it is important to be able to express the standard elementary functions by means of power series. The question becomes, given a function f and a point c, can we find coefficients $\{a_n\}$ for which $f(x) = \sum_{n=0}^{\infty} a_n(x-c)^n$? The answer lies in the theory of Taylor series, which is the focus of the last two sections of this chapter. Because the partial sums $\sum_{n=0}^{N} a_n(x-c)^n$ of a power series are polynomials, our study of Taylor series will lead to powerful techniques for approximating functions by polynomials.

8.1 Series

The idea of adding together finitely many numbers is a simple and familiar one. We are so used to the properties of finite sums that we tend not to think about them. Now we will learn about sums of *infinitely* many numbers, and we will have to be a bit more careful. As the next example shows, contradictions can arise if we do not establish certain ground rules.

▶ **EXAMPLE 1** Group the terms of $-1+1-1+1-1+1-1+\cdots$ in two ways to produce different results.

Solution If we group the terms as $(-1+1)+(-1+1)+(-1+1)+\cdots$, then the sum seems to be $0+0+0+\cdots$, which ought to be 0. On the other hand, if we group the terms as $-1+(1-1)+(1-1)+(1-1)+\cdots$, then the sum seems to be $-1+0+0+0+\cdots$, which ought to be -1. Apparently the associative law of addition fails for this infinite sum. ◀

Example 1 points to the need for a precise definition that establishes what it means to add infinitely many numbers. With a carefully conceived definition, sums such as $-1+1-1+1-1+1-1+\cdots$ are not given any meaning, and ambiguities such as that of Example 1 do not arise. The key idea in the precise definition of an infinite sum is the notion of a limit of a sequence. It is that concept that we review next.

Limits of Infinite Sequences–A Review

Recall that an *infinite sequence* $\{a_n\}_{n=1}^{\infty}$ is a list $a_1, a_2, a_3, \ldots$ of terms indexed by the positive integers. For example, 1, 1/2, 1/4, 1/8, 1/16,... is the infinite sequence $\{1/2^{n-1}\}_{n=1}^{\infty}$ of nonnegative powers of 1/2. Often it is convenient to begin a sequence with a first index that differs from 1. Thus $\{1/2^n\}_{n=0}^{\infty}$ is another way to write the sequence $1, 1/2, 1/4, 1/8, 1/16, \ldots$.

We say that a sequence $\{a_n\}_{n=1}^{\infty}$ converges to a real number ℓ (or has limit ℓ) if, for each $\epsilon > 0$, there is an integer N such that $|a_n - \ell| < \epsilon$ for all $n \geq N$. In this case, we write $\lim_{n\to\infty} a_n = \ell$, or, equivalently, $a_n \to \ell$ as $n \to \infty$. A sequence that has a limit is said to be *convergent*. If a sequence does not converge, then we say that it *diverges*, and we call it a *divergent sequence*. Some basic convergent sequences are

$$\lim_{n\to\infty} r^n = 0 \quad \text{if } r \text{ is a constant with } |r| < 1, \tag{8.1.1}$$

$$\lim_{n\to\infty} b^{1/n} = 1 \quad \text{if } b \text{ is a positive constant}, \tag{8.1.2}$$

and

$$\lim_{n\to\infty} \frac{1}{n^p} = 0 \quad \text{if } p \text{ is a positive constant}. \tag{8.1.3}$$

There are a few rules that, when used with these basic limits, enable us to easily evaluate the limits of many common sequences. For example, if $\{a_n\}_{n=1}^{\infty}$ and $\{b_n\}_{n=1}^{\infty}$ are convergent sequences, and if α is any real number, then

$$\lim_{n\to\infty} c_n = \beta, \quad \text{if } c_n = \beta \text{ for all } n, \tag{8.1.4}$$

$$\lim_{n\to\infty} (\alpha \cdot a_n) = \alpha \cdot \lim_{n\to\infty} a_n, \tag{8.1.5}$$

$$\lim_{n\to\infty} (a_n + b_n) = \lim_{n\to\infty} a_n + \lim_{n\to\infty} b_n, \tag{8.1.6}$$

$$\lim_{n\to\infty} (a_n - b_n) = \lim_{n\to\infty} a_n - \lim_{n\to\infty} b_n, \tag{8.1.7}$$

$$\lim_{n\to\infty} (a_n \cdot b_n) = (\lim_{n\to\infty} a_n) \cdot (\lim_{n\to\infty} b_n), \tag{8.1.8}$$

and

$$\lim_{n\to\infty} \frac{a_n}{b_n} = \frac{\lim_{n\to\infty} a_n}{\lim_{n\to\infty} b_n} \quad \text{if} \quad \lim_{n\to\infty} b_n \neq 0. \tag{8.1.9}$$

▶ **EXAMPLE 2** Let

$$a_n = \frac{2n^4 + 3\sqrt{n} - 5}{3n^4 + 10}.$$

Use limit formula (8.1.3) to calculate $\lim_{n\to\infty} a_n$.

Solution We divide each summand in the numerator and denominator of a_n by the highest power, n^4, obtaining

$$\lim_{n\to\infty} a_n = \lim_{n\to\infty} \frac{2 + 3\sqrt{n}/n^4 - 5/n^4}{3 + 10/n^4}.$$

Observing that

$$\lim_{n\to\infty}\left(2 + 3\frac{\sqrt{n}}{n^4} - \frac{5}{n^4}\right) = \lim_{n\to\infty}(2) + \lim_{n\to\infty}\frac{3}{n^{7/2}} - \lim_{n\to\infty}\left(\frac{5}{n^4}\right)$$
$$= 2 + 3\cdot\lim_{n\to\infty}\frac{1}{n^{7/2}} - 5\cdot\lim_{n\to\infty}\left(\frac{1}{n^4}\right) = 2 + 3\cdot 0 - 5\cdot 0 = 2$$

and

$$\lim_{n\to\infty}\left(3 + \frac{10}{n^4}\right) = \lim_{n\to\infty}(3) + \lim_{n\to\infty}\frac{10}{n^4} = 3 + 10\cdot\lim_{n\to\infty}\frac{1}{n^4} = 3 + 10\cdot 0 = 3 \neq 0,$$

we use limit formula (8.1.9) to conclude that $\lim_{n\to\infty} a_n = 2/3$. ◀

Example 2 has a generalization that is used frequently. If $p_1, p_2, \ldots, p_k$ and $q_1, q_2, \ldots, q_\ell$ are constants with $p_1 > p_2 > \ldots > p_k \geq 0$ and $q_1 > q_2 > \ldots > q_\ell \geq 0$, and if $c_1, c_2, \ldots, c_k$ and $d_1, d_2, \ldots, d_\ell$ are constants with $c_1 \neq 0$ and $d_1 \neq 0$, then

$$\lim_{n\to\infty} \frac{c_1 n^{p_1} + c_2 n^{p_2} + \cdots + c_k n^{p_k}}{d_1 n^{q_1} + d_2 n^{q_2} + \cdots + d_\ell n^{q_\ell}} = \begin{cases} c_1/d_1 & \text{if } p_1 = q_1 \\ 0 & \text{if } p_1 < q_1 \end{cases}. \tag{8.1.10}$$

In words, if the highest power in the numerator is the *same* as the highest power in the denominator, then the limit is the ratio of the coefficients of the highest powers. If the highest power in the numerator is *less* than the highest power in the denominator, then the limit is 0.

In another important technique for calculating $\lim_{n\to\infty} a_n$, we treat the index variable n as a continuous variable x that tends to infinity. Then we may apply tools from calculus, such as l'Hôpital's Rule. The next two examples illustrate this idea.

▶ **EXAMPLE 3** Let $a_n = n/e^{\lambda n}$ where λ is a positive constant. Calculate $\lim_{n\to\infty} a_n$.

Solution Because $\lambda > 0$, we have $\lim_{n\to\infty} e^{\lambda n} = \infty$ and $\lim_{n\to\infty} a_n$ has the indeterminate form $\frac{\infty}{\infty}$. An application of l'Hôpital's Rule shows that

$$\lim_{x\to\infty} \frac{x}{e^{\lambda x}} = \lim_{x\to\infty} \frac{\frac{d}{dx}x}{\frac{d}{dx}e^{\lambda x}} = \lim_{x\to\infty} \frac{1}{\lambda e^{x}} = 0.$$

If we restrict the approach of x to infinity through integer values n, then we obtain $\lim_{n\to\infty} a_n = 0$. ◀

▶ **EXAMPLE 4** Show that

$$\lim_{n\to\infty} n^{1/n} = 1. \tag{8.1.11}$$

Solution In Example 10 of Section 4.7, we used l'Hôpital's Rule to show that $\lim_{x\to\infty} x^{1/x} = 1$. If we restrict the approach of x to infinity through integer values n, then we obtain limit formula (8.1.11). ◀

The Definition of an Infinite Series

If $a_1, a_2, a_3, \ldots$ is an infinite sequence, then the (formal) expression $a_1 + a_2 + a_3 + \cdots$ is said to be an *infinite series*. It is convenient to extend the sigma notation

$$\sum_{n=M}^{N} a_n = a_M + a_{M+1} + \cdots + a_{N-1} + a_N$$

for finite sums that we introduced in Section 5.1 of Chapter 5. To signify an infinite series, we simply use the symbol ∞ for the upper limit of summation. Thus we let $\sum_{n=1}^{\infty} a_n$ denote the infinite series $a_1 + a_2 + a_3 + \cdots$. For instance, if $a_n = 1/2^{n-1}$, then $\sum_{n=1}^{\infty} a_n$ denotes the infinite series $1 + 1/2 + 1/4 + 1/8 + 1/16 + \cdots$.

Example 1 suggests that care is needed if we are to assign a value to the infinite series $\sum_{n=1}^{\infty} a_n$. Nevertheless, there is a quite natural way to define and interpret infinite series. Consider the decimalization of 1/3, namely 0.3333... (ad infinitum). In reality, the decimal number 0.3333... symbolizes the infinite series $\sum_{n=1}^{\infty} a_n$ with $a_n = 3/10^n$. If we need a decimal expansion that is accurate to two decimal places, then we use $a_1 + a_2 = 0.33$. If we require five decimal digits of 1/3, then we use $a_1 + a_2 + a_3 + a_4 + a_5 = 0.33333$, and so on. By taking "partial sums" of the infinite series $\sum_{n=1}^{\infty} a_n$, we obtain approximations that become closer and closer to the exact value of $\sum_{n=1}^{\infty} a_n$. This familiar example suggests the method by which we define the "sum" of an infinite series in general.

DEFINITION If $a_1, a_2, a_3, \ldots$ is an infinite sequence, then we say that the expression

$$S_N = \sum_{n=1}^{N} a_n = a_1 + a_2 + \cdots + a_N$$

is a *partial sum* of the infinite series $\sum_{n=1}^{\infty} a_n$. The sequence $\{S_N\}_{N=1}^{\infty}$ is said to be the sequence of partial sums of the infinite series $\sum_{n=1}^{\infty} a_n$.

▶ **EXAMPLE 5** Compute the sequence of partial sums of the infinite series $\sum_{n=1}^{\infty}(-1)^n$.

Solution According to the definition, $S_1 = \sum_{n=1}^{1}(-1)^n = -1$, $S_2 = \sum_{n=1}^{2}(-1)^n = (-1)^1 + (-1)^2 = -1+1=0$, $S_3 = \sum_{n=1}^{3}(-1)^n = (-1)^1+(-1)^2+(-1)^3 = -1$, and so on. The sequence of partial sums of $\sum_{n=1}^{\infty}(-1)^n$ is $-1, 0, -1, 0, -1, 0, \ldots$ ◀

▶ **EXAMPLE 6** Find a formula for the partial sum $S_N = \sum_{n=1}^{N}\frac{1}{2^{n-1}}$ of the infinite series $\sum_{n=1}^{\infty}\frac{1}{2^{n-1}}$.

Solution For small values of N, ordinary addition yields the values of S_N. For example, we have $S_1 = 1/2^0 = 1$, $S_2 = 1/2^0 + 1/2^1 = 3/2$, $S_3 = 1/2^0 + 1/2^1 + 1/2^2 = 7/4$, and so on. More efficiently, whenever we calculate one partial sum, we can obtain the next partial sum by adding the next term of the series. Thus $S_4 = S_3 + 1/2^3 = 7/4 + 1/8 = 15/8$. To calculate the partial sum

$$S_N = \sum_{n=1}^{N}\frac{1}{2^{n-1}} = 1 + \frac{1}{2} + \frac{1}{2^2} + \frac{1}{2^3} + \cdots + \frac{1}{2^{N-1}}$$

that corresponds to a general value of N, we use equation

$$1 + r + r^2 + \cdots + r^{N-1} = \frac{r^N - 1}{r - 1}, \quad r \neq 1 \tag{8.1.12}$$

from Theorem 4 of Section 2.5 in Chapter 2. If we specify $r = 1/2$ in equation (8.1.12), we see that

$$S_N = \sum_{n=1}^{N}\left(\frac{1}{2}\right)^{n-1} = 1 + \frac{1}{2} + \frac{1}{2^2} + \frac{1}{2^3} + \cdots + \frac{1}{2^{N-1}} = \frac{\left(\frac{1}{2}\right)^N - 1}{\left(\frac{1}{2}\right) - 1} = \frac{1 - \frac{1}{2^N}}{\frac{1}{2}} = 2\left(1 - \frac{1}{2^N}\right) = 2 - \frac{1}{2^{N-1}}.$$

This general formula agrees with the specific calculations of S_N for small values of N that we performed at the beginning of the solution. ◀

Convergence of Infinite Series

The meaning that we give to the "sum" of an infinite series $\sum_{n=1}^{\infty} a_n$ is determined by the behavior of its sequence $\{S_N\}$ of partial sums as N tends to infinity.

DEFINITION Let $\{S_N\}$ be the sequence of partial sums of an infinite series $\sum_{n=1}^{\infty} a_n$. If $\lim_{N\to\infty} S_N = S$, then we say that the infinite series $\sum_{n=1}^{\infty} a_n$ *converges* to S. In this case, when we refer to $\sum_{n=1}^{\infty} a_n$, we mean the number S, and we write $\sum_{n=1}^{\infty} a_n = S$. We call S the *sum* or *value* of the infinite series. If the sequence $\{S_N\}$ does not converge, then we say that the series *diverges*. A series that converges is said to be *convergent*; otherwise, we say that it is *divergent*.

INSIGHT The terms "convergent" and "divergent" suggest an analogy between infinite series $\sum_{n=1}^{\infty} a_n$ and improper integrals $\int_1^{\infty} f(x)\,dx$. Indeed, the approaches to the two concepts are similar. To calculate $\int_1^{\infty} f(x)\,dx$, we compute $\int_1^{N} f(x)\,dx$ for large values of N and take the limit of $\int_1^{N} f(x)\,dx$ as N tends to infinity. Analogously, to calculate

$\sum_{n=1}^{\infty} a_n$, we compute the partial sum $S_N = a_1 + a_2 + a_3 + \cdots + a_N$ for large integer values N and take the limit of S_N as N tends to infinity. We define $\sum_{n=1}^{\infty} a_n$ to be the limit $S = \lim_{N\to\infty} S_N = \lim_{N\to\infty}(a_1 + a_2 + a_3 + \cdots + a_N)$, if it exists. Although we call S the "sum" of the series, it is not a sum in the conventional sense of the word: By definition, it is a limit.

Examples

▶ **EXAMPLE 7** Does the series $\sum_{n=1}^{\infty} \frac{1}{2^{n-1}}$ converge? If so, to what limit?

Solution In Example 6, we calculated the Nth partial sum $S_N = 2 - 1/2^{N-1}$ of this infinite series. We may now pass to the limit:

$$\lim_{N\to\infty} S_N = \lim_{N\to\infty}\left(2 - \frac{1}{2^{N-1}}\right) = 2.$$

By definition, we say that the series $\sum_{n=1}^{\infty} \frac{1}{2^{n-1}}$ converges to 2. ◀

▶ **EXAMPLE 8** Discuss convergence for the series $\sum_{n=1}^{\infty} (-1)^n = -1 + 1 - 1 + 1 - 1 + \cdots$.

Solution If N is even then the partial sum $S_N = \sum_{n=1}^{N} (-1)^n = -1 + 1 - \cdots - 1 + 1$ comprises an equal number of $+1$'s and -1's. Therefore $S_N = 0$ for N even. If N is odd, then there is one more -1 than $+1$ in the sum $S_N = \sum_{n=1}^{N} (-1)^n$ and $S_N = -1$. Therefore the sequence $\{S_N\}_{N=1}^{\infty}$ of partial sums is just the sequence $-1, 0, -1, 0, -1, 0, \ldots$ which does not converge. We conclude that the series $\sum_{n=1}^{\infty} (-1)^n$ does not converge. ◀

INSIGHT Example 8 sheds some light on the apparent contradiction that we encountered in Example 1. When we first investigated the series $\sum_{n=1}^{\infty} (-1)^n = -1 + 1 - 1 + 1 - 1 + \cdots$, we formed two different groupings of its summands. Those groupings suggested different values, -1 and 0, for the series. Now we understand that -1 is the limit of the odd partial sums, and 0 is the limit of the even partial sums. The partial sums taken together do not have a limit, and the series therefore does not converge. The lesson is that we should not attempt to perform arithmetic operations on series that do not converge.

◀ **A LOOK BACK** It is important to understand the difference between a sequence, which is a list of numbers, and a series, which is a sum (or a limit of a sum) of numbers. A sequence $a_1, a_2, a_3, \ldots$ is used to create an infinite series $a_1 + a_2 + a_3 + \cdots$. In turn, this series gives rise to a second sequence, the sequence $a_1, a_1 + a_2, a_1 + a_2 + a_3, \ldots$ of partial sums of the infinite series. We determine whether we can assign a value to the infinite series $\sum_{n=1}^{\infty} a_n$ by examining the limiting behavior of its sequence of partial sums.

A Telescoping Series

It is relatively rare to find an explicit formula for the partial sums of an infinite series. In the next example, we employ a useful algebraic "trick" to evaluate the partial sums.

▶ **EXAMPLE 9** Discuss convergence for the series $\sum_{n=1}^{\infty} \frac{1}{n(n+1)}$.

Solution If you use a calculator to compute some partial sums, then you will probably be convinced that the series converges. We can justify this conclusion by writing each summand in its partial fraction decomposition:

$$\frac{1}{n(n+1)} = \frac{1}{n} - \frac{1}{n+1}.$$

As a result, we have

$$\begin{aligned} S_N &= \frac{1}{1\cdot 2} + \frac{1}{2\cdot 3} + \frac{1}{3\cdot 4} + \cdots + \frac{1}{N\cdot(N+1)} \\ &= \left(\frac{1}{1} - \frac{1}{2}\right) + \left(\frac{1}{2} - \frac{1}{3}\right) + \left(\frac{1}{3} - \frac{1}{4}\right) + \cdots + \left(\frac{1}{N-1} - \frac{1}{N}\right) + \left(\frac{1}{N} - \frac{1}{N+1}\right). \end{aligned}$$

Almost everything cancels, and we have $S_N = 1 - 1/(N+1)$. Clearly $\lim_{N\to\infty} S_N = 1$. Therefore the series $\sum_{n=1}^{\infty} \frac{1}{n(n+1)}$ converges to 1. ◀

INSIGHT If we set $f(n) = 1/n$, then we may write the series of Example 9 in the form

$$\sum_{n=1}^{\infty} \Big(f(n) - f(n+1)\Big).$$

In general, a series that may be expressed this way is said to be a *collapsing* or *telescoping* series. That is because each partial sum collapses to two summands:

$$\begin{aligned} \sum_{n=1}^{N} \Big(f(n) - f(n+1)\Big) &= \Big(f(1) - f(2)\Big) + \Big(f(2) - f(3)\Big) + \Big(f(3) - f(4)\Big) + \cdots \\ &\quad + \Big(f(N-1) - f(N)\Big) + \Big(f(N) - f(N+1)\Big) \\ &= f(1) - f(N+1). \end{aligned}$$

Such a series converges if and only if $\lim_{N\to\infty} f(N) = 0$.

The Harmonic Series

The infinite series $\sum_{n=1}^{\infty} 1/n$ comes up often in mathematics and physics. It is called the "harmonic series" because of its role in the theory of acoustics: If the natural frequencies at which a given body vibrates are $\rho_1, \rho_2, \ldots$ and if $\rho_n/\rho_{n+1} = 1/(n+1)$ for all n, then the frequencies are said to form a sequence of harmonics. Although the harmonic series is divergent, you might not think so if you calculate a large number of partial sums using a computer. The divergence is taking place so slowly that it is difficult to be certain what is actually happening. For example, if we sum the first 1 million terms, then we obtain $S_{1\,000\,000} = 14.3927$ (rounded to four decimal places). If we add the next 1,000 terms to this, then we obtain $S_{1\,001\,000} = 14.3937$—the change is barely noticeable despite the additional 1,000 terms! If you sum the first 10^{43} terms of the harmonic series, then the result is a value that is still less than 100. To be certain of the divergence, we need a mathematical proof.

THEOREM 1 The harmonic series $\sum_{n=1}^{\infty} \frac{1}{n}$ is divergent.

Proof. Notice that

$$\begin{aligned}
S_1 &= 1 = \frac{2}{2} \\
S_2 &= 1 + \frac{1}{2} = \frac{3}{2} \\
S_4 &= 1 + \frac{1}{2} + \left(\frac{1}{3} + \frac{1}{4}\right) \\
&\geq 1 + \frac{1}{2} + \left(\frac{1}{4} + \frac{1}{4}\right) \geq 1 + \frac{1}{2} + \frac{1}{2} = \frac{4}{2} \\
S_8 &= 1 + \frac{1}{2} + \left(\frac{1}{3} + \frac{1}{4}\right) + \left(\frac{1}{5} + \frac{1}{6} + \frac{1}{7} + \frac{1}{8}\right) \\
&\geq 1 + \frac{1}{2} + \left(\frac{1}{4} + \frac{1}{4}\right) + \left(\frac{1}{8} + \frac{1}{8} + \frac{1}{8} + \frac{1}{8}\right) \\
&= \frac{5}{2}.
\end{aligned}$$

In general, this argument shows that $S_{2^k} \geq (k+2)/2$. The sequence of S_N's is increasing because the series contains only positive terms. The fact that the partial sums $S_1, S_2, S_4, S_8, \ldots$ increase without bound shows that the entire sequence of partial sums must increase without bound. We conclude that the series diverges. ■

Basic Properties of Series

We now present some elementary properties of series that are similar to the properties of sequences that you learned in Chapter 2.

THEOREM 2 Suppose that $\sum_{n=1}^{\infty} a_n$ converges to A, that $\sum_{n=1}^{\infty} b_n$ converges to B, that $\sum_{n=1}^{\infty} c_n$ diverges, and that λ is a constant. Then

a. $\sum_{n=1}^{\infty} (a_n + b_n)$ converges to $A + B$, and $\sum_{n=1}^{\infty} (a_n - b_n)$ converges to $A - B$.
b. $\sum_{n=1}^{\infty} (\lambda a_n)$ converges to λA.
c. $\sum_{n=1}^{\infty} (\lambda c_n)$ diverges if $\lambda \neq 0$.

▶ **EXAMPLE 10** Discuss convergence of the series $\displaystyle\sum_{n=1}^{\infty} \left(\frac{5}{n(n+1)} - \frac{3}{2^{n-1}}\right)$.

Solution Example 9 combined with Theorem 2b tells us that $\sum_{n=1}^{\infty} 5 \cdot \dfrac{1}{n(n+1)}$ converges to 5. Similarly, Example 7 and Theorem 2b tell us that $\sum_{n=1}^{\infty} 3 \cdot \dfrac{1}{2^{n-1}}$ converges to $3 \cdot 2$, or 6. Hence, by Theorem 2a, we may conclude that the given series converges to $5 - 6$, or -1. ◀

▶ **EXAMPLE 11** Let $a_n = \dfrac{2^n - n}{n2^n}$. Does the series $\sum_{n=1}^{\infty} a_n$ converge?

Solution We may write the summand a_n as

$$a_n = \frac{2^n - n}{n2^n} = \frac{2^n}{n2^n} - \frac{n}{n2^n} = \frac{1}{n} - \frac{1}{2^n}.$$

Let $b_n = 1/2^n$. Because $b_n = \lambda \cdot \dfrac{1}{2^{n-1}}$ with $\lambda = 1/2$, we know from Example 7 and Theorem 2b that $\sum_{n=1}^{\infty} b_n$ converges. If the series $\sum_{n=1}^{\infty} a_n$ were to converge, then so would the series $\sum_{n=1}^{\infty} (a_n + b_n)$ by Theorem 2a. But $a_n + b_n = 1/n$, and we know that the harmonic series diverges (Theorem 1). Therefore $\sum_{n=1}^{\infty} a_n$ must diverge. ◀

INSIGHT The method used to solve Example 10 shows that if $\sum_{n=1}^{\infty} a_n$ converges and $\sum_{n=1}^{\infty} b_n$ diverges, then $\sum_{n=1}^{\infty} (a_n + b_n)$ diverges.

Series of Powers (Geometric Series)

We now consider the series $\sum_{n=0}^{\infty} r^n = 1 + r + r^2 + r^3 + \cdots$ of all nonnegative integer powers of a fixed number r. We call this series a *geometric series* with ratio r. Notice that this series begins with index value $n = 0$. Occasionally, it will be useful to begin such series, and others, with a first summation index value that is an integer M not equal to 1.

THEOREM 3 *If $|r| < 1$, then the geometric series with ratio r converges and*

$$\sum_{n=0}^{\infty} r^n = \frac{1}{1-r}. \tag{8.1.13}$$

In general, for any M in $\mathbb{Z}$, if $|r| < 1$, then $\sum_{n=M}^{\infty} r^n$ converges and

$$\sum_{n=M}^{\infty} r^n = \frac{r^M}{1-r}. \tag{8.1.14}$$

If $1 \le |r|$, then the series $\sum_{n=0}^{\infty} r^n$ and $\sum_{n=M}^{\infty} r^n$ diverge.

Proof. Let $S_N = \sum_{n=0}^{N-1} r^n = 1 + r + r^2 + \cdots + r^{N-1}$ denote the Nth partial sum of the geometric series $\sum_{n=0}^{\infty} r^n$. If $1 \le r$, then it is clear that S_N satisfies $S_N = 1 + r + r^2 + \cdots + r^{N-1} \ge 1 + 1 + 1 + \cdots + 1 = N$. It follows that the sequence $\{S_N\}$ cannot converge to any real number ℓ. In other words, the series $\sum_{n=0}^{\infty} r^n$ is divergent for $1 \le r$. We put aside the case $r \le -1$ for now. (In Section 8.2, we will see how to prove this divergence without analyzing partial sums.)

Now suppose that $|r| < 1$. Then, using formula (8.1.12), we have

$$S_N = \sum_{n=0}^{N-1} r^n = 1 + r + r^2 + \cdots + r^{N-1} = \frac{r^N - 1}{r - 1}.$$

Because $|r| < 1$, we have $\lim_{N\to\infty} r^N = 0$ and

$$\sum_{n=0}^{\infty} r^n = \lim_{N\to\infty} S_N = \frac{0-1}{r-1} = \frac{1}{1-r},$$

which is equation (8.1.13). Finally, if $|r| < 1$ and if M is any integer, then we deduce equation (8.1.14) from (8.1.13) as follows:

$$\begin{aligned}\sum_{n=M}^{\infty} r^n &= \lim_{N\to\infty} \sum_{n=M}^{M+N} r^n = \lim_{N\to\infty} \left(r^M + r^{M+1} + r^{M+2} + \cdots + r^{M+N}\right) \\ &= r^M \lim_{N\to\infty} \left(1 + r + r^2 + \cdots + r^N\right) = r^M \cdot \frac{1}{1-r} = \frac{r^M}{1-r}.\end{aligned}$$

■

▶ **EXAMPLE 12** At a certain aluminum recycling plant, the recycling process turns x pounds of used aluminum into $0.9x$ pounds of new aluminum. Including the initial quantity, how much usable aluminum will 100 pounds of virgin aluminum ultimately yield, if we assume that it is continually returned to the same recycling plant?

Solution We begin with 100 pounds of aluminum. When it is returned to the plant as scrap and recycled, $0.9 \cdot 100$ pounds of new aluminum results. When that aluminum is returned to the plant as scrap and recycled, $0.9 \cdot (0.9 \cdot 100) = 0.9^2 \cdot 100$ pounds of new aluminum results. This process continues forever. Writing the initial quantity of aluminum as $0.9^0 \cdot 100$, and using formula (8.1.13) and Theorem 2b, we see that the total amount of aluminum created is

$$0.9^0 \cdot 100 + 0.9 \cdot 100 + 0.9^2 \cdot 100 + \cdots = \sum_{n=0}^{\infty} (0.9^n \cdot 100) = 100 \sum_{n=0}^{\infty} 0.9^n = 100 \cdot \frac{1}{1 - 0.9} = 1000.$$

Therefore 1000 pounds of usable aluminum is ultimately generated by an initial 100 pounds of virgin aluminum. ◀

▶ **EXAMPLE 13** Express the repeating decimal 1212.121212121... as a ratio of two integers.

Solution There are several ways we can apply formula (8.1.14) with $r = 1/100$. For example, choosing $M = 1$, we have

$$\begin{aligned} 1212.121212121\ldots &= 1212 + 0.12 + 0.0012 + 0.000012 + \cdots \\ &= 1212 + 12 \cdot \left(\frac{1}{100}\right)^1 + 12 \cdot \left(\frac{1}{100}\right)^2 + 12 \cdot \left(\frac{1}{100}\right)^3 + \cdots \\ &= 1212 + 12 \cdot \sum_{n=1}^{\infty} \left(\frac{1}{100}\right)^n \\ &= 1212 + 12 \cdot \frac{(1/100)}{1 - (1/100)} \\ &= 1212 + 12 \cdot \frac{1}{99} \\ &= \frac{40000}{33}. \end{aligned}$$

Alternatively, we can use formula (8.1.14) with $r = 1/100$ and $M = -1$ as follows:

$$\begin{aligned} 1212.121212121\ldots &= 1200 + 12 + 0.12 + 0.0012 + 0.000012 + \cdots \\ &= 12 \cdot \left(\frac{1}{100}\right)^{-1} + 12 \cdot \left(\frac{1}{100}\right)^0 + 12 \cdot \left(\frac{1}{100}\right)^1 + 12 \cdot \left(\frac{1}{100}\right)^2 + 12 \cdot \left(\frac{1}{100}\right)^3 + \cdots \\ &= 12 \cdot \sum_{n=-1}^{\infty} \left(\frac{1}{100}\right)^n \\ &= 12 \cdot \frac{(1/100)^{-1}}{1 - (1/100)} \\ &= 12 \cdot \frac{10000}{99} \\ &= \frac{40000}{33}. \end{aligned}$$ ◀

QUICK QUIZ

1. Which of the numbers 9/8, 10/8, 11/8, 12/8, is a partial sum of $\sum_{n=1}^{\infty} n/2^n$?
2. True or false: The sum of two convergent series is also convergent.
3. What is the value of $\sum_{n=0}^{\infty} 5/3^n$?
4. Does $\sum_{n=1}^{\infty} 5/n$ converge or diverge?

Answers

1. 11/8 2. True 3. 15/2 4. Diverge

EXERCISES

Problems for Practice

In each of Exercises 1–20, evaluate $\lim_{n\to\infty} a_n$ for the given sequence $\{a_n\}$.

1. $a_n = \dfrac{2n^2 - 1 + n^3}{5n^3 + n + 2}$
2. $a_n = \dfrac{3n^2 + n + 4}{2n^3 + 1}$
3. $a_n = \dfrac{2^{n+1} + 5}{2^n + 3}$
4. $a_n = \dfrac{3^{2n} + 2}{9^{n+1} + 1}$
5. $a_n = \dfrac{3^n + 2 \cdot 5^n}{2^n + 3 \cdot 5^n}$
6. $a_n = \dfrac{2^n + 5^n}{7^n}$
7. $a_n = \dfrac{6n + \cos(n)}{3n + 2 - \sin(n^2)}$
8. $a_n = \dfrac{n2^n + 2^n + 1}{n2^n + 1}$
9. $a_n = \dfrac{n2^n + 3^n + 1}{n2^n + 3^n + 2}$
10. $a_n = \dfrac{\ln(n)}{n}$
11. $a_n = \dfrac{\ln^2(n)}{\sqrt{n}}$
12. $a_n = ne^{-2n}$
13. $a_n = n^2/e^n$
14. $a_n = \displaystyle\sum_{k=1}^{n} \frac{1}{k(k+1)}$
15. $a_n = \sqrt{n^2 + 3n} - n$
16. $a_n = \arcsin\left(\dfrac{n}{n+1}\right)$
17. $a_n = \arctan(n)$
18. $a_n = n\sin(1/n)$
19. $a_n = \left(\dfrac{n+1}{n}\right)^n$
20. $a_n = \left(\dfrac{n-1}{n}\right)^n$

In each of Exercises 21–28, a series $\sum_{n=1}^{\infty} a_n$ is given. Calculate the first five partial sums of the series. That is, calculate $S_N = \sum_{n=1}^{N} a_n$ for $N = 1, 2, 3, 4, 5$.

21. $\sum_{n=1}^{\infty} 1/n$
22. $\sum_{n=1}^{\infty} 2^n/n!$
23. $\sum_{n=1}^{\infty} (2^{-n} + 1)$
24. $\displaystyle\sum_{n=1}^{\infty} \left(\frac{1}{n} - \frac{1}{n+1}\right)$
25. $\sum_{n=1}^{\infty} 2^n/3^n$
26. $\sum_{n=1}^{\infty} 1/n^2$
27. $\sum_{n=1}^{\infty} (-1)^{n+1}/n^2$
28. $\sum_{n=1}^{\infty} n^2/2^n$

In each of Exercises 29–44, find the sum of the given series.

29. $\sum_{n=1}^{\infty} (3/7)^n$
30. $\sum_{n=1}^{\infty} (2/3)^{2n}$
31. $\sum_{n=0}^{\infty} (-2/3)^n$
32. $\sum_{n=1}^{\infty} 8^{-n}$
33. $\sum_{n=0}^{\infty} 9(0.1)^n$
34. $\sum_{n=4}^{\infty} (0.2)^n$
35. $\sum_{n=-3}^{\infty} (1/5)^n$
36. $\sum_{n=1}^{\infty} 7^{(-n/3)}$
37. $\displaystyle\sum_{n=2}^{\infty} \frac{1}{2^{n/2}}$
38. $\sum_{n=-1}^{\infty} (2/3)^{2n+1}$
39. $\sum_{n=3}^{\infty} (0.1)^n/(0.2)^{n+2}$
40. $\sum_{n=3}^{\infty} (2^{-n} + 3^{-n})$
41. $\sum_{n=1}^{\infty} (2^{-n} \cdot 3^{-n} + 7^{-n})$
42. $\sum_{n=1}^{\infty} (5^{-n} \cdot 3^{-n} \cdot 4^n)$
43. $\displaystyle\sum_{n=1}^{\infty} \frac{2^{n+1}}{5^{n-1}}$
44. $\displaystyle\sum_{n=1}^{\infty} \frac{3^{n-1}}{4^{n+1}}$

In Exercises 45–50, write each of the given repeating decimals as a constant times a geometric series (the geometric series will contain powers of 0.1). Use the formula for the sum of a geometric series to express the repeating decimal as a rational number.

45. 0.8888888 ...
46. 0.131313131313 ...
47. 0.017017017 ...
48. 0.983983983 ...
49. 0.1221212121 ...
50. 0.31323232 ...

In Each of Exercises 51–56, the partial sum $S_N = \sum_{n=1}^{N} a_n$ of an infinite series $\sum_{n=1}^{\infty} a_n$ is given. Determine the value of the infinite series.

51. $S_N = 2 - 1/N^2$
52. $S_N = (N+1)/(N+4)$
53. $S_N = (3N+2)/(N+1)$
54. $S_N = 2N^2/(N^2+2)$
55. $S_N = (3N+1)/(2N+4)$
56. $S_N = 2 - 2^{N+1}/3^{N+2}$

Further Theory and Practice

In each of Exercises 57–62, prove that the given series diverges by showing that the Nth partial sum satisfies $S_N \geq k \cdot N$ for some positive constant k.

57. $\sum_{n=1}^{\infty} (1.01)^n$

58. $\displaystyle\sum_{n=1}^{\infty} \frac{n}{n+1}$

59. $\displaystyle\sum_{n=1}^{\infty} \frac{n}{2n+3}$

60. $\displaystyle\sum_{n=1}^{\infty} \frac{n}{\sqrt{n^2+1}}$

61. $\displaystyle\sum_{n=1}^{\infty} \frac{4^n}{4^n+2^n}$

62. $\displaystyle\sum_{n=1}^{\infty} \frac{n+10}{10n}$

In each of Exercises 63–68, calculate the Nth partial sum S_N of the given series in closed form. Sum the series by finding $\lim_{N\to\infty} S_N$.

63. $\displaystyle\sum_{n=1}^{\infty} \left(\frac{1}{n} - \frac{1}{n+1}\right)$

64. $\displaystyle\sum_{n=1}^{\infty} \left(\frac{n+1}{n+2} - \frac{n}{n+1}\right)$

65. $\displaystyle\sum_{n=1}^{\infty} \left(\frac{2n-2}{n^3} - \frac{2n}{(n+1)^3}\right)$

66. $\displaystyle\sum_{n=1}^{\infty} \left(\frac{1}{n^2} - \frac{1}{(n+1)^2}\right)$

67. $\displaystyle\sum_{n=1}^{\infty} \left(\frac{1}{\sqrt{n}} - \frac{1}{\sqrt{n+1}}\right)$

68. $\displaystyle\sum_{n=1}^{\infty} \Big(\arctan(n+1) - \arctan(n)\Big)$

In each of Exercises 69–72, use partial fractions to calculate the Nth partial sum S_N of the given series in closed form. Sum the series by finding $\lim_{N\to\infty} S_N$.

69. $\displaystyle\sum_{n=1}^{\infty} \frac{1}{n(n+2)}$

70. $\displaystyle\sum_{n=1}^{\infty} \frac{1}{n(n+4)}$

71. $\displaystyle\sum_{n=1}^{\infty} \frac{1}{(2n+1)(2n+3)}$

72. $\displaystyle\sum_{n=1}^{\infty} \frac{2n+1}{(n^2+n)^2}$

73. Calculate the Nth partial sum of $\sum_{n=1}^{\infty} \ln\left(\frac{n}{n+1}\right)$. Show that the series is divergent.

74. Show that

$$\sum_{n=2}^{N} \ln\left(1 - \frac{1}{n^2}\right) = \ln((N-1)!) + \ln((N+1)!) - 2\ln(N!) - \ln(2),$$

and use this formula to sum $\sum_{n=2}^{\infty} \ln\left(1 - \frac{1}{n^2}\right)$.

75. Suppose that r is a constant greater than 1. Calculate the Nth partial sum of $\sum_{n=1}^{\infty} \frac{r^n}{(r^n-1)(r^{n+1}-1)}$, and use the formula to evaluate the infinite series.

76. The Fibonacci sequence $\{f_n\}$ is recursively defined by $f_0 = f_1 = 1$ and $f_n = f_{n-1} + f_{n-2}$ for $n \geq 2$. Show that

$$\frac{1}{f_n f_{n+2}} = \frac{1}{f_n f_{n+1}} - \frac{1}{f_{n+1} f_{n+2}},$$

and use this formula to sum $\sum_{n=0}^{\infty} 1/(f_n f_{n+2})$.

77. It is known that $\sum_{n=1}^{\infty} 1/n^2 = \pi^2/6$. What are the values of the convergent series

$$1/2^2 + 1/4^2 + 1/6^2 + 1/8^2 + \cdots$$

and

$$1 + 1/3^2 + 1/5^2 + 1/7^2 + \cdots ?$$

78. It is known that $\sum_{n=1}^{\infty} \frac{n}{2^n} = 2$ and $\sum_{n=1}^{\infty} \frac{1}{n2^n} = \ln(2)$. Supposing that a, b, and c are constants, evaluate $\sum_{n=1}^{\infty} \frac{an^2+cn+b}{n2^n}$.

79. It is known that $\sum_{n=1}^{\infty} \frac{1}{n^2} = \frac{1}{6}\pi^2$ and $\sum_{n=1}^{\infty} \frac{1}{n^4} = \frac{1}{90}\pi^4$. Let $f(x,y) = \sum_{n=1}^{\infty} \frac{x + y \cdot (\pi n)^2}{n^4}$. Describe the solution set $\{(x,y) \in \mathbb{R}^2 : f(x,y) = 0\}$.

80. The infinite series $\sum_{n=1}^{\infty} nr^{n-1}$ arises in genetics and other fields. Notice that the partial $S_N = \sum_{n=1}^{N} nr^{n-1}$ is equal to $\frac{d}{dr}\sum_{n=1}^{N} r^n$. Use this observation to derive a closed form expression for S_N. Show that $\sum_{n=1}^{\infty} nr^{n-1}$ converges for $|r| < 1$, and determine its value.

81. Prove that $\sum_{n=1}^{\infty} \ln(1/n)$ diverges.

82. Use the inequality $x/2 < \ln(1+x)$ for $x \in (0,1)$ to prove that $\sum_{n=1}^{\infty} \ln(1+1/n)$ diverges.

83. A tournament ping-pong ball bounces to 2/3 of its original height when it is dropped from a height of 25 cm or less onto a hard surface. How far will the ball travel if dropped from a height of 20 cm and allowed to bounce forever?

84. Suppose that every dollar that we spend gives rise (through wages, profits, etc.) to 90 cents for someone else to spend. That 90 cents will generate a further 81 cents for spending, and so on. How much spending will result from the purchase of a \$16,000 automobile, the car included? (This phenomenon is known as the *multiplier effect.*)

85. Suppose that the national savings rate is 5%. Of every dollar earned, 5¢ is saved and 95¢ is spent. Of the 95¢ that goes into the hands of others, 5% will be saved and 95% spent, and so on. How much spending will be generated by a 100 billion dollar (\$$10^{11}$) tax cut?

86. A heart patient must take a 0.5 mg daily dose of a medication. Each day, his body eliminates 90% of the medication present. The amount S_N of the medicine that is present after N days is a partial sum of which infinite series? Approximately what amount of the medicine is maintained in the patient's body after a long period of treatment?

87. A particular pollutant breaks down in water as follows: If the mass of the pollutant at the beginning of the time interval $[t, t+T]$ is M, then the mass is reduced to Me^{-kT} by the end of the interval. Suppose that at intervals 0, T, $2T$, $3T$, ... a factory discharges an amount M_0 of the pollutant into a holding tank containing water. After a long period, the mixture from this tank is fed into a river. If the quantity of pollutant released into the river is required to be no greater than Q, then, in terms of k, T, and Q, how large can M_0 be if the factory is compliant?

88. Sketch the graphs of $y = x^n$, $n = 1, 2, 3, \ldots$ in one viewing window $[0,1] \times [0,1]$. Find the area between two consecutive graphs $y = x^{n-1}$ and $y = x^n$. Use your calculation to calculate $\sum_{n=1}^{\infty} 1/(n \cdot (n+1))$ by a geometric argument.

89. If r is a *fixed* number in $(0,1)$, then $\sum_{n=1}^{\infty} r^n$ converges. However, if $\{r_n\}$ is a sequence of numbers in $(0,1)$, then $\sum_{n=1}^{\infty} r_n^n$ may diverge. Prove the divergence of this series for $r_n = 1 - 1/n$.

90. Show that

$$2 \cdot 3 \cdot 6 \cdot 8 \cdots (2N) = N!2^N.$$

91. Show that

$$1 \cdot 3 \cdot 5 \cdot 7 \cdots (2N-3)(2N-1) = \frac{(2N)!}{N!2^N}.$$

92. Mr. Woodman has an unlimited supply of 1 inch long, 3/16 inch thick dominoes. He stacks N of them so that for each $2 \le n \le N$, the nth domino, counting from the *bottom* of the stack, protrudes $1/(2(N-n+1))$ inch over the end of the $(n-1)^{\text{st}}$ domino (see Figure 1). Show that the center of mass of Mr. Woodman's tower of dominoes lies over the bottom domino (and so the stack will not fall). Deduce that by using a sufficiently large number N, Mr. Woodman can make his stack span, from left to right, any given distance. About how many dominoes would it take to span the 10 foot length of his playroom? How high would the tower be?

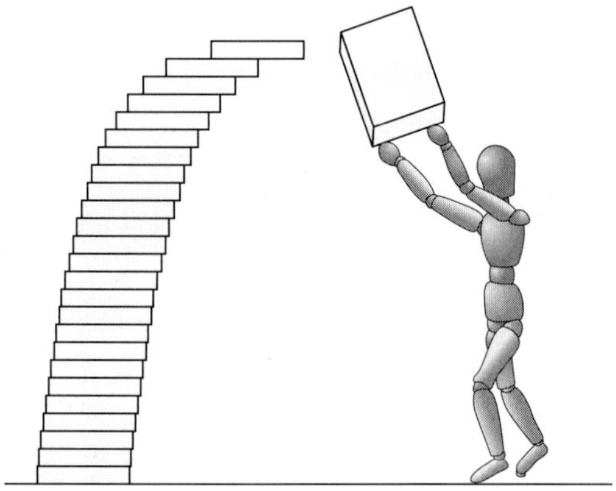

▲ Figure 1

Calculator/Computer Exercises

In each of Exercises 93–97, a convergent series is given. Estimate the value ℓ of the series by calculating its partial sums S_N for $N = 1, 2, 3, \ldots$. Round your evaluations to four decimal places and stop when three consecutive rounded partial sums agree. (This procedure does *not* ensure that the last partial sum calculated agrees with ℓ to four decimal places. The error that results when a partial sum is used to approximate an infinite series is called a *truncation error*. Methods of estimating truncation errors will be discussed in later sections.)

93. $\displaystyle\sum_{n=1}^{\infty} \frac{e^{-n}}{n}$

94. $\displaystyle\sum_{n=1}^{\infty} \frac{e^n}{n!}$

95. $\sum_{n=1}^{\infty} \left(\frac{1.1}{n}\right)^n$

96. $\sum_{n=1}^{\infty} \frac{2^n + n^2}{10^n}$

97. $\sum_{n=1}^{\infty} \frac{2n+1}{n^5+n+1}$

98. Let a and b be positive numbers, and set $r = 1/(1 + a^{-b})$. Observe that $0 < r < 1$, and therefore $\sum_{n=1}^{\infty} r^n$ is convergent. Let S denote the value of this infinite series.

a. Let $\tau(N) = \sum_{n=N+1}^{\infty} r^n$. Show that

$$\tau(N) = \frac{a^{b(N+1)}}{(a^b+1)^N}.$$

b. For what constants α and β is $\ln(\tau(N)) = \alpha \ln(a) - \beta \ln(a^b + 1)$?

c. Calculate $\ln(\tau(N))$ for $a = 10$, $b = 50$, and $N = 10^6$. By about how much does the millionth partial sum S_{10^6} of

$$\sum_{n=1}^{\infty} \left(\frac{1}{1+10^{-50}}\right)^n$$

differ from the full sum? (The difference *is* very large. The number N of terms that must be summed for $S - S_N$ to be less than 0.1 is around 1.17×10^{52}.)

8.2 The Divergence Test and the Integral Test

In the preceding section, you learned that the convergence of a series $\sum_{n=1}^{\infty} a_n$ depends on the behavior of the Nth partial sum $S_N = \sum_{n=1}^{N} a_n = a_1 + a_2 + \cdots + a_{N-1} + a_N$ as N tends to infinity. The problem is that it is often difficult or impossible to find an explicit formula for S_N. Fortunately, in many cases, we do not need such an explicit formula. Beginning in this section, we will develop "tests" that permit us to decide whether an infinite series converges or diverges without having to directly analyze the sequence of partial sums.

The Divergence Test

To show that a series $\sum_{n=1}^{\infty} a_n$ converges, we must show that the sequence $\{S_N\}$ of partial sums converges (either directly or by using one of the theorems that will be presented in this and subsequent sections). We may wonder whether the convergence of $\{a_n\}$ itself tells us anything about the convergence of $\sum_{n=1}^{\infty} a_n$. There are three possibilities:

a. $\lim_{n\to\infty} a_n$ does not exist.
b. $\lim_{n\to\infty} a_n$ exists and this limit is nonzero.
c. $\lim_{n\to\infty} a_n = 0$.

The next theorem tells us that, in cases a and b, the series $\sum_{n=1}^{\infty} a_n$ diverges.

THEOREM 1 **(The Divergence Test)** If the summands a_n of an infinite series $\sum_{n=1}^{\infty} a_n$ do not tend to 0, then the series $\sum_{n=1}^{\infty} a_n$ diverges.

Proof. The statement, "A implies B," is logically equivalent to the statement, "not B implies not A." This equivalence is known as the *law of the contrapositive*. Using this law with $A =$ "$\lim_{n\to\infty} a_n \neq 0$" and $B =$ "$\sum_{n=1}^{\infty} a_n$ diverges," we will prove the Divergence Test in the form, $\sum_{n=1}^{\infty} a_n$ converges implies $\lim_{n\to\infty} a_n = 0$.

We may always write a partial sum $S_N = a_1 + a_2 + \cdots + a_{N-1} + a_N$ in terms of the preceding partial sum S_{N-1}:

$$S_N = (a_1 + a_2 + \cdots + a_{N-1}) + a_N = S_{N-1} + a_N.$$

This equation may also be written as $a_N = S_N - S_{N-1}$. Notice that if we have $\lim_{N\to\infty} S_N = \ell$, then we also have $\lim_{N\to\infty} S_{N-1} = \ell$. As a result, if the infinite series $\sum_{n=1}^{\infty} a_n$ converges to ℓ, then

$$\lim_{N\to\infty} a_N = \lim_{N\to\infty} (S_N - S_{N-1}) = \ell - \ell = 0. \quad \blacksquare$$

Notice that the Divergence Test tells us that certain infinite series *diverge*, but it *never* tells us that a particular infinite series converges. The next two examples show how the Divergence Test can be used and how it should not be used.

▶ **EXAMPLE 1** What does the Divergence Test tell us about the geometric series $\sum_{n=0}^{\infty} r^n$?

Solution Let $a_n = r^n$. If $|r| < 1$, then $\lim_{n\to\infty} a_n = 0$. In this case, the hypothesis of the Divergence Test is not satisfied. The Divergence Test therefore yields no information about the series $\sum_{n=0}^{\infty} r^n$ when $|r| < 1$. Of course, we do know from Section 8.1 that the series $\sum_{n=0}^{\infty} r^n$ converges for $|r| < 1$.

Now suppose that $1 \le |r|$. Then, $1 \le |r^n|$, and $\lim_{n\to\infty} a_n \ne 0$. In this case, the hypothesis of the Divergence Test *is* satisfied. Thus the Divergence Test tells us that $\sum_{n=0}^{\infty} r^n$ diverges when $1 \le |r|$. (This application of the Divergence Test completes the proof of Theorem 3, Section 8.1.) ◀

▶ **EXAMPLE 2** What does the Divergence Test tell us about the two series $\sum_{n=1}^{\infty} 1/n$ and $\sum_{n=1}^{\infty} 1/(n^2 + n)$?

Solution The answer is, "Not a thing!" The Divergence Test can only be applied to infinite series $\sum_{n=1}^{\infty} a_n$ for which $\lim_{n\to\infty} a_n \ne 0$. That is not the case with either of these series. We do know from Section 8.1 that $\sum_{n=1}^{\infty} 1/n$ diverges and $\sum_{n=1}^{\infty} 1/(n^2 + n)$ converges (Theorem 1 and Example 9, respectively). But the Divergence Test does not provide us with an alternative justification for either of these conclusions. ◀

INSIGHT It is a common error to conclude that an infinite series converges because its terms tend to 0. Such reasoning is faulty and the conclusion may be *false*: the harmonic series shows that a series $\sum_{n=1}^{\infty} a_n$ may be divergent even though $\lim_{n\to\infty} a_n = 0$. In the theory of infinite series, it often happens that one particular test is inconclusive, which only means that we must resort to a different line of reasoning.

Series with Nonnegative Terms

Some series have both positive and negative terms and converge because the terms tend to cancel each other out. The situation is more straightforward when a series has summands that are all nonnegative. Suppose that $a_n \ge 0$ for each n. Let $S_N = \sum_{n=1}^{N} a_n$. Notice that $S_N = S_{N-1} + a_N$, and therefore $S_{N-1} \le S_N$. Thus the sequence of partial sums is increasing: $0 \le S_1 \le S_2 \le S_3 \le \ldots$. There are now only two possibilities. Either the S_N's increase without bound, or they do not. If they increase without bound, then the series diverges. On the other hand, if the S_N's are bounded above, then by the Monotone Convergence Property (Section 2.6 of Chapter 2), the S_N's converge to a limit ℓ. In this case, $\sum_{n=1}^{\infty} a_n = \ell$. We state this observation as a theorem.

THEOREM 2 Suppose that $0 \le a_n$ for each n. Let $S_N = \sum_{n=1}^{N} a_n$. If there is a real number U such that $S_N \le U$ for all N, then the infinite series $\sum_{n=1}^{\infty} a_n$ converges.

Theorem 2 does not tell us the value of a convergent series $\sum_{n=1}^{\infty} a_n$ but it is extremely useful nonetheless. Until now, we have had to find an explicit formula for the partial sum S_N in order to determine the convergence of $\sum_{n=1}^{\infty} a_n$. Theorem 2 allows us to determine convergence with less precise information about the partial sums. Here is an example of how Theorem 2 can be used.

▶ **EXAMPLE 3** Discuss convergence for the series $\sum_{n=1}^{\infty} 1/n^n$.

Solution The summands $1/n^n$ are all positive. Theorem 2 tells us that $\sum_{n=1}^{\infty} 1/n^n$ is convergent if there is a real number U such that $S_N \le U$ for all N. In fact, $U = 3/2$ will do:

$$\begin{aligned} S_N &= 1 + \frac{1}{2^2} + \frac{1}{3^3} + \frac{1}{4^4} + \cdots + \frac{1}{N^N} \\ &\le 1 + \frac{1}{2^2} + \frac{1}{2^3} + \frac{1}{2^4} + \cdots + \frac{1}{2^N} \\ &= \left(1 + \frac{1}{2^1} + \frac{1}{2^2} + \frac{1}{2^3} + \frac{1}{2^4} + \cdots + \frac{1}{2^N}\right) - \frac{1}{2^1} \\ &< \sum_{n=0}^{\infty} \left(\frac{1}{2}\right)^n - \frac{1}{2} \\ &= 2 - \frac{1}{2} \qquad \text{by equation (8.1.13) with } r = 1/2. \end{aligned}$$

To recapitulate, the infinite series $\sum_{n=1}^{\infty} 1/n^n$ is convergent because its terms are positive and its partial sums are bounded above. ◀

INSIGHT In Example 3, we are able to conclude that the infinite series $\sum_{n=1}^{\infty} 1/n^n$ converges without actually finding its sum ℓ. Because our analysis shows that $S_N \le 3/2$ for all N, we deduce that $\sum_{n=1}^{\infty} 1/n^n = \lim_{N\to\infty} S_N \le 3/2$. If we need a more precise estimate, then we must calculate partial sums. Using a computer, we calculate $S_9 = 1.291285997$ (to ten significant digits) and $S_{10} = 1.291285997$ as well. With a little more calculation, we see that adding more terms changes only those digits that are beyond the ninth decimal place.

▶ **EXAMPLE 4** Give an example of a divergent infinite series $\sum_{n=1}^{\infty} a_n$ whose partial sums $S_N = \sum_{n=1}^{N} a_n$ are bounded.

Solution According to Theorem 2, the summands a_n cannot all be nonnegative. In fact, we have already encountered an infinite series that has the properties we seek: $\sum_{n=1}^{\infty} (-1)^n$. We have observed that the partial sums of this series are all either -1 or 0 (Section 8.1, Example 5) and the series diverges (Section 8.1, Example 8). The lesson is that the hypothesis $a_n \ge 0$ cannot be omitted from Theorem 2. ◀

The Tail End of a Series

Whether or not a given series converges does not depend on the first million or so terms. For suppose that we are given the series $\sum_{n=1}^{\infty} a_n$ and that we are able to ascertain that $\sum_{n=10^6+1}^{\infty} a_n$ converges. This means that there is a real number ℓ such that $\lim_{N\to\infty} \sum_{n=10^6+1}^{N} a_n = \ell$. But this limit allows us to see that the partial sums of the full series converge:

$$\lim_{N\to\infty}\sum_{n=1}^{N} a_n = \lim_{N\to\infty}\left(\sum_{n=1}^{10^6} a_n + \sum_{n=10^6+1}^{N} a_n\right) = \sum_{n=1}^{10^6} a_n + \lim_{N\to\infty}\sum_{n=10^6+1}^{N} a_n = \sum_{n=1}^{10^6} a_n + \ell.$$

Of course, there is nothing special about the number 10^6. If M is any positive integer, then we refer to $\sum_{n=M+1}^{\infty} a_n$ as a *tail* of $\sum_{n=1}^{\infty} a_n$. The full series $\sum_{n=1}^{\infty} a_n$ is convergent if and only if the tail $\sum_{n=M+1}^{\infty} a_n$ is convergent. In this case, the two infinite series $\sum_{n=1}^{\infty} a_n$ and $\sum_{n=M+1}^{\infty} a_n$ differ only by the partial sum $\sum_{n=1}^{M} a_n$, which, comprising finitely many summands, is finite.

▶ **EXAMPLE 5** Let $a_n = 1/2^{n-1}$ if $n < 1000$, and $a_n = 1/n$ for $n \geq 1000$. Does the series $\sum_{n=1}^{\infty} a_n$ converge?

Solution The series must diverge. For if $\sum_{n=1}^{\infty} a_n$ were convergent, then the tail $\sum_{n=1000}^{\infty} a_n$, or $\sum_{n=1000}^{\infty} 1/n$, would be convergent. But this tail of $\sum_{n=1}^{\infty} a_n$ is also a tail of $\sum_{n=1}^{\infty} 1/n$. The convergence of $\sum_{n=1000}^{\infty} 1/n$ would therefore imply the convergence of $\sum_{n=1}^{\infty} 1/n$. That is false (Section 8.1, Theorem 1). ◀

▶ **EXAMPLE 6** Does the series $\sum_{n=10}^{\infty} \frac{1}{n(n+1)}$ converge? If so, to what number?

Solution Yes: the given series is a tail of the convergent series $\sum_{n=1}^{\infty} \frac{1}{n(n+1)}$ (Section 8.1, Example 9), and so it too is convergent. The difference between the series $\sum_{n=1}^{\infty} \frac{1}{n(n+1)}$ and the given series is just the finite sum $\sum_{n=1}^{9} \frac{1}{n(n+1)}$, which we calculate to be 9/10. Because we know that $\sum_{n=1}^{\infty} \frac{1}{n(n+1)} = 1$ from Example 9 of Section 8.1, we see that

$$\sum_{n=10}^{\infty} \frac{1}{n(n+1)} = \sum_{n=1}^{\infty} \frac{1}{n(n+1)} - \sum_{n=1}^{9} \frac{1}{n(n+1)} = 1 - \frac{9}{10} = \frac{1}{10}. \quad ◀$$

The Integral Test

We have already noted the analogy between infinite series and improper integrals over an infinite interval. We develop that idea further to obtain a convergence test for infinite series of positive terms.

THEOREM 3 **(The Integral Test)** Let f be a positive, continuous, decreasing function on the interval $[1, \infty)$. Then the infinite series $\sum_{n=1}^{\infty} f(n)$ converges if and only if the improper integral $\int_1^{\infty} f(x)\,dx$ converges. In this case, we have

$$\int_1^{\infty} f(x)\,dx \leq \sum_{n=1}^{\infty} f(n) \leq f(1) + \int_1^{\infty} f(x)\,dx. \tag{8.2.1}$$

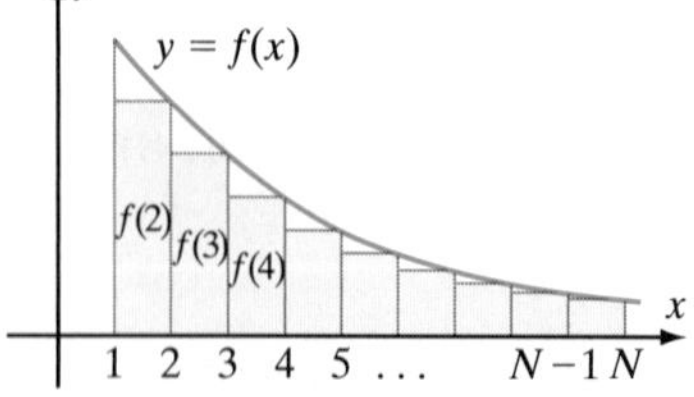

▲ Figure 1 The area of the shaded rectangles is the series $\sum_{n=2}^{N} f(n)$.

Proof. Look at Figure 1. Because the base of each inscribed rectangle has length 1, the area of each rectangle is equal to its height. Thus the area of the first rectangle is $f(2)$, the area of the second box is $f(3)$, and so on. All the boxes lie under the graph. Therefore

$$\sum_{n=2}^{N} f(n) \leq \int_1^{N} f(x)\,dx. \tag{8.2.2}$$

If the improper integral $\int_1^\infty f(x)\,dx$ is convergent, then the numbers $\int_1^N f(x)\,dx$ increase to a (finite) limit as N tends to infinity. This limit is denoted by $\int_1^\infty f(x)\,dx$. Using inequality (8.2.2) we deduce that

$$\sum_{n=1}^{N} f(n) = f(1) + \sum_{n=2}^{N} f(n) \leq f(1) + \int_1^N f(x)\,dx \leq f(1) + \int_1^\infty f(x)\,dx.$$

Thus the partial sums of $\sum_{n=1}^\infty f(n)$ are bounded above by $f(1) + \int_1^\infty f(x)\,dx$, and, as a result, the series $\sum_{n=1}^\infty f(n)$ is convergent and bounded above by $f(1) + \int_1^\infty f(x)\,dx$.

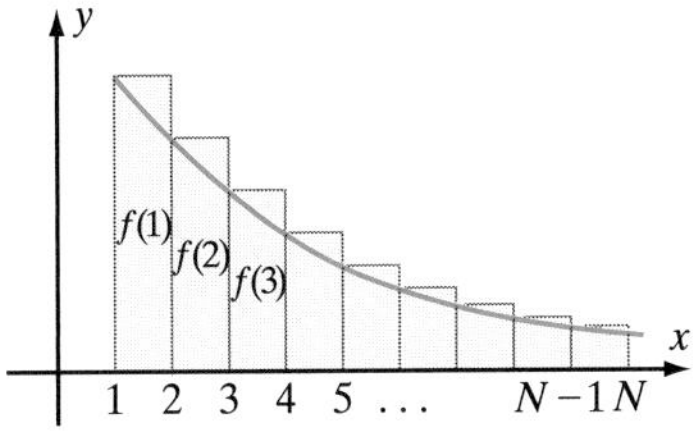

▲ Figure 2 The area of the shaded rectangles is the series $\sum_{n=1}^{N-1} f(n)$.

Now look at the superscribed rectangles in Figure 2. Clearly

$$\int_1^N f(x)\,dx \leq \sum_{n=1}^{N-1} f(n). \tag{8.2.3}$$

If $\sum_{n=1}^\infty f(n)$ is convergent, then inequality (8.2.3) tells us that the numbers $\int_1^N f(x)\,dx$ are bounded above by $\sum_{n=1}^\infty f(n)$. As a result, $\int_1^\infty f(x)\,dx$ is convergent and bounded above by $\sum_{n=1}^\infty f(n)$. ■

The crux of the proof of the Integral Test is in the geometry. Remember the pictures, and you will remember why the test works. Here are several examples.

▶ **EXAMPLE 7** Show that the series $\sum_{n=1}^{\infty} \frac{1}{1+n^2}$ converges, and estimate its value

Solution We apply the Integral Test to the continuous, positive, decreasing function $f(x) = 1/(1+x^2)$. Because

$$\int_1^\infty f(x)\,dx = \int_1^\infty \frac{1}{1+x^2}\,dx = \lim_{N\to\infty} \arctan(x)\Big|_{x=1}^{x=N} = \lim_{N\to\infty}\left(\arctan(N) - \frac{\pi}{4}\right) = \frac{\pi}{2} - \frac{\pi}{4} = \frac{\pi}{4},$$

we conclude that $\int_1^\infty f(x)\,dx$ is convergent. According to the Integral Test, the series $\sum_{n=1}^\infty f(n) = \sum_{n=1}^\infty 1/(1+n^2)$ is also convergent. The Integral Test also provides an estimate of the value of the series. Because $\int_1^\infty 1/(1+x^2)\,dx = \pi/4$, line (8.2.1) tells us that

$$0.785 \approx \frac{\pi}{4} = \int_1^\infty \frac{1}{1+x^2}\,dx \leq \sum_{n=1}^{\infty} \frac{1}{1+n^2} \leq \frac{1}{1+1^2} + \int_1^\infty \frac{1}{1+x^2}\,dx = \frac{1}{2} + \frac{\pi}{4} \approx 1.285.$$

In fact, a computer calculation shows that $\sum_{n=1}^\infty 1/(1+n^2) = 1.076\ldots$, which is indeed in the interval $[0.785, 1.285]$ that we determined. It is sometimes useful to have such estimates even when they are not particularly accurate. For the series of Example 7, we can improve on the lower bound 0.785 obtained from the Integral Test by simply summing three terms $\sum_{n=1}^{3} 1/(1+n^2) = 4/5 = 0.8$. ◀

INSIGHT The Integral Test is stated in terms of a function f. In practice, we apply the Integral Test when presented with an infinite series $\sum_{n=1}^\infty a_n$. Example 7 shows how to define an appropriate function f starting from the series: In the expression for a_n, replace the summation index n with x, and the formula for $f(x)$ results. Of course, before applying the Integral Test, we must be sure that f is positive, continuous, and decreasing.

▶ **EXAMPLE 8** Show that the series $\sum_{n=1}^{\infty} \dfrac{n}{e^n}$ converges, and estimate its value.

Solution The function $f(x) = x/e^x$ is positive, continuous, and decreasing for $x \geq 1$ because for these values of x, we have $f'(x) = (e^x - xe^x)/e^{2x} = (1-x)/e^x \leq 0$. Next, we will investigate the improper integral of $f(x)$. We have, by an application of integration by parts with $u = x$ and $dv = e^{-x}dx$,

$$\int_1^{\infty} f(x)\,dx = \lim_{N\to\infty} \int_1^N xe^{-x}\,dx = \lim_{N\to\infty} \left((-xe^{-x})\Big|_{x=1}^{x=N} + \int_1^N e^{-x}\,dx \right) = \lim_{N\to\infty} (-xe^{-x} - e^{-x})\Big|_{x=1}^{x=N}.$$

Therefore

$$\int_1^{\infty} f(x)\,dx = \lim_{N\to\infty} (-Ne^{-N} - e^{-N}) - (-e^{-1} - e^{-1}) = -\lim_{N\to\infty} Ne^{-N} + \frac{2}{e} = \frac{2}{e},$$

where we have used Example 3 of Section 8.1 to conclude that $\lim_{N\to\infty} Ne^{-N} = 0$. Because $\int_1^{\infty} f(x)\,dx$ converges and equals $2/e$, it follows from the Integral Test that $\sum_{n=1}^{\infty} n/e^n$ also converges. Line (8.2.1) tells us that

$$0.7358 \approx \frac{2}{e} = \int_1^{\infty} \frac{x}{e^x}\,dx \leq \sum_{n=1}^{\infty} \frac{n}{e^n} \leq \frac{1}{e^1} + \int_1^{\infty} \frac{x}{e^x}\,dx = \frac{1}{e} + \frac{2}{e} = \frac{3}{e} \approx 1.1036.$$

Using more advanced techniques, it can be shown that the value of the given series is $e/(e-1)^2 = 0.92067\ldots$, which indeed lies between our lower and upper estimates. ◀

INSIGHT The limit formula

$$\lim_{x\to\infty} \frac{x^p}{e^{\lambda x}} = 0 \quad (0 < \lambda), \tag{8.2.4}$$

which holds for any constant p, is frequently useful. For $p = 1$, equation (8.2.4) is proved by applying l'Hôpital's Rule, as in Example 3 of Section 8.1. In general, equation (8.2.4) is proved by repeatedly applying l'Hôpital's Rule.

p-Series

If p is a fixed number then the infinite series $\sum_{n=1}^{\infty} 1/n^p$ is called a p-series. Our next theorem completely describes how the convergence of the p-series depends on the value of p.

THEOREM 4 Fix a real number p. The series $\sum_{n=1}^{\infty} \dfrac{1}{n^p}$ converges if $1 < p$; it diverges if $p \leq 1$.

Proof. If p is not positive, then the terms of the series do not tend to 0. According to the Divergence Test, the series must diverge. If p is positive, then $f(x) = 1/x^p$ is positive, continuous, and decreasing on the interval $[1, \infty)$. We may therefore apply the Integral Test. There are three cases to consider. If $0 < p < 1$, then

$$\int_1^{\infty} \frac{1}{x^p}\,dx = \lim_{N\to\infty} \frac{1}{1-p} \cdot x^{1-p}\Big|_1^N = \infty.$$

The Integral Test tells us that the series diverges. If $p = 1$, then the series $\sum_{n=1}^{\infty} 1/n^p$ is the harmonic series, which diverges. (The Integral Test can also be used to demonstrate this divergence). Finally, if $p > 1$, then

$$\int_1^{\infty} \frac{1}{x^p}\, dx = \lim_{N\to\infty} \int_1^N x^{-p}\, dx = \lim_{N\to\infty} \frac{x^{1-p}}{1-p}\Bigg|_1^N = \frac{1}{1-p} \lim_{N\to\infty} \left(\frac{1}{N^{p-1}} - 1\right) = \frac{1}{p-1}.$$

The Integral Test then tells us that the series converges. ■

▶ **EXAMPLE 9** Determine whether $\sum_{n=1}^{\infty} \frac{n+2}{n^{3/2}}$ is convergent.

Solution The series $\sum_{n=1}^{\infty} 1/n^{3/2}$ is a p-series with $p = 3/2 > 1$. Therefore it is convergent. If the given series were convergent, then Theorem 2 of Section 8.1 (with $a_n = (n+2)/n^{3/2}$, $b_n = 1/n^{3/2}$, and $\lambda = -2$) would tell us that

$$\sum_{n=1}^{\infty} (a_n + 2b_n) = \sum_{n=1}^{\infty} \left(\frac{n+2}{n^{3/2}} - 2\frac{1}{n^{3/2}}\right) = \sum_{n=1}^{\infty} \frac{n}{n^{3/2}} = \sum_{n=1}^{\infty} \frac{1}{n^{1/2}}$$

is convergent. But $\sum_{n=1}^{\infty} 1/n^{1/2}$ is a p-series with $p = 1/2 \leq 1$. Therefore it is divergent. We conclude that the given series must be divergent; otherwise, we would have a contradiction. ◀

An Extension

It is often convenient and sometimes necessary to apply the Integral Test with an initial summation index M greater than 1. We first apply the Integral Test to see that $\sum_{n=M}^{\infty} a_n$ and $\int_M^{\infty} f(x)\, dx$ have the same convergence behavior. Then we note that the full series $\sum_{n=1}^{\infty} a_n$ shares the behavior of its tail.

THEOREM 5 Suppose that f is a positive, continuous, decreasing function on the interval $[M, \infty)$. Let $\{a_n\}$ be a sequence with $a_n = f(n)$ for all n with $M \leq n$. Then, the infinite series $\sum_{n=1}^{\infty} a_n$ converges if and only if the improper integral $\int_M^{\infty} f(x)\, dx$ converges.

▶ **EXAMPLE 10** Determine whether the series $\sum_{n=2}^{\infty} \frac{1}{n\ln(n)}$ converges.

Solution This series begins with $n = 2$; notice that the summand is not even defined for $n = 1$. Because x and $\ln(x)$ are increasing functions of x, so is the product $x\ln(x)$. The reciprocal $f(x) = 1/(x\ln(x))$ of this product is therefore decreasing. Because f is also positive and continuous on the interval $[2, \infty)$ we may apply the Integral Test. Using the substitution $u = \ln(x)$, $du = (1/x)dx$, we calculate

$$\int_2^{\infty} f(x)\, dx = \lim_{N\to\infty} \int_2^N \frac{1}{x\ln(x)}\, dx = \lim_{N\to\infty} \int_{\ln(2)}^{\ln(N)} \frac{1}{u}\, du = \lim_{N\to\infty} \ln(|u|)\Bigg|_{u=\ln(2)}^{u=\ln(N)} = \lim_{N\to\infty} \Big(\ln(\ln(N)) - \ln(\ln(2))\Big) = \infty.$$

The divergence of the improper integral $\int_2^{\infty} f(x)\, dx$ implies the divergence of the infinite series $\sum_{n=2}^{\infty} f(n) = \sum_{n=2}^{\infty} 1/\big(n\ln(n)\big)$. ◀

▶ **EXAMPLE 11** Use the Integral Test to show that the series $\sum_{n=1}^{\infty} n^2 e^{-n/10}$ converges.

Solution The given series is equal to $\sum_{n=1}^{\infty} f(n)$ for $f(x) = x^2 e^{-x/10}$. In this case, f has an increasing factor, namely x^2. To apply the Integral Test, we must be sure that f itself is decreasing. The plot of f in Figure 3 suggests that $f(x)$ increases for $x < 20$ and decreases for $20 < x$. We can confirm this last observation by showing that $f'(x) < 0$ for $20 < x$:

$$f'(x) = 2xe^{-x/10} + x^2\left(-\frac{1}{10}\right)e^{-x/10} = \frac{1}{10}xe^{-x/10}(20 - x) < 0 \quad \text{for } 20 < x.$$

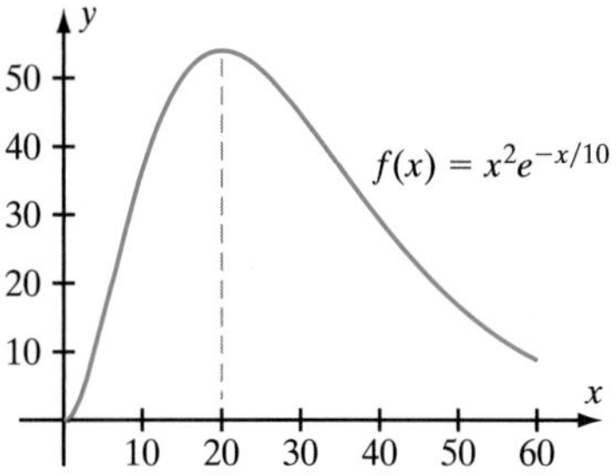

▲ Figure 3

We may therefore apply the Integral Test over the interval $[20, \infty)$. By integrating by parts twice and using equation (8.2.4) with both $p = 2$ and $p = 1$, we calculate

$$\begin{aligned}\int_{20}^{\infty} f(x)\,dx &= \lim_{N\to\infty} \int_{20}^{N} x^2 e^{-x/10}\,dx \\ &= \lim_{N\to\infty} \left(-10x^2 e^{-\frac{1}{10}x} - 200xe^{-\frac{1}{10}x} - 2000e^{-\frac{1}{10}x}\right)\Bigg|_{x=20}^{x=N} \\ &= \frac{10000}{e^2}.\end{aligned}$$

It follows that $\int_{20}^{\infty} f(x)\,dx$ is convergent, and, as a result, $\sum_{n=20}^{\infty} f(n)$ is convergent. Therefore $\sum_{n=1}^{\infty} n^2 e^{-n/10} = \sum_{n=1}^{\infty} f(n) = \sum_{n=1}^{19} f(n) + \sum_{n=20}^{\infty} f(n)$ is also convergent. ◀

◀ **A LOOK BACK** The Integral Test works by comparing the partial sums of a series with the area under a curve. It is often natural to compare the partial sums of one series with those of a series whose convergence is known. The next section is devoted to this powerful technique.

QUICK QUIZ

1. True or false: $\sum_{n=1}^{\infty} 1/\sqrt{n}$ converges because $\lim_{n\to\infty}(1/\sqrt{n}) = 0$.
2. True or false: The Divergence Test cannot be applied to the series $\sum_{n=1}^{\infty} 1/n^2$.
3. True or false: $\sum_{n=1}^{\infty} 1/n^{3/2}$ converges.
4. True or false: A series with bounded partial sums is convergent.

Answers

1. False 2. True 3. True 4. False

EXERCISES

Problems for Practice

▶ In Exercises 1–20, state what conclusion, if any, may be drawn from the Divergence Test. ◀

1. $\sum_{n=1}^{\infty} ne^{-n}$

2. $\displaystyle\sum_{n=1}^{\infty} \frac{1}{\sqrt{1+\sqrt{n}}}$

3. $\displaystyle\sum_{n=1}^{\infty} \frac{n^2}{n^2+1}$

4. $\displaystyle\sum_{n=1}^{\infty} \frac{3^n}{3^n+4}$

5. $\displaystyle\sum_{n=1}^{\infty} \frac{1}{1+1/n}$

6. $\sum_{n=1}^{\infty} \frac{\ln(n)}{n}$

7. $\sum_{n=1}^{\infty} \frac{3^n}{4^n + 3}$

8. $\sum_{n=1}^{\infty} \frac{1}{1 + \ln(n)}$

9. $\sum_{n=1}^{\infty} \frac{3^n + 5^n}{8^n}$

10. $\sum_{n=1}^{\infty} \frac{5^n}{2^n + 3^n}$

11. $\sum_{n=1}^{\infty} \frac{2^n + 1}{2^n + n^2}$

12. $\sum_{n=1}^{\infty} \frac{n^{1/4}}{\sqrt{1 + \sqrt{n}}}$

13. $\sum_{n=1}^{\infty} \frac{2n}{3n^2 + 1}$

14. $\sum_{n=1}^{\infty} \frac{2n^2}{3n^2 + n + 1}$

15. $\sum_{n=1}^{\infty} n^{1/n}$

16. $\sum_{n=1}^{\infty} \frac{n!}{n^n}$

17. $\sum_{n=1}^{\infty} (\pi/2 - \arctan(n))$

18. $\sum_{n=1}^{\infty} \cos(1/n)$

19. $\sum_{n=1}^{\infty} \sin(1/n)$

20. $\sum_{n=1}^{\infty} (n/(n+1) - 1/(n+2))$

In each sentence in Exercises 21–26, use either the word "may" or the word "must" to fill in the blank so that the completed sentence is correct. Explain your answer by referring to a theorem or example.

21. A series with summands tending to 0__converge.

22. A series that converges __ have summands that tend to 0.

23. If a series diverges, then the Divergence Test __ succeed in proving the divergence.

24. If the partial sums of an infinite series are bounded, then the series __ converge.

25. If a series diverges, then its terms __ diverge.

26. If the partial sums of an infinite series diverge, then the series __ diverge.

In each of Exercises 27–46, use the Integral Test to determine whether the given series converges or diverges. Before you apply the test, be sure that the hypotheses are satisfied.

27. $\sum_{n=1}^{\infty} e^{-n}$

28. $\sum_{n=1}^{\infty} \frac{n}{n^2 + 1}$

29. $\sum_{n=1}^{\infty} \frac{1}{n^2 + 4}$

30. $\sum_{n=1}^{\infty} \frac{1}{n(n+2)}$

31. $\sum_{n=1}^{\infty} \frac{1}{n + 3}$

32. $\sum_{n=1}^{\infty} n^2 e^{-n^3}$

33. $\sum_{n=1}^{\infty} \frac{2n^2}{n^3 + 4}$

34. $\sum_{n=2}^{\infty} \frac{1}{n \ln(n)}$

35. $\sum_{n=1}^{\infty} \frac{1}{(n+3)^{5/4}}$

36. $\sum_{n=2}^{\infty} \frac{1}{n^2 - n}$

37. $\sum_{n=8}^{\infty} \frac{e^n}{(1 + e^n)^2}$

38. $\sum_{n=1}^{\infty} \frac{n}{10^n}$

39. $\sum_{n=2}^{\infty} \frac{3}{n^2 + n}$

40. $\sum_{n=4}^{\infty} \frac{1}{n \ln^4(n)}$

41. $\sum_{n=1}^{\infty} \frac{\ln(n)}{n}$

42. $\sum_{n=1}^{\infty} \frac{1}{n^2 + 2n + 1}$

43. $\sum_{n=1}^{\infty} n e^{-2n}$

44. $\sum_{n=1}^{\infty} n 2^{-n}$

45. $\sum_{n=2}^{\infty} \frac{1}{n\sqrt{n^2 - 1}}$

46. $\sum_{n=3}^{\infty} \frac{1}{n \ln^{3/2}(n)}$

In each of Exercises 47–54, use known facts about p-series to determine whether the given series converges or diverges.

47. $\sum_{n=1}^{\infty} \frac{1}{\sqrt{n}}$

48. $\sum_{n=1}^{\infty} \frac{1}{n^{(\pi - e)}}$

49. $\sum_{n=1}^{\infty} \frac{\sqrt{3}}{n^{\sqrt{2}}}$

50. $\sum_{n=1}^{\infty} \frac{7n - \sqrt{2}}{4n^2}$

51. $\sum_{n=1}^{\infty} \frac{\sqrt{n} + 5}{n^2}$

52. $\sum_{n=1}^{\infty} \frac{n^3+1}{n^4}$

53. $\sum_{n=1}^{\infty} \frac{n^2+11}{n^4}$

54. $\sum_{n=1}^{\infty} \frac{2n^{1/8}+3n^{1/16}+1}{n^{5/4}}$

Further Theory and Practice

In Exercises 55–68, state what conclusion, if any, may be drawn from theDivergence Test.

55. $\sum_{n=1}^{\infty} \left(\frac{1}{n}\right)^{1/n}$

56. $\sum_{n=1}^{\infty} (\sin(1/n) - 1/n)$

57. $\sum_{n=1}^{\infty} (1+1/n)^n$

58. $\sum_{n=1}^{\infty} (\cos(1/n) - \sec(1/n))$

59. $\sum_{n=1}^{\infty} \frac{\ln(n^3)}{n}$

60. $\sum_{n=1}^{\infty} (\sqrt{n^2+3} - n)$

61. $\sum_{n=1}^{\infty} ((1+1/n)^n - n)$

62. $\sum_{n=1}^{\infty} \frac{\sin(n)}{n}$

63. $\sum_{n=1}^{\infty} n\sin(1/n)$

64. $\sum_{n=1}^{\infty} \frac{(n!)^2}{(2n)!}$

65. $\sum_{n=1}^{\infty} \frac{1}{\arctan(n)}$

66. $\sum_{n=1}^{\infty} \tan(\pi/3 - 1/n)$

67. $\sum_{n=1}^{\infty} (\operatorname{arcsec}(n) - \arctan(n))$

68. $\sum_{n=1}^{\infty} \left(\frac{n-1}{n}\right)^n$

In Exercises 69–74, use the Integral Test to determine whether the given series converges.

69. $\sum_{n=1}^{\infty} \frac{n^2}{2^n}$

70. $\sum_{n=1}^{\infty} \frac{\ln(n)}{n^2}$

71. $\sum_{n=1}^{\infty} \frac{1}{1+n^{1/2}}$

72. $\sum_{n=1}^{\infty} \ln\left(\frac{n+3}{n}\right)$

73. $\sum_{n=1}^{\infty} \ln\left(\frac{n^2+1}{n^2}\right)$

74. $\sum_{n=1}^{\infty} \frac{\arctan(n)}{1+n^2}$

In each of Exercises 75–78, verify that the given function f satisfies the hypotheses of the Integral Test and that $\int_1^{\infty} f(x)\,dx$ is convergent. Use line (8.2.1) to find an interval in which $\sum_{n=1}^{\infty} f(n)$ lies.

75. $f(x) = x/(3+x^2)^{5/2}$

76. $f(x) = x^2/(1+x^3)^3$

77. $f(x) = \dfrac{\exp(x)}{(1+\exp(x))^3}$

78. $f(x) = \dfrac{\exp(x)}{1+\exp(2x)}$

79. Show that $\sum_{n=1}^{\infty} (1/2)^{n^2}$ is convergent.

80. Derive the inequality $\sqrt{n}/4^n < 1/2^n$, which holds for every positive integer n. Prove that the infinite series $\sum_{n=1}^{\infty} \frac{\sqrt{n}}{4^n}$ is convergent by showing that its partial sums are bounded by 1.

81. Notice that

$$n! = n \cdot (n-1) \ldots 4 \cdot 3 \cdot 2 \cdot 1 > 2 \cdot 2 \ldots 2 \cdot 2 \cdot 2 \cdot 1$$

for $n \geq 3$. Use this observation to find a number U that is an upper bound of the infinite series $\sum_{n=1}^{\infty} 1/n!$ (thereby showing that this series is convergent).

82. Determine whether $\sum_{n=2}^{\infty} \int_n^{2n} x^{-2}\,dx$ converges.

83. Determine whether $\sum_{n=2}^{\infty} \int_n^{2n} e^{-x}\,dx$ converges.

84. Use the Integral Test to prove that the series

$$\sum_{n=20}^{\infty} \frac{1}{n\ln(n)\ln(\ln(n))}$$

diverges.

85. Define $f(x) = \sin^2(\pi x) + 1/x^2$ for $x \geq 1$. Notice that f is positive and continuous. Show that $\int_1^{\infty} f(x)\,dx$ diverges, yet $\sum_{n=1}^{\infty} f(n)$ converges. Explain why the conclusion of the Integral Test is not valid.

86. For which values of p is $\sum_{n=1}^{\infty} \left(\frac{n}{n^2+1}\right)^p$ convergent?

87. Let $\{d_n\}$ be a sequence of positive numbers. Let $a_n = \min(d_n/2, 1/2^n)$. Show that $\sum_{n=1}^{\infty} a_n$ converges.

88. Prove that if $0 \leq a_n$ and $\sum_{n=1}^{\infty} a_n$ converges then there exist numbers b_n with $a_n < b_n$ such that $\sum_{n=1}^{\infty} b_n$ still converges.

89. Prove that if $0 > c_n$ and $\sum_{n=1}^{\infty} c_n$ diverges, then there exist numbers d_n satisfying $0 < d_n < c_n$ and such that $\sum_{n=1}^{\infty} d_n$ diverges. This exercise shows that there is no "largest convergent series" and no "smallest divergent series."

Let $H_N = \sum_{n=1}^{N} 1/n$ denote the partial sums of the harmonic series. Let $A_N = H_N - \ln(N)$. Exercises 90–93 establish that $\lim_{N\to\infty} A_N$ exists. This important number, usually denoted by γ, is called the *Euler-Mascheroni constant.* Its value is 0.5772156649 to ten decimal places.

90. Sketch $y = 1/x$. Use the ideas of the Integral Test to illustrate the geometric significance of A_N. Draw a

staircase-shaped region whose area is H_N. Shade the region (comprising $N-1$ "triangular" regions and one rectangle) whose area is A_N.

91. Let a_n denote the area of the "triangular" region above $y = 1/x$, below $y = 1/n$, and between $x = n$ and $x = n+1$. Your sketch should show that

$$A_N = \sum_{n=1}^{N-1} a_n + \frac{1}{N}.$$

By comparison with a circumscribed rectangle, show that

$$0 < a_n < \left(\frac{1}{n} - \frac{1}{n+1}\right).$$

92. Show that the partial sums of the series $\sum_{n=1}^{\infty} a_n$ are bounded by the value of the convergent series

$$\sum_{n=1}^{\infty}\left(\frac{1}{n} - \frac{1}{n+1}\right) = 1.$$

Deduce that the limit $\gamma = \lim_{N\to\infty} A_N$ exists and is a number between 1/2 and 1.

93. This exercise presents an alternative, analytic approach to the limit $\lim_{N\to\infty} A_N$. Let $f(x) = x + \ln(1-x)$ for $0 \le x < 1$. Show that $f'(x) < 0$ for $0 < x < 1$. Deduce that $f(x) < 0$ for $0 < x < 1$. Show that $A_{N+1} - A_N = f(1/(N+1))$. Deduce that $\{A_N\}$ is a positive decreasing sequence, hence convergent.

94. Suppose $\{a_n\}$ and f satisfy the hypotheses of the Extended Integral Test (Theorem 5). Show that

$$\sum_{n=1}^{M-1} a_n + \int_M^{\infty} f(x)\,dx \le \sum_{n=1}^{\infty} a_n$$

and

$$\sum_{n=1}^{\infty} a_n \le \sum_{n=1}^{M} a_n + \int_M^{\infty} f(x)\,dx.$$

Use these estimates together with the integral formula $\int_M^{\infty} x^2 e^{-x/4}\,dx = e^{-M/4}(4M^2 + 32M + 128)$ to find an interval in which $\sum_{n=1}^{\infty} n^2 e^{-n/4}$ lies.

Calculator/Computer Exercises

Suppose that f is positive, continuous, and decreasing on $[1, \infty)$. Let

$$L(M) = \sum_{n=1}^{M-1} f(n) + \int_M^{\infty} f(x)\,dx$$

and

$$U(M) = f(M) + L(M).$$

Then, according to the result of Exercise 94,

$$L(M) < \sum_{n=1}^{\infty} f(n) < U(M).$$

In each of Exercises 95–98, for the given function f, calculate the least value M such that $U(M) - L(M) < 5 \times 10^{-4}$. State $\sum_{n=1}^{\infty} f(n)$ to three decimal places.

95. $f(x) = 1/x^3$

96. $f(x) = \dfrac{1}{x^2(x+1)}$

97. $f(x) = x/(3 + x^2)^{5/2}$

98. $f(x) = x^2/(1 + x^3)^3$

8.3 The Comparison Tests

In the preceding section, we studied the convergence of a series $\sum_{n=1}^{\infty} a_n$ of nonnegative terms by comparing the series with an improper integral. The idea was to show the boundedness of the increasing sequence of partial sums. In this section, we will see that the same goal can often be reached by comparing the given infinite series with another series.

The Comparison Test for Convergence

Suppose that $\sum_{n=1}^{\infty} b_n$ is a convergent series of nonnegative terms with $\sum_{n=1}^{\infty} b_n = \ell$. Because the partial sums $\sum_{n=1}^{N} b_n$ form an increasing sequence, these partial sums must increase to ℓ. In particular, each partial sum satisfies $\sum_{n=1}^{N} b_n \le \ell$. If $\sum_{n=1}^{\infty} a_n$ is another series satisfying $0 \le a_n \le b_n$ for every n, then the partial sums for this series satisfy

$$\sum_{n=1}^{N} a_n \le \sum_{n=1}^{N} b_n \le \ell.$$

Thus the partial sums of the series $\sum_{n=1}^{\infty} a_n$ are increasing (because the a_n's are nonnegative) and bounded above by ℓ. By the Monotone Convergence Property (Section 2.6 in Chapter 2), we conclude that the sequence of partial sums of $\sum_{n=1}^{\infty} a_n$ converge. We summarize in the following theorem.

THEOREM 1 **(The Comparison Test for Convergence)** Let $0 \leq a_n \leq b_n$ for every n. If the series $\sum_{n=1}^{\infty} b_n$ converges, then the series $\sum_{n=1}^{\infty} a_n$ also converges.

▶ **EXAMPLE 1** Show that the series $\sum_{n=1}^{\infty} \dfrac{1}{\sqrt{n}(n+\sqrt{n}+1)}$ converges.

Solution Observe that $\sqrt{n}(n+\sqrt{n}+1) = n\sqrt{n}+n+\sqrt{n} > n\sqrt{n} = n^{3/2}$ for every positive n. We conclude that $\dfrac{1}{\sqrt{n}(n+\sqrt{n}+1)} < \dfrac{1}{n^{3/2}}$ for every positive n. Also, $\sum_{n=1}^{\infty} 1/n^{3/2}$ is a p-series that is convergent (because $p = 3/2 > 1$). By the Comparison Test for Convergence, the given series $\sum_{n=1}^{\infty} \dfrac{1}{\sqrt{n}(n+\sqrt{n}+1)}$ converges. ◀

INSIGHT When the Comparison Test for Convergence is applied to a series $\sum_{n=1}^{\infty} a_n$, a *second* series $\sum_{n=1}^{\infty} b_n$ must be found. How does one go about finding such a series for comparison? In Example 1, look at the denominator $\sqrt{n}(n+\sqrt{n}+1)$, or $n^{3/2} + n + n^{1/2}$, of the general term of the series. Compare the contribution of each summand to the total. When n is 10,000, for instance, the term $n^{3/2}$, which is 1,000,000, is comparable to $n^{3/2} + n + n^{1/2}$, which is 1,010,100. This sort of rounding is encountered all the time in the reporting of sales figures or economic data. In summary, discarding all but the largest power of n in the denominator of $\dfrac{1}{\sqrt{n}(n+\sqrt{n}+1)}$ results in the simpler expression $1/n^{3/2}$, which is the comparison used in Example 1.

▶ **EXAMPLE 2** Discuss convergence for the series $\sum_{n=1}^{\infty} \dfrac{1}{(3n-2)^2}$.

Solution Notice that $3n-2 = n+2(n-1) \geq n$ for all $n \geq 1$. Therefore $(3n-2)^2 \geq n^2$ so that $1/(3n-2)^2 \leq 1/n^2$. Finally, $\sum_{n=1}^{\infty} 1/n^2$ is a convergent p-series (with $p = 2 > 1$). By the Comparison Test for Convergence, we conclude that $\sum_{n=1}^{\infty} 1/(3n-2)^2$ converges. ◀

INSIGHT It is often possible to come to an intuitive decision about the convergence of a series without doing a precise comparison. In Example 2, we see that for large n, the summand $1/(3n-2)^2$ is about the same size as $1/(3n)^2$. Such a rough comparison is enough for us to predict the convergence of $\sum_{n=1}^{\infty} 1/(3n-2)^2$ by comparison with $(1/9)\sum_{n=1}^{\infty} 1/n^2$. The Limit Comparison Test, which will appear later in this section, will make this idea precise.

▶ **EXAMPLE 3** Discuss convergence for the series $\sum_{n=1}^{\infty} \dfrac{\sin^2(2n+5)}{n^4+8n+6}$.

Solution Observe that

$$0 \le \frac{\sin^2(2n+5)}{n^4+8n+6} \le \frac{1}{n^4+8n+6} \le \frac{1}{n^4}.$$

Also, $\sum_{n=1}^{\infty} 1/n^4$ is a p-series that is convergent (because $p = 4 > 1$). By the Comparison Test for Convergence, the given series converges. ◄

An Extension of the Comparison Test For Convergence

Because the convergence or divergence of a series does not depend on the first several terms, we can restate the Comparison Test for Convergence so that it applies even when there are a finite number of exceptions to the termwise comparison $0 \le a_n \le b_n$. We say that a property P holds "for all sufficiently large n" if there is an integer M such that P holds for all $n \ge M$.

THEOREM 2 **(The Comparison Test for Convergence)** Let $0 \le a_n \le b_n$ for all sufficiently large n. If the series $\sum_{n=1}^{\infty} b_n$ converges, then the series $\sum_{n=1}^{\infty} a_n$ also converges.

► **EXAMPLE 4** Determine whether the series $\displaystyle\sum_{n=10}^{\infty} \frac{n^{10}}{4^n}$ converges.

Solution This series begins with large terms—the first is $10^{10}/4^{10}$, or 9536.7. However, the exponential in the denominator eventually catches up to the polynomial in the numerator, exceeds it, and then continues to grow substantially faster than the numerator. We therefore expect the series to converge. To justify our intuition, we can reason as follows:

$$\frac{n^{10}}{4^n} = \left(\frac{n^{10}}{2^n}\right)\left(\frac{1}{2}\right)^n = \left(\frac{n^{10}}{e^{n\ln(2)}}\right)\left(\frac{1}{2}\right)^n = \left(\frac{n}{e^{n\ln(2)/10}}\right)^{10}\left(\frac{1}{2}\right)^n = \left(\frac{n}{e^{\lambda n}}\right)^{10}\left(\frac{1}{2}\right)^n,$$

where $\lambda = \ln(2)/10$. Example 3 of Section 8.1 shows that $\lim_{n\to\infty} \dfrac{n}{e^{\lambda n}} = 0$. Therefore there is an N such that $\dfrac{n}{e^{\lambda n}} < 1$ for all $n \ge N$. Thus if $a_n = n^{10}/4^n$ and $b_n = (1/2)^n$, then we have

$$a_n = \frac{n^{10}}{4^n} \le \left(\frac{1}{2}\right)^n = b_n$$

for all $n \ge N$. We compare $\sum_{n=N}^{\infty} a_n$ to the series $\sum_{n=N}^{\infty} b_n$, which, being a geometric series with positive ratio 1/2 (less than 1), is convergent. We conclude that the given series that begins with summation index 10 converges as well. ◄

The Comparison Test for Divergence

We can now reverse our reasoning to obtain a comparison test for divergence.

THEOREM 3 **(The Comparison Test for Divergence)** Let $0 \le b_n \le a_n$ for all sufficiently large n. If the series $\sum_{n=1}^{\infty} b_n$ diverges, then the series $\sum_{n=1}^{\infty} a_n$ also diverges.

Proof. Suppose that there is an M such that $b_n \le a_n$ for all $M \le n$. If $\sum_{n=1}^{\infty} a_n$ converged, then the Comparison Test for Convergence would imply that $\sum_{n=1}^{\infty} b_n$ converged, which would contradict the hypothesis. Thus $\sum_{n=1}^{\infty} a_n$ must diverge. ■

▶ **EXAMPLE 5** Show that the series $\sum_{n=1}^{\infty} \frac{\ln(n+4)}{n}$ diverges.

Solution Observe that $1 < \ln(1+4) \le \ln(n+4)$ for every $n \ge 1$. It follows that $1/n < \ln(n+4)/n$ for $n \ge 1$. Because the harmonic series $\sum_{n=1}^{\infty} 1/n$ diverges, the Comparison Test for Divergence allows us to conclude that the given series diverges as well. ◀

▶ **EXAMPLE 6** Analyze the series $\sum_{n=2}^{\infty} \frac{1}{\sqrt{n\ln(n)}}$.

Solution Notice that $\sqrt{y} < y$ if $1 < y$. We may apply this observation to $y = n\ln(n)$, which is greater than 1 for every $n \ge 2$. Thus each term of the given series satisfies $\sqrt{n\ln(n)} < n\ln(n)$ and $1/(n\ln(n)) < 1/\sqrt{n\ln(n)}$. In Example 10 of Section 8.2, we proved that the series $\sum_{n=2}^{\infty} 1/(n\ln(n))$ diverges. By the Comparison Theorem for Divergence, the given series also must diverge. ◀

▶ **EXAMPLE 7** Use the Comparison Theorem for Divergence to study the series $\sum_{n=1}^{\infty} \frac{1}{\sqrt{\sqrt{n}+3}}$

Solution A rough analysis tells us that we are dealing with terms that are comparable to $1/n^{1/4}$ for large n. Because the p-series with $p = 1/4$ is divergent, we predict that the given series diverges. We can make this informal observation precise by noting that, for $n \ge 9$, we have $n^{1/2} \ge 3$, or $2n^{1/2} = n^{1/2} + n^{1/2} \ge n^{1/2} + 3$, or $(2n^{1/2})^{1/2} \ge (n^{1/2}+3)^{1/2}$, or

$$\frac{1}{\sqrt{2}} \cdot \frac{1}{n^{1/4}} \le \frac{1}{\sqrt{\sqrt{n}+3}}.$$

The series $(1/\sqrt{2}) \sum_{n=1}^{\infty} 1/n^{1/4}$ diverges (because it is a multiple of a p-series with $p = 1/4 < 1$). Therefore the given series diverges by the Comparison Test for Divergence. ◀

An Advanced Example

The next example is more sophisticated than the preceding examples. As you read the solution, pay particular attention to the way the logarithm is handled. The basic principle to keep in mind is that logarithmic growth is very slow. In fact,

$$\lim_{x\to\infty} \frac{\ln(x)}{x^q} = 0 \tag{8.3.1}$$

for *any* positive number q. To derive limit formula (8.3.1), we apply l'Hôpital's Rule as follows:

$$\lim_{x\to\infty} \frac{\frac{d}{dx}\ln(x)}{\frac{d}{dx}x^q} = \frac{1}{q} \lim_{x\to\infty} \frac{1/x}{x^{q-1}} = \frac{1}{q} \lim_{x\to\infty} \frac{1}{x^q} = 0.$$

Limit formula (8.3.1) tells us that $\ln(x)/x^q$ becomes, and remains, less than 1. In other words, for any positive q, there is an integer M such that

$$\ln(x) \le x^q \quad (M \le x). \tag{8.3.2}$$

▶ **EXAMPLE 8** Does the series $\sum_{n=3}^{\infty} \frac{\ln(n)}{n^2}$ converge?

Solution This series of nonnegative summands termwise exceeds the series $\sum_{n=3}^{\infty} 1/n^2$. But this last series *converges*. Thus the Comparison Test for Divergence does not tell us anything about the convergence or divergence of the series $\sum_{n=3}^{\infty} \ln(n)/n^2$ based on the comparison we have made. We must try another approach. This time, we will analyze the size of $\ln(n)/n^2$ more carefully. The idea is that we can use some of the growth in the denominator to counteract the logarithmic growth in the numerator. We write

$$\frac{\ln(n)}{n^2} = \frac{\ln(n)}{n^q} \cdot \frac{1}{n^{2-q}}$$

for a positive q that will be chosen appropriately. If $1 < 2 - q$, or $q < 1$, then we may make a comparison with the convergent p-series $\sum_{n=3}^{\infty} (1/n^{2-q})$. Choosing $q = 1/2$—any positive value less than 1 would do just as well—we see from inequality (8.3.2) that there is an integer M such that

$$\frac{\ln(n)}{n^2} = \frac{\ln(n)}{n^{1/2}} \cdot \frac{1}{n^{3/2}} \leq 1 \cdot \frac{1}{n^{3/2}} = \frac{1}{n^{3/2}}$$

for $n \geq M$. Because the infinite series $\sum_{n=M}^{\infty} 1/n^{3/2}$ is a convergent p-series ($p = 3/2 > 1$), it follows from the Comparison Test for Convergence that the series $\sum_{n=M}^{\infty} \frac{\ln(n)}{n^2}$ converges. Because the given series $\sum_{n=3}^{\infty} \frac{\ln(n)}{n^2}$ has a convergent tail, it must also be convergent. ◀

INSIGHT Example 8 reminds us that the comparison tests are *not* "if and only if" statements. If we find a *divergent* series $\sum_{n=1}^{\infty} b_n$ such that $a_n \leq b_n$, then the Comparison Test for Convergence tells us nothing about $\sum_{n=1}^{\infty} a_n$ because not all the hypotheses of the test are satisfied. Likewise, if we find a *convergent* series $\sum_{n=1}^{\infty} b_n$ such that $a_n \geq b_n$, then the Comparison Test for Divergence tells us nothing about $\sum_{n=1}^{\infty} a_n$.

The Limit Comparison Test

We close this section with a variant of the comparison tests called the Limit Comparison Test. In practice, you may find this new test easier to apply than the Comparison Tests from Theorems 1 and 2. The idea behind this new test is also easy to understand. If $\sum_{n=1}^{\infty} a_n$ and $\sum_{n=1}^{\infty} b_n$ are series of positive terms and if

$$\ell = \lim_{n\to\infty} \frac{a_n}{b_n} \text{ exists} \quad \text{and} \quad 0 < \ell < \infty,$$

then for large n, we have the approximation $a_n \approx \ell \cdot b_n$. Because the terms of the tail series of $\sum_{n=1}^{\infty} a_n$ and $\sum_{n=1}^{\infty} b_n$ are approximately proportional, we expect the two series to have the same convergence properties. We state this as a theorem.

THEOREM 4 **(The Limit Comparison Test)** Let $\sum_{n=1}^{\infty} a_n$ and $\sum_{n=1}^{\infty} b_n$ be series of positive terms. Suppose that $\ell = \lim_{n\to\infty} a_n/b_n$ exists and $0 < \ell < \infty$. Then $\sum_{n=1}^{\infty} a_n$ converges if and only if $\sum_{n=1}^{\infty} b_n$ converges.

▶ **EXAMPLE 9** Does the series $\sum_{n=1}^{\infty} \frac{n}{2n^3 - 4n + 6}$ converge?

Solution As discussed in the *Insight* following Example 1, we neglect the lower order terms $4n$ and 6 in the denominator. The nth summand a_n is approximately $n/(2n^3)$, or $1/(2n^2)$ for large values of n. This suggests that we apply the Limit Comparison Test with the series $\sum_{n=1}^{\infty} b_n = \sum_{n=1}^{\infty} 1/(2n^2)$. We calculate

$$\lim_{n\to\infty} \frac{a_n}{b_n} = \lim_{n\to\infty} \frac{n/(2n^3 - 4n + 6)}{1/(2n^2)} = \lim_{n\to\infty} \frac{2n^3}{2n^3 - 4n + 6} = \lim_{n\to\infty} \frac{2}{2 - 4n/n^2 + 6/n^3} = 1.$$

Because the limit is positive and finite and because the simpler series $\sum_{n=1}^{\infty} 1/(2n^2)$ converges, we conclude that the original series converges as well. ◀

INSIGHT The Limit Comparison Test allows us to compare series without having to do a term-by-term comparison. In Example 10, we are able to compare the series $\sum_{n=1}^{\infty} n/(2n^3 - 4n + 6)$ with the series $\sum_{n=1}^{\infty} 1/(2n^2)$. In doing so, we do not have to worry about proving the inequality $n/(2n^3 - 4n + 6) \le 1/(2n^2)$ as we would with the Comparison Test for Convergence. (In fact, for $n > 1$, this inequality is false.)

▶ **EXAMPLE 10** Show that the series $\sum_{n=1}^{\infty} \frac{2^n + 1}{n2^n + 1}$ diverges.

Solution The nth summand $a_n = (2^n + 1)/(n2^n + 1)$ may be written as $a_n = (1 + 2^{-n})/(n + 2^{-n})$. It follows that $a_n \approx 1/n$ for large n. We apply the Limit Comparison Test using the infinite series $\sum_{n=1}^{\infty} b_n = \sum_{n=1}^{\infty} 1/n$ for comparison:

$$\lim_{n\to\infty} \frac{a_n}{b_n} = \lim_{n\to\infty} \frac{(2^n + 1)/(n2^n + 1)}{1/n} = \lim_{n\to\infty} \frac{n2^n + n}{n2^n + 1} = \lim_{n\to\infty} \frac{1 + 1/2^n}{1 + 1/(n2^n)} = 1.$$

Because this limit is finite and positive and because the harmonic series $\sum_{n=1}^{\infty} 1/n$ diverges, we conclude that the given series diverges. ◀

QUICK QUIZ

1. True or false: The Comparison Tests yield no information about the series $\sum_{n=1}^{\infty} a_n$ when $0 \le b_n \le a_n$ for all n and $\sum_{n=1}^{\infty} b_n$ is convergent.
2. True or false: If a series $\sum_{n=1}^{\infty} a_n$ of positive terms converges, then $\sum_{n=1}^{\infty} \frac{n+1}{n} \cdot a_n$ also converges.
3. True or false: It follows from the Limit Comparison Test that if $\sum_{n=1}^{\infty} a_n$ is a divergent series of positive terms and if $\{\lambda_n\}$ is a sequence of positive numbers with a finite limit, then $\sum_{n=1}^{\infty} \lambda_n \cdot a_n$ is also divergent.
4. For what values of p is $\sum_{n=1}^{\infty} \frac{n+1}{n^p}$ convergent?

Answers

1. True 2. True 3. False 4. $2 < p$

EXERCISES

Problems for Practice

In each of Exercises 1–16, use the Comparison Test for Convergence to show that the given series converges. State the series that you use for comparison and the reason for its convergence.

1. $\sum_{n=1}^{\infty} \frac{n}{n^3+1}$
2. $\sum_{n=1}^{\infty} \frac{5n-4}{n^{9/4}}$
3. $\sum_{n=1}^{\infty} \frac{1}{(n+\sqrt{2})^2}$
4. $\sum_{n=1}^{\infty} \frac{1}{(\sqrt{n}+\sqrt{2})^4}$
5. $\sum_{n=1}^{\infty} \frac{2+\sin(n)}{n^4}$
6. $\sum_{n=1}^{\infty} \frac{2n-1}{ne^n}$
7. $\sum_{n=1}^{\infty} \frac{n+2}{2n^{5/2}+3}$
8. $\sum_{n=1}^{\infty} \frac{n^2+2n+10}{2n^4}$
9. $\sum_{n=1}^{\infty} \frac{2^n}{n3^n}$
10. $\sum_{n=1}^{\infty} (1/3)^{n^2}$
11. $\sum_{n=1}^{\infty} \frac{1}{2n\sqrt{n}-1}$
12. $\sum_{n=1}^{\infty} \frac{\sqrt{n}}{n^2+1}$
13. $\sum_{n=1}^{\infty} \frac{n^3}{n^{4.01}+1}$
14. $\sum_{n=1}^{\infty} \frac{2}{2^n+1/2^n}$
15. $\sum_{n=1}^{\infty} \frac{1}{n!}$
16. $\sum_{n=1}^{\infty} \frac{1}{n^n}$

In each of Exercises 17–28, use the Comparison Test for Divergence to show that the given series diverges. State the series that you use for comparison and the reason for its divergence.

17. $\sum_{n=1}^{\infty} \frac{1}{2n-1}$
18. $\sum_{n=1}^{\infty} \frac{1}{n+\sqrt{n}}$
19. $\sum_{n=2}^{\infty} \frac{1}{\ln(n)}$
20. $\sum_{n=1}^{\infty} \frac{1+\ln(n)}{3n-2}$
21. $\sum_{n=1}^{\infty} \frac{2n+5}{n^2+1}$
22. $\sum_{n=1}^{\infty} \frac{n+2}{2n^{3/2}+3}$
23. $\sum_{n=1}^{\infty} \frac{1}{\sqrt{10+n^2}}$
24. $\sum_{n=1}^{\infty} \frac{2+\sin(n)}{\sqrt{n}}$
25. $\sum_{n=1}^{\infty} \frac{3^n+1}{3^n+2^n}$
26. $\sum_{n=1}^{\infty} \frac{2^n}{n^2}$
27. $\sum_{n=1}^{\infty} \frac{3^n+n}{\sqrt{n}3^n+1}$
28. $\sum_{n=1}^{\infty} \frac{2^n+1}{2^n+n^2}$

In Exercises 29–48, use the Limit Comparison Test to determine whether the given series converges or diverges.

29. $\sum_{n=1}^{\infty} \frac{3n}{2n^2-1}$
30. $\sum_{n=1}^{\infty} \frac{\sqrt{n+1}}{n^2}$
31. $\sum_{n=1}^{\infty} \frac{\sqrt{n^2+1}}{n^2}$
32. $\sum_{n=1}^{\infty} \frac{(n-1)^2}{(n+1)^4}$
33. $\sum_{n=1}^{\infty} \frac{10n+3}{100n^3-99}$
34. $\sum_{n=1}^{\infty} \frac{n^2+2}{(n^2+1)^2}$
35. $\sum_{n=1}^{\infty} \frac{\sqrt{n}+1}{2n^2-n}$
36. $\sum_{n=1}^{\infty} \frac{3}{2^n+3}$
37. $\sum_{n=1}^{\infty} \frac{2^n+11}{3^n-1}$

38. $\sum_{n=1}^{\infty} \frac{\arctan(n)}{n^2}$

39. $\sum_{n=1}^{\infty} \frac{\sqrt{1+n^2}}{(1+n^8)^{1/4}}$

40. $\sum_{n=1}^{\infty} \frac{n^{1/3}}{\sqrt{1+n^{5/2}}}$

41. $\sum_{n=1}^{\infty} \frac{2^n}{1+2^{2n}}$

42. $\sum_{n=1}^{\infty} \sin(1/n)$

43. $\sum_{n=1}^{\infty} \frac{\sin(1/n)}{n}$

44. $\sum_{n=1}^{\infty} \frac{n+\ln(n)}{n^3}$

45. $\sum_{n=1}^{\infty} \frac{1}{n+\ln(n)}$

46. $\sum_{n=1}^{\infty} \frac{2^n+n^3}{3^n+n^2}$

47. $\sum_{n=1}^{\infty} \frac{7^n}{10^n+n^{10}}$

48. $\sum_{n=1}^{\infty} \left(\frac{2n+5}{3n+7}\right)^n$

Further Theory and Practice

In each of Exercises 49–72, use a Comparison Test to determine whether the given series converges or diverges.

49. $\sum_{n=1}^{\infty} \frac{2n+1}{n^2}$

50. $\sum_{n=1}^{\infty} \frac{2+\sin(n)}{\sqrt{n}}$

51. $\sum_{n=1}^{\infty} \frac{n+2}{2n^{3/2}+3}$

52. $\sum_{n=1}^{\infty} \frac{5^n+6}{7^n}$

53. $\sum_{n=1}^{\infty} \left(\frac{n+2}{n^2+1}\right)^2$

54. $\sum_{n=1}^{\infty} \sqrt{\frac{n+2}{n^3+1}}$

55. $\sum_{n=1}^{\infty} \frac{1+1/n}{n}$

56. $\sum_{n=1}^{\infty} \frac{\ln(n^3)}{n^3}$

57. $\sum_{n=2}^{\infty} \frac{n}{\ln(n)^3}$

58. $\sum_{n=1}^{\infty} \frac{\arctan(n)}{n}$

59. $\sum_{n=1}^{\infty} \frac{100^n}{n^n}$

60. $\sum_{n=1}^{\infty} \frac{\tan(1/n)}{n}$

61. $\sum_{n=1}^{\infty} \sin(1/n^2)$

62. $\sum_{n=1}^{\infty} (1-\cos(1/n))$

63. $\sum_{n=1}^{\infty} \left(\frac{1}{2}-\frac{1}{3n}\right)^n$

64. $\sum_{n=1}^{\infty} \left(\frac{2}{5}+\frac{\sin(n)}{3}\right)^n$

65. $\sum_{n=1}^{\infty} \frac{2^n+3^n}{7^n+5^n}$

66. $\sum_{n=1}^{\infty} \sqrt{\sin(1/n^3)}$

67. $\sum_{n=1}^{\infty} \frac{n^3}{(n^3+n^2+1)^2}$

68. $\sum_{n=1}^{\infty} \left(\frac{4}{n}-\frac{12}{3n+1}\right)$

69. $\sum_{n=1}^{\infty} \left(\sqrt{n^4+2\sqrt{n}}-\sqrt{n^4+\sqrt{n}}\right)$

70. $\sum_{n=1}^{\infty} \left(\sec\left(\frac{1}{\sqrt{n}}\right)-\cos\left(\frac{1}{\sqrt{n}}\right)\right)$

71. $\sum_{n=2}^{\infty} \frac{\sqrt{n^2+1}}{(n\ln(n))^2}$

72. $\sum_{n=2}^{\infty} \left(\frac{1+\ln(n^2)}{\ln(n^4)}\right)^n$

73. Let $\sum_{n=1}^{\infty} a_n$ and $\sum_{n=1}^{\infty} b_n$ be series of positive terms. Suppose that $\lim_{n\to\infty} a_n/b_n = 0$. Show that if $\sum_{n=1}^{\infty} b_n$ converges, then $\sum_{n=1}^{\infty} a_n$ also converges.

74. Let $\sum_{n=1}^{\infty} a_n$ and $\sum_{n=1}^{\infty} b_n$ be series of positive terms. Suppose that $\lim_{n\to\infty} a_n/b_n = \infty$. Show that if $\sum_{n=1}^{\infty} b_n$ diverges, then $\sum_{n=1}^{\infty} a_n$ also diverges.

75. Suppose that $\{a_n\}$ is a sequence of positive numbers. Show that if $\sum_{n=1}^{\infty} a_n$ converges, then $\sum_{n=1}^{\infty} a_n^2$ also converges.

76. Suppose that $\{a_n\}$ and $\{c_n\}$ are sequences of positive numbers. Suppose that $c_n \to \ell$ where ℓ is a real number. Show that if $\sum_{n=1}^{\infty} a_n$ converges, then $\sum_{n=1}^{\infty} a_n c_n$ also converges.

77. Let p be any number. Show that $\sum_{n=1}^{\infty} n^p r^n$ is convergent if $0 < r < 1$.

78. Let q be any number. Show that $\sum_{n=1}^{\infty} \ln^q(n)/n^p$ is convergent if $1 < p$.

79. Show that $\sum_{n=1}^{\infty} n!/n^n$ is convergent.

Calculator/Computer Exercises

80. Let $S = \sum_{n=1}^{\infty}\left(\frac{\ln(n)}{n^{1/3}}\right)^n$. A brute-force computer calculation shows that the 1000th partial sum satisifes $S_{1000} = 941.010575563\ldots$. As a consequence of limit formula (8.3.1), there is a positive integer M such that $\frac{\ln(n)}{n^{1/3}} < 0.9$ for $n \geq M$. Find this value of M, and obtain an upper bound $S_{M-1} + \sum_{n=M}^{\infty}(0.9)^n$ for S.

81. Find a positive integer M such that $n^{10} \cdot 0.99^n < 0.995^n$ for all $n \geq M$. Prove that the series $\sum_{n=1}^{\infty} n^{10} \cdot 0.99^n$ is convergent.

8.4 Alternating Series

In this section, we study series with terms that alternate between positive and negative values. Because the partial sums of such series do not form an increasing sequence, the Integral Test and the Comparison Tests do not apply. In this section, we will develop a different test for handling these so-called alternating series.

The Alternating Series Test

The idea of cancellation motivates the next convergence test, which is concerned with the series we consider in this section. These series have alternating signs $(\ldots, +, -, +, -, \ldots)$, and are called *alternating series*. It is convenient to account for the sign changes by writing the series in the form $\sum_{n=1}^{\infty}(-1)^{n-1}a_n = a_1 - a_2 + a_3 - \ldots$ where $0 < a_n$ for all n. If instead, the signs of the even terms are positive, then we write the series as $\sum_{n=1}^{\infty}(-1)^n a_n = -a_1 + a_2 - a_3 + \cdots$ where $0 < a_n$ for all n. Because this last series is just the negative of $\sum_{n=1}^{\infty}(-1)^{n-1}a_n$, each statement that we make about one of the series will hold for the other.

THEOREM 1 **(Alternating Series Test)** Let $\{a_n\}$ be a sequence of nonnegative numbers that satisfy the following two conditions:

a. $a_1 \geq a_2 \geq a_3 \geq \ldots$
b. $\lim_{n\to\infty} a_n = 0$

Then $\sum_{n=1}^{\infty}(-1)^{n-1}a_n$ converges. Furthermore, the limit $\ell = \sum_{n=1}^{\infty}(-1)^{n-1}a_n$ lies between each pair of consecutive partial sums $S_N = \sum_{n=1}^{N}(-1)^{n-1}a_n$ and $S_{N+1} = S_N + (-1)^N a_{N+1}$. In particular, the limit ℓ satisfies the inequality

$$|\ell - S_N| \leq a_{N+1}. \tag{8.4.1}$$

Proof. First observe that

$$\begin{aligned} S_1 &= a_1 \\ &\geq a_1 + (-a_2 + a_3) = S_3 \\ &\geq a_1 + (-a_2 + a_3) + (-a_4 + a_5) = S_5 \\ &\geq a_1 + (-a_2 + a_3) + (-a_4 + a_5) + (-a_6 + a_7) = S_7 \ldots. \end{aligned}$$

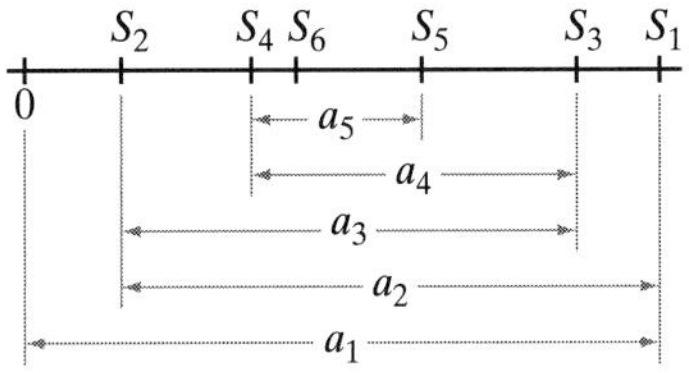

▲ Figure 1

Therefore the odd partial sums form a decreasing sequence. In a similar way, we see that the even partial sums form an increasing sequence. The sequence of even partial sums lies to the left of the sequence of odd partial sums (see Figure 1).

Therefore the two monotone sequences are bounded, hence convergent. Let ℓ denote the limit of the odd partial sums. For any positive integer N, $S_{2N} - S_{2N-1} = a_{2N} \to 0$. Hence the even and odd partial sums tend to the same limit ℓ. This means that the series converges. Because ℓ lies between each pair of consecutive partial sums S_N and $S_N + a_{N+1}$, inequality (8.4.1) follows. ■

Some Examples The Alternating Series Test applies to series in which the terms (1) have alternating signs and (2) decrease in absolute value to 0. Always be sure to verify these hypotheses when applying the test.

▶ **EXAMPLE 1** Apply Theorem 1 to the series $\sum_{n=1}^{\infty} (-1)^{n-1}\frac{1}{n}$.

Solution Set $a_n = 1/n$. Then, $a_1 \geq a_2 \geq a_3 \geq \ldots$ and $a_n \to 0$. Thus the series has the form of an alternating series as in Theorem 1. The Alternating Series Test applies, and $\sum_{n=1}^{\infty} (-1)^{n-1}/n$ converges. ◀

INSIGHT Notice that if we remove the minus signs from the series in Example 1, then we obtain the harmonic series, which is divergent. The minus signs cause enough cancellation to occur so that the series converges. Theorem 1 does not tell us the value ℓ of $\sum_{n=1}^{\infty} (-1)^{n-1}/n$, but it does allow us to make an estimate. If we want to evaluate ℓ with an error less than 0.01, then we find the first index n for which $1/n < 0.01$. This value is 101. By inequality (8.4.1), we can be sure that

$$\left|\ell - \sum_{n=1}^{100} (-1)^{n-1}/n\right| \leq \frac{1}{101} < 0.01.$$

With a computer, we can calculate $\sum_{n=1}^{100} (-1)^{n-1}/n \approx 0.6882$. This number is our approximation. In fact, it is known that $\sum_{n=1}^{\infty} (-1)^{n-1}/n$ is exactly equal to $\ln(2)$, or $0.693\ldots$.

▶ **EXAMPLE 2** Analyze the series $\sum_{n=1}^{\infty} (-1)^n \frac{1}{\sqrt{n}}$.

Solution Because $a_n = 1/\sqrt{n}$ decreases to 0, the Alternating Series Test may be applied to the series $\sum_{n=1}^{\infty} (-1)^{n-1}/\sqrt{n}$, which is the negative of the given series. Therefore the original series also converges. ◀

▶ **EXAMPLE 3** Does the series $\sum_{n=1}^{\infty} (-1)^n \frac{1}{1+1/n}$ converge?

Solution The given series is indeed alternating. However, the terms do *not* tend to 0. Therefore the Alternating Series Test does *not* apply; we may draw no conclusion from that test. However, the Divergence Test *does* apply; it tells us that the series diverges because the terms do not tend to 0. ◀

Because the first terms of a series do not affect convergence, we may be able to deduce that a series converges by applying the Alternating Series Test to an appropriate tail. The next example illustrates this idea.

▶ **EXAMPLE 4** Show that the series

$$\sum_{n=1}^{\infty} (-1)^{n-1} \frac{\ln(n^4)}{\sqrt{n}}$$

converges.

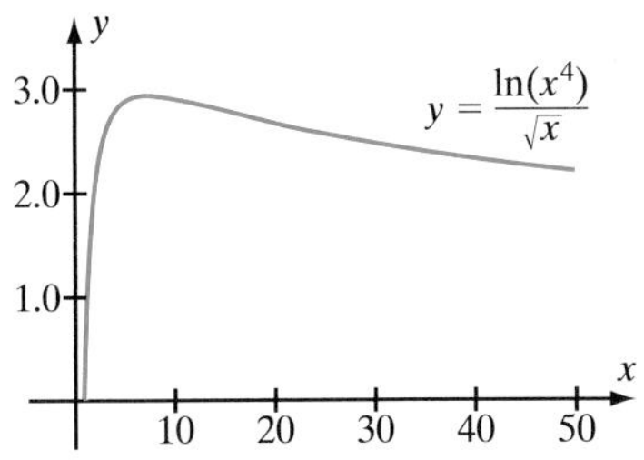

▲ Figure 2

Solution A plot of $\ln(x^4)/\sqrt{x}$ (Figure 2) indicates that this expression is not decreasing for all positive x, but it does appear to decrease from some point forward. To be sure, we differentiate the expression. We find that

$$\frac{d}{dx}\left(\frac{\ln(x^4)}{\sqrt{x}}\right) = 4\frac{d}{dx}\left(\frac{\ln(x)}{x^{1/2}}\right) = 4\left(\frac{x^{1/2}\cdot x^{-1} - \frac{1}{2}x^{-1/2}\ln(x)}{x}\right) = 2\cdot\frac{2-\ln(x)}{x^{3/2}}.$$

This expression is negative for $\ln(x) > 2$, or $x > e^2 \approx 7.39$. It follows that the sequence $a_n = \ln(n^4)/\sqrt{n}$ decreases for $n \geq 8$. Also notice that

$$\lim_{n\to\infty} a_n = \lim_{n\to\infty} \frac{\ln(n^4)}{\sqrt{n}} = 4\lim_{n\to\infty}\frac{\ln(n)}{n^{1/2}} \overset{(8.3.1)}{=} 0.$$

Thus the Alternating Series Test does apply to the tail series $\sum_{n=8}^{\infty}(-1)^{n-1} \ln(n^4)/\sqrt{n}$. Because the tail converges, the full series also converges.

Absolute Convergence

It is useful to distinguish between series that converge because of cancellation and series that converge because of size. Then we can learn something by comparing the two. These considerations motivate the following definition.

DEFINITION Let $\sum_{n=1}^{\infty} a_n$ be a series, possibly containing terms of both positive and negative sign. If the series $\sum_{n=1}^{\infty} |a_n|$ of absolute values converges, then we say that the series $\sum_{n=1}^{\infty} a_n$ converges *absolutely*.

▶ **EXAMPLE 5** Does the series $\sum_{n=1}^{\infty}(-1)^{n-1}/n$ converge absolutely?

Solution The given series converges by the Alternating Series Test, as noted in Example 1. But the series $\sum_{n=1}^{\infty}|(-1)^{n-1}/n|$ is the harmonic series, $\sum_{n=1}^{\infty} 1/n$, which diverges. Therefore the series $\sum_{n=1}^{\infty}(-1)^{n-1}/n$ does *not* converge absolutely. ◀

▶ **EXAMPLE 6** Does the series $\sum_{n=1}^{\infty}(-1/2)^n$ converge? Does it converge absolutely?

Solution The given series is a geometric series with ratio $-1/2$, which is less than 1 in absolute value. Therefore it is convergent. For the same reason, the series of absolute values, $\sum_{n=1}^{\infty}|(-1/2)^n| = \sum_{n=1}^{\infty}(1/2)^n$, converges. Therefore the given series converges absolutely. ◀

INSIGHT Our intuition tells us that absolute convergence is "better" than ordinary convergence. Example 5 suggests that it is harder for a series to converge absolutely than it is for a series to simply converge. By making the even-indexed terms of the divergent series $1+1/2+1/3+1/4+\cdots$ negative, the resulting series $1-1/2+1/3-1/4+\cdots$ converges thanks to the partial cancellation among consecutive terms with opposite signs. By contrast, an absolutely convergent series converges without this aid. The next theorem makes this intuition substantive.

THEOREM 2 If the series $\sum_{n=1}^{\infty}|a_n|$ converges, then the series $\sum_{n=1}^{\infty}a_n$ converges. In other words, absolute convergence implies convergence.

Proof. Let $b_n = a_n + |a_n|$. Then, $0 \le b_n \le 2|a_n|$. By hypothesis, $\sum_{n=1}^{\infty}|a_n|$ converges so that $\sum_{n=1}^{\infty}2|a_n|$ converges. Using the Comparison Test for Convergence, we conclude that $\sum_{n=1}^{\infty}b_n$ also converges. However, $a_n = b_n - |a_n|$. By Theorem 2a of Section 1, we conclude that $\sum_{n=1}^{\infty}a_n$ converges, as required. ■

▶ **EXAMPLE 7** Test the series $\displaystyle\sum_{n=1}^{\infty}\frac{\sin(n)}{2^n}$ for convergence.

Solution Notice that $\sin(n)$ is sometimes positive and sometimes negative, but it does *not* alternate in sign. The pattern of signs begins $+,+,+,-,-,-,+,+,+,-,-,-\ldots$, but the pattern is less simple than these initial terms suggest: $\sin(22)$, $\sin(23)$, $\sin(24)$, $\sin(25)$ produce four consecutive negative terms. Therefore the Alternating Series Test cannot be used. The Comparison and Integral Tests do not apply because the series has both positive and negative terms. In short, *not one* of the tests for convergence that we have learned in previous sections can be applied to the given series. But we *can* apply the Comparison Test for Convergence to the series $\sum_{n=1}^{\infty}|\sin(n)/2^n|$ of absolute values. We have $0 \le |\sin(n)/2^n| \le 1/2^n$, and because $\sum_{n=1}^{\infty}1/2^n$ is a convergent geometric series, we can conclude that $\sum_{n=1}^{\infty}|\sin(n)/2^n|$ converges. Theorem 2 then tells us that the original series $\sum_{n=1}^{\infty}\sin(n)/2^n$ converges. ◀

The sum of two different absolutely convergent series remains absolutely convergent. To see why, suppose that $\sum_{n=0}^{\infty}a_n$ and $\sum_{n=0}^{\infty}b_n$ are absolutely convergent. Notice that $|a_n+b_n| \le |a_n|+|b_n|$ for each index n by the triangle inequality. Therefore

$$\sum_{n=1}^{N}|a_n+b_n| \le \sum_{n=1}^{N}|a_n| + \sum_{n=1}^{N}|b_n| \le \sum_{n=1}^{\infty}|a_n| + \sum_{n=1}^{\infty}|b_n| < \infty.$$

It follows that the partial sums of the series $\sum_{n=1}^{\infty}|a_n+b_n|$ are bounded and that the series is convergent. We record this and some other simple facts as the following theorem.

THEOREM 3 Let $\sum_{n=0}^{\infty}a_n$ and $\sum_{n=0}^{\infty}b_n$ be absolutely convergent series. Then,

a. $\sum_{n=0}^{\infty}(a_n+b_n)$ and $\sum_{n=0}^{\infty}(a_n-b_n)$ are absolutely convergent.
b. $\sum_{n=0}^{\infty}\lambda a_n$ converges absolutely for any real constant λ.

Conditional Convergence

We have seen that the alternating harmonic series $\sum_{n=1}^{\infty}(-1)^{n+1}/n$ is convergent (Example 1) but not absolutely convergent (Example 5). This suggests that there is an interesting class of series that converge but do not converge absolutely.

DEFINITION If a series $\sum_{n=1}^{\infty} a_n$ converges but does not converge absolutely, then we say that the series converges *conditionally*.

▶ **EXAMPLE 8** For what positive values of p is the alternating p-series $\sum_{n=1}^{\infty}(-1)^{n+1}/n^p$ conditionally convergent?

Solution For $p>1$, the series $\sum_{n=1}^{\infty}|(-1)^{n+1}/n^p|$ is just the convergent p-series $\sum_{n=1}^{\infty} 1/n^p$. Therefore $\sum_{n=1}^{\infty}(-1)^{n+1}/n^p$ is absolutely convergent for $1<p$. By definition, an absolutely convergent series is not conditionally convergent.

Now suppose that $0<p\leq 1$. The series $\sum_{n=1}^{\infty}|(-1)^{n+1}/n^p|$ is the divergent p-series $\sum_{n=1}^{\infty} 1/n^p$. Thus $\sum_{n=1}^{\infty}(-1)^{n+1}/n^p$ is *not* absolutely convergent. Because $a_n = 1/n^p$ decreases to 0 as n tends to infinity, the series $\sum_{n=1}^{\infty}(-1)^{n+1}/n^p$ *is* convergent by the Alternating Series Test. We conclude that $\sum_{n=1}^{\infty}(-1)^{n+1}/n^p$ is conditionally convergent for $0<p\leq 1$ and for no other positive values of p. ◀

QUICK QUIZ

1. True or false: If $1 - 1/3 + 1/5 - 1/7 + 1/9$ is used to approximate $\sum_{n=1}^{\infty}(-1)^{n-1}/(2n-1)$, then the error is less than 0.1.
2. For what values of p is the series $\sum_{n=1}^{\infty}(-1)^n/n^p$ conditionally convergent?
3. True or false: If $\sum_{n=1}^{\infty} a_n$ is conditionally convergent, then $\sum_{n=1}^{\infty}|a_n|$ is divergent.
4. True or false: Every series is either absolutely convergent or conditionally convergent or divergent.

Answers

1. True 2. $0<p\leq 1$ 3. True 4. True

EXERCISES

Problems for Practice

▶ In each of Exercises 1–12, the given series may be shown to converge by using the Alternating Series Test. Show that the hypotheses of the Alternating Series Test are satisfied. ◀

1. $\displaystyle\sum_{n=1}^{\infty}\frac{(-1)^n}{n^3+1}$

2. $\displaystyle\sum_{n=1}^{\infty}\frac{(-1)^n}{\sqrt{n}}$

3. $\displaystyle\sum_{n=1}^{\infty}\frac{(-1)^n}{\sqrt{n^2+1}}$

4. $\displaystyle\sum_{n=2}^{\infty}\frac{(-1)^n}{\ln(n)}$

5. $\displaystyle\sum_{n=1}^{\infty}\frac{(-1)^{n+1}}{(2/3)^n}$

6. $\displaystyle\sum_{n=1}^{\infty}\frac{(-1)^{n+1}}{n!}$

7. $\displaystyle\sum_{n=1}^{\infty}\frac{(-4/5)^n}{n+2}$

8. $\displaystyle\sum_{n=1}^{\infty}(-1)^{n+1}\frac{n}{n^2+1}$

9. $\displaystyle\sum_{n=2}^{\infty}(-1)^{n+1}\frac{n^2}{e^n}$

10. $\displaystyle\sum_{n=1}^{\infty}(-1)^n\left(1+\frac{1}{n}\right)\ln\left(1+\frac{1}{n}\right)$

11. $\sum_{n=2}^{\infty} \cos(\pi n)\sin(\pi/n)$

12. $\sum_{n=1}^{\infty} (-1)^n(\pi/2 - \arctan(n))$

In each of Exercises 13–36, determine whether the given series converges absolutely, converges conditionally, or diverges.

13. $\sum_{n=1}^{\infty} \frac{(-1)^n}{n^3+1}$

14. $\sum_{n=1}^{\infty} \frac{(-1)^n}{\sqrt{n}}$

15. $\sum_{n=1}^{\infty} (-1)^n \frac{n}{n^2+1}$

16. $\sum_{n=1}^{\infty} (-1)^n \frac{n}{\sqrt{n^2+1}}$

17. $\sum_{n=2}^{\infty} \frac{(-1)^n}{\ln(n)}$

18. $\sum_{n=1}^{\infty} (-1)^n \frac{n^{2/3}}{\sqrt{n^3+e}}$

19. $\sum_{n=1}^{\infty} (-1)^n \frac{\sqrt{\sqrt{n}+1}}{\sqrt{n^3+1}}$

20. $\sum_{n=1}^{\infty} (-1)^{n+1}(2/3)^n$

21. $\sum_{n=1}^{\infty} (-1)^{n+1} \frac{(n+\ln(n))}{n^{3/2}}$

22. $\sum_{n=2}^{\infty} (-1)^n \frac{n}{\ln^3(n)}$

23. $\sum_{n=1}^{\infty} \sin(n)e^{-n}$

24. $\sum_{n=1}^{\infty} (-1)^n \frac{\ln^3(n)}{n^{1/3}}$

25. $\sum_{n=1}^{\infty} (-1)^n \left(\frac{\ln(n)}{n}\right)^3$

26. $\sum_{n=1}^{\infty} \frac{(-1)^n}{1+\sin(n)}$

27. $\sum_{n=1}^{\infty} (-1)^n \left(\frac{1+1/n}{2}\right)^n$

28. $\sum_{n=1}^{\infty} (-1)^n \left(\frac{en+1}{\pi n+1}\right)^n$

29. $\sum_{n=1}^{\infty} (-1)^n(1+1/n)$

30. $\sum_{n=1}^{\infty} (-1)^n(1+1/n)^n$

31. $\sum_{n=1}^{\infty} (-1)^n(1-1/n)^n$

32. $\sum_{n=1}^{\infty} (-1)^n \frac{\arctan(n)}{2n^4-n}$

33. $\sum_{n=1}^{\infty} (-1)^n \frac{3^n}{5^{n^2}}$

34. $\sum_{n=2}^{\infty} (-1)^n \frac{\ln(n)}{\ln(n^2)}$

35. $\sum_{n=3}^{\infty} (-1)^n \frac{1}{n\sqrt{\ln(n)}}$

36. $\sum_{n=3}^{\infty} (-1)^n \frac{1}{n\ln^2(n)}$

In each of Exercises 37–42, a convergent alternating series $S = \sum_{n=1}^{\infty} (-1)^{n+1}a_n$ is given. Use inequality (8.4.1) to find a value of N such that the partial sum $S_N = \sum_{n=1}^{N} (-1)^{n+1}a_n$ approximates the given infinite series to within 0.01. That is, find N so that $|S - S_N| \leq 0.01$ holds.

37. $\sum_{n=1}^{\infty} (-1)^{n+1} \frac{1}{\sqrt{n}}$

38. $\sum_{n=1}^{\infty} (-1)^{n+1} \frac{1}{n^3}$

39. $\sum_{n=1}^{\infty} (-1)^{n+1} \frac{4}{2n-1}$

40. $\sum_{n=1}^{\infty} (-1)^{n+1} \frac{1}{n!}$

41. $\sum_{n=1}^{\infty} (-1)^{n+1} \frac{1}{n^2+15n}$

42. $\sum_{n=1}^{\infty} (-1)^{n+1} \frac{1}{\ln(n+1)}$

Further Theory and Practice

In each of Exercises 43–46, find a value of M for which the Alternating Series Test may be applied to the tail $\sum_{n=M}^{\infty} (-1)^n a_n$ of the given series $\sum_{n=1}^{\infty} (-1)^n a_n$.

43. $\sum_{n=1}^{\infty} \frac{(-1)^n}{n^2-20n+101}$

44. $\sum_{n=1}^{\infty} \frac{(-1)^n}{n^4-32n^2+400}$

45. $\sum_{n=1}^{\infty} (-1)^n \frac{n^5}{2^n}$

46. $\sum_{n=1}^{\infty} (-1)^n \frac{\ln(n^{10})}{\sqrt{n}}$

In each of Exercises 47–50, calculate the given alternating series to three decimal places.

47. $\sum_{n=1}^{\infty} \frac{(-1)^n}{n!}$

48. $\sum_{n=1}^{\infty} (-1)^n(-1)^n \left(\frac{1+1/n}{10}\right)^n$

49. $\sum_{n=1}^{\infty} (-1)^n \dfrac{n!}{(2n)!}$

50. $\sum_{n=1}^{\infty} (-1)^n n^{-2n}$

Calculator/Computer Exercises

In each of Exercises 51–56, an equation is given that expresses the value of an alternating series. For the given d, use the Alternating Series Test to determine a partial sum that is within $5 \times 10^{-(d+1)}$ of the value of the infinite series. Verify that the asserted accuracy is achieved.

51. $\sum_{n=1}^{\infty} (-1)^{n+1} \dfrac{1}{n} = \ln(2)$ $\quad d = 2$

52. $\sum_{n=1}^{\infty} (-1)^{n+1} \dfrac{4}{2n-1} = \pi$ $\quad d = 2$

53. $\sum_{n=1}^{\infty} (-1)^{n+1} \dfrac{1}{n!} = 1 - \dfrac{1}{e}$ $\quad d = 5$

54. $\sum_{n=1}^{\infty} (-1)^{n+1} \dfrac{1}{(2n-1)!} = \sin(1)$ $\quad d = 5$

55. $\sum_{n=0}^{\infty} (-1)^{n} \dfrac{1}{(2n)!} = \cos(1)$ $\quad d = 5$

56. $\sum_{n=0}^{\infty} (-1)^{n} \dfrac{(\pi/3)^{2n}}{(2n)!} = \dfrac{1}{2}$ $\quad d = 5$

In each of Exercises 57–60, find a value of M for which the Alternating Series Test may be applied to the tail $\sum_{n=M}^{\infty} (-1)^n a_n$ of the given series $\sum_{n=1}^{\infty} (-1)^n a_n$.

57. $\sum_{n=1}^{\infty} (-1)^n \dfrac{9n^2+13}{n^3+55n+60}$

58. $\sum_{n=1}^{\infty} (-1)^n n^4 e^{-n} \ln(n)$

59. $\sum_{n=1}^{\infty} (-1)^n \dfrac{100n^{9/4}+n}{150+n^{5/2}}$

60. $\sum_{n=1}^{\infty} (-1)^n \dfrac{n^3}{\exp(\sqrt{n/100})}$

8.5 The Ratio and Root Tests

There are two powerful and easy-to-use tests that tell when a series converges absolutely: the Ratio Test and the Root Test. We will discuss them in this section. Bear in mind, however, that they are of no use in telling whether a series converges conditionally.

The Ratio Test

Suppose that $\sum_{n=1}^{\infty} a_n$ is a series of positive terms for which each term a_{n+1} is approximately equal to a fixed multiple L of its predecessor a_n. That is, suppose that $a_{n+1} \approx L \cdot a_n$ for every n. For $n = 1$ and $n = 2$, we have $a_2 \approx L \cdot a_1$ and $a_3 \approx L \cdot a_2$. Because $a_3 \approx L \cdot a_2 \approx L \cdot (L \cdot a_1)$, we conclude that $a_3 \approx L^2 \cdot a_1$. Similarly, $a_4 \approx L \cdot a_3 \approx L \cdot (L^2 \cdot a_1) = L^3 \cdot a_1$. By continuing this process, we obtain the estimate $a_n \approx L^{n-1} \cdot a_1$ for each positive integer n.

Now, if the approximate ratio L is less than 1, then the series $\sum_{n=1}^{\infty} a_n$ will converge by comparison with the convergent geometric series $a_1 \sum_{n=1}^{\infty} L^{n-1}$. On the other hand, if the approximate ratio L is greater than 1, then $a_{n+1} > a_n$, and the terms of the series do not tend to 0. In this case, the series diverges by the Divergence Test. These are the ideas behind the next test.

THEOREM 1 **(The Ratio Test)** Let $\sum_{n=1}^{\infty} a_n$ be a series. Suppose that

$$\lim_{n\to\infty} \left| \frac{a_{n+1}}{a_n} \right| = L.$$

a. If $L < 1$, then the series converges absolutely.
b. If $L > 1$, then the series diverges.
c. If $L = 1$, then the Ratio Test gives no information.

Take particular notice that the Ratio Test yields no information when $L = 1$. This point will be made clearer in the examples. Also remember that the limit of the ratio $|a_{n+1}/a_n|$ might not even exist. In this case, we cannot use Theorem 1. Finally, notice that the test only depends on the *limit* of the ratio $|a_{n+1}/a_n|$. The outcome does *not* depend on the first million or so terms of the series, and that is the way it should be: The convergence or divergence of a series depends only on its tail.

Now we look at some examples that illustrate when the Ratio Test gives convergence and when it gives divergence.

▶ **EXAMPLE 1** Use the Ratio Test to analyze the series $\sum_{n=1}^{\infty} \frac{2^n}{n^{10}}$.

Solution We set $a_n = 2^n/n^{10}$. Then, as $\to \infty$,

$$L = \lim_{n\to\infty}\left|\frac{a_{n+1}}{a_n}\right| = \lim_{n\to\infty}\frac{2^{n+1}/(n+1)^{10}}{2^n/n^{10}} = \lim_{n\to\infty}\frac{2^{n+1}}{(n+1)^{10}}\cdot\frac{n^{10}}{2^n} = \lim_{n\to\infty} 2\left(\frac{n}{n+1}\right)^{10} = 2\cdot 1^{10} = 2.$$

Because $L > 1$, the Ratio Test says that the series diverges. (The Divergence Test also can be used to reach this conclusion.) ◀

▶ **EXAMPLE 2** Discuss the convergence or divergence of the series $\sum_{n=1}^{\infty} \frac{\sqrt{n+1}}{2^n}$.

Solution In this example, we let $a_n = \sqrt{n+1}/2^n$. Then,

$$L = \lim_{n\to\infty}\left|\frac{a_{n+1}}{a_n}\right| = \lim_{n\to\infty}\frac{\sqrt{n+2}/2^{n+1}}{\sqrt{n+1}/2^n} = \lim_{n\to\infty}\left(\frac{2^n}{2^{n+1}}\cdot\sqrt{\frac{n+2}{n+1}}\right) = \frac{1}{2}\lim_{n\to\infty}\sqrt{\frac{1+2/n}{1+1/n}} = \frac{1}{2}.$$

Part a of the Ratio Test then tells us that the series converges. (As a matter of style the adjective "absolute" is generally not used in discussions about the convergence of a series of positive terms a_n. Doing so would add no information because $\sum_{n=1}^{\infty}|a_n| = \sum_{n=1}^{\infty} a_n$.) ◀

▶ **EXAMPLE 3** Use the Ratio Test to show that the series $\sum_{n=0}^{\infty} \frac{x^n}{n!}$ converges for every real number x.

Solution Recall that $0! = 1,\ 1! = 1$, and $n! = n\cdot(n-1)\cdot(n-2)\cdots 3\cdot 2\cdot 1$. When using the Ratio Test, it is easiest to spot cancellation when a larger factorial is written in terms of a smaller one. Thus

$$\begin{aligned}(n+1)! &= (n+1)\cdot n\cdot(n-1)\cdot(n-2)\cdots 3\cdot 2\cdot 1\\ &= (n+1)\cdot n!.\end{aligned}$$

Let $a_n = x^n/n!$, so that

$$\left|\frac{a_{n+1}}{a_n}\right| = \frac{\dfrac{|x|^{n+1}}{(n+1)!}}{\dfrac{|x|^n}{n!}} = \frac{|x|\cdot n!}{(n+1)!} = \frac{|x|}{n+1}.$$

As $n \to \infty$, the limit of this expression is 0. Therefore $L = \lim_{n\to\infty} |a_{n+1}/a_n|$ exists and is less than 1. By part a of the Ratio Test, we conclude that the series converges. ◀

INSIGHT The Ratio Test often works well when the general term a_n of a series factors into powers and factorials. In such cases, the ratio $|a_{n+1}/a_n|$ can simplify greatly due to cancellation, as Example 3 illustrates.

▶ **EXAMPLE 4** Test the series $\displaystyle\sum_{n=1}^{\infty} \frac{n^n}{n!}$ for convergence.

Solution Set $a_n = n^n/n!$. Then

$$\begin{aligned}\left|\frac{a_{n+1}}{a_n}\right| &= \frac{(n+1)^{n+1}/(n+1)!}{n^n/n!}\\ &= \left(\frac{n+1}{n}\right)^n\\ &= \left(1+\frac{1}{n}\right)^n.\end{aligned}$$

This last expression converges to the number e when $n \to \infty$, as we know from Section 2.6 in Chapter 2. Because $e = 2.718\ldots > 1$, the series diverges. Exercise 64 contains some further remarks about this example. ◀

▶ **EXAMPLE 5** Apply the Ratio Test to the harmonic series $\sum_{n=1}^{\infty} 1/n$.

Solution Setting $a_n = 1/n$, we have

$$\lim_{n\to\infty}\left|\frac{a_{n+1}}{a_n}\right| = \lim_{n\to\infty}\frac{1/(n+1)}{1/n} = \lim_{n\to\infty}\frac{n}{n+1} = 1.$$

Therefore the Ratio Test gives no information whatever. Of course we already know from independent reasoning that the harmonic series diverges. ◀

▶ **EXAMPLE 6** Apply the Ratio Test to the series $\sum_{n=1}^{\infty} 1/n^3$.

Solution Set $a_n = 1/n^3$. Then,

$$\lim_{n\to\infty}\left|\frac{a_{n+1}}{a_n}\right| = \lim_{n\to\infty}\frac{1/(n+1)^3}{1/n^3} = \lim_{n\to\infty}\frac{n^3}{(n+1)^3} = \lim_{n\to\infty}\left(\frac{n}{n+1}\right)^3 = 1^3 = 1.$$

Therefore the Ratio Test gives no information. Of course we already know from independent reasoning that this series is a convergent p-series. ◀

INSIGHT Examples 5 and 6 together explain what we mean when we say that the Ratio Test gives *no* information when the limit L is equal to 1. Under these circumstances, the series could diverge (Example 5), or the series could converge (Example 6). There is no way to tell when $L = 1$; you must use another convergence test to find out.

The Root Test The next test, the Root Test, has a similar flavor to the Ratio Test. Sometimes it is easier to apply the Ratio Test than it is to apply the Root Test (and vice versa). Thus even though the tests look rather similar, you should be well-versed in using both of them.

THEOREM 2 Let $\sum_{n=1}^{\infty} a_n$ be a series. Suppose that $\lim_{n\to\infty} |a_n|^{1/n} = L$.

a. If $L < 1$, then the series converges absolutely.
b. If $L > 1$, then the series diverges.
c. If $L = 1$, then the test gives no information.

The Root Test is based on a comparison with a tail of the geometric series $\sum_{n=1}^{\infty} L^n$. If $\lim_{n\to\infty} |a_n|^{1/n} = L$, then there is an N such that $|a_n| \approx L^n$ for all $n \geq N$. Thus we compare $\sum_{n=N}^{\infty} |a_n|$ with the geometric series $\sum_{n=N}^{\infty} L^n$, which is convergent when $L < 1$ and divergent when $L > 1$. Once again, take particular note that when $L = 1$, the Root Test gives no information. Also, the limit L might not even exist. In this case, we cannot use Theorem 2. Finally, the Root Test does not depend on the first million or so terms of the series—only on the tail.

When applying the Root Test, the limit formula

$$\lim_{n\to\infty} n^{1/n} = 1 \tag{8.5.1}$$

is often useful. We have considered this limit already (Example 4 of Section 8.1), but let us give a quick reminder of its proof. We use the basic formula $\lim_{x\to\infty} \dfrac{\ln(x)}{x} = 0$, which is limit formula (8.3.1) with $q = 1$. It then follows that

$$\lim_{n\to\infty} n^{1/n} = \lim_{n\to\infty} \exp\left(\ln(n^{1/n})\right) = \lim_{n\to\infty} \exp\left(\frac{\ln(n)}{n}\right) = \exp\left(\lim_{n\to\infty} \frac{\ln(n)}{n}\right) = e^0 = 1.$$

Examples Now we look at some examples that will illustrate when the Root Test gives convergence, when it gives divergence, and when it gives no information.

▶ **EXAMPLE 7** Analyze the series $\displaystyle\sum_{n=1}^{\infty} \frac{n}{(n^2+6)^n}$

Solution We apply the Root Test with $a_n = n/(n^2+6)^n$. Then, $|a_n|^{1/n} = n^{1/n}/(n^2+6)$. The numerator of this expression tends to 1 by equation (8.5.1). Because the denominator becomes unbounded, we have $L = \lim_{n\to\infty} |a_n|^{1/n} = 0$. Because this limit is less than 1, the Root Test tells us that the series converges. ◀

INSIGHT Notice that

$$\frac{n}{(n^2+6)^n} \leq \frac{n}{n^{2n}} = \frac{1}{n^{2n-1}} \leq \frac{1}{n^3}.$$

Thus the series from Example 7 is termwise dominated by the convergent p-series $\sum_{n=1}^{\infty} 1/n^3$. We can therefore deduce that the given series converges by applying the Comparison Test for Convergence. The Root Test, however, proves to be more straightforward to apply in this example.

▶ **EXAMPLE 8** Apply the Root Test to the series $\sum_{n=3}^{\infty} \dfrac{1}{\ln^n(n)}$.

Solution When $a_n = 1/\ln^n(n)$, we have $|a_n|^{1/n} = 1/\ln(n)$. This expression tends to 0 as $n \to \infty$. Therefore $L = 0 < 1$, and the Root Test says that the series converges. ◀

INSIGHT Try using the Ratio Test or the Integral Test on the series of Example 8. Each of those tests results in a difficult calculation. The Root Test is an obvious choice for Example 8 because a_n is a nth power. The Comparison Test for Convergence might also be used. For example, $\ln(n)$ is greater than 2 for $n \geq 8$. Therefore $1/\ln^n(n) < 1/2^n$ for $n \geq 8$. We deduce that the given series is convergent by comparison with the convergent geometric series $\sum_{n=8}^{\infty} 1/2^n$.

▶ **EXAMPLE 9** Apply the Root Test to the series $\sum_{n=1}^{\infty} \dfrac{(-2)^n}{n^{10}}$.

Solution With $a_n = (-2)^n/n^{10}$, we have

$$\lim_{n\to\infty} |a_n|^{1/n} = \lim_{n\to\infty} \frac{2}{n^{10/n}} = \frac{2}{(\lim_{n\to\infty} n^{1/n})^{10}} \overset{(8.5.1)}{=} \frac{2}{1} > 1.$$

We conclude, using the Root Test, that the series diverges. ◀

▶ **EXAMPLE 10** Let p be positive. Apply the Root Test to the p-series $\sum_{n=1}^{\infty} 1/n^p$

Solution If $a_n = 1/n^p$, then

$$L = \lim_{n\to\infty} |a_n^{1/n}| = \lim_{n\to\infty} 1/n^{p/n} = \frac{1}{(\lim_{n\to\infty} n^{1/n})^p} \overset{(8.5.1)}{=} 1.$$

Whatever the value of p, the Root Test is inconclusive. ◀

INSIGHT Example 10 explains what we mean when we say that the Root Test gives no information when the limit L is equal to 1. Under these circumstances, the series could diverge (a p-series with $p \leq 1$) or the series could converge (a p-series with $p > 1$). There is no way to tell from the Root Test; you must use another test to find out.

QUICK QUIZ

1. True or false: The Ratio Test fails if applied to a series $\sum_{n=1}^{\infty} a_n$ for which $\lim_{n\to\infty} |a_{n+1}/a_n| = \infty$.
2. True or false: The Ratio Test always fails when applied to a conditionally convergent series.
3. What is $\lim_{n\to\infty} n^{1/n}$?
4. Determine whether $\sum_{n=1}^{\infty} 3^{2n}/(-10)^n$ converges absolutely, converges conditionally, or diverges.

Answers

1. False 2. True 3. 1 4. converges absolutely

EXERCISES

Problems for Practice

In each of Exercises 1–12, use the Ratio Test to determine the convergence or divergence of the given series.

1. $\sum_{n=1}^{\infty} \frac{n}{e^n}$
2. $\sum_{n=1}^{\infty} \frac{n^2}{2^n}$
3. $\sum_{n=1}^{\infty} \frac{2^n}{n^3}$
4. $\sum_{n=1}^{\infty} \frac{10^n}{n!}$
5. $\sum_{n=1}^{\infty} \frac{n^{100}}{n!}$
6. $\sum_{n=1}^{\infty} \frac{n!}{11^n (n+12)^{13}}$
7. $\sum_{n=1}^{\infty} \frac{n!}{n3^n}$
8. $\sum_{n=1}^{\infty} \frac{\sqrt{n!}}{n^5 \cdot 7^n}$
9. $\sum_{n=1}^{\infty} \frac{3^n + n}{2^n + n^3}$
10. $\sum_{n=1}^{\infty} \frac{3n + n^2}{3^n}$
11. $\sum_{n=1}^{\infty} \frac{2^n \sqrt{n}}{3^n}$
12. $\sum_{n=1}^{\infty} \frac{2^n + n^2}{3^n + n}$

In each of Exercises 13–24, verify that the Ratio Test yields no information about the convergence of the given series. Use other methods to determine whether the series converges absolutely, converges conditionally, or diverges.

13. $\sum_{n=2}^{\infty} (-1)^n \frac{\ln(n)}{n}$
14. $\sum_{n=2}^{\infty} (-1)^n \frac{1}{\ln(n^2)}$
15. $\sum_{n=1}^{\infty} (-1)^n \frac{\ln(n)}{n^2}$
16. $\sum_{n=1}^{\infty} (-1)^n \cdot \frac{n+1}{n^3+1}$
17. $\sum_{n=1}^{\infty} (-1)^n \cdot \left(1 + \frac{1}{n}\right)$
18. $\sum_{n=1}^{\infty} (-1)^n \frac{\sqrt{n}}{n+4}$
19. $\sum_{n=1}^{\infty} (-1)^n \frac{e^n + \ln(n)}{e^n \cdot n^2}$
20. $\sum_{n=1}^{\infty} (-1)^n \frac{2^n + n}{2^n + n^3}$
21. $\sum_{n=1}^{\infty} \frac{(-3)^n}{n^3 + 3^n}$
22. $\sum_{n=1}^{\infty} (-1)^n \cdot \frac{n!}{n! + 2^n}$
23. $\sum_{n=1}^{\infty} (-1)^n \arctan(n)$
24. $\sum_{n=1}^{\infty} (-1)^n \sin\left(\frac{1}{n}\right)$

In each of Exercises 25–34, use the Root Test to determine the convergence or divergence of the given series.

25. $\sum_{n=1}^{\infty} n^{-n/2}$
26. $\sum_{n=1}^{\infty} \frac{2^{3n}}{3^{2n}}$
27. $\sum_{n=1}^{\infty} \frac{n}{2^n}$
28. $\sum_{n=1}^{\infty} \frac{10^n}{n^{10}}$
29. $\sum_{n=1}^{\infty} \frac{n^{100}}{(1+n)^n}$
30. $\sum_{n=2}^{\infty} \frac{n^7}{\ln^n(n)}$
31. $\sum_{n=3}^{\infty} \left(\frac{\ln(n)}{\ln(n^2-4)}\right)^n$
32. $\sum_{n=1}^{\infty} \left(\frac{37}{n}\right)^n$
33. $\sum_{n=1}^{\infty} \left(\frac{n^2+7n+13}{2n^2+1}\right)^n$
34. $\sum_{n=1}^{\infty} \left(1 - \frac{1}{n}\right)^{n^2}$

Further Theory and Practice

In Exercises 35–40, use the Ratio Test to determine the convergence or divergence of the given series.

35. $\sum_{n=1}^{\infty} \frac{(2n)!}{(3n)!}$

36. $\sum_{n=1}^{\infty} \frac{(n!)^2}{(2n)!}$

37. $\sum_{n=1}^{\infty} \frac{(2n)!}{(n! \cdot 2^n)}$

38. $\sum_{n=1}^{\infty} \frac{n!}{2^{n^n}}$

39. $\sum_{n=1}^{\infty} \frac{n+3^n}{n^3+2^n}$

40. $\sum_{n=1}^{\infty} \frac{2^n}{1+\ln^n(n)}$

In each of Exercises 41–44, use the Root Test to determine the convergence or divergence of the given series.

41. $\sum_{n=1}^{\infty} \frac{4^n}{(n^{1/n}+2)^n}$

42. $\sum_{n=1}^{\infty} (n^{1/n}+1/2)^n$

43. $\sum_{n=1}^{\infty} \frac{2^n}{1+\ln^n(n)}$

44. $\sum_{n=1}^{\infty} \frac{\exp(n)}{1+\ln^n(n)}$

In each of Exercises 45–60, determine whether the series converges absolutely, converges conditionally, or diverges. The tests that have been developed in Section 5 are not the most appropriate for some of these series. You may use any test that has been discussed in this chapter.

45. $\sum_{n=1}^{\infty} (-1)^n \frac{1}{\sqrt{n+10}}$

46. $\sum_{n=1}^{\infty} (-1)^n \frac{1}{n\sqrt{n+10}}$

47. $\sum_{n=1}^{\infty} (-1)^n \frac{n!}{3^n}$

48. $\sum_{n=1}^{\infty} (-3/4)^n n$

49. $\sum_{n=1}^{\infty} (-1)^n \frac{n}{n^2-11}$

50. $\sum_{n=2}^{\infty} (-1)^n \frac{1}{n\ln^3(n)}$

51. $\sum_{n=2}^{\infty} (-1)^n \frac{n^{1/n}}{\ln(n)}$

52. $\sum_{n=1}^{\infty} (-1)^n \ln(n^{1/n})$

53. $\sum_{n=1}^{\infty} (-1)^n \left(\frac{n^{1/n}}{1+1/n}\right)$

54. $\sum_{n=1}^{\infty} (-1)^n \left(\frac{n^2+2n+2}{3n^3+7}\right)$

55. $\sum_{n=1}^{\infty} (-1)^n \frac{e^{1/n}}{n^e}$

56. $\sum_{n=1}^{\infty} (-1)^n \ln(2+1/n)$

57. $\sum_{n=1}^{\infty} (-1)^n \ln(1+1/n)$

58. $\sum_{n=1}^{\infty} (-1)^n n 2^{-n^2}$

59. $\sum_{n=1}^{\infty} (-1)^n \frac{1}{(1+1/n)^n}$

60. $\sum_{n=1}^{\infty} (-1)^n \frac{\ln(n)}{\sqrt{n}}$

61. Let p be a fixed real number (possibly negative). Show that the Ratio Test results in the same conclusion when applied to both $\sum_{n=1}^{\infty} n^p a_n$ and $\sum_{n=1}^{\infty} a_n$. Explain how this observation can save work in the application of the Ratio Test. Deduce that the Ratio Test is inconclusive when applied to all p-series.

62. Suppose that p is a polynomial. Show that the Ratio Test results in the same conclusion when applied to both $\sum_{n=1}^{\infty} p(n)a_n$ and $\sum_{n=1}^{\infty} a_n$.

63. Let p be a fixed real number (possibly negative). Show that the Ratio Test results in the same conclusion when applied to both $\sum_{n=2}^{\infty} a_n \ln^p(n)$ and $\sum_{n=2}^{\infty} a_n$.

64. In Example 4, the Ratio Test is used to show that $\sum_{n=1}^{\infty} n^n/n!$ diverges. Use the Divergence Test to reach the same conclusion. However, the Divergence Test says nothing about $\sum_{n=1}^{\infty} n!/n^n$. Use the calculation of Example 4 to show that this series converges.

Calculator/Computer Exercises

Let $\{a_n\}$ be a sequence of positive numbers. In a course on mathematical analysis, one learns that if the two limits $\lim_{n\to\infty} a_{n+1}/a_n$ and $\lim_{n\to\infty} a_n^{1/n}$ exist, then they are equal. In each of Exercises 65–68, produce a plot that illustrates the equality of these two limits. Your plot should include a horizontal line that is the asymptote of the points $\{(n, a_{n+1}/a_n)\}$ and $\{(n, {a_n}^{1/n})\}$.

65. $a_n = 1/n^2$

66. $a_n = n/2^n$

67. $a_n = e^n/\ln(1+n)$

68. $a_n = n!/n^n$

8.6 Introduction to Power Series

The simplest functions that we know are those that are of the form $x \mapsto x^n$ where n is a nonnegative integer: The constant function $f(x) = x^0 = 1$, the functions $g(x) = x$, $h(x) = x^2$, and so on. These functions are easy to understand and simple to compute. We use them to form polynomials $a_0 + a_1 x + \cdots + a_N x^N$, which are also straightforward to handle. Now that we understand the concept of "infinite series," it is natural to create functions by summing infinitely many powers of x.

DEFINITION An expression of the form $a_0 + a_1 x + a_2 x^2 + a_3 x^3 + \cdots$, where the a_n's are constants, is called a *power series* in x. It is convenient to denote the constant term a_0 by $a_0 x^0$ so that we may use sigma notation to express the power series $a_0 + a_1 x + a_2 x^2 + a_3 x^3 + \cdots$ as $\sum_{n=0}^{\infty} a_n x^n$.

▶ **EXAMPLE 1** Are the infinite series $\sum_{n=0}^{\infty} (\sqrt{x})^n$, $\sum_{n=3}^{\infty} (2x)^n$ and $\sum_{n=0}^{\infty} (x^2)^n$ power series in x?

Solution In the definition of a power series, each power of x that appears is a nonnegative integer. Therefore the series $\sum_{n=0}^{\infty} (\sqrt{x})^n = \sum_{n=0}^{\infty} x^{n/2} = 1 + \sqrt{x} + x + x\sqrt{x} + \cdots$ is *not* a power series. The series $\sum_{n=3}^{\infty} (2x)^n = \sum_{n=3}^{\infty} 2^n x^n$ may be written as $0 + 0 \cdot x + 0 \cdot x^2 + 2^3 \cdot x^3 + 2^4 \cdot x^4 + \cdots$. Therefore $\sum_{n=3}^{\infty} (2x)^n$ *is* a power series even though it begins with the index $n = 3$. Similarly,

$$\sum_{n=0}^{\infty} (x^2)^n = 1 + x^2 + x^4 + \cdots = 1 + 0 \cdot x + x^2 + 0 \cdot x^3 + 1 \cdot x^4 + \cdots$$

is a power series. ◀

Notice that the power series $a_0 + a_1 x + a_2 x^2 + \cdots$ converges for $x = 0$ because it reduces to its initial term a_0 when $x = 0$. Thus the set on which the series $\sum_{n=0}^{\infty} a_n x^n$ converges is not empty; it contains the point 0 at the least. We may therefore regard $\sum_{n=0}^{\infty} a_n x^n$ as a function of x. The domain of this function is the (nonempty) set of x for which the series converges.

▶ **EXAMPLE 2** For what values of x does the power series $f(x) = \sum_{n=0}^{\infty} x^n$ converge?

Solution Compare $\sum_{n=0}^{\infty} x^n$ with the left side of equation (8.1.2). The only difference is that here we are using x instead of r. Thus we see that $f(x)$ is a geometric series with ratio x. As we know from Section 1, it converges when $|x| < 1$ and diverges when $|x| \geq 1$. Thus the function f is defined for $x \in (-1, 1)$. From our study of geometric series, we know that $f(x) = 1/(1-x)$ for these values of x. ◀

Radius and Interval of Convergence

The set of values at which a power series converges always has a special form. Suppose, for example, that the power series $\sum_{n=0}^{\infty} a_n x^n$ converges at some $x = t$ and that $|s| < |t|$. Then, $\lim_{n\to\infty} |a_n t^n| = 0$ by the Divergence Test (Section 8.2). In particular, we have $|a_n t^n| \leq 1$ for sufficiently large n. For such n, and for $r = |s|/|t|$, we see that $r < 1$ and $|a_n s^n| = r^n \cdot |a_n t^n| \leq r^n$. We conclude that $\sum_{n=0}^{\infty} |a_n s^n|$ converges by by comparison with the convergent geometric series $\sum_{n=0}^{\infty} r^n$. In summary, we have proved that if a power series $\sum_{n=0}^{\infty} a_n x^n$ converges at some value of x, then it converges absolutely at all points that have a smaller absolute value. It follows that the set of points where the series converges is an interval centered at the origin. This interval can have infinite length, finite positive length, or zero length (see Figure 1). We state this as the following theorem.

THEOREM 1 Let $\sum_{n=0}^{\infty} a_n x^n$ be a power series. Then precisely one of the following statements holds:

a. The series converges absolutely for every real x.
b. There is a positive number R such that the series converges absolutely for $|x| < R$ and diverges for $|x| > R$.
c. The series converges only at $x = 0$.

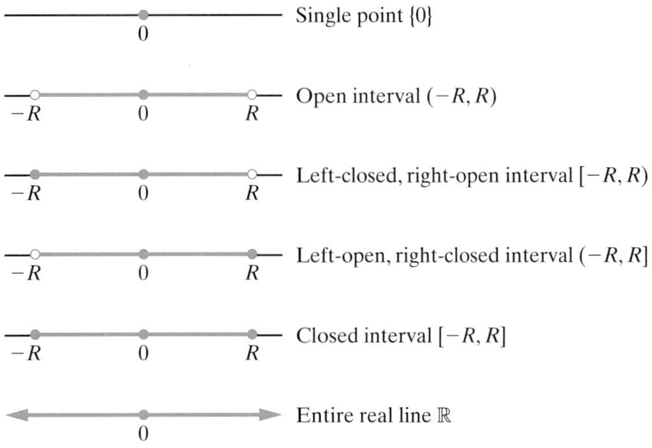

▲ Figure 1 Intervals of convergence.

Whichever the case, there is a quantity R that is either a nonnegative real number or $+\infty$ such that the power series $\sum_{n=0}^{\infty} a_n x^n$ converges absolutely for $|x| < R$ and diverges for $|x| > R$. In case a of Theorem 1, $R = \infty$ and in case c, $R = 0$. In all cases, we say that R is the *radius of convergence* of the power series. We now learn how to compute this quantity.

THEOREM 2 Suppose that the limit $\ell = \lim_{n\to\infty} |a_n|^{1/n}$ exists (as a nonnegative real number or ∞). Let R denote the radius of convergence of the power series $\sum_{n=0}^{\infty} a_n x^n$. If $0 < \ell < \infty$, then $R = 1/\ell$. If $\ell = 0$, then $R = \infty$, and if $\ell = \infty$, then $R = 0$.

Proof. The Root Test (Section 8.5) applied to $\sum_{n=0}^{\infty} a_n x^n$ tells us the series converges absolutely when $L = \lim_{n\to\infty} |a_n x^n|^{1/n} < 1$ and diverges when $L = \lim_{n\to\infty} |a_n x^n|^{1/n} > 1$. Notice that $L = |x| \cdot \lim_{n\to\infty} |a_n|^{1/n} = |x| \cdot \ell$. Thus the series converges absolutely for

$$|x| \cdot \ell < 1 \tag{8.6.1}$$

and diverges for

$$|x| \cdot \ell > 1. \tag{8.6.2}$$

If $\ell = 0$, then inequality (8.6.1) is satisfied for all x. Consequently R equals ∞. If $\ell = \infty$, then inequality (8.6.2) is satisfied for all $x \neq 0$, and therefore $R = 0$. If ℓ is finite and positive, then the series converges for $|x| < 1/\ell$ and diverges for $|x| > 1/\ell$. This means that $R = 1/\ell$. ■

DEFINITION The set of points at which a power series $\sum_{n=1}^{\infty} a_n x^n$ converges is called the *interval of convergence*.

If $R = \infty$, then the interval of convergence is the entire real line. If $R = 0$, then the interval of convergence is the single point $\{0\}$. When $0 < R < \infty$, the series converges on the interval $(-R, R)$ but it may or may not converge at the endpoints $x = R$ and $x = -R$. To determine the interval of convergence, each endpoint must be tested separately by substituting the values $x = R$ and $x = -R$ into the series. Thus when R is positive and finite, the interval of convergence will have the form $[-R, R]$ or $(-R, R]$ or $[-R, R)$ or $(-R, R)$, as shown earlier in Figure 1.

▶ **EXAMPLE 3** Find the interval of convergence for the power series $\displaystyle\sum_{n=0}^{\infty} \left(\frac{x}{3}\right)^n$.

Solution With $a_n = 1/3^n$, we calculate

$$\ell = \lim_{n\to\infty} |a_n|^{1/n} = \lim_{n\to\infty} \left(\left|\frac{1}{3^n}\right|^{1/n}\right) = \lim_{n\to\infty} \frac{1}{3} = \frac{1}{3}.$$

The radius of convergence R is the reciprocal of ℓ. Thus $R = 3$. This tells us that the series converges (absolutely) when $|x| < 3$ and diverges when $|x| > 3$. To completely determine the interval of convergence, we must investigate each endpoint. When $x = 3$, the series becomes $\sum_{n=0}^{\infty} 1$, which diverges. When $x = -3$, the series becomes $\sum_{n=0}^{\infty} (-1)^n$, which also diverges. The interval of convergence is therefore $(-3, 3)$. As a function of x, the domain of the power series $\sum_{n=0}^{\infty} x^n/3^n$ is the open interval $(-3, 3)$. ◀

▶ **EXAMPLE 4** Find the interval of convergence for the series $\sum_{n=1}^{\infty} x^n/n$.

Solution We set $a_n = 1/n$ and calculate

$$\ell = \lim_{n\to\infty} |a_n|^{1/n} = \lim_{n\to\infty} \left|\frac{1}{n}\right|^{1/n} = \lim_{n\to\infty} \frac{1}{n^{1/n}} \overset{(8.5.1)}{=} 1.$$

The radius of convergence R is given by $R = 1/\ell = 1$. Thus the series converges (absolutely) for $|x| < 1$ and diverges when $|x| > 1$. When $x = 1$ the series becomes $\sum_{n=1}^{\infty} 1/n$; this is the harmonic series, which diverges. When $x = -1$, the series becomes $\sum_{n=1}^{\infty} (-1)^n/n$, which is convergent by the Alternating Series Test. We conclude that the interval of convergence is $[-1, 1)$. The power series $\sum_{n=1}^{\infty} x^n/n$ defines a function of x with domain $[-1, 1)$. ◀

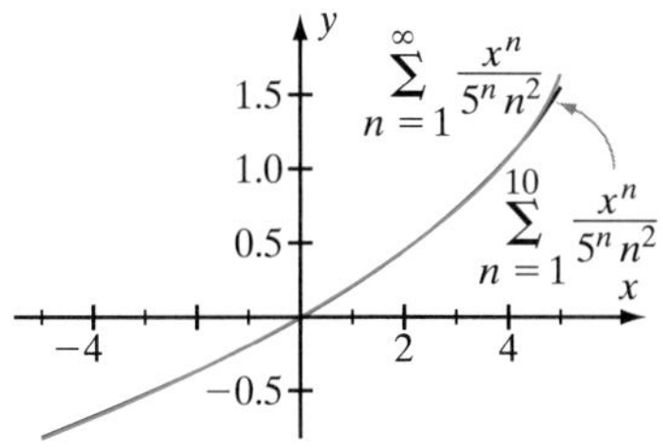

▲ Figure 2

INSIGHT We know from Theorem 1 that convergence is absolute at any point of the interval of convergence other than an endpoint. Nevertheless, the definition of "interval of convergence" refers only to convergence—not absolute convergence. If $R \in (0, \infty)$ is the radius of convergence of a power series $\sum_{n=0}^{\infty} a_n x^n$ and if the series converges when $x = R$, then R is in the interval of convergence whether the convergence is conditional or absolute. The same statement holds for $-R$. Thus in Example 4, the endpoint -1 belongs to the interval of convergence of $\sum_{n=1}^{\infty} x^n/n$ even though $\sum_{n=1}^{\infty} x^n/n$ converges conditionally when $x = -1$.

▶ **EXAMPLE 5** Find the radius and interval of convergence for the power series $\displaystyle\sum_{n=1}^{\infty} \frac{x^n}{5^n \cdot n^2}$.

Solution Let $a_n = 1/(5^n \cdot n^2)$. Then the radius of convergence R is given by

$$R = \left(\lim_{n\to\infty} |a_n|^{1/n}\right)^{-1} = \left(\lim_{n\to\infty} \left|\frac{1}{5^n \cdot n^2}\right|^{1/n}\right)^{-1} = \left(\lim_{n\to\infty} \frac{1}{5 \cdot n^{2/n}}\right)^{-1} = 5 \lim_{n\to\infty} (n^{1/n})^2 \overset{(8.5.1)}{=} 5.$$

Thus the series converges (absolutely) when $|x| < 5$ and diverges when $|x| > 5$. When $x = 5$, the series becomes $\sum_{n=1}^{\infty} 1/n^2$, which is a convergent p-series. When $x = -5$, the series becomes $\sum_{n=1}^{\infty} (-1)^n/n^2$, which, by the Alternating Series Test, also converges. We conclude that the interval of convergence is $[-5, 5]$. The power series $\sum_{n=1}^{\infty} x^n/(5^n \cdot n^2)$ therefore defines a function with domain $[-5, 5]$. This function is plotted in Figure 2. The graph of the approximating partial sum, $\sum_{n=1}^{10} x^n/(5^n \cdot n^2)$, is also shown in the figure. The approximating curve and the original curve are so close that they are difficult to distinguish. ◀

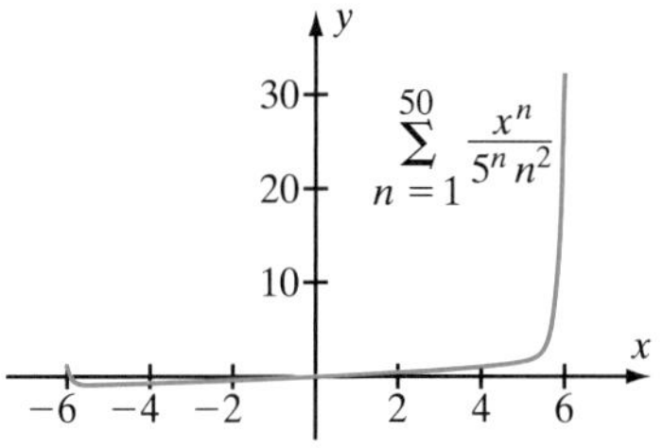

▲ Figure 3a

INSIGHT Graphing some partial sums of a power series can provide clues to its convergence properties. Example 5 shows that $[-5, 5]$ is the interval of convergence of $\sum_{n=1}^{\infty} x^n/(5^n \cdot n^2)$. Figure 3 suggests the divergence outside this interval. In Figure 3a, the partial sum $\sum_{n=1}^{50} x^n/(5^n \cdot n^2)$ is plotted over the interval $[-6, 6]$. In Figure 3b, the partial sum $\sum_{n=1}^{100} x^n/(5^n \cdot n^2)$ is plotted. Notice the scales on the vertical axes.

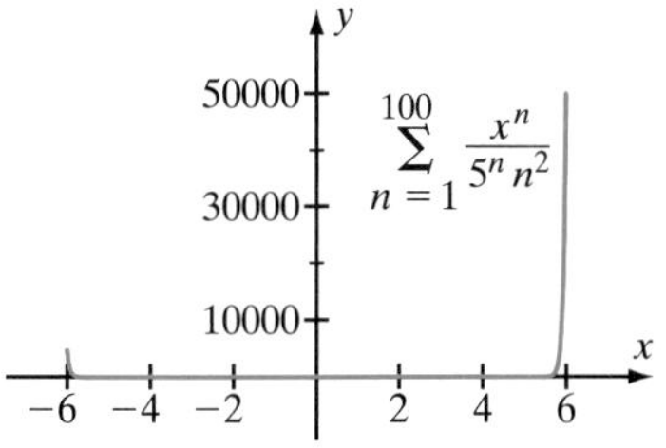

▲ Figure 3b

Sometimes it is convenient to use the Ratio Test when calculating the radius of convergence. The next example will illustrate this technique.

▶ **EXAMPLE 6** Find the interval of convergence for the power series $\displaystyle\sum_{n=0}^{\infty} \frac{x^n}{n!}$ by using the Ratio Test.

Solution According to the Ratio Test (Section 8.5), the series $\sum_{n=1}^{\infty} u_n$ converges absolutely if $\lim_{n\to\infty} |u_{n+1}/u_n| < 1$ and diverges if $\lim_{n\to\infty} |u_{n+1}/u_n| > 1$. We set $u_n = x^n/n!$ and calculate

$$\lim_{n\to\infty} \left|\frac{u_{n+1}}{u_n}\right| = \lim_{n\to\infty} \left|\frac{\dfrac{x^{n+1}}{(n+1)!}}{\dfrac{x^n}{n!}}\right| = \lim_{n\to\infty} \left|\frac{n!}{(n+1)!} \cdot \frac{x^{n+1}}{x^n}\right| = \lim_{n\to\infty} \left|\frac{n!}{(n+1) \cdot n!} \cdot x\right| = \lim_{n\to\infty} \left|\frac{x}{n+1}\right| = 0.$$

In particular, the limit is *always* less than 1. Hence, the Ratio Test tells us that the series converges absolutely for every real x. In other words, $R = \infty$, and the interval of convergence is $(-\infty, \infty)$. ◀

INSIGHT Remember that a ratio of factorials can be simplified by peeling off factors from the larger factorial until both numerator and denominator involve the same factorial (which is then canceled). Example 6 illustrates this technique, as does the next example.

▶ **EXAMPLE 7** Find the radius and interval of convergence for the series $\sum_{n=1}^{\infty} \frac{(2n)!}{n!} x^n$.

Solution We set $u_n = (2n)!x^n/n!$ and calculate

$$\lim_{n\to\infty} \left| \frac{u_{n+1}}{u_n} \right| = \lim_{n\to\infty} \left| \frac{\frac{(2n+2)!}{(n+1)!} x^{n+1}}{\frac{(2n)!}{n!} x^n} \right| = \lim_{n\to\infty} \left| \frac{(2n+2)!}{(2n)!} \cdot \frac{n!}{(n+1)!} \cdot x \right|.$$

Because $(n+1)! = (n+1) \cdot n!$ and $(2n+2)! = (2n+2)(2n+1) \cdot (2n)!$, we have

$$\lim_{n\to\infty} \left| \frac{u_{n+1}}{u_n} \right| = \lim_{n\to\infty} \left| \frac{(2n+2)(2n+1) \cdot (2n)!}{(2n)!} \cdot \frac{n!}{(n+1) \cdot n!} \cdot x \right| = \lim_{n\to\infty} \left| \frac{(2n+2)(2n+1)}{(n+1)} \cdot x \right| = \lim_{n\to\infty} |2(2n+1)x|.$$

This limit is less than 1 only when $x = 0$. It exceeds 1 (in fact the limit is infinite) for any nonzero value of x. Therefore $R = 0$, and the interval of convergence is the set $\{0\}$ that contains only the origin. ◀

The calculations that we have performed in Examples 6 and 7 can be generalized to yield another formula for the radius of convergence. We state this formula now and indicate its derivation in Exercise 65.

THEOREM 3 Let R be the radius of convergence of the power series $\sum_{n=0}^{\infty} a_n x^n$. Suppose that

$$\ell = \lim_{n\to\infty} \frac{|a_{n+1}|}{|a_n|}$$

exists (as a nonnegative real number or ∞). If $0 < \ell < \infty$, then $R = 1/\ell$. If $\ell = 0$, then $R = \infty$. If $\ell = \infty$, then $R = 0$.

INSIGHT Each of Theorems 2 and 3 gives a formula for the radius of convergence. Which should be used? Naturally, when both limits exist, they must have the same value. In courses on Advanced Calculus, it is shown that, of the two, the formula based on the Root Test has wider scope. However, the formula based on the Ratio Test is frequently easier to calculate. Therefore the Ratio Test is often tried first.

Power Series about an Arbitrary Base Point

Often, it is useful to consider series that consist of powers of the form $(x-c)^n$, where c is *any* fixed constant. These series will have a radius of convergence R just as before, but now the interval of convergence will be centered at the base point c. (Until now, we have considered only power series centered at 0.) We will still have to test the endpoints separately to determine the exact interval of convergence.

DEFINITION Suppose that $\{a_n\}$ is a sequence of constants and that c is a real number. An expression of the form

$$S = a_0 + a_1(x-c) + a_2(x-c)^2 + a_3(x-c)^3 + \cdots$$

is called a *power series* in x *with base point (or center)* c.

It is convenient to denote the constant term a_0 of S by $a_0(x-c)^0$ so that we may write S as $\sum_{n=0}^{\infty} a_n(x-c)^n$. Notice that a power series with base point 0 is exactly the sort of power series with which we have already been working. The next theorem expresses the content of Theorems 1, 2, and 3 for an arbitrary base point.

THEOREM 4 Let $\{a_n\}_{n=0}^{\infty}$ be a sequence for which $\ell = \lim_{n\to\infty} |a_n|^{1/n}$ exists as a nonnegative real number or ∞. Then ℓ is also given by the formula $\ell = \lim_{n\to\infty} \dfrac{|a_{n+1}|}{|a_n|}$, if this limit exists. If $0 < \ell < \infty$, set $R = 1/\ell$. If $\ell = 0$, set $R = \infty$. If $\ell = \infty$, set $R = 0$. Then $\sum_{n=0}^{\infty} a_n(x-c)^n$ converges absolutely for $\{|x-c| < R\}$ and diverges for $|x-c| > R$. In particular, if $R = \infty$, then the series converges absolutely for every real x. If $R = 0$, then the series converges only at $x = c$.

As with the case in which $c = 0$, we call R the *radius of convergence* of the power series. The set on which the series converges is called the *interval of convergence*. If $R = \infty$, then the interval of convergence is the entire real line. If $R = 0$, then the interval of convergence is the single point $\{c\}$. When $0 < R < \infty$, the series may or may not converge at the endpoints $x = c - R$ and $x = c + R$. To determine the interval of convergence, each endpoint must be tested separately by substituting the values $x = c - R$ and $x = c + R$ into the series. Thus when R is positive and finite, the interval of convergence will have the form $[c-R, c+R]$ or $(c-R, c+R]$ or $[c-R, c+R)$ or $(c-R, c+R)$. Figure 4 illustrates the possibilities.

▶ **EXAMPLE 8** Calculate the interval of convergence for the power series $\sum_{n=0}^{\infty} (x-2)^n$.

Solution Because $\sum_{n=0}^{\infty} (x-2)^n = \sum_{n=0}^{\infty} a_n(x-2)^n$ with $a_n = 1$ for all n, we calculate $\ell = \lim_{n\to\infty} |a_n|^{1/n} = \lim_{n\to\infty} 1^{1/n} = 1$. We set $R = 1/\ell = 1$. Theorem 4 tells us that the series converges (absolutely) for every x that satisfies $|x-2| < 1$ and diverges for every x such that $|x-2| > 1$. The points $x = 2 + 1 = 3$ and $x = 2 - 1 = 1$ must be tested separately. When $x = 3$, the series becomes $\sum_{n=0}^{\infty} 1^n$, which diverges. When $x = 1$, the series becomes $\sum_{n=0}^{\infty} (-1)^n$, which also diverges. Thus the interval of convergence is the open interval $(1, 3)$. ◀

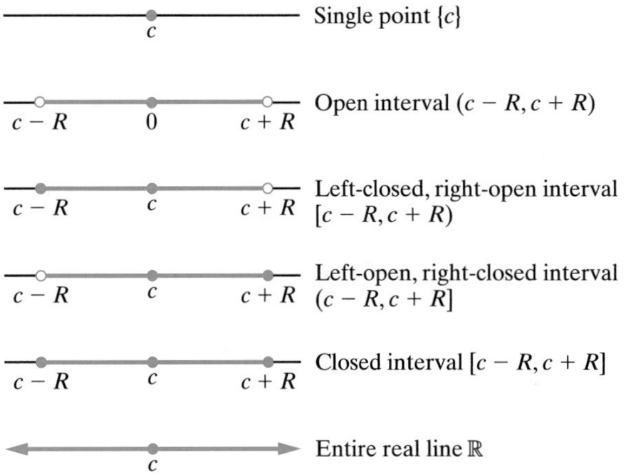

▲ Figure 4 Intervals of convergence centered at c.

The final example involves several elements of calculating the interval of convergence of a power series. We first enumerate the steps.

Steps for Calculating an Interval of Convergence

1. Write the power series in the standard form $\sum_{n=0}^{\infty} a_n(x-c)^n$. In particular, the coefficient of x in the power $(x-c)^n$ should be 1, and the base point c should be *subtracted* from x. To complete step 1, identify the center c of the interval of convergence and the coefficient a_n.
2. Calculate $\ell = \lim_{n\to\infty} |a_n|^{1/n}$ or $\ell = \lim_{n\to\infty} |a_{n+1}/a_n|$. To complete step 2, identify the radius of convergence R.
   If $\ell = 0$, then $R = \infty$, and the interval of convergence is $(-\infty, \infty)$. If $\ell = \infty$, then $R = 0$, and the interval of convergence is $\{0\}$. In both of these cases, the calculation is complete. If $0 < \ell < \infty$, then $R = 1/\ell$. Steps 3 and 4 are necessary in this case.
3. Substitute $x = c - R$ in $\sum_{n=0}^{\infty} a_n(x-c)^n$ and determine the convergence of the resulting series, $\sum_{n=0}^{\infty} (-1)^n a_n R^n$. At the conclusion of this step, it will be known if the interval of convergence does or does not contain its left endpoint $c - R$.
4. Substitute $x = c + R$ in $\sum_{n=0}^{\infty} a_n(x-c)^n$, and determine the convergence of the resulting series, $\sum_{n=0}^{\infty} a_n R^n$. At the conclusion of this step, it will be known if the interval of convergence does or does not contain its right endpoint $c + R$.

▶ **EXAMPLE 9** Determine the interval of convergence of the series $\displaystyle\sum_{n=0}^{\infty} \frac{(-1)^n}{5n+1}(2x+5)^n$.

Solution The first step is to write this power series in the standard form $\sum_{n=0}^{\infty} a_n(x-c)^n$. We have

$$\sum_{n=0}^{\infty} (-1)^n \frac{(2x+5)^n}{5n+1} = \sum_{n=0}^{\infty} \frac{(-1)^n 2^n}{5n+1}\left(x+\frac{5}{2}\right)^n = \sum_{n=0}^{\infty} \frac{(-2)^n}{5n+1}\left(x-\left(-\frac{5}{2}\right)\right)^n.$$

Now we can identify the base point $c = -5/2$ and the coefficients $a_n = (-2)^n/(5n+1)$. For step 2, we calculate

$$\ell = \lim_{n\to\infty} \frac{|a_{n+1}|}{|a_n|} = \lim_{n\to\infty} \frac{\left|\dfrac{(-2)^{n+1}}{5(n+1)+1}\right|}{\left|\dfrac{(-2)^n}{5n+1}\right|} = \lim_{n\to\infty} \frac{(5n+1)\cdot 2^{n+1}}{(5n+6)\cdot 2^n} = 2\lim_{n\to\infty}\frac{5n+1}{5n+6} = 2\lim_{n\to\infty}\frac{5+1/n}{5+6/n} = 2.$$

Therefore $R = 1/\ell = 1/2$. The endpoints of the interval of convergence are $c - R = -5/2 - 1/2 = -3$ and $c + R = -5/2 + 1/2 = -2$. When $x = -3$, the series becomes $\sum_{n=0}^{\infty} 1/(5n+1)$, which diverges by the Integral Test (or by the Limit Comparison Test using the harmonic series for comparison). At this point, the end of step 3, we know that the interval of convergence is either $(-3,-2)$ or $(-3,-2]$. When $x = -2$, the series becomes $\sum_{n=0}^{\infty} (-1)^n/(5n+1)$, which converges by the Alternating Series Test. Therefore the interval of convergence is $(-3,-2]$. ◄

Addition and Scalar Multiplication of Power Series

The first result tells us how to add or subtract two power series *with the same base point* and how to multiply a power series by a scalar. It is a straightforward application of Theorem 3, Section 9.5 of Chapter 9.

THEOREM 5 Let λ be any real number. Suppose that $\sum_{n=0}^{\infty} a_n(x-c)^n$ and $\sum_{n=0}^{\infty} b_n(x-c)^n$ are power series that converge absolutely on an open interval I. Then

a. The power series $\sum_{n=0}^{\infty} (a_n + b_n)(x-c)^n$ and $\sum_{n=0}^{\infty} (a_n - b_n)(x-c)^n$ also converge absolutely on I; moreover,

$$\sum_{n=0}^{\infty} a_n(x-c)^n + \sum_{n=0}^{\infty} b_n(x-c)^n = \sum_{n=0}^{\infty} (a_n + b_n)(x-c)^n$$

and

$$\sum_{n=0}^{\infty} a_n(x-c)^n - \sum_{n=0}^{\infty} b_n(x-c)^n = \sum_{n=0}^{\infty} (a_n - b_n)(x-c)^n.$$

b. The power series $\sum_{n=0}^{\infty} \lambda a_n(x-c)^n$ converges absolutely on I and

$$\lambda \sum_{n=0}^{\infty} a_n(x-c)^n = \sum_{n=0}^{\infty} \lambda a_n(x-c)^n.$$

► **EXAMPLE 10** On what open interval does $\sum_{n=0}^{\infty}\left(5 - \frac{1}{3^n}\right)(x-2)^n$ converge?

Solution The series $\sum_{n=0}^{\infty} (1/3)^n x^n$ has radius of convergence 3 (as shown in Example 3). Because the calculation of a radius of convergence depends only on the coefficients $\{a_n\}$ of $\sum_{n=0}^{\infty} a_n(x-c)^n$ and not on the center, it follows that $\sum_{n=0}^{\infty} (1/3)^n(x-2)^n$ also has radius of convergence 3. Therefore $\sum_{n=0}^{\infty} (1/3)^n(x-2)^n$ converges absolutely when x is in $(2-3, 2+3)$, or $(-1,5)$. By Example 8, the series $\sum_{n=0}^{\infty} (x-2)^n$ converges absolutely when $x \in (1,3)$. Therefore by Theorem 5b, so does the series $\sum_{n=0}^{\infty} 5(x-2)^n$. By Theorem 5a, $\sum_{n=0}^{\infty} (5 - (1/3)^n)(x-2)^n$ is equal to the difference $\sum_{n=0}^{\infty} 5(x-2)^n - \sum_{n=0}^{\infty} (1/3)^n (x-2)^n$ and converges absolutely on the intersection $(1,3)$ of the two intervals. ◀

Differentiation and Antidifferentiation of Power Series

A function that has derivatives of all orders on an open interval is said to be *infinitely differentiable*. The next theorem says that a power series that converges on an open interval is infinitely differentiable. Furthermore, the power series may be differentiated termwise and integrated termwise.

THEOREM 6 Suppose that a power series $\sum_{n=0}^{\infty} a_n(x-c)^n$ has a positive radius of convergence R. Let I denote the interval $(c-R, c+R)$ and define a function f on I by $f(x) = \sum_{n=0}^{\infty} a_n(x-c)^n$. Then,

a. The function f is infinitely differentiable on I and

$$f'(x) = \sum_{n=1}^{\infty} n a_n (x-c)^{n-1}.$$

The power series for f' has radius of convergence R.

b. The power series

$$F(x) = \sum_{n=0}^{\infty} \frac{a_n}{n+1}(x-c)^{n+1}$$

has radius of convergence R. The function F satisfies the equation $F(x) = \int_c^x f(t)\,dt$ for all x in I. In particular, $F'(x) = f(x)$ on I. The indefinite integral of $f(x)$ is given by $\int f(x)\,dx = F(x) + C$ where C is an arbitrary constant.

▶ **EXAMPLE 11** Calculate the derivative and the indefinite integral of the power series $f(x) = \sum_{n=1}^{\infty} \sqrt{n} \cdot x^n$ for x in the interval of convergence $I = (-1, 1)$.

Solution By Theorem 6a, $f'(x) = \sum_{n=1}^{\infty} n\sqrt{n}\, x^{n-1}$ for x in I. Although this formula correctly represents $f'(x)$ as a power series, we generally prefer to express a power series in standard form (in which the power of the variable is equal to the index of summation). If we change the index of summation by setting $m = n-1$, then as n ranges over the integers from 1 to ∞, m ranges over the integers from 0 to ∞, and $f'(x) = \sum_{m=0}^{\infty} (m+1)\sqrt{m+1}\, x^m$.

Also, according to Theorem 6, the indefinite integral of $f(x)$ is

$$\int f(x)\,dx = \sum_{n=1}^{\infty} \left(\frac{\sqrt{n}}{n+1}\right) x^{n+1} + C.$$

To write this power series in standard form, we change the index of summation to $m = n+1$. Then, as n ranges over the integers from 1 to infinity, m ranges over the integers from 2 to infinity, and

$$\int f(x)\,dx = C + \sum_{m=2}^{\infty} \left(\frac{\sqrt{m-1}}{m}\right) x^m. \quad ◀$$

INSIGHT If we explicitly write out several terms of $f(x)$, $f'(x)$, and $\int f(x)\,dx$, then our term-by-term differentiation and integration becomes more evident:

$$\begin{aligned} f(x) &= x + \sqrt{2}x^2 + \sqrt{3}x^3 + \cdots, \\ f'(x) &= 1 + 2\sqrt{2}x + 3\sqrt{3}x^2 + \cdots, \\ \int f(x)\,dx &= C + \frac{1}{2}x^2 + \frac{\sqrt{2}}{3}x^3 + \frac{\sqrt{3}}{4}x^3 + \cdots. \end{aligned}$$

QUICK QUIZ

1. True or false: If $\lim_{n\to\infty}\left|\frac{a_{n+1}}{a_n}\right| = 2$, then the radius of convergence of $\sum_{n=0}^{\infty} a_n x^n$ is 2.
2. True or false: If $\lim_{n\to\infty} |a_n|^{1/n} = 2$, then the radius of convergence of $\sum_{n=0}^{\infty} a_n x^n$ is 1/2.
3. What is the center of the interval of convergence of $\sum_{n=0}^{\infty} (3x+2)^n/n!$?
4. If $f(x) = \sum_{n=0}^{\infty} x^n/(n+2)$, what is the coefficient of x^4 in the power series for $f'(x)$?

Answers

1. False 2. True 3. $-2/3$ 4. 5/7

EXERCISES

Problems for Practice

In Exercises 1–10, use Theorem 2 and, where necessary, limit formula (8.5.1) to calculate the radius of convergence R. Determine the interval of convergence I by checking the endpoints.

1. $\sum_{n=0}^{\infty} (2x)^n$

2. $\sum_{n=0}^{\infty} \left(\frac{-x}{3}\right)^n$

3. $\sum_{n=1}^{\infty} nx^n$

4. $\sum_{n=0}^{\infty} \frac{10}{(n+1)^2} x^n$

5. $\sum_{n=1}^{\infty} (-1)^{n+1} n^3 x^n$

6. $\sum_{n=1}^{\infty} \frac{(4x)^n}{n}$

7. $\sum_{n=0}^{\infty} \frac{n}{5^n} x^n$

8. $\sum_{n=0}^{\infty} \left(\frac{2nx}{3n+1}\right)^n$

9. $\sum_{n=1}^{\infty} n^n x^n$

10. $\sum_{n=1}^{\infty} \left(\frac{-x}{n}\right)^n$

In Exercises 11–20, use Theorem 3 to calculate the radius of convergence R. Determine the interval of convergence I by checking the endpoints.

11. $\sum_{n=0}^{\infty} (-1)^n \frac{3}{\sqrt{n^2+1}} x^n$

12. $\sum_{n=0}^{\infty} e^n x^n$

13. $\sum_{n=0}^{\infty} \frac{n^2}{n^3+1} \cdot x^n$

14. $\sum_{n=3}^{\infty} \frac{3n+5}{n-2} \cdot x^n$

15. $\sum_{n=0}^{\infty} \frac{3^n}{n^3+1} x^n$

16. $\sum_{n=0}^{\infty} \frac{1+2^n}{1+3^n} \cdot x^n$

17. $\sum_{n=0}^{\infty} (-1)^n \frac{n}{\sqrt{n+1}} x^n$

18. $\sum_{n=1}^{\infty} (-1)^{n+1} \frac{x^n}{\sqrt{n} \cdot 10^n}$

19. $\sum_{n=2}^{\infty} \frac{2^{2n}}{\ln(n)} x^n$

20. $\sum_{n=0}^{\infty} \frac{(2x)^n}{n!}$

▶ In each of Exercises 21–30, use Theorem 4 to calculate the radius of convergence R. Determine the interval of convergence I by checking the endpoints. ◀

21. $\sum_{n=1}^{\infty} \frac{(x+6)^n}{\sqrt{n}}$

22. $\sum_{n=1}^{\infty} \frac{(\pi - x)^n}{n}$

23. $\sum_{n=0}^{\infty} \frac{(x+4)^n}{n^5+1}$

24. $\sum_{n=0}^{\infty} \frac{(x-3)^n}{2n+3}$

25. $\sum_{n=0}^{\infty} (2x+1)^n$

26. $\sum_{n=0}^{\infty} \frac{(3x+2)^n}{n+\sqrt{n}+1}$

27. $\sum_{n=0}^{\infty} (-1)^n \frac{n+1}{3n+1}(x+5)^n$

28. $\sum_{n=0}^{\infty} \left(\frac{x}{2} - \frac{1}{3}\right)^n$

29. $\sum_{n=1}^{\infty} \left(\frac{x+1}{n}\right)^n$

30. $\sum_{n=0}^{\infty} \frac{2^n+n}{3^n+n}\left(x+\frac{1}{2}\right)^n$

▶ In each of Exercises 31–34, find the open interval on which the given power series converges absolutely. ◀

31. $\sum_{n=0}^{\infty} (2^n - 2^{-n})(x-1)^n$

32. $\sum_{n=1}^{\infty} \left(\frac{1}{n^3} - \frac{1}{3^n}\right)(x+1)^n$

33. $\sum_{n=0}^{\infty} (3^n + 2^n)(x+2)^n$

34. $\sum_{n=0}^{\infty} \left(5^n + \frac{4^n}{n!}\right)(x-3)^n$

▶ In each of Exercises 35–38, the given power series $\sum_{n=0}^{\infty} a_n(x-c)^n$ defines a function f. State the partial sum $\sum_{n=0}^{4} b_n(x-c)^n$ of the power series for $f'(x)$, and determine the interval I of convergence of $f'(x)$. ◀

35. $\sum_{n=1}^{\infty} n\,x^n$

36. $\sum_{n=1}^{\infty} \frac{x^n}{n^3}$

37. $\sum_{n=1}^{\infty} \frac{(x-1)^n}{n^{3/2}}$

38. $\sum_{n=1}^{\infty} \frac{2^n}{n^2}\left(x+\frac{1}{2}\right)^n$

▶ In each of Exercises 39–42, the given power series $\sum_{n=0}^{\infty} a_n(x-c)^n$ defines a function f. Calculate the partial sum $\sum_{n=1}^{4} b_n(x-c)^n$ of the power series for $F(x) = \int_c^x f(t)\,dt$, and determine the interval I of convergence of $F(x)$. ◀

39. $\sum_{n=0}^{\infty} \sqrt{n+1}\,x^n$

40. $\sum_{n=0}^{\infty} \frac{2^n}{n+1}x^n$

41. $\sum_{n=0}^{\infty} \frac{1}{3^n}(x-2)^n$

42. $\sum_{n=0}^{\infty} \frac{\sqrt{n+1}}{2^n}(x+2)^n$

Further Theory and Practice

▶ In each of Exercises 43–48, find the radius of convergence of the given power series. ◀

43. $\sum_{n=0}^{\infty} n(n+1)(n+2)(2x+3)^n$

44. $\sum_{n=1}^{\infty} (1/n - 1/2^n)(x+1)^n$

45. $\sum_{n=1}^{\infty} \frac{(n!)^2}{(2n)!}x^n$

46. $\sum_{n=1}^{\infty} \left(\left(\frac{n+1}{n}\right)^n x\right)^n$

47. $\sum_{n=1}^{\infty} \frac{(n!)^2}{(n^2)!}(nx)^n$

48. $\sum_{n=0}^{\infty} \frac{10^{2^n}}{(2^n)^{10}}x^n$

▶ Each power series in Exercises 49–52, comprises only even powers $(x-c)^{2n}$. Make the substitution $t = (x-c)^2$, find the interval of convergence of the series in t, and use it to find the interval of convergence of the original series. ◀

49. $\sum_{n=0}^{\infty} (-1)^n(x+1)^{2n}/9^n$

50. $\sum_{n=1}^{\infty} 4^n(x+3)^{2n}/n^2$

51. $\sum_{n=0}^{\infty} 4^n\,(2x-1)^{2n}$

52. $\sum_{n=1}^{\infty} n\,(3x)^{2n}/2^n$

▶ Each power series in Exercises 53–56 comprises only odd powers $(x-c)^{2n+1}$. Factor out $(x-c)^1$ to produce a series consisting of even powers. Follow the instructions to Exercises 49–52 to find the interval of convergence of the given series. ◀

53. $\sum_{n=1}^{\infty} (2x)^{2n+1}/(n+1)$

54. $\sum_{n=1}^{\infty} n(x-1)^{2n+1}/25^n$

55. $\sum_{n=0}^{\infty} (x+e)^{2n+1}/3^n$

56. $\sum_{n=0}^{\infty} 2^n(x+\sqrt{2})^{2n+1}$

▶ Suppose that α and β are positive real numbers with $\alpha < \beta$. In each of Exercises 57–60, find a power series whose interval of convergence is precisely the given interval. ◀

57. (α, β)

58. $(\alpha, \beta]$

59. $[\alpha, \beta)$
60. $[\alpha, \beta]$

61. Is it possible for a power series to have interval of convergence $(-\infty, 1)$? Could it have interval of convergence $(0, \infty)$? Explain your answers.

62. What is the radius of convergence of $\sum_{n=1}^{\infty} n!\,(x/n)^n$?

63. Let $a_n = \sum_{k=1}^{n} 1/k^2$. What is the radius of convergence of $\sum_{n=1}^{\infty} a_n x^n$?

64. Let $b_n = \sum_{k=1}^{n} 1/\sqrt{k}$. What is the radius of convergence of $\sum_{n=1}^{\infty} b_n x^n$?

65. Suppose that $a_n \neq 0$ for $n \geq 0$. Show that if $\ell = \lim_{n\to\infty} |a_{n+1}|/|a_n|$ exists (as a nonnegative number or ∞), then $R = 1/\ell$ is the radius of convergence of $\sum_{n=0}^{\infty} a_n x^n$. *Hint:* Follow Examples 6 and 7, and apply the Ratio Test to the series $\sum_{n=0}^{\infty} u_n$ where $u_n = a_n x^n$.

66. Let $a_1 = 1$, $a_2 = 0$, $a_3 = 3^{-3}$, $a_4 = a_5 = 0$, $a_6 = 6^{-6}$, $a_7 = a_8 = a_9 = 0$, $a_{10} = 10^{-10}$, and, in general, $a_N = N^{-N}$ if $N = n(n+1)/2$ for some positive integer n and $a_N = 0$ otherwise. Calculate the radius of convergence of $\sum_{n=0}^{\infty} a_n x^n$

67. For each odd positive integer n, let $a_n = n$. For each even positive integer n, let $a_n = 2a_{n-1}$. Calculate the radius of convergence of $\sum_{n=1}^{\infty} a_n x^n$.

Calculator/Computer Exercises

68. We know the validity of the equation $\sum_{n=0}^{\infty} (-1)^n x^n = 1/(1+x)$ for $-1 < x < 1$. In this exercise, we will explore this equation by graphing partial sums $S_N(x) = \sum_{n=0}^{N} (-1)^n x^n$ of $\sum_{n=0}^{\infty} (-1)^n x^n$ for several values of N.

a. Plot $f(x) = 1/(1+x)$ for $-0.6 < x < 0.6$. In the same window, first add the plot of $S_2(x)$, then the plot of $S_4(x)$, and finally $S_6(x)$. You will notice the increasing accuracy as N increases. In this window, the graphs of $S_{20}(x)$ and $f(x)$ would not be distinguishable because the functions agree to four decimal places.

b. Repeat part a but plot over the interval $0.6 < x < 0.95$. Notice that the convergence of $S_N(x)$ is less rapid as x moves away from the center 0 of the interval of convergence. If you are working with a computer algebra system, replace the plot of $S_6(x)$ with a plot of $S_{50}(x)$. You will see that, for x in the interval of convergence but near an endpoint, $S_N(x)$ can be used to accurately approximate $f(x)$, but a large value of N may be required.

c. Repeat part a but plot over the interval $-0.98 < x < -0.6$. Here the problem is that the function $f(x)$ has a singularity at the left endpoint -1 of the interval of convergence of the power series.

69. We know the validity of the equation $\sum_{n=0}^{\infty} x^n = 1/(1-x)$ for $-1 < x < 1$. In this exercise, we will explore this equation by graphing partial sums $S_N(x) = \sum_{n=0}^{N} x^n$ of $\sum_{n=0}^{\infty} x^n$ for several values of N.

a. Plot $f(x) = 1/(1-x)$ for $-0.6 < x < 0.6$. In the same window, first add the plot of $S_2(x)$, then the plot of $S_4(x)$, and finally $S_6(x)$. You will notice the increasing accuracy as N increases. In this window, the graphs of $S_{20}(x)$ and $f(x)$ would not be distinguishable because the functions agree to four decimal places.

b. Repeat part a but plot over the interval $-0.95 < x < -0.6$. Notice that the convergence of $S_N(x)$ is less rapid as x moves away from the center 0 of the interval of convergence. If you are working with a computer algebra system, replace the plot of $S_6(x)$ with a plot of $S_{50}(x)$. You will see that, for x in the interval of convergence but near an endpoint, $S_N(x)$ can be used to accurately approximate $f(x)$, but a large value of N may be required.

c. Repeat part a but plot over the interval $0.6 < x < 0.98$. Here the problem is that the function $f(x)$ has a singularity at the right endpoint 1 of the interval of convergence of the power series.

70. It is known that the equation $\sum_{n=0}^{\infty} nx^n = x/(x-1)^2$ holds for $-1 < x < 1$. In this exercise, we will explore this equation by graphing partial sums $S_N(x) = \sum_{n=0}^{N} nx^n$ of $\sum_{n=0}^{\infty} nx^n$ for several values of N.

a. Plot $f(x) = x/(x-1)^2$ for $-0.5 < x < 0.5$. In the same window, first add the plot of $S_2(x)$, then the plot of $S_4(x)$, and finally $S_8(x)$. You will notice the increasing accuracy as N increases.

b. Repeat part a but plot over the interval $-0.8 < x < -0.5$. Notice that the convergence of $S_N(x)$ is less rapid as x moves away from the center 0 of the interval of convergence. If you are working with a computer algebra system, replace the plot of $S_8(x)$ with a plot of $S_{20}(x)$. You will see that, for x in the interval of convergence but near an endpoint, $S_N(x)$ can be used to accurately approximate $f(x)$, but a large value of N may be required.

c. Repeat part a but plot over the interval $0.5 < x < 0.9$. If you are working with a computer algebra system, replace the plot of $S_8(x)$ with a plot of $S_{25}(x)$.

71. We know that $\sum_{n=0}^{\infty} (-1)^n t^n = 1/(1+t)$ for $-1 < t < 1$. Therefore for $-1 < x < 1$, we have

$$\ln(1+x) = \int_0^x \frac{1}{1+t}\,dt = \int_0^x \sum_{n=0}^{\infty} (-1)^n t^n\,dt.$$

In this exercise, we will investigate graphically the approximation that results when an infinite series is truncated and then integrated term by term.

a. Plot $f(x) = \ln(1+x)$ for $-0.5 \leq x \leq 1$. In the same window, first add the plot of $S_N(x) = \int_0^x \sum_{n=0}^{N} (-1)^n t^n\,dt$ for $N = 2$, then the plot of $S_4(x)$, and finally $S_6(x)$. You will notice the increasing accuracy as N increases. In this window, the graphs of $S_{50}(x)$ and $f(x)$ would be distinguishable only at the rightmost few pixels.

b. Repeat part a but plot over the interval $-0.99 < x < -0.5$. Here the problem is that the function $f(x)$ has a singularity at the left endpoint -1 of the interval of convergence of the power series. Still, accuracy in the interval of convergence can be attained by increasing the value of N. If you are working with a computer algebra system, replace the plot of $S_6(x)$ with a plot of $S_{50}(x)$.

8.7 Representing Functions by Power Series

A natural question to ask is, if a power series defines a function, can we always say what function it is? The answer is, *No*. Most functions, those defined by power series included, do not have familiar names. A more fruitful question is, can we write the functions with which we are familiar as power series? The answer to this question is, with certain restrictions, *Yes*. Beginning with this section, we will study how to go back and forth between functions and power series.

Power Series Expansions of Some Standard Functions

For each fixed x, the power series $\sum_{n=0}^{\infty} x^n$ can be regarded as a geometric series. Because we know that this series converges to $1/(1-x)$ if x belongs to the interval $(-1, 1)$, we have

$$\frac{1}{1-x} = \sum_{n=0}^{\infty} x^n = 1 + x + x^2 + x^3 + \cdots \qquad (-1 < x < 1).$$

This formula can be used to find power series representations for many functions that are related to $1/(1-x)$. To facilitate substitution, we rewrite this equation with a different variable:

$$\frac{1}{1-u} = \sum_{n=0}^{\infty} u^n = 1 + u + u^2 + u^3 + \cdots \qquad (-1 < u < 1). \qquad (8.7.1)$$

▶ **EXAMPLE 1** Express $\dfrac{x^3}{(4+x^2)}$ by means of a power series with base point 0.

Solution The key idea is to rewrite the denominator of the given function in the form $1-u$ and then use formula (8.7.1). We accomplish this manipulation by writing

$$\frac{1}{4+x^2} = \frac{1}{4\big(1-(-x^2/4)\big)} = \frac{1}{4}\frac{1}{(1-u)},$$

where $u = -x^2/4$. If $-2 < x < 2$, then $-1 < u < 1$. Substituting $u = -x^2/4$ in equation (8.7.1), we obtain

$$\frac{1}{4+x^2} = \frac{1}{4}\frac{1}{(1-u)} \overset{(8.7.1)}{=} \frac{1}{4}\sum_{n=0}^{\infty} u^n = \frac{1}{4}\sum_{n=0}^{\infty}\left(-\frac{x^2}{4}\right)^n \qquad (-2 < x < 2).$$

We conclude that

$$\frac{x^3}{4+x^2} = \frac{x^3}{4}\sum_{n=0}^{\infty}\left(-\frac{x^2}{4}\right)^n = \sum_{n=0}^{\infty}(-1)^n\left(\frac{1}{4}\right)^{n+1} x^{2n+3} \qquad (-2 < x < 2).$$

If we write out several terms, then this equation becomes

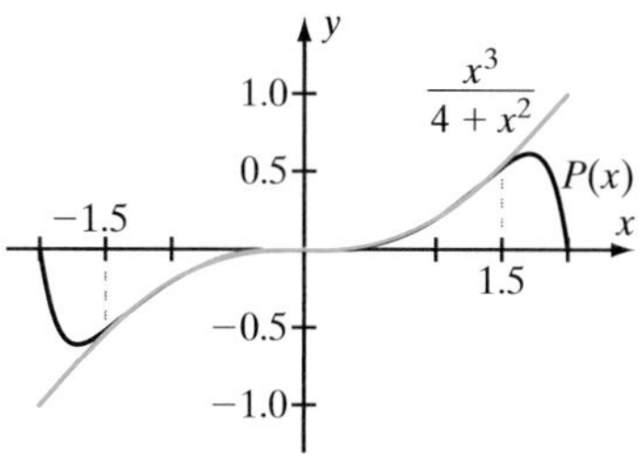

Figure 1

$$\frac{x^3}{4+x^2} = \underbrace{\frac{1}{4}x^3 - \frac{1}{16}x^5 + \frac{1}{64}x^7 - \frac{1}{256}x^9 + \frac{1}{1024}x^{11} - \frac{1}{4096}x^{13}}_{P(x)} + \cdots \qquad -2 < x < 2.$$

Figure 1 shows that the degree 13 polynomial $P(x)$ that we obtain by truncating the infinite series is quite a good approximation to $x^3/(4+x^2)$ for x not too far away from the center of the interval of convergence. In the figure, the plots of $x^3/(4+x^2)$ and $P(x)$ are barely distinguishable for $-1.5 < x < 1.5$. ◀

▶ **EXAMPLE 2** Find a power series representation for the function $F(x) = \ln(1+x)$.

Solution The relationship of F to formula (8.7.1) becomes apparent after differentiation:

$$F'(x) = \frac{d}{dx}\ln(1+x) = \frac{1}{1+x} = \frac{1}{1-(-x)} = \frac{1}{1-u} \quad \text{where} \quad u = -x.$$

For $-1 < x < 1$, we also have $-1 < u < 1$. We may therefore apply formula (8.7.1) to obtain

$$F'(x) = \frac{1}{1-u} \overset{(8.7.1)}{=} \sum_{n=0}^{\infty} u^n = \sum_{n=0}^{\infty} (-x)^n = \sum_{n=0}^{\infty} (-1)^n x^n \qquad (-1 < x < 1).$$

Next, we integrate both sides of this equation. Theorem 5b of Section 8.6 tells us that the right-hand side may be integrated termwise. The result is

$$F(x) = \sum_{n=0}^{\infty} (-1)^n \frac{x^{n+1}}{n+1} + C \qquad (-1 < x < 1),$$

where C is a constant of integration. By setting $x = 0$ in this equation, we obtain $F(0) = C$. Because $F(0) = \ln(1+0) = 0$, we conclude that $C = 0$. Thus we have derived the formula

$$\ln(1+x) = \sum_{n=0}^{\infty} (-1)^n \frac{x^{n+1}}{n+1} \qquad (-1 < x < 1).$$

It is not wrong to leave the series in this form, but we prefer to simplify the exponent to which x is raised. By changing the index of summation to $m = n+1$, we obtain

$$\ln(1+x) = \sum_{m=1}^{\infty} \frac{(-1)^{m-1}}{m} x^m \qquad (-1 < x < 1). \qquad (8.7.2a)$$ ◀

INSIGHT It is usually a good idea to write out the first several terms of a new series as a visual aid. For instance, equation (8.7.2a) becomes

$$\ln(1+x) = x - \frac{1}{2}x^2 + \frac{1}{3}x^3 - \frac{1}{4}x^4 + \frac{1}{5}x^5 - \frac{1}{6}x^6 + \frac{1}{7}x^7 - \cdots \quad \text{for } -1 < x < 1. \quad (8.7.2b)$$

▶ **EXAMPLE 3** Use formula (8.7.2b) to calculate ln(1.1) with an error no greater than 0.0001.

Solution Formula (8.7.2b) tells us that

$$\ln(1.1) = \ln(1+0.1) = 0.1 - \frac{1}{2}(0.1)^2 + \frac{1}{3}(0.1)^3 - \frac{1}{4}(0.1)^4 + \cdots.$$

We can estimate this quantity by truncating the alternating infinite series. According to inequality (8.4.1) of the Alternating Series Test, the error that results is less than the first term that is omitted. Therefore

$$\ln(1+0.1) = 0.1 - \frac{1}{2}(0.1)^2 + \frac{1}{3}(0.1)^3 + \varepsilon$$

where ε is an error that is less than $\frac{1}{4}(0.1)^4 = 0.000025$ in absolute value. The required approximation is

$$\ln(1.1) \approx 0.1 - \frac{1}{2}(0.1)^2 + \frac{1}{3}(0.1)^3 = 0.09533\ldots$$

A check with a calculator will verify the indicated accuracy. ◀

▶ **EXAMPLE 4** Find a power series expansion for $F(x) = \arctan(x)$ that is valid for $-1 < x < 1$.

Solution We use the same idea that was the basis of Example 2. Our starting point is

$$F'(x) = \frac{d}{dx}\arctan(x) = \frac{1}{1+x^2} = \frac{1}{1-u} \quad \text{where} \quad u = -x^2.$$

Notice that $-1 < u \leq 0$ for $-1 < x < 1$. For these values of u, the geometric series $\sum_{n=0}^{\infty} u^n$ converges to $1/(1-u)$. Therefore

$$F'(x) = \frac{1}{1-u} \overset{(8.7.1)}{=} \sum_{n=0}^{\infty} u^n = \sum_{n=0}^{\infty}(-x^2)^n = \sum_{n=0}^{\infty}(-1)^n x^{2n} \qquad (-1 < x < 1).$$

Now we find antiderivatives of both sides. The result is

$$\arctan(x) = \sum_{n=0}^{\infty} \frac{(-1)^n x^{2n+1}}{2n+1} + C \qquad (-1 < x < 1).$$

Setting $x = 0$ shows that $C = \arctan(0) = 0$. Therefore

$$\arctan(x) = \sum_{n=0}^{\infty} \frac{(-1)^n x^{2n+1}}{2n+1} \qquad (-1 < x < 1), \qquad (8.7.3a)$$

or, in a form that is more easily recognized and remembered,

$$\arctan(x) = x - \frac{1}{3}x^3 + \frac{1}{5}x^5 - \frac{1}{7}x^7 + \frac{1}{9}x^9 - \cdots \quad \text{for } -1 < x < 1. \qquad (8.7.3b)$$ ◀

INSIGHT Power series representations (8.7.2b) and (8.7.3b) for $\ln(1+x)$ and $\arctan(x)$ are valid for $-1<x<1$. The Alternating Series Test tells us that the series in these formulas converge when $x=1$ as well. Because each side of equations (8.7.2b) and (8.7.3b) makes sense for $x=1$, it is plausible to think that these equations remain valid for $x=1$. That is,

$$\ln(2) = 1 - \frac{1}{2} + \frac{1}{3} - \frac{1}{4} + \frac{1}{5} - \frac{1}{6} + \frac{1}{7} - \cdots \qquad (8.7.4)$$

and, because $\arctan(1) = \pi/4$,

$$\frac{\pi}{4} = 1 - \frac{1}{3} + \frac{1}{5} - \frac{1}{7} + \frac{1}{9} - \cdots. \qquad (8.7.5)$$

In fact, both of these formulas are true. Proving them requires a delicate interchange of limits that we will omit. Although formulas (8.7.4) and (8.7.5) are interesting, they are not very efficient from a computational viewpoint. For instance, after summing the first 100 terms that appear on the right side of formula (8.7.5), we get

$$\frac{\pi}{4} \approx 1 - \frac{1}{3} + \frac{1}{5} - \frac{1}{7} + \frac{1}{9} - \frac{1}{11} + \cdots - \frac{1}{199} = 0.7828982\ldots.$$

This results in the approximation $\pi \approx 4 \times 0.7828982\ldots = 3.131\ldots$, which is correct to only one decimal place!

Thus far, the base point used in all examples of this section has been 0. At times, other base points may be desirable or, as in the next example, necessary.

▶ **EXAMPLE 5** Expand $1/x$ and $1/x^2$ in powers of $x-1$ for $|x-1|<1$.

Solution First notice that, on applying formula (8.7.1) with $u = 1-x$, we have

$$\frac{1}{x} = \frac{1}{1-(1-x)} \overset{(8.7.1)}{=} \sum_{n=0}^{\infty} (1-x)^n = \sum_{n=0}^{\infty} (-1)^n (x-1)^n, \quad |x-1|<1.$$

Differentiating both sides of this equation, we obtain $-1/x^2 = \sum_{n=1}^{\infty} n(-1)^n (x-1)^{n-1}$, or $1/x^2 = \sum_{n=1}^{\infty} n(-1)^{n-1}(x-1)^{n-1}$ for $|x-1|<1$. To write this power series in standard form, we set $m = n-1$. As n runs through the integers from 1 to infinity, the index m runs through the integers from 0 to infinity. This change of summation index gives us

$$\frac{1}{x^2} = \sum_{m=0}^{\infty} (-1)^m (m+1)(x-1)^m \quad \text{for} \quad |x-1|<1.$$

If we write out several terms, then we obtain

$$\frac{1}{x^2} = 1 - 2(x-1) + 3(x-1)^2 - 4(x-1)^3 + \cdots \quad \text{for} \quad |x-1|<1. \qquad (8.7.6)$$ ◀

A LOOK FORWARD ▶ Every power series representation that we have obtained so far has been derived from the identity $1/(1-x) = \sum_{n=0}^{\infty} x^n$ for $|x|<1$. It is surprising, perhaps, that functions such as $\ln(1+x)$ and $\arctan(x)$ can be treated using this approach. These examples should not, however, lead us to overestimate the scope of the techniques that have been presented. Power series representations of other important functions such as e^x, $\sin(x)$, and $\cos(x)$ require the more general methods that we develop in the next section.

The Relationship between the Coefficients and Derivatives of a Power Series

Theorem 6 of Section 8.7 tells us that, if a power series $\sum_{n=0}^{\infty} a_n(x-c)^n$ converges on an open interval I centered at c, then the function $f(x)$ defined by $f(x) = \sum_{n=0}^{\infty} a_n(x-c)^n$ is infinitely differentiable on I. In other words, the nth derivative $f^{(n)}(c)$ of f at c exists for every positive integer n. The next theorem tells us that the coefficient a_n can be expressed in terms of $f^{(n)}(c)$. To avoid having to treat the index $n=0$ as a special case, it will be convenient to understand $f^{(0)}(c)$ to mean $f(c)$.

THEOREM 1 Suppose that the power series $\sum_{n=0}^{\infty} a_n(x-c)^n$ converges on an open interval I centered at c. Let f be the function defined on I by $f(x) = \sum_{n=0}^{\infty} a_n(x-c)^n$. Then

$$a_n = \frac{f^{(n)}(c)}{n!}, \quad n = 0, 1, 2, 3, \ldots \tag{8.7.7}$$

and

$$f(x) = \sum_{n=0}^{\infty} \frac{f^{(n)}(c)}{n!} \cdot (x-c)^n. \tag{8.7.8}$$

Proof. Because $a_0 = f(c) = f^{(0)}(c) = f^{(0)}(c)/0!$, equation (8.7.7) holds when $k=0$. Now suppose that k is a positive integer. Observe that

$$\frac{d^k}{dx^k}(x-c)^n = \begin{cases} 0 & \text{if } n<k \\ k! & \text{if } n=k. \\ n(n-1)\cdots(n-k+1)\cdot(x-c)^{n-k} & \text{if } n>k \end{cases}$$

From the last line of this equation, we see that

$$\left.\frac{d^k}{dx^k}(x-c)^n\right|_{x=c} = n\cdot(n-1)\cdots(n-k+1)\cdot(c-c)^{n-k} = 0$$

when $n>k$. By dividing the power series for $f(x)$ into three parts,

$$f(x) = \underbrace{\sum_{n=0}^{k-1} a\,(x-c)^n}_{n<k} + \underbrace{a_k(x-c)^k}_{n=k} + \underbrace{\sum_{n=k+1}^{\infty} a_n(x-c)^n}_{n>k},$$

and differentiating k times term by term, we see that

$$f^{(k)}(x) = 0 + k!\cdot a_k + \sum_{n=k+1}^{\infty} n(n-1)\cdots(n-k+1)\cdot a_n(x-c)^{n-k}$$

and

$$f^{(k)}(c) = 0 + k!\cdot a_k + 0.$$

Equation (8.7.7) follows. Formula (8.7.8) is then obtained by substituting the value equation (8.7.7) gives for a_n into $f(x) = \sum_{n=0}^{\infty} a_n(x-c)^n$. ■

▶ **EXAMPLE 6** Let $f(x) = \sum_{n=0}^{\infty} (-1)^n (n+1)(x-1)^n$. What are $f'(1)$, $f''(1)$, and $f'''(1)$?

Solution The coefficients $\{a_n\}$ of the power series defining f are given by $a_n = (-1)^n (n+1)$ for $0 \le n < \infty$. The base point is $c = 1$. Because

$$\ell = \lim_{n\to\infty} \left| \frac{a_{n+1}}{a_n} \right| = \lim_{n\to\infty} \frac{n+2}{n+1} = \lim_{n\to\infty} \frac{1+2/n}{1+1/n} = 1,$$

we see that the radius of convergence R is given by $R = 1/\ell = 1$. All that is needed to apply Theorem 1 is that $R > 0$. We may therefore use formula (8.7.7), which tells us that $f^{(k)}(c) = k! \cdot a_k$. Consequently, on setting $k = 1$, 2, and 3, we have $f'(1) = 1! \cdot a_1 = 1! \cdot (-1)^1 (1+1) = -2$, $f''(1) = 2! \cdot a_2 = 2! \cdot (-1)^2 (2+1) = 6$, and $f'''(1) = 3! \cdot a_3 = 3! \cdot (-1)^3 (3+1) = -24$. ◀

INSIGHT Notice that the formula $f^{(k)}(c) = k! \cdot a_k$ permits the calculation of $f^{(k)}(c)$ without the benefit of a closed form for $f(x)$. However, equation (8.7.6) shows that the power series that defines $f(x)$ in Example 6 is equal to $1/x^2$. In other words, the function $f(x)$ of Example 6 is equal to $1/x^2$. We may therefore check the results of Example 6 by directly calculating $f'(x) = -2/x^3$, $f''(x) = 6/x^{-4}$, and $f'''(x) = -24/x^5$. On evaluation at $x = 1$, we obtain $f'(1) = -2$, $f''(1) = 6$, and $f'''(1) = -24$, in agreement with the values obtained by using the formula $f^{(k)}(c) = k! \cdot a_k$.

In Example 6, the coefficient a_k of a power series representing a function f is used to calculate $f^{(k)}(c)$. In the next example, equation (8.7.7) is applied in the opposite direction: $f^{(k)}(c)$ is used to calculate a_k.

▶ **EXAMPLE 7** The error function $x \mapsto \mathrm{erf}(x)$ is defined by

$$\mathrm{erf}(x) = \frac{2}{\sqrt{\pi}} \int_0^x e^{-t^2}\, dt, \qquad -\infty < x < \infty.$$

As its name suggests, the error function arises in the analysis of observational errors. The error function also plays an important role in the theory of refraction and the theory of heat conduction. It is known that the error function has a convergent power series representation $\sum_{n=0}^{\infty} a_n x^n$. Calculate the partial sum $\sum_{n=0}^{3} a_n x^n$.

Solution Let $f(x) = \mathrm{erf}(x)$. By the Fundamental Theorem of Calculus, we have $f'(x) = (2/\sqrt{\pi})e^{-x^2}$. We calculate $f^{(2)}(x) = -(4/\sqrt{\pi})xe^{-x^2}$ and $f^{(3)}(x) = -(4/\sqrt{\pi})e^{-x^2} + (8/\sqrt{\pi})x^2 e^{-x^2}$. It follows that $f(0) = 0$, $f^{(1)}(0) = 2/\sqrt{\pi}$, $f^{(2)}(0) = 0$, and $f^{(3)}(0) = -4/\sqrt{\pi}$. We obtain the requested partial sum by setting $c = 0$ in equation (8.7.8), omitting all terms with index n greater than 3, and substituting in the computed values of $f(0)$, $f^{(1)}(0)$, $f^{(2)}(0)$, and $f^{(3)}(0)$:

$$\sum_{n=0}^{3} a_n x^n = \sum_{n=0}^{3} \frac{f^{(n)}(0)}{n!}(x-0)^n = 0 + \frac{(2/\sqrt{\pi})}{1!}x + \frac{0}{2!}x^2 + \frac{(-4/\sqrt{\pi})}{3!}x^3 = \frac{2}{\sqrt{\pi}}x - \frac{2}{3\sqrt{\pi}}x^3.$$

Figure 2 reveals that, even though this partial sum has only two nonzero terms, it is still a very good approximation to the error function for values of x that are close to the base point 0. ◀

An Application to Differential Equations

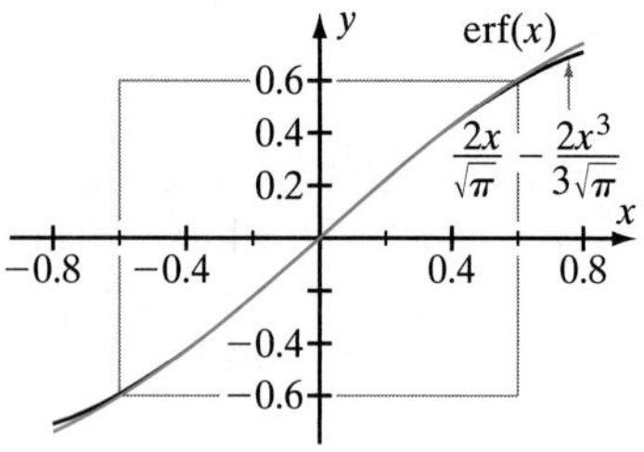

▲ Figure 2

Equation (8.7.8) tells us that, if we fix a base point, then the power series expansion of a function, if it has one, is unique Thus if we use one method or formula to calculate a power series representation $f(x) = \sum_{n=0}^{\infty} a_n (x-c)^n$ and if we use a second method or formula to obtain another power series representation $f(x) = \sum_{n=0}^{\infty} b_n (x-c)^n$, then $a_n = b_n$ for every n (because each is equal to $f^{(n)}(c)/n!$). The next example shows how to use this uniqueness to solve a differential equation.

▶ **EXAMPLE 8** Find a power series solution of the initial value problem $\dfrac{dy}{dx} = y - x,\ y(0) = 2$.

Solution Suppose that $y(x) = \sum_{n=0}^{\infty} a_n x^n$. We differentiate term-by-term to obtain $\dfrac{dy}{dx} = \sum_{n=1}^{\infty} n a_n x^{n-1}$. In order for y to satisfy the given differential equation, we must have

$$\underbrace{\sum_{n=1}^{\infty} n a_n x^{n-1}}_{y'} = \underbrace{\sum_{n=0}^{\infty} a_n x^n}_{y} - x.$$

It is convenient to make the change of summation index $m = n - 1$ for the series on the left side. Because n runs from 1 to infinity, the new index m runs from 0 to infinity. Consequently, we have

$$\sum_{m=0}^{\infty} (m+1) a_{m+1} x^m = \sum_{n=0}^{\infty} a_n x^n - x.$$

Renaming the index of summation of the series on the left, we obtain

$$\sum_{n=0}^{\infty} (n+1) a_{n+1} x^n = \sum_{n=0}^{\infty} a_n x^n - x.$$

By separating the terms that correspond to $n = 0$ and $n = 1$, we may write this equation as

$$a_1 + 2a_2 x + \sum_{n=2}^{\infty} (n+1) a_{n+1} x^n = a_0 + (a_1 - 1)x + \sum_{n=2}^{\infty} a_n x^n.$$

The uniqueness theorem that we discussed allows us to match up coefficients and deduce that $a_1 = a_0$, $2a_2 = a_1 - 1$, and $(n+1)a_{n+1} = a_n$ for $n \geq 2$. The initial condition $y(0) = 2$ tells us that $a_0 = 2$. It follows that $a_1 = 2$, and $a_2 = (a_1 - 1)/2 = (2-1)\ /2 = 1/2$. Substituting $n = 2$ in the equation $(n+1)a_{n+1} = a_n$, we obtain $a_3 = a_2/3 = 1/3!$. Substituting $n = 3$ in the equation $(n+1)_{an+1} = a_n$, we obtain $a_4 = a_3/4 = 1/4!$. Continuing in this way, we obtain $a_n = 1/n!$ for $n \geq 2$. It follows that, on its interval of convergence,

$$y(x) = 2 + 2x + \sum_{n=2}^{\infty} \frac{x^n}{n!} \tag{8.7.9}$$

is the solution of the given initial value problem. Because Example 6 of Section 8.6 shows that the power series on the right side of (8.7.9) converges on the entire real line, we can be sure that equation (8.7.9) provides the complete solution of the given initial value problem. ◄

Taylor Series and Polynomials

Theorem 6 of Section 8.7 tells us that, if we hope to represent a function f by a power series on an open interval I containing c, then f must be infinitely differentiable on I. If f does have this property, then we may form the power series

$$T(x) = \sum_{n=0}^{\infty} \frac{f^{(n)}(c)}{n!}(x-c)^n. \tag{8.7.10}$$

We call $T(x)$ the *Taylor series* of f with base point c. A Taylor series with base point 0 is also called a *Maclaurin series*. Theorem 1 of Section 8.7 tells us that $T(x)$ is the *only* power series with base point c that can represent f on an open interval centered at c. Whether it does represent f or not requires analysis. The first question is, does Taylor series (8.7.10) converge at points other than the base point? That question can be resolved by calculating the radius of convergence in a standard way. However, even if a Taylor series $T(x)$ generated by a function f has a positive radius of convergence, it is not automatic that $T(x)$ is equal to $f(x)$. We will devote the next section to this question. For now, we will concern ourselves only with the partial sums of Taylor series.

DEFINITION If a function f is N times continuously differentiable on an interval containing c, then

$$\begin{aligned} T_N(x) &= \sum_{n=0}^{N} \frac{f^{(n)}(c)}{n!}(x-c)^n \\ &= \frac{f^{(0)}(c)}{0!}(x-c)^0 + \frac{f^{(1)}(c)}{1!}(x-c)^1 + \frac{f^{(2)}(c)}{2!}(x-c)^2 + \frac{f^{(3)}(c)}{3!}(x-c)^3 + \cdots \\ &\quad + \frac{f^{(N)}(c)}{N!}(x-c)^N \end{aligned}$$

is called the *Taylor polynomial of order N and base point c* for the function f.

THEOREM 2 Suppose that f is N times continuously differentiable. Then

$$T_N(c) = f(c), \quad T_N'(c) = f'(c), \quad T_N''(c) = f''(c),$$
$$T_N^{(3)}(c) = f^{(3)}(c), \ldots T_N^{(N)}(c) = f^{(N)}(c).$$

Proof. Fix an integer k between 0 and N. By dividing T_N into three parts,

$$T_N(x) = \underbrace{\sum_{n=0}^{k-1} \frac{f^{(n)}(c)}{n!}(x-c)^n}_{n<k} + \underbrace{\frac{f^{(k)}(c)}{k!}(x-c)^k}_{n=k} + \underbrace{\sum_{n=k+1}^{N} \frac{f^{(n)}(c)}{n!}(x-c)^n}_{n>k},$$

we see, as in Theorem 1, that

$$\frac{d^k}{dx^k} T_N(x)\bigg|_{x=c} = 0 + \frac{f^{(k)}(c)}{k!} \cdot k! + 0 = f^{(k)}(c).$$

Thus $T_N^{(k)}(c) = f^{(k)}(c)$. ■

INSIGHT Because $0! = 1$, $f^{(0)}(c) = f(c)$, and $(x-c)^0 = 1$, the first term,

$$\frac{f^{(0)}(c)}{0!} \cdot (x-c)^0,$$

of a Taylor polynomial always simplifies to $f(c)$. The longer, unsimplified, notation serves to emphasize that the first term fits the pattern of all the other terms.

▶ **EXAMPLE 9** Compute the Taylor polynomials of order one, two, and three for the function $f(x) = e^{2x}$ expanded with base point $c = 0$.

Solution First of all,

$$\begin{aligned} f(0) &= 1 \\ f^{(1)}(0) &= 2 \cdot e^{2x}\big|_{x=0} = 2 \\ f^{(2)}(0) &= 4 \cdot e^{2x}\big|_{x=0} = 4 \\ f^{(3)}(0) &= 8 \cdot e^{2x}\big|_{x=0} = 8. \end{aligned}$$

Therefore

$$T_1(x) = \frac{f(0)}{0!} \cdot (x-0)^0 + \frac{f^{(1)}(0)}{1!} \cdot (x-0)^1 = 1 + 2x,$$

$$T_2(x) = \frac{f(0)}{0!} \cdot (x-0)^0 + \frac{f^{(1)}(0)}{1!} \cdot (x-0)^1 + \frac{f^{(2)}(0)}{2!} \cdot (x-0)^2 = 1 + 2x + 2x^2$$

and

$$T_3(x) = \frac{f(0)}{0!} \cdot (x-0)^0 + \frac{f^{(1)}(0)}{1!} \cdot (x-0)^1 + \frac{f^{(2)}(0)}{2!} \cdot (x-0)^2 + \frac{f^{(3)}(0)}{3!} \cdot (x-0)^3 = 1 + 2x + 2x^2 + \frac{4}{3}x^3.$$

See Figure 3. ◀

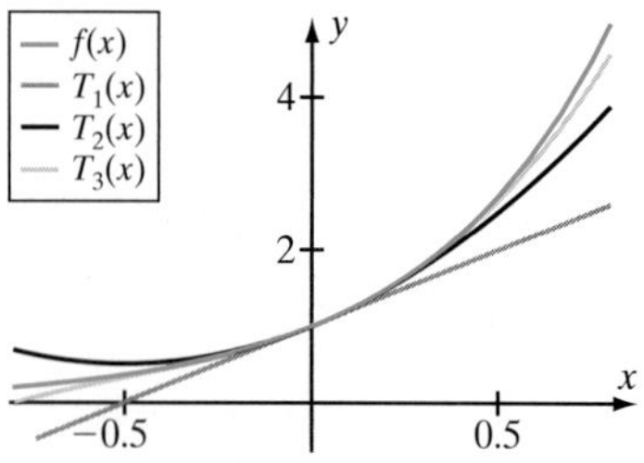

▲ Figure 3

INSIGHT If $T_N(x)$ has been computed, then we need not start from scratch in computing the next Taylor polynomial $T_{N+1}(x)$; we have only to compute and add one more term:

$$T_{N+1}(x) = T_N(x) + \frac{f^{(N+1)}(c)}{(N+1)!}(x-c)^{N+1}. \tag{8.7.11}$$

Thus in Example 9, $T_2(x) = T_1(x) + 2x^2 = (1 + 2x) + 2x^2$ and $T_3(x) = T_2(x) + \frac{4}{3}x^3 = (1 + 2x + 2x^2) + \frac{4}{3}x^3$.

▶ **EXAMPLE 10** Find the Taylor polynomial of order two for the function $f(x) = \arctan(x)$ about the point $c = -1$.

Solution We begin by calculating

$$f(-1) = -\frac{\pi}{4}$$

$$f^{(1)}(-1) = \left.\frac{1}{1+x^2}\right|_{x=-1} = \frac{1}{2}$$

$$f^{(2)}(-1) = \left.\frac{-2x}{(1+x^2)^2}\right|_{x=-1} = \frac{1}{2}.$$

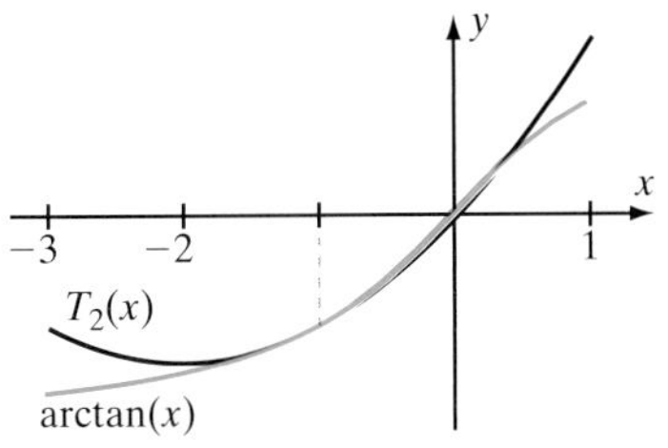

▲ Figure 4

Therefore

$$T_2(x) = \frac{-\pi/4}{0!}(x+1)^0 + \frac{1/2}{1!}(x+1)^1 + \frac{1/2}{2!}(x+1)^2 = -\frac{\pi}{4} + \frac{1}{2}(x+1) + \frac{1}{4}(x+1)^2.$$

The plots of f and T_2 are shown in Figure 4. ◀

INSIGHT Notice that the Taylor polynomial for arctan (x) that is calculated in Example 10 does *not* agree with the first few terms of the Taylor series for arctan (x) in formula (8.7.3b). The reason is that $c = 0$ is the base point of expansion (8.7.3b), whereas $c = -1$ is the base point used in Example 10.

QUICK QUIZ

1. If $f(x) = \sum_{n=1}^{\infty} x^n/n^2$, what is $f'''(0)$?
2. What is the coefficient of $(x-9)^2$ in the Taylor series of $\sqrt{x}$ with base point 9?
3. What is the coefficient of x^3 in the Maclaurin series of e^{-x}?
4. What is the order 4 Taylor polynomial of $1/(1+x^2)$ with base point 0?

Answers

1. 2/3 2. −1/216 3. −1/6 4. $1 - x^2 + x^4$

EXERCISES

Problems for Practice

▶ In each of Exercises 1–10, express the given function as a power series in x with base point 0. Calculate the radius of convergence R. ◀

1. $\dfrac{1}{1-2x}$

2. $\dfrac{x}{1+x}$

3. $\dfrac{1}{2-x}$

4. $\dfrac{1}{1+x^2}$

5. $\dfrac{1}{4+x}$

6. $\dfrac{1}{9-x^2}$

7. $\dfrac{x^2+1}{1-x^2}$

8. $\dfrac{x}{1+x^2}$

9. $\dfrac{x^3}{1+x^4}$
10. $\dfrac{1}{16-x^4}$

▶ In each of Exercises 11–16, use formula (8.7.2) to express the given function as a power series in x with base point 0. State the radius of convergence R. ◀

11. $\ln(1+4x)$
12. $\ln(5+2x)$
13. $\ln(1+x^2)$
14. $x\ln(1+x^2)$
15. $\int_0^x \ln(1+t^2)\,dt$
16. $\int_0^x t\ln(1+t)\,dt$

▶ In each of Exercises 17–20, use the equation (8.7.3) to express the given function as a power series in x with base point 0. State the radius of convergence R. ◀

17. $\arctan(x^2)$
18. $\int_0^x \arctan(t)\,dt$
19. $\int_0^x \arctan(t^2)\,dt$
20. $\int_0^x t\arctan(t)\,dt$

▶ In each of Exercises 21–26, a function f and a point c are given. Use the equation

$$\frac{1}{1-(t-c)} = \sum_{n=0}^{\infty}(t-c)^n, \quad |t-c|<1,$$

together with some algebra, to express $f(x)$ as a power series with base point c. State the radius of convergence R. ◀

21. $f(x) = \dfrac{1}{6-x}$ $\quad c=5$
22. $f(x) = \dfrac{1}{x}$ $\quad c=1$
23. $f(x) = \dfrac{1}{x+1}$ $\quad c=-3$
24. $f(x) = \dfrac{1}{x-4}$ $\quad c=1$
25. $f(x) = \dfrac{1}{2x+5}$ $\quad c=-1$
26. $f(x) = \dfrac{1}{x}$ $\quad c=2$

▶ In Exercises 27–38, compute the Taylor polynomial T_N of the given function f with the given base point c and given order N. ◀

27. $f(x)=\cos(x)$ $\quad N=4$ $\quad c=\pi/3$
28. $f(x)=\sin(x)$ $\quad N=5$ $\quad c=\pi/2$
29. $f(x)=\sqrt{x}$ $\quad N=3$ $\quad c=4$
30. $f(x)=e^{-x}$ $\quad N=5$ $\quad c=0$
31. $f(x)=5x^2+\dfrac{1}{x}$ $\quad N=4$ $\quad c=1$
32. $f(x)=\sin(3x)$ $\quad N=4$ $\quad c=\pi/6$
33. $f(x)=1/x^3$ $\quad N=3$ $\quad c=2$
34. $f(x)=x^2+1+1/x^2$ $\quad N=5$ $\quad c=-1$
35. $f(x)=\ln(3x+7)$ $\quad N=4$ $\quad c=-2$
36. $f(x)=e^{x+2}/x$ $\quad N=2$ $\quad c=-2$
37. $f(x)=\tan(x)$ $\quad N=3$ $\quad c=0$
38. $f(x)=\sec(x)$ $\quad N=3$ $\quad c=0$

Further Theory and Practice

▶ In each of Exercises 39–42, compute the Taylor polynomial $T_3(x)$ of the given function f with the given base point c. ◀

39. $f(x)=\dfrac{x}{x^2+1}$ $\quad c=1$
40. $f(x)=x\cdot e^x$ $\quad c=0$
41. $f(x)=(x+5)^{1/3}$ $\quad c=3$
42. $f(x)=(1+x)^{1/2}$ $\quad c=2$

▶ For each function f given in Exercises 43–46, use the equation

$$\frac{d}{dx}\left(\frac{1}{1-x}\right) = \frac{1}{(1-x)^2}$$

to find the power series of $f(x)$ with base point 0. ◀

43. $f(x)=1/(1-x)^2$
44. $f(x)=1/(1-2x)^2$
45. $f(x)=(2+x)/(1-x)^2$
46. $f(x)=1/(1-x)^3$

▶ In each of Exercises 47–52, find the sum of the given series in closed form. State the radius of convergence R. ◀

47. $\sum_{n=0}^{\infty} x^{n+2}$
48. $\sum_{n=1}^{\infty} n\,x^{n+1}$
49. $\sum_{n=1}^{\infty} n(n+1)x^n$
50. $\sum_{n=2}^{\infty} n(n-1)x^{n-2}$
51. $\sum_{n=1}^{\infty} n\,x^{2n+1}$
52. $\sum_{n=0}^{\infty} n(x+2)^n$

▶ In each of Exercises 53–58, use the Uniqueness Theorem to determine the coefficients $\{a_n\}$ of the solution $y(x)=\sum_{n=0}^{\infty} a_n x^n$ of the given initial value problem. ◀

53. $dy/dx=y$ $\quad y(0)=1$
54. $dy/dx=2y$ $\quad y(0)=3$
55. $dy/dx=x+y$ $\quad y(0)=1$
56. $dy/dx=2x-y$ $\quad y(0)=1$
57. $dy/dx=1+x+y$ $\quad y(0)=0$
58. $dy/dx=x+xy$ $\quad y(0)=0$

▶ If $f(x)=\sum_{n=0}^{\infty} a_n x^n$ and $g(x)=\sum_{n=0}^{\infty} b_n x^n$ converge on $(-R,R)$, then we may formally multiply the series as though they were polynomials. That is, if $h(x)=f(x)g(x)$, then

$$h(x)=\sum_{n=0}^{\infty}\left(\sum_{k=0}^{n} a_k b_{n-k}\right)x^n.$$

The product series, which is called the *Cauchy product*, also converges on $(-R,R)$. Exercises 59–62 concern the Cauchy product. ◀

59. Suppose $|x| < 1$. Calculate the power series of $h(x) = 1/(1-x)^2$ with base point 0 by using the method of Exercise 43. Using $f(x) = g(x) = 1/(1-x) = \sum_{n=0}^{\infty} x^n$, verify the Cauchy product formula for $h = f \cdot g$ up to the x^6 term.

60. Suppose $|x| < 1$. Calculate the power series of $h(x) = 1/(1-x^2)$ with base point 0 by substituting $t = x^2$ the equation $1/(1-t) = \sum_{n=0}^{\infty} t^n$. Let $f(x) = 1/(1-x)$ and $g(x) = f(-x)$. Verify the Cauchy product formula for $h = f \cdot g$ up to the x^8 term.

61. The secant function has a known power series expansion that begins

$$\sec(x) = 1 + \frac{1}{2}x^2 + \frac{5}{24}x^4 + \frac{61}{720}x^6 \cdots.$$

The sine function has a known power series expansion that begins

$$\sin(x) = x - \frac{x^3}{3!} + \frac{x^5}{5!} - \frac{x^7}{7!} \cdots.$$

The tangent function has a known power series expansion that begins

$$\tan(x) = x + \frac{1}{3}x^3 + \frac{2}{15}x^5 + \frac{17}{315}x^7 + \cdots.$$

Verify the Cauchy product formula for $\tan(x) = \sin(x)\sec(x)$ up to the x^7 term.

62. Suppose that the series $\sum_{n=0}^{\infty} a_n x^n$ converges on $(-R, R)$ to a function $f(x)$ and that $|f(x)| \geq k > 0$ on that interval for some positive constant k. Then, $1/f(x)$ also has a convergent power series expansion on $(-R, R)$. Compute its coefficients in terms of the a_n's. *Hint:* Set

$$\frac{1}{f(x)} = g(x) = \sum_{n=0}^{\infty} b_n x^n.$$

Use the equation $f(x) \cdot g(x) = 1$ to solve for the b_n's.

Calculator/Computer Exercises

63. Let $f(x) = (x+1)^4/(x^4+1)$. It is known that f has a power series expansion of the form

$$f(x) = 1 + 4x + 6x^2 + 4x^3 - 4x^5 - 6x^6 - 4x^7 + 4x^9 + \cdots.$$

a. Plot the central difference quotient approximation $D_0 f(x, 10^{-5})$ of $f'(x)$ for $-0.5 < x < 0.5$. Use the given power series to find a degree 8 polynomial approximation of $f'(x)$. Add the plot of this polynomial to the viewing rectangle.

b. Repeat part a but plot for $-3/4 < x < 3/4$. Notice that the approximation becomes less accurate as the distance betwen x and 0, the base point of the given series, increases.

64. Let $f(x) = \sqrt{5 + x^2}$. It is known that f has a power series expansion of the form

$$f(x) = 3 + \frac{2}{3}(x-2) + \frac{5}{54}(x-2)^2 - \frac{5}{243}(x-2)^3 + \frac{55}{17496}(x-2)^4 + \cdots.$$

Plot the central difference quotient approximation $D_0 f(x, 10^{-5})$ of $f'(x)$ for $0 < x < 4$. Use the given power series to find a degree 3 polynomial approximation $T(x)$ of $f'(x)$. Plot the central difference quotient approximation $D_0 f(x, 10^{-5})$ of $f'(x)$ for $0 < x < 4$. To this viewing window, add the plot of $T(x)$.

65. Consider the initial value problem

$$\frac{dy}{dx} = x^2 + y, \qquad y(0) = 1.$$

a. Calculate the power series expansion $y(x) = \sum_{n=0}^{\infty} a_n x^n$ of the solution up to the x^7 term.

b. Using the coefficients you have calculated, plot $S_3(x) = \sum_{n=0}^{3} a_n x^n$ in the viewing rectangle $[-3,3] \times [-11,44]$.

c. The exact solution to the initial value problem is $y(x) = 3e^x - x^2 - 2x - 2$, as can be determined using the methods of Section 7.7 in Chapter 7. Add the plot of the exact solution to the viewing window. From the two plots, we see that the approximation is fairly accurate for $-1 \leq x \leq 1$, but the accuracy decreases outside this subinterval.

d. When a partial sum $S_N(x)$ is used to approximate an infinite series, an increase in the value of N requires more computation, but improved accuracy is the reward. To see the effect in this example, replace the graph of $S_3(x)$ with that of $S_7(x)$.

66. Consider the initial value problem

$$\frac{dy}{dx} = 2 - x - y, \qquad y(0) = 1.$$

a. Calculate the power series expansion $y(x) = \sum_{n=0}^{\infty} a_n x^n$ of the solution up to the x^7 term.

b. Using the coefficients you have calculated, plot $S_3(x) = \sum_{n=0}^{3} a_n x^n$ in the viewing rectangle $[-2,2] \times [-10,1.7]$.

c. The exact solution to the initial value problem is $y(x) = 3 - x - 2e^{-x}$, as can be determined using the methods of Section 7.7 (in Chapter 7). Add the plot of the exact solution to the viewing window. From the two plots, we see that the approximation is fairly accurate for $-1 \leq x \leq 1$, but the accuracy decreases outside this subinterval.

d. To see the improvement in accuracy that results from using more terms in a partial sum, replace the graph of $S_3(x)$ with that of $S_7(x)$.

8.8 Taylor Series

In the preceding section, we associated Taylor series and Taylor polynomials to a differentiable function f. In this section, we explore two basic, related questions: How well do the Taylor polynomials approximate f, and how can we tell when a Taylor series converges to f?

Taylor's Theorem

If f is defined on an open interval centered at c, and if $T_N(x)$ is the order N Taylor polynomial of f with base point c, then we define a function R_N by

$$R_N(x) = f(x) - T_N(x). \tag{8.8.1}$$

Rearranging this equation, we obtain

$$f(x) = T_N(x) + R_N(x). \tag{8.8.2}$$

From this viewpoint, $R_N(x)$ is said to be a *remainder term*, or *error term*: it is what we must add to the approximation $T_N(x)$ to obtain the exact value of $f(x)$. The key to understanding how the Taylor polynomial T_N can be used to approximate f is to obtain a formula for $R_N(x)$ that we can analyze and estimate. Such a formula is provided by Taylor's Theorem.

THEOREM 1 ***(Taylor's Theorem)*** For any natural number N, suppose that f is N + *1* times continuously differentiable on an open interval I centered at c. Define $R_N(x)$ by equation (8.8.1). If x is a point in I, then there is a number s between c and x such that

$$R_N(x) = \frac{f^{(N+1)}(s)}{(N+1)!}(x-c)^{N+1}. \tag{8.8.3}$$

By combining equations (8.8.2) and (8.8.3), we obtain

$$f(x) = T_N(x) + \frac{f^{(N+1)}(s)}{(N+1)!}(x-c)^{N+1}. \tag{8.8.4}$$

Compare this equation with equation (8.7.11):

$$T_{N+1}(x) = T_N(x) + \frac{f^{(N+1)}(c)}{(N+1)!}(x-c)^{N+1}.$$

We see that the remainder $R_N(x)$ is similar to the term that must be added to the order N Taylor polynomial $T_N(x)$ to obtain the Taylor polynomial $T_{N+1}(x)$ of one order higher. The difference is that the derivative $f^{(N+1)}$ that appears in the formula for R_N is not evaluated at the base point c, as it is in T_{N+1}, but at some other, unspecified point s.

Because the order 0 Taylor polynomial $T_0(x)$ of f is the constant function $f(c)$, equation (8.8.4) with $N = 0$ tells us that there is a point s between c and x such that

$$f(x) = T_0(x) + \frac{f^{(1)}(s)}{1!} \cdot (x-c)^1 = f(c) + f'(s) \cdot (x-c),$$

or

$$f'(s) = \frac{f(x) - f(c)}{x - c}.$$

This is precisely the assertion of the Mean Value Theorem. For this reason, Taylor's Theorem may be seen as a generalization of the Mean Value Theorem. A proof is outlined in Exercise 80 for $N = 1$ and in Exercise 81 for the general case.

▶ **EXAMPLE 1** For $f(x) = \sqrt{x}$ and $c = 1$, calculate the error $R_1(4/9)$ that results when $T_1(4/9)$ is used to approximate $f(4/9)$. Verify formula (8.8.3) for $N = 1$ and $x = 4/9$.

Solution We calculate

$$f(x) = 1, \quad f^{(1)}(c) = \frac{d}{dx}\sqrt{x}\bigg|_{x=c} = \frac{1}{2\sqrt{x}}\bigg|_{x=1} = \frac{1}{2}, \quad \text{and} \quad f^{(2)}(c) = \frac{d^2}{dx^2}\sqrt{x}\bigg|_{x=c} = -\frac{1}{4x^{3/2}}\bigg|_{x=1} = -\frac{1}{4}.$$

Therefore we have

$$T_1(x) = f(c) + \frac{f^{(1)}(c)}{1!}(x-c)^1 = 1 + \frac{1}{2}(x-1),$$

which results in the approximation

$$T_1\left(\frac{4}{9}\right) = 1 + \frac{1}{2}\left(\frac{4}{9} - 1\right) = \frac{13}{18}$$

of $f(4/9) = 2/3$. The error $R_1(4/9)$ is given by

$$R_1\left(\frac{4}{9}\right) = f\left(\frac{4}{9}\right) - T_1\left(\frac{4}{9}\right) = \frac{2}{3} - \frac{13}{18} = -\frac{1}{18}. \tag{8.8.5}$$

Because $f^{(2)}(x) = -1/(4x^{3/2})$, equation (8.8.3) tells us that

$$R_1(x) = \frac{f^{(2)}(s)}{2!}(x-1)^2 = -\frac{1}{8\,s^{3/2}}(x-1)^2$$

for some s between 1 and x. To verify this equation for $x = 4/9$, we must show that

$$-\frac{1}{18} \overset{(8.8.5)}{=} R_1\left(\frac{4}{9}\right) = -\frac{1}{8\,s^{3/2}}\left(\frac{4}{9} - 1\right)^2 = -\frac{25}{648} \cdot \frac{1}{s^{3/2}}$$

has a solution s in the interval $(4/9, 1)$. We solve the equation for s, obtaining $s = (25/36)^{2/3} = 0.784\ldots$, which is, indeed, in $(4/9, 1)$. ◀

▶ **EXAMPLE 2** Calculate the order 7 Taylor polynomial $T_7(x)$ with base point 0 of $\sin(x)$. If $T_7(x)$ is used to approximate $\sin(x)$ for $-1 \le x \le 1$, what accuracy is guaranteed by Taylor's Theorem?

Solution We first calculate

$$\begin{aligned}
f(0) &= \sin(0) = 0\\
f^{(1)}(0) &= \cos(x)|_{x=0} = 1\\
f^{(2)}(0) &= -\sin(x)|_{x=0} = 0\\
f^{(3)}(0) &= -\cos(x)|_{x=0} = -1\\
f^{(4)}(0) &= \sin(x)|_{x=0} = 0\\
f^{(5)}(0) &= \cos(x)|_{x=0} = 1\\
f^{(6)}(0) &= -\sin(x)|_{x=0} = 0\\
f^{(7)}(0) &= -\cos(x)|_{x=0} = -1\\
f^{(8)}(x) &= \sin(x).
\end{aligned}$$

Therefore

$$\begin{aligned}
T_7(x) &= \frac{0}{0!}\cdot(x-0)^0 + \frac{1}{1!}\cdot(x-0)^1 + \frac{0}{2!}\cdot(x-0)^2 + \frac{(-1)}{3!}\cdot(x-0)^3\\
&\quad + \frac{0}{4!}\cdot(x-0)^4 + \frac{1}{5!}\cdot(x-0)^5 + \frac{0}{6!}\cdot(x-0)^6 + \frac{(-1)}{7!}\cdot(x-0)^7\\
&= x - \frac{x^3}{3!} + \frac{x^5}{5!} - \frac{x^7}{7!}.
\end{aligned}$$

The remainder term is given by

$$R_7(x) = \frac{f^{(8)}(s)}{8!}\cdot(x-0)^8 = \frac{\sin(s)}{8!}\cdot x^8.$$

Thus

$$\sin(x) = x - \frac{x^3}{3!} + \frac{x^5}{5!} - \frac{x^7}{7!} + \frac{\sin(s)}{8!}\cdot x^8$$

for some value of s between 0 and x. Because $|\sin(s)|$ can be no greater than 1 and $1/8! < 0.000025$, we can make the estimate

$$|R_7(x)| = \left|\frac{\sin(s)}{8!}\cdot x^8\right| \le \frac{|\sin(s)|}{8!} < 0.000025 \quad \text{for} \quad |x| \le 1.$$

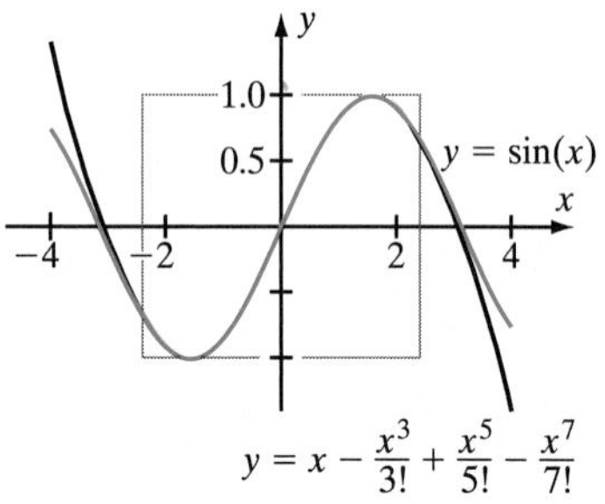

▲ Figure 1

Therefore $\sin(x)$ and $x - x^3/3! + x^5/5! - x^7/7!$ agree to four decimal places when $|x| \le 1$. As Figure 1 shows, $x - x^3/3! + x^5/5! - x^7/7!$ is a good approximation of $\sin(x)$ on an even larger interval. ◀

▶ **EXAMPLE 3** Calculate the order 4 Taylor polynomial $T_4(x)$ with base point $\pi/4$ of $\sin(x)$. If $T_4(x)$ is used to approximate $\sin(x)$ for $0 \le x \le \pi/2$, what accuracy is guaranteed by Taylor's Theorem?

Solution Using the derivatives calculated in the preceding example, we have $f(\pi/4) = 1/\sqrt{2}$, $f^{(1)}(\pi/4) = \cos(\pi/4) = 1/\sqrt{2}$, $f^{(2)}(\pi/4) = -\sin(\pi/4) = -1/\sqrt{2}$, $f^{(3)}(\pi/4) = -\cos(\pi/4) = -1/\sqrt{2}$, $f^{(4)}(\pi/4) = \sin(\pi/4) = 1/\sqrt{2}$, and $f^{(5)}(x) = \cos(x)$. Therefore

$$T_4(x) = \frac{1/\sqrt{2}}{0!}\left(x - \frac{\pi}{4}\right)^0 + \frac{1/\sqrt{2}}{1!}\left(x - \frac{\pi}{4}\right)^1 - \frac{1/\sqrt{2}}{2!}\left(x - \frac{\pi}{4}\right)^2 - \frac{1/\sqrt{2}}{3!}\left(x - \frac{\pi}{4}\right)^3 + \frac{1/\sqrt{2}}{4!}\left(x - \frac{\pi}{4}\right)^4,$$

or

$$T_4(x) = \frac{1}{\sqrt{2}} + \frac{1}{\sqrt{2}}\left(x - \frac{\pi}{4}\right) - \frac{1}{2!\sqrt{2}}\left(x - \frac{\pi}{4}\right)^2 - \frac{1}{3!\sqrt{2}}\left(x - \frac{\pi}{4}\right)^3 + \frac{1}{4!\sqrt{2}}\left(x - \frac{\pi}{4}\right)^4.$$

According to formula (8.8.3), the remainder is given by

$$R_4(x) = \frac{f^{(5)}(s)}{5!}\left(x - \frac{\pi}{4}\right)^5 = \frac{\cos(s)}{5!}\left(x - \frac{\pi}{4}\right)^5,$$

where s is between $\pi/4$ and x. Because $|x - \pi/4| \le \pi/2 - \pi/4 = \pi/4 = 0.785398\ldots$ and $\cos(s) \le 1$, we have

$$|R_4(x)| = \frac{|f^{(5)}(s)|}{5!}\left|x - \frac{\pi}{4}\right|^5 \le \frac{1}{5!}(0.785398\ldots)^5 < 0.0025.$$

Thus for x in the interval $[0, \pi/2]$, $T_4(x)$ approximates $\sin(x)$ to two decimal places. ◀

INSIGHT As we observed in the preceding section, Taylor polynomials that are calculated at distinct base points may look quite different. The Taylor polynomials of $\sin(x)$ involve only odd powers of x when $c = 0$ is used as the base point (Example 1). That is not the case when the base point is $\pi/4$ (Example 2).

Estimating the Error Term

The following theorem formalizes the method of estimation employed in Examples 1 and 2.

THEOREM 2 *Let f be a function that is N + 1 times continuously differentiable on an open interval I centered at c. For each x in I, let J_x denote the closed interval with endpoints x and c. Thus $J_x = [c, x]$ if $c \le x$ and $J_x = [x, c]$ if $x < c$. Let*

$$M = \max_{t \in J_x} |f^{(N+1)}(t)|. \tag{8.8.6}$$

Then the error term $R_N(x)$ defined by equation (8.8.1) satisfies

$$|R_N(x)| \le \frac{M}{(N+1)!} \cdot |x - c|^{N+1}. \tag{8.8.7}$$

Proof. Taylor's Theorem expresses the remainder $R_N(x)$ in terms of a derivative $f^{(N+1)}(s)$ that is evaluated at an undetermined point s between c and x. We bound the remainder by using the largest possible value for this derivative. Because s belongs to J_x, we have

$$|R_N(x)| = \left|\frac{f^{(N+1)}(s)}{(N+1)!} \cdot (x - c)^{N+1}\right| = \frac{|f^{(N+1)}(s)|}{(N+1)!} \cdot |x - c|^{N+1} \le \frac{M}{(N+1)!} \cdot |x - c|^{N+1}. \quad ■$$

▶ **EXAMPLE 4** Use the third order Taylor polynomial of e^x with base point 0 to approximate $e^{-0.1}$. Estimate your accuracy.

Solution If $f(x) = e^x$, then $f^{(k)}(x) = e^x$ for all natural numbers k. Thus $f^{(k)}(0) = 1$ for $k = 0, 1, 2, 3$, and

$$T_3(x) = \sum_{k=0}^{3} \frac{f^{(k)}(0)}{k!} x^k = 1 + x + \frac{1}{2!}x^2 + \frac{1}{3!}x^3.$$

Our approximation is therefore $T_3(-0.1) = 1 + (-0.1) + (-0.1)^2/2 + (-0.1)^3/6 = 0.90483\ldots$. Next, to estimate the accuracy of this approximation, we calculate

$$M = \max_{-0.1 \le t \le 0} \left| \frac{d^4}{dt^4} e^t \right| = \max_{-0.1 \le t \le 0} e^t = e^0 = 1.$$

Therefore

$$|R_3(-0.1)| \le \frac{1}{4!} \cdot |-0.1 - 0|^4 = 4.166\ldots \times 10^{-6} < 5 \times 10^{-6}.$$

Our error estimate guarantees that our approximation is accurate to five decimal places. A check with a calculator shows this to be correct. ◀

INSIGHT When $|f^{(N+1)}(t)|$ is a monotonic function on the interval J_x between c and x, as in Example 3, the maximum $M = \max_{t \in J_x} |f^{(N+1)}(t)|$ occurs at an endpoint of J. That is, M is the larger of $|f^{(N+1)}(c)|$ and $|f^{(N+1)}(x)|$.

Achieving a Desired Degree of Accuracy

To approximate the value of $f(x)$ by the value $T_N(x)$ of a Taylor polynomial, we must select a base point c and an order N. The base point c is chosen to facilitate computation. Another important consideration is that, if $|x - c|$ is small, then the factor $|x - c|^{N+1}$ of $|R_N(x)|$ will be small. Choosing a large value of N is also an obvious strategy for producing a small error term. However, it is not advantageous to choose a value of N that is larger than necessary. To minimize the amount of calculation, we desire the smallest N that gives the required accuracy. Studying several examples is the best way to understand the method of choosing values of c and N that are appropriate for attaining a particular degree of accuracy.

▶ **EXAMPLE 5** Compute $\ln(1.2)$ to an accuracy of four decimal places.

Solution Let $f(x) = \ln(x)$. Saying that we require four decimal places of accuracy means that we want the error term to be less than $5 \cdot 10^{-5}$. Thus we require $|R_N(1.2)| < 5 \cdot 10^{-5}$. We can arrange for this estimate to be true by choosing c sensibly and by making N sufficiently large. We take $c = 1$, which is reasonably close to 1.2, and note that, for $f(x) = \ln(x)$, we have $f^{(1)}(x) = x^{-1}$, $f^{(2)}(x) = -1 \cdot x^{-2} = -1! \cdot x^{-2}$, $f^{(3)}(x) = 2 \cdot x^{-3} = 2! \cdot x^{-3}$, $f^{(4)}(x) = -3 \cdot 2 \cdot x^{-4} = -3! \cdot x^{-4}$, $f^{(5)}(x) = 4! \cdot x^{-5}$, and, in general,

$$f^{(n)}(x) = (-1)^{n-1}(n-1)! \cdot x^{-n} \quad n = 1, 2, 3, \ldots . \tag{8.8.8}$$

Thus

$$\frac{f^{(n)}(c)}{n!} = \frac{(-1)^{n-1}(n-1)!\cdot c^{-n}}{n!} = \frac{(-1)^{n-1}(n-1)!\cdot 1^{-n}}{n!} = \frac{(-1)^{n-1}}{n}$$

and

$$T_N(1.2) = \ln(c) + \sum_{n=1}^{N} \frac{f^{(n)}(c)}{n!}(1.2-c)^n = \ln(1) + \sum_{n=1}^{N} \frac{(-1)^{n-1}}{n}(1.2-1)^n = \sum_{n=1}^{N} \frac{(-1)^{n-1}(0.2)^n}{n}. \quad (8.8.9)$$

To estimate the error, we note that $f^{(N+1)}(t) = (-1)^N\, N!\cdot t^{-(N+1)}$ by formula (8.8.8). Therefore

$$M = \max_{1\le t\le 1.2}\left|f^{(N+1)}(t)\right| = \max_{1\le t\le 1.2}\left|(-1)^{N+2}N!\cdot t^{-(N+1)}\right| = N! \max_{1\le t\le 1.2}\frac{1}{t^{N+1}} = N!$$

(because the maximum of $1/t^{N+1}$ for $1\le t\le 1.2$ occurs at $t=1$). Thus

$$|R_N(1.2)| \le M\cdot\frac{|1.2-1|^{N+1}}{(N+1)!} = N!\cdot\frac{(0.2)^{N+1}}{(N+1)!} = \frac{(0.2)^{N+1}}{N+1}.$$

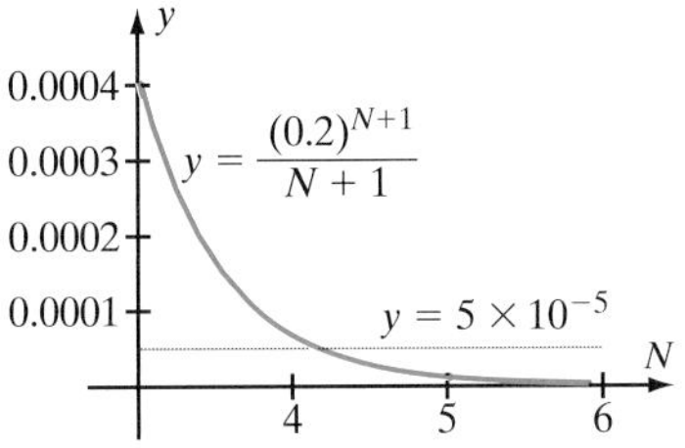

▲ Figure 2

To ensure that the error is less than $5\cdot 10^{-5}$, it suffices to choose N large enough so that $(0.2)^{N+1}/(N+1) < 5\cdot 10^{-5}$. Substituting in a few values for N, we see that the inequality will first be true when $N=5$. We could also determine this value of N by plotting $(0.2)^{N+1}/(N+1)$, as in Figure 2.

In conclusion, the formula $\ln(1.2) = T_5(1.2) + R_5(1.2)$ and the inequality $|R_5(1.2)| < 5\cdot 10^{-5}$ assure us that $\ln(1.2)$ and $T_5(1.2)$ agree to four decimal places. Substituting $N=5$ in line (8.8.9), we calculate

$$\ln(1.2) \approx T_5(1.2) = \sum_{n=1}^{5}\frac{(-1)^{n-1}(0.2)^n}{n} = 0.1823\ldots \quad ◀$$

▶ **EXAMPLE 6** In Example 1 of Section 3.9 of Chapter 3, we used the differential approximation to calculate three decimal places of $\sqrt{4.1}$. How many decimal places of accuracy are obtained by using a second order Taylor polynomial to approximate $\sqrt{4.1}$?

Solution Let $f(x) = \sqrt{x}$ and $c = 4$. We calculate $f^{(1)}(x) = \frac{1}{2}x^{-1/2}$, $f^{(2)}(x) = -\frac{1}{4}x^{-3/2}$, and $f^{(3)}(x) = \frac{3}{8}x^{-5/2}$. Using the first two of these derivatives, we obtain

$$T_2(4.1) = \frac{f^{(0)}(4)}{0!}(4.1-4)^0 + \frac{f^{(1)}(4)}{1!}(4.1-4)^1 + \frac{f^{(2)}(4)}{2!}(4.1-4)^2 = 2 + \frac{(4.1-4)}{4} - \frac{(4.1-4)^2}{64},$$

or

$$T_2(4.1) = 2 + \frac{(0.1)}{4} - \frac{(0.1)^2}{64} = 2.02484375.$$

Using the last calculated derivative, $f^{(3)}(x) = (3/8)x^{-5/2}$, we have

$$M = \max_{4\le t\le 4.1}\left|f^{(3)}(t)\right| = \max_{4\le t\le 4.1}\left|\frac{3}{8}t^{-5/2}\right| = \frac{3}{8}4^{-5/2} = \frac{3}{256}.$$

Therefore

$$|R_2(4.1)| \le \frac{M}{3!} \cdot (4.1-4)^3 = \frac{3}{256 \cdot 3!} \cdot (0.1)^3 = 1.953125 \times 10^{-6} < 5 \times 10^{-6}.$$

Thus $T_2(4.1) = 2.02484375$ approximates $\sqrt{4.1}$ with an error less that 5×10^{-6}. In other words, our approximation is accurate to five decimal places. ◄

Taylor Series Expansions of the Common Transcendental Functions

Suppose that f is infinitely differentiable on an open interval centered at c. We can then form the Taylor series

$$T(x) = \sum_{n=0}^{\infty} \frac{f^{(n)}(c)}{n!}(x-c)^n \qquad (8.8.10)$$

of f. (Recall that the term *Maclaurin series* is also used when $c = 0$.) Let R denote the radius of convergence of this series and suppose that $R > 0$. From Section 8.7, we know that $T(x)$ is the only power series with base point c that can represent f on the open interval $I = (c-R, c+R)$. Whether $T(x)$ is $f(x)$ or another function is the subject of the next theorem.

THEOREM 3 Suppose that f is infinitely differentiable on an interval containing points c and x. Let $R_N(x)$ be defined by equation *(8.8.1)*. Then

$$T(x) = \lim_{N\to\infty} \sum_{n=0}^{N} \frac{f^{(n)}(c)}{n!}(x-c)^n$$

exists if and only if $\lim_{N\to\infty} R_N(x)$ *exists and*

$$f(x) = \sum_{n=0}^{\infty} \frac{f^{(n)}(c)}{n!}(x-c)^n$$

if and only if

$$\lim_{N\to\infty} R_N(x) = 0.$$

Proof. We let N tend to infinity in equation (8.8.2). The result is

$$f(x) = \lim_{N\to\infty} \left(T_N(x) + R_N(x)\right) = \lim_{N\to\infty} \left(\sum_{n=0}^{N} \frac{f^{(n)}(c)}{n!}(x-c)^n + R_N(x) \right),$$

from which both assertions follow. ■

In Section 8.7, we calculated power series expansions of $\ln(1+x)$, $\arctan(x)$, and a few other functions. Now we will use Theorem 3 to do the same for some other basic functions. Before we attack some concrete examples, we must first compute the limit of a certain sequence that will occur in most of our examples.

THEOREM 4 If b is a fixed positive constant, then

$$\lim_{k\to\infty} \frac{b^k}{k!} = 0. \qquad (8.8.11)$$

Proof. Because b is fixed, we may choose an integer ℓ that exceeds b. Notice that b^k and $k!$ each have k factors. For $k > \ell$, we have

$$\frac{b^k}{k!} = \frac{b}{k} \cdot \underbrace{\left(\frac{b}{(k-1)} \cdot \frac{b}{(k-2)} \cdots \frac{b}{\ell} \right)}_{\text{number less than 1}} \cdot \underbrace{\left(\frac{b}{(\ell-1)} \cdot \frac{b}{(\ell-2)} \cdots \frac{b}{2} \cdot \frac{b}{1} \right)}_{\text{fixed number}}.$$ ■

Because the first factor on the right tends to 0 as k tends to infinity, the proposition follows.

▶ **EXAMPLE 7 (The Power Series Expansion of the Exponential Function)** Show that e^x is represented by its Maclaurin series on the entire real line:

$$e^x = \sum_{n=0}^{\infty} \frac{x^n}{n!} = 1 + x + \frac{1}{2!}x^2 + \frac{1}{3!}x^3 + \frac{1}{4!}x^4 + \cdots \quad (-\infty < x < \infty). \tag{8.8.12}$$

Solution By observing that $\frac{d^n}{dx^n} e^x = e^x$ and $\frac{d^n}{dx^n} e^x|_{x=0} = e^x|_{x=0} = 1$ for every n, we see that the Taylor polynomial $T_N(x)$ with base point 0 is $\sum_{n=0}^{N} x^n/n!$, and, the Taylor series is $\sum_{n=0}^{\infty} x^n/n!$. In Example 3 of Section 8.5, we used the Ratio Test to show that this Taylor series is convergent for every real x. Now fix a value of x. Let J denote the interval between 0 and x. Then,

$$M = \max_{t \in J} \left| \frac{d^{N+1}}{dt^{N+1}} e^t \right| = \max_{t \in J} (e^t) \le e^{|x|}.$$

Theorem 2 tells us that

$$|R_N(x)| \le \frac{M}{(N+1)!} \cdot |x|^{N+1} \le e^{|x|} \cdot \frac{|x|^{N+1}}{(N+1)!}.$$

Because x is fixed, Theorem 4 says that the right hand side of this inequality tends to 0 as $N \to \infty$. We conclude that $\lim_{N\to\infty} R_N(x) = 0$. This is precisely what is needed to show that e^x is equal to its Maclaurin series for every value of x. ◀

INSIGHT Read Example 7 again to identify the main steps. You will notice that we use the Taylor expansion of f to generate a power series that we *hope* converges to f. We check that it *does* converge to f by verifying that the remainder term tends to 0. It is not enough simply to verify that the Taylor series converges for each x. We want to know that the series converges back to the function f that we began with. To check this, we *must* use the error term.

◀ **A LOOK BACK** In Example 8 of Section 8.7, we proved that $y(x) = 2 + 2x + \sum_{n=2}^{\infty} x^n/n!$ is the solution of the initial value problem $dy/dx = y - x$, $y(0) = 2$. We may now simplify the expression for $y(x)$ as follows:

$$y(x) = 2 + 2x + \sum_{n=2}^{\infty} \frac{x^n}{n!} = 2 + 2x + \sum_{n=0}^{\infty} \frac{x^n}{n!} - 1 - x = 1 + x + \left(\sum_{n=0}^{\infty} \frac{x^n}{n!} \right) = 1 + x + e^x.$$

In this explicit form, we can easily verify that $y(x)$ does satisfy the differential equation $dy/dx = y - x$.

▶ **EXAMPLE 8** Express

$$f(x) = xe^{-2x^2}$$

as a Maclaurin series.

Solution Calculating the Maclaurin series directly from formula (8.8.10) is not appealing—the derivatives $f^{(n)}(0)$ become more and more complicated as n increases. Instead, we use the known Maclaurin series (8.8.12). It is convenient to begin with u as the variable:

$$e^u = \sum_{n=0}^{\infty} \frac{u^n}{n!}.$$

Now substitute $u = -2x^2$ to obtain

$$e^{-2x^2} = \sum_{n=0}^{\infty} \frac{(-2x^2)^n}{n!} = \sum_{n=0}^{\infty} \frac{(-2)^n x^{2n}}{n!}.$$

Finally,

$$xe^{-2x^2} = x\sum_{n=0}^{\infty} \frac{(-2)^n x^{2n}}{n!} = \sum_{n=0}^{\infty} \frac{(-2)^n x^{2n+1}}{n!} = x - 2x^3 + 2x^5 - \frac{4}{3}x^7 + \frac{2}{3}x^9 - \cdots.$$ ◀

▶ **EXAMPLE 9** **(The Power Series Expansion of the Sine Function)** Show that $\sin(x)$ is represented by its Maclaurin series on the entire real line:

$$\sin(x) = x - \frac{x^3}{3!} + \frac{x^5}{5!} - \frac{x^7}{7!} + \frac{x^9}{9!} - \frac{x^{11}}{11!} + \cdots \qquad -\infty < x < \infty. \tag{8.8.13}$$

Solution The calculations of Example 2 show that the right side of equation (8.8.13) is indeed the Maclaurin series of $\sin(x)$. We must still show that this Maclaurin series converges to $\sin(x)$. Let x be fixed, let J be the interval with endpoints 0 and x, and let $M = \max_{t\in J}\left|\frac{d^{N+1}}{dt^{N+1}}\sin(t)\right|$. Because $\frac{d^{N+1}}{dt^{N+1}}\sin(t)$ is $\pm\sin(t)$ or $\pm\cos(t)$, we may conclude that $M_{N+1} \le 1$. Theorem 2 tells us that

$$|R_N(x)| \le \frac{M}{(N+1)!}|x|^{N+1} \le \frac{|x|^{N+1}}{(N+1)!}.$$

Using Theorem 4, we conclude that $\lim_{N\to\infty} R_N(x) = 0$, which is exactly what is required for equation (8.8.13) to hold. ◀

INSIGHT Not every property of a function is quickly detected from its power series. However, because odd powers of x appear in the Maclaurin series of $\sin(x)$, we can easily see that $\sin(x)$ is an odd function: $\sin(-x) = -\sin(x)$. Because we know that $\cos(x)$ is an even function, that is, $\cos(-x) = \cos(x)$ for all x, we expect the Maclaurin series of $\cos(x)$ to involve only even powers of x. Indeed, a calculation that is analogous to that of Example 9 shows that

$$\cos(x) = 1 - \frac{x^2}{2!} + \frac{x^4}{4!} - \frac{x^6}{6!} + \frac{x^8}{8!} - \frac{x^{10}}{10!} + \cdots \qquad -\infty < x < \infty. \tag{8.8.14}$$

The Binomial Series

Fix a nonzero real number α. Let us expand the function $f(x) = (1+x)^\alpha$ about the point $c = 0$. We calculate $f^{(1)}(x) = \alpha(1+x)^{\alpha-1}$, $f^{(2)}(x) = \alpha(\alpha-1)(1+x)^{\alpha-2}$, $f^{(3)}(x) = \alpha(\alpha-1)(\alpha-2)(1+x)^{\alpha-3}$, and so on. In general, $f^{(n)}(x) = \alpha(\alpha-1)(\alpha-2)\cdots(\alpha-(n-1))(1+x)^{\alpha-n}$ for $n = 0, 1, 2, \ldots$. Substituting $x = 0$, we obtain

$$f^{(n)}(0) = \alpha(\alpha-1)(\alpha-2)\cdots(\alpha-(n-2))(\alpha-(n-1)).$$

The Maclaurin series of f is therefore

$$\sum_{n=0}^{\infty} \frac{\alpha(\alpha-1)(\alpha-2)\cdots(\alpha-(n-2))(\alpha-(n-1))}{n!} x^n.$$

There is a useful notation that can be used to simplify the unwieldy appearance of this Taylor series. For any real number α and nonnegative integer n, we define the expression $\binom{\alpha}{n}$ by $\binom{\alpha}{0} = 1$ and, for a positive integer n,

$$\binom{\alpha}{n} = \frac{\alpha(\alpha-1)(\alpha-2)\cdots(\alpha-(n-2))(\alpha-(n-1))}{n!}. \tag{8.8.15}$$

Notice that, when α is a positive integer, equation (8.8.15) is just the same as the usual definition of the binomial coefficient $\binom{\alpha}{n}$. Using equation (8.8.15), the Maclaurin series associated to the function $f(x) = (1+x)^\alpha$ can be written as $\sum_{n=0}^{\infty} \binom{\alpha}{n} x^n$. This series is known as *Newton's binomial series*. Its coefficients are known as *binomial coefficients*, even when α is not a positive integer.

A simple application of the Ratio Test shows that, for every value of α, the binomial series converges for $|x| < 1$. For these values of x, the Maclaurin series converges to the function f:

$$(1+x)^\alpha = \sum_{n=0}^{\infty} \binom{\alpha}{n} x^n \qquad -1 < x < 1. \tag{8.8.16}$$

A proof of this assertion for general α and x in $(-1, 1)$ is explored in Exercise 79.

▶ **EXAMPLE 10** Calculate the binomial series for $\alpha = 1/2$.

Solution We have

$$\begin{aligned}
\binom{1/2}{0} &= 1 \\
\binom{1/2}{1} &= \frac{1/2}{1!} = \frac{1}{2} \\
\binom{1/2}{2} &= \frac{(1/2)\cdot(-1/2)}{2!} = -\frac{1}{8} \\
\binom{1/2}{3} &= \frac{(1/2)\cdot(-1/2)\cdot(-3/2)}{3!} = \frac{1}{16} \\
\binom{1/2}{4} &= \frac{(1/2)\cdot(-1/2)\cdot(-3/2)\cdot(-5/2)}{4!} = -\frac{5}{128}. \\
&\cdots
\end{aligned}$$

We conclude that

$$(1+x)^{1/2} = 1 + \frac{1}{2}x - \frac{1}{8}x^2 + \frac{1}{16}x^3 - \frac{5}{128}x^4 + \cdots. \tag{8.8.17}$$ ◀

Using Taylor Series to Approximate

If a convergent Taylor series is an alternating series, then inequality (8.4.1) can be used to estimate the remainder. The next two examples illustrate this point.

▶ **EXAMPLE 11** Approximate $\sqrt{1.02}$ to five decimal places.

Solution We may use the binomial series expansion given by equation (8.8.16) to obtain

$$(1+0.02)^{1/2} = 1 + \frac{0.02}{2} - \frac{(0.02)^2}{8} + \frac{(0.02)^3}{16} - \frac{(0.02)^4}{128} + \cdots.$$

As this is an alternating series, the truncation error is less than the first term that is omitted. The first term that is less than 5×10^{-6}, the maximum allowable error, is $(0.02)^3/16 = 5.0 \times 10^{-7}$. Our approximation is therefore $1 + 0.02/2 - (0.02)^2/8 = 1.00995$. A calculator evaluation yields $\sqrt{1.02} = 1.0099504\ldots$, which confirms our analysis. ◀

▶ **EXAMPLE 12** Calculate $\int_0^{0.3} \cos(u^2)du$ to six decimal places.

Solution By substituting $x = u^2$ in equation (8.8.14), we obtain

$$\cos(u^2) = 1 - \frac{u^4}{2!} + \frac{u^8}{4!} - \frac{u^{12}}{6!} + \frac{u^{16}}{8!} - \frac{u^{20}}{10!} + \cdots.$$

Recall that we may integrate a power series term by term in its interval of convergence. Doing so gives us

$$\int_0^{0.3} \cos(u^2)du = \int_0^{0.3} \left(1 - \frac{u^4}{2!} + \frac{u^8}{4!} - \frac{u^{12}}{6!} + \frac{u^{16}}{8!} - \frac{u^{20}}{10!} + \cdots\right)du$$

$$= \left(u - \frac{u^5}{(2!)\cdot 5} + \frac{u^9}{(4!)\cdot 9} - \frac{u^{13}}{(6!)\cdot 13} + \cdots\right)\Bigg|_0^{0.3}.$$

It follows that

$$\int_0^{0.3} \cos(u^2)du = 0.3 - \frac{(0.3)^5}{(2!)\cdot 5} + \frac{(0.3)^9}{(4!)\cdot 9} - \frac{(0.3)^{13}}{(6!)\cdot 13} + \cdots.$$

By calculating the first few terms of the alternating series on the right side, we find that the third term, $(0.3)^9/(4!\cdot 9)$, or 9.1125×10^{-8}, is the first that is less than 5×10^{-7}. If we omit it and all subsequent terms, then the error that results is less than 9.1125×10^{-8} and certainly less than 5×10^{-7}. Thus the approximation

$$\int_0^{0.3} \cos(u^2)du \approx 0.3 - \frac{(0.3)^5}{(2!)\cdot 5} = 0.299757$$

is correct to six decimal places. Software for numerical integration gives us $\int_0^{0.3} \cos(u^2)\,du = 0.29975709\ldots$, which confirms our estimate. ◀

Using Taylor Polynomials to Evaluate Indeterminate Forms

If a Taylor series of f with base point c is known, then it is often an efficient tool for calculating an indeterminate form involving f at c. This method is often simpler to implement than l'Hôpital's Rule. The next example will illustrate.

▶ **EXAMPLE 13** Evaluate

$$\lim_{x\to 0}\frac{\arctan(x^2)}{1-\cos(x)}.$$

Solution The requested limit is indeterminate of form $\frac{0}{0}$. From (8.7.3b) and (8.8.14), we have

$$\lim_{x\to 0}\frac{\arctan(x^2)}{1-\cos(x)} = \lim_{x\to 0}\frac{(x^2)-\dfrac{1}{3}(x^2)^3+\dfrac{1}{5}(x^2)^5-\cdots}{\dfrac{x^2}{2!}-\dfrac{x^4}{4!}+\dfrac{x^6}{6!}-\cdots} = \lim_{x\to 0}\frac{x^2\left(1-\dfrac{1}{3}x^4+\dfrac{1}{5}x^8-\cdots\right)}{x^2\left(\dfrac{1}{2}-\dfrac{x^2}{4!}+\dfrac{x^4}{6!}-\cdots\right)}$$

$$= \lim_{x\to 0}\frac{1-\dfrac{1}{3}x^4+\dfrac{1}{5}x^8-\cdots}{\dfrac{1}{2}-\dfrac{x^2}{4!}+\dfrac{x^4}{6!}-\cdots} = \frac{1}{1/2} = 2.$$

QUICK QUIZ

1. What is the coefficient of x^{12} in the Maclaurin series of $\sin(x)$?
2. What is the coefficient of x^6 in the Maclaurin series of $\cos(x)$?
3. What is the coefficient of x^n in the Maclaurin series of e^x?
4. What is the order 2 Taylor polynomial with base point 0 of $1/\sqrt{1+x}$?

Answers

1. 0 2. $-1/720$ 3. $1/n!$ 4. $1-x/2+3x^2/8$

EXERCISES

Problems for Practice

▶ In each of Exercises 1–12, approximate $f(x)$ at the indicated value of x by using a Taylor polynomial of given order N and base point c. Use formula (8.8.3) of Theorem 1 to express the error resulting from this approximation in terms of a number s between c and x. ◀

1. $f(x)=e^x$	$x=0.1$	$N=4$	$c=0$
2. $f(x)=e^{2x}$	$x=-0.01$	$N=3$	$c=0$
3. $f(x)=\sin(x)$	$x=0.3$	$N=5$	$c=0$
4. $f(x)=\sin(x)$	$x=1.6$	$N=5$	$c=\pi/2$
5. $f(x)=\cos(x)$	$x=0.5$	$N=2$	$c=\pi/6$
6. $f(x)=\cos(x)$	$x=-0.85$	$N=3$	$c=-\pi/4$
7. $f(x)=\ln(x)$	$x=3$	$N=3$	$c=e$
8. $f(x)=\ln(x)$	$x=1.2$	$N=4$	$c=1$
9. $f(x)=\arctan(x)$	$x=1.5$	$N=2$	$c=\sqrt{3}$
10. $f(x)=1/\sqrt{x}$	$x=3.9$	$N=3$	$c=4$
11. $f(x)=\sqrt{1+x}$	$x=3.2$	$N=2$	$c=3$
12. $f(x)=\sqrt{x}$	$x=24.4$	$N=3$	$c=25$

▶ In each of Exercises 13–22, a function f, an order N, a base point c, and a point x_0 are given. The interval with endpoints c and x_0 is denoted by J. ◀

a. Calculate the approximation $T_N(x_0)$ of $f(x_0)$.

b. By showing that $M=\max_{t\in J}|f^{(N+1)}(t)|$ is the greater of $|f^{(N+1)}(c)|$ and $|f^{(N+1)}(x_0)|$, and by using inequality (8.8.7) of Theorem 2, find an upper bound for the absolute error $|R_N(x_0)|=|f(x_0)-T_N(x_0)|$ that results from the approximation of part a.

13. $f(x)=\sqrt{1+x}$	$N=3$	$J=[c, x_0]=[0, 0.4]$
14. $f(x)=x+\sin(x)$	$N=3$	$J=[x_0, c]=[\pi 6, \pi/4]$
15. $f(x)=\sin(x^2-1)$	$N=2$	$J=[x_0, c]=[0.9, 1]$
16. $f(x)=x+e^x$	$N=3$	$J=[c, x_0]=[0, 0.8]$
17. $f(x)=x^2+e^{1+x}$	$N=2$	$J=[x_0, c]=[-1.2, -1]$
18. $f(x)=x/(1+x)$	$N=2$	$J=[c, x_0]=[0, 0.5]$
19. $f(x)=\ln(x)$	$N=2$	$J=[x_0, c]=[2.5, e]$
20. $f(x)=\ln(1+x^2)$	$N=2$	$J=[c, x_0]=[0, 0.4]$
21. $f(x)=\arctan(x)$	$N=2$	$J=[c, x_0]=[0, 0.4]$
22. $f(x)=1/(1+x^2)$	$N=2$	$J=[c, x_0]=[0, 0.2]$

▶ In each of Exercises 23–34, derive the Maclaurin series of the given function $f(x)$ by using a known Maclaurin series. ◀

23. $f(x)=\sin(2x)$
24. $f(x)=x\cos(x)$
25. $f(x)=x^2\sin(x/2)$
26. $f(x)=x^3+\cos(x^2)$

27. $f(x) = 5\cos(2x) - 4\sin(3x)$
28. $f(x) = e^{1-x}$
29. $f(x) = \exp(-x^2)$
30. $f(x) = x^2/(1-x)$
31. $f(x) = x/(1+x^2)$
32. $f(x) = \ln(1+x^2)$
33. $f(x) = \sqrt{1+x^3}$
34. $f(x) = 1/\sqrt{1-x^2}$

In Exercises 35–42, use formula (8.8.10) with the given base point c to compute the Taylor series of the given function f.

35. $f(x) = \sqrt{x}$ $c = 1$ (See Exercise 91, Section 8.1)
36. $f(x) = \sqrt{x}$ $c = 4$ (See Exercise 91, Section 8.1)
37. $f(x) = xe^x$ $c = 1$
38. $f(x) = 1/x$ $c = 2$
39. $f(x) = \ln(x)$ $c = 1$
40. $f(x) = 2^x$ $c = 2$
41. $f(x) = 3^{-x}$ $c = -3$
42. $f(x) = x\ln(x)$ $c = 1$

In each of Exercises 43–48, write Newton's binomial series for the given expression up to and including the x^3 term.

43. $(1+x)^{3/4}$
44. $(1+x)^4$
45. $1/\sqrt{1+x}$
46. $1/(1+x)$
47. $1/(1+x)^{3/2}$
48. $1/(1+x)^2$

In each of Exercises 49–54, use Taylor series to calculate the given limit.

49. $\lim_{x\to 0} \dfrac{\sin(x) - x}{x^3}$
50. $\lim_{x\to 0} \dfrac{1-\cos(x)}{x^2}$
51. $\lim_{x\to 0} \dfrac{\ln(1+x) - x}{x^2}$
52. $\lim_{x\to 0} \dfrac{\arctan(x) - x}{x^3}$
53. $\lim_{x\to 0} \dfrac{1-e^{2x}}{x}$
54. $\lim_{x\to 0} \dfrac{e^x - e^{-x}}{x}$

Further Theory and Practice

In each of Exercises 55–60, use Taylor series to calculate the given limit.

55. $\lim_{x\to 0} \dfrac{\cos(x) - \exp(-x^2)}{x^2}$
56. $\lim_{x\to 0} \dfrac{e^{3x} - e^{-3x} - 6x}{x^3}$
57. $\lim_{x\to 0} \dfrac{\arctan(x) - \sin(x)}{x(1-\cos(x))}$
58. $\lim_{x\to 0} \dfrac{\sin(x) - x}{(1-\cos(x))\cdot\ln(1+x)}$
59. $\lim_{x\to 0} \dfrac{\cos^2(x) - \exp(x^2)}{x\sin(x)}$
60. $\lim_{x\to 0} \dfrac{\ln(1+2x^2)}{e^{3x} - 3x - \cos(x)}$

61. Use Half Angle Formula (1.6.10) to find the Maclaurin series of $\sin^2(x)$.
62. Use Half Angle Formula (1.6.11) to find the Maclaurin series of $\cos^2(x)$.
63. Use the Maclaurin series found in Exercise 61 to evaluate

$$\lim_{x\to 0} \frac{\sin^2(x) - \sin(x^2)}{x^4}.$$

64. Use the Maclaurin series for $\exp(x)$ to calculate

$$\lim_{x\to 0} \frac{\sinh(3x) - \tanh(3x)}{x(\cosh(2x) - \operatorname{sech}(2x))}.$$

65. Suppose that f'' is continuous on an open interval centered at c. Use Taylor's Theorem to show that

$$\lim_{h\to 0} \frac{f(c+h) - 2f(c) + f(c-h)}{h^2} = f''(c).$$

In each of Exercises 66–71, evaluate the given series (exactly).

66. $\dfrac{\pi}{4} - \dfrac{\pi^3}{4^3\cdot 3!} + \dfrac{\pi^5}{4^5\cdot 5!} - \dfrac{\pi^7}{4^7\cdot 7!} + \cdots$
67. $1 - \dfrac{\pi^2}{3^2\cdot 2!} + \dfrac{\pi^4}{3^4\cdot 4!} - \dfrac{\pi^6}{3^6\cdot 6!} + \dfrac{\pi^8}{3^8\cdot 8!} - \cdots$
68. $1 - \dfrac{\pi^2}{2^2\cdot 3!} + \dfrac{\pi^4}{2^4\cdot 5!} - \dfrac{\pi^6}{2^6\cdot 7!} + \cdots$
69. $1 + \dfrac{1}{1!} + \dfrac{1}{2!} + \dfrac{1}{3!} + \dfrac{1}{4!} + \cdots$
70. $1 - \dfrac{1}{1!} + \dfrac{1}{2!} - \dfrac{1}{3!} + \dfrac{1}{4!} - \cdots$
71. $\dfrac{2}{1!} + \dfrac{4}{2!} + \dfrac{8}{3!} + \dfrac{16}{4!} + \dfrac{32}{5!} + \cdots$

72. Find the Maclaurin series of $\cosh(x)$ and $\sinh(x)$.
73. Use the Maclaurin series of $\cos(x)$, namely $1 + u$ where $u = \sum_{m=1}^{\infty} (-1)^n x^{2n}/(2n)!$, and the Maclaurin series $\sum_{m=1}^{\infty} (-1)^{m+1} u^m/m$ of $\ln(1+u)$, to show that the Maclaurin series of $\ln(\cos(x))$ is $-x^2/2 - x^4/12 - x^6/45 - \ldots$. Use this result to obtain the Maclaurin series of $\tan(x)$ up to and including the x^5 term.
74. Derive the Maclaurin series expansion

$$\ln\left(\frac{1+x}{1-x}\right) = 2x + \frac{2x^3}{3} + \frac{2x^5}{5} + \frac{2x^7}{7} + \cdots,$$

which is valid for $-1 < x < 1$. Use this expansion to estimate $\ln(2)$ to five decimal place accuracy. The series used here is more rapidly convergent than the more familiar series

$$\ln(2) = 1 - \frac{1}{2} + \frac{1}{3} - \frac{1}{4} + \frac{1}{5} - \cdots.$$

75. Show that

$$\binom{-1/2}{n} = \frac{(-1)^n(2n)!}{2^{2n}(n!)^2}.$$

Using this formula, state the Maclaurin series of $(1+u)^{-1/2}$, and derive the Maclaurin series of $\arcsin(x)$.

76. Show that the Taylor series of $f(x) = x^2e^x$ with base point 1 is

$$\sum_{n=0}^{\infty} \frac{(n^2+n+1)}{n!} \cdot e \cdot (x-1)^n.$$

77. Let $f(t) = \tan(t/3) + 2\sin(t/3)$. Snell's Inequality, discovered in 1621, states that $t \le f(t)$ for $0 < t < 3\pi/2$. The approximation $t \approx f(t)$ is remarkably good for $0 < t < \pi/2$. Calculate the Taylor polynomial of degree 5 for f about $c = 0$. Use it to explain the accuracy of the approximation.

78. The inequality

$$\frac{3\sin(t)}{2+\cos(t)} < t \quad (\text{for all } t > 0)$$

was discovered by Nicholas Cusa in 1458. Calculate the Taylor polynomial of degree 5 for $f(t) = 3\sin(t)/(2+\cos(t))$ about $c = 0$. Use it to explain Cusa's inequality for small positive values of t.

79. Fix a nonzero real number α. Define $f(x) = (1+x)^\alpha$ and

$$g(x) = \sum_{n=0}^{\infty} \binom{\alpha}{n} x^n.$$

Our goal is to show that $f(x) = g(x)$ for $x \in (-1, 1)$. We know from the text that the series certainly converges on that interval. Now complete the following steps:

a. Prove that

$$(n+1) \cdot \binom{\alpha}{n+1} + n \cdot \binom{\alpha}{n} = \alpha \cdot \binom{\alpha}{n}.$$

b. Prove that

$$(1+x) \cdot g'(x) = \alpha \cdot g(x), \qquad x \in (-1, 1).$$

c. Prove that

$$(1+x) \cdot f'(x) = \alpha \cdot f(x), \qquad x \in (-1, 1).$$

d. Notice that $f(0) = g(0) = 1$.

e. Solve the equation $(1+x)y' = \alpha \cdot y$ by rewriting it as

$$\frac{y'}{y} = \frac{\alpha}{1+x}.$$

f. Find the unique solution to the equation $(1+x)y' = \alpha \cdot y$ satisfying $y(0) = 1$. Conclude that $f = g$.

80. Suppose that f is twice continuously differentiable in an open interval I that is centered at c. Let x be any fixed point in I. Consider the function

$$\phi(t) = f(x) - f(t) - (x-t)f'(t) - \left(\frac{x-t}{x-c}\right)^2 \left(f(x) - T_1(x)\right)$$

for $c < t < x$. Apply Rolle's Theorem to φ to deduce that there is a s between c and x for which $\varphi^{(1)}(s) = 0$. Conclude that $f(x) = T_1(x) + \frac{f^{(2)}(s)}{2}(x-c)^2$ for some s between c and x.

81. Suppose that f is N times continuously differentiable in an open interval I that is centered at c. Let x be any fixed point in I. Define the function

$$\rho_N(t) = f(x) - \sum_{n=0}^{N} \frac{f^{(j)}(t)}{n!}(x-t)^n$$

for $c < t < x$. Let

$$\varphi(t) = \rho_N(t) - \left(\frac{x-t}{x-c}\right)^{N+1} \rho_N(c).$$

Note that $\varphi(c) = \varphi(x) = 0$. Apply Rolle's Theorem to φ to deduce that there is a s between c and x for which $\varphi^{(1)}(s) = 0$. Deduce Taylor's Theorem from this fact.

Calculator/Computer Exercises

In each of Exercises 82–85, use an alternating Maclaurin series to approximate the given expression to four decimal places.

82. $\cos(0.2)$
83. $\sin(0.3)$
84. $\exp(-0.2)$
85. $\arctan(0.15)$

86. Use a Taylor polynomial with base point $c = e^3$ to approximate $\ln(20)$ to five decimal places. Do not use a calculator to evaluate any value of $\ln(x)$, but you may use a calculator for arithmetic with the number e and its powers.

In each of Exercises 87–90, a function f, a base point c, and a number x_0 are given. In the ty-plane, plot the horizontal line $y = f(x_0) - T_3(x_0)$ and the graph of $y = f^{(4)}(t) \cdot (x_0 - c)^4/4!$ for t between c and x_0. Use your graph to determine a value of s for which equation (8.8.3) holds.

87. $s = f(x) = \sqrt{3+x}$	$c = 1$	$x_0 = 1.4$
88. $f(x) = x2^{-x}$	$c = 2$	$x_0 = 1.7$
89. $f(x) = \sin(x/2) - 3\cos(x)$	$c = \pi$	$x_0 = 3.5$
90. $f(x) = \dfrac{x^2+1}{x+2}$	$c = 3$	$x_0 = 2.6$

In each of Exercises 91–94 a function f, a base point c, and a point x_0 are given. Plot $y = |f^{(3)}(t)|$ for t between c and x_0. Use your plot to estimate the quantity M of Theorem 2. Then use your value of M to obtain an upper bound for the absolute error $|R_2(x_0)| = |f(x_0) - T_2(x_0)|$ that results when $f(x_0)$ is approximated by the order 2 Taylor polynomial with base point c.

91. $f(x) = x\sin(x)$ $\qquad c = \pi/6 \qquad x_0 = 0.6$

92. $f(x) = \ln(2x^2 - 1)$ $c = 1$ $x_0 = 1.3$

93. $f(x) = \sqrt{16 + x^2}$ $c = 3$ $x_0 = 2.4$

94. $f(x) = e^{\sin(x)}$ $c = 0$ $x_0 = 0.4$

▶ In Exercises 95–98, use a Taylor polynomial to calculate the given integral to five decimal places. ◀

95. $\int_0^{1/2} e^{-x2}\, dx$

96. $\int_0^{\pi/4} \frac{\sin(x)}{x}\, dx$

97. $\displaystyle\int_0^{1/3} \frac{1}{1+x^5}\, dx$

98. $\int_0^{1/2} \sqrt{1+x^3}\, dx$

99. Plot

$$f(x) = \frac{\sin(\tan(x)) - \tan(\sin(x))}{\arcsin(\arctan(x)) - \arctan(\arcsin(x))}$$

for $0.01 < x < 0.02$. If your plot did not come out correctly, then the following considerations may help. The numerator of $f(x)$ has Maclaurin series

$$-\frac{x^7}{30} - \frac{29x^9}{756} + \cdots$$

and the denominator has Maclaurin series

$$-\frac{x^7}{30} + \frac{13x^9}{756} + \cdots.$$

Use these two series to plot f in the given interval.

Summary of Key Topics in Chapter 8

Series (Section 8.1)

An infinite sum $\sum_{n=1}^{\infty} a_n$ is called a series. (Sometimes it is convenient to begin a series at an index other than $n = 1$.) The expression $S_N = \sum_{n=1}^{N} a_n$ is called the Nth partial sum of the series. The series is said to converge to a limit ℓ if $S_N \to \ell$ as $N \to \infty$.

Some Special Series (Section 8.1, Section 8.2)

The harmonic series $\sum_{n=1}^{\infty} 1/n$ diverges.

The geometric series $\sum_{n=M}^{\infty} r^n$ with ratio r converges if and only if $|r| < 1$. When $|r| < 1$, the series sums to $r^M/(1-r)$. Here M can be any integer.

The p-series $\sum_{n=1}^{\infty} 1/n^p$ converges if and only if $p > 1$.

Properties of Series (Section 8.1)

If $\sum_{n=1}^{\infty} a_n$ converges to A and $\sum_{n=1}^{\infty} b_n$ converges to B, then

a. $\sum_{n=1}^{\infty} (a_n + b_n) = A + B$, $\sum_{n=1}^{\infty} (a_n - b_n) = A - B$.
b. $\sum_{n=1}^{\infty} \lambda a_n = \lambda A$ for any real constant λ.

The Divergence Test (Section 8.2)

If $\sum_{n=1}^{\infty} a_n$ converges, then $a_n \to 0$. Equivalently, if the summands of a series do not tend to 0 then the series cannot converge.

The Integral Test (Section 8.2)

Let f be a positive, continuous, decreasing function with domain $\{x : 1 \le x < \infty\}$. Then $\int_1^{\infty} f(x)\, dx$ converges if and only if $\sum_{n=1}^{\infty} f(n)$ converges.

The Comparison Tests (Section 8.3)

If $0 \le a_n \le b_n$ and $\sum_{n=1}^{\infty} b_n$ converges, then $\sum_{n=1}^{\infty} a_n$ converges.

If $0 \le b_n \le a_n$ and $\sum_{n=1}^{\infty} b_n$ diverges, then $\sum_{n=1}^{\infty} a_n$ diverges.

Let $\sum_{n=1}^{\infty} a_n$ and $\sum_{n=1}^{\infty} b_n$ be series of positive terms. Suppose that $\lim_{n\to\infty} a_n/b_n$ exists as a (finite) positive number. Then $\sum_{n=1}^{\infty} a_n$ converges if and only if $\sum_{n=1}^{\infty} b_n$ converges.

Alternating Series Test (Section 8.4)

If $a_1 \geq a_2 \geq a_3 \geq \ldots$ and if $\lim_{n\to\infty} a_n = 0$, then $\sum_{n=1}^{\infty}(-1)^{n+1}a_n$ converges.

Series with Positive and Negative Terms (Section 8.4)

A series $\sum_{n=1}^{\infty} a_n$ is said to *converge absolutely* if $\sum_{n=1}^{\infty}|a_n|$ converges.

If a series converges absolutely, then it converges.

A series that is convergent but not absolutely convergent is called *conditionally convergent*.

The Ratio and Root Tests (Section 8.5)

The Ratio Test: Let $\sum_{n=1}^{\infty} a_n$ satisfy $\lim_{n\to\infty}|a_{n+1}/a_n| = \ell$. If $\ell < 1$, then the series converges absolutely. If $\ell > 1$, then the series diverges. If $\ell = 1$, then no conclusion is possible from the Ratio Test.

The Root Test: Let $\sum_{n=1}^{\infty} a_n$ satisfy $\lim_{n\to\infty}|a_n|^{1/n} = \ell$. If $\ell < 1$, then the series converges absolutely. If $\ell > 1$, then the series diverges. If $\ell = 1$, then no conclusion is possible from the Root Test.

Power Series and Radius of Convergence (Section 8.6)

An expression

$$\sum_{n=0}^{\infty} a_n(x-c)^n$$

is called a power series with *base point* c (or *center* c). We set $\ell = \lim_{n\to\infty}|a_n|^{1/n}$, provided that this limit exists as a finite or infinite number. If it is convenient, then the formula $\ell = \lim_{n\to\infty}|a_{n+1}|/|a_n|$ may be used instead. The number $R = 1/\ell$ is called the *radius of convergence* of the power series (with the understanding that $R = 0$ when $\ell = \infty$, and $R = \infty$ when $\ell = 0$). The power series will converge absolutely for $|x-c| < R$ and diverge for $|x-c| > R$. The *interval of convergence* of a power series is the set of points at which the series is convergent. If R is a positive finite number, then the interval of convergence will be $(c-R, c+R)$, $[c-R, c+R)$, $(c-R, c+R]$, or $[c-R, c+R]$. If R is infinite, then the interval of convergence is the entire real line. If $R = 0$, then the interval of convergence is just the point c.

Basic Properties of Power Series (Section 8.6)

Theorem: In the interval of convergence,

$$\sum_{n=0}^{\infty}(a_n+b_n)x^n = \sum_{n=0}^{\infty}a_n + \sum_{n=0}^{\infty}b_n, \quad \sum_{n=0}^{\infty}(a_n-b_n)x^n = \sum_{n=0}^{\infty}a_n - \sum_{n=0}^{\infty}b_n,$$

$$\text{and} \quad \sum_{n=0}^{\infty}\lambda a_n x^n = \lambda\sum_{n=0}^{\infty}a_n x^n \quad \text{for every } \lambda \in \mathbb{R}.$$

Theorem: Power series may be differentiated and integrated term by term in the interior of the interval of convergence.

Theorems 1 and 2 may be used to generate convergent power series expansions for several functions.

Expanding Functions in Power Series (Section 8.7)

The geometric series with ratio x can be regarded as a power series for its sum:

$$\frac{1}{1-x} = \sum_{n=0}^{\infty} x^n = 1 + x + x^2 + x^3 + \cdots \quad \text{for } -1 < x < 1.$$

This basic power series leads to other power series expansions through substitution and integration. Examples include

$$\ln(1+x) = \sum_{n=1}^{\infty} \frac{(-1)^{n-1}}{n} x^n = x - \frac{1}{2}x^2 + \frac{1}{3}x^3 - \frac{1}{4}x^4 + \frac{1}{5}x^5 - \frac{1}{6}x^6 + \frac{1}{7}x^7 - \cdots \quad \text{for } -1 < x < 1$$

and

$$\arctan(x) = \sum_{n=0}^{\infty} \frac{(-1)^n x^{2n+1}}{2n+1} = x - \frac{1}{3}x^3 + \frac{1}{5}x^5 - \frac{1}{7}x^7 + \frac{1}{9}x^9 - \cdots \quad \text{for } -1 < x < 1.$$

The Relationship between the Coefficients and Derivatives of a Power Series (Section 8.7)

If a power series $\sum_{n=0}^{\infty} a_n \,(x-c)^n$ converges on an open interval I centered at c, then the function f defined by $f(x) = \sum_{n=0}^{\infty} a_n \,(x-c)^n$ is infinitely differentiable on I and

$$a_n = \frac{f^{(n)}(c)}{n!}$$

for every natural number n.

Taylor Series and Taylor Polynomials (Section 8.7)

If f is infinitely differentiable on an interval centered about a point c, then the power series

$$T(x) = \sum_{n=0}^{\infty} \frac{f^{(n)}(c)}{n!} \,(x-c)^n$$

can be defined. It is called the *Taylor series* of f with base point c. A Taylor series with base point 0 is also called a *Maclaurin series*. If a power series converges to f, then it must be a Taylor series. The converse is not always true: A Taylor series $T(x)$ of f need not converge to $f(x)$ if x is not the base point. A polynomial of the form

$$T_N(x) = \sum_{n=0}^{N} \frac{f^{(n)}(c)}{n!} \,(x-c)^n$$

is called an order N Taylor polynomial of f with base point c.

Taylor's Theorem with Remainder (Section 8.8)

If f is $N+1$ times continuously differentiable on an interval I centered at c, then, for any $x \in I$, we have

$$f(x) = \underbrace{\sum_{n=0}^{N} \frac{f^{(n)}(c)}{n!} \cdot (x-c)^n}_{T_N(x)} + \underbrace{\frac{f^{(N+1)}(s)}{(N+1)!} \cdot (x-c)^{N+1}}_{R_N(x)}$$

for some s between c and x. We call R_N the *remainder term* or *error term* of order N.

Estimating the Remainder Term (Section 8.8)

The remainder term R_N satisfies the estimate

$$|R_N(x)| \le M \cdot \frac{|x-c|^{N+1}}{(N+1)!}$$

where M is the maximum value of $|f^{(N+1)}(t)|$ for t between c and x. The Taylor series $T(x)$ of f converges to $f(x)$ if and only if $R_N(x) \to 0$ as $N \to \infty$.

Taylor's theorem together with the error estimate may be used to approximate the values of many transcendental functions to any desired degree of accuracy.

Taylor Series Representations of Common Transcendental Functions (Section 8.8)

Many familiar functions have convergent power series representations. For example,

$$\sin(x) = \sum_{n=0}^{\infty} \frac{(-1)^n}{(2n+1)!} x^{2n+1} = x - \frac{1}{3!}x^3 + \frac{1}{5!}x^5 - \frac{1}{7!}x^7 + \cdots, \quad -\infty < x < \infty$$

$$\cos(x) = \sum_{n=0}^{\infty} \frac{(-1)^n}{(2n)!} x^{2n} = 1 - \frac{1}{2!}x^2 + \frac{1}{4!}x^4 - \frac{1}{6!}x^6 + \cdots, \quad -\infty < x < \infty$$

$$e^x = \sum_{n=0}^{\infty} \frac{1}{n!} x^n = 1 + x + \frac{1}{2!}x^2 + \frac{1}{3!}x^3 + \frac{1}{4!}x^4 + \frac{1}{5!}x^5 + \frac{1}{6!}x^6 + \cdots, \quad -\infty < x < \infty.$$

The function $f(x) = (1+x)^\alpha$ has a particularly interesting power series representation called the binomial series:

$$(1+x)^\alpha = \sum_{n=0}^{\infty} \binom{\alpha}{n} x^n \quad \text{for} \quad |x| < 1$$

where

$$\binom{\alpha}{n} = \frac{\alpha \cdot (\alpha-1) \cdots (\alpha-n+1)}{n!}.$$

Review Exercises for Chapter 8

1. Calculate the first five partial sums $S_1, \ldots, S_5$ of the series $\sum_{n=1}^{\infty} \frac{6n}{(n+1)}$.

2. Calculate the first five partial sums $S_1, \ldots, S_5$ of the series $\sum_{n=1}^{\infty} \frac{n-2}{n!}$.

3. Evaluate $\sum_{n=0}^{\infty} (3/5)^n$.

4. Evaluate $\sum_{n=1}^{\infty} 4(2/3)^n$.

5. Evaluate $\sum_{n=2}^{\infty} 3^{n+1}/7^{n-1}$.

6. Evaluate $\sum_{n=2}^{\infty} (-1/4)^{n+1}$.

7. Express $0.6313636363\ldots$ as the ratio of two integers.

8. Express $0.183181818\ldots$ as the ratio of two integers.

In Exercises 9–44, determine if the given series converges conditionally, converges absolutely, or diverges.

9. $\sum_{n=0}^{\infty} (-1)^n \frac{\sqrt{1+\sqrt{n}}}{(1+\sqrt{n})^2}$

10. $\sum_{n=0}^{\infty} (-1)^n \frac{2^n}{\ln(2+n)}$

11. $\sum_{n=1}^{\infty} (-1)^n \frac{1}{1+(1/n)^2}$

12. $\sum_{n=0}^{\infty} (-1)^n \frac{1}{2+\sin(n)}$

13. $\sum_{n=0}^{\infty} \frac{(-1)^n}{\sqrt[3]{n^2+1}}$

14. $\sum_{n=0}^{\infty} (-1)^n \frac{n^2}{2^n}$

15. $\sum_{n=2}^{\infty}(-1)^{n+1}\dfrac{n^2}{\ln^4(n)}$

16. $\sum_{n=0}^{\infty}(-1)^{n}\dfrac{1}{1+\sqrt{n}}$

17. $\sum_{n=0}^{\infty}(-1)^{n}\dfrac{1}{1+\sqrt{n}+n\sqrt{n}}$

18. $\sum_{n=0}^{\infty}(-1)^{n}\dfrac{5^n}{n!}$

19. $\sum_{n=10}^{\infty}(-1)^{n}\dfrac{1}{n\ln^2(n)}$

20. $\sum_{n=0}^{\infty}(-1)^{n}\dfrac{7^n}{2^{3n}}$

21. $\sum_{n=0}^{\infty}(-1)^{n+1}\dfrac{\sqrt{n}}{n-\sqrt{2}}$

22. $\sum_{n=1}^{\infty}(-1)^{n+1}\dfrac{\ln(n)}{n}$

23. $\sum_{n=1}^{\infty}(-1)^{n}\dfrac{n^5}{(1+n^6)^{14/13}}$

24. $\sum_{n=0}^{\infty}(-1)^{n}\dfrac{2^n}{n^2+2^n}$

25. $\sum_{n=2}^{\infty}(-1)^{n+1}\dfrac{\ln^2(n)}{n^{3/2}}$

26. $\sum_{n=1}^{\infty}(-1)^{n+1}\sin^2(1/n)$

27. $\sum_{n=1}^{\infty}(-1)^{n+1}n\sin^2(1/n)$

28. $\sum_{n=1}^{\infty}\dfrac{(-1)^n}{n^n}$

29. $\sum_{n=0}^{\infty}\dfrac{(-1)^n}{n+(1.3)^n}$

30. $\sum_{n=0}^{\infty}(-1)^{n}\dfrac{n^{2/3}}{1+n\sqrt{n}}$

31. $\sum_{n=1}^{\infty}(-1)^{n}\dfrac{\arctan(n)}{n\sqrt{n}}$

32. $\sum_{n=0}^{\infty}(-1)^{n}\dfrac{(1.4)^n}{n^3+(1.2)^n}$

33. $\sum_{n=1}^{\infty}\dfrac{(-1)^n}{n^{1/n}}$

34. $\sum_{n=1}^{\infty}(-1)^{n+1}(\sqrt{n^2+3n}-n)$

35. $\sum_{n=1}^{\infty}(-1)^{n}\left(\dfrac{n}{n+1}\right)^{n^2}$

36. $\sum_{n=1}^{\infty}(-1)^{n}\dfrac{\arccos(1/n)}{n}$

37. $\sum_{n=2}^{\infty}(-1)^{n+1}\dfrac{1}{\ln(n^2)}$

38. $\sum_{n=1}^{\infty}(-1)^{n}\sin\left(\dfrac{1}{n}\right)\csc\left(\dfrac{2}{n}\right)$

39. $\sum_{n=0}^{\infty}(-1)^{n}\left(\dfrac{n+3}{2n+1}\right)^{n}$

40. $\sum_{n=0}^{\infty}(-1)^{n}\dfrac{2\cdot 4\cdot 6\cdots(2n)}{(n!)^2}$

41. $\sum_{n=1}^{\infty}(-1)^{n}\dfrac{n!\cdot 2^{2n-1}}{(2n-1)!}$

42. $\sum_{n=0}^{\infty}(-1)^{n}\dfrac{1\cdot 3\cdot 5\cdots(2n+1)}{2\cdot 5\cdot 8\cdots(3n+2)}$

43. $\sum_{n=0}^{\infty}(-1)^{n+1}\dfrac{(n!)^3}{(3n)!}$

44. $\sum_{n=1}^{\infty}(-1)^{n}\dfrac{n!}{n^n}$

In each of Exercises 45–50, examine the given series and use either the word *conclusive* or the word *inconclusive* to complete each of the following two sentences. (a) *The Divergence Test applied to this series is* __. (b) *The Ratio Test applied to this series is* __.

45. $\sum_{n=1}^{\infty}(-1)^{n}\dfrac{n^2+1}{n^2+n}$

46. $\sum_{n=1}^{\infty}(-1)^{n}\dfrac{n}{n^2+1}$

47. $\sum_{n=1}^{\infty}(-1)^{n}\dfrac{n}{n^3+1}$

48. $\sum_{n=1}^{\infty}(-1)^{n}\dfrac{1}{\ln(n+1)}$

49. $\sum_{n=1}^{\infty}(-1)^{n}\dfrac{5^n}{n^2}$

50. $\sum_{n=1}^{\infty}(-1)^{n}\left(\dfrac{2}{n}\right)^{n}$

51. Estimate $\sum_{n=1}^{\infty}\dfrac{(-1)^{n+1}}{4n^3}$ with an error no greater than 0.002.

52. Estimate $\sum_{n=0}^{\infty}\dfrac{(-2)^n}{(2n)!}$ with an error no greater than 0.0005.

53. Use the Integral Test to determine lower and upper estimates for $\sum_{n=1}^{\infty}\dfrac{1}{n^{3/2}}$.

54. Use the Integral Test to determine lower and upper estimates for $\sum_{n=1}^{\infty}\dfrac{2n+1}{(n^2+n)^2}$.

In Exercises 55–66, determine the interval of convergence of the given power series.

55. $\sum_{n=0}^{\infty} (3x+4)^n$

56. $\sum_{n=1}^{\infty} \frac{(-1)^n}{\sqrt{n}} \left(\frac{x}{3}\right)^n$

57. $\sum_{n=0}^{\infty} \frac{3n}{(n+1)^3} x^n$

58. $\sum_{n=1}^{\infty} (1+1/n)^n x^n$

59. $\sum_{n=0}^{\infty} (-1)^n \frac{x^n}{\sqrt{n^{4/3}+1}}$

60. $\sum_{n=1}^{\infty} (-1)^{n+1} \frac{3^n}{n^3} x^n$

61. $\sum_{n=1}^{\infty} (-1)^{n+1} \frac{(x+1/2)^n}{\sqrt{n}}$

62. $\sum_{n=0}^{\infty} \frac{(2-x)^n}{2n+3}$

63. $\sum_{n=1}^{\infty} \frac{(2x+3)^n}{3^n \cdot n}$

64. $\sum_{n=0}^{\infty} (-1)^n \frac{(x+1)^n}{3n+1}$

65. $\sum_{n=2}^{\infty} \frac{(-1)^n}{3^n \cdot \ln^2(n)} (3x-1)^n$

66. $\sum_{n=0}^{\infty} \frac{2^n+n}{3^n+n} \left(x+\frac{1}{2}\right)^n$

In Exercises 67–72, use the Maclaurin series for $1/(1-x)$ together with algebra to find the Maclaurin series of the given function.

67. $\frac{2}{3+x}$

68. $\frac{1}{4-x^2}$

69. $\frac{27}{81-x^4}$

70. $\frac{x^3}{1-x}$

71. $\frac{x^2}{1+x^2}$

72. $\frac{x}{1+2x}$

In each of Exercises 73–76, approximate $f(x)$ at the indicated value of x by using a Taylor polynomial of order 2 and given base point c. Use formula (8.8.3) of Theorem 1 to express the error resulting from this approximation in terms of a number s between c and x.

73. $f(x) = x\,2^{1-x}$	$x = 5/4$	$c = 1$
74. $f(x) = x\,e^{2x}$	$x = 1/2$	$c = 0$
75. $f(x) = 3 - \cos(x)$	$x = \pi/2$	$c = \pi/3$
76. $f(x) = \sqrt{2}\sin(x)$	$x = \pi/8$	$c = \pi/4$

In each of Exercises 77–80, a function f, a base point c, and a point x_0 are given.

a. Calculate the approximation $T_3(x_0)$ of $f(x_0)$.

b. Use inequality (8.8.7) to find an upper bound for the absolute error $|R_3(x_0)| = |f(x_0) - T_3(x_0)|$ that results from the approximation of part (*a*).

77. $f(x) = \sqrt{4+x}$	$[c, x_0] = [0, 0.41]$
78. $f(x) = 1/x$	$[x_0, c] = [1/4, 1/2]$
79. $f(x) = x + \exp(2x)$	$[c x_0] = [0, 1/2]$
80. $f(x) = \ln(x)$	$[x_0, c] = [e - 1/2, e]$

In Exercises 81–84, use a Taylor polynomial to calculate the given integral with an error less than 10^{-3}.

81. $\int_0^{0.2} \sin(x^2)\, dx$

82. $\int_0^{0.2} \sqrt{1 + \frac{1}{2}x^3}\, dx$

83. $\int_0^{0.2} \frac{1}{1+x^3}\, dx$

84. $\int_0^{0.1} e^{-2x^2}\, dx$

In Exercises 85–87, use power series to calculate the requested limit.

85. $\lim_{x\to 0} \frac{\exp(x) + \exp(-x) - 2}{x^2}$.

86. $\lim_{x\to 0} \frac{2 - 2\cos(x) - x^2}{x^4}$.

87. $\lim_{x\to 0} \frac{x\arctan(x^2)}{x - \sin(x)}$.

88. $\lim_{x\to 0} \frac{\exp(2x) - \exp(-2x)}{\exp(3x) - \exp(-3x)}$.

Infinite Series in Ancient Greece

The concept of infinite series, like many of the mathematical ideas that we study today, originated in ancient Greece. Zeno's Arrow paradox, which was posed in the 5th century B.C.E., arises from decomposing the number 1 into infinitely many summands by means of the geometric series, $1 = 1/2 + 1/4 + 1/8 + 1/16 + \cdots$. Two hundred years later, Archimedes, who worked precisely and confidently with infinite processes, used the formula $1 + 1/4 + 1/16 + 1/64 + 1/256 + \cdots = 4/3$ to calculate the area of a parabolic sector.

Nicole Oresme

With the decline of Greek mathematics, the study of infinite series was put on hold. Progress resumed in the Middle Ages. Nicole Oresme (ca. 1320–1382) was one of the greatest of the medieval philosophers. Trained in theology, Oresme abandoned an academic career to rise through the ranks of the clergy, obtaining a Bishop's chair in 1377. Along the way, he served as financial advisor to France's King Charles V. In response to a request from Charles, Oresme translated several of Aristotle's works from Latin into French. These translations are considered to have had an important influence on the development of the French language.

Oresme's own writings were both extensive and varied. He is commonly held to be the leading economist of the Middle Ages—one can make a case for a similar status in physics. But it is Oresme's work in mathematics that is best known today. It was Oresme who first demonstrated that an infinite series can be divergent even though its terms converge to 0. (The proof of the divergence of the harmonic series that is found in Section 8.1 is that of Oresme.) Oresme also found a clever way to show that $\sum_{n=1}^{\infty} \frac{n}{2^{n-1}} = 4$.

Oresme turned out to be something of an isolated genius in medieval mathematics. He had no immediate followers to develop his theories and extend his mathematical work. Not all of Oresme's work sank into oblivion, but his work on infinite series certainly did.

Mādhavan

The Indian mathematician Mādhavan (1350–1425), a near-contemporary of Oresme, also discovered remarkable facts about infinite series. Mādhavan's work was even more overlooked than that of Oresme and remained largely unknown until its recognition midway through the 20th century. Two examples will suffice to show Mādhavan's depth and originality:

$$\frac{\pi}{\sqrt{12}} = \sum_{n=0}^{\infty} (-1)^n \frac{1}{(2n+1)3^n} = 1 - \frac{1}{3\times 3} + \frac{1}{5\times 3^2} - \frac{1}{7\times 3^3} + \cdots$$

and

$$\frac{\pi}{4} = \sum_{n=0}^{\infty} (-1)^n \frac{1}{(2n+1)} = 1 - \frac{1}{3} + \frac{1}{5} - \frac{1}{7} + \cdots.$$

The simplicity of this second formula, which relates the circular ratio π to the odd integers, is striking. With the methods learned in Chapter 8, we know how to derive such formulas, but historians do not know how Mādhavan himself obtained them—mathematicians who followed Mādhavan recorded only the statements of his main theorems. Nothing of Mādhavan's own mathematical writing has survived.

Infinite Series in the 17th Century

The study of infinite series began to flourish in the 17th century. Pietro Mengoli (1626–1686) found the value of the alternating harmonic series,

$$\ln(2) = \sum_{n=1}^{\infty} (-1)^{n+1}\frac{1}{n} = 1 - \frac{1}{2} + \frac{1}{3} + \frac{1}{4} - \frac{1}{5} + \cdots,$$

as did several other mathematicians. Mengoli also reproved Oresme's forgotten theorem concerning the divergence of the harmonic series. Mādhavan's work did not become known in the West until modern times, but his infinite series for $\pi/4$ was rediscovered in the second half of the 17th century. In a 1676 letter to Newton, Leibniz communicated the formula $\pi/4 = 1 - 1/3 + 1/5 - 1/7 + \ldots$, the right side of which is now called Leibniz's series. Newton, who had previously obtained similar results, responded with an astonishing example of scientific one-upmanship. Providing only the sketchiest of hints, Newton sent Leibniz an analogous but much less transparent series involving π and the odd natural numbers:

$$\frac{\pi}{2\sqrt{2}} = \frac{1}{1} + \frac{1}{3} - \frac{1}{5} - \frac{1}{7} + \frac{1}{9} + \frac{1}{11} - \frac{1}{13} - \frac{1}{15} + \frac{1}{17} + \frac{1}{19} - \frac{1}{21} - \frac{1}{23} + \cdots.$$

Euler

A contemporary of the 18th century mathematician, Leonhard Euler, once remarked that Euler calculated as effortlessly as "men breathe, as eagles sustain themselves in the air." Euler was a master of both convergent and divergent infinite series. One of his earliest successes was the evaluation of a series that eluded his predecessors:

$$\frac{\pi^2}{6} = \sum_{n=1}^{\infty} \frac{1}{n^2} = 1 + \frac{1}{4} + \frac{1}{9} + \frac{1}{16} + \frac{1}{25} + \cdots.$$

In fact, Euler discovered a method to evaluate all p-series $\zeta(p) = \sum_{n=1}^{\infty} n^{-p}$ for which p is a positive even integer. More precisely, Euler was able to determine rational numbers q_k such that $\zeta(2k) = q_k \pi^{2k}$ for each positive integer k. For example, $q_1 = 1/6$, $\zeta(2) = \pi^2/6$, $q_2 = 1/90$, $\zeta(4) = \pi^4/90$, and $q_3 = 1/945$, $\zeta(6) = \pi^6/945$. The numerators of q_1 through q_5 are all 1, but that pattern ends with q_6. Euler calculated explicitly as far as $q_{13} = 1315862/11094481976030578125$.

The function $\zeta(p)$ is called the *zeta function*, after the Greek letter by which it is usually denoted. Given our knowledge of $\zeta(2)$, $\zeta(4)$, $\zeta(6)$, and so on, it is natural to wonder about $\zeta(3)$. To this day, the p-series $\sum_{n=1}^{\infty} 1/n^{2k+1}$ with odd exponents remain mysterious. In 1978, Roger Apéry (1916–1994) created a sensation when he proved that $\zeta(3)$ is irrational. Despite a great deal of effort since then, we do not know the exact value of $\zeta(3)$ or of any other odd power p-series.

An Infinite Series of Ramanujan

The formula of Mādhavan and Leibniz can be used to calculate the digits of π, but it is a very inefficient procedure. Sum 1,000 terms of Leibniz's series, and you do not even get the third decimal place of π for your effort. The grade school approximation 22/7 is very nearly as accurate. Compare that with the formula

$$\frac{1}{\pi} = \frac{\sqrt{8}}{9801} \sum_{n=0}^{\infty} \frac{(4n)!(1103 + 26390n)}{(n!)^4 (396)^{4n}}$$

that the Indian mathematician, Srinivasa Ramanujan (1887–1920), found early in the 20th century. Sum just the first two terms of this series, and you obtain 0.3183098861837906, every digit of which agrees with that of $1/\pi$. Ever since Ramanujan's untimely death, mathematicians have been at work analyzing and proving the mysterious equations that filled his notebooks.

The Early History of Power Series

Before the general theory of Taylor series was developed, many particular power series expansions were discovered and rediscovered. For example, in 1624, Henry Briggs (1561–1631) derived the power series formula

$$(1 + x)^{1/2} = 1 + \frac{1}{2}x - \frac{1}{8}x^2 + \frac{1}{16}x^3 - \frac{5}{128}x^4 + \cdots$$

in the course of calculating logarithms. This formula of Briggs is a special case of the binomial series. The Maclaurin series of $\ln(1 + x)$

$$\ln(1 + x) = x - \frac{1}{2}x^2 + \frac{1}{3}x^3 - \frac{1}{4}x^4 + \frac{1}{5}x^5 - \frac{1}{6}x^6 + \frac{1}{7}x^7 - \cdots,$$

was independently discovered by Johann Hudde (1628–1704) in 1656, by Sir Isaac Newton in 1665, and by Nicolaus Mercator (c. 1619–1687) in 1668.

James Gregory and Sir Isaac Newton

Newton derived the binomial series as well as several other power series in 1665. These series played an important role in his first approach to calculus. Early in the 1670s, Newton began to correspond with the Scottish mathematician, James Gregory. By 1671, Gregory had obtained the Maclaurin series for $\arcsin(x)$, $\arctan(x)$, $\tan(x)$, and $\sec(x)$. Until then, the series that had been discovered—Newton's binomial series and the power series for the logarithm, the arcsine, and the arctangent, in particular—had all been derived by special techniques. An analysis of Gregory's private papers, however, reveal that Gregory hit upon the idea of Taylor series in the early 1670s and used the construction to derive the power series expansions of $\tan(x)$ and $\sec(x)$. It is likely that Gregory, thinking that he had rediscovered a method already known to Newton, withheld publication of his discovery of Taylor series in order to cede the privilege of first publication to Newton. In fact, Newton did not discover Taylor series until 1691 and that discovery, hidden in his private papers, did not come to light until 1976!

Brook Taylor (1685–1731)

Brook Taylor was born into a well-to-do family with connections to the nobility. Although he studied law at Cambridge, his scientific career was already under way by the time he graduated. Encouraged by contacts at the Royal Society to develop and communicate his discoveries, Taylor privately disclosed the series that now bears his name in a letter of 1712. He first publicly communicated his work on series in his book, *The Method of Increments*, which appeared in 1715. At the time, the unpublished work of Gregory and Newton remained unknown. Johann Bernoulli and Abraham De Moivre had published anticipations of Taylor series in 1694 and 1708, but they did not deny Taylor's claim to priority. Moreover, Taylor's immediate successors, including Maclaurin and Euler, credited him not only for introducing Taylor series, but also for demonstrating the uses to which they can be put. The term *Taylor series* has been traced to 1786, after which it quickly became established. By the mid-1800s, one writer commented, "A single analytical formula in the *Method of Increments* has conferred a celebrity on its author, which the most voluminous works have not often been able to bestow." It is ironic, then, that Taylor series were known to Gregory fourteen years before Taylor was born and the error term, which is a crucial component of Taylor's Theorem in practice, was introduced by Lagrange several years after Taylor's death.

Colin Maclaurin (1698–1746)

Colin Maclaurin was born a minister's son. Orphaned at the age of nine, he became the ward of an uncle, who was also a minister. The precocious boy entered the University of Glasgow just two years later. The plan was that he too would become a minister. At the age of twelve, however, Maclaurin discovered the *Elements* of Euclid and was led astray. Within a few days he mastered the first six books of Euclid. His interest in mathematics did not wane thereafter. Three years later, at the age of fifteen, he defended his thesis on *Gravity* and graduated from university.

The young Maclaurin had to bide his time for a few years, but he was still only nineteen when, in 1717, he was appointed Professor of Mathematics at the University of Aberdeen. In the next two years, he forged a warm friendship with Sir Isaac Newton, a bond that Maclaurin described as his "greatest honour and happiness." On Newton's recommendation, the University of Edinburgh offered Maclaurin a professorship in 1725. It came in the nick of time—Aberdeen was about to dismiss Maclaurin for being absent without leave for nearly three years. His teaching at Edinburgh had happier results. One student wrote that "He made mathematics a fashionable study."

Maclaurin's research also received the highest praise. Lagrange described Maclaurin's study of tides as "a masterpiece of geometry, comparable to the most beautiful and ingenious results that Archimedes left to us." Maclaurin's *Treatise of Fluxions*, which appeared in 1742, is considered his major work. He conceived it as an answer to the criticisms of Bishop Berkeley that have been discussed in *Genesis and Development 2 and 3*. Although *Maclaurin series* are to be found in the *Treatise of Fluxions*, Maclaurin himself attributed them to Taylor. Thus the terminology *Maclaurin series* is not historically justified. Nonetheless, it is convenient—it saves us from having to say "Taylor series expanded about 0." Furthermore, it is appropriate to remember this great mathematician in some way. Ideas such as the Integral Test that Maclaurin did introduce in his *Treatise* do not bear his name.

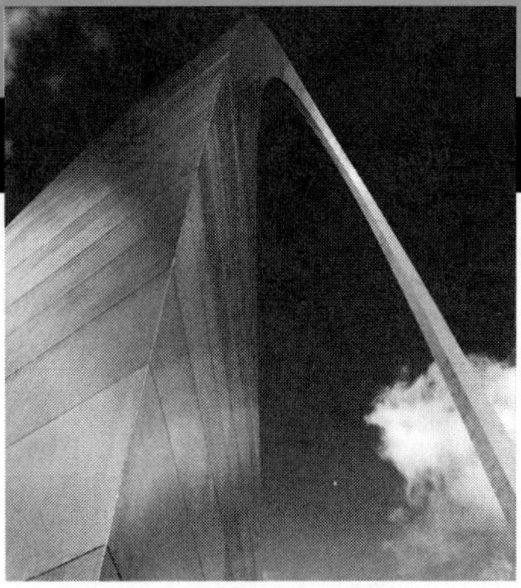

CHAPTER 9

Vectors

PREVIEW

In this chapter, we lay the geometric foundation for all of our coming work with functions of two or more variables. Those studies will require us to have a mathematical model of three-dimensional space. Just as we use a real number as the mathematical model of a point on a line, and an ordered pair of real numbers as the mathematical model of a point in the plane, so we use an ordered triple (x, y, z) as the mathematical model of a point in three-dimensional space. Equations among these space variables describe curves and surfaces in space. In this chapter, we will study the simplest such objects, lines and planes, in detail. As an aid to our investigations, we introduce a new concept, the *vector*, that will be a vital tool in every chapter to follow.

A vector may be thought of as an arrow that has a length and a direction, but whose initial point is of no importance. If we reposition the arrow while preserving its direction, then it still represents the same vector. We will learn how to add vectors and how to perform a number of other algebraic operations with them. The interplay between the geometry and algebra of vectors is the perfect device for understanding the relationship between the geometry of lines and planes and their Cartesian equations.

The vector construction is ideal for capturing many concepts in both algebra and geometry. But the use of vectors extends to other sciences as well. Many fundamental quantities in physics are understood by means of their direction and magnitude. Force, velocity, and acceleration are examples. Using the length of a vector to represent magnitude, we can model these physical quantities by means of vectors.

9.1 Vectors in the Plane

For many purposes in calculus and physics, we need a concept that simultaneously contains the notions of direction and magnitude. For instance, force, velocity, and acceleration are all quantities that have both direction and magnitude. In this section, we will develop a mathematical tool, the *vector*, for handling these concepts.

DEFINITION A line segment $\overline{AB}$ between two points A and B is said to be a *directed line segment* if one endpoint is considered to be the initial point of the line segment and the other endpoint the terminal point. We denote the directed line segment with initial point A and terminal point B by $\overrightarrow{AB}$. The same pair of points determines a second directed line segment, $\overrightarrow{BA}$, which is said to be *opposite* in direction to $\overrightarrow{AB}$ (see Figure 1).

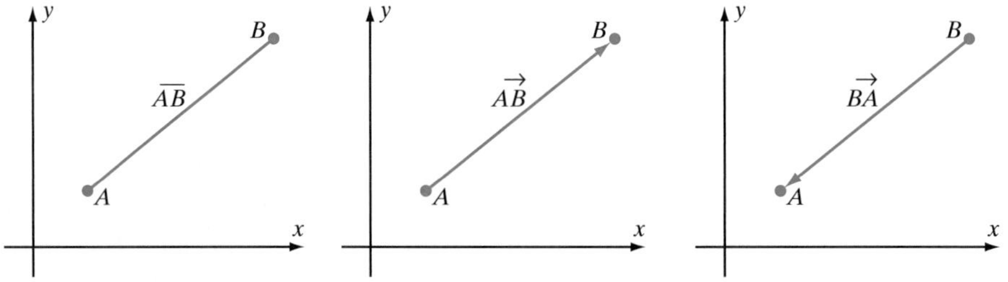

▲ Figure 1 Line segment $\overline{AB}$, directed line segment $\overrightarrow{AB}$, and directed line segment $\overrightarrow{BA}$

The directed line segment $\overrightarrow{AB}$ may be thought of as a straight path from A to B. Its direction is indicated by an arrow, as in Figure 1. We often find it useful to parameterize $\overrightarrow{AB}$ so that its direction is "respected." In such a parameterization, the initial point of the interval of paramaterization corresponds to A, and the endpoint of the interval of parameterization corresponds to B. Example 1 demonstrates this technique.

▶ **EXAMPLE 1** Suppose that $A = (5, 6)$ and $B = (9, 14)$. Find a parameterization $x = f(t), y = g(t), 0 \le t \le 1$ of $\overrightarrow{AB}$ that respects its direction.

Solution We are to find functions f and g so that the point $(f(t), g(t))$ traverses the line segment from A to B as t increases from 0 to 1. From the initial point A to the terminal point B, the x-displacement is $9 - 5$, or 4. The y-displacement is $14 - 6$, or 8. We use these displacements together with the coordinates of A to define $f(t) = 5 + 4t$ and $g(t) = 6 + 8t$. Then, at the endpoints, we have $(f(0), g(0)) = (5, 6) = A$ and $(f(1), g(1)) = (5 + 4, 6 + 8) = B$. Thus $x = f(t)$, $y = g(t)$, $0 \le t \le 1$ parameterizes a curve that begins at A and ends at B. To see that this parameterized curve is a line segment, we use the equation $x = f(t) = 5 + 4t$ to obtain $t = (x - 5)/4$. It follows that $y = g(t) = 6 + 8t = 6 + 8(x - 5)/4 = 6 + 2(x - 5)$, or $y = 2x - 4$, which is the equation of a line. Thus the equations $x = 5 + 4t$, $y = 6 + 8t$, $0 \le t \le 1$ parameterize the directed line segment $\overrightarrow{AB}$. ◀

INSIGHT The calculation of Example 1 may be carried out with any two points $A = (x_0, y_0)$ and $B = (x_1, y_1)$. The result is that the equations

$$x = x_0 + t(x_1 - x_0),\ y = y_0 + t(y_1 - y_0)\quad (0 \le t \le 1) \tag{9.1.1}$$

parameterize the directed line segment $\overrightarrow{AB}$.

A directed line segment is determined by its initial point, direction, and length. Given a directed line segment $\overrightarrow{AB}$, it is important to be able to construct directed line segments with the same length and direction as $\overrightarrow{AB}$ but with other initial points. The next example shows how this is done.

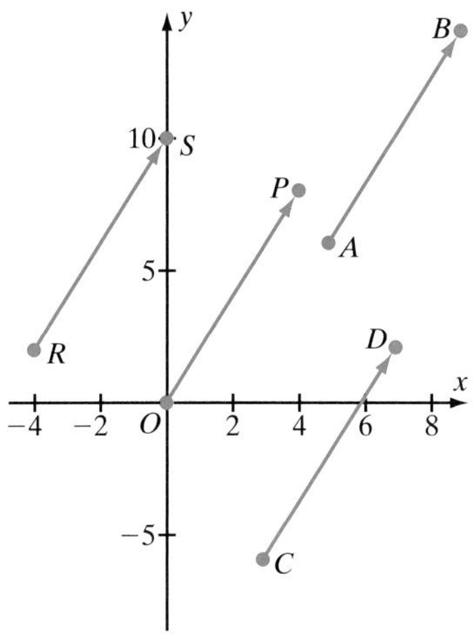

Figure 2 Directed line segments with the same length and direction as $\overrightarrow{AB}$.

▶ **EXAMPLE 2** Let $A = (5, 6)$ and $B = (9, 14)$ as in Example 1. Let O denote the origin $(0, 0)$. If $C = (3, -6)$ and $S = (0, 10)$, then find points D, P, and R so that the directed line segments, $\overrightarrow{OP}$, $\overrightarrow{CD}$ and $\overrightarrow{RS}$ have the same length and direction as $\overrightarrow{AB}$.

Solution The length and direction of $\overrightarrow{AB}$ are both determined by its x-displacement $9 - 5 = 4$ and y-displacement $14 - 6 = 8$. If we add these displacements to the coordinates of the initial points O and C, then we obtain the terminal points $P = (0 + 4, 0 + 8) = (4, 8)$ and $D = (3 + 4, -6 + 8) = (7, 2)$. If we subtract these displacements from the coordinates of the terminal point S, then we obtain the initial point $R = (0 - 4, 10 - 8) = (-4, 2)$. In this way, $\overrightarrow{OP}$, $\overrightarrow{CD}$, and $\overrightarrow{RS}$ have the same x- and y-increments as $\overrightarrow{AB}$ and therefore the same length and direction (see Figure 2.) ◀

Vectors

Imagine that you are exerting a force of constant magnitude to push a stalled automobile (Figure 3). If the street is straight, then the direction of the force **F** that you exert is also constant. An arrow may be used to represent the direction of the force **F**, as in Figure 3. We can use the length of the arrow to represent the magnitude of **F**. The initial point of the arrow can be used to represent the point of application of **F**. As the car moves, the position of the arrow changes, but the direction and length do not. We can therefore regard the force **F** as a directed line segment that (a) has a fixed direction, (b) has a fixed length, and (c) can be applied at any point. These considerations suggest creating a mathematical object that captures the direction and length of a directed line segment but which disregards its initial point. Such an object is called a *vector.*

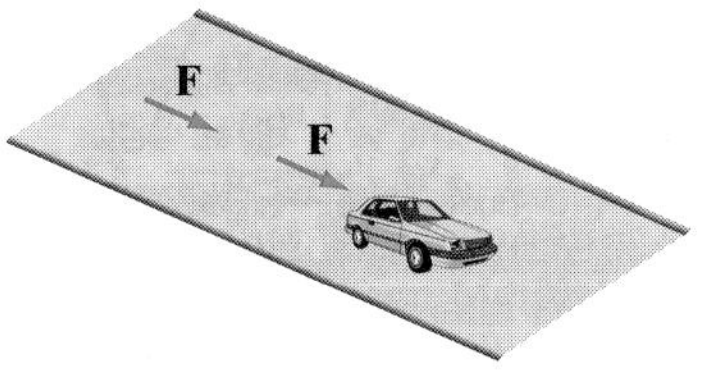

Figure 3

DEFINITION Let A and B be points in the plane. The collection of all directed line segments having the same length and direction as $\overrightarrow{AB}$ is said to be a *vector.* Every directed line segment in this collection is said to *represent* the vector.

To help distinguish between vectors and ordinary numbers, we often refer to real numbers as *scalars.* In textbooks, boldface type such as **v** is frequently used to denote a vector. When rendered by hand, the vector **v** is often denoted by $\vec{v}$. Informally, we can think of a vector as a "floating" arrow whose initial point can be chosen in any convenient way. Figure 2 illustrates four representations of a *single* vector. In practice, we often refer to a vector by one of the directed line segments that represent it. This can lead to no confusion and is often very helpful. Thus we may refer to the vector that appears in Figure 2 as vector $\overrightarrow{AB}$, vector $\overrightarrow{PQ}$, vector $\overrightarrow{ON}$, or vector $\overrightarrow{RS}$. Even though these four directed line segments are all different, they all represent the same vector.

When a particular directed line segment $\overrightarrow{P_0P_1}$ is used to represent a vector $\mathbf{v}$ (as in Figure 4), the x-displacement $v_1 = x_1 - x_0$ and the y-displacement $v_2 = y_1 - y_0$ do not depend on the choice of initial point P_0. We can therefore use the notation $\langle v_1, v_2\rangle$ to unambiguously denote $\mathbf{v}$. Thus $\mathbf{v} = \langle v_1, v_2\rangle$. The quantities v_1 and v_2 are said to be the *components* or *entries* of $\langle v_1, v_2\rangle$. Notice that, if $P = (x, y)$ and $O = (0,0)$, then vector $\overrightarrow{OP}$ is equal to $\langle x - 0, y - 0\rangle$ or $\langle x, y\rangle$. Thus every point $P = (x, y)$ gives rise to the vector $\langle x, y\rangle$ that is represented by the directed line segment from the origin to P (Figure 5). The vector $\overrightarrow{OP} = \langle x, y\rangle$ is sometimes called the *position vector* of $P = (x, y)$.

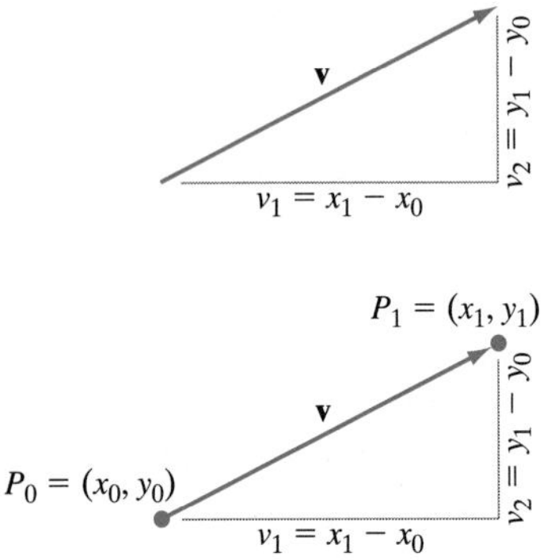

▲ Figure 4 The components v_1 and v_2 of vector $\mathbf{v}$ are the same in each directed line segment representation.

▶ **EXAMPLE 3** Let $A = (5,6)$ and $B = (9,14)$ as in Examples 1 and 2. Let $\mathbf{v}$ be the vector that is represented by $\overrightarrow{AB}$. Write $\mathbf{v}$ in terms of its components. If $\mathbf{v}$ is represented by the directed line segment $\overrightarrow{OP}$ with the origin O as initial point, then what is the terminal point P?

Solution If $\mathbf{v} = \langle a, b\rangle$, then the components a and b of $\mathbf{v}$ are given by $a = 9 - 5 = 4$ and $b = 14 - 6 = 8$. Thus $\mathbf{v} = \langle 4,8\rangle$. The *point* $P = (4,8)$ is the terminal point of the directed line segment that begins at the origin O and that represents $\mathbf{v} = \langle 4,8\rangle$. Refer again to Figure 2. ◀

Vector Algebra

In this subsection, we introduce some algebraic operations that can be performed with vectors.

DEFINITION The *sum* $\mathbf{v} + \mathbf{w}$ of two vectors $\mathbf{v} = \langle v_1, v_2\rangle$ and $\mathbf{w} = \langle w_1, w_2\rangle$ is formed by adding the vectors componentwise:

$$\mathbf{v} + \mathbf{w} = \langle v_1, v_2\rangle + \langle w_1, w_2\rangle = \langle v_1 + w_1, v_2 + w_2\rangle.$$

We can interpret vector addition geometrically by first drawing a directed line segment $\overrightarrow{AB}$ that represents $\mathbf{v}$ and then drawing a directed line segment $\overrightarrow{BC}$ that represents $\mathbf{w}$. Notice (as shown in Figure 6) that the initial point of the second directed line segment is the terminal point of the first. The sum $\mathbf{v} + \mathbf{w}$ is then equal to vector $\overrightarrow{AC}$. In other words, $\overrightarrow{AB} + \overrightarrow{BC} = \overrightarrow{AC}$. Figure 7 shows that the sum $\mathbf{v} + \mathbf{w}$ is the diagonal of the parallelogram determined by $\mathbf{v}$ and $\mathbf{w}$. It follows that $\mathbf{v} + \mathbf{w} = \mathbf{w} + \mathbf{v}$, as is also evident from the algebraic definition.

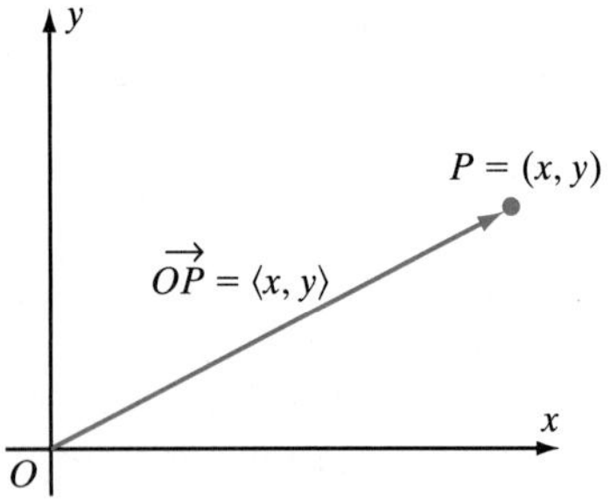

▲ Figure 5 The position vector of point $P = (x, y)$

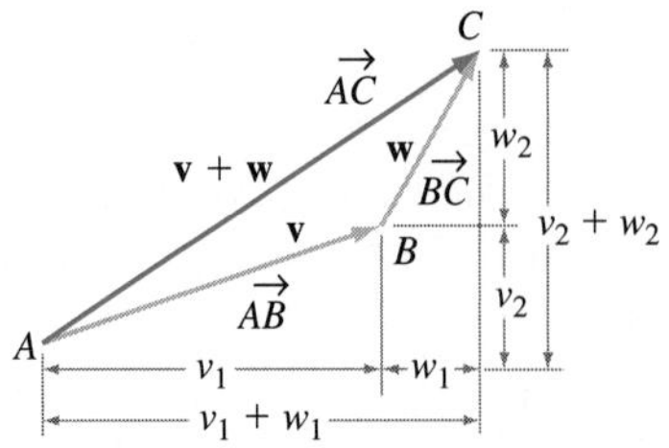

▲ Figure 6

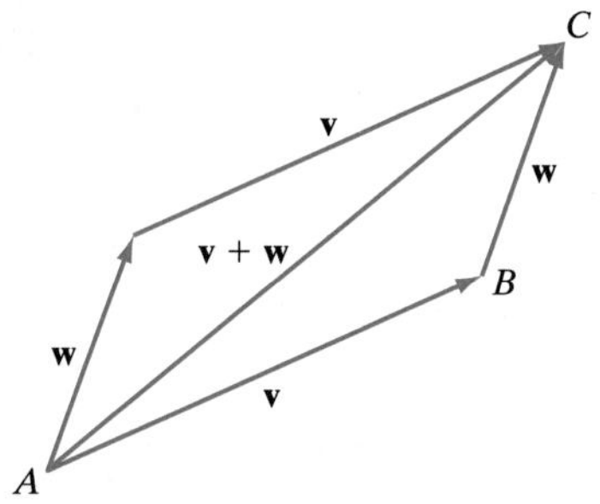

▲ Figure 7

▶ **EXAMPLE 4** Add the vectors $\mathbf{v} = \langle -3, 9\rangle$ and $\mathbf{w} = \langle 1, 8\rangle$.

Solution We have $\mathbf{v} + \mathbf{w} = \langle -3+1,\ 9+8\rangle = \langle -2,\ 17\rangle$. ◀

Next we define the vector analogue of the scalar 0.

DEFINITION The *zero vector* $\mathbf{0}$ is the vector both of whose components are 0. That is, $\mathbf{0} = \langle 0, 0\rangle$.

The zero vector is the identity for vector addition. This means that, if $\mathbf{v}$ is any vector, then $\mathbf{v} + \mathbf{0} = \mathbf{0} + \mathbf{v} = \mathbf{v}$.

DEFINITION If $\mathbf{v} = \langle v_1, v_2\rangle$ is a vector, and λ is a real number, then we define the *scalar multiplication* of $\mathbf{v}$ by λ to be

$$\lambda\mathbf{v} = \langle \lambda v_1, \lambda v_2\rangle.$$

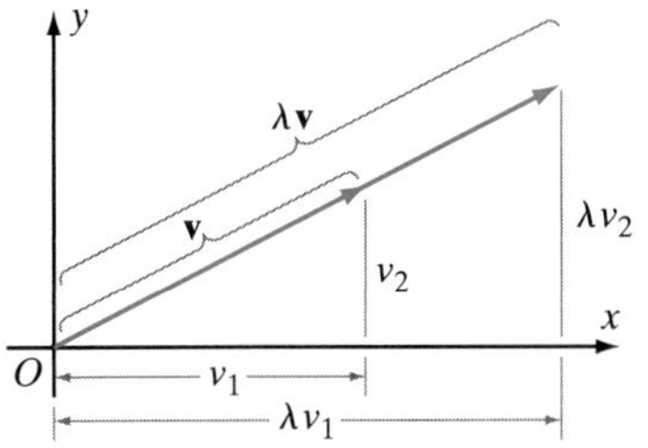

▲ Figure 8a $\lambda > 0$

Geometrically, we think of scalar multiplication as producing a vector that is "parallel" to $\mathbf{v}$. One may use the idea of similar triangles to check this idea, as Figure 8 suggests. If $\lambda > 0$, then $\lambda\mathbf{v}$ and $\mathbf{v}$ have the same direction. If $\lambda < 0$, then $\lambda\mathbf{v}$ and $\mathbf{v}$ have opposite directions. These intuitive notions will be made more precise when a formal definition of *direction* is given later.

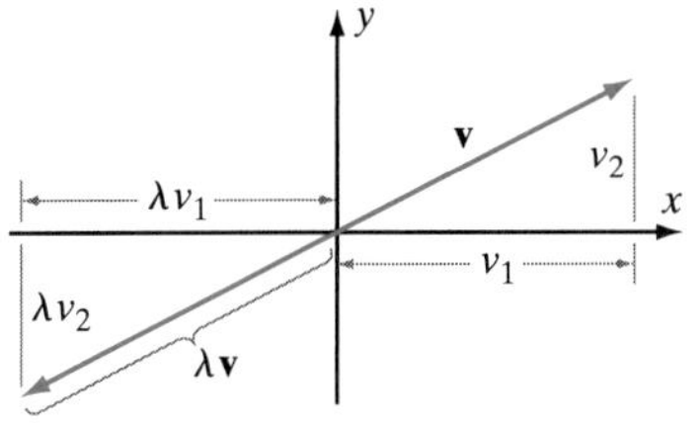

▲ Figure 8b $\lambda < 0$

▶ **EXAMPLE 5** If $\mathbf{v} = \langle 2, -1\rangle$, then calculate $2\mathbf{v}$ and $-3\mathbf{v}$.

Solution We have $2\mathbf{v} = \langle 2\cdot 2,\ 2\cdot(-1)\rangle = \langle 4, -2\rangle$ and $-3\mathbf{v} = \langle (-3)\cdot 2, (-3)\cdot(-1)\rangle = \langle -6,\ 3\rangle$. ◀

Vector addition and scalar multiplication can be used together to define *vector subtraction.* If $\mathbf{v}$ and $\mathbf{w}$ are given vectors, then the expression $\mathbf{v} - \mathbf{w}$ is interpreted to mean $\mathbf{v} + ((-1)\mathbf{w})$. See Figure 9, which features the parallelogram that is generated by $\mathbf{v}$ and $\mathbf{w}$. This same parallelogram has been used in Figure 7 to visualize the sum $\mathbf{v} + \mathbf{w}$, which is represented by the directed diagonal $\overrightarrow{AC}$. In Figure 9, notice that $\mathbf{v} - \mathbf{w}$ is represented by the *other* directed diagonal, $\overrightarrow{DB}$, of the parallelogram. To understand why, focus on the three sides of the upper-right triangle, which show that $\overrightarrow{DB} = \overrightarrow{DC} + \overrightarrow{CB} = \mathbf{v} + (-\mathbf{w}) = \mathbf{v} - \mathbf{w}$. Furthermore, by concentrating on the three sides of the lower-left triangle of Figure 9, we see that $\mathbf{v} - \mathbf{w}$ is the vector that we add to $\mathbf{w}$ to obtain $\mathbf{v}$.

A simple calculation shows that subtraction of vectors is performed componentwise, just like addition of vectors. Thus if $\mathbf{v} = \langle v_1, v_2\rangle$ and $\mathbf{w} = \langle w_1, w_2\rangle$, then

$$\mathbf{v} - \mathbf{w} = \langle v_1, v_2\rangle + (-1)\langle w_1, w_2\rangle = \langle v_1, v_2\rangle + \langle -w_1, -w_2\rangle = \langle v_1 - w_1, v_2 - w_2\rangle.$$

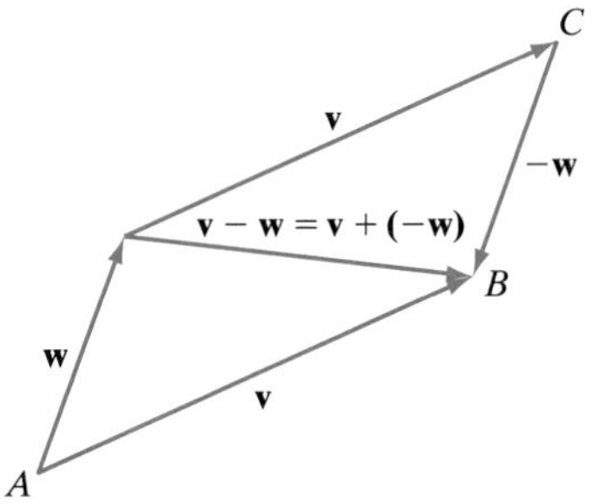

▲ Figure 9

▶ **EXAMPLE 6** Calculate the difference $\mathbf{v} - \mathbf{w}$ of the vectors $\mathbf{v} = \langle -6, 12\rangle$ and $\mathbf{w} = \langle 19, -7\rangle$.

Solution We have $\mathbf{v} - \mathbf{w} = \langle -6 - 19, 12 - (-7)\rangle = \langle -25,\ 19\rangle$. ◀

The Length (or Magnitude) of a Vector

If $\mathbf{v} = \langle v_1, v_2 \rangle$ is a vector, then its *length* is defined to be

$$\|\mathbf{v}\| = \sqrt{(v_1)^2 + (v_2)^2}.$$

This quantity is also called the *magnitude* of $\mathbf{v}$. If $\overrightarrow{AB}$ is a directed line segment that represents $\mathbf{v}$, then $\|\mathbf{v}\|$ is just the distance between the points A and B. The length of a nonzero vector is positive. The zero vector $\mathbf{0}$ is the only vector that has length equal to 0.

▶ **EXAMPLE 7** Let $O = (0,0)$ denote the origin. Calculate $\|\overrightarrow{PQ}\|$, $\|\overrightarrow{OP}\|$, and $\|\overrightarrow{OQ}\|$, for $P = (-9,6)$ and $Q = (2,1)$.

Solution Because $\overrightarrow{PQ} = \langle 2-(-9),\ 1-6\rangle = \langle 11, -5\rangle$, $\overrightarrow{OP} = \langle -9,\ 6\rangle$, and $\overrightarrow{OQ} = \langle 2,1\rangle$, we have $\|\overrightarrow{PQ}\|, = \sqrt{11^2 + (-5)^2} = \sqrt{146}$, $\|\overrightarrow{OP}\| = \sqrt{(-9)^2 + 6^2} = \sqrt{117}$, and $\|\overrightarrow{OQ}\| = \sqrt{2^2 + 1^2} = \sqrt{5}$. ◀

THEOREM 1 If $\mathbf{v}$ is a vector and λ is a scalar, then

$$\|\lambda\mathbf{v}\| = |\lambda|\,\|\mathbf{v}\|. \tag{9.1.2}$$

In words: The length of $\lambda\mathbf{v}$ is the absolute value of λ times the length of $\mathbf{v}$.

Proof. Equation (9.1.2) is easily derived from the definitions of length and scalar multiplication:

$$\|\lambda\mathbf{v}\| = \|\langle\lambda v_1, \lambda v_2\rangle\| = \sqrt{(\lambda v_1)^2 + (\lambda v_2)^2} = \sqrt{\lambda^2}\sqrt{(v_1)^2 + (v_2)^2} = |\lambda|\,\|\mathbf{v}\|. \quad ■$$

▶ **EXAMPLE 8** Verify equation (9.1.2) for $\mathbf{v} = \langle 4, 12\rangle$ and $\lambda = -5$.

Solution We calculate

$$|\lambda|\,\|\mathbf{v}\| = |-5|\sqrt{4^2 + 12^2} = 5\sqrt{160} = 20\sqrt{10}$$

and

$$\|\lambda\mathbf{v}\| = \|(-5)\langle 4,12\rangle\| = \|\langle -20, -60\rangle\| = \sqrt{(-20)^2 + (-60)^2} = 20\sqrt{(-1)^2 + (-3)^2} = 20\sqrt{10}. \quad ◀$$

Unit Vectors and Directions

If the length of $\mathbf{u} = \langle u_1, u_2\rangle$ is 1, that is, if $\|\mathbf{u}\| = 1$, then $\mathbf{u}$ is called a *unit vector*. Observe that $\mathbf{u} = \langle u_1, u_2\rangle$ is a unit vector if and only if the point (u_1, u_2) lies on the unit circle (Figure 10). A unit vector $\mathbf{u}$ therefore has the form

$$\mathbf{u} = \langle\cos(\alpha), \sin(\alpha)\rangle \tag{9.1.3}$$

for some angle α in $[0, 2\pi)$. Sometimes we refer to a unit vector $\mathbf{u}$ as a *direction vector* or simply a *direction*. For any nonzero vector $\mathbf{v} = \langle v_1, v_2\rangle$, the vector

$$\mathbf{u} = \frac{1}{\|\mathbf{v}\|}\mathbf{v} \tag{9.1.4}$$

is a unit vector because, according to equation (9.1.2),

$$\|\mathbf{u}\| = \left\|\frac{1}{\|\mathbf{v}\|}\mathbf{v}\right\| = \left|\frac{1}{\|\mathbf{v}\|}\right|\|\mathbf{v}\| = \frac{1}{\|\mathbf{v}\|}\|\mathbf{v}\| = 1.$$

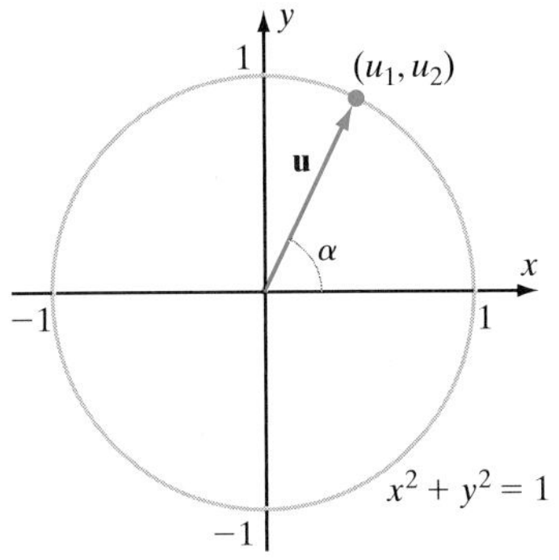

▲ Figure 10 A unit vector $\mathbf{u} = \langle u_1, u_2 \rangle$

We refer to the direction vector **u** defined by equation (9.1.4) as the *direction* of **v** and write

$$\mathbf{dir}(\mathbf{v}) = \frac{1}{\|\mathbf{v}\|}\mathbf{v} \quad (\mathbf{v} \neq \mathbf{0}). \tag{9.1.5}$$

We do not define the direction of the zero vector. By rearranging equation (9.1.5), we see that every nonzero vector **v** can be expressed as the scalar multiplication of the direction vector of **v** by the magnitude of **v**:

$$\mathbf{v} = \|\mathbf{v}\|\,\mathbf{dir}(\mathbf{v}). \tag{9.1.6}$$

The right side of equation (9.1.6) is sometimes said to be the *polar form* of **v**.

▶ **EXAMPLE 9** Let $P = (-1, 2)$ and $Q = (2, 1)$. What is the direction **u** of the vector **v** represented by $\overrightarrow{PQ}$? What angle α, as given by formula (9.1.3), does **u** make with the positive x-axis?

Solution We calculate $\mathbf{v} = \langle 2-(-1),\ 1-2 \rangle = \langle 3, -1 \rangle$ and $\|\mathbf{v}\| = \sqrt{3^2 + (-1)^2} = \sqrt{10}$. The unit vector

$$\mathbf{u} = \frac{1}{\|\mathbf{v}\|}\mathbf{v} = \frac{1}{\sqrt{10}}\langle 3, -1 \rangle = \left\langle \frac{3}{\sqrt{10}}, \frac{-1}{\sqrt{10}} \right\rangle$$

is the direction of **v**. Figure 11 illustrates both **v** and its direction **u**. The angle α that $\mathbf{u} = \langle \cos(\alpha), \sin(\alpha) \rangle$ makes with the positive x-axis is the value of α between $3\pi/2$ and 2π such that $\cos(\alpha) = 3/\sqrt{10}$ and $\sin(\alpha) = -1/\sqrt{10}$. That is, $\alpha = 2\pi - \arctan(1/3)$. With the aid of a calculator, we find that $\alpha \approx 5.96$ radians. ◀

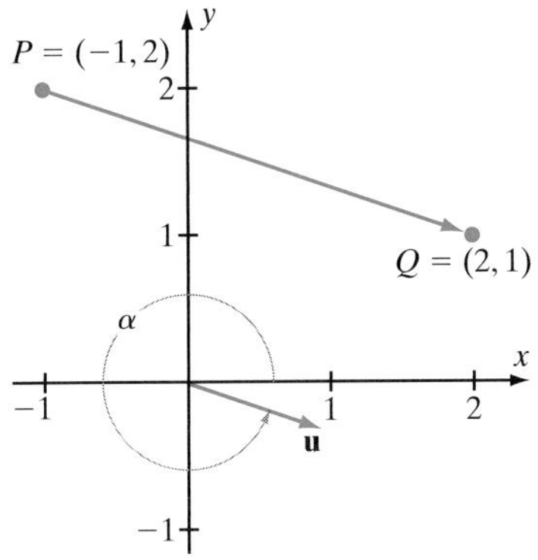

▲ Figure 11

DEFINITION Let **v** and **w** be nonzero vectors. We say that **v** and **w** have the *same direction* if $\mathbf{dir}(\mathbf{v}) = \mathbf{dir}(\mathbf{w})$. We say that **v** and **w** are *opposite in direction* if $\mathbf{dir}(\mathbf{v}) = -\mathbf{dir}(\mathbf{w})$. We say that **v** and **w** are *parallel* if either (a) **v** and **w** have the same direction or (b) **v** and **w** are opposite in direction. Although the zero vector **0** does not have a direction, it is conventional to say that **0** is parallel to every vector.

Notice that nonzero vectors **v** and **w** are parallel if and only if $\mathbf{dir}(\mathbf{v}) = \pm\mathbf{dir}(\mathbf{w})$. Our next theorem provides us with a simple algebraic condition for recognizing parallel vectors.

THEOREM 2 Vectors **v** and **w** are parallel if and only if at least one of the following two equations holds: (a) $\mathbf{v} = \mathbf{0}$ or (b) $\mathbf{w} = \lambda\mathbf{v}$ for some scalar λ. Moreover, if **v** and **w** are both nonzero and $\mathbf{w} = \lambda\mathbf{v}$, then **v** and **w** have the same direction if $0 < \lambda$ and opposite directions if $\lambda < 0$.

Proof. We begin by supposing that equation (a) or equation (b) is true. If $\mathbf{v} = \mathbf{0}$, then **v** and **w** are parallel by convention. Now suppose that $\mathbf{v} \neq \mathbf{0}$ and $\mathbf{w} = \lambda\mathbf{v}$ for some scalar λ. If $\lambda = 0$, then $\mathbf{w} = \mathbf{0}$, in which case **v** and **w** are parallel. If $\lambda \neq 0$, then the following three facts hold: neither **v** nor **w** is **0**, both $\|\mathbf{v}\|$ and $\|\mathbf{w}\|$ are nonzero, and both **dir**(**v**) and **dir**(**w**) are defined. Consequently, we have

$$\mathbf{dir}(\mathbf{w}) = \frac{1}{\|\mathbf{w}\|}\mathbf{w} = \frac{1}{\|\lambda\mathbf{v}\|}(\lambda\mathbf{v}) = \left(\frac{\lambda}{|\lambda|}\right)\frac{1}{\|\mathbf{v}\|}\mathbf{v} = \left(\frac{\lambda}{|\lambda|}\right)\mathbf{dir}(\mathbf{v}) = \pm\mathbf{dir}(\mathbf{v}).$$

This equation tells us that **v** and **w** are parallel. Furthermore, because $\lambda/|\lambda| = +1$ when $\lambda > 0$ and $\lambda/|\lambda| = -1$ when $\lambda < 0$, we see that **v** and **w** have the same direction if $\lambda > 0$ and opposite directions if $\lambda < 0$.

To prove the converse, we suppose that **v** and **w** are parallel and that $\mathbf{v} \neq \mathbf{0}$. We must prove that $\mathbf{w} = \lambda\mathbf{v}$ for some scalar λ. If $\mathbf{w} = \mathbf{0}$, then $\mathbf{w} = \lambda\mathbf{v}$ for $\lambda = 0$. We may therefore assume that **w**, in addition to **v**, is nonzero. In this case, both **dir**(**v**) and **dir**(**w**) are defined. Moreover, because **v** and **w** are parallel, we have $\mathbf{dir}(\mathbf{v}) = \pm\mathbf{dir}(\mathbf{w})$. Finally, noting that $\|\mathbf{v}\| \neq 0$ and using equation (9.1.5), we calculate

$$\mathbf{w} = \|\mathbf{w}\|\mathbf{dir}(\mathbf{w}) = \pm\|\mathbf{w}\|\mathbf{dir}(\mathbf{v}) = \pm\frac{\|\mathbf{w}\|}{\|\mathbf{v}\|}\mathbf{v} = \lambda\mathbf{v}$$

for $\lambda = \pm\|\mathbf{w}\|/\|\mathbf{v}\|$. ■

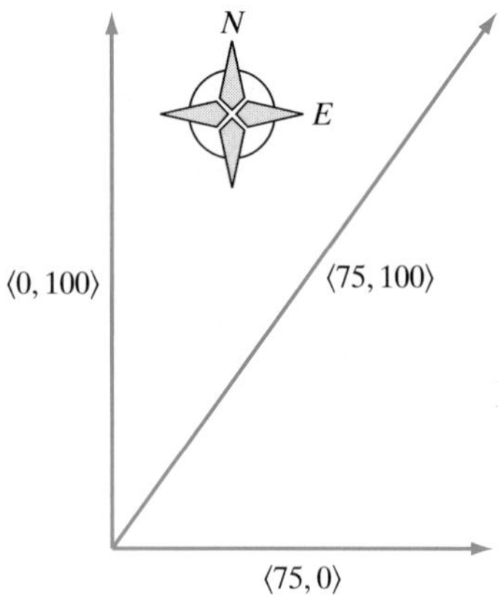

▲ Figure 12

▶ **EXAMPLE 10** For what value of a are the vectors $\langle a, -1\rangle$ and $\langle 3, 4\rangle$ parallel?

Solution For the two given nonzero vectors to be parallel, it is necessary and sufficient for $\langle 3, 4\rangle = \lambda\,\langle a, -1\rangle$ for some scalar λ. This vector equation is equivalent to the two scalar equations, $\lambda a = 3$ and $-\lambda = 4$. We obtain $a = 3/\lambda$, or $a = -3/4$. ◀

An Application to Physics

As discussed earlier, *force* is a quantity that is naturally described by the vector concept. If a force of magnitude F is applied in direction $\langle\cos(\alpha), \sin(\alpha)\rangle$, then the force is written as the vector $\langle F\cos(\alpha), F\sin(\alpha)\rangle$. The *principle of superposition* says that, if two forces act on a body, then the *resultant force* is obtained by adding the vectors corresponding to the two given forces.

▶ **EXAMPLE 11** Two workers are each pulling on a rope attached to a dead tree stump. One pulls in the northerly direction with a force of 100 pounds and the other in the easterly direction with a force of 75 pounds. Compute the resultant force that is applied to the tree stump.

Solution Look at Figure 12. The force vector for the first worker is $\langle 0, 100\rangle$ and that for the second worker is $\langle 75, 0\rangle$. The resultant force is the sum of these, or $\langle 75, 100\rangle$. The resultant force vector is shown in Figure 12. The associated directed line segment has magnitude (length) equal to $(75^2 + 100^2)^{1/2} = 125$. ◀

The Special Unit Vectors i and j

It is common to let **i** denote the vector $\langle 1, 0\rangle$ and **j** denote the vector $\langle 0, 1\rangle$, as shown in Figure 13. If $\mathbf{v} = \langle a, b\rangle$ is any vector, then we may express **v** in terms of **i** and **j** as follows:

$$\mathbf{v} = \langle a, b\rangle = a\langle 1, 0\rangle + b\langle 0, 1\rangle = a\mathbf{i} + b\mathbf{j}.$$

▶ **EXAMPLE 12** Suppose that $\mathbf{v} = \langle 3, -5\rangle$ and $\mathbf{w} = \langle 2, 4\rangle$. Express **v**, **w**, and $\mathbf{v} + \mathbf{w}$ in terms of **i** and **j**.

Solution We have $\mathbf{v} = 3\mathbf{i} - 5\mathbf{j}$ and $\mathbf{w} = 2\mathbf{i} + 4\mathbf{j}$. We can calculate

$$\mathbf{v} + \mathbf{w} = \langle 3, -5\rangle + \langle 2, 4\rangle = \langle 5, -1\rangle = 5\mathbf{i} - \mathbf{j}.$$

Alternatively, we can express **v** and **w** in terms of **i** and **j**, as we have already done, and add these expressions:

$$\mathbf{v}+\mathbf{w}=(3\mathbf{i}-5\mathbf{j})+(2\mathbf{i}+4\mathbf{j})=(3+2)\mathbf{i}+(-5+4)\mathbf{j}=5\mathbf{i}-\mathbf{j}. \blacktriangleleft$$

The Triangle Inequality

A glance back at ΔABC in Figure 7 suggests that, if **v** and **w** are nonparallel vectors, then the length of $\mathbf{v}+\mathbf{w}$ is never greater than the sum of the lengths of **v** and **w**. In other words,

$$\|\mathbf{v}+\mathbf{w}\| \le \|\mathbf{v}\| + \|\mathbf{w}\|. \tag{9.1.7}$$

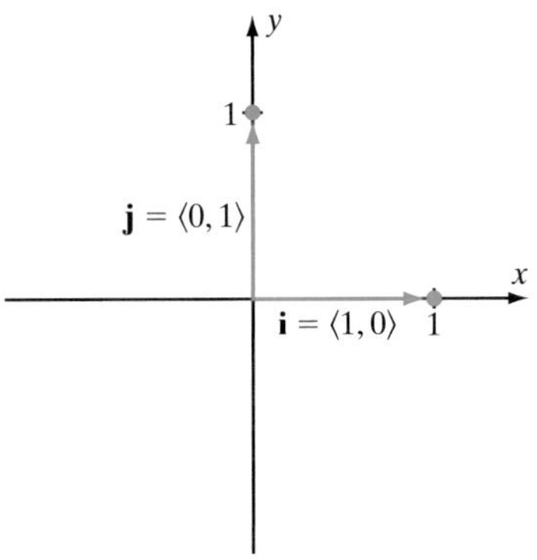

▲ Figure 13

In fact, this inequality, known as the *Triangle Inequality*, is just a vector interpretation of the familiar theorem in Euclidean geometry that states that the sum of the lengths of any two sides of a triangle is greater than the length of the third side. Inequality (9.1.7) is also true when **v** and **w** are parallel, for, in this case, we may write one of the vectors, say **w**, as a scalar multiple of the other: $\mathbf{v}=\lambda\mathbf{w}$. We then use equation (9.1.2) together with the Triangle Inequality for scalars to obtain

$$\begin{aligned}\|\mathbf{v}+\mathbf{w}\| &= \|\mathbf{v}+\lambda\mathbf{v}\| = \|(1+\lambda)\mathbf{v}\| = |1+\lambda|\|\mathbf{v}\| \le (1+|\lambda|)\|\mathbf{v}\| \\ &= \|\mathbf{v}\|+|\lambda|\|\mathbf{v}\| = \|\mathbf{v}\|+\|\lambda\mathbf{v}\| = \|\mathbf{v}\|+\|\mathbf{w}\|. \blacktriangleleft\end{aligned}$$

▶ **EXAMPLE 13** Verify the Triangle Inequality for the vectors $\mathbf{v}=\langle -3, 4\rangle$ and $\mathbf{w}=\langle 8,6\rangle$.

Solution After calculating $\|\mathbf{v}\|+\|\mathbf{w}\|=\sqrt{(-3)^2+4^2}+\sqrt{8^2+6^2}=5+10=15$, we find that

$$\|\mathbf{v}+\mathbf{w}\| = \|\langle 5,10\rangle\| = \sqrt{125} \approx 11.18 < 15 = \|\mathbf{v}\|+\|\mathbf{w}\|. \blacktriangleleft$$

We conclude this section by stating the distributive, associative, and commutative laws for scalar multiplication and vector addition. If **u**, **v**, **w** are vectors and λ, μ are scalars then

a. $\mathbf{v}+\mathbf{w}=\mathbf{w}+\mathbf{v}$
b. $\mathbf{u}+(\mathbf{v}+\mathbf{w})=(\mathbf{u}+\mathbf{v})+\mathbf{w}$
c. $\lambda(\mu\mathbf{v})=(\lambda\mu)\mathbf{v}$
d. $\lambda(\mathbf{v}+\mathbf{w})=\lambda\mathbf{v}+\lambda\mathbf{w}$
e. $(\lambda+\mu)\mathbf{v}=\lambda\mathbf{v}+\mu\mathbf{v}$

QUICK QUIZ

1. If $P=(2,-1)$, then for what point Q does $\overrightarrow{PQ}$ represent $\langle -3,-2\rangle$?
2. What is the magnitude of $\langle -3, 4\rangle$?
3. What is the direction vector of $\langle 12, -5\rangle$?
4. If $\langle a, b\rangle$ is opposite in direction to $\langle -4, 3\rangle$ and has length 2, then what is a?

Answers

1. $(-1,-3)$ 2. 5 3. $\langle 12/13, -5/13\rangle$ 4. $8/5$

EXERCISES

Problems for Practice

▶ In each of Exercises 1–8, sketch $\overrightarrow{PQ}$, and write it in the form $\langle a, b\rangle$. ◀

1. $P = (5, 8)$ $Q = (4, 4)$
2. $P = (1, 1)$ $Q = (3, 9)$
3. $P = (0, 0)$ $Q = (-5, 1)$
4. $P = (1, 1)$ $Q = (2, 2)$
5. $P = (-1, 8)$ $Q = (4, -3)$
6. $P = (0, 1)$ $Q = (0, 0)$
7. $P = (-5, 6)$ $Q = (1, 0)$
8. $P = (2, 2)$ $Q = (2, -2)$

▶ In each of Exercises 9–12, calculate the lengths of $\overrightarrow{PQ}$ and $\overrightarrow{RS}$. Also determine whether these vectors are parallel. ◀

9. $P = (1, 2)$ $Q = (2, 0)$ $R = (3, 6)$ $S = (6, 0)$
10. $P = (1, 3)$ $Q = (3, 1)$ $R = (1, 2)$ $S = (6, 3)$
11. $P = (3, 1)$ $Q = (-6, 4)$ $R = (7, -2)$ $S = (5, 5)$
12. $P = (1, 1)$ $Q = (-7, 5)$ $R = (1, 0)$ $S = (17, -8)$

▶ In each of Exercises 13–16, calculate the indicated vector given that $\mathbf{v} = \langle 4, 2\rangle$ and $\mathbf{w} = \langle 1, -3\rangle$. For each exercise, draw a sketch of **v**, **w**, and the vector you have calculated. ◀

13. $2\mathbf{v} + \mathbf{w}$
14. $(1/2)\mathbf{v} + 3\mathbf{w}$
15. $\mathbf{v} - 2\mathbf{w}$
16. $-3\mathbf{v} + 2\mathbf{w}$

▶ In each of Exercises 17–20, two points P and Q are given. Determine (a) the vector **v** that is represented by $\overrightarrow{PQ}$, (b) the length of **v**, (c) the vector that has the same length as **v** but is in the opposite direction of **v**, (d) the direction vector of **v**, and (e) a unit vector that is in the opposite direction of **v**. ◀

17. $P = (5, -7)$ $Q = (0, 5)$
18. $P = (-7, 1)$ $Q = (-14)$
19. $P = (-3, 7)$ $Q = (1, -9)$
20. $P = (3, 2)$ $Q = (1, 1)$

▶ In each of Exercises 21–24, vectors **v** and **w** are given. Express **v**, **w**, $-4\mathbf{v}$, $3\mathbf{v} - 2\mathbf{w}$, and $4\mathbf{v} + 7\mathbf{j}$ in terms of the unit vectors **i** and **j**. ◀

21. $\mathbf{v} = \langle -7, 2\rangle$ $\mathbf{w} = \langle -2, 9\rangle$
22. $\mathbf{v} = \langle 0, 3\rangle$ $\mathbf{w} = \langle -5, 0\rangle$
23. $\mathbf{v} = \langle 6, -2\rangle$ $\mathbf{w} = \langle 9, -3\rangle$
24. $\mathbf{v} = \langle 1/3, 1\rangle$ $\mathbf{w} = \langle -3/2, 1/2\rangle$

▶ In each of Exercises 25–28, calculate the resultant force $\mathbf{F} = \mathbf{F}_1 + \mathbf{F}_2$. What is the magnitude of **F**? ◀

25. $\mathbf{F}_1 = \langle 3, 0\rangle$ $\mathbf{F}_2 = \langle 0, 4\rangle$
26. $\mathbf{F}_1 = \langle 1, 2\rangle$ $\mathbf{F}_2 = \langle 2, 2\rangle$
27. $\mathbf{F}_1 = \mathbf{i} - 2\mathbf{j}$ $\mathbf{F}_2 = 2\mathbf{i} + \mathbf{j}$
28. $\mathbf{F}_1 = \mathbf{i} - \mathbf{j}$ $\mathbf{F}_2 = -5\mathbf{i} + 13\mathbf{j}$

▶ In each of Exercises 29–32, determine the direction vector **u** that makes the given positive angle α with the positive x-axis. ◀

29. $\pi/6$
30. $3\pi/4$
31. π
32. $5\pi/3$

▶ In each of Exercises 33–36, write the given vector **v** in the form $\lambda\mathbf{u}$ where λ is a positive scalar, and **u** is a direction vector. ◀

33. $\mathbf{v} = \langle 6, -5/2\rangle$
34. $\mathbf{v} = \langle -\sqrt{11}, 5\rangle$
35. $\mathbf{v} = 3\mathbf{i} - 2\mathbf{j}$
36. $\mathbf{v} = -\dfrac{1}{2}\mathbf{i} - \dfrac{\sqrt{7}}{2}\mathbf{j}$

▶ In each of Exercises 37–40, write the given vector **v** in the form $\langle a, b\rangle$. ◀

37. $2(5\mathbf{i} + 3\mathbf{j}) - 3(\mathbf{i} - \mathbf{j})$
38. $4(\mathbf{i} - 2\mathbf{j}) + 5(\mathbf{i} + 2\mathbf{j})$
39. $3\mathbf{j} - 2(-3\mathbf{i} + \mathbf{j}) - 2\mathbf{i}$
40. $5\mathbf{j} + 2(-\mathbf{j} + 4\mathbf{i}) - 3(2\mathbf{j} + \mathbf{i})$

Further Theory and Practice

▶ In Exercises 41–54, determine the value of a from the given information about $\mathbf{v} = \langle a, b\rangle$. ◀

41. $a > b = 3$ and $\|\mathbf{v}\| = 7$
42. $a < b = 5$ and $\|\mathbf{v}\| = 13$
43. **v** is parallel to $\langle 5, 3b\rangle$
44. **v** is parallel to $\langle 7, -b/2\rangle$
45. **v** is opposite in direction to $\mathbf{w} = \langle 2, c\rangle$ and $\|\mathbf{v}\| = 3\|\mathbf{w}\|$
46. **v** is opposite in direction to $\mathbf{w} = \langle -3, c\rangle$ and $\|\mathbf{v}\| = \dfrac{1}{6}\|\mathbf{w}\|$
47. both a and b are positive, and **v** is a diagonal of the parallelogram generated by $\langle 3, 5\rangle$ and $\langle -1, 7\rangle$
48. both a and b are negative, and **v** is a diagonal of the parallelogram generated by $\langle 4, 7\rangle$ and $\langle -1, 8\rangle$
49. both a and b are positive, and **v** is a diagonal of the parallelogram generated by $\langle -7, 6\rangle$ and $\langle 3, 7\rangle$
50. both a and b are negative, and **v** is a diagonal of the parallelogram generated by $\langle 2, 9\rangle$ and $\langle -5, 3\rangle$
51. $\mathbf{dir}(\mathbf{v}) = \mathbf{dir}(\mathbf{w})$ and $\|\mathbf{v}\| = 3$ for $\mathrm{w} = \langle 1, \sqrt{3}\rangle$
52. $\mathbf{dir}(\mathbf{v}) = \mathbf{dir}(2\mathbf{w})$ and $\|\mathbf{v}\| = 1/2$ for $\mathrm{w} = \langle 2, 1\rangle$
53. $\mathbf{dir}(\mathbf{v}) = \mathbf{dir}(-\mathbf{w})$ and $\|\mathbf{v}\| = 10$ for $\mathrm{w} = \langle 3, -4\rangle$
54. $\mathbf{dir}(\mathbf{v}) = -\mathbf{dir}(\mathbf{w})$ and $\|\mathbf{v}\| = 1/\sqrt{2}$ for $\mathrm{w} = \langle -\sqrt{3}, \sqrt{5}\rangle$
55. Mr. and Mrs. Woodman are pulling on ropes tied to a heavy wagon. Refer to Figure 14. If Mr. Woodman pulls with a strength of 100 pounds, then how hard must

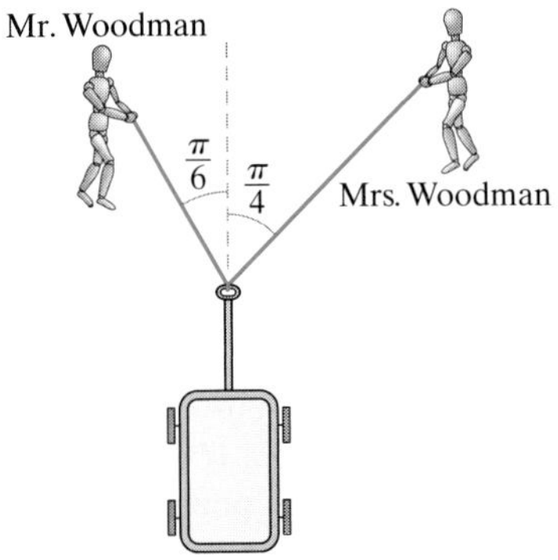

▲ Figure 14

Mrs. Woodman pull so that the wagon moves along the dotted line? What is the magnitude of the resultant force?

56. Three forces $\mathbf{F}_1$, $\mathbf{F}_2$, and $\mathbf{F}_3$ are applied to a point mass. Suppose that $\mathbf{F}_1 = 100\,\mathbf{j}$ newtons and that $\mathbf{F}_2$ has magnitude 120 newtons applied in the direction $\langle 3/5, 4/5\rangle$. What must $\mathbf{F}_3$ be if the point mass is to remain at rest?

57. A boat maintains a straight course across a 500 m wide river at a rate of 50 m/min. The current pulls the boat down river at a rate of 20 m/min. When the boat docks on the other shore, how far down river will it have been carried?

58. A river is 1 mile wide and flows south with a current of 3 miles per hour. What speed and heading should a motorboat adopt in order to cross the river in 10 minutes and reach a point on the opposite bank due east of its point of departure?

59. Bjarne, Leif, and Sammy are towing their vessel. The forces that they exert are directed along the tow lines, as indicated in Figure 15, which also provides the magnitudes of their forces. (Note that the actual force vectors are not drawn in Figure 15.) What is the resultant force?

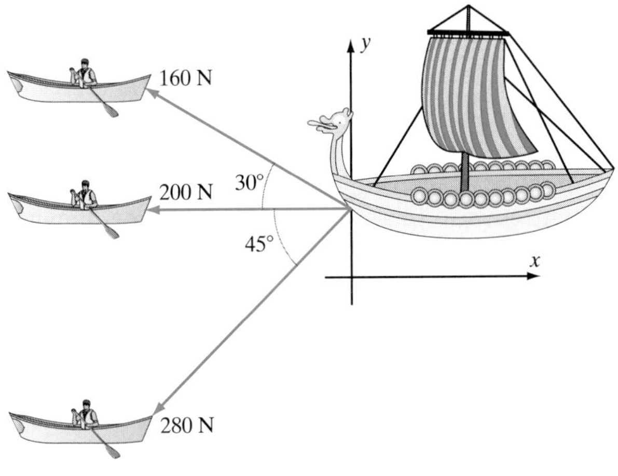

▲ Figure 15

60. Let $P = (p_1, p_2)$ and $Q = (q_1, q_2)$ be distinct points in the plane. Let M be the midpoint of the segment joining P and Q. Use the relationship $\overrightarrow{OM} = \overrightarrow{OP} + \frac{1}{2}\overrightarrow{PQ}$ to determine the Cartesian coordinates of M.

61. Let $\mathbf{v}$ and $\mathbf{w}$ be two given nonzero vectors. Prove that there is a unique number λ such that $\mathbf{v} + \lambda\mathbf{w}$ is as short as possible. In other words, the function from $\mathbb{R}$ to $\mathbb{R}$ defined by $\lambda \mapsto \|\mathbf{v} + \lambda\mathbf{w}\|$ attains an absolute minimum value.

62. Let m and b be constants. Show that $\mathbf{v}$ is parallel to $\overrightarrow{PQ}$ for every pair of points P and Q on the line $y = mx + b$ if and only if $\mathbf{v} = \lambda,\ \langle 1, m\rangle$ for some scalar λ.

63. Verify that $\mathbf{w} = \langle -b, a\rangle$ is perpendicular to $\mathbf{v} = \langle a, b\rangle$. Describe all vectors that are perpendicular to $\mathbf{v}$.

64. Let $\mathbf{v}$ and $\mathbf{w}$ be two nonzero vectors in the plane that are *not* parallel. Let $\mathbf{u}$ be any other vector. Prove that there are unique scalars λ and μ such that

$$\mathbf{u} = \lambda\mathbf{v} + \mu\mathbf{w}.$$

(If $\mathbf{v} = \mathbf{i}$ and $\mathbf{w} = \mathbf{j}$, then λ and μ are the entries of vector $\mathbf{u}$. In general, we can think of $\mathbf{v}$ and $\mathbf{w}$ generating a coordinate system with λ and μ the entries of vector $\mathbf{u}$ in this new coordinate system.)

▶ In Exercises 65 and 66, develop a formula for the distance of a point to a line. ◀

65. Suppose that A and B are not 0. Consider the line ℓ whose Cartesian equation is $Ax + By + D = 0$. Suppose that $P_0 = (x_0, y_0)$ does not lie on ℓ. Show that $\mathbf{n} = \langle A, B\rangle$ is a vector that is perpendicular to ℓ. Let $Q_x = (x, -Ax/B - D/B)$ be a point on l. Calculate $\overrightarrow{P_0Q_x}$. For what value ξ of x is $\overrightarrow{P_0Q_\xi}$ parallel to $\mathbf{n}$?

66. For the value of ξ determined in Exercise 65, show that

$$\|\overrightarrow{P_0Q_\xi}\| = \frac{|Ax_0 + By_0 + D|}{\sqrt{A^2 + B^2}}.$$

What does it mean to state that this is the distance of P_0 to ℓ?

67. Let $P = (p_1, p_2)$ and $Q = (q_1, q_2)$ be distinct points in the plane. Suppose that $0 < t < 1$. What are the Cartesian coordinates of the point P_t on line segment $\overline{PQ}$ whose distance from P is $t \cdot \|\overrightarrow{PQ}\|$? Use vectors to find an elegant solution to this problem.

▶ Let A, B, and C denote any three distinct points that are not collinear. Let α, β, and γ denote the midpoints of line segments $\overline{BC}$, $\overline{AC}$, and $\overline{AB}$, respectively. The line segments $\overline{A\alpha}$, $\overline{B\beta}$, and $\overline{C\gamma}$ are called *medians* of ΔABC. Exercises 68–70 concern these medians and their common point of intersection, the *centroid* of ΔABC. ◀

68. Prove that $\overrightarrow{A\alpha} + \overrightarrow{B\beta} + \overrightarrow{C\gamma} = \overrightarrow{\mathbf{0}}$.

69. Let M be point of intersection of $\overline{A\alpha}$ and $\overline{B\beta}$. Show that $\overrightarrow{AM} = (2/3)\overrightarrow{A\alpha}$ and $\overrightarrow{BM} = (2/3)\,\overrightarrow{B\beta}$. Symmetrically, the same relationship holds when the medians $\overline{B\beta}$ and $\overline{C\gamma}$ are selected. Deduce that all three medians intersect in one point. It is called the *centroid* of ΔABC.

70. Express the coordinates of the centroid of ΔABC in terms of the coordinates of the vertices A, B, and C.

71. Let P_t be the point (t^2, t^3) for $t > 0$. Find the value of t such that $\overrightarrow{P_tP_{t+3}}$ has the same direction as $\langle 1, 7\rangle$.

72. In his first season in Major League Baseball (1941), Stan Musial's batting average was 0.426. That year, Joe DiMaggio, already an established star, hit for 0.357. The next year, Musial's average was 0.315 and DiMaggio's was 0.305. Because Musial's average exceeded DiMaggio's in both of those years, it seems obvious that his combined average over the two years must have exceeded that of DiMaggio. Not so fast! Define the *slope* of $\langle a, b\rangle$ to be b/a if $a \neq 0$. We can represent Musial's number of at bats in 1941 (47) and hits (20) by the vector $\mathbf{m}_1 = \langle 47, 20\rangle$. The slope of this vector, 20/47, or 0.426, is Musial's batting average for 1941. The vector $\mathbf{m}_2$ representing his data for 1942 is $\mathbf{m}_2 = \langle 467, 147\rangle$. The corresponding vectors for Joe DiMaggio are $\mathbf{d}_1 = \langle 541, 193\rangle$ and $\mathbf{d}_2 = \langle 610, 186\rangle$. Calculate the combined average for Musial over the two-year period. Do the same for DiMaggio—you will find that, contrary to expectation, it is *greater* than Musial's. The phenomenon at work here is known as *Simpson's Paradox* in statistics. Explain geometrically how this paradox can occur by representing the combined averages that you calculated as the slopes of the diagonals of two parallelograms. Observe that a diagonal of parallelogram P_2 may have greater slope than the corresponding diagonal of P_1 even if the edges of P_1 have greater slopes than the the corresponding edges of P_2.

Calculator/Computer Exercises

73. Plot $y = e^x$ and $y = 2 - x^2$ for $-2 \le x \le 1$. Let P and Q be the points of intersection of the two curves in the second and first quadrants respectively. What is vector $\overrightarrow{PQ}$?

74. Let $f(x) = x^2 + 2x + 2$. Let P_x be the point $(x, f(x))$. Plot $y = \|\overrightarrow{OP_x}\|$. What is the point on the graph of f that is closest to the origin?

75. For $-2 \le t \le 3$ let Q_t, denote the point $(t, e^{-t} + t^2)$. Let $f(t) = \|\overrightarrow{PQ_t}\|$ where $P = (1, 4)$. Plot $f(t)$. At what value of t is $f(t)$ minimized?

76. Suppose that for each point $P_t = (t, t^2)$ on the curve $y = x^2$, a point $Q_t = (\xi(t), \eta(t))$ is found such that (a) $\xi(t) > t$, (b) $\overrightarrow{P_tQ_t}$ is a unit vector, and (c) $\overrightarrow{P_tQ_t}$ is tangent to the curve. Plot the curve whose parametric equations are $x = \xi(t)$, $y = \eta(t)$. What point on this parametric curve is farthest below the y-axis?

77. Suppose that, for each point $P_t = (t, t^3)$ on the curve $y = x^3$, a point $Q_t = (\xi(t), \eta(t))$ is found such that (a) $\xi(t) > t$, (b) $\overrightarrow{P_tQ_t}$ is a unit vector, and (c) $\overrightarrow{P_tQ_t}$ is tangent to the curve. Plot the curve whose parametric equations are $x = \xi(t)$, $y = \eta(t)$. What are its x-and y-intercepts? To what values of t do these intercepts correspond?

9.2 Vectors in Three-Dimensional Space

We are familiar with locating points on the line by using one coordinate and points in the plane by using two coordinates. Now we turn to three-dimensional space, where locating points requires three coordinates.

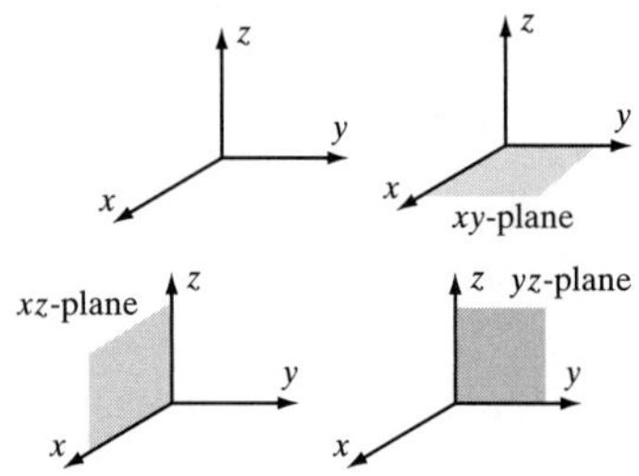

▲ Figure 1

Figure 1 exhibits three mutually perpendicular axes: the x-axis, the y-axis, and the z-axis. The yz-plane contains the y- and z-axes. It coincides with the plane of this page. The xy- and xz-planes are defined analogously. They emerge perpendicularly from this page. A point with spatial coordinates $(x, y, 0)$ lies in the xy-plane. In general, the third coordinate z of a point (x, y, z) indicates the *height* of the point above or below the xy-plane. Figure 2 exhibits a point (x, y, z) with $z > 0$: It lies z units above the point $(x, y, 0)$ in the xy-plane. Notice how we use dotted lines in our picture to create a sense of perspective so that we can picture a three-dimensional concept on two-dimensional paper. Points (x, y, z) with $z < 0$ lie below the xy-plane.

The coordinates x, y, and z of the point $P = (x, y, z)$ are called the *Cartesian coordinates* of P. An equation involving Cartesian coordinates is said to be a *Cartesian equation.* The terms *rectangular coordinates* and *rectangular equation* are sometimes used instead. The set of all spatial points is denoted by $\mathbb{R}^3$ to signify that three coordinates are needed to describe such points.

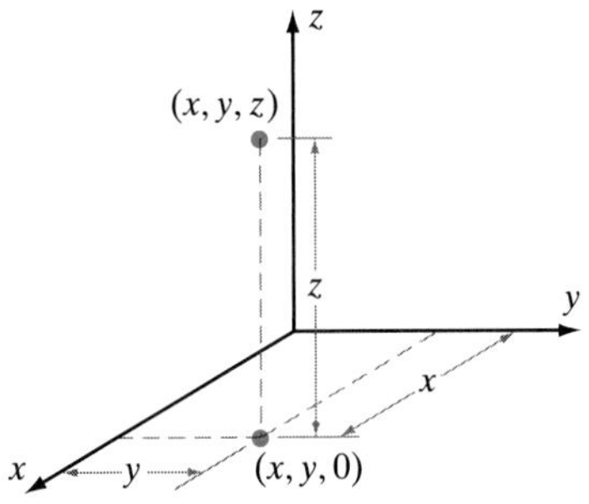

▲ Figure 2

Just as the coordinate axes in the plane divide the plane into four *quadrants*, so do the xy-plane, the xz-plane, and the yz-plane divide space into eight *octants*. The octant that contains points with all three coordinates positive is called the *first octant*. It is sometimes helpful to think of the xy-plane as the floor of a room. Think of the origin $O = (0,0,0)$ as a corner of the floor and of the positive z-axis as the join of two walls (the two walls are the xz-plane and the yz-plane). Look at Figure 3, which depicts the first octant from this point of view. Let $P = (x_0, y_0, z_0)$ be a point with all three coordinates positive. Then the first two coordinates of P tell which position $P' = (x_0, y_0, 0)$ on the floor the point P lies over. The last coordinate tells how high above the floor P is.

▶ **EXAMPLE 1** Sketch the points $(3,2,5)$, $(2,3,-3)$, and $(-1,-2,1)$.

Solution These points are graphed in Figure 4. Some polygonal paths connecting these points to the origin are also plotted for perspective. For example, to reach the point $(-1,-2,1)$ from $O = (0,0,0)$, we may go 2 units in the negative y-direction, arriving at the point $(0,-2,0)$, then 1 unit in the negative x-direction, arriving at the point $(-1,-2,0)$, and finally 1 unit in the positive z-direction, finishing at $(-1,-2,1)$. ◀

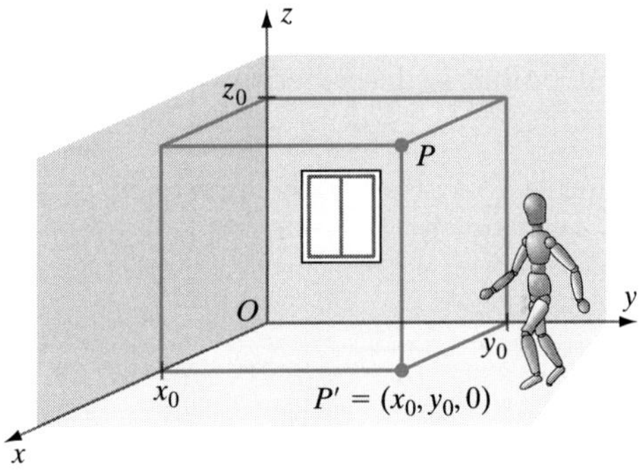

▲ Figure 3 A point $P = (x_0, y_0, z_0)$ in the first octant and the projection of P to the xy-plane

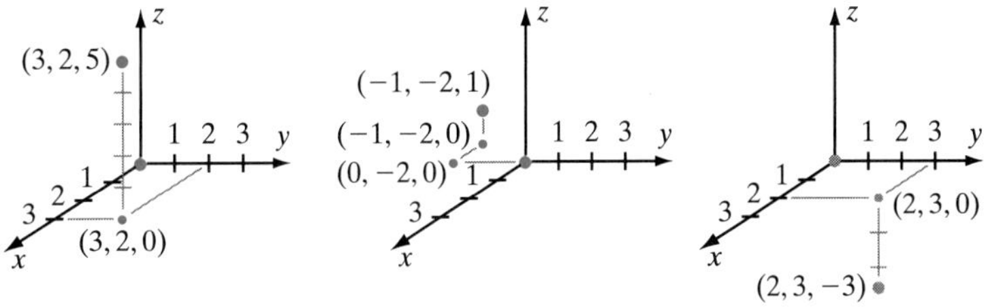

▲ Figure 4

Distance

Two distinct points in space determine a box with sides parallel to the coordinate axes (see Figure 5). The length of a main diagonal of this box coincides with the *distance d* between these two points. We have labeled length, width, and height of the box as α, β, and γ. One main diagonal with length d is drawn while δ is the length of a minor diagonal that is shown. Notice that, by the Pythagorean Theorem, $\delta^2 = \beta^2 + \gamma^2$ and $d^2 = \alpha^2 + \delta^2$. Substituting the first of these equations into the second yields $d^2 = \alpha^2 + \beta^2 + \gamma^2$. Translating this calculation into coordinates gives us a formula for the distance between two points in space:

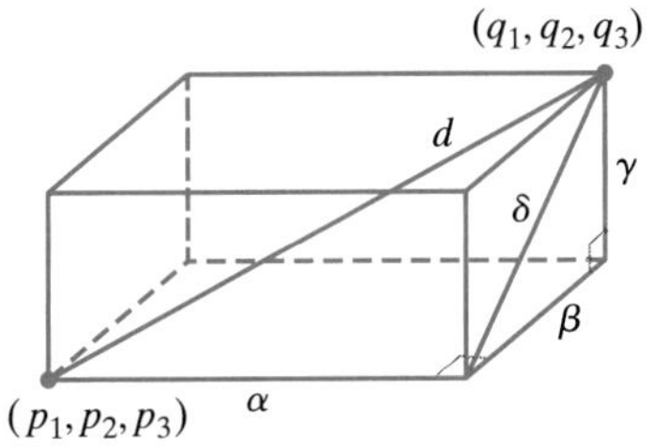

▲ Figure 5

THEOREM 1 The distance $d(P,Q)$ between the points $P = (p_1, p_2, p_3)$ and $Q = (q_1, q_2, q_3)$ is given by

$$d(P,Q) = \sqrt{(q_1 - p_1)^2 + (q_2 - p_2)^2 + (q_3 - p_3)^2}. \tag{9.2.1}$$

▶ **EXAMPLE 2** Calculate the distance between the points $(4,-9,7)$ and $(2,1,4)$.

Solution According to Theorem 1, the required distance is

$$\sqrt{(2-4)^2+(1-(-9))^2+(4-7)^2}=\sqrt{(-2)^2+(10)^2+(-3)^2}=\sqrt{113}. \quad ◀$$

If $P_0=(x_0, y_0, z_0)$ is a fixed point in space and if $r>0$ is a fixed number, then the collection of all points with distance r from P_0 is the *sphere* with center P_0 and radius r. From Theorem 1, we see that the set of points that satisfy the equation

$$\sqrt{(x-x_0)^2+(y-y_0)^2+(z-z_0)^2}=r,$$

or

$$(x-x_0)^2+(y-y_0)^2+(z-z_0)^2=r^2 \tag{9.2.2}$$

is the sphere with center $P_0=(x_0, y_0, z_0)$ and radius r. Notice that the left side of equation (9.2.2) is quadratic in the three space variables x, y, and z. The coefficients of x^2, y^2, and z^2 are all 1. There are no mixed products xy, xz, or yz.

▶ **EXAMPLE 3** Determine what set of points is described by the equation $x^2+y^2+z^2-6x+4y+6=0$.

Solution We complete the squares for x and y (because of the presence of the linear terms $-6x$ and $4y$): The equation $x^2+y^2+z^2-6x+4y+6=0$ is equivalent to $(x^2-6x)+(y^2+4y)+z^2=-6$ or

$$\left(x^2-6x+(-6/2)^2\right)+\left(y^2+4y+(4/2)^2\right)+z^2=-6+(-6/2)^2+(4/2)^2.$$

After simplification, we have $(x-3)^2+(y+2)^2+z^2=7$, or $(x-3)^2+(y-(-2))^2+(z-0)^2=7$. We see that the given equation describes a sphere with center $(3,-2,0)$ and radius $\sqrt{7}$ (see Figure 6). ◀

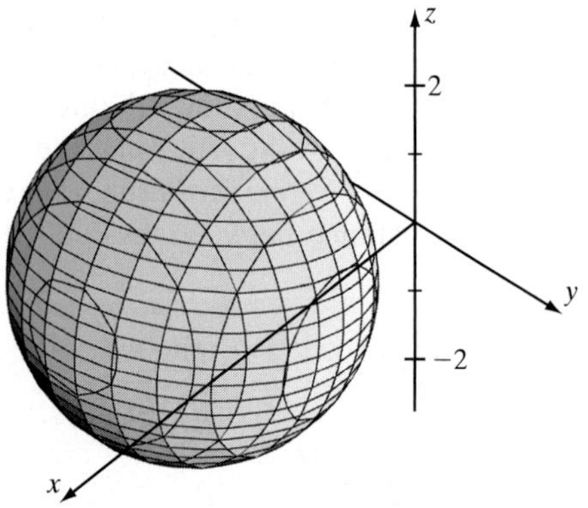

▲ Figure 6 The sphere $(x-3)^2+(y-(-2))^2+(z-0)^2=7$

DEFINITION Let $P_0 = (x_0, y_0, z_0)$ be a point in space, and let r be a positive number. The set

$$\{P \in \mathbb{R}^3 : d(P, P_0) < r\},$$

or, equivalently,

$$\{(x, y, z) : (x - x_0)^2 + (y - y_0)^2 + (z - z_0)^2 < r^2\},$$

is the set of all points inside the sphere $(x - x_0)^2 + (y - y_0)^2 + (z - z_0)^2 = r^2$. This set is called the *open ball with center P_0 and radius r*. The set

$$\{P \in \mathbb{R}^3 : d(P, P_0) \leq r\},$$

or, equivalently,

$$\{(x, y, z) : (x - x_0)^2 + (y - y_0)^2 + (z - z_0)^2 \leq r^2\},$$

is the preceding open ball plus its boundary, the sphere. We call this set the *closed ball with center P_0 and radius r.*

▶ **EXAMPLE 4** Describe the set of points S that satisfy the equation $x^2 + y^2 + z^2 + 80 + 8y + 16z < 6x$.

Solution This set can be identified by rearranging the given inequality as $(x^2 - 6x) + (y^2 + 8y) + (z^2 + 16z) < -80$ and completing the squares in each set of parentheses:

$$\left(x^2 - 6x + \left(\frac{6}{2}\right)^2\right) + \left(y^2 + 8y + \left(\frac{8}{2}\right)^2\right) + \left(z^2 + 16z + \left(\frac{16}{2}\right)^2\right)$$
$$< -80 + \left(\frac{6}{2}\right)^2 + \left(\frac{8}{2}\right)^2 + \left(\frac{16}{2}\right)^2.$$

This inequality simplifies to $(x - 3)^2 + (y + 4)^2 + (z + 8)^2 < 9$ or $(x - 3)^2 + (y - (-4))^2 + (z - (-8))^2 < 3^2$. From this last inequality, we see that the given set consists of those points whose distance from the point $(3, -4, -8)$ is less than 3. In other words, S is the open ball with center $(3, -4, -8)$ and radius 3. ◀

Vectors in Space

A *vector* in three-dimensional space is an ordered triple $\langle v_1, v_2, v_3 \rangle$. This is the algebraic definition of vector and is the one that we use in calculations. But there is a geometric interpretation as well. Like a vector in the plane, a vector in space can be represented by a directed line segment that specifies magnitude and direction. We say that a directed line segment $\overrightarrow{PQ}$ with initial point P and terminal point Q *represents* the vector $\mathbf{v} = \langle v_1, v_2, v_3 \rangle$ if the points $P = (p_1, p_2, p_3)$ and $Q = (q_1, q_2, q_3)$ satisfy

$$v_1 = q_1 - p_1, \quad v_2 = q_2 - p_2, \quad \text{and} \quad v_3 = q_3 - p_3. \tag{9.2.3}$$

If $P = (x, y, z)$ is a point in space, then the vector $\overrightarrow{OP} = \langle x, y, z \rangle$ is sometimes called the *position vector* of P.

▶ **EXAMPLE 5** Let $P = (3, -1, 4)$, $Q = (2, 1, 0)$, $R = (-6, 9, 7)$, and $S = (-7, 11, 3)$. What vectors are represented by the directed line segments $\overrightarrow{PQ}$, $\overrightarrow{PR}$, and $\overrightarrow{QS}$? Do any of these directed line segments represent the same vector?

Solution The vector represented by $\overrightarrow{PQ}$ is $\langle 2-3, 1-(-1), 0-4\rangle$ or $\langle -1, 2, -4\rangle$, the vector represented by $\overrightarrow{PR}$ is $\langle -6-3, 9-(-1), 7-4\rangle$ or $\langle -9, 10, 3\rangle$, and the vector represented by $\overrightarrow{QS}$ is $\langle -7-2, 11-1, 3-0\rangle$ or $\langle -9, 10, 3\rangle$. Thus $\overrightarrow{PQ}$ and $\overrightarrow{PR}$ represent different vectors. However, $\overrightarrow{PR}$ and $\overrightarrow{QS}$ represent the same vector, $\langle -9, 10, 3\rangle$. ◀

Vector Operations

Addition. If $\mathbf{v} = \langle v_1, v_2, v_3\rangle$ and $\mathbf{w} = \langle w_1, w_2, w_3\rangle$, then we define

$$\mathbf{v} + \mathbf{w} = \langle v_1 + w_1, v_2 + w_2, v_3 + w_3\rangle.$$

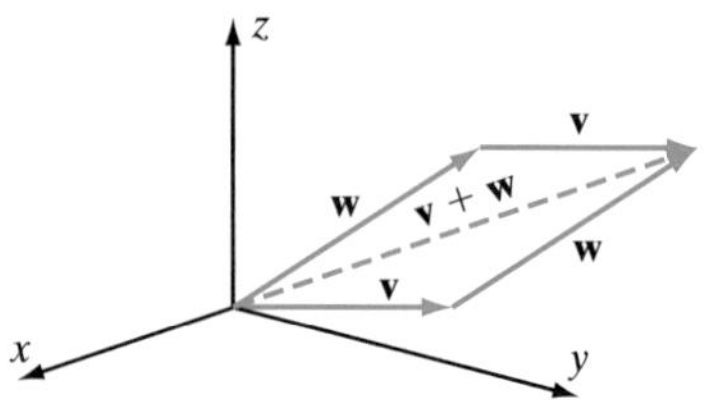

▲ Figure 7

As in the plane, the geometric interpretation of vector addition is obtained by using the parallelogram rule. The sum $\mathbf{v} + \mathbf{w}$, indicated by the dotted directed line segment in Figure 7, is the diagonal of the parallelogram determined by $\mathbf{v}$ and $\mathbf{w}$.

▶ **EXAMPLE 6** Suppose that $\mathbf{v} = \langle 3, 0, 1\rangle$ and $\mathbf{w} = \langle 0, 4, 2\rangle$. Calculate $\mathbf{v} + \mathbf{w}$, and sketch the three vectors.

Solution We calculate $\mathbf{v} + \mathbf{w} = \langle 3+0, 0+4, 1+2\rangle = \langle 3, 4, 3\rangle$. The three vectors are drawn in Figure 8. ◀

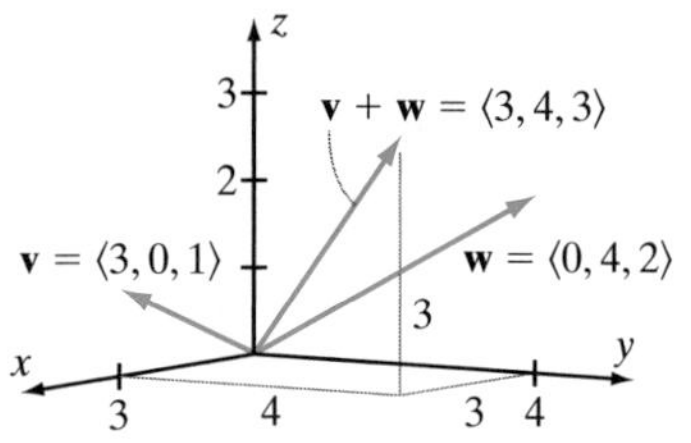

▲ Figure 8

Scalar Multiplication. If $\mathbf{v} = \langle v_1, v_2, v_3\rangle$ is a vector and λ is a real number (i.e., a *scalar*) then we define the *scalar multiplication* of $\mathbf{v}$ by λ to be

$$\lambda\mathbf{v} = \langle \lambda v_1, \lambda v_2, \lambda v_3\rangle.$$

Geometrically, we think of scalar multiplication as producing a vector that is parallel to $\mathbf{v}$ but with different length when $|\lambda| \neq 1$. An argument with similar triangles establishes that two nonzero vectors have the same direction if one is a positive scalar multiple of the other. They have opposite directions if one is a negative scalar multiple of the other. For now, these notions are intuitive generalizations of facts that we already know for vectors in the plane. They will be made more precise in the discussion that follows later in this section.

▶ **EXAMPLE 7** Suppose that $\mathbf{v} = \langle 2, -1, 1\rangle$. Calculate $3\mathbf{v}$ and $-4\mathbf{v}$.

Solution We see that $3\mathbf{v} = \langle 6, -3, 3\rangle$ and $-4\mathbf{v} = \langle -8, 4, -4\rangle$. ◀

Subtraction. If $\mathbf{v}$ and $\mathbf{w}$ are given vectors, then the expression $\mathbf{v} - \mathbf{w}$ is interpreted to mean $\mathbf{v} + ((-1)\mathbf{w})$. If $\mathbf{v} = \langle v_1, v_2, v_3\rangle$ and $\mathbf{w} = \langle w_1, w_2, w_3\rangle$, then

$$\mathbf{v} - \mathbf{w} = \mathbf{v} + (-1)\mathbf{w} = \langle v_1 - w_1, v_2 - w_2, v_3 - w_3\rangle.$$

▶ **EXAMPLE 8** Suppose that $\mathbf{v} = \langle -6, 12, 5\rangle$ and $\mathbf{w} = \langle 19, -7, 4\rangle$. Calculate $\mathbf{v} - \mathbf{w}$ and $\mathbf{w} - 3\mathbf{v}$.

Solution We see that $\mathbf{v} - \mathbf{w} = \langle -6-19, 12-(-7), 5-4\rangle = \langle -25, 19, 1\rangle$ and $\mathbf{w} - 3\mathbf{v} = \langle 19 - 3(-6), -7 - 3\cdot 12, 4 - 3\cdot 5\rangle = \langle 37, -43, -11\rangle$. ◀

The Length of a Vector

If $\mathbf{v} = \langle v_1, v_2, v_3 \rangle$ is a vector, then the *length* or *magnitude* of $\mathbf{v}$ is defined to be

$$\|\mathbf{v}\| = \sqrt{(v_1)^2 + (v_2)^2 + (v_3)^2}. \tag{9.2.4}$$

If $\overrightarrow{PQ}$ is a directed line segment that represents $\mathbf{v}$, then $\|\mathbf{v}\|$ is just the distance between the points P and Q, as can be seen from formulas (9.2.1) and (9.2.3). If $\mathbf{v}$ is a vector and λ is a scalar, then the magnitudes of vectors $\mathbf{v}$ and $\lambda\mathbf{v}$ are related by

$$\|\lambda\mathbf{v}\| = |\lambda|\,\|\mathbf{v}\|. \tag{9.2.5}$$

Thus the length of $\lambda\mathbf{v}$ is $|\lambda|$ times the length of $\mathbf{v}$.

▶ **EXAMPLE 9** Let O denote the origin. Suppose that $P = (-9, 6, -3)$ and $Q = (2, 1, 7)$. Calculate $\|\overrightarrow{PQ}\|$, $\|\overrightarrow{OP}\|$, and $\|\overrightarrow{OQ}\|$.

Solution We find that

$$\|\overrightarrow{PQ}\| = \sqrt{(2-(-9))^2 + (1-6)^2 + (7-(-3))^2} = \sqrt{11^2 + 5^2 + 10^2} = \sqrt{246}.$$

Also $\|\overrightarrow{OP}\| = \sqrt{(-9)^2 + 6^2 + (-3)^2} = \sqrt{126}$ and $\|\overrightarrow{OQ}\| = \sqrt{2^2 + 1^2 + 7^2} = \sqrt{54}$. ◀

Unit Vectors and Directions

A vector with length one is called a *unit vector*. We sometimes refer to a unit vector as a *direction vector* or a *direction*.

▶ **EXAMPLE 10** Is there a value of r for which $\mathbf{u} = \langle -1/3, 2/3, r \rangle$ is a unit vector? Is there a value of s for which $\mathbf{v} = \langle -4/5, 4/5, s \rangle$ is a unit vector?

Solution We set $\|\mathbf{u}\| = 1$ and attempt to solve for r. This gives us $(-1/3)^2 + (2/3)^2 + r^2 = 1$ or $r^2 = 1 - 1/9 - 4/9 = 4/9$. Thus if $r = 2/3$ or $r = -2/3$, then $\mathbf{u}$ is a unit vector. On the other hand, $\|\mathbf{v}\| = \sqrt{(-4/5)^2 + (4/5)^2 + s^2} = \sqrt{32/25 + s^2} > \sqrt{1 + s^2}$. Because this inequality is strict, we see that $\|\mathbf{v}\| > 1$ for every value of s. Therefore vector $\mathbf{v}$ cannot be a unit vector for any value of s. ◀

If $\mathbf{v}$ is any nonzero vector, then we may form the vector

$$\mathbf{dir}(\mathbf{v}) = \frac{1}{\|\mathbf{v}\|}\,\mathbf{v} \qquad (\mathbf{v} \neq \mathbf{0}), \tag{9.2.6}$$

which, according to equation (9.2.5) with $\lambda = 1/\|\mathbf{v}\|$, has length 1. We call $\mathbf{dir}(\mathbf{v})$ the *direction* of $\mathbf{v}$. If we multiply each side of equation (9.2.6) by $\|\mathbf{v}\|$, then we obtain

$$\mathbf{v} = \|\mathbf{v}\|\ \mathbf{dir}(\mathbf{v}) \qquad (\mathbf{v} \neq \mathbf{0}). \tag{9.2.7}$$

The right side of equation (9.2.7) is sometimes said to be the *polar form* of $\mathbf{v}$.

DEFINITION Let $\mathbf{v}$ and $\mathbf{w}$ be nonzero vectors. We say that $\mathbf{v}$ and $\mathbf{w}$ have the *same direction* if $\mathbf{dir}(\mathbf{v}) = \mathbf{dir}(\mathbf{w})$. We say that $\mathbf{v}$ and $\mathbf{w}$ are *opposite* in direction if $\mathbf{dir}(\mathbf{v}) = -\mathbf{dir}(\mathbf{w})$. We say that $\mathbf{v}$ and $\mathbf{w}$ are *parallel* if either (a) $\mathbf{v}$ and $\mathbf{w}$ have the same direction or (b) $\mathbf{v}$ and $\mathbf{w}$ are opposite in direction. Although the zero vector $\mathbf{0}$ does not have a direction, it is conventional to say that $\mathbf{0}$ is parallel to every vector.

Notice that nonzero vectors **v** and **w** are parallel if and only if $\mathbf{dir}(\mathbf{v}) = \pm\mathbf{dir}(\mathbf{w})$. Our next theorem provides us with a simple algebraic condition for recognizing parallel vectors.

THEOREM 2 Vectors **v** and **w** are parallel if and only if at least one of the following two equations holds: (a) $\mathbf{v} = \mathbf{0}$ or (b) $\mathbf{w} = \lambda\mathbf{v}$ for some scalar λ. Moreover, if **v** and **w** are both nonzero and $\mathbf{w} = \lambda\mathbf{v}$, then **v** and **w** have the same direction if $\lambda > 0$ and opposite directions if $\lambda < 0$.

The proof of this theorem is the same as that of its planar analogue (Theorem 2 of Section 1).

▶ **EXAMPLE 11** Suppose that $\mathbf{v} = \langle 4, 3, 1\rangle$ and $\mathbf{w} = \langle 2, b, c\rangle$. Are there values of b and c for which **v** and **w** are parallel?

Solution If $\mathbf{v} = \lambda\mathbf{w}$ for some scalar λ then $\langle 4,3,1\rangle = \lambda\langle 2, b, c\rangle = \langle 2\lambda, \lambda b, \lambda c\rangle$. It follows that $4 = 2\lambda$ or $\lambda = 2$. Therefore $3 = \lambda b = 2b$ and $1 = \lambda c = 2c$. We conclude that $\mathbf{v} = \langle 4,3,1\rangle$ and $\mathbf{w} = \langle 2, b, c\rangle$ are parallel if and only if $b = 3/2$ and $c = 1/2$. ◀

The Special Unit Vectors i, j, and k

Let **i** denote the vector $\langle 1,0,0\rangle$, **j** the vector $\langle 0,1,0\rangle$, and **k** the vector $\langle 0,0,1\rangle$. These are the unit vectors that give the positive directions of the coordinate axes, as shown in Figure 9. If $\mathbf{v} = \langle v_1, v_2, v_3\rangle$ is any vector, then we may express **v** in terms of **i**, **j**, and **k** as follows: $\mathbf{v} = v_1\mathbf{i} + v_2\mathbf{j} + v_3\mathbf{k}$.

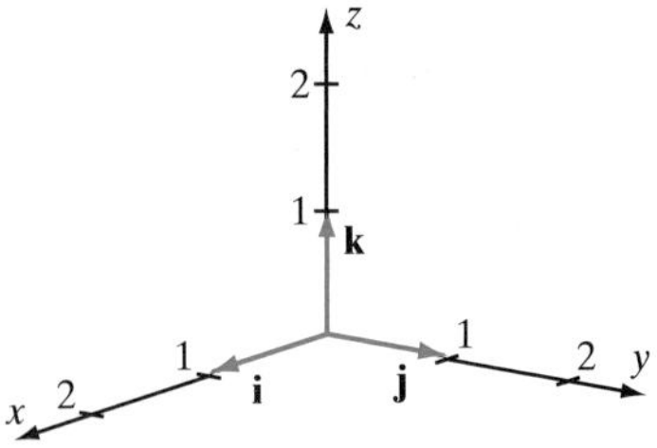

▲ Figure 9

▶ **EXAMPLE 12** For what values of b and c are the vectors $\mathbf{v} = 9\mathbf{i} + b\mathbf{j} - \mathbf{k}$ and $\mathbf{w} = -3\mathbf{i} + 2\mathbf{j} + c\mathbf{k}$ opposite in direction?

Solution The vectors $\mathbf{v} = 9\mathbf{i} + b\mathbf{j} - \mathbf{k}$ and $\mathbf{w} = -3\mathbf{i} + 2\mathbf{j} + c\mathbf{k}$ are opposite in direction if and only if $\mathbf{v} = \lambda\mathbf{w}$ for some negative scalar λ. The vector equation $\mathbf{v} = \lambda\mathbf{w}$ is equivalent to the three simultaneous scalar equations $9 = -3\lambda$, $b = 2\lambda$, and $-1 = c\lambda$. The first scalar equation tells us that $\lambda = -3$, which indeed is negative. The second and third scalar equations give us $b = 2(-3) = -6$ and $c = -1/(-3) = 1/3$. Substituting these values for b and c in the formulas for **v** and **w**, we obtain the vectors $9\mathbf{i} - 6\mathbf{j} - \mathbf{k}$ and $-3\mathbf{i} + 2\mathbf{j} + (1/3)\mathbf{k}$, which are indeed opposite in direction. ◀

The Triangle Inequality

The Triangle Inequality,

$$\|\mathbf{v} + \mathbf{w}\| \le \|\mathbf{v}\| + \|\mathbf{w}\|, \tag{9.2.8}$$

which we learned for planar vectors in Section 1, holds for spatial vectors as well. The proof given in Section 1 for planar vectors also applies in space. In the case that **v** and **w** are nonparallel, we see from Figure 7 earlier that equation (9.2.8) is just a vector interpretation of the theorem in Euclidean geometry that states that the sum of the lengths of any two sides of a triangle is greater than the length of the third side.

▶ **EXAMPLE 13** Verify the Triangle Inequality for the vectors $\mathbf{v} = \langle 2, -2, -1\rangle$ and $\mathbf{w} = \langle 3, -6, -2\rangle$.

Solution After calculating $\|\mathbf{v}\| + \|\mathbf{w}\| = \sqrt{2^2 + (-2)^2 + (-1)^2} + \sqrt{3^2 + (-6)^2 + (-2)^2} = 3 + 7 = 10$, we find that

$$\|\mathbf{v} + \mathbf{w}\| = \|\langle 5, -8, -3\rangle\| = \sqrt{5^2 + (-8)^2 + (-3)^2} = \sqrt{98} < \sqrt{100} = 10 = \|\mathbf{v}\| + \|\mathbf{w}\|.$$ ◀

We conclude this section by stating the distributive, associative, and commutative laws for scalar multiplication and vector addition. If **u**, **v**, **w** are vectors in space and λ, μ are scalars, then

a. $\mathbf{v} + \mathbf{w} = \mathbf{w} + \mathbf{v}$
b. $\mathbf{u} + (\mathbf{v} + \mathbf{w}) = (\mathbf{u} + \mathbf{v}) + \mathbf{w}$
c. $\lambda(\mu\mathbf{v}) = (\lambda\mu)\mathbf{v}$
d. $\lambda(\mathbf{v} + \mathbf{w}) = \lambda\mathbf{v} + \lambda\mathbf{w}$
e. $(\lambda + \mu)\mathbf{v} = \lambda\mathbf{v} + \mu\mathbf{v}$

Equations a through e may be routinely verified.

QUICK QUIZ

1. What is the distance between the points $(3, 1, 5)$ and $(1, -1, 6)$?
2. If $P = (-2, -1, 2)$ and $Q = (5, 3, 6)$, then what is the length of the vector represented by the directed line segment $\overrightarrow{PQ}$?
3. What is the direction of $\langle 8, -1, -4\rangle$?
4. Write $2\mathbf{k} - 2(\mathbf{i} + 4\mathbf{j} - 7\mathbf{k}) + 3(-\mathbf{i} + 4\mathbf{j} - 5\mathbf{k})$ in the form $\langle a, b, c\rangle$.

Answers

1. 3   2. 9   3. $\langle 8/9, -1/9, -4/9\rangle$ 4. $\langle -5, 4, 1\rangle$

EXERCISES

Problems for Practice

▶ In each of Exercises 1–4, graph the given point P. Also graph the point that lies directly above or below P in the xy-plane. ◀

1. $P = (3, 2, 1)$
2. $P = (2, 0, 5)$
3. $P = (-4, 1, -2)$
4. $P = (3, -1, 2)$

▶ In Exercises 5–8, calculate the distance between the given pair of points. ◀

5. $(2, 0, -3)$ $(8, 3, 6)$
6. $(-2, -3, -5)$ $(4, 2, 0)$
7. $(4, 4, 1)$ $(1, 1, 4)$
8. $(-4, 2, 1)$ $(8, 0, 4)$

▶ In each of Exercises 9–14, determine the center and radius of each sphere whose Cartesian equation is given. ◀

9. $x^2 + y^2 + z^2 + 2x + 4y + 6z = 8$
10. $3x^2 + 3y^2 + 3z^2 - 12x + 6y = 0$
11. $4x^2 + 4y^2 + 4z^2 + 4x + 4y + 4z = 2$
12. $-x^2 - y^2 - z^2 = 6x - 8y + 12z - 4$
13. $x^2 + y^2 + z^2 = x$
14. $-4x^2 - 4y^2 - 4z^2 = -2x - 2y - 2z - 6$

▶ In each of Exercises 15–20, explain how to recognize that the given Cartesian equation is not the equation of a sphere. ◀

15. $x^2 + y^2 + 2z^2 + 2x + 4z = 8$
16. $2x^2 - 2y^2 + 2z^2 = 1$
17. $x^2 + y^2 + z^2 + 4xy = 2$
18. $x^2 + y^2 + z^2 + 2z + 2 = 0$
19. $(x - y)^2 + z^2 = 1$
20. $x^2 + y^2 = 2x + 4y - 7 - z^2$

▶ In each of Exercises 21–26, find an equation or inequality $P(x, y, z)$ such that the given set is described by $\{(x, y, z) : P(x, y, 9z)\}$. For example, the origin is the set $\{(x, y, z) : x^2 + y^2 + z^2 = 0\}$, and the set of points above the xy-plane is $\{(x, y, z) : 0 < z\}$. ◀

21. The open ball with center $(3, -2, 6)$ and radius 4.
22. The closed ball with center $(1, 2, 9)$ and radius 1.
23. The set of points whose distance to the point $(\pi, \pi, -\pi)$ is π.

24. The set of points whose distance to the point $(1,-2,0)$ is not greater than 3.
25. The set of points outside the closed ball with center $(2,1,0)$ and radius 2.
26. The set of points outside the open ball with center $(-3,-1,-5)$ and radius 5.

▶ Points P, Q, R, and S are given in each of Exercises 27–30, Calculate the lengths of vectors $\overrightarrow{PQ}$ and $\overrightarrow{RS}$. Also determine if the two vectors are parallel. ◀

27. $P=(1,2,-1)$ $Q=(2,0,-7)$ $R=(2,6,9)$ $S=(4,2,-3)$
28. $P=(-1,3,6)$ $Q=(4,1,5)$ $R=(1,1,0)$ $S=(1,3,5)$
29. $P=(0,1,4)$ $Q=(-1,3,6)$ $R=(6,-3,0)$ $S=(1,1,2)$
30. $P=(1,1,2)$ $Q=(-7,5,4)$ $R=(1,0,6)$ $S=(19,-12,0)$

▶ In each of Exercises 31–38, let $\mathbf{v}=\langle 3,2,6\rangle$ and $\mathbf{w}=\langle 4,-1,8\rangle$. Calculate the specified vector. ◀

31. $-3\mathbf{w}$
32. $3\mathbf{v}-4\mathbf{w}$
33. $-4\mathbf{v}+3\mathbf{w}$
34. $(1/2)\mathbf{w}-(1/3)\mathbf{v}$
35. $2\mathbf{dir}(\mathbf{v})$
36. $21\mathbf{dir}(\mathbf{v})-18\mathbf{dir}(\mathbf{w})$
37. $\mathbf{v}+3\mathbf{k}$
38. $\mathbf{w}-4\mathbf{i}+7\mathbf{k}$

▶ In each of Exercises 39–42, points P and Q are given. State (a) the vector $\mathbf{v}$ which is represented by $\overrightarrow{PQ}$, (b) the vector that has the same length as $\mathbf{v}$ and direction opposite to that of $\mathbf{v}$, (c) the length of $\mathbf{v}$, (d) the direction $\mathbf{dir}(\mathbf{v})$ of $\mathbf{v}$, and (e) the vector with length 12 that has the same direction as $\mathbf{v}$. ◀

39. $P=(1,2,-1)$ $Q=(3,-1,5)$
40. $P=(-2,2,3)$ $Q=(4,-4,-4)$
41. $P=(3,1,-3)$ $Q=(4,0,-5)$
42. $P=(9,8,-1)$ $Q=(5,12,6)$

▶ In each of Exercises 43–48, let $\mathbf{v}=\langle 3,-4,1\rangle$ and $\mathbf{w}=\langle -5,2,0\rangle$. Express the given vector in the form $a\mathbf{i}+b\mathbf{j}+c\mathbf{k}$. ◀

43. $-5\mathbf{v}$
44. $4\mathbf{w}$
45. $\mathbf{w}+2\mathbf{v}$
46. $2\mathbf{v}-\mathbf{w}$
47. $\mathbf{v}-3\mathbf{i}$
48. $2\mathbf{w}+4\mathbf{j}$

▶ In each of Exercises 49–52, the two given vectors determine a parallelogram $\mathcal{P}$. Calculate the vectors with positive first entries that represent the diagonals of $\mathcal{P}$. ◀

49. $\langle 3,-2,1\rangle$, $\langle 1,-4,2\rangle$
50. $\langle -1,1,0\rangle$, $\langle 4,0,1\rangle$
51. $-\mathbf{i}+\mathbf{j}+\mathbf{k}$, $2\mathbf{j}-\mathbf{k}$
52. $-2\mathbf{i}+3\mathbf{k}$, $-\mathbf{j}+\mathbf{k}$

Further Theory and Practice

53. Let $P_0=(x_0, y_0, z_0)$ be a point in space. Which point in the xy-plane is nearest to P_0? Which point in the yz-plane is nearest to P_0? How about the xz-plane?

▶ In each of Exercises 54–61, find an equation or inequality $P(x,y,z)$ such that the given set is described by $\{(x,y,z) : P(x, y,z)\}$. See the instructions to Exercises 21–26 for examples. ◀

54. The yz-coordinate plane.
55. The xz-coordinate plane.
56. The plane that is parallel to the xy-plane and that passes through the point (π, π^2, π^3).
57. The open half space that contains all points that lie on the same side of the xz-coordinate plane as the point $(1,1,1)$.
58. The half space that comprises the yz-plane as well as all points that are on the same side of the yz-plane as the point $(1,1,1)$.
59. The set of points whose distance to the xy-coordinate plane is greater than 5.
60. The set of points that lie in one or more of the three coordinate planes. (In other words, the union of the three coordinate planes.)
61. The set of points that are farther from the origin than from $(2,-1,-2)$

62. What are the lengths of the diagonals of the parallelogram determined by the vectors $\langle 1,1,0\rangle$ and $\langle 0,1,2\rangle$?
63. Give a geometric description of the set of all vectors of the form

$$\langle\cos(\theta)\sin(\phi), \sin(\theta)\sin(\phi), \cos(\phi)\rangle \quad 0\le\theta<2\pi\,, 0\le\phi\le\pi.$$

64. Consider the vectors $\mathbf{u}=\langle 1,3,4\rangle$, $\mathbf{v}=\langle -2,1,6\rangle$, and $\mathbf{w}=\langle 0,1,5\rangle$. If $\mathbf{m}$ is any other vector in space, then show that there is a unique triple of numbers a, b, c such that

$$\mathbf{m}=a\mathbf{u}+b\mathbf{v}+c\mathbf{w}.$$

65. A student walking at the rate of 4 feet per second crosses a 16 foot high pedestrian bridge. A car passes directly underneath traveling at constant speed 40 feet per second. How fast is the distance between the student and the car changing 2 seconds later?
66. Suppose that P, Q, and R are three noncollinear points. Show that there are three points A, B, and C, any one of which together with P, Q, and R forms a parallelogram. Show that

$$\overrightarrow{OA}+\overrightarrow{OB}+\overrightarrow{OC}=\overrightarrow{OP}+\overrightarrow{OQ}+\overrightarrow{OR}.$$

67. On which sphere do the four points $(5,2,3)$, $(1,6,-1)$, $(3,-2,5)$, and $(-1,2,-3)$ lie?
68. What is the minimum value of $d(P,Q)$ if P is a point on the sphere $x^2+10x+y^2-12y+z^2-2z+37=0$ and Q is a point on the sphere $x^2-12x+y^2+8y+z^2-6z+45=0$?
69. Atomic particles may carry a positive charge (like a proton) or a negative charge (like an electron). A charged

particle exerts a force on every other charged particle. Suppose that P is the location of a charge p and Q is the location of a charge q. Let r be the distance between P and Q. Coulomb's Law states that the electric force exerted by the particle at P on the particle at Q is given by

$$\mathbf{F} = \frac{pq}{4\pi\varepsilon_0 r^2}\mathbf{u}$$

where $\mathbf{u}$ is the unit vector in the direction of $\overrightarrow{PQ}$ and ε_0 is a positive constant, the numerical value of which depends on the units of measurement. Discuss the physical meaning of the sign of the coefficient of $\mathbf{u}$ in this equation. In particular, explain how the principle "Like charges repel and opposite charges attract" is captured by the formula for the force.

70. Describe the intersections of the surface $z = x^2 + y^2/4$ with the coordinate planes and with the planes $z = h$ where h is a constant. Sketch the surface.

71. Describe the intersections of the surface $x^2 + y^2/9 + z^2/16 = 1$ with the coordinate planes. Sketch the surface.

72. Describe the intersections of the surface $z = \sqrt{9x^2 + y^2}$ with the coordinate planes and with the planes $z = h$ where h is a constant. Sketch the surface.

Suppose that $\mathbf{v}_1, \mathbf{v}_2, \ldots, \mathbf{v}_N$ are vectors and $\lambda_1, \lambda_2, \ldots, \lambda_N$ are scalars. The vector $\mathbf{w} = \lambda_1\mathbf{v}_1 + \lambda_2\mathbf{v}_2 + \ldots + \lambda_N\mathbf{v}_N$ is said to be a *linear combination* of the vectors $\mathbf{v}_1, \mathbf{v}_2, \ldots, \mathbf{v}_N$. If every vector in $\mathbb{R}^3$ is a linear combination of $\mathbf{v}_1, \mathbf{v}_2, \ldots, \mathbf{v}_N$, then we say that the vectors $\mathbf{v}_1, \mathbf{v}_2, \ldots, \mathbf{v}_N$ *span* or *generate* $\mathbb{R}^3$. The vectors $\mathbf{i}$, $\mathbf{j}$, $\mathbf{k}$, for example, span $\mathbb{R}^3$. In each of Exercises 73–76, determine whether the three given vectors span $\mathbb{R}^3$.

73. $\mathbf{i}+\mathbf{j},\ \mathbf{j}+\mathbf{k},\ \mathbf{i}+\mathbf{k}$

74. $\mathbf{i}+\mathbf{j}+\mathbf{k},\ \mathbf{i}+\mathbf{j},\ \mathbf{i}-\mathbf{j}$

75. $\mathbf{i}+\mathbf{j}+\mathbf{k},\ \mathbf{i}+\mathbf{j},\ -\mathbf{i}-\mathbf{j}+2\,\mathbf{k}$

76. $\mathbf{i}+2\mathbf{j}+3\mathbf{k},\ \mathbf{i}-2\mathbf{j}+3\mathbf{k},\ \mathbf{i}-3\mathbf{k}$

Calculator/Computer Exercises

77. For what values of t, to four decimal places, are the lengths of $\langle 2t, t, t\rangle$, and $\langle t^2, 1-t, 1\rangle$ equal?

78. For what values, to six decimal places, of s and t in $(0,1)$ do the vectors $\langle s^2-t, t^2+s, st+1\rangle$ and $\langle 1,2,3\rangle$ have the same direction?

79. A particle's position P_t at time t is the point $(t-\cos(t), \sin(t), t\sin(t))$. How close to the origin does the particle come?

80. At time $t \geq 0$, a particle's position P_t is the point $(t, t\cos(t), t\sin(t))$. Does the vector $\overrightarrow{P_tP_{t+1}}$ ever have length 2?

9.3 The Dot Product and Applications

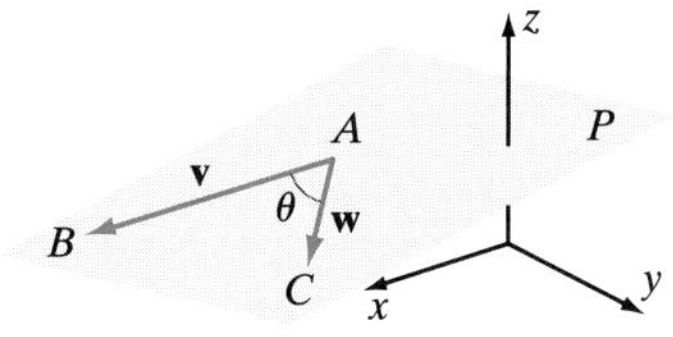

Figure 1

If two nonzero vectors $\mathbf{v}$ and $\mathbf{w}$ have the same direction, then the angle between them is 0. If the two vectors are in opposite directions, then the angle between them is π. In all other cases, we may use the two vectors to form a plane $\mathcal{P}$ by selecting directed line segment representatives $\overrightarrow{AB}$ and $\overrightarrow{AC}$ that have the same initial point (see Figure 1). The angle between $\mathbf{v}$ and $\mathbf{w}$ is understood to be the angle $\theta \in (0,\pi)$ between $\overrightarrow{AB}$ and $\overrightarrow{AC}$ in plane $\mathcal{P}$. It does not depend on the initial point A that is selected. In this section, we will study an algebraic operation that can be used to calculate the angle between two vectors.

The Algebraic Definition of the Dot Product

In Sections 1 and 2, we discussed the addition of vectors and the multiplication of vectors by scalars. We now introduce a useful way to form the product of two vectors.

DEFINITION The *dot product* $\mathbf{v}\cdot\mathbf{w}$ of two vectors $\mathbf{v}$ and $\mathbf{w}$ is the sum of the products of corresponding entries of $\mathbf{v}$ and $\mathbf{w}$. That is, if $\mathbf{v} = \langle v_1, v_2\rangle$ and $\mathbf{w} = \langle w_1, w_2\rangle$ are planar vectors, then $\mathbf{v}\cdot\mathbf{w}$ is defined by

$$\mathbf{v}\cdot\mathbf{w} = v_1w_1 + v_2w_2.$$

If $\mathbf{v} = \langle v_1, v_2, v_3\rangle$ and $\mathbf{w} = \langle w_1, w_2, w_3\rangle$ are spatial vectors, then $\mathbf{v}\cdot\mathbf{w}$ is defined by

$$\mathbf{v}\cdot\mathbf{w} = v_1w_1 + v_2w_2 + v_3w_3.$$

Notice that the dot product of two vectors is a scalar, or real number. For that reason, the quantity $\mathbf{v}\cdot\mathbf{w}$ is sometimes called the *scalar product* of $\mathbf{v}$ and $\mathbf{w}$. This terminology should not be confused with *scalar multiplication*, which we have discussed in the preceding two sections. When we create a vector $\lambda\mathbf{v}$ by scalar multiplication, we multiply *one* vector $\mathbf{v}$ by a scalar λ. The result of scalar multiplication is a *vector*. We *never* use a dot to signify this type of multiplication. By contrast, the scalar product is an operation between *two* vectors, and the result is a *scalar*. We *always* use a dot to signify this type of multiplication.

▶ **EXAMPLE 1** Let $\mathbf{u}=\langle 2,3,-1\rangle$, $\mathbf{v}=\langle 4,6,-2\rangle$, and $\mathbf{w}=\langle -2,-1,-7\rangle$. Calculate the dot products $\mathbf{u}\cdot\mathbf{v}$, $\mathbf{u}\cdot\mathbf{w}$, and $\mathbf{v}\cdot\mathbf{w}$.

Solution We have

$$\begin{aligned}\mathbf{u}\cdot\mathbf{v} &= (2)(4)+(3)(6)+(-1)(-2)=28\\ \mathbf{u}\cdot\mathbf{w} &= (2)(-2)+(3)(-1)+(-1)(-7)=0\\ \mathbf{v}\cdot\mathbf{w} &= (4)(-2)+(6)(-1)+(-2)(-7)=0.\end{aligned}$$

In the next theorem, we collect several simple algebraic rules for working with the dot product.

THEOREM 1 Suppose that $\mathbf{u}$, $\mathbf{v}$, and $\mathbf{w}$ are vectors and that λ is a scalar. The dot product satisfies the following elementary properties:

a. $\mathbf{u}\cdot(\mathbf{v}+\mathbf{w})=\mathbf{u}\cdot\mathbf{v}+\mathbf{u}\cdot\mathbf{w}$
b. $\mathbf{v}\cdot\mathbf{w}=\mathbf{w}\cdot\mathbf{v}$
c. $(\lambda\mathbf{v})\cdot\mathbf{w}=\mathbf{v}\cdot(\lambda\mathbf{w})=\lambda(\mathbf{v}\cdot\mathbf{w})$.

INSIGHT Theorem 1c tells us that the value of the expression $\lambda\mathbf{v}\cdot\mathbf{w}$ is unambiguous: Both of the two possible interpretations, $(\lambda\mathbf{v})\cdot\mathbf{w}$ and $\lambda(\mathbf{v}\cdot\mathbf{w})$, result in the same number. Nevertheless, we will generally insert parentheses for clarity. A similar remark applies to the expression $\mathbf{v}\cdot\lambda\mathbf{w}$. Because we do not form the dot product of a vector $\mathbf{v}$ and a scalar λ, we cannot interpret $\mathbf{v}\cdot\lambda\mathbf{w}$ as $(\mathbf{v}\cdot\lambda)\mathbf{w}$. Thus we can understand $\mathbf{v}\cdot\lambda\mathbf{w}$ only as $\mathbf{v}\cdot(\lambda\mathbf{w})$. By including the parentheses, we make this meaning instantly evident.

▶ **EXAMPLE 2** Let $\mathbf{u}=\langle 3,2\rangle$ and $\mathbf{v}=\langle 4,-5\rangle$. Calculate $(\mathbf{u}\cdot\mathbf{v})\,\mathbf{u}+(\mathbf{v}\cdot\mathbf{u})\mathbf{v}$.

Solution Using Theorem 1b and then 1a, we have

$$\begin{aligned}(\mathbf{u}\cdot\mathbf{v})\mathbf{u}+(\mathbf{v}\cdot\mathbf{u})\mathbf{v} &= (\mathbf{u}\cdot\mathbf{v})\mathbf{u}+(\mathbf{u}\cdot\mathbf{v})\mathbf{v}=(\mathbf{u}\cdot\mathbf{v})(\mathbf{u}+\mathbf{v})\\ &= ((3)(4)+(2)(-5))\langle 3+4,2-5\rangle=2\langle 7,-3\rangle=\langle 14,-6\rangle.\end{aligned}$$ ◀

A Geometric Formula for the Dot Product

Because $\mathbf{v}\cdot\mathbf{v}=(v_1)^2+(v_2)^2+(v_3)^2$, we can relate the dot product $\mathbf{v}\cdot\mathbf{v}$ to the length of $\mathbf{v}$ by rewriting equation (9.2.4) as

$$\|\mathbf{v}\|=\sqrt{\mathbf{v}\cdot\mathbf{v}}, \tag{9.3.1}$$

or

$$\|\mathbf{v}\|^2=\mathbf{v}\cdot\mathbf{v}. \tag{9.3.2}$$

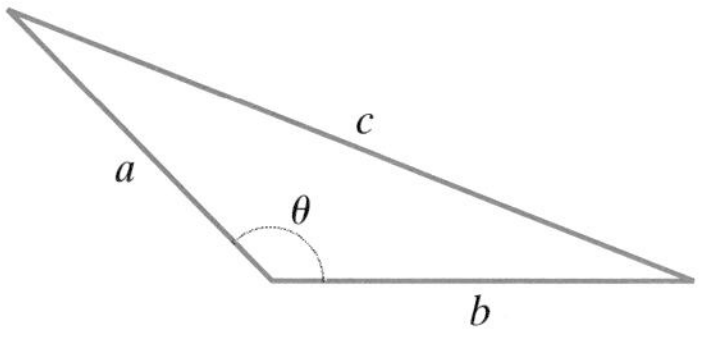

▲ Figure 2 Law of Cosines: $c^2 = a^2 + b^2 - 2ab\cos(\theta)$

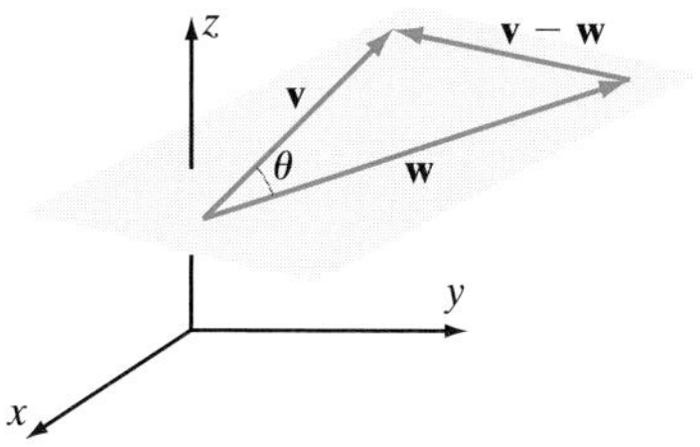

▲ Figure 3

We will use this formula to develop a method for calculating the angle between two vectors and for calculating projections. To that end, recall that the Law of Cosines says that if the sides of a triangle measure a, b, c and if θ is the angle between the first two sides, as in Figure 2, then

$$c^2 = a^2 + b^2 - 2ab\cos(\theta). \tag{9.3.3}$$

Supposing that $\mathbf{v}$ and $\mathbf{w}$ are not parallel, we apply the Law of Cosines to the triangle determined by $\mathbf{v}$, $\mathbf{w}$, and $\mathbf{v} - \mathbf{w}$ in Figure 3. This trigonometric formula tells us that

$$\|\mathbf{v} - \mathbf{w}\|^2 = \|\mathbf{v}\|^2 + \|\mathbf{w}\|^2 - 2\|\mathbf{v}\|\,\|\mathbf{w}\|\cos(\theta). \tag{9.3.4}$$

Using equation (9.3.2) and the rules of Theorem 1, we may expand the left side of equation (9.3.4), obtaining

$$\|\mathbf{v} - \mathbf{w}\|^2 = (\mathbf{v} - \mathbf{w}) \cdot (\mathbf{v} - \mathbf{w}) = \mathbf{v} \cdot \mathbf{v} - \mathbf{v} \cdot \mathbf{w} - \mathbf{w} \cdot \mathbf{v} + \mathbf{w} \cdot \mathbf{w} = \|\mathbf{v}\|^2 - 2\mathbf{v} \cdot \mathbf{w} + \|\mathbf{w}\|^2.$$

Substituting this expression for $\|\mathbf{v} - \mathbf{w}\|^2$ into the left side of equation (9.3.4), we obtain

$$\|\mathbf{v}\|^2 - 2\mathbf{v} \cdot \mathbf{w} + \|\mathbf{w}\|^2 = \|\mathbf{v}\|^2 + \|\mathbf{w}\|^2 - 2\|\mathbf{v}\|\,\|\mathbf{w}\|\cos(\theta).$$

Canceling like terms from each side results in the identity $-2\mathbf{v} \cdot \mathbf{w} = -2\|\mathbf{v}\| \cdot \|\mathbf{w}\| \cdot \cos(\theta)$, or

$$\mathbf{v} \cdot \mathbf{w} = \|\mathbf{v}\|\,\|\mathbf{w}\|\cos(\theta). \tag{9.3.5}$$

THEOREM 2 Let $\mathbf{v}$ and $\mathbf{w}$ be nonzero vectors. Then, the angle θ between $\mathbf{v}$ and $\mathbf{w}$ satisfies the equation

$$\cos(\theta) = \frac{\mathbf{v} \cdot \mathbf{w}}{\|\mathbf{v}\|\,\|\mathbf{w}\|}. \tag{9.3.6}$$

Proof. If $\mathbf{v}$ and $\mathbf{w}$ are not parallel, then equation (9.3.5) holds. We obtain equation (9.3.6) by dividing by $\|\mathbf{v}\|\,\|\mathbf{w}\|$, which is not 0. If $\mathbf{v}$ and $\mathbf{w}$ are parallel and have the same direction, then $\theta = 0$ and $\mathbf{w} = \lambda\mathbf{v}$ for some positive scalar λ. The left side of equation (9.3.6) is cos(0), or 1. Using Theorem 1c, we calculate

$$\|\mathbf{v}\|\,\|\mathbf{w}\| = \|\mathbf{v}\|\,\|\lambda\mathbf{v}\| \overset{(9.2.5)}{=} |\lambda|\,\|\mathbf{v}\|\,\|\mathbf{v}\| = \lambda\,\|\mathbf{v}\|^2 \overset{(9.3.2)}{=} \lambda(\mathbf{v} \cdot \mathbf{v}) = \mathbf{v} \cdot (\lambda\mathbf{v}) = \mathbf{v} \cdot \mathbf{w}, \tag{9.3.7}$$

which shows that the right side of equation (9.3.6) is also 1. Finally, if $\mathbf{v}$ and $\mathbf{w}$ are parallel and have the opposite direction, then $\theta = \pi$ and $\mathbf{w} = \lambda\mathbf{v}$ for some negative scalar λ. The left side of equation (9.3.6) is $\cos(\pi)$, or -1. Using Theorem 1c, we calculate

$$\|\mathbf{v}\|\,\|\mathbf{w}\| = \|\mathbf{v}\|\,\|\lambda\mathbf{v}\| \overset{(9.2.5)}{=} |\lambda|\,\|\mathbf{v}\|\,\|\mathbf{v}\| = -\lambda\,\|\mathbf{v}\|^2 \overset{(9.3.2)}{=} -\lambda(\mathbf{v} \cdot \mathbf{v}) = -\mathbf{v} \cdot (\lambda\mathbf{v}) = -\mathbf{v} \cdot \mathbf{w}, \tag{9.3.8}$$

which shows that the right side of equation (9.3.6) is also -1. ■

▶ **EXAMPLE 3** Calculate the angle between the two vectors $\mathbf{v} = \langle 2, 2, 4\rangle$ and $\mathbf{w} = \langle 2, -1, 1\rangle$.

Solution If θ is the angle between $\mathbf{v}$ and $\mathbf{w}$, then Theorem 2 tells us that

$$\begin{aligned}\cos(\theta) &= \frac{\mathbf{v}\cdot\mathbf{w}}{\|\mathbf{v}\|\,\|\mathbf{w}\|}\\ &= \frac{\langle 2,2,4\rangle\cdot\langle 2,-1,1\rangle}{\|\langle 2,2,4\rangle\|\,\|\langle 2,-1,1\rangle\|}\\ &= \frac{2\cdot 2+2\cdot(-1)+4\cdot 1}{\sqrt{2^2+2^2+4^2}\cdot\sqrt{2^2+(-1)^2+1^2}}\\ &= \frac{6}{\sqrt{24}\cdot\sqrt{6}}\\ &= \frac{1}{2}.\end{aligned}$$

Therefore θ is $\pi/3$. ◄

Equation (9.3.5) gives rise to an important inequality. Because $|\cos(\theta)|\le 1$ for every angle θ, we have

$$|\mathbf{v}\cdot\mathbf{w}|\le\|\mathbf{v}\|\,\|\mathbf{w}\|. \tag{9.3.9}$$

This relationship, called the *Cauchy-Schwarz Inequality*, can also be deduced from the identity

$$\|\mathbf{v}\|^2\|\mathbf{w}\|^2-(\mathbf{v}\cdot\mathbf{w})^2=(v_1w_2-w_1v_2)^2+(v_1w_3-w_1v_3)^2+(v_2w_3-w_2v_3)^2 \tag{9.3.10}$$

for $\mathbf{v}=\langle v_1,\,v_2,\,v_3\rangle$ and $\mathbf{w}=\langle w_1,\,w_2,\,w_3\rangle$. See Exercise 59.

► **EXAMPLE 4** Verify that the two vectors $\mathbf{v}=\langle 2,2,4\rangle$ and $\mathbf{w}=\langle 12,13,24\rangle$ satisfy the Cauchy-Schwarz Inequality.

Solution We calculate $\mathbf{v}\cdot\mathbf{w}=(2)(12)+(2)(13)+(4)(24)=146$ and

$$\|\mathbf{v}\|\,\|\mathbf{w}\|=\sqrt{2^2+2^2+4^2}\sqrt{12^2+13^2+24^2}=(2\sqrt{6})(\sqrt{889})=146.068\ldots,$$

which is greater than $|\mathbf{v}\cdot\mathbf{w}|$, as the Cauchy-Schwarz Inequality asserts. ◄

DEFINITION Let θ be the angle between nonzero vectors $\mathbf{v}$ and $\mathbf{w}$. If $\theta=\pi/2$, then we say that the vectors $\mathbf{v}$ and $\mathbf{w}$ are *orthogonal* or *mutually perpendicular*. Although the zero vector $\mathbf{0}$ has no direction, it is conventional to say that $\mathbf{0}$ is orthogonal to every vector.

Our next theorem serves two important purposes. The definition of orthogonality of two vectors is geometric, as it should be. However, because the definition does not involve the entries of the vectors, it does not allow us to easily verify orthogonality. Theorem 3 provides us with a simple algebraic criterion for doing so. By contrast, our treatment of parallel vectors has been, up until now, algebraic and easy to apply. What it has lacked has been a transparent geometric foundation that refers to the angle between the two vectors. Theorem 3 fills in this gap.

THEOREM 3 Let $\mathbf{v}$ and $\mathbf{w}$ be any vectors Then:

a. $\mathbf{v}$ and $\mathbf{w}$ are orthogonal if and only if $\mathbf{v} \cdot \mathbf{w} = 0$.
b. $\mathbf{v}$ and $\mathbf{w}$ are parallel if and only if $|\mathbf{v} \cdot \mathbf{w}| = \|\mathbf{v}\| \, \|\mathbf{w}\|$.
c. If $\mathbf{v}$ and $\mathbf{w}$ are nonzero, and if θ is the angle between them, then $\mathbf{v}$ and $\mathbf{w}$ are parallel if and only if $\theta = 0$ or $\theta = \pi$. In this case, $\mathbf{v}$ and $\mathbf{w}$ have the same direction if $\theta = 0$ and opposite directions if $\theta = \pi$.

Proof. Recall that a vector is $\mathbf{0}$ if and only if its length is 0. With this in mind, we see from equation (9.3.5) that $\mathbf{v} \cdot \mathbf{w} = 0$ if and only if $\mathbf{v} = \mathbf{0}$, $\mathbf{w} = \mathbf{0}$, or $\theta = \pi/2$. Statement a is an immediate consequence of this observation.

To prove statement b, recall from Theorem 2, Section 9.2 that $\mathbf{v}$ and $\mathbf{w}$ are parallel if and only if $\mathbf{v} = \mathbf{0}$ or $\mathbf{w} = \lambda\mathbf{v}$ for some scalar λ. Thus statement b asserts that $\mathbf{v} = \mathbf{0}$ or $\mathbf{w} = \lambda\mathbf{v}$ for some scalar λ if and only if $|\mathbf{v} \cdot \mathbf{w}| = \|\mathbf{v}\| \, \|\mathbf{w}\|$. It is evident that $\mathbf{v} = \mathbf{0}$ implies $|\mathbf{v} \cdot \mathbf{w}| = \|\mathbf{v}\| \, \|\mathbf{w}\|$: Both sides are 0 in this case. That $\mathbf{w} = \lambda\mathbf{v}$ implies $|\mathbf{v} \cdot \mathbf{w}| = \|\mathbf{v}\| \, \|\mathbf{w}\|$ has been proved in lines (9.3.7) and (9.3.8). For the last step of the proof of statement b, we must show that if $|\mathbf{v} \cdot \mathbf{w}| = \|\mathbf{v}\| \, \|\mathbf{w}\|$ and $\mathbf{v} \neq \mathbf{0}$, then $\mathbf{w} = \lambda\mathbf{v}$. In this case, we write $\mathbf{v} = \langle v_1, v_2, v_3 \rangle$, $\mathbf{w} = \langle w_1, w_2, w_3 \rangle$, and use equation (9.3.10) to deduce that

$$(v_1 w_2 - w_1 v_2)^2 + (v_1 w_3 - w_1 v_3)^2 + (v_2 w_3 - w_2 v_3)^2 = 0,$$

or, equivalently,

$$v_1 w_2 = w_1 v_2, \quad v_1 w_3 = w_1 v_3, \quad \text{and} \quad v_2 w_3 = w_2 v_3. \tag{9.3.11}$$

Because $\mathbf{v} \neq \mathbf{0}$, at least one of the entries v_1, v_2, v_3 is nonzero. Whichever entry is not 0, the argument is the same. Suppose, for example, $v_2 \neq 0$. Let $\lambda = w_2/v_2$. Then $w_2 = \lambda v_2$. Using this equation with the first equation of line (9.3.11), we obtain $v_1(\lambda v_2) = w_1 v_2$, or $w_1 = \lambda v_1$. Similarly, using the last equation of line (9.3.11), we obtain $v_2 w_3 = (\lambda v_2) v_3$, or $w_3 = \lambda v_3$. Thus

$$\mathbf{w} = \langle w_1, w_2, w_3 \rangle = \langle \lambda v_1, \lambda v_2, \lambda v_3 \rangle = \lambda \langle v_1, v_2, v_3 \rangle = \lambda\mathbf{v}.$$

Finally, from formula (9.3.6), we see that $|\mathbf{v} \cdot \mathbf{w}| = \|\mathbf{v}\| \, \|\mathbf{w}\|$ holds for nonzero vectors $\mathbf{v}$ and $\mathbf{w}$ if and only if $|\cos(\theta)| = 1$, which is to say $\theta = 0$ or $\theta = \pi$. Statement c now follows from statement b. ■

► **EXAMPLE 5** Consider the vectors $\mathbf{u} = \langle 2, 3, -1 \rangle$, $\mathbf{v} = \langle 4, 6, -2 \rangle$, and $\mathbf{w} = \langle -2, -1, -7 \rangle$ of Example 1. Are any of these vectors orthogonal? Parallel?

Solution From the dot products $\mathbf{u} \cdot \mathbf{w} = 0$ and $\mathbf{v} \cdot \mathbf{w} = 0$ that we calculated in Example 1, we conclude that $\mathbf{u}$ and $\mathbf{v}$ *are both* orthogonal to $\mathbf{w}$. Because $\mathbf{u} \cdot \mathbf{v} = 28 \neq 0$, we can tell that $\mathbf{u}$ is *not* orthogonal to $\mathbf{v}$. Indeed, the equation

$$\|\mathbf{u}\| \, \|\mathbf{v}\| = \sqrt{2^2 + 3^2 + (-1)^2}\sqrt{4^2 + 6^2 + (-2)^2} = \sqrt{14 \cdot 56} = \sqrt{784} = 28 = |\mathbf{u} \cdot \mathbf{v}|$$

tells that $\mathbf{u}$ and $\mathbf{v}$ are parallel (by Theorem 3b). Since $\|\mathbf{u}\| \, \|\mathbf{v}\|$ actually equals $+\mathbf{u} \cdot \mathbf{v}$, we may further observe that $\mathbf{u}$ and $\mathbf{v}$ have the same direction. Notice that the equality $\mathbf{v} = 2\mathbf{u}$ is another way to reach this conclusion. ◄

Projection

One of the most powerful constructions in geometry is the *projection*. Figure 4a shows two nonzero vectors $\mathbf{v}$ and $\mathbf{w}$ represented by directed line segments that

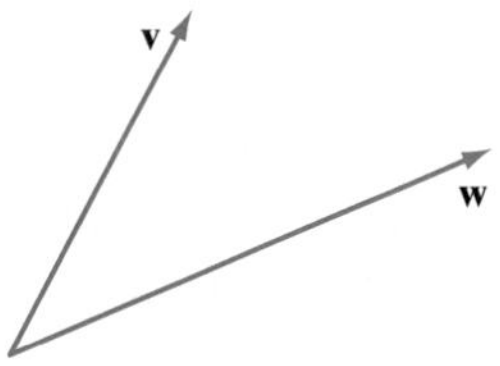

▲ Figure 4a

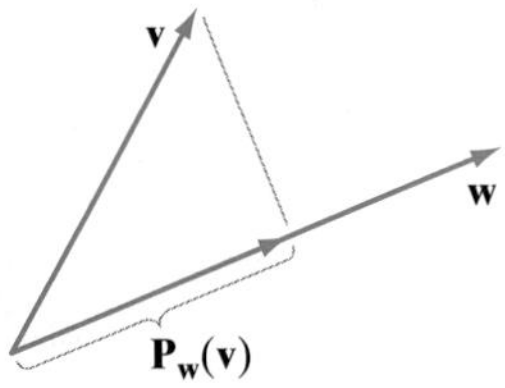

▲ Figure 4b

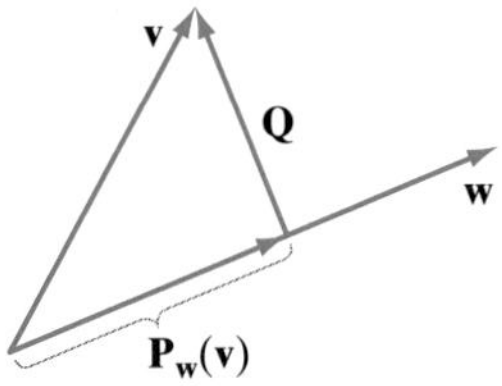

▲ Figure 4c

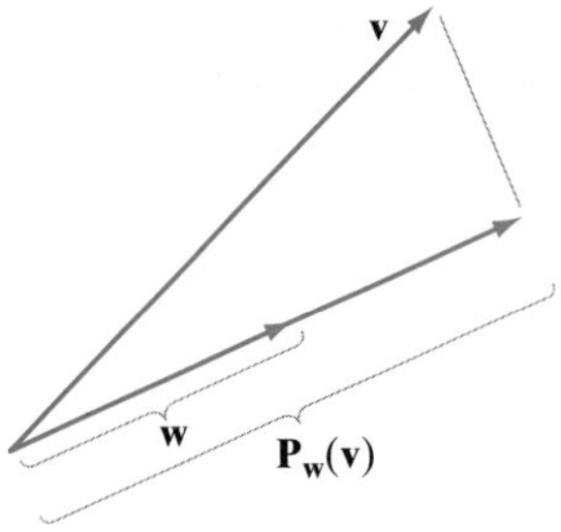

▲ Figure 5a

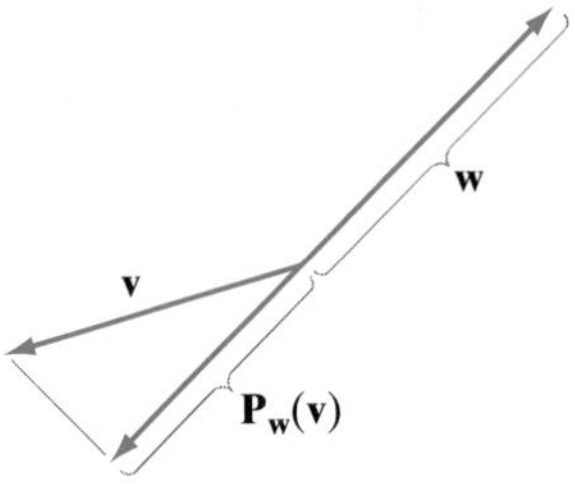

▲ Figure 5b

share the same initial point. The projection $\mathbf{P_w(v)}$ of $\mathbf{v}$ onto $\mathbf{w}$ is shown in Figure 4b. The main thing to notice in Figure 4c is that the projection of $\mathbf{v}$ onto $\mathbf{w}$ is parallel to $\mathbf{w}$ and determines, along with $\mathbf{v}$ and $\mathbf{Q}$, a right triangle. Figure 5 shows two other projections that will aid your geometric understanding of the concept. The next theorem provides us with an analytic expression for $\mathbf{P_w(v)}$ as a scalar multiple of the direction, $\mathbf{dir(w)} = (1/\|\mathbf{w}\|)\mathbf{w}$, of $\mathbf{w}$.

THEOREM 4 If $\mathbf{v}$ and $\mathbf{w}$ are nonzero vectors, then the projection $\mathbf{P_w(v)}$ of $\mathbf{v}$ onto $\mathbf{w}$ is given by

$$\mathbf{P_w(v)} = \left(\frac{\mathbf{v}\cdot\mathbf{w}}{\mathbf{w}\cdot\mathbf{w}}\right)\mathbf{w} \tag{9.3.12}$$

or, equivalently,

$$\mathbf{P_w(v)} = (\mathbf{v}\cdot\mathbf{dir(w)})\mathbf{dir(w)}. \tag{9.3.13}$$

The length of $\mathbf{P_w(v)}$ is given by

$$\|\mathbf{P_w(v)}\| = |\mathbf{v}\cdot\mathbf{dir(w)}|, \tag{9.3.14}$$

or, equivalently,

$$\|\mathbf{P_w(v)}\| = \frac{|\mathbf{v}\cdot\mathbf{w}|}{\|\mathbf{w}\|}. \tag{9.3.15}$$

Proof. Let $\mathbf{Q}$ denote the vector orthogonal to $\mathbf{w}$ that is depicted in Figure 4c. We have

$$\mathbf{P_w(v)} + \mathbf{Q} = \mathbf{v}. \tag{9.3.16}$$

Because $\mathbf{P_w}(\mathbf{v})$ is parallel to $\mathbf{w}$, we may write

$$\mathbf{P_w}(\mathbf{v}) = c\mathbf{w} \tag{9.3.17}$$

for some scalar c. Substituting this expression for $\mathbf{P_w}(\mathbf{v})$ into equation (9.3.16), we arrive at $c\mathbf{w} + \mathbf{Q} = \mathbf{v}$. Taking the dot product of each side of this equation with $\mathbf{w}$ and using the distributive law for the dot product gives us $c\mathbf{w}\cdot\mathbf{w} + \mathbf{Q}\cdot\mathbf{w} = \mathbf{v}\cdot\mathbf{w}$, or $(\mathbf{w}\cdot\mathbf{w})c + 0 = \mathbf{v}\cdot\mathbf{w}$. Thus $c = (\mathbf{w}\cdot\mathbf{w})^{-1}\mathbf{v}\cdot\mathbf{w}$. With this value for c, equation (9.3.17) becomes

$$\mathbf{P_w}(\mathbf{v}) = \left(\frac{\mathbf{v}\cdot\mathbf{w}}{\mathbf{w}\cdot\mathbf{w}}\right)\mathbf{w} = \left(\frac{\mathbf{v}\cdot\mathbf{w}}{\|\mathbf{w}\|^2}\right)\mathbf{w} = \left(\mathbf{v}\cdot\left(\frac{1}{\|\mathbf{w}\|}\mathbf{w}\right)\right)\frac{1}{\|\mathbf{w}\|}\mathbf{w} = \left(\mathbf{v}\cdot\mathbf{dir}(\mathbf{w})\right)\mathbf{dir}(\mathbf{w}),$$

which establishes formulas (9.3.12) and (9.3.13). By equating the lengths of the vectors on each side of formula (9.3.13), using equation (9.2.5), and noting that $\|\mathbf{dir}(\mathbf{w})\| = 1$, we see that

$$\|\mathbf{P_w}(\mathbf{v})\| = |\mathbf{v}\cdot\mathbf{dir}(\mathbf{w})|\,\|\mathbf{dir}(\mathbf{w})\| = |\mathbf{v}\cdot\mathbf{dir}(\mathbf{w})| = \left|\mathbf{v}\cdot\left(\frac{1}{\|\mathbf{w}\|}\mathbf{w}\right)\right| = \frac{|\mathbf{v}\cdot\mathbf{w}|}{\|\mathbf{w}\|},$$

which establishes both (9.3.14) and (9.3.15). ■

In formulas (9.3.12) and (9.3.13), notice that (a) the expressions $\left(\frac{\mathbf{v}\cdot\mathbf{w}}{\mathbf{w}\cdot\mathbf{w}}\right)$ and $(\mathbf{v}\cdot\mathbf{dir}(\mathbf{w}))$ are *scalars,* and (b) each of these scalars multiplies a *vector*. The result of these scalar multiplications, the projection $\mathbf{P_w}(\mathbf{v})$ of $\mathbf{v}$ onto $\mathbf{w}$, is a *vector*. Because equation (9.3.16) decomposes $\mathbf{v}$ as a sum of two orthogonal vectors, $\mathbf{P_w}(\mathbf{v})$ is also called the *orthogonal projection* of $\mathbf{v}$ in the direction of $\mathbf{w}$. The scalar $\mathbf{v}\cdot\mathbf{dir}(\mathbf{w})$, which may also be written as $\left(\frac{\mathbf{v}\cdot\mathbf{w}}{\|\mathbf{w}\|}\right)$, is called the *component* of $\mathbf{v}$ in the direction of $\mathbf{w}$. The absolute value of the component of $\mathbf{v}$ in the direction of $\mathbf{w}$ is the length of the projection $\mathbf{P_w}(\mathbf{v})$.

▶ **EXAMPLE 6** Let $\mathbf{v} = \langle 2, 1, -1\rangle$ and $\mathbf{w} = \langle 1, -2, 2\rangle$. Calculate the projection of $\mathbf{v}$ onto $\mathbf{w}$, the projection of $\mathbf{w}$ onto $\mathbf{v}$, and the lengths of these projections. Also calculate the component of $\mathbf{v}$ in the direction of $\mathbf{w}$ and the component of $\mathbf{w}$ in the direction of $\mathbf{v}$.

Solution To efficiently calculate $\mathbf{P_w}(\mathbf{v})$ and the other quantities associated with this projection, we determine two scalars: $\|\mathbf{w}\|$ and $\mathbf{v}\cdot\mathbf{dir}(\mathbf{w})$. We first calculate $\|\mathbf{w}\| = \sqrt{1^2 + (-2)^2 + 2^2} = 3$ and $\mathbf{dir}(\mathbf{w}) = \frac{1}{\|\mathbf{w}\|}\mathbf{w} = \langle 1/3, -2/3, 2/3\rangle$. The component of $\mathbf{v}$ in the direction of $\mathbf{w}$ is $\mathbf{v}\cdot\mathbf{dir}(\mathbf{w})$, or $(2)(1/3) + (1)(-2/3) + (-1)(2/3)$, which simplifies to $-2/3$. The absolute value of this quantity is the length of the projection. That is, $\|\mathbf{P_w}(\mathbf{v})\| = |-2/3| = 2/3$. Formula (9.3.13) gives us the projection $\mathbf{P_w}(\mathbf{v})$ itself:

$$\mathbf{P_w}(\mathbf{v}) = \left(\mathbf{v}\cdot\mathbf{dir}(\mathbf{w})\right)\mathbf{dir}(\mathbf{w}) = -\frac{2}{3}\left\langle\frac{1}{3}, -\frac{2}{3}, \frac{2}{3}\right\rangle = \left\langle -\frac{2}{9}, \frac{4}{9}, -\frac{4}{9}\right\rangle.$$

For $\mathbf{P_v}(\mathbf{w})$, we calculate $\|\mathbf{v}\| = \sqrt{2^2 + 1^2 + (-1)^2} = \sqrt{6}$ and $\mathbf{dir}(\mathbf{v}) = \frac{1}{\|\mathbf{v}\|}\mathbf{v} = \langle 2/\sqrt{6}, 1/\sqrt{6}, -1/\sqrt{6}\rangle$. The component of $\mathbf{w}$ in the direction of $\mathbf{v}$ is $\mathbf{w}\cdot\mathbf{dir}(\mathbf{v})$, or

$(1)(2/\sqrt{6}) + (-2)(1/\sqrt{6}) + (2)(-1/\sqrt{6})$, which simplifies to $-2/\sqrt{6}$. The absolute value, $2/\sqrt{6}$, of this quantity is the length of $\|\mathbf{P_v}(\mathbf{w})\|$. Formula (9.3.13) gives us the projection $\mathbf{P_v}(\mathbf{w})$ itself:

$$\mathbf{P_v}(\mathbf{w}) = \left(\mathbf{w} \cdot \mathbf{dir}(\mathbf{v})\right) \mathbf{dir}(\mathbf{v}) = -\frac{2}{\sqrt{6}}\left\langle \frac{2}{\sqrt{6}}, \frac{1}{\sqrt{6}}, -\frac{1}{\sqrt{6}} \right\rangle = \left\langle -\frac{2}{3}, -\frac{1}{3}, \frac{1}{3} \right\rangle.$$

As we could have anticipated from the geometry of projections, the two projections $\mathbf{P_w}(\mathbf{v})$ and $\mathbf{P_v}(\mathbf{w})$ are different. ◀

INSIGHT For theoretical reasons, it is useful to have a variety of formulas. Thus equation (9.3.15) allows us to easily prove that

$$\|\mathbf{P_w}(\lambda\mathbf{v})\| = |\lambda|\,\|\mathbf{P_w}(\mathbf{v})\| \quad \text{and} \quad \|\mathbf{P}_{\lambda\mathbf{w}}(\mathbf{v})\| = \|\mathbf{P_w}(\mathbf{v})\|.$$

As a practical matter, however, Example 6 shows that formula (9.3.13) contains all the information that pertains to the projection of one vector onto another.

Projection onto a unit vector $\mathbf{u}$ is simple. Setting $\mathbf{w} = \mathbf{u}$ in equations (9.3.13) and (9.3.14) and noting that $\mathbf{dir}(\mathbf{u}) = \mathbf{u}$, we obtain the simplifications

$$\mathbf{P_u}(\mathbf{v}) = (\mathbf{v} \cdot \mathbf{u})\,\mathbf{u} \quad \text{and} \quad \|\mathbf{P_u}(\mathbf{v})\| = |\mathbf{v} \cdot \mathbf{u}|. \tag{9.3.18}$$

Also, the component of $\mathbf{v}$ in the direction of $\mathbf{u}$ is just the dot product $\mathbf{v} \cdot \mathbf{u}$.

▶ **EXAMPLE 7** Let $\mathbf{v} = \langle \sqrt{2}, 8, -2 \rangle$ and $\mathbf{u} = \langle 1/\sqrt{2}, 1/2, -1/2 \rangle$. Calculate the projection $\mathbf{P_u}(\mathbf{v})$ of $\mathbf{v}$ onto $\mathbf{u}$. What is the component of $\mathbf{v}$ in the direction of $\mathbf{u}$?

Solution To calculate $\mathbf{P_u}(\mathbf{v})$, we proceed as in Example 6 and determine the two scalars $\|\mathbf{u}\|$ and $\mathbf{v} \cdot \mathbf{dir}(\mathbf{u})$. Thus $\|\mathbf{u}\| = \sqrt{(1/\sqrt{2})^2 + (1/2)^2 + (-1/2)^2} = 1$ and $\mathbf{v} \cdot \mathbf{u} = \langle \sqrt{2}, 8, -2 \rangle \langle 1/\sqrt{2}, 1/2, -1/2 \rangle = 1 + 4 + 1 = 6$. Because $\mathbf{u}$ is a unit vector, the component of $\mathbf{v}$ in the direction of $\mathbf{u}$ is $\mathbf{v} \cdot \mathbf{u}$, or 6, and

$$\mathbf{P_u}(\mathbf{v}) = (\mathbf{v} \cdot \mathbf{u})\,\mathbf{u} = 6\left\langle \frac{1}{\sqrt{2}}, \frac{1}{2}, -\frac{1}{2} \right\rangle = \left\langle \frac{6}{\sqrt{2}}, \frac{6}{2}, -\frac{6}{2} \right\rangle = \left\langle 3\sqrt{2}, 3, -3 \right\rangle. \quad ◀$$

Projection and the Standard Basis Vectors

As noted in Section 2, it is often useful to express vectors in terms of the three unit vectors $\mathbf{i} = \langle 1, 0, 0 \rangle$, $\mathbf{j} = \langle 0, 1, 0 \rangle$, and $\mathbf{k} = \langle 0, 0, 1 \rangle$.

▶ **EXAMPLE 8** Let $\mathbf{v} = \langle 2, -6, 12 \rangle$. Calculate $\mathbf{P_i}(\mathbf{v})$, $\mathbf{P_j}(\mathbf{v})$, and $\mathbf{P_k}(\mathbf{v})$, and express $\mathbf{v}$ as a linear combination of these projections.

Solution Because $\mathbf{i}$, $\mathbf{j}$, and $\mathbf{k}$ are unit vectors, we use the projection formula found in line (9.3.18):

$$\mathbf{P_i}(\mathbf{v}) = (\mathbf{v} \cdot \mathbf{i})\,\mathbf{i} = (\langle 2, -6, 12 \rangle \cdot \langle 1, 0, 0 \rangle)\mathbf{i} = 2\mathbf{i} = \langle 2, 0, 0 \rangle,$$

$$\mathbf{P_j}(\mathbf{v}) = (\mathbf{v} \cdot \mathbf{j})\,\mathbf{j} = (\langle 2, -6, 12 \rangle \cdot \langle 0, 1, 0 \rangle)\mathbf{j} = -6\mathbf{j} = \langle 0, -6, 0 \rangle,$$

and

$$\mathbf{P_k}(\mathbf{v}) = (\mathbf{v} \cdot \mathbf{k})\,\mathbf{k} = (\langle 2, -6, 12 \rangle \cdot \langle 0, 0, 1 \rangle)\mathbf{k} = 12\mathbf{k} = \langle 0, 0, 12 \rangle.$$

Observe that

$$\mathbf{v} = \langle 2, -6, 12\rangle = \langle 2, 0, 0\rangle + \langle 0, -6, 0\rangle + \langle 0, 0, 12\rangle = \mathbf{P_i}(\mathbf{v}) + \mathbf{P_j}(\mathbf{v}) + \mathbf{P_k}(\mathbf{v}) \quad ◄$$

INSIGHT If we repeat the calculations of Example 8 with a general vector $\mathbf{v} = \langle a, b, c\rangle$, then we find that the projections of $\mathbf{v}$ on $\mathbf{i}$, $\mathbf{j}$, and $\mathbf{k}$ give the displacements of $\mathbf{v}$ in the x-, y-, and z-directions, respectively: $\mathbf{P_i}(\mathbf{v}) = a\mathbf{i}$, $\mathbf{P_j}(\mathbf{v}) = b\mathbf{j}$, and $\mathbf{P_k}(\mathbf{v}) = c\mathbf{k}$. In particular,

$$\mathbf{v} = \mathbf{P_i}(\mathbf{v}) + \mathbf{P_j}(\mathbf{v}) + \mathbf{P_k}(\mathbf{v}). \tag{9.3.19}$$

Direction Cosines and Direction Angles

Suppose that $\mathbf{u} = \langle u_1, u_2, u_3\rangle = u_1\mathbf{i} + u_2\mathbf{j} + u_3\mathbf{k}$ is a unit vector. Because $u_1^2 + u_2^2 + u_3^2 = 1$, it follows that the numbers u_1, u_2, u_3 all lie between -1 and 1. Thus there are unique numbers α, β, γ between 0 and π such that $u_1 = \cos(\alpha)$, $u_2 = \cos(\beta)$, and $u_3 = \cos(\gamma)$. As a result,

$$\mathbf{u} = \cos(\alpha)\mathbf{i} + \cos(\beta)\mathbf{j} + \cos(\gamma)\mathbf{k}.$$

The numbers α, β, and γ are called the *direction angles* for $\mathbf{u}$, and the entries u_1, u_2, u_3 are called the *direction cosines* of $\mathbf{u}$. Notice that, by Theorem 2:

α is the angle that $\mathbf{u}$ makes with the positive x-axis.
β is the angle that $\mathbf{u}$ makes with the positive y-axis.
γ is the angle that $\mathbf{u}$ makes with the positive z-axis.

Refer to Figure 6.

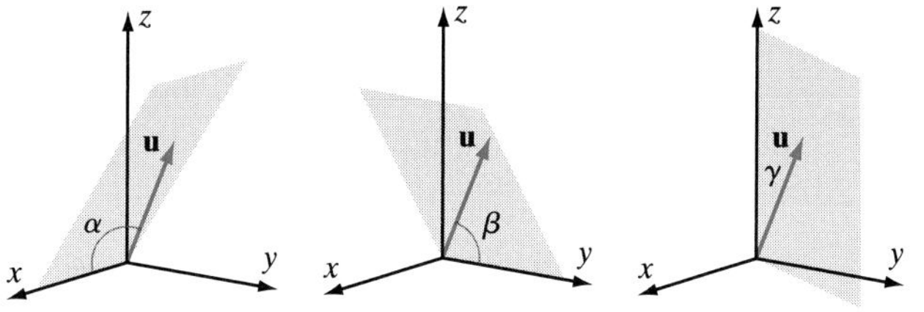

▲ Figure 6 $\mathbf{u} = \cos(\alpha)\mathbf{i} + \cos(\beta)\mathbf{j} + \cos(\gamma)\mathbf{k}$.

If $\mathbf{v}$ is *any* nonzero vector, then the direction angles and direction cosines for $\mathbf{v}$ are understood to be the direction angles and cosines for the unit vector $(1/\|\mathbf{v}\|)\mathbf{v}$. The direction angles for $\mathbf{v}$ are the angles that $\mathbf{v}$ makes with the positive x-, y-, and z-axes.

► **EXAMPLE 9** Calculate the direction cosines and direction angles for the vector $\mathbf{v} = \langle 0, -3\sqrt{3}, 3\rangle$

Solution The associated unit vector is

$$\mathbf{u} = \frac{1}{\|\mathbf{v}\|}\mathbf{v} = \frac{1}{\sqrt{0^2 + (-3\sqrt{3})^2 + 3^2}}\langle 0, -3\sqrt{3}, 3\rangle = \frac{1}{6}\langle 0, -3\sqrt{3}, 3\rangle = \left\langle 0, -\frac{\sqrt{3}}{2}, \frac{1}{2}\right\rangle.$$

The direction cosines for $\mathbf{v}$ are therefore $\cos(\alpha) = 0, \cos(\beta) = -\sqrt{3}/2$, and $\cos(\gamma) = 1/2$. It follows that $\alpha = \pi/2$, $\beta = 5\pi/6$, and $\gamma = \pi/3$. ◄

Applications

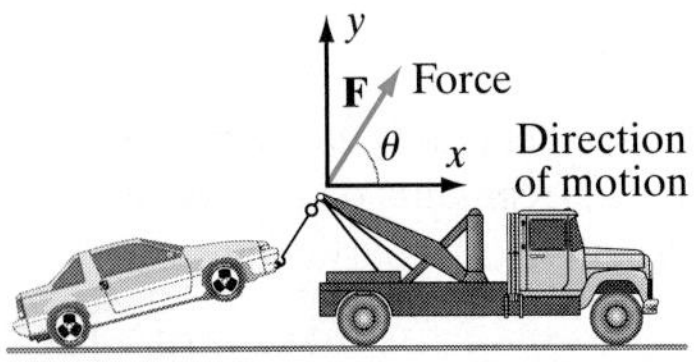

▲ Figure 7

If a constant force F is applied along the line of motion to move an object a distance d, then, as we learned in Chapter 8, the *work* performed is $W = Fd$. However, in many applications, the force $\mathbf{F}$ is a vector that is not applied in the direction of motion. Imagine a truck towing a car (Figure 7). The force is applied in a direction that makes an angle θ with the direction of motion. Let $\mathbf{u}$ be a unit vector along the direction of motion. Let $g = \mathbf{F} \cdot \mathbf{u} = \|F\|\cos(\theta)$ be the component of $\mathbf{F}$ in the direction $\mathbf{u}$. Then the work performed in moving the body in the figure a distance d is $W = gd$.

▶ **EXAMPLE 10** A tow truck pulls a disabled vehicle a total of 20,000 feet. To keep the vehicle in motion, the truck must apply a constant force of 3,000 pounds. The hitch is set up so that the force is exerted at an angle of 30° with the horizontal. How much work is performed?

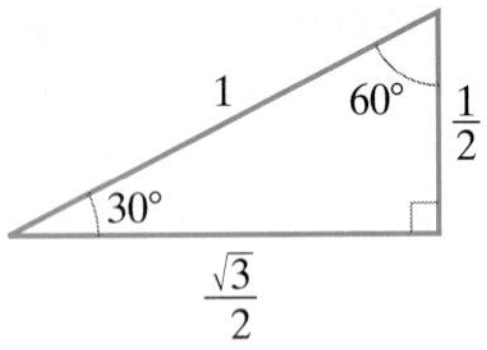

▲ Figure 8a

Solution We use a $30° - 60° - 90°$ triangle, as shown in Figure 8, panel (a), to resolve the force vector $\mathbf{F}$ as a sum of two orthogonal vectors, one of which is in the direction $\mathbf{i}$ of motion. We find that the direction of $\mathbf{F}$ is given by $\mathbf{v} = (\sqrt{3}/2)\mathbf{i} + (1/2)\mathbf{j}$ (see panel (b) of Figure 8). Because the force vector $\mathbf{F}$ has magnitude 3,000 and direction $\mathbf{v}$, we have

$$\mathbf{F} = \|\mathbf{F}\|\mathbf{v} = 3000\left(\frac{\sqrt{3}}{2}\mathbf{i} + \frac{1}{2}\mathbf{j}\right) = 1500\sqrt{3}\mathbf{i} + 1500\mathbf{j},$$

as is indicated in Figure 8, panel (c). The component of vector $\mathbf{F}$ in the direction $\mathbf{i}$ of motion is $g = 1500\sqrt{3}$. As a result, the work performed is

$$W = gd = 1500\sqrt{3} \cdot 20,000 = 3\sqrt{3} \cdot 10^7 \text{foot-pounds.} \quad ◀$$

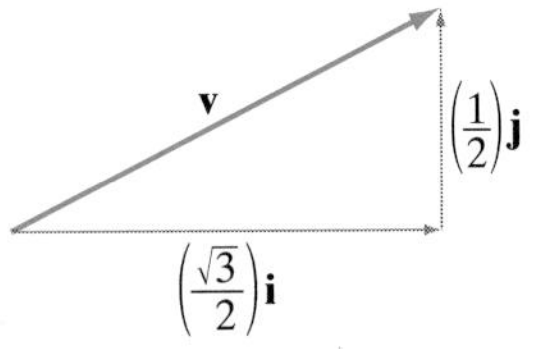

▲ Figure 8b
$\mathbf{v} = (\sqrt{3}/2)\mathbf{i} + (1/2)\mathbf{j}$

▶ **EXAMPLE 11** The force of the wind is given by the vector $\mathbf{F} = \langle 2, 1, 3 \rangle$. In navigation, it is useful to resolve this force into its component in the direction of motion and its component perpendicular to the direction of motion. If $\langle 1, 1, -4 \rangle$ represents the direction of motion, then perform this resolution.

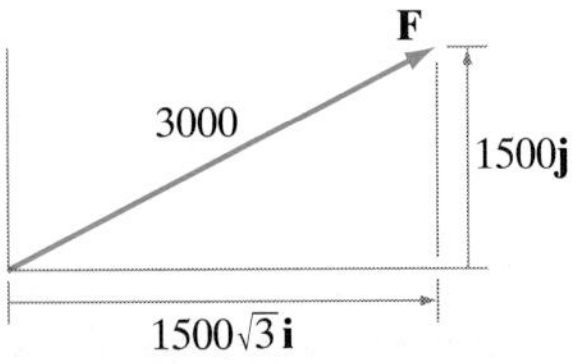

▲ Figure 8c
$\mathbf{F} = 1500\sqrt{3}\mathbf{i} + 1500\mathbf{j}$

Solution The component of $\mathbf{F}$ in the direction $\mathbf{v} = \langle 1, 1, -4 \rangle$ is just the projection $\mathbf{P}_\mathbf{v}(\mathbf{F})$. We calculate

$$\begin{aligned} \mathbf{P}_\mathbf{v}(\mathbf{F}) &= \left(\frac{\mathbf{F} \cdot \mathbf{v}}{\|\mathbf{v}\|^2}\right)\mathbf{v} \\ &= \frac{-9}{18}\langle 1, 1, -4 \rangle \\ &= \left\langle -\frac{1}{2}, -\frac{1}{2}, 2 \right\rangle. \end{aligned}$$

Then we write

$$\begin{aligned} \langle 2, 1, 3 \rangle = \mathbf{F} &= \mathbf{P}_\mathbf{v}(\mathbf{F}) + \big(\mathbf{F} - \mathbf{P}_\mathbf{v}(\mathbf{F})\big) \\ &= \left\langle -\frac{1}{2}, -\frac{1}{2}, 2 \right\rangle + \left\langle \frac{5}{2}, \frac{3}{2}, 1 \right\rangle. \end{aligned}$$

Notice that the two vectors into which we have decomposed $\mathbf{F}$ are orthogonal. The first is a multiple of $\mathbf{v}$, while the second is perpendicular to $\mathbf{v}$. This is the resolution that we desire. ◀

QUICK QUIZ

1. Calculate $\langle 1,2,-1\rangle \cdot \langle 3,4,2\rangle$.
2. Use the arccosine to express the angle between $\langle 6,3,2\rangle$ and $\langle 4,-7,4\rangle$.
3. For what value of a are $\langle 1,a,-1\rangle$ and $\langle 3,4,a\rangle$ perpendicular?
4. Calculate the projection of $\langle 3,1,-1\rangle$ onto $\langle 8,4,1\rangle$.

Answers

1. 9 2. arccos (11/63) 3. −1 4. $\langle 8/3, 4/3, 1/3\rangle$

EXERCISES

Problems for Practice

In each of Exercises 1–6, calculate the dot product of the given vectors.

1. $\langle 3,-2,4\rangle$ $\langle 2,1,6\rangle$
2. $\langle -6,-3,-5\rangle$ $\langle -8,1,1\rangle$
3. $\langle 0,4,0\rangle$ $\langle 5,1,0\rangle$
4. $\langle 0,-2,6\rangle$ $\langle -4,2,5\rangle$
5. $\langle 1,4,9\rangle$ $\langle 2,3,-5\rangle$
6. $\langle -2,-10,4\rangle$ $\langle -8,0,2\rangle$

In each of Exercises 7–12, calculate the angle between the given pair of vectors.

7. $\langle 1,1,0\rangle$ $\langle 0,1,1\rangle$
8. $\langle 2,-2,2\rangle$ $\langle -1,1,5\rangle$
9. $\langle 3,0,4\rangle$ $\langle 0,\sqrt{7},-5\rangle$
10. $\langle 3,4,7\rangle$ $\langle 18,24,5\rangle$
11. $\langle 2,-1,9\rangle$ $\langle -4,1,1\rangle$
12. $\langle 1,1,2\rangle$ $\langle 2,-1,1\rangle$

In each of Exercises 13–16, determine whether the two given vectors are orthogonal. Give a reason for your answer.

13. $\langle -3,1,5\rangle$ $\langle 4,-2,3\rangle$
14. $\langle 0,-6,7\rangle$ $\langle 8,14,12\rangle$
15. $\langle 2,-5,8\rangle$ $\langle -2,4,3\rangle$
16. $\langle 1,6,5\rangle$ $\langle -8,3,-2\rangle$

In each of Exercises 17–22, find $\mathbf{P_v}(\mathbf{w})$ and $\mathbf{P_w}(\mathbf{v})$. State the component of $\mathbf{w}$ in the direction of $\mathbf{v}$ and the component of $\mathbf{v}$ in the direction of $\mathbf{w}$.

17. $\mathbf{v}=\langle -3,7,2\rangle$ $\mathbf{w}=\langle 1,4,-4\rangle$
18. $\mathbf{v}=\langle 0,0,1\rangle$ $\mathbf{w}=\langle -9,7,5\rangle$
19. $\mathbf{v}=\langle 2,4,6\rangle$ $\mathbf{w}=\langle 4,8,12\rangle$
20. $\mathbf{v}=\langle 2,1,9\rangle$ $\mathbf{w}=\langle 1,0,1\rangle$
21. $\mathbf{v}=\langle 1,0,1\rangle$ $\mathbf{w}=\langle 0,1,0\rangle$
22. $\mathbf{v}=\langle 2,4,6\rangle$ $\mathbf{w}=\langle -2,-4,-6\rangle$

In each of Exercises 23–28, find the direction of the given vector. Then find the direction angles for the vector.

23. $\langle 6,-2\sqrt{3},0\rangle$
24. $\langle 0,3\sqrt{3},-9\rangle$
25. $\langle 0,1,0\rangle$
26. $\langle 3\sqrt{2},-3,3\rangle$
27. $\langle -6,0,6\rangle$
28. $\langle 0,1,1\rangle$

In each of Exercises 29–32, calculate the dot product of the given vectors and their lengths. Verify that the Cauchy-Schwarz Inequality holds for the pair.

29. $\langle 2,-4,4\rangle$ $\langle -2,1,-2\rangle$
30. $\langle 1,2,3\rangle$ $\langle 1,3,2\rangle$
31. $\langle 2,0,1\rangle$ $\langle 4,\sqrt{3},1\rangle$
32. $\langle 2,1,-2\rangle$ $\langle 4,1,-8\rangle$

In each of Exercises 33–36, calculate the projection of the given vector $\mathbf{v}$ onto the given direction $\mathbf{u}$. Verify that $\mathbf{P_u}(\mathbf{v})$ and $\mathbf{v}-\mathbf{P_u}(\mathbf{v})$ are orthogonal.

33. $\mathbf{v}=\langle 4,1,-8\rangle$ $\mathbf{u}=\langle 2/3,2/3,-1/3\rangle$
34. $\mathbf{v}=\langle 4,2,-3\rangle$ $\mathbf{u}=\langle 3/5,0,-4/5\rangle$
35. $\mathbf{v}=\langle \sqrt{12},0,\sqrt{48}\rangle$ $\mathbf{u}=\langle 1/\sqrt{3},1/\sqrt{3},-1/\sqrt{3}\rangle$
36. $\mathbf{v}=\langle 3,2,-1\rangle$ $\mathbf{u}=\langle 1/9,-4/9,8/9\rangle$

In each of Exercises 37–40, find all values of s for which the two given vectors are orthogonal.

37. $\langle 3,2,-1\rangle$ $\langle s,1,-4\rangle$
38. $\langle s,5,-12\rangle$ $\langle 1,s,2\rangle$
39. $\langle 3,1,1\rangle$ $\langle -7,s^2,5\rangle$
40. $\langle s,-5,1\rangle$ $\langle s,s,6\rangle$

Further Theory and Practice

41. Find a unit vector $\mathbf{u}$ that is perpendicular to both $\mathbf{v}=\mathbf{i}+2\mathbf{j}+\mathbf{k}$ and $\mathbf{w}=\mathbf{i}-\mathbf{j}$. Show that any other vector that is perpendicular to both $\mathbf{v}$ and $\mathbf{w}$ is parallel to $\mathbf{u}$.

42. Show that if $\mathbf{v}$ is any vector, then

$$\mathbf{v}=(\mathbf{v}\cdot\mathbf{i})\mathbf{i}+(\mathbf{v}\cdot\mathbf{j})\mathbf{j}+(\mathbf{v}\cdot\mathbf{k})\mathbf{k}.$$

43. Prove that the vectors $\mathbf{a}=\langle \sqrt{3}/2, 1/(2\sqrt{2}), -1/(2\sqrt{2})\rangle$, $\mathbf{b}=\langle 0,1/\sqrt{2},1/\sqrt{2}\rangle$, and $\mathbf{c}=\langle 1/2,-\sqrt{6}/4,\sqrt{6}/4\rangle$ are each perpendicular to the other two and are of unit length.

44. If **v** is any vector and if **a**, **b**, and **c** are the three vectors of Exercise 43, then prove that

$$\mathbf{v} = (\mathbf{v}\cdot\mathbf{a})\mathbf{a} + (\mathbf{v}\cdot\mathbf{b})\mathbf{b} + (\mathbf{v}\cdot\mathbf{c})\mathbf{c}.$$

45. Mrs. Woodman pulls a railroad car along a track using a rope that makes a 30° angle with the track (Figure 9). Calculate how much work is performed if she exerts a force of 200 pounds and succeeds in pulling the car 1000 feet.

▲ Figure 9

46. Show that the direction cosines for any vector **v** satisfy

$$\cos^2(\alpha) + \cos^2(\beta) + \cos^2(\gamma) = 1.$$

If **v** lies in the xy-plane, then to what familiar identity does this equation reduce?

47. Prove the Parallelogram Law:

$$\|\mathbf{v}+\mathbf{w}\|^2 + \|\mathbf{v}-\mathbf{w}\|^2 = 2\|\mathbf{v}\|^2 + \|\mathbf{w}\|^2.$$

Give a geometric interpretation for this equality.

48. Prove the Polarization Formula:

$$\|\mathbf{v}+\mathbf{w}\|^2 - \|\mathbf{v}-\mathbf{w}\|^2 = 4\mathbf{v}\cdot\mathbf{w}.$$

49. In each of the following, find a real number λ such that $\mathbf{P_u}(\mathbf{v}) = \lambda\mathbf{w}$:

a. $\mathbf{u} = \langle 3,6,9\rangle$ $\quad\mathbf{v} = \langle 1,-2,5\rangle$ $\quad\mathbf{w} = \langle 1,2,3\rangle$

b. $\mathbf{u} = \langle -2,0,4\rangle$ $\quad\mathbf{v} = \langle 1,-4,6\rangle$ $\quad\mathbf{w} = \langle -1,0,2\rangle$

c. $\mathbf{u} = \langle 2,10,-6\rangle$ $\quad\mathbf{v} = \langle -3,4,8\rangle$ $\quad\mathbf{w} = \langle 1,5,-3\rangle$

d. $\mathbf{u} = \langle 12,0,-16\rangle$ $\quad\mathbf{v} = \langle 1,1,3\rangle$ $\quad\mathbf{w} = \langle 3,0,-4\rangle$

50. Prove that for any two vectors **v** and **w**, the angle between $\mathbf{u} = \|\mathbf{w}\|\mathbf{v} + \|\mathbf{v}\|\mathbf{w}$ and **v** equals the angle between **u** and **w**.

51. Let $\overrightarrow{OP}$ be the position vector for an arbitrary point $P = (x,y,z)$ in space. Using vector methods, find an equation involving $\overrightarrow{OP}$, **j**, the dot product, and norm to describe the cone with vertex at the origin, forming an angle of 60° at the vertex, and axis of symmetry the y-axis.

52. Let $\overrightarrow{OP}$ be the position vector for an arbitrary point $P = (x,y,z)$ in space. Using vector methods, find an equation involving $\overrightarrow{OP}$, **k**, the dot product, and norm to describe the cone with vertex at the origin, forming an angle of 45° at the vertex, and axis of symmetry the z-axis.

53. Show that the points $(2,1,4)$, $(5,3,2)$, and $(7,4,6)$ are the vertices of a right triangle.

54. Suppose that P, Q, and R are vertices of a cube with $\overline{PQ}$ the diagonal of the cube and $\overline{PR}$ the diagonal of a face of the cube. What is the angle between $\overrightarrow{PQ}$ and $\overrightarrow{PR}$?

55. Suppose that a and b are nonnegative. When the Cauchy-Schwarz Inequality is applied to $\mathbf{v} = \sqrt{a}\,\mathbf{i} + \sqrt{b}\,\mathbf{j}$ and $\mathbf{w} = \sqrt{b}\,\mathbf{i} + \sqrt{a}\,\mathbf{j}$, what inequality results?

56. Let v_1, v_2, and v_3 be any numbers. Find a suitable vector **w** such that the Cauchy-Schwarz Inequality applied to $\mathbf{v} = \langle v_1, v_2, v_3\rangle$ and **w** results in the inequality

$$(v_1 + v_2 + v_3)^2 \le 3(v_1^2 + v_2^2 + v_3^2).$$

57. Use the Cauchy-Schwarz Inequality to prove the Triangle Inequality:

$$\|\mathbf{v}+\mathbf{w}\| \le \|\mathbf{v}\| + \|\mathbf{w}\|.$$

58. Suppose that a, b, and c are positive numbers. The equation $x/a + y/b + z/c = 1$ represents a plane. What are the intercepts A, B, and C with the x-axis, y-axis, and z-axis, respectively? Calculate the cosines of the three angles of ΔABC. What arccosine identity can be deduced from these angles?

59. Verify the identity

$$\begin{aligned}&(v_1^2 + v_2^2 + v_3^2)(w_1^2 + w_2^2 + w_3^2) - (v_1w_1 + v_2w_2 + v_3w_3)^2 \\ &= (v_1w_2 - w_1v_2)^2 + (v_1w_3 - w_1v_3)^2 + (v_2w_3 - w_2v_3)^2.\end{aligned}$$

Let $\mathbf{v} = \langle v_1, v_2, v_3\rangle$ and $\mathbf{w} = \langle w_1, w_2, w_3\rangle$. Deduce that

$$\begin{aligned}\|\mathbf{v}\|^2\|\mathbf{w}\|^2 - (\mathbf{v}\cdot\mathbf{w})^2 &= (v_1w_2 - w_1v_2)^2 + (v_1w_3 - w_1v_3)^2 \\ &\quad + (v_2w_3 - w_2v_3)^2.\end{aligned}$$

Prove the Cauchy-Schwarz Inequality from this identity.

60. Suppose that $\mathbf{v} = \langle v_1, v_2, v_3\rangle$ and $\mathbf{w} = \langle w_1, w_2, w_3\rangle$ are parallel. Use Theorem 3b to deduce that

$$\|\mathbf{v}\|^2\|\mathbf{w}\|^2 - (\mathbf{v}\cdot\mathbf{w})^2 = 0.$$

Use the second identity of Exercise 59 to conclude that

$$(v_1w_2 - w_1v_2)^2 + (v_1w_3 - w_1v_3)^2 + (v_2w_3 - w_2v_3)^2 = 0,$$

which implies the following three equations:

$$v_1w_2 = w_1v_2,\quad v_1w_3 = w_1v_3,\quad \text{and}\quad v_2w_3 = w_2v_3.$$

If one entry v_i of **v** is nonzero, then show that $\mathbf{w} = (w_i/v_i)\,\mathbf{v}$. Deduce that in any event, one of **v** and **w** is a scalar multiple of the other.

61. Let V denote the set of planar vectors. Let $T : V \to V$ be defined by

$$T(\langle a,b\rangle) = \langle a\cos(\theta) - b\sin(\theta), a\sin(\theta) + b\cos(\theta)\rangle.$$

Show that the angle between $\langle a,b\rangle$ and $T(\langle a,b\rangle)$ is θ. Also show that $\|T(\langle a,b\rangle)\| = \|\langle a,b\rangle\|$.

Calculator/Computer Exercises

62. Calculate the angles of the triangle with vertices $(1,1,2)$, $(2,0,3)$, and $(1,2,-5)$. What is the sum of these numbers?
63. Calculate the angles of the triangle with vertices $(1,1,2)$, $(2,0,3)$, and $(1,2,-5)$. What is the sum of these numbers?
64. For what value(s) of a are the vectors $\langle a^2, 2a, -3\rangle$ and $\langle a, 3a, 1\rangle$ orthogonal?
65. Plot $f(t) = \langle e^{-t}, t, 1\rangle \cdot \langle 1, 2t, 3\rangle$ for $-2 \le t \le 2$. At what value of t does $f(t)$ attain a minimum?
66. At time $t \ge 0$, a particle's position P_t is the point $(t, t\cos(t), t\sin(t))$. Is the vector $\overrightarrow{P_tP_{t+1}}$ ever perpendicular to $\langle 1,2,3\rangle$?

9.4 The Cross Product and Triple Product

If we are given two vectors $\mathbf{v}$ and $\mathbf{w}$ in space, then it is frequently useful to find a nonzero vector that is perpendicular to both of them. Of course such a vector is not unique: If $\mathbf{n}$ is one such vector, then, for any scalar c, the vector $c\mathbf{n}$ is also perpendicular to both $\mathbf{v}$ and $\mathbf{w}$. In this section, we present an algebraic method for producing a vector that is not only perpendicular to both $\mathbf{v}$ and $\mathbf{w}$ but also has a length with particular geometric significance.

The Cross Product of Two Spatial Vectors

In the preceding section, we learned a product of two vectors, the dot product, that results in a scalar. We now define a product of two spatial vectors that results in another vector.

DEFINITION If $\mathbf{v} = \langle v_1, v_2, v_3\rangle$ and $\mathbf{w} = \langle w_1, w_2, w_3\rangle$, then we define their *cross product* $\mathbf{v}\times\mathbf{w}$ to be

$$\mathbf{v}\times\mathbf{w} = (v_2w_3 - w_2v_3)\,\mathbf{i} - (v_1w_3 - w_1v_3)\,\mathbf{j} + (v_1w_2 - w_1v_2)\,\mathbf{k}. \qquad (9.4.1)$$

Formula (9.4.1) for the cross product is somewhat odd-looking and cannot be easily committed to memory. We will soon present an alternative approach to calculating cross products, but first we verify that $\mathbf{v}\times\mathbf{w}$ performs an important geometric job.

THEOREM 1 If $\mathbf{v}$ and $\mathbf{w}$ are vectors, then $\mathbf{v}\times\mathbf{w}$ is orthogonal to both $\mathbf{v}$ and $\mathbf{w}$.

Proof. Recall that "orthogonal" is a synonym for "perpendicular." Let $\mathbf{v} = \langle v_1, v_2, v_3\rangle$ and $\mathbf{w} = \langle w_1, w_2, w_3\rangle$. We calculate the dot product of $\mathbf{v}$ with $\mathbf{v}\times\mathbf{w}$:

$$\begin{aligned}\mathbf{v}\cdot(\mathbf{v}\times\mathbf{w}) &= \langle v_1, v_2, v_3\rangle \cdot \langle (v_2w_3 - w_2v_3), -(v_1w_3 - w_1v_3), (v_1w_2 - w_1v_2)\rangle \\ &= v_1(v_2w_3 - w_2v_3) - v_2(v_1w_3 - w_1v_3) + v_3(v_1w_2 - w_1v_2) \\ &= 0.\end{aligned}$$

The verification that $\mathbf{w}\cdot(\mathbf{v}\times\mathbf{w}) = 0$ is similar. ■

▶ **EXAMPLE 1** Let $\mathbf{v} = \langle 2, -1, 3\rangle$ and $\mathbf{w} = \langle 5, 4, -6\rangle$. Calculate $\mathbf{v}\times\mathbf{w}$. Verify that $\mathbf{v}\times\mathbf{w}$ is orthogonal to both $\mathbf{v}$ and $\mathbf{w}$.

Solution We calculate

$$\begin{aligned}\mathbf{v}\times\mathbf{w} &= \langle 2,-1,3\rangle \times \langle 5,4,-6\rangle \\ &= ((-1)\cdot(-6)-4\cdot 3)\mathbf{i} - (2\cdot(-6)-5\cdot 3)\mathbf{j} + (2\cdot 4 - 5\cdot(-1))\mathbf{k} \\ &= \langle -6,27,13\rangle .\end{aligned}$$

We may now use the dot product to verify that $\mathbf{v}\times\mathbf{w}$ is perpendicular to both $\mathbf{v}$ and $\mathbf{w}$:

$$\mathbf{v}\cdot(\mathbf{v}\times\mathbf{w}) = \langle 2,-1,3\rangle \cdot \langle -6,27,13\rangle = 2\cdot(-6)+(-1)\cdot 27 + 3\cdot 13 = 0$$

and

$$\mathbf{w}\cdot(\mathbf{v}\times\mathbf{w}) = \langle 5,4,-6\rangle \cdot \langle -6,27,13\rangle = 5\cdot(-6)+4\cdot 27 + (-6)\cdot 13 = 0. \quad \blacktriangleleft$$

The Relationship between Cross Products and Determinants

There is a nice way to remember the formula for a cross product using the language of determinants. The determinant is a procedure for calculating a number based on the entries of a square array (or square *matrix*) of numbers such as

$$\underbrace{\begin{bmatrix} a & b \\ c & d \end{bmatrix}}_{2\times 2\ \text{array}} \quad \text{or} \quad \underbrace{\begin{bmatrix} \alpha & \beta & \gamma \\ a & b & c \\ d & e & f \end{bmatrix}}_{3\times 3\ \text{array}}.$$

By "square," we mean that the number of rows in the array equals the number of columns. In the general theory of determinants, the number of rows and columns may be any positive integer. For our purposes, we need only consider 2×2 and 3×3 arrays.

The *determinant* of a 2×2 array $\begin{bmatrix} a & b \\ c & d \end{bmatrix}$ is defined by

$$M = \begin{bmatrix} a & b \\ c & d \end{bmatrix}$$

$\det(M) = ad - bc$

▲ Figure 1

$$\det\left(\begin{bmatrix} a & b \\ c & d \end{bmatrix}\right) = ad - bc. \tag{9.4.2}$$

Notice that the determinant is the alternating sum of the product of diagonal entries indicated in Figure 1.

The determinant of a 3×3 array is defined by

$$\det\left(\begin{bmatrix} \alpha & \beta & \gamma \\ a & b & c \\ d & e & f \end{bmatrix}\right) = \alpha\det\left(\begin{bmatrix} b & c \\ e & f \end{bmatrix}\right) - \beta\det\left(\begin{bmatrix} a & c \\ d & f \end{bmatrix}\right) + \gamma\det\left(\begin{bmatrix} a & b \\ d & e \end{bmatrix}\right). \tag{9.4.3}$$

This definition is said to be an *expansion along the first row* of the array. The right side of equation (9.4.3) is an alternating sum of three products, one for each entry in the first row. The factor of each entry is the determinant of the 2×2 array obtained by crossing out the row and column of the entry. Figure 2 illustrates this procedure. Notice the signs, $+,-,+$, that appear on the right side of formula (9.4.3). This pattern of alternating sums is the reason for describing the determinant as an "alternating sum."

An equivalent scheme for calculating the determinant of a 3×3 array is sometimes used. As Figure 3 shows, two copies of the 3×3 array are placed side-by-side. The products of the six indicated diagonals are then computed. The determinant is the sum of the products that arise from the diagonals that slope down to the right minus the sum of the products that arise from the diagonals that slope down to the left.

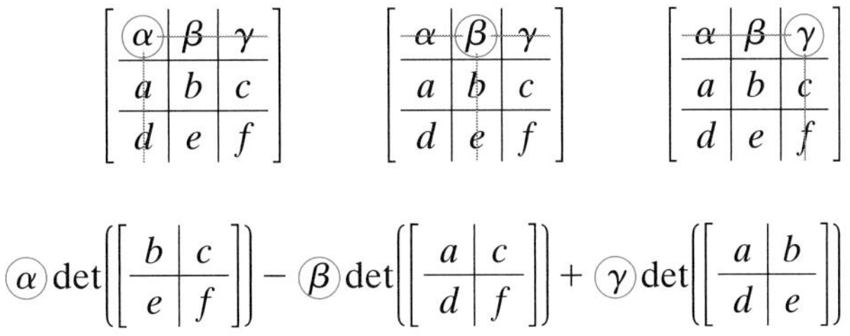

9.4 The Cross Product and Triple Product

$$\alpha \det\left(\begin{bmatrix} b & c \\ e & f \end{bmatrix}\right) - \beta \det\left(\begin{bmatrix} a & c \\ d & f \end{bmatrix}\right) + \gamma \det\left(\begin{bmatrix} a & b \\ d & e \end{bmatrix}\right)$$

▲ Figure 2

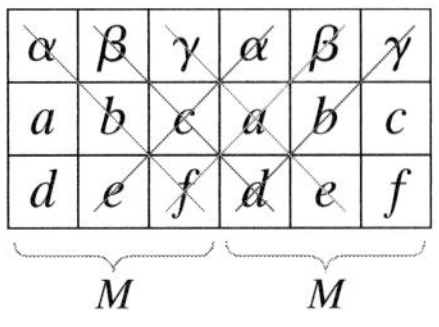

▲ Figure 3 $\det(M) = \alpha bf + \beta cd + \gamma ae - \alpha ce - \beta af - \gamma bd$

If $\mathbf{v} = \langle v_1, v_2, v_3 \rangle$ and $\mathbf{w} = (w_1, w_2, w_3\rangle$, then a simple calculation shows that

$$\mathbf{v} \times \mathbf{w} = \det\left(\begin{bmatrix} \mathbf{i} & \mathbf{j} & \mathbf{k} \\ v_1 & v_2 & v_3 \\ w_1 & w_2 & w_3 \end{bmatrix}\right). \tag{9.4.4}$$

Here the determinant is expanded just as if $\mathbf{i}, \mathbf{j}, \mathbf{k}$ were scalars. When we do so, we obtain

$$\begin{aligned} \mathbf{v} \times \mathbf{w} &= \det\left(\begin{bmatrix} v_2 & v_3 \\ w_2 & w_3 \end{bmatrix}\right)\mathbf{i} - \det\left(\begin{bmatrix} v_1 & v_3 \\ w_1 & w_3 \end{bmatrix}\right)\mathbf{j} + \det\left(\begin{bmatrix} v_1 & v_2 \\ w_1 & w_2 \end{bmatrix}\right)\mathbf{k} \\ &= (v_2 w_3 - w_2 v_3)\mathbf{i} - (v_1 w_3 - w_1 v_3)\mathbf{j} + (v_1 w_2 - w_1 v_2)\mathbf{k}, \end{aligned}$$

which is the formula for the cross product, as given by equation (9.4.1).

▶ **EXAMPLE 2** Use a determinant to calculate the cross product of $\mathbf{v} = \langle 2, -1, 6 \rangle$ and $\mathbf{w} = \langle -3, 4, 1 \rangle$.

Solution We write

$$\begin{aligned} \mathbf{v} \times \mathbf{w} &= \det\left(\begin{bmatrix} \mathbf{i} & \mathbf{j} & \mathbf{k} \\ 2 & -1 & 6 \\ -3 & 4 & 1 \end{bmatrix}\right) \\ &= \mathbf{i}((-1)\cdot 1 - 4 \cdot 6) - \mathbf{j}(2 \cdot 1 - (-3) \cdot 6) + \mathbf{k}(2 \cdot 4 - (-3) \cdot (-1)) \\ &= -25\mathbf{i} - 20\mathbf{j} + 5\mathbf{k}. \end{aligned}$$

Thus $\mathbf{v} \times \mathbf{w} = \langle -25, -20, 5 \rangle$. ◀

INSIGHT Recall that the dot product $\mathbf{v} \cdot \mathbf{w}$ of two vectors is a *scalar;* however, the *cross product* of two vectors is another *vector*. Texts that call $\mathbf{v} \cdot \mathbf{w}$ the *scalar product* often call $\mathbf{v} \times \mathbf{w}$ the *vector product*. We will not use this alternative terminology.

Algebraic Properties of the Cross Product

If $\mathbf{u}, \mathbf{v}$, and $\mathbf{w}$ are vectors and λ and μ are scalars, then

$$\begin{aligned} \mathbf{0} \times \mathbf{v} &= \mathbf{0}, \\ \mathbf{u} \times \mathbf{0} &= \mathbf{0}, \\ \mathbf{u} \times (\mathbf{v} + \mathbf{w}) &= (\mathbf{u} \times \mathbf{v}) + (\mathbf{u} \times \mathbf{w}), \\ (\mathbf{v} + \mathbf{w}) \times \mathbf{u} &= (\mathbf{v} \times \mathbf{u}) + (\mathbf{w} \times \mathbf{u}), \\ \mathbf{v} \times \mathbf{w} &= -\mathbf{w} \times \mathbf{v}, \\ (\lambda\mathbf{v}) \times (\mu\mathbf{w}) &= (\lambda\mu)(\mathbf{v} \times \mathbf{w}). \end{aligned}$$

These algebraic properties for the cross product can be verified by routine calculation. The properties of distribution are as expected. However, pay particular attention to the law $\mathbf{v} \times \mathbf{w} = -\mathbf{w} \times \mathbf{v}$, which shows that the cross product is *not* a commutative operation. In fact, because reversing the order of the operands of $\mathbf{v} \times \mathbf{w}$ changes the sign of the product, the cross product is said to be *anticommutative.*

THEOREM 2 If $\mathbf{v}$ is any vector, then $\mathbf{v} \times \mathbf{v} = \mathbf{0}$. More generally, if $\mathbf{u}$ and $\mathbf{v}$ are parallel vectors, then $\mathbf{u} \times \mathbf{v} = \mathbf{0}$. That is, the cross product of any two parallel vectors is the zero vector.

Proof. The anticommutative property of the cross product states that $\mathbf{v} \times \mathbf{w} = -\mathbf{w} \times \mathbf{v}$ for any vectors $\mathbf{v}$ and $\mathbf{w}$. It follows that, if $\mathbf{w} = \mathbf{v}$, then $\mathbf{v} \times \mathbf{v} = -\mathbf{v} \times \mathbf{v}$. By adding $\mathbf{v} \times \mathbf{v}$ to each side of this equation, we conclude that $2\mathbf{v} \times \mathbf{v} = \mathbf{0}$, or $\mathbf{v} \times \mathbf{v} = \mathbf{0}$. Next, suppose that $\mathbf{u}$ and $\mathbf{v}$ are parallel. According to Theorem 2 of Section 2, either $\mathbf{u} = \mathbf{0}$ or $\mathbf{v} = \lambda\mathbf{u}$ for some scalar λ. In the first case, $\mathbf{u} \times \mathbf{v} = \mathbf{0} \times \mathbf{v} = \mathbf{0}$, as we have already observed. In the second case, we use the equation $\mathbf{u} \times \mathbf{u} = \mathbf{0}$, which we proved as the first part of this theorem, and obtain $\mathbf{u} \times \mathbf{v} = \mathbf{u} \times (\lambda\mathbf{u}) = \lambda\,(\mathbf{u} \times \mathbf{u}) = \lambda\mathbf{0} = \mathbf{0}$. ■

▶ **EXAMPLE 3** Give an example to show that the cross product does not satisfy a cancellation property. In other words, the equality $\mathbf{v} \times \mathbf{w} = \mathbf{v} \times \mathbf{u}$ does not imply that $\mathbf{w} = \mathbf{u}$. Give an example to show that the cross product does not satisfy the associative property. In other words, $(\mathbf{u} \times \mathbf{v}) \times \mathbf{w} \neq \mathbf{u} \times (\mathbf{v} \times \mathbf{w})$ in general.

Solution A counterexample to the cancellation property is given by the parallel vectors $\mathbf{v} = \langle 1,0,0\rangle$, $\mathbf{w} = \langle 2,0,0\rangle$, and $\mathbf{u} = \langle 3,0,0\rangle$. We have $\mathbf{v} \times \mathbf{w} = \mathbf{0} = \mathbf{v} \times \mathbf{u}$ yet $\mathbf{w} \neq \mathbf{u}$. A counterexample to the associative property is given by $\mathbf{u} = \langle 1,0,0\rangle$, $\mathbf{v} = \langle 1,0,0\rangle$, and $\mathbf{w} = \langle 0,1,0\rangle$. Because $\mathbf{u} = \mathbf{v}$, we have $(\mathbf{u} \times \mathbf{v}) \times \mathbf{w} = \mathbf{0} \times \mathbf{w} = \mathbf{0}$. On the other hand, we calculate $\mathbf{v} \times \mathbf{w} = \langle 0,0,1\rangle$ and $\mathbf{u} \times (\mathbf{v} \times \mathbf{w}) = \langle 1,0,0\rangle \times \langle 0,0,1\rangle = \langle 0,-1,0\rangle \neq \mathbf{0}$. ◀

▶ **EXAMPLE 4** Use Theorem 2 to determine whether there is any nonzero vector of the form $\mathbf{v} = \langle a, b, b^2\rangle$ that is parallel to $\mathbf{u} = \langle 1,2,3\rangle$.

Solution We calculate $\mathbf{u} \times \mathbf{v} = \langle 2b^2 - 3b, 3a - b^2, b - 2a\rangle$. Theorem 2 tells us that if $\mathbf{u}$ and $\mathbf{v}$ are parallel, then the entries of this cross product must all be 0. That is, $2b^2 - 3b = 0$, $3a - b^2 = 0$, and $b - 2a = 0$. The last of these three equations gives us $b = 2a$. If we substitute this value for b into the second equation, then we obtain $3a - 4a^2 = 0$, or $a(3/4 - a) = 0$. Because $b = 2a$, the solution $a = 0$ implies $b = 0$, which results in $\mathbf{v}$ being the zero vector. Therefore the only admissible solution is $a = 3/4$. This leads to $b = 3/2$. We note that this value of b also satisfies the first equation, $2b^2 - 3b = 0$. Thus Theorem 2 tells us that $\mathbf{v} = \langle 3/4, 3/2, 9/4\rangle$ is the only nonzero vector that can be a solution. Because $\mathbf{v} = (3/4)\langle 1,2,3\rangle = (3/4)\,\mathbf{u}$, we see that $\mathbf{v}$ *is* parallel to $\mathbf{u}$. ◀

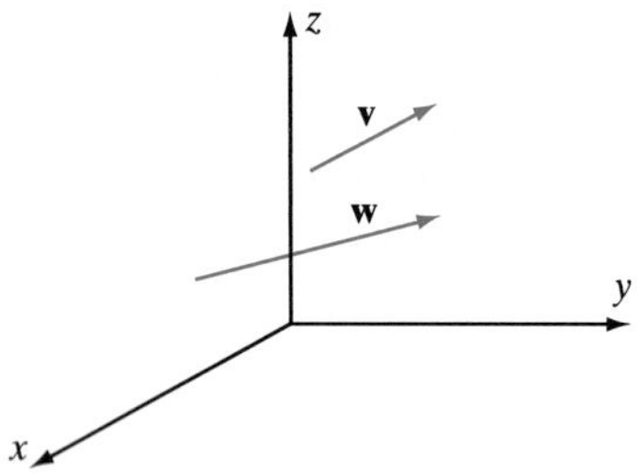

▲ Figure 4

A Geometric Understanding of the Cross Product

Now we would like to develop a geometric way to think of the cross product. If we have a picture of nonparallel vectors $\mathbf{v}$ and $\mathbf{w}$ (Figure 4), then how can we visualize the cross product $\mathbf{v} \times \mathbf{w}$? The first step is to use $\mathbf{v}$ and $\mathbf{w}$ to determine a plane. If P_0 is a point in space, then we may represent $\mathbf{v}$ and $\mathbf{w}$ by directed line segments that

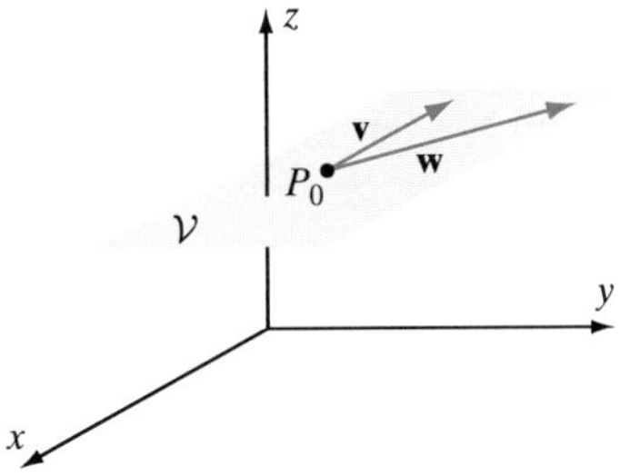

▲ Figure 5

have P_0 as their common initial point. In doing so, we obtain a plane $\mathcal{V}$ through P_0 that contains **v** and **w** (see Figure 5).

Next, we observe that, for any point P_0 on a plane $\mathcal{V}$ in space, there is a unique line ℓ through P_0 that is perpendicular to $\mathcal{V}$ (see Figure 6). We say that ℓ is *the normal line* to $\mathcal{V}$ through P_0. If we do not refer to the point of intersection P_0 of ℓ and $\mathcal{V}$, then we use the indefinite article and say that ℓ is *a normal line* to $\mathcal{V}$. A nonzero vector **n** is *normal* to $\mathcal{V}$ if **n** can be represented by a directed line segment $\overrightarrow{P_1P_2}$ where P_1 and P_2 are distinct points on a normal line ℓ to $\mathcal{V}$ (see Figure 7).

If **n** is normal to a plane $\mathcal{V}$ that is determined by **v** and **w**, then we say that **n** is a *normal vector* for the pair **v** and **w**. If **n** is also a unit vector, then it is said to be a *unit normal vector* for **v** and **w**. A pair of nonparallel vectors always has two unit normal vectors: If **n** is one of them, then $-\mathbf{n}$ is the other (see Figure 8). With the following right-hand rule, we will associate one of these two unit normal vectors with the ordered pair $(\mathbf{v}, \mathbf{w})$ and call it the *standard* unit normal vector for $(\mathbf{v}, \mathbf{w})$. The negative of this vector is the standard unit normal vector for the ordered pair $(\mathbf{w}, \mathbf{v})$.

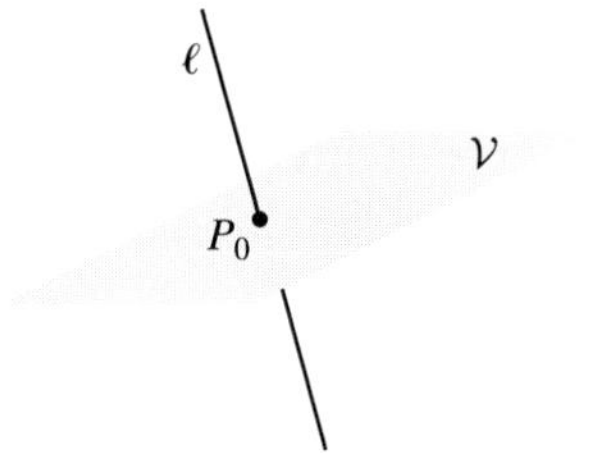

▲ Figure 6

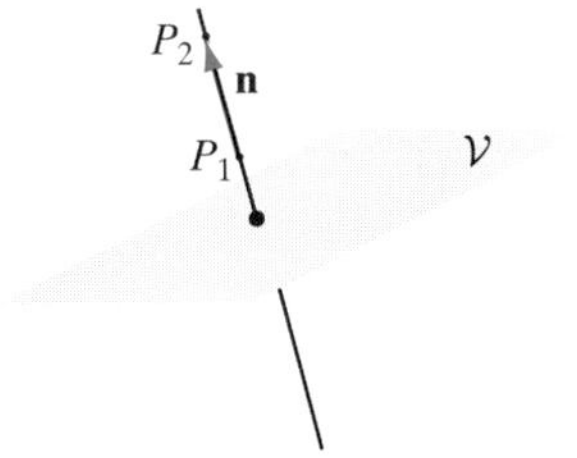

▲ Figure 7

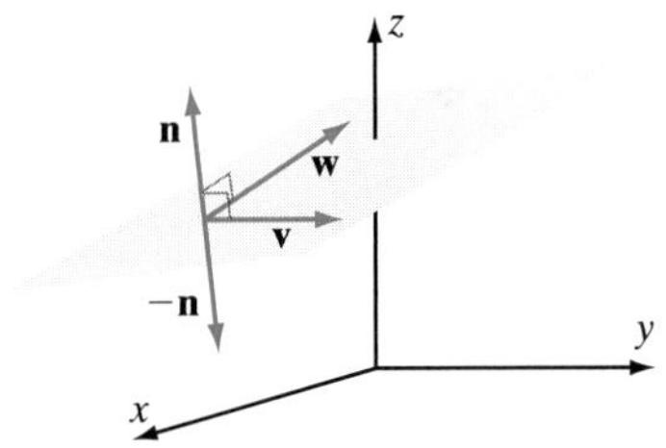

▲ Figure 8

The Right-Hand Rule for Finding the Direction of the Standard Unit Normal

Let $(\mathbf{v}, \mathbf{w})$ be an ordered pair of two nonparallel vectors. Represent **v** and **w** by directed line segments with a common initial point. Point the fingers of your right hand along the first vector **v** and then curl them toward the second vector **w**, as in Figure 9. The direction that the thumb points during this process is the direction of the standard unit normal vector for the vectors **v**, **w**. Look again at Figure 8. The direction of vector **n** is the standard unit normal for the ordered pair $(\mathbf{v}, \mathbf{w})$.

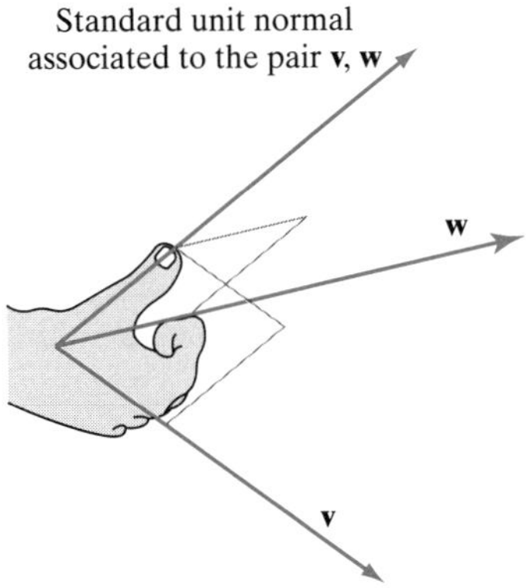

▲ Figure 9

▶ **EXAMPLE 5** Find the standard unit normal for the pairs $(\mathbf{i}, \mathbf{j})$ and $(\mathbf{j}, \mathbf{k})$ and $(\mathbf{k}, \mathbf{i})$. Find also the standard unit normal for the pairs $(\mathbf{j}, \mathbf{i})$ and $(\mathbf{k}, \mathbf{j})$ and $(\mathbf{i}, \mathbf{k})$.

Solution If you curl the fingers of your right hand from **i** toward **j**, then your thumb will point in the direction of the positive z-axis. Thus the standard unit normal for the pair $(\mathbf{i}, \mathbf{j})$ is **k**. Similar reasoning shows that the standard unit normal for $(\mathbf{j}, \mathbf{i})$ is $-\mathbf{k}$. We leave it as an exercise to check that the standard unit normal for the pair $(\mathbf{j}, \mathbf{k})$ is **i**, and the standard unit normal for the pair $(\mathbf{k}, \mathbf{i})$ is **j**. In addition, the standard unit normal for $(\mathbf{k}, \mathbf{j})$ is $-\mathbf{i}$, and the standard unit normal for $(\mathbf{i}, \mathbf{k})$ is $-\mathbf{j}$. ◀

INSIGHT According to the right-hand rule, the standard unit normal vector for the ordered pair $(\mathbf{v}, \mathbf{w})$ points in the opposite direction to the standard unit normal vector for the ordered pair $(\mathbf{w}, \mathbf{v})$. Example 4 provides particular cases of this general fact.

If $\mathbf{v} \times \mathbf{w} \neq \mathbf{0}$, then we may form the unit vector $\mathbf{dir}(\mathbf{v} \times \mathbf{w}) = \|\mathbf{v} \times \mathbf{w}\|^{-1}(\mathbf{v} \times \mathbf{w})$. Because $\mathbf{v} \times \mathbf{w}$ is perpendicular to both $\mathbf{v}$ and $\mathbf{w}$, the vector $\mathbf{dir}(\mathbf{v} \times \mathbf{w})$ must be the standard unit normal for either the ordered pair $(\mathbf{v}, \mathbf{w})$ or the ordered pair $(\mathbf{w}, \mathbf{v})$. Our next theorem tells us which. It also specifies the length of $\mathbf{v} \times \mathbf{w}$ and tells us precisely when the equation $\mathbf{v} \times \mathbf{w} = \mathbf{0}$ holds.

THEOREM 3 Let $\mathbf{v}$ and $\mathbf{w}$ be vectors. Then:

a. $\|\mathbf{v} \times \mathbf{w}\|^2 = \|\mathbf{v}\|^2 \|\mathbf{w}\|^2 - (\mathbf{v} \cdot \mathbf{w})^2$.
b. If $\mathbf{v}$ and $\mathbf{w}$ are nonzero, then $\|\mathbf{v} \times \mathbf{w}\| = \|\mathbf{v}\|\,\|\mathbf{w}\| \sin(\theta)$ where $\theta \in [0, \pi]$ denotes the angle between $\mathbf{v}$ and $\mathbf{w}$.
c. $\mathbf{v}$ and $\mathbf{w}$ are parallel if and only if $\mathbf{v} \times \mathbf{w} = \mathbf{0}$.
d. If $\mathbf{v}$ and $\mathbf{w}$ are not parallel, then $\mathbf{v} \times \mathbf{w}$ points in the direction of the standard unit normal for the pair $(\mathbf{v}, \mathbf{w})$. In particular, $\mathbf{dir}(\mathbf{v} \times \mathbf{w}) = \|\mathbf{v} \times \mathbf{w}\|^{-1}(\mathbf{v} \times \mathbf{w})$ is the standard unit normal vector for the pair $(\mathbf{v}, \mathbf{w})$.

Proof. Part a is an identity among the entries of $\mathbf{v} = \langle v_1, v_2, v_3 \rangle$ and $\mathbf{w} = \langle w_1, w_2, w_3 \rangle$. We have

$$\begin{aligned} \|\mathbf{v} \times \mathbf{w}\|^2 &= (v_2 w_3 - w_2 v_3)^2 + (v_1 w_3 - w_1 v_3)^2 + (v_1 w_2 - w_1 v_2)^2 \\ &= (v_1^2 + v_2^2 + v_3^2)(w_1^2 + w_2^2 + w_3^2) - (v_1 w_1 + v_2 w_2 + v_3 w_3)^2 \\ &= \|\mathbf{v}\|^2 \|\mathbf{w}\|^2 - (\mathbf{v} \cdot \mathbf{w})^2. \end{aligned}$$

The second equality in this chain is not obvious, but it may be verified by multiplying everything out.

If neither $\mathbf{v}$ nor $\mathbf{w}$ is the zero vector, then the scalars $\|\mathbf{v}\|$ and $\|\mathbf{w}\|$ are nonzero. In this case, the identity of part a becomes

$$\|\mathbf{v} \times \mathbf{w}\|^2 = \|\mathbf{v}\|^2 \|\mathbf{w}\|^2 - (\mathbf{v} \cdot \mathbf{w})^2 = \|\mathbf{v}\|^2 \|\mathbf{w}\|^2 \left(1 - \frac{(\mathbf{v} \cdot \mathbf{w})^2}{\|\mathbf{v}\|^2 \|\mathbf{w}\|^2}\right) = \|\mathbf{v}\|^2 \|\mathbf{w}\|^2 (1 - \cos^2(\theta)),$$

where the last equality is obtained by using formula (9.3.6). It follows that $\|\mathbf{v} \times \mathbf{w}\|^2 = \|\mathbf{v}\|^2 \|\mathbf{w}\|^2 \sin^2(\theta)$ and, on taking the square root, $\|\mathbf{v} \times \mathbf{w}\| = \|\mathbf{v}\|\,\|\mathbf{w}\|\,|\sin(\theta)|$. Because $0 \leq \theta \leq \pi$, it follows that $0 \leq \sin(\theta)$. Therefore $|\sin(\theta)| = \sin(\theta)$ and $\|\mathbf{v} \times \mathbf{w}\| = \|\mathbf{v}\|\,\|\mathbf{w}\| \sin(\theta)$, as asserted in part b.

Theorem 3 of Section 3 tells us that vectors $\mathbf{v}$ and $\mathbf{w}$ are parallel if and only if $|\mathbf{v} \cdot \mathbf{w}| = \|\mathbf{v}\|\,\|\mathbf{w}\|$. Using the identity of part a, we conclude that $\mathbf{v}$ and $\mathbf{w}$ are parallel if and only if $\|\mathbf{v} \times \mathbf{w}\| = 0$. Because $\mathbf{0}$ is the only zero length vector, assertion c follows.

For part d, we will limit our verification to a few special cases. Notice that $\mathbf{i} \times \mathbf{j} = \mathbf{k}$, $\mathbf{j} \times \mathbf{k} = \mathbf{i}$, and $\mathbf{k} \times \mathbf{i} = \mathbf{j}$. Based on our observations in Example 4, we conclude that, for the basic vectors $\mathbf{i}, \mathbf{j}$, and $\mathbf{k}$, the operation of cross product produces the standard unit normal. Geometric reasoning can be used to show that the cross product $\mathbf{v} \times \mathbf{w}$ always points in the direction of the standard unit normal for the vectors $\mathbf{v}, \mathbf{w}$. ■

▶ **EXAMPLE 6** Let $\mathbf{v} = \langle 1, -3, 2 \rangle$ and $\mathbf{w} = \langle 1, -1, 4 \rangle$. What is the standard unit normal vector for $(\mathbf{v}, \mathbf{w})$?

Solution We calculate the vector

$$\begin{aligned}\mathbf{v}\times\mathbf{w} &= \big((-3)\cdot 4-(-1)\cdot 2\big)\mathbf{i}-(1\cdot 4-1\cdot 2)\mathbf{j}+\big(1\cdot(-1)-1\cdot(-3)\big)\mathbf{k}\\ &= -10\mathbf{i}-2\mathbf{j}+2\mathbf{k}\end{aligned}$$

and its length

$$\|\mathbf{v}\times\mathbf{w}\| = \sqrt{(-10)^2+(-2)^2+2^2} = \sqrt{108} = \sqrt{(36)(3)} = 6\sqrt{3}.$$

Part d of Theorem 2 tells us that

$$\mathbf{dir}(\mathbf{v}\times\mathbf{w}) = \|\mathbf{v}\times\mathbf{w}\|^{-1}(\mathbf{v}\times\mathbf{w}) = \frac{1}{6\sqrt{3}}(-10\mathbf{i}-2\mathbf{j}+2\mathbf{k}) = -\frac{5}{3\sqrt{3}}\mathbf{i}-\frac{1}{3\sqrt{3}}\mathbf{j}+\frac{1}{3\sqrt{3}}\mathbf{k}$$

is the vector we seek. ◄

Cross Products and the Calculation of Area

Now we learn a connection between cross products and areas of triangles and parallelograms. To be specific, if **v** and **w** are two nonparallel vectors that are represented by directed line segments $\overrightarrow{OP}$ and $\overrightarrow{OQ}$ with common initial point O, then we speak of triangle OPQ as "the triangle determined by the vectors **v** and **w**." See Figure 10 for this triangle. The parallelogram determined by **v** and **w** is shown in Figure 11. Notice that, when we declare OPQ to be the triangle determined by **v** and **w**, the third side $\overrightarrow{QP}$ represents $\mathbf{v}-\mathbf{w}$. Had we positioned **v** so that its initial point coincided with the terminal point Q of **w**, then we would have obtained triangle OQR with third side $\mathbf{v}+\mathbf{w}$. Notice that the areas of triangles OPQ and OQR are the same because each is half the area of the parallelogram. Thus had we declared OQR as the triangle determined by the vectors **v** and **w** instead of OPQ, we would have been speaking of a triangle with the same area.

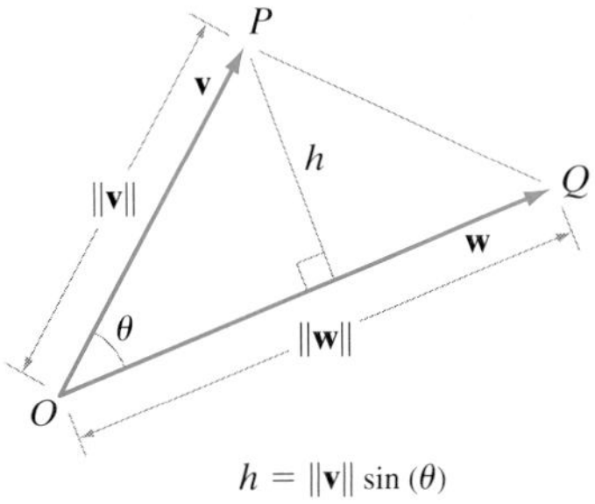

▲ Figure 10

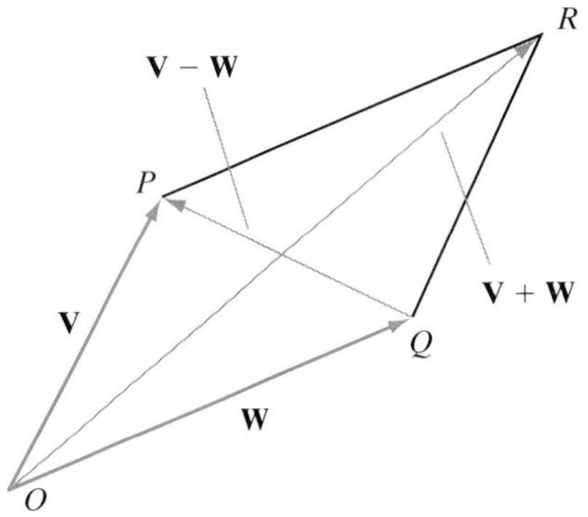

▲ Figure 11

THEOREM 4 Suppose that **v** and **w** are nonparallel vectors. The area of the triangle determined by **v** and **w** is $\frac{1}{2}\|\mathbf{v}\times\mathbf{w}\|$. The area of the parallelogram determined by **v** and **w** is $\|\mathbf{v}\times\mathbf{w}\|$.

Proof. As we see from Figure 10, the altitude from vertex P to base OQ has length $\|\mathbf{v}\|\sin(\theta)$. The area of triangle OPQ, namely half the product of its base and height, is therefore $\frac{1}{2}\|\mathbf{w}\|\,\|\mathbf{v}\|\sin(\theta)$. This quantity is $\frac{1}{2}\|\mathbf{v}\times\mathbf{w}\|$ by Theorem 3b. As discussed, the second assertion follows from the first because the area of the parallelogram determined by **v** and **w** is twice the area of the triangle that the two vectors determine. ■

▶ **EXAMPLE 7** Find the area of the triangle determined by the vectors $\mathbf{v} = \langle 0, 2, 1\rangle$ and $\mathbf{w} = \langle 3, 1, -1\rangle$.

Solution We calculate that

$$\frac{1}{2}\left\|\mathbf{v}\times\mathbf{w}\right\| = \frac{1}{2}\left\|-3\mathbf{i}+3\mathbf{j}-6\mathbf{k}\right\| = \frac{1}{2}\sqrt{54} = \frac{3}{2}\sqrt{6}. \blacktriangleleft$$

▶ **EXAMPLE 8** Calculate the area of the parallelogram determined by the vectors $\mathbf{v} = \langle -2, 1, 3\rangle$ and $\mathbf{w} = \langle 1, 0, 4\rangle$.

Solution The required area is $\left\|\mathbf{v}\times\mathbf{w}\right\| = \left\|4\mathbf{i}+11\mathbf{j}-\mathbf{k}\right\| = \sqrt{138}.$ ◀

A Physical Application of the Cross Product

Experimental evidence teaches us that if a magnetic field **M** acts on a charged particle with charge s, and if the charged particle has velocity **v**, then the resultant force **F** that is exerted is

$$\mathbf{F} = (s\mathbf{v})\times\mathrm{M}.$$

▶ **EXAMPLE 9** In the picture tube for an oscilloscope, a magnetic field is used to control the path of ions that transmit the image to the screen. If s is the charge of the particle, **v** is its velocity, and **M** is the magnetic field, then the force **F** exerted on the particle is $\mathbf{F} = (s\mathbf{v})\times\mathbf{M}$. Suppose that the velocity vector for the ions is $\langle c, -c, 0\rangle$, where c is a physical constant. The magnetic field will have the form $\langle a, 1, 1\rangle$, where the value of a will be varied to force the ions to go in different directions. If we want to exert a force on a positively charged ion in the direction $\langle c/2, c/2, c/2\rangle$, then how should we select a?

Solution We need to solve the equation $\lambda\langle c/2, c/2, c/2\rangle = s\langle c, -c, 0\rangle \times \langle a, 1, 1\rangle$. The constant λ on the left side allows us to adjust for length. Calculating the cross product $\langle c, -c, 0\rangle \times \langle a, 1, 1\rangle$, we find that

$$\frac{\lambda}{2}\langle c, c, c\rangle = s\langle -c, -c, c+ca\rangle.$$

We choose $\lambda = -2s$ so that the first two entries on either side match up. For this vector equation to hold, the third entries $\lambda c/2$ and $s(c+ca)$ must also be equal. We must therefore choose a so that $(-2s)\,c/2 = s(c+ca)$. This forces us to set $a = -2$. ◀

The Triple Scalar Product

We have just seen that the cross product is useful for finding the areas of triangles and parallelograms. We now develop this idea further by introducing a product that involves three vectors.

> **DEFINITION** If $\mathbf{u}, \mathbf{v}$, and $\mathbf{w}$ are given vectors, then we define their *triple scalar product* to be the number $(\mathbf{u}\times\mathbf{v})\cdot\mathbf{w}$.

Notice that the triple scalar product $(\mathbf{u}\times\mathbf{v})\cdot\mathbf{w}$ of three vectors $\mathbf{u}, \mathbf{v}$, and **w** is a *scalar* because it is the dot product of the two vectors $\mathbf{u}\times\mathbf{v}$ and **w**. The parentheses in the triple scalar product are not really necessary: The association $\mathbf{u}\times(\mathbf{v}\cdot\mathbf{w})$

makes no sense because the cross product of vector **u** with scalar **v** · **w** is not defined. However, the expression **u** · (**v** × **w**) *is* defined, and, by expanding both it and the triple scalar product (**u** × **v**) · **w** in terms of the entries of **u**, **v**, and **w**, we find that

$$(\mathbf{u} \times \mathbf{v}) \cdot \mathbf{w} = \mathbf{u} \cdot (\mathbf{v} \times \mathbf{w}). \tag{9.4.5}$$

In other words, to compute the triple scalar product of **u**, **v**, and **w**, we write down the vectors *in the given order*, we insert the operations · and × in either order, and we then associate in the only way possible (see Figure 12).

$$\underbrace{\mathbf{u}\ \mathbf{v}\ \mathbf{w}}_{\text{Three vectors, in order}} \Rightarrow \underbrace{\begin{array}{c}\mathbf{u} \times \mathbf{v} \cdot \mathbf{w} \\ \text{or} \\ \mathbf{u} \cdot \mathbf{v} \times \mathbf{w}\end{array}}_{\text{Insert · and ×}} \Rightarrow \underbrace{\begin{array}{c}(\mathbf{u} \times \mathbf{v}) \cdot \mathbf{w} \\ \text{or} \\ \mathbf{u} \cdot (\mathbf{v} \times \mathbf{w})\end{array}}_{\text{Associate}}$$

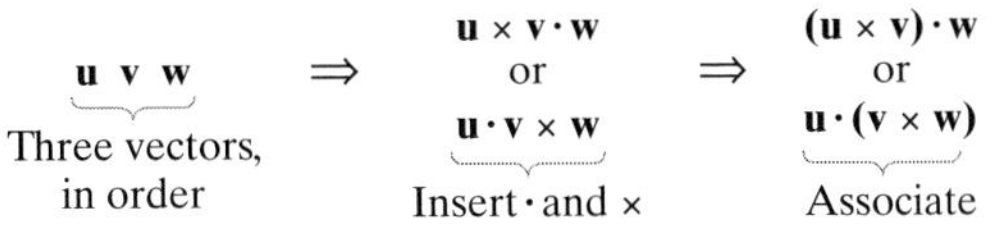

▲ **Figure 12** The triple scalar product of *u*, *v*, and *w*

▶ **EXAMPLE 10** Calculate the triple scalar product of $\mathbf{u} = \langle 2, -1, 4 \rangle$, $\mathbf{v} = \langle 7, 2, 3 \rangle$, and $\mathbf{w} = \langle -1, 1, 2 \rangle$ in two different ways.

Solution Because $\mathbf{u} \times \mathbf{v} = \langle -11, 22, 11 \rangle$, we may calculate the triple scalar product as

$$(\mathbf{u} \times \mathbf{v}) \cdot \mathbf{w} = \langle -11, 22, 11 \rangle \cdot \langle -1, 1, 2 \rangle = 11 + 22 + 22 = 55.$$

Because $\mathbf{v} \times \mathbf{w} = \langle 1, -17, 9 \rangle$, an alternative calculation is

$$\mathbf{u} \cdot (\mathbf{v} \times \mathbf{w}) = \langle 2, -1, 4 \rangle \cdot \langle 1, -17, 9 \rangle = 2 + 17 + 36 = 55. \quad \blacktriangleleft$$

The next theorem shows that we can compute a triple scalar product without first calculating a cross product.

THEOREM 5 The triple scalar product of $\mathbf{u} = \langle u_1, u_2, u_3 \rangle$, $\mathbf{v} = \langle v_1, v_2, v_3 \rangle$, and $\mathbf{w} = \langle w_1, w_2, w_3 \rangle$ is given by the formula

$$(\mathbf{u} \times \mathbf{v}) \cdot \mathbf{w} = \det\left(\begin{bmatrix} u_1 & u_2 & u_3 \\ v_1 & v_2 & v_3 \\ w_1 & w_2 & w_3 \end{bmatrix}\right). \tag{9.4.6}$$

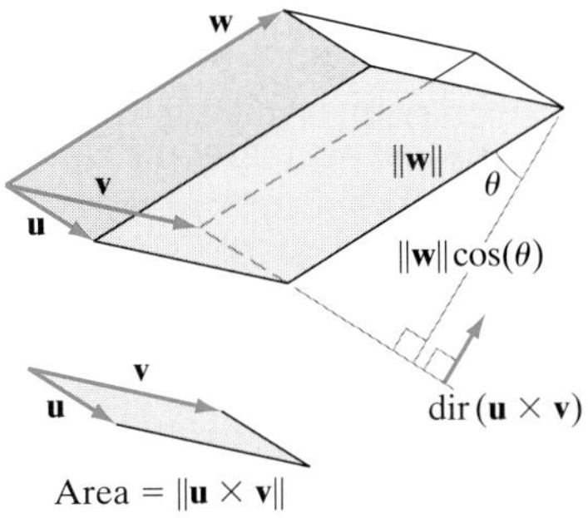

▲ **Figure 13**

Proof. Because $(\mathbf{u} \times \mathbf{v}) \cdot \mathbf{w} = \mathbf{u} \cdot (\mathbf{v} \times \mathbf{w})$ and

$$\mathbf{v} \times \mathbf{w} = \det\left(\begin{bmatrix} v_2 & v_3 \\ w_2 & w_3 \end{bmatrix}\right)\mathbf{i} - \det\left(\begin{bmatrix} v_1 & v_3 \\ w_1 & w_3 \end{bmatrix}\right)\mathbf{j} + \det\left(\begin{bmatrix} v_1 & v_2 \\ w_1 & w_2 \end{bmatrix}\right)\mathbf{k},$$

we have

$$(\mathbf{u} \cdot \mathbf{v}) \times \mathbf{w} = u_1 \det\left(\begin{bmatrix} v_2 & v_3 \\ w_2 & w_3 \end{bmatrix}\right) - u_2 \det\left(\begin{bmatrix} v_1 & v_3 \\ w_1 & w_3 \end{bmatrix}\right) + u_3 \det\left(\begin{bmatrix} v_1 & v_2 \\ w_1 & w_2 \end{bmatrix}\right) = \det\left(\begin{bmatrix} u_1 & u_2 & u_3 \\ v_1 & v_2 & v_3 \\ w_1 & w_2 & w_3 \end{bmatrix}\right). \quad \blacksquare$$

To understand the geometric significance of the triple scalar product, we look at Figure 13. The solid region determined by **u**, **v**, and **w** is called a "parallelepiped." By the right-hand rule, $\mathbf{u} \times \mathbf{v}$ points upward in the figure and is perpendicular to the base of the parallelepiped. Moreover, $\|\mathbf{u} \times \mathbf{v}\|$ is the area of the base. The height of the parallelepiped is $\|\mathbf{w}\|\cos(\theta)$, where θ is the angle between $\mathbf{w}$ and $\mathbf{u}\times\mathbf{v}$ (as shown in Figure 13). We may always take θ to be between 0 and $\pi/2$. We calculate the volume of the parallelepiped as follows:

$$\begin{aligned}(\text{volume of parallelepiped}) &= (\text{area of base})\cdot(\text{height})\\ &= \|\mathbf{u}\times\mathbf{v}\|\cdot\|\mathbf{w}\|\cos\theta\\ &= |(\mathbf{u}\times\mathbf{v})\cdot\mathbf{w}|.\end{aligned}$$

Because the volume of the parallelepiped is obviously independent of the order in which we take the vectors **u**, **v**, and **w**, we deduce that the absolute value of the triple product, taken in any order, gives the same answer. We record the results of this investigation as a theorem:

THEOREM 6 Let $\mathbf{u},\mathbf{v},\mathbf{w}$ be vectors. The volume of the parallelepiped determined by these vectors is given by any of the expressions

$$|(\mathbf{u}\times\mathbf{v})\cdot\mathbf{w}| = |(\mathbf{u}\times\mathbf{w})\cdot\mathbf{v}| = |(\mathbf{v}\times\mathbf{w})\cdot\mathbf{u}| = \left|\det\left(\begin{bmatrix} u_1 & u_2 & u_3\\ v_1 & v_2 & v_3\\ w_1 & w_2 & w_3\end{bmatrix}\right)\right|. \quad (9.4.7)$$

In case $\mathbf{u}\times\mathbf{v}$ and $\mathbf{w}$ point to the same side of the plane determined by **u** and **v** (the vectors form a right-hand system), then $(\mathbf{u}\times\mathbf{v})\cdot\mathbf{w}$ is already positive and equals the volume of the parallelepiped.

▶ **EXAMPLE 11** Use the determinant to calculate the volume of the parallelepiped determined by the vectors $\langle -3,2,5\rangle$, $\langle 1,0,3\rangle$, and $\langle 3,-1,-2\rangle$.

Solution The required volume is

$$\left|\det\left(\begin{bmatrix} -3 & 2 & 5\\ 1 & 0 & 3\\ 3 & -1 & -2\end{bmatrix}\right)\right| = \left|-3\det\left(\begin{bmatrix} 0 & 3\\ -1 & -2\end{bmatrix}\right) - 2\det\left(\begin{bmatrix} 1 & 3\\ 3 & -2\end{bmatrix}\right) + 5\det\left(\begin{bmatrix} 1 & 0\\ 3 & -1\end{bmatrix}\right)\right|$$
$$= |-3(3) - 2(-11) + 5(-1)| = 8. \quad ◀$$

Vectors are said to be *coplanar* when they lie in the same plane. Our next theorem gives us a simple way to tell when three vectors are coplanar:

THEOREM 7 Three vectors $\mathbf{u} = \langle u_1,u_2,u_3\rangle$, $\mathbf{v} = \langle v_1,v_2,v_3\rangle$, and $\mathbf{w} = \langle w_1,w_2,w_3\rangle$ are coplanar if and only if $\mathbf{u}\cdot(\mathbf{v}\times\mathbf{w}) = 0$. Equivalently, they are coplanar if and only if

$$\det\left(\begin{bmatrix} u_1 & u_2 & u_3\\ v_1 & v_2 & v_3\\ w_1 & w_2 & w_3\end{bmatrix}\right) = 0.$$

Proof. According to Theorem 6, this determinant is 0 if and only if the parallelepiped determined by the three vectors has 0 volume. But this means that the three vectors are coplanar. ■

▶ **EXAMPLE 12** Show that $\mathbf{u} = \langle 3,1,1\rangle, \mathbf{v} = \langle 1,2,0\rangle$, and $\mathbf{w} = \langle 1,-3,1\rangle$ are coplanar.

Solution We calculate

$$\det\left(\begin{bmatrix} 3 & 1 & 1 \\ 1 & 2 & 0 \\ 1 & -3 & 1 \end{bmatrix}\right) = 3\det\left(\begin{bmatrix} 2 & 0 \\ -3 & 1 \end{bmatrix}\right) - 1\cdot\det\left(\begin{bmatrix} 1 & 0 \\ 1 & 1 \end{bmatrix}\right) + 1\cdot\det\left(\begin{bmatrix} 1 & 2 \\ 1 & -3 \end{bmatrix}\right) = 6 - 1 - 5 = 0. \quad ◀$$

Triple Vector Products

It is reasonable for us to ask, "Is there a triple product of spatial vectors that results in a vector?" An affirmative answer is contained in the next definition.

DEFINITION If $\mathbf{u}, \mathbf{v}$, and $\mathbf{w}$ are given spatial vectors, then each of the vectors $\mathbf{u}\times(\mathbf{v}\times\mathbf{w})$ and $(\mathbf{u}\times\mathbf{v})\times\mathbf{w}$ is said to be a *triple vector product* of $\mathbf{u}, \mathbf{v}$, and $\mathbf{w}$.

INSIGHT Although the triple scalar product $\mathbf{u}\cdot\mathbf{v}\times\mathbf{w}$ does not require parentheses, a triple vector product *does*. The expression $\mathbf{u}\times\mathbf{v}\times\mathbf{w}$ is ambiguous because its meaning depends on how the terms are associated. In general, $(\mathbf{u}\times\mathbf{v})\times\mathbf{w} \neq \mathbf{u}\times(\mathbf{v}\times\mathbf{w})$, as we observed in Example 3.

Our final example, which involves a triple vector product, illustrates several geometric properties of the cross product.

▶ **EXAMPLE 13** Let $\mathbf{v}$ and $\mathbf{w}$ be perpendicular spatial vectors. Show that

$$\mathbf{v}\times(\mathbf{v}\times\mathbf{w}) = -(\mathbf{v}\cdot\mathbf{v})\,\mathbf{w} \tag{9.4.8}$$

and

$$\mathbf{w}\times(\mathbf{v}\times\mathbf{w}) = (\mathbf{w}\cdot\mathbf{w})\,\mathbf{v}. \tag{9.4.9}$$

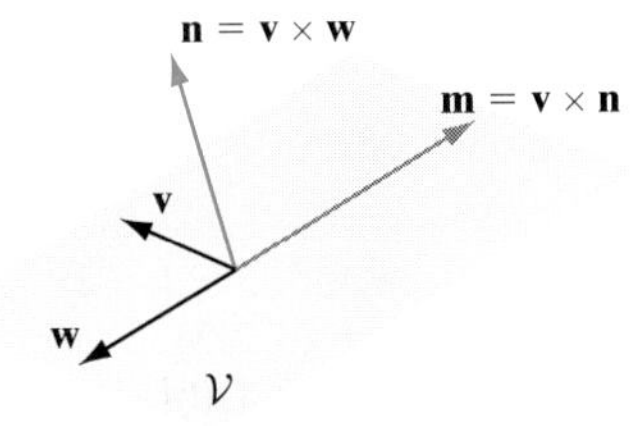

▲ Figure 14

Solution If either $\mathbf{v}$ or $\mathbf{w}$ is $\mathbf{0}$, then the asserted equation holds because each side is $\mathbf{0}$. Suppose that $\mathbf{v}$ and $\mathbf{w}$ are both nonzero. Then $\mathbf{n} = \mathbf{v}\times\mathbf{w}$ is normal to the plane $\mathcal{V}$ determined by $\mathbf{v}$ and $\mathbf{w}$. Let $\mathbf{m} = \mathbf{v}\times\mathbf{n}$. Then $\mathbf{m}$ is perpendicular to $\mathbf{v}$ and $\mathbf{n}$. Because $\mathbf{m}$ is perpendicular to $\mathbf{n}$, a normal to plane $\mathcal{V}$, it follows that $\mathbf{m}$ lies in $\mathcal{V}$. Because $\mathbf{m}$, $\mathbf{v}$, and $\mathbf{w}$ all lie in $\mathcal{V}$, and $\mathbf{m}$ and $\mathbf{w}$ are both perpendicular to $\mathbf{v}$, it follows that $\mathbf{m}$ and $\mathbf{w}$ are parallel. Thus there is a scalar λ such that $\mathbf{m} = \lambda\mathbf{w}$. If we apply the right-hand rule a first time to find the direction of $\mathbf{n} = \mathbf{v}\times\mathbf{w}$ and a second time to find the direction of $\mathbf{m} = \mathbf{v}\times\mathbf{n}$, then we see that $\mathbf{m}$ and $\mathbf{w}$ have oppositive directions—use Figure 14 to help you see this. It follows that $\lambda<0$ and $\lambda = -|\lambda|$. Finally, by applying Theorem 3b, first to the perpendicular pair $\mathbf{v}, \mathbf{n}$ and then to the perpendicular pair $\mathbf{v}, \mathbf{w}$, we obtain

$$|\lambda|\,\|\mathbf{w}\| = \|\lambda\mathbf{w}\| = \|\mathbf{m}\| = \|\mathbf{v}\times\mathbf{n}\| = \|\mathbf{v}\|\,\|\mathbf{n}\| = \|\mathbf{v}\|\,\|\mathbf{v}\times\mathbf{w}\| = \|\mathbf{v}\|\,\|\mathbf{v}\|\,\|\mathbf{w}\| = \|\mathbf{v}\|^2\,\|\mathbf{w}\| = (\mathbf{v}\cdot\mathbf{v})\,\|\mathbf{w}\|.$$

We conclude that $|\lambda| = \mathbf{v}\cdot\mathbf{v}$ and $\lambda = -|\lambda| = -(\mathbf{v}\cdot\mathbf{v})$. Thus

$$\mathbf{v}\times(\mathbf{v}\times\mathbf{w}) = \mathbf{v}\times\mathbf{n} = \mathbf{m} = \lambda\mathbf{w} = -(\mathbf{v}\cdot\mathbf{v})\mathbf{w},$$

which is equation (9.4.8). We obtain equation (9.4.9) from this equation (with **v** and **w** interchanged) as follows:

$$\mathbf{w} \times (\mathbf{v} \times \mathbf{w}) = -\mathbf{w} \times (\mathbf{w} \times \mathbf{v}) \overset{(9.4.8)}{=} (\mathbf{w} \cdot \mathbf{w})\,\mathbf{v}.$$

Formulas (9.4.8) and (9.4.9) can also be proved by direct calculation with the entries of **v** and **w**. ◀

QUICK QUIZ

1. Calculate $\langle 2,1,2\rangle \times \langle 1,-2,-1\rangle$.
2. Find the area of the parallelogram determined by $\langle 2,1,-2\rangle$ and $\langle 1,1,0\rangle$.
3. Find the standard unit normal vector for the ordered pair $(\langle 2,1,-2\rangle, \langle 1,1,0\rangle)$.
4. True or false: (a) $\mathbf{v}\times\mathbf{w} = \mathbf{w}\times\mathbf{v}$; (b) $\mathbf{u}\times(\mathbf{v}\times\mathbf{w}) = (\mathbf{u}\times\mathbf{v})\times\mathbf{w}$; (c) $\mathbf{u}\cdot\mathbf{v}\times\mathbf{w} = \mathbf{u}\times\mathbf{v}\cdot\mathbf{w}$?

Answers

1. $\langle 3,4,-5\rangle$ 2. 3 3. $\langle 2/3, -2/3, 1/3\rangle$ 4. (a) False; (b) False; (c) True

EXERCISES

Problems for Practice

▶ In each of Exercises 1–8, compute $\mathbf{v}\times\mathbf{w}$. Verify that **v** and **w** are perpendicular to $\mathbf{v}\times\mathbf{w}$ by showing that $\mathbf{v}\cdot(\mathbf{v}\times\mathbf{w})$ and $\mathbf{w}\cdot(\mathbf{v}\times\mathbf{w})$ are both 0. ◀

1. $\mathbf{v} = \langle 3,3,-5\rangle$ $\mathbf{w} = \langle -1,3,-2\rangle$
2. $\mathbf{v} = \langle 0,1,1\rangle$ $\mathbf{w} = \langle 1,1,0\rangle$
3. $\mathbf{v} = \langle 4,-3,6\rangle$ $\mathbf{w} = \langle 2,1,2\rangle$
4. $\mathbf{v} = \langle 2,0,-2\rangle$ $\mathbf{w} = \langle 2,0,2\rangle$
5. $\mathbf{v} = \langle -4,-2,3\rangle$ $\mathbf{w} = \langle 7,1,-5\rangle$
6. $\mathbf{v} = \langle -9,4,6\rangle$ $\mathbf{w} = \langle 0,0,1\rangle$
7. $\mathbf{v} = \langle 1,6,8\rangle$ $\mathbf{w} = \langle 1,2,8\rangle$
8. $\mathbf{v} = \langle 2,3,3\rangle$ $\mathbf{w} = \langle -2,-2,3\rangle$

▶ In each of Exercises 9–14, find the standard unit normal vector associated to the given ordered pair of vectors. ◀

9. $\langle 2,1,3\rangle$ $\langle -3,-2,-5\rangle$
10. $\langle 0,1,1\rangle$ $\langle 1,1,0\rangle$
11. $\langle 3,-1,1\rangle$ $\langle 1,1,1\rangle$
12. $\langle -2,-4,3\rangle$ $\langle 2,2,-1\rangle$
13. $\langle 1,2,2\rangle$ $\langle 2,-1,1\rangle$
14. $\langle 0,1/2,1\rangle$ $\langle 1/2,0,1\rangle$

▶ In each of Exercises 15–18, find the area of the triangle determined by the two given vectors. ◀

15. $\langle 2,1,2\rangle$ $\langle 3,2,3\rangle$
16. $\langle 1/3,0,-1\rangle$ $\langle 1,3,3\rangle$
17. $\langle -3,4,1\rangle$ $\langle -2,0,1\rangle$
18. $\langle 1,3,4\rangle$ $\langle 0,-3,-4\rangle$

▶ In each of Exercises 19–22, calculate the area of the parallelogram determined by the two given vectors. ◀

19. $\langle 2,1,1\rangle$ $\langle 0,-3,1\rangle$
20. $\langle -1,3,2\rangle$ $\langle -1,1,1\rangle$
21. $\langle 1,0,2\rangle$ $\langle 1,-3,0\rangle$
22. $\langle 1,3,4\rangle$ $\langle 1,1,2\rangle$

▶ In each of Exercises 23–26, calculate the triple vector products $\mathbf{u}\times(\mathbf{v}\times\mathbf{w})$ and $(\mathbf{u}\times\mathbf{v})\times\mathbf{w}$. (In each exercise, notice that $\mathbf{u}\times(\mathbf{v}\times\mathbf{w}) \neq (\mathbf{u}\times\mathbf{v})\times\mathbf{w}$.) ◀

23. $\mathbf{u} = \langle -4,1,0\rangle$ $\mathbf{v} = \langle 2,-1,-3\rangle$ $\mathbf{w} = \langle 3,-2,-3\rangle$
24. $\mathbf{u} = \langle 0,4,-3\rangle$ $\mathbf{v} = \langle 3,0,-2\rangle$ $\mathbf{w} = \langle -4,-3,-2\rangle$
25. $\mathbf{u} = \langle 1,1,2\rangle$ $\mathbf{v} = \langle 2,1,2\rangle$ $\mathbf{w} = \langle -3,4,-3\rangle$
26. $\mathbf{u} = \langle 2,-3,-2\rangle$ $\mathbf{v} = \langle -3,1,-5\rangle$ $\mathbf{w} = \langle 2,1,0\rangle$

▶ In each of Exercises 27–30, calculate $\mathbf{u}\times\mathbf{v}$, $\mathbf{v}\times\mathbf{w}$, and the triple scalar product $(\mathbf{u}\times\mathbf{v})\cdot\mathbf{w}$. Verify that

$$(\mathbf{u}\times\mathbf{v})\cdot\mathbf{w} = \mathbf{u}\cdot(\mathbf{v}\times\mathbf{w}). ◀$$

27. $\mathbf{u} = \langle 3,0,1\rangle$ $\mathbf{v} = \langle 2,-1,-3\rangle$ $\mathbf{w} = \langle -1,-3,2\rangle$
28. $\mathbf{u} = \langle 0,4,-3\rangle$, $\mathbf{v} = \langle 2,0,-6\rangle$, $\mathbf{w} = \langle -3,-1,-2\rangle$
29. $\mathbf{u} = \langle 1,1,-2\rangle$ $\mathbf{v} = \langle 2,1,-2\rangle$ $\mathbf{w} = \langle 3,2,3\rangle$
30. $\mathbf{u} = \langle 4,3,-2\rangle$ $\mathbf{v} = \langle 2,1,-5\rangle$ $\mathbf{w} = \langle 2,1,0\rangle$

▶ In each of Exercises 31–34, use a determinant to calculate the triple scalar product $(\mathbf{u}\times\mathbf{v})\cdot\mathbf{w}$. ◀

31. $\mathbf{u} = \langle 1,-2,4\rangle$ $\mathbf{v} = \langle 2,0,1\rangle$ $\mathbf{w} = \langle 3,1,1\rangle$
32. $\mathbf{u} = \langle 2,1,2\rangle$ $\mathbf{v} = \langle 2,2,3\rangle$ $\mathbf{w} = \langle 3,3,-5\rangle$
33. $\mathbf{u} = \langle 1,1,1\rangle$ $\mathbf{v} = \langle 3,0,2\rangle$ $\mathbf{w} = \langle -2,-3,3\rangle$
34. $\mathbf{u} = \langle -5,-1,0\rangle$ $\mathbf{v} = \langle 1,0,3\rangle$ $\mathbf{w} = \langle 0,2,2\rangle$

▶ In each of Exercises 35–38, use the triple scalar product to verify that the three given vectors are coplanar. ◀

35. $\mathbf{u} = \langle 1,-2,4\rangle$ $\mathbf{v} = \langle 2,0,1\rangle$ $\mathbf{w} = \langle 5,-2,6\rangle$

36. $\mathbf{u} = \langle 2,1,2\rangle$ $\quad \mathbf{v} = \langle 2,2,3\rangle$ $\quad \mathbf{w} = \langle 6,4,7\rangle$
37. $\mathbf{u} = \langle 1,1,1\rangle$ $\quad \mathbf{v} = \langle 3,0,2\rangle$ $\quad \mathbf{w} = \langle 8,2,6\rangle$
38. $\mathbf{u} = \langle -5, -1, 0\rangle$ $\quad \mathbf{v} = \langle 1,0,3\rangle$ $\quad \mathbf{w} = \langle -6, -1, -3\rangle$

In each of Exercises 39–42, vectors $\mathbf{u}, \mathbf{v}$, and $\mathbf{w}$ are given. Calculate the volume of the parallelepiped that they determine.

39. $\mathbf{u} = \langle 2,1,1\rangle$ $\quad \mathbf{v} = \langle 1,1,-1\rangle$ $\quad \mathbf{w} = \langle 1, -1, -1\rangle$
40. $\mathbf{u} = \langle 2,1,-2\rangle$ $\quad \mathbf{v} = \langle 3,1,-1\rangle$ $\quad \mathbf{w} = \langle 1,-1,1\rangle$
41. $\mathbf{u} = \langle 3,2,1\rangle$ $\quad \mathbf{v} = \langle 1,2,3\rangle$ $\quad \mathbf{w} = \langle 1,4,-2\rangle$
42. $\mathbf{u} = \langle 5,2,3\rangle$ $\quad \mathbf{v} = \langle 4,-2,3\rangle$ $\quad \mathbf{w} = \langle 3,-1,2\rangle$

Further Theory and Practice

In each of Exercises 43–46, find scalars s and t for which $\mathbf{u} \times (\mathbf{v} \times \mathbf{w}) = s\mathbf{v} + t\mathbf{w}$.

43. $\mathbf{u} = \langle 1,-2,4\rangle$ $\quad \mathbf{v} = \langle 2,0,1\rangle$ $\quad \mathbf{w} = \langle 5,-3,2\rangle$
44. $\mathbf{u} = \langle 2,1,2\rangle$ $\quad \mathbf{v} = \langle 2,2,3\rangle$ $\quad \mathbf{w} = \langle 1,1,1\rangle$
45. $\mathbf{u} = \langle 1,1,-1\rangle$ $\quad \mathbf{v} = \langle 3,0,2\rangle$ $\quad \mathbf{w} = \langle 4,2,1\rangle$
46. $\mathbf{u} = \langle 5,-1,0\rangle$ $\quad \mathbf{v} = \langle 1,0,3\rangle$ $\quad \mathbf{w} = \langle -2, -1, 7\rangle$

In each of Exercises 47–50, vectors $\mathbf{v}$ and $\mathbf{w}$ are given. Calculate $\|\mathbf{v} \times \mathbf{w}\|^2$ and verify that this quantity equals $\|\mathbf{v}\|^2 \|\mathbf{w}\|^2 - (\mathbf{v} \cdot \mathbf{w})^2$, as asserted by Theorem 3a.

47. $\mathbf{v} = \langle 2,1,2\rangle$ $\quad \mathbf{w} = \langle 1, -2, -2\rangle$
48. $\mathbf{v} = \langle 3,-1,1\rangle$ $\quad \mathbf{w} = \langle 1,-1,1\rangle$
49. $\mathbf{v} = \langle 2,3,1\rangle$ $\quad \mathbf{w} = \langle 2,3,-1\rangle$
50. $\mathbf{v} = \langle 4,0,1\rangle$ $\quad \mathbf{w} = \langle 7,1,2\rangle$

In each of Exercises 51–54, vectors $\mathbf{v}$ and $\mathbf{w}$ are given. Let θ be the angle between $\mathbf{v}$ and $\mathbf{w}$. Calculate (a) $\mathbf{v} \times \mathbf{w}$, (b) $\sin(\theta)$, (c) $\mathbf{v} \cdot \mathbf{w}$, and (d) $\cos(\theta)$. Verify that the values of $\sin(\theta)$ and $\cos(\theta)$ are consistent.

51. $\mathbf{v} = \langle 2,1,1\rangle$ $\quad \mathbf{w} = \langle 1,2,-1\rangle$
52. $\mathbf{v} = \langle 1,0,-2\rangle$ $\quad \mathbf{w} = \langle 2,1,-1\rangle$
53. $\mathbf{v} = \langle 2, -2,1\rangle$ $\quad \mathbf{w} = \langle 8,4,1\rangle$
54. $\mathbf{v} = \langle \sqrt{2}, \sqrt{3}, 2\rangle$ $\quad \mathbf{w} = \langle 2\sqrt{2}, 0, 1\rangle$

55. Suppose that $\mathbf{u}, \mathbf{v}, \mathbf{w}$ are spatial vectors. Prove that

$$\mathbf{u} \times (\mathbf{v} + \mathbf{w}) = (\mathbf{u} \times \mathbf{v}) + (\mathbf{u} \times \mathbf{w}).$$

56. Suppose that $\mathbf{u}, \mathbf{v}, \mathbf{w}$ are spatial vectors. Use the result of the preceding exercise to deduce that

$$(\mathbf{v} + \mathbf{w}) \times \mathbf{u} = (\mathbf{v} \times \mathbf{u}) + (\mathbf{w} \times \mathbf{u}).$$

57. Suppose that $\mathbf{v}$ and $\mathbf{w}$ are spatial vectors and that λ and μ are scalars. Prove that

$$(\lambda\mathbf{v}) \times (\mu\mathbf{w}) = (\lambda\mu)(\mathbf{v} \times \mathbf{w}).$$

58. Suppose that $\mathbf{v}$ and $\mathbf{w}$ are spatial vectors. Prove that

$$(\mathbf{u} \times \mathbf{v}) \times \mathbf{w} = \mathbf{w} \times (\mathbf{v} \times \mathbf{u}).$$

59. Suppose that $\mathbf{w}$ is a fixed spatial vector. Find all vectors $\mathbf{v}$ that satisfy the equation $\mathbf{v} \times \mathbf{w} = \mathbf{v} + \mathbf{w}$.

The cross product can be interpreted physically in terms of *moment of force.* Namely, let $\mathbf{F}$ be a force (vector) applied at a point Q in space, and let P be another point in space. The *moment* or *torque* $\boldsymbol{\tau}$ of the force $\mathbf{F}$ at the point Q about the point P is defined to be

$$\boldsymbol{\tau} = \overrightarrow{PQ} \times \mathbf{F}.$$

See Figure 15. Notice that $\boldsymbol{\tau}$ is a vector. If a certain body that can pivot about P extends to the point Q, and if $\mathbf{F}$ is applied at Q, then $\|\boldsymbol{\tau}\|$ measures the tendency $\mathbf{F}$ to make the body rotate about P. The direction of $\boldsymbol{\tau}$ gives the axis of rotation. A bolt at point P will be driven by a wrench in the direction of $\boldsymbol{\tau}$ (see Figure 15). Exercises 60–62 concern torque.

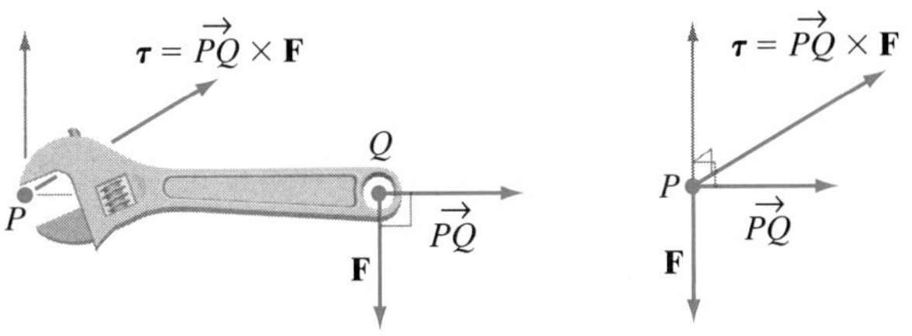

▲ Figure 15

60. A diver who weighs 520 newtons stands at the end of a rigid diving board: It is *not* a spring board. The diving board is 2 meters long and extends in an upward direction from the edge of the pool at a 30 degree angle with the horizontal. The weight of the diver results in a torque about the point at which the diving board is attached. What is the magnitude of this torque?

61. An 8-inch wrench is used to drive a bolt at point P. A force $\mathbf{F}$ is applied to the end of the handle (point Q). If $\|\mathbf{F}\| = 60$ pounds and the angle between and $\overrightarrow{PQ}$ is $90°$, then what is the magnitude (in foot-pounds) of the torque that is produced?

62. Repeat Exercise 61 but suppose that the angle of application is $120°$.

63. Two distinct points P and Q determine a line ℓ. Suppose that R is a point not on ℓ. The distance $d(R, \ell)$ of R to ℓ is defined to be the minimum value of $\|\overrightarrow{RT}\|$ as T varies over all points of ℓ. Let S be the point on ℓ such that $\overrightarrow{RS}$ and $\overrightarrow{PQ}$ are mutually perpendicular. Show that $d(R, \ell) = \|\overrightarrow{RS}\|$. Use the cross product to calculate $d(R, \ell)$ in terms of the given points P, Q, and R.

64. Prove the converse to equation (9.4.8): If $\mathbf{v} \times (\mathbf{v} \times \mathbf{w}) = -(\mathbf{v} \cdot \mathbf{v}), \mathbf{w}$, then $\mathbf{v}$ and $\mathbf{w}$ are perpendicular.

65. Let s and t be scalars. Show that $\mathbf{u}, \mathbf{v}$, and $s\mathbf{u} + t\mathbf{v}$ are coplanar.

66. The *Gram determinant* of two spatial vectors $\mathbf{v}$ and $\mathbf{w}$ is defined to be

$$G = \det\left(\begin{bmatrix} \mathbf{v}\cdot\mathbf{v} & \mathbf{v}\cdot\mathbf{w} \\ \mathbf{w}\cdot\mathbf{v} & \mathbf{w}\cdot\mathbf{w} \end{bmatrix}\right).$$

Show that $\mathbf{v}$ and $\mathbf{w}$ are parallel if and only if $G = 0$.

67. Prove the Lagrange Identity:

$$(\mathbf{v}\times\mathbf{w})\cdot(\mathbf{p}\times\mathbf{q}) = (\mathbf{v}\cdot\mathbf{p})(\mathbf{w}\cdot\mathbf{q}) - (\mathbf{v}\cdot\mathbf{q})(\mathbf{w}\cdot\mathbf{p}).$$

68. Prove the Jacobi Identity:

$$(\mathbf{u}\times\mathbf{v})\times\mathbf{w} + (\mathbf{v}\times\mathbf{w})\times\mathbf{u} + (\mathbf{w}\times\mathbf{u})\times\mathbf{v} = 0.$$

Calculator/Computer Exercises

69. Let $P = (x, y, z)$, $P_0 = (1, 2, 3)$, $\mathbf{v} = \langle 2, 1, -4\rangle$, and $\mathbf{w} = \langle 5, -2, 1\rangle$. Write the equation $\overrightarrow{P_0P}\cdot(\mathbf{v}\times\mathbf{w}) = 0$ as a Cartesian equation in x, y, and z. Plot the solution set S of the equation. What geometric figure results? What is its relationship to the vector $\mathbf{v}\times\mathbf{w}$?

70. Consider the curve that is parameterized by $t \mapsto (5 + \cos(t), 3 + \sin(t), 5)$, $0 \le t \le 2\pi$. Find the point P on the curve that is closest to the origin. Find a vector $\mathbf{v}$ that is tangent to the curve at P. Calculate $\|\mathbf{v}\times\overrightarrow{OP}\|$ and $\|\mathbf{v}\|\cdot\|\overrightarrow{OP}\|$. What significance for $\mathbf{v}$ and $\overrightarrow{OP}$ do these calculations have?

71. Suppose that $\mathbf{v}_t = \langle 1 - t^2, t, 4 - t^2\rangle$ and $\mathbf{w} = \langle 1, 2, 3\rangle$. Plot

$$f(t) = \|\mathbf{v_t}\times\mathbf{w}\| - \mathbf{v_t}\cdot\mathbf{w}$$

for $-3 \le t \le 3$. What is the minimum value of f?

72. Suppose that $\mathbf{v}_t = \langle t, t^2, 1 - t^2\rangle$ and $\mathbf{w}_t = \langle 1 - t^2, t, t^2\rangle$. Plot $f(t) = \|\mathbf{v}_t\times\mathbf{w}_t\|^2$ for $-1 \le t \le 1$. What are the local and global minima of f?

9.5 Lines and Planes in Space

When we learned to graph with Cartesian coordinates in the two-dimensional plane, we found that the set of points satisfying one linear equation is a line, while the set of points satisfying two linear equations is (usually) a point. The philosophy here is that each equation "removes a degree of freedom" or takes away a dimension.

The same philosophy works in three dimensions: The set of points in space satisfying one linear equation will be a plane, and the set of points in space satisfying two linear equations will (usually) be a line. Notice that we had to add the word "usually" because sometimes two equations have no solution or have too many solutions.

Cartesian Equations of Planes in Space

We can describe a line in the plane by giving a point through which the line passes and the direction or slope of the line. We want to use the same idea for determining a plane $\mathcal{V}$ in space. However, we specify the "direction" of a plane in a rather indirect way. Namely, we give a nonzero vector $\mathbf{n}$ that is perpendicular to $\mathcal{V}$. By this, we mean that $\mathbf{n}$ is perpendicular to every vector that lies in $\mathcal{V}$. Recall that such a vector $\mathbf{n}$ is said to be a *normal vector* for $\mathcal{V}$. All normal vectors for $\mathcal{V}$ are then of the form $\lambda\mathbf{n}$ for a nonzero scalar λ. Notice that a normal vector determines the "tilt" of a plane in space, but not its position. Any two parallel planes will have the same normal vectors (see Figure 1). Finally, any given point in space lies in exactly one of these parallel planes. That is the geometric reasoning that underlies the next theorem.

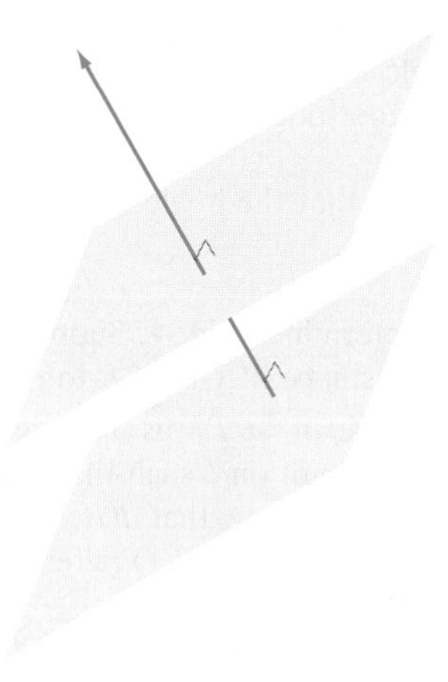

▲ Figure 1 Parallel planes have the same normal vectors.

THEOREM 1 Let $\mathcal{V}$ be a plane in space. Suppose that $\mathbf{n} = \langle A, B, C\rangle$ is a normal vector for $\mathcal{V}$ and that $P_0 = (x_0, y_0, z_0)$ is a point on $\mathcal{V}$. Then,

$$A(x - x_0) + B(y - y_0) + C(z - z_0) = 0 \qquad (9.5.1)$$

is a Cartesian equation for $\mathcal{V}$. This equation may be written in the alternative form

$$Ax + By + Cz = D \tag{9.5.2}$$

where $D = Ax_0 + By_0 + Cz_0$.

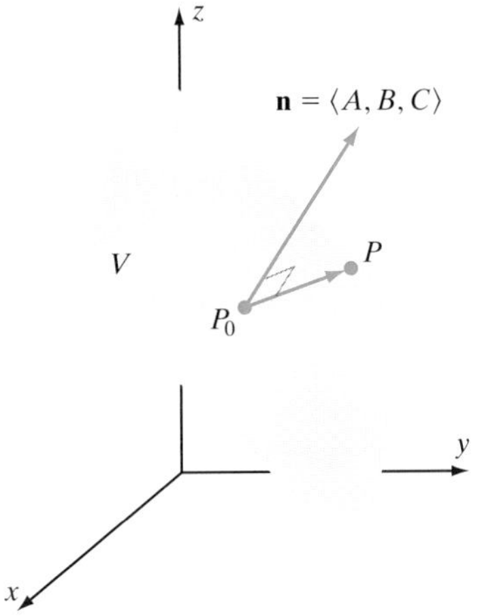

Figure 2 $\mathbf{n} \cdot \overrightarrow{P_0P} = 0$

Proof. Figure 2 shows a plane $\mathcal{V}$ and a nonzero vector $\mathbf{n} = \langle A, B, C\rangle$ that is normal to $\mathcal{V}$. Figure 2 also shows a fixed point $P_0 = (x_0, y_0, z_0)$ in $\mathcal{V}$. Now let $P = (x, y, z)$ be a general point in the plane. The key geometric fact that we need is that the vector $\overrightarrow{P_0P}$ lies in the plane. This is true precisely when the vector $\overrightarrow{P_0P}$ is perpendicular to $\mathbf{n}$. In other words, P lies on $\mathcal{V}$ if and only if $\mathbf{n} \cdot \overrightarrow{P_0P} = 0$. In coordinates, the vector $\overrightarrow{P_0P}$ is given by $\langle x - x_0, y - y_0, z - z_0\rangle$, and the equation $\mathbf{n} \cdot \overrightarrow{P_0P} = 0$ becomes $\langle A, B, C\rangle \cdot \langle x - x_0, y - y_0, z - z_0\rangle = 0$. By expanding the dot product we obtain $A(x - x_0) + B(y - y_0) + C(z - z_0) = 0$, which is equation (9.5.1). Equation (9.5.2) results from this one when the products on the left are expanded and the constant terms are brought to the right side. ∎

▶ **EXAMPLE 1** Determine a Cartesian equation for the plane that has normal vector $\mathbf{n} = \langle -2, 7, 3\rangle$ and passes through the point $P_0 = (-5, -2, 1)$.

Solution Equation (9.5.2) tells us that the required equation has the form $-2x + 7y + 3z = D$ where $D = -2x_0 + 7y_0 + 3z_0$ for any point (x_0, y_0, z_0) in the plane. We are given one such point, P_0. Using its coordinates, we calculate $D = -2(-5) + 7(-2) + 3(1) = -1$. Consequently, $-2x + 7y + 3z = -1$ is a Cartesian equation for the plane. ◀

The chain of reasoning that we have used to determine a Cartesian equation for a given plane can be reversed. We state this observation as a theorem.

THEOREM 2 Suppose that at least one of the coefficients A, B, C is nonzero. Then, the solution set of the equation $A(x - x_0) + B(y - y_0) + C(z - z_0) = 0$ is the plane that has $\langle A, B, C\rangle$ as a normal vector and passes through the point (x_0, y_0, z_0). The solution set of the equation $Ax + By + Cz = D$ is a plane that has $\langle A, B, C\rangle$ as a normal vector.

▶ **EXAMPLE 2** Find a vector that is normal to, and two points that lie in, the plane $\mathcal{V}$ whose Cartesian equation is $z = 3y - 2x$.

Solution By writing the given equation in the standard form $2x - 3y + z = 0$, we may conclude that the vector $\langle 2, -3, 1\rangle$ is a normal vector to $\mathcal{V}$. To find a point $P_0 = (x_0, y_0, z_0)$ in $\mathcal{V}$, we may choose two of the entries of P_0 in an arbitrary way and use the equation to solve for the third entry. For example, if we arbitrarily set $x_0 = 7$ and $y_0 = 5$, then we obtain $z_0 = 3y_0 - 2x_0 = (3)(5) - (2)(7) = 1$. Thus the point $(7, 5, 1)$ is on $\mathcal{V}$. Similarly, if we (arbitrarily) set $y_0 = 3$ and $z_0 = 1$, then we have $2x_0 - 3(3) + (1) = 0$, or $x_0 = 4$. Therefore the point $(4, 3, 1)$ is also on $\mathcal{V}$. ◀

INSIGHT If the three space variables x, y, z have no equations or restrictions imposed on them, then they have three "degrees of freedom" and generate all of three dimensional space. In general, each new equation imposed on x, y, z removes one degree of freedom. We therefore expect the solution set of one Cartesian equation among the variables x, y, z to be a two-dimensional surface. Theorem 2 tells us that, if the equation is linear, then the two-dimensional surface is a plane.

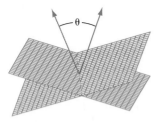

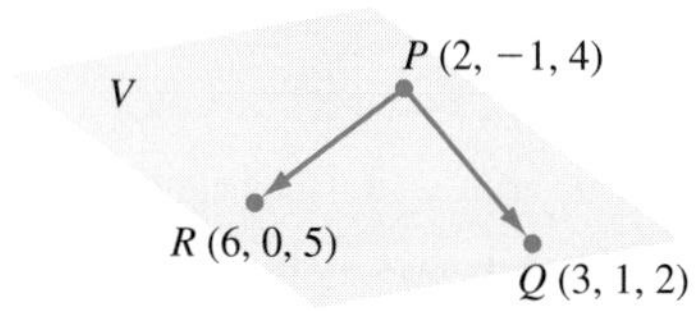

▲ Figure 3

It is intuitively clear that any three points that are not on the same straight line determine a plane; imagine that the three points are three fingertips and balance your notebook on the fingertips—that is the plane we seek. To use equation (9.5.1) or (9.5.2), we must be able to obtain a normal vector from the three points. The next example shows how this is done.

▶ **EXAMPLE 3** Find an equation for the plane $\mathcal{V}$ passing through the points $P = (2, -1, 4), Q = (3, 1, 2)$, and $R = (6, 0, 5)$.

Solution To determine a normal vector to $\mathcal{V}$, we notice that the vectors $\overrightarrow{PQ} = \langle 3-2, 1-(-1), 2-4\rangle = \langle 1, 2, -2\rangle$ and $\overrightarrow{PR} = \langle 6-2, 0-(-1), 5-4\rangle = \langle 4, 1, 1\rangle$ lie in the plane (see Figure 3). The cross product of these two vectors will be perpendicular to both of them, hence perpendicular to $\mathcal{V}$. Thus a normal vector will be $\overrightarrow{PQ} \times \overrightarrow{PR} = 4\mathbf{i} - 9\mathbf{j} - 7\mathbf{k}$. According to formula (9.5.2), the desired equation has the form $4x - 9y - 7z = D$. Each point in the plane satisfies this equation, yielding the same value for D on the right side. We are given three such points. By choosing one, say R, we obtain $D = 4(6) - 9(0) - 7(5) = -11$. Thus $4x - 9y - 7z = -11$ is a Cartesian equation for $\mathcal{V}$.

INSIGHT In working the type of problem illustrated by Example 3, there is a lot of choice. We started by choosing two pairs of points: P, Q and P, R. We might have chosen P, Q and Q, R (or P, R and Q, R) equally well. We used each pair of points to determine a vector in the plane, but we could just as well have chosen the opposite vector. For example, in taking the cross product to find a normal vector, we could have used $\overrightarrow{QP}$ instead of $\overrightarrow{PQ}$. After we found a normal vector, we used equation (9.5.2) to obtain a Cartesian equation for the plane $\mathcal{V}$. We might have used equation (9.5.1) instead. The coordinates of any one of the three given points could then have been used for (x_0, y_0, z_0). To illustrate with just one of these combinations, $\overrightarrow{RQ} \times \overrightarrow{QP}$ results in the normal $-4\mathbf{i} + 9\mathbf{j} + 7\mathbf{k}$. Using this normal in equation (9.5.1) and choosing $P = (2, -1, 4)$ for (x_0, y_0, z_0), we obtain $-4(x-2) + 9(y-(-1)) + 7(z-4) = 0$, or $-4x + 9y + 7z - 11 = 0$, which is equivalent to the equation obtained in Example 3.

▲ Figure 4

We define the *angle between two planes* to be the angle θ between their normals, as in Figure 4. We avoid ambiguity by always selecting the angle θ such that $0 \leq \theta < \pi$.

▶ **EXAMPLE 4** Find the angle between the plane with Cartesian equation $x - y - z = 7$ and the plane with Cartesian equation $-x + y - 3z = 6$.

Solution As we have noticed, $\langle 1, -1, -1\rangle$ is a normal vector for the first plane, and $\langle -1, 1, -3\rangle$ is a normal for the second plane. The angle θ between the two given planes satisfies

$$\begin{aligned}\cos(\theta) &= \frac{\langle 1,-1,-1\rangle \cdot \langle -1,1,-3\rangle}{\|\langle 1,-1,-1\rangle\|\,\|\langle -1,1,-3\rangle\|}\\ &= \frac{1}{\sqrt{3}\sqrt{11}}\\ &= \frac{1}{\sqrt{33}}.\end{aligned}$$

Thus using a calculator, we see that the angle between the two planes is about 1.4 radians. ◀

Parametric Equations of Planes in Space

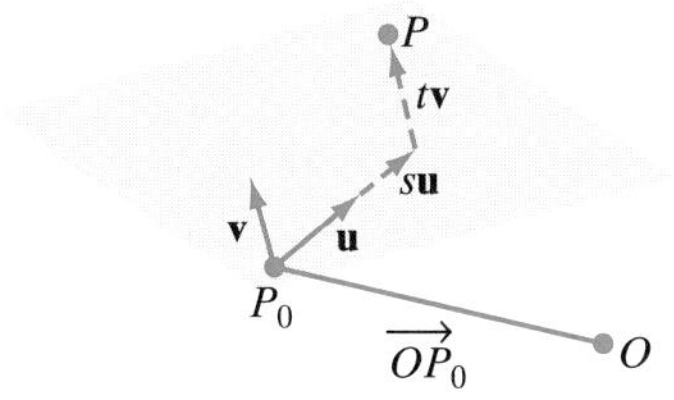

▲ Figure 5 $\overrightarrow{OP} = \overrightarrow{OP_0} + s\mathbf{u} + t\mathbf{v}$.

Suppose that $\mathbf{u} = \langle u_1, u_2, u_3\rangle$ and $\mathbf{v} = \langle v_1, v_2, v_3\rangle$ are nonparallel vectors that lie in a plane $\mathcal{V}$. Let $P_0 = (x_0, y_0, z_0)$ be any fixed point that lies in $\mathcal{V}$. We can think of P_0 as an "origin" in the plane. Any point $P = (x, y, z)$ in the plane can be reached from P_0 by adding a scalar multiple of **u** and then a scalar multiple of **v**, as illustrated in Figure 5. The position vector $\overrightarrow{OP}$ can then be obtained by vector addition as

$$\overrightarrow{OP} = \overrightarrow{OP_0} + s\mathbf{u} + t\mathbf{v}.$$

If we write out this vector equation in coordinates, then we get three scalar parametric equations for the plane:

$$\begin{aligned}x &= x_0 + su_1 + tv_1\\ y &= y_0 + su_2 + tv_2\\ z &= z_0 + su_3 + tv_3.\end{aligned}$$

The parameters s and t may be thought of as coordinates in the plane $\mathcal{V}$: Each point of $\mathcal{V}$ is determined by specifying the values of s and t. We state these observations as a theorem.

THEOREM 3 If $P_0 = (x_0, y_0, z_0)$ is a point in a plane $\mathcal{V}$, and if $\mathbf{u} = \langle u_1, u_2, u_3\rangle$ and $\mathbf{v} = \langle v_1, v_2, v_3\rangle$ are any two nonparallel vectors that are perpendicular to a normal vector for $\mathcal{V}$, then $\mathcal{V}$ consists precisely of those points (x, y, z) with coordinates that satisfy the vector equation

$$\langle x, y, z\rangle = \langle x_0, y_0, z_0\rangle + s\,\mathbf{u} + t\,\mathbf{v}. \qquad (9.5.3)$$

When written coordinatewise, equation (9.5.3) yields parametric equations for $\mathcal{V}$:

$$\begin{aligned}x &= x_0 + su_1 + t\,v_1\\ y &= y_0 + su_2 + t\,v_2\\ z &= z_0 + su_3 + t\,v_3\end{aligned}$$

It should be noted that parametric equations for planes are not unique. Other choices for P_0, **u** and **v** will result in a different parameterization.

▶ **EXAMPLE 5** Find parametric equations for the plane $\mathcal{V}$ whose Cartesian equation is $3x - y + 2z = 7$.

Solution We must find one point $P_0 = (x_0, y_0, z_0)$ and two nonparallel vectors $\mathbf{u}$ and $\mathbf{v}$ in $\mathcal{V}$. To find P_0, we may select any values for x_0 and y_0 and use the Cartesian equation $3x - y + 2z = 7$ to solve for z_0. For example, if we (arbitrarily) set $x_0 = 1$ and $y_0 = 0$, then we obtain $3(1) - (0) + 2z_0 = 7$, or $z_0 = 2$. Thus $P_0 = (1, 0, 2)$ is a point on $\mathcal{V}$. The vectors $\mathbf{u}$ and $\mathbf{v}$ can be any nonparallel vectors that are perpendicular to the normal vector $\mathbf{n} = \langle 3, -1, 2\rangle$. For example, we may take $\mathbf{u} = \langle 0, 2, 1\rangle$ and $\mathbf{v} = \langle 2, 0, -3\rangle$, as you can verify by calculating $\mathbf{u} \cdot \mathbf{n} = 0$ and $\mathbf{v} \cdot \mathbf{n} = 0$. With these choices, the parametric equations become $x = 1 + 0s + 2t$, $y = 0 + 2s + 0t, z = 2 + s - 3t$. We may (and should) verify these parametric equations by substituting them into the Cartesian equation and checking that the resulting equation holds for all s and t. Here the substitutions yield the equation $3(1 + 2t) - (2s) + 2(2 + s - 3t) \equiv 7$, which we see is true for every s and t by simplifying the left side. Having verified that $x = 1 + 2t, y = 2s, z = 2 + s - 3t$ are valid parametric equations for $\mathcal{V}$, we should remark that alternative choices lead to equally valid yet quite different parametric equations. See Exercise 92. ◀

INSIGHT The cross product is useful for producing *one* vector that is orthogonal to *two* given vectors. The solution of Example 5 illustrates a simple procedure for finding *two* nonparallel vectors that are orthogonal to *one* given vector $\mathbf{n} = \langle A, B, C\rangle$. To do that, we may replace one component of $\mathbf{n}$ by 0, interchange the other two, and change the sign of one of them. Thus $\langle 0, C, -B\rangle$, $\langle C, 0, -A\rangle$, and $\langle B, -A, 0\rangle$ are all orthogonal to $\mathbf{n}$.

Parametric Equations of Lines in Space

A line ℓ in space can be described by a point that it passes through and a vector that is parallel to it. Look at Figure 6. Let $P_0 = (x_0, y_0, z_0)$ be the point, and let $\mathbf{m} = \langle a, b, c\rangle$ the vector parallel to ℓ. An arbitrary point $P = (x, y, z)$ lies on ℓ if and only if the vector $\overrightarrow{P_0P}$ is parallel to $\mathbf{m}$. This happens if and only if $\overrightarrow{P_0P}$ is a multiple of $\mathbf{m}$. That is, $\overrightarrow{P_0P}$ is parallel to $\mathbf{m}$ if and only if, $\overrightarrow{P_0P} = t\,\mathbf{m}$ for some scalar t. In coordinates, this last equation becomes $\langle x - x_0, y - y_0, z - z_0\rangle = t\langle a, b, c\rangle$, or $\langle x, y, z\rangle - \langle x_0, y_0, z_0\rangle = \langle ta, tb, tc\rangle$. By adding vector $\langle x_0, y_0, z_0\rangle$ to each side, we obtain

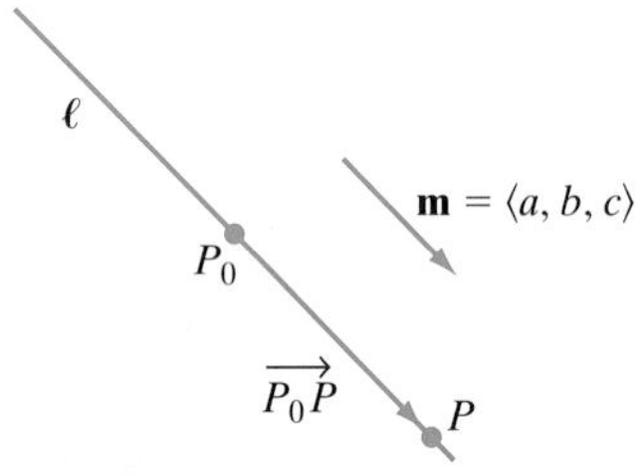

▲ Figure 6

$$\langle x, y, z\rangle = \langle x_0 + ta, y_0 + tb, z_0 + tc\rangle.$$

Notice that t is a scalar and will play the role of a parameter. After we match up coordinates, the parametric equations for a line in space become

$$\begin{aligned} x &= x_0 + ta \\ y &= y_0 + tb \\ z &= z_0 + tc \end{aligned}$$

Let us summarize our observations with a theorem.

THEOREM 4 **(Parametric Equations for a Line)** The line in space that passes through the point $P_0 = (x_0, y_0, z_0)$ and is parallel to the vector $\mathbf{m} = \langle a, b, c\rangle$ has equation

$$\overrightarrow{P_0P} = t\,\mathbf{m}.$$

Here, $P = (x, y, z)$ is a variable point on the line. In coordinates, the equation may be written as three parametric equations:

$$\begin{aligned} x &= x_0 + ta \\ y &= y_0 + tb \\ z &= z_0 + tc. \end{aligned}$$

▶ **EXAMPLE 6** Give three points that lie on the line ℓ whose parametric equations are

$$\begin{aligned} x &= -5 + 3t \\ y &= 7 - 8t \\ z &= 1 + 2t. \end{aligned}$$

Does the point $(1, 2, 3)$ lie on ℓ? How about the point $(-11, 23, -3)$? Give parametric equations for the line through the origin that is parallel to ℓ.

Solution Each value assigned to t will generate a point on ℓ. For instance, $t = 0$ gives $x = -5, y = 7, z = 1$, or the point $(-5, 7, 1)$. The value $t = 3$ gives $x = 4, y = -17$, $z = 7$, or the point $(4, -17, 7)$. The value $t = -1$ gives $x = -8, y = 15, z = -1$, or the point $(-8, 15, -1)$. To determine whether the point $(1, 2, 3)$ lies on the line, we need to determine whether some value of t, when substituted into the parametric equations, yields this point. Thus we need to find a *single* t that satisfies all three equations

$$\begin{aligned} 3t - 5 &= 1 \\ -8t + 7 &= 2 \\ 2t + 1 &= 3 \end{aligned}$$

simultaneously. The first equation gives $t = 2$, the second gives $t = 5/8$. We need look no further: There is no single value of t which satisfies all three equations. Thus the point $(1, 2, 3)$ does not lie on the line.

We have better luck with the point $(-11, 23, -3)$. We endeavor to find a (single) t by solving

$$\begin{aligned} 3t - 5 &= -11 \\ -8t + 7 &= 23 \\ 2t + 1 &= -3. \end{aligned}$$

The first equation gives $t = -2$, and so does the second, and so does the third. Thus substituting $t = -2$ into our parametric equations yields the point $(-11, 23, -3)$, and this point therefore lies on the line. Finally, the parametric equations tell us that the vector $\langle 3, -8, 2 \rangle$ is parallel to ℓ. We now use Theorem 4 with $P_0 = (x_0, y_0, z_0) = (0, 0, 0)$ to see that the line parameterized by $x = 0 + 3t, y = 0 - 8t, z = 0 + 2t$ is parallel to ℓ and passes through the origin. ◀

Cartesian Equations of Lines in Space

As our next theorem shows, Cartesian equations of a line are obtained by eliminating the parameter from a parameterization.

THEOREM 5 Suppose that a line ℓ passes through the point $P_0 = (x_0, y_0, z_0)$ and is parallel to the vector $\mathbf{m} = \langle a, b, c \rangle$ with a, b, and c nonzero. Then ℓ is the solution set of the following system of equations:

$$\frac{x - x_0}{a} = \frac{y - y_0}{b} = \frac{z - z_0}{c}. \qquad (9.5.4)$$

Proof. According to Theorem 4, we may parameterize ℓ by the equations $x = x_0 + ta, y = y_0 + tb, z = z_0 + tc$. After solving for the parameter t in each of these equations, we have $t = (x - x_0)/a, t = (y - y_0)/b$, and $t = (z - z_0)/c$. We equate these three expressions for t to obtain line (9.5.4). ■

The equations of line (9.5.4) are known as *symmetric equations* for ℓ. For the cases in which one or two of a, b, c are zero, see Exercise 96.

▶ **EXAMPLE 7** Line ℓ of Example 6 is parameterized by $x = -5 + 3t$, $y = 7 - 8t, z = 1 + 2t$. Find Cartesian equations for ℓ. Is the point $(1, -9, 5)$ on ℓ? How about the point $(4, -1, 7)$?

Solution We solve for t in each of the parametric equations, obtaining $t = (x + 5)/3$, $t = (y - 7)/(-8)$ and $t = (z - 1)/2$. We now equate these three expressions for t to obtain the symmetric form of ℓ:

$$\frac{x+5}{3} = \frac{y-7}{(-8)} = \frac{z-1}{2}. \tag{9.5.5}$$

A point $P = (x, y, z)$ lies on ℓ if and only if all three of these quantities are equal. For example, the point $(1, -9, 5)$ lies on the line because each quotient in line (9.5.5) becomes 2 when we substitute $x = 1, y = -9$, and $z = 5$. However, the point $(4, -1, 7)$ is not on the line because when we substitute $x = 4, y = -1, z = 5$ into line (9.5.5), the first quantity becomes 3 and the second becomes 1. ◀

Just as two points in the plane determine a straight line, so do two points in space. The next example shows how to find the symmetric form of a line that passes through two given points.

▶ **EXAMPLE 8** Let $P = (2, -3, 5)$ and $Q = (-6, 1, 12)$. Write Cartesian equations for the line ℓ passing through P and Q.

Solution The symmetric form of a line requires a point on the line and a vector $\mathbf{m}$ that is parallel to the line. The piece of information we are initially missing is $\mathbf{m}$, but the vector $\overrightarrow{PQ} = \langle -6 - 2, 1 - (-3), 12 - 5\rangle = \langle -8, 4, 7\rangle$ will certainly do the job. We can use either P or Q as the point we take to be (x_0, y_0, z_0) in line (9.5.4). If we choose P, then the symmetric form for ℓ is

$$\frac{x-2}{(-8)} = \frac{y+3}{4} = \frac{z-5}{7}. \quad ◀$$

▶ **EXAMPLE 9** Convert the symmetric equations

$$\frac{x+1}{3} = \frac{y}{2} = \frac{2z-3}{6}$$

to parametric form.

Solution We set each of these three equal quantities equal to t. We obtain

$$\frac{x+1}{3} = t, \qquad \frac{y}{2} = t, \qquad \frac{2z-3}{6} = t$$

or, equivalently,

$$x = 3t - 1, \qquad y = 2t, \qquad z = 3t + \frac{3}{2}.$$

These are parametric equations for the given line. ◀

▶ **EXAMPLE 10** Find symmetric equations for the line ℓ that is perpendicular to the plane $3x - 7y + 4z = 2$ and that passes through the point $(1, 4, 5)$.

Solution Notice that the vector $\mathbf{m} = \langle 3, -7, 4 \rangle$ is normal to the plane. So the line we seek will be parallel to $\mathbf{m}$. Because the line passes through $(1, 4, 5)$, its parametric equations will be

$$\begin{aligned} x &= 1 + 3t \\ y &= 4 - 7t \\ z &= 5 + 4t. \end{aligned}$$

To pass to symmetric equations for ℓ, we solve for t in each equation and then equate the resulting expressions for t. Thus

$$\frac{x-1}{3} = \frac{y-4}{-7} = \frac{z-5}{4},$$

are symmetric equations for ℓ. ◀

INSIGHT Symmetric equations for a line ℓ describe ℓ by means of *two* linear equations. Each of these equations is, taken by itself, the equation of a plane. Therefore symmetric equations of a line permit us to visualize the line as the intersection of two planes. The Cartesian equations in Example 10 may be rewritten as

$$\frac{x-1}{3} = \frac{y-4}{-7} \quad \text{and} \quad \frac{y-4}{-7} = \frac{z-5}{4}.$$

Rewriting these equations as

$$7x + 3y = 19 \quad \text{and} \quad 4y + 7z = 51,$$

we see that *our line is the set of points that lie in the intersection of two planes.* These ideas are illustrated in Figure 7.

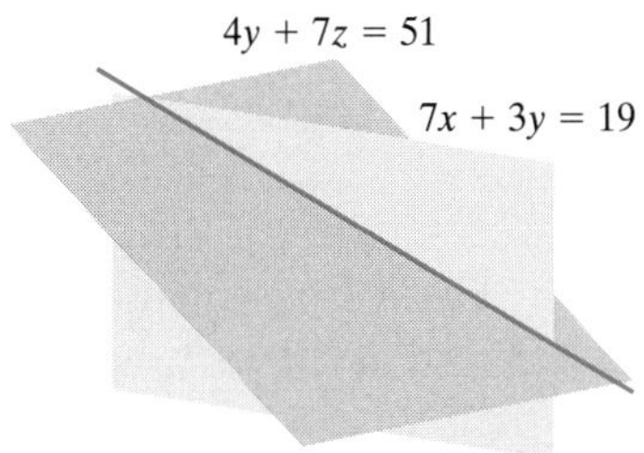

▲ Figure 7 The line $\frac{x-1}{3} = \frac{y-4}{-7} = \frac{z-5}{4}$ as an intersection of two planes.

▶ **EXAMPLE 11** Find parametric equations of the line of intersection of the two planes

$$x - 2y + z = 4 \qquad \text{and} \qquad 2x + y - z = 3.$$

Solution To do so, we set $x = t$ and substitute this into the given equations:

$$-2y + z = 4 - t \qquad \text{and} \qquad y - z = 3 - 2t.$$

We may solve these equations simultaneously for y and z in terms of t:

$$\begin{aligned} y &= 3t - 7 \\ z &= 5t - 10. \end{aligned}$$

If we combine these equations with the equation $x = t$ with which we began our computation, then we see that

$$\begin{aligned} x &= t \\ y &= 3t - 7 \\ z &= 5t - 10 \end{aligned}$$

are parametric equations for the line of intersection. Solving for t in each of the parametric equations and then equating the expressions for t yields $x = (y + 7)/3 = (z + 10)/5$ for symmetric equations. ◀

INSIGHT We could have done the last example by setting either $y = t$ or $z = t$. These choices would have resulted in different parameterizations for the same line. For clarity, let's set $y = s$ (instead of t) and see what happens: We obtain

$$x + z = 4 + 2s$$

and

$$2x - z = 3 - s.$$

Solving these simultaneously, for x and z in terms of t, yields

$$x = \frac{s}{3} + \frac{7}{3}$$
$$z = \frac{5}{3}s + \frac{5}{3}.$$

This leads to parametric equations

$$x = \frac{s}{3} + \frac{7}{3}$$
$$y = s$$
$$z = \frac{5}{3}s + \frac{5}{3}.$$

These three equations describe the very same line as in the last example, but they look quite different from the parametric equations that we found there. What is the relationship between these two parameterizations? The answer is that the change of variable $s = 3t - 7$ transforms one parametrization into the other.

▶ **EXAMPLE 12** Let $\mathcal{V}$ be the plane whose Cartesian equation is $2x + y - 2z = 10$. Let ℓ be the line that is perpendicular to $\mathcal{V}$ and that passes through $P = (3, -8, 3)$. Find the point R at which ℓ intersects $\mathcal{V}$.

Solution For the first step, we follow the method of Example 10 to parameterize ℓ. Because $\mathbf{m} = \langle 2, 1, -2 \rangle$ is orthogonal to $\mathcal{V}$, ℓ has parameterization $x = 3 + 2t$, $y = -8 + t, z = 3 - 2t$. We substitute the parametric equations for x, y, and z into the equation of the plane. This gives us $2(3 + 2t) + (-8 + t) - 2(3 - 2t) = 10$, or $-8 + 9t = 10$. Solving for t, we obtain $t = 2$. The coordinates of R are therefore $x = 3 + 2(2) = 7, y = -8 + 2 = -6$, and $z = 3 - 2(2) = -1$. Thus $R = (7, -6, -1)$ is the point of intersection that we seek. ◀

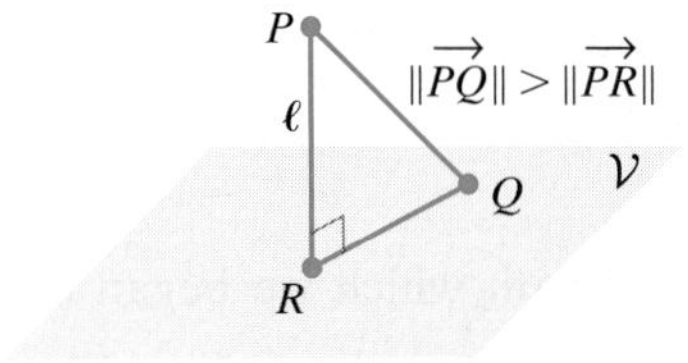

▲ Figure 8 R is the point on $\mathcal{V}$ that is closest to P

INSIGHT In Example 12, the distance between P and R is $\sqrt{(7-3)^2 + (-6-(-8))^2 + (-1-3)^2}$, or 6. Using the Theorem of Pythagoras, we see that R is the point on $\mathcal{V}$ that is closest to P (see Figure 8). Later in this section, we will discuss an efficient method of calculating this distance without first locating the point R of intersection.

▶ **EXAMPLE 13** Determine whether the lines parameterized by

$$x = 3t - 7$$
$$y = -2t + 5$$
$$z = t + 1$$

and

$$\begin{aligned} x &= -3t + 2 \\ y &= t + 1 \\ z &= 2t - 2 \end{aligned}$$

intersect.

Solution First of all, let us note that we cannot expect that the point of intersection, if there is one, will correspond to the same value of t for each line. (Imagine two cars crossing each other's path at an intersection. When all goes well the two cars pass through the same point but *at different times*.) We overcome this complication by using a different parameter, say s, for the first line. Now we equate the expressions for x, y, and z. We have

$$\begin{aligned} 3s - 7 &= -3t + 2 \\ -2s + 5 &= t + 1 \\ s + 1 &= 2t - 2. \end{aligned}$$

Notice that there are three equations in only two unknowns. Solving the first two equations simultaneously gives $s = 1$ and $t = 2$. There will be a point of intersection if and only if these values for s and t also satisfy the third equation. That is the case in this example. The value $s = 1$ corresponds to the point $(-4, 3, 2)$ on the first line; the value $t = 2$ corresponds to the same point on the second line. This is the only point of intersection. ◄

INSIGHT When we determine the (line of) intersection of two planes, we solve two equations in three unknowns. Such a system generally has a solution (corresponding to the fact that two nonparallel planes in space intersect). However, when we determine the (point of) intersection of two lines, we solve three equations in two unknowns. Such a system generally does *not* have a solution (corresponding to the fact that two lines in space generally do not intersect). Much of the best mathematics arises from the elegant fashion in which algebra reinforces geometry. The situation of intersecting planes and intersecting lines is a simple instance of this principle.

Calculating Distance

If S_1 and S_2 are two sets of points in space, then we define the distance between S_1 and S_2 to be the minimum, if it exists, of all distances $d(P_1, P_2)$ where P_1 belongs to S_1 and P_2 belongs to S_2. We now use vector methods to find formulas for the distances between various geometric figures.

THEOREM 6 Suppose that $P = (x_0, y_0, z_0)$ is a point and that $\mathcal{V}$ is a plane. Let $\mathbf{n} = \langle A, B, C\rangle$ be a normal vector for $\mathcal{V}$ and let $Q = (x_1, y_1, z_1)$ be any point on $\mathcal{V}$. Then the distance between P and $\mathcal{V}$ is the length $\|\mathbf{P_n}(\overrightarrow{PQ})\|$ of the orthogonal projection of $\overrightarrow{PQ}$ in the direction of $\mathbf{n}$. That is, the distance between P and $\mathcal{V}$ is equal to

$$\frac{|\overrightarrow{PQ} \cdot \mathbf{n}|}{\|\mathbf{n}\|}, \tag{9.5.6}$$

or, equivalently,

$$\frac{|Ax_0 + By_0 + Cz_0 - D|}{\sqrt{A^2 + B^2 + C^2}} \tag{9.5.7}$$

where $D = Ax_1 + By_1 + Cz_1$.

Proof. Let R be the intersection of $\mathcal{V}$ with the line ℓ through P that is normal to $\mathcal{V}$, as in Figure 8. According to Pythagoras's Theorem, R is the point on $\mathcal{V}$ that is closest to P. The distance of P to $\mathcal{V}$ is therefore $\|\overrightarrow{PR}\|$. Because $\mathbf{n}$ and $\overrightarrow{PR}$ are parallel, this length is precisely the length of the orthogonal projection of $\overrightarrow{PQ}$ in the direction of $\mathbf{n}$. Formula (9.5.6) follows from equation (9.3.15). Formula (9.5.7) is then obtained from (9.5.6) by observing that $\|\mathbf{n}\| = \sqrt{A^2 + B^2 + C^2}$ and

$$\begin{aligned}|\overrightarrow{PQ} \cdot \mathbf{n}| &= |\langle x_1 - x_0, y_1 - y_0, z_1 - z_0\rangle \cdot \langle A, B, C\rangle| \\ &= |Ax_1 + By_1 + Cz_1 - (Ax_0 + By_0 + Cz_0)| = |Ax_0 + By_0 + Cz_0 - D|. \quad ■\end{aligned}$$

The proof of Theorem 6 demonstrates how powerful the concept of projection can be. Elementary geometry tells us that the distance we seek is the length $\|\overrightarrow{PR}\|$, but this quantity depends on a point R that has not been given. By calculating a projection, we are able to find the length $\|\overrightarrow{PR}\|$ without first finding the point R.

▶ **EXAMPLE 14** Find the distance between the point $P = (3, -8, 3)$ and the plane $\mathcal{V}$ whose Cartesian equation is $2x + y - 2z = 10$. (The point and the plane of this example are the same as those of Example 12.)

Solution From the given Cartesian equation for $\mathcal{V}$, we see that $\mathbf{n} = \langle 2, 1, -2\rangle$ is a normal for $\mathcal{V}$. If $Q = (x_1, y_1, z_1)$ is any point on $\mathcal{V}$, then $D = 2x_1 + y_1 - 2z_1 = 10$. According to formula (9.5.7), the distance between P and $\mathcal{V}$ is

$$\frac{|2(3) + (-8) - 2(3) - 10|}{\sqrt{2^2 + 1^2 + (-2)^2}},$$

or 6. This answer agrees with the distance that we calculated in the *Insight* following Example 12. That computation was considerably more laborious because it required the calculation of the point R at which $\mathcal{V}$ and the line normal to $\mathcal{V}$ through P intersect. ◀

The next theorem, which gives a method for calculating the distance of a point to a line, is a nice application of vector projection, the vector triple product, and the scalar triple product.

THEOREM 7 Suppose that $P = (x_0, y_0, z_0)$ is a point and that ℓ is a line in space that is parallel to $\mathbf{m}$. Let $Q = (x_1, y_1, z_1)$ be any point on ℓ. Let $\mathbf{n} = \mathbf{m} \times (\mathbf{m} \times \overrightarrow{PQ})$. Then, the distance between P and ℓ is the length $\|\mathbf{P}_{\mathbf{n}}(\overrightarrow{PQ})\|$ of the orthogonal projection of $\overrightarrow{PQ}$ in the direction of $\mathbf{n}$. That is, the distance between P and ℓ is equal to

$$\frac{|\overrightarrow{PQ} \cdot \mathbf{n}|}{\|\mathbf{n}\|}. \tag{9.5.8}$$

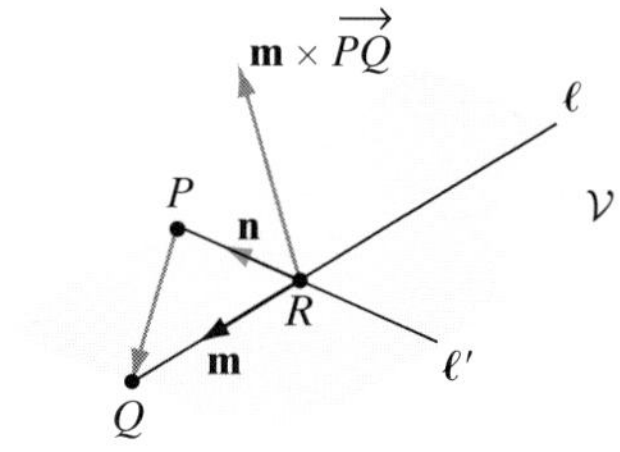

▲ Figure 9

Proof. Let $\mathcal{V}$ be the plane containing both P and ℓ, let ℓ' be the line in $\mathcal{V}$ that passes through P and that is perpendicular to ℓ, and let R be the point of intersection of ℓ and ℓ' (see Figure 9). Reasoning as in Theorem 6, we see that R is the point on ℓ that is closest to P. The distance of P to ℓ is therefore $\|\overrightarrow{PR}\|$. Our next step is to show that $\overrightarrow{PR}$ and $\mathbf{n}$ are parallel. To do this, we need to see that $\mathbf{n}$ is perpendicular to ℓ and lies in $\mathcal{V}$. Because $\mathbf{n} = \mathbf{m} \times (\mathbf{m} \times \overrightarrow{PQ})$, we know that $\mathbf{n}$ is perpendicular to the first operand of the cross product, $\mathbf{m}$, which is to say $\mathbf{n}$ is perpendicular to ℓ. We also know that $\mathbf{n}$ is perpendicular to the second operand of the cross product, $\mathbf{m} \times \overrightarrow{PQ}$. But $\mathbf{m}$ and $\overrightarrow{PQ}$ are nonparallel vectors in $\mathcal{V}$. Therefore $\mathbf{m} \times \overrightarrow{PQ}$ is orthogonal to $\mathcal{V}$. Because $\mathbf{n}$ is perpendicular to a normal for $\mathcal{V}$, it follows that $\mathbf{n}$ lies in $\mathcal{V}$. Thus $\mathbf{n}$ and $\overrightarrow{PR}$ are parallel, and the length $\|\overrightarrow{PR}\|$ is precisely the length of the orthogonal projection of $\overrightarrow{PQ}$ in the direction of $\mathbf{n}$. Formula (9.5.8) therefore follows from equation (9.3.15). ■

► **EXAMPLE 15** Find the distance between $P = (5, 3, 3)$ and the line ℓ with symmetric equations $x - 1 = (y + 8)/(-4) = (z - 7)/3$.

Solution Writing the term $x - 1$ as $(x - 1)/1$ (to attain the exact form recorded in line (9.5.4) in which each expression is a quotient), we see that $Q = (1, -8, 7)$ is a point on ℓ, and $\mathbf{m} = \langle 1, -4, 3 \rangle$ is parallel to ℓ. We calculate $\overrightarrow{PQ} = \langle -4, -11, 4 \rangle$, $\mathbf{m} \times \overrightarrow{PQ} = \langle 17, -16, -27 \rangle$, $\mathbf{n} = \mathbf{m} \times (\mathbf{m} \times \overrightarrow{PQ}) = \langle 156, 78, 52 \rangle = 26\langle 6, 3, 2 \rangle$,

$$\frac{1}{\|\mathbf{n}\|}\mathbf{n} = \frac{1}{26\|\langle 6,3,2\rangle\|} 26\,\langle 6,3,2\rangle = \frac{1}{7}\langle 6,3,2\rangle, \text{ and } \frac{|\overrightarrow{PQ}\cdot\mathbf{n}|}{\|\mathbf{n}\|}$$

$$= \frac{|(-4)(6) + (-11)(3) + 4(2)|}{7} = \frac{|-49|}{7} = 7.$$

Thus the distance between P and ℓ is 7. In Exercise 93, you are asked to verify this result by using the methods of Chapter 4. ◄

Our next theorem provides a method for calculating the distance between two lines.

THEOREM 8 Suppose that $\mathbf{m}_1$ and $\mathbf{m}_2$ are nonparallel vectors, that ℓ_1 is a line through $P_1 = (x_1, y_1, z_1)$ parallel to $\mathbf{m}_1$, and that ℓ_2 is a line through $P_2 = (x_2, y_2, z_2)$ parallel to $\mathbf{m}_2$. Let $\mathbf{n} = \mathbf{m}_1 \times \mathbf{m}_2$. Then, the distance between ℓ_1 and ℓ_2 is the length $\|\mathbf{P}_n(\overrightarrow{P_1P_2})\|$ of the orthogonal projection of $\overrightarrow{P_1P_2}$ in the direction of $\mathbf{n}$. That is, the distance between ℓ_1 and ℓ_2 is equal to

$$\frac{|\overrightarrow{P_1P_2}\cdot\mathbf{n}|}{\|\mathbf{n}\|}. \tag{9.5.9}$$

Proof. Let R_1 and R_2 be the points on ℓ_1 and ℓ_2, respectively, that are closest to each other. Then, $\overrightarrow{R_1R_2}$ is perpendicular to both ℓ_1 and ℓ_2. It follows that $\overrightarrow{R_1R_2}$ is parallel to $\mathbf{n}$, and the length $\|\overrightarrow{R_1R_2}\|$ is precisely the length of the orthogonal projection of $\overrightarrow{P_1P_2}$ in the direction of $\mathbf{n}$. Formula (9.5.9) therefore follows from equation (9.3.15). ■

▶ **EXAMPLE 16** Let ℓ_1 and ℓ_2 have symmetric equations

$$x = y + 1 = \frac{z-3}{-4} \qquad \text{and} \qquad \frac{x-9}{2} = \frac{y-2}{-1} = \frac{z+3}{-2}.$$

respectively. Find the distance between ℓ_1 and ℓ_2.

Solution From the symmetric equations for ℓ_1 we see that ℓ_1 passes through $P_1 = (0, -1, 3)$ and is parallel to $\mathbf{m}_1 = \langle 1, 1, -4\rangle$. From the symmetric equations for ℓ_2 we see that ℓ_2 passes through $P_2 = (9, 2, -3)$ and is parallel to $\mathbf{m}_2 = \langle 2, -1, -2\rangle$. We calculate $\overrightarrow{P_1P_2} = \langle 9, 3, -6\rangle = 3\,\langle 3, 1, -2\rangle$, $\mathbf{n} = \mathbf{m}_1 \times \mathbf{m}_2 = \langle -6, -6, -3\rangle = -3\,\langle 2, 2, 1\rangle$, and

$$\frac{|\overrightarrow{P_1P_2} \cdot \mathbf{n}|}{\|\mathbf{n}\|}. = \frac{3^2 |\langle 3, 1, -2\rangle \cdot \langle 2, 2, 1\rangle|}{3\sqrt{2^2 + 2^2 + 1^2}} = (3)(2) + (1)(2) + (-2)(1) = 6.$$

Thus the distance between ℓ_1 and ℓ_2 is 6. With considerably more labor, the points $R_1 = (1, 0, -1)$ on ℓ_1 and $R_2 = (5, 4, 1)$ on ℓ_2 that are closest can be located. Vector projection allows us to avoid such calculations. ◀

Formulas for Triple Vector Products

If $\mathbf{v}$ and $\mathbf{w}$ are nonparallel vectors represented by directed line segments with a common initial point, then they determine a plane $\mathcal{V}$ with $\mathbf{v} \times \mathbf{w}$ as a normal. Because the triple vector product $\mathbf{u} \times (\mathbf{v} \times \mathbf{w})$ is perpendicular to the second operand $\mathbf{v} \times \mathbf{w}$, we see that $\mathbf{u} \times (\mathbf{v} \times \mathbf{w})$ lies in $\mathcal{V}$. In our work on the parametric equations of planes, we observed that every vector in $\mathcal{V}$ can be written as $s\mathbf{v} + t\mathbf{w}$ for some scalars s and t. The consequence is that there are scalars s and t such that $\mathbf{u} \times (\mathbf{v} \times \mathbf{w}) = s\mathbf{v} + t\mathbf{w}$. Our next theorem provides specific formulas for these scalars. A shorter proof is discussed in Exercise 94; the argument uses tedious algebraic identities involving the nine entries of $\mathbf{u}, \mathbf{v}$, and $\mathbf{w}$.

THEOREM 9 The triple vector products $\mathbf{u} \times (\mathbf{v} \times \mathbf{w})$ and $(\mathbf{u} \times \mathbf{v}) \times \mathbf{w}$ satisfy

$$\mathbf{u} \times (\mathbf{v} \times \mathbf{w}) = (\mathbf{u} \cdot \mathbf{w})\,\mathbf{v} - (\mathbf{u} \cdot \mathbf{v})\,\mathbf{w} \tag{9.5.10}$$

and

$$(\mathbf{u} \times \mathbf{v}) \times \mathbf{w} = -(\mathbf{v} \cdot \mathbf{w})\,\mathbf{u} + (\mathbf{u} \cdot \mathbf{w})\,\mathbf{v}. \tag{9.5.11}$$

Proof. We prove the first equation and leave (9.5.11) as Exercise 95.

If $\mathbf{v}$ and $\mathbf{w}$ are parallel, then $\mathbf{v} \times \mathbf{w} = \mathbf{0}$, and either $\mathbf{v} = \mathbf{0}$ or $\mathbf{w} = \lambda\mathbf{v}$. For the left side of (9.5.10), we have $\mathbf{u} \times (\mathbf{v} \times \mathbf{w}) = \mathbf{u} \times \mathbf{0} = \mathbf{0}$. For the right side of (9.5.10), we have either $(\mathbf{u} \cdot \mathbf{w})\,\mathbf{v} - (\mathbf{u} \cdot \mathbf{v})\,\mathbf{w} = (\mathbf{u} \cdot \mathbf{w})\,\mathbf{0} - (\mathbf{u} \cdot \mathbf{0})\,\mathbf{w} = \mathbf{0}$ or $(\mathbf{u} \cdot \mathbf{w})\,\mathbf{v} - (\mathbf{u} \cdot \mathbf{v})\,\mathbf{w} = (\mathbf{u} \cdot \lambda\mathbf{v})\,\mathbf{v} - (\mathbf{u} \cdot \mathbf{v})\,\lambda\mathbf{v} = \lambda(\mathbf{u} \cdot \mathbf{v})\,(\mathbf{v} - \mathbf{v}) = \mathbf{0}$. Thus equation (9.5.10) holds when $\mathbf{v}$ and $\mathbf{w}$ are parallel.

Now suppose that $\mathbf{v}$ and $\mathbf{w}$ are perpendicular. Let $\mathcal{V}$ be the plane they determine. We may write $\mathbf{u} = \alpha\mathbf{v} + \beta\mathbf{w} + \mathbf{n}$ where $\mathbf{n}$ is orthogonal to $\mathcal{V}$ and therefore parallel to $\mathbf{v} \times \mathbf{w}$. Then (9.5.10) will hold for $\mathbf{u}$ if it holds for each summand. Starting with $\alpha\mathbf{v}$ and noting that $\alpha\mathbf{v} \cdot \mathbf{w} = 0$, we have

$$\alpha\mathbf{v} \times (\mathbf{v} \times \mathbf{w}) \overset{(9.4.8)}{=} -\alpha(\mathbf{v} \cdot \mathbf{v})\,\mathbf{w} = -(\alpha\mathbf{v} \cdot \mathbf{v})\,\mathbf{w} = (\alpha\mathbf{v} \cdot \mathbf{w})\,\mathbf{v} - (\alpha\mathbf{v} \cdot \mathbf{v})\,\mathbf{w},$$

which is (9.5.10) for the vector $\alpha\mathbf{v}$. Next, working with $\beta\mathbf{w}$ and noting that $(\beta\mathbf{w})\cdot\mathbf{v} = \mathbf{0}$, we have

$$\beta\mathbf{w}\times(\mathbf{v}\times\mathbf{w}) \overset{(9.4.9)}{=} \beta(\mathbf{w}\cdot\mathbf{w})\,\mathbf{v} = (\beta\mathbf{w}\cdot\mathbf{w})\,\mathbf{v} = (\beta\mathbf{w}\cdot\mathbf{w})\,\mathbf{v} - (\beta\mathbf{w}\cdot\mathbf{v})\,\mathbf{w},$$

which is (9.5.10) for the vector $\beta\mathbf{w}$. Finally, because $\mathbf{n}\times(\mathbf{v}\times\mathbf{w}) = \mathbf{0}, \mathbf{n}\cdot\mathbf{w} = \mathbf{0}$, and $\mathbf{n}\cdot\mathbf{v} = 0$, we have

$$\mathbf{n}\times(\mathbf{v}\times\mathbf{w}) = \mathbf{0} = (\mathbf{0})\,\mathbf{v} - (\mathbf{0})\,\mathbf{w} = (\mathbf{n}\cdot\mathbf{w})\,\mathbf{v} - (\mathbf{n}\cdot\mathbf{v})\,\mathbf{w},$$

which is (9.5.10) for the vector $\mathbf{n}$.

Now we make no assumptions about $\mathbf{v}$. As in equation (9.3.16), we write $\mathbf{v} = \mathbf{P}_{\mathbf{w}}(\mathbf{v}) + \mathbf{Q}$ where $\mathbf{P}_{\mathbf{w}}(\mathbf{v})$ is parallel to $\mathbf{w}$ and $\mathbf{Q}$ is perpendicular to $\mathbf{w}$. Because equation (9.5.10) has been proved for both summands of $\mathbf{v}$, it holds for $\mathbf{v}$ as well. ■

▶ **EXAMPLE 17** Let $\mathbf{u} = \langle 2, -1, 3\rangle, \mathbf{v} = \langle 3, 2, -2\rangle$, and $\mathbf{w} = \langle -1, 0, 2\rangle$. Calculate $\mathbf{u}\times(\mathbf{v}\times\mathbf{w})$, and verify equation (9.5.10).

Solution We have

$$\mathbf{v}\times\mathbf{w} = \det\left(\begin{bmatrix} \mathbf{i} & \mathbf{j} & \mathbf{k} \\ 3 & 2 & -2 \\ -1 & 0 & 2 \end{bmatrix}\right) = 4\mathbf{i} - 4\mathbf{j} + 2\mathbf{k}$$

and

$$\mathbf{u}\times(\mathbf{v}\times\mathbf{w}) = \det\left(\begin{bmatrix} \mathbf{i} & \mathbf{j} & \mathbf{k} \\ 2 & -1 & 3 \\ 4 & -4 & 2 \end{bmatrix}\right) = 10\mathbf{i} + 8\mathbf{j} - 4\mathbf{k}.$$

Also, $\mathbf{u}\cdot\mathbf{w} = (2)(-1) + (-1)(0) + (3)(2) = 4$ and $\mathbf{u}\cdot\mathbf{v} = (2)(3) + (-1)(2) + (3)(-2) = -2$. Therefore

$$(\mathbf{u}\cdot\mathbf{w})\,\mathbf{v} - (\mathbf{u}\cdot\mathbf{v})\,\mathbf{w} = 4(3\mathbf{i} + 2\mathbf{j} - 2\mathbf{k}) - (-2)(-\mathbf{i} + 0\mathbf{j} + 2\mathbf{k}) = (12-2)\mathbf{i} + (8-0)\mathbf{j} + (-8+4)\mathbf{k} = 10\mathbf{i} + 8\mathbf{j} - 4\mathbf{k},$$

which agrees with our direct calculation of $\mathbf{u}\times(\mathbf{v}\times\mathbf{w})$. ◀

QUICK QUIZ

1. If $\mathbf{n} = \langle 1, \beta, \gamma\rangle$ is orthogonal to the plane with Cartesian equation $2x - 5y + 6z = 11$, then what is γ?
2. The point $(3, 15, -2)$ is on a plane that has $\langle 3, 1, 6\rangle$ as a normal and Cartesian equation $x + By + Cz = D$. What is D?
3. If $x = (3 - 2y)/8 = 1 + 2z$ are Cartesian equations for a line ℓ and $\langle 3, b, c\rangle$ is parallel to ℓ, then what is b?
4. What is the distance of the plane $x + y + 2z = 2$ to the origin?

Answers

1. 3 2. 4 3. −12 4. $2/\sqrt{6}$

EXERCISES

Problems for Practice

In each of Exercises 1–4, a Cartesian equation that describes a plane $\mathcal{V}$ is given. Using only the given equation, describe all normals to $\mathcal{V}$. Then, verify that the given points P, Q, and R are on $\mathcal{V}$. Calculate $\mathbf{n} = \overrightarrow{PQ} \times \overrightarrow{PR}$, and verify that $\mathbf{n}$ is normal to $\mathcal{V}$.

1. $x - 3y + 4z = 2$ $P = (2,0,0)$ $Q = (1,1,1)$ $R = (5,1,0)$
2. $z = 5 - 7x + 11y$ $P = (0,0,5)$ $Q = (1,0,-2)$ $R = (2,1,2)$
3. $3x - 4z = 1$ $P = (3,0,2)$ $Q = (7,1,5)$ $R = (-5,2,-4)$
4. $x = 2y$ $P = (0,0,0)$ $Q = (-4,-2,3)$ $R = (2,1,1)$

In each of Exercises 5–8, describe all normals to the plane that is parameterized by the given equations.

5. $x = 3 + 2s + t$ $y = -1 + 2s - t$ $z = 2 + s - 3t$
6. $x = 1 + s + t$ $y = 2 + 2s + 2t$ $z = 3 + 2s + 3t$
7. $x = 2s$ $y = 1 - t$ $z = s - 2t$
8. $x = 2s + t$ $y = 2s$ $z = -3t$

In each of Exercises 9–12, find a Cartesian equation for the plane with the given normal vector $\mathbf{n}$ and passing through the given point P.

9. $\mathbf{n} = \langle -3,7,9 \rangle$ $P = (1,4,6)$
10. $\mathbf{n} = \langle 9,-3,7 \rangle$ $P = (6,0,-9)$
11. $\mathbf{n} = \langle 2,2,5 \rangle$ $P = (-5,9,2)$
12. $\mathbf{n} = \langle 7,1,1 \rangle$ $P = (2,0,0)$

In each of Exercises 13–16, find a Cartesian equation for the plane determined by the three given points.

13. $(0,1,3)$ $(3,-9,5)$ $(4,1,6)$
14. $(1,1,8)$ $(0,1,2)$ $(1,9,5)$
15. $(2,-6,8)$ $(1,-9,-4)$ $(-5,1,4)$
16. $(6,1,3)$ $(8,2,5)$ $(-5,4,6)$

In Exercises 17–20, find parametric equations for the line that passes through the given point P_0 and that is parallel to the vector $\mathbf{m}$

17. $P_0 = (2,1,9)$ $\mathbf{m} = \langle -3,1,7 \rangle$
18. $P_0 = (7,-11,3)$ $\mathbf{m} = \langle 4,6,-8 \rangle$
19. $P_0 = (0,1,0)$ $\mathbf{m} = \langle 2,-1,1 \rangle$
20. $P_0 = (2,2,-5)$ $\mathbf{m} = \langle 1,-1,-3 \rangle$

In Exercises 21–24, find symmetric equations for the line that passes through the given point P_0 and that is parallel to the given vector:

21. $P_0 = (2,1,5)$ $\langle -5,8,2 \rangle$
22. $P_0 = (-4,-2,-1)$ $\langle -6,-3,-7 \rangle$
23. $P_0 = (0,1,0)$ $\langle 1,1,9 \rangle$
24. $P_0 = (-5,-3,-9)$ $\langle 9,3,-1 \rangle$

In Exercises 25–28, find parametric equations for the line perpendicular to the given plane and passing through the given point.

25. $x - 3y + 7z = 6$ $(7,-4,6)$
26. $3x + 5y + z = 9$ $(2,3,8)$
27. $x - y + z/3 = 0$ $(-1,-1,-8)$
28. $x/2 + y + 2z = 1$ $(0,2,1)$

In each of Exercises 29–32, find symmetric equations for the line that passes through the two given points.

29. $(1,0,1)$ $(2,1,0)$
30. $(1,1,-1)$ $(-1,-1,1)$
31. $(1,2,1)$ $(3,1,0)$
32. $(2,2,1)$ $(0,0,0)$

In each of Exercises 33–36, find, in parametric form, the line of intersection of the two given planes.

33. $x - 3y + 5z = 2$ $2x + 6y + 3z = 4$
34. $x - 3y = 4$ $y + 8z = 6$
35. $4x + 6y - z = 1$ $4x + z = 0$
36. $x + 2y - z = 1$ $2x - y - z = 2$

In each of Exercises 37–40, find symmetric equations for the line of intersection of the two given planes.

37. $x - y + z = 2$ $x + y + 3z = 6$
38. $x - 3y = 1$ $y + 8z = 1$
39. $2x + y - z = 2$ $4x - z = 3$
40. $x + 2y - z = 0$ $2x - y - z = 1$

In each of Exercises 41–44, find the point of intersection of the given plane and the given line.

41. $x - 3y + 5z = 0$ $x = 2t + 6$ $y = 6t + 4$ $z = -t - 3$
42. $2x + 3y - 5z = 16$ $x = t - 4$ $y = -3t + 2$ $z = 4t + 1$
43. $x + y = 4$ $x = -t$ $y = 2t + 6$ $z = -2t + 3$
44. $x + y - 3z = 5$ $x = 4 - t$ $y = 2t + 6$ $z = 2t + 5$

In each of Exercises 45–48, find the point of intersection of the given plane and the given line.

45. $x - 3y + 5z = 19$ $x = y + 1 = z/2$
46. $2x + 3y - 5z = 0$ $(x-1)/2 = (y+1)/4 = 2z - 5$
47. $x + y = 4$ $(x+1)/2 = y + 1 = 3z + 2$
48. $x + y - 2z = 5$ $(x-2)/2 = (y+1)/3 = z - 1$

In each of Exercises 49–52, find the cosine of the angle between the two given planes.

49. $4x - 3z = 4$ $2x + y + 2z = 2$
50. $2x - 2y + z = 1$ $x - 2y + 2z = 5$
51. $-5x - 3y - 4z = 0$ $x + y + z = 1$
52. $8z - x - 4y = 1$ $z - x - y = 5$

▶ In each of Exercises 53–56, find the distance between the given point and the given plane: ◀

53.	$(1,0,1)$	$2x+y+z=9$
54.	$(-2,1,1)$	$2x-2y+z=13$
55.	$(1,1,1)$	$2x+2y+z=14$
56.	$(0,0,0)$	$x-6y+4z=3$

▶ In each of Exercises 57–60, determine whether the given point lies on the given line. ◀

57.	$(-7,6,2)$	$x=3t-1$	$y=-t+4$	$z=2t+6$
58.	$(-4,7,8)$	$x=-3t+5$	$y=t+4$	$z=5t-7$
59.	$(1,5,1)$	$x=t$	$y=5$	$z=-3t+1$
60.	$(21,4,-4)$	$x=6t-3$	$y=t$	$z=-t$

▶ In each of Exercises 61–64, determine whether the pairs of lines intersect; if they do, find the point(s) of intersection. ◀

61.	$x=3t-5$	$y=-4t-5$	$z=t+1$
	$x=t+1$	$y=t-6$	$z=t+5$
62.	$x=-t+6$	$y=t+3$	$z=4t-6$
	$x=2t-4$	$y=t-5$	$z=-3t+4$
63.	$x=t-1$	$y=-2t+14$	$z=2t+3$
	$x=3t+7$	$y=-t+4$	$z=-2t+11$
64.	$x=t+4$	$y=-2t+1$	$z=-3t-3$
	$x=t-2$	$y=t+7$	$z=2t+6$

▶ In each of Exercises 65–68, find the distance between the given point and the line with given symmetric equations. ◀

65.	$(0,0,0)$	$x-1=y-2=z-3$
66.	$(0,-1,2)$	$x=y=-z$
67.	$(1,0,-1)$	$x=1-y=2z$
68.	$(1,0,-1)$	$x-2=y-2=z/2$

▶ In each of Exercises 69–72, verify equation (9.5.10) for the triple vector product. ◀

69.	$\mathbf{u}=\langle 1,-2,4\rangle$	$\mathbf{v}=\langle 2,0,1\rangle$	$\mathbf{w}=\langle 3,1,1\rangle$
70.	$\mathbf{u}=\langle 2,1,2\rangle$	$\mathbf{v}=\langle 2,2,3\rangle$	$\mathbf{w}=\langle 3,3,-5\rangle$
71.	$\mathbf{u}=\langle 1,1,1\rangle$	$\mathbf{v}=\langle 3,0,2\rangle$	$\mathbf{w}=\langle -2,-3,3\rangle$
72.	$\mathbf{u}=\langle -5,-1,0\rangle$	$\mathbf{v}=\langle 1,0,3\rangle$	$\mathbf{w}=\langle 0,2,2\rangle$

Further Theory and Practice

▶ In each of Exercises 73–76, find a Cartesian equation for the plane that is parallel to the given plane and that passes through the point $(1,2,3)$. ◀

73. $3x+4y+z=0$
74. $z=2x$
75. $(x-1)+2(y-2)+3(z-3)=1$
76. $z=2x-3y$

▶ In each of Exercises 77–80, a Cartesian equation for a plane is given. Calculate the intercepts of the plane with the three coordinate axes. Sketch the part of the plane that lies in the first octant. ◀

77. $2x+3y+z=6$
78. $x+4y+6z=12$
79. $3x+5y+15z=15$
80. $2x+3y+9z=18$

81. Determine whether the following pairs of planes are parallel:

a.	$x-2y+4z=7$	$2x-4y+8z=5$
b.	$x+y+5z=1$	$3x+3y+15z=-3$
c.	$4x-6y+2z=1$	$x-y+5z=8$
d.	$x+4y-6z=-5$	$-2x-8y+12z=1$
e.	$x+y+z=0$	$2x+2y+2z=9$

82. Find a formula for the distance between the two planes with Cartesian equations $Ax+By+Cz=D_1$ and $Ax+By+Cz=D_2$.

83. Suppose that you are given parametric equations of a line ℓ and a Cartesian equation for a plane $\mathcal{V}$. How do you tell if ℓ lies in $\mathcal{V}$? If ℓ does not lie in $\mathcal{V}$, how do you tell whether or not ℓ and $\mathcal{V}$ intersect (without attempting to solve the equations simultaneously)? In each of parts a to d, apply your answer to each given line and plane. If the line and plane do intersect, find the point of intersection.
a. $x-5y+2;\ z=8;\ x=-3t+4;\ y=-t-6;\ z=-t+2$
b. $3x-7y+2;\ z=-22;\ x=3t-6;\ y=t+2;\ z=-t+5$
c. $x+y-2;\ z=10;\ x=4t;\ y=t+6;\ z=-3t+9$
d. $-3x-2y+6;\ z=1;\ x=2t+1;\ y=3t-4;\ z=2t+8$

▶ In each of Exercises 84–87, find the distance between the lines with given symmetric equations. ◀

84.	$x=1-y=z+1$	$x=y=z$
85.	$x-1=y/2=z$	$x=y=z$
86.	$x-1=y-1=z$	$2x=y-1=z-1$
87.	$x-1=y+1=-z$	$x=y=2z$

▶ In each of Exercises 88–91, a point Q and a Cartesian equation for a plane $\mathcal{V}$ are given. Compute the distance of Q to $\mathcal{V}$ as follows: First find the line through Q that is perpendicular to $\mathcal{V}$; next, find the point R of intersection of this line and $\mathcal{V}$; finally, calculate the distance from Q to R to obtain the distance between Q and $\mathcal{V}$. ◀

88.	$Q=(-6,-3,13)$	$2x+y-3z=2$
89.	$Q=(8,-2,4)$	$x-y+z=4$
90.	$Q=(-1,5,3)$	$-3x+y-4z=2$
91.	$Q=(2,-3,6)$	$2x-5y+7z=4$

92. Verify that $x=2+s+t, y=-1+s-t, z=-s-2t$ are equally valid parametric equations for the plane $\mathcal{V}$ of Example 5 even though they are obviously quite different from the ones found in the solution.

93. In Example 15, vector methods are used to show that the distance between $P=(5,3,3)$ and the line ℓ with symmetric equations $x-1=(y+8)/(-4)=(z-7)/3$ is 7.

Obtain this result by parameterizing the line, expressing the distance between P and a point on the line in terms of the parameter, and using calculus to minimize the distance function.

94. Writing $\mathbf{u} = \langle a, b, c\rangle$, $\mathbf{v} = \langle d, e, f\rangle$, and $\mathbf{w} = \langle g, h, p\rangle$, explicitly calculate each side of vector equation (9.5.10). Verify that corresponding entries of each side are equal.

95. Derive equation (9.5.11) by applying (9.5.10) with the vectors appropriately permuted to the right side of $(\mathbf{u} \times \mathbf{v}) \times \mathbf{w} = -\mathbf{w} \times (\mathbf{u} \times \mathbf{v})$.

96. Let ℓ be a line whose direction $\mathbf{m} = \langle a, b, c\rangle$ has one or two zero entries. Show that ℓ has the Cartesian equations

$$\frac{x - x_0}{a} = \frac{y - y_0}{b}, \quad z = z_0 \quad \text{when} \quad \mathbf{m} = \langle a, b, 0\rangle,$$

$$\frac{x - x_0}{a} = \frac{z - z_0}{c}, \quad y = y_0 \quad \text{when} \quad \mathbf{m} = \langle a, 0, c\rangle,$$

$$\frac{y - y_0}{b} = \frac{z - z_0}{c}, \quad x = x_0 \quad \text{when} \quad \mathbf{m} = \langle 0, b, c\rangle$$

and

$$y = y_0, \ z = z_0 \quad \text{when} \quad \mathbf{m} = \langle a, 0, 0\rangle,$$

$$x = x_0, \ z = z_0 \quad \text{when} \quad \mathbf{m} = \langle 0, b, 0\rangle,$$

$$x = x_0, \ y = y_0 \quad \text{when} \quad \mathbf{m} = \langle 0, 0, c\rangle.$$

Calculator/Computer Exercises

97. Plot the sphere with center $(2, 1, 2)$ and radius 3. Verify that $P_0 = (3, 3, 4)$ is a point on this sphere. Add to your plot the plane that passes through P_0 and that is tangent to the sphere.

98. Plot the plane $2x + y + z = 5$. Add to this plot a sphere that is tangent to this plane at the point $(1, 2, 1)$ and that has radius 1.

99. The line ℓ that is parallel to $\langle -1, -2, 3\rangle$ and passes through the point $(1, 1, 0)$ intercepts the solution set S of the Cartesian equation $z = x^4 + y^4$ in two points. Find them.

Summary of Key Topics in Chapter 9

Points in Space (Section 11.2)

A point in space is located with three coordinates: (x, y, z). The distance between two points (a_1, b_1, c_1) and (a_2, b_2, c_2) is

$$\sqrt{(a_1 - a_2)^2 + (b_1 - b_2)^2 + (c_1 - c_2)^2}.$$

It follows that the equation of a sphere with center (a, b, c) and radius $r > 0$ is

$$(x - a)^2 + (y - b)^2 + (z - c)^2 = r^2.$$

The sets

$$\{(x, y, z) : (x - a)^2 + (y - b)^2 + (z - c)^2 < r^2\}$$

and

$$\{(x, y, z) : (x - a)^2 + (y - b)^2 + (z - c)^2 \le r^2\}$$

are called, respectively, the *open* and *closed balls* with center (a, b, c) and radius r.

Vectors (Sections 11.1, 11.2)

A vector in two dimensions is an ordered pair $\mathbf{v} = \langle v_1, v_2\rangle$. A vector in three dimensions is an ordered triple $\mathbf{w} = \langle w_1, w_2, w_3\rangle$. If $P = (p_1, p_2)$ and $Q = (q_1, q_2)$ are points in the plane, and if $\mathbf{v} = \langle v_1, v_2\rangle$ where $v_1 = q_1 - p_1$ and $v_2 = q_2 - p_2$, then we say that the *directed line segment* $\overrightarrow{PQ}$ with initial point P and endpoint Q represents the vector $\mathbf{v}$. A similar definition is used in three dimensions.

If $\mathbf{v} = \langle v_1, v_2, v_3\rangle$ and $\mathbf{w} = \langle w_1, w_2, w_3\rangle$ are vectors, then

$$v + w = \langle v_1 + w_1, v_2 + w_2, v_3 + w_3\rangle.$$

Geometrically, two vectors are added by following the displacement of the first by the displacement of the second. The zero vector is defined by $\mathbf{0} = \langle 0, 0, 0\rangle$. If λ is a real number, then

$$\lambda\mathbf{v} = \langle \lambda v_1, \lambda v_2, \lambda v_3\rangle.$$

The most important attributes of a vector $\mathbf{v} = \langle v_1, v_2, v_3 \rangle$ are its *magnitude* (or *length*)

$$\|\mathbf{v}\| = \sqrt{(v_1)^2 + (v_2)^2 + (v_3)^2}$$

and, if $\mathbf{v} \neq \mathbf{0}$, its *direction*

$$\mathbf{dir}(\mathbf{v}) = \frac{1}{\|\mathbf{v}\|}\langle v_1, v_2, v_3 \rangle.$$

If λ is positive, then $\lambda\mathbf{v}$ points in the same direction as $\mathbf{v}$: $\mathbf{dir}(\lambda\mathbf{v}) = \mathbf{dir}(\mathbf{v})$. If λ is negative, then $\lambda\mathbf{v}$ points in the direction opposite to $\mathbf{v}$: $\mathbf{dir}(\lambda\mathbf{v}) = -\mathbf{dir}(\mathbf{v})$. In both cases, the length of $\|\lambda\mathbf{v}\|$ is obtained by dilating the length of $\|\mathbf{v}\|$ by a factor of $\|\lambda\|$: $\|\lambda\mathbf{v}\| = |\lambda|\|\mathbf{v}\|$. The Triangle Inequality tells us that the length of a sum of vectors is less than the sum of the lengths of the summands:

$$\|\mathbf{v} + \mathbf{w}\| \leq \|\mathbf{v}\| + \|\mathbf{w}\|.$$

Dot Product (Section 11.3)

If $\mathbf{v} = \langle v_1, v_2, v_3 \rangle$ and $\mathbf{w} = \langle w_1, w_2, w_3 \rangle$ are vectors, then their dot product is

$$\mathbf{v} \cdot \mathbf{w} = v_1 w_1 + v_2 w_2 + v_3 w_3.$$

The dot product is commutative ($\mathbf{v} \cdot \mathbf{w} = \mathbf{w} \cdot \mathbf{v}$) and distributes over addition ($\mathbf{u} \cdot (\mathbf{v} + \mathbf{w}) = \mathbf{u} \cdot \mathbf{v} + \mathbf{u} \cdot \mathbf{w}$ and $(\mathbf{u} + \mathbf{v}) \cdot \mathbf{w} = \mathbf{u} \cdot \mathbf{w} + \mathbf{v} \cdot \mathbf{w}$). The dot product commutes with scalar multiplication ($(\lambda\mathbf{v}) \cdot \mathbf{w} = \mathbf{w} \cdot (\lambda\mathbf{v}) = \lambda(\mathbf{v} \cdot \mathbf{w})$). The length of a vector is related to the dot product by the equation

$$\mathbf{v} \cdot \mathbf{v} = \|\mathbf{v}\|^2.$$

If $\mathbf{v}$ and $\mathbf{w}$ are nonzero vectors, then

$$\cos(\theta) = \frac{\mathbf{v} \cdot \mathbf{w}}{\|\mathbf{v}\|\,\|\mathbf{w}\|},$$

where $\theta \in [0, \pi]$ is the angle between the two vectors.

Two vectors $\mathbf{v}$ and $\mathbf{w}$ are orthogonal (that is, mutually perpendicular) if and only if $\mathbf{v} \cdot \mathbf{w} = 0$. They are parallel if and only if $\|\mathbf{v}\|\,\|\mathbf{w}\| = |\mathbf{v} \cdot \mathbf{w}|$, or, equivalently, if and only if $\mathbf{v} = \lambda\mathbf{w}$ or $\mathbf{w} = \lambda\mathbf{v}$ for some scalar λ. By convention, $\mathbf{0}$ is both orthogonal and parallel to every vector.

Projection (Section 11.3)

The projection of a vector $\mathbf{v}$ onto a vector $\mathbf{w}$ is

$$\mathbf{P}_{\mathbf{w}}(\mathbf{v}) = \left(\frac{\mathbf{v} \cdot \mathbf{w}}{\|\mathbf{w}\|^2}\right)\mathbf{w}.$$

The component of $\mathbf{v}$ in the direction of $\mathbf{w}$ is $\mathbf{v} \cdot \mathbf{w}/\|\mathbf{w}\|$. The length of the projection is $|\mathbf{v} \cdot \mathbf{w}|/\|\mathbf{w}\|$.

Direction Vectors (Section 11.3)

If $\mathbf{v}$ is any nonzero vector, then the vector $(1/\|\mathbf{v}\|)\mathbf{v}$ is a unit vector pointing in the same direction. A unit vector is called a *direction*. The unit vectors

$$\begin{aligned} \mathbf{i} &= \langle 1,0,0\rangle \\ \mathbf{j} &= \langle 0,1,0\rangle \\ \mathbf{k} &= \langle 0,0,1\rangle \end{aligned}$$

are called the *standard basis vectors*. Any vector $\mathbf{u} = \langle u_1, u_2, u_3\rangle$ can be written uniquely in the form

$$\mathbf{u} = u_1\mathbf{i} + u_2\mathbf{j} + u_3\mathbf{k}.$$

If $\mathbf{u}$ is a unit vector, then we can write

$$\mathbf{u} = \cos(\alpha)\mathbf{i} + \cos(\beta)\mathbf{j} + \cos(\gamma)\mathbf{k}.$$

The coefficients of $\mathbf{i}, \mathbf{j}$, and $\mathbf{k}$ are called the *direction cosines* for $\mathbf{u}$, and the angles α, β, and γ are called the *direction angles*.

Cross Product (Section 11.4)

The *cross product* of two vectors $\mathbf{v} = \langle v_1, v_2, v_3\rangle$ and $\mathbf{w} = \langle w_1, w_2, w_3\rangle$ is defined to be the vector

$$\mathbf{v} \times \mathbf{w} = (v_2w_3 - w_2v_3)\mathbf{i} - (v_1w_3 - w_1v_3)\mathbf{j} + (v_1w_2 - w_1v_2)\mathbf{k}.$$

The geometric interpretation of cross product is that

$$\mathbf{v} \times \mathbf{w} = \|\mathbf{v}\|\ \|\mathbf{w}\|\sin(\theta)\,\mathbf{n},$$

where θ is the angle between the two vectors, and $\mathbf{n}$ is the standard unit normal to $\mathbf{v}$ and $\mathbf{w}$ determined by the right-hand rule. We may use determinant notation to express the cross product as

$$\mathbf{v} \times \mathbf{w} = \det\left(\begin{bmatrix} \mathbf{i} & \mathbf{j} & \mathbf{k} \\ v_1 & v_2 & v_3 \\ w_1 & w_2 & w_3 \end{bmatrix}\right).$$

The cross product is anticommutative ($\mathbf{v} \times \mathbf{w} = -\mathbf{w} \times \mathbf{v}$) and distributes over addition ($\mathbf{u} \times (\mathbf{v} + \mathbf{w}) = \mathbf{u} \times \mathbf{v} + \mathbf{u} \times \mathbf{w}$ and $(\mathbf{u} + \mathbf{v}) \times \mathbf{w} = \mathbf{u} \times \mathbf{w} + \mathbf{v} \times \mathbf{w}$). The area of the triangle determined by the two vectors $\mathbf{v}$ and $\mathbf{w}$ is $\|\mathbf{v} \times \mathbf{w}\|/2$, and the area of the paralellogram they determine is $\|\mathbf{v} \times \mathbf{w}\|$.

Lines and Planes (Section 11.5)

The plane with normal vector $\mathbf{N} = \langle A, B, C\rangle$ and passing through the point $P_0 = (x_0, y_0, z_0)$ has equation

$$A(x - x_0) + B(y - y_0) + C(z - z_0) = 0.$$

A line in space that passes through $P_0 = (x_0, y_0, z_0)$ and is parallel to a vector $\mathbf{m} = \langle a, b, c\rangle$ is given parametrically by

$$\begin{aligned} x &= x_0 + ta \\ y &= y_0 + tb \\ z &= z_0 + tc. \end{aligned}$$

By solving each of these equations for t and equating, we can express the line in terms of x, y, and z only. This is called the *symmetric* form for the line. If a, b, and c are all nonzero, then the symmetric equations are

$$\frac{x - x_0}{a} = \frac{y - y_0}{b} = \frac{z - z_0}{c}.$$

Triple Scalar Product (Section 11.4) If $\mathbf{u}, \mathbf{v}$, and $\mathbf{w}$ are vectors, then the volume of the parallelepiped which they determine is given by the triple product $|(\mathbf{u} \times \mathbf{v}) \cdot \mathbf{w}|$. The product may be taken in any order. The vectors $\mathbf{u}, \mathbf{v}, \mathbf{w}$ are coplanar if and only if $(\mathbf{u} \times \mathbf{v}) \cdot \mathbf{w} = 0$.

Review Exercises for Chapter 9

In Exercises 1–4, calculate the distance between the given pair of points.

1. $(-1, 0, -3)$ $(6, 4, 1)$
2. $(-3, -4, 3)$ $(4, 2, -3)$
3. $(5, 1, -2)$ $(6, -7, 2)$
4. $(-1, 2, 1)$ $(1, 6, -3)$

In each of Exercises 5–8, determine the center and radius of the sphere whose Cartesian equation is given.

5. $x^2 + y^2 + z^2 - 4x + 8y + 2z = 15$
6. $x^2 + y^2 + z^2 + 12x - 6y + 4z = 0$
7. $4x^2 + 4y^2 + 4z^2 + 4x + 4y - 4z = 1$
8. $2 - x^2 - y^2 - z^2 = \sqrt{3}x - y + 2z$

In each of Exercises 9–12, points P and Q are given. Write the vector $\mathbf{v}$ represented by $\overrightarrow{PQ}$ in the form $\|\mathbf{v}\|\,\mathbf{dir}(\mathbf{v})$.

9. $P = (-1, 2, -2)$ $Q = (5, 8, 1)$
10. $P = (1, -1, 4)$ $Q = (0, 1, 2)$
11. $P = (2, 1, 2)$ $Q = (4, -1, 0)$
12. $P = (1, -11, 3)$ $Q = (12, -1, 1)$

In each of Exercises 13–16, let $\mathbf{v} = \langle 4, 2, 4 \rangle$ and $\mathbf{w} = \langle 2, 2, -1 \rangle$. Calculate the specified vector.

13. $\|\mathbf{v}\|\,\mathbf{i} + \|\mathbf{w}\|\,\mathbf{j} + (\mathbf{v} \cdot \mathbf{w})\,\mathbf{k}$
14. $(\mathbf{v} \cdot \mathbf{w})\,\mathbf{v} - (\mathbf{v} \cdot \mathbf{i})\,\mathbf{w}$
15. $\|\mathbf{w}\|\,\mathbf{v} - (\mathbf{v} \cdot \mathbf{v})\,\mathbf{w}$
16. $(\mathbf{v} \cdot \mathbf{k})\,\mathbf{j} + 6\,\mathbf{dir}(\mathbf{w})$
17. What vector of length 3 has the same direction as $\langle 10, 10, 5 \rangle$?
18. What vector of length 5 is opposite in direction to $\langle 11, 10, 2 \rangle$?
19. For what value of a is $\langle a, 3, 1 \rangle$ perpendicular to $\langle 4, 5, 13 \rangle$?
20. If $\langle a, b, -12 \rangle$ is parallel to $\langle 27, 5, 9 \rangle$, then what is a?

In each of Exercises 21–24, calculate the dot product of the given vectors.

21. $\langle 2, -3, 4 \rangle$ $\langle 2, 4, 5 \rangle$
22. $\langle -3, 3, -5 \rangle$ $\langle -8, 1, 6 \rangle$
23. $\langle \sqrt{2}, \sqrt{3}, -1 \rangle$ $\langle 3\sqrt{2}, 3, 3 + 3\sqrt{3} \rangle$
24. $2\mathbf{i} + 3\mathbf{j} + 4\mathbf{k}$ $3\mathbf{i} - 4\mathbf{j} + 5\mathbf{k}$
25. Find all values of s such that $\langle 3, s, s \rangle$ and $\langle 2, s, -5 \rangle$ are perpendicular.
26. What are the lengths of the parallelograms determined by $\langle 2, -1, -3 \rangle$ and $\langle 3, 2, -1 \rangle$?

In each of Exercises 27–30, use the arccosine to express the angle between the given pair of vectors.

27. $\langle 1, 1, -1 \rangle$ $\langle -1, 1, 1 \rangle$
28. $\langle 2, 2, -2 \rangle$ $\langle 1, -1, -1 \rangle$
29. $\langle 1, 1, 4 \rangle$ $\langle 3, 0, 3 \rangle$
30. $\langle 4, -2, 2 \rangle$ $\langle 1, 2, 1 \rangle$

In each of Exercises 31–34, calculate the projection of the given vector $\mathbf{v}$ onto the given vector $\mathbf{w}$. Verify that $\mathbf{P_w}(\mathbf{v})$ and $\mathbf{v} - \mathbf{P_w}(\mathbf{v})$ are mutually perpendicular.

31. $\mathbf{v} = \langle 4, 2, 6 \rangle$ $\mathbf{w} = \langle 2, 2, 1 \rangle$
32. $\mathbf{v} = \langle 20, 4, 9 \rangle$ $\mathbf{w} = \langle 9, 2, 6 \rangle$
33. $\mathbf{v} = \langle 75, 30, 21 \rangle$ $\mathbf{w} = \langle 7, 4, 4 \rangle$
34. $\mathbf{v} = \langle 10, 3, -20 \rangle$ $\mathbf{w} = \langle 2, 6, -3 \rangle$

In each of Exercises 35–38, calculate the dot product of the given vectors and their lengths. Verify that the Cauchy-Schwarz Inequality holds for the pair.

35. $\mathbf{v} = \langle 6, -2, 3 \rangle$ $\mathbf{w} = \langle 8, -1, 4 \rangle$
36. $\mathbf{v} = \langle 7, 4, 4 \rangle$ $\mathbf{w} = \langle 7, 6, 6 \rangle$
37. $\mathbf{v} = \langle 10, 10, 5 \rangle$ $\mathbf{w} = \langle 11, 10, 2 \rangle$
38. $\mathbf{v} = \langle 3, \sqrt{2}, 5 \rangle$ $\mathbf{w} = \langle 3, 3\sqrt{2}, 13 \rangle$

In each of Exercises 39–42, calculate the cross product $\mathbf{v} \times \mathbf{w}$.

39. $\mathbf{v} = \langle -4, 1, 0 \rangle$ $\mathbf{w} = \langle 3, -2, -3 \rangle$
40. $\mathbf{v} = \langle 3, 1, -2 \rangle$ $\mathbf{w} = \langle -1, 3, 2 \rangle$
41. $\mathbf{v} = \langle 2, 1, 2 \rangle$ $\mathbf{w} = \langle -2, 2, 2 \rangle$
42. $\mathbf{v} = \langle 3, 1, -5 \rangle$ $\mathbf{w} = \langle 2, 1, 1 \rangle$

In each of Exercises 43–46, find the area of the triangle determined by the two given vectors.

43. $\langle 4, 2, 4 \rangle$ $\langle 1, -1, 1 \rangle$
44. $\langle 2, 2, -1 \rangle$ $\langle 4, 6, 2 \rangle$
45. $\langle 2, 2, 3 \rangle$ $\langle 2, 2, -2 \rangle$
46. $\langle 3, 1, 3 \rangle$ $\langle 3, 2, 1 \rangle$

In each of Exercises 47–50, calculate the area of the parallelogram determined by the two given vectors.

47. $\langle 4, 2, 1 \rangle$ $\langle 1, 1, 3 \rangle$
48. $\langle 5, 1, 1 \rangle$ $\langle 1, 1, -1 \rangle$
49. $\langle 3, 1, -1 \rangle$ $\langle 1, 2, 2 \rangle$
50. $\langle 4, 1, -2 \rangle$ $\langle 2, 0, -3 \rangle$

In each of Exercises 51–54, vectors **u**, **v**, and **w** are given. Calculate the triple scalar product $(\mathbf{u} \times \mathbf{v}) \cdot \mathbf{w}$.

51. $\mathbf{u} = \langle 2, 1, 1\rangle$ $\mathbf{v} = \langle 1, -1, -1\rangle$ $\mathbf{w} = \langle 1, 1, -2\rangle$
52. $\mathbf{u} = \langle 3, 1, 2\rangle$ $\mathbf{v} = \langle 1, 1, -4\rangle$ $\mathbf{w} = \langle 2, 1, 2\rangle$
53. $\mathbf{u} = \langle 1, 1, -3\rangle$ $\mathbf{v} = \langle 1, -1, 17\rangle$ $\mathbf{w} = \langle 2, 1, 5\rangle$
54. $\mathbf{u} = \langle 2, 3, -6\rangle$ $\mathbf{v} = \langle 1, 1, 3\rangle$ $\mathbf{w} = \langle 3, 2, 5\rangle$

In each of Exercises 55–58, vectors **u**, **v**, and **w** are given. Calculate the volume of the parallelepiped that they determine.

55. $\mathbf{u} = \langle 3, 1, 2\rangle$ $\mathbf{v} = \langle 1, 1, -4\rangle$ $\mathbf{w} = \langle 2, 1, 2\rangle$
56. $\mathbf{u} = \langle 2, 1, 1\rangle$ $\mathbf{v} = \langle 1, -1, -1\rangle$ $\mathbf{w} = \langle 1, 1, -2\rangle$
57. $\mathbf{u} = \langle 2, 3, -6\rangle$ $\mathbf{v} = \langle 1, 1, 3\rangle$ $\mathbf{w} = \langle 3, 2, 5\rangle$
58. $\mathbf{u} = \langle 1, 1, -3\rangle$ $\mathbf{v} = \langle 1, -1, 17\rangle$ $\mathbf{w} = \langle 2, 1, 5\rangle$

59. For what values of x is

$$\langle 1, x, x\rangle \times (\langle x, x, 1\rangle \times \langle x, 1, x\rangle) = 0?$$

In each of Exercises 60–63, a Cartesian equation that describes a plane $\mathcal{V}$ is given. If $\langle 1, B, C\rangle$ is normal to $\mathcal{V}$, then what is B?

60. $2x - 6y + z = 3$
61. $z = 2 - 5x + 7y$
62. $3x - 4z = 1$
63. $x = 2y$

In each of Exercises 64–67, $\langle A, B, 1\rangle$ is normal to the plane described by the given parametric equations. What is A?

64. $x = 1 + 3s + 4t$ $y = -1 + 2s + 3t$ $z = s + t$
65. $x = -1 + s - t$ $y = 2s + t$ $z = 3 + 3s + 2t$
66. $x = 1 + 2s$ $y = 2 + 3t$ $z = s - t$
67. $x = s + 2t$ $y = 2s$ $z = 3t$

In each of Exercises 68–71, $Ax + By + Cz = 1$ is a Cartesian equation for the plane with the given normal vector **n** and passing through the given point P. What is A?

68. $\mathbf{n} = \langle 3, 2, -1\rangle$ $P = (1, 1, -4)$
69. $\mathbf{n} = \langle 10, 3, 2\rangle$ $P = (2, -5, 5)$
70. $\mathbf{n} = \langle 6, 5, 1\rangle$ $P = (3, -2, 7)$
71. $\mathbf{n} = \langle 16, -2, -3\rangle$ $P = (-1, -1, 0)$

In each of Exercises 72–75, find a Cartesian equation for the plane determined by the three given points.

72. $(1, 1, 1)$ $(2, 3, 1)$ $(3, 1, 4)$
73. $(4, 2, 1)$ $(5, 3, 2)$ $(6, 1, 0)$
74. $(1, 2, 1)$ $(2, 1, 3)$ $(2, 2, 0)$
75. $(6, 1, 3)$ $(8, 2, 3)$ $(7, 1, 5)$

In Exercises 76–79, find parametric equations for the line that passes through the given point P_0 and that is parallel to the vector **m**.

76. $P_0 = (0, 1, -1)$ $\mathbf{m} = \langle -5, 2, 4\rangle$
77. $P_0 = (1, 3, 5)$ $\mathbf{m} = \langle 1, 0, 3\rangle$
78. $P_0 = (1, -1, 2)$ $\mathbf{m} = \langle 1, -1, 2\rangle$
79. $P_0 = (0, 0, 1)$ $\mathbf{m} = \langle 2, -2, -3\rangle$

In Exercises 80–83, find symmetric equations for the line that passes through the given point P_0 and that is parallel to the given vector:

80. $P_0 = (3, 0, -5)$ $\langle -3, 7, 2\rangle$
81. $P_0 = (-2, 2, -1)$ $\langle 1/2, 1, -1\rangle$
82. $P_0 = (1, 6, -1)$ $\langle 1, -1, -2/3\rangle$
83. $P_0 = (-2, -3, -5)$ $\langle 1/3, 2/3, -2/3\rangle$

In each of Exercises 84–87, find symmetric equations for the line that passes through the two given points.

84. $(1, -1, 1)$ $(2, 3, 0)$
85. $(1, 1, -1)$ $(-1, 0, 1)$
86. $(-1, -2, 7)$ $(3, 1, 4)$
87. $(1/3, 4/3, 2/3)$ $(1, 2, 1)$

In each of Exercises 88–91, find, in parametric form, the line of intersection of the two given planes.

88. $2x + y + z = 6$ $x + y + 2z = 6$
89. $x + y + 2z = 4$ $x - y + z = 2$
90. $x + 2y - z = 0$ $x + z = 4$
91. $x + 2y - z = 2$ $5x + y + z = 4$

In each of Exercises 92–97, find the point of intersection of the given plane and the given line.

92. $x - y + 2z = 0$ $x = 2t - 6$ $y = -3t + 7$ $z = -t - 1$
93. $x + y - z = 2$ $x = 2t - 4$ $y = 4t + 2$ $z = 2t + 2$
94. $x + y + 2z = 4$ $(x + 3)/2 = y + 2 = (z - 12)/(-3)$
95. $x = -1 + s + 3t,\ y = -10 + 2s + t,\ z = -2 + s;$ $x = y + 1 = z/2$
96. $\langle x, y, z\rangle = \langle 1, 0, 1\rangle + s\langle 0, 1, 1\rangle + t\langle 1, 0, 2\rangle;$ $\langle x, y, z\rangle = \langle 6, 5, 13\rangle + u\langle 1, 2, 3\rangle$
97. $x = s + t, y = -s, z = 4 - t;\ (x - 3)/2 = y/2 = (z + 11)/4$

98. Find the distance between the point $(1, 1, 2)$ and the plane that has Cartesian equation $2x - y + z = 5$.

99. Find the distance between $P = (2, 0, 0)$ and the line with symmetric equations $x = y = 2z$.

The roots of vector calculus can be traced to 1799. In that year, both Carl Friedrich Gauss and a Norwegian surveyor, Caspar Wessel (1745–1818), "identified" the complex numbers $\mathbb{C}$ with the plane $\mathbb{R}^2$. They did this by associating each point (x, y) of the plane with the complex number $x + y\sqrt{-1}$. Gauss and Wessel used this identification to transfer the geometry of the plane to the field of complex numbers. This geometric interpretation of complex numbers gave rise to a new mathematical subject, the calculus of complex-valued functions of a complex variable.

In fact, the identification of $\mathbb{R}^2$ with $\mathbb{C}$ is a two-way street: We can also use it to transfer the algebraic structure of the complex number system to the plane. For example, the multiplication of complex numbers,

$$(x + y\sqrt{-1})(x' + y'\sqrt{-1}) = (xx' - yy') + (xy' + yx')\sqrt{-1},$$

can be used to define a product of points in the plane:

$$(x, y) \odot (x', y') = (xx' - yy', xy' + yx').$$

The search for an analogous algebraic structure that could be imposed on the points of three-dimensional space led to the creation of vector calculus in 1843. The breakthrough was attained more or less simultaneously by two mathematicians, Hermann Grassmann and Sir William Rowan Hamilton, working independently.

Hermann Günther Grassmann and Vector Calculus

Hermann Günther Grassmann (1809–1877) was born and raised in Stettin, a city that is now known by its Polish name, Szczecin. Grassmann acquired his education in Berlin, where he studied philology and theology but received no mathematical training at the university level. After returning to Stettin, he became a high school teacher and father.

The demands of rearing eleven children notwithstanding, Grassmann maintained his interest in languages and also carried out a program of original research in mathematics and physics. He first published his work on vector calculus in a book that appeared in 1844. Almost no notice was taken of it during his lifetime. After selling only a few copies, his publisher rendered the remaining inventory of 600 copies into waste paper. With that failure, Grassmann's hopes for landing a university position were dashed. Spurned in mathematics, he returned to the study of languages, a field in which he finally earned recognition. At the time of his death, and for some time thereafter, his Sanskrit-to-German translation of the *Rig-Veda* appeared to be his primary legacy. In fact, Grassmann's mathematical work did not attract any serious attention until nearly fifty years after his death.

One of Grassmann's most original ideas was the creation an anticommutative algebra of vectors that is now called "exterior algebra." In the 20th century, this algebra has played a prominent role in geometry and analysis. As Grassmann developed his theory of exterior algebra, he discovered not only the cross product but also the dot product, the vector norm, orthogonal projection, and several sophisticated concepts such as the "Hodge star operator" that were rediscovered and introduced into mainstream mathematics at a much later date. Frustrated by the apathy with which his research was greeted, Grassmann did not waver in his belief of its importance. As he expressed it,

> I remain completely confident that the labor I have expended on the science presented here and which has demanded a significant part of my life as well as the most strenuous application of my powers, will not be lost.... I know and feel obliged to state (though I run the risk of seeming arrogant) that even if this work should again remain unused for another seventeen years or even longer, without entering into the actual development of science, still that time will come when it will be brought forth from the dust of oblivion and when ideas now dormant will bring forth fruit.

Sir William Rowan Hamilton and Vector Calculus

In stark contrast to Grassmann, Sir William Rowan Hamilton was the best known scientist of his day. Even before his talents in mathematics and physics became evident, Hamilton attracted attention for his precocious ability to learn languages. By five years of age, he was already proficient in Latin, Greek, and Hebrew. As a teenager he was versed in the rudiments of thirteen languages. All this knowledge was the result of self-study under the tutelage of an uncle. Hamilton received his only formal schooling when he attended university.

Hamilton's interest in mathematics was kindled by the reading of a Latin copy of *Euclid* when he was ten years old. When he was sixteen, Hamilton detected a hitherto unsuspected error in Laplace's treatise on celestial mechanics. He published original mathematical research one year later and made his name in physics over the next ten years by applying the *calculus of variations* to discover fundamental principles of mechanics and optics. The Nobel laureate Erwin Schrödinger summarized Hamilton's influence as follows

> The central conception of all modern physics is the '"Hamiltonian." If you wish to apply modern theory to any particular problem, you must start with putting the problem "in Hamiltonian form."

On the basis of his theoretical investigations, Hamilton predicted the phenomenon of conical refraction. In the words of a prominent scientist of the time, this phenomenon was "unheard of and without analogy." The experimental verification of conical refraction created a sensation and brought fame to Hamilton beyond scientific circles.

Hamilton's introduction of vector calculus was a by-product of his creation of the *quaternions* Q, an algebraic system that may be identified with $\mathbb{R}^4$. A typical quaternion is of the form

$$t + x\mathbf{i} + y\mathbf{j} + z\mathbf{k}.$$

Quaternions of the form $x\mathbf{i} + y\mathbf{j} + z\mathbf{k}$ may be identified with points of three-dimensional space $\mathbb{R}^3$ by the one-to-one correspondence $(x, y, z) \to x\mathbf{i} + y\mathbf{j} + z\mathbf{k}$. At first, Hamilton referred to these spatial quaternions as *triplets*.

Addition and subtraction of quaternions was no problem. Finding a suitable multiplication proved difficult. The stumbling block was Hamilton's initial reluctance to abandon the commutative law: $\mathbf{pq} = \mathbf{qp}$ $(\mathbf{p},\mathbf{q} \in Q)$. The answer came to him on October 16, 1843. Shortly before his death in 1865, Hamilton described the circumstances in a letter to his son Archibald:

> Every morning in the early part of [October 1843], on my coming down from breakfast your (then) little brother William Edwin, and yourself, used to ask me "Well, Papa, can you multiply triplets?" Whereto I was always obliged to reply, with a sad shake of the head: "No. I can only add and subtract them."
>
> But on the sixteenth day of the same month... I was walking... along the Royal Canal... An electric current seemed to close; and a spark flashed forth, the herald... of many long years to come of definitely directed thought and work... Nor could I resist the impulse to cut with a knife on a stone of Brougham Bridge... the fundamental formula with the symbols **i**, **j**, **k**; namely
>
> $$\mathbf{i}^2 = \mathbf{j}^2 = \mathbf{k}^2 = \mathbf{ijk} = -1,$$
>
> which contains the Solution of the problem, but of course, as an inscription, has long since mouldered away. [Since 1954, a plaque on Brougham Bridge has commemorated the vanished engraving.]

Hamilton retained the *associative* law of multiplication: $(\mathbf{pq})\mathbf{r} = \mathbf{p}(\mathbf{qr})$ $\quad(\mathbf{p}, \mathbf{q}, \mathbf{r} \in \mathbf{Q})$; he even coined the term. Therefore no bracketing is necessary in the relation $\mathbf{ijk} = -1$ that Hamilton carved in stone. But Hamilton recognized that he would have to give up the commutative property of multiplication. To see how to multiply two quaternions, imagine the face of a clock with **i**, **j**, and **k** positioned at 12, 4, and 8 o'clock, respectively (Figure 1). A clockwise product of any two adjacent terms results in the third term; a counterclockwise product results in the negative of the remaining term. Thus $\mathbf{ij} = \mathbf{k} = -\mathbf{ji}$. Two general quaternions are then multiplied by the ordinary rules of arithmetic except that the order of factors must be preserved. For example,

$$\begin{aligned}(2+3\mathbf{j})(5\mathbf{i}+6\mathbf{j}+7\mathbf{k}) &= 10\mathbf{i}+12\mathbf{j}+14\mathbf{k}+15\mathbf{ji}+18\mathbf{j}^2+21\mathbf{jk}\\ &= -18+(10+21)\mathbf{i}+12\mathbf{j}+(14-15)\mathbf{k}\\ &= -18+31\mathbf{i}+12\mathbf{j}-\mathbf{k}.\end{aligned}$$

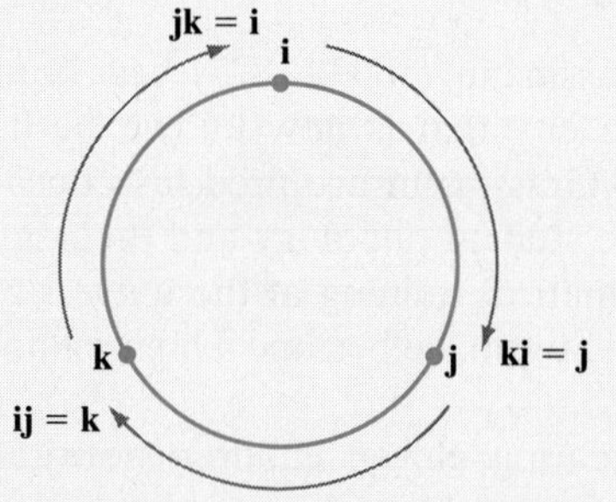

▲ Figure 1

By 1846, Hamilton was calling the triplet $x\mathbf{i} + y\mathbf{j} + z\mathbf{k}$ a *vector*. This terminology was derived from *radius vector*, a term that was already used in analytic geometry. At the same time, he wrote that "the algebraically real part [namely, the quantity t in the quaternion

$t + x\mathbf{i} + y\mathbf{j} + z\mathbf{k}$] may receive… all values contained on the one scale of progression of number from negative to positive infinity; we shall call it therefore the *scalar* part." In his notation, he wrote the quaternion $\mathbf{q} = t + x\mathbf{i} + y\mathbf{j} + z\mathbf{k}$ as $-S.\mathbf{q} + \mathbf{V}.\mathbf{q}$ with $-\mathrm{S}.\mathbf{q}$, the scalar component of $\mathbf{q}$, equal to t and $\mathbf{V}.\mathbf{q}$, the vector component of $\mathbf{q}$, equal to $x\mathbf{i} + y\mathbf{j} + z\mathbf{k}$.

Hamilton also introduced the *scalar product* (i.e., the dot product) and the *vector product*, (i.e., the cross product). These products arise simultaneously when the scalar and vector parts of a quaternionic product of two *vectors* are separated. In other words, if **u** and **v** are vectors (quaternions **u** and **v** with $S.\mathbf{u} = S.\mathbf{v} = 0$), then $\mathbf{uv} = -S.\mathbf{uv} + \mathbf{V}.\mathbf{uv}$ with $S.\mathbf{uv} = \mathbf{u} \cdot \mathbf{v}$ and $\mathbf{V}.\mathbf{uv} = \mathbf{u} \times \mathbf{v}$.

Despite his professional success, Hamilton's personal life was a shambles. At the age of twenty, Hamilton was rejected in love by Catherine Disney, the sister of a friend. He never really recovered from this unrequited first love. A second courtship six years later led to a second rejection. Hamilton finally proposed to a woman who was too timid to reject him but who was also too timid for very much else. Throughout his marriage, he remained in love with Catherine Disney but was torn by guilt over it. Depression and drink became frequent problems. In 1848, Catherine initiated a correspondence with Hamilton, but her conscience was not up to it: She soon confessed to her husband and attempted suicide. Hamilton responded to the stress by drinking more.

By 1853, Hamilton had collected his work on quaternions and vectors into a book which he titled *Lectures on Quaternions*. It ran to more than 800 pages. The famous astronomer, Sir John Herschel, complained that it would "take any man… half a lifetime to digest." In the United States, the mathematician Thomas Hill, who later become president of Harvard, reviewed the book in extremely favorable terms. He concluded that "in quaternions …there is as much real promise of benefit to mankind as in any event of [Queen] Victoria's reign." Hill lamented that "Perhaps not fifty men on this side of the Atlantic have seen [Hamilton's book], certainly not five have read it." In his review, he admitted that he himself was not actually among that handful.

After Catherine's death in 1853, Hamilton's life was marked by grief, reclusiveness, concentrated work at irregular intervals, and a struggle with alcoholism. As one biographer put it, "When thirsty, he would visit the locker, and the one blemish in the man's personal character is that these latter visits were sometimes paid too often." It was during this last phase of his work, when he was organizing his final thoughts on vectors and quaternions, that Hamilton discovered the theorem on square arrays (or matrices) that is fundamental to matrix algebra.

Although the theory of quaternions never assumed the central importance that Hamilton expected, his influence on mathematics has been as enduring as his influence on physics. The study of noncommutative algebraic systems that he initiated was taken up by first-rate mathematicians such as Arthur Cayley (1821–1895) and William Kingdon Clifford (1845–1879). Their work gave rise to the subject of *noncommutative algebra*. To this day noncommutative algebra continues to be an important branch of mathematical research.

CHAPTER 10

Vector-Valued Functions

PREVIEW

Until now, all of the functions $x \mapsto f(x)$ that we have studied have been scalar-valued: We evaluate f at a real number x, and we obtain a real number $f(x)$ as the result. In this chapter, we will study the calculus of *vector-valued* functions $t \mapsto \mathbf{r}(t)$. Here we evaluate a function $\mathbf{r}$ at a real number t, and we obtain a *vector* $\mathbf{r}(t)$ as the result. By writing out the vector $\mathbf{r}(t)$ in terms of its components $\langle x(t), y(t), z(t) \rangle$, we see that we can identify the value of $\mathbf{r}(t)$ with the point $(x(t), y(t), z(t))$ in space. As t varies in the domain of $\mathbf{r}$, the values $\mathbf{r}(t)$ trace out a curve in three-dimensional space. By developing the calculus of vector-valued functions, we will be able to analyze such space curves. For example, if we differentiate each component of $\mathbf{r}(t)$, then we obtain a vector that is tangent to the space curve. That much is analogous to the scalar-valued theory that we already know. But the geometry of curves in three-dimensional space is very rich, and there will be much that is new in this chapter.

Early in the 16th century, Johannes Kepler deduced the three laws of planetary motion that now bear his name. They were among the first precise quantitative laws known to science. What was lacking, however, was any theory that provided a reason for those laws. Half a century later, Isaac Newton demonstrated the power of calculus, then in its infancy, by deducing Kepler's Laws from basic physical principles. Chapter 10 concludes by applying the calculus of vector-valued functions to derive Kepler's Laws.

10.1 Vector-Valued Functions–Limits, Derivatives, and Continuity

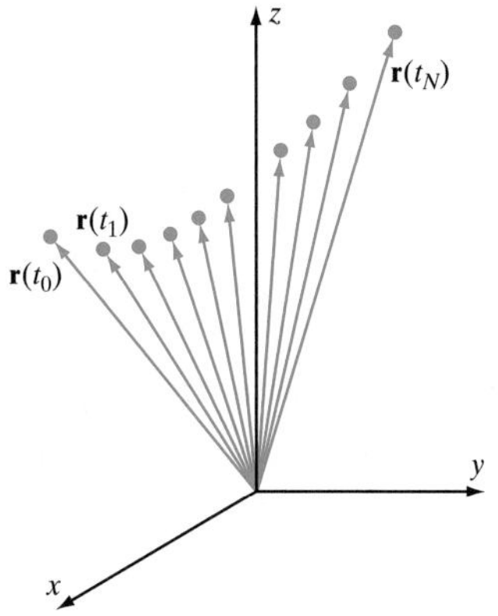

▲ Figure 1a

A function whose range consists of vectors is called a *vector-valued function*. For example, the formula $\mathbf{r}(t) = \cos(t)\mathbf{i} + \sin(t)\mathbf{j} + t\mathbf{k}$ defines a vector-valued function $\mathbf{r}$ that depends on a single variable t. Such functions will be the focus of our study in this chapter. One reason for our interest in this matter is that a vector-valued function $\mathbf{r}$ can be used to describe a curve $\mathcal{C}$ in space. To do so, we draw $\mathbf{r}(t)$ as a directed line segment with initial point at the origin. Figure 1a shows such directed line segments for several values, $t_0, t_1, \ldots, t_N$, in the domain of $\mathbf{r}$. The curve $\mathcal{C}$ is the collection of terminal points of the vectors $\mathbf{r}(t)$ as t runs through all values of the domain of $\mathbf{r}$, as Figure 1b illustrates. We say that the curve $\mathcal{C}$ is *parameterized* by $\mathbf{r}$. We refer to the vector-valued function $\mathbf{r}$ as a *parametric curve*.

▶ **EXAMPLE 1** Describe the curve that is parameterized by the vector-valued function $\mathbf{r}(t) = (5 - t)\mathbf{i} + (1 + 2t)\mathbf{j} - 3t\mathbf{k}$.

Solution From Section 11.5, we know that the equations $x = 5 - t$, $y = 1 + 2t$, $z = -3t$ are parametric equations of the line L that (a) is parallel to the vector $\langle -1, 2, -3 \rangle$ and (b) passes through the point $(5, 1, 0)$. This straight line is therefore the curve described by $\mathbf{r}$. ◀

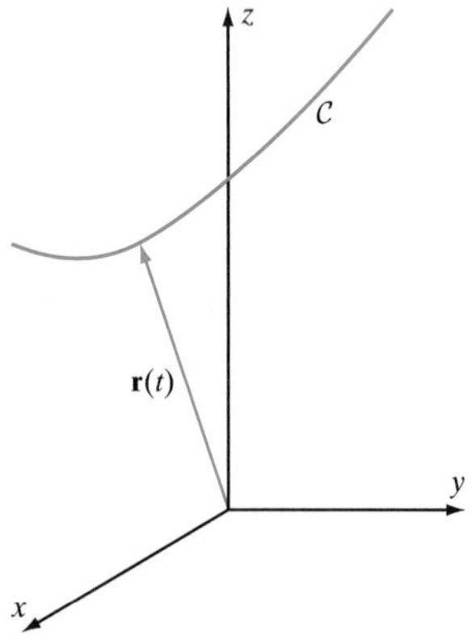

▲ Figure 1b

Suppose that a curve $\mathcal{C}$ is parameterized by $t \mapsto \mathbf{r}(t)$. If we think of the variable t as "time," then we can imagine a particle tracing the curve so that its position at time t is the terminal point of $\mathbf{r}(t)$. With this interpretation of the curve, we say that $\mathbf{r}(t)$ is the *position vector* of the particle at time t. If we write the position vector as $\mathbf{r}(t) = x(t)\mathbf{i} + y(t)\mathbf{j} + z(t)\mathbf{k}$, or, equivalently, $\mathbf{r}(t) = \langle x(t), y(t), z(t) \rangle$, then $(x(t), y(t), z(t))$ is the position of the particle at time t, and $x = x(t)$, $y = y(t)$, $z = z(t)$ are parametric equations for the motion. In this context, we often call $\mathcal{C}$ a *trajectory*.

▶ **EXAMPLE 2** The position of a particle moving through space is given by $\mathbf{r}(t) = \cos(t)\mathbf{i} + \sin(t)\mathbf{j} + t\mathbf{k}$. What is the position of the body at $t = 0$, $t = \pi/2$, $t = \pi$, $t = 3\pi/2$ and $t = 2\pi$? Describe the curve $\mathcal{C}$ along which the particle moves.

Solution At time 0, the body has position vector

$$\mathbf{r}(0) = \cos(0)\mathbf{i} + \sin(0)\mathbf{j} + 0\mathbf{k} = \mathbf{i} = \langle 1, 0, 0 \rangle.$$

This means that, at time $t = 0$, the particle is at the point $(1, 0, 0)$ in space. At time $t = \pi/2$, the particle has position vector

$$\mathbf{r}\left(\frac{\pi}{2}\right) = \cos\left(\frac{\pi}{2}\right)\mathbf{i} + \sin\left(\frac{\pi}{2}\right)\mathbf{j} + \frac{\pi}{2}\mathbf{k} = \left\langle 0, 1, \frac{\pi}{2} \right\rangle.$$

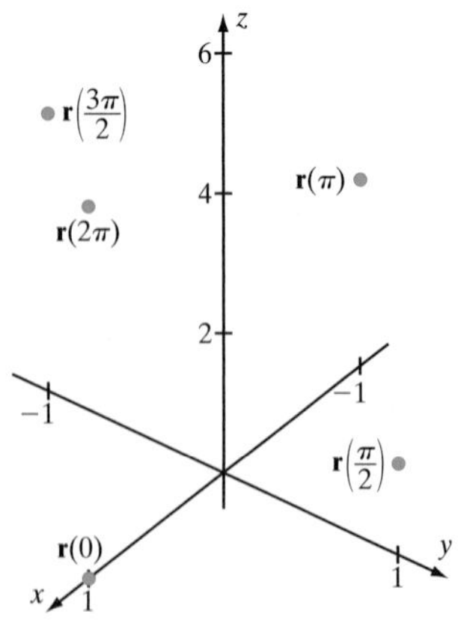

▲ Figure 2

Thus at $t = \pi/2$, the particle is at the point $(0, 1, \pi/2)$ in space. Similarly, we find that the particle is at the points $(-1, 0, \pi)$, $(0, -1, 3\pi/2)$, and $(1, 0, 2\pi)$ at times $t = \pi$, $t = 3\pi/2$, and $t = 2\pi$, respectively. These five points are plotted in Figure 2; they do not, however, give us a very good idea of the particle's path!

To better understand the trajectory, we observe that the values $x = \cos(t)$, $y = \sin(t)$, $z = t$ satisfy the equation $x^2 + y^2 = 1$. Now the set $\{(x, y, 0) : x^2 + y^2 = 1\}$ is a circle in the xy-plane of xyz-space, as is shown in Figure 3. The coordinates of every point above or below that circle also satisfy the equation $x^2 + y^2 = 1$ because the equation imposes no requirement on the z-coordinate. Therefore in xyz-space, the graph of $x^2 + y^2 = 1$ is a cylinder, as can also be seen in Figure 3. Because curve $\mathcal{C}$ lies on this cylinder, we can picture how $\mathcal{C}$ passes through the five plotted points. See Figure 4 for the arc of $\mathcal{C}$ that joins the five plotted points and Figure 5 for the arc of $\mathcal{C}$ that is obtained by plotting $\mathbf{r}(t)$ for values of t between -3π and 3π. We call $\mathcal{C}$ a *helix*. ◀

If we study a particle moving through space, then it stands to reason that we will want to calculate its velocity and its acceleration. Therefore we will need to calculate derivatives. Before we learn to differentiate vector-valued functions, we must first define their limits.

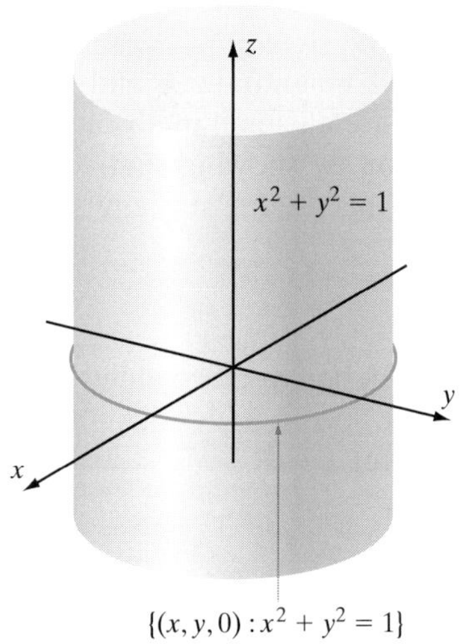

▲ Figure 3

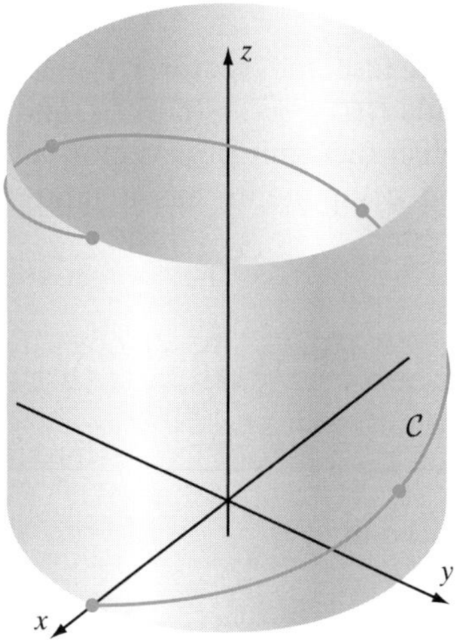

▲ Figure 4

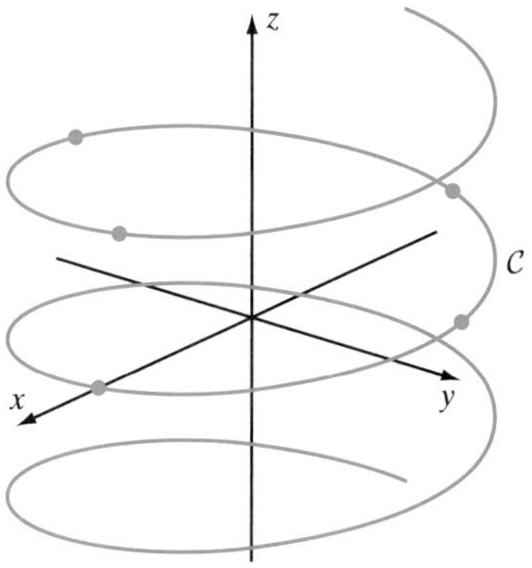

▲ Figure 5

Limits of Vector-Valued Functions

If $\mathbf{L}$ is a vector and $\mathbf{r}$ is a vector-valued function, then we say that the limit of $\mathbf{r}(t)$ is $\mathbf{L}$ as t tends to c, or $\mathbf{r}(t)$ tends to $\mathbf{L}$ as t tends to c, if we can make $\mathbf{r}(t)$ be arbitrarily close to $\mathbf{L}$ by taking t sufficiently close to c. The following definition states this idea precisely.

DEFINITION We say that $\mathbf{r}(t) = r_1(t)\mathbf{i} + r_2(t)\mathbf{j} + r_3(t)\mathbf{k}$ *converges* to the vector $\mathbf{L} = \langle L_1, L_2, L_3 \rangle$ as t tends to c if, for any $\varepsilon > 0$, there is a $\delta > 0$ such that

$$0 < |t - c| < \delta \quad \text{implies} \quad \|\mathbf{r}(t) - \mathbf{L}\| < \varepsilon.$$

If $\mathbf{r}(t)$ converges to $\mathbf{L}$ as t tends to c, then we write

$$\lim_{t \to c} \mathbf{r}(t) = \mathbf{L},$$

and we say that $\mathbf{L}$ is the *limit* of $\mathbf{r}(t)$ as t tends to c.

Is there anything new in this definition? It looks just like the rigorous definition of limit in Section 2.2 in Chapter 2. The difference is that, because our function is vector-valued, we measure distance in the range by using the length $\|\ \ \|$ instead of the absolute value $|\ \ |$. Figure 6 illustrates the geometry behind the definition.

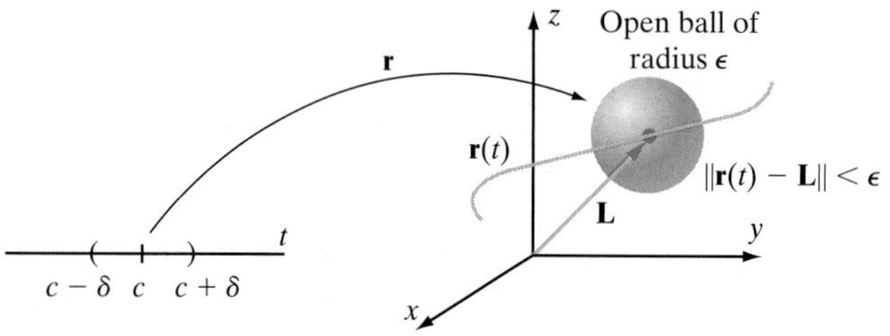

▲ **Figure 6** The function **r** maps all points within δ of c to an open ball of radius ε centered at the terminal point of **L**.

Because the vector $\mathbf{r}(t) - \mathbf{L}$ has a small magnitude if and only if its components $r_1(t) - L_1$, $r_2(t) - L_2$, and $r_3(t) - L_3$ are all small in absolute value, we can establish the limit of a vector-valued function by showing that each of its scalar-valued components has a limit. We state this useful strategy as the following theorem.

THEOREM 1 Let $\mathbf{r}(t) = r_1(t)\mathbf{i} + r_2(t)\mathbf{j} + r_3(t)\mathbf{k}$ be a vector-valued function. Then

$$\lim_{t\to c} \mathbf{r}(t) = L_1\mathbf{i} + L_2\mathbf{j} + L_3\mathbf{k}$$

if and only if

$$\lim_{t\to c} \mathbf{r}_1(t) = L_1, \quad \lim_{t\to c} \mathbf{r}_2(t) = L_2, \quad \textit{and} \quad \lim_{t\to c} \mathbf{r}_3(t) = L_3.$$

We mention in passing that, if the limit of a vector-valued function exists, then it is unique. This assertion follows from reasoning that is similar to that used in Chapter 2 to see that the limit of a scalar-valued function is unique when it exists. Now we can legitimately calculate limits in the most convenient way, and the resulting answer will have the right physical significance.

▶ **EXAMPLE 3** Let

$$\mathbf{r}(t) = (t^2 - 4t)\mathbf{i} + \left(\frac{t^2-9}{t-3}\right)\mathbf{j} + \frac{\sin(\pi t)}{t-3}\mathbf{k}.$$

What is $\lim_{t\to 3}\mathbf{r}(t)$?

Solution We calculate

$$\lim_{t\to 3}(t^2 - 4t) = -3 \quad \text{and} \quad \lim_{t\to 3}\frac{t^2-9}{t-3} = \lim_{t\to 3}(t+3) = 6.$$

To calculate the limit of the third component, notice that $\lim_{t\to 3}\sin(\pi t) = \sin(3\pi) = 0$ and $\lim_{t\to 3}(t-3) = 0$. To resolve the indeterminate form $0/0$, we apply l'Hôpital's Rule as follows:

$$\lim_{t\to 3}\frac{\sin(\pi t)}{t-3} = \lim_{t\to 3}\frac{\frac{d}{dt}\sin(\pi t)}{\frac{d}{dt}(t-3)} = \lim_{t\to 3}\frac{\pi\cos(\pi t)}{1} = \pi\cos(3\pi) = -\pi.$$

Thus according to Theorem 1, we have $\lim_{t\to 3}\mathbf{r}(t) = -3\mathbf{i} + 6\mathbf{j} - \pi\mathbf{k}$. ◄

The next theorem collects a number of results about calculating limits of vector-valued functions.

THEOREM 2 Let f and g be vector-valued functions, let ϕ be a scalar-valued function, let c be a real number, and let λ be a constant. Assume that $\lim_{t\to c}\mathbf{f}(t)$, $\lim_{t\to c}\mathbf{g}(t)$, and $\lim_{t\to c}\phi(t)$ exist. Then,

a. $\lim_{t\to c}(\mathbf{f}+\mathbf{g})(t) = \lim_{t\to c}\mathbf{f}(t) + \lim_{t\to c}\mathbf{g}(t)$
b. $\lim_{t\to c}(\mathbf{f}-\mathbf{g})(t) = \lim_{t\to c}\mathbf{f}(t) - \lim_{t\to c}\mathbf{g}(t)$
c. $\lim_{t\to c}(\mathbf{f}\cdot\mathbf{g})(t) = \left(\lim_{t\to c}\mathbf{f}(t)\right)\cdot\left(\lim_{t\to c}\mathbf{g}(t)\right)$
d. $\lim_{t\to c}\left(\lambda\mathbf{f}(t)\right) = \lambda\lim_{t\to c}\mathbf{f}(t)$
e. $\lim_{t\to c}(\mathbf{f}\times\mathbf{g})(t) = \left(\lim_{t\to c}\mathbf{f}(t)\right)\times\left(\lim_{t\to c}\mathbf{g}(t)\right)$
f. $\lim_{t\to c}(\phi(t)\mathbf{f}(t)) = \left(\lim_{t\to c}\phi(t)\right)\left(\lim_{t\to c}\mathbf{f}(t)\right)$.

INSIGHT These formulas are similar to those in Theorem 2 of Section 2.4 in Chapter 2. However, you should notice two things. First, the "$\mathbf{f}\cdot\mathbf{g}$" in part c now signifies a dot product: It *must* because $\mathbf{f}(t)$ is a vector and $\mathbf{g}(t)$ is a vector. Second, the "$\mathbf{f}\times\mathbf{g}$" in part e represents the cross product of **f** and **g**. Third, there is no result about the limit of a quotient, simply because we may not take the quotient of two vectors.

► **EXAMPLE 4** Calculate $\lim_{t\to 4}\left((t^2\mathbf{i} - 3t\mathbf{j})\cdot(4\mathbf{i} - \sqrt{t}\mathbf{j} + 5t\mathbf{k})\right)$.

Solution We calculate

$$\lim_{t\to 4}\left((t^2\mathbf{i} - 3t\mathbf{j} + 0\mathbf{k})\cdot(4\mathbf{i} - \sqrt{t}\mathbf{j} + 5t\mathbf{k})\right) = \lim_{t\to 4}\left(4t^2 + 3t\sqrt{t} + 0\right) = 64 + 24 = 88.$$

According to Theorem 2c, we obtain the same answer if we first compute the limits of the terms that appear in the dot product:

$$\lim_{t\to 4}(t^2\mathbf{i} - 3t\mathbf{j})\cdot\lim_{t\to 4}(4\mathbf{i} - \sqrt{t}\mathbf{j} + 5t\mathbf{k}) = (16\mathbf{i} - 12\mathbf{j} + 0\mathbf{k})\cdot(4\mathbf{i} - 2\mathbf{j} + 20\mathbf{k}) = 64 + 24 = 88. \quad ◄$$

Continuity The definition of continuity for a vector-valued function is just the same as the definition given in Chapter 2 for a scalar-valued function. The essence of the definition is that the actual value of the function at c agrees with the value the function approaches as $t \to c$. To be precise, see the following definition.

DEFINITION We say that a vector-valued function $\mathbf{r}$ is *continuous* at a point c of its domain if

$$\lim_{t \to c} \mathbf{r}(t) = \mathbf{r}(c).$$

If $\mathbf{r}$ is not continuous at a point c in its domain, then we say that $\mathbf{r}$ is *discontinuous* at c.

Because of Theorem 1, we find that we can test for continuity componentwise, as the following theorem observes.

THEOREM 3 Suppose that c belongs to the domain of a vector-valued function $\mathbf{r}$. Then, $\mathbf{r}$ is continuous at c if and only if each component function of $\mathbf{r}$ is continuous at c.

▶ **EXAMPLE 5** Discuss the continuity properties of the vector-valued function $\mathbf{r}$ that is defined on the set $\mathcal{D} = \{t \in \mathbb{R} : t \neq 2\}$ by the formula $\mathbf{r}(t) = \cos(t)\mathbf{i} + \ln(|t-2|)\mathbf{j} + t^3\mathbf{k}$.

Solution The functions $t \mapsto \cos(t)$ and $t \mapsto t^3$ are defined for every real t. Both of these functions are continuous everywhere. The function $t \mapsto \ln(|t-2|)$ is defined and continuous at all real $t \neq 2$. From these observations, we use Theorem 2 to conclude that the function $\mathbf{r}$ is continuous at all points in its domain $\mathcal{D}$. ◀

INSIGHT We do not say that $\mathbf{r}(t)$ is discontinuous at $t = 2$. That is because $\mathbf{r}$ is undefined at 2; we discuss continuity and discontinuity only at points of a function's domain.

The next theorem collects a number of results that are often convenient for establishing continuity.

THEOREM 4 Let λ be a constant. Suppose that ϕ is a scalar-valued function and that $\mathbf{f}$ and $\mathbf{g}$ are vector-valued functions. Suppose that these three functions are continuous at a common value c of their domains. Then,

a. $\mathbf{f} + \mathbf{g}$ *and* $\mathbf{f} - \mathbf{g}$ *are continuous at* c
b. $\mathbf{f} \cdot \mathbf{g}$ *is continuous at* c
c. $\lambda\mathbf{f}$ *is continuous at* c
d. $\mathbf{f} \times \mathbf{g}$ *is continuous at* c
e. $\phi\mathbf{f}$ *is continuous at* c.

INSIGHT These facts about continuous functions are similar to those in Section 2.3 of Chapter 2 about scalar-valued continuous functions. Two things should be noted. First, there is no statement about quotients of continuous vector-valued functions because we do not have a way to form the quotient of two vectors. Second, there are three statements about products: one for the dot product, one for cross product, and one for scalar multiplication. Just as in Section 2.3, the proofs of these rules are immediate applications of the rules for limits.

▶ **EXAMPLE 6** Discuss the continuity properties of the functions $\mathbf{f}(t) = |t|\mathbf{i} - 3\mathbf{j} + t^3\mathbf{k}$ and $\mathbf{g}(t) = \frac{1}{t+1}\mathbf{i} - \sec(t)\mathbf{k}$ at the value 0. Next, discuss the continuity properties of $\mathbf{f} \cdot \mathbf{g}$ and $\mathbf{f} \times \mathbf{g}$ at 0.

Solution Both $\mathbf{f}$ and $\mathbf{g}$ are continuous at $t = 0$ because all of their component functions are. Therefore $(\mathbf{f} \cdot \mathbf{g})(t)$ and $(\mathbf{f} \times \mathbf{g})(t)$ are continuous at 0 by Theorem 4b and 4d. ◀

INSIGHT Of course we can always calculate

$$(\mathbf{f} \cdot \mathbf{g})(t) = \frac{|t|}{t+1} - t^3\sec(t) \quad \text{and} \quad (\mathbf{f} \times \mathbf{g})(t) = 3\sec(t)\mathbf{i} + \left(|t|\sec(t) + \frac{t^3}{t+1}\right)\mathbf{j} + \frac{3}{t+1}\mathbf{k}.$$

The first of these, $\mathbf{f} \cdot \mathbf{g}$, is a scalar-valued function that can be shown to be continuous at $t = 0$ using the methods of Chapter 2. Theorem 3 can be used to show the continuity of the second function $\mathbf{f} \times \mathbf{g}$ at $t = 0$ because each of its components is continuous at $t = 0$.

Derivatives of Vector-Valued Functions

Now that we understand limits, it is a simple matter to define derivatives.

DEFINITION Suppose that $\mathbf{r}$ is a vector-valued function that is defined on an open interval that contains c. If the limit

$$\lim_{\Delta t \to 0} \frac{1}{\Delta t}\big(\mathbf{r}(c + \Delta t) - \mathbf{r}(c)\big)$$

exists, then we call this limit the *derivative* of the function $\mathbf{r}$ at the point c, and we denote this quantity by the symbols

$$\mathbf{r}'(c) \quad \text{and} \quad \left.\frac{d\mathbf{r}}{dt}\right|_{t=c}.$$

The notation $\dot{\mathbf{r}}(c)$ is also used, especially when t represents time. The process of calculating $\mathbf{r}'$ is called *differentiation* of $\mathbf{r}$. If the derivative $\mathbf{r}'(c)$ exists, then $\mathbf{r}$ is said to be *differentiable* at c.

INSIGHT Notice how similar the definition of $\mathbf{r}'(c)$ is to the definition of the derivative that is given in Section 3.2 of Chapter 3. The difference, of course, is that we are now considering vector-valued functions, so our limit process is a bit different. The quantity $\mathbf{r}(c + \Delta t) - \mathbf{r}(c)$ is a vector, and the product $\frac{1}{\Delta t}(\mathbf{r}(c + \Delta t) - \mathbf{r}(c))$ signifies the operation of scalar multiplication. The result of this operation is a vector. Thus the derivative $\mathbf{r}'(c)$ is a vector.

By Theorem 1, the process of calculating a limit may be performed by calculating the limit in each component separately. It follows that a vector-valued function may be differentiated by differentiating each component separately, *provided that all of the components are differentiable.*

THEOREM 5 A vector-valued function $\mathbf{r}(t) = r_1(t)\mathbf{i} + r_2(t)\mathbf{j} + r_3(t)\mathbf{k}$ is differentiable at $t = c$ if and only if each component function of $\mathbf{r}$ is differentiable at c. In this case,

$$\mathbf{r}'(c) = r_1'(c)\mathbf{i} + r_2'(c)\mathbf{j} + r_3'(c)\mathbf{k}.$$

▶ **EXAMPLE 7** Let $\mathbf{r}(t) = e^{2t}\mathbf{i} + |t|\mathbf{j} - \cos(t)\mathbf{k}$. For what values of t is $\mathbf{r}$ differentiable? What is $\mathbf{r}'(t)$ at these values?

Solution We calculate $\mathbf{r}'(t)$ by differentiating each component:

$$\mathbf{r}'(t) = \left(\frac{d}{dt}e^{2t}\right)\mathbf{i} + \left(\frac{d}{dt}|t|\right))\mathbf{j} - \left(\frac{d}{dt}\cos(t)\right)\mathbf{k} = 2e^{2t}\mathbf{i} + \left(\frac{d}{dt}|t|\right)\mathbf{j} + \sin(t)\mathbf{k}.$$

We see that $\mathbf{r}(t)$ is *not* differentiable at $t = 0$ because the second component, $|t|$, is not. However, for $t > 0$, we have $|t| = t$ and $\frac{d}{dt}|t| = 1$. It follows that $\mathbf{r}(t)$ is differentiable for $t > 0$, and $\mathbf{r}'(t) = 2e^{2t}\mathbf{i} + \mathbf{j} + \sin(t)\mathbf{k}$. For $t < 0$, we have $|t| = -t$ and $\frac{d}{dt}|t| = -1$. It follows that $\mathbf{r}(t)$ is differentiable for $t < 0$ and $\mathbf{r}'(t) = 2e^{2t}\mathbf{i} - \mathbf{j} + \sin(t)\mathbf{k}$. ◀

The derivative of a vector-valued function is another vector-valued function. Therefore we can differentiate the derivative function to create a second derivative, and so on. Just as for real-valued functions, the second derivative of $\mathbf{r}(t)$ is denoted by

$$\mathbf{r}''(t) \quad \text{or} \quad \frac{d^2\mathbf{r}}{dt^2}.$$

The notation $\ddot{\mathbf{r}}(t)$ is also used, especially when t represents time.

▶ **EXAMPLE 8** Let $\mathbf{r}(t) = (3/t - t)\mathbf{i} + \cos(\pi t)\mathbf{j} + e^{-t}\mathbf{k}$. Calculate $\mathbf{r}'$ and $\mathbf{r}''$. What is $\mathbf{r}''(1)$?

Solution We have $\mathbf{r}'(t) = (-3/t^2 - 1)\mathbf{i} - \pi\sin(\pi t)\mathbf{j} - e^{-t}\mathbf{k}$ and $\mathbf{r}''(t) = (6/t^3)\mathbf{i} - \pi^2\cos(\pi t)\mathbf{j} + e^{-t}\mathbf{k}$. In particular, $\mathbf{r}''(1) = 6\mathbf{i} + \pi^2\mathbf{j} + (1/e)\mathbf{k}$. ◀

The next theorem gathers together several useful differentiation rules.

THEOREM 6 Let λ be a constant. Suppose that the vector-valued functions $\mathbf{f}$ and $\mathbf{g}$ and the scalar-valued function ϕ are defined on an open interval containing the point c. Suppose further that $\mathbf{f}'(c)$, $\mathbf{g}'(c)$, and $\phi'(c)$ exist. Then

a. $(\mathbf{f}+\mathbf{g})'(c)$ exists and
$$(\mathbf{f}+\mathbf{g})'(c)=\mathbf{f}'(c)+\mathbf{g}'(c);$$

b. $(\mathbf{f}-\mathbf{g})'(c)$ exists and
$$(\mathbf{f}-\mathbf{g})'(c)=\mathbf{f}'(c)-\mathbf{g}'(c);$$

c. $(\mathbf{f}\cdot\mathbf{g})'(c)$ exists and
$$(\mathbf{f}\cdot\mathbf{g})'(c)=\mathbf{f}'(c)\cdot\mathbf{g}(c)+\mathbf{f}(c)\cdot\mathbf{g}'(c);$$

d. $(\lambda\mathbf{f})'(c)$ exists and
$$(\lambda\mathbf{f})'(c)=\lambda(\mathbf{f}'(c))$$

e. $(\mathbf{f}\times\mathbf{g})'(c)$ exists and
$$(\mathbf{f}\times\mathbf{g})'(c)=\mathbf{f}'(c)\times\mathbf{g}(c)+\mathbf{f}(c)\times\mathbf{g}'(c).$$

f. $(\phi\mathbf{f})'(c)$ exists and
$$(\phi\mathbf{f})'(c)=\phi'(c)\mathbf{f}(c)+\phi(c)\mathbf{f}'(c).$$

g. If ψ is a scalar-valued differentiable function on an open interval I that contains the point a, if $\psi(a)=c$, and if the composition $\mathbf{f}\circ\psi$ makes sense on I, then $\mathbf{f}\circ\psi$ is differentiable at a, and
$$(\mathbf{f}\circ\psi)'(a)=\psi'(a)\mathbf{f}'\big(\psi(a)\big).$$

INSIGHT These differentiation rules are similar to those of Chapter 3 for functions with scalar values. However, two points should be noted. First, there is no quotient rule because we do not have a way to form the quotient of two vectors. Second, we have three product rules. Rule c is a product rule for the dot product. Rule e is a product rule for the cross product. Rule f is a product rule for scalar multiplication. Care should be taken to distinguish among these three rules.

Take particular note that it is important on the right-hand side of formula e that the order of $\mathbf{f}$ and $\mathbf{g}$ match the order of $\mathbf{f}$ and $\mathbf{g}$ on the left-hand side. (Remember that the cross product is an anticommutative operation: Interchanging the operands of a cross product changes the sign of the result.) Also notice the form of the Chain Rule stated in part g. On the right side of the formula, $\psi'(a)$ is a scalar and $\mathbf{f}'(\psi(a))$ is a vector. Thus the right side of the formula is the scalar multiplication of a vector.

▶ **EXAMPLE 9** Let $\mathbf{f}(t) = \cos(t)\mathbf{j} - \ln(t)\mathbf{k}$ and $\mathbf{g}(t) = t^2\mathbf{i} - t^{-2}\mathbf{j} + t\mathbf{k}$. Calculate $(\mathbf{f} \cdot \mathbf{g})'(t)$.

Solution We apply formula Theorem 6c to obtain

$$\begin{aligned}(\mathbf{f} \cdot \mathbf{g})'(t) &= \mathbf{f}'(t) \cdot \mathbf{g}(t) + \mathbf{f}(t) \cdot \mathbf{g}'(t)\\ &= \left(-\sin(t)\mathbf{j} - \frac{1}{t}\mathbf{k}\right) \cdot (t^2\mathbf{i} - t^{-2}\mathbf{j} + t\mathbf{k}) + \big(\cos(t)\mathbf{j} - \ln(t)\mathbf{k}\big) \cdot (2t\mathbf{i} + 2t^{-3}\mathbf{j} + \mathbf{k})\\ &= \left((0)(t^2) + \big(-\sin(t)\big)(-t^{-2}) + \left(-\frac{1}{t}\right)(t)\right) + \Big((0)(2t) + \big(\cos(t)\big)(2t^{-3}) + \big(-\ln(t)\big)(1)\Big)\\ &= t^{-2}\sin(t) - 1 + 2t^{-3}\cos(t) - \ln(t).\end{aligned}$$ ◀

INSIGHT We are not obliged to use Theorem 6 in Example 9. As an alternative, we can first calculate $(\mathbf{f} \cdot \mathbf{g})(t) = (0)(t^2) + \cos(t)(-t^{-2}) + \big(-\ln(t)\big)(t) = -t^{-2}\cos(t) - t\ln(t)$ and then differentiate:

$$\begin{aligned}(\mathbf{f} \cdot \mathbf{g})'(t) &= \frac{d}{dt}(-t^{-2}\cos(t) - t\ln(t)) = \frac{d}{dt}(-t^{-2}\cos(t)) + \frac{d}{dt}(-t\ln(t))\\ &= (2t^{-3}\cos(t) + t^{-2}\sin(t)) + \left(-\ln(t) - t\left(\frac{1}{t}\right)\right) = 2t^{-3}\cos(t) + t^{-2}\sin(t) - \ln(t) - 1.\end{aligned}$$

▶ **EXAMPLE 10** Define $\mathbf{f}(t) = \cos(t)\mathbf{i} - \sin(t)\mathbf{j}$, $\mathbf{g}(t) = t^3\mathbf{j} + t^{-1}\mathbf{k}$, $\phi(t) = t^2$, and $\lambda = 5$. Calculate $(\phi\mathbf{f} + \lambda\mathbf{g})'(t)$.

Solution We use several of our differentiation rules:

$$\begin{aligned}(\phi\mathbf{f} + \alpha\mathbf{g})'(t) &= (\phi\mathbf{f})'(t) + (\alpha\mathbf{g})'(t) && \text{Theorem 6a}\\ &= \phi'(t)\mathbf{f}(t) + \phi(t)\mathbf{f}'(t) + \alpha\big(\mathbf{g}'(t)\big) && \text{Theorem 6d, 6f}\\ &= 2t\big(\cos(t)\mathbf{i} - \sin(t)\mathbf{j}\big) + t^2\big(-\sin(t)\mathbf{i} - \cos(t)\mathbf{j}\big) + 5(3t^2\mathbf{j} - t^{-2}\mathbf{k})\\ &= \big(2t\cos(t) - t^2\sin(t)\big)\mathbf{i} + \big(-2t\sin(t) - t^2\cos(t) + 15t^2\big)\mathbf{j} - 5t^{-2}\mathbf{k}.\end{aligned}$$

As an alternative, we can first calculate $(\phi\mathbf{f} + \lambda\mathbf{g})(t) = t^2\cos(t)\mathbf{i} + (-t^2\sin(t) + 5t^3)\mathbf{j} + 5t^{-1}\mathbf{k}$. We can then directly differentiate this expression. As an exercise, verify that this second method results in the same formula for $(\phi\mathbf{f} + \lambda\mathbf{g})'(t)$. ◀

▶ **EXAMPLE 11** Let $\mathbf{f}(t) = \tan(t)\mathbf{i} - \cos(t)\mathbf{k}$ and $\mathbf{g}(t) = \sin(t)\mathbf{j}$. Calculate $(\mathbf{f} \times \mathbf{g})'(t)$.

Solution By Theorem 6e,

$$\begin{aligned}(\mathbf{f} \times \mathbf{g})'(t) &= \mathbf{f}'(t) \times \mathbf{g}(t) + \mathbf{f}(t) \times \mathbf{g}'(t)\\ &= \big(\sec^2(t)\mathbf{i} + \sin(t)\mathbf{k}\big) \times \sin(t)\mathbf{j} + \big(\tan(t)\mathbf{i} - \cos(t)\mathbf{k}\big) \times \cos(t)\mathbf{j}\\ &= \big(-\sin^2(t)\mathbf{i} + \sec^2(t)\sin(t)\mathbf{k}\big) + \big(\cos^2(t)\mathbf{i} + \sin(t)\mathbf{k}\big)\\ &= \big(\cos^2(t) - \sin^2(t)\big)\mathbf{i} + \big(\sin(t) + \sec^2(t)\sin(t)\big)\mathbf{k}.\end{aligned}$$

Verify this answer by first calculating the cross product $\mathbf{f} \times \mathbf{g}$ and then differentiating. ◀

We next turn to a statement relating differentiability and continuity of a vector-valued function. The proof illustrates how we often analyze a vector-valued function by reducing the investigation to its scalar-valued components.

THEOREM 7 If $\mathbf{f}$ is a vector-valued function that is differentiable at c, then $\mathbf{f}$ is continuous at c.

Proof. If $\mathbf{f}(t) = f_1(t)\mathbf{i} + f_2(t)\mathbf{j} + f_3(t)\mathbf{k}$ is differentiable at c, then the scalar-valued expressions $f_1(t)$, $f_2(t)$, and $f_3(t)$ are all differentiable at c. Each of the functions f_1, f_2, f_3 is therefore continuous at c by Theorem 1 of Section 3.2 in Chapter 3. Theorem 3 of this section now tells us that $\mathbf{f}$ is continuous at c. ■

Antidifferentiation

If $\mathbf{f}(t) = f_1(t)\mathbf{i} + f_2(t)\mathbf{j} + f_3(t)\mathbf{k}$ is continuous, then we may consider the antiderivative

$$\begin{aligned}\mathbf{F}(t) &= \int \mathbf{f}(t)dt \\ &= \left(\int f_1(t)dt\right)\mathbf{i} + \left(\int f_2(t)dt\right)\mathbf{j} + \left(\int f_3(t)dt\right)\mathbf{k} + \mathbf{C}.\end{aligned}$$

Here it should be clearly understood that each of the antidifferentiations in the $\mathbf{i}$, $\mathbf{j}$, and $\mathbf{k}$ components will have a constant of integration. Therefore the constant of integration $\mathbf{C}$ is a vector of the form $\mathbf{C} = C_1\mathbf{i} + C_2\mathbf{j} + C_3\mathbf{k}$.

▶ **EXAMPLE 12** Find the antiderivative $\mathbf{F}(t)$ of $\mathbf{f}(t) = 3t^2\mathbf{i} - 4t\mathbf{j} + 8\mathbf{k}$ that satisfies the equation $\mathbf{F}(1) = 2\mathbf{i} - 3\mathbf{j} + 2\mathbf{k}$.

Solution We have

$$\begin{aligned}\mathbf{F}(t) &= \int \mathbf{f}(t)dt \\ &= t^3\mathbf{i} - 2t^2\mathbf{j} + 8t\mathbf{k} + \mathbf{C}.\end{aligned}$$

Now the additional condition that has been given tells us that $2\mathbf{i} - 3\mathbf{j} + 2\mathbf{k} = \mathbf{F}(1) = \mathbf{i} - 2\mathbf{j} + 8\mathbf{k} + \mathbf{C}$. Solving for $\mathbf{C}$ gives $\mathbf{C} = \mathbf{i} - \mathbf{j} - 6\mathbf{k}$. Therefore $\mathbf{F}(t) = (t^3 + 1)\mathbf{i} - (2t^2 + 1)\mathbf{j} + (8t - 6)\mathbf{k}$. ◀

QUICK QUIZ

1. What planar curve is described by $\mathbf{r}(t) = \cos(t)\mathbf{i} + \sin(t)\mathbf{j}$, $0 \le t < 2\pi$?
2. If $\mathbf{f}(\pi) = \langle 2, 0, 3\rangle$, $\mathbf{f}'(\pi) = \langle 2, 3, 0\rangle$, $\mathbf{g}(\pi) = \langle 0, 1, 4\rangle$, and $\mathbf{g}'(\pi) = \langle -1, 0, 1\rangle$, then what is $(\mathbf{f} \cdot \mathbf{g})'(\pi)$?
3. Referring to $\mathbf{f}$ and $\mathbf{g}$ of the preceeding question, what is $(\mathbf{f} \times \mathbf{g})'(\pi)$?
4. What vector-valued function $\mathbf{F}$ satisfies $\mathbf{F}'(t) = \langle 3, \sin(t), 2\exp(t)\rangle$ and $\mathbf{F}(0) = \langle 5, 5, 5\rangle$?

Answers

1. The unit circle, traversed counterclockwise 2. 4 3. $\langle 12, -13, 2\rangle$
4. $\mathbf{F}(t) = \langle 3t + 5, 6 - \cos(t), 3 + 2\exp(t)\rangle$

EXERCISES

Problems for Practice

In each of Exercises 1–6, describe and sketch the curve that is defined by the given vector-valued function.

1. $\mathbf{r}(t) = t\mathbf{i} + t^2\mathbf{j}$
2. $\mathbf{r}(t) = t^2\mathbf{i} + t^2\mathbf{j}$
3. $\mathbf{r}(t) = t\mathbf{i} + t^2\mathbf{j} + 2\mathbf{k}$
4. $\mathbf{r}(t) = t^2\mathbf{i} + t^2\mathbf{j} + 3\mathbf{k}$
5. $\mathbf{r}(t) = \mathbf{i} + \cos(t)\mathbf{j} + \sin(t)\mathbf{k}$
6. $\mathbf{r}(t) = t\mathbf{i} + \cos(t)\mathbf{j} + \sin(t)\mathbf{k}$

In each of Exercises 7–10, the position vector $\mathbf{r}(t)$ of a particle is given. Describe the particle's trajectory.

7. $\mathbf{r}(t) = t^2\mathbf{i} + t^2\mathbf{j} + \mathbf{k}$ $\quad -1 \le t \le 2$
8. $\mathbf{r}(t) = t^2\mathbf{i} + t^4\mathbf{j}$ $\quad -1 \le t \le 2$
9. $\mathbf{r}(t) = t\mathbf{i} + t\mathbf{j} + \sin(t)\mathbf{k}$ $\quad 0 \le t \le 2\pi$
10. $\mathbf{r}(t) = \cos(t^2)\mathbf{i} + \sin(t^2)\mathbf{j} + tk$ $\quad \sqrt{2\pi} \le t \le 2\sqrt{\pi}$

In each of Exercises 11–14, a vector-valued function $\mathbf{r}$ and a point c are given. Calculate the limit of $\mathbf{r}(t)$ as $t \to c$.

11. $\mathbf{r}(t) = \cos(\pi t)\mathbf{i} + 2^t\mathbf{j}$ $\quad c = -1$
12. $\mathbf{r}(t) = \ln(1 + t^2)\mathbf{i} + \mathbf{j}$ $\quad c = 0$
13. $\mathbf{r}(t) = \langle t^2, t^t, e^{2\ln(t-1)} \rangle$ $\quad c = 2$
14. $\mathbf{r}(t) = \dfrac{t-1}{t^2-1}\mathbf{i} + \dfrac{t+1}{t^2+1}\mathbf{j}$ $\quad c = 1$

In each of Exercises 15–20, calculate $\mathbf{r}'(t)$.

15. $\mathbf{r}(t) = t^{1/2}\mathbf{i} - 3\mathbf{j} + t^2\mathbf{k}$
16. $\mathbf{r}(t) = t^{3/2}\mathbf{i} - t^2\mathbf{j} - 5\sqrt{t}\mathbf{k}$
17. $\mathbf{r}(t) = \tan(t)\mathbf{i} - \cot(t)\mathbf{j} + \sec(t)\mathbf{k}$
18. $\mathbf{r}(t) = (1/t)\mathbf{i} - t\mathbf{j} + \ln(t)\mathbf{k}$
19. $\mathbf{r}(t) = t\mathbf{i} + (t+5)^{-1}\mathbf{j} + (t-7)^{-2}\mathbf{k}$
20. $\mathbf{r}(t) = \langle \sqrt{1-t^2}, t/\sqrt{1-t^2}, \arcsin(t) \rangle$

In each of Exercises 21–26, calculate $\mathbf{r}''(t)$.

21. $\mathbf{r}(t) = t^2\mathbf{i} - t^{3/2}\mathbf{j} + (t+1)^{5/2}\mathbf{k}$
22. $\mathbf{r}(t) = (t^2+1)^{-1}\mathbf{i} - \sec(t)\mathbf{k}$
23. $\mathbf{r}(t) = e^{-t}\mathbf{i} - \ln(t^2 + t^4)\mathbf{j} + \mathbf{k}$
24. $\mathbf{r}(t) = \sec(t)\mathbf{j} - \cot(t)\mathbf{k}$
25. $\mathbf{r}(t) = t^{-1/3}\mathbf{i} - (1+t^2)^{-1/2}\mathbf{j} + t^{1/4}\mathbf{k}$
26. $\mathbf{r}(t) = \arctan(t)\mathbf{i} - t\sin(t)\mathbf{j} + \arcsin(t)\mathbf{k}$

In each of Exercises 27–30, calculate the indefinite integral

27. $\int (t^2\mathbf{i} - t^{-1/2}\mathbf{j} + \cos(2t)\mathbf{k})dt$
28. $\int (e^{3t}\mathbf{i} - e^{-2t}\mathbf{j} + te^t\mathbf{k})dt$
29. $\int (\ln(t)\mathbf{i} - \cos^2(t)\mathbf{j} + \tan(t)\mathbf{k})dt$
30. $\int (\sqrt{t}\mathbf{i} - t^{-1/3}\mathbf{j} + \sec^2(t)\mathbf{k})dt$

In each of Exercises 31–34, find the antiderivative $\mathbf{F}(t)$ for $\mathbf{f}(t)$ that satisfies the additional condition that is given.

31. $\mathbf{f}(t) = 2t\mathbf{i} - 3t^2\mathbf{j} + 4t^3\mathbf{k}$ $\quad \mathbf{F}(2) = 5\mathbf{i} - 2\mathbf{j} + \mathbf{k}$
32. $\mathbf{f}(t) = \sqrt{t}\mathbf{i} - t^2\mathbf{j} + \mathbf{k}$ $\quad \mathbf{F}(2) = 5\mathbf{i} - 2\mathbf{j}$
33. $\mathbf{f}(t) = \cos(t)\mathbf{i} - \sin(2t)\mathbf{j} + \sec(t)\tan(t)\mathbf{k}$
$\quad \mathbf{F}(0) = 3\mathbf{i} + (3/2)\mathbf{j} - 2\mathbf{k}$
34. $\mathbf{f}(t) = \ln(t)\mathbf{j}$ $\quad \mathbf{F}(e) = -2\mathbf{i} + \mathbf{j} - 4\mathbf{k}$

In each of Exercises 35–48, suppose that $\mathbf{f}(t) = e^t\mathbf{i} - \cos(t)\mathbf{j} + \ln(1+t^2)\mathbf{k}$, $\mathbf{g}(t) = t^3\mathbf{i} - t\mathbf{k}$, $\phi(t) = 1 + 5t$, and $\lambda = 3$. Calculate the derivative of the given function.

35. $\mathbf{r}(t) = \mathbf{f}(\lambda t)$
36. $\mathbf{r}(t) = \phi(t)^\lambda \mathbf{f}(t)$
37. $\mathbf{r}(t) = \mathbf{g}(\phi(t))$
38. $\mathbf{r}(t) = \lambda\mathbf{f}(t) - \phi(t)\mathbf{g}(t)$
39. $\mathbf{r}(t) = (\phi(t)\mathbf{k}) \times \mathbf{f}(t) \cdot (\lambda\mathbf{j})$
40. $\mathbf{r}(t) = \mathbf{g}(\lambda t) - \lambda\mathbf{g}(t)$
41. $\mathbf{r}(t) = \mathbf{g}(1/t^3)$
42. $\psi(t) = \mathbf{f}(t) \cdot \mathbf{g}(t)$
43. $\psi(t) = \|\mathbf{g}(t)\|^2$
44. $\mathbf{r}(t) = \mathbf{f}(t) \times (\phi(t)\mathbf{f}(t))$
45. $\psi(t) = \phi(\mathbf{f}(t) \cdot \mathbf{g}(t))$
46. $\mathbf{r}(t) = t\mathbf{j} \times \mathbf{g}(t)$
47. $\mathbf{r}(t) = (\phi(t)\mathbf{i} + \lambda\mathbf{k}) \times \mathbf{g}(t)$
48. $\mathbf{r}(t) = \mathbf{f}(t) \times \mathbf{g}(t)$

Further Theory and Practice

In each of Exercises 49–52, suppose that $\mathbf{f}$ and $\mathbf{g}$ are the functions of Exercises 35–48. Calculate the derivative of the given function.

49. $\mathbf{r}(t) = (\mathbf{i} \cdot \mathbf{f}(t))(\mathbf{i} \times \mathbf{g}(t))$
50. $\mathbf{r}(t) = \big((\phi(t)\mathbf{j} + \lambda\mathbf{k}) \cdot \mathbf{g}(t)\big)\mathbf{f}(t)$
51. $\psi(t) = \mathbf{f}(t) \cdot (\mathbf{f}(t) \times \mathbf{g}(t))$
52. $\mathbf{r}(t) = (\phi(t)\mathbf{j} \times \mathbf{g}(t)) \times \mathbf{g}(t) - \phi(t)\mathbf{j} \times (\mathbf{g}(t) \times \mathbf{g}(t))$

In each of Exercises 53–56, calculate the given limit.

53. $\lim_{t\to 0}\left(|t|^t\mathbf{i} + \cos^{1/t}(t)\mathbf{j} + (1-t)^t\mathbf{k}\right)$
54. $\lim_{t\to 0}\left(\dfrac{\sin(t)}{t}\mathbf{i} + \dfrac{t}{\cos(t)}\mathbf{j} + \dfrac{\tan(t)}{t}\mathbf{k}\right)$
55. $\lim_{t\to \pi}\left\langle \dfrac{\sin(t)}{t-\pi}, \dfrac{\ln(t^2)}{\cos(t)}, |\sec(t)| \right\rangle$
56. $\lim_{t\to 0}\left((1+t)^{1/t}\mathbf{i} + (1+2t)^{1/t}\mathbf{j} + (1-t)^{1/t}\mathbf{k}\right)$

Determine where each of the functions in Exercises 57–60 is continuous.

57. $\mathbf{r}(t) = \phi(t)\mathbf{i} - 7t^2\mathbf{j}$, where

$$\phi(t) = \begin{cases} 1 & \text{if } t<5 \\ -1 & \text{if } t\geq 5 \end{cases}$$

58. $\mathbf{r}(t) = \tan(t)\mathbf{i} - t^3\mathbf{j} + \phi(t)\mathbf{k}$ where

$$\phi(t) = \begin{cases} \sin(1/t) & \text{if } t\neq 0 \\ 1 & \text{if } t=0 \end{cases}$$

59. $\mathbf{r}(t) = \phi(t)\mathbf{i} + \ln(1+t^2)\mathbf{j} + |t|\mathbf{k}$ where

$$\phi(t) = \begin{cases} \sin(t)/t & \text{if } t>0 \\ \cos(t) & \text{if } t\leq 0. \end{cases}$$

60. $\mathbf{r}(t) = \mathbf{i} + \phi(t)\mathbf{j} + (1/(1+|t|))|\mathbf{k}$ where

$$\phi(t) = \begin{cases} t^3 - 3t - 1 & \text{if } t<2 \\ 3t-5 & \text{if } t\geq 2 \end{cases}$$

61. Suppose that $\mathbf{r}(t) = p(t)\mathbf{i} + q(t)\mathbf{j} + s(t)\mathbf{k}$ where p, q, and s are polynomials. Prove that there exists a positive integer N such that the Nth derivative of $\mathbf{r}$ is identically the zero vector. What is the least N that will suffice?

62. Verify that the function $\mathbf{r}(t) = \cos(t)\mathbf{i} - \sin(t)\mathbf{k}$ satisfies the differential equation

$$\frac{d^2}{dt}\mathbf{r}(t) + \mathbf{r}(t) = 0.$$

63. Verify that the curve described by $\mathbf{r}(t) = \cos^2(t)\mathbf{i} + \cos(t)\sin(t)\mathbf{j} + \sin(t)\mathbf{k}$ lies on the sphere $x^2 + y^2 + z^2 = 1$.

64. Let $\mathcal{S}$ be the surface that consists of all points (x, y, z) that satisfy the equation $x^2 + y^2 = z^2$. What are the intersections of $\mathcal{S}$ with horizontal planes $z = h$? What is the intersection of $\mathcal{S}$ with the yz-plane $x = 0$? Describe $\mathcal{S}$. Describe the space curve $\mathcal{C}$ defined by $\mathbf{r}(t) = t\cos(t)\mathbf{i} + t\sin(t)\mathbf{j} + t\mathbf{k}$ and verify that $\mathcal{C}$ lies on $\mathcal{S}$.

65. Give an example of a vector-valued function that is continuous at a point c but is not differentiable at c

66. Formulate a definition of right-hand limit and a definition of left-hand limit for vector-valued functions. Prove that these limits may be calculated componentwise.

67. Suppose that $\mathbf{r}(t)$ is a vector-valued function. Prove that $\lim_{t\to c}\mathbf{r}(t) = \mathbf{L}$ can hold for at most one value of $\mathbf{L}$. *Hint:* Use the analogous theorem for scalar-valued functions.

68. Suppose that $\mathbf{r}(t)$ is a vector-valued function satisfying $\lim_{t\to c}\mathbf{r}(t) = \mathbf{L}$. Prove that $\lim_{t\to c}\|\mathbf{r}(t)\| = \|\mathbf{L}\|$. Prove that the converse is false.

69. Suppose that $t \mapsto \mathbf{r}(t)$ is continuous at $t = c$. Prove that $t \mapsto \|\mathbf{r}(t)\|$ is continuous at $t = c$. Prove that the converse is false.

70. Suppose $\mathbf{r}(c) \neq 0$. Show that the differentiability of $t \mapsto \mathbf{r}(t)$ at $t = c$ implies the differentiability of $t \mapsto \|\mathbf{r}(t)\|$ at $t = c$. Prove that the converse is false.

71. Let $\mathbf{f}$ be a vector-valued function for which

$$\|\mathbf{f}(s) - \mathbf{f}(t)\| \leq |s-t|$$

for all s and t. Prove that $\mathbf{f}$ is continuous.

Calculator/Computer Exercises

In each of Exercises 72–79, a vector-valued function $\mathbf{r}(t) = r_1(t)\mathbf{i} + r_2(t)\mathbf{j}$ and a value t_0 in the domain of $\mathbf{r}$ are given. Plot the planar curve associated with $\mathbf{r}$. To your plot, add the straight line through $\mathbf{r}(t_0)$ with slope $r_2'(t_0)/r_1'(t_0)$.

Use the Chain Rule to explain the relationship of this quantity to dy/dx at $\mathbf{r}(t_0)$.

72. $\mathbf{r}(t) = (5 + 2\cos(t))\mathbf{i} + (3 + \sin(t))\mathbf{j}$, $0 \leq t \leq 2\pi$; $t_0 = \pi/3$

73. $\mathbf{r}(t) = (t^2 + t)\mathbf{i} + (t^2 - t)\mathbf{j}$, $-2 \leq t \leq 2$; $t_0 = -1$

74. $\mathbf{r}(t) = \tan(t)\mathbf{i} + \sec(t)\mathbf{j}$, $-\pi/4 < t < \pi/4$; $t_0 = \pi/6$

75. $\mathbf{r}(t) = 3t/(1+t^3)\mathbf{i} + 3t^2/(1+t^3)\mathbf{j}$, $-30 < t < -2$ and $-1/2 < t < 30$ (arcs of the *Folium of Descartes*); $t_0 = 2$

76. $\mathbf{r}(t) = (e^t + e^{-t})\mathbf{i} + (e^t - e^{-t})\mathbf{j}$, $-2 \leq t \leq 2$; $t_0 = 1$

77. $\mathbf{r}(t) = (t^3 - t)\mathbf{i} + (t^2 - 1)\mathbf{j}$, $-5/4 \leq t \leq 5/4$; $t_0 = -5/6$

78. $\mathbf{r}(t) = 2t/(1+t^2)\mathbf{i} + (1-t^2)/(1+t^2)\mathbf{j}$, $-\infty < t < \infty$; $t_0 = 4$

79. $\mathbf{r}(t) = (t - \sin(t))\mathbf{i} + (1 - \cos(t))\mathbf{j}$, $0 \leq t \leq 4\pi$ (two arches of the *Cycloid*); $t_0 = 3\pi/2$

In each of Exercises 80 and 81, plot the space curve associated with the given vector-valued function $\mathbf{r}$.

80. $\mathbf{r}(t) = \cos^2(t)\mathbf{i} + \cos(t)\sin(t)\mathbf{j} + \sin(t)\mathbf{k} \quad 0 \leq t \leq 2\pi$

81. $\mathbf{r}(t) = (\cos(t)/t)\mathbf{i} + (\sin(t)/t)\mathbf{j} + (1/t)\mathbf{k} \quad 4\pi \leq t \leq 12\pi$

10.2 Velocity and Acceleration

Imagine a particle traveling through space with position vector $\mathbf{r}(t) = r_1(t)\mathbf{i} + r_2(t)\mathbf{j} + r_3(t)\mathbf{k}$. In Figure 1, we see that the vector $\mathbf{r}(t+\Delta t) - \mathbf{r}(t)$ is the displacement of the particle's position from $\mathbf{r}(t)$ to $\mathbf{r}(t+\Delta t)$: It represents how far and in what direction the particle moves from time t to time $t + \Delta t$. Therefore

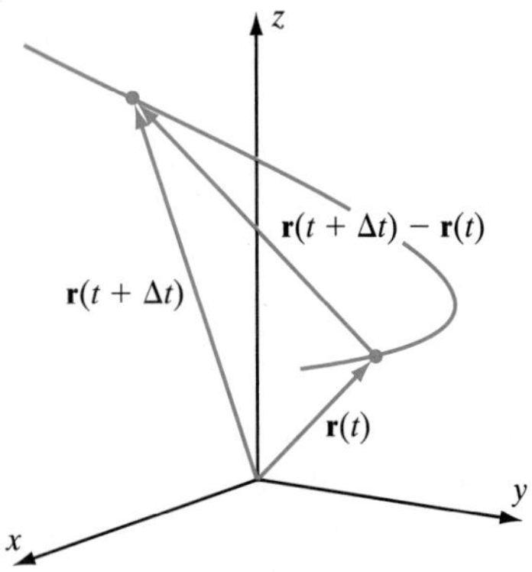

▲ Figure 1

$$\frac{1}{\Delta t}(\mathbf{r}(t+\Delta t)-\mathbf{r}(t))$$

represents the average change in position over the time interval $[t, t+\Delta t]$ or *average velocity* of the particle over the time interval $[t, t+\Delta t]$. Following the line of reasoning used in Chapter 3, we define the *instantaneous velocity* of the particle at time t to be

$$\mathbf{v}(t)=\mathbf{r}'(t)=\lim_{\Delta t\to 0}\frac{1}{\Delta t}(\mathbf{r}(t+\Delta t)-\mathbf{r}(t))$$

provided that this limit exists.

INSIGHT Our definition of instantaneous velocity is similar to the one we adopted for scalar-valued functions in Section 3.1 of Chapter 3. Notice that the expression that defines average velocity and that appears in the formula for instantaneous velocity does *not* contain the quotient of two vectors; it is actually the scalar multiplication of the vector $\mathbf{r}(t+\Delta t)-\mathbf{r}(t)$ by the number $1/\Delta t$. Thus average velocity is a vector and instantaneous velocity is a vector. This is an important point to keep in mind.

How could we have been so clumsy in Chapter 3 to have thought that velocity was a number, when now it turns out to be a vector? Notice that, in space, a vector has three components. In the plane, a vector has two components. But in one dimension, a vector has just one component; in other words, in one dimension a vector is just a number. Therefore in one dimension, there is no need to use vector language. Now that we are working in space, there is definitely a need.

The advantage of treating instantaneous velocity as a vector is that the direction of the vector is the instantaneous direction of motion of the moving body. If we wish to ignore direction and merely talk about magnitude, then we can proceed as follows: $\|\mathbf{r}(t+\Delta t)-\mathbf{r}(t)\|$ is the *magnitude* of the displacement vector over the time interval $[t, t+\Delta t]$. It represents distance traveled, without regard to direction traveled. Therefore

$$\left\|\frac{1}{\Delta t}(\mathbf{r}(t+\Delta t)-\mathbf{r}(t))\right\|$$

represents the average rate of change of distance traveled over the time interval $[t, t+\Delta t]$. This number is always non-negative. We define the *instantaneous speed* to be

$$v(t)=\lim_{\Delta t\to 0}\left\|\frac{1}{\Delta t}(\mathbf{r}(t+\Delta t)-\mathbf{r}(t))\right\|$$

provided that this limit exists. Notice that $v(t)=\|\mathbf{v}(t)\|=\|\mathbf{r}'(t)\|$. Now we will summarize what we have learned.

DEFINITION Suppose that a particle moving through space has position vector $\mathbf{r}(t)$. If $\mathbf{r}$ is differentiable at t, then the *instantaneous velocity* of the particle at time t is the vector $\mathbf{v}(t) = \mathbf{r}'(t)$. We refer to $\mathbf{v}(t) = \mathbf{r}'(t)$ as the *velocity vector* for short. The nonnegative number $v(t) = \|\mathbf{v}(t)\| = \|\mathbf{r}'(t)\|$ is the *instantaneous speed* of the particle. If the instantaneous speed is positive, that is, if $\mathbf{r}'(t) \neq \mathbf{0}$, then the direction vector $\dfrac{1}{\|\mathbf{r}'(t)\|}\mathbf{r}'(t)$ is said to be the *instantaneous direction of motion* of the particle.

▶ **EXAMPLE 1** Let the motion of a particle in space be given by $\mathbf{r}(t) = \cos(t)\mathbf{i} + \sin(t)\mathbf{j} + t\mathbf{k}$. Sketch the curve of motion on a set of axes. Calculate the velocity vector for any t. What value does the velocity have at time $t = \pi/2$? What is the speed at this time? Add the velocity vector $\mathbf{v}(\pi/2)$ to your sketch, representing it by a directed line segment whose initial point is the terminal point of $\mathbf{r}(t)$.

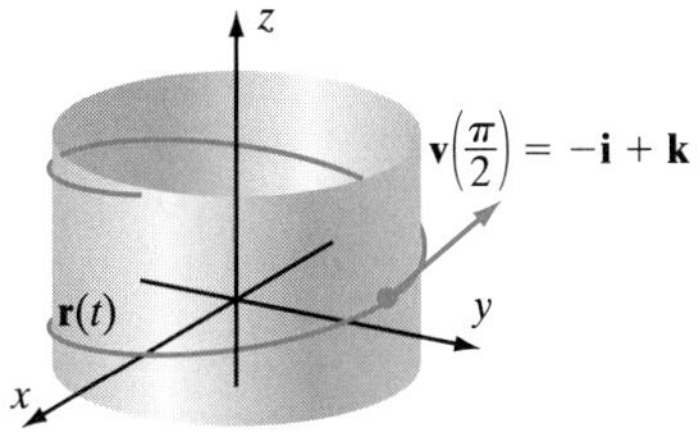

▲ Figure 2

Solution We have studied the curve of motion, a helix, in Example 2 of Section 10.1. Recall that the **i** and **j** components represent motion around a circle, and the **k** component represents a simple vertical motion. The velocity vector for any time t is $\mathbf{v}(t) = \mathbf{r}'(t) = -\sin(t)\mathbf{i} + \cos(t)\mathbf{j} + \mathbf{k}$. We therefore have $\mathbf{v}(\pi/2) = -\mathbf{i} + \mathbf{k}$. The speed of the particle at time $t = \pi/2$ is $v(t) = \|\mathbf{v}(\pi/2)\| = \sqrt{(-1)^2 + 0^2 + 1^2} = \sqrt{2}$. The curve of motion and this velocity vector are sketched in Figure 2. ◀

INSIGHT Figure 2 suggests that the velocity vector $\mathbf{r}'(\pi/2)$ is tangent to the curve of motion at the point $\mathbf{r}(\pi/2)$. Imagine that the curve is a roller-coaster track and that the particle is a roller-coaster car. Suppose that the coaster loses its grip on the track and hurtles off into space. If we ignore air resistance and the effect of gravity, then the car will continue to move in the direction it was moving at the instant that it left the track; that direction is the direction of the instantaneous velocity vector. These ideas are developed in the remainder of this section and also in Section 10.5.

The Tangent Line to a Curve in Space

Let $\mathcal{C}$ be a space curve that is parameterized by $t \mapsto \mathbf{r}(t)$. Suppose that $P_0 = \mathbf{r}(t_0)$ is a point on $\mathcal{C}$ at which $\mathbf{r}'(t_0)$ exists. If Δt is sufficiently small, then the secant line through the two points $\mathbf{r}(t_0)$ and $\mathbf{r}(t_0 + \Delta t)$ can be used as an approximation to the (as yet undefined) tangent line to $\mathcal{C}$ at P_0. Figure 3 illustrates this idea. As you look at panels (a), (b), and (c), notice that the approximation improves as Δt decreases to 0. The secant line in panel (c) is very close to our intuitive conception of the tangent line, which is shown in panel (d).

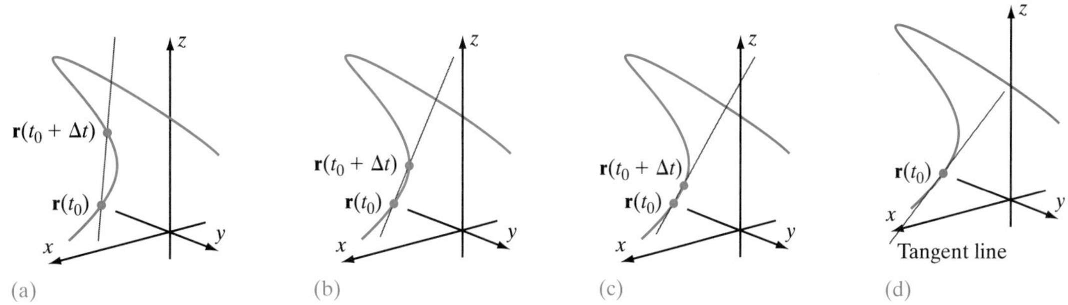

▲ Figure 3

Notice that the displacement vector $\mathbf{r}(t_0+\Delta t)-\mathbf{r}(t_0)$ captures the direction of the secant line through the points $\mathbf{r}(t_0)$ and $\mathbf{r}(t_0+\Delta t)$. We would like to let Δt tend to 0 but, when we do so, the displacement vector $\mathbf{r}(t_0+\Delta t)-\mathbf{r}(t_0)$ approaches $\mathbf{0}$ and its direction becomes undefined. However, if Δt is positive, then the average velocity

$$\frac{1}{\Delta t}\big(\mathbf{r}(t_0+\Delta t)-\mathbf{r}(t_0)\big)$$

has both (a) the same direction as $\mathbf{r}(t_0+\Delta t)-\mathbf{r}(t_0)$ and (b) a typically nonzero limit $\mathbf{r}'(t_0)$ as Δt tends to 0. These ideas suggest the following definition for the tangent line to $\mathcal{C}$ at the point P_0.

DEFINITION Let P_0 be a point on a space curve $\mathcal{C}$. Suppose that $\mathbf{r}$ is a parameterization of $\mathcal{C}$ with $P_0=\mathbf{r}(t_0)$. If $\mathbf{r}'(t_0)$ exists and is not the zero vector, then the *tangent line* to $\mathcal{C}$ at the point $P_0=\mathbf{r}(t_0)$ is the line through P_0 that is parallel to vector $\mathbf{r}'(t_0)$. The tangent line is parameterized by $u\mapsto\mathbf{r}(t_0)+u\mathbf{r}'(t_0)$.

INSIGHT Why do we introduce a new parameter u to describe the tangent line? Notice that the curve described by $t\mapsto\mathbf{r}(t)$ and its tangent line $u\mapsto\mathbf{r}(t_0)+u\mathbf{r}'(t_0)$ pass through the point $P_0=\mathbf{r}(t_0)$ at the values $t=t_0$ and $u=0$, respectively. It is therefore prudent to use different parameters for $\mathcal{C}$ and its tangent line.

▶ **EXAMPLE 2** Consider the curve $\mathcal{C}$ defined by the vector-valued function $\mathbf{r}(t)=\langle t\cos(t), t\sin(t), t\rangle$. What are parametric equations for the tangent line to $\mathcal{C}$ at the point $P_0=(0,\pi/2,\pi/2)$?

Solution We calculate

$$\mathbf{r}'(t)=\left\langle\frac{d}{dt}(t\cos(t)),\frac{d}{dt}(t\sin(t)),\frac{d}{dt}t\right\rangle=\langle\cos(t)-t\sin(t),\sin(t)+t\cos(t),1\rangle.$$

Therefore $\mathbf{r}'(\pi/2)=\langle-\pi/2,1,1\rangle$. The tangent line to $\mathcal{C}$ at $\mathbf{r}(\pi/2)=\langle 0,\pi/2,\pi/2\rangle$ is therefore parameterized by

$$u\mapsto\mathbf{r}\left(\frac{\pi}{2}\right)+u\mathbf{r}'\left(\frac{\pi}{2}\right)=\left\langle 0,\frac{\pi}{2},\frac{\pi}{2}\right\rangle+u\left\langle-\frac{\pi}{2},1,1\right\rangle=\left\langle-\frac{\pi}{2}u,\frac{\pi}{2}+u,\frac{\pi}{2}+u\right\rangle.$$

The parametric equations for the tangent line are

$$x=-\frac{\pi}{2}u,\quad y=\frac{\pi}{2}+u,\quad z=\frac{\pi}{2}+u. \quad ◀$$

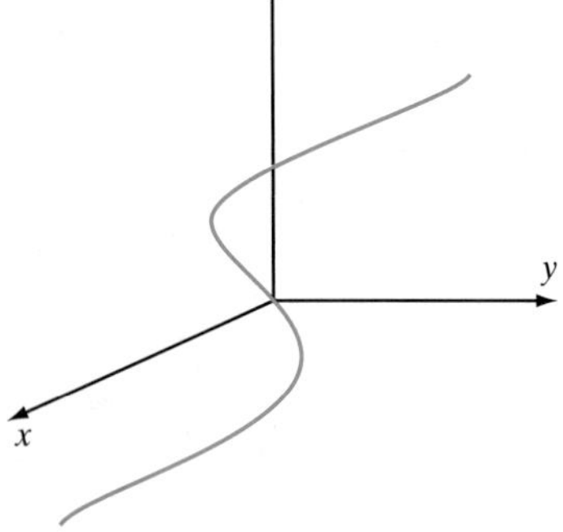

$\mathbf{r}(t)=\langle t\cos(t), t\sin(t), t\rangle$

▲ Figure 4a

INSIGHT A plot of $\mathcal{C}$ may be found in Figure 4a. A static, two-dimensional image of a space curve often does not impart a clear understanding of how the curve twists in space. To help visualize the curve $\mathcal{C}$ of Example 2, we have plotted the solution set $\mathcal{S}$ of the equation $x^2+y^2=z^2$. The graph of $\mathcal{S}$ comprises the two cones that are shown in

Figure 4b. (In Section 11.2 in Chapter 11, you will learn techniques for sketching such surfaces.) Because the entries of $\mathbf{r}(t) = \langle t\cos(t), t\sin(t), t\rangle$ satisfy the equation

$$\underbrace{(t\cos(t))^2}_{x^2} + \underbrace{(t\sin(t))^2}_{y^2} = \underbrace{t^2}_{z^2}$$

we see that $\mathcal{C}$ lies on $\mathcal{S}$. Superimposing the plots of $\mathcal{C}$ and $\mathcal{S}$ in Figure 4c reveals that $\mathcal{C}$ winds around $\mathcal{S}$. Finally, the tangent line that we have calculated in Example 2 has been added in Figure 4d.

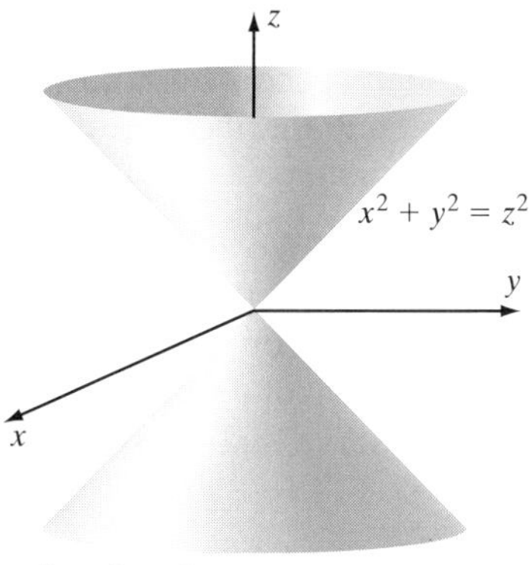

▲ Figure 4b

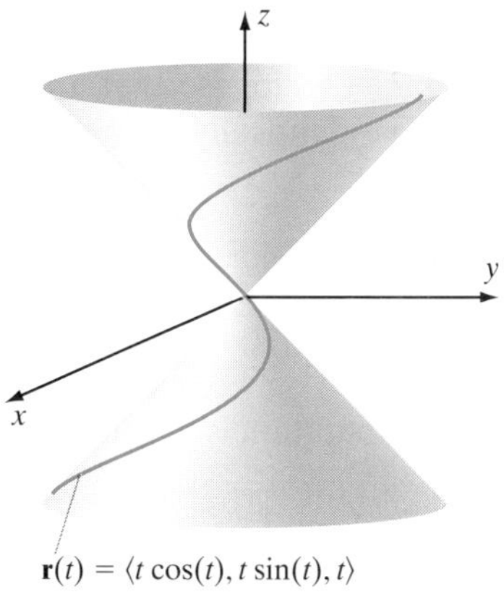

▲ Figure 4c

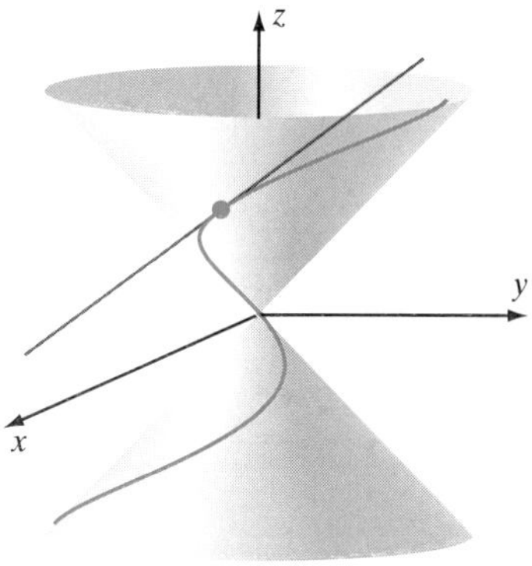

▲ Figure 4d

Acceleration

As we might expect, acceleration is obtained by differentiating velocity. Acceleration is again represented by a vector:

DEFINITION If $\mathbf{r}(t)$ is the position vector of a body moving through space with velocity $\mathbf{v}(t) = \mathbf{r}'(t)$, then the *instantaneous acceleration* of the body at time is $\mathbf{a}(t) = \mathbf{v}'(t)$, provided that this derivative exists. Equivalently, we may define $\mathbf{a}(t) = \mathbf{r}''(t)$, provided that this second derivative exists.

INSIGHT While the velocity vector points in the direction of motion of the moving body, the acceleration vector. usually does *not*. In fact, Newton's Second Law tells us that force $\mathbf{F}$ is equal to mass m times acceleration $\mathbf{a}$. That is,

$$\mathbf{F} = m\mathbf{a}.$$

Now mass is a positive scalar, and force and acceleration are vectors. So Newton's Second Law tells us that the acceleration vector points in the same direction as the force being applied to the body. The next example illustrates this point strikingly.

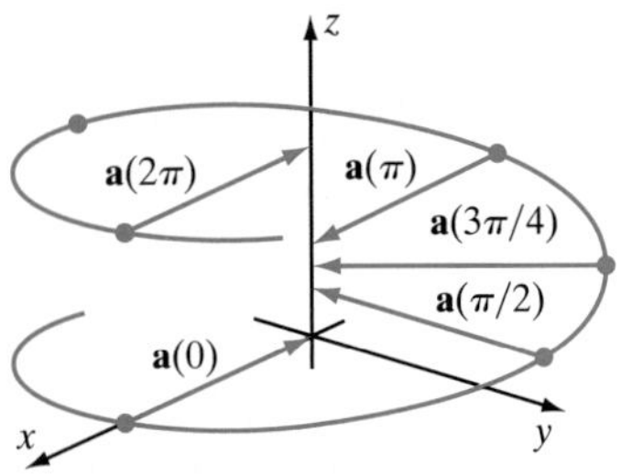

▲ Figure 5
$\mathbf{r}(t) = \cos(t)\mathbf{i} + \sin(t)\mathbf{j} + t\mathbf{k}$

▶ **EXAMPLE 3** Let $\mathbf{r}(t) = \cos(t)\mathbf{i} + \sin(t)\mathbf{j} + t\mathbf{k}$ as in Example 1. Calculate the acceleration $\mathbf{a}(t)$. Evaluate the acceleration at $t = 0$, $\pi/2$, $3\pi/4$, π, and 2π.

Solution We have $\mathbf{v}(t) = -\sin(t)\mathbf{i} + \cos(t)\mathbf{j} + \mathbf{k}$ and $\mathbf{a}(t) = -\cos(t)\mathbf{i} - \sin(t)\mathbf{j}$. In particular, $\mathbf{a}(0) = -\mathbf{i}$, $\mathbf{a}(\pi/2) = -\mathbf{j}$, $\mathbf{a}(3\pi/4) = (1/\sqrt{2})\mathbf{i} + (-1/\sqrt{2})\mathbf{j}$, $\mathbf{a}(\pi) = \mathbf{i}$, and $\mathbf{a}(2\pi) = -\mathbf{i}$. These acceleration vectors are exhibited in Figure 5. In each instance, the acceleration vector points horizontally and *into* the helix. That is also the direction of the force that is required to hold the particle on the spiral path. ◀

Because differentiation of a vector-valued function is performed componentwise, it follows that antidifferentiation is also performed componentwise. We saw this phenomenon at the end of Section 10.1. Now we use this idea in a physical example.

▶ **EXAMPLE 4** Suppose that a particle moves through space with acceleration vector that is identically $\mathbf{0}$. What does that tell us about the motion of the particle?

Solution Let $\mathbf{r}(t)$ be the position vector of the particle at time t. We are given that $\mathbf{v}'(t) = \mathbf{a}(t) = 0\mathbf{i} + 0\mathbf{j} + 0\mathbf{k}$. Antidifferentiating each side of this equation componentwise yields $\mathbf{v}(t) = c_1\mathbf{i} + c_2\mathbf{j} + c_3\mathbf{k}$, where c_1, c_2, c_3 are constants. Notice here that we are using only the simple fact that the antiderivative of $\mathbf{0}$ is a constant. Although our computation is not finished, we can already observe that the speed of the particle is constant: $\|\mathbf{v}(t)\| = \sqrt{c_1^2 + c_2^2 + c_3^2}$ for all t.

Next, we work with the equations $\mathbf{r}'(t) = \mathbf{v}(t)$ and $\mathbf{v}(t) = c_1\mathbf{i} + c_2\mathbf{j} + c_3\mathbf{k}$. We equate to obtain $\mathbf{r}'(t) = c_1\mathbf{i} + c_2\mathbf{j} + c_3\mathbf{k}$. Again we antidifferentiate componentwise, this time obtaining

$$\mathbf{r}(t) = (c_1t + d_1)\mathbf{i} + (c_2t + d_2)\mathbf{j} + (c_3t + d_3)\mathbf{k}.$$

Here we have used the fact that the most general antiderivative of the constant c_1 is $c_1t + d_1$, and so on. Of course, d_1, d_2, d_3 are new constants of integration. Our calculation is complete. We have learned that $\mathbf{r}(t) = \langle d_1, d_2, d_3\rangle + t\langle c_1, c_2, c_3\rangle$, which is the parameterization of a line. In summary, we have shown that if a particle moves through space with an acceleration vector that is identically $\mathbf{0}$, then the particle travels with constant speed along a line. ◀

INSIGHT Example 4 is essentially Newton's First Law: A body moves at constant speed along a straight line unless a force acts upon it. In particular, Example 4 tells us that, whenever a body travels along a path that is not a straight line, there must be some acceleration (and hence some force) present.

▶ **EXAMPLE 5** Show that the acceleration vector of a particle moving through space is always perpendicular to the velocity vector if and only if the particle travels at constant speed.

Solution Let $\mathbf{r}(t)$, $\mathbf{v}(t) = \mathbf{r}'(t)$, and $\mathbf{v}'(t) = \mathbf{r}''(t)$, respectively, denote the position vector, the velocity vector, and the acceleration vector of the particle at time t. Let $v(t) = \|\mathbf{v}(t)\| = \sqrt{\mathbf{v}(t)\cdot\mathbf{v}(t)}$ denote the speed of the particle. Then, $v(t)^2 = \mathbf{v}(t)\cdot\mathbf{v}(t)$ and

$$\frac{d}{dt}\left(v(t)^2\right) = \frac{d}{dt}\left(\mathbf{v}(t)\cdot\mathbf{v}(t)\right) = \mathbf{v}(t)\cdot\frac{d}{dt}\mathbf{v}(t) + \left(\frac{d}{dt}\mathbf{v}(t)\right)\cdot\mathbf{v}(t) = 2\mathbf{v}(t)\cdot\mathbf{v}'(t).$$

Of course $\frac{d}{dt}\left(v(t)^2\right) = 0$ if and only if $v(t)^2$ is constant. We conclude that $v(t)^2$ is constant if and only if $\mathbf{v}(t)\cdot\mathbf{v}'(t) = 0$ for all t. It follows that $v(t)$ is constant if and only if $\mathbf{v}(t)$ and $\mathbf{v}'(t)$ are perpendicular. ◀

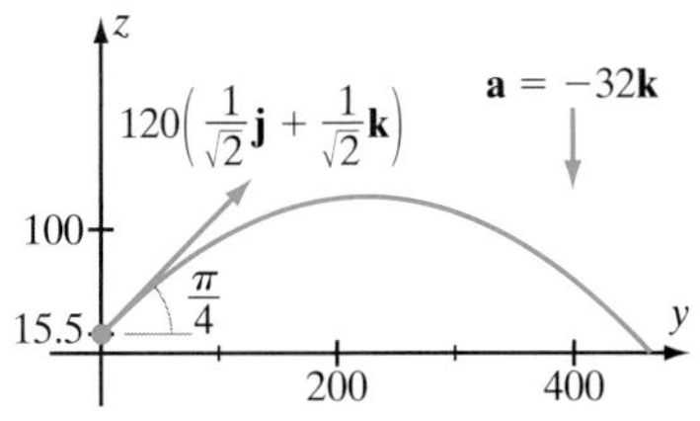

▲ Figure 6

▶ **EXAMPLE 6** An arrow is shot into the air from a height of $15\frac{1}{2}$ feet at a speed of 120 feet per second. The initial angle of the arrow from the horizontal is $\pi/4$ (see Figure 6). Assuming that the acceleration due to gravity is 32 ft/sec^2, determine how far the arrow travels horizontally.

Solution The initial position vector is $0\mathbf{i} + 0\mathbf{j} + (31/2)\mathbf{k}$. The motion takes place entirely in the yz-plane (there is no side-to-side motion). Writing the initial velocity as the scalar multiplication of its magnitude and direction vector, we have $120((1/\sqrt{2})\mathbf{j} + (1/\sqrt{2})\mathbf{k}) = 0\mathbf{i} + 60\sqrt{2}\mathbf{j} + 60\sqrt{2}\mathbf{k}$. The acceleration in the problem is just that due to gravity: $\mathbf{a} = -32\mathbf{k}$.

We begin with the acceleration and work backward: $\mathbf{a}(t) = -32\mathbf{k} = 0\mathbf{i} + 0\mathbf{j} - 32\mathbf{k}$. Antidifferentiating in each component separately (don't forget the constants of integration!) gives $\mathbf{v}(t) = c_1\mathbf{i} + c_2\mathbf{j} + (-32t + c_3)\mathbf{k}$. Now we have

$$0\mathbf{i} + 60\sqrt{2}\mathbf{j} + 60\sqrt{2}\mathbf{k} = \mathbf{v}(0) = c_1\mathbf{i} + c_2\mathbf{j} + (-32\cdot 0 + c_3)\mathbf{k}.$$

This means that $c_1 = 0, c_2 = 60\sqrt{2}$, and $c_3 = 60\sqrt{2}$. We conclude that $\mathbf{v}(t) = 60\sqrt{2}\mathbf{j} + (-32t + 60\sqrt{2})\mathbf{k}$. Antidifferentiating once again gives

$$\mathbf{r}(t) = d_1\mathbf{i} + (60\sqrt{2}t + d_2)\mathbf{j} + (-16t^2 + 60\sqrt{2}t + d_3)\mathbf{k}.$$

But then

$$0\mathbf{i} + 0\mathbf{j} + (31/2)\mathbf{k} = \mathbf{r}(0) = d_1\mathbf{i} + d_2\mathbf{j} + d_3\mathbf{k}.$$

We conclude that $d_1 = 0, d_2 = 0$, and $d_3 = 31/2$. To summarize, the trajectory of the arrow is given by

$$\mathbf{r}(t) = 60\sqrt{2}t\mathbf{j} + (-16t^2 + 60\sqrt{2}t + 31/2)\mathbf{k}.$$

The motion of the arrow ceases when the arrow hits the ground, which is when the $\mathbf{k}$ component of $\mathbf{r}$ is 0 . We solve the quadratic equation $-16t^2 + 60\sqrt{2}t + 31/2 = 0$ to obtain

$$t = \frac{-60\sqrt{2}\pm\sqrt{(-60\sqrt{2})^2 - 4(-16)(31/2)}}{-32} = \frac{-60\sqrt{2}\pm\sqrt{8192}}{-32} = \frac{-60\sqrt{2}\pm 64\sqrt{2}}{-32}.$$

The negative solution $(-60\sqrt{2} + 64\sqrt{2})/(-32) = -\sqrt{2}/8$ has no physical relevance. For $t = (-60\sqrt{2} - 64\sqrt{2})/(-32) = 31\sqrt{2}/8$, the $\mathbf{j}$ component of $\mathbf{r}$ is $(60\sqrt{2})\cdot(31\sqrt{2}/8) = 465$. Thus the horizontal distance traveled by the arrow is 465 feet. ◀

The Physics of Baseball

Many experts consider the fastball to be the most important pitch in baseball. If the pitcher imparts a certain backspin to the ball when it is released, then the ball can appear to abruptly sink shortly before it reaches the plate. The amount by which it sinks can be several inches. Because the sweet spot of a standard baseball bat has diameter 2.75 inches, and the batter must get his "fix" on the ball during the first 30 feet of the ball's 60-foot journey to the plate, the "sinking fastball" can certainly fool a batter.

Let us use the ideas we have learned to understand why the fastball actually sinks. Suppose that the pitcher stands at the origin of coordinates and throws in the direction of the positive x-axis with an initial velocity vector $\mathbf{v}_0 = 130\mathbf{i} - \mathbf{k}$. Here the units are distance measured in feet and time measured in seconds. The speed of 130 feet per second is about 88.6 miles per hour, a plausible speed for a fastball thrown by a good pitcher.

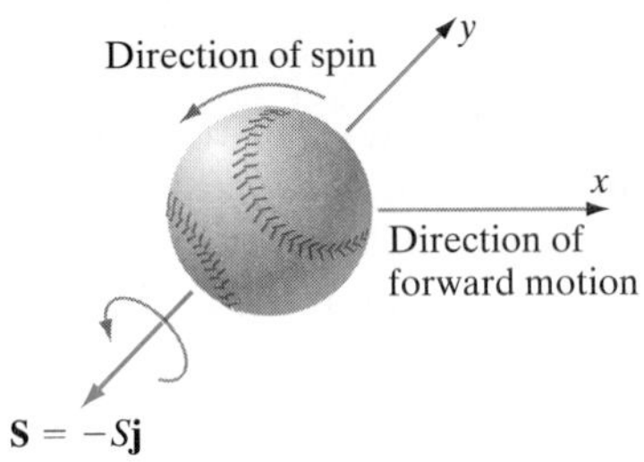

▲ Figure 7

Assume in addition that the pitcher releases the ball at a height of 5.5 feet, so that the initial position is $\mathbf{r}_0 = 5.5\mathbf{k}$.

Now the interesting part of this analysis is the acceleration. Of course there is a component of acceleration that is due to gravity; this component contributes $-32\mathbf{k}$. Because the ball is given an initial backward spin, there is also a component called "spin acceleration." Use the right-hand rule to see that the so-called spin vector $\mathbf{S}$ for the ball is pointing in the direction of the negative y-axis (Figure 7). We shall write $\mathbf{S} = -S\mathbf{j}$.

According to the theory of aerodynamics, the spin causes a difference in air pressure on the sides of the ball and results in a contribution of *spin acceleration* that is given by

$$\mathbf{a}^s = c\mathbf{S} \times \mathbf{v}_0.$$

Here c is a physical constant that is determined through experimentation. Let us calculate this cross product:

$$\mathbf{S} \times \mathbf{v}_0 = \det\left(\begin{bmatrix} \mathbf{i} & \mathbf{j} & \mathbf{k} \\ 0 & -S & 0 \\ v_1 & v_2 & v_3 \end{bmatrix}\right) = (-Sv_3)\mathbf{i} + (Sv_1)\mathbf{k}.$$

Noting that $v_1 = 130$ and $v_3 = -1$, we have thus determined that $\mathbf{a}^s = c(S\mathbf{i} + 130S\mathbf{k})$ and

$$\mathbf{a} = \mathbf{a}^s - 32\mathbf{k} = c(S\mathbf{i} + 130S\mathbf{k}) - 32\mathbf{k} = cS\mathbf{i} + (130cS - 32)\mathbf{k}.$$

Typically, the amount of spin placed on the ball is about $S = 2.5$ revolutions per second. Experimental evidence suggests that $c = 0.08$ is plausible. Thus we arrive at

$$\mathbf{a} = 0.2\mathbf{i} - 6\mathbf{k}.$$

As usual, we integrate this last equation to obtain

$$\mathbf{v} = 0.2t\mathbf{i} - 6t\mathbf{k} + \mathbf{C}.$$

The value of $\mathbf{C}$ is determined by setting $\mathbf{v}(0) = \mathbf{v}_0$. We find that $\mathbf{C} = 130\mathbf{i} - \mathbf{k}$. Hence

$$\mathbf{v} = (130 + 0.2t)\mathbf{i} + (-1 - 6t)\mathbf{k}.$$

Integrating a second time yields

$$\mathbf{r}(t) = (130t + 0.1t^2)\mathbf{i} + (-t - 3t^2)\mathbf{k} + \mathbf{D}.$$

The value of the constant vector $\mathbf{D}$ is determined by setting $\mathbf{f}(0) = \mathbf{r}_0$. We find that $\mathbf{D} = 5.5\mathbf{k}$. The final result is then

$$\mathbf{r}(t) = (130t + 0.1t^2)\mathbf{i} + (5.5 - t - 3t^2)\mathbf{k}.$$

What is the meaning of this formula for the motion of the ball? It is subtle. The distance of the pitcher's mound from home plate is 60 feet. To find out when the ball crosses the plate, we solve $130t + 0.1t^2 = 60$ to obtain $t \approx 0.46$. Suppose that the batter gets his fix on the ball at the halfway point. Solving $130t + 0.1t^2 = 30$, we obtain $t \approx 0.23$ seconds. At that time the height of the ball is $5.5 - 0.23 - 3 \cdot (0.23)^2$. Thus, the batter sees that, halfway through its course, the ball drops $0.23 + 3 \cdot (0.23)^2 = 0.3887$ feet. He would expect it to drop a similar amount on the second half of its journey. However, at time $t = 0.46$, the ball has height $(5.5 - 0.46 - 3 \cdot (0.46)^2)$. The ball has dropped a total of $0.46 + 3 \cdot (0.46)^2 = 1.0948$ feet *instead* of the $0.3887 + 0.3887 = 0.7774$ feet that the batter expected. So, in effect, the ball has "sunk" $1.0948 - 0.7774 = 0.3174$ feet or 3.81 inches. Refer to Figure 8 to see the path of the pitched ball. It is worth noting that the vertical component of the trajectory of the ball is parabolic. Thus the ball does *not* travel a linear path; it instead traces a parabola.

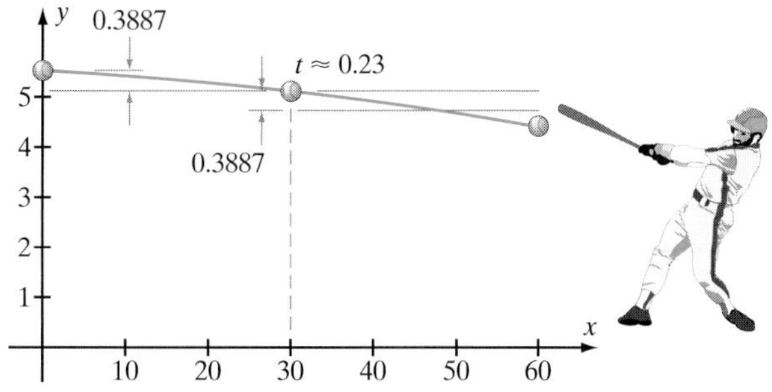

Figure 8 The expected drop after the halfway point is 0.3887 ft.

QUICK QUIZ

1. If $\mathbf{r}(t) = \langle t, t^2, t^3\rangle$, what is $\mathbf{r}'(2)$?
2. Find symmetric equations for the tangent line to $\mathbf{r}(t) = \langle t, 2t^2, 2 + t^2\rangle$ at $t = 1$.
3. A particle's position is given by $\mathbf{r}(t) = \langle t, \sqrt{4+t}, e^{-2t}\rangle$. What is its velocity when $t = 0$?
4. A particle's position is given by $\mathbf{r}(t) = \langle e^{-t}, \cos(t), t^2\rangle$. What is its acceleration when $t = 0$?

Answers

1. $\langle 1, 4, 12\rangle$ 2. $x - 1 = (y-2)/4 = (z-3)/2$ 3. $\langle 1, 1/4, -2\rangle$ 4. $\langle 1, -1, 2\rangle$

EXERCISES

Problems for Practice

For each position function $\mathbf{r}$ in Exercises 1–8, calculate the corresponding velocity $\mathbf{v}$, speed v, and acceleration $\mathbf{a}$.

1. $\mathbf{r}(t) = t\mathbf{i} + t^2\mathbf{j} + t^3\mathbf{k}$
2. $\mathbf{r}(t) = (2 - t^2)\mathbf{i} - 5t^4\mathbf{k}$
3. $\mathbf{r}(t) = (1 - t^2)\mathbf{i} + 2t\mathbf{j} + (1/(1 + t^2))\mathbf{k}$
4. $\mathbf{r}(t) = e^t\mathbf{i} + e^{-t}\mathbf{j} - e^{-2t}\mathbf{k}$
5. $\mathbf{r}(t) = t\mathbf{i} - 5\mathbf{j} + e^t\mathbf{k}$
6. $\mathbf{r}(t) = \ln(|t|)\mathbf{i} + t^{-1}\mathbf{j} - t^{-2}\mathbf{k}$
7. $\mathbf{r}(t) = -\cos(t^2)\mathbf{i} + \sin(t^2)\mathbf{j} + t^3\mathbf{k}$
8. $\mathbf{r}(t) = t^{1/2}\mathbf{i} + t^{1/3}\mathbf{j} + t^{1/4}\mathbf{k}$

In each of Exercises 9–12, a body traveling through space near Earth's surface has constant acceleration due to gravity given by $-32\mathbf{k}$. The initial velocity $\mathbf{v}_0 = \mathbf{v}(0)$ and initial position $\mathbf{r}_0 = \mathbf{r}(0)$ are given in each problem. Using antidifferentiation, compute a formula for the position $\mathbf{r}(t)$.

9. $\mathbf{v}_0 = 3\mathbf{i} - 2\mathbf{j} + \mathbf{k}$	$\mathbf{r}_0 = 2\mathbf{i} - 5\mathbf{k}$
10. $\mathbf{v}_0 = 4\mathbf{i} - 7\mathbf{j}$	$\mathbf{r}_0 = 2\mathbf{j} + 8\mathbf{k}$
11. $\mathbf{v}_0 = \mathbf{i} + \mathbf{j} + \mathbf{k}$	$\mathbf{r}_0 = 3\mathbf{i} - 2\mathbf{j} - \mathbf{k}$
12. $\mathbf{v}_0 = \mathbf{k}$	$\mathbf{r}_0 = \mathbf{j}$

In Exercises 13–22, determine parametric equations for the tangent line to the curve described by the given function $\mathbf{r}$ at the given point P_0.

13. $\mathbf{r}(t) = t\mathbf{i} + t^2\mathbf{j} + t^3\mathbf{k}$	$P_0 = (1, 1, 1)$
14. $\mathbf{r}(t) = (2 - t^2)\mathbf{i} - (1 - t^2)\mathbf{j} - 5t\mathbf{k}$	$P_0 = (1, 0, -5)$
15. $\mathbf{r}(t) = e^t\mathbf{i} + e^{-t}\mathbf{j} - e^{-2t}\mathbf{k}$	$P_0 = (1, 1, -1)$
16. $\mathbf{r}(t) = \ln(1 + t)\mathbf{i} + 1/(1 + t)\mathbf{j} - 1/(1 - t)^2\mathbf{k}$	$P_0 = (0, 1, -1)$
17. $\mathbf{r}(t) = t\mathbf{i} - \mathbf{j} + e^{t/2}\mathbf{k}$	$P_0 = (2, -1, e)$
18. $\mathbf{r}(t) = -\cos(t)\mathbf{i} + \sin(t)\mathbf{j} + t\mathbf{k}$	$P_0 = (1, 0, \pi)$
19. $\mathbf{r}(t) = t\mathbf{i} - 5\mathbf{j} + (5/t)\mathbf{k}$	$P_0 = (5, -5, 1)$
20. $\mathbf{r}(t) = (5 - t^2)\mathbf{i} + 2t\mathbf{j} + (5/(1 + t^2))\mathbf{k}$	$P_0 = (1, -4, 1)$
21. $\mathbf{r}(t) = (3 + t)/(1 + t^2)\mathbf{i} + t^2\mathbf{j} + \sin(\pi t)\mathbf{k}$	$P_0 = (1, 4, 0)$
22. $\mathbf{r}(t) = e^t\mathbf{i} + te^t\mathbf{j} + t^2e^t\mathbf{k}$	$P_0 = (1, 0, 0)$

In each of Exercises 23–32, determine symmetric equations for the tangent line to the curve described by the given function $\mathbf{r}$ at the given point P_0.

23. $\mathbf{r}(t) = t\mathbf{i} - t^2\mathbf{j} + t^3\mathbf{k}$	$P_0 = (-2, -4, -8)$
24. $\mathbf{r}(t) = t\mathbf{i} - 5\mathbf{j} + e^t\mathbf{k}$	$P_0 = (2, -5, e^2)$
25. $\mathbf{r}(t) = t\mathbf{i} - 5\mathbf{j} + (5/t)\mathbf{k}$	$P_0 = (5, -5, 1)$
26. $\mathbf{r}(t) = (7 - t^2)\mathbf{i} + 2\mathbf{j} + (t + 1/t)\mathbf{k}$	$P_0 = (6, 2, -2)$
27. $\mathbf{r}(t) = (2/t)\mathbf{i} + 2t\mathbf{j} + \mathbf{k}$	$P_0 = (1, 4, 1)$
28. $\mathbf{r}(t) = 4\arctan(t)\mathbf{i} - \cos(\pi t)\mathbf{j} + (1/t)\mathbf{k}$	$P_0 = (\pi, 1, 1)$
29. $\mathbf{r}(t) = -\cos(t^2)\mathbf{i} + \sin(t^2)\mathbf{j} + t\mathbf{k}$	$P_0 = (1, 0, \sqrt{\pi})$
30. $\mathbf{r}(t) = e^t\mathbf{i} + e^{2t}\mathbf{j} + e^{3t}\mathbf{k}$	$P = (1, 1, 1)$
31. $\mathbf{r}(t) = (5 - t^2)\mathbf{i} + 2t\mathbf{j} + (5/(1 + t^2))\mathbf{k}$	$P_0 = (1, -4, 1)$
32. $\mathbf{r}(t) = \ln(1 + t^2)\mathbf{i} + \arcsin(t)\mathbf{j} + \sec(t)\mathbf{k}$	$P_0 = (0, 0, 1)$

Further Theory and Practice

33. An arrow is shot into the air with initial height 4 ft, initial trajectory forming an angle of $\pi/6$ radians with the horizontal, and initial speed 100 ft/s. Assuming that the acceleration due to gravity is constantly $-32\mathbf{k}$, compute the function $\mathbf{r}(t)$ describing the motion of the arrow. After how many seconds does the arrow hit the ground? How far does the arrow travel horizontally? What is the greatest height that the arrow achieves?
34. A rock is thrown from a 500-foot cliff, with a downward trajectory forming an initial angle of $\pi/6$ radians with the horizontal. The initial speed is 50 ft/s. The acceleration due to gravity is constantly $-32\mathbf{k}$. Derive a formula for the position $\mathbf{r}(t)$ of the rock. When does the rock hit the ground? How far does it travel horizontally?
35. A roller coaster is speeding along a track; its motion describes the path
$$\mathbf{r}(t) = -5t^2\mathbf{i} + t^3\mathbf{j} + (4t + 30)\mathbf{k}.$$
At time $t = 2$, the roller coaster leaves the track. Ignoring the effects of gravity and drag, determine where the roller coaster will be at time $t = 5$.
36. An arrow is shot into the air from an initial height of 6 ft, and with trajectory making initial angle of $\pi/3$ radians with the horizontal. If it were a calm day, the force exerted by the bow would have launched the arrow with initial speed 120 ft/s. However, a sudden gust of wind reduces the *horizontal* component of the arrow's initial velocity by 20 ft/s (the wind blows horizontally, and dies out after its instantaneous gusting). The acceleration due to gravity is constantly $-32\mathbf{k}$. What is the velocity of the arrow when it strikes the ground? How far does the arrow travel horizontally before it strikes the ground?
37. A body moving through space has constant acceleration vector and zero initial velocity. What can you say about the path traveled by the body (give a geometric description)?
38. A body in motion through space has the property that its velocity vector always equals its acceleration vector. What can you say about the motion? Can you write a formula for it?
39. A body travels through space in such a way that its position vector is always perpendicular to its velocity vector. Show that this implies that the motion is on the surface of a sphere.
40. Redo Exercise 35, but this time take into account the effect of gravity.
41. Redo Exercise 36 if the arrow is shot downward from a height of 500 feet with an initial angle of 60 degrees from the horizontal.
42. A particle moves along the curve that is parameterized by $x = 1 + t$, $y = 2t^2$, $z = 6t + t^2$. Its vertical speed is 16 at all

times. What is the velocity vector of the particle when it passes through the point $(2,2,7)$?

43. Let $P_0 = (0,0,2r)$ denote the north pole of the sphere $x^2+y^2+(z-r)^2 = r^2$. Let P_1 be any other point on the sphere (see Figure 9). Suppose that a rigid straight line joins P_0 to P_1 and that a particle initially at rest is allowed to slide down the line from P_0 to P_1 under a constant force equal to $-mg\,\mathbf{k}$. Show that the elapsed time does not depend on the choice of the point P_1.

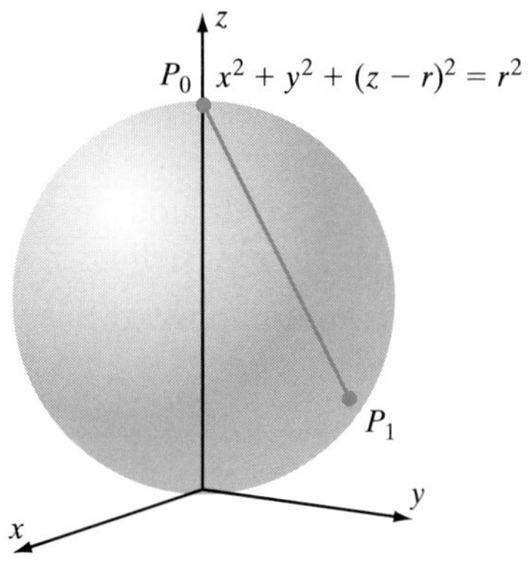

▲ Figure 9

44. Let $\mathbf{r}_0$ and $\mathbf{v}_0$ be constant vectors. Let ω be a positive constant. Show that

$$\mathbf{r}(t) = \cos(\omega t)\mathbf{r}_0 + \frac{\sin(\omega t)}{\omega}\mathbf{v}_0$$

satisfies the initial value problem

$$\frac{d^2}{dt^2}\mathbf{r}(t) + \omega^2\mathbf{r}(t) = 0, \quad \mathbf{r}(0) = \mathbf{r}_0, \quad \mathbf{r}'(0) = \mathbf{v}_0.$$

45. Let $\mathbf{r}(t)$ denote the position at time t of a body moving through space. Suppose that $\mathbf{r}(t)$ is parallel to $\mathbf{a}(t) = \mathbf{r}''(t)$ for all t. Prove that $\mathbf{r}(t) \times \mathbf{r}'(t)$ is a constant vector.

46. To each point $P_t = (\cos(t), \sin(t), t)$ with $t \geq 0$, we associate another point Q_t as follows: $\overline{P_tQ_t}$ is tangent to the helix traced by P_t, $\|\overrightarrow{P_tQ_t}\| = \sqrt{2}t$, and $\overrightarrow{P_tQ_t} \cdot \mathbf{k} \geq 0$. Parameterize the curve traced by Q_t.

47. Show that the speed of a moving particle is decreasing if the angle between its velocity and acceleration vectors is always acute. Show that the speed of the particle is decreasing if the angle is always obtuse.

48. Let $\mathbf{r}(t) = (t^3 - t)\mathbf{i} + (t^2 - 1)\mathbf{j}$. Plot the points $\mathbf{r}(-2)$, $\mathbf{r}(-1)$, $\mathbf{r}(-1/2)$, $\mathbf{r}(0)$, $\mathbf{r}(1/2)$, $\mathbf{r}(1)$, $\mathbf{r}(2)$. Sketch the curve. Describe the tangent lines to the curve at the point $(0,0)$.

49. Let $\mathcal{C}$ be the curve with parameterization $\mathbf{r}(t) = \langle t, t^2, t^3\rangle$ for $t > 0$. Find a point P_0 on $\mathcal{C}$ for which the tangent line to $\mathcal{C}$ at P_0 passes through the point $(0, -1/2, -1/\sqrt{2})$.

50. Show that the curves $t \mapsto \langle t^3 + t^2 - 3t + 1, t^2, 3t - 1\rangle$ and $t \mapsto \langle 3t+1, 2t, t^2 + t - 1\rangle$ intersect at two points. At each of these points, what is the angle between the tangent lines to the curves?

Calculator/Computer Exercises

▶ In each of Exercises 51–58, plot the planar curve associated with the given vector-valued function $\mathbf{r}(t) = r_1(t)\mathbf{i} + r_2(t)\mathbf{j}$. Calculate $\mathbf{r}'(t_0)$ at the given value of t_0 and add the tangent line to the curve at $\mathbf{r}(t_0)$ to your plot. ◀

51. $\mathbf{r}(t) = (5 + 2\cos(t))\mathbf{i} + (3 + \sin(t))\mathbf{j}$, $0 \leq t \leq 2\pi$, $t_0 = \pi/3$

52. $\mathbf{r}(t) = (t^2 + t)\mathbf{i} + (t^2 - t)\mathbf{j}$, $-2 \leq t \leq 2$, $t_0 = 1$

53. $\mathbf{r}(t) = \tan(t)\mathbf{i} + \sec(t)\mathbf{j}$, $-\pi/3 \leq t \leq \pi/3$, $t_0 = \pi/4$

54. $\mathbf{r}(t) = 3t/(1+t^3)\mathbf{i} + 3t^2/(1+t^3)\mathbf{j}$, $-0.6 \leq t \leq 15$, $t_0 = 1/3$

55. $\mathbf{r}(t) = (e^t + e^{-t})\mathbf{i} + (e^t - e^{-t})\mathbf{j}$, $-3 \leq t \leq 3$, $t_0 = 2$

56. $\mathbf{r}(t) = (t^3 - t)\mathbf{i} + (t^2 - 1)\mathbf{j}$, $-2 \leq t \leq 2$, $t_0 = 1/2$

57. $\mathbf{r}(t) = 2t/(1+t^2)\mathbf{i} + (1-t^2)/(1+t^2)\mathbf{j}$, $-20 \leq t \leq 20$, $t_0 = 1/2$

58. $\mathbf{r}(t) = (t - \sin(t))\mathbf{i} + (1 - \cos(t))\mathbf{j}$, $0 \leq t \leq 2\pi$, $t_0 = \pi/6$

59. Plot the planar curve $\mathbf{r}(t) = 3t/(1+t^3)\mathbf{i} + 3t^2/(1+t^3)\mathbf{j}$, $t > -1/2$. It is an arc of the *Folium of Descartes*. Find all points at which the curve has a tangent line parallel to the line $y = x$. Find all points at which the curve has a horizontal tangent line. Find all points at which the curve has a vertical tangent line.

60. Plot $\mathbf{r}(t) = \cos(3t)\cos(t)\mathbf{i} + \sin(2t)\sin(t)\mathbf{j}$ for $0 < t < \pi/2$. Find the point P_0 where the curve crosses itself, and add the two tangents at P_0 to your plot.

61. Plot $\mathbf{r}(t) = \langle t^2 - 1, t^4 - 1, (t-1)\ln(2+t)\rangle$ for $-6/5 \leq t \leq 6/5$. To your plot, add the tangent lines at the point $(0,0,0)$.

62. The curve $\mathbf{r}(t) = \langle t\cos(t), t\sin(t), t\rangle$, $0 \leq t \leq \pi/6$ has one tangent line that intersects the vertical line $x = 1$, $y = 1$. What is the intersection point? Plot the curve, its tangent line, and the vertical line.

10.3 Tangent Vectors and Arc Length

In Section 10.2, we studied the derivative $\mathbf{r}'$ of a vector-valued function $\mathbf{r}$. Our focus was on the velocity and acceleration of the motion that $\mathbf{r}$ defines. We also learned how to use the derivative to obtain tangent lines to the graph of $\mathbf{r}$. In this section, we will investigate the properties of tangents in greater depth. We will also learn an application to arc length.

Unit Tangent Vectors

To effectively study the geometry of a space curve $\mathcal{C}$, we must use a parameterization $\mathbf{r}$ that provides us with information about the tangent lines of $\mathcal{C}$. This forces us to eliminate certain parameterizations from our discussion. As an example of what we must avoid, consider the vector-valued function $\mathbf{r}(t) = \langle\cos(t^3), \sin(t^3), 0\rangle$, which parameterizes the circle $\mathcal{C} = \{(x, y, 0) : x^2 + y^2 = 1\}$. Notice that the derivative $\mathbf{r}'(t) = \langle -3t^2\sin(t^3), 3t^2\cos(t^3), 0\rangle$ vanishes when $t = 0$. Because $\mathbf{r}'(0) = \mathbf{0}$, we obtain no information about the direction of the tangent line to $\mathcal{C}$ at the point $\mathbf{r}(0)$. Our next definition excludes such inconvenient parameterizations.

DEFINITION Suppose that $\mathbf{r}(t) = r_1(t)\mathbf{i} + r_2(t)\mathbf{j} + r_3(t)\mathbf{k}$ is continuous for $a \leq t \leq b$. We say that $\mathbf{r}$ is a *smooth* parameterization of the curve it defines if

a. the scalar-valued functions r_1, r_2, r_3 are all twice continuously differentiable on (a, b), and
b. $\mathbf{r}'(t) \neq \mathbf{0}$ for every t in (a, b).

If we can divide the interval $[a, b]$ into finitely many subintervals $[a, x_1]$, $[x_1, x_2]$, ..., $[x_{N-1}, b]$ such that the restriction of $\mathbf{r}$ to each subinterval is smooth, then we say that $\mathbf{r}$ is *piecewise smooth.*

Until Chapter 13, we will restrict our attention to smooth parameterizations. A moving particle whose position vector $\mathbf{r}$ is a smooth parameterization of its path has, at each instant of time, a differentiable velocity vector $\mathbf{r}'(t)$, a positive speed $v(t) = \|\mathbf{r}'(t)\|$, and a continuous acceleration vector $\mathbf{a}(t) = \mathbf{r}''(t)$. For such a parameterization $\mathbf{r}$ of a curve $\mathcal{C}$, we define the *unit tangent vector* to $\mathcal{C}$ at the point $P = \mathbf{r}(t)$ to be the vector

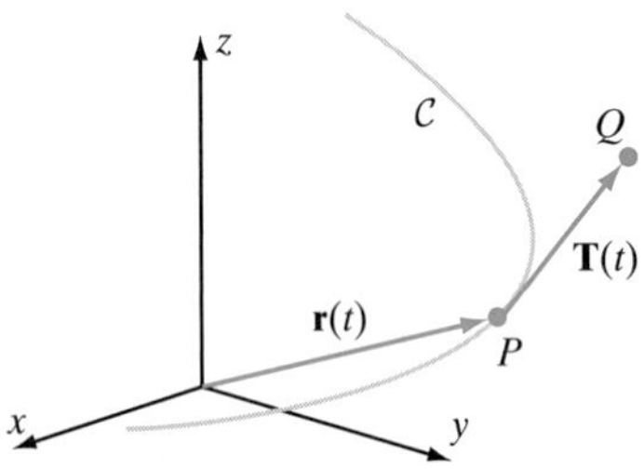

▲ Figure 1

$$\mathbf{T}(t) = \left(\frac{1}{\|\mathbf{r}'(t)\|}\right)\mathbf{r}'(t). \tag{10.3.1}$$

As its name implies, $\mathbf{T}(t)$ is a unit vector. Furthermore, if $\mathbf{T}(t)$ is represented by a directed line segment $\overrightarrow{PQ}$ with initial point P, then $\overrightarrow{PQ}$ is tangent to $\mathcal{C}$ at P (see Figure 1). Notice that a particle with position vector $\mathbf{r}(t)$ traces $\mathcal{C}$ in the direction indicated by $\mathbf{T}(t)$. Finally observe that, by rewriting equation (10.3.1), we may express the velocity vector $\mathbf{r}'(t)$ as the scalar product of the unit tangent vector with the speed:

$$\mathbf{r}'(t) = \|\mathbf{r}'(t)\|\mathbf{T}(t) = v(t)\mathbf{T}(t). \tag{10.3.2}$$

▶ **EXAMPLE 1** Let $\mathbf{r}(t) = e^t\mathbf{i} + e^{-t}\mathbf{j} + \sqrt{2}t\,\mathbf{k}$. Calculate the unit tangent vector $\mathbf{T}(t)$. What are the unit tangent vectors at $P_0 = \mathbf{r}(0)$ and $P_1 = \mathbf{r}(\ln(2))$?

Solution Differentiating each component of $\mathbf{r}(t)$, we see that $\mathbf{r}'(t) = e^t\mathbf{i} - e^{-t}\mathbf{j} + \sqrt{2}\mathbf{k}$ is a tangent vector to the curve defined by $\mathbf{r}$ at the point $\mathbf{r}(t)$. The length of $\mathbf{r}'(t)$ is given by

$$\|\mathbf{r}'(t)\| = \sqrt{(e^t)^2 + (-e^{-t})^2 + \left(\sqrt{2}\right)^2} = \sqrt{e^{2t} + e^{-2t} + 2} = \sqrt{(e^t + e^{-t})^2} = e^t + e^{-t}.$$

The unit tangent vector $\mathbf{T}(t)$ is therefore

$$\mathbf{T}(t) = \frac{1}{e^t + e^{-t}}\left(e^t\mathbf{i} - e^{-t}\mathbf{j} + \sqrt{2}\mathbf{k}\right) = \frac{e^t}{e^t + e^{-t}}\mathbf{i} - \frac{e^{-t}}{e^t + e^{-t}}\mathbf{j} + \frac{\sqrt{2}}{e^t + e^{-t}}\mathbf{k}.$$

In particular, we have $\mathbf{T}(0) = (1/2)\mathbf{i} - (1/2)\mathbf{j} + (\sqrt{2}/2)\mathbf{k}$ and

$$\mathbf{T}(\ln(2)) = \frac{2}{2+1/2}\mathbf{i} - \frac{1/2}{2+1/2}\mathbf{j} + \frac{\sqrt{2}}{2+1/2}\mathbf{k} = \frac{4}{5}\mathbf{i} - \frac{1}{5}\mathbf{j} + \frac{2\sqrt{2}}{5}\mathbf{k}.$$

These two unit tangent vectors are plotted in Figure 2. ◄

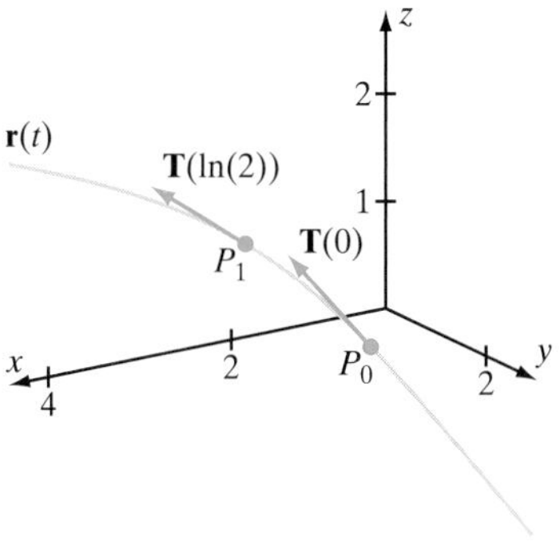

▲ Figure 2

► **EXAMPLE 2** Let $\mathbf{r}(t) = -t\mathbf{i} + t^2\mathbf{j} + (t^3 - 5t)\mathbf{k}$. The point $P_0 = (-1, 1, -4)$ corresponds to $t = 1$ and lies on this curve. Using the unit tangent vector at this point, write symmetric equations for the tangent line to the curve at P_0.

Solution The velocity vector at $t = 1$ is

$$\mathbf{r}'(1) = (-\mathbf{i} + 2t\mathbf{j} + (3t^2 - 5)\mathbf{k})\big|_{t=1} = -\mathbf{i} + 2\mathbf{j} - 2\mathbf{k},$$

and the speed is $\|\mathbf{r}'(1)\| = \sqrt{(-1)^2 + 2^2 + (-2)^2} = 3$. The unit tangent vector is therefore

$$\mathbf{T}(1) = \frac{1}{\|\mathbf{r}'(1)\|}\mathbf{r}'(1) = \frac{1}{3}(-\mathbf{i} + 2\mathbf{j} - 2\mathbf{k}) = -\frac{1}{3}\mathbf{i} + \frac{2}{3}\mathbf{j} - \frac{2}{3}\mathbf{k}.$$

The tangent line we seek is

$$\frac{x - (-1)}{(-1/3)} = \frac{y - 1}{(2/3)} = \frac{z - (-4)}{(-2/3)},$$

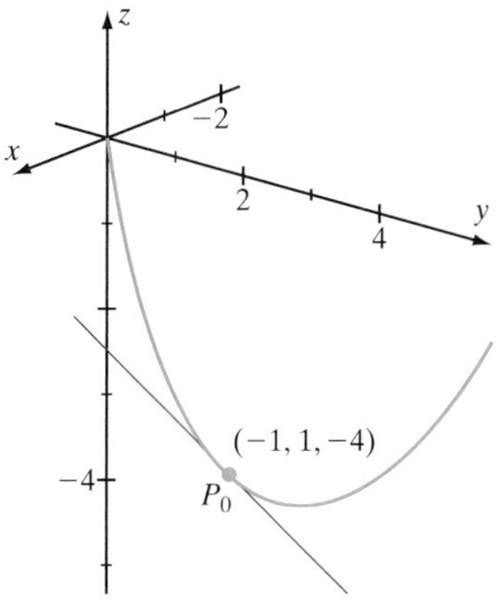

▲ Figure 3

or $2(x + 1) = 1 - y = z + 4$. Figure 3 shows the plot of $\mathbf{r}(t)$ for $0 \le t \le 2$. The tangent line at P_0 has also been plotted. ◄

Arc Length

If a curve $\mathcal{C}$ is defined by a smooth parameterization $\mathbf{r}(t)$ for $a \le t \le b$, then we may ask, What is the *length* of $\mathcal{C}$ from $\mathbf{r}(a)$ to $\mathbf{r}(b)$? An intuitive way to think about length is this: Lay a piece of string along the curve from $\mathbf{r}(a)$ to $\mathbf{r}(b)$. Mark the string where it touches $\mathbf{r}(a)$ and $\mathbf{r}(b)$, straighten the string out, and measure the distance between the marks. That is the length. To obtain a method for actually calculating length from the function $\mathbf{r}$, we follow the scheme that we used in Section 7.2 of Chapter 7 to compute the lengths of planar curves.

We divide the interval $[a, b]$ into small subintervals by a uniform partition

$$a = t_0 < t_1 < \cdots < t_N = b.$$

Let $\Delta t = (b - a)/N$ be the common width of the subintervals of this partition. Set $P_j = \mathbf{r}(t_j) = (x_j, y_j, z_j)$. The basic idea is that the length of the curve $\mathbf{r}$ is

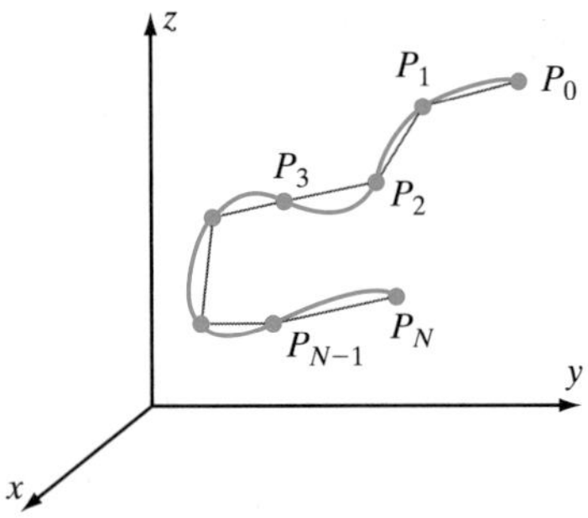

▲ Figure 4

approximated by the sum of the lengths of the directed line segments $\overrightarrow{P_0P_1}$, $\overrightarrow{P_1P_2}$, $\overrightarrow{P_2P_3}, \ldots, \overrightarrow{P_{N-1}P_N}$ (see Figure 4). Thus the length of the curve is approximately given by

$$S_N = \sum_{j=1}^{N} \|\overrightarrow{P_{j-1}P_j}\| = \sum_{j=1}^{N} \sqrt{(x_j - x_{j-1})^2 + (y_j - y_{j-1})^2 + (z_j - z_{j-1})^2}.$$

The Mean Value Theorem tells us that

$$\begin{aligned} x_j - x_{j-1} &= r_1(t_j) - r_1(t_{j-1}) = r_1{}'(\rho_j)\Delta t, \\ y_j - y_{j-1} &= r_2(t_j) - r_2(t_{j-1}) = r_2{}'(\sigma_j)\Delta t, \\ z_j - z_{j-1} &= r_3(t_j) - r_3(t_{j-1}) = r_3{}'(\tau_j)\Delta t, \end{aligned}$$

for some numbers ρ_j, σ_j, τ_j in the subinterval (t_{j-1}, t_j). Then we can rewrite our formula for S_N as

$$S_N = \sum_{j=1}^{N} \sqrt{r_1'(\rho_j)^2 + r_2'(\sigma_j)^2 + r_3'(\tau_j)^2}\,\Delta t.$$

On the one hand, this sum approximates the length of the curve. On the other hand, S_N approaches the Riemann integral $\int_a^b \|\mathbf{r}'(t)\|\, dt$ as N tends to infinity. We are led to the following definition.

DEFINITION Let $\mathcal{C}$ be a curve with smooth parameterization $t \mapsto \mathbf{r}(t)$, $a \le t \le b$. The *arc length* of $\mathcal{C}$ is $\int_a^b \|\mathbf{r}'(t)\|\, dt$.

By restricting our attention to a smooth parameterization $\mathbf{r}$, which has the property that $\mathbf{r}'$ is continuous, we ensure that the function $t \mapsto \|\mathbf{r}'(t)\|$ is continuous. This property in turn assures the existence of the Riemann integral $\int_a^b \|\mathbf{r}'(t)\|\, dt$. It should be noted that the formula for arc length can be used to find the arc length between *any* two points of a curve. If $P_1 = \mathbf{r}(t_1)$ and $P_2 = \mathbf{r}(t_2)$ with $t_1 < t_2$, then $\int_{t_1}^{t_2} \|\mathbf{r}'(t)\|\, dt$ is the arc length between P_1 and P_2.

▶ **EXAMPLE 3** Let $\mathbf{r}(t) = \sin(2t)\mathbf{i} - \cos(2t)\mathbf{j} + \sqrt{5}t\mathbf{k}$. Calculate the arc length of that portion of the curve between $(0, -1, 0)$ and $(1, 0, \sqrt{5}\pi/4)$.

Solution Differentiating $\mathbf{r}(t)$, we obtain $\mathbf{r}'(t) = 2\cos(2t)\mathbf{i} + 2\sin(2t)\mathbf{j} + \sqrt{5}\mathbf{k}$. The point $(0, -1, 0)$ corresponds to $t = 0$, and the point $(1, 0, \sqrt{5}\pi/4)$ corresponds to $t = \pi/4$. The arc length of the portion of the curve between the point $(0, -1, 0)$ and $(1, 0, \sqrt{5}\pi/4)$ is therefore

$$\begin{aligned} \int_0^{\pi/4} \|\mathbf{r}'(t)\|\, dt &= \int_0^{\pi/4} \sqrt{\big(2\cos(2t)\big)^2 + \big(2\sin(2t)\big)^2\mathbf{j} + (\sqrt{5})^2 dt} \\ &= \int_0^{\pi/4} \sqrt{4\big(\cos^2(2t) + \sin^2(2t)\big) + 5dt} \\ &= \int_0^{\pi/4} \sqrt{9}dt \\ &= 3\pi/4. \end{aligned}$$ ◀

Reparameterization It is important to bear in mind that many different vector-valued functions can describe the same curve in space. Imagine, for example, a railroad line between the city of New York and Boston. Each time a train travels this track from New York to Boston, it traces the same curve $\mathcal{C}$. Yet, if the train travels at different speeds on different trips, then different parameterizations of $\mathcal{C}$ result.

To make this idea concrete, let us consider a specific example. The vector-valued functions

$$\mathbf{r}(t) = \cos(t)\mathbf{i} + \sin(t)\mathbf{j} + t\mathbf{k}, \quad 0 \le t \le 2\pi$$

and

$$\mathbf{p}(s) = \cos(2s)\mathbf{i} + \sin(2s)\mathbf{j} + 2s\mathbf{k}, \quad 0 \le s \le \pi$$

describe the same curve in space. To understand why notice that, as s runs through the interval $[0, \pi]$, the expression $t = 2s$ runs through the interval $[0, 2\pi]$. Furthermore, for every s in $[0, \pi]$, we have

$$\mathbf{p}(s) = \mathbf{p}\left(\frac{t}{2}\right) = \cos\left(2\left(\frac{t}{2}\right)\right)\mathbf{i} + \sin\left(2\left(\frac{t}{2}\right)\right)\mathbf{j} + 2\left(\frac{t}{2}\right)\mathbf{k} = \cos(t)\mathbf{i} + \sin(t)\mathbf{j} + t\mathbf{k} = \mathbf{r}(t).$$

This equation shows that $\mathbf{p}$ and $\mathbf{r}$ describe the same points in space (albeit at different values of their domains). The functions $\mathbf{r}$ and $\mathbf{p}$ are simply different parameterizations of the same curve. Notice that the invertible function $\psi : [0, \pi] \to [0, 2\pi]$ defined by $\psi(s) = 2s = t$ affects the way we pass from one parameterization to the other: Given a value of s in $[0, \pi]$, we have $\mathbf{p}(s) = \mathbf{r}(\psi(s)) = \mathbf{r}(t)$. Equivalently, given a value of t in $[0, 2\pi]$, we have $\mathbf{r}(t) = \mathbf{p}(\psi^{-1}(t)) = \mathbf{p}(s)$. Finally, observe that ψ is *increasing*. As a result, $\mathbf{r}$ and $\mathbf{p}$ parameterize the curve in the same direction. That is, as t increases from 0 to 2π, and as s increases from 0 to π, the points $\mathbf{r}(t)$ and $\mathbf{p}(s)$ both trace the curve from the initial point $\mathbf{r}(0) = \mathbf{p}(0) = (1, 0, 0)$ to the terminal point $\mathbf{r}(2\pi) = \mathbf{p}(\pi) = (1, 0, 2\pi)$.

In general, passing from one vector-valued function $\mathbf{r}$ to another function $\mathbf{p}$ that describes the same curve $\mathcal{C}$ in the same direction is called *reparameterization*. If the domain of $\mathbf{r}$ is $[a, b]$ and the domain of $\mathbf{p}$ is $[c, d]$, then the reparameterization $\mathbf{p}$ is the composition of $\mathbf{r}$ with a continuously differentiable increasing function $\psi : [c, d] \to [a, b]$. That is,

$$\mathbf{p}(s) = (\mathbf{r} \circ \psi)(s) = \mathbf{r}(\psi(s)), \quad c \le s \le d.$$

▶ **EXAMPLE 4** Suppose that $\mathbf{r}$ is a smooth parameterization of a curve $\mathcal{C}$ with $P = \mathbf{r}(t)$. Suppose also that $\mathbf{p} = \mathbf{r} \circ \psi$ is a reparameterization with $\psi(s) = t$. Show that the tangent line to $\mathcal{C}$ at the point $\mathbf{p}(s) = \mathbf{r}(t) = P$ will be the same whichever parameterization we use.

Solution The parameterization $\mathbf{r}$ yields $\mathbf{r}'(t)$ as a tangent vector to $\mathcal{C}$ at P. The reparameterization $\mathbf{p}$ yields $\mathbf{p}'(s)$ as a tangent vector to $\mathcal{C}$ at P. We relate the two vectors by means of Theorem 6g of Section 10.1:

$$\mathbf{p}'(s) = (\mathbf{r} \circ \psi)'(s) = \psi'(s)\mathbf{r}'(\psi(s)) = \psi'(s)\mathbf{r}'(t). \tag{10.3.3}$$

Equation (10.3.3) tells us that the vectors $\mathbf{p}'(s)$ and $\mathbf{r}'(t)$ are parallel. It follows that the tangent line to C at the point $\mathbf{p}(s) = \mathbf{r}(t) = P$ will be the same whichever parameterization we use. ◀

INSIGHT Suppose that a differentiable, increasing function ψ is used to obtain a reparameterization $\mathbf{p} = \mathbf{r} \circ \psi$. From equation (10.3.3), we see that $\mathbf{p}'(s) = \mathbf{0}$ at any s for which $\psi'(s) = 0$. Thus if $\mathbf{p}$ is a smooth parameterization, then $\psi'(s) > 0$ must hold for all s.

The next theorem shows that the arc length of a curve does not depend on the chosen parameterization of the curve.

THEOREM 1 Suppose that C is a curve with smooth parameterization $t \mapsto \mathbf{r}(t)$, $a \le t \le b$. Let $\psi : [c, d] \to [a, b]$ be a continuously differentiable increasing function, and let $\mathbf{p} = \mathbf{r} \circ \psi$ be a reparameterization of C. Then the arc length of C when computed using the reparameterization $\mathbf{p}$ is equal to the arc length of C when computed using the parameterization $\mathbf{r}$.

Proof. We compute the arc length of C using the reparameterization $\mathbf{p}$. Using equation (10.3.3), we have

$$\int_c^d \|\mathbf{p}'(s)\|\, ds = \int_c^d \|\psi'(s)\mathbf{r}'(\psi(s))\|\, ds = \int_c^d \psi'(s)\|\mathbf{r}'(\psi(s))\|\, ds.$$

Now we make the substitution $\tau = \psi(s)$, $d\tau = \psi'(s)ds$. The last integral then becomes $\int_a^b \|\mathbf{r}'(\tau)\|\, d\tau$, which is the arc length of C as determined by the parameterization $\mathbf{r}$. ■

Parameterizing a Curve by Arc Length

Suppose that the vector-valued function $\mathbf{r}$ defined on the interval $[a, b]$ is a smooth parameterization of C. Let $L = \int_a^b \|\mathbf{r}'(\tau)\|\, d\tau$ denote the (total) arc length of C. The *arc length function* σ of $\mathbf{r}$ is the increasing function from $[a, b]$ to $[0, L]$, which is defined by

$$\sigma(t) = \int_a^t \|\mathbf{r}'(\tau)\|\, d\tau, \quad a \le t \le b. \tag{10.3.4}$$

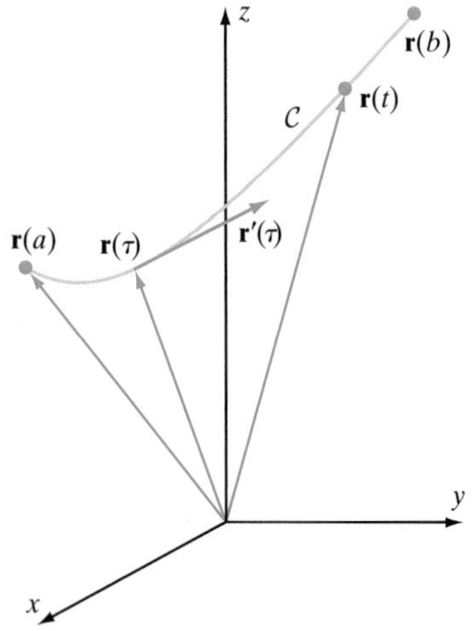

▲ Figure 5 $\sigma(t) = \int_a^t \|\mathbf{r}'(\tau)\| d\tau$ is the arc length of C from $\mathbf{r}(a)$ to $\mathbf{r}(t)$.

For every t in the domain of the parameterization, $\sigma(t)$ measures the distance along C from the initial point $\mathbf{r}(a)$ to the point $\mathbf{r}(t)$ (see Figure 5). When we apply the Fundamental Theorem of Calculus to equation (10.3.4), we obtain

$$\sigma'(t) = \|\mathbf{r}'(t)\|. \tag{10.3.5}$$

If we think of a particle with position vector $\mathbf{r}(t)$ tracing the curve, then equation (10.3.5) has a satisfying interpretation: The instantaneous rate of change of distance along the curve with respect to time (that is, $\sigma'(t)$) is equal to the speed of the particle (that is, $\|\mathbf{r}'(t)\|$).

For some types of problems, it is convenient to reparameterize C by arc length. We will use the arc length function σ of $\mathbf{r}$ to accomplish this task. As a consequence of our assumption that C is smooth, we have $\mathbf{r}'(t) \neq \mathbf{0}$ for all t. From equation (10.3.5), we deduce that $\sigma'(t) > 0$. This inequality implies that σ is a strictly increasing function. Therefore σ has an inverse function σ^{-1}, which is also increasing. We let $\mathbf{p}$ be the reparameterization of C defined by $\mathbf{p}(s) = r(\sigma^{-1}(s))$ for $0 \leq s \leq L$, where L is the arc length of C. Using Theorem 1 with $\psi = \sigma^{-1}$ and $[c, d] = [0, s]$, we see that the arc length along C between $\mathbf{p}(0)$ and $\mathbf{p}(s)$ is

$$\int_0^s \|\mathbf{p}'(u)\|\,du = \int_a^{\sigma^{-1}(s)} \|\mathbf{r}'(\tau)\|\,d\tau = \sigma(\sigma^{-1}(s)) = s. \qquad (10.3.6)$$

This means that the point $\mathbf{p}(s)$ is situated s units of arc length along C from its initial point—note the equality of the parameter s of $\mathbf{p}(s)$ and the distance. Clearly, this property of the reparameterization $\mathbf{p}$ is very special. It is the basis for the following definition.

DEFINITION Let L be the length of a curve C. A parameterization $\mathbf{p}(s)$, $0 \leq s \leq L$, of C is called the *arc length parameterization* of C if the arc length between $\mathbf{p}(0)$ and $\mathbf{p}(s)$ is equal to s for every s in the interval $[0, L]$. We also say that $\mathbf{p}$ parameterizes C with respect to arc length.

The next theorem tells us how to recognize an arc length parameterization of a curve without calculating arc length.

THEOREM 2 Let $s \mapsto \mathbf{p}(s)$, $0 \leq s \leq L$ be the arc length parameterization of a curve C. Then, $\|\mathbf{p}'(s)\| = 1$ for all s. Moreover, the velocity vector $\mathbf{p}'(s)$ is equal to the unit tangent vector $\mathbf{T}(s)$ for every s:

$$\mathbf{T}(s) = \mathbf{p}'(s). \qquad (10.3.7)$$

Conversely, if $t \mapsto \mathbf{r}(t)$, $0 \leq t \leq b$ is a smooth parameterization of a curve C such that $\|\mathbf{r}'(t)\| = 1$ for all t, then $\mathbf{r}$ is the arc length parameterization of C, and b is the length of C.

Proof. Suppose that $s \mapsto \mathbf{p}(s)$, $0 \leq s \leq L$ is the arc length parameterization of C. The identity $\int_0^s \|\mathbf{p}'(u)\|\,du = s$ then holds by definition. If we differentiate each side of this equation with respect to s, then, according to the Fundamental Theorem of Calculus, we have

$$\|\mathbf{p}'(s)\| = \frac{d}{ds}\int_0^s \|\mathbf{p}'(u)\|\,du = \frac{d}{ds}s = 1,$$

which is the first assertion of Theorem 2. Applying formula (10.3.1) with parameterization $\mathbf{p}$ and substituting $\|\mathbf{p}'(s)\| = 1$, we obtain

$$\mathbf{T}(s) = \left(\frac{1}{\|\mathbf{p}'(s)\|}\right)\mathbf{p}'(s) = \mathbf{p}'(s),$$

which is formula (10.3.7).

Conversely, if $t \mapsto \mathbf{r}(t), 0 \le t \le b$, such that $\|\mathbf{r}'(t)\| = 1$ for all t, then the arc length along $\mathcal{C}$ between $\mathbf{r}(0)$ and $\mathbf{r}(t)$ is $\int_0^t \|\mathbf{r}'(\tau)\| d\tau = \int_0^t 1\, d\tau = t$. This is precisely the requirement for $\mathbf{r}$ to be the arc length parametrization of $\mathcal{C}$. Finally, we see that b is the arc length of $\mathcal{C}$ because $b = \int_0^b \|\mathbf{r}'(\tau)\| d\tau$, and the right side of this equation is the arc length. ■

INSIGHT Suppose that we observe the path $\mathcal{C}$ of a moving particle, beginning at time $s = 0$. Theorem 2 tells us that the particle's position vector is the arc length parameterization of $\mathcal{C}$ if and only if the particle moves with constant speed 1.

▶ **EXAMPLE 5** Are $\mathbf{r}(t) = \cos(2t)\mathbf{i} - \sin(2t)\mathbf{k}$, $0 \le t \le \pi$ and $\mathbf{p}(u) = \cos(u)\mathbf{i} - \sin(u)\mathbf{k}$, $0 \le u \le 2\pi$ arc length parameterizations of the curves that they define?

Solution The function $\mathbf{r}$ is *not* an arc length parameterization, for $\mathbf{r}'(t) = -2\sin(2t)\mathbf{i} - 2\cos(2t)\mathbf{k}$ and $\|\mathbf{r}'(t)\| = 2 \ne 1$. On the other hand, $\mathbf{p}'(u) = -\sin(u)\mathbf{i} - \cos(u)\mathbf{k}$ so that $\|\mathbf{p}'(u)\| = 1$ for all u. Therefore the function $\mathbf{p}(u) = \cos(u)\mathbf{i} - \sin(u)\mathbf{k}$ *is* an arc length parameterization. Notice that the functions $\mathbf{r}$ and $\mathbf{p}$ describe the same set of points in space. Indeed, we obtain $\mathbf{p}$ from $\mathbf{r}$ by composing with the increasing function $\psi(u) = u/2$: $\mathbf{p}(u) = (\mathbf{r} \circ \psi)(u) = \mathbf{r}(\psi(u)) = \mathbf{r}(u/2)$. ◀

▶ **EXAMPLE 6** Reparameterize the curve $\mathbf{r}(t) = \langle \cos(t), \sin(t), t \rangle$, $0 \le t \le 2\pi$ so that it is parameterized with respect to arc length.

Solution We calculate $\mathbf{r}'(t) = \langle -\sin(t), \cos(t), 1 \rangle$ and $\|\mathbf{r}'(t)\| = \sqrt{\sin^2(t) + \cos^2(t) + 1^2} = \sqrt{2}$. Therefore the arc length function of $\mathbf{r}$ is given by $s = \sigma(t) = \int_0^t \|\mathbf{r}'(\tau)\| d\tau = \int_0^t \sqrt{2}\, d\tau = \sqrt{2}t$. Then $t = s/\sqrt{2} = \sigma^{-1}(s)$ and

$$\mathbf{p}(s) = \mathbf{r}(\sigma^{-1}(s)) = \mathbf{r}(s/\sqrt{2}) = \left\langle \cos\left(\frac{s}{\sqrt{2}}\right), \sin\left(\frac{s}{\sqrt{2}}\right), \frac{s}{\sqrt{2}} \right\rangle, \quad 0 \le s \le 2\sqrt{2}\pi$$

is the required parameterization by arc length. ◀

Unit Normal Vectors

Suppose that $t \mapsto \mathbf{r}(t)$ is a smooth parameterization of a curve $\mathcal{C}$. Fix a point $P = \mathbf{r}(t)$ on $\mathcal{C}$. Figure 6 shows the plane through point P that is perpendicular to the tangent vector to $\mathcal{C}$ at P. It is called the *normal plane* of $\mathcal{C}$ at P. Although every vector in the normal plane is perpendicular to $\mathbf{T}(t)$, we will see that two of these normal vectors are of particular interest. Our starting point is the identity $\|\mathbf{T}(t)\|^2 = 1$, which we may write as $\mathbf{T}(t) \cdot \mathbf{T}(t) = 1$. If we differentiate both sides of this last equation with respect to t, then we obtain $\mathbf{T}'(t) \cdot \mathbf{T}(t) + \mathbf{T}(t) \cdot \mathbf{T}'(t) = 0$ or $2\mathbf{T}(t) \cdot \mathbf{T}'(t) = 0$. The resulting equation,

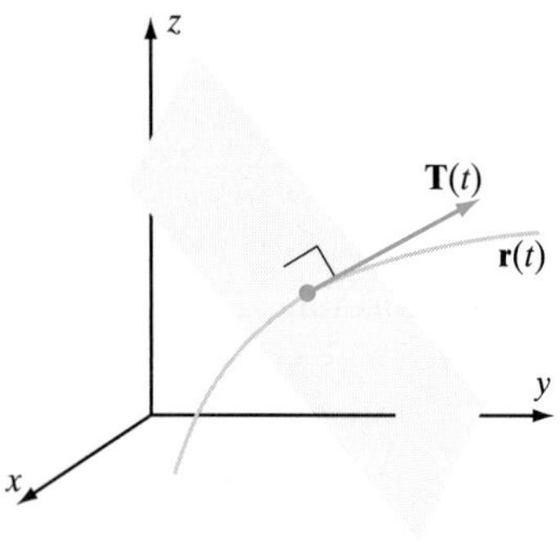

▲ Figure 6

$$\mathbf{T}(t) \cdot \mathbf{T}'(t) = 0, \tag{10.3.8}$$

tells us that the vector $\mathbf{T}'(t)$ *is perpendicular to* the unit tangent vector $\mathbf{T}(t)$. Therefore $\mathbf{T}'(t)$ lies in the normal plane of $\mathcal{C}$ at P. Although the vector $\mathbf{T}(t)$ has unit length, its derivative $\mathbf{T}'(t)$ generally does not have this property. If $\mathbf{T}'(t) \ne \mathbf{0}$, then we obtain a unit vector by scalar multiplying $\mathbf{T}'(t)$ by $1/\|\mathbf{T}'(t)\|$.

DEFINITION If $t \mapsto \mathbf{r}(t)$ is a smooth parameterization of a curve $\mathcal{C}$ with $\mathbf{T}'(t) \neq \mathbf{0}$, then the vector

$$\mathbf{N}(t) = \frac{1}{\|\mathbf{T}'(t)\|}\mathbf{T}'(t) \tag{10.3.9}$$

is called the *principal unit normal* to $\mathcal{C}$ at $\mathbf{r}(t)$.

INSIGHT If $s \mapsto \mathbf{p}(s)$ is the arc length parameterization of $\mathcal{C}$, then we can use equation (10.3.7) to write the principal unit normal as

$$\mathbf{N}(s) = \frac{1}{\|\mathbf{p}''(s)\|}\mathbf{p}''(s). \tag{10.3.10}$$

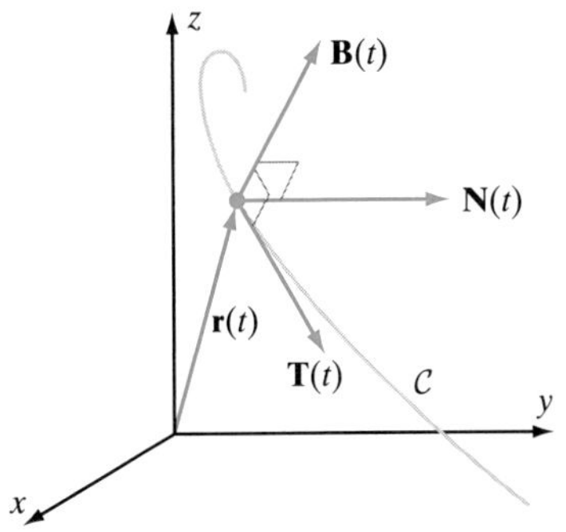

▲ Figure 7 $\mathbf{B}(t) = \mathbf{T}(t) \times \mathbf{N}(t)$

The cross product

$$\mathbf{B}(t) = \mathbf{T}(t) \times \mathbf{N}(t) \tag{10.3.11}$$

of the unit tangent vector and the principal unit normal vector is called the *binormal* vector to $\mathcal{C}$ at the point $\mathbf{r}(t)$. Our work in Section 9.4 shows that the binormal vector is a unit vector that is perpendicular to both $\mathbf{T}(t)$ and $\mathbf{N}(t)$ (see Figure 7).

INSIGHT Imagine a particle with position vector $\mathbf{p}(s)$ tracing a curve $\mathcal{C}$ with unit speed. Formula (10.3.10) tells us that $\mathbf{N}(s)$ is the direction of the particle's acceleration vector, $\mathbf{p}''(s)$. Therefore $\mathbf{N}(s)$ *points to the side to which* $\mathcal{C}$ *curves*. (Refer again to Figure 7.) This observation will be important in the next section.

▶ **EXAMPLE 7** Let $\mathcal{C}$ be the curve defined by $\mathbf{r}(t) = t\mathbf{i} + (t^3/3)\mathbf{j} + (1 - t^2)/\sqrt{2}\mathbf{k}$. Calculate the principal unit normal and the binormal vector of $\mathcal{C}$ at the point $P = (1, 1/3, 0)$.

Solution Observe that $\mathbf{r}'(t) = \mathbf{i} + t^2\mathbf{j} - \sqrt{2}t\mathbf{k}$ and

$$\|\mathbf{r}'(t)\| = \sqrt{1^2 + (t^2)^2 + \left(-\sqrt{2}t\right)^2} = \sqrt{1 + 2t^2 + t^4} = 1 + t^2.$$

It follows that

$$\mathbf{T}(t) = \frac{1}{1+t^2}\mathbf{i} + \frac{t^2}{1+t^2}\mathbf{j} - \frac{\sqrt{2}t}{1+t^2}\mathbf{k}$$

and

$$\mathbf{T}'(t) = \left(-\frac{2t}{(1+t^2)^2}\right)\mathbf{i} + \left(\frac{2t}{(1+t^2)^2}\right)\mathbf{j} + \sqrt{2}\left(\frac{t^2-1}{(1+t^2)^2}\right)\mathbf{k}.$$

The point P corresponds to $t = 1$, and for this value of t we have

$$\mathbf{T}(1) = \frac{1}{2}\mathbf{i} + \frac{1}{2}\mathbf{j} - \frac{\sqrt{2}}{2}\mathbf{k}$$

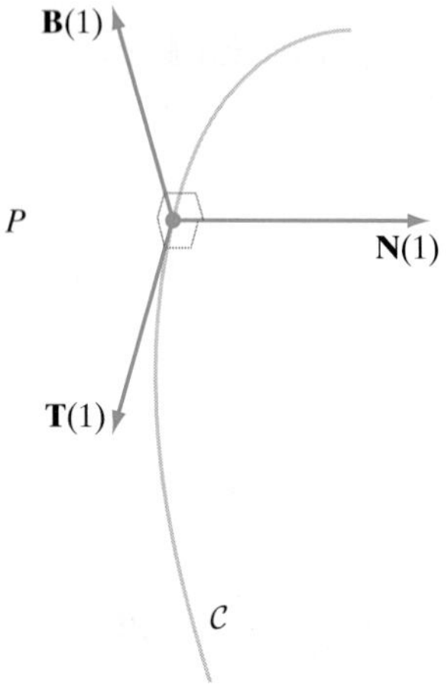

▲ Figure 8 $\mathbf{T}(1)$, $\mathbf{N}(1)$, and $\mathbf{B}(1)$ are mutually perpendicular.

and $\mathbf{T}'(1) = (-1/2)\mathbf{i} + (1/2)\mathbf{j} + 0\mathbf{k} = (1/2)(-\mathbf{i}+\mathbf{j})$. The required principal unit normal vector $\mathbf{N}(1)$ is the direction of $\mathbf{T}'(1)$, which is the direction of the vector $-\mathbf{i}+\mathbf{j}$. It is quickly calculated to be

$$\mathbf{N}(1) = -\frac{1}{\sqrt{2}}\mathbf{i} + \frac{1}{\sqrt{2}}\mathbf{j}.$$

Therefore

$$\mathbf{B}(1) = \mathbf{T}(1) \times \mathbf{N}(1) = \left(\frac{1}{2}\mathbf{i} + \frac{1}{2}\mathbf{j} - \frac{\sqrt{2}}{2}\mathbf{k}\right) \times \left(-\frac{1}{\sqrt{2}}\mathbf{i} + \frac{1}{\sqrt{2}}\mathbf{j}\right)$$

$$= \det\left(\begin{bmatrix} \mathbf{i} & \mathbf{j} & \mathbf{k} \\ 1/2 & 1/2 & -\sqrt{2}/2 \\ -1/\sqrt{2} & 1/\sqrt{2} & 0 \end{bmatrix}\right) = \frac{1}{2}\mathbf{i} + \frac{1}{2}\mathbf{j} + \frac{\sqrt{2}}{2}\mathbf{k}.$$

In Figure 8, the vectors $\mathbf{T}(1)$, $\mathbf{N}(1)$, and $\mathbf{B}(1)$ have been added to a plot of $\mathbf{r}(t)$ for $0 \le t \le 2$. ◄

INSIGHT Like the vectors $\mathbf{i}$, $\mathbf{j}$, and $\mathbf{k}$, the vectors $\mathbf{T}(t)$, $\mathbf{N}(t)$, and $\mathbf{B}(t)$ are orthogonal and have length 1. Like $\mathbf{i}$, $\mathbf{j}$, and $\mathbf{k}$, the vectors $\mathbf{T}(t)$, $\mathbf{N}(t)$, and $\mathbf{B}(t)$ form a *right-handed system*. Unlike $\mathbf{i}$, $\mathbf{j}$, and $\mathbf{k}$, the vectors $\mathbf{T}(t)$, $\mathbf{N}(t)$, and $\mathbf{B}(t)$ are not fixed directions; they change as $\mathbf{r}(t)$ traverses the curve $\mathcal{C}$. The triple $(\mathbf{T}, \mathbf{N}, \mathbf{B})$ is said to be a *moving frame*. It is particularly useful for studying the geometry of space curves and the trajectories of moving bodies.

QUICK QUIZ

1. For $\mathbf{r}(t) = t^3\mathbf{i} + t^2\mathbf{j} + t^6\mathbf{k}$, what is the unit tangent vector $\mathbf{T}(1)$?
2. What integral represents the arc length of the plane curve parametrized by $\mathbf{r}(t) = t\mathbf{i} + e^t\mathbf{j} + \frac{1}{2}e^{2t}\mathbf{k}$, $0 \le t \le 1$?
3. If $\mathbf{r}$ is an arc length parameterization of a curve of length 6, what is $\|\mathbf{r}'(4)\|$?
4. If, at a point on a curve, $\langle 1/\sqrt{2}, 0, 1/\sqrt{2}\rangle$ is the unit tangent vector, and $\langle -1/\sqrt{2}, 0, 1/\sqrt{2}\rangle$ is the principal unit normal vector, then what is the binormal vector?

Answers

1. $\langle 3/7, 2/7, 6/7\rangle$ 2. $\int_0^1 \sqrt{1+e^{2t}+e^{4t}}\,dt$ 3. 1 4. $\langle 0, -1, 0\rangle$

EXERCISES

Problems for Practice

► In each of Exercises 1–8, a parameterization $\mathbf{r}(t)$ of a curve is given. Calculate the tangent vector $\mathbf{r}'(t)$ at $\mathbf{r}(t)$, the unit tangent vector $\mathbf{T}(t)$ at $\mathbf{r}(t)$, and the tangent line at the given point P_0. ◄

1. $\mathbf{r}(t) = t\mathbf{i} - t^2\mathbf{j} + t^3\mathbf{k}$ $\quad P_0 = (1, -1, 1)$

2. $\mathbf{r}(t) = -\mathbf{j} + t\mathbf{k}$ $\quad P_0 = (0, -1, 5)$

3. $\mathbf{r}(t) = t\mathbf{i} - t^{1/2}\mathbf{k}$ $\quad P_0 = (4, 0, -2)$

4. $\mathbf{r}(t) = e^t\mathbf{i} + e^{2t}\mathbf{j} + e^{3t}\mathbf{k}$ $\quad P_0 = (3, 9, 27)$

5. $\mathbf{r}(t) = (e^t + t^2)\mathbf{i} + 3\mathbf{j}$ $\quad P_0 = (1, 3, 0)$

6. $\mathbf{r}(t) = \ln(t)\mathbf{i} - t\ln(t)\mathbf{j} + t\mathbf{k}$ $\quad P_0 = (1, -e, e)$

7. $\mathbf{r}(t) = \cos(t)\mathbf{i} + \sin(t)\mathbf{j} + t\mathbf{k}$ $\quad P_0 = (0, 1, \pi/2)$

8. $\mathbf{r}(t) = 2\ln(\cos(t))\mathbf{i} + 2\ln(\sin(t))\mathbf{j} + 4t\mathbf{k}$, $P_0 = (-\ln(2), -\ln(2), \pi)$

In each of Exercises 9–14, calculate the arc length of the curve described by the given vector-valued function.

9. $\mathbf{r}(t) = \sin(t^2)\mathbf{i} - \cos(t^2)\mathbf{j} + t^2\mathbf{k}$ $\quad 0 \le t \le \sqrt{\pi}$
10. $\mathbf{r}(t) = e^{4t}\mathbf{i} + 8t\mathbf{j} + 2e^{-4t}\mathbf{k}$ $\quad 0 \le t \le \ln(2)$
11. $\mathbf{r}(t) = \cos^3(t)\mathbf{i} + 3\mathbf{j} - \sin^3(t)\mathbf{k}$ $\quad \pi \le t \le 3\pi$
12. $\mathbf{r}(t) = \sin(3t)\mathbf{i} - \cos(3t)\mathbf{j} + t^{3/2}\mathbf{k}$ $\quad \pi/2 \le t \le \pi$
13. $\mathbf{r}(t) = (1+t)^{3/2}\mathbf{i} + (1-t)^{3/2}\mathbf{j} + t^{3/2}\mathbf{k}$ $\quad 0 \le t \le 1$
14. $\mathbf{r}(t) = \sin(t)\mathbf{i} + \cos(t)\mathbf{j} + \ln(\cos(t))\mathbf{k}$ $\quad 0 \le t \le \pi/4$

In each of Exercises 15–18, determine whether the given function is an arc length parameterization of the curve it defines.

15. $\mathbf{r}(t) = t\mathbf{i} - (t^2/2)\mathbf{j} + (t^3/3)\mathbf{k}$
16. $\mathbf{r}(t) = \cos(t/\sqrt{2})\mathbf{i} + \sin(t/\sqrt{2})\mathbf{j} + (t/\sqrt{2})\mathbf{k}$
17. $\mathbf{r}(t) = \dfrac{3}{5}\cos(t)\mathbf{i} - \dfrac{4}{5}\cos(t)\mathbf{j} + \sin(t)\mathbf{k}$
18. $\mathbf{r}(t) = \cos^2(t)\mathbf{i} + \cos(t)\sin(t)\mathbf{j} + \sin(t)\mathbf{k}$

In each of Exercises 19–22, a parameterization $\mathbf{r}(t)$ of a curve is given. Calculate the arc length between the two specified points:

19. $\mathbf{r}(t) = t^2\mathbf{i} - \ln(t)\mathbf{j} + 2t\mathbf{k}$; $(1, 0, 2)$ and $(e^2, -1, 2e)$
20. $\mathbf{r}(t) = (\sin(t) - t\cos(t))\mathbf{i} + t^2\mathbf{j} - (t\sin(t) + \cos(t))\mathbf{k}$; $\mathbf{r}(0)$ and $\mathbf{r}(2\pi)$
21. $\mathbf{r}(t) = e^t\mathbf{i} + e^{-t}\mathbf{j} + \sqrt{2}t\mathbf{k}$; $(1, 1, 0)$ and $(e, 1/e, \sqrt{2})$
22. $\mathbf{r}(t) = \cos(t)\mathbf{i} + \sin(t)\mathbf{j} - t^{3/2}\mathbf{k}$; $(1, 0, 0)$ and $(0, 1, -(\pi/2)^{3/2})$

In each of Exercises 23–30, a parameterization $\mathbf{r}$ of a curve $\mathcal{C}$ and a point P_0 are specified. Give symmetric equations for the line through P_0 with direction vector equal to the principal unit normal vector to $\mathcal{C}$ at P_0. This line is called the *normal line* to $\mathcal{C}$ at P_0.

23. $\mathbf{r}(t) = t\mathbf{i} - t^2\mathbf{j} + t^3\mathbf{k}$ $\quad P_0 = (1, -1, 1)$
24. $\mathbf{r}(t) = t^2\mathbf{i} + 2t^2\mathbf{j} + 3t^3\mathbf{k}$ $\quad P_0 = (1, 2, -3)$
25. $\mathbf{r}(t) = \cos(t)\mathbf{i} + \sin(t)\mathbf{j} + t\mathbf{k}$ $\quad P_0 = (0, 1, \pi/2)$
26. $\mathbf{r}(t) = t\mathbf{i} + (2/t)\mathbf{j} + t^2\mathbf{k}$ $\quad P_0 = (2, 1, 4)$
27. $\mathbf{r}(t) = e^t\mathbf{i} + e^{-t}\mathbf{j} - t^2\mathbf{k}$ $\quad P_0 = (1, 1, 0)$
28. $\mathbf{r}(t) = e^t\mathbf{i} + te^t\mathbf{j} + (t + e^t)\mathbf{k}$ $\quad P_0 = (1, 0, 1)$
29. $\mathbf{r}(t) = \sqrt{t}\mathbf{i} + (4/t)\mathbf{j} + (4/\sqrt{t})\mathbf{k}$ $\quad P_0 = (2, 1, 2)$
30. $\mathbf{r}(t) = \ln(t)\mathbf{i} + \arctan(t)\mathbf{j} + (1/t)\mathbf{k}$ $\quad P_0 = (0, \pi/4, 1)$

In each of Exercises 31–38, a parameterization $\mathbf{r}(t)$ of a curve is given. Find the Cartesian equation for the normal plane to the curve at the given point P_0.

31. $\mathbf{r}(t) = t\mathbf{i} - t^2\mathbf{j} + t^3\mathbf{k}$ $\quad P_0 = (1, -1, 1)$
32. $\mathbf{r}(t) = t^2\mathbf{i} + 2t^2\mathbf{j} + 3t^3\mathbf{k}$ $\quad P_0 = (1, 2, -3)$
33. $\mathbf{r}(t) = t\mathbf{i} + (2/t)\mathbf{j} + t^2\mathbf{k}$ $\quad P_0 = (2, 1, 4)$
34. $\mathbf{r}(t) = \cos(t)\mathbf{i} + \sin(t)\mathbf{j} + t\mathbf{k}$ $\quad P_0 = (0, 1, \pi/2)$
35. $\mathbf{r}(t) = e^t\mathbf{i} + e^{-t}\mathbf{j} - t^2\mathbf{k}$ $\quad P_0 = (1, 1, 0)$
36. $\mathbf{r}(t) = e^t\mathbf{i} + e^{2t}\mathbf{j} + e^{3t}\mathbf{k}$ $\quad P_0 = (3, 9, 27)$
37. $\mathbf{r}(t) = \sqrt{t}\mathbf{i} + (4/t)\mathbf{j} + (t^2 - 15)\mathbf{k}$ $\quad P_0 = (2, 1, 1)$
38. $\mathbf{r}(t) = \ln(t)\mathbf{i} + \arctan(t)\mathbf{j} + (1/t)\mathbf{k}$ $\quad P_0 = (0, \pi/4, 1)$

Further Theory and Practice

In each of Exercises 39–42, a curve is parameterized by the given function $\mathbf{r}(t)$ for t in the interval $[0, 1]$. Find an explicit formula for the arc length function $\sigma(t)$ that is defined by equation (10.3.4). Calculate the inverse of σ by setting $s = \sigma(t)$ and solving for t as a function of s—in other words, find the function σ^{-1} such that $t = \sigma^{-1}(s)$. Now substitute $t = \sigma^{-1}(s)$ into the given formula for $\mathbf{r}(t)$ to obtain $\mathbf{p}(s) = \mathbf{r}(\sigma^{-1}(s))$. The reparameterized curve will be parameterized according to arc length.

39. $\mathbf{r}(t) = t\mathbf{i} + t\mathbf{j} + t^{3/2}\mathbf{k}$
40. $\mathbf{r}(t) = \cosh(t)\mathbf{i} - \sinh(t)\mathbf{j} + t\mathbf{k}$
41. $\mathbf{r}(t) = e^t\mathbf{i} + e^{-t}\mathbf{j} + \sqrt{2}t\mathbf{k}$
42. $\mathbf{r}(t) = \cos^3(t)\mathbf{i} + \sin^3(t)\mathbf{k}$

In each of Exercises 43–50, a parameterization $\mathbf{r}(t)$ of a planar curve is given. Find Cartesian equations for the tangent and normal lines to the curve at $\mathbf{r}(t_0)$. (The normal line is defined in the instructions to Exercises 23–30.)

43. $\mathbf{r}(t) = (5 + 2\cos(t))\mathbf{i} + (3 + \sin(t))\mathbf{j}$ $\quad t_0 = \dfrac{\pi}{3}$
44. $\mathbf{r}(t) = (t^2 + t)\mathbf{i} + (t^2 - t)\mathbf{j}$ $\quad t_0 = 1$
45. $\mathbf{r}(t) = \tan(t)\mathbf{i} + \sec(t)\mathbf{j}$ $\quad t_0 = \dfrac{\pi}{4}$
46. $r(t) = \dfrac{3t}{1+t^3}\mathbf{i} + \dfrac{3t^2}{1+t^3}\mathbf{j}$ $\quad t_0 = 1$
47. $\mathbf{r}(t) = (e^t + e^{-t})\mathbf{i} + (e^t - e^{-t})\mathbf{j}$ $\quad t_0 = 0$
48. $\mathbf{r}(t) = (t^3 - t)\mathbf{i} + (t^2 - 1)\mathbf{j}$ $\quad t_0 = \dfrac{1}{2}$
49. $\mathbf{r}(t) = \dfrac{2t}{1+t^2}\mathbf{i} + \dfrac{1-t^2}{1+t^2}\mathbf{j}$ $\quad t_0 = \dfrac{1}{2}$
50. $\mathbf{r}(t) = (t - \sin(t))\mathbf{i} + (1 - \cos(t))\mathbf{j}$ $\quad t_0 = \dfrac{\pi}{2}$

In each of Exercises 51–56, a parameterization $\mathbf{r}(t)$ of a space curve is given. Calculate the moving frame $(\mathbf{T}, \mathbf{N}, \mathbf{B})$ at $\mathbf{r}(t_0)$.

51. $\mathbf{r}(t) = \cos(t)\mathbf{i} + \sin(2t)\mathbf{j} + t\mathbf{k}$ $\quad t_0 = \pi/4$
52. $\mathbf{r}(t) = (t^2 + t)\mathbf{i} + (t^2 - t)\mathbf{j} + t^2\mathbf{k}$ $\quad t_0 = 1$
53. $\mathbf{r}(t) = \tan(t)\mathbf{i} + \sec(t)\mathbf{j} + t\mathbf{k}$ $\quad t_0 = \pi/4$
54. $\mathbf{r}(t) = t\mathbf{i} + t^2\mathbf{j} + t^3\mathbf{k}$ $\quad t_0 = 1$
55. $\mathbf{r}(t) = e^t\mathbf{i} + e^{-t}\mathbf{j} + (e^t + e^{-t})\mathbf{k}$ $\quad t_0 = \ln(2)$
56. $\mathbf{r}(t) = (t^3 - t)\mathbf{i} + (t^2 - 1)\mathbf{j} + t^2\mathbf{k}$ $\quad t_0 = 1$

57. Given any positive number N, describe a way to define a piecewise smooth curve parameterized by $\mathbf{r}(t) = x(t)\mathbf{i} + y(t)\mathbf{j}$ that has length greater than N and components that satisfy $0 \le x(t), y(t) \le 1$.
58. Show that the tangent lines to the helix parameterized by $\mathbf{r}(t) = \langle\cos(t), \sin(t), t/2\rangle$ make a constant angle with $\mathbf{k}$. Show that the normal lines intersect the z-axis.

Calculator/Computer Exercises

In each of Exercises 59–62, use Simpson's Rule to approximate to two decimal places—that is, with an error no greater than 0.005—the arc length of the curve defined by the given vector-valued function **r** over the given interval.

59. $\mathbf{r}(t) = t^2\mathbf{i} - t^4\mathbf{j} + t^{-1}\mathbf{k}$ $\quad 1 \le t \le 4$

60. $\mathbf{r}(t) = t\mathbf{i} - e^{-t}\mathbf{j} + t^5\mathbf{k}$ $\quad 0.4 \le t \le 0.9$

61. $\mathbf{r}(t) = \cos(t)\mathbf{i} + \sin(2t)\mathbf{j} + \sqrt{t}\mathbf{k}$ $\quad 1 \le t \le 4$

62. $\mathbf{r}(t) = \ln(t)\mathbf{i} + (2/t)\mathbf{j} + t^{3/2}\mathbf{k}$ $\quad 1 \le t \le 2$

10.4 Curvature

In this section, we learn what the second derivative $\mathbf{r}''$ of a vector-valued function $\mathbf{r}$ can tell us about the geometry of the curve $\mathcal{C}$ defined by $\mathbf{r}$. To be specific, we will see that $\mathbf{r}''$ can be used to measure the way $\mathcal{C}$ deviates from a straight line. The geometric tool for this study is the concept of *curvature*.

What does it mean to say that a circle is curved while a line is not? One answer is that, as we move along a circle, the unit tangent vector changes direction, while if we move along a line, the unit tangent vector does not change direction. We define the curvature to be the magnitude of the instantaneous rate of change of direction of the unit tangent vector with respect to arc length.

DEFINITION Suppose that $s \mapsto \mathbf{p}(s)$ is a smooth arc length parameterization of a curve $\mathcal{C}$. The quantity

$$\kappa(s) = \left\| \frac{d\mathbf{T}(s)}{ds} \right\| = \left\| \frac{d}{ds}\mathbf{p}'(s) \right\| = \|\mathbf{p}''(s)\| \tag{10.4.1}$$

is called the *curvature* of $\mathcal{C}$ at the point $\mathbf{r}(s)$.

INSIGHT The definition of curvature involves differentiation with respect to arc length. That choice is influenced by geometric considerations, but there is a subtle motivation as well. As we saw in the preceding section, there are always many ways to parameterize a curve. The concept of curvature should depend only on the geometry of a curve and not on the choice of parameterization. By referring to the arc length parameterization, we avoid potential ambiguities that might arise from different choices of parameterization. Theorem 1 of Section 10.3 shows that any two parameterizations of $\mathcal{C}$ will give rise to the same arc length reparameterization.

▶ **EXAMPLE 1** Let $\mathbf{r}(s) = \rho\cos(s/\rho)\mathbf{i} + \rho\sin(s/\rho)\mathbf{j}$ for some positive constant ρ. Calculate the curvature at each value of s.

Solution The curve $\mathcal{C}$ described by $\mathbf{r}$ is a circle in the xy-plane, $z = 0$. The radius of $\mathcal{C}$ is ρ, and the center of $\mathcal{C}$ is $(0, 0, 0)$. Notice that $\mathbf{r}'(s) = (-\sin(s/\rho))\mathbf{i} + (\cos(s/\rho))\mathbf{j}$, so that $\|\mathbf{r}'(s)\| = \sqrt{\sin^2(s/\rho) + \cos^2(s/\rho)} = 1$ for all s. We deduce that $\mathbf{r}(s)$ is an arc length parameterization by Theorem 2 of the preceding section. Finally,

$$\mathbf{r}''(s) = -\frac{1}{\rho}\cos\left(\frac{s}{\rho}\right)\mathbf{i} - \frac{1}{\rho}\sin\left(\frac{s}{\rho}\right)\mathbf{j}$$

and therefore

$$\kappa(s) = \|\mathbf{r}''(s)\| = \sqrt{\left(\frac{-\cos(s/\rho)}{\rho}\right)^2 + \left(\frac{-\sin(s/\rho)}{\rho}\right)^2} = \frac{1}{\rho}\sqrt{\cos^2\left(\frac{s}{\rho}\right) + \sin^2\left(\frac{s}{\rho}\right)} = \frac{1}{\rho}. \quad ◀$$

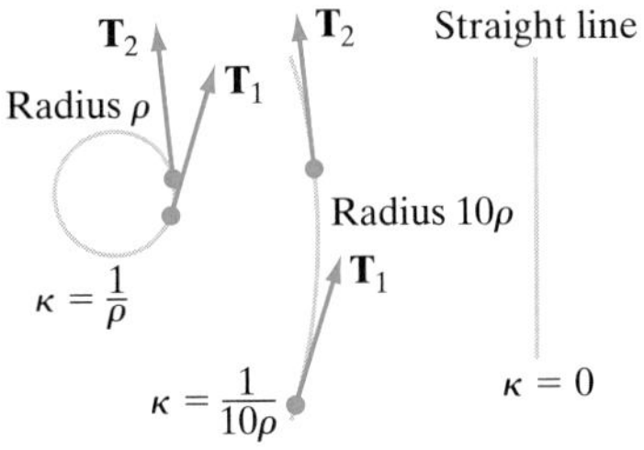

▲ Figure 1

INSIGHT Example 1 shows that our notion of curvature is a sensible one because it says that the curvature of a circle of radius ρ is $1/\rho$ at all points of the circle. In particular, the curvature of a circle of large radius is small (the vector $\mathbf{T}$ changes slowly with respect to arc length), while the curvature of a circle of small radius is large (the vector $\mathbf{T}$ changes rapidly with respect to arc length). See Figure 1, in which circles with radii ρ and 10ρ are plotted. The same pair of tangent vectors, T_1 and T_2, appears on each circle. The change of direction shown on the small circle is the same as that shown on the large circle, yet it is attained with a much smaller change of arc length. The rate of change of direction with respect to arc length is therefore greater on the smaller circle.

Calculating Curvature Without Reparameterizing

As we have discussed, there are good reasons for defining curvature by means of formula (10.4.1). However, this approach does have the drawback of requiring the arc length parameterization of the curve, which is often tedious or difficult to calculate. It is therefore desirable to have an alternative expression for curvature that does not refer to arc length.

THEOREM 1 Suppose that $\mathbf{r}$ is a smooth parameterization of a curve $\mathcal{C}$. Then,

$$\kappa_{\mathbf{r}}(t) = \frac{\|\mathbf{r}'(t) \times \mathbf{r}''(t)\|}{\|\mathbf{r}'(t)\|^3} \tag{10.4.2}$$

is the curvature of $\mathcal{C}$ at the point $\mathbf{r}(t)$. The curvature at $\mathbf{r}(t)$ may also be expressed as

$$\kappa_{\mathbf{r}}(t) = \frac{\sqrt{(\|\mathbf{r}'(t)\| \cdot \|\mathbf{r}''(t)\|)^2 - (\mathbf{r}'(t) \cdot \mathbf{r}''(t))^2}}{\|\mathbf{r}'(t)\|^3}. \tag{10.4.3}$$

Proof. Let $\sigma(t) = \int_0^t \|\mathbf{r}'(\tau)\| d\tau$ denote the arc length function for the parameterization $\mathbf{r}$. Using equation (10.3.5) to express the speed $\|\mathbf{r}'(t)\|$ as $d\sigma(t)/dt$, we may rewrite equation (10.3.2) as

$$\mathbf{r}'(t) = \frac{d\sigma(t)}{dt}\mathbf{T}(t). \tag{10.4.4}$$

When we differentiate each side of equation (10.4.4) with respect to time, we obtain

$$\mathbf{r}''(t) = \frac{d^2\sigma(t)}{dt^2}\mathbf{T}(t) + \frac{d\sigma(t)}{dt}\mathbf{T}'(t). \tag{10.4.5}$$

Substituting the right sides of equations (10.4.4) and (10.4.5) into the expression $\mathbf{r}'(t) \times \mathbf{r}''(t)$ and noting that $\mathbf{T}(t) \times \mathbf{T}(t) = \mathbf{0}$, we obtain

$$\mathbf{r}'(t) \times \mathbf{r}''(t) = \left(\frac{d\sigma(t)}{dt}\right)^2 \mathbf{T}(t) \times \mathbf{T}'(t). \tag{10.4.6}$$

Recall that $\mathbf{T}(t)$ and $\mathbf{T}'(t)$ are perpendicular by equation (10.3.8). Therefore $\|\mathbf{T}(t) \times \mathbf{T}'(t)\| = \|\mathbf{T}(t)\|\,\|\mathbf{T}'(t)\| = \|\mathbf{T}'(t)\|$ by Theorem 3b of Section 9.4. Comparing the magnitudes of each side of equation (10.4.6), we deduce that

$$\|\mathbf{r}'(t) \times \mathbf{r}''(t)\| = \left(\frac{d\sigma(t)}{dt}\right)^2 \|\mathbf{T}'(t)\|,$$

or

$$\|\mathbf{T}'(t)\| = \frac{\|\mathbf{r}'(t) \times \mathbf{r}''(t)\|}{\|\mathbf{r}'(t)\|^2}. \tag{10.4.7}$$

Now let $s = \sigma(t)$ be the arc length of $\mathcal{C}$ between the points $\mathbf{r}(a)$ and $\mathbf{r}(t)$. Our work in Section 10.3 shows that the arc length reparameterization is $\mathbf{p}(s) = \mathbf{r}(\sigma^{-1}(s)) = \mathbf{r}(t)$. Observe that

$$\mathbf{p}'(s) = \mathbf{T}(t) \tag{10.4.8}$$

because each side is the unit tangent at the same point on the curve. As a result, we have

$$\mathbf{T}'(t) = \frac{d}{dt}\mathbf{T}(t) = \frac{ds}{dt}\frac{d}{ds}\mathbf{T}(t) \overset{(10.4.8)}{=} \|\mathbf{r}'(t)\| \frac{d}{ds}\mathbf{p}'(s) = \|\mathbf{r}'(t)\|\mathbf{p}''(s). \tag{10.4.9}$$

Finally, the curvature at the point in question is, by definition (10.4.1),

$$\kappa(s) = \|\mathbf{p}''(s)\| \overset{(10.4.9)}{=} \frac{\|\mathbf{T}'(t)\|}{\|\mathbf{r}'(t)\|} \overset{(10.4.7)}{=} \frac{\|\mathbf{r}'(t) \times \mathbf{r}''(t)\|}{\|\mathbf{r}'(t)\|^3},$$

which is formula (10.4.2) for $\kappa_{\mathbf{r}}(t)$. Equation (10.4.3) can be derived from (10.4.2) by elementary but tedious algebraic manipulations (Exercise 55). ■

INSIGHT Each formula for curvature has advantages and disadvantages. Equation (10.4.1) is simple, natural, and easy to remember. However, it is often useless for computation. Formula (10.4.2) is more complicated than (10.4.1), but it is usually the easier formula to use because it does not require reparameterization. The downside is that formula (10.4.2) does not have any obvious connection with the concept of curvature and certainly seems to be dependent on the choice of parameterization. Formula (10.4.3) shares these drawbacks and has the further disadvantage of being less memorable. The upside is that formula (10.4.3) does not require a cross product calculation.

▶ **EXAMPLE 2** Let $\mathcal{C}$ be the helix parameterized by $\mathbf{r}(t) = \cos(t)\mathbf{i} + \sin(t)\mathbf{j} + t\mathbf{k}$ for $0 \le t \le 4\pi$. Calculate the arc length function σ for $\mathbf{r}$. Use σ to find the arc length reparameterization $\mathbf{p}(s)$ of $\mathcal{C}$. Using formula (10.4.1), show that the curvature is $1/2$ at each point of $\mathcal{C}$ for all s. Verify that formulas (10.4.2) and (10.4.3) both yield this same constant value for the curvature.

Solution Observe that $\mathbf{r}'(t) = -\sin(t)\mathbf{i} + \cos(t)\mathbf{j} + \mathbf{k}$ and $\|\mathbf{r}'(t)\| = \sqrt{(-\sin(t))^2 + (\cos(t))^2 + 1^2} = \sqrt{2}$. It follows that the arc length function σ for the curve is $\sigma(t) = \int_0^t \|\mathbf{r}'(\tau)\|\,d\tau = \int_0^t \sqrt{2}\,d\tau = \sqrt{2}t$. Therefore if $\sigma(t) = s$, then $\sqrt{2}t = s$, or $t = s/\sqrt{2}$. As we learned in Section 10.3, this implies that

$$\mathbf{p}(s) = \mathbf{r}(\sigma^{-1}(s)) = \mathbf{r}\left(\frac{s}{\sqrt{2}}\right) = \cos\left(\frac{s}{\sqrt{2}}\right)\mathbf{i} + \sin\left(\frac{s}{\sqrt{2}}\right)\mathbf{j} + \frac{s}{\sqrt{2}}\mathbf{k}$$

is the arc length parameterization of $\mathcal{C}$. We have

$$\mathbf{p}'(s) = -\frac{1}{\sqrt{2}}\sin\left(\frac{s}{\sqrt{2}}\right)\mathbf{i} + \frac{1}{\sqrt{2}}\cos\left(\frac{s}{\sqrt{2}}\right)\mathbf{j} + \frac{1}{\sqrt{2}}\mathbf{k}$$

and

$$\mathbf{p}''(s) = -\frac{1}{2}\cos\left(\frac{s}{\sqrt{2}}\right)\mathbf{i} - \frac{1}{2}\sin\left(\frac{s}{\sqrt{2}}\right)\mathbf{j}.$$

Therefore

$$\kappa(s) = \|\mathbf{p}''(s)\| = \sqrt{\left(-\frac{1}{2}\cos\left(\frac{s}{\sqrt{2}}\right)\right)^2 + \left(-\frac{1}{2}\sin\left(\frac{s}{\sqrt{2}}\right)\right)^2} = \frac{1}{2}\sqrt{\cos^2\left(\frac{s}{\sqrt{2}}\right) + \sin^2\left(\left(\frac{s}{\sqrt{2}}\right)\right)} = \frac{1}{2}.$$

To use formula (10.4.2), we calculate

$$\mathbf{r}'(t) \times \mathbf{r}''(t) = (-\sin(t)\mathbf{i} + \cos(t)\mathbf{j} + \mathbf{k}) \times (-\cos(t)\mathbf{i} - \sin(t)\mathbf{j}) = \det\left(\begin{bmatrix} \mathbf{i} & \mathbf{j} & \mathbf{k} \\ -\sin(t) & \cos(t) & 1 \\ -\cos(t) & -\sin(t) & 0 \end{bmatrix}\right),$$

which gives us $\mathbf{r}'(t) \times \mathbf{r}''(t) = \sin(t)\mathbf{i} - \cos(t)\mathbf{j} + \mathbf{k}$. Therefore $\|\mathbf{r}'(t) \times \mathbf{r}''(t)\| = \sqrt{\sin^2(t) + (-\cos(t))^2 + 1^2} = \sqrt{2}$. Formula (10.4.2) then tells us that curvature at $\mathbf{r}(t)$ is equal to

$$\frac{\|\mathbf{r}'(t) \times \mathbf{r}''(t)\|}{\|\mathbf{r}'(t)\|^3} = \frac{\sqrt{2}}{(\sqrt{2})^3} = \frac{1}{2},$$

which agrees with the result of our first calculation. Using formula (10.4.3), we find that the curvature at $\mathbf{r}(t)$ is

$$\frac{\sqrt{\left(\|\mathbf{r}'(t)\|\cdot\|\mathbf{r}''(t)\|\right)^2-(\mathbf{r}'(t)\cdot\mathbf{r}''(t))^2}}{\|\mathbf{r}'(t)\|^3}=\frac{\sqrt{(\sqrt{2}\cdot 1)^2-((-\sin(t)\mathbf{i}+\cos(t)\mathbf{j}+\mathbf{k})\cdot(-\cos(t)\mathbf{i}-\sin(t)\mathbf{j}))^2}}{(\sqrt{2})^3}$$

$$=\frac{\sqrt{2-0^2}}{2\sqrt{2}}=\frac{1}{2},$$

which agrees with the previously obtained value. ◀

In Example 2, the curvature calculation is feasible whether or not Theorem 1 is used. The next example better illustrates the power of Theorem 1.

▶ **EXAMPLE 3** Let $\mathcal{C}$ be the curve parameterized by $\mathbf{r}(t)=e^t\mathbf{i}+e^{-t}\mathbf{j}+t\mathbf{k}$. Find the curvature $\kappa_{\mathbf{r}}(t)$ at point $\mathbf{r}(t)$.

Solution We calculate $\mathbf{r}'(t)=e^t\mathbf{i}-e^{-t}\mathbf{j}+\mathbf{k}$. Notice that, because the arc length function for $\mathbf{r}$ is

$$\sigma(t)=\int_0^t\|\mathbf{r}'(\tau)\|\,d\tau=\int_0^t\|e^\tau\mathbf{i}-e^{-\tau}\mathbf{j}+\mathbf{k}\|\,d\tau=\int_0^t(e^{2\tau}+e^{-2\tau}+1)^{1/2}\,d\tau,$$

it is not practical to find the arc length parameterization of $\mathcal{C}$. We appeal to Theorem 1 instead. Because $\mathbf{r}''(t)=e^t\mathbf{i}+e^{-t}\mathbf{j}$, it follows that

$$\mathbf{r}'(t)\times\mathbf{r}''(t)=(e^t\mathbf{i}-e^{-t}\mathbf{j}+\mathbf{k})\times(e^t\mathbf{i}+e^{-t}\mathbf{j})=\det\left(\begin{bmatrix}\mathbf{i}&\mathbf{j}&\mathbf{k}\\ e^t&-e^{-t}&1\\ e^t&e^{-t}&0\end{bmatrix}\right)=-e^{-t}\mathbf{i}+e^t\mathbf{j}+2\mathbf{k}.$$

Therefore the curvature is

$$\kappa_{\mathbf{r}}(t)=\frac{\|\mathbf{r}'(t)\times\mathbf{r}''(t)\|}{\|\mathbf{r}'(t)\|^3}=\frac{\|-e^{-t}\mathbf{i}+e^t\mathbf{j}+2\mathbf{k}\|}{\|e^t\mathbf{i}-e^{-t}\mathbf{j}+\mathbf{k}\|^3}=\frac{\sqrt{e^{2t}+e^{-2t}+4}}{(e^{2t}+e^{-2t}+1)^{3/2}}.$$ ◀

The Osculating Circle

Let $s\mapsto\mathbf{p}(s)$ be a smooth arc length parameterization of a curve $\mathcal{C}$. Recall that formula (10.3.10) expresses the principal unit normal vector as $\mathbf{N}(s)=\|\mathbf{p}''(s)\|^{-1}\mathbf{p}''(s)$. Using formula (10.4.1) for the curvature, we can rewrite this equation for $\mathbf{N}(s)$ in the form

$$\mathbf{p}''(s)=\kappa(s)\mathbf{N}(s). \tag{10.4.10}$$

Equation (10.4.10) states that, if a particle with position vector $\mathbf{p}(s)$ traces $\mathcal{C}$ with unit speed, then the direction of its acceleration vector $\mathbf{p}''(s)$ is the principal unit normal $\mathbf{N}(s)$, and the magnitude of its acceleration is the curvature $\kappa(s)$.

If $\kappa(s)\neq 0$, then we let $\rho(s)=1/\kappa(s)$. By multiplying each side of equation (10.4.10) by $\rho(s)$, we obtain

$$\mathbf{N}(s)=\rho(s)\mathbf{p}''(s). \tag{10.4.11}$$

The quantity $\rho(s)$ is known as the *radius of convergence* of $\mathcal{C}$ at $\mathbf{p}(s)$. The reason for this terminology can be understood from the case in which $\mathcal{C}$ is a circle. We know from Example 1 that a circle of radius ρ has constant curvature $\kappa(s) = 1/\rho$. Thus for a circle of radius ρ, we have $\rho(s) = 1/\kappa(s) = \rho$. In other words, the radius of curvature for a circle is the radius of the circle.

In general, we can approximate $\mathcal{C}$ near the point $\mathbf{p}(s)$ by a circle that is tangent to $\mathbf{T}(s)$ at $\mathbf{p}(s)$ and that has the same curvature, $\kappa(s)$, at $\mathbf{p}(s)$ as $\mathcal{C}$. The radius of this approximating circle is $\rho(s)$, and the center is a distance $\rho(s)$ from $\mathbf{p}(s)$. If $\mathcal{C}$ were actually a circle, then the direction of acceleration, $\mathbf{v}'(s) = \mathbf{p}''(s)$, would point to the center of the circle (Example 5, Section 10.2). Therefore by equation (10.4.11), the vector from $\mathbf{p}(s)$ to the center of our approximating circle should have direction $\mathbf{N}(s)$. Finally, if $\mathcal{C}$ were a circle, then it would lie in a plane; that plane would contain its velocity and acceleration vectors. Therefore our approximating circle should lie in the plane determined by the vectors $\mathbf{T}(s)$ and $\mathbf{N}(s)$. We summarize all of these considerations with a precise definition of the approximating circle.

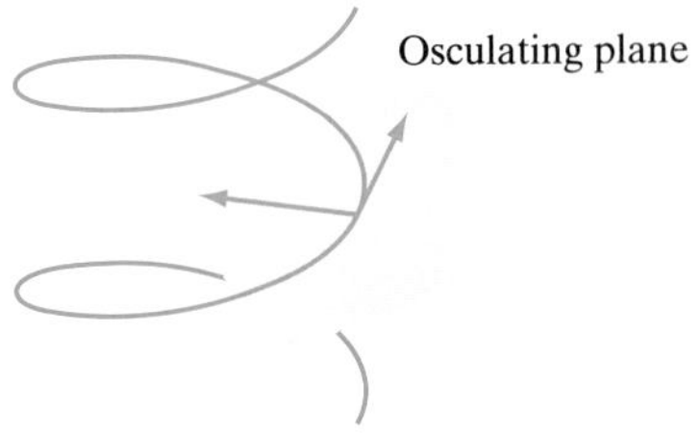

▲ Figure 2a

DEFINITION Let $\mathbf{p}$ be a smooth arc length parameterization of a space curve $\mathcal{C}$. If $\kappa(s) > 0$, then the *osculating circle* (or the *circle of curvature*) of $\mathcal{C}$ at $\mathbf{p}(s)$ is the unique circle satisfying the following conditions:

a. The circle has radius $\rho(s) = 1/\kappa(s)$.
b. The circle has center $\mathbf{p}(s) + \rho(s)\mathbf{N}(s)$. This point is called the *center of curvature* of $\mathcal{C}$ at $\mathbf{p}(s)$.
c. The circle lies in the plane determined by the vectors $\mathbf{N}(s)$ and $\mathbf{T}(s)$. This plane is called the *osculating plane* of $\mathcal{C}$ at $\mathbf{p}(s)$.

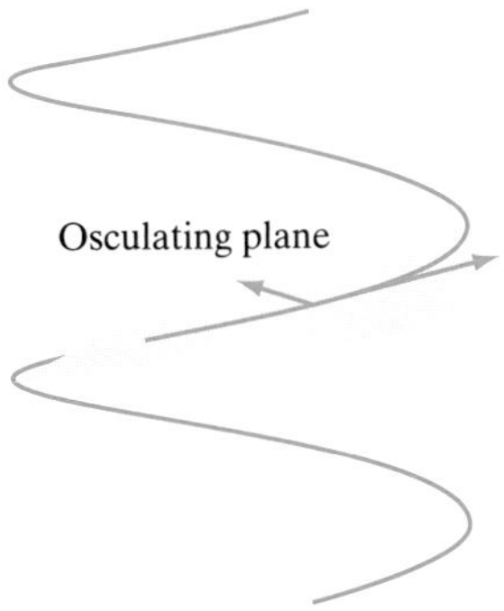

▲ Figure 2b

INSIGHT By construction, the osculating circle has the same principal unit normal vector and tangent vector as the curve $\mathcal{C}$ at the point of contact $\mathbf{p}(s)$. It therefore has the same binormal vector as well. The osculating circle and the curve $\mathcal{C}$ just touch at the point of contact $\mathbf{p}(s)$ because they share the same tangent line. That is the reason for the name: "Osculating" is a synonym for "kissing." An osculating plane for a helix is shown in Figure 2a. The unit tangent vector and the principal unit normal vector are also graphed. A different view of the same curve and osculating plane appears in Figure 2b. It shows how well the osculating plane "fits" the curve near the point of contact. The osculating circle is added in Figure 3.

Although we have used the arc length parameterization to simplify the discussion, it should be noted that all of the ingredients needed for the osculating plane and osculating circle can be calculated without reparameterization. The next example will illustrate this point.

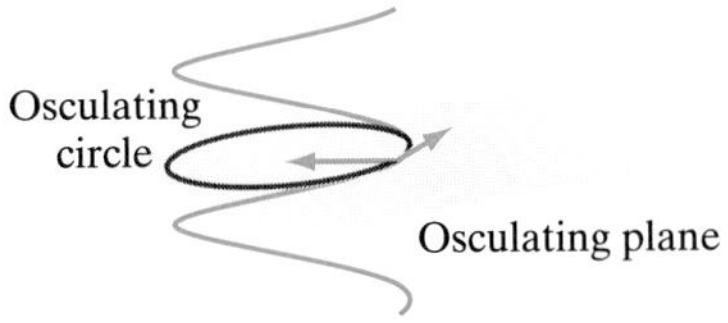

▲ Figure 3

▶ **EXAMPLE 4** Let $\mathcal{C}$ be the curve parameterized by $\mathbf{r}(t) = t^2\mathbf{i} + t\mathbf{j} + t\mathbf{k}$. Find the curvature $\kappa_{\mathbf{r}}(t)$ at the point $\mathbf{r}(t)$. Also determine the radius of curvature, principal unit normal, and center of the osculating circle at the point $\mathbf{r}(t)$. What is the Cartesian equation of the osculating plane at $\mathbf{r}(t)$?

Solution We calculate $\mathbf{r}'(t) = 2t\mathbf{i} + \mathbf{j} + \mathbf{k}$ and $\mathbf{r}''(t) = 2\mathbf{i}$. It follows that

$$\mathbf{r}'(t) \times \mathbf{r}''(t) = (2t\mathbf{i} + \mathbf{j} + \mathbf{k}) \times (2\mathbf{i}) = \det\left(\begin{bmatrix} \mathbf{i} & \mathbf{j} & \mathbf{k} \\ 2t & 1 & 1 \\ 2 & 0 & 0 \end{bmatrix}\right) = 2\mathbf{j} - 2\mathbf{k}.$$

Therefore the curvature at the point $\mathbf{r}(t)$ is

$$\kappa_{\mathbf{r}}(t) = \frac{\|\mathbf{r}'(t) \times \mathbf{r}''(t)\|}{\|\mathbf{r}'(t)^3\|} = \frac{\|2\mathbf{j} - 2\mathbf{k}\|}{\|2t\mathbf{i} + \mathbf{j} + \mathbf{k}\|^3} = \frac{2\sqrt{2}}{(2+4t^2)^{3/2}} = \frac{1}{(1+2t^2)^{3/2}}.$$

In particular, the radius of curvature at $\mathbf{r}(t)$ is $(1+2t^2)^{3/2}$. The unit tangent vector at $\mathbf{r}(t)$ is

$$\mathbf{T}(t) = \frac{1}{\|\mathbf{r}'(t)\|}\mathbf{r}'(t) = \frac{1}{\sqrt{2+4t^2}}(2t\mathbf{i} + \mathbf{j} + \mathbf{k}) = \frac{2t}{\sqrt{2+4t^2}}\mathbf{i} + \frac{1}{\sqrt{2+4t^2}}\mathbf{j} + \frac{1}{\sqrt{2+4t^2}}\mathbf{k}.$$

We calculate

$$\mathbf{T}'(t) = \frac{4}{(2+4t^2)^{3/2}}\mathbf{i} - \frac{4t}{(2+4t^2)^{3/2}}\mathbf{j} - \frac{4t}{(2+4t^2)^{3/2}}\mathbf{k} = \frac{4}{(2+4t^2)^{3/2}}(\mathbf{i} - t\mathbf{j} - t\mathbf{k}).$$

The direction of $\mathbf{T}'(t)$ is the principal unit normal vector. This direction is the same as the direction of $\mathbf{i} - t\mathbf{j} - t\mathbf{k}$. It follows that

$$\mathbf{N}(t) = \frac{1}{\sqrt{1+2t^2}}\mathbf{i} - \frac{t}{\sqrt{1+2t^2}}\mathbf{j} - \frac{t}{\sqrt{1+2t^2}}\mathbf{k}$$

is the principal unit normal vector at $\mathbf{r}(t)$. The center of curvature is therefore

$$\begin{aligned} \mathbf{r}(t) + \rho(t)\mathbf{N}(t) &= (t^2\mathbf{i} + t\mathbf{j} + t\mathbf{k}) + (1+2t^2)^{3/2}\left(\frac{1}{\sqrt{1+2t^2}}\mathbf{i} - \frac{t}{\sqrt{1+2t^2}}\mathbf{j} - \frac{t}{\sqrt{1+2t^2}}\mathbf{k}\right) \\ &= (t^2\mathbf{i} + t\mathbf{j} + t\mathbf{k}) + \left((1+2t^2)\mathbf{i} - t(1+2t^2)\mathbf{j} - t(1+2t^2)\mathbf{k}\right) \\ &= (1+3t^2)\mathbf{i} - 2t^3\mathbf{j} - 2t^3\mathbf{k}. \end{aligned}$$

The binormal vector

$$\mathbf{B}(t) = \mathbf{T}(t) \times \mathbf{N}(t) = \det\left(\begin{bmatrix} \mathbf{i} & \mathbf{j} & \mathbf{k} \\ 2t/\sqrt{2+4t^2} & 1/\sqrt{2+4t^2} & 1/\sqrt{2+4t^2} \\ 1/\sqrt{1+2t^2} & -t/\sqrt{1+2t^2} & -t/\sqrt{1+2t^2} \end{bmatrix}\right) = \frac{\sqrt{2}}{2}(\mathbf{j} - \mathbf{k})$$

is perpendicular to the osculating plane. Because the osculating plane passes through the point $\mathbf{r}(t) = t^2\mathbf{i} + t\mathbf{j} + t\mathbf{k}$, its equation is

$$0(x - t^2) + \frac{\sqrt{2}}{2}(y - t) + \left(-\frac{\sqrt{2}}{2}\right)(z - t) = 0,$$

which simplifies to $y = z$. The curve, its osculating plane, and osculating circle are shown in Figure 4. ◀

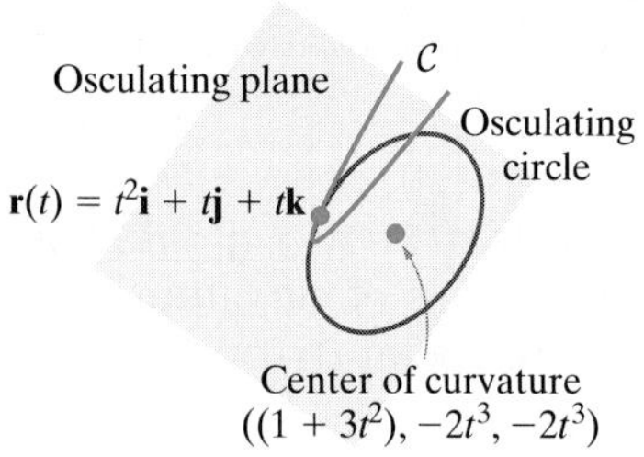

▲ Figure 4

INSIGHT In Example 4, the entire curve $\mathbf{r}(t) = t^2\mathbf{i} + t\mathbf{j} + t\mathbf{k}$ is contained in the plane $y = z$. It is therefore not surprising that the plane that contains the curve turns out to be the osculating plane at each point.

Planar Curves

The theory that we have developed for space curves can also be applied to curves in the xy-plane. Such a curve $\mathcal{C}$ might be the graph $y = f(x)$ of a function f, or it might be given parametrically by functions $x(t)$ and $y(t)$: $\mathcal{C} = \{(x, y) : x = x(t), y = y(t), a \leq t \leq b\}$. In fact, the graph $y = f(x)$ of a function can be realized parametrically by means of the equations $x = x(t), y = y(t)$ where $x(t) = t$, and $y(t) = f(t)$. We can imagine the xy-plane to be the plane $z = 0$ in xyz-space. Then the planar curve can be realized by the vector-valued function $\mathbf{r}(t) = x(t)\mathbf{i} + y(t)\mathbf{j} + 0\mathbf{k}$, and our formula for the curvature of space curves applies. The next theorem will make this explicit.

THEOREM 2 Suppose that $x(t)$ and $y(t)$ are twice differentiable functions that define a planar curve $\mathcal{C}$ by means of the parametric equations $x = x(t)$, $y = y(t)$. Then, at any point $P = ((x(t), y(t))$ for which the velocity vector $\langle x'(t), y'(t)\rangle$ is not $\mathbf{0}$, the curvature of $\mathcal{C}$ is given by

$$\kappa(t) = \frac{|x'(t)y''(t) - y'(t)x''(t)|}{\left((x'(t))^2 + (y'(t))^2\right)^{3/2}}. \tag{10.4.12}$$

In particular, the curvature of the graph of $y = f(x)$ at the point $(x, f(x))$ is

$$\kappa(x) = \frac{|f''(x)|}{\left(1 + (f'(x))^2\right)^{3/2}}. \tag{10.4.13}$$

Proof. With the xy-plane realized as the plane $z = 0$ in xyz-space and with $\mathcal{C}$ parameterized by the equation $\mathbf{r}(t) = x(t)\mathbf{i} + y(t)\mathbf{j} + 0\mathbf{k}$, we have

$$\mathbf{r}'(t) \times \mathbf{r}''(t) = \det\left(\begin{bmatrix} \mathbf{i} & \mathbf{j} & \mathbf{k} \\ x'(t) & y'(t) & 0 \\ x''(t) & y''(t) & 0 \end{bmatrix}\right) = \left(x'(t)y''(t) - y'(t)x''(t)\right)\mathbf{k}.$$

It follows that $\|\mathbf{r}'(t)\| = \left((x'(t))^2 + (y'(t))^2\right)^{1/2}$ and $\|\mathbf{r}'(t) \times \mathbf{r}''(t)\| = |x'(t)y''(t) - y'(t)x''(t)|$. Formula (10.4.12) results when these values are substituted into equation (10.4.2). If $x(t) = t$ and $y(t) = f(t)$, then $x'(t) = 1$, $x''(t) = 0$, $y'(t) = f'(t)$, and formula (10.4.12) reduces to (10.4.13) ■

▶ **EXAMPLE 5** The planar curve defined by $\mathbf{r}(t) = \left(t - \sin(t)\right)\mathbf{i} + \left(1 - \cos(t)\right)\mathbf{j}$ is called a *cycloid*. Calculate its curvature at the point P that corresponds to $t = \pi/3$. Repeat for $t = 3\pi$.

Solution Let $x(t) = t - \sin(t)$ and $y(t) = 1 - \cos(t)$. Then $x'(t) = 1 - \cos(t)$, $x''(t) = \sin(t)$, $y'(t) = \sin(t)$, and $y''(t) = \cos(t)$. Formula (10.4.12) becomes

$$\kappa(t) = \frac{|x'(t)y''(t) - y'(t)x''(t)|}{\left((x'(t))^2 + (y'(t))^2\right)^{3/2}} = \frac{|(1-\cos(t))\cos(t) - \sin(t)\sin(t)|}{\left((1-\cos(t))^2 + (\sin(t))^2\right)^{3/2}},$$

which simplifies to

$$\kappa(t) = \frac{|\cos(t) - \cos^2(t) - \sin^2(t)|}{\left(1 - 2\cos(t) + \cos^2 t + \sin^2(t)^2\right)^{3/2}} = \frac{|\cos(t) - 1|}{2^{3/2}(1-\cos(t))^{3/2}} = \frac{\sqrt{2}}{4\sqrt{1-\cos(t)}}.$$

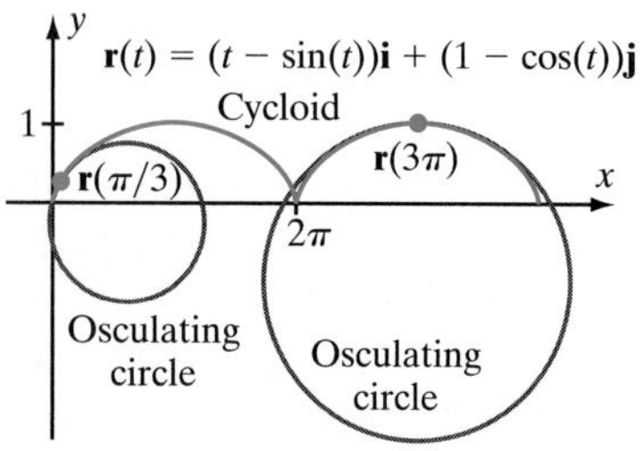

▲ Figure 5

When $t = \pi/3$, the curvature is $\kappa(\pi/3) = \sqrt{2}/(4\sqrt{1-\cos(\pi/3)}) = 1/2$. Similarly, we find that $\kappa(3\pi) = \sqrt{2}/(4\sqrt{1-\cos(3\pi)}) = \sqrt{2}/(4\sqrt{1-(-1)}) = 1/4$. Our calculations tell us that the osculating circles at $\mathbf{r}(\pi/3)$ and $\mathbf{r}(3\pi)$ have radii 2 and 4, respectively. The cycloid is plotted for $0 \le t \le 4\pi$ in Figure 5, which also shows the osculating circles at the points corresponding to $t = \pi/3$ and $t = 3\pi$. ◀

▶ **EXAMPLE 6** Find the circle of curvature of the curve $y = \sin(2x)$ at the point $P = (\pi/4, 1)$.

Solution We set $f(x) = \sin(2x)$ and notice that $f'(\pi/4) = 2\cos(\pi/2) = 0$ and $f''(\pi/4) = -4\sin(\pi/2) = -4$. Then, according to formula (10.4.13), we have

$$\kappa(\pi/4) = \frac{4}{(1+0^2)^{3/2}} = 4.$$

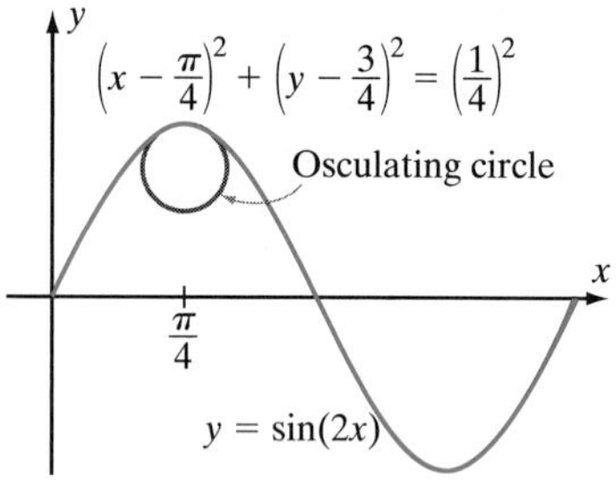

▲ Figure 6

The radius of curvature is therefore 1/4. Because $f'(\pi/4) = 0$, the graph of $y = \sin(2x)$ has a horizontal tangent. The principal unit normal therefore points vertically. Because the graph of $y = \sin(2x)$ is concave down at P and the principal unit normal points in the direction to which the curve bends, we conclude that the principal unit normal points downward. The center of curvature is the point $(\pi/4, 1 - 1/4)$, and the circle of curvature has Cartesian equation $(x - \pi/4)^2 + (y - 3/4)^2 = (1/4)^2$. The graph of f together with this osculating circle is shown in Figure 6. ◀

QUICK QUIZ

1. The position $\mathbf{p}(s)$ of a moving particle is such that $\mathbf{p}'(s) = \sin(\pi s)\mathbf{i} + \frac{1}{\sqrt{2}}\cos(\pi s)\mathbf{j} - \frac{1}{\sqrt{2}}\cos(\pi s)\mathbf{k}$. What is the curvature of the trajectory of the particle at $s = 2$?
2. As a particle passes through point P, its speed is 2, the magnitude of its acceleration is 6, and its acceleration is perpendicular to its velocity. What is the curvature of the trajectory of the particle at P?
3. If $t \mapsto \mathbf{r}(t)$ is a smooth parameterization of a curve $\mathcal{C}$ and if $\kappa_{\mathbf{r}}(3) > 0$, then what unit vectors are perpendicular to the osculating plane of $\mathcal{C}$ at $\mathbf{r}(3)$?
4. If the curvature at a point P on $\mathcal{C}$ is $5/4$, then what is the radius of the osculating circle of $\mathcal{C}$ at P?

Answers

1. π 2. $3/2$ 3. $\pm\mathbf{B}(3)$ 4. $4/5$

EXERCISES

Problems for Practice

In each of Exercises 1–6, an arc length parameterization of a curve is given. Verify this assertion and then calculate a. $\mathbf{T}(s)$; b. $\mathbf{N}(s)$; and c. $\kappa(s)$.

1. $\mathbf{r}(s) = \sin(s)\mathbf{i} - \cos(s)\mathbf{j} + 3\mathbf{k}$
2. $\mathbf{r}(s) = (3 - \cos(s))\mathbf{i} - (6 + \sin(s))\mathbf{j}$
3. $\mathbf{r}(s) = \cos(s/\sqrt{2})\mathbf{i} - \sin(s/\sqrt{2})\mathbf{j} + (s/\sqrt{2})\mathbf{k}$
4. $\mathbf{r}(s) = \sin(s/4)\mathbf{i} + \cos(s/4)\mathbf{j} - (\sqrt{15}/4)s\mathbf{k}$
5. $\mathbf{r}(s) = (3\sin(s)/5)\mathbf{i} + \cos(s)\mathbf{j} + (4\sin(s)/5)\mathbf{k}$
6. $\mathbf{r}(s) = \dfrac{2\sqrt{2}}{3}s\mathbf{i} - \dfrac{1}{3}\sin(s)\mathbf{j} + \dfrac{1}{3}\cos(s)\mathbf{k}$
7. For the arc length parameterization given in Exercise 3, calculate the radius of curvature $\rho(s)$ and the center of the osculating circle at each point of the curve.
8. For the arc length parameterization given in Exercise 4, calculate the radius of curvature $\rho(s)$ and the center of the osculating circle at each point of the curve.
9. For the arc length parameterization given in Exercise 5, calculate the radius of curvature $\rho(s)$ and the center of the osculating circle at each point of the curve.

In each of Exercises 10–15, a parameterization of a planar curve is given. Calculate the curvature at each point of the curve

10. $\mathbf{r}(t) = \cos^2(t^2)\mathbf{i} + \sin^2(t^2)\mathbf{j}$
11. $\mathbf{r}(t) = e^t\mathbf{i} - e^{2t}\mathbf{j}$
12. $\mathbf{r}(t) = t^2\mathbf{i} - t^3\mathbf{j}$
13. $\mathbf{r}(t) = \ln(|t|)\mathbf{i} - (1/t)\mathbf{j}$
14. $\mathbf{r}(t) = t^{1/2}\mathbf{i} + t^{3/2}\mathbf{j}$
15. $\mathbf{r}(t) = \tan(t)\mathbf{i} + \sec(t)\mathbf{j}$

In each of Exercises 16–21, a function f is given. Calculate the curvature at every point on the graph (in the xy-plane) of the equation $y = f(x)$.

16. $f(x) = x^3$
17. $f(x) = x^{1/2}$
18. $f(x) = \ln(|x|)$
19. $f(x) = \ln(\sec(|x|))$
20. $f(x) = (e^x + e^{-x})/2$
21. $f(x) = x^3/6 + 1/(2x)$

In each of Exercises 22–27, a parameterization of a space curve is given. At each point of the curve, calculate the radius of curvature and the center of curvature.

22. $\mathbf{r}(t) = t\mathbf{i} + t\mathbf{j} + t^2\mathbf{k}$
23. $\mathbf{r}(t) = t^2\mathbf{i} + t^2\mathbf{j} + (1 + 2t)\mathbf{k}$
24. $\mathbf{r}(t) = \sqrt{t}\mathbf{i} + (2 - t)\mathbf{j} + 3t\mathbf{k}$
25. $\mathbf{r}(t) = e^t\mathbf{i} + t\mathbf{j} + e^t\mathbf{k}$
26. $\mathbf{r}(t) = \ln(t)\mathbf{i} + t\mathbf{j} + 2t\mathbf{k}$
27. $\mathbf{r}(t) = t^4\mathbf{i} + t\mathbf{j} + 2t\mathbf{k}$

In each of Exercises 28–32, a parameterization of a space curve is given. Find the Cartesian equation of its osculating plane at the given point P_0.

28. $\mathbf{r}(t) = (2 - t^2)\mathbf{i} + 2t^3\mathbf{j} + t^2\mathbf{k}$ $\quad P_0 = (1, 2, 1)$
29. $\mathbf{r}(t) = e^t\mathbf{i} + t\mathbf{j} + e^{2t}\mathbf{k}$ $\quad P_0 = (1, 0, 1)$
30. $\mathbf{r}(t) = \sqrt{t}\mathbf{i} + (4/t)\mathbf{j} + (t^2/4)\mathbf{k}$ $\quad P_0 = (2, 1, 4)$
31. $\mathbf{r}(t) = (t^2 + t)\mathbf{i} + (t^2 - t)\mathbf{j} + t^2\mathbf{k}$ $\quad P_0 = (2, 6, 4)$
32. $\mathbf{r}(t) = \sin(t)\mathbf{i} + \cos(2t)\mathbf{j} + 6t\mathbf{k}$ $\quad P_0 = (1/2, 1/2, \pi)$

Further Theory and Practice

Given a curve $\mathcal{C}$, the locus of its centers of curvature is called the *evolute* of $\mathcal{C}$. In each of Exercises 33–40, find the evolute of the given plane curve.

33. $\mathbf{r}(t) = (2 - t^2)\mathbf{i} + 2t^3\mathbf{j}$
34. $\mathbf{r}(t) = t\mathbf{i} + \cos(t)\mathbf{j}$
35. $\mathbf{r}(t) = t\mathbf{i} + e^t\mathbf{j}$
36. $\mathbf{r}(t) = 2\cos(t)\mathbf{i} + \sin(t)\mathbf{j}$
37. $\mathbf{r}(t) = (t - \sin(t))\mathbf{i} + (1 - \cos(t))\mathbf{j}$
38. $\mathbf{r}(t) = t\cos(t)\mathbf{i} + t\sin(t)\mathbf{j}$
39. $\mathbf{r}(t) = t\mathbf{i} + \ln(\sec(t))\mathbf{j}$
40. $\mathbf{r}(t) = e^t\cos(t)\mathbf{i} + e^t\sin(t)\mathbf{j}$

In each of Exercises 41–44, find the point at which the graph $y = f(x)$ of the given function f has the greatest curvature.

41. $f(x) = x^2$
42. $f(x) = e^x$
43. $f(x) = \ln(2x^2)$
44. $f(x) = x^4/16 + 1/(2x^2) \quad x > 0$

In each of Exercises 45–48, find a parameterization for the curve that is the graph of the given equation. Find the curvature at the specified point.

45. $x^2/a^2 + y^2/b^2 = 1 \quad (a/\sqrt{2}, b/\sqrt{2})$
46. $x^3 + y^3 = 4xy$; $(2, 2)$ *Hint:* Choose parameter $t = y/x$, divide the cubic equation by x^3, and find x in terms of t.
47. $x^{2/3} + y^{2/3} = 1$; $(1/8, 3\sqrt{3}/8)$ *Hint:* Take advantage of the identity $\cos^2(t) + \sin^2(t) = 1$.
48. $x(x + y)^2 = 9 + (x + y)^2$; $(2, 1)$ *Hint:* Try the parameter $t = x + y$ and solve for x in terms of t.
49. Prove that a curve with curvature 0 at every point is a line.

In Exercises 50–53, treat the concept of *torsion.*

50. Suppose that $\mathbf{p}(s)$ is an arc length parameterization of a curve. Prove that there is a scalar-valued function $s \mapsto \tau(s)$ such that the unit tangent vector $\mathbf{T}(s)$, the principal unit normal $\mathbf{N}(s)$, and the binormal $\mathbf{B}(s)$ satisfy the equations:

$$\begin{aligned} \mathbf{T}'(s) &= \kappa(s)\mathbf{N}(s) \\ \mathbf{N}'(s) &= -\kappa(s)\mathbf{T}(s) + \tau(s)\mathbf{B}(s) \\ \mathbf{B}'(s) &= -\tau(s)\mathbf{N}(s). \end{aligned}$$

The function $s \mapsto \tau(s)$ is called the *torsion* of the curve. The three equations are collectively known as the *Frenet Formulas*.

51. Show that the torsion is given by the formula

$$\tau(s) = \frac{(\mathbf{p}'(s) \times \mathbf{p}''(s)) \cdot \mathbf{p}'''(s)}{\|\mathbf{p}'(s) \times \mathbf{p}''(s)\|^2}.$$

52. Compute the torsion of the helix $\mathbf{p}(s) = \sin(s/\sqrt{2})\mathbf{i} + \cos(s/\sqrt{2})\mathbf{j} + (s/\sqrt{2})\mathbf{k}$.

53. The torsion of a curve measures the extent to which the curve is twisting out of the osculating plane. Prove that a curve with 0 torsion lies in a plane. Conversely, if a curve is planar, it has 0 torsion.

54. Let $p(x)$ be a polynomial of degree 2 or greater. Prove that there is a point on the plane curve $y = p(x)$ where the curvature is greatest.

55. Prove formula (10.4.3).

56. Prove that

$$\kappa_{\mathbf{r}}(t) = \frac{\left\| \|\mathbf{r}'(t)\|^2 \mathbf{r}''(t) - (\mathbf{r}'(t) \cdot \mathbf{r}''(t))\mathbf{r}'(t) \right\|}{\|\mathbf{r}'(t)\|^4}.$$

Calculator/Computer Exercises

In each of Exercises 57–62, a parameterization of a plane curve C is given. Plot the locus of the centers of curvature of C. The resulting plane curve is called the *evolute* of C.

57. $\mathbf{r}(t) = (2 - t^2)\mathbf{i} + 2t^3\mathbf{j}$
58. $\mathbf{r}(t) = t\mathbf{i} + \cos(t)\mathbf{j}$
59. $\mathbf{r}(t) = t\mathbf{i} + e^t\mathbf{j}$
60. $\mathbf{r}(t) = 2\cos(t)\mathbf{i} + \sin(t)\mathbf{j}$
61. $\mathbf{r}(t) = (t - \sin(t))\mathbf{i} + (1 - \cos(t))\mathbf{j}$
62. $\mathbf{r}(t) = t\cos(t)\mathbf{i} + t\sin(t)\mathbf{j}$

In each of Exercises 63–66, a parameterization of a space curve C is given. Plot the locus of the centers of curvature of C. The resulting space curve is called the *evolute* of C.

63. $\mathbf{r}(t) = \cos(t)\mathbf{i} + \sin(t)\mathbf{j} + t\mathbf{k}$
64. $\mathbf{r}(t) = t\mathbf{i} + t^2\mathbf{j} + t^3\mathbf{k}$
65. $\mathbf{r}(t) = \cos(t)\mathbf{i} + \sin(t)\mathbf{j} + t^2\mathbf{k}$
66. $\mathbf{r}(t) = (1 + t^2)\cos(t)\mathbf{i} + (1 + t^2)\sin(t)\mathbf{j} + t^2\mathbf{k}$

10.5 Applications of Vector-Valued Functions to Motion

Suppose that $\mathbf{r}$ is a smooth parameterization of a space curve C. Then, the second derivative $\mathbf{r}''$ has a physical interpretation: It is the acceleration of a moving particle that has $\mathbf{r}$ as its position vector. There is also a geometric significance to $\mathbf{r}''$ that shows up in the formula for curvature. We begin this section by establishing a connection between the two roles $\mathbf{r}''$ plays. Let us try to anticipate the relationship. When we go around a sharp curve in a car or a roller coaster, we feel a strong force in the outward direction. By Newton's second law, strong force means strong acceleration. So there is a correlation between large curvature and the acceleration vector having a large normal component.

To make these considerations precise, we let $\mathbf{T}(t)$ and $\mathbf{N}(t)$ denote the unit tangent and principal unit normal vectors at the point $\mathbf{r}(t)$. The velocity and acceleration vectors are given by $\mathbf{v}(t) = \mathbf{r}'(t)$ and $\mathbf{a}(t) = \mathbf{v}'(t) = \mathbf{r}''(t)$, respectively. The magnitude of the velocity vector is the speed, $v(t) = \|\mathbf{r}'(t)\|$. Let $\sigma(t) = \int_0^t \|\mathbf{r}'(\tau)\|\, d\tau$ be the arc length function of $\mathbf{r}$. Then, $d\sigma/dt$ is another way to express the speed.

Recall equation (10.4.5),

$$\mathbf{r}''(t) = \frac{d^2\sigma(t)}{dt^2}\mathbf{T}(t) + \frac{d\sigma(t)}{dt}\mathbf{T}'(t),$$

which decomposes the acceleration vector into orthogonal vectors in the osculating plane. We rewrite this equation as

$$\mathbf{a}(t) = \frac{dv(t)}{dt}\mathbf{T}(t) + v(t)\mathbf{T}'(t). \tag{10.5.1}$$

Also recall formula (10.3.9),

$$\mathbf{N}(t) = \frac{1}{\|\mathbf{T}'(t)\|}\mathbf{T}'(t),$$

which we rewrite as

$$\mathbf{T}'(t) = \|\mathbf{T}'(t)\|\mathbf{N}(t). \tag{10.5.2}$$

Substituting equation (10.5.2) into (10.5.1), we obtain

$$\mathbf{a}(t) = \frac{dv(t)}{dt}\mathbf{T}(t) + v(t)\|\mathbf{T}'(t)\|\mathbf{N}(t). \tag{10.5.3}$$

Using formula (10.4.7) to substitute for $\|\mathbf{T}'(t)\|$ in equation (10.5.3), we get

$$\mathbf{a}(t) = \frac{dv(t)}{dt}\mathbf{T}(t) + v(t)\frac{\|\mathbf{r}'(t) \times \mathbf{r}''(t)\|}{\|\mathbf{r}'(t)\|^2}\mathbf{N}(t),$$

or

$$\mathbf{a}(t) = \frac{dv(t)}{dt}\mathbf{T}(t) + v(t)\|\mathbf{r}'(t)\|\frac{\|\mathbf{r}'(t) \times \mathbf{r}''(t)\|}{\|\mathbf{r}'(t)\|^3}\mathbf{N}(t). \tag{10.5.4}$$

Formula (10.4.2) tells us that the ratio preceding $\mathbf{N}(t)$ in (10.5.4) is $\kappa_{\mathbf{r}}(t)$. Furthermore, $\|\mathbf{r}'(t)\|$ is an alternative way to express the speed $v(t)$. Substituting these equivalent expressions into equation (10.5.4), we obtain

$$\mathbf{a}(t) = \frac{dv(t)}{dt}\mathbf{T}(t) + v(t)^2\kappa_{\mathbf{r}}(t)\mathbf{N}(t). \tag{10.5.5}$$

Let us summarize these computations with the following theorem.

THEOREM 1 Suppose that $t \mapsto \mathbf{r}(t)$ is a smooth parameterization of a space curve $\mathcal{C}$. Let $v(t) = \|\mathbf{r}'(t)\|$ denote the speed of a particle moving along the curve with position vector $\mathbf{r}(t)$. Then, the acceleration vector $\mathbf{a}(t) = \mathbf{r}''(t)$ can be decomposed as the sum of two vectors, one with direction $\mathbf{T}(t)$, the unit tangent to $\mathcal{C}$ at $\mathbf{r}(t)$, and the other with direction $\mathbf{N}(t)$, the principal unit normal to $\mathcal{C}$ at $\mathbf{r}(t)$. The decomposition has the form

$$\mathbf{a}(t) = a_T(t)\mathbf{T}(t) + a_N(t)\mathbf{N}(t) \tag{10.5.6}$$

where

$$a_T(t) = \frac{dv(t)}{dt} \quad \textit{and} \quad a_N(t) = v(t)^2\kappa_{\mathbf{r}}(t). \tag{10.5.7}$$

The scalar $a_T(t)$ is called the *tangential component of acceleration*, and the scalar $a_N(t)$ is called the *normal component of acceleration.* Observe that the normal component of acceleration cannot be negative. Equation (10.5.6), which expresses acceleration as a linear combination of the unit tangent and principal unit normal directions, contains a great deal of information. The component of force in the direction of $\mathbf{N}(t)$ that is required to hold a moving body on its curved trajectory is called the *centripetal force.* By Newton's Second Law, force is mass times acceleration. Thus centripetal force is the mass of the body times $a_N(t)\mathbf{N}(t)$, a vector that is called the *centripetal or normal acceleration.* Its magnitude $a_N(t)$ depends on the speed and the curvature (but not on the change of speed). The vector $a_T(t)\mathbf{T}(t)$ is called the *tangential acceleration.* Its magnitude $|a_T|$ depends on the rate of change of speed with respect to time. If the path of the moving body becomes more curved, then the normal component of the acceleration increases in magnitude. For the body to follow such a path, the normal component of force must be increasing. Furthermore, if the speed of the body increases, then the normal component of acceleration also increases. Once again, we infer that such a trajectory requires a force with increasing normal component

When calculating centripetal acceleration, the following theorem is often useful.

THEOREM 2 The tangential and normal components of acceleration satisfy

a. $a_T(t) = \mathbf{a}(t) \cdot \mathbf{T}(t)$

b. $a_N(t) = \mathbf{a}(t) \cdot \mathbf{N}(t)$

c. $\|\mathbf{a}(t)\|^2 = (a_T(t))^2 + (a_N(t))^2$.

Proof. Each of the three formulas is obtained by calculating a dot product and using the equations $\mathbf{T}(t) \cdot \mathbf{T}(t) = 1$, $\mathbf{N}(t) \cdot \mathbf{N}(t) = 1$, and $\mathbf{T}(t) \cdot \mathbf{N}(t) = 0$. Thus part a is obtained by taking the dot product of each side of equation (10.5.6) with $\mathbf{T}(t)$. The result is $\mathbf{a}(t) \cdot \mathbf{T}(t) = a_T(t)\mathbf{T}(t) \cdot \mathbf{T}(t) + a_N(t)\mathbf{N}(t) \cdot \mathbf{T}(t) = a_T(t) + 0$, which is part a. Part b is obtained in a similar way: We take the dot product of each side of equation (10.5.6) with $\mathbf{N}(t)$. Finally,

$$\begin{aligned}
\|\mathbf{a}(t)\|^2 &= \mathbf{a}(t) \cdot \mathbf{a}(t) \\
&= \big(a_T(t)\mathbf{T}(t) + a_N(t)\mathbf{N}(t)\big) \cdot \big(a_T(t)\mathbf{T}(t) + a_N(t)\mathbf{N}(t)\big) \\
&= \big(a_T(t)\big)^2\mathbf{T}(t) \cdot \mathbf{T}(t) + 2a_T(t)a_N(t)\mathbf{T}(t) \cdot \mathbf{N}(t) + \big(a_N(t)\big)^2\mathbf{N}(t) \cdot \mathbf{N}(t) \\
&= \big(a_T(t)\big)^2 + \big(a_N(t)\big)^2.
\end{aligned}$$

■

▶ **EXAMPLE 1** Let $\mathbf{r}(t) = \sin(t)\mathbf{i} - \cos(t)\mathbf{j} - (t^2/2)\mathbf{k}$. Calculate $a_T(t)$ and $a_N(t)$. Express $\mathbf{a}(t)$ as a linear combination of $\mathbf{T}(t)$ and $\mathbf{N}(t)$.

Solution We have $\mathbf{r}'(t) = \cos(t)\mathbf{i} + \sin(t)\mathbf{j} - t\mathbf{k}$ and $\mathbf{a}(t) = \mathbf{r}''(t) = -\sin(t)\mathbf{i} + \cos(t)\mathbf{j} - \mathbf{k}$. Therefore

$$\frac{ds}{dt} = v(t) = \|\mathbf{r}'(t)\| = \sqrt{\cos^2(t) + \sin^2(t) + t^2} = \sqrt{1 + t^2}.$$

It follows that

$$a_T(t) = \frac{d^2s}{dt^2} = \frac{d}{dt}\sqrt{1+t^2} = \frac{t}{\sqrt{1+t^2}}.$$

Also, $\|\mathbf{a}(t)\|^2 = \|-\sin(t)\mathbf{i} + \cos(t)\mathbf{j} - \mathbf{k}\|^2 = \left(-\sin(t)\right)^2 + \left(\cos(t)\right)^2 + (-1)^2 = 2.$ According to Theorem 2,

$$a_N(t) = \sqrt{\|\mathbf{a}(t)\|^2 - \left(a_T(t)\right)^2} = \sqrt{2 - \frac{t^2}{1+t^2}} = \sqrt{\frac{2+t^2}{1+t^2}}.$$

In conclusion, equation (10.5.6) becomes

$$\mathbf{a}(t) = \frac{t}{\sqrt{1+t^2}}\mathbf{T}(t) + \sqrt{\frac{2+t^2}{1+t^2}}\mathbf{N}(t). \qquad (10.5.8)$$ ◄

INSIGHT Equation (10.5.6) can be used in conjunction with Theorem 2b as an alternative calculation of the principal unit normal vector $\mathbf{N}(t)$. For instance, in Example 1, we can use equation (10.5.8) to solve for $\mathbf{N}(t)$:

$$\mathbf{N}(t) = \sqrt{\frac{1+t^2}{2+t^2}}\left(\mathbf{a}(t) - \frac{t}{\sqrt{1+t^2}}\mathbf{T}(t)\right).$$

The values for vectors $\mathbf{a}(t)$ and $\mathbf{T}(t)$ can then be substituted into this formula to yield $\mathbf{N}(t)$.

Central Force Fields

Let O be a fixed point in space. We will use it as the origin of our coordinate axes. For each point $P \neq O$, let $r_P = \|\overrightarrow{OP}\|$ and $\mathbf{u}_P = \mathbf{dir}\left(\overrightarrow{OP}\right)$. We say that a force $\mathbf{F}$ is a *central force field* if there is a scalar-valued function f of a real variable such that $\mathbf{F}(P) = f(r_P)\mathbf{u}_P$ for every $P \neq O$ (see Figure 1). If $f(r) > 0$, then the force is directed away from O. If $f(r) < 0$, then the force is directed toward O and is said to be *attractive.* As an example of an attractive central force field, consider the sun as a point mass that is fixed at a point O in space. If a planet is at point P, then, according to Newton's Universal Law of Gravitation, the gravitational force exerted by the sun on the planet is equal to

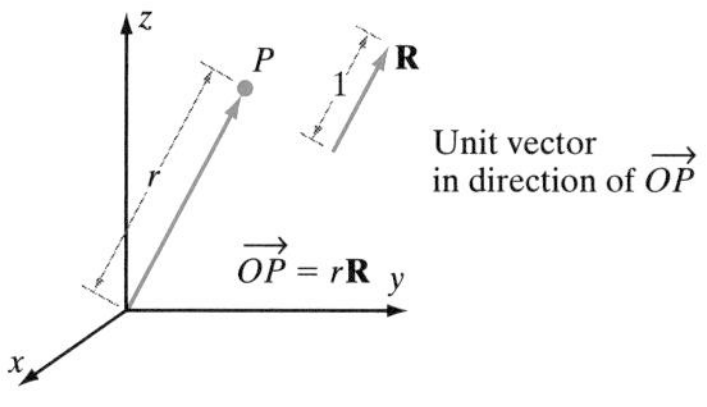

▲ Figure 1 A central force field $\mathbf{F}(P) = f(r)\mathbf{R}$ acts in the direction of the position vector $\overrightarrow{OP}$ if $f(r) > 0$ and in the direction opposite to the position vector if $f(r) < 0$.

$$\mathbf{F}(P) = -\frac{GMm}{r_P^2}\mathbf{u}_P \qquad (10.5.9)$$

where G is a universal constant (i.e., the same for all planets), M is the mass of the sun, m is the mass of the planet, r_P is the distance of the planet to the sun, and $\mathbf{u}_P$ is the direction of $\overrightarrow{OP}$. By setting $f(r) = -GMm/r^2$, we see that the force of gravity is an attractive central force field.

► **EXAMPLE 2** Suppose that $a > b > 0$. A force $\mathbf{F}$ acts on a particle of mass m in such a way that the particle moves in the xy-plane with position described by $\mathbf{r}(t) = a\cos(t)\mathbf{i} + b\sin(t)\mathbf{j}$. Show that $\mathbf{F}$ is a central force field.

Solution According to Newton's Second Law of Motion, $\mathbf{F} = m\mathbf{a}(t)$ and therefore,

$$\mathbf{F} = m\mathbf{r}''(t) = m\big(-a\cos(t)\mathbf{i} - b\sin(t)\mathbf{j}\big) = -m\mathbf{r}(t) = -m\|\mathbf{r}(t)\| \operatorname{dir}\big(\mathbf{r}(t)\big).$$

This formula for $\mathbf{F}$ shows that it is a central force field directed toward the origin. ◀

A key fact about central force fields is that they always give rise to trajectories that lie in a plane.

THEOREM 3 If a particle moving in space is subject only to a central force field, then the particle's trajectory lies in a plane.

Proof. Let $\mathbf{r}(t)$ be the position vector of the particle, and let m be its mass. The force on the particle can be written as $\mathbf{F} = f(r)\mathbf{u}$ where $r = \|\mathbf{r}(t)\|$ and $\mathbf{u} = (1/r)\mathbf{r}(t)$. Notice that $\mathbf{r}(t) \times \mathbf{F} = 0$ because $\mathbf{r}(t)$ and $\mathbf{u}$ have the same direction. Newton's Second Law of Motion, force equals mass times acceleration, allows us to conclude that $\mathbf{r}(t) \times (m\mathbf{r}''(t)) = 0$, or $\mathbf{r}(t) \times \mathbf{r}''(t) = 0$. We can write this as

$$\mathbf{r}(t) \times \frac{d}{dt}\mathbf{r}'(t) = \mathbf{0}.$$

From this last equation, we deduce that

$$\frac{d}{dt}\big(\mathbf{r}(t) \times \mathbf{r}'(t)\big) = \mathbf{r}'(t) \times \mathbf{r}'(t) + \mathbf{r}(t) \times \frac{d}{dt}\mathbf{r}'(t) = \mathbf{0} + \mathbf{0} = \mathbf{0}.$$

It follows that

$$\mathbf{r}(t) \times \mathbf{r}'(t) = \mathbf{c} \qquad (10.5.10)$$

for some constant vector $\mathbf{c}$. This tells us that $\mathbf{r}(t)$ is perpendicular to $\mathbf{c}$ whatever the value of t may be. In other words, $\mathbf{r}(t)$ is in a plane that is normal to $\mathbf{c}$. ■

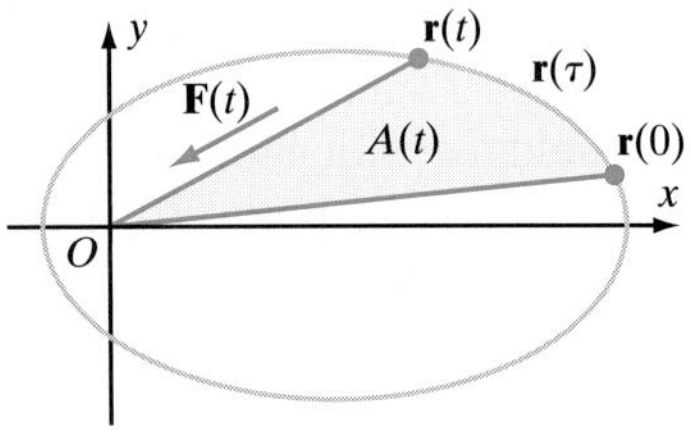

▲ Figure 2

When we study the trajectory of a body under a central force field, the motion is planar, so it is convenient to assume that the motion is in the xy-plane. As the body moves along its trajectory, its position vector sweeps out a region in the plane of motion (see Figure 2). Let $A(t)$ denote the area of the region swept out by the position vector $\mathbf{r}(\tau)$ for $0 \le \tau \le t$.

THEOREM 4 If a moving particle is subject only to a central force field, then the particle's position vector sweeps out a region whose area $A(t)$ has a constant rate of change with respect to t.

Proof. Let Δt be an increment of time, $\Delta\mathbf{r}(t)$ the corresponding increment of position, and ΔA the increment of area swept out (Figure 3). We see that ΔA is

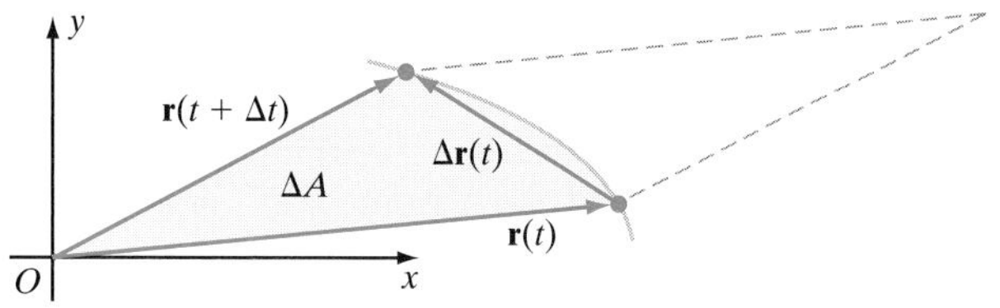

Figure 3

approximately equal to half the area of the parallelogram determined by the vectors $\mathbf{r}(t)$ and $\mathbf{r}(t+\Delta t) = \mathbf{r}(t) + \Delta\mathbf{r}(t)$. The area of this parallelogram is $\|\mathbf{r}(t) \times \mathbf{r}(t+\Delta t)\| = \|\mathbf{r}(t) \times (\mathbf{r}(t) + \Delta\mathbf{r}(t))\| = \|\mathbf{r}(t) \times \Delta\mathbf{r}(t)\|$. Thus

$$\frac{\Delta A}{\Delta t} \approx \frac{1}{2}\frac{\|\mathbf{r}(t) \times \Delta\mathbf{r}(t)\|}{\Delta t} = \frac{1}{2}\left\|\mathbf{r}(t) \times \frac{\Delta\mathbf{r}(t)}{\Delta t}\right\|.$$

Letting $\Delta t \to 0$ gives

$$A'(t) = \lim_{\Delta t \to 0}\frac{\Delta A}{\Delta t} = \frac{1}{2}\left\|\mathbf{r}(t) \times \lim_{\Delta t \to 0}\frac{\Delta\mathbf{r}(t)}{\Delta t}\right\| = \frac{1}{2}\|\mathbf{r}(t) \times \mathbf{r}'(t)\| \overset{(10.5.10)}{=} \frac{1}{2}\|\mathbf{c}\|.$$

We conclude that area $A(t)$ is swept out at a constant rate. This is Kepler's second law of planetary motion. ∎

Ellipses

For the remainder of our work in this section, we will need an in-depth understanding of one particular planar curve, the ellipse. Fix two distinct points F_1 and F_2 in the xy-plane. Let c denote half of the distance between the two points. Suppose that a is a fixed positive constant that is greater than c. The set of points $P = (x, y)$ with the property that the distance from P to F_1 plus the distance from P to F_2 equals $2a$ is called an *ellipse* (see Figure 4).

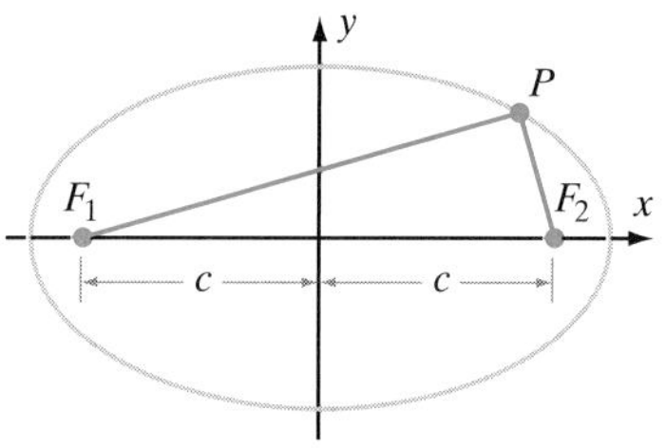

Figure 4 $\|\overline{PF_1}\| + \|\overline{PF_2}\| = 2a$

Each of the points F_1 and F_2 is called a *focus* of the ellipse. Together they are called the *foci* (plural of "focus") of the ellipse. The midpoint of line segment $\overline{F_1F_2}$ is called the *center* of the ellipse. The chord of the ellipse passing through the two foci is called the *major axis* of the ellipse. The chord perpendicular to the major axis and passing through the center is called the *minor axis* of the ellipse. These features are shown in Figure 5.

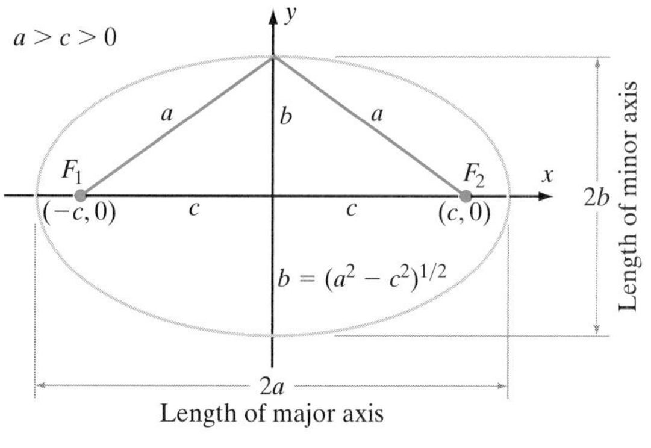

Figure 5

Let us derive the formula for an ellipse with major axis along the x-axis and center at the origin. Let c be a positive constant. We place the foci of the ellipse at points $F_1 = (-c, 0)$ and $F_2 = (c, 0)$, symmetrically situated on the x-axis. Let a be a number that exceeds c. A point (x, y) lies on the ellipse provided that

$$\underbrace{\left((x-(-c))^2 + (y-0)^2\right)^{1/2}}_{\text{distance to } F_1} + \underbrace{\left((x-c)^2 + (y-0)^2\right)^{1/2}}_{\text{distance to } F_2} = 2a,$$

or

$$\left((x-(-c))^2 + (y-0)^2\right)^{1/2} = 2a - \left((x-c)^2 + (y-0)^2\right)^{1/2}.$$

We square both sides and simplify to obtain

$$a^2 - cx = a(x^2 - 2cx + c^2 + y^2)^{1/2}.$$

Squaring both sides again and simplifying gives

$$(a^2 - c^2)x^2 + a^2y^2 = a^2(a^2 - c^2).$$

Because $a > c > 0$, it follows that each of the coefficients appearing in this equation is positive. Let us simplify matters by setting

$$b = (a^2 - c^2)^{1/2}. \tag{10.5.11}$$

We then have $b^2x^2 + a^2y^2 = a^2b^2$ or, equivalently,

$$\frac{x^2}{a^2} + \frac{y^2}{b^2} = 1$$

with $a > b > 0$. This last equation is the standard form for an ellipse with foci on the x-axis and center at the origin. By definition, the major axis will be a segment in the x-axis. Because the x-intercepts of the ellipse are $(\pm a, 0)$ (just set $y = 0$ to find these), we see that the major axis is the segment connecting $(-a, 0)$ to $(a, 0)$. Similarly the minor axis is the segment connecting $(0, -b)$ to $(0, b)$ (see Figure 5).

If we were to repeat the preceding calculation with foci $(0, -c)$ and $(0, c)$ and $b = \sqrt{a^2 - c^2}$, then we would obtain the equation

$$\frac{x^2}{b^2} + \frac{y^2}{a^2} = 1.$$

In general, an ellipse with center (h, k), major axis length $2a$, minor axis length $2b$, and axes parallel to the coordinate axes has equation

$$\underbrace{\frac{(x-h)^2}{a^2} + \frac{(y-k)^2}{b^2} = 1}_{\text{foci at } (h\pm c, k)} \quad \text{or} \quad \underbrace{\frac{(x-h)^2}{b^2} + \frac{(y-k)^2}{a^2} = 1}_{\text{foci at } (h, k\pm c)}, \qquad (b^2 + c^2 = a^2). \tag{10.5.12}$$

▶ **EXAMPLE 3** Suppose that $a > b > 0$. Show that the curve described by $\mathbf{r}(t) = a\cos(t)\mathbf{i} + b\sin(t)\mathbf{j}$, $0 \le t \le 2\pi$, is an ellipse. Where are the foci located?

Solution If $x = a\cos(t)$ and $y = b\sin(t)$, then $x^2/a^2 + y^2/b^2 = \cos^2(t) + \sin^2(t) = 1$. Therefore $\mathbf{r}(t)$ describes an ellipse. Although the half-distance c between the two foci does not explicitly appear in the equation $x^2/a^2 + y^2/b^2 = 1$, we can use the formula $b = (a^2 - c^2)^{1/2}$ to determine that $c = (a^2 - b^2)^{1/2}$. Because $a > b$, the major axis of the ellipse lies along the x-axis. The foci are therefore located at $\left(-\sqrt{a^2 - b^2}, 0\right)$ and $\left(\sqrt{a^2 - b^2}, 0\right)$. ◀

DEFINITION Let c denote the half-distance between the foci of an ellipse. Let a denote half the length of the major axis of the ellipse. The quantity $e = c/a$ is called the *eccentricity* of the ellipse.

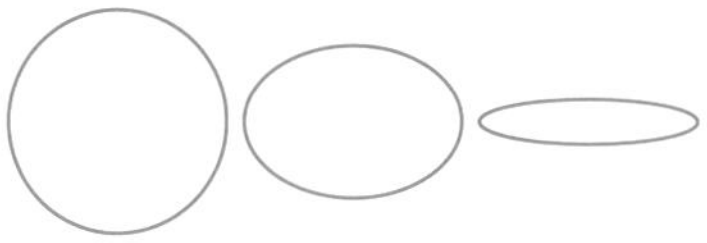

Eccentricity 0.1 Eccentricity 0.75 Eccentricity 0.98

▲ Figure 6

It is traditional to use the letter e for the eccentricity of an ellipse. In this context, the letter e has nothing to do with the base of the natural logarithm. Notice that the eccentricity e of an ellipse is a nonnegative number that is less than 1. The nearer that e is to 0, the more the ellipse will look like a circle. The closer that e is to 1, the more the ellipse will look like a line segment (see Figure 6).

▶ **EXAMPLE 4** If the length of the major axis of an ellipse is double the length of its minor axis, then what is the eccentricity of the ellipse?

Solution We suppose that $2a$ is the length of the major axis, and $2b$ is the length of the minor axis. We are given that $b = a/2$. Therefore

$$e = \frac{c}{a} = \frac{\sqrt{a^2 - b^2}}{a} = \frac{\sqrt{a^2 - (a/2)^2}}{a} = \sqrt{1 - 1/4} = \frac{\sqrt{3}}{2} \approx 0.866.$$

Until now, we have described an ellipse by means of its two foci and the length $2a$ of its major axis. It is often convenient to have an alternative characterization of an ellipse. ◀

THEOREM 5 Let $\mathcal{C}$ be a closed curve in the plane. Then, $\mathcal{C}$ is an ellipse if and only if there is a real number e with $0 < e < 1$, a point F, and a line $\mathcal{D}$ such that $\mathcal{C}$ is the locus of all points P that satisfy $\|\overline{PF}\| = e \cdot |P\mathcal{D}|$. Here, $|\overline{PF}|$denotes the distance between the points P and F, $|P\mathcal{D}|$ the distance of P to the line $\mathcal{D}$. In other words, $|P\mathcal{D}|$ is the length of a perpendicular dropped from P to $\mathcal{D}$. If $\mathcal{C}$ is the ellipse defined by the equation $|\overline{PF}| = e \cdot |P\mathcal{D}|$, then

a. The eccentricity of $\mathcal{C}$ is e.
b. $\mathcal{C}$ lies on one side of $\mathcal{D}$.
c. F is a focus of $\mathcal{C}$, the focus closest to $\mathcal{D}$.
d. The major axis of $\mathcal{C}$ is perpendicular to $\mathcal{D}$.

Proof. Rather than give a general proof, we will sketch the key algebraic procedure for establishing this theorem. Consider the ellipse $x^2/a^2 + y^2/b^2 = 1$ with foci$(-c, 0)$ and $F = (c, 0)$. The distance $|\overline{PF}|$ between F and a point $P = (x, y)$ on the ellipse is given by

$$|\overline{PF}| = \sqrt{(x-c)^2 + y^2} = \sqrt{(x-c)^2 + b^2\left(1 - \frac{x^2}{a^2}\right)}$$

$$= \sqrt{(x-c)^2 + (a^2 - c^2)\left(1 - \frac{x^2}{a^2}\right)} \overset{\text{Verify!}}{=} \sqrt{\frac{(a^2 - xc)^2}{a^2}} = \frac{|a^2 - xc|}{a}.$$

If we take the vertical line $x = a^2/c$ to be $\mathcal{D}$, then we have

$$|\overline{PF}| = c\frac{|a^2/c - x|}{a} = e\left|\frac{a^2}{c} - x\right| = e|P\mathcal{D}|.$$

The steps of this algebraic argument may be reversed to show that the locus of points P such that $|\overline{PF}| = e|P\mathcal{D}|$ is an ellipse. ■

DEFINITION If an ellipse is defined by the equation $|\overline{PF}| = e|P\mathcal{D}|$ then the line $\mathcal{D}$ is called a *directrix* for the ellipse.

▶ **EXAMPLE 5** Suppose that $F = (0, 0)$ and $\mathcal{D}$ is the line $x = 2$. What is the Cartesian equation for the locus of points $P = (x, y)$ for which $|\overline{PF}| = (1/2)|P\mathcal{D}|$? What are the lengths of the major and minor axes? Where are the foci located?

Solution If $P = (x, y)$ and $F = (0, 0)$, then $|\overline{PF}| = \sqrt{x^2 + y^2}$. The distance of the point (x, y) from the line $x = 2$ is $|2 - x|$. The equation $|\overline{PF}| = (1/2)|P\mathcal{D}|$ becomes

$$\sqrt{x^2 + y^2} = \frac{1}{2}|2 - x|.$$

Squaring each side results in the equation $x^2 + y^2 = (2 - x)^2/4$. If we expand the right side of this last equation and bring the terms involving x to the left side, then we obtain $3x^2/4 + x + y^2 = 1$ or $x^2 + 4x/3 + 4y^2/3 = 4/3$. We now complete the square on the left side: $\left(x^2 + 4x/3 + (2/3)^2\right) + 4y^2/3 = 4/3 + (2/3)^2$ or $(x + 2/3)^2 + 4y^2/3 = 16/9$ Finally, writing this last equation in standard form (10.5.12), we obtain

$$\frac{(x + 2/3)^2}{(4/3)^2} + \frac{y^2}{(2/\sqrt{3})^2} = 1.$$

This equation tells us that the center of the ellipse is at $(-2/3, 0)$, the length of the major axis is $2a = 2(4/3) = 8/3$, and the length of the minor axis is $2b = 2(2/\sqrt{3}) =$

$4/\sqrt{3}$. One focus is $F = (0, 0)$. The distance c between focus $F = (0, 0)$ and center $(-2/3, 0)$ is $2/3$. The second focus is also situated on the major axis a distance c from the center; it is therefore the point $(-4/3, 0)$ (see Figure 7). ◀

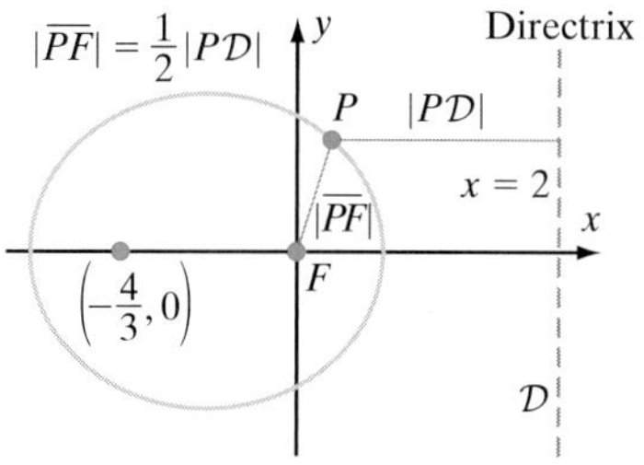

▲ Figure 7

INSIGHT In Example 5, there are several other ways to calculate the value of c. For instance, we may use the formula $c = \sqrt{a^2 - b^2}$ to obtain $c = \sqrt{16/9 - 4/3} = 2/3$. Alternatively, the equation $|\overline{PF}| = (1/2)|P\mathcal{D}|$ tells us that the eccentricity e equals $1/2$. It follows that $c = e \cdot a = (1/2)(4/3) = 2/3$.

▶ **EXAMPLE 6** Let e and d be positive constants with $e < 1$. Suppose that $\mathbf{U}$ and $\mathbf{V}$ are orthogonal direction vectors in space. Represent $\mathbf{U}$ and $\mathbf{V}$ by directed line segments that have the origin O as their common initial point. These directed line segments determine a unique plane $\mathcal{P}$. Figure 8a shows a curve $\mathcal{C}$ that lies in $\mathcal{P}$. Suppose that, for every point P on $\mathcal{C}$, the distance r of $\mathcal{P}$ to the origin O and the angle θ that $\overrightarrow{OP}$ makes with $\mathbf{U}$ are related by the equation

$$r = \frac{ed}{1 + e\cos(\theta)}.$$

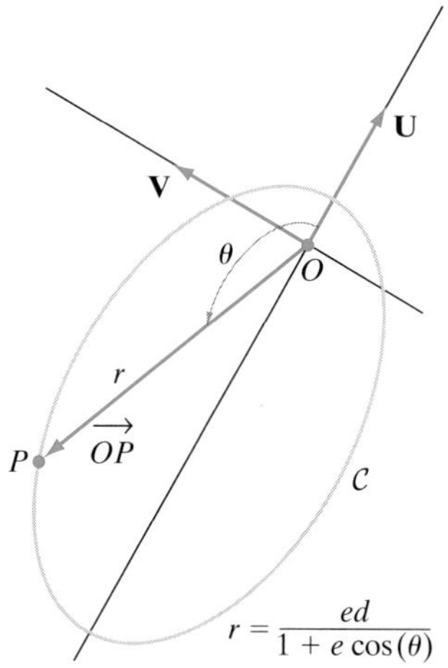

▲ Figure 8a

Show that $\mathcal{C}$ is an ellipse with one focus at O and eccentricity e.

Solution Let us write (u, v) as the coordinates in $\mathcal{P}$ of the point whose position vector is $u\mathbf{U} + v\mathbf{V}$ (see Figure 8b). In these coordinates, elementary geometry shows that $r = \sqrt{u^2 + v^2}$ and $u = r\cos(\theta)$. On cross-multiplication, the displayed equation becomes $\sqrt{u^2 + v^2} + eu = ed$. After isolating the radical and squaring, we obtain

$$u^2 + v^2 = (ed - eu)^2 = e^2d^2 - 2e^2du + e^2u^2$$

or $(1 - e^2)u^2 + 2e^2du + v^2 = e^2d^2$. Division by the nonzero coefficient of u^2 results in the equation

$$u^2 + \frac{2e^2d}{1 - e^2}u + \frac{v^2}{1 - e^2} = \frac{e^2d^2}{1 - e^2}.$$

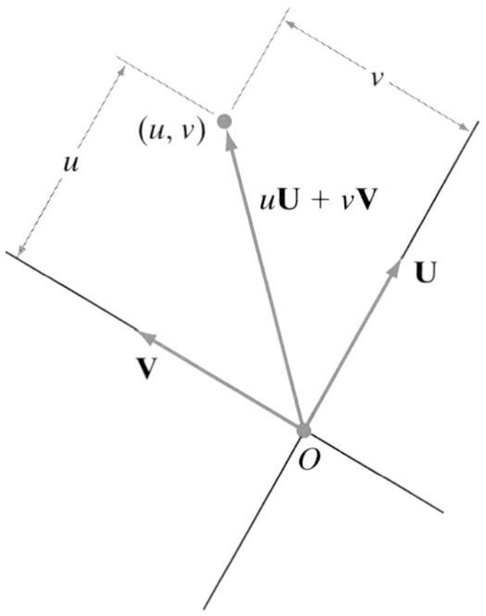

▲ Figure 8b

We complete the square by adding the square of one half of the coefficient of u to both sides of the equation:

$$\left(u^2 + \frac{2e^2d}{1 - e^2}u + \left(\frac{e^2d}{1 - e^2}\right)^2\right) + \frac{v^2}{1 - e^2} = \frac{e^2d^2}{1 - e^2} + \left(\frac{e^2d}{1 - e^2}\right)^2.$$

This equation simplifies to

$$\left(u + \frac{e^2d}{1 - e^2}\right)^2 + \frac{v^2}{1 - e^2} = \frac{e^2d^2}{(1 - e^2)^2},$$

which may be written in standard form as

$$\frac{\left(u+\dfrac{e^2 d}{1-e^2}\right)^2}{\left(\dfrac{ed}{1-e^2}\right)^2}+\frac{v^2}{\left(\dfrac{ed}{\sqrt{1-e^2}}\right)^2}=1.$$

Because $0<1-e^2<1$, we deduce that $1-e^2<\sqrt{1-e^2}$. From this inequality, it follows that $ed/(1-e^2)>ed/\sqrt{1-e^2}$. Thus the major axis has direction $\mathbf{U}$. The distance between focus and center is

$$\sqrt{\left(\frac{ed}{1-e^2}\right)^2-\left(\frac{ed}{\sqrt{1-e^2}}\right)^2}=\frac{e^2 d}{1-e^2}.$$

Consequently the eccentricity of the ellipse is the quotient of this distance $e^2 d/(1-e^2)$ by half the length of the major axis, or $ed/(1-e^2)$. The value of this quotient is simply e. ◀

Applications to Planetary Motion

In ancient times, it was believed that the planets traveled in circular orbits about Earth. The 16th century astronomer Copernicus (1473–1543) argued that the planets orbit the sun, but he believed that the paths of motion were circles. Early in the 17th century, Johannes Kepler (1571–1630), by studying many detailed planetary observations of Tycho Brahe (1546–1601), was able to devise three simple and elegant laws of planetary motion.

Kepler's Laws of Planetary Motion

I. The orbit of each planet is an ellipse with the sun at one focus.
II. The line segment from the center of the sun to the center of an orbiting planet sweeps out area at a constant rate.
III. The square of the period of revolution of a planet is proportional to the cube of the length of the major axis of its elliptical orbit, with the same constant of proportionality for any planet.

Kepler's First Law is that each planet travels in an ellipse. It turns out that the eccentricities of the ellipses that arise in the orbits of the planets are very small, so that the orbits are nearly circles, but they are definitely *not* circles. That is the importance of Kepler's First Law.

Kepler's Second Law tells us that when the planet is at its aphelion (farthest from the sun), then it is traveling relatively slowly, whereas at its perihelion (nearest point to the sun), it is traveling relatively rapidly—Figure 9.

Kepler's Third Law allows us to calculate the length of a year on any given planet from knowledge of the shape of its orbit.

Let us see how to derive Kepler's Laws from Newton's inverse square law of gravitational attraction (10.5.9). To keep matters as simple as possible, we will assume that our solar system contains a fixed sun at the origin and just one planet. The problem of analyzing interactions of gravity among three or more bodies is incredibly complicated and is still not thoroughly understood.

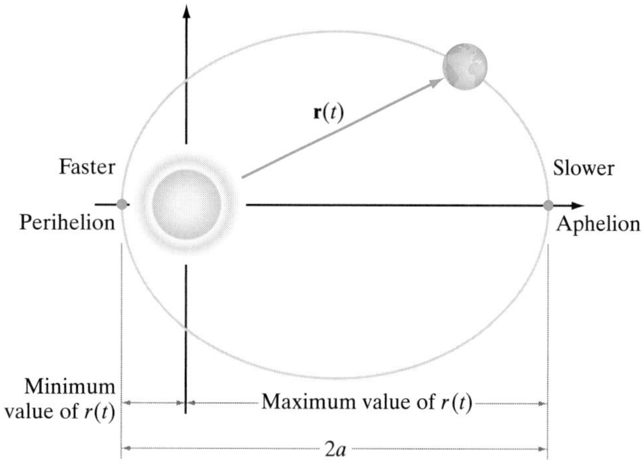

▲ Figure 9

We denote the position of the planet at time t by $\mathbf{r}(t)$. We can write this position vector as $\mathbf{r}(t) = r(t)\mathbf{u}(t)$ where $\mathbf{u}(t)$ is a unit vector pointing in the same direction as $\mathbf{r}(t)$, and $r(t)$ is a positive scalar representing the length of $\mathbf{r}(t)$. Because the gravitation force is a central force field, we know from Theorem 3 that the motion of the planet is contained in a plane. To be specific, equation (10.5.10) tells us that there is a vector $\mathbf{c}$ such that $\mathbf{r}(t) \times \mathbf{r}'(t) = \mathbf{c}$.

If $\mathbf{F}$ is force, m is the mass of the planet, and $\mathbf{a}(t) = \mathbf{r}''(t)$ is the acceleration of the planet, then Newton's Second Law of Motion says that $\mathbf{F} = m\mathbf{r}''(t)$. We can also use Newton's Law of Gravitation, formula (10.5.9), to express the force. By equating the two different expressions for the force, we conclude that $m\mathbf{r}''(t) = -\big(GMm/r(t)^2\big)\mathbf{u}(t)$ or, equivalently,

$$\mathbf{r}''(t) = -\frac{GM}{r(t)^2}\mathbf{u}(t).$$

Therefore

$$\mathbf{r}''(t) \times \mathbf{c} = -\frac{GM}{r(t)^2}\mathbf{u}(t) \times \mathbf{c} \overset{(10.5.10)}{=} -\frac{GM}{r(t)^2}\mathbf{u}(t) \times \big(\mathbf{r}(t) \times \mathbf{r}'(t)\big) = -\frac{GM}{r(t)}\mathbf{u}(t) \times \big(\mathbf{u}(t) \times \mathbf{r}'(t)\big).$$

Now

$$\mathbf{r}'(t) = \frac{d}{dt}\big(r(t)\mathbf{u}(t)\big) = r'(t)\mathbf{u}(t) + r(t)\mathbf{u}'(t).$$

Therefore

$$\mathbf{r}''(t) \times \mathbf{c} = -\frac{GM}{r(t)}\mathbf{u}(t) \times \Big(\mathbf{u}(t) \times \big(r'(t)\mathbf{u}(t) + r(t)\mathbf{u}'(t)\big)\Big) = -GM\mathbf{u}(t) \times \big(\mathbf{u}(t) \times \mathbf{u}'(t)\big). \quad (10.5.13)$$

According to formula (9.5.10), the vector triple product in equation (10.5.13) may be written as

$$\mathbf{u}(t) \times \big(\mathbf{u}(t) \times \mathbf{u}'(t)\big) = \big(\mathbf{u}(t) \cdot \mathbf{u}'(t)\big)\mathbf{u}(t) - \big(\mathbf{u}(t) \cdot \mathbf{u}(t)\big)\mathbf{u}'(t). \qquad (10.5.14)$$

Because $\mathbf{u}(t) \cdot \mathbf{u}'(t) = 0$ and $\mathbf{u}(t) \cdot \mathbf{u}(t) = 1$, equation (10.5.14) simplifies to

$$\mathbf{u}(t) \times \big(\mathbf{u}(t) \times \mathbf{u}'(t)\big) = -\mathbf{u}'(t).$$

If we substitute this result into equation (10.5.13), then we obtain

$$\mathbf{r}''(t) \times \mathbf{c} = GM\,\mathbf{u}'(t).$$

By integrating each side of this last equation, we see that

$$\mathbf{r}'(t) \times \mathbf{c} = GM\,\mathbf{u}(t) + \mathbf{h}$$

for some (constant) vector $\mathbf{h}$. Taking the dot product of each side with $\mathbf{r}(t)$ results in

$$\mathbf{r}(t) \cdot \big(\mathbf{r}'(t) \times \mathbf{c}\big) = GM\,\mathbf{r}(t) \cdot \mathbf{u}(t) + \mathbf{r}(t) \cdot \mathbf{h} = GM\,r(t) + \mathbf{r}(t) \cdot \mathbf{h} = GM\,r(t) + r(t)\|\mathbf{h}\|\cos\big(\theta(t)\big),$$

where $\theta(t)$ is the angle between $\mathbf{r}(t)$ and $\mathbf{h}$. After we set $e = \|\mathbf{h}\|/(GM)$, our equation becomes

$$\mathbf{r}(t) \cdot \big(\mathbf{r}'(t) \times \mathbf{c}\big) = GMr(t)\Big(1 + e\cos\big(\theta(t)\big)\Big). \qquad (10.5.15)$$

But, according to formula (9.4.5), we may write the triple scalar product on the left side of equation (10.5.15) as

$$\mathbf{r}(t) \cdot \big(\mathbf{r}'(t) \times \mathbf{c}\big) \overset{(9.4.5)}{=} \big(\mathbf{r}(t) \times \mathbf{r}'(t)\big) \cdot \mathbf{c} \overset{(10.5.10)}{=} \mathbf{c} \cdot \mathbf{c} = \|\mathbf{c}\|^2.$$

It follows that

$$\|\mathbf{c}\|^2 = GMr(t)\Big(1 + e\cos\big(\theta(t)\big)\Big). \qquad (10.5.16)$$

Observe that the left side of equation (10.5.16) is positive. We conclude that $e < 1$ and

$$r(t) = \frac{\|\mathbf{c}\|^2/(GM)}{1 + e\cos\big(\theta(t)\big)}.$$

Because Example 6 shows that this equation describes an ellipse, the derivation of Kepler's First Law is complete.

Kepler's Second Law about the area function $A(t)$ is a special case of Theorem 4.

The proof of Kepler's Third Law begins with an observation from Figure 9. The length $2a$ of the major axis of the elliptical orbit is equal to the maximum value of $r(t)$ plus the minimum value of $r(t)$. From the equation for the ellipse, we see that these occur respectively when $\cos\big(\theta(t)\big)$ is -1 and when $\cos\big(\theta(t)\big)$ is 1. Thus

$$2a = \frac{\|\mathbf{c}\|^2/(GM)}{1 - e} + \frac{\|\mathbf{c}\|^2/(GM)}{1 + e} = \frac{2\|\mathbf{c}\|^2}{GM(1 - e^2)}.$$

We conclude that

$$\|\mathbf{c}\| = \sqrt{aGM(1-e^2)}.$$

Now recall from the proof of Theorem 4 that the area function A satisfies $A'(t) = \|\mathbf{c}\|/2$. Then, by antidifferentiating, we find that $A(t) = \|\mathbf{c}\|t/2$. (The constant of integration is 0 because $A(0) = 0$.). It follows that

$$A(t) = \frac{t}{2}\sqrt{aGM(1-e^2)}.$$

Let T be the time it takes to sweep out one orbit. In other words, T is the year for the particular planet. Because the area inside an ellipse with major axis $2a$ and eccentricity e is $\pi a^2\sqrt{1-e^2}$, we have

$$\pi a^2\sqrt{1-e^2} = A(T) = \frac{T}{2}\sqrt{aGM(1-e^2)}.$$

Solving for T, we obtain

$$T = \frac{2\pi}{\sqrt{GM}}a^{3/2}$$

or

$$\frac{T^2}{a^3} = \frac{4\pi^2}{GM}.$$

This is Kepler's Third Law.

QUICK QUIZ

1. For two points P and Q on the trajectory of a particle, the curvature at Q is twice that at P. If the speed of the particle at Q is three times its speed at P, then by what factor is its normal component of acceleration at Q greater than it is at P?
2. At a point $\mathbf{r}(t_0)$ in its trajectory, the acceleration of a particle is $\mathbf{r}''(t_0) = \langle 6, 6, 7\rangle$, and the normal component $a_N(t_0)$ of its acceleration is $4\sqrt{6}$. What is the rate of change of the particle's speed?
3. The major axis of an ellipse in the xy-plane is the line segment with endpoints $(-1, -2)$ and $(-1, 8)$. The eccentricity of the ellipse is $4/5$. What is the right endpoint of the minor axis?
4. In a solar system, Auric's year is 64 times that of Stavro. By what factor is Auric's major axis greater than that of Stavro?

Answers

1. 18 2. 5 3. $(2, 3)$ 4. 16

EXERCISES

Problems for Practice

▸ In each of Exercises 1–8, the position vector $\mathbf{r}$ of a particle is given. Calculate $\mathbf{r}'(t)$, $\mathbf{r}''(t)$, $\mathbf{T}(t)$, $\mathbf{N}(t)$, $\kappa_{\mathbf{r}}(t)$, and the tangential and normal components, a_T and a_N, of acceleration for the planar motion. ◂

1. $\mathbf{r}(t) = t\mathbf{i} - t^2\mathbf{k}$
2. $\mathbf{r}(t) = t^2\mathbf{i} - t^3\mathbf{j}$
3. $\mathbf{r}(t) = (t - t^2)\mathbf{i} + (t + t^2)\mathbf{j}$
4. $\mathbf{r}(t) = \ln(t)\mathbf{i} + t^{1/2}\mathbf{j}$
5. $\mathbf{r}(t) = (t - \sin(t))\mathbf{i} + (1 - \cos(t))\mathbf{j}$
6. $\mathbf{r}(t) = 5\cos(3t + \pi/6)\mathbf{i} + 5\sin(3t + \pi/6)\mathbf{j}$
7. $\mathbf{r}(t) = t^{3/2}\mathbf{i} + t^{1/2}\mathbf{j}$
8. $\mathbf{r}(t) = \cos^3(t)\mathbf{i} + \sin^3(t)\mathbf{j}$

▸ In each of Exercises 9–16, the position vector $\mathbf{r}$ of a particle is given. Calculate $\mathbf{r}'(t)$, $\mathbf{r}''(t)$, $\mathbf{T}(t)$, $\mathbf{N}(t)$, $\kappa_{\mathbf{r}}(t)$, and the tangential and normal components, a_T and a_N, of acceleration for the motion in space. ◂

9. $\mathbf{r}(t) = t\mathbf{i} + t^2\mathbf{j} + t^3\mathbf{k}$
10. $\mathbf{r}(t) = (t - t^2)\mathbf{i} + (t + t^2)\mathbf{j} + t^2\mathbf{k}$
11. $\mathbf{r}(t) = (1 + t)^{3/2}\mathbf{i} + (1 - t)^{3/2}\mathbf{j} + t\mathbf{k}$
12. $\mathbf{r}(t) = t^{1/2}\mathbf{i} + t^{-1/2}\mathbf{j} + t^{-1}\mathbf{k}$
13. $\mathbf{r}(t) = e^t\mathbf{i} + e^{-t}\mathbf{j} + \sqrt{2}t\mathbf{k}$
14. $\mathbf{r}(t) = e^t\mathbf{i} + te^t\mathbf{j} + 2e^t\mathbf{k}$
15. $\mathbf{r}(t) = \cos(t)\mathbf{i} + \sin(t)\mathbf{j} + t^2\mathbf{k}$
16. $\mathbf{r}(t) = \ln(|t|)\mathbf{i} + t^{-1}\mathbf{j} + t\mathbf{k}$

▸ In each of Exercises 17–20, the position vector $\mathbf{r}$ of a particle is given. Calculate $\mathbf{a}(t) = \mathbf{r}''(t)$. Calculate the tangential and normal components, a_T and a_N, of the decomposition $\mathbf{a}(t) = a_T\mathbf{T}(t) + a_N\mathbf{N}(t)$ without calculating $\mathbf{T}(t)$ and $\mathbf{N}(t)$. ◂

17. $\mathbf{r}(t) = (t + 2)\mathbf{i} + t^2\mathbf{j} + 2t\mathbf{k}$
18. $\mathbf{r}(t) = (t^2 + 1)\mathbf{i} + (t^2 - 1)\mathbf{j} + t^2\mathbf{k}$
19. $\mathbf{r}(t) = e^t\mathbf{i} + e^{-t}\mathbf{j} + t\mathbf{k}$
20. $\mathbf{r}(t) = \cos(t)\mathbf{i} + \sin(t)\mathbf{j} + \ln(\cos(t))\mathbf{k}$

▸ In each of Exercises 21–30, determine the Cartesian equation of the ellipse that is described. ◂

21. center $(0, 0)$, major axis 6, minor axis 4, foci on y-axis
22. foci $(0, 3)$ and $(0, -3)$, major axis 10
23. foci $(4, 0)$ and $(-4, 0)$, minor axis 6
24. foci $(5, 0)$ and $(-5, 0)$, eccentricity $5/13$
25. foci $(1, 14)$ and $(1, -10)$, eccentricity $12/13$
26. foci $(2, 4)$, center $(2, 1)$, minor axis 8
27. center $(1, 2)$, focus $(1, 10)$; eccentricity 0.8
28. center $(1, 2)$, focus $(-5, 2)$, minor axis 8
29. directrix $x = 6$, focus nearest the directrix $(2, 0)$, eccentricity $1/2$
30. directrix $y = 4$, focus nearest the directrix $(1, 1)$, eccentricity $3/5$

Further Theory and Practice

31. A particle moves at constant speed with position vector $\mathbf{r}$. If $\mathbf{r}'' \neq \mathbf{0}$, then what is the relationship of $\mathbf{r}''$ to $\mathbf{r}'$?
32. A particle moves with position vector $\mathbf{r}$. If $\mathbf{r}' \cdot \mathbf{r}'' = 0$, then what can be said about the speed $\|\mathbf{r}'\|$?
33. Prove Huygens's law of centripetal force for uniform circular motion: If a body of mass m moves with constant speed v along a circle of radius r, then the centripetal force $\mathbf{F}$ acting on the body has magnitude $|\mathbf{F}| = mv^2/r$.
34. A car weighing 3,000 pounds races at a constant speed of 100 mph around a circular track that has radius 1 mi. Calculate the centripetal force, in pounds, needed to hold the car on the track.
35. A man-made satellite orbits Earth in an elliptical path with major axis 1850 km. The orbit of Earth's moon has semi-major axis 354, 400 km and period 27.322 days. Calculate the time for one orbit of the man-made satellite.
36. It is known that the eccentricity of Earth's orbit about the sun is about 0.0167, and the semi-minor axis of the elliptical orbit is about 92, 943, 235 mi. Calculate the semi-major axis. Of course, it is also known that the length of time for one orbit is one Earth year (by definition). Using Kepler's Third Law and this other information, calculate GM where M is the sun's mass.
37. Earth's mass is known to be 5.976×10^{27} grams, and the gravitational constant G equals 6.637×10^{-8} $\text{cm}^3/(g \cdot \text{sec}^2)$. If a man-made satellite orbits Earth, with each orbit taking 15 h, then what is the length of the major axis of the elliptical orbit?
38. The sun's mass is 2×10^{33} grams, and the gravitational constant is given in Exercise 37. Use Kepler's Third Law to calculate the length, in kilometers, of the semi-major axis of Earth's orbit around the sun.
39. A man-made satellite orbits Earth. It is known that, at a certain moment in its orbit, the satellite's distance from Earth's center is 4,310 mi, and its speed is 900 mi/h. Moreover, astronomical observations determine that, at that moment, the angle between the position vector and the velocity vector for this satellite is $\pi/3$. Give a numerical value for $\dfrac{dA}{dt}$.
40. Show that the area inside an ellipse with major axis $2a$ and minor axis $2b$ is πab. Show that the area can also be expressed as $\pi a^2\sqrt{1 - e^2}$, where e is the eccentricity of the ellipse.

▸ Earth's semi-major axis is 149, 597, 887 km, and its year is 365.256 days. Use this information as needed in Exercises 41–44. ◂

41. The eccentricity of Mercury's orbit is 0.2056. Its year is 87.97 Earth days. Calculate Mercury's perihelion distance to the sun.

42. Pluto's orbit of the sun lasts 248.08 Earth years. Its perihelion distance to the sun is 4,436,824,613 km. Determine the eccentricity of the orbit.

43. The eccentricity of Jupiter's orbit is 0.04839. At its farthest, Jupiter is 816,081,455 km from the sun. Determine the period of Jupiter's orbit.

44. Saturn's nearest and farthest distances to the sun are 1,349,467,375 km and 1,503,983,449 km, respectively. Determine the duration of its orbit.

Exercises 45–50 outline an alternative demonstration that planets follow elliptic orbits. We assume that the motion is planar. Suppose that the sun is at the origin of the plane of motion and that the x axis passes through the aphelion of the planet, as in Figure 10. Let $r = r(t)$ denote the distance of the planet to the sun, and let θ be the angle between the position vector of the planet and the positive x axis. This angle is called the *true anomaly* in astronomy. We assume that $r^2\dot{\theta} = C$ where C is a constant, and $\dot{\theta}$ is the derivative of the true anomaly with respect to time—as we will see in Chapter 12, this equation expresses Kepler's second law. Let M and m denote the masses of the sun and planet, respectively, and let G denote the gravitational constant.

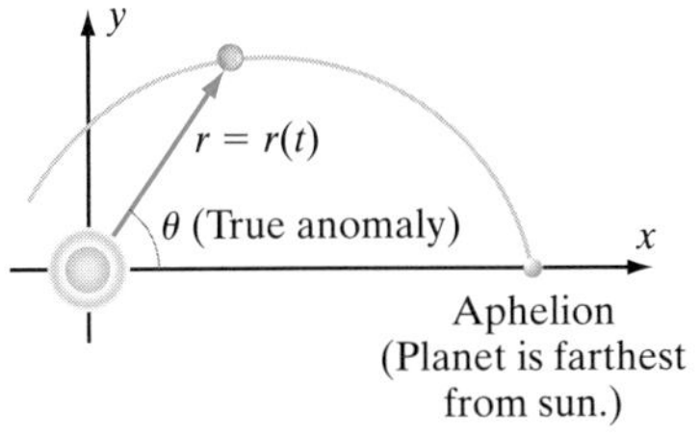

Figure 10

45. After resolving gravitational acceleration into the x and y directions, use the Chain Rule and the equation $r^2\dot{\theta} = C$ to show that

$$\frac{d\dot{x}}{d\theta} = -\frac{GM}{C}\cos(\theta)$$

and

$$\frac{d\dot{y}}{d\theta} = -\frac{GM}{C}\sin(\theta).$$

46. The hodograph of a motion $t \mapsto \langle x(t), y(t)\rangle$ is the parameterized velocity curve $t \mapsto \langle \dot{x}(t), \dot{y}(t)\rangle$. Integrate the equations of the preceding exercise to show that

$$\dot{x} = -\frac{GM}{C}\sin(\theta) + a$$

and

$$\dot{y} = \frac{GM}{C}\cos(\theta) + b$$

for some constants a and b. Deduce that the hodograph of planetary motion is a circle.

47. Substitute $x = r\cos(\theta)$ and $y = r\sin(\theta)$ into the equations of Exercise 46. Eliminate $\dot{r}$ from the resulting equations to obtain

$$r \cdot \dot{\theta} = \frac{GM}{C} - a\sin(\theta) + b\cos(\theta).$$

Deduce that

$$r = \frac{C}{GM/C - a\sin(\theta) + b\cos(\theta)}.$$

48. Use the Addition Formula for the cosine to show that

$$r = \frac{ed}{1 + e\cos(\theta)}$$

for some constants e and d.

49. Show that

$$\frac{d\dot{x}}{dt} = -\frac{GM}{r^3}x$$

and

$$\frac{d\dot{y}}{dt} = -\frac{GM}{r^3}y.$$

Then, show that

$$\frac{d}{dt}(\dot{x})^2 = -\frac{GM}{r^3}\frac{d}{dt}x^2$$

and

$$\frac{d}{dt}(\dot{y})^2 = -\frac{GM}{r^3}\frac{d}{dt}y^2.$$

50. Use the equations of the preceding exercise to show that there is a constant E such that

$$\frac{1}{2}m(\dot{x}^2 + \dot{y}^2) = \frac{GMm}{r} + E.$$

(In Chapter 13, we will learn that this equation expresses the conservation of total energy.)

51. Suppose that the acceleration $\mathbf{a}(t)$ of a smooth trajectory is differentiable. Show that if $\|\mathbf{a}(t_0)\| = a_N(t_0) = A$ with $A \neq 0$, then the graphs of $t \mapsto \|\mathbf{a}(t)\|$ and $t \mapsto a_N(t)$ are tangent to each other at the point (t_0, A).

Calculator/Computer Exercises

In each of Exercises 52–57, the trajectory $t \mapsto \mathbf{r}(t)$ of a moving particle is given. Plot $t \mapsto \|\mathbf{a}(t)\|$, $t \mapsto a_T(t)$, and $t \mapsto a_N(t)$ in the specified viewing window W before answering the questions about the acceleration.

52. $\mathbf{r}(t) = t^4\mathbf{i} + (2+t)\mathbf{j} + t^2\mathbf{k}$, $W = [0,1] \times [0,12.2]$ At what point on the curve defined by $\mathbf{r}$ are the normal and tangential components of acceleration equal? At what point does the normal component of acceleration have a local minimum?

53. $\mathbf{r}(t) = t^3\mathbf{i} + e^t\mathbf{j} + t^2\mathbf{k}$, $W = [-1,1] \times [-1,5]$ At what point on the curve defined by $\mathbf{r}$ are the normal and tangential components of acceleration equal? At what point do the graphs of normal acceleration and magnitude of acceleration touch? What condition must be satisfied at such a point of contact?

54. $\mathbf{r}(t) = 1/(1+t^2)\mathbf{i} + t/(1+t^2)\mathbf{j} + t^3\mathbf{k}$, $W = [-1,1] \times [-6,6]$ Over what time intervals is the speed decreasing? At what points do the graphs of normal acceleration and magnitude of acceleration touch? What condition must be satisfied at such a point of contact?

55. $\mathbf{r}(t) = \cos(t)\mathbf{i} + \sin(2t)\mathbf{j} + (\cos(3t) + \sin(5t))\mathbf{k}$, $W = [1/4,1] \times [-30,35]$ On what interval is the particle slowing down? On what interval are both components of acceleration decreasing?

56. $\mathbf{r}(t) = (t+\cos(t))\mathbf{i} + (t-\sin(t))\mathbf{j} + t\mathbf{k}, W = [0,\pi] \times [-3/4,1]$ On what interval is tangential acceleration greater than normal acceleration? Find the points at which $|a_T|$ has a local extremum. Show that a_N has a local extremum at each of these points. Explain why this behavior happens for $\mathbf{r}$ but cannot be expected to happen for other trajectories.

57. $\mathbf{r}(t) = \exp(\sin(t))\mathbf{i} + \exp(\cos(t))\mathbf{j} + \ln(2+\cos(t))\mathbf{k}$, $W = [0,2\pi] \times [-7/4,3]$ Where does the absolute maximum of normal acceleration occur? Where does the absolute maximum value A of the magnitude of acceleration occur? (The quantities a_T, a_N, and A are required for Exercise 58.)

58. Let $a_T(t)$, $a_N(t)$, and A be as in Exercise 57. Plot the planar curve $t \mapsto a_T(t)\mathbf{i} + a_N(t)\mathbf{j}$, $0 < t < 2\pi$. To your plot, add the semicircle of radius A that is centered at the origin and lies in the upper half-plane. Explain the relationship between the two plotted curves.

59. Calculate the tangential component $a_T(t)$ and normal component $a_N(t)$ of acceleration for the space curve $\mathbf{r}(t) = 1/(1+t^2)\mathbf{i} + t/(1+t^2)\mathbf{j} + t^3\mathbf{k}$ (the curve of Exercise 54). Plot the planar curve $t \mapsto a_T(t)\mathbf{i} + a_N(t)\mathbf{j}$ in the viewing window $[-0.3,0.3] \times [1.9,2.7]$. Explain how the loop that you see pertains to the particle's movement. Add vertical tangent lines to the plot, and explain how they are obtained.

60. Calculate the tangential component $a_T(t)$ and normal component $a_N(t)$ of acceleration for the space curve $\mathbf{r}(t) = t^2\mathbf{i} + 1/(1+t^2)\mathbf{j} + (t+t^2)\mathbf{k}$. Plot the planar curve $t \mapsto a_T(t)\mathbf{i} + a_N(t)\mathbf{j}, 3/8 < t < 9/8$. Find the two values t_0 and t_1 for which $a_T(t_0) = a_T(t_1)$ and $a_N(t_0) = a_N(t_1)$.

Summary of Key Topics in Chapter 10

A vector-valued function (in space) of a real variable has the form $\mathbf{r}(t) = r_1(t)\mathbf{i} + r_2(t)\mathbf{j} + r_3(t)\mathbf{k}$. For each t, we think of $\mathbf{r}(t)$ as the position vector of a point in space. We represent $\mathbf{r}(t)$ by a directed line segment with the origin as its initial point. As t runs through the domain of $\mathbf{r}$, the endpoints of the directed line segments representing $\mathbf{r}(t)$ form a curve $\mathcal{C}$ in space. When t represents time, we can think of the curve $\mathcal{C}$ as the path of a particle moving through space. We say that $\mathbf{r}$ is a *parameterization of* $\mathcal{C}$.

If $\mathbf{L}$ is a vector, then we say that

$$\lim_{t \to c} \mathbf{r}(t) = \mathbf{L}$$

if, for any $\varepsilon > 0$, there is a $\delta > 0$ such that

$$0 < |t-c| < \delta \quad \text{implies} \quad \|\mathbf{r}(t) - \mathbf{L}\| < \varepsilon.$$

We say that $\mathbf{r}$ is continuous at $t = c$ if c is in the domain of $\mathbf{r}$ and

$$\lim_{t \to c} \mathbf{r}(t) = \mathbf{r}(c).$$

We define the derivative of $\mathbf{r}$ at $t = c$ to be

$$\lim_{\Delta t \to 0} \frac{1}{\Delta t}\big(\mathbf{r}(c + \Delta t) - \mathbf{r}(c)\big),$$

provided that the limit exists.

A vector-valued function is continuous at c if and only if all component functions r_j are continuous at c. It is differentiable at c if and only if all component functions r_j are differentiable at c.

Properties of Limits
(Section 10.1)

Limits of vector-valued functions satisfy the following familiar properties: If $\mathbf{f}(t)$ and $\mathbf{g}(t)$ are vector-valued functions and if $\lim_{t \to c}\mathbf{f}$ and $\lim_{t \to c}\mathbf{g}$ exist, then

a. $\lim_{t \to c}(\mathbf{f} \pm \mathbf{g})(t) = \lim_{t \to c}\mathbf{f}(t) \pm \lim_{t \to c}\mathbf{g}(t)$
b. $\lim_{t \to c}(\mathbf{f} \cdot \mathbf{g})(t) = (\lim_{t \to c}\mathbf{f}(t)) \cdot (\lim_{t \to c}\mathbf{g}(t))$
c. $\lim_{t \to c}(\lambda \mathbf{f}(t)) = \lambda(\lim_{t \to c}\mathbf{f}(t))$ for any constant λ.

It follows that the set of continuous functions is closed under addition and under multiplication by scalars.

If $\mathbf{f}$ and $\mathbf{g}$ are vector-valued functions that are continuous at c, then the scalar-valued function $\mathbf{f} \cdot \mathbf{g}$, and the vector-valued function $\mathbf{f} \times \mathbf{g}$, will also be continuous at c. If $\mathbf{f}$ and $\mathbf{g}$ are differentiable at c, then $\mathbf{f} \cdot \mathbf{g}$ and $\mathbf{f} \times \mathbf{g}$ will also be differentiable at c. We have the product rules

$$(\mathbf{f} \cdot \mathbf{g})'(t) = \mathbf{f}'(t) \cdot \mathbf{g}(t) + \mathbf{f}(t) \cdot \mathbf{g}'(t)$$

and

$$(\mathbf{f} \times \mathbf{g})'(t) = \big(\mathbf{f}'(t) \times \mathbf{g}(t)\big) + \big(\mathbf{f}(t) \times \mathbf{g}'(t)\big).$$

There is also a sum-difference rule for differentiation and a rule for multiplication by scalars. If ϕ is a scalar-valued differentiable function, then $\phi\mathbf{f}$ makes sense, is differentiable, and

$$(\phi\mathbf{f})'(t) = \phi'(t)\mathbf{f}(t) + \phi(t)\mathbf{f}'(t).$$

Finally, if ψ is a differentiable scalar valued function and if $\mathbf{r} \circ \psi$ makes sense, then $\mathbf{r} \circ \psi$ is differentiable and

$$(\mathbf{r} \circ \psi)'(t) = \psi'(t)\mathbf{r}'\big(\psi(t)\big).$$

Velocity and Acceleration
(Section 10.2)

If $\mathbf{r}$ is a differentiable, vector-valued function of a real variable, and if we think of $\mathbf{r}$ as describing the motion of a body through space, then $\mathbf{r}'(t)$ is *velocity* at time t, and $\mathbf{r}''(t)$ is *acceleration* at time t. Speed is a scalar-valued quantity and is given by

$$v(t) = \|\mathbf{r}'(t)\|.$$

If $\mathbf{r}'(t)$ and $\mathbf{r}''(t)$ exist, if $\mathbf{r}''(t)$ is continuous, and if $\mathbf{r}'(t)$ is never the 0 vector, then we say that $\mathbf{r}$ is a *smooth parameterization* of the curve it defines.

Tangent Vectors and Arc Length (Section 10.3)

If $\mathbf{r}$ is a smooth parameterization of a curve C, then the vector $\mathbf{r}'(t)$ can be interpreted as the tangent vector to C at the point $\mathbf{r}(t)$. The *unit tangent vector* is then

$$\mathbf{T}(t) = \frac{1}{\|\mathbf{r}'(t)\|}\mathbf{r}'(t).$$

The tangent line to C at the point $\mathbf{r}(t)$ is the line passing through the point $\mathbf{r}(t)$ with direction $\mathbf{T}(t)$.

If $t \mapsto \mathbf{r}(t), a \le t \le b$ is a smooth parameterization of a curve C, then the arc length of C is

$$\int_a^b \|\mathbf{r}'(\tau)\|\, d\tau.$$

The arc length of C does not depend on the choice of parameterization.

A vector function $s \mapsto \mathbf{p}(s), 0 \le s \le L$ is said to *parameterize* a curve C *with respect to arc length* if $\mathbf{p}$ is a parameterization of C such that, for all s in the domain of parameterization, the arc length from $\mathbf{p}(0)$ to $\mathbf{p}(s)$ equals s. A smooth parameterization $t \mapsto \mathbf{r}(s), 0 \le s \le L$ of C is the arc length parameterization of C if and only if $\|\mathbf{r}'(s)\| = 1$ for all s. Otherwise, the reparameterization $\mathbf{p}(s) = \mathbf{r}(\sigma^{-1}(s))$, where

$$\sigma(t) = \int_a^t \|\mathbf{r}'(\tau)\|\, d\tau,$$

is the arc length parameterization.

Curvature (Section 10.4)

If C is a curve that is parameterized according to arc length by $\mathbf{p}$, then the *curvature* of C at $\mathbf{p}(s)$ is defined to be

$$\kappa(s) = \|\mathbf{T}'(s)\| = \|\mathbf{p}''(s)\|.$$

The vector $\mathbf{p}''(s)$ is always perpendicular to $\mathbf{T}(s)$. We define the principal unit normal to $\mathbf{p}$ at t to be

$$\mathbf{N}(s) = \frac{1}{\|\mathbf{p}''(s)\|}\mathbf{p}''(s).$$

We have

$$\mathbf{p}''(s) = \kappa(s)\mathbf{N}(s).$$

The *radius of curvature* at $\mathbf{p}(s)$ is defined to be $\rho(s) = 1/\kappa(s)$ if $\kappa(s) \neq 0$. The *osculating plane* at $\mathbf{p}(s)$ is the plane through $\mathbf{p}(s)$ that contains the vectors $\mathbf{T}$ and $\mathbf{N}$.

The *osculating circle* (or *circle of curvature*) at the point $\mathbf{p}(s)$ is the unique circle with radius $\rho(s)$, center $\mathbf{p}(s) + \rho(s)\mathbf{N}(s)$, and lying in the osculating plane.

If $\mathbf{r}$ is *any* smooth parameterization of a curve, then the curvature at $\mathbf{r}(t)$ is given by

$$\kappa_{\mathbf{r}}(t) = \frac{\|\mathbf{r}'(t) \times \mathbf{r}''(t)\|}{\|\mathbf{r}'(t)\|^3}.$$

For a planar curve that is parameterized by $\mathbf{r}(t) = x(t)\mathbf{i} + y(t)\mathbf{j}$, then the curvature at $\mathbf{r}(t)$ is given by

$$\kappa_{\mathbf{r}}(t) = \frac{|x'(t)y''(t) - y'(t)x''(t)|}{(x'(t)^2 + y'(t)^2)^{3/2}}.$$

If the planar curve is the graph of the equation $y = f(x)$, then the curvature at $(x, f(x))$ is

$$\kappa(x) = \frac{|f''(x)|}{\left(1 + (f'(x))^2\right)^{3/2}}.$$

Tangential and Normal Components of Acceleration (Section 10.5)

If $\mathbf{r}(t)$ represents a motion through space, then the acceleration vector $\mathbf{r}''(t)$ can be decomposed as

$$\mathbf{r}''(t) = \left(\frac{d}{dt}v(t)\right)\mathbf{T}(t) + v(t)^2\kappa_{\mathbf{r}}(t)\mathbf{N}(t)$$

where $v(t) = \|\mathbf{r}'(t)\|$ is the speed of the motion.

Kepler's Three Laws of Planetary Motion (Section 10.5)

A planet travels in an elliptical orbit with the sun at one focus. The rate at which area is swept out by the ray from the sun to the moving planet is constant. The square of the period of revolution is proportional to the cube of the major axis of the elliptical orbit, with constant of proportionality independent of the particular planet.

Review Exercises for Chapter 10

In Exercises 1–3, sketch the space curve defined by the given vector-valued function.

1. $\mathbf{r}(t) = t\mathbf{i} + 3t\mathbf{j} + \sin(t)\mathbf{k}$ $\quad 0 \le t \le 2\pi$
2. $\mathbf{r}(t) = t^2\mathbf{i} + t\mathbf{j} + 2\mathbf{k}$ $\quad 0 \le t \le 2$
3. $\mathbf{r}(t) = t\mathbf{i} + \sin(t)\mathbf{j} + \cos(t)\mathbf{k}$ $\quad 0 \le t \le 4\pi$

In Exercises 4–6, a vector-valued function $\mathbf{r}$ and a point c are given. Calculate the limit of $\mathbf{r}(t)$ as $t \to c$.

4. $\mathbf{r}(t) = \sin(t)\mathbf{i} - \cos(t)\mathbf{j} + t\mathbf{k}$ $\quad c = \pi$
5. $\mathbf{r}(t) = \langle e^t, t^2, \ln(t-1)\rangle$ $\quad c = 2$
6. $\mathbf{r}(t) = \dfrac{1-1}{1+t}\mathbf{i} - \dfrac{2}{1+t^2}\mathbf{j} + t^2\mathbf{k}$ $\quad c = 2$

In Exercises 7–9, calculate $\mathbf{r}'(t)$.

7. $\mathbf{r}(t) = t^{1/3}\mathbf{i} - 3t^2\mathbf{j} + 2t\mathbf{k}$
8. $\mathbf{r}(t) = \langle \sin(t), \tan(t), e^t\rangle$

9. $\mathbf{r}(t) = \frac{1}{t}\mathbf{i} - \frac{t}{t+1}\mathbf{j} + t^{3/2}\mathbf{k}$

In Exercises 10–12, calculate the indicated indefinite integral.

10. $\int \left(t^2\mathbf{i} + \cos(t)\mathbf{j} - \sin(t)\mathbf{k}\right)dt$

11. $\int \left(e^t\mathbf{i} - e^{2t}\mathbf{j} + e^{3t}\mathbf{k}\right)dt$

12. $\int \left(\sqrt[3]{t}\mathbf{i} - \sqrt{t}\mathbf{j} + \sqrt[4]{t}\mathbf{k}\right)dt$

In Exercises 13–15, find the antiderivative $\mathbf{G}(t)$ of $g(t)$ that satisfies the additional given condition.

13. $\mathbf{g}(t) = t^2\mathbf{i} - t^4\mathbf{j} + t^6\mathbf{k}$ $\quad \mathbf{G}(1) = 2\mathbf{i} - 3\mathbf{j} + \mathbf{k}$

14. $\mathbf{g}(t) = \sqrt{t}\mathbf{i} - t^3\mathbf{j} + t\mathbf{k}$ $\quad \mathbf{G}(0) = 4\mathbf{i} + 7\mathbf{j} - \mathbf{k}$

15. $\mathbf{g}(t) = \sqrt{t+1}\mathbf{i} + t^3\mathbf{j} - 3t\mathbf{k}$ $\quad \mathbf{G}(2) = 2\mathbf{i} - \mathbf{j} + 4\mathbf{k}$

In Exercises 16–18, let $\mathbf{f}(t) = e^{3t}\mathbf{i} - \cos(t)\mathbf{j} + t^2\mathbf{k}$, $\varphi(t) = t^3 + t$, and $\lambda = 4$. In each problem, calculate the derivative of $\mathbf{r}$.

16. $\mathbf{r}(t) = \mathbf{f}(\varphi(t))$

17. $\mathbf{r}(t) = \varphi(t)\mathbf{f}(t)$

18. $\mathbf{r}(t) = \lambda\mathbf{f}(t) + 3\lambda\varphi(t)\mathbf{f}(t)$

In Exercises 19–21, calculate $r''(t)$.

19. $\mathbf{r}(t) = e^{2t}\mathbf{i} - e^{3t}\mathbf{j} + \tan(t)\mathbf{k}$

20. $\mathbf{r}(t) = \ln(t+2)\mathbf{i} + \frac{t}{t+1}\mathbf{j} - \sqrt{t}\mathbf{k}$

21. $\mathbf{r}(t) = \arcsin(t)\mathbf{i} - \arccos(t)\mathbf{j} + \tan(t)\mathbf{k}$

In Exercises 22–24, calculate for each motion $\mathbf{r}$ the corresponding velocity $\mathbf{v}$, speed v, and acceleration $\mathbf{a}$.

22. $\mathbf{r}(t) = t^3\mathbf{i} - \sqrt{t}\mathbf{j} + t^{-1/4}\mathbf{k}$

23. $\mathbf{r}(t) = e^{2t}\mathbf{i} - e^{4t}\mathbf{j} + e^{-3t}\mathbf{k}$

24. $\mathbf{r}(t) = \ln(t)\mathbf{i} + e^t\mathbf{j} - 5t^5\mathbf{k}$

In each of Exercises 25–27, a vector-valued function $\mathbf{r}$ and a point P_0 are given. Determine parametric equations for the line that is tangent at P_0 to the curve described by $\mathbf{r}$.

25. $\mathbf{r}(t) = 3t^2\mathbf{i} - 2t^3\mathbf{j} + t^4\mathbf{k}$ $\quad P_0 = (3, -2, 1)$

26. $\mathbf{r}(t) = \frac{1}{1-t}\mathbf{i} - \frac{t}{1+t}\mathbf{j} + t^3\mathbf{k}$ $\quad P_0 = (-1, 2/3, 8)$

27. $\mathbf{r}(t) = \cos(\pi t)\mathbf{i} + \ln(1+t)\mathbf{j} - \sin(\pi t)\mathbf{k}$ $\quad P_0 = (-1, \ln(2), 0)$

In each of Exercises 28–30, a vector-valued function $\mathbf{r}$ and a point P_0 are given. Determine symmetric equations for the line that is tangent at P_0 to the curve described by $\mathbf{r}$.

28. $\mathbf{r}(t) = \frac{2}{t}\mathbf{i} - \frac{t}{1+t}\mathbf{j} + \sqrt{t}\mathbf{k}$ $\quad P_0 = (2, -1/2, 1)$

29. $\mathbf{r}(t) = t^4\mathbf{i} - t^3\mathbf{j} + t^2\mathbf{k}$ $\quad P_0 = (16, -8, 4)$

30. $\mathbf{r}(t) = \sqrt[3]{t}\mathbf{i} + \sqrt[5]{t}\mathbf{j} - \sqrt[4]{t}\mathbf{k}$ $\quad P_0 = (1, 1, -1)$

In Exercises 31–33, a body is falling near the Earth's surface with constant acceleration due to gravity given by $\mathbf{a}(t) = -32\mathbf{k}$. The initial height $\mathbf{r}_0 = \mathbf{r}(0)$ and initial velocity $\mathbf{v}_0 = \mathbf{v}(0)$ are given in each problem. Using antidifferentation, calculate an explicit formula for $\mathbf{r}(t)$.

31. $\mathbf{r}_0 = \mathbf{i} - 2\mathbf{k}$ $\quad \mathbf{v}_0 = -\mathbf{i} + 4\mathbf{j} - \mathbf{k}$

32. $\mathbf{r}_0 = 3\mathbf{i} - 4\mathbf{j} + 2\mathbf{k}$ $\quad \mathbf{v}_0 = \mathbf{i} - \mathbf{j}$

33. $\mathbf{r}_0 = \mathbf{i} - \mathbf{j} + 2\mathbf{k}$ $\quad \mathbf{v}_0 = \mathbf{j}$

In each of Exercises 34–36, a vector-valued function $\mathbf{r}$ is given, as well as a point P_0 that lies on the curve that $\mathbf{r}$ parameterizes. At, P_0 calculate the tangent vector $\mathbf{r}'$, the unit tangent vector $\mathbf{T}$, and the tangent line.

34. $\mathbf{r}(t) = (e^t + e^{-t})\mathbf{i} + (e^t - e^{-t})\mathbf{j} + e^{2t}\mathbf{k}$ $\quad P_0 = (2, 0, 1)$

35. $\mathbf{r}(t) = -t^2\mathbf{i} + t^4\mathbf{j} - t^3\mathbf{k}$ $\quad P_0 = (-4, 16, -8)$

36. $\mathbf{r}(t) = \sin(\pi t)\mathbf{i} - \cos(\pi t)\mathbf{j} + t\mathbf{k}$ $\quad P_0 = (0, -1, 0)$

In Exercises 37–39, calculate the arc length of that portion of the curve that is parametrized by the given function $\mathbf{r}$ over the specified interval.

37. $\mathbf{r}(t) = \cos(t^2)\mathbf{i} - t^2\mathbf{j} + \sin(t^2)\mathbf{k}$ $\quad 0 \le t \le \sqrt{2\pi}$

38. $\mathbf{r}(t) = \cos(5t)\mathbf{i} - \sin(5t)\mathbf{j} + t^{5/2}\mathbf{k}$ $\quad \pi \le t \le 2\pi$

39. $\mathbf{r}(t) = e^{2t}\mathbf{i} - 2\sqrt{2}t\mathbf{j} + e^{-2t}\mathbf{k}$ $\quad 0 \le t \le \ln(2)$

In each of Exercises 40–42, a vector-valued function $\mathbf{r}$ is given, as well as a point P_0 that lies on the curve that $\mathbf{r}$ parameterizes. Give parametric equations for the normal line through P_0. This is the line that passes through P_0 that has the principal unit normal vector to the curve at P_0 for its direction.

40. $\mathbf{r}(t) = t^3\mathbf{i} - t^2\mathbf{j} + 3t\mathbf{k}$ $\quad P_0 = (27, -9, 9)$

41. $\mathbf{r}(t) = \sin(3t)\mathbf{i} - \cos(3t)\mathbf{j} + 3t\mathbf{k}$ $\quad P_0 = (0, 1, 3\pi)$

42. $\mathbf{r}(t) = e^{2t}\mathbf{i} - e^{-2t}\mathbf{j} + t\mathbf{k}$ $\quad P_0 = (1, -1, 0)$

In each of Exercises 43–45, a function $\mathbf{r}$ and a value t_0 are given. At $\mathbf{r}(t_0)$, find parametric equations for the tangent and normal lines to the curve described by $\mathbf{r}$. (Refer to the instructions to Exercises 40–42 for the definition of the normal line.)

43. $\mathbf{r}(t) = t\mathbf{i} - t^2\mathbf{j} + 3t\mathbf{k}$ $\quad t_0 = 0$

44. $\mathbf{r}(t) = \cos(3t)\mathbf{i} + \sin(3t)\mathbf{j} + \tan(3t)\mathbf{k}$ $\quad t_0 = \pi$

45. $\mathbf{r}(t) = e^t\mathbf{i} + e^{-t}\mathbf{j} + \sqrt{2}t\mathbf{k}$ $\quad t_0 = 0$

In each of Exercises 46–48, a parametrized space curve is given. Calculate the moving frame $(\mathbf{T}, \mathbf{N}, \mathbf{B})$ at the point $t = t_0$.

46. $\mathbf{r}(t) = \sec(2t)\mathbf{i} - \tan(2t)\mathbf{j} + 4t\mathbf{k}$ $\quad t_0 = 2\pi$

47. $\mathbf{r}(t) = \cos(2t)\mathbf{i} - \sin(2t)\mathbf{j} + 2t\mathbf{k}$ $\quad t_0 = \pi$

48. $\mathbf{r}(t) = e^{2t}\mathbf{i} - e^{-2t}\mathbf{j} + (e^{2t} - e^{-2t})\mathbf{k}$ $\quad t_0 = 0$

In each of Exercises 49–51, a function $\mathbf{r}$ is given, as well as a point P_0 on the curve parameterized by $\mathbf{r}$. Find a Cartesian equation for the normal plane to the curve at P_0.

49. $\mathbf{r}(t) = t^3\mathbf{i} - 3t^4\mathbf{j} - 4t^5\mathbf{k}$ $\quad P_0 = (1, -3, -4)$

50. $\mathbf{r}(t) = \sqrt{t}\mathbf{i} - t^{-1/3}\mathbf{j} + t^{1/4}\mathbf{k}$ $\quad P_0 = (1, -1, 1)$

51. $\mathbf{r}(t) = \cos(t)\mathbf{i} - \sin(t)\mathbf{j} + t\mathbf{k} \quad P_0 = (-1, 0, \pi)$

In each of Exercises 52–54, verify that the given parametrization $\mathbf{p}$ is an arc length parametrization. Then calculate $\mathbf{T}(s)$, $\mathbf{N}(s)$, and $\kappa(s)$.

52. $\mathbf{p}(s) = 3\mathbf{i} - \sin(s)\mathbf{j} - \cos(s)\mathbf{k}$
53. $\mathbf{p}(s) = 3\cos(s)/5\mathbf{i} - 4\cos(s)/5\mathbf{j} + \sin(s)\mathbf{k}$
54. $\mathbf{p}(s) = (3 - \sin(s))\mathbf{i} + (6 + \cos(s))\mathbf{j}$

In each of Exercises 55–57, a vector-valued function $\mathbf{r}$ is given, as well as a value t_0 in its domain. For the space curve parameterized by $\mathbf{r}$, calculate the radius of curvature and the center of curvature at $\mathbf{r}(t_0)$.

55. $\mathbf{r}(t) = e^{-t}\mathbf{i} - e^{t}\mathbf{j} + \sqrt{2}t\mathbf{k} \quad t_0 = 0$
56. $\mathbf{r}(t) = t^2\mathbf{i} - t\mathbf{j} + t\mathbf{k} \quad t_0 = 1$
57. $\mathbf{r}(t) = (1 - 2t)\mathbf{i} + t^2\mathbf{j} - t\mathbf{k} \quad t_0 = 0$

In each of Exercises 58–60, calculate the curvature at each point of the graph of $y = f(x)$.

58. $f(x) = x^4$
59. $f(x) = e^x - e^{-x}$
60. $f(x) = 3x^2 - 4x$

In each of Exercises 61–63, a vector-valued function $\mathbf{r}$ is given. Calculate the curvature at $\mathbf{r}(t)$ of the planar curve that $\mathbf{r}$ parameterizes.

61. $\mathbf{r}(t) = \cos(t^2)\mathbf{i} - \sin(t^2)\mathbf{j}$
62. $\mathbf{r}(t) = e^{-t}\mathbf{i} + e^{t}\mathbf{j}$
63. $\mathbf{r}(t) = t^{3/2}\mathbf{i} - t^{1/2}\mathbf{j}$

In each of Exercises 64–66, a vector-valued function $\mathbf{r}$ is given, as well as a point P_0 that lies on the space curve parameterized by $\mathbf{r}$. Calculate a Cartesian equation of the osculating plane at P_0.

64. $\mathbf{r}(t) = t\mathbf{i} - e^{t}\mathbf{j} + e^{2t}\mathbf{k} \quad P_0 = (0, -1, 1)$
65. $\mathbf{r}(t) = \cos(2t)\mathbf{i} + \sin(2t)\mathbf{j} + 2t\mathbf{k} \quad P_0 = (1, 0, 2\pi)$
66. $\mathbf{r}(t) = t^{1/3}\mathbf{i} - t^2\mathbf{j} + 3t\mathbf{k} \quad P_0 = (2, -64, 24)$

In each of Exercises 67–69, a vector-valued function $\mathbf{r}$ is given, as well as a point t_0 in its domain. For the planar motion that $\mathbf{r}$ describes, calculate $\mathbf{r}'(t_0)$, $\mathbf{r}''(t_0)$, $\mathbf{T}(t_0)$, $\mathbf{N}(t_0)$, $\kappa_{\mathbf{r}}(t_0)$, and the tangential and normal components, a_T and a_N, of acceleration at $\mathbf{r}(t_0)$.

67. $\mathbf{r}(t) = t\mathbf{i} - t^2\mathbf{j} \quad t_0 = 0$
68. $\mathbf{r}(t) = \ln(2t + 1)\mathbf{i} + \sqrt{t}\mathbf{j} \quad t_0 = 1$
69. $\mathbf{r}(t) = e^{t}\mathbf{i} + e^{-t}\mathbf{j} \quad t_0 = 0$

In each of Exercises 70–72, a vector-valued function $\mathbf{r}$ is given, as well as a point t_0 in its domain. For the spatial motion that $\mathbf{r}$ describes, calculate $\mathbf{r}'(t_0)$, $\mathbf{r}''(t_0)$, $\mathbf{T}(t_0)$, $\mathbf{N}(t_0)$, $\kappa_{\mathbf{r}}(t_0)$, and the tangential and normal components, a_T and a_N, of acceleration at $\mathbf{r}(t_0)$.

70. $\mathbf{r}(t) = t^2\mathbf{i} - t^3\mathbf{j} + t\mathbf{k} \quad t_0 = 0$
71. $\mathbf{r}(t) = e^{t}\mathbf{i} - e^{-t}\mathbf{j} + \sqrt{2}t\mathbf{k} \quad t_0 = 0$
72. $\mathbf{r}(t) = t\mathbf{i} - \ln(t)\mathbf{j} + t^{-1}\mathbf{k} \quad t_0 = 1$

In each of Exercises 73–75, the parametrization $\mathbf{r}(t)$ of a motion in space is given. Calculate $\mathbf{a}(t) = \mathbf{r}''(t)$. Calculate the tangential and normal components, a_T and a_N, of the decomposition $\mathbf{a}(t) = a_T\mathbf{T}(t) + a_N\mathbf{N}(t)$ *without calculating* $\mathbf{T}(t)$ and $\mathbf{N}(t)$.

73. $\mathbf{r}(t) = (t^2 - t)\mathbf{i} + (t^2 + t)\mathbf{j} - t\mathbf{k}$
74. $\mathbf{r}(t) = e^{-2t}\mathbf{i} + e^{2t}\mathbf{j} + 3t\mathbf{k}$
75. $\mathbf{r}(t) = (t - 1)\mathbf{i} + (t + 1)\mathbf{j} + t^2\mathbf{k}$

In each of Exercises 76–78, write the Cartesian equations of the ellipse that is described in words.

76. foci at $(4, 0)$ and $(-4, 0)$, major axis 6
77. center $(0, 0)$, major axis 8, minor axis 6, foci on the x-axis
78. foci $(2, 6)$, $(2, 10)$, eccentricity $3/4$

Tycho Brahe

On the night of November 11, 1572, Tycho Brahe, a young Danish nobleman with a hobbyist's interest in astronomy, cast his eyes toward the constellation Cassiopeia. To his astonishment, he sighted a new star, one that was much brighter than any other. Tycho was well aware that the appearance of a new star in the firmament, a *nova* as he called it, was extraordinary. He characterized his discovery as "the greatest wonder that has ever shown itself in the whole of nature since the beginning of the world." At the very least, Tycho understood that the spectacle he witnessed contradicted the ancient astronomy of Aristotle.

We now know that Tycho observed a cataclysmic explosion that is called a type I supernova. This celestial event is the final evolutionary stage of a white dwarf star. It occurs when the core of the star collapses, releasing an enormous quantity of energy that blows the star to bits. In our galaxy, such supernovae occur very infrequently—only three have ever been observed. The first was recorded in 1054 A.D. by astronomers in China, Korea, and Japan. Tycho observed the second.

Tycho Brahe was born in his family's ancestral seat, Knutstorp Castle, in 1546. At the age of two he was abducted by a paternal uncle and childless aunt, who brought him up. Tycho studied at the University of Copenhagen and at several renowned universities in the Germanic territories. It was at Basel that he learned the new theories of Copernicus, which he did not entirely accept. While he was a student in Rostok, Tycho engaged in a quarrel that escalated into a sword fight. During the battle Tycho received a blow that left a diagonal slash across his forehead and hacked off a substantial chunk of his nose. (A nasal prosthetic can be seen in several subsequent portraits.)

After his return to the Kingdom of Denmark, Tycho's social standing afforded him the leisure and wealth to pursue his interest in astronomy. In time, the King of Denmark, Frederick II, offered Tycho a choice of several fiefdoms. Tycho demurred. He wrote to a friend "I am displeased with society here. Among people of my own class I waste much time." While staying in the castle made famous by Shakespeare's Hamlet, Frederick conceived a way of retaining Denmark's leading scholar. From his window, he spotted the island of Hven. He granted it to Tycho and awarded him a pension so that he could build and operate an observatory there. Tycho took possession of the island in 1576. For the next twenty-one years, Tycho watched and recorded the planets.

Tycho's fortune took a turn for the worse when his royal patron died. Frederick's son and successor, Christian IV, was not inclined to continue the pension that his father had granted Tycho. Nor would he agree to honor his father's pledge to allow Tycho's children to inherit Hven. (Because Tycho had married a commoner, his children did not have automatic rights of inheritance.) Tycho abandoned Hven for Copenhagen, where he started over. When Christian had him dismantle his new observatory because it obstructed the view from the royal palace, Tycho departed Denmark for good.

Johannes Kepler

When Tycho began his exile in 1597, Johannes Kepler was twenty-six years old and teaching in Graz. His teaching duties were light enough that he had time to ponder the mechanics of the solar system. He published his first book on the planets, the *Mysterium Cosmographicum*, in 1596. Though this work is more flight-of-fancy than science, it reveals Kepler's early interest in the three mathematical questions that he would eventually answer with his laws of planetary motion.

By nature, Kepler was more a mathematician than a stargazer. He did not have the instruments of Tycho, and his eyesight was poor. Tycho, to whom Kepler sent a copy of his book, commented that the observations of Copernicus on which Kepler relied were not sufficiently accurate to be the basis of the conclusions Kepler reached. Moreover, Tycho did not think highly of Kepler's inclination to theorize. Tycho chided Kepler that the force behind the motion of the planets could only be established "a posteriori, after the motions have been definitely established, and not a priori as you would do." Kepler was frustrated. "I did not wish to be discouraged, but to be taught." He was aware that in order to make progress, he needed Tycho's planetary data, data that Tycho did not share with other scholars. The problem was how to gain access. Referring to Tycho's wealth of data, Kepler wrote "Like most rich men Tycho does not know how to make proper use of his riches. Therefore one must take pains to wring his treasures from him."

As it happened, fate brought the two men together. At the same time royal hostility was driving Tycho from his native Denmark, religious persecution was driving Kepler from his post in Graz. Tycho made his way to Prague, the capital of the Holy Roman Empire, in 1599. Kepler arrived shortly afterwards. Although Tycho maintained a tight grip on his secrets at first, he eventually allowed Kepler to work on the orbit of Mars. Its eccentricity had caused great difficulties for all the circular motion theories that Ptolemy, Copernicus, and Tycho had put forth.

In 1601, Kepler became Tycho's salaried assistant. Only a few days later, Tycho attended a banquet where, according to Kepler's account, he "drank a little over-generously and experienced pressure on his bladder. He felt less concern for the state of his health than for etiquette," which required guests to remain seated at the dinner table. On the basis of this document, Tycho's death has been traditionally attributed to a burst bladder. Other reports, originating soon after Tycho's death, raised the suspicion of heavy metal poisoning. The modern verdict is that Tycho died of kidney failure brought on by enlargement of the prostate.

Tycho was laid to rest in the famous Týn Church of Prague. During the religious turmoil of the 1620s, many graves were desecrated. According to legend, Tycho's corpse was removed from its burial site during one of those desecrations. In 1901, the 300th anniversary of his interment, Tycho's crypt was refurbished. That restoration provided an opportunity to investigate the contents of the crypt. It was opened and the grave-robbing story debunked: The male skeleton found there had a cranial wound that was consistent with the injury Tycho suffered during his sword fight. In the late 20th century, samples of Tycho's hair, stored since 1901, were analyzed. The results indicate a very high level of mercury present. There is no evidence, however, of criminal poisoning. It seems likely that Tycho ingested the mercury as a remedy for his urinary difficulties.

Kepler's Laws

Kepler succeeded Tycho in the position of Imperial Mathematician. By compensating Tycho's heirs, the emperor was able to ensure that Tycho's observations were placed at Kepler's disposal. Those measurements were all Kepler had to work with—there were no established physical theories such as gravitation to guide him, no advanced mathematical tools such as analytic geometry and calculus to aid in calculation, no tools such as logarithms to ease the burden of numerical computation. After four years of intense mathematical labor, Kepler discovered the first two of the planetary laws that now carry his name. He published them in his *Astronomia Nova* of 1609.

For the most part, Kepler left remarkably candid accounts of the steps that precipitated his discoveries. In the case of his third law, however, Kepler was unusually silent. He did record the date, March 8, 1618, when the idea first came to him. His initial efforts to confirm his theory were not successful, but on May 15, 1618, he was able to write "A fresh assault overcame the darkness of my reason ... I feel carried away and possessed by an unutterable rapture over the divine spectacle of the heavenly harmony ... I write a book for the present time, or for posterity. It is all the same to me. It may wait a hundred years for its readers, as God has also waited six thousand years for an onlooker."

Such moments of elation were brief. As one scientist has remarked, Kepler's standard biography has passages that can bring its readers to tears. Poverty, religious persecution, war, smallpox, typhus, plague—Kepler's life was an unrelenting struggle filled with hardship and sorrow. In his 58th year, he had a premonition of death and composed his own epitaph:

> *I used to measure the heavens, now I measure the shadows of the Earth.*
>
> *Although my mind was heaven-bound, the shadow of my body lies here.*

A few months after penning this distich, Kepler took ill and died. The peace that eluded Kepler in life, eluded his mortal remains as well. Scarcely a few years passed before war ravaged the churchyard of St. Peter's, Regensburg, obliterating every trace of Kepler's burial site.

Sir Isaac Newton

The idea of a gravitational force originated in the early 16th century. By the last half of the 17th century, the foremost question of natural philosophy had become, *How could Kepler's laws of planetary motion be derived from a theory of gravitation?* The first step was to deduce the form of the gravity law. This was a problem with which Kepler wrestled unsuccessfully for thirty years. Ironically, his third law became one of the two keys that together were used to unlock the secrets of gravity. Christiaan Huygens (1629–1695) provided the second when, in 1659, he published the law of centrifugal force for uniform circular motion.

If a body of mass m moves with constant speed v along a circle of radius r, then its *centrifugal* force $\mathbf{F}$ is directed away from the center of the circle with magnitude

$$|\mathbf{F}| = \frac{mv^2}{r}.$$

To prevent a planet from flying out of its orbit, the sun must exert a gravitational force $-\mathbf{F}$ that exactly counterbalances the planet's centrifugal force. Now imagine planetary motion in an elliptic orbit with semi-major axis r and nearly 0 eccentricity—an orbit that for all practical purposes is a circle of radius r. According to Kepler's third law, there is a constant k such that the period T satisfies

$$T^2 = kr^3.$$

On the other hand, a body traveling around a circle of radius r with constant speed v completes one orbit in time

$$T = \frac{2\pi r}{v} \quad \left(\text{Time} = \frac{\text{Distance}}{\text{Rate}}\right).$$

By substituting this value of T in Kepler's third law, we see that $4\pi^2 r^2/v^2 = kr^3$ or $v^2 = 4\pi^2/(kr)$. If we now substitute this formula for v into Huygens's formula for centrifugal force, we find

$$|\mathbf{F}| = \frac{mv^2}{r} = \frac{4\pi^2 m}{k} \cdot \frac{1}{r^2}.$$

The fall of an apple caused Newton to reflect on the nature of gravity, but it was by these considerations that he deduced, in 1666, the law of gravity. Independently, Sir Edmond Halley, Sir Christopher Wren, and Robert Hooke came to the same conclusion several years later.

The next step was to show that the inverse square law *implies* Kepler's laws. During a visit to Cambridge in 1684, Halley asked Newton if he knew the curve that an inverse square law would entail. Without hesitation, Newton answered that it would be an ellipse. Three months later, he sent Halley a short manuscript that derived the three laws of Kepler from the inverse square law of gravitation. For the next fifteen months, Newton went into seclusion, devoting himself to setting out a general science of dynamics that included the Universal Law of Gravitation as well as Kepler's laws. The result was Newton's *Principia*, the most important scientific treatise ever written.

Publication was always an anxious process for Newton. In 1672, he was forced to defend a paper on optics against several criticisms, most notably from Robert Hooke, who managed to question both Newton's conclusions and Newton's priority. Newton found the business tiresome. Shortly thereafter, he declined to publish a more complete treatment of optics, remarking that with further use of the press "I shall not enjoy my former serene liberty." Many years later, when Newton looked back on the ensuing period of silence, he explained, "I began for the sake of a quiet life to decline correspondencies by Letters about Mathematical & Philosophical matters finding [that they] tend to disputes and controversies."

The *Principia* interrupted Newton's quiet life, embroiling him in a new dispute with his old nemesis, Robert Hooke. As the *Principia* neared completion, Hooke began to stir things up. Halley, who was overseeing the publication of the *Principia*, communicated the problem to Newton: "Mr Hook has some pretensions upon the invention of y[e] rule of the decrease of Gravity ... He sais you had the notion from him ... Mr Hook seems to expect you should make some mention of him in the preface." Newton was aggravated: "Philosophy is such an impertinently litigious Lady that a man had as good be engaged in Law suits as have to do with her. I found it so formerly and now I no sooner come near her again but she gives me warning." Newton responded to the baseless charge of plagiarism by deleting some references to Hooke that were already in the manuscript. In private correspondence, he referred to his antagonist as an "ignoramus."

After completing the *Principia*, Newton rapidly lost interest in mathematical research. A number of documents leave no doubt that he suffered a complete mental breakdown in 1693. The cause remains uncertain, but mercury poisoning (resulting from carefree handling of substances in chemical experiments) is a plausible candidate. Samples of Newton's hair, taken from four preserved locks, were analyzed in 1979. Although elevated levels of mercury, antimony, arsenic, gold, and lead were measured, a conclusive diagnosis is not possible 300 years after the illness. What is not in question is that in the remaining thirty-four years of his life, Newton's scientific activity was largely confined to polishing the exposition of earlier work. He retired from academic life in 1696, serving at the Mint until his death in 1727.

In 1820, the apple tree that set Newton to think about the nature of gravity succumbed to disease and was felled. Scions were taken and the line lives on, both in England and the United States. The fruit is pear-shaped and, it is said, without flavor.

Many unexpected details concerning Newton's life came to light in the 20th century. Newton's interests in alchemy, theology, and biblical chronology had long been known. However, the *extent* of those interests remained hidden until 1936 when a large portion of his estate was put up for auction. Of the volumes in Newton's personal library, 3% and 7% pertained to physics and mathematics, respectively, whereas 8% and 27.5% concerned alchemy and theology. Of Newton's personal papers, about 1 million words were devoted to scientific subjects compared to 2 million words on alchemy, theology, and chronology. These revelations led the great economist John Maynard Keynes (1883–1946) to say of Newton, "Not the first of the age of Reason. He was the last of the magicians."

CHAPTER 11

Functions of Several Variables

PREVIEW

Many of the quantities that we study in mathematics and other fields depend on two or more variables. For example, the output of a factory depends on the amount of capital that is allocated to labor and the amount that is allocated to equipment. The air pressure at a point in the atmosphere depends on the altitude of the point and also on the temperature at the point. The height above sea level of a point on the earth's surface depends on the longitude and latitude of the point. Functions of two or more variables are, indeed, commonplace. To analyze these functions, we need to develop appropriate tools of calculus. That is the purpose of this chapter. We will ask and answer the same types of questions that we posed for functions of one variable.

We begin by learning how to plot a function f of two variables: Its graph is a surface in three-dimensional space. At a point at which the function is well behaved, its graph will have a tangent plane that is composed of tangent lines. The mathematical concept that is the key to obtaining the tangent planes to the graph of f is the *partial derivative*.

If f is a function of variables x and y, then an increment Δx in x or an increment Δy in y will generally cause a change in the value of f. The numerator of the quotient

$$\frac{f(x_0 + \Delta x, y_0) - f(x_0, y_0)}{\Delta x}$$

represents the change of $f(x, y_0)$ as x varies from x_0 to $x_0 + \Delta x$ while the variable y is held constant at y_0. The denominator Δx represents the change in x. The quotient therefore represents the *average* rate of change of $f(x, y)$ as x varies from x_0 to $x_0 + \Delta x$ with y being held constant at y_0. The limit

$$\frac{\partial f}{\partial x}(x_0, y_0) = \lim_{\Delta x \to 0} \frac{f(x_0 + \Delta x, y_0) - f(x_0, y_0)}{\Delta x}$$

represents the instantaneous rate of change of f with respect to x at (x_0, y_0). This quantity is called the *partial derivative of f with respect to x*. Similarly, if we allow

y to vary while holding x fixed at x_0, then we obtain the *partial derivative of f with respect to y*:

$$\frac{\partial f}{\partial y}(x_0, y_0) = \lim_{\Delta y \to 0} \frac{f(x_0, y_0 + \Delta y) - f(x_0, y_0)}{\Delta y}.$$

As in the one-variable case, the derivative is essential for understanding and answering many of the questions that arise in the analysis of functions of many variables. As an application, we will learn how to use the partial derivatives of f to identify and classify the local maxima and minima of f.

11.1 Functions of Several Variables

The first nine chapters of this text have been concerned with scalar-valued functions that depend on one real variable. Such a function f is evaluated at a real number x in its domain, and the result of the evaluation is a real number $f(x)$. Chapter 10 is devoted to vector-valued functions that depend on one real variable. Such a function $\mathbf{r}$ is evaluated at a real number t in its domain, and the result of the evaluation is a vector $\mathbf{r}(t)$. Now we begin the study of functions that are evaluated at two or more real numbers. In this chapter, we will discuss the differential calculus of scalar-valued functions of two or more real variables.

DEFINITION Suppose that $\mathcal{D}$ is a set of ordered pairs of real numbers and that $\mathcal{R}$ is a set of real numbers. We say that f is a *(scalar-valued) function of two variables* with domain $\mathcal{D}$ and range $\mathcal{R}$ if, for every ordered pair (x, y) in $\mathcal{D}$, there is associated a unique real number in $\mathcal{R}$; we denote this number by $f(x, y)$. The number $f(x, y)$ is said to be the *image* of the point (x, y) under f. We also say that f is *evaluated* at (x, y), or that f *maps* (x, y) to the value $f(x, y)$. The notation $(x, y) \mapsto f(x, y)$ is often used; the arrow-like symbol is read as "maps to." A function F of three variables is similarly defined: The only difference is that the domain of F is a set of ordered triples. The notation $(x, y, z) \mapsto F(x, y, z)$ is used for a function F of three variables.

▶ **EXAMPLE 1** Express the surface area A and volume V of a rectangular box as functions of the side lengths.

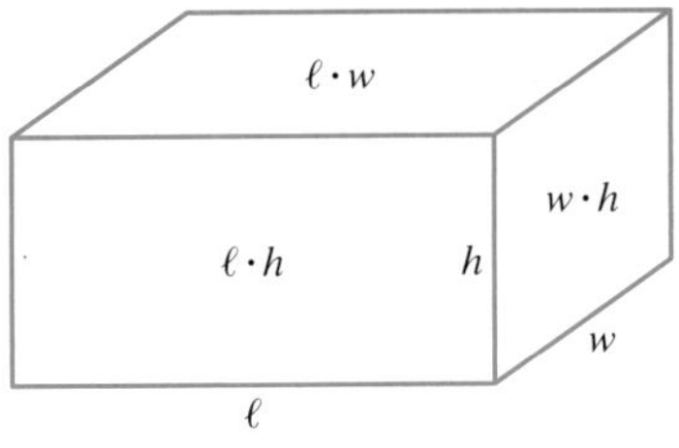

▲ Figure 1

Solution Suppose that the side lengths of the box are ℓ, w, and h (see Figure 1). Then the top and bottom have area ℓw, the left and right sides have area wh, and the front and back have area ℓh. Thus the surface area is $A(\ell, w, h) = 2\ell w + 2wh + 2\ell h$. The volume is, of course, $V(\ell, w, h) = \ell wh$. Each function has for its domain the set of ordered triples (ℓ, w, h) with all three entries positive. ◀

It is often convenient to interpret an expression to be a function. When this is done, the domain of the function is taken to be the largest set of points for which the expression makes sense and evaluates to a real number. For example, the expression $\sqrt{25 - (x^2 + y^2)}$ defines the function $(x, y) \mapsto \sqrt{25 - (x^2 + y^2)}$ whose domain consists of all ordered pairs (x, y) with $x^2 + y^2 \le 25$. We employ the same convention when we use an equation such as $f(x, y) = \sqrt{25 - (x^2 + y^2)}$ to define a function without explicitly stating its domain. As in the one-variable case, it is convenient to think of a function as an input-output machine (Figure 2). The domain is thought of as the set of input values. Evaluation is the process of getting a unique output value from an input value that is fed into the machine.

▶ **EXAMPLE 2** Let a be any constant. Discuss the domains of the functions $f(x, y) = x^2 + y^2$, $g(x, y) = a/(x^2 + y^2)$ and $h(x, y, z) = z/(x^2 + y^2)$.

Solution The expression $x^2 + y^2$ is defined for all values of x and y. Therefore the domain of f is the set of all pairs (x, y) of real numbers. For the function g, we observe that, whatever the value of a, the expression $a/(x^2 + y^2)$ is defined if x and y

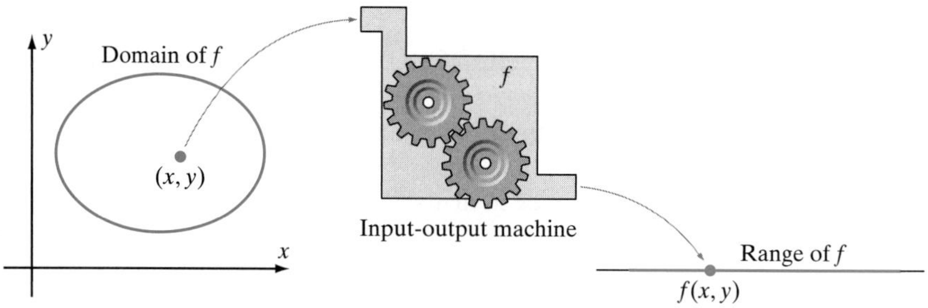

▲ Figure 2

are not both 0 and undefined if $x = y = 0$. Therefore the domain of g is the set of all ordered pairs (x, y) other than $(0, 0)$. For the same reason, the domain of h is the set of all ordered triples (x, y, z) for which x and y are not both 0. The set that is excluded from the domain of h is the z-axis in xyz-space. In other words, the domain of h is the set of all points not on the z-axis. ◄

INSIGHT Observe the distinction between functions g and h in Example 2. When we define $g(x, y) = a/(x^2 + y^2)$, we regard a as a parameter of g, not a variable. Although a may have any value, its value does not change when we consider $g(x, y)$. Indeed, the notation $g(x, y)$ tells us to treat a as a constant. On the other hand, the notation $h(x, y, z) = z/(x^2 + y^2)$ tells us that the numerator z of $h(x, y, z)$ is a variable. Even though the values $g(x, y)$ and $h(x, y, a)$ are the same, the functions g and h are quite different.

Combining Functions

Much of the elementary material that we have learned about functions in Chapter 1 still applies here. For instance, we can add, subtract, multiply, and divide functions of two variables to form new functions. If λ is a constant and if f and g are functions with a common domain $\mathcal{D}$, then we define the functions λf, $f + g$, $f - g$, fg, and $\left(\frac{f}{g}\right)$ on $\mathcal{D}$ by

a. $(\lambda f)(x, y) = \lambda f(x, y)$.
b. $(f + g)(x, y) = f(x, y) + g(x, y)$.
c. $(f - g)(x, y) = f(x, y) - g(x, y)$.
d. $(fg)(x, y) = f(x, y)g(x, y)$.
e. $\left(\frac{f}{g}\right)(x, y) = \frac{f(x, y)}{g(x, y)}$ provided $g(x, y) \neq 0$.

There are similar formulas for functions of three variables.

► **EXAMPLE 3** Let $f(x, y) = 2x + 3y^2$, $g(x, y) = 5 + x^3 y$, and $h(x, y, z) = xyz^2$. Compute $(f + 3g)(1, 2)$, $(fg)(1, 2)$, $\left(\frac{f}{g}\right)(1, 2)$, and $\left(\frac{1}{9}h^2\right)(1, 2, 3)$.

Solution We have $f(1, 2) = 2(1) + 3(2)^2 = 14$ and $g(1, 2) = 5 + (1)^3\, 2 = 7$. Therefore

$$(f + 3g)(1, 2) = f(1, 2) + 3g(1, 2) = 14 + 3(7) = 35,$$

$$(fg)(1, 2) = f(1, 2)g(1, 2) = (14)(7) = 98,$$

$$\left(\frac{f}{g}\right)(1,2) = \frac{f(1,2)}{g(1,2)} = \frac{14}{7} = 2,$$

and

$$\left(\frac{1}{9}h^2\right)(1,2,3) = \frac{1}{9}\left(h(1,2,3)\right)^2 = \frac{1}{9}\left((1)(2)(3)^2\right)^2 = 36. \quad ◀$$

Now we want to discuss composition of functions. For this, a schematic diagram is essential (see Figure 3). Consider a scalar-valued function $(x,y) \mapsto f(x,y)$ of two variables. If g is a function of one variable, and if the values of f lie in the domain of g, then we may consider the expression $g(f(x,y))$ for any point (x,y) in the domain of f. We write

$$(g \circ f)(x,y) = g\big(f(x,y)\big), \tag{11.1.1}$$

and call the function $g \circ f$ defined by (11.1.1) the *composition* of g with f. The domain of $g \circ f$ is the same as the domain of f.

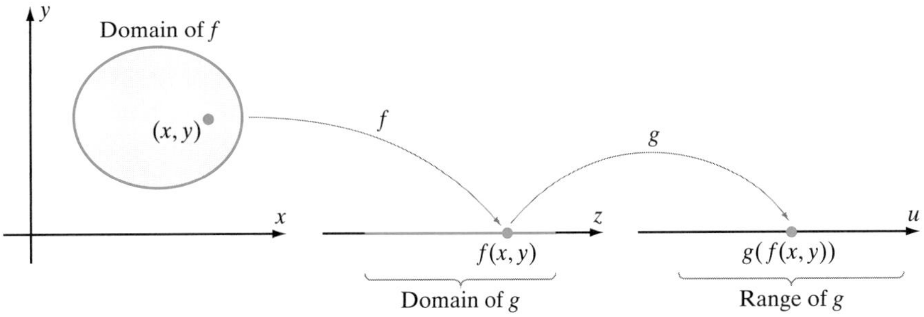

▲ Figure 3 Schematic diagram of $(g \circ f)(x,y) = g(f(x,y))$

▶ **EXAMPLE 4** Let g be defined on the set of all nonnegative numbers by the formula $g(u) = \sqrt{u}$. Suppose that f is defined on the set $\mathcal{D} = \{(x,y) : x^2 + y^2 \le 25\}$ by the formula $f(x,y) = 25 - (x^2 + y^2)$. Show that the composition $g \circ f$ is defined on $\mathcal{D}$, and calculate $(g \circ f)(2, \sqrt{5})$. If $h(u) = 1/\sqrt{9-u}$, then what is the largest subset of $\mathcal{D}$ on which $h \circ f$ is defined?

Solution The composition $(g \circ f)(x,y) = g\big(f(x,y)\big)$ is defined precisely when the values $f(x,y)$ are in the domain of g. In other words, $(g \circ f)(x,y)$ is defined provided $f(x,y) \ge 0$. This inequality holds if and only if $x^2 + y^2 \le 25$. Because the coordinates of every ordered pair (x,y) in the domain $\mathcal{D}$ of f satisfies this inequality, the composition $g \circ f$ can be formed. In particular, for (x,y) in $\mathcal{D}$, we have

$$(g \circ f)(x,y) = g\big(f(x,y)\big) = \sqrt{f(x,y)} = \sqrt{25 - (x^2 + y^2)}.$$

To calculate $(g \circ f)(2, \sqrt{5})$, we first observe that, because $2^2 + (\sqrt{5})^2 = 9$ is not greater than 25, the point $(2, \sqrt{5})$ is in the domain of f, and

$$(g \circ f)(2, \sqrt{5}) = \sqrt{25 - \left(2^2 + (\sqrt{5})^2\right)} = \sqrt{25 - 9} = 4.$$

For

$$(h \circ f)(x, y) = \frac{1}{\sqrt{9 - f(x, y)}} = \frac{1}{\sqrt{9 - \left(25 - (x^2 + y^2)\right)}} = \frac{1}{\sqrt{(x^2 + y^2) - 16}}$$

to be defined, we must have $x^2 + y^2 > 16$. Therefore the largest subset of $\mathcal{D}$ on which $h \circ f$ is defined is $\{(x, y) : 16 < x^2 + y^2 \leq 25\}$. This set is sketched in Figure 4. ◄

Graphing Functions of Several Variables

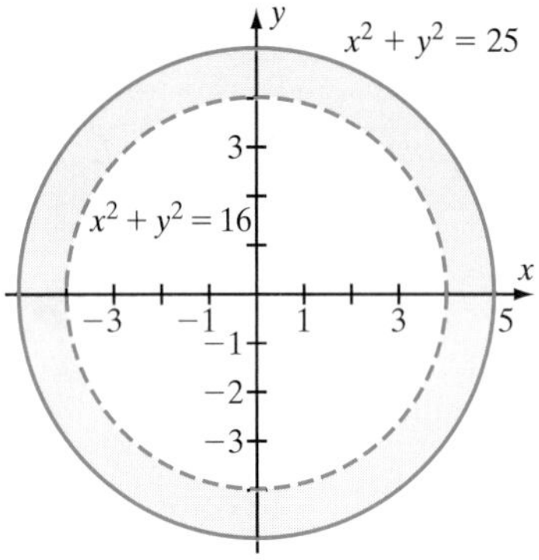

▲ Figure 4 The set $\{(x, y) : 16 < x^2 + y^2 \leq 25\}$

When we graph a scalar-valued function of one variable, we use the two-dimensional plane: One dimension is for the domain of the function, and the other is for the range. Now we will learn to graph scalar-valued functions of two variables. This will be done in three-dimensional space because we need two dimensions for the domain of the function and one dimension for the range. To be specific, the graph of the function $(x, y) \mapsto f(x, y)$ consists of all points (x, y, z) such that (x, y) is in the domain of f, and $z = f(x, y)$. In other words, we plot a point on the graph of f by locating an element (x, y) of the domain of f in the xy-plane and then, in space, going up or down $f(x, y)$ units.

Some caution is in order here. In spite of all the sophisticated techniques that we know for graphing functions of one variable, we are often tempted to just plot points and connect them with a curve. Sometimes we can get away with this because curves are often fairly simple. But now we are working in three dimensions. Our graph will be a *surface*, and it is virtually impossible to plot some points by hand and "connect them with a surface."

In fact, we need a new idea to graph functions of two variables. The idea is that a surface in three-dimensional space can be described by its two-dimensional slices. For instance, suppose that every horizontal slice of a surface is a circle of radius 1 with center lying on the z-axis (Figure 5). The only thing that this surface could be is a cylinder (Figure 6). On the other hand, suppose that every horizontal slice of a surface is a circle with center lying on the z-axis, but that the radius of the circle increases with the height. Then, you may have a mental picture of the object in Figure 7a or perhaps the one in Figure 7b (there are other possibilities as well). We need to understand how to use horizontal slices effectively.

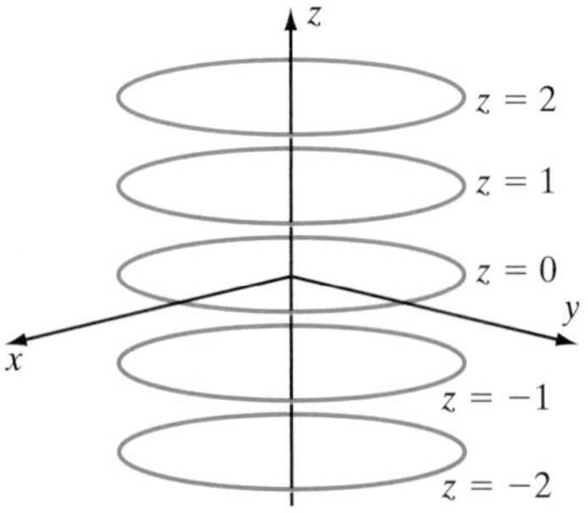

▲ Figure 5 Horizontal slices of a surface $z = f(x, y)$

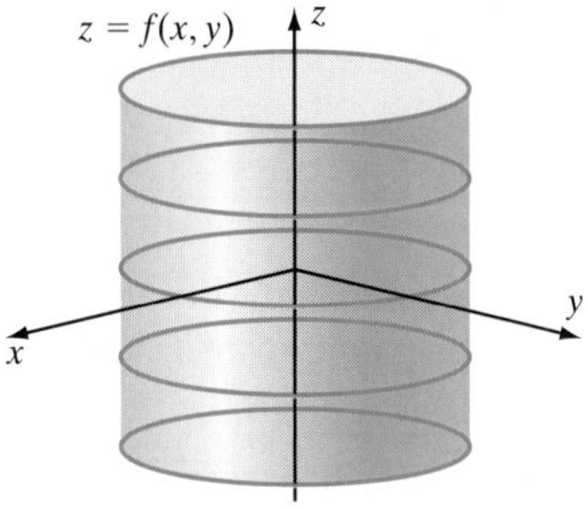

▲ Figure 6

▲ Figure 7a

▲ Figure 7b

DEFINITION Let $(x, y) \mapsto f(x, y)$ be a function of two variables. If c is a constant, then we call the set $L_c = \{(x, y) : f(x, y) = c\}$ a *level set* of f.

Notice that, by definition, a level set of a function of x and y is a subset of the xy-plane. Level sets of functions of two variables are often curves; the terminology *level curve* is used in such cases. The terms *contour, contour curve, contour line*, and *isoline* are also used.

If $L_c = \{(x, y) : f(x, y) = c\}$ is a level set of a function f, then the set $\{(x, y, c) : (x, y) \in L_c\}$ is the intersection of the graph of f with the horizontal plane $z = c$. This means that the level curves of f are obtained by projecting the horizontal slices of the graph of f into the xy-plane (see Figure 8).

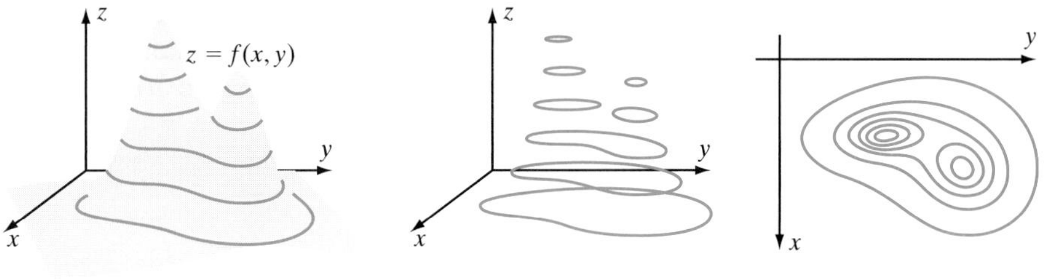

▲ Figure 8 Graph of $z = f(x, y)$ with six horizontal slices; the six horizontal slices; the level curves corresponding to the slices

▶ **EXAMPLE 5** Let $f(x, y) = x^2 + y^2 + 4$. Calculate and graph the level sets that correspond to horizontal slices at heights 20, 13, 5, 4, and 2.

Solution The level set that corresponds to the slice at height 20 is the set $\{(x, y) : x^2 + y^2 + 4 = 20\} = \{(x, y) : x^2 + y^2 = 16\}$. This is a circle of radius 4 centered at the origin of the xy-plane. The level set that corresponds to the slice at height 13 is $\{(x, y) : x^2 + y^2 + 4 = 13\} = \{(x, y) : x^2 + y^2 = 9\}$. This set is the circle of radius 3 centered at the origin of the xy-plane. Similarly, we find that the level set that corresponds to the slice at height 5 is the circle of radius 1 centered at the origin of the xy-plane. The level set that corresponds to the slice at height 4 is $\{(x, y) : x^2 + y^2 + 4 = 4\} = \{(x, y) : x^2 + y^2 = 0\} = \{(0, 0)\}$. This level set is just a single point. The graphs of these level sets appear in Figure 9. The level set that corresponds to the slice at height 2 is

$$\{(x, y) : x^2 + y^2 + 4 = 2\} = \{(x, y) : x^2 + y^2 = -2\} = \emptyset \quad \text{(the empty set)}.$$

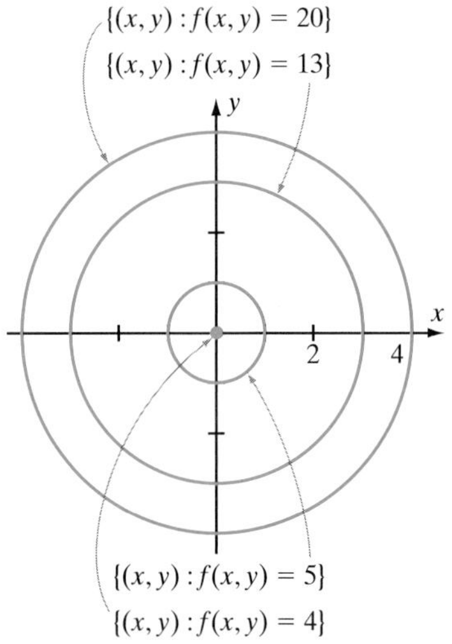

▲ Figure 9

There are no points in this level set. In fact, all the level sets that correspond to slices at heights less than 4 are empty. ◀

The level sets of a function tell us what the horizontal slices of its graph are. These in turn tell us a great deal about the graph of the function.

▶ **EXAMPLE 6** Let $f(x, y) = x^2 + y^2 + 4$, as in Example 5. Use the level sets calculated in Example 5 to plot the horizontal slices at heights 20, 13, 5, 4, and 2. Then plot the graph of f.

Solution As we calculated in Example 5, the level set that corresponds to the slice $z = 20$ is $\{(x, y) : x^2 + y^2 = 16\}$. This means that the points $\{(x, y, 20) : x^2 + y^2 = 16\}$

lie on the graph of f. This set is a circle with center $(0, 0, 20)$ and radius 4. It is located at height 20. Likewise, the slice of the graph of f at height 13 is the circle with center $(0, 0, 13)$ and radius 3. And the slice of the graph at height 5 is the circle with center $(0, 0, 5)$ and radius 1. Finally, the slice of the graph at height 4 is just the point $(0, 0, 4)$. It is important to note that, at heights below 4, the slice of the graph is the empty set. This means that the graph does not extend below the plane $z = 4$.

All this information is amalgamated into a picture in Figure 10a. How do we go from a few horizontal slices to the plot of the entire surface? This is a critical step. We piece together the horizontal slices by taking a slice in another direction: Imagine slicing the figure with the yz-plane. The shape of this last slice will tell us a lot about the shape of the figure. The yz-plane has equation $x = 0$. Now substituting $x = 0$ into $z = f(x, y)$ results in $z = y^2 + 4$, which is the equation of a parabola in the yz-plane (Figure 10b). Because all the nonempty level sets of f are circles with centers at (0, 0), it follows that the graph of f has rotational symmetry about the z-axis. Therefore the slice we have taken in the yz-plane suffices to confirm that the graph should be as shown in Figure 10c. ◄

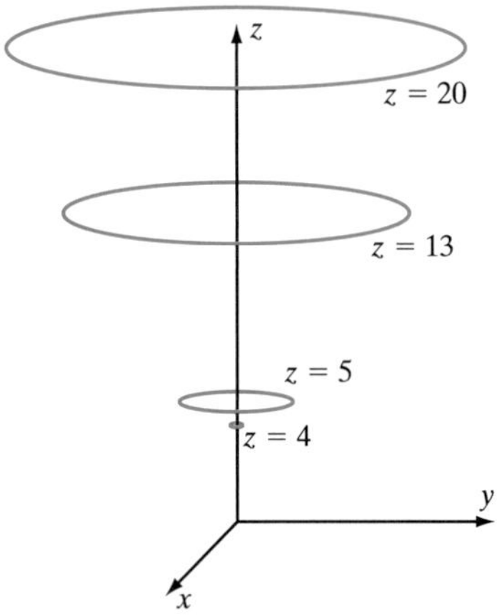

▲ Figure 10a Four horizontal slices of $z = x^2 + y^2 + 4$

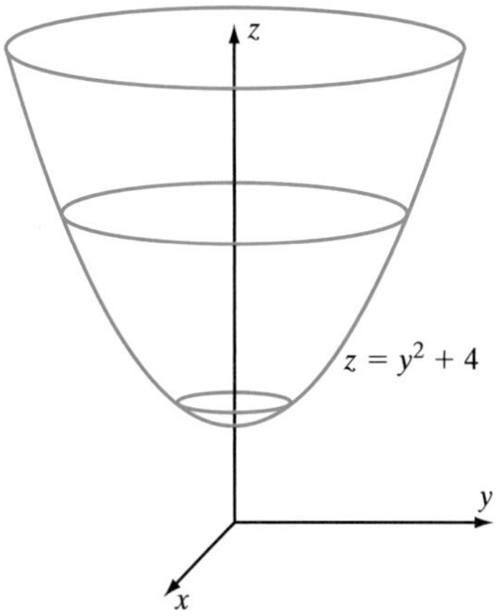

▲ Figure 10b The slice with the yz-plane is added to the plot

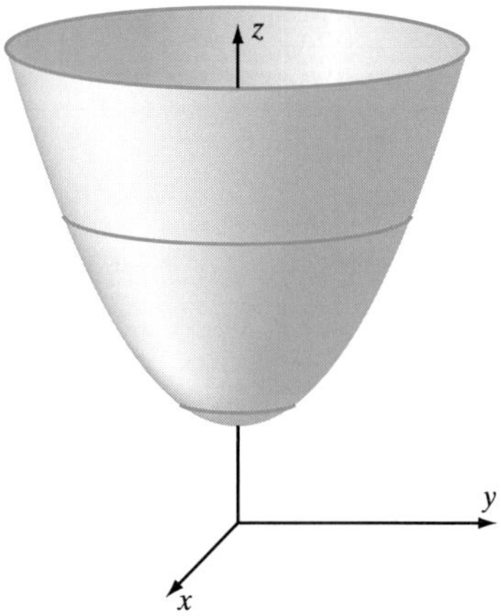

▲ Figure 10c The plot of $z = x^2 + y^2 + 4$

INSIGHT Our new procedure for graphing has several steps. We first must draw several two-dimensional level sets. Then, we must merge these into a three-dimensional picture. One or more slices perpendicular to the xy-plane are used in this last step. Symmetries about a plane or rotational symmetry about an axis provide important shortcuts when present.

▶ **EXAMPLE 7** Sketch the graph of $f(x, y) = x^2 + y$.

Solution Intersecting the graph of f with the plane $z = c$ results in a horizontal slice of the form $\{(x, y, c) : x^2 + y = c\}$. Several of these slices are exhibited in Figure 11a. Notice that they are all parabolas that are symmetric in the yz-plane. When we slice the graph of f with the yz-plane by setting $x = 0$, we see that the vertices of the parabolas lie on the line $x = 0, z = f(0, y) = y$. The plot appears in Figure 11b.

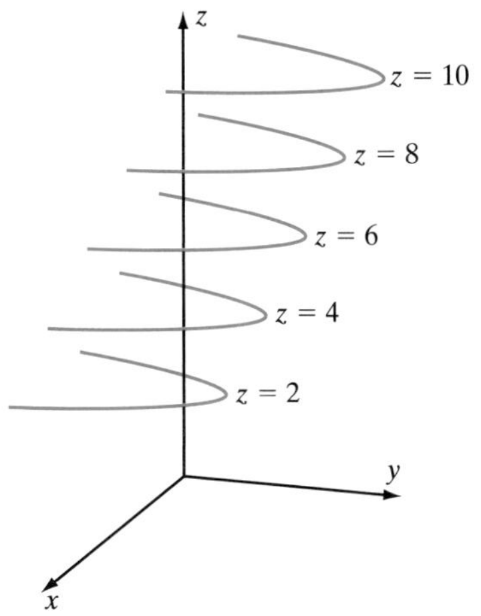

▲ Figure 11a Horizontal slices of $f(x, y) = x^2 + y$

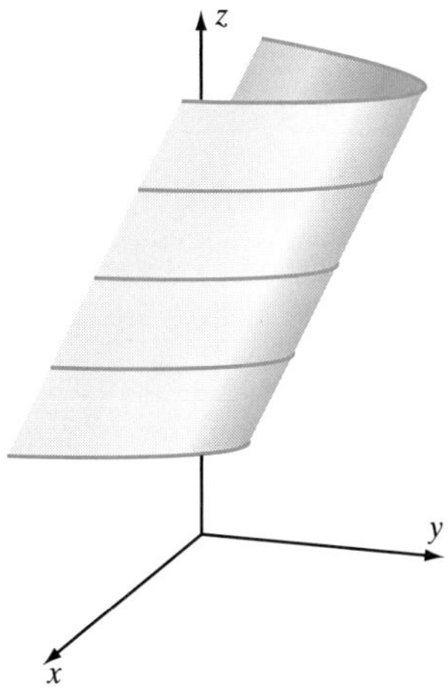

▲ Figure 11b $f(x, y) = x^2 + y$

INSIGHT When you graph a function $(x, y) \mapsto f(x, y)$, you may use any slices that are convenient. In Example 7, if we slice the graph of f with planes $y = c$ that are parallel to the xz-plane, then we obtain a family of parabolas of the form $y = c$, $z = x^2 + c$ (Figure 12a). These vertical slices can be pieced together to obtain the plot of f (Figure 12b). By comparing Figures 11b and 12b, we can see that different slices can provide different perspectives on a surface.

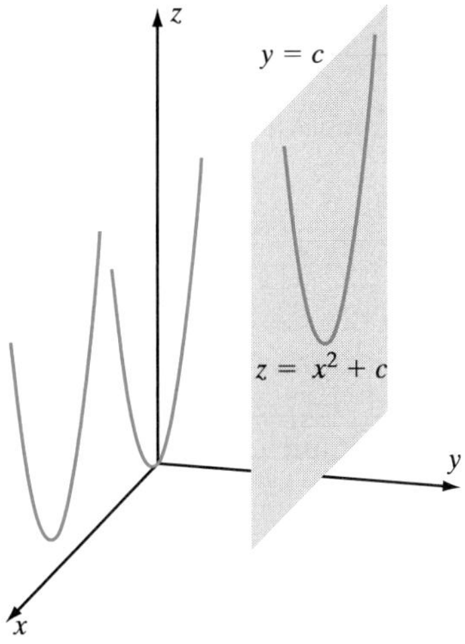

▲ Figure 12a Vertical slices of $f(x, y) = x^2 + y$

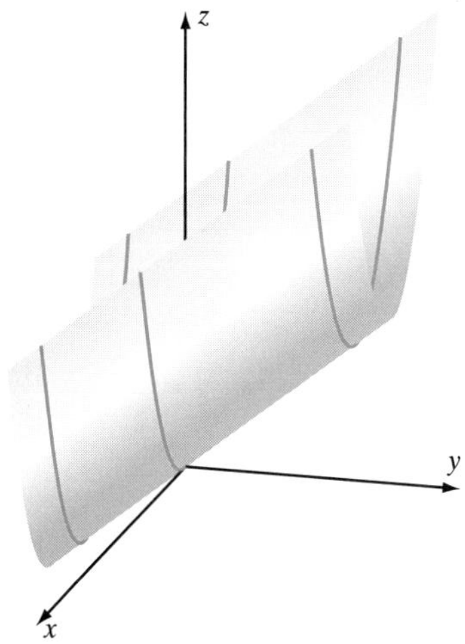

▲ Figure 12b $f(x, y) = x^2 + y$

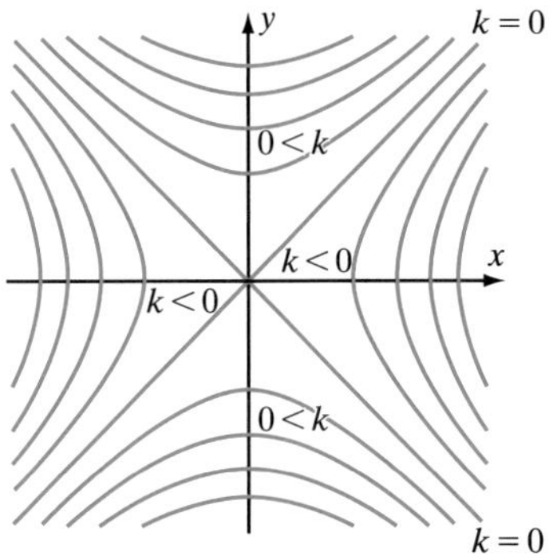

▲ Figure 13
The family of $y^2 - x^2 = k$, where k is a constant

▶ **EXAMPLE 8** Sketch the graph of $f(x, y) = y^2 - x^2$.

Solution Several level sets are exhibited in Figure 13. Notice that they are all hyperbolas. These hyperbolas change orientation at height $c = 0$. In fact, $c = 0$ corresponds to the level set $y^2 - x^2 = 0$ or $(y - x)(y + x) = 0$, which is a pair of crossed lines. The level sets are hyperbolas that open sideways (in the xy-plane) when $c < 0$ and are hyperbolas that open up and down (in the xy-plane) when $c > 0$. We use the slice corresponding to $x = 0$, which is a parabola, and the slice corresponding to $y = 0$, which is also a parabola, to complete our idea of the graph. The plot of f is shown in Figure 14. ◀

More on Level Sets

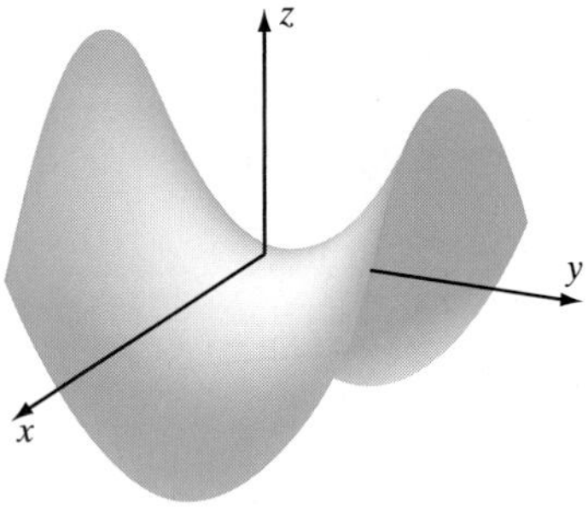

▲ Figure 14 The graph of $z = y^2 - x^2$

Level sets occur in many applications of mathematics. They are frequently used in geography, geology, and meteorology. For example, the contours that we see on a topographic map are level sets of the *height function*. Thus if h is the function that assigns to each point on the map the height above sea level at that point, then the curves $h(x, y) = c$ that are drawn on the map give us an idea of where the mountains, valleys, and ridges are located. An example is shown in Figure 15. A contour that represents constant elevation is called an *isohypse*. A curve that represents a constant depth of a body of water is called an *isobath*.

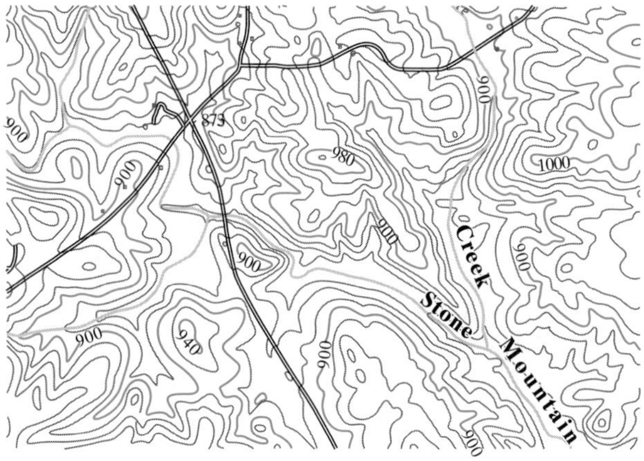

▲ Figure 15 Topographic map of Stone Mountain, Georgia

In newspapers, we often see maps featuring level sets of the temperature function. That is, let $T(x, y)$ be the function that assigns to each point on the map the temperature in degrees Fahrenheit at that point. The curves $T(x, y) = c$, called *isothermal curves* or *isotherms* for short, give us an idea of the changes in weather, the location of cold pockets, and so on (see Figure 16). Likewise, *isobars* are the level sets of the barometric pressure function, *isohyets* are level sets of the rainfall function, and so on.

We have seen that the graph $z = f(x, y)$ of a function of two variables will generally be a surface. Although we cannot graph a function F of three variables, we *can* graph its level sets $\{(x, y, z) : F(x, y, z) = c\}$. The level sets of a function of three variables are generally surfaces (and are called *level surfaces*). A simple example illustrates this idea.

▶ **EXAMPLE 9** Discuss the level sets of the function $F(x, y, z) = x^2 + y^2 + z^2$.

Solution Remember that this is a function of *three variables*: We may substitute values of z independently of the values of x and y. We therefore do not expect the level sets of F to be curves (as we would for a function of two variables). The level

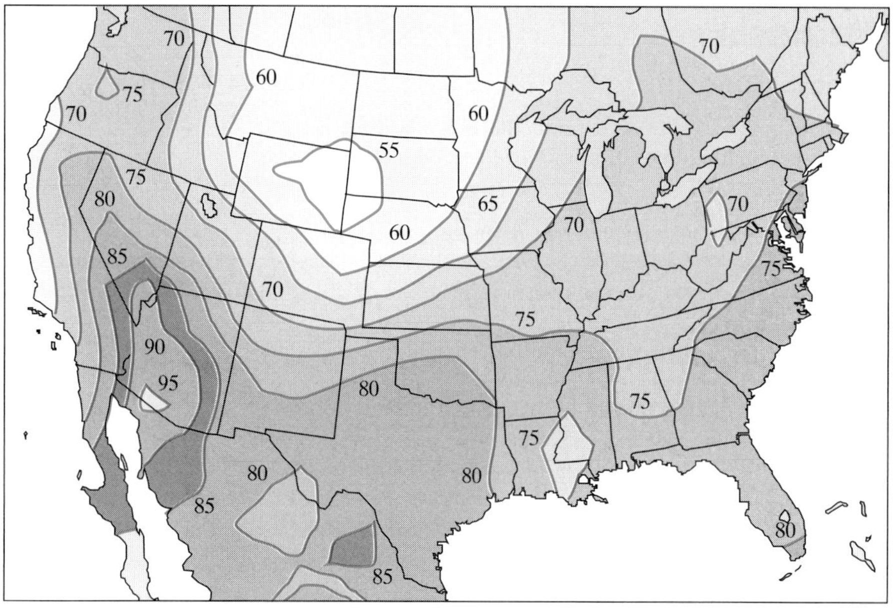

▲ Figure 16 Isotherms 13 August 2002 12 AM EDT

sets have the form $x^2 + y^2 + z^2 = c$. For $c > 0$, the level set is the sphere with center 0 and radius $\sqrt{c}$; for $c = 0$, the level set is a single point; for $c < 0$, the level set is the empty set. ◀

◀ **A LOOK BACK** To understand the different types of functions that have been studied, keep in mind the geometry that each type of function describes. The graphs of scalar-valued functions of one variable are planar curves that meet the Vertical Line Test. Vector-valued functions of one variable describe curves in the plane or in space. The graphs of scalar-valued functions of two variables are surfaces in space. The level sets of scalar-valued functions of three variables describe surfaces in three-dimensional space.

QUICK QUIZ

1. Describe the domain of $f(x, y) = \dfrac{\ln(3 + 2x - x^2)}{\sqrt{4 - y^2}}$.
2. When the graph of $f(x, y) = x - y^2$ is sliced with planes that are parallel to the yz-plane, what curves result?
3. Describe the level sets of $f(x, y) = x^2 - y^2$.
4. Describe the level sets of $F(x, y, z) = x + 2y - 3z$.

Answers

1. The open rectangle $(-1, 3) \times (-2, 2)$ 2. Parabolas that open downward
3. Hyperbolas 4. Planes perpendicular to $\langle 1, 2, -3 \rangle$

EXERCISES

Problems for Practice

In each of Exercises 1–8, express the quantity that is described as a function of two or more variables. Describe the domain.

1. The volume $V(h,r)$ of a cylinder in terms of its height h and its radius r
2. The surface area $S(h,r)$ of a cylinder in terms of its height h and its radius r
3. The volume $V(h,r)$ of a cone in terms of its height h and its radius r
4. The area $A(r,\theta)$ of a circular sector of radius r and interior angle θ
5. The area $A(a,b)$ under the graph of $y=x^2$ and over the interval $[a,b]$
6. The surface area $S(x,y,V)$ of a rectangular box interms of two side lengths, x and y, and its volume V
7. The mass $m(M,\tau,T)$ of a radioactive substance as a function of the initial mass M, the half-life τ, and the amount of time T that has elapsed since the initial measurement
8. The magnitude $a_N(\kappa,v)$ of centripetal acceleration of a moving body at a moment when its speed is v and the curvature of its trajectory is κ

In each of Exercises 9–12, let $f(x,y)=e^{x-y}$, $g(x,y)=x^2-y^3$, $h(x,y,z)=xy^{z+2}$, $k(x,y,z)=y/(x+z)$, and calculate the given quantity.

9. $(2f+g)(5,2)$
10. $(fg)(3,2)$
11. $(h-k/3)(1/2,3,0)$
12. $(h/k)(3,4,-1)$

In each of Exercises 13–18, let $f(x,y)=3xy/(x^2+y^2)$, $g(x,y)=x^2+2y$, $\phi(t)=\sqrt{t}$, and calculate the indicated quantity.

13. $(\phi\circ f)(3,4)$
14. $(\phi\circ g)(4,10)$
15. $(\phi\circ(2f+11g))(1,1)$
16. $((\phi\circ f)g)(1,2)$
17. $((2\phi)\circ(fg))(3,4)$
18. $(((2\phi)\circ f)((2\phi)\circ g))(3,4)$

In each of Exercises 19–28, sketch the level sets of f that correspond to horizontal slices at heights -6, -4, -2, 0, 2, 4, and 6.

19. $f(x,y)=3x$
20. $f(x,y)=x+2y$
21. $f(x,y)=3x-2y$
22. $f(x,y)=4x^2-y$
23. $f(x,y)=1+x-y^2$
24. $f(x,y)=x^2+y^2/4$
25. $f(x,y)=1-x^2/9-y^2$
26. $f(x,y)=x^2-2x+y^2$
27. $f(x,y)=x^2-y^2$
28. $f(x,y)=2+y^2-x^2$

In each of Exercises 29–38, a function f of two variables is given. Sketch several level sets of f. Then, sketch the graph of $z=f(x,y)$.

29. $f(x,y)=1+x$
30. $f(x,y)=6-2x-3y$
31. $f(x,y)=x^2+y^2-1$
32. $f(x,y)=2-(x^2+y^2)$
33. $f(x,y)=\sqrt{x^2+y^2}$
34. $f(x,y)=3-\sqrt{x^2+y^2}$
35. $f(x,y)=\sqrt{1-x^2-y^2}$
36. $f(x,y)=1+\sqrt{4-x^2-y^2}$
37. $f(x,y)=y^2+2x+x^2$
38. $f(x,y)=x^2+y^2/4+6$

In each of Exercises 39–44, sketch level sets corresponding to heights -8, -4, 0, 4, and 8 for the given function F.

39. $F(x,y,z)=x+y+z$
40. $F(x,y,z)=10-x-y$
41. $F(x,y,z)=12-z^2$
42. $F(x,y,z)=x^2+y^2+z^2-8$
43. $F(x,y,z)=x^2+2x+y^2+4y+z^2+6z-25$
44. $F(x,y,z)=x^2+y^2-12$

Further Theory and Practice

45. Let $f(x,y)=xy/\sqrt{x^2+y^2}$. Calculate $f(f(x,y),f(x,y))$.
46. Match the given functions to the correct set of level curves in Figure 17.
    a. $f(x,y)=x^2-y$
    b. $f(x,y)=xy$
    c. $f(x,y)=|x+y|$
    d. $f(x,y)=2x^2+y^2$
    e. $f(x,y)=x^2-y^2$
    f. $f(x,y)=(2x-y)^2$
47. How do the level sets of $f_1(x,y)=x-y$ and $f_2(x,y)=(x-y)^2$ differ?
48. Calculate the time $T(h,v_0)$ until ground impact of an object propelled downward from a height h with initial speed v_0. (Assume gravity is the only force present.)
49. Calculate the x-intercept $a(h,c)$ of the tangent line to the curve $x\mapsto h+e^x$ at $x=c$.

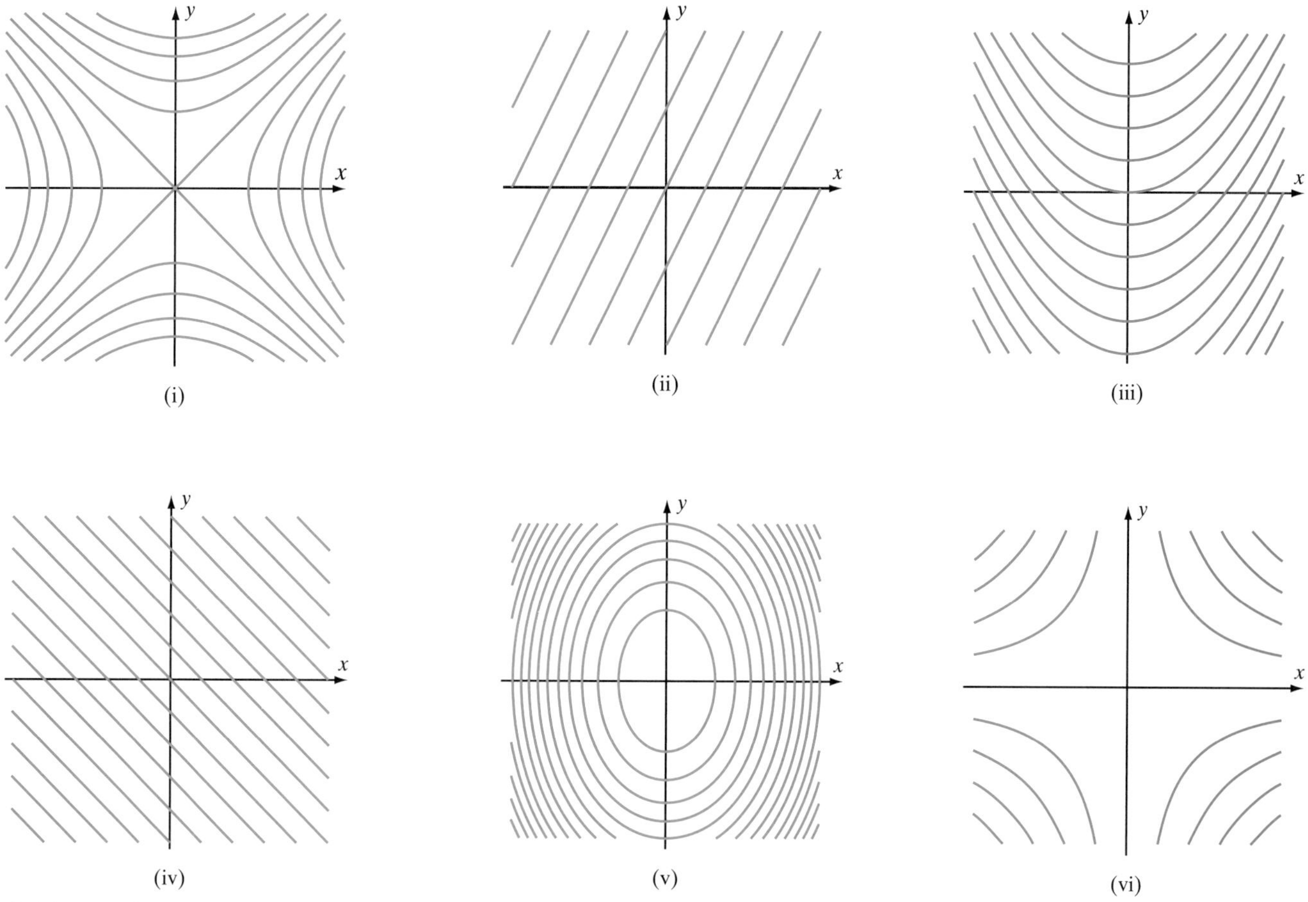

▲ Figure 17

50. Calculate the y-intercept $b(k,c)$ of the tangent line to the curve $y = k/x$ at $x = c$.

51. Let $f(x,y) = x^2 + y^2$ and $g(x,y) = 4 - 2x^2 - 2y^2$. Sketch the set $S = \{(x,y) : f(x,y) > g(x,y)\}$.

52. Let $f(x,y) = x^2 + y^2$. Sketch the set $S = \{(x,y) : f(x,y) \geq f(x,y)^2\}$.

53. Let $f(x,y) = x^2 - y + 4$. Sketch the set $S = \{(x,y) : 2 \leq f(x,y) < 4\}$.

▶ In each of Exercises 54–59, describe the level sets of the given function f. ◀

54. $f(x,y) = \ln(xy)$.

55. $f(x,y) = \sin(y - x^2)$.

56. $f(x,y) = y\sin(x)$

57. $f(x,y) = \sin(x^2 + y^2)$

58. $f(x,y) = y + \sin(x)$

59. $f(x,y) = y^2/(1 + x^2)$

60. Figure 18 shows typical level curves of a function f. Answer the following questions about f, stating the reasons for your answers.

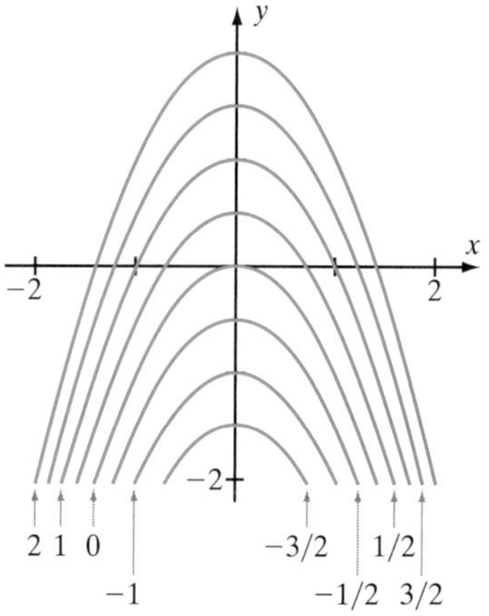

▲ Figure 18 Level curves of f

a. Does $f(0,1) = f(1,0)$?

b. Does $f(1/2,0) = f(0,1/2)$?

c. Which is greater: $f(-1,1)$ or $f(1/2,1)$?

d. As t increases, does $f(-1/2-t, -3/2+t)$ increase or decrease?

e. On what interval is $f(x,-1)$ an increasing function of x?

f. On what interval is $f(-1/2, y)$ an increasing function of y?

61. Express the function $f(x,y)=\sum_{j=0}^{\infty}(x/y)^j$ without using sigma-notation. What is its domain?

62. What is the domain of the function $f(x,y)=\sum_{n=1}^{\infty} n^{x+y}$? Sketch its level sets.

63. Give an example of a function $f(x,y)$ such that for each fixed y_0, the graph of the function $x \mapsto f(x,y_0)$ is a parabola, while for each fixed x_0, the function $y \mapsto f(x_0,y)$ is a cubic.

64. Parameterize the level curves of $f(x,y)=x^2+2y^2$.

65. Parameterize the level curves of $f(x,y)=x^2-2xy+2y^2$.

Calculator/Computer Exercises

In each of Exercises 66–69, a function f and a viewing window R are given. Plot the level curves of the function f in the given viewing window. Use the horizontal slices $z=k/4$ for $k=-3,-2,\ldots,3$.

66. $f(x,y)=x^3+x^2y^4$ $\quad R=[-1,1]\times[-3,3]$

67. $f(x,y)=x^2+xy+y^3$ $\quad R=[-1.2,1]\times[-1,1]$

68. $f(x,y)=(1-x+y)/(1+x^2+y^2)$ $\quad R=[-6,4]\times[-4,6]$

69. $f(x,y)=2xy^2+x^4y$ $\quad R=[-2,2]\times[-2,2]$

11.2 Cylinders and Quadric Surfaces

You already know from your experience with functions of one variable that it requires some practice to feel comfortable with graphing. We have to do a lot of graphing and encounter a great many different functions and pictures before we can visualize ideas readily. The purpose of this section is to provide that vital experience for functions of two variables.

Some of the graphs in this section will *not* be the graphs of functions. Part of what we will learn is to distinguish which graphs come from functions. We begin now with the simplest graphs that are encountered in the theory of several variables.

Cylinders

An equation in three-dimensional space has a graph that is said to be a *cylinder* if one (or more) of the variables x, y, z does not appear in the equation. The justification for the term "cylinder" is apparent from the graph of the surface $\{(x,y,z): x^2+y^2=1\}$. In Example 2 of Section 12.1 of Chapter 12, we deduced that the plot of this equation in xyz-space is a cylinder (in the ordinary sense of the word). The next examples show that the mathematical meaning of "cylinder" is quite a bit more general.

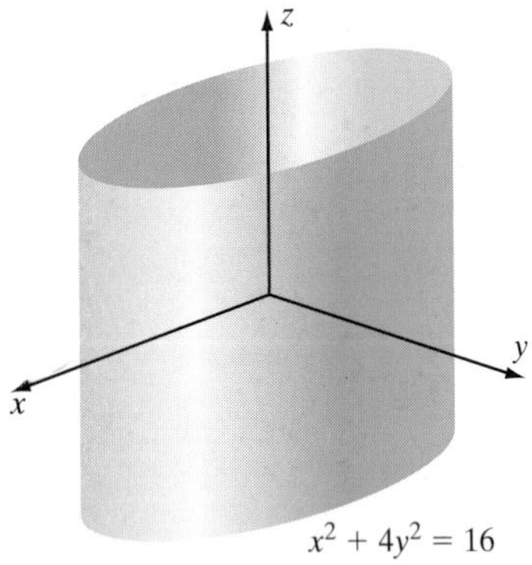

▲ Figure 1 Elliptic cylinder

▶ **EXAMPLE 1** Sketch the set of points in three-dimensional space satisfying the equation $x^2+4y^2=16$.

Solution We exploit the missing variable by taking slices that are perpendicular to its axis. In this case, z is missing, so we calculate the equations of some level sets $z=c$:

$$z=-1 \qquad x^2+4y^2=16$$
$$z=0 \qquad x^2+4y^2=16$$
$$z=2 \qquad x^2+4y^2=16.$$

Notice that all the level sets of the graph are the same ellipse. The reason for this is simple: The variable z does not appear in the equation that we are graphing. Substituting in different values of z has no effect. Now we sketch the graph. We plot the ellipse in the xy-plane and form the cylindrical surface with this ellipse (and its translates up and down) as cross-sections (see Figure 1). ◀

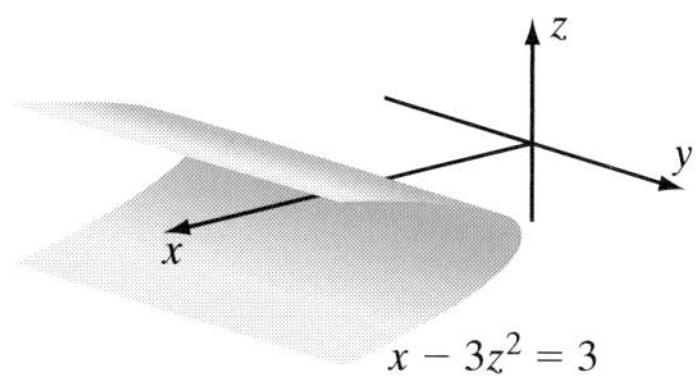

▲ Figure 2

▶ **EXAMPLE 2** Sketch the graph of $x - 3z^2 = 3$.

Solution The variable y is missing, so we know that the graph will be a cylinder and that we should use slices obtained by setting y equal to a constant. We also know that no matter what value c we take for y, the level set will be the parabola $x = 3 + 3z^2$. The resulting graph is sketched in Figure 2.

Quadric Surfaces

Quadratic equations provide a rich variety of examples on which to hone our technique at sketching graphs. We will only discuss equations of the form

$$Ax^2 + By^2 + Cz^2 + Dx + Ey + Fz = G. \tag{11.2.1}$$

The graphs of equations with terms such as xy, yz, xz in them are obtained by rotating the graphs of equations of the form (11.2.1).

Ellipsoids

An *ellipsoid* is the set of points in space satisfying an equation of the form

$$\frac{x^2}{\alpha^2} + \frac{y^2}{\beta^2} + \frac{z^2}{\gamma^2} = 1 \tag{11.2.2}$$

with α, β, γ all positive constants. The distinguishing feature of such an equation is that all of its intersections with planes parallel to the coordinate planes are ellipses. More precisely, if we set $x = c$, then we obtain the set

$$\frac{y^2}{\beta^2} + \frac{z^2}{\gamma^2} = 1 - \frac{c^2}{\alpha^2}.$$

This is an ellipse if $|c| < \alpha$, a point (which may be regarded as a degenerate ellipse) if $|c| = \alpha$, and empty otherwise. Similarly, if we set $y = c$, then we obtain the set

$$\frac{x^2}{\alpha^2} + \frac{z^2}{\gamma^2} = 1 - \frac{c^2}{\beta^2}.$$

This is an ellipse if $|c| < \beta$, a point if $|c| = \beta$, and empty otherwise. The intersections with the planes $z = c$ are similar. It is worth noting that equation (11.2.2) is unchanged when x is replaced by $-x$, or y by $-y$, or z by $-z$. Thus the graph of equation (11.2.2) will be symmetric with respect to each coordinate plane—that is, the graph looks the same on both sides of each coordinate plane. We put this information together to obtain the plot shown in Figure 3. Notice that, when α, β, and γ are distinct, the three axes of the ellipsoid have different lengths. When they are all equal, the ellipsoid is, in fact, a sphere.

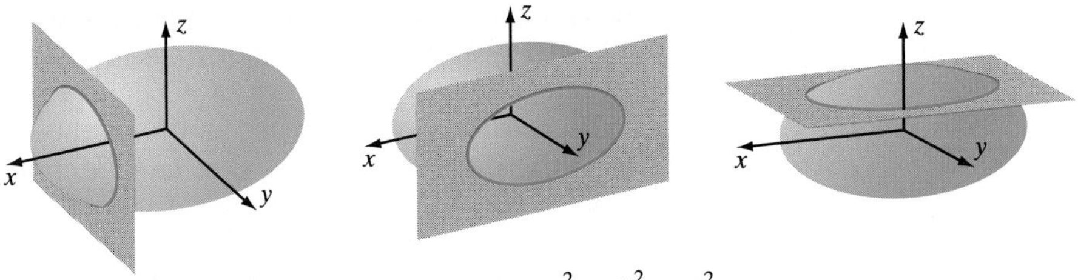

▲ Figure 3 Intersections of the ellipsoid $\frac{x^2}{\alpha^2} + \frac{y^2}{\beta^2} + \frac{z^2}{\gamma^2} = 1$, with planes parallel to the coordinate planes, are ellipses.

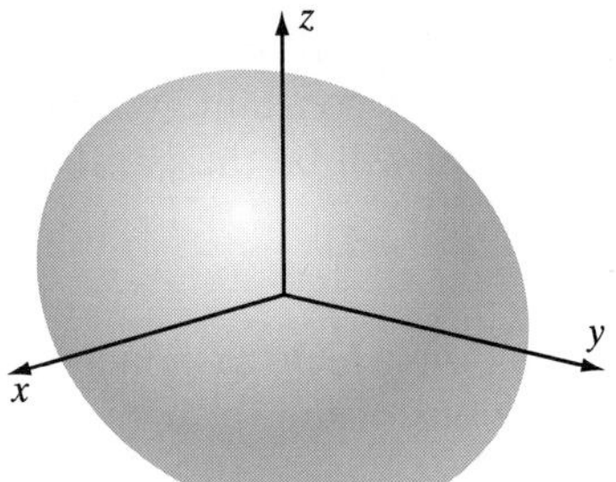

$4x^2 + y^2 + 2z^2 = 4$

▲ Figure 4

▶ **EXAMPLE 3** Sketch the set of points satisfying the equation $4x^2 + y^2 + 2z^2 = 4$.

Solution We divide through by 4 to write the equation in standard form:

$$\frac{x^2}{1^2} + \frac{y^2}{2^2} + \frac{z^2}{(\sqrt{2})^2} = 1.$$

According to our discussion, this is the equation of an ellipsoid. Its graph is sketched in Figure 4. Observe that it is easy to read, from the standard form of the equation, that the intercepts of the surface are $(\pm 1, 0, 0)$, $(0, \pm 2, 0)$, and $(0, 0, \pm\sqrt{2})$. ◀

Elliptic Cones

Elliptic cones arise from equations of the form

$$\frac{x^2}{\alpha^2} + \frac{y^2}{\beta^2} = \frac{z^2}{\gamma^2} \quad \text{or} \quad \frac{y^2}{\beta^2} + \frac{z^2}{\gamma^2} = \frac{x^2}{\alpha^2} \quad \text{or} \quad \frac{z^2}{\gamma^2} + \frac{x^2}{\alpha^2} = \frac{y^2}{\beta^2}$$

with α, β, and γ positive. We will analyze the first of these equations. The remaining two can be handled in a similar way.

When $z = c$, the level set of $x^2/\alpha^2 + y^2/\beta^2 = z^2/\gamma^2$ has the form

$$\frac{x^2}{\alpha^2} + \frac{y^2}{\beta^2} = \frac{c^2}{\gamma^2}.$$

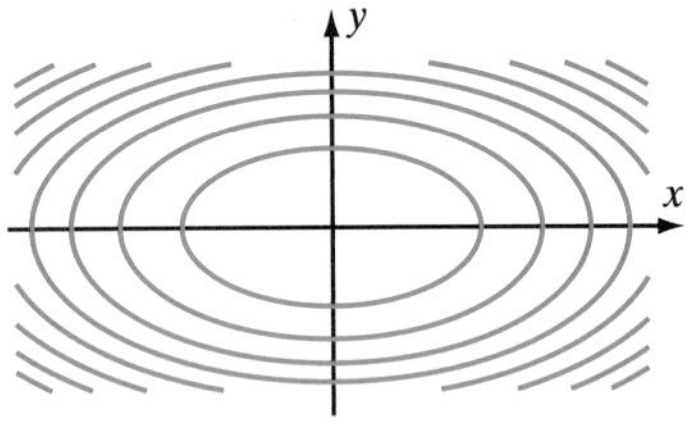

▲ Figure 5 Some level sets of $\frac{x^2}{\alpha^2} + \frac{y^2}{\beta^2} = \frac{c^2}{\gamma^2}$

The graph of this equation is an ellipse whose axes have ratio $\alpha : \beta$. As $|c|$ becomes larger, the ellipse becomes larger, but the ratio of the axes remains the same. Some level curves are sketched in Figure 5.

It remains to see how the level sets stack up. To do this, we take some slices parallel to the xz- and yz-planes. Setting $x = 0$, that is, slicing with the yz-coordinate plane, we obtain $y^2/\beta^2 = z^2/\gamma^2$ or

$$y = \pm\frac{\beta}{\gamma}z,$$

which is a pair of lines. Likewise, setting $y = 0$, that is, slicing with the xz-coordinate plane, we obtain the lines

$$x = \pm\frac{\alpha}{\gamma}z.$$

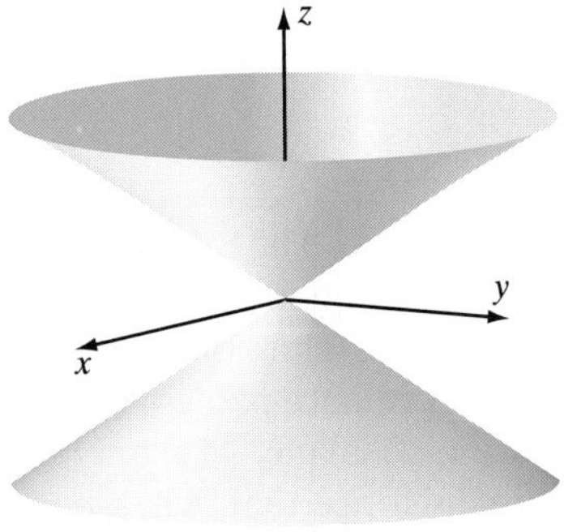

$\frac{x^2}{\alpha^2} + \frac{y^2}{\beta^2} = \frac{z^2}{\gamma^2}$

▲ Figure 6 An elliptic cone

Finally, we notice that the graph will be symmetric in all three coordinate planes. Taking into account all this information, we exhibit the graph in Figure 6.

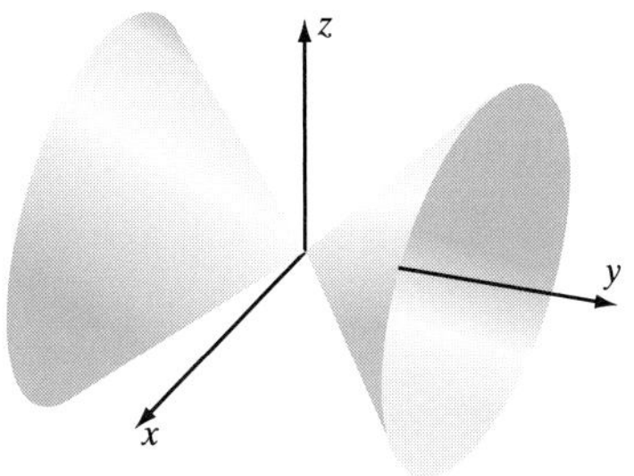

▲ Figure 7 The elliptic cone $x^2 + 2z^2 = 2y^2$

▶ **EXAMPLE 4** Sketch the set of points satisfying the equation $x^2 + 2z^2 = 2y^2$.

Solution We divide through by 2 to write the equation in standard form:

$$\frac{x^2}{(\sqrt{2})^2} + \frac{z^2}{1^2} = \frac{y^2}{1^2}.$$

According to our discussion, this surface is an elliptic cone. Its graph is sketched in Figure 7. ◀

Elliptic Paraboloids

Elliptic paraboloids arise from equations of the form

$$\frac{x^2}{\alpha^2} + \frac{y^2}{\beta^2} = \frac{z}{\gamma} \quad \text{or} \quad \frac{y^2}{\beta^2} + \frac{z^2}{\gamma^2} = \frac{x}{\alpha} \quad \text{or} \quad \frac{z^2}{\gamma^2} + \frac{x^2}{\alpha^2} = \frac{y}{\beta}.$$

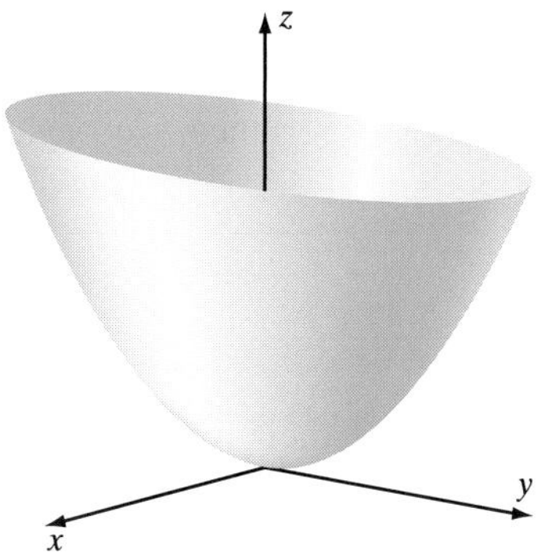

▲ Figure 8 $\dfrac{x^2}{\alpha^2} + \dfrac{y^2}{\beta^2} = \dfrac{z}{\gamma}$; Elliptic paraboloid

As usual, we assume that $\alpha, \beta, \gamma > 0$. We will discuss only the first of these equations, the others yielding to a similar treatment. As in the case of the elliptic cone, the level set of $x^2/\alpha^2 + y^2/\beta^2 = z/\gamma$ obtained by setting $z = c$ is the ellipse

$$\frac{x^2}{\alpha^2} + \frac{y^2}{\beta^2} = \frac{c}{\gamma}.$$

As $c > 0$ becomes larger, the ellipses get larger. For, $c = 0$, the level set consists only of the origin. The level sets for values $c < 0$ are empty.

Next we look at slices parallel to the xz- and yz-planes to see how to stack the ellipses. Setting $x = 0$, that is, slicing with the yz-coordinate plane, we obtain $y^2/\beta^2 = z/\gamma$ or

$$z = \gamma \frac{y^2}{\beta^2},$$

which is an upward-opening parabola. Likewise, setting $y = 0$, that is, slicing with the xz-coordinate plane, we obtain $x^2/\alpha^2 = z/\gamma$ or

$$z = \gamma \frac{x^2}{\alpha^2}.$$

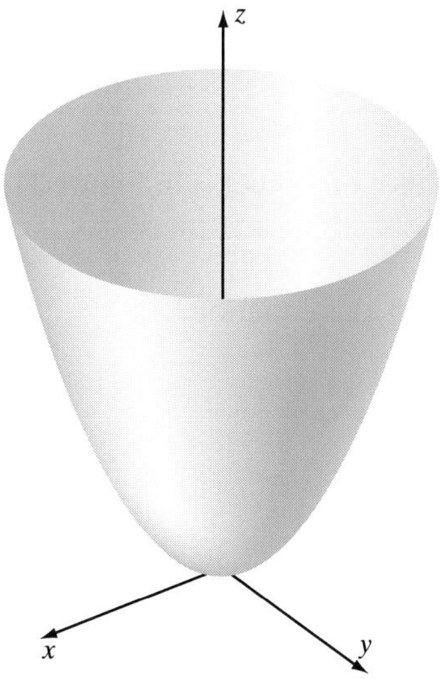

▲ Figure 9 $\dfrac{x^2}{\alpha^2} + \dfrac{y^2}{\alpha^2} = \dfrac{z}{\gamma}$; Circular paraboloid

This, too, is an upward-opening parabola. Finally, we notice that the graph will be symmetric in the xz-coordinate plane (replacing y by $-y$ results in no change) and in the yz-coordinate plane (replacing x by $-x$ results in no change). However, it will *not* be symmetric in the xy-coordinate plane. The resulting sketch appears in Figure 8.

INSIGHT The special case of the elliptic paraboloid when $\alpha = \beta$ is called the *circular paraboloid*. For this surface, the level sets are circles. The sketch is in Figure 9. Circular paraboloids are of interest because they are the model for the shape of radio, television, and radar antennas. The reflecting properties of paraboloids are also crucial to the construction of reflecting telescopes.

▶ **EXAMPLE 5** Sketch the set of points satisfying the equation $x^2 + 2z^2 = y$.

Solution We divide through by 2 to put the equation in standard form:

$$\frac{x^2}{(\sqrt{2})^2} + \frac{z^2}{1^2} = \frac{y}{2}.$$

According to our discussion, the locus of this equation is an elliptic paraboloid. Its graph is shown in Figure 10. ◀

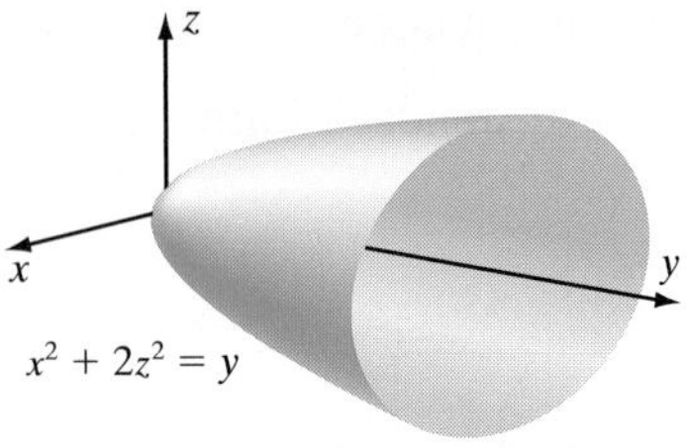

▲ Figure 10 Elliptic paraboloid

Hyperboloids of One Sheet

The hyperboloid of one sheet arises from the equations

$$\frac{x^2}{\alpha^2} + \frac{y^2}{\beta^2} - \frac{z^2}{\gamma^2} = 1 \quad \text{or} \quad \frac{y^2}{\beta^2} + \frac{z^2}{\gamma^2} - \frac{x^2}{\alpha^2} = 1 \quad \text{or} \quad \frac{z^2}{\gamma^2} + \frac{x^2}{\alpha^2} - \frac{y^2}{\beta^2} = 1.$$

We assume that α, β, γ are positive. Again, we will discuss only the first of these equations, the others two being similar. Following previous experience, we begin by looking at level sets, the ellipses

$$\frac{x^2}{\alpha^2} + \frac{y^2}{\beta^2} = \frac{c^2}{\gamma^2} + 1.$$

These ellipses increase in size as $|c|$ increases but the ratio of their axes remains constant. To see how the ellipses stack up, we look at slices parallel to the xz- and yz-planes. We slice with the yz-plane by setting $x = 0$ and obtain

$$\frac{y^2}{\beta^2} - \frac{z^2}{\gamma^2} = 1.$$

This is a hyperbola. Also, we slice by the xz-plane, setting $y = 0$, to obtain

$$\frac{x^2}{\alpha^2} - \frac{z^2}{\gamma^2} = 1.$$

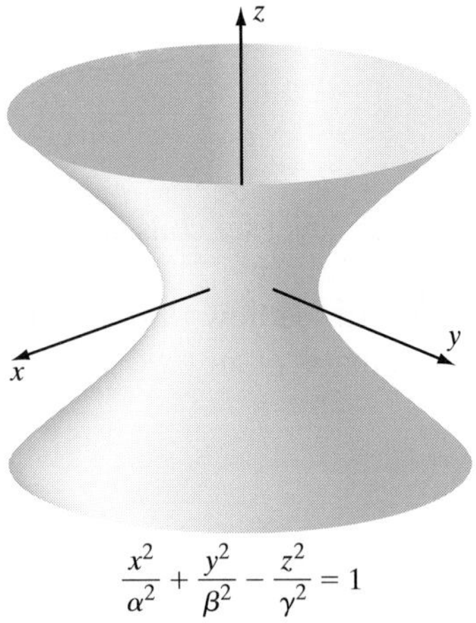

▲ Figure 11 Hyperboloid of one sheet

This too is a hyperbola. The picture of our hyperboloid of one sheet is in Figure 11. The phrase "one sheet" means simply that the surface is all of one piece.

▶ **EXAMPLE 6** Sketch the set of points satisfying the equation

$$3x^2 - \frac{1}{2}y^2 + z^2 = 1.$$

Solution We put the equation in standard form:

$$\frac{x^2}{(1/\sqrt{3})^2} + \frac{z^2}{1^2} - \frac{y^2}{(\sqrt{2})^2} = 1.$$

According to our discussion, this surface is a hyperboloid of one sheet. Its graph is shown in Figure 12. ◀

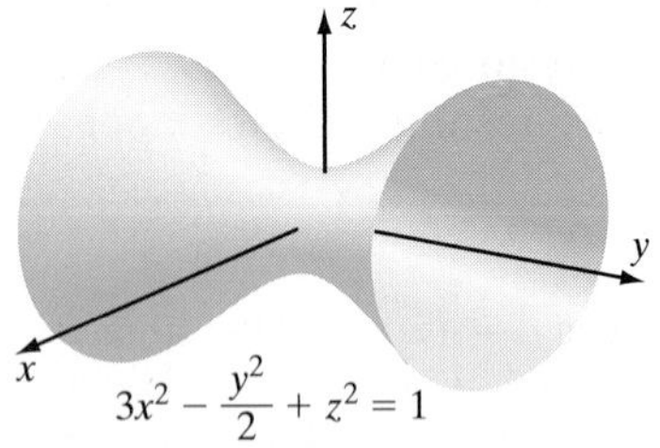

▲ Figure 12 Hyperboloid of one sheet

Hyperboloids of Two Sheets

Consider the equation

$$\frac{z^2}{\gamma^2} - \frac{x^2}{\alpha^2} - \frac{y^2}{\beta^2} = 1.$$

We begin by considering the level sets of the graph of this equation. Setting $z = c$, we obtain

$$\frac{x^2}{\alpha^2} + \frac{y^2}{\beta^2} = \frac{c^2}{\gamma^2} - 1.$$

If $|c| \geq \gamma$, then this level set is an ellipse. The ellipses obtained in this way increase in size with $|c|$, but the ratio of their axes remains constant. The level set corresponding to $c = \gamma$ contains only the point $(0, 0, \gamma)$. The level set corresponding to $c = -\gamma$ contains only the point $(0, 0, -\gamma)$. If $|c| < \gamma$, then the level set is empty.

We now slice by the coordinate planes. The slice with the yz-plane is a hyperbola

$$\frac{z^2}{\gamma^2} - \frac{y^2}{\beta^2} = 1.$$

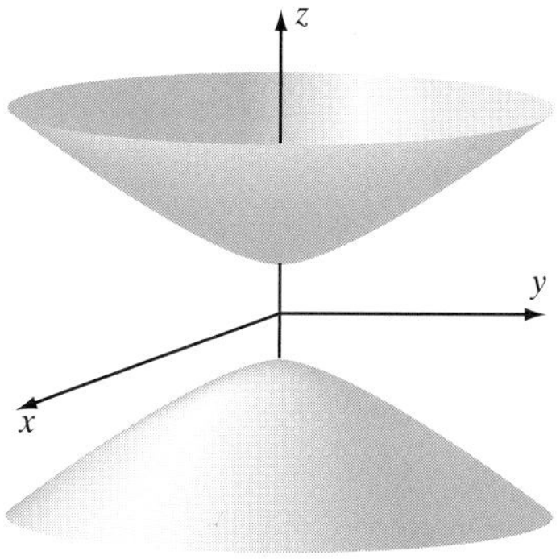

$$\frac{z^2}{\gamma^2} - \frac{x^2}{\alpha^2} - \frac{y^2}{\beta^2} = 1$$

▲ Figure 13 Hyperboloid of two sheets

Similarly, the slice with the xz-plane is the hyperbola

$$\frac{z^2}{\gamma^2} - \frac{x^2}{\alpha^2} = 1.$$

The picture of our hyperboloid of two sheets is in Figure 13. By analogous reasoning, we see that the equations

$$\frac{x^2}{\alpha^2} - \frac{y^2}{\beta^2} - \frac{z^2}{\gamma^2} = 1 \quad \text{and} \quad \frac{y^2}{\beta^2} - \frac{z^2}{\gamma^2} - \frac{x^2}{\alpha^2} = 1$$

are also hyperboloids of two sheets.

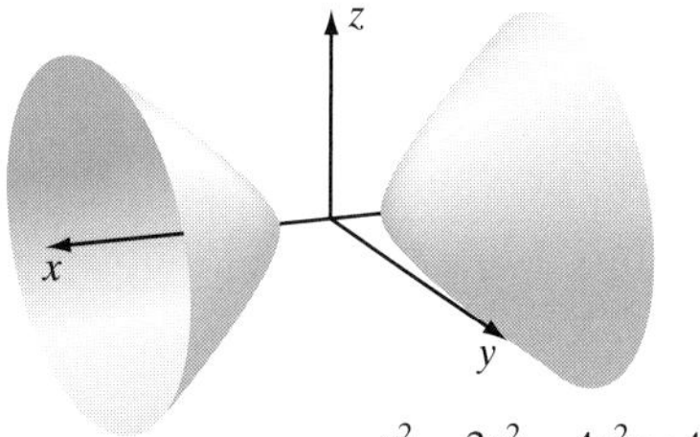

$x^2 - 2z^2 - 4y^2 = 4$

▲ Figure 14 Hyperboloid of two sheets

▶ **EXAMPLE 7** Sketch the set of points satisfying $x^2 - 2z^2 - 4y^2 = 4$.

Solution We divide by 4 to put the equation in standard form:

$$\frac{x^2}{2^2} - \frac{y^2}{1^2} - \frac{z^2}{(\sqrt{2})^2} = 1.$$

According to our discussion, this surface is a hyperboloid of two sheets. Its graph is shown in Figure 14. ◀

The Hyperbolic Paraboloid

Finally, we consider the equation

$$\frac{x^2}{\alpha^2} - \frac{y^2}{\beta^2} = \frac{z}{\gamma}.$$

As usual, α, β, γ are positive. As long as $c \neq 0$, the level set corresponding to $z = c$ is the hyperbola

$$\frac{x^2}{\alpha^2} - \frac{y^2}{\beta^2} = \frac{c}{\gamma}.$$

When $c = 0$, the level set is $x^2/\alpha^2 - y^2/\beta^2 = 0$ or $x = \pm\alpha y/\beta$, which is a pair of lines. Figure 15 shows the family of level sets for $c<0$, $c=0$, and $c>0$. The slices corresponding to $x=0$ and $y=0$ are parabolas. The final picture is in Figure 16. The resulting surface is commonly known as a *saddle surface*. It is of particular interest when we solve maximum-minimum problems because the origin looks like a local minimum in the x-direction and looks like a local maximum in the y-direction.

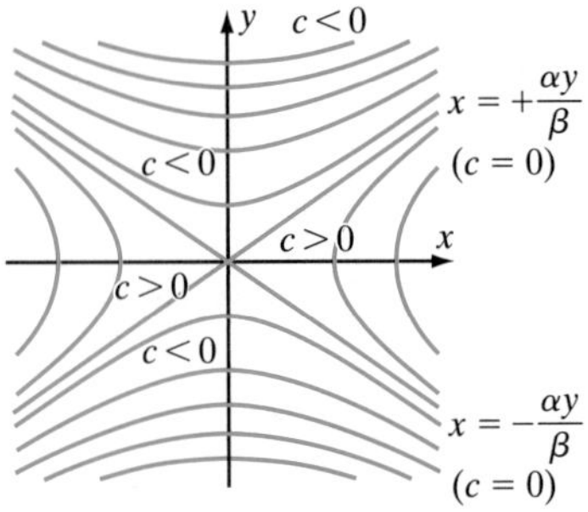

▲ Figure 15 Level curves of $\frac{x^2}{\alpha^2} - \frac{y^2}{\beta^2} = \frac{z}{\gamma}$

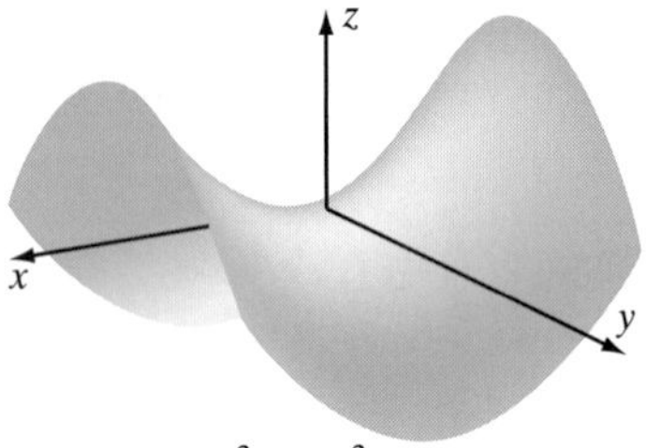

▲ Figure 16 Hyperbolic paraboloid (saddle surface)

► **EXAMPLE 8** Sketch the set of points satisfying the equation

$$z = 2y^2 - 4x^2.$$

Solution We divide through by 4 to put the equation in standard form:

$$\frac{y^2}{(\sqrt{2})^2} - \frac{x^2}{1^2} = \frac{z}{4}.$$

According to our discussion, the surface is a hyperbolic paraboloid. The graph is shown in Figure 17. ◄

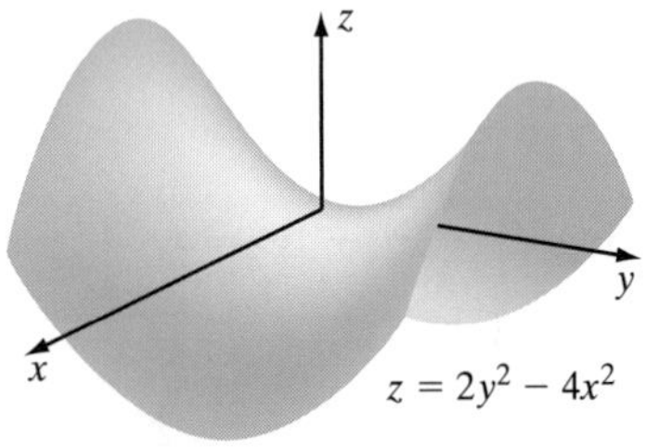

▲ Figure 17 Hyperbolic paraboloid

Recognizing the Graph of a Function

Many of the quadric surfaces that we have discussed in this section are *not* the graphs of functions, while a few of them are. How do we tell which are which?

A function $(x, y) \mapsto f(x, y)$ has the crucial property that, for each (x_0, y_0) in its domain, there is precisely one value z_0 such that $z_0 = f(x_0, y_0)$. This means that, if (x_0, y_0) is in the domain of f, and we cut the graph with the vertical line through the point $(x_0, y_0, 0)$, then the line will intersect the graph just once. This is the analogue of the Vertical Line Test for functions of one variable:

A surface in space is the graph of a function $z = f(x, y)$ if and only if every vertical line intersects the surface *at most* once.

► **EXAMPLE 9** Is the set of points satisfying the equation $x^2 + 3y^2 + z^2 = 12$ the graph of a function of x and y? What about the surface described by $x^2 + 3y^2 + z = 12$?

Solution The first surface is *not* the graph of a function. For If we slice it with the line $x = 0, y = 1$, we see that the equation becomes $z^2 = 9$ or $z = \pm 3$. Thus both the points $(0, 1, 3)$ and $(0, 1, -3)$ lie on the graph and have xy-coordinates $(0, 1)$. By contrast, the surface $x^2 + 3y^2 + z = 12$ *is* the graph of a function because when $x = x_0, y = y_0$ are specified, then z must be

$$z = 12 - (x_0)^2 - (y_0)^2$$

in order for (x_0, y_0, z) to lie on the surface. For the second surface, z is uniquely determined by x and y. For the first, *it is not.* ◄

We have studied six kinds of quadric surfaces. It is a good idea to go through the following table, noting similarities and differences between the equations of these surfaces. Note: Only one example of each type of quadric surface is listed. For each type, permuting the variables x, y, and z results in a quadric surface of the same type.

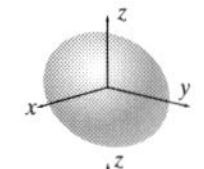	$\frac{x^2}{\alpha^2}+\frac{y^2}{\beta^2}+\frac{z^2}{\gamma^2}=1$	ellipsoid
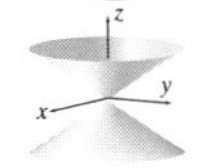	$\frac{x^2}{\alpha^2}+\frac{y^2}{\beta^2}=\frac{z^2}{\gamma^2}$	elliptic cone
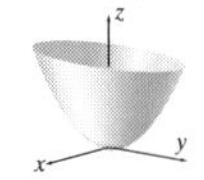	$\frac{x^2}{\alpha^2}+\frac{y^2}{\beta^2}=\frac{z}{\gamma}$	elliptic paraboloid
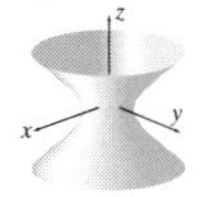	$\frac{x^2}{\alpha^2}+\frac{y^2}{\beta^2}-\frac{z^2}{\gamma^2}=1$	hyperboloid of one sheet
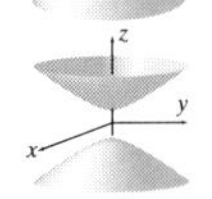	$\frac{z^2}{\gamma^2}-\frac{x^2}{\alpha^2}-\frac{y^2}{\beta^2}=1$	hyperboloid of two sheets
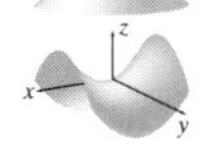	$\frac{x^2}{\alpha^2}-\frac{y^2}{\beta^2}=\frac{z}{\gamma}$	hyperbolic paraboloid

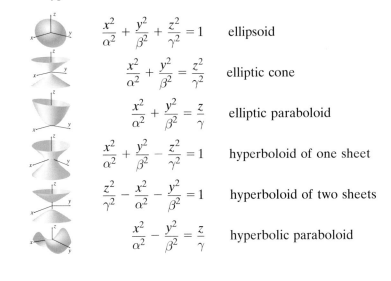

QUICK QUIZ

1. The graphs of which of the following equations are cylinders in space? a) $x=y^2+z^2$; b) $y=y^2+z^2$; c) $e^{x+2z}=y$; d) $e^{x+2z}=z^2$
2. The graphs of which of the following equations are cones in space? a) $x^2=y^2-2z^2$; b) $x^2=y^2+2z^2$; c) $x=y^2+z^2$; d) $x=y^2-z^2$
3. The graphs of which of the following equations are hyperboloids of one sheet in space? a) $x^2-y^2+2z^2=-1$; b) $x^2-y^2+2z^2=0$; c) $x^2-y^2+2z^2=1$; d) $x^2-y^2-2z^2=1$
4. The graphs of which of the following equations are hyperbolic paraboloids in space? a) $x=y^2-2z^2$; b) $x=y^2+2z^2$; c) $z^2+y^2=x^2$; d) $z+y^2=x^2$

Answers

1. b, d 2. b 3. c 4. a, d

EXERCISES

Problems for Practice

In each of Exercises 1–14, the equation of a cylinder in xyz-space is given. Sketch its plot.

1. $x+y=5$
2. $4x^2+z^2=4$
3. $4x^2+z=4$
4. $y^2+2z^2=4$
5. $x^2+z=-4$
6. $y-x^2=9$
7. $x^2+y=4$
8. $z^2+y^2=2y$
9. $x^2-2y^2=1$
10. $4x^2-2y^2=0$
11. $y^2-4z^2=1$
12. $z=\sin(y)$
13. $x-2xy=1$
14. $z^2-y^2=1$

In each of Exercises 15–30, sketch some slices, and then give a complete plot of the quadric surface that is the graph of the given equation. Identify each quadric surface as to type (cylinder, paraboloid, etc.), and determine if the surface is the graph of a function of x and y. ◀

15. $x^2 - 4y^2 + 9z^2 = 36$
16. $x^2 + y^2 = z^2$
17. $y - x^2 = z^2$
18. $x^2 - 9z^2 - 4y^2 = 16$
19. $4x^2 + y^2 + 9z^2 = 25$
20. $2x^2 + 4z^2 = y$
21. $x^2 + y^2 + z = 4$
22. $y^2 + x^2 - z = 4$
23. $4x^2 + y^2 = z$
24. $4x^2 + y^2 + 16z^2 = 16$
25. $y^2 - z^2 - x^2 = 1$
26. $z^2 - 4x^2 = y$
27. $y^2 + 8z^2 = x^2$
28. $z^2 - 4x^2 - 9y^2 = 36$
29. $y^2 - x - z^2 = 0$
30. $x^2 - y^2 - 2z^2 = 6$

Further Theory and Practice

31. True or false: If every level curve of a surface is a quadratic equation, then the surface is a quadric surface. Give reasons for your answer.

32. If a particle bounces off a smooth surface, then the angle of the bounce is determined just as though it bounces off the tangent plane at that point: The incoming angle with the tangent plane is the same as the outgoing angle. Now consider a circular paraboloid $z = x^2 + y^2$. Show that there is a point $P = (0, 0, p)$ on the positive z- axis with the property that any particle falling vertically through the paraboloid and striking the wall of the paraboloid will bounce to the point P. Thus P plays the role of a focal point. This is why parabolic mirrors and antennas are important. (Hint: Reduce the problem to a two-dimensional problem with a slice, and use the tangent line to see how the particle bounces.)

33. Prove that the circular paraboloid is the only surface that has circular symmetry in the x and y variables and that has the property described in Exercise 32.

▶ The projection in the xy-plane of a space curve C is the set $\{(x, y) : (x, y, z) \in C \text{ for some } z\}$. In Exercises 34 and 35, find the projection in the xy-plane of the given space curve C. ◀

34. C is the intersection of the elliptic paraboloids $z = x^2 + 2y^2 - 8$ and $z = 12 - 4x^2 - 3y^2$.

35. C is the intersection of the hyperbolic paraboloid $z = x^2/2 - y^2$ and the elliptic paraboloid $z = 1 - x^2/2 - 5y^2/4$.

Calculator/Computer Exercises

▶ In each of Exercises 36–39, plot the given quadric surface. ◀

36. $x^2 + 2xy + 2y^2 + z^2 = 1$
37. $x^2 + 2xy + 2y^2 - z^2 = 0$
38. $x^2 + 2xy + 2y^2 - z^2 = 1$
39. $x^2 + 2xy + 2y^2 - z^2 = -1$

40. Plot $z = x^2 + y^2$ and $2x + 2y + z = 2$ for $-4 \le x \le 3$, $-4 \le y \le 3$. Parameterize the curve of intersection. Add the graph of this curve to your plot.

41. Plot $z^2 = 4(x^2 + y^2)$ for $0 \le z \le 6$. Plot the plane $z = \sqrt{2}(1 - x)$ for $-6/\sqrt{2} + 1 \le x \le 1$, $-3 \le y \le 3$. Display the two plots in the same viewing box. Parameterize the curve of intersection, graph it, and add it to your previous plot.

11.3 Limits and Continuity

Now that we have treated functions and graphs, we can discuss limits. Recall from Chapter 2 that the limit at c of a function of *one variable* is the value that we *expect* the function to take at c. The same philosophy will prevail for functions of several variables, but there is an important difference. In one variable, the domain is linear, and we decide what we "expect" the function will do at c by approaching c from the left and from the right. In two variables, however, the domain is planar, and we decide what we "anticipate" the function will do at c by approaching c along *all* possible paths that tend to c (see Figure 1). In a sense, then, it is more difficult for a function of two variables to have a limit.

If $P_1 = (x_1, y_1)$ and $P_2 = (x_2, y_2)$ are points in the plane, then it is convenient to denote the distance between them by $d(P_1, P_2)$. This distance is the length of the

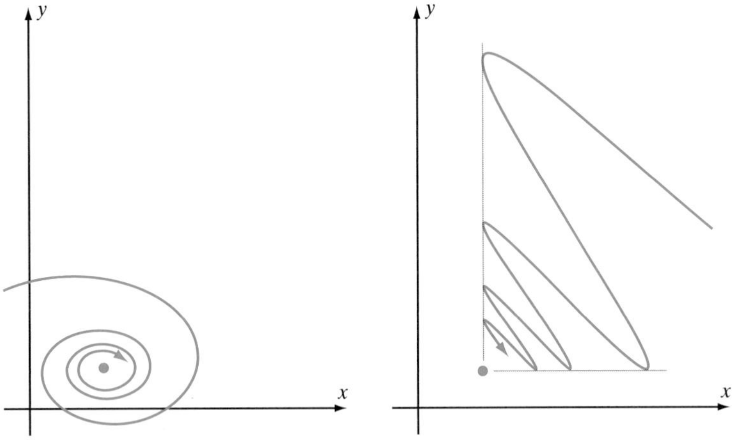

▲ Figure 1 Two possible paths that tend to a planar point

directed line segment $\overrightarrow{P_1P_2}$, or, equivalently, the magnitude of the vector that $\overrightarrow{P_1P_2}$ represents:

$$d(P_1,P_2) = \left\| \overrightarrow{P_1P_2} \right\| = \sqrt{(x_2 - x_1)^2 + (y_2 - y_1)^2}.$$

If $P_0 = (x_0, y_0)$ is a point in the plane and r is a positive number, then $D(P_0, r)$ denotes the open disk of positive radius r that is centered at P_0. That is, $D(P_0, r) = \{P \in \mathbb{R}^2 : d(P, P_0) < r\}$. Because of the strict inequality, the boundary of the disk is not included in $D(P_0, r)$ (see Figure 2).

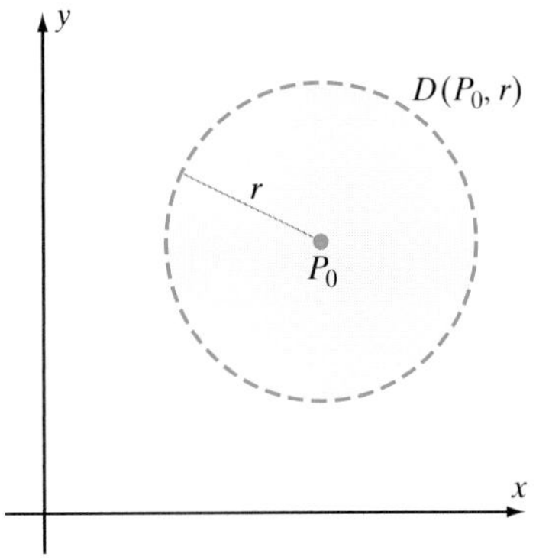

▲ Figure 2

Limits

In many of the constructions of calculus, we must investigate the limit of a function at a point that is not in its domain. For example, we learned in Chapter 3 that the formula $\dfrac{d}{dx}\sin(x) = \cos(x)$ depends entirely on establishing the limit $\lim_{x\to 0} \dfrac{\sin(x)}{x} = 1$. Yet the point 0 is not in the domain of the expression $\dfrac{\sin(x)}{x}$ that is under consideration. To handle similar situations in the plane, it is convenient to refer to the "punctured" disk $D_*(P_0, r)$ formed from the open disk $D(P_0, r)$ by removing the center: $D_*(P_0, r) = \{P \in \mathbb{R}^2 : 0 < d(P, P_0) < r\}$ (see Figure 3).

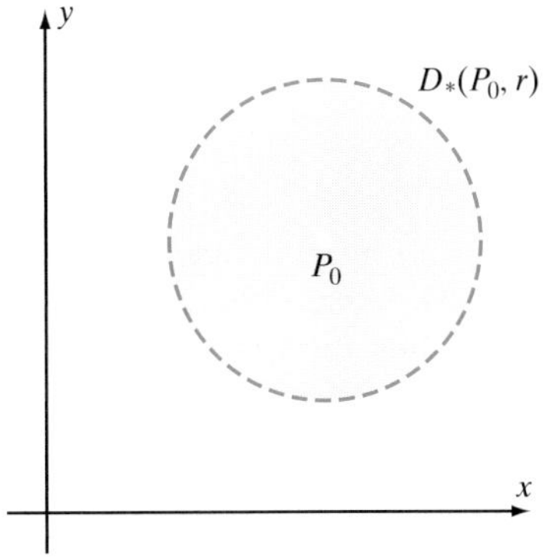

▲ Figure 3

DEFINITION Suppose that f is a function of two variables. Let $P_0 = (x_0, y_0)$ be a fixed point in the plane such that every punctured disk $D_*(P_0, r)$ intersects the domain of f. We say that the real number ℓ is the *limit* of $f(P)$ as $P = (x, y)$ approaches (or tends to) P_0, and we write

$$\lim_{P\to P_0} f(P) = \ell \quad \text{or, equivalently,} \quad \lim_{(x,y)\to(x_0,y_0)} f(x,y) = \ell,$$

if, for any $\varepsilon > 0$, there is a $\delta > 0$ such that $|f(P) - \ell| < \varepsilon$ for all points P in the domain of f with $0 < d(P, P_0) < \delta$.

Notice that it is *not* required that P_0 be in the domain of f in order for us to consider $\lim_{P\to P_0} f(P)$.

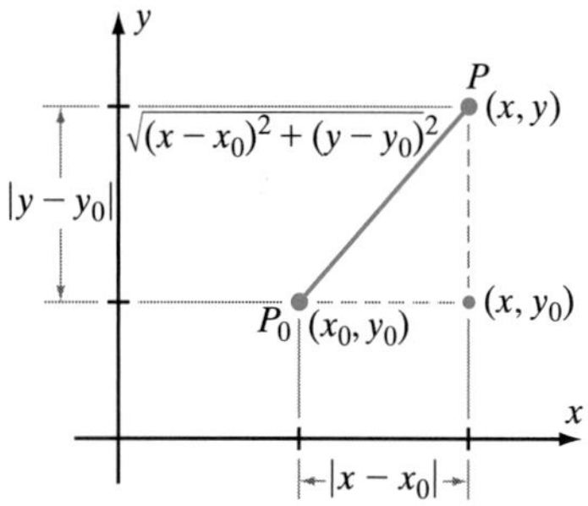

▲ Figure 4

INSIGHT Compare this definition of limit for a function of two variables with the definition given in Section 2.2 of Chapter 2 for a function of one variable. The principal difference is that the distance between $P=(x,y)$ and $P_0=(x_0,y_0)$ is measured using $d(P,P_0)=\sqrt{(x-x_0)^2+(y-y_0)^2}$ rather than $|\ |$. What does this mean? Because the hypotenuse of the right triangle in Figure 4 is at least as long as each leg, we have $|x-x_0|\le d(P,P_0)$ and $|y-y_0|\le d(P,P_0)$. These inequalities tell us that when P is close to P_0, the variable x is close to x_0, *and*, simultaneously, the variable y is close to y_0. Also, the Triangle Inequality tells us that

$$d(P,P_0)=\left\|\overrightarrow{P_0P}\right\|=\|(x-x_0)\mathbf{i}+(y-y_0)\mathbf{j}\|\le\|(x-x_0)\mathbf{i}\|+\|(y-y_0)\mathbf{j}\|=|x-x_0|+|y-y_0|.$$

From this inequality, we see that, when both $|x-x_0|$ and $|y-y_0|$ are small, the distance $d(P,P_0)$ must also be small. In summary, $d(P,P_0)=\sqrt{(x-x_0)^2+(y-y_0)^2}$ is small precisely when x is close to x_0 *and*, simultaneously, y is close to y_0.

▶ **EXAMPLE 1** Define $f(x,y)=x^2+y^2$. Verify that $\lim_{(x,y)\to(0,0)} f(x,y)=0$.

Solution Let $\ell=0$, $P=(x,y)$, and $P_0=(0,0)$. Suppose that $\varepsilon>0$. Then,

$$|f(P)-\ell|=|f(x,y)-0|=|x^2+y^2|.$$

Thus $|f(P)-\ell|<\varepsilon$ if $|x^2+y^2|<\varepsilon$. But this last inequality holds when $d(P,P_0)=\sqrt{(x-0)^2+(y-0)^2}<\sqrt{\varepsilon}$. Unwinding these observations, we see that we should set $\delta=\sqrt{\varepsilon}$. It follows that if $0<d(P,P_0)<\delta$ then $|f(P)-\ell|<\varepsilon$. ◀

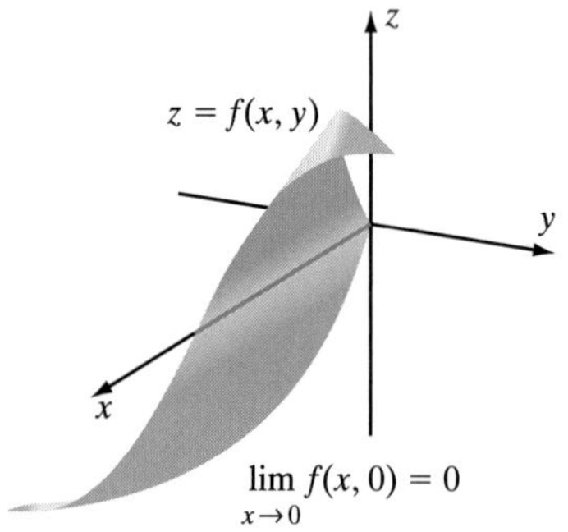

▲ Figure 5

▶ **EXAMPLE 2** Define

$$f(x,y)=\begin{cases}\dfrac{xy}{x^2+y^2} & \text{if } (x,y)\ne(0,0)\\ 0 & \text{if } (x,y)=(0,0).\end{cases}$$

Discuss the limiting behavior of $f(x,y)$ as $(x,y)\to(0,0)$.

Solution Notice that $\lim_{x\to0} f(x,0)=\lim_{x\to0} 0=0$. So $f(x,y)$ approaches 0 as (x,y) approaches the origin along the x-axis (see Figure 5). Also, $\lim_{y\to0} f(0,y)=\lim_{y\to0} 0=0$. Thus $f(x,y)$ approaches 0 as (x,y) approaches the origin along the y-axis. However, we find a different limiting behavior when we consider points $(x,y)=(t,t)$ that approach the origin diagonally as $t\to0$. In this case, we have

$$\lim_{t\to0} f(t,t)=\lim_{t\to0}\frac{t^2}{t^2+t^2}=\lim_{t\to0}\frac{1}{2}=\frac{1}{2}.$$

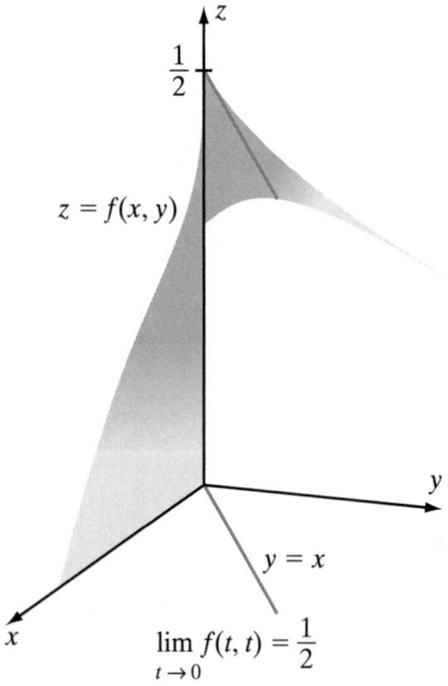

▲ Figure 6

See Figure 6. In summary, if (x,y) is sufficiently close to the origin on either axis then $f(x,y)$ is near 0, but if (x,y) is near the origin on the line $y=x$, then $f(x,y)$ is near 1/2. For the limit of $f(x,y)$ to exist at the origin, the values of f must approach the same number no matter how the point (x,y) tends to the origin. Hence $\lim_{(x,y)\to(0,0)} f(x,y)$ does *not* exist. ◀

INSIGHT Although the function f in Example 2 happens to have been defined at $P_0 = (0,0)$, we do not actually use the defined value $f(P_0)$ when we investigate the existence of $\lim_{P \to P_0} f(P)$. That is because the definition of "limit" refers to the punctured disk $D_*(P_0, \delta)$ from which the point P_0 has been removed. Our analysis would have been exactly the same had $f(P_0)$ been assigned a different value or, indeed, had $f(P_0)$ not been defined at all.

Rules for Limits

Fortunately, there are a number of rules about limits of functions that often make it easy in practice to determine limits. These rules are similar to those that we learned in Chapter 2 for functions of one variable. We record the new rules here for reference.

THEOREM 1 Suppose that f and g are functions of two variables such that

$$\lim_{(x,y)\to(x_0,y_0)} f(x,y) \quad \text{and} \quad \lim_{(x,y)\to(x_0,y_0)} g(x,y)$$

exist. Then,

a. $\lim_{(x,y)\to(x_0,y_0)} (f+g)(x,y) = \lim_{(x,y)\to(x_0,y_0)} f(x,y) + \lim_{(x,y)\to(x_0,y_0)} g(x,y)$.

b. $\lim_{(x,y)\to(x_0,y_0)} (f-g)(x,y) = \lim_{(x,y)\to(x_0,y_0)} f(x,y) - \lim_{(x,y)\to(x_0,y_0)} g(x,y)$.

c. $\lim_{(x,y)\to(x_0,y_0)} (fg)(x,y) = \left(\lim_{(x,y)\to(x_0,y_0)} f(x,y)\right)\left(\lim_{(x,y)\to(x_0,y_0)} g(x,y)\right)$.

d. $\lim_{(x,y)\to(x_0,y_0)} \left(\dfrac{f}{g}\right)(x,y) = \dfrac{\lim_{(x,y)\to(x_0,y_0)} f(x,y)}{\lim_{(x,y)\to(x_0,y_0)} g(x,y)}$ provided that $\lim_{(x,y)\to(x_0,y_0)} g(x,y) \neq 0$.

e. $\lim_{(x,y)\to(x_0,y_0)} \left(\lambda f(x,y)\right) = \lambda \lim_{(x,y)\to(x_0,y_0)} f(x,y)$ for any constant λ.

▶ **EXAMPLE 3** Define $f(x,y) = (x+y+1)/(x^2-y^2)$. What is the limiting behavior of f as (x,y) tends to $(1,2)$?

Solution Because

$$\lim_{(x,y)\to(1,2)} (x^2-y^2) = \lim_{(x,y)\to(1,2)} x^2 - \lim_{(x,y)\to(1,2)} y^2 = 1-4 = -3,$$

which is nonzero, we have

$$\lim_{(x,y)\to(1,2)} \frac{x+y+1}{x^2-y^2} = \frac{\lim_{(x,y)\to(1,2)}(x+y+1)}{\lim_{(x,y)\to(1,2)}(x^2-y^2)} = \frac{\lim_{(x,y)\to(1,2)} x + \lim_{(x,y)\to(1,2)} y + 1}{-3} = \frac{1+2+1}{-3} = -\frac{4}{3}.$$ ◀

▶ **EXAMPLE 4** Evaluate the limit

$$\lim_{(x,y)\to(3,-2)} \frac{2x^2+5xy+3y^2}{2x+3y}.$$

Solution Notice that, as $(x, y)\to(3, -2)$, both the numerator and the denominator tend to 0. Using our experience from Chapter 2, we factor the numerator to see if anything cancels. Our limit problem then becomes

$$\lim_{(x,y)\to(3,-2)} \frac{2x^2+5xy+3y^2}{2x+3y} = \lim_{(x,y)\to(3,-2)} \frac{(2x+3y)(x+y)}{2x+3y} = \lim_{(x,y)\to(3,-2)} (x+y) = 1. \blacktriangleleft$$

Continuity

The definition of continuity for a function of two variables is no different from the one in Section 2.3 in Chapter 2 for functions of one variable, except that we now use our new notion of limit. The philosophy is the same: We require that the value that we *anticipate* f will take at P_0 be the value that f *actually* takes:

DEFINITION Suppose that f is a function of two variables that is defined at a point $P_0 = (x_0, y_0)$. If $f(x, y)$ has a limit as (x, y) approaches (x_0, y_0), and if

$$\lim_{(x,y)\to(x_0,y_0)} f(x, y) = f(x_0, y_0),$$

then we say that f is *continuous* at P_0. If f is not continuous at a point in its domain, then we say that f is *discontinuous* there.

Notice that we discuss the continuity or discontinuity of a function only at a point of its domain. Continuity of f at P_0 means

a. $f(P_0)$ is defined,

b. $\lim_{P\to P_0} f(P)$ exists, and

c. $f(P_0) = \lim_{P\to P_0} f(P)$.

The function $f(x, y) = x^2 + y^2$ from Example 1 *is* continuous at $(0, 0)$ because $\lim_{(x,y)\to(0,0)} f(x, y) = 0 = f(0, 0)$. However, the function

$$f(x,y) = \begin{cases} \dfrac{xy}{x^2+y^2} & \text{if } (x,y) \neq (0,0) \\ 0 & \text{if } (x,y) = (0,0) \end{cases}$$

from Example 2 is *not* continuous at $(0, 0)$ because $\lim_{(x,y)\to(0,0)} f(x, y)$ does not exist, so it certainly cannot equal $f(0, 0)$.

▶ **EXAMPLE 5** Suppose that

$$f(x,y) = \begin{cases} \dfrac{(x^2-y^2)^2}{x^2+y^2} & \text{if } (x,y) \neq (0,0) \\ 1 & \text{if } (x,y) = (0,0). \end{cases}$$

Is f continuous at $(0, 0)$?

Solution Observe that, for $(x, y) \neq (0, 0)$, we have

$$0 \leq |f(x, y)| = \frac{(x^2 - y^2)^2}{x^2 + y^2} \leq \frac{(x^2 + y^2)^2}{x^2 + y^2} = x^2 + y^2.$$

In Example 1, we proved that $\lim_{(x,y)\to(0,0)}(x^2 + y^2) = 0$. It follows that $f(x, y)$ also tends to 0 as $(x, y)\to(0, 0)$. Therefore

$$\lim_{(x,y)\to(0,0)} f(x, y) = 0 \neq 1 = f(0, 0).$$

We conclude that f is not continuous at $(0, 0)$. ◄

Rules for Continuity

Our first rule for continuity tells us that a continuous function of one variable gives rise to a pair of continuous functions of two variables.

THEOREM 2 Suppose that φ is defined on an interval (a, b) and continuous at a point c inside the interval. Let $f(x, y) = \varphi(x)$ for $a < x < b$, $-\infty < y < \infty$ and let $g(x, y) = \varphi(y)$ for $-\infty < x < \infty, a < y < b$. Then, for every y_0, the function f is continuous at (c, y_0), and, for every x_0, the function g is continuous at (x_0, c).

Proof. We will prove that f is continuous at $P_0 = (c, y_0)$. Continuity of g is handled in the same way. Let ε be an arbitrary positive number. Because φ is continuous at c, there is a positive number δ such that $|\varphi(x) - \varphi(c)| < \varepsilon$ if $|x - c| < \delta$. Now, if $P = (x, y)$ satisfies $d(P, P_0) < \delta$, then we have

$$|x - c| = \sqrt{(x - c)^2} \leq \sqrt{(x - c)^2 + (y - y_0)^2} = d(P, P_0) < \delta.$$

It follows that

$$|f(P) - f(P_0)| = |f(x, y) - f(c, y_0)| = |\varphi(x) - \varphi(c)| < \varepsilon$$

if $d(P, P_0) < \delta$, which establishes the continuity of f at P_0. ■

We will use Theorem 2 in conjunction with the following rules for combining functions of two variables, which are similar to those that we learned in Chapter 2 for functions of one variable.

THEOREM 3 Suppose that f and g are functions that are continuous at $P_0 = (x_0, y_0)$. Then, $f + g, f - g$, and fg are also continuous at P_0. If $g(P_0) \neq 0$, then f/g is also continuous at P_0. If ϕ is a function of one variable that is continuous at $f(P_0)$, then $\phi \circ f$ is continuous at P_0.

► **EXAMPLE 6** Discuss the continuity of

$$U(x, y) = y^3 \sin(x) - \frac{\cos(xy^2)}{(2x - y)^2}.$$

Solution We first observe that $\mathcal{D} = \{(x,y) : y \neq 2x\}$ is the domain of U. Suppose that $P_0 = (x_0, y_0)$ is a point in $\mathcal{D}$. Because $x \mapsto \sin(x)$ and $y \mapsto y^3$ are continuous functions of one variable, we know from Theorem 2 that $f_0(x,y) = y^3$ and $g_0(x,y) = \sin(x)$ are continuous at P_0. Theorem 3 then tells us that $f(x,y) = f_0(x,y)\, g_0(x,y) = y^3 \sin(x)$ is also continuous at P_0. By similar reasoning, we see that the continuous single-variable functions $x \mapsto x$, $x \mapsto 2x$, and $y \mapsto y^2$ can be combined to form the continuous functions $f_1(x,y) = xy^2$ and $f_2(x,y) = 2x - y$. Because $\phi_1(u) = \cos(u)$ and $\phi_2(u) = u^2$ are continuous, Theorem 3 tells us that $g_1(x,y) = (\phi_1 \circ f_1)(x,y) = \phi_1(f_1(x,y)) = \cos(xy^2)$ and $g_2(x,y) = (\phi_2 \circ f_2)(x,y) = \phi_2(f_2(x,y)) = (2x-y)^2$ are continuous at P_0. Because $g_2(P_0) \neq 0$, we conclude from Theorem 3 that

$$g(x,y) = \frac{g_1(x,y)}{g_2(x,y)} = \frac{\cos(xy^2)}{(2x-y)^2}$$

is continuous at P_0. Finally, by Theorem 3, $U(x,y) = f(x,y) - g(x,y)$ is also continuous at P_0. In summary, U is continuous at every point in its domain. ◀

A function that is continuous at every point in its domain, such as the function U of Example 6, is said to be a *continuous* function.

Functions of Three Variables

We conclude by noting that the entire discussion of this section applies to functions of three variables x, y, z, with no change (except for notation). In particular, all of the rules and remarks are still true.

▶ **EXAMPLE 7** Show that $V(x,y,z) = z^3 \cos(xy^2)$ is a continuous function.

Solution The domain of V is all of xyz-space because the expression $z^3\cos(xy^2)$ makes sense for all values of x, y, and z. Therefore we must show that V is continuous at every point (x,y,z). In Example 6, we observed that the function $(x,y) \mapsto \cos(xy^2)$ of two variables is continuous at every point (x,y). The other factor of $V(x,y,z)$, the function $z \mapsto z^3$ of one variable, is also continuous. The extension of Theorem 2 to three variables tells us that the functions $f(x,y,z) = z^3$ and $g(x,y,z) = \cos(xy^2)$ are continuous at every point (x,y,z). The extension of Theorem 3 to three variables then tells us that the product $V(x,y,z) = f(x,y,z) \cdot g(x,y,z)$ is also continuous at every point (x,y,z). That is, V is a continuous function. ◀

QUICK QUIZ

1. True or false: If f is not defined at P_0, then $\lim_{P \to P_0} f(P)$ does not exist.
2. What is $\lim_{(x,y)\to(3,3)} \dfrac{x^2 - y^2}{x - y}$?
3. True or false: If ϕ and ψ are functions of one variable that are continuous at a and b, respectively, then $f(x,y) = \phi(x)\psi(y)$ is continuous at (a,b).
4. True or false: If f is continuous at $(1,4)$, if $\lim_{x\to 1} f(x, 4x) = 5$, and if $\ell = \lim_{y \to 4} f(\tan(\pi/y), y)$, then $\ell = 5$.

Answers

1. False 2. 6 3. True 4. True

EXERCISES

Problems for Practice

In each of Exercises 1–14, calculate the indicated limit.

1. $\lim_{(x,y)\to(2,3)} x(2xy-3)^2$
2. $\lim_{(x,y)\to(-1,6)} \sin\left(\frac{\pi xy}{4}\right)$
3. $\lim_{(x,y)\to(4,2)} \sqrt{x^2y-x}$
4. $\lim_{(x,y)\to(1,-1)} \frac{x^2-y^2}{x+y}$
5. $\lim_{(x,y)\to(-1,4)} \ln(y^2+xy)$
6. $\lim_{(x,y)\to(4,-8)} x^{-1/2}y^{1/3}$
7. $\lim_{(x,y)\to(1/4,1/4)} \frac{\arccos(x-y)}{e^{x-y}}$
8. $\lim_{(x,y)\to(\pi/2,2)} \arctan\left(\sqrt{y}\cos\left(\frac{x}{2}\right)\right)$
9. $\lim_{(x,y)\to(0,\pi/2)} \sin(y\cos(x)) + \cos(x+\cos(y))$
10. $\lim_{(x,y)\to(0,0)} \frac{\sin(x+y)}{x+y}$
11. $\lim_{(x,y)\to(1,2)} \frac{(x-1)(y^2-4)}{(y-2)(x^2-1)}$
12. $\lim_{(x,y)\to(0,3)} \frac{y\sin(x)}{x(y^2+1)}$
13. $\lim_{(x,y)\to(6,3)} \frac{x^2-xy-2y^2}{x-2y}$
14. $\lim_{(x,y)\to(0,0)} \frac{1-\exp(x^2+y^2)}{x^2+y^2}$

In each of Exercises 15–20, the given expression does *not* have a limit at $(0,0)$. Let (x, y) approach the origin along two different paths, as in Example 2, to verify the nonexistence of the limit.

15. y/x
16. $y(x^2+y^2)^{-1/2}$
17. $xy^2/|xy^2|$
18. $(x^2+y^2)/(x^2-y^2)$
19. $(x+y)/(x-y)$
20. $(x+y)/(x+2y)$

Further Theory and Practice

In each of Exercises 21–24, determine whether the given function is continuous at $(0,0)$.

21.
$$f(x,y)=\begin{cases}\dfrac{\sin(x^2+y^2)}{x^2+y^2} & (x,y)\neq(0,0)\\ 1 & (x,y)=(0,0)\end{cases}$$

22.
$$f(x,y)=\begin{cases}\dfrac{(x^2+y^2)^2}{x^4+y^4} & (x,y)\neq(0,0)\\ 1 & (x,y)=(0,0)\end{cases}$$

23.
$$f(x,y)=\begin{cases}\dfrac{y^2-2x^2}{x^2+y^2} & (x,y)\neq(0,0)\\ 1 & (x,y)=(0,0)\end{cases}$$

24.
$$f(x,y)=\begin{cases}\dfrac{x^4+3y^4}{x^2+y^2} & (x,y)\neq(0,0)\\ 0 & (x,y)=(0,0)\end{cases}$$

25. If ϕ and ψ are continuous functions, then $(x,y)\mapsto\phi(x)+\psi(y)$ is continuous, as can be deduced from Theorems 2 and 3. Instead, give a direct, ε-δ, proof.
26. If f is continuous at (x_0,y_0), show that $\phi(x)=f(x,y_0)$ is continuous at x_0 and that $\psi(y)=f(x_0,y)$ is continuous at y_0.
27. Suppose that $\phi(x)=f(x,y_0)$ is continuous at x_0 and $\psi(y)=f(x_0,y)$ is continuous at y_0. Is it true that f is continuous at (x_0,y_0)? Give a reason for your answer.
28. Calculate $\lim_{(x,y)\to(1,0)}(1+xy)^{1/(xy)}$.
29. Calculate $\lim_{(x,y)\to(0,0)} x^2\ln(x^2+y^2)$.
30. Let $f(x,y)=2xy/(x^2+y^2)$. Calculate $\lim_{y\to0} f(0,y)$. Also, for any slope m with $-\infty<m<\infty$, calculate $\lim_{x\to0} f(x,mx)$. Show that $f(x,y)$ has a limit as (x,y) approaches $(0,0)$ along any straight line approaching the origin, and that the limit can be any number in the interval $[-1,1]$.
31. Let $f(x,y)=xy^2/(x^2+y^4)$. This function does not have a limit as (x, y) tends to $(0,0)$. However, you cannot detect this by looking only at straight-line paths approaching the origin. Calculate $\lim_{y\to0} f(0,y)$. Also, for any slope $-\infty<m<\infty$, calculate $\lim_{x\to0} f(x,mx)$. Deduce that if (x,y) approaches the origin along any straight line, then the limit of $f(x,y)$ is 0. However, show that $f(x,y)$ has a different limit as (x,y) approaches $(0,0)$ along the parabolic path $x=y^2$.
32. Let $f(x,y)=xy^3/(x^2+y^6)$. Prove that along any straight-line path approaching the origin, the limit of $f(x,y)$ is 0. However, show that there is another path along which $f(x,y)$ has a different limit as (x,y) approaches $(0,0)$. Therefore $f(x,y)$ does not have a limit as (x, y) tends to $(0,0)$.

Calculator/Computer Exercises

33. Plot $f(x,y)=xy/(x^2+y^2)$ for (x,y) in a rectangle centered at $(0,0)$. Use your plot to verify visually that f does not have a limit at $(0,0)$.

34. Plot $f(x, y) = (x^2 - y^2)/(x^2 + y^2)$ for (x, y) in a rectangle centered at $(0, 0)$. Use your plot to verify visually that f does not have a limit at $(0, 0)$.

35. Plot $f(x, y) = (x + y)^2/(x^2 + y^2)$ for (x, y) in a rectangle centered at $(0, 0)$. Use your plot to verify visually that f does not have a limit at $(0, 0)$.

36. Plot $f(x, y) = xy^2/(x^2 + y^4)$ for (x, y) in a rectangle centered at $(0, 0)$. Use your plot to verify visually that f does not have a limit at $(0, 0)$. Note, however, that the limit *does* exist and equals 0 along straight lines through the origin.

11.4 Partial Derivatives

In this section, we will learn how to determine the rate at which a function f of two variables changes. To do so, we will need an appropriate notion of differentiation. Because there are *two* independent variables, we shall develop a procedure for differentiating in each variable separately. The geometric idea behind these new differentiation techniques is simple: We take slices of the graph of f. You are already familiar with the strategy of intersecting the graph of $z = f(x, y)$ with planes. Up until now, we have relied primarily on horizontal planes as an aid in plotting. Such planes, however, are of no use for our current purposes because the values of $f(x, y)$ do not change on the level curves that they produce. Instead, we will now intersect the graph of $z = f(x, y)$ with planes that are parallel to the xz- and yz-coordinate planes.

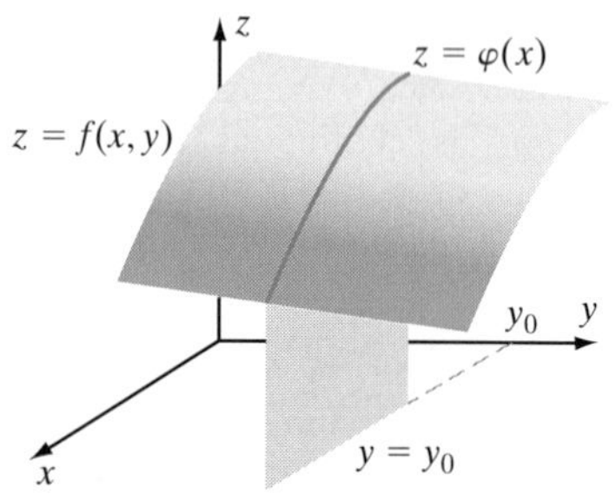

▲ Figure 1

Let $P_0 = (x_0, y_0)$ be a point in the xy-plane. Suppose that f is a function whose domain contains a disk centered at P_0. If we fix y_0 and only allow x to vary, then we obtain a function $\varphi(x) = f(x, y_0)$ of one variable. The graph of $z = \varphi(x)$ can be realized by intersecting the graph of $z = f(x, y)$ with the plane $y = y_0$ as shown in Figure 1. Ordinarily, the graph of $z = \varphi(x)$ would be plotted in the xz-plane, which is the plane $y = 0$ in xyz-space. Here it is realized in the plane $y = y_0$, which is parallel to the xz-plane.

Because φ is a function of one variable, we can differentiate it in the ordinary manner. Not only that, the number

$$\varphi'(x_0) = \lim_{\Delta x \to 0} \frac{\varphi(x_0 + \Delta x) - \varphi(x_0)}{\Delta x} = \lim_{\Delta x \to 0} \frac{f(x_0 + \Delta x, y_0) - f(x_0, y_0)}{\Delta x}$$

has its usual interpretation—the instantaneous rate of change of φ at x_0. Because φ is just the function f in which the y variable has been fixed at y_0, the value of $\varphi'(x_0)$ is the instantaneous rate of change of f at (x_0, y_0) as x changes and $y = y_0$ remains fixed.

In an analogous manner, we fix $x = x_0$, define $\psi(y) = f(x_0, y)$, and use

$$\psi'(y_0) = \lim_{\Delta y \to 0} \frac{\psi(y_0 + \Delta y) - \psi(y_0)}{\Delta y} = \lim_{\Delta y \to 0} \frac{f(x_0, y_0 + \Delta y) - f(x_0, y_0)}{\Delta y}$$

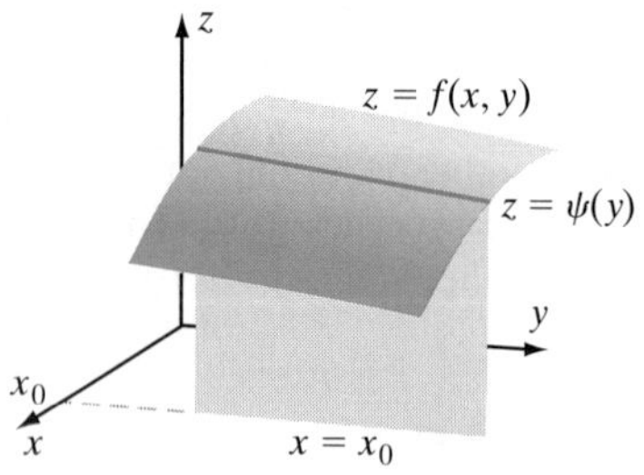

▲ Figure 2

to tell us the instantaneous rate of change of f at (x_0, y_0) as y changes and $x = x_0$ remains fixed. See Figure 2 for the geometry. These considerations give us a way to define differentiation for a function of two variables.

DEFINITION Let $P_0 = (x_0, y_0)$ be a point in the xy-plane. Suppose that f is a function that is defined on a disk $D(P_0, r)$. We say that f is *differentiable* with respect to x at P_0 if

$$\lim_{\Delta x \to 0} \frac{f(x_0 + \Delta x, y_0) - f(x_0, y_0)}{\Delta x}$$

exists. We call this limit the *partial derivative* of f with respect to x at the point P_0 and denote it by

$$\frac{\partial f}{\partial x}(P_0).$$

Similarly, we say that f is *differentiable* with respect to y at P_0 if

$$\lim_{\Delta y \to 0} \frac{f(x_0, y_0 + \Delta y) - f(x_0, y_0)}{\Delta y}$$

exists. We call this limit the *partial derivative* of f with respect to y at the point P_0 and denote it by

$$\frac{\partial f}{\partial y}(P_0).$$

▶ **EXAMPLE 1** Define $f(x, y) = xy^2$. Calculate the partial derivatives with respect to x and with respect to y. What are $\dfrac{\partial f}{\partial x}(1,3)$ and $\dfrac{\partial f}{\partial y}(1,3)$? Compare these values with the derivatives $\left.\dfrac{d\varphi}{dx}\right|_{x=1}$ and $\left.\dfrac{d\psi}{dy}\right|_{y=3}$ where φ and ψ are the functions of one variable that are defined by $\varphi(x) = f(x, 3)$ and $\psi(y) = f(1, y)$.

Solution When we calculate $\partial f/\partial x$, we differentiate the expression $f(x, y)$ with respect to x, treating y, and any expression depending only on y, as if it were a constant. In the case of xy^2, that means we differentiate, with respect to x, the product of the constant y^2 and the variable x. The result is y^2 times 1. Thus $\partial f/\partial x = y^2$. Similarly, when we calculate $\partial f/\partial y$, we think of x as a constant and differentiate the expression $f(x, y)$ with respect to y. In other words, we differentiate with respect to y, the product of the constant x and the expression y^2. The result is the constant x times $2y$. Thus $\partial f/\partial y = 2xy$. By substituting $x = 1$ and $y = 3$, we obtain

$$\frac{\partial f}{\partial x}(1,3) = y^2\big|_{x=1, y=3} = 9 \quad \text{and} \quad \frac{\partial f}{\partial y}(1,3) = 2xy\big|_{x=1, y=3} = 6.$$

Next, consider the function $\varphi(x) = f(x, 3) = (x)(3)^2 = 9x$. We calculate $\varphi'(x) = 9$. Therefore $\varphi'(1) = 9$, which is the same as $\partial f/\partial x$ evaluated at $(1, 3)$. Similarly, $\psi(y) = f(1, y) = (1)(y^2) = y^2$ and

$$\psi'(3) = \frac{d}{dy}(y^2)\bigg|_{y=3} = 2y\bigg|_{y=3} = 6, \text{ which equals } \frac{\partial f}{\partial y}(1,3). \blacktriangleleft$$

All of our familiar rules for differentiation—the Product Rule, the Quotient Rule, the Chain Rule, and so on—are still valid for partial derivatives because partial differentiation is just the differentiation that we already know, applied one variable at a time.

▶ **EXAMPLE 2** Calculate the partial derivatives $\partial f/\partial x$ and $\partial f/\partial y$ for $f(x,y) = \sqrt{x^3 + y^2}$.

Solution Applying the Chain Rule, we obtain

$$\frac{\partial f}{\partial x}(x,y) = \frac{\partial}{\partial x}(x^3+y^2)^{1/2} = \frac{1}{2}(x^3+y^2)^{-1/2}\frac{\partial}{\partial x}(x^3+y^2) = \frac{3}{2}x^2(x^3+y^2)^{-1/2}$$

and

$$\frac{\partial f}{\partial y}(x,y) = \frac{\partial}{\partial y}(x^3+y^2)^{1/2} = \frac{1}{2}(x^3+y^2)^{-1/2}\frac{\partial}{\partial y}(x^3+y^2) = y(x^3+y^2)^{-1/2}. \blacktriangleleft$$

There are several common notations for the partial derivatives of a function $(x,y) \mapsto f(x,y)$ of two variables. Thus

$$\frac{\partial f}{\partial x}, \quad D_x f, \quad f_x, \quad f_1, \quad \text{and} \quad D_1 f$$

are all used for the partial derivative of f with respect to x. Similarly, $\partial f/\partial y$, $D_y f$, f_y, f_2, and $D_2 f$ all stand for the partial derivative of f with respect to y.

▶ **EXAMPLE 3** Calculate f_x and f_y for the function f defined by

$$f(x,y) = \ln(x)e^{x\cos(y)}.$$

Solution We calculate f_x using the Product Rule and Chain Rule:

$$f_x(x,y) = \left(\frac{\partial}{\partial x}\ln(x)\right)e^{x\cos(y)} + \ln(x)\frac{\partial}{\partial x}e^{x\cos(y)} = \frac{1}{x}e^{x\cos(y)} + \ln(x)\left(e^{x\cos(y)}\cos(y)\right).$$

When we calculate a partial derivative with respect to y, we treat x as a constant. Therefore $\ln(x)$ is treated as a constant when calculating f_y: We do not need the Product Rule for this calculation:

$$f_y(x,y) = \ln(x)\frac{\partial}{\partial y}\left(e^{x\cos(y)}\right) = \ln(x)e^{x\cos(y)}\frac{\partial}{\partial y}(x\cos(y)) = -\ln(x)e^{x\cos(y)}x\sin(y). \blacktriangleleft$$

▶ **EXAMPLE 4** A string in the xy-plane vibrates up and down in the y-direction and has endpoints that are fixed at $(0,0)$ and $(1,0)$. Suppose that the displacement of the string at point x and time t is given by $y(x,t) = \sin(\pi x)\sin(2t)$. What is the instantaneous rate of change of y with respect to time at the point $x = 1/4$?

Solution We calculate

$$\frac{\partial}{\partial t}y(x,t) = \frac{\partial}{\partial t}\big(\sin(\pi x)\sin(2t)\big) = \sin(\pi x)\frac{\partial}{\partial t}\sin(2t) = 2\sin(\pi x)\cos(2t).$$

Therefore the instantaneous rate of change with respect to time of the string's displacement from equilibrium at the point $x = 1/4$ is $2\sin(\pi/4)\cos(2t)$ or $\sqrt{2}\cos(2t)$. ◄

INSIGHT Example 4 demonstrates that the partial derivative with respect to a variable can be interpreted as the *rate of change* with respect to that variable. Many physical quantities depend on several parameters, so that we need to have this new multivariate way to measure rates of change. For instance, if heat is being applied to the ends of a rod, then the temperature T at a point x on the rod depends both on x and on the time t. The partial derivatives $\partial T/\partial x$ and $\partial T/\partial t$ tell us about the rates of change when one of the two independent variables is held constant.

Functions of Three Variables

The notion of partial derivative also applies to functions of three (or more) variables. If $F(x, y, z)$ is an expression involving three independent variables x, y, and z, then we calculate $\partial F/\partial x$ by holding y and z constant and differentiating with respect to x. The partial derivatives in y and z are calculated similarly.

► **EXAMPLE 5** Calculate the partial derivatives of $F(x, y, z) = xz\sin(y^2z)$ with respect to x, with respect to y, and with respect to z.

Solution When we hold y and z constant, the expression $z\sin(y^2z)$ is a constant and $f(x, y, z)$ is this constant times x. Therefore $\partial F/\partial x = z\sin(y^2z)$. Next, hold x and z constant to calculate the partial derivative with respect to y:

$$\begin{aligned}
\frac{\partial F}{\partial y}(x,y,z) &= \frac{\partial}{\partial y}\big(xz\sin(y^2z)\big) \\
&= xz\frac{\partial}{\partial y}\big(\sin(y^2z)\big) && \text{(the constant slides outside the derivative)} \\
&= xz\cos(y^2z)\frac{\partial}{\partial y}(y^2z) && \text{(using the Chain Rule)} \\
&= xz^2\cos(y^2z)\frac{\partial}{\partial y}(y^2) && \text{(the constant slides outside the derivative)} \\
&= 2xyz^2\cos(y^2z).
\end{aligned}$$

We calculate $\partial F/\partial z$ in a similar way, but for this partial derivative we require the Product Rule:

$$\frac{\partial F}{\partial z}(x,y,z) = \frac{\partial}{\partial z}\big(xz\sin(y^2z)\big) = x\frac{\partial}{\partial z}\big(z\sin(y^2z)\big) = x\sin\big(y^2z\big) + xz\frac{\partial}{\partial z}\big(\sin(y^2z)\big).$$

Continuing the calculation with an application of the Chain Rule, we have

$$\frac{\partial F}{\partial z}(x,y,z) = x\sin(y^2z) + xz\cos(y^2z)\frac{\partial}{\partial z}(y^2z) = x\sin(y^2z) + xy^2z\cos(y^2z). \quad \blacktriangleleft$$

For a function $(x,y,z)\mapsto F(x,y,z)$ of three variables, there are several alternative notations for the partial derivative with respect to x:

$$\frac{\partial F}{\partial x}, \quad D_xF, \quad F_x, \quad F_1, \quad D_1F$$

Similarly, $\partial F/\partial y$, D_yF, F_y, F_2, and D_2F can be used interchangeably for the partial derivative with respect to y. Likewise, $\partial F/\partial z$, D_zF, F_z, F_3, and D_3F all refer to the partial derivative with respect to z.

Higher Partial Derivatives

Because the partial derivative of a given function $f(x,y)$ is a new function, we may attempt to differentiate that function again. The result, if it exists, is called a *second partial derivative*. Some examples are

$$\frac{\partial^2 f}{\partial x^2} \text{ which means } \frac{\partial}{\partial x}\left(\frac{\partial}{\partial x}f\right);$$

$$\frac{\partial^2 f}{\partial y^2} \text{ which means } \frac{\partial}{\partial y}\left(\frac{\partial}{\partial y}f\right);$$

$$\frac{\partial^2 f}{\partial x \partial y} \text{ which means } \frac{\partial}{\partial x}\left(\frac{\partial}{\partial y}f\right);$$

$$\frac{\partial^2 f}{\partial y \partial x} \text{ which means } \frac{\partial}{\partial y}\left(\frac{\partial}{\partial x}f\right).$$

The last two of these higher order partial derivatives involve differentiation with respect to both x and y. They are therefore sometimes called *mixed partial derivatives*.

Of course we can also use the other partial derivative notations for second derivatives. For example, $\partial^2 f/\partial x^2$ can be written as

$$D_{xx}f \quad \text{or} \quad f_{xx} \quad \text{or} \quad f_{1,1} \quad \text{or} \quad D_{1,1}f,$$

and $\dfrac{\partial^2 f}{\partial y \partial x}$ may be written as

$$D_{xy}f \quad \text{or} \quad f_{xy} \quad \text{or} \quad f_{1,2} \quad \text{or} \quad D_{1,2}f,$$

and so on.

▶ **EXAMPLE 6** Calculate all the second partial derivatives of $f(x,y) = xy - y^3 + x^2y^4$.

Solution We have

$$\frac{\partial^2 f}{\partial x^2}(x,y) = \frac{\partial}{\partial x}\left(\frac{\partial f}{\partial x}(x,y)\right) = \frac{\partial}{\partial x}(y + 2xy^4) = 2y^4,$$

$$\frac{\partial^2 f}{\partial y^2} = \frac{\partial}{\partial y}\left(\frac{\partial f}{\partial y}(x,y)\right) = \frac{\partial}{\partial y}(x - 3y^2 + 4x^2y^3) = -6y + 12x^2y^2,$$

$$\frac{\partial^2 f}{\partial x \partial y} = \frac{\partial}{\partial x}\left(\frac{\partial f}{\partial y}(x,y)\right) = \frac{\partial}{\partial x}(x - 3y^2 + 4x^2y^3) = 1 + 8xy^3, \text{ and}$$

$$\frac{\partial^2 f}{\partial y \partial x} = \frac{\partial}{\partial y}\left(\frac{\partial f}{\partial x}(x,y)\right) = \frac{\partial}{\partial y}(y + 2xy^4) = 1 + 8xy^3. \quad ◀$$

INSIGHT A noteworthy occurrence may be observed in Example 6. The first order partial derivatives for $f(x,y) = xy - y^3 + x^2y^4$ turned out to be

$$\frac{\partial f}{\partial x}(x,y) = y + 2xy^4 \quad \text{and} \quad \frac{\partial f}{\partial y}(x,y) = x - 3y^2 + 4x^2y^3.$$

At first glance, these expressions do not appear to have much in common. Yet, when we used $\frac{\partial f}{\partial x}$ and $\frac{\partial f}{\partial y}$ to calculate the mixed partial derivatives $\frac{\partial^2 f}{\partial y \partial x}$ and $\frac{\partial^2 f}{\partial x \partial y}$, we discovered that

$$\frac{\partial^2 f}{\partial x \partial y} = 1 + 8xy^3 = \frac{\partial^2 f}{\partial y \partial x}.$$

That is, differentiating first with respect to y and then with respect to x results in the same expression that is obtained by differentiating first with respect to x and then with respect to y. Is this a coincidence, or a particular case of something that is true all the time? The answer is that if we restrict our attention to a class of reasonably nice functions, then the order of differentiation is, indeed, irrelevant. The class of "nice" functions that we will study is described in the next definition.

DEFINITION A function $(x,y) \mapsto f(x,y)$ is called *continuously differentiable* on an open disk $D(P_0, r)$ if f and its partial derivatives

$$f_x \quad \text{and} \quad f_y$$

are continuous on $D(P_0, r)$. The function f is called *twice continuously differentiable* on $D(P_0, r)$ if f and its partial derivatives

$$f_x, \quad f_y, \quad f_{xx}, \quad f_{yy}, \quad f_{xy}, \quad f_{yx}$$

are all continuous on $D(P_0, r)$.

THEOREM 1 If $(x, y) \mapsto f(x, y)$ is twice continuously differentiable on an open disk $D(P_0, r)$, then its mixed partial derivatives are equal in the disk:

$$f_{xy}(x, y) = f_{yx}(x, y) \quad (x, y) \in D(P_0, r). \tag{11.4.1}$$

INSIGHT The proof of Theorem 1 is discussed in Exercise 78. Does Theorem 1 mean that we are going to have to sit down and check that every function we encounter is twice continuously differentiable? Fortunately, the answer is "No." The functions that we encounter in practice are built up using algebraic operations and composition involving functions that are obviously twice continuously differentiable. For the functions that we will be using in the rest of this book, the real meaning of Theorem 1 is that we *can* differentiate in any order we like.

Concluding Remarks about Partial Differentiation

The derivatives that we learned about in Chapter 3, namely derivatives of functions of one variable, are sometimes called *ordinary derivatives* to distinguish them from the partial derivatives that we discuss in this chapter. When we compute an ordinary derivative, we write d/dx, but when we compute a partial derivative, we write $\partial/\partial x$ or $\partial/\partial y$ (or one of the other notations that we have discussed). Sometimes a partial derivative can turn out to be an ordinary derivative. For example, if g is a function of one variable, then

$$\frac{\partial^2}{\partial x \partial y}\left(g(x)y + 4\sin(x)\right) = \frac{\partial}{\partial x}\left(\frac{\partial}{\partial y}\left(g(x)y + 4\sin(x)\right)\right) = \frac{d}{dx}g(x).$$

In some applications, it is necessary to calculate derivatives of order higher than two. For example,

$$\frac{\partial}{\partial x}\left(\frac{\partial^2 f}{\partial x \partial y}\right)$$

is a third derivative obtained by applying $\dfrac{\partial}{\partial x}$ to the already familiar quantity $\dfrac{\partial^2 f}{\partial x \partial y}$. We write the resulting partial derivative as $\dfrac{\partial^3 f}{\partial x^2 \partial y}$. In general, the expression

$$\frac{\partial^m f}{\partial x^j \partial y^k},$$

where $j + k = m$, denotes a derivative of order m: Differentiation in x is applied j times, and differentiation in y is applied k times. To guarantee that the differentiations may be applied in any order with the same result, one usually assumes that the function is m times continuously differentiable (all partial derivatives up to and including order m exist and are continuous).

▶ **EXAMPLE 7** Calculate $\dfrac{\partial^3 f}{\partial y^3}$ and $\dfrac{\partial^3 f}{\partial x^2 \partial y}$ for the function $f(x,y) = xy - y^3 + x^2y^4$ of Example 6.

Solution Using the calculation $f_{yy}(x,y) = -6y + 12x^2y^2$ from Example 6, we obtain

$$\frac{\partial^3 f}{\partial y^3} = \frac{\partial}{\partial y}\left(\frac{\partial^2 f}{\partial y^2}(x,y)\right) = \frac{\partial}{\partial y}(-6y + 12x^2y^2) = -6 + 24x^2y.$$

Using the calculation $f_{xx}(x,y) = 2y^4$ from Example 6, we obtain

$$\frac{\partial^3 f}{\partial x^2 \partial y}(x,y) = \frac{\partial}{\partial y}\left(\frac{\partial^2 f}{\partial x^2}(x,y)\right) = \frac{\partial}{\partial y}(2y^4) = 8y^3.$$

Alternatively, we might have used the calculation $f_{xy}(x,y) = 1 + 8xy^3$ from Example 6 as our starting point for the last calculation. Because f is three times continuously differentiable, the result is the same:

$$\frac{\partial^3 f}{\partial x^2 \partial y}(x,y) = \frac{\partial}{\partial x}\left(\frac{\partial^2 f}{\partial x \partial y}\right) = \frac{\partial}{\partial x}(1 + 8xy^3) = 8y^3. \quad ◀$$

QUICK QUIZ

1. True or false: If $\varphi(x) = f(x,2)$ is differentiable, then $\varphi'(3) = \dfrac{\partial f}{\partial x}(3,2)$.
2. Calculate $f_x(3,-1)$ and $f_y(3,-1)$ for $f(x,y) = x^2y^3 + 4x + 10y^2$.
3. Calculate $f_{yy}(5,3)$ and $f_{xy}(5,3)$ for $f(x,y) = x^2\ln(y)$.
4. Calculate $\dfrac{\partial^3 f}{\partial x \partial y^2}$ for the function $f(x,y) = xy - y^3 + x^2y^4$ of Example 6.

Answers

1. True 2. $f_x(3,-1) = -2$ and $f_y(3,-1) = 7$ 3. $f_{yy}(5,3) = -25/9$ and $f_{xy}(5,3) = 10/3$ 4. $24xy^2$

EXERCISES

Problems for Practice

▶ In each of Exercises 1–10, a function f and a point $P_0 = (x_0, y_0)$ are given. Calculate $\varphi(x) = f(x, y_0)$, $\psi(y) = f(x_0, y)$, $\varphi'(x_0)$, and $\psi'(y_0)$. Then, calculate $(\partial f/\partial x)(x_0, y_0)$ and $(\partial f/\partial y)(x_0, y_0)$. Verify that $\varphi'(x_0) = (\partial f/\partial x)(x_0, y_0)$ and $\psi'(y_0) = (\partial f/\partial y)(x_0, y_0)$. ◀

1. $f(x,y) = 2x + y^3$ $\quad P_0 = (5,1)$
2. $f(x,y) = \pi + \sin(x) - \cos(y)$ $\quad P_0 = (\pi/4, \pi/3)$
3. $f(x,y) = 7 - 3y^4$ $\quad P_0 = (2,-1)$
4. $f(x,y) = 1 + 3xy^2 - y$ $\quad P_0 = (1,2)$
5. $f(x,y) = \cos(xy^2)$ $\quad P_0 = (\pi, 1/2)$
6. $f(x,y) = xe^y$ $\quad P_0 = (1,0)$
7. $f(x,y) = x\ln(xy)$ $\quad P_0 = (1,e)$
8. $f(x,y) = \dfrac{x^2}{x+2y}$ $\quad P_0 = (4,2)$
9. $f(x,y) = x^y$ $\quad P_0 = (2,2)$
10. $f(x,y) = \dfrac{2x-y}{1+3x+3y}$ $\quad P_0 = (3,-2)$

▶ In each of Exercises 11–20, a function $(x,y) \mapsto f(x,y)$ is given. Calculate f_{xx}, f_{yy}, f_{xy}, and f_{yx}. Verify that the mixed partial derivatives are equal. ◀

11. $f(x,y) = x^2y - xy + y^5$
12. $f(x,y) = \dfrac{x}{x+y}$
13. $f(x,y) = x^2 - x/y$
14. $f(x,y) = \sqrt{x^2 + 2y}$
15. $f(x,y) = \dfrac{\sin(x)}{\cos(y)}$
16. $f(x,y) = \sin(xy)$
17. $f(x,y) = \ln(xy - y^2)$

18. $f(x,y) = y^2 \ln(x+y)$
19. $f(x,y) = \cos(x)\sin(y)$
20. $f(x,y) = \ln(x + e^y)$

For each of the functions in Exercises 21–30, calculate $D_y f$, $D_{yy} f$, $D_{yx} f$, f_{12}, and f_{11}. Verify that $D_{yx} f = f_{12}$.

21. $f(x,y) = x\sec(y)$
22. $f(x,y) = \dfrac{x}{1-xy}$
23. $f(x,y) = \sin(x^3 + 2y)$
24. $f(x,y) = \tan(x^2 + y^2)$
25. $f(x,y) = \left(\sin(x) - \cos(y)\right)^5$
26. $f(x,y) = e^{x-2y}$
27. $f(x,y) = (x+3y)^4$
28. $f(x,y) = 1/(4+xy^2)^3$
29. $f(x,y) = \sqrt{1+3x^2+y^4}$
30. $f(x,y) = 3y^2 - \sin(x\sqrt{y})$

In each of Exercises 31–40, a function $(x,y,z) \mapsto f(x,y,z)$ is given. Calculate f_x, f_y, f_z, f_{xy}, f_{xz}, f_{yz}, f_{xx}, f_{yy}, and f_{zz}.

31. $f(x,y,z) = 2x^3 + y^5 z^8$
32. $f(x,y,z) = x^3 y - 3z\ln(y)$
33. $f(x,y,z) = (x+2y+3z)^5$
34. $f(x,y,z) = \sin(x+yz)$
35. $f(x,y,z) = e^{xy-yz}$
36. $f(x,y,z) = x^2/(y^2+z^2)$
37. $f(x,y,z) = y\ln(xz)$
38. $f(x,y,z) = \ln(x^3 y + 2z)$
39. $f(x,y,z) = (x^2+y^3+z^4)^3$
40. $f(x,y,z) = e^{z\cos(x)\sin(y)}$

In each of Exercises 41–44, a function $(x,y,z) \mapsto f(x,y,z)$ is given. Calculate f_{xyz}, f_{yxz}, f_{xzy}, f_{zxy}, f_{yzx}, and f_{zyx}. (All six calculations should result in the same final expression but will involve different intermediate results.)

41. $f(x,y,z) = xy^2z^3$
42. $f(x,y,z) = xy + 2xz$
43. $f(x,y,z) = \dfrac{3+y\cos(z)}{x}$
44. $f(x,y,z) = (x+2y+3z)^{-2}$

Further Theory and Practice

In each of Exercises 45–50, a function $(x,y) \mapsto f(x,y)$ is given. Calculate f_{xx}, f_{yy}, f_{xy}, and f_{yx}. Verify that the mixed partial derivatives are equal.

45. $f(x,y) = \dfrac{x-y}{x+y}$
46. $f(x,y) = e^{xy}\cos(x)$
47. $f(x,y) = \tan(x^2 y)$
48. $f(x,y) = x\sqrt{x^2+4y}$
49. $f(x,y) = \arctan(x+2y)$
50. $f(x,y) = x^y$

In each of Exercises 51–54, an expression $f(x,y)$ and a point $P_0 = (x_0, y_0, z_0)$ are given. The plane $x = x_0$ (respectively, $y = y_0$) intersects the graph of $z = f(x,y)$ in a curve $\mathcal{C}_1$ (respectively, $\mathcal{C}_2$) with tangent line ℓ_1 (respectively, ℓ_2) at P_0. Find the y-coordinate (respectively, x-coordinate) of the point of intersection of ℓ_1 (respectively, ℓ_2) with the xy-plane.

51. $x^3 y - 2xy^3$	$(2,1,4)$
52. $2x/(x+y)$	$(3,-2,6)$
53. $x^2\sqrt{1+2x+y^2}$	$(2,-2,12)$
54. $5y + \frac{1}{9}x\ln(x^2-y^3)$	$(3,2,10)$

55. Express the volume V. of a cone as a function of height h. and radius r. What is the rate of change of volume with respect to radius? With respect to height?

56. Express the surface area A of a box in terms of length x, width y, and height z. What is the rate of change of surface area with respect to length?

57. The pressure P, volume V, and temperaure T of a sample of an ideal gas are related by the equation $PV = kT$ for a constant k. Express each variable, P, V, and T, as a function of the other two variables. Then show that

$$\frac{\partial P}{\partial V}\frac{\partial V}{\partial T}\frac{\partial T}{\partial P} = -1.$$

(Notice that the partial differentials cannot be formally canceled.)

58. Show that $u(x,y) = x\sin(x/y)$ satisfies the equation $xu_x(x,y) + yu_y(x,y) = u(x,y)$.

59. Show that $u(x,y) = \ln(e^x + e^y)$ satisfies the equation $u_x(x,y) + u_y(x,y) = 1$.

60. Show that $u(x,y) = \arctan(y/x)$, $u(x,y) = \ln(y) - \ln(x)$, and $u(x,y) = 2xy/(x^2+y^2)$ all satisfy the equation $xu_x(x,y) + yu_y(x,y) = 0$.

The partial differential equation

$$\frac{\partial^2 u}{\partial x^2} + \frac{\partial^2 u}{\partial y^2} = 0$$

is called *Laplace's equation*. It describes, for example, the steady state heat distribution in a thin metal plate. A function $u(x,y)$ that satisfies Laplace's equation is said to be *harmonic*. Exercises 61–67 concern solutions of Laplace's equation.

61. Which of the following functions is harmonic?
a. $u(x,y) = x^2 - y^2$
b. $u(x,y) = xy$
c. $u(x,y) = e^x\cos(y)$
d. $u(x,y) = x^2 + y^2$
e. $u(x,y) = x^3$
f. $u(x,y) = e^{-x}\sin(y)$

62. Verify that the function

$$P(x,y) = \frac{1}{2\pi}\frac{1-(x^2+y^2)}{(x-1)^2+y^2}$$

is harmonic. It is called the *Poisson kernel* for the open unit disk $D(0,1)$.

63. Verify that the function

$$P(x,y) = \frac{1}{\pi}\frac{y}{x^2+y^2}$$

is harmonic. It is called the *Poisson kernel* for the upper half-plane $\{(x,y): y>0\}$.

64. Determine all nonzero solutions of Laplace's equation having the form $u(x,y) = Ax^2 + Bxy + Cy^2$. These are the *spherical harmonics* of degreee 2.

65. Suppose that A and E are given constants. Determine B and C so that $u(x,y) = Ax^3 + Bx^2y + Cxy^2 + Ey^3$ satisfies Laplace's equation. In this case, we call u a spherical harmonic of degree 3.

66. Let (x_0, y_0) be any fixed point in the plane. Show that $u(x,y) = \ln\left((x-x_0)^2 + (y-y_0)^2\right)$ is harmonic.

67. Let $v_n(x,y) = (x^2+y^2)^n$. Show that

$$\left(\frac{\partial^2}{\partial x^2} + \frac{\partial^2}{\partial y^2}\right)v_n(x,y) = \frac{4n^2}{x^2+y^2}v_n(x,y).$$

Deduce that $v_n(x,y)$ cannot be harmonic for any $n \neq 0$.

68. Show that $v_n(x,y,z) = (x^2+y^2+z^2)^n$ satisfies

$$\left(\frac{\partial^2}{\partial x^2} + \frac{\partial^2}{\partial y^2} + \frac{\partial^2}{\partial z^2}\right)v_n(x,y,z) = \frac{2n(2n+1)}{x^2+y^2+z^2}v_n(x,y,z).$$

69. If $u(x,t)$ denotes the temperature at position x in a long thin rod at time t, then u satisfies the *heat equation*

$$\frac{\partial u}{\partial t} = \alpha^2\frac{\partial^2 u}{\partial x^2}.$$

The number α is a constant determined by the heat conductivity properties of the rod. Show that the functions

$$u_n(x,t) = \exp(-n^2\alpha^2 t)\sin(nx), \quad n \in \mathbb{Z},$$

satisfy the heat equation.

70. Let $u(x,t)$ describe the height above or below the x axis at time t of the point x on a vibrating string with ends fixed at $x=0$ and $x=2\pi$. Using principles of elementary mechanics, it can be shown that u satisfies the *wave equation*

$$\frac{\partial^2 u}{\partial t^2} = c^2\frac{\partial^2 u}{\partial x^2}$$

where c is a constant that depends on the properties of the string. Show that the functions

$$u_m(x,t) + A_m\sin(cmt)\sin(mx), \quad m \in \mathbb{Z},$$

satisfy the wave equation.

Suppose that $f_x(x,y) = 2x$ and $f_y(x,y) = 3y^2$. If we integrate the first equation with respect to x, then we obtain $f(x,y) = x^2 + g(y)$. Notice that the constant of integration is a function of y because y is held constant when f_x is calculated. If we differentiate this last equation with respect to y, holding x fixed, we obtain $f_y(x,y) = g'(y)$. But we are given $f_y(x,y) = 3y^2$. It follows that $g'(y) = 3y^2$ and $g(y) = y^3 + C$. Here C is a constant of integration. If we substitute this expression for $g(y)$ into our formula for $f(x,y)$, then we see that $f(x,y) = x^2 + y^3 + C$. In each of Exercises 71–74, use the method just outlined to find a function $f(x,y)$ that satisfies the given equations.

71. $f_x(x,y) = 2xy$ $\quad$ $f_y(x,y) = x^2 + 1$
72. $f_x(x,y) = 2x + 2$ $\quad$ $f_y(x,y) = 2y$
73. $f_x(x,y) = 2x$ $\quad$ $f_y(x,y) = 1 - 2y$
74. $f_x(x,y) = 3(x+y)^2 - 1$ $\quad$ $f_y(x,y) = 3(x+y)^2 + 2$

75. Let $P(x,y) = e^{-xy}\dfrac{\partial^{n+m}}{\partial x^n \partial y^m}e^{xy}$. Show that $P(x,y)$ is a polynomial of total degree $m+n$ in x and y. In fact,

$$P(x,y) = \sum_{k=0}^{\mu} k!\binom{n}{k}\binom{m}{k}x^{m-k}y^{n-k}$$

where μ is the minimum of m and n.

76. Define

$$f(x,y) = \begin{cases} \dfrac{xy(x^2-y^2)}{x^2+y^2} & \text{if} \quad (x,y) \neq (0,0) \\ 0 & \text{if} \quad (x,y) = (0,0). \end{cases}$$

Calculate f_x and f_y. (Note: To calculate these at $(0,0)$, you have to resort to the *definition* of partial derivative, using a limit, and you have to calculate the limit by hand.) Now calculate $f_{xy}(0,0)$ and $f_{yx}(0,0)$. They are unequal. Explain why Theorem 1 is not contradicted.

77. In Chapter 3, we observed that if a function of one variable is differentiable at a point, then it is continuous at that point. However, for a function of two variables, continuity does not follow from the existence of the partial derivatives. For the function

$$f(x,y) + \begin{cases} \dfrac{xy}{x^2+y^2} & \text{if } (x,y) \neq (0,0) \\ 0 & \text{if } (x,y) = 0 \end{cases}$$

prove that f_x and f_y exist at $(0,0)$, yet f is discontinuous at $(0,0)$.

78. Complete the following outline to obtain a proof of Theorem 1.

a. Fix a base point (x_0, y_0). Expand f in a Taylor expansion of order 1 *in the x-variable* about the point (x_0, y).

b. Expand the Taylor polynomial P_1 from part a in a first-order Taylor expansion *in the y-variable* about the point y_0.

c. Repeat steps a and b with the roles of x and y reversed.

d. Subtract the two formulas for $f(x,y)$ that you obtained in parts b and c, canceling like terms.

e. Divide through by $(x-x_0)\cdot(y-y_0)$, and let $(x,y)\to(0,0)$ to obtain the result.

Calculator/Computer Exercises

In Exercises 79 and 80, use numerical differentiation to calculate $f_x(P_0)$ and $f_y(P_0)$ for the given function f and the given point P_0.

79. $f(x,y) = (\sqrt{x} + e^{y-x})/\sqrt{3+x^2+y^4}$, $P_0 + (4,3)$

80. $f(x,y) = \sqrt{x^2\cos^2(y)+y^4}/\sqrt{1+y^2\cos^2(x)+x^4}$, $P_0 = (\pi/4, 3\pi/4)$

81. Use numerical differentiation to calculate $f_{xx}(P_0)$ and $f_{yy}(P_0)$ for the function f and the point P_0 of Exercise 79.

82. Use numerical differentiation to calculate $f_{xx}(P_0)$ and $f_{yy}(P_0)$ for the function f and the point P_0 of Exercise 80.

11.5 Differentiability and the Chain Rule

Partial derivatives tell us how a multivariable function changes one variable at a time. In practice, it often happens that two or more of the independent variables are simultaneously changing. In this section, we will learn how to analyze the rate of change of a function when more than one of its variables is changing. Throughout this discussion, the notation $a \cdot b$ will be used to signify the product of scalars; there are *no* dot products of vectors in this section.

Let us review what we know for a function ϕ of one variable that is differentiable at c. By definition of $\phi'(c)$, the quantity

$$\varepsilon = \frac{\phi(x)-\phi(c)}{x-c} - \phi'(c)$$

tends to 0 as x tends to c. After multiplying by $x-c$ and isolating $\phi(x)$, we obtain

$$\phi(x) = \phi(c) + \phi'(c)\cdot(x-c) + \varepsilon\cdot(x-c) \qquad (11.5.1)$$

where $\lim_{x\to c}\varepsilon = 0$. Equation (11.5.1) is the essence of the differential approximation. The sum $\phi'(c)\cdot(x-c)+\varepsilon\cdot(x-c)$ measures how far $\phi(x)$ is from $\phi(c)$. Because each of these two summands has the factor $x-c$, each summand is small when x is close to c. However, the term $\varepsilon\cdot(x-c)$ has the additional small factor ε. Consequently, as $x\to c$, the expression $\varepsilon\cdot(x-c)$ becomes negligible compared to the other terms on the right side of equation (11.5.1). Now we will develop similar ideas that are key to the concept of differentiability of functions of two or more variables. For simplicity, we will discuss the case of a function of two variables—functions of three or more variables are treated analogously.

DEFINITION Suppose that $P_0 = (x_0, y_0)$ is the center of an open disk $D(P_0, r)$ on which a function f of two variables is defined. Suppose that $f_x(P_0)$ and $f_y(P_0)$ both exist. We say that f is *differentiable* at the point P_0 if we can express $f(x, y)$ by the formula

$$\begin{aligned} f(x,y) = f(P_0) &+ f_x(P_0)\cdot(x-x_0) + f_y(P_0)\cdot(y-y_0) + \varepsilon_1\cdot(x-x_0) \\ &+ \varepsilon_2\cdot(y-y_0), \end{aligned} \qquad (11.5.2)$$

where

$$\lim_{(x,y)\to(x_0,y_0)} \varepsilon_1 = 0 \quad \text{and} \quad \lim_{(x,y)\to(x_0,y_0)} \varepsilon_2 = 0.$$

We say that f is *differentiable* on a set if it is differentiable at each point of the set.

It is clear that each of the last four summands on the right side of equation (11.5.2) tends to 0 as (x, y) tends to (x_0, y_0). Therefore if f is differentiable at P_0, we have $\lim_{(x,y)\to(x_0,y_0)} f(x, y) = f(P_0)$. We state this observation as the following theorem.

THEOREM 1 If f is differentiable at P_0, then f is continuous at P_0.

It is reasonable to wonder whether the existence of both partial derivatives at a point is enough to ensure the differentiability of the function at the point. As Example 1 will show, the answer is, *No!* In fact, Example 1 shows that the existence of both partial derivatives at a point does *not* even ensure the continuity of the function at the point.

▶ **EXAMPLE 1** Show that the function f defined by

$$f(x,y) = \begin{cases} \dfrac{xy}{x^2+y^2} & \text{if} \quad (x,y) \neq (0,0) \\ 0 & \text{if} \quad (x,y) = (0,0) \end{cases}$$

is not differentiable at the origin even though both partial derivatives $f_x(0, 0)$ and $f_y(0, 0)$ exist.

Solution We have considered this function in Example 2 of Section 11.3. We observed that $\lim_{(x,y)\to(0,0)} f(x, y)$ does not exist. Therefore f is not continuous at $(0, 0)$. In view of Theorem 1, this rules out differentiability at $(0, 0)$. On the other hand,

$$\lim_{\Delta x\to 0} \frac{f(0+\Delta x, 0) - f(0,0)}{\Delta x} = \lim_{\Delta x\to 0} \frac{1}{\Delta x}\cdot\left(\frac{(\Delta x)\cdot 0}{(\Delta x)^2 + 0^2} - 0\right) = \lim_{\Delta x\to 0} \frac{0}{\Delta x} = 0.$$

Therefore $f_x(0, 0)$ exists and is 0. An analogous argument shows that $f_y(0, 0)$ exists and is 0 as well. ◀

INSIGHT In Example 1, we must use the definitions of the partial derivatives f_x and f_y to calculate $f_x(0,0)$ and $f_y(0,0)$. We cannot apply the Quotient Rule to $xy/(x^2+y^2)$ at the point $(0,0)$ because the denominator would be 0.

Example 1 shows that we cannot take differentiability for granted; it must be checked. Unfortunately, equation (11.5.2) is difficult to work with. Our next theorem provides us with what we need: sufficient conditions for differentiability that are usually easy to verify.

THEOREM 2 Suppose that $P_0=(x_0,y_0)$ is the center of an open disk $D(P_0,r)$ on which a function f of two variables is defined. If both $f_x(x,y)$ and $f_y(x,y)$ exist and are continuous on $D(P_0,r)$, then f is differentiable at P_0. In other words, if f is continuously differentiable at P_0 (as defined in Section 11.4), then f is differentiable at P_0.

Proof. We start with the identity

$$f(x,y)=f(x_0,y_0)+\big(f(x,y)-f(x_0,y)\big)+\big(f(x_0,y)-f(x_0,y_0)\big) \tag{11.5.3}$$

and apply the Mean Value Theorem to the function of one variable $x\mapsto f(x,y)$. Because we are holding y fixed, the derivative of this function with respect to x is the partial derivative f_x. The Mean Value Theorem tells us that there is a point P_1 on the line segment joining (x_0,y) to (x,y) such that $f(x,y)-f(x_0,y)=f_x(P_1)\cdot(x-x_0)$. See Figure 1. A similar application of the Mean Value Theorem tells us that there is a point P_2 on the line segment joining (x_0,y_0) to (x_0,y) such that $f(x_0,y)-f(x_0,y_0)=f_y(P_2)\cdot(y-y_0)$. We can therefore rewrite equation (11.5.3) as

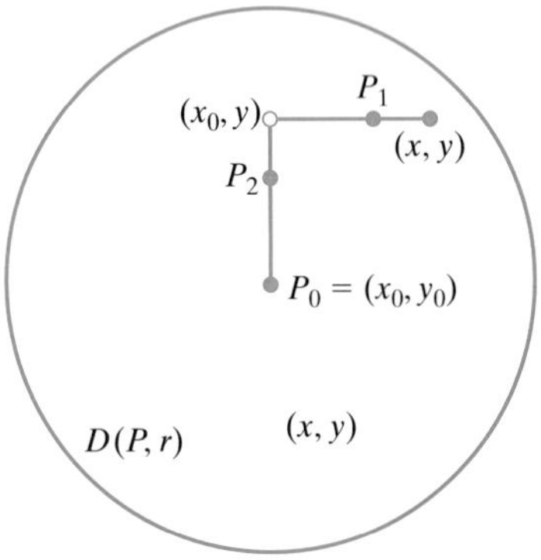

▲ Figure 1

$$f(x,y)=f(x_0,y_0)+f_x(P_1)\cdot(x-x_0)+f_y(P_2)\cdot(y-y_0). \tag{11.5.4}$$

Let $\varepsilon_1=f_x(P_1)-f_x(P_0)$ and $\varepsilon_2=f_y(P_2)-f_y(P_0)$. Then, $f_x(P_1)=f_x(P_0)+\varepsilon_1$ and $f_y(P_2)=f_y(P_0)+\varepsilon_2$. If we substitute these formulas into equation (11.5.4), then we obtain

$$f(x,y)=f(x_0,y_0)+f_x(P_0)\cdot(x-x_0)+f_y(P_0)\cdot(y-y_0)+\varepsilon_1\cdot(x-x_0)+\varepsilon_2\cdot(y-y_0).$$

Furthermore, P_1 and P_2 both tend to P_0 as (x,y) does (refer to Figure 1). By the assumed continuity of f_x and f_y, we have

$$\lim_{(x,y)\to(x_0,y_0)}\varepsilon_1=\lim_{(x,y)\to(x_0,y_0)}\big(f_x(P_1)-f_x(P_0)\big)=0 \quad\text{and}\quad \lim_{(x,y)\to(x_0,y_0)}\varepsilon_2=\lim_{(x,y)\to(x_0,y_0)}\big(f_y(P_2)-f_y(P_0)\big)=0.$$

In conclusion, we have established equation (11.5.2) with ε_1 and ε_2 having the required limit properties. This proves the differentiability of f at P_0. ■

▶ **EXAMPLE 2** Show that $f(x,y)=y/(1+x^2)$ is differentiable on the entire xy-plane.

Solution We have $f_x(x,y) = -2xy/(1+x^2)^2$ and $f_y(x,y) = 1/(1+x^2)$. Because these partial derivatives exist and are continuous at each point of the plane, we may apply Theorem 2 to obtain the desired conclusion. ◄

The Chain Rule for a Function of Two Variables Each Depending on Another Variable

Suppose that $z = f(x,y)$ is a function of the two variables x and y. Suppose also that $x = \rho(s)$ and $y = \sigma(s)$ are functions of another variable s. The situation is represented schematically in Figure 2. We see that z depends depends on x and y, each of which in turn depends on s. So z is, indirectly, a function of s, and we may ask for the value of dz/ds. Naturally, the answer is given in terms of a Chain Rule.

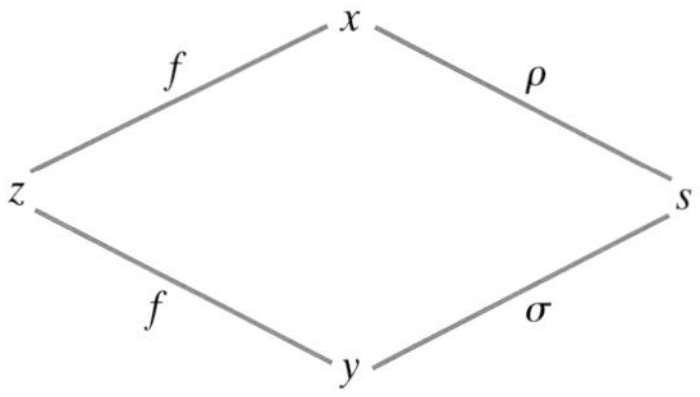

$z = f(x,y),\ x = \rho(s),\ y = \sigma(s)$

▲ Figure 2

THEOREM 3 Let $z = f(x,y)$ be a differentiable function of x and y. Suppose that

$$x = \rho(s) \quad \text{and} \quad y = \sigma(s)$$

are differentiable functions of s. Then $z = f\big(\rho(s), \sigma(s)\big)$ is a differentiable function of s and

$$\frac{dz}{ds} = \frac{\partial f}{\partial x}\frac{d\rho}{ds} + \frac{\partial f}{\partial y}\frac{d\sigma}{ds}. \tag{11.5.5}$$

When written entirely in terms of variables, equation (11.5.5) takes the form

$$\frac{dz}{ds} = \frac{\partial z}{\partial x}\frac{dx}{ds} + \frac{\partial z}{\partial y}\frac{dy}{ds}. \tag{11.5.6}$$

Proof. Let $x = \rho(s+\Delta s)$, $y = \sigma(s+\Delta s)$, $x_0 = \rho(s)$ and $y_0 = \sigma(s)$. Then equation (11.5.2) can be written as

$$f\big(\rho(s+\Delta s), \sigma(s+\Delta s)\big) - f\big(\rho(s), \sigma(s)\big) = \left(\frac{\partial f}{\partial x}(x,y) + \varepsilon_1\right) \cdot \big(\rho(s+\Delta s) - \rho(s)\big) + \left(\frac{\partial f}{\partial y}(x,y) + \varepsilon_2\right) \cdot \big(\sigma(s+\Delta s) - \sigma(s)\big).$$

Dividing each term by Δs, we obtain

$$\frac{f\big(\rho(s+\Delta s), \sigma(s+\Delta s)\big) - f\big(\rho(s), \sigma(s)\big)}{\Delta s} = \left(\frac{\partial f}{\partial x}(x,y) + \varepsilon_1\right) \cdot \left(\frac{\rho(s+\Delta s) - \rho(s)}{\Delta s}\right) + \left(\frac{\partial f}{\partial y}(x,y) + \varepsilon_2\right) \cdot \left(\frac{\sigma(s+\Delta s) - \sigma(s)}{\Delta s}\right).$$

On letting Δs tend to 0 the left side of this equation tends to dz/ds. The quotients on the right side tend to $d\rho/ds$ and $d\sigma/ds$ while the coefficients ε_1 and ε_2 both tend to 0. The limiting form of the equation is therefore

$$\frac{dz}{ds} = \frac{\partial f}{\partial x}\frac{d\rho}{ds} + \frac{\partial f}{\partial y}\frac{d\sigma}{ds},$$

as asserted. ■

▶ **EXAMPLE 3** Define $z = f(x,y) = x^2 + y^3$, $x = \sin(s)$, and $y = \cos(s)$. Calculate dz/ds.

Solution The Chain Rule tells us that

$$\frac{dz}{ds} = \frac{\partial z}{\partial x}\frac{dx}{ds} + \frac{\partial z}{\partial y}\frac{dy}{ds} = (2x)\cos(s) + (3y^2)\big(-\sin(s)\big).$$

This is not a satisfactory form for our answer because it uses the x and y variables to express a rate of change with respect to s. Therefore we substitute for x and y. The result is

$$\frac{dz}{ds} = 2\sin(s)\cos(s) + 3\cos^2(s)\big(-\sin(s)\big) = \sin(s)\cos(s)\,\big(2 - 3\cos(s)\big).$$ ◀

INSIGHT An alternative approach to Example 3 is to substitute the formulas for x and y in terms of s directly into the formula for z. By doing so, we express

$$z = f(x,y) = x^2 + y^3 = \big(\sin(s)\big)^2 + \big(\cos(s)\big)^3$$

explicitly as a function of s so that we may differentiate it directly. (Verify that this differentiation yields the same answer obtained in Example 3.) In spite of the fact that the Chain Rule is not absolutely necessary for calculations such as those of Example 3, you will find that it is an excellent device for organizing complicated differentiation problems. In the next few sections, you will also see that the Chain Rule is essential for certain theoretical developments.

The Chain Rule for a Function of Two Variables Each Depending on Two Other Variables

Frequently we are faced with a function $(x,y) \mapsto f(x,y)$ for which each variable x and y depends on more than one independent variable. To begin, let us suppose that $x = \rho(s,t)$ and $y = \sigma(s,t)$. The Chain Rule for this setup is as follows in Theorem 4.

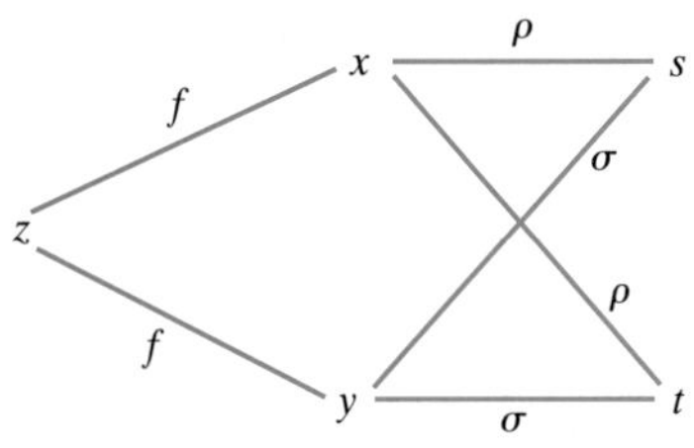

$z = f(x,y),\ x = \rho(s,t),\ y = \sigma(s,t)$

▲ Figure 3

THEOREM 4 Let $z = f(x,y)$ be a differentiable function of x and y. Furthermore, assume that

$$x = \rho(s,t) \quad \text{and} \quad y = \sigma(s,t)$$

are differentiable functions of s and t (as represented in Figure 3). Then, the composition $z = f\big(\rho(s,t), \sigma(s,t)\big)$ is a differentiable function of s and t. Furthermore,

$$\frac{\partial z}{\partial s} = \frac{\partial f}{\partial x}\frac{\partial \rho}{\partial s} + \frac{\partial f}{\partial y}\frac{\partial \sigma}{\partial s} \tag{11.5.7}$$

and

$$\frac{\partial z}{\partial t} = \frac{\partial f}{\partial x}\frac{\partial \rho}{\partial t} + \frac{\partial f}{\partial y}\frac{\partial \sigma}{\partial t}. \tag{11.5.8}$$

When written entirely in terms of variables, equations (11.5.7) and (11.5.8) take the form

$$\frac{\partial z}{\partial s} = \frac{\partial z}{\partial x}\frac{\partial x}{\partial s} + \frac{\partial z}{\partial y}\frac{\partial y}{\partial s} \qquad (11.5.9)$$

and

$$\frac{\partial z}{\partial t} = \frac{\partial z}{\partial x}\frac{\partial x}{\partial t} + \frac{\partial z}{\partial y}\frac{\partial y}{\partial t}. \qquad (11.5.10)$$

The derivations of equations (11.5.7) and (11.5.8) are essentially the same as the derivation of equation (11.5.5). Notice that the form of the Chain Rule when there are two independent variables is similar to the form when there is just one variable: We simply treat each of the variables s and t one at a time.

▶ **EXAMPLE 4** Consider the functions $z = f(x, y) = \ln(x + y^2)$, $x = se^t$, and $y = te^{-s}$. Calculate $\partial z/\partial s$ and $\partial z/\partial t$.

Solution In this example, x and y depend both on s and on t. Let us first differentiate z with respect to s. According to formula (11.5.9), we have

$$\frac{\partial z}{\partial s} = \frac{\partial z}{\partial x}\frac{\partial x}{\partial s} + \frac{\partial z}{\partial y}\frac{\partial y}{\partial s} = \left(\frac{1}{x + y^2}\right)e^t + \left(\frac{2y}{x + y^2}\right)\left(t(-e^{-s})\right).$$

Finally, we substitute for x and y to obtain

$$\frac{\partial z}{\partial s} = \left(\frac{1}{se^t + t^2e^{-2s}}\right)e^t + \left(\frac{2t \cdot e^{-s}}{se^t + t^2e^{-2s}}\right)(-te^{-s}).$$

A similar calculation using formula (11.5.10) yields

$$\frac{\partial z}{\partial t} = \underbrace{\left(\frac{1}{s\,e^t + t^2e^{-2s}}\right)}_{\frac{\partial z}{\partial x}}\underbrace{(s \cdot e^t)}_{\frac{\partial x}{\partial t}} + \underbrace{\left(\frac{2te^{-s}}{se^t + t^2e^{-2s}}\right)}_{\frac{\partial z}{\partial y}}\underbrace{(e^{-s})}_{\frac{\partial y}{\partial t}}. \quad ◀$$

The Chain Rule for a Function of Three Variables Each Depending on Another Variable

Theorems 3 and 4 can be stated for functions of any number of variables. We consider the case of three variables next.

THEOREM 5 Suppose that $w = f(x, y, z)$ is a function of x, y, and z, and that these variables are functions of the variable s. That is, suppose that there are functions ρ, σ, and τ such that $x = \rho(s)$, $y = \sigma(s)$, and $z = \tau(s)$. If the functions f, ρ, σ, and τ are differentiable, then $w = f\big(\rho(s), \sigma(s), \tau(s)\big)$ is a differentiable function of s and

$$\frac{dw}{ds} = \frac{\partial f}{\partial x}\frac{d\rho}{ds} + \frac{\partial f}{\partial y}\frac{d\sigma}{ds} + \frac{\partial f}{\partial z}\frac{d\tau}{ds}.$$

Schematically, we may write this as

$$\frac{dw}{ds} = \frac{\partial w}{\partial x}\frac{dx}{ds} + \frac{\partial w}{\partial y}\frac{dy}{ds} + \frac{\partial w}{\partial z}\frac{dz}{ds}. \qquad (11.5.11)$$

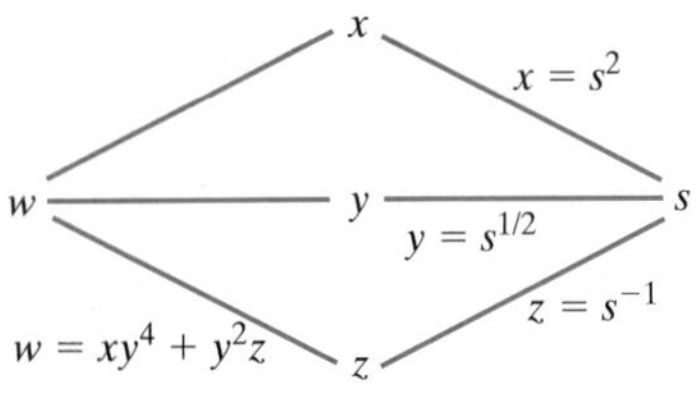

▲ Figure 4

▶ **EXAMPLE 5** Suppose that $w = xy^4 + y^2z$, $x = s^2$, $y = s^{1/2}$, and $z = s^{-1}$. Calculate dw/ds.

Solution Refer to Figure 4. According to Theorem 5, we have

$$\frac{dw}{ds} = \frac{\partial w}{\partial x}\frac{dx}{ds} + \frac{\partial w}{\partial y}\frac{dy}{ds} + \frac{\partial w}{\partial z}\frac{dz}{ds} = y^4(2s) + (4xy^3 + 2yz)\left(\frac{1}{2}s^{-1/2}\right) + y^2(-s^{-2}).$$

Substituting for x, y, z and simplifying gives

$$\frac{dw}{ds} = \left(s^{1/2}\right)^4(2s) + \left(2s^2s^{3/2} + s^{1/2}s^{-1}\right)s^{-1/2} + \left(s^{1/2}\right)^2(-s^{-2}) = 4s^3.$$

You may verify this result by substituting the expressions for x, y, and z into w. It turns out that $w = s^4 + 1$ and the calculation $dw/ds = 4s^3$ follows immediately. ◀

The Chain Rule for a Function of Three Variables Each Depending on Two Other Variables

THEOREM 6 Suppose that $w = f(x, y, z)$ is a function of x, y, and z, and that these variables are functions of the variables s and t. That is, suppose that there are functions ρ, σ, and τ such that $x = \rho(s, t)$, $y = \sigma(s, t)$, and $z = \tau(s, t)$. If the functions f, ρ, σ, and τ are differentiable, then $w = f(\rho(s, t), \sigma(s, t), \tau(s, t))$ is a differentiable function of s and t. Furthermore,

$$\frac{\partial w}{\partial s} = \frac{\partial f}{\partial x}\frac{\partial \rho}{\partial s} + \frac{\partial f}{\partial y}\frac{\partial \sigma}{\partial s} + \frac{\partial f}{\partial z}\frac{\partial \tau}{\partial s},$$

and

$$\frac{\partial w}{\partial t} = \frac{\partial f}{\partial x}\frac{\partial \rho}{\partial t} + \frac{\partial f}{\partial y}\frac{\partial \sigma}{\partial t} + \frac{\partial f}{\partial z}\frac{\partial \tau}{\partial t}.$$

Schematically, we may write this as

$$\frac{\partial w}{\partial s} = \frac{\partial f}{\partial x}\frac{\partial x}{\partial s} + \frac{\partial f}{\partial y}\frac{\partial y}{\partial s} + \frac{\partial f}{\partial z}\frac{\partial z}{\partial s}, \tag{11.5.12}$$

and

$$\frac{\partial w}{\partial t} = \frac{\partial f}{\partial x}\frac{\partial x}{\partial t} + \frac{\partial f}{\partial y}\frac{\partial y}{\partial t} + \frac{\partial f}{\partial z}\frac{\partial z}{\partial t}. \tag{11.5.13}$$

▶ **EXAMPLE 6** Suppose that $w = xy + z^2$, $x = 2s - t$, $y = 3t$, and $z = s$. Calculate $\partial w/\partial s$ and $\partial w/\partial t$.

Solution Theorem 6 tells us that

$$\frac{\partial w}{\partial s} = \frac{\partial w}{\partial x}\frac{\partial x}{\partial s} + \frac{\partial w}{\partial y}\frac{\partial y}{\partial s} + \frac{\partial w}{\partial z}\frac{\partial z}{\partial s} = (y)(2) + (x)(0) + (2z)(1) = (3t)(2) + (2s)(1) = 6t + 2s$$

and

$$\frac{\partial w}{\partial t} = \frac{\partial w}{\partial x}\frac{\partial x}{\partial t} + \frac{\partial w}{\partial y}\frac{\partial y}{\partial t} + \frac{\partial w}{\partial z}\frac{\partial z}{\partial t} = (y)(-1) + (x)(3) + (2z)(0) = (3t)(-1) + (2s - t)(3) = -6t + 6s.$$

These formulas can also be obtained without the Chain Rule by substituting the formulas $x = 2s - t$, $y = 3t$, and $z = s$ into the equation for w. The result is $w = (2s - t)(3t) + s^2 = 6st - 3t^2 + s^2$, from which $\partial w/\partial s$ and $\partial w/\partial t$ are easily calculated. ◀

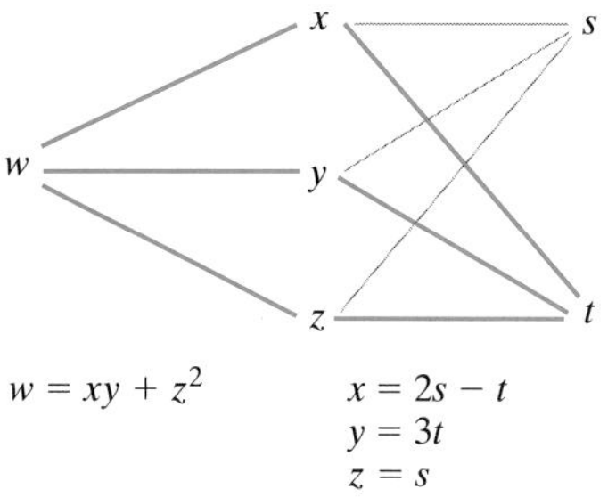

▲ Figure 5

INSIGHT A schematic diagram can help you keep all terms straight, even in complicated examples. Figure 5 shows how this works in Example 6. There are three "paths" from w to t, which tells us that the formula for $\partial w/\partial t$ has three summands. Each path has two segments, which tells us that each summand is the product of two factors. Each factor is obtained by differentiating the variable that appears on the left of the segment with respect to the variable that appears on the right.

◀ **A LOOK BACK AND A LOOK FORWARD** ▶ Theorems 3–6 are special cases of a general Chain Rule that includes equations (11.5.6) and (11.5.9)–(11.5.13). We will not state the general Chain Rule formula because it is best developed using the language of matrices. However, in Section 11.6, we will introduce notation that enables us to concisely unify several of our formulas. In the meantime, it is a good idea to review the formulas that we have learned and to notice that they follow the common pattern discussed in the preceding *Insight*. An understanding of that pattern will help you to remember all forms of the Chain Rule without having to memorize several special cases. Furthermore, you will be able to generalize the Chain Rule to situations that have not been explicitly considered. For example, suppose that V is a function of w, x, y, and z, and that each of these variables depends on q, r, s, t, and u. Then, we may calculate the partial derivative of V with respect to any of the variables q, r, s, t, and u. The first formula is

$$\frac{\partial V}{\partial q} = \frac{\partial V}{\partial w}\frac{\partial w}{\partial q} + \frac{\partial V}{\partial x}\frac{\partial x}{\partial q} + \frac{\partial V}{\partial y}\frac{\partial y}{\partial q} + \frac{\partial V}{\partial z}\frac{\partial z}{\partial q},$$

and the other four are obtained by replacing q with r, s, t, and u.

Taylor's Formula in Several Variables

The basic ideas connected with Taylor's formula for functions of one variable have been developed in Chapter 8. Now we want to consider an analogous formula for functions of two variables. Whereas the Taylor formula in one variable was specified on an interval centered at the point about which the formula was expanded, we now formulate the theorem on a rectangle $I = \{(x, y) : |x - x_0| < r_1, |y - y_0| < r_2\}$ that is centered about the base point of expansion (x_0, y_0). We will state only the order 1 and order 2 cases of the multivariable Taylor's Theorem.

THEOREM 7 Suppose that $P = (x, y)$ is a point in a rectangle I that is centered at $P_0 = (x_0, y_0)$. Set $h = x - x_0$ and $k = y - y_0$. If f is twice continuously differentiable on I, then

$$f(P) = f(P_0) + f_x(P_0)h + f_y(P_0)k + R_1(P) \qquad (11.5.14)$$

where

$$R_1(P) = \frac{1}{2!}\left(f_{xx}(Q_1)h^2 + 2f_{xy}(Q_1)hk + f_{yy}(Q_1)k^2\right)$$

for some point Q_1 on the line segment between P_0 and P. If f is three times continuously differentiable on I, then

$$f(P) = f(P_0) + f_x(P_0)h + f_y(P_0)k + \frac{1}{2!}\Big(f_{xx}(P_0)h^2 + 2f_{xy}(P_0)hk + f_{yy}(P_0)k^2\Big) + R_2(P) \qquad (11.5.15)$$

where

$$R_2(P) = \frac{1}{3!}\Big(f_{xxx}(Q_2)h^3 + 3f_{xxy}(Q_2)h^2k + 3f_{xyy}(Q_2)hk^2 + f_{yyy}(Q_2)k^3\Big)$$

for some point Q_2 on the line segment between P_0 and P.

Proof. We parameterize the line segment between P_0 and P by $P_t = (x_0 + th, y_0 + tk)$ for $0 \le t \le 1$. We may apply the one-variable Taylor Theorem with base point 0 to $F(t) = f(P_t) = f(x_0 + th, y_0 + tk)$. Using $N = 1$ in equation (8.8.4) yields a derivation of formula (11.5.14). Let us instead set $N = 2$ in equation (8.8.4) and prove (11.5.15). The result is

$$F(t) = F(0) + \frac{F'(0)}{1!}t + \frac{F''(0)}{2!}t^2 + \frac{F'''(s)}{3!}t^3 \qquad (11.5.16)$$

for some number s between 0 and t. The next step is to express each derivative on the right side of equation (11.5.16) in terms of derivatives of f. We use the Chain Rule (Theorem 3) to calculate

$$F'(t) = f_x(P_t)\frac{d}{dt}(x_0 + th) + f_y(P_t)\frac{d}{dt}(y_0 + tk),$$

or

$$F'(t) = f_x(P_t)h + f_y(P_t)k. \qquad (11.5.17)$$

To calculate $F''(t)$, we apply the Chain Rule to equation (11.5.17), obtaining

$$F''(t) = \big(f_{xx}(P_t)h + f_{xy}(P_t)k\big)h + \big(f_{yx}(P_t)h + f_{yy}(P_t)k\big)k,$$

or

$$F''(t) = f_{xx}(P_t)h^2 + 2f_{xy}(P_t)hk + f_{yy}(P_t)k^2. \qquad (11.5.18)$$

One further application of the Chain Rule (Theorem 3) yields

$$F'''(t) = f_{xxx}(P_t)h^3 + 3f_{xxy}(P_t)h^2k + 3f_{xyy}(P_t)hk^2 + f_{yyy}(P_t)k^3. \qquad (11.5.19)$$

On substituting $t = 0$ into equations (11.5.17) and (11.5.18), we obtain

$$F'(0) = f_x(P_0)h + f_y(P_0)k \quad \text{and} \quad F''(0) = f_{xx}(P_0)h^2 + 2f_{xy}(P_0)hk + f_{yy}(P_0)k^2. \qquad (11.5.20)$$

On substituting $t = s$ into equation (11.5.19), we obtain

$$F'''(s) = f_{xxx}(P_s)h^3 + 3f_{xxy}(P_s)h^2k + 3f_{xyy}(P_s)hk^2 + f_{yyy}(P_s)k^3. \qquad (11.5.21)$$

Substituting formulas (11.5.20) and (11.5.21) into equation (11.5.16), letting Q_2 denote P_s, and setting $t = 1$ yields equation (11.5.15). ■

The terms $R_1(P)$ and $R_2(P)$ on the right sides of equations (11.5.14) and (11.5.15) are called *remainder* terms, as in the one-variable theory. When we discard $R_1(P)$ from the right side of (11.5.14), we obtain the order 1 Taylor polynomial $T_1(x, y)$ of $f(x, y)$ with base point $P_0 = (x_0, y_0)$. That is,

$$T_1(x, y) = f(P_0) + f_x(P_0)h + f_y(P_0)k, \qquad (11.5.22)$$

where we continue to let $h = x - x_0$ and $k = y - y_0$. Similarly, if we discard $R_2(P)$ from the right side of equation (11.5.15), then we are left with the order 2 Taylor polynomial of $f(x, y)$ with base point $P_0 = (x_0, y_0)$:

$$T_2(x, y) = f(P_0) + f_x(P_0)h + f_y(P_0)k + \frac{1}{2!}\left(f_{xx}(P_0)h^2 + 2f_{xy}(P_0)hk + f_{yy}(P_0)k^2\right). \qquad (11.5.23)$$

As in Section 8.8, we use these Taylor polynomials, T_1 and T_2, together with the estimates of the remainder terms, R_1 and R_2, to approximate f near the base point.

▶ **EXAMPLE 7** Find a quadratic polynomial $T_2(x, y)$ that approximates the function $f(x, y) = \cos(x)\cos(y)$ near the origin.

Solution We choose $(x_0, y_0) = (0, 0)$ for the base point in formula (11.5.23). Given this choice of base point, the increments $h = x - x_0$ and $k = y - y_0$ simplify to x and y, respectively. Equation (11.5.23) becomes

$$T_2(x, y) = f(0, 0) + \left(f_x(0, 0)x + f_y(0, 0)y\right) + \frac{1}{2!}\left(f_{xx}(0, 0)x^2 + 2f_{xy}(0, 0)xy + f_{yy}(0, 0)y^2\right). \quad (11.5.24)$$

We calculate

$$\begin{aligned}
f(x, y) &= \cos(x)\cos(y)\\
f(0, 0) &= \cos(x)\cos(y)\big|_{(0,0)} = 1\\
f_x(0, 0) &= -\sin(x)\cos(y)\big|_{(0,0)} = 0\\
f_y(0, 0) &= -\cos(x)\sin(y)\big|_{(0,0)} = 0\\
f_{xx}(0, 0) &= -\cos(x)\cos(y)\big|_{(0,0)} = -1\\
f_{xy}(0, 0) &= \sin(x)\sin(y)\big|_{(0,0)} = 0\\
f_{yy}(0, 0) &= -\cos(x)\cos(y)\big|_{(0,0)} = -1.
\end{aligned}$$

Substituting these values into equation (11.5.24), we obtain the approximation

$$\cos(x)\cos(y) \approx T_2(x, y) = 1 - \frac{1}{2}(x^2 + y^2). \quad ◀$$

▶ **EXAMPLE 8** Use Taylor's formula to find a quadratic polynomial $T_2(x,y)$ that approximates the function $f(x,y) = xy + 3/x + 9/y$ near the point $(1,3)$.

Solution We begin by calculating the first and second order partial derivatives of $f(x,y)$: $f_x(x,y) = y - 3/x^2$, $f_y(x,y) = x - 9/y^2$, $f_{xy}(x,y) = 1$, $f_{xx}(x,y) = 6/x^3$, and $f_{yy}(x,y) = 18/y^3$. Thus at the base point $P_0 = (1,3)$, we have $f_x(1,3) = 3 - 3/1^2 = 0$, $f_y(1,3) = 1 - 9/3^2 = 0$, $f_{xy}(1,3) = 1$, $f_{xx}(1,3) = 6/1^3 = 6$, and $f_{yy}(1,3) = 18/3^3 = 2/3$. The required degree 2 Taylor polynomial is therefore

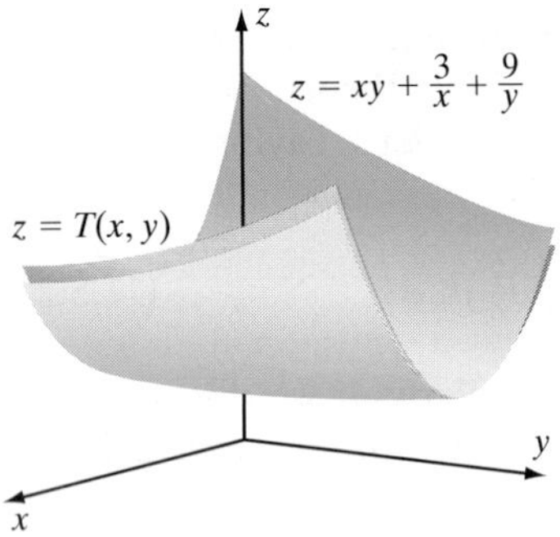

▲ Figure 6

$$f(1,3) + \big(f_x(1,3)(x-1) + f_y(1,3)(y-3)\big) + \frac{1}{2!}\Big(f_{xx}(1,3)(x-1)^2 + 2f_{xy}(1,3)(x-1)(y-3) + f_{yy}(1,3)(y-3)^2\Big)$$

or

$$\begin{aligned} T(x,y) &= 9 + \frac{1}{2!}\Big(6(x-1)^2 + 2(x-1)(y-3) + \frac{2}{3}(y-3)^2\Big) \\ &= 9 + 3(x-1)^2 + (x-1)(y-3) + \frac{1}{3}(y-3)^2. \end{aligned}$$

Figure 6 shows the graphs of f and T in the viewing box $[0.9, 1.1] \times [2.9, 3.1] \times [8.985, 9.05]$.

INSIGHT It is important to have analytic techniques for determining the extrema of functions of several variables. When we graph a function of one variable in a small window containing a maximum or minimum value, we expect to have visual evidence of the extremum. However, the extrema of functions of two variables can be difficult to spot in a graph. It turns out that the Taylor polynomial $T(x,y)$ of Example 8 has an absolute minimum at the base point $(1,3)$, a fact that is not particularly evident from Figure 6. However, if we complete the square in the expression for $T(x,y)$, then we obtain

$$\begin{aligned} T(x,y) &= 9 + 3\Big((x-1)^2 + \frac{1}{3}(x-1)(y-3) + \frac{1}{36}(y-3)^2 + \Big(\frac{1}{9} - \frac{1}{36}\Big)(y-3)^2\Big) \\ &= 9 + 3\Big(x - 1 + \frac{1}{6}(y-3)\Big)^2 + \frac{1}{4}(y-3)^2, \end{aligned}$$

which shows us that $T(x,y)$ is the sum of $T(1,3) = 9$ and two nonnegative quantities. We conclude that T has an absolute minimum value of 9 at the point $(1,3)$. Because $f(x,y)$ differs from $T(x,y)$ by a remainder term that is small when the point (x,y) is close to the base point $(1,3)$, we expect f to also have a local minimum at the point $(1,3)$, and it does. We will return to these ideas in Section 11.8.

QUICK QUIZ

1. True or false: If both $f_x(P_0)$ and exist $f_y(P_0)$, then f is continuous at P_0.
2. True or false: If f is differentiable at P_0, then f is continuous at P_0.
3. If $f_x(-1,2) = 3$, $f_y(-1,2) = -5$, and $z = f(7t-8, 2t)$, then what is dz/dt when $t = 1$?
4. Give a quadratic polynomial that approximates the function $x/(1+y)$ near the origin.

Answers

1. False 2. True 3. 11 4. $x - xy$

EXERCISES

Problems for Practice

In each of Exercises 1–10, calculate dz/ds. Do this by two methods:

a. Use the Chain Rule.

b. Substitute the formulas for x and y into the formula for z, and calculate the derivative directly.

1. $z = f(x, y) = x^2y - y^3$ $\quad x = \rho(s) = s^3 - 1$ $\quad y = \sigma(s) = s^{-1}$

2. $z = f(x, y) = x\sin(y)$ $\quad x = \rho(s) = e^s$ $\quad y = \sigma(s) = e^{-2s}$

3. $z = f(x, y) = \dfrac{x}{x^2 + y^2}$ $\quad x = \rho(s) = \cos(s)$ $\quad y = \sigma(s) = \sin(s)$

4. $z = f(x, y) = xe^y$ $\quad x = \rho(s) = s^2 - s$ $\quad y = \sigma(s) = s^3 + s$

5. $z = f(x, y) = \ln(x^3 + y)$ $\quad x = \rho(s) = s^{1/2}$ $\quad y = \sigma(s) = s^{3/2}$

6. $z = f(x, y) = \tan(x^2y)$ $\quad x = \rho(s) = s \cdot e^s$ $\quad y = \sigma(s) = s^2e^s$

7. $z = f(x, y) = e^{(2x-3y)}$ $\quad x = \rho(s) = \cos(s)$ $\quad y = \sigma(s) = \sin(s)$

8. $z = f(x, y) = x(y - 1)$ $\quad x = \rho(s) = \tan(s)$ $\quad y = \sigma(s) = \cot(s)$

9. $z = f(x, y) = \dfrac{x + y}{x - y}$ $\quad x = \rho(s) = \cos(s^2)$ $\quad y = \sigma(s) = \sin(s^2)$

10. $z = f(x, y) = xy^2$ $\quad x = \rho(s) = s \cdot \ln(s)$ $\quad y = \sigma(s) = s^3$

In each of Exercises 11–18, use the Chain Rule to calculate

$$\frac{\partial z}{\partial s} \text{ and } \frac{\partial z}{\partial t}.$$

11. $z = f(x, y) = x^2 - y^7$ $\quad x = \rho(s, t) = \sin(3st)$ $\quad y = \sigma(s, t) = \cos(3st)$

12. $z = f(x, y) = \ln(3x - y^2)$ $\quad x = \rho(s, t) = s^2 - 4t^3$ $\quad y = \sigma(s, t) = s^3t^{-3}$

13. $z = f(x, y) = e^{3xy}$ $\quad x = \rho(s, t) = s/t$ $\quad y = \sigma(s, t) = t/s$

14. $z = f(x, y) = x/y$ $\quad x = \rho(s, t) = \tan(3st)$ $\quad y = \sigma(s, t) = \cot(3st)$

15. $z = f(x, y) = \sqrt{x^2 - y^2}$ $\quad x = \rho(s, t) = (s^2 - t^2)^{-1}$ $\quad y = \sigma(s, t) = (s^2 + t^2)^{-1}$

16. $z = f(x, y) = \dfrac{1}{x + y}$ $\quad x = \rho(s, t) = e^{s-t}$ $\quad y = \sigma(s, t) = e^{8t-5s}$

17. $z = f(x, y) = \exp(2x - y)$ $\quad x = \rho(s, t) = \ln(s - t)$ $\quad y = \sigma(s, t) = \ln(2s + 3t)$

18. $z = f(x, y) = \dfrac{x + 3y}{y - 5x}$ $\quad x = \rho(s, t) = e^{5st}$ $\quad y = \sigma(s, t) = e^{-4st}$

In each of Exercises 19–26, calculate dw/ds using the Chain Rule.

19. $w = f(x, y, z) = xyz^2$ $\quad x = \rho(s) = e^s$ $\quad y = \sigma(s) = s^3$ $\quad z = \tau(s) = (s - 2)^3$

20. $w = f(x, y, z) = \dfrac{x^2 + y}{z}$ $\quad x = \rho(s) = \cos(s)$ $\quad y = \sigma(s) = \sin(s)$ $\quad z = \tau(s) = \tan(s)$

21. $w = f(x, y, z) = \sqrt{x^2 - y^3 + z}$ $\quad x = \rho(s) = 3s^3$ $\quad y = \sigma(s) = 2s^2$ $\quad z = \tau(s) = \ln(s)$

22. $w = f(x, y, z) = yz\sin(x)$ $\quad x = \rho(s) = s^5$ $\quad y = \sigma(s) = s^7$ $\quad z = \tau(s) = s^9 + 3$

23. $w = f(x, y, z) = \sin(x^3y^2z)$ $\quad x = \rho(s) = e^s$ $\quad y = \sigma(s) = s^{-2}$ $\quad z = \tau(s) = s^7$

24. $w = f(x, y, z) = \ln(x + y^2 + z^3)$ $\quad x = \rho(s) = s^2$ $\quad y = \sigma(s) = (s^2 + 3)^{1/2}$ $\quad z = \tau(s) = (s^2 + 8)^{1/3}$

25. $w = f(x, y, z) = \dfrac{x^2}{y^2 + z^2}$ $\quad x = \rho(s) = 8^s$ $\quad y = 2^s$ $\quad z = \tau(s) = 4^s$

26. $w = f(x, y, z) = x^3\ln(z^4 + y)$ $\quad x = \rho(s) = \sin^3(s)$ $\quad y = \sigma(s) = s^{-4}$ $\quad z = \tau(s) = s^{-1}$

In each of Exercises 27–32, for the given function f and using the given base point P_0, calculate the order 2 Taylor polynomial.

27. $f(x, y) = y\ln(x) - \sin(xy)$ $\quad P_0 = (1, \pi)$

28. $f(x, y) = \cos(x^2y)$ $\quad P_0 = (0, 0)$

29. $f(x, y) = 1/(1 + xy)$ $\quad P_0 = (0, 0)$

30. $f(x, y) = e^{x+2y}$ $\quad P_0 = (0, 0)$

31. $f(x, y) = \tan(2x - y)$ $\quad P_0 = (\pi/4, \pi/4)$

32. $f(x, y) = \sqrt{2 + xy}$ $\quad P_0 = (1, 2)$

Further Theory and Practice

33. Suppose that $z = h(x, y)$, $x = \rho(s, t)$, $y = \sigma(s, t)$, $s = \alpha(u, v)$, and $t = \beta(u, v)$. State a Chain Rule for $\dfrac{\partial z}{\partial u}$ and $\dfrac{\partial z}{\partial v}$.

34. A rectangular box has sides that, due to temperature fluctuations, undergo changes in dimension. At a given moment, the length increases at a rate of 0.01 cm/s, the width decreases at a rate of 0.02 cm/s, and the height increases at a rate of 0.005 cm/s. At this particular moment, the length is 15 cm, the width is 8 cm, and the height is 5 cm. Is the volume increasing or decreasing at this moment?

35. In Exercise 34, is the surface area increasing or decreasing?

36. Given that $z = f(x, y), x = \rho(s, t)$, and $y = \sigma(s, t)$, find formulas for $(\partial^2 z)/(\partial s^2)$, $(\partial^2 z)/(\partial t^2)$, and $(\partial^2 z)/(\partial s \partial t)$.

37. Let ϕ be a continuously differentiable function of one variable. Suppose that f is a differentiable function of two variables. Set $g(x, y) = \phi(f(x, y))$. Prove that

$$g_x(x,y) = \phi'(f(x,y))\, f_x(x,y)$$

and

$$g_y(x,y) = \phi'(f(x,y))\, f_y(x,y).$$

Draw a schematic diagram for this version of the Chain Rule. Use the function $g(x,y) = \sin(x^2 \ln(y))$ to illustrate this Chain Rule.

38. Let ϕ and ψ be continuously differentiable functions of one variable. Suppose that for each x and t, there is a unique value $f(x,t)$ for which $f(x,t) = t + x\phi(f(x,t))$. Let $v(x,t) = \psi(f(x,t))$. Show that

$$\frac{\partial v}{\partial x} = \phi(f(x,t))\frac{\partial v}{\partial t}.$$

39. Let a and b be constants. Suppose that ϕ is a continuously differentiable function of one variable. Show that $u(x,y) = \phi(ax+by)$ satisfies the equation

$$b\frac{\partial u}{\partial x}(x,y) = a\frac{\partial u}{\partial y}(x,y).$$

40. Let ϕ be a continuously differentiable function of one variable. Show that $u(x,y) = \phi(xy)$ satisfies the equation

$$x\frac{\partial u}{\partial x}(x,y) = y\frac{\partial u}{\partial y}(x,y).$$

41. Let ϕ be continuously differentiable functions of one variable. Show that $v(x,y) = \phi(x^2+y^2)$ satisfies

$$y\frac{\partial v}{\partial x}(x,y) = x\frac{\partial v}{\partial y}(x,y).$$

42. If $y(x,t)$ represents the displacement of a vibrating string at time t and position x, then y satisfies the wave equation

$$\frac{\partial^2 y}{\partial x^2}(x,t) = \frac{1}{c^2}\frac{\partial^2 y}{\partial t^2}(x,t)$$

where c is a positive constant that depends on the tension in the string. Show that if ϕ and ψ are any twice differentiable functions of a single variable, then

$$y(x,t) = \phi(x+ct) + \psi(x-ct)$$

is a solution of the wave equation. Notice that

$$\psi((x+c\Delta t) - c(t+\Delta t)) = \psi(x-ct).$$

What physical interpretation of the term $\psi(x-ct)$ results from this observation? What is the interpretation of $\phi(x+ct)$?

43. Let $\xi = x+ct, \eta = x-ct$. Suppose that y is a twice continuously differentiable function of two variables such that $y_{xx}(x,t) = (1/c^2)y_{tt}(x,t)$. Show that

$$v(\xi,\eta) = y\left(\frac{\xi+\eta}{2}, \frac{\xi-\eta}{2c}\right)$$

satisfies $v_{\xi\eta}(\xi,\eta) = 0$. Solve this differential equation and deduce that

$$y(x,t) = \phi(x+ct) + \psi(x-ct)$$

for some twice continuously differentiable functions ϕ and ψ.

44. Let $x = r\cos(\theta)$ and $y = r\sin(\theta)$. (Here r is the distance of $P = (x,y)$ to the origin, and θ is the angle that the position vector $\overrightarrow{OP}$ makes with the positive x-axis.) Given a twice continuously differentiable function u of two variables, let $v(r,\theta) = u(r\cos(\theta), r\sin(\theta))$. If u satisfies Laplace's equation $u_{xx}(x,y) + u_{yy}(x,y) = 0$, then what partial differential equation does v satisfy?

45. Let c be a constant, and f be a differentiable function of two variables. Suppose that the equation $f(x,y) = c$ defines y implicitly as a differentiable function of x. Show that $y'(x) = -f_x(x,y)/f_y(x,y)$ at any point at which the denominator is nonzero.

46. Let c be a constant, and F be a differentiable function of three variables. Suppose that the equation $F(x,y,z) = c$ defines z implicitly as a differentiable function of x and y. Show that

$$\frac{\partial z}{\partial x}(x,y) = -\frac{F_x(x,y,z)}{F_z(x,y,z)} \quad \text{and} \quad \frac{\partial z}{\partial y}(x,y) = -\frac{F_y(x,y,z)}{F_z(x,y,z)}$$

at any point at which the denominator is nonzero.

47. Let $f(x,y)$ be a continuously differentiable function of two variables. Calculate the derivative with respect to x of $f(f(x,y), f(x, f(x,y)))$.

48. Any point $P = (x,y)$ with $x > 0$ and $y > 0$ may be written as $(r\cos(\theta), r\sin(\theta))$ where $r = \|OP\|$ and θ is the (acute) angle between $\overrightarrow{OP}$ and $\overrightarrow{OP'}$, where $P' = (x,0)$. If $f(x,y)$ has first quadrant for its domain, let $u(r,\theta) = f(r\cos(\theta), r\sin(\theta))$. Prove the following formulas (in which each partial derivative of u is evaluated at (r,θ) and each partial derivative of f is evaluated at $(r\cos(\theta), r\sin(\theta))$):

a. $u_r = \cos(\theta)f_x + \sin(\theta)f_y$
b. $u_{rr} = \cos^2(\theta)f_{xx} + 2\cos(\theta)\sin(\theta)f_{xy} + \sin^2(\theta)f_{yy}$
c. $u_\theta = -r\sin(\theta)f_x + r\cos(\theta)f_y$
d. $u_{\theta\theta} = -ru_r + r^2(\sin^2(\theta)f_{xx} - 2\sin(\theta)\cos(\theta)f_{xy} + \cos^2(\theta)f_{yy})$
e. $u_{rr} + \frac{1}{r}u_r + \frac{1}{r^2}u_{\theta\theta} = f_{xx} + f_{yy}$

Formula e expresses the Laplacian $f_{xx} + f_{yy}$ of f in terms of the *polar coordinates* r and θ of P. Polar coordinates will be studied in greater detail in Chapter 12.

Calculator/Computer Exercises

In each of Exercises 49–52, functions $F(x,y)$, $\rho(s,t)$, and $\sigma(s,t)$ are given. Let $f(s,t) = F(\rho(s,t), \sigma(s,t))$. For the given point (s_0, t_0), set $x_0 = \rho(s_0,t_0)$, and $y_0 = \sigma(s_0,t_0)$. Use central difference quotients to evaluate $f_s(s_0,t_0)$ and $f_t(s_0,t_0)$. Also, use central difference quotients to evaluate $\rho_s(s_0,t_0)$, $\rho_t(s_0,t_0)$, $\sigma_s(s_0,t_0)$, $\sigma_t(s_0,t_0)$, $F_x(x_0,y_0)$, and $F_y(x_0,y_0)$. Then, use these values together with the Chain Rule to obtain alternative evaluations of $f_s(s_0,t_0)$ and $f_t(s_0,t_0)$.

49. $F(x,y) = (x+y)/(x+1)$, $\rho(s,t) = \sqrt{st+t}$, $\sigma(s,t) = 2^{s-t}$, $(s_0, t_0) = (3,1)$

50. $F(x,y) = xe^{2y-x}$, $\rho(s,t) = s/(1+t)$, $\sigma(s,t) = s^t$, $(s_0, t_0) = (2,0)$

51. $F(x,y) = \ln(1+x^4+y^2)$, $\rho(s,t) = (2s+t)^{1/3}$, $\sigma(s,t) = \tan(s+t)$, $(s_0, t_0) = (1,-1)$

52. $F(x,y) = \sin(2x+6y)$, $\rho(s,t) = \arcsin(s/t)$, $\sigma(s,t) = \arcsin\left((2s-1)/t^2\right)$, $(s_0, t_0) = (1, -\sqrt{2})$

11.6 Gradients and Directional Derivatives

Recall that a *direction* is another word for a *unit vector*. In this section, we want to discuss the notion of differentiating a function of several variables in a given direction. For motivation, look at the topographic map in Figure 1. The level sets are for the height function f (measured in feet above sea level). They are plotted at 20 foot increments. Imagine that you are standing at the point P that is the origin of the superimposed coordinate axes. The level curve that passes through P tells us that P is situated 1040 feet above sea level. If you look east, the height is increasing; So the derivative of f in that direction should be positive. If you look south or north (along the level curve), the height is not changing, so the derivatives of f in those directions should be 0. If you look west, the height is decreasing, so the derivative in that direction should be negative.

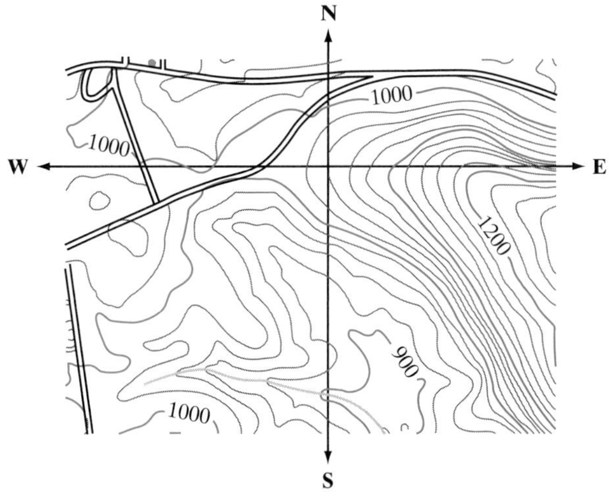

Figure 1 Stone Mountain, Georgia

At what rate does your elevation change if you set out from P in a given direction? In which direction does elevation increase most rapidly? Decrease most rapidly? This section introduced methods for answering these questions.

The Directional Derivative

Let $\mathbf{u} = u_1\mathbf{i} + u_2\mathbf{j}$ be a direction vector and $P_0 = (x_0, y_0)$ a point in the xy-plane. Let $L_\mathbf{u}$ denote the line in the xy-plane that passes through P_0 and has direction $\mathbf{u}$. We parameterize $L_\mathbf{u}$ by the vector-valued function $\mathbf{r}(t) = \overrightarrow{OP_0} + t\mathbf{u}$. Because $\|\mathbf{r}'(t)\| = \|\mathbf{u}\| = 1$ for every t, we note that $t \mapsto \mathbf{r}(t)$ is the arc length parameterization of $L_\mathbf{u}$ that passes through P_0 at $t = 0$ (see Figure 2).

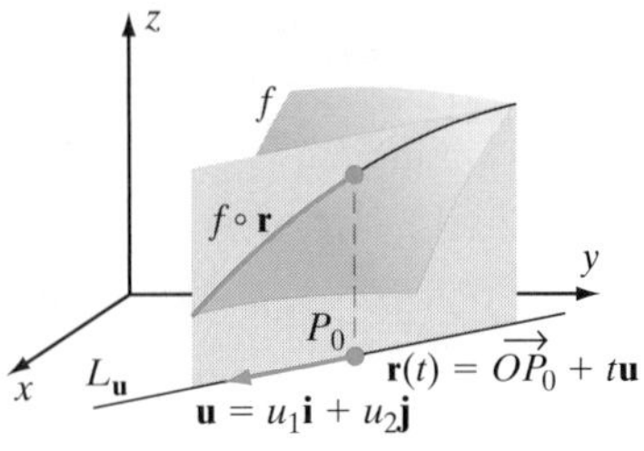

▲ Figure 2

Let $(x, y) \mapsto f(x, y)$ be a differentiable function defined on a disk centered at P_0. The first step toward calculating the rate of change of f at the point P_0 in the direction $\mathbf{u} = u_1\mathbf{i} + u_2\mathbf{j}$ is to restrict f to $L_\mathbf{u}$ by forming the composition $f \circ \mathbf{r}$:

$$(f \circ \mathbf{r})(t) = f(\overrightarrow{OP_0} + t\mathbf{u}) = f(x_0 + tu_1, y_0 + tu_2).$$

This is a scalar-valued function of one variable. Its plot can be obtained by intersecting the surface $z = f(x, y)$ with the plane that contains the line $L_\mathbf{u}$ and that is perpendicular to the xy-plane (refer to Figure 2). The rate of change that we seek is obtained from the derivative of $f \circ \mathbf{r}$.

DEFINITION The *directional derivative* of the function f in the direction $\mathbf{u} = u_1\mathbf{i} + u_2\mathbf{j}$ at the point P_0 is defined to be

$$D_\mathbf{u} f(P_0) = \frac{d}{dt} f(\overrightarrow{OP_0} + t\mathbf{u})\Big|_{t=0} = \frac{d}{dt} f(x_0 + tu_1, y_0 + tu_2)\Big|_{t=0}.$$

Now that we have defined the directional derivative, how can we calculate it? We use the Chain Rule to differentiate the function $z = f \circ \mathbf{r}$ with respect to t:

$$\frac{dz}{dt}\Big|_{t=0} = \frac{\partial f}{\partial x}(P_0)\left(\frac{dx}{dt}\Big|_{t=0}\right) + \frac{\partial f}{\partial y}(P_0)\left(\frac{dy}{dt}\Big|_{t=0}\right) = \frac{\partial f}{\partial x}(P_0)u_1 + \frac{\partial f}{\partial y}(P_0)u_2.$$

Now we can summarize what we have found.

THEOREM 1 Let P_0 be a point in the plane, $\mathbf{u} = u_1\mathbf{i} + u_2\mathbf{j}$ a unit vector, and f a differentiable function on a disk centered at P_0. Then, the directional derivative of f at P_0 in the direction $\mathbf{u}$ is given by the formula

$$D_\mathbf{u} f(P_0) = \frac{\partial f}{\partial x}(P_0)u_1 + \frac{\partial f}{\partial y}(P_0)u_2.$$

INSIGHT Notice that, if $\mathbf{u} = \mathbf{i}$, then $u_1 = 1$ and $u_2 = 0$. Therefore $D_\mathbf{i} f(P_0) = f_x(P_0)$. Similarly, $D_\mathbf{j} f(P_0) = f_y(P_0)$. This tells us that the partial derivatives $f_x(P_0)$ and $f_y(P_0)$ are themselves directional derivatives: They correspond to the directions of the positive x-axis and positive y-axis, respectively. From this point of view, the directional derivative may be regarded as a generalization of the concept of the partial derivative.

▶ **EXAMPLE 1** Let $f(x, y) = 1 - x^2 y$. Let $\mathbf{v} = (\sqrt{2}/2)\mathbf{i} + (\sqrt{2}/2)\mathbf{j}$ and $\mathbf{u} = (1/2)\mathbf{i} - (\sqrt{3}/2)\mathbf{j}$. Calculate $D_\mathbf{v} f(1, 3)$ and $D_\mathbf{u} f(1, 3)$. What is $D_{-\mathbf{u}} f(1, 3)$?

Solution To begin, notice that $\mathbf{u}$ and $\mathbf{v}$ are unit vectors, as required by the definition of the directional derivative. We calculate that

$$\frac{\partial f}{\partial x}(1, 3) = -2xy\Big|_{(1,3)} = -6 \quad \text{and} \quad \frac{\partial f}{\partial y}(1, 3) = -x^2\Big|_{(1,3)} = -1.$$

Therefore

$$D_{\mathbf{v}}f(1,3) = \frac{\partial f}{\partial x}(1,3)v_1 + \frac{\partial f}{\partial y}(1,3)v_2 = (-6)\frac{\sqrt{2}}{2} + (-1)\frac{\sqrt{2}}{2} = -7\frac{\sqrt{2}}{2}.$$

Likewise,

$$D_{\mathbf{u}}f(1,3) = \frac{\partial f}{\partial x}(1,3)u_1 + \frac{\partial f}{\partial y}(1,3)u_2 = (-6)\frac{1}{2} + (-1)\left(-\frac{\sqrt{3}}{2}\right) = -\left(3 - \frac{\sqrt{3}}{2}\right).$$

Similarly,

$$D_{-\mathbf{u}}f(1,3) = \frac{\partial f}{\partial x}(1,3)(-u_1) + \frac{\partial f}{\partial y}(1,3)(-u_2) = -\left(\frac{\partial f}{\partial x}(1,3)u_1 + \frac{\partial f}{\partial y}(1,3)u_2\right) = -D_{\mathbf{u}}f(1,3) = 3 - \frac{\sqrt{3}}{2}.$$

INSIGHT The last computation in Example 1 shows that $D_{-\mathbf{u}}f(P_0) = -D_{\mathbf{u}}f(P_0)$. There is a simple geometric explanation for this fact. Remember that the sign of a derivative indicates whether the function increases (positive derivative) or decreases (negative derivative). Now think of a mountain path while looking at Figure 2. Follow the path downward. Relative to the ground, your path has a heading **u**. Because you are descending, your height function has a negative rate of change. Now follow the same path upward. You do this by reversing your ground direction from **u** to $-\mathbf{u}$. The slope of the path is the same but, because you are now ascending, the rate of change of height reverses in sign.

▶ **EXAMPLE 2** Let $f(x,y) = 1 + 2x + y^3$. What is the directional derivative of f at $P = (2,1)$ in the direction from P to $Q = (14,6)$?

Solution We calculate $f_x(x,y) = 2$ and $f_y(x,y) = 3y^2$. Therefore $f_x(P) = 2$ and $f_y(P) = 3$. The next step is to compute the required direction. The initial candidate, $\overrightarrow{PQ} = 12\mathbf{i} + 5\mathbf{j}$, has length $\sqrt{12^2 + 5^2}$, which simplifies to 13. Because $\overrightarrow{PQ}$ is not a unit vector, we do not use it in the formula for the directional derivative. Instead, we use the *unit* vector in the direction of P to Q, namely $\mathbf{u} = (1/13)\overrightarrow{PQ} = (12/13)\mathbf{i} + (5/13)\mathbf{j}$. The required directional derivative is

$$D_{\mathbf{u}}f(P) = (2)\left(\frac{12}{13}\right) + (3)\left(\frac{5}{13}\right) = \frac{24+15}{13} = 3. \quad ◀$$

The Gradient

Before proceeding, we introduce a notation that we will use to organize our calculations.

DEFINITION Let f be a differentiable function of two variables. The *gradient function* of f is the vector-valued function ∇f defined by

$$\nabla f(P) = \left(\frac{\partial f}{\partial x}(P)\right)\mathbf{i} + \left(\frac{\partial f}{\partial y}(P)\right)\mathbf{j}.$$

In this notation, ∇f is a vector-valued function created from the scalar-valued function f and $\nabla f(P)$ is the function ∇f evaluated at a point P. We say that $\nabla f(P)$ is the *gradient* of f at the point P. The notation $\text{grad}(f)$ is an alternative to ∇f.

We can use the gradient to rewrite the directional derivative as the dot product

$$D_{\mathbf{u}}f(P) = \nabla f(P) \cdot \mathbf{u}. \tag{11.6.1}$$

Notice that if $\nabla f(P) = \mathbf{0}$, that is, if $f_x(P) = 0$ and $f_y(P) = 0$, then $D_u f(P) = 0$ for all unit vectors $\mathbf{u}$. In other words, if the gradient at P is $\mathbf{0}$, then the rate of change of f at P is 0 in all directions.

▶ **EXAMPLE 3** Let $f(x, y) = x\sin(y)$. Calculate $\nabla f(x, y)$. If $\mathbf{u} = (-3/5)\mathbf{i} + (4/5)\mathbf{j}$, then what is $D_{\mathbf{u}}f(2, \pi/6)$?

Solution We have

$$\nabla f(x, y) = \left(\frac{\partial f}{\partial x}(x, y)\right)\mathbf{i} + \left(\frac{\partial f}{\partial y}(x, y)\right)\mathbf{j} = \sin(y)\mathbf{i} + x\cos(y)\mathbf{j}.$$

Therefore $\nabla f(2, \pi/6) = \sin(\pi/6)\mathbf{i} + 2\cos(\pi/6)\mathbf{j} = (1/2)\mathbf{i} + \sqrt{3}\mathbf{j}$ and

$$D_{\mathbf{u}}f\left(2, \frac{\pi}{6}\right) = \nabla f\left(2, \frac{\pi}{6}\right) \cdot \mathbf{u} = \left(\frac{1}{2}\mathbf{i} + \sqrt{3}\mathbf{j}\right) \cdot \left(-\frac{3}{5}\mathbf{i} + \frac{4}{5}\mathbf{j}\right) = -\frac{3}{10} + \frac{4\sqrt{3}}{5}.$$ ◀

The Directions of Greatest Increase and Decrease

Suppose now that the point P_0 and the function f are fixed. Which direction $\mathbf{u}$ will result in the greatest possible value for $D_{\mathbf{u}}f(P_0)$? Which direction $\mathbf{u}$ will result in the least possible value for $D_{\mathbf{u}}f(P_0)$? Because $D_{\mathbf{u}}f(P_0)$ measures the rate of change of f at P_0 in the direction $\mathbf{u}$, we can pose our questions in this way: What is the direction of greatest increase for the function f at P_0? What is the direction of greatest decrease for the function f at P_0? Theorem 2 answers these questions.

THEOREM 2 Suppose that f is a differentiable function for which $\nabla f(P_0) \neq \mathbf{0}$. Then, $D_{\mathbf{u}}f(P_0)$ is maximal when the unit vector $\mathbf{u}$ is the direction of the gradient $\nabla f(P_0)$. For this choice of $\mathbf{u}$, the directional derivative is $D_{\mathbf{u}}f(P_0) = \|\nabla f(P_0)\|$. Also, $D_{\mathbf{u}}f(P_0)$ is minimal when $\mathbf{u}$ is opposite in direction to $\nabla f(P_0)$. For this choice of $\mathbf{u}$, the directional derivative is $D_{\mathbf{u}}f(P_0) = -\|\nabla f(P_0)\|$.

Proof. To begin, fix an *arbitrary* direction $\mathbf{u}$. Because $\nabla f(P_0) \neq \mathbf{0}$, there is a well-defined angle θ between the nonzero vectors $\nabla f(P_0)$ and $\mathbf{u}$. Expressing the dot product of these vectors in terms of θ, we have

$$D_{\mathbf{u}}f(P_0) = \nabla f(P_0) \cdot \mathbf{u} = \|\nabla f(P_0)\| \, \|\mathbf{u}\|\cos(\theta) = \|\nabla f(P_0)\|\cos(\theta).$$

Remember that f and P_0 are fixed. We are trying to select a direction $\mathbf{u}$ to make $D_{\mathbf{u}}f(P_0)$ as big or as small as possible. In other words, we want to determine θ so that $\cos(\theta)$ is as big or as small as possible. Clearly, $\theta = 0$ gives the greatest value for $\cos(\theta)$, namely 1. This value of θ occurs when $\mathbf{u}$ is the direction of the gradient $\nabla f(P_0)$. For this choice of $\mathbf{u}$, we have $D_{\mathbf{u}}f(P_0) = \|\nabla f(P_0)\|(1) = \|\nabla f(P_0)\|$.

Also observe that $\theta = \pi$ results in the least value of $\cos(\theta)$, namely -1. This value of θ occurs when $\mathbf{u}$ is the opposite direction of the gradient $\nabla f(P_0)$. For this choice of $\mathbf{u}$, we have $D_{\mathbf{u}}f(P_0) = \|\nabla f(P_0)\|(-1) = -\|\nabla f(P_0)\|$. ■

▶ **EXAMPLE 4** At the point $P_0 = (-2, 1)$, what is the direction that results in the greatest increase for $f(x, y) = x^2 + y^2$, and what is the direction of greatest decrease? What are the greatest and least values of the directional derivative at P_0?

Solution The gradient of f at (x, y) is $\nabla f(x, y) = 2x\mathbf{i} + 2y\mathbf{j}$. At P_0, this gradient is equal to $-4\mathbf{i} + 2\mathbf{j}$. The direction of greatest increase is therefore

$$\mathbf{u} = \frac{1}{\|\nabla f(P_0)\|}\nabla f(P_0) = \frac{1}{\sqrt{20}}(-4\mathbf{i} + 2\mathbf{j}) = -\frac{2}{\sqrt{5}}\mathbf{i} + \frac{1}{\sqrt{5}}\mathbf{j}.$$

The corresponding greatest value for the directional derivative at P_0 is

$$D_{\mathbf{u}}f(P_0) = \|\nabla f(P_0)\| = \sqrt{20} = 2\sqrt{5}.$$

The direction of greatest decrease is just the negative $(2/\sqrt{5})\mathbf{i} + (-1/\sqrt{5})\mathbf{j}$ of the vector $\mathbf{u}$. The corresponding least value for the directional derivative at P_0 is $-\|\nabla f(P_0)\| = -2\sqrt{5}$. ◄

The Gradient and Level Curves

We turn to another important application of the gradient. Suppose that f is a function that is differentiable at a point P_0 in the plane. Let c denote the value $f(P_0)$, and let $\mathbf{r} = \langle r_1, r_2\rangle$ be an arc length parameterization of the level curve $\mathcal{C} = \{(x, y) : f(x, y) = c\}$ with $\mathbf{r}(s_0) = P_0$. Then, $\mathbf{T} = \mathbf{r}'(s_0)$ is the unit tangent vector to $\mathcal{C}$ at P_0. Because $\mathbf{r}(s)$ is a point on $\mathcal{C}$ for all s, we see that $(f \circ \mathbf{r})(s) = f(\mathbf{r}(s)) = c$ is constant. It follows that $(f \circ \mathbf{r})'(s) = 0$ for all s. In particular, for $s = s_0$, we have

$$D_{\mathbf{T}}f(P_0) = \nabla f(P_0) \cdot \mathbf{T} = \nabla f(P_0) \cdot \mathbf{r}'(s_0) = \frac{\partial f}{\partial x}(P_0)r_1'(s_0) + \frac{\partial f}{\partial y}(P_0)r_2'(s_0) \overset{\text{Chain Rule}}{=} (f \circ \mathbf{r})'(s_0) = 0.$$

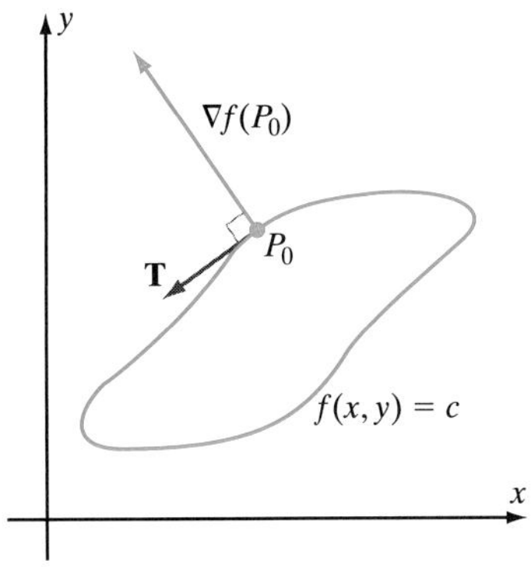

▲ Figure 3

We gather together some conclusions that can be drawn from this equation:

THEOREM 3 Suppose that f is differentiable at P_0. Let $\mathbf{T}$ be a unit tangent vector to the level curve of f at P_0. Then:

a. $D_{\mathbf{T}}f(P_0) = 0$.
b. $\nabla f(P_0)$ is perpendicular to $\mathbf{T}$.

In words, part a tells us that the instantaneous rate of change of f at P_0 is 0 in the direction of the tangent to the level curve. Part b tells us that the gradient $\nabla f(P_0)$ is normal to the level curve of f at P_0 (see Figure 3).

INSIGHT In combination, Theorems 2 and 3 tell us that a direction perpendicular to a level curve of f at P is either the direction in which f has the greatest increase at P or the direction in which f has the greatest decrease at P. This principle is actually quite familiar to us. Imagine ascending the ramp shown in Figure 4. If we begin at point P, then our instinct is to walk toward Q, in the direction that is perpendicular to the level curves, rather than toward another point such as R.

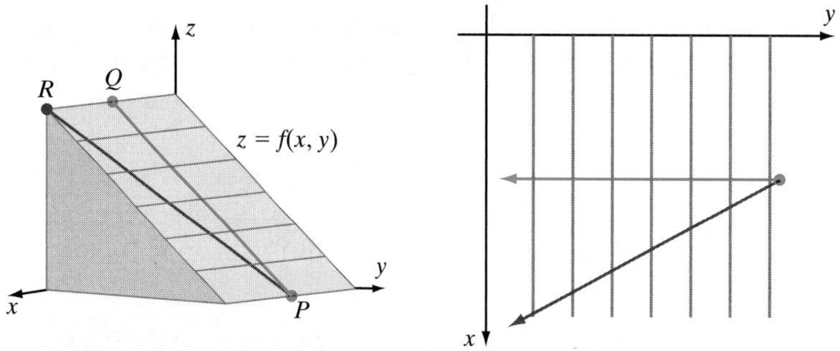

▲ Figure 4 The direction perpendicular to the level curves of f is the direction of greatest increase or the direction of greatest decrease.

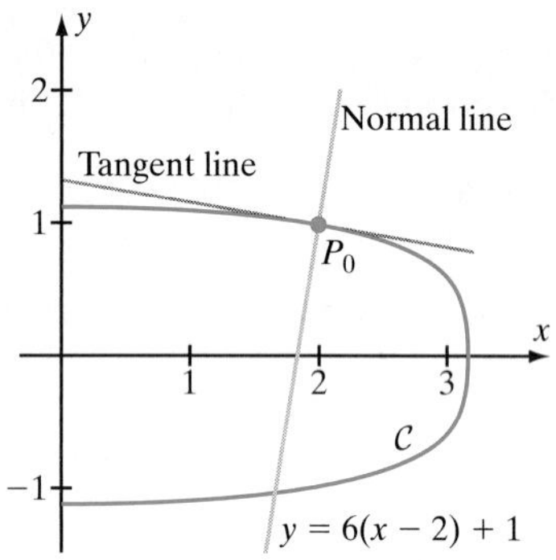

▲ Figure 5 A normal line to the graph of $x^2 + 6y^4 = 10$

▶ **EXAMPLE 5** Consider the curve $\mathcal{C}$ in the xy-plane that is the graph of the equation $x^2 + 6y^4 = 10$. Find the line that is normal to the curve at the point $(2, 1)$.

Solution Let $P_0 = (2, 1)$. We may regard $\mathcal{C}$ as the level curve $f(x, y) = 10$ where $f(x, y) = x^2 + 6y^4$. We calculate

$$\nabla f(P_0) = \left(2x\mathbf{i} + 24y^3\mathbf{j}\right)_{(x,y)=(2,1)} = 4\mathbf{i} + 24\mathbf{j} = 4(\mathbf{i} + 6\mathbf{j}).$$

According to Theorem 3, this gradient vector is normal to $\mathcal{C}$ at P_0. We may therefore write the normal line as $(x-2)/1 = (y-1)/6$ or, equivalently, $y = 6(x-2) + 1$ (see Figure 5). ◀

Functions of Three or More Variables

All of the ideas introduced in this section are also valid for functions $(x, y, z) \mapsto F(x, y, z)$ of three variables. The gradient of such a function is

$$\nabla F = \frac{\partial F}{\partial x}\mathbf{i} + \frac{\partial F}{\partial y}\mathbf{j} + \frac{\partial F}{\partial z}\mathbf{k}.$$

The directional derivative of F at the point $P_0 = (x_0, y_0, z_0)$ in the direction $\mathbf{u} = \langle u_1, u_2, u_3 \rangle$ is

$$D_{\mathbf{u}}F(P_0) = \nabla F(P_0) \cdot \mathbf{u}.$$

The direction that results in the greatest rate of increase of the function f at the point P_0 is

$$\mathbf{u} = \frac{1}{\|\nabla F(P_0)\|}\nabla F(P_0).$$

With this choice of direction $\mathbf{u}$, we have $D_{\mathbf{u}}F(P_0) = \|\nabla F(P_0)\|$. The direction that results in the greatest rate of decrease of the function F at the point P_0 is

$$-\mathbf{u} = -\frac{1}{\|\nabla F(P_0)\|}\nabla F(P_0).$$

With this choice of direction, we have $D_{-\mathbf{u}}F(P_0) = -\|\nabla F(P_0)\|$.

The level set $\{(x, y, z) : F(x, y, z) = c\}$ is a surface. At the point P_0, the gradient $\nabla F(P_0)$ is normal to the level surface of F that passes through P_0. We will discuss this point further in the next section.

▶ **EXAMPLE 6** Find the directions of greatest rate of increase and greatest rate of decrease for the function $F(x,y,z) = xyz$ at the point $(-1,2,1)$.

Solution To do this, we calculate $\nabla F(x,y,z) = yz\mathbf{i} + xz\mathbf{j} + xy\mathbf{k}$ and $\nabla F(-1,2,1) = 2\mathbf{i} - \mathbf{j} - 2\mathbf{k}$. Then, the direction of greatest rate of increase is

$$\frac{1}{\|\nabla F(-1,2,1)\|}\nabla F(-1,2,1) = \frac{1}{\sqrt{2^2 + (-1)^2 + (-2)^2}}(2\mathbf{i} - \mathbf{j} - 2\mathbf{k}) = \frac{2}{3}\mathbf{i} - \frac{1}{3}\mathbf{j} - \frac{2}{3}\mathbf{k}.$$

The direction of greatest rate of decrease at $(-1,2,1)$ is then

$$-\frac{1}{\|\nabla F(-1,2,1)\|}\nabla F(-1,2,1) = -\frac{2}{3}\mathbf{i} + \frac{1}{3}\mathbf{j} + \frac{2}{3}\mathbf{k}. \quad ◀$$

The Gradient and the Chain Rule

We conclude this section by observing that the gradient gives us another way to think about the Chain Rule. From this perspective, the multivariable Chain Rule will look a little bit more like the one-variable Chain Rule. Let us illustrate this point of view for functions of two variables. Namely, if $z = f(x,y)$ where $x = \rho(t)$ and $y = \sigma(t)$, then we can think of $\mathbf{r}(t) = \rho(t)\mathbf{i} + \sigma(t)\mathbf{j}$ as the parameterization of a curve. The Chain Rule tells us that

$$\frac{dz}{dt} = \frac{\partial f}{\partial x}\frac{d\rho}{dt} + \frac{\partial f}{\partial y}\frac{d\sigma}{dt},$$

which we can write as

$$\frac{dz}{dt} = (\nabla f) \cdot \left(\frac{d\mathbf{r}}{dt}\right). \qquad (11.6.2)$$

We see that the gradient allows us to write the Chain Rule in the more familiar form of a product, provided we interpret the "·" as the dot product. Notice that this last notation for the Chain Rule does not look any different when more independent variables are present. For instance, if $w = F(x,y,z)$ where $x = \sigma_1(t)$, $y = \sigma_2(t), z = \sigma_3(t)$, then we can think of $\mathbf{r}(t) = \sigma_1(t)\mathbf{i} + \sigma_2(t)\mathbf{j} + \sigma_3(t)\mathbf{k}$ as the parameterization of a curve in xyz-space. In this case, the Chain Rule is

$$\frac{dw}{dt} = \frac{\partial F}{\partial x}\frac{d\sigma_1}{dt} + \frac{\partial F}{\partial y}\frac{d\sigma_2}{dt} + \frac{\partial F}{\partial z}\frac{d\sigma_3}{dt},$$

which can be written as

$$\frac{dw}{dt} = (\nabla F) \cdot \left(\frac{d\mathbf{r}}{dt}\right). \qquad (11.6.3)$$

QUICK QUIZ

1. Suppose that $\mathbf{u} = (3/5)\mathbf{i} + (4/5)\mathbf{j}$ and $f(x,y) = 10x + y^5$. What is $D_{\mathbf{u}}f(2,-1)$?
2. For what vector $\mathbf{u}$ is $D_{\mathbf{u}}f(P_0) = f_y(P_0)$?
3. Suppose that $f(x,y) = 1 + 2x + y^3$. What is $\nabla f(5,4)$?
4. Suppose that $f_x(P_0) = 2\sqrt{5}$ and $f_y(P_0) = 4$. What is the greatest directional derivative of f at P_0?

Answers

1. 10 2. $\mathbf{j}$ 3. $\langle 2, 48\rangle$ 4. 6

EXERCISES

Problems for Practice

Calculate the gradient of each of the functions in Exercises 1–10.

1. $f(x,y) = x\sin(y)$
2. $f(x,y) = x\ln(x+3y)$
3. $f(x,y) = \tan(xy^2)$
4. $f(x,y,z) = xy^2/z^3$
5. $f(x,y,z) = \cos(xy)\sin(yz)$
6. $f(x,y) = x^{3/2}y^{-5/2} - xy$
7. $f(x,y) = \dfrac{xy^2}{\cos(y)}$
8. $f(x,y) = \sqrt{y}\cot(x+y)$
9. $f(x,y,z) = z\sin(xy)$
10. $f(x,y,z) = xy^2\sin(yz^2)$

In each of Exercises 11–22, calculate the directional derivative of f at P_0 in the given direction $\mathbf{u}$.

11. $f(x,y) = x^2y - y^3x$, $P_0 = (2,-3)$, $\mathbf{u} = \langle -6/10, 8/10\rangle$
12. $f(x,y) = x\ln(x^2+y^2)$, $P_0 = (2,4)$, $\mathbf{u} = \langle 4/5, 3/5\rangle$
13. $f(x,y) = \cos(x^2-y^2)$, $P_0 = (\pi^{1/2}, \pi^{1/2}/2)$, $\mathbf{u} = \langle 1/\sqrt{2}, -1/\sqrt{2}\rangle$
14. $f(x,y) = \dfrac{x+y}{x-y}$, $P_0 = (4,7)$, $\mathbf{u} = \langle \sqrt{3}/2, -1/2\rangle$
15. $f(x,y) = e^{4x-7y}$, $P_0 = (0, \ln(2))$, $\mathbf{u} = \langle 1, 0\rangle$
16. $f(x,y) = \tan(x/y)$, $P_0 = (\pi/4, 1)$, $\mathbf{u} = \langle -12/13, 5/13\rangle$
17. $f(x,y) = \dfrac{1}{x^2+y^2}$, $P = (2,2)$, $\mathbf{u} = \langle 0, 1\rangle$
18. $f(x,y) = (x-y)\ln(x+y)$, $P_0 = (1,1)$, $\mathbf{u} = \left\langle \sqrt{3/2}, -1/2\right\rangle$
19. $f(x,y) = \dfrac{\sin(xy)}{x+y}$, $P_0 = (0,1)$, $\mathbf{u} = \langle 1/\sqrt{2}, -1/\sqrt{2}\rangle$
20. $f(x,y) = e^{xy}e^{-y}$, $P_0 = (1, \ln 2)$, $\mathbf{u} = \langle 0, -1\rangle$
21. $f(x,y) = \sin^3(x+y)$, $P_0 = (\pi/3, -\pi/6)$, $\mathbf{u} = \langle -1, 0\rangle$
22. $f(x,y) = (x^2+y^3)^4$, $P_0 = (1,1)$, $\mathbf{u} = \langle -1/2, \sqrt{3}/2\rangle$

In each of Exercises 23–32, find the following:

a. the direction $\mathbf{u}$ of greatest rate of increase for the function f at the point P_0
b. the directional derivative of f at P_0 in the direction of $\mathbf{u}$
c. the direction $\mathbf{v}$ of greatest rate of decrease for f at the point P_0
d. the directional derivative of f at P_0 in the direction of $\mathbf{v}$

23. $f(x,y) = x^3y^2 - yx$ $\quad P_0 = (2, 1/2)$
24. $f(x,y) = \cos(x-y^2)$ $\quad P_0 = (\pi, \pi^{1/2}/2)$
25. $f(x,y) = \dfrac{\ln(x+y)}{x+y}$ $\quad P_0 = (1,1)$
26. $f(x,y) = x^2\tan(y)$ $\quad P = (1, \pi/4)$
27. $f(x,y) = \dfrac{\sin(x+y)}{\cos(x-y)}$ $\quad P_0 = (\pi/2, \pi/3)$
28. $f(x,y) = \tan(x^2y)$ $\quad P_0 = (1, \pi/3)$
29. $f(x,y) = e^{-x}\cos(y)$ $\quad P_0 = (1, \pi)$
30. $f(x,y) = \dfrac{x}{x-y}$ $\quad P_0 = (4,1)$
31. $f(x,y) = (x^2+y)^5$ $\quad P_0 = (1,1)$
32. $f(x,y) = \sec(x-y)$ $\quad P_0 = (\pi/2, \pi/3)$

In each of Exercises 33–38, find the directional derivative of the function F at the point P_0 in the direction $\mathbf{u}$.

33. $F(x,y,z) = xyz^3$, $P_0 = (2,1,-3)$, $\mathbf{u} = \langle 2/3, -2/3, 1/3\rangle$
34. $F(x,y,z) = \dfrac{x}{y+z}$, $P_0 = (4,2,2)$, $\mathbf{u} = \langle 3/5, \sqrt{7}/5, -3/5\rangle$
35. $F(x,y,z) = y\sin(xz)$, $P_0 = (1/2, 3, \pi)$, $\mathbf{u} = \langle \sqrt{3}/8, -6/8, 5/8\rangle$
36. $F(x,y,z) = \dfrac{x}{yz}$, $P_0 = (6,3,2)$, $\mathbf{u} = \langle 5/7, -2\sqrt{2}/7, -4/7\rangle$
37. $F(x,y,z) = \ln(xy+z^3)$, $P_0 = (1,1,1)$, $\mathbf{u} = \langle -1/2, 3/4, \sqrt{3}/4\rangle$
38. $F(x,y,z) = \dfrac{\sin(x-z)}{\cos(y+z)}$, $P_0 = (\pi/6, \pi/3, \pi/2)$, $\mathbf{u} = \langle 1/2, -1/2, 1/\sqrt{2}\rangle$

In each of Exercises 39–42, find the following:

a. the direction $\mathbf{u}$ of greatest rate of increase for the function F at the point P_0
b. the directional derivative of F at P_0 in the direction of $\mathbf{u}$
c. the direction $\mathbf{v}$ of greatest rate of decrease for F at the point P_0
d. the directional derivative of F at P_0 in the direction of $\mathbf{v}$

39. $F(x,y,z) = x^2(y^3 - zy)$ $\quad P_0 = (2,-1,4)$
40. $F(x,y,z) = \dfrac{\ln(x-y)}{z}$ $\quad P_0 = (2,1,3)$

41. $F(x,y,z) = \sin(x^2 - y + z)$ $\quad P_0 = (0, \pi, 2\pi)$

42. $F(x,y,z) = \sin(2x) + 12\cos(yz)$ $\quad P_0 = (\pi/6, 1, \pi/6)$

Further Theory and Practice

43. Let $f(x,y)$ be a continuously differentiable function, and suppose that $\nabla f(P) = 0$. What does this say about the direction of greatest increase or greatest decrease of f at P? Give examples.

In each of Exercises 44–47, find the directional derivative of the given function f at P in the direction from P to Q.

44. $f(x,y) = x^2 + xy$ $\quad P = (1,2)$ $\quad Q = (3,4)$

45. $f(x,y) = x\sqrt{1+xy}$ $\quad P = (2,4)$ $\quad Q = (5,8)$

45. $f(x,y) = \exp(8x/y^2 + y/x)$ $\quad P = (1, -2)$ $\quad Q = (-5,6)$

47. $f(x,y) = \tan(xy^2)$ $\quad P = (\pi, 1/2)$ $\quad Q = (-3\pi, 1/2 + 3\pi)$

In each of Exercises 48–51, find $\dfrac{\partial f}{\partial x}(P)$ and $\dfrac{\partial f}{\partial y}(P)$ using the given information.

48. $\mathbf{u} = \langle 0,1 \rangle$, $\mathbf{v} = \langle 1/\sqrt{2}, 1/\sqrt{2} \rangle$, $D_{\mathbf{u}}f(P) = 2\sqrt{2}$, $D_{\mathbf{v}}f(P) = 3\sqrt{2}$

49. $\mathbf{u} = \langle 3/5, 4/5 \rangle$, $\mathbf{v} = \langle 1/\sqrt{2}, 1/\sqrt{2} \rangle$, $D_{\mathbf{u}}f(P) = 1$, $D_{\mathbf{v}}f(P) = \sqrt{2}$

50. $\mathbf{u} = \langle 5/13, 12/13 \rangle$, $\mathbf{v} = \langle -12/13, 5/13 \rangle$, $D_{\mathbf{u}}f(P) = 19$, $\mathbf{D}_{\mathbf{v}}f(P) = 22$

51. $\mathbf{u} = \langle 1/2, -\sqrt{3}/2 \rangle$, $\mathbf{v} = \langle 1/\sqrt{2}, 1/\sqrt{2} \rangle$, $D_{\mathbf{u}}f(P) = 3$, $D_{\mathbf{v}}f(P) = 0$

52. Let $f(x,y) = x^2 - y^3$. At the point $P = (3,7)$, which direction(s) $\mathbf{u}$ will have the property that $D_{\mathbf{u}}f(P) = 8$?

53. Let $f(x,y) = xy - y^4$ and $P = (8, -2)$. Explain why there is no direction $\mathbf{u}$ such that $D_{\mathbf{u}}f(P) = 50$.

54. You are standing at the bottom, $(0,0,0)$, of a hole that is shaped like the graph of $f(x,y) = x^4 + y^2 + x^2y^6$. You need to climb out of the hole. What is the direction of steepest ascent? What is the direction of slowest ascent?

55. A heat-seeking missile can only sense heat in its immediate vicinity. In an effort to find the source of the heat, the missile always moves in the direction of greatest increase in temperature. Suppose that the temperature at any point in space is given by $T(x,y,z) = 1000 - 10(x-90)^2 - 2(y-12)^4 - z^2$, and the missile is currently at $(100, 15, 8)$. In what direction will it move next?

56. Laplace's equation for a function of two variables is given by

$$\Delta f(x,y) = \left(\frac{\partial^2}{\partial x^2} + \frac{\partial^2}{\partial y^2}\right) f(x,y) = 0.$$

Let $\mathbf{u}$ and $\mathbf{v}$ be any pair of orthogonal unit vectors in the plane. Prove that Laplace's equation is also given by

$$D_{\mathbf{u}}(D_{\mathbf{u}}f) + D_{\mathbf{v}}(D_{\mathbf{v}}f)(x,y) = 0.$$

57. For a vector-valued function $\mathbf{f}(x,y) = f_1(x,y)\mathbf{i} + f_2(x,y)\mathbf{j}$, we define the divergence of $\mathbf{f}$ to be

$$\operatorname{div}\mathbf{f}(x,y) = \frac{\partial f_1}{\partial x}(x,y) + \frac{\partial f_2}{\partial y}(x,y).$$

Show that if φ is a twice continuously differentiable function, then

$$\operatorname{div}(\operatorname{grad}(\varphi)) = \Delta\varphi.$$

58. Let f be a continuously differentiable function defined on a planar region. Suppose that P is a point in the domain of f. Show that the set of all possible values for the directional derivative of f at P forms an interval. What are the endpoints of this interval?

59. Let f be a twice continuously differentiable function with P_0 in its domain. Let $\mathbf{u} = u_1\mathbf{i} + u_2\mathbf{j}$ and $\mathbf{v} = v_1\mathbf{i} + v_2\mathbf{j}$ be directions. Give formulas for $D_{\mathbf{u}}(D_{\mathbf{u}}f)(P_0)$, and $D_{\mathbf{v}}(D_{\mathbf{v}}f)(P_0)$, and $D_{\mathbf{u}}(D_{\mathbf{v}}f)(P_0)$. What does $D_{\mathbf{u}}(D_{\mathbf{u}}f)(P_0)$ tell us about the concavity of the graph of f?

Calculator/Computer Exercises

60. In the viewing window $[-1, 3.2] \times [-2.2, 2.2]$, plot the level curve of $f(x,y) = 4x + y - x^2/8 - x^4/8 + 4y^2 - y^4$ at height 1. Calculate and sketch the directions of the gradients at the six points of intersection with the coordinate axes.

61. In the viewing window $[-1.25, 4.5] \times [-5, 0.75]$, plot the level curves $L_c = \{(x,y) : f(x,y) = c\}$ of $f(x,y) = 1 - x^2y/2 - x^2 - y^2 - 2y$ for $c = -1/2, 1/2, 3/2, 5/2$, and $7/2$. Calculate and sketch the directions of the gradients at the following points.

a. $(3, -3/2)$ on $L_{-1/2}$

b. $(2, -2 - \sqrt{2}/2)$ and $(2, -2 + \sqrt{2}/2)$ on $L_{1/2}$

c. $(1, -1)$ and $(1, -3/2)$ on $L_{3/2}$

d. $(3, -3)$ and $(3, -7/2)$ on $L_{5/2}$

e. $(4, -5 + \sqrt{26}/2)$ on $L_{7/2}$

11.7 Tangent Planes

If f is a differentiable function of one variable, then its derivative at a point x tells us the slope of the tangent line at that point. Now we will learn that the gradient of a function of two variables tells us about the tangent lines to the graph of f. We will see that the gradient plays the role of a "total derivative."

First we must address an important question: *What* geometric object should the tangent to the graph of a function of two variables be? Look at Figure 1, which exhibits the graph of $f(x, y) = 1 + x^2 + y^2$. There are a great many tangent lines at the point $(0, 0, 1)$, as the figure illustrates. The union of these tangent lines suggests that a *tangent plane* is the geometric object we seek (see Figure 2).

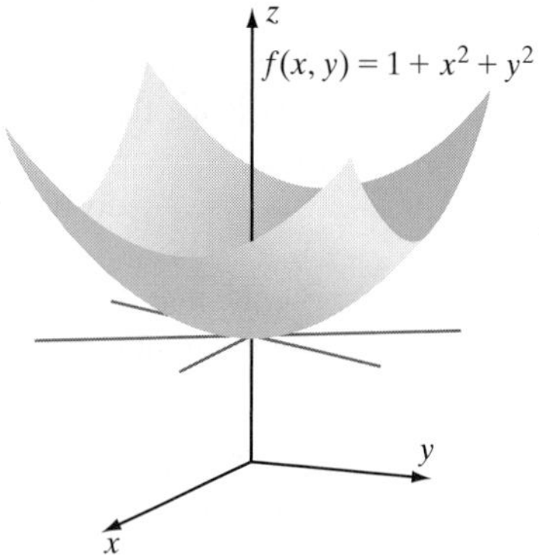

▲ Figure 1 Some tangent lines at the point $(0, 0, 1)$

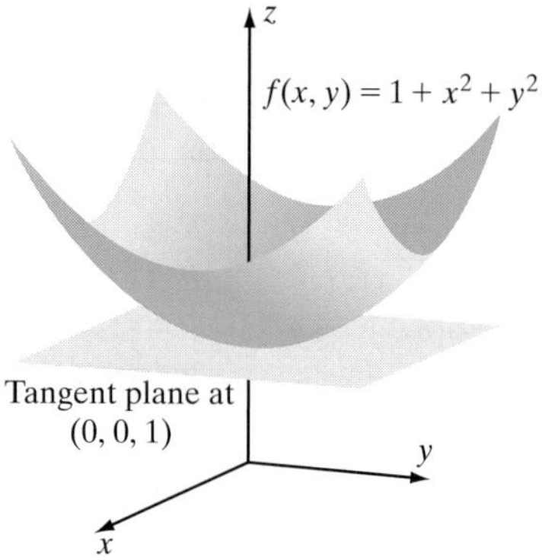

▲ Figure 2 The union of all tangent lines at a point is a tangent plane.

Now consider an arbitrary differentiable function f of two variables. Fix a point $P_0 = (x_0, y_0)$ in the domain of f. Also fix two nonparallel directions $\mathbf{u} = u_1\mathbf{i} + u_2\mathbf{j}$ and $\mathbf{v} = v_1\mathbf{i} + v_2\mathbf{j}$ in the xy-plane. Then, the curves $\mathcal{C}_1$ and $\mathcal{C}_2$ that are parameterized by

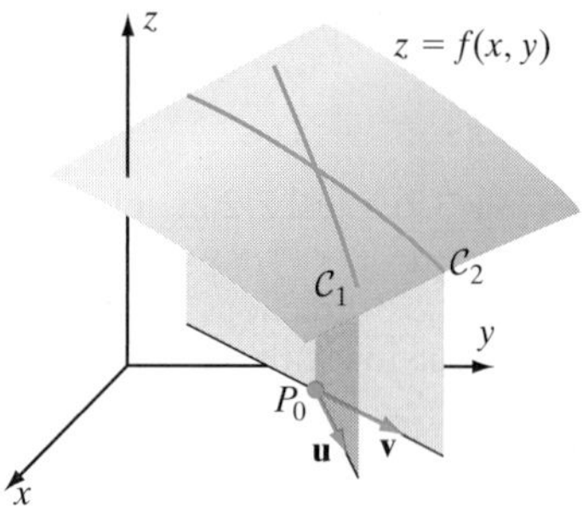

▲ Figure 3

$$r_{\mathbf{u}}(t) = (x_0 + tu_1)\mathbf{i} + (y_0 + tu_2)\mathbf{j} + f(x_0 + tu_1, y_0 + tu_2)\mathbf{k}$$

and

$$r_{\mathbf{v}}(t) = (x_0 + tv_1)\mathbf{i} + (y_0 + tv_2)\mathbf{j} + f(x_0 + tv_1, y_0 + tv_2)\mathbf{k}$$

are subsets of the graph of f. Each passes through the point P_0 when $t = 0$. The geometry is shown in Figure 3. We know that the vectors $r_{\mathbf{u}}'(0)$ and $r_{\mathbf{v}}'(0)$ are tangent to $\mathcal{C}_1$ and $\mathcal{C}_2$, respectively. We use the Chain Rule to calculate these two tangent vectors:

$$r_{\mathbf{u}}'(0) = u_1\mathbf{i} + u_2\mathbf{j} + \left(\frac{d}{dt}f(x_0 + tu_1, y_0 + tu_2)\Big|_{t=0}\right)\mathbf{k} = u_1\mathbf{i} + u_2\mathbf{j} + D_{\mathbf{u}}f(P_0)\mathbf{k}$$

and

$$r_{\mathbf{v}}'(0) = v_1\mathbf{i} + v_2\mathbf{j} + \left(\frac{d}{dt}f(x_0 + tv_1, y_0 + tv_2)\Big|_{t=0}\right)\mathbf{k} = v_1\mathbf{i} + v_2\mathbf{j} + D_{\mathbf{v}}f(P_0)\mathbf{k}.$$

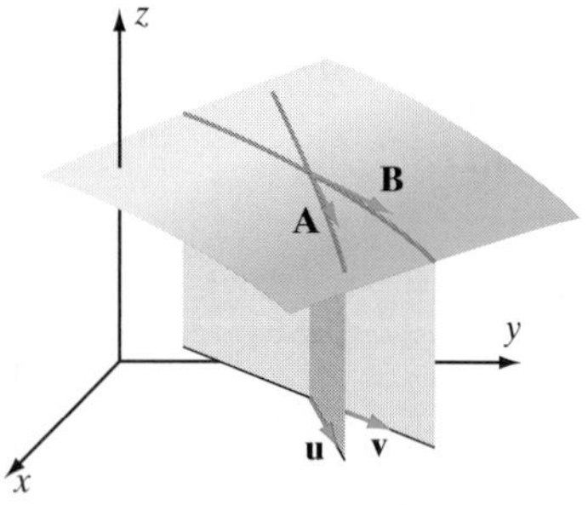

▲ Figure 4 Tangent vectors $\mathbf{A} = u_1\mathbf{i} + u_2\mathbf{j} + D_{\mathbf{u}}f(P_0)\mathbf{k}$ and $\mathbf{B} = v_1\mathbf{i} + v_2\mathbf{j} + D_{\mathbf{v}}f(P_0)\mathbf{k}$

See Figure 4. Let us use the cross product to calculate a vector $\mathbf{N}$ that is normal to both of these tangent vectors. We obtain

$$\mathbf{N} = (u_1\mathbf{i} + u_2\mathbf{j} + D_{\mathbf{u}}f(P_0)\mathbf{k}) \times (v_1\mathbf{i} + v_2\mathbf{j} + D_{\mathbf{v}}f(P_0)\mathbf{k}) = \det\left(\begin{bmatrix} \mathbf{i} & \mathbf{j} & \mathbf{k} \\ u_1 & u_2 & D_{\mathbf{u}}f(P_0) \\ v_1 & v_2 & D_{\mathbf{v}}f(P_0) \end{bmatrix}\right)$$

or, equivalently,

$$\mathbf{N} = \big(u_2 D_{\mathbf{v}} f(P_0) - v_2 D_{\mathbf{u}} f(P_0)\big)\mathbf{i} - \big(u_1 D_{\mathbf{v}} f(P_0) - v_1 D_{\mathbf{u}} f(P_0)\big)\mathbf{j} + (u_1 v_2 - u_2 v_1)\mathbf{k}.$$

After substituting $D_{\mathbf{u}} f(P_0) = u_1 f_x(P_0) + u_2 f_y(P_0)$ and $D_{\mathbf{v}} f(P_0) = v_1 f_x(P_0) + v_2 f_y(P_0)$ into the formula for $\mathbf{N}$, we find that the resulting expression simplifies to

$$\mathbf{N} = c\big(f_x(P_0)\mathbf{i} + f_y(P_0)\mathbf{j} - \mathbf{k}\big)$$

where c is the scalar $u_2 v_1 - v_2 u_1$. We have discovered a remarkable fact: No matter which two directions $\mathbf{u}$ and $\mathbf{v}$ we choose, any normal to the corresponding tangent vectors is a multiple of the vector $f_x(P_0)\mathbf{i} + f_y(P_0)\mathbf{j} - \mathbf{k}$. Because the tangent lines are all perpendicular to the vector $f_x(P_0)\mathbf{i} + f_y(P_0)\mathbf{j} - \mathbf{k}$, they must form a plane that has this vector as its normal. These observations allow us to make the following definition.

DEFINITION If f is a differentiable function of two variables and (x_0, y_0) is in its domain, then the *tangent plane* to the graph of f at $(x_0, y_0, f(x_0, y_0))$ is the plane that passes through the point $(x_0, y_0, f(x_0, y_0))$ and that is normal to the vector $f_x(x_0, y_0)i + f_y(x_0, y_0)\mathbf{j} - \mathbf{k}$. We say that the vector $f_x(x_0, y_0)\mathbf{i} + f_y(x_0, y_0)\mathbf{j} - \mathbf{k}$ is *normal* to the graph of f at the point $(x_0, y_0, f(x_0, y_0))$.

Using equation (9.5.1), we see that

$$f_x(x_0, y_0)(x - x_0) + f_y(x_0, y_0)(y - y_0) - \big(z - f(x_0, y_0)\big) = 0, \tag{11.7.1}$$

or, equivalently,

$$z = f(x_0, y_0) + f_x(x_0, y_0)(x - x_0) + f_y(x_0, y_0)(y - y_0) \tag{11.7.2}$$

is a Cartesian equation of the tangent plane to the graph of f at the point $\big(x_0, y_0, f(x_0, y_0)\big)$.

INSIGHT When we use equations (11.7.1) and (11.7.2), we calculate the derivatives $f_x(x, y)$ and $f_y(x, y)$, and then we evaluate them at the point (x_0, y_0). Consequently, the coefficients $f_x(x_0, y_0)$ and $f_y(x_0, y_0)$ are constants. When our calculation of equation (11.7.1) or (11.7.2) is complete, each of the variables x and y appears at most once, has power 1, and is multiplied by a constant. Equation (11.7.2), for example, has the form

$$\underset{\text{variable}}{\underset{\uparrow}{z}} = \underbrace{f(x_0, y_0)}_{\text{constant}} + \underbrace{f_x(x_0, y_0)}_{\text{constant}}\,(\underset{\text{variable}}{\underset{\uparrow}{x}} - \underbrace{x_0}_{\text{constant}}) + \underbrace{f_y(x_0, y_0)}_{\text{constant}}\,(\underset{\text{variable}}{\underset{\uparrow}{y}} - \underbrace{y_0}_{\text{constant}})$$

▶ **EXAMPLE 1** Find a Cartesian equation of the tangent plane to the graph of $f(x, y) = 2x - 3xy^3$ at the point $(2, -1, 10)$.

Solution We calculate $f_x(2, -1) = (2 - 3y^3)|_{(2,-1)} = 5$ and $f_y(2, -1) = (-9xy^2)|_{(2,-1)} = -18$. Also, $f(2, -1) = 10$. According to equation (11.7.1), $5(x - 2) + (-18)(y - (-1)) - (z - 10) = 0$, or $5x - 18y - z = 18$, is a Cartesian equation of the tangent plane to the graph of f at $(2, -1, 10)$. ◀

At this point in the theory, the variables x, y, and z do not have symmetric roles in the formula for the tangent plane to the graph of $z = f(x, y)$. A partial explanation is given by the fact that z is a dependent variable, whereas x and y are independent variables. However, a more geometric explanation is provided by the following discussion of level surfaces.

Level Surfaces

In Section 11.6, we learned that the gradient $\nabla f(x_0, y_0)$ of a function of two variables is perpendicular to the level curve of F that passes through the point (x_0, y_0). An analogous principle holds true for a function F of three variables. To understand why, fix a point $Q_0 = (x_0, y_0, z_0)$ in the domain of F. Let $\mathcal{S} = \{(x, y, z) : F(x, y, z) = c\}$ denote the level surface of F that passes through Q_0. We will show that $\nabla F(Q_0)$ is perpendicular to $\mathcal{S}$ at Q_0 by demonstrating that it is perpendicular to all curves that lie in $\mathcal{S}$ and that pass through Q_0. Indeed, suppose that $\mathbf{r} = r_1\mathbf{i} + r_2\mathbf{j} + r_3\mathbf{k}$ is a parameterization of such a curve $\mathcal{C}$ with $\mathbf{r}(t_0) = Q_0$. Then

$$\frac{d}{dt} F\big(\mathbf{r}(t)\big)\Big|_{t=t_0} = \nabla F(Q_0) \cdot \mathbf{r}'(t_0) \qquad (11.7.3)$$

by equation (11.6.3). Because $\mathcal{C}$ is a subset of the level surface $\mathcal{S}$, it follows that $F(\mathbf{r}(t)) = c$ for all values of t. The left side of equation (11.7.3) is therefore 0. Thus $\nabla F(Q_0) \cdot \mathbf{r}'(t_0) = 0$, which implies that $\nabla F(Q_0)$ is perpendicular to the velocity vector $\mathbf{r}'(t_0)$. Because this velocity vector is tangent to $\mathcal{C}$ at Q_0, we conclude that $\nabla F(Q_0)$ is perpendicular to $\mathcal{C}$. We state the result of this analysis as a theorem.

THEOREM 1 If F is a differentiable function of three variables, then $\nabla F(x_0, y_0, z_0)$ is perpendicular to the level surface of F at (x_0, y_0, z_0).

We can now tie together the idea of tangent plane with our new insight into level surfaces. Suppose that $(x, y) \mapsto f(x, y)$ is a function of two variables, and we want to know a normal to the tangent plane to the graph at the point $Q_0 = \big(x_0, y_0, f(x_0, y_0)\big)$. Define

$$F(x, y, z) = f(x, y) - z. \qquad (11.7.4)$$

Then the graph of f is just the set $\{(x, y, z) : F(x, y, z) = 0\}$. That is, the graph of f is a level surface of F. Therefore $\nabla F(Q_0)$ is perpendicular to the graph of f. But

$$\nabla F(Q_0) = \langle f_x(x_0, y_0), f_y(x_0, y_0), -1\rangle.$$

Thus we have rediscovered the fact that $\langle f_x(x_0, y_0), f_y(x_0, y_0), -1\rangle$ is a normal to the graph of f at the point $\big(x_0, y_0, f(x_0, y_0)\big)$.

▶ **EXAMPLE 2** Find the tangent plane to the surface $x^2 + 4y^2 + 8z^2 = 13$ at the point $(1, -1, 1)$.

Solution We do *not* need to solve for z as a function of x and y. Instead, we define the function of three variables $F(x, y, z) = x^2 + 4y^2 + 8z^2$. Then, our surface is just the level set $F(x, y, z) = 13$. By Theorem 1, a normal to the surface $F(x, y, z) = 13$ at the point $(1, -1, 1)$ is given by

$$\nabla F(1, -1, 1) = (2x\mathbf{i} + 8y\mathbf{j} + 16z\mathbf{k})\big|_{(x,y,z) = (1,-1,1)} = 2\mathbf{i} - 8\mathbf{j} + 16\mathbf{k}.$$

Because the tangent plane also passes through the point $(x_0, y_0, z_0) = (1, -1, 1)$, we may write the equation of the plane as $2(x-1) - 8(y-(-1)) + 16(z-1) = 0$. This equation simplifies to $2x - 8y + 16z = 26$. ◀

▶ **EXAMPLE 3** Let $\mathcal{S}$ be the surface defined by the equation $z = \sqrt{\dfrac{5 + x^2 + 2y^4}{y^2 + x^4}}$. Find the tangent plane to $\mathcal{S}$ at the point $(1, -1, 2)$.

Solution Although z is given explicitly as a function of x and y, the calculation of the derivatives $\partial z/\partial x$ and $\partial z/\partial y$ involves a lengthy application of the Quotient Rule. We may avoid this difficulty by observing that $\mathcal{S}$ is a subset of a level surface of a function that is easy to differentiate. Indeed, every point (x, y, z) of $\mathcal{S}$ satisfies the equation $z^2 = (5 + x^2 + 2y^4)/(y^2 + x^4)$, or $5 + x^2 + 2y^4 - (y^2 + x^4)z^2 = 0$. Therefore $\mathcal{S}$ is a subset of a level set of $F(x, y, z) = 5 + x^2 + 2y^4 - (y^2 + x^4)z^2$. We calculate $\nabla F(x, y, z) = (2x - 4x^3z^2)\mathbf{i} + (8y^3 - 2yz^2)\mathbf{j} - 2(y^2 + x^4)z\mathbf{k}$. A normal to $\mathcal{S}$ at the point $(1, -1, 2)$ is therefore given by

$$\nabla F(1, -1, 2) = ((2x - 4x^3z^2)\mathbf{i} + (8y^3 - 2yz^2)\mathbf{j} - 2(y^2 + x^4)z\mathbf{k})\big|_{(x,y,z)\,=\,(1,-1,2)} = -14\mathbf{i} - 8\mathbf{k}.$$

Because the tangent plane we seek passes through the point $(x_0, y_0, z_0) = (1, -1, 2)$, we may write its equation as $(-14)(x-1) + (0)(y-(-1)) + (-8)(z-2) = 0$. This equation simplifies to $7x + 4z = 15$. ◀

Normal Lines

The normal line to a surface at a point $Q_0 = (x_0, y_0, f(x_0, y_0))$ on the surface is the line with direction vector given by a normal to the surface at Q_0 and passing through Q_0.

▶ **EXAMPLE 4** Find symmetric equations for the normal line to the graph of $f(x, y) = -y^2 - x^3 + xy^2$ at the point $(1, 4, -1)$.

Solution We will solve this problem in two ways. A normal to the graph at Q_0 is

$$\langle f_x(x, y), f_y(x, y), -1\rangle\big|_{x=1,y=4} = \langle -3x^2 + y^2,\ -2y + 2xy, -1\rangle\big|_{x=1,y=4} = \langle 13, 0, -1\rangle.$$

Because the desired line also passes through $Q_0 = (1, 4, f(1, 4)) = (1, 4, -1)$, we may write the line parametrically as

$$x = 1 + 13t, \quad y = 4, \quad z = -1 + (-1)t.$$

Solving for t in terms of x and also in terms of z, and then equating the expressions for t leads to the symmetric equations

$$\frac{x-1}{13} = \frac{z+1}{-1}, \quad y = 4.$$

As an alternative, we can obtain the same result by defining

$$F(x, y, z) = f(x, y) - z = -y^2 - x^3 + xy^2 - z.$$

Then the level set $\{(x, y, z) : F(x, y, z) = 0\}$ is precisely the graph of f. Therefore a normal to the graph of f is

$$\nabla F(1, 4, -3) = \left((-3x^2 + y^2)\mathbf{i} + (-2y + 2xy)\mathbf{j} - 1\mathbf{k}\right)\big|_{x=1, y=4, z=f(1,4)} = 13\mathbf{i} - \mathbf{k}.$$

The rest of the computation continues exactly as in the first solution. ◀

Numerical Approximations Using the Tangent Plane

As we learned in Section 3.9, the tangent line can be used as a means for approximating a differentiable function of one variable. Now we apply the same idea to functions of two variables. Let $\mathcal{T}$ be the tangent plane to the graph of a differentiable function $(x, y) \mapsto f(x, y)$ at the point $(x_0, y_0, f(x_0, y_0))$. Equation (11.7.2) tells us that $\mathcal{T}$ is the graph of the function

$$L(x, y) = f(P_0) + f_x(P_0)(x - x_0) + f_y(P_0)(y - y_0) \qquad (11.7.5)$$

where $P_0 = (x_0, y_0)$. Recall that the terms $f(P_0)$, $f_x(P_0)$, and $f_y(P_0)$ are just constants. After we have calculated these three fixed numbers, we need only perform two multiplications and a few additions to compute $L(x, y)$ for any x and y. What is significant is that this particularly simple function gives a good approximation to $f(x, y)$ for x and y close to x_0 and y_0, respectively, as can be seen from equation (11.5.2), which states that

$$f(x, y) = f(P_0) + f_x(P_0)(x - x_0) + f_y(P_0)(y - y_0) + \varepsilon_1 \cdot (x - x_0) + \varepsilon_2 \cdot (y - y_0).$$

Notice that the first three summands on the right side exactly constitute the formula for $L(x, y)$. We conclude that

$$f(x, y) = L(x, y) + \varepsilon_1 \cdot (x - x_0) + \varepsilon_2 \cdot (y - y_0).$$

Finally, recall that ε_1 and ε_2 both tend to 0 as x and y tend to x_0 and y_0, respectively. This means that $f(x, y) - L(x, y)$ is the sum of two terms, $\varepsilon_1 \cdot (x - x_0)$ and $\varepsilon_2 \cdot (y - y_0)$, each of which becomes small compared to the increments $k = x - x_0$ and $k = y - y_0$ as x and y tend to x_0 and y_0. We summarize our analysis in the following theorem.

THEOREM 2 Let f be a differentiable function of two variables on a rectangular region centered at $P_0 = (x_0, y_0)$. Let L be the linear function of two variables defined by

$$L(x, y) = f(P_0) + f_x(P_0)(x - x_0) + f_y(P_0)(y - y_0).$$

Then, for every point (x, y) near the fixed point P_0, we have the approximation $f(x, y) \approx L(x, y)$. The difference $f(x, y) - L(x, y)$ between the exact value $f(x, y)$ and the approximating value $L(x, y)$ is equal to $\varepsilon_1 \cdot (x - x_0) + \varepsilon_2 \cdot (y - y_0)$ where ε_1 and ε_2 both tend to 0 as x and y tend to x_0 and y_0, respectively.

DEFINITION The expression $L(x, y)$ defined by equation (11.7.5) is called the *linear approximation* (or the *tangent plane approximation*) to $f(x, y)$ at the point P_0.

Today with calculators and computers so common, we can easily compute the expression $f(x, y)$ when it is given by an explicit formula that involves x and y. What is the point of an approximation? The answer is that $f(x, y)$ may involve complicated combinations and compositions of expressions that are difficult to analyze. Because the expression $L(x, y)$ is either a constant or a degree 1 polynomial in x and y, it is always easy to work with.

It is natural to ask how good the approximation given by the theorem is. After all, we do not have exact expressions for the terms ε_1 and ε_2 that appear in the error $f(x, y) - L(x, y)$. We will use the multivariable Taylor's Theorem to estimate this difference. If we set $h = x - x_0$ and $k = y - y_0$, then formula (11.7.5) becomes

$$L(x, y) = f(P_0) + f_x(P_0)h + f_y(P_0)k = T_1(h, k),$$

where $T_1(h, k)$ is the order 1 Taylor polynomial of f with base point P_0, as given by equation (11.5.22). Thus if f is twice continuously differentiable, then equation (11.5.14) can be written as $f(x, y) = L(x, y) + R_1(x, y)$, or

$$f(x, y) - L(x, y) = \frac{1}{2!}\left(f_{xx}(Q_1)h^2 + 2f_{xy}(Q_1)hk + f_{yy}(Q_1)k^2\right)$$

for some point Q_1 on the line segment between P_0 and (x, y). It follows that

$$|f(x, y) - L(x, y)| \le \frac{M}{2}\left(|x - x_0| + |y - y_0|\right)^2 \qquad (11.7.6)$$

for (x, y) in the rectangular region $R = \{(x, y) : |x - x_0| \le r_1, |y - y_0| \le r_2\}$, where M is any number that is larger than the absolute values of all the second derivatives $|f_{xx}|, |f_{yy}|, |f_{xy}|$ in R.

► **EXAMPLE 5** Let f be defined by $f(x, y) = \sqrt{x^2 + y^4}$. Using $P_0 = (3, 2)$, what is the tangent plane approximation $L(2.9, 2.1)$ of $f(2.9, 2.1)$?

Solution We use formula (11.7.5) with $(x, y) = (2.9, 2.1)$ and $(x_0, y_0) = (3, 2)$. Because

$$f(3, 2) = 5, \quad f_x(3, 2) = \left.\frac{x}{\sqrt{x^2 + y^4}}\right|_{(3,2)} = \frac{3}{5}, \quad \text{and} \quad f_y(3, 2) = \left.\frac{2y^3}{\sqrt{x^2 + y^4}}\right|_{(3,2)} = \frac{16}{5},$$

we have

$$\begin{aligned} f(2.9, 2.1) \approx L(2.9, 2.1) &= f(3, 2) + f_x(3, 2)(2.9 - 3) + f_y(3, 2)(2.1 - 2) \\ &= 5 + \left(\frac{3}{5}\right)(-0.1) + \left(\frac{16}{5}\right)(0.1) = 5.26. \end{aligned}$$

The true value of $f(2.9, 2.1)$ is $5.278\ldots$, so our approximation is accurate to one decimal place. ◄

INSIGHT If we assume that the rectangular region R on which Example 5 takes place is $R = \{(x, y) : |x - 3| \leq 0.1, |y - 2| \leq 0.1\}$, then we may show that the second derivatives of f are not larger than $M = 10$ in absolute value. Thus estimate (11.7.6) for the maximal size of the error tells us that it cannot be larger than

$$\frac{M}{2}(|-0.1| + |0.1|)^2 = 0.2.$$

In fact, the accuracy that we actually achieved is even better than this prediction. The error bound given by estimate (11.7.6) is a "worst case" value.

A Restatement of Theorem 2 Using Increments

The linear approximation is often stated using the notation of increments. Let us write $x = x_0 + \Delta x$, where Δx is a small increment. Similarly, we write $y = y_0 + \Delta y$. Continuing with this theme, we let $f(x, y) = f(x_0, y_0) + \Delta f$. Then the approximation $f(x, y) \approx L(x, y) = f(P_0) + f_x(P_0)(x - x_0) + f_y(P_0)(y - y_0)$, or $f(x, y) - f(P_0) \approx f_x(P_0)(x - x_0) + f_y(P_0)(y - y_0)$, becomes

$$\Delta f \approx f_x(P_0)\Delta x + f_y(P_0)\Delta y. \qquad (11.7.7)$$

This estimate is sometimes recorded as an equation among differentials:

$$df = f_x(P_0)dx + f_y(P_0)dy. \qquad (11.7.8)$$

The quantity on the right side of equation (11.7.8) is called the *total differential* of f at the point P_0. In more advanced courses, a precise meaning is given to the differentials in equation (11.7.8) so that equality does hold.

▶ **EXAMPLE 6** A machinist is making a bearing that is in the shape of a small cone. It is being manufactured from a costly titanium-vanadium alloy. The bearing is to have base of radius 3 millimeters and height 6 millimeters. The manufacturer can only allow errors of 1% in the amount of material used in each bearing. How large an error is allowable in the dimensions of the bearing?

Solution The volume of the cone is $V(r, h) = \pi r^2 h/3$. The maximum allowable error is 1% of volume or

$$\text{Allowable error} = (0.01) \cdot \frac{1}{3}\pi r^2 h\Big|_{r=3,h=6} = 0.18\pi.$$

According to Theorem 2,

$$\Delta V \approx V_r(3,6)\Delta r + V_h(3,6)\Delta h = \frac{2}{3}\pi r h\Big|_{r=3,h=6}\Delta r + \frac{1}{3}\pi r^2\Big|_{r=3,h=6}\Delta h = 12\pi\Delta r + 3\pi\Delta h.$$

It is therefore required that Δr and Δh satisfy $12\pi\Delta r + 3\pi\Delta h \leq 0.18\pi$ or

$$4\Delta r + \Delta h \leq 0.06.$$

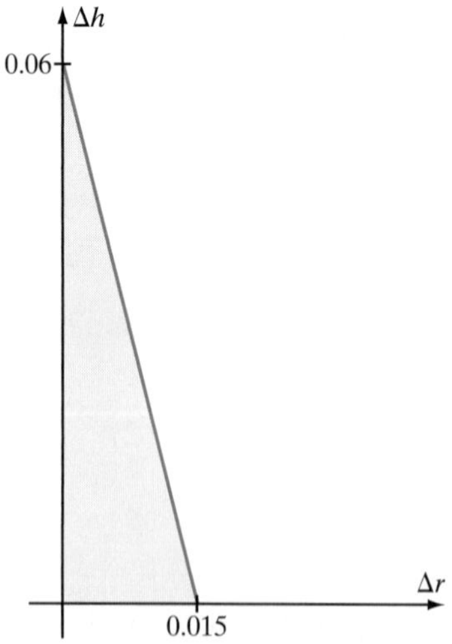

▲ Figure 5

How do we interpret this inequality? See Figure 5 for a schematic of Δr and Δh. If a given bearing has *no error* in the height, then the formula reduces to $\Delta r \leq 0.015$.

Thus a *maximum error* of 0.015 millimeter can be allowed in the radius of the bearing if there is no error in the height. Similarly, if there is no error in the radius, then the formula becomes $\Delta h \leq 0.06$. We see that a maximum error of 0.06 millimeters in the height can be allowed if there is no error in the radius. It is more realistic to suppose that the machinist can keep the height error down to 0.01 millimeters and ask how much tolerance can be allowed in the radius. To apply our error estimate, we set $\Delta h = 0.01$ to obtain $4\Delta r + 0.01 \leq 0.06$ or $\Delta r \leq 0.0125$. Thus the greatest variation in radius that we can allow is 0.0125 millimeters. ◀

Functions of Three or More Variables

The idea of linear approximation applies to functions of three or more variables. If $(x, y, z) = (x_0 + \Delta x, y_0 + \Delta y, z_0 + \Delta z)$ is near the point $Q_0 = (x_0, y_0, z_0)$, then the linear approximation of $F(x, y, z)$ is

$$F(x, y, z) \approx F(Q_0) + F_x(Q_0)(x - x_0) + F_y(Q_0)(y - y_0) + F_z(Q_0)(z - z_0).$$

Stated another way, the difference ΔF between the values of F at (x, y, z) and $Q_0 = (x_0, y_0, z_0)$ is approximately

$$\Delta F \approx F_x(Q_0)\Delta x + F_y(Q_0)\Delta y + F_z(Q_0)\Delta z.$$

Using the notation of differentials, this approximation is written as

$$dF = F_x(Q_0)dx + F_y(Q_0)dy + F_z(Q_0)dz, \tag{11.7.9}$$

and the right side is called the *total differential* of F at Q_0.

▶ **EXAMPLE 7** The initial mass of a bacterial sample is measured to be $m = 3$ mg. The mass grows according to the exponential law me^{kt}. One hour after the initial measurement, a second measurement $M = 3.3$ mg is taken. The two measurements allow the researchers to determine the growth constant: $k = \ln(3.3/3) = \ln(1.1)$. It is therefore predicted that 8 hours after the initial measurement, the mass will be $3e^{8\ln(1.1)}$ mg, or 6.43 mg. By about how much can the actual mass exceed this figure if the two mass measurements were as much as 0.05 mg off and the time measurement $t = 1$ hr was as much as 0.001 hr in error?

Solution If an exact initial mass measurement m was made, and if an exact mass measurement M was made exactly T hours after the first, then we would have $M = me^{kT}$ or $k = \dfrac{1}{T}\ln\left(\dfrac{M}{m}\right)$. The mass W at a fixed instant of time t hours after the initial measurement would be

$$W(m, M, T) = me^{kt} = m\exp\left(\frac{t}{T}\ln\left(\frac{M}{m}\right)\right) = m\exp\left(\ln\left(\left(\frac{M}{m}\right)^{t/T}\right)\right) = m\left(\frac{M}{m}\right)^{t/T}.$$

In particular, the mass would be

$$W(m, M, T) = m\left(\frac{M}{m}\right)^{8/T}$$

8 hours after the initial measurement. Using the values $m_0 = 3$, $M_0 = 3.3$, and $T_0 = 1$, we have $W(3, 3.3, 1) = 6.43$, as given in the statement of the problem. We calculate

$$W_m(3.0, 3.3, 1) = \frac{\partial}{\partial m}\left(m\left(\frac{M}{m}\right)^{8/T}\right)\Bigg|_{m=3,M=3.3,T=1} = \left(\frac{M}{m}\right)^{8/T} - 8\frac{(M/m)^{8/T}}{T}\Bigg|_{m=3,M=3.3,T=1} = -15.005,$$

$$W_M(3.0, 3.3, 1) = \frac{\partial}{\partial M}\left(m\left(\frac{M}{m}\right)^{8/T}\right)\Bigg|_{m=3,M=3.3,T=1} = 8m\frac{(M/m)^{8/T}}{TM}\Bigg|_{m=3,M=3.3,T=1} = 15.59,$$

and

$$W_T(3.0, 3.3, 1) = \frac{\partial}{\partial T}\left(m\left(\frac{M}{m}\right)^{8/T}\right)\Bigg|_{m=3,M=3.3,T=1} = -8m\frac{(M/m)^{8/T}}{T^2}\ln\left(\frac{M}{m}\right)\Bigg|_{m=3,M=3.3,T=1} = -4.903.$$

The error ΔW is approximately

$$\begin{aligned}\Delta W &\approx \frac{\partial W}{\partial m}(3, 3.3, 1)\Delta m + \frac{\partial W}{\partial M}(3, 3.3, 1)\Delta M + \frac{\partial W}{\partial T}(3, 3.3, 1)\Delta T \\ &= (-15.005)\Delta m + (15.590)\Delta M + (-4.903)\Delta T.\end{aligned}$$

Now the given bounds for the measurement errors are $-0.05 \le \Delta m \le 0.05$, $-0.05 \le \Delta M \le 0.05$, and $-0.001 \le \Delta T \le 0.001$. In the worst case, the signs of Δm, ΔM, and ΔT will cause the three summands of ΔW to have the same sign. When this happens, the individual measurement errors will accumulate and not cancel. Thus if $\Delta m = -0.05$, $\Delta M = +0.05$, and $\Delta T = -0.001$, then

$$\Delta W \approx (-15.005)(-0.05) + (15.590)(0.05) + (-4.903)(-0.001) \approx 1.535.$$

Because $1.535/6.43 \approx 0.238$, our error analysis tells us that, in the worst case, the error can be about 24% of the predicted value. ◀

◀ **A LOOK BACK** In Section 3.9 of Chapter 3, we learned the linear approximation $\Delta f \approx f'(c)\Delta x$ of a function f of one variable. Equation (3.9.7), $df = f'(c)dx$, expresses this linear approximation in the language of differentials. Equations (11.7.8) and (11.7.9) are the two-variable and three-variable analogues of equation (3.9.7).

QUICK QUIZ

1. What is a normal vector to the graph of $z = x^2 + 2y$ at $(3, -2, 5)$?
2. What is the Cartesian equation $Ax + By - z = D$ of the plane that is tangent to the graph of $z = x^2 + 2y$ at the point $(3, -2, 5)$?
3. What is the Cartesian equation $x + By + Cz = D$ of the plane that is tangent to the graph of $xz + 2y^2 = 4$ at the point $(2, -1, 1)$?
4. If $f(x, y) = \sqrt{x + 3y}$, what is the tangent plane approximation to $f(3.3, 1.8)$ if the base point $(3, 2)$ is used?

Answers

1. $\langle 6, 2, -1\rangle$ 2. $6x + 2y - z = 9$ 3. $x - 4y + 2z = 8$ 4. 2.95

EXERCISES

Problems for Practice

In each of Exercises 1–12, a differentiable function $(x, y) \mapsto f(x, y)$ is given, as is a point P_0 in the domain of f. Calculate a normal vector to the graph of f at the point $(P_0, f(P_0))$. Then, write the equation of the tangent plane to the graph of f at the point $(P_0, f(P_0))$.

1. $f(x, y) = xy - y^3 + x^2$ $\quad P_0 = (1, 4)$
2. $f(x, y) = \sin(x - y)$ $\quad P_0 = (\pi/2, \pi/3)$
3. $f(x, y) = \ln(2 + x^2 + y^2)$ $\quad P_0 = (1, 2)$
4. $f(x, y) = e^{2x-3y}$ $\quad P_0 = (\ln(2), \ln(3))$
5. $f(x, y) = \dfrac{x}{x + y}$ $\quad P_0 = (2, 1)$
6. $f(x, y) = \sin(x)\cos(y)$ $\quad P_0 = (\pi/3, \pi/4)$
7. $f(x, y) = xye^{x+y}$ $\quad P_0 = (1, 2)$
8. $f(x, y) = \dfrac{\sin(x)}{\cos(y)}$ $\quad P_0 = (\pi/3, \pi/3)$
9. $f(x, y) = \dfrac{x^2 + y^2}{x + y}$ $\quad P_0 = (1, 1)$
10. $f(x, y) = (x + y + 1)^{1/3}$ $\quad P_0 = (6, 1)$
11. $f(x, y) = (x - y)^{-1/3}$ $\quad P_0 = (9, 1)$
12. $f(x, y) = \cot(x/y)$ $\quad P_0 = (\pi/2, 2)$

In each of Exercises 13–20, find a normal to the surface at the point Q_0. Then, find the equation of the tangent plane to the surface at the point Q_0.

13. $x^2 - 4y^2 + z^2 = -14$ $\quad Q_0 = (1, 2, 1)$
14. $xyz = 8$ $\quad Q_0 = (4, 1, 2)$
15. $\dfrac{y}{x + z} = 2$ $\quad Q_0 = (1, 8, 3)$
16. $xy\sin(z) = 5$ $\quad Q_0 = (5, 2, \pi/6)$
17. $\ln(1 + x + y + z) = 3$ $\quad Q_0 = (e^2 + 2, e^3 - e^2, -3)$
18. $(x + y)(z - y^2) = -14$ $\quad Q_0 = (-2, 4, 9)$
19. $(x + y + z^2)^6 = 1$ $\quad Q_0 = (3, -3, 1)$
20. $\cos(xyz) = 0$ $\quad Q_0 = (4, \pi, 1/8)$

In Exercises 21–26, write symmetric equations for the normal line to the graph of the given function f at the given point.

21. $f(x, y) = x^3 - y^5$ $\quad (2, 1, 7)$
22. $f(x, y) = \sqrt{x + y - 2}$ $\quad (2, 1, 1)$
23. $f(x, y) = xe^{x+y}$ $\quad (2, -2, 2)$
24. $f(x, y) = \dfrac{y^2}{x - y}$ $\quad (1, 2, -4)$
25. $f(x, y) = \cos(x) - \sin(y)$ $\quad (0, \pi, 1)$
26. $f(x, y) = x\sin(y)$ $\quad (\pi/2, \pi/2, \pi/2)$

In each of Exercises 27–30, calculate the linear approximation to $f(P)$ using the tangent plane at $(P_0, f(P_0))$.

27. $f(x, y) = \sqrt{x^4 + y^2}$ $\quad P = (2.1, 2.9)$ $\quad P_0 = (2, 3)$
28. $f(x, y) = x/\sqrt{1 + x^2 + y^2}$ $\quad P = (1.9, 2.1)$ $\quad P_0 = (2, 2)$
29. $f(x, y) = x\cos(2x - y)$ $\quad P = (0.9, 2.2)$ $\quad P_0 = (1, 2)$
30. $f(x, y) = \ln(1 + \sqrt{x} - y)$ $\quad P = (9.3, 3.2)$ $\quad P_0 = (9, 3)$

31. A triangular plot of land is to have base 1000 feet and height 500 feet. Suppose that surveying errors are kept within 1%. Use the total differential to estimate the greatest error in area.
32. A rectangular beam with cross section of width w and thickness t is known to have stiffness $\alpha \cdot w^3 \cdot t$, where α is a physical constant. The beam is cut to have dimensions 10 cm (width) and 5 cm (thickness). Assuming neither cut is more than 1% off, use the total differential to estimate how much the stiffness of the beam can vary.
33. A rectangular box is to be built with length 6 centimeters, width 4 centimeters, and height 3 centimeters. Suppose that measurement error can be kept accurate to half of 1%. Use the total differential to estimate the greatest error in volume.
34. A machinist's jig is to be made in the shape of an isosceles triangle. The base is to measure 12 centimeters and the legs 8 centimeters. However, the man building the jig is a novice, and his measurements are off by 0.1 centimeter. Estimate the resulting error in the area of the jig.

Further Theory and Practice

In each of Exercises 35–38, a function f and a point P_0 are given. Also given is an incorrect equation for the tangent plane to the graph of f at P_0. Explain what error has been made, and state the correct equation for the tangent plane.

35. $f(x, y) = 1 + 3x^2 + 2y^3$ $\quad P_0 = (1, -1)$
    $z = 6(x - 1) + 6(y + 1)$
36. $f(x, y) = 5 + x^2 + y^3$ $\quad P_0 = (2, 1)$
    $z = 10 + 2x(x - 2) + 3y^2(y - 1)$
37. $f(x, y) = 3 + y + xy$ $\quad P_0 = (0, -3)$
    $z = yx + (x + 1)(y + 3)$
38. $f(x, y) = 7x + 4/y$ $\quad P_0 = (1, 2)$ $\quad z = 9 + 7x - y$

39. Let θ be any number. Show that $\mathbf{r}_\theta(t) = t\cos(\theta)\mathbf{i} + t\sin(\theta)\mathbf{j} + t\mathbf{k}$ is a curve that passes through the origin and that lies on the surface $z^2 = x^2 + y^2$. Show that the zero vector is the only vector that is perpendicular to $\mathbf{r}'_\theta(0)$ for every θ. Use this observation to discuss the concept of normal vector and tangent plane to the surface $z^2 = x^2 + y^2$. Sketch the surface. What, if anything, does Theorem 1 tell us when we apply it to the differentiable function $F(x, y, z) = x^2 + y^2 - z^2$ at the origin?
40. Define the angle between two surfaces at a point of intersection to be the angle between their tangent planes, which is the same thing as the angle between their normals. What is the angle between the surfaces $x^2 + y^2 +$

$z^2 = 16$ and $x - y + 2z = 0$ at their points of intersection? (Hint: There are many points of intersection.)

41. Refer to Exercise 40 for terminology. Calculate the angle between the surfaces $x^2 - 3xz + 4y = 0$ and $y^2 + 3xyz - 8x = 0$ at the point $(0, 0, 0)$.

42. Calculate the angle between the surface $z = x^2 - 4y$ and the line

$$x = 2t, y = -t + 1, z = 4t + 12$$

at their two points of intersection. (Hint: This is the same as the angle between the direction vector for the line and the tangent plane to the surface.)

43. Calculate the angle between the line $x = 3t + 6$, $y = 4t + 8$, $z = 5t + 5$ and the sphere $x^2 + y^2 + z^2 = 25$ at their two points of intersection.

44. Fix a twice continuously differentiable function $(x, y) \mapsto f(x, y)$, a point $P_0 = (x_0, y_0)$ in the domain of f, and a direction $\mathbf{u} = u_1\mathbf{i} + u_2\mathbf{j}$. Consider the function $\phi(t) = f(P_0 + t\mathbf{u})$. Write out the first order Taylor Expansion of ϕ about $t = 0$. What do the zero- and first-order terms that you see in this expansion have to do with the equation for the tangent plane to the graph of f at P_0?

45. A solid, round bead (with no hole drilled through it) of radius 4 mm is to be made of an alloy that is 60% gold. If the mass of the gold that is used is not to be greater than 1% more than intended, then approximately what errors in size and alloy percentage can be permitted?

46. Let S be any sphere and ℓ any line that intersects that sphere in two points. Show that the angle between the line and the sphere is the same at the two points of intersection.

In each of Exercises 47–56, a function $f(x, y)$ and a point P_0 on the graph of the equation $z = f(x, y)$ are given. By expressing the graph of f near P_0 as the solution set of an equation of the form $F(x, y, z) = 0$, find a Cartesian equation for the tangent plane to the graph of f at P_0. (One way to do this is to use equation (11.7.4). Instead, follow the method of Example 3 to avoid tedious differentiations.)

47. $f(x, y) = (x + y + 2xy^2)/(x - y)$ $\quad (3, 1, 5)$

48. $f(x, y) = 4(x + y)/(x^2 + 2y)$ $\quad (2, 3, 2)$

49. $f(x, y) = (2x^3 + x^2y - xy^3)^{1/3}$ $\quad (-1, 2, 2)$

50. $f(x, y) = (1 + x^6 + y^4)^{1/4}$ $\quad (2, -2, 3)$

51. $f(x, y) = \sqrt{(x + 2y)/(x - y)}$ $\quad (2, 1, 2)$

52. $f(x, y) = \left(\dfrac{y^2 - 3x}{x^2 - y}\right)^{1/3}$ $\quad (1, 2, -1)$

53. $f(x, y) = \dfrac{2 + \ln(y + 2)}{1 + \ln(x)}$ $\quad (1, -1, 2)$

54. $f(x, y) = \dfrac{4\exp(x - 3y)}{\ln(x - 2y)}$ $\quad (3, 1, 4)$

55. $f(x, y) = \arctan\left(\dfrac{y - x}{3x - y}\right)$ $\quad (1, 2, \pi/4)$

56. $f(x, y) = xy^2/(x - y) + 10x^2y/(x + y)$ $\quad (2, 3, 6)$

Calculator/Computer Exercises

In each of Exercises 57–60, plot the given function f in a suitable viewing box containing the given point P_0. On the same set of axes, plot the tangent plane at the point P_0. Zoom in on the given point until it becomes apparent that the tangent plane closely approximates the graph of f.

57. $f(x, y) = 3xy^2/(1 + x^2 + y^4)$ $\quad P_0 = (2, 1, 1)$

58. $f(x, y) = (3x - y)^2/\sqrt{6 + x^2 + y^4}$ $\quad P_0 = (3, 1, 16)$

59. $f(x, y) = (x^2 - y)e^{2x-y}$ $\quad P_0 = (2, 3, e)$

60. $f(x, y) = 17\ln(1 + x^2)/(1 + x^2y^4)$ $\quad P_0 = (1, 2, \ln(2))$

11.8 Maximum-Minimum Problems

Throughout this section, we assume that f is defined on a planar set $\mathcal{G}$. We further assume that each point in $\mathcal{G}$ is the center of a disk of positive radius that is entirely contained in $\mathcal{G}$. Such a set $\mathcal{G}$ is said to be *open*. A point P_0 in $\mathcal{G}$ is called a *local* (or *relative*) *minimum* for f if it is the center of a disk $D(P_0, r)$ with positive radius r such that

$$f(P_0) \leq f(x, y) \quad \text{for all } (x, y) \in D(P_0, r).$$

If $f(P_0) \leq f(x, y)$ for all (x, y) in $\mathcal{G}$, then P_0 is called an *absolute* (or *global*) *minimum* for f. Similarly, a point P_0 in $\mathcal{G}$ is called a *local* (or *relative*) *maximum* for f if it is the center of a disk $D(P_0, r)$ with positive radius r such that

$$f(P_0) \geq f(x, y) \quad \text{for all } (x, y) \in D(P_0, r).$$

If $f(P_0) \geq f(x, y)$ for all (x, y) in $\mathcal{G}$, then P_0 is called an *absolute* (or *global*) *maximum* for f. A point that is either a local minimum or a local maximum is said to be a *local extremum*.

The Analogue of Fermat's Theorem

Pictorially, we expect a local minimum $P_0 = (x_0, y_0)$ to correspond to a point $(x_0, y_0, f(P_0))$ where the graph of f looks like the plot in Figure 1a. We expect a local maximum to correspond to a point $(x_0, y_0, f(P_0))$ where the graph of f looks like the plot in Figure 2a. Figures 1b and 2b suggest that if P_0 is a local extremum of f, then the tangent plane to the graph of f at $(x_0, y_0, f(P_0))$ will be parallel to the xy-plane. But this would happen only if the normal vector to the tangent plane, $\langle f_x(P_0), f_y(P_0), -1\rangle$, were a multiple of $\langle 0, 0, 1\rangle$. In other words, the geometry indicates that $f_x(P_0) = 0$ and $f_y(P_0) = 0$ must both hold if P_0 is to be a local extremum.

Let us confirm these conjectures, supposing that both $f_x(P_0)$ and $f_y(P_0)$ exist. Notice that if $P_0 = (x_0, y_0)$ is a local minimum for f, then the one-variable function $\varphi(x) = f(x, y_0)$ has a local minimum at $x = x_0$. It follows from the one variable theory that $\varphi'(x_0) = 0$. But $\varphi'(x_0) = f_x(x_0, y_0)$, as we have seen in Section 11.4. It follows that $f_x(x_0, y_0) = 0$ if (x_0, y_0) is a local minimum for f. By similar reasoning with $\psi(y) = f(x_0, y)$, we have $f_y(x_0, y_0) = \psi'(y_0) = 0$. An analogous argument shows that at a local maximum (x_0, y_0), it is also the case that both $f_x(x_0, y_0) = 0$ and $f_y(x_0, y_0) = 0$. We summarize our findings in the following theorem.

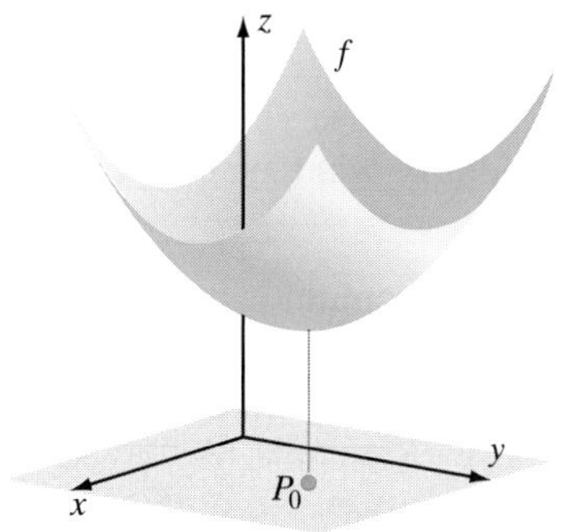

▲ Figure 1a Local minimum at $P_0 = (x_0, y_0)$

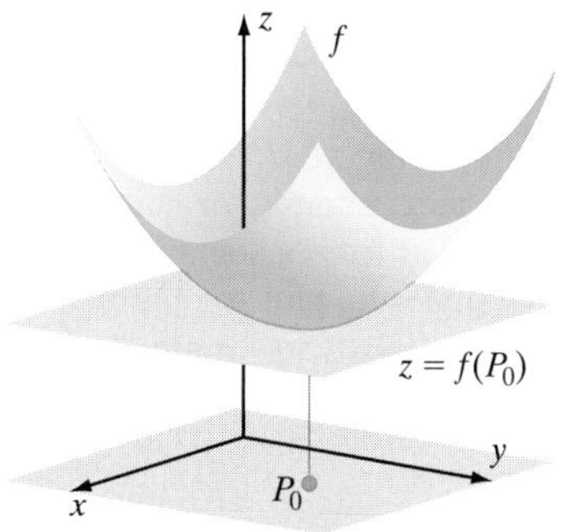

▲ Figure 1b Horizontal tangent plane at $(x_0, y_0, f(P_0))$

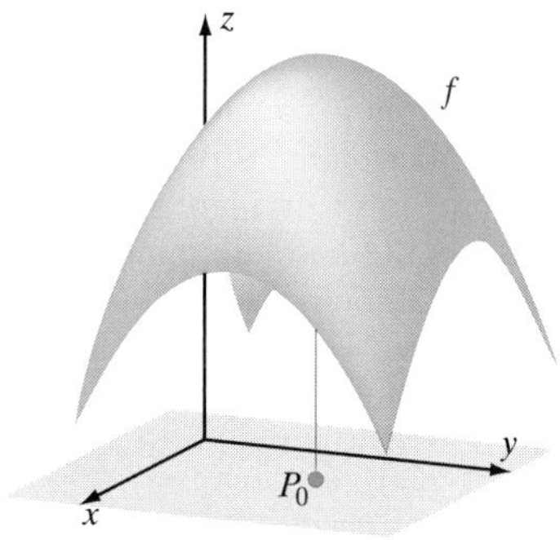

▲ Figure 2a Local maximum at $P_0 = (x_0, y_0)$

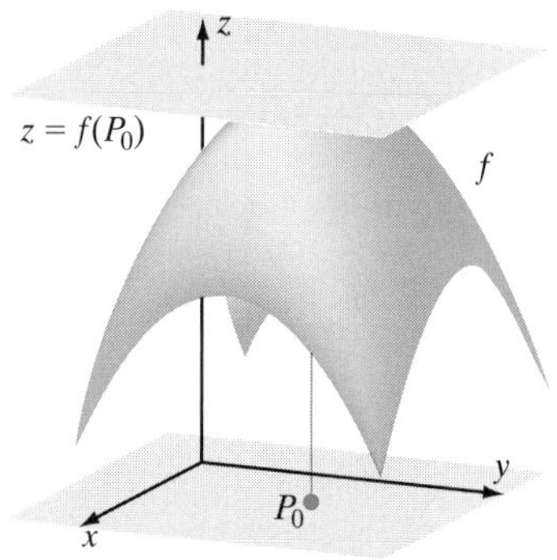

▲ Figure 2b Horizontal tangent plane at $(x_0, y_0, f(p_0))$

THEOREM 1 Suppose that $(x, y) \mapsto f(x, y)$ is a function of two variables defined on an open set $\mathcal{G}$, that P_0 is a point in $\mathcal{G}$, and that both partial derivatives $f_x(P_0)$ and $f_y(P_0)$ exist.

a. If P_0 is a local minimum for f, then $f_x(P_0) = 0$ and $f_y(P_0) = 0$.
b. If P_0 is a local maximum for f, then $f_x(P_0) = 0$ and $f_y(P_0) = 0$.

In short, at a local extremum P_0, we have $\nabla f(P_0) = \mathbf{0}$.

DEFINITION Suppose that $(x, y) \mapsto f(x, y)$ is a function of two variables defined on an open set $\mathcal{G}$ and that P_0 is a point in $\mathcal{G}$. We say that P_0 is a *critical point* for f if either

a. both partial derivatives $f_x(P_0)$ and $f_y(P_0)$ exist and $\nabla f(P_0) = \mathbf{0}$, or
b. f fails to have either a partial derivative with respect to x at P_0 or a partial derivative with respect to y at P_0.

From Theorem 1, we see that the critical points of f are the only points at which f might have a local extremum.

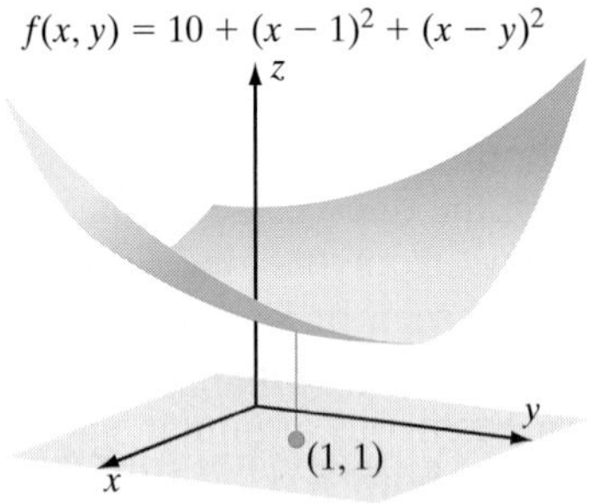

▲ Figure 3a

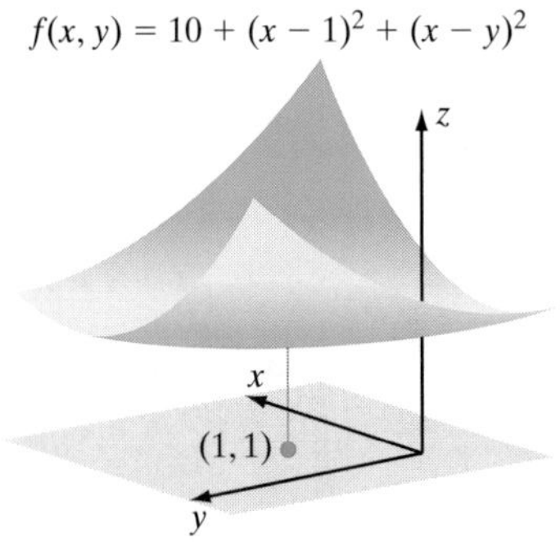

▲ Figure 3b

▶ **EXAMPLE 1** Let $f(x, y) = 10 + (x-1)^2 + (x-y)^2$. Use Theorem 1 to locate all points that might be local extrema for f. Identify what type of critical points these are.

Solution Because f is everywhere differentiable, a local extremum can occur only at a point for which $\nabla f(x, y) = \mathbf{0}$. We calculate $\nabla f(x, y) = \langle 2(x-1) + 2(x-y), -2(x-y) \rangle$. We must therefore determine all points (x, y) where *both* $2(x-1) + 2(x-y) = 0$ and $-2(x-y) = 0$. The second of these equations tells us that $y = x$ and, when we substitute this information into the first equation, we obtain $x = 1$. Therefore $(1, 1)$ is the only critical point of f. According to Theorem 1, $(1, 1)$ is the only possible point at which f might have a local extremum. Theorem 1 does *not* tell us that $(1, 1)$ is a local extremum—it merely tells us that $(1, 1)$ is the only point that *could* be a local extremum. However, we may notice that $f(x, y)$ is the sum of 10 and two nonnegative quantities. It follows that $f(x, y) \geq f(1, 1) = 10$ for all values of x and y. Therefore f has an absolute minimum at the point $(1, 1)$. ◀

INSIGHT Theorem 1 is analogous to Fermat's Theorem from Chapter 4. It gives us a way to find that are *candidates* for local maxima and minima. After we isolate the candidates, our next job is to identify them as local maxima, local minima, or neither. Figure 3a illustrates the absolute minimum of Example 1, viewed from a standard perspective. As Figure 3b demonstrates, it may be possible to rotate the plot to obtain a more convincing view. Nevertheless, it is desirable to have an analytic procedure for identifying a local extremum. Later in this section, we will learn the analogue of the Second Derivative Test for Local Extrema.

Saddle Points

As in the one variable situation, a critical point need not be a local extremum. For example, the point $P_0 = (0, 0)$ is a critical point for $f(x, y) = 1 + x^3 + y^2$. However, if r is any positive number, then the disk $D(P_0, r)$ contains the points $P_1 = (-r/2, 0)$ and $P_2 = (r/2, 0)$. The inequalities

$$f(P_1) = 1 - \frac{r^3}{8} < 1 = f(P_0) < 1 + \frac{r^3}{8} = f(P_2)$$

rule out P_0 as either a local minimum or a local maximum (see Figure 4). The most important instance of what a critical point can be, besides a local maximum or a local minimum, is illustrated in the next example.

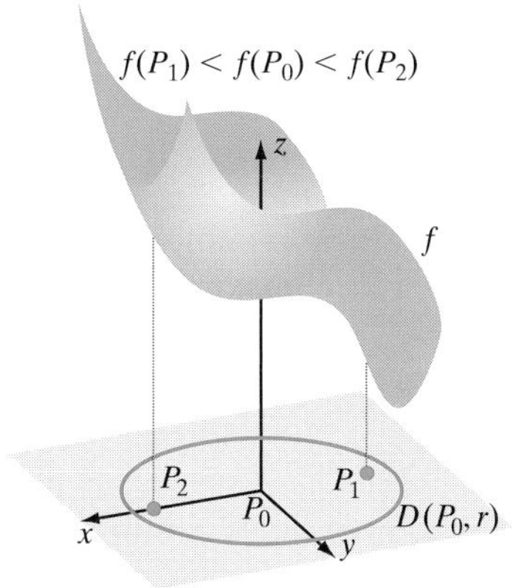

▲ Figure 4 Critical point P_0 that is not a local extremum

▶ **EXAMPLE 2** Locate and analyze the critical points of $f(x, y) = x^2 - y^2$.

Solution Refer to Section 11.2 for details on plotting the graph of f, which appears in Figure 5a. Because $\nabla f(x, y) = \langle 2x, -2y \rangle$, the only critical point is $(0, 0)$. But this point is neither a local maximum nor a local minimum. If we restrict attention to the x-direction by setting $y = 0$, then we have $f(x, 0) = x^2$. Thus the curve that is formed by intersecting the graph of f and the plane $y = 0$ is the parabola that consists of the points $(x, 0, x^2)$. A minimum occurs when $x = 0$ (see Figure 5b). If we restrict attention to the y direction by setting $x = 0$, then we have $f(0, y) = -y^2$. The curve that is formed by intersecting the graph of f and the plane $x = 0$ is a parabola that has maximum value when $y = 0$. It follows that $(0, 0)$ is neither a local maximum nor a local minimum for the function $f(x, y)$ of two variables in the sense introduced in the beginning of this section. Our surface, concave up in one direction and concave down in the perpendicular direction, looks like a saddle. ◀

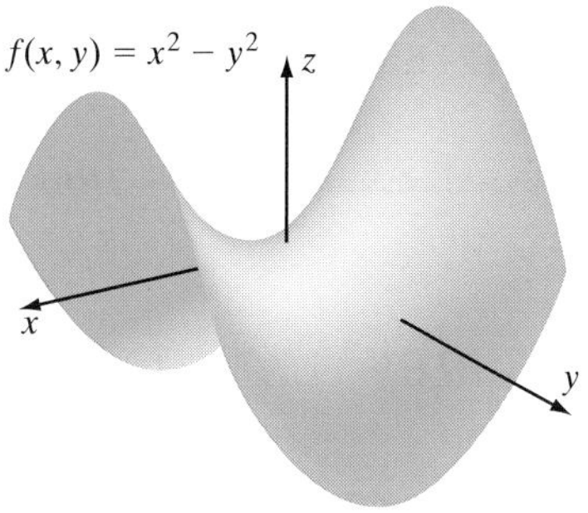

▲ Figure 5a Saddle surface

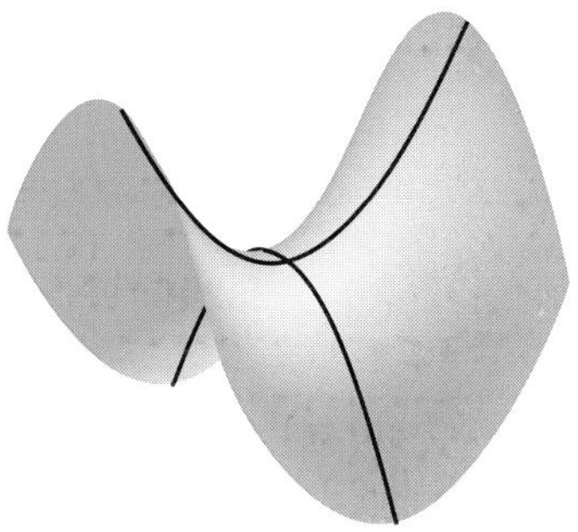

▲ Figure 5b

DEFINITION Suppose that P_0 is a critical point for the function f. We call P_0 a *saddle point* for f if f has neither a maximum value at P_0 nor a minimum value at P_0. In particular, if there are two smooth curves $\mathcal{C}_1$ and $\mathcal{C}_2$ in the domain of f that pass through P_0, if $f(P_0) < f(P)$ for all P on $\mathcal{C}_1$ with $P \neq P_0$, and if $f(P_0) > f(P)$ for all P on $\mathcal{C}_2$ with $P \neq P_0$, then P_0 is a saddle point for f.

The Second Derivative Test

The phenomenon of saddle points makes the analysis of critical points in the plane somewhat more complicated than in the one variable case. The following construction is frequently useful for streamlining the analysis.

DEFINITION Let $(x, y) \mapsto f(x, y)$ be a twice continuously differentiable function. The scalar-valued function defined by

$$\mathcal{D}(f, P_0) = \det\left(\begin{bmatrix} f_{xx}(P_0) & f_{xy}(P_0) \\ f_{yx}(P_0) & f_{yy}(P_0) \end{bmatrix}\right) = \frac{\partial^2 f}{\partial x^2}(P_0)\frac{\partial^2 f}{\partial y^2}(P_0) - \left(\frac{\partial^2 f}{\partial x \partial y}(P_0)\right)^2 \quad (11.8.1)$$

is called the *discriminant* of f.

THEOREM 2 **(Second Derivative Test)** Let $(x,y)\mapsto f(x,y)$ be a twice continuously differentiable function, and suppose that P_0 is a critical point for f. Let $\mathcal{D}(f,P_0)$ be the discriminant of f at P_0.

a. If $\mathcal{D}(f,P_0)>0$, then $f_{xx}(P_0)$ and $f_{yy}(P_0)$ have the same sign.
   i. If $f_{xx}(P_0)$ and $f_{yy}(P_0)$ are both positive, then P_0 is a local minimum for f.
   ii. If $f_{xx}(P_0)$ and $f_{yy}(P_0)$ are both negative, then P_0 is a local maximum for f.

b. If $\mathcal{D}(f,P_0)<0$, then P_0 is a saddle point for f.

c. If $\mathcal{D}(f,P_0)=0$, then we can draw no conclusion from the discriminant.

Proof. For the proof that we give, we will assume that f is three times continuously differentiable, which is a bit more than the stated hypothesis. Let $A=f_{xx}(P_0)$, $B=f_{xy}(P_0)$, $C=f_{yy}(P_0)$, $h=x-x_0$, and $k=y-y_0$. We use Taylor's Theorem to expand f about the base point P_0, noting that the first powers of $h=(x-x_0)$ and $k=(y-y_0)$ have coefficients $f_x(P_0)$ and $f_y(P_0)$, which are both zero. With this simplification, equation (11.5.15) can be written as

$$f(x,y)-f(P_0)=\frac{1}{2}\left(Ah^2+2Bhk+Ck^2\right)+R_2(x,y), \tag{11.8.2}$$

where

$$R_2(x,y)=\frac{1}{3!}\left(f_{xxx}(Q_2)h^3+3f_{xxy}(Q_2)h^2k+3f_{xyy}(Q_2)hk^2+f_{yyy}(Q_2)k^3\right)$$

for some point Q_2 between P_0 and (x,y). Notice that each summand of $\frac{1}{2}(Ah^2+2Bhk+Ck^2)$ is a monomial of degree 2 in h and k, whereas each summand of R_2 is a monomial of degree 3 in h and k. As a result, $R_2(x,y)$ is negligible compared to $\frac{1}{2}(Ah^2+2Bhk+Ck^2)$ when the point (x,y) is restricted to a disk $D(P_0,r)$ of sufficiently small radius. (Imagine, for example, that $h=k=10^{-3}$. Then, $h^2=hk=k^2=10^{-6}$, whereas $h^3=h^2k=hk^2=k^3=10^{-9}$.) From this observation and equation (11.8.2), it follows that the sign of $f(x,y)-f(P_0)$ is the same as the sign of $\frac{1}{2}(Ah^2+2Bhk+Ck^2)$ when (x,y) is in $D(P_0,r)$. This observation is the key to both parts a and b.

The hypothesis of part a is that $AC-B^2>0$. Because $AC>B^2\geq 0$, we see that A and C have the same sign. For now, let us assume that both $A>0$ and $C>0$. Then we can form the expressions $\sqrt{A}$ and $\sqrt{C}$. By expanding the left side of the inequality $\left(\sqrt{A}|h|-\sqrt{C}|k|\right)^2\geq 0$, we obtain $Ah^2-2\sqrt{A}\sqrt{C}|h||k|+Ck^2\geq 0$, or $Ah^2+Ck^2\geq 2\sqrt{A}\sqrt{C}|h||k|$. Using this inequality together with the inequality $\sqrt{AC}-|B|>0$, which follows from the hypothesis $AC-B^2=\mathcal{D}(f,P_0)>0$, we obtain

$$\frac{1}{2}(Ah^2+2Bhk+Ck^2)\geq\sqrt{A}\sqrt{C}|h|\,|k|+Bhk\geq\sqrt{A}\sqrt{C}|h|\,|k|-|B|\,|h|\,|k|=\left(\sqrt{AC}-|B|\right)|h|\,|k|\geq 0.$$

Because $f(x,y)-f(P_0)$ and $\frac{1}{2}(Ah^2+2Bhk+Ck^2)$ have the same sign for (x,y) in $D(P_0,r)$, we conclude that $f(x,y)-f(P_0)\geq 0$ hence $f(x,y)\geq f(P_0)$ for (x,y) in $D(P_0,r)$. In other words, f has a local minimum at P_0. This proves case i of part a.

If f satisfies the hypotheses of case ii of part a, then $g = -f$ satisfies the hypotheses of case i of part a. From what we have already established, this means that g has a local minimum at P_0. That is, $g(x,y) \geq g(P_0)$ for (x,y) in some disk $D(P_0, r)$. For these points (x,y), it follows that $f(x,y) = -g(x,y) \leq -g(P_0) = f(P_0)$. In other words, f has a local maximum at P_0.

Part b is proved by showing that, when $\mathcal{D}(f, P_0) < 0$, there is a direction $\langle h_1, k_1 \rangle$ for which $Ah_1^2 + 2Bh_1k_1 + Ck_1^2 < 0$ and a direction $\langle h_2, k_2 \rangle$ for which $Ah_2^2 + 2Bh_2k_2 + Ck_2^2 > 0$. Only elementary algebra is needed to do this, but the formulas appear unmotivated when they are not developed from a more advanced point of view. Students who study quadratic forms in a linear algebra class will encounter the missing details. ■

$f(x,y) = 2x^2 + 3xy + 4y^2 - 5x + 2y + 3$

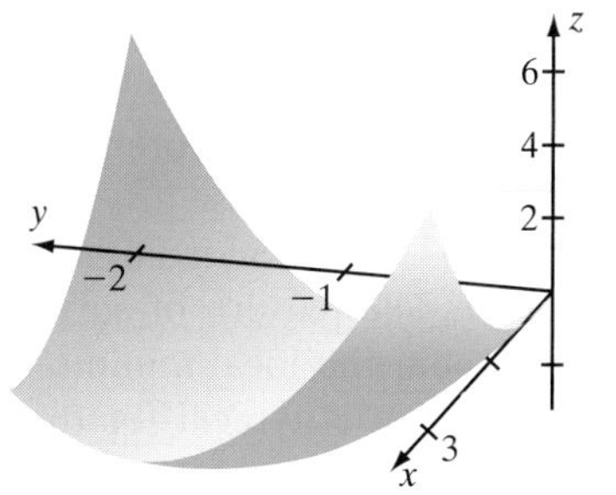

▲ Figure 6a

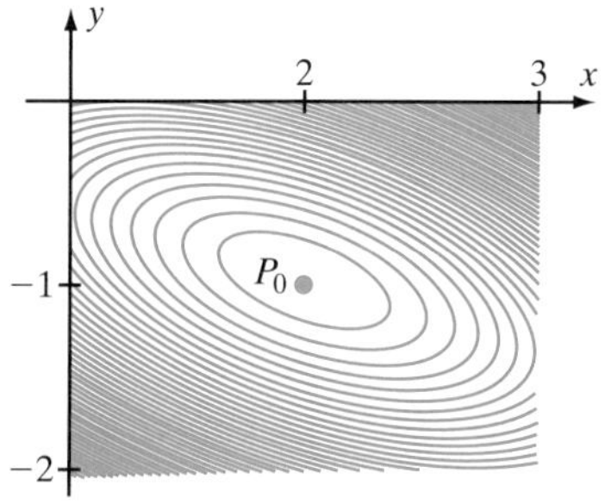

▲ Figure 6b Level curves of Figure 6a. Local minimum at $P_0 = (2, -1)$

▶ **EXAMPLE 3** Locate all local maxima, local minima, and saddle points for the function

$$f(x,y) = 2x^2 + 3xy + 4y^2 - 5x + 2y + 3.$$

Solution We calculate that $f_x(x,y) = 4x + 3y - 5$, $f_y(x,y) = 3x + 8y + 2$, $f_{xx}(x,y) = 4$, $f_{yy}(x,y) = 8$, and $f_{xy}(x,y) = f_{yx}(x,y) = 3$. The critical points are determined by the simultaneous solution of the equations

$$f_x(x,y) = 4x + 3y - 5 = 0 \quad \text{and} \quad f_y(x,y) = 3x + 8y + 2 = 0.$$

Solving simultaneously, we find that $x = 2$ and $y = -1$. That is, $P_0 = (2, -1)$ is the only critical point. Because

$$\mathcal{D}(f, P_0) = \det\left(\begin{bmatrix} f_{xx}(P_0) & f_{xy}(P_0) \\ f_{yx}(P_0) & f_{yy}(P_0) \end{bmatrix}\right) = \det\left(\begin{bmatrix} 4 & 3 \\ 3 & 8 \end{bmatrix}\right) = 23 > 0,$$

we see that we are in Theorem 2a. Because $f_{xx}(2,-1) > 0$ and $f_{yy}(2,-1) > 0$, we know that case i applies. Therefore the point $(2,-1)$ is a local minimum. Figure 6a shows a plot of f. Several level curves near the minimum are plotted in Figure 6b. The pattern of curves enclosing the local extremum is typical. ◀

INSIGHT The key to success with maximum-minimum problems is to be systematic. The Second Derivative Test requires several pieces of information, and the only way to keep track of them all is to follow the recipe given in Example 3.

▶ **EXAMPLE 4** Locate and identify the critical points of the function $f(x,y) = 2x^3 - 2y^3 - 4xy + 5$.

Solution We calculate that

$$f_x = 6x^2 - 4y, \quad f_y = -6y^2 - 4x, \quad f_{xx} = 12x, \quad f_{yy} = -12y, \quad \text{and} \quad f_{xy} = f_{yx} = -4.$$

The critical points are the simultaneous solutions of the equations

$$6x^2 - 4y = 0 \quad \text{and} \quad -6y^2 - 4x = 0.$$

The first equation yields $y = (3/2)x^2$, which we may substitute into the second equation to obtain $-6((3/2)x^2)^2 - 4x = 0$ or, equivalently, $27x^4 + 8x = 0$. We factor this as $x(27x^3 + 8) = 0$. It follows that $x = 0$ and $x = -2/3$ are the only possible

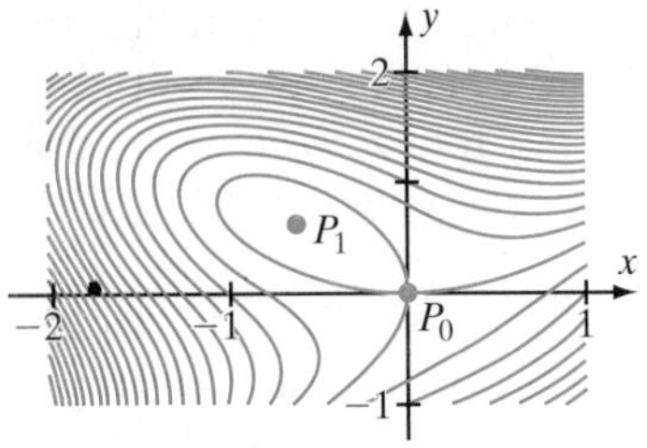

▲ Figure 7 Saddle at $P_0 = (0,0)$, Local maximum at $P_1 = (-2/3, 2/3)$

solutions. We return to the rewritten first equation, $y = (3/2)x^2$, to find that the value of y corresponding to $x = 0$ is 0, and the value of y corresponding to $x = -2/3$ is $2/3$. Thus our critical points are $P_0 = (0,0)$ and $P_1 = (-2/3, 2/3)$. To identify them, we need to examine the discriminant. Now for $P = (x, y)$, we have

$$\mathcal{D}(f,P) = \det\left(\begin{bmatrix} f_{xx}(P) & f_{xy}(P) \\ f_{yx}(P) & f_{yy}(P) \end{bmatrix}\right) = \det\left(\begin{bmatrix} 12x & -4 \\ -4 & -12y \end{bmatrix}\right) = -144xy - 16.$$

We see that $\mathcal{D}(f, P_0) = -16 < 0$ and conclude that $P_0 = (0,0)$ is a saddle point. Next, we consider $P_1 = (-2/3, 2/3)$. Because $\mathcal{D}(f, P_1) = 48 > 0$, Theorem 2a applies. We calculate $f_{xx}(-2/3, 2/3) = -8$ and $f_{yy}(-2/3, 2/3) = -8$. These are both negative and so we find ourselves in case ii. That is, $(-2/3, 2/3)$ is a local maximum. Level curves near the critical points are shown in Figure 7. The pattern of curves that enclose the local maximum is similar to what we saw near the local minimum of Example 3. The "crossing" that appears at the saddle point is typical for that type of critical point. ◀

Applied Maximum-Minimum Problems

Certainly part of the interest of learning to find local maxima and minima is their use in solving practical problems.

▶ **EXAMPLE 5** A rectangular box, with a top, is to hold 20 cubic inches. The material used to make the top and bottom costs 2 cents per square inch, while the material used to make the front and back and the sides costs 3 cents per square inch. What dimensions will yield the most economical box?

Solution To solve this problem, we follow the general scheme for solving maximum-minimum problems that was introduced in Section 4.3 of Chapter 4. We introduce the function that we wish to minimize, namely *cost*. We use variable z to denote the height of the box, x to represent the width of its front and back, and y to denote the depth of the box. Then the following information becomes clear:

a. Front and back each have area xz square inches and cost $3xz$ cents.
b. Sides each have area yz square inches and cost $3yz$ cents.
c. Top and bottom each have area xy square inches and cost $2xy$ cents.

It follows that the total cost is given by

$$\text{cost} = 2(3xz) + 2(3yz) + 2(2xy) = 6xz + 6yz + 4xy.$$

Notice that the factors of two appear because there are two of each item.

If either $x = 0$ or $y = 0$, then the box degenerates and can hold no powder. So we will assume in what follows that $x > 0$ and $y > 0$. Just as in the problems in Section 4.3 of Chapter 4, we now use information in the problem to eliminate one of the variables. Namely, Volume $= xyz = 20$, or $z = 20/(xy)$. Substituting this information into the formula for C gives

$$C(x, y) = \frac{120}{y} + \frac{120}{x} + 4xy, \quad (0 < x, y < \infty)$$

as the function we are required to minimize. Now the methods developed in this section can be applied to analyze $C(x, y)$. We have

$$\frac{\partial C}{\partial x}(x, y) = -\frac{120}{x^2} + 4y, \quad \frac{\partial C}{\partial y}(x, y) = -\frac{120}{y^2} + 4x,$$

and

$$\frac{\partial^2 C}{\partial x^2}(x,y) = \frac{240}{x^3}, \quad \frac{\partial^2 C}{\partial y^2} = \frac{240}{y^3}, \quad \frac{\partial^2 C}{\partial x \partial y}(x,y) = \frac{\partial^2 C}{\partial y \partial x}(x,y) = 4.$$

The critical points are the simultaneous solutions of

$$-\frac{120}{x^2} + 4y = 0 \quad \text{and} \quad -\frac{120}{y^2} + 4x = 0.$$

We solve the first equation for y to obtain $y = 30/x^2$. Substituting this into the second equation gives

$$-x^4 + 30x = 0.$$

As a result, we find that $x = 0$ or $x = 30^{1/3}$. But $x = 0$ does not correspond to a point in the domain of $C(x,y)$. Thus the only critical point is $P_0 = (30^{1/3}, 30^{1/3})$. We analyze this critical point by using the discriminant,

$$\mathcal{D}(f, P_0) = \det\left(\begin{bmatrix} 240/x^3 & 4 \\ 4 & 240/y^3 \end{bmatrix}\right)\Bigg|_{x=y=30^{1/3}} = \frac{57600}{x^3\,y^3} - 16\Bigg|_{x=y=30^{1/3}} = 48.$$

Therefore P_0 is either a local maximum or a local minimum. But, because $C_{xx}(30^{1/3}, 30^{1/3})$ and $C_{yy}(30^{1/3}, 30^{1/3})$ are both positive, the point P_0 is a local minimum. Next, observe that if either x or y tends to 0, then $C(x,y) = 120/y + 120/x + 4xy$ tends to infinity. This is not surprising. If, for example, y tends to 0, then $xz = 20/y$ tends to infinity. So the areas of the front and back of the box become infinite, and the cost of material does as well. Because $C(x,y) \to \infty$ as (x,y) tends to a point on the boundary of its domain, we deduce that the single local extremum in the domain of C must be an absolute minimum. Thus the optimal dimensions are

$$x = 30^{1/3}, \quad y = 30^{1/3}, \quad z = 20 \cdot 30^{-2/3}. \quad ◀$$

We now consider a strategy that often eases the computational burden of locating extrema. Suppose that ϕ is an increasing function of one variable and that f is a differentiable function of two variables. Set $g = \phi \circ f$. Because increasing functions preserve inequalities, we have $g(P_0) = \phi(f(P_0)) \le \phi(f(P)) = g(P)$ if and only if $f(P_0) \le f(P)$. Thus P_0 is a local minimum for g if and only if it is a local minimum for f. The same reasoning is true for local maxima. We can benefit from this observation as follows: By composing f with a suitable increasing function ϕ, we may create a function that is easier to analyze. The next example illustrates this idea.

▶ **EXAMPLE 6** Find the point on the plane $3x + 2y + z = 6$ that is nearest to the origin.

Solution The function to be minimized is the distance to the origin, $f(x,y,z) = \sqrt{x^2 + y^2 + z^2}$. Because the partial derivatives of f are ratios with the term $\sqrt{x^2 + y^2 + z^2}$ in the denominator, we see that it is simpler to minimize the *square* of the distance to the origin: $f(x,y,z)^2 = x^2 + y^2 + z^2$. (Here we have composed f with the increasing function $\phi(u) = u^2$.) We make a substitution to eliminate one of the variables. Because we are only considering points in the plane $3x + 2y + z = 6$, we see that $z = 6 - 2y - 3x$. Substituting this expression for z into $x^2 + y^2 + z^2$ yields the function we want to minimize:

$$g(x, y) = x^2 + y^2 + (6 - 2y - 3x)^2 = 10x^2 + 12xy + 5y^2 - 36x - 24y + 36.$$

Now

$$g_x(x, y) = 20x + 12y - 36, \quad g_y(x, y) = 12x + 10y - 24$$

and

$$g_{xx}(x, y) = 20, \quad g_{yy}(x, y) = 10, \quad g_{xy}(x, y) = g_{yx}(x, y) = 12.$$

The only critical point is the simultaneous solution of the equations

$$20x + 12y - 36 = 0 \quad \text{and} \quad 12x + 10y - 24 = 0,$$

which is $P_0 = (9/7, 6/7)$. We calculate that

$$\mathcal{D}(g, P_0) = \det\left(\begin{bmatrix} 20 & 12 \\ 12 & 10 \end{bmatrix}\right) = (20)(10) - 12^2 = 56 > 0.$$

Thus our critical point is either a relative maximum or minimum. Because $g_{xx}(9/7, 6/7)$ and $g_{yy}(9/7, 6/7)$ are both positive, the point P_0 is a relative minimum. The corresponding point on the plane is

$$x = \frac{9}{7}, \quad y = \frac{6}{7}, \quad z = 6 - 2y - 3x = \frac{3}{7}.$$

Thus $(9/7, 6/7, 3/7)$ is the point in the plane that is nearest to the origin. ◀

Least Squares Lines

The following table lists the weights and pulse rates of ten land mammals. The column labeled x represents the natural logarithms of the weights, and the column labeled y represents the natural logarithms of the pulse rates.

Mammal	**Weight (g)**	**Pulse Rate (1/min)**	x	y
Mouse	25	670	3.2189	6.5073
Rat	200	420	5.2983	6.0403
Guinea Pig	300	300	5.7038	5.7038
Rabbit	2000	205	7.6009	5.3230
Dog (small)	5000	120	8.5172	4.7875
Dog (large)	30000	85	10.309	4.4427
Sheep	50000	70	10.820	4.2485
Human	70000	72	11.156	4.2767
Horse	450000	38	13.017	3.6376
Elephant	3000000	48	14.914	3.8712

Source: A. J. Clark, Comparative Physiology of the Heart, Macmillan, 1972.

Figure 8 shows a "scatter plot" of the ten data points (x, y). The plotted points certainly suggest a straight line relationship, but what line should we use? The ten data points give rise to forty-five pairs of points, each of which determines a straight line. Whichever line we select, it will not pass through the other eight points. To choose between these possible lines, and other candidate lines, which may not pass through any of the points, we need a measure of how well a line ℓ "fits" the data. To each data point not on ℓ, there is an error. If we quantify this error and sum over all data points (having made sure that all errors are taken to be positive so that there is no cancellation when we sum), then we obtain a measure of how well our line ℓ fits the data. The line for which the total error is minimized is called a *best-fitting line.*

The first question, then, is how do we measure the amount by which a line $y = mx + b$ misses a data point (x_j, y_j)? One possibility is to use the absolute value $|y_j - (mx_j + b)|$ of the vertical distance that separates the point from the line. The absolute value is necessary so that errors on different sides of the line do not cancel. However, summing the absolute errors results in a function that is not differentiable. That prevents us from using calculus to identify the minimum. Instead, we use the *sum of the squares of the errors*:

$$SSE(m, b) = \sum_{j=1}^{N} \left(y_j - (mx_j + b)\right)^2. \qquad (11.8.3)$$

Although this function seems complicated, it is simply a quadratic in the variables m and b. The best-fitting line that we obtain when we minimize SSE is called a *least-squares line*. The term *regression line* is also used (see Figure 9).

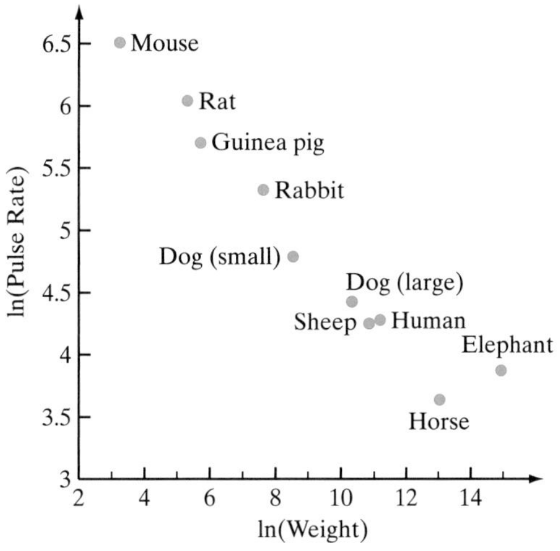

▲ Figure 8

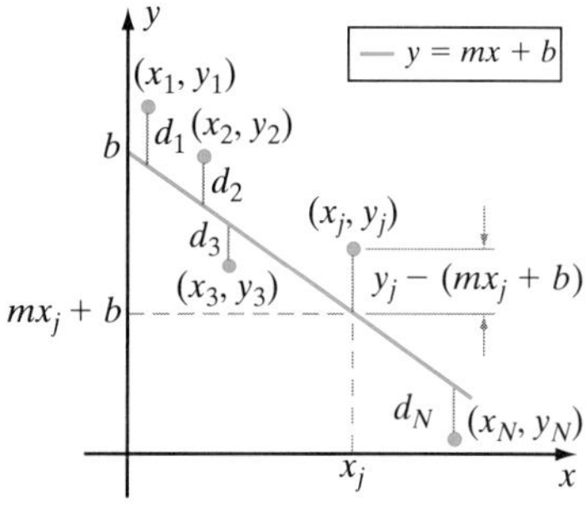

▲ Figure 9
The least squares line minimizes $d_1^2 + d_2^2 + \cdots + d_N^2$.

THEOREM 3 Suppose that N is an integer greater than or equal to 2. Given N observations $(x_1, y_1), (x_2, y_2), \ldots, (x_N, y_N)$, the least squares line is $y = mx + b$ where

$$m = \frac{N\sum_{j=1}^{N} x_j y_j - \left(\sum_{j=1}^{N} x_j\right)\left(\sum_{j=1}^{N} y_j\right)}{N\sum_{j=1}^{N} x_j^2 - \left(\sum_{j=1}^{N} x_j\right)^2} \tag{11.8.4}$$

and

$$b = \frac{1}{N}\left(\sum_{j=1}^{N} y_j - m\sum_{j=1}^{N} x_j\right). \tag{11.8.5}$$

INSIGHT It is common to use vector notation to simplify the appearance of these formulas. Thus $\mathbf{x}\cdot\mathbf{y}$ is used to denote the (dot product) sum $\sum_{j=1}^{N} x_j y_j$, $\mathbf{x}\cdot\mathbf{x}$ is used for the sum $\sum_{j=1}^{N} x_j^2$, and $\overline{x}$ and $\overline{y}$ are used for the averages $\left(\sum_{j=1}^{N} x_j\right)/N$ and $\left(\sum_{j=1}^{N} y_j\right)/N$. Then, the formulas become

$$m = \frac{\mathbf{x}\cdot\mathbf{y} - N\,\overline{x}\overline{y}}{\mathbf{x}\cdot\mathbf{x} - N\,\overline{x}\overline{x}} \quad \text{and} \quad b = \overline{y} - m\overline{x}. \tag{11.8.6}$$

Proof. Because

$$\frac{\partial}{\partial m}\left(y_j - (mx_j + b)\right)^2 = -2x_j(y_j - mx_j - b) \quad \text{and} \quad \frac{\partial}{\partial b}\left(y_j - (mx_j + b)\right)^2 = -2(y_j - mx_j - b),$$

we have

$$\frac{\partial}{\partial m} SSE(m,b) = -2\sum_{j=1}^{N} x_j(y_j - mx_j - b) = -2\sum_{j=1}^{N} x_j y_j + 2m\sum_{j=1}^{N} x_j^2 + 2b\sum_{j=1}^{N} x_j$$

and

$$\frac{\partial}{\partial b} SSE(m,b) = -2\sum_{j=1}^{N} (y_j - mx_j - b) = -2\sum_{j=1}^{N} y_j + 2m\sum_{j=1}^{N} x_j + 2Nb.$$

To find the critical point, we set $SSE_m(m,b) = 0$ and $SSE_b(m,b) = 0$. Using the vector notation discussed previously, we obtain

$$-2\mathbf{x}\cdot\mathbf{y} + 2m\mathbf{x}\cdot\mathbf{x} + 2Nb\overline{x} = 0 \tag{11.8.7}$$

and

$$-2N\,\overline{y} + 2mN\,\overline{x} + 2Nb = 0. \tag{11.8.8}$$

Equation (11.8.8) simplifies to $b = \overline{y} - m\overline{x}$, which is one of the required equations. If we replace b with $\overline{y} - m\overline{x}$ in equation (11.8.7) and solve for m, then we obtain the other required equation after a little algebraic simplification. An application of the Second Derivative Test, outlined in Exercise 37, shows that the critical point we have found is a local minimum. From the nature of the function SSE, we can reason that the critical point corresponds to an absolute minimum. ■

▶ **EXAMPLE 7** Use the ten observations in the table given previously to formulate an explicit relationship between the body weight and the pulse rate of land mammals.

Solution We calculate

$$\mathbf{x}\cdot\mathbf{y} = \ln(25)\ln(670) + \ln(200)\ln(420) + \ln(300)\ln(300) + \cdots + \ln(3000000)\ln(48) = 411.283,$$

$$\mathbf{x}\cdot\mathbf{x} = \ln(25)^2 + \ln(200)^2 + \ln(300)^2 + \cdots + \ln(3000000)^2 = 940.96,$$

$$\overline{x} = \frac{1}{10}\Big(\ln(25) + \ln(200) + \ln(300) + \cdots + \ln(3000000)\Big) = 9.056,$$

and

$$\overline{y} = \frac{1}{10}\Big(\ln(670) + \ln(420) + \ln(300) + \cdots + \ln(48)\Big) = 4.88.$$

The equation of the least squares line is $y = mx + b$ where

$$m = \frac{\mathbf{x}\cdot\mathbf{y} - 10\overline{x}\,\overline{y}}{\mathbf{x}\cdot\mathbf{x} - \overline{x}\,\overline{x}} = \frac{411.283 - 10(9.056)(4.88)}{940.96 - 10(9.056)(9.056)} = -0.254$$

and

$$b = \overline{y} - m\overline{x} = 4.88 - (-0.254)(9.056) = 7.18.$$

In Figure 10, we have superimposed this line on the scatter plot of the data. Plots such as this are often called log-log plots because we plotted the log of the dependent variable p, the pulse rate (measured in beats per minute), as a function of the log of the independent variable w, the body weight (measured in grams). In terms of these variables, the equation that we have found is $\ln(p) = -0.254\ln(w) + 7.18$ or

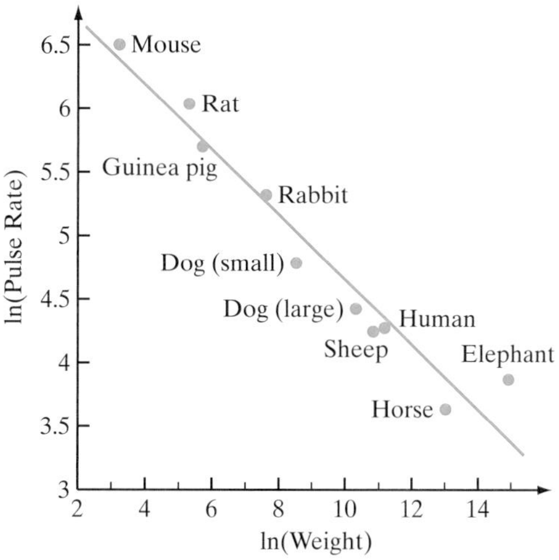

▲ Figure 10

$$p = \frac{1312.91}{w^{0.254}}.$$

Roughly speaking, pulse rate is inversely proportional to the fourth root of weight.

The first "mouse-to-elephant curve," such as the line in Figure 10, was discovered by the biologist Max Kleiber in the 1930s. In 1947, Kleiber found that metabolic rate is proportional to the $(3/4)^{\text{th}}$ power of weight. It has been a puzzling problem in physiology to convincingly explain the powers that appear in these laws. The mathematical theory of fractals is the basis of an explanation that has recently been advanced. ◀

◀ **A LOOK BACK** In Section 1.3 of Chapter 1, we used elementary algebra to find the least squares line that passes through a specified point. That task is a one-variable problem in which only the slope m must be determined. In practice, however, it rarely is the case that there is a preferred observation through which to pass the best-fitting line. Now that we are able to minimize a function of two variables, we can determine the least squares line without making any assumption about the line.

QUICK QUIZ

Suppose that $f(x, y)$ is twice continuously differentiable on an open disk centered at P_0. What conclusion about the behavior of $f(x, y)$ at P_0 can be drawn from the given information?

1. $f_{xx}(P_0) = 4$, $f_{yy}(P_0) = 9$, $f_{xy}(P_0) = 1$
2. $f_x(P_0) = 0$, $f_y(P_0) = 0$, $f_{xx}(P_0) = 4$, $f_{yy}(P_0) = 9$, $f_{xy}(P_0) = 6$
3. $f_x(P_0) = 0$, $f_y(P_0) = 0$, $f_{xx}(P_0) = 4$, $f_{yy}(P_0) = 9$, $f_{xy}(P_0) = 5$
4. $f_x(P_0) = 0$, $f_y(P_0) = 0$, $f_{xx}(P_0) = 4$, $f_{yy}(P_0) = 9$, $f_{xy}(P_0) = 7$

Answers

1. No local maximum 2. No conclusion 3. Local minimum 4. Saddle point

EXERCISES

Problems for Practice

▶ In each of Exercises 1–20, find all the critical points of the given function f. For each critical point P, determine whether P is a local minimum, a local maximum, or a saddle point of f. ◀

1. $f(x, y) = -2x^2 - 2y^2 + 4x + 8y + 7$
2. $f(x, y) = x^2 - 4xy + y^2 + 6x + 8y + 6$
3. $f(x, y) = x^3 - y^3 - 6xy + 8$
4. $f(x, y) = x\cos(y)$
5. $f(x, y) = 3x^2 + 2y^3 - 6xy$
6. $f(x, y) = 3xy + 3/x + 2/y$
7. $f(x, y) = (3 + x + y)xy$
8. $f(x, y) = x^2 + y^2 + 4x - 8y + 7$
9. $f(x, y) = 2x^2 + 4y^2 - 12x - 24y + 6$
10. $f(x, y) = x^2 + 2y^2 - 4xy + 6x + 12y + 9$
11. $f(x, y) = (x - 2)(y + 1)x$
12. $f(x, y) = x^2 - 3y^2 + 6xy + 3x + 5y + 4$
13. $f(x, y) = xy^2 + yx^2 + 8xy + 6$
14. $f(x, y) = 3y^2 + 6x^2 - 2xy + 4x + 3y + 7$
15. $f(x, y) = y^3 + 3x^2y + 3x^2 - 15y + 6$
16. $f(x, y) = -2y^3 + 3x^2 + 6y^2 + 6xy$
17. $f(x, y) = x^4 + y^4 - 8xy$
18. $f(x, y) = \sin(x) + \cos(y)$
19. $f(x, y) = \ln(1 + x^2 + y^2)$
20. $f(x, y) = y^2 - 4xy + x^2 + 5y - 2x + 7$

21. Find three positive numbers the sum of which is 100 and whose product is as large as possible.
22. Which point on the plane $x + 2y + 2z = 6$ is the nearest to the origin?
23. A box in the shape of a rectangular parallelepiped is made of three different types of material. The material for the top costs 2 cents per square inch, that for the sides costs 1/2 cent per square inch, and that for the bottom costs 1 cent per square inch. If the box is to hold 64 cubic inches, then what dimensions will result in the cheapest box?
24. Find the point(s) on the surface $z = x^2 + 8y^2$ that are nearest to the point $(0, 0, 1)$
25. A triangle is to enclose area 100. What dimensions for the triangle result in the least perimeter?
26. Which point on the sphere

$$(x-4)^2 + (y-2)^2 + (z-3)^2 = 1$$

is nearest to the origin?
27. The following table records several paired values of automobile mileage (x) measured in thousands of miles and hydrocarbon emissions per mile (y) measured in grams:

x	5.013	10.124	15.060	24.899	44.862
y	0.270	0.277	0.282	0.310	0.345

Plot these points. Determine the regression line for this data. If this pattern continues for higher mileage cars, about how many grams of hydrocarbons per mile would a car with 100000 miles emit?
28. The following table displays total annual American income (x) and consumption (y) in billions of dollars for the years 1985–1991.

	Income (x)	Consumption (y)
1985	3325.3	2629.0
1986	3526.2	2797.4
1987	3776.6	3009.4
1988	4070.8	3296.1
1989	4384.3	3523.1
1990	4679.8	3748.4
1991	4834.4	3887.7

Plot these points. Determine the regression line for this data.
29. In the accompanying table, x represents cigarette consumption per adult (in 100s) for the year 1962 for eight countries. The variable y represents mortality per 100,000 due to heart disease in 1962.

	x	y
Australia	32.2	238.1
Belgium	17.0	118.1
Canada	33.5	211.6
West Germany	18.9	150.3
Ireland	27.7	187.3
Netherlands	18.14	124.7
Great Britain	27.9	194.1
United States	39	256.9

Plot the eight points (x, y).
Find the least squares line $y = mx + b$ that could be used to model the relationship between cigarette consumption and mortality due to heart disease.

Further Theory and Practice

30. A 20-inch piece of wire is to be cut into three pieces. From one piece is made a square and from another is made a rectangle with length equal to twice its width. From the third is made an equilateral triangle. How should the wire be cut so that the sum of the three areas is a maximum?
31. A radioactive isotope of gold is used to diagnose arthritis. Let $y(t)$ denote the blood serum gold at time t (measured in days) as a fraction of the initial dose at time $t = 0$. Measured values of $y(t)$ for $t = 1, 2, \ldots 6$ are as follows:

t	Serum gold
0	1.0
1	0.91
2	0.77
3	0.66
4	0.56
5	0.49
6	0.43
7	0.38
8	0.34

Plot the points $(t, \ln(y))$. Find a least squares line of the form $\ln(y) = -mt$. Use the model to predict the fraction of the initial dose that remains in the blood serum after 10 days.
32. If $P_1 = (x_1, y_1, z_1)$, $P_2 = (x_2, y_2, z_2)$, $\ldots$, $P_N = (x_N, y_N, z_N)$ are points in space, then their "fit" to the plane $z = Ax + By + D$ is measured by

$$S = \sum_{j=1}^{N} \left(z_j - (Ax_j + By_j + D)\right)^2.$$

The plane of best fit is that plane that minimizes the expression S. Find, in terms of $P_1, \ldots, P_N$, the plane of best fit.

33. Consider a quadratic function of the form

$$f(x,y) = Ax^2 + Cy^2 + Dx + Ey + F.$$

Explain how, by completing squares, one can immediately identify any local maxima or minima.

34. Apply the method of Exercise 33 to analyze the following quadratic functions:

a. $f(x,y) = x^2 + 3y^2 - 6x + 8y - 7$
b. $f(x,y) = -3x^2 + 6x - 8y - 5y^2 + 6$
c. $f(x,y) = 4x^2 - 7y^2 + 4x - 7y + 3$
d. $f(x,y) = 5x^2 + 8y^2 - 10x + 24y + 2$

35. If $F(x,y,z) = 0$ is a surface in space, P is a point that does not lie on the surface, and Q is the point on the surface that is nearest to P, then prove that the line through P and Q is orthogonal to the surface at the point Q.

36. Let $f(x,y) = (x^2 - y)(3x^2 - y) = 3x^4 - 4x^2y + y^2$.

a. Show that $(0,0)$ is a critical point. Show that the Second Derivative Test is inconclusive, however.

b. Plot $y = x^2$ and $y = 3x^2$ in the xy-plane. These two curves divide the plane into three regions. The function f takes positive values on two of the regions, negative values on the other. Label these regions $+$ and $-$ accordingly.

c. By referring to your plot in part b, show that the restriction of f to any straight line through $(0,0)$ has a minimum there.

d. Surprisingly, in view of part c, $(0,0)$, is a saddle point for f. Show this by using your plot to find a parabolic curve $\mathcal{C}$ such that f restricted to $\mathcal{C}$ has a maximum value at $(0,0)$.

37. Let $SSE(m,b)$ be defined by equation (13.17).

a. Show that for any point P,

$$\mathcal{D}(SSE, P) = 4\left((N-1)\sum_{j=1}^{N} x_j^2 - 2 \sum_{1 \le i < j \le N} x_i x_j \right).$$

b. Show that

$$(N-1)\sum_{j=1}^{N} x_j^2 = \sum_{i=1}^{N} \sum_{j=i+1}^{N} (x_i^2 + x_j^2).$$

c. Deduce that

$$\mathcal{D}(SSE, P) = 4 \sum_{1 \le i < j \le N} (x_i - x_j)^2 > 0.$$

Then apply the Second Derivative Test to show that formulas (13.18) and (13.19) define a point at which SSE has a local minimum.

Calculator/Computer Exercises

38. Plot $x^4 + y^2 - 10xy + y$ for $3 \le x \le 4$ and $16 \le y \le 18$. Identify the coordinates of the local minimum that you see in the plot. Use the Second Derivative Test to verify this local minimum.

39. Plot $x^4 - y^5 + x^2y + x$ for $-1 \le x \le 0$ and $0 \le y \le 1$. Identify the coordinates of the saddle point that you see in the plot. Use the Second Derivative Test to verify that this critical point is not an extremum.

40. Plot $y^3 - xy - x^5$ for $0 \le x \le 1$, $-1 \le y \le 0$. Identify the coordinates of the local maximum that you see in the plot. Use the Second Derivative Test to verify this local maximum.

41. Plot $\ln(1 + x^4) - \sin(y)$ for $-1 \le x \le 1$ and $1 \le y \le 2$. Identify the coordinates of a critical point that you see in the plot. Verify that the Second Derivative Test is inconclusive at this critical point. What does your plot tell you about the behavior?

11.9 Lagrange Multipliers

Many times, when we extremize a function $f(x,y)$, we are not interested in the entire domain of the function f. In these applications, we constrain the point (x,y) to lie on a curve $\mathcal{C}$ in the domain of f. To allow for the greatest generality, we assume that the constraint curve $\mathcal{C}$ has the form $g(x,y) = c$ for some constant c. For example, every point on a mountain has a height $f(x,y)$ above its ground coordinates (x,y). But when we walk across the mountain and speak of the highest point of our walk, we are only interested in the points along the path that we follow. This situation is illustrated in Figure 1a with a path whose ground coordinates satisfy the equation $y = c$ for some constant c. Here we take $g(x,y) = y$ so that the constraint curve has the form $g(x, y) = c$. The problem is to maximize the expression $f(x, y)$ subject to the constraint that (x, y) must satisfy the equation $g(x, y) = c$.

In this section, we study an elegant mathematical procedure for determining the extrema of a function subject to a constraint. We begin by finding what might be called critical points for $f(x, y)$ subject to the constraint $g(x, y) = c$. These points are analogous to the critical points that we studied in earlier parts of this book. That is, we locate the critical points by finding the points where derivatives involving f and g are 0. These are the potential extrema. We then examine the critical points to determine the extremum that we seek.

Lagrange Multipliers—A Geometric Approach

The example of a mountain path will provide you with a good geometric understanding of the technique. Look again at Figure 1a and imagine yourself traversing the path from point A to point B. By following this path, you are constraining your x and y coordinates to the curve $g(x, y) = c$ in the xy-plane. As you start out from point A toward point B, you begin to ascend the mountain. Your projection into the xy-plane moves along the curve $g(x, y) = c$ from point A_0 toward point B_0, *crossing* the level curves of f (as seen in Figure 1b). That crossing behavior corresponds to passing from one level to another on the mountain. At the moment you pass through the high point P along your path, you break the pattern of passing to a higher level. That suggests that the curve $g(x, y) = c$ will touch but not cross the level curve of f at the corresponding point P_0 in the xy-plane. Indeed, Figure 1b shows that the curve $g(x, y) = c$ is tangent at P_0 to the level curve of f through P_0. The same holds true even when the constraint curve is not a straight line (Figure 2).

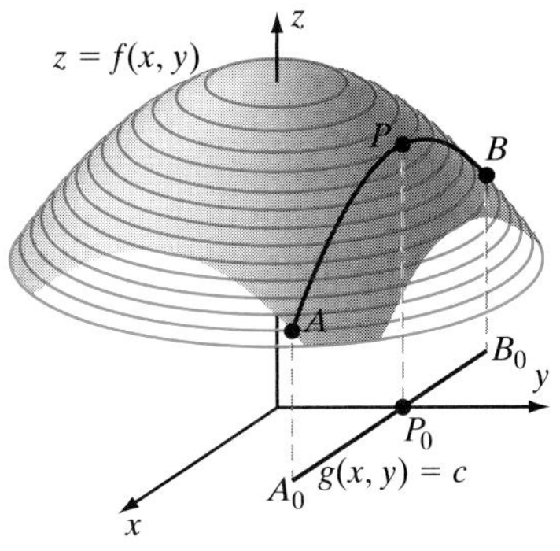

▲ Figure 1a

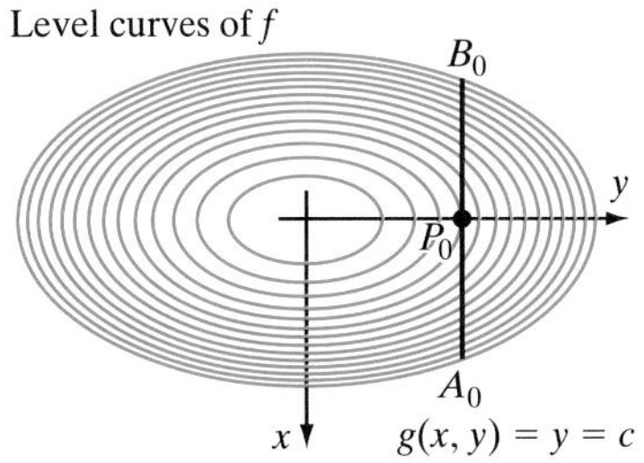

▲ Figure 1b

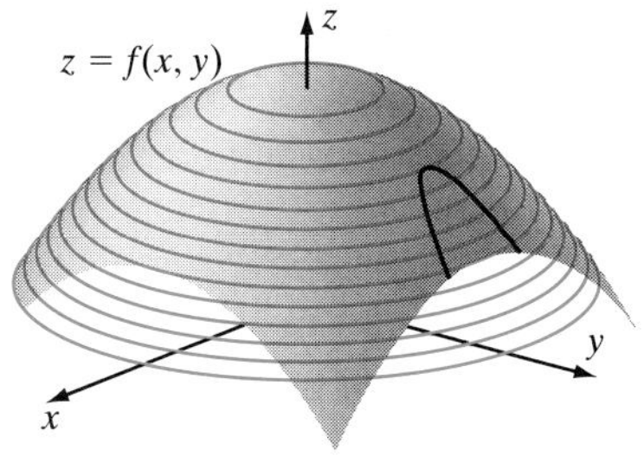

▲ Figure 2a

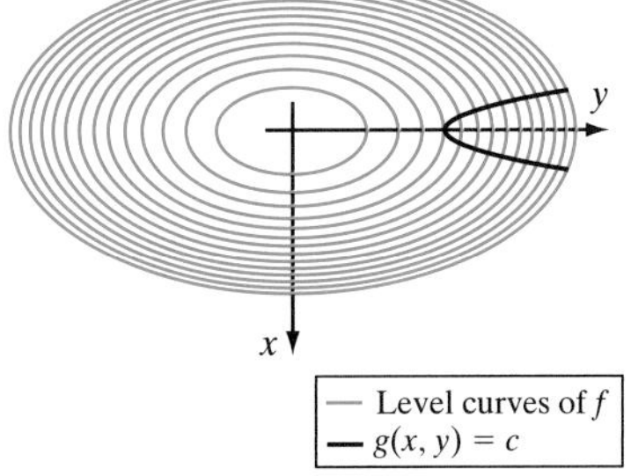

▲ Figure 2b

What we need, then, is an algebraic condition that tells us where $g(x, y) = c$ is tangent to the level curve of f through (x, y). Now two curves are tangent at a point of intersection P_0 if and only if their normal vectors at the point are parallel. But we already know that $\nabla g(P_0)$ is normal to the curve $g(x, y) = c$ at P_0 and that $\nabla f(P_0)$ is normal to the level curve of f at P_0. So we want to find all points P_0 at which the vectors $\nabla g(P_0)$ and $\nabla f(P_0)$ are parallel. According to Theorem 2 of Section 9.1 in Chapter 9, the two vectors $\nabla g(P_0)$ and $\nabla f(P_0)$ are parallel if and only if either

$$\nabla f(P_0) = \lambda \nabla g(P_0)$$

for some scalar λ or

$$\nabla g(P_0) = 0.$$

The critical points of $f(x, y)$ subject to the constraint $g(x, y) = c$ are the places on the constraint curve at which one of these two equations holds. The factor λ is known as *a Lagrange multiplier*. The procedure for finding critical points by solving the equation $\nabla f(x, y) = \lambda \nabla g(x, y)$ is called the *method of Lagrange multipliers*.

INSIGHT It is important to understand that a critical point of f in the sense of Section 11.8 is not, in general, a critical point of $f(x, y)$ subject to the constraint $g(x, y) = c$. The surface $z = f(x, y)$ that is plotted Figures 1 and 2 has a horizontal tangent plane only at the point $(0, 0)$. That point is the only critical point of $f(x, y)$, just as P_0 is the only critical point of $f(x, y)$ subject to the constraint $g(x, y) = c$. There is no connection between the set of critical points of f and the set of critical points of f subject to the constraint $g(x, y) = c$.

Let us see how the method of Lagrange multipliers is applied in practice.

▶ **EXAMPLE 1** Find the point on the hyperbola $x^2 - y^2 = 4$ that is nearest to the point $(0, 2)$.

Solution Our first job is to identify the function that we are maximizing or minimizing and to identify the constraint function. The constraint function is easy. We are only interested in points on the curve $x^2 - y^2 = 4$. So, let $g(x, y) = x^2 - y^2$. Then, our constraint is $g(x, y) = 4$. We calculate $\nabla g(x, y) = \langle 2x, -2y\rangle$. This vector is 0 only at the point $(0, 0)$, which is not on the constraint curve. It is therefore not a critical point.

We turn to the Lagrange multiplier equation $\nabla f = \lambda \nabla g$ for a function f that we will now specify. Rather than minimize the distance to $(0, 2)$, let us employ the strategy discussed in Section 11.8 and minimize the square of the distance to $(0, 2)$, namely $f(x, y) = x^2 + (y - 2)^2$. The reason we want to do this is because the expression $x^2 + (y - 2)^2$ is simpler than $\sqrt{x^2 + (y - 2)^2}$. We *can* do this because the expressions $x^2 + (y - 2)^2$ and $\sqrt{x^2 + (y - 2)^2}$ are both minimized at the same point. According to the method of Lagrange multipliers, we find the critical points for this problem by solving the equation $\nabla f(x, y) = \lambda \nabla g(x, y)$, or $\langle 2x, 2y - 4\rangle) = \lambda\langle 2x, -2y\rangle)$. Equating components, we convert this vector equation into the two scalar equations,

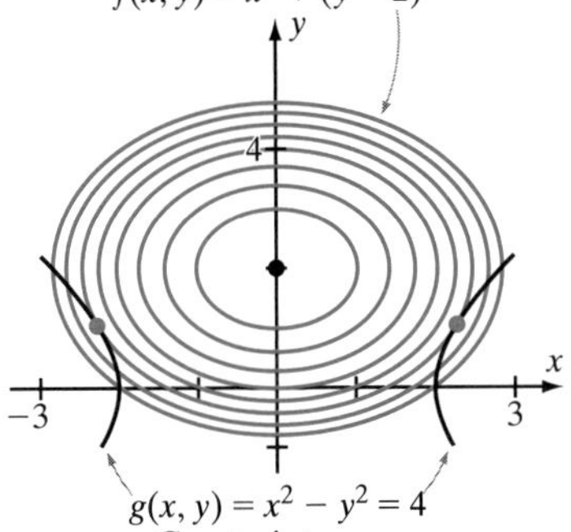

▲ Figure 3 Critical points at $(\sqrt{5}, 1)$ and $(-\sqrt{5}, 1)$

$$2x = \lambda 2x \quad \text{and} \quad 2y - 4 = \lambda(-2y),$$

which we need to solve for x, y and possibly λ. The first equation can be written as $2x(\lambda - 1) = 0$, from which we see that either $x = 0$ or $\lambda = 1$. But $x = 0$ is not possible for points on the constraint curve $x^2 - y^2 = 4$. Therefore $\lambda = 1$. Substituting this into the second equation gives $2y - 4 = (1)(-2y)$, or $4y - 4 = 0$, or $y = 1$. The points on the curve $x^2 - y^2 = 4$ with $y = 1$ are $(\sqrt{5}, 1)$ and $(-\sqrt{5}, 1)$. These are the critical points for our problem. We know from common sense that the problem of finding the nearest point to $(0, 2)$ has a solution. Which of the critical points is it? Symmetry considerations (see Figure 3) show that our two critical points are equidistant from $(0, 2)$. So both $(\sqrt{5}, 1)$ and $(-\sqrt{5}, 1)$ are solutions to our constrained extremal problem. ◀

INSIGHT Notice that, in the process of applying the method of Lagrange multipliers in Example 1, we have solved for λ. However, the actual value of λ is not a part of our final solution of the problem. Although the parameter λ is part of the technique, and its value is often crucial to the solution process, the value of λ is usually of no interest. In some applications in economics and other constrained extremum problems, the actual value of λ does have significance. But, in general, λ is merely a proportionality factor between two parallel vectors and does not have any intrinsic importance.

Why the Method of Lagrange Multipliers Works

Let us say a few more words about why the Lagrange multiplier technique works. Look at Figure 4, which illustrates the level curve $g(x, y) = c$. Our job is to find the maxima and minima of f among those points that lie on this curve. Say that we are standing at a point P_1 on the curve, and looking for a maximum. If we are not already at a maximum, then we should move to a different point in our search. In which direction should we go? Ideally, we should go in the direction of $\nabla f(P_1)$ because that is the direction of most rapid increase of f. The trouble is that $\nabla f(P_1)$ probably points out of the curve (as happens in Figure 4). The next best thing is that we can move in the direction of the projection of $\nabla f(P_1)$ onto the tangent to the curve (as shown in Figure 4). Thus we move a little bit in that direction to a point P_2, and repeat our program: If we are not already at a maximum, then we move in the direction of the projection of $\nabla f(P_2)$ onto the tangent to the curve at P_2. When we find the maximum P' we seek, this procedure breaks down: There is no direction to move that will increase the value of f. Thus $\nabla f(P')$ must have zero projection into the tangent to the curve. In other words, at a maximum P', the vector $\nabla f(P')$ is normal to the curve. But $\nabla g(P')$ is normal to the curve $g(x, y) = c$ at P', so $\nabla f(P')$ and $\nabla g(P')$ are parallel. Thus either $\nabla g(P') = 0$ or $\nabla f(P') = \lambda \nabla g(P')$ for some constant λ. The same reasoning works at a minimum. This explains why we declare places at which $\nabla f(P') = \lambda \nabla g(P')$ to be critical points.

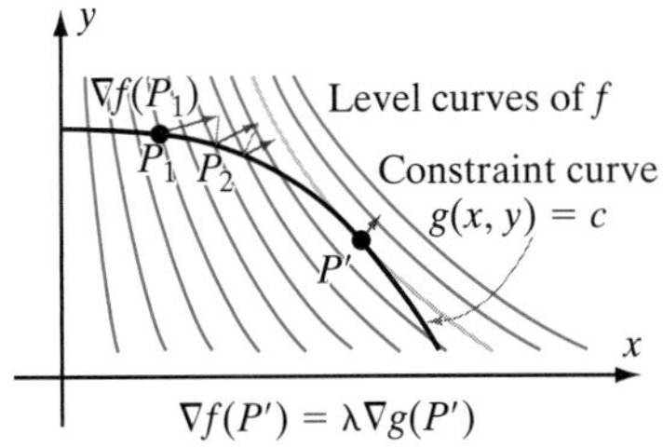

▲ Figure 4

We state the results of our investigation as a theorem. A rigorous proof is outlined in Exercise 45.

THEOREM 1 **(Method of Lagrange Multipliers).** Suppose that $(x, y) \mapsto f(x, y)$ and $(x, y) \mapsto g(x, y)$ are differentiable functions. Let c be a constant. If f has an extreme value at a point P' on the constraint curve $g(x, y) = c$, then either $\nabla g(P') = 0$ or there is a constant λ such that $\nabla f(P') = \lambda \nabla g(P')$.

INSIGHT As noted, the points P' for which $\nabla g(P') = \mathbf{0}$ must be considered alongside the solutions of the Lagrange multiplier equation $\nabla f(P') = \lambda \nabla g(P')$ as potential extrema. Also, the method of Lagrange multipliers fails at points P' where either $\nabla f(P')$ or $\nabla g(P')$ is undefined. As with the theory of extrema of one variable, these situations must be handled with *ad hoc* techniques. We will not treat them here.

▶ **EXAMPLE 2** Find the maximum and minimum values of the function $f(x, y) = 2x^2 - y^2$ on the ellipse $x^2 + 2(y-1)^2 = 2$.

Solution Let $g(x, y) = x^2 + 2(y-1)^2$. Our job is to maximize the function f subject to the constraint $g(x, y) = 2$. Notice that $\nabla g(x, y) = \langle 2x, 4(y-1)\rangle$, which is $\mathbf{0}$ only at the point $(0, 1)$. Because this point is not on the constraint curve $g(x, y) = 2$, it is not a critical point for our problem. The only critical points will be the solutions of the equation $\nabla f(x, y) = \lambda \nabla g(x, y)$ that lie on the constraint curve. Therefore we want to solve $\langle 4x, -2y\rangle = \lambda\langle 2x, 4(y-1)\rangle$ or

$$4x = \lambda 2x \qquad (11.9.1)$$

and

$$-2y = \lambda 4(y-1). \qquad (11.9.2)$$

Equation (11.9.1) is equivalent to $2x(2-\lambda) = 0$, which holds when either $\lambda = 2$ or $x = 0$.

a. If $x = 0$, then the corresponding points on the curve $g(x, y) = 2$ are $(0, 0)$ and $(0, 2)$. We still must determine whether equation (11.9.2) is satisfied at either of these two points. In fact, by taking $\lambda = 0$, equation (11.9.2) holds for the point $(0, 0)$. Similarly, we find that the point $(0, 2)$ satisfies equation (11.9.2) for $\lambda = -1$. Therefore $(0, 0)$ and $(0, 2)$ are both critical points.
b. If $\lambda = 2$, then equation (11.9.2) becomes $-2y = 8(y-1)$, or $-10y = -8$, or $y = 4/5$. The corresponding points on the curve $g(x, y) = 2$ are $(4\sqrt{3}/5, 4/5)$ and $(-4\sqrt{3}/5, 4/5)$.

We have found four critical points:

$$(0, 0), \quad (0, 2), \quad (4\sqrt{3}/5, 4/5), \quad \text{and} \quad (-4\sqrt{3}/5, 4/5).$$

The maximum we seek will occur at one (or more) of these four points. The values of $f(x, y)$ at these four points are $0, -4, 80/25$, and $80/25$, respectively. We conclude that the maximum value of f, when restricted to the curve $g(x, y) = 2$, is $80/25$. This maximum value is attained at the points $(4\sqrt{3}/5, 4/5)$, and $(-4\sqrt{3}/5, 4/5)$. The minimum value of f when restricted to the curve $g(x, y) = 2$ is -4. This minimum value is attained at the point $(0, 2)$. Figure 5 shows the location of the three extrema on the constraint curve. The critical point $(0, 0)$ yields neither a maximum nor a minimum. There is no contradiction in this: As in the extremization topics that were treated earlier, a critical point is a *candidate* for an extremum, but it may turn out not to actually be one. ◀

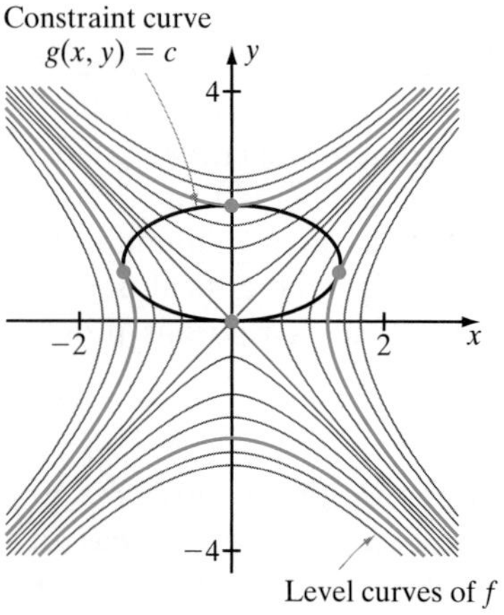

▲ Figure 5

INSIGHT The Lagrange multiplier equation $\nabla f(x, y) = \lambda \nabla g(x, y)$ is a vector equation that is equivalent to two scalar equations—equations (11.9.1) and (11.9.2) in Example 2. These *two* equations involve *three* unknowns: x, y, and λ. When solving for three variables, we expect to work with three equations. It is the constraint equation $g(x, y) = c$ that provides the third equation.

A Strategy for Solving Lagrange Multiplier Equations

The algebra in Lagrange multiplier problems is often tricky. If we are not careful, then we may easily overlook or lose solutions. The following strategy can be helpful in looking for solutions of $\nabla f(x, y) = \lambda \nabla g(x, y)$ where $\nabla g(x, y) \neq \mathbf{0}$. The vector equation $\nabla f(x, y) = \lambda \nabla g(x, y)$ is equivalent to the two scalar equations

$$f_x(x, y) = \lambda g_x(x, y) \tag{11.9.3}$$

and

$$f_y(x, y) = \lambda g_y(x, y). \tag{11.9.4}$$

Because $\nabla g(x, y) \neq \mathbf{0}$, we have $g_x(x, y) \neq 0$ or $g_y(x, y) \neq 0$. Suppose, for example, that $g_x(x, y) \neq 0$. On multiplying each side of (11.9.4) by $g_x(x, y)$, we obtain $f_y(x, y) g_x(x, y) = \lambda g_x(x, y) g_y(x, y)$, or, in view of equation (11.9.3),

$$f_y(x, y) g_x(x, y) = f_x(x, y) g_y(x, y). \tag{11.9.5}$$

Equation (11.9.5) together with the constraint equation, $g(x, y) = c$, makes for two equations in the two unknowns x and y. It is a heuristic principle of algebra that, when the number of unknowns equals the number of equations, the system will have a finite number of solutions.

▶ **EXAMPLE 3** Maximize $x + 2y$ subject to the constraint $x^2 + y^2 = 5$.

Solution Let $f(x, y) = x + 2y$ and $g(x, y) = x^2 + y^2$. Then, $\nabla f(x, y) = \langle 1, 2 \rangle$ and $\nabla g(x, y) = \langle 2x, 2y \rangle$. Because $\nabla g(x, y) = \mathbf{0}$ only at the point $(0, 0)$, which is not on the constraint curve, we may assume that $\nabla g(x, y) \neq \mathbf{0}$. Then we may use equation (11.9.5), which gives us $(2)(2x) = (1)(2y)$, or $y = 2x$. Substituting this into the constraint, we obtain $x^2 + (2x)^2 = 5$, or $5x^2 = 5$, or $x = \pm 1$. There are therefore two critical points: $(1, 2)$ and $(-1, -2)$. The maximum of $f(x, y) = x + 2y$ clearly occurs for positive values of the variables and is $f(1, 2)$, or 5. ◀

A Cautionary Example

Our next example emphasizes that it is important for us to know that an extremum exists before we leap to any conclusion about a candidate that is located by the method of Lagrange multipliers.

▶ **EXAMPLE 4** What are the extreme values of $f(x, y) = x^2 + 2y + 16$, subject to the constraint $(x + y)^2 = 1$?

Solution Here, $g(x, y) = (x + y)^2$ and $\nabla g(x, y) = 2\langle x + y, x + y \rangle$. We observe that $\nabla g(x, y) = \mathbf{0}$ if and only if $x + y = 0$. The constraint $(x + y)^2 = 1$ rules out such points. We proceed to the vector equation $\nabla f = \lambda \nabla g$, which gives us the two scalar equations $2x = 2\lambda(x + y)$ and $2 = 2\lambda(x + y)$. Because the right sides of these two equations are the same, we see that $2x = 2$, or $x = 1$. Substituting this value of x

into the constraint equation $(x+y)^2=1$, we find that $y=0$ or $y=-2$. Thus $(1,0)$ and $(1,-2)$ are the only critical points for $f(x,y)$ subject to the constraint $(x+y)^2=1$. Because $17=f(1,0)>f(1,-2)=13$, it is tempting to think that we have located a maximum at $(1,0)$. That conclusion would be wrong, however. Notice that the constraint condition tells us that either $x+y=1$ or $x+y=-1$. Therefore on the constraint curve, either $y=1-x$ or $y=-1-x$. Because $f(x,1-x)=x^2+2(1-x)+16=(x-1)^2+17$ and $f(x,-1-x)=x^2+2(-1-x)+16=(x-1)^2+13$, we see that, subject to the given constraint, f can assume arbitrarily large values. Therefore f does not have an extreme value at the point $(1,0)$. On the other hand, our formulas for $f(x,1-x)$ and $f(x,-1-x)$ show that $f(x,y)\geq 13=f(1,-2)$ at points on the constraint curve. Therefore subject to the constraint, f has a minimum value of 13, which occurs at $(1,-2)$ (see Figure 6). ◀

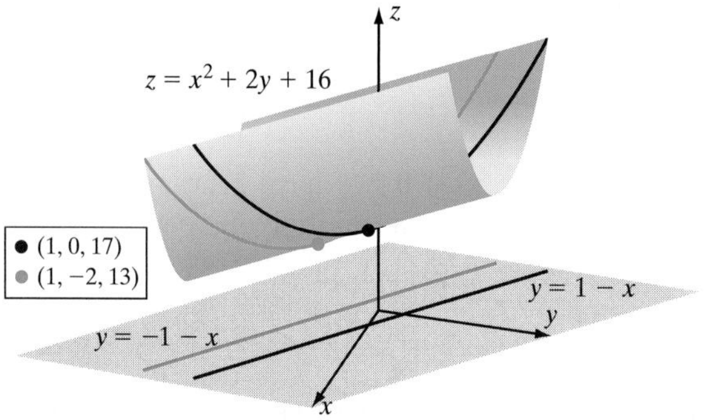

▲ Figure 6

Lagrange Multipliers and Functions of Three Variables

The method of Lagrange multipliers works in much the same way for functions of three variables. If we want to extremize $F(x,y,z)$ subject to the constraint $G(x,y,z)=c$, then we must find all points $P=(x,y,z)$ on the constraint curve for which the vector equation $\nabla F(P)=\lambda\nabla G(P)$ holds. Any point P where $\nabla G(P)=\mathbf{0}$ or where $\nabla F(P)$ does not exist or where $\nabla G(P)$ does not exist is also included among the critical points.

▶ **EXAMPLE 5** The temperature at a point (x,y,z) in space is given by $F(x,y,z)=8x-4y+2z$. Find the maximum and minimum temperatures on the sphere $x^2+y^2+z^2=21$.

Solution Let $G(x,y,z)=x^2+y^2+z^2$. Thus we are finding extrema of f subject to the constraint $G(x,y,z)=21$. First we calculate $\nabla G(x,y,z)=\langle 2x,2y,2z\rangle$). From this, we see that $\nabla G(x,y,z)=\mathbf{0}$ holds only at the origin, a point that may be disregarded because it does not satisfy the constraint equation. We find the critical points by setting $\nabla F(x,y,z)=\lambda\nabla G(x,y,z)$. Writing out the gradients more explicitly gives $\langle 8,-4,2\rangle=\lambda\langle 2x,2y,2z\rangle$ or

$$8=\lambda 2x \quad\text{and}\quad -4=\lambda 2y \quad\text{and}\quad 2=\lambda 2z.$$

The first and third equations yield

$$\frac{4}{x}=\lambda=\frac{1}{z} \quad\text{or}\quad x=4z.$$

The second and third equations yield

$$\frac{-2}{y} = \lambda = \frac{1}{z} \quad \text{or} \quad y = -2z.$$

We may substitute these into the constraint equation $G(x, y, z) = 21$. The result of these substitutions is $(4z)^2 + (-2z)^2 + z^2 = 21$. Therefore $z = \pm 1$. If $z = 1$, then $x = 4$ and $y = -2$, so we have found the critical point $(4, -2, 1)$. Likewise, $z = -1$ yields the critical point $(-4, 2, -1)$. Because $F(4, -2, 1) = 42$ and $F(-4, 2, -1) = -42$, we conclude the following: The minimum value of F on the surface $G(x, y, z) = 21$ is -42, and this value is assumed at the point $(-4, 2, -1)$; the maximum value of F on the surface $G(x, y, z) = 21$ is 42, and this value is assumed at the point $(4, -2, 1)$. ◀

Extremizing a Function Subject to Two Constraints

It is sometimes necessary to find extreme values of an expression $F(x, y, z)$ on the curve that is formed by the intersection of two surfaces $G(x, y, z) = c_1$ and $H(x, y, z) = c_2$. In this case, we use two Lagrange multipliers λ and μ. The Lagrange multiplier equation is then

$$\nabla F(P) = \lambda \nabla G(P) + \mu \nabla H(P). \tag{11.9.6}$$

By extracting components, we obtain from this vector equation *three* scalar equations in the *five* unknowns x, y, z, λ, and μ. The other two equations that we need are the constraint equations $G(x, y, z) = c_1$ and $H(x, y, z) = c_2$. The reason that this procedure works is discussed in Exercise 48. However, the algebra required to solve the resulting system of five equations in five unknowns is usually tedious and often complicated.

▶ **EXAMPLE 6** Let $\mathcal{C}$ be the curve that results when the plane $2x + y + z = 11/5$ intersects the surface $x^2 + y^2 + z = 1$. What point on $\mathcal{C}$ has the largest z value?

Solution Because $z = 1 - (x^2 + y^2) \le 1$ for each point (x, y, z) on $\mathcal{C}$, we deduce that there is a maximum value of z on $\mathcal{C}$. Figure 7, which shows the two surfaces, also makes it clear that our problem has a solution. It appears from the figure that there is also a minimum value of z. We should therefore expect two solutions of the Lagrange multiplier equations. The function to maximize is $F(x, y, z) = z$. The constraints are $G(x, y, z) = 2x + y + z = 11/5$ and $H(x, y, z) = x^2 + y^2 + z = 1$. Equation (11.9.6) becomes $\langle 0, 0, 1 \rangle = \lambda \langle 2, 1, 1 \rangle + \mu \langle 2x, 2y, 1 \rangle$ or

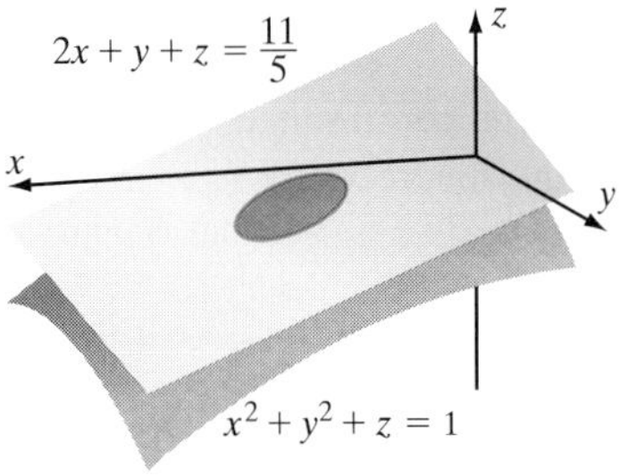

▲ Figure 7

$$0 = 2\lambda + 2\mu x \quad \text{and} \quad 0 = \lambda + 2\mu y \quad \text{and} \quad 1 = \lambda + \mu.$$

The third equation gives us $\lambda = 1 - \mu$. Substituting this into $0 = 2\lambda + 2\mu x$ and simplifying, we obtain $\mu(1 - x) = 1$. This equation tells us that $\mu \ne 0$ and $1 - x = 1/\mu$. Similarly, replacing λ with $1 - \mu$ in $0 = \lambda + 2\mu y$, we obtain $\mu(1 - 2y) = 1$. Therefore $1 - x = 1/\mu = 1 - 2y$ or $x = 2y$. We can now eliminate x from the two constraints: $G(2y, y, z) = 2(2y) + y + z = 11/5$ or $z = 11/5 - 5y$ and $H(2y, y, z) = (2y)^2 + y^2 + z = 1$. Replacing z with $11/5 - 5y$ in this last equation gives us $5y^2 + (11/5 - 5y) = 1$ or $5y^2 - 5y + 6/5 = 0$. Using the Quadratic Formula, we have

$$y = \frac{-(-5) \pm \sqrt{(-5)^2 - 4(5)(6/5)}}{2(5)} = \frac{5 \pm 1}{10} = \frac{2}{5}, \frac{3}{5}.$$

But we have found that $z = 11/5 - 5y$. Therefore the largest value of z on the curve of intersection is $11/5 - 5(2/5) = 1/5$. The least value is $11/5 - 5(3/5) = -4/5$. ◄

QUICK QUIZ

1. True or false: If f and g are differentiable functions of x and y, then an extreme value of $f(x, y)$ subject to the constraint $g(x, y) = c$ must occur at a point P_0 for which $\nabla f(P_0) = \lambda \nabla g(P_0)$.
2. True or false: If two critical points arise in the solution of a constrained extremum problem, then a maximum occurs at one of the points and a minimum occurs at the other.
3. Maximize, if possible, $x^2 + y^2$ subject to the constraint $x + 2y = 5$.
4. Minimize, if possible, $x^2 + y^2$ subject to the constraint $x + 2y = 5$.

Answers

1. False 2. False 3. No maximum 4. 5

EXERCISES

Problems for Practice

► In each of Exercises 1–12, find the extrema of $f(x, y)$ subject to the constraint $g(x, y) = c$. ◄

1. $f(x, y) = 2x - 3y + 6$	$g(x, y) = x^2 + 2y^2 = 4$
2. $f(x, y) = 3x - 4y$	$g(x, y) = 4x^2 + y^2 = 7$
3. $f(x, y) = (x + 1)^2 + y^2$	$g(x, y) = x^2 + 4y^2 = 16$
4. $f(x, y) = x^2 - y^2$	$g(x, y) = x^2 + (y - 2)^2 = 9$
5. $f(x, y) = 3xy^2 - 24$	$g(x, y) = x^2 + y^2 = 16$
6. $f(x, y) = xy$	$g(x, y) = x^2 + 9y^2 = 18$
7. $f(x, y) = x + y^2$	$g(x, y) = x^2 + y^2 = 9$
8. $f(x, y) = y^2 - x^2$	$g(x, y) = y^2 + 2x^2 = 4$
9. $f(x, y) = 4x^2 + 4y^2$	$g(x, y) = x^4 + y^4 = 16$
10. $f(x, y) = \sin^2(x) + \sin^2(y)$	$g(x, y) = x + y = \pi$
11. $f(x, y, z) = xyz$	$g(x, y, z) = x^2 + y^2 + z^2 = 4$
12. $f(x, y, z) = 4x - 7y + 6z$	$g(x, y, z) = x^2 + 7y^2 + 12y^2 = 84$

► In each of Exercises 13–25, use the method of Lagrange multipliers to solve the stated problem. ◄

13. Minimize the function $f(x, y) = x^2 + 2y^2 + 9$ subject to the constraint $2x - 6y = 5$.

14. Maximize the function $f(x, y) = x - y^2$ subject to the constraint $2x + y^2 = 4$.

15. Minimize $x^2 + y^2 + 4y$ subject to the constraint $x^2 + 2y^2 = 8$.

16. Maximize the product xy subject to the condition that $6x^2 + y^2 = 8$.

17. Maximize $x^3 + 2y$ subject to the condition that $x^2 + y^2 = 4/3$.

18. A window is to be constructed in the shape of a rectangle surmounted by an isosceles triangle. Building codes require the total area of the window to be 6 square feet, but the material used in the frame is very costly. What dimensions will minimize the perimeter of the window?

19. Minimize the surface area $2\pi r^2 + 2\pi rh$ of a cylindrical can of height h and radius r, subject to the constraint that the volume $\pi r^2 h$ of the cylinder is equal to a fixed positive constant V_0.

20. Minimize $x^2 - y^2$ subject to the constraint $x^2 + 2y^2 = 8$.

21. Maximize $x^4 + y^4 + z^4$ on the sphere $x^2 + y^2 + z^2 = 12$.

22. Find the point on the plane $x - 3y + 5z = 6$ that is nearest to the origin.

23. Find the point on the ellipsoid $x^2 + 2y^2 + 4z^2 = 4$ that is nearest to $(1, 0, 0)$.

24. The temperature of any point (x, y, z) in space is given by

$$T(x, y, z) = 2x - 6y + 5z.$$

Find the greatest and least temperatures on the surface $x^2 + 6y^2 + 4z^2 = 24$.

25. Maximize the product xy^2z subject to the constraint $x^2 + 3y^2 + 2z^2 = 64$.

Further Theory and Practice

26. By using the method of Lagrange multipliers, find two level curves of $f(x,y) = 5x + 4y$ that are tangent to the hyperbola $x^2 - y^2 = 1$. Does an extreme value occur at either point of tangency?

27. What is the largest possible y-coordinate of a point on the curve $3x^2 + 2xy + 3y^2 = 24$?

28. The temperature on an ellipsoid $2x^2 + y^2 + 2z^2 = 8$ is given by the formula

$$T(x,y,z) = 8z^2 + 4xy - 12y + 200.$$

What are the locations of the hottest and coldest points on the ellipsoid?

29. A certain county consists of the region $\{(x,y) : 4x^2 + 2y^2 \le 16\}$. The altitude at any point of this county is given by $a(x,y) = 80x - 70y + 150$. What are the highest and lowest points in this county?

30. If α, β, γ are the angles of a triangle, then how large can $\sin(\alpha)\sin(\beta)\sin(\gamma)$ be?

31. Some cylindrical cans, such as containers of juice concentrate, are manufactured with metal tops and bottoms but cardboard sides. Suppose that the cost per unit area of the metal is k times that of the cardboard. Minimize the cost of materials of such a can if its volume is to equal a fixed positive constant V_0.

32. A capsule has the shape of a cylinder that is "capped" at each end by a hemisphere. The cylinder has height $h > 0$ and radius $r > 0$. Is it possible to maximize or minimize the surface area $4\pi r^2 + 2\pi rh$ subject to the constraint that the volume of the capsule is equal to a fixed positive constant V_0?

In each of Exercises 33–38, constraint equations $G(x,y,z) = c_1$ and $H(x,y,z) = c_2$ are given. As in Example 6, use two Lagrange multipliers to find the requested extreme value of the given function $F(x,y,z)$.

33. $x + y + z = 6$, $x + 2y - 3z = 12$; minimize $F(x,y,z) = x^2 + y^2 + z^2$

34. $x + y + z = 6$, $x + 3y - z = 12$; minimize $F(x,y,z) = x^2 + y^2$

35. $x + 2y + 3z = 6$, $5(x^2 + y^2 + z^2) = 14$; maximize $F(x,y,z) = z$

36. $2x + y + z = 2$, $x^2 + y^2/4 + z^2 = 1$; maximize $F(x,y,z) = x$

37. $x + 2y + z = 10$, $x^2 + y^2 - z = 0$; maximize $F(x,y,z) = x^2 + y^2 + z^2$

38. $x^2 + 2y^2 - z = 0$, $2x^2 + y^2 + z = 27$; maximize $F(x,y,z) = z$

In economics, a *utility function* $f(x,y)$ quantifies the satisfaction a consumer derives from x units of one item and y units of a second item. The level curves of f are called *indifference curves*. The consumer's *budget line* is the line segment in the first quadrant of the xy plane that represents how many of the two items can be purchased for a total amount T. Exercises 39 and 40 concern these concepts.

39. A Cobb-Douglas utility function has the form $f(x,y) = Cx^p y^q$ where C, p, and q are positive constants. If the first item costs A per unit and the second costs B per unit, and if the consumer has a total amount T that he can spend on the two items, then for what values of x and y is the consumer's utility maximized?

40. Using $A = 15$, $B = 10$, and $T = 120$, plot the consumer's budget line in the viewing window $[0,12]) \times [0,12])$. In this viewing window, add the plots (or sketches) of several indifference curves of the Cobb-Douglas utility function described in the preceding exercise. Use the values $C = 1$, $p = 3/4$, and $q = 1/4$. Include the indifference curve on which the consumer's maximum utility lies. What is the relationship of this indifference curve to the budget line?

41. A company allocates an amount T to produce an item. The available capital can be divided into labor costs and plant costs. Let $f(x,y)$ be the number of units of an item that can be produced with x units of labor and y units of plant expenditures. Suppose that the unit costs of these outlays are a and b, respectively. Show that when production is maximized, $f_x/f_y = a/b$. This equation can be stated as the following economic principle: *When labor and plant costs are allocated optimally, then the ratio of their marginal productivities is equal to the ratio of their unit costs.*

42. Suppose that f and g are differentiable functions of two variables. Suppose also that for every c there is a unique point $(x(c), y(c))$ that maximizes f subject to the constraint $g(x,y) = c$. Assume that $\nabla g(x(c), y(c)) \neq \vec{\mathbf{0}}$ and let $\lambda(c)$ be the Lagrange multiplier defined by $\nabla f(x(c), y(c)) = \lambda(c) \nabla g(x(c), y(c))$. Let $\mathbf{r}(c) = \langle x(c), y(c) \rangle$. Set $M(c) = f(x(c), y(c))$, the maximum value of f subject to the constraint $g(x,y) = c$.

a. Show that $1 = \nabla g\big(x(c), y(c)\big) \cdot \mathbf{r}'(c)$.

b. Show that $\lambda(c) = \nabla f\big(x(c), y(c)\big) \cdot \mathbf{r}'(c)$.

c. Deduce that $\lambda(c) = M'(c)$.

43. Let $f(x,y) = xy^{3/4}$ be a production function (in the sense of Exercise 41). Suppose that the equation $15x + 12y = c$ describes the allocation of labor and plant costs. For an arbitrary positive constant c, find the point $(x(c), y(c))$ on the constraint curve at which f is maximized. Calculate the maximum value $M(c)$ and the Lagrange multiplier $\lambda(c)$. According to Exercise 42, $\lambda(c)$ is approximately equal to the marginal productivity $M(c+1) - M(c)$. Verify the approximation $\lambda(c) \approx M(c+1) - M(c)$ for $c = 10000$.

44. Suppose that A and B are positive constants, that $f(x,y) = Ax + By$, and that $g(x,y) = x^2 + y^2$. Let $(x(c), y(c))$ be the point on the constraint curve $g(x,y) = c$ at which f is maximized. Let $M(c) = f(x(c), y(c))$. Calculate the Lagrange multiplier $\lambda(c)$, and verify that $\lambda(c) = M'(c)$.

45. *The Milkmaid Problem:* A river, in the shape of a smooth curve, flows near a house H and a barn B. Each morning, a milkmaid leaves the house, fills a bucket of water at a point R on the river, and then goes to the barn. The distance of this walk is $|\overline{HR}|) + |\overline{RB}|)$. If the point R_0 minimizes this distance, then show that there is an ellipse with foci B and H and that is tangent to the river at R_0 (see Figure 8).

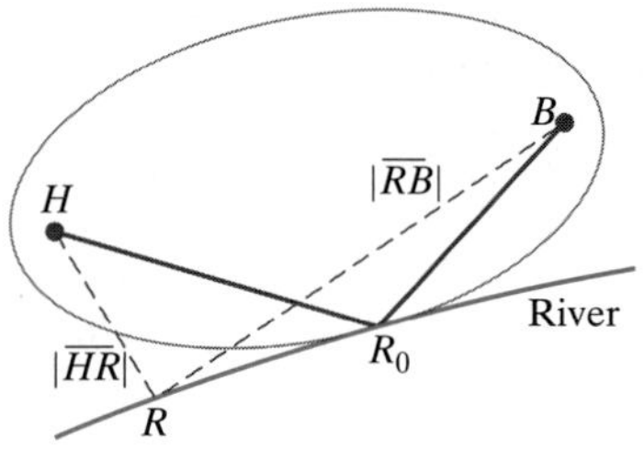

▲ Figure 8

46. Suppose that $a, b, c, \alpha, \beta, \gamma$ are constants that satisfy the inequalities $a^2 + b^2 + c^2 > 0$ and $\alpha\beta\gamma \neq 0$. Let $f(x, y, z) = ax + by + cz$ be a linear function, and let

$$g(x, y, z) = \alpha^2(x - e)^2 + \beta^2(y - f)^2 + \gamma^2(z - g)^2.$$

Prove that the extrema of f on the region

$$\mathcal{R} = \{(x, y, z) : g(x, y, z) \leq 1\}$$

will always occur on the boundary of $\mathcal{R}$. Prove that there will always be exactly one maximum and one minimum, and they will occur at diametrically opposite points of the region $\mathcal{R}$.

47. Complete the following outline to obtain a more rigorous treatment of the method of Lagrange Multipliers. We will assume that ∇f and ∇g exist at the critical points and that ∇g does not vanish at those points. We seek extrema of $f(x, y)$ subject to the constraint $g(x, y) = c$.

a. Let $P_0 = (x_0, y_0)$ satisfy $g(P_0) = c$, and let $\mathbf{r}(t) = x(t)\mathbf{i} + y(t)\mathbf{j}$ be a parametrization of a curve lying in $g(x, y) = c$ that passes through P_0. Say that $\mathbf{r}(t_0) = P_0$.

b. If P_0 is an extremum for f subject to the constraint $g = c$, then the function of one variable $f(\mathbf{r}(t))$ has an extremum at $t = t_0$.

c. We have

$$\frac{d}{dt} f(\mathbf{r}(t))\Big|_{t=t_0} = 0.$$

d. We conclude that

$$\nabla f(P_0) \perp \mathbf{r}'(t_0).$$

e. Because $\mathbf{r}$ was an arbitrary curve in the level set $g = c$, we conclude that $\nabla f(P_0)$ is normal to the level set $g = c$ at P_0.

f. Deduce that $\nabla f(P_0) = \lambda \nabla g(P_0)$ for some $\lambda \in \mathbb{R}$.

48. Suppose that surfaces $G(x, y, z) = c_1$ and $H(x, y, z) = c_2$ have the following two properties: they intersect in a smooth curve $\mathcal{C}$ containing a point $P_0 = (x_0, y_0, z_0)$, and they do not have the same tangent plane at P_0. Let $\mathbf{r}(t) = x(t)\mathbf{i} + y(t)\mathbf{j} + z(t)\mathbf{k}$ be a parametrization of $\mathcal{C}$ with $\mathbf{r}(t_0) = P_0$. Suppose further that P_0 is an extremum for $F(x, y, z)$ subject to the constraint $G(x, y, z) = c_1$ and $H(x, y, z) = c_2$. Show that $t \mapsto F(\mathbf{r}(t))$ has an extremum at $t = t_0$. Deduce that

$$\frac{d}{dt} F(\mathbf{r}(t))\Big|_{t=t_0} = 0$$

and therefore $\nabla F(P_0) \perp \mathbf{r}'(t_0)$. Conclude that $\nabla F(P_0)$ is in the plane that is normal to $\mathcal{C}$ at P_0. Show that every vector in this normal plane has the form $\lambda \nabla G(P_0) + \mu \nabla H(P_0)$ for some scalars λ and μ.

49. Maximize $x_1 + x_2 + \cdots + x_N$ subject to the constraint $x_1^2 + x_2^2 + \cdots + x_N^2 = 1$. Deduce the inequality

$$\frac{x_1 + x_2 + \cdots + x_N}{N} \leq \sqrt{\frac{x_1^2 + x_2^2 + \cdots + x_N^2}{N}}.$$

50. Suppose $x_1, x_2, \ldots, x_N$ are positive numbers. Minimize $x_1 + x_2 + \cdots + x_N$ subject to the constraint $x_1 x_2 \cdots x_N = 1$. Deduce the *Arithmetic-Geometric Mean Inequality:*

$$(x_1 x_2 \cdots x_N)^{1/N} \leq \frac{x_1 + x_2 + \cdots + x_N}{N}.$$

Calculator/Computer Exercises

51. Find the largest value of $x\exp(x^2 - xy)$ subject to the constraint $x^2 + y^2 = 1$.

52. Find the point on the curve $x\exp(x^2 - xy) = 1$ that is closest to the origin.

53. Find the minimum value of $(1 + x^2 + xy)/(1 + x^2 + y^4)$ subject to the constraint $x^2 + y^2/4 = 1$.

54. Minimize $\sqrt{x^2 + y^2} + \sqrt{(x - 2)^2 + y^2}$ subject to the constraint $xy = 2$.

55. Find the extreme values of $x + y + z$ subject to the constraints $z^3 + z = e^x$ and $x^2 + y^2 = 1$.

Summary of Key Topics in Chapter 11

Functions of Several Variables
(Section 11.1)

A function of two variables has as its domain a set of ordered pairs of numbers; a function of three variables has as its domain a set of ordered triples. In this chapter, these functions will be scalar-valued. Functions of several variables may be added, subtracted, and multiplied in the usual way. They may be divided as long as we do not divide by zero. If f is a scalar-valued function of several variables, ϕ is a scalar-valued function of one variable, and the values of f lie in the domain of ϕ, then we may form the composition $\phi \circ f$ to obtain a new scalar-valued function of several variables.

Functions of two variables are graphed in three-dimensional space. The graph of f is a surface consisting of the points $\{(x, y, f(x, y)) : (x, y) \in \text{domain of } f\}$. In general, a surface inspace is the graph of a function if and only if no vertical line intersects the surface more than once. We form the graph by sketching level curves

$$\{(x, y) : f(x, y) = c\}$$

in the plane and then amalgamating them into the final graph. Level curves arise in applications of mathematics as isotherms, isobars, isohyets, and so on.

Cylinders
(Section 11.2)

A cylinder is the set of points in space satisfying an equation that is missing one variable. Such surfaces have uniform sections, or level curves, and are easy to sketch.

Quadric Surfaces
(Section 11.2)

The quadric surfaces are classified as follows:

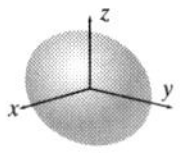

$$\frac{x^2}{\alpha^2} + \frac{y^2}{\beta^2} + \frac{z^2}{\gamma^2} = 1$$ ellipsoid

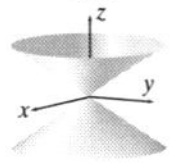

$$\frac{x^2}{\alpha^2} + \frac{y^2}{\beta^2} = \frac{z^2}{\gamma^2}$$ elliptic cone

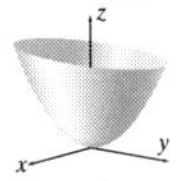

$$\frac{x^2}{\alpha^2} + \frac{y^2}{\beta^2} = \frac{z}{\gamma}$$ elliptic paraboloid

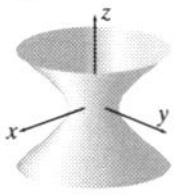

$$\frac{x^2}{\alpha^2} + \frac{y^2}{\beta^2} - \frac{z^2}{\gamma^2} = 1$$ hyperboloid of one sheet

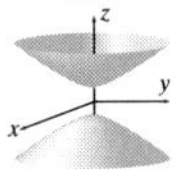

$$\frac{z^2}{\gamma^2} - \frac{x^2}{\alpha^2} - \frac{y^2}{\beta^2} = 1$$ hyperboloid of two sheets

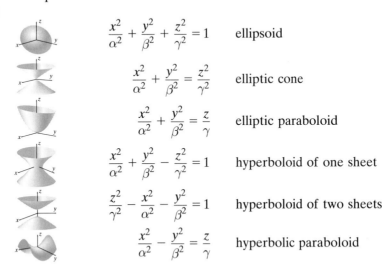

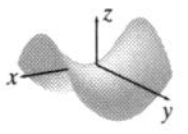

$$\frac{x^2}{\alpha^2} - \frac{y^2}{\beta^2} = \frac{z}{\gamma}$$ hyperbolic paraboloid

Limits and Continuity
(Section 11.3)

If $P = (p_1, p_2)$ is a point in the plane, let $D(P_0, r)$ be the disk of radius $r > 0$ centered at P_0. Let $D_*(P_0, r)$ be the same disk but with the center removed. Let $d(P_0, P)$ denote the distance between the points P and P_0. If the domain of $f(x, y)$ contains $D_*(P_0, r)$, then we say that

$$\lim_{(x,y)\to P_0} f(x,y) = \ell$$

if, for any $\varepsilon > 0$, there is a $\delta > 0$ such that

$$0 < d(P_0, P) < \delta \quad \text{implies} \quad |f(x,y) - \ell| < \varepsilon.$$

We say that f is continuous at $P_0 = (x_0, y_0)$ if the domain of f contains P_0 and

$$\lim_{(x,y)\to P_0} f(x,y) = f(x_0, y_0).$$

The standard limit rules apply to these new notions of limit; the standard results about continuity also hold.

Partial Derivatives
(Section 11.4)

If the domain of f contains a disc $D(P_0, r)$, then we define the partial derivatives of f at $P = (x_0, y_0)$ to be

$$\frac{\partial f}{\partial x}(P_0) = \lim_{\Delta x \to 0} \frac{f(x_0 + \Delta x, y_0) - f(x_0, y_0)}{\Delta x}$$

and

$$\frac{\partial f}{\partial y}(P_0) = \lim_{\Delta y \to 0} \frac{f(x_0, y_0 + \Delta y) - f(x_0, y_0)}{\Delta y},$$

provided these limits exist. The partial differentiation process can be iterated to obtain higher partial derivatives. We will usually work with functions with continuous derivatives, called continuously differentiable functions. The familiar differentiation rules also hold for partial differentiation.

Other notations for the partial derivatives of f are f_x, f_1, f_y, f_2, f_{xy}, f_{12}, $D_x f$, $D_2 f$, and so on.

All of these ideas can be adapted to functions of three variables.

The Chain Rule
(Section 11.5)

If $z = f(x,y)$ and in turn $x = \rho(s)$ and $y = \sigma(s)$, all functions continuously differentiable, then

$$\frac{dz}{ds} = \frac{\partial z}{\partial x}\frac{dx}{ds} + \frac{\partial z}{\partial y}\frac{dy}{ds}.$$

A similar Chain Rule applies when $x = \rho(z,t)$ and $y = \sigma(s,t)$:

$$\frac{\partial z}{\partial s} = \frac{\partial z}{\partial x}\frac{\partial x}{\partial s} + \frac{\partial z}{\partial y}\frac{\partial y}{\partial s}$$

and

$$\frac{\partial z}{\partial t} = \frac{\partial z}{\partial x}\frac{\partial x}{\partial t} + \frac{\partial z}{\partial y}\frac{\partial y}{\partial t}.$$

There is an analogous Chain Rule when f is a function of three or more variables.

Gradients and Directional Derivatives (Section 11.6)

If $f(x, y)$ is a continuously differentiable function, then the gradient of f is

$$\operatorname{grad} f(x, y) = \nabla f(x, y) = f_x(x, y)\mathbf{i} + f_y(x, y)\mathbf{j}.$$

If $\mathbf{u}$ is a unit vector, then the directional derivative of f in the direction $\mathbf{u}$ at the point (x, y) is

$$D_{\mathbf{u}} f(x, y) = \nabla f(x, y) \cdot \mathbf{u}.$$

If P_0 is fixed, then the greatest directional derivative of f at P_0 occurs when $\mathbf{u}$ is the direction of $\nabla f(P_0)$. In this case, $D_{\mathbf{u}} f(P_0) = \|\nabla f(P_0)\|$. Also the least directional derivative of f at P_0 occurs when $\mathbf{u}$ is the direction of $-\nabla f(P_0)$. In this case, $D_{\mathbf{u}} f(P_0) = -\|\nabla f(P_0)\|$.

All of these ideas can be adapted to functions of three variables.

Normal Vectors and Tangent Planes (Section 11.7)

If $f(x, y)$ is a continuously differentiable function, and $P_0 = (x_0, y_0)$ is a point in its domain, then

$$f_x(P_0)\mathbf{i} + f_y(P_0)\mathbf{j} - \mathbf{k}$$

is a normal vector to the graph at $(P_0, f(P_0))$. The tangent plane to the graph at $(P_0, f(P_0))$ will be the plane passing through $(P_0, f(P_0))$ and normal to this vector. It will have equation

$$f_x(P_0)(x - x_0) + f_y(P_0)(y - y_0) - \big(z - f(x_0, y_0)\big) = 0.$$

For any continuously differentiable function f of several variables, the gradient of f is perpendicular to the level surfaces of f.

Numerical Approximation (Section 11.7)

Because the tangent plane geometrically approximates a graph near the point of contact, we may approximate $f(x, y)$ for (x, y) near $P_0 = (x_0, y_0)$ by the function

$$L(x, y) = f(P_0) + f_x(P_0)(x - x_0) + f_y(P_0)(y - y_0).$$

The error, or rate of approximation, can be estimated using the second derivatives of f. This approximation is useful in numerical calculations and in error analysis.

Critical Points (Section 11.8)

Local maxima and minima are defined as for functions of one variable. A point at which a function has a local minimum when approached from one direction, and a local maximum when approached from another direction, is called a saddle point. If $f(x, y)$ is continuously differentiable, then a critical point is a point where the gradient vanishes. Let $P_0 = (x_0, y_0)$ be a critical point. If f is twice continuously differentiable, then define

$$\mathcal{D}(f, P_0) = \det\left(\begin{bmatrix} f_{xx}(P_0) & f_{xy}(P_0) \\ f_{yx}(P_0) & f_{yy}(P_0) \end{bmatrix}\right) = f_{xx}(P_0) \cdot f_{yy}(P_0) - \big(f_{xy}(P_0)\big)^2.$$

a. If $\mathcal{D}(f, P_0) > 0$, $f_{xx}(P_0) > 0$, and $f_{yy}(P_0) > 0$, then P_0 is a local minimum for f.

b. If $\mathcal{D}(f, P_0) > 0$, $f_{yy}(P_0) < 0$, and $f_{yy}(P_0) < 0$, then P_0 is a local maximum for f.

c. If $\mathcal{D}(f, P_0) < 0$, then P_0 is a saddle point for f.

d. If $\mathcal{D}(f, P_0) = 0$, then we can draw no conclusion.

The ability to find and identify critical points is an aid in graphing and enables us to solve applied problems about extrema.

Lagrange Multipliers (Section 11.9)

To find the critical points for the problem of extremizing $f(x, y)$ subject to the constraint $g(x, y) = c$, we solve the equation

$$\nabla f(x, y) = \lambda \nabla g(x, y).$$

This procedure will locate those critical points for the problem at which both ∇f and ∇g exist and $\nabla g \neq \mathbf{0}$.

Review Exercises for Chapter 11

In Exercises 1–3, express the quantity described as a function of two or more variables. Give an explicit description of the domain of this function.

1. The surface area A of a cylinder in terms of its height h and its radius r.

2. The time T required for a round trip of a total distance $2L$ if the average speed of the first leg of the trip is s and the average speed for the return is v.

3. The perimeter P of a right triangle in terms of the lengths ℓ_1 and ℓ_2 of its two legs.

In Exercises 4–6, let $f(x, y) = xy/(x^2 + y^2)$, $g(x, y) = \sin(x \cdot y)$, $h(x, y) = x^2 y^3$. Calculate the specified quantity.

4. $(f + 3g)(4, 6)$

5. $(f \cdot g)(1, 3)$

6. $(f/(g + h + 10))(0, 2)$

In Exercises 7–9, let $f(x, y) = (x - y)/(x^2 + 1)$, $g(x, y) = x^2 - y^3$, and $\varphi(t) = t^{3/2}$. Calculate the specified quantity.

7. $(\varphi \circ g)(4, 2)$

8. $((\varphi \circ f) \cdot (\varphi \circ g))(-4, 2)$

9. $(2\varphi \circ f)(0, -3)$

In Exercises 10–12, sketch (on a set of planar axes) the level sets of the function f that correspond to values -8, -4, 0, 4, 8.

10. $f(x, y) = y^2 + 6x^2$

11. $f(x, y) = x^2 - 4y^2$

12. $f(x, y) = x - 5y$

In each of Exercises 13–15, a function g of two variables is given. Sketch four level sets of g, and then sketch the graph in space of $z = g(x, y)$.

13. $g(x, y) = x^2 + 2y^2 + 1$

14. $g(x, y) = 6 - y$

15. $g(x, y) = 2 - (x^2 + y^2)$

In each of Exercises 16–18, sketch the level set of the function $g(x, y, z)$ of three variables that corresponds to level 5.

16. $g(x, y, z) = x - y$

17. $g(x, y, z) = x^2 + 2y^2 + 4z^2 + 4$

18. $g(x, y, z) = 6 - x^2 - z^2$

In each of Exercises 19–21, sketch the given cylinder on axes in three-dimensional space.

19. $x^2 - y = 4$

20. $y^2 - z^2 = 1$

21. $x^2 + 4z^2 = 2$

In each of Exercises 22–24, sketch the graph of the given equation and identify the type of quadric surface you have drawn.

22. $z - y^2 = x^2$

23. $y^2 + 4x^2 = z$

24. $x^2 - y^2 + z^2 = 9$

In each of Exercises 25–27, calculate the indicated limit.

25. $\lim_{(x,y)\to(1,1)} \dfrac{x^2 - 2x + 1}{xy - 2x - y + 2}$

26. $\lim_{(x,y)\to(2,3)} \sqrt{xy^2 - y^3}$

27. $\lim_{(x,y)\to(3,5)} \dfrac{\sin(5x - 3y)}{5x - 3y}$

▶ In each of Exercises 28–30, the provided function does *not* have a limit at the point $(0,0)$. Calculate the limit of the function along two different paths, and derive two different answers, to demonstrate that the limit does not exist. ◀

28. $\dfrac{x^2 - 2y^2}{x^2 + 4y^2}$

29. $\dfrac{x - y}{x + y}$

30. $\dfrac{x}{\sqrt{x^2 + y^2}}$

▶ In each of Exercises 31–33, calculate $\partial f/\partial x$ and $\partial f/\partial y$ at the point P. ◀

31. $f(x,y) = \cos(yx^2)$ $P = (2, \pi)$

32. $f(x,y) = \ln(x^2 - xy)$ $P = (2, 4)$

33. $f(x,y) = xy^2 e^{x+y}$ $P = (0, 1)$

▶ In each of Exercises 34–36, a function f of the two variables x and y is given. Calculate f_{xx}, f_{yy}, f_{xy}, and f_{yx}. In each case verify that $f_{xy} = f_{yx}$. ◀

34. $f(x,y) = \tan(xy^2)$

35. $f(x,y) = \dfrac{\sin(y)}{\cos(x)}$

36. $f(x,y) = e^{xy}\sin(x^2 - y)$

▶ In each of Exercises 37–39, a function $f(x,y,z)$ of three variables is given. Calculate $f_x, f_y, f_z, f_{xy}, f_{yz}, f_{xyz}$, and f_{zz}. ◀

37. $f(x,y,z) = x^3yz - \sin(yz)$

38. $f(x,y,z) = \dfrac{xz}{x^2 + y^2}$

39. $f(x,y,z) = xze^{y^2x}$

▶ In each of Exercises 40–42, a function $f(x,y,z)$ of three variables is given. Calculate $f_{xyz}, f_{yxz}, f_{zyx}, f_{zxy}, f_{yzx}, f_{yxz}$. All six calculations should yield the same final answer. ◀

40. $f(x,y,z) = \dfrac{xyz}{x^2 + y^3 + z^4}$

41. $f(x,y,z) = z\sin(xy)$

42. $f(x,y,z) = y^2x\ln(x^2 - z^2)$

▶ In each of Exercises 43–45, calculate dz/ds by (i) using the Chain Rule and (ii) substituting the formulas for x and y into the formula for z and calculating the derivative directly. ◀

43. $z = f(x,y) = x^2 - y^2x, x = \rho(s) = s^2 - s, y = \sigma(s) = 1/s$

44. $z = f(x,y) = \dfrac{x^2 - y}{x + y}, x = \rho(s) = s + e^s, y = \sigma(s) = s^2$

45. $z = f(x,y) = (x + y)(y - x), x = \rho(s) = s\cos(s),$
$y = \sigma(s) = s\sin(s)$

▶ In each of Exercises 46–48, use the Chain Rule to calculate $\partial z/\partial s$ and $\partial z/\partial t$. ◀

46. $z = f(x,y) = xy\sin(x^2 - y^2), x = \rho(s,t) = s\cos(t),$
$y = \sigma(s,t) = t\sin(s)$

47. $z = f(x,y) = \ln(x - y^2), x = \rho(s,t) = se^t, y = \sigma(s,t) = te^s$

48. $z = f(x,y) = x/y, x = \rho(s,t) = \ln(s + t), y = \sigma(s,t) = e^{s+t}$

▶ In each of Exercises 49–51, use the Chain Rule to calculate dw/ds. ◀

49. $w = f(x,y,z) = \dfrac{z}{x - y}, x = \rho(s) = s\ln(s), y = \sigma(s) = se^s,$
$z = \tau(s) = s\sin(s)$

50. $w = f(x,y,z) = xyz\tan(x), x = \rho(s) = s/(s+1),$
$y = \sigma(s) = s(s+1),\ z = \tau(s) = s^2$

51. $w = f(x,y,z) = \ln(x - y^2 + z^3), x = \rho(s) = \ln(s)/s,$
$y = \sigma(s) = s\ln(s), z = \tau(s) = e^{2s}$

▶ In each of Exercises 52–54, give the Taylor polynomial of order 2 about the point P. ◀

52. $f(x,y) = \sin(x^2y)$ $P = (\sqrt{\pi}, 2)$

53. $f(x,y) = \ln(x - y^2)$ $P = (2, 1)$

54. $f(x,y) = e^{x-2y}$ $P = (0, 0)$

▶ In each of Exercises 55–57, calculate the gradient of the given function. ◀

55. $f(x,y) = x^2\sin(xy)$

56. $f(x,y) = \cos(xy)\sin(y^2x)$

57. $g(x,y,z) = xyze^{xz - y^2}$

▶ In each of Exercises 58–60, calculate the directional derivative of the function f in the direction $\mathbf{u}$ at the point P. ◀

58. $f(x,y) = xy^2\sin(x - y)$ $P = (1, 4)$ $\mathbf{u} = \langle 3/5, -4/5\rangle$

59. $f(x,y) = e^y\ln(x - y)$ $P = (1, 0)$ $\mathbf{u} = \langle 1/\sqrt{2}, -1/\sqrt{2}\rangle$

60. $f(x,y) = xy^2e^{xy}$ $P = (0, 0)$ $\mathbf{u} = \langle 6/10, -8/10\rangle$

▶ In each of Exercises 61–63, calculate these four quantities:

- the direction $\mathbf{u}$ of greatest increase of the function f at the point P;
- the directional derivative of f at P in the direction $\mathbf{u}$;
- the direction $\mathbf{v}$ of greatest decrease of the function f at the point P;
- the directional derivative of f at P in the direction $\mathbf{v}$. ◀

61. $f(x,y) = xy^3 - x^4$ $P = (1, 2)$

62. $f(x,y) = \ln(xy - y^3)$ $P = (3, 1)$

63. $f(x,y) = \sin(x - y^2)$ $P = (2\pi, \sqrt{\pi})$

▶ In each of Exercises 64–66, find the directional derivative of the function $g(x,y,z)$ at the point P in the direction $\mathbf{v}$. ◀

64. $f(x,y,z) = x^2 \sin(yz), P = (1,0,1),$
$\mathbf{v} = \langle 3/5, -\sqrt{7}/5, -3/5 \rangle$

65. $f(x,y,z) = y^2 z(1+x^2), P = (1,1,5), \mathbf{v} = \langle -2/3, 1/3, 2/3 \rangle$

66. $f(x,y,z) = xze^{yz}, P = (0,0,0), \mathbf{v} = \langle -1/\sqrt{2}, 1/2, -1/2 \rangle$

In each of Exercises 67–69, calculate these four quantities:

- the direction $\mathbf{u}$ of greatest increase of the function F at the point P;
- the directional derivative of F at P in the direction $\mathbf{u}$;
- the direction $\mathbf{v}$ of greatest decrease of the function F at the point P;
- the directional derivative of F at P in the direction $\mathbf{v}$.

67. $F(x,y,z) = xy(z^3 - zx)$ $\quad P = (2,1,4)$

68. $F(x,y,z) = \sin(x^2 + y^2 + z^4)$ $\quad P = (0, \sqrt{\pi}, 0)$

69. $F(x,y,z) = \ln(x^2 + 2yz + 1)$ $\quad P = (1,1,1)$

In each of Exercises 70–72, calculate a normal vector to the graph of the function $f(x,y)$ at the point P. Then write the equation of the tangent plane to the graph of f at the point P.

70. $f(x,y) = y(x^2 - y^2)$ $\quad P = 2,3)$

71. $f(x,y) = \sin(x^2 - y^2)$ $\quad P = (\sqrt{\pi}, 0)$

72. $f(x,y) = \ln(xy + 1)$ $\quad P = (0,0)$

In each of Exercises 73–75, find a normal vector to the given surface in three-dimensional space at the point Q. Then find the equation of the tangent plane to the surface at Q.

73. $x^2 + 4y^2 + 8z^2 = 13$ $\quad Q = (1,1,1)$

74. $(x^2 + y^2 + z)^2 = 36$ $\quad Q = (1,2,1)$

75. $\ln(1 + x^2 + z^2) = y$ $\quad Q = (1, \ln(2), 0)$

In each of Exercises 76–78, find parametric equations for the normal line to the graph of the given function at the given point.

76. $f(x,y) = \dfrac{x^2}{x^2 - y^2}$ $\quad P = (2,1,4/3)$

77. $f(x,y) = xy\tan(x+y)$ $\quad P = (\pi/2, 0, 0)$

78. $f(x,y) = \ln(xy + x^2)$ $\quad P = (1,1,\ln(2))$

In each of Exercises 79–81, calculate the linear approximation (given by the tangent plane at $(P_0, f(P_0))$) to $f(P)$.

79. $f(x,y) = \sqrt{y^3 + x^2}$ $\quad P = (3.9, 5.9) \quad P_0 = (4,6)$

80. $f(x,y) = y\sin((2y - x)\pi)$ $\quad P = (0.9, 1.9) \quad P_0 = (1,2)$

81. $f(x,y) = \ln(\sqrt{x} + y)$ $\quad P = (1.1, 0.1) \quad P_0 = (1,0)$

82. A cylindrical can is to be built with radius 3 inches and height 6 inches. Suppose that the error in measurement can be kept accurate to 0.3%. Use the total differential to estimate the greatest possible error in volume.

In each of Exercises 83–85, find all the critical points of the function f. Determine whether each critical point is a local minimum, a local maximum, a saddle point, or none of these.

83. $f(x,y) = 4x^2 + 6y^2 - 6x + 8y - 4$

84. $f(x,y) = 3xy - 2/x - 4/y$

85. $f(x,y) = (x-4)(y+1)y$

86. Find the point(s) on the surface $z = xy$ that are nearest to $(1,1,1)$.

87. Find three positive numbers whose product is 100 and whose sum is as small as possible.

In each of Exercises 88–90, find the extrema of the function $f(x,y)$ subject to the constraint given by the function $g(x,y)$.

88. $f(x,y) = \cos^2(x) + \cos^2(y), g(x,y) = x - y = \pi$

89. $f(x,y) = (y+2)^2 + x^2, g(x,y) = y^2 + 6x^2 = 36$

90. $f(x,y) = xy^2 z, g(x,y) = x^2 + y^2 + z^2 = 9$

In each of Exercises 91–93, use the method of Lagrange Multipliers to solve the problem.

91. Minimize the function $f(x,y) = 4x^2 + 8y^2 + 4$ subject to the constraint $x - 4y = 8$.

92. Maximize the quantity xy^2 subject to the constraint $y^2 + 2x^2 = 4$.

93. Minimize $x^2 - 3y^2$ subject to the constraint $5x^2 + y^2 = 25$.

The two founders of calculus, Newton and Leibniz, both had occasion to employ partial derivatives. For example, in a letter written to l'Hôpital in 1694, Leibniz introduced the total differential of a function of two variables. Although Leibniz wrote δf and vf for the partial derivatives $\partial f/\partial x$ and $\partial f/\partial y$, respectively, the practice of using notation to distinguish between ordinary and partial derivatives did not take hold among the mathematicians who followed. In fact, the notation, $\partial f/\partial x$ and $\partial f/\partial y$, introduced by Carl Gustav Jacobi in 1841, did not receive widespread acceptance until the end of the 19th century.

Clairaut's Theorem

Today, we often refer to the identity $f_{xy}(x, y) = f_{yx}(x, y)$ as *Clairaut's Theorem*. Alexis Claude Clairaut (1713–1765), the son of a mathematics teacher and the only one of twenty siblings to reach adulthood, was a mathematical prodigy who published his first paper at the age of thirteen. When only sixteen, he initiated the study of space curves, and, as a result, he became the youngest member ever appointed to the prestigious Paris Academy of Sciences. In addition to geometry, Clairaut undertook important research in differential equations, the calculus of variations, mechanics, and astronomy. He participated in an expedition to Lapland to measure one degree of longitude, with the aim of verifying Newton's assertion that Earth is an oblate sphere. Clairaut's scientific activity continued unabated until his early death, but, not to the exclusion of all other interests. As one contemporary mathematician wrote, "Occupied by dinner parties and late-night discussions, burdened by unrestrained womanizing, and content to mix pleasure with business, Clairaut lost his repose, his health, and, finally, at the age of 52, his life."

Before Clairaut, the equality of mixed partial derivatives had been observed in particular cases and was tacitly assumed to be true in general. In 1740, Clairaut published a derivation of the equation $f_{xy}(x, y) = f_{yx}(x, y)$, as did Euler. However, when the standards of mathematical rigor tightened up half a century later, both proofs were judged inadequate. A succession of eminent mathematicians, including Lagrange in 1797 and Cauchy in 1823, advanced alternative demonstrations that were equally flawed. In 1867, the Finnish mathematician Leonard Lorenz Lindelöf (1827–1908) published a critique of the existing faulty proofs. Reinforcing the need for a valid proof, he also exhibited a nondifferentiable function f for which $f_{xy}(x, y)$ and $f_{yx}(x, y)$ exist but are unequal. At last, in 1873, Hermann Amandus Schwarz (1843–1921) conclusively established the validity of Clairaut's Theorem under appropriate assumptions on f. Thereafter, work on Clairaut's Theorem was directed toward relaxing the hypotheses. Well-known mathematicians such as Camille Jordan (1838–1922), Ulisse Dini (1845–1918), Axel Harnack (1851–1888), and Giuseppe Peano (1858–1932) all contributed to extending Clairaut's Theorem to a wider class of functions.

The Vibrating String

A large part of the mathematical enterprise of the 18th century was devoted to applying calculus to natural phenomena. One of the problems that first attracted attention was the motion of a vibrating string (Figure 1). If $y(x, t)$ represents the vertical displacement of the string at point x $(0 \le x \le L)$ and time t, then it can be shown that

$$\frac{\partial^2 y}{\partial t^2} = c^2 \frac{\partial^2 y}{\partial x^2},$$

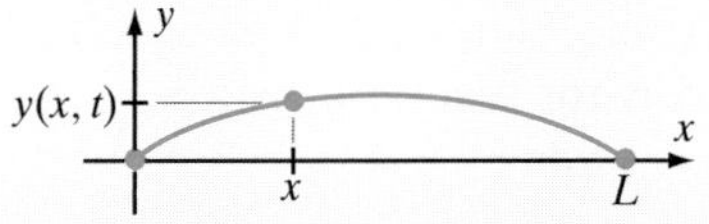

▲ Figure 1 A snapshot of a vibrating string at time t

where c^2 is the tension in the string divided by the mass density of the string. This partial differential equation is called the *one-dimensional wave equation*. It appeared for the first time in a paper that Jean Le Rond d'Alembert (1717–1783) published in 1747. In the same article, d'Alembert also argued that any solution of the one-dimensional wave equation can be expressed in the form

$$y(x, t) = f_1(x - ct) + f_2(x + ct) \qquad (1)$$

where f_1 and f_2 are twice-differentiable functions of one variable. Letting $\Delta x = c\Delta t$, we observe that $f_1((x + \Delta x) - c(t + \Delta t)) = f_1(x - ct)$. This means that the displacement of the string at a point $c\Delta t$ to the right of x is the same displacement that existed at x at a time Δt units earlier. We conclude that f_1 represents a wave that travels to the right with speed c (see Figure 2). Similarly, $f_2((x-\Delta x)+c(t+\Delta t)) = f_2(x+ct)$, so f_2 represents a wave that travels to the left with speed c.

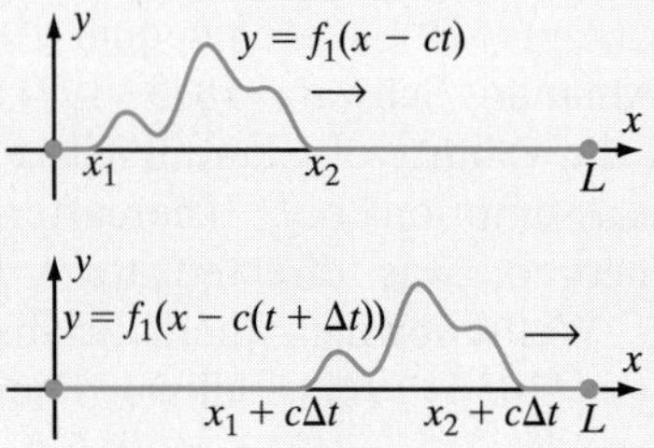

▲ Figure 2 Snapshots at times t and $t+\Delta t$ of a traveling wave on a vibrating string

Six years later, Daniel Bernoulli (1700–1782), son of Johann Bernoulli, discovered an alternative method of solving the wave equation. His idea was to represent the solution as a superposition of fundamental waves (Figure 3). Following that approach, Bernoulli found that the general solution of the wave equation has the form

$$y(x,t) = \sum_{n=0}^{\infty} \left(a_n \cos\left(\frac{n\pi ct}{L}\right) + b_n \sin\left(\frac{n\pi ct}{L}\right)\right) \sin\left(\frac{n\pi x}{L}\right). \quad (2)$$

Now it often happens that different methods lead to different forms of an answer. Usually these differences can be easily reconciled. In the case of the vibrating string, however, there was no obvious way to relate the solutions of d'Alembert and Bernoulli. Calculus was left in something of a crisis for several decades.

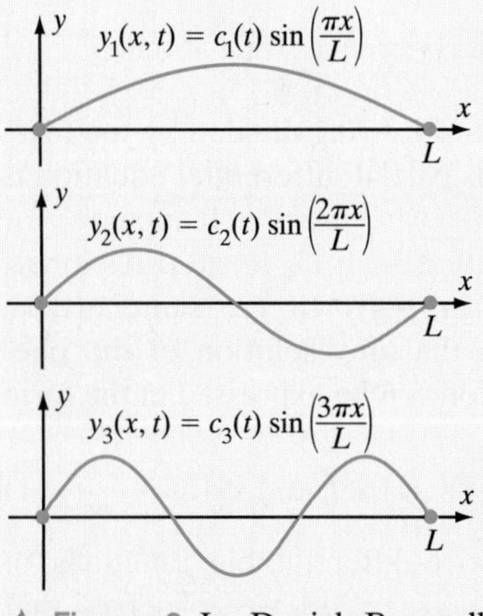

▲ Figure 3 In Daniel Bernoulli's solution of the wave equation, every vibration is the superposition of waves that have fundamental frequencies.

Other Equations of Mathematical Physics

As the 18th century progressed, mathematicians used partial differential equations to study acoustics, hydrodynamics, elasticity, and celestial mechanics. One particular type of expression appeared repeatedly. If y depends on only one space variable x, if z depends on only two space variables x and y, and if u depends on all three space variables x, y, and z, then we write,

$$\Delta y = \frac{\partial^2 y}{\partial x^2}, \Delta z = \frac{\partial^2 z}{\partial x^2} + \frac{\partial^2 z}{\partial y^2}, \text{ and } \Delta u = \frac{\partial^2 u}{\partial x^2} + \frac{\partial^2 u}{\partial y^2} + \frac{\partial^2 u}{\partial z^2}.$$

Using this notation, d'Alembert's one-dimensional wave equation takes the form $y_{tt} = c^2 \Delta y$. Similarly, the two-dimensional wave equation, which Euler introduced in a 1764 study of the vibrating circular membrane (Figure 4), has the form $z_{tt} = c^2 \Delta z$.

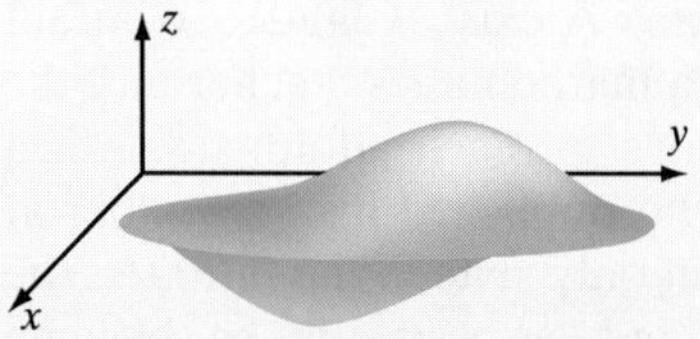

▲ Figure 4 A vibrating circular membrane

Although the expression Δu first appeared in a 1752 paper of Euler that concerned fluid flow, Pierre-Simon Laplace (1749–1827) used the equation $\Delta u = 0$ so extensively in his work on celestial mechanics that Δu is now called the *Laplacian* of u. One partial differential equation involving the Laplacian, the heat equation $\Delta u = k u_t$, has had a particularly important impact on mathematical analysis. This equation, so-named because it describes the conduction of heat, was derived by Jean Baptiste Joseph Fourier (1768–1830) in 1807. Although Fourier's method of solution did not gain immediate acceptance, he undertook a more thorough exposition, titled *Théorie Analytique de la Chaleur*, which he published in 1822.

When Fourier's book came out, the controversy over the vibrating string was still very much alive even though the original participants in the dispute were long dead. Fourier realized that his method could finally "resolve all the difficulties that the analysis employed by Daniel Bernoulli presented." To that end, consider a wave $y(x,t) = f(x-ct)$ such that $y(0,t) = y(L,t) = 0$ for all t. We have, on the one hand, $y(x,0) = f(x)$. On the other hand, Bernoulli's solution tells us that we can express $y(x, 0)$ by substituting $t = 0$ into the right side of equation (2). This leads to the equation

$$f(x) = \sum_{n=1}^{\infty} a_n \sin\left(\frac{n\pi x}{L}\right). \tag{3}$$

Before Fourier's work on heat conduction, most mathematicians did not believe that a general function could admit such an expansion as a trigonometric series. Fourier was convinced that such skepticism would vanish once a concrete relationship between f and the coefficients a_n was made. As he phrased the matter, "Geometers only admit that which they cannot dispute. Of all the derivations of Bernoulli's solution, the most complete is that which consists of actually resolving an arbitrary function into such a series [as equation (3) above] and assigning the values of the coefficients." The methods Fourier brought to bear on the heat equation were exactly the right tools for determining the coefficients a_n in equation (3):

$$a_n = \frac{2}{L}\int_0^L f(x)\sin\left(\frac{n\pi x}{L}\right)dx. \tag{4}$$

Fourier emphasized that his solution is "applicable to the case where the initial figure of the string is that of a triangle or a trapezoid, or is such that only one part of the string is set in motion while the other parts blend with the axis." In other words, he asserted that his method remains valid for a function that is defined by different analytic expressions on different parts of its domain. Figure 5 shows the graph of a function f on the interval $[0, 10]$. This function is precisely of the type that Fourier mentioned: partly trapezoidal and partly "blended" with the axis. The trigonometric polynomial $T_5(x) = \sum_{n=1}^{5} a_n \sin(n\pi x/10)$, with coefficients specified by equation (4), is also shown in Figure 5. Notice that $T_5(x)$ already captures the general shape of the graph of f even though it is composed of a small number of terms.

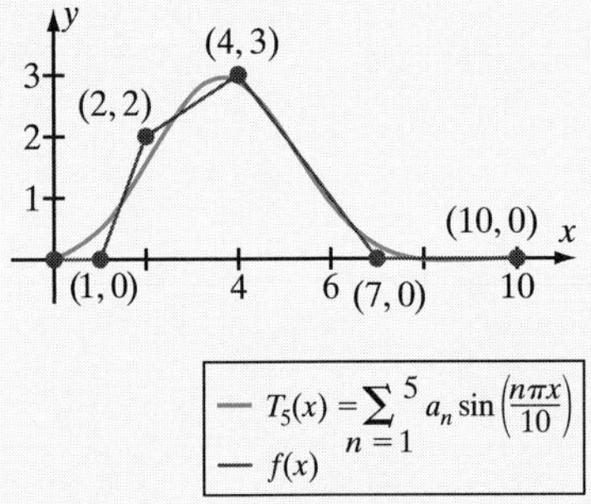

▲ Figure 5

There would be little point displaying an analogous superposition of the graphs of $f(x)$ and $T_{100}(x) = \sum_{n=1}^{100} a_n \sin(n\pi x/10)$ because no differences would be discernible. Indeed, look at Figure 6, which shows the plot of the error function $y = f(x) - T_{100}(x)$. The graph suggests that the convergence of the trigonometric series is slowest at the points $x = 1$, $x = 2$, $x = 4$, and $x = 7$. These are the points at which f is not differentiable. Even so, the maximum error is only about 0.02.

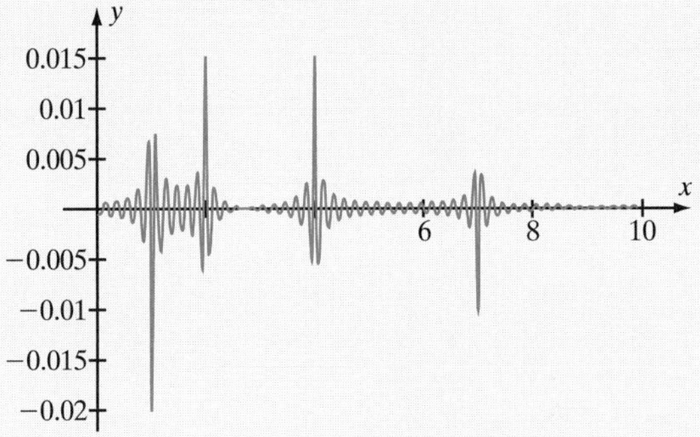

▲ Figure 6

The repercussions of Fourier's work touched nearly every aspect of mathematical analysis. The modern definition of a function as a correspondence of real numbers (as opposed to an analytic expression) ensued in large measure from Fourier's influence. The introduction of Fourier series prompted Niels Henrik Abel (1802–1829) and Peter Gustav Lejeune Dirichlet (1805–1859) to clarify the meaning of convergence of infinite series. By 1850, Fourier series were being used in number theory, far afield from mathematical physics. Because many natural functions of number theory are considerably more irregular than the garden-variety functions of physics, Georg Friedrich Bernhard Riemann (1826–1866) thought it prudent to examine the meaning of equation (4) in cases where f has numerous discontinuities. The Riemann integral was the result of his investigations. In 1871, Georg Cantor (1845–1918) studied the uniqueness of representation by trigonometric series. The question he sought to answer was, for what sets X is it true that

$$\sum_{n=1}^{\infty} a_n \sin\left(\frac{n\pi x}{L}\right) = \sum_{n=1}^{\infty} b_n \sin\left(\frac{n\pi x}{L}\right), \quad x \in X$$

implies $a_n = b_n$ for all n? It was through these studies that Cantor came to create the subject of set theory.

CHAPTER 12

Multiple Integrals

PREVIEW

In the preceding chapter, we learned how to differentiate functions of two or more variables. Now we will learn how to integrate such functions. To be specific, given a region $\mathcal{R}$ of the xy-plane and a continuous function f defined on $\mathcal{R}$, we will define the integral of f over $\mathcal{R}$. This integral is denoted by $\iint_{\mathcal{R}} f(x, y)\, dA$ and is called a *double integral*. Analogously, if F is a continuous function of three variables that is defined on a solid region $\mathcal{U}$ in xyz-space, then we will define the *triple integral* $\iiint_{\mathcal{U}} F(x, y, z)\, dV$. The main theorem of this chapter tells us that multiple integrals can be evaluated by iterating the familiar process of integration, one variable at a time.

In Chapter 7, we studied certain physical applications of the integral: calculating centers of mass of planar regions, volumes of solids of revolution, and so on. Because we had only single integrals available to us at the time, the scope of our discussion was necessarily limited. For example, when we calculated volumes, we treated only solids with symmetries. We exploited the symmetry to express volume by means of a single integral. In this chapter, we will take a second look at some of the physical interpretations of the integral. As we will see, double and triple integrals allow us to handle these concepts in greater generality. Moreover, we will be able to treat some more substantial problems in Chapter 13 using multiple integrals.

12.1 Double Integrals Over Rectangular Regions

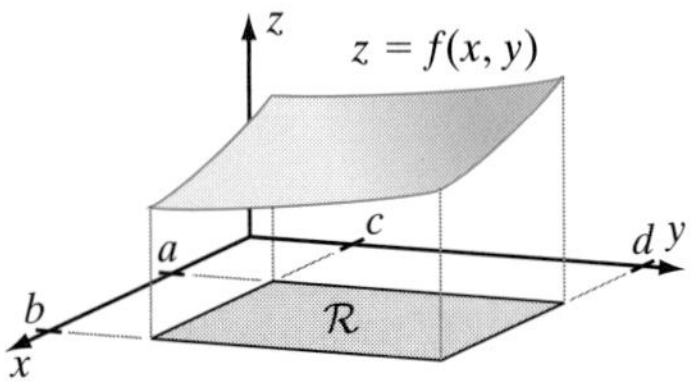

▲ Figure 1

Suppose that $(x, y) \mapsto f(x, y)$ is a positive, continuous function whose domain contains the rectangle $\mathcal{R} = \{(x, y) : a \le x \le b, c \le y \le d\}$. How should we calculate the volume of the solid that lies under the graph of f and above rectangle $\mathcal{R}$? See Figure 1.

We begin by partitioning the domain of the function. A convenient way to do this is to partition the domain of the x variable:

$$a = x_0 < x_1 < x_2 < \cdots < x_{N-1} < x_N = b$$

and also to partition the domain of the y variable:

$$c = y_0 < y_1 < y_2 < \cdots < y_{N-1} < y_N = d.$$

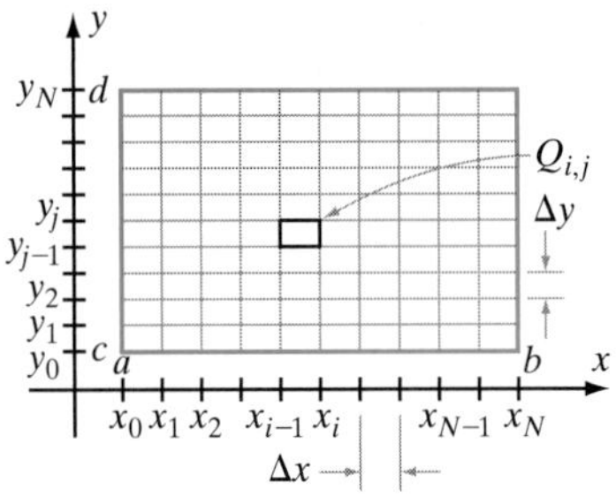

▲ Figure 2

For simplicity, we use only uniform partitions, with the same number N of intervals in both the x-direction and the y-direction. This partition breaks up the rectangle $\mathcal{R}$ into smaller rectangles of equal sizes, as shown in Figure 2. Specifically, to each pair x_{i-1}, x_i of consecutive points in the x-partition and each pair y_{j-1}, y_j of consecutive points in the y-partition, there corresponds a subrectangle $Q_{i,j}$. The side lengths of $Q_{i,j}$ are $\Delta x = x_i - x_{i-1} = (b-a)/N$ and $\Delta y = y_j - y_{j-1} = (d-c)/N$.

Over each subrectangle $Q_{i,j}$, we erect a box with height $f(\xi_{i,j})$, where $\xi_{i,j}$ is some point in $Q_{i,j}$ (see Figure 3). This box has volume

$$V_{i,j} = f(\xi_{i,j})\ \Delta x \Delta y = f(\xi_{i,j})\ \Delta A.$$

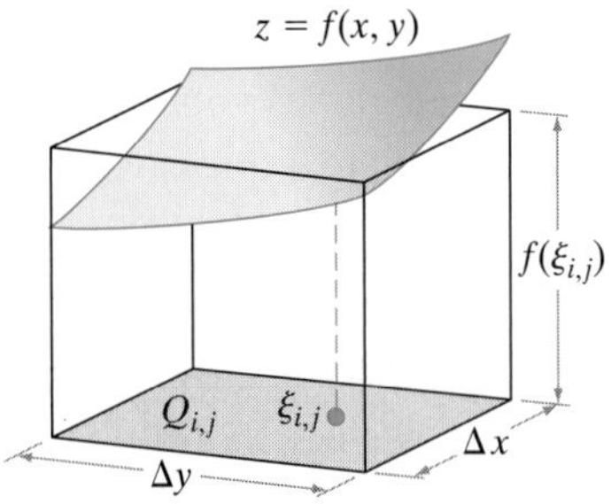

▲ Figure 3 The volume $V_{i,j} = f(\xi_{i,j}) \Delta x \Delta y$ of the box of height $f(\xi_{i,j})$ and base area $\Delta A = \Delta x \Delta y$ approximates the volume of the solid under the graph of $z = f(x, y)$ and over the subrectangle $Q_{i,j}$

Here the symbol ΔA denotes the increment $(\Delta x)(\Delta y)$ of (planar) area. The volume V of the solid under the graph of $z = f(x, y)$ and over rectangle $\mathcal{R}$ is then approximated by the sum of the volumes of these boxes. That is,

$$V \approx \sum_{i=1}^{N} \sum_{j=1}^{N} V_{i,j} = \sum_{i=1}^{N} \sum_{j=1}^{N} f(\xi_{i,j})\ \Delta A. \tag{12.1.1}$$

Approximation (12.1.1) is illustrated for $N = 2$ in Figure 4a and for $N = 4$ in Figure 4b. In Figure 4c, with $N = 16$, we have omitted the graph of $z = f(x, y)$. To see how well the tops of the 16^2 boxes approximate the graph of f—compare Figure 4c with Figure 1.

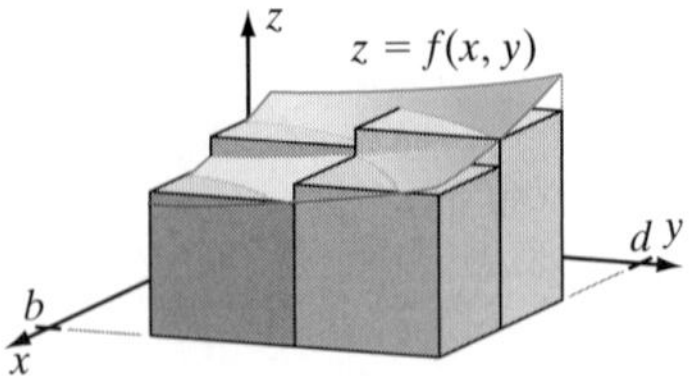

▲ Figure 4a

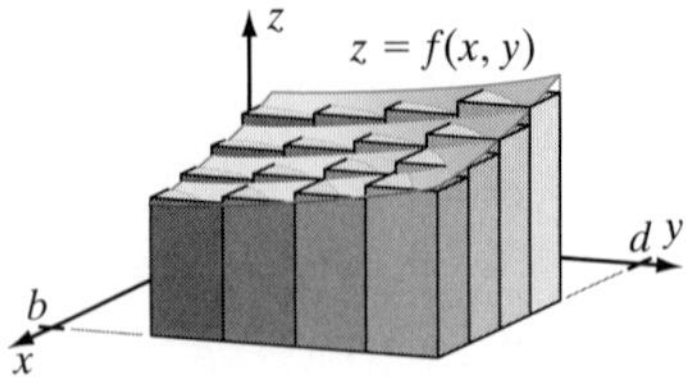

▲ Figure 4b

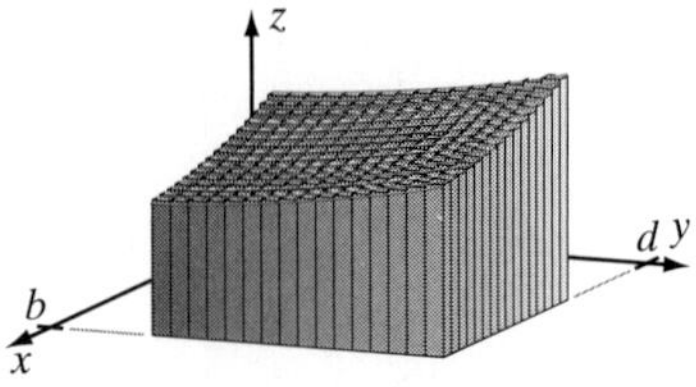

▲ Figure 4c

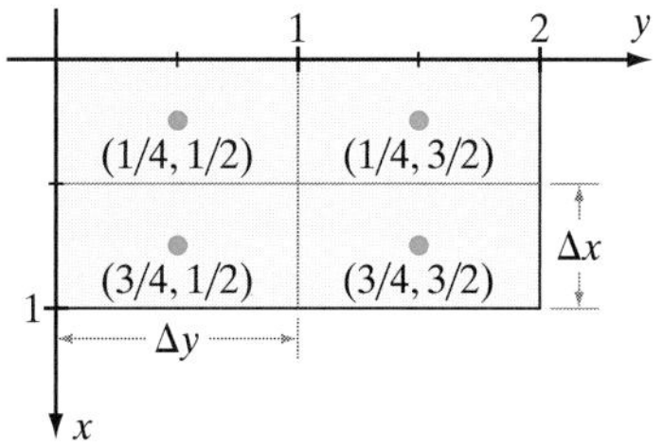

▲ Figure 5 The rectangle $[0, 1] \times [0, 2]$ with two equal subdivisions of each side: The midpoints of the subrectangles constitute the choice of points $\{\xi_{i,j}\}$.

▶ **EXAMPLE 1** Use approximation (12.1.1) with $N = 2$ to estimate the volume of the solid that lies under the surface $z = 6 - x^2 - y^2$ and above the rectangle $[0, 1] \times [0, 2]$ in the xy-plane. Use the midpoint of each subrectangle for the choice of points $\{\xi_{ij}\}$.

Solution Figure 5 illustrates the four subrectangles that arise from the uniform partitions of the x-interval $[0, 1]$ and y-interval $[0, 2]$ with $N = 2$. The midpoints of these rectangles are also shown. The increment of area is $\Delta A = \Delta x \Delta y = (1/2)(1) = 1/2$. We use $f(x, y) = 6 - x^2 - y^2$ in approximation (12.1.1) to obtain

$$V \approx \left(6 - \left(\frac{1}{4}\right)^2 - \left(\frac{1}{2}\right)^2\right)\left(\frac{1}{2}\right) + \left(6 - \left(\frac{1}{4}\right)^2 - \left(\frac{3}{2}\right)^2\right)\left(\frac{1}{2}\right)$$
$$+ \left(6 - \left(\frac{3}{4}\right)^2 - \left(\frac{1}{2}\right)^2\right)\left(\frac{1}{2}\right) + \left(6 - \left(\frac{3}{4}\right)^2 - \left(\frac{3}{2}\right)^2\right)\left(\frac{1}{2}\right) = \frac{71}{8}.$$

Figure 6a illustrates the solid. Figure 6b shows the four approximating boxes. ◀

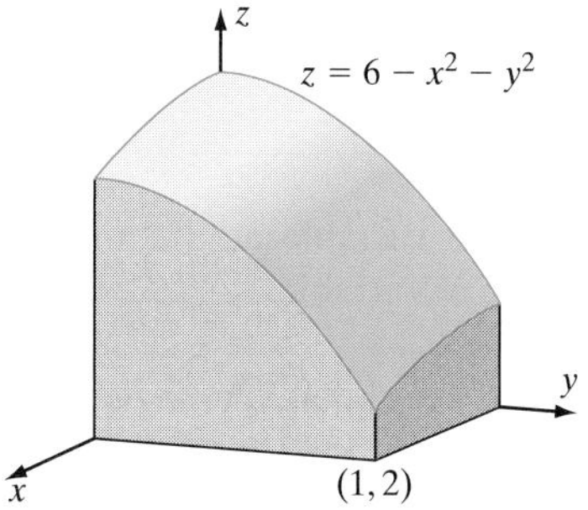

▲ Figure 6a

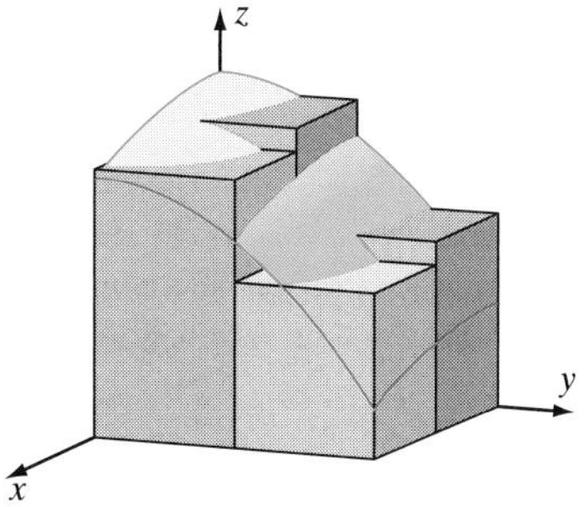

▲ Figure 6b

We call the double sum in line (12.1.1) a *Riemann sum*. As we have seen, these sums can be used to estimate the volume under the graph of f when f is a positive function. However, line (12.1.1) makes perfect sense for any continuous function. Furthermore, it can be shown that the Riemann sums of any continuous function f tend to a limit as N tends to infinity. We may therefore make the following definition:

DEFINITION Let f be a continuous function that is defined on the rectangle $\mathcal{R} = \{(x, y) : a \le x \le b, c \le y \le d\}$. For each positive integer N, let $\{x_0, x_1, \ldots, x_N\}$ and $\{y_0, y_1, \ldots, y_N\}$ be uniform partitions of the intervals $[a, b]$ and $[c, d]$, respectively. Set $\Delta x = (b - a)/N$, $\Delta y = (d - c)/N$, and $\Delta A = (\Delta x)(\Delta y)$. Let ξ_{ij} be any point in subrectangle $Q_{ij} = [x_{i-1}, x_i] \times [y_{j-1}, y_j]$. Then, the *double integral* of f over $\mathcal{R}$ is defined to be the limit of the Riemann sums

$$\sum_{i=1}^{N} \sum_{j=1}^{N} f(\xi_{i,j})\ \Delta A$$

as N tends to infinity. We denote the double integral of f over the rectangle $\mathcal{R}$ by the symbol $\iint_{\mathcal{R}} f(x, y)\, dA$. Thus

$$\iint_{\mathcal{R}} f(x, y)\, dA = \lim_{N \to \infty} \sum_{i=1}^{N} \sum_{j=1}^{N} f(\xi_{i,j})\ \Delta A = \lim_{N \to \infty} \sum_{i=1}^{N} \sum_{j=1}^{N} f(\xi_{i,j})\ \Delta x \Delta y. \quad (12.1.2)$$

Iterated Integrals

Now that we understand what quantity the double integral $\iint_{\mathcal{R}} f(x, y)\, dA$ signifies, we need to know how to calculate it. The technique is to integrate the integrand $f(x, y)$ one variable at a time. The idea is similar to partial differentiation: One variable of $f(x, y)$ is held constant while we apply an operation of calculus to the

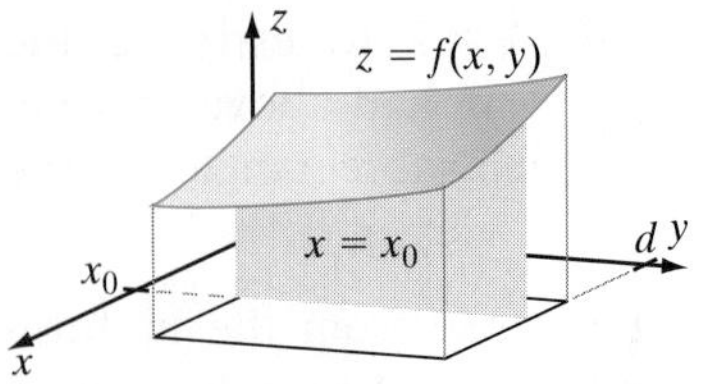

▲ Figure 7 The area of the shaded planar region is $\int_c^d f(x_0, y)\,dy$.

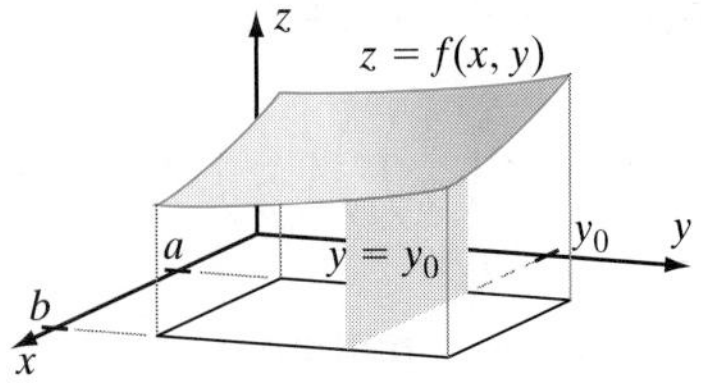

▲ Figure 8 The area of the shaded planar region is $\int_a^b f(x, y_0)\,dx$.

other variable. We continue to suppose that f is continuous on the rectangle $\mathcal{R} = \{(x,y) : a \le x \le b, c \le y \le d\} = [a,b] \times [c,d]$. If we fix the value $x = x_0$ in the interval $[a,b]$, then we can form the integral $\int_c^d f(x_0, y)\,dy$. If $y \mapsto f(x_0, y)$ happens to be positive, then the integral $\int_c^d f(x_0, y)\,dy$ represents the area of the region in the $x = x_0$ plane that lies under the slice $z = f(x_0, y)$ and above the xy-plane (see Figure 7). Similarly, if we fix the value $y = y_0$ in the interval $[c, d]$, then we can form the integral $\int_a^b f(x, y_0)\,dx$. See Figure 8 for the area interpretation of this integral when $x \mapsto f(x, y_0)$ is a positive function.

▶ **EXAMPLE 2** Let $f(x,y) = 6 - x^2 - y^2$ for $0 \le x \le 1$ and $0 \le y \le 2$. For each fixed x in the interval $[0,1]$, calculate $\mathcal{I}(x) = \int_0^2 f(x,y)\,dy$. For each fixed y in the interval $[0,2]$, calculate $\mathcal{J}(y) = \int_0^1 f(x,y)\,dx$.

Solution In the integral that defines $\mathcal{I}(x)$, we treat the x variable as a constant. Thus the antiderivative of $6 - x^2 - y^2$ with respect to the y variable is $(6 - x^2)y - y^3/3$. Therefore we have

$$\mathcal{I}(x) = \int_0^2 (6 - x^2 - y^2)\,dy = \left((6 - x^2)y - \frac{y^3}{3}\right)\Bigg|_{y=0}^{y=2} = 2(6 - x^2) - \frac{8}{3} = \frac{28}{3} - 2x^2.$$

Similarly, in the integral that defines $\mathcal{J}(y)$, we treat the y variable as a constant. Thus the antiderivative of $6 - x^2 - y^2$ with respect to the x variable is $(6 - y^2)x - x^3/3$. Therefore we have

$$\mathcal{J}(y) = \int_0^1 (6 - x^2 - y^2)\,dx = \left((6 - y^2)x - \frac{x^3}{3}\right)\Bigg|_{x=0}^{x=1} = (6 - y^2) - \frac{1}{3} = \frac{17}{3} - y^2. \quad ◀$$

Using the continuity of f, it can be shown that the functions $\mathcal{I}(x) = \int_c^d f(x,y)\,dy$ and $\mathcal{J}(y) = \int_a^b f(x,y)\,dx$ are continuous for $a \le x \le b$ and $c \le y \le d$, respectively. In particular, each of these two functions can be integrated over its domain. Thus the quantities

$$\int_a^b \left(\int_c^d f(x,y)\,dy\right)dx = \int_a^b \mathcal{I}(x)\,dx \quad \text{and} \quad \int_c^d \left(\int_a^b f(x,y)\,dx\right)dy = \int_c^d \mathcal{J}(y)\,dy$$

are well-defined.

DEFINITION The expressions $\int_a^b \left(\int_c^d f(x,y)\,dy\right)dx$ and $\int_c^d \left(\int_a^b f(x,y)\,dx\right)dy$ are called *iterated integrals*.

In the iterated integral $\int_a^b \left(\int_c^d f(x,y)\,dy\right)dx$, we first calculate the inner integral $\int_c^d f(x,y)\,dy$. For this computation, we treat the variable x as if it were a constant. Because $\int_c^d f(x,y)\,dy$ is a definite integral in the y variable, its value does not depend on y, but it does depend on x. We integrate this function of x to complete the calculation of the iterated integral. We handle the integral $\int_c^d \left(\int_a^b f(x,y)\,dx\right)dy$ in an analogous way.

▶ **EXAMPLE 3** Let $f(x,y)=6-x^2-y^2$ for $0\le x\le 1$ and $0\le y\le 2$, as in Example 2. Calculate the iterated integrals $\int_0^1\left(\int_0^2 f(x,y)\,dy\right)dx$ and $\int_0^2\left(\int_0^1 f(x,y)\,dx\right)dy$.

Solution Let $\mathcal{I}(x)=\int_0^2 f(x,y)\,dy$ and $\mathcal{J}(y)=\int_0^1 f(x,y)\,dx$. In Example 2, we showed that $\mathcal{I}(x)=28/3-2x^2$ and $\mathcal{J}(y)=17/3-y^2$. Therefore

$$\int_0^1\left(\int_0^2 f(x,y)\,dy\right)dx=\int_0^1 \mathcal{I}(x)\,dx=\int_0^1\left(\frac{28}{3}-2x^2\right)dx=\left(\frac{28}{3}x-\frac{2}{3}x^3\right)\Big|_0^1=\frac{26}{3}$$

and

$$\int_0^2\left(\int_0^1 f(x,y)\,dx\right)dy=\int_0^2 \mathcal{J}(y)\,dy=\int_0^2\left(\frac{17}{3}-y^2\right)dy=\left(\frac{17}{3}y-\frac{y^3}{3}\right)\Big|_0^2=\frac{26}{3}.\quad ◀$$

INSIGHT In Example 3, the integrands of the integrals $\int_0^1 \mathcal{I}(x)\,dx$ and $\int_0^2 \mathcal{J}(y)\,dy$ differ, as do the intervals of integration. Nevertheless, Example 3 demonstrates that the two integrals have the same value. In other words, the two iterated integrals $\int_0^1\left(\int_0^2 f(x,y)\,dy\right)dx$ and $\int_0^2\left(\int_0^1 f(x,y)\,dx\right)dy$ are equal. In the next subsection, we will see that these expressions are equal because they both represent the volume of the solid that lies under the graph of f and over the rectangle $[0,1]\times[0,2]$.

Using Iterated Integrals to Calculate Double Integrals

Next we will show that, when f is continuous and positive, the iterated integrals $\int_a^b\left(\int_c^d f(x,y)\,dy\right)dx$ and $\int_c^d\left(\int_a^b f(x,y)\,dx\right)dy$ both equal the volume of the solid $\mathcal{U}$ that lies under the graph of f and above the rectangle $\mathcal{R}=[a,b]\times[c,d]$. Let $\mathcal{I}(x)=\int_c^d f(x,y)\,dy$ and $\mathcal{J}(y)=\int_a^b f(x,y)\,dx$. Partition the interval $[a,b]$ into N equal subintervals by means of the points $a=x_0, x_1, \ldots, x_{N-1}, x_N=b$. As before, we set $\Delta x=(b-a)/N$. Figure 9 shows that, when Δx is small, the quantity $\left(\int_c^d f(x_i,y)\,dy\right)\Delta x$ is a good approximation to the volume of the section of $\mathcal{U}$ that lies between the planes $x=x_{i-1}$ and $x=x_i$. It follows that the sum $\sum_{i=1}^N\left(\int_c^d f(x_i,y)\,dy\right)\Delta x$ is a good approximation to the volume of $\mathcal{U}$ when Δx is small. Therefore

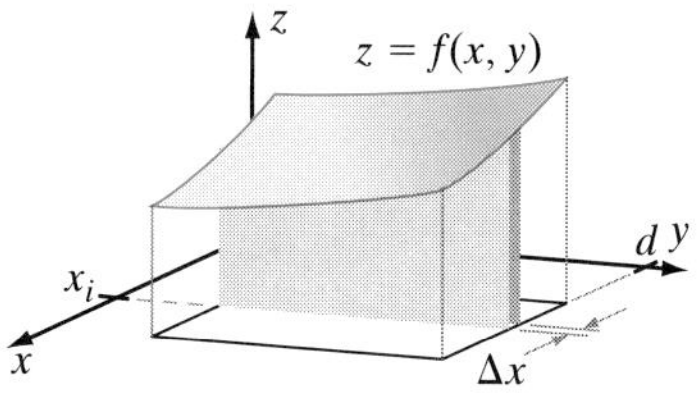

▲ Figure 9 The area of the shaded solid is approximately $(\int_c^d f(x_i,y)dy)\Delta x$.

$$\lim_{N\to\infty}\sum_{i=1}^N\left(\int_c^d f(x_i,y)\,dy\right)\Delta x=\iint_{\mathcal{R}} f(x,y)\,dA. \tag{12.1.3}$$

On the other hand, the sum $\sum_{i=1}^N\left(\int_c^d f(x_i,y)\,dy\right)\Delta x$ is a Riemann sum for the Riemann integral $\int_a^b \mathcal{I}(x)\,dx$. Therefore

$$\lim_{N\to\infty}\sum_{i=1}^N\left(\int_c^d f(x_i,y)\,dy\right)\Delta x=\int_a^b \mathcal{I}(x)\,dx=\int_a^b\left(\int_c^d f(x,y)\,dy\right)dx. \tag{12.1.4}$$

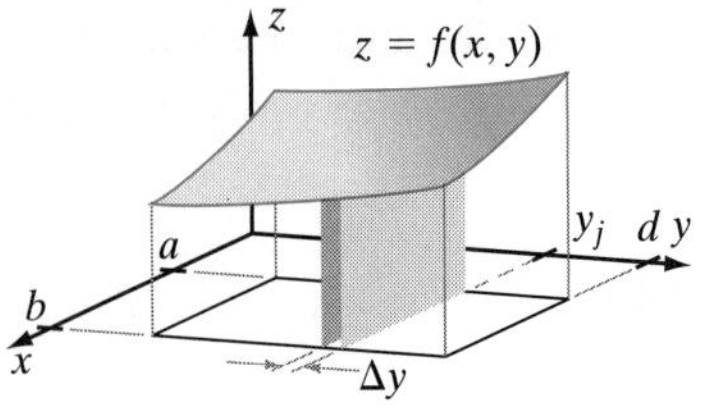

▲ Figure 10 The area of the shaded solid is approximately $\left(\int_a^b f(x,y_j)dx\right)\Delta y$.

From equations (12.1.3) and (12.1.4), we conclude that $\iint_{\mathcal{R}} f(x,y)\, dA = \int_a^b \left(\int_c^d f(x,y)\, dy\right) dx$.

Figure 10 reveals an analogous situation for the other iterated integral. If we partition the interval $[c,d]$ into N equal subintervals by means of the points $c = y_0, y_1, \ldots, y_{N-1}, y_N = d$, and we set $\Delta y = (d-c)/N$, then $\left(\int_a^b f(x,y_j)\, dx\right)\Delta y$ is a good approximation to the volume of the section of $\mathcal{U}$ that lies between the planes $y = y_{j-1}$ and $y = y_j$. It follows that the sum $\sum_{j=1}^{N} \left(\int_a^b f(x,y_j)\, dx\right)\Delta y$ is a good approximation to the volume of $\mathcal{U}$, and it is also a Riemann sum for $\int_c^d \mathcal{J}(y)\, dy$. Letting N tend to infinity, we obtain

$$\lim_{N\to\infty} \sum_{j=1}^{N} \left(\int_a^b f(x,y_j)\, dx\right)\Delta y = \iint_{\mathcal{R}} f(x,y)\, dA$$

and

$$\lim_{N\to\infty} \sum_{j=1}^{N} \left(\int_a^b f(x,y_j)\, dx\right)\Delta y = \int_c^d \mathcal{J}(y)\, dy = \int_c^d \left(\int_a^b f(x,y)\, dx\right) dy.$$

It follows that $\iint_{\mathcal{R}} f(x,y)\, dA = \int_c^d \left(\int_a^b f(x,y)\, dx\right) dy$.

Although, our discussion has assumed that f is positive, the equalities we have obtained are valid for any continuous function, as our next theorem asserts.

THEOREM 1 Let f be continuous on the rectangle $\mathcal{R} = \{(x,y) : a \le x \le b, c \le y \le d\} = [a,b] \times [c,d]$. Then, the double integral $\iint_{\mathcal{R}} f(x,y)\, dA$ exists and

$$\iint_{\mathcal{R}} f(x,y)\, dA = \int_a^b \left(\int_c^d f(x,y)\, dy\right) dx = \int_c^d \left(\int_a^b f(x,y)\, dx\right) dy.$$

INSIGHT If you think back to the method of disks in Section 7.1 of Chapter 7, then you will realize we were using a similar idea because we would calculate the areas of slices in the y variable and then integrate out in the x variable, or vice versa.

▶ **EXAMPLE 4** Use Theorem 1 to calculate the integral of the function $f(x,y) = \cos(x)\sin(y)$ over the rectangle $\mathcal{R} = \{(x,y) : \pi/6 \le x \le \pi/2, \pi/4 \le y \le \pi/3\}$.

Solution By Theorem 1,

$$\iint_{\mathcal{R}} f(x,y)\, dA = \int_{\pi/6}^{\pi/2} \left(\int_{\pi/4}^{\pi/3} \cos(x)\sin(y)\, dy\right) dx.$$

We evaluate an iterated integral by working from the inside out. Notice that the inner integral is an integral in the y variable. Just as in the theory of partial differentiation, we treat the x variable as a constant. Thus the antiderivative of $\cos(x)\sin(y)$ with respect to the y variable is $\cos(x)(-\cos(y))$, or $-\cos(x)\cos(y)$. We have

$$\int_{\pi/4}^{\pi/3} \cos(x)\sin(y)\,dy = -\cos(x)\cos(y)\Big|_{y=\pi/4}^{y=\pi/3} = -\cos(x)\cos\left(\frac{\pi}{3}\right) + \cos(x)\cos\left(\frac{\pi}{4}\right)$$

$$= -\frac{1}{2}\cos(x) + \frac{\sqrt{2}}{2}\cos(x) = \left(\frac{\sqrt{2}-1}{2}\right)\cos(x).$$

To finish the calculation of the given double integral, we now do a straightforward integral in one variable. The result is

$$\iint_{\mathcal{R}} f(x,y)\,dA = \int_{\pi/6}^{\pi/2} \left(\frac{\sqrt{2}-1}{2}\right)\cos(x)\,dx$$

$$= \left(\frac{\sqrt{2}-1}{2}\right)\sin(x)\Big|_{x=\pi/6}^{x=\pi/2} = \left(\frac{\sqrt{2}-1}{2}\right)\cdot 1 - \left(\frac{\sqrt{2}-1}{2}\right)\frac{1}{2} = \frac{\sqrt{2}-1}{4}. \blacktriangleleft$$

▶ **EXAMPLE 5** Integrate the function $f(x,y) = x^2y - y^3 + 2x$ over the rectangle $\mathcal{R} = \{(x,y) : 1 \le x \le 3,\ -1 \le y \le 0\}$.

Solution According to Theorem 1, we have

$$\iint_{\mathcal{R}} f(x,y)\,dA = \int_{-1}^{0} \left(\int_{1}^{3} (x^2y - y^3 + 2x)\,dx\right)dy.$$

We evaluate the inside integral in the x variable first. When calculating the antiderivative in the x variable, we treat y as a constant. Thus

$$\int_1^3 (x^2y - y^3 + 2x)\,dx = \left(y\frac{x^3}{3} - y^3x + x^2\right)\Big|_{x=1}^{x=3} = (9y - 3y^3 + 9) - \left(\frac{1}{3}y - y^3 + 1\right) = \frac{26}{3}y - 2y^3 + 8.$$

Therefore

$$\iint_{\mathcal{R}} f(x,y)\,dA = \int_{-1}^{0} \left(\frac{26}{3}y - 2y^3 + 8\right)dy = \left(\frac{13}{3}y^2 - \frac{1}{2}y^4 + 8y\right)\Big|_{y=-1}^{y=0} = 0 - \left(\frac{13}{3} - \frac{1}{2} - 8\right) = \frac{25}{6}. \blacktriangleleft$$

INSIGHT Theorem 1 gives us a choice. We can calculate $\iint_{\mathcal{R}} f(x,y)\,dA$ by integrating first in the y variable and then in the x variable, as in Example 4. Alternatively, we can integrate first with respect to x and then with respect to y, as in Example 5.

▶ **EXAMPLE 6** Consider the integral $\iint_{\mathcal{R}} x\cos(xy)\,dA$, where $\mathcal{R}=\{(x,y): 0\le x\le 1, 0\le y\le \pi\}$. Use Theorem 1 to convert this to an iterated integral. Evaluate the iterated integral by performing the y integration first.

Solution The antiderivative of $x\cos(xy)$ with respect to y is $\sin(xy)$. Our integral is therefore

$$\iint_{\mathcal{R}} x\cos(xy)\,dA = \int_0^1 \left(\int_0^{\pi} x\cos(xy)\,dy\right)dx = \int_0^1 \left(\sin(xy)\Big|_{y=0}^{y=\pi}\right)dx$$
$$= \int_0^1 \big(\sin(\pi x)-\sin(0)\big)\,dx = -\frac{1}{\pi}\cos(\pi x)\Big|_{x=0}^{x=1} = -\frac{1}{\pi}(-1-1) = \frac{2}{\pi}. \quad ◀$$

INSIGHT In Example 6, we could have integrated with respect to x first. However, then we would have had to use the method of integration by parts. Because we chose to do the y integration first, our calculation was much simpler.

QUICK QUIZ

1. Estimate the volume under the graph of $f(x,y)=2x+3y^2$ and above the rectangle $[0,2]\times[2,6]$. Use a Riemann sum with $N=2$. For the choice of points $\{\xi_{ij}\}$, use the midpoint of each subrectangle.
2. The expression

$$\sum_{i=1}^{100}\sum_{j=1}^{100}\left(3+\frac{i}{50}\right)\left(4+\frac{j}{100}\right)^2\left(\frac{1}{50}\right)\left(\frac{1}{100}\right)$$

is a Riemann sum for the double integral $\iint_{\mathcal{R}}(xy^2)\,dA$. The upper-right vertex of each subrectangle Q_{ij} of $\mathcal{R}$ is used as the point ξ_{ij}. What is $\mathcal{R}$?
3. Calculate $\mathcal{I}(x)=\int_2^6(2x+3y^2)\,dy$ and $\mathcal{J}(y)=\int_0^2(2x+3y^2)\,dx$.
4. Evaluate $\iint_{\mathcal{R}}(2x+3y^2)\,dA$, where $\mathcal{R}=\{(x,y): 0\le x\le 2, 2\le y\le 6\}$.

Answers

1. 424 2. $[3,5]\times[4,5]$ 3. $\mathcal{I}(x)=8x+208$; $\mathcal{J}(y)=6y^2+4$ 4. 432

EXERCISES

Problems for Practice

▶ In each of Exercises 1–4, approximate the volume of the solid region that lies under the graph of the given function f and above the given rectangle R in the xy-plane. Use a Riemann sum with $N=2$ and, for the choice of points $\{\xi_{ij}\}$, the midpoints of the four subrectangles of R. ◀

1. $f(x,y)=x+2y$ $\quad R=[0,2]\times[1,5]$
2. $f(x,y)=1+6xy^2$ $\quad R=[0,1]\times[1,2]$
3. $f(x,y)=3x^2+2y$ $\quad R=[-1,3]\times[0,2]$
4. $f(x,y)=y/x$ $\quad R=[3,7]\times[2,4]$

▶ In each of Exercises 5–10, calculate $\mathcal{I}(x)=\int_c^d f(x,y)\,dy$ and $\mathcal{J}(y)=\int_a^b f(x,y)\,dx$ for the given function f and the given rectangle $R=[a,b]\times[c,d]$ in the xy-plane. ◀

5. $f(x,y)=x+2y$ $\quad R=[0,2]\times[1,5]$
6. $f(x,y)=1+6xy^2$ $\quad R=[0,1]\times[1,2]$
7. $f(x,y)=3x^2+2y$ $\quad R=[-1,3]\times[-2,2]$
8. $f(x,y)=y/x$ $\quad R=[3,7]\times[2,4]$
9. $f(x,y)=y\exp(xy^2)$ $\quad R=[0,1]\times[0,1]$
10. $f(x,y)=\sin(x+2y)$ $\quad R=[0,2\pi]\times[0,\pi]$

In each of Exercises 11–26, evaluate the double integral by converting it into an iterated integral.

11. $\iint_{\mathcal{R}}(x^2-y)\,dA$, $\mathcal{R}=\{(x,y): -2\le x\le 5, 1\le y\le 4\}$

12. $\iint_{\mathcal{R}}(\cos(x)-\sin(y))\,dA$, $\mathcal{R}=\{(x,y): 0\le x\le \pi/2 - \pi/3\le y\le \pi\}$

13. $\iint_{\mathcal{R}} x\cos(y)\,dA$, $\mathcal{R}=\{(x,y): 0\le x\le \pi, 0\le y\le \pi\}$

14. $\iint_{\mathcal{R}}(x+2y)\,dA$, $\mathcal{R}=\{(x,y): 0\le x\le 4,\ -2\le y\le 0\}$

15. $\iint_{\mathcal{R}} e^{x-y}\,dA$, $\mathcal{R}=\{(x,y): 0\le x\le 1,\ -2\le y\le 0\}$

16. $\iint_{\mathcal{R}}\sin(x)\tan(y)\,dA$, $\mathcal{R}=\{(x,y): 0\le x\le \pi/4,\ -\pi/3\le y\le 0\}$

17. $\displaystyle\iint_{\mathcal{R}}\left(\frac{\cos(x)}{y}-\frac{\cos(y)}{x}\right)dA$, $\mathcal{R}=\{(x,y): \pi/2\le x\le \pi, \pi\le y\le 2\pi\}$

18. $\iint_{\mathcal{R}}(x\,e^y - y\,e^x)\,dA$, $\mathcal{R}=\{(x,y): 0\le x\le 1, 1\le y\le 2\}$

19. $\iint_{\mathcal{R}}(y\sqrt{x}-x\sqrt{y})\,dA$, $\mathcal{R}=\{(x,y): 1\le x\le 4, 1\le y\le 4\}$

20. $\displaystyle\iint_{\mathcal{R}}\frac{x^2}{1+y^2}\,dA$, $\mathcal{R}=\{(x,y): 0\le x\le 2, 1\le y\le \sqrt{3}\}$

21. $\displaystyle\iint_{\mathcal{R}}\left(\frac{x}{y^2}+\frac{y}{x^3}\right)dA$, $\mathcal{R}=\{(x,y): -3\le x\le -1,\ -4\le y\le -2\}$

22. $\iint_{\mathcal{R}} e^x\cos(y)\,dA$, $\mathcal{R}=\{(x,y): 0\le x\le 1, 0\le y\le \pi/2\}$

23. $\iint_{\mathcal{R}}\cos^2(x)\sin^2(y)\,dA$, $\{(x,y): \pi/2\le x\le \pi, 0\le y\le \pi\}$

24. $\displaystyle\iint_{\mathcal{R}}\frac{y}{\sqrt{1-x^2}}\,dA$, $\mathcal{R}=\{(x,y): 0\le x\le 1/2,\ -1\le y\le 2\}$

25. $\iint_{\mathcal{R}}\cos^3(x)\sin(y)\,dA$, $\mathcal{R}=\{(x,y): 0\le x\le \pi/2, 0\le y\le \pi\}$

26. $\iint_{\mathcal{R}}(x+y)^3\,dA$, $\mathcal{R}=\{(x,y): 1\le x\le 2, 2\le y\le 3\}$

In each of Exercises 27–30, evaluate the given double integral twice by performing iterated integrations in both orders. Your answer should come out the same either way.

27. $\iint_{\mathcal{R}} x^3y^2\,dA$, $\mathcal{R}=\{(x,y): 0\le x\le 1, 0\le y\le 1\}$

28. $\iint_{\mathcal{R}}(x+y)^4\,dA$, $\mathcal{R}=\{(x,y): 0\le x\le 1, 1\le y\le 2\}$

29. $\displaystyle\iint_{\mathcal{R}}\frac{1}{x+y}\,dA$, $\mathcal{R}=\{(x,y): 0\le x\le 2, 1\le y\le e\}$

30. $\iint_{\mathcal{R}}\sin(2x+y)\,dA$, $\mathcal{R}=\{(x,y): -\pi/4\le x\le \pi, \pi/6\le y\le \pi\}$

Further Theory and Practice

In each of Exercises 31–34, evaluate the given double integral.

31. $\iint_{\mathcal{R}}\ln(xy)\,dA$, $\mathcal{R}=\{(x,y): 1\le x\le e, 1\le y\le e\}$

32. $\iint_{\mathcal{R}}\ln(x^y)\,dA$, $\mathcal{R}=\{(x,y): 1\le x\le e, 1\le y\le 2e\}$

33. $\iint_{\mathcal{R}} xy\ln(y)\,dA$, $\mathcal{R}=\{(x,y): 1\le x\le e, 1\le y\le e\}$

34. $\iint_{\mathcal{R}} y^2e^xe^y\,dA$, $\mathcal{R}=\{(x,y): -1\le x\le 1,\ -2\le y\le 0\}$

In each of Exercises 35–38, evaluate the given iterated integral.

35. $\int_0^1\left(\int_0^1 xe^{xy}\,dy\right)dx$

36. $\int_0^{1/2}\left(\int_0^{\pi} x\sin(xy)\,dy\right)dx$

37. $\int_1^4\left(\int_1^2 2x^2y\ln(x^2y^2)\,dy\right)dx$

38. $\int_0^{\pi}\left(\int_0^1 xy\cos(xy^2)\,dy\right)dx$

In each of Exercises 39–42, calculate $\iint_{\mathcal{R}} f(x,y)\,dA$ for the given function f and the given rectangle $\mathcal{R}$.

39. $f(x,y)=x^2\cos(xy)$, $\mathcal{R}=\{(x,y): -1\le x\le 1, -1\le y\le 1\}$

40. $f(x,y)=y^2\sin(xy)\cos(xy)$, $\mathcal{R}=\{(x,y): 0\le x\le 1, 0\le y\le \pi/2\}$

41. $f(x,y)=xy\ln(xy)\,dA$, $\mathcal{R}=\{(x,y): 1\le x\le e, 1\le y\le e\}$

42. $(x,y)=y\exp(x^2+y^2)$, $\mathcal{R}=\{(x,y): 0\le x\le 1,\ -1\le y\le 1\}$

43. Suppose that ϕ and ψ are continuous functions of one variable on the intervals $[a,b]$ and $[c,d]$ respectively. Show that if $f(x,y)=\phi(x)\,\psi(y)$, then

$$\int_a^b\left(\int_c^d f(x,y)\,dy\right)dx=\left(\int_a^b\phi(x)\,dx\right)\left(\int_c^d\psi(y)\,dy\right).$$

Let $\mathcal{R}$ denote the rectangle $[0,1]\times[0,2]$ in the xy-plane. Given that

$$\int_0^1 x^2(1-x^2)^{3/2}\,dx=\frac{\pi}{32}$$

and

$$\int_0^2 y^2(4-y^2)^{1/2}dy=\pi,$$

what is $\iint_{\mathcal{R}}(xy)^2\sqrt{(1-x^2)^3(4-y^2)}\,dA$?

44. Suppose that f has continuous partial derivatives on an open set that contains the rectangle $\mathcal{R}=[a,b]\times[c,d]$ in the xy-plane. Evaluate $\iint_{\mathcal{R}} f_{xy}(x,y)\,dA$.

In each of Exercises 45–48, functions $f(x,y)$ and $g(x,y)$ are given, along with a rectangle $\mathcal{R}$ in the xy-plane. Calculate the volume of the solid $\mathcal{U}$ that is bounded above by $z=f(x,y)$, below by $z=g(x,y)$, and that is above $\mathcal{R}$. In other words, what is the volume of

$$\mathcal{U}=\{(x,y,z): (x,y)\in\mathcal{R}, g(x,y)\le z\le f(x,y)\}?$$

45. $f(x,y)=1+x$, $g(x,y)=y$ $\quad\mathcal{R}=[1,2]\times[0,1]$

46. $f(x,y)=xy$, $g(x,y)=y$ $\quad\mathcal{R}=[1,3]\times[1,2]$

47. $f(x,y)=2x+y$, $g(x,y)=x+2y$ $\quad\mathcal{R}=[1,2]\times[0,1]$

48. $f(x,y)=e^x+y$, $g(x,y)=xy$ $\quad\mathcal{R}=[0,1]\times[0,1]$

Calculator/Computer Exercises

By using a calculator or computer, we may use approximation (12.1) with a large value of N to estimate a double integral $\iint_{\mathcal{R}} f(x,y)\,dA$. Figure 11 illustrates how this can be done using the computer algebra system Maple. In the illustration, the integrand is given by $f(x,y)=\sqrt{25-9x^2-y^2}$, and the rectangle of integration is $\mathcal{R}=[0,1]\times[0,4]$. Each side of the rectangle has been divided into $N=100$ equal subintervals, and the midpoints of the subrectangles have been used for the choice of points $\{\xi_{ij}\}$. In Exercises 49–52, approximate the double integral $\iint_{\mathcal{R}} f(x,y)\,dA$ for the given function f and rectangle $\mathcal{R}$. Use $N=50$, and take the midpoints of the subrectangles for the choice of points $\{\xi_{ij}\}$. (In each of these exercises, the suggested approximation is accurate to at least three decimal places.)

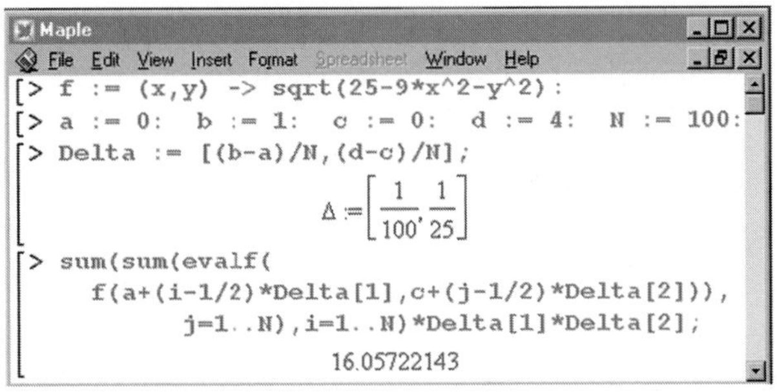

▲ Figure 11

49. $f(x,y)=\cos(\sqrt{1+x}+y)$ $\quad\mathcal{R}=[-1,1]\times[0,1]$

50. $f(x,y)=\sqrt{1+x+y^3}$ $\quad\mathcal{R}=[1,3]\times[0,1]$

51. $f(x,y)=\exp(-x^2-y^2)$ $\quad\mathcal{R}=[-1,1]\times[-1,1]$

52. $f(x,y)=(1+x+y)/(1+x^2+y^4)$ $\quad\mathcal{R}=[0,2]\times[0,1]$

12.2 Integration Over More General Regions

Many integration problems are formulated over planar regions other than rectangles. We need a technique for defining an integral over the interior of a triangle, a circle, an ellipse, or over even more complicated sets in the plane. We will restrict attention to regions bounded by a finite number of smooth curves. To begin, we need to know how to partition such a set. These are the considerations that we treat in the present section.

Planar Regions Bounded by Finitely Many Curves

Let $\mathcal{R}$ be a closed region in the plane that is contained in some bounded rectangle Q and that is bounded by finitely many continuously differentiable curves (see Figure 1). Suppose that $Q=\{(x,y): a\le x\le b, c\le y\le d\}$. Then, we can partition the rectangle Q in the usual way by partitioning the intervals $[a,b]$ and $[c,d]$:

$$a=x_0<x_1<x_2<\cdots<x_{N-1}<x_N=b$$

and

$$c=y_0<y_1<y_2<\cdots<y_{N-1}<y_N=d.$$

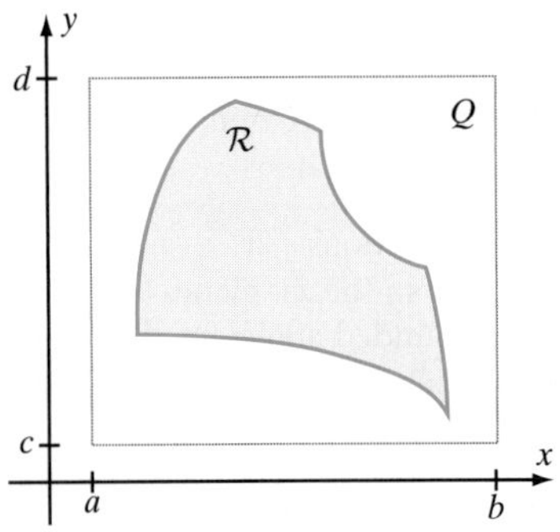

▲ Figure 1

For each pair x_{i-1}, x_i in the partition of $[a,b]$ and y_{j-1}, y_j in the partition of $[c,d]$ there is a small subrectangle $Q_{i,j}$ as shown in Figure 2. The only rectangles $Q_{i,j}$ relevant for integration over the region $\mathcal{R}$ are the ones that are completely contained in $\mathcal{R}$. Clearly the rectangles that lie completely outside $\mathcal{R}$ are irrelevant; the ones that overlap the boundary of $\mathcal{R}$ make a negligible error contribution when the subrectangles of the partition are small.

If f is a continuous function defined on $\mathcal{R}$, then we consider Riemann sums

$$\sum_{Q_{i,j}\subset\mathcal{R}} f(\xi_{i,j})\Delta x\Delta y=\sum_{Q_{i,j}\subset\mathcal{R}} f(\xi_{i,j})\Delta A,$$

where $\xi_{i,j}$ is any point selected from $Q_{i,j}$ for each i,j

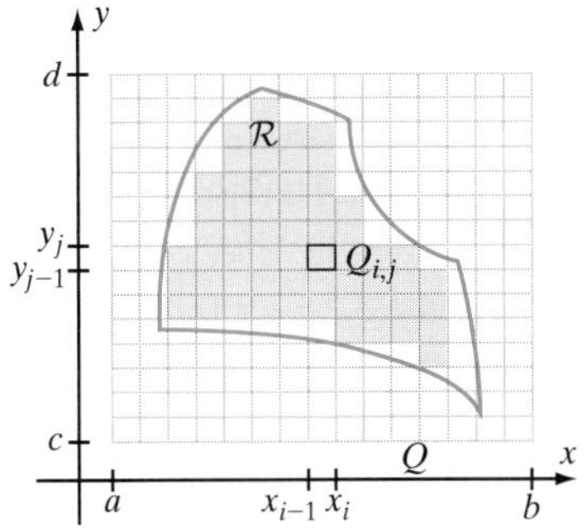

▲ Figure 2

A theorem from mathematical analysis guarantees that, if we let N tend to infinity, then these Riemann sums tend to a limit. We call this limit the *integral* of f over the region $\mathcal{R}$, and we denote it by

$$\iint_{\mathcal{R}} f(x,y)\, dA.$$

The practical issue that we now need to address is how to evaluate integrals over general regions $\mathcal{R}$. To keep things from getting too complicated, we restrict attention to two types of regions: the *x-simple regions*, which are regions of the form

$$\mathcal{R} = \{(x,y) : c \le y \le d, \alpha_1(y) \le x \le \alpha_2(y)\}$$

for some continuously differentiable functions $\alpha_1(y) \le \alpha_2(y)$ on $[c,d]$ (see Figure 3), and the *y-simple regions*, which are regions of the form

$$\mathcal{R} = \{(x,y) : a \le x \le b, \beta_1(x) \le y \le \beta_2(x)\}$$

for some continuously differentiable functions $\beta_1(x) \le \beta_2(x)$ on $[a,b]$ (see Figure 4). Notice that the first type of region is called x-simple because it is spanned from left to right by horizontal line segments; a similar reason motivates the name of the second type of region.

The following two theorems give us a method for evaluating integrals over x-simple and y-simple regions. They tell us that double integrals over simple regions can be evaluated by iterated integrals (just as with the rectangles of Section 12.1).

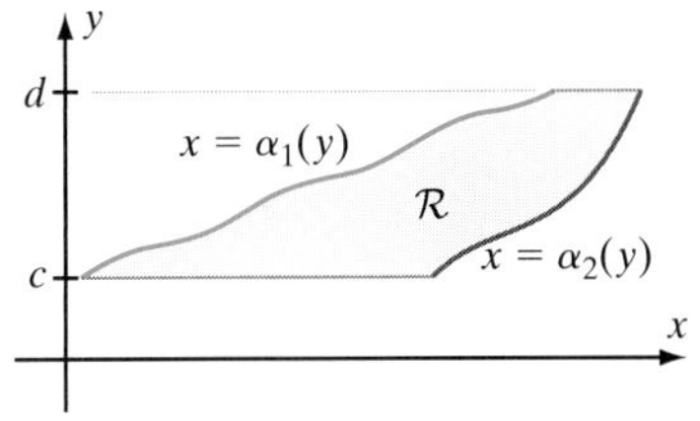

▲ Figure 3 The region $\mathcal{R} = \{(x,y) : c \le y \le d, \alpha_1(y) \le x \le \alpha_2(y)\}$ is x-simple.

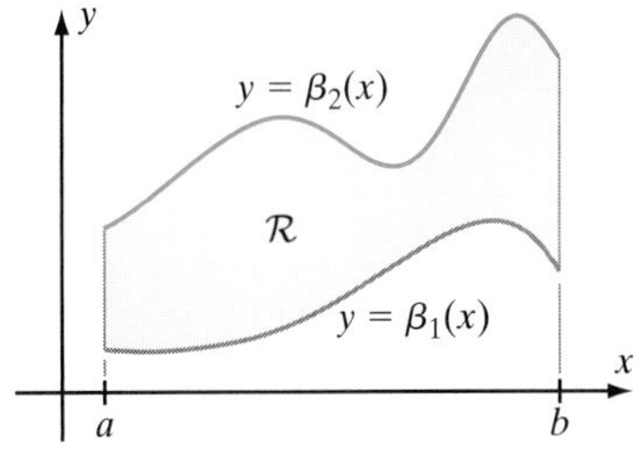

▲ Figure 4 The region $\mathcal{R} = \{(x,y) : a \le x \le b, \beta_1(x) \le y \le \beta_2(x)\}$ is y-simple.

THEOREM 1 Suppose that $\alpha_1(y) \le \alpha_2(y)$ for $y \in [c,d]$. Let

$$\mathcal{R} = \{(x,y) : c \le y \le d, \alpha_1(y) \le x \le \alpha_2(y)\}$$

be the corresponding x-simple region. If f is a continuous function on $\mathcal{R}$, then

$$\iint_{\mathcal{R}} f(x,y)\, dA = \int_c^d \left(\int_{\alpha_1(y)}^{\alpha_2(y)} f(x,y)\, dx \right) dy.$$

THEOREM 2 Suppose that $\beta_1(x) \le \beta_2(x)$ for $y \in [c,d]$. Let

$$\mathcal{R} = \{(x,y) : a \le x \le b, \beta_1(x) \le y \le \beta_2(x)\}$$

be the corresponding y-simple region. If f is a continuous function on $\mathcal{R}$, then

$$\iint_{\mathcal{R}} f(x,y)\, dA = \int_a^b \left(\int_{\beta_1(x)}^{\beta_2(x)} f(x,y) dy \right) dx.$$

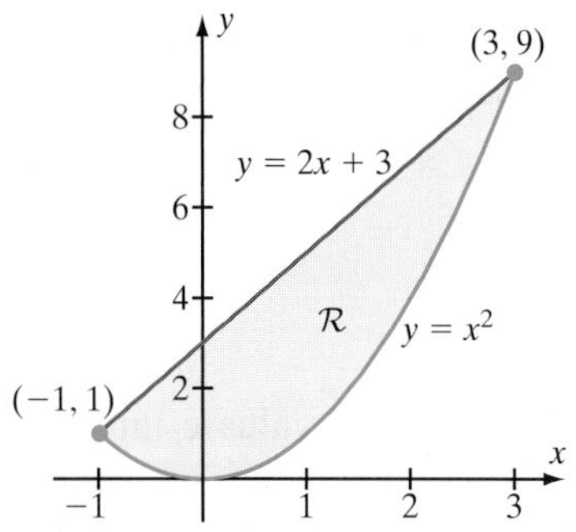

▲ Figure 5 The region $\mathcal{R}$ is y-simple.

You should think back to the method of disks that we used to calculate the volumes of solids of revolution. Recall that in those integrations, we would calculate the areas of slices in the y variable and then integrate out in the x variable, or vice versa. In Theorem 1, we are, in effect, taking slices in the x direction and then integrating out in the y variable. In Theorem 2, we are taking slices in the y direction and then integrating out in the x variable.

▶ **EXAMPLE 1** Integrate the function $f(x,y) = 3xy$ over the region $\mathcal{R}$ between the parabola $y = x^2$ and the line $y = 2x + 3$.

Solution To find the points of intersection of the parabola and the line, we solve $x^2 = y = 2x + 3$, or $x^2 - 2x - 3 = 0$. The left side of this equation factors to yield $(x+1)(x-3) = 0$, from which we see that $(-1,1)$ and $(3,9)$ are the points of intersection. Figure 5 reveals that $\mathcal{R} = \{(x,y) : -1 \le x \le 3, x^2 \le y \le 2x+3\}$ is a y-simple region. According to Theorem 2, we have

$$\iint_{\mathcal{R}} 3xy\, dA = \int_{-1}^{3} \left(\int_{x^2}^{2x+3} 3xy\, dy \right) dx = \int_{-1}^{3} \left(\frac{3}{2} xy^2 \Big|_{y=x^2}^{y=2x+3} \right) dx$$
$$= \frac{3}{2} \int_{-1}^{3} x\left((2x+3)^2 - (x^2)^2 \right) dx.$$

On expanding the integrand, we obtain

$$\iint_{\mathcal{R}} 3xy\, dA = \frac{3}{2} \int_{-1}^{3} (-x^5 + 4x^3 + 12x^2 + 9x)\, dx = \frac{3}{2}\left(-\frac{1}{6}x^6 + x^4 + 4x^3 + \frac{9}{2}x^2 \right)\Big|_{x=-1}^{x=3} = \frac{3}{2}\left(108 - \frac{4}{3} \right) = 160.$$ ◀

▶ **EXAMPLE 2** Integrate the function $f(x,y) = 2x - 4y$ over the region $\mathcal{R}$ between the curves $x = y^2$ and $x = y^3$ (see Figure 6).

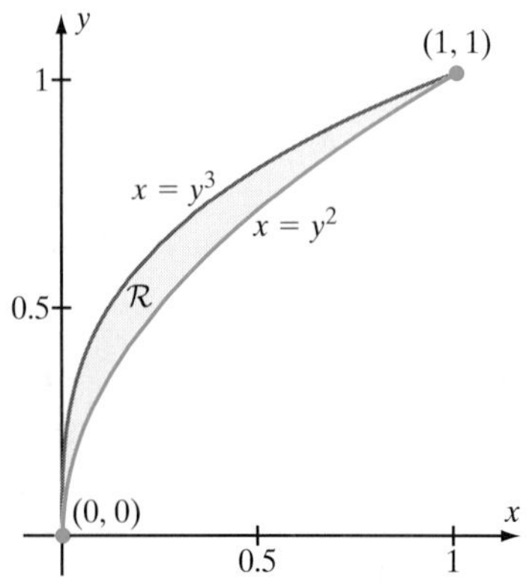

▲ Figure 6 The region $\mathcal{R}$ is x-simple and y-simple.

Solution We first notice that the curves intersect at the points $(0,0)$ and $(1,1)$. The region $\mathcal{R} = \{(x,y) : 0 \le y \le 1,\ y^3 \le x \le y^2\}$ is x-simple. According to Theorem 1, the value of the integral is

$$\iint_{\mathcal{R}} (2x - 4y)\, dA = \int_0^1 \left(\int_{y^3}^{y^2} (2x - 4y)\, dx \right) dy = \int_0^1 \left((x^2 - 4xy)\Big|_{x=y^3}^{x=y^2} \right) dy$$
$$= \int_0^1 \left((y^4 - 4y^3) - (y^6 - 4y^4) \right) dy.$$

The last integrand simplifies to $-y^6 + 5y^4 - 4y^3$ and therefore

$$\iint_{\mathcal{R}} (2x - 4y)\, dA = \int_0^1 (-y^6 + 5y^4 - 4y^3)\, dy = \left(-\frac{1}{7}y^7 + y^5 - y^4 \right)\Big|_{y=0}^{y=1} = -\frac{1}{7}.$$ ◀

INSIGHT Notice that the region $\mathcal{R}$ in Example 2 is also y-simple; indeed,

$$\mathcal{R} = \{(x,y) : 0 \le x \le 1,\ x^{1/2} \le y \le x^{1/3}\}.$$

From this point of view, the integral of f over $\mathcal{R}$ is calculated as follows:

$$\iint_{\mathcal{R}} (2x-4y)\, dA = \int_0^1 \left(\int_{x^{1/2}}^{x^{1/3}} (2x-4y)dy \right) dx = 2\int_0^1 (xy-y^2)\Big|_{y=x^{1/2}}^{y=x^{1/3}} dx$$

$$= \int_0^1 (x^{4/3}-x^{2/3}-x^{3/2}+x)\, dx = \left(\frac{6}{7}x^{7/3} - \frac{6}{5}x^{5/3} - \frac{4}{5}x^{5/2} + x^2 \right)\Big|_0^1 = -\frac{1}{7}.$$

Changing the Order of Integration

Sometimes an iterated integral that is difficult to calculate becomes much easier if the order of integration is reversed. Theorems 1 and 2 taken together guarantee that if a region is both x-simple and y-simple, then we get the same answer no matter which order we choose to do the integration.

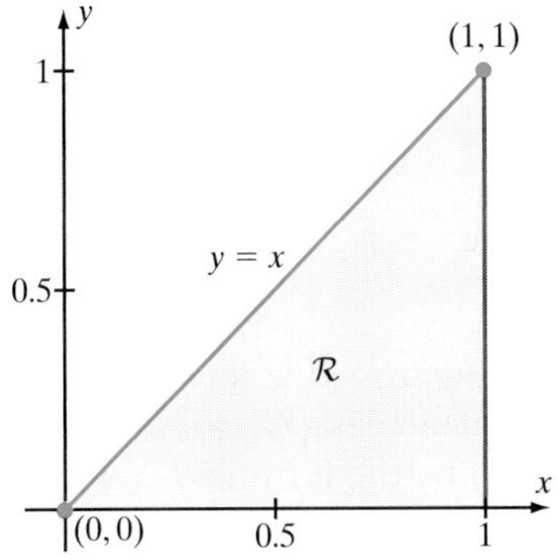

▲ Figure 7 The region $\mathcal{R}$ is x-simple and y-simple.

▶ **EXAMPLE 3** Calculate the integral $\int_0^1\int_y^1 6\,y\,e^{x^3}\,dxdy$.

Solution The inner integral with respect to x is not one that we can do with the techniques that we have learned. Let us instead change the order of integration in an attempt to reformulate the problem. A glance at Figure 7 shows that the region $\mathcal{R}$ of integration is not only x-simple but also y-simple:

$$\mathcal{R} = \{(x,y) : 0 \le x \le 1, 0 \le y \le x\}.$$

Therefore the given iterated integral may be rewritten as

$$\int_0^1 \left(\int_0^x 6\,y\,e^{x^3}\,dy \right) dx,$$

an integral that we easily evaluate as follows:

$$\int_0^1 \left(\int_0^x 6ye^{x^3}\,dy \right) dx = \int_0^1 \left(3y^2e^{x^3}\Big|_{y=0}^{y=x} \right) dx = \int_0^1 3x^2e^{x^3}\,dx = e^{x^3}\Big|_{x=0}^{x=1} = e-1. \quad ◀$$

INSIGHT When you are changing the order of integration, there is no substitute for drawing a good picture. It tells you immediately the new limits of integration and the order in which they should appear.

Part of our work in switching the order of integration in Example 3 was to re-express the boundary curves in terms of the x variable instead of the y variable. Sometimes, the domain of a double integral must be broken up if it is to be evaluated in a certain order. However, it may be possible to avoid breaking up the integral if we evaluate in the opposite order. Here is an example.

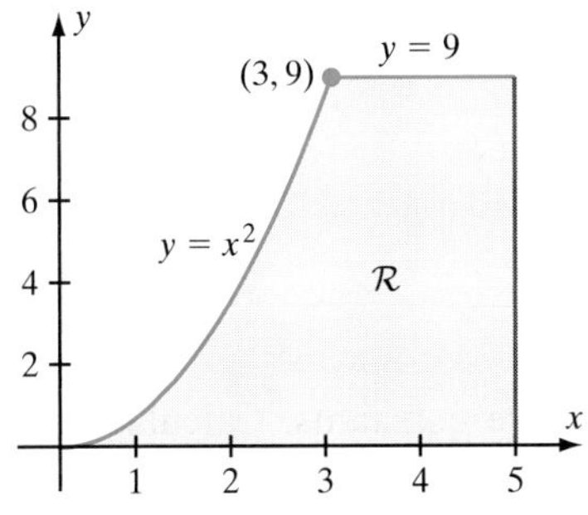

▲ Figure 8

▶ **EXAMPLE 4** Evaluate the integral $\iint_{\mathcal{R}} 4xy\,dA$ where $\mathcal{R}$ is the region shown in Figure 8.

Solution Notice that the region of integration is the union of two y-simple regions. If we choose to apply Theorem 2 and make y the inside variable, then we must note that for points (x, y) in $\mathcal{R}$ with $0 \le x \le 3$, the ordinate y satisfies $0 \le y \le x^2$, whereas when $3 \le x \le 5$, the ordinate y satisfies $0 \le y \le 9$. Because of the different formulas for y at the upper limits of integration, namely $y = x^2$ over the interval $[0, 3]$ and $y = 9$ over the interval $[3, 5]$, we must write the given double integral as a sum of two iterated integrals,

$$\iint_{\mathcal{R}} 4xy\, dA = \int_0^3 \left(\int_0^{x^2} 4xy\, dy \right) dx + \int_3^5 \left(\int_0^9 4xy\, dy \right) dx. \qquad (12.2.1)$$

On the other hand, $\mathcal{R}$ is x-simple. In fact, $\mathcal{R} = \{(x, y) : 0 \le y \le 9, \sqrt{y} \le x \le 5\}$. Therefore by Theorem 1,

$$\begin{aligned} \iint_{\mathcal{R}} 4xy\, dA &= \int_0^9 \left(\int_{\sqrt{y}}^5 4xy\, dx \right) dy = \int_0^9 \left(2x^2 y \Big|_{x=\sqrt{y}}^{x=5} \right) dy \\ &= \int_0^9 (50y - 2y^2)\, dy = \left(25y^2 - \frac{2}{3} y^3 \right) \Big|_0^9 = (25)(81) - \frac{2}{3} 9^3 = 1539. \end{aligned}$$

You may verify this answer by showing that the first integral on the right side of equation (12.2.1) evaluates to 243 and the second to 1296, for a sum of 1539. ◄

The Area of a Planar Region

Suppose that $\mathcal{R}$ is a bounded region in the plane, as in Figure 9. Fix a rectangle Q that contains $\mathcal{R}$ and partition Q into N^2 congruent subrectangles $\{Q_{i,j}\}$. Next, let f be the function that is identically 1 on $\mathcal{R}$. That is, define f by the formula $f(x, y) = 1$ for every point (x, y) in $\mathcal{R}$. Then, whatever choice of points $\{\xi_{i,j}\}$ is made, the Riemann sum for f over $\mathcal{R}$ is equal to

$$\sum_{Q_{i,j} \subseteq \mathcal{R}} 1\, \Delta A. \qquad (12.2.2)$$

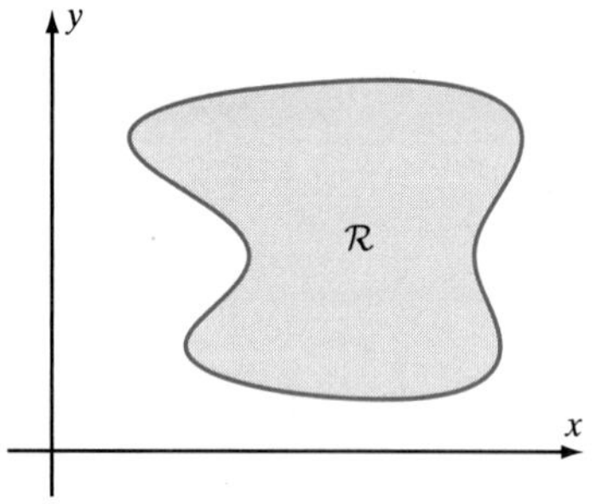

▲ **Figure 9** A bounded region in the plane.

This quantity is just the sum of the areas of the small rectangles $Q_{i,j}$ that are contained in $\mathcal{R}$ (see Figure 10). When the number N^2 of subrectangles of the partition is large, formula (12.2.2) gives a good approximation to the area of $\mathcal{R}$. It can be shown that, if the boundary of $\mathcal{R}$ is a piecewise smooth curve, then the Riemann sums in line (12.2.2) tend to a limit as N tends to infinity. We denote this limit by $\iint_{\mathcal{R}} 1\, dA$ and make the following definition.

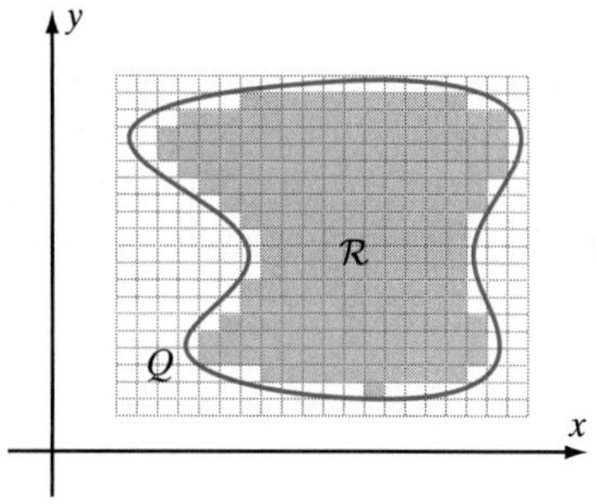

▲ **Figure 10** The area of the shaded region is the Riemann sum $\sum_{Q_{i,j} \subseteq R} 1 \Delta A$.

> **DEFINITION** If $\mathcal{R}$ is a region in the plane bounded by finitely many continuously differentiable curves, then the area of $\mathcal{R}$ is defined to be
>
> $$\iint_{\mathcal{R}} 1\, dA.$$

► **EXAMPLE 5** Suppose that a and b are positive constants. Calculate the area $\mathcal{A}$ inside the ellipse $x^2/a^2 + y^2/b^2 = 1$.

Solution The given ellipse is a y-simple region:

$$\mathcal{R} = \left\{(x,y) : -a \le x \le a, -b\sqrt{1-x^2/a^2} \le y \le b\sqrt{1-x^2/a^2}\right\}.$$

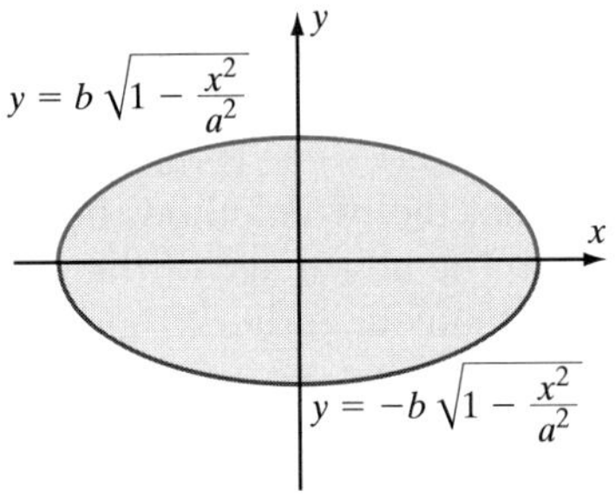

▲ **Figure 11** The ellipse $\frac{x^2}{a^2} + \frac{y^2}{b^2} = 1$ is a y-simple region.

See Figure 11. We have

$$\mathcal{A} = \iint_{\mathcal{R}} 1\, dA = \int_{-a}^{a} \left(\int_{-b\sqrt{1-x^2/a^2}}^{b\sqrt{1-x^2/a^2}} 1\, dy \right) dx = \int_{-a}^{a} \left(b\sqrt{1-\frac{x^2}{a^2}} - \left(-b\sqrt{1-\frac{x^2}{a^2}} \right) \right) dx,$$

or

$$\mathcal{A} = 2b \int_{-a}^{a} \sqrt{1 - \frac{x^2}{a^2}}\, dx. \tag{12.2.3}$$

In Section 6.3 of Chapter 6, we learned that this type of integral can be handled by the change of variables $x = a\sin(\theta)$, $dx = a\cos(\theta)d\theta$ which, using formula (6.2.4), leads to

$$\mathcal{A} = 2ab \int_{-\pi/2}^{\pi/2} \sqrt{1-\sin^2(\theta)}\cos(\theta)\, d\theta = 2ab \int_{-\pi/2}^{\pi/2} \cos^2(\theta)\, d\theta = 2ab \left(\frac{\theta}{2} + \frac{\sin(2\theta)}{4} \right) \Bigg|_{-\pi/2}^{\pi/2} = \pi ab. \quad ◀$$

INSIGHT It sometimes happens that an integrand can be successfully treated by different substitutions. If the change of variable $x = a\,u$, $dx = a\,du$ is made in line (12.2.3), then the area computation goes as follows:

$$\mathcal{A} = 2b \int_{-a}^{a} \sqrt{1-\frac{x^2}{a^2}}\, dx = 2ab \int_{-1}^{1} \sqrt{1-u^2}\, du$$

$$= 2ab\ (\text{area of the upper half of the unit circle}) = 2ab\left(\frac{\pi}{2}\right) = \pi ab.$$

▶ **EXAMPLE 6** Suppose that $\mathcal{R}_1$ is the x-simple region given by $\mathcal{R}_1 = \{(x,y) : c \le y \le d, \alpha_1(y) \le x \le \alpha_2(y)\}$. Suppose that $\mathcal{R}_2$ is the y-simple region given by $\mathcal{R}_2 = \{(x,y) : a \le x \le b, \beta_1(x) \le y \le \beta_2(x)\}$. Using the double integral definition of planar area, express the areas of $\mathcal{R}_1$ and $\mathcal{R}_2$ as "single" (as opposed to double) Riemann integrals.

Solution According to Theorem 1, the area of $\mathcal{R}_1$ is given by

$$\iint_{\mathcal{R}} 1\, dA = \int_c^d \left(\int_{\alpha_1(y)}^{\alpha_2(y)} 1\, dx \right) dy = \int_c^d \left(x \Big|_{x=\alpha_1(y)}^{x=\alpha_2(y)} \right) dy = \int_c^d (\alpha_2(y) - \alpha_1(y))dy.$$

Similarly, using Theorem 2, the area of $\mathcal{R}_2$ is given by

$$\iint_{\mathcal{R}} 1\, dA = \int_a^b \left(\int_{\beta_1(x)}^{\beta_2(x)} 1\, dy \right) dx = \int_a^b \left(y \Big|_{y=\beta_1(x)}^{y=\beta_2(x)} \right) dx = \int_a^b (\beta_2(x) - \beta_1(x))dx. \quad ◀$$

A LOOK BACK In Section 5.7 of Chapter 5, we learned to express the area between two curves using a single integral. Example 6 shows that, in such a situation, the definition of area by means of a double integral leads to the very same formula that our earlier method prescribes. Of course it is essential that the two methods be consistent when they can be applied to the same region. The advantage of the double integral definition is that it has wider applicability than our earlier definition.

QUICK QUIZ

1. The triangular region $\mathcal{R}$ with vertices $(0,0)$, $(2,6)$, and $(3,6)$ is x-simple. Specify constants c and d, and functions $\alpha_1(y)$ and $\alpha_2(y)$, for which $\mathcal{R} = \{(x,y): c \le y \le d, \alpha_1(y) \le x \le \alpha_2(y)\}$.
2. Let $\mathcal{R}$ be the region bounded by the vertical line segment with endpoints $O=(0,0)$ and $P=(0,1)$, the horizontal line segment with endpoints $P=(0,1)$ and $Q=(\pi/2,1)$, and the arc of $y=\sin(x)$ between O and Q. The area of $\mathcal{R}$ is $\int_0^1 \int_0^{\alpha(y)} 1\,dx\,dy$ for what function $\alpha(y)$?
3. For what constant b and function $\beta(x)$ is $\int_1^2 \int_{y^2}^{3y-2} f(x,y)\,dx\,dy = \int_1^b \int_{1+(x-1)/3}^{\beta(x)} f(x,y)\,dydx$?
4. Evaluate the integral $\iint_{\mathcal{R}} 2x^3y\,dA$, where $\mathcal{R} = \{(x,y) : 0 \le x \le 1, 0 \le y \le x^2\}$.

Answers

1. $c=0$, $d=6$, $\alpha_1(y) = y/3$, $\alpha_2(y) = y/2$ 2. $\alpha(y) = \arcsin(y)$
3. $b=4$, $\beta(x) = \sqrt{x}$ 4. $1/8$

EXERCISES

Problems for Practice

In each of Exercises 1–6, sketch the region $\mathcal{R}$ that is described. Decide whether $\mathcal{R}$ is x-simple, y-simple, or both x-simple and y-simple. If $\mathcal{R}$ is x-simple (respectively, y-simple), state the functions $y \mapsto \alpha_1(y)$ and $y \mapsto \alpha_2(y)$ (respectively, $x \mapsto \beta_1(x)$ and $x \mapsto \beta_2(x)$) whose graphs form part of its boundary.

1. The region between the curves $x = -y^2 + 2$ and $-2y - 1$
2. The region between the curves $y = x$ and $y = x^2$
3. The region between curves $x = y^2$ and $x = -2y^2 - 3y + 18$
4. The region between the curves $y = x^2$ and $y = 8x^{1/2}$
5. The region above the curve $y = x^{3/2}$, below the curve $y = 6x - 1$, and between $x = 1$ and $x = 3$
6. The region below $y = 2$, above $y = -1$, and between the curves $y = x^3$ and $y = x + 8$

In each of Exercises 7–14, evaluate the given iterated integral.

7. $\int_{-1}^{1} \int_{(x+1)^2}^{8-(x+1)^2} 3x\,dydx$

8. $\int_1^{64} \int_{x^{1/2}}^{x^{1/3}} (y/x)\,dydx$

9. $\int_0^2 \int_{x/4}^{4x} (2x-1)\sqrt{y}\,dydx$

10. $\int_0^1 \int_y^{\sqrt{y}} 2xe^y\,dxdy$

11. $\int_{\pi/2}^{\pi} \int_0^{\sin(x)} 4x\,dydx$

12. $\int_1^4 \int_y^{y+1} 4x\,dxdy$

13. $\int_{\pi/4}^{\pi/2} \int_0^{\sin(y)} 5\,dxdy$

14. $\int_1^3 \int_{x-1}^{x+1} y^{1/2}(x+1)\,dydx$

In each of Exercises 15–22, calculate the integral of the given function f over the given region $\mathcal{R}$.

15. $f(x,y) = 6y - 4x$; $\mathcal{R}$ is the region between the curves $y = 2x$ and $y = x^2$.
16. $f(x,y) = \cos(x)$; $\mathcal{R}$ is the region between the curves $y = x$ and $y = 2x$, $0 \le x \le \pi$.
17. $f(x,y) = xe^y$; $\mathcal{R}$ is the region above $y = 0$, below $y = 2x - 6$, and between $x = 4$ and $x = 7$.
18. $f(x,y) = 9\sqrt{y}/4$; $\mathcal{R}$ is the region below $y = 4$, above $y = 1$, and between $x = y^3$ and $x = y^2 + 48$.
19. $f(x,y) = y\sqrt{x}$; $\mathcal{R}$ is the region between $x = 3$ and $x = 4$, above $y = x^2 - 3x$ and below $y = x^2 + 8$.
20. $f(x,y) = y^2/x^2$; $\mathcal{R}$ is the region above $y = -5$, below $y = -3$, and between $x = y^2$ and $x = 5$.
21. $f(x,y) = e^x$; $\mathcal{R}$ is the region between $x = y$ and $x = 6y$, $0 \le x \le 1$.

22. $f(x,y)=y\sin(x)$; $\mathcal{R}$ is the region bounded by $y=\cos(x)$ and $y=\sin(x)$, $0\le x\le \pi/4$.

In each of Exercises 23–26, switch the order of integration to make the integral manageable and then calculate the iterated integral.

23. $\displaystyle\int_0^{\pi/2}\int_x^{\pi/2}\frac{\sin(y)}{y}\,dydx$

24. $\displaystyle\int_1^2\int_1^{x^2}\frac{x}{y}\,dydx$

25. $\displaystyle\int_0^2\int_{x^2}^4\frac{e^y}{\sqrt{y}}\,dydx$

26. $\int_0^1\int_y^1 \exp(-x^2)\,dxdy$

Further Theory and Practice

In each of Exercises 27–34, calculate the given integral.

27. $\int_0^{\sqrt{\pi}}\int_x^{\sqrt{\pi}}\sin(y^2)\,dydx$

28. $\int_0^1\int_0^{\arccos(y)}\cos^4(x)\,dxdy$

29. $\displaystyle\int_0^1\int_{\sqrt{y}}^1\frac{y(1-x^2)^2}{x^3}\,dxdy$

30. $\int_0^2\int_0^{\sqrt{4-x^2}}(4-y^2)^{3/2}\,dydx$

31. $\iint_{\mathcal{R}}(x-y)\,dA$ where $\mathcal{R}$ is the region bounded by $y=\cos(x)$ and $y=\sin(x)$, $0\le x\le\pi/4$.

32. $\iint_{\mathcal{R}}(y+\sec^2(x))\,dA$ where $\mathcal{R}$ is the region bounded by $y=\tan(x)$ for $0\le x\le\pi/4$, $y=-x$ for $-1\le x\le 0$, and $y=1$ for $-1\le x\le\pi/4$

33. $\int_0^1\left(\int_y^1 x^2\sin(xy)\,dx\right)dy$

34. $\int_0^3\left(\int_0^{\sqrt{9-y^2}}\sqrt{9-x^2}\,dx\right)dy$

In each of Exercises 35–44, sketch the set $\mathcal{R}$ that is the union of the two regions of integration. Reverse the order of integrations and write the given sum as one iterated integral over $\mathcal{R}$. Calculate this iterated integral.

35. $\int_{-1}^0\int_{-x}^1 y\,dydx+\int_0^1\int_{x^2}^1 y\,dydx$

36. $\int_0^1\int_0^y xy\,dxdy+\int_1^2\int_0^{2-y}xy\,dxdy$

37. $\int_{-1}^0\int_{-y}^1 xy\,dxdy+\int_0^1\int_{\sqrt{y}}^1 xy\,dxdy$

38. $\int_0^1\int_0^x xy\,dydx+\int_1^3\int_0^1 xy\,dydx$

39. $\int_0^1\int_{-\sqrt{y}}^{\sqrt{y}}x\,dxdy+\int_1^4\int_{y-2}^{\sqrt{y}}x\,dxdy$

40. $\int_0^1\int_1^2 1dydx+\int_1^2\int_1^{x^2-4x+5}1dydx$

41. $\int_{-\sqrt{2}}^{-1}\int_{-\sqrt{2-y^2}}^{\sqrt{2-y^2}}x\,dxdy+\int_{-1}^1\int_y^1 x\,dxdy$

42. $\int_0^1\int_{-\sqrt{y}}^{\sqrt{y}}y\,dxdy+\int_1^2\int_{-\sqrt{2-y}}^{\sqrt{2-y}}y\,dxdy$

43. $\int_0^{1/\sqrt{2}}\int_0^{\arcsin(y)}y\,dxdy+\int_{1/\sqrt{2}}^1\int_0^{\arccos(y)}y\,dxdy$

44. $\int_0^1\int_{-1}^{-y}|x|\,dxdy+\int_0^1\int_y^1|x|\,dxdy$

In each of Exercises 45–54, calculate $\iint_{\mathcal{R}}f(x,y)\,dA$ for the given function f and region $\mathcal{R}$.

45. $f(x,y)=ax+by$; $\mathcal{R}$ is the triangle with vertices $(0,0)$, $(1,3)$, $(10,4)$.

46. $f(x,y)=x^2+y$; $\mathcal{R}$ is the parallelogram with vertices $(-1,0)$, $(0,0)$, $(1,1)$, $(0,1)$.

47. $f(x,y)=x+y$; $\mathcal{R}$ is the region in the first quadrant that is below $y=1+x^2$ and to the left of $y=5(x-1)$.

48. $f(x,y)=x^2y$; $\mathcal{R}$ is the region bounded above by $y=4+2x-x^2$ and below by $y=x^2$.

49. $f(x,y)=x+y$; $\mathcal{R}$ is that part of the unit circle that lies in the first quadrant.

50. $f(x,y)=x\sqrt{y}$; $\mathcal{R}$ is the region above $y=1-x$ and inside the unit circle.

51. $f(x,y)=y\sin(x)$; $\mathcal{R}$ is the region in the first quadrant that is bounded above by $y=1-\cos(x)$, and that is bounded on the right by $x=\pi/2$.

52. $f(x,y)=1/x^3$; $\mathcal{R}=\{(x,y): 1\le x\le 2, x\sin(1/x)\le y\le 2\}$.

53. $f(x,y)=|y-x|$; $\mathcal{R}$ is the region in the first quadrant that is bounded above by $y=2x$, and on the right by $x=1$.

54. $f(x,y)=2y/(2-x)$; $\mathcal{R}$ is the region in the first quadrant that is bounded above by $y=\sqrt{4-x^2}$ and on the right by $x=1$.

55. Show that the error function

$$\operatorname{erf}(x)=\frac{2}{\sqrt{\pi}}\int_0^x\exp(-y^2)\,dy$$

satisfies

$$\int_0^z\operatorname{erf}(x)\,dx=z\operatorname{erf}(z)-\frac{1}{\sqrt{\pi}}(1-\exp(-z^2)).$$

(Hint: Write the left side as an iterated integral and change the order of integration.) Deduce that

$$\int_0^z(1-\operatorname{erf}(x))\,dx=z(1-\operatorname{erf}(z))+\frac{1}{\sqrt{\pi}}(1-\exp(-z^2)).$$

Given that $\lim_{z\to\infty}\operatorname{erf}(z)=1$, use l'Hôpital's Rule to evaluate the convergent improper integral $\int_0^\infty(1-\operatorname{erf}(x))\,dx$.

Calculator/Computer Exercises

In each of Exercises 56–59, a function f and a region $\mathcal{R}$ are given. Calculate $\iint_{\mathcal{R}}f(x,y)\,dA$ by writing the double integral as an iterated integral and finding the outer limits of integration.

56. $f(x,y)=1+x+xy$; $\mathcal{R}$ is the region that is bounded above by $y=1-x^4$ and below by $y=x^3$.

57. $f(x,y)=xy$; $\mathcal{R}$ is the region that is bounded above by $y=1+3x$ and below by $y=e^x$.

58. $f(x, y) = x$; $\mathcal{R}$ is the region in the first quadrant that is bounded above by $y = 4 - x^{1/8}$ and below by $y = 1 + \sqrt{x}$.

59. $f(x, y) = x$; $\mathcal{R}$ is the region in the first quadrant that is bounded above by $y = 1/(1 + x^2)$ and below by $y = x^{10}$.

In each of Exercises 60–63, a function f and a region $\mathcal{R}$ are given. Approximate $\iint_{\mathcal{R}} f(x, y)\, dA$ to two decimal places by writing the double integral as an iterated integral for which the inner integration can be evaluated exactly. Use Simpson's Rule for the outer integration.

60. $f(x, y) = \sqrt{1 - x^4}$; $\mathcal{R}$ is that part of the unit circle that lies in the first quadrant .

61. $f(x, y) = \sqrt{x^2 + y}$; $\mathcal{R}$ is that part of the unit circle that lies in the first quadrant.

62. $f(x, y) = \sqrt{x + y}$; $\mathcal{R}$ is the region in the first quadrant that is bounded above by $y = x^3$ and on the right by $x = 1$.

63. $f(x, y) = \exp(-y^2)$; $\mathcal{R}$ is the region in the first quadrant that is bounded on the left by $y = \tan(x)$ for $0 \le x \le \pi/4$, and on the right by $y = \sqrt{2}\cos(x)$ for $\pi/4 \le x \le \pi/2$.

The iterated integral $I = \int_a^b \left(\int_{\beta_1(x)}^{\beta_2(x)} f(x, y)\, dy\right) dx$ may be approximated by using an appropriate Trapezoidal Rule. Let $\Delta x = (b - a)/N$, $x_i = a + i\Delta x$ $0 \le i \le N$, $(\Delta y)_i = (\beta_2(x_i) - \beta_1(x_i))/N$, and $y_{ij} = \beta_1(x_i) + j(\Delta y)_i$ $0 \le j \le N$. Then,

$$I \approx \frac{\Delta x}{2} \sum_{i=0}^{N} \frac{(\Delta y)_i}{2} \sum_{j=0}^{N} c_{ij} f(x_i, y_{ij})$$

where $c_{ij} = 1$ if $\{i, j) \subset \{0, N)$, $c_{ij} = 4$ if $\{i, j) \subset \{1, 2, \ldots, N - 1)$, and $c_{ij} = 2$ otherwise (i.e., if exactly one of i and j is 0 or N). In Exercises 64–67, approximate the given integral using $N = 4$. Calculate the integral exactly. What is the approximation error?

64. $\int_0^1 \int_0^x (x - y)\, dy dx$

65. $\int_{-1}^1 \int_0^{1-x^2} y\, dy dx$

66. $\int_{\pi/6}^{\pi/4} \int_0^{\tan(x)} \cot(x)\, dy dx$

67. $\int_0^1 \int_{x^2}^x \sqrt{x}\, dy dx$

12.3 Calculation of Volumes of Solids

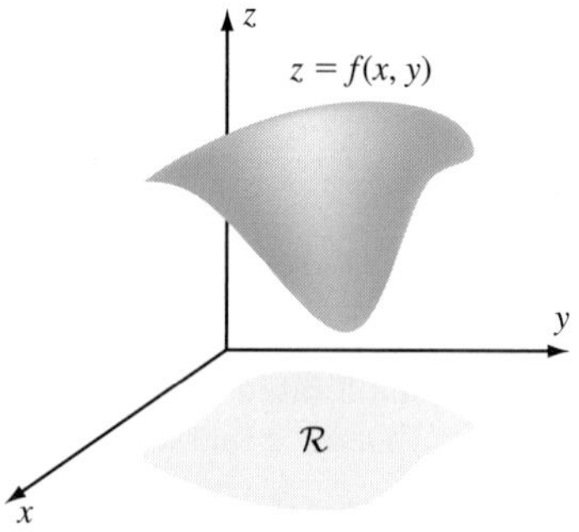

▲ Figure 1a The graph of f over a planar region $\mathcal{R}$

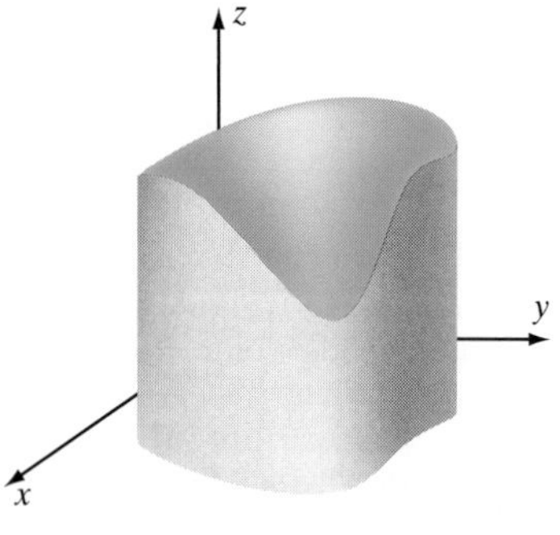

▲ Figure 1b The solid $\mathcal{U}$ that lies under the graph of f and over the planar region $\mathcal{R}$

In this section, we use double integrals to calculate volumes of solids in three-dimensional space. To be specific, given a nonnegative continuous function f defined on a region $\mathcal{R}$ of the xy-plane, we are interested in determining the volume of the solid that is above the region $\mathcal{R}$ and below the graph of f (see Figure 1). Recall that we motivated integration of functions of two variables by volume considerations. We now use the integral to precisely define what we mean by the volume of a solid.

DEFINITION Suppose that f is a nonnegative, continuous function defined on a region $\mathcal{R}$ of the xy-plane. Let $\mathcal{U}$ be the three-dimensional solid that lies above $\mathcal{R}$ and below the graph of f. The volume of $\mathcal{U}$ is defined to be

$$\iint_{\mathcal{R}} f(x, y)\, dA.$$

▶ **EXAMPLE 1** Figure 2 illustrates the three-dimensional solid $\mathcal{U}$ that lies below the plane $f(x, y) = 2x - 6y$ and above the planar region

$$\mathcal{R} = \{(x, y) : 2 \le x \le 5, -3 \le y \le -1\}.$$

Find the volume of $\mathcal{U}$.

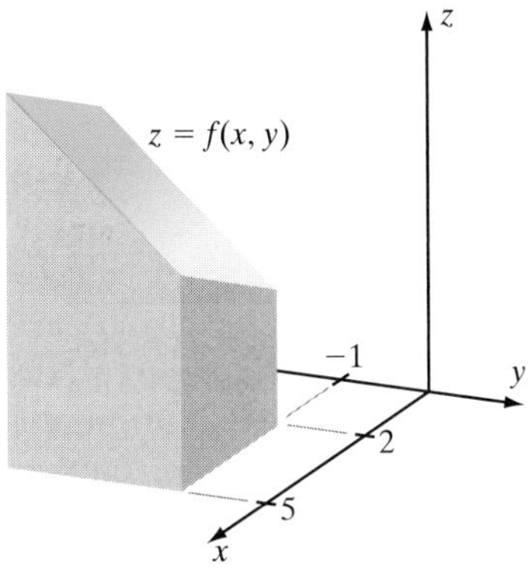

▲ Figure 2 The solid that lies under the graph of $f(x, y) = 2x - 6y$ and over a planar rectangle

Solution Because $x > 0$ and $y < 0$ on $\mathcal{R}$, we have $f(x, y) > 0$ on $\mathcal{R}$. The volume of $\mathcal{U}$ is therefore

$$\iint_{\mathcal{R}} f(x,y)\,dA = \int_{-3}^{-1}\left(\int_{2}^{5}(2x-6y)\,dx\right)dy = \int_{-3}^{-1}(x^2-6xy)\Big|_{x=2}^{x=5}dy$$
$$= \int_{-3}^{-1}(21-18y)\,dy = (21y-9y^2)\Big|_{-3}^{-1} = 114.$$

Sometimes a little geometric analysis is required before we can calculate a volume as shown in the next example. ◀

▶ **EXAMPLE 2** Calculate the volume below the plane $8x + 4y + 2z = 16$, above the xy-plane, and in the first octant.

Solution Solving for z, we think of the plane as the graph of the function $z = f(x, y) = 8 - 4x - 2y$. We notice that $f(x, y) = 0$ when (x, y) lies on the line $4x + 2y = 8$ in the xy-plane. Therefore if $\mathcal{R}$ is the region in the xy-plane that is bounded by the x-axis, the y-axis, and the line $4x + 2y = 8$, then our problem is to find the volume of the solid that is below the graph of f and above $\mathcal{R}$ (see Figure 3). The region $\mathcal{R}$ is y-simple:

$$\mathcal{R} = \{(x, y) : 0 \le x \le 2, 0 \le y \le 4 - 2x\}.$$

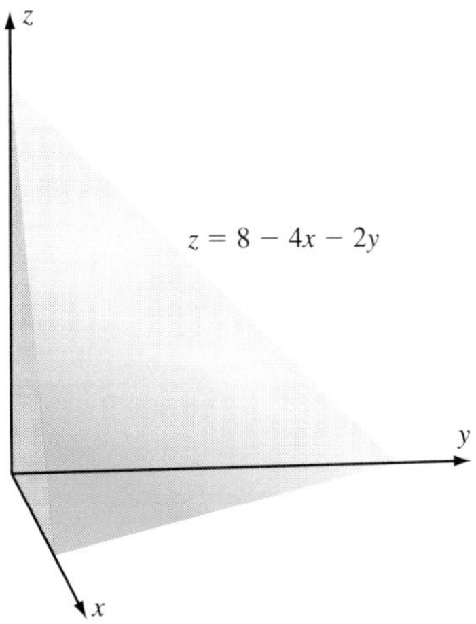

▲ Figure 3a The solid in the first octant that lies below the plane $z = 8 - 4x - 2y$

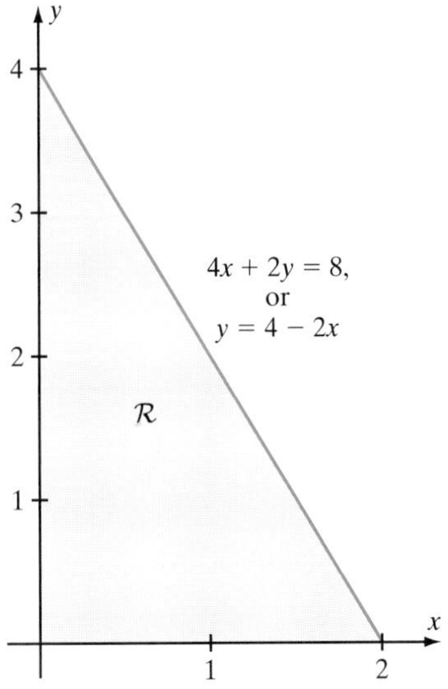

▲ Figure 3b

The required volume $\iint_{\mathcal{R}} f(x, y)\,dA$ is therefore

$$\int_0^2\left(\int_0^{4-2x}(8-4x-2y)\,dy\right)dx = \int_0^2\left((8y-4xy-y^2)\Big|_{y=0}^{y=4-2x}\right)dx$$
$$= \int_0^2 (16-16x+4x^2)\,dx$$

$$= \left(\left(16x - 8x^2 + \frac{4}{3}x^3 \right) \Big|_0^2 \right)$$

$$= \frac{32}{3}. \blacktriangleleft$$

The Volume Between Two Surfaces

So far, we have discussed the volume of a region lying below the graph of a continuous function and above a region in the xy-plane. Now let us treat the volume below the graph of a continuous function f, above the graph of a continuous function g, and over a region $\mathcal{R}$ in the xy-plane. In this instance, the relevant Riemann sums would be for the function $f - g$, and the integral representing the desired volume is

$$\iint_{\mathcal{R}} (f(x,y) - g(x,y))\, dA. \tag{12.3.1}$$

The geometry is illustrated in Figure 4. The following is an example.

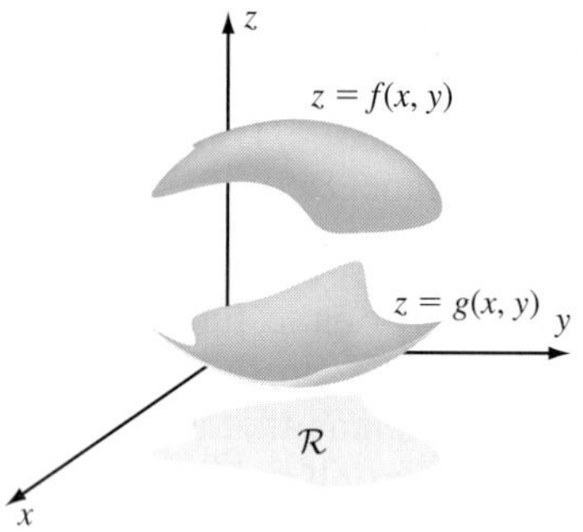

▲ Figure 4a The graphs of f and g over a planar region $\mathcal{R}$

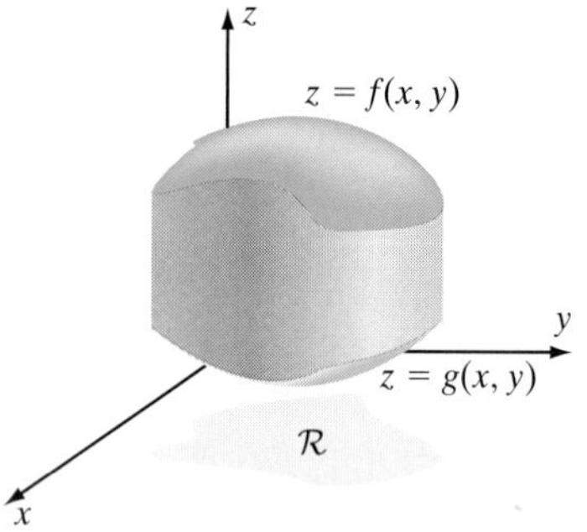

▲ Figure 4b The solid $\mathcal{U}$ that lies under the graph of f, above the graph of g, and over the planar region $\mathcal{R}$

▶ **EXAMPLE 3** Let $\mathcal{U}$ be the solid that is bounded below by the paraboloid that is the graph of $g(x,y) = 8 + x^2 + y^2$ and above by the plane that is the graph of $f(x,y) = 2x + 2y + 15$. Set up (but do not calculate) the integral for finding the volume of $\mathcal{U}$.

Solution The two surfaces intersect when $8 + x^2 + y^2 = 2x + 2y + 15$ or, equivalently, $(x^2 - 2x) + (y^2 - 2y) = 7$. After completing the squares, we obtain $(x-1)^2 + (y-1)^2 = 3^2$. Thus the solid we want to study "casts a shadow in the xy-plane"on the disk $\mathcal{R}$ that is bounded by the circle with center $(1,1)$ and radius 3. Figure 5 shows the two boundary surfaces of $\mathcal{U}$ and the region $\mathcal{R}$ of integration. The boundary of the region $\mathcal{R}$ of integration may be written as $y - 1 = \pm(9 - (x-1)^2)^{1/2}$ or, equivalently, $y = 1 \pm (8 + 2x - x^2)^{1/2}$. We write $\mathcal{R}$ as the y-simple region given by

$$\mathcal{R} = \left\{ (x,y) : -2 \le x \le 4, 1 - \sqrt{8 + 2x - x^2} \le y \le 1 + \sqrt{8 + 2x - x^2} \right\}.$$

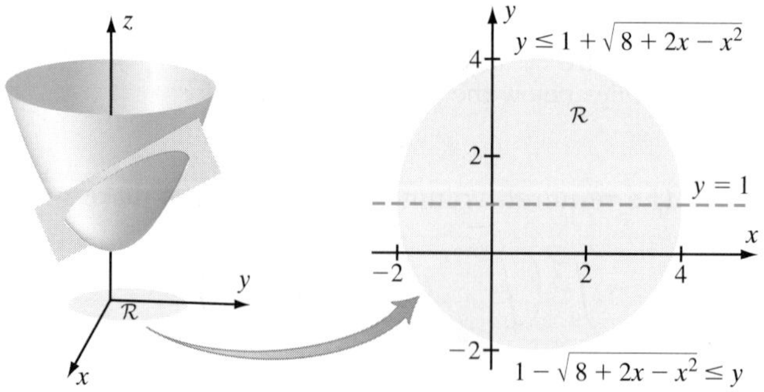

▲ Figure 5 The paraboloid $z = 8 + x^2 + y^2$ and the plane $z = 2x + 2y + 15$

The desired volume is therefore

$$\iint_{\mathcal{R}} (f(x,y) - g(x,y))\, dA = \int_{-2}^{4} \int_{1-\sqrt{8+2x-x^2}}^{1+\sqrt{8+2x-x^2}} ((2x+2y+15) - (8+x^2+y^2))\, dydx.$$ ◀

▶ **EXAMPLE 4** Find the volume V of the solid $\mathcal{U}$ that is below the plane $z = x + 4y - 7$, above the plane $z = x + 2y - 4$, and bounded on the sides by the parabolic cylinder $x = (y-4)^2 + 3$ and the plane $x + 2y = 11$.

Solution The graph of $\mathcal{U}$ and its projection $\mathcal{R}$ in the xy-plane appear in Figure 6a. Observe that $\mathcal{R}$, the region of integration, is the set of points in the xy-plane between the parabola $x = (y-4)^2 + 3$ and the line $x + 2y = 11$. For the points of intersection, we solve $11 - 2y = x = (y-4)^2 + 3$, or $y^2 - 6y + 8 = 0$. The roots $y = 2$ and $y = 4$ are the ordinates of the points $(7, 2)$ and $(3, 4)$ of intersection. The plot of $\mathcal{R}$ in the xy-plane (Figure 6b) shows that $\mathcal{R}$ is x-simple: $\mathcal{R} = \{(x,y) : 2 \le y \le 4, (y-4)^2 + 3 \le x \le 11 - 2y\}$. Because the plane $z = x + 4y - 7$ forms the upper boundary of $\mathcal{U}$, and the plane $z = x + 2y - 4$ forms the lower boundary, we calculate the volume of $\mathcal{U}$ using formula (12.3.1) with $f(x,y) = x + 4y - 7$ and $g(x,y) = x + 2y - 4$. Thus

$$V = \iint_{\mathcal{R}} \Big((x+4y-7) - (x+2y-4)\Big)\, dA = \int_2^4 \int_{(y-4)^2+3}^{11-2y} (2y-3)\,dxdy = \int_2^4 (2y-3)x\Big|_{x=(y-4)^2+3}^{x=11-2y}\, dy.$$

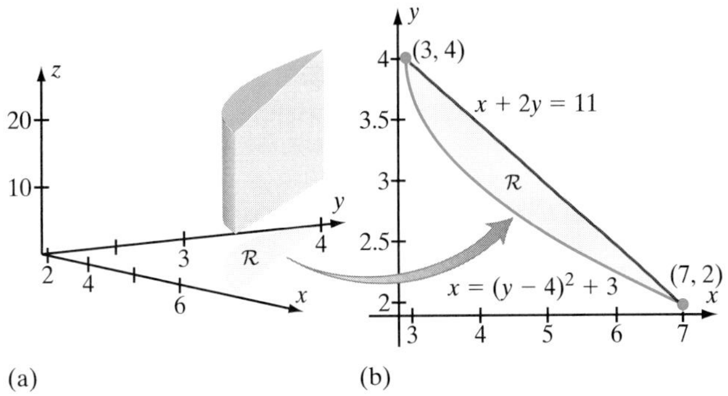

▲ Figure 6

On evaluating, expanding, and simplifying this integrand, we obtain

$$(2y-3)x\Big|_{x=(y-4)^2+3}^{x=11-2y} = (2y-3)\Big((11-2y) - \big((y-4)^2+3\big)\Big) = -2y^3 + 15y^2 - 34y + 24.$$

Therefore

$$V = \int_2^4 (-2y^3 + 15y^2 - 34y + 24)\, dy = \Big(-\frac{1}{2}y^4 + 5y^3 - 17y^2 + 24y\Big)\Big|_2^4 = 16 - 12 = 4.$$ ◀

▶ **EXAMPLE 5** Calculate the volume V of the solid $\mathcal{U}$ enclosed by the cylinders $x^2+y^2=a^2$ and $x^2+z^2=a^2$.

Solution Figure 7a shows the two cylinders. We can use symmetry to simplify the geometry of the problem. The left side of Figure 7b shows the part of the intersection of the two cylinders that lies in the first octant. We will compute the volume V_0 of this smaller solid $\mathcal{U}_0$ and multiply by 8 to obtain V. Because $\mathcal{U}_0$ lies over the quarter-disk shown in the right side of Figure 7b, the region of integration is $\mathcal{R}=\{(x,y): 0\le x\le a, 0\le y\le \sqrt{a^2-x^2}\}$, which is y-simple. The cylinder $x^2+z^2=a^2$ is the upper surface of $\mathcal{U}_0$. By solving for z, we realize this surface as the graph of $f(x,y)=\sqrt{a^2-x^2}$. The lower surface of $\mathcal{U}_0$ is the xy-plane, which we realize as $g(x,y)=0$. Therefore

$$V=8V_0=8\int_0^a\int_0^{\sqrt{a^2-x^2}}(f(x,y)-g(x,y))\,dydx=8\int_0^a\left(\int_0^{\sqrt{a^2-x^2}}\sqrt{a^2-x^2}\,dy\right)dx.$$

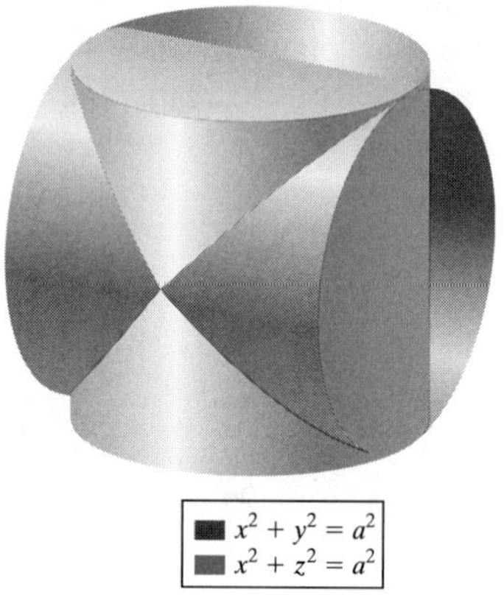

▲ Figure 7a

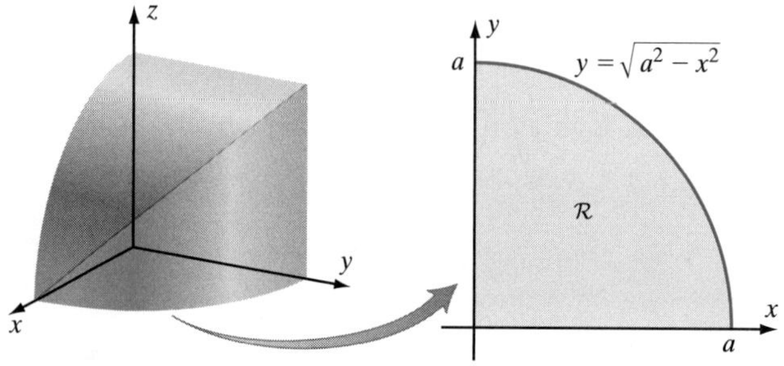

▲ Figure 7b

Finally, we obtain

$$V=8\int_0^a\left((\sqrt{a^2-x^2})\,y\right)\Big|_{y=0}^{y=\sqrt{a^2-x^2}}dx=8\int_0^a(a^2-x^2)\,dx=8\left(a^2x-\frac{x^3}{3}\right)\Big|_0^a=\frac{16}{3}a^3.$$

QUICK QUIZ

1. Let V denote the volume of the solid that lies above the region $\mathcal{R}$ in the xy-plane, which is bounded above by $z=6+x^2-y^2$ and bounded below by $z=1+x^2+2y$. Then, V is equal to $\iint_{\mathcal{R}} h(x,y)\,dA$ for what function $h(x,y)$?
2. Calculate the volume of the solid that lies below the graph of $f(x,y)=2x+3y^2$ and over the rectangle $[0,1]\times[0,1]$ in the xy-plane.
3. Calculate the volume of the solid that lies below the graph of $f(x,y)=1+x$, above the graph of $g(x,y)=2x$, and over the rectangle $[0,1]\times[0,1]$ in the xy-plane.
4. Let $\mathcal{R}$ be the region in the first quadrant of the xy-plane bounded by the curves $y=x^2$ and $y=x$. Calculate the volume of the solid that lies above $\mathcal{R}$, which is bounded above by $z=13x$ and below by $z=x$.

Answers

1. $5-y^2-2y$ 2. 2 3. 1/2 4. 1

EXERCISES

Problems for Practice

▶ In each of Exercises 1–8, calculate the volume of the solid that lies below the graph of $f(x, y) = 5 + x$ and over the given region in the xy-plane. ◀

1. The region bounded by the curves $y = x^2 + x$ and $y = 2x^2 + 2x - 12$

2. The region bounded by the curves $x = y^2 + y + 2$ and $x = 2y^2 + 4y + 4$

3. The region bounded by the curves $y = x^2 - 2$ and $y = x$

4. The region bounded by the curves $x = y^4$ and $x = 16$

5. The region bounded by the curves $y = 5x^2 - 7x + 3$ and $y = x^3$

6. The region bounded by the curves $y = |x|$ and $y = 2$

7. The region bounded by the curves $x + y = 2$ and $y = x^2 - 4$

8. The region bounded by the curves $y = x^2$ and $y = 5 - (x - 1)^2$

▶ In each of Exercises 9–23, calculate the volume of the solid $\mathcal{U}$ that lies below the graph of $f(x, y)$ and over the region $\mathcal{R}$ in the xy-plane. ◀

9. $f(x, y) = x - 2y$;
$\mathcal{R} = \{(x, y) : 8 \le x \le 10, 2 \le y \le 4\}$.

10. $f(x, y) = x^2 + y^2$;
$\mathcal{R} = \{(x, y) : 2 \le x \le 4, 1 \le y \le 6\}$.

11. $f(x, y) = x^2 y^2$;
$\mathcal{R} = \{(x, y) : -4 \le x \le -2, -5 \le y \le -2\}$.

12. $f(x, y) = \sin(x)$;
$\mathcal{R} = \{(x, y) : 0 \le x \le \pi, 1 \le y \le 3\}$.

13. $f(x, y) = 10x - 3y + 40$;
$\mathcal{R} = \{(x, y) : 3 \le x \le 6, 3x \le y \le 2x + 8\}$.

14. $f(x, y) = x^{-3/2} y^{1/2}$;
$\mathcal{R} = \{(x, y) : 1 \le y \le 4, y^2 \le x \le y^4\}$.

15. $f(x, y) = x - y$;
$\mathcal{R} = \{(x, y) : 3 \le x \le 5, -6x \le y \le -2x\}$

16. $f(x, y) = x/y^2$;
$\mathcal{R} = \{(x, y) : -3 \le y \le -2, y^2 \le x \le y^4\}$

17. $f(x, y) = x + 2$;
$\mathcal{R}$ is the region between the curves $y = x^2 + x$ and $y = 3x + 8$.

18. $f(x, y) = 12 + x + y$;
$\mathcal{R}$ is the region between the curves $y = -x^2 + 4$ and $y = x^2 + 2x - 8$.

19. $f(x, y) = e^x$;
$\mathcal{R}$ is the region bounded by the curves $y = 2x + 3$; $y = x$; and $x = 2$.

20. $f(x, y) = x/y$;
$\mathcal{R}$ is the region below $y = 1$ and above $y = \exp(x^2 - x)$.

21. $f(x, y) = x + 2y$;
$\mathcal{R}$ is the region bounded by the curves $y = \sqrt{x}$, $x = 0$, $y = x + 2$, and $x = 4$.

22. $f(x, y) = 8 - 2x + y$;
$\mathcal{R}$ is the region between the curves $y = x^3$ and $y = x^3 + x^2 - 2x - 3$.

23. $f(x, y) = \sin(y)$;
$\mathcal{R}$ is the region bounded by the curves $y = x + 1$, $y = -x + 1$, and $x = 1$.

▶ In each of Exercises 24–30, calculate the volume of the solid $\mathcal{U}$. ◀

24. $\mathcal{U}$ lies below the graph of $x + 2y + 3z = 6$ and in the first octant.

25. $\mathcal{U}$ lies below the graph of $z = x + 2y^2 + 3$, above the graph of $z = x + y$, and over the region $\mathcal{U}$ in the xy-plane that lies between $x = y^2$ and $x = -y^2 + 2$.

26. $\mathcal{U}$ lies below the graph of $z = 3x + y + 7$, above the graph of $z = -x - 2y - 4$, and over the region $\mathcal{U}$ in the xy-plane that is cut off by the line $x + 3y = 12$ in the first quadrant.

27. $\mathcal{U}$ is the solid that is bounded above by the cylinder $y^2 + z^2 = 49$, below by the xy-plane, and laterally by the cylinder $x^2 + y^2 = 49$.

28. $\mathcal{U}$ is the solid that is bounded by the two paraboloids $z = x^2 + y^2 + 1$ and $z = -x^2 - y^2 + 19$.

29. $\mathcal{U}$ is the solid above the graph of $z = x - y + 1$, below the graph of $z = 3x + 5y + 7$, and bounded on the sides by the vertical planes $x = y, x = -y$, and $y = -4x + 4$.

30. $\mathcal{U}$ is the region bounded above by the plane $3x - 2y - z = -8$, below by the plane $z = -2x - 3y - 6$, and with vertical sides given by the parabolic cylinders $x = y^2$ and $x = -y^2 - 8y + 10$.

Further Theory and Practice

31. Let $\mathcal{R}$ be the smaller region inside the circle $x^2 + y^2 - 4y + 8x = 6$ that is cut off by the line $y = x$. Calculate the volume of the solid that lies above $\mathcal{R}$ and below the plane $z = 2x + 8$.

32. Let $\mathcal{R}$ be the bounded region between the line $y = x + 12$ and the left branch of the hyperbola $x^2 - 4y^2 = 84$. Calculate the volume of the solid that lies over $\mathcal{R}$ and under the plane $z = -x$.

33. Find the volume of the solid that lies above the planar disk $(x + 4)^2 + y^2 \le 1$ and below the plane $z = x + 4$.

34. Find the volume of the solid that lies below the graph of $z = 1/(x^2 + y^2)$ and that is bounded laterally by the cylinder set $y = |x|$ and the planes $y = 2$ and $y = 8$.

35. Find the volume of the solid that is bounded by the two paraboloids $z = x^2 + y^2 + 3x + 6y - 2$ and $z = -x^2 - y^2 - 5x + 2y + 6$.

36. Find the volume in the first octant that is below both the plane $z = 2x + y$ and the plane $z = 2 - x - y$.

37. Find the volume in the first octant that is inside the cylinder $x^2 + y^2 = 1$ and below the surface $z = xy$.

38. Find the volume of the solid that is bounded by the planes $z=0$, $y=0$, $z=1-x+y$ and by the parabolic cylinder $y=1-x^2$.
39. Find the volume of the solid that is bounded above by the parabolic cylinder $z=1-y^2$ and below by the parabolic cylinder $z=x^2$.
40. Find the volume of the infinite solid that lies below the surface $z=\exp(-x^2)$ and above the planar region $\{(x,y,0) : x \geq |y|\}$.
41. Find the volume of the solid that is bounded above by the surface $z=1/(x+y)$, and that lies over the region in the xy-plane that is between $y=1/x$ and $y=1$ for $1 \leq x \leq 2$.
42. Find the volume of the solid that is bounded by the paraboloid $z=1+x^2+y^2$ and by the planes $x=0$, $y=0$, $z=0$, and $x+2y=2$.
43. Find the volume of the solid region that is bounded above by $z=xy$, below by $z=0$, and laterally by the plane $y=x$ and by the cylinder set $y=x^3$.
44. Suppose that $a<b$. Let $\mathcal{R}$ be the triangle with vertices (a,a), (b,a), and (b,b). Let u and v be continuously differentiable functions of one variable on $[a,b]$. Calculate $I=\iint_{\mathcal{R}} u'(x)v'(y)\, dA$ by using iterated integrals. Next, reverse the order of integration. and calculate I again. Equate the two expressions for I, and solve for $\int_a^b u(y)v'(y)\, dy$. What formula have you found?

Calculator/Computer Exercises

45. Let $\mathcal{R}$ be the region in the first quadrant of the xy-plane that lies between the curves $y=10x-1-6x^2$ and $y=x^3-3x^2+3x$. Find the volume of the solid that lies above $\mathcal{R}$ and below the plane $z=x+y$.
46. Let $\mathcal{R}$ be the region in the first quadrant of the xy-plane that lies between the curves $y=e^x$ and $y=1+3x$. Find the volume of the solid that lies above $\mathcal{R}$ and below the plane $z=10-y$.
47. Let $\mathcal{R}$ be the region in the xy-plane that lies between the curves $y=2-x^3$ and $y=1+x^4$. Find the volume of the solid that lies above $\mathcal{R}$ and below the plane $z=y$.
48. Let $\mathcal{R}$ be the region in the xy-plane that is bounded by the lines $x=0$, $y=0$, and $y=1-x$. Use the Trapezoidal Rule approximation with $N=4$ to estimate the volume of the solid region that lies above $\mathcal{R}$ and below $z=1/(1+xy)$. (The Trapezoidal Rule approximation for double integrals is discussed in the headnote to Exercises 65–68 of Section 12.2.)

12.4 Polar Coordinates

You are familiar with the idea of locating a point in the plane with two real numbers: In the Cartesian setup, one of these numbers corresponds to the x-displacement and the other to the y-displacement. These represent signed distances to perpendicular axes. In certain situations, however, it is natural to express the location of a point P using the distance $r=\|\overrightarrow{OP}\|$ to a fixed point O, which is taken to be the origin. For example, we found it convenient to refer to the coordinate r when we analyzed central force fields in Section 10.5 of Chapter 10. Of course the distance r alone does not determine the location of the point P. For that, we will need information that tells us the direction of $\overrightarrow{OP}$. The angle that $\overrightarrow{OP}$ makes with a fixed half-line emanating from O makes a convenient choice. This fixed half-line is called the *polar axis*. We will choose the positive x-axis of Cartesian coordinates to be the polar axis, as is conventional.

The Polar Coordinate System

Suppose that r is a positive number and that θ is any real number. Then there is a unique point P in the plane such that P is a distance r from the origin and such that the angle between $\overrightarrow{OP}$ and the polar axis measures θ radians. The angle is measured counterclockwise when θ is positive (Figure 1a) and clockwise when θ is negative (Figure 1b). We say that r and θ are *polar coordinates* for the point P. As with Cartesian coordinates, we write polar coordinates as an ordered pair, (r,θ). We must understand from the context which coordinate system is being used. Thus in Figure 2, point P has polar coordinates $(2,1)$, whereas point Q has

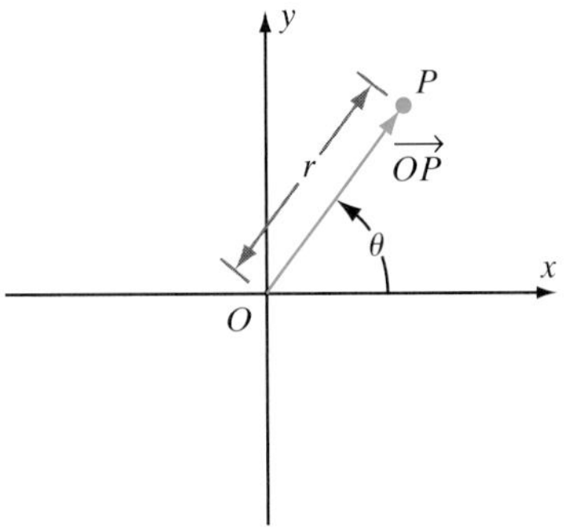

P has polar coordinates (r, θ) with $\theta > 0$

▲ Figure 1a

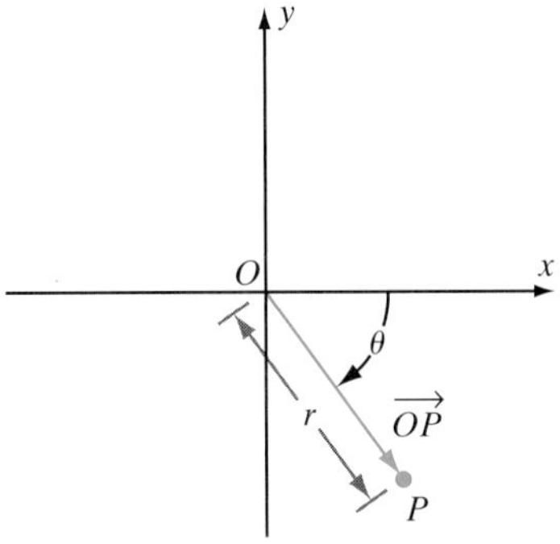

P has polar coordinates (r, θ) with $\theta < 0$

▲ Figure 1b

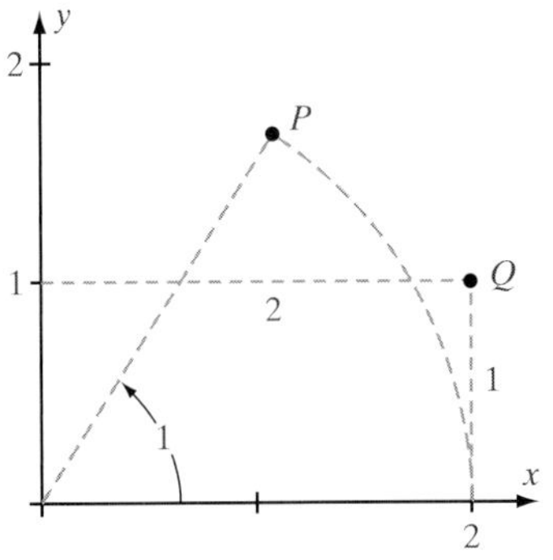

▲ Figure 2

Cartesian coordinates $(2, 1)$. We must know the coordinate system to which the ordered pair $(2, 1)$ refers before we can know which planar point has $(2, 1)$ for its coordinates.

In polar coordinates, once you have specified the radial coordinate r of a point P, then you have already restricted the point to lie on the circle of radius r centered at the origin (see Figure 3). The angular coordinate θ specifies which point on the circle P is. Alternatively, once you have specified the angular coordinate θ, then you have restricted the point to lie on a certain ray. The radial coordinate r then tells you how far out on that ray the point lies. These ideas are also shown in Figure 3.

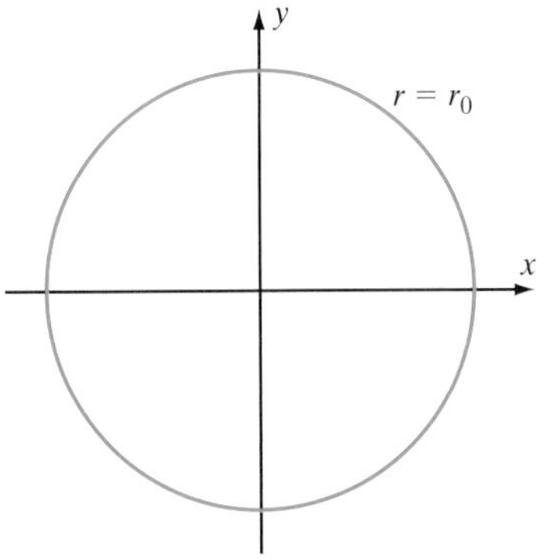

▲ Figure 3a The circle $r = r_0$

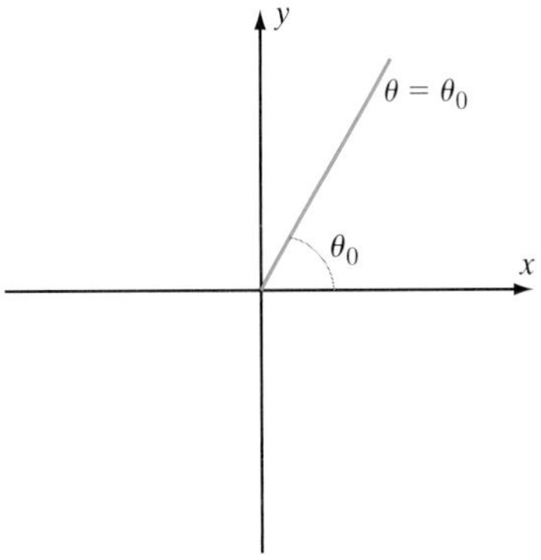

▲ Figure 3b The ray $\theta = \theta_0$

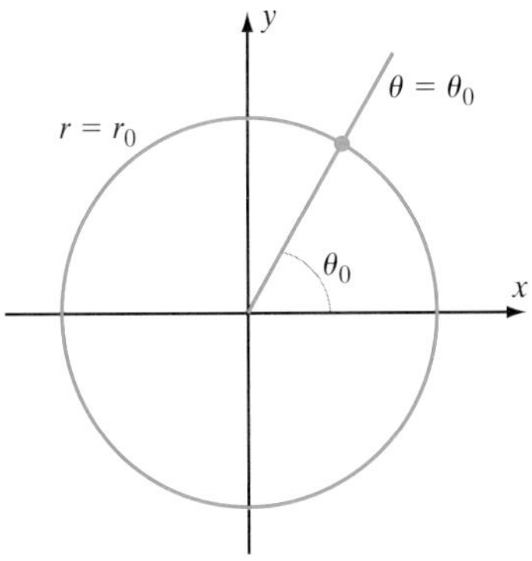

▲ Figure 3c The point whose polar coordinates are given by $r = r_0$ and $\theta = \theta_0$

In polar coordinates, the number 0 is also a possible value for the radial coordinate r. For any θ, the point in the plane with polar coordinates $(0, \theta)$ is the origin O because it is the only point a distance 0 from O. Notice that the polar coordinates of the origin are ambiguous: The radial coordinate r must be 0, but the angular coordinate θ may have any value.

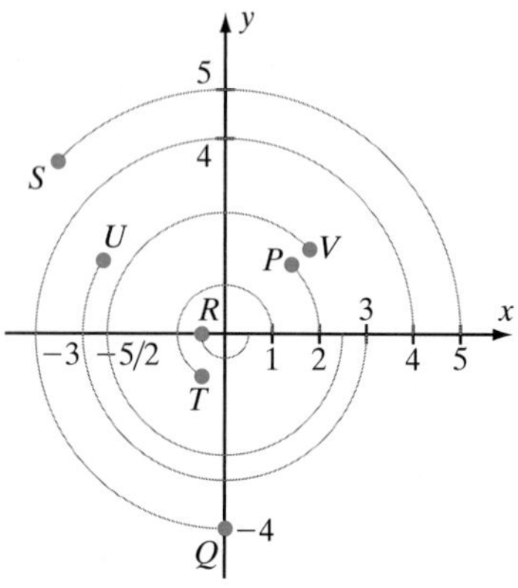

▲ Figure 4
$P = (2, \pi/4)$ $Q = (4, 3\pi/2)$
$R = (1/2, -\pi)$ $S = (5, 3\pi/4)$
$T = (1, 4\pi/3)$ $U = (3, -7\pi/6)$
$V = (5/2, -7\pi/4)$

▶ **EXAMPLE 1** Plot the points whose polar coordinates are tabulated.

Point	P	Q	R	S	T	U	V
Polar coordinates	$(2, \pi/4)$	$(4, 3\pi/2)$	$(1/2, -\pi)$	$(5, 3\pi/4)$	$(1, 4\pi/3)$	$(3, -7\pi/6)$	$(5/2, -7\pi/4)$

Solution The points are exhibited in Figure 4. ◀

INSIGHT In Example 1, compare the first point P and the last point V. They lie on the same ray emanating from the origin. That is because their angular coordinates differ by an integer multiple of 2π: $\pi/4 = -7\pi/4 + 2\pi$. As you can imagine from Figure 4, if the radial coordinates of points P and V had been equal, then the points would have coincided despite their different angular coordinates. The polar coordinates of a point in the plane are *not* unique (unlike its rectangular coordinates). Indeed, the polar coordinates $(r_0, \theta_0 + 2n\pi)$ represent the same point in the plane for all integer values of n. Thus every point in the plane can be specified in infinitely many different ways in the polar coordinate system.

Negative Values of the Radial Variable

It is often convenient to allow the radial coordinate r to be negative. Suppose that $r < 0$ and that θ is any number. Let P' be the point with polar coordinates $(|r|, \theta)$. Then the polar coordinates (r, θ) refer to the point P that is a distance $|r|$ from the origin and that lies on the ray opposite to $\overrightarrow{OP'}$. In other words, because $|r| = -r$, the ordered pairs (r, θ) and $(-r, \theta)$ are polar coordinates for points P and P' that are reflections of each other through the origin. Another way to think of this relationship is that the ordered pairs (r, θ) and $(-r, \theta + \pi)$ are polar coordinates for the same point. See Figure 5 in which P has polar coordinates (r, θ) with $r < 0$.

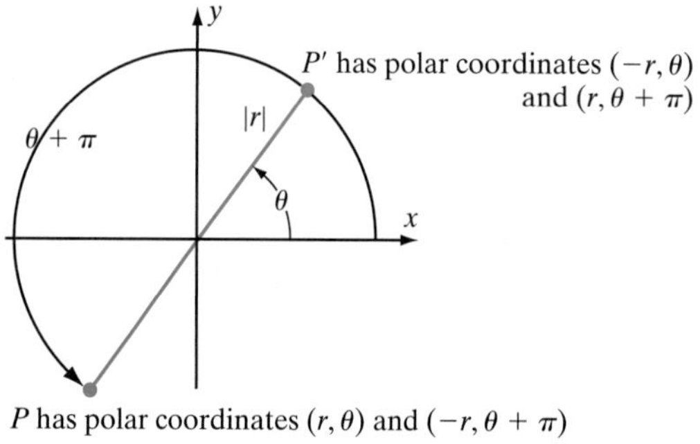

▲ Figure 5 Polar coordinates with $r < 0$

▶ **EXAMPLE 2** Give all possible polar coordinates of the point P that is a distance 4 from the origin and for which $\overrightarrow{OP}$ makes an angle of $5\pi/6$ with the positive x-axis.

Solution In effect, we have been told that $(4, 5\pi/6)$ is a pair of polar coordinates for P. However, we may add any multiple of 2π to the angle and obtain another valid set of polar coordinates for this point. Therefore

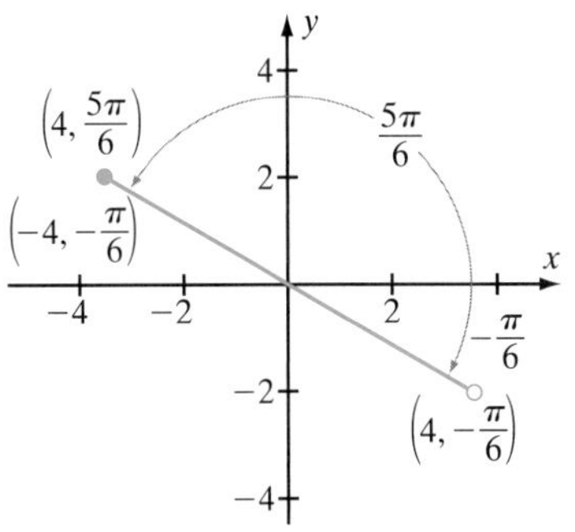

▲ Figure 6

$$\ldots(4, -19\pi/6),\ (4, -7\pi/6),\ (4, 5\pi/6),\ (4, 17\pi/6),\ (4, 29\pi/6)\ldots$$

are all polar coordinates for P. We summarize this by saying that

$$(4, 5\pi/6 + 2n\pi), \quad n \in \mathbb{Z}$$

are polar coordinates for P. Is this a complete list of all possible polar coordinates for P? No, because we can allow r to be negative. Thus P may be thought of as the point $(-4, -\pi/6)$. See Figure 6. Again, we can add any multiple of 2π to the angle. Therefore the pairs

$$(-4, -\pi/6 + 2n\pi), \quad n \in \mathbb{Z}$$

are also valid polar coordinates for P. Because P must have radial coordinate ± 4 and must have angular coordinate either $5\pi/6 + 2n\pi$ or $-\pi/6 + 2n\pi$, we have accounted for all possible polar coordinate representations of P. ◀

Relating Polar Coordinates to Rectangular Coordinates

Now we discuss the connection between polar coordinates and Cartesian (or rectangular) coordinates. Look at Figure 7. The point Q has polar coordinates (r, θ) and rectangular coordinates (x, y). By examining the right triangle in the figure, we see that

$$x = r\cos(\theta) \quad \text{and} \quad y = r\sin(\theta). \tag{12.4.1}$$

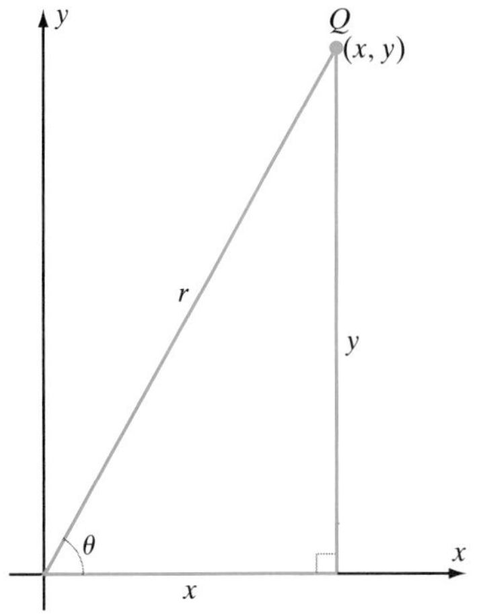

▲ Figure 7

As expected, the rectangular coordinates x and y of a point are uniquely determined by its polar coordinates r and θ. We also have equations that we can use to obtain polar coordinates r and θ for the point with rectangular coordinates (x, y):

$$r^2 = x^2 + y^2 \quad \text{and, if} \quad x \neq 0, \quad \tan(\theta) = \frac{y}{x}. \tag{12.4.2}$$

As we expect from our earlier discussion, we see that these equations do not determine r and θ uniquely. The first formula in line (12.4.2) tells us the magnitude $|r| = \sqrt{x^2 + y^2}$ of the radial coordinate r but not its sign. The second formula in line (12.4.2) is satisfied by infinitely many values of the angular coordinate θ.

▶ **EXAMPLE 3** Compute the rectangular coordinates of the point P whose polar coordinates are $(4, 5\pi/6)$.

Solution Observe that $r = 4$ and $\theta = 5\pi/6$ so $x = 4\cos(5\pi/6) = -2\sqrt{3}$ and $y = 4\sin(5\pi/6) = 2$. Thus P has the ordered pair $(-2\sqrt{3}, 2)$ for its Cartesian coordinates. ◀

INSIGHT Notice that P is the same point in Examples 2 and 3. In Example 2, we determined infinitely many ways to express P in polar coordinates. Whichever of these representations we choose to work with, the conversion to rectangular coordinates will be the ordered pair $(-2\sqrt{3}, 2)$ that we obtained in Example 3. For instance, the polar coordinates $(-4, -\pi/6)$ lead to $x = (-4)\cos(-\pi/6) = -2\sqrt{3}$ and $y = (-4)\sin(-\pi/6) = 2$.

▶ **EXAMPLE 4** Calculate all possible polar coordinates for the point Q with rectangular coordinates $(-5, 5)$.

Solution We know that $x = -5, y = 5$. Therefore $r^2 = x^2 + y^2 = 25 + 25 = 50$. If we take $r = +\sqrt{50} = 5\sqrt{2}$, then the identities $x = r\cos(\theta)$ and $y = r\sin(\theta)$ yield

$$\cos(\theta) = -\frac{1}{\sqrt{2}} \quad \text{and} \quad \sin(\theta) = \frac{1}{\sqrt{2}}.$$

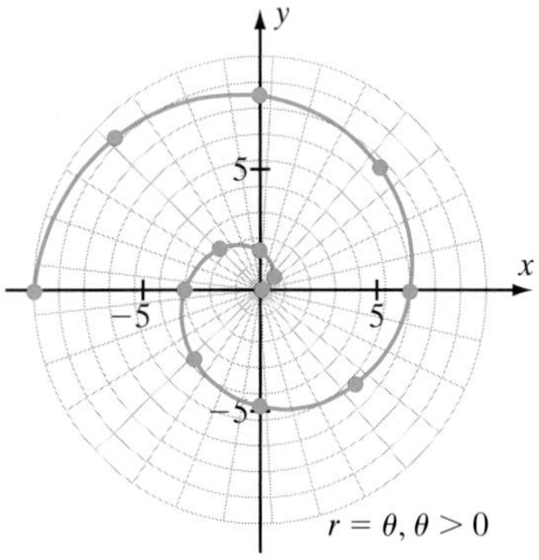

▲ Figure 8

It follows that $\theta = 3\pi/4 + 2n\pi$, $n \in \mathbb{Z}$. So we have found the polar coordinates $(5\sqrt{2}, 3\pi/4 + 2n\pi)$, $n \in \mathbb{Z}$. However, r could also have the value $-\sqrt{50}$, or $-5\sqrt{2}$. In this case, the equations $x = r\cos(\theta)$ and $y = r\sin(\theta)$ yield

$$\cos(\theta) = \frac{1}{\sqrt{2}} \quad \text{and} \quad \sin(\theta) = -\frac{1}{\sqrt{2}}.$$

It follows that $\theta = 7\pi/4 + 2n\pi$, $n \in \mathbb{Z}$. So we have found the polar coordinates $(-5\sqrt{2},\ 7\pi/4 + 2n\pi)$, $n \in \mathbb{Z}$. This accounts for all possible polar coordinates for Q. ◀

Graphing in Polar Coordinates

Graphing in polar coordinates is different from graphing in rectangular coordinates. We mention two new features in particular. The first is that the pictorial notions we have developed for graphing in rectangular coordinates—slope, x- and y-intercepts, and so on no longer apply. The second is that the ambiguity in the polar coordinates of a point means that we have to be careful not to forget part of the graph. Because of these twists, we begin with the most elementary concepts.

▶ **EXAMPLE 5** Sketch the graph of $r = \theta$.

Solution The graph of $y = x$ in *rectangular* coordinates is a line, but do not expect to see anything like that now. In Figure 8, we have plotted a few points corresponding to $\theta = 0, \pi/4, \pi/2, 3\pi/4, \pi$, and so on. We notice that, as θ increases from $\theta = 0$, so does r increase. Therefore it is logical to connect the points with a smooth curve. Notice that the resulting curve is a spiral that grows with rotation in the counterclockwise direction. The graph so far is not complete because we have not considered negative values of θ. We plot some points corresponding to $\theta = -\pi/4, -\pi/2, -3\pi/4, -\pi$, and so on and connect them as we did for $\theta > 0$. Thus the complete graph consists of two spirals, one expanding with clockwise rotation and the other expanding with counterclockwise rotation (Figure 9). The curve we have graphed is called a *Spiral of Archimedes* (as is any curve whose equation in polar coordinates is $r = a\theta$ where a is a constant). ◀

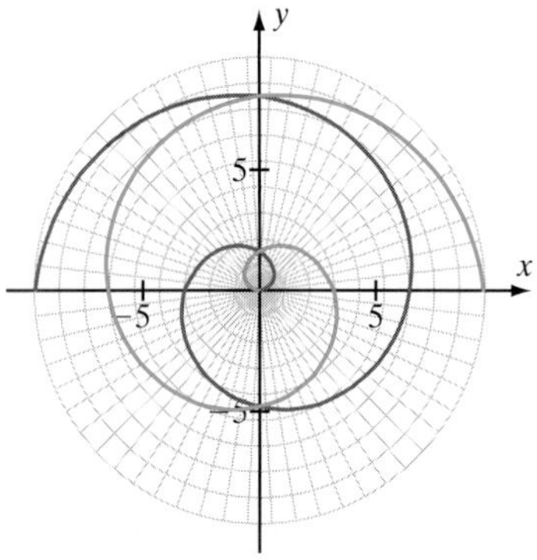

▲ Figure 9 Spiral of Archimedes $r = \theta$

▶ **EXAMPLE 6** Describe the graphs of $r\cos(\theta) = 2$ and $r\sin(\theta) = -1$.

Solution The easiest thing to do is to convert these equations to rectangular coordinates: They become $x = 2$ and $y = -1$. The graphs are lines parallel to the rectangular coordinate axes. ◀

▶ **EXAMPLE 7** Sketch the graph of $r = 2\sin(\theta)$ by two different methods.

Solution For the first method, we create a table of values.

θ	0	$\pi/6$	$\pi/4$	$\pi/3$	$\pi/2$	$2\pi/3$	$3\pi/4$	$5\pi/6$	π	$7\pi/6$	$5\pi/4$	$4\pi/3$	$\pi/2$	$5\pi/3$	$7\pi/4$	$11\pi/6$	2π
r	0	1	$\sqrt{2}$	$\sqrt{3}$	2	$\sqrt{3}$	$\sqrt{2}$	1	0	-1	$-\sqrt{2}$	$-\sqrt{3}$	-2	$-\sqrt{3}$	$-\sqrt{2}$	-1	0

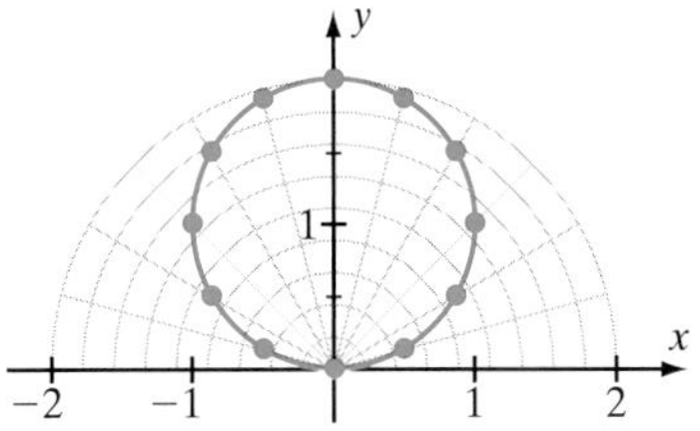

▲ Figure 10a

The corresponding points are plotted in Figure 10a. These points are connected, *in increasing order of* θ, by a smooth curve. We see that a circular loop beginning and ending at the origin is traced when θ increases from 0 to π. This loop is retraced when θ increases from π to 2π.

An alternative method to sketch this graph is to multiply both sides of the equation by r to obtain $r^2 = 2r\sin(\theta)$. Then, it is easy to convert to rectangular coordinates: $x^2 + y^2 = 2y$. By completing the square, we obtain the equation $x^2 + (y-1)^2 = 1$ of the circle with center $(0, 1)$ and radius 1. ◀

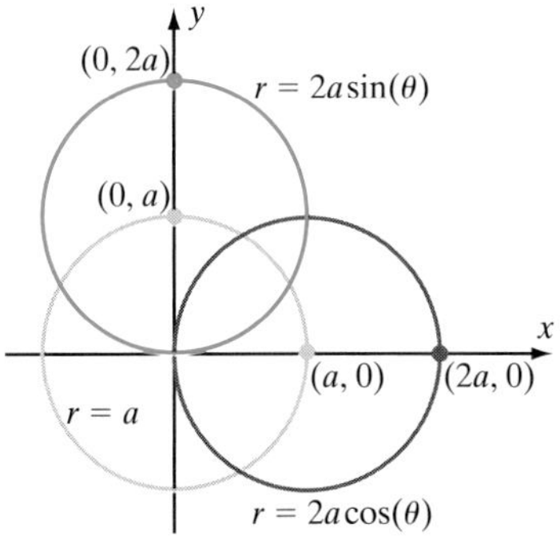

▲ Figure 10b

INSIGHT The graph of $r = a$ is a circle of radius a. But $r = a$ is not the only polar equation that has a circle of radius a as its graph. When generalized, the calculations of Example 7 show that the graph of $r = 2a\sin(\theta)$ is a circle of radius a. In a similar manner, we may show that the graph of $r = 2a\cos(\theta)$ is also a circle of radius a. Figure 10b shows these three circles in one viewing window.

▶ **EXAMPLE 8** Sketch $r = 3\big(1 + \cos(\theta)\big)$.

Solution Notice that, as θ increases from 0 to $\pi/2$, the values of $\cos(\theta)$ decrease from 1 to 0. Therefore r, being equal to $3\big(1 + \cos(\theta)\big)$, decreases from 6 to 3. Likewise, as θ increases from $\pi/2$ to π, the values of r decrease from 3 to 0. This information is indicated in Figure 11. We now have the upper half of the curve. We could continue this type of analysis to get the lower half of the curve. Instead, we notice that, because $\cos(-\theta) = \cos(\theta)$, the formula for r does not change when θ is replaced by $-\theta$. So, for every point (r, θ) on the graph, there is a corresponding point $(r, -\theta)$, which means that the graph of $r = 3\big(1 + \cos(\theta)\big)$ is symmetric about the horizontal axis. The complete sketch appears in Figure 12. A curve whose polar equation has the form $r = a\big(1 + \cos(\theta)\big)$ or $r = a(1 - \cos(\theta))$ or $r = a\big(1 + \sin(\theta)\big)$ or $r = a\big(1 - \sin(\theta)\big)$ is called a *cardioid* (because the locus is shaped like a heart). ◀

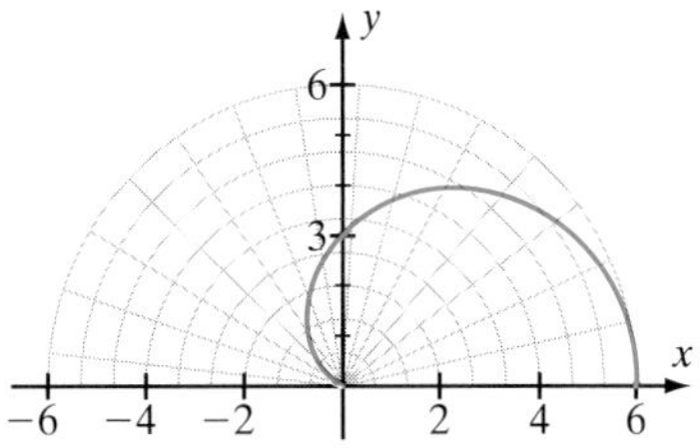

$r = 3(1 + \cos(\theta)), 0 \le \theta \le \pi$

▲ Figure 11

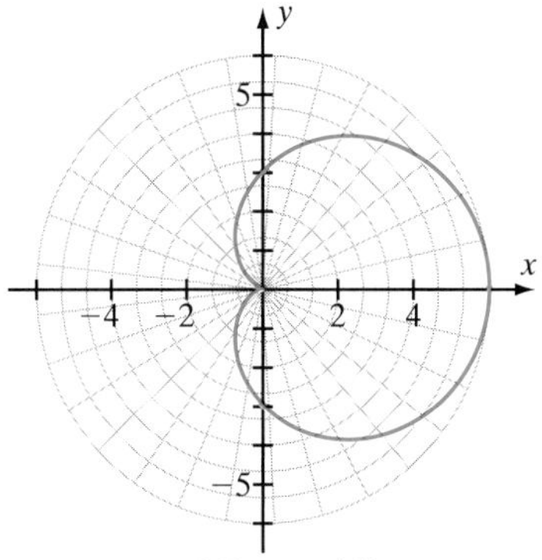

$r = 3(1 + \cos(\theta))$

▲ Figure 12

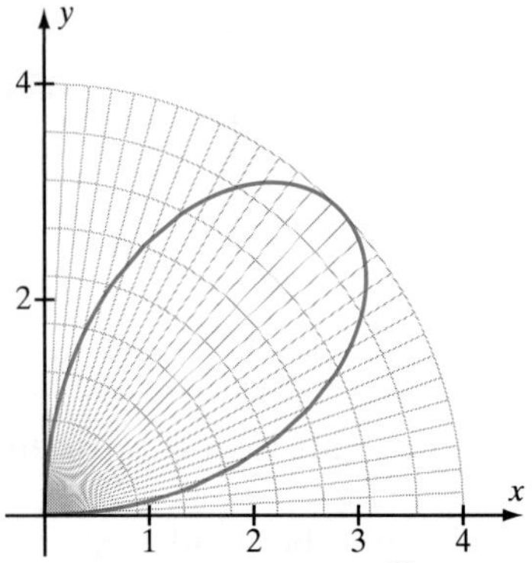

$r = 4\sin(2\theta), 0 \le \theta \le \frac{\pi}{2}$

▲ Figure 13

▶ **EXAMPLE 9** Sketch the graph of $r = 4\sin(2\theta)$.

Solution When θ increases from 0 to $\pi/4$, we see that r increases from 0 to 4. When θ then passes from $\pi/4$ to $\pi/2$, the value of r decreases monotonically back to 0. The corresponding sketch is in Figure 13. Figure 14 shows the added information obtained from considering $\pi/2 \le \theta \le \pi$: The value of r decreases from 0 to -4 and then increases again back to 0. Notice that although the values of θ correspond to the second quadrant, this portion of the plot appears in the diagonally opposite quadrant, the fourth quadrant, because r is negative.

Now let us see what we can learn from symmetry. Notice that

$$\sin\big(2(\theta + \pi)\big) = \sin(2\theta).$$

So to each point (r, θ) on the curve, there is a corresponding point $(r, \theta + \pi)$. This tells us that the graph is symmetric with respect to reflection through the origin. Glancing at Figure 14 and taking this symmetry into account, we complete our sketch as in Figure 15. This curve is called a *four-leafed rose.* ◀

▶ **EXAMPLE 10** Sketch the curve $r^2 = -\cos(2\theta)$.

Solution We notice that r is undefined when $\cos(2\theta)$ is positive, so certain values of θ must be ruled out. For instance, r is undefined for $0 < \theta < \pi/4$. However, when θ increases from $\pi/4$ to $\pi/2$, the values of $\cos(2\theta)$ decrease from 0 to -1. Therefore the values of r^2 increase from 0 to 1. This will give rise to *two branches* of our curve: one corresponding to r increasing from 0 to 1 and the other corresponding to r decreasing from 0 to -1. These are illustrated in Figure 16. Once again, notice the phenomenon of plotting that part of a curve with $r < 0$: It appears as the reflection through the origin of the arc that would have been plotted had the values of r been positive.

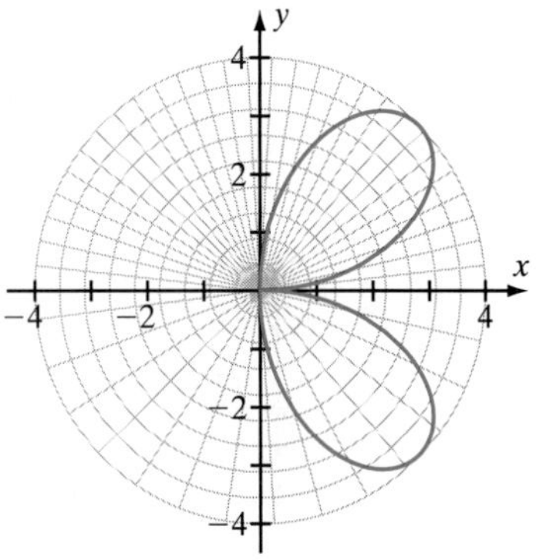

$r = 4\sin(2\theta), 0 \le \theta \le \pi$

▲ Figure 14

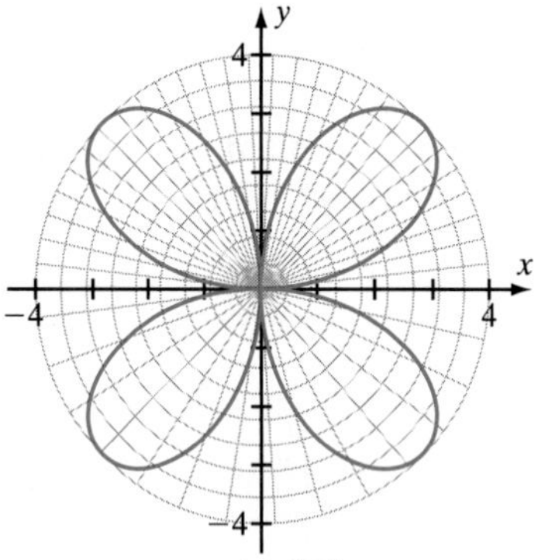

$r = 4\sin(2\theta)$

▲ Figure 15

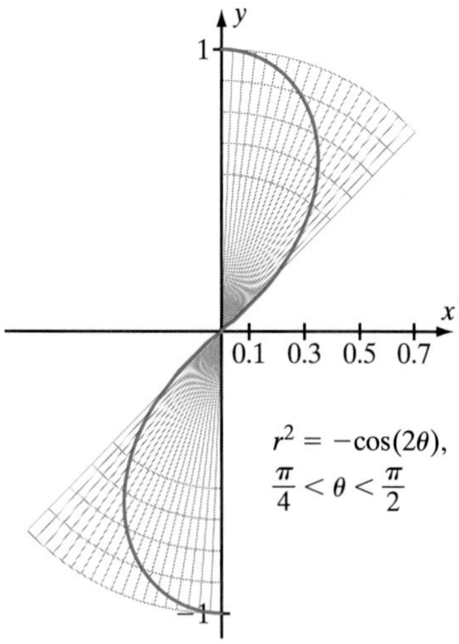

▲ Figure 16

Next we observe that $\cos(2(\pi - \theta)) = \cos(2\theta)$; hence the equation $r^2 = -\cos(2\theta)$ is unchanged under the map $\theta \mapsto \pi - \theta$. It follows that for each point (r, θ) on the curve, there is a corresponding point $(r, \pi - \theta)$ that is also on the curve. This property tells us that the graph is symmetric with respect to the y-axis (see Figure 17). The graph of $r^2 = -\cos(2\theta)$ is called a *lemniscate*.

INSIGHT To be sure that we have not forgotten any part of the graph in the previous example, we can perform a check. We have already considered $0 < \theta < \pi/2$. For $\pi/2 \leq \theta \leq 3\pi/4$, the quantity $-\cos(2\theta)$ decreases from 1 to 0, which is indicated on Figure 17. For $3\pi/4 < \theta < \pi$, $-\cos(2\theta)$ is negative so that r is undefined. That is also indicated by our figure. If θ is between π and 2π, then we may write $\theta = \theta_0 + \pi$ with θ_0 between 0 and π. Remembering that $(-r, \theta_0)$ and $(r, \theta_0 + \pi)$ are polar coordinates for the same point, and observing that $r^2 = -\cos(2\theta)$ if and only if $(-r)^2 = -\cos(2\theta) = -\cos(2\theta_0 + 2\pi) = -\cos(2\theta)$, we see that every point on the curve that has polar coordinates (r, θ), or $(r, \theta_0 + \pi)$, is a point on the curve with polar coordinates $(-r, \theta_0)$. In other words, the picture that is drawn as θ ranges from 0 to π is repeated as θ ranges from π to 2π.

Symmetry Principles in Graphing

We have seen that symmetry can be a powerful tool in graphing polar equations. We summarize here the basic tests for symmetry.

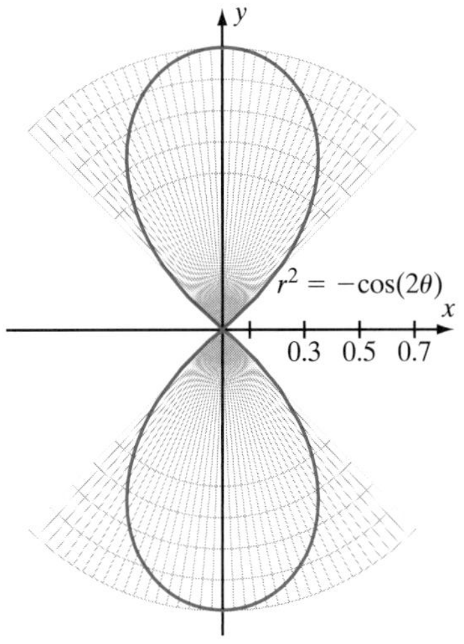

▲ Figure 17

Symmetry with respect to	Test
x-axis	Polar equation is unchanged when θ is replaced with $-\theta$.
y-axis	Polar equation is unchanged when θ is replaced with $\pi - \theta$.
origin	Polar equation is unchanged when r is replaced by $-r$, or polar equation is unchanged when θ is replaced by $\theta + \pi$.

▶ **EXAMPLE 11** Test the equation $r = 3 - \sin(2\theta)$ for symmetry.

Solution Because

$$\sin\big(2(\pi - \theta)\big) = \sin(2\pi - 2\theta) = \sin(-2\theta) = -\sin(2\theta),$$

the test for symmetry in the y-axis fails. Because $\sin(2(-\theta)) \neq \sin(2\theta)$, the test for symmetry in the x-axis fails. Finally, $\sin(2(\theta + \pi)) = \sin(2\theta + 2\pi) = \sin(2\theta)$, so the equation is unaltered if θ is replaced by $\theta + \pi$. We conclude that there is symmetry in the origin. The curve is sketched in Figure 18. ◀

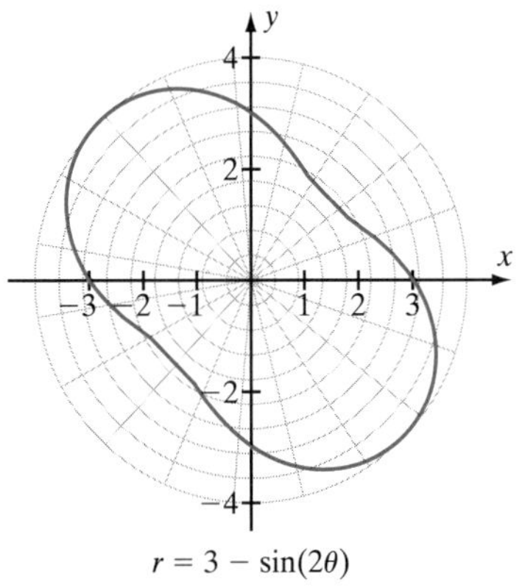

▲ Figure 18

INSIGHT Notice that there are two tests for symmetry in the origin. The one that we used in Example 11 is to replace θ by $\theta + \pi$. The other test is to replace r by $-r$. The curve in Example 11 *fails* this second test. Again, the fact that a single point has several different polar representations is causing confusion. Always remember that our symmetry tests are *sufficient* conditions for symmetry, not necessary conditions. This means that if a curve passes the test, then it is certainly symmetric; if it does not pass the test, it may or may not be symmetric. In practice, you should just apply the tests that you know and use to advantage whatever information they give you.

QUICK QUIZ

1. What are the polar coordinates of the point with rectangular coordinates $(1, \sqrt{3})$?
2. What are the rectangular coordinates of the point with polar coordinates $(-2, \pi/2)$?
3. Describe the curve that has polar equation $r = 1 + 2\theta$ for $0 < \theta$.
4. If f is an even function, then what symmetry does the curve defined by $r = f(\theta)$ have?

Answers

1. $(2, \pi/3)$ 2. $(0, -2)$ 3. A spiral 4. Symmetry with respect to the x-axis

EXERCISES

Problems for Practice

In each of Exercises 1–8, plot the four points whose polar coordinates are given.

1. $(3, \pi/4), (-3, \pi/4), (3, -\pi/4), (-3, -\pi/4)$
2. $(2, 5\pi/6), (-2, 5\pi/6), (2, -5\pi/6), (-2, -5\pi/6)$
3. $(0, \pi/2), (0, \pi), (0, -\pi), (0, -\pi/2)$
4. $(6, 9\pi/4), (-6, 9\pi/4), (6, -9\pi/4), (-6, -9\pi/4)$
5. $(4, -7\pi/2), (-4, -7\pi/2), (4, 7\pi/2), (-4, 7\pi/2)$
6. $(1, \pi/3), (-1, \pi/3), (1, -\pi/3), (-1, -\pi/3)$
7. $(2, 7\pi/3), (-2, 7\pi/3), (2, -7\pi/3), (-2, -7\pi/3)$
8. $(10, 0), (-10, 0), (6, 0), (0, 0)$

In each of Exercises 9–16, the polar coordinates of a point are given. What are its rectangular coordinates?

9. $(4, \pi/4)$
10. $(7, 9\pi/2)$
11. $(0, \pi)$
12. $(-8, -3\pi/4)$
13. $(-4, -17\pi/6)$
14. $(-3, 5\pi/6)$
15. $(-6, \pi/3)$
16. $(3, 0)$

In each of Exercises 17–22, give all possible polar coordinates for the point that is given in rectangular coordinates.

17. $(4, -4)$
18. $(-4\sqrt{2}, -4\sqrt{2})$
19. $(0, -9)$
20. $(-5, -5)$
21. $(-1, 0)$
22. $(-8\sqrt{3}, 8)$

In each of Exercises 23–32, test the given equation for symmetry in the x-axis, the y-axis, and the origin.

23. $r^2 = 3\sin(\theta)$
24. $r = 4 - 2\tan(2\theta)$
25. $r = 3$
26. $\theta = \pi/2$
27. $r = 1 + \cos^2(\theta)$
28. $r\sin(\theta) = \cos(\theta)$
29. $r = 6 - \cos(3\theta)$
30. $r = 4 + \sin(3\theta)$
31. $r = \sin(\theta) - \cos(\theta)$
32. $r^2 = 1 - 2\sin(\theta)$

In Exercises 33–44, sketch the curve that is described by the given equation in polar coordinates. (The name of each curve has been provided in parentheses.)

33. $r = 1 - \cos(\theta)$ (cardioid)
34. $r = 3 + 3\sin(\theta)$ (cardioid)
35. $r = 3\cos(2\theta)$ (four-leafed rose)
36. $r^2 = 9\cos(2\theta)$ (lemniscate)
37. $r = 4 - 2\sin(\theta)$ (limaçon)
38. $r = 2\sin(3\theta)$ (three-leafed rose)
39. $r = -1 + 3\cos(\theta)$ (limaçon)
40. $r^2 = -3\sin(2\theta)$ (lemniscate)
41. $r = 3\sin(\theta) - 2\cos(\theta)$ (circle)
42. $r = -3\sin(5\theta)$ (five-leafed rose)
43. $r = 3 - 3\sin(\theta)$ (cardioid)
44. $r = -2\theta$ (spiral of Archimedes)

Further Theory and Practice

45. Give all possible polar coordinates for the points that have rectangular coordinates: a)$(3, \pi/4)$; b)$(-6, 7\pi/2)$; and c) $(\pi/4, -8\pi/3)$.
46. Give rectangular coordinates for the points that have polar coordinates a) $(2, -6)$; b) $(-4, 7)$; c) $(\pi/3, 9)$; and d) $(7\pi/3, -6)$.
47. Give a formula in polar coordinates for the distance between two points in the plane with polar coordinates (r_1, θ_1) and (r_2, θ_2). What is the polar equation of a circle of radius 1 with center given by $(3, \pi/4)$ in polar coordinates?
48. A regular N-gon(a convex polygon with N equal sides) is inscribed in a circle of radius r that is centered at the origin. If one of the vertices is on the positive x-axis, then what are the polar coordinates of the vertices?

In each of Exercises 49–54, find all points of intersection of the two polar curves that are given.

49. $r = \cos(\theta)$ and $r = 1 - \cos(\theta)$
50. $r = 1 + \sin(\theta)$ and $r = 3(1 - \sin(\theta))$
51. $r = 1$ and $r = \tan(\theta)$
52. $r = 3\sec(\theta)$ and $r = 4\cos(\theta)$
53. $r = 1 + \cos(\theta)$ and $r^2 = (1/2)\cos(\theta)$
54. $r = 4 - 5\sin(\theta)$ and $r = 3\sin(\theta)$

55. Find each ordered pair (a, b) that represents the same point in the plane in both polar and rectangular coordinates.

56. Suppose that f is a continuously differentiable scalar-valued function of one variable. If $\mathcal{C}$ is a curve given in polar coordinates by the equation $r = f(\theta)$ for $\alpha \leq \theta \leq \beta$, then the arc length of C is

$$\int_{\alpha}^{\beta} \sqrt{f(\theta)^2 + f'(\theta)^2}\, d\theta.$$

Prove this formula. Hint: Describe the curve in terms of the position vector $\mathbf{r}(\theta) = \langle x(\theta), y(\theta)\rangle$, and use the formula for arc length that was developed in Chapter 10.

Refer to Exercise 56 for the arc length formula in polar coordinates. In each of Exercises 57–60, calculate the arclength of the given curve.

57. $r = e^{\theta}$ $\quad -\pi/2 \leq \theta \leq \pi/2$
58. $r = a(1-\cos(\theta))$ $\quad 0 \leq \theta \leq 2\pi$
59. $r = \sin^3(\theta/3)$ $\quad 0 \leq \theta \leq \pi/2$
60. $r = \sec(\theta)$ $\quad 0 \leq \theta \leq \pi/4$

In each of Exercises 61–66, sketch the graph of the given polar curve.

61. $r = \exp(\theta/5),\ -3\pi \leq \theta \leq 2\pi$
62. $r = 2\sin^2(\theta/2)$
63. $r^2 = -3\sin(4\theta)$
64. $r^{1/3} = \sin(\theta)$
65. $r^{1/2} = \sin(\theta)$
66. $r^3 = \sin^2(\theta)$.

67. Sketch the graphs of $r = \theta^2$ and $\theta = r^2$.

68. Explain how and why the graphs of $r^2 = \sin^2(\theta)$ and $r = \sin(\theta)$ differ.

69. In a θr-plane, plot the Cartesian equation $r = \theta + \cos(\theta,$ $0 \leq \theta \leq 2\pi$. Then, plot this equation by interpreting it as a polar equation in the xy-plane.

70. In a θr-plane, plot the Cartesian equation

$$r = \frac{2+\theta^2}{1+\theta^2}, \quad 0 \leq \theta \leq 4\pi.$$

Then, plot this equation by interpreting it as a polar equation in the xy-plane.

Calculator/Computer Exercises

71. Plot the polar curves $r = \sin(j\theta)$ for $j = 1, 2, 3, 4, 5$. Describe the graphs of $r = \sin(50\,\theta)$ and $r = \sin(51\theta)$. Suppose that j is a positive integer. For what positive integer values of k does the graph of $r = \sin(k\,\theta)$ have more petals than the graph of $r = \sin(j\theta)$?

72. Plot the polar curves $r = j + \sin(\theta)$ for $j = 0, 1, 2, 3, 5, 10, 20, 30$. How does changing the additive constant affect the graph?

73. Plot the polar curves $r = \sin(\theta + j)$ for $j = 0, 1, 2, 3, 5, 10, 20, 30$. How does changing the additive constant affect the graph?

74. Plot Fay's Butterfly $r = e^{\cos(\theta)} - 2\cos(4\theta) + \sin^5(\theta/12)$ for $0 \leq \theta \leq 2\pi$. Repeat, using the range $0 \leq \theta \leq 4\pi$ and again for $0 \leq \theta \leq 6\pi$. Continue. For what range $0 \leq \theta \leq 2N\pi$ do you obtain the complete plot without any repetition?

12.5 Integrating in Polar Coordinates

The theory of area in Cartesian coordinates rests on the basic notion that the area of any planar region can be thought of as a sum of areas of thin rectangles, up to a relatively small error. This is sensible because the Cartesian coordinate system is a rectangular one—the axes are perpendicular, and the notion of rectangular area *fits* the notion of rectangular coordinates. Now we are dealing with polar coordinates, and there is no longer a standard pair of perpendicular directions. In fact, we have a radial direction and an angular direction, which are qualitatively quite different from the information coming from rectangular coordinates. As a result, the building blocks we use for area are not rectangles. Instead, they are *sectors*.

A *sector* is a portion of a circle cut off by two rays: look at Figure 1. The area A of the sector depends on the radius r of the circle and the angle θ between the two rays. In fact, by comparing the sector with the full circle and using proportionality, we deduce that $\frac{A}{\pi r^2} = \frac{\theta}{2\pi}$, or

$$A = \frac{1}{2}r^2\theta. \qquad (12.5.1)$$

Our method for calculating area in polar coordinates will be determined by this formula.

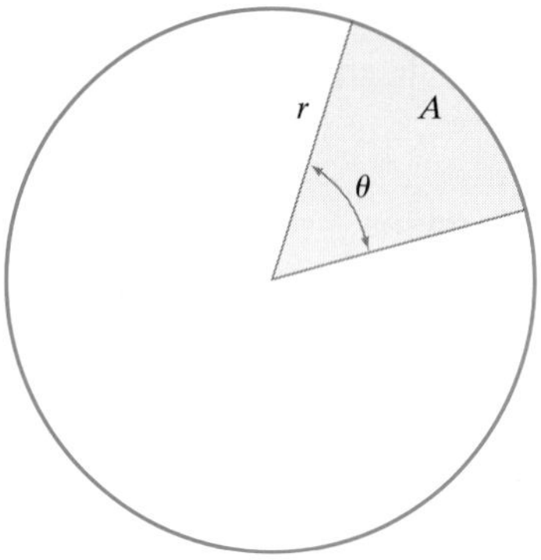

▲ Figure 1 The area of the sector is $A = \frac{1}{2}r^2\theta$.

Areas of More General Regions

We consider a region $\mathcal{R}$ in the xy-plane that is contained inside a polar curve $r = \varphi(\theta)$. More precisely, if we set

$$\mathcal{S} = \{(r, \theta) : 0 \le r \le \varphi(\theta),\ \alpha \le \theta \le \beta\} \qquad (12.5.2)$$

for a given nonnegative function $\varphi(\theta)$ defined for θ between α and β, then we consider the region $\mathcal{R}$ in the xy-plane defined by

$$\mathcal{R} = \{(x, y) : x = r\cos(\theta), y = r\sin(\theta) \text{ for some } (r, \theta) \text{ in } \mathcal{S}\}. \qquad (12.5.3)$$

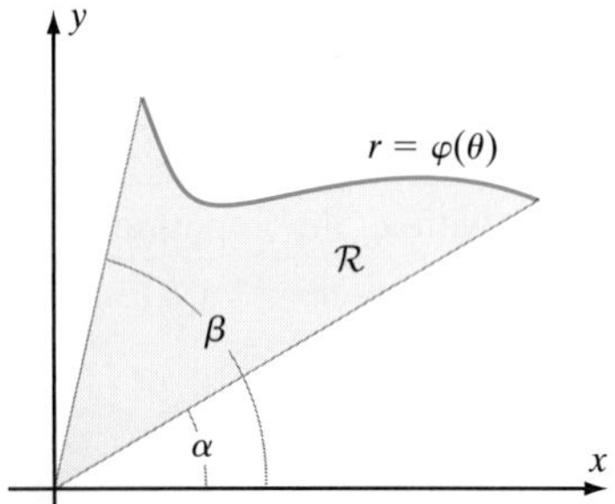

▲ Figure 2

Notice that the description of $\mathcal{R}$ that appears in equation (12.5.3) is in terms of the Cartesian coordinates (x, y) of the points of $\mathcal{R}$. The set $\mathcal{S}$ that appears in equation (12.5.2) describes $\mathcal{R}$ in terms of the polar coordinates of the points of $\mathcal{R}$. To avoid ambiguities, we usually restrict α and β to differ by not more than 2π. The region is shown in Figure 2.

We estimate the area of the region $\mathcal{R}$ by forming a uniform partition,

$$\alpha = \theta_0 < \theta_1 < \theta_2 < \cdots < \theta_{N-1} < \theta_N = \beta,$$

of the interval $[\alpha, \beta]$, which is the domain of φ. The angular increment $\Delta\theta$ satisfies

$$\Delta\theta = \theta_j - \theta_{j-1} = \frac{\beta - \alpha}{N}$$

for all $1 \le j \le N$. Figure 3 shows how the arc corresponding to $\alpha \le \theta \le \beta$ is broken into subarcs by the partition. Now each subarc determines (not a rectangle but) a sector. If S_j is the area of the jth sector, then the sum of the areas of the sectors,

$$\sum_{j=1}^{N} S_j,$$

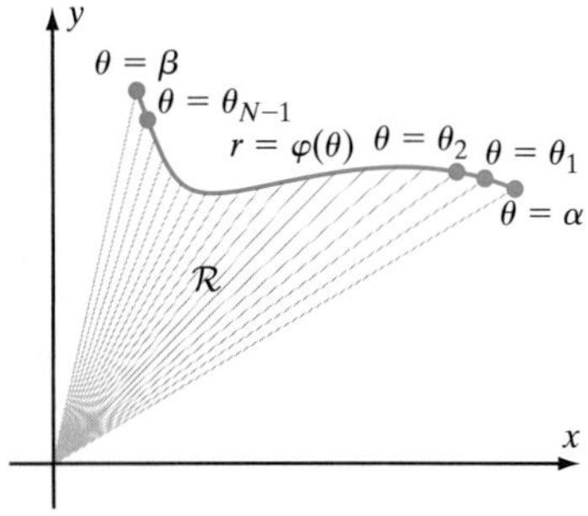

▲ Figure 3

approximates the area of the region $\mathcal{R}$.

In order to calculate this sum, we need to know the area of the jth sector S_j. The angle S_j forms at the origin is $\Delta\theta$. The radius is approximately given by $r_j = \varphi(\xi_j)$, where ξ_j is any number chosen from the interval $[\theta_{j-1}, \theta_j]$. See Figure 4. Thus according to formula (12.5.1), the area of the jth sector is approximately

$$S_j \approx \frac{1}{2}\varphi(\xi_j)^2\Delta\theta.$$

The area of the region $\mathcal{R}$ is therefore approximately

$$\sum_{j=1}^{N} \frac{1}{2} \varphi(\xi_j)^2 \Delta\theta.$$

However, this sum is a Riemann sum for the integral

$$\int_{\alpha}^{\beta} \frac{1}{2} \varphi(\theta)^2 \, d\theta.$$

We summarize our findings in a theorem.

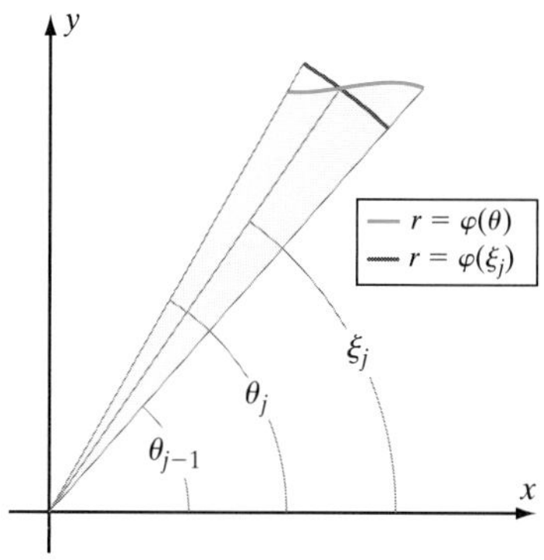

▲ Figure 4

THEOREM 1 Suppose that φ is a nonnegative function on the interval $[\alpha, \beta]$, where $\beta - \alpha \leq 2\pi$. Let $\mathcal{S} = \{(r, \theta) : 0 \leq r \leq \varphi(\theta),\ \alpha \leq \theta \leq \beta\}$ and let $\mathcal{R}$ be the region in the xy-plane that is described in polar coordinates by $\mathcal{S}$. Then, the area of $\mathcal{R}$ is given by the integral

$$\frac{1}{2} \int_{\alpha}^{\beta} \varphi(\theta)^2 \, d\theta. \qquad (12.5.4)$$

▶ **EXAMPLE 1** Find the area inside the curve $r = 2\sin(\theta)$.

Solution In Section 12.4, we learned that the curve $r = 2\sin(\theta)$, shown in Figure 5, is a circle of radius 1 that is traced out once as θ varies from 0 to π. Therefore we expect to obtain $\pi = \pi(1^2)$ as our answer. That is exactly what we obtain when we apply Theorem 1 using the correct interval of integration, $[0, \pi]$. Thus we have

$$\text{Area} = \frac{1}{2} \int_0^{\pi} (2\sin(\theta))^2 \, d\theta = 2 \int_0^{\pi} \sin^2(\theta) \, d\theta \overset{(6.2.8)}{=} 2 \cdot \frac{\pi}{2} = \pi. \quad ◀$$

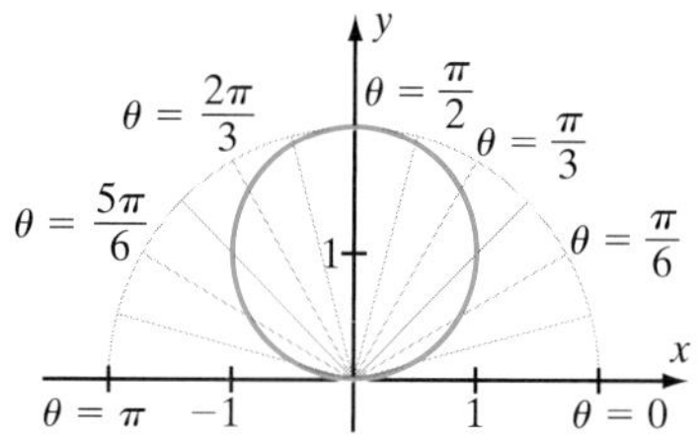

▲ Figure 5 The curve described by the equation $r = 2\sin(\theta)$ in polar coordinates

INSIGHT Example 1 illustrates an important point. Because of the ambiguity built into polar coordinates, the circle in Example 1 is generated twice as θ ranges from 0 to 2π. If we attempt to calculate the area inside it by integrating from 0 to 2π, we count the same area twice and obtain the incorrect value 2π as the answer. As a result of this observation, we must begin the solution to each area problem with a brief analysis to be sure we are avoiding duplication of area. A sketch will usual play a vital role in this analysis.

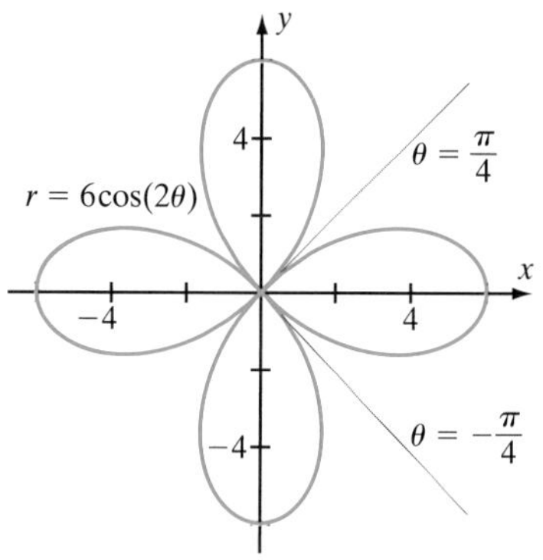

▲ Figure 6

▶ **EXAMPLE 2** Find the area inside one leaf of the rose $r = 6\cos(2\theta)$.

Solution By the periodicity of the cosine function, each of the leaves has the same area, so we choose for convenience the leaf on the right (look at Figure 6). We must integrate from $\theta = -\pi/4$ to $\theta = \pi/4$. By Theorem 1, the area of the leaf is

$$A = \frac{1}{2}\int_{-\pi/4}^{\pi/4} (6\cos(2\theta))^2\, d\theta = 18\int_{-\pi/4}^{\pi/4} \cos^2(2\theta)\, d\theta.$$

To handle this integrand, we use the method we learned in Section 6.2 of Chapter 6. Replacing θ with 2θ in formula (6.2.2), we obtain

$$\cos^2(2\theta) = \frac{1}{2}\big(1 + \cos(4\theta)\big).$$

It follows that the area A of one leaf of the rose is

$$A = 18\int_{-\pi/4}^{\pi/4} \frac{(1+\cos(4\theta))}{2}\, d\theta = 9\left(\theta + \frac{\sin(4\theta)}{4}\right)\Bigg|_{-\pi/4}^{\pi/4} = 9\left(\left(\frac{\pi}{4} - \left(-\frac{\pi}{4}\right)\right) + \left(\frac{\sin(\pi)}{4} - \frac{\sin(-\pi)}{4}\right)\right) = \frac{9\pi}{2}. \quad ◀$$

▶ **EXAMPLE 3** Find the area A of the region $\mathcal{R}$ in the xy-plane that is inside the polar curve $r = 4 + \cos(\theta)$ and outside the polar curve $r = 2 + \cos(\theta)$.

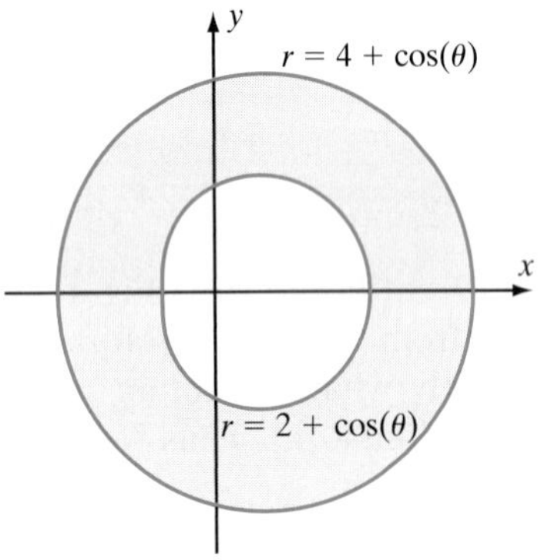

▲ Figure 7

Solution The boundary curves of $\mathcal{R}$ are called *limaçons* (see Figure 7). To avoid duplicating our calculations, we will work with a general limaçon $r = c + \cos(\theta)$, $0 \le \theta \le 2\pi$ where c is a constant greater than 1. The area $A(c)$ inside this limaçon is

$$\begin{aligned} A(c) &= \frac{1}{2}\int_0^{2\pi} (c + \cos(\theta))^2\, d\theta = \frac{1}{2}\int_0^{2\pi} \left(c^2 + 2c\cos(\theta) + \cos^2(\theta)\right) d\theta \\ &= \frac{1}{2}\int_0^{2\pi} c^2\, d\theta + c\int_0^{2\pi} \cos(\theta)\, d\theta + \frac{1}{2}\int_0^{2\pi} \cos^2(\theta)\, d\theta. \end{aligned}$$

The first summand on the right side evaluates to πc^2, the middle summand evaluates to 0, and the last summand evaluates to $\pi/2$ by formula (6.2.9). Thus

$$A(c) = \pi c^2 + 0 + \frac{\pi}{2} = \pi c^2 + \frac{\pi}{2}. \tag{12.5.5}$$

The area A of $\mathcal{R}$ is obtained by subtracting the area inside the limaçon $r = 2 + \cos(\theta)$ from the area inside the limaçon $r = 4 + \cos(\theta)$. Taking $c = 4$ and $c = 2$ in formula (12.5.5), we have

$$A = A(4) - A(2) = \left(\pi(4)^2 + \frac{\pi}{2}\right) - \left(\pi(2)^2 + \frac{\pi}{2}\right) = 12\pi. \quad ◀$$

INSIGHT Example 3 illustrates the calculation of the area of a region bounded by two polar curves. In general, if $\varphi_0(\theta) \le \varphi_1(\theta)$ for $\alpha \le \theta \le \beta$, then the formula

$$\frac{1}{2}\int_\alpha^\beta \varphi_1(\theta)^2\, d\theta - \frac{1}{2}\int_\alpha^\beta \varphi_0(\theta)^2\, d\theta$$

gives the area between the two curves $r=\varphi_0(\theta)$ and $r=\varphi_1(\theta)$ for $\alpha\leq\theta\leq\beta$. This difference of integrals is equivalent to

$$\frac{1}{2}\int_{\alpha}^{\beta}\left(\varphi_1(\theta)^2-\varphi_0(\theta)^2\right)\,d\theta. \tag{12.5.6}$$

Notice the error that we have avoided in Example 3: It would be a mistake to subtract first and then square—the integral $\frac{1}{2}\int_{\alpha}^{\beta}(\varphi_1(\theta)-\varphi_0(\theta))^2\,d\theta$ has nothing to do with the area between the two curves.

Using Iterated Integrals to Calculate Area in Polar Coordinates

Suppose that $\mathcal{R}$ is a planar region given in polar coordinates by

$$\mathcal{S}=\{(r,\theta):\alpha\leq\theta\leq\beta,0\leq r\leq\varphi(\theta)\}$$

where $\beta-\alpha\leq 2\pi$. Although formula (12.5.4) suffices for the calculation of the area of $\mathcal{R}$, it is instructive to derive an area formula that uses iterated integrals. Starting from equation (12.5.4), we proceed as follows:

$$\text{Area of }\mathcal{R}=\frac{1}{2}\int_{\alpha}^{\beta}\varphi^2(\theta)\,d\theta=\int_{\alpha}^{\beta}\frac{1}{2}r^2\Big|_{r=0}^{r=\varphi(\theta)}\,d\theta=\int_{\alpha}^{\beta}\int_{0}^{\varphi(\theta)}r\,drd\theta.$$

It is just as easy to calculate the area of a region $\mathcal{R}$ that is bounded by two polar curves. Suppose that $\mathcal{S}=\{(r,\theta):\alpha\leq\theta\leq\beta,\ \varphi_0(\theta)\leq r\leq\varphi_1(\theta)\}$ and that $\mathcal{R}$ is the region in the xy-plane that is described in polar coordinates by $\mathcal{S}$, as in Figure 8. Then the area of $\mathcal{R}$ is clearly

$$\int_{\alpha}^{\beta}\int_{0}^{\varphi_1(\theta)}r\,drd\theta-\int_{\alpha}^{\beta}\int_{0}^{\varphi_0(\theta)}r\,drd\theta=\int_{\alpha}^{\beta}\left(\int_{0}^{\varphi_1(\theta)}r\,dr-\int_{0}^{\varphi_0(\theta)}r\,dr\right)d\theta=\int_{\alpha}^{\beta}\int_{\varphi_0(\theta)}^{\varphi_1(\theta)}r\,dr\,d\theta.$$

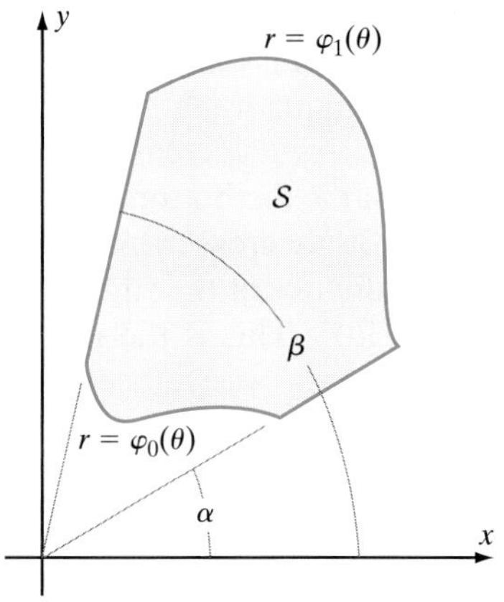

▲ Figure 8

We state our observations as a theorem.

THEOREM 2 Suppose that $0\leq\beta-\alpha\leq 2\pi$ and $0\leq\varphi_0(\theta)\leq\varphi_1(\theta)$ for all θ. Let $\mathcal{S}=\{(r,\theta):\alpha\leq\theta\leq\beta,\ \varphi_0(\theta)\leq r\leq\varphi_1(\theta)\}$. Let $\mathcal{R}$ be the region in the xy-plane that is described in polar coordinates by $\mathcal{S}$. That is, $\mathcal{R}=\{(x,y): x=r\cos(\theta), y=r\sin(\theta)\text{ for some }(r,\theta)\text{ in }\mathcal{S}\}$. Then, the area of $\mathcal{R}$ is given by

$$\iint_{\mathcal{R}}1\,dxdy=\int_{\alpha}^{\beta}\int_{\varphi_0(\theta)}^{\varphi_1(\theta)}r\,drd\theta. \tag{12.5.7}$$

▶ **EXAMPLE 4** Calculate the area of the planar region lying between the curves $r=\cos(\theta)$ and $r=1+\sin(\theta)$ for θ ranging between 0 and $\pi/4$.

Solution Notice that, for this range of θ, we have $\cos(\theta) \le 1 + \sin(\theta)$. See Figure 9. According to Theorem 2, this area is

$$A = \int_0^{\pi/4} \int_{\cos(\theta)}^{1+\sin(\theta)} r\,dr\,d\theta = \int_0^{\pi/4} \left(\frac{1}{2} r^2 \Big|_{r=\cos(\theta)}^{r=1+\sin(\theta)} \right) d\theta = \frac{1}{2} \int_0^{\pi/4} \left(1 + 2\sin(\theta) + \sin^2(\theta) - \cos^2(\theta)\right) d\theta.$$

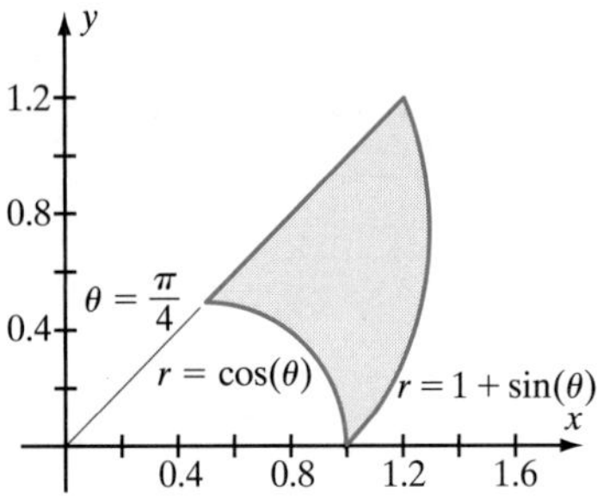

▲ Figure 9

We could proceed as in previous problems in this section by using the half-angle formulas for sine and cosine. Here, however, it is simpler to exploit the combination of $\cos^2(\theta)$ and $\sin^2(\theta)$ that has arisen. We use the Double Angle Formula for the cosine: $\cos(2\theta) = \cos^2(\theta) - \sin^2(\theta)$. As a result,

$$A = \frac{1}{2} \int_0^{\pi/4} \left(1 + 2\sin(\theta) - \cos(2\theta)\right) d\theta = \left(\frac{\theta}{2} - \cos(\theta) - \frac{\sin(2\theta)}{4} \right) \Bigg|_0^{\pi/4}$$

$$= \left(\frac{\pi}{8} - \frac{\sqrt{2}}{2} - \frac{1}{4} \right) - (0 - 1 - 0) = \frac{3}{4} + \frac{\pi}{8} - \frac{\sqrt{2}}{2}. \quad ◀$$

Integrating Functions in Polar Coordinates

Because $\iint_{\mathcal{R}} 1\,dA$ and $\iint_{\mathcal{R}} 1\,dxdy$ represent the area of a planar region $\mathcal{R}$ when we are working in rectangular coordinates, we sometimes use the expression *element of area* to describe dA and $dxdy$. Theorem 2 tells us that, in polar coordinates, the element of area is $r\,drd\theta$. Suppose that $\mathcal{R}$ is a planar region that is described in terms of polar coordinates by a set $\mathcal{S}$. Let $(r, \theta) \mapsto g(r, \theta)$ be a continuous function defined on $\mathcal{S}$. That is, g is a function of the polar coordinates of the points of $\mathcal{R}$. Then we can integrate g over $\mathcal{S}$ using the element of area $r\,drd\theta$. That is, we can consider the integral $\iint_{\mathcal{S}} g(r, \theta)\, r\,dr\,d\theta$. Observe that the element of area in polar coordinates contributes a nontrivial factor, namely r, to the integrand.

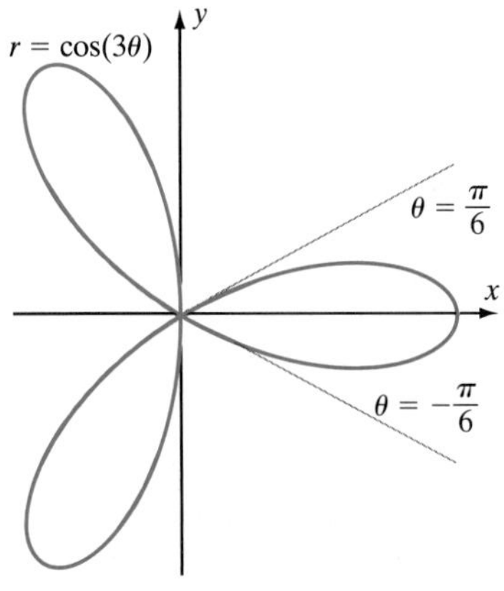

▲ Figure 10

▶ **EXAMPLE 5** Let $\mathcal{R}$ be the region in the plane that is enclosed by the rightmost petal of the three-leafed rose given in polar coordinates by $r = \cos(3\theta)$. Let $\mathcal{S}$ be the set of polar coordinates of the points of $\mathcal{R}$. Integrate $g(r, \theta) = r\sin^2(3\theta)$ over $\mathcal{S}$ using the element of area in polar coordinates.

Solution The plot of $r = \cos(3\theta)$ appears in Figure 10. We need to find the range of θ for the rightmost loop. Because the origin is the "initial" point and "endpoint" of the loop, we must find the values of θ in the first and fourth quadrants for which $\cos(3\theta) = 0$. These are $\theta = \pi/6$ and $\theta = -\pi/6$, respectively. As θ increases from $-\pi/6$ to 0, the value of $r = \cos(3\theta)$ increases from 0 to 1. That is the bottom half of the petal. As θ increases from 0 to $\pi/6$, the value of $r = \cos(3\theta)$ decreases from 1 to 0, generating the top half of the petal. The set of polar coordinates of the rightmost petal is therefore $\mathcal{S} = \{(r, \theta) : -\pi/6 \le \theta \le \pi/6,\ 0 \le r \le \cos(3\theta)\}$. This is the region of integration. The requested integral is

$$\iint_{\mathcal{S}} g(r, \theta)\, r\,dr\,d\theta = \int_{-\pi/6}^{\pi/6} \int_0^{\cos(3\theta)} r\sin^2(3\theta)\, r\,drd\theta = \int_{-\pi/6}^{\pi/6} \left(\sin^2(3\theta) \int_0^{\cos(3\theta)} r^2 dr \right) d\theta$$

$$= \frac{1}{3} \int_{-\pi/6}^{\pi/6} \sin^2(3\theta) \cos^3(3\theta)\, d\theta.$$

We have learned the technique for evaluating such integrals in Chapter 6. We write $\cos^3(3\theta) = \cos^2(3\theta)\cos(3\theta) = \left(1 - \sin^2(3\theta)\right)\cos(3\theta)$ and make the substitution $u = \sin(3\theta), du = 3\cos(3\theta)\, d\theta$. The integral becomes

$$\begin{aligned} \frac{1}{3}\int_{-\pi/6}^{\pi/6} \sin^2(3\theta)\cos^3(3\theta)\, d\theta &= \frac{1}{3}\int_{-\pi/6}^{\pi/6} \sin^2(3\theta)\left(1-\sin^2(3\theta)\right)\cos(3\theta)\, d\theta \\ &= \frac{1}{9}\int_{-1}^{1} u^2(1-u^2)\, du \\ &= \frac{1}{9}\int_{-1}^{1} (u^2-u^4)\, du \\ &= \frac{4}{135}. \end{aligned}$$ ◀

Sometimes an integration problem originally presented in rectangular coordinates can be tackled more easily in polar coordinates. We cannot just write $x = r\cos(\theta)$ and $y = r\sin(\theta)$ and use $f(r\cos(\theta), r\sin(\theta))$ instead of $f(x,y)$. We must take into account the elements of area, which are different in the two coordinate systems. In other words, we must remember that the element of area $dxdy$ in rectangular coordinates becomes $r\, drd\theta$ in polar coordinates. It follows that, when we convert an integration problem from rectangular to polar coordinates, the iterated integral $\iint_{\mathcal{R}} f(x,y)\, dxdy$ becomes $\iint_{\mathcal{S}} f(r\cos(\theta), r\sin(\theta) r\, drd\theta$. We will give a more formal treatment of the change from rectangular coordinates to polar coordinates later in this section. For now, we state the resulting theorem and turn to some examples.

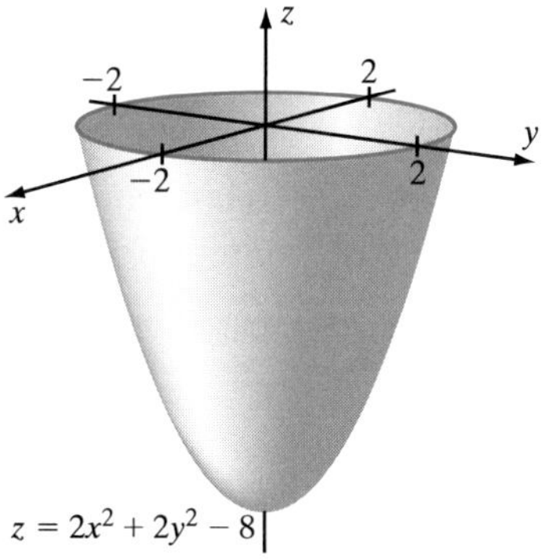

▲ Figure 11

THEOREM 3 Let $(x,y) \mapsto f(x,y)$ be a continuous function that is defined on a planar region $\mathcal{R}$. Suppose that $\mathcal{R}$ can be described in polar coordinates by $\mathcal{S} = \{(r,\theta) : \alpha \le \theta \le \beta,\ \varphi_0(\theta) \le r \le \varphi_1(\theta)\}$ where $0 \le \beta - \alpha \le 2\pi$ and $0 \le \varphi_0(\theta) \le \varphi_1(\theta)$ for all θ. Then, the integral of f over $\mathcal{R}$ is given by

$$\iint_{\mathcal{R}} f(x,y)\, dA = \int_{\alpha}^{\beta}\int_{\varphi_0(\theta)}^{\varphi_1(\theta)} f\left(r\cos(\theta), r\sin(\theta)\right)\, r\, drd\theta. \qquad (12.5.8)$$

▶ **EXAMPLE 6** Calculate the volume V of the solid bounded by the paraboloid $z = 2x^2 + 2y^2 - 8$ and the xy-plane.

Solution Look at Figure 11. The paraboloid intersects the xy-plane in the circle whose equation is $2x^2 + 2y^2 - 8 = 0$, or $x^2 + y^2 = 4$. Thus we calculate the volume by integrating the height function $f(x,y) = 0 - (2x^2 + 2y^2 - 8) = 8 - 2(x^2 + y^2)$ over the disk $\mathcal{S}$ that is bounded by the circle with radius 2 that is centered at the origin of the xy-plane. We can accomplish this by integrating the function $f\big(r\cos(\theta), r\sin(\theta)\big) = 8 - 2\left(r^2\cos^2(\theta) + r^2\sin^2(\theta)\right) = 8 - 2r^2$ over the region $\mathcal{S} = \{(r,\theta) : 0 \le \theta \le 2\pi, 0 \le r \le 2\}$ using the polar element of area. We obtain

$$V = \int_0^{2\pi}\int_0^2 (8-2r^2)\, r\, drd\theta = \int_0^{2\pi}\int_0^2 (8r - 2r^3)\, drd\theta = \int_0^{2\pi}\left(4r^2 - \frac{2}{4}r^4\right)\Big|_{r=0}^{r=2} d\theta = \int_0^{2\pi} 8\, d\theta = 16\pi.$$ ◀

INSIGHT Rectangular coordinates can also be used to calculate the volume of the solid in Example 6. The required integral is $V = \int_{-2}^{2} \int_{-\sqrt{4-x^2}}^{\sqrt{4-x^2}} (8 - 2(x^2 + y^2))\, dy\, dx$, which, with some effort, does evaluate to 16π. The calculation is decidedly simpler in polar coordinates because of the symmetry about the z-axis. Problems involving circular symmetry are often best solved using polar coordinates.

▶ **EXAMPLE 7** Set up the integral for calculating the volume V of the solid $\mathcal{U}$ that consists of the points inside the lower branch of the cone $(z-3)^2 = x^2 + y^2$, above the xy-plane, and inside the circular cylinder $x^2 + (y-1)^2 = 1$.

Solution Look at Figure 12. Notice that the solid $\mathcal{U}$ lies entirely inside the cylinder, while its upper boundary is determined by the lower branch of the cone, and its lower boundary by the xy-plane. The equation for the lower half of the cone is $z = 3 - (x^2 + y^2)^{1/2}$. Therefore we must integrate the function $f(x, y) = 3 - (x^2 + y^2)^{1/2}$ over the disk that is bounded by the circle $x^2 + (y-1)^2 = 1$. This circle is the graph of the polar coordinate equation $r = 2\sin(\theta)$ for $0 \le \theta \le \pi$, as we have seen in Example 7 of Section 12.4. Therefore we can calculate V by using polar coordinates, integrating the function $f(r\cos(\theta), r\sin(\theta)) = 3 - r$ over the region $\mathcal{S} = \{(r, \theta) : 0 \le r \le \pi,\ 0 \le r \le 2\sin(\theta)\}$. The volume of $\mathcal{U}$ is

$$V = \int_0^{\pi} \int_0^{2\sin(\theta)} (3 - r)\, r\, dr d\theta = \int_0^{\pi} \int_0^{2\sin(\theta)} (3r - r^2)\, dr d\theta.$$

Therefore

$$V = \int_0^{\pi} \left(\frac{3}{2} r^2 - \frac{1}{3} r^3 \right) \Big|_{r=0}^{r=2\sin(\theta)} d\theta = \int_0^{\pi} \left(6\sin^2(\theta) - \frac{8}{3}\sin^3(\theta) \right) d\theta.$$

The interested reader may verify that the value of V is $3\pi - 32/9$. (Use equation (6.2.8) for the integration of $\sin^2(\theta)$ and refer to Section 6.2 of Chapter 6 for the integration of $\sin^3(\theta)$.)

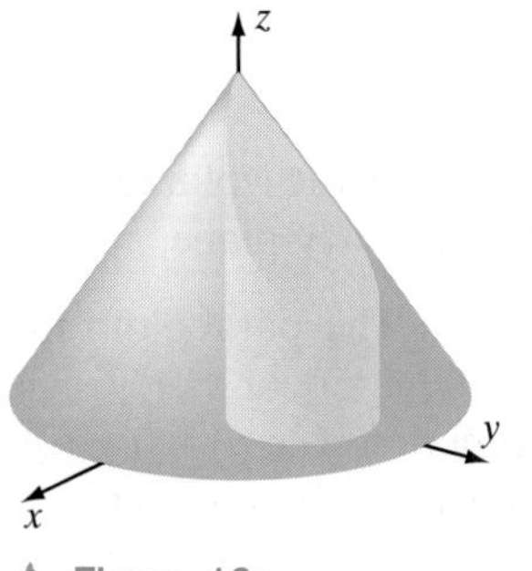

▲ Figure 12a

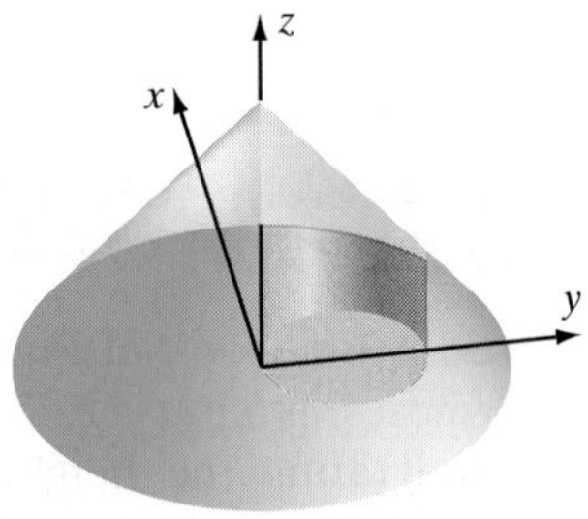

▲ Figure 12b Viewed from below

Change of Variable and the Jacobian

Our discussion of the element of area in polar coordinates suggests an important question: How does a change of variable affect a double integral? That is the problem that we now investigate. Suppose that $\mathcal{S}$ is a planar region on which two differentiable functions ϕ and ψ are defined. These are the functions that we use to effect the change of variables on $\mathcal{S}$. Let $\mathcal{R}$ be the region defined by

$$\mathcal{R} = \{(x, y) : x = \phi(u, v) \quad \text{and} \quad y = \psi(u, v) \text{ for some point } (u, v) \text{ in } \mathcal{S}\}.$$

Figure 13 shows that we can imagine ϕ and ψ transforming the region $\mathcal{S}$ into the region $\mathcal{R}$. In other words, we define a function $T : \mathcal{S} \to \mathbb{R}^2$ by the equation $T(u, v) = (\phi(u, v), \psi(u, v))$ and define $\mathcal{R}$ to be the image of $\mathcal{S}$ under T. That is,

$$\mathcal{R} = \{(x, y) : (x, y) = T(u, v) = (\phi(u, v), \psi(u, v)) \text{ for some point } (u, v) \text{ in } \mathcal{S}\}.$$

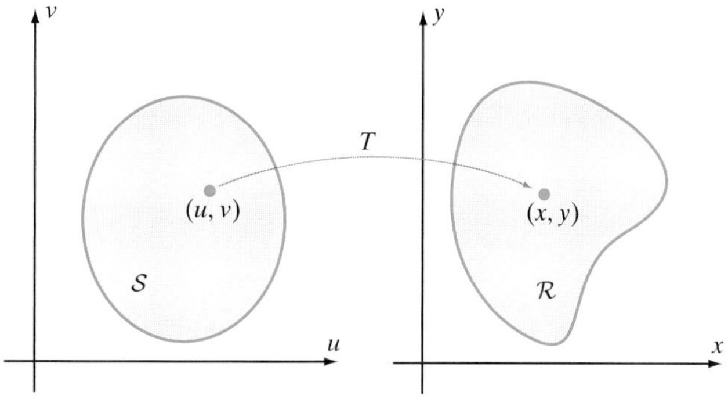

▲ Figure 13 The equation $(x, y) = (\phi(u, v), \psi(u, v))$ transfoms the region $\mathcal{S}$ in to the region $\mathcal{R}$.

We assume that T is one-to-one so that, given a point (x, y) in $\mathcal{R}$, there is a unique point (u, v) in $\mathcal{S}$ for which $T(u, v) = (x, y)$. In other words, T is an invertible function from $\mathcal{S}$ onto $\mathcal{R}$.

Because the components, ϕ and ψ, of T are differentiable, we can form the matrix

$$D(T)(u, v) = \begin{bmatrix} \phi_u(u, v) & \phi_v(u, v) \\ \psi_u(u, v) & \psi_v(u, v) \end{bmatrix},$$

which is called the *Jacobian matrix* of T. The determinant of this square matrix,

$$J_T(u, v) = \det\left(\begin{bmatrix} \phi_u(u, v) & \phi_v(u, v) \\ \psi_u(u, v) & \psi_v(u, v) \end{bmatrix}\right) = \phi_u(u, v)\psi_v(u, v) - \phi_v(u, v)\psi_u(u, v) = (\phi_u\psi_v - \phi_v\psi_u)(u, v) \qquad (12.5.9)$$

is called the *Jacobian determinant* (sometimes abbreviated to just the *Jacobian*) of T. The classical notation $\dfrac{\partial(x, y)}{\partial(u, v)}$ for the Jacobian determinant is still encountered frequently.

The next example will reveal that the Jacobian determinant has been present all along in the integrations that we have been performing.

▶ **EXAMPLE 8** Let $\mathcal{R}$ be a region in the plane. Let $\mathcal{S}$ be the set of polar coordinates of points of $\mathcal{R}$. Let $T : \mathcal{S} \to \mathcal{R}$ be defined by $T(r, \theta) = (r\cos(\theta), r\sin(\theta))$. Show that

$$\iint_{\mathcal{R}} 1 \, dxdy = \iint_{\mathcal{S}} 1 \, |J_T(r, \theta)| \, drd\theta. \qquad (12.5.10)$$

Solution We calculate the Jacobian matrix of T as follows:

$$D(T)(r,\theta) = \begin{bmatrix} \frac{\partial}{\partial r}(r\cos(\theta)) & \frac{\partial}{\partial \theta}(r\cos(\theta)) \\ \frac{\partial}{\partial r}(r\sin(\theta)) & \frac{\partial}{\partial \theta}(r\sin(\theta)) \end{bmatrix} = \begin{bmatrix} \cos(\theta) & -r\sin(\theta) \\ \sin(\theta) & r\cos(\theta) \end{bmatrix}.$$

The Jacobian determinant is given by

$$J_T(r,\theta) = \det\left(\begin{bmatrix} \cos(\theta) & -r\sin(\theta) \\ \sin(\theta) & r\cos(\theta \end{bmatrix}\right) = \cos(\theta)r\cos(\theta) - (-r\sin(\theta))\sin(\theta) = r\big(\cos^2(\theta) + \sin^2(\theta)\big) = r.$$

Using this result together with Theorem 2, we have

$$\iint_{\mathcal{R}} 1\, dxdy = \iint_{\mathcal{S}} r\, drd\theta = \iint_{\mathcal{S}} 1\, |J_T(r,\theta)|\, drd\theta.$$

The next example shows that the Jacobian determinant plays a role in area calculations even when polar coordinates are not involved. ◀

▶ **EXAMPLE 9** For a point (u,v) in $\mathcal{S} = \{(u,v): 0 \le u \le 1, 0 \le v \le 1\}$, let $\phi(u,v) = 2u + v$, $\psi(u,v) = u - v$, and $T(u,v) = (2u + v, u - v)$. Show that T is invertible. What is the image $\mathcal{R}$ of $\mathcal{S}$ under T? Calculate the area of $\mathcal{R}$ and show that

$$\iint_{\mathcal{R}} 1\, dx\, dy = \iint_{\mathcal{S}} 1|J_T(u,v)|\, dudv. \tag{12.5.11}$$

Solution If P is the point in $\mathcal{R}$ with coordinates $(x,y) = (2u + v, u - v)$, then we can solve the equations $x = 2u + v, y = u - v$ simultaneously for u and v. We find $u = (x + y)/3$ and $v = (x - 2y)/3$. Because this solution is unique, we have shown that $T: \mathcal{S} \to \mathcal{R}$ is invertible. If we write the position vector $\langle x, y\rangle$ of P as $\langle 2u + v, u - v\rangle$, or $u\,\langle 2,1\rangle + v\,\langle 1,-1\rangle$, then we see that, as u and v take on all values between 0 and 1, the point P runs through all points in the parallelogram $\mathcal{R}$ that has sides $\langle 2,1\rangle$ and $\langle 1,-1\rangle$. See Figure 14. If we write these two sides as spatial vectors that have 0 as the third coordinate, then we calculate

$$\text{Area of } \mathcal{R} = \big\|(2\mathbf{i} + 1\mathbf{j} + 0\mathbf{k}) \times (1\mathbf{i} + (-1)\mathbf{j} + 0\mathbf{k})\big\| = \big\|0\mathbf{i} + 0\mathbf{j} + (-3)\mathbf{k}\big\| = 3.$$

Also,

$$D(T)(u,v) = \begin{bmatrix} \frac{\partial}{\partial u}\phi(u,v) & \frac{\partial}{\partial v}\phi(u,v) \\ \frac{\partial}{\partial u}\psi(u,v) & \frac{\partial}{\partial v}\psi(u,v) \end{bmatrix} = \begin{bmatrix} \frac{\partial}{\partial u}(2u + v) & \frac{\partial}{\partial v}(2u + v) \\ \frac{\partial}{\partial u}(u - v) & \frac{\partial}{\partial v}(u - v) \end{bmatrix} = \begin{bmatrix} 2 & 1 \\ 1 & -1 \end{bmatrix}.$$

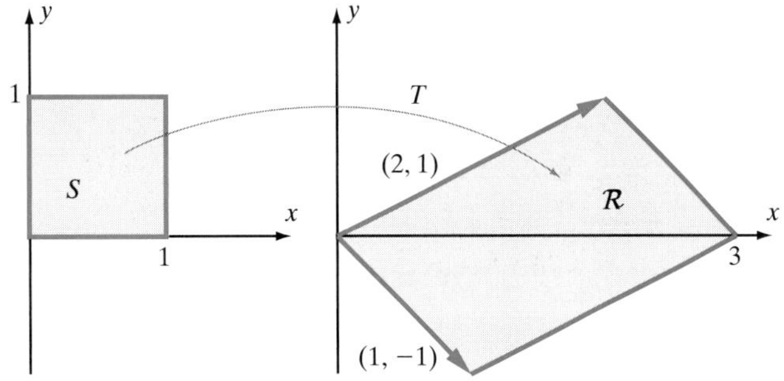

▲ Figure 14

Therefore $J_T(u, v) = (2)(-1) - (1)(1) = -3$. Because the area of $\mathcal{S}$ is 1, we have

$$\iint_{\mathcal{R}} 1\, dxdy = 3 = (3)(1) = 3\iint_{\mathcal{S}} 1\, dudv = \iint_{\mathcal{S}} 1|J_T(u, v)|\, dudv,$$

which is equation (12.5.11). ◄

Theorem 4 shows that equations (12.5.10) and (12.5.11) are special cases of a general change of variable formula. As you read Theorem 4, you may find it helpful to refer to Figure 15 for a schematic diagram of the functions $T, f, f \circ T$, and J_T that appear in the theorem.

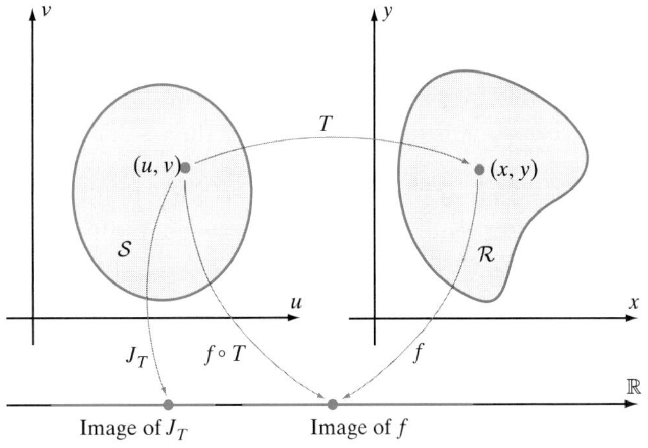

▲ Figure 15

THEOREM 4 Suppose that $T(u, v) = (\phi(u, v), \psi(u, v))$ is a one-to-one transformation of a region $\mathcal{S}$ in the uv-plane onto a region $\mathcal{R}$ in the xy-plane. Suppose that the component functions ϕ, ψ of T are continuously differentiable and that f is a continuous function on $\mathcal{R}$. Then

$$\iint_{\mathcal{R}} f(x,y)\,dxdy = \iint_{\mathcal{S}} f(\phi(u,v),\psi(u,v))\,|J_T(u,v)|\,dudv, \tag{12.5.12}$$

where

$$J_T(u,v) = \det\left(\begin{bmatrix} \phi_u(u,v) & \phi_v(u,v) \\ \psi_u(u,v) & \psi_v(u,v) \end{bmatrix}\right) = (\phi_u\psi_v - \psi_u\phi_v)(u,v)$$

is the Jacobian determinant of T.

Proof. Consider a partition of $\mathcal{R}$. Let the lengths and widths of the rectangles in this partition be Δx and Δy, respectively. Let $\xi_{i,j}$ be a point in the i,j^{th} rectangle in the partition of $\mathcal{R}$. These choices determine a Riemann sum $\sum f(\xi_{i,j})\Delta x\Delta y$ for f over $\mathcal{R}$. Let $\eta_{i,j}$ be the point of $\mathcal{S}$ such that $T(\eta_{i,j}) = \xi_{i,j}$. Our job is to show that the Riemann sum $\sum f(T(\eta_{i,j}))\,|J_T(\eta_{i,j})|\Delta u\Delta y$ over $\mathcal{S}$ corresponds to $\sum f(\xi_{i,j})\Delta x\Delta y$.

Suppose that Δu and Δv are small increments in u and v. Then, from formula (11.7.7), we have $\Delta x \approx \phi_u\Delta u + \phi_v\Delta v$ and $\Delta y \approx \psi_u\Delta u + \psi_v\Delta v$. Interpreted geometrically, this means that a small horizontal displacement in the xy-coordinates, which we may represent by the vector $\langle \Delta x, 0\rangle$, corresponds to a displacement in the uv-coordinates given by the vector $\langle \phi_u\Delta u, \phi_v\Delta v\rangle$. Similarly, a small vertical displacement in the xy-coordinates, which we may represent by the vector $\langle 0, \Delta y\rangle$, corresponds to a displacement in the uv-coordinates given by the vector $\langle \psi_u\Delta u, \psi_v\Delta v\rangle$.

In Figure 16, we see that the rectangle with sides Δx and Δy in the xy-plane corresponds to a parallelogram that is determined by the vectors $\langle \phi_u\Delta u, \phi_v\Delta v\rangle$ and $\langle \psi_u\Delta u, \psi_v\Delta v\rangle$ in the uv-plane. Therefore the element of area $\Delta x\Delta y$ in the xy-plane corresponds to the area ΔA of the parallelogram that is determined by the vectors $\langle \phi_u\Delta u, \phi_v\Delta v\rangle$ and $\langle \psi_u\Delta u, \psi_v\Delta v\rangle$ in the uv-plane. To calculate ΔA, we think of the uv-plane as a coordinate plane in three-dimensional space. Then, as we learned in Section 9.4 of Chapter 9, ΔA is equal to the length of the cross product of the two vectors $\langle \phi_u\Delta u, \phi_v\Delta v, 0\rangle$ and $\langle \psi_u\Delta u, \psi_v\Delta v, 0\rangle$. As a result, we have

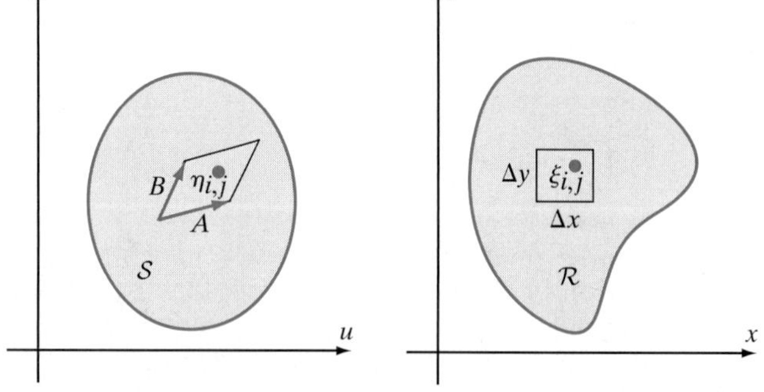

▲ **Figure 16** The parallelogram determined by vectors $\vec{A} = \langle \phi_u\Delta u, \phi_v\Delta v\rangle$ and $\vec{B} = \langle \psi_u\Delta u, \psi_v\Delta v\rangle$ corresponds to the rectangle in $\mathcal{R}$ with sides Δx and Δy.

$$\Delta A = \left\| \det\left(\begin{bmatrix} \mathbf{i} & \mathbf{j} & \mathbf{k} \\ \phi_u \Delta u & \phi_v \Delta v & 0 \\ \psi_u \Delta u & \psi_v \Delta v & 0 \end{bmatrix} \right) \right\| = \| (\phi_u \psi_v - \psi_u \phi_v) \Delta u \Delta v \mathbf{k} \| = |\phi_u \psi_v - \psi_u \phi_v| \Delta u \Delta v.$$

From this computation, we see that the Riemann sum $\sum f(\xi_{i,j}) \Delta x \Delta y$ for f over $\mathcal{R}$ corresponds to the Riemann sum in uv-coordinates given by

$$\sum f(T(\eta_{i,j})) |(\phi_u \psi_v - \psi_u \phi_v)(\eta_{i,j})| \Delta u \Delta v.$$

(Remember that, in the algebra of functions, $(\phi_u \psi_v - \psi_u \phi_v)(\eta)$ means $\phi_u(\eta)\psi_v(\eta) - \psi_u(\eta)\phi_v(\eta)$.) Passing to the limit, we obtain equation (12.5.12).

We have already used Theorem 4 several times. Indeed, Theorem 3 is the special case of Theorem 4 in which T transforms polar coordinates to Cartesian coordinates. We conclude this section with an application of Theorem 4 that is new to us.

▶ **EXAMPLE 10** Evaluate the integral

$$\int_0^3 \int_{x/3}^{(x/3)+2} \left(\frac{x+y}{2} \right) dy dx$$

using the transformation $u = x/3$, $v = (x+y)/2$.

Solution In this problem, the region $\mathcal{R}$ is given by $\{(x, y) : x/3 \le y \le x/3 + 2, 0 \le x \le 3\}$. We begin by rewriting the transformation in a form that fits the theorem. The first equation is equivalent to $x = 3u$. The second equation is equivalent to $y = -x + 2v$, or $y = -3u + 2v$. Our transformation T is therefore $x = \phi(u, v) = 3u$, $y = \psi(u, v) = -3u + 2v$. To find the region $\mathcal{S}$ that corresponds to $\mathcal{R}$, notice that $0 \le x \le 3$ corresponds to $0 \le u \le 1$ and $(x/3) \le y \le (x/3) + 2$ corresponds to $u \le 2v - 3u \le u + 2$. This last chain of inequalities simplifies, by adding $3u$ to all parts, to $4u \le 2v \le 4u + 2$, or $2u \le v \le 2u + 1$. Thus $\mathcal{S} = \{(u, v) : 2u \le v \le 2u + 1, 0 \le u \le 1\}$. See Figure 17. Because the Jacobian determinant of the transformation is

$$J_T(u, v) = (\phi_u \psi_v - \psi_u \phi_v)(u, v) = (3)(2) - (-3)(0) = 6,$$

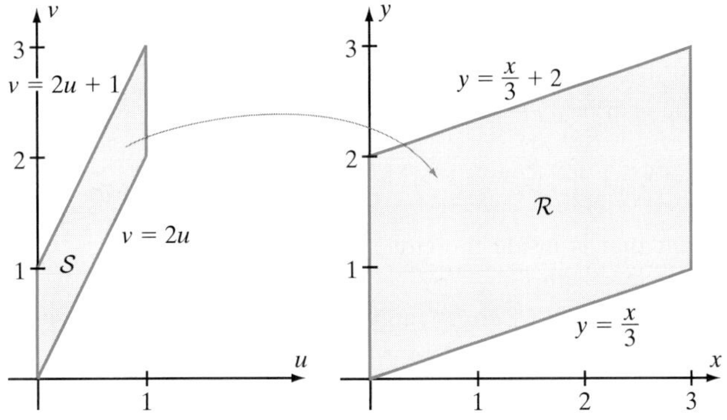

▲ Figure 17 The transformation $x = \phi(u, v) = 3u$, $y = \psi(u, v) = -3u + 2v$

the integral becomes

$$\int_0^3 \int_{x/3}^{(x/3)+2} \left(\frac{x+y}{2}\right) dydx = \int_0^1 \int_{2u}^{2u+1} 6v\, dvdu = 6 \int_0^1 \left(\frac{v^2}{2}\Big|_{v=2u}^{v=2u+1} \right) du$$
$$= 3 \int_0^1 (4u+1)\, du = 3\left((2u^2+u)\Big|_0^1\right) = 9. \blacktriangleleft$$

QUICK QUIZ

1. What is the area enclosed by the polar curve $r = \sqrt{\cos(\theta)}$, $-\pi/2 \le \theta \le \pi/2$?
2. What is the element of area in polar coordinates?
3. Use polar coordinates to calculate the integral of $f(x, y) = x^2 + y^2$ over the region that lies outside the unit circle and inside the circle $x^2 + y^2 = 4$.
4. What is the Jacobian determinant $J_T(u, v)$ of the transformation $T(u, v) = (u^3 + 4v, 3u + 2v)$?

Answers

1. 1 2. $r\,drd\theta$ 3. $15\pi/2$ 4. $6u^2 - 12$

EXERCISES

Problems for Practice

In each of Exercises 1–6, use formula (12.5.2) to evaluate the area enclosed by the given curve.

1. $r = 1 - \sin(\theta)$
2. $r = 7 - 3\cos(\theta)$
3. $r = 4\sin(3\theta)$
4. $r = 5\sin(2\theta)$
5. $r^2 = -\sin(\theta)$
6. $r^2 = \cos(\theta)$

In each of Exercises 7–12, use formula (12.5.4) to find the area of the region that is described.

7. The region between the graphs of $r = 6 - \sin(\theta)$ and $r = 1 + \cos(\theta)$.
8. The region in the first quadrant between the graphs of $r = 3\sin(2\theta)$ and $r = 4 - \cos(\theta)$.
9. The region enclosed by the curves $r = \sin(\theta)$ and $r = \cos(\theta)$.
10. The "lune" outside the cardiod $r = 1 + \sin(\theta)$ and inside the circle $r = 3\sin(\theta)$.
11. The region in the first quadrant that is inside the circle $r = \sqrt{3}\cos(\theta)$ and outside the circle $r = \sin(\theta)$.
12. The region that is inside the circle $r = \sqrt{3}\cos(\theta)$ and outside the cardioid $r = 1 + \sin(\theta)$.

In each of Exercises 13–18, calculate the area of the given region $\mathcal{R}$ using a double integral in polar coordinates.

13. $\mathcal{R} = \{(r, \theta) : -\pi/4 \le \theta \le \pi/4, 2 \le r \le 4\cos(\theta)\}$
14. $\mathcal{R}$ is one leaf of the rose $r = 3\sin(4\theta)$.
15. $\mathcal{R}$ is the region inside the circle with center the origin and radius 5 and outside the limaçon $r = 3 + \cos(\theta)$.
16. $\mathcal{R} = \{(r, \theta) : \pi/6 \le \theta \le \pi/4, \sin(\theta) \le r \le \cos(\theta)\}$
17. $\mathcal{R}$ is the region outside the cardioid $r = 1 + \cos(\theta)$ and inside the limaçon $r = 4 - \sin(\theta)$.
18. $\mathcal{R}$ is the region inside the circle $r = 4\cos(\theta)$ and outside the right hand leaf of the rose $r = \cos(2\theta)$.

In each of Exercises 19–26, integrate the given function f over the given planar region $\mathcal{R}$, using a double integral in polar coordinates.

19. $f(r, \theta) = \cos(\theta)$; $\mathcal{R}$ is the region inside the circle $r = \cos(\theta)$.
20. $f(r, \theta) = \sin^2(\theta)$; $\mathcal{R}$ is the region inside the limaçon $r = 3 - \cos(\theta)$.
21. $f(r, \theta) = \cos(\theta)$; $\mathcal{R}$ is the region inside the leftmost petal of the rose $r = \cos(2\theta)$.
22. $f(r, \theta) = \theta$; $\mathcal{R} = \{(r, \theta) : 0 \le \theta \le \pi/2, 0 \le r \le \cos^{1/2}(\theta)\}$
23. $f(r, \theta) = \sin(\theta)$; $\mathcal{R}$ is the region inside the circle $r = \sin(\theta)$.
24. $f(r, \theta) = r$; $\mathcal{R} = \{(r, \theta) : 0 \le \theta \le \pi/2, 0 \le r \le \sin(\theta)\}$
25. $f(r, \theta) = 3r - 10$; $\mathcal{R}$ is the region outside the circle $r = 3$ and inside the limaçon $r = 4 - \cos(\theta)$.
26. $f(r, \theta) = \sin(\theta)$; $\mathcal{R}$ is the region in the first quadrant that is inside the circle $r = 1$ and outside the cardioid $r = 1 - \cos(\theta)$.

In each of Exercises 27–38, use a double integral in polar coordinates to determine the volume of the solid.

27. The solid bounded by the surface $z = -x^2 - y^2 + 4$ and the xy-plane
28. The solid bounded by cylinder $(x-2)^2 + y^2 = 4$, the paraboloid $z = x^2 + y^2 + 6$, and the xy-plane
29. The solid bounded by the paraboloids $z = 3x^2 + 3y^2 - 7$ and $z = -x^2 - y^2 + 9$
30. The solid bounded by the paraboloids $z = 2x^2 + 2y^2 + 4$ and $z = -4x^2 - 4y^2 - 4$ and by the cylinder $x^2 + y^2 = 16$
31. The solid bounded by the upper nappe of the cone $(z+3)^2 = 4x^2 + 4y^2$ and by the paraboloid $z = 5 - x^2 - y^2$
32. The solid bounded by the paraboloids $z = 4x^2 + 4y^2 + 2$ and $z = 7 - x^2 - y^2$
33. The solid bounded by the hyperboloid $z^2 - x^2 - y^2 = 1$ and by the plane $z = 4$
34. The solid bounded by the upper nappe of the cone $z^2 = 4x^2 + 4y^2$ and by the plane $z = 12$
35. The solid bounded by the cylinder $x^2 + y^2 = 4$ and by the hyperboloid $z^2 - x^2 - y^2 = 4$
36. The solid bounded by the the sphere $x^2 + y^2 + z^2 = 20$ and by the paraboloid $z = x^2 + y^2$
37. The solid outside the cone $z^2 = x^2 + y^2$ and inside the cylinder $x^2 + y^2 = 4$
38. The solid bounded by the cylinder $x^2 + (y-1)^2 = 1$ and the sphere $x^2 + y^2 + z^2 = 4$

In each of Exercises 39–44, perform the indicated change of variable in the double integral and then evaluate the integral.

39. $\int_0^2 \int_{y^2/4}^{y/2} (x-y)^2 \, dxdy$; $u = (x-y)/2, v = y/2$
40. $\int_0^4 \int_0^{\sqrt{16-x^2}} xy \, dydx$; $x = r\cos(2\theta), y = r\sin(2\theta)$
41. $\int_1^5 \int_3^9 y^2/x^2 \, dxdy$; $u = 1/x, v = y$
42. $\int_0^4 \int_{y/2}^{(y/2)+1} (2x-y)/2 \, dxdy$; $u = (2x-y)/2, v = y/2$
43. $\int_0^1 \int_3^4 \sqrt{4-xy} \, dydx$; $u = xy, v = y$
44. $\int_0^{16} \int_0^{\sqrt{x}} \sqrt{y/x} \, dydx$; $u^2 = x, v^2 = y$

Further Theory and Practice

In each of Exercises 45–52, find the area of the region that is described.

45. The region between the graphs of $r = 3\sin(2\theta)$ and $r = 4 - \cos(\theta)$
46. The region enclosed by both $r = 1 + \sin(\theta)$ and $r = 3\sin(\theta)$
47. The region inside the cardioid $r = 1 - \cos(\theta)$ and also inside the circle $r = \cos(\theta)$
48. The region inside both the circle $r = \sqrt{3}\cos(\theta)$ and the cardioid $r = 1 + \sin(\theta)$
49. The region inside both the rose $r = \cos(2\theta)$ and the circle $r = 1/2$
50. The region between the two loops of $r = \sqrt{3} - 2\cos(\theta)$
51. The region inside the smaller loop of $r = 1 - 2\sin(\theta)$
52. The region in the first and second quadrants bounded by the x-axis, the spiral arc $r = \theta$ for $0 \le \theta \le \pi$, and the spiral arc $r = \theta$ for $2\pi \le \theta \le 3\pi$

Perform the integrations in Exercises 53–56 by converting them to polar coordinates.

53. $\int_{-2}^{2} \int_{-\sqrt{4-x^2}}^{\sqrt{4-x^2}} (x^2 + y^2 + 1)^{-1/2} \, dydx$
54. $\int_0^1 \int_0^{(1-x^2)^{1/2}} \cos(x^2 + y^2) \, dydx$
55. $\int_{-3}^{0} \int_{-\sqrt{9-y^2}}^{0} (x^2 + y^2)^{1/3} \, dxdy$
56. $\int_{-3}^{3} \int_0^{\sqrt{9-y^2}} (1 + x^2 + y^2)^{3/2} \, dxdy$

57. Integrate the function $f(r,\theta) = \theta$ over the region $\mathcal{R} = \{(r,\theta) : 0 \le \theta \le \pi/2, 0 \le r \le \cos(\theta)\}$.
58. Integrate the function $f(r,\theta) = \cos(\theta)$ over the region $\mathcal{R} = \{(r,\theta) : 0 \le \theta \le \pi, 0 \le r \le \theta\}$.
59. Calculate the area in the first quadrant bounded by the lines $y = x$, $y = 2x$ and by the curves $xy = 2$ and $xy = 4$. After setting up the integral, simplify matters by using the change of variables $u = xy, v = y/x$.
60. The convergent improper integral

$$I = \int_{-\infty}^{\infty} e^{(-x^2)} \, dx$$

arises in probability, statistics, error analysis, and many other topics. It cannot be evaluated by the techniques of Chapter 6. Instead, write

$$\begin{aligned} I^2 &= \left(\int_{-\infty}^{\infty} e^{(-x^2)} dx\right)\left(\int_{-\infty}^{\infty} e^{(-y^2)} dy\right) \\ &= \int_{-\infty}^{\infty}\left(\int_{-\infty}^{\infty} e^{(-x^2)} dx\right) e^{(-y^2)} dy \\ &= \int_{-\infty}^{\infty}\int_{-\infty}^{\infty} e^{-(x^2+y^2)} \, dxdy. \end{aligned}$$

Convert to polar coordinates to evaluate this iterated integral. Then, solve for I.
61. Use polar coordinates to determine for which real values of α the improper integral

$$\int_{-\infty}^{\infty}\int_{\infty}^{\infty} \frac{1}{(1 + x^2 + y^2)^{\alpha}} \, dxdy$$

converges. Here, you should understand this integral to *mean*

$$\lim_{R \to \infty} \int_{-R}^{R}\int_{-R}^{R} \frac{1}{(1 + x^2 + y^2)^{\alpha}} \, dxdy.$$

62. Prove that, for *any* real value of α, the improper integral

$$\int_{-\infty}^{\infty}\int_{-\infty}^{\infty} \frac{\exp(-(x^2 + y^2))}{(1 + x^2 + y^2)^{\alpha}} \, dxdy$$

converges.

Calculator/Computer Exercises

63. Find the value θ_0 of θ in $[0, \pi/2]$ for which $\theta = 3\cos^2(\theta)$. Calculate the area of the region in the first quadrant that is outside the curve $r = \theta$ and inside the curve $r = 3\cos^2(\theta)$ for $0 \le \theta \le \theta_0$.

64. Find the value θ_0 of θ in $[0, \pi/2]$ for which $2 + \theta = e^\theta$. Calculate the area of the region in the first quadrant that is between the curves $r = 2 + \theta$ and $r = e^\theta$ for $0 \le \theta \le \theta_0$.

▶ If an iterated integral $\int_\alpha^\beta \int_{\phi_0(\theta)}^{\phi_1(\theta)} f(r\cos(\theta), r\sin(\theta))\, r\, dr d\theta$ arises from integration in polar coordinates, then we may estimate its value by applying any approximation technique for double integrals. We do so by using the product $f(r\cos(\theta), r\sin(\theta))\, r$ as the integrand in the approximation formula. In Exercises 65 and 66, use the Trapezoidal Rule with $N = 4$ to approximate the given integral. (Refer to the instructions for Exercise 64–67 from Section 12.2 for a discussion of the Trapezoidal Rule for double integrals.) ◀

65. The integral of $f(r, \theta) = 1/(1 + \theta + r^3)$ over the region bounded by the y-axis and the spiral $r = \theta$, $0 \le \theta \le \pi/2$

66. The integral of $f(r, \theta) = \sqrt{\pi + \theta + \sqrt{r}}$ over the half-disk $0 \le r \le 1, -\pi/2 \le \theta \le \pi/2$

12.6 Triple Integrals

In this section we discuss the theory of the triple integral. In fact, integration can be done in any number of dimensions. We concentrate here on three dimensions because that is sufficient to study the basic properties of bodies in space.

The Concept of the Triple Integral

Let $\mathcal{U}$ be a solid in space and f a continuous function defined on $\mathcal{U}$. We want to develop a concept of integrating f over $\mathcal{U}$. This will parallel the development in Section 12.2, so we will be brief.

Suppose that $\mathcal{U}$ is contained in a box $Q = \{(x, y, z) : a \le x \le b, c \le y \le d, e \le z \le f\}$. We partition each of the intervals $[a, b]$, $[c, d]$, and $[e, f]$ into N equal subintervals, giving rise to N^3 small boxes $Q_{j,k,\ell}$. We select a point $\xi_{j,k,\ell}$ in each $Q_{j,k,\ell}$ and form the Riemann sum

$$\sum_{Q_{j,k,\ell} \subset \mathcal{U}} f(\xi_{j,k,\ell})\, \Delta x \Delta y \Delta z,$$

or

$$\sum_{Q_{j,k,\ell} \subset \mathcal{U}} f(\xi_{j,k,\ell})\, \Delta V,$$

where $\Delta V = (\Delta x)(\Delta y)(\Delta z)$ is the increment of volume. As in Section 12.2, we sum only those boxes that are contained in $\mathcal{U}$. If the geometry of $\mathcal{U}$ is relatively simple (a concept that is made precise in the upcoming Theorem 1), then it is known that the Riemann sums will have a limiting value as N tends to infinity. The limiting value of these Riemann sums is called the *triple Riemann integral*. This integral is denoted by the symbol

$$\iiint_{\mathcal{U}} f(x, y, z)\, dV.$$

The most important fact for us about triple integrals is that we have a means of evaluating them when the solid $\mathcal{U}$ has a regular form. The next theorem tells us how to calculate triple integrals in terms of iterated integrals.

THEOREM 1 Suppose that $\alpha_1, \alpha_2, \gamma_1$, and γ_2 are continuously differentiable functions of one variable. Suppose that β_1 and β_2 are continuously differentiable functions of two variables. Assume that $\mathcal{U}_1$ and $\mathcal{U}_2$ are solids in space given by

$$\mathcal{U}_1 = \{(x, y, z) : a \le x \le b, \alpha_1(x) \le y \le \alpha_2(x), \beta_1(x, y) \le z \le \beta_2(x, y)\},$$

and

$$\mathcal{U}_2 = \{(x, y, z) : c \le y \le d, \gamma_1(y) \le x \le \gamma_2(y), \beta_1(x, y) \le z \le \beta_2(x, y)\}.$$

If f is a continuous function defined on $\mathcal{U}_1$, then

$$\iiint_{\mathcal{U}_1} f(x, y, z)\, dV = \int_a^b \int_{\alpha_1(x)}^{\alpha_2(x)} \int_{\beta_1(x,y)}^{\beta_2(x,y)} f(x, y, z)\, dzdydx.$$

If f is a continuous function defined on $\mathcal{U}_2$, then

$$\iiint_{\mathcal{U}_2} f(x, y, z)\, dV = \int_c^d \int_{\gamma_1(y)}^{\gamma_2(y)} \int_{\beta_1(x,y)}^{\beta_2(x,y)} f(x, y, z)\, dzdxdy.$$

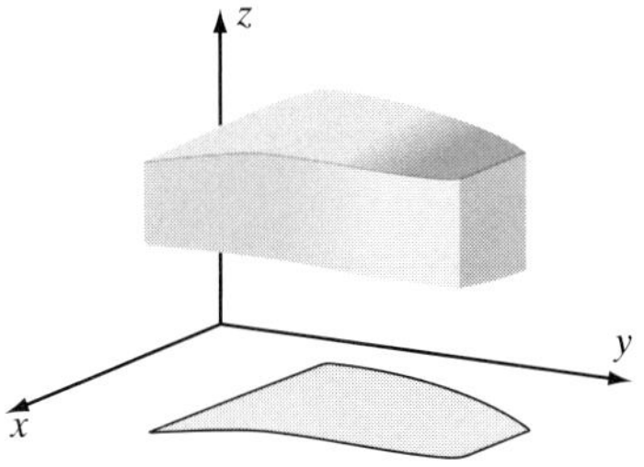

▲ Figure 1

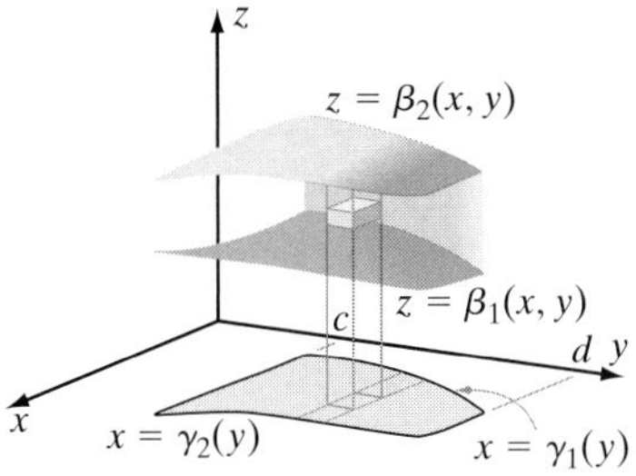

▲ Figure 2

Solids of the type described in Theorem 1 are called *z-simple*. Figure 1 shows a z-simple solid of the form $\mathcal{U} = \{(x, y, z) : c \le y \le d, \gamma_1(y) \le x \le \gamma_2(y), \beta_1(x, y) \le z \le \beta_2(x, y)\}$. The cut-away image of $\mathcal{U}$ that is given in Figure 2 reveals a typical box of volume ΔV that arises in the definition of $\iiint_{\mathcal{U}} f(x, y, z)\, dV$. In the iterated integral $\int_c^d \int_{\gamma_1(y)}^{\gamma_2(y)} \int_{\beta_1(x,y)}^{\beta_2(x,y)} f(x, y, z)\, dzdxdy$, the inner integral, namely $\int_{\beta_1(x,y)}^{\beta_2(x,y)} f(x, y, z)\, dz$, integrates up the vertical column that is shown in Figure 2. This result is then integrated, by means of $\int_{\gamma_1(y)}^{\gamma_2(y)} \ldots\, dx$, along the strip that is depicted in Figure 2. The result of this calculation is then integrated over all such strips from $y = c$ to $y = d$.

An analogue of Theorem 1 is true for x-simple and y-simple solids, which are defined similarly. You will see all three of these types of solids in the examples that follow.

► **EXAMPLE 1** Integrate the function $f(x, y, z) = 4x - 12z$ over the solid $\mathcal{U} = \{(x, y, z) : 1 \le x \le 2, x \le y \le 2x, y - x \le z \le y\}$.

Solution By Theorem 1, we have

$$\iiint_{\mathcal{U}} f(x, y, z)\, dV = \int_1^2 \int_x^{2x} \int_{y-x}^{y} (4x - 12z)\, dzdydx.$$

We begin by evaluating the inside integral, and then we work our way outward. The last line equals

$$\int_1^2 \int_x^{2x} (4xz - 6z^2)\Big|_{z=y-x}^{z=y} dydx$$
$$= \int_1^2 \int_1^2 \int_x^{2x} (4xy - 6y^2) - \left(4x(y-x) - 6(y-x)^2\right) dydx$$
$$= \int_1^2 \int_x^{2x} (10x^2 - 12xy)\, dydx$$
$$= \int_1^2 (10x^2y - 6xy^2)\Big|_{y=x}^{y=2x} dx$$
$$= \int_1^2 (-8x^3)\, dx$$
$$= -2x^4\Big|_1^2$$
$$= -30. \blacktriangleleft$$

If we take the function f that is being integrated to be the constant function $f(x, y, z) = 1$, then the Riemann sums

$$\sum_{Q_{j,k,\ell} \subset \mathcal{R}} f(\xi_{j,k,\ell})\Delta V = \sum_{Q_{j,k,\ell} \subset \mathcal{R}} \Delta V$$

form a good approximation to the volume of $\mathcal{U}$. This observation motivates the following definition:

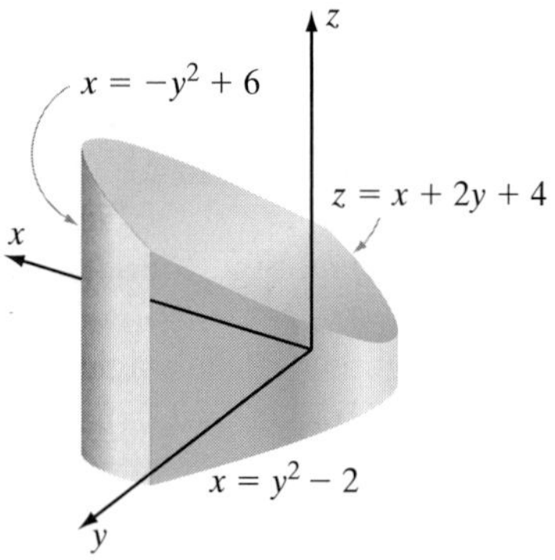

▲ Figure 3

DEFINITION The volume of a solid $\mathcal{U}$ is defined to be $\iiint_{\mathcal{U}} 1\, dV$ when the integral exists.

Here is an example:

▶ **EXAMPLE 2** Compute the volume V of the three-dimensional solid $\mathcal{U}$ that is bounded laterally by the parabolic cylinders $x = y^2 - 2$ and $x = -y^2 + 6$, above by the plane $z = x + 2y + 4$, and below by the plane $z = -x - 4$.

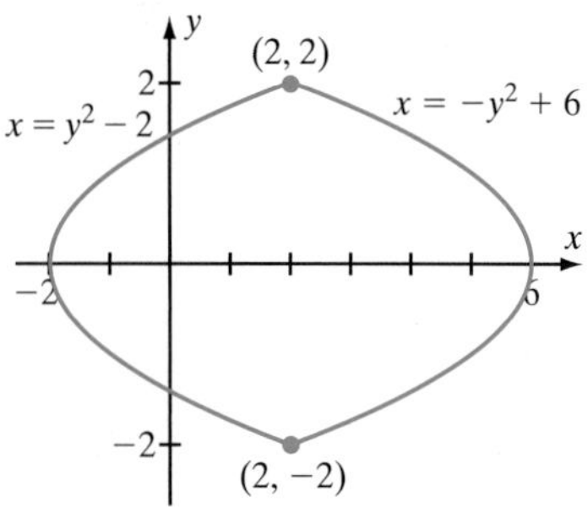

▲ Figure 4

Solution The solid $\mathcal{U}$ is shown in Figure 3. Observe that the graphs of $x = y^2 - 2$ and $x = -y^2 + 6$ are vertical sides of $\mathcal{U}$. The projection of $\mathcal{U}$ into the xy-plane is the region bounded by the parabolas $x = y^2 - 2$ and $x = -y^2 + 6$. These parabolas, shown in Figure 4, intersect at the points $(2, 2)$ and $(2, -2)$. From Figure 4, we see that the y-variable ranges between -2 and 2. So our outer integral will be

$$\int_{-2}^{2} \ldots dy.$$

Now, for each given values of y, Figure 4 shows that x ranges from $y^2 - 2$ to $-y^2 + 6$. Thus the next integral is the integral in the x-variable, and we have

$$\int_{-2}^{2} \int_{y^2-2}^{-y^2+6} \ldots dxdy.$$

Finally, for each choice of x and y, the least value of z is $-x - 4$ (because $\mathcal{U}$ is bounded below by the plane $z = -x - 4$), and the greatest value of z is $x + 2y + 4$

(because $\mathcal{U}$ is bounded above by the plane $z = x + 2y + 4$). Our formula for the volume finally takes the form

$$V = \int_{-2}^{2} \int_{y^2-2}^{-y^2+6} \int_{-x-4}^{x+2y+4} 1 \, dzdxdy.$$

Working from the inside out, it is now straightforward to evaluate this integral: It equals

$$\begin{aligned} V &= \int_{-2}^{2} \int_{y^2-2}^{-y^2+6} \big((x+2y+4) - (-x-4)\big) \, dxdy \\ &= \int_{-2}^{2} \int_{y^2-2}^{-y^2+6} (2x+2y+8) \, dxdy = \int_{-2}^{2} (x^2 + 2(y+4)x)\Big|_{x=y^2-2}^{x=-y^2+6} dy. \end{aligned}$$

After some simplification, we obtain

$$V = \int_{-2}^{2} (96 + 16y - 24y^2 - 4y^3) \, dy = (96y + 8y^2 - 8y^3 - y^4)\Big|_{-2}^{2} = 256. \quad ◀$$

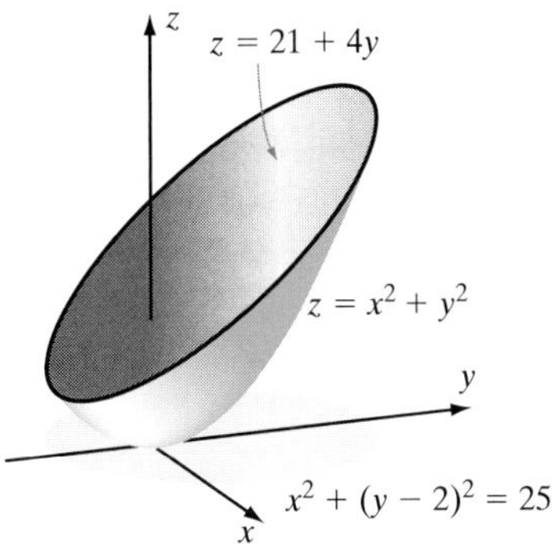

▲ Figure 5

INSIGHT Take careful note of the analysis that we used in Example 2 to set up the triple integral. This is the key to success in triple integral problems: Work from the outside in to set up the integral, but work from the inside out to evaluate the integral.

▶ **EXAMPLE 3** Set up, but do not evaluate, the integral of $(x, y, z) \mapsto f(x, y, z)$ over the solid $\mathcal{U}$ bounded by the paraboloid $z = x^2 + y^2$ and the plane $z = 21 + 4y$. Treat $\mathcal{U}$ as a z-simple solid.

Solution The graph of $\mathcal{U}$ appears in Figure 5. When we consider $\mathcal{U}$ as a z-simple solid, we think of x and y as the outside variables of integration and z as the inside variable of integration. Let $\mathcal{D}$ be the projection of $\mathcal{U}$ to the xy-plane. If the point $(x, y, 0)$ belongs to $\mathcal{D}$, then we integrate z from $x^2 + y^2$ to $21 + 4y$. In particular, $x^2 + y^2 \leq 21 + 4y$ and, on the boundary of $\mathcal{D}$, $x^2 + y^2 = 21 + 4y$, or, equivalently, $x^2 + (y-2)^2 = 25$. Thus $\mathcal{D}$ is the disk in the xy-plane that is bounded by the circle with center $(0, 2)$ and radius 5 (see Figure 6). We see that y ranges from -3 to 7. We conclude that the integral will have the form

$$\int_{-3}^{7} \dots dy.$$

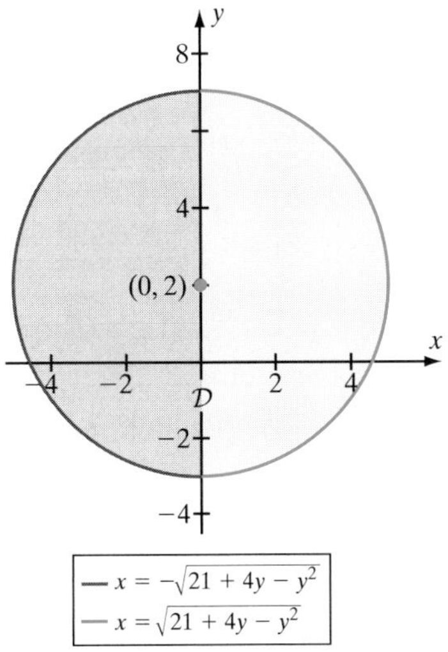

▲ Figure 6

For each choice of y in this interval of integration, x ranges from $-\sqrt{25-(y-2)^2}$, or $-\sqrt{21+4y-y^2}$, to $\sqrt{21+4y-y^2}$. Therefore our integral will have the form

$$\int_{-3}^{7} \int_{-\sqrt{21+4y-y^2}}^{\sqrt{21+4y-y^2}} \dots \, dxdy.$$

After y and x are chosen, z ranges from $x^2 + y^2$ to $21 + 4y$. Our integral takes the form

$$\int_{-3}^{7}\int_{-\sqrt{21+4y-y^2}}^{\sqrt{21+4y-y^2}}\int_{x^2+y^2}^{21+4y} f(x,y,z)\ dzdxdy. \quad ◀$$

▶ **EXAMPLE 4** Set up the integral from Example 3, treating $\mathcal{U}$ as an x-simple solid.

Solution Refer to Figure 5 again and imagine the projection $\mathcal{P}$ of $\mathcal{U}$ into the yz-plane. In this case, the projection $\mathcal{P}$ is actually the intersection of $\mathcal{U}$ with the yz-plane. Setting $x = 0$, we see that $\mathcal{P}$ is the region of the yz-plane that is bounded below by the parabola $z = y^2$ and above by the line $z = 21 + 4y$ (see Figure 7). Our analysis proceeds as follows: First, the variable y ranges from $y = -3$ to $y = 7$. We conclude that the integral will have the form

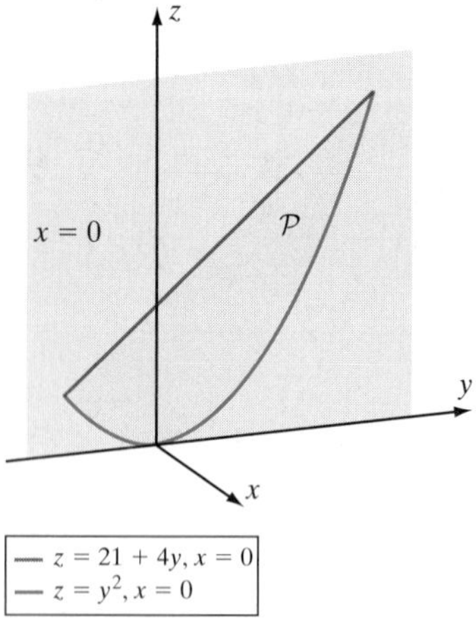

▲ Figure 7

$$\int_{-3}^{7} \ldots dy.$$

Next, for each choice of y in the interval of integration, the z variable ranges from $z = y^2$ to $z = 21 + 4y$. Therefore our integral will have the form

$$\int_{-3}^{7}\int_{y^2}^{21+4y} \ldots dzdy.$$

Finally, for each choice of y and z, the variable x ranges from the side of the paraboloid $z = x^2 + y^2$ on which $x = -\sqrt{z - y^2}$ to the other side of the paraboloid, on which $x = \sqrt{z - y^2}$. Our integral therefore takes the form

$$\int_{-3}^{7}\int_{y^2}^{21+4y}\int_{-\sqrt{z-y^2}}^{\sqrt{z-y^2}} f(x,y,z)\ dxdzdy. \quad ◀$$

QUICK QUIZ

1. Let $\mathcal{U}$ be the solid in the first octant that lies underneath the plane $3x + 4y + 2z = 12$. Set up (but do not evaluate) an iterated integral for the volume of $\mathcal{U}$ using the volume element $dzdydx$.
2. Let $\mathcal{U}$ be the solid ball with center $(0,0,3)$ and radius 1. Express $\iiint_{\mathcal{U}} f(x,y,z)dV$ by means of an integrated integral with volume element $dzdydx$ (but do not evaluate the integral).
3. Calculate $\iiint_{\mathcal{U}} 12xy^2z^3 dV$, where $\mathcal{U} = \{(x,y,z) : 1 \le x \le 2,\ -1 \le y \le 1,\ 0 \le z \le 1\}$.
4. Let $\mathcal{U}$ be the solid that lies below the paraboloid $z = 9 - x^2 - y^2$ and above the plane $z = 5$. Set up the triple integral $\iiint_{\mathcal{U}} f(x,y,z)dV$ as an iterated integral.

Answers

1. $\int_0^4 \int_0^{3-3x/4} \int_0^{6-3x/2-2y} 1\ dzdydx$ 2. $\int_{-1}^{1}\int_{-\sqrt{1-x^2}}^{\sqrt{1-x^2}}\int_{3-\sqrt{1-x^2-y^2}}^{3+\sqrt{1-x^2-y^2}} f(x,y,z)dzdydx$

3. 3 4. $\int_{-2}^{2}\int_{-\sqrt{4-x^2}}^{\sqrt{4-x^2}}\int_{5}^{9-x^2-y^2} f(x,y,z)dzdydx$

EXERCISES

Problems for Practice

Evaluate the given iterated integral in each of Exercises 1–10.

1. $\int_2^4 \int_x^{x+1} \int_x^{x+y} (6x - 3z)\, dzdydx$
2. $\int_{-2}^2 \int_0^{2+y} \int_{y-x}^{y} 3x\, dzdxdy$
3. $\int_1^4 \int_{-z}^{z} \int_{y-z}^{y+z} (x+3)\, dxdydz$
4. $\int_3^5 \int_2^z \int_0^{\sqrt{z^2-x^2}} 6y\, dydxdz$
5. $\int_e^{e^2} \int_{1/(2y)}^{1/y} \int_1^y (4/z)\, dzdxdy$
6. $\int_0^3 \int_y^{2y} \int_z^{z+y} 8xyz\, dxdzdy$
7. $\int_0^{\pi/2} \int_0^x \int_x^z 2\cos(y)\, dydzdx$
8. $\int_{-3}^{-2} \int_0^{2y} \int_0^z 16y \exp(-z^2)\, dxdzdy$
9. $\int_0^\pi \int_{-y}^{\sqrt{\pi}} \int_{-x}^0 4y\cos(x^2)\, dzdxdy$
10. $\int_0^{\pi/2} \int_{-5}^3 \int_0^{\cos(y)} 3\sqrt{1+\sin(y)}\, dxdzdy$

In each of Exercises 11–18, use a triple integral to express the volume of the solid that is bounded by the given surfaces in space. Evaluate the volume by means of iterated integrals.

11. $z = x^2 + y^2,\ z = 8,\ z = 2$
12. $z = x^2 + y^2,\ z = 8 - x^2 - y^2$
13. $y = x^2,\ z = 1 - y,\ z = -1 + y$
14. $x + 2y + 4z = 8,\ y = x/2,\ x = 0,\ z = 0$
15. $x^2 + y^2 = 4,\ z = x + y,\ z = -6 - x$
16. $y = x^2,\ z = -y + 4,\ z = 0$
17. $z = -x^2 - y^2 + 9,\ z = 2x + 4y - 2$
18. $y = x^2 + z^2 - 3,\ y = -x^2 - z^2 + 5$

In each of Exercises 19–24, evaluate the given iterated integral. Describe the solid $\mathcal{U}$ over which the integration is performed.

19. $\int_0^2 \int_0^{3-3x/2} \int_0^{6-3x-2y} x\, dzdydx$
20. $\int_0^6 \int_0^6 \int_0^{xy/6} z\, dzdydx$
21. $\int_0^1 \int_0^{\sqrt{1-x^2}} \int_0^{\sqrt{1-x^2-y^2}} xz\, dzdydx$
22. $\int_0^1 \int_0^{\sqrt{1-x^2}} \int_0^{1-y^2} x\, dzdydx$
23. $\int_0^1 \int_0^x \int_0^{\sqrt{1-x^2}} z\, dzdydx$
24. $\int_0^1 \int_0^{\sqrt{1-y^2}} \int_0^{\sqrt{x^2+y^2}} x(1+y)\, dzdxdy$

Further Theory and Practice

25. The iterated integral of Exercise 19 represents a triple integral $\iiint_{\mathcal{U}} x\, dV$ over a solid $\mathcal{U}$ that is z-simple, as we can see from the inner variable of integration. However, the solid $\mathcal{U}$ is also y-simple, and, reflecting this property, $\iiint_{\mathcal{U}} x\, dV$ can also be expressed as an iterated integral with y the inner variable of integration: Both

$$\int_0^2 \int_0^{6-3x} \int_0^{3-3x/2-z/2} x\, dydzdx$$

and

$$\int_0^6 \int_0^{2-z/3} \int_0^{3-3x/2-z/2} x\, dydxdz$$

do the job. Examine each iterated integral in Exercises 20–24. If the integral represents a triple integral over a solid that is y-simple, then express the integral as an iterated integral with y the inner variable of integration.

Draw a sketch of the domain of integration of each iterated integral in Exercises 26–31.

26. $\int_0^2 \int_x^{2x+1} \int_x^y f(x,y,z)\, dzdydx$
27. $\int_{-1}^1 \int_y^{y+3} \int_{x+2}^{x+4} f(x,y,z)\, dzdxdy$
28. $\int_{-1}^0 \int_z^4 \int_0^{2x+4} f(x,y,z)\, dydxdz$
29. $\int_{-3}^3 \int_{-\sqrt{9-y^2}}^{\sqrt{9-y^2}} \int_{-x-y-5}^{x+y+5} f(x,y,z)\, dzdxdy$
30. $\int_{-1}^1 \int_{-\sqrt{4-4x^2}}^{\sqrt{4-4x^2}} \int_{-4-x-z}^{2+x+z} f(x,y,z)\, dydzdx$
31. $\int_1^4 \int_y^6 \int_y^{x^2} f(x,y,z)\, dzdxdy$

32. Suppose that a, b, and c are positive constants. Use a triple integral to evaluate the volume inside an ellipsoid of the form $x^2/a^2 + y^2/b^2 + z^2/c^2 = 1$.
33. Use a triple integral to give a formula for the volume of the solid bounded by the paraboloid $z = x^2 + y^2$ and the plane $z = ax + by + c, a > 0, b > 0, c > 0$.
34. Let a, b, and c be positive. Use a triple integral to find a formula for the volume of the solid in the first octant cut off by the plane $x/a + y/b + z/c = 1$.
35. Integrate $f(x,y,z) = x^2 + y^2 + z^2$ over the solid that is bounded above by the cone $z = m\sqrt{x^2+y^2}$ and below by the paraboloid $z = x^2 + y^2$.
36. Use a triple integral to find the volume of the solid bounded by the cone $z^2 = x^2 + y^2$ and the planes $z = a$ and $z = b$, $0 < a < b$.
37. Calculate $\int_0^1 \int_x^1 \int_0^x z\sin(y^4)\, dzdydx$.

In Exercises 38 and 39, a solid $\mathcal{U}$ is described. Evaluate $\iiint_{\mathcal{U}} (xy + z)\, dV$.

38. $\mathcal{U}$ is the solid in the first octant that is bounded by the three coordinate planes, by the plane $x = 1$, and by the plane $3y + 2z = 6$.

39. $\mathcal{U}$ is the solid in the first octant that is bounded by the three coordinate planes, by the plane $z = 2 + 7x$, and by the parabolic cylinder $y = 4 - x^2$.

In Exercises 40 and 41, a solid $\mathcal{U}$ is described. Write $\iiint_{\mathcal{U}} f(x, y, z)\, dV$ in each of the six possible orders of integration.

40. $\mathcal{U}$ is the solid in the first octant that is bounded by the three coordinate planes and by the plane $5x + 3y + 2z = 30$.

41. $\mathcal{U}$ is the solid in the first octant that is bounded by the planes $x = 0$, $x = y$, $y = 1$, $z = 0$, and $z = 1$.

42. Write the iterated integral

$$\int_{-4}^{4} \int_{9}^{25-x^2} \int_{-\sqrt{25-x^2-y}}^{\sqrt{25-x^2-y}} (x^2 - yz)\, dzdydx$$

in all possible orders.

43. Write the iterated integral

$$\int_{-1}^{0} \int_{2}^{37} \int_{x+1}^{\sqrt{x+1}} y\sqrt{1-x}\, dydzdx$$

in all possible orders.

44. Suppose that a, b, and c are positive constants with $b > a$. Integrate $x^3\sqrt{c^2 - y^2}$ over the solid that is bounded by the parabolic cylinder $y = x^2$ and by the planes $x = 0$, $y = c$, $z = by$, and $z = ay$.

45. Suppose that a, b, c are positive constants. Let $\mathcal{U}$ be the solid that is bounded below by the xy-plane, above by the graph of $\sqrt{x/a} + \sqrt{y/b} + \sqrt{z/c} = 1$, and by the xz- and yz-planes on the side. Show that the volume of $\mathcal{U}$ is $abc/90$. Also show that

$$\iiint_{\mathcal{U}} xyz\, dV = \frac{a^2b^2c^2}{277200}.$$

Calculator/Computer Exercises

46. Let $\mathcal{R}$ be the region of the xy-plane that is bounded above by $y = 3 - x^2$ and below by $y = 1 + (x + 1)^3$. Let $\mathcal{U}$ be the solid that is bounded below by $\mathcal{R}$, above by the plane $z = y$, and laterally by the cylinder sets $y = 3 - x^2$ and $y = 1 + (x + 1)^3$. Calculate $\iiint_{\mathcal{U}} x^2yz\, dV$.

47. Let $\mathcal{R}$ be the region in the first quadrant of the xy-plane that is bounded above by $y = \sin(x)$ and below by $y = 1/(1 + x^2)$. Let $\mathcal{U}$ be the solid in the first octant that is bounded below by $\mathcal{R}$, above by the plane $z = x$, and laterally by the cylinder sets $y = \sin(x)$ and $y = 1/(1 + x^2)$. Calculate $\iiint_{\mathcal{U}} z\, dV$.

48. Let $\mathcal{U}$ be the solid in the first octant that is bounded below by the plane $z = x$, above by the plane $z = 2$, and laterally by the cylinder sets $y = e^x$ and $y = 6x - x^2 - 1$. Calculate $\iiint_{\mathcal{U}} z\, dV$.

12.7 Physical Applications

Multiple integrals are useful for calculating a number of physical quantities. In this section, we learn how to calculate the mass of a body, the center of mass of a body, the first moment of a body about an axis, and the moment of inertia of a body about an axis.

Mass

Imagine a thin metal plate (sometimes called a *lamina*) made of material that varies in composition from point to point. We think of the plate as occupying a region $\mathcal{R}$ in the xy-plane and assume that it has mass density given by a continuous function $\delta(x, y)$ at the point (x, y). In this context, "mass density" at a point (x, y) means mass per unit area at the point (x, y). This physical concept is defined like a partial derivative. To be specific, let $M(x, y, h)$ be the mass of a square of side length h that is centered at the point (x, y). To obtain the density at (x, y), we divide this mass by the area of the square and take the limit as h tends to 0:

$$\delta(x, y) = \lim_{h \to 0^+} \frac{M(x, y, h)}{h^2}. \tag{12.7.1}$$

We now answer the following question: If the mass density of a plate is known at each point, then how can we calculate the mass of the entire plate?

As usual, we think of the plate as lying inside a rectangle $Q = \{(x,y): a \le x \le b, c \le y \le d\}$. Partitioning the intervals $[a,b]$ and $[c,d]$ into N subintervals each, we obtain rectangles $Q_{i,j}$. We consider only the rectangles that lie entirely inside $\mathcal{R}$. If $\xi_{i,j} = (x_i, y_j)$ is a point in $Q_{i,j}$, then the mass $M_{i,j}$ of the square $Q_{i,j}$ is approximately

$$M_{i,j} = \delta(\xi_{i,j})\Delta x \Delta y = \delta(\xi_{i,j})\Delta A.$$

Notice that we estimate the mass by multiplying $\delta(\xi_{i,j})$, which represents mass per unit of area, by the area of the small rectangle. Limit formula (12.7.1) justifies this approximation. The total mass of the plate is about

$$\sum_{Q_{i,j} \subset \mathcal{R}} M_{i,j} = \sum_{Q_{i,j} \subset \mathcal{R}} \delta(\xi_{i,j})\Delta A.$$

Letting the number of subrectangles of the partition tend to infinity leads us to define the mass M of the plate by

$$M = \iint_{\mathcal{R}} \delta(x,y)\, dA. \tag{12.7.2}$$

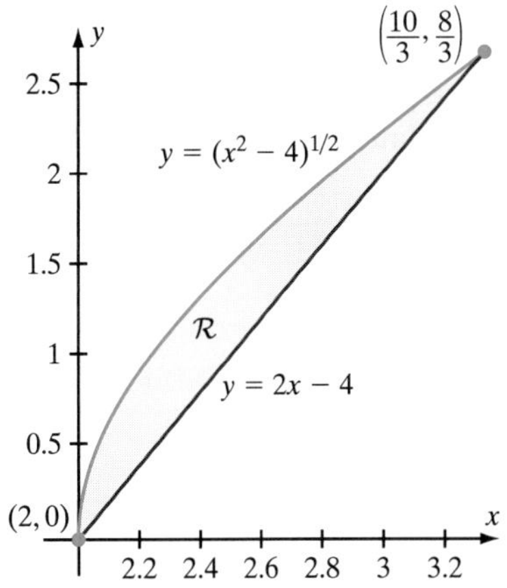

▲ Figure 1

▶ **EXAMPLE 1** Suppose that a thin plate $\mathcal{R}$ is in the shape shown in Figure 1. The bounding curves are $y = (x^2 - 4)^{1/2}$ and $y = 2x - 4$. Assume also that the density of $\mathcal{R}$ at the point (x,y) is $\delta(x,y) = 4x$. Find the mass of $\mathcal{R}$.

Solution The curves intersect when $x = 2$ and $x = 10/3$. We may write $\mathcal{R}$ as a y-simple region as follows:

$$\mathcal{R} = \{(x,y) : 2 \le x \le 10/3, 2x - 4 \le y \le (x^2 - 4)^{1/2}\}.$$

Then the mass of $\mathcal{R}$ is

$$\begin{aligned} M &= \iint_{\mathcal{R}} \delta(x,y)\, dA \\ &= \int_2^{10/3} \int_{2x-4}^{(x^2-4)^{1/2}} 4x\, dy dx \\ &= \int_2^{10/3} 4x\Big((x^2-4)^{1/2} - (2x-4)\Big)\, dx \\ &= \left(\frac{4}{3}(x^2-4)^{3/2} - \frac{8}{3}x^3 + 8x^2\right)\Bigg|_2^{10/3} \\ &= \frac{384}{81}. \end{aligned}$$

The mass of the plate is 384/81, or 128/27. ◀

First Moments

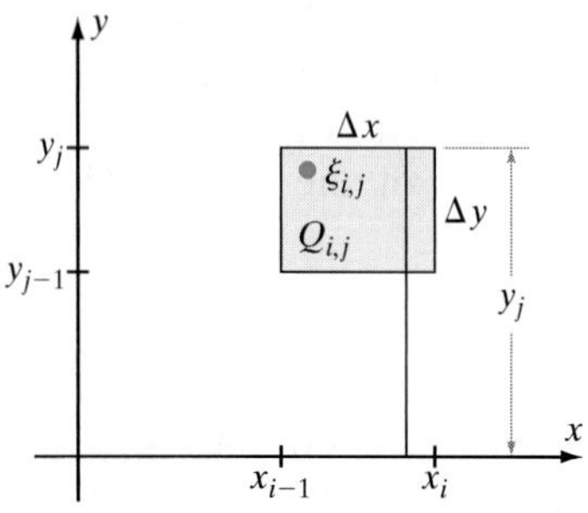

▲ Figure 2

Next we consider the first moments of a lamina about each axis. For a point mass, the *first moment* about an axis is the product of the mass times the distance to the axis. For each i and j, the first moment of the small rectangle $Q_{i,j}$ about the x-axis is approximately $y_j \cdot \delta(\xi_{i,j})\Delta A$ where $\xi_{i,j}$ is a point arbitrarily chosen from Q_{ij}. Look at Figure 2. There are two approximations in estimating this first moment. The different points that make up $Q_{i,j}$ do not have constant distance to the x-axis, but, if Q_{ij} is small, then the distances of the constituent points do not vary greatly, and y_j can be used as an approximation. Also, we approximate the mass of Q_{ij} by $\delta(\xi_{i,j})\Delta A$, as usual. Summing over i and j, we obtain

$$\sum_{Q_{i,j}\subseteq\mathcal{R}} y_j\delta(\xi_{i,j})\Delta A$$

as an approximation to the first moment of the plate about the x-axis. Letting the number of subrectangles tend to infinity leads us to define the *first moment* of the plate with respect to the x-axis, which we write as $y = 0$, to be

$$M_{y=0} = \iint_{\mathcal{R}} y\ \delta(x, y)\ dA. \tag{12.7.3}$$

Similar computations lead to the definition that the first momentof the plate with respect to the y-axis, which we write as $x = 0$, is

$$M_{x=0} = \iint_{\mathcal{R}} x\ \delta(x, y)\ dA. \tag{12.7.4}$$

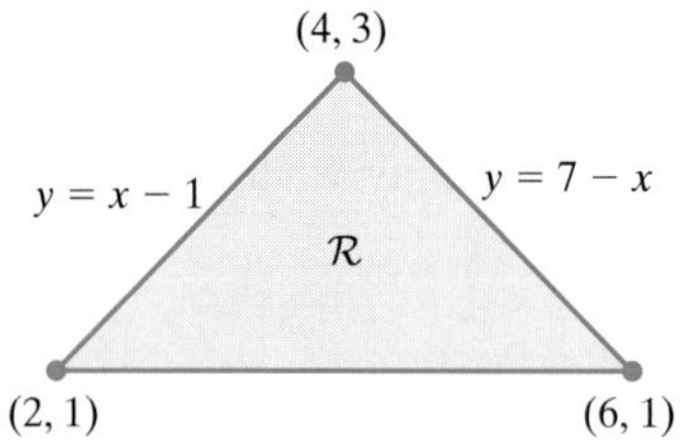

▲ Figure 3

▶ **EXAMPLE 2** A metal plate $\mathcal{R}$ sitting in the xy-plane is shown in Figure 3. Its density at a point (x, y) is $\delta(x, y) = 6x + 9y$. Calculate its first moment with respect to the x-axis.

Solution First, write the region $\mathcal{R}$ in x-simple form: $\mathcal{R} = \{(x, y) : 1 \le y \le 3,\ y + 1 \le x \le -y + 7\}$. We have

$$\begin{aligned} M_{y=0} &= \iint_{\mathcal{R}} y\delta(x, y)\, dA \\ &= \int_1^3 \int_{y+1}^{-y+7} y(6x + 9y)\, dxdy \\ &= \int_1^3 (3yx^2 + 9y^2x)\Big|_{x=y+1}^{x=-y+7} dy \\ &= \int_1^3 (-18y^3 + 6y^2 + 144y)\, dy \\ &= \left(-\frac{9}{2}y^4 + 2y^3 + 72y^2\right)\Big|_1^3 \\ &= 268. \end{aligned}$$

We conclude that the first moment of $\mathcal{R}$ with respect to the x-axis is 268 units. For example, if δ is measured in kilograms per square meter, and x and y are measured in meters, then our answer is measured in meter-kilograms. ◀

Center of Mass

Suppose that M, $M_{x=0}$, and $M_{y=0}$ denote the mass and first moments of a plate. Define the numbers $\overline{x}$ and $\overline{y}$ by

$$\overline{x} = \frac{M_{x=0}}{M} \quad \text{and} \quad \overline{y} = \frac{M_{y=0}}{M}. \tag{12.7.5}$$

Referring to the definition of $M_{x=0}$, we see that $\overline{x}$ is the average value of the x-coordinate function on the plate, weighted according to the varying density function $\delta(x, y)$. Similarly, $\overline{y}$ is the average y value on the plate, weighted according to the mass density. The *center of mass* (or *center of gravity*) of the plate is now defined to be the point

$$\overline{c} = (\overline{x}, \overline{y}).$$

It is not difficult to see where the definition of "center of gravity" comes from. The point $\overline{c}$ is determined by the property that the tendency of the plate to rotate about $\overline{c}$ in any given direction is balanced exactly by its tendency to rotate in the opposite direction. The physical significance of the center of gravity is that the plate would balance on the point of a pin placed at $\overline{c}$. When the density function is not mentioned explicitly, then we assume that mass is "uniformly distributed" (i.e., $\delta(x, y) = \delta_0$, a constant), and we speak of the "centroid" of a region. In this case, notice that the constant density δ_0 can be brought outside each of the integrals that define M, $M_{x=0}$, and $M_{y=0}$. The constant δ_0 then cancels in the ratios $\overline{x} = M_{x=0}/M$ and $\overline{y} = M_{y=0}/M$. That is why the value of a uniform density is often not specified.

▶ **EXAMPLE 3** Consider a plate $\mathcal{R}$ in the xy-plane bounded above by the curve $y = 12 - 3x^2$ and below by the x-axis. Assume that the plate has constant density $\delta(x, y) = 1$. Calculate the center of mass of the plate.

Solution By setting $y = 0$ in $y = 12 - 3x^2$ and solving for x, we see that the curve that forms the upper boundary of $\mathcal{R}$ crosses the x-axis at $x = \pm 2$. Thus $\mathcal{R}$ is the y-simple region $\{(x, y) : -2 \le x \le 2, 0 \le y \le 12 - 3x^2\}$. The mass is therefore

$$M = \iint_{\mathcal{R}} \delta(x, y)\, dA = \int_{-2}^{2} \int_{0}^{12-3x^2} 1\, dydx = \int_{-2}^{2} (12 - 3x^2)\, dx = 32.$$

Further, we know that the first moment with respect to the x-axis is

$$\begin{aligned} M_{y=0} &= \iint_{\mathcal{R}} y\, \delta(x, y)\, dA = \int_{-2}^{2} \int_{0}^{12-3x^2} y\, dydx = \int_{-2}^{2} \frac{y^2}{2} \bigg|_{y=0}^{y=12-3x^2} dx \\ &= \int_{-2}^{2} \left(\frac{9}{2}x^4 - 36x^2 + 72 \right) dx = \frac{768}{5}. \end{aligned}$$

Similarly, the first moment with respect to the y-axis is

$$M_{x=0} = \iint_{\mathcal{R}} x\, \delta(x, y)\, dA = \int_{-2}^{2} \int_{0}^{12-3x^2} x\, dydx = \int_{-2}^{2} (12 - 3x^2)\, x\, dx = 0.$$

In physical terms, the first moment with respect to the y-axis is equal to 0 because the plate is symmetric with respect to the y-axis. We conclude that

$$\overline{x} = \frac{M_{x=0}}{M} = \frac{0}{32} = 0 \quad \text{and} \quad \overline{y} = \frac{M_{y=0}}{M} = \frac{768/5}{32} = \frac{24}{5}.$$

Therefore the center of gravity is $\overline{c} = (\overline{x}, \overline{y}) = (0, 24/5)$. ◀

Moment of Inertia

The *second moments*, or *moments of inertia*, of a plate $\mathcal{R}$ about the coordinate axes are defined as follows:

$$I_{y=0} = \iint_{\mathcal{R}} y^2 \delta(x,y)\, dA$$

$$I_{x=0} = \iint_{\mathcal{R}} x^2 \delta(x,y)\, dA.$$

Here $I_{y=0}$ is the moment of inertia about the x-axis (i.e., about the line $y = 0$), and $I_{x=0}$ is the moment of inertia about the y-axis (the line $x = 0$). Notice that moments of inertia differ from first moments in that we use the *square* of the distance to the axis. Moments of inertia can also be defined about points. For example, the second moment about the origin $(0, 0)$, or the *polar moment of inertia*, I_0, is defined by

$$I_0 = I_{x=0} + I_{y=0} = \iint_{\mathcal{R}} (x^2 + y^2)\, \delta(x,y)\, dA.$$

Second moments are used, for instance, to measure the kinetic energy of rotation of a spinning body. These matters are studied in more detail in physics texts.

▶ **EXAMPLE 4** Calculate the second moment $I_{y=0}$ of the lamina with constant density $\delta = 3$ that occupies the region bounded by the curves $y = -x^2 + 4$ and $y = 0$.

Solution The x limits of integration are -2 and 2. Thus according to the definition,

$$\begin{aligned}
I_{y=0} &= \iint_{\mathcal{R}} y^2 \cdot 3\, dA \\
&= \int_{-2}^{2} \int_{0}^{-x^2+4} 3y^2\, dy dx \\
&= \int_{-2}^{2} y^3 \Big|_{0}^{-x^2+4} dx \\
&= \int_{-2}^{2} (-x^2 + 4)^3\, dx \\
&= \int_{-2}^{2} (-x^6 + 12x^4 - 48x^2 + 64)\, dx \\
&= \left(-\frac{x^7}{7} + \frac{12x^5}{5} - 16x^3 + 64x \right) \Big|_{-2}^{2} \\
&= \frac{4096}{35}.
\end{aligned}$$ ◀

Mass, First Moment, Moment of Inertia, and Center of Mass in Three Dimensions

The reasoning that we used to derive double integral formulas for mass, first moment, moment of inertia, and center of mass in two dimensions can also be used to derive triple integral formulas for these quantities in three dimensions. If $\mathcal{U}$ is a solid in space and if the density of the solid at the point (x, y, z) is $\delta(x, y, z)$, then an argument as in the first part of this section leads us to define the *mass* of $\mathcal{U}$ to be

$$M = \iiint_{\mathcal{U}} \delta(x, y, z)\, dV.$$

Similarly, the first moments with respect to the xy-plane, yz-plane, and xz-plane are given by

$$M_{z=0} = \iiint_{\mathcal{U}} z\delta(x, y, z)\, dV, M_{x=0} = \iiint_{\mathcal{U}} x\delta(x, y, z)\, dV, \text{ and } M_{y=0} = \iiint_{\mathcal{U}} y\delta(x, y, z)\, dV.$$

The coordinates of the center of mass of the body are then given by

$$\overline{x} = \frac{M_{x=0}}{M},$$
$$\overline{y} = \frac{M_{y=0}}{M},$$
$$\overline{z} = \frac{M_{z=0}}{M}.$$

The center of mass is the point

$$\overline{c} = (\overline{x}, \overline{y}, \overline{z}).$$

As in two dimensions, when the density function is not mentioned explicitly, we assume that mass is uniformly distributed, and we speak of the *centroid* of a region.

Finally, if $\mathcal{U}$ is a solid with density δ, then its *second moments*, or *moments of inertia*, about the x-axis, the y-axis, and the z-axis are respectively given by

$$I_x = \iiint_{\mathcal{U}} (y^2 + z^2)\delta(x, y, z)\, dV,$$

$$I_y = \iiint_{\mathcal{U}} (x^2 + z^2)\delta(x, y, z)\, dV,$$

and

$$I_z = \iiint_{\mathcal{U}} (x^2 + y^2)\delta(x, y, z)\, dV.$$

Notice that the factors $y^2 + z^2$, $x^2 + z^2$, and $x^2 + y^2$ of these moments of inertia are the squares of the distances of the point (x, y, z) to the x-axis, the y-axis, and the z-axis, respectively.

▶ **EXAMPLE 5** Assuming that the body

$$\mathcal{U} = \{(x, y, z) : -2 \le y \le 0, y + 3 \le x \le y + 5, 2x + y + 4 \le z \le 2y + 14\}$$

has a uniform mass distribution $\delta(x, y, z) = 3$, calculate its first moment with respect to the xz-plane.

Solution According to the definition, the first moment of $\mathcal{U}$ with respect to the xz-plane is

$$\begin{aligned}
M_{y=0} &= \iint_{\mathcal{R}} 3y \, dV \\
&= \int_{-2}^{0} \int_{y+3}^{y+5} \int_{2x+y+4}^{2y+14} 3y \, dz dx dy \\
&= \int_{-2}^{0} \int_{y+3}^{y+5} 3yz \Big|_{z=2x+y+4}^{z=2y+14} dx dy \\
&= \int_{-2}^{0} \int_{y+3}^{y+5} (3y^2 - 6xy + 30y) \, dx dy \\
&= \int_{-2}^{0} -6y^2 + 12y \, dy \\
&= -40. \quad ◀
\end{aligned}$$

INSIGHT A negative first moment simply means that the solid is positioned so that a coordinate of its center of mass is negative. By contrast, the second moments of a solid cannot be negative, no matter where it is situated relative to the coordinate axes.

▶ **EXAMPLE 6** Assume that the solid body

$$\mathcal{U} = \{(x, y, z) : -1 \le z \le 1, 1 \le y \le 2, 0 \le x \le z + y\}$$

has density function $\delta(x, y, z) = 2x + y$. Compute the center of mass of $\mathcal{U}$.

Solution First, we have

$$\begin{aligned}
M &= \iint_{\mathcal{R}} \delta(x, y, z) \, dV \\
&- \int_{-1}^{1} \int_{1}^{2} \int_{0}^{z+y} (2x + y) \, dx dy dz \\
&= \int_{-1}^{1} \int_{1}^{2} (x^2 + yx) \Big|_{0}^{z+y} dy dz \\
&= \int_{-1}^{1} \int_{1}^{2} (z^2 + 3zy + 2y^2) \, dy dz
\end{aligned}$$

$$= \int_{-1}^{1} \left(z^2 y + \frac{3}{2} z y^2 + \frac{2}{3} y^3 \right) \Big|_{y=1}^{y=2} dz$$

$$= \int_{-1}^{1} z^2 + \frac{9}{2} z + \frac{14}{3} dz$$

$$= 10.$$

Similar calculations show that

$$M_{x=0} = \int_{-1}^{1} \int_{0}^{1} \int_{0}^{z+y} x(2x+y)\, dxdydz = \frac{45}{4},$$

$$M_{y=0} = \int_{-1}^{1} \int_{0}^{1} \int_{0}^{z+y} y(2x+y)\, dxdydz = 16,$$

$$M_{z=0} = \int_{-1}^{1} \int_{0}^{1} \int_{0}^{z+y} z(2x+y)\, dxdydz = 3.$$

Therefore

$$\overline{x} = \frac{M_{x=0}}{M} = \frac{45/4}{10} = \frac{9}{8},$$

$$\overline{y} = \frac{M_{y=0}}{M} = \frac{16}{10} = \frac{8}{5},$$

$$\overline{z} = \frac{M_{z=0}}{M} = \frac{3}{10}.$$

We conclude that the center of mass of $\mathcal{U}$ is the point $(9/8, 8/5, 3/10)$. ◀

QUICK QUIZ

1. A planar region $\mathcal{R}$ has mass density $\delta(x,y) = 3x^2y^3$. What function do we integrate over $\mathcal{R}$ to calculate the first moment about the x-axis?
2. For a planar region $\mathcal{R}$, $M_{x=0} = 72$, $M_{y=0} = 24$, and $\overline{x} = 12$. What is $\overline{y}$?
3. What is the polar moment of inertia of the unit disk $\{(x,y) : 0 \le x^2 + y^2 \le 1\}$ if its mass density is $\delta(x,y) = 1 + x^2 + y^2$?
4. What is the moment of inertia of the solid cylinder $\mathcal{U} = \{(x,y,z) : 1 \le z \le 2, 0 \le x^2 + y^2 \le 1\}$ about the z-axis if its mass density is $\delta(x,y,z) = 2z$?

Answers

1. $3x^2y^4$ 2. 4 3. $5\pi/6$ 4. $3\pi/2$

EXERCISES

Problems for Practice

▶ In Exercises 1–4, use a double integral to find the mass of the planar region $\mathcal{R}$ with density function $\delta(x,y)$. ◀

1. $\mathcal{R}=\{(x,y): 0\le x\le 2, x+1\le y\le 2x+3\}$, $\delta(x,y)=3x+4y$
2. $\mathcal{R}=\{(x,y): -2\le x\le 2, 0\le y\le 4-x^2\}$, $\delta(x,y)=x+3$
3. $\mathcal{R}=\{(x,y): 0\le x\le 1, 1-x^2\le y\le 2(1-x^2)\}$, $\delta(x,y)=12xy$
4. $\mathcal{R}=\{(x,y): 0\le y\le 2, y\le x\le 4-y\}$, $\delta(x,y)=3y$

▶ In Exercises 5–12, find the first moment of the region $\mathcal{R}$ with density function $\delta(x,y)$ about the indicated axis. ◀

5. $\mathcal{R}=\{(x,y): 0\le x\le 1, 1-x^3\le y\le 2(1-x^3)\}$, $\delta(x,y)=12xy$, y-axis
6. $\mathcal{R}=\{(x,y): 0\le y\le 2, y\le x\le 4-y\}$, $\delta(x,y)=3y$, x-axis
7. $\mathcal{R}=\{(x,y): 1\le x\le 2, x\le y\le 5-x\}$, $\delta(x,y)=12x$, x-axis
8. $\mathcal{R}=\{(x,y): 1\le y\le 4, 1\le x\le 2/\sqrt{y}\}$, $\delta(x,y)=12/\sqrt{y}$, x-axis
9. $\mathcal{R}=\{(x,y): 1\le y\le 4, \sqrt{y}\le x\le y^2\}$, $\delta(x,y)=10$, x-axis
10. $\mathcal{R}=\{(x,y): 0\le x\le 1, 0\le y\le x^2\}$, $\delta(x,y)=\sqrt{x}$, y-axis
11. $\mathcal{R}=\{(x,y): -3\le x\le -1, 2\le y\le 4\}$, $\delta(x,y)=-x/y$, y-axis
12. $\mathcal{R}=\{(x,y): -1\le y\le 0, 0\le x\le y+1\}$, $\delta(x,y)=\sqrt{|y|}$, y-axis

▶ In each of Exercises 13–18, find the center of mass of the planar region $\mathcal{R}$ with given density function $\delta(x,y)$. ◀

13. $\mathcal{R}=\{(x,y): 0\le x\le 1, x\le y\le \sqrt{x}\}$, $\delta(x,y)=x+6$
14. $\mathcal{R}$ is the region between the parabolas $y=-x^2+4$ and $y=x^2-4$, $\delta(x,y)=3-x$
15. $\mathcal{R}=\{(x,y): -1\le y\le 1, 2\le x\le y+3\}$, $\delta(x,y)=x(y+1)$
16. $\mathcal{R}=\{(x,y): 3\le x\le 4, -2\le y\le 1\}$, $\delta(x,y)=x+y+2$
17. $\mathcal{R}=\{(x,y): 1\le y\le 4, 1\le x\le 2/\sqrt{y}\}$, $\delta(x,y)=12/\sqrt{y}$
18. $\mathcal{R}=\{(x,y): 1\le x\le 2, x\le y\le 5-x\}$, $\delta(x,y)=12x$

▶ In Exercises 19–24, calculate the mass of the solid $\mathcal{U}$ with density function $\delta(x,y,z)$: ◀

19. $\mathcal{U}=\{(x,y,z): 0\le x\le 1, 0\le y\le 2, 0\le z\le xy\}$, $\delta(x,y,z)=x+1$
20. $\mathcal{U}=\{(x,y,z): 0\le y\le 1, y\le x\le 4, y\le z\le x\}$, $\delta(x,y,z)=\sqrt{x}+\sqrt{y}$
21. $\mathcal{U}=\{(x,y,z): 0\le z\le 2, z\le y\le 3, y\le x\le y+1\}$, $\delta(x,y,z)=x+z$
22. $\mathcal{U}=\{(x,y,z): 0\le x\le 1, 0\le y\le 1, 0\le z\le x+y\}$, $\delta(x,y,z)=x+y+z$
23. $\mathcal{U}=\{(x,y,z): 0\le x\le 1, x^2\le y\le x, y\le z\le 1\}$, $\delta(x,y,z)=12x$
24. $\mathcal{U}=\{(x,y,z): 0\le x\le 1, 0\le y\le \sqrt{1-x^2}, x^2+y^2\le z\le 2-x^2-y^2\}$, $\delta(x,y,z)=60y$

▶ In each of Exercises 25–30, calculate the mass of the solid $\mathcal{U}$, the first moments of $\mathcal{U}$ about the three coordinate planes, and the center of mass of $\mathcal{U}$. ◀

25. $\mathcal{U}=\{(x,y,z): -1\le x\le 1, -\sqrt{1-x^2}\le y\le \sqrt{1-x^2}, 0\le z\le 12\}$, $\delta(x,y,z)=z$
26. $\mathcal{U}=\{(x,y,z): 0\le x\le 1, 0\le y\le 2, 0\le z\le 1+2x+3y\}$, $\delta(x,y,z)=3$
27. $\mathcal{U}=\{(x,y,z): -1\le x\le 0, x\le y\le 1, x+y\le z\le 2\}$, $\delta(x,y,z)=x+3$
28. $\mathcal{U}=\{(x,y,z): 1\le z\le 2, z^2\le y\le 4, 2\le x\le 2z\}$, $\delta(x,y,z)=\sqrt{y}$
29. $\mathcal{U}=\{(x,y,z): 1\le z\le 2, 0\le x\le z, x\le y\le z\}$ $\delta(x,y,z)=y-x$
30. $\mathcal{U}=\{(x,y,z): x^2+y^2\le 4, 0\le z\le 4\}$ $\delta(x,y,z)=x^2+y^2+z^2$

▶ In each of Exercises 31–34, calculate the second moments of the planar region $\mathcal{R}$ with density function $\delta(x,y)$ about each coordinate axis. ◀

31. $\mathcal{R}=\{(x,y): 0\le x\le 2, 0\le y\le 4-x^2\}$, $\delta(x,y)=x+1$
32. $\mathcal{R}=\{(x,y): 1\le x\le 2, x\le y\le 2x\}$, $\delta(x,y)=y$
33. $\mathcal{R}=\{(x,y): 0\le x\le 1, 1-x\le y\le 2(1-x)\}$, $\delta(x,y)=x+y$
34. $\mathcal{R}=\{(x,y): 1\le x\le 2, x\le y\le 2\}$, $\delta(x,y)=x$

▶ In each of Exercises 35–38, calculate the polar moment of inertia of the planar region $\mathcal{R}$ with density function $\delta(x,y)$. ◀

35. $\mathcal{R}=\{(x,y): 0\le x\le 1, 0\le y\le 2\}$, $\delta(x,y)=6x$
36. $\mathcal{R}=\{(x,y): 1\le x\le 2, x\le y\le 2\}$, $\delta(x,y)=5x$
37. $\mathcal{R}=\{(x,y): 0\le x\le 1, 1\le y\le 2-x\}$, $\delta(x,y)=6y/(x^2+y^2)$
38. $\mathcal{R}=\{(x,y): 1\le x\le 2, x\le y\le 2\}$, $\delta(x,y)=xy$

▶ In each of Exercises 39–42, calculate the moments of inertia with respect to the three coordinate axes. ◀

39. $\mathcal{U} = \{(x, y, z) : 3 \le x \le 5, 1 \le y \le 4, 1 \le z \le 2\}$
$\delta(x, y, z) = xy$

40. $\mathcal{U} = \{(x, y, z) : -1 \le y \le 2, 0 \le x \le 2 - y, 0 \le z \le x\}$,
$\delta(x, y, z) = y + 2$

41. $\mathcal{U} = \{(x, y, z) : 0 \le x \le 3, 1 \le y \le 2, x + y \le z \le x + y + 1\}$,
$\delta(x, y, z) = 4$

42. $\mathcal{U} = \{(x, y, z) : 1 \le z \le 2, 1 \le x \le z, 1 \le y \le x\}$
$\delta(x, y, z) = x$

Further Theory and Practice

In each of Exercises 43–46, find the center of mass of the planar region $\mathcal{R}$ with given density function $\delta(x, y)$.

43. $\mathcal{R} = \{(x, y) : 0 \le x \le \pi, 0 \le y \le \sin(x)\}$, $\delta(x, y) = 8y$

44. $\mathcal{R} = \{(x, y) : 0 \le x \le 1, e^x \le y \le e\}$, $\delta(x, y) = 1/y$

45. $\mathcal{R} = \{(x, y) : 0 \le x \le 1, 0 \le y \le e^x\}$, $\delta(x, y) = 8x$

46. $\mathcal{R} = \{(x, y) : 1 \le x \le e, 0 \le y \le \ln(x)\}$, $\delta(x, y) = 72x$

47. Give an example of a planar region $\mathcal{R}$ such that the center of gravity $\mathcal{R}$ is not an element of $\mathcal{R}$.

48. Suppose that a region $\mathcal{R}$ in the plane consists of each point that lies in one or more of the sets

$$\{(x, y) : x^2 + y^2 \le 1\},$$

$$\{(x, y) : (x - 4)^2 + (y - 3)^2 \le 1\},$$

$$\{(x, y) : x^2 + (y - 12)^2 \le 4\}.$$

That is, $\mathcal{R}$ is the union of the three sets. Assume that the mass distribution is uniform in this region. Find the center of mass.

49. A solid $\mathcal{U}$ rotates about the x-axis with constant angular speed ω. Express the solid's kinetic energy $KE_{\mathcal{U}}$ in terms of ω and the moment of inertia I_x of $\mathcal{U}$. Let $M_{\mathcal{U}}$ denote the mass of $\mathcal{U}$. The *radius of gyration* of $\mathcal{U}$ about the x-axis is the distance ρ from the x-axis of a point-mass that has mass $M_{\mathcal{U}}$ and kinetic energy $KE_{\mathcal{U}}$. Find a formula for ρ in terms of I_x and $M_{\mathcal{U}}$. (Verify that your formula has the correct dimensions.)

50. Pappus's Theorem says that if a region in the right half of the xy-plane (i.e., in the set $\{(x, y, 0): x > 0\}$, is rotated about the y-axis, then the volume of the resulting solid is the area of the region times the circumference of the circle traversed by the centroid of the region. Prove Pappus's Theorem.

51. Apply Pappus's Theorem (as stated in Exercise 50) to calculate the volume obtained when a disk of radius r and center $(R, 0)$ with $r < R$ is rotated about the y-axis.

Calculator/Computer Exercises

52. Let $\mathcal{R}$ be the region of the xy-plane that is bounded above by $y = 3 - x^2$ and below by $y = 1 + (x + 1)^3$. Calculate the center of mass of $\mathcal{R}$ given that its mass density is $\delta(x, y) = x + 3y$.

53. Let $\mathcal{R}$ be the region in the first quadrant of the xy-plane that is bounded above by $y = 1 + x + x^2$ and below by $y = \exp(x)$. Calculate the center of mass of $\mathcal{R}$ given that its mass density is $\delta(x, y) = 1 + x$.

54. Let $\mathcal{R}$ be the region in the first quadrant of the xy-plane that is bounded above by $y = 1 + x + 2x^2$ and below by $y = 1 + x^3$. The mass density of $\mathcal{R}$ is $\delta(x, y) = 1 + x\sqrt{y}$. Calculate the second moments of $\mathcal{R}$ about the x- and y-axes.

55. Let $\mathcal{R}$ be the square in the xy-plane with vertices $(0, 0)$, $(0, 1)$, $(1, 1)$, and $(1, 0)$. The mass density of $\mathcal{R}$ is $\delta(x, y) = 2 + \sin(x^2) + \exp(-y^4)$. Approximate the center of mass of $\mathcal{R}$.

12.8 Other Coordinate Systems

Earlier in this chapter, we learned that polar coordinates make the equations of planar curves with circular symmetry much easier to study. A similar situation occurs when we are doing calculus in space. In this section, we learn about two special coordinate systems for handling surfaces and solids with special symmetries.

Cylindrical Coordinates

The cylindrical coordinate system is created by using polar coordinates in the xy-plane and the standard Euclidean z-coordinate in the vertical direction. The idea is illustrated in Figure 1. Notice that we locate the point P in space by first looking at the projection P' of P into the xy-plane. We denote by r the distance of P' from the origin. We denote by θ the angle by which we rotate the positive x-axis

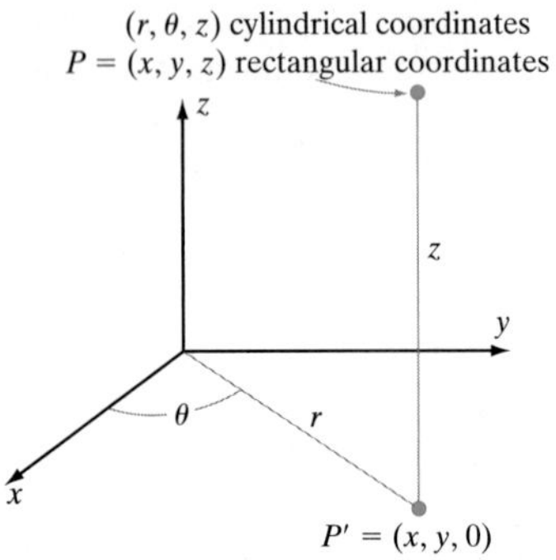

▲ Figure 1

counterclockwise until it lies over the ray $\overrightarrow{OP'}$. The third coordinate is simply z, the signed distance of P from P'. We write the cylindrical coordinates for P as (r, θ, z). The formulas that we learned in Section 12.4 for converting from rectangular to polar coordinates and back are valid in the context of cylindrical coordinates:

$$x = r\cos(\theta), \quad y = r\sin(\theta), \quad x^2 + y^2 = r^2, \quad \text{and,} \quad \text{if } x \neq 0, \quad \tan(\theta) = \frac{y}{x}.$$

▶ **EXAMPLE 1** The point P has rectangular coordinates $(1, 1, 4)$. What are its cylindrical coordinates?

Solution The projection of P into the xy-plane is $P' = (1, 1, 0)$, as shown in Figure 2. Then, we see that $r = \sqrt{2}$ and $\theta = \pi/4$. Also, $z = 4$. Therefore the cylindrical coordinates for P are $(\sqrt{2}, \pi/4, 4)$. ◀

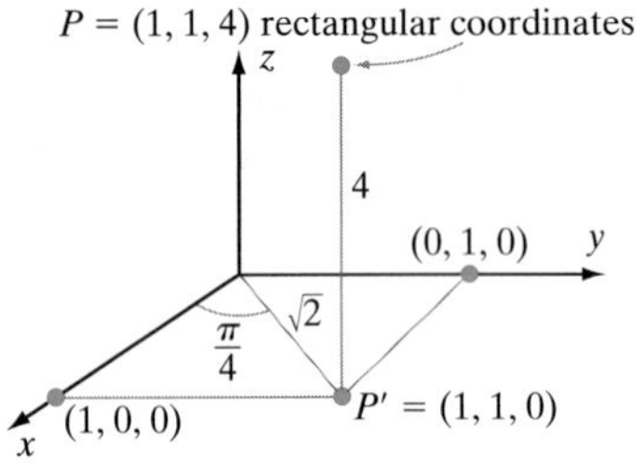

▲ Figure 2

We can calculate many integrals easily by using cylindrical coordinates. To do so, we replace the rectangular element of area $dxdy$ with the polar element of area $r\,drd\theta$. So our element of volume, which is $dV = dxdydz$ in rectangular coordinates, becomes

$$dV = r\,drd\theta dz$$

in cylindrical coordinates.

▶ **EXAMPLE 2** Integrate the function $f(x, y, z) = x^2 + y^2$ over the solid $\mathcal{U}$ that is inside the cylinder $x^2 + y^2 = 4$, below the plane $z = 3$, and above the plane $z = -2$.

Solution The solid $\mathcal{U}$ is drawn in Figure 3. It is convenient to convert the problem to cylindrical coordinates. Then $f(r\cos(\theta), r\sin(\theta), z) = r^2$ and the solid over which we wish to integrate becomes

$$\mathcal{U} = \{(r, \theta, z) : 0 \le r \le 2, 0 \le \theta \le 2\pi, \ -2 \le z \le 3\}.$$

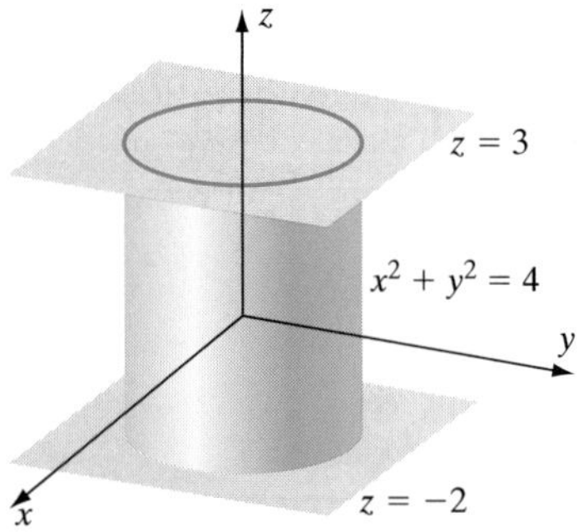

▲ Figure 3

Bearing in mind that the volume element is $r\,drd\theta dz$, we find that our integral is

$$\iiint_{\mathcal{U}} f(x, y, z)\, dV = \int_{-2}^{3}\int_{0}^{2\pi}\int_{0}^{2} (r^2)\, r\,drd\theta dz$$

$$= \int_{-2}^{3}\int_{0}^{2\pi} \left.\frac{r^4}{4}\right|_{r=0}^{r=2} d\theta dz = \frac{16}{4}(2\pi)(3 - (-2)) = 40\pi. \quad ◀$$

▶ **EXAMPLE 3** Calculate the volume of the solid $\mathcal{U}$ that lies above the cone $z = \sqrt{x^2 + y^2}$ and below the hemisphere $z = 1 + \sqrt{1 - x^2 - y^2}$.

Solution The solid $\mathcal{U}$ appears in Figure 4. We observe that, in cylindrical coordinates, the cone may be written as $z = r$, and the hemisphere may be written as $z = 1 + \sqrt{1 - r^2}$. From Figure 4, we see that the projection of $\mathcal{U}$ in the xy-plane is a disk $\mathcal{D}$. The boundary of $\mathcal{D}$ is the projection of the circle that is the intersection of the cone and the hemisphere. Because the z-coordinate of each point on this circle

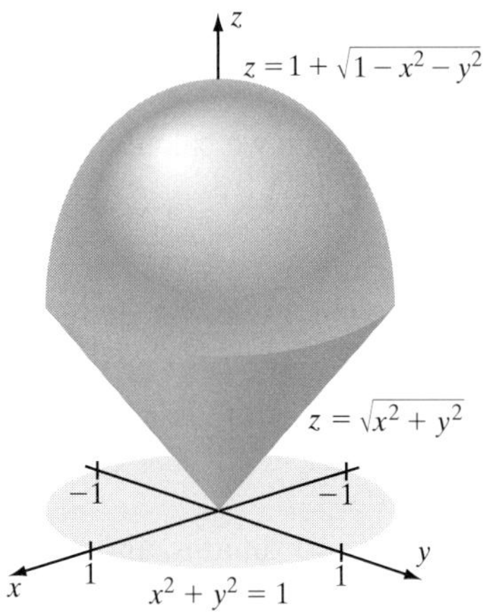

▲ Figure 4

of intersection is given both by r and by $1+\sqrt{1-r^2}$, we set $r=1+\sqrt{1-r^2}$. We solve this equation by rewriting it as $r-1=\sqrt{1-r^2}$, or $(r-1)^2=1-r^2$, which simplifies to $2r(r-1)=0$. The solution is $r=0$ or $r=1$, which means that $\mathcal{D}$ consists of the points with cylindrical coordinates that satisfy $z=0$ and the inequalities $0\le\theta\le 2\pi$ and $0\le r\le 1$. These inequalities tell us the limits of integration for the outer and middle integrals. To determine the limits of integration for the inner integral, we observe that a point with cylindrical coordinates (r,θ,z) lies in $\mathcal{U}$ if and only if $r\le z\le 1+\sqrt{1-r^2}$. Thus the volume of $\mathcal{U}$ is

$$\iiint_{\mathcal{U}} 1\,dV = \int_0^{2\pi}\int_0^1\int_r^{1+\sqrt{1-r^2}} (1)\,r\,dz dr d\theta = \int_0^{2\pi}\int_0^1 \left(1+\sqrt{1-r^2}-r\right) r\,dr d\theta.$$

Hence we have

$$\iiint_{\mathcal{U}} 1\,dV = \int_0^{2\pi} \left(\frac{r^2}{2}-\frac{(1-r^2)^{3/2}}{3}-\frac{r^3}{3}\right)\Bigg|_{r=0}^{r=1} d\theta$$
$$= 2\pi\left(\left(\frac{1}{2}-0-\frac{1}{3}\right)-\left(0-\frac{1}{3}-0\right)\right)=\pi. \quad ◀$$

Spherical Coordinates

$P = (x, y, z)$ rectangular coordinates
$P = (r, \theta, z)$ cylindrical coordinates
$P = (\rho, \phi, \theta)$ spherical coordinates

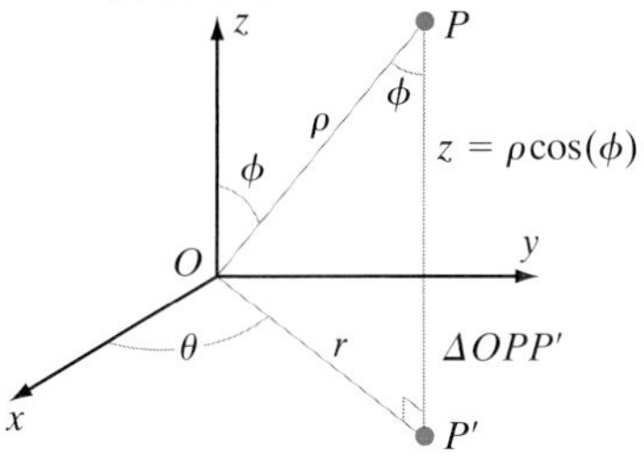

$P' = (x, y, 0)$ rectangular coordinates
$P' = (r, \theta, 0)$ cylindrical coordinates
$P' = (r, \frac{\pi}{2}, \theta)$ spherical coordinates

▲ Figure 5

For calculations involving objects in three-dimensional space that are symmetrical in the origin, spherical coordinates are often most convenient. Referring to Figure 5, we see that, in this new coordinate system, we locate a point P according to its distance ρ from the origin, the angle ϕ that the ray $\overrightarrow{OP}$ makes with the positive z-axis, and the angle θ that its projected ray $\overrightarrow{OP'}$ makes in the xy-plane with the positive x-axis. We write the spherical coordinates for P as (ρ,ϕ,θ).

Every point in space can be located with spherical coordinates (ρ,ϕ,θ) satisfying $0\le\rho$, $0\le\phi\le\pi$, and $0\le\theta<2\pi$. By convention, we limit the values of the spherical coordinates to these ranges. Notice that θ and ϕ have different ranges. We can use the point P_0 with rectangular coordinates $(0,-1,0)$ to understand why we need not ever take ϕ to be greater than π. If convention permitted, we could describe P_0 by the spherical coordinates $\rho=1$, $\phi=3\pi/2$, and $\theta=\pi/2$. But such a description is unnecessary because we can also describe P_0 by the spherical coordinates $\rho=1$, $\phi=\pi/2$, and $\theta=3\pi/2$ (see Figure 6).

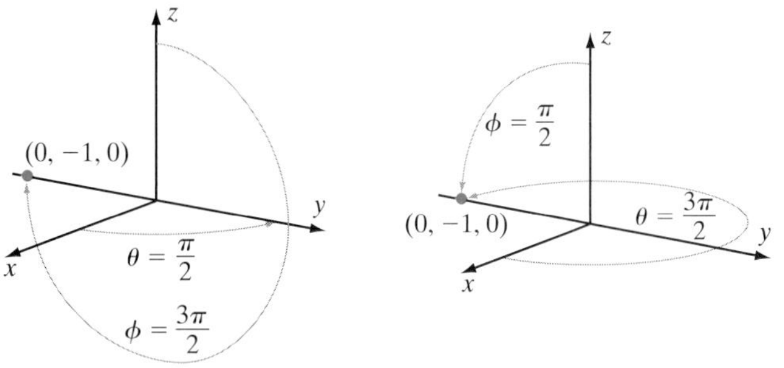

▲ Figure 6

▶ **EXAMPLE 4** What are spherical coordinates for the point P_0 with rectangular coordinates $(1, 1, \sqrt{6})$?

Solution Using the formula for the distance between two points, we calculate that the distance of P_0 from the origin is $\sqrt{(1-0)^2 + (1-0)^2 + (\sqrt{6}-0)^2}$, or $\sqrt{8}$. Thus $\rho = \sqrt{8} = 2\sqrt{2}$. Because the projection of P_0 into the xy-plane is $P_0' = (1, 1, 0)$, we see that $\theta = \pi/4$. Next we use the equation $z = \rho\cos(\phi)$, which can be deduced from Figure 5. This equation tells us that the spherical coordinate ϕ of P_0 satisfies $\sqrt{6} = 2\sqrt{2}\cos(\phi)$, or $\cos(\phi) = \sqrt{6}/(2\sqrt{2}) = \sqrt{3}/2$. It follows that $\phi = \pi/6$. In summary, the spherical coordinates of P_0 are $(2\sqrt{2}, \pi/6, \pi/4)$. ◀

Converting from spherical coordinates to rectangular coordinates and back again requires some calculation. The formula for the distance between the point with Cartesian coordinates (x, y, z) and the origin gives us

$$\rho = \sqrt{x^2 + y^2 + z^2}. \tag{12.8.1}$$

Written in terms of cylindrical coordinates, equation (12.8.1) becomes

$$\rho = \sqrt{r^2 + z^2}. \tag{12.8.2}$$

As can be seen from right triangle OPP' in Figure 5, we have

$$z = \rho\cos(\phi), \tag{12.8.3}$$

and

$$r = \rho\sin(\phi). \tag{12.8.4}$$

Equations (12.8.1)–(12.8.4) allow us to pass back and forth between spherical and cylindrical coordinates. If we substitute equation (12.8.4) into the equations $x = r\cos(\theta)$ and $y = r\sin(\theta)$, then we obtain

$$x = \rho\cos(\theta)\sin(\phi) \tag{12.8.5}$$

and

$$y = \rho\sin(\theta)\sin(\phi). \tag{12.8.6}$$

Equations (12.8.3), (12.8.5), and (12.8.6) tell us the Cartesian coordinates of a point with spherical coordinates (ρ, ϕ, θ). Observe that these formulas imply that

$$\begin{aligned} x^2 + y^2 + z^2 &= \rho^2\cos^2(\theta)\sin^2(\phi) + \rho^2\sin^2(\theta)\sin^2(\phi) + \rho^2\cos^2(\phi) \\ &= \rho^2\sin^2(\phi)\big(\cos^2(\theta) + \sin^2(\theta)\big) + \rho^2\cos^2(\phi) \\ &= \rho^2, \end{aligned}$$

which confirms that ρ is the radial coordinate in space, analogous to the coordinate r in polar coordinates in the plane.

▶ **EXAMPLE 5** A point P_0 in space has spherical coordinates given by $(2, 2\pi/3, 5\pi/6)$. Calculate its rectangular and cylindrical coordinates.

Solution Denote the rectangular coordinates of P_0 by (x_0, y_0, z_0). Using equations (12.8.3), (12.8.5), and (12.8.6), we calculate

$$x_0 = 2\cos\left(\frac{5\pi}{6}\right)\sin\left(\frac{2\pi}{3}\right) = 2\left(-\frac{\sqrt{3}}{2}\right)\left(\frac{\sqrt{3}}{2}\right) = -\frac{3}{2}$$

$$y_0 = 2\sin\left(\frac{5\pi}{6}\right)\sin\left(\frac{2\pi}{3}\right) = 2\left(\frac{1}{2}\right)\left(\frac{\sqrt{3}}{2}\right) = \frac{\sqrt{3}}{2}$$

$$z_0 = 2\cos\left(\frac{2\pi}{3}\right) = 2\left(-\frac{1}{2}\right) = -1.$$

Thus the rectangular coordinates of P_0 are $(-3/2, \sqrt{3}/2, -1)$. It is now easy to convert to cylindrical coordinates (r_0, θ_0, z_0). The cylindrical coordinate z_0 is the same as the rectangular coordinate z_0, namely -1. The cylindrical coordinate θ_0 is the same as the spherical coordinate θ_0, namely $5\pi/6$. Finally,

$$r_0 = \sqrt{x_0^2 + y_0^2} = \sqrt{\left(-\frac{3}{2}\right)^2 + \left(\frac{\sqrt{3}}{2}\right)^2} = \sqrt{3}.$$

So the cylindrical coordinates of P_0 are $(\sqrt{3}, 5\pi/6, -1)$. ◀

Naturally we are interested in performing integrations in spherical coordinates. We need to determine the volume element in this new coordinate system. Figure 7 exhibits an increment of volume swept out by increments $\Delta\rho$, $\Delta\phi$, and $\Delta\theta$ of the coordinates. This solid is approximately a box, with sides that are determined by arcs of circles and a radial height $\Delta\rho$. One circular edge is an arc of a circle that is parallel to the xy-plane and that has radius $\rho\sin(\phi)$. Because this edge subtends an angle $\Delta\theta$, it has arc length $\rho\sin(\phi)\,\Delta\theta$. Another circular edge of the box is an arc of a circle of radius ρ. This edge subtends an angle $\Delta\phi$ and therefore has arc length $\rho\Delta\phi$. The height of the box in the radial direction is $\Delta\rho$. Thus the product

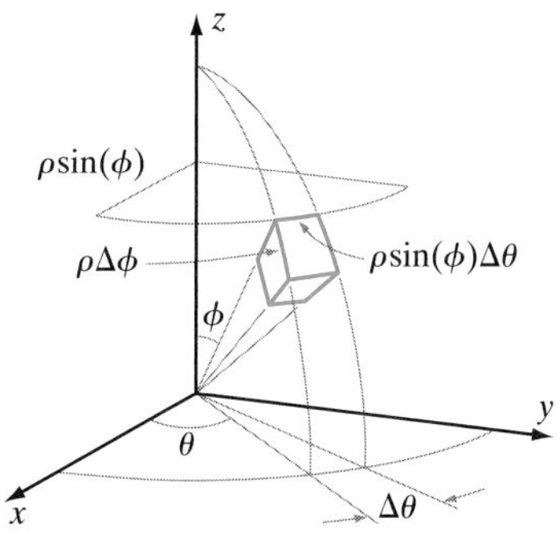

▲ Figure 7 A small "box" in spherical coordinates

$$\big(\rho\sin(\phi)\Delta\theta\big)(\rho\Delta\phi)(\Delta\rho) = \rho^2\sin(\phi)\,\Delta\theta\Delta\phi\Delta\rho$$

approximates the volume of the small box. We conclude, by using familiar limiting arguments, that the volume element for spherical coordinates is

$$dV = \rho^2\sin(\phi)\,d\rho\,d\phi\,d\theta. \qquad (12.8.7)$$

INSIGHT Together, Example 8 and Theorem 4 of Section 12.5 show how to apply a change of variable formula in the plane to the transformation $T(r,\theta) = (r\cos(\theta), r\sin(\theta))$ to obtain the polar element of area $dA = |J_T(r,\theta)|dr d\theta = r\,dr d\theta$. Although we will not derive it, there is an analogous change of variable formula for space. The spherical coordinate transformation $T(\rho,\phi,\theta) = (\rho\cos(\theta)\sin(\phi), \rho\sin(\theta)\sin(\phi), \rho\cos(\phi))$ has Jacobian matrix

$$D(T)(\rho,\phi,\theta) = \begin{bmatrix} \frac{\partial}{\partial\rho}(\rho\cos(\theta)\sin(\phi)) & \frac{\partial}{\partial\phi}(\rho\cos(\theta)\sin(\phi)) & \frac{\partial}{\partial\theta}(\rho\sin(\phi)\cos(\theta)) \\ \frac{\partial}{\partial\rho}(\rho\sin(\theta)\sin(\phi)) & \frac{\partial}{\partial\phi}(\rho\sin(\theta)\sin(\phi)) & \frac{\partial}{\partial\theta}(\rho\sin(\phi)\sin(\theta)) \\ \frac{\partial}{\partial\rho}(\rho\cos(\phi)) & \frac{\partial}{\partial\phi}(\rho\cos(\phi)) & \frac{\partial}{\partial\theta}(\rho\cos(\phi)) \end{bmatrix},$$

or

$$D(T)(\rho,\phi,\theta) = \begin{bmatrix} \cos(\theta)\sin(\phi) & \rho\cos(\theta)\cos(\phi) & -\rho\sin(\theta)\sin(\phi) \\ \sin(\theta)\sin(\phi) & \rho\sin(\theta)\cos(\phi) & \rho\cos(\theta)\sin(\phi) \\ \cos(\phi) & -\rho\sin(\phi) & 0 \end{bmatrix}.$$

From this, we calculate the Jacobian determinant to be

$$J_T(\rho,\phi,\theta) = \det\big(D(T)(\rho,\phi,\theta)\big) = \rho^2\sin(\phi).$$

These considerations provide an alternative approach to formula (12.8.7).

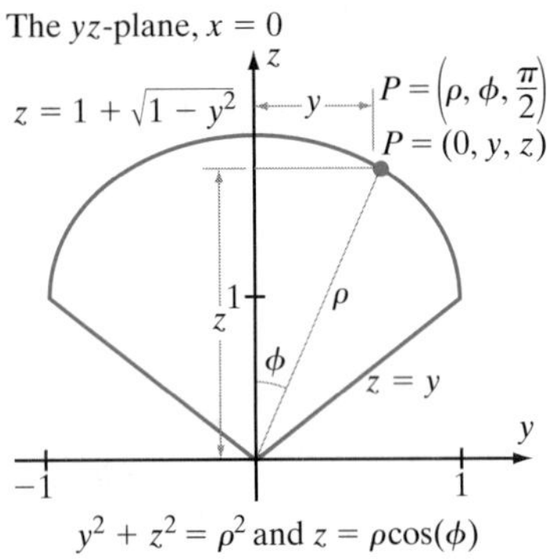

▲ Figure 8

▶ **EXAMPLE 6** Use spherical coordinates to calculate the volume V of the solid $\mathcal{U}$ that lies above the cone $z=\sqrt{x^2+y^2}$ and below the hemisphere $z=1+\sqrt{1-x^2-y^2}$.

Solution This solid is pictured in Figure 4, We have already calculated its volume to be π by using cylindrical coordinates in Example 3.

The intersection of $\mathcal{U}$ with the yz-plane is shown in Figure 8. Because $x=0$ in this plane, the equation of the hemisphere is $z=1+\sqrt{1-y^2}$ or $(z-1)^2=1-y^2$. After expanding the left side of this equation, we get $y^2+z^2=2z$. But, as Figure 8 shows, in the yz-plane the relationships between y, z, ρ, and ϕ are $y^2+z^2=\rho^2$ and $z=\rho\cos(\phi)$. Therefore the equation of the intersection of the hemisphere with the yz-plane is $\rho^2=2\rho\cos(\phi)$ or $\rho=2\cos(\phi)$. It follows that the solid $\mathcal{U}$ is described in spherical coordinates by the inequalities $0\le\theta\le 2\pi$, $0\le\phi\le\pi/4$, $0\le\rho\le 2\cos(\phi)$. According to Theorem 1,

$$V = \int_0^{2\pi}\int_0^{\pi/4}\int_0^{2\cos(\phi)} (1)\ \rho^2\sin(\phi)\ d\rho d\phi d\theta.$$

We evaluate this integral as follows:

$$V = \frac{1}{3}\int_0^{2\pi}\int_0^{\pi/4} 8\cos^3(\phi)\sin(\phi)\,d\phi\,d\theta = \frac{8}{3}\int_0^{2\pi}\left(-\frac{\cos^4(\phi)}{4}\Bigg|_{\phi=0}^{\phi=\pi/4}\right)d\theta$$

$$= 2\pi\left(\frac{8}{3}\right)\left(\frac{1}{4}-\frac{(1/\sqrt{2})^4}{4}\right) = \pi. \quad ◀$$

▶ **EXAMPLE 7** Calculate the center of gravity of the solid in the last example, assuming a uniform mass distribution $\delta(x, y, z) = 1$. (Refer to Section 12.7 for terminology and notation.)

Solution It follows from symmetry considerations that $\bar{x} = \bar{y} = 0$. So, we must calculate $\bar{z}$. Now, substituting formula (12.8.3) for z, 1 for $\delta(x, y, z)$, and formula (12.8.7) for dV, we obtain

$$M_{z=0} = \iiint_{\mathcal{U}} z\delta(x,y,z)\,dV = \int_0^{2\pi}\int_0^{\pi/4}\int_0^{2\cos(\phi)} (\rho\cos(\phi))(1)\rho^2\sin(\phi)\,d\rho\,d\phi\,d\theta$$

$$= \int_0^{2\pi}\int_0^{\pi/4} \cos(\phi)\sin(\phi)\left(\int_0^{2\cos(\phi)} \rho^3 d\rho\right)\,d\phi\,d\theta.$$

Therefore

$$M_{z=0} = \int_0^{2\pi}\left(\int_0^{\pi/4}\cos(\phi)\sin(\phi)\left(\frac{1}{4}\rho^4\Big|_{\rho=0}^{\rho=2\cos(\phi)}\right)\,d\phi\right)d\theta = 4\int_0^{2\pi}\left(\int_0^{\pi/4}\cos(\phi)\sin(\phi)\cos^4(\phi)\,d\phi\right)d\theta.$$

Because the inner integral does not depend on θ, we have

$$M_{z=0} = 2\pi\cdot 4\int_0^{\pi/4}\cos^5(\phi)\sin(\phi)\,d\phi.$$

We make the substitution $u = \cos(\phi)$, $du = -\sin(\phi)\,d\phi$, obtaining

$$M_{z=0} = 8\pi\int_1^{1/\sqrt{2}}(-u^5)\,du = -\frac{8\pi}{6}u^6\Big|_1^{1/\sqrt{2}} = -\frac{8\pi}{6}\left(\frac{1}{8}-1\right) = \frac{7}{6}\pi.$$

From Examples 3 and 6, we know that the volume of the solid is π. The mass is therefore $M = \iiint_{\mathcal{U}}\delta(x,y,z)\,dV = \iiint_{\mathcal{U}} 1\,dV = \pi$. We conclude that

$$\bar{z} = \frac{M_{z=0}}{M} = \frac{7\pi/6}{\pi} = \frac{7}{6}.$$

Therefore the center of gravity is $(0, 0, 7/6)$. ◀

QUICK QUIZ

1. What are the cylindrical coordinates of the point with rectangular coordinates $(1, 2, 3)$?
2. What are the spherical coordinates of the point with rectangular coordinates $(1, 2, 3)$?
3. What are the rectangular coordinates of the point that has $(2\sqrt{6}, \pi/4, \pi/3)$ for spherical coordinates?
4. What is the element of volume in spherical coordinates?

Answers

1. $(\sqrt{5}, \arctan(2), 3)$ 2. $(\sqrt{14}, \arccos(3/\sqrt{14}), \arctan(2))$
3. $(\sqrt{3}, 3, 2\sqrt{3})$ 4. $\rho^2\sin(\phi)\,d\rho\,d\phi\,d\theta$

EXERCISES

Problems for Practice

In each of Exercises 1–8, a point P is given in rectangular coordinates. Give cylindrical coordinates for P.

1. $P = (2\sqrt{3}, 2, 5)$
2. $P = (4, 0, 6)$
3. $P = (-2, 2, -1)$
4. $P = (\sqrt{2}, -\sqrt{2}, 2)$
5. $P = (0, -3, -2)$
6. $P = (5, 5, 0)$
7. $P = (1, -\sqrt{3}, -2)$
8. $P = (0, 0, -3)$

In each of Exercises 9–16, rectangular coordinates are given for a point P. Give spherical coordinates for P.

9. $P = (1, 1, \sqrt{2})$
10. $P = (-1, \sqrt{3}, 2)$
11. $P = (4, 0, 4)$
12. $P = (0, 0, 3)$
13. $P = (1, 1, \sqrt{6})$
14. $P = (2, 0, -2\sqrt{3})$
15. $P = (-4, 4, -4\sqrt{6})$
16. $P = (\sqrt{3}, -3, \sqrt{12})$

In each of Exercises 17–24, rectangular, spherical, or cylindrical coordinates are given for the point P. Calculate coordinates in the other two coordinate systems.

17. $P = (3, \pi/3, 2\pi/3)$ spherical
18. $P = (1, 5\pi/6, 7\pi/4)$ spherical
19. $P = (2, -2, -2\sqrt{2})$ rectangular
20. $P = (0, \sqrt{3}, -1)$ rectangular
21. $P = (2, -3\pi/4, -2)$ cylindrical
22. $P = (6, -7\pi/4, -6)$ cylindrical
23. $P = (3, 0, -3)$ rectangular
24. $P = (5, 3\pi/4, 2\pi/3)$ spherical

In each of Exercises 25–30, integrate the function f over the solid $\mathcal{U}$ using cylindrical coordinates.

25. $f(x,y,z) = x^2 + y^2$; $\mathcal{U}$ is the region below the paraboloid $z = -x^2 - y^2$ and above the plane $z = -4$.
26. $f(x,y,z) = x + y$; $\mathcal{U}$ is the solid in the first octant bounded by $z = x^2 + y^2$ and $z = \sqrt{x^2 + y^2}$.
27. $f(x,y,z) = \sqrt{x^2 + y^2}$; $\mathcal{U}$ is the solid bounded by the cylinder $x^2 + y^2 = 4$ and by $z = x^2 + y^2 + 1, z = -3$.
28. $f(x,y,z) = 1/(1 + x^2 + y^2)$; $\mathcal{U}$ is the solid bounded by the paraboloids $z = x^2 + y^2 + 2$ and $z = 10 - x^2 - y^2$.
29. $f(x,y,z) = 1/(x^2 + y^2)$; $\mathcal{U}$ is the solid bounded by the cylinders $x^2 + y^2 = 2$ and $x^2 + y^2 = 4$ and by the planes $z = 8$ and $z = -2$.
30. $f(x,y,z) = xy$; $\mathcal{U}$ is the solid inside the cone $z^2 = x^2 + y^2$, outside the cylinder $x^2 + y^2 = 9$, and between $z = -5$ and $z = 5$.

In each of Exercises 31–38, integrate the function f over the solid $\mathcal{U}$ using spherical coordinates.

31. $f(x,y,z) = 5$; $\mathcal{U}$ is the solid inside the sphere $x^2 + y^2 + z^2 = 9$.
32. $f(x,y,z) = x^2 + y^2 + z^2$; $\mathcal{U}$ is the solid outside the cone $z^2 = x^2 + y^2$ and inside the sphere $x^2 + y^2 + z^2 = 8$.
33. $f(x,y,z) = (x^2 + y^2 + z^2)^{1/2}$; $\mathcal{U}$ is the solid inside the sphere $x^2 + y^2 + z^2 = 9$ and outside the sphere $x^2 + y^2 + z^2 = 4$.
34. $f(x,y,z) = \cos\left((x^2 + y^2 + z^2)^{3/2}\right)$; $\mathcal{U}$ is the solid inside the sphere $x^2 + y^2 + z^2 = 9$ and above the xy-plane.
35. $f(x,y,z) = z$; $\mathcal{U}$ is the solid below the surface $z = \sqrt{x^2 + y^2}$ and inside the sphere $x^2 + y^2 + z^2 = 4$.
36. $f(x,y,z) = x^2 + y^2$; $\mathcal{U}$ is the region outside the cone $z^2 = x^2 + y^2$ and inside the sphere $x^2 + y^2 + z^2 = 4$.
37. $f(x,y,z) = (x^2 + y^2 + z^2)^{-1/2}$; $\mathcal{U}$ is the solid between the spheres $x^2 + y^2 + z^2 = 2$ and $x^2 + y^2 + z^2 = 3$.
38. $f(x,y,z) = (1 + x^2 + y^2 + z^2)^{-2}$; $\mathcal{U}$ is the solid above the cone $z = \sqrt{x^2 + y^2}$ and below the sphere $x^2 + y^2 + z^2 = 3$.

Further Theory and Practice

39. Use cylindrical coordinates to calculate the volume outside the cone $z^2 = x^2 + y^2$ and inside the cylinder $x^2 + y^2 = 4$.
40. A point P has rectangular coordinates $(3, 5\pi/6, 2\pi/3)$. What are the cylindrical coordinates of P? What are the spherical coordinates of P?
41. A point P has spherical coordinates $(3, 2, 5)$. What are the rectangular coordinates of P? What are the cylindrical coordinates of P?
42. Determine all points that have spherical coordinates equal to their rectangular coordinates.
43. Determine all points that have cylindrical coordinates equal to their rectangular coordinates.
44. Find formulas for converting directly from cylindrical to spherical coordinates and vice versa.
45. Sketch the graph of the surface $\rho = \cos(\phi)$ in spherical coordinates.
46. Sketch the graph of $\phi = \pi/3$ in spherical coordinates.
47. Sketch the graph of $z = 2 + r^2$ in cylindrical coordinates.
48. Sketch the graph of $\phi = \theta$ in spherical coordinates.

Calculator/Computer Exercises

49. Let $\mathcal{R}$ be the planar region in the first quadrant that lies outside the curve $r = \sin(\theta)^3$ and inside the curve

$r = 3 - 2\cos(\theta)$. Let $\mathcal{U}$ be the solid that lies above $\mathcal{R}$ and below $z = y$. Calculate

$$\iiint_{\mathcal{U}} (x^2 + y^2)^{1/3} dV.$$

50. Let $V(\alpha)$ be the volume of that part of the unit ball that consists of points with spherical coordinates $0 \leq \theta \leq \alpha$ and $0 \leq \phi \leq \alpha$. For what value of α is $V(\alpha)$ equal to one-fourth the volume of the unit ball?

51. Let a be a constant between 0 and 1. The spherical cap of height a of the unit ball is the solid

$$\mathcal{U}_a = \{(x, y, z) : x^2 + y^2 + z^2 \leq 1, 1 - a \leq z\}.$$

For what a is the volume of $\mathcal{U}_a$ equal to one-fourth the volume of the unit ball?

Summary of Key Topics in Chapter 12

Double Integrals Over Rectangular Regions (Section 12.1)

The double integral of a continuous function f over a rectangular region

$$\mathcal{R} = \{(x, y) : a \leq x \leq b, c \leq y \leq d\}$$

is defined to be the limit of a Riemann sum over the product of partitions of $[a, b]$ and $[c, d]$. The integral is written

$$\iint_{\mathcal{R}} f(x, y)\, dA.$$

This double integral is evaluated by computing either of the iterated integrals

$$\int_a^b \left(\int_c^d f(x, y)\, dy \right) dx \quad \text{or} \quad \int_c^d \left(\int_a^b f(x, y)\, dx \right) dy.$$

Integration Over More General Planar Regions (Section 12.2)

If $\mathcal{R}$ is a planar region bounded by finitely many smooth curves and contained in a bounded rectangle $Q = \{(x, y) : a \leq x \leq b, c \leq y \leq d\}$, then for each positive integer N, we partition Q into N^2 subrectangles by dividing each interval $[a, b]$ and $[c, d]$ into N equal subintervals. To integrate a continuous function f over $\mathcal{R}$, we form Riemann sums using only subrectangles of the partition that lie completely in $\mathcal{R}$. The limit of the Riemann sums, as the number of subrectangles tends to infinity, gives the double integral; we denote the integral by

$$\iint_{\mathcal{R}} f(x, y)\, dA.$$

Simple Regions (Section 12.2)

A region $\mathcal{R}$ is x-simple if it has the form

$$\mathcal{R} = \{(x, y) : c \leq y \leq d,\ \alpha_1(y) \leq x \leq \alpha_2(y)\}.$$

It is y-simple if it has the form

$$\mathcal{R} = \{(x, y) : a \leq x \leq b,\ \beta_1(x) \leq y \leq \beta_2(x)\}.$$

The integral of a continuous f over an x-simple region $\mathcal{R}$ is calculated with the iterated integral

$$\int_c^d \left(\int_{\alpha_1(y)}^{\alpha_2(y)} f(x,y)\, dx \right) dy.$$

The integral of a continuous f over a y-simple region $\mathcal{R}$ is calculated with the iterated integral

$$\int_a^b \left(\int_{\beta_1(x)}^{\beta_2(x)} f(x,y)\, dy \right) dx.$$

If a region is both x-simple and y-simple, then the integral may be evaluated in either order; this observation is often useful in simplifying integration problems.

Area and Volume
(Section 12.3)

The area of a bounded region $\mathcal{R}$ whose boundary consists of finitely many smooth curves is given by

$$\iint_{\mathcal{R}} 1\, dA.$$

The volume under the graph of a continuous function $z = f(x,y)$ and above a region $\mathcal{R}$ in the xy-plane is given by

$$\iint_{\mathcal{R}} f(x,y)\, dA.$$

The volume under the graph of a function $z = f(x,y)$ and above the graph of $z = g(x,y)$, both defined on the region $\mathcal{R}$ in the xy-plane, is given by

$$\iint_{\mathcal{R}} (f(x,y) - g(x,y))\, dA.$$

Polar Coordinates
(Section 12.5)

A planar point P is located by polar coordinates (r, θ), where r is the distance of P from the origin, and θ is the signed angle that the ray $\overrightarrow{OP}$ makes with the positive x-axis. In the case $r < 0$, the point with polar coordinates (r, θ) is the reflection in the origin of the point with polar coordinates $(|r|, \theta)$.

Polar coordinates (r, θ) are related to Cartesian coordinates of a point P by the formulas

$$x = r\cos(\theta), \quad y = r\sin(\theta), \quad r^2 = x^2 + y^2, \quad \text{and, if } x \neq 0, \quad \tan(\theta) = \frac{y}{x}.$$

Graphing in Polar Coordinates
(Section 12.4)

To graph a curve $r = \varphi(\theta)$, plotting points is a vital tool. Also we take note of intervals of increase of $\varphi(\theta)$ and intervals of decrease of $\varphi(\theta)$. We test for symmetry as follows:

$$\text{symmetry in the } x\text{-axis if:} \quad \varphi(\theta) = \varphi(-\theta)$$

symmetry in the y-axis if: $\varphi(\theta) = \varphi(\pi - \theta)$

symmetry in the origin if: $\varphi(\theta) = \varphi(\theta + \pi)$

If the curve is defined implicitly by an equation of the form $\Psi(r, \theta) = 0$, then the curve is symmetric about the origin if $\Psi(-r, \theta) = \Psi(r, \theta)$.

Area in Polar Coordinates (Section 12.5)

If $\varphi(\theta) \geq 0$ for $\theta \in [\alpha, \beta]$, if $\beta - \alpha \leq 2\pi$, and if $\mathcal{R}$ is plannar region described in polar coordinates by $\mathcal{S} = \{(r, \theta) : 0 \leq r \leq \varphi(\theta), \alpha \leq \theta \leq \beta\}$, then the area of $\mathcal{R}$ is given by

$$\frac{1}{2} \int_{\alpha}^{\beta} \varphi^2(\theta) d\theta.$$

If $0 \leq \varphi_0(\theta) \leq \varphi_1(\theta)$ for $\theta \in [\alpha, \beta]$ and if $\mathcal{R}$ is the plannar region described in polar coordinates by $\mathcal{S} = \{(r, \theta) : \varphi_0(\theta) \leq r \leq \varphi_1(\theta), \alpha \leq \theta \leq \beta\}$, then the area of $\mathcal{R}$ is given by

$$\frac{1}{2} \int_{a}^{b} \left(\varphi_1^2(\theta) - \varphi_0^2(\theta)\right) d\theta.$$

Double Integrals in Polar Coordinates (Section 12.5)

If, a plannar region $\mathcal{R}$ is described in polar coordinates by $\mathcal{S} = \{(r, \theta) : \alpha \leq \theta \leq \beta, \varphi_0(\theta) \leq r \leq \varphi_1(\theta)\}$, with $\beta - \alpha \leq 2\pi$ and $0 \leq \varphi_0(\theta) \leq \varphi_1(\theta)$ for all θ, then the area of $\mathcal{R}$ is given by

$$\int_{\alpha}^{\beta} \int_{\varphi_0(\theta)}^{\varphi_1(\theta)} r \, dr d\theta$$

If $f(r, \theta)$ is a continuous function on $\mathcal{R}$, then the integral of f over $\mathcal{R}$ is given by

$$\int_{\alpha}^{\beta} \int_{\varphi_0(\theta)}^{\varphi_1(\theta)} f(r, \theta) \, r \, dr d\theta.$$

Integrals in polar coordinates can also be used to calculate volumes.

Change of Variable in Double Integrals (Section 12.5)

Let $\mathcal{S}$ be a region in the uv-plane bounded by finitely many continuously differentiable curves. Suppose that $T(u, v) = (\varphi(u, v), \psi(u, v))$ is a transformation of $\mathcal{S}$ to a region $\mathcal{R}$ in the xy-plane that is also bounded by finitely many continuously differentiable curves. We assume that T is one-to-one and onto and that the component functions ϕ, ψ are continuously differentiable.

If $\mathcal{S}$ and $\mathcal{R}$ and the transformation T are as just described and if f is a continuous function on the region $\mathcal{R}$, then

$$\iint_{\mathcal{R}} f(x, y) \, dxdy = \iint_{\mathcal{S}} f(\phi(u, v), \psi(u, v)) |(\phi_u \psi_v - \psi_u \phi_v)(u, v)| \, dudv.$$

The expression

$$J_T(u,v) = \det\left(\begin{bmatrix} \phi_u(u,v) & \phi_v(u,v) \\ \psi_u(u,v) & \psi_v(u,v) \end{bmatrix}\right) = (\phi_u\psi_v - \psi_u\phi_v)(u,v)$$

that appears in the change of variable formula is called the *Jacobian determinant* of the transformation T (or sometimes simply the *Jacobian*).

Triple Integrals (Section 12.6)

For a solid $\mathcal{U}$ in space contained in a cube

$$Q = \{(x,y,z) : a \le x \le b, c \le y \le d, e \le z \le f\},$$

and a continuous function $f(x,y,z)$ on $\mathcal{U}$, the Riemann sums for f over $\mathcal{U}$ are defined using a product of partitions of the intervals $[a,b], [c,d], [e,f]$. The triple integral is a limit of the Riemann sums whenever the limit exists. The integral is denoted

$$\iiint_{\mathcal{U}} f(x,y,z)\,dV.$$

When $\mathcal{U}$ is a solid of the form

$$\mathcal{U} = \{(x,y,z) : a \le x \le b, \alpha_1(x) \le y \le \alpha_2(x), \beta_1(x,y) \le z \le \beta_2(x,y)\},$$

then the triple integral may be evaluated as the iterated integral

$$\int_a^b \int_{\alpha_1(x)}^{\alpha_2(x)} \int_{\beta_1(x,y)}^{\beta_2(x,y)} f(x,y,z)\,dz dy dx.$$

The roles of x, y, and z may be permuted in this last formula when they are similarly permuted in the definition of $\mathcal{R}$.

Mass, Moment of Inertia, and Center of Mass (Section 12.7)

If a planar plate (or lamina) $\mathcal{R}$ has density $\delta(x,y)$ at the point (x,y), then the mass of the plate is given by

$$M = \iint_{\mathcal{R}} \delta(x,y)\,dA.$$

The *first moment* of the plate with respect to the x-axis (the axis with Cartesian equation $y=0$) is

$$M_{y=0} = \iint_{\mathcal{R}} y\delta(x,y)\,dA,$$

and the first moment with respect to the y-axis (the axis with Cartesian equation $x=0$) is

$$M_{x=0} = \iint_{\mathcal{R}} x\delta(x, y)\, dA.$$

The respective *second moments*, or *moments of inertia*, are

$$I_{y=0} = \iint_{\mathcal{R}} y^2\delta(x, y)\, dA$$

and

$$I_{x=0} = \iint_{\mathcal{R}} x^2\delta(x, y)\, dA.$$

The *center of mass* (or *center of gravity*) of the plate is the point $\overline{c} = (\overline{x}, \overline{y})$ where

$$\overline{x} = \frac{M_{x=0}}{M}$$

and

$$\overline{y} = \frac{M_{y=0}}{M}.$$

Mass, moments, and centers of mass are defined similarly for three-dimensional solids using triple integrals.

Cylindrical Coordinates (Section 12.8)

For a point P in three-dimensional space, the *cylindrical coordinates* of P have the form (r, θ, z), where r and θ are the polar coordinates of the projection P' of P in the xy-plane, and z is the vertical signed distance of the point from the xy-plane. Cylindrical coordinates are related to rectangular coordinates by the formulas

$$x = r\cos(\theta), \quad y = r\sin(\theta), \quad x^2 + y^2 = r^2.$$

The volume element dV in cylindrical coordinates is $r\, dzdrd\theta$.

Spherical Coordinates (Section 12.8)

For a point P in three-dimensional space, spherical coordinates have the form (ρ, ϕ, θ), where ρ is the distance of the point from the origin, ϕ is the angle that the ray $\overrightarrow{OP}$ makes with the positive z-axis, and θ is angular polar coordinate of the projection P' of P in the xy-plane. Spherical coordinates are related to rectangular coordinates by the formulas

$$x = \rho\cos(\theta)\sin(\phi), \qquad y = \rho\sin(\theta)\sin(\phi), \qquad z = \rho\cos(\phi)$$

and

$$\rho = \sqrt{x^2 + y^2 + z^2}.$$

When symmetry about the z-axis is present, it is often easiest to integrate using cylindrical coordinates. The volume element in cylindrical coordinates is $r\,dr d\theta dz$. When symmetry about the origin is present, it is often easiest to integrate using spherical coordinates. The volume element in spherical coordinates is $\rho^2 \sin(\phi)\, d\rho d\phi d\theta$.

Review Exercises for Chapter 12

In Exercises 1–3, use a Riemann sum with $N = 2$, and $\{\xi_{i,j}\}$ the midpoints of the four resulting subrectangles of $\mathcal{R}$, to approximate the volume of the solid that lies under the graph of f and above the rectangle $\mathcal{R}$ in the xy-plane.

1. $f(x, y) = 3x - y$ $\qquad \mathcal{R} = [0, 3] \times [1, 2]$
2. $f(x, y) = xy^2 - x$ $\qquad \mathcal{R} = [1, 2] \times [0, 1]$
3. $f(x, y) = x/y$ $\qquad \mathcal{R} = [2, 6] \times [4, 8]$

In each of Exercises 4–6, calculate $\mathcal{I}(x) = \int_c^d f(x, y)\, dy$ and $\mathcal{J}(y) = \int_a^b f(x, y)\, dx$ for the given function f and the rectangle $\mathcal{R} = [a, b] \times [c, d]$ in the xy-plane.

4. $f(x, y) = xe^{xy^2}$ $\qquad \mathcal{R} = [0, 2] \times [0, 2]$
5. $f(x, y) = \sin(2x - y)$ $\qquad \mathcal{R} = [0, \pi] \times [0, 2\pi]$
6. $f(x, y) = yx^2 - x$ $\qquad \mathcal{R} = [0, 3] \times [0, 2]$

In each of Exercises 7–9, evaluate the double integral by converting it to an iterated integral.

7. $\iint_{\mathcal{R}} y^3 - x\, dA$ $\qquad \mathcal{R} = [1, 4] \times [-2, 1]$
8. $\iint_{\mathcal{R}} \sin(x)\cos(y)\, dA$ $\qquad \mathcal{R} = [\pi, 3\pi/2] \times [\pi/2, \pi]$
9. $\iint_{\mathcal{R}} e^{x+y}\, dA$ $\qquad \mathcal{R} = [0, 1] \times [0, 2]$

In each of Exercises 10–12, evaluate the double integral by converting to an iterated integral in the order $\iint_{\mathcal{R}} dxdy$ and also by converting to an iterated integral in the order $\iint_{\mathcal{R}} dydx$. The result of each calculation should of course give the same answer.

10. $\iint_{\mathcal{R}} y^3 x\, dA$ $\qquad \mathcal{R} = [0, 1] \times [1, 3]$
11. $\iint_{\mathcal{R}} \cos(x - y)\, dA$ $\qquad \mathcal{R} = [0, \pi/2] \times [\pi/2, 3\pi/2]$
12. $\iint_{\mathcal{R}} e^{2x-y}\, dA$ $\qquad \mathcal{R} = [0, 2] \times [0, 1]$

In each of Exercises 13–15, draw a sketch of the region $\mathcal{R}$. State whether the region is x-simple, y-simple, or both. If $\mathcal{R}$ is x-simple, then exhibit the functions $y \mapsto \alpha_1(y)$ and $y \mapsto \alpha_2(y)$ that describe the boundary of $\mathcal{R}$. Perform the analogous task when $\mathcal{R}$ is y-simple.

13. The region between the curves $y = x^2 - 2$ and $y = x + 1$
14. The region between the curves $x = y^2$ and $x = y$
15. The region between the curves $x = y^2$ and $x = 2y^{1/2}$

In each of Exercises 16–18, evaluate the iterated integral.

16. $\int_{-1}^{1} \int_{x^2}^{x^4+6} xy^2 - y\, dydx$
17. $\int_0^4 \int_{x/3}^{3x} yx^{1/3}\, dydx$
18. $\displaystyle\int_1^{27} \int_{y^{1/4}}^{y^{1/3}} \frac{y}{x}\, dxdy$

In each of Exercises 19–21, calculate the integral of the function f over the region $\mathcal{R}$.

19. $f(x, y) = 5y - 3x$, $\mathcal{R}$ is the region bounded by $x = y^2$ and $x = 2y$.
20. $f(x, y) = \sqrt{x}\sqrt{y}$, $\mathcal{R}$ is the region bounded on the left and right by $x = 2$ and $x = 4$ and above and below by $y = x^2 + 4$ and $y = x^2 - 2$.
21. $f(x, y) = x$, $\mathcal{R}$ is the region bounded by $y = \cos(x)$ and $y = \sin(x)$ for $0 \le x \le \pi/4$

In each of Exercises 22–24, switch the order of integration to make the iterated integral calculable, and then perform the integration.

22. $\int_0^{\pi/2} \int_y^{\pi/2} \cos(x)/x\, dxdy$
23. $\int_0^1 \int_y^1 e^{-x^2}\, dxdy$
24. $\displaystyle\int_1^4 \int_1^{y^2} \frac{y}{x}\, dxdy$

In each of Exercises 25–27, calculate the volume of the solid that lies below the graph of $z = f(x, y)$ and over the region $\mathcal{R}$ in the xy-plane.

25. $f(x, y) = x^2 + 2y^2 + 4$, $\mathcal{R} = \{(x, y) : -1 \le x \le 1, -2 \le y \le 2\}$
26. $f(x, y) = 5 - x + y$, $\mathcal{R} = \{(x, y) : x^2 + y^2 \le 1\}$
27. $f(x, y) = x^2 - y^2 + 8$, $\mathcal{R} = \{(x, y) : |x| \le y \le 4\}$

In each of Exercises 28–30, calculate the volume of the solid $\mathcal{U}$.

28. $\mathcal{U}$ is bounded by the paraboloids $y = z^2 + x^2 + 4$ and $y = -z^2 - x^2 - 2$.

29. $\mathcal{U}$ is bounded on the sides by the cylinder $x^2+4y^2=4$, above by $z=x+y+6$, and below by $z=0$.
30. $\mathcal{U}$ is lies below the graph of $x-3y+2z=6$ and in the fourth octant.

In each of Exercises 31–33, plot the given points on a polar coordinate graph.

31. $(2,\pi/4)$, $(-4,2\pi/3)$, $(6,-\pi)$, $(1,3\pi/2)$
32. $(-5,9\pi/2)$, $(-3,-2\pi/3)$, $(4,11\pi/3)$, $(-2,7\pi/2)$
33. $(4,0)$, $(-4,0)$, $(0,7)$, $(0,-7)$

In each of Exercises 34–36, you are given a point in polar coordinates. Give the rectangular coordinates of the point.

34. $(2,5\pi/4)$
35. $(-3,-9\pi/2)$
36. $(5,11\pi/4)$

In each of Exercises 37–39, you are given rectangular coordinates for a point. Give all possible polar coordinates for that same point.

37. $(7\sqrt{2},-7\sqrt{2})$
38. $(4,\pi)$
39. $(\pi,4)$

In each of Exercises 40–42, test the given polar equation for symmetry in the x-axis, in the y-axis, and in the origin.

40. $r^2=3\cos(\theta)$
41. $r=6+3\cot(\theta)$
42. $r=4-\cos(3\theta)$

In each of Exercises 43–45, sketch the polar curve.

43. $r=2-2\sin(\theta)$
44. $r=2\cos(\theta)$
45. $r=3\theta$

In each of Exercises 46–48, use formula (12.11) to calculate the area enclosed by the given curve.

46. $r=1+\cos(\theta)$
47. $r^2=\cos(\theta)$
48. $r=3\cos(2\theta)$

In each of Exercises 49–51, use formula (12.12) to calculate the area bounded by the two given curves.

49. The region outside the curve $r=1+\cos(\theta)$ and inside the circle $r=3\cos(\theta)$
50. The region in the second quadrant that is inside the circle $r=\sqrt{2}\sin(\theta)$ and outside the circle $r=\sin(\theta)$
51. The region inside the circle $r=3\sin(\theta)$ and outside the circle $r=\sin(\theta)$

In each of Exercises 52–54, calculate the area of the region $\mathcal{R}$ by using a double (or iterated) integral in polar coordinates.

52. $\mathcal{R}$ is one leaf of the rose $r=5\cos(4\theta)$
53. $\mathcal{R}$ is the region inside the circle $r=\sqrt{3}\sin(\theta)$ and outside the rose $r=\sin(2\theta)$.
54. $\mathcal{R}=\{(r,\theta):\pi/6\le\theta\le\pi/4, 1\le r\le 2\sin(\theta)\}$

In each of Exercises 55–57, use a double (or iterated) polar coordinate integral to integrate the function f over the planar region $\mathcal{R}$.

55. $f(r,\theta)=\sin(\theta)$, $\mathcal{R}$ is the region inside the circle $r=2\sin(\theta)$
56. $f(r,\theta)=\theta$, $\mathcal{R}$ is the region inside the limaçon $r=4-\sin(\theta)$
57. $f(r,\theta)=\cos(\theta)$, $\mathcal{R}$ is the region in the second quadrant that is inside the circle $r=3$ and outside the cardioid $r=1-\sin(\theta)$

In each of Exercises 58–60, use a double integral in polar coordinates to calculate the volume of the indicated solid.

58. The solid bounded by the surface $z=-2x^2-2y^2+8$ and the plane $z=1$
59. The solid bounded by the upper nappe of the cone $z^2=2x^2+2y^2$ and the plane $z=10\sqrt{2}$.
60. The solid bounded by the cylinder $(x-1)^2+y^2=1$ and the sphere $x^2+y^2+z^2=16$

In each of Exercises 61–63, perform the indicated change of variable in the double integral. Then evaluate the integral.

61. $\int_0^4\int_0^{\sqrt{y}}\sqrt{x/y}\,dx\,dy$ $\quad u^2=x \quad v^2=y$
62. $\int_0^1\int_2^3\sqrt{3+xy}\,dx\,dy$ $\quad v=xy \quad u=x$
63. $\displaystyle\int_1^4\int_2^3\frac{x^2}{y^2}\,dy\,dx$ $\quad v=1/y \quad u=x$

In each of Exercises 64–66, evaluate the iterated integral.

64. $\int_1^3\int_{y+1}^{2y+4}\int_y^{x+y}3y-xy\,dz\,dx\,dy$
65. $\int_1^e\int_{2x}^{3x}\int_1^z xy\,dy\,dz\,dx$
66. $\int_0^\pi\int_z^{\sqrt{\pi}}\int_y^0 4z\sin(y^2)\,dx\,dy\,dz$

In each of Exercises 67–69, use a triple integral to express the volume of the solid that is bounded by the given surfaces in space. Then determine the volume by calculating the integral.

67. $z=4x^2+4y^2, z=1, z=16$
68. $x=y^4, z=-x+6, z=0$
69. $z=2x^2+2y^2-4, z=x^2+y^2+4$

In each of Exercises 70–72, evaluate the iterated integral. Describe in words the solid $\mathcal{U}$ over which the integration is calculated.

70. $\int_0^4\int_0^4\int_0^{yz/4}x\,dx\,dy\,dz$
71. $\int_0^2\int_0^z\int_0^{\sqrt{4-z^2}}y\,dy\,dx\,dz$
72. $\int_0^1\int_0^{1-y/2}\int_0^{4-2x-y}y\,dz\,dx\,dy$

In each of Exercises 73–75, use a double integral to find the mass of the planar region $\mathcal{R}$ with density function $\delta(x, y)$.

73. $\mathcal{R} = \{(x, y) : 0 \le y \le 3, y + 2 \le x \le 3y + 5\}, \delta(x, y) = x + 3y$

74. $\mathcal{R} = \{(x, y) : -2 \le x \le 3, 0 \le y \le 9 - x^2\}, \delta(x, y) = 4 + y$

75. $\mathcal{R} = \{(x, y) : -1 \le y \le 1, -y^2 - 2 \le x \le y^2 + 1\}$, $\delta(x, y) = x^2 + y$

In each of Exercises 76–78, find the first moment of the region $\mathcal{R}$ with density function $\delta(x, y)$ about the indicated axis.

76. $\mathcal{R} = \{(x, y) : -2 \le y \le -1, 1 \le x \le 4\}, \delta(x, y) = -y/x$, x-axis

77. $\mathcal{R} = \{(x, y) : 0 \le y \le 2, 0 \le x \le 2y^2\}, \delta(x, y) = \sqrt[3]{y}$, x-axis

78. $\mathcal{R} = \{(x, y) : 2 \le x \le 3, x^{1/3} \le y \le x^{1/2}\}, \delta(x, y) = 2$, y-axis

In each of Exercises 79–81, find the center of mass of the planar region $\mathcal{R}$ with density function $\delta(x, y)$.

79. $\mathcal{R} = \{(x, y) : 1 \le y \le 2, \sqrt{y} \le x \le y\}, \delta(x, y) = y + 1$

80. $\mathcal{R} = \{(x, y) : 1 \le x \le 2, 3 \le y \le 5\}, \delta(x, y) = 12 - x - y$

81. $\mathcal{R} = \{(x, y) : -2 \le x \le 1, x + 2 \le y \le 2x + 5\}, \delta(x, y) = y$

In each of Exercises 82–84, calculate the mass of the solid body $\mathcal{U}$ having density function $\delta(x, y, z)$.

82. $\mathcal{U} = \{(x, y, z) : 1 \le x \le 3, 2 \le y \le 4, x + y \le z \le 8\}$, $\delta(x, y, z) = y + x + 4$

83. $\mathcal{U} = \{(x, y, z) : 0 \le z \le 3, 1 \le x \le 4, 2 \le y \le 5\}$, $\delta(x, y, z) = \sqrt{x}$

84. $\mathcal{U} = \{(x, y, z) : 1 \le x \le 4, 2 \le z \le 3, 0 \le y \le x + 2z\}, \delta(x, y, z) = 2x + y + 3z$

In each of Exercises 85–87, calculate the three first moments and the center of mass of the given solid in space.

85. $\mathcal{U} = \{(x, y, z) : 0 \le x \le 1, 0 \le y \le 2, 0 \le z \le y + 1\}$, $\delta(x, y, z) = y + 5$

86. $\mathcal{U} = \{(x, y, z) : x^2 + 4y^2 \le 4, 0 \le z \le 2\}$, $\delta(x, y, z) = x^2 + y^2 + 2z^2$

87. $\mathcal{U} = \{(x, y, z) : 2 \le z \le 3, 1 \le x \le 2, 1 \le y \le x\}$, $\delta(x, y, z) = z$

In each of Exercises 88–90, a point P is given in rectangular coordinates. Calculate cylindrical coordinates for P.

88. $P = (2, 0, 3)$

89. $P = (-1, 2, \sqrt{3})$

90. $P = (-2, \sqrt{2}, -\sqrt{2})$

In each of Exercises 91–93, a point P is given in rectangular coordinates. Calculate spherical coordinates for P.

91. $P = (0, 0, 5)$

92. $P = (\sqrt{3}, \sqrt{3}, 3\sqrt{2})$

93. $P = (-3, 3\sqrt{3}, 6)$

In each of Exercises 94–96, either rectangular or spherical or cylindrical coordinates are specified for the point P. Calculate coordinates of the point in the other two coordinate systems.

94. $P = (5, \pi/4, 3\pi/4)$ spherical

95. $P = (-2\sqrt{2}, 2\sqrt{2}, 4)$ rectangular

96. $P = (8, -\pi/3, -8)$ cylindrical

In each of Exercises 97–99, integrate the function f over the solid $\mathcal{U}$ using cylindrical coordinates.

97. $f(x, y, z) = x^2 + y^2$, $\mathcal{U}$ is the region below the paraboloid $z = 4 - 2x^2 - 2y^2$ and above the plane $z = 2$.

98. $f(x, y, z) = 1/(x^2 + y^2)$, $\mathcal{U}$ is the region bounded by the cylinders $x^2 + y^2 = 8$ and $x^2 + y^2 = 16$ and by the planes $z = 6$ and $z = 1$

99. $f(x, y, z) = x^2 + y^2$, $\mathcal{U}$ is the solid inside the cylinder $x^2 + y^2 = 1$ and bounded by $z = x^2 + y^2$ and $z = (x^2 + y^2)^{1/4}$

In each of Exercises 100–102, integrate the function f over the solid $\mathcal{U}$ using spherical coordinates.

100. $f(x, y, z) = 8, \mathcal{U} = \{(x, y, z) : x^2 + y^2 + z^2 \le 4\}$

101. $f(x, y, z) = x^2 + y^2 + z^2$, $\mathcal{U}$ is the solid inside the sphere $\{(x, y, z) : x^2 + y^2 + z^2 = 4\}$ and above the xy-plane

102. $f(x, y, z) = (1 + x^2 + y^2 + z^2)^{-1}$, $\mathcal{U}$ is the solid above the cone $z = 4\sqrt{x^2 + y^2}$ and below the sphere $x^2 + y^2 + z^2 = 16$

The double integral was introduced by Euler in 1769. Four years later, Lagrange used triple integrals to solve a problem of mathematical physics. Both of these mathematicians discovered the method for changing variables in an integral. In particular, Lagrange obtained the formula for integration in spherical coordinates and used it to study gravitational attraction. The general theory of changing variables in a multiple integral was developed by the Ukrainian mathematician, Mikhail Vasilevich Ostrogradskii (1801–1862), in 1836. However, Ostrogradskii's work did not receive wide dissemination in the West. Thus when Carl Gustav Jacobi covered much the same ground in 1841, he received credit for the "Jacobian," a determinant that already had appeared in the neglected papers of Ostrogradskii.

Exotic Functions

When 19th century mathematicians enlarged the concept of function to include any correspondence of real numbers, they opened the door to some exceedingly complicated functions. For some time, it was believed that the assumption of continuity would be enough to exclude the functions that are beyond the realm of differential calculus. Indeed, it is difficult to imagine a continuous function that has a graph without *any* smoothness. However, many efforts by notable mathematicians such as André Marie Ampère (1775–1836) failed to demonstrate that a continuous function must be differentiable on a "substantial" set. Eventually, some mathematicians began to suspect that continuity does not necessarily entail any differentiability.

The first "everywhere-continuous nowhere-differentiable" function was conceived by Bernhard Bolzano (1781–1848) in 1830. Because communication beyond the established mathematical centers was then very spotty, Bolzano's work remained unknown until 1922. Around 1860, the Swiss mathematician, Charles Cellérier (1818–1890), also constructed an everywhere-continuous nowhere-differentiable function,

$$C(x) = \sum_{n=1}^{\infty} \frac{\sin(2^n x)}{2^n}.$$

However, Cellérier's function was posthumously published after a delay of thirty years, and by that time, his method of construction had become old hat.

Indeed, at more or less the same time that Cellérier was concocting his novel function, Bernhard Riemann (1826–1866) was thinking along the same lines. Some time before 1861, Riemann defined and studied the everywhere-continuous function

$$R(x) = \sum_{n=1}^{\infty} \frac{\sin(n^2 x)}{n^2}.$$

Although Riemann's function is differentiable at some points, it is nondifferentiable at infinitely many points of any open interval. The graph of R over the interval $[0, 1]$ is shown in Figure 1. The plotted curve does not seem too irregular in some places, such as the small rectangle drawn in Figure 1, but when we zoom in, as in Figure 2, we find that the apparent smoothness is an illusion.

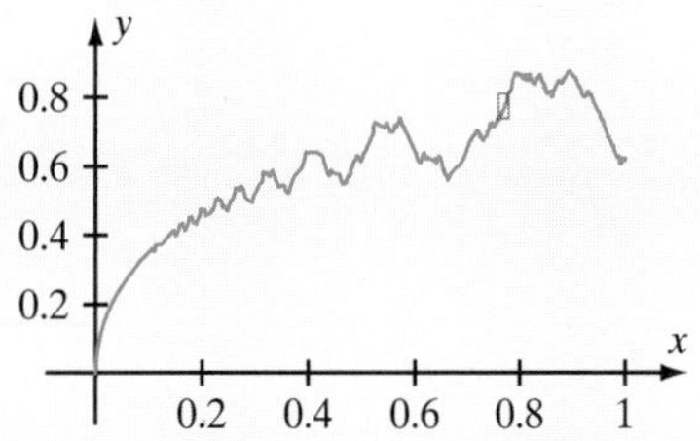

▲ Figure 1 The graph of Riemann's function $R(x) = \sum_{n=1}^{\infty} \frac{\sin(n^2 x)}{n^2}$.

There is no record that any mathematician other than Karl Weierstrass (1815–1897) was aware of Riemann's example. Weierstrass took Riemann's construction a step further and, in 1872, announced that

$$W(x) = \sum_{n=0}^{\infty} \frac{\cos(3^n \pi x)}{2^n}$$

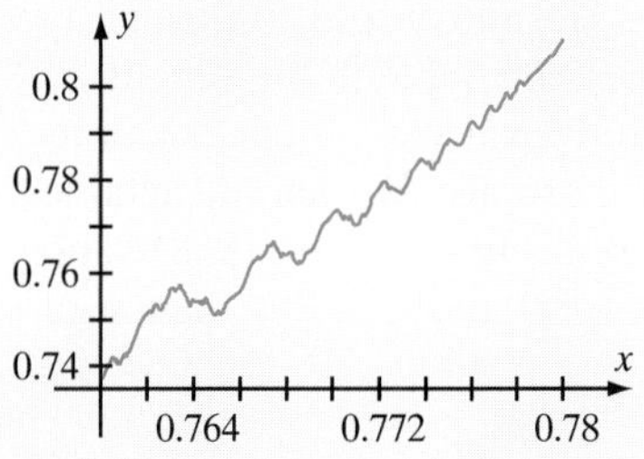

▲ Figure 2 A detail of the graph of Riemann's function R, obtained by zooming in on the small viewing window shown in Figure 1

is an everywhere-continuous nowhere-differentiable function. Weierstrass's function is plotted over the interval $[0, 1]$ in Figure 3. A detail appears in Figure 4.

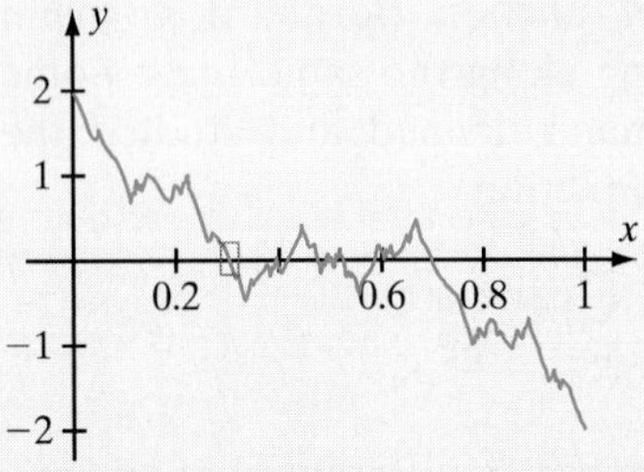

▲ Figure 3 The graph of Weierstrass's everywhere-continuous nowhere-differentiable function on the interval [0, 1]

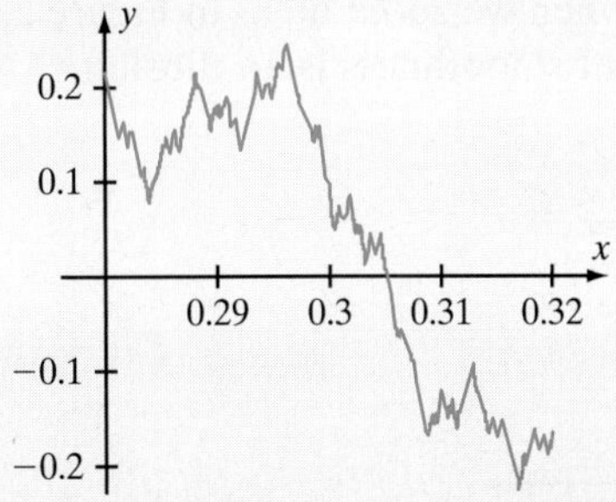

▲ Figure 4 A detail of the graph of Weierstrass's function, obtained by zooming in on the small rectangle drawn in Figure 3.

In 1873, Gaston Darboux (1842–1917) defined yet another nondifferentiable continuous function,

$$G(x) = \sum_{n=1}^{\infty} \frac{\sin(x\ (n+1)!)}{n!},$$

and it was far from the last. By the end of the 19th century, such pathological functions had become so commonplace that Charles Hermite wrote, "I turn away with fright and horror from this lamentable plague of functions that do not have derivatives." Such reactions notwithstanding, the very existence of such intractable functions prompted Ulisse Dini (1845–1918) and others to undertake a deeper study of the theory of differentiation.

The Lebesgue Integral

As badly behaved as they are for differential calculus, the continuous functions that we have been discussing pose no difficulties for the Riemann integral. However, as mathematicians became familiar with ever more troublesome functions, they discovered some undesirable limitations of the Riemann integral. Above all, the Riemann integral does not "respect" some of the most fundamental processes of calculus.

Let's consider one situation in which the Riemann integral comes up short. Set $S_1 = \{0, 1\}$, $S_2 = \{0, 1/2, 1\}$, $S_3 = \{0, 1/3, 1/2, 2/3, 1\}$, and, in general,

$$S_n = S_{n-1} \cup \left\{\frac{1}{n}, \frac{2}{n}, \frac{3}{n}, \ldots, \frac{n-1}{n}\right\}.$$

We define f_n on the interval $[0, 1]$ by the formula

$$f_n(x) = \begin{cases} 0 & \text{if } x \in S_n \\ 1 & \text{if } x \notin S_n \end{cases}.$$

The graphs of f_4 and f_5 are shown in Figure 5. In general, the graph of f_n is obtained from the graph of f_{n-1}

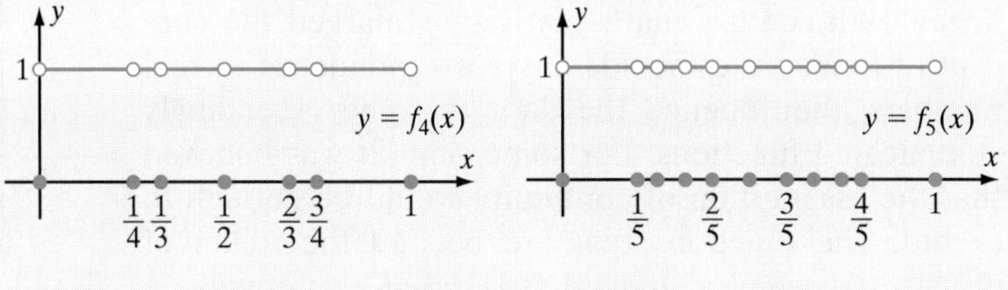

▲ Figure 5

by moving at most $n-1$ points from the line $y = 1$ to the line $y = 0$. Next, define the function f on $[0, 1]$ by

$$f(x) = \begin{cases} 0 & \text{if } x \text{ is a rational number} \\ 1 & \text{if } x \text{ is an irrational number.} \end{cases}$$

It is not hard to deduce that $f_n(x) \to f(x)$ for every x. Moreover, this convergence is as nice as can be: The values $\{f_n(x)\}$ decrease to $f(x)$, and the graphs of all the functions are contained in the square $[0, 1] \times [0, 1]$. Under these conditions, it is highly desirable that the limit function f have an integral and that we can interchange limit and integral:

$$\lim_{n\to\infty} \int_0^1 f_n(x)\,dx = \int_0^1 \left(\lim_{n\to\infty} f_n(x)\right) dx = \int_0^1 f(x)\,dx.$$

In the case of the sequence we have just constructed, however, the limit function f does not have a Riemann integral. No matter how many subintervals of $[0, 1]$ we take, some of the Riemann sums of f will be 1, and some will be 0.

Double integrals highlight another weakness of the Riemann integral. In a first course in calculus, we limit our attention to *simple* planar regions. These regions are enclosed by a finite number of well-behaved curves. By the 1880s, serious attempts were underway to understand the integral $\iint_{\mathcal{R}} f(x, y)\, dA$ when $\mathcal{R}$ is a very irregular set of points in the plane. Simple, intuitive concepts of *area* proved to be inadequate for the irregular planar point sets being considered. Therefore Camille Jordan (1838–1922) and Giuseppe Peano (1858–1932) initiated a program of *measuring* such complicated point sets.

Iterated Riemann integrals can also be problematic. In the 1870s, mathematicians discovered that changing the order of integration does not always result in the same answers for some discontinuous integrands of two variables. Additional concerns arose. As we have learned, the boundary of a planar region plays an important role when we calculate a double integral by means of an iterated integral. For a simple planar region, the boundary is "infinitely thin" and therefore has zero area. Yet, in 1890, Peano constructed a continuous curve that passes through every point of the square $[0, 1] \times [0, 1]$. Such *space-filling curves* added to the urgency of coming to a better understanding of multiple integrals.

Jordan's theory of measure did not prove to be the right instrument for resolving all the sore points that beset the theory of integration. However, in the 1890s, Emile Borel (1871–1956) published refinements to Jordan's theory, and those refinements guided the way to a successful resolution. That step was accomplished by Henri Lebesgue (1875–1941), who introduced a new integral in his doctoral thesis, which was published in 1902. Within a few years of its introduction, the *Lebesgue integral* became one of the primary tools of mathematical analysis. It revitalized the study of trigonometric series, paved the way for the study of function spaces, and played a fundamental role in the axiomatic treatment of probability theory.

CHAPTER 13

Vector Calculus

PREVIEW

Imagine a fluid flowing through a planar region. Because velocity has both magnitude and direction, the velocity of the flow at various points can be indicated by vectors. Suppose that we want to concentrate our attention on the flow in a certain region. In this chapter, we will learn to relate the flow through the boundary of the region to the flow in the interior of the region. In the end, we will see that these physical considerations are related to a remarkable multidimensional version of the Fundamental Theorem of Calculus.

This chapter begins by developing the mathematical tools that are the basis of vector calculus. It is plain that a number of new ideas are needed. After we have learned the necessary tools in the first few sections, we will come to the theorems—Green's Theorem, Stokes's Theorem, and the Divergence Theorem—that give precise formulations of the relationship between the behavior of a vector field on the boundary of a region in the plane or in space and its behavior in the interior.

First we need functions that describe flows. Such a function would have to assign a direction and a magnitude to each point in the region; in other words, it would have to be a vector-valued function on a planar or spatial region. Next, we will need to develop certain integrals that involve vector-valued functions. They will enable us to integrate over curves and surfaces.

Also, if we are going to use information on the boundary curve or surface to obtain information inside the curve or surface, then we need a mathematical way to tell the inside from the outside: This distinction will be closely connected to the idea of "orienting" a surface.

Finally, we will need vector operations that help to measure the flow properties of a vector-valued function. Together with the gradient (which we have already seen in Chapter 11), we will use two new operations called *divergence* and *curl*.

13.1 Vector Fields

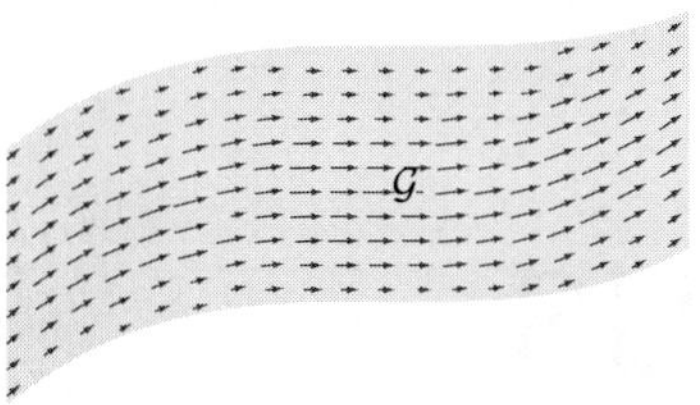

▲ Figure 1

Suppose that a fluid passes through a region $\mathcal{G}$ in the plane or in space. At each point P, we may use a vector $\mathbf{v}(P)$ to represent the magnitude and direction of the velocity of the flow. Several of these vectors are sketched in Figure 1. By associating a vector $\mathbf{v}(P)$ to each point P in $\mathcal{G}$, we are in effect creating a vector-valued function with domain $\mathcal{G}$. Such a function is said to be a *vector field* on the region $\mathcal{G}$. Let us make these concepts precise.

DEFINITION We say that a planar set $\mathcal{G}$ is *open* if each of its points is the center of a disk with positive radius that is contained in $\mathcal{G}$. A set $\mathcal{G}$ in three-dimensional space is *open* if each of its points is the center of a ball with positive radius that is contained in $\mathcal{G}$. A set $\mathcal{G}$ in the plane or in space is said to be *path-connected* if every pair of points in $\mathcal{G}$ can be connected by a continuous curve that lies entirely in $\mathcal{G}$. A set that is both open and path-connected is called a *region*.

Figure 2 shows a set that is both open and path-connected: It is a region. Figure 3 shows an open set $\mathcal{G}$ that comprises two components. This set is not a region because it is not path-connected: A point in one component cannot be connected to a point in the other component by means of a continuous curve that lies entirely within $\mathcal{G}$. In Figure 4, a new set $\mathcal{S}$ is formed by connecting the two components of $\mathcal{G}$ with a line segment $\mathcal{L}$. The set $\mathcal{S}$ *is* path-connected, but it is still *not* a region because it is not an open set: Each point of $\mathcal{L}$ is in $\mathcal{S}$ but is not the center of any disk that is contained in $\mathcal{S}$.

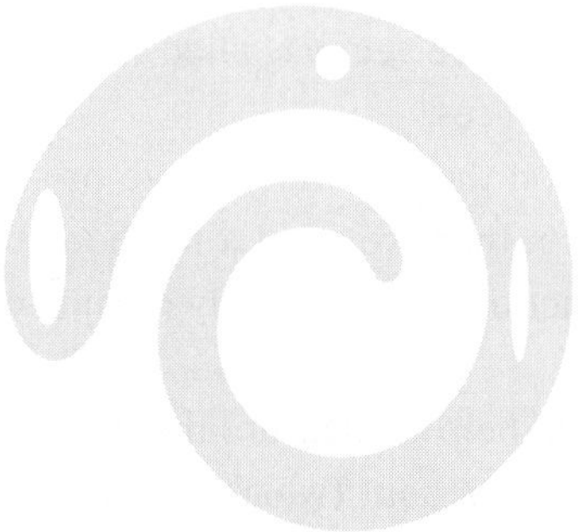

▲ Figure 2 A set that is open and path-connected: It is a region.

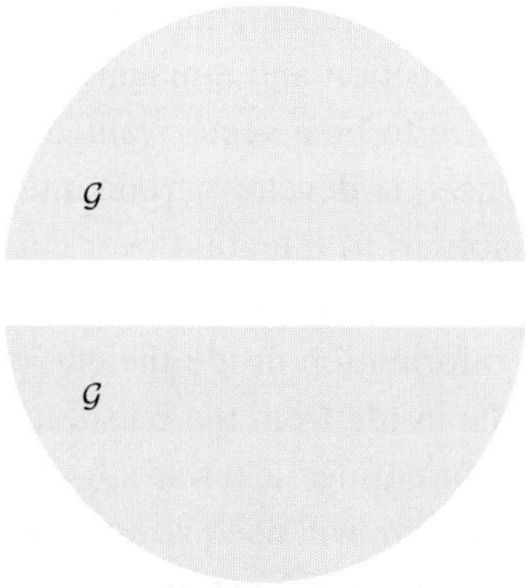

▲ Figure 3 The set $\mathcal{G}$ comprises two components. $\mathcal{G}$ is open but not path-connected; it is *not* a region.

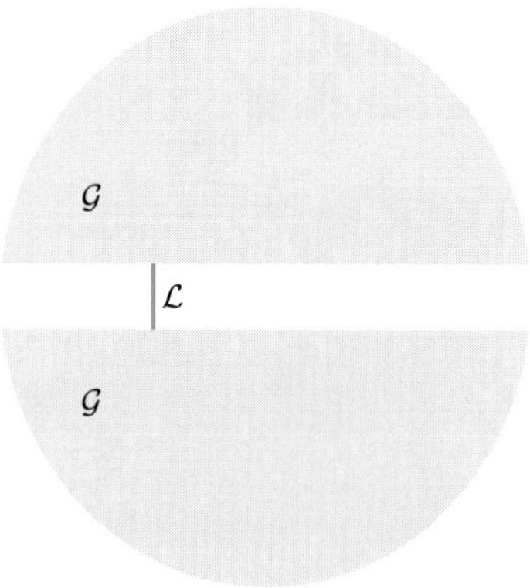

▲ Figure 4 The set $\mathcal{S} = \mathcal{G} \cup \mathcal{L}$ is path-connected but not open; it is *not* a region.

DEFINITION A *vector field* $\mathbf{v}$ on a region $\mathcal{G}$ is a function that assigns to each point of $\mathcal{G}$ a vector. If $\mathcal{G}$ is a planar region, then the values of $\mathbf{v}$ are planar vectors. That is, $\mathbf{v}$ has the form $\mathbf{v}(x, y) = v_1(x, y)\mathbf{i} + v_2(x, y)\mathbf{j}$ for each point (x, y) in $\mathcal{G}$. If $\mathcal{G}$ is a region in space, then the values of $\mathbf{v}$ are vectors in space, and $\mathbf{v}$ has the form $\mathbf{v}(x, y, z) = v_1(x, y, z)\mathbf{i} + v_2(x, y, z)\mathbf{j} + v_3(x, y, z)\mathbf{k}$ for each point (x, y, z) in $\mathcal{G}$.

INSIGHT In Chapter 10, we used vector-valued functions $t \mapsto \mathbf{r}(t)$ of a scalar variable t to parameterize curves. A vector field $P \mapsto \mathbf{v}(P)$ is also a vector-valued function. It differs from a curve in that its domain consists not of scalars but of points P that have two or three coordinates.

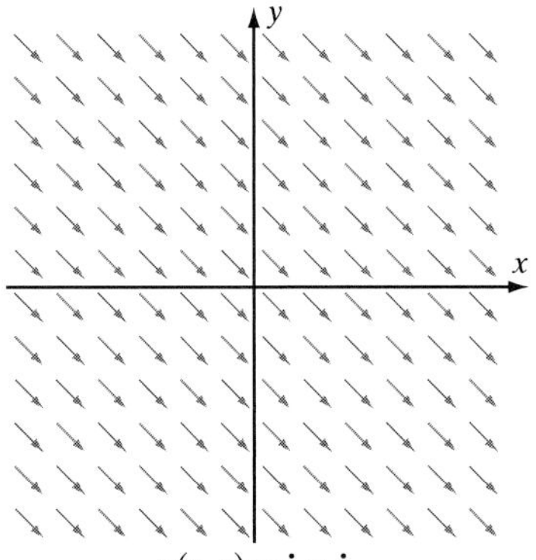

$\mathbf{v}(x, y) = \mathbf{i} - \mathbf{j}$

▲ Figure 5

▶ **EXAMPLE 1** Let $\mathcal{G}$ denote the entire plane (which is a two-dimensional region). Sketch the vector fields $\mathbf{v}(x, y) = \mathbf{i} - \mathbf{j}$ and $\mathbf{w}(x, y) = x\mathbf{i} - (y/4)\mathbf{j}$, which are defined on $\mathcal{G}$.

Solution The vector field $\mathbf{v}$ assigns to each point (x,y) in the plane the vector $\mathbf{i} - \mathbf{j}$. It is a constant vector field. A sketch is indicated in Figure 5. Note that we make a sketch by drawing *some* of the vectors (if we drew them all, the picture would just be solid ink). Now consider the vector field $\mathbf{w}$. This is a vector field that assigns to the point (x, y) in the plane the vector $x\mathbf{i} - (y/4)\mathbf{j}$. Some of the values of this vector field are indicated in Table 1.

(x, y)	$(1,0)$	$(0,1)$	$(-1,0)$	$(0,-1)$	$(1,1)$	$(-1,1)$	$(1,-1)$	$(-1,-1)$	$(2,2)$	$(-2,2)$
$w(x, y)$	$\mathbf{i}$	$-\frac{1}{4}\mathbf{j}$	$-\mathbf{i}$	$\frac{1}{4}\mathbf{j}$	$\mathbf{i} - \frac{1}{4}\mathbf{j}$	$-\mathbf{i} - \frac{1}{4}\mathbf{j}$	$\mathbf{i} + \frac{1}{4}\mathbf{j}$	$-\mathbf{i} + \frac{1}{4}\mathbf{j}$	$2\mathbf{i} - \frac{1}{2}\mathbf{j}$	$-2\mathbf{i} - \frac{1}{2}\mathbf{j}$

Table 1

These values of the vector field $\mathbf{w}$, and a few others, are shown in Figure 6. ◀

In previous chapters, we have seen that it is convenient to deal with functions that are continuously differentiable. These functions have much nicer properties than arbitrary functions. The same is true for vector fields. As a result, we will deal primarily with continuously differentiable vector fields. We call a planar vector field $\mathbf{v}(x, y) = v_1(x, y)\mathbf{i} + v_2(x, y)\mathbf{j}$ or a spatial vector field $\mathbf{w}(x, y, z) = w_1(x, y, z)\mathbf{i} + w_2(x, y, z)\mathbf{j} + w_3(x, y, z)\mathbf{k}$ *continuously differentiable* if each of the component functions v_1, v_2 or w_1, w_2, w_3 is continuously differentiable.

Vector Fields in Physics

Vector fields arise naturally in many physical contexts. For example, gravitational, magnetic, and electrostatic forces generate vector fields that are called force fields. Another example from physics that will be important for us is the velocity vector field.

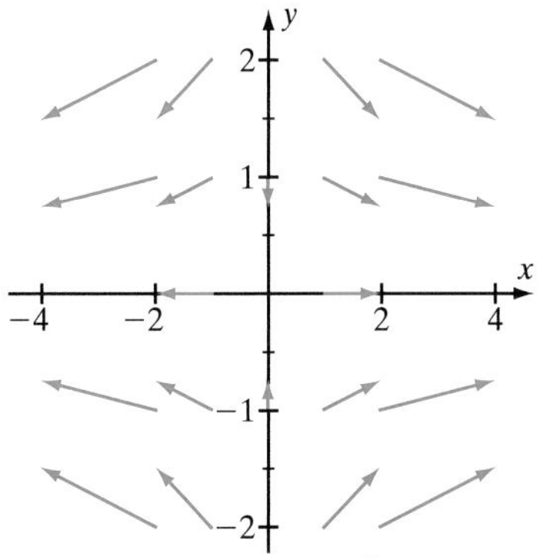

$\mathbf{w}(x, y) = x\mathbf{i} - (y/4)\mathbf{j}$

▲ Figure 6

▶ **EXAMPLE 2** Let $\mathbf{v}$ denote the velocity of a wind that blows through a wind tunnel parallel to the sides of the tunnel. Suppose that, at each point $P = (x, y, z)$, the magnitude of $\mathbf{v}(P)$ is proportional to the height of P above the ground. Assume that the ground is at height $z = 0$ and that the vector $\mathbf{i}$ points down the tunnel in the direction that the wind is blowing (see Figure 7). Describe the velocity as a vector field.

Solution We let c be the constant of proportionality. At $P = (x, y, z)$, the magnitude of $\mathbf{v}(P)$ is cz, and the direction of $\mathbf{v}(P)$ is $\mathbf{i}$. The vector field we seek is therefore $\mathbf{v}(x, y, z) = cz\mathbf{i}$. ◀

▶ **EXAMPLE 3** Let ρ be a positive constant. Suppose that a solid ball $\mathcal{B} = \{(x, y, z) : x^2 + y^2 + z^2 \le \rho^2\}$ is rotating about the z-axis with constant angular

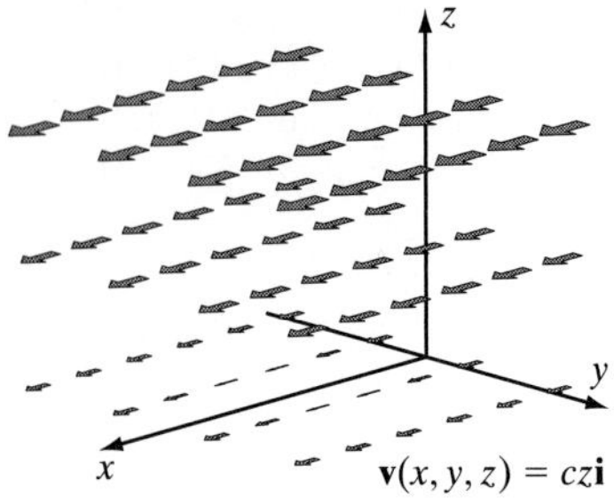

▲ Figure 7

speed ω. Looking down at the xy-plane from the point $(0, 0, 2\rho)$ on the positive z-axis, the rotation of $\mathcal{B}$ is counterclockwise. Show that there is a vector $\mathbf{\Omega}$ such that the velocity vector field $P \mapsto \mathbf{v}(P)$ is given by $\mathbf{v}(P) = \mathbf{\Omega} \times \overrightarrow{OP}$ for each point $P = (x, y, z)$ in $\mathcal{B}$. (The vector $\mathbf{\Omega}$ is called the angular velocity.)

Solution Each point $P = (x, y, z)$ in $\mathcal{B}$ traces out a circle $\mathcal{C}_P$ in a horizontal plane. The center of the circle is the point $A = (0, 0, z)$ on the z-axis. The radius vector from A to P is $\mathbf{r} = \overrightarrow{AP} = \langle x - 0, y - 0, z - z \rangle = \langle x, y, 0 \rangle$. Because P traces out $\mathcal{C}_P$, the velocity vector $\mathbf{v}(P)$ at $P = (x, y, z)$ is tangent to $\mathcal{C}_P$. Because $\mathcal{C}_P$ is a circle, we deduce that $\mathbf{v}(P)$ is perpendicular to the radius vector $\mathbf{r} = \langle x, y, 0 \rangle$ of $\mathcal{C}_P$ (see Figure 8). It follows that the direction, $\mathbf{dir}(\mathbf{v}(P))$, of $\mathbf{v}(P)$ is equal to $\|\mathbf{r}\|^{-1}\langle -y, x, 0 \rangle$. Letting $v(P)$ denote the magnitude of $\mathbf{v}(P)$, we obtain

$$\mathbf{v}(P) \overset{(9.2.7)}{=} \|\mathbf{v}(P)\| \, \mathbf{dir}\big(\mathbf{v}(P)\big) = v(P)\|\mathbf{r}\|^{-1}\langle -y, x, 0 \rangle.$$

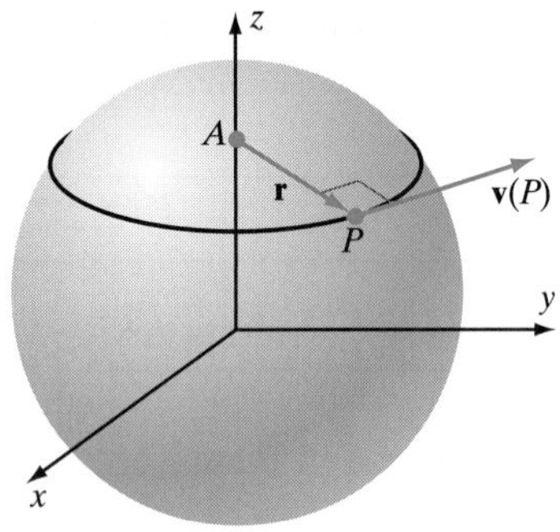

▲ Figure 8 The velocity vector $\mathbf{v}(P)$ at $P = (x, y, z)$ is perpendicular to vector $\mathbf{r} = \langle x, y, 0 \rangle$.

To determine $v(P)$, notice that the arc length Δs that P traces during a time interval Δt is equal to the product of the radius and the central angle subtended by the traced arc. That is, $\Delta s = \|\mathbf{r}\|(\omega \Delta t)$. This equation tells us that

$$v(P) = \frac{ds}{dt} = \lim_{\Delta t \to 0} \frac{\Delta s}{\Delta t} = \lim_{\Delta t \to 0} \frac{\|\mathbf{r}\|(\omega \Delta t)}{\Delta t} = \omega \|\mathbf{r}\|.$$

Substituting this value for $v(P)$ into the formula for $\mathbf{v}(P)$, we obtain

$$\mathbf{v}(P) = (\omega\|\mathbf{r}\|)\|\mathbf{r}\|^{-1}\langle -y, x, 0 \rangle = \omega\langle -y, x, 0 \rangle = -\omega y\mathbf{i} + \omega x\mathbf{j} + 0\mathbf{k}.$$

We may now verify that $\mathbf{\Omega} = \omega\mathbf{k}$ is a vector that satisfies the requirements:

$$\mathbf{\Omega} \times \overrightarrow{OP} = \det\left(\begin{bmatrix} \mathbf{i} & \mathbf{j} & \mathbf{k} \\ 0 & 0 & \omega \\ x & y & z \end{bmatrix}\right) = -\omega y\mathbf{i} + \omega x\mathbf{j} + 0\mathbf{k} = \mathbf{v}(P). \quad ◀$$

▶ **EXAMPLE 4** A mass is located at the point $P_0 = (x_0, y_0, z_0)$. It exerts a gravitational attraction $\mathbf{F}(P)$ at each point $P \neq P_0$ in space. At P, the direction of the gravitational force is from P toward P_0 and, according to Newton's law of gravitation, the magnitude of the force is proportional to the reciprocal of the square of the distance of P to P_0. Using α as the proportionality constant, express $\mathbf{F}(P)$ as a vector field.

Solution Let $\mathbf{r} = \overrightarrow{OP}$ and $\mathbf{r}_0 = \overrightarrow{OP_0}$. The distance of $P = (x, y, z)$ to P_0 is

$$\|\mathbf{r} - \mathbf{r}_0\| = \Big((x - x_0)^2 + (y - y_0)^2 + (z - z_0)^2\Big)^{1/2}.$$

Newton's law tells us that the magnitude of the gravitational force vector is $\alpha\|\mathbf{r} - \mathbf{r}_0\|^{-2}$. The gravitational force vector points in the same direction as $\overrightarrow{PP_0} = \mathbf{r}_0 - \mathbf{r}$. Because the direction vector of $\overrightarrow{PP_0}$ *is* $(-1/\|\mathbf{r} - \mathbf{r}_0\|)(\mathbf{r} - \mathbf{r}_0)$, we conclude that

$$\mathbf{F}(P) = \|\mathbf{F}(P)\| \, \mathbf{dir}\big(\mathbf{F}(P)\big) = \alpha\|\mathbf{r} - \mathbf{r}_0\|^{-2}\left(\frac{-1}{\|\mathbf{r} - \mathbf{r}_0\|}\right)(\mathbf{r} - \mathbf{r}_0) = -\frac{\alpha}{\|\mathbf{r} - \mathbf{r}_0\|^3}(\mathbf{r} - \mathbf{r}_0).$$

In component form, the force is given by

$$\mathbf{F}(x,y,z) = -\frac{\alpha(x-x_0)}{((x-x_0)^2+(y-y_0)^2+(z-z_0)^2)^{3/2}}\mathbf{i} - \frac{\alpha(y-y_0)}{\left((x-x_0)^2+(y-y_0)^2+(z-z_0)^2\right)^{3/2}}\mathbf{j} - \frac{\alpha(z-z_0)}{\left((x-x_0)^2+(y-y_0)^2+(z-z_0)^2\right)^{3/2}}\mathbf{k}. \blacktriangleleft$$

Integral Curves (Streamlines)

If $\mathbf{F}$ is a vector field on a region $\mathcal{G}$, and $t \mapsto \mathbf{r}(t)$ is a vector-valued function whose values lie in $\mathcal{G}$, then we say that the curve $\mathcal{C}$ described by $\mathbf{r}$ is an *integral curve* of $\mathbf{F}$ if

$$\mathbf{r}'(t) = \mathbf{F}(\mathbf{r}(t)). \tag{13.1.1}$$

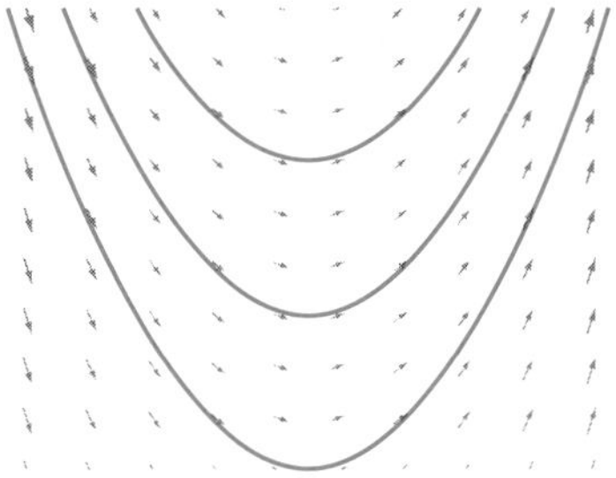

▲ Figure 9 A vector field and some of its integral curves

Equation (13.1.1) tells us that for each point $P = \mathbf{r}(t)$ of the integral curve $\mathcal{C}$, the vector $\mathbf{F}(P)$ is tangent to $\mathcal{C}$ at P. This behavior can be seen in Figure 9, which shows a vector field and three of its integral curves. Integral curves are often called by other names, such as *field lines*. (A curve that passes through a vector field is often called a "line," whether or not it is straight.) If we think of $\mathbf{r}$ as the position vector of a point mass moving along $\mathcal{C}$, then $\mathbf{F}$ may be interpreted as a velocity field. In this case, the integral curves are called *flow lines* or *streamlines*. As equation (13.1.1) suggests, determining the integral curves of a vector field usually amounts to solving a differential equation or a system of differential equations.

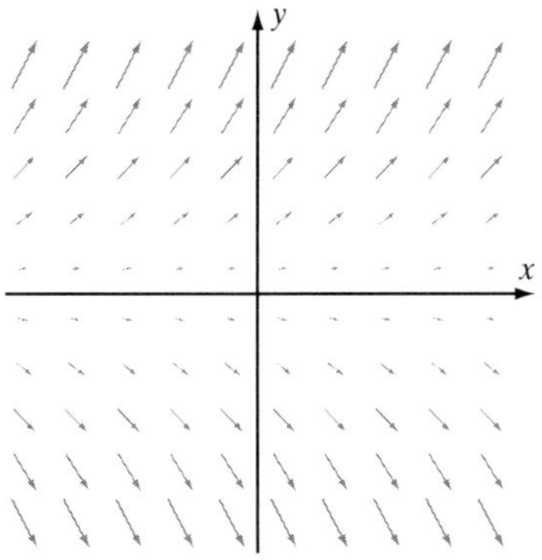

$\mathbf{F}(x, y) = \mathbf{i} + y\mathbf{j}$

▲ Figure 10

▶ **EXAMPLE 5** Determine the field lines of the vector field $\mathbf{F}(x,y) = \mathbf{i} + y\mathbf{j}$ that is shown in Figure 10.

Solution We seek $\mathbf{r}(t) = x(t)\mathbf{i} + y(t)\mathbf{j}$ such that

$$x'(t)\mathbf{i} + y'(t)\mathbf{j} = \mathbf{r}'(t) = \mathbf{F}(\mathbf{r}(t)) = \mathbf{i} + y(t)\mathbf{j} = 1\mathbf{i} + y(t)\mathbf{j}.$$

By equating the $\mathbf{i}$ and $\mathbf{j}$ components, we obtain two differential equations: $x'(t) = 1$ and $y'(t) = y(t)$. The first of these differential equations can be integrated to give $x = t + C_1$ for some constant C_1. The second differential equation, $y'(t) = y(t)$, is an equation of exponential growth: Each of its solutions is of the form $y = C_2e^t$ for some constant C_2. (To derive this result, apply the technique for solving separable differential equations, as discussed in Section 7.6 of Chapter 7. Alternatively, set $\alpha = 0$ and $\beta = -1$ in differential equation (7.7.4) and its solution (7.7.5).) Combining the two solutions $x = t + C_1$ and $y = C_2e^t$ that we have found, we see that

$$y = C_2e^t = C_2e^{x-C_1} = (C_2e^{-C_1})e^x = Ce^x$$

for the constant $C = C_2e^{-C_1}$. Thus the field lines of $\mathbf{F}(x,y) = \mathbf{i} + y\mathbf{j}$ are the graphs of the equations $y = Ce^x$ where C is a constant. Several of these field lines are shown in Figure 11. ◀

▶ **EXAMPLE 6** Let ω be a positive constant. The differential equation

$$\frac{d^2}{dt^2}u(t) + \omega^2 u(t) = 0 \tag{13.1.2}$$

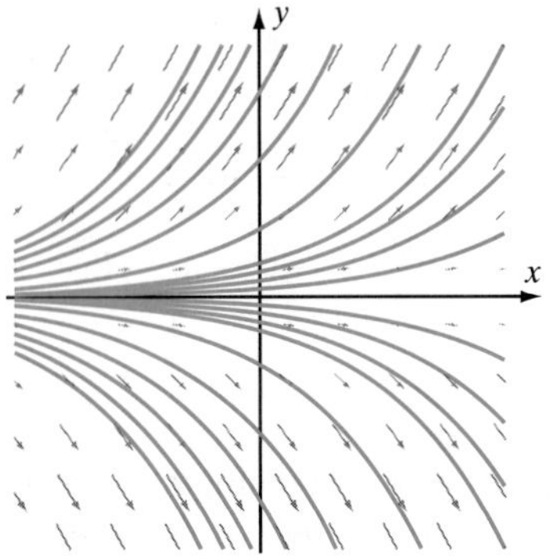

▲ Figure 11 Field lines of $\mathbf{F}(x, y) = \mathbf{i} + y\mathbf{j}$

is called the harmonic oscillator equation. It is known that $u(t)$ is a solution of this equation if and only if $u(t) = A\cos(\omega t) + B\sin(\omega t)$ for some constants A and B. Use this fact to determine the integral curves of the vector field $\mathbf{F}(x, y) = -\omega y\mathbf{i} + \omega x\mathbf{j}$.

Solution We seek $\mathbf{r}(t) = x(t)\mathbf{i} + y(t)\mathbf{j}$ such that

$$x'(t)\mathbf{i} + y'(t)\mathbf{j} = \mathbf{r}'(t) = \mathbf{F}(\mathbf{r}(t)) = -\omega y(t)\mathbf{i} + \omega x(t)\mathbf{j}.$$

By equating **i** and **j** components, we obtain the following system of two differential equations:

$$x'(t) = -\omega y(t) \tag{13.1.3}$$

and

$$y'(t) = \omega x(t). \tag{13.1.4}$$

By differentiating equation (13.1.3) with respect to t and then using equation (13.1.4) to substitute for $y'(t)$, we obtain

$$x''(t) = -\omega y'(t) = -\omega\big(\omega x(t)\big) = -\omega^2 x(t),$$

which may be rewritten as $x''(t) + \omega^2 x(t) = 0$. In other words, $x(t)$ is a solution of the harmonic oscillator equation. From the stated information, it follows that

$$x(t) = A\cos(\omega t) + B\sin(\omega t) \tag{13.1.5}$$

for some constants A and B. By differentiating this formula for $x(t)$, we find that

$$x'(t) = -A\omega\sin(\omega t) + B\omega\cos(\omega t). \tag{13.1.6}$$

Now rewrite equation (13.1.3) as $y(t) = (-1/\omega)x'(t)$, and, into this equation for $y(t)$, substitute formula (13.1.6) for $x'(t)$. The result is

$$y(t) = -B\cos(\omega t) + A\sin(\omega t). \tag{13.1.7}$$

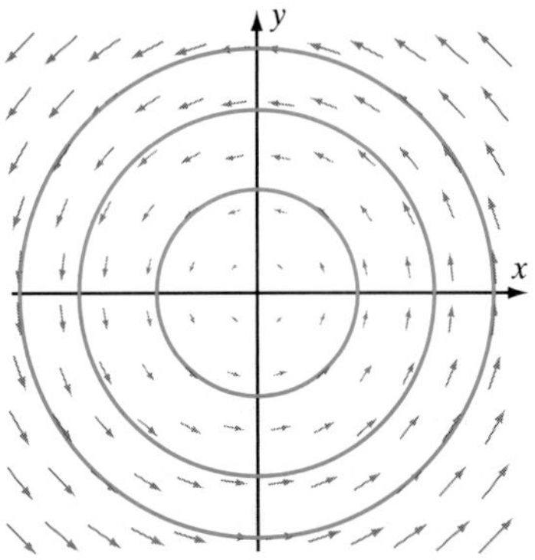

▲ Figure 12 The vector field $\mathbf{F}(x, y) = -\omega y\mathbf{i} + \omega x\mathbf{j}$ with three of its integral curves

Putting together formulas (13.1.5) and (13.1.7), we conclude that the integral curves of the vector field $\mathbf{F}(x, y) = -\omega y\mathbf{i} + \omega x\mathbf{j}$ have the form

$$\mathbf{r}(t) = \langle x(t), y(t)\rangle = \langle A\cos(\omega t) + B\sin(\omega t),\ -B\cos(\omega t) + A\sin(\omega t)\rangle.$$

In the case that B is 0, it is clear that the integral curve $\mathbf{r}(t) = A\langle\cos(\omega t), \sin(\omega t)\rangle$ describes the counterclockwise traversal of a circle of radius $|A|$. In general, the calculation

$$\begin{aligned}\|\mathbf{r}(t)\|^2 &= \big(A\cos(\omega t) + B\sin(\omega t)\big)^2 + \big(-B\cos(\omega t) + A\sin(\omega t)\big)^2 \\ &= A^2\big(\cos^2(\omega t) + \sin^2(\omega t)\big) + B^2\big(\cos^2(\omega t) + \sin^2(\omega t)\big) + 2AB\cos(\omega t)\sin(\omega t) - 2AB\cos(\omega t)\sin(\omega t) \\ &= A^2 + B^2\end{aligned}$$

shows that $\mathbf{r}$ parameterizes a circle of radius $\sqrt{A^2 + B^2}$ (see Figure 12). ◄

Gradient Vector Fields and Potential Functions

In Section 11.6 of Chapter 11, we introduced the gradient operator. When applied to the scalar-valued function $u(x, y)$, the gradient results in the vector-valued function $\nabla u(x, y) = u_x(x, y)\mathbf{i} + u_y(x, y)\mathbf{j}$, and, when applied to the scalar-valued function $u(x, y, z)$, the gradient results in the vector-valued function $\nabla u(x, y, z) = u_x(x, y, z)\mathbf{i} + u_y(x, y, z)\mathbf{j} + u_z(x, y, z)\mathbf{k}$. Notice that $(x, y) \mapsto \nabla u(x, y)$ is a planar vector field, and $(x, y, z) \mapsto \nabla u(x, y, z)$ is a spatial vector field. Thus whether u is a scalar-valued function of two or three variables, its gradient ∇u is a vector field.

▶ **EXAMPLE 7** Let the temperature at a point in the plane be given by $T(x, y) = x^2 + 6y^2$. Calculate the gradient of T, and give a physical interpretation.

Solution The gradient vector field of the temperature function is $\nabla T(x, y) = 2x\mathbf{i} + 12y\mathbf{j}$. Physically, this vector field tells us that if we are standing at a point P in the plane, then $\nabla T(P)$ points in the direction of most rapid increase of temperature at P. The magnitude of $\nabla T(P)$ indicates the rate of increase of the function T at the point P in the direction of $\nabla T(P)$. ◀

DEFINITION We say that a vector field $\mathbf{F}$ on a region $\mathcal{G}$ is a *gradient vector field* if there is a scalar-valued function u on $\mathcal{G}$ for which $\mathbf{F}(P) = \nabla u(P)$ at each point P of $\mathcal{G}$. A gradient vector field is also called a *conservative vector field*. If V is a scalar-valued function on $\mathcal{G}$ such that $\mathbf{F}(P) = -\nabla V(P)$ for each point P in $\mathcal{G}$, then we say that V is a *potential function* for $\mathbf{F}$.

A few remarks are in order concerning the term *potential function*, which is a bit ambiguous. If u is a scalar-valued function, and we set $V = -u$, then

$$\mathbf{F}(P) = \nabla u(P) \tag{13.1.8}$$

if and only if

$$\mathbf{F}(P) = -\nabla V(P). \tag{13.1.9}$$

In other words, given a vector field $\mathbf{F}$, equation (13.1.8) has a solution u if and only if equation (13.1.9) has a solution V. This means that a vector field $\mathbf{F}$ is conservative (equation (13.1.8) can be solved for u) if and only if $\mathbf{F}$ has a potential function (equation (13.1.9) can be solved for V). Of the two equivalent equations, mathematicians tend to work with (13.1.8). In the mathematical literature, a scalar function u that solves equation (13.1.8) is often called a potential function for the vector field $\mathbf{F}$. In physical applications, however, a function V that solves equation (13.1.9) has an important significance—it is the function used to define *potential energy*. The next example develops this background.

▶ **EXAMPLE 8** Suppose that a force field $\mathbf{F}$ has a potential function V in a region $\mathcal{G}$. Suppose that, due to the action of $\mathbf{F}$, a particle of mass m moves through $\mathcal{G}$ with a continuously differentiable position vector $\mathbf{r}(t)$, $a \le t \le b$. Show that

$$\frac{1}{2}m\|\mathbf{r}'(t)\|^2 + V\big(\mathbf{r}(t)\big) = C \tag{13.1.10}$$

for some scalar constant C.

Solution According to Newton's law, force equals mass times acceleration. We therefore have $\mathbf{F}(\mathbf{r}(t)) = m\mathbf{r}''(t)$. Because $\mathbf{F}$ has a potential function V, we may rewrite the force equation as $-\nabla V(\mathbf{r}(t)) = m\mathbf{r}''(t)$, or

$$m\mathbf{r}''(t) + \nabla V\big(\mathbf{r}(t)\big) = \mathbf{0}. \tag{13.1.11}$$

If we form the dot product of each side of equation (13.1.11) with $\mathbf{r}'(t)$, then we obtain

$$m\mathbf{r}'(t) \cdot \mathbf{r}''(t) + \nabla V\big(\mathbf{r}(t)\big) \cdot \mathbf{r}'(t) = 0. \tag{13.1.12}$$

Now $\|\mathbf{r}'(t)\|^2 = \mathbf{r}'(t) \cdot \mathbf{r}'(t)$. Therefore

$$\frac{d}{dt}\|\mathbf{r}'(t)\|^2 = \frac{d}{dt}\left(\mathbf{r}'(t) \cdot \mathbf{r}'(t)\right) = \left(\frac{d}{dt}\mathbf{r}'(t)\right) \cdot \mathbf{r}'(t) + \mathbf{r}'(t) \cdot \left(\frac{d}{dt}\mathbf{r}'(t)\right) = \mathbf{r}''(t) \cdot \mathbf{r}'(t) + \mathbf{r}'(t) \cdot \mathbf{r}''(t) = 2\mathbf{r}'(t) \cdot \mathbf{r}''(t),$$

which allows us to write equation (13.1.12) as

$$\frac{1}{2}m\frac{d}{dt}\|\mathbf{r}'(t)\|^2 + \nabla V\big(\mathbf{r}(t)\big) \cdot \mathbf{r}'(t) = 0. \tag{13.1.13}$$

Next we look at the second summand on the left side of equation (13.1.13). The Chain Rule, in the form of equation (11.6.3), tells us that

$$\frac{d}{dt}V\big(\mathbf{r}(t)\big) = \nabla V\big(\mathbf{r}(t)\big) \cdot \mathbf{r}'(t).$$

Taking this equation into account, we rewrite (13.1.13) as

$$\frac{1}{2}m\frac{d}{dt}\|\mathbf{r}'(t)\|^2 + \frac{d}{dt}V\big(\mathbf{r}(t)\big) = 0,$$

or

$$\frac{d}{dt}\left(\frac{1}{2}m\|\mathbf{r}'(t)\|^2 + V\big(\mathbf{r}(t)\big)\right) = 0.$$

It follows that

$$\frac{1}{2}m\|\mathbf{r}'(t)\|^2 + V\big(\mathbf{r}(t)\big) = C,$$

which is equation (13.1.10). ◄

Suppose that, as in Example 8, a conservative force field $\mathbf{F}$ acts on a particle that has mass m and position vector $\mathbf{r}(t)$. If V_0 is a potential function for $\mathbf{F}$, and if k is any scalar constant, then, because $\nabla(V_0 + k) = \nabla V_0 + \nabla k = -\mathbf{F} + \mathbf{0} = -\mathbf{F}$, we see that $V = V_0 + k$ is also a potential function for $\mathbf{F}$. After a choice of potential function V has been fixed, the quantity $V(P)$ is said to be the *potential energy* of the mass at point $P = \mathbf{r}(t)$. The quantity $\frac{1}{2}m\|\mathbf{r}'(t)\|^2$, that is, half the product of the mass and the square of the speed, is defined to be the *kinetic energy* of the mass at $P = \mathbf{r}(t)$. Equation (13.1.10) states that the total mechanical energy of the particle—kinetic energy plus potential energy—is constant. For this reason, equation

(13.1.10) is called the *Law of Conservation of Mechanical Energy* for the conservative force **F**.

▶ **EXAMPLE 9** Suppose that **F** is the planar force field given by $\mathbf{F}(x, y) = -mg\mathbf{j}$. This is the force field for a particle of mass m that falls with constant acceleration $\mathbf{a} = -g\mathbf{j}$. Show that **F** is conservative. Find the potential function V that satisfies $V(x, 0) = 0$ for all x. Verify the Law of Conservation of Mechanical Energy for a particle of mass m that is dropped from the point (x_0, y_0).

Solution If $u(x, y) = -mgy + C$ and $V(x, y) = mgy + C$ for any constant C, then we have $\mathbf{F} = \nabla u$ and $\mathbf{F} = -\nabla V$. Therefore **F** is a conservative force field. If we take $C = 0$, then $V(x, y) = mgy$ satisfies $V(x, 0) = 0$ for all x.

Suppose that $y_0 > 0$. If the particle is dropped from the point (x_0, y_0) and its position at time t is $\mathbf{r}(t)$, then we have

$$\mathbf{a}(t) = \mathbf{r}''(t) = -g\mathbf{j},$$

$$\mathbf{v}(t) = \mathbf{r}'(t) = \int_0^t (-g\mathbf{j})\, d\tau = -\,gt\,\mathbf{j},$$

and

$$\mathbf{r}(t) = x_0\,\mathbf{i} + y_0\,\mathbf{j} + \int_0^t \mathbf{r}'(\tau)\, d\tau = x_0\,\mathbf{i} + y_0\,\mathbf{j} + \int_0^t (-g\tau\,\mathbf{j})\, d\tau = x_0\,\mathbf{i} + \left(y_0 - \frac{1}{2}gt^2\right)\mathbf{j}.$$

At time t, the kinetic energy of the particle is

$$\frac{1}{2}m\|\mathbf{r}'(t)\|^2 = \frac{1}{2}m\|-gt\mathbf{j}\|^2 = \frac{1}{2}mg^2t^2,$$

and the potential energy of the particle is

$$V(\mathbf{r}(t)) = V\left(x_0, y_0 - \frac{1}{2}gt^2\right) = mg\left(y_0 - \frac{1}{2}gt^2\right) = mgy_0 - \frac{1}{2}mg^2t^2.$$

The total mechanical energy is

$$\frac{1}{2}m\|\mathbf{r}'(t)\|^2 + V(\mathbf{r}(t)) = \frac{1}{2}mg^2t^2 + \left(mgy_0 - \frac{1}{2}mg^2t^2\right) = mgy_0,$$

which does not depend on time, as asserted by the Law of Conservation of Mechanical Energy. ◀

Finding Potential Functions

In Section 13.3, we will learn how to *recognize* a conservative vector field. The next two examples show how to *find* a potential function for a vector field that is known to be conservative.

▶ **EXAMPLE 10** The vector field $\mathbf{F}(x, y) = (y - 3x^2)\mathbf{i} + (x + \sin(y))\mathbf{j}$ is known to be conservative. Find a scalar-valued function u for which $\mathbf{F} = \nabla u$.

Solution We want u to satisfy the vector equation

$$(y - 3x^2)\mathbf{i} + (x + \sin(y))\mathbf{j} = \mathbf{F}(x, y) = \nabla u(x, y) = u_x(x, y)\mathbf{i} + u_y(x, y)\mathbf{j}.$$

This vector equation is equivalent to the two simultaneous scalar equations

$$u_x(x, y) = y - 3x^2 \tag{13.1.14}$$

and

$$u_y(x, y) = x + \sin(y). \tag{13.1.15}$$

In order to remove the partial derivative from u in equation (13.1.14), we antidifferentiate each side of the equation $u_x(x, y) = y - 3x^2$ with respect to the x variable. Because we treat y as a constant when we calculate $u_x(x, y)$, we treat y as a constant when we integrate to undo the process. As a result, the constant of integration that we obtain *can* depend on y. Thus antidifferentiating each side of equation (13.1.14) yields

$$u(x, y) = xy - x^3 + \phi(y), \tag{13.1.16}$$

where the constant of integration is some unknown function $\phi(y)$. To verify our assertion concerning the constant of integration, observe that partial differentiating the right side of equation (13.1.16) with respect to x does result in the right side of equation (13.1.14). Our goal now is to determine the expression $\phi(y)$ that has arisen in the integration. To do so, we differentiate each side of equation (13.1.16) with respect to y, obtaining

$$u_y(x, y) = x + \phi'(y). \tag{13.1.17}$$

Notice that the expression $\phi'(y)$ is an ordinary derivative because ϕ depends only on y. Comparing equation (13.1.17) with equation (13.1.15), we see that $\phi'(y) = \sin(y)$. After antidifferentiating with respect to y, we obtain $\phi(y) = -\cos(y) + C$. Observe that, in this last antidifferentiation, which we performed to remove an *ordinary* derivative, the constant of integration C is a numerical constant. By substituting our formula for $\phi(y)$ into equation (13.1.16), we obtain

$$u(x, y) = xy - x^3 - \cos(y) + C. \quad ◀$$

▶ **EXAMPLE 11** The vector field $\mathbf{F}(x, y, z) = y^2\mathbf{i} + (2xy + z)\mathbf{j} + (y + 3)\mathbf{k}$ is known to be conservative. Find a scalar-valued function u for which $\mathbf{F} = \nabla u$.

Solution We want to find a function u whose partial derivatives satisfy the equations

$$\text{(a) } u_x(x, y, z) = y^2, \quad \text{(b) } u_y(x, y, z) = 2xy + z, \quad \text{and} \quad \text{(c) } u_z(x, y, z) = y + 3.$$

We antidifferentiate the first of these equations with respect to the x variable. The result is $u(x, y, z) = xy^2 + \phi(y, z)$. Notice that the constant of integration $\phi(y, z)$ is constant only with respect to x. It can depend on y and z.

Next, we determine the expression $\phi(y, z)$. To do so, we differentiate with respect to y each side of the equation $u(x, y, z) = xy^2 + \phi(y, z)$. This differentiation yields the equation $u_y(x, y, z) = 2xy + \phi_y(y, z)$. Comparing this last equation with equation (b), we see that $2xy + \phi_y(y,z) = 2xy + z$, or $\phi_y(y,z) = z$. We antidifferentiate this last equation with respect to y, noting that the constant of integration can depend on z. The result is $\phi(y,z) = yz + \psi(z)$. We substitute this formula for $\phi(y,z)$ into our equation for u, obtaining

$$u(x,y,z) = xy^2 + yz + \psi(z). \qquad (13.1.18)$$

Our next task is to identify the unknown expression $\psi(z)$. To do so, we take the partial derivative of each side of equation (13.1.18) with respect to z. The result is $u_z(x,y,z) = y + \psi'(z)$. Comparing the right side of this last equation with the right side of equation (c), we see that $y + \psi'(z) = y + 3$, or $\psi'(z) = 3$. It follows that $\psi(z) = 3z + C$ where C is a numerical constant. We conclude our computation by substituting our expression for $\psi(z)$ into equation (13.1.18), obtaining

$$u(x,y,z) = xy^2 + yz + 3z + C. \quad ◀$$

QUICK QUIZ

1. Is $\{(x,y,z) : 0 < x, 0 < y, 0 < z\}$ a region? If the answer is *No*, then why not?
2. Is $\{(x,y,z) : 0 < xyz)$ a region? If the answer is *No*, then why not?
3. True or false: If a particle is in motion due to a conservative force field, then the kinetic energy of the particle is conserved.
4. The vector field $\mathbf{F}(x,y) = y\mathbf{i} + (x+3)\mathbf{j}$ is known to be conservative. Find a scalar-valued function u for which $\mathbf{F} = \nabla u$.

Answers

1. Yes 2. Not path-connected 3. False 4. $u(x,y) = xy + 3y + C$

EXERCISES

Problems for Practice

▶ In each of Exercises 1–8, sketch the given vector field in the plane by drawing the vectors that are assigned to several points. ◀

1. $\mathbf{F}(x,y) = 3\mathbf{i} + \mathbf{j}$
2. $\mathbf{F}(x,y) = x\mathbf{i} + y\mathbf{j}$
3. $\mathbf{F}(x,y) = 2y\mathbf{i} - 3\mathbf{j}$
4. $\mathbf{F}(x,y) = 3x\mathbf{i} + x\mathbf{j}$
5. $\mathbf{F}(x,y) = y\mathbf{i} - 5x\mathbf{j}$
6. $\mathbf{F}(x,y) = x^2\mathbf{i} + y^2\mathbf{j}$
7. $\mathbf{F}(x,y) = x^2\mathbf{i} - y^2\mathbf{j}$
8. $\mathbf{F}(x,y) = \mathbf{i} + \sin(y)\mathbf{j}$

▶ In each of Exercises 9–12, determine an explicit formula for the spatial vector field that is described. ◀

9. A wind blows down a 10 foot high hall with vertical walls and width 8 feet. The speed of the wind at a point is proportional to the distance of the point from a line running down the middle of the floor. The line down the middle of the floor is the y-axis, with the positive direction being the direction that the wind blows. Write the velocity vector field.
10. A cylindrical tank is 5 meters high with a radius of 4 meters. Water circulates about the tank's central axis, the line segment $[0,5]$ on the z-axis, in such a way that the flow is clockwise as you look down on it, and the speed at any point is proportional to the distance of the point from the center of the tank. Write the velocity vector field.
11. In the tank described in the preceding exercise, water circulates about the tank's central axis in such a way that the flow is clockwise as you look down on it, and the speed at any point is proportional to the distance of the point from the central axis of the tank. Write the velocity vector field.
12. Particles circulate in space in such a way that rotation takes place about the y-axis. Each particle rotates in a plane perpendicular to the y-axis so that its oriented axis of rotation, determined by the right-hand rule, is the vector $\mathbf{j}$. The speed of any particle is proportional to the square of its distance from the y-axis. Write the velocity vector field.

In each of Exercises 13–20, write the gradient vector field associated with the given scalar-valued function.

13. $u(x,y) = xy - y^2 + 3x$
14. $u(x,y) = \sin(xy^2)$
15. $u(x,y) = \ln(x^3 - y^2)$
16. $u(x,y) = e^{2y-5x}$
17. $u(x,y,z) = x^2y^3z$
18. $u(x,y,z) = \dfrac{\sin(xy)}{\cos(z)}$
19. $u(x,y,z) = \dfrac{\ln(x-y)}{\ln(z)}$
20. $u(x,y,z) = e^{\cos(x)}e^{\sin(y)}e^z$

In each of Exercises 21–26, the given vector field $\mathbf{F}(x,y) = M(x,y)\mathbf{i} + N(x,y)\mathbf{j}$ is conservative. Find an expression $u(x,y)$ such that $\mathbf{F} = \nabla u$.

21. $\mathbf{i} + \mathbf{j}$
22. $x\mathbf{i} + y\mathbf{j}$
23. $y\mathbf{i} + x\mathbf{j}$
24. $y\mathbf{i} + (x+2)\mathbf{j}$
25. $(y^3 + 2xy)\mathbf{i} + (3xy^2 + x^2)\mathbf{j}$
26. $2e^{2x-y}\mathbf{i} - e^{2x-y}\mathbf{j}$

In each of Exercises 27–32, the given vector field $\mathbf{F}(x,y,z) = M(x,y,z)\mathbf{i} + N(x,y,z)\mathbf{j} + R(x,y,z)\mathbf{k}$ is conservative. Find an expression $u(x,y,z)$ such that $\mathbf{F} = \nabla u$.

27. $2x\mathbf{i} + \mathbf{j} + 0\mathbf{k}$
28. $\mathbf{i} + \mathbf{j} + 2z\mathbf{k}$
29. $2xyz^3\mathbf{i} + (x^2z^3 + 5)\mathbf{j} + 3x^2yz^2\mathbf{k}$
30. $y\mathbf{i} + (x + 2z)\mathbf{j} + 2y\mathbf{k}$
31. $(z^3 - 2xy^2)\mathbf{i} + (z^2 - 2x^2y)\mathbf{j} + (3xz^2 + 2yz)\mathbf{k}$
32. $(\sin(y) - z\sin(x))\mathbf{i} + x\cos(y)\mathbf{j} + \cos(x)\mathbf{k}$

In each of Exercises 33–38, find the integral curves $t \mapsto \mathbf{r}(t)$ of the given vector field $\mathbf{F}(x,y) = M(x,y)\mathbf{i} + N(x,y)\mathbf{j}$.

33. $\mathbf{F}(x,y) = \mathbf{i} + \mathbf{j}$
34. $\mathbf{F}(x,y) = \mathbf{i} + 2\mathbf{j}$
35. $\mathbf{F}(x,y) = x\mathbf{i} + 2\mathbf{j}$
36. $\mathbf{F}(x,y) = x\mathbf{i} - y\mathbf{j}$
37. $\mathbf{F}(x,y) = 2y\mathbf{i} + \mathbf{j}$
38. $\mathbf{F}(x,y) = \mathbf{i} - x\mathbf{j}$

Further Theory and Practice

In each of Exercises 39–42, the given vector field $\mathbf{F}(x,y) = M(x,y)\mathbf{i} + N(x,y)\mathbf{j}$ is conservative. Find an expression $u(x,y)$ such that $\mathbf{F} = \nabla u$.

39. $(e^{2x} + 2xe^{2x})\mathbf{i} + \mathbf{j}$
40. $\left(\dfrac{y}{1+x^2y^2}\right)\mathbf{i} + \left(\dfrac{x}{1+x^2y^2}\right)\mathbf{j}$
41. $y^2\cos(xy)\mathbf{i} + \big(\sin(xy) + xy\cos(xy)\big)\mathbf{j}$
42. $\left(\ln(x+y) + \dfrac{x}{x+y}\right)\mathbf{i} + \dfrac{x}{x+y}\mathbf{j}$

In each of Exercises 43–46, the given vector field $\mathbf{F}(x,y,z) = M(x,y,z)\mathbf{i} + N(x,y,z)\mathbf{j} + R(x,y,z)\mathbf{k}$ is conservative. Find an expression $u(x,y,z)$ such that $\mathbf{F} = \nabla u$.

43. $z(y+z)^{-1}\mathbf{i} - xz(y+z)^{-2}\mathbf{j} + xy(y+z)^{-2}\mathbf{k}$
44. $(y^2+z^2)^{-1}\mathbf{i} - 2xy(y^2+z^2)^{-2}\mathbf{j} - 2xz(y^2+z^2)^{-2}\mathbf{k}$
45. $x^{y-1}y\mathbf{i} + x^y\ln(x)\mathbf{j} + \ln(z)\mathbf{k}$
46. $\ln(y)^z\mathbf{i} + (xz/y)\ln(y)^{z-1}\mathbf{j} + x\ln(y)^z\ln(\ln(y))\mathbf{k}$

47. Show that $\mathbf{F}(x,y) = y\mathbf{i} + 2x\mathbf{j}$ is not the gradient vector field of any continuously differentiable function.

48. On separate sets of axes, sketch the following vector fields in a neighborhood of the origin in the plane:
a. $\mathbf{F}(x,y) = x\mathbf{i} + y\mathbf{j}$
b. $\mathbf{F}(x,y) = -x\mathbf{i} - y\mathbf{j}$
c. $\mathbf{F}(x,y) = -x\mathbf{i} + y\mathbf{j}$
d. $\mathbf{F}(x,y) = x\mathbf{i} - y\mathbf{j}$
e. $\mathbf{F}(x,y) = y\mathbf{i} + x\mathbf{j}$
f. $\mathbf{F}(x,y) = -y\mathbf{i} + x\mathbf{j}$
g. $\mathbf{F}(x,y) = y\mathbf{i} - x\mathbf{j}$
h. $\mathbf{F}(x,y) = -y\mathbf{i} - x\mathbf{j}$
Each of these exhibits qualitatively different behavior near the origin, which is the point where $\mathbf{F}$ vanishes. It can be proved, using advanced techniques, that typical vector fields satisfying a certain nondegeneracy condition and vanishing at the origin exhibit one of these eight modes of behavior.

49. Let $r(x,y,z) = \|x\mathbf{i} + y\mathbf{j} + z\mathbf{k}\|$. Calculate $\nabla(1/r)$ at all points where the derivative exists.

50. Coulomb's Law says that the force $\mathbf{F}$ acting on a charge q at position $\mathbf{r}$ because of a charge Q at the origin is

$$\mathbf{F} = \frac{\varepsilon Qq}{\|\mathbf{r}\|^3}\mathbf{r}$$

where ε is a positive constant that depends on the physical units that are used. Show that $V = \varepsilon Qq/\|\mathbf{r}\|$ is a potential function for $\mathbf{F}$. If $Qq > 0$, then $\mathbf{F}$ is said to be *repulsive*, while if $Qq < 0$, then $\mathbf{F}$ is said to be *attractive*. Explain why this terminology is appropriate.

51. At the point (x,y), the elevation of the terrain is $h(x,y) = -x^2 + 3xy + 2y^2 + 100$ feet. Sketch the vector field that shows the direction of quickest descent at each point.

52. At the point (x,y) on a sheet of aluminum, the temperature is $T(x,y) = 150 - xy + 4y^3$. Sketch the vector field that shows at each point the direction of greatest increase in temperature.

If r_1 and r_2 are distinct real numbers, then all solutions of the differential equation

$$\frac{d^2}{dt^2}u(t) - (r_1 + r_2)\frac{d}{dx}u(t) + r_1r_2u(t) = 0$$

have the form $u(t) = A\exp(r_1 t) + B\exp(r_2 t)$ for some constants A and B. In each of Exercises 53–56, use this fact to determine the integral curves $t \mapsto \mathbf{r}(t)$ of the given vector field $\mathbf{F}(x, y) = M(x, y)\mathbf{i} + N(x, y)\mathbf{j}$.

53. $\mathbf{F}(x, y) = (3x + y)\mathbf{i} + 2y\mathbf{j}$
54. $\mathbf{F}(x, y) = 2y\mathbf{i} + (3y - x)\mathbf{j}$
55. $\mathbf{F}(x, y) = (3x - 4y)\mathbf{i} - x\mathbf{j}$
56. $\mathbf{F}(x, y) = (3x + 2y)\mathbf{i} + (12x - 7y)\mathbf{j}$

57. Find a continuous vector field $\mathbf{F}(x, y) = M(x, y)\mathbf{i} + N(x, y)\mathbf{j}$ such that, at each point (x, y) on the circle $\mathcal{C} = \{(x, y) : x^2 + y^2 = 1\}$, the vector $\mathbf{F}(x, y)$ satisfies $\|F(x, y)\| = 1$, and $\mathbf{F}(x, y)$ is tangent to $\mathcal{C}$. According to the "Combing the Hairy Sphere" Theorem, there is no such continuous vector field on the unit sphere in three-dimensional space.

Calculator/Computer Exercises

58. Draw the given planar vector fields in the viewing window $[-1, 1) \times [-1, 1)$:
a. $(x + y)\mathbf{i} + (1 + xy)\mathbf{j}$
b. $xy\mathbf{i} + (1 + y - x)\mathbf{j}$
c. $(x + y)\mathbf{i} + (x - y^2)\mathbf{j}$
d. $(x^2 - y^2)\mathbf{i} + (x^2 + y^2)\mathbf{j}$

13.2 Line Integrals

In Chapters 5 through 11, the integrations we performed were over subintervals of the real line. In Chapter 12, we learned to integrate over regions in the plane and over solids in space. In this section, we will study how to integrate over *curves* in the plane and in space. By expanding the types of sets over which we can integrate, we will be able to learn new applications of the integral. For instance, if we apply a force at each point of a curve, then we need to integrate the force along the curve to calculate the work performed. Similarly, if a thin wire takes the shape of a curve, then we can integrate a mass density over the curve to obtain the total mass of the wire. In this section, we study the mathematical ideas that are the basis of these and other applications.

Work Along a Curved Path

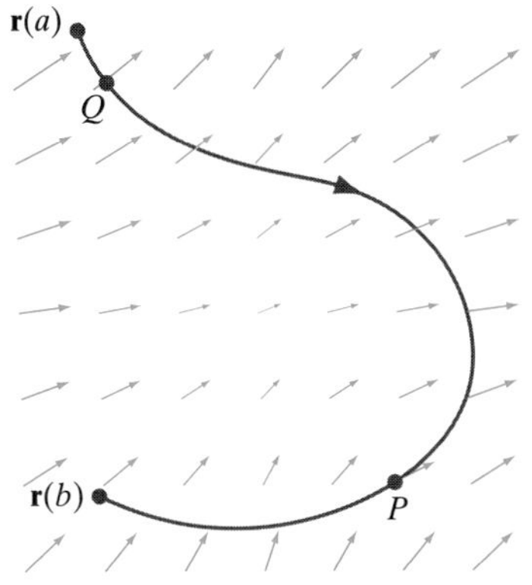

▲ Figure 1

In Section 7.5 of Chapter 7, we developed a method for computing the work performed in moving a mass along a line segment. Now we would like to calculate work along a curved path. To be specific, we suppose that $\mathbf{F}$ is a continuous force field in a region of the plane or in space. If a particle moves through this force field with position vector $\mathbf{r}(t)$ for $a \le t \le b$, then how much work is performed? See Figure 1. The interesting feature here is that sometimes the force might be directly opposed to the path (point P in Figure 1), whereas at other times it will be skew to the path (point Q in Figure 1). In other words, we do not assume that there is any relationship between the direction of the force, $\mathbf{dir}(\mathbf{F}(P))$, and the direction of the curve, $\mathbf{T}(t) = \mathbf{dir}(\mathbf{r}'(t))$, at a point $P = \mathbf{r}(t)$ of the curve.

We estimate the work W by partitioning the domain $[a, b]$ of $\mathbf{r}$ into N subintervals $I_j = [t_{j-1}, t_j]$ with equal lengths $\Delta t = (b - a)/N$. For each j, we let W_j denote the work performed as t passes from t_{j-1} to t_j. If ξ_j is an arbitrary point in I_j, then we have the approximation

$$W_j \approx \big(\text{component of Force in direction of motion at } \mathbf{r}(\xi_j)\big) \cdot \big(\text{distance between } \mathbf{r}(t_{j-1}) \textbf{ and } \mathbf{r}(t_j)\big).$$

Using the first equation of line (9.3.8) to evaluate the component of $\mathbf{F}$ along the tangent vector to the curve at $\mathbf{r}(\xi_j)$, we obtain

$$W_j \approx \Big(\mathbf{F}\big(\mathbf{r}(\xi_j)\big) \cdot \mathbf{T}(\xi_j)\Big)\|\mathbf{r}(t_j) - \mathbf{r}(t_{j-1})\| = \left(\mathbf{F}\big(\mathbf{r}(\xi_j)\big) \cdot \frac{1}{\|\mathbf{r}'(\xi_j)\|}\mathbf{r}'(\xi_j)\right)\|\mathbf{r}(t_j) - \mathbf{r}(t_{j-1})\|.$$

Because we can approximate the velocity vector $\mathbf{r}'(\xi_j)$ by the difference quotient

$$\frac{1}{\Delta t}\big(\mathbf{r}(t_j) - \mathbf{r}(t_{j-1})\big),$$

it follows that $\|\mathbf{r}(t_j) - \mathbf{r}(t_{j-1})\| \approx \|\mathbf{r}'(\xi_j)\|\Delta t$. When we substitute this last expression into our approximation for W_j, we obtain $W_j \approx \big(\mathbf{F}\big(\mathbf{r}(\xi_j)\big) \cdot \mathbf{r}'(\xi_j)\big)\Delta t$. Therefore the work W performed in traversing the entire curve from $\mathbf{r}(a)$ to $\mathbf{r}(b)$ is approximately

$$\sum_{j=1}^{N} \mathbf{F}\Big(\mathbf{r}(\xi_j)\Big) \cdot \mathbf{r}'(\xi_j)\Delta t,$$

which we notice is a Riemann sum. As N increases, the approximation becomes more accurate, and the sum tends to the integral of $\mathbf{F}\big(\mathbf{r}(t)\big) \cdot \mathbf{r}'(t)$ over the interval $[a, b]$. This observation leads to the definition of work.

DEFINITION The work done by a force $\mathbf{F}$ along a parameterized curve $t \mapsto \mathbf{r}(t)$, $a \le t \le b$, is

$$\int_a^b \mathbf{F}\big(\mathbf{r}(t)\big) \cdot \mathbf{r}'(t)\,dt. \tag{13.2.1}$$

▶ **EXAMPLE 1** What is the work done by the planar force field $\mathbf{F}(x, y) = \langle y, 1 + x\rangle$ moving a particle along the path $\mathbf{r}(t) = \langle t^3, t^2\rangle$, $0 \le t \le 1$? If the particle retraces its path with trajectory $\tilde{\mathbf{r}}(t) = \big\langle (2-t)^3, (2-t)^2\big\rangle$, $1 \le t \le 2$, then what is the work done by $\mathbf{F}$ during this "return trip"?

Solution We have $\mathbf{F}(\mathbf{r}(t)) = \mathbf{F}(t^3, t^2) = \langle t^2, 1 + t^3\rangle$ and $\mathbf{r}'(t) = \langle 3t^2, 2t\rangle$. Therefore according to formula (13.2.1), the work performed is equal to

$$\begin{aligned}\int_0^1 \mathbf{F}\big(\mathbf{r}(t)\big) \cdot \mathbf{r}'(t)\,dt &= \int_0^1 \langle t^2, 1 + t^3\rangle \cdot \langle 3t^2, 2t\rangle\,dt \\ &= \int_0^1 \big(3t^4 + 2t(1 + t^3)\big)\,dt = \int_0^1 (5t^4 + 2t)\,dt = (t^5 + t^2)\Big|_0^1 = 2.\end{aligned}$$

The work performed along the return path is

$$\begin{aligned}\int_1^2 \mathbf{F}\big(\tilde{\mathbf{r}}(t)\big) \cdot \tilde{\mathbf{r}}'(t)\,dt &= \int_1^2 \big\langle (2-t)^2, 1 + (2-t)^3\big\rangle \cdot \big\langle -3(2-t)^2, -2(2-t)\big\rangle\,dt \\ &= -\int_1^2 \big(5(2-t)^4 + 2(2-t)\big)dt\end{aligned}$$

and therefore

$$\int_1^2 \mathbf{F}(\tilde{\mathbf{r}}(t)) \cdot \tilde{\mathbf{r}}'(t)\, dt = \left((2-t)^5 + (2-t)^2\right)\Big|_1^2 = -2. \quad \blacktriangleleft$$

INSIGHT Work is a scalar quantity that can be positive, negative, or zero. In Example 1, the values of work associated with the paths traced by $\mathbf{r}$ and $\tilde{\mathbf{r}}$ are opposite in sign. The physical meaning of "negative work" will be explored in Section 13.3. However, we can understand right away why the work done by $\mathbf{F}$ over $\tilde{\mathbf{r}}$ is negative. Observe that the functions $\mathbf{r}$ and $\tilde{\mathbf{r}}$ both trace the graph of $y = x^{2/3}$, $0 \le x \le 1$ but in opposite directions (Figure 2). We can see from the plot that the angles between the vectors $\mathbf{F}(\mathbf{r}(t))$ and $\mathbf{r}'(t)$ are acute. The integrand $\mathbf{F}(\mathbf{r}(t)) \cdot \mathbf{r}'(t)$ is therefore nonnegative and $\int_0^1 \mathbf{F}(\mathbf{r}(t)) \cdot \mathbf{r}'(t)dt > 0$. On the other hand, $\tilde{\mathbf{r}}'$ points in the opposite direction of $\mathbf{r}'$, the angles between $\mathbf{F}(\tilde{\mathbf{r}}(t))$ and $\tilde{\mathbf{r}}'(t)$ are obtuse, and the integrand $\mathbf{F}(\tilde{\mathbf{r}}(t)) \cdot \tilde{\mathbf{r}}'(t)$ is negative. It follows that $\int_0^1 \mathbf{F}(\tilde{\mathbf{r}}(t)) \cdot \tilde{\mathbf{r}}'(t)dt < 0$.

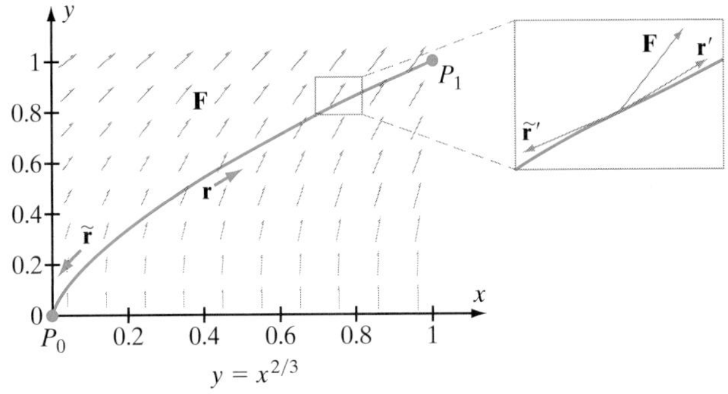

▲ Figure 2

Line Integrals

Let us recall that a *smooth parametric curve* is a continuously differentiable vector-valued function $t \mapsto \mathbf{r}(t), a \le t \le b$ such that $\mathbf{r}'(t) \ne \mathbf{0}$. Of course, when we use the term "curve" in its everyday sense, we really mean the geometric set of values, $\mathcal{C} = \{\mathbf{r}(t) : a \le t \le b\}$, which $\mathbf{r}$ assumes as t varies over its domain. In fact, $\mathcal{C}$ has an additional attribute. Because $\mathbf{r}$ imparts to $\mathcal{C}$ an initial point $\mathbf{r}(a)$, a terminal point $\mathbf{r}(b)$, and a direction for traversing $\mathcal{C}$ from $\mathbf{r}(a)$ to $\mathbf{r}(b)$, we refer to $\mathcal{C}$ as a *directed* or *oriented* curve. Whenever we speak of a parameterization of a directed curve $\mathcal{C}$, we imply that the parameterization respects the direction of $\mathcal{C}$.

We now want to define the integral of a vector field $\mathbf{F}$ over a directed curve $\mathcal{C}$. Our calculation of work motivates the definition of the integral of $\mathbf{F}$ over $\mathcal{C}$.

DEFINITION Let $\mathcal{C}$ be a directed curve (in the plane or in space) with parameterization $t \mapsto \mathbf{r}(t)$, $a \le t \le b$. Suppose that $\mathbf{F}$ is a continuous vector field defined on $\mathcal{C}$. Then, the *line integral* (or *path integral*) of $\mathbf{F}$ over $\mathcal{C}$ is denoted by the symbol $\int_{\mathcal{C}} \mathbf{F} \cdot d\mathbf{r}$ and defined to be

$$\int_{\mathcal{C}} \mathbf{F} \cdot d\mathbf{r} = \int_a^b \mathbf{F}(\mathbf{r}(t)) \cdot \mathbf{r}'(t)\, dt. \tag{13.2.2}$$

If $\mathcal{C}$ is a piecewise-smooth directed curve that is the union of smooth arcs $\mathcal{C}_1, \mathcal{C}_2, \ldots, \mathcal{C}_n$ (as in Figure 3), then $\int_{\mathcal{C}} \mathbf{F} \cdot d\mathbf{r}$ is defined by

$$\int_{\mathcal{C}} \mathbf{F} \cdot d\mathbf{r} = \int_{\mathcal{C}_1} \mathbf{F} \cdot d\mathbf{r} + \int_{\mathcal{C}_2} \mathbf{F} \cdot d\mathbf{r} + \ldots + \int_{\mathcal{C}_n} \mathbf{F} \cdot d\mathbf{r},$$

where each summand on the right side of the equation is defined by formula (13.2.2). (In each of the expressions $\int_{\mathcal{C}_i} \mathbf{F} \cdot d\mathbf{r}$, it is implicit that $\mathbf{r}$ is a parameterization of the particular curve $\mathcal{C}_i$ that appears in the expression. In other words, the symbol $d\mathbf{r}$ signifies different parameterizations in the different summands.)

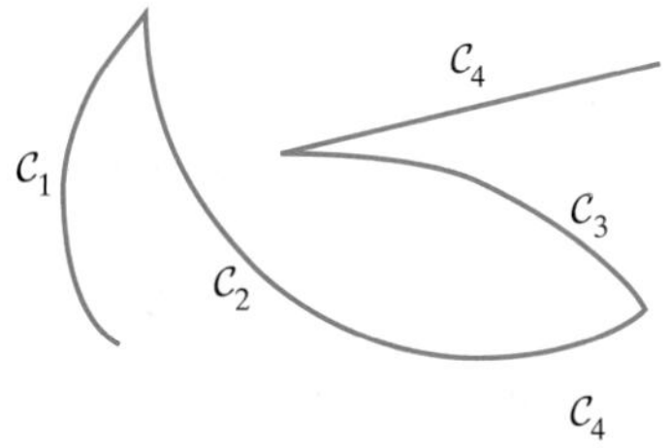

▲ Figure 3

If $\mathbf{F}$ is a force field, then equation (13.2.1) shows that the line integral $\int_{\mathcal{C}} \mathbf{F} \cdot d\mathbf{r}$ has a physical interpretation as the work performed in traversing $\mathcal{C}$ from initial point to terminal point. However, we may form the line integral $\int_{\mathcal{C}} \mathbf{F} \cdot d\mathbf{r}$ of *any* continuous vector field $\mathbf{F}$ defined on the directed curve $\mathcal{C}$. The vector field $\mathbf{F}$ does not have to be a force field, and the line integral $\int_{\mathcal{C}} \mathbf{F} \cdot d\mathbf{r}$ does not have to represent work. Notice that the value of a line integral is a scalar: The dot product notation $\mathbf{F} \cdot d\mathbf{r}$ serves to remind us of this fact.

▶ **EXAMPLE 2** Let $\mathcal{C}$ be the planar directed curve parameterized by $\mathbf{r}(t) = \langle t, -t^2 \rangle, 0 \le t \le 1$. Calculate $\int_{\mathcal{C}} \mathbf{F} \cdot d\mathbf{r}$ for $\mathbf{F}(x,y) = e^x\mathbf{i} + xy\mathbf{j}$.

Solution By definition, this line integral equals $\int_0^1 \mathbf{F}(\mathbf{r}(t)) \cdot \mathbf{r}'(t)dt$. Now

$$\mathbf{F}(\mathbf{r}(t)) = \mathbf{F}(t, -t^2) = e^t\mathbf{i} + t(-t^2)\mathbf{j} = e^t\mathbf{i} - t^3\mathbf{j}.$$

and $\mathbf{r}'(t) = 1\mathbf{i} - 2t\mathbf{j}$. Therefore the line integral equals

$$\int_0^1 (e^t\mathbf{i} - t^3\mathbf{j}) \cdot (1\mathbf{i} - 2t\mathbf{j})\, dt = \int_0^1 (e^t + 2t^4)\, dt = \left(e^t + \frac{2}{5}t^5\right)\Bigg|_{t=0}^{t=1} = e - \frac{3}{5}. \quad ◀$$

As we know from Chapter 10, a directed curve $\mathcal{C}$ may be parameterized in several different ways. Formula (13.2.2) seems to imply that the line integral $\int_{\mathcal{C}} \mathbf{F} \cdot d\mathbf{r}$ of $\mathbf{F}$ over $\mathcal{C}$ depends on the parameterization $\mathbf{r}$ of $\mathcal{C}$ that is used in the calculation. The next theorem demonstrates that the value of the line integral $\int_{\mathcal{C}} \mathbf{F} \cdot d\mathbf{r}$ does not actually depend on the particular parameterization of $\mathcal{C}$ that is chosen (provided that the parameterization respects the direction of $\mathcal{C}$).

THEOREM 1 Let $t \mapsto \mathbf{r}(t),\ a \le t \le b$ be a continuously differentiable parametrization for a directed curve $\mathcal{C}$ in the plane or in space. Suppose that $\mathbf{F}$ is a continuous vector field whose domain contains $\mathcal{C}$. Let $s \mapsto \mathbf{p}(s), 0 \le s \le L$ be the arc length parameterization of $\mathcal{C}$ with $\mathbf{p}(0) = \mathbf{r}(a)$ and $\mathbf{p}(L) = \mathbf{r}(b)$. Let $\mathbf{T}(s) = \mathbf{p}'(s)$ denote the unit tangent vector to $\mathcal{C}$ at the point $\mathbf{p}(s)$. Then the line integral of $\mathbf{F}$ over $\mathcal{C}$ is given by the equation

$$\int_{\mathcal{C}} \mathbf{F} \cdot d\mathbf{r} = \int_0^L \mathbf{F}(\mathbf{p}(s)) \cdot \mathbf{T}(s)\, ds. \qquad (13.2.3)$$

In particular, because the right side of equation (13.2.3) does not involve the parameterization $\mathbf{r}$, the line integral of $\mathbf{F}$ over $\mathcal{C}$ does not depend on the chosen parameterization of $\mathcal{C}$.

Proof. The derivation of equation (13.2.3) is very similar to the proof of Theorem 1, Section 10.3 in Chapter 10. If $s = \int_a^t \|\mathbf{r}'(\tau)\|\,d\tau$, then

$$\mathbf{p}(s) = \mathbf{r}(t), \quad \mathbf{T}(s) = \frac{1}{\|\mathbf{r}'(t)\|}\mathbf{r}'(t), \quad \text{and} \quad \frac{ds}{dt} = \|\mathbf{r}'(t)\|.$$

Making the substitution $s = \int_a^t \|\mathbf{r}'(\tau)\|\,d\tau$, $ds = \|\mathbf{r}'(t)\|\,dt$ in the integral on the right side of equation (13.2.3), we obtain

$$\int_0^L \mathbf{F}(\mathbf{p}(s)) \cdot \mathbf{T}(s)\,ds = \int_a^b \mathbf{F}(\mathbf{r}(t)) \cdot \left(\frac{1}{\|\mathbf{r}'(t)\|}\mathbf{r}'(t)\right)\|\mathbf{r}'(t)\|\,dt = \int_a^b \mathbf{F}(\mathbf{r}(t)) \cdot \mathbf{r}'(t)\,dt = \int_\mathcal{C} \mathbf{F} \cdot d\mathbf{r}. \quad ■$$

▶ **EXAMPLE 3** Let $\mathcal{C}$ be the quarter circle in the xy-plane that is parameterized by $\mathbf{r}(t) = t\mathbf{i} + \sqrt{1-t^2}\mathbf{j}$, $0 \le t \le 1$. Verify Theorem 1 for the vector field $\mathbf{F}(x, y) = x\mathbf{i} + xy\mathbf{j}$.

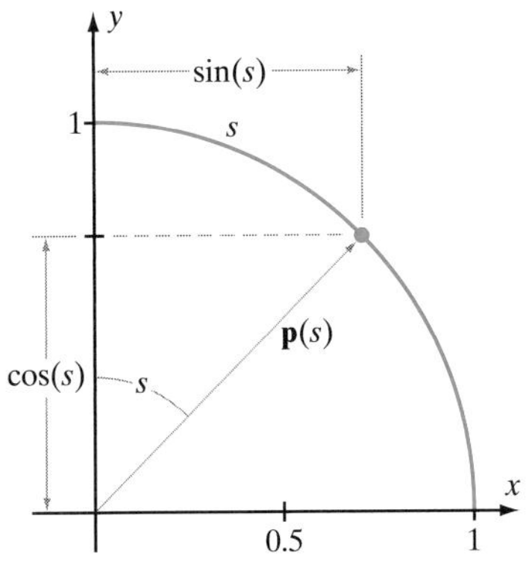

▲ Figure 4

Solution Because $\mathbf{F}(\mathbf{r}(t)) = \mathbf{F}(t, \sqrt{1-t^2}) = t\mathbf{i} + t\sqrt{1-t^2}\mathbf{j}$ and $\mathbf{r}'(t) = \mathbf{i} - (t/\sqrt{1-t^2})\mathbf{j}$, we have $\mathbf{F}(\mathbf{r}(t)) \cdot \mathbf{r}'(t) = t - t^2$. Therefore

$$\int_\mathcal{C} \mathbf{F} \cdot d\mathbf{r} = \int_0^1 \mathbf{F}(\mathbf{r}(t)) \cdot \mathbf{r}'(t)\,dt = \int_0^1 (t - t^2)\,dt = \left(\frac{t^2}{2} - \frac{t^3}{3}\right)\Bigg|_0^1 = \frac{1}{6}.$$

In this example, the arc length parameterization of $\mathcal{C}$ can be established from geometric principles: The length of an arc of a circle of radius a is a times the angle at the center that is subtended by the arc. As Figure 4 shows, this implies that $\mathbf{p}(s) = \sin(s)\mathbf{i} + \cos(s)\mathbf{j}$, $0 \le s \le \pi/2$, is the arc length parameterization of $\mathcal{C}$. Thus $L = \pi/2$, $\mathbf{F}(\mathbf{p}(s)) = \sin(s)\mathbf{i} + \sin(s)\cos(s)\mathbf{j}$, $\mathbf{T}(s) = \mathbf{p}'(s) = \cos(s)\mathbf{i} - \sin(s)\mathbf{j}$, and $\mathbf{F}(\mathbf{p}(s)) \cdot \mathbf{T}(s) = \sin(s)\cos(s) - \sin^2(s)\cos(s)$. We therefore have

$$\int_0^L \mathbf{F}(\mathbf{p}(s)) \cdot \mathbf{T}(s)\,ds = \int_0^{\pi/2} (\sin(s)\cos(s) - \sin^2(s)\cos(s))\,ds = \left(\frac{\sin^2(s)}{2} - \frac{\sin^3(s)}{3}\right)\Bigg|_0^{\pi/2} = \frac{1}{6}. \quad ◀$$

For every directed curve $\mathcal{C}$ from P_0 to P_1 there is a directed curve, often denoted by $-\mathcal{C}$, that is *opposite* to $\mathcal{C}$. It consists of the same points as $\mathcal{C}$ but has P_1 as its initial point and P_0 as its terminal point. The line integrals of a field $\mathbf{F}$ over the two directed curves $\mathcal{C}$ and $-\mathcal{C}$ are related by the formula

$$\int_\mathcal{C} \mathbf{F} \cdot d\mathbf{r} = -\int_{-\mathcal{C}} \mathbf{F} \cdot d\tilde{\mathbf{r}}, \tag{13.2.4}$$

where $\mathbf{r}$ and $\tilde{\mathbf{r}}$ denote parameterizations of $\mathcal{C}$ and $-\mathcal{C}$, respectively.

▶ **EXAMPLE 4** Let $t \mapsto \mathbf{r}(t), a \le t \le b$ be a parameterization of a directed curve $\mathcal{C}$. Prove that $t \mapsto \tilde{\mathbf{r}}(t) = \mathbf{r}(a+b-t), a \le t \le b$ is a parameterization of the opposite curve $-\mathcal{C}$. Verify equation (13.2.4).

Solution As t increases from a to b, the expression $a+b-t$ decreases from b to a. It follows that the values $\tilde{\mathbf{r}}(t) = \mathbf{r}(a+b-t)$, $a \le t \le b$, are the points of $\mathcal{C}$ traversed from $\mathbf{r}(b)$ to $\mathbf{r}(a)$. Because $\tilde{\mathbf{r}}'(t) = -\mathbf{r}'(a+b-t)$, we have

$$\int_{-\mathcal{C}} \mathbf{F} \cdot d\tilde{\mathbf{r}} = \int_a^b \mathbf{F}(\tilde{\mathbf{r}}(t)) \cdot \tilde{\mathbf{r}}'(t)\,dt = -\int_a^b \mathbf{F}(\mathbf{r}(a+b-t)) \cdot \mathbf{r}'(a+b-t)\,dt.$$

By making the change of variable $u = a+b-t$, $du = (-1)dt$ in this last integral, we obtain

$$\int_{-\mathcal{C}} \mathbf{F} \cdot d\tilde{\mathbf{r}} = \int_b^a \mathbf{F}(\mathbf{r}(u)) \cdot \mathbf{r}'(u)\,\frac{du}{dt}\,dt = \int_b^a \mathbf{F}(\mathbf{r}(u)) \cdot \mathbf{r}'(u)\,du = -\int_a^b \mathbf{F}(\mathbf{r}(u)) \cdot \mathbf{r}'(u)\,du = -\int_{\mathcal{C}} \mathbf{F} \cdot d\mathbf{r}. \quad ◀$$

INSIGHT Equation (13.2.4) is a generalization of the formula $\int_b^a f(t)dt = -\int_a^b f(t)dt$, from one-variable calculus. Just as on the real axis, reversing the direction of integration in a line integral changes the sign of the integral's value.

Dependence on Path

We have observed that, if $\mathcal{C}$ is a directed curve from P_0 to P_1, then the line integral $\int_{\mathcal{C}} \mathbf{F} \cdot d\mathbf{r}$ does not depend on the parameterization of $\mathcal{C}$. It is reasonable to wonder if the line integral depends on the path at all. That is, will a different path $\mathcal{C}_*$ between P_0 and P_1 lead to a line integral $\int_{\mathcal{C}_*} \mathbf{F} \cdot d\mathbf{r}_*$ with the same value as $\int_{\mathcal{C}} \mathbf{F} \cdot d\mathbf{r}$? The next example shows that the answer to this question is generally *No*.

▶ **EXAMPLE 5** Let $\mathbf{F}(x,y) = -y\mathbf{i} + x\mathbf{j}$. Set $P_0 = (1,0)$ and $P_1 = (-1,0)$. Consider two paths from P_0 to P_1: $\mathcal{C}$ parameterized by

$$\mathbf{r}(t) = \cos(t)\mathbf{i} + \sin(t)\mathbf{j}, \quad 0 \le t \le \pi$$

and $\mathcal{C}_*$ parameterized by

$$\mathbf{r}_*(t) = \cos(t)\mathbf{i} - \sin(t)\mathbf{j}, \quad 0 \le t \le \pi.$$

Is the line integral of $\mathbf{F}$ over $\mathcal{C}$ equal to the line integral of $\mathbf{F}$ over $\mathcal{C}_*$?

Solution We calculate

$$\begin{aligned}\int_{\mathcal{C}} \mathbf{F} \cdot d\mathbf{r} = \int_0^\pi \mathbf{F}(\mathbf{r}(t)) \cdot \mathbf{r}'(t)\,dt &= \int_0^\pi (-\sin(t)\mathbf{i} + \cos(t)\mathbf{j}) \cdot (-\sin(t)\mathbf{i} + \cos(t)\mathbf{j})\,dt \\ &= \int_0^\pi (\sin^2(t) + \cos^2(t))\,dt = \pi.\end{aligned}$$

However,

$$\begin{aligned}\int_{\mathcal{C}_*} \mathbf{F} \cdot d\mathbf{r}_* &= \int_0^\pi \mathbf{F}(\tilde{\mathbf{r}}(t)) \cdot \mathbf{r}_*'(t)\,dt = \int_0^\pi (\sin(t)\mathbf{i} + \cos(t)\mathbf{j}) \cdot (-\sin(t)\mathbf{i} - \cos(t)\mathbf{j})\,dt \\ &= -\int_0^\pi (\sin^2(t) + \cos^2(t))\,dt = -\pi.\end{aligned}$$

The line integrals have different values. ◀

INSIGHT Figure 5 suggests why it is plausible that the two line integrals of Example 5 should be different. The line integral over $\mathcal{C}$ is counterclockwise along the upper half of the unit circle: It is in the direction of the vector field $\mathbf{F}$. However, the line integral over $\mathcal{C}_*$ is clockwise along the lower half of the unit circle: It is *against* the direction of the vector field. Thus we would anticipate the values of the integrals to have opposite signs. It makes sense that the value of a line integral over a path from P_0 to P_1 ought to depend on the path between the points. But it turns out that line integrals of conservative vector fields *are* independent of path. Section 13.3 treats these ideas in depth.

Other Notation for the Line Integral

If $\mathbf{F} = M(x,y,z)\mathbf{i} + N(x,y,z)\mathbf{j} + R(x,y,z)\mathbf{k}$ is a vector field defined on a directed curve $\mathcal{C}$ that is parameterized by $\mathbf{r}(t) = \langle x(t), y(t), z(t)\rangle, a \le t \le b$, then the line integral $\int_{\mathcal{C}} \mathbf{F}\cdot d\mathbf{r} = \int_a^b \mathbf{F}(\mathbf{r}(t))\cdot \mathbf{r}'(t)dt$ becomes

$$\int_{\mathcal{C}} \mathbf{F}\cdot d\mathbf{r} = \int_a^b \left(M\big(x(t), y(t), z(t)\big)\frac{dx}{dt} + N\big(x(t), y(t), z(t)\big)\frac{dy}{dt} + R\big(x(t), y(t), z(t)\big)\frac{dz}{dt}\right) dt \qquad (13.2.5)$$

when expanded. Using the formalism of calculus,

$$\frac{dx}{dt}dt = dx, \quad \frac{dy}{dt}dt = dy, \quad \text{and} \quad \frac{dz}{dt}dt = dz,$$

we may write equation (13.2.5) more compactly as

$$\int_{\mathcal{C}} \mathbf{F}\cdot d\mathbf{r} = \int_{\mathcal{C}} M\,dx + N\,dy + R\,dz. \qquad (13.2.6)$$

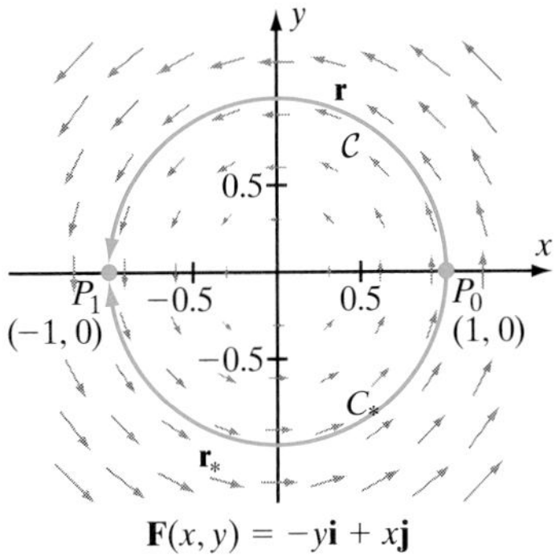

▲ Figure 5

▶ **EXAMPLE 6** Let $\mathcal{C}$ be the directed curve that is parameterized by $\mathbf{r}(t) = \langle 1-t, 2-t^2, 1\rangle$, $-1 \le t \le 2$. Calculate $\int_{\mathcal{C}} xy\,dx + z\,dy + x\,dz$.

Solution We use formula (13.2.6) with $x(t) = 1-t$, $y(t) = 2-t^2$, and $z(t) = 1$. Then $dx/dt = -1$, $dy/dt = -2t$, and $dz/dt = 0$. As a result, we have

$$\begin{aligned}\int_{\mathcal{C}} xy\,dx + z\,dy + x\,dz &= \int_{-1}^{2} \left((1-t)(2-t^2)\frac{dx}{dt} + 1\frac{dy}{dt} + (1-t)\frac{dz}{dt}\right)dt \\ &= \int_{-1}^{2} \Big((1-t)(2-t^2)(-1) + (1)(-2t) + (1-t)(0)\Big)dt.\end{aligned}$$

On expanding the integrand, we obtain

$$\begin{aligned}\int_{\mathcal{C}} xy\,dx + z\,dy + x\,dz &= \int_{-1}^{2} (-2+t^2-t^3)\,dt = \left(-2t + \frac{t^3}{3} - \frac{t^4}{4}\right)\Bigg|_{-1}^{2} \\ &= \left(-4 + \frac{8}{3} - \frac{16}{4}\right) - \left(2 - \frac{1}{3} - \frac{1}{4}\right) = -\frac{27}{4}.\end{aligned}$$ ◀

Closed Curves

If a directed curve $\mathcal{C}$ is parameterized by $t \mapsto \mathbf{r}(t)$ for $a \le t \le b$ and if $\mathbf{r}(a) = \mathbf{r}(b)$, then we say that the curve $\mathcal{C}$ is *closed*. A closed curve begins and ends at the same point. We often write

$$\oint_C \mathbf{F} \cdot d\mathbf{r} \qquad \text{or} \qquad \oint_C M\,dx + N\,dy + R\,dz$$

for the line integral of a vector field $\mathbf{F} = \langle M, N, R \rangle$ over a closed, directed curve $\mathcal{C}$. The small circle in the middle of the integral sign reminds us that the curve is closed.

▶ **EXAMPLE 7** Calculate $\oint_{\mathcal{C}} (x+y)\,dx + z\,dy + x^2 dz$, where $\mathcal{C}$ comprises the line segment from $P_0 = (1,0,0)$ to $P_1 = (0,1,0)$, the line segment from P_1 to $P_2 = (0,0,2)$, and the line segment from P_2 *to* P_0.

Solution Let $\mathcal{C}_1$ be the line segment from $P_0 = (1,0,0)$ to $P_1 = (0,1,0)$. We may parameterize it by $\mathbf{r}_1(t) = \overrightarrow{OP_0} + t(\overrightarrow{P_0P_1}) = \langle 1-t, t, 0 \rangle$, $0 \le t \le 1$. Then $dx/dt = -1$, $dy/dt = 1$, and $dz/dt = 0$. We have

$$\int_{\mathcal{C}_1} (x+y)\,dx + z\,dy + x^2 dz = \int_0^1 \left(((1-t)+t)\frac{dx}{dt} + 0\frac{dy}{dt} + (1-t)^2\frac{dz}{dt} \right) dt = \int_0^1 (-1+0+0)\,dt = -1.$$

Similarly, we parameterize the directed line segment $\mathcal{C}_2$ from $P_1 = (0,1,0)$ to $P_2 = (0,0,2)$ by $\mathbf{r}_2(t) = \overrightarrow{OP_1} + t(\overrightarrow{P1\,P_2}) = \langle 0, 1-t, 2t \rangle$, $0 \le t \le 1$. Then $dx/dt = 0$, $dy/dt = -1$, $dz/dt = 2$, and

$$\int_{\mathcal{C}_2} (x+y)\,dx + z\,dy + x^2 dz = \int_0^1 \left(\left(0 + (1-t)\right)\frac{dx}{dt} + 2t\frac{dy}{dt} + 0^2\frac{dz}{dt} \right) dt = \int_0^1 (0 - 2t + 0)\,dt = -1.$$

Finally, we parameterize the directed line segment $\mathcal{C}_3$ from $P_2 = (0,0,2)$ to $P_0 = (1,0,0)$ by $\mathbf{r}_3(t) = \overrightarrow{OP_2} + t(\overrightarrow{P_2P_0}) = \langle 0,0,2 \rangle + t\langle 1,0,-2 \rangle = \langle t, 0, 2-2t \rangle, 0 \le t \le 1$. Then, $dx/dt = 1$, $dy/dt = 0$, $dz/dt = -2$, and

$$\int_{\mathcal{C}_3} (x+y)\,dx + z\,dy + x^2 dz = \int_0^1 \left((t+0)\frac{dx}{dt} + (2-2t)\frac{dy}{dt} + t^2\frac{dz}{dt} \right) dt = \int_0^1 (t + 0 - 2t^2)\,dt = -\frac{1}{6}.$$

It follows that $\oint_{\mathcal{C}} (x+y)\,dx + z\,dy + x^2 dz = -1 + (-1) + (-1/6) = -13/6$. ◀

Other Applications of Line Integrals

Suppose that $\mathcal{C}$ is a directed curve parameterized by arc length $s \mapsto \mathbf{p}(s)$, $0 \le s \le L$. Then equation (13.2.3) tells us that

$$\int_{\mathcal{C}} \mathbf{F} \cdot d\mathbf{r} = \int_0^L f(\mathbf{p}(s))\,ds, \tag{13.2.7}$$

where $f(\mathbf{p}(s)) = \mathbf{F}(\mathbf{p}(s)) \cdot \mathbf{T}(s)$. Now the integral on the right side of equation (13.2.7) makes sense for any scalar-valued function f whose domain contains $\mathcal{C}$. For example, if $f(x,y,z)$ is identically 1, then integral (13.2.7) is the arc length of $\mathcal{C}$, a topic that we studied in Chapter 10.

We obtain a new application of the line integral by supposing that $f(x,y,z)$ is the mass density of a thin wire that has the shape $\mathcal{C}$. If we divide $\mathcal{C}$ into N arcs of equal length $\Delta s = L/N$ and choose a point $\mathbf{p}(s_j)$ in the jth arc, then the mass of the wire is approximately $\sum_{j=1}^{N} f(\mathbf{p}(s_j))\Delta s$. The approximation tends to become more accurate as N tends to infinity. At the same time,

$$\lim_{N\to\infty} \sum_{j=1}^{N} f(\mathbf{p}(s_j))\Delta s = \int_0^L f(\mathbf{p}(s))\, ds.$$

Thus when f is a mass density, the right side of equation (13.2.7) represents the total mass of the curve. Of course similar interpretations can be made when f is another type of density.

The right side of equation (13.2.7) is often written as $\int_{\mathcal{C}} f\, ds$. The proof of Theorem 1 shows that, in general, if $\mathbf{r}$ is any parameterization of $\mathcal{C}$, not necessarily the arc length parameterization, then

$$\int_{\mathcal{C}} f\, ds = \int_a^b f(\mathbf{r}(t))\, \|\mathbf{r}'(t)\|\, dt. \tag{13.2.8}$$

▶ **EXAMPLE 8** Suppose that a wire has the shape of the curve $\mathcal{C}$ that is parameterized by $\mathbf{r}(t) = \langle 3t, 4t, t^2\rangle$, $0 \le t \le 6$. The charge density of the wire is given by $q(x, y, z) = 9x - y + \sqrt{z}$. What is the total charge held by the wire?

Solution We use formula (13.2.8). Observe that $\mathbf{r}'(t) = \langle 3, 4, 2t\rangle$ and $\|\mathbf{r}'(t)\| = \sqrt{3^2 + 4^2 + (2t)^2} = \sqrt{25 + 4t^2}$. The total charge is therefore given by

$$\int_{\mathcal{C}} q\, ds = \int_0^6 q(\mathbf{r}(t))\, \|\mathbf{r}'(t)\| dt = \int_0^6 \left(9(3t) - (4t) + \sqrt{t^2}\right)\sqrt{25 + 4t^2}\, dt = \int_0^6 24t\sqrt{25 + 4t^2}\, dt.$$

Making the change of variable $u = 25 + 4t^2$, $du = 8t\, dt$, and noting the limits of integration $25 + 4(0)^2$, or 25, and $25 + 4(6)^2$, or 169, for u, we have

$$\int_{\mathcal{C}} q\, ds = \int_0^6 24t\sqrt{25 + 4t^2}\, dt = 3\int_{25}^{169} u^{1/2}\, du = 2u^{3/2}\Big|_{25}^{169} = 2(13^3 - 5^3) = 4144.$$ ◀

QUICK QUIZ

1. What is the work done by the planar force field $\mathbf{F}(x, y) = \langle y, x\rangle$ moving a particle along the path $\mathbf{r}(t) = \langle t, t^2\rangle$, $1 \le t \le 2$?
2. True or false: The work performed by a force field $\mathbf{F}$ over a directed curve $\mathcal{C}$ does not depend on the choice of parameterization of $\mathcal{C}$ provided that the parameterization respects the direction of $\mathcal{C}$.
3. What is $\int_{\mathcal{C}} 2x\, dx + 2y\, dy + 5\sqrt{xz}\, dz$ if $\mathcal{C}$ is the curve parameterized by $\mathbf{r}(t) = \langle t, t^2, t^3\rangle$, $0 \le t \le 1$?
4. If $\mathcal{C}$ is the curve that is parameterized by $\mathbf{r}(t) = \langle 3t, 2t^2\rangle$, $0 \le t \le 1$ and if $f(x, y) = 8x$, then what is $\int_{\mathcal{C}} f\, ds$?

Answers

1. 7 2. True 3. 5 4. 49

EXERCISES

Problems for Practice

▶ In each of Exercises 1–8, calculate $\int_C \mathbf{F}\cdot d\mathbf{r}$ for the given planar vector field $\mathbf{F}$ and the given parametric curve $\mathbf{r}$. ◀

1. $\mathbf{F}(x,y)=xy\mathbf{i}+y^2\mathbf{j},\ \mathbf{r}(t)=\cos(t)\mathbf{i}+\sin(t)\mathbf{j}$ $\quad 0\le t\le \pi/3$
2. $\mathbf{F}(x,y)=y^2\mathbf{i}-x^2\mathbf{j},\ \mathbf{r}(t)=\sin(t)\mathbf{i}+\cos(t)\mathbf{j}$ $\quad 0\le t\le \pi$
3. $\mathbf{F}(x,y)=\sin(x)\mathbf{i}-\cos(y)\mathbf{j},\ \mathbf{r}(t)=t^2\mathbf{i}+t^3\mathbf{j}$ $\quad 0\le t\le \sqrt{\pi}$
4. $\mathbf{F}(x,y)=\ln(y)\mathbf{i}-e^x\mathbf{j},\ \mathbf{r}(t)=\ln(t)\mathbf{i}+t^3\mathbf{j}$ $\quad 1\le t\le e$
5. $\mathbf{F}(x,y)=xy\mathbf{i}+x^2y^2\mathbf{j},\ \mathbf{r}(t)=t^{1/2}\mathbf{i}+t^{-3/2}\mathbf{j}$ $\quad 1\le t\le 4$
6. $\mathbf{F}(x,y)=y^{1/2}\mathbf{i}+x^{3/2}\mathbf{j},\ \mathbf{r}(t)=t^2\mathbf{i}+t^4\mathbf{j}$ $\quad -1\le t\le 2$
7. $\mathbf{F}(x,y)=(x^2+1)^{-1}\mathbf{j}-xy\mathbf{i},\ \mathbf{r}(t)=t\mathbf{i}+t^2\mathbf{j}$ $\quad -4\le t\le -1$
8. $\mathbf{F}(x,y)=y\mathbf{i}-x\mathbf{j},\ \mathbf{r}(t)=\cos(t)\mathbf{i}+\sin(t)\mathbf{j}$ $\quad -\pi/2\le t\le \pi/2$

▶ In each of Exercises 9–16, calculate $\int_C \mathbf{F}\cdot d\mathbf{r}$ for the given spatial vector field $\mathbf{F}$ and the given parametric curve $\mathbf{r}$. ◀

9. $\mathbf{F}(x,y,z)=xz\mathbf{i}-yx\mathbf{j}+xy\mathbf{k},\ \mathbf{r}(t)=t\mathbf{i}+t^2\mathbf{j}+t^3\mathbf{k},\ -1\le t\le 1$
10. $\mathbf{F}(x,y,z)=y\mathbf{i}-\sin(4z)\mathbf{j}+x\mathbf{k},\ \mathbf{r}(t)=\cos(t)\mathbf{i}+\sin(t)\mathbf{j}+(t/4)\mathbf{k},$ $-3\pi\le t\le -\pi/2$
11. $\mathbf{F}(x,y,z)=\cos(y)\mathbf{i}+\sin(z)\mathbf{j}+x\mathbf{k},\ \mathbf{r}(t)=\cos(t)\mathbf{i}+t\mathbf{j}+t\mathbf{k},$ $\pi/4\le t\le \pi/2$
12. $\mathbf{F}(x,y,z)=(x/y)\mathbf{i}-(y/z)\mathbf{j}+xz\mathbf{k},$ $\mathbf{r}(t)=t^3\mathbf{i}+t^2\mathbf{j}+(1/t)\mathbf{k},\ -2\le t\le -1$
13. $\mathbf{F}(x,y,z)=z\mathbf{i}+\mathbf{k},\ \mathbf{r}(t)=t\mathbf{i}+\mathbf{j}+\cos^2(t)\mathbf{k},\ 0\le t\le \pi$
14. $\mathbf{F}(x,y,z)=y\mathbf{i}+(x-z)\mathbf{j}+z\mathbf{k},$ $\mathbf{r}(t)=\sin(t)\mathbf{i}+\cos(t)\mathbf{j}+\sin(t)\mathbf{k},\ 0\le t\le 2\pi$
15. $\mathbf{F}(x,y,z)=\dfrac{x}{x^2+y^2}\mathbf{k},\ \mathbf{r}(t)=\sin(t)\mathbf{i}+\cos(t)\mathbf{j}+t\mathbf{k},\ 0\le t\le \pi$
16. $\mathbf{F}(x,y,z)=yz\mathbf{i}+\mathbf{j}+\dfrac{1}{1+y^2}\mathbf{k},\ \mathbf{r}(t)=t\mathbf{i}+\tan(t)\mathbf{j}+\tan(t)\mathbf{k},$ $0\le t\le \pi/4$

▶ In each of Exercises 17–20, calculate the work performed by a force $\mathbf{F}$ in moving a particle along the parametric curve $\mathbf{r}$. ◀

17. $\mathbf{F}(x,y)=xy\mathbf{i}+x^3\mathbf{j},\ \mathbf{r}(t)=t^{1/2}\mathbf{i}+t^{1/4}\mathbf{j}$ $\quad 1\le t\le 16$
18. $\mathbf{F}(x,y)=3y^2\mathbf{i}-2x^3y\mathbf{j},\ \mathbf{r}(t)=\cos(t)\mathbf{i}+\sin(t)\mathbf{j}$ $\quad 0\le t\le \pi$
19. $\mathbf{F}(x,y,z)=x^2\mathbf{i}+y^2\mathbf{j}+z^2\mathbf{k},$ $\mathbf{r}(t)=(1+t^2)\mathbf{i}+t^2\mathbf{j}+(2+\sin(\pi t))\mathbf{k}$ $\quad 0\le t\le 1$
20. $\mathbf{F}(x,y,z)=(x+y)\mathbf{i}-(y-x)\mathbf{j}+4z\mathbf{k},$ $\mathbf{r}(t)=\sin(3t)\mathbf{i}+\cos(3t)\mathbf{j}+3t\mathbf{k}$ $\quad 0\le t\le 2\pi$

▶ In each of Exercises 21–24, calculate the given line integral over the directed curve C that is parameterized by $\mathbf{r}(t)=t^2\mathbf{i}+t^3\mathbf{j}+t^4\mathbf{k},\ 0\le t\le 1$. ◀

21. $\int_C xyz\,dx$
22. $\int_C dx+dy+dz$
23. $\int_C xz\,dx+yz\,dy+xy\,dz$
24. $\int_C xe^z\,dx+yz\,dy$

▶ In each of Exercises 25–28, calculate the given line integral $\oint_C M\,dx+N\,dy$ where C is the triangle with vertices $P_0=(0,1),\ P_1=(2,1),\ P_2=(3,4)$ with counterclockwise orientation. ◀

25. $\oint_C x\,dy$
26. $\oint_C x\,dx+dy$
27. $\oint_C (2x+y)\,dx+y\,dy$
28. $\oint_C y\,dx-x\,dy$

▶ In each of Exercises 29–32, calculate the line integral $\int_C f\,ds$ for the given function f and parametric curve $\mathbf{r}$. ◀

29. $f(x,y)=x^2+y^2$ $\quad \mathbf{r}(t)=\sin(t)\mathbf{i}+\cos(t)\mathbf{j}$ $\quad 0\le t\le 2\pi$
30. $f(x,y)=y/x$ $\quad \mathbf{r}(t)=t\mathbf{i}+t^2\mathbf{j}$ $\quad 1\le t\le 2$
31. $f(x,y)=x+\sqrt{y}$ $\quad \mathbf{r}(t)=t^2\mathbf{i}+t^4\mathbf{j}$ $\quad 1\le t\le 2$
32. $f(x,y)=ye^x$ $\quad \mathbf{r}(t)=t\mathbf{i}+e^t\mathbf{j}$ $\quad 0\le t\le 1$

Further Theory and Practice

▶ In each of Exercises 33–36, a vector field $\mathbf{F}$ is given. Calculate $\int_{C_i}\mathbf{F}\cdot d\mathbf{r_i}$ for $i=1,2,3$ where C_1 is parameterized by

$$\mathbf{r}_1(t)=\langle\cos(2t),1-2t/\pi\rangle,0\le t\le \pi,$$

C_2 is parameterized by

$$\mathbf{r}_2(t)=\langle 1+\sin(2\pi t),\cos(\pi t)\rangle,0\le t\le 1,$$

and C_3 is parameterized by

$$\mathbf{r}_3(t)=\langle 1,1-2t\rangle,0\le t\le 1.$$

Notice that all three paths begin at $(1,1)$ and terminate at $(1,-1)$. Verify that, for the given $\mathbf{F}$, the values of the line integrals are the same for all three paths. (The vector fields of Exercises 33–36 are "path-independent." Section 13.3 is devoted to this type of vector field.) ◀

33. $\mathbf{F}(x,y)=3\mathbf{i}-2\mathbf{j}$
34. $\mathbf{F}(x,y)=\mathbf{i}+y\mathbf{j}$
35. $\mathbf{F}(x,y)=x\mathbf{i}+y\mathbf{j}$
36. $\mathbf{F}(x,y)=y\mathbf{i}+x\mathbf{j}$

37. Let $\mathbf{F}(x,y)=2y\mathbf{i}+x\mathbf{j}$. Calculate $\int_{C_i}\mathbf{F}\cdot d\mathbf{r}_i$ for $i=1,2,3$ where C_1, C_2, and C_3 are the parametric curves described in the instructions to Exercises 33–36. The values will not all be the same in this case.
38. Calculate $\int_C xe^z\,dx+yz\,dy+xe^y\,dz$ over the directed curve C that is parameterized by $\mathbf{r}(t)=t^2\mathbf{i}+t^3\mathbf{j}+t^4\mathbf{k},\ 0\le t\le 1$.

▶ In each of Exercises 39–42, a closed curve C with counterclockwise orientation is specified. Calculate

$$\oint_C \frac{-y\,dx+x\,dy}{x^2+y^2}.$$ ◀

39. $\mathcal{C}$ is the circle $x^2 + y^2 = a^2$.

40. $\mathcal{C}$ is the square with vertices $(\pm 1, \pm 1)$.

41. $\mathcal{C}$ is the arc of the circle $x^2 + y^2 = 2$, from $(1, -1)$ to $(1, 1)$ followed by the line segment from $(1, 1)$ to $(1, -1)$.

42. $\mathcal{C}$ is the arc of the circle $x^2 + y^2 = 2$, from $(-1, -1)$ to $(-1, 1)$ followed by the line segment from $(-1, 1)$ to $(-1, -1)$.

43. Calculate $\oint_{\mathcal{C}} xyz\,dz$ where $\mathcal{C}$ is the closed curve of intersection of the cylinder $x^2 + y^2 = 1$ and the plane $z = 1 + x$, oriented counterclockwise when viewed from above.

44. Calculate $\oint_{\mathcal{C}} x\,dx + x^2\,dy + (y - z)\,dz$ where $\mathcal{C}$ is the arc of the curve of intersection of the cylinders $x^2 + y^2 = 1$ and $x^2 + z^2 = 1$ that lies in the first octant and that begins at the point $(1, 0, 0)$ and terminates at $(0, 1, 1)$.

Calculator/Computer Exercises

45. Let a and b denote the real roots of $4 + 2x - 3x^4$ with $a < b$. Calculate $\oint_{\mathcal{C}} dx + x^2\,dy$ where $\mathcal{C}$ is the piecewise-smooth closed curve that comprises the line segment from $(a, 0)$ to $(b, 0)$ and the arc of the graph of $y = 4 + 2x - 3x^4$ from $(b, 0)$ to $(a, 0)$.

46. Let P denote the point of intersection of the graphs of $y = x^{1/3}$ and $y = 1 - x^2$ in the first quadrant. Calculate $\oint_{\mathcal{C}} y^3\,dx + x\,dy$ where $\mathcal{C}$ is the piecewise-smooth closed curve that comprises the arc of the graph of $y = x^{1/3}$ from $(0, 0)$ to P, the arc of the graph of $y = 1 - x^2$ from P to $(0, 1)$, and the line segment $(0, 1)$ to $(0, 0)$.

47. Let $\mathcal{C}$ be the planar curve whose equation is $\mathbf{r}(t) = \langle t, t^2 \rangle, 0 \leq t \leq 1$. Calculate $\int_{\mathcal{C}} xe^y\,ds$.

48. Let $\mathcal{C}$ be the spatial curve whose equation is $\mathbf{r}(t) = \langle t, t^2, t^3 \rangle, 0 \leq t \leq 1$. Calculate $\int_{\mathcal{C}} \sqrt{x + y + z}\,ds$.

49. Let $\mathcal{C}$ be the planar curve whose equation in polar coordinates is $r = \theta, 0 \leq \theta \leq 2\pi$. Calculate $\int_{\mathcal{C}} x^2\,ds$.

50. Calculate $\oint_{\mathcal{C}} (x^2 + y^2)\,ds$ where $\mathcal{C}$ is the ellipse $x^2 + 2y^2 = 1$.

13.3 Conservative Vector Fields and Path Independence

In the preceding section, we learned that to compute the line integral $\int_{\mathcal{C}} \mathbf{F} \cdot d\mathbf{r}$ of a vector field $\mathbf{F}$ over a directed curve $\mathcal{C}$, we use a parameterization $t \mapsto \mathbf{r}(t), a \leq t \leq b$, of $\mathcal{C}$ and evaluate

$$\int_a^b \mathbf{F}(\mathbf{r}(t)) \cdot \mathbf{r}'(t)\,dt.$$

We observed that this integral does not depend on the choice of parameterization $\mathbf{r}$. However, the integrand *does* depend on the values $\mathbf{F}(\mathbf{r}(t))$ that $\mathbf{F}$ assumes along $\mathcal{C}$. Therefore if we were to use two different paths to connect a given pair of points, we would expect to obtain two line integrals with different evaluations. Indeed, Example 5 of the preceding section confirms this expectation: If we connect a pair of points by a path $\mathcal{C}$, then the value of $\int_{\mathcal{C}} \mathbf{F} \cdot d\mathbf{r}$ *can* depend on $\mathcal{C}$. There is, however, an important class of vector fields with line integrals that are *independent* of the path. In this section, we will study and characterize such vector fields.

DEFINITION Suppose that $\mathbf{F}$ is a vector field defined on a region $\mathcal{G}$. We say that $\mathbf{F}$ is *path-independent* on $\mathcal{G}$ if, for any two points P_0 and P_1 in $\mathcal{G}$, the line integral $\int_{\mathcal{C}} \mathbf{F} \cdot d\mathbf{r}$ has the same value for all directed curves $\mathcal{C}$ in $\mathcal{G}$ with initial point P_0 and terminal point P_1.

Our first theorem characterizes path-independent vector fields by means of their line integrals over *closed* curves.

THEOREM 1 Let $\mathbf{F}$ be a vector field on a region $\mathcal{G}$. Then $\mathbf{F}$ is path-independent on $\mathcal{G}$ if and only if $\oint_{\mathcal{C}} \mathbf{F} \cdot d\mathbf{r} = 0$ for every closed curve $\mathcal{C}$ in $\mathcal{G}$.

Proof. If $\mathcal{C}_1$ and $\mathcal{C}_2$ are two directed curves from P_0 to P_1, then the union of $\mathcal{C}_1$ and $-\mathcal{C}_2$ is a closed curve $\mathcal{C}$ (see Figure 1). On the other hand, if $\mathcal{C}$ is a closed curve with initial and terminal point P_0, then we may (arbitrarily) choose another point P_1 on $\mathcal{C}$ and write $\mathcal{C} = -\mathcal{C}_2 \cup \mathcal{C}_1$ where $\mathcal{C}_1$ is the path along $\mathcal{C}$ from P_0 to P_1 and $-\mathcal{C}_2$ is the path along $\mathcal{C}$ from P_1 to P_0. Then

$$\oint_{\mathcal{C}} \mathbf{F} \cdot d\mathbf{r} = \int_{\mathcal{C}_1} \mathbf{F} \cdot d\mathbf{r} + \int_{-\mathcal{C}_2} \mathbf{F} \cdot d\mathbf{r} \overset{(13.2.4)}{=} \int_{\mathcal{C}_1} \mathbf{F} \cdot d\mathbf{r} - \int_{\mathcal{C}_2} \mathbf{F} \cdot d\mathbf{r}.$$

Thus

$$\oint_{\mathcal{C}} \mathbf{F} \cdot d\mathbf{r} = 0 \quad \text{if and only if} \quad \int_{\mathcal{C}_1} \mathbf{F} \cdot d\mathbf{r} = \int_{\mathcal{C}_2} \mathbf{F} \cdot d\mathbf{r}.$$

■

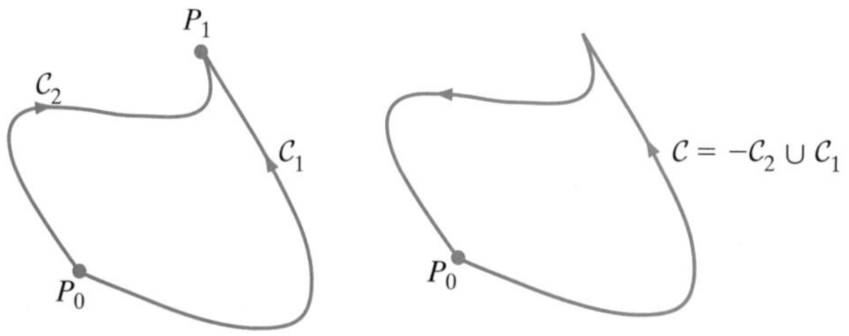

▲ Figure 1

At first glance, it might seem that few vector fields could have the remarkable property of path-independence. The next theorem produces an abundant supply of path-independent vector fields, however.

THEOREM 2 Suppose that u is a continuously differentiable scalar-valued function on a region $\mathcal{G}$ in the plane or in space. Let $\mathbf{F}$ be the gradient vector field defined by $\mathbf{F} = \nabla u$. Let $\mathcal{C}$ be a piecewise-smooth directed curve in $\mathcal{G}$ that has P_0 as its initial point and P_1 as its terminal point. Then

$$\int_{\mathcal{C}} \mathbf{F} \cdot d\mathbf{r} = u(P_1) - u(P_0). \tag{13.3.1}$$

In particular, the value of the line integral $\int_{\mathcal{C}} \mathbf{F} \cdot d\mathbf{r}$ depends on the initial and terminal points P_0 and P_1 but not on the path $\mathcal{C}$ that joins them. In other words, a gradient vector field is path-independent.

Proof. Suppose that $\mathcal{C}$ is a smooth curve. Let $t \mapsto \mathbf{r}(t)$, $a \le t \le b$, be any parameterization of $\mathcal{C}$. Then, using the Chain Rule, we see that

$$\begin{aligned}\int_C \mathbf{F}\cdot d\mathbf{r} &= \int_a^b \mathbf{F}(\mathbf{r}(t))\cdot\mathbf{r}'(t)\,dt\\ &= \int_a^b \nabla u(\mathbf{r}(t))\cdot\mathbf{r}'(t)\,dt\\ &= \int_a^b \frac{d}{dt}(u(\mathbf{r}(t)))\,dt\\ &= u(\mathbf{r}(t))\Big|_a^b\\ &= u(\mathbf{r}(b))-u(\mathbf{r}(a))\\ &= u(P_1)-u(P_0).\end{aligned}$$

The line integral depends on P_0, P_1, and u alone, as asserted. This proves the theorem in the smooth case.

Now suppose that $\mathcal{C}$ is a continuous, directed curve comprising two smooth pieces, $\mathcal{C}_1$ from P_0 to P' and $\mathcal{C}_2$ from P' to P_1. According to the smooth case that we have just proved,

$$\int_{\mathcal{C}} \mathbf{F}\cdot d\mathbf{r} = \int_{\mathcal{C}_1} \mathbf{F}\cdot d\mathbf{r}_1 + \int_{\mathcal{C}_2} \mathbf{F}\cdot d\mathbf{r}_2 = (u(P')-u(P_0)) + (u(P_1)-u(P')) = u(P_1)-u(P_0).$$

This proves the theorem in the case that the piecewise-smooth curve is made up of two smooth pieces. The case in which $\mathcal{C}$ consists of an arbitrary number of smooth pieces is proved in the same way. ■

INSIGHT It is worth noting the similarity between equation (13.3.1) and the Fundamental Theorem of Calculus. If $\mathbf{F}(x,y,z) = \nabla u(x,y,z)$, then u can be thought of as an antiderivative of the first component of $\mathbf{F}$ with respect to x, the second component of $\mathbf{F}$ with respect to y, and the third component of $\mathbf{F}$ with respect to z. Theorem 1 tells us that the line integral of $\mathbf{F}$ over $\mathcal{C}$ can be calculated by evaluating the antiderivative u at the end points of $\mathcal{C}$.

Recall from Section 13.1 that we say that a gradient vector field $\mathbf{F} = \nabla u$ is conservative and that $V = -u$ is a potential function for $\mathbf{F}$. Theorem 2 states that a conservative vector field is path-independent. An equivalent formulation is that a vector field with a potential function is path-independent.

▶ **EXAMPLE 1** Consider the vector field $\mathbf{F}(x,y,z) = yz\mathbf{i} + xz\mathbf{j} + xy\mathbf{k}$. Show that $\mathbf{F}$ is path independent. Calculate the line integral of $\mathbf{F}$ from $P_0 = (0,-1,2)$ to $P_1 = (2,1,4)$.

Solution Observe that $\mathbf{F} = \nabla u$ where $u(x,y,z) = xyz$. Therefore $\mathbf{F}$ is a conservative vector field. Equation (13.3.1) tells us that, if $\mathcal{C}$ is any directed curve from P_0 to P_1, then

$$\int_{\mathcal{C}} \mathbf{F}\cdot d\mathbf{r} = u(P_1) - u(P_0) = (2)(1)(4) - (0)(-1)(2) = 8.$$

Notice that we have evaluated $\int_{\mathcal{C}} \mathbf{F}\cdot d\mathbf{r}$ without actually having calculated a line integral. Let's verify our result by calculating $\int_{\mathcal{C}} \mathbf{F}\cdot d\mathbf{r}$ as a line integral. For simplicity, we take $\mathcal{C}$ to be a straight-line path joining P_0 to P_1. We can then parameterize $\mathcal{C}$ by

$$\mathbf{r}(t) = \overrightarrow{OP_0} + t\,\overrightarrow{P_0P_1} = \langle 0,-1,2\rangle + t\langle 2,2,2\rangle = \langle 2t,-1+2t,2+2t\rangle, \quad 0 \le t \le 1.$$

With these choices, we have

$$\mathbf{F}(\mathbf{r}(t)) = (-1+2t)(2+2t)\mathbf{i} + 2t(2+2t)\mathbf{j} + 2t(-1+2t)\mathbf{k} \quad \text{and} \quad \mathbf{r}'(t) = 2\mathbf{i}+2\mathbf{j}+2\mathbf{k}.$$

We calculate $\mathbf{F}(\mathbf{r}(t)) \cdot \mathbf{r}'(t) = (-1+2t)(2+2t)(2) + 2t(2+2t)2 + 2t(-1+2t)2 = -4+8t+24t^2$. Therefore the line integral of $\mathbf{F}$ from P_0 to P_1 is

$$\int_{\mathcal{C}} \mathbf{F}\cdot d\mathbf{r} = \int_0^1 (-4+8t+24t^2)\,dt = 8. \quad ◀$$

A Characterization of Path-Independent Vector Fields

Theorem 2 tells us that conservative vector fields are path-independent. The next theorem tells us that there are no other path-independent vector fields.

THEOREM 3 Suppose that $\mathbf{F}$ is a continuous vector field on a region $\mathcal{G}$ in the plane or in space. If $\mathbf{F}$ is path-independent, then $\mathbf{F} = \nabla u$ for some continuously differentiable function u.

Proof. Let $\mathbf{F} = \langle M, N, R\rangle$ be a continuous, path-independent vector field in a spatial region $\mathcal{G}$. (The planar case is proved in the same way.) Fix a point P_0 in $\mathcal{G}$. For each point $P = (x,y,z)$ in $\mathcal{G}$, set

$$u(x,y,z) = \int_{\mathcal{C}_P} \mathbf{F}\cdot d\mathbf{r},$$

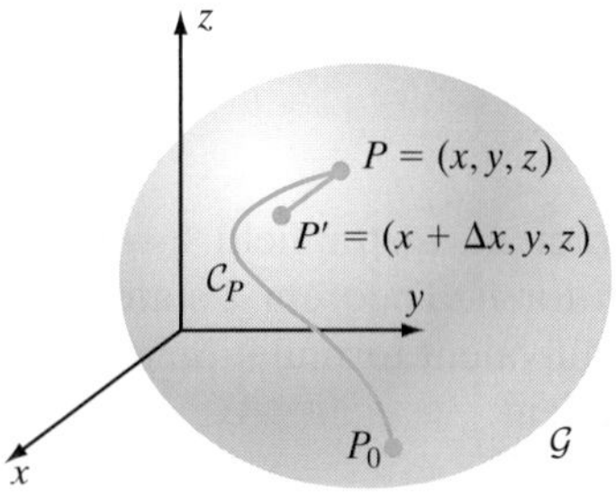

$\mathcal{C}_{P'} = \mathcal{C}_P \cup \overrightarrow{PP'}$

▲ Figure 2

where $\mathcal{C}_P$ is *any* curve that connects P_0 to P and lies entirely in $\mathcal{G}$. Such a curve exists because $\mathcal{G}$ is a region (and therefore path-connected). Moreover, because $\mathbf{F}$ is path independent, the value of $u(x,y,z)$ will be the same no matter what connecting path $\mathcal{C}_P$ we choose. This tells us that there is no ambiguity in the definition of $u(x,y,z)$. For sufficiently small Δx, we can choose the path $\mathcal{C}_{P'}$ from P_0 to $P' = (x+\Delta x, y, z)$ to be the piecewise-smooth curve that comprises $\mathcal{C}_P$ and the directed line segment $\overrightarrow{PP'}$ (Figure 2). We can parameterize this segment by $\tilde{\mathbf{r}}(t) = \langle x + t\Delta x, y, z\rangle, \quad 0 \le t \le 1.$ Because $\tilde{\mathbf{r}}'(t) = \langle \Delta x, 0, 0\rangle$ it follows that $\mathbf{F}(\tilde{\mathbf{r}}(t)) \cdot \tilde{\mathbf{r}}'(t) = M(x+t\Delta x, y, z)\Delta x$ and

$$\int_{\overrightarrow{PP'}} \mathbf{F}\cdot d\tilde{\mathbf{r}} = \int_0^1 M(x+t\Delta x, y, z)\Delta x\,dt = \Delta x \int_0^1 M(x+t\Delta x, y, z)\,dt.$$

Thus

$$\begin{aligned} u(x+\Delta x, y, z) - u(x,y,z) &= \left(\int_{\mathcal{C}_P} \mathbf{F}\cdot d\mathbf{r} + \int_{\overrightarrow{PP'}} \mathbf{F}\cdot d\tilde{\mathbf{r}}\right) - \int_{\mathcal{C}_P} \mathbf{F}\cdot d\mathbf{r} \\ &= \Delta x \int_0^1 M(x+t\Delta x, y, z)\,dt \end{aligned}$$

and

$$u_x(x,y,z) = \lim_{\Delta x \to 0} \frac{u(x+\Delta x, y, z) - u(x,y,z)}{\Delta x} = \lim_{\Delta x \to 0} \int_0^1 M(x + t\Delta x, y, z)\, dt.$$

In courses on real analysis, it is shown that this limit can be taken inside the integral:

$$\lim_{\Delta x \to 0} \int_0^1 M(x + t\Delta x, y, z)\, dt = \int_0^1 \lim_{\Delta x \to 0} M(x + t\Delta x, y, z)\, dt.$$

From this equality, it follows that

$$u_x(x,y,z) = \int_0^1 \lim_{\Delta x \to 0} M(x + t\Delta x, y, z)\, dt = \int_0^1 M(x,y,z)\, dt = M(x,y,z).$$

Similar arguments show that $u_y(x,y,z) = N(x,y,z)$ and $u_z(x,y,z) = R(x,y,z)$. Together these three equations show that $\mathbf{F} = \nabla u$. ■

▶ **EXAMPLE 2** Let $\mathbf{F}$ be a path-independent force field on a region $\mathcal{G}$. If we move a particle through the field from one point P_0 in $\mathcal{G}$ to another point P_1 in $\mathcal{G}$, then how does the work that we do relate to the change in potential energy of the particle?

Solution To move a particle from P_0 to P_1 along a path $\mathcal{C}$, we must apply a force opposite to $\mathbf{F}$. The work W that we do is therefore given by $W = \int_{\mathcal{C}} (-\mathbf{F}) \cdot d\mathbf{r}$. According to Theorem 2, there is a scalar-valued function u for which $\mathbf{F} = \nabla u$. Theorem 1 tells us that

$$W = -\int_{\mathcal{C}} \mathbf{F} \cdot d\mathbf{r} = -\big(u(P_1) - u(P_0)\big) = V(P_1) - V(P_0) \qquad (13.3.2)$$

where $V = -u$. Because V is a potential function for $\mathbf{F}$, the potential energy at each point P of $\mathcal{G}$ may be defined to be $V(P)$, as discussed in Section 13.1. Observe that the change in potential energy between any two points in $\mathcal{G}$ does not depend on the choice of potential function that is used to define potential energy. Indeed, equation (13.3.2) tells us that the work we do is equal to the change in the particle's potential energy. ◀

▶ **EXAMPLE 3** Suppose that $V(x,y) = mgy$ and that $\mathbf{F}(x,y) = -mg\mathbf{j} = -\nabla V(x,y)$ is the conservative planar force field of Example 9, Section 13.1. If we move a particle of mass m from the point $P_0 = (x_0, y_0)$ to the point $P_1 = (x_1, y_1)$, then how much work have we done?

Solution The simplest solution is to use Example 2, which tells us that the work W we have done is the change of potential energy, $V(P_1) - V(P_0)$, or $mgy_1 - mgy_0$. We can also obtain this result by calculating a line integral. The force we must apply is $\tilde{\mathbf{F}}(x,y) = mg\mathbf{j} = \nabla V(x,y)$, which is conservative. Because $\tilde{\mathbf{F}}$ is path-independent (Theorem 2), we may calculate $W = \int_{\mathcal{C}} \tilde{\mathbf{F}} \cdot d\mathbf{r}$ using any path joining P_0 to P_1. We will use the piecewise-smooth path $\mathcal{C}$ formed from the horizontal line segment $\mathcal{C}_h$ parameterized by $t \mapsto \langle x_0 + t(x_1 - x_0), y_0 \rangle$, $0 \le t \le 1$, followed by the vertical line segment $\mathcal{C}_v$ parameterized by $t \mapsto \langle x_1, y_0 + t(y_1 - y_0) \rangle$, $0 \le t \le 1$. We have

$$\int_{\mathcal{C}_h} \tilde{\mathbf{F}} \cdot d\mathbf{r} = \int_0^1 \tilde{\mathbf{F}}\big(x_0 + t(x_1 - x_0), y_0\big) \cdot \frac{d}{dt}\langle x_0 + t(x_1 - x_0), y_0 \rangle \, dt$$

$$= \int_0^1 \langle 0, mg \rangle \cdot \langle x_1 - x_0, 0 \rangle \, dt = \int_0^1 0 \, dt = 0$$

and

$$\int_{\mathcal{C}_v} \tilde{\mathbf{F}} \cdot d\mathbf{r} = \int_0^1 \tilde{\mathbf{F}}\big(x_1, y_0 + t(y_1 - y_0)\big) \cdot \frac{d}{dt}\langle x_1, y_0 + t(y_1 - y_0) \rangle \, dt$$

$$= \int_0^1 \langle 0, mg \rangle \cdot \langle 0, y_1 - y_0 \rangle dt = \int_0^1 mg(y_1 - y_0) dt = mg(y_1 - y_0).$$

Therefore

$$W = \int_{\mathcal{C}_h} \tilde{\mathbf{F}} \cdot d\mathbf{r} + \int_{\mathcal{C}_v} \tilde{\mathbf{F}} \cdot d\mathbf{r} = 0 + mg(y_1 - y_0) = mgy_1 - mgy_0. \quad ◀$$

Closed Vector Fields

Because it is not usually feasible to check path-independence directly, it is important to have a practical method for recognizing when a vector field is conservative. Our next theorem is a first step toward this goal: it provides us with an easily verifiable condition that will help us identify planar vector fields that are *not* conservative.

THEOREM 4 Let $\mathbf{F}(x, y) = M(x, y)\mathbf{i} + N(x, y)\mathbf{j}$ be a continuously differentiable vector field in a region $\mathcal{G}$. If $\boldsymbol{F}$ is conservative, then

$$\frac{\partial M}{\partial y} = \frac{\partial N}{\partial x} \tag{13.3.3}$$

at each point in $\mathcal{G}$.

Proof. Because $\mathbf{F}$ is conservative and continuously differentiable, there is a twice continuously differentiable function u such that

$$M\mathbf{i} + N\mathbf{j} = \mathbf{F} = \nabla u = \frac{\partial u}{\partial x}\mathbf{i} + \frac{\partial u}{\partial y}\mathbf{j}.$$

In other words, $M = \partial u / \partial x$ and $N = \partial u / \partial y$. It follows that

$$\frac{\partial}{\partial y}(M) = \frac{\partial}{\partial y}\left(\frac{\partial}{\partial x}u\right) = \frac{\partial}{\partial x}\left(\frac{\partial}{\partial y}u\right) = \frac{\partial}{\partial x}(N). \quad ■$$

A vector field $\mathbf{F}(x,y) = M(x,y)\mathbf{i} + N(x,y)\mathbf{j}$ that satisfies equation (13.3.3) at each point of a region $\mathcal{G}$ is said to be *closed* on $\mathcal{G}$. Theorem 4 says that every conservative vector field is closed. A logically equivalent reformulation is that any vector field that is not closed is not conservative.

▶ **EXAMPLE 4** Is the vector field $\mathbf{F}(x,y) = \left(x^2 - \sin(y)\right)\mathbf{i} + \left(y^3 + \cos(x)\right)\mathbf{j}$ conservative?

Solution Notice that

$$\frac{\partial}{\partial y}\left(x^2 - \sin(y)\right) = -\cos(y) \qquad \text{and} \qquad \frac{\partial}{\partial x}\left(y^3 + \cos(x)\right) = -\sin(x).$$

Because these expressions are unequal, Theorem 4 guarantees that $\mathbf{F}$ is not a conservative vector field. ◀

As we can see from Example 4, Theorem 4 gives us a handy way to tell when a planar vector field is *not* conservative: If the vector field does not satisfy equation (13.3.3), then the vector field is not conservative. However, Theorem 4 does *not* say anything about a planar vector field that does satisfy equation (13.3.3). In particular, Theorem 4 does *not* assert that a closed vector field is conservative. Indeed, the next example shows that a closed vector field need not be conservative.

▶ **EXAMPLE 5** Show that the vector field

$$\mathbf{F}(x,y) = \left(\frac{-y}{x^2+y^2}\right)\mathbf{i} + \left(\frac{x}{x^2+y^2}\right)\mathbf{j}$$

is closed but not conservative on the region $\mathcal{G} = \{(x,y) : (x,y) \neq (0,0)\}$.

Solution Observe that $\mathbf{F}$ is defined and differentiable on $\mathcal{G} = \{(x,y) : (x,y) \neq (0,0)\}$. To verify equation (13.3.3) on $\mathcal{G}$, we set $M(x,y) = -y/(x^2+y^2)$, $N(x,y) = x/(x^2+y^2)$, and calculate

$$\frac{\partial M}{\partial y} = \frac{\partial}{\partial y}\left(\frac{-y}{x^2+y^2}\right) = -\frac{x^2-y^2}{(x^2+y^2)^2} = \frac{\partial}{\partial x}\left(\frac{x}{x^2+y^2}\right) = \frac{\partial N}{\partial x}.$$

Thus $\mathbf{F}$ is closed on $\mathcal{G}$. To show that $\mathbf{F}$ is not conservative on $\mathcal{G}$, it suffices to observe that $\mathbf{F}$ is not path-independent. Set $P_0 = (1,0)$, and $P_1 = (-1,0)$, and consider two paths from P_0 to P_1: $\mathcal{C}$ parameterized by

$$\mathbf{r}(t) = \cos(t)\mathbf{i} + \sin(t)\mathbf{j}, \quad 0 \le t \le \pi,$$

and $\tilde{\mathcal{C}}$ parameterized by

$$\tilde{\mathbf{r}}(t) = \cos(t)\mathbf{i} - \sin(t)\mathbf{j}, \quad 0 \le t \le \pi.$$

We calculate that

$$\begin{aligned}\int_{\mathcal{C}} \mathbf{F}\cdot d\mathbf{r} &= \int_0^{\pi} \mathbf{F}(\mathbf{r}(t))\cdot \mathbf{r}'(t)dt = \int_0^{\pi}\left(-\frac{\sin(t)}{1}\mathbf{i} + \frac{\cos(t)}{1}\mathbf{j}\right)\cdot\left(-\sin(t)\mathbf{i} + \cos(t)\mathbf{j}\right)dt\\ &= \int_0^{\pi}\left(\sin^2(t) + \cos^2(t)\right)dt = \pi.\end{aligned}$$

However,

$$\int_{\tilde{\mathcal{C}}} \mathbf{F}\cdot d\tilde{\mathbf{r}} = \int_0^{\pi} \mathbf{F}(\tilde{\mathbf{r}}(t))\cdot\tilde{\mathbf{r}}'(t)\,dt = \int_0^{\pi}\left(\frac{\sin(t)}{1}\mathbf{i}+\frac{\cos(t)}{1}\mathbf{j}\right)\cdot(-\sin(t)\mathbf{i}-\cos(t)\mathbf{j})\,dt$$
$$= -\int_0^{\pi}(\sin^2(t)+\cos^2(t))\,dt = -\pi.$$

Our calculations show that **F** is not path-independent. We use Theorem 2 to conclude that **F** is not conservative. ◀

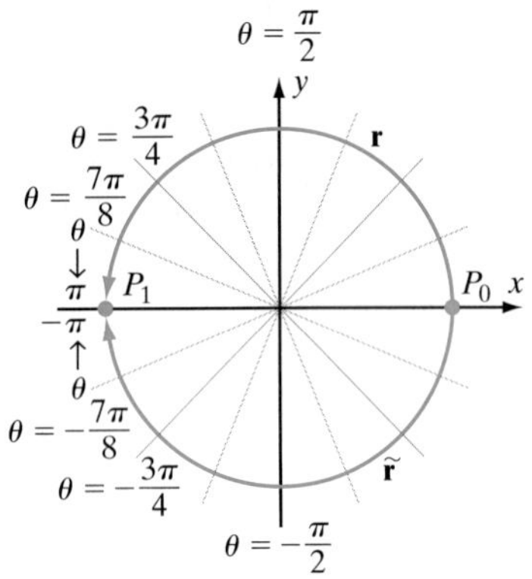

▲ Figure 3

INSIGHT If **F** is the vector field of Example 5 and if $\mathcal{G}$ is the region of Example 5, then it is instructive to investigate why we cannot find a differentiable function u such that $\nabla u(x,y) = \mathbf{F}(x,y)$ on $\mathcal{G}$. Using the method presented in Example 10 of Section 13.1 to solve the equations $u_x(x,y) = -y/(x^2+y^2)$ and $u_y(x,y) = x/(x^2+y^2)$, we integrate the equation $u_x(x,y) = -y/(x^2+y^2)$ with respect to x and find that $u(x,y) = \arctan(y/x) + \phi(y)$. If we substitute this expression for $u(x,y)$ into the equation $u_y(x,y) = x/(x^2+y^2)$, then we obtain $\phi'(y) = 0$, or $\phi(y) = C$ for some constant C. Thus if $\nabla u(x,y) = F(x,y)$, then $u(x,y) = \arctan(y/x) + C$. Now, on the region $\tilde{\mathcal{G}} = \{(x,y) : x > 0\}$, the function u is differentiable and $\mathbf{F}(x,y) = \nabla u(x,y)$. As a result, **F** is conservative on $\tilde{\mathcal{G}}$. Although we cannot extend the expression $\arctan(y/x) + C$ to a larger region because it is not defined for $x = 0$, we can exploit an alternative formulation. Notice that $\arctan(y/x)$ is the polar coordinate $\theta \in (-\pi, \pi)$ of each point (x,y) that does not lie on the negative x-axis $\{(x,0) : x \le 0\}$. This observation allows us to define $u(x,y) = \theta + C$ on the region $\mathcal{G}_*$ that we obtain by removing the negative x-axis from the plane. But we cannot extend the definition of u from $\mathcal{G}_*$ to all of $\mathcal{G}$. To understand why, notice that the polar angle θ increases from 0 to π as we traverse the curve $\mathcal{C}$ from $P_0 = (1,0)$ to $P_1 = (-1,0)$. See Figure 3. On the other hand, as we traverse the curve $\tilde{\mathcal{C}}$ from $P_0 = (1,0)$ to $P_1 = (-1,0)$, the polar angle θ decreases from 0 to $-\pi$. Consequently, there is no continuous and unambiguous way to define $u(x,y)$ on $\mathcal{G} = \{(x,y) : (x,y) \ne (0,0)\}$.

DEFINITION A planar or spatial region $\mathcal{G}$ is called *simply connected* if any closed curve in it can be continuously deformed to a point in $\mathcal{G}$ while staying within the region.

Roughly speaking, a planar region is simply connected if it has no holes in it. Figure 4 illustrates some regions that are *not* simply connected: Any closed curve that encircles a hole in one of these regions *cannot* be continuously deformed to a point while staying within the region. By contrast, the regions in Figure 5 *are* simply connected.

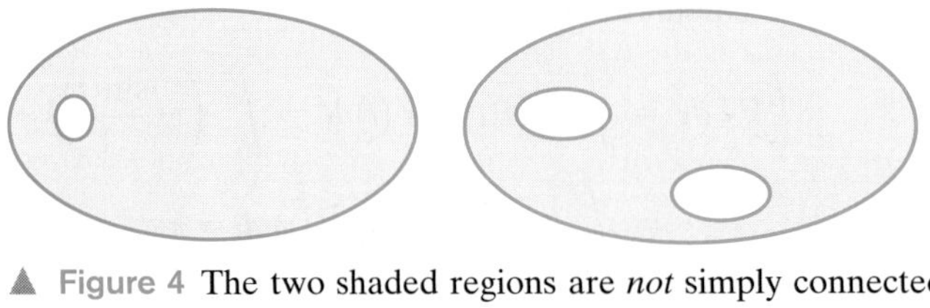

▲ Figure 4 The two shaded regions are *not* simply connected.

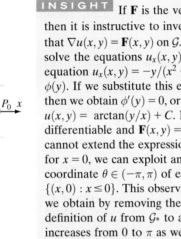

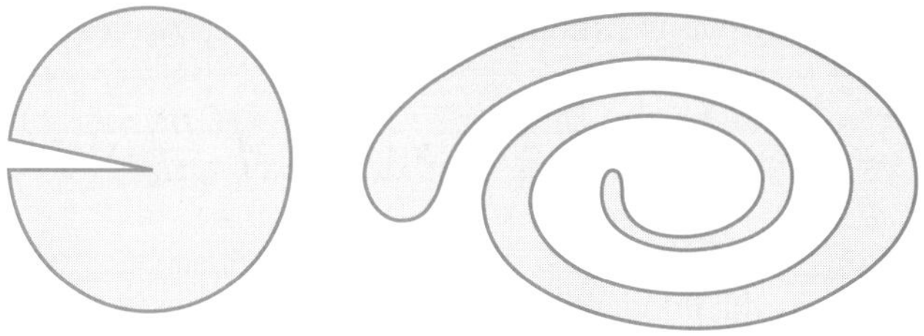

▲ Figure 5 The two shaded regions *are* simply connected.

If we remove one point from a simply connected planar region, then the resulting punctured region will not be simply connected. For instance, the region $\mathcal{G} = \{(x, y) : (x, y) \neq (0, 0)\}$ of Example 5 is not simply connected. That example shows that a closed vector field need not be conservative on a region that is punctured. It is a surprising fact that a closed vector field on a simply connected region *is* conservative. That is, we have the following theorem.

THEOREM 5 Let $\mathbf{F} = M(x, y)\mathbf{i} + N(x, y)\mathbf{j}$ be a continuously differentiable vector field on a simply connected planar region, $\mathcal{G}$. If $\partial M/\partial y = \partial N/\partial x$ on $\mathcal{G}$, then there is a twice continuously differentiable function u on $\mathcal{G}$ such that $\nabla u = \mathbf{F}$. In short, a closed vector field on a simply connected region is conservative.

Proof. We will outline the proof for a rectangle $\mathcal{G} = \{(x, y) : a < x < b, c < y < d\}$. For simplicity, we assume that the origin lies in $\mathcal{G}$. We define

$$u(x, y) = \int_0^x M(s, y)\, ds + \int_0^y N(0, t)\, dt.$$

Notice that, because our region $\mathcal{G}$ is a rectangle, we are integrating along paths that lie entirely in $\mathcal{G}$, and these integrals are guaranteed to make sense.

We assert that u satisfies $\nabla u = \mathbf{F}$. Let's see why. First we compute u_x. The second expression in the formula for u does not even depend on x, so we have

$$u_x(x, y) = \frac{\partial}{\partial x} \int_0^x M(s, y) ds.$$

The Fundamental Theorem of Calculus tells us that this last expression is $M(x, y)$, which is just what we want. Next we calculate u_y. We have

$$u_y(x, y) = \frac{\partial}{\partial y} \int_0^x M(s, y)\, ds + \frac{\partial}{\partial y} \int_0^y N(0, t)\, dt = \frac{\partial}{\partial y} \int_0^x M(s, y)\, ds + N(0, y).$$

It is a deep but plausible result that we may pass the differentiation under the integral sign to obtain

$$u_y(x, y) = \int_0^x \frac{\partial M}{\partial y}(s, y)\, ds + N(0, y).$$

Now we use the hypothesis that $\partial M/\partial y = \partial N/\partial x$ to obtain

$$u_y(x,y) = \int_0^x \frac{\partial N}{\partial x}(s,y)\,ds + N(0,y).$$

By the Fundamental Theorem of Calculus, we obtain $u_y(x,y) = \big(N(x,y) - N(0,y)\big) + N(0,y) = N(x,y)$ as desired. ■

▶ **EXAMPLE 6** Let $M(x,y) = \sin(y) + y\sin(x)$ and $N(x,y) = x\cos(y) - \cos(x) + 1$. Is the vector field $\mathbf{F}(x,y) = M(x,y)\mathbf{i} + N(x,y)\mathbf{j}$ conservative on the rectangle $\mathcal{G} = \{(x,y) : -1 < x < 1, -1 < y < 1\}$? If it is, then find a function u such that $\mathbf{F} = \nabla u$.

Solution We notice that $\mathcal{G}$ is simply connected and

$$\frac{\partial M}{\partial y}(x,y) = \cos(y) + \sin(x) = \frac{\partial N}{\partial x}(x,y).$$

Our vector field $\mathbf{F}$ passes the test: It is closed on $\mathcal{G}$. By Theorem 5, $\mathbf{F}$ is conservative.

Let us find a function u on $\mathcal{G}$ whose gradient is $\mathbf{F}$. Proceeding as in Example 10 of Section 13.1, we first study the equation

$$\frac{\partial u}{\partial x} = M(x,y) = \sin(y) + y\sin(x).$$

This equation tells us that u is an antiderivative with respect to x of the right-hand side. Therefore $u(x,y) = x\sin(y) - y\cos(x) + \phi(y)$, where the constant of integration $\phi(y)$ is independent of x but possibly dependent on y. We differentiate this formula for u, with respect to y, obtaining

$$\frac{\partial u}{\partial y} = x\cos(y) - \cos(x) + \phi'(y).$$

Because $\nabla u = \mathbf{F}$, we also know that

$$\frac{\partial u}{\partial y} = N(x,y) = x\cos(y) - \cos(x) + 1.$$

On equating our two expressions for $\partial u/\partial y$, we obtain $\phi'(y) = 1$. It follows that $\phi(y) = y + C$, where C is a numerical constant. Putting this information together yields $u(x,y) = x\sin(y) - y\cos(x) + y + C$. ◀

Vector Fields in Space

Theorems 4 and 5 pertain to planar vector fields, but they have analogues that apply to vector fields in space. Our next theorem states a necessary condition for a spatial vector field to be conservative.

THEOREM 6 If $\mathbf{F}(x,y,z) = M(x,y,z)\mathbf{i} + N(x,y,z)\mathbf{j} + R(x,y,z)\mathbf{k}$ is a continuously differentiable conservative vector field on a region $\mathcal{G}$, then all three of the following equalities hold on $\mathcal{G}$:

$$\frac{\partial M}{\partial y} = \frac{\partial N}{\partial x}, \quad \frac{\partial M}{\partial z} = \frac{\partial R}{\partial x}, \quad \frac{\partial N}{\partial z} = \frac{\partial R}{\partial y}. \tag{13.3.4}$$

To understand what Theorem 6 says, let's put it in words: If $\mathbf{F}(x,y,z)$ is conservative, then the derivative of the first component of $\mathbf{F}$ with respect to the second variable equals the derivative of the second component of $\mathbf{F}$ with respect to the first variable; the derivative of the first component of $\mathbf{F}$ with respect to the third variable equals the derivative of the third component of $\mathbf{F}$ with respect to the first variable; the derivative of the second component of $\mathbf{F}$ with respect to the third variable equals the derivative of the third component of $\mathbf{F}$ with respect to the second variable. A vector field $\mathbf{F}(x,y,z) = M(x,y,z)\mathbf{i} + N(x,y,z)\mathbf{j} + R(x,y,z)\mathbf{k}$ that satisfies the three equations of line (13.3.4) at every point of a region $\mathcal{G}$ is said to be a *closed* vector field on $\mathcal{G}$. Theorem 6 states that every conservative vector field on a spatial region is closed. Equivalently, a vector field that is not closed is not conservative.

▶ **EXAMPLE 7** Is the vector field in space given by $\mathbf{F}(x,y,z) = xz\mathbf{i} + yz\mathbf{j} + (y^2/2)\mathbf{k}$ a conservative vector field?

Solution Let's write $\mathbf{F}(x,y,z) = M(x,y,z)\mathbf{i} + N(x,y,z)\mathbf{j} + R(x,y,z)\mathbf{k}$, where $M(x,y,z) = xz$, $N(x,y,z) = yz$, and $R(x,y,z) = y^2/2$. We notice that the partial derivative of the first component of $\mathbf{F}$ with respect to the third variable z does not equal the partial derivative of the third component of $\mathbf{F}$ with respect to the first variable x:

$$\frac{\partial M}{\partial z} = x \neq 0 = \frac{\partial R}{\partial x}.$$

Because one of the equations of line (13.3.4) is not satisfied, the vector field $\mathbf{F}$ cannot be conservative. ◀

INSIGHT Whereas a planar vector field must satisfy only one equation to be conservative, a spatial vector field must satisfy three. The vector field $\mathbf{F}(x,y,z) = M(x,y,z)\mathbf{i} + N(x,y,z)\mathbf{j} + R(x,y,z)\mathbf{k}$ of Example 6 *does* satisfy two of the three necessary equations to be conservative, namely $M_y = 0 = N_x$ and $N_z = y = R_y$, but that does *not* suffice.

Theorem 6 has a converse that will work on geometrically simple regions.

THEOREM 7 Suppose that $\mathbf{F}(x,y,z) = M(x,y,z)\mathbf{i} + N(x,y,z)\mathbf{j} + R(x,y,z)\mathbf{k}$ is a continuously differentiable vector field on a simply connected region $\mathcal{G}$ in space. If $\mathbf{F}$ satisfies all three equations of line (13.3.4), then there is a twice continuously differentiable function u on $\mathcal{G}$ such that $\nabla u = \mathbf{F}$. In short, a closed vector field on a simply connected region in space is conservative.

▶ **EXAMPLE 8** Is the vector field $\mathbf{F}(x,y,z) = yz^2\mathbf{i} + (xz^2 - z)\mathbf{j} + (2xyz - y)\mathbf{k}$ conservative on the box $\mathcal{G} = \{(x,y,z) : 0 < x < 2, 0 < y < 2, 0 < z < 2\}$? If it is, find a function u for which $\mathbf{F} = \nabla u$.

Solution First we check whether $\mathbf{F}$ is closed. Because

$$\frac{\partial}{\partial y}(yz^2) = z^2 = \frac{\partial}{\partial x}(xz^2 - z), \qquad \frac{\partial}{\partial z}(yz^2) = 2yz = \frac{\partial}{\partial x}(2xyz - y),$$

and

$$\frac{\partial}{\partial z}(xz^2 - z) = 2xz - 1 = \frac{\partial}{\partial y}(2xyz - y),$$

we conclude that $\mathbf{F}$ is closed. By Theorem 7, it is conservative. Thus $\mathbf{F} = \nabla u$ for some function u. Let us find u. First, u is the antiderivative in the x variable of yz^2. It follows that

$$u(x,y,z) = xyz^2 + \phi(y,z). \tag{13.3.5}$$

Here the constant of integration does not depend on x, but it can depend on y and z. Now we differentiate each side of equation (13.3.5) with respect to y, treating x and z as constants:

$$\frac{\partial u}{\partial y} = xz^2 + \frac{\partial}{\partial y}\phi(y,z).$$

But the equality of the second components of the vector equation $\nabla u = \mathbf{F}$ also gives us

$$\frac{\partial u}{\partial y} = xz^2 - z.$$

We conclude that

$$xz^2 + \frac{\partial}{\partial y}\phi(y,z) = xz^2 - z$$

or, equivalently,

$$\frac{\partial}{\partial y}\phi(y,z) = -z.$$

It follows that

$$\phi(y,z) = -yz + \psi(z),$$

where now our constant of integration is independent of x and y but may depend on z. If we substitute our formula for $\phi(y,z)$ into equation (13.2.5), then we obtain

$$u(x,y,z) = xyz^2 - yz + \psi(z). \tag{13.3.6}$$

Finally, when we differentiate each side of equation (13.3.6) with respect to z, we obtain

$$\frac{\partial u}{\partial z} = 2xyz - y + \psi'(z).$$

But the equality of the third components of the vector equation $\nabla u = \mathbf{F}$ tells us that

$$\frac{\partial u}{\partial z} = 2xyz - y.$$

By equating the two expressions for $\partial u/\partial z$, we see that $\psi'(z) = 0$. We conclude that $\psi(z) = C$ for some numerical constant C. Substituting this into equation (13.3.6), we arrive at our final formula, $u(x, y, z) = xyz^2 - yz + C$. ◄

Summary of Principal Ideas

Let's summarize the key ideas that we have learned so far in this section. Let $\mathbf{F}$ be a continuously differentiable vector field either on a planar or spatial region $\mathcal{G}$. Then the following three properties are equivalent—if $\mathbf{F}$ possesses any one of them, then $\mathbf{F}$ possesses all three:

a. $\oint_C \mathbf{F} \cdot d\mathbf{r} = 0$ for every closed curve $\mathcal{C}$ in $\mathcal{G}$.

b. $\mathbf{F}$ is conservative: $\mathbf{F} = \nabla u$ for some u.

c. $\mathbf{F}$ is path-independent: $\int_{C_1} \mathbf{F} \cdot d\mathbf{r} = \int_{C_2} \mathbf{F} \cdot d\mathbf{r}$ for any two directed curves in $\mathcal{G}$ that share the same initial point and the same terminal point.

   If $\mathbf{F}$ possesses any one (and hence all) of properties a, b, or c, then $\mathbf{F}$ also possesses the following property:

d. $\mathbf{F}$ is closed: $\mathbf{F}$ satisfies equation (13.3.3) if planar, the three equations of line (13.3.4) if spatial.

If $\mathcal{G}$ is simply connected, and $\mathbf{F}$ possesses property d, then $\mathbf{F}$ possesses any one (and hence all) of properties a, b, and c. (The schematic diagram of implications and equivalences in Figure 6 may help you to organize these relationships.)

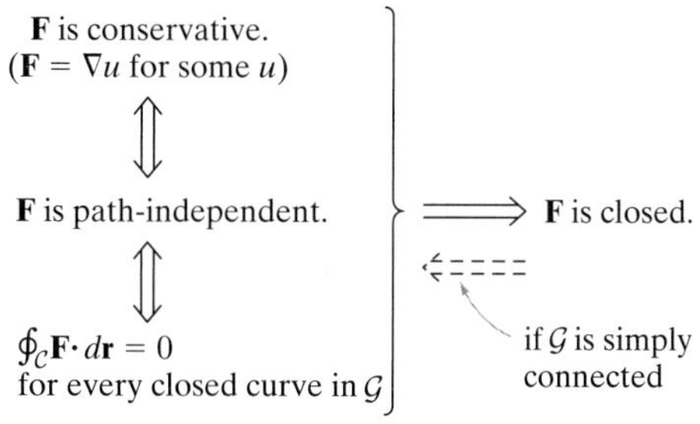

▲ Figure 6

QUICK QUIZ

1. True or false: If $M(u)$ and $N(u)$ are polynomials of one variable, then $\oint_C M(x)\,dx + N(y)\,dy = 0$ for every closed path in the plane.
2. True or false: For a vector field on a region, the properties of being path-independent and being closed are equivalent, even if the region is not simply connected.

3. True or false: For a vector field on a region, the properties of being path-independent and being conservative are equivalent even if the region is not simply connected.
4. True or false: If the first quadrant is removed from the plane, then the region that remains is not simply connected.

Answers

1. True 2. False 3. True 4. False

EXERCISES

Problems for Practice

▶ In each of Exercises 1–12, a planar vector field $\mathbf{F}$ is given. Determine if $\mathbf{F}$ is closed on the square $Q = \{(x,y) : -1<x<1,\ -1<y<1\}$. If $\mathbf{F}$ is closed, then find a potential function $V(x,y)$. ◀

1. $\mathbf{F}(x,y) = (x+\pi)\mathbf{i} + \mathbf{j}$
2. $\mathbf{F}(x,y) = (x^2 - y^2)\mathbf{i} + (2xy - y^3)\mathbf{j}$
3. $\mathbf{F}(x,y) = y^3\mathbf{i} + (3y^2x - 7y^6)\mathbf{j}$
4. $\mathbf{F}(x,y) = 2x\ln(y+2)\mathbf{i} - \dfrac{x^2}{y+2}\mathbf{j}$
5. $\mathbf{F}(x,y) = (y^3 - xy^2)\mathbf{i} - (x^2y^2 + x)\mathbf{j}$
6. $\mathbf{F}(x,y) = \dfrac{y}{(x+1)^2}\mathbf{i} + \dfrac{x}{x+1}\mathbf{j}$
7. $\mathbf{F}(x,y) = (x-2)^{-2}\mathbf{i} + (y-1/2)^2\mathbf{j}$
8. $\mathbf{F}(x,y) = x\cos(xy)\mathbf{i} + y\sin(xy)\mathbf{j}$
9. $\mathbf{F}(x,y) = 2x\sin(x^2 - y)\mathbf{i} - \sin(x^2 - y)\mathbf{j}$
10. $\mathbf{F}(x,y) = \dfrac{x}{1+x^2+y^2}\mathbf{i} + \dfrac{y}{1+x^2+y^2}\mathbf{j}$
11. $\mathbf{F}(x,y) = -\dfrac{y}{x^2+y^2}\mathbf{i} + \dfrac{x}{x^2+y^2}\mathbf{j}$
12. $\mathbf{F}(x,y) = \left(\dfrac{y}{3+x} - 2\ln(3+y)\right)\mathbf{i} + \left(\ln(3+x) - \dfrac{2x}{3+y}\right)\mathbf{j}$

▶ In each of Exercises 13–22, a spatial vector field $\mathbf{F}$ is given. Determine if $\mathbf{F}$ is closed on the cube $Q = \{(x,y,z) : -1<x,y,z<1\}$. If $\mathbf{F}$ is closed, then find a potential function $V(x,y,z)$. ◀

13. $\mathbf{F}(x,y,z) = (yz+1)\mathbf{i} + (xz+2)\mathbf{j} + (xy+3)\mathbf{k}$
14. $\mathbf{F}(x,y,z) = z^3\mathbf{i} - 2yz\mathbf{j} + (3xz^2 - y^2)\mathbf{k}$
15. $\mathbf{F}(x,y,z) = \dfrac{z}{2+x+y}\mathbf{i} + \dfrac{z}{2+x+y}\mathbf{j} + (y + \ln(2+x+y))\mathbf{k}$
16. $\mathbf{F}(x,y,z) = 3x^2y^2z^3\mathbf{i} + 2x^3yz^3\mathbf{j} + 3x^3y^2z\mathbf{k}$
17. $\mathbf{F}(x,y,z) = -\dfrac{z^2}{(x+y-4)^2}(\mathbf{i}+\mathbf{j}) + \dfrac{2z}{x+y-4}\mathbf{k}$
18. $\mathbf{F}(x,y,z) = (3+y^2)\mathbf{i} + (2xy + z^3)\mathbf{j} + (3yz^2 + 8z)\mathbf{k}$
19. $\mathbf{F}(x,y,z) = x\cos(z)\mathbf{i} - z\sin(y)\mathbf{j} + y\sin(x)\mathbf{k}$
20. $\mathbf{F}(x,y,z) = z\sin(y)\mathbf{i} + xz\cos(y)\mathbf{j} + z^2\sin(y)\mathbf{k}$
21. $\mathbf{F}(x,y,z) = z\sec^2(x+y)\mathbf{i} + z\sec^2(x+y)\mathbf{j} + \tan(x+y)\mathbf{k}$
22. $\mathbf{F}(x,y,z) = (xy + z + 2)^{-1}(y\mathbf{i} + x\mathbf{j} + \mathbf{k})$

▶ In each of Exercises 23–28, a vector field $\mathbf{F}$ is given, as are the initial point P_0 and the terminal point P_1 of a directed curve C. Verify that $\mathbf{F}$ is closed in a simply connected region that contains P_0 and P_1. Then, calculate the line integral $\int_C \mathbf{F}\cdot d\mathbf{r}$. ◀

23. $\mathbf{F}(x,y) = \dfrac{5}{(5+xy)^2}\mathbf{i} - \dfrac{x^2}{(5+xy)^2}\mathbf{j}$, $P_0 = (1,-1)$, $P_1 = (2,2)$
24. $\mathbf{F}(x,y) = (1 + \ln(xy))\mathbf{i} + (1 + x/y)\mathbf{j}$, $P_0 = (1,1)$, $P_1 = (\sqrt{2}, \sqrt{2})$
25. $\mathbf{F}(x,y) = (y^2 + ye^{xy})\mathbf{i} + (2xy + xe^{xy})\mathbf{j}$, $P_0 = (0,0)$, $P_1 = (-1,1)$
26. $\mathbf{F}(x,y,z) = (y^2 + z^3)\mathbf{i} + (2xy + 2yz^3)\mathbf{j} + (3y^2z^2 + 3xz^2)\mathbf{k}$, $P_0 = (1,-1,-1)$, $P_1 = (-1,1,0)$
27. $\mathbf{F}(x,y,z) = e^{y+2z}(\mathbf{i} + x\mathbf{j} + 2x\mathbf{k})$, $P_0 = (0,0,0)$, $P_1 = (-1,2,1)$
28. $\mathbf{F}(x,y,z) = (3x^2z + 1/x)\mathbf{i} + (1/z - 1/y)\mathbf{j} + (x^3 - y/z^2)\mathbf{k}$, $P_0 = (1,1,1)$, $P_1 = (4,2,2)$

▶ In each of Exercises 29–32 a conservative vector field $\mathbf{F}$ and three points P_0, P_1, and P_2 are given. Verify that ◀

$$\int_{\overrightarrow{P_0P_1}} \mathbf{F}\cdot d\mathbf{r} + \int_{\overrightarrow{P_1P_2}} \mathbf{F}\cdot d\mathbf{r} + \int_{\overrightarrow{P_2P_0}} \mathbf{F}\cdot d\mathbf{r} = 0.$$

29. $\mathbf{F}(x,y) = (y^2 + 2y)\mathbf{i} + (2xy + 2x)\mathbf{j}$, $P_0 = (0,0)$, $P_1 = (0,1)$, $P_2 = (1,1)$
30. $\mathbf{F}(x,y) = (2x - y)\mathbf{i} - x\mathbf{j}$, $P_0 = (1,1)$, $P_1 = (3,4)$, $P_2 = (2,5)$
31. $\mathbf{F}(x,y,z) = 2xy\mathbf{i} + x^2\mathbf{j} + 2z\mathbf{k}$, $P_0 = (0,0,1)$, $P_1 = (0,1,1)$, $P_2 = (1,1,1)$
32. $\mathbf{F}(x,y,z) = yz^2\mathbf{i} + xz^2\mathbf{j} + 2xyz\mathbf{k}$, $P_0 = (0,0,0)$, $P_1 = (-1,2,1)$, $P_2 = (0,2,0)$

Further Theory and Practice

33. If $\mathbf{F}(x,y,z) = yz\mathbf{i} + xz\mathbf{j} + xy\mathbf{k}$ and if C is any directed curve from $P_0 = (0,-1,2)$ to $P_1 = (2,1,4)$, then $\int_C \mathbf{F}\cdot d\mathbf{r} = 8$, as shown in Example 1. Verify this equation for the parameterized curve

$$\mathbf{r}(t) = \langle 12t^2 - 10t, 8t^3 - 8t^2 + 2t - 1, 4t^2 - 2t + 2\rangle$$

defined for $0 \le t \le 1$.

The gravitational force that a point mass m at the origin O exerts on a unit point mass at $P = (x, y, z) \ne O$ is $\mathbf{F} = -(Gmr^{-3})\overrightarrow{OP}$ where $\mathbf{r} = \|\overrightarrow{OP}|$. In Exercises 34 and 35 let $P_0 = (x_0, y_0, z_0) \ne O$ be a fixed point and let V be a fixed choice of potential function for $\mathbf{F}$. Let W_P be the work done in moving a unit mass from P_0 to $P = (x, y, z) \ne O$.

34. Calculate the work W_P. For what points Q is $W_Q = W_P$?

35. Show that $W_\infty = \lim_{P\to\infty} W_P$ exists in the following sense: For any $\varepsilon > 0$, there is an R such that $|W_P - W_\infty| < \varepsilon$ for all P with $\|\overrightarrow{OP}\| > R$. (Frequently in physics, a potential function is chosen so that W_∞ has a convenient value for a point P_0 of interest.)

36. Consider the square $Q = \{(x, y) : |x| < 1, |y| < 1\}$. Let $L_c = \{(0, y) : y > c\}$ $\mathcal{G} = \{(x, y) \in Q : (x, y) \notin L_c\}$. For what values of c is the region $\mathcal{G}$ simply connected?

37. Consider the square $Q = \{(x, y) : |x| < 1, |y| < 1\}$. Let $L_{c,d} = \{(0, y) : c < y < d\}$ and $\mathcal{G} = \{(x, y) \in Q : (x, y) \notin L_{c,d}\}$. Describe the pairs (c, d) for which the region $\mathcal{G}$ is simply connected.

38. Is the solid $\mathcal{U} = \{(x, y, z) : 1 < x^2 + y^2 + z^2 < 4\}$ simply connected? What about the solid $\mathcal{V} = \{(x, y, z) : 1 < x^2 + y^2, x^2 + y^2 + z^2 < 4\}$?

39. Sketch two simply connected spatial regions the intersection of which is not a simply connected region.

40. Verify that

$$\mathbf{F}(x, y) = \frac{2}{\sqrt{\pi}} \exp(-(x - y)^2)(\mathbf{i} - \mathbf{j})$$

is conservative. Calculate $\int_C \mathbf{F} \cdot d\mathbf{r}$ for a curve that joins $(0, 0)$ to $(1, 1)$.

Calculator/Computer Exercises

41. Verify that $\mathbf{F}(x, y) = \sin(\pi xy^2)(x^{-1}\mathbf{i} + 2y^{-1}\mathbf{j})$ is conservative in the first quadrant. Calculate $\int_C \mathbf{F} \cdot d\mathbf{r}$ on a directed curve $\mathcal{C}$ from $(1, 1/2)$ to $(1/2, 1)$.

42. The vector field $\mathbf{F}(x, y) = 2\pi^{-1/2}\exp(-(x - y)^2)(\mathbf{i} - \mathbf{j})$ is conservative—see Exercise 40. Calculate $\int_C \mathbf{F} \cdot d\mathbf{r}$ on a directed curve $\mathcal{C}$ from $(0, 0)$ to $(1, 2)$.

13.4 Divergence, Gradient, and Curl

In this section, we will learn about the differential operators that will be needed to formulate the fundamental theorems of vector analysis. Each of the constructs that we will see has an interesting physical interpretation.

Divergence of a Vector Field

Suppose that $\mathbf{F}(x, y) = M(x, y)\mathbf{i} + N(x, y)\mathbf{j}$ is a differentiable vector field that is defined on a region $\mathcal{G}$ of the plane. The scalar-valued function $(x, y) \mapsto M_x(x, y) + N_y(x, y)$ is said to be the *divergence* of $\mathbf{F}$. We denote this function by $\mathrm{div}(\mathbf{F})$:

$$\mathrm{div}(\mathbf{F}) = \frac{\partial M}{\partial x} + \frac{\partial N}{\partial y}. \tag{13.4.1}$$

Similarly, if $\mathbf{F}(x, y, z) = M(x, y, z)\mathbf{i} + N(x, y, z)\mathbf{j} + R(x, y, z)\mathbf{k}$ is a differentiable vector field on a region $\mathcal{G}$ in space, then the *divergence* of $\mathbf{F}$ is defined to be

$$\mathrm{div}(\mathbf{F}) = \frac{\partial M}{\partial x} + \frac{\partial N}{\partial y} + \frac{\partial R}{\partial z}. \tag{13.4.2}$$

If $\mathrm{div}(\mathbf{F})(P) = 0$ at every point P of $\mathcal{G}$, then we say that $\mathbf{F}$ is *solenoidal* on $\mathcal{G}$. Gauss's Law for magnetism states that a magnetic field $\mathbf{B}$ is solenoidal.

▶ **EXAMPLE 1** Let $\mathbf{F}(x, y, z) = xy\mathbf{i} + yz\mathbf{j} - 2xz\mathbf{k}$ and $\mathbf{G}(x, y, z) = 2xy\mathbf{i} + (z - y^2)\mathbf{j} + xy\mathbf{k}$. Calculate $\operatorname{div}(\mathbf{F})$ and $\operatorname{div}(\mathbf{G})$. Is either vector field solenoidal on $\mathbb{R}^3$?

Solution We have

$$\operatorname{div}(\mathbf{F})(x, y, z) = \frac{\partial}{\partial x}(xy) + \frac{\partial}{\partial y}(yz) + \frac{\partial}{\partial z}(-2xz) = y + z - 2x$$

and

$$\operatorname{div}(\mathbf{G})(x, y, z) = \frac{\partial}{\partial x}(2xy) + \frac{\partial}{\partial y}(z - y^2) + \frac{\partial}{\partial z}(xy) = 2y - 2y + 0 = 0.$$

Thus $\mathbf{G}$ is solenoidal on $\mathbb{R}^3$. Although $\operatorname{div}(\mathbf{F})(x, y, z)$ equals 0 at each point of the plane $y + z - 2x = 0$, these are the only points at which $\operatorname{div}(\mathbf{F})$ vanishes. Consequently, $\mathbf{F}$ is not solenoidal on $\mathbb{R}^3$. ◀

▶ **EXAMPLE 2** Suppose that a point mass at the origin exerts a gravitational field

$$\mathbf{F}(x, y, z) = \frac{\lambda}{\|\mathbf{r}\|^3}\mathbf{r}, \quad (x, y, z) \neq (0, 0, 0)$$

where λ is a proportionality constant and $\mathbf{r}$ is the position vector $x\mathbf{i} + y\mathbf{j} + z\mathbf{k}$ of the point (x, y, z). Show that $\mathbf{F}$ is solenoidal on $\mathcal{G} = \{(x, y, z) : (x, y, z) \neq (0, 0, 0)\}$.

Solution We calculate

$$\begin{aligned}
\frac{\partial}{\partial x}\frac{x}{(x^2 + y^2 + z^2)^{3/2}} &= \frac{1}{(x^2 + y^2 + z^2)^3}\left((x^2 + y^2 + z^2)^{3/2} - x\frac{\partial}{\partial x}(x^2 + y^2 + z^2)^{3/2}\right) \\
&= \frac{1}{(x^2 + y^2 + z^2)^3}\left((x^2 + y^2 + z^2)^{3/2} - 3x^2(x^2 + y^2 + z^2)^{1/2}\right) \\
&= \frac{(x^2 + y^2 + z^2)^{1/2}}{(x^2 + y^2 + z^2)^3}\left((x^2 + y^2 + z^2) - 3x^2\right) \\
&= \frac{(y^2 + z^2 - 2x^2)}{(x^2 + y^2 + z^2)^{5/2}}.
\end{aligned}$$

By symmetry, we deduce that

$$\begin{aligned}
\operatorname{div}(\mathbf{F})(x, y, z) &= \lambda\left(\frac{(y^2 + z^2 - 2x^2)}{(x^2 + y^2 + z^2)^{5/2}} + \frac{(x^2 + z^2 - 2y^2)}{(x^2 + y^2 + z^2)^{5/2}} + \frac{(x^2 + y^2 - 2z^2)}{(x^2 + y^2 + z^2)^{5/2}}\right) \\
&= \lambda\left(\frac{0}{(x^2 + y^2 + z^2)^{5/2}}\right) = 0. \quad ◀
\end{aligned}$$

What is the significance of the divergence? Imagine that $\mathbf{F}(x, y, z) = M(x, y, z)\mathbf{i} + N(x, y, z)\mathbf{j} + R(x, y, z)\mathbf{k}$ represents the velocity of a fluid flow.

Recalling from Section 11.6 of Chapter 11 that $\partial/\partial x$ is the directional derivative $D_{\mathbf{i}}$ in the **i** direction, $\partial/\partial y$ is the directional derivative $D_{\mathbf{j}}$ in the **j** direction, and $\partial/\partial z$ is the directional derivative $D_{\mathbf{k}}$ in the **k** direction, we have

$$\operatorname{div}(\mathbf{F}) = \frac{\partial M}{\partial x} + \frac{\partial N}{\partial y} + \frac{\partial R}{\partial z} = \frac{\partial}{\partial x}(\mathbf{F}\cdot\mathbf{i}) + \frac{\partial}{\partial y}(\mathbf{F}\cdot\mathbf{j}) + \frac{\partial}{\partial z}(\mathbf{F}\cdot\mathbf{k}) = D_{\mathbf{i}}(\mathbf{F}\cdot\mathbf{i}) + D_{\mathbf{j}}(\mathbf{F}\cdot\mathbf{j}) + D_{\mathbf{k}}(\mathbf{F}\cdot\mathbf{k}).$$

Thus div(**F**) sums the rate of change of the **i** component of **F** in the **i** direction, the rate of change of the **j** component of **F** in the **j** direction, and the rate of change of the **k** component of **F** in the **k** direction. If, at a point $P = (x, y, z)$, the divergence is positive, then the fluid is tending to flow outward from P. If div(**F**) is negative at P, then the fluid is tending to flow inward at P. Thus the divergence measures the tendency of a vector field to expand or contract. These ideas are illustrated in the next two examples.

▶ **EXAMPLE 3** Define $\mathbf{F}(x,y) = x\mathbf{i} + y\mathbf{j}$, $\mathbf{G}(x,y) = -x\mathbf{i} - y^3\mathbf{j}$, and $\mathbf{H}(x,y) = x^2\mathbf{i} - y^2\mathbf{j}$. Calculate the divergence of each vector field at the origin, and relate your answer to the physical properties of the flow that the vector field represents.

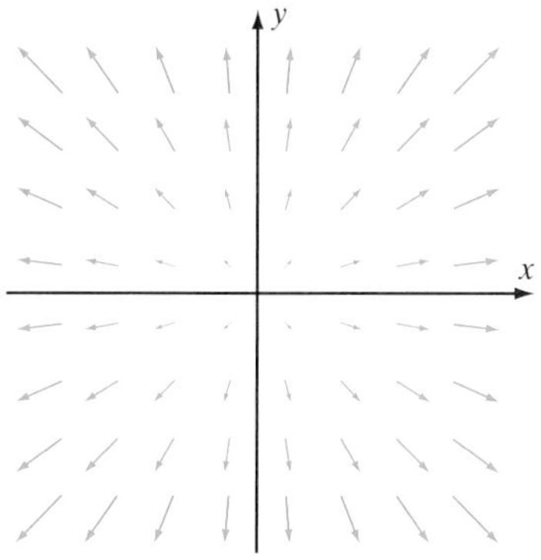

$\mathbf{F}(x, y) = x\mathbf{i} + y\mathbf{j}$

▲ Figure 1

Solution We begin with **F**. Look at Figure 1. The flow at the origin seems to be outward from 0. We calculate that $\operatorname{div}(\mathbf{F})(0,0) = 1 + 1 = 2$, which, being positive, is the indication of an outward flow, as expected. Next, we calculate that $\operatorname{div}(\mathbf{G})(x,y) = -1 - 3y^2$. Therefore $\operatorname{div}(\mathbf{G})(0,0) = -1$, which, being negative, is the indication of an inward flow. Indeed, Figure 2 shows that the flow of **G** at the origin is inward. Finally, $\operatorname{div}(\mathbf{H})(x,y) = 2x - 2y = 2(x - y)$, which is 0 at the origin. The (net) flow at the origin is neither outward nor inward, as the plot of vector field **H** suggests (Figure 3). ◀

▶ **EXAMPLE 4** Define $\mathbf{F}(x,y) = 2xy\mathbf{i} + (1 + 2xy - x^2)\mathbf{j}$. Figure 4 represents a sketch of the vector field $(x,y) \mapsto (1/2)\mathbf{dir}(\mathbf{F}(x,y))$, which shows us the directions of **F** scaled to a convenient size. The points $P = (0,1)$, $Q = (0,-1)$, and $R = (1,-1)$ are also plotted. Use Figure 4 to determine the sign of $\operatorname{div}(\mathbf{F})(P)$ and $\operatorname{div}(\mathbf{F})(Q)$.

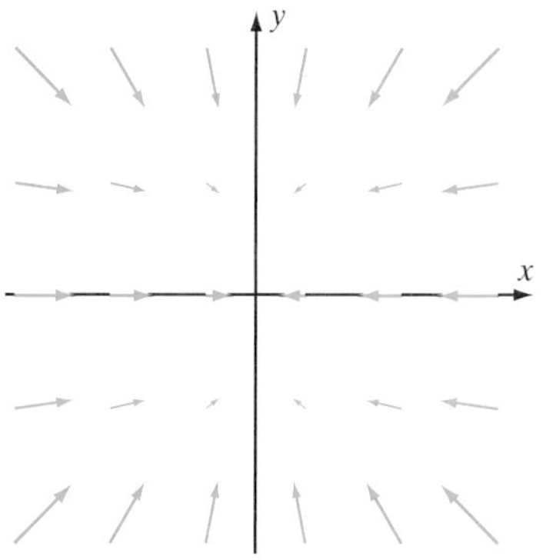

$\mathbf{G}(x, y) = -x\mathbf{i} - y^3\mathbf{j}$

▲ Figure 2

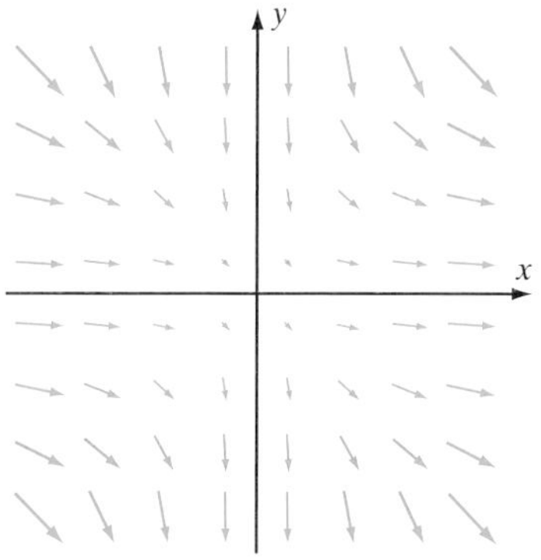

$\mathbf{H}(x, y) = x^2\mathbf{i} - y^2\mathbf{j}$

▲ Figure 3

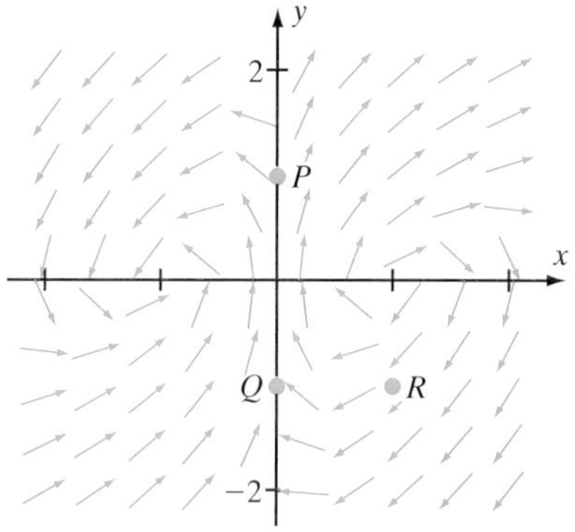

$\frac{1}{2}\mathbf{dir}(2xy\mathbf{i} + (1 + 2xy - x^2)\mathbf{j})$

▲ Figure 4

Confirm your deductions by calculating these values exactly. What can you say about $\mathrm{div}(\mathbf{F})(R)$?

Solution Because the flow at $P = (0, 1)$ is outward, we deduce that $\mathrm{div}(\mathbf{F})(P) > 0$. Similarly, the inward flow at $Q = (0, -1)$ tells us that $\mathrm{div}(\mathbf{F})(Q) < 0$. However, the figure is less revealing at R (and nearby points). Although the net flow does not appear to be strongly inward or strongly outward at such points, the visual evidence is not sufficiently accurate to determine whether the divergence is exactly 0 or merely a small positive or small negative value. To be sure, we calculate

$$\mathrm{div}(\mathbf{F})(x, y) = \frac{\partial}{\partial x}(2xy) + \frac{\partial}{\partial y}(1 + 2xy - x^2) = 2y + 2x.$$

Thus $\mathrm{div}(\mathbf{F})(P) = 2 > 0$ and $\mathrm{div}(\mathbf{F})(Q) = -2 < 0$, confirming our deductions. We also discover that $\mathrm{div}(\mathbf{F})(R) = 0$. The points just above R have a small positive divergence, and the points just below have a small negative divergence. ◀

Many terms related to divergence have been coined to describe fluid flow. Suppose that the vector field $\mathbf{v}$ represents the velocity of a fluid in a region $\mathcal{G}$. Then $\mathbf{v}$ (and the fluid) is called *incompressible* if $\mathrm{div}(\mathbf{v}) = 0$ at each point of $\mathcal{G}$. After we learn the Divergence Theorem in Section 13.8, we will understand that, if a fluid is incompressible in a region $\mathcal{G}$, then the quantity of fluid entering $\mathcal{G}$ equals the quantity exiting $\mathcal{G}$. There are circumstances in which the fluid can be compressed so that the net flow into the region exceeds the net flow out. This situation corresponds to $\mathrm{div}(\mathbf{F}) < 0$ on $\mathcal{G}$. We then call the vector field (and the fluid) *compressible*. Finally, if $\mathrm{div}(\mathbf{F}) > 0$ on $\mathcal{G}$, then the vector field and the fluid are called *expanding*.

The Curl of a Vector Field

If $\mathbf{F}(x, y) = M(x, y)\mathbf{i} + N(x, y)\mathbf{j}$ is a vector field in the plane, then we define its *curl* to be the *spatial* vector field

$$\mathbf{curl}(\mathbf{F})(x, y) = \left(\frac{\partial N}{\partial x} - \frac{\partial M}{\partial y}\right)\mathbf{k}. \qquad (13.4.3)$$

If $\mathbf{F}(x, y, z) = M(x, y, z)\mathbf{i} + N(x, y, z)\mathbf{j} + R(x, y, z)\mathbf{k}$ is a vector field in space, then we define its *curl* to be the vector field

$$\mathbf{curl}(\mathbf{F})(x, y, z) = \left(\frac{\partial R}{\partial y} - \frac{\partial N}{\partial z}\right)\mathbf{i} - \left(\frac{\partial R}{\partial x} - \frac{\partial M}{\partial z}\right)\mathbf{j} + \left(\frac{\partial N}{\partial x} - \frac{\partial M}{\partial y}\right)\mathbf{k}. \qquad (13.4.4)$$

Formula (13.4.4) can be symbolically written as

$$\mathbf{curl}(\mathbf{F}) = \det\left(\begin{bmatrix} \mathbf{i} & \mathbf{j} & \mathbf{k} \\ \dfrac{\partial}{\partial x} & \dfrac{\partial}{\partial y} & \dfrac{\partial}{\partial z} \\ M & N & R \end{bmatrix}\right),$$

a form which should be regarded as a memory aid.

▶ **EXAMPLE 5** Define $\mathbf{F}(x, y, z) = y^2 z\mathbf{i} - x^3\mathbf{j} + xy\mathbf{k}$. Calculate $\mathbf{curl}(\mathbf{F})$.

Solution We have

$$\frac{\partial R}{\partial y} - \frac{\partial N}{\partial z} = x - 0, \quad \frac{\partial R}{\partial x} - \frac{\partial M}{\partial z} = y - y^2, \quad \text{and} \quad \frac{\partial N}{\partial x} - \frac{\partial M}{\partial y} = -3x^2 - 2yz.$$

Therefore $\mathbf{curl}(\mathbf{F})(x, y, z) = x\mathbf{i} - (y - y^2)\mathbf{j} + (-3x^2 - 2yz)\mathbf{k}$. ◀

What is the geometric or physical meaning of the curl of a vector field? The vector field **curl** (**F**) measures the tendency of the vector field to "curl" or "rotate." More specifically, curl **F** points along the *axis of rotation* of **F** (determined by the right-hand rule), and the length of curl(**F**) measures the amount of curling. This is best seen by studying a simple example.

▶ **EXAMPLE 6** Sketch the vector field $\mathbf{F}(x, y, z) = -y\mathbf{i} + x\mathbf{j} + 0\mathbf{k}$ and its curl.

Solution Set $M = -y$, $N = x$, and $R = 0$. We calculate $\partial R/\partial y = 0$, $\partial N/\partial z = 0$, $\partial R/\partial x = 0$, $\partial M/\partial z = 0$, $\partial N/\partial x = 1$, and $\partial M/\partial y = -1$. Substituting these values into formula (13.4.4), we get $\mathbf{curl}(\mathbf{F})(x, y, z) = 2\mathbf{k}$. Both **F** and **curl(F)** are sketched in Figure 5. We see that the vector field **F** curls counterclockwise about the origin in the xy-plane. If you curl your right hand in the direction of the curl, then your thumb points in the direction **k**. Thus **curl**(**F**) coincides with our physical perception of the curl of a flow. If the vector field were rotating in the opposite direction—say that $\mathbf{H} = y\mathbf{i} - x\mathbf{j}$—then we would expect the curl to have opposite sign. And, indeed, $\mathbf{curl}(\mathbf{H}) = -2\mathbf{k}$. ◀

The notation **rot**(**F**) (abbreviation for rotation of **F**) is sometimes found in older texts instead of **curl**(**F**). In physics, a vector field for which $\mathbf{curl}(\mathbf{F}) = \mathbf{0}$ at all points is called *irrotational*. Comparing equation (13.3.3) with equation (13.4.3) in the planar case, and line (13.3.4) with equation (13.4.4) in the spatial case, we see that $\mathbf{curl}(\mathbf{F}) = \mathbf{0}$ on a region if and only if **F** is closed on the region. From our work in Section 13.3, it follows that, if **F** is conservative, or, equivalently, path-independent, then $\mathbf{curl}(\mathbf{F}) = \mathbf{0}$. Also, *on a simply connected region*, $\mathbf{curl}(\mathbf{F}) = \mathbf{0}$ if and only if **F** is conservative. This information is recorded in the diagram in Figure 6. It will be of use when we study Stokes's Theorem in Section 13.7.

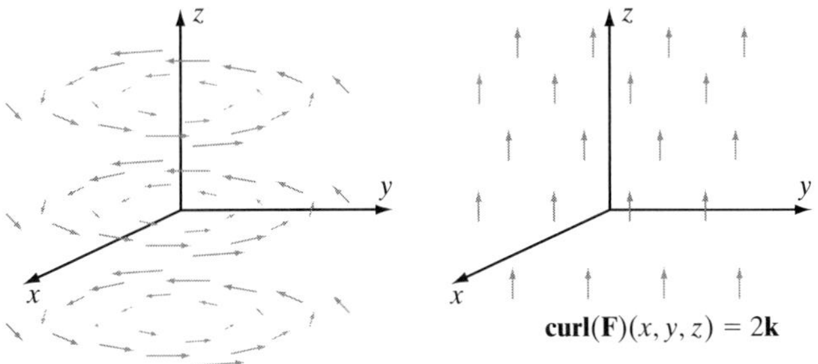

$\mathbf{F}(x, y, z) = -y\mathbf{i} + x\mathbf{j} + 0\mathbf{k}$ $\quad$ $\mathbf{curl}(\mathbf{F})(x, y, z) = 2\mathbf{k}$

▲ Figure 5

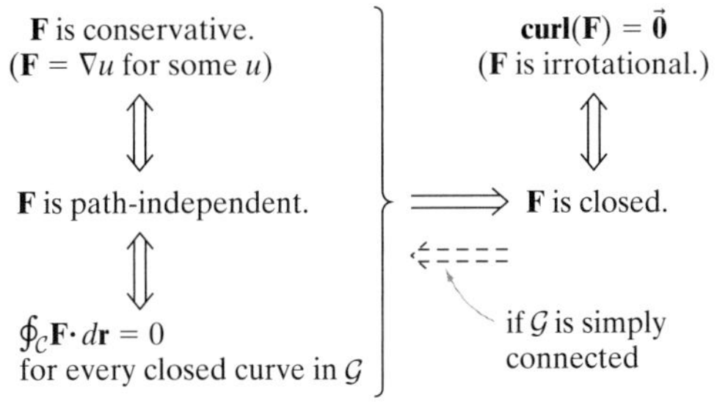

▲ Figure 6

▶ **EXAMPLE 7** Verify that the vector field $\mathbf{F}(x, y, z) = 2x\mathbf{i} + (z^2 - 2y)\mathbf{j} + 2yz\mathbf{k}$ is irrotational on all of three-dimensional space.

Solution We must show that $\mathbf{curl}(\mathbf{F})(x,y,z)=\mathbf{0}$ for all x,y,z. Setting $M(x,y,z)=2x$, $N(x,y,z)=z^2-2y$, and $R(x,y,z)=2yz$, we have

$$\frac{\partial R}{\partial y}-\frac{\partial N}{\partial z}=2z-(2z)=0,\quad \frac{\partial R}{\partial x}-\frac{\partial M}{\partial z}=0-0=0,\quad \text{and}\quad \frac{\partial N}{\partial x}-\frac{\partial M}{\partial y}=0-0=0. \quad ◀$$

The Del Notation Let us introduce the formal notation

$$\nabla=\frac{\partial}{\partial x}\mathbf{i}+\frac{\partial}{\partial y}\mathbf{j}+\frac{\partial}{\partial z}\mathbf{k}.$$

This is not a vector, but it is a convenient notation in calculations. We are already accustomed to writing the gradient of a function u as ∇u. The gradient of a function u may also be written as $\mathbf{grad}(u)$. Now we note that the divergence of a vector field $\mathbf{F}=M\mathbf{i}+N\mathbf{j}+R\mathbf{k}$ may be written as

$$\operatorname{div}(\mathbf{F})=\nabla\cdot\mathbf{F}. \tag{13.4.5}$$

This means that, if we treat ∇ algebraically as though it were a vector, then

$$\nabla\cdot\mathbf{F}=\left(\frac{\partial}{\partial x}\mathbf{i}+\frac{\partial}{\partial y}\mathbf{j}+\frac{\partial}{\partial z}\mathbf{k}\right)\cdot(M\mathbf{i}+N\mathbf{j}+R\mathbf{k})=\frac{\partial M}{\partial x}+\frac{\partial N}{\partial y}+\frac{\partial R}{\partial z}=\operatorname{div}(\mathbf{F}).$$

Thus a field $\mathbf{F}$ is solenoidal on a region if and only if $\nabla\cdot\mathbf{F}=0$ on the region. Stated in this manner, Gauss's law for magnetism becomes $\nabla\cdot\mathbf{B}=0$, which is one of Maxwell's equations for classical electromagnetism.

In a similar vein, we may write the curl of the vector field $\mathbf{F}=M\mathbf{i}+N\mathbf{j}+R\mathbf{k}$ as

$$\mathbf{curl}(\mathbf{F})=\nabla\times\mathbf{F}. \tag{13.4.6}$$

This means that, if we treat ∇ algebraically as though it were a vector, then

$$\nabla\times\mathbf{F}=\det\left(\begin{bmatrix}\mathbf{i}&\mathbf{j}&\mathbf{k}\\ \frac{\partial}{\partial x}&\frac{\partial}{\partial y}&\frac{\partial}{\partial z}\\ M&N&R\end{bmatrix}\right)=\left(\frac{\partial R}{\partial y}-\frac{\partial N}{\partial z}\right)\mathbf{i}-\left(\frac{\partial R}{\partial x}-\frac{\partial M}{\partial z}\right)\mathbf{j}+\left(\frac{\partial N}{\partial x}-\frac{\partial M}{\partial y}\right)\mathbf{k}.$$

In physics books, it is quite common to see divergence and curl expressed using the notation ∇. We call this the *del notation*. The symbol ∇ is also referred to as *nabla*.

Probably the most important partial differential operator of mathematical physics is the *Laplace operator*, which is often denoted by the symbol $\triangle$. Applied to a twice continuously differentiable scalar-valued function u of x and y, this operator is given by the formula

$$\triangle u=\frac{\partial^2 u}{\partial x^2}+\frac{\partial^2 u}{\partial y^2}.$$

For a twice continuously differentiable scalar-valued function v of x, y, and z, the Laplace operator (or laplacian) is given by

$$\triangle v=\frac{\partial^2 v}{\partial x^2}+\frac{\partial^2 v}{\partial y^2}+\frac{\partial^2 v}{\partial z^2}.$$

We must be careful not to confuse the laplacian $\triangle v$ with its look-alike, Δv, which represents a small increment in the value of v. Generally, we can infer which meaning is intended from the context. It is often convenient to define the laplacian of a vector-valued function $\mathbf{F} = \langle M, N, R\rangle$ by applying $\triangle$ to the components of $\mathbf{F}$:

$$\triangle\mathbf{F} = \langle \triangle M, \triangle N, \triangle R\rangle.$$

Sometimes we write the laplacian as $\nabla \cdot \nabla$ because

$$\triangle v = \frac{\partial^2 v}{\partial x^2} + \frac{\partial^2 v}{\partial y^2} + \frac{\partial^2 v}{\partial z^2} = \frac{\partial}{\partial x}\frac{\partial v}{\partial x} + \frac{\partial}{\partial y}\frac{\partial v}{\partial y} + \frac{\partial}{\partial z}\frac{\partial v}{\partial z} = \nabla \cdot (\nabla v). \tag{13.4.7}$$

The notation $\nabla \cdot \nabla$ is often contracted to ∇^2.

Many important differential equations in mathematical physics involve the laplacian. For example, if the edges of a metal plate are heated up, then the second law of thermodynamics implies that the steady state heat distribution $h(x, y)$ in the plate satisfies the equation $\triangle h = 0$. A function u that satisfies $\triangle u = 0$ is called *harmonic*. Harmonic functions are important objects of study in mathematical analysis, physics, and engineering.

▶ **EXAMPLE 8** Show that the functions $u(x, y) = x^2 - y^2$ and $v(x, y, z) = e^{(x+y)}\cos(\sqrt{2}z)$ are harmonic.

Proof. We have

$$\frac{\partial^2 u}{\partial x^2} + \frac{\partial^2 u}{\partial y^2} = 2 + (-2) = 0$$

and

$$\frac{\partial^2 v}{\partial x^2} + \frac{\partial^2 v}{\partial y^2} + \frac{\partial^2 v}{\partial z^2} = e^{(x+y)}\cos\left(\sqrt{2}z\right) + e^{(x+y)}\cos\left(\sqrt{2}z\right) + \left(-\left(\sqrt{2}\right)^2 e^{(x+y)}\cos\left(\sqrt{2}z\right)\right) = 0. \quad \blacksquare$$

Identities Involving div, curl, grad, and $\triangle$

The next theorem gathers together some basic identities that involve the operators we have been studying. Several other identities may be found in the exercises for this section.

THEOREM 1 Let u be a twice continuously differentiable scalar-valued function on a region $\mathcal{G}$ in the plane or in space. Let $\mathbf{F}$ be a twice continuously differentiable vector field on $\mathcal{G}$. Then

a. $\operatorname{div}\left(\mathbf{grad}(u)\right) = \triangle u$,
b. $\operatorname{div}\left(\mathbf{curl}(F)\right) = 0$,
c. $\mathbf{curl}\left(\mathbf{grad}(u)\right) = 0$,
d. $\mathbf{curl}\left(\mathbf{curl}(\mathbf{F})\right) = \mathbf{grad}\left(\operatorname{div}(\mathbf{F})\right) - \triangle\mathbf{F}$,
e. $\operatorname{div}(u\mathbf{F}) = u\operatorname{div}(\mathbf{F}) + \mathbf{grad}(u) \cdot \mathbf{F}$.

Proof. The first assertion is simply a restatement of equation (13.4.7). Part b follows from the equality of the mixed partial derivatives of the components of $\mathbf{F} = \langle M, N, R \rangle$:

$$\operatorname{div}(\mathbf{curl}(\mathbf{F})) = \frac{\partial}{\partial x}\left(\frac{\partial R}{\partial y} - \frac{\partial N}{\partial z}\right) + \frac{\partial}{\partial y}\left(\frac{\partial M}{\partial z} - \frac{\partial R}{\partial x}\right) + \frac{\partial}{\partial z}\left(\frac{\partial N}{\partial x} - \frac{\partial M}{\partial y}\right)$$

or

$$\operatorname{div}(\mathbf{curl}(\mathbf{F})) = \left(\frac{\partial^2 R}{\partial x \partial y} - \frac{\partial^2 R}{\partial y \partial x}\right) + \left(\frac{\partial^2 N}{\partial z \partial x} - \frac{\partial^2 N}{\partial x \partial z}\right) + \left(\frac{\partial^2 M}{\partial y \partial z} - \frac{\partial^2 M}{\partial z \partial y}\right) = 0.$$

Part c asserts that the conservative vector field $\mathbf{grad}(u)$ is closed, a fact which we established in Section 13.3. Part d can be proved by a brute force calculation that is similar in spirit to part b. Finally, part e is obtained as follows:

$$\begin{aligned}
\operatorname{div}(u\mathbf{F}) &= \frac{\partial(uM)}{\partial x} + \frac{\partial(uN)}{\partial y} + \frac{\partial(uR)}{\partial z} \\
&= \left(u\frac{\partial M}{\partial x} + M\frac{\partial u}{\partial x}\right) + \left(u\frac{\partial N}{\partial y} + N\frac{\partial u}{\partial y}\right) + \left(u\frac{\partial R}{\partial z} + R\frac{\partial u}{\partial z}\right) \\
&= u\left(\frac{\partial M}{\partial x} + \frac{\partial N}{\partial y} + \frac{\partial R}{\partial z}\right) + \left(M\frac{\partial u}{\partial x} + N\frac{\partial u}{\partial y} + R\frac{\partial u}{\partial z}\right) \\
&= u\operatorname{div}(\mathbf{F}) + (\nabla u) \cdot \mathbf{F}.
\end{aligned}$$

■

▶ **EXAMPLE 9** For $\mathbf{F}(x, y, z) = \langle y, z^3, \cos(x) + e^z \rangle$, verify the identity $\mathbf{curl}(\mathbf{curl}(\mathbf{F})) = \mathbf{grad}(\operatorname{div}(\mathbf{F})) - \triangle \mathbf{F}$.

Solution We have

$$\begin{aligned}
\mathbf{curl}(\mathbf{F}) &= \left(\frac{\partial}{\partial y}(\cos(x) + e^z) - \frac{\partial}{\partial z}z^3\right)\mathbf{i} + \left(\frac{\partial}{\partial z}y - \frac{\partial}{\partial x}(\cos(x) + e^z)\right)\mathbf{j} + \left(\frac{\partial}{\partial x}z^3 - \frac{\partial}{\partial y}y\right)\mathbf{k} \\
&= -3z^2\mathbf{i} + \sin(x)\mathbf{j} + (-1)\mathbf{k}
\end{aligned}$$

and therefore

$$\begin{aligned}
\mathbf{curl}(\mathbf{curl}(\mathbf{F})) &= \left(\frac{\partial}{\partial y}(-1) - \frac{\partial}{\partial z}\sin(x)\right)\mathbf{i} + \left(\frac{\partial}{\partial z}(-3z^2) - \frac{\partial}{\partial x}(-1)\right)\mathbf{j} + \left(\frac{\partial}{\partial x}\sin(x) - \frac{\partial}{\partial y}(-3z^2)\right)\mathbf{k} \\
&= -6z\mathbf{j} + \cos(x)\mathbf{k}.
\end{aligned}$$

On the other hand,

$$\mathbf{grad}(\operatorname{div}(\mathbf{F})) = \mathbf{grad}\left(\frac{\partial}{\partial x}y + \frac{\partial}{\partial y}z^3 + \frac{\partial}{\partial z}(\cos(x) + e^z)\right) = \mathbf{grad}(e^z) = e^z\mathbf{k}$$

and

$$\triangle \mathbf{F} = (\triangle y)\mathbf{i} + (\triangle z^3)\mathbf{j} + \Big(\triangle\,(\cos(x) + e^z)\Big)\mathbf{k} = 6z\mathbf{j} + (-\cos(x) + e^z)\mathbf{k}.$$

Therefore $\mathbf{grad}\,(\mathrm{div}(\mathbf{F})) - \triangle \mathbf{F} = e^z\mathbf{k} - \Big(6z\mathbf{j} + (-\cos(x) + e^z)\mathbf{k}\Big) = -6z\mathbf{j} + \cos(x)\mathbf{k}$, which equals $\mathbf{curl}(\mathbf{curl}(\mathbf{F}))$. ◀

QUICK QUIZ

1. Calculate the divergence of $x^2yz\mathbf{i} + 7y\mathbf{j} - x^y\mathbf{k}$ at the point $(1, 2, -1)$.
2. Calculate the curl of $x^2\mathbf{i} + (12/z)\mathbf{j} - y^3\mathbf{k}$ at the point $(-4, 3, 2)$.
3. Calculate $\nabla \times (\nabla u)$ for $u(x, y, z) = x^2y^3 + z^4$.
4. True or false: $\mathbf{F}$ is conservative implies $\mathbf{curl}(\mathbf{F}) = \mathbf{0}$.

Answers

1. 3 2. $-24\mathbf{i}$ 3. $\mathbf{0}$ 4. True

EXERCISES

Problems for Practice

▶ In each of Exercises 1–10, calculate the divergence of the given planar vector field $\mathbf{F}$. ◀

1. $\mathbf{F}(x, y) = 3x^2y^3\,\mathbf{i} - xy^4\,\mathbf{j}$
2. $\mathbf{F}(x, y) = (5x + 2y)\,\mathbf{i} + (5x - 2y)\,\mathbf{j}$
3. $\mathbf{F}(x, y) = x^2\sin(y)\,\mathbf{i} + \big(x^2 + \sin(y)\big)\,\mathbf{j}$
4. $\mathbf{F}(x, y) = \tan^2(y^3)\,\mathbf{i} + \big(\cos^2(x) + 2y^3\big)\,\mathbf{j}$
5. $\mathbf{F}(x, y) = \cos(x)\,\mathbf{i} - \sin^2(xy)\,\mathbf{j}$
6. $\mathbf{F}(x, y) = \ln(x + y)\,\mathbf{i} + e^{xy}\,\mathbf{j}$
7. $\mathbf{F}(x, y) = \dfrac{x}{x + y}\,\mathbf{i} - \dfrac{2y}{x - y}\,\mathbf{j}$
8. $\mathbf{F}(x, y) = xe^y\,\mathbf{i} - ye^x\,\mathbf{j}$
9. $\mathbf{F}(x, y) = \tan(x/y)\,\mathbf{j} + \cot(y/x)\,\mathbf{i}$
10. $\mathbf{F}(x, y) = (x^2 + y^2)\,e^{x+y}(\mathbf{i} + \mathbf{j})$

▶ In each of Exercises 11–20, calculate the divergence of the given spatial vector field $\mathbf{F}$. ◀

11. $\mathbf{F}(x, y, z) = \langle 2y - 3z, 4z - 5x, 6x + 7z\rangle$
12. $\mathbf{F}(x, y, z) = \langle xz - xy, xy - yz, yz - xz\rangle$
13. $\mathbf{F}(x, y, z) = (x + 2y + 3z)\langle 1, 2, 3\rangle$
14. $\mathbf{F}(x, y, z) = \langle z\ln(x), yz/x, z^2/x\rangle$
15. $\mathbf{F}(x, y, z) = x^2\,\mathbf{i} + y^2\mathbf{j} + z^2\,\mathbf{k}$
16. $\mathbf{F}(x, y, z) = x^3y\,\mathbf{i} - y^2zx\,\mathbf{j} + xyz\,\mathbf{k}$
17. $\mathbf{F}(x, y, z) = e^{x-z}\,\mathbf{i} + e^{z-y}\,\mathbf{j} - e^{y-x}\,\mathbf{k}$
18. $\mathbf{F}(x, y, z) = x^2\sin(y)\,(\mathbf{i} - \mathbf{j} + \mathbf{k})$
19. $\mathbf{F}(x, y, z) = \cos(xy)\,\mathbf{i} - \sin(yz)\,\mathbf{j} + \cos(y)\sin(x)\,\mathbf{k}$
20. $\mathbf{F}(x, y, z) = x^y\,\mathbf{i} + y^z\,\mathbf{j} - x^z\,\mathbf{k}$

▶ In each of Exercises 21–26, calculate the curl of the given vector field $\mathbf{F}$. ◀

21. $\mathbf{F}(x, y, z) = xz\,\mathbf{i} - y^3\,\mathbf{j} + xyz\,\mathbf{k}$
22. $\mathbf{F}(x, y, z) = \cos^2(z)\,\mathbf{i} - \sin(xy)\,\mathbf{j} + x^4z\,\mathbf{k}$
23. $\mathbf{F}(x, y, z) = \ln(x + z)\,\mathbf{i} - e^{yz}\,\mathbf{j} + xy\,\mathbf{k}$
24. $\mathbf{F}(x, y, z) = z^2\,\mathbf{i} - xz\,\mathbf{j} + x^3\,\mathbf{k}$
25. $\mathbf{F}(x, y, z) = \tan(xy)\,\mathbf{i} + \cos(xy)\,\mathbf{j} - \sin(xy)\,\mathbf{k}$
26. $\mathbf{F}(x, y, z) = xyz\,\mathbf{i} + (xyz)^2\,\mathbf{j} + (xyz)^3\,\mathbf{k}$

▶ In each of Exercises 27–32, calculate the curl of the given vector field $\mathbf{F}$. State whether $\mathbf{F}$ is closed or not. ◀

27. $\mathbf{F}(x, y, z) = y^2z\,\mathbf{i} + 2xyz\,\mathbf{j} + xy^2\,\mathbf{k}$
28. $\mathbf{F}(x, y, z) = (y/x)\,\mathbf{i} + (z/y)\,\mathbf{j} + (x/z)\,\mathbf{k}$
29. $\mathbf{F}(x, y, z) = \cos(yz)\,\mathbf{i} - xz\sin(yz)\,\mathbf{j} - xy\sin(yz)\,\mathbf{k}$
30. $\mathbf{F}(x, y, z) = \dfrac{y}{x + z}\,\mathbf{i} + \ln(x + z)\,\mathbf{j} + \dfrac{y}{x + z}\,\mathbf{k}$
31. $\mathbf{F}(x, y, z) = x\cos(y)\,\mathbf{i} - z\sin(x)\,\mathbf{j} + (xy)\sin(z)\,\mathbf{k}$
32. $\mathbf{F}(x, y, z) = e^x\,\mathbf{i} - e^{x-z}\,\mathbf{j} + e^{y-x}\,\mathbf{k}$

▶ In each of Exercises 33–38, a scalar-valued function u is given. Calculate $\mathrm{div}(\mathbf{grad}(u))$. ◀

33. $u(x, y) = x^2 - y^2$
34. $u(x, y) = e^x\cos(y)$
35. $u(x, y, z) = x/(z - y^2)$
36. $u(x, y, z) = y\sin(x - z)$
37. $u(x, y, z) = x(y + 2z)$
38. $u(x, y, z) = (2x - z)/(yz)$

▶ In each of Exercises 39–42, a vector-valued function $\mathbf{F}$ is given. Calculate $\mathbf{grad}(\mathrm{div}(\mathbf{F}))$. ◀

39. $\mathbf{F}(x, y) = \cos(x)\,\mathbf{i} + \sin(y)\,\mathbf{j}$
40. $\mathbf{F}(x, y, z) = \langle x/y, -x/y^2\rangle$
41. $\mathbf{F}(x, y, z) = xyz\,\mathbf{i} + yz^2\,\mathbf{j} + xz\,\mathbf{k}$
42. $\mathbf{F}(x, y, z) = \langle x^2, y^3, z^4\rangle$

Further Theory and Practice

▶ In each of Exercises 43–46, calculate $\mathbf{curl}(\mathbf{curl}(\mathbf{F}))$ for the given vector field $\mathbf{F}$. ◀

43. $\mathbf{F}(x,y,z) = \langle xyz, xyz, xyz \rangle$
44. $\mathbf{F}(x,y,z) = \langle x^2+y^2, y^2+z^2, xy^2 \rangle$
45. $\mathbf{F}(x,y,z) = (x/y)\,\mathbf{i} + (z/y)\,\mathbf{j} + (x^2-y^2)\,\mathbf{k}$
46. $\mathbf{F}(x,y,z) = \cos(y)\,\mathbf{i} + \sin(x+2z)\,\mathbf{j} + \sin(x)\cos(y)\,\mathbf{k}$

Suppose that g, h are twice continuously differentiable scalar-valued functions and that $\mathbf{F}$ and $\mathbf{G}$ are continuously differentiable vector fields. Prove each of the identities in Exercises 47–52.

47. $\nabla(gh) = g\nabla(h) + h\nabla(g)$
48. $\nabla\cdot(g\mathbf{F}) = (\nabla g)\cdot\mathbf{F} + g(\nabla\cdot\mathbf{F})$
49. $\nabla\times(g\mathbf{F}) = (\nabla g)\times\mathbf{F} + g(\nabla\times\mathbf{F})$
50. $\nabla\cdot(\mathbf{F}\times\mathbf{G}) = (\nabla\times\mathbf{F})\cdot\mathbf{G} - \mathbf{F}\cdot(\nabla\times\mathbf{G})$
51. $\nabla\times(g\nabla h - h\nabla g) = 2\nabla g\times\nabla h$
52. $\nabla\times(g\nabla h + h\nabla g) = \mathbf{0}$

53. Let $\mathbf{r} = \langle x,y,z \rangle$. Define $\mathbf{F}(x,y,z) = \lambda\|\mathbf{r}\|^{-3}\mathbf{r}$ as in Example 3. Verify that $\nabla\times\mathbf{F} = \mathbf{0}$ at all points of space except for the origin.

Harmonic functions on the plane have a remarkable mean value property: The average value of a harmonic function on a circle is equal to the value of the function at the center. To be precise, if u is harmonic, then, for any $r>0$,

$$u(x_0,y_0) = \frac{1}{2\pi}\int_0^{2\pi} u\big(x_0 + r\cos(t), y_0 + r\sin(t)\big)\,dt.$$

In Exercises 54 and 55, verify that the given function u is harmonic, and then verify the mean value property at the given point (x_0,y_0).

54. $u(x,y) = x^2 - y^2$ $\quad (3,2)$
55. $u(x,y) = x^3 - 3xy^2 + 1$ $\quad (2,1)$

56. Suppose that u is a twice continuously differentiable scalar-valued function. Set $\mathbf{r}(x,y,z) = x\,\mathbf{i} + y\,\mathbf{j} + z\,\mathbf{k}$. Prove that $\mathbf{r}\times\nabla u$ is solenoidal.

57. Prove that $\nabla\times(\boldsymbol{\omega}\times\mathbf{r}) = 2\boldsymbol{\omega}$ when $\boldsymbol{\omega}$ is a constant vector and $\mathbf{r}(x,y,z) = x\,\mathbf{i} + y\,\mathbf{j} + z\,\mathbf{k}$.

58. Prove that $\mathbf{F}\times\mathbf{G}$ is solenoidal when both spatial vector fields $\mathbf{F}$ and $\mathbf{G}$ are irrotational.

59. Suppose that the curl of a spatial vector field $\mathbf{F}$ has a potential function V. Prove that V is harmonic and $\triangle\mathbf{F} = \mathbf{grad}\big(\mathrm{div}(\mathbf{F})\big)$.

60. Theorem 1 tells us that the curl of any twice-continuously differentiable vector field on a region $\mathcal{G}$ is solenoidal on $\mathcal{G}$. The converse is also true, provided $\mathcal{G}$ has certain properties. In this exercise, assume that $\mathcal{G}$ is an open ball containing the origin O. Let $\mathbf{F}$ be a differentiable vector field such that $\mathrm{div}(\mathbf{F}) = 0$ on $\mathcal{G}$. For each $P = (x,y,z) \in \mathcal{G}$, let $\mathbf{r}_P$ be the parameterization of $\overrightarrow{OP}$ given by $\mathbf{r}_P(t) = \langle tx, ty, tz \rangle, 0 \le t \le 1$. Define

$$\mathbf{G}(P) = \int_0^1 t\,\mathbf{F}\big(\mathbf{r}_P(t)\big)\times\mathbf{r}_P'(t)\,dt.$$

Show that $\mathbf{F} = \mathrm{curl}(\mathbf{G})$.

Calculator/Computer Exercises

In Exercises 61 and 62, verify the mean value property of the given harmonic function u at the given point (x_0,y_0). (See the instructions for Exercises 54 and 55). Use $r=1$.

61. $u(x,y) = e^x\cos(y)$ $\quad (x_0,y_0) = (1,3)$
62. $u(x,y) = \ln(x^2+y^2)$ $\quad (x_0,y_0) = (3,4)$

13.5 Green's Theorem

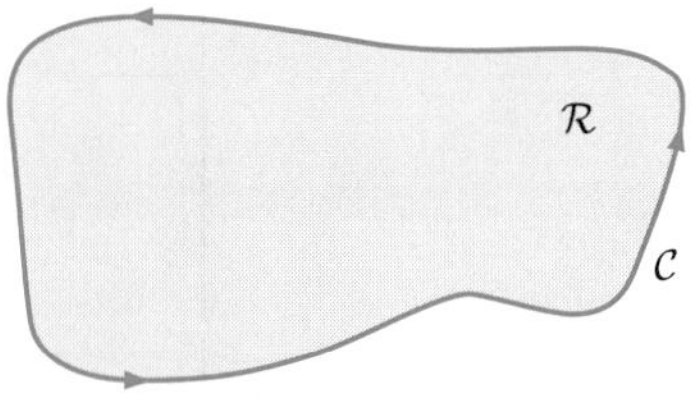

Figure 1

In this section, we learn our first theorem that relates the behavior of a flow at the boundary of a region to the behavior in the interior. There will be several interesting applications. Everything in this section takes place in the two-dimensional plane. Throughout this section (and also in the remaining sections), we will write dA for the element of area in the plane.

We will begin our study with a simply connected bounded region $\mathcal{R}$ in the plane. Its boundary consists of a single closed curve C that we parameterize by a vector-valued function $t \mapsto \mathbf{r}(t), a \le t \le b$ (see Figure 1). We assume four things about $\mathbf{r}$:

a. $\mathbf{r}$ is piecewise-smooth.
b. $\mathbf{r}(a) = \mathbf{r}(b)$ (C is *closed*).
c. $\mathbf{r}(t_1) \ne \mathbf{r}(t_2)$ for $a \le t_1 < t_2 \le b$.
d. As $\mathbf{r}(t)$ traverses C with increasing t, the enclosed region $\mathcal{R}$ lies to the left.

The first condition states that C is made up of finitely many smooth arcs. The second condition says that $\mathbf{r}$ begins and ends at the same point. The third condition

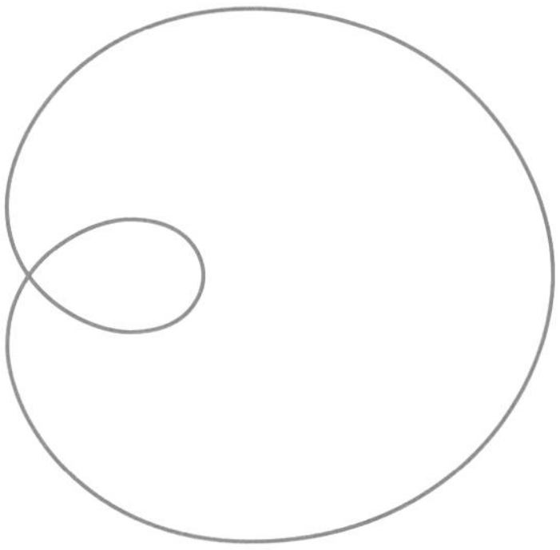

▲ Figure 2 A curve that intersects itself is *not* simple.

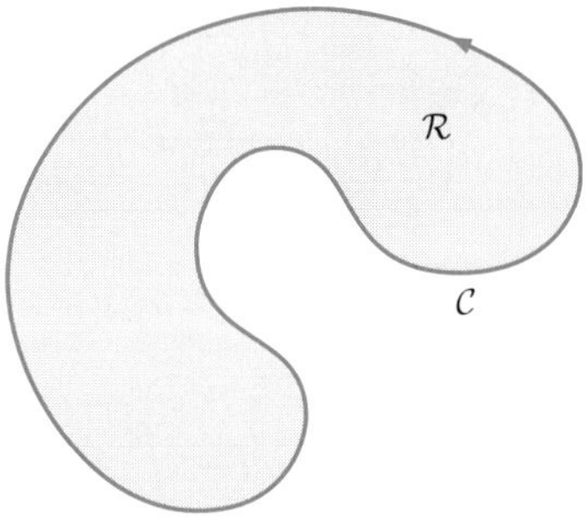

▲ Figure 3a Positive orientation of boundary

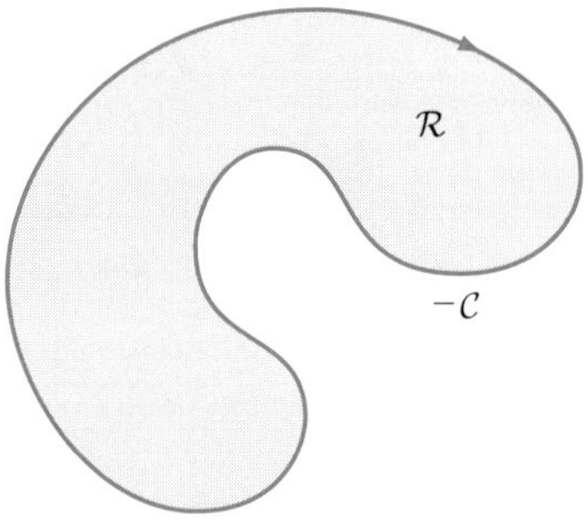

▲ Figure 3b Negative orientation of boundary

states that $\mathcal{C}$ does not cross itself. It rules out curves like the one in Figure 2. A closed curve that satisfies condition c is said to be *simple*. Condition d says that $\mathcal{C}$ is oriented as in Figure 3a, *not* as in Figure 3b. This is said to be the *positive* orientation of $\mathcal{C}$. If we regard the xy-plane as the $z = 0$ plane of xyz-space, then the right- hand-rule for cross products tells us that $\mathbf{k} \times \mathbf{r}'$ points into the region $\mathcal{R}$ when $\mathcal{C}$ is positively oriented.

THEOREM 1 **(Green's Theorem)**. Suppose that $\mathcal{R}$ is a simply connected region with boundary $\mathcal{C}$ that is parameterized by a vector-valued function $\mathbf{r}$ with the four properties previously listed. If $\mathbf{F}(x,y) = M(x,y)\mathbf{i} + N(x,y)\mathbf{j}$ is a continuously differentiable vector field on a region containing $\mathcal{R}$ and its boundary, then

$$\oint_{\mathcal{C}} \mathbf{F} \cdot d\mathbf{r} = \iint_{\mathcal{R}} \left(\frac{\partial N}{\partial x} - \frac{\partial M}{\partial y} \right) dA. \qquad (13.5.1)$$

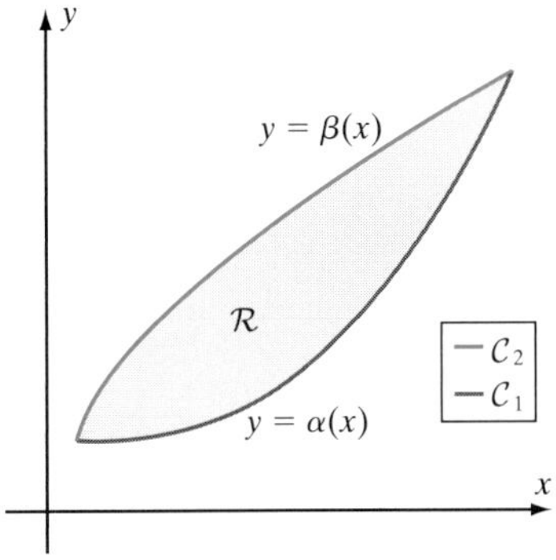

▲ Figure 4 $\mathcal{R}$ is both x-simple and y-simple.

Proof. A number of subtleties arise in the proof of Green's Theorem in its full generality, and these subtleties are a proper topic for a course in advanced calculus. Therefore we will settle for a proof of Green's Theorem in a special case. We assume that our region is both x-simple and y-simple as in Figure 4, with boundaries consisting of the graphs of two functions $y = \alpha(x)$ and $y = \beta(x)$. We can parameterize the lower curve $\mathcal{C}_1$ by $\mathbf{r}_1(t) = \langle t, \alpha(t) \rangle, a \le t \le b$, and the upper curve $\mathcal{C}_2$ by $\mathbf{r}_2(t) = \langle b + a - t, \beta(b + a - t) \rangle, a \le t \le b$. Now, we begin with the right side of the formula in Green's Theorem:

$$\iint_{\mathcal{R}} \left(\frac{\partial N}{\partial x} - \frac{\partial M}{\partial y} \right) dA = \iint_{\mathcal{R}} \frac{\partial N}{\partial x}\, dA - \iint_{\mathcal{R}} \frac{\partial M}{\partial y}\, dA.$$

We can calculate the second double integral on the right side by expressing it as an iterated integral and using the Fundamental Theorem of Calculus:

$$\int_a^b \int_{\alpha(x)}^{\beta(x)} \frac{\partial M}{\partial y}\, dy dx = \int_a^b M(x,y)\Big|_{y=\alpha(x)}^{y=\beta(x)} dx = \int_a^b M\big(x, \beta(x)\big)\, dx - \int_a^b M\big(x, \alpha(x)\big)\, dx. \qquad (13.5.2)$$

Now,

$$\begin{aligned}
\oint_C M\mathbf{i}\cdot d\mathbf{r} &= \int_{C_1} M\mathbf{i}\cdot d\mathbf{r}_1 + \int_{C_2} M\mathbf{i}\cdot d\mathbf{r}_2 \\
&= \int_a^b \left(M(t,\alpha(t))\frac{dt}{dt} + 0\frac{d\alpha(t)}{dt}\right) dt \\
&\quad + \int_a^b \left(M(b+a-t,\beta(b+a-t))\frac{d(b+a-t)}{dt} + 0\frac{d\beta(b+a-t)}{dt}\right) dt \\
&= \int_a^b M(t,\alpha(t))dt - \int_a^b M(b+a-t,\beta(b+a-t))\,dt \\
&= \int_a^b M(t,\alpha(t))dt + \int_b^a M(x,\beta(x))dx, \text{(after substitution } x=b+a-t, dx=-dt) \\
&= \int_a^b M(x,\alpha(x))dx - \int_a^b M(x,\beta(x))dx \\
&= -\iint_{\mathcal{R}} \frac{\partial M}{\partial y}dA \quad \text{by (13.5.2).}
\end{aligned}$$

This is half of the job. A similar calculation shows that

$$\int_C N\mathbf{j}\cdot d\mathbf{r} = \iint_{\mathcal{R}} \frac{\partial N}{\partial x}dA.$$

Adding the last two equations results in the identity that is Green's Theorem. ■

INSIGHT Expressed in component form, Green's Theorem says that

$$\oint_C M\,dx + N\,dy = \iint_{\mathcal{R}} \left(\frac{\partial N}{\partial x} - \frac{\partial M}{\partial y}\right) dA. \tag{13.5.3}$$

It is instructive to compare equation (13.5.3) with the Fundamental Theorem of Calculus,

$$f(b) - f(a) = \int_a^b f'(t)\,dt,$$

which relates the values of f on the boundary of an interval $[a, b]$, namely the points a and b, to the integral of the derivative of f on the interval. Observe that the boundary orientation is important in the Fundamental Theorem of Calculus: The term $f(b)$ appears with a " + " and the term $f(a)$ appears with a " − ". It is plain that Green's Theorem is a two-dimensional version of the Fundamental Theorem of Calculus, in that it relates the integral of **F** on the oriented boundary of a region $\mathcal{R}$ to the integral of a certain derivative of **F** on the region.

► **EXAMPLE 1** Verify Green's Theorem for $\mathbf{F}(x,y) = -3y\mathbf{i} + 6x\mathbf{j}$ and $\mathcal{R} = \{(x,y) : x^2 + y^2 < 1\}$.

Solution The boundary of $\mathcal{R}$ is the unit circle $\mathcal{C}$ traversed counterclockwise when positively oriented. We may use the parameterization $\mathbf{r}(t) = \cos(t)\mathbf{i} + \sin(t)\mathbf{j}$, $0 \le t \le 2\pi$. We have

$$\oint_{\mathcal{C}} \mathbf{F} \cdot d\mathbf{r} = \int_0^{2\pi} \left(-3\sin(t)\frac{d}{dt}\cos(t) + 6\cos(t)\frac{d}{dt}\sin(t)\right) dt = \int_0^{2\pi} \left(3\sin^2(t) + 6\cos^2(t)\right) dt.$$

But $\int_0^{2\pi} \sin^2(t)\,dt = \int_0^{2\pi} \cos^2(t)\,dt = \pi$ by formula (6.2.9). It follows that $\oint_{\mathcal{C}} \mathbf{F} \cdot d\mathbf{r} = 9\pi$. Having calculated the left side of equation (13.5.1) for the data of this example, we turn to the right side. We have

$$\iint_{\mathcal{R}} \left(\frac{\partial N}{\partial x} - \frac{\partial M}{\partial y}\right) dA = \iint_{\mathcal{R}} \left(\frac{\partial(6x)}{\partial x} - \frac{\partial(-3y)}{\partial y}\right) dA = \iint_{\mathcal{R}} 9\,dA = 9 \times \text{ (area of unit disk)} = 9\pi.$$

Thus we have verified Green's theorem for this particular example. ◀

At first Example 1 may seem pointless. Now we have two ways of calculating something instead of one. That's just twice as much work! But this example was just for mechanical practice, to help us to understand the *components* of Green's Theorem. It is not the way that we usually use Green's Theorem. The next theorem shows one of the really striking applications.

THEOREM 2 Suppose that $\mathcal{R}$, $\mathcal{C}$, and $\mathbf{r}$ satisfy the hypotheses of Green's Theorem. Then we have

$$(\text{Area of } \mathcal{R}) = \frac{1}{2}\oint_{\mathcal{C}} -y\,dx + x\,dy. \tag{13.5.4}$$

Proof. For the vector field $\mathbf{F} = M\mathbf{i} + N\mathbf{j} = (-y/2)\mathbf{i} + (x/2)\mathbf{j}$, we have $N_x - M_y = (1/2) - (-1/2) = 1$. Green's Theorem tells us that

$$\oint_{\mathcal{C}} (-y/2)\,dx + (x/2)\,dy = \iint_{\mathcal{R}} (N_x - M_y)\,dA = \iint_{\mathcal{R}} (1)\,dA.$$

This, of course, is just the area of $\mathcal{R}$. ■

▶ **EXAMPLE 2** Use formula (13.5.4) to calculate the area A inside the ellipse $x^2/a^2 + y^2/b^2 = 1$.

Solution We can parameterize the ellipse $\mathcal{C}$ by $\mathbf{r}(t) = a\cos(t)\mathbf{i} + b\sin(t)\mathbf{j}, 0 \le t \le 2\pi$. Then

$$A = \frac{1}{2}\oint_{\mathcal{C}} (-y)\,dx + x\,dy = \frac{1}{2}\int_0^{2\pi} \left((-b\sin(t))\frac{d}{dt}a\cos(t) + (a\cos(t))\frac{d}{dt}b\sin(t)\right) dt.$$

By simplifying the integrand, we obtain

$$A = \frac{1}{2}\int_0^{2\pi} ab\left(\cos^2(t) + \sin^2(t)\right)dt = \frac{1}{2}\int_0^{2\pi} ab \cdot 1\, dt = \pi ab. \blacktriangleleft$$

Theorem 5 of Section 13.3 asserts that a closed vector field on a simply connected planar region is conservative. This is a deep result whose proof we have outlined for rectangular regions. Now we can use Green's Theorem to obtain a more general proof.

▶ **EXAMPLE 3** Use Green's Theorem to show that a closed vector field $\mathbf{F} = \langle M, N\rangle$ on a simply connected planar region $\mathcal{G}$ is conservative.

Solution Suppose that $\mathbf{F} = \langle M, N\rangle$ is closed on $\mathcal{G}$ and that $\mathcal{C}$ is a simple closed curve in $\mathcal{G}$. Let $\mathcal{R}$ denote the region enclosed by $\mathcal{C}$. Then,

$$\oint_{\mathcal{C}} \mathbf{F}\cdot d\mathbf{r} = \iint_{\mathcal{R}} \left(\frac{\partial N}{\partial x} - \frac{\partial M}{\partial y}\right) dA = \iint_{\mathcal{R}} 0\, dA = 0.$$

It follows from Theorems 1 and 3 of Section 17.3 that $\mathbf{F}$ is conservative. ◀

A Vector Form of Green's Theorem

Green's Theorem has many variants, each of which reveals new information. In particular, there is a formulation of Green's Theorem for each of the differential operators div, curl, and $\triangle$ that we studied in Section 13.4. For example, for $\mathbf{F} = M\mathbf{i} + N\mathbf{j}$, we have $\mathbf{curl}(\mathbf{F}) = (N_x - M_y)\mathbf{k}$, and, therefore, $N_x - M_y = \mathbf{curl}(\mathbf{F})\cdot\mathbf{k}$. We can therefore state Green's Theorem as

$$\oint_{\mathcal{C}} \mathbf{F}\cdot d\mathbf{r} = \iint_{\mathcal{R}} \mathbf{curl}(\mathbf{F})\cdot\mathbf{k}\, dA. \tag{13.5.5}$$

Notice that $\mathbf{k}$ is a unit normal to the region $\mathcal{R}$. The direction of this unit normal (as opposed to the *other* unit normal to $\mathcal{R}$, namely $-\mathbf{k}$) is determined by the right-hand rule: Curl your fingers of your right hand in the positive orientation of $\mathcal{C}$, and your thumb points in the direction of $\mathbf{k}$. Equation (13.5.5) tells us that we can compute the line integral of $\mathbf{F}$ over a simple closed curve $\mathcal{C}$ by integrating the component of $\mathbf{curl}(\mathbf{F})$ in the direction of the outward unit normal $\mathbf{k}$ to the enclosed region $\mathcal{R}$, the integration being performed over $\mathcal{R}$.

Next we will develop a divergence form of Green's Theorem. Suppose that $\mathcal{R}$, $\mathcal{C}$, $t \mapsto \mathbf{r}(t) = x(t)\mathbf{i} + y(t)\mathbf{j}$, and $\mathbf{F}(x, y) = M(x, y)\mathbf{i} + N(x, y)\mathbf{j}$ all satisfy the hypotheses of Green's Theorem. Corresponding to each point $\mathbf{r}(t)$ of $\mathcal{C}$, we define the unit vector

$$\mathbf{n}\big(\mathbf{r}(t)\big) = \frac{y'(t)}{\sqrt{x'(t)^2 + y'(t)^2}}\mathbf{i} + \frac{(-x'(t))}{\sqrt{x'(t)^2 + y'(t)^2}}\mathbf{j} = \frac{1}{\|\mathbf{r}'(t)\|}\big(y'(t)\mathbf{i} - x'(t)\mathbf{j}\big).$$

Notice that

$$\mathbf{r}'(t)\cdot\mathbf{n}\big(\mathbf{r}(t)\big) = \frac{1}{\|\mathbf{r}'(t)\|}\Big(x'(t)y'(t) + y'(t)\big(-x'(t)\big)\Big) = 0.$$

Because $\mathbf{r}'(t)$ is tangent to $\mathcal{C}$ at the point $\mathbf{r}(t)$, we deduce that $\mathbf{n}\big(\mathbf{r}(t)\big)$ is normal to $\mathcal{C}$ at $\mathbf{r}(t)$. By observing the signs of $x'(t)$ and $y'(t)$ as $\mathbf{r}(t)$ traverses $\mathcal{C}$ in its positive orientation, we see that $\mathbf{n}(\mathbf{r}(t))$ is the *outward* unit normal to $\mathcal{C}$ at $\mathbf{r}(t)$.

To obtain the divergence form of Green's Theorem, we apply formula (13.5.1) to the vector field $\tilde{\mathbf{F}}(x,y) = -N(x,y)\mathbf{i} + M(x,y)\mathbf{j}$. We obtain

$$\oint_{\mathcal{C}} \tilde{\mathbf{F}} \cdot d\mathbf{r} = \iint_{\mathcal{R}} \left(\frac{\partial M}{\partial x} + \frac{\partial N}{\partial y}\right) dxdy = \iint_{\mathcal{R}} \operatorname{div}(\mathbf{F})\, dA. \qquad (13.5.6)$$

Now let us look at the line integral on the left side of equation (13.5.6). We have

$$\begin{aligned}\oint_{\mathcal{C}} \tilde{\mathbf{F}} \cdot d\mathbf{r} &= \int_a^b \Big(-N\big(x(t),y(t)\big)x'(t) + M\big(x(t),y(t)\big)y'(t)\Big)dt \\ &= \int_a^b \left(M\big(x(t),y(t)\big)\frac{y'(t)}{\|\mathbf{r}'(t)\|} + N\big(x(t),y(t)\big)\frac{\big(-x'(t)\big)}{\|\mathbf{r}'(t)\|}\right)\|\mathbf{r}'(t)\|\, dt \\ &= \int_a^b (\mathbf{F}\cdot\mathbf{n})\big(\mathbf{r}(t)\big)\|\mathbf{r}'(t)\|\, dt.\end{aligned}$$

But formula (13.2.8) allows us to identify this last expression as the line integral $\oint_{\mathcal{C}} \mathbf{F} \cdot \mathbf{n}\, ds$. Thus we have

$$\oint_{\mathcal{C}} \mathbf{F} \cdot \mathbf{n}\, ds = \iint_{\mathcal{R}} \operatorname{div}(\mathbf{F})\, dA. \qquad (13.5.7)$$

This form of Green's Theorem is sometimes called the "Divergence Theorem in two dimensions." The line integral on the left side of equation (13.5.7) is called the *flux* of vector field $\mathbf{F}$ across $\mathcal{C}$. If $\mathbf{F}$ is the velocity field of a fluid, the flux is the sum total of the scalar component of the fluid's velocity in the direction of the outward normal along the boundary curve.

▶ **EXAMPLE 4** Let $\mathbf{F}(x,y) = x\mathbf{i} + y\mathbf{j}$ and $\mathcal{R} = \{(x,y) : x^2 + y^2 < 1\}$. Assume that $\mathbf{F}$ represents the velocity of the flow of a fluid in the region. Explain what Green's Theorem tells us about this fluid flow.

Solution Let $\mathcal{C}$ denote the positively oriented unit circle. The integral

$$\int_{\mathcal{C}} \mathbf{F} \cdot \mathbf{n}\, ds$$

measures the fluid flow across the boundary of $\mathcal{R}$. If the integral is positive, then more fluid is flowing out of the region than into it (indicating that there is a *source* in the region); if the integral is negative, then more fluid is flowing into the region than out of it (indicating that there is a *sink* in the region). According to equation (13.5.6), we have

$$\int_{\mathcal{C}} \mathbf{F} \cdot \mathbf{n}\, ds = \iint_{\mathcal{R}} \operatorname{div}(\mathbf{F})\, dA = \iint_{\mathcal{R}} (1+1)\, dA = 2(\text{area of } \mathcal{R}) = 2\pi.$$

Because this number is positive, we see that the overall fluid flow is out of the region. A glance at the plot of $\mathbf{F}$ in Figure 5 confirms our conclusion. ◀

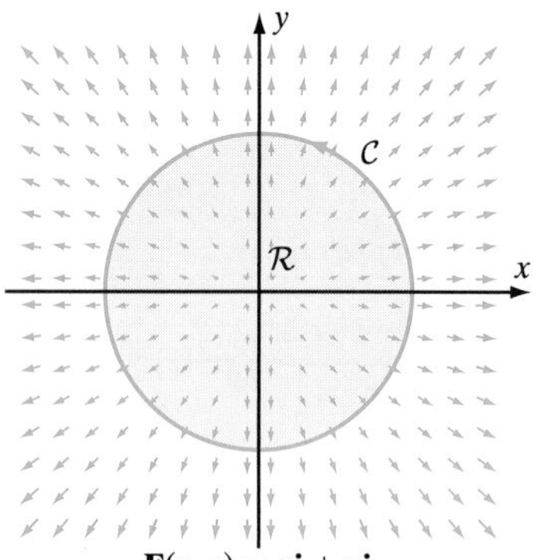

$\mathbf{F}(x, y) = x\mathbf{i} + y\mathbf{j}$

▲ Figure 5

Green's Theorem for More General Regions

Now we want to discuss Green's Theorem on some regions different from those specified in Theorem 1. For example, the boundary of the region in Figure 6 consists of two smooth, simple closed curves. This new feature poses no problem. We orient *each* curve so that the region $\mathcal{R}$ is to its left as it is traversed (as in Figure 7). Then we calculate the line integral by integrating over each curve, one at a time, and adding up the results.

THEOREM 3 Let $\mathcal{R}$ be a region in the plane bounded by one or more continuously differentiable curves $\mathcal{C}_1, \ldots, \mathcal{C}_N$ with parameterizations $\mathbf{r}_1, \mathbf{r}_2, \ldots, \mathbf{r}_N$. Assume that these curves are oriented so that the region $\mathcal{R}$ is on the left as each curve is traversed. If $\mathbf{F}(x, y) = M(x, y)\mathbf{i} + N(x, y)\mathbf{j}$ is a continuously differentiable vector field on a neighborhood of $\mathcal{R}$ and its boundary, then

$$\sum_{j=1}^{N} \int_{\mathcal{C}_j} \mathbf{F} \cdot d\mathbf{r}_j = \iint_{\mathcal{R}} \left(\frac{\partial N}{\partial x} - \frac{\partial M}{\partial y} \right) dA. \tag{13.5.8}$$

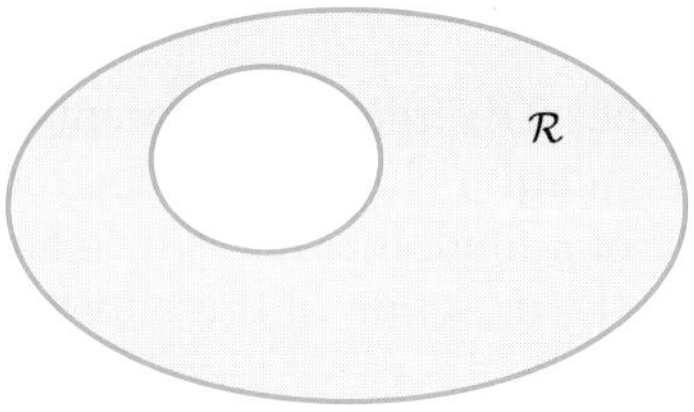

▲ Figure 6

Proof. Rather than give a complete proof, we will simply indicate how to prove Green's Theorem for a region that is not simply connected. Look again at the region $\mathcal{R}$ shown in Figure 6. Orient the boundary as in Figure 7. Connect the two curves by line segments as in Figure 8. Doing so creates two simply connected subregions. By applying Green's Theorem to each subregion, we obtain

$$\begin{aligned} \iint_{\mathcal{R}} \left(\frac{\partial N}{\partial x} - \frac{\partial M}{\partial y} \right) dA &= \iint_{\mathcal{R}_1} \left(\frac{\partial N}{\partial x} - \frac{\partial M}{\partial y} \right) dA + \iint_{\mathcal{R}_2} \left(\frac{\partial N}{\partial x} - \frac{\partial M}{\partial y} \right) \\ &= \int_{\mathcal{C}_1} \mathbf{F} \cdot d\mathbf{r}_1 + \int_{\mathcal{C}_2} \mathbf{F} \cdot d\mathbf{r}_2 = \int_{\mathcal{C}} \mathbf{F} \cdot d\mathbf{r}. \end{aligned}$$

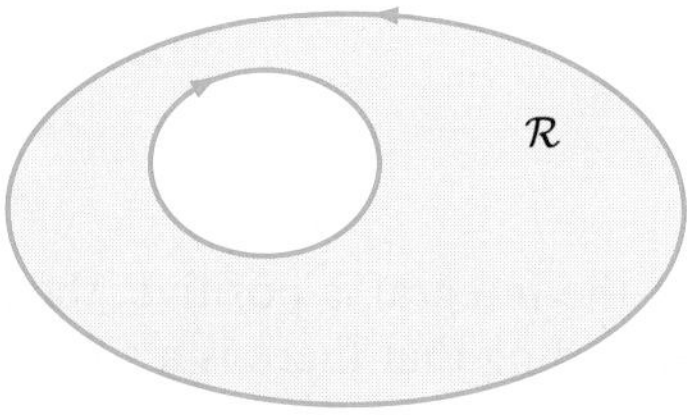

▲ Figure 7

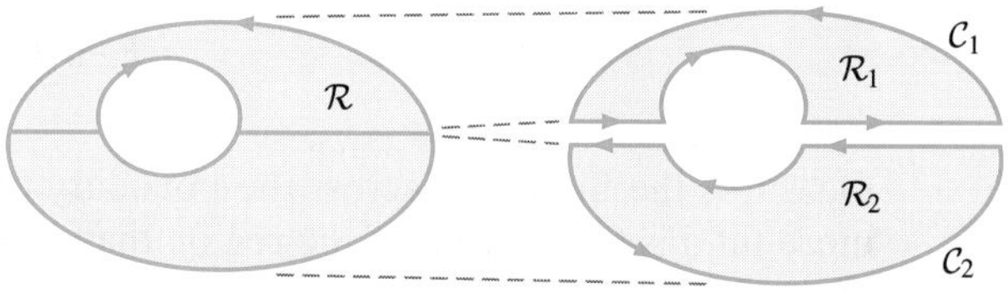

▲ Figure 8

To obtain the last equality in this chain that we notice when we traverse the inserted line segments in $\mathcal{C}_1$ and $\mathcal{C}_2$, we travel in opposite directions, so these portions of the line integrals cancel when we add $\int_{\mathcal{C}_1} \mathbf{F} \cdot d\mathbf{r}_1$ and $\int_{\mathcal{C}_2} \mathbf{F} \cdot d\mathbf{r}_2$. ■

▶ **EXAMPLE 5** Consider the region $\mathcal{R}$ (with piecewise-smooth boundary) that is shown in Figure 9. Verify equation (13.5.8) for $\mathbf{F} = \mathbf{i} + x^2\mathbf{j}$.

Solution Here $M = 1$, $N = x^2$, and $N_x - M_y = 2x$. We have

$$\iint_{\mathcal{R}} \left(\frac{\partial N}{\partial x} - \frac{\partial M}{\partial y}\right) dA = \int_0^1 \int_{1-x}^{\sqrt{1-x^2}} 2x\, dy dx = \int_0^1 2xy\Big|_{y=1-x}^{y=\sqrt{1-x^2}} dx = \int_0^1 \left(2x\sqrt{1-x^2} - 2x(1-x)\right) dx.$$

Making the substitution $u = 1 - x^2$, $du = -2x\, dx$, we calculate

$$\int_0^1 \left(2x\sqrt{1-x^2}\right) dx = -\int_1^0 \sqrt{u}\, du = -\frac{2}{3}u^{3/2}\Big|_{u=1}^{u=0} = -0 - \left(-\frac{2}{3}\right) = \frac{2}{3}.$$

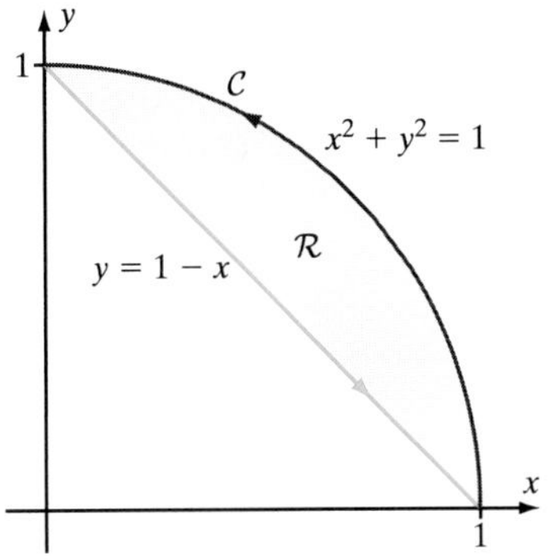

▲ Figure 9

Because

$$\int_0^1 (-2x(1-x))dx = \int_0^1 (2x^2 - 2x)\, dx = \frac{2}{3}x^3 - x^2\Big|_{x=0}^{x=1} = -\frac{1}{3},$$

we conclude that

$$\iint_{\mathcal{R}} \left(\frac{\partial N}{\partial x} - \frac{\partial M}{\partial y}\right) dA = \frac{2}{3} + \left(-\frac{1}{3}\right) = \frac{1}{3}.$$

To calculate the left side of equation (13.5.8), we parameterize the boundary of $\mathcal{R}$ by using $\mathbf{r}_1(t) = \cos(t)\mathbf{i} + \sin(t)\mathbf{j}, 0 \le t \le \pi/2$ for the circular arc C_1 and $\mathbf{r}_2(t) = t\mathbf{i} + (1-t)\mathbf{j}, 0 \le t \le 1$ for the line segment C_2. Then

$$\int_{C_1} \mathbf{F} \cdot d\mathbf{r} = \int_0^{\pi/2} \left(1\frac{d}{dt}\cos(t) + \cos^2(t)\frac{d}{dt}\sin(t)\right) dt = \int_0^{\pi/2} (-\sin(t) + \cos^3(t))\, dt.$$

An antiderivative of $\cos^3(t)$, namely $\sin(t) - \sin^3(t)/3$, is calculated in Example 5 of Section 6.2 in Chapter 6. It follows that

$$\int_{C_1} \mathbf{F} \cdot d\mathbf{r} = \cos(t) + \left(\sin(t) - \frac{\sin^3(t)}{3}\right)\Big|_{t=0}^{t=\pi/2} = \left(1 - \frac{1}{3}\right) - 1 = -\frac{1}{3}.$$

Finally, we have

$$\int_{C_2} \mathbf{F} \cdot d\mathbf{r} = \int_0^1 \left(1\frac{d}{dt}t + t^2\frac{d}{dt}(1-t)\right) dt = \int_0^1 (1 - t^2)\, dt = \left(t - \frac{t^3}{3}\right)\Big|_{t=0}^{t=1} = \frac{2}{3}.$$

Thus

$$\int_C \mathbf{F} \cdot d\mathbf{r} = \int_{C_1} \mathbf{F} \cdot d\mathbf{r} + \int_{C_2} \mathbf{F} \cdot d\mathbf{r} = \left(-\frac{1}{3}\right) + \frac{2}{3} = \frac{1}{3},$$

which is the same number we obtained when we evaluated the right side of equation (13.5.8). ◀

▶ **EXAMPLE 6** Let $\mathcal{R}$ be the region bounded by the circles $x^2 + y^2 = 1$ and $x^2 + y^2 = 4$. Verify Green's Theorem for the vector field $\mathbf{F} = -y^3\mathbf{i} + 2\mathbf{j}$ defined on $\mathcal{R}$.

Solution Here, $M = -y^3$, $N = 2$, and $N_x - M_y = 3y^2$. Using polar coordinates and formula (6.2.9), we have

$$\iint_{\mathcal{R}} \left(\frac{\partial N}{\partial x} - \frac{\partial M}{\partial y}\right) dA = \iint_{\mathcal{R}} 3y^2 dA = \int_0^{2\pi}\int_1^2 3r^2 \sin^2(\theta)\, r\, dr d\theta$$

$$= 3\int_0^{2\pi} \sin^2(\theta) \frac{r^4}{4}\bigg|_{r=1}^{r=2} d\theta = \frac{45}{4}\int_0^{2\pi} \sin^2(\theta)\, d\theta = \frac{45}{4}\pi.$$

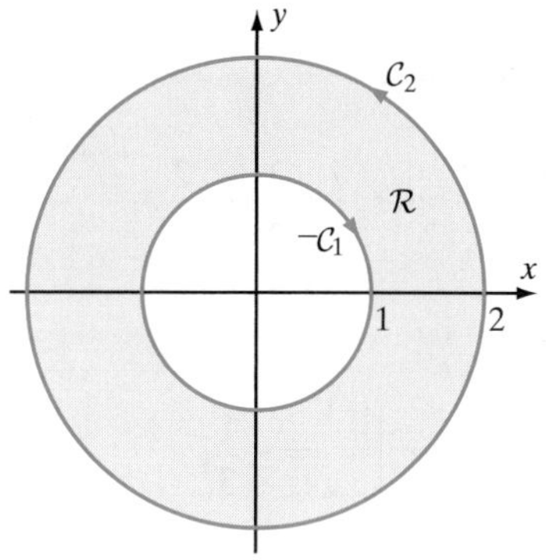

▲ Figure 10

Next we integrate over the oriented boundary $\mathcal{C}$ of $\mathcal{R}$. If we let $\mathcal{C}_a$ denote the circle of radius a that is centered at the origin and that has counterclockwise orientation, then we observe that $\mathcal{C} = \mathcal{C}_2 \cup (-\mathcal{C}_1)$ and

$$\int_{\mathcal{C}} \mathbf{F} \cdot d\mathbf{r} = \int_{\mathcal{C}_2} \mathbf{F} \cdot d\mathbf{r}_2 - \int_{\mathcal{C}_1} \mathbf{F} \cdot d\mathbf{r}_1. \tag{13.5.9}$$

See Figure 10. We parameterize $\mathcal{C}_a$ by $\mathbf{r}(t) = a\cos(t)\mathbf{i} + a\sin(t)\mathbf{j}$, $0 \le t \le 2\pi$ and calculate

$$\int_{\mathcal{C}_a} F \cdot d\mathbf{r} = \int_0^{2\pi} \left(-a^3 \sin^3(t)\frac{d}{dt}a\cos(t) + 2\frac{d}{dt}a\sin(t)\right) dt = \int_0^{2\pi} (a^4 \sin^4(t) + 2a\cos(t))\, dt,$$

or

$$\int_{\mathcal{C}_a} \mathbf{F} \cdot d\mathbf{r} = a^4 \int_0^{2\pi} \sin^4(t)\, dt + 2a\underbrace{\int_0^{2\pi} \cos(t) dt}_{0}$$

We can calculate $\int_0^{2\pi} \sin^4(t) dt$ using formula (6.2.11) with $n = 4$. We obtain

$$\int_{\mathcal{C}_a} \mathbf{F} \cdot d\mathbf{r} = a^4 \int_0^{2\pi} \sin^4(t) dt = a^4 \left(-\frac{1}{4}\sin^3(t)\cos(t)\bigg|_{t=0}^{t=2\pi} + \frac{3}{4}\int_0^{2\pi} \sin^2(t) dt\right) \overset{(6.2.9)}{=} \frac{3}{4}\pi a^4. \tag{13.5.10}$$

We conclude by using formula (13.5.10) with $a = 1$ and $a = 2$ to evaluate the two summands of the right side of (13.5.9). We obtain

$$\int_{\mathcal{C}} \mathbf{F} \cdot d\mathbf{r} = \int_{\mathcal{C}_2} \mathbf{F} \cdot d\mathbf{r}_2 - \int_{\mathcal{C}_1} \mathbf{F} \cdot d\mathbf{r}_1 = \frac{3}{4}\pi \cdot 2^4 - \frac{3}{4}\pi \cdot 1^4 = \frac{45}{4}\pi,$$

which agrees with our calculation of $\iint_{\mathcal{R}}(N_x - M_y)dA$. ◀

QUICK QUIZ

1. True or false: If a simple closed curve $\mathcal{C}$ is part of the boundary of a region, then the positive orientation of $\mathcal{C}$ is always counterclockwise.
2. If $\mathcal{C}$ is a positively oriented simple closed curve that encloses a region of area 3, and if $\mathbf{F}(x, y) = 5y\mathbf{i} - 2x\mathbf{j}$, then what is the value of $\oint_{\mathcal{C}} \mathbf{F} \cdot d\mathbf{r}$?

3. If $\mathcal{C}$ is a positively oriented simple closed curve that encloses a region of area 3, if $\mathbf{F}(x,y) = 7x\mathbf{i} - 5y\mathbf{j}$, and if $\mathbf{n}$ is the outward unit normal along $\mathcal{C}$, then what is the value of $\oint_{\mathcal{C}} \mathbf{F} \cdot \mathbf{n}\, ds$?
4. Calculate $\oint_{\mathcal{C}} -y\,dx + x\,dy$ where $\mathcal{C}$ is the triangle with vertices $(0,0)$, $(6,6)$, and $(4,6)$, traversed counterclockwise.

Answers

1. False 2. −21 3. 6 4. 12

EXERCISES

Problems for Practice

In each of Exercises 1–6, calculate both sides of the formula in Green's Theorem, and verify that they are equal.

1. $\mathbf{F}(x,y) = y^2\mathbf{i} - 3x\mathbf{j}$, $\mathbf{r}(t) = \sin(t)\mathbf{i} - \cos(t)\mathbf{j}$, $\pi \le t \le 3\pi$.
2. $\mathbf{F}(x,y) = xy\mathbf{i} - x^2\mathbf{j}$, $\mathbf{r}(t) = \cos(t)\mathbf{i} + 3\sin(t)\mathbf{j}$, $0 \le t \le 2\pi$.
3. $\mathbf{F}(x,y) = (y-2x)\mathbf{i} + (3x-4y)\mathbf{j}$, $\mathbf{r}(t) = \cos(2t)\mathbf{i} + \sin(2t)\mathbf{j}$, $0 \le t \le \pi$.
4. $\mathbf{F}(x,y) = xy^2\mathbf{i} - xy\mathbf{j}$, $\mathbf{r}(t) = \sin(2t)\mathbf{i} - \cos(2t)\mathbf{j}$, $\pi/2 \le t \le 3\pi/2$.
5. $\mathbf{F}(x,y) = y\mathbf{i} + 2x\mathbf{j}$, $\mathbf{r}(t) = \sin(t)\mathbf{i} - \cos(t)\mathbf{j}$, $-\pi \le t \le \pi$.
6. $\mathbf{F}(x,y) = y\mathbf{i} + y^2\mathbf{j}$, $\mathbf{r}(t) = \cos(4t)\mathbf{i} + \sin(4t)\mathbf{j}$, $-\pi/2 \le t \le 0$.

In each of Exercises 7–10, determine the region $\mathcal{R}$ bounded by the oriented curve $\mathcal{C}$ whose parameterization $\mathbf{r}$ is given. Then, use Green's Theorem to calculate $\int_{\mathcal{C}} \mathbf{F} \cdot d\mathbf{r}$ by integrating over $\mathcal{R}$.

7. $\mathbf{F}(x,y) = xy\mathbf{i} - x^2\mathbf{j}$, $\mathbf{r}(t) = \cos(t)\mathbf{i} + \sin(t)\mathbf{j}$, $0 \le t \le 2\pi$.
8. $\mathbf{F}(x,y) = 3y\mathbf{i} - 4x\mathbf{j}$, $\mathbf{r}(t) = 3\sin(t)\mathbf{i} - 4\cos(t)\mathbf{j}$, $\pi \le t \le 3\pi$
9. $\mathbf{F}(x,y) = -y^2\mathbf{i} + x^3\mathbf{j}$, $\mathbf{r}(t) = \sin(t)\mathbf{i} - \cos(t)\mathbf{j}$, $-\pi \le t \le \pi$
10. $\mathbf{F}(x,y) = (y-x)\mathbf{i} + (x-3y)\mathbf{j}$, $\mathbf{r}(t) = 4\cos(t)\mathbf{i} + 3\sin(t)\mathbf{j}$, $0 \le t \le 2\pi$

In each of Exercises 11–16, the given region has a piecewise-smooth boundary. Apply Green's Theorem to evaluate the line integral of $\mathbf{F}$ around the positively oriented boundary curve.

11. $\{(x,y) : -1 \le x \le 1, -2 \le y \le 2\}$, $\mathbf{F}(x,y) = xy^2\mathbf{i} - x^3y^2\mathbf{j}$
12. the triangle with vertices $(\pi/4, 0)$, $(0, \pi/4)$, and $(0,0)$, $\mathbf{F} = \cos(y)\mathbf{i} + \sin(x)\mathbf{j}$
13. the square with vertices $(1,0)$, $(0,1)$, $(-1,0)$, and $(0,-1)$, $\mathbf{F}(x,y) = \ln(3+y)\mathbf{i} - xy\mathbf{j}$
14. the rectangle with vertices $(-2,-1)$, $(2,-1)$, $(2,1)$, and $(-2,1)$, $\mathbf{F}(x,y) = \sin^3(y)\mathbf{i} - \cos^3(x)\mathbf{j}$
15. the region bounded by $y = x^2$ and $y = 4x + 5$, $\mathbf{F}(x,y) = x^2y\mathbf{i} + xy\mathbf{j}$
16. $\{(x,y) : 0 \le y \le \sqrt{4-x^2},\ -2 \le x \le 2\}$, $\mathbf{F}(x,y) = e^x\mathbf{i} - xy\mathbf{j}$

In each of Exercises 17–22, use formula (13.5.4) to calculate the area of the given region.

17. The triangle with vertices $(2,5)$, $(3,-6)$, and $(4,1)$
18. The trapezoid with vertices $(4,1)$, $(-2,1)$, $(3,3)$, and $(-1,3)$
19. The region bounded by the parabola $y = -2x^2 + 6$ and the line $y = 4x$
20. $\{(x,y) : 1/3 \le y \le 1/x, 1/2 \le x \le 3\}$
21. The area inside the circle $x^2 + y^2 = 1$ and lying above the line $y = (x+1)/\sqrt{3}$
22. The region obtained by removing the disc with boundary $x^2 + y^2 = 4$ from the triangle with vertices $(-14,16)$, $(8,12)$, and $(1,-20)$

In each of Exercises 23–25, compute the flux of $\mathbf{F}$ across the boundary of the given region.

23. $\mathbf{F}(x,y) = xy^2\mathbf{i} - x\mathbf{j}$, $\mathcal{R} = \{(x,y) : 0 < |x| < y < 1\}$
24. $\mathbf{F}(x,y) = \dfrac{y}{x^2+1}\mathbf{i} - \dfrac{3x}{y^2+1}\mathbf{j}$, $\mathcal{R} = \{(x,y) : 0 < x < 2,\ -1 < y < 2\}$
25. $\mathbf{F}(x,y) = xy^{7/2}\mathbf{i} - x^{5/2}y\mathbf{j}$, $\mathcal{R} = \{(x,y) : 0 < x < y < 1\}$

Further Theory and Practice

26. Use formula (13.5.4) to calculate the area of the region bounded by $\mathbf{r}(t) = \langle \cos(t) - 2\sin(t), \cos(t) \rangle$, $0 \le t \le 2\pi$.
27. Use formula (13.5.4) to calculate the area inside the cardioid that is described by the polar equation $\mathbf{r} = 2 + 2\cos(\theta)$.
28. Use formula (13.5.4) to calculate the area inside the limaçon that is described by the polar equation $\mathbf{r} = 5 - 3\sin(\theta)$.
29. Calculate $\displaystyle\oint_{\mathcal{C}} \left(y + \arcsin\left(\frac{x}{2}\right)\right) dx + (2xy + \ln(1+y^4))\, dy$ where $\mathcal{C}$ is the closed counterclockwise path consisting of the graphs $y = x^2, 0 \le x \le 1$ and $y = \sqrt{x}, 0 \le x \le 1$.
30. Calculate $\oint_{\mathcal{C}} (x - y\sec^2(x))\, dx + \arctan(y^2)\, dy$ where $\mathcal{C}$ is the closed counterclockwise path consisting of the graphs $y = \tan(x)^2$, $0 \le x \le \pi/4$ and $y = \tan(x)^4$, $0 \le x \le \pi/4$.

31. Sketch the curve $\mathcal{C}$ parameterized by $\mathbf{r}(t) = \langle \sin(t), \sin(2t) \rangle, 0 \le t \le \pi$. Use formula (13.5.4) to calculate the area of the region enclosed by $\mathcal{C}$.

32. Sketch the curve $\mathcal{C}$ parameterized by $\mathbf{r}(t) = \langle t - t^2, t - t^3 \rangle$, $0 \le t \le 1$. Use formula (13.5.4) to calculate the area of the region enclosed by $\mathcal{C}$.

33. The Folium of Descartes defined by $x^3 + y^3 = 3xy$ is shown in Figure 11. The loop in the folium is a simple closed curve that can be parameterized by

$$\mathbf{r}(t) = \frac{3t}{1+t^3}\mathbf{i} + \frac{3t^2}{1+t^3}\mathbf{j}, \quad 0 \le t < \infty.$$

Although the interval of the parameter is unbounded, formula (13.5.4) does yield the correct area of the region enclosed by the loop. Calculate the area.

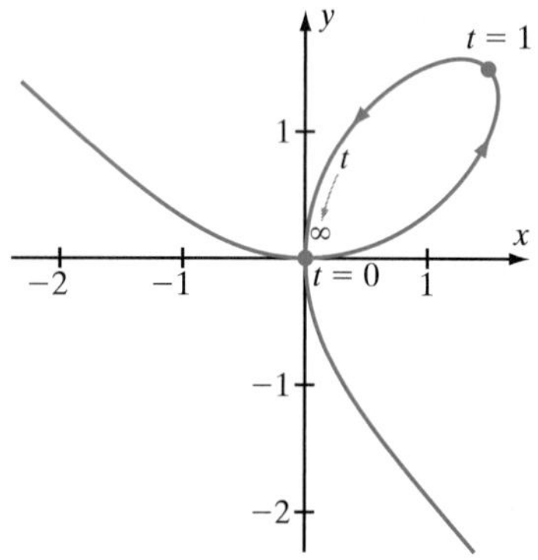

$\mathbf{r}(t) = \frac{3t}{1+t^3}\mathbf{i} + \frac{3t^2}{1+t^3}\mathbf{j}$

▲ Figure 11

34. Suppose that u and v are twice continuously differentiable functions in a neighborhood of a region $\mathcal{R}$ and its boundary curve $\mathcal{C}$. Suppose that $\mathcal{R}$ and $\mathcal{C}$ are as in the statement of Green's Theorem. Prove the following identities involving Laplace's operator, $\triangle = \partial^2/\partial x^2 + \partial^2/\partial y^2$.

a. Green's First Identity

$$\iint_{\mathcal{R}} (u \triangle v + \nabla u \cdot \nabla v) dA = \int_{\mathcal{C}} (u v_x dy - u v_y dx).$$

b. Green's Second Identity

$$\iint_{\mathcal{R}} (u \triangle v - v \triangle u) dA = \int_{\mathcal{C}} (u v_x - v u_x) dy + (v u_y - u v_y) dx$$

Hint: Use part a, by reversing the roles of u and v.

35. Use Green's Second Identity (from the preceding exercise) to show that

$$\int_{\mathcal{C}} u_y\, dx - u_x dy = 0$$

if u is harmonic ($\triangle u = 0$ at all points).

Calculator/Computer Exercises

36. Plot the simple closed curve that is parameterized by $\mathbf{r}(t) = \sin(2t)\mathbf{i} - \sin^2(3t)\mathbf{j}$, $\pi/12 < t < 5\pi/12$. Calculate the area that is enclosed.

▶ In each of Exercises 37–40, verify Green's Theorem for the given vector field $\mathbf{F}$ and region $\mathcal{R}$. ◀

37. $\mathbf{F}(x,y) = \sin(x + y^2)\mathbf{i}$, $\mathcal{R} = \{(x,y) : x^2 < y < x, 0 < x < 1\}$

38. $\mathbf{F}(x,y) = \exp(yx^2)\mathbf{i} + x\exp(xy^2)\mathbf{j}$,
$\mathcal{R} = \{(x,y) : 0 < y < 1 - x^2,\ -1 < x < 1\}$

39. $\mathbf{F}(x,y) = \dfrac{x^2}{1+y^4}\mathbf{i} + \ln(1+x^4)\mathbf{j}$,
$\mathcal{R} = \{(x,y) : 0 < y < \sin(x), 0 < x < \pi\}$

40. $\mathbf{F}(x,y) = (1 + x^2 + y^2)^{-1/2}\mathbf{i} + \sqrt{1 + x^2 + y^2}\mathbf{j}$,
$\mathcal{R} = \left\{(x,y) : (x-1)^2 < y < 1 - x + x^2 - x^5, 0 < x < 1\right\}$

13.6 Surface Integrals

In this section, we develop the final concept that will be needed for the principal theorems of vector analysis. That is the concept of the surface integral, which enables us to integrate a function that is defined on a surface. We begin with the problem of calculating the area of a surface that is the graph of a function.

Suppose that a surface $\mathcal{S}$ is given as the graph of a function $z = f(x,y)$ over a region $\mathcal{R}$ in the xy-plane as in Figure 1. How do we calculate the surface area of $\mathcal{S}$? As you might suspect, we can develop a method for calculating surface area by using a double integral of a certain expression over the region $\mathcal{R}$.

The Integral for Surface Area

Suppose that a surface $\mathcal{S}$ is the graph of a function f over a bounded region $\mathcal{R}$ in the xy-plane, as in Figure 1. Let $Q = \{(x,y) : a < x < b, c < y < d\}$ be a rectangle that contains $\mathcal{R}$. We partition the x-interval $[a,b]$ as

$$a = x_0 < x_1 < x_2 < \cdots < x_{N-1} < x_N = b \quad \text{where } x_i = a + i\Delta x,\ \ \Delta x = (b-a)/N,$$

and we partition the y-interval $[c,d]$ as

$$c = y_0 < y_1 < y_2 < \cdots < y_{N-1} < y_N = d \quad \text{where } y_j = c + j\Delta y,\ \ \Delta y = (d-c)/N.$$

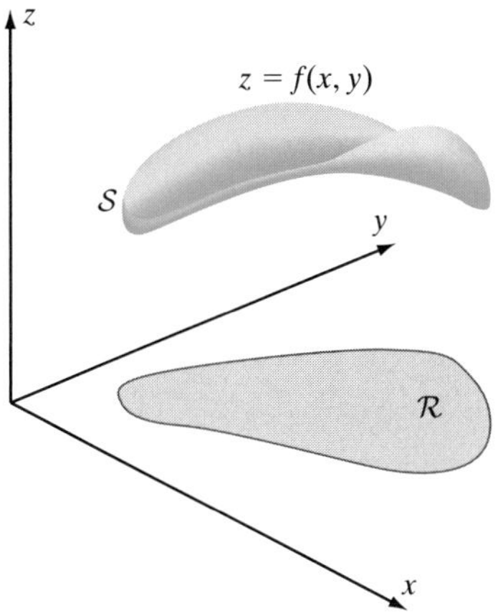

▲ Figure 1

Corresponding to each pair x_{i-1}, x_i in the partition of the x-variable and each pair y_{j-1}, y_j in the partition of the y-variable, there is a small subrectangle $Q_{i,j}$ in the xy-plane. The only relevant rectangles $Q_{i,j}$ for our calculation are the ones that lie entirely in $\mathcal{R}$ (see Figure 2). We want to approximate the area $A_{i,j}$ of the small piece of surface lying above $Q_{i,j}$. Examine Figure 3. We estimate $A_{i,j}$ by the area of a parallelogram that closely approximates the graph of f over $Q_{i,j}$. To do this, we select a point ξ_{ij} inside $Q_{i,j}$ and use the parallelogram that is tangent to $\mathcal{S}$ at the point over ξ_{ij} and that has the vectors $\mathbf{v} = \langle \Delta x, 0, f_x(\xi_{ij})\Delta x\rangle$ and $\mathbf{w} = \langle 0, \Delta y, f_y(\xi_{ij})\Delta y\rangle$ for its sides. Refer to Figure 4. We know from Section 9.4 in Chapter 9 that the area of this parallelogram is

$$\|\mathbf{v}\times\mathbf{w}\| = \left\| \det\left(\begin{bmatrix} \mathbf{i} & \mathbf{j} & \mathbf{k} \\ \Delta x & 0 & f_x(\xi_{ij})\Delta x \\ 0 & \Delta y & f_y(\xi_{ij})\Delta y \end{bmatrix}\right)\right\| = \left\|(-f_x(\xi_{ij})\Delta x\Delta y)\mathbf{i} + \left(-f_y(\xi_{ij})\Delta x\Delta y\right)\mathbf{j} + (\Delta x\Delta y)\mathbf{k}\right\|$$

$$= \sqrt{1 + f_x(\xi_{ij})^2 + f_y(\xi_{ij})^2}\,\Delta x\Delta y.$$

If we sum over rectangles $Q_{i,j}$ that lie in $\mathcal{R}$, then we get an approximation $\sum_{Q_{i,j}\subset\mathcal{R}} \sqrt{1 + f_x(\xi_{ij})^2 + f_y(\xi_{ij})^2}\,\Delta x\Delta y$ to the surface area of $\mathcal{S}$. As the number N

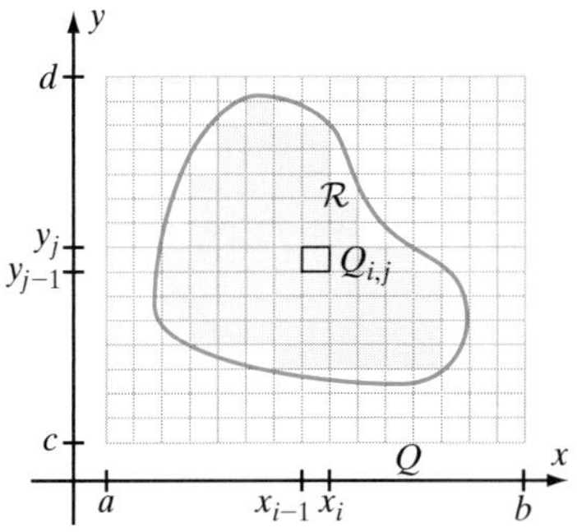

▲ Figure 2

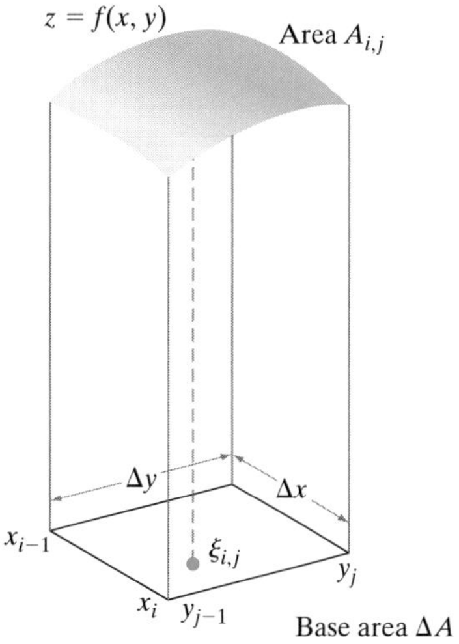

▲ Figure 3

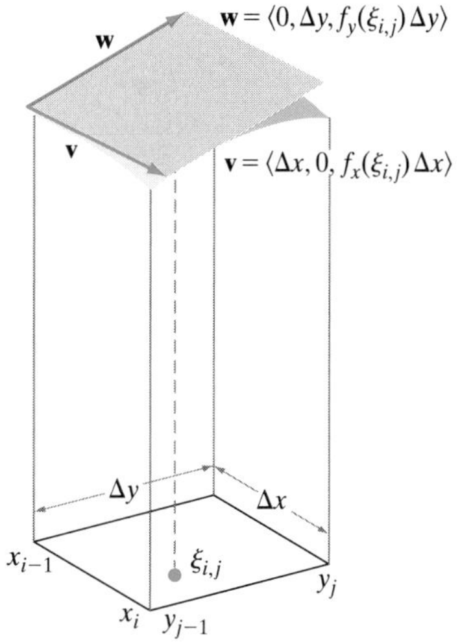

▲ Figure 4

increases, our approximation improves. On the other hand, our approximation is a Riemann sum for the double integral $\iint_{\mathcal{R}} \sqrt{1+f_x(x,y)^2+f_y(x,y)^2}\,dA$. We therefore make the following definition.

DEFINITION Let $\mathcal{S}$ be the graph of a continuously differentiable function $(x,y)\mapsto f(x,y)$ defined on a region $\mathcal{R}$ in the xy-plane. We refer to

$$dS=\sqrt{1+\Big(\frac{\partial}{\partial x}f(x,y)\Big)^2+\Big(\frac{\partial}{\partial y}f(x,y)\Big)^2}\,dA$$

as the *element of surface area* on $\mathcal{S}$. The *surface area* of the graph of f over $\mathcal{R}$ is defined by

$$\text{Surface area of } \mathcal{S}=\iint_{\mathcal{S}} 1\,dS=\iint_{\mathcal{R}}\sqrt{1+\Big(\frac{\partial}{\partial x}f(x,y)\Big)^2+\Big(\frac{\partial}{\partial y}f(x,y)\Big)^2}\,dA. \quad (13.6.1)$$

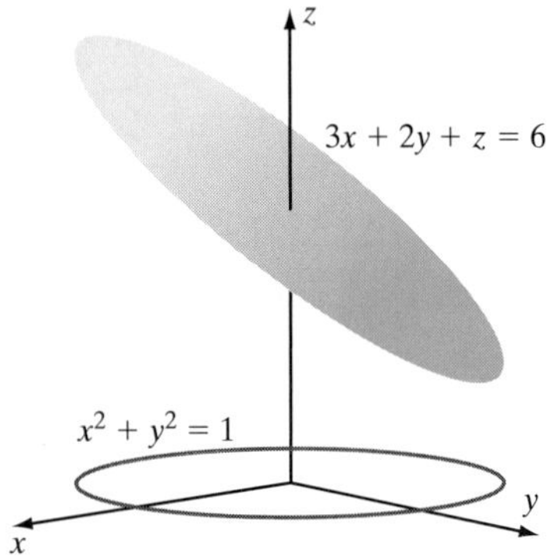

▲ Figure 5

▶ **EXAMPLE 1** Find the area of the portion of the plane $3x+2y+z=6$ that lies over the interior of the circle $x^2+y^2=1$ in the xy-plane.

Solution The surface whose area we are computing is *not itself a disk*. It is the interior of an ellipse (see Figure 5). But we can do the area calculation without determining the exact nature of this ellipse. We need only notice that the plane is the graph of the function $z=f(x,y)=6-3x-2y$. We need to calculate the surface area of the graph over the region $\mathcal{R}=\{(x,y): x^2+y^2<1\}$. According to the definition, the required surface area is

$$\iint_{\mathcal{R}}\sqrt{1+(f_x)^2+(f_y)^2}dA=\iint_{\mathcal{R}}\sqrt{1+(-3)^2+(-2)^2}\,dA=\sqrt{14}\iint_{\mathcal{R}}dA=\sqrt{14}\cdot\pi\cdot 1^2=\sqrt{14}\pi. \quad ◀$$

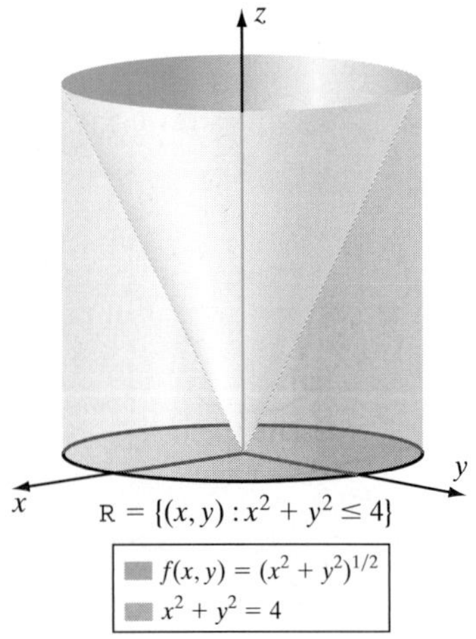

▲ Figure 6

▶ **EXAMPLE 2** Calculate the surface area of that portion of the graph of $z=\sqrt{x^2+y^2}$ that lies inside the cylinder $x^2+y^2=4$.

Solution See Figure 6. To determine the surface area of the graph of $z=\sqrt{x^2+y^2}$, we let $f(x,y)=(x^2+y^2)^{1/2}$ and calculate

$$\sqrt{1+f_x(x,y)^2+f_y(x,y)^2}=\sqrt{1+\frac{x^2}{x^2+y^2}+\frac{y^2}{x^2+y^2}}=\sqrt{\frac{(x^2+y^2)+x^2+y^2}{x^2+y^2}}$$
$$=\sqrt{\frac{2(x^2+y^2)}{x^2+y^2}}=\sqrt{2}\,.$$

The region $\mathcal{R}$ that we integrate over is the disk in the xy-plane that is inside the cylinder $\mathcal{R}$. The required surface area is therefore

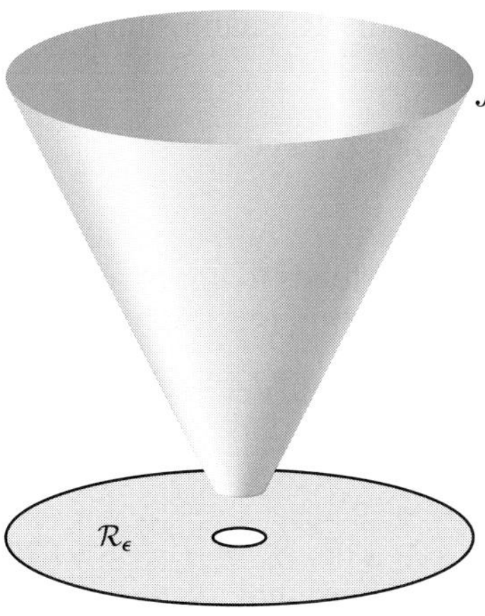

$\mathcal{R}_\epsilon = \{(x, y) : \epsilon^2 < x^2 + y^2 < 4\}$

▲ Figure 7

$$\iint_{\mathcal{R}} \sqrt{1 + (f_x)^2 + (f_y)^2}\,dA = \iint_{\mathcal{R}} \sqrt{2}\,dA = \sqrt{2} \cdot (\text{Area of } \mathcal{R}) = \sqrt{2} \cdot (\pi \cdot 2^2) = 4\pi\sqrt{2}. \quad ◀$$

INSIGHT Strictly speaking, the definition of surface area that we have given does not apply to the conical surface in Example 2. The corner on the graph of f at the origin means that the first derivatives of f do not exist there. However, we may justify what we have done by calculating the area of the part of the graph over the ring $\mathcal{R}_\varepsilon = \{(x, y) : \varepsilon^2 < x^2 + y^2 < 4\}$ and then letting ε tend to 0 (see Figure 7).

When a new formula is developed, it is always a good idea to verify that the results it yields are consistent with those that are obtained by existing methods (when those methods can be applied). For the conical surface of Example 2, we can apply formula (7.2.7) for the surface area of the frustum of a cone. In that formula, $A = 2\pi\left(\frac{r+R}{2}\right)s$, we set $r = 0$ (the radius at the vertex), $R = 2$ (the radius of the base), and $s = 2\sqrt{2}$ (the slant height). We obtain $2\pi\left(\frac{0+2}{2}\right)2\sqrt{2}$, or $4\pi\sqrt{2}$, which agrees with the value obtained in Example 2.

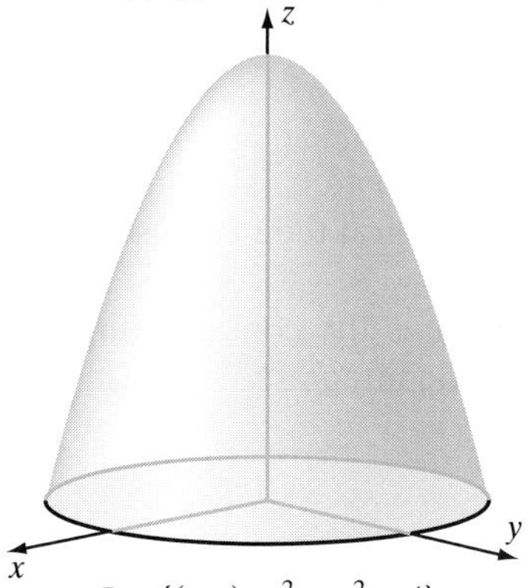

$\mathcal{R} = \{(x, y) : x^2 + y^2 \le 4\}$

▲ Figure 8

▶ **EXAMPLE 3** Calculate the surface area of that part of the graph of $f(x, y) = 4 - x^2 - y^2$ that lies above the xy-plane.

Solution Look at Figure 8. We are required to calculate the surface area of the graph of f over the region $\mathcal{R} = \{(x, y) : x^2 + y^2 \le 4\}$:

$$\iint_{\mathcal{R}} \sqrt{1 + \left(\frac{\partial}{\partial x} f(x, y)\right)^2 + \left(\frac{\partial}{\partial y} f(x, y)\right)^2}\,dA$$

$$= \iint_{\mathcal{R}} \sqrt{1 + (-2x)^2 + (-2y)^2}\,dA = \iint_{\mathcal{R}} \sqrt{1 + 4(x^2 + y^2)}\,dA.$$

This integration is easiest if we use polar coordinates. The region of integration is given by $\mathcal{R} = \{(r, \theta) : 0 \le \theta \le 2\pi, 0 \le r \le 2\}$ in polar coordinates. The integral becomes

$$\int_0^{2\pi} \int_0^2 \sqrt{1 + 4r^2}\, r\,dr\,d\theta = \int_0^{2\pi} \frac{1}{12}(1 + 4r^2)^{3/2}\Big|_{r=0}^{r=2} d\theta = \int_0^{2\pi} \left(\frac{17^{3/2} - 1}{12}\right) d\theta = \left(\frac{17^{3/2} - 1}{6}\right)\pi. \quad ◀$$

Integrating a Function Over a Surface

Once again, consider the surface $\mathcal{S}$ that is shown in Figure 1. Suppose that, at each point (x, y, z) of $\mathcal{S}$, we know the weight density $\varphi(x, y, z)$ of the surface. In order to calculate the total weight of the surface, we repeat the procedure that we used to find surface area. That is, we enclose the region $\mathcal{R}$ in a rectangle Q, we subdivide each side of Q into N equal length subintervals (refer to Figure 2), and we choose a point $\xi_{i,j}$ in each of the resulting small rectangles $Q_{i,j}$ that are entirely contained within $\mathcal{R}$. The weight of that part of $\mathcal{S}$ that lies over $Q_{i,j}$ (as shown earlier in Figure 3) is approximated by

$$\underbrace{\varphi(\xi_{i,j}, f(\xi_{i,j}))}_{\text{weight per unit area}} \cdot \underbrace{\sqrt{1 + f_x(\xi_{i,j})^2 + f_y(\xi_{i,j})^2} \cdot \Delta x \cdot \Delta y}_{\text{approximate surface area of patch over } Q_{i,j}}.$$

Summing over all the subrectangles $Q_{i,j}$ in $\mathcal{R}$, we obtain the Riemann sum

$$\sum_{i,j} \varphi\left(\xi_{i,j}, f(\xi_{i,j})\right) \sqrt{1 + f_x(\xi_{i,j})^2 + f_y(\xi_{i,j})^2} \cdot \Delta x \Delta y$$

as an approximation of the weight of $\mathcal{S}$. This approximation tends to improve as the number N of subdivisions increases. Because our approximation is a Riemann sum for the double integral $\iint_{\mathcal{R}} \varphi(x, y, f(x, y)) \sqrt{1 + f_x(x, y)^2 + f_y(x, y)^2}\, dA$, we make the following definition.

DEFINITION Let $\mathcal{S}$ be the graph of a continuously differentiable function f defined on a region $\mathcal{R}$ in the xy-plane. If φ is a continuous function on $\mathcal{S}$, then the *surface integral* of φ over $\mathcal{S}$, denoted by $\iint_{\mathcal{S}} \varphi\, dS$, is given by

$$\iint_{\mathcal{S}} \varphi\, dS = \iint_{\mathcal{R}} \varphi(x, y, f(x, y)) \sqrt{1 + \left(\frac{\partial}{\partial x} f(x, y)\right)^2 + \left(\frac{\partial}{\partial y} f(x, y)\right)^2}\, dA. \quad (13.6.2)$$

Formula (13.6.2) is often used when φ is a density function (such as weight density, mass density, charge density, etc.). The surface integral then represents the total value of the quantity whose density is represented by φ. Of course formula (13.6.2) has meaning for any continuous function φ. If φ is the constant function that is identically 1, then the surface integral reduces to the integral for surface area given by formula (13.6.1).

INSIGHT It is worth taking a moment to understand the roles of the two functions that appear in surface integral (13.6.2). The function $(x, y) \mapsto f(x, y)$ is a function of two variables. It defines the surface $\mathcal{S}$ and gives rise to the factor $\sqrt{1 + f_x(x, y)^2 + f_y(x, y)^2}$ that is part of the element of surface area. The function $(x, y, z) \mapsto \varphi(x, y, z)$ is a function of three variables that is defined on a region of space that includes the surface $\mathcal{S}$. This is the function we integrate. Because the integration is over $\mathcal{S}$, the value $\varphi(x, y, z)$ at a general point (x, y, z) does *not* appear in formula (13.6.2), as it does in the triple integrals $\iiint_{\mathcal{U}} \varphi(x, y, z)\, dV$ of Section 12.6 in Chapter 12 (where we integrate φ over a solid $\mathcal{U}$). Instead, when we integrate φ over the surface $\mathcal{S}$ that is the graph of $z = f(x, y)$, we evaluate φ only on $\mathcal{S}$. Therefore only the values $\varphi(x, y, f(x, y))$ are used in formula (13.6.2).

▶ **EXAMPLE 4** Let the surface $\mathcal{S}$ be the graph of $f(x, y) = x^2 + y^2$ for (x, y) in the planar region $\mathcal{R} = \{(x, y) : x^2 + y^2 < 2\}$. Calculate the surface integral of $\varphi(x, y, z) = z - x^2$ over the surface $\mathcal{S}$.

Solution The element of surface area is

$$dS = \sqrt{1 + f_x(x, y)^2 + f_y(x, y)^2}\, dA = \sqrt{1 + (2x)^2 + (2y)^2}\, dA = \sqrt{1 + 4(x^2 + y^2)}\, dA$$

and

$$\varphi(x,y,f(x,y)) = \varphi(x,y,z)|_{z=f(x,y)} = (z-x^2)|_{z=f(x,y)} = f(x,y) - y^2 = (x^2+y^2) - x^2 = y^2.$$

The surface integral therefore equals

$$\iint_{\mathcal{S}} \varphi\, dS = \iint_{\mathcal{R}} y^2\sqrt{1+4(x^2+y^2)}\, dA = \int_0^{2\pi}\int_0^{\sqrt{2}} r^2\sin^2(\theta)(1+4r^2)^{1/2} r\, dr\, d\theta,$$

where the last integral has been obtained by converting to polar coordinates. We compute this last integral by making the substitution $u = 1+4r^2$, $du = 8r\,dr$. With this substitution, $r = 0$ corresponds to $u = 1+4(0)^2 = 1$, and $r = \sqrt{2}$ corresponds to $u = 1+4\left(\sqrt{2}\right)^2 = 9$. Because the equation $u = 1+4r^2$ gives us $r^2 = (u-1)/4$, we obtain

$$\iint_{\mathcal{S}} \varphi\, dS = \int_0^{2\pi}\int_1^9 \left(\frac{u-1}{4}\right)\sin^2(\theta)u^{1/2}\frac{1}{8}\, du\, d\theta = \frac{1}{32}\int_0^{2\pi}\sin^2(\theta)\left(\int_1^9\left(u^{3/2}-u^{1/2}\right)du\right)d\theta. \quad (13.6.3)$$

The inner integral involving u is routinely determined:

$$\int_1^9\left(u^{3/2}-u^{1/2}\right)du = \left(\frac{2}{5}u^{5/2} - \frac{2}{3}u^{3/2}\right)\Bigg|_1^9 = \frac{1192}{15}.$$

Substituting this value into equation (13.6.3) and using formula (6.2.9), we obtain

$$\iint_{\mathcal{S}} \varphi\, dS = \frac{1}{32}\cdot\frac{1192}{15}\int_0^{2\pi}\sin^2(\theta)\, d\theta = \frac{1}{32}\cdot\frac{1192}{15}\pi = \frac{149}{60}\pi. \quad ◀$$

▶ **EXAMPLE 5** Integrate the function $\varphi(x,y,z) = xy - 3z$ over the surface $\mathcal{S}$ given by the graph of $f(x,y) = xy$ over the region $\mathcal{R} = \{(x,y) : 0<x<1, 0<y<2\}$.

Solution The element of surface area is

$$dS = \sqrt{1+f_x(x,y)^2+f_y(x,y)^2}\, dA = (1+y^2+x^2)^{1/2}\, dx\, dy$$

and

$$\varphi\big(x,y,f(x,y)\big) = \varphi(x,y,z)|_{z=f(x,y)} = xy - 3z|_{z=f(x,y)} = xy - 3f(x,y) = xy - 3xy = -2xy.$$

Thus

$$\iint_{\mathcal{S}} \varphi\, dS = \int_0^2\int_0^1 -2xy(1+y^2+x^2)^{1/2}\, dx\, dy = -2\int_0^2 y\left(\int_0^1 x(1+y^2+x^2)^{1/2}\, dx\right)dy. \quad (13.6.4)$$

To calculate the inner integral, we make the substitution $u = 1+y^2+x^2$, $du = 2x\,dx$. After finding the antiderivative with respect to u, we will resubstitute for x and use the original limits of integration. Thus

$$\int_0^1 x(1+y^2+x^2)^{1/2}\,dx = \frac{1}{2}\int_{x=0}^{x=1} u^{1/2}\,du = \frac{1}{2}\left(\frac{2}{3}u^{3/2}\right)\Bigg|_{x=0}^{x=1} = \frac{1}{3}(1+y^2+x^2)^{3/2}\Bigg|_{x=0}^{x=1} = \frac{1}{3}\left((2+y^2)^{3/2}-(1+y^2)^{3/2}\right).$$

On substituting the result of this integration into (13.6.4), we obtain

$$\iint_{\mathcal{S}} \varphi\,dS = -\frac{2}{3}\int_0^2 y\left((2+y^2)^{3/2}-(1+y^2)^{3/2}\right)dy = -\frac{2}{3}\int_0^2 y(2+y^2)^{3/2}\,dy + \frac{2}{3}\int_0^2 y(1+y^2)^{3/2}\,dy. \qquad (13.6.5)$$

The two integrals on the right side of equation (13.6.5) are quite similar and can be handled by one integration, if performed in appropriate generality. We replace the constants 2 and 1 that are added to y^2 with a general constant a and consider $\int_0^2 y(a+y^2)^{3/2}\,dy$. We then make the change of variable $u = a + y^2$, $du = 2y\,dy$, and note that $y = 0$ corresponds to $u = a$ and $y = 2$ corresponds to $u = a + 4$. We have

$$\int_0^2 y(a+y^2)^{3/2}\,dy = \frac{1}{2}\int_a^{a+4} u^{3/2}\,du = \frac{1}{2}\cdot\frac{2}{5}u^{5/2}\Bigg|_{u=a}^{u=a+4} = \frac{1}{5}\left((a+4)^{5/2}-a^{5/2}\right).$$

Substituting $a = 2$ and $a = 1$ into this formula results in

$$\int_0^2 y(2+y^2)^{3/2}\,dy = \frac{1}{5}\left((2+4)^{5/2}-2^{5/2}\right) = \frac{1}{5}\left(6^{5/2}-2^{5/2}\right)$$

and

$$\int_0^2 y(1+y^2)^{3/2}\,dy = \frac{1}{5}\left((1+4)^{5/2}-1^{5/2}\right) = \frac{1}{5}\left(5^{5/2}-1\right).$$

Substituting the results of these two integrations into the right side of (13.6.5) yields

$$\iint_{\mathcal{S}} \varphi\,dS = -\frac{2}{3}\left(\frac{1}{5}(6^{5/2}-2^{5/2})\right) + \frac{2}{3}\left(\frac{1}{5}(5^{5/2}-1)\right) = \frac{2}{15}\left(2^{5/2}+5^{5/2}-6^{5/2}-1\right).$$

An Application We now give an application of the concept of surface integral to the determination of a center of mass.

▶ **EXAMPLE 6** Assuming that it has uniform mass distribution δ, determine the center of mass of the upper half of the sphere $x^2 + y^2 + z^2 = a^2$.

Solution Refer to Section 12.7 in Chapter 12 for the concept of center of mass. The surface to be studied is $\mathcal{S} = \{(x,y,z) : x^2+y^2+z^2 = a^2, z > 0\}$. By symmetry considerations, it is clear that $\overline{x} = 0$ and $\overline{y} = 0$. Thus we need only calculate $\overline{z}$. We think of the surface as the graph of the function $f(x,y) = \sqrt{a^2-x^2-y^2}$ over the region $\mathcal{R} = \{(x,y) : x^2+y^2 < a^2\}$. We see that

$$M_{z=0} = \iint_{\mathcal{R}} \delta\cdot z\cdot\sqrt{1+f_x(x,y)^2+f_y(x,y)^2}\,dA.$$

Now

$$f_x(x,y) = \frac{-x}{\sqrt{a^2 - x^2 - y^2}} \quad \text{and} \quad f_y(x,y) = \frac{-y}{\sqrt{a^2 - x^2 - y^2}}.$$

It follows that

$$\sqrt{1 + f_x(x,y)^2 + f_y(x,y)^2} = \sqrt{1 + \left(\frac{x^2}{a^2 - x^2 - y^2}\right) + \left(\frac{y^2}{a^2 - x^2 - y^2}\right)} = \frac{a}{\sqrt{a^2 - x^2 - y^2}}. \tag{13.6.6}$$

Therefore

$$M_{z=0} = \iint_{\mathcal{R}} \delta \cdot z \cdot \frac{a}{\sqrt{a^2 - x^2 - y^2}}\, dA = a \iint_{\mathcal{R}} \frac{\sqrt{a^2 - x^2 - y^2}}{\sqrt{a^2 - x^2 - y^2}}\, dA = a\delta \iint_{\mathcal{R}} 1\, dA = a\delta(\text{area of } \mathcal{R}) = \pi a^3 \delta.$$

We also need to calculate the mass of S. In Example 7 of Section 7.2 in Chapter 7, we learned that the surface area of a sphere of radius a is $4\pi a^2$. The surface area of the hemisphere S is therefore $2\pi a^2$. The mass M of S is therefore $2\pi a^2 \delta$. It follows that

$$\overline{z} = \frac{M_{z=0}}{M} = \frac{\pi a^3 \delta}{2\pi a^2 \delta} = \frac{a}{2}.$$

Therefore the center of mass is $(0, 0, a/2)$. ◄

INSIGHT It is worth pausing to observe the difference between the calculation of Example 6 and the calculations found in Section 12.7 of Chapter 12. In that section, the geometric objects we treated were either planar regions or three-dimensional solids. For example, using the formulas of Section 12.7, we are able to calculate the center of mass of the *solid* half-ball $\mathcal{U}$ defined by the inequalities $x^2 + y^2 + z^2 \le a^2$, $0 \le z$. If $\mathcal{U}$ has uniform density δ, then its mass is $\frac{2}{3}\pi a^3 \delta$. Using cylindrical coordinates, we calculate $\overline{z}$ as follows:

$$\overline{z} = \frac{M_{z=0}}{M} = \frac{1}{2\pi a^3 \delta/3} \int_0^{2\pi} \int_0^a \int_0^{\sqrt{a^2 - r^2}} \delta z\, dz\, r\, dr\, d\theta = \frac{3}{4\pi a^3} \int_0^{2\pi} \int_0^a (a^2 - r^2) r\, dr\, d\theta = \frac{3}{8} a.$$

The calculation involves a triple integral, and the value of the z-coordinate of the center of mass is now different.

The Element of Area for a Surface Given Parametrically

In many applications, it is most convenient to describe a surface parametrically. Because a surface is a two-dimensional geometric figure in three-dimensional space, a parameterization specifies the three space coordinates, x, y, and z, in terms of two parameters. We will use u and v as generic parameters and indicate the dependence of x, y, and z on u and v by writing $x(u, v)$, $y(u, v)$, $z(u, v)$. Using these three functions of u and v, we obtain a vector-valued function $\mathbf{r}(u, v) = \langle x(u, v), y(u, v), z(u, v)\rangle$. Notice that we use the letter $\mathbf{r}$ as the position vector for a point on a parameterized surface, just as we do for a point on a parameterized curve. Of course, we know that $\mathbf{r}(t)$ refers to a point on a curve because there is one parameter, and $\mathbf{r}(u,v)$ refers to a point on a surface because there are two parameters.

We now briefly describe how to calculate the element of area on a surface S that is parameterized by a vector-valued function $\mathbf{r}(u,v)=\langle x(u,v), y(u,v), z(u,v)\rangle$, where u and v range over a parameter set $\mathcal{R}$ in the plane (see Figure 9). Fix values u_0 and v_0. Figure 10 shows a small rectangle Q with vertex at (u_0, v_0), width Δu, and height Δv. The patch on S that is the image, $\mathbf{r}(Q)$, of Q is also shown. In analogy with our preceding investigation, we want to find the area of a parallelogram that approximates $\mathbf{r}(Q)$. The functions

$$u \mapsto \langle x(u,v_0), y(u,v_0), z(u,v_0)\rangle \qquad \text{and} \qquad v \mapsto \langle x(u_0,v), y(u_0,v), z(u_0,v)\rangle$$

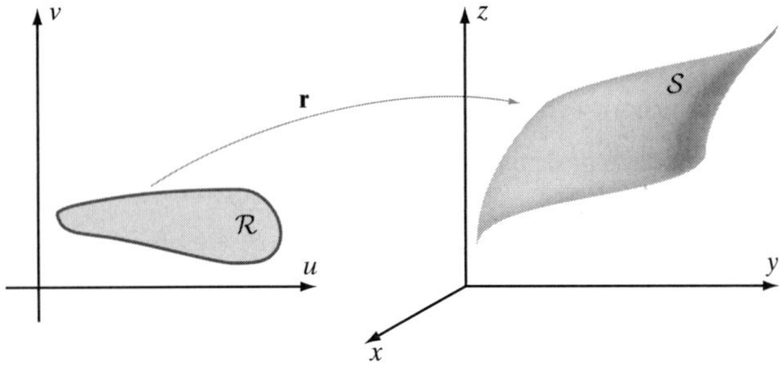

▲ Figure 9

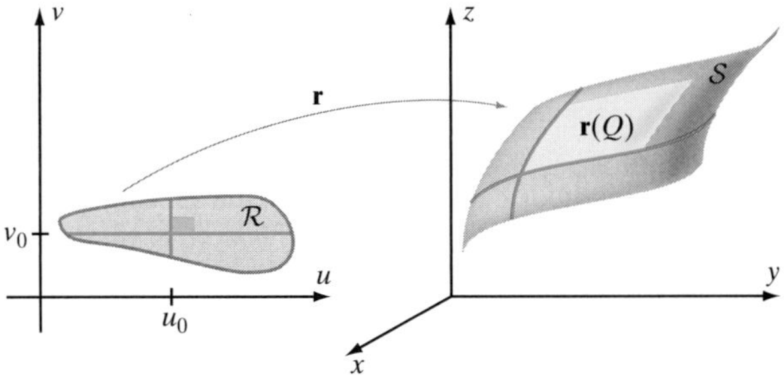

▲ Figure 10 Rectangle Q of width Δu and height Δv

describe curves on the surface S that pass through the point $P_0=\mathbf{r}(u_0,v_0)$. The velocity vectors $\mathbf{r}_u(u_0,v_0)$ and $\mathbf{r}_v(u_0,v_0)$ are tangent to S at P_0, as are their scalar multiples $(\Delta u)\mathbf{r}_u(u_0,v_0)$ and $(\Delta v)\mathbf{r}_v(u_0,v_0)$. (Here the subscripts u and v denote partial differentiation with respect to u and v, respectively). By reasoning similar to that used at the beginning of the section, we find that $\mathbf{r}(Q)$ is closely approximated by the area of the parallelogram determined by the vectors $(\Delta u)\mathbf{r}_u(u_0,v_0)$ and $(\Delta v)\mathbf{r}_v(u_0,v_0)$ as shown in Figure 11. That area is just $\|(\Delta u)\mathbf{r}_u(u_0,v_0)\times(\Delta v)\mathbf{r}_v(u_0,v_0)\|$, or $\|(\mathbf{r}_u\times\mathbf{r}_v)(u_0,v_0)\|\Delta u\Delta v$. Adding these expressions, we obtain a Riemann sum for the integral $\iint_{\mathcal{R}}\|(\mathbf{r}_u\times\mathbf{r}_v)(u,v)\|\,du dv$. We are therefore led to the following definition of surface area.

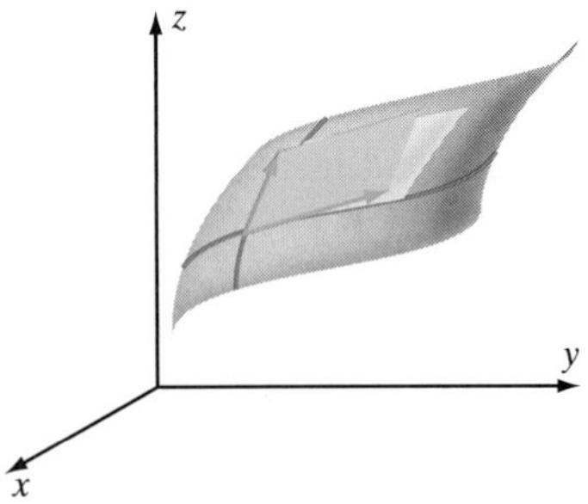

▲ Figure 11 The vectors $\Delta u(\mathbf{r}_u(u_0,v_0))$, $\Delta v(\mathbf{r}_v(u_0,v_0))$, and the parallelogram that they determine

DEFINITION Let $\mathbf{r}(u,v)=\langle x(u,v), y(u,v), z(u,v)\rangle$ be a continuously differentiable vector-valued function with domain $\mathcal{R}$ in the plane. Let S be the surface that is the image of $\mathcal{R}$ under $\mathbf{r}$. Then, the surface area of S is given by the formula

$$\text{Surface area of } \mathcal{S} = \iint_{\mathcal{R}} \|(\mathbf{r}_u\times\mathbf{r}_v)(u,v)\|\,du dv. \qquad (13.6.7)$$

We refer to the expression $\|(\mathbf{r}_u\times\mathbf{r}_v)(u_0,v_0)\|\,du dv$ as the *element of area* for the surface $\mathcal{S}$.

▶ **EXAMPLE 7** If a is a positive constant, and if we set $\rho = a$ in equations (12.8.3), (12.8.5), and (12.8.6) for the spherical coordinates of the point (x, y, z), then we see that

$$\mathbf{r}(\theta, \phi) = \langle a\cos(\theta)\sin(\phi), a\sin(\theta)\sin(\phi), a\cos(\phi)\rangle, 0 \le \theta \le 2\pi, 0 \le \phi \le \pi$$

parameterizes the sphere with radius a and center at the origin. Use this parameterization to determine the surface area of a sphere of radius $a > 0$.

Solution We have

$$\mathbf{r}_\theta(\theta, \phi) = a\langle -\sin(\theta)\sin(\phi), \cos(\theta)\sin(\phi), 0\rangle \quad \text{and} \quad \mathbf{r}_\phi(\theta, \phi) = a\langle \cos(\theta)\cos(\phi), \sin(\theta)\cos(\phi), -\sin(\phi)\rangle .$$

Thus

$$(\mathbf{r}_\theta \times \mathbf{r}_\phi)(\theta, \phi) = a^2 \det\left(\begin{bmatrix} \mathbf{i} & \mathbf{j} & \mathbf{k} \\ -\sin(\theta)\sin(\phi) & \cos(\theta)\sin(\phi) & 0 \\ \cos(\theta)\cos(\phi) & \sin(\theta)\cos(\phi) & -\sin(\phi) \end{bmatrix}\right),$$

or

$$(\mathbf{r}_\theta \times \mathbf{r}_\phi)(\theta, \phi) = -a^2\big(\cos(\theta)\sin^2(\phi)\mathbf{i} + \sin(\theta)\sin^2(\phi)\mathbf{j} + \big(\sin^2(\theta)\sin(\phi)\cos(\phi) + \cos^2(\theta)\cos(\phi)\sin(\phi)\big)\mathbf{k}\big).$$

On simplification, we have

$$(\mathbf{r}_\theta \times \mathbf{r}_\phi)(\theta, \phi) = -a^2\sin(\phi)\big(\cos(\theta)\sin(\phi)\mathbf{i} + \sin(\theta)\sin(\phi)\mathbf{j} + \cos(\phi)\mathbf{k}\big)$$

and so

$$\|(\mathbf{r}_\theta \times \mathbf{r}_\phi)(\theta, \phi)\| = a^2\sin(\phi)\sqrt{\cos^2(\theta)\sin^2(\phi) + \sin^2(\theta)\sin^2(\phi) + \cos^2(\phi)} = a^2\sin(\phi).$$

It follows that the surface area of a sphere of radius a is

$$\int_0^{2\pi}\int_0^{\pi} a^2\sin(\phi)\,d\phi\,d\theta = a^2\int_0^{2\pi}\big(-\cos(\phi)\big)\Big|_{\phi=0}^{\phi=\pi} d\theta = a^2\int_0^{2\pi}\big(-(-1) - (-1)\big)\,d\theta = 4\pi a^2,$$

which agrees with the value obtained in Example 7 of Section 7.2 in Chapter 7. ◀

Surface Integrals Over Parameterized Surfaces

Having already determined the element of surface area for a parametrically defined surface S, we may proceed directly to the definition of a surface integral on S.

DEFINITION Let $\mathbf{r}(u, v) = \langle x(u, v), y(u, v), z(u, v)\rangle$ be a continuously differentiable vector-valued function with domain $\mathcal{R}$ in the plane. Let S be the surface that is the image of $\mathcal{R}$ under $\mathbf{r}$. If φ is a continuous function on S, then the surface integral of φ over S is denoted by $\iint_S \varphi\, dS$ and is given by the formula

$$\iint_S \varphi\, dS = \iint_{\mathcal{R}} \varphi(\mathbf{r}(u, v))\,\|(\mathbf{r}_u \times \mathbf{r}_v)(u, v)\|\,du\,dv.$$

▶ **EXAMPLE 8** Integrate the function $\varphi(x,y,z) = z - xy$ over the surface S that is parameterized by

$$\mathbf{r}(u,v) = \langle u+v, u-v, 3u-2v\rangle, \quad 0 \le u \le 1, 0 \le v \le 2.$$

Solution We calculate that $\mathbf{r}_u(u,v) = \langle 1,1,3\rangle$ and $\mathbf{r}_v(u,v) = \langle 1,-1,-2\rangle$. Then

$$\|(\mathbf{r}_u \times \mathbf{r}_v)(u,v)\| = \|\mathbf{i} + 5\mathbf{j} - 2\mathbf{k}\| = \sqrt{30}.$$

The integral that we want to evaluate is

$$\iint_S \varphi\, dS = \int_0^2 \int_0^1 \varphi(u+v, u-v, 3u-2v)\sqrt{30}\, du\, dv$$

$$= \sqrt{30}\int_0^2 \int_0^1 \left((3u-2v) - (u+v)(u-v)\right) du\, dv.$$

Performing the necessary algebra, we find that

$$\begin{aligned}\iint_S \varphi\, dS &= \sqrt{30}\int_0^2 \left(\frac{3}{2}u^2 - 2uv - \frac{1}{3}u^3 + uv^2\right)\Bigg|_{u=0}^{u=1} dv \\ &= \sqrt{30}\int_0^2 \left(\frac{7}{6} - 2v + v^2\right) dv \\ &= \sqrt{30}\left(\frac{7}{6}v - v^2 + \frac{1}{3}v^3\right)\Bigg|_0^2 \\ &= \sqrt{30}. \quad ◀\end{aligned}$$

QUICK QUIZ

1. What is the element of area for a surface that is the graph of $f(x,y) = 2x + y^2$ for (x,y) in a region of the xy-plane?
2. Calculate $\iint_S \varphi\, dS$ where $\varphi(x,y,z) = z\sqrt{2+y^2}$ and S is the graph of $f(x,y) = x + y^2/2$ for (x,y) in the unit square $[0,1]\times[0,1]$.
3. What is the element of surface area for a surface that is the graph of $\mathbf{r}(u,v) = \langle 2u, -v, u+v\rangle$ for (u,v) in a region of the uv-plane?

Answers

1. $\sqrt{5+4y^2}\,dx\,dy$ 2. $8/5$ 3. $3\,du\,dv$

EXERCISES

Problems for Practice

▶ In each of Exercises 1–12, calculate the surface area of the graph of the function f over the region $\mathcal{R}$ in the xy-plane. ◀

1. $f(x,y) = 3x - 4y + 8$, $\mathcal{R} = \{(x,y) : 2 \le x \le 5, -3 \le y \le -2\}$
2. $f(x,y) = x^2 + y^2 + 6$, $\mathcal{R} = \{(x,y) : x^2 + y^2 < 16\}$
3. $f(x,y) = (x^2 + y^2)^{1/2}$, $\mathcal{R} = \{(x,y) : 1 < x^2 + y^2 < 9\}$
4. $f(x,y) = 2\sqrt{2}x + (3/2)y^2$, $\mathcal{R} = \{(x,y) : 0 \le y \le 3, 0 \le x \le 2y\}$
5. $f(x,y) = x^2 - y^2 + 3$, $\mathcal{R} = \{(x,y) : x^2 + y^2 < 4\}$
6. $f(x,y) = xy$, $\mathcal{R} = \{(x,y) : x^2 + y^2 < 9\}$
7. $f(x,y) = x^2/2$, $\mathcal{R} = \{(x,y) : 2 \le x \le 6, 0 \le y \le x\}$
8. $f(x,y) = 2y^2 - 2x^2$, $\mathcal{R} = \{(x,y) : 16 < x^2 + y^2 < 25\}$
9. $f(x,y) = x + 6y$, $\mathcal{R} = \{(x,y) : 1 \le x \le 4, 2 \le y \le 2x\}$
10. $f(x,y) = x^2 - \sqrt{2}y$, $\mathcal{R} = \{(x,y) : 3 \le x \le 7, 0 \le y \le x\}$
11. $f(x,y) = 2x^2 - 2y^2 + 2$, $\mathcal{R} = \{(x,y) : 4 < x^2 + y^2 < 9\}$
12. $f(x,y) = \sqrt{1-x^2}$, $\mathcal{R} = \{(x,y) : -1 < x < 1, 0 < y < 1\}$

▶ In each of Exercises 13–20, calculate the area of the given surface. ◀

13. The cone with base of radius 5 and height 7 (do not count the area of the base)
14. The portion of the sphere $x^2 + y^2 + z^2 = 9$ that lies above the plane $z = 2$
15. The portion of the cone $z^2 = x^2 + y^2$ that lies between $z = 4$ and $z = 10$
16. The portion of the surface $z = xy$ that lies within the cylinder $x^2 + y^2 = 16$
17. The portion of the surface $z = (2/3)(x^{3/2} + y^{3/2})$ that lies over the region $\mathcal{R} = \{(x, y) : 1 \le x \le 3, 2 \le y \le 5\}$
18. The portion of the surface $z = (2/3)x^{3/2} + y + 2$ that lies over the region $\mathcal{R} = \{(x, y) : 1 \le x \le 4, 1 \le y \le 4\}$
19. The portion of the surface $3x - 3y + z = 12$ that lies over the interior of the ellipse $x^2 + 4y^2 = 4$
20. The portion of the surface $2x + 4y + z = 11$ that lies inside the paraboloid $z = x^2 + y^2$

In each of Exercises 21–28, integrate the function $\phi(x, y, z)$ over the surface given by the graph of f over the region $\mathcal{R}$.

21. $\phi(x, y, z) = x - 2y + 3z$, $f(x, y) = x^2 + y^2 + 8$, $\mathcal{R} = \{(x, y) : x^2 + y^2 < 6\}$
22. $\phi(x, y, z) = 3x^2 - 4y^2 + 7z$, $f(x, y) = x - 3y$, $\mathcal{R} = \{(x, y) : |x| < 2, |y| < 3\}$
23. $\phi(x, y, z) = (3x^2 + 3y^2 + z + 1)^{1/2}$, $f(x, y) = x^2 + y^2$, $\mathcal{R} = \{(x, y) : 1 < x^2 + y^2 < 4\}$
24. $\phi(x, y, z) = x^2 - y^2 + z^2$, $f(x, y) = 2x - 2y$, $\mathcal{R} = \{(x, y) : x^2 + y^2 < 1\}$
25. $\phi(x, y, z) = y^2 + z$, $f(x, y) = 8 - y^2$, $\mathcal{R} = \{(x, y) : 0 < x < y < \sqrt{2}\}$
26. $\phi(x, y, z) = \sin(x) + \sin(z)$, $f(x, y) = 9x - 4y$, $\mathcal{R} = \{(x, y) : 0 < x < \pi, 0 < y < \pi\}$
27. $\phi(x, y, z) = x - xy + z$, $f(x, y) = 3x - 7y$, $\mathcal{R} = \{(x, y) : 1 < 2y < x < 4\}$
28. $\phi(x, y, z) = z$, $f(x, y) = \sqrt{x^2 + y^2}$, $\mathcal{R} = \{(x, y) : 0 < x^2 + y^2 < 1\}$

In each of Exercises 29–32, write the integral that gives the surface area of the parametrically defined surface. Do *not* evaluate the integral.

29. $x = u^2 - v, y = u + v^2, z = v, 0 \le u \le 1, 0 \le v \le 4$
30. $x = \cos(u), y = \sin(v), z = u - v$, $\pi/6 \le u \le \pi/3, \pi/4 \le v \le 3\pi/4$
31. $x = e^{u+v}, y = e^{u-v}, z = u, 0 \le u \le 1, -1 \le v \le 1$
32. $x = u/(u + v), y = v/(u + v), z = u + v, 1 \le u \le 2, 1 \le v \le 2$

In each of Exercises 33–36, integrate the function ϕ over the parametrically defined surface.

33. $x = 2u - v, y = v + 2u, z = v - u$, $0 \le u \le 2, 0 \le v \le 3, \phi(x, y, z) = x + y + z$
34. $x = u + 3v, y = 4u, z = -3v$, $-2 \le u \le 1, -1 \le v \le 0, \phi(x, y, z) = \sin(\pi(x - y + z)/3)$
35. $x = v, y = u, z = u + v$, $3 \le u \le 5, 2 \le v \le 3, \phi(x, y, z) = \sqrt{(z - x)y}$
36. $x = u^2, y = v, z = u + 3v$, $\sqrt{2} \le u \le \sqrt{3}, 1 \le v \le 3, \phi(x, y, z) = z - 3y$

Further Theory and Practice

37. Let f be a positive, continuously differentiable function of one variable. Let S be the surface of revolution that results when the graph of f over $[a, b]$, realized as the space curve $\{(x, f(x), 0) : a \le x \le b\}$, is rotated about the x-axis. Parameterize S and obtain a formula for its surface area.
38. Suppose that $b > a > 0$. Let S be the surface that results when the circle $\{(0, y, z) : (y - b)^2 + z^2 = a^2\}$ is rotated about the z-axis—S is called a *torus*. Calculate the surface area of S.
39. Let S be the graph of a function $f(x, y)$ over a planar region $\mathcal{R}$. For each $(x, y) \in \mathcal{R}$, let $\mathbf{n}(x, y)$ be the unit upward normal to S at the point $(x, y, f(x, y))$. Let $\gamma(x, y)$ be the angle between $\mathbf{k}$ and $\mathbf{n}$. Give an argument to demonstrate that the surface area of S is given by

$$\iint_{\mathcal{R}} \sec(\gamma)\, dA,$$

and the integral of a continuous function $\phi(x, y, z)$ over M is given by

$$\iint_{\mathcal{R}} \phi(x, y, f(x, y)) \sec(\gamma)\, dA.$$

40. Define a concept of average temperature over a surface. If the temperature of any point in space is given by

$$T(x, y, z) = 3x - 8y + z,$$

then find the average temperature over the upper hemisphere of the unit sphere centered at the origin.
41. Use an improper integral to integrate the function

$$\phi(x, y, z) = \frac{1}{(x^2 + y^2 + 1)^3}$$

over the surface that is the graph of $f(x, y) = x^2 + y^2/2$.
42. Use an improper integral to integrate the function $\phi(x, y) = ze^{-(x^2+y^2)}$ over the surface that is the graph of the function $f(x, y) = x^2 - y^2$.
43. Assuming that it has uniform mass distribution, find the center of gravity of the cone that is the graph of $z = 2\sqrt{x^2 + y^2}$ $(0 \le x^2 + y^2 \le 4)$.
44. Assuming that it has uniform mass distribution, find the center of gravity of the portion of the sphere $x^2 + y^2 + z^2 = 4$ that lies above the plane $z = -1$.
45. Assuming that it has uniform mass distribution, find the center of gravity of the portion of the paraboloid $z = 4 - x^2 - y^2$ that lies above the xy-plane.

46. Let F be a continuously differentiable function of three variables. Let $\mathcal{R}$ be a region in the xy-plane. Suppose that for each point $(x,y) \in \mathcal{R}$, there is exactly one value of z such that $F(x,y,z)=0$. Further assume that $\partial F/\partial z$ is not 0 at each such point. Prove that the area $\mathcal{A}$ of the portion of the graph of $\{(x,y,z) : F(x,y,z)=0\}$ lying over $\mathcal{R}$ is given by

$$\mathcal{A} = \iint_{\mathcal{R}} \frac{\|\nabla F\|}{|\nabla F \cdot \mathbf{k}|}\, dA.$$

47. Use the formula developed in Exercise 46 to calculate the surface area of a hemisphere of radius a.
48. Use the formula developed in Exercise 46 to calculate the surface area of a cone with base of radius r and height h.
49. A *prolate spheroid* is an ellipsoid with equal shorter axes. Calculate the surface area of the prolate spheroid $x^2/a^2 + y^2/a^2 + z^2/c^2 = 1$ where $0 < a < c$.
50. An *oblate spheroid* is an ellipsoid with equal longer axes. Calculate the surface area of the oblate spheroid $x^2/a^2 + y^2/c^2 + z^2/c^2 = 1$ where $0 < a < c$.
51. Let a be a positive constant. Verify that $\mathbf{r}(\theta,z) = \langle a\cos(\theta), a\sin(\theta), z\rangle, -\infty < z < \infty, 0 \leq \theta < 2\pi$ is a parameterization of the circular cylinder $x^2 + y^2 = a^2$. Deduce that the element of surface area dS is $a\,dz\,d\theta$ when this parameterization is used. (Symmetrically, the parameterization $\mathbf{r}(\theta,y) = \langle a\cos(\theta), y, a\sin(\theta)\rangle, -\infty < y < \infty$, $0 \leq \theta < 2\pi$ yields $a\,dy\,d\theta$ for the element of surface area of the circular cylinder $x^2 + z^2 = a^2$ and the parameterization $\mathbf{r}(\theta,x) = \langle x, a\cos(\theta), a\sin(\theta)\rangle, -\infty < x < \infty, 0 \leq \theta < 2\pi$ yields $a\,dx\,d\theta$ for the element of surface area of the circular cylinder $y^2 + z^2 = a^2$.)

In each of Exercises 52–54, calculate the specified surface integral by using the parameterization described in Exercise 51.

52. The surface integral of $\phi(x,y,z) = x + 2z$ over that part of the cylinder $x^2 + y^2 = 4$ that lies in the first octant and under the plane $z=3$.
53. The surface integral of $\phi(x,y,z) = xyz$ over that part of the cylinder $x^2 + z^2 = 16$ that lies in the first octant between the planes $y = 2$ and $y = 4$.
54. The surface integral of $\phi(x,y,z) = y^2 z$ over that part of the cylinder $y^2 + z^2 = 9$ that lies in the first octant between the planes $x = 0$ and $x = 2$.

Calculator/Computer Exercises

55. Calculate the surface area of that portion of the graph of $f(x,y) = \exp(x^2)$ that lies over the interior of the square $\{(x,y) : -1 < x < 1, -1 < y < 1\}$.
56. Calculate the surface area of that portion of the graph of $f(x,y) = \cos(\sqrt{y})$ that lies over the interior of the rectangle $\{(x,y) : 0 < x < 1, 0 < y < \pi^2\}$.
57. Calculate the surface area of that portion of the graph of $f(x,y) = 2 - x^2y^4$ that lies over the interior of the square $\{(x,y) : -1 < x < 1, -1 < y < 1\}$.
58. Calculate the surface area of that portion of the graph of $f(x,y) = x^2 + y^2$ that lies over the interior of the ellipse $\{(x,y) : x^2 + 4y^2 < 4\}$.

13.7 Stokes's Theorem

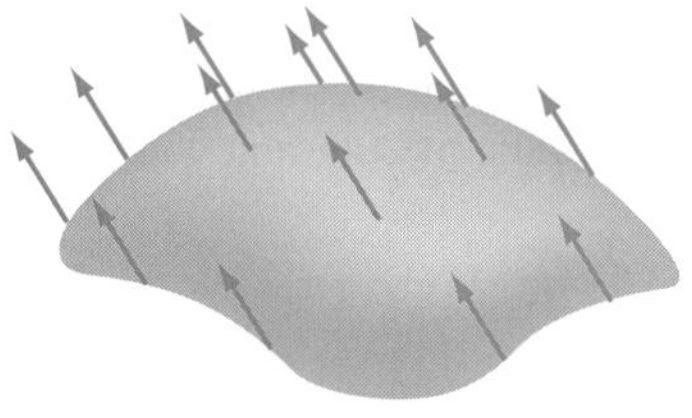

When we studied Green's Theorem in Section 13.5, we used the notion of planar flows to gain a physical understanding of the theorem. Now imagine a flow through a surface (see Figure 1). We may ask how the flow through the interior of the surface is related to the flow at the boundary of the surface. This is the subject of Stokes's Theorem, which generalizes Green's Theorem from planar regions to surfaces in space.

Orientable Surfaces and Their Boundaries

A crucial idea in formulating and using Stokes's Theorem is that of orientation. Intuitively, we want to say two things about the orientation of a surface and its boundary. The first of our requirements is that we may continuously assign to each point of S a "preferred" unit normal vector.

DEFINITION A continuous vector field $P \mapsto \mathbf{n}(P)$ defined on a smooth surface S is said to be an *orientation* of S if, at each point P of S, the vector $\mathbf{n}(P)$ is a unit normal to S at P. We say that a surface S in space is *orientable* if it possesses an orientation. An orientable surface together with a fixed choice of orientation is called an *oriented surface*.

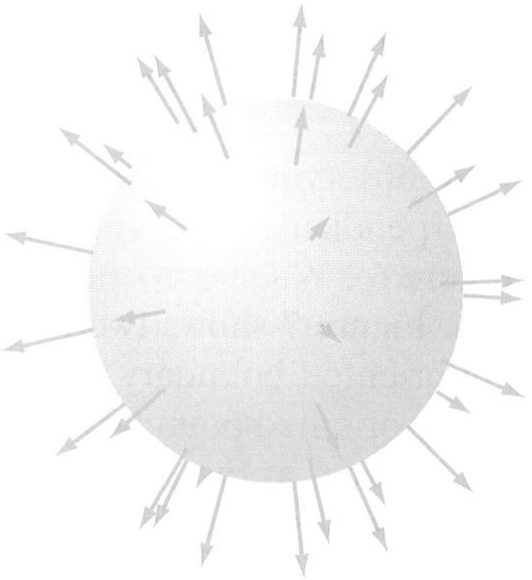

▲ Figure 2 The outward orientation of the sphere

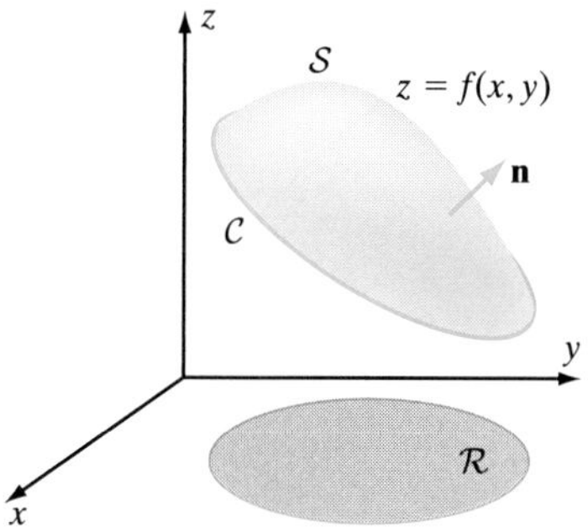

▲ Figure 3

▲ Figure 4

INSIGHT If $P \mapsto \mathbf{n}(P)$is an orientation of S, then so is the opposite vector field $P \mapsto -\mathbf{n}(P)$. We therefore say that an orientable surface is *two-sided*. If S is oriented, then we say that the side out of which the orientation $\mathbf{n}$ points is the *positive* side, and the side out of which $-\mathbf{n}$ points is the negative side.

▶ **EXAMPLE 1** Suppose that $a > 0$. What are the two orientations of the sphere $x^2 + y^2 + z^2 = a^2$?

Solution By elementary geometry, we know that the position vector $\langle x, y, z\rangle$ is perpendicular to the sphere at the point (x, y, z). Because $\|\langle x, y, z\rangle\| = \sqrt{x^2 + y^2 + z^2} = a$, it follows that the function $(x, y, z) \mapsto \langle x/a, y/a, z/a\rangle$ is a continuous unit normal on the sphere. It is the *outward* unit normal (see Figure 2). The opposite orientation of the sphere is given by the *inward* unit normal $(x, y, z) \mapsto -\langle x/a, y/a, z/a\rangle$. ◀

INSIGHT A surface with no boundary is said to be *closed*. The sphere is an example. With such a surface, we can speak of an inward normal or an outward normal vector field. When a surface is the graph of a function, we may speak of an *upward* normal vector field (as shown in Figure 3) or a *downward* normal vector field.

Although familiar surfaces such as the sphere are orientable, nonorientable surfaces *do* exist. For example, August Ferdinand Möbius (1790–1868) discovered a "one-sided" surface, now called the *Möbius strip*, that is not orientable (see Figure 4). Notice that, when you move a unit normal continuously around a Möbius strip, returning to your original location (but on the opposite side), it winds up opposite to its initial position. You can make your own Möbius strip by taking a strip of paper, giving one end a 180° twist, and then joining the ends.

Suppose now that S is a smooth, bounded surface with orientation $P \mapsto \mathbf{n}(P)$ and with closed curve $\mathcal{C}$ as its boundary (see Figure 5a). We want to use the orientation of S to induce a particular direction on $\mathcal{C}$. Imagine yourself standing at a point P on $\mathcal{C}$, perpendicular to S with your foot-to-head direction given by $\mathbf{n}(P)$. There are two directions in which you may walk around $\mathcal{C}$. Choose the direction for which S lies to your *left* (see Figure 5b). We say that this direction around $\mathcal{C}$ is

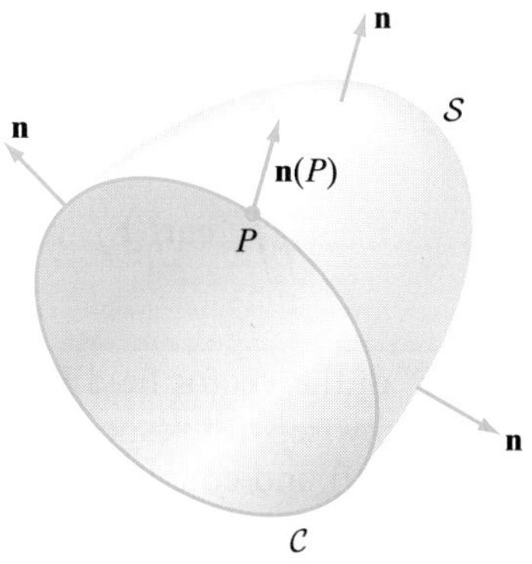

▲ Figure 5a

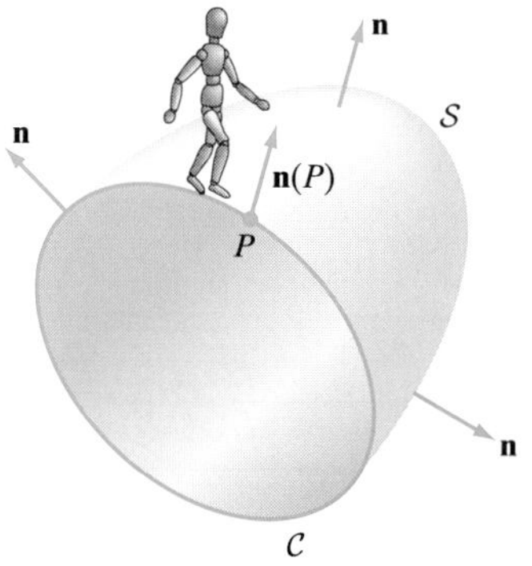

▲ Figure 5b $\mathcal{C}$ is oriented so that $\mathcal{S}$ lies to the left when $\mathcal{C}$ is traversed in the direction of the orientation.

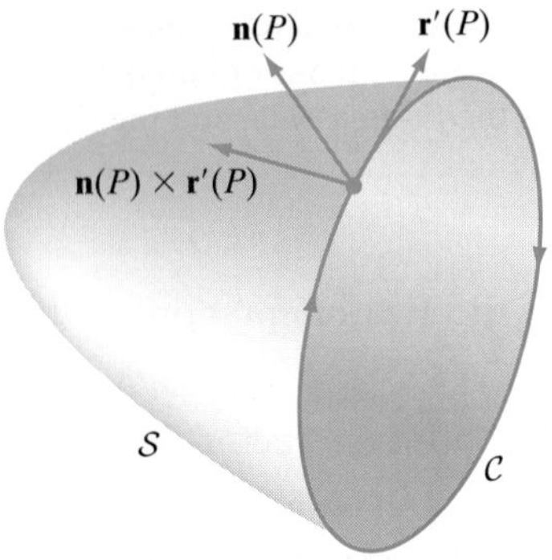

▲ Figure 6a $\mathbf{n}(P) \times \mathbf{r}'(P)$ points *into* $\mathcal{S}$: $\mathcal{S}$ and $\mathcal{C}$ are consistently oriented.

induced by the oriented surface S. If $\mathcal{C}$ is a parameterized curve, we say that $\mathcal{C}$ is *oriented consistently* with S if the orientation of the parameterization is the same as the one induced by S.

These notions can be formulated more precisely by observing that, if $\mathbf{r}$ is a parameterization of $\mathcal{C}$, then the vector $\mathbf{n}(P) \times \mathbf{r}'(P)$ is perpendicular to $\mathcal{C}$ and tangent to S. It points either away from S or into S (see Figure 6). When $\mathcal{C}$ is oriented consistently with S, the vector $\mathbf{n}(P) \times \mathbf{r}'(P)$ points *into* S. Figure 7 shows two possible orientations of a disk in space with a consistently oriented boundary curve. Figure 8 shows another oriented surface with a consistently oriented boundary that comprises two disconnected pieces. We will discuss such surfaces at the end of this section. For now, just note that the definition of *consistent orientation* can be applied to each component of a disconnected boundary.

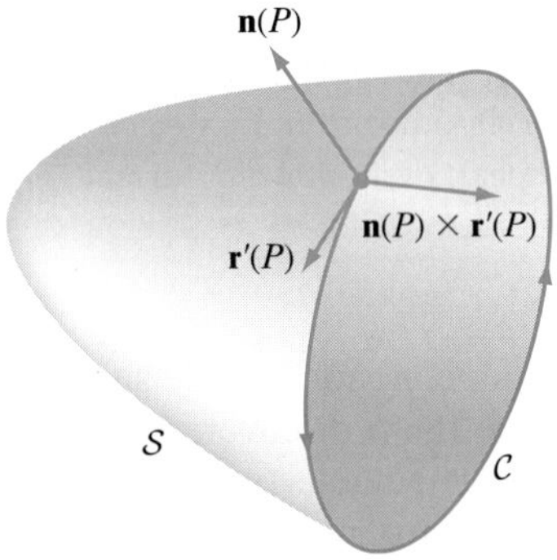

▲ Figure 6b $\mathbf{n}(P) \times \mathbf{r}'(P)$ points *away from* $\mathcal{S}$: $\mathcal{S}$ and $\mathcal{C}$ are *not* consistently oriented.

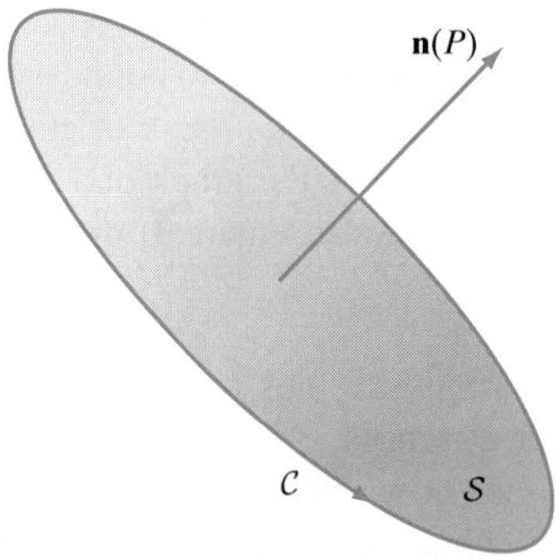

▲ Figure 7a Upward-pointing orientation of $\mathcal{S}$ with consistent boundary orientation

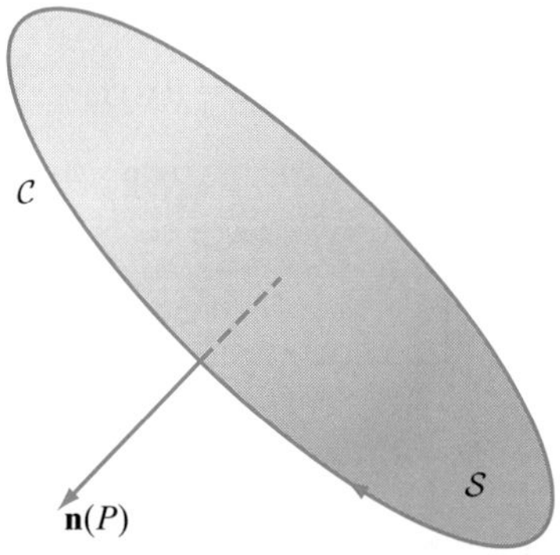

▲ Figure 7b Downward-pointing orientation of $\mathcal{S}$ with consistent boundary orientation

The Component of Curl in the Normal Direction

Recall that the curl operator, when applied to a vector field $\mathbf{F}$, can be interpreted as a measure of the vorticity of the vector field about a given point. In our calculations, we saw that $\mathbf{curl}(\mathbf{F})$ points in the direction of the axis of rotation. For example, if $\mathbf{F}(x, y) = M(x, y)\mathbf{i} + N(x, y)\mathbf{j}$ is a planar vector field, then equation (13.4.3) defines $\mathbf{curl}(\mathbf{F})$ to be a multiple of $\mathbf{k}$. In particular, the curl points out of the plane in which the circulation is taking place. These observations suggest that, if we want to measure the magnitude of the curl of a vector field as it relates to a surface S, then we should consider the component of the curl in the direction normal to the surface. We therefore use $\mathbf{curl}(\mathbf{F}(P)) \cdot \mathbf{n}(P)$ to measure how much the vector field is curling in the surface at P. The surface integral

$$\iint_{\mathcal{S}} \mathbf{curl}(\mathbf{F}) \cdot \mathbf{n}\, dS \tag{13.7.1}$$

represents the "total curl" of the vector field $\mathbf{F}$ on S. Bear in mind that, in general, the vector field $\mathbf{F}$ curls clockwise with respect to $\mathbf{n}(P)$ at some points (the points P at which $\mathbf{curl}(\mathbf{F}(P)) \cdot \mathbf{n}(P) < 0$) and counterclockwise at other points (the points at which $\mathbf{curl}(\mathbf{F}(P)) \cdot \mathbf{n}(P) > 0$). So, surface integral (13.7.1) takes into account certain cancellations.

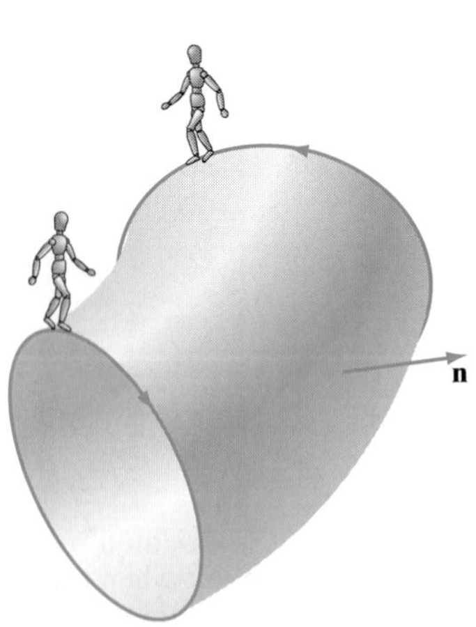

▲ Figure 8

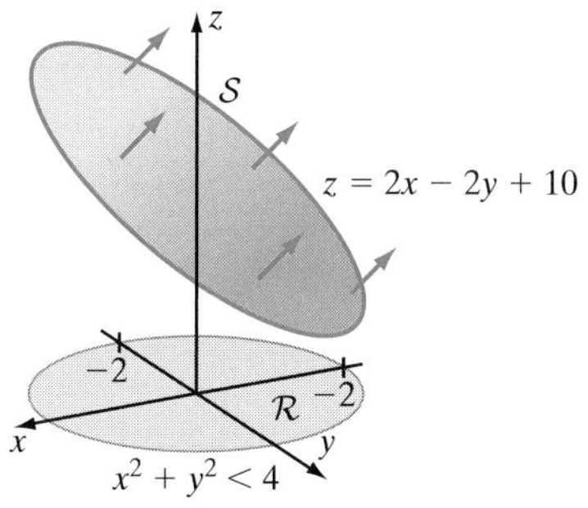

▲ Figure 9

▶ **EXAMPLE 2** Let S be the graph of $z = 2x - 2y + 10$ over the circular region $\mathcal{R} = \{(x, y) : x^2 + y^2 < 4\}$. Orient S with the upward pointing unit normal $\mathbf{n}$ (see Figure 9). Define $\mathbf{F}(x, y, z) = (y - z)\mathbf{i} - (x + y)\mathbf{j} + 2x\mathbf{k}$. Calculate $\iint_S \mathbf{curl}(\mathbf{F}) \cdot \mathbf{n}\, dS$.

Solution Our surface is just that portion of the plane $2x - 2y - z = -10$ that lies above the interior of the circle centered at the origin and with radius 2. We use the coefficients that appear in the equation of the plane, $2x - 2y - z = -10$, to obtain a vector $2\mathbf{i} - 2\mathbf{j} - 1\mathbf{k}$ that is normal to S. An upward normal at any point is then given by $-2\mathbf{i} + 2\mathbf{j} + 1\mathbf{k}$ (we know this is upward because the $\mathbf{k}$ component is positive). Dividing by the length of this vector, which is $\sqrt{(-2)^2 + 2^2 + 1^2}$, or 3, we obtain $\mathbf{n}(x, y, z) = (-2/3)\mathbf{i} + (2/3)\mathbf{j} + (1/3)\mathbf{k}$. Next we calculate

$$\mathbf{curl}(\mathbf{F}) = \left(\frac{\partial}{\partial y}(2x) - \frac{\partial}{\partial z}(-(x+y))\right)\mathbf{i} - \left(\frac{\partial}{\partial x}(2x) - \frac{\partial}{\partial z}(y - z)\right)\mathbf{j}$$
$$+ \left(\frac{\partial}{\partial x}(-(x+y)) - \frac{\partial}{\partial y}((y - z))\right)\mathbf{k} = 0\mathbf{i} - 3\mathbf{j} - 2\mathbf{k}.$$

and

$$\mathbf{curl}(\mathbf{F}) \cdot \mathbf{n} = (0\mathbf{i} - 3\mathbf{j} - 2\mathbf{k}) \cdot \left(-\frac{2}{3}\mathbf{i} + \frac{2}{3}\mathbf{j} + \frac{1}{3}\mathbf{k}\right) = -\frac{8}{3}.$$

Now the element of area dS on S is

$$dS = \sqrt{1 + f_x(x, y)^2 + f_y(x, y)^2}\, dA = \sqrt{1 + 2^2 + (-2)^2}\, dA = 3\, dA.$$

Putting all this information together, and noting that $\mathcal{R}$ is a disk of radius 2, we have

$$\iint_S \mathbf{curl}(\mathbf{F}) \cdot \mathbf{n}\, dS = \iint_{\mathcal{R}} \left(-\frac{8}{3} \cdot 3\right) dA = -8 \iint_{\mathcal{R}} 1\, dA = -8 \cdot (\text{area of } \mathcal{R}) = -8 \cdot (\pi \cdot 2^2) = -32\pi. \quad ◀$$

Stokes's Theorem

Because the line integral $\oint_C \mathbf{F} \cdot d\mathbf{r}$ sums the tangential component of $\mathbf{F}$, namely $\mathbf{F} \cdot \mathbf{r}'$, over C, we can regard it as a measure of the tendency of $\mathbf{F}$ to flow about the boundary of S. It therefore makes good physical sense for this line integral over the boundary of S to be related to the surface integral $\iint_S \mathbf{curl}(\mathbf{F}) \cdot \mathbf{n}\, dS$. Indeed, Green's Theorem, in the form of equation (13.5.5), states that

$$\oint_C \mathbf{F} \cdot d\mathbf{r} = \iint_S \mathbf{curl}(\mathbf{F}) \cdot \mathbf{n}\, dS \qquad (13.7.2)$$

when S is a planar region with boundary C and upward pointing orientation $\mathbf{n} = \mathbf{k}$. The next example, which pertains to a surface that is not contained in the xy-plane, provides further evidence of this relationship.

▶ **EXAMPLE 3** As in Example 2, let $\mathbf{F}(x, y, z) = (y - z)\mathbf{i} - (x + y)\mathbf{j} + 2x\mathbf{k}$, and let S be the graph of $z = 2x - 2y + 10$ over the circular region $\mathcal{R} = \{(x, y) : x^2 + y^2 < 4\}$. Let $\mathbf{n}$ be the upward pointing unit normal on S, and orient the boundary C of S consistently. Verify equation (13.7.2).

Solution In Example 2, we proved that the surface integral over S on the right side of equation (13.7.2) is -32π. We now calculate the line integral over $\mathcal{C}$ on the left side. The orientation of $\mathcal{C}$ that is consistent with S is counterclockwise when viewed from above. The vector-valued function

$$\mathbf{r}(t) = 2\cos(t)\mathbf{i} + 2\sin(t)\mathbf{j} + \big(2(2\cos(t)) - 2(2\sin(t)) + 10\big)\mathbf{k}, 0 \le t \le 2\pi$$

parameterizes $\mathcal{C}$ with the proper orientation. We calculate

$$\begin{aligned}\mathbf{F}(\mathbf{r}(t)) &= \big(2\sin(t) - (2(2\cos(t)) - 2(2\sin(t)) + 10)\big)\mathbf{i} - \big((2\cos(t)) + (2\sin(t))\big)\mathbf{j} + 2(2\cos(t))\mathbf{k}\\ &= \big(6\sin(t) - 4\cos(t) - 10\big)\mathbf{i} - \big(2\cos(t) + 2\sin(t)\big)\mathbf{j} + 4\cos(t)\mathbf{k},\end{aligned}$$

and

$$\mathbf{r}'(t) = -2\sin(t)\mathbf{i} + 2\cos(t)\mathbf{j} + \big(-4\sin(t) - 4\cos(t)\big)\mathbf{k}.$$

After some simplification, we obtain

$$\mathbf{F}(\mathbf{r}(t)) \cdot \mathbf{r}'(t) = -12\sin(t)\cos(t) + 20\sin(t) - 8\cos^2(t) - 12 = -6\sin(2t) + 20\sin(t) - 8\cos^2(t) - 12.$$

Therefore using formula (6.2.9) to evaluate the definite integral of $\cos^2(t)$, we have

$$\oint_{\mathcal{C}} \mathbf{F} \cdot d\mathbf{r} = \int_0^{2\pi} \big(-6\sin(2t) + 20\sin(t) - 8\cos^2(t) - 12\big)dt = 0 + 0 - 8 \cdot \pi - 12 \cdot (2\pi) = -32\pi.$$

This equation together with the result of Example 2 establishes the validity of equation (13.7.2) for $\mathbf{F}$ and S. ◄

Our next theorem tells us that equation (13.7.2) holds quite generally.

THEOREM 1 **(Stokes's Theorem).** Suppose that $(x, y, z) \mapsto \mathbf{F}(x, y, z) = M(x, y, z)\mathbf{i} + N(x, y, z)\mathbf{j} + R(x, y, z)\mathbf{k}$ is a continuously differentiable vector field defined on an oriented surface S and its boundary $\mathcal{C}$. Let $P \mapsto \mathbf{n}(P)$ denote the orientation of S, and let $\mathbf{r}$ be a parameterization of $\mathcal{C}$ that orients $\mathcal{C}$ consistently with $\mathcal{S}$. Then

$$\oint_{\mathcal{C}} \mathbf{F} \cdot d\mathbf{r} = \iint_S \mathbf{curl}(\mathbf{F}) \cdot \mathbf{n}\, dS.$$

Proof. Our scheme is to reduce Stokes's Theorem to a calculation involving Green's Theorem. For ease of calculation, we will assume that our surface S is the graph of a continuously differentiable function f over a bounded planar region $\mathcal{R}$. We suppose that the boundary of $\mathcal{R}$ is a continuously differentiable curve $\mathcal{C}_0$ with a parametrization $t \mapsto \mathbf{r}_0(t) = \xi(t)\mathbf{i} + \eta(t)\mathbf{j}$, $a \le t \le b$, that gives $\mathcal{C}_0$ a counterclockwise orientation when viewed from above. The vector-valued function $\mathbf{r}(t) = \langle \mathbf{r}_0(t), f(\mathbf{r}_0(t))\rangle$, $a \le t \le b$ is then a parameterization of $\mathcal{C}$ that orients $\mathcal{C}$ consistently with the upward pointing orientation $P \mapsto \mathbf{n}(P)$ of S. Figure 10 illustrates the geometry.

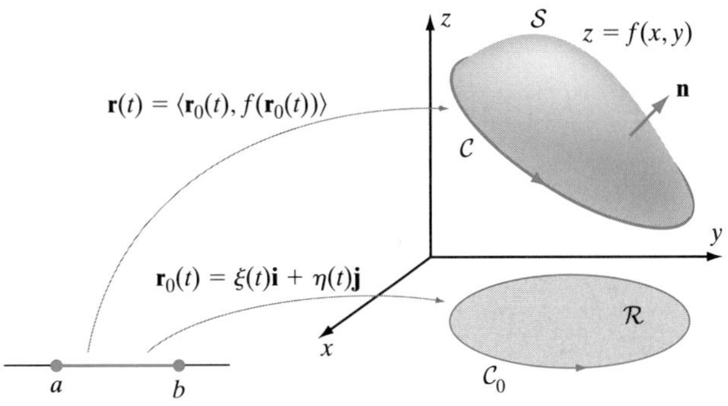

▲ Figure 10

Let us begin by rewriting the surface integral. The element of area on S is

$$dS = \sqrt{1 + f_x(x,y)^2 + f_y(x,y)^2}\, dA.$$

Because S is the graph of f, a normal to S at each point $P = (x, y, f(x,y))$ is given by $f_x(x,y)\mathbf{i} + f_y(x,y)\mathbf{j} - 1\mathbf{k}$, as we learned in Section 11.7 of Chapter 11. Of course an upward normal is then $-f_x(x,y)\mathbf{i} - f_y(x,y)\mathbf{j} + 1\mathbf{k}$ (because the $\mathbf{k}$ component is then positive). Thus

$$\mathbf{n}(P) = \frac{1}{\sqrt{1 + f_x(x,y)^2 + f_y(x,y)^2}}\big(-f_x(x,y)\mathbf{i} - f_y(x,y)\mathbf{j} + 1\mathbf{k}\big).$$

Using these expressions for dS and $\mathbf{n}(P)$, and using formula (13.4.4) for $\mathbf{curl}(\mathbf{F})$, we obtain

$$\begin{aligned}\iint_{\mathcal{S}} \mathbf{curl}(\mathbf{F}) \cdot \mathbf{n}\, dS &= \iint_{\mathcal{R}} \mathbf{curl}(\mathbf{F}) \cdot \big(-f_x(x,y)\mathbf{i} - f_y(x,y)\mathbf{j} + 1\mathbf{k}\big)\, dA \\ &= \iint_{\mathcal{R}} \big(-(R_y - N_z)f_x - (M_z - R_x)f_y + (N_x - M_y)\big)\, dA. \qquad (13.7.3)\end{aligned}$$

In the last integration, each partial derivative of M, N, and R is evaluated at the points $(x, y, f(x,y))$ of S. This is an explicit expression for the right-hand side of the formula in Stokes's Theorem.

Now we look at the line integral portion of Stokes's Theorem. We have

$$\oint_{\mathcal{C}} \mathbf{F} \cdot d\mathbf{r} = \int_a^b \mathbf{F}\big(\mathbf{r}(t)\big) \cdot \mathbf{r}'(t)dt = \int_a^b M\big(\mathbf{r}(t)\big)\xi'(t) + N\big(\mathbf{r}(t)\big)\eta'(t) + R\big(\mathbf{r}(t)\big)\Big(\frac{d}{dt}f\big(\xi(t), \eta(t)\big)\Big)dt.$$

Using the Chain Rule, we obtain

$$\oint_{\mathcal{C}} \mathbf{F} \cdot d\mathbf{r} = \int_a^b \Big(M\big(\mathbf{r}(t)\big)\xi'(t) + N\big(\mathbf{r}(t)\big)\eta'(t) + R\big(\mathbf{r}(t)\big)f_x\big(\mathbf{r}_0(t)\big)\xi'(t) + R\big(\mathbf{r}(t)\big)f_y\big(\mathbf{r}_0(t)\big)\eta'(t)\Big)dt,$$

or

$$\oint_{\mathcal{C}} \mathbf{F} \cdot d\mathbf{r} = \int_a^b \Big(\big(M(\mathbf{r}(t)) + R(\mathbf{r}(t)) f_x(\mathbf{r}_0(t)) \big) \xi'(t) + \big(N(\mathbf{r}(t)) + R(\mathbf{r}(t)) f_y(\mathbf{r}_0(t)) \big) \eta'(t) \Big) dt$$
$$= \oint_{\mathcal{C}_0} \tilde{\mathbf{F}} \cdot d\mathbf{r}_0,$$

where

$$\tilde{\mathbf{F}}(x,y) = \big(M(x,y,f(x,y)) + R(x,y,f(x,y)) f_x(x,y)\big)\mathbf{i} + \big(N(x,y,f(x,y)) + R(x,y,f(x,y)) f_y(x,y)\big)\mathbf{j}.$$

We may apply Green's Theorem to the planar line integral $\oint_{\mathcal{C}_0} \tilde{\mathbf{F}} \cdot d\mathbf{r}_0$ that has arisen. Doing so gives us

$$\oint_{\mathcal{C}_0} \tilde{\mathbf{F}} \cdot d\mathbf{r}_0 = \iint_{\mathcal{R}} \Big(\frac{\partial}{\partial x} \big(N(x,y,f(x,y)) + R(x,y,f(x,y)) f_y(x,y) \big)$$
$$- \frac{\partial}{\partial y} \big(M(x,y,f(x,y)) + R(x,y,f(x,y)) f_x(x,y) \big) \Big) \, dA. \quad (13.7.4)$$

If these differentiations are written out, using the Product Rule and the Chain Rule, then

$$\frac{\partial}{\partial x}(N + R \cdot f_y) = N_x + N_z f_x + R_x f_y + R_z f_x f_y + R f_{yx}$$

and

$$\frac{\partial}{\partial y}(M + R \cdot f_x) = M_y + M_z f_y + R_y f_x + R_z f_y f_x + R f_{xy}.$$

If we substitute these expressions into the right side of equation (13.7.4), then the integral reduces to the right side of equation (13.7.3), which establishes Stokes's Theorem. ■

Although the proof given here is limited, Stokes's theorem is not limited to surfaces that are the graphs of functions. In practice, it is often convenient to treat a general surface by breaking it into pieces, each of which can be realized as the graph of a function. In addition to its theoretical importance, Stokes's Theorem can be used to advantage whenever the computation of one side of equation (13.7.2) is substantially easier than the computation of the other side.

▶ **EXAMPLE 4** Let $\mathcal{S}$ be the graph of $f(x,y) = 4x - 8y + 5$ over $\mathcal{R} = \{(x,y) : (x-1)^2 + 9(y-3)^2 < 36\}$. Let $\mathbf{n}(P)$ be the upward unit normal at each point P of $\mathcal{S}$. Consider the vector field $\mathbf{F}(x,y,z) = -3z\mathbf{i} + (x+y)\mathbf{j} + y\mathbf{k}$. Let $\mathcal{C}$ be the boundary of $\mathcal{S}$ with consistent orientation. Evaluate $\oint_{\mathcal{C}} \mathbf{F} \cdot d\mathbf{r}$.

Solution To calculate the given line integral directly would involve an awkward parameterization of the boundary of $\mathcal{S}$. With Stokes's Theorem, matters become much simpler. We calculate

$$dS = \sqrt{1 + f_x(x,y)^2 + f_y(x,y)^2}\,dA = \sqrt{1 + 4^2 + (-8)^2} = 9\,dA$$

and

$$\mathbf{curl}(\mathbf{F}) = \det\left(\begin{bmatrix} \mathbf{i} & \mathbf{j} & \mathbf{k} \\ \dfrac{\partial}{\partial x} & \dfrac{\partial}{\partial y} & \dfrac{\partial}{\partial z} \\ -3z & x+y & y \end{bmatrix}\right) = \mathbf{i} - 3\mathbf{j} + \mathbf{k}.$$

Finally, because our surface is contained in the plane $4x - 8y - z = -5$, the vector $-4\mathbf{i} + 8\mathbf{j} + 1\mathbf{k}$ is an upward normal (the coefficient of $\mathbf{k}$ is positive) and $\mathbf{n}(x,y,z) = (-4/9)\mathbf{i} + (8/9)\mathbf{j} + (1/9)\mathbf{k}$. Thus

$$\oint_{\mathcal{C}} \mathbf{F}\cdot d\mathbf{r} = \iint_{\mathcal{S}} \mathbf{curl}(\mathbf{F})\cdot\mathbf{n}\,dS = \iint_{\mathcal{R}} (\mathbf{i} - 3\mathbf{j} + \mathbf{k})\cdot\left(-\frac{4}{9}\mathbf{i} + \frac{8}{9}\mathbf{j} + \frac{1}{9}\mathbf{k}\right)9\,dA$$

$$= \iint_{\mathcal{R}} (-27)\,dA = -27\cdot(\text{area of } \mathcal{R}).$$

But $\mathcal{R}$ is an ellipse with semi-major axis 6 and semi-minor axis 2. Therefore it has area $6\cdot 2\cdot\pi$, or 12π. We conclude that $\oint_{\mathcal{C}}\mathbf{F}\cdot d\mathbf{r} = -27\cdot 12\pi = -324\pi$. ◀

▶ **EXAMPLE 5** Consider the surface $\mathcal{S}$ consisting of the part of the sphere of radius 3 centered at the origin that lies above the plane $z = \sqrt{5}$. Let $P \mapsto \mathbf{n}(P)$ denote the upward pointing normal on $\mathcal{S}$. Define $\mathbf{F}(x,y,z) = -y\mathbf{i} + xz\mathbf{j} + y^2\mathbf{k}$. Calculate $\iint_{\mathcal{S}}\mathbf{curl}(\mathbf{F})\cdot\mathbf{n}\,dS$.

Solution The intersection of the level plane $z = \sqrt{5}$ and the sphere $x^2 + y^2 + z^2 = 9$ is the circle $\{(x,y,z) : x^2 + y^2 = 4, z = \sqrt{5}\}$. See Figure 11. The surface $\mathcal{S}$ is therefore the graph of the function $f(x,y) = (9 - x^2 - y^2)^{1/2}$ over the region $\mathcal{R} = \{(x,y) : x^2 + y^2 < 4\}$. The counterclockwise-oriented boundary curve $\mathcal{C}$ can be parameterized by

$$\mathbf{r}(t) = 2\cos(t)\mathbf{i} + 2\sin(t)\mathbf{j} + \sqrt{5}\mathbf{k}, \quad 0 \le t \le 2\pi.$$

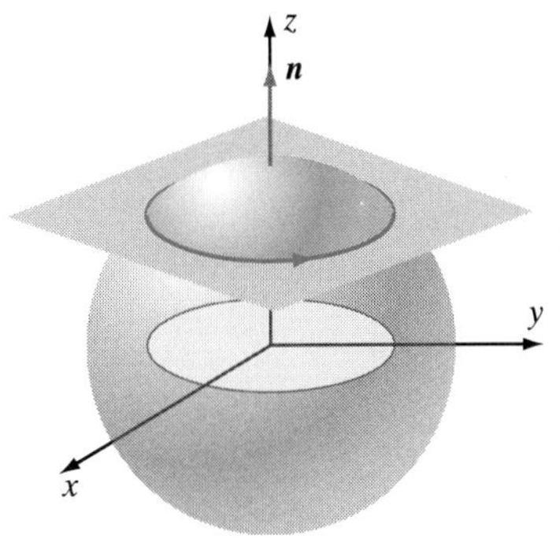

▲ Figure 11

We have

$$\mathbf{F}\big(\mathbf{r}(t)\big) = -2\sin(t)\mathbf{i} + 2\sqrt{5}\cos(t)\mathbf{j} + 4\sin^2(t)\mathbf{k} \quad\text{and}\quad \mathbf{r}'(t) = -2\sin(t)\mathbf{i} + 2\cos(t)\mathbf{j} + 0\mathbf{k}.$$

Therefore $\mathbf{F}\big(\mathbf{r}(t)\big)\cdot\mathbf{r}'(t) = 4\sin^2(t) + 4\sqrt{5}\cos^2(t)$ and, by Stokes's Theorem,

$$\iint_{\mathcal{S}} \mathbf{curl}(\mathbf{F})\cdot\mathbf{n}\,dS = \oint_{\mathcal{C}} \mathbf{F}\cdot d\mathbf{r} = \int_0^{2\pi} \mathbf{F}\big(\mathbf{r}(t)\big)\cdot\mathbf{r}'(t)\,dt$$

$$= \int_0^{2\pi} 4\sin^2(t)dt + \int_0^{2\pi} 4\sqrt{5}\cos^2(t)dt \overset{(6.2.9)}{=} (4 + 4\sqrt{5})\pi. \quad ◀$$

Stokes's Theorem on a Region with a Piecewise-Smooth Boundary

In many applications, the boundary of the surface $\mathcal{S}$ is not given by a single curve but rather by finitely many continuously differentiable curves. Stokes's Theorem is also valid in that setting.

The oriented surface $\mathcal{S}$ is the image of a continuously differentiable, vector-valued function $(u, v) \mapsto \mathbf{r}(u, v)$ defined on a bounded region $\mathcal{R}$ of the uv-plane. The boundary of $\mathcal{S}$ consists of finitely many continuously differentiable curves $\mathcal{C}_1, \ldots, \mathcal{C}_N$ having parameterizations $\mathbf{r}_1, \ldots, \mathbf{r}_N$, each of which is consistent with the orientation $P \mapsto \mathbf{n}(P)$ of $\mathcal{S}$ (see Figure 12). Now the statement of Stokes's Theorem is that if $\mathbf{F}$ is a continuously differentiable vector field on a region of space that contains $\mathcal{S}$ and its boundary, then

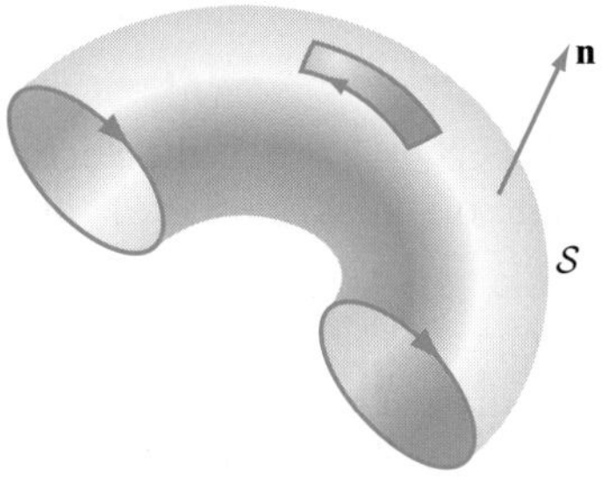

▲ Figure 12

$$\iint_{\mathcal{S}} \mathbf{curl}(\mathbf{F}) \cdot \mathbf{n}\, dS = \sum_{j=1}^{N} \int_{\mathcal{C}_j} \mathbf{F} \cdot d\mathbf{r}_j. \tag{13.7.5}$$

(Because the curve $\mathcal{C}_j$ might not be closed, we do not use the symbol $\oint_{\mathcal{C}_j}$. If $\mathcal{C}_j$ does happen to be closed, there is no harm in using the more general notation $\int_{\mathcal{C}_j}$.)

▶ **EXAMPLE 6** Let $\mathcal{S}$ be the graph of $f(x, y) = 4x - 8y + 30$ over the rectangle $\mathcal{R} = \{(x, y) : -2 < x < 3, 0 < y < 2\}$ (as shown in Figure 13). Consider the vector field $\mathbf{F}(x, y, z) = -x^2\mathbf{i} + xz\mathbf{j} + yx\mathbf{k}$. Use Stokes's Theorem to evaluate $\oint_{\mathcal{C}} \mathbf{F} \cdot d\mathbf{r}$ where $\mathcal{C}$ is the boundary of $\mathcal{S}$, oriented in the counterclockwise direction when viewed from above.

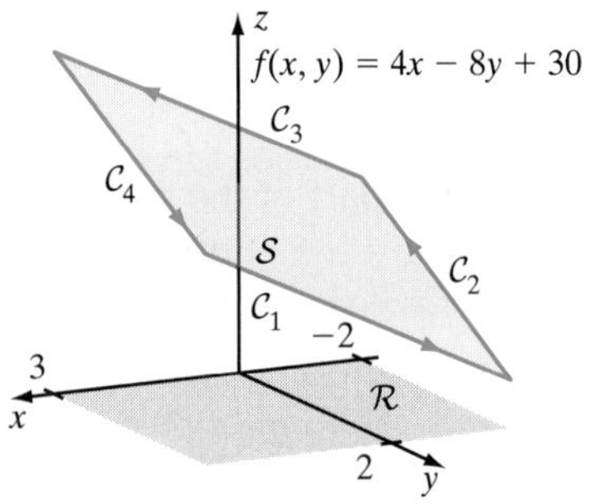

▲ Figure 13

Solution We could certainly calculate the required line integral directly, but to do so would involve parameterizing four separate line segments and evaluating four separate integrals. Using Stokes's Theorem, the task becomes much simpler. We have

$$dS = \sqrt{1 + f_x(x, y)^2 + f_y(x, y)^2}\, dA = \sqrt{1 + 4^2 + (-8)^2} = 9\, dA.$$

Also

$$\mathbf{curl}(\mathbf{F})(x, y, z) = \det\left(\begin{bmatrix} \mathbf{i} & \mathbf{j} & \mathbf{k} \\ \dfrac{\partial}{\partial x} & \dfrac{\partial}{\partial y} & \dfrac{\partial}{\partial z} \\ -x^2 & xz & yx \end{bmatrix}\right) = 0\mathbf{i} - y\mathbf{j} + z\mathbf{k}.$$

Finally, because our surface is $4x - 8y - z = -30$, the vector $-4\mathbf{i} + 8\mathbf{j} + 1\mathbf{k}$ is an upward normal and $\mathbf{n}(P) = (-4/9)\mathbf{i} + (8/9)\mathbf{j} + (1/9)\mathbf{k}$. Thus on $\mathcal{S}$, we have

$$\begin{aligned} \mathbf{curl}(\mathbf{F}) \cdot \mathbf{n} &= (0\mathbf{i} - y\mathbf{j} + z\mathbf{k}) \cdot \left(-\frac{4}{9}\mathbf{i} + \frac{8}{9}\mathbf{j} + \frac{1}{9}\mathbf{k}\right) = \frac{1}{9}(z - 8y) \\ &= \frac{1}{9}\big((4x - 8y + 30) - 8y\big) = \frac{1}{9}(4x - 16y + 30). \end{aligned}$$

We conclude that

$$\oint_{\mathcal{C}} \mathbf{F} \cdot d\mathbf{r} = \sum_{j=1}^{4} \int_{\mathcal{C}_j} \mathbf{F} \cdot d\mathbf{r}_j = \iint_{\mathcal{S}} \text{curl}(\mathbf{F}) \cdot \mathbf{n}\, dS = \frac{1}{9} \iint_{\mathcal{R}} (4x - 16y + 30)\, 9\, dA = \int_0^2 \int_{-2}^{3} (4x - 16y + 30)\, dx\, dy.$$

For the inner integration, we have

$$\int_{-2}^{3} (4x - 16y + 30)\, dx = (2x^2 - 16xy + 30x)\Big|_{x=-2}^{x=3} = -80y + 160.$$

Finally,

$$\oint_{\mathcal{C}} \mathbf{F} \cdot d\mathbf{r} = \int_0^2 (-80y + 160)\, dy = (-40y^2 + 160y)\Big|_{y=0}^{y=2} = -160 + 320 = 160. \blacktriangleleft$$

Example 6 involved a line integral over a piecewise-smooth boundary comprising several smooth curves. The next example concerns a surface that has a boundary consisting of disconnected curves.

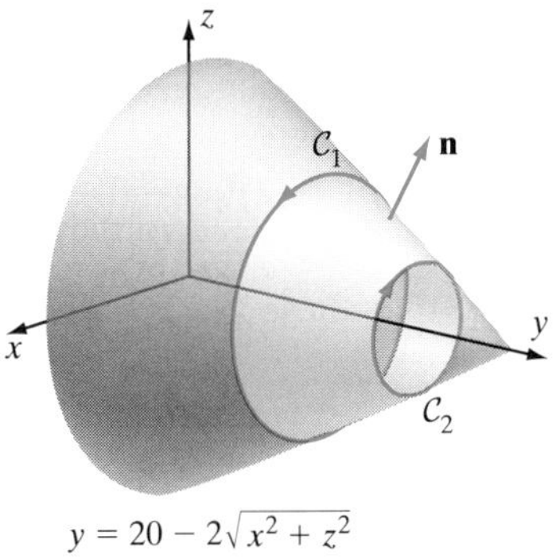

▲ Figure 14

▶ **EXAMPLE 7** Let $\mathcal{S}$ be the frustum of the cone $y = 20 - 2\sqrt{x^2 + z^2}$ that lies between the planes $y = 8$ and $y = 14$. Suppose that $\mathcal{S}$ and its boundary curves are consistently oriented as shown in Figure 14. Consider the vector field $\mathbf{F}(x, y, z) = z\mathbf{i} + 1\mathbf{j} + z^2\mathbf{k}$. Verify Stokes's Theorem for $\mathbf{F}$ and $\mathcal{S}$.

Solution Notice that $\mathcal{S}$ is not the graph of a function. We may use polar coordinates in the xz-plane as an aid in parameterizing $\mathcal{S}$. If $x = r\cos(\theta)$ and $z = r\sin(\theta)$, then on $\mathcal{S}$ we have $y = 20 - 2\sqrt{x^2 + z^2} = 20 - 2r$, or $r = 10 - y/2$. Therefore the vector-valued function

$$\mathbf{r}(\theta, y) = \left(10 - \frac{y}{2}\right)\cos(\theta)\mathbf{i} + y\mathbf{j} + \left(10 - \frac{y}{2}\right)\sin(\theta)\mathbf{k}, \quad 0 \le \theta \le 2\pi, 8 \le y \le 14,$$

parameterizes $\mathcal{S}$. We calculate the partial derivatives

$$\mathbf{r}_\theta(\theta, y) = -\left(10 - \frac{y}{2}\right)\sin(\theta)\mathbf{i} + 0\mathbf{j} + \left(10 - \frac{y}{2}\right)\cos(\theta)\mathbf{k}, \quad \mathbf{r}_y(\theta, y) = -\frac{1}{2}\cos(\theta)\mathbf{i} + 1\mathbf{j} - \frac{1}{2}\sin(\theta)\mathbf{k}.$$

A normal vector at a point $P = (x, y, z)$ on $\mathcal{S}$ is then given by

$$\mathbf{N} = \mathbf{r}_\theta(\theta, y) \times \mathbf{r}_y(\theta, y) = \det\left(\begin{bmatrix} \mathbf{i} & \mathbf{j} & \mathbf{k} \\ -(10 - y/2)\sin(\theta) & 0 & (10 - y/2)\cos(\theta) \\ -(1/2)\cos(\theta) & 1 & -(1/2)\sin(\theta) \end{bmatrix}\right)$$

$$= \left(-10 + \frac{y}{2}\right)\cos(\theta)\mathbf{i} + \left(-5 + \frac{y}{4}\right)\mathbf{j} + \left(-10 + \frac{y}{2}\right)\sin(\theta)\mathbf{k}.$$

If P is in the upper half of the yz-plane, then $x = r\cos(\theta) = 0$ so $\theta = \pi/2$, and the $\mathbf{k}$-component of $\mathbf{N}$ is $(-10 + y/2)\sin(\pi/2) = y/2 - 10 < 0$. Therefore $\mathbf{n} = -(1/\|\mathbf{N}\|)\mathbf{N}$ gives the correct orientation of $\mathcal{S}$. We calculate

$$\mathbf{curl}(\mathbf{F}) = \det\left(\begin{bmatrix} \mathbf{i} & \mathbf{j} & \mathbf{k} \\ \dfrac{\partial}{\partial x} & \dfrac{\partial}{\partial y} & \dfrac{\partial}{\partial z} \\ z & 1 & z^2 \end{bmatrix}\right) = 0\mathbf{i} + 1\mathbf{j} + 0\mathbf{k} = \mathbf{j}.$$

Therefore

$$\mathbf{curl}(\mathbf{F})\cdot\mathbf{n}\,dS=\mathbf{j}\cdot\left(-\left(\frac{1}{\|\mathbf{N}\|}\mathbf{N}\right)\right)\|N\|\,dA=\left(5-\frac{y}{4}\right)dyd\theta.$$

Thus

$$\iint_{\mathcal{S}}\mathbf{curl}(\mathbf{F})\cdot\mathbf{n}\,dS=\int_0^{2\pi}\int_8^{14}\left(5-\frac{y}{4}\right)dy\,d\theta=\int_0^{2\pi}\left(5y-\frac{y^2}{8}\right)\Big|_8^{14}d\theta$$

$$=2\pi\left(\left(70-\frac{14^2}{8}\right)-\left(40-\frac{8^2}{8}\right)\right)=27\pi.$$

To compute the right side of equation (13.7.5), we must calculate two line integrals. Figure 14 shows the two components $\mathcal{C}_1$ and $\mathcal{C}_2$ of the boundary. Let

$$\mathbf{r}_1(\theta)=\mathbf{r}(-\theta,8)=6\cos(\theta)\mathbf{i}+8\mathbf{j}-6\sin(\theta)\mathbf{k},\quad \mathbf{r}_2(\theta)=\mathbf{r}(\theta,14)=3\cos(\theta)\mathbf{i}+14\mathbf{j}+3\sin(\theta)\mathbf{k}$$

be parameterizations of $\mathcal{C}_1$ and $\mathcal{C}_2$, respectively. (Verify that these parameterizations orient $\mathcal{C}_1$ and $\mathcal{C}_2$ consistently with $\mathcal{S}$.) We obtain

$$\mathbf{F}(\mathbf{r}_1(\theta))\cdot\mathbf{r}_1'(\theta)=\left(-6\sin(\theta)\mathbf{i}+1\mathbf{j}+36\sin^2(\theta)\mathbf{k}\right)\cdot\left(-6\sin(\theta)\mathbf{i}+0\mathbf{j}-6\cos(\theta)\mathbf{k}\right)=36\sin^2(\theta)-216\sin^2(\theta)\cos(\theta).$$

Thus

$$\int_{\mathcal{C}_1}\mathbf{F}\cdot d\mathbf{r}_1=\int_0^{2\pi}\left(36\sin^2(\theta)-216\sin^2(\theta)\cos(\theta)\right)d\theta\overset{(6.2.9)}{=}36\pi-0=36\pi.$$

Similarly, we have

$$\mathbf{F}(\mathbf{r}_2(\theta))\cdot\mathbf{r}_2'(\theta)=\left(3\sin(\theta)\mathbf{i}+1\mathbf{j}+9\sin^2(\theta)\mathbf{k}\right)\cdot\left(-3\sin(\theta)\mathbf{i}+0\mathbf{j}+3\cos(\theta)\mathbf{k}\right)=-9\sin^2(\theta)+27\sin^2(\theta)\cos(\theta)$$

and

$$\int_{\mathcal{C}_2}\mathbf{F}\cdot d\mathbf{r}_2=\int_0^{2\pi}\left(-9\sin^2(\theta)+27\sin^2(\theta)\cos(\theta)\right)d\theta\overset{(6.2.9)}{=}-9\pi+0=-9\pi.$$

Thus

$$\int_{\mathcal{C}_1}\mathbf{F}\cdot d\mathbf{r}_1+\int_{\mathcal{C}_2}\mathbf{F}\cdot d\mathbf{r}_2=36\pi-9\pi=27\pi=\iint_{\mathcal{S}}\text{curl}(\mathbf{F})\cdot\mathbf{n}\,dS,$$

which is the assertion of Stokes's Theorem. ◄

INSIGHT Notice that, because of cancellation, we did not need to calculate $\|\mathbf{N}\|$ in Example 7. In general, when a surface $\mathcal{S}$ is parameterized by $(u, v) \mapsto \mathbf{r}(u, v)$, $(u, v) \in \mathcal{R}$, we have

$$\mathbf{n} = \pm \frac{1}{\|\mathbf{r}_u \times \mathbf{r}_v\|} \mathbf{r}_u \times \mathbf{r}_v \quad \text{and} \quad dS = \|\mathbf{r}_u \times \mathbf{r}_v\| \, du dv.$$

Therefore the scalar $\|\mathbf{r}_u \times \mathbf{r}_v\|$ cancels when $\mathbf{curl}(\mathbf{F}) \cdot \mathbf{n}\, dS$ is calculated, resulting in the formula

$$\iint_{\mathcal{S}} \mathbf{curl}(\mathbf{F}) \cdot \mathbf{n}\, dS = \pm \iint_{\mathcal{R}} \mathbf{curl}(\mathbf{F}) \cdot (\mathbf{r}_u \times \mathbf{r}_v)\, du dv. \tag{13.7.6}$$

THEOREM 2 Let $\mathcal{S}$ be an oriented surface with piecewise-smooth boundary such as we have been discussing. Let the boundary curves be $\mathcal{C}_1, \ldots, \mathcal{C}_N$ with consistent parameterizations $\mathbf{r}_1, \ldots, \mathbf{r}_N$. Suppose that $\mathbf{F}$ is a continuously differentiable irrotational vector field on a neighborhood of $\mathcal{S}$ and its boundary. Then

$$\sum_{j=1}^{N} \int_{\mathcal{C}_j} \mathbf{F} \cdot d\mathbf{r}_j = 0. \tag{13.7.7}$$

Proof. As discussed in Section 13.4, saying that $\mathbf{F}$ is irrotational means that $\mathbf{curl}(\mathbf{F}) = 0$ at each point of $\mathcal{S}$. We apply Stokes's Theorem to obtain

$$\sum_{j=1}^{N} \int_{\mathcal{C}_j} \mathbf{F} \cdot d\mathbf{r}_j = \iint_{\mathcal{S}} \mathbf{curl}(\mathbf{F}) \cdot \mathbf{n}\, dS = \iint_{\mathcal{S}} 0\, dS = 0.$$ ■

If the irrotational vector field $\mathbf{F}$ represents a flow, then we see that the flow around the boundary integrates to 0. Though this is to be expected, there can be some subtleties involved. Remember that, if $\mathcal{S}$ is not simply connected, then the property "$\mathbf{F}$ is irrotational" is weaker than the property "$\mathbf{F}$ is conservative." If $\mathbf{F}$ is irrotational but not conservative, then a summand $\int_{\mathcal{C}_j} \mathbf{F} \cdot d\mathbf{r}_j$ in equation (13.7.7) can be nonzero even if the path $\mathcal{C}_j$ is closed. The next example will illustrate this point.

▶ **EXAMPLE 8** Verify equation (13.7.7) for the irrotational vector field

$$\mathbf{F}(x, y, z) = \left(\frac{-y}{x^2 + y^2}\right)\mathbf{i} + \left(\frac{x}{x^2 + y^2}\right)\mathbf{j} + 0\mathbf{k}$$

on the planar surface $\mathcal{S} = \{(x, y, z) : z = 0, 1 \le x^2 + y^2 \le 4\}$ oriented by an upward pointing normal.

Solution It is a routine matter to verify that **F** is irrotational. Let a be any positive constant. The vector-valued function $\mathbf{r}(t) = a\cos(t)\mathbf{i} + a\sin(t)\mathbf{j} + 0\mathbf{k}, 0 \le t \le 2\pi$, parameterizes a counterclockwise circle $\mathcal{C}$ in the xy-plane of radius a and center $(0,0,0)$. We have $\mathbf{F}(\mathbf{r}(t)) = -(1/a)\sin(t)\mathbf{i} + (1/a)\cos(t)\mathbf{j} + 0\mathbf{k}$ and $\mathbf{r}'(t) = -a\sin(t)\mathbf{i} + a\cos(t)\mathbf{j} + 0\mathbf{k}$. Therefore $\mathbf{F}(\mathbf{r}(t)) \cdot \mathbf{r}'(t) = \sin^2(t) + \cos^2(t)$ and

$$\int_{\mathcal{C}} \mathbf{F} \cdot d\mathbf{r} = \int_0^{2\pi} (\sin^2(t) + \cos^2(t))dt = \int_0^{2\pi} 1dt = 2\pi.$$

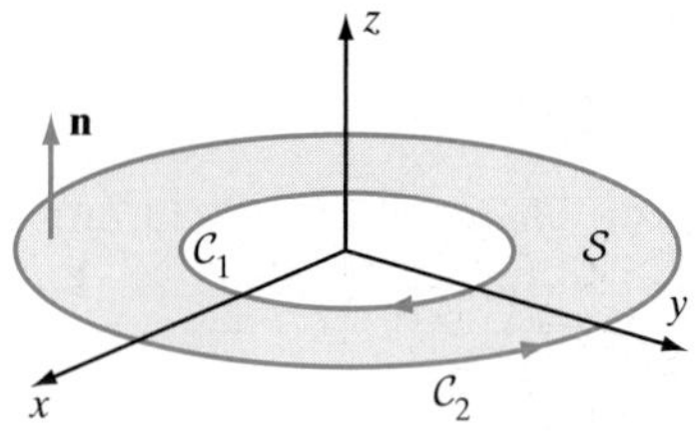

$\{(x, y, z) : z = 0, 1 \le x^2 + y^2 \le 4\}$

▲ Figure 15

Notice that the value of this line integral does not depend on the radius a. If $\mathcal{C}_1$ and $\mathcal{C}_2$ are the components of the boundary indicated in Figure 15, then $\mathcal{C}_1$ has clockwise orientation, $\mathcal{C}_2$ has counterclockwise orientation, and

$$\int_{\mathcal{C}_1} \mathbf{F} \cdot d\mathbf{r}_1 + \int_{\mathcal{C}_2} \mathbf{F} \cdot d\mathbf{r}_2 = (-2\pi) + (2\pi) = 0. \quad ◀$$

An Application

The background material on curl that is presented in Section 13.4 provides ample evidence that Stokes's Theorem is important in physics. Here we provide just one example.

▶ **EXAMPLE 9** Imagine that **B** is a magnetic field in space. Let the surface $\mathcal{S}$ together with its continuously differentiable boundary curve $\mathcal{C}$ be the graph of a continuously differentiable function $z = f(x, y)$. Suppose that $\mathcal{S}$ is given the upward unit normal **n** and that $\mathcal{C}$ is parameterized by **r** with consistent orientation. What does Stokes's Theorem tell us about the flow of **B** around $\mathcal{C}$?

Solution By Stokes's theorem,

$$\oint_{\mathcal{C}} \mathbf{B} \cdot d\mathbf{r} = \iint_{\mathcal{S}} \mathbf{curl}(\mathbf{B}) \cdot \mathbf{n}\, dS.$$

Let us normalize this equation so that we can introduce some standard physical terminology. Let c denote the speed of light (about 186,000 miles per second). The vector

$$\mathbf{J} = \frac{c}{4\pi}\mathbf{curl}(\mathbf{B})$$

is called the *current density* for the magnetic field. It measures the current per unit area of the surface. Then the total current I for the field on the surface $\mathcal{S}$ is

$$I = \iint_{\mathcal{S}} \mathbf{J} \cdot \mathbf{n}\, dS.$$

In summary, we see that

$$\oint_{\mathcal{C}} \mathbf{B} \cdot d\mathbf{r} = \frac{4\pi}{c} I. \tag{13.7.8}$$

Thus the flow of **B** around the boundary of $\mathcal{S}$ is a physical constant times the total current I. Equation (13.7.8) is known as Ampère's Law, after its discoverer André Marie Ampère (1775–1836). It is geometrically obvious that a closed curve $\mathcal{C}$ bounds infinitely many different surfaces $\mathcal{S}$. It is remarkable that Ampère's Law does not depend on which surface $\mathcal{S}$ we are considering. ◀

QUICK QUIZ

1. True or false: For every surface $\mathcal{S}$ in space, we can choose a unit normal $\mathbf{n}(P)$ to $\mathcal{S}$ at every point P on $\mathcal{S}$ such that $P \mapsto \mathbf{n}(P)$ is continuous on $\mathcal{S}$.
2. Suppose that a surface $\mathcal{S}$ is oriented by a unit normal vector field $P \mapsto \mathbf{n}(P)$ and that a boundary curve $\mathcal{C}$ is oriented consistently. Suppose that you are standing at a point P on $\mathcal{C}$ at the edge of $\mathcal{S}$, that $\mathbf{n}(P)$ is the direction of the vector pointing from your feet to your head, and that you are facing in the direction of the tangent vector to $\mathcal{C}$ at P. On what side of you does $\mathcal{S}$ lie: right or left?
3. Stokes's theorem concerns an equation of the form $\oint_{\mathcal{C}} \mathbf{F} \cdot d\mathbf{r} = \iint_{\mathcal{S}} \varphi \, dS$ where φ is a scalar-valued function that involves a unit normal vector field $\mathbf{n}$ on $\mathcal{S}$ that is used to orient $\mathcal{S}$ and $\mathcal{C}$ consistently. What is φ?
4. True or false: Green's Theorem in the plane is a special case of Stokes's Theorem.

Answers

1. False 2. left 3. $\mathbf{curl}(\mathbf{F}) \cdot \mathbf{n}$ 4. True

EXERCISES

Problems for Practice

▶ In each of Exercises 1–12, a planar region $\mathcal{R}$, a vector field $\mathbf{F}$, and a scalar-valued function f are given. Let $\mathcal{S}$ denote the graph of f over the region $\mathcal{R}$, and let $\mathbf{n}$ be the upward unit normal of $\mathcal{S}$. Calculate the surface integral of $\mathbf{curl}(\mathbf{F}) \cdot \mathbf{n}$ over $\mathcal{S}$ directly. Then, verify the value you obtained by orienting the boundary of $\mathcal{S}$ consistently and using Stoke's Theorem. ◀

1. $\mathbf{F}(x,y,z) = z^3\mathbf{i} - xy\mathbf{j} + xz\mathbf{k}$, $f(x,y) = 2x - y$, $\mathcal{R} = \{(x,y) : |x| < 2, |y| < 1\}$
2. $\mathbf{F}(x,y,z) = (y-z)\mathbf{i} - (x+z)\mathbf{j} + (x+y)\mathbf{k}$, $f(x,y) = -5x + 2y$, $\mathcal{R} = \{(x,y) : 1 < x < 2, 0 < y < 2\}$
3. $\mathbf{F}(x,y,z) = e^z\mathbf{i} + 2\mathbf{j} - y\mathbf{k}$, $f(x,y) = x - 2y$, $\mathcal{R} = \{(x,y) : \ln(2) < x < \ln(4), 0 < y < 1\}$
4. $\mathbf{F}(x,y,z) = z\mathbf{i} - z^2\mathbf{j} + x^2\mathbf{k}$, $f(x,y) = 4x - y$, $\mathcal{R} = $ (the triangle with vertices $(1,0), (0,1)$, and $(-1,0)$)
5. $\mathbf{F}(x,y,z) = xz\mathbf{i} - x\mathbf{j} + yz\mathbf{k}$, $f(x,y) = (1 - x^2 - y^2)^{1/2}$, $\mathcal{R} = \{(x,y) : x^2 + y^2 < 1\}$
6. $\mathbf{F}(x,y,z) = \sin(z)\mathbf{i} - \cos(x)\mathbf{j} + \sin(y)\mathbf{k}$, $f(x,y) = 2x - 3y$, $\mathcal{R} = \{(x,y) : |x| < 3\pi, 0 < y < \pi/2\}$
7. $\mathbf{F}(x,y,z) = 7z\mathbf{i} - 5x\mathbf{j} + 3y\mathbf{k}$, $f(x,y) = x^2$, $\mathcal{R} = \{(x,y) : 0 < x < 3, 1 < y < 2\}$
8. $\mathbf{F}(x,y,z) = 2y\mathbf{i} + (x+y)\mathbf{j} + yz\mathbf{k}$, $f(x,y) = (1 - x^2 - y^2)^2$, $\mathcal{R} = \{(x,y) : x^2 + y^2 < 1\}$
9. $\mathbf{F}(x,y,z) = x\mathbf{i} + (y - z^2)\mathbf{j} + \mathbf{k}$, $f(x,y) = x + y + 3$, $\mathcal{R} = \{(x,y) : |x| < y < 1\}$
10. $\mathbf{F}(x,y,z) = y\mathbf{i} - x\mathbf{j} + \mathbf{k}$, $f(x,y) = 2x - 3y + 1$, $\mathcal{R} = \{(x,y) : |y| < x < 1\}$
11. $\mathbf{F}(x,y,z) = z\mathbf{i} - x\mathbf{k}$, $f(x,y) = x + 3y$, $\mathcal{R} = \{(x,y) : |x| < 1, |y| < 1\}$
12. $\mathbf{F}(x,y,z) = xy\mathbf{i} + zx\mathbf{j} - xy\mathbf{k}$, $f(x,y) = x^2 - y^2$, $\mathcal{R} = \{(x,y) : 0 < y < 1 - x^2\}$

▶ In each of Exercises 13–18, use Stokes's Theorem to evaluate $\iint_{\mathcal{S}} \mathbf{curl}(\mathbf{F}) \cdot \mathbf{n} \, dS$ for the given vector field $\mathbf{F}$ and oriented surface $\mathcal{S}$. ◀

13. $\mathbf{F}(x,y,z) = y\mathbf{i} + z^2\mathbf{j} + \sqrt{1+z^4}\mathbf{k}$; $\mathcal{S} = \{(x,y,z) : x^2 + y^2 + (z-4)^2 = 25, 0 < z\}$ oriented by $P \mapsto \mathbf{n}(P)$ with $\mathbf{n}(0,0,9) = \mathbf{k}$
14. $\mathbf{F}(x,y,z) = x\mathbf{i} + x\mathbf{j} + \exp(xyz)\mathbf{k}$; $\mathcal{S} = \{(x,y,z) : z = \sqrt{x^2+y^2}, z < 1\}$ oriented by the upward-pointing unit normal vector field
15. $\mathbf{F}(x,y,z) = zx\mathbf{j}$; $\mathcal{S} = \{(x,y,z) : x^2 + y^2 = 9, -2 < z < 5\}$ oriented by $P \mapsto \mathbf{n}(P)$ with $\mathbf{n}(3,0,0) = -\mathbf{i}$
16. $\mathbf{F}(x,y,z) = 2x\mathbf{i} + (z+1)y\mathbf{j} + \mathbf{k}$; $\mathcal{S}$ is the union of the cylinder $\{(x,y,z) : x^2 + y^2 = 4, 0 < z < 1\}$, the disk $\{(x,y,0) : x^2 + y^2 \le 4\}$, and the annulus $\{(x,y,1) : 1 < x^2 + y^2 \le 4\}$, oriented by $P \mapsto \mathbf{n}(P)$ with $\mathbf{n}(0,0,0) = -\mathbf{k}$

17. $\mathbf{F}(x,y,z) = y^2\mathbf{i} + x\mathbf{j} + \sin(z^2)\mathbf{k}$; $\mathcal{S} = \Delta PQT \cup \Delta QRT \cup \Delta RST \cup \Delta SPT$ where $P = (0,0,0)$, $Q = (2,0,0)$, $R = (2,2,0)$, $S = (0,2,0)$, $T = (1,1,2)$, oriented by the upward-pointing unit normal vector field

18. $\mathbf{F}(x,y,z) = (3x+y)\mathbf{i} - z\mathbf{j} + y^2\mathbf{k}$; $\mathcal{S}$ is the union of the cylinder $\{(x,y,z) : x^2+y^2 = 4, 2 < z < 4\}$ and the hemisphere $\{(x,y,z) : x^2+y^2+(z-2)^2 = 4, z \leq 2\}$ oriented by $P \mapsto \mathbf{n}(P)$ with $\mathbf{n}(0,0,0) = -\mathbf{k}$

Further Theory and Practice

In each of Exercises 19–30, an oriented surface $\mathcal{S}$ and a vector field $\mathbf{F}$ are given. Calculate $\iint_{\mathcal{S}} \mathbf{curl}(\mathbf{F}) \cdot \mathbf{n}\, dS$ directly. Then, verify the value you obtained by orienting the boundary of $\mathcal{S}$ consistently, and using Stokes's Theorem.

19. $\mathbf{F}(x,y,z) = y^3\mathbf{i} + z^3\mathbf{j} + x^3\mathbf{k}$. $\mathcal{S}$ is the hemisphere $\{(x,y,z) : x^2+y^2+z^2 = 1, y > 0\}$ oriented by $P \mapsto \mathbf{n}(P)$ with $\mathbf{n}(0,1,0) = \mathbf{j}$.

20. $\mathbf{F}(x,y,z) = z\mathbf{i} + x\mathbf{j} + y\mathbf{k}$. $\mathcal{S} = \{(x,y,z) : y^2+z^2 = 1, 0 < x < 1, 0 < z\}$ oriented by $P \mapsto \mathbf{n}(P)$ with $\mathbf{n}(1/2,0,1) = \mathbf{k}$.

21. $\mathbf{F}(x,y,z) = y\mathbf{i} - x\mathbf{j} + 3\mathbf{k}$. $\mathcal{S} = \left\{(x,y,z) : z = \sqrt{x^2+y^2}, 0 < z < 2\right\}$ oriented by $P \mapsto \mathbf{n}(P)$ with $\mathbf{n}(P) \cdot \mathbf{k} < 0$.

22. $\mathbf{F}(x,y,z) = -y\mathbf{i} + x\mathbf{j} + z\mathbf{k}$. $\mathcal{S} = \{(x,y,z) : x^2+y^2+z^2 = 4, 0 < x, 0 < y, 0 < z\}$ oriented by $P \mapsto \mathbf{n}(P)$ with $\mathbf{n}(P) \cdot \mathbf{k} > 0$.

23. $\mathbf{F}(x,y,z) = z\mathbf{i} + y\mathbf{j} + z\mathbf{k}$. $\mathcal{S} = \{(x,y,z) : z = x^2+y^2, z < x+y\}$ oriented by $P \mapsto \mathbf{n}(P)$ with $\mathbf{n}(P) \cdot \mathbf{k} < 0$.

24. $\mathbf{F}(x,y,z) = -xz\mathbf{j}$. $\mathcal{S} = \{(x,y,z) : x^2+y^2+z^2 = 2, -1 < z\}$ oriented by $P \mapsto \mathbf{n}(P)$ with $\mathbf{n}(0,0,\sqrt{2}) = \mathbf{k}$.

25. $\mathbf{F}(x,y,z) = xy^2\mathbf{j} - xz\mathbf{k}$. $\mathcal{S} = \{(x,y,z) : x^2+y^2+z^2 = 25, 3 < z < 4\}$ oriented by $P \mapsto \mathbf{n}(P)$ with $\mathbf{n}(P) \cdot \mathbf{k} > 0$.

26. $\mathbf{F}(x,y,z) = yz\mathbf{i}$. $\mathcal{S} = \{(x,y,z) : x^2+y^2+z^2 = 25, 0 < z < 4\}$ oriented by $P \mapsto \mathbf{n}(P)$ with $\mathbf{n}(P) \cdot \mathbf{k} > 0$.

27. $\mathbf{F}(x,y,z) = x^3\mathbf{i} - z\mathbf{j} + y\mathbf{k}$. $\mathcal{S} = \{(x,y,z) : x^2 = y^2+z^2, 1 < x < 4\}$ oriented by $P \mapsto \mathbf{n}(P)$ with $\mathbf{n}(P) \cdot \mathbf{i} < 0$.

28. $\mathbf{F}(x,y,z) = x\mathbf{i} - xyz^3\mathbf{j} + z^2\mathbf{k}$. $\mathcal{S} = \{(x,y,z) : x^2+y^2 = 1, 0 < z < 1\}$ oriented by $P \mapsto \mathbf{n}(P)$ with $\mathbf{n}(1,0,1/2) \cdot \mathbf{i} < 0$.

29. $\mathbf{F}(x,y,z) = y^2\mathbf{i} - xz\mathbf{j} + x\mathbf{k}$. $\mathcal{S} = \{(x,y,z) : x^2+y^2 = 4, -2 < z < 3\}$ oriented by $P \mapsto \mathbf{n}(P)$ with $\mathbf{n}(2,0,0) = \mathbf{i}$.

30. $\mathbf{F}(x,y,z) = yz\mathbf{i} - xyz\mathbf{j} + x^2\mathbf{k}$. $\mathcal{S} = \{(x,y,z) : x^2+y^2 = 4, -1 < z < 1\}$ oriented by $P \mapsto \mathbf{n}(P)$ with $\mathbf{n}(2,0,0) = \mathbf{i}$.

31. Let $\mathcal{S} = \{(x,y,z) : x^2+y^2+(z-1)^2 = 2, 0 < z\}$ oriented by $P \mapsto \mathbf{n}(P)$ with $\mathbf{n}(0,0,1+\sqrt{2}) = \mathbf{k}$. Let $\mathbf{F}(x,y,z) = y\cos(xz)\mathbf{i} - xy\sin(z^2)\mathbf{j} + \exp(x^2)\mathbf{k}$. Calculate $\iint_{\mathcal{S}} \mathrm{curl}(\mathbf{F}) \cdot \mathbf{n}\, dS$ by using Stokes's Theorem twice. Convert the required surface integral to a line integral, and then convert that to a surface integral over which integration is easy.

32. Let $\mathcal{S} = \{(x,y,z) : z = 1 - x^2/4 - y^2, 0 < z\}$ oriented by $P \mapsto \mathbf{n}(P)$ with $\mathbf{n}(0,0,1) = \mathbf{k}$. Let $\mathbf{F}(x,y,z) = y\exp(x\sin(z))\mathbf{i} - x\ln(1+z^2)\mathbf{j} + \sin(x^2)\mathbf{k}$. Calculate $\iint_{\mathcal{S}} \mathbf{curl}(\mathbf{F}) \cdot \mathbf{n}\, dS$ by using Stokes's Theorem twice. Convert the required surface integral to a line integral, and then convert that to a surface integral over which integration is easy.

33. Consider the surface $\mathcal{S}$, which is the unit sphere in space centered at the origin. Let $\mathbf{n}$ denote the outward unit normal on $\mathcal{S}$. Use Stokes's Theorem to show that if $\mathbf{F}$ is a continuously differentiable vector field on a neighborhood of $\mathcal{S}$, then

$$\iint_{\mathcal{S}} \mathbf{curl}(\mathbf{F}) \cdot \mathbf{n}\, dS = 0.$$

In each of Exercises 34–37, $\mathcal{S}$ is that part of the cylinder $x^2+y^2 = 9$ between the planes $z = 0$ and $z = 2$. Let $\mathbf{n}$ be the unit outward normal on $\mathcal{S}$. For the given vector field $\mathbf{F}$, use the parameterization of Exercise 51 from Section 13.6 to calculate $\iint_{\mathcal{S}} \mathbf{curl}(\mathbf{F}) \cdot \mathbf{n}\, dS$. Then, orient the boundary of $\mathcal{S}$ consistently and verify your answer by using Stokes's Theorem.

34. $\mathbf{F}(x,y,z) = yz^2\,\mathbf{i}$

35. $\mathbf{F}(x,y,z) = 2yz\,\mathbf{i} + xz\mathbf{j}$

36. $\mathbf{F}(x,y,z) = \langle -7y, -xz, z\rangle$

37. $\mathbf{F}(x,y,z) = \langle x, 2xz/(x^2+y^2), y\rangle$

Calculator/Computer Exercises

In each of Exercises 38–41, an oriented surface $\mathcal{S}$ and a vector field $\mathbf{F}$ are given. Orient the boundary of $\mathcal{S}$ consistently, and verify Stokes's Theorem.

38. $\mathbf{F}(x,y,z) = z^3\mathbf{i} + 3y\mathbf{j}$. $\mathcal{S} = \{(x,y,z) : z = x^2+2y, (x-2)^2 < y < \ln(x)\}$ oriented by $P \mapsto \mathbf{n}(P)$ with $\mathbf{n}(2,1/2,5) = \mathbf{k}$.

39. $\mathbf{F}(x,y,z) = xe^{-y}\mathbf{j}$. $\mathcal{S} = \{(x,y,z) : z = \exp(x^2+y^2), x^2+y^2 < 1\}$ oriented by $P \mapsto \mathbf{n}(P)$ with $\mathbf{n}(0,0,1) = -\mathbf{k}$.

40. $\mathbf{F}(x,y,z) = \ln(2+y)\mathbf{i}$. $\mathcal{S} = \{(x,y,z) : z = 1 - x^2/4 - y^2, 0 < z\}$ oriented by $P \mapsto \mathbf{n}(P)$ with $\mathbf{n}(0,0,1) = \mathbf{k}$.

41. $\mathbf{F}(x,y,z) = \exp(z^2)\mathbf{j}$. $\mathcal{S} = \{(x,y,z) : x+y+z = 1, 0 < x, 0 < y, 0 < z\}$ oriented by $P \mapsto \mathbf{n}(P)$ with $\mathbf{n} \cdot \mathbf{k} > 0$.

13.8 The Divergence Theorem

The Fundamental Theorem of Calculus relates the boundary data of a function at the endpoints of an interval with information about the derivative of the function on the interval:

$$F(b) - F(a) = \int_a^b F'(t)\,dt.$$

This is a *one-dimensional* theorem.

Green's Theorem relates the boundary data of a vector field on a planar region $\mathcal{R}$ with information about certain derivatives of the vector field $M\mathbf{i} + N\mathbf{j}$ in the interior:

$$\oint_C M\,dx + N\,dy = \iint_{\mathcal{R}} \left(\frac{\partial N}{\partial x} - \frac{\partial M}{\partial y} \right) dA.$$

This is a *two-dimensional* theorem. Stokes's Theorem is a similar two-dimensional theorem for surfaces. Now we will learn a three-dimensional theorem in the same vein—the Divergence Theorem. As with Stokes's Theorem, we must begin by considering the question of orientation.

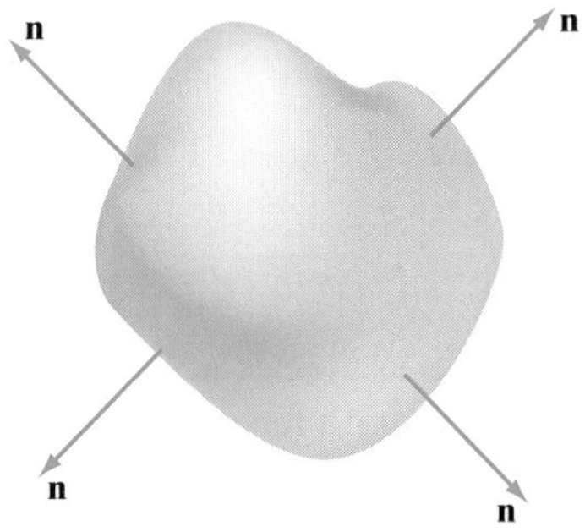

▲ Figure 1

Suppose that $\mathcal{U}$ is a bounded, three-dimensional solid in space whose boundary $\mathcal{S}$ consists of one or more continuously differentiable surfaces. To each point (x, y, z) on $\mathcal{S}$, associate the unit normal vector that points *out* of the solid (see Figure 1). Call this vector field $\mathbf{n}$. Suppose that $\mathbf{F}$ is a continuously differentiable vector field on a neighborhood of $\mathcal{U}$ and its boundary. At a point $P = (x, y, z)$ on $\mathcal{S}$, the quantity $\mathbf{F}(x, y, z) \cdot \mathbf{n}(x, y, z)$ is the normal component of $\mathbf{F}$. If $\mathbf{F}$ represents a flow and if $\mathbf{F}(P) \cdot \mathbf{n} > 0$, then the flow at P is outward, while if $\mathbf{F}(P) \cdot \mathbf{n} < 0$, then the flow at P is inward. The surface integral

$$\iint_{\mathcal{S}} \mathbf{F} \cdot \mathbf{n}\,dS$$

represents the overall normal component of flow at the boundary. We call this quantity the *flux* of the vector field $\mathbf{F}$ across the boundary. If the flux is positive, then, overall, the flow is outward; if the flux is negative, then, overall, the flow is inward.

The Divergence Theorem

It makes good physical sense that the flux of a vector field $\mathbf{F}$ can also be obtained by summing its divergence over the interior. This is in fact what the Divergence Theorem tells us.

THEOREM 1 **(The Divergence Theorem)** With $\mathcal{U}$, $\mathcal{S}$, $\mathbf{n}$, and $\mathbf{F}$ as described previously, we have

$$\iint_{\mathcal{S}} \mathbf{F} \cdot \mathbf{n}\,dS = \iiint_{\mathcal{U}} \operatorname{div}(\mathbf{F})\,dV. \tag{13.8.1}$$

A proof of the Divergence Theorem in its full generality is far beyond the scope of this book. We therefore supply only a proof of a relatively simple case. Because even that simplified derivation is quite lengthy, we defer it to the end of the section. For now, we will concentrate on understanding the Divergence Theorem in particular examples.

▶ **EXAMPLE 1** Let $\mathcal{U}$ be the unit ball in space. The boundary surface $\mathcal{S}$ is the unit sphere $x^2 + y^2 + z^2 = 1$. Verify the Divergence Theorem for the vector field $\mathbf{F}(x, y, z) = 3x\mathbf{i} + 3y\mathbf{j} + 3z\mathbf{k}$ defined on $\mathcal{U}$.

Solution First, $\text{div}(\mathbf{F}) = 3 + 3 + 3 = 9$. Therefore

$$\iiint_{\mathcal{U}} \text{div}(\mathbf{F})dV = 9 \cdot \ (\text{volume of ball of radius } 1) = 9 \cdot \left(\frac{4}{3}\pi \cdot 1^3\right) = 12\pi.$$

Turning our attention to the left side of equation (13.8.1), we observe that $\mathbf{n}(x, y, z) = x\mathbf{i} + y\mathbf{j} + z\mathbf{k}$ is the unit outward pointing normal at the point (x, y, z) of $\mathcal{S}$. Therefore on $\mathcal{S}$, where $x^2 + y^2 + z^2 = 1$, we have

$$\mathbf{F} \cdot \mathbf{n} = (3x\mathbf{i} + 3y\mathbf{j} + 3z\mathbf{k}) \cdot (x\mathbf{i} + y\mathbf{j} + z\mathbf{k}) = 3(x^2 + y^2 + z^2) = 3.$$

It follows that

$$\iint_{\mathcal{S}} \mathbf{F} \cdot \mathbf{n}dS = \iint_{\mathcal{S}} 3dS = 3 \cdot \ (\text{surface area of the unit sphere}) = 3 \cdot (4\pi \cdot 1^2) = 12\pi,$$

as required. ◀

▶ **EXAMPLE 2** Let $\mathcal{U}$ be the cube centered at the origin with faces parallel to the coordinate planes and side length 2: $\mathcal{U} = \{(x, y, z) : -1 < x < 1, -1 < y < 1, -1 < z < 1\}$. Verify the Divergence Theorem for $\mathbf{F}(x, y, z) = 2x^2\mathbf{i} - y^2\mathbf{j} + z\mathbf{k}$ defined on $\mathcal{U}$.

Solution We have

$$\iiint_{\mathcal{U}} \text{div}(\mathbf{F})dV = \iiint_{\mathcal{U}} (4x - 2y + 1)dV = \int_{-1}^{1}\int_{-1}^{1}\int_{-1}^{1} (4x - 2y + 1)\,dz\,dy\,dx = 2\int_{-1}^{1}\int_{-1}^{1} (4x - 2y + 1)\,dy\,dx.$$

Continuing,

$$\begin{aligned} 2\int_{-1}^{1}\int_{-1}^{1} (4x - 2y + 1)\,dy\,dx &= 2\int_{-1}^{1} (4xy - y^2 + y)\Big|_{y=-1}^{y=1} dx \\ &= 2\int_{-1}^{1} \big((4x - 1 + 1) - (-4x - 1 - 1)\big)\,dx \\ &= 2\int_{-1}^{1} (8x + 2)\,dx \\ &= 2(4x^2 + 2x)\Big|_{x=-1}^{x=1} \\ &= 8. \end{aligned}$$

Let $\mathcal{S}$ be the boundary of $\mathcal{U}$. It consists of the six faces of the cube: top and bottom, front and back, left and right. On the top, the unit normal field is $\mathbf{n}(x, y, z) = \mathbf{k}$, the element of area is $dxdy$, and $z = 1$. Thus the surface integral of $\mathbf{F} \cdot \mathbf{n}$ over the top of the cube is

$$\int_{-1}^{1} \int_{-1}^{1} (2x^2\mathbf{i} - y^2\mathbf{j} + 1\mathbf{k}) \cdot \mathbf{k}\, dx\, dy = \int_{-1}^{1} \int_{-1}^{1} 1\, dx\, dy = 4.$$

Similarly, we see that the normal to the bottom is $-\mathbf{k}$, the element of area is $dxdy$, and $z = -1$. Therefore the surface integral of $\mathbf{F} \cdot \mathbf{n}$ over the bottom of the cube is

$$\int_{-1}^{1} \int_{-1}^{1} (2x^2\mathbf{i} - y^2\mathbf{j} - 1\mathbf{k}) \cdot (-\mathbf{k})\, dx\, dy = \int_{-1}^{1} \int_{-1}^{1} 1\, dx\, dy = 4.$$

In the same fashion, it can be calculated that the integral over the front is 8, the integral over the back is -8, the integral over the left side is 4, and the integral over the right side is -4.

Summing these values, we see that

$$\iint_{\mathcal{S}} \mathbf{F} \cdot \mathbf{n}\, dS = 4 + 4 + 8 - 8 + 4 - 4 = 8,$$

which verifies the Divergence Theorem. From the positive value of the flux, we conclude that the net flow is *out* of the cube. ◀

Of course there is no point in calculating the same quantity twice. Examples 1 and 2 are just for practice. The next example is a more typical application of the Divergence Theorem.

▶ **EXAMPLE 3** Use the Divergence Theorem to calculate the flux of $\mathbf{F}(x, y, z) = -x^3\mathbf{i} - y^3\mathbf{j} + 3z^2\mathbf{k}$ across the boundary $\mathcal{S}$ of

$$\mathcal{U} = \{(x, y, z) : x^2 + y^2 < 16, 0 < z < 5\}.$$

Solution Observe that $\mathcal{U}$ is a cylinder with radius 4 and height 5. The flux of $\mathbf{F}$ across $\mathcal{S}$ is defined to be $\iint_{\mathcal{S}} \mathbf{F} \cdot \mathbf{n}\, dS$. In order to evaluate this integral, we would have to break up the calculation into an integral over the top, an integral over the bottom, and an integral over the cylindrical side. It is therefore considerably easier to use the Divergence Theorem to calculate the flux. Thus

$$\iint_{\mathcal{S}} \mathbf{F} \cdot \mathbf{n}\, dS = \iiint_{\mathcal{U}} \operatorname{div}(\mathbf{F})\, dV = \iiint_{\mathcal{U}} \left(\frac{\partial}{\partial x}(-x^3) + \frac{\partial}{\partial y}(-y^3) + \frac{\partial}{\partial z}3z^2\right) dV = \iiint_{\mathcal{U}} (-3x^2 - 3y^2 + 6z) dV.$$

Introducing cylindrical coordinates, we may rewrite this integral as

$$\int_{0}^{5} \int_{0}^{2\pi} \int_{0}^{4} (-3r^2 + 6z)\, r\, dr\, d\theta\, dz,$$

the evaluation of which is straightforward. We have

$$\iint_{\mathcal{S}} \mathbf{F} \cdot \mathbf{n}\, dS = \int_0^5 \int_0^{2\pi} \left(-\frac{3}{4}r^4 + 3zr^2\right)\Big|_{r=0}^{r=4} d\theta dz$$

$$= \int_0^5 \int_0^{2\pi} (-192 + 48z)\, d\theta dz = 2\pi \int_0^5 (-192 + 48z)\, dz = -720\pi.$$

The flux is negative, and we conclude that, overall, the flow is *into* the cylinder. ◀

Some Applications The Divergence Theorem has many important uses in several branches of physics. We turn to a few representative applications.

▶ **EXAMPLE 4** Let $\mathcal{S}$ be the boundary of a solid that contains the origin in its interior. Denote the outward unit normal of $\mathcal{S}$ by $\mathbf{n}$. Suppose that a point charge at the origin exerts a gravitational field

$$\mathbf{F}(x, y, z) = \frac{k}{\|\mathbf{r}\|^3}\mathbf{r}, \quad (x, y, z) \neq (0, 0, 0)$$

where k is a proportionality constant, and $\mathbf{r}$ is the position vector $x\mathbf{i} + y\mathbf{j} + z\mathbf{k}$ of the point (x, y, z). Show that

$$\iint_{\mathcal{S}} \mathbf{F} \cdot \mathbf{n}\, dS = 4\pi k. \tag{13.8.2}$$

Solution Let us first calculate $\int\int_{\mathcal{S}} \mathbf{F} \cdot \mathbf{n}\, dS$ when $\mathcal{S}$ is the sphere Σ_a of radius a centered at the origin, oriented with outward pointing unit normal. (The symbol Σ is often used to denote a sphere, particularly when it cannot be confused with its familiar role as summation symbol.) At any point (x, y, z) of Σ_a, the outward unit normal is $(1/a)\mathbf{r}$. Therefore

$$\mathbf{F} \cdot \mathbf{n} = \frac{k}{a^3}\mathbf{r} \cdot \frac{1}{a}\mathbf{r} = \frac{k}{a^4}\|\mathbf{r}\|^2 = \frac{k}{a^2}$$

and

$$\iint_{\Sigma_a} \mathbf{F} \cdot \mathbf{n}\, dS = \iint_{\Sigma_a} \frac{k}{a^2}\, dS = \frac{k}{a^2}(\text{surface area of sphere of radius } a)$$

$$= \frac{k}{a^2}(4\pi a^2) = 4\pi k. \tag{13.8.3}$$

Equation (13.8.3) is a special case of equation (13.8.2). We will use this special case, together with the Divergence Theorem, to derive equation (13.8.2) in general. Of course, we cannot apply the Divergence Theorem to $\mathbf{F}$ on $\mathcal{U}$ because $\mathbf{F}$ is not differentiable at the origin. However, $\mathbf{F}$ is differentiable on the solid $\mathcal{U}_a$ which is obtained by removing a small ball centered at the origin and having radius a from $\mathcal{U}$. Let $\mathcal{S}_a$ denote the boundary of $\mathcal{U}_a$ with outward normal $\mathbf{n}$. Because $\text{div}(\mathbf{F}) = 0$ at each point other than the origin (by Example 4 of Section 13.4), we have

$$\iint_{\mathcal{S}_a} \mathbf{F} \cdot \mathbf{n} \, dS = \iiint_{\mathcal{U}_a} \operatorname{div}(\mathbf{F}) \, dV = \iiint_{\mathcal{U}_a} 0 \, dV = 0. \tag{13.8.4}$$

But $\mathcal{S}_a$ comprises $\mathcal{S}$ and the sphere $-\Sigma_a$ of radius a centered at the origin—the minus sign signifies that the sphere is oriented with the inward pointing normal. Taking these observations into account in equation (13.8.4), we obtain

$$0 = \iint_{\mathcal{S}_a} \mathbf{F} \cdot \mathbf{n} \, dS = \iint_{\mathcal{S}} \mathbf{F} \cdot \mathbf{n} \, dS + \iint_{-\Sigma_a} \mathbf{F} \cdot \mathbf{n} \, dS = \iint_{\mathcal{S}} \mathbf{F} \cdot \mathbf{n} \, dS - \iint_{\Sigma_a} \mathbf{F} \cdot \mathbf{n} \, dS = \iint_{\mathcal{S}} \mathbf{F} \cdot \mathbf{n} \, dS - 4\pi k,$$

from which equation (13.8.2) immediately follows. ◀

INSIGHT Equation (13.8.2) is known as *Gauss's Law*. Because the force between two charged particles has the same mathematical form as Newton's Law of Gravitation, Gauss's Law is also applicable to the theory of electrostatics.

There are several variants of the Divergence Theorem. In the next example, we examine one that is often useful in applications.

▶ **EXAMPLE 5** Suppose that $\mathcal{U}$, $\mathcal{S}$, and $\mathbf{n}$ are as in the statement of the Divergence Theorem. Suppose that f is a continuously differentiable scalar-valued function on an open set containing $\mathcal{U}$ and its boundary. Prove that

$$\iint_{\mathcal{S}} f \mathbf{n} \, dS = \iiint_{\mathcal{U}} \nabla f \, dV. \tag{13.8.5}$$

Solution First of all, observe that the indicated equation is a *vector* equation. The integrations are performed componentwise. Let $\mathbf{c}$ be a constant vector. Define the vector field $\mathbf{F} = f(x,y,z)\mathbf{c}$. Theorem 1e, Section 13.4, tells us that $\operatorname{div}(\mathbf{F}) = f(x,y,z)\operatorname{div}(\mathbf{c}) + (\nabla f) \cdot \mathbf{c}$. But $\operatorname{div}(\mathbf{c}) = 0$ because $\mathbf{c}$ is constant. Therefore $\operatorname{div}(\mathbf{F}) = (\nabla f) \cdot \mathbf{c}$. The Divergence Theorem tells us that

$$\iiint_{\mathcal{U}} (\nabla f) \cdot \mathbf{c} \, dV = \iiint_{\mathcal{U}} \operatorname{div}(\mathbf{F}) \, dV = \iint_{\mathcal{S}} \mathbf{F} \cdot \mathbf{n} \, dS = \iint_{\mathcal{S}} (f\mathbf{c}) \cdot \mathbf{n} \, dS.$$

Therefore

$$\left(\iiint_{\mathcal{U}} \nabla f \, dV - \iint_{\mathcal{S}} f \mathbf{n} \, dS \right) \cdot \mathbf{c} = 0$$

for every constant vector $\mathbf{c}$. Because the zero vector is the only vector perpendicular to every constant vector, we conclude that

$$\iiint_{\mathcal{U}} \nabla f \, dV - \iint_{\mathcal{S}} f \mathbf{n} \, dS = \mathbf{0},$$

from which the required identity follows. ◀

Next we use the Divergence Theorem to derive a famous law of physics, *Archimedes's Law of Buoyancy*.

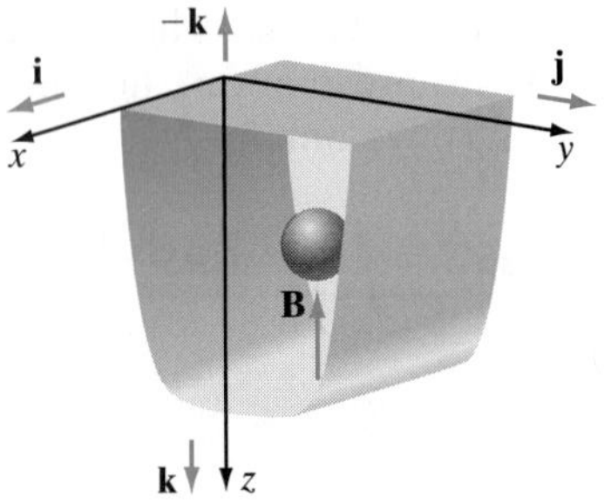

▲ Figure 2

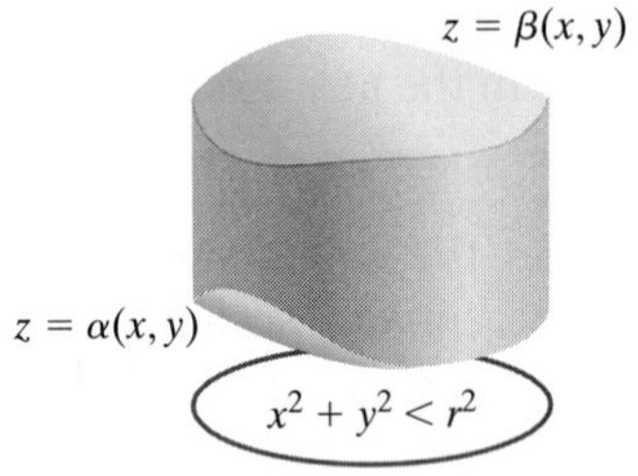

$\mathcal{U} = \{(x, y, z) : \alpha(x, y) < z < \beta(x, y), x^2 + y^2 < r^2\}$

▲ Figure 3

▶ **EXAMPLE 6** Let $\mathcal{U}$ be a solid body with piecewise-smooth boundary $\mathcal{S}$. The body is immersed in water having constant mass density μ. For convenience, we use a coordinate system so that the positive z-axis points down into the water. The surface of the water corresponds to the plane $z = 0$ (see Figure 2). The pressure exerted by the water at depth z is $p = \mu g z$, where g is the gravitational acceleration (we arranged for the positive z-axis to be the downward direction so that this formula for pressure would have no minus sign in it). Use the Divergence Theorem to show that the buoyant force exerted on the body in the direction $-\mathbf{k}$ has magnitude equal to the weight of the fluid displaced by the body. This statement is known as Archimedes's Law of Buoyancy.

Solution Let $\mathbf{n}$ denote the outward unit normal to $\mathcal{S}$. Let $\mathbf{B}$ be the buoyant force. By Newton's Third Law,

$$\mathbf{B} = -\iint_{\mathcal{S}} p\mathbf{n}\, dS = -\iiint_{\mathcal{U}} \nabla p\, dV, \tag{13.8.6}$$

the second equality resulting from equation (13.8.5). But, because $p = \mu g z$, we have $\nabla p = \mu g \mathbf{k}$. This makes sense because the direction of greatest change for the pressure function is down, which is in the $\mathbf{k}$ direction in the coordinate system that we have chosen. If we replace p and ∇p in line (13.8.6) with $\mu g z$ and $\mu g \mathbf{k}$, respectively, then we obtain

$$\mathbf{B} = -\iiint_{\mathcal{U}} \mu g \mathbf{k}\, dV = -\left(g \iiint_{\mathcal{U}} \mu\, dV\right)\mathbf{k}.$$

Because of the minus sign, we see that the vector $\mathbf{B}$ is directed upward. (Remember, $\mathbf{k}$ points downward in our coordinate system.) By definition of density, the value of the triple integral is just the total mass of water displaced. The quantity in the parentheses is therefore the weight of the displaced water.

Proof of the Divergence Theorem

Here we will prove the Divergence Theorem in the case that $\mathcal{U}$ is the z-simple solid between the graphs of two functions $\alpha(x,y)$ and $\beta(x,y)$ over a disk $\mathcal{D} = \{(x,y) : x^2 + y^2 < r^2\}$, that is, $\mathcal{U} = \{(x,y,z) : \alpha(x,y) < z < \beta(x,y), x^2 + y^2 < r^2\}$. Look at Figure 3. The boundary $\mathcal{S}$ of $\mathcal{U}$ consists of three pieces. The upper boundary is the surface $z = \beta(x,y), x^2 + y^2 < r^2$; the lower boundary is $z = \alpha(x,y), x^2 + y^2 < r^2$; the boundary on the side is the cylindrical surface $x^2 + y^2 = r^2, \alpha(x,y) < z < \beta(x,y)$.

Let us write our continuously differentiable vector field $\mathbf{F}$ in component form: $\mathbf{F} = M\mathbf{i} + N\mathbf{j} + R\mathbf{k}$. Our job is to prove equation (13.8.1), which we may write as

$$\iint_{\mathcal{S}} (M\mathbf{i} + N\mathbf{j} + R\mathbf{k}) \cdot \mathbf{n}\, dS = \iiint_{\mathcal{U}} (M_x + N_y + R_z)\, dV, \quad \text{or}$$

$$\iint_{\mathcal{S}} M\mathbf{i} \cdot \mathbf{n} dS + \iint_{\mathcal{S}} N\mathbf{j} \cdot \mathbf{n}\, dS + \iint_{\mathcal{S}} R\mathbf{k} \cdot \mathbf{n}\, dS = \iiint_{\mathcal{U}} M_x\, dV + \iiint_{\mathcal{U}} N_y\, dV + \iiint_{\mathcal{U}} R_z\, dV. \tag{13.8.7}$$

We will show that the third summand on the left equals the third summand on the right. The equalities of the first summands and of the second summands can be obtained by similar calculations.

Let's begin with the surface integral $\iint_{\mathcal{S}} R\mathbf{k}\cdot\mathbf{n}\,dS$. It consists of an integration over the top, an integration over the bottom, and an integration over the side. For the top, $\mathbf{n}$ points upward and

$$\mathbf{n}(x,y,z) = \frac{1}{\left(1+\beta_x(x,y,z)^2+\beta_y(x,y,z)^2\right)^{1/2}}\left(-\beta_x(x,y,z)\mathbf{i}-\beta_y(x,y,z)\mathbf{j}+1\mathbf{k}\right).$$

Notice that $\mathbf{k}\cdot\mathbf{n} = \left(1+\beta_x(x,y,z)^2+\beta_y(x,y,z)^2\right)^{-1/2}$ as a result. Also, we have

$$dS = \left(1+\beta_x(x,y,z)^2+\beta_y(x,y,z)^2\right)^{1/2} dA,$$

and so $\mathbf{k}\cdot\mathbf{n}\,dS = dA$. The upshot of these observations is that

$$\iint_{\text{top of } \mathcal{S}} R\mathbf{k}\cdot\mathbf{n}\,dS = \iint_{\mathcal{D}} R\big(x,y,\beta(x,y)\big)\,dA. \tag{13.8.8}$$

For the bottom, the analysis is the same except that the normal points downward (and of course we use $z=\alpha(x,y)$ instead of $z=\beta(x,y)$). The result is

$$\iint_{\text{bottom of } \mathcal{S}} R\mathbf{k}\cdot\mathbf{n}\,dS = -\iint_{\mathcal{D}} R\big(x,y,\alpha(x,y)\big)\,dA. \tag{13.8.9}$$

For the side, $\mathbf{n}$ is horizontal. Therefore $\mathbf{k}\cdot\mathbf{n}=0$. We conclude that

$$\iint_{\text{side of } \mathcal{S}} R\mathbf{k}\cdot\mathbf{n}\,dS = 0. \tag{13.8.10}$$

Summing equations (13.8.8), (13.8.9), and (13.8.10), we obtain

$$\iint_{\mathcal{S}} R\mathbf{k}\cdot\mathbf{n}\,dS = \iint_{\mathcal{D}} \Big(R\big(x,y,\beta(x,y)\big) - R\big(x,y,\alpha(x,y)\big)\Big)\,dA. \tag{13.8.11}$$

We turn our attention now to the last summand on the right side of equation (13.8.7), applying the Fundamental Theorem of Calculus to the integrand in the z-variable to obtain

$$\begin{aligned}\iiint_{\mathcal{U}} R_z\,dV &= \iint_{\mathcal{D}}\left(\int_{\alpha(x,y)}^{\beta(x,y)} R_z(x,y,z)\,dz\right)dA \\ &= \iint_{\mathcal{D}} R(x,y,z)\Big|_{z=\alpha(x,y)}^{z=\beta(x,y)} dA = \iint_{\mathcal{D}}\Big(R\big(x,y,\beta(x,y)\big)-R\big(x,y,\alpha(x,y)\big)\Big)\,dA.\end{aligned}$$

Because the double integral that terminates this chain of equalities is the right side of equation (13.8.11), we have proved the desired equality.

QUICK QUIZ

1. Suppose that $\mathbf{n}$ is the outward unit normal on the boundary $\mathcal{S}$ of a solid. The flux of a vector field $\mathbf{F}$ across $\mathcal{S}$ is $\iint_{\mathcal{S}}\varphi\,dS$ for what scalar-valued function on $\mathcal{S}$?
2. The Divergence Theorem states that the flux of a vector field $\mathbf{F}$ across a surface $\mathcal{S}$ that bounds a solid $\mathcal{U}$ is equal to $\iiint_{\mathcal{U}} u\,dV$ for some scalar-valued function u on $\mathcal{U}$. What is u?

3. Let $\mathbf{n}$ be the unit outward normal on $\mathcal{S}$, the boundary of $\mathcal{U} = \{(x, y, z) : 0 \le x^2 + y^2 \le 9, 1 \le z \le 3)$. Calculate $\iint_{\mathcal{S}} \mathbf{F} \cdot \mathbf{n}\, dS$ for $\mathbf{F}(x, y, z) = \langle x, 2y, z^2 \rangle$

Answers

1. $\mathbf{F} \cdot \mathbf{n}$ 2. $\text{div}(\mathbf{F})$ 3. 126π

EXERCISES

Problems for Practice

► In each of Exercises 1–20, calculate the flux of the given vector field $\mathbf{F}(x, y, z)$ across the boundary $\mathcal{S}$ of the given solid $\mathcal{U}$ in space by calculating the surface integral directly. Then, use the Divergence Theorem to verify your answer. ◄

1. $\mathbf{F}(x, y, z) = 5\mathbf{k}$, $\mathcal{U} = \{(x, y, z) : x^2 + y^2 < 9, \ -1 < z < 2\}$

2. $\mathbf{F}(x, y, z) = z\mathbf{i}$, $\mathcal{U} = \{(x, y, z) : (x-1)^2 + y^2 < 1, 0 < z < 3\}$

3. $\mathbf{F}(x, y, z) = z\mathbf{k}$, $\mathcal{U} = \{(x, y, z) : x^2 + y^2 < 1, 0 < z < 1\}$

4. $\mathbf{F}(x, y, z) = x\mathbf{i} + y\mathbf{j} + z\mathbf{k}$, $\mathcal{U} = \{(x, y, z) : 0 < z < \sqrt{1 - x^2 - y^2}\}$

5. $\mathbf{F}(x, y, z) = x\mathbf{i}$, $\mathcal{U} = \{(x, y, z) : x^2 + y^2 < 1, 0 < z < 1\}$

6. $\mathbf{F}(x, y, z) = 3xz\mathbf{i} - 2yz\mathbf{j}$, $\mathcal{U} = \{(x, y, z) : x^2 + z^2 < 25, 3 < z, 0 < y < 2\}$

7. $\mathbf{F}(x, y, z) = 7x\mathbf{i} + y\mathbf{j} - 2z\mathbf{k}$, $\mathcal{U} = \{(x, y, z) : x + y + z < 1, 0 < x, 0 < y, 0 < z\}$

8. $\mathbf{F}(x, y, z) = \mathbf{i} + 2\mathbf{j} + 3z\mathbf{k}$, $\mathcal{U} = \{(x, y, z) : \sqrt{x^2 + y^2} < z < 1\}$

9. $\mathbf{F}(x, y, z) = y^2\mathbf{i} - yz\mathbf{j} + xz\mathbf{k}$, $\mathcal{U} = \{(x, y, z) : |x| < 2, |y| < 2, |z| < 1\}$

10. $\mathbf{F}(x, y, z) = \dfrac{x}{z+2}\mathbf{i} + \dfrac{y}{x+2}\mathbf{j} + \dfrac{z}{y+2}\mathbf{k}$, $\mathcal{U} = \{(x, y, z) : |x| < 1, |y| < 1, 0 < z < 1\}$

11. $\mathbf{F}(x, y, z) = xz\mathbf{i} - yz\mathbf{j} + xz\mathbf{k}$, $\mathcal{U} = \{(x, y, z) : 0 < z < x^2 - y^2, 0 < x < 1\}$

12. $\mathbf{F}(x, y, z) = (x - z)\mathbf{i} + (y + z)\mathbf{j} + z\mathbf{k}$, $\mathcal{U} = \{(x, y, z) : 0 < z < 6 - x^2 - y^2, x^2 + y^2 < 4\}$

13. $\mathbf{F}(x, y, z) = (x - z)\mathbf{i} + y^2\mathbf{j} + (x + z)\mathbf{k}$, $\mathcal{U} = \{(x, y, z) : x^2 + z^2 < 9, 0 < x, 0 < z, 1 < y < 2\}$

14. $\mathbf{F}(x, y, z) = x^2\mathbf{i} - 4xy\mathbf{j} - 6xz\mathbf{k}$, $\mathcal{U} = \{(x, y, z) : 0 < x < 1, 0 < y < 1, 0 < z < -x - 4y + 8\}$

15. $\mathbf{F}(x, y, z) = (y - 2z)\mathbf{i} + (y - 5x)\mathbf{j} - 9y\mathbf{k}$, $\mathcal{U} = \{(x, y, z) : 4 < x < 5, \ -2 < z < 3, x + 2z < y < 8 + 2x + 3z\}$

16. $\mathbf{F}(x, y, z) = z\mathbf{i} + 3y^2\mathbf{j} + \mathbf{k}$, $\mathcal{U}$ is the solid bounded by the planes $x + y = 1$, $-x + y = 1$, $y = -2$, $z = -3$, and $z = 2x + y + 9$.

17. $\mathbf{F}(x, y, z) = xy\mathbf{i} + y^2\mathbf{j} + z\mathbf{k}$, $\mathcal{U} = \{(x, y, z) : x^2 + y^2 < 1, x^2 + y^2 + z^2 < 4\}$

18. $\mathbf{F}(x, y, z) = xz\mathbf{i} - yz\mathbf{j} + x^2z\mathbf{k}$, $\mathcal{U} = \{(x, y, z) : x^2 + y^2 < z^2 < 9\}$

19. $\mathbf{F}(x, y, z) = (x^2 - 2xz)\mathbf{i} - (3 - y^2)\mathbf{j} + z^2\mathbf{k}$, $\mathcal{U} = \{(x, y, z) : x^2 + y^2 < z < 4, 0 < y\}$

20. $\mathbf{F} = x^2\mathbf{j} - y^2\mathbf{k}$, $\mathcal{U} = \{(x, y, z) : y^2 + 4z^2 < x < 4\}$

Further Theory and Practice

21. Let $R = \sqrt{x^2 + y^2 + z^2}$, and set $\mathbf{F}(x, y, z) = R^{-3}(x\mathbf{i} + y\mathbf{j} + z\mathbf{k})$. For $0 < \alpha < \beta$, calculate the flux of $\mathbf{F}$ across the boundary of the solid $\mathcal{U} = \{(x, y, z) : \alpha^2 \le x^2 + y^2 + z^2 \le \beta^2\}$.

22. Use the result of Exercise 21 to show that the flux of the vector field $\mathbf{F}$ across any sphere $\Sigma_r = \{(x, y, z) : x^2 + y^2 + z^2 = r\}$ is independent of r. (This statement is less precise than Gauss's Law.)

23. Use the results of Exercises 21 and 22, together with the Divergence Theorem, to see that if $\mathcal{U}$ is *any* solid with no holes that contains the origin, then the flux of $\mathbf{F}$ across the boundary of $\mathcal{U}$ is the same as the flux across any Σ_r. (This statement is again less precise than Gauss's Law.)

24. Suppose that $\mathbf{F}(x, y, z) = (x - yz)\mathbf{i} + (y - xy)\mathbf{j} + (xz - z)\mathbf{k}$. Let

$$\mathcal{U} = \{(x, y, z) : x^2 + 4y^2 < 8, \ -2 < z < \phi(x, y)\},$$

for some continuously differentiable function $\phi(x, y)$. If the flux of $\mathbf{F}$ through the side of this solid is known to be 9 then what is the total flux through the top and bottom?

25. Suppose that $\mathbf{F}$ is a twice continuously differentiable vector field and that $\mathbf{G} = \mathbf{curl}(\mathbf{F})$. Prove that the total flux of $\mathbf{G}$ across the boundary of a solid such as the one in the statement of the Divergence Theorem is 0.

26. Let $\mathbf{F}$ be a constant vector field and $\mathcal{U}$ a solid in space that is bounded by finitely many continuously differentiable surfaces. What is the total flux of $\mathbf{F}$ across the boundary of $\mathcal{U}$?

27. Let $\mathcal{U}$ be the solid in space between the two spheres $\{(x, y, z) : x^2 + y^2 + z^2 = 1\}$ and $\{(x, y, z) : x^2 + y^2 + z^2 = 4\}$. Apply the Divergence Theorem to evaluate the flux of the vector field $\mathbf{F} = x\mathbf{i} - y^2\mathbf{j} + xz\mathbf{k}$ across the boundary of $\mathcal{U}$.

28. Suppose that $\mathbf{F}$ is a continuously differentiable vector field defined on all of space and assume that $\text{div}(\mathbf{F}) = 0$ at all points. Suppose that $\mathcal{S}$ is a surface in the upper half-space $\{(x, y, z) : 0 < z)$ and that the boundary of $\mathcal{S}$ is a simple closed curve $\mathcal{C}$ in the xy-plane. Let $\mathcal{U}$ be the solid that is bounded above by $\mathcal{S}$ and below by the xy-plane. Suppose that $\mathcal{S}'$ is a surface in $\mathcal{U}$ that, like $\mathcal{S}$, has $\mathcal{C}$ for its boundary. Let $\mathcal{S}$ and $\mathcal{S}'$ be oriented by their upward normal vector fields. Prove that

$$\iint_{\mathcal{S}} \mathbf{F} \cdot \mathbf{n} dS = \iint_{\mathcal{S}'} \mathbf{F} \cdot \mathbf{n} dS.$$

29. Let $\mathbf{F}(x, y, z) = (1/3)(x\mathbf{i} + y\mathbf{j} + z\mathbf{k})$, and let $\mathcal{U}$ be a solid in space bounded by finitely many continuously differentiable surfaces. Prove that the flux of $\mathbf{F}$ across the boundary of $\mathcal{U}$ equals the volume of $\mathcal{U}$.

30. Let $\mathcal{U}$ be a solid in space bounded by finitely many continuously differentiable surfaces. Let u be a twice continuously differentiable function whose domain contains a neighborhood of $\mathcal{U}$ and its boundary $\mathcal{S}$. Assume that u is harmonic, that is, $u_{xx} + u_{yy} + u_{zz} = 0$. Prove that $\iint_{\mathcal{S}} \nabla u \cdot \mathbf{n} dS = 0$.

31. With $\mathcal{U}$, $\mathcal{S}$, $\mathbf{n}$, and u as in Exercise 30, prove that

$$\iint_{\mathcal{S}} (\nabla u \cdot \mathbf{n}) u \, dS = \iiint_{\mathcal{U}} |\nabla u|^2 dV.$$

32. Let $\mathcal{U}$ be a solid in space. Assume that the boundary $\mathcal{S}$ of $\mathcal{U}$ is piecewise continuously differentiable. Let $\mathbf{F}$ be a continuously differentiable vector field with domain that contains both $\mathcal{U}$ and $\mathcal{S}$. At any smooth point of $\mathcal{S}$, let $\mathbf{n}$ be the outward unit normal. Prove that

$$\iint_{\mathcal{S}} \mathbf{n} \times \mathbf{F} dS = \iiint_{\mathcal{U}} \nabla \times \mathbf{F} \, dV.$$

33. We now present a bogus proof that all magnetic fields are 0. (This nugget is taken from G. Arfken, *Am. J. Physics* 27 (1959), p. 526.) The mistake is purely a mathematical one. Your job is to fill in the details of the argument and then determine where the error lies. One of Maxwell's laws says that if $\mathbf{B}$ is a magnetic field, then $\text{div}(\mathbf{B}) = \mathbf{0}$. (Accept this fact—it is not the problem!) The Divergence Theorem then says that

$$\iint_{\mathcal{S}} \mathbf{B} \cdot \mathbf{n} dS = 0.$$

Also, because $\text{div}(\mathbf{B}) = 0$, there is a vector field $\mathbf{F}$ such that $\mathbf{B} = \mathbf{curl}(\mathbf{F})$. Combining all this information yields the equation

$$\iint_{\mathcal{S}} \mathbf{curl}(\mathbf{F}) \cdot \mathbf{n} dS = 0.$$

Now we apply Stokes's Theorem to see that

$$\oint_{\mathcal{C}} \mathbf{F} \cdot d\mathbf{r} = \iint_{\mathcal{S}} \mathbf{curl}(\mathbf{F}) \cdot \mathbf{n} dS = 0.$$

Thus $\mathbf{F}$ is path independent. We conclude that $\mathbf{F} = \mathbf{grad}(u)$ for some function u. But then

$$B = \mathbf{curl}(\mathbf{grad}(u)) = \vec{\mathbf{0}}$$

by Theorem 1c, Section 13.4. Of course this is absurd. Where is the error?

In each of Exercises 34–37, calculate the vector $\iiint_{\mathcal{U}} \nabla f \, dV$ for the given solid $\mathcal{U}$ and scalar-valued function f. Then, verify the value you obtained by calculating the left side of equation (13.8.5).

34. $f(x, y, z) = 2y + z^2, \mathcal{U} = \{(x, y, z) : -1 < x, y, z < 1\}$
35. $f(x, y, z) = xz, \mathcal{U} = \{(x, y, z) : 0 < z < 1 - x^2 - y^2\}$
36. $f(x, y, z) = x + 2y + 3z, \mathcal{U} = \{(x, y, z) : x^2 + y^2 < 9, 0 < z < 2\}$
37. $f(x, y, z) = 3z, \mathcal{U} = \{(x, y, z) : x^2 + y^2 + z^2 < 1\}$

Calculator/Computer Exercises

In each of Exercises 38–41, use the Divergence Theorem to calculate the flux of the given vector field $\mathbf{F}(x, y, z)$ across the boundary $\mathcal{S}$ of the given solid $\mathcal{U}$.

38. $\mathbf{F}(x, y, z) = x(1 + z)\sin(y^2)\mathbf{i}$,
$\mathcal{U} = \{(x, y, z) : |x| < 2, |y| < 2, |z| < 1\}$
39. $\mathbf{F}(x, y, z) = \ln(y^2 + z^2)\mathbf{i} + \ln(x^2 + z^2)\mathbf{j} + z^2\ln(2 + y)\mathbf{k}$,
$\mathcal{U} = \{(x, y, z) : 1 < z < 2 - x^2 - y^2\}$
40. $\mathbf{F}(x, y, z) = x\exp(-y^2)\mathbf{i} + y^2\exp(-x^2)\mathbf{j}$,
$\mathcal{U} = \{(x, y, z) : 0 < x, y, z < 1\}$
41. $\mathbf{F}(x, y, z) = (1 + z)^y\mathbf{i} + (1 + x)^z\mathbf{j} + e^z\arctan(1 + x)\mathbf{k}$,
$\mathcal{U} = \{(x, y, z) : 0 < z < 1 - x^2 - y^2\}$

Summary of Key Topics in Chapter 13

A vector field in the plane is a function $\mathbf{F}(x, y) = M(x, y)\mathbf{i} + N(x, y)\mathbf{j}$ of two real variables with vector values in the plane. A vector field in space has the form $\mathbf{F}(x, y, z) = M(x, y, z)\mathbf{i} + N(x, y, z)\mathbf{j} + R(x, y, z)\mathbf{k}$. We usually assume that the coefficient functions are continuously differentiable. Vector fields are useful for describing force fields and the velocity of a flow. They also arise when the gradient operator is applied to a scalar-valued function.

Line Integrals (Section 13.2)

If $\mathbf{F}$ is a vector field and if $t \mapsto \mathbf{r}(t), a \leq t \leq b$ parameterizes a directed curve $\mathcal{C}$ over the interval $[a, b]$, then the line integral of $\mathbf{F}$ over $\mathbf{r}$ is

$$\int_{\mathcal{C}} \mathbf{F} \cdot d\mathbf{r} = \int_a^b \mathbf{F}\big(\mathbf{r}(t)\big) \cdot \mathbf{r}'(t)\, dt.$$

If $\mathbf{F}$ is a force field, this integral represents work performed in traversing the curve. The line integral does not depend on the choice of the parameterization (as long as the parameterization traverses $\mathcal{C}$ in its direction).

Conservative and Path Independent Vector Fields (Section 13.3)

A vector field is *conservative* if it is the gradient of a scalar-valued function. If $\mathbf{F}$ is conservative, then there is an integration procedure for finding a function u such that $\nabla u = \mathbf{F}$. We then say that $V = -u$ is a potential function for $\mathbf{F}$. If $\mathbf{F}$ is a conservative force field, then a potential function can be used to define potential energy. A vector field $\mathbf{F}$ is conservative in a region $\mathcal{R}$ if and only if

$$\int_{\mathcal{C}} \mathbf{F} \cdot d\mathbf{r} = \int_{\tilde{\mathcal{C}}} \mathbf{F} \cdot d\tilde{\mathbf{r}}$$

for every two directed curves $\mathcal{C}$ and $\tilde{\mathcal{C}}$ in $\mathcal{R}$ that begin at the same point and end at the same point. A vector field with this property is called *path-independent.*

A curve $\mathcal{C}$ parameterized by $t \mapsto \mathbf{r}(t)$, $t \in [a, b]$ is said to be closed if $\mathbf{r}(a) = \mathbf{r}(b)$. When we write the path integral over a closed curve, we often use the special integral sign

$$\oint_{\mathcal{C}} \mathbf{F} \cdot d\mathbf{r}.$$

Closed Vector Fields (Section 13.3)

A planar vector field $\mathbf{F} = M\mathbf{i} + N\mathbf{j}$ is closed if

$$\frac{\partial M}{\partial y} = \frac{\partial N}{\partial x}.$$

Conservative vector fields are closed. If the domain of a closed vector field is simply connected, then the vector field is conservative. On regions with holes, this last assertion can be false.

In space, all these statements are still correct, with the definition of a closed vector field $\mathbf{F} = M\mathbf{i} + N\mathbf{j} + R\mathbf{k}$ being

$$\frac{\partial M}{\partial y} = \frac{\partial N}{\partial x}, \quad \frac{\partial M}{\partial z} = \frac{\partial R}{\partial x}, \quad \frac{\partial N}{\partial z} = \frac{\partial R}{\partial y},$$

(all three equations must be satisfied).

Divergence (Section 13.4)

If $\mathbf{F} = M\mathbf{i} + N\mathbf{j}$ is a planar vector field, then its divergence is defined to be

$$\operatorname{div}(\mathbf{F}) = \frac{\partial M}{\partial x} + \frac{\partial N}{\partial y}.$$

For a spatial vector field $\mathbf{F} = M\mathbf{i} + N\mathbf{j} + R\mathbf{k}$, the divergence is

$$\operatorname{div}(\mathbf{F}) = \frac{\partial M}{\partial x} + \frac{\partial N}{\partial y} + \frac{\partial R}{\partial z}.$$

Curl (Section 13.4)

If $\mathbf{F}(x, y) = M(x, y)\mathbf{i} + N(x, y)\mathbf{j}$ is a planar vector field, then the curl of $\mathbf{F}$ is defined to be

$$\mathbf{curl}(\mathbf{F})(x, y, z) = \left(\frac{\partial}{\partial x} N(x, y) - \frac{\partial}{\partial y} M(x, y)\right)\mathbf{k}.$$

If $\mathbf{F} = M\mathbf{i} + N\mathbf{j} + R\mathbf{k}$ is a spatial vector field, then the curl of $\mathbf{F}$ is defined to be

$$\mathbf{curl}(\mathbf{F}) = \left(\frac{\partial R}{\partial y} - \frac{\partial N}{\partial z}\right)\mathbf{i} - \left(\frac{\partial R}{\partial x} - \frac{\partial M}{\partial z}\right)\mathbf{j} + \left(\frac{\partial N}{\partial x} - \frac{\partial M}{\partial y}\right)\mathbf{k}.$$

The Del Notation (Section 13.4)

The del operator is

$$\nabla = \frac{\partial}{\partial x}\mathbf{i} + \frac{\partial}{\partial y}\mathbf{j} + \frac{\partial}{\partial z}\mathbf{k}.$$

The nabla symbol ∇ has already been used in Chapter 11 to give an alternative notation, ∇u, for the gradient $\mathbf{grad}(u)$, of a scalar-valued function u. Using the del notation, we may write $\operatorname{div}(\mathbf{F})$ as $\nabla \cdot \mathbf{F}$ and $\mathbf{curl}(\mathbf{F})$ as $\nabla \times \mathbf{F}$.

Operations with Div and Curl (Section 13.4)

A spatial vector field $\mathbf{F} = M\mathbf{i} + N\mathbf{j} + R\mathbf{k}$ is closed if and only if $\mathbf{curl}(\mathbf{F}) = \mathbf{0}$. In physics, such a vector field is called *irrotational.* A closely related fact is that $\operatorname{div}(\mathbf{curl}(\mathbf{F})) = 0$ for any twice continuously differentiable vector field $\mathbf{F}$.

Green's Theorem (Section 13.5)

This result says that if $\mathbf{F}(x, y) = M(x, y)\mathbf{i} + N(x, y)\mathbf{j}$ is a continuously differentiable vector field defined on a neighborhood of a region $\mathcal{R}$ and its piecewise continuously differentiable boundary curve $\mathcal{C}$, then

$$\int_{\mathcal{C}} \mathbf{F} \cdot d\mathbf{r} = \iint_{\mathcal{R}} \left(\frac{\partial N}{\partial x} - \frac{\partial M}{\partial y}\right) dA.$$

This result relates the flux of a flow across the boundary to the flow across the interior. The parameterization $\mathbf{r}$ of $\mathcal{C}$ should have the property that the vector obtained by rotating $\mathbf{r}'$ counterclockwise by 90° points into $\mathcal{R}$. If $\mathbf{n}$ is the unit *outward* pointing normal vector field on the boundary $\mathcal{C}$, then Green's Theorem may be rewritten as

$$\int_{\mathcal{C}} \mathbf{F}\cdot\mathbf{n}\,ds = \iint_{\mathcal{R}} \operatorname{div}(\mathbf{F})\,dA.$$

Surface Area and Surface Integrals (Section 13.6)

If $\mathcal{R}$ is a region in the plane and if $f(x,y)$ is a continuously differentiable function on $\mathcal{R}$, then the element of surface area on $\mathcal{S} = \{(x,y,f(x,y)) : (x,y) \in \mathcal{R}\}$ is

$$dS = \sqrt{1 + f_x(x,y)^2 + f_y(x,y)^2}\,dA$$

where dA is the element of area in the xy-plane. Thus the area of $\mathcal{S}$ is given by

$$A = \iint_{\mathcal{R}} \sqrt{1 + f_x(x,y)^2 + f_y(x,y)^2}\,dA.$$

If $\varphi(x,y,z)$ is a continuous function defined on the surface $\mathcal{S}$, then the surface integral of φ over $\mathcal{S}$ is denoted by $\iint_{\mathcal{S}} \varphi\,dS$ and defined by

$$\iint_{\mathcal{R}} \varphi(x,y,f(x,y))\sqrt{1 + f_x(x,y)^2 + f_y(x,y)^2}\,dA.$$

If a surface $\mathcal{S}$ in space is given parametrically by the vector-valued function $(u,v) \mapsto \mathbf{r}(u,v)$, $(u,v) \in \mathcal{R}$, then the element of surface area is $\|\mathbf{r}_u \times \mathbf{r}_v\|\,dA$ where dA is the element of area in the uv-plane. Thus the area of the surface is

$$A = \iint_{\mathcal{R}} \|\mathbf{r}_u \times \mathbf{r}_v\|\,dA.$$

If φ is a continuous function defined on the surface, then the integral of φ over the surface is

$$\iint_{\mathcal{S}} \varphi\,dS = \iint_{\mathcal{R}} \varphi(\mathbf{r}(u,v))\,\|(\mathbf{r}_u \times \mathbf{r}_v)(u,v)\|\,du\,dv.$$

Stokes's Theorem (Section 13.7)

Let $\mathcal{S}$ be a surface in space with boundary curve $\mathcal{C}$. We say that $\mathcal{S}$ is *oriented* if it is possible to make a continuous assignment of a unit normal vector $\mathbf{n}(P)$ to each point P of $\mathcal{S}$. We say that the boundary $\mathcal{C}$ of $\mathcal{S}$ is *oriented consistently* with $\mathcal{S}$ if it is parameterized by $\mathbf{r}$ so that, at each boundary point, $\mathbf{n} \times \mathbf{r}'$ points into the region $\mathcal{S}$. If $\mathcal{S}$ and $\mathcal{C}$ are consistently oriented, and if $\mathbf{F}$ is a continuously differentiable vector field defined on $\mathcal{S}$ and its boundary, then Stokes's Theorem states that

$$\oint_{\mathcal{C}} \mathbf{F}\cdot d\mathbf{r} = \iint_{\mathcal{S}} \mathbf{curl}(\mathbf{F})\cdot\mathbf{n}\,dS.$$

The Divergence Theorem (Section 13.8) If $\mathcal{U}$ is a solid in space bounded by finitely many continuously differentiable surfaces, if the boundary of $\mathcal{U}$ is $\mathcal{S}$, if $\mathbf{n}$ is the unit outward pointing normal vector field on $\mathcal{S}$, and if $\mathbf{F}$ is a continuously differentiable vector field on an open set that contains $\mathcal{U}$ and its boundary $\mathcal{S}$, then the Divergence Theorem says that

$$\iint_{\mathcal{S}} \mathbf{F} \cdot \mathbf{n}\, dS = \iiint_{\mathcal{U}} \operatorname{div}(\mathbf{F})\, dV.$$

Review Exercises for Chapter 13

In Exercises 1–3, sketch the given vector field in the plane.

1. $\mathbf{F}(x, y) = x\mathbf{i} - y\mathbf{j}$
2. $\mathbf{F}(x, y) = 4y\mathbf{i} + 3x\mathbf{j}$
3. $\mathbf{F}(x, y) = y\mathbf{i} - x\mathbf{j}$

In each of Exercises 4–6, give an explicit formula for the spatial vector field that is described in words.

4. Particles move in space away from the origin with speed equal to the cube of the distance to the origin. Write the velocity vector field.
5. Particles move in a wind tunnel parallel to a line running down the center of the tunnel. Think of that line as the x-axis. The velocity of a particle is equal to the square of its distance to the line plus its height from the floor (which is at $z = -5$). The positive direction on the x-axis is *opposite* to the direction in which the wind blows. Write the velocity vector field.
6. Particles move outward inside the cone $z^2 = x^2 + y^2$ (inside both nappes!) at a speed equal to the square of the distance of the point from the origin. Write the velocity vector field.

In each of Exercises 7–9, write the gradient vector field that is induced by the given scalar-valued function u.

7. $u(x, y) = xy^2 - yx^3 + x$
8. $u(x, y) = \ln(x^2 - y^3)$
9. $u(x, y) = \sin(x^3 y)$

In each of Exercises 10–12, the given vector field $\mathbf{F}(x, y) = M(x, y)\mathbf{i} + N(x, y)\mathbf{j}$ is conservative. Find a potential function $u(x, y)$ so that $\nabla u(x, y) = F(x, y)$.

10. $\mathbf{F}(x, y) = \sin(y)\mathbf{i} + x\cos(y)\mathbf{j}$
11. $\mathbf{F}(x, y) = xy\mathbf{i} + \dfrac{x^2}{2}\mathbf{j}$
12. $\mathbf{F}(x, y) = \ln(y)\mathbf{i} + \dfrac{x}{y}\mathbf{j}$

In each of Exercises 13–15, the vector field $\mathbf{F}(x, y, z) = M(x, y)\mathbf{i} + N(x, y)\mathbf{j} + R(x, y)\mathbf{k}$ is conservative. Find a potential function $u(x, y, z)$ such that $\nabla u(x, y, z) = \mathbf{F}(x, y, z)$.

13. $\mathbf{F}(x, y, z) = z\mathbf{i} + 1\mathbf{j} + x\mathbf{k}$
14. $\mathbf{F}(x, y, z) = \cos(y)\mathbf{i} - x\sin(y)\mathbf{j} + 1\mathbf{k}$
15. $\mathbf{F}(x, y, z) = y\ln(z)\mathbf{i} + x\ln(z)\mathbf{j} + \dfrac{xy}{z}\mathbf{k}$

In each of Exercises 16–18, find the integral curves $t \mapsto \mathbf{r}(t)$ of the vector field $\mathbf{F}(xy) = M(x, y)\mathbf{i} + N(x, y)\mathbf{j}$.

16. $\mathbf{F}(\mathbf{t}) = \mathbf{i} - \mathbf{j}$
17. $\mathbf{F}(t) = 2x\mathbf{i} - 3\mathbf{j}$
18. $\mathbf{F}(t) = 3z\mathbf{i} - 3y\mathbf{j}$

In each of Exercises 19–21, calculate $\int_{\mathcal{C}} \mathbf{F} \cdot d\mathbf{r}$ for the planar vector field $\mathbf{F}$ and the parametric curve $\mathbf{r}$.

19. $\mathbf{F}(x, y) = y^2\mathbf{i} - xy\mathbf{j}$, $\mathbf{r}(t) = \sin(t)\mathbf{i} - \cos(t)\mathbf{j}, 0 \le t \le \pi/2$
20. $\mathbf{F}(x, y) = x^{1/3}\mathbf{i} - y^{1/4}\mathbf{j}, \mathbf{r}(t) = t^3\mathbf{i} + t^2\mathbf{j}, 1 \le t \le 2$
21. $\mathbf{F}(x, y) = \sin(x)\mathbf{i} + \cos(y)\mathbf{j}, \mathbf{r}(t) = t\mathbf{i} + t^2\mathbf{j}, \pi/2 \le t \le \pi$

In each of Exercises 22–24, calculate $\int_{\mathcal{C}} \mathbf{F} \cdot d\mathbf{r}$ for the spatial vector field $\mathbf{F}$ and the parametric curve $\mathbf{r}$.

22. $\mathbf{F}(x, y, z) = xz\mathbf{i} - y^2x\mathbf{j} + x^2y\mathbf{k}, \mathbf{r}(t) = t^2\mathbf{i} - t\mathbf{k} + t^3\mathbf{k},\ 0 \le t \le 1$
23. $\mathbf{F}(x, y, z) = \sin(x)\mathbf{i} + z\mathbf{j} - \cos(z)\mathbf{k}, \mathbf{r}(t) = t\mathbf{i} - t\mathbf{j} + t^2\mathbf{k}$, $0 \le t \le \pi$
24. $\mathbf{F}(x, y, z) = (y/z)\mathbf{i} - (x/y)\mathbf{j} + yz\mathbf{k}, \mathbf{r}(t) = (1/t)\mathbf{i} + t^2\mathbf{j} - t\mathbf{k}$, $1 \le t \le 2$

In each of Exercises 25–27, calculate the work performed by a force $\mathbf{F}$ in moving a particle along the parametrized curve $\mathbf{r}$.

25. $\mathbf{F}(x, y) = y^2\mathbf{i} - x^2\mathbf{j}, \mathbf{r}(t) = \sin(t)\mathbf{i} - \cos(t)\mathbf{j}, 0 \le t \le \pi$
26. $\mathbf{F}(x, y, z) = (x - y + z)\mathbf{i} + (y + 2x - z)\mathbf{j} + (z + x)\mathbf{k}, \mathbf{r}(t) = \cos(2t)\mathbf{i} - \sin(2t)\mathbf{j} + 2t\mathbf{k}, \pi/2 \le t \le \pi$
27. $\mathbf{F}(x, y) = xy^2\mathbf{i} - yx^2\mathbf{j}, \mathbf{r}(t) = t^{1/3}\mathbf{i} - t^{1/2}\mathbf{j}, 1 \le t \le 3$

In each of Exercises 28–30, calculate the indicated line integral, where C is the triangle with vertices (1, 2), (3, 5), and (4, 1) and with clockwise orientation.

28. $\oint_{\mathcal{C}} xy^2\, dx$
29. $\oint_{\mathcal{C}} y\, dx - x\, dy$

30. $\oint_C (3y - x)dx - (x + y)\,dy$

In each of Exercises 31–33, calculate the line integral $\oint_C f\,ds$ for the given function f and the parametric curve $\mathcal{C}$ parametrized by $\mathbf{r}(t)$.

31. $f(x, y) = x^2 - y^3, \mathbf{r}(t) = \sin(t)\mathbf{i} + \cos(t)\mathbf{j}, 0 \le t \le \pi/4$
32. $f(x, y) = xe^{2y}, \mathbf{r}(t) = e^{-t}\mathbf{i} + t\mathbf{j}, 0 \le t \le 1$
33. $f(x, y) = 3x/y, \mathbf{r}(t) = t^2\mathbf{i} - t\mathbf{j}, 2 \le t \le 4$

In each of Exercises 34–36, determine whether the given vector field is closed on the region $Q = \{(x, y) : -1 < x < 1, -2 < y < 2\}$. If it is closed, then find a potential function.

34. $\mathbf{F}(x, y) = (y - 1)\mathbf{i} + x\mathbf{j}$
35. $\mathbf{F}(x, y) = y^2\mathbf{i} - x^2\mathbf{j}$
36. $\mathbf{F}(x, y) = x\cos(y)\mathbf{i} - (x^2/2)\cos(y)\mathbf{j}$

In each of Exercises 37–39, determine whether the given vector field is closed on the cube $Q = \{(x, y, z) : -1 < x < 1, -1 < y < 1, -1 < z < 1\}$. If it is closed, then find a potential function.

37. $\mathbf{F}(x, y, z) = y\mathbf{i} - z\mathbf{j} + x\mathbf{k}$
38. $\mathbf{F}(x, y, z) = (x^2 - y^2)\mathbf{i} + (y^2 - z^2)\mathbf{j} + (x^2 - z^2)\mathbf{k}$
39. $\mathbf{F}(x, y, z) = (yz - 4)\mathbf{i} + (xz + 6)\mathbf{j} + (xy + 2)\mathbf{k}$

In each of Exercises 40–42, a vector field $\mathbf{F}(x, y)$ is given. The initial point P_0 and the terminal point P_1 for a directed curve $\mathcal{C}$ are specified. But the curve is not. Check that $\mathbf{F}$ is closed in a simply connected region containing P_0 and P_1. Then calculate the line integral $\int_C \mathbf{F} \cdot d\mathbf{r}$ for some (hence any) curve $\mathcal{C}$ that connects P_0 to P_1.

40. $\mathbf{F}(x, y) = \ln(xy)\mathbf{i} + \frac{x}{y}\mathbf{j}, P_0 = (1, 1), P_1 = (2, 2)$
41. $\mathbf{F}(x, y, z) = y^2z\mathbf{i} + 2xyz\mathbf{j} + xy^2\mathbf{k}, P_0 = (2, 1, 3), P_1 = (4, 2, 5)$
42. $\mathbf{F}(x, y, z) = -y^2z\sin(xy^2)\mathbf{i} - 2xyz\sin(xy^2)\mathbf{j} + \cos(xy^2)\mathbf{k}$, $P_0 = (0, 0, 0), P_1 = (2, 1, 3)$

In each of Exercises 43–45, a conservative vector field $\mathbf{F}$ and three points P_0, P_1, P_2 are given. Verify that

$$\int_{\overrightarrow{P_0P_1}} \mathbf{F} \cdot d\mathbf{r} + \int_{\overrightarrow{P_1P_2}} \mathbf{F} \cdot d\mathbf{r} + \int_{\overrightarrow{P_2P_0}} F \cdot d\mathbf{r} = 0.$$

43. $\mathbf{F}(x, y) = y^2\mathbf{i} + 2xy\mathbf{j}, P_0 = (1, 1), P_1 = (2, 2), P_2 = (3, 3)$
44. $\mathbf{F}(x, y, z) = 2xyz^3\mathbf{i} + x^2z^3\mathbf{j} + 3x^2yz^2\mathbf{k}$, $P_0 = (0, 0, 1), P_1 = (0, 1, 0), P_2 = (1, 0, 0)$
45. $\mathbf{F}(x, y) = e^y\mathbf{i} + xe^y\mathbf{j}, P_0 = (2, 3), P_1 = (4, 6), P_2 = (1, 5)$

In each of Exercises 46–48, calculate the divergence of the planar vector field $\mathbf{F}$.

46. $\mathbf{F}(x, y) = \cos(xy)\mathbf{i} - \sin(x, y)\mathbf{j}$
47. $\mathbf{F}(x, y) = \ln(x^2 - y)\mathbf{i} + \ln(y^2 - x)\mathbf{j}$
48. $\mathbf{F}(x, y) = (x^2 - y^3)\mathbf{i} + (y^2 - x)\mathbf{j}$

In each of Exercises 49–51, calculate the divergence of the spatial vector field $\mathbf{F}$.

49. $\mathbf{F}(x, y, z) = e^{x+2z}\mathbf{i} + e^{z-2y}\mathbf{j} - e^{x+y}\mathbf{k}$
50. $\mathbf{F}(x, y, z) = (x^2 - y^3)\mathbf{i} + (y^4 - xz)\mathbf{j} + (z^3 - x)\mathbf{k}$
51. $\mathbf{F}(x, y, z) = \sin(xy)\mathbf{i} - \cos(yz)\mathbf{j} + \tan(xyz)\mathbf{k}$

In each of Exercises 52–54, calculate the curl of the vector field $\mathbf{F}$.

52. $\mathbf{F}(x, y, z) = z^2yz\mathbf{i} - x^2z\mathbf{j} + y^2x\mathbf{k}$
53. $\mathbf{F}(x, y, z) = \ln(x - y)\mathbf{i} + e^{y+z}\mathbf{j} - xyz\mathbf{k}$
54. $\mathbf{F}(x, y, z) = \frac{x}{y + z}\mathbf{i} + \frac{y}{x + z}\mathbf{j} - \frac{z}{x + y}\mathbf{k}$

In each of Exercises 55–57, calculate the curl of the vector field $\mathbf{F}$. State whether $\mathbf{F}$ is closed.

55. $\mathbf{F}(x, y, z) = xy\mathbf{i} - yz\mathbf{j} + xyz\mathbf{k}$
56. $\mathbf{F}(x, y, z) = e^{yz}\mathbf{i} + xze^{yz}\mathbf{j} + xye^{yz}\mathbf{k}$
57. $\mathbf{F}(x, y, z) = \sin(xyz)\mathbf{i} + \cos(yz)\mathbf{j} + \sin(xy)\mathbf{k}$

In each of Exercises 58–60, a scalar-valued function u is given. Calculate div(**grad**(u)).

58. $u(x, y) = x^3 - xy^2$
59. $u(x, y, z) = y^2\sin(z^2 - x^2)$
60. $u(x, y) = xy\ln(y^2 - x^2)$

In each of Exercises 61–63, calculate both sides of the fomula in Green's Theorem and verify that they are equal.

61. $\mathbf{F}(x, y) = x^2\mathbf{i} - 3y\mathbf{j}, \mathbf{r}(t) = \cos(t)\mathbf{i} - \sin(t)\mathbf{j}, -\pi \le t \le \pi$
62. $\mathbf{F}(x, y) = xy\mathbf{i} - yx^3\mathbf{j}, \mathbf{r}(t) = \cos(2t)\mathbf{i} - \sin(2t)\mathbf{j}, \pi \le t \le 2\pi$
63. $\mathbf{F}(x, y) = x^2y\mathbf{i} + y^3\mathbf{j}, \mathbf{r}(t) = \sin(8t)\mathbf{i} - \cos(8t)\mathbf{j}, 0 \le t \le \pi/4$

In each of Exercises 64–66, determine the region that is bounded by the oriented curve $\mathcal{C}$. Then use Green's Theorem to calculate $\int_C \mathbf{F} \cdot d\mathbf{r}$.

64. $\mathbf{F}(x, y) = xy\mathbf{i} - x^2\mathbf{j}, \mathbf{r}(t) = \sin(t)\mathbf{i} - \cos(t)\mathbf{j}, 0 \le t \le 2\pi$
65. $\mathbf{F}(x, y) = (x - 3y)\mathbf{i} + (y - 4x)\mathbf{j}, \mathbf{r}(t) = \cos(2t)\mathbf{i} + \sin(2t)\mathbf{j}$, $\pi \le t \le 2\pi$
66. $\mathbf{F}(x, y) = y^2\mathbf{i} - x^2\mathbf{j}, \mathbf{r}(t) = -\sin(t)\mathbf{i} - \cos(t)\mathbf{j}$, $-\pi/2 \le t \le 3\pi/2$

In each of Exercises 67–69, the given region $\mathcal{R}$ has piecewise smooth boundary. Apply Green's Theorem to evaluate the line integral of $\mathbf{F}$ around the positively oriented boundary curve.

67. $\mathcal{R} = \{(x, y) : -1 \le x \le 1, -2 \le y \le 2\}$, $\mathbf{F}(x, y) = y^2x\mathbf{i} - x^3y^2\mathbf{j}$
68. $\mathcal{R}$ is the triangle with vertices $(0, 0)$, $(1, 0)$, $(0, 1)$, and $\mathbf{F}(x, y) = \sin(y)\mathbf{i} - \cos(x)\mathbf{j}$
69. R is the square with vertices $(2, 0)$, $(0, 2)$, $(0, 0)$, $(2, 2)$, and $\mathbf{F}(x, y) = e^{x-y}\mathbf{i} + e^{x+y}\mathbf{j}$

In each of Exercises 70–72, use formula (15.36) to calculate the area of the given region.

70. The triangle with vertices $(2,0)$, $(0,2)$, $(1,1)$
71. The region inside the circle $x^2+y^2=8$ and lying above the line $y=x$
72. The region bounded by the parabola $y=x^2$ and the line $y=x+1$

In each of Exercises 73–75, compute the flux of the vector field $\mathbf{F}$ across the boundary of the region $\mathcal{R}$.

73. $\mathbf{F}(x,y)=y^2x\mathbf{i}-x^2y\mathbf{j}, \mathcal{R}=\{(x,y): |x|<1, |y|<2\}$
74. $\mathbf{F}(x,y)=yx^{3/2}\mathbf{i}-xy^{3/2}\mathbf{j}, \mathcal{R}=\{(x,y): 0<y<x<1\}$
75. $\mathbf{F}(x,y)=(x/y)\mathbf{i}-(y/x)\mathbf{j}, \mathcal{R}=\{(x,y): 2<x<4, 4<y<6\}$

In each of Exercises 76–78, calculate the surface area of the graph of the function $f(x, y)$ over the region $\mathcal{R}$ in the xy-plane.

76. $f(x,y)=x^2+y^2+4, \mathcal{R}=\{(x,y): x^2+y^2<9\}$
77. $f(x,y)=3x-6y, \mathcal{R}=\{(x,y): -2<x<2, -3<y<3\}$
78. $f(x,y)=4x^2-4y^2, \mathcal{R}=\{(x,y): 4<x^2+y^2<9\}$

In each of Exercises 79–81, calculate the area of the given surface.

79. The cone with base of radius 3 and height 5 (not including the base itself)
80. The portion of the surface $z=xy$ that lies inside the cylinder $x^2+y^2=1$
81. The part of the surface $z=2x+2y+2$ that lies inside the paraboloid $z=x^2+y^2$

In each of Exercises 82–84, integrate the function $\varphi(x,y,z)$ over the surface which is the graph of the function f over the region $\mathcal{R}$.

82. $\varphi(x,y,z)=\cos(y)+\cos(z), f(x,y)=4x-3y,$ $\mathcal{R}=\{(x,y): 0<x<2y<2\}$
83. $\varphi(x,y,z)=z-xz+y, f(x,y)=x+4y,$ $\mathcal{R}=\{(x,y): -1<x<1, -3<y<3\}$
84. $\varphi(x,y,z)=y, f(x,y)=\sqrt{x^2+y^2}, \mathcal{R}=\{(x,y): x^2+y^2<4\}$

In each of Exercises 85–87, write the integral that gives the surface area of the parametrically defined surface. Do *not* attempt to evaluate the integral.

85. $x=u^2-v^2, y=u^2+v^2, z=uv, 0\le u\le 2, 1\le v\le 3$
86. $x=e^{u-v}, y=e^{v-3u}, z=v, 0\le u\le 1, 2\le v\le 4$
87. $x=\sin(uv), y=\cos(v), z=u+v, 0\le u\le \pi/2,$ $\pi/3\le v\le 2\pi/3$

In each of Exercises 88–90, integrate the function φ over the parametrically defined surface.

88. $x=u+v, y=u-v, z=3v-u, 0\le u\le 3,$ $1\le v\le 2, \varphi(x,y,z)=x-y+z$
89. $x=v, y=u, z=2u-v,$ $1\le u\le 4, 2\le v\le 3, \varphi(x,y,z)=y+2z$
90. $x=3u-4v, y=u+v, z=u-v, -2\le u\le -1,$ $-2\le v\le 1, \varphi(x,y,z)=4x-3z$

In each of Exercises 91–93, a planar region $\mathcal{R}$, a vector field $\mathbf{F}$ on $\mathcal{R}$, and a scalar-valued function f on $\mathcal{R}$ are given. Let $\mathcal{S}$ denote the graph of f. Calculate the surface integral of $\mathbf{curl}(\mathbf{F})\cdot\mathbf{n}$ over $\mathcal{S}$ in two different ways:

- Perform the integration directly as a surface integral.
- Use Stokes's Theorem to reduce the problem to a line integral.

91. $\mathbf{F}(x,y,z)=y^2\mathbf{i}-zy\mathbf{j}+yz\mathbf{k}, f(x,y)=x-3y,$ $\mathcal{R}=\{(x,y): |x|<1, |y|<2\}$
92. $\mathbf{F}(x,y,z)=y\mathbf{i}-z^2\mathbf{j}+x^2\mathbf{k}, f(x,y)=y-3x,$ $\mathcal{R}=\{(x,y): 1<x<2, 4<y<5\}$
93. $\mathbf{F}(x,y,z)=e^y\mathbf{i}-x\mathbf{j}+5\mathbf{k}, f(x,y)=y+4x,$ $\mathcal{R}=\{(x,y): |x|<2, |y|<3\}$

In each of Exercises 94–96, use Stokes's Theorem to evaluate $\iint_{\mathcal{S}}\mathbf{curl}(\mathbf{F})\cdot\mathbf{n}\,dS$ for the vector field $\mathbf{F}$ and the oriented surface $\mathcal{S}$.

94. $\mathbf{F}(x,y,z)=3z\mathbf{i}-xy\mathbf{j}+y\mathbf{k}, \mathcal{S}=\{(x,y,z): x^2+y^2=4,$ $2\le z\le 4\}$, oriented by $P\mapsto \mathbf{n}(P)$ with $\mathbf{n}(2,0,0)=\mathbf{i}$
95. $\mathbf{F}(x,y,z)=\cos(y)\mathbf{i}-z\mathbf{j}+x\mathbf{k}$, $\mathcal{S}$ is the interior of the triangle with vertices $(1,0,0)$, $(0,1,0)$, $(1,1,0)$ oriented by the upward-pointing unit normal vector field.
96. $\mathbf{F}(x,y,z)=yx\mathbf{k}, \mathcal{S}=\{(x,y,z): y^2+z^2=1, 1\le x\le 2\},$ oriented by $P\mapsto \mathbf{n}(P)$ with $\mathbf{n}(0,0,1)=\mathbf{k}$

In each of Exercises 97–101, calculate the flux of the given vector field $\mathbf{F}(x, y, z)$ across the boundary S of the given solid U in space by the following two methods:

- calculating the surface integral directly
- using the Divergence Theorem

97. $\mathbf{F}(x,y,z)=4z\mathbf{k}, \mathcal{U}=\{(x,y,z): x^2+y^2<4, 2<z<6\}$
98. $\mathbf{F}(x,y,z)=(z-3y)\mathbf{i}+(4x+y)\mathbf{j}-(y-x)\mathbf{k},$ $\mathcal{U}=\{(x,y,z): 0<z<4-x^2-y^2, x^2+y^2<2\}$
99. $\mathbf{F}(x,y,z)=yz\mathbf{i}-xy\mathbf{j}+yz\mathbf{k},$ $\mathcal{U}=\{(x,y,z): x^2+y^2+z^2<4\}$
100. $\mathbf{F}(x,y,z)=y^2\mathbf{j}+x^2\mathbf{k}, \mathcal{U}=\{(x,y,z): x^2+z^2<y<2\}$
101. $\mathbf{F}(x,y,z)=(x/z)\mathbf{i}-(y/x)\mathbf{j}+(z/y)\mathbf{k},$ $\mathcal{U}=\{(x,y,z): 1<x<4, 2<y<3, 3<z<4\}$

Chapter 13 has been devoted to multivariable analogues—Green's Theorem, Stokes's Theorem, and the Divergence Theorem—of the Fundamental Theorem of Calculus. In fact, each of these theorems is a special case of a single, very general formula that can be written in a remarkably concise form:

$$\int_M df = \int_{\partial\mathcal{M}} f. \qquad (1)$$

This *Genesis and Development* will discuss the background and history of equation (1).

The Beginning of Potential Theory

In 1752, while investigating a question in fluid flow, Euler introduced the partial differential equation

$$\Delta V = \frac{\partial^2 V}{\partial x^2} + \frac{\partial^2 V}{\partial y^2} + \frac{\partial^2 V}{\partial z^2} \equiv 0. \qquad (2)$$

Nine years later, Lagrange encountered equation (2) in his own research on fluid dynamics. Lagrange's work on celestial mechanics in 1773 provided him the occasion for a deeper study of equation (2). He noticed that the force

$$\mathbf{F}(P) = -MG\frac{1}{\|\overrightarrow{P_0P}\|^3}\overrightarrow{P_0P}$$

that a mass M at point P_0 exerts on a unit mass at point P can be expressed (using modern vector notation) as $\mathbf{F}(P) = -\nabla V(P)$ where $V = -MG\|\overrightarrow{P_0P}\|^{-1}$. Lagrange also observed that the potential function V satisfies equation (2). In a series of papers that followed, Lagrange laid the foundation for a theory of the potential.

In 1782, Laplace, then only twenty-four years old, used the theory of potential functions to solve a problem about planetary motion that had baffled Newton, Euler, and Lagrange: the "great inequality of Jupiter and Saturn." According to all observations available at that time, the average angular velocity of Jupiter was increasing, whereas that of Saturn was decreasing. The implication was that Jupiter would eventually be pulled into the sun and that Saturn would fly off from the solar system. Not knowing how to reconcile these observations with his theory of gravitation, Newton believed that divine intervention was occasionally necessary to keep the planets in their place. Laplace was able to show that the planets could not have "secular" (or constant direction) accelerations due to their mutual attractions. Laplace deduced that the inequality of Jupiter and Saturn was not secular but periodic of very long period. Observation has borne out the truth of his deductions. His use of equation (2) became so well known that it was named *Laplace's equation*.

Laplace ranks as one of the most important mathematicians of his era. As a professor at the École Militaire, Laplace had the opportunity to be the examiner of the young Napoléon Bonaparte. After the French Revolution, Napoléon appointed Laplace to France's new cabinet (Minister of the Interior) but dismissed him after only six weeks. Napoléon is alleged to have commented that "Laplace saw no question from its true point of view; he sought subtleties everywhere, had only doubtful ideas, and finally carried the spirit of the infinitely small into administration." Regardless, Laplace's reputation as a scientist continued to grow with the publication of his *Traité de Mécanique Céleste,* which appeared in five huge volumes from 1798 to 1825. In addition to the scientific accolades that Laplace received, he was made a marquis.

George Green

George Green (1793–1841) was the self-taught son of a baker and miller. He used the top floor of his father's mill as his study. Inspired by Laplace's mathematical success in celestial mechanics, Green attempted to establish a mathematical foundation for electricity and magnetism. The results of his research appeared in a paper that was published by subscription in 1828. Green announced his viewpoint with these words:

> ... by confining the attention solely on that peculiar function on whose differentials they all depend, I was induced to try whether it would be possible to discover any general relations, existing between this function and the quantities of electricity in the bodies producing it.

Green coined the name *potential function* for the "peculiar function." Because there were only fifty-two subscribers to Green's publication, his work had no influence at first. Five years after its publication, in 1833, Green became an undergraduate at Cambridge at thirty-nine years of age. He obtained a B.A. degree and stayed on at Cambridge for a few years until poor health forced his return to Green's Mill in 1840. Green died in obscurity one year later. It was not until the republication of his paper in 1846 that Green's research received the attention it deserved. His work re-established the significance of the potential function when progress in the theory created by Lagrange and Laplace was in danger of coming to a halt. Green also discovered formulas that are equivalent to what we now call Green's Theorem. Additionally, he introduced a construct, now called the *Green's function*, that is fundamental to potential theory.

Today, *Green's Mill and Centre* is a tourist attraction. What was once a windmill is now very much part of the 20th century. Included in the exhibits is a weather satellite receiving system that displays data live from orbiting and geostationary satellites. Among the memorabilia that may be purchased there is a pencil with Green's Theorem inscribed on the side.

William Thomson and George Stokes

Only a few weeks before Green's death, William Thomson (1824–1907) entered Cambridge. Thomson's mathematical interests had already been formed by his reading of Fourier's *Théorie Analytique de la Chaleur* and Laplace's *Mécanique Céleste*, which he had acquired at the age of fifteen. As an undergraduate, Thomson encountered a reference to Green's work on potential theory but was not able to locate a copy until after his graduation in 1845. Thomson recognized at once the importance of this work and caused it to be republished the next year. He furthered the application of potential theory to electromagnetic theory and developed his own integral formulas. Indeed, Thomson discovered Stokes's Theorem.

George Gabriel Stokes (1819–1903) was Lucasian professor at Cambridge and a leading expert in fluid dynamics. Stokes's Theorem, in the form

$$\int (\alpha dx + \beta dy + \gamma dz) = \pm \iint \left\{ \ell\left(\frac{d\beta}{dz} - \frac{d\gamma}{dy}\right) + m\left(\frac{d\gamma}{dx} - \frac{d\alpha}{dz}\right) + n\left(\frac{d\alpha}{dy} - \frac{d\beta}{dx}\right) \right\} dS,$$

first appeared in a postscript to an 1850 letter that Thomson sent to Stokes. Stokes liked Thomson's equation enough to include it on several examinations at Cambridge, the first of which was given in 1854. Thomson's theorem eventually came to be associated with, and named after, Stokes.

Although Thomson published many mathematical papers, he is better known as a physicist, thanks to his fundamental contributions to thermodynamics, electricity, and magnetism. Thomson also made his mark as a teacher. At the University of Glasgow, he established the first teaching laboratory in Great Britain. His *Treatise on Natural Philosophy*, written with P. G. Tait, first appeared in 1867 and remained the leading physics textbook for decades.

In the mid 1850s, Thomson served on the board of directors of a consortium of British industrialists who were planning to lay an underwater telegraph cable from Ireland to Newfoundland. Over Thomson's objections, the first two attempts were based on unsound electrical practices and failed. The third attempt, which followed Thomson's design, proved successful. Thomson's role in saving an enormous capital investment made him a hero in British financial circles. A knighthood had long been the appropriate reward for such a substantial contribution, and Thomson became Sir William. He later acquired the title Baron Kelvin of Largs, and you will usually find him referred to as Lord Kelvin in physics texts.

As the twentieth century approached, the aging Lord Kelvin rejected many of the scientific theories that were beginning to take hold. In biology, he was a staunch opponent of Darwin, and, in physics, he rejected the atomic theory of matter. His scientific conservatism manifested itself in mathematics as well. Although we now state Kelvin's best known mathematical theorem in the language of vector analysis, Kelvin's own opinion was that

> "vector" is a useless survival, or offshoot, from quaternions, and has never been of the slightest use to any creature.

The Divergence Theorem was first published by Mikhail Vasilievich Ostrogradsky (1801–1862) in 1831. Ostrogradsky's ambition had been to become an officer in the military. Instead, because the life of an officer was too expensive for someone of modest means, he entered university in preparation for a career in civil service. After graduating from the University of Kharkov, Ostrogradsky traveled to Paris, where he learned Fourier's theory of heat conduction. Research on the partial differential equation that describes heat conduction in three dimensions led Ostrogradsky to the Divergence Theorem. A few years later, in 1839, Gauss discovered the form of the Divergence Theorem that is named for him.

The Fundamental Theorem of Calculus: An Advanced Point of View

Long after the separate discoveries of Green's Theorem, Stokes's Theorem, and the Divergence Theorem, mathematicians came to understand that these formulas are all special cases of one general theorem, a theorem that also contains the Fundamental Theorem of Calculus of one variable. To understand what is involved, we must first recast the Fundamental Theorem of Calculus into an appropriate form.

The equation

$$\int_a^b f'(x)dx = f(b) - f(a),$$

contains two basic objects, the function f and the interval $\mathcal{M} = [a, b]$. There are also two operations at work on them. One is the operation d of differentiation on the function. We have already used the notation df for $f'(x)dx$. The other operation is a geometric one: To the oriented interval $\mathcal{M}$, we associate its boundary $\partial\mathcal{M} = \{a, b\}$ with orientation induced from $\mathcal{M}$. When the interval is given its standard left-to-right orientation, $\partial\mathcal{M}$ is oriented by assigning a " + " at b and a " − " at a.

Recall that an integral over a region is the limit of Riemann sums associated with decompositions of the region. When that region contains only two points, there is no need to take limits. In view of the preceding discussion, we may write $f(b) - f(a)$ as $\int_{\partial\mathcal{M}} f$ and reformulate the Fundamental Theorem of Calculus as

$$\int_{\mathcal{M}} df = \int_{\partial\mathcal{M}} f.$$

A theory of integration has been worked out over very general oriented regions. The typical "differentials" ω that are integrated in three-dimensional space vary in form according to the dimension of the geometric object $\mathcal{M}$ over which they are being integrated. They may be written in the following forms:

Differental Form ω	Dimension of $\mathcal{M}$	Description of $\mathcal{M}$
$f_1 dx + f_2 dy + f_3 dz$	1	curve
$g_1\, dx \wedge dy + g_2\, dy \wedge dz + g_3\, dz \wedge dx$	2	surface
$h\, dx \wedge dy \wedge dz$	3	solid

Here $f_1, f_2, f_3, g_1, g_2, g_3$, and h are all real valued functions of x, y, and z. The basic differentials dx, dy, and dz are provided with an associative algebra structure in which multiplication is denoted by the symbol "∧" and the following rules apply: $dx \wedge dx = dy \wedge dy = dz \wedge dz = 0$, $dy \wedge dx = -dx \wedge dy$, $dz \wedge dy = -dy \wedge dz$, and $dx \wedge dz = -dz \wedge dx$. The integration of $\mathbf{F} \cdot \mathbf{n}$ on an oriented surface $\mathcal{S}$ with normal $\mathbf{n}$ is written as

$$\iint_{\mathcal{S}} F_1 dy \wedge dz + F_2 dz \wedge dx + F_3 dx \wedge dy$$

in this notation instead of $\iint_{\mathcal{S}} \mathbf{F} \cdot \mathbf{n}\, dS$.

Notice that, in a general differential form, the coefficients of the basic differentials are functions. We apply d to any function f to obtain the total differential $df = (\partial f/\partial x)dx + (\partial f/\partial y)dy + (\partial f/\partial z)dz$. The differential operator d is applied to a differential form by applying d to each coefficient and then using the calculus of differentials just described. For example

$$\begin{aligned} &d(xy^2 z\, dx \wedge dy + \sin(x)\, dy \wedge dz) \\ &\quad = (y^2 z\, dx + 2xyz\, dy + xy^2 dz) \wedge dx \wedge dy \\ &\qquad + \cos(x)\, dx \wedge dy \wedge dz = \left(xy^2 + \cos(x)\right) dx \wedge dy \wedge dz. \end{aligned}$$

Using these rules of differentiation, we can express the Fundamental Theorem of Calculus, Green's Theorem, Stokes's Theorem, and the Divergence Theorem—and many other related integral equations—in the form

$$\int_{\partial\mathcal{M}} \omega = \int_{\mathcal{M}} d\omega$$

by specifying a differential form ω that is appropriate for the theorem. A formal verification of the Divergence Theorem,

$$\iiint_M \operatorname{div}(\mathbf{F}) dxdydz = \iint_{\partial M} \mathbf{F} \cdot \mathbf{n}\, dS,$$

runs as follows: Because $\mathbf{F}\cdot\mathbf{n}\,dS = F_1\,dy\wedge dz + F_2\,dz\wedge dx + F_3\,dx\wedge dy$, the calculus of differentials gives

$$\begin{aligned} d(\mathbf{F}\cdot\mathbf{n}dS) &= \left(\frac{\partial F_1}{\partial x}dx + \frac{\partial F_1}{\partial y}dy + \frac{\partial F_1}{\partial z}dz\right)\wedge dy\wedge dz \\ &\quad + \left(\frac{\partial F_2}{\partial x}dx + \frac{\partial F_2}{\partial y}dy + \frac{\partial F_2}{\partial z}dz\right)\wedge dz\wedge dx \\ &\quad + \left(\frac{\partial F_3}{\partial x}dx + \frac{\partial F_3}{\partial y}dy + \frac{\partial F_3}{\partial z}dz\right)\wedge dx\wedge dy \\ &= \left(\frac{\partial F_1}{\partial x} + \frac{\partial F_2}{\partial y} + \frac{\partial F_3}{\partial z}\right)dxdydz \\ &= \mathrm{div}(\mathbf{F}). \end{aligned}$$

A complete understanding of the calculus of differential forms requires concepts from linear algebra, mathematical analysis, and differential geometry. These subjects, like the theory of differential equations, make natural follow-ups to your study of calculus.

Table of Integrals

Integrals Involving x^p

1. $\int \frac{1}{x}\,dx = \ln(|x|) + C$

2. $\int x^p\,dx = \frac{1}{p+1}x^{p+1} + C, \qquad (p \neq -1)$

3. $\int \frac{1}{x^2}\,dx = -\frac{1}{x} + C$

4. $\int \frac{1}{\sqrt{x}}\,dx = 2\sqrt{x} + C$

Integrals Involving $a + bx$

5. $\int \frac{1}{a+bx}\,dx = \frac{1}{b}\ln(|a+bx|) + C$

6. $\int (a+bx)^p\,dx = \frac{1}{(p+1)b}(a+bx)^{p+1} + C, \qquad (p \neq -1)$

7. $\int \frac{x}{a+bx}\,dx = \frac{1}{b^2}\left((a+bx) - a\ln(|a+bx|)\right) + C$

8. $\int \frac{x}{(a+bx)^2}\,dx = \frac{1}{b^2}\left(\ln(|a+bx|) + \frac{a}{(a+bx)}\right) + C$

9. $\int \frac{x}{(a+bx)^p}\,dx = \frac{1}{b^2}\left(\frac{a}{(p-1)(a+bx)^{p-1}} - \frac{1}{(p-2)(a+bx)^{p-2}}\right) + C, \qquad (p \neq 1, 2)$

10. $\int \frac{x^2}{a+bx}\,dx = \frac{1}{b^3}\left(\frac{1}{2}(a+bx)^2 - 2a(a+bx) + a^2\ln(|a+bx|)\right) + C$

11. $\int \frac{x^2}{(a+bx)^2}\,dx = \frac{1}{b^3}\left((a+bx) - 2a\ln(|a+bx|) - \frac{a^2}{a+bx}\right) + C$

12. $\int \frac{x^2}{(a+bx)^3}\,dx = \frac{1}{b^3}\left(\ln(|a+bx|) + \frac{2a}{a+bx} - \frac{a^2}{2(a+bx)^2}\right) + C$

13. $\int \frac{x^2}{(a+bx)^p}\,dx = \frac{1}{b^3}\left(-\frac{a^2}{(p-1)(a+bx)^{p-1}} + \frac{2a}{(p-2)(a+bx)^{p-2}} - \frac{1}{(p-3)(a+bx)^{p-3}}\right) + C, \qquad (p \neq 1, 2, 3)$

14. $\int \frac{x^3}{a+bx}\,dx = \frac{1}{b^4}\left(\frac{1}{3}(a+bx)^3 - \frac{3a}{2}(a+bx)^2 + 3a^2(a+bx) - a^3\ln(|a+bx|)\right) + C$

15. $\displaystyle\int \frac{x^3}{(a+bx)^2}\,dx = \frac{1}{b^4}\left(\frac{1}{2}(a+bx)^2 - 3a(a+bx) + 3a^2\ln(|a+bx|) + \frac{a^3}{a+bx}\right) + C$

16. $\displaystyle\int \frac{x^3}{(a+bx)^3}\,dx = \frac{1}{b^4}\left((a+bx) - 3a\ln(|a+bx|) - \frac{3a^2}{a+bx} + \frac{a^3}{2(a+bx)^2}\right) + C$

17. $\displaystyle\int \frac{x^3}{(a+bx)^4}\,dx = \frac{1}{b^4}\left(\ln(|a+bx|) + \frac{3a}{a+bx} - \frac{3a^2}{2(a+bx)^2} + \frac{a^3}{3(a+bx)^3}\right) + C$

18. $$\int \frac{x^3}{(a+bx)^p}\,dx = \frac{1}{b^4}\left(\frac{a^3}{(p-1)(a+bx)^{p-1}} - \frac{3a^2}{(p-2)(a+bx)^{p-2}} + \frac{3a}{(p-3)(a+bx)^{p-3}} - \frac{1}{(p-4)(a+bx)^{p-4}}\right) + C, \quad (p \neq 1,2,3,4)$$

19. $\displaystyle\int \frac{1}{x(a+bx)}\,dx = -\frac{1}{a}\ln\left(\left|\frac{a+bx}{x}\right|\right) + C$

20. $\displaystyle\int \frac{1}{x(a+bx)^2}\,dx = -\frac{1}{a^2}\left(\ln\left(\left|\frac{a+bx}{x}\right|\right) + \frac{bx}{a+bx}\right) + C$

21. $\displaystyle\int \frac{1}{x^2(a+bx)^2}\,dx = -b\left(\frac{1}{a^2(a+bx)} + \frac{1}{a^2bx} - \frac{2}{a^3}\ln\left(\left|\frac{a+bx}{x}\right|\right)\right) + C$

Integrals Involving $x+a$ and $x+b$, $a \neq b$

22. $\displaystyle\int \frac{x+a}{x+b}\,dx = x + (a-b)\ln(x+b) + C$

23. $\displaystyle\int \frac{1}{(x+a)(x+b)}\,dx = \frac{1}{a-b}\ln\left(\left|\frac{x+b}{x+a}\right|\right) + C$

24. $\displaystyle\int \frac{x}{(x+a)(x+b)}\,dx = \frac{1}{a-b}\left(a\ln(|x+a|) - b\ln(|x+b|)\right) + C$

25. $\displaystyle\int \frac{1}{(x+a)(x+b)^2}\,dx = -\frac{1}{(a-b)(x+b)} + \frac{1}{(a-b)^2}\ln\left(\left|\frac{x+a}{x+b}\right|\right) + C$

26. $\displaystyle\int \frac{x}{(x+a)(x+b)^2}\,dx = \frac{b}{(a-b)(x+b)} - \frac{a}{(a-b)^2}\ln\left(\left|\frac{x+a}{x+b}\right|\right) + C$

27. $\displaystyle\int \frac{x^2}{(x+a)(x+b)^2}\,dx = -\frac{b^2}{(a-b)(x+b)} + \frac{a^2}{(a-b)^2}\ln(|x+a|) + \frac{b^2-2ab}{(a-b)^2}\ln(|x+b|) + C$

28. $\displaystyle\int \frac{1}{(x+a)^2(x+b)^2}\,dx = -\frac{1}{(a-b)^2}\left(\frac{1}{x+a} + \frac{1}{x+b}\right) + \frac{2}{(a-b)^3}\ln\left(\left|\frac{x+a}{x+b}\right|\right) + C$

29. $\displaystyle\int \frac{x}{(x+a)^2(x+b)^2}\,dx = \frac{1}{(a-b)^2}\left(\frac{a}{x+a} + \frac{b}{x+b}\right) - \frac{(a+b)}{(a-b)^3}\ln\left(\left|\frac{x+a}{x+b}\right|\right) + C$

30. $\displaystyle\int \frac{x^2}{(x+a)^2(x+b)^2}\,dx = -\frac{1}{(a-b)^2}\left(\frac{a^2}{x+a} + \frac{b^2}{x+b}\right) + \frac{(2ab)}{(a-b)^3}\ln\left(\left|\frac{x+a}{x+b}\right|\right) + C$

Integrals Involving $a^2 + x^2$

31. $\displaystyle\int \frac{1}{a^2+x^2}\,dx = \frac{1}{a}\arctan\left(\frac{x}{a}\right) + C$

32. $\displaystyle\int \frac{1}{(a^2+x^2)^2}\,dx = \frac{x}{2a^2(a^2+x^2)} + \frac{1}{2a^3}\arctan\left(\frac{x}{a}\right) + C$

33. $\displaystyle\int \frac{1}{(a^2+x^2)^3}\,dx = \frac{x}{4a^2(a^2+x^2)^2} + \frac{3x}{8a^4(a^2+x^2)} + \frac{3}{8a^5}\arctan\left(\frac{x}{a}\right) + C$

34. $\displaystyle\int \frac{1}{(a^2+x^2)^4}\,dx = \frac{x}{6a^2(a^2+x^2)^3} + \frac{5x}{24a^4(a^2+x^2)^2} + \frac{5x}{16a^6(a^2+x^2)} + \frac{5}{16a^7}\arctan\left(\frac{x}{a}\right) + C$

35. $\displaystyle\int \frac{x^2}{a^2+x^2}\,dx = x - a\arctan\left(\frac{x}{a}\right) + C$

36. $\displaystyle\int \frac{x^2}{(a^2+x^2)^2}\,dx = -\frac{x}{2(a^2+x^2)} + \frac{1}{2a}\arctan\left(\frac{x}{a}\right) + C$

37. $\displaystyle\int \frac{x^2}{(a^2+x^2)^3}\,dx = -\frac{x}{4(a^2+x^2)^2} + \frac{x}{8a^2(a^2+x^2)} + \frac{1}{8a^3}\arctan\left(\frac{x}{a}\right) + C$

38. $\displaystyle\int \frac{x^2}{(a^2+x^2)^4}\,dx = -\frac{x}{6(a^2+x^2)^3} + \frac{x}{24a^2(a^2+x^2)^2} + \frac{x}{16a^4(a^2+x^2)} + \frac{1}{16a^5}\arctan\left(\frac{x}{a}\right) + C$

39. $\displaystyle\int \frac{1}{x(a^2+x^2)}\,dx = \frac{1}{2a^2}\ln\left(\frac{x^2}{a^2+x^2}\right) + C$

40. $\displaystyle\int \frac{1}{x(a^2+x^2)^2}\,dx = \frac{1}{2a^2(a^2+x^2)} + \frac{1}{2a^4}\ln\left(\frac{x^2}{a^2+x^2}\right) + C$

41. $\displaystyle\int \frac{1}{x^2(a^2+x^2)}\,dx = -\frac{1}{a^2x} - \frac{1}{a^3}\arctan\left(\frac{x}{a}\right) + C$

42. $\displaystyle\int \frac{1}{x^2(a^2+x^2)^2}\,dx = -\frac{1}{a^4x} - \frac{x}{2a^4(a^2+x^2)} - \frac{3}{2a^5}\arctan\left(\frac{x}{a}\right) + C$

Integrals Involving $a^2 - x^2$

43. $\displaystyle\int \frac{1}{a^2-x^2}\,dx = \frac{1}{2a}\ln\left(\left|\frac{x+a}{x-a}\right|\right) + C$

44. $\displaystyle\int \frac{1}{a^2-x^2}\,dx = \frac{1}{a}\tanh^{-1}\left(\frac{x}{a}\right) + C$

45. $\displaystyle\int \frac{1}{(a^2-x^2)^2}\,dx = \frac{x}{2a^2(a^2-x^2)} + \frac{1}{4a^3}\ln\left(\left|\frac{a+x}{a-x}\right|\right) + C$

46. $\displaystyle\int \frac{1}{(a^2-x^2)^3}\,dx = \frac{x}{4a^2(a^2-x^2)^2} + \frac{3x}{8a^4(a^2-x^2)} + \frac{3}{16a^5}\ln\left(\left|\frac{a+x}{a-x}\right|\right) + C$

47. $\displaystyle\int \frac{1}{(a^2-x^2)^4}\,dx = \frac{x}{6a^2(a^2-x^2)^3} + \frac{5x}{24a^4(a^2-x^2)^2} + \frac{5x}{16a^6(a^2-x^2)} + \frac{5}{32a^7}\ln\left(\left|\frac{a+x}{a-x}\right|\right) + C$

48. $\displaystyle\int \frac{x^2}{a^2-x^2}\,dx = -x + \frac{a}{2}\ln\left(\left|\frac{a+x}{a-x}\right|\right) + C$

49. $\displaystyle\int \frac{x^2}{(a^2-x^2)^2}\,dx = \frac{x}{2(a^2-x^2)} - \frac{1}{4a}\ln\left(\left|\frac{a+x}{a-x}\right|\right) + C$

50. $\displaystyle\int \frac{x^2}{(a^2-x^2)^3}\,dx = \frac{x}{4(a^2-x^2)^2} - \frac{x}{8a^2(a^2-x^2)} - \frac{1}{16a^3}\ln\left(\left|\frac{a+x}{a-x}\right|\right) + C$

51. $\displaystyle\int \frac{x^2}{(a^2-x^2)^4}\,dx = \frac{x}{6(a^2-x^2)^3} - \frac{x}{24a^2(a^2-x^2)^2} - \frac{x}{16a^4(a^2-x^2)} - \frac{1}{32a^5}\ln\left(\left|\frac{a+x}{a-x}\right|\right) + C$

52. $\displaystyle\int \frac{1}{x(a^2-x^2)}\,dx = \frac{1}{2a^2}\ln\left(\left|\frac{x^2}{a^2-x^2}\right|\right) + C$

53. $\displaystyle\int \frac{1}{x(a^2-x^2)^2}\,dx = \frac{1}{2a^2(a^2-x^2)} + \frac{1}{2a^4}\ln\left(\left|\frac{x^2}{a^2-x^2}\right|\right) + C$

54. $\displaystyle\int \frac{1}{x^2(a^2-x^2)}\,dx = -\frac{1}{a^2x} + \frac{1}{2a^3}\ln\left(\left|\frac{a+x}{a-x}\right|\right) + C$

55. $\displaystyle\int \frac{1}{x^2(a^2-x^2)^2}\,dx = -\frac{1}{a^4x} + \frac{x}{2a^4(a^2-x^2)} + \frac{3}{4a^5}\ln\left(\left|\frac{a+x}{a-x}\right|\right) + C$

Integrals Involving $a^4 \pm x^4$

56. $\displaystyle\int \frac{1}{a^4+x^4}\,dx = \frac{1}{4a^3\sqrt{2}}\ln\left(\frac{x^2+ax\sqrt{2}+a^2}{x^2-ax\sqrt{2}+a^2}\right) + \frac{1}{2a^3\sqrt{2}}\arctan\left(\frac{ax\sqrt{2}}{a^2-x^2}\right) + C$

57. $\displaystyle\int \frac{x}{a^4+x^4}\,dx = \frac{1}{2a^2}\arctan\left(\frac{x^2}{a^2}\right) + C$

58. $\displaystyle\int \frac{x^2}{a^4+x^4}\,dx = -\frac{1}{4a\sqrt{2}}\ln\left(\frac{x^2+ax\sqrt{2}+a^2}{x^2-ax\sqrt{2}+a^2}\right) + \frac{1}{2a\sqrt{2}}\arctan\left(\frac{ax\sqrt{2}}{a^2-x^2}\right) + C$

59. $\displaystyle\int \frac{1}{a^4-x^4}\,dx = \frac{1}{4a^3}\ln\left(\left|\frac{a+x}{a-x}\right|\right) + \frac{1}{2a^3}\arctan\left(\frac{x}{a}\right) + C$

60. $\displaystyle\int \frac{x}{a^4-x^4}\,dx = \frac{1}{4a^2}\ln\left(\left|\frac{a^2+x^2}{a^2-x^2}\right|\right) + C$

61. $\displaystyle\int \frac{x^2}{a^4-x^4}\,dx = \frac{1}{4a}\ln\left(\left|\frac{a+x}{a-x}\right|\right) - \frac{1}{2a}\arctan\left(\frac{x}{a}\right) + C$

Integrals Involving $\sqrt{x}$ and $a^2 \pm b^2x$

62. $\displaystyle\int \frac{1}{(a^2+b^2x)\sqrt{x}}\,dx = \frac{2}{ab}\arctan\left(\frac{b\sqrt{x}}{a}\right) + C$

63. $\displaystyle\int \frac{1}{(a^2+b^2x)^2\sqrt{x}}\,dx = \frac{\sqrt{x}}{a^2(a^2+b^2x)} + \frac{1}{a^3b}\arctan\left(\frac{b\sqrt{x}}{a}\right) + C$

64. $\displaystyle\int \frac{\sqrt{x}}{a^2+b^2x}\,dx = \frac{2\sqrt{x}}{b^2} - \frac{2a}{b^3}\arctan\left(\frac{b\sqrt{x}}{a}\right) + C$

65. $\displaystyle\int \frac{\sqrt{x}}{(a^2+b^2x)^2}\,dx = -\frac{\sqrt{x}}{b^2(a^2+b^2x)} + \frac{1}{ab^3}\arctan\left(\frac{b\sqrt{x}}{a}\right) + C$

66. $\displaystyle\int \frac{x^{3/2}}{a^2+b^2x}\,dx = \frac{2x^{3/2}}{3b^2} - \frac{2a^2\sqrt{x}}{b^4} + \frac{2a^3}{b^5}\arctan\left(\frac{b\sqrt{x}}{a}\right) + C$

67. $\displaystyle\int \frac{x^{3/2}}{(a^2+b^2x)^2}\,dx = \frac{2x^{3/2}}{b^2(a^2+b^2x)} + \frac{3a^2\sqrt{x}}{b^4(a^2+b^2x)} - \frac{3a}{b^5}\arctan\left(\frac{b\sqrt{x}}{a}\right) + C$

68. $\displaystyle\int \frac{1}{(a^2-b^2x)\sqrt{x}}\,dx = \frac{1}{ab}\ln\left(\left|\frac{a+b\sqrt{x}}{a-b\sqrt{x}}\right|\right) + C$

69. $\displaystyle\int \frac{1}{(a^2-b^2x)^2\sqrt{x}}\,dx = \frac{\sqrt{x}}{a^2(a^2-b^2x)} + \frac{1}{2a^3b}\ln\left(\left|\frac{a+b\sqrt{x}}{a-b\sqrt{x}}\right|\right) + C$

70. $\displaystyle\int \frac{\sqrt{x}}{a^2-b^2x}\,dx = -\frac{2\sqrt{x}}{b^2} + \frac{a}{b^3}\ln\left(\left|\frac{a+b\sqrt{x}}{a-b\sqrt{x}}\right|\right) + C$

71. $\displaystyle\int \frac{\sqrt{x}}{(a^2-b^2x)^2}\,dx = \frac{\sqrt{x}}{b^2(a^2-b^2x)} - \frac{1}{2ab^3}\ln\left(\left|\frac{a+b\sqrt{x}}{a-b\sqrt{x}}\right|\right) + C$

72. $$\int \frac{x^{3/2}}{a^2-b^2x}\,dx = -\frac{2x^{3/2}}{3b^2} - \frac{2a^2\sqrt{x}}{b^4} + \frac{a^3}{b^5}\ln\left(\left|\frac{a+b\sqrt{x}}{a-b\sqrt{x}}\right|\right) + C$$

73. $$\int \frac{x^{3/2}}{(a^2-b^2x)^2}\,dx = -\frac{2b^2x^{3/2}}{b^4(a^2-b^2x)} + \frac{3a^2\sqrt{x}}{b^4(a^2-b^2x)} - \frac{3a}{2b^5}\ln\left(\left|\frac{a+b\sqrt{x}}{a-b\sqrt{x}}\right|\right) + C$$

Integrals Involving $\sqrt{a+bx}$

74. $$\int \frac{1}{x\sqrt{a+bx}}\,dx = \frac{1}{\sqrt{a}}\ln\left(\frac{\sqrt{a+bx}-\sqrt{a}}{\sqrt{a+bx}+\sqrt{a}}\right) + C$$

75. $$\int \frac{x}{\sqrt{a+bx}}\,dx = \frac{2}{b^2}\left(\frac{1}{3}(a+bx)^{3/2} - a\sqrt{a+bx}\right) + C$$

76. $$\int \frac{x^2}{\sqrt{a+bx}}\,dx = \frac{2}{b^3}\left(\frac{1}{5}(a+bx)^{5/2} - \frac{2a}{3}(a+bx)^{3/2} + a^2\sqrt{a+bx}\right) + C$$

77. $$\int \frac{1}{x(a+bx)^{3/2}}\,dx = \frac{2}{a\sqrt{a+bx}} + \frac{1}{a^{3/2}}\ln\left(\frac{\sqrt{a+bx}-\sqrt{a}}{\sqrt{a+bx}+\sqrt{a}}\right) + C$$

78. $$\int \frac{x}{(a+bx)^{3/2}}\,dx = \frac{2}{b^2}\left(\sqrt{a+bx} + \frac{a}{\sqrt{a+bx}}\right) + C$$

79. $$\int \frac{x^2}{(a+bx)^{3/2}}\,dx = \frac{2}{b^3}\left(\frac{1}{3}(a+bx)^{3/2} - 2a\sqrt{a+bx} - \frac{a^2}{\sqrt{a+bx}}\right) + C$$

80. $$\int \frac{\sqrt{a+bx}}{x}\,dx = 2\sqrt{a+bx} + \sqrt{a}\ln\left(\frac{\sqrt{a+bx}-\sqrt{a}}{\sqrt{a+bx}+\sqrt{a}}\right) + C$$

81. $$\int x\sqrt{a+bx}\,dx = \frac{2}{b^2}\left(\frac{1}{5}(a+bx)^{5/2} - \frac{a}{3}(a+bx)^{3/2}\right) + C$$

82. $$\int x^2\sqrt{a+bx}\,dx = \frac{2}{b^3}\left(\frac{1}{7}(a+bx)^{7/2} - \frac{2a}{5}(a+bx)^{5/2} + \frac{a^2}{3}(a+bx)^{3/2}\right) + C$$

83. $$\int \frac{1}{\sqrt{1+x}\sqrt{a+bx}}\,dx = \frac{2}{\sqrt{b}}\ln\left(\sqrt{a+bx} + \sqrt{b}\sqrt{1+x}\right) + C$$

84. $$\int \frac{\sqrt{1+x}}{\sqrt{a+bx}}\,dx = \frac{\sqrt{1+x}\sqrt{a+bx}}{b} - \frac{(a-b)}{b^{3/2}}\ln\left(\sqrt{a+bx} + \sqrt{b}\sqrt{1+x}\right) + C$$

85. $$\int \sqrt{1+x}\sqrt{a+bx}\,dx = \frac{(a-b)}{4b}\sqrt{1+x}\sqrt{a+bx} + \frac{1}{2}(1+x)^{3/2}\sqrt{a+bx} - \frac{(a-b)^2}{4b^{3/2}}\ln\left(\sqrt{a+bx} + \sqrt{b}\sqrt{1+x}\right) + C$$

86. $$\int \frac{\sqrt{a+bx}}{\sqrt{1+x}}\,dx = \sqrt{1+x}\sqrt{a+bx} + \frac{(a-b)}{\sqrt{b}}\ln\left(\sqrt{a+bx} + \sqrt{b}\sqrt{1+x}\right) + C$$

Integrals Involving $\sqrt{a^2+x^2}$

87. $$\int \frac{1}{(a^2+x^2)^{3/2}}\,dx = \frac{x}{a^2\sqrt{a^2+x^2}} + C$$

88. $$\int \frac{1}{\sqrt{a^2+x^2}}\,dx = \ln\left(x + \sqrt{a^2+x^2}\right) + C$$

89. $$\int \sqrt{a^2+x^2}\,dx = \frac{1}{2}x\sqrt{a^2+x^2} + \frac{1}{2}a^2\ln\left(x + \sqrt{a^2+x^2}\right) + C$$

90. $$\int (a^2+x^2)^{3/2}\,dx = \frac{1}{4}x(a^2+x^2)^{3/2} + \frac{3}{8}a^2x\sqrt{a^2+x^2} + \frac{3}{8}a^4\ln\left(x + \sqrt{a^2+x^2}\right) + C$$

91. $\displaystyle\int \frac{x^2}{(a^2+x^2)^{3/2}}\,dx = -\frac{x}{\sqrt{a^2+x^2}} + \ln\left(x+\sqrt{a^2+x^2}\right) + C$

92. $\displaystyle\int \frac{x^2}{\sqrt{a^2+x^2}}\,dx = \frac{1}{2}x\sqrt{a^2+x^2} - \frac{1}{2}a^2\ln\left(x+\sqrt{a^2+x^2}\right) + C$

93. $\displaystyle\int x^2\sqrt{a^2+x^2}\,dx = \frac{1}{4}x(a^2+x^2)^{3/2} - \frac{1}{8}a^2x\sqrt{a^2+x^2} - \frac{1}{8}a^4\ln\left(x+\sqrt{a^2+x^2}\right) + C$

94. $\displaystyle\int \frac{1}{x(a^2+x^2)^{3/2}}\,dx = \frac{1}{a^2\sqrt{a^2+x^2}} - \frac{1}{a^3}\ln\left(\left|\frac{a+\sqrt{a^2+x^2}}{x}\right|\right) + C$

95. $\displaystyle\int \frac{1}{x\sqrt{a^2+x^2}}\,dx = -\frac{1}{a}\ln\left(\left|\frac{a+\sqrt{a^2+x^2}}{x}\right|\right) + C$

96. $\displaystyle\int \frac{\sqrt{a^2+x^2}}{x}\,dx = \sqrt{a^2+x^2} - a\ln\left(\left|\frac{a+\sqrt{a^2+x^2}}{x}\right|\right) + C$

97. $\displaystyle\int \frac{(a^2+x^2)^{3/2}}{x}\,dx = \frac{1}{3}(a^2+x^2)^{3/2} + a^2\sqrt{a^2+x^2} - a^3\ln\left(\left|\frac{a+\sqrt{a^2+x^2}}{x}\right|\right) + C$

Integrals Involving $\sqrt{a^2-x^2}$

98. $\displaystyle\int \frac{1}{(a^2-x^2)^{3/2}}\,dx = \frac{x}{a^2\sqrt{a^2-x^2}} + C$

99. $\displaystyle\int \frac{1}{\sqrt{a^2-x^2}}\,dx = \arcsin\left(\frac{x}{a}\right) + C$

100. $\displaystyle\int \sqrt{a^2-x^2}\,dx = \frac{1}{2}x\sqrt{a^2-x^2} + \frac{1}{2}a^2\arcsin\left(\frac{x}{a}\right) + C$

101. $\displaystyle\int (a^2-x^2)^{3/2}\,dx = \frac{1}{4}x(a^2-x^2)^{3/2} + \frac{3}{8}a^2x\sqrt{a^2-x^2} + \frac{3}{8}a^4\arcsin\left(\frac{x}{a}\right) + C$

102. $\displaystyle\int \frac{x^2}{(a^2-x^2)^{3/2}}\,dx = \frac{x}{\sqrt{a^2-x^2}} - \arcsin\left(\frac{x}{a}\right) + C$

103. $\displaystyle\int \frac{x^2}{\sqrt{a^2-x^2}}\,dx = -\frac{1}{2}x\sqrt{a^2-x^2} + \frac{1}{2}a^2\arcsin\left(\frac{x}{a}\right) + C$

104. $\displaystyle\int x^2\sqrt{a^2-x^2}\,dx = -\frac{1}{4}x(a^2-x^2)^{3/2} + \frac{a^2}{8}x\sqrt{a^2-x^2} + \frac{a^4}{8}\arcsin\left(\frac{x}{a}\right) + C$

105. $\displaystyle\int \frac{1}{x(a^2-x^2)^{3/2}}\,dx = -\frac{1}{a^3}\ln\left(\left|\frac{a+\sqrt{a^2-x^2}}{x}\right|\right) + \frac{1}{a^2\sqrt{a^2-x^2}} + C$

106. $\displaystyle\int \frac{1}{x\sqrt{a^2-x^2}}\,dx = -\frac{1}{a}\ln\left(\left|\frac{a+\sqrt{a^2-x^2}}{x}\right|\right) + C$

107. $\displaystyle\int \frac{\sqrt{a^2-x^2}}{x}\,dx = \sqrt{a^2-x^2} - a\ln\left(\left|\frac{a+\sqrt{a^2-x^2}}{x}\right|\right) + C$

108. $\displaystyle\int \frac{(a^2-x^2)^{3/2}}{x}\,dx = \frac{1}{3}(a^2-x^2)^{3/2} + a^2\sqrt{a^2-x^2} - a^3\ln\left(\left|\frac{a+\sqrt{a^2-x^2}}{x}\right|\right) + C$

Integrals Involving $\sqrt{x^2-a^2}$

109. $\displaystyle\int \frac{1}{(x^2-a^2)^{3/2}}\,dx = -\frac{x}{a^2\sqrt{x^2-a^2}} + C$

110. $\int \frac{1}{\sqrt{x^2 - a^2}}\, dx = \ln\left(x + \sqrt{x^2 - a^2}\right) + C$

111. $\int \sqrt{x^2 - a^2}\, dx = \frac{1}{2} x\sqrt{x^2 - a^2} - \frac{1}{2} a^2 \ln\left(x + \sqrt{x^2 - a^2}\right) + C$

112. $\int (x^2 - a^2)^{3/2}\, dx = \frac{1}{4} x(x^2 - a^2)^{3/2} - \frac{3}{8} a^2 x\sqrt{x^2 - a^2} + \frac{3}{8} a^4 \ln\left(x + \sqrt{x^2 - a^2}\right) + C$

113. $\int \frac{x^2}{(x^2 - a^2)^{3/2}}\, dx = -\frac{x}{\sqrt{x^2 - a^2}} + \ln\left(x + \sqrt{x^2 - a^2}\right) + C$

114. $\int \frac{x^2}{\sqrt{x^2 - a^2}}\, dx = \frac{1}{2} x\sqrt{x^2 - a^2} + \frac{1}{2} a^2 \ln\left(x + \sqrt{x^2 - a^2}\right) + C$

115. $\int x^2\sqrt{x^2 - a^2}\, dx = \frac{1}{4} x(x^2 - a^2)^{3/2} + \frac{1}{8} a^2 x\sqrt{x^2 - a^2} - \frac{1}{8} a^4 \ln\left(x + \sqrt{x^2 - a^2}\right) + C$

116. $\int \frac{1}{x(x^2 - a^2)^{3/2}}\, dx = -\frac{1}{a^3} \operatorname{arcsec}\left(\frac{x}{a}\right) - \frac{1}{a^2\sqrt{x^2 - a^2}} + C, \quad 0 < x$

117. $\int \frac{1}{x\sqrt{x^2 - a^2}}\, dx = \frac{1}{a} \operatorname{arcsec}\left(\frac{x}{a}\right) + C, \quad 0 < x$

118. $\int \frac{\sqrt{x^2 - a^2}}{x}\, dx = \sqrt{x^2 - a^2} - a \operatorname{arcsec}\left(\frac{x}{a}\right) + C, \quad 0 < x$

119. $\int \frac{(x^2 - a^2)^{3/2}}{x}\, dx = \frac{1}{3}(x^2 - a^2)^{3/2} - a^2\sqrt{x^2 - a^2} + a^3 \operatorname{arcsec}\left(\frac{x}{a}\right) + C, \quad 0 < x$

Integrals Involving sin(x)

120. $\int \sin(x)\, dx = -\cos(x) + C$

121. $\int \sin^2(x)\, dx = \frac{1}{2} x - \frac{1}{2} \cos(x)\sin(x) + C$

122. $\int \sin^3(x)\, dx = -\frac{1}{3} \sin^2(x)\cos(x) - \frac{2}{3} \cos(x) + C$

123. $\int \sin^3(x)\, dx = \frac{1}{3} \cos^3(x) - \cos(x) + C$

124. $\int \sin^4(x)\, dx = -\frac{1}{4} \sin^3(x)\cos(x) - \frac{3}{8} \cos(x)\sin(x) + \frac{3}{8} x + C$

125. $\int x\sin(x)\, dx = \sin(x) - x\cos(x) + C$

126. $\int x^2\sin(x)\, dx = 2x\sin(x) - (x^2 - 2)\cos(x) + C$

127. $\int x^3\sin(x)\, dx = 3(x^2 - 2)\sin(x) - (x^3 - 6x)\cos(x) + C$

128. $\int \frac{1}{1 + \sin(x)}\, dx = -\frac{2}{1 + \tan(x/2)} + C$

129. $\int \frac{1}{1 - \sin(x)}\, dx = \frac{2}{1 - \tan(x/2)} + C$

130. $\int \frac{x}{1 + \sin(x)}\, dx = \ln\left(1 + \sin(x)\right) - \frac{x\cos(x)}{1 + \sin(x)} + C$

131. $\int \frac{x}{1 - \sin(x)}\, dx = \ln\left(1 - \sin(x)\right) + \frac{x\cos(x)}{1 - \sin(x)} + C$

132. $\int \frac{1}{a + b\sin(x)}\, dx = \frac{2}{\sqrt{a^2 - b^2}} \arctan\left(\frac{a\tan(x/2) + b}{\sqrt{a^2 - b^2}}\right) + C, \quad b < a$

133. $\displaystyle\int \frac{1}{a+b\sin(x)}\,dx = \frac{1}{\sqrt{b^2-a^2}}\ln\left(\left|\frac{a\tan(x/2)+b-\sqrt{b^2-a^2}}{a\tan(x/2)+b+\sqrt{b^2-a^2}}\right|\right)+C, \qquad a<b$

134. $\displaystyle\int \frac{1}{1+\sin^2(x)}\,dx = \frac{1}{\sqrt{2}}\arctan\left(\sqrt{2}\tan(x)\right)+C$

135. $\displaystyle\int \frac{1}{a^2-b^2\sin^2(x)}\,dx = \frac{1}{a\sqrt{a^2-b^2}}\arctan\left(\frac{\sqrt{a^2-b^2}\tan(x)}{a}\right)+C, \qquad b<a$

136. $\displaystyle\int \frac{1}{a^2-b^2\sin^2(x)}\,dx = \frac{1}{2a\sqrt{b^2-a^2}}\ln\left(\left|\frac{\sqrt{b^2-a^2}\tan(x)+a}{\sqrt{b^2-a^2}\tan(x)-a}\right|\right)+C, \qquad a<b$

137. $\displaystyle\int \sin(ax)\sin(bx)\,dx = \frac{1}{2(a-b)}\sin((a-b)x)-\frac{1}{2(a+b)}\sin((a+b)x)+C, \qquad a\neq b$

138. $\displaystyle\int \sin(ax^2)\,dx = \sqrt{\frac{\pi}{2a}}\,\text{FresnelS}\left(\sqrt{\frac{2a}{\pi}}x\right)+C$

139. $\displaystyle\int \sqrt{x}\sin(ax)\,dx = -\frac{1}{a}\sqrt{x}\cos(ax)+\sqrt{\frac{\pi}{2a^3}}\,\text{FresnelC}\left(\sqrt{\frac{2a}{\pi}}x\right)+C$

140. $\displaystyle\int \frac{\sin(ax)}{\sqrt{x}}\,dx = \sqrt{\frac{2\pi}{a}}\,\text{FresnelS}\left(\sqrt{\frac{2a}{\pi}}x\right)+C$

141. $\displaystyle\int \frac{\sin(ax)}{x}\,dx = \text{Si}(ax)+C$

Integrals Involving cos(*x*)

142. $\int \cos(x)\,dx = \sin(x)+C$

143. $\displaystyle\int \cos^2(x)\,dx = \frac{1}{2}x+\frac{1}{2}\cos(x)\sin(x)+C$

144. $\displaystyle\int \cos^3(x)\,dx = \frac{1}{3}\cos^2(x)\sin(x)+\frac{2}{3}\sin(x)+C$

145. $\displaystyle\int \cos^3(x)\,dx = \sin(x)-\frac{1}{3}\sin^3(x)+C$

146. $\displaystyle\int \cos^4(x)\,dx = \frac{3}{8}x+\frac{1}{4}\cos^3(x)\sin(x)+\frac{3}{8}\cos(x)\sin(x)+C$

147. $\int x\cos(x)\,dx = \cos(x)+x\sin(x)+C$

148. $\int x^2\cos(x)\,dx = (x^2-2)\sin(x)+2x\cos(x)+C$

149. $\int x^3\cos(x)\,dx = (x^3-6x)\sin(x)+(3x^2-6)\cos(x)+C$

150. $\displaystyle\int \frac{1}{1+\cos(x)}\,dx = \tan\left(\frac{x}{2}\right)+C$

151. $\displaystyle\int \frac{1}{1-\cos(x)}\,dx = -\cot\left(\frac{x}{2}\right)+C$

152. $\displaystyle\int \frac{x}{1+\cos(x)}\,dx = x\tan\left(\frac{x}{2}\right)+2\ln\left(\left|\cos\left(\frac{x}{2}\right)\right|\right)+C$

153. $\displaystyle\int \frac{x}{1-\cos(x)}\,dx = -x\cot\left(\frac{x}{2}\right)+2\ln\left(\left|\sin\left(\frac{x}{2}\right)\right|\right)+C$

154. $\displaystyle\int \frac{1}{a+b\cos(x)}\,dx = \frac{2}{\sqrt{a^2-b^2}}\arctan\left(\frac{(a-b)\tan(x/2)}{\sqrt{a^2-b^2}}\right)+C, \qquad b<a$

155. $\displaystyle\int \frac{1}{a+b\cos(x)}\,dx = \frac{1}{\sqrt{b^2-a^2}}\ln\left(\left|\frac{(b-a)\tan(x/2)+\sqrt{b^2-a^2}}{(b-a)\tan(x/2)-\sqrt{b^2-a^2}}\right|\right)+C, \qquad a<b$

156. $\displaystyle\int \frac{1}{1+\cos^2(x)}\,dx = \frac{1}{\sqrt{2}}\arctan\left(\frac{1}{\sqrt{2}}\tan(x)\right)+C$

157. $\displaystyle\int \frac{1}{a^2-b^2\cos^2(x)}\,dx = \frac{1}{a\sqrt{a^2-b^2}}\arctan\left(\frac{a\tan(x)}{\sqrt{a^2-b^2}}\right)+C, \qquad b<a$

158. $\displaystyle\int \frac{1}{a^2-b^2\cos^2(x)}\,dx = \frac{1}{2a\sqrt{b^2-a^2}}\ln\left(\left|\frac{a\tan(x)-\sqrt{b^2-a^2}}{a\tan(x)+\sqrt{b^2-a^2}}\right|\right)+C, \qquad a<b$

159. $\displaystyle\int \cos(ax)\cos(bx)\,dx = \frac{1}{2(a-b)}\sin\big((a-b)x\big)+\frac{1}{2(a+b)}\sin\big((a+b)x\big)+C, \qquad a\neq b$

160. $\displaystyle\int \cos(ax^2)\,dx = \sqrt{\frac{\pi}{2a}}\,\mathrm{FresnelC}\left(\sqrt{\frac{2a}{\pi}}x\right)+C$

161. $\displaystyle\int \sqrt{x}\cos(ax)\,dx = \frac{1}{a}\sqrt{x}\sin(ax)-\sqrt{\frac{\pi}{2a^3}}\,\mathrm{FresnelS}\left(\sqrt{\frac{2a}{\pi}}x\right)+C$

162. $\displaystyle\int \frac{\cos(ax)}{\sqrt{x}}\,dx = \sqrt{\frac{2\pi}{a}}\,\mathrm{FresnelC}\left(\sqrt{\frac{2a}{\pi}}x\right)+C$

163. $\displaystyle\int \frac{\cos(ax)}{x}\,dx = \mathrm{Ci}(ax)+C$

Integrals Involving sin(x) and cos(x)

164. $\displaystyle\int \sin^2(x)\cos^2(x)\,dx = \frac{1}{8}x-\frac{1}{4}\cos^3(x)\sin(x)+\frac{1}{8}\cos(x)\sin(x)+C$

165. $\displaystyle\int \sin^2(ax)\cos^2(bx)\,dx = \frac{1}{8}\left(2x-\frac{1}{a}\sin(2ax)+\frac{1}{b}\sin(2bx)-\frac{1}{2(a-b)}\sin\big(2(a-b)x\big)-\frac{1}{2(a+b)}\sin\big(2(a+b)x\big)\right)$
$\displaystyle +C, \quad a\neq b$

166. $\displaystyle\int \frac{1}{\sin(x)\cos(x)}\,dx = \ln\big(|\tan(x)|\big)+C$

167. $\displaystyle\int \frac{1}{\sin^2(x)\cos(x)}\,dx = -\csc(x)+\ln\big|\big(\sec(x)+\tan(x)\big)\big|+C$

168. $\displaystyle\int \frac{1}{\sin(x)\cos^2(x)}\,dx = \sec(x)-\ln\big(|\csc(x)+\cot(x)|\big)+C$

169. $\displaystyle\int \frac{1}{\sin^2(x)\cos^2(x)}\,dx = \sec(x)\csc(x)-2\cot(x)+C$

170. $\displaystyle\int \frac{1}{\sin^2(x)\cos^2(x)}\,dx = \tan(x)-\cot(x)+C$

171. $\displaystyle\int \frac{1}{\sin^3(x)\cos(x)}\,dx = -\frac{1}{2}\csc^2(x)+\ln\big(|\tan(x)|\big)+C$

172. $\displaystyle\int \frac{1}{\sin(x)\cos^3(x)}\,dx = \frac{1}{2}\sec^2(x)+\ln\big(|\tan(x)|\big)+C$

173. $\displaystyle\int \frac{\sin(x)}{\cos(x)}\,dx = \ln\big(|\sec(x)|\big)+C$

174. $\displaystyle\int \frac{\sin^2(x)}{\cos(x)}\,dx = -\sin(x)+\ln\big(|\sec(x)+\tan(x)|\big)+C$

175. $\displaystyle\int \frac{\sin^3(x)}{\cos(x)}\,dx = -\frac{1}{2}\sin^2(x) - \ln\left(|\cos(x)|\right) + C$

176. $\displaystyle\int \frac{\cos(x)}{\sin(x)}\,dx = \ln\left(|\sin(x)|\right) + C$

177. $\displaystyle\int \frac{\cos^2(x)}{\sin(x)}\,dx = \cos(x) - \ln\left(|\csc(x) + \cot(x)|\right) + C$

178. $\displaystyle\int \frac{\cos^3(x)}{\sin(x)}\,dx = \frac{1}{2}\cos^2(x) + \ln\left(|\sin(x)|\right) + C$

179. $\displaystyle\int \frac{\sin^2(x)}{\cos^2(x)}\,dx = \tan(x) - x + C$

180. $\displaystyle\int \frac{\sin^3(x)}{\cos^2(x)}\,dx = \cos(x) + \sec(x) + C$

181. $\displaystyle\int \frac{\cos^2(x)}{\sin^2(x)}\,dx = -\cot(x) - x + C$

182. $\displaystyle\int \frac{\cos^3(x)}{\sin^2(x)}\,dx = -\sin(x) - \csc(x) + C$

183. $\displaystyle\int \frac{1}{a^2\cos^2(x) + b^2\sin^2(x)}\,dx = \frac{1}{ab}\arctan\left(\frac{b}{a}\tan(x)\right) + C$

184. $\displaystyle\int \frac{1}{a^2\cos^2(x) - b^2\sin^2(x)}\,dx = \frac{1}{2ab}\ln\left(\left|\frac{b\tan(x) + a}{b\tan(x) - a}\right|\right) + C$

Integrals Involving tan(*x*)

185. $\int \tan(x)\,dx = \ln\left(|\sec(x)|\right) + C$

186. $\int \tan^2(x)\,dx = \tan(x) - x + C$

187. $\displaystyle\int \tan^3(x)\,dx = \frac{1}{2}\tan^2(x) + \ln\left(|\cos(x)|\right) + C$

188. $\displaystyle\int \tan^4(x)\,dx = \frac{1}{3}\tan^3(x) - \tan(x) + x + C$

Integrals Involving cot(*x*)

189. $\int \cot(x)\,dx = -\ln\left(|\csc(x)|\right) + C$

190. $\int \cot^2(x)\,dx = -\cot(x) - x + C$

191. $\displaystyle\int \cot^3(x)\,dx = -\frac{1}{2}\cot^2(x) + \ln\left(|\csc(x)|\right) + C$

192. $\displaystyle\int \cot^4(x)\,dx = -\frac{1}{3}\cot^3(x) + \cot(x) + x + C$

Integrals Involving sec(*x*)

193. $\int \sec(x)\,dx = \ln\left(|\sec(x) + \tan(x)|\right) + C$

194. $\int \sec^2(x)\,dx = \tan(x) + C$

195. $\int \sec^3(x)\,dx = \frac{1}{2}\sec(x)\tan(x) + \frac{1}{2}\ln(|\sec(x) + \tan(x)|) + C$

196. $\int \sec^4(x)\,dx = \frac{1}{3}\tan^3(x) + \tan(x) + C$

Integrals Involving tan(x) and sec(x)

197. $\int \sec(x)\tan(x)\,dx = \sec(x) + C$

198. $\int \sec^2(x)\tan(x)\,dx = \frac{1}{2}\sec^2(x) + C$

199. $\int \sec(x)\tan^2(x)\,dx = \frac{1}{2}\sec(x)\tan(x) - \frac{1}{2}\ln(|\sec(x) + \tan(x)|) + C$

200. $\int \sec^2(x)\tan^2(x)\,dx = \frac{1}{3}\tan^3(x) + C$

201. $\int \sec^3(x)\tan(x)\,dx = \frac{1}{3}\sec^3(x) + C$

202. $\int \sec(x)\tan^3(x)\,dx = \frac{1}{3}\sec^3(x) - \sec(x) + C$

Integrals Involving csc(x)

203. $\int \csc(x)\,dx = -\ln(|\csc(x) + \cot(x)|) + C$

204. $\int \csc^2(x)\,dx = -\cot(x) + C$

205. $\int \csc^3(x)\,dx = -\frac{1}{2}\csc(x)\cot(x) - \frac{1}{2}\ln(|\csc(x) + \cot(x)|) + C$

206. $\int \csc^4(x)\,dx = -\frac{1}{3}\cot^3(x) - \cot(x) + C$

Integrals Involving cot(x) and csc(x)

207. $\int \csc(x)\cot(x)\,dx = -\csc(x) + C$

208. $\int \csc^2(x)\cot(x)\,dx = -\frac{1}{2}\csc^2(x) + C$

209. $\int \csc(x)\cot^2(x)\,dx = -\frac{1}{2}\cot(x)\csc(x) + \frac{1}{2}\ln(|\csc(x) + \cot(x)|) + C$

210. $\int \csc^2(x)\cot^2(x)\,dx = -\frac{1}{3}\cot^3(x) + C$

211. $\int \csc^3(x)\cot(x)\,dx = -\frac{1}{3}\csc^3(x) + C$

212. $\int \csc(x)\cot^3(x)\,dx = -\frac{1}{3}\csc^3(x) + \csc(x) + C$

Integrals Involving e^x

213. $\int e^{bx}\,dx = \frac{1}{b}e^{bx} + C$

214. $\int a^x\,dx = \frac{1}{\ln(a)}a^x + C, \qquad a \neq 1$ (preceding formula with $b = \ln(a)$)

215. $\int xe^{ax}\,dx = \frac{1}{a}\left(x - \frac{1}{a}\right)e^{ax} + C$

216. $\int x^2e^{ax}\,dx = \frac{1}{a^3}(a^2x^2 - 2ax + 2)e^{ax} + C$

217. $\int \frac{1}{a + b\exp(x)}\,dx = \frac{1}{a}\ln\left(\frac{\exp(x)}{a + b\exp(x)}\right) + C$

218. $\int \exp(-ax^2)\,dx = \frac{1}{2}\operatorname{erf}(\sqrt{a}x) + C$

219. $\int_{-\infty}^{\infty} \exp(-ax^2)\,dx = \sqrt{\frac{\pi}{a}}$

220. $\int \sqrt{x}\exp(-ax)\,dx = -\frac{1}{a}\sqrt{x}\exp(-ax) + \frac{1}{2}\sqrt{\frac{\pi}{a^3}}\operatorname{erf}(\sqrt{ax}) + C$

221. $\int_0^{\infty} \sqrt{x}\exp(-ax)\,dx = \frac{1}{2}\sqrt{\frac{\pi}{a^3}}$

222. $\int \frac{1}{\sqrt{x}}\exp(-ax)\,dx = \sqrt{\frac{\pi}{a}}\operatorname{erf}(\sqrt{ax}) + C$

223. $\int_0^{\infty} \frac{1}{\sqrt{x}}\exp(-ax)\,dx = \sqrt{\frac{\pi}{a}}$

224. $\int_0^{\infty} x^n\exp(-ax)\,dx = \frac{n!}{a^{n+1}} \quad (n = 1, 2, 3, \ldots)$

225. $\int x^2\exp(-ax^2)\,dx = -\frac{1}{2a}x\exp(-ax^2) + \frac{1}{4}\sqrt{\frac{\pi}{a^3}}\operatorname{erf}(\sqrt{a}x) + C$

226. $\int e^{ax}\sin(bx)\,dx = \frac{1}{a^2 + b^2}\big(a\sin(bx) - b\cos(bx)\big)e^{ax} + C$

227. $\int e^{ax}\cos(bx)\,dx = \frac{1}{a^2 + b^2}\big(a\cos(bx) + b\sin(bx)\big)e^{ax} + C$

Integrals Involving ln(*x*)

228. $\int \ln(x)\,dx = x\ln(x) - x + C$

229. $\int \ln(ax)\,dx = x\ln(ax) - x + C$

230. $\int x\ln(ax)\,dx = \frac{1}{2}x^2\ln(ax) - \frac{1}{4}x^2 + C$

231. $\int x^2\ln(ax)\,dx = \frac{1}{3}x^3\ln(ax) - \frac{1}{9}x^3 + C$

232. $\int x^b\ln(ax)\,dx = \frac{1}{b+1}x^{b+1}\ln(ax) - \frac{1}{(b+1)^2}x^{b+1} + C$

233. $\int \frac{\ln(ax)}{x^b}\,dx = -\frac{1}{b-1}\frac{\ln(ax)}{x^{b-1}} - \frac{1}{(b-1)^2}\frac{1}{x^{b-1}} + C$

234. $\int \ln^2(ax)\,dx = x\ln^2(ax) - 2x\ln(ax) + 2x + C$

235. $\int \ln(a + bx)\,dx = \frac{1}{b}(a + bx)\ln(a + bx) - x + C$

236. $\int x\ln(a + bx)\,dx = -\frac{(a^2 - b^2x^2)}{2b^2}\ln(a + bx) + \frac{a}{2b}x - \frac{1}{4}x^2 + C$

237. $\int \ln(a^2 + x^2)\,dx = x\ln(a^2 + x^2) - 2x + 2a\arctan\left(\frac{x}{a}\right) + C$

238. $\int x\ln(a^2 + x^2)\,dx = \frac{1}{2}(a^2 + x^2)\ln(a^2 + x^2) - \frac{1}{2}(a^2 + x^2) + C$

239. $\int x^2\ln(a^2 + x^2)\,dx = \frac{1}{3}x^3\left(\ln(a^2 + x^2) - \frac{2}{3}\right) + \frac{2a^2}{3}x - \frac{2a^3}{3}\arctan\left(\frac{x}{a}\right) + C$

Integrals Involving Special Functions

240. $\int \mathrm{FresnelC}(x)\,dx = x\,\mathrm{FresnelC}(x) - \frac{1}{\pi}\sin\left(\frac{1}{2}\pi x^2\right) + C$

241. $\int \mathrm{FresnelS}(x)\,dx = x\,\mathrm{FresnelS}(x) + \frac{1}{\pi}\cos\left(\frac{1}{2}\pi x^2\right) + C$

242. $\int \mathrm{erf}(x)\,dx = x\,\mathrm{erf}(x) + \frac{1}{\sqrt{\pi}}\exp(-x^2) + C$

243. $\int x\,\mathrm{erf}(x)\,dx = \frac{1}{2}x^2\mathrm{erf}(x) + \frac{1}{2\sqrt{\pi}}x\exp(-x^2) - \frac{1}{4}\mathrm{erf}(x) + C$

244. $\int \mathrm{erf}(\sqrt{x})\,dx = x\,\mathrm{erf}(\sqrt{x}) + \frac{1}{\sqrt{\pi}}\sqrt{x}e^{-x} - \frac{1}{2}\mathrm{erf}(\sqrt{x}) + C$

245. $\int \mathrm{Si}(x)\,dx = \mathrm{Si}(x) + \cos(x) + C$

246. $\int \mathrm{Ci}(x)\,dx = x\,\mathrm{Ci}(x) - \sin(x) + C$

Answers to Selected Exercises

Section 1.1

1. 213/100 **3.** 23/99

5. 4996/999 **7.** 0.025

9. $1.\overline{6}$ (The overline identifies the repeating block.)

11. 0.72 **13.** $[1, 3]$

15. $(1, \infty)$ **17.** $[-14, 6]$

19.
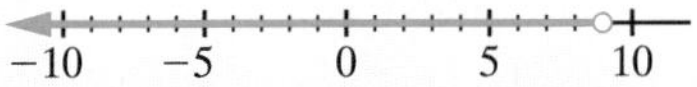

21.
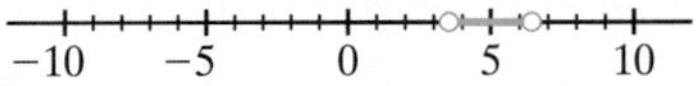

23.
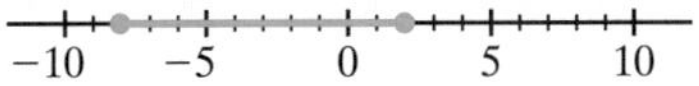

25. $\{x : |x - 1| \leq 2\}$

27. $\{x : |x - 1| < \pi + 1\}$

31. Both approximations are rational numbers, but π is not.

33. $[34.155, 34.845]$; $\{x : |x - 34.5| \leq 0.345\}$

35.

37.
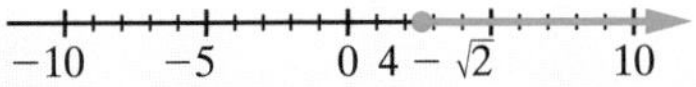

39.
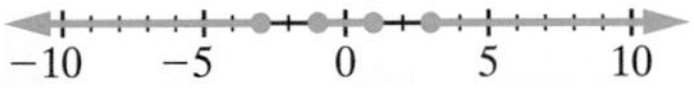

41.
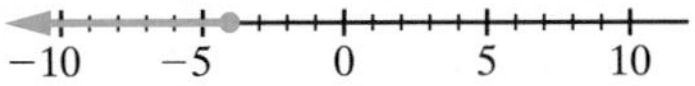

43.
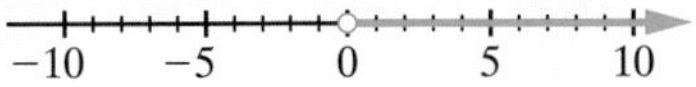

45. $\{-3\}$

47. $\{x : 1 < x \leq 2\sqrt{2}\}$

49. $\{s : -13 < s < -5/3\}$

51. $\{w : |w + 1/2| < 1/2\}$

59. $[0.4485, 0.4495]$

61. $[0.000499, 0.001499]$

63. 4.001

65. Calculated values of $\alpha \cdot \beta$ using $N = 12$, 14, 16, 18, and 20 digits are tabulated:

N	$\alpha \cdot \beta$
12	34562960.8
14	−345629.6
16	0.000000
18	0.000000
20	1.0369

67.

x	**Relative Error**
10^{10}	0.5
10^{12}	9.901×10^{-3}
10^{14}	9.999×10^{-5}
10^{16}	9.99999×10^{-7}
10^{18}	9.9999999×10^{-9}

x	Relative Error
-10^{12}	0.0101
-10^{14}	1.0001×10^{-4}
-10^{16}	1.000001×10^{-6}
-10^{18}	$1.00000001 \times 10^{-8}$
-10^{20}	$1.000000000 \times 10^{-10}$

69.

x	Relative Error
10^3	1.031
10^6	0.0345
10^9	0.00003
10^{12}	0.3×10^{-7}
10^{15}	0.3×10^{-10}

x	Relative Error
-10^3	0.9709
-10^6	0.0322
-10^9	0.00003
-10^{12}	0.3×10^{-7}
-10^{15}	0.3×10^{-10}

Section 1.2

1.

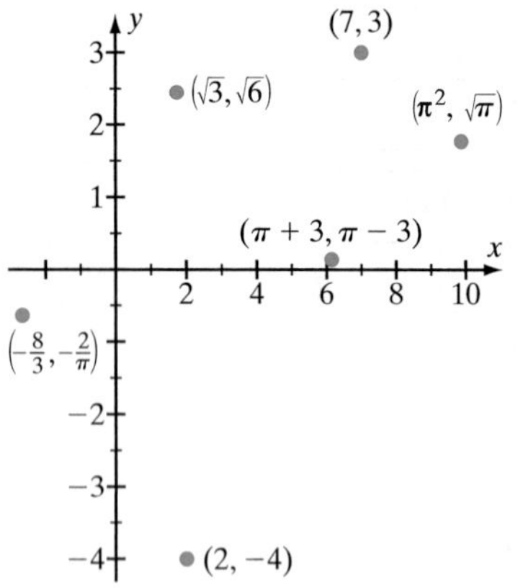

3.

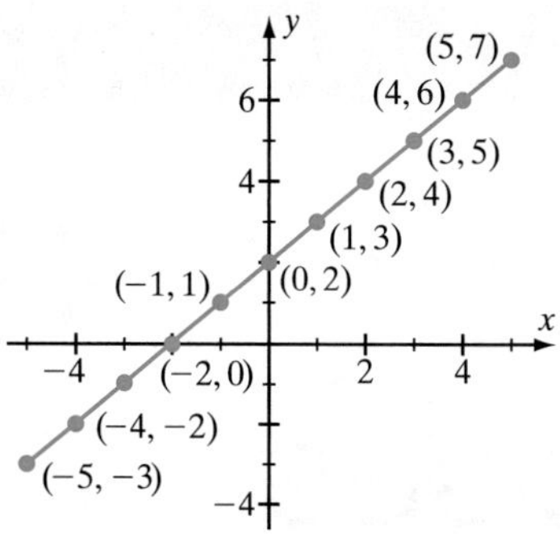

5. $|\overline{AB}| = 2\sqrt{13}, |\overline{AC}| = \sqrt{130}, |\overline{BC}| = \sqrt{170}$

7. Center: $(1, 3)$; radius: 3

9. Center: $(0, -5)$; radius: $\sqrt{2}$

11. Center: $(0, 1/2)$; radius: $1/2$

13. Center: $(-2, 1)$; radius: $\sqrt{17/3}$

15. Center: $(4, -1)$; radius: $\sqrt{17}$

17. $(x + 3)^2 + (y - 5)^2 = 36$

19. $(x + 4)^2 + (y - \pi)^2 = 25$

21. Vertex: $(0, -3)$; axis of symmetry: $x = 0$

23. Vertex: $(1, 1)$; axis of symmetry: $x = 1$

25. Vertex: $(-3, 9/2)$; axis of symmetry: $x = -3$

27. Ellipse; standard form: $x^2/(1/2)^2 + y^2 = 1$; center: $(0, 0)$

29. Ellipse; standard form: $(x + 1/2)^2/2^2 + y^2/(2/3)^2 = 1$; center: $(-1/2, 0)$

31. Ellipse; standard form: $(x + 1)^2/7^2 + (y + 3)^2/(7/2)^2 = 1$; center: $(-1, -3)$

33. Hyperbola; standard form: $x^2/(10/\sqrt{2})^2 - (y + 1)^2/(10/\sqrt{3})^2 = 1$; center: $(0, -1)$

35.

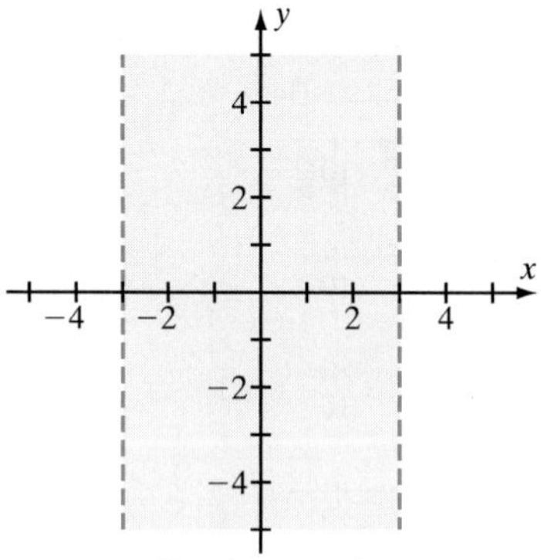

$\{(x, y) : |x| < 3\}$

37.

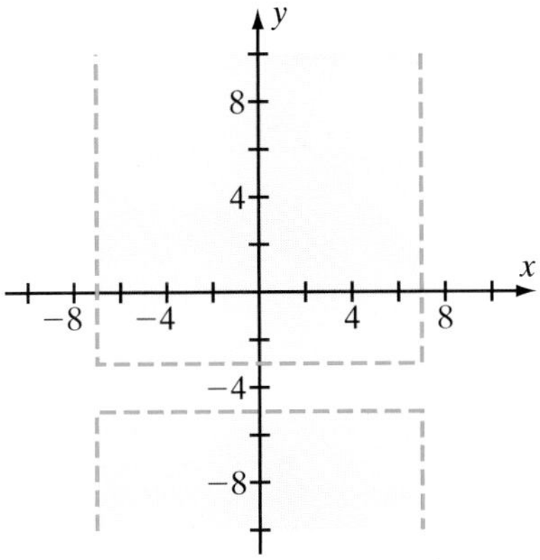

$\{(x, y) : |x| < 7, |y + 4| > 1\}$

39.

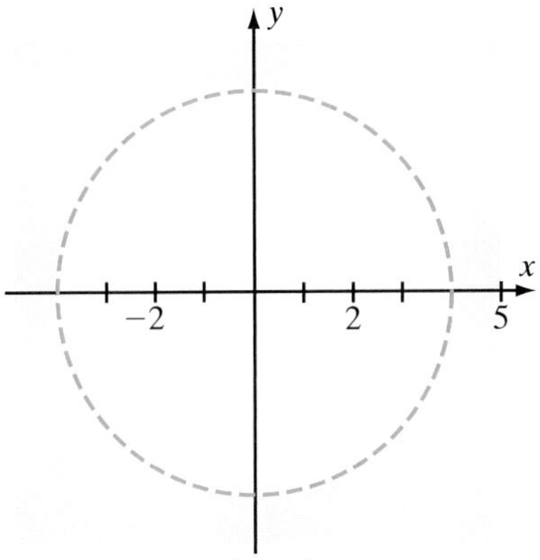

$\{(x, y) : x^2 + y^2 > 16\}$

41.

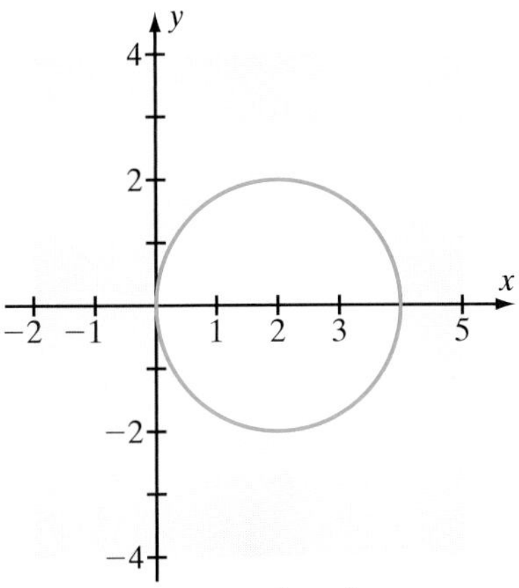

$\{(x, y) : (x - 2)^2 + y^2 \geq 4\}$

43.

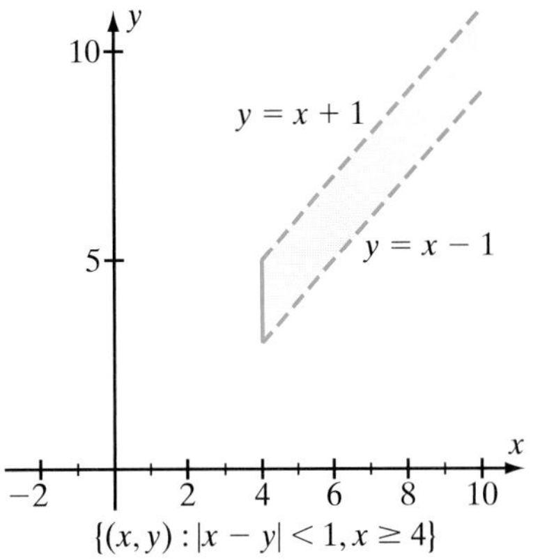

$\{(x, y) : |x - y| < 1, x \geq 4\}$

45.

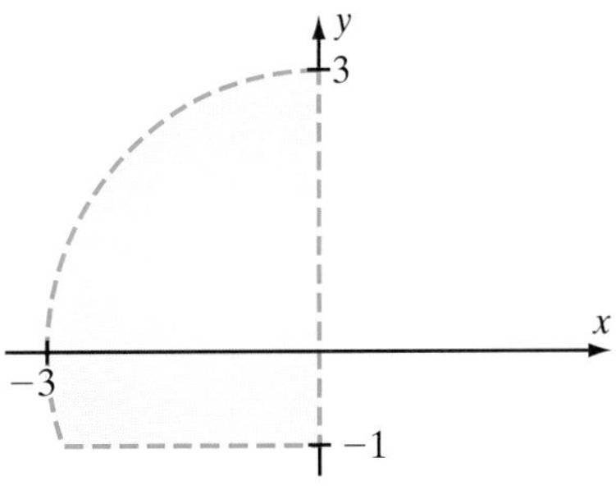

$\{(x, y) : x^2 + y^2 < 9, x < 0, y > -1\}$

47. $(3, -1)$ **49.** $(0, 0)$

51. Yes: It is a circle centered at $(1/5, -3/10)$ with radius $\sqrt{133}/10$.

53. No: It is the equation of a hyperbola.

55. $81/2$ **57.** $1/2$

59. $a = -7$, $b = 1$

61.

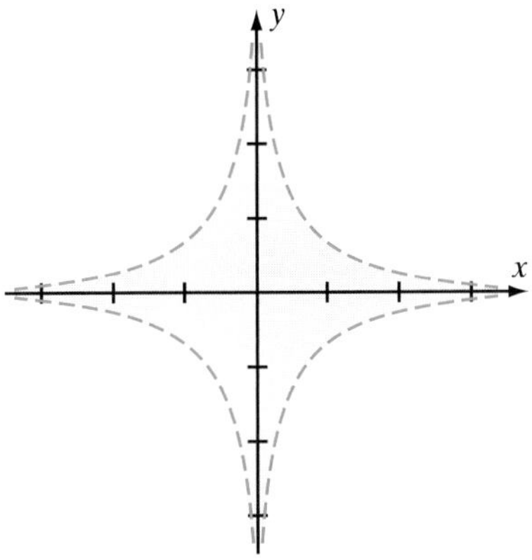

$\{(x, y) : |x| \cdot |y| < 1\}$

63.

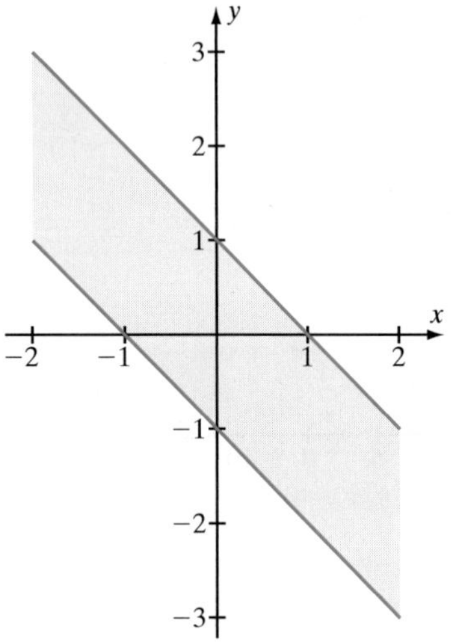

$\{(x, y) : |x + y| \leq 1\}$

65.

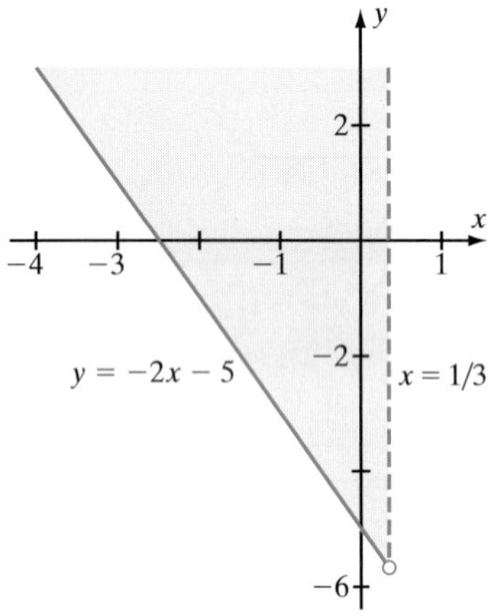

$\{(x, y) : 3x - 1 < 0 \leq y + 2x + 5\}$

67.

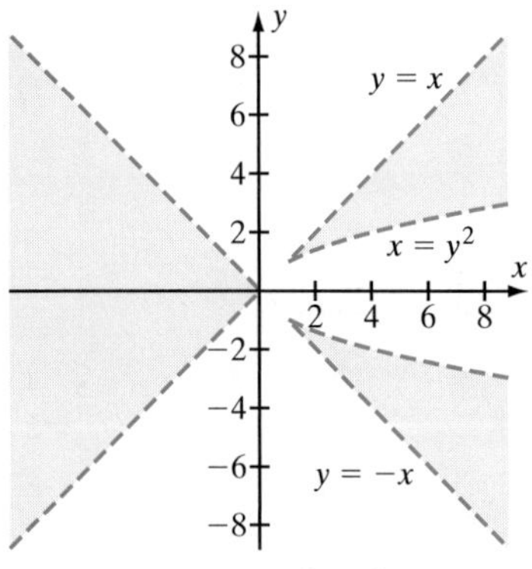

$\{(x, y) : x < y^2 < x^2\}$

69.

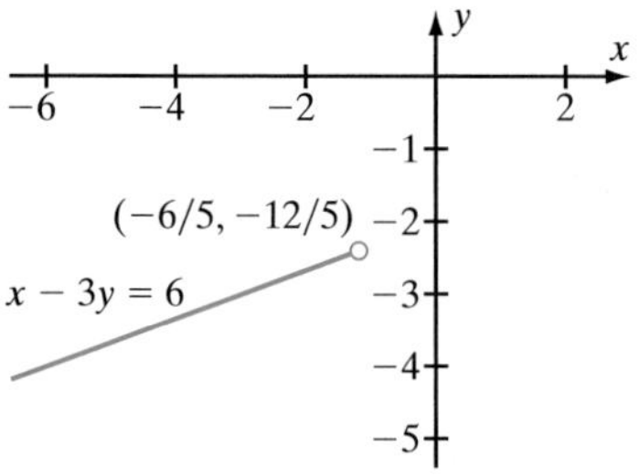

$\{(x, y) : y > 2x, x - 3y = 6\}$

73. $(x + 1)^2 + y^2 = 5^2$

75.

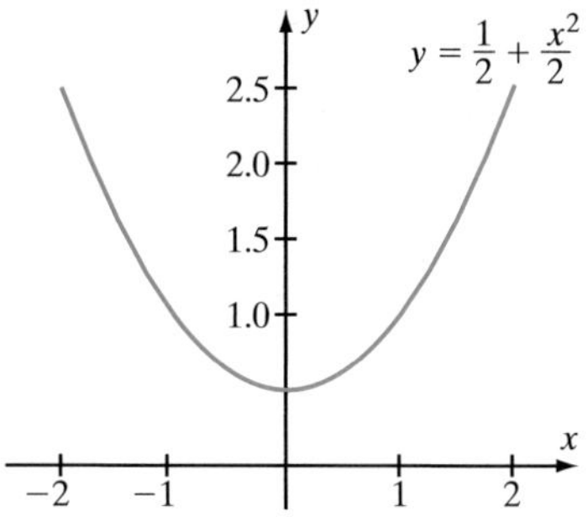

77.

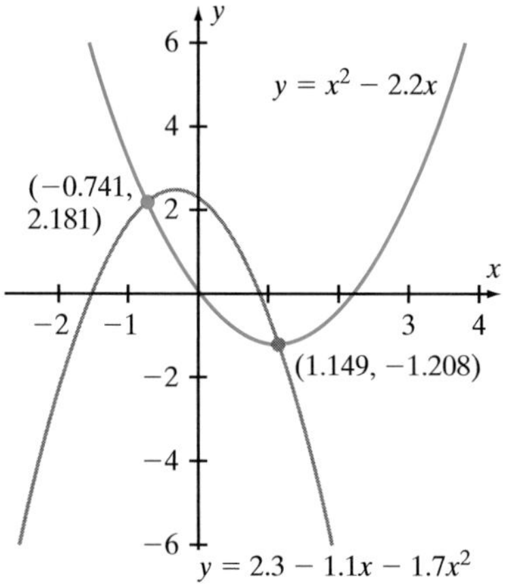

79.

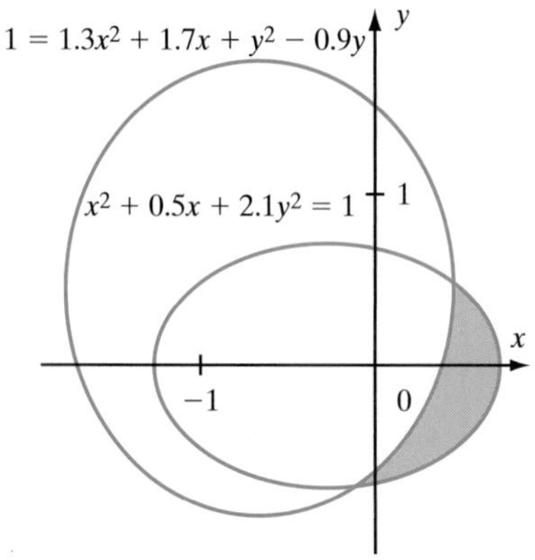

81.

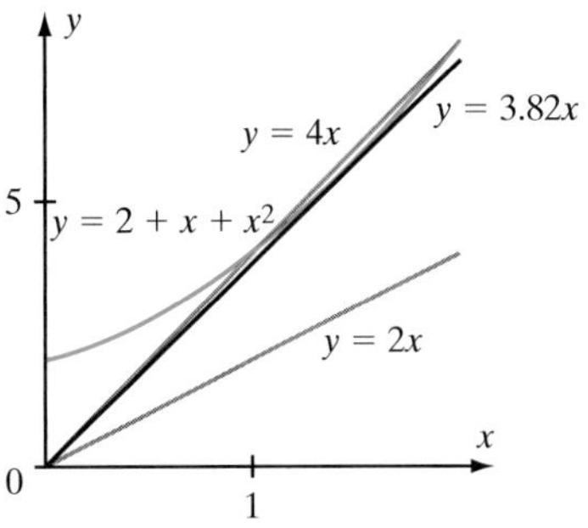

Section 1.3

1.

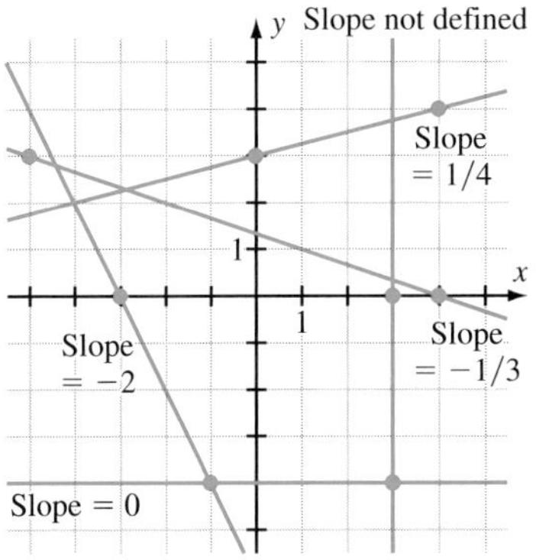

3.

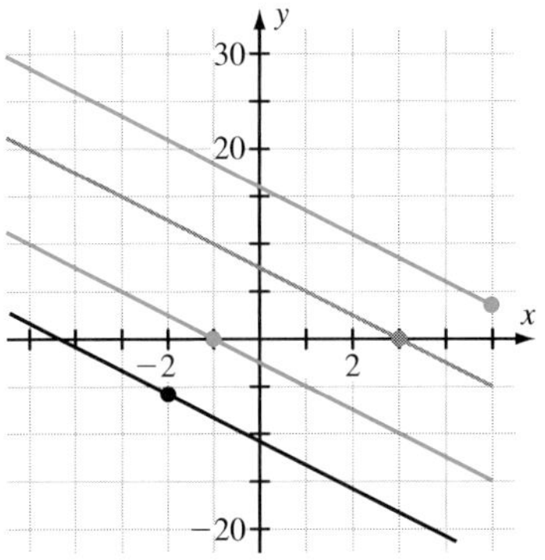

5. $y = 5(x + 3) + 7$

7. $y = -(x + \sqrt{2})$

9. $y = (-11/4)(x - 2) + 7$ or $y = (-11/4)(x - 6) - 4$

11. $y = (1/4)(x + 7) - 4$ or $y = (1/4)(x - 9)$

13. $y = -4x + 9$

15. $y = \sqrt{2}x - \sqrt{3}$

17. $y = 3x + 12$

19. $y = 3x + 6$

21. $y = 3x + 1$

23. $y = -4x - 1$

25. $x/(-2) + y/6 = 1$

27. $x/(-1) + y/3 = 1$

29. $y = (-1/2)x - 3/2$

31. $y = 2x - 3$

33. $y = 2x - 4$

35. $y = (3/2)x - 1/2$

37. x-intercept: 2/3; y-intercept: 1/2; slope: −3/4

39. x-intercept: 3; y-intercept: −1; slope: 1/3

41.

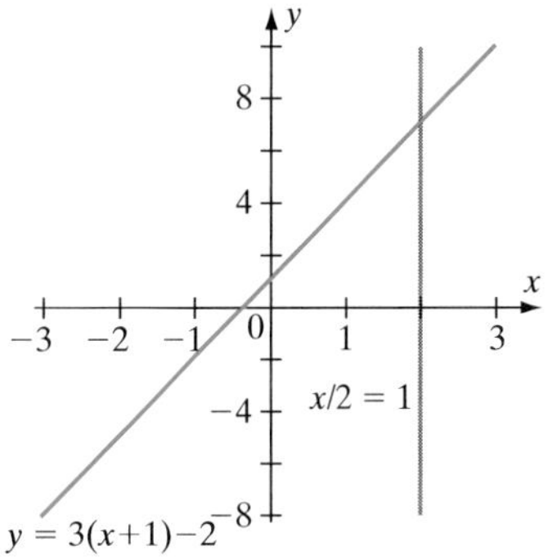

43. The figure for 41 has this plot.

45.

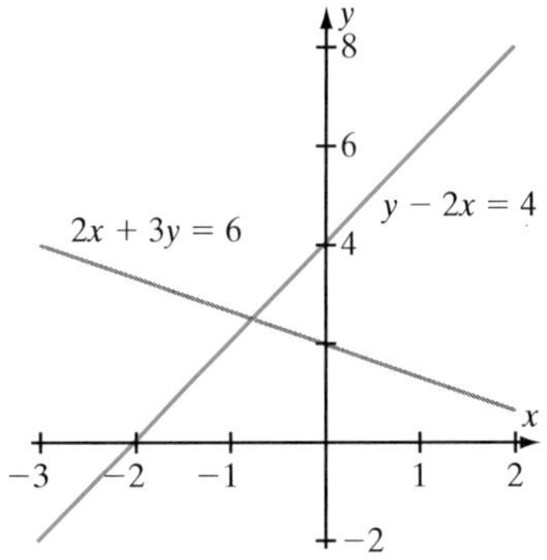

47. The figure for 45 has this plot.

49. No; no

51. $y = 3x - 29$

53. All points $(a, 2a - 17)$ with $a \neq 5$; for example, $(0, -17)$

55. All points $(a, 1 - 3a)$ with $a \neq -2$; for example, $(1, -2)$

57. One example is $(a, b) = (1, 0)$ and $(c, d) = (3, 4)$.

59. $x = 1$, $y = 2$

61. $x = -4$, $y = 6$

63. (3, 1)

67. −1999997 and 2000003

69. Equation of budget line: $x \cdot p_X + y \cdot p_Y = C$, $0 \leq x$, and $0 \leq y$; x-intercept: C/p_X; y-intercept: C/p_Y; slope: $-p_X/p_Y$. The budget line L' that corresponds to C' is parallel to the budget line L that corresponds to C. L' lies above L if $C' > C$; L' lies below L if $C' < C$.

75. Slope ≈ (1.002001 − .998001)/(1.001 − .999), or 2.0 (See accompanying figure.)

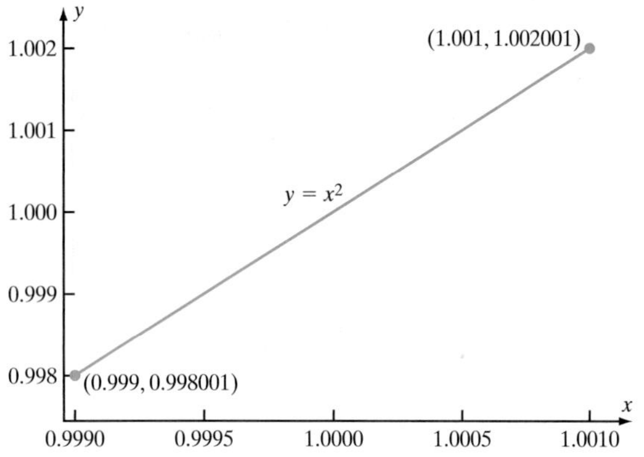

77. Slope ≈ (0.001999998 − (−0.001999998))/(0.001 − (−0.001)), or 1.999998 (See accompanying figure.)

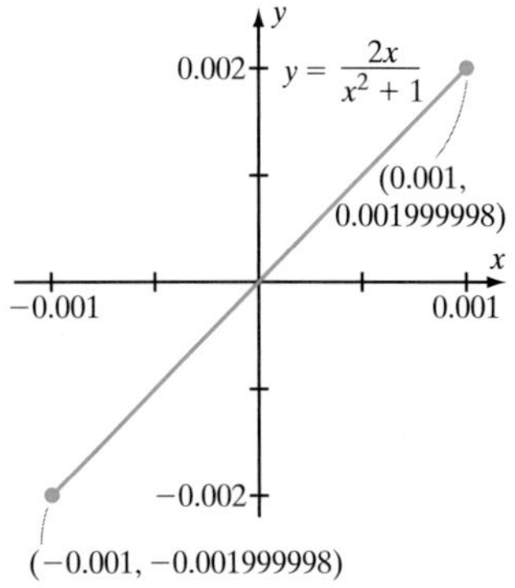

79. Slope ≈ 0 (See accompanying figure.)

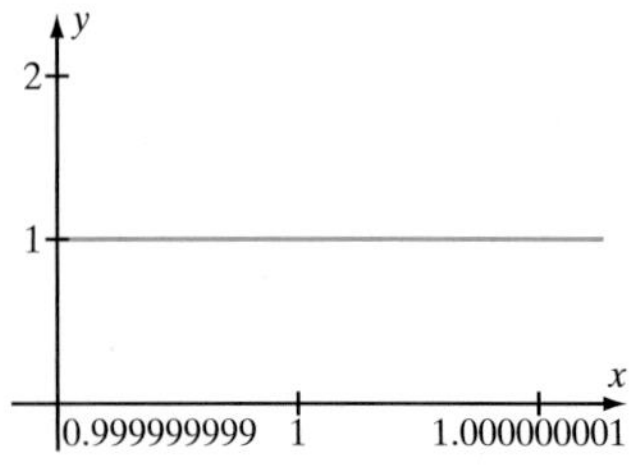

81.

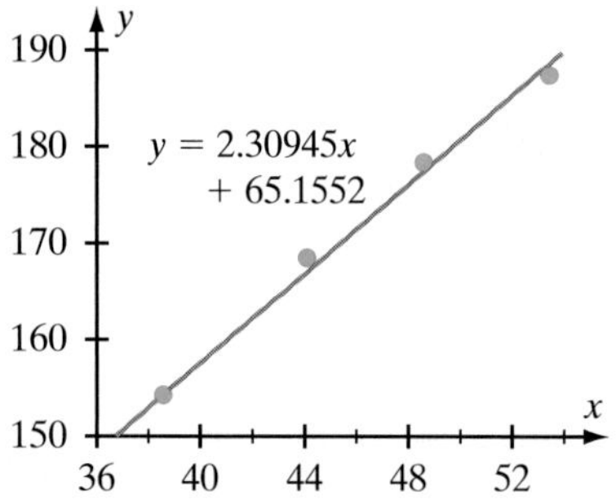

83. $Q = (3/4, \sqrt{3}/2)$; line ℓ has equation $y = (1/\sqrt{3})(x - 3/4) + \sqrt{3}/2$

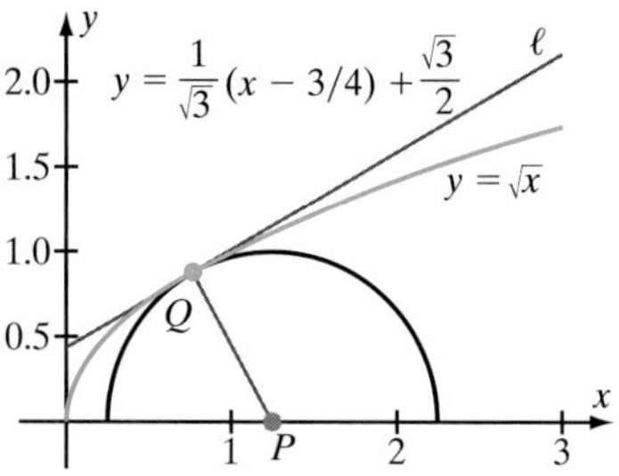

85. $m \approx 4.76923$ minimizes the sum of the squared errors; $m \approx 4.6666667$ minimizes the sum of the absolute errors

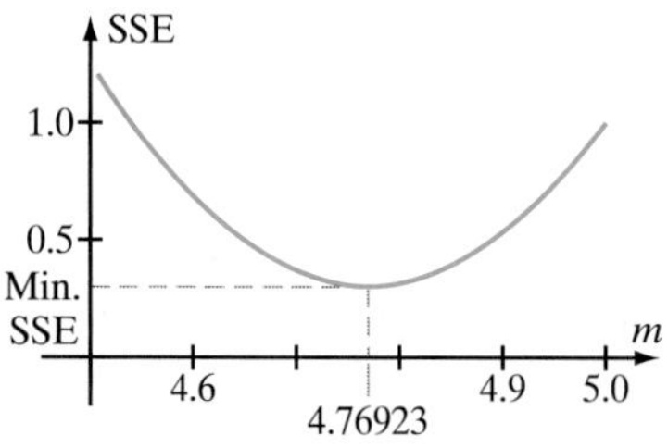

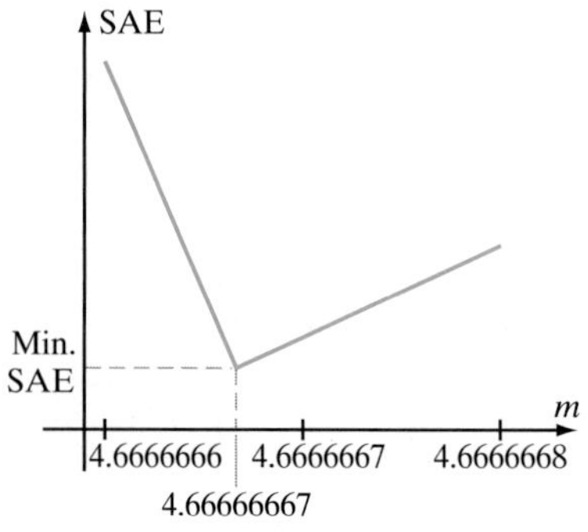

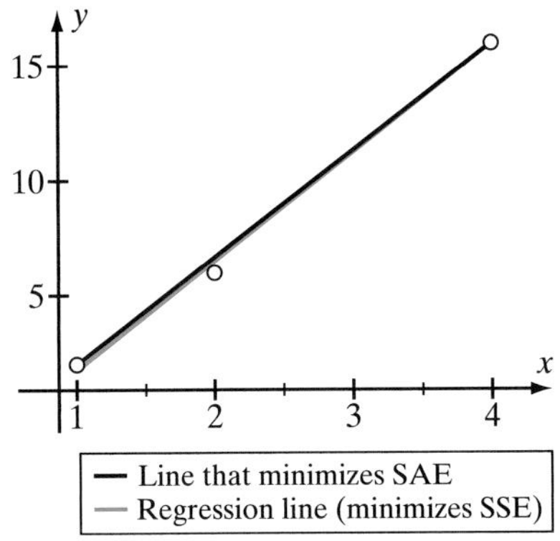

Section 1.4

1. $\{x \in \mathbb{R} : x \neq -1\}$

3. $\{x \in \mathbb{R} : x \leq -\sqrt{2} \text{ or } x \geq \sqrt{2}\}$ (or, equivalently, $\{x \in \mathbb{R} : \sqrt{2} \leq |x|\}$)

5. $\{x \in \mathbb{R} : x \neq -1 \text{ and } x \neq 1\}$

7. $\mathbb{R}$

9.

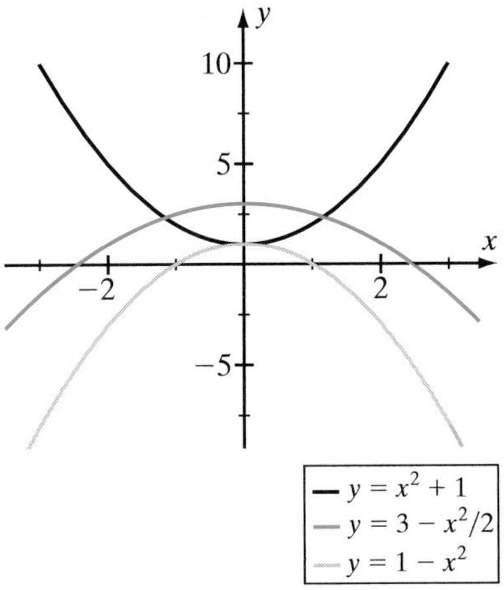

11. The figure for Exercise 9 has this plot.

13. The figure for Exercise 9 has this plot.

15.

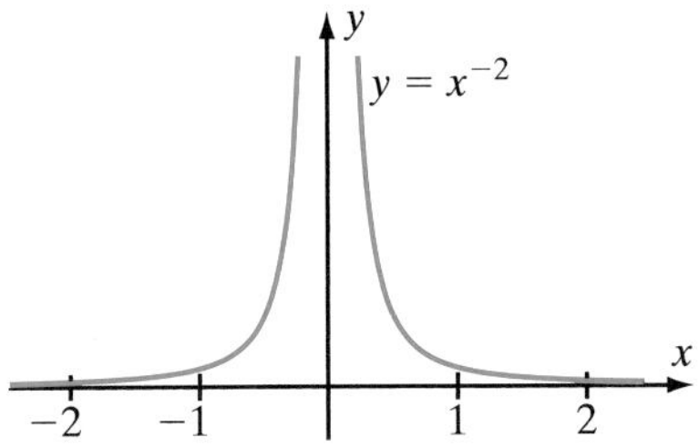

17.

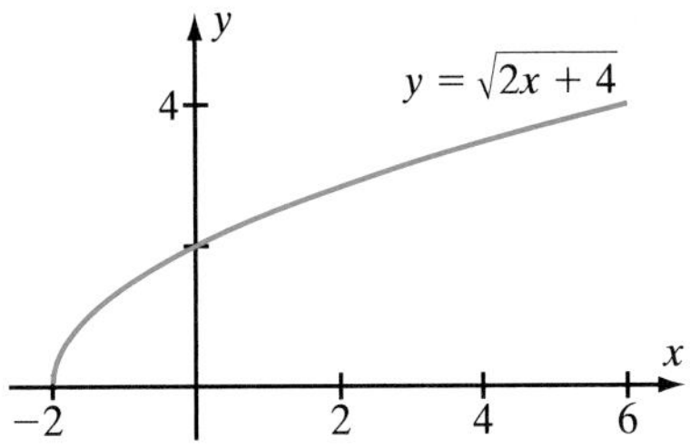

19.

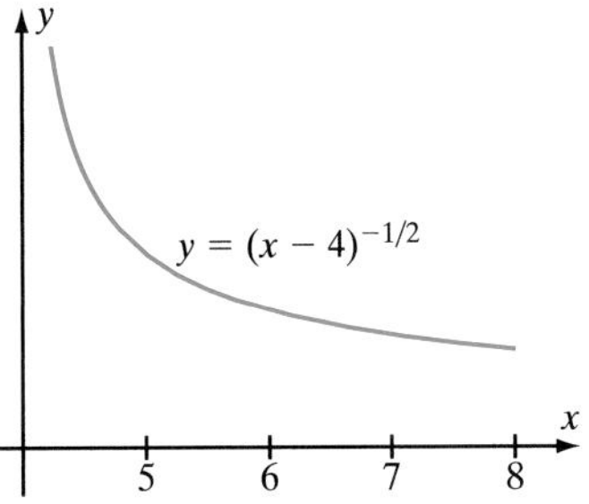

21.

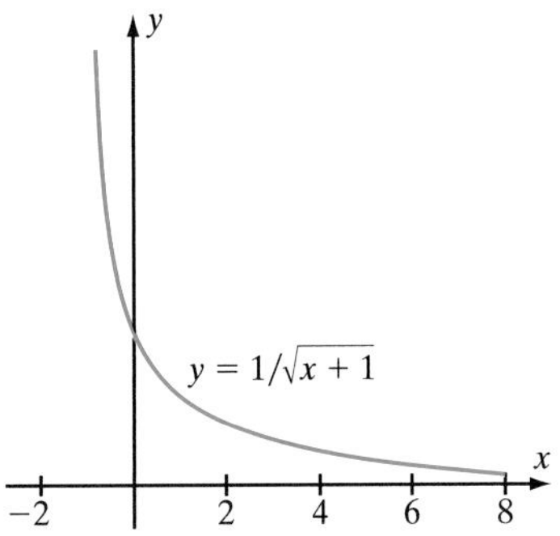

23.

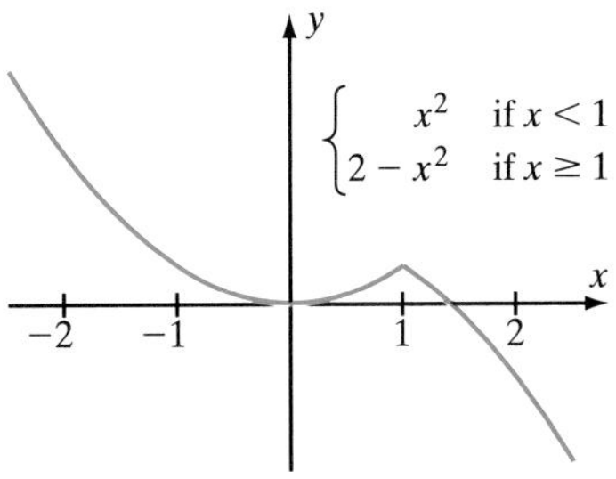

25. $s(\alpha) = (\pi/180)\, r\alpha$

27. $A(r, \alpha) = (\pi/360)\, r^2\alpha$

29. $A = -(r+s)$; $B = rs$

31. $$f(x) = \begin{cases} 1 + 2x & \text{if } 0 \leq x \leq 1 \\ 3(3-x)/2 & \text{if } 1 < x \leq 3 \\ x - 3 & \text{if } 3 < x \leq 4 \\ 1 & \text{if } 4 < x \leq 5 \end{cases}$$

33. $m(x) = \begin{cases} 0.10 \text{ if } 0 < x < 8025 \\ 0.15 \text{ if } 8025 < x < 32550, \\ 0.25 \text{ if } 32550 < x < 50000 \end{cases}$ $\quad A(x) = T(x)$

35. $f_1 = 2 \quad f_n = 2 \cdot f_{n-1}$ for $2 \le n$

37. $f_1 = 1 \quad f_n = n \cdot f_{n-1}$ for $2 \le n$

39. $f_1 = 1 \quad f_n = n + f_{n-1}$ for $2 \le n$

41. $\lfloor x \rfloor = \begin{cases} \text{Int}(x) & \text{if } x > 0 \text{ or } x \in \mathbb{Z} \\ \text{Int}(x) - 1 & \text{if } x < 0 \end{cases}$

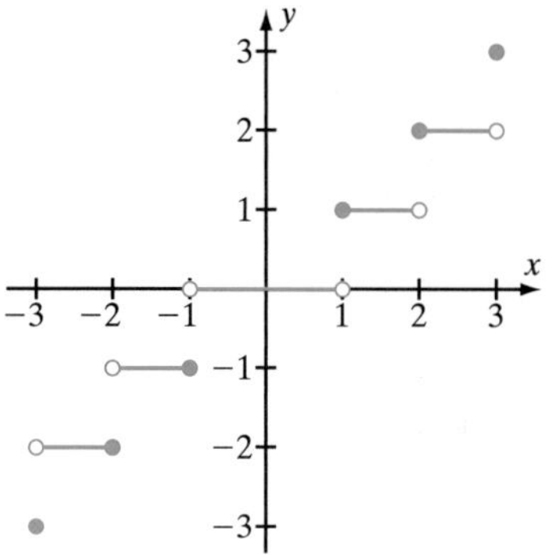

$y = \text{Int}(x)$ (also called $y = \text{trunc}(x)$)

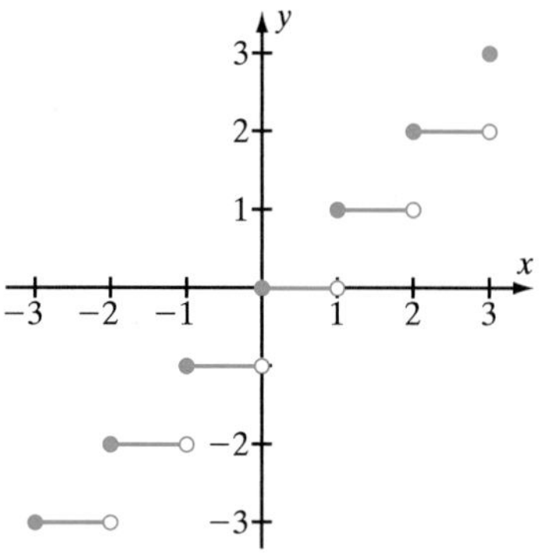

$y = \lfloor x \rfloor$

43. $I(P, m, n) = 12mn - P$

45. 1, 1, 2, 5, 14, 42, 132, 429, 1430

47. $R(x) = x(0 < x)(x \le 1) + (1 < x)$

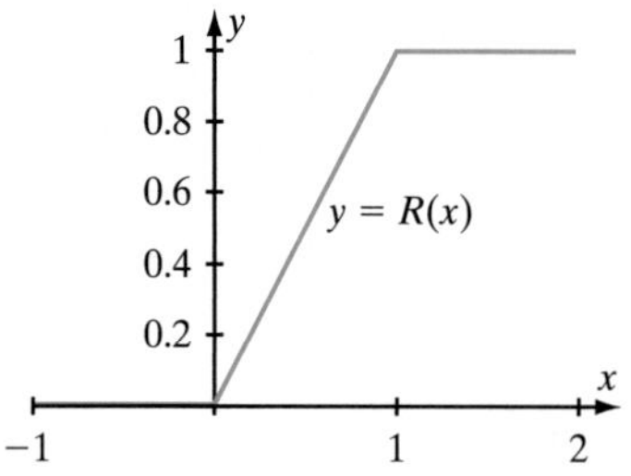

49. $b(x) = x(0 < x)(x \le 1) + (2 - x), (1 < x)(x \le 2)$

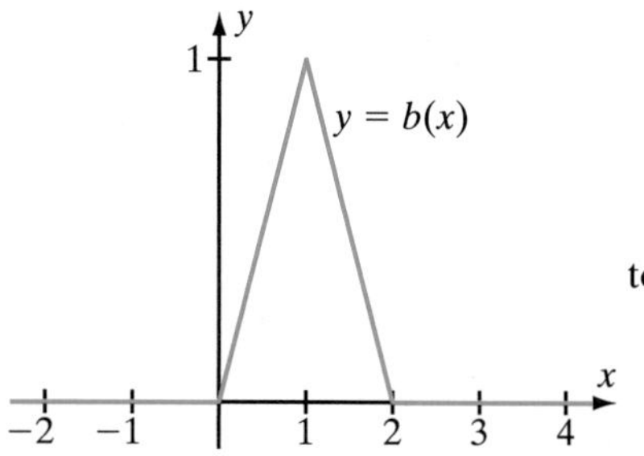

top of right column

51.

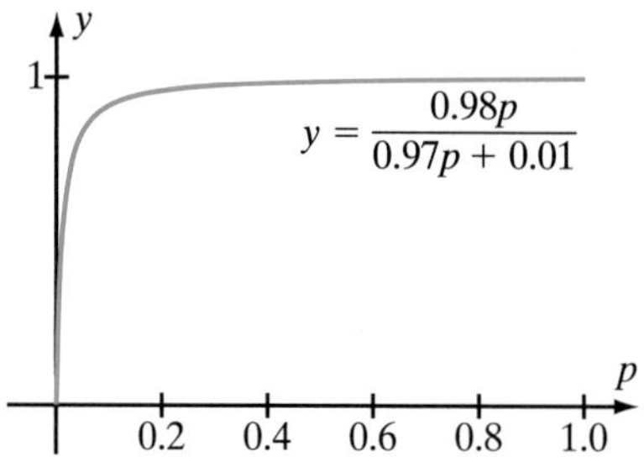

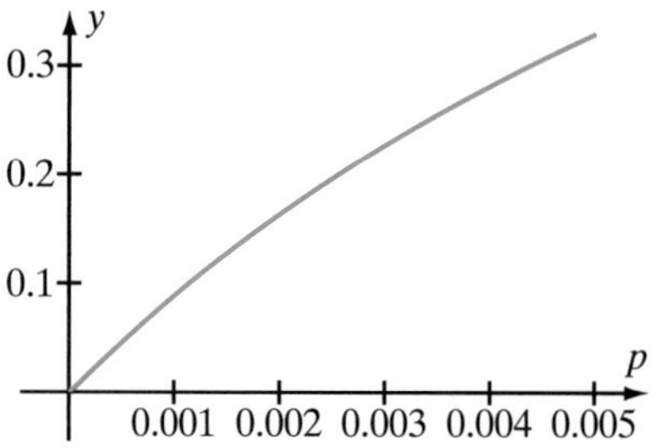

The probability of a true positive becomes very small as x approaches 0. Routine screening is not advisable if the disease is rare. $y = R(x)$

53. **a.** $x_0 = 4$;
b.

h	10^{-3}	10^{-4}	10^{-5}
$F(h, 3.99)$	1.5031×10^{-3}	1.437×10^{-4}	1.44×10^{-5}
$F(h, x_0)$	7.1972×10^{-5}	7.1997×10^{-7}	7.200×10^{-9}

c.

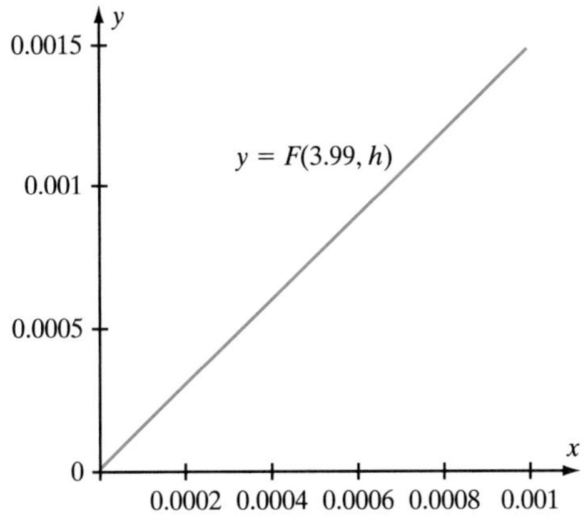

d.

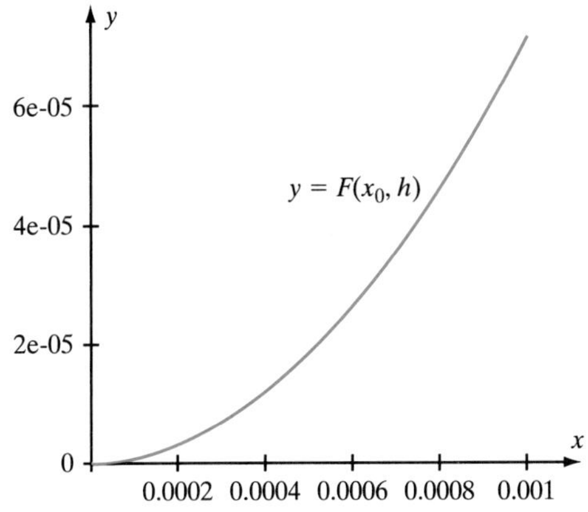

57. $p_1 = 2.828427124$, $p_2 = 3.061467458$
$p_3 = 3.121445152$, $p_4 = 3.136548492$
$p_5 = 3.140331158$, $p_6 = 3.141277254$
$p_7 = 3.141513803$, $p_8 = 3.141572945$

Section 1.5

1. $(x^3 - x^2 + 6x - 4)/(x - 1)$,

3. $(x - 2)/(x - 3)$

5. $2x^3 - 7x^2 + 10x - 30$

7. $(x - 2)^2/(x - 3)^2 + 5$

9. $-x$

11. $f(x) = x - 2$, $g(x) = x^2$

13. $f(x) = x^3 + 3x$, $g(x) = x^4$

15. $g(x) = 3x^2 + 1$

17. $g(x) = x/(x^2 + 2)$

19. 3

21. 1/2

23. $(x + 5)(x - 1)$

25. $(x + 2)(x - 2)(x^2 + 2x + 2)$

27. $f^{-1}(t) = \sqrt{t - 1}$

29. $f^{-1}(t) = (-1 + \sqrt{1 + 4t})/2$

31. $f^{-1}(t) = (t + 27)^{1/3}$

33. $f^{-1}(t) = \sqrt{t/(t - 1)}$

35. $f^{-1}(t) = t^2 - 3$

37. Not onto

39. **a.** Invertible

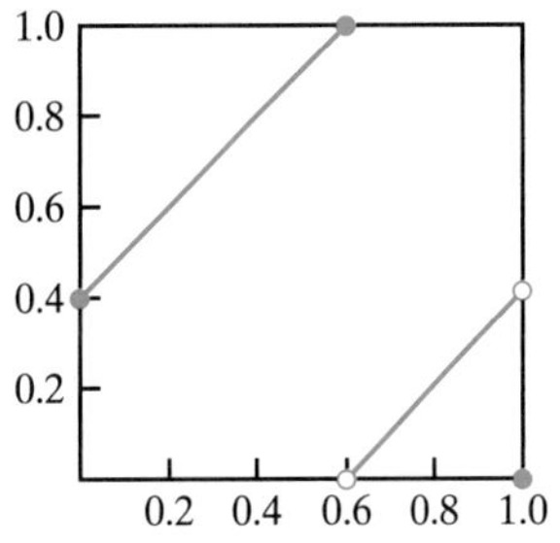

b. Not invertible **c.** Not invertible **d.** Not invertible
e. Invertible

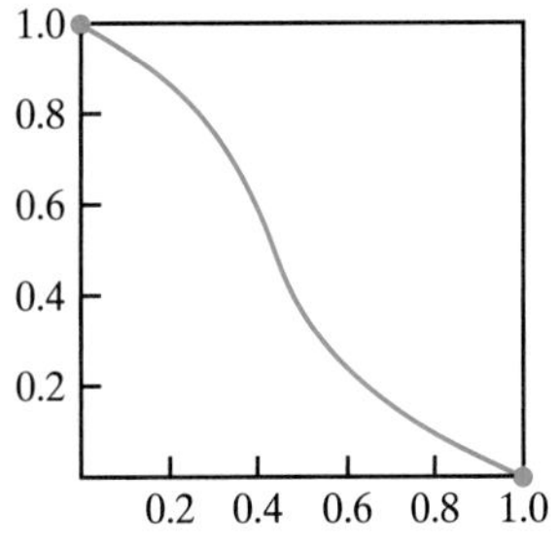

f. Invertible

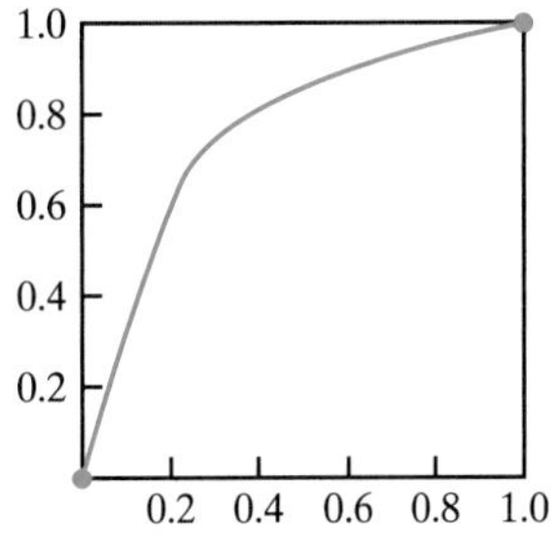

41. $h = -2$; horizontal translation 2 units to the *right*

43. $h = -1$; horizontal translation 1 unit to the *right*

45. $h = 2$; $k = -12$; The graph of g is obtained by translating the graph of f 2 units to the left and 12 units down.

47. $h = 3$; $k = -2$; The graph of g is obtained by translating the graph of f 3 units to the left and 2 units down.

49. The line $y = 3x - 7$

51. The parabola $x = 3y^2 + 1$

53. $\deg(p \cdot q) = \deg(p) + \deg(q)$; $\deg(p \circ q) = \deg(p) \cdot \deg(q)$; $\deg(p \circ q) = \deg(q \circ p)$; $\deg(p \pm q) \leq \max(\deg(p), \deg(q))$

55. $\deg(f) = n^2$; every root of p is a root of f; $q(x) = x^2 - x - 6$.

61. $g(x) = x^2 + 2$

63. $g(x) = 2x + 18$

65. $f(x) = x - 4$

67. $f(x) = (x^2 - 1)^{1/3}$

69. $f(x) = (x - 3)^2$

71. $f(x) = (1 - x^3)/(x^2 + 1)$

73. True for $p = 0$ and $p = 2$

75. $\phi(x) = \pi$; $\psi(x) = \pi^{1/5}$; $\mu(x) = x^{1/5} + \pi$; $\lambda(x) = (x + \pi)^{1/5}$

77.

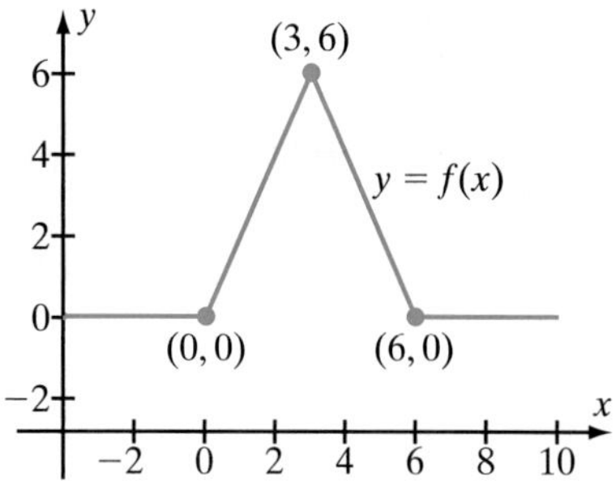

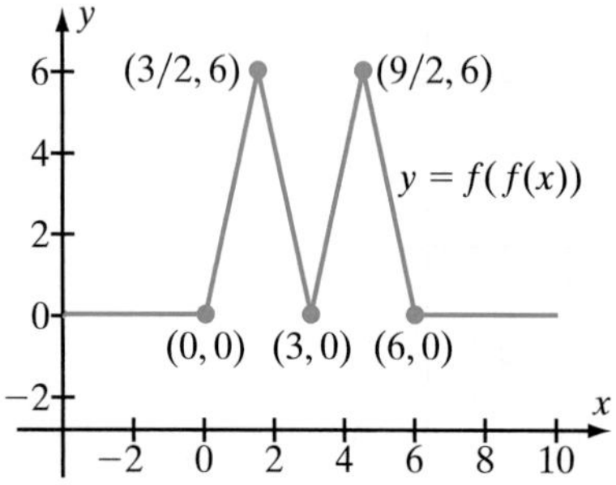

79. $G(x) = x^p$, $F(x) = px$, $H(x) = \log_{10}(p) + x$

81. $x = y^{2/3}$

83. $y = 3 \cdot 2^x$

85. $f \circ S = f$; $g \circ S = -g$

87.

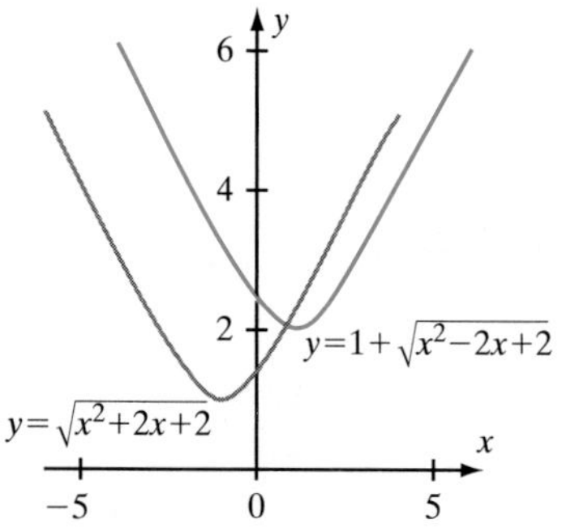

89.

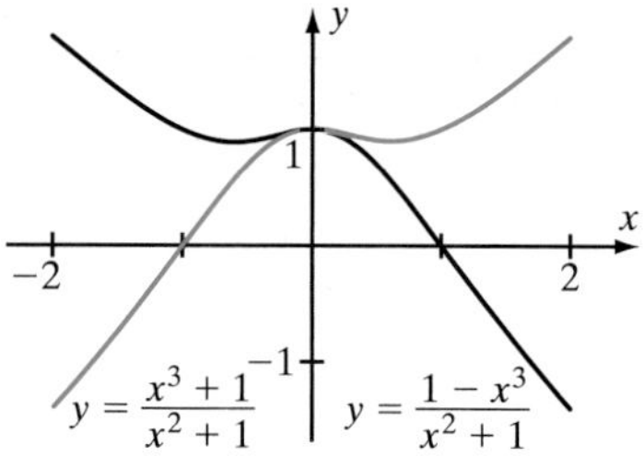

91.

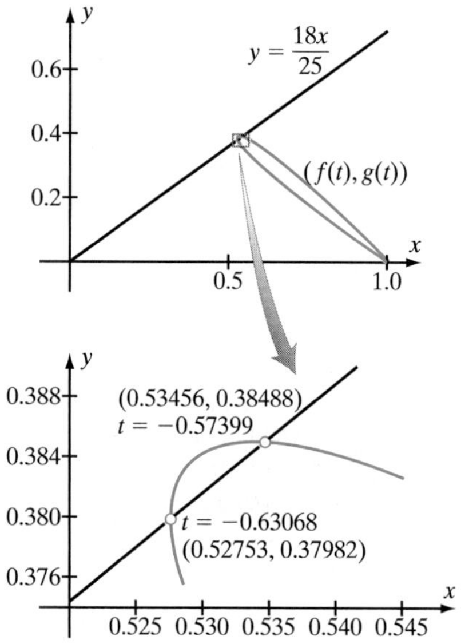

93.

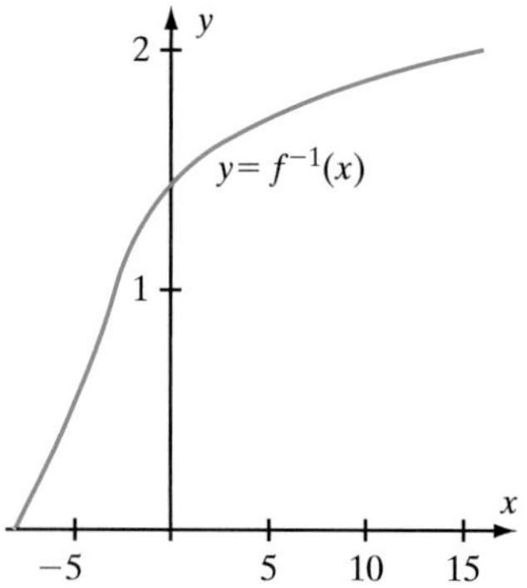

Section 1.6

1. $\sin(\pi/6) = 1/2, \cos(\pi/6) = \sqrt{3}/2, \tan(\pi/6) = \sqrt{3}/3$, $\csc(\pi/6) = 2, \sec(\pi/6) = 2\sqrt{3}/3, \cot(\pi/6) = \sqrt{3}$

3. $\sin(2\pi/3) = \sqrt{3}/2, \cos(2\pi/3) = -1/2$, $\tan(2\pi/3) = -\sqrt{3}, \csc(2\pi/3) = 2\sqrt{3}/3$, $\sec(2\pi/3) = -2, \cot(2\pi/3) = -\sqrt{3}/3$

5. $\sqrt{3}/4$ 7. $(\sqrt{3}+1)/2$ 9. 3 11. 1

13. -1 15. $2\sqrt{2}/3$ 17. 24/25 19. $\sqrt{25/26}$

21. sin, cos, tan, csc, sec, cot

23. tan, cot

25.

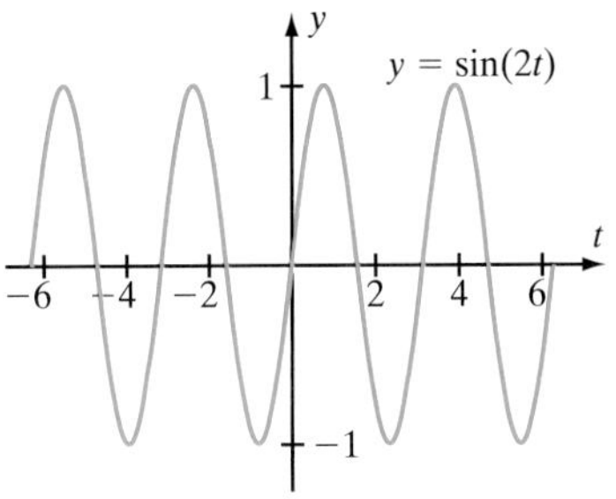

27.

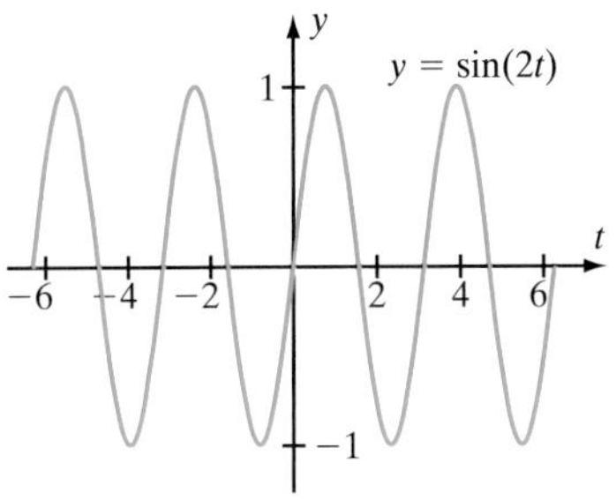

29. $x = \cos(t)$, $y = \sin(t)$

31. $x = \sin(t)$, $y = \cos(t)$

33. $r\theta$

35. The first equation holds only for $\theta = n\pi/2$ where n is an integer. The second equation is an identity: It holds for *every* value of θ.

45. $\sin(7\pi/12) = \frac{1}{2}\sqrt{2+\sqrt{3}}, \cos(7\pi/12) = -\frac{1}{2}\sqrt{2-\sqrt{3}}$, $\tan(7\pi/12) = -2-\sqrt{3}$, $\csc(7\pi/12) = 2\sqrt{2-\sqrt{3}}, \sec(7\pi/12) = -2\sqrt{2+\sqrt{3}}$, $\cot(7\pi/12) = -2+\sqrt{3}$

51. $C = \sqrt{A^2+B^2}$; ϕ satisfies $\sin(\phi) = A/C$ and $\cos(\phi) = B/C$

55. The line segment $x/a + y/b = 1$ traversed from $(a,0)$ to $(0,b)$ and then back to $(a,0)$.

57. $y = ab/x,\ 0 < x < \infty$

59. 2π 61. π 63. 2π

65. $T(t) = 27.5 + 23.5\sin(\pi t/6 - \pi/2)$.

73. $T(t) = 59 + 19\sin(\pi t/6 - 7\pi/12)$

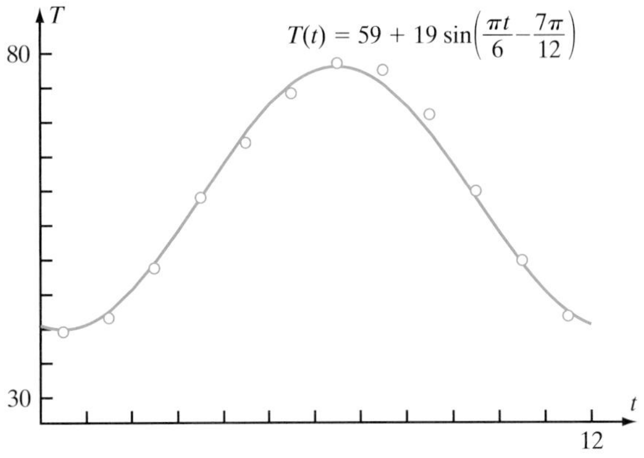

75. $\omega = 7$; the frequency of the modulated amplitude is 1

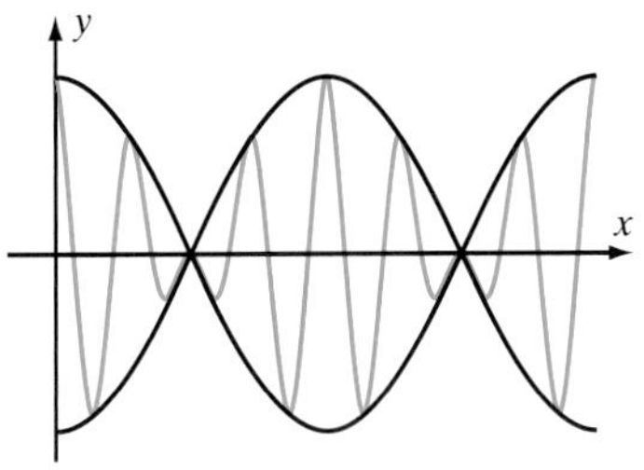

77.

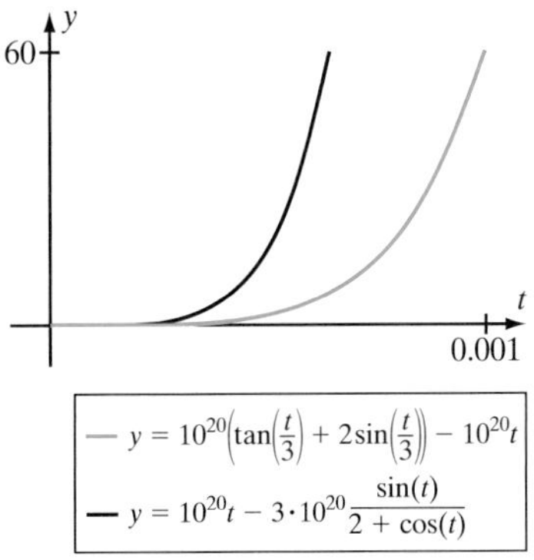

REVIEW EXERCISES FOR CHAPTER 1

1. $(-11, 1)$ 3. $[0, 3]$

5. $(-\infty, -1/6) \cup (1/2, \infty)$

7. $(-\infty, -2] \cup (0, 3]$

9. 13 **11.** (3,−7/2)

13. circle; axes of symmetry: $x = 1$, $y = -1$; radius: $\sqrt{2}$

15. ellipse; axes of symmetry: $x = -1$, $y = 2$

17. $y = 3x - 6$

19. $y = -x - 4$

21. $y = -6(x + 2) + 1$

23. $y = 2(x + 2) + 1$

25. $x/4 + y/(-3) = 1$

27. $x/(3/2) + y/3 = 1$

29. $(2, 3)$

31. $(-1, -8)$ and $(4, 7)$

33. $(-\infty, \infty)$

35. $(-\infty, 5) \cup (5, \infty)$

37. $(-\infty, -2) \cup (-2, \infty)$

39. $[-5, -3) \cup (-3, 3) \cup (3, 5]$

41. $1/2, 2/3, 3/4, 4/5, 5/6$

43. $4, 8, 14, 24, 42$

45. $7, 15, 31, 63, 127$

47. 16 **49.** 41

51. $4x + 12\sqrt{x} + 9$

53. $-x^2/(2 + x^2)$

55. $2x^4 + 4x^2 + 5$

57. $2\sqrt{1 + x} + 3$

59. $-(3x + 4)/(2x + 3)$

61. $T = [1/5, 1/2]$; $f^{-1}(t) = 1/t - 1$

63. $T = [2, 5]$; $f^{-1}(t) = -1 + \sqrt{t - 1}$

65. Onto but not one-to-one

67. One-to-one but not onto

69. Translate left 2 units, down 1

71. Translate right 3 units, down 9

73. The line segment from $(-1/3, 0)$ to $(0, 1/2)$

75. The line segment from $(1, 0)$ to $(0, 1)$

77. $\sin(A) = 3/\sqrt{13}, \cos(A) = 2/\sqrt{13}, \tan(A) = 3/2,$
$\cot(A) = 2/3, \sec(A) = \sqrt{13}/2, \csc(A) = \sqrt{13}/3$

79. $\sin(A) = 1/\sqrt{10}, \cos(A) = 3/\sqrt{10},$
$\tan(A) = 1/3, \cot(A) = 3, \sec(A) = \sqrt{10}/3, \csc(A) = \sqrt{10}$

81. $\sin(\alpha + \pi/2) = 5/13, \cos(\alpha + \pi/2) = -12/13,$
$\tan(\alpha + \pi/2) = -5/12, \cot(\alpha + \pi/2) = -12/5,$
$\sec(\alpha + \pi/2) = -13/12, \csc(\alpha + \pi/2) = 13/5$

83. 2 **85.** $-2\sqrt{2}$ **87.** $3\sqrt{7}/8$ **89.** $-24/7$

Section 2.1

1. 5 **3.** 57 **5.** −1

7. −4 **9.** 10 **11.** −1/14

13. Does not have a limit.

15. $\lim_{x \to 2} f(x) = -2$

17. $\lim_{x \to 5} f(x) = 7$

19. 0.01 **21.** 0.002 **23.** 0.2 **25.** 0.0019

27. $\lim_{x \to 2^-} f(x) = 0 \neq 4 = \lim_{x \to 2^+} f(x)$

29. **a.** The domains of g, f, h, and k are, respectively, $\{x \in \mathbb{R} : x \neq 1\}, \mathbb{R}, \mathbb{R}$, and $\mathbb{R}$. **b.** f and h
c. $\lim_{x \to 1} g(x) = \lim_{x \to 1} f(x) = \lim_{x \to 1} h(x) = \lim_{x \to 1} k(x) = 2$

31. **a.** The domains of f, g, k, and h are, respectively, $\mathbb{R}, \mathbb{R}, \mathbb{R}$, and $\{x \in \mathbb{R} : x \neq 0\}$. **b.** f and k
c. $\lim_{x \to 0} g(x) = \lim_{x \to 0} f(x) = \lim_{x \to 0} h(x) = \lim_{x \to 0} k(x) = 0$

33. $\lim_{x \to 0^+} H(x) = 1$; $\lim_{x \to 0^-} H(x) = 0$; $\lim_{x \to 0} H(x)$ does not exist

35. 3.28×10^{-3}

37. **a.** Does not exist. **b.** 0 **c.** 1 **d.** −1

39. See Figure for Exercise 41, Section 1.4;
$\lim_{x \to n^+} \lfloor x \rfloor = 1 + \lim_{x \to n^-} \lfloor x \rfloor \neq \lim_{x \to n^-} \lfloor x \rfloor$

41. **a.** 3 **b.** 2.5 **c.** 2.1 **d.** $2 + h$ e. 2

43. **a.** 4 **b.** 0.1 **c.** 0.01 **d.** ε

45. **a.** 8 **b.** 0.1 **c.** 0.01 **d.** ε

47. $6 + h$; 6

49. $h + 4$; 4

51. 0

53. $|x - 2| < 0.0095$

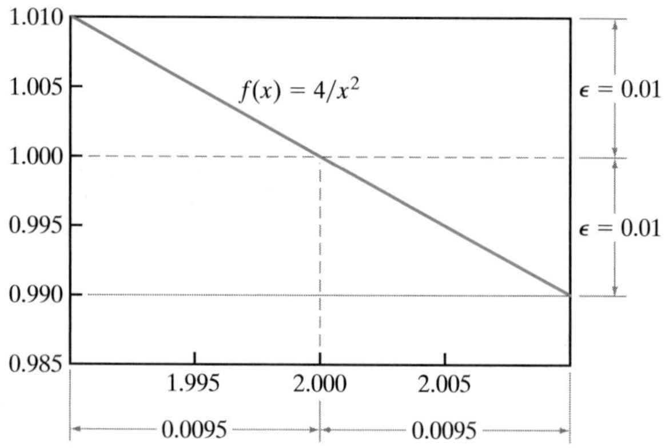

55. $|x - 10| < 0.00046$

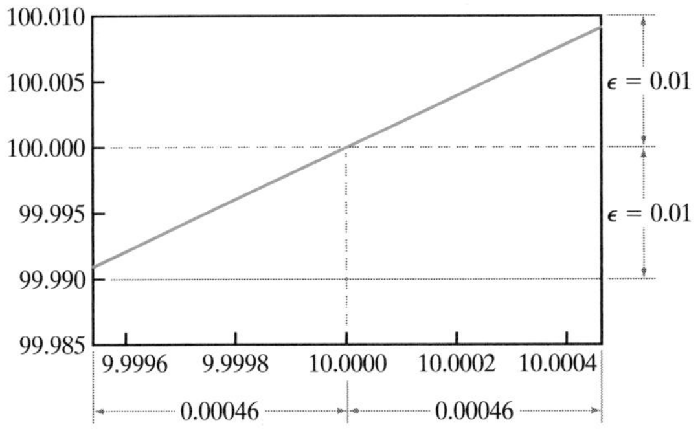

57. $\delta = 0.5166368178$ (or smaller)

59. 7.82×10^{-4} mm

61. a.

x	$f(x)$
$\pi/2 - 1/10$	-0.99833
$\pi/2 - 1/100$	-0.99998
$\pi/2 - 1/1000$	-0.99999

b. -1 **c.** $\delta = 0.24$ will do

d.

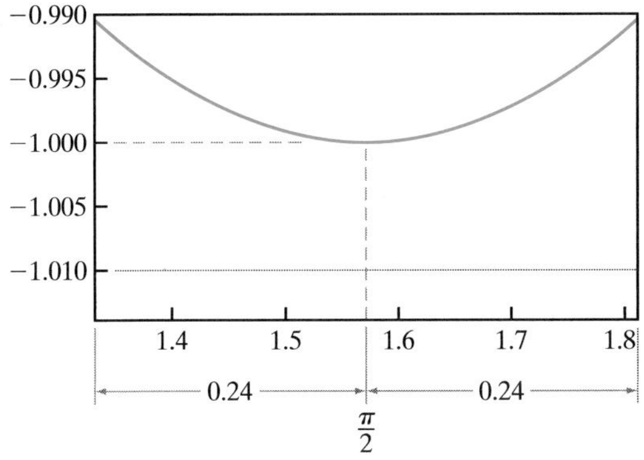

63. a.

x	$f(x)$
1/10	0.16658
1/100	0.16666
1/1000	0.16666

b. 1/6 **c.** $\delta = 1.05$ will do

d.

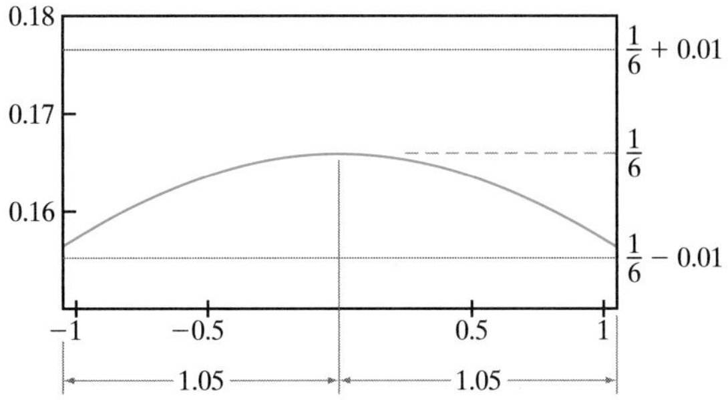

65. a. 4 **b.** $h < 0.01662$ will do.
c. $h < 0.0016662$ will do.

67. a. 3 **b.** $h < 0.45187$ will do.
c. $h < 0.044205$ will do.

69. a. $\bar{v}(t) = \left(\sin(t + 0.0001) - \sin(t - 0.0001)\right)/0.0002$

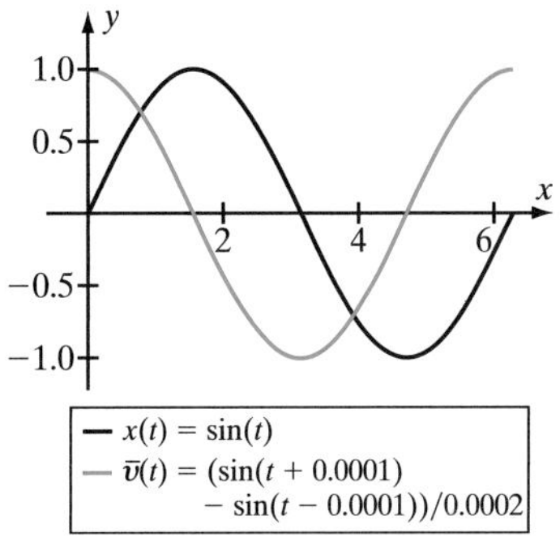

b. $\bar{v}(t)$ is the greatest near $t = 0$ and 2π; $\bar{v}(t)$ is the most negative near $t = \pi$
c. $\bar{v}(t)$ is zero near $t = \pi/2$ and $3\pi/2$
d. $\bar{v}(t) < 0$ for t in $I = [\pi/2, 3\pi/2]$; $x(t)$ decreases on I; $\bar{v}(t) < 0$ suggests that x is decreasing.
e. $\cos(t)$

Section 2.2

1. 6 **3.** 4 **5.** $\sqrt{2}$ **7.** 0

9. $\lim_{x \to 5^-} f(x) = 15 \neq 4 = \lim_{x \to 5^+} f(x)$

11. $\lim_{x \to 5^-} f(x) = 0 \neq -4 = \lim_{x \to 5^+} f(x)$

13. $\lim_{x \to 4^-} f(x) = 17 = \lim_{x \to 4^+} f(x)$

15. $\lim_{x \to 4^-} f(x) = 64 = \lim_{x \to 4^+} f(x)$

17. 14 **19.** -5 **21.** 2 **23.** $-4/5$

25. 1/2 **27.** 6 **29.** -2

31. $(\pi - \sqrt[3]{\pi})/(\sqrt{\pi} - 3)$

33. 0 **35.** 1 **37.** 3

39. $\lim_{x \to 2} f(x) = 2$

41. $\lim_{x \to 0} f(x) = 25$

43. $\lim_{x \to 1^+} f(x) = 0$ and $\lim_{x \to 2^-} f(x) = 0$

45. $\lim_{x \to (-2)^+} f(x) = 2$ and $\lim_{x \to 2^-} f(x) = 0$

Neither $\lim_{x \to 0^+} f(x)$ nor $\lim_{x \to 0^-} f(x)$ exist

47. −1 **49.** 28 **51.** 36 **53.** −8

55. Does not exist.

57. 3 **59.** 2 **61.** 0 **63.** $\pi/180$

65. 6 **67.** 4 **69.** 1 **71.** −1/8

73. 3/2 **75.** −1 **77.** 3/2 **79.** 1

87. 0.00062 will do.

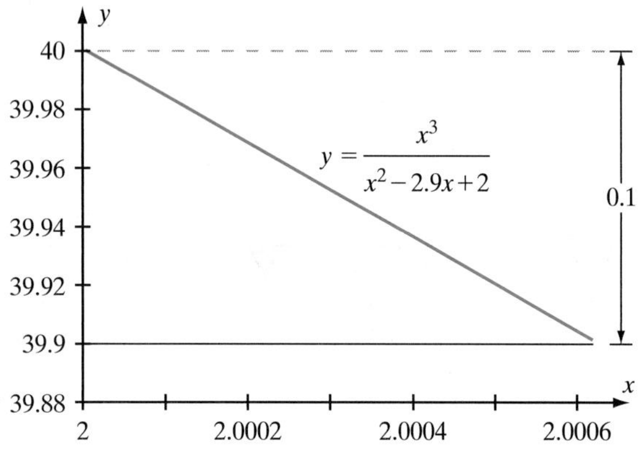

89. 0.046 will do.

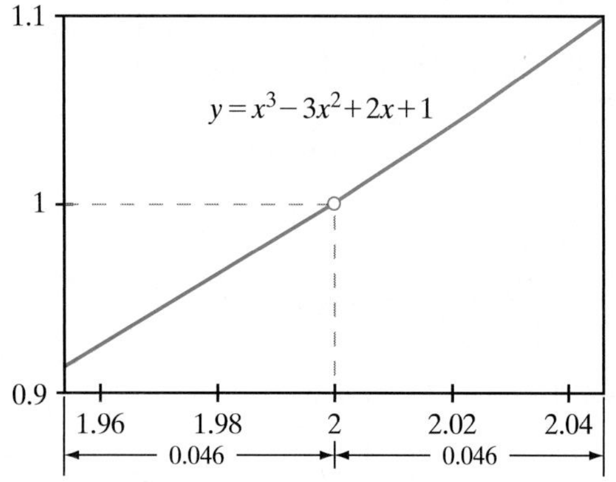

91. From the graph: approximately 0.01745 (See Exercise 63 for exact limit.)

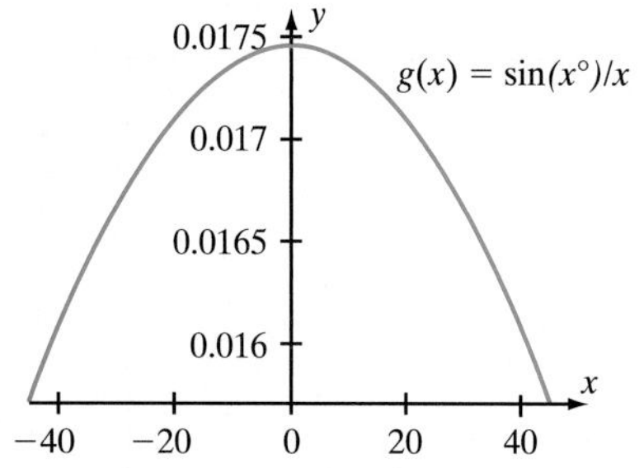

93. $\lim_{x \to 1} f(x) = 2$

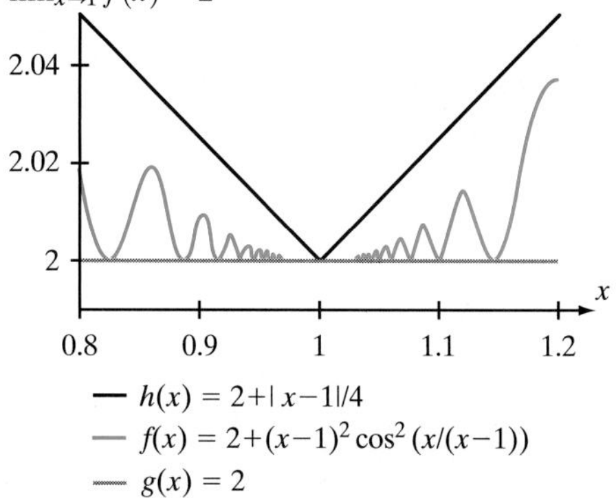

95. $\lim_{x \to 0} f(x) = 0$

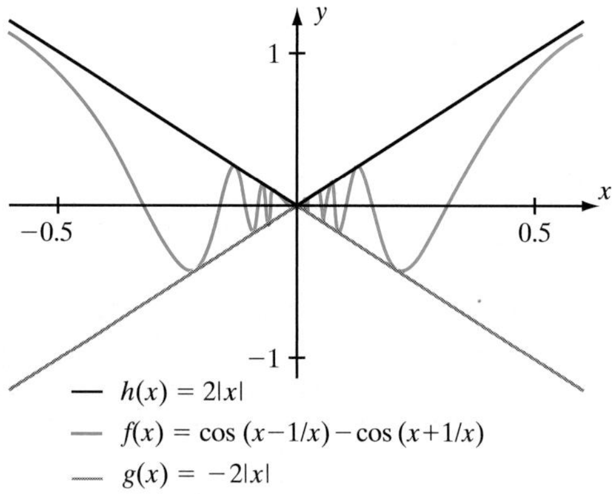

Section 2.3

1. −1, 1, 2 **3.** 1, 2

5. f is continuous at all values of its domain, which is $\mathbb{R}$.

7. f is continuous at all values of its domain, which is $\{x \in \mathbb{R} : x \neq -1\}$.

9. f is continuous at all values of its domain, which is $\mathbb{R}$.

11. f is continuous at all values of its domain, which is $\{x \in \mathbb{R} : x \neq 3, x \neq -4\}$.

13. The domain of f is $\mathbb{R}$; f is continuous at each point of $\{x \in \mathbb{R} : x \neq 3\}$.

15. f is continuous at all values of its domain, which is $\mathbb{R}$.

17. f is continuous at all values of its domain, which is $\mathbb{R}$.

19. The domain of f is $\mathbb{R}$; f is continuous at each point of $\{x \in \mathbb{R} : x \neq -1\}$.

21. The domain of f is $\mathbb{R}$; f is continuous at each point of $\{x \in \mathbb{R} : x \neq 0\}$.

23. f is continuous at all values of its domain, which is $\mathbb{R}$.

25. $F(2) = 11$

27. $F(2) = 4$

29. Left continuous on $(-\infty, \infty)$; right continuous on $\{x \in \mathbb{R} : x \neq 5\}$; continuous on $\{x \in \mathbb{R} : x \neq 5\}$

31. Left continuous on $\{x \in \mathbb{R} : x \neq 3\}$; right continuous on $(-\infty, \infty)$; continuous on $\{x \in \mathbb{R} : x \neq 3\}$

33.

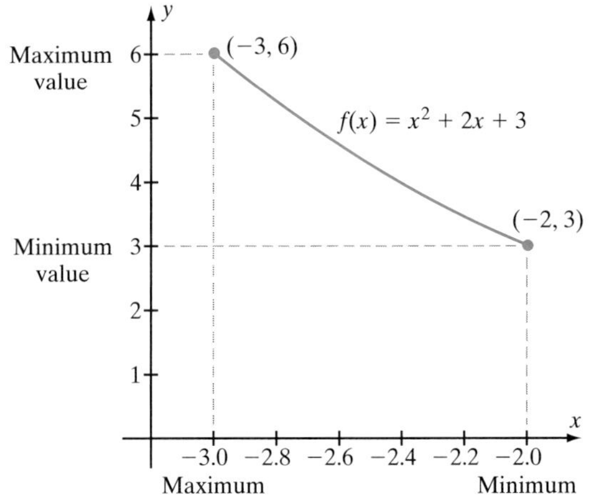

35.

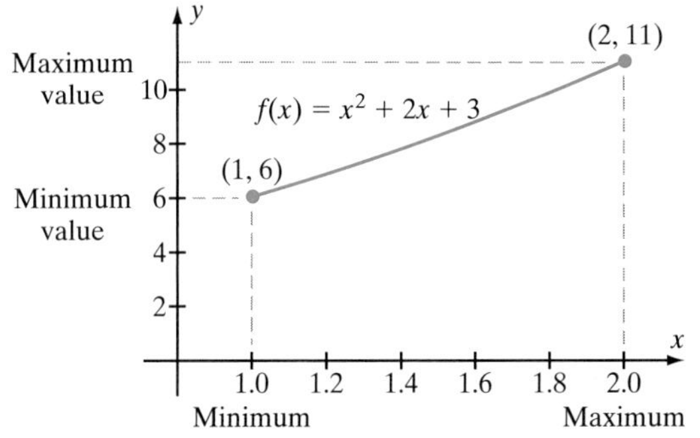

37.

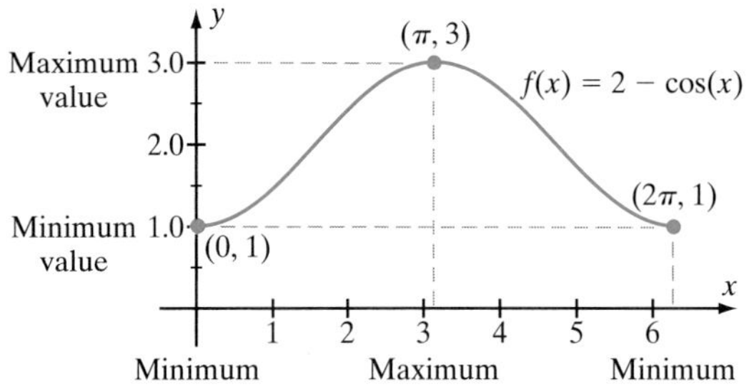

39. $-1/2, \sqrt{2}/2$

41. $1/4, 49/96$

43. $a = -7/2, b = 9/2$

45. $a = 25/6, b = -7/2$

47. Discontinuous at each odd integer

49. Discontinuous at each integer $n \neq k + 6m$ for some integer m and $k = \pm 1$

53. $\lim_{x \to 1} f(x), \lim_{x \to 1} g(x), \lim_{x \to 1} h(x)$, and $\lim_{x \to 1} k(x)$ all exist and equal 2; g and h are continuous at 1, while k is discontinuous at 1; g and h are identical.

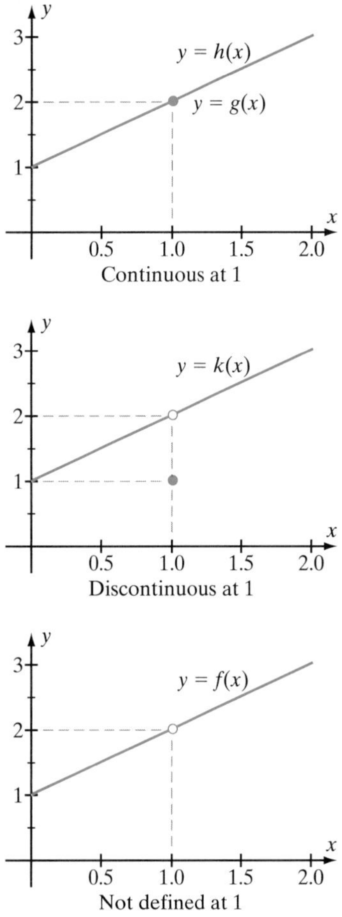

$$T(x) = \begin{cases} 0.08x & \text{if } 0 < x \le 8025 \\ 642 + 0.12(x - 8025) & \text{if } 8025 < x \le 32550 \\ 3585 + 0.16(x - 32550) & \text{if } 32550 < x \le 78850 \\ 10993 + 0.20(x - 78850) & \text{if } 78850 < x \le 164550 \\ 28133 + 0.25(x - 164550) & \text{if } 164550 < x \le 357700 \\ 76420.50 + 0.30(x - 357700) & \text{if } 357700 < x \end{cases}$$

55. **a.** 2 **b.** -1; **c.** -1; **d.** 0

65. False

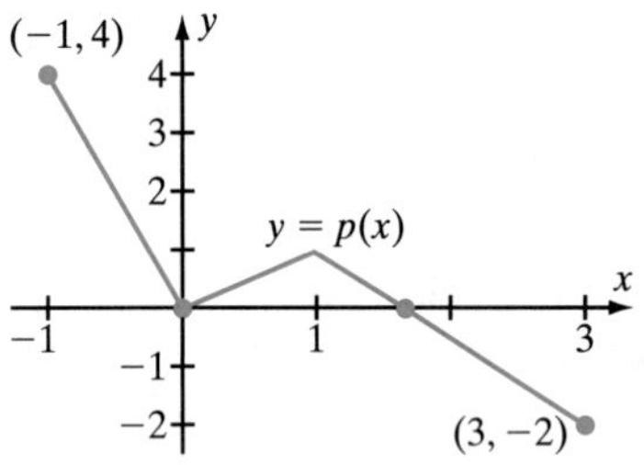

67. False (The figure for answer 65 illustrates this as well.)

69. False

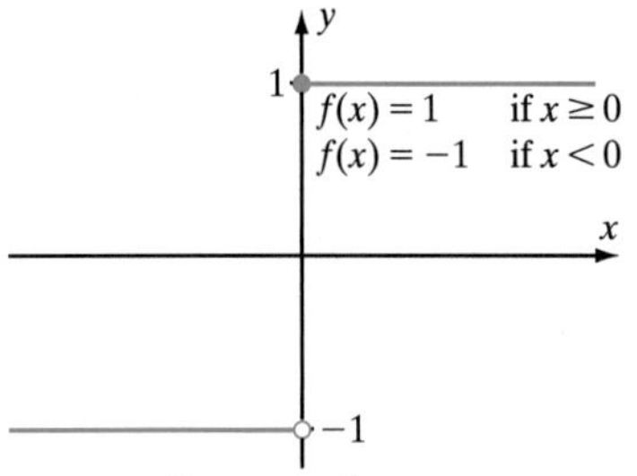

f is not continuous

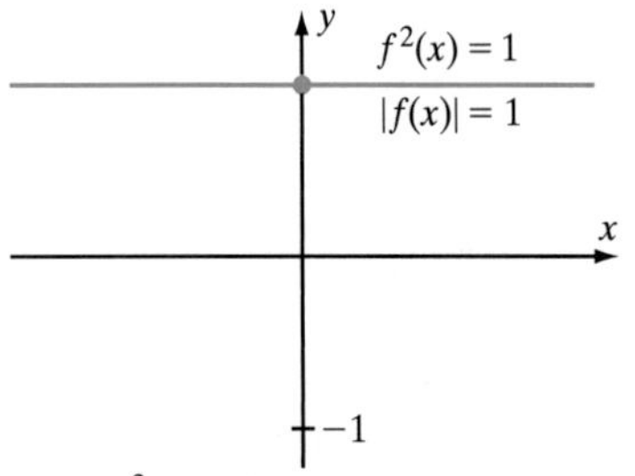

f^2 and $|f|$ are continuous

71. False (The figure for answer 69 illustrates this as well.)

73. True; otherwise, the Extreme Value Theorem would be contradicted.

83. 4.3017

85. No continuous extension

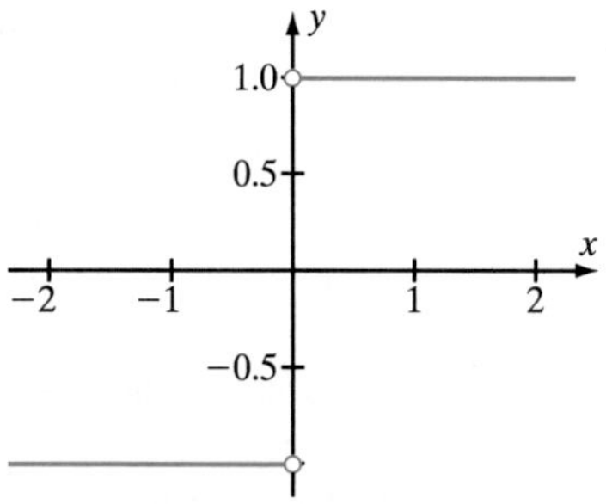

$f(x) = x/|x|$

87. A continuous extension F is obtained by setting $F(2) = -8$.

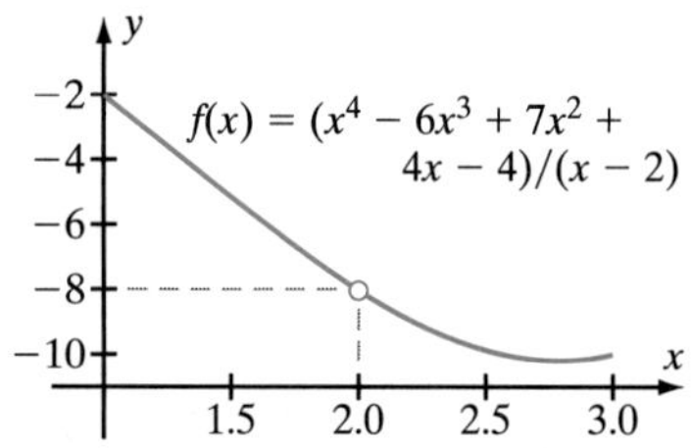

89. For each $\gamma \ge 10$, the equation $p(x) = \gamma$ has at least one solution x; the equation has no solution when $\gamma < 10$. When $\gamma = 20$, the solutions are 0.7267 and 4.1132.

91. Minimum at $x = -1.8301$; maximum at $x = 0.9369$

Section 2.4

1. ∞	**3.** 0	**5.** 1	**7.** 1	**9.** ∞
11. 1/3	**13.** ∞	**15.** ∞	**17.** $-\infty$	**19.** $-\infty$

21. Vertical asymptote: $x = 7$; horizontal asymptote: $y = 1$

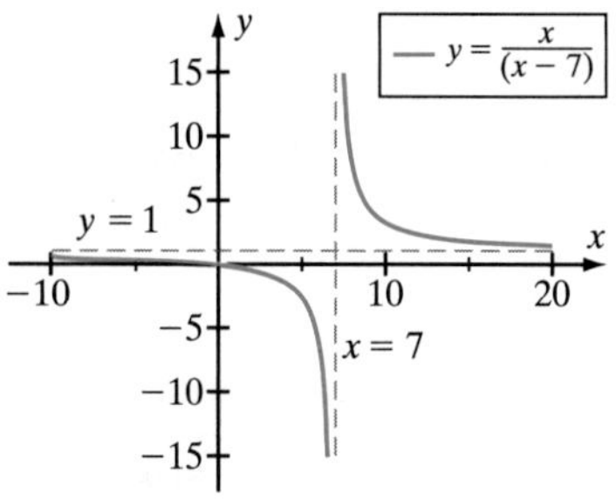

23. Vertical asymptote: $x = 0$; horizontal asymptote: $y = 0$

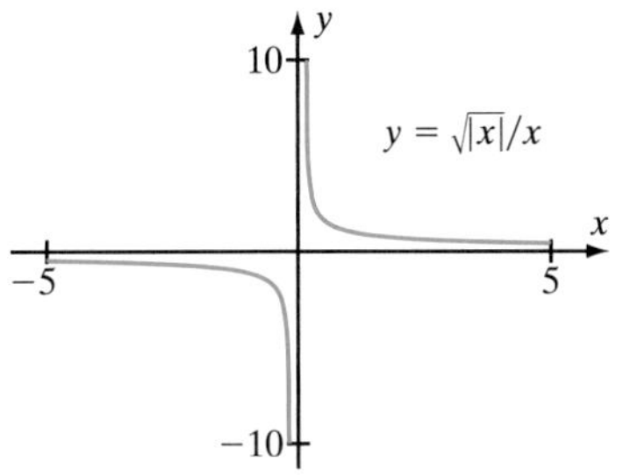

25. Horizontal asymptote: $y = 1$

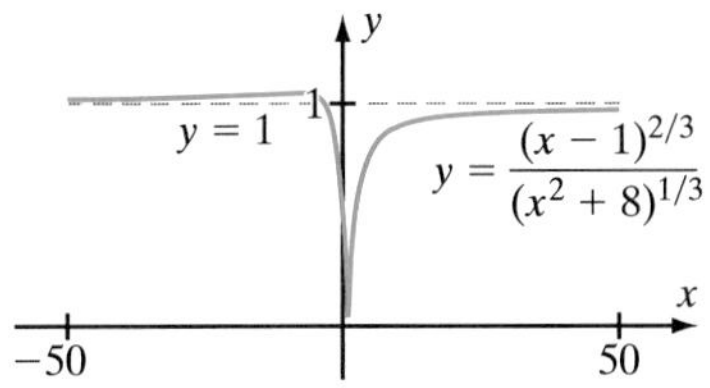

27. Horizontal asymptote: $y = 0$

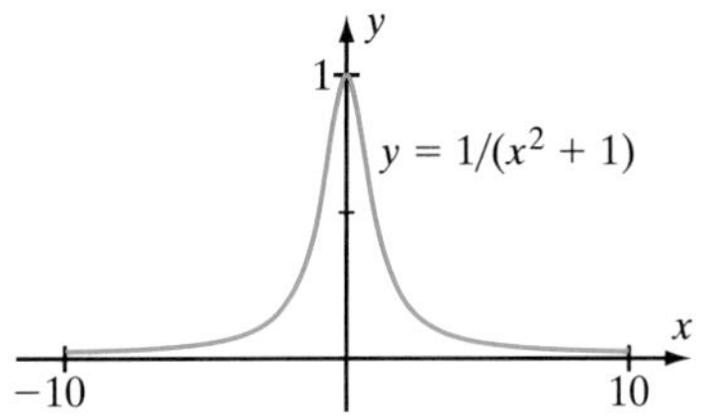

29. Vertical asymptote: $x = 0$; horizontal asymptote: $y = 0$

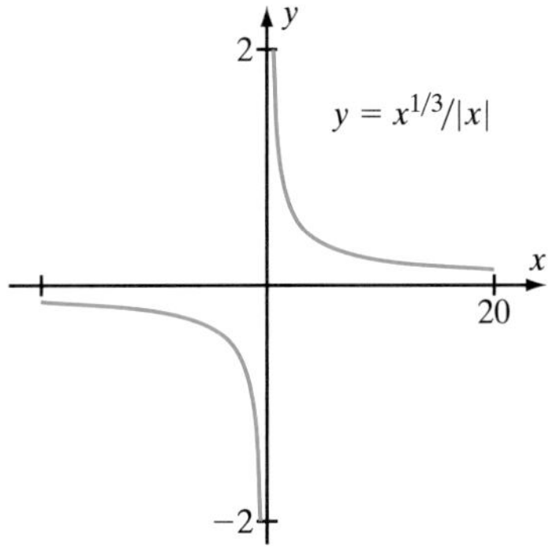

31. Vertical asymptotes: $x = -4$, $x = -1$, $x = 0$; horizontal asymptote: $y = 0$

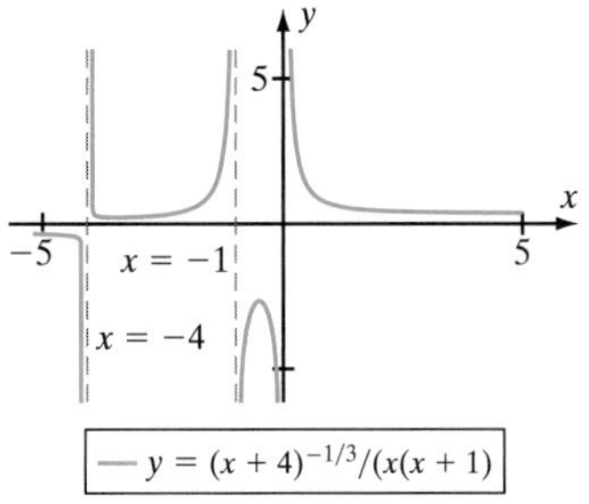

33. Vertical asymptote: $x = 0$

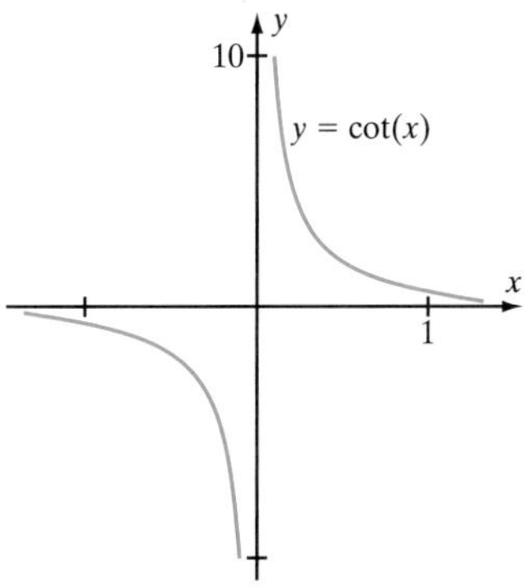

41. Vertical asymptotes: $x = n\pi$ for every nonzero integer n

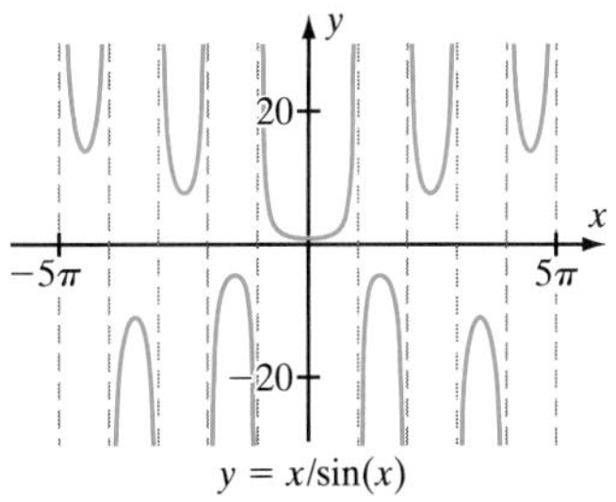

43. Vertical asymptote: $x = 0$; horizontal asymptote: $y = 0$

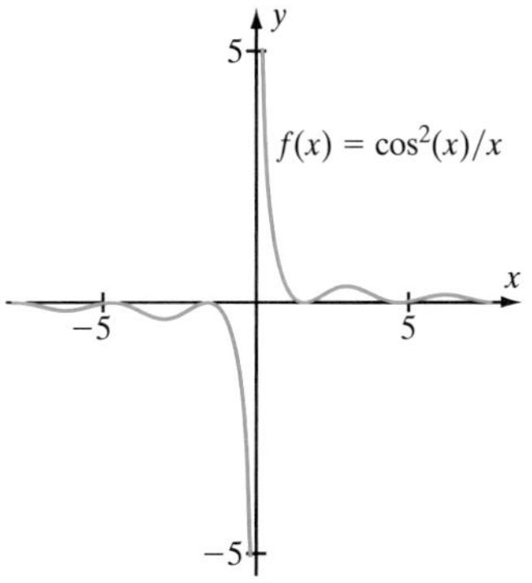

45. Vertical asymptotes: $x = n\pi$ for every integer n

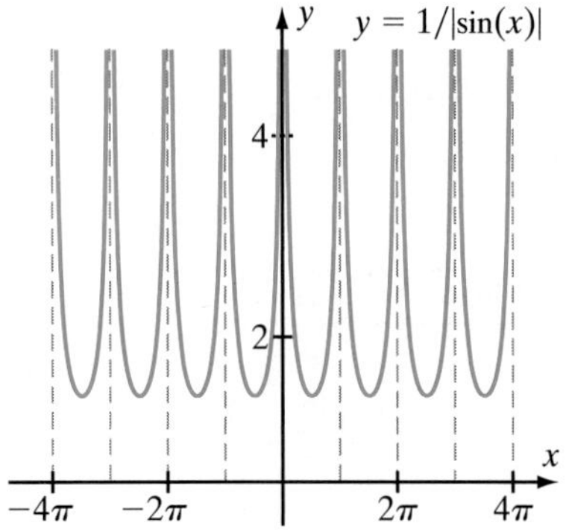

47. $x = 0$ is *not* a vertical asymptote.

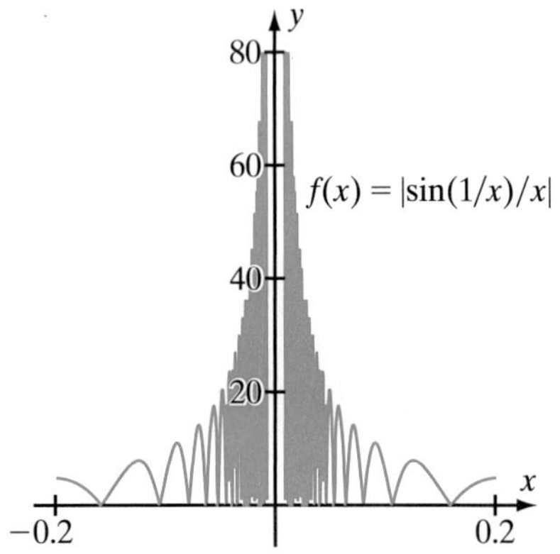

49. $f(10^{80}) = 10^{-20}$ and $\lim_{x\to\infty} f(x) = \infty$;
$g(10^{80}) = 10^{20}$ and $\lim_{x\to\infty} g(x) = 0$

51. Vertical asymptote:

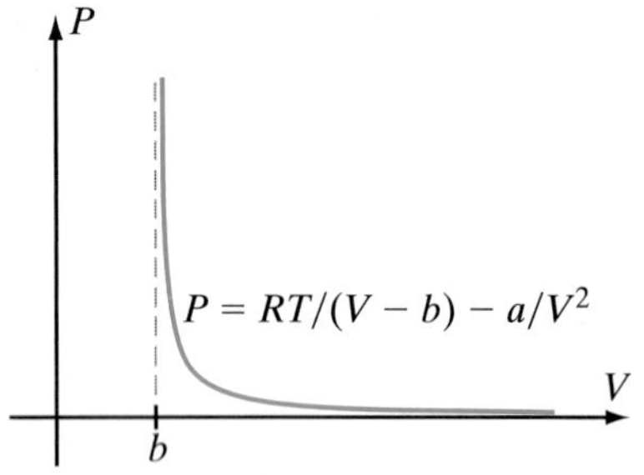

57. $y = x^2 - 3x + 9$ is a parabolic asymptote of $F(x)$;
$y = x^2 - 5x + 25$ is a parabolic asymptote of $G(x)$;
$y = x^2 + 1$ is a parabolic asymptote of $H(x)$.

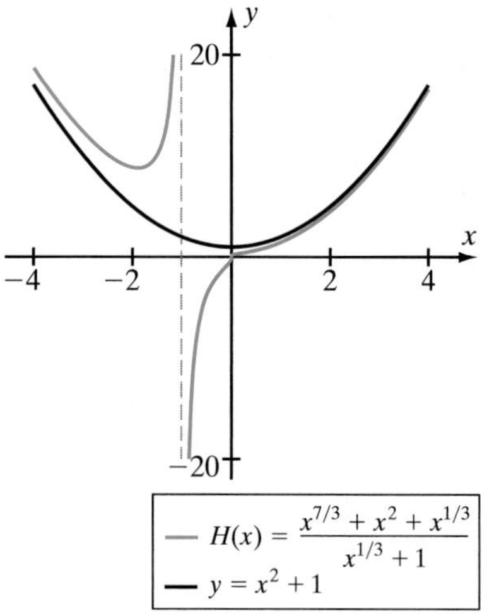

59. $x > (500 + \sqrt{249999})^2 \approx 999997.999999$

61.

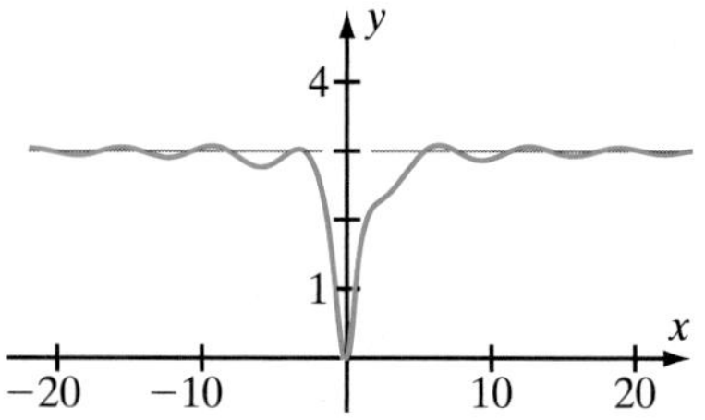

$f(x) = (3x^2 + x\cos(x))/(x^2 + 1)$

Graphed in the viewing window.
$[-25, 25] \times [0, 4]$

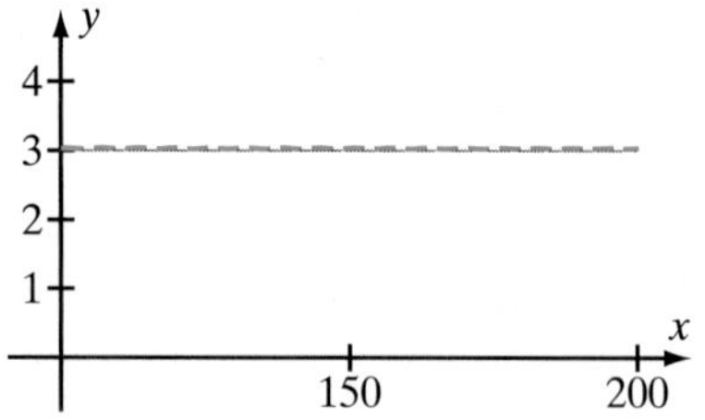

$f(x) = (3x^2 + x\cos(x))/(x^2 + 1)$

Graphed in the viewing window.
$[100, 200] \times [0, 4]$

63.

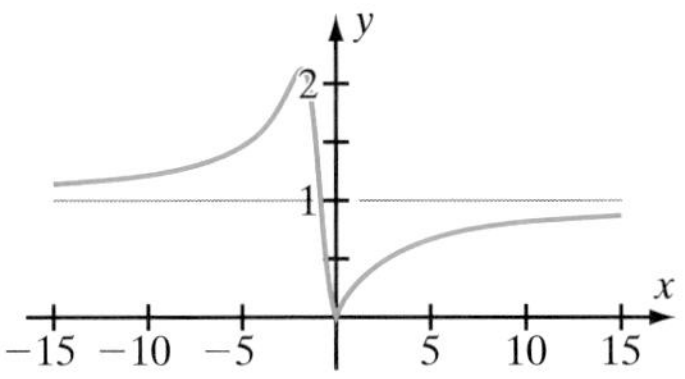

$f(x) = \sqrt{x^4 + x\sin(x)}/(x^2 + 2x + 2)$

Graphed in the viewing window.
$[-15, 15] \times [0, 2]$

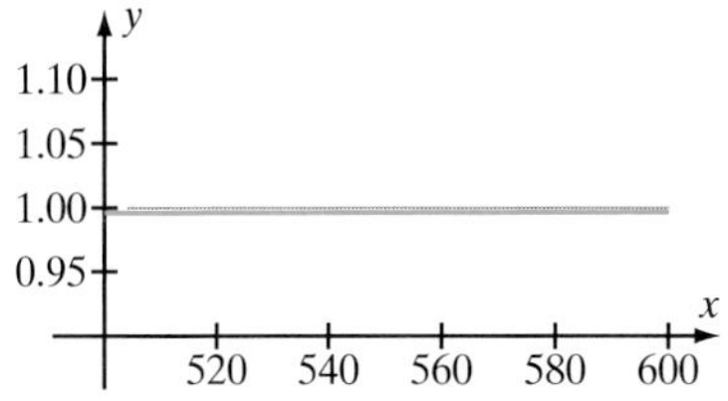

$f(x) = \sqrt{x^4 + x\sin(x)}/(x^2 + 2x + 2)$

Graphed in the viewing window.
$[500, 600] \times [0.9, 1.1]$

65.

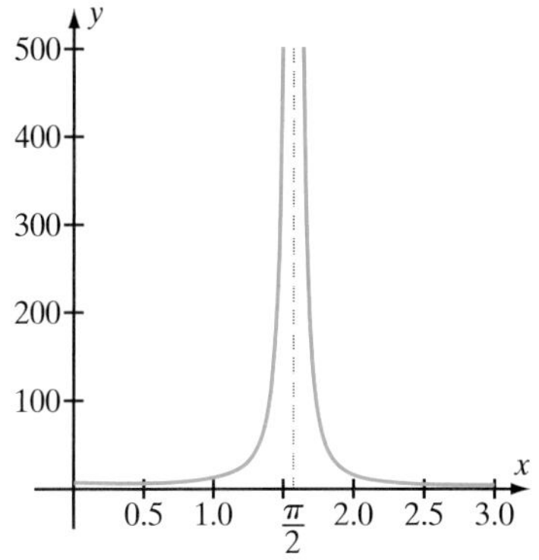

$f(x) = (4 + \cos(x))/(\sin(x) + \cos(2x))$

Section 2.5

1. 0 **3.** 3/2 **5.** 0

7. Does not converge.

9. 0

11. Does not converge.

13. −1 **15.** 1 **17.** −6 **19.** 1/2

21. 6 **23.** 0 **25.** 1/2 **27.** 0

29. 3 **31.** $\sqrt{3}$ **33.** 1/4 **35.** 0

37. 1 **39.** 0 **41.** 1/3 **43.** 1/6

45. 3/2 **47.** $3/(3-\sqrt{3})$ **49.** 10/9 **51.** 0

53. 2 **55.** 1/2 **57.** 16 **59.** 608911/4950

77. $\lim_{n\to\infty} n^{1/n} = 1$

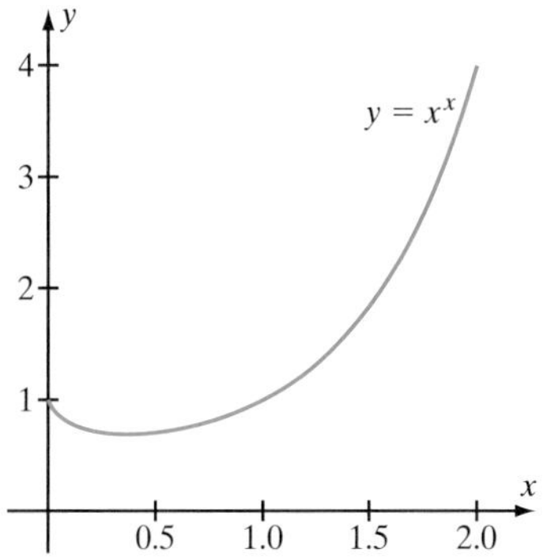

Section 2.6

1. $2^{\sqrt{3}}$ **3.** 2^{π} **5.** 11 **7.** 9

9. −3 **11.** 3 **13.** $\dfrac{2}{3} - \dfrac{1}{3}x$ **15.** 1

17. $-2x$ **19.** $3x$ **21.** 9 **23.** −3/2 **25.** $\log_6(2)$

27.

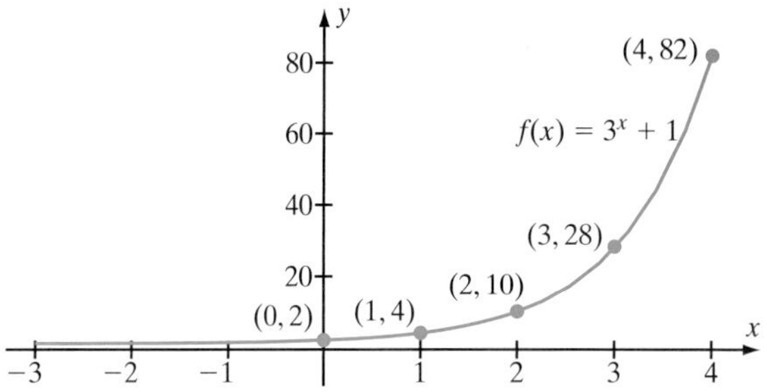

29.

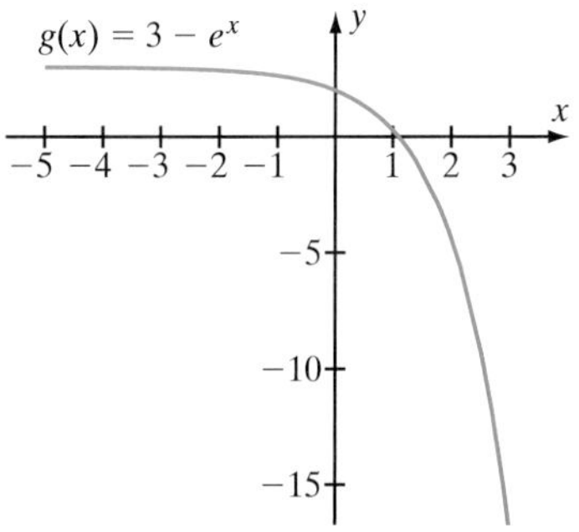

31.

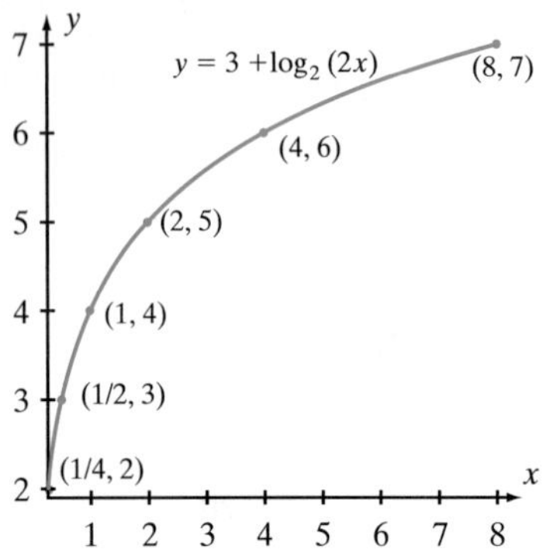

33. e **35.** $\exp(e)$ **37.** $e^{3/4}$ **39.** $e^{-\pi}$

41. $e^{-\pi/4}$ **43.** 8 **45.** 0 **47.** 1

49. e^3 **51.** 1

53. Vertical asymptote: $x = 0$; horizontal asymptotes: $y = 1$ and $y = -1$

55. Vertical asymptote: $x = 3$; horizontal asymptotes: $y = 0$ and $y = 4$

57. **a.** 1060 **b.** 1060.90 **c.** 1061.36 **d.** 1061.83 **e.** 1061.84

59. **a.** 5829.57 **b.** 5841.02 **c.** 5846.91 **d.** 5852.84 **e.** 5852.90

61. 17510 **63.** 7211 **65.** 3.707 g **67.** 15.77 g

69. $40\ln(2)/\ln(3/2) \approx 68.38$ yr

71. $100(\ln(2))/\ln(7/5) \approx 206$ yr

73.

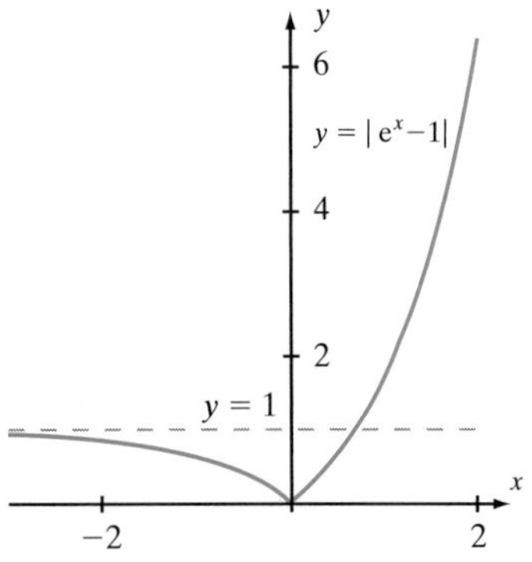

75.

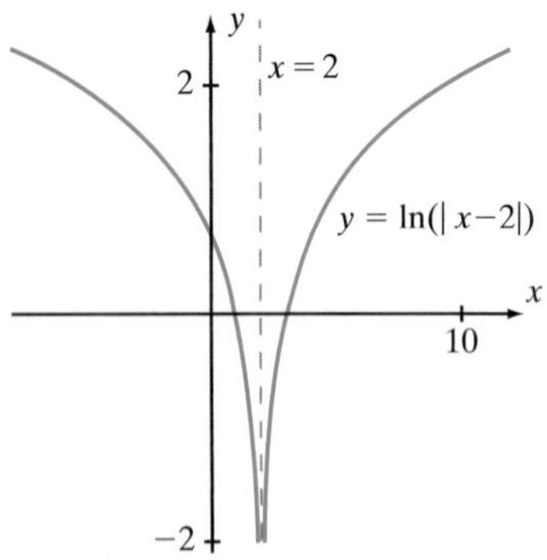

77. e **79.** e^6 **83.** \$44693

85. $\$1000(e^{-2/25} + e^{-3/25} + e^{-4/25})$, or about \$2662.18

87. Exact doubling time: $100\ln(2)/r$; 69 better approximates $100\ln(2)$

89. $y = 0$, $y = M$

91. $\lim_{t\to\infty} C(t) = \alpha/\beta$; horizontal asymptote : $y = \alpha/\beta$

93. α **95.** 63.84% **97.** 19996 yr

REVIEW EXERCISES FOR CHAPTER 2

1. Does not exist

3. -2 **5.** -12 **7.** $-1/10$ **9.** 6

11. Does not exist

13. -1 **15.** 1 **17.** 2 **19.** 4

21. 0 **23.** 7/3 **25.** 0 **27.** 3

29. Does not exist

31. 32 **33.** 4 **35.** 1 **37.** $1/\sqrt{2}$

39. -1 **41.** 9 **43.** ∞

45. Does not exist

47. 0 **49.** 1 **51.** $1/e$ **53.** 0

55. 0.01 **57.** 0.000500024 **59.** 1 **61.** -2

63. Horizontal asymptote: $y = 1$; vertical asymptote $x = -3$

65. Horizontal asymptote: $y = 0$

67. Horizontal asymptote: $y = 3$; vertical asymptote $x = -2$

69. 9 **70.** e^3 **71.** 64 **73.** $x + 2$

75. -2 **77.** $\log_{9/2}(18)$

79. $f(x) = \begin{cases} 0 & \text{if } x \text{ is not an integer} \\ 1 & \text{if } x \text{ is an integer} \end{cases}$

80. 3 **81.** 1 **83.** 0 **85.** 0

87. 3.97 more weeks

89. $1620\ln(5)/\ln(2) \approx 3761.5$ yr

91. **a.** 0.775 **b.** 0.802

93. 0.524 **95.** 5001 yr

Section 3.1

1. 0 **3.** 42 **5.** 6 **7.** 7

9. -2 **11.** -4 **13.** Forward **15.** Backward

17. 9 **19.** $-25/2$ **21.** 6 **23.** -6

25. $y = 20(x-5) + 50$

27. $y = 12(x+2) - 7$

29. $y = 20x - 20$

31. $y = 8x - 3$

33. $y = -(1/20)(x-5) + 50$

35. $y = -(1/12)\ (x+2) - 7$

37. $y = x/6 + 19/6$

39. $y = -x/28 + 85/14$

41. 6 **43.** 0.05 cents

45. 5 **47.** 16 **49.** $-5/4$

51. $y = 12(x-2) + 13$

53. $y = 12(x+3) + 38$

55. $A:10;\ B:3;\ C:-1;\ D:-3;\ E:0;\ F:20$

57. 64π in

59. Positive: $-1/2 < t$; negative: $t < -1/2$

61. $y = 18x/2^{1/3}$

63. 1, 5

65. $(1,-6)$, $(1/3, 4/27)$

67. $-\sqrt{2/3},\ \sqrt{2/3},\ -\sqrt{2},\ \sqrt{2}$

69. 2

71. a and b have the same sign.

77. **a.** i. -8 ii. 6.4 iii. 7.84 iv. 7.984 **b.** 8

Section 3.2

1. $\cos(3) - \sin(3)$

3. 2 **5.** 592

7. $3\cos(c) - 4\sin(c)$

9. $36 - \sin(2)$

11. $-\pi - 1/(\pi x^2)$

13. $12x^2 + 12x$

15. 3 **17.** $-\sqrt{2}$ and $\sqrt{2}$

19.

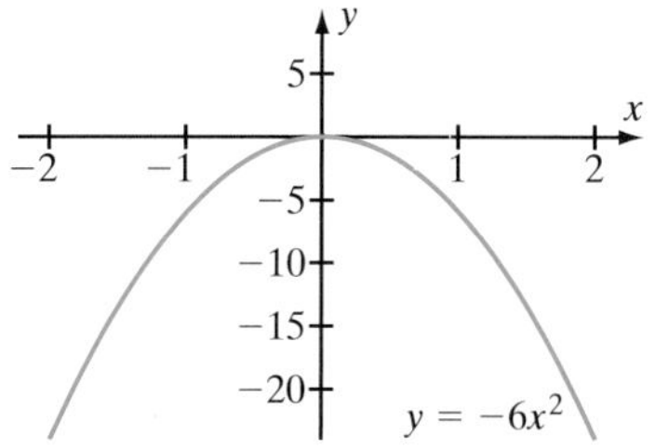

21.

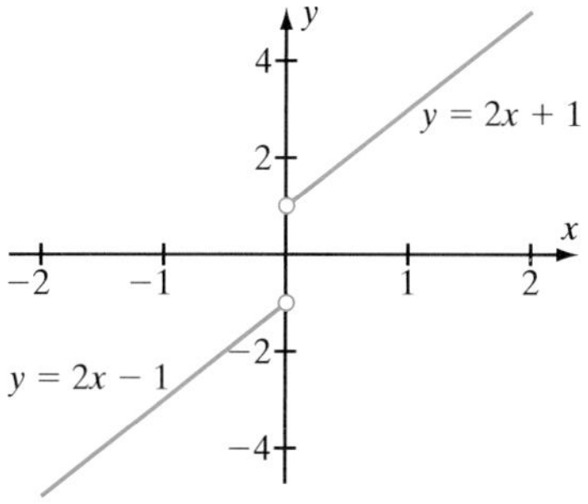

23. $y = \frac{1}{2}\left(x - \frac{\pi}{3}\right) + \frac{1}{2}\sqrt{3}$

25. $y = \sqrt{2}(x - 3\pi/4) - 3\sqrt{2}$

27. $1/(2\sqrt{x})$

29. $3/(2\sqrt{3x+7})$

33. 0 **35.** -32

37. $-5x + 16$

39. $-1/(1+x)^2$

41. $-x^{-1/2}(1+\sqrt{x})^{-2}/2$

43. No **45.** Yes

47. $f(x) = \sqrt{x}, c = 4$

49. $f(x) = 1/x,\ c = 5$

55. $(1-c^2)/(1+c^2)^2$

59. $E(p) = -pq'(p)/q(p)$

67.

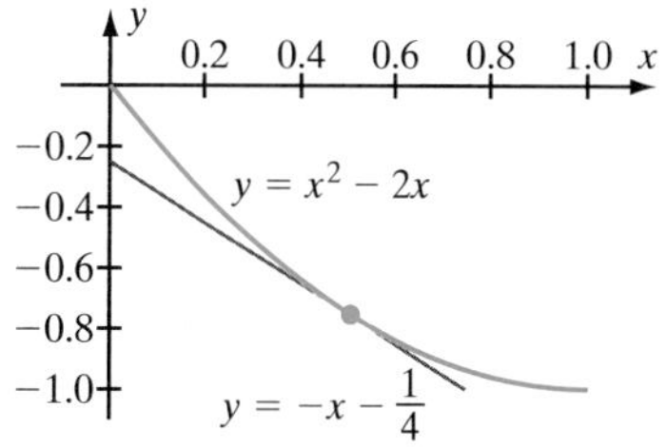

69.

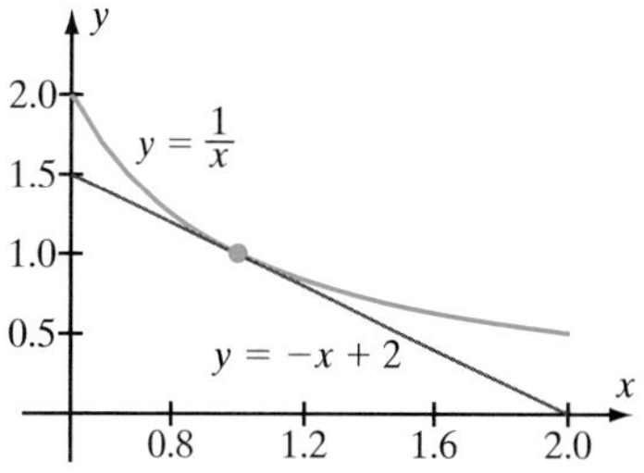

71. $f'(c) \approx 0.22727273$

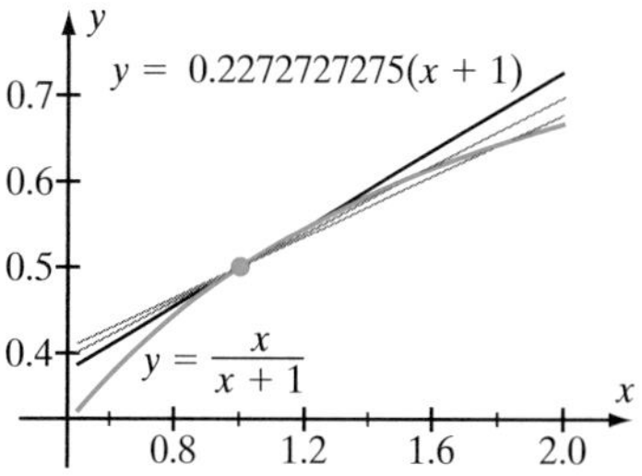

73. $f'(c) \approx 0.7019192$

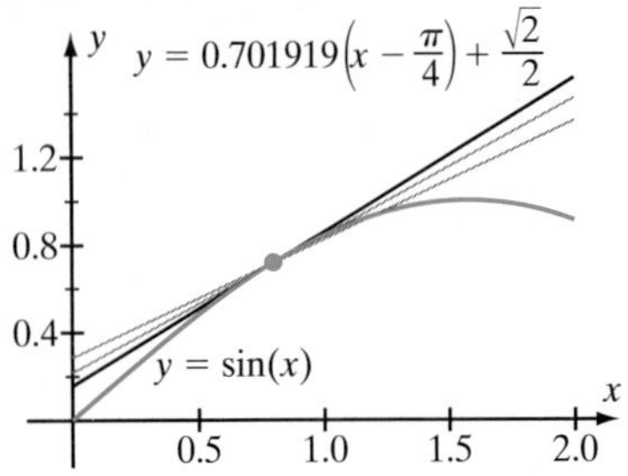

79. $f'(1)$ does *not* exist

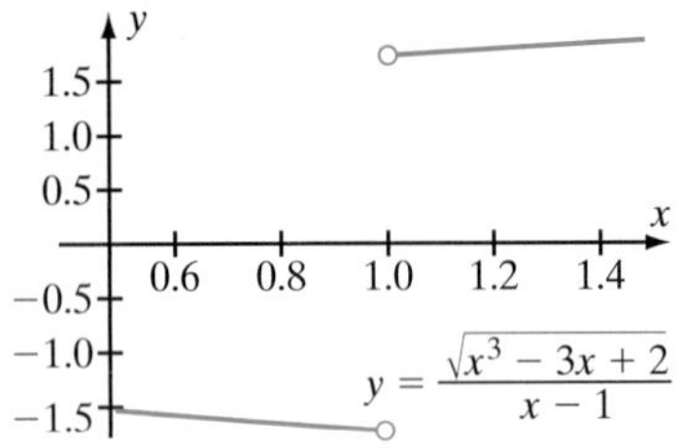

81. $f'(\pi/2)$ does *not* exist

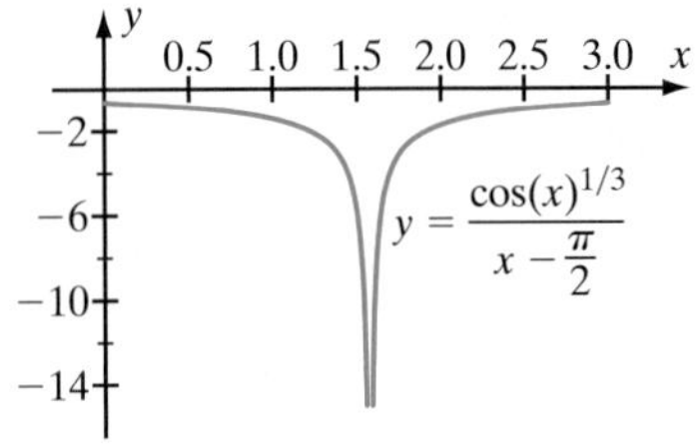

83. $f'(0) = 0$

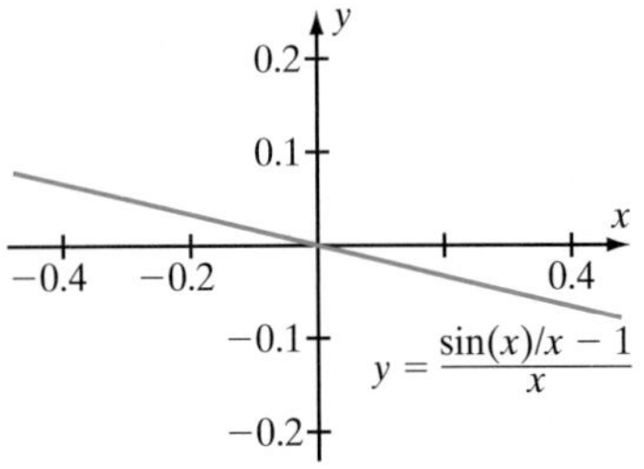

85.

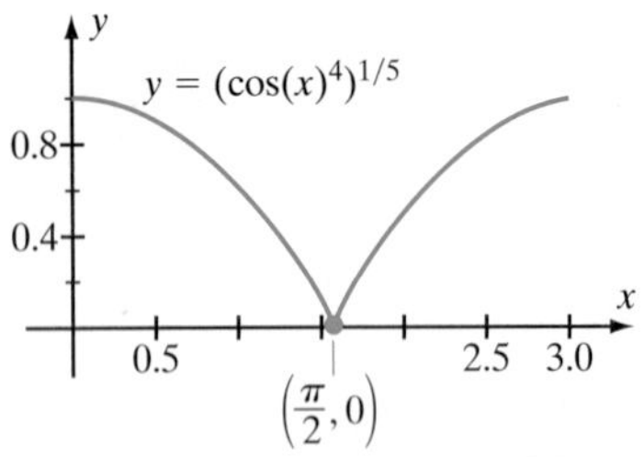

87.

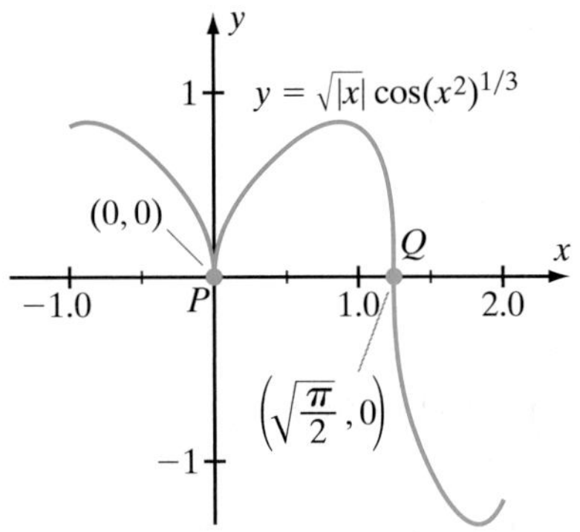

89. $f'(x) = 2\cos(2x)$

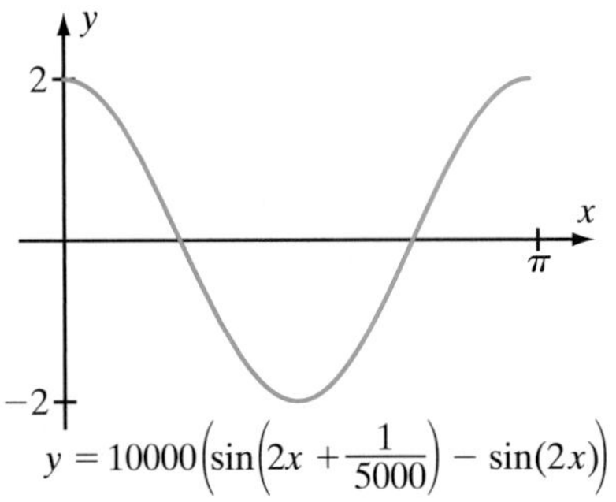

Section 3.3

1. $15x^2 - 6x$

3. $3x^2/\pi - \pi\sin(x)$

5. $\frac{4}{5}\cos(x) + \frac{3}{5}\sin(x) + 5$

7. $2x(x + 3) + x^2$, or $3x^2 + 6x$

9. $3x^2\cos(x) - x^3\sin(x)$

11. $2\sin(x)\cos(x)$

13. $\cos^2(x) - \sin^2(x)$

15. $7(2x + 1)\sin(x) + 7(x^2 + x)\cos(x)$

17. $-\sin(x)/x^2 + \cos(x)/x$

19. $-1/(1 + x)^2$

21. $-12x/(3x^2 + 4)^2$

23. $-(1 - 3\sin(x))/(x + 3\cos(x))^2$

25. $-27(x^2 - \sin(x))/(3\cos(x) + x^3)^2$

27. $1/(1+x)^2$ **29.** $2/(1+x)^2$ **31.** $4/x^3$

33. $\dfrac{x\cos(x)-\sin(x)}{x^2}$

35. $\dfrac{2x\cos(x)+(x^2+7)\sin(x)}{\cos^2(x)}$

37. $2\sin(x)/(1+\cos(x))^2$

39. $\dfrac{x(x+1)\cos(x)+\sin(x)}{(x+1)^2}$

41. $6x^5+6x^2+4x^3+2$

43. $2\sin(x)+\sin^2(x)\sec(x)\tan(x)$

45. $y=(5-x)/2$

47. $y=6-3x$

49. $y=x$

51. $\sec(x)\ (1+x\tan(x))$

53. $-2\sec(x)\tan(x)/(1+2\sec(x))^2$

55. 5.0 **57.** 20

59. $(f\cdot g)'(2)=22$, $(f/g)'(2)=29/32$

61. $(f\cdot g)'(-1)=-32$, $(f^2)'(-1)=20$, $(g^2)'(-1)=-64$

63. $2x$ **65.** $-6/x^7$

67. $\sin(x)\cos(x)+x(\cos^2(x)-\sin^2(x))$

69. $6x(1+x^2)^2$

71. **a.** $2/(c+1)^2$ **b.** $y=2(x+c^2)/(c+1)^2$
c. $4=2(2+c^2)/(c+1)^2$
d. $y=2x$ and $y=2x/9+32/9$

73. $A=4$, $B=-1$

81. **b.** 0.62% decrease
c. $E(70)\ \approx 7/3$, $E(90)\approx\ 39/4$, $E(80)\ \approx 102/25$

83.

n	Forward	Backward	Center
1	.48943	−.48943	0.00000
2	.04934	−.04934	0.00000
3	.00493	−.00493	0.00000
4	.00049	−.00049	0.00000
5	.00005	−.00005	0.00000

85.

n	Forward	Backward	Center
1	.24279	.24401	.24342
2	.24337	.24349	.24343
3	.24342	.24343	.24343
4	.24344	.24343	.24344
5	.24340	.24340	.24350

87. 0.86603, $h=1/1000$ suffices.

89. −1.41421, $h=1/1000$ suffices.

91.

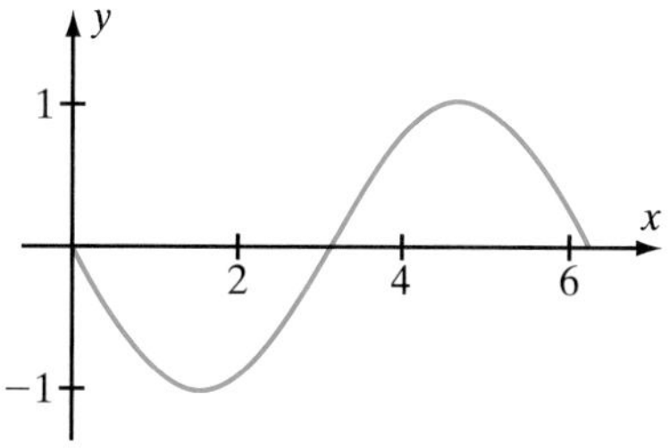

93. $L(2.70)\ \approx 0.99322$, $L(2.73)\approx 1.00435$.

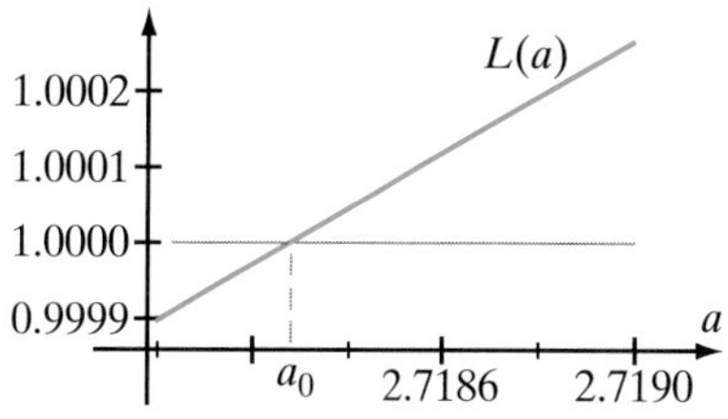

$a_0\approx 2.718$ and $\dfrac{d}{dx}a_0^x|_{x=0}\approx L(2.718)=1$.

Section 3.4

1. $80x^9+30x^{-6}$

3. $10x^{2/3}-15x^{-2/5}$

5. $2+e^x$

7. $\left(x-\dfrac{3}{2}\right)x^{-5/2}e^x$

9. $-9x^{-10}\cot\ (x)\ -x^{-9}\csc^2(x)$

11. $\dfrac{1}{2}x^{-3/2}\sec(x)\left(2x\tan(x)-1\right)$

13. $-\csc(x)\cot(x)-\csc^2(x)$

15. $-\cot^2(x)\csc\ (x)-\csc^3(x)$

17. $\tan^2(x)\sec(x)+\sec^3(x)$

19. $e^x(\sin(x)+\cos(x))$

21. $-21e^x/(5+7e^x)^2$

23. $-\dfrac{\sec(x)(1+\sec(x))}{\tan^2(x)}$

25. $x(x+2e^x-xe^x)/(x+e^x)^2$

27. $\dfrac{x^{-2/3}-1}{3(1+x^{2/3})^2}$

29. $y = x + 4$

31. $y = -(x - \pi)/\pi^3$

33. $y = 14(x - \pi/3) + 2\sqrt{3}$

35. $y = 8x + 4\pi x - \pi^2$

37. $2e^{2x}$

39. $-2\csc^2(x)\cot(x)$

41. $x^7 - 2x^2 + 6x + C$, where C is any constant

43. $x^9/9 + x^6 - 3x^2/2 + C$, where C is any constant

45. $2\sqrt{x} + C$, where C is any constant

47. $8\cot(x) + C$, where C is any constant

51. $\tan(x) - x + C$, where C is any constant

53. -7 **55.** 63 **59.** 1 **63.** 0 and 1

65. $\sqrt{p/(2-p)}$ **71. a.** $bp/(a - bp)$ **b.** b

73. $\tanh'(x) = \operatorname{sech}^2(x)$, $\operatorname{sech}'(x) = -\tanh(x)\operatorname{sech}(x)$

75. + 77.

Interval where f increases	Interval where f decreases	Point at which f has a horizontal tangent
75. $[-0.4805, 0.5]$	$[-1.3, -0.4805]$	-0.4805
77. $[0, 4.9936]$	$[4.9936, 7]$	4.9936
Interval where $f' > 0$	**Interval where $f' < 0$**	**Point(s) at which $f' = 0$**
75. $[-0.4805, 0.5]$	$[-1.3, -.4805]$	-0.4805
77. $[0, 4.9936]$	$[4.9936, 7]$	4.9936

79. $g_k'(0) = k$

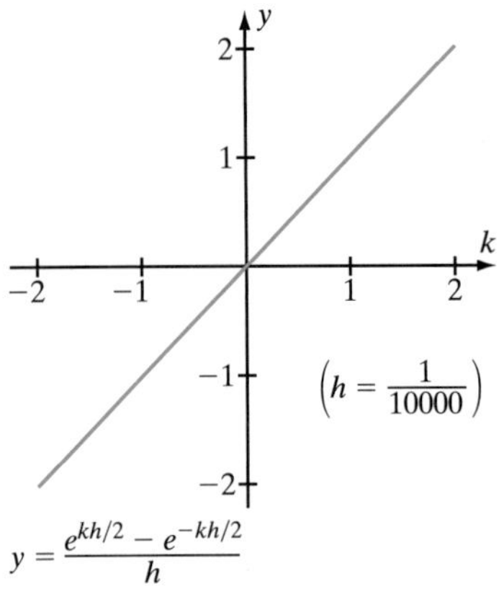

81. a. $(\mu^4\tau_0 p_0 - \tau_0 p_0 + 2\mu^2\Delta)/(p + \mu^2 p_0)^2$

e.

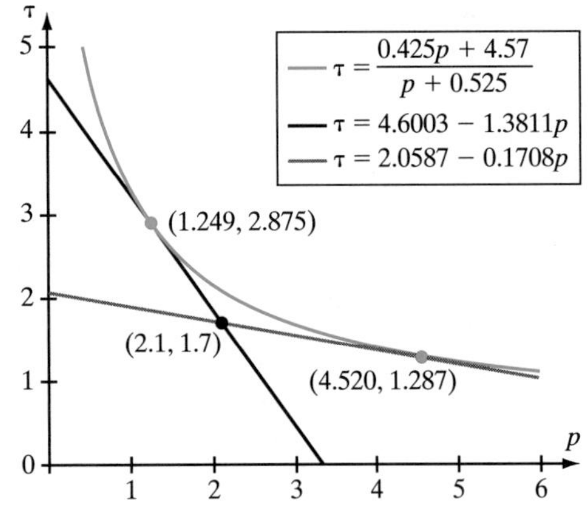

Section 3.5

1. $3\cos(3x)$

3. $5e^{5x}$

5. $(1/x^2)\sin(1/x)$

7. $(2x + 3)\cos(x^2 + 3x)$

9. $-\sin(x)\sec(\cos(x))\tan(\cos(x))$

11. $-7\sec^2(1 - 7x)$

13. $(1 + x/2)/(1 + x)^{3/2}$

15. $-2x\csc(x^2 + 4)(\csc(x^2 + 4) + \cot(x^2 + 4))$

17. $(2/x^3)\exp(-1/x^2)$

19. $3(2 + \cos(x))(2x + \sin(x))^2$

21. $x/\sqrt{4 + x^2}$

23. $3\sin^2(x)\cos(x)$

25. $6e^x/\sqrt{1 + e^x}$

27. $(f \circ g)'(x) = 2x + 7$; $(g \circ f)'(x) = (2x^3 + 7x)/\sqrt{x^4 + 7x^2}$

29. $(f \circ g)'(x) = 3x^2/(x^3 + 1)^2$; $(g \circ f)'(x) = 3x^2/(1 + x)^4$

31. $2^x \ln(2)$ **33.** $2(3^{2x})\ln(3)$

35. $8^x \ln(8) - 8x^7$

37. $-(1/x^2)(1 + e)^{1/x}\ln(1 + e)$

39. $(3^x \ln(3) - (3^x + 2)\ln(5))/5^x$

41. 32 **43.** 4

45. $5(1 + 2x)^4/\sqrt{1 + (1 + 2x)^5}$

47. $-6\cos^2(2x)\sin(2x)$

49. $3\sec^2(\sqrt{3x + 2})/\sqrt{3x + 2}$

51. $5/36$ **53.** $9\sqrt{10}/40$

59. $-0.001213\ \mathrm{yr}^{-1}$, 0, 0

61. $8\sqrt{2}\,\mathrm{in}^2/\mathrm{min}$

63. $(1-2t)f'(t^2); f'(t) - g'(\sqrt{t}) = (1-2\sqrt{t})f'(t)$

65. $10\sqrt{7/x}\sec^5(\sqrt{7x})\tan(\sqrt{7x})$

67. $\sec^2(\sqrt{x})/\sqrt{x} + 3\sec^2(x)/\sqrt{\tan(x)}$

69. $x(5x^3+1)(25x^3+30x-1)/(x^2+1)^{3/2}$

71. $4\pi^2 x\sin\big(\pi\sin(\pi x^2)\big)\cos\big(\pi\sin(\pi x^2)\big)\cos(\pi x^2)$

73. $3\sqrt{(\sqrt{2x-1}+\sqrt{2x+1})/(4x^2-1)}$

75. Yes: $c = 1/3$

77. **a.** P_0, M

b. $P_0 M^2 (M-P_0)ke^{-kMt}/\big(P_0 + (M-P_0)e^{-kMt}\big)^2$

c. 0

79. **a.** $\sqrt{g/\kappa}$

b. $4g\exp(-2t\sqrt{g\kappa})/\Big(1+\exp(-2t\sqrt{g\kappa})\Big)^2$

c. 0

81. $\cosh'(ax) = a\sinh(ax)$, $\sinh'(ax) = a\cosh(ax)$

85. 0.38677, 0.26795, 1

87. -1.921×10^{-11} g/yr

89.

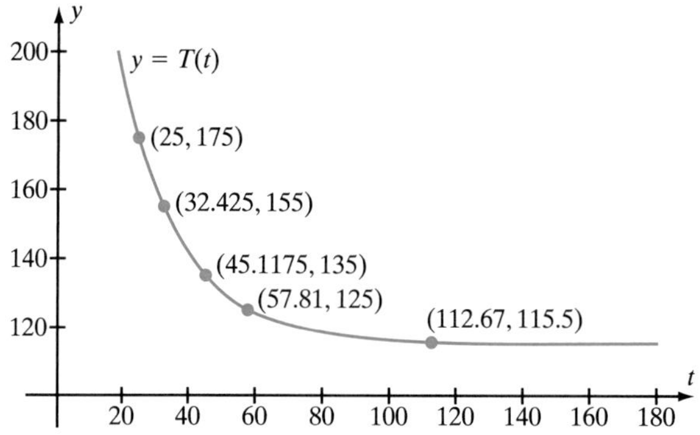

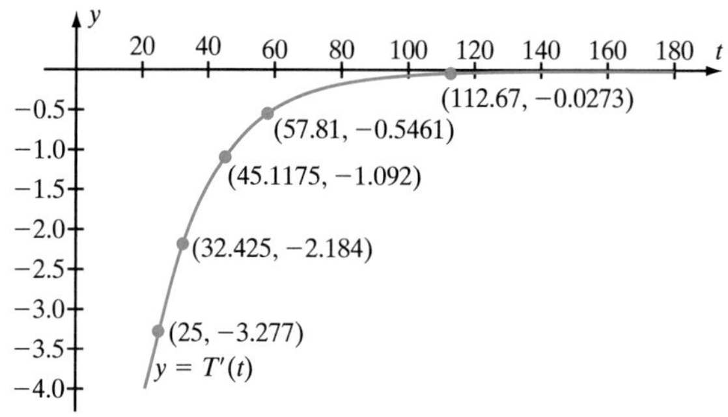

91. v is everywhere continuous; v is differentiable at all points except at $t = 8$. v' is continuous and differentiable at all points of its domain, which does not include the point $t = 8$.

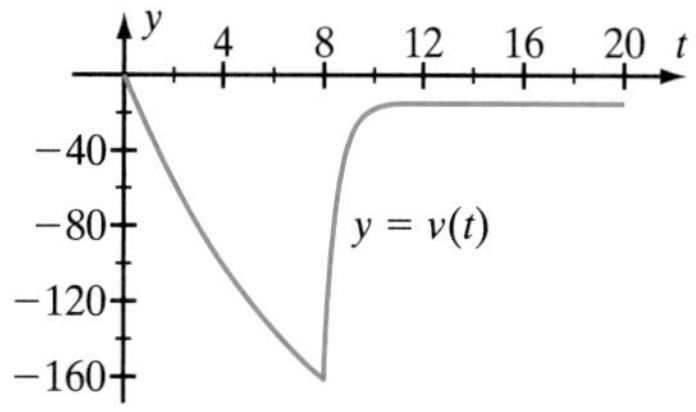

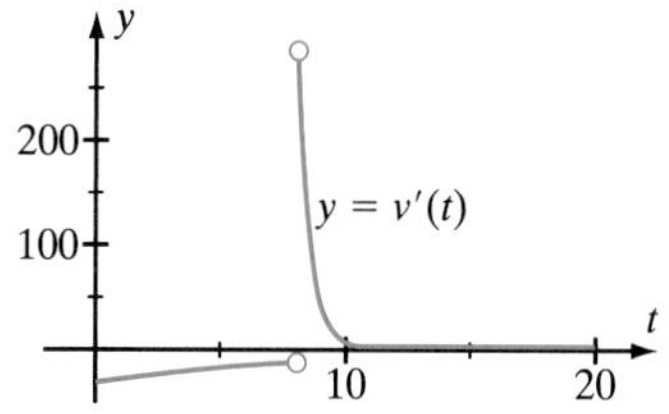

Section 3.6

1. 1/2 **3.** 1/3 **5.** 1/6 **7.** 4/5

9. 1/3 **11.** $-2/t^3$

13. $(t-2)^{-4/5}/5$

15. $1/(2\sqrt{1+t})$

17. $t/\sqrt{t^2-9}$

19. $1/2 + t/(2\sqrt{t^2-4})$, or $(t^2 + t\sqrt{t^2-4} - 2)/(t^2 + t\sqrt{t^2-4} - 4)$

21. $2\ln(t)/t$

23. $2^t\ln(2)$

25. $1/\big(3t\ln(2)(\log_2(t))^{2/3}\big)$

27. $1/x$ **29.** $\ln(x)$

31. $(1-\ln(x))/x^2$

33. $3/\big(x\sqrt{\ln(x)}\big)$

35. $\tan(x)$

37. $1/(x\ln(x))$

39. $\big(1+\exp(x)\big)/\big(x+\exp(x)\big)$

41. $\dfrac{\ln(3)}{\ln(2)}\,\dfrac{3^{\log_2(x)}}{x}$

43. $5/\big((5x+3)\ln(10)\big)$

45. $2^x\Big(\ln(1+4x) + 4/\big((1+4x)\ln(2)\big)\Big)$

47. $3x^2\log_{10}(6-x) - x^3/\big((6-x)\ln(10)\big)$

49. $1/\big(x(5)\ln(x)\big)$

51. $\dfrac{4(2x+1)(x^2+x)^3(x^3+x^2-1)^3}{(x^2+2)^2} + \dfrac{3(3x^2+2x)(x^2+x)^4(x^3+x^2-1)^2}{(x^2+2)^2} - \dfrac{4x(x^2+x)^4(x^3+x^2-1)^3}{(x^2+2)^3}$

53. $3x^{3x}(\ln(x)+1)$

55. $(2\ln(x)+1)x^{x^2+1}$

57. $2/x$

59. $-1/(2x)$

61. $s=-54(t-1/3)+6$

63. $s=(t-3)/5+1$

65. $f^{-1}(5)=2,\ (f^{-1})'(5)=1/14$

67. $f^{-1}(3/2)=1,\ (f^{-1})'(3/2)=1$

69. $f^{-1}(0)=1,\ (f^{-1})'(0)=1$

71. $f^{-1}(9)=64,\ (f^{-1})'(9)=128\ln(2)/3$

73. $1/7$

75. $2x^{\ln(x)}\ln(x)/x$

77. $(\log_2(x))^x(\ln(\log_2(x))+1/(\ln(2)\log_2(x)))$

79. $2\ln(x)/x$

81. $((g\circ f)^{-1})'(u)=1/(g'(t)f'(s))$ where $t=f(s)$ and $u=g(t)$.

83. $c=e,\ \gamma=1$

85. **a.** length / time **c.** $y=C$ **d.** domain: $[0,\infty)$, image: $(C,c(0)]$; **e.** domain: $(C,c(0)]$, image: $[0,\infty)$, $c^{-1}(s)=-\dfrac{V}{kA}\ln\left(\dfrac{s-C}{c(0)-C}\right)$ *f.* $c'(t)=(CkA/V)\exp(-kAt/V)$, $(c^{-1})'(s)=-V\cdot(kA(s-C))^{-1}$

87. $p=p_0-\dfrac{p_0}{q_0\,E(p_0)}(q-q_0)$

89. 0.3065

91. 0.02701

93.

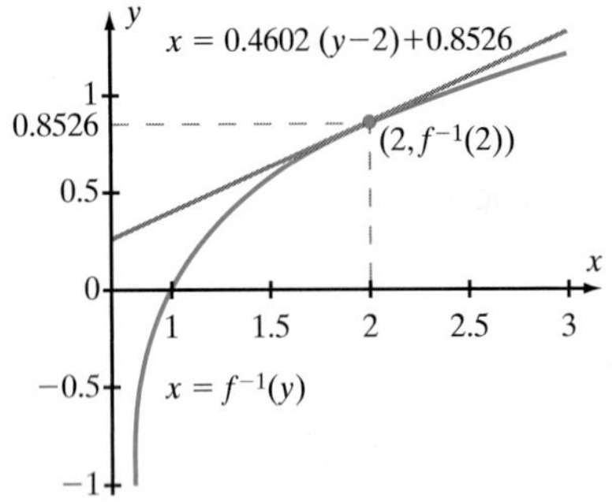

Graph of f^{-1} for $f(x)=1+x\exp(x)/2$

95.

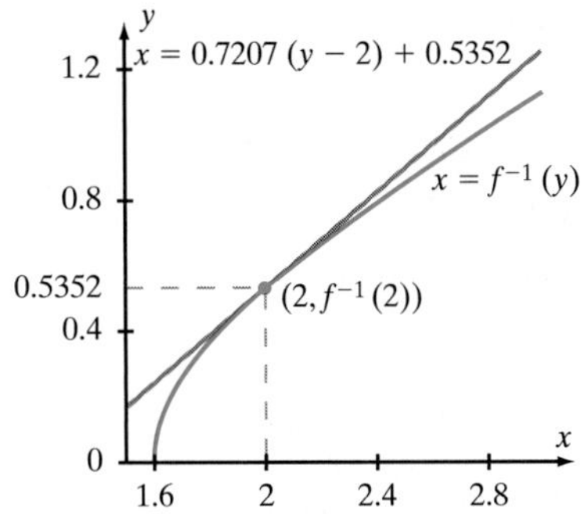

Graph of f^{-1} for $f(x)=\sin^2(x)-\cos(x)+2.6$

Section 3.7

1. $10x-6,\ 10,\ 0$

3. $-1/x^2,\ 2/x^3,\ -6/x^4$

5. $1/x,\ -1/x^2,\ 2/x^3$

7. $3/\sqrt{6x+5},\ -9/(6x+5)^{3/2},\ 81/(6x+5)^{5/2}$

9. $-3\sin(3x)$, $-9\cos(3x)$, $27\sin(3x)$

11. $2e^{2x},\ 4e^{2x},\ 8e^{2x}$

13. $1/(x+1),\ -1/(x+1)^2,\ 2/(x+1)^3$

15. $-2/(x-1)^2,\ 4/(x-1)^3,\ -12/(x-1)^4$

17. $90x(x^2+1)^{2/3},\ 30(x^2+1)^{-1/3}(7x^2+3),\ 40x(x^2+1)^{-4/3}(7x^2+9)$

19. $\cos(2x)-2x\sin(2x),\ -4\sin(2x)-4x\cos(2x),\ -12\cos(2x)+8x\sin(2x)$

21. $2^x\ln(2),\ 2^x\ln^2(2),\ 2^x\ln^3(2)$

23. $e^x(\sin(x)+\cos(x)),\ 2e^x\cos(x),\ 2e^x(\cos(x)-\sin(x))$

25. $1/(x\ln(3)),\ -1/(x^2\ln(3)),\ 2/(x^3\ln(3))$

27. $\cos(x)$

29. 696

31. $-360x^2+840x$

33. $64\sin(4x+3)$

35. $\pi/2$ **37.** $0\ \mathrm{m/s^2}$ **39.** $-16\ \mathrm{m/s^2}$

41. $0\ \mathrm{m/s^2}$ **43.** $6\ \mathrm{m/s^2}$

45. $38\ \mathrm{m/s^2},\ 74\ \mathrm{m/s^2}$

47. $128-32t$ ft/s, $-32\ \mathrm{ft/s^2}$

49. 324 ft **51.** -176 ft/s

53. $(-6-3t)\ \mathrm{ft/s^2}$

59. x^3-4x^2+2x+2

61. $2\cos(x)-4x\sin(x)-x^2\cos(x)$

63. $(2\ln(x) - 3)/x^3$

65. $x^{-3/2}(x^2 + x - 1/4)e^x$

69. $P_0(x) = 1,\ P_1(x) = x,\ P_2(x) = (3x^2 - 1)/2,$
$P_3(x) = (5x^3 - 3x)/2,\ P_4(x) = (35x^4 - 30x^2 + 3)/8$

71. $H_0(x) = 1,\ H_1(x) = 2x,\ H_2(x) = 4x^2 - 2,$
$H_3(x) = 8x^3 - 12x, H_4(x) = 16x^4 - 48x^2 + 12$

79.

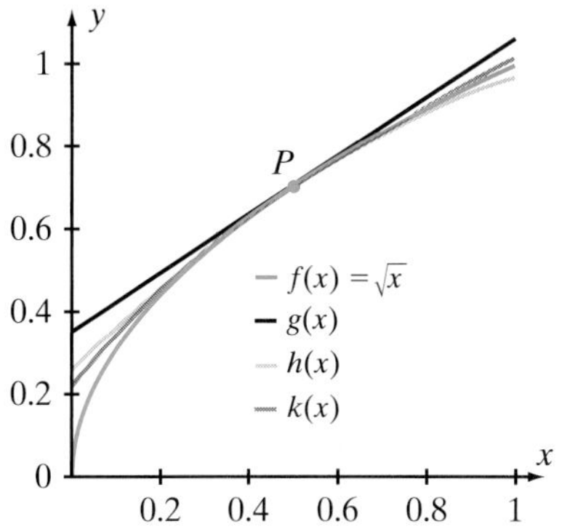

81.

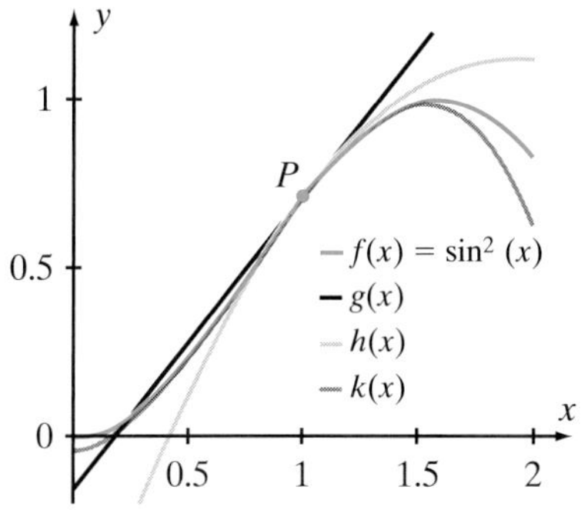

83. -0.0061 **85.** -0.0069

Section 3.8

1. $6y^{1/2}(dy/dx)$

3. $2y^{-1/2} - xy^{-3/2}(dy/dx)$

5. $e^y + xe^y(dy/dx)$

7. $-8/5$ **9.** $-1/16$ **11.** $-5/12$ **13.** $-1/5$

15. $y = 13/7 - 3x/14$

17. $y = \sqrt{3}(x - \pi/3)/2 + 1/4$

19. $y = x/2$

21. $y = x/4 - 1$

23. $y = 12/5 - 2x/5$

25. $y = -7(x - 2) + 2$

27. $y = -(x - \pi/4) + \pi/4$

29. $-1/4$; $-3/32$

31. $-9/8$; $135/128$

33. -1; -2

35. $y = 3x/2 - 1$

37. $y = x/7 - 32/63$

39. $y = (3 + 2\sqrt{2})(x - \pi/4 - \sqrt{2}/2) + \pi/4 + \sqrt{2}/2$

41. $y = 124x/45 - 16/15$

43. $y = -\sqrt{3}e^{(1-\sqrt{3})/2}\,(x - e^{\sqrt{3}/2}) + e^{1/2}$

45. $y = \ln(27)(x - 5)/\ln(4) + 5$

51. 1

53. $4(\ln(2) - 1)/(2\ln(2) - 1)$

55. a^2

57. $-A\ x^{-7/2}\,y^3\,(b + 3x/2)(y - 2)^{-1}\exp(b/x)$

59. **a.** -0.289 **b.** 0.503

61. -1.0006

63. -2.862

65. $y_0 = -10.6657$; $y = -1.00018x - 0.6639$

67. $y_0 = 4.4599$; $y = -3.9162x - 3.4214$

69.

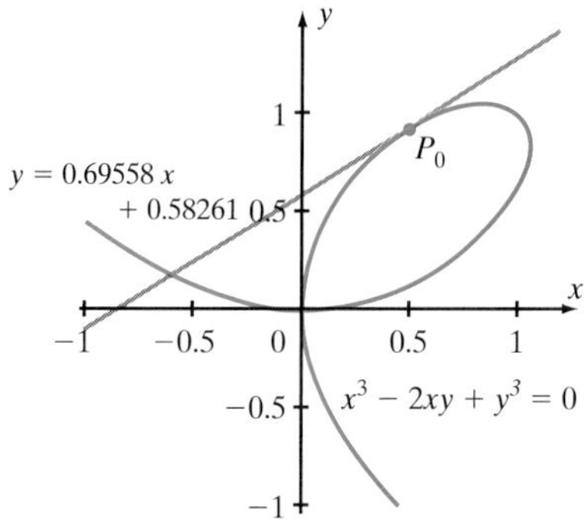

71.

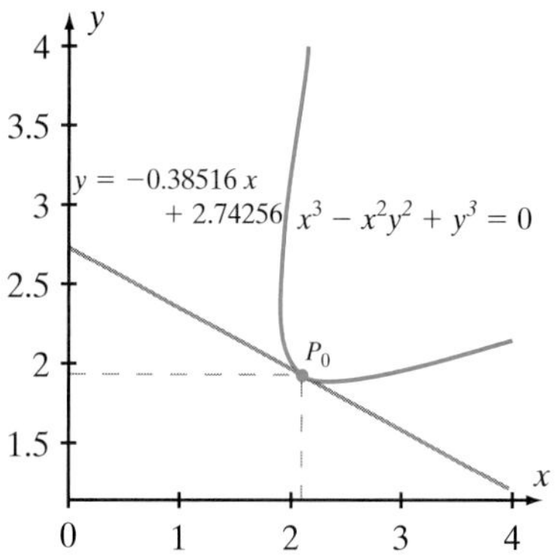

73.

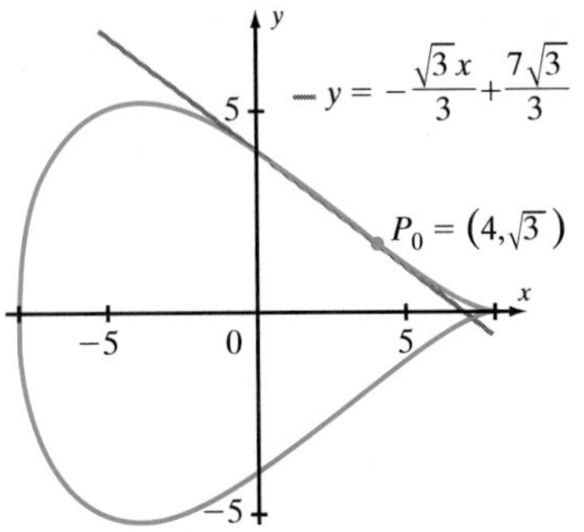

Tear Drop Curve $x = 8\cos(t)$, $y = 8\sin(t)\,(\sin(t/2))^2$

75.

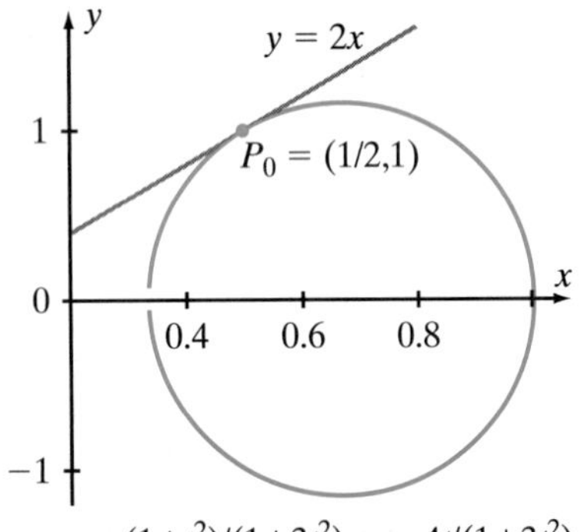

$x = (1+t^2)/(1+3t^2)$, $y = 4t/(1+3t^2)$

77.

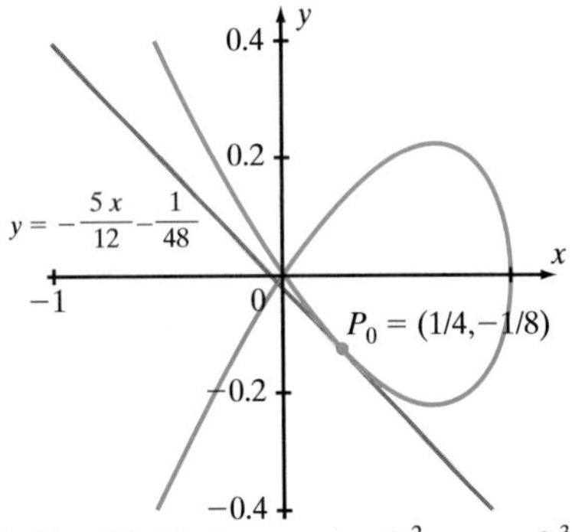

Cubic of Tschirnhaus $x = 1 - 3t^2$, $y = t - 3t^3$

79. $(4^{1/3}, 0)$

81. $q'(4) = -0.4317$, $q(4.2) = 4.0680$, $q(3.8) = 4.2408$, $q'(4) \approx D_0 q(4, 0.4) = -0.432$

Section 3.9

1. 1.975 **3.** 0.4985417 **5.** 4.05

7. 3.825 **9.** 0.48891

11. $2.6 - \pi/2 \approx 1.0292$

13. 1.138968

15. $2 + 20/\exp(3) \approx 2.995741$

17. 0.83

19. $1 + 0.13\ln(16) \approx 1.3604$

21. 0.9627

23. $1.2 - \pi/4 \approx 0.4146$

25. $c = 25$; estimate: 49/10; error: $= 1.02 \times 10^{-3}$

27. $c = 9$; estimate: 2.02917; error: $= 7.5 \times 10^{-4}$

29. $c = 60$; estimate: $1/2 + \sqrt{3}\pi/72$; error: $= 0.002$

31. $1/3 - 2x/9$ **33.** $x/8$ **35.** $1 - x$

37. $1/2 - \sqrt{3}(x - \pi/3)/2$

39. $\frac{10}{3}\ln\left(\frac{5}{3}\right)\left(x - \frac{1}{2}\right) + \frac{5}{3}$

41. 1

43. $1 + p(a - b)x$

47. 2.6 **49.** 3/2

51. Slope: −0.26124; 3235 units at \$6.80; 3287 units at \$6.60

53. 365987

55. One-step approximation: 3.03333; three-step approximation: 3.03309; exact: 3.03296...

57. One-step approximation: 1.10364; three-step approximation: 1.10109; exact: 1.09861...

59. $y_1 = 2.6$; $z_1 = 2.68$; exact solution: $4\exp(0.2) - 2.2$(about 2.6856)

61. $y_1 = 1.5$; $z_1 = 1.8828125$; exact solution: $5/32 + 11/(4\sqrt{e})$(about 1.82421)

63. Approximate demand: 9989; exact demand: 10000; relative error: 0.11%

65. Approximate demand: 3072693; exact demand: 3200000; relative error: 3.98%

Section 3.10

1. $\pi/2$ **3.** $-\pi/2$ **5.** $\pi/3$ **7.** $\pi/4$

9. $-\pi/3$ **11.** $\pi/4$ **13.** $-\pi/4$

15. $-\pi/4$ **17.** $5\pi/6$

19. $-1/\sqrt{2x - x^2}$

21. $\arcsin(x) + x/\sqrt{1 - x^2}$

23. $-1/\left(\ln(2)\sqrt{1 - x^2}\arccos(x)\right)$

25. $e^x/(1 + e^{2x})$

27. $\arctan(x) + x/(x^2 + 1)$

29. $2x\,\mathrm{arcsec}(x) + x/\sqrt{x^2 - 1}$

31. $-2/(x\sqrt{x^4 - 1})$

33. $-1/(2\sqrt{x}(x+1))$

35. $3\cosh(3x)$

37. $\coth(x)$

39. $(x\sinh(x) - \cosh(x))x^2$

41. $\operatorname{sech}^2(x)/\tanh(x)$

43. $\operatorname{sech}^2(\tan x)\sec^2(x)$

45. $-\operatorname{sech}(-\sqrt{x})\tanh(\sqrt{x})/(2\sqrt{x})$

47. $5/\sqrt{1+25x^2}$

49. $\cosh(x)/(1+\sinh^2(x))$

51. $-\csc(x)$

53. $-\operatorname{csch}(x)$

55. $\arcsin(x) = \pi/6 + (2/\sqrt{3})(x - 1/2)$

57. $\pi/3 + (x - \sqrt{3})/4$

77. $p(u, v) = u^2 + v^4 - 6v^2 - 2uv^2 - 6u + 9$

79. $p(u, v) = u^2 + v^6 - 2uv^3 - u$

81. $f'(2.1) \approx 0.0887$

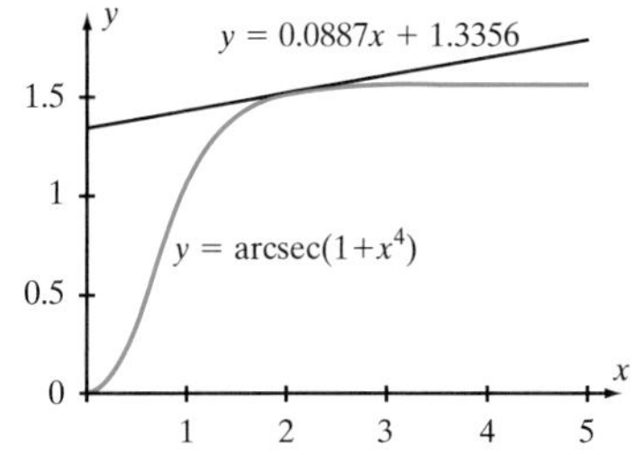

83. $f'(4.5) \approx 0.47213$

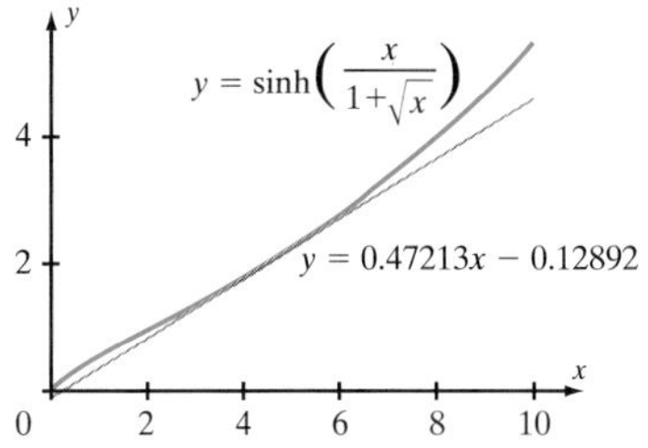

85. $f'(2.5) \approx 0.34815$

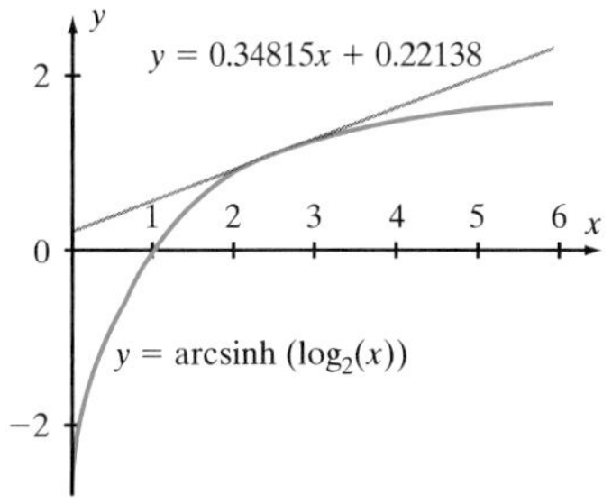

87.

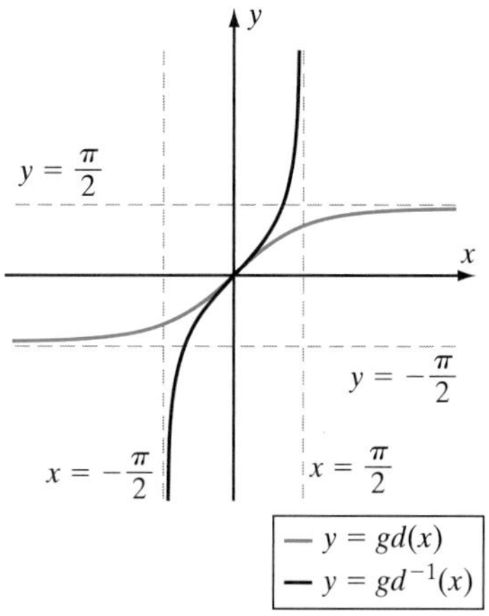

REVIEW EXERCISES FOR CHAPTER 3

1. $y = 12x + 16$

3. $y = (-3/32)x + 9/2$

5. $y = (6 + \pi\sqrt{3})x - \pi\sqrt{3}/6$

7. $y = (7/3)x - 2$

9. $y = (3/e^2)x - 2/e$

11. $y = -3\ln(2)x + 3$

13. $\frac{1}{2}x^{-1/2}$

15. $-(x+6)/x^3$

17. $3x^2 - 3/x^4$

19. $1 + 2/x^2$

21. $-9x^2/(2+x^3)^4$

23. $\cos(x)$

25. $-2\sin(2x + \pi/4)$

27. $2\sec(2x)\tan(2x)$

29. $x^{-2}\csc^2(1/x)$

31. $30x\sin^2(5x^2)\cos(5x^2)$

33. $3x^2\sin(x^2) + 2x^4\cos(x^2)$

35. $(1 + \sin(x) + \cos(x))/(1+\cos(x))^2$

37. $-3x\sin(3x^2)/\sqrt{1+\cos(3x^2)}$

39. $(1+3x)\,e^{3x}$

41. $(x^2+x+1)\,e^x/(x+1)^2$

43. $-2^{-x}\ln(2)$

45. $1/x - 1/x^2$

47. $x^2(1 + 3\ln(x))$

49. $1/(x\ln(3x))$

51. $2x\log_2(x+2) + x^2/((x+2)\ln(2)$

53. $\ln(30)2^x \cdot 3^x \cdot 5^x$

55. $-2/(x^2+4)$ **57.** -1

59. $2x\sinh(x^2)$

61. $e^x/\sqrt{1+e^{2x}}$

63. $60-32t$; -32

65. $-40t/(1+t^2)^2$; $40(3t^2-1)/(1+t^2)^3$

67. $5e^{-5t}$; $-25e^{-5t}$

69. $1/15$ **71.** 1 **73.** $1/(2e)$

75. $1/(1+\exp(y))$

77. $2x/(1-2y)$

79. $y^2/(x(y^2-1))$

81. $1/(1+3y^2)$; $-6y/(1+3y^2)^3$

83. $2\sin(x)/\cos(y)$; $2\sec^2(y)\,(\cos(y)\cos(x)+2\sin^2(x)\tan(y))$

85. $y=4x-6$

87. $y=\sqrt{3}x+(2-\pi\sqrt{3}/3)$

89. $y=-x+4$

91. 5.825 **93.** 21.3

95. $f'(c)\approx 2.214; f''(c)\approx 4.144$

97. $f'(c)\approx 27.605; f''(c)\approx 30.5$

Section 4.1

1. 60 **3.** 1 **5.** 3 **7.** -2 **9.** -3

11. $s_0=-4$

13. $v_0=-2$

15. $s_0=-24$

17. $s_0=1/2$; tangent line: $y=(x+13)/6$

19. $v_0=-1/2$; tangent line: $y=4x-2$

21. $s_0=-32/23$; tangent line: $y=(16x-2)/23$

23. 10 in^2/min

25. 27 cm^2/s

27. 3

29. 0.48 ft/s

31. $1/(384\pi)$ cm/min

33. $81\pi/8$ **35.** 115 ft/s

37. 1 **39.** -8

41. $120/\pi$ cm/min

43. 0.15 Ω/s

45. $(1/4, 1/2)$

47. $(1/2)$ cm/min

51. 405π cm^3/min

53. 2.349 cm/s

Section 4.2

1. Local minimum for $c=2$

3. Local maximum for $c=0$

5. Local minimum for $c=(4n+3)\pi/8$, integer n; local maximum for $c=(4n+1)\pi/8$, integer n

7. 6 **9.** $-1, 0, 1$ **11.** $-5, 1$ **13.** 1

15. 0 **17.** 1 **19.** e **21.** 3

23. 1 **25.** 4/9 **27.** $(-4+\sqrt{13})/3$ **29.** 0

31. 1/31 **33.** 5 **35.** $5x+C$

37. $x^3/3+\pi x+C$

39. $\sin(x)+C$

41. 15 **43.** 11 **45.** $2+e^3$ **47.** 3/2

49. $3\pi/4$ **51.** $\pi/4$ **53.** $1+\sqrt{3}$ **55.** $\sqrt{19}$

57. $f'(5)$ does not exist.

59. $f'(0)$ does not exist.

91.

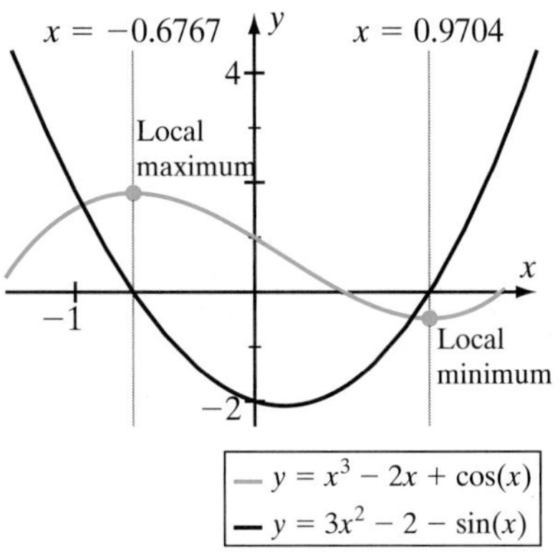

93.

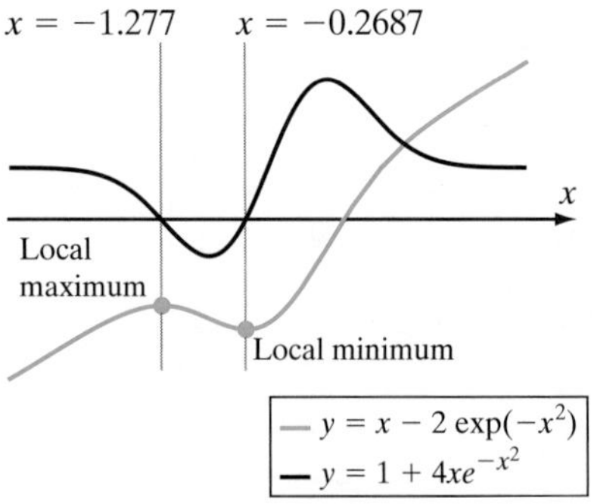

95.

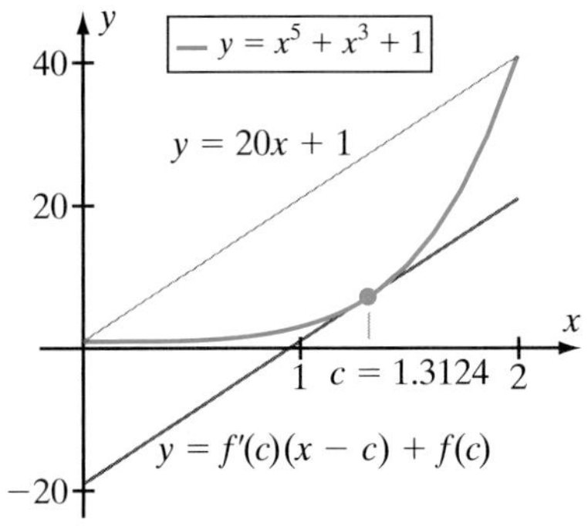

97.

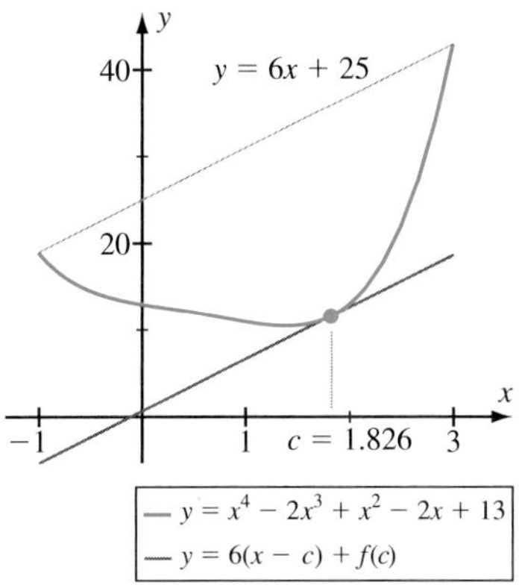

Section 4.3

1. Decreasing: $(-\infty, -1/2)$; increasing: $(-1/2, \infty)$; $c = -1/2$: (absolute) minimum

3. Decreasing: $(-5, 3)$; increasing: $(-\infty, -5) \cup (3, \infty)$; $c = -5$: local maximum; $c = 3$: local minimum

5. Decreasing: $(-\infty, -3) \cup (-2, 0)$; increasing: $(-3, -2) \cup (0, \infty)$; $c = -3$: local minimum; $c = -2$: local maximum; $c = 0$: local minimum

7. Decreasing: $(-5, 5)$; increasing: $(-\infty, -5) \cup (5, \infty)$; $c = -5$: local maximum; $c = 5$: local minimum

9. Decreasing: $(-\infty, 1) \cup (1, \infty)$

11. Decreasing: $(-2, -4/3)$; increasing: $(-\infty, -2) \cup (-4/3, \infty)$; $c = -2$: local maximum; $c = -4/3$: local minimum

13. Decreasing: $(-3, -2) \cup (-2, -1)$; increasing: $(-\infty, -3) \cup (-1, \infty)$; $c = -3$: local maximum; $c = -1$: local minimum

15. Decreasing : $(-3/2, 0) \cup (0, 3/2)$; increasing: $(-\infty, -3/2) \cup (3/2, \infty)$; $c = -3/2$: local maximum; $c = 3/2$: local minimum

17. Increasing: $(-\infty, \infty)$

19. Decreasing: $(-\infty, 0)$; increasing: $(0, \infty)$; $c = 0$: local minimum

21. Decreasing: $(-\infty, 0) \cup (2, \infty)$; increasing: $(0, 2)$; $c = 0$: local minimum; $c = 2$: local maximum

23. Decreasing: $\left(-\infty, -\ln(\ln(2))/\ln(2)\right)$; increasing: $\left(\ln(\ln(2))/\ln(2), \infty\right)$; $c = -\ln(\ln(2))/\ln(2)$: local minimum

25. Decreasing: $(0, 1/e)$; increasing: $(1/e, \infty)$; $c = 1/e$: local minimum

27. Decreasing: $(0, 1) \cup (1, e)$; increasing : (e, ∞); $c = e$: local minimum

29. Local minimum at $-2/5$

31. No extremum at 0; local minimum at 2

33. Local maximum at $-1/27$; no extremum at 0; local minimum at $1/27$

35. No extremum at 0; local minimum at 8

37. Local maximum at -5; local minimum at -1; no extremum at 1

39. Local maximum at $-1/3$; local minimum at 3; no extremum at 4;

41. Decreasing: $(-1, 0)$ and $(3/2, 2)$; increasing: $(-3/2, -1)$, $(0, 3/2)$, and $(2, 5/2)$

43. Local minima at $-\sqrt{\dfrac{\ln(2/\ln(2))}{\ln(2)}}$ and $\sqrt{\dfrac{\ln(2/\ln(2))}{\ln(2)}}$; local maximum at 0

45. No local extremum at $8/3$

47. Local maximum at $\sqrt{3}/9$

49. Local minimum at $\sqrt{5} - 2$

51. Local minimum at $-\dfrac{1}{2}\ln(3)$

53. Local minimum at $1/\sqrt{3}$

77. Decreasing on $(-\infty, 2.33344)$; increasing on $(2.33344, \infty)$

79. Decreasing on $(-\infty, -1.3363)$ and $(0.3449, 2.8134)$; increasing on $(-1.3363, 0.3449)$ and $(2.8134, \infty)$

81. Decreasing on $(-\infty, -4.0471)$ and $(-0.46916, 0.83853)$; increasing on $(-4.0471, -0.46916)$ and $(0.83853, \infty)$

83. Decreasing on $(0.36835, 215.9861)$; increasing on $(-\infty, 0.36835)$ and $(215.9861, \infty)$

85. Local maximum at -0.724068; local minimum at 0.4628463

87. Local minimum at 2.04289; local maxima at 0.50638 and 2.95695

Section 4.4

1. 25 m (each side)
3. $(2^{-1/3}, 2^{1/6})$
5. 10/3 in (each side)
7. -2
9. $\sqrt{3}/3$
11. Height = width = $2 + 4\sqrt{5}$ in
13. Base: $4\sqrt{3}/3$; height: 8/3
15. Side length of base: $2 \cdot 9^{1/3}$ m
17. $6\sqrt{33(6+\sqrt{3})}$ m
19. $5\pi/3$ in^3
21. 100
23. 25
25. $-1/4$, 2
27. $-1/2$, 31/2
29. -3, 9
31. 12, 13
33. 1, $e-1$
35. The positive integer in $(0, \alpha/\beta)$ nearest $\alpha/(2\beta)$
41. $(b+\sqrt{a^2+b^2})/a$
43. Circumference of circle: $\pi L/(\pi+4)$
45. Minimum area: $\pi L^2/(6\pi^2+8\pi+8)$; maximum area: $L^2/(4\pi)$
47. $(b/a)^{1/4}$
49. $(2b)^{1/6}$
53. $(r/R)^4$
55. 5000 m^2
57. $8/\pi$
59. $\ell/6$
63. -0.2228589, 1.059793
65. $10(2-2^{2/3}+2^{1/3})/3$ cm from the stronger source
67. $\frac{1}{2}\arccos\left(gh/(v^2+gh)\right)$
69. 1.029
71. 0.64

Section 4.5

1. Concave up: $(0,\infty)$; concave down: no interval; inflection points: none; critical points: 4; local minimum value at $x=4$; local maximum value: none
3. Concave up: $(0,9)$; concave down: $(9,\infty)$; inflection point: $(9,4)$; critical points: 3; local minimum value at $x=3$; local maximum value: none
5. Concave up: $(-\infty, -3\cdot 2^{1/3})$ and $(0,\infty)$; concave down: $(-3\cdot 2^{1/3}, 0)$; inflection point: $(-3\cdot 2^{1/3}, 0)$; critical points: 3; local minimum value at $x=3$; local maximum value: none
7. Concave up: $(-3,\infty)$; concave down: $(-\infty,-3)$; inflection point: $(-3,132)$; critical points: $-7, 1$; local minimum value at $x=1$; local maximum value at $x=-7$
9. Concave up: $(1/2,\infty)$; concave down: $(-\infty,1/2)$; inflection point: $(1/2,-11/2)$; critical points: -1, 2; local minimum value at $x=2$; local maximum value at $x=-1$
11. Concave up: $(-\infty,(1-\sqrt{7})/2)$ and $((1+\sqrt{7})/2,\infty)$; concave down: $((1-\sqrt{7})/2,\ (1+\sqrt{7})/2)$; inflection points: $((1-\sqrt{7})/2,\ -71/4+5\sqrt{7})$ and $((1+\sqrt{7})/2,\ -71/4-5\sqrt{7})$; critical points: $-3/2$, 0, 3; local minimum values at $x=-3/2$ and $x=3$; local maximum value at $x=0$
13. Concave up: $(-1,0)$ and $(1,\infty)$; concave down: $(-\infty,-1)$ and $(0,1)$; inflection points: $(-1,-16)$, $(0,-8)$, and $(1,0)$; critical points: -1, 1; local minimum values: none; local maximum values: none
15. Concave up: $(-3,0)$ and $(3,\infty)$; concave down: $(-\infty,-3)$ and $(0,3)$; inflection points: $(-3,-1/4)$, $(0,0)$, and $(3,1/4)$; critical points: $-\sqrt{3}$ and $\sqrt{3}$; local minimum value at $x=-\sqrt{3}$; local maximum value at $x=\sqrt{3}$
17. Concave up: $(-\sqrt{3}/3,\sqrt{3}/3)$; concave down: $(-\infty,-\sqrt{3}/3)$ and $(\sqrt{3}/3,\infty)$; inflection points: $(-\sqrt{3}/3,-1/2)$ and $(\sqrt{3}/3,-1/2)$; critical point: 0; local minimum value at $x=0$; local maximum values: none
19. Concave up: $(\pi/2,\ 3\pi/2)$; concave down: $(0,\ \pi/2)$ and $(3\pi/2,\ 2\pi)$; inflection points: $(\pi/2,\ \pi/2)$ and $(3\pi/2,\ 3\pi/2)$; critical points: $\pi/2$; local minimum values: none; local maximum values: none
21. Concave up: $(0,\pi)$; concave down: $(-\pi,0)$; inflection point: $(0,0)$; critical points: none; local minimum values: none; local maximum values: none
23. Concave up: $(0,\pi/4)$ and $(3\pi/4,\pi)$; concave down: $(\pi/4,\ 3\pi/4)$; inflection points: $(\pi/4,1/2)$ and $(3\pi/4,1/2)$; critical points: $\pi/2$; local minimum value: none; local maximum value at $x=\pi/2$
25. Concave up: $(-\infty,\infty)$; concave down: nowhere; inflection points: none; critical points: 0; local minimum value at $x=0$; local maximum values: none
27. Concave up: $(0,\infty)$; concave down: nowhere; inflection points: none; critical points: 1; local minimum value at $x=1$; local maximum values: none
29. Concave up: $(0,\infty)$; concave down: nowhere; inflection points: none; critical points: 2; local minimum value at $x=2$; local maximum values: none
31. Concave up: $(-\infty,-5/2)$ and $(0,\infty)$; concave down: $(-5/2,0)$; inflection points: $(-5/2, f(-5/2))$ and $(0,0)$; critical points: 5/4, 0; local minimum value at $x=5/4$; local maximum values: none
33. Concave up: $(-\infty,\infty)$; concave down: nowhere; inflection points: none; critical points: $\ln(\ln(2))/\ln(2)$; local minimum value at $x=\ln(\ln(2))/\ln(2)$; local maximum values: none

35. Concave up: $(\sqrt{2}/2, 1)$; concave down: $(-1, \sqrt{2}/2)$; inflection point: $(\sqrt{2}/2, \pi/4 - 1/2)$; critical point: $\sqrt{2}/2$; local minimum values: none; local maximum value: none

37. Local maximum at 0; local minimum at 1

39. Local maximum at -1; local minimum at 1

41. Local maximum at -2; local minimum at 2

43. $-1, 0$ **45.** none **47.** $-5, 5$

49. Inflection points: $(-2a/\sqrt{3}, 3a/2)$ and $(2a/\sqrt{3}, 3a/2)$; concave up: $(-\infty, -2a/\sqrt{3})$ and $(2a/\sqrt{3}, \infty)$; concave down: $(-2a/\sqrt{3}, 2a/\sqrt{3})$

51. Concave up for all x

53. Inflection point: $(m^{-1}(c), c)$; concave up on $(0, m^{-1}(c))$and concave down on $(m^{-1}(c), \infty)$

57. Concave down on $(0, 2\pi)$; local maximum at π

59. Concave up on $(-\infty, -\sqrt{3}/3)$ and $(\sqrt{3}/3, \infty)$; concave down on $(-\sqrt{3}/3, \sqrt{3}/3)$

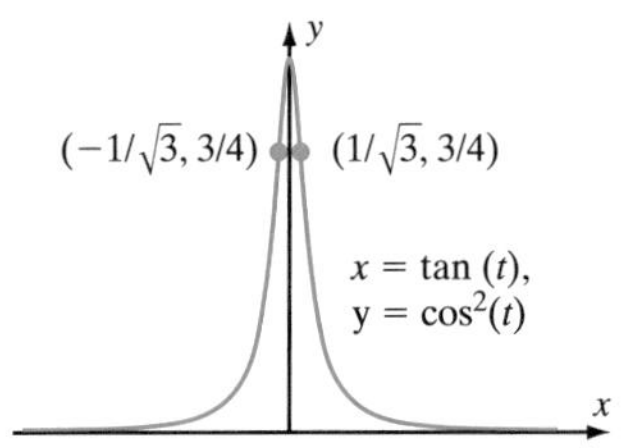

61. $|\cos(x)|/\left(1+\sin^2(x)\right)^{3/2}$

63. $2|x|^3/(x^4+1)^{3/2}$

65. $1/\sqrt[4]{5}$ **67.** $-\dfrac{1}{2}\ln(2)$

73. Critical point: 1.27846 local maximum; inflection points: $(-2.39936, -2.19968)$ and $(2.39936, 0.19968)$

75. Critical points: -0.80778 local minimum and 6.14167 local maximum; inflection points: $(-0.34378, -2.43554)$, $(0.53289, -1.72839)$, and $(5.01770, 346.83)$

77. Critical points: -1.44403 local maximum and 0.56043 local minimum; inflection points: $(-1.82524, 1.07320)$, $(-0.75879, 0.84663)$, and $(1.22627, 0.59430)$

79. Critical point: 0.70871 local maximum; inflection points: $(-0.40297, -0.52219)$ and $(1.44141, 0.21890)$

Section 4.6

1. Local maximum: $P = (-1, 12)$; local minimum: $R = (3, -20)$; point of inflection: $Q = (1, -4)$; increasing: $(-\infty, -1) \cup (3, \infty)$; decreasing: $(1, 3)$; concave up: $(1, \infty)$; concave down: $(-\infty, 1)$

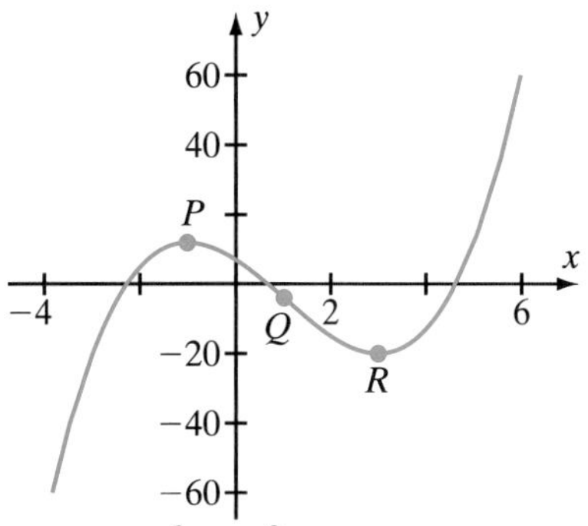

$f(x) = x^3 - 3x^2 - 9x + 7$

3. Local maximum: $P = (0, 4)$; local minimum: $R = (2, 0)$; point of inflection: $Q = (1, 2)$; increasing: $(-\infty, 0) \cup (2, \infty)$; decreasing: $(0, 2)$; concave up: $(1, \infty)$; concave down: $(-\infty, 1)$

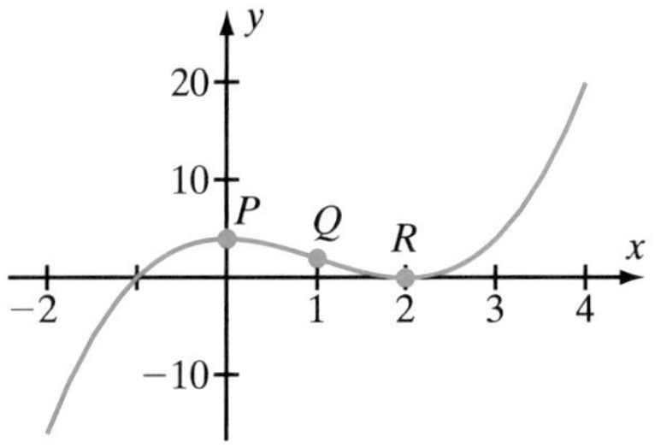

5. Local maximum: $S = (2, 1/4)$; local minimum: $Q = (-2, -1/4)$; points of inflection: $P = (-2\sqrt{3}, -\sqrt{3}/8)$, $R = (0, 0)$, and $T = (2\sqrt{3}, \sqrt{3}/8)$; increasing: $(-2, 2)$; decreasing: $(-\infty, -2) \cup (2, \infty)$; concave up: $(-2\sqrt{3}, 0) \cup (2\sqrt{3}, \infty)$; concave down: $(-\infty, -2\sqrt{3}) \cup (0, 2\sqrt{3})$; horizontal asymptote: $y = 0$

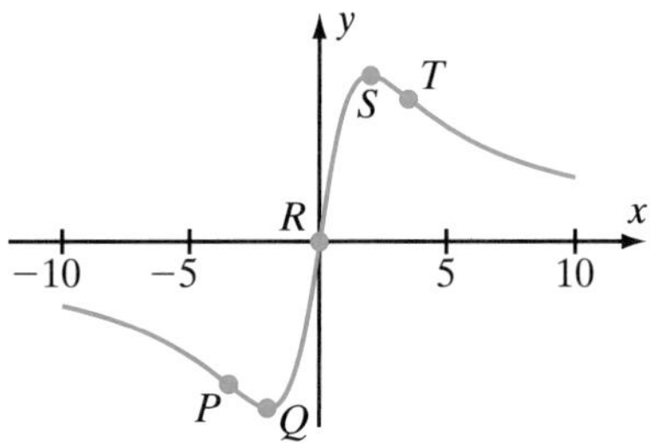

7. Local maximum: $P = (0, -1)$; increasing: $(-\infty, -2) \cup (-2, 0)$; decreasing: $(0, 2) \cup (2, \infty)$; concave up: $(-\infty, 2) \cup (2, \infty)$; concave down: $(-2, 2)$; vertical asymptotes $x = -2$ and $x = 2$; horizontal asymptote: $y = 1$

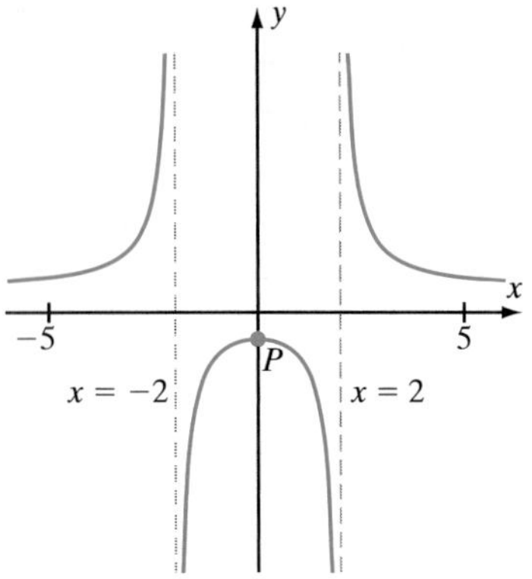

9. Local minimum: $P = (2^{-4/3}, f(2^{-4/3}))$; increasing: $(2^{-4/3}, \infty)$; decreasing: $(0, 2^{-4/3})$; concave up: $(0, \infty)$

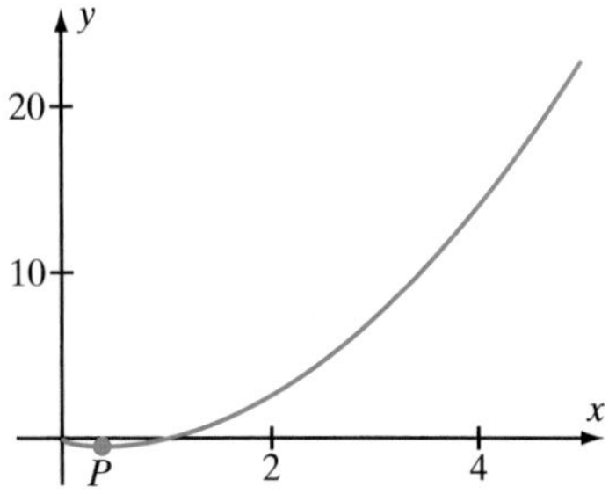

11. Points of inflection: $(n\pi, -n\pi)$ for every integer n; concave down: $(2k\pi, (2k+1)\pi)$ for every integer k; concave up: $((2k-1)\pi, 2k\pi)$ for every integer k

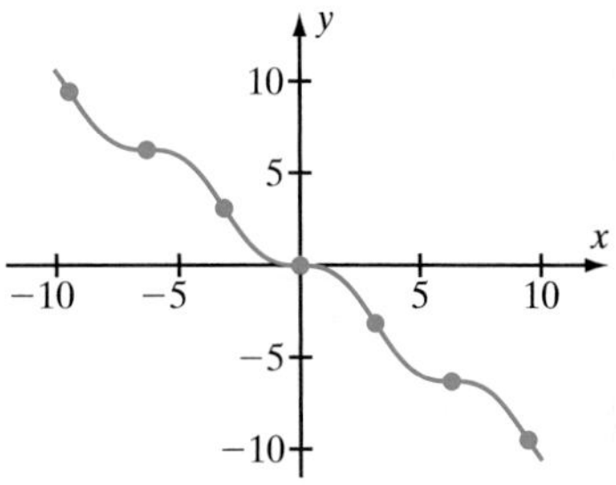

13. Local maximum: $P = (-27/8, 4/27)$; point of inflection: $(-(9/5)^3, 100/729)$; increasing: $(-\infty, -27/8)$; decreasing: $(-27/8, 0) \cup (0, \infty)$; concave up: $(-\infty, -(9/5)^3) \cup (0, \infty)$; concave down: $(-(9/5)^3, 0)$; vertical asymptote: $x = 0$; horizontal asymptote: $y = 0$

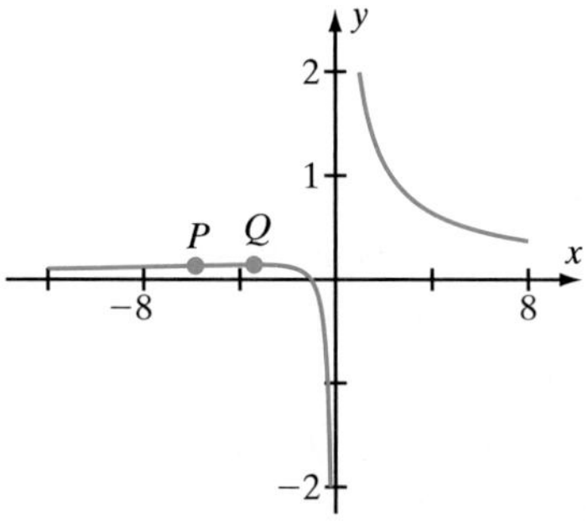

15. Local maximum: $Q = (3/2, f(3/2))$; points of inflection: $P = (3/2 - 9\sqrt{5}/10, f(3/2 - 9\sqrt{5}/10))$, $O = (0,0)$, and $R = (3/2 + 9\sqrt{5}/10, f(3/2 + 9\sqrt{5}/10))$; increasing: $(-\infty -3) \cup (-3, 3/2)$; decreasing: $(3/2, \infty)$; concave up: $(-\infty, -3) \cup (3/2 - 9\sqrt{5}/10, 0) \cup (3/2 + 9\sqrt{5}/10, \infty)$; concave down: $(-3, 3/2 - 9\sqrt{5}) \cup (0, 3/2 + 9\sqrt{5})$; vertical asymptote $x = -3$; horizontal asymptote: $y = 0$

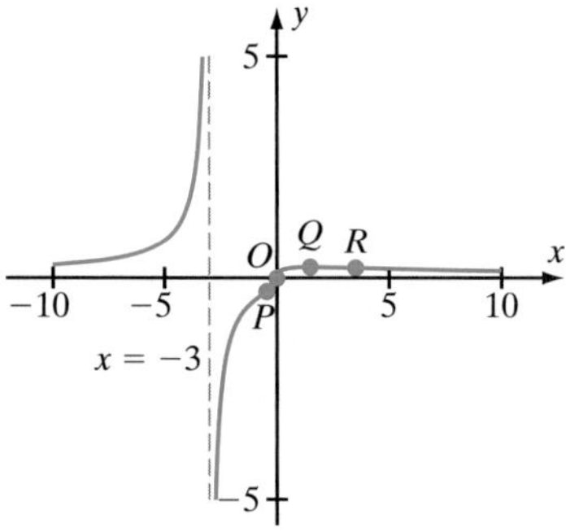

17. Local minimum: $P = (-5/6, f(-5/6))$; points of inflection: $Q = (-5/3, f(-5/3))$ and $R = (-1, 0)$; increasing: $(-5/3, \infty)$; decreasing: $(-\infty, -5/3)$; concave up: $(-\infty, -5/3) \cup (-1, \infty)$; concave down: $(-5/3, -1)$

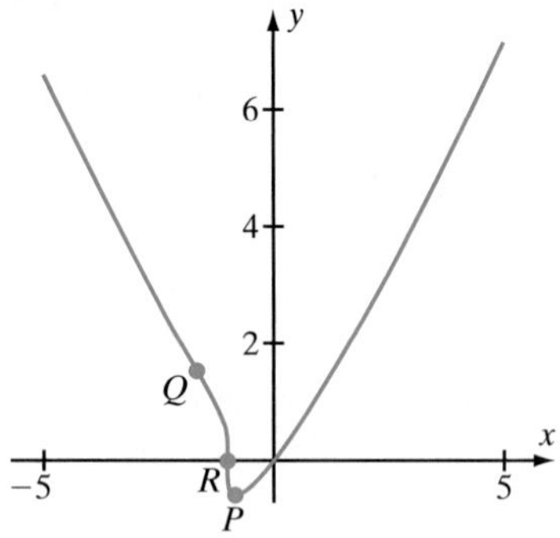

19. Local minimum: $P = (1/2, -3 \cdot 2^{-4/3})$; points of inflection: $Q = (0,0)$ and $R = (-1, 3)$; increasing: $(1/2, \infty)$; decreasing: $(-\infty, 1/2)$; concave up: $(-\infty, -1) \cup (0, \infty)$; concave down: $(-1, 0)$

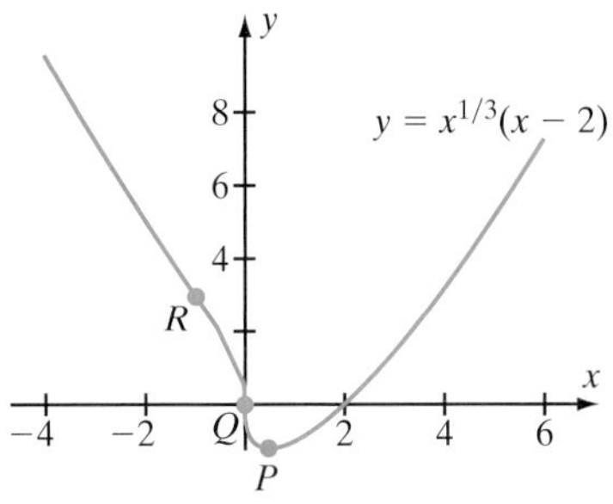

21. Local maximum: $P = (1, 1/2)$; point of inflection: $Q = \left(1 + 2/\sqrt{3}, f(1 + 2/\sqrt{3})\right)$; increasing: $(0, 1)$; decreasing: $(1, \infty)$; concave up: $(1 + 2/\sqrt{3}, \infty)$; concave down: $(0, 1 + 2/\sqrt{3})$

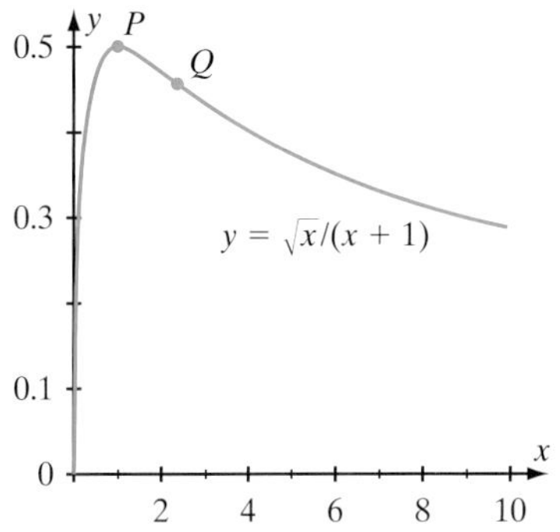

23. Local minimum: $P = (2, 12)$; point of inflection: $Q = (-2^{4/3}, 0)$; increasing: $(2, \infty)$; decreasing: $(-\infty, 0)$ and $(0, 2)$; concave up: $(-\infty, -2^{4/3}) \cup (0, \infty)$; concave down: $(-2^{4/3}, 0)$

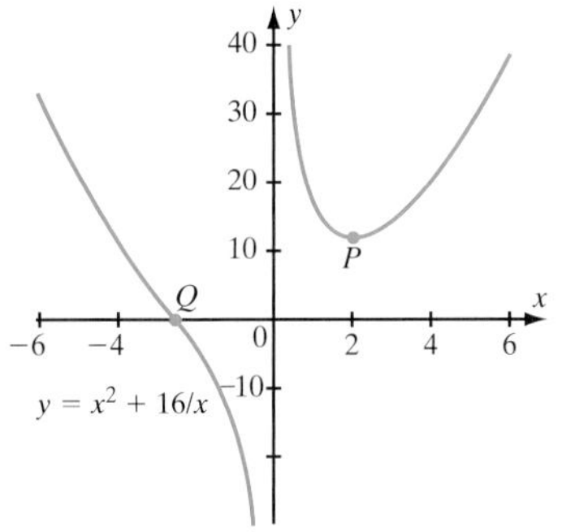

25. Local maximum: $Q = (-2, f(-2))$; points of inflection: $P = \left(-2 - 6\sqrt{5}/5, f(-2 - 6\sqrt{5}/5)\right)$, $R = (0, 0)$, $S = \left(-2 + 6\sqrt{5}/5, f(-2 + 6\sqrt{5}/5)\right)$; increasing: $(-\infty, -2)$; decreasing: $(-2, 4) \cup (4, \infty)$; concave up: $(-\infty, -2 - 6\sqrt{5}/5) \cup (0, -2 + 6\sqrt{5}/5) \cup (4, \infty)$; concave down: $(-2 - 6\sqrt{5}/5, 0) \cup (-2 + 6\sqrt{5}/5, 4)$; vertical asymptote $x = 4$; horizontal asymptote: $y = 0$

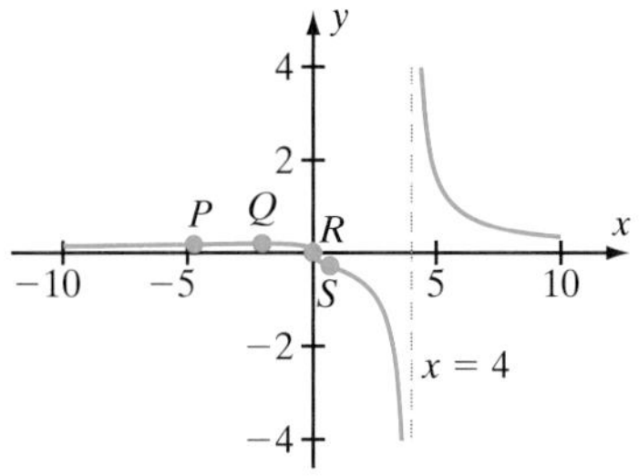

27. Local maxima: $P = (-\sqrt{6}/3, 2\sqrt{3}/9)$ and $T = (\sqrt{6}/3, 2\sqrt{3}/9)$; local minimum: $R = (0, 0)$; points of inflection: $Q = \left(-\sqrt{27 - 3\sqrt{33}}/6, f(-\sqrt{27 - 3\sqrt{33}})\right)$ and $S = \left(\sqrt{27 - 3\sqrt{33}}/6, f(\sqrt{27 - 3\sqrt{33}})\right)$; increasing: $[-1, -\sqrt{6}/3) \cup (0, \sqrt{6}/3)$; decreasing: $(-\sqrt{6}/3, 0) \cup (\sqrt{6}/3, 1]$; concave up: $(-\sqrt{27 - 3\sqrt{33}}/6, \sqrt{27 - 3\sqrt{33}}/6)$; concave down: $[-1, -\sqrt{27 - 3\sqrt{33}}/6) \cup (\sqrt{27 - 3\sqrt{33}}/6, 1]$

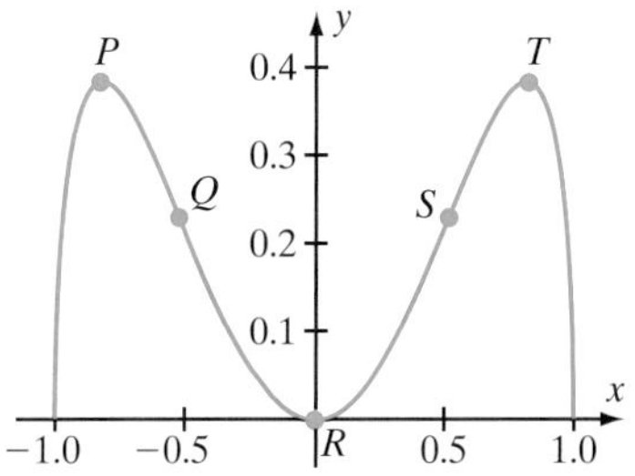

29. Local maximum: $P = (-9/2, \sqrt{33}/3)$; point of inflection: $Q = (-6, 1/\sqrt{6})$; increasing: $(-\infty, -9/2)$; decreasing: $(-9/2, 0)$ and $(0, \infty)$; concave up: $(-\infty, -6)$ and $(0, \infty)$; concave down: $(-6, 0)$; vertical asymptote $x = 0$; horizontal asymptotes: $y = -1$ and $y = 1$

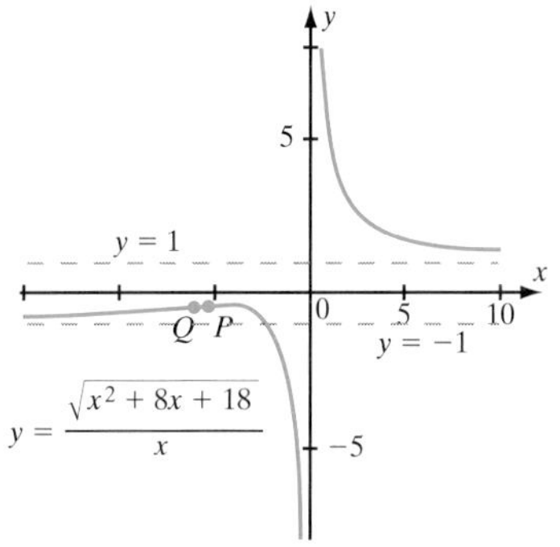

31. Local minimum: $Q = (-2, 0)$; local maximum: $R = (1/3, f(1/3))$; point of inflection: $P = (-13/3, f(-13/3))$;

increasing: $(-\infty, -2) \cup (1/3, \infty)$; decreasing: $(-2, 1/3)$; concave up: $(-13/3, -2) \cup (-2, \infty)$; concave down: $(-\infty, -13/3)$

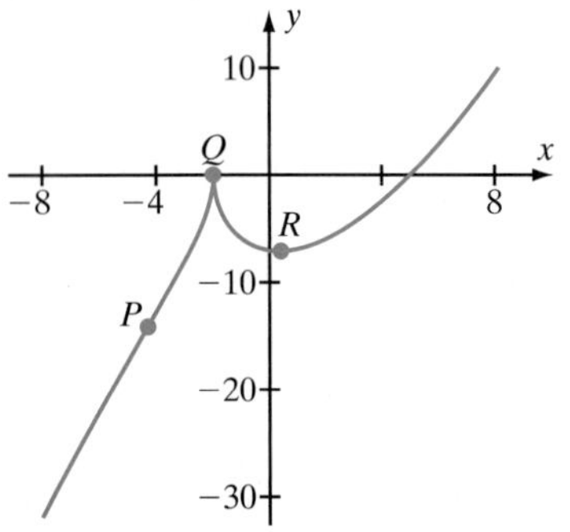

33. Local minimum: $Q = (2^{-1/3}, f(2^{-1/3}))$; point of inflection: $P = (-1, 1)$; increasing: $(2^{-1/3}, \infty)$; decreasing: $(-\infty, 0) \cup (0, 2^{-1/3})$; concave up: $(-\infty, -1) \cup (0, \infty)$; concave down: $(-1, 0)$; vertical asymptote: $x = 0$

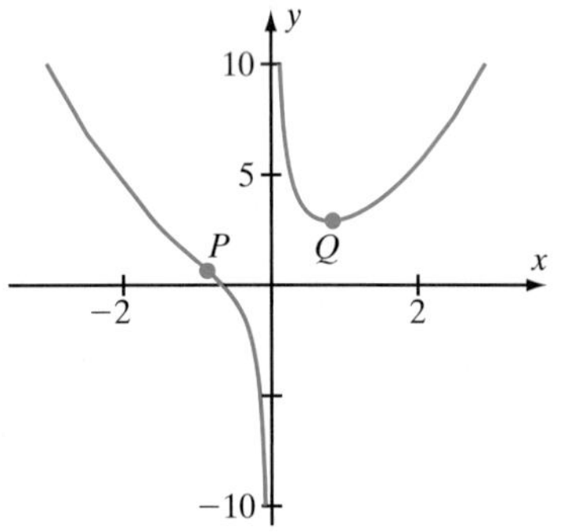

35. No local extrema; no points of inflection (concavity changes at points $-\pi/2 + 2k\pi$ for integral k, but these points are not in the domain of f); increasing: on its entire domain; concave up: intervals of the form $(-\pi/2 + 2k\pi, \pi/2 + 2k\pi)$ for integral k; concave down: intervals of the form: $(\pi/2 + 2k\pi, 3\pi/2 + 2k\pi)$ for integral k; vertical asymptotes: lines of the form $x = \pi/2 + 2k\pi$ for integral k

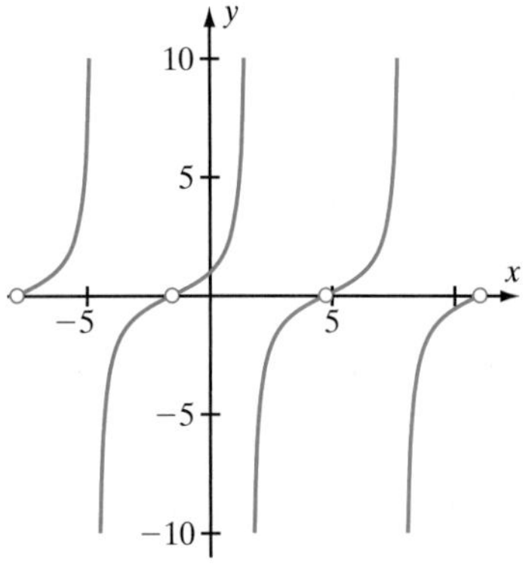

37. Local minima: $P = (-1, 0)$ and $R = (0, 0)$; local maximum: $Q = (-1/3, 4/27)$; points of inflection: $s = (-2/3, 2/27)$ and $R = (0, 0)$; increasing: $(-1, -1/3) \cup (0, \infty)$; decreasing: $(-\infty, -1) \cup (-1/3, 0)$; concave up: $(-\infty, -2/3) \cup (0, \infty)$; concave down: $(-2/3, 0)$

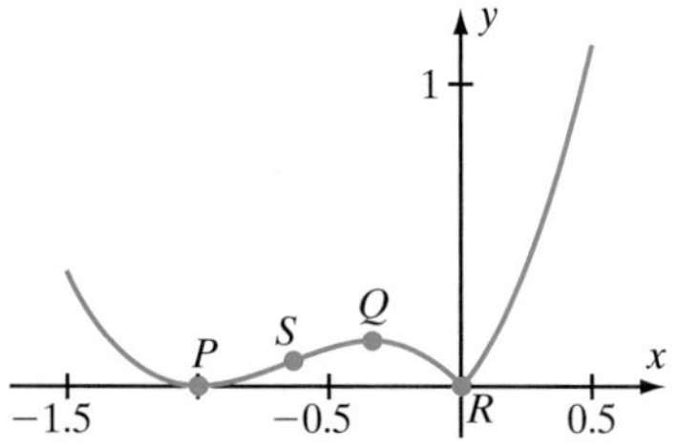

39. Local minimum: $Q = \left(-14/13 - 3\sqrt{3}/13, f(-14/13 - 3\sqrt{3}/13)\right)$; local maximum: $R = \left(-14/13 + 3\sqrt{3}/13, f(-14/13 + 3\sqrt{3}/13)\right)$; point of inflection: $P = (-2, -4)$; concave up: $(-2, -1) \cup (1, \infty)$; concave down: $(-\infty, -2) \cup (-1, 1)$; vertical asymptotes: $x = -1$ and $x = 1$

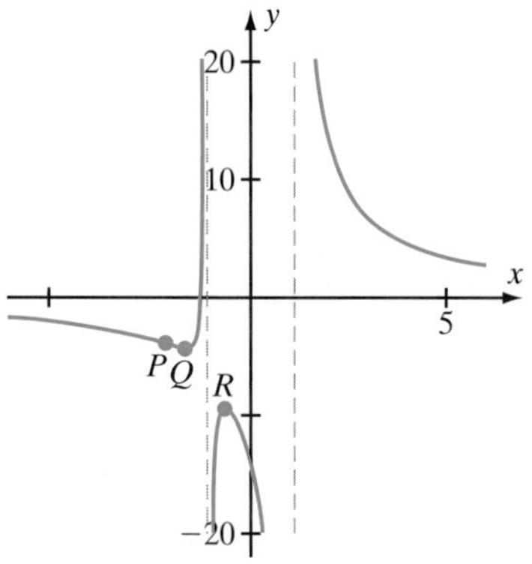

41. Increasing: $(-\infty, 0) \cup (0, \infty)$; concave up: $(-\infty, 0)$; concave down: $(0, \infty)$; vertical asymptote: $x = 0$; skew-asymptote: $y = x/2$

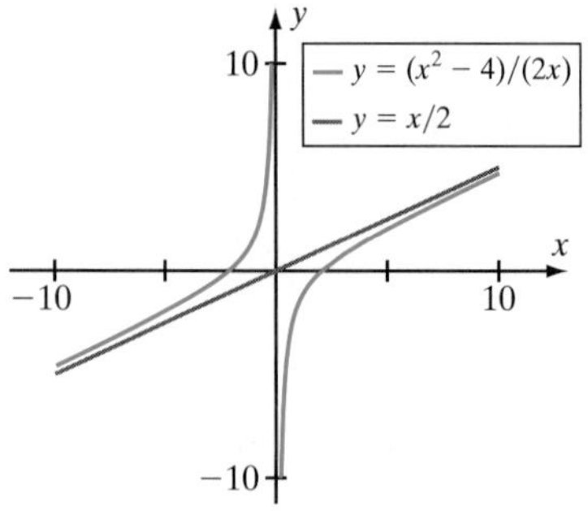

45. Global minimum: $P = (-4.39926, -125.33442)$; points of inflection: $Q = (-2.94338, -76.22328)$ and

$R = (-0.05662, -4.05450)$; increasing: $(-4.39926, \infty)$; decreasing: $(-\infty, -4.39926)$; concave up: $(-\infty, \infty)$ and $(-0.05662, \infty)$; concave down: $(-2.94338, -0.05662)$

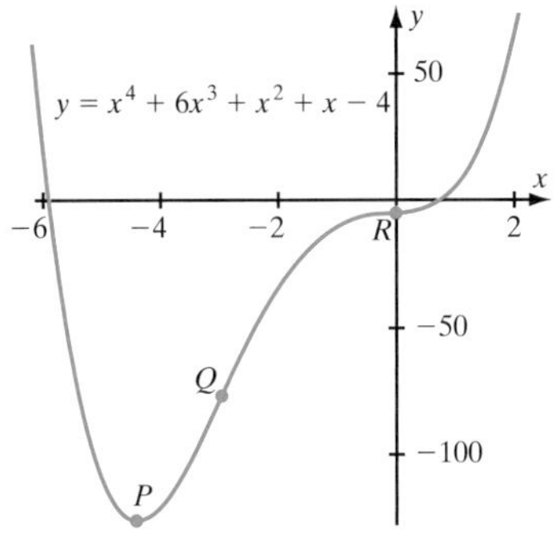

47. Local maxima: $P = (-1.6012, f(-1.6012))$ and $T = (0.39915, f(0.39915))$; local minima: $R = (-1.1839, f(-1.1839))$ and $V = (1.5859, f(1.5859))$; points of inflection: $Q = (-1.4117, f(-1.4117))$, $S = (-0.31451, f(-0.31451))$ and $U = (1.1262, f(1.1262))$; increasing: $(-\infty, -1.6012) \cup (-1.1839, 0.39915) \cup (1.5859, \infty)$; decreasing: $(-1.6012, -1.1839) \cup (0.39915, 1.5859)$; concave up: $(-1.4117, -0.31451) \cup (1.1262, \infty)$: concave down: $(-\infty, -1.4117) \cup (-0.31451, 1.1262)$

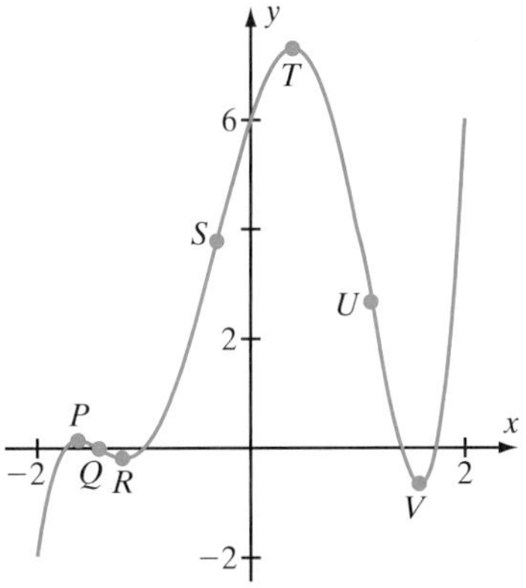

49. Local maximum: $P = (0.92388, 1.20711)$; local minimum: $R = (-0.38268, -0.20711)$; point of inflection: $Q = (0.51201, 0.70196)$; increasing: $(-0.38268, 0.92388)$; decreasing: $(-1, -0.38268)$ and $(0.92388, 1)$; concave up: $(-1, 0.51201)$: concave down: $(0.51201, 1)$

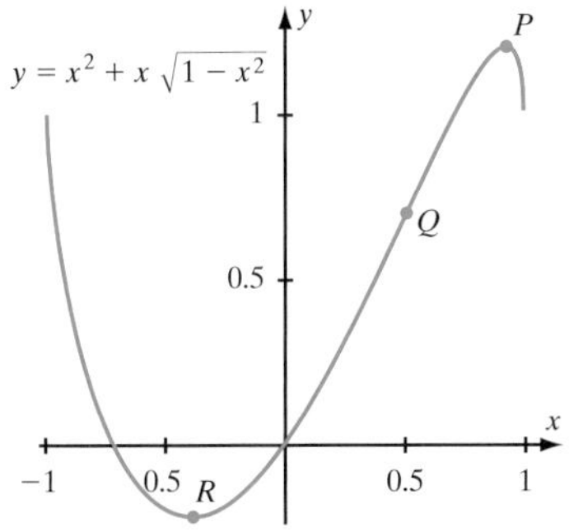

51. Global minima: $P = (-2, 0)$ and $V = (2, 0)$; global maximum: $S = (0, 2)$; points of inflection $Q = (-1.7321, f(-1.7321))$, $R = (-0.58288, f(-0.58288))$, $T = (0.58288, f(0.58288))$, and $U = (1.7321, f(1.7321))$; increasing: $(-2, 0)$; decreasing: $(0, 2)$; concave up: $(-1.7321, -0.58288) \cup (0.58288, 1.7321)$: concave down: $(-2, -1.7321) \cup (-0.58288, 0.58288) \cup (1.7321, 2)$

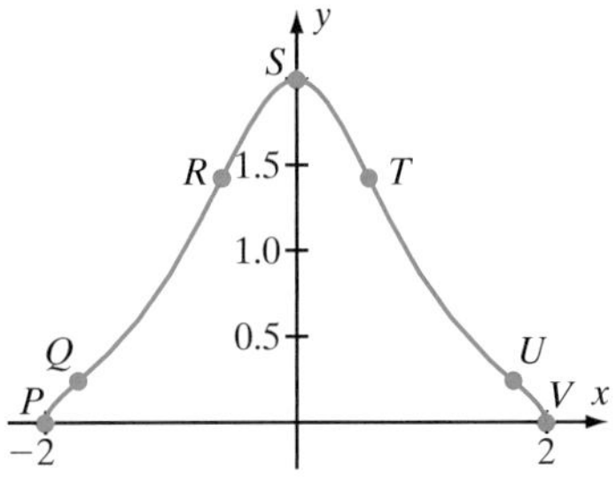

53. Local maximum: $P = (-1.49714, 9.06986)$; local minimum: $Q = (1.97537, -3.00630)$; points of inflection: $R = (-2.58587, 6.34340)$, $S = (0, 15/4)$, and $T = (3.39777, -1.12363)$; increasing: $(-\infty, -1.49714)$ and $(1.97537, \infty)$; decreasing: $(-1.49714, 1.97537)$; concave up: $(-\infty, -2.58587)$ and $(0, 3.39777)$: concave down: $(-2.58587, 0)$ and $(3.39777, \infty)$; skew asymptote: $y = x$

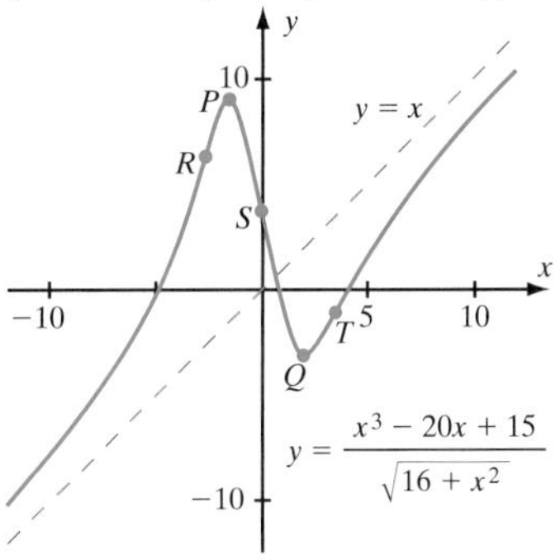

Section 4.7

1. -1
3. $1/5$
5. $-1/8$
7. $1/4$
9. Does not exist.
11. 2
13. $1/3$
15. 1
17. 0
19. 1
21. $4/\pi^2$
23. $-4/3$
25. 0
27. 0
29. 1
31. $-4/\pi$
33. 1
35. $1/e^{-5}$
37. 1
39. 1
41. -1
43. 0
45. Does not exist.
47. 0
49. -1
51. $-1/2$
53. 1
55. 1
57. $1/2$
59. $-1/2$
61. $1/e^2$
63. e^3
65. e^x

67. $a(a^2 - 1)/(2b(b^2 - 1))$

69. 0 **71.** 1/3 **73.** 2

75. $1/e$

77.

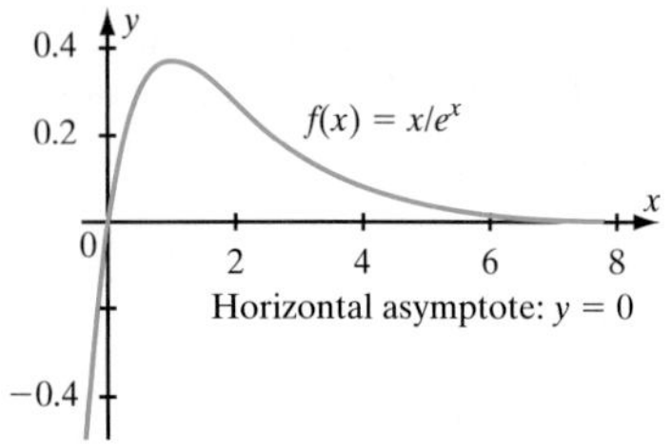

Horizontal asymptote: $y = 0$

79.

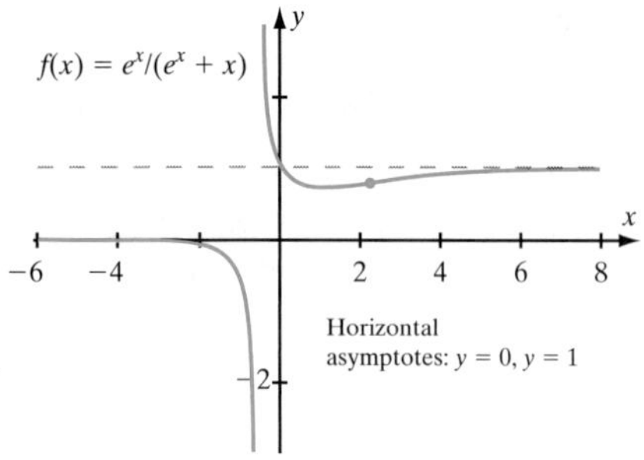

81.

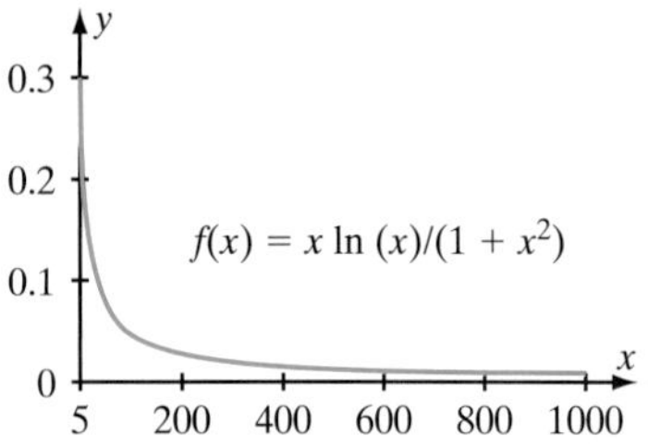

Horizontal asymptote: $y = 0$

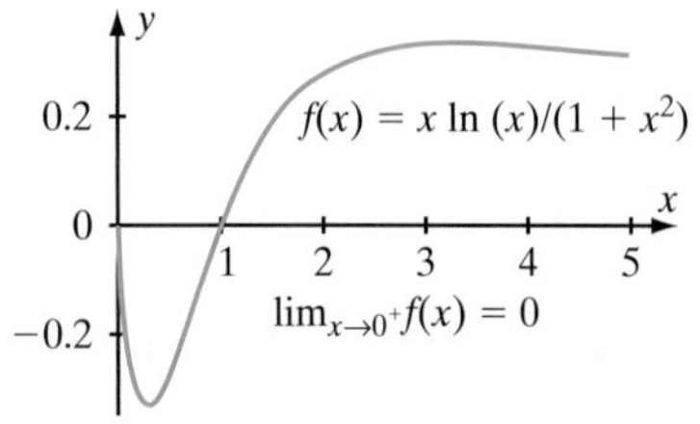

89. The limit is 1/2.

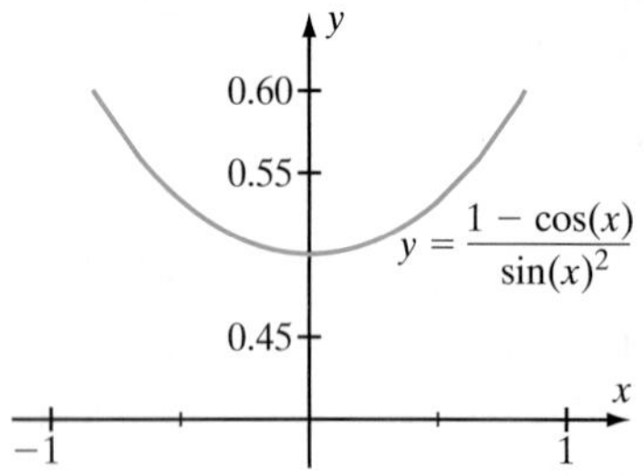

91. The limit is 1.

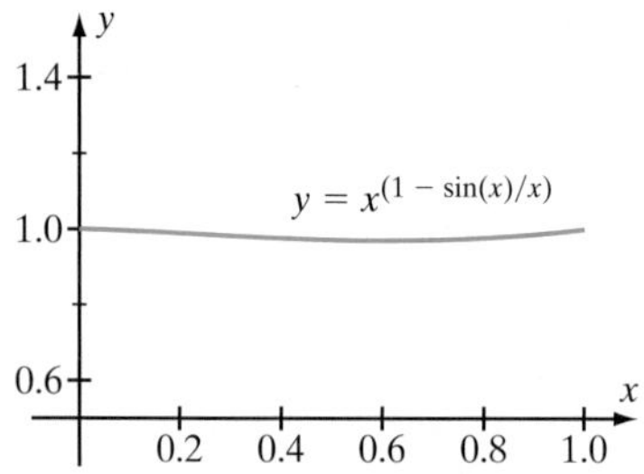

93.

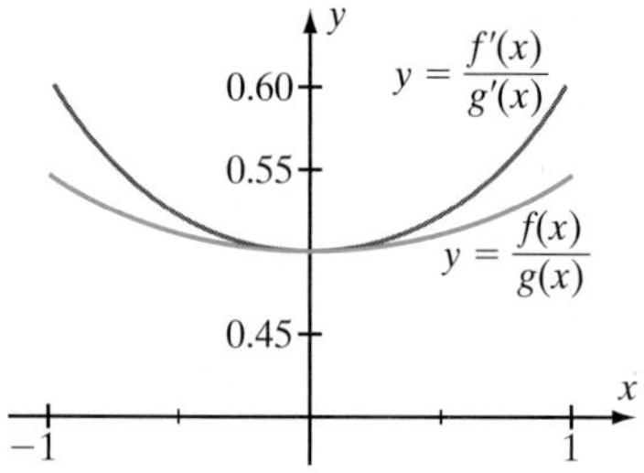

$f(x) = 1 - \cos(x), g(x) = x\sin(x)$

95.

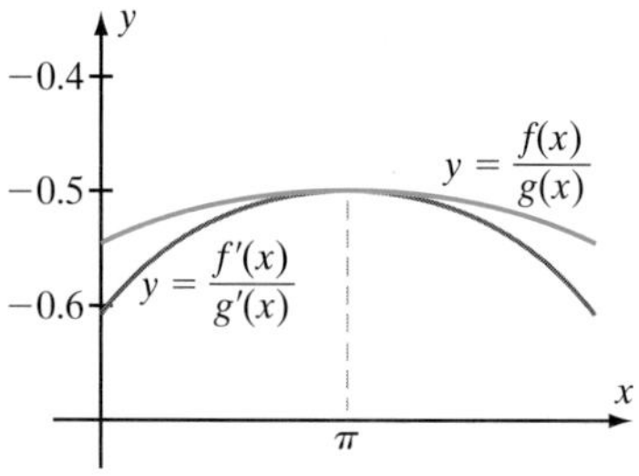

$f(x) = 1 + \cos(x), g(x) = (x - \pi)\sin(x)$

Section 4.8

1. $x_2 = 2.5$, $x_3 = 2.45$

3. $x_2 = 4.6667$, $x_3 = 4.5641$

5. $x_2 = 1.9$, $x_3 = 1.8883$

7. $x_2 = 1.6667$, $x_3 = 1.5911$

9. $x_2 = 2.45$, $x_3 = 2.4495$

11. $x_2 = 3.5$, $x_3 = 3.1786$

13. 1 whatever value of x_1 is chosen

29. 0.995598772 **31.** 0.62902

33. 12 **35.** 11

37. $M = 4$, 4, 3, and 4 for Exercises 33, 34, 35, and 36, respectively

39. 1.080

41. **a.** 7.75%; **b.** 5.82%

43. The second: It has an effective yield of 6.50%, whereas the first has an effective yield of 6.25%.

45. The second: It has an effective yield of 7.37%, whereas the first has an effective yield of 7.00%.

49. 1.02876, occuring at 0.51282

Section 4.9

1. $\frac{1}{3}x^3 - \frac{5}{2}x^2 + C$

3. $e^x + C$

5. $\frac{2}{3}(x+2)^{3/2} + C$

7. $-1/x - 1/(6x^6) + C$

9. $\frac{1}{3}(x+1)^3 + C$

11. $\frac{1}{e}\exp(ex)$

13. $-\frac{3}{4}x^{-4/3} - 12x^{1/3} + C$

15. $\frac{3}{4}\sin(4x) + x^2 + C$

17. $\frac{1}{8}\tan(8x) + C$

19. $\frac{1}{12}(3x-2)^4 + C$

21. 1 **23.** $-3/2$ **25.** $3e - 4$

27. $4\sqrt{10}$ m/s

29. 155.5 ft; -91.61 ft/s

31. $121/54$ ft/s^2

33. 4.738 s

35. $\frac{1}{2}\sin^2(x) + C$, or $-\frac{1}{2}\cos^2(x) + C$, or $-\frac{1}{4}\cos(2x) + C$

37. $\tan(x) - x + C$

39. $\frac{1}{2}x + \frac{1}{4}\sin(2x) + C$

41. $\sec(x) + C$

43. $\frac{1}{\ln^2(2)}2^{x\ln 2} + C$

45. Local maximum at -3; local minimum at 2

47. $G(f(x)) + C$

49. $\sin(x^2+7) + C$

51. $\frac{1}{2}\exp(x^2) + C$

55. $\exp(-x)/\sqrt{\pi x}$

57. 420.9 ft; 95 ft/s

59. Impact at 17 mi/hr

61. 484 ft

63. $3825/4$ ft; $-60\sqrt{17}$ ft/s

65. 120 s

67. $H - kgt + k^2 g(1 - e^{-t/k})$

69. 1839.6 yr

71. $T\infty = 40^\circ C; \tau_1 = 25.54$ s; $\tau_2 = 20.27$s; $\tau_3 = 14.39$ s

73. $v(t) = 8.855(1 + \exp(4.427t))^{-1} - 4.427$; $v_\infty = -4.427$; $v(1.717) = 0.999\, v_\infty$

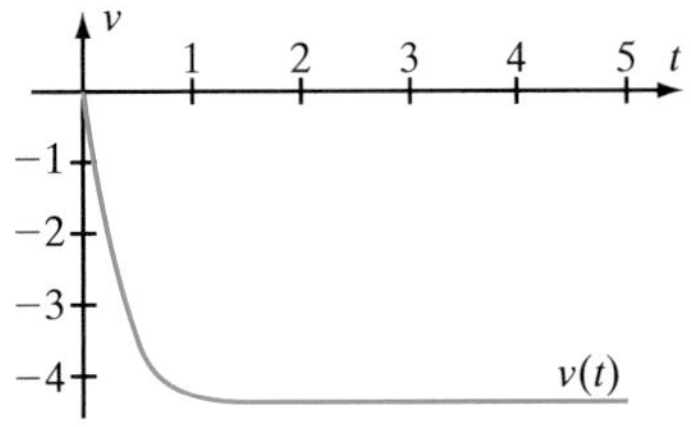

REVIEW EXERCISES FOR CHAPTER 4

1. 26/5 **3.** 6

5. $3/(32\pi)$ cm/min

7. 4/15 m/min

9. -105

11. 1/2 ft/s

13. $c = \sqrt{7}; f'(c) = 21$

15. $c = -9; f'(c) = -1/81$

17. -1 **19.** $e - 1$ **21.** $8 - \ln(2)$ **23.** 11

25. Local maximum at -2; local minimum at 2

27. Local minimum at $16^{1/3}$

29. Local maximum at -2; local minimum at 2

31. Local minimum at 1; local maximum at e^2

33. Local maximum at 5

35. Local maximum at 1; local minimum at 2

37. No extremum at 0; local minimum at 1/2; no extremum at 1

39. -14, 67 **41.** $-e$, $1/e$

43. $10/(4+\pi)$ **47.** $\frac{3}{8}2^{1/3}$

49. $4\sqrt{3}$ in^3

51. **a.** $(-\infty,-1)$ and $(1,\infty)$ **b.** $(-1,1)$ **c.** $-1, 1$

53. **a.** $(-\infty,3)$ **b.** $(3,\infty)$ **c.** 3

55. **a.** $(0,6)$ **b.** $(6,\infty)$ **c.** 6

57. **a.** $(e^{5/6},\infty)$ **b.** $(0,e^{5/6})$ **c.** $e^{5/6}$

59. **a.** $(-3, 0)$ and $(0, \infty)$ **b.** $(-\infty, -3)$ **c.** -3

61. 2 **63.** -3 **65.** -2 **67.** 0

69. 4 **71.** e^{-3} **73.** 1/2

75. 3.181725815, 3.146284845

77. 6.823500280, 6.854079461

79. $\sec(x) + C$

81. $(3x+1)^3/9 + C$

83. $x^4/4 - x^2/2 + C$

85. $3x^{7/3}/7 - 6x^{2/3} + C$

87. $\frac{1}{3}\tan(3x) + C$

89. $\frac{2}{9}x^{9/2} + \frac{1}{2}x^4 + C$

Section 5.1

1. 30 **3.** -108

5. 73/6 **7.** 118

9. 19/4 **11.** $\sum_{j=1}^{5}(j+1)$

13. $\sum_{j=1}^{6}(5+4j)$ **15.** $\sum_{j=4}^{8} 1/j$

17. 650 **19.** 806

21. 676 **23.** $(e^{20}-1)/(e-1)$

25. ln(720) **27.** 23

29. $(2-\log_2(3))/4$ **31.** 3π

33. 197/60 **35.** $\pi^2/2$

37. 14 **39.** $\pi(2+\sqrt{3})/24$

41. 17/6 **43.** 581/1200

45. 16/3 **47.** $p = 5050$; $q = 100!$

49. $N(2N-1)(2N+1)/3$ **65.** $ab - A$

69. 0.16 **71.** 1.57

73. 1.94 **75.** 0.30

Section 5.2

1. a. 11/6 **b.** ln(7)

3. a. 72 **b.** 224/3

5. a. 378 **b.** 696

7. a. 42 **b.** 37

9. a. $2(e^2+1+e^{-2})$ **b.** $e^2 - e^{-4}$

11. a. $\pi/3$ **b.** $-\sqrt{3}/2$

13. lower: 45/64; upper: 365/64

15. lower: $-\pi$; upper: $\pi/3$

17. $2\sqrt{2}$ **19.** 5 **21.** 9/2 **23.** 13/2

25. 11 **27.** 14/3 **29.** 1 **31.** 1

33. 195/4 **35.** $e-1$ **37.** -1 **39.** 7/ln(2)

41. $1-\sqrt{3}/3$ **43.** 21/2

45. $\mathcal{R}(f,\mathcal{L}_N) = 1$; $\mathcal{R}(f,\mathcal{S}) = \sqrt{e}$; $\mathcal{R}(f,\mathcal{U}_N) = e$

47. $\mathcal{R}(f,\mathcal{L}_N) = 7/12$; $\mathcal{R}(f,\mathcal{S}) = 24/35$; $\mathcal{R}(f,\mathcal{U}_N) = 5/6$

49. $\mathcal{R}(f,\mathcal{L}_N) = \ln(6)$; $\mathcal{R}(f,\mathcal{S}) = \ln(105/8)$; $\mathcal{R}(f,\mathcal{U}_N) = \ln(24)$

51. 2/3 **53.** 1/2

55. $1 - 1/e$

57. $\mathcal{R}(f,\mathcal{L}_3) = 11 < 33/2 = \int_0^3 f(x)\,dx < 22 = \mathcal{R}(f,\mathcal{U}_3)$

59. $s_1 = 10^{1/3}$, $s_2 = 68^{1/3}$

61. $s_1 = 1/(2\ln(3/2))$, $s_2 = 1/(2\ln(4/3))$

71. 0.375 **73.** 3.14195155

75. $\mathcal{R}(f,\mathcal{L}_{50}) = 8.035809924$; $\mathcal{R}(f,\mathcal{U}_{50}) = 8.244835710$

77. $\mathcal{R}(f,\mathcal{L}_{50}) = 1.893551751$; $\mathcal{R}(f,\mathcal{U}_{50}) = 1.908235239$

Section 5.3

1. 4 **3.** 8 **5.** 10 **7.** 75

9. -64 **11.** -18 **13.** -5

15. $f_{\text{avg}} = 4$; $c = 3$

17. $f_{avg} = 14/9$; $c = 196/81$

19. $f_{\text{avg}} = 41$; $c = \sqrt{3}$

21. 18 **23.** 8/9 **25.** $2/\pi$ **27.** $4/\pi$

29. $A = 9$; $B = 18$

31. $A = 4/5$; $B = 4/3$

33. $\int_a^b f(x)\,dx = 0$; $\int_a^b g(x)\,dx = -1$

35. $\int_a^b f(x)\,dx = -5$; $\int_a^b g(x)\,dx = -1$

37. $6 - 5\ln(3)$

39. $3 - 1/e$

41. 4/15 **43.** $6e - 9$ **45.** No

49. $A = 3/5$; $B = 3$

51. $A = 2/\sqrt{e}$; $B = 4/e$

53. $F(x) = x^3 + x - 2$

55. $F(x) = 1/x - x^2$

57. $F(x) = 2x - 2$

59. $|\int_0^1 (\pi x - 3)\,dx| = 3 - \pi/2 = 1.429\ldots$;
$\int_0^1 |\pi x - 3|\,dx = 9/\pi - 3 + \pi/2 = 1.435\ldots$

61. $|\int_{-2}^1 (1-9x^2)\,dx| = 24$;
$\int_{-2}^1 |1-9x^2|dx = 224/9 = 24.888\ldots$

65. $A = 6/17$; $B = 2.8406\ldots$

67. $A = \pi/2$; $B = 5.2413\ldots$

69. $f_{\text{avg}} = \pi/4 + 2/\pi$; $c = 0.74445\ldots$

71. $f_{\text{avg}} = \frac{52}{3} + \ln(3)$; $c = 4.18032\ldots$

Section 5.4

1. $(e^6-1)/3$ **3.** $6 - 3\sqrt{3}$

5. 1 **7.** $6 - 2\sqrt{3}$

9. $\pi/8$

11. $x^3 + x + 2$

13. $3x^{4/3} - 48$

15. $(e^x - e^{-x})/2$

17. $\sec(x) - \sqrt{2}$

19. $10x^{3/2} + 6x^{5/2} - 16$

21. $(2x+1)(x+2)$

23. $x^2 + 3$

25. $\tan(x^2)$

27. $x\sqrt{1+x}$

29. $-\cos(4x)$

31. $-\cot(x)$

33. $-\sqrt{2 - \sin^2(x)}$

35. $2\sin(4x) - \sin(2x)$

37. $2x\cot(x^2) - \cot(x)$

39. $x(e^x - e^{-x})$

41. $\sqrt{2x-1}/(2\sqrt{x}) - 2\sqrt{8x^2-1}$

43. $F'(x) = x(x-1)$; $F''(x) = 2x - 1$; increasing: $(-\infty, 0) \cup (1, \infty)$; decreasing: $(0,1)$; concave up: $(1/2, \infty)$; concave down: $(-\infty, 1/2)$

45. $F'(x) = x\ln(x)$; $F''(x) = 1 + \ln(x)$; increasing: $(1, \infty)$; decreasing: $(0,1)$; concave up: $(1/e, \infty)$; concave down: $(0, 1/e)$

51.

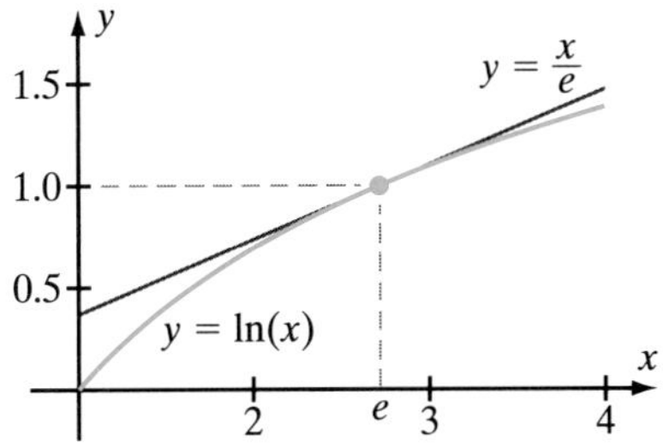

53.

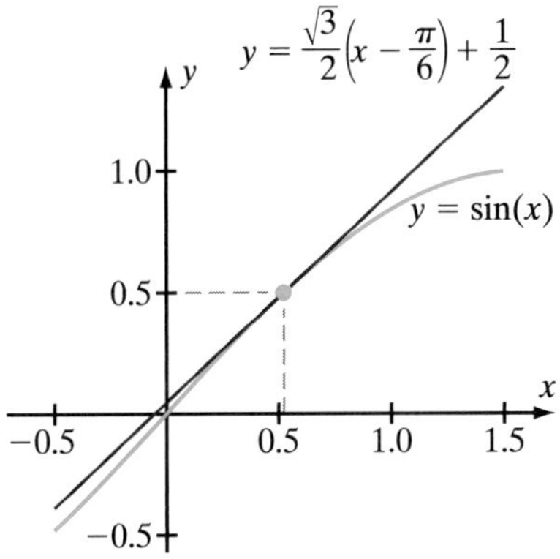

55. $f(t) = \begin{cases} 2 & \text{if } 0 \le t \le 1 \\ 0 & \text{if } 1 < t \le 2 \\ -1 & \text{if } 2 < t \le 3 \end{cases}$

59. $f(g(h(x))) \cdot g'(h(x)) \cdot h'(x)$

65. $1 - 2xF(x)$

71.

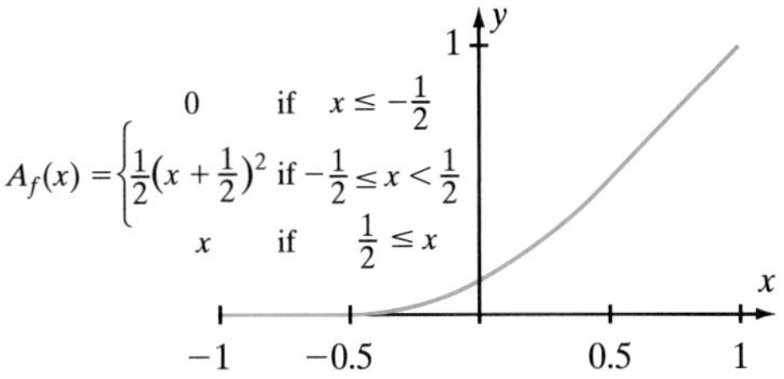

75. Decreasing: $(-\infty, -0.71)$; increasing: $(-0.71, \infty)$; concave up: $(-\infty, \infty)$

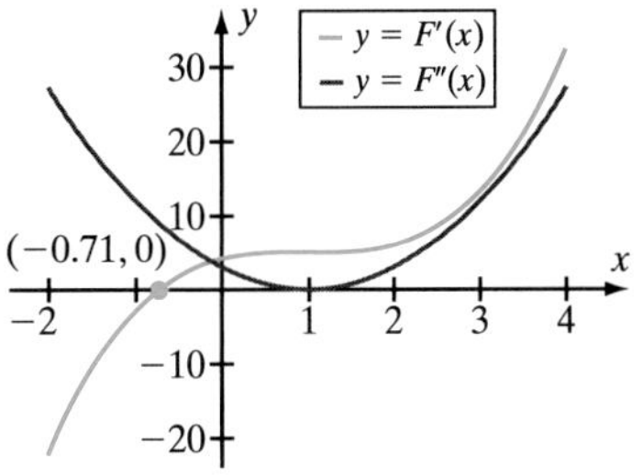

$F(x) = \int_{-3}^{x} (t^3 - 3t^2 + 3t + 4)\,dt$

77. Decreasing: $(-\infty, -1) \cup (0, 1)$; increasing: $(-1, 0) \cup (1, \infty)$; concave down: $(-\infty, -3.2143) \cup (-0.46081, 0.67513)$; concave up: $(-3.2143, -0.46081) \cup (0.67513, \infty)$

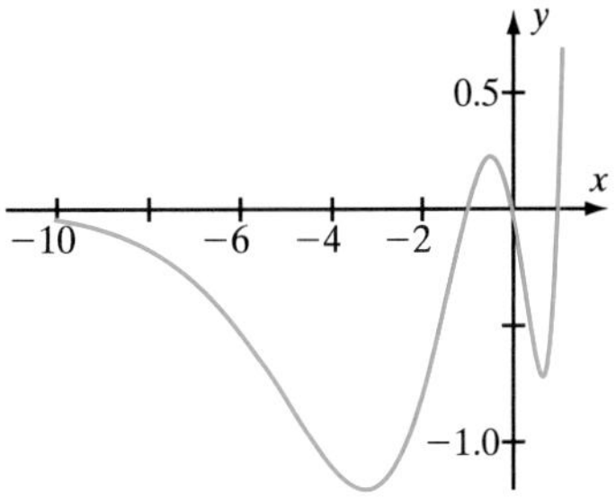

$y = F'(x)$ where $F(x) = \int_0^x (t^3 - t)\exp(t)\,dt$

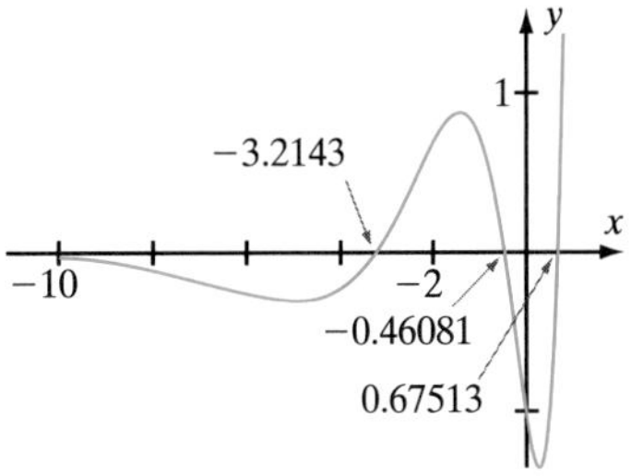

$y = F''(x)$ where $F(x) = \int_0^x (t^3 - t)\exp(t)\,dt$

79. The temperature reaches at least $37^\circ C$ on about $365 - N(37) \approx 6$ days.

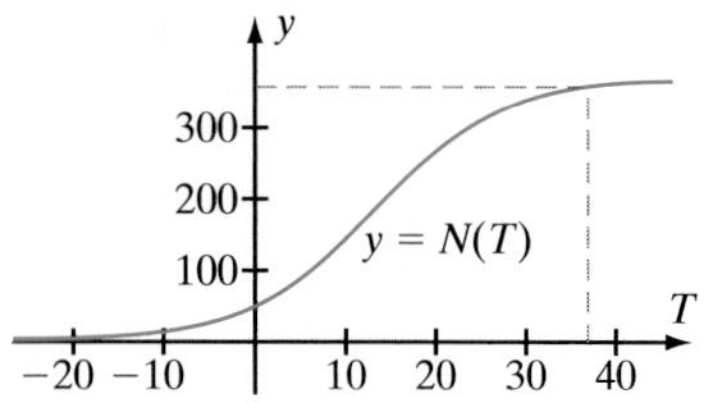

81.

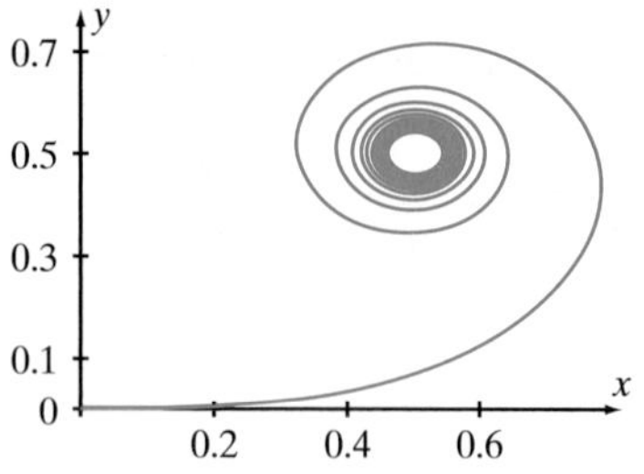

Arc of Cornu Spiral

83. $F'(4) = 5.0000001\ldots$; $f(2) = 5$

85. $F'(4) = 2.0000001\ldots$; $f(4) = 2$

Section 5.5

1. -4

3. 3

5. 4

7. 8

9. 12

11. $12 + \ln(2)$

13. $1/x$

15. $2/(2x + 3)$

17. $-1/(2(1 - x))$

19. $-x^{-1}(4 + \ln(x))^{-2}$

21. $2x^{-1}\ln(x)$

23. $2^x(1 + x\ln(2))$

25. $2^{\ln(x)}\ln(2)/x$

27. $-1/(x\ln(10))$

29. $1/(x\ln(2))$

31. $1/\ln(2)$

33. $1 + 9/(10\ln(10))$

35. First derivative: $2^x\ln(2) - 1$; second derivative: $2^x\ln^2(2)$; local minimum at $-\ln(\ln(2))/\ln(2)$

37. First derivative: $10^{-x}(1 - x\ln(10))$; second derivative: $10^{-x}(x\ln(10) - 2)\ln(10)$; local maximum at $1/\ln(10)$

39. First derivative: $(1 + \ln(x))/\ln(10)$; second derivative: $1/(x\ln(10))$; local minimum at $1/e$

41. First derivative: $2\ln(x)/x$; second derivative: $2(1 - \ln(x))/x^2$; local minimum at 1

43. $F'(x) = 1/(x\ln(x))$; $F''(x) = -1/(x^2\ln(x)) - 1/(x^2\ln^2(x))$

45. $F'(x) = \ln(x)/x^2$; $F''(x) = (3 - 2(1 + \ln(x)))/x^3$

47. $F'(x) = \ln(\ln(x))/x$; $F''(x) = (1/\ln(x) - \ln(\ln(x)))/x^2$

53. $A = 1$, $B = -1$

75. Minimum: $1.36517\ldots$ at $x = 4.92155\ldots$

77. Local maximum at $-3.09654\ldots$; local minimum at $-0.788845\ldots$

Section 5.6

1. $u = 3x$, $du = 3\,dx$; $-\cos(3x)/3 + C$

3. $u = x^8 + 1$, $du = 8x^7dx$; $-2/(x^8 + 1)^4 + C$

5. $u = x^3 - 5$, $du = 3x^2dx$; $4(x^3 - 5)^{5/2} + C$

7. $u = \cos(s)$, $du = -\sin(s)\,ds$; $-\cos^5(s)/5 + C$

9. $u = \pi/t$, $du = (-\pi/t^2)$, dt; $(1/\pi)\cos(\pi/t) + C$

11. $(x^2 + 1)^8 + C$

13. $\sqrt{x^2 + 1} + C$

15. $\sec(x) + C$

17. $2\sin^6(2x) + C$

19. $24\sqrt{1 + x^2} + C$

21. $-2\cos(\sqrt{x}) + C$

23. $32/3$

25. $52/3$

27. 3

29. $-15/8$

31. -2

33. $7/3$

35. $6\ln(3) - 4$

37. $\frac{1}{4}\ln(2)$

39. $4\ln(1 + \sqrt{2})$

41. $\ln(2)/(4\pi)$

43. $(4/\pi)\ln(1 + \sqrt{2})$

45. $\frac{1}{9}\ln\left(\left|\frac{4t + 3}{t + 3}\right|\right) + C$

47. $t - 4\arctan\left(\frac{t + 3}{4}\right) + C$

49. $\frac{1}{8}\ln\left(\frac{\exp(2t)}{4 + \exp(2t)}\right) + C$

51. $2\sqrt{t} - 4\ln(|2 + \sqrt{t}|) + C$

53. $\frac{1}{5}(x - 1)(2x + 3)^{3/2} + C$

55. $2(x - 6)\sqrt{x + 3}/3 + C$

57. $\sqrt{6 + x^2}(x^4 - 8x^2 + 96)/5 + C$

59. $2(x - 5)^{3/2}(3x + 20)/15 + C$

61. $\arcsin(1/2\sin(x)) + C$

63. $2\ln^2(x) + C$

65. $\ln(|\ln(x)|) + 2/\ln(x) + C$

67. $\arctan^2(x)/2 + C$

69. $\arctan(e^x) + C$

71. $-1/(1 + e^x) + C$

73. $\arcsin(x/2) - (2x^2 + 19)\sqrt{4 - x^2} + C$

75. $-\sqrt{1 - e^{2x}} + C$

77. $-\frac{1}{2}\ln\left(\frac{\sqrt{4 + \cos(t)} - 2}{\sqrt{4 + \cos(t)} + 2}\right) + C$

79. $-3\ln\left(\frac{2 + \sqrt{4 + t^4}}{t^2}\right) + C$, or $\frac{3}{2}\ln\left(\frac{\sqrt{4 + t^4} - 2}{\sqrt{4 + t^4} + 2}\right) + C$

81. $\sec(t) - \arctan(\sec(t)) + C$

91. $-\frac{1}{2}\left(\frac{1}{\alpha + \beta}\cos((\alpha + \beta)x) + \frac{1}{\alpha - \beta}\cos((\alpha - \beta)x)\right) + C$

93. $\ln(t^2+1) - 1/(t+1) + C$

95. 0.585096...

97. 1.95265...

Section 5.7

1. $4 - \sqrt{2} - \sqrt{3}$

3. 80 **5.** 29/4 **7.** 101/48 **9.** 1137/8

11. 49 **13.** 95/6 **15.** $3 - \sqrt{2}$ **17.** 28

19. 36 **21.** 9/2 **23.** 36 **25.** 32/3

27. 1/2 **29.** 8 **31.** 7 **33.** 1/2

35. $16\sqrt{2}/3$

37. $2\sqrt{2}$

39. $(3-e)/2$

41. $\ln(2)/2 - 1/4$

43. 4 **45.** 64/3 **47.** 1/12

49. $\int_0^2 (4-y^2)\,dy$

51. $\int_0^{\pi/2} (1-\sin(y))\,dy$

53. $\int_0^2 (2y - y^2)\,dy$

55. $\int_0^3 (y-(-y))\,dy$

57. $\int_1^2 (x-1)\,dx + \int_2^3 (3-x)\,dx$

59. $\int_{-1}^1 ((2-x^2) - |x|)dx$

61. $\int_0^1 ((3-y)-(1+y))\,dy$

63. $\int_0^1 (y-(-y))\,dy + \int_1^2 (\sqrt{2-y} - (-\sqrt{2-y}))\,dy$

65. $\int_0^{\sqrt{2}} ((4-y^2)-y^2)dy$

67. $\int_0^{\pi/4} (\cos(y) - \sin(y))dy$

69. $\int_0^2 \left(\sqrt{1-y/2} - (-(1-y/2)^2)\right)dy$

71. $\sqrt{5}/6$

73. 17.102649... **75.** 0.4011197...

77. 0.737141... **79.** 1.2895159...

Section 5.8

1. 17/2, 9, 26/3

3. $(\sqrt{7}+\sqrt{11})/2, (2+3\sqrt{2})/2, (2+5\sqrt{2})/3$

5. $\ln(15/4)$, $\ln(2) + \ln(3)/2$, $(4\ln(2) + \ln(3))/3$

7. $(-\pi/2)\cos(\pi/8)$, $-\pi(2+\sqrt{2})/8$, $-\pi(4+\sqrt{2})/12$

9. 36, 81/2, 75/2

11. a. 7.752..., **b.** 8.112..., **c.** $6 + \ln(7) \approx 7.945...$

13. a. 2.872..., **b.** 2.895..., **c.** $3(5^{2/3}-1)/2 \approx 2.886...$

15. a. −163.56, **b.** −152.76, **c.** −156

17. a. 22.3963..., **b.** 22.8968..., **c.** $(\exp(4) - \exp(9/4))/2 \approx 22.555...$

19. a. 682.475..., **b.** 682.605..., **c.** 682.5

21. a. 17.315..., **b.** 17.349..., **c.** $52/3 \approx 17.333...$

23. 9/25, or 0.36

25. 6/25, or 0.24

27. 91/375, or 0.2426...

29. 6.7 L/min

31. 255, 18

33. 10 **35.** 163 **37.** 4 **39.** 28

43. $A = 4/3$; $\mathcal{M}_2 = \sqrt{2}$; $|A - \mathcal{M}_2| = 0.08088...$;
$\mathcal{S}_2 = 2/3$; $|A - \mathcal{S}_2| = 0.66666...$;
$\mathcal{M}_4 = (1+\sqrt{3})/2$; $|A - \mathcal{M}_4| = 0.03269...$;
$\mathcal{S}_4 = (1+2\sqrt{2})/3$; $|A - \mathcal{S}_4| = 0.05719...$

47. $(5000 - 3034)/5000 = 983/2500 \approx 0.3932$

49. Midpoint: 254.4 m^3; Trapezoidal: 253.62 m^3; Simpson's: 253.88 m^3

51. $\mathcal{M}_2 = 164 \le \mathcal{S}_4 = 616/3 \le \mathcal{T}_2 = 288$

53. $\mathcal{M}_3 = 11/6 \approx 1.83 \le \mathcal{S}_6 = 617/315 \approx 1.96 \le \mathcal{T}_3 = 232/105 \approx 2.21$

55. 22/3 km ≈ 7.66667 km **57.** 7, 3.8766...

59. 37, 0.6036... **61.** 9.0373

63. 13.9829 **65.** 1.89831

67. 3.21047 **69.** 0.6827

71. 55625/18, or 3090.277...

73. a. approximate: 364.996..., exact: 365 **b.** 3 to 4 **c.** 39.7%

REVIEW EXERCISES FOR CHAPTER 5

1. $\sum_{n=1}^{25} n^2$ **3.** $\sum_{n=2}^{100} (1/n)$ **5.** 216

7. 43/30 **9.** $\pi/2$ **11.** 138

13. 52 **15.** $-\ln(2)$ **17.** 13

19. $3(\sqrt{3}-1)/2$ **21.** 76/3 **23.** 12

25. $8/\ln(3)$ **27.** $\sqrt{2}-1$

29. $f_{avg} = 3/7$; $c = 7\sqrt{21}/9$

31. $f_{avg} = 14/9$; $c = 196/81$

33. 3/2 **35.** 7 **37.** −1/2

39. $6\ln(2)$ **41.** −24 **43.** 6

45. 18 **47.** 3 **49.** 12

51. $-\frac{1}{2}\csc^2(x) + C$

53. $12\sqrt{2+x^4}+C$

55. $-\dfrac{1}{2}\ln(3+2\cos(x))+C$

57. $\dfrac{3}{2}(\sqrt{x}+1)^{4/3}+C$

59. $-2/\ln(x)+C$

61. $\dfrac{1}{5}(6x-1)(4x+1)^{3/2}+C$

63. $\dfrac{2}{3}(x-4)\sqrt{x+2}$

65. $\dfrac{1}{3}\arctan^3(x)+C$

67. $\dfrac{1}{2}(\arcsin^2(x))+C$

69. $2\sqrt{1-x^2}+\arcsin(x)+C$

71. $8\ln(2)$ **73.** $\dfrac{3}{2}\ln(3)$ **75.** $49/3$

77. $125/6$ **79.** $608/5$ **81.** $71/6$

83. $148/3$ **85.** 16, 4, 12

87. 57/64, 101/105, 292/315

89. $\alpha=37.318\ldots, \beta=37.363\ldots$

91. $\alpha=11.223\ldots, \beta=11.286\ldots$

93. 48.6 km^2, 52.9 km^2

95. 53.9 m

97. $\dfrac{2}{2x+3}+\ln\left(\left|\dfrac{2x+1}{2x+3}\right|\right)+C$

99. $\ln\left(\left|\dfrac{\exp(t)+2}{\exp(t)-2}\right|\right)-\exp(t)+C$

Section 6.1

1. $(x-1)e^x+C$

3. $3e^{x/3}(2x-1)+C$

5. $\sin(2x)-(2x+1)\cos(2x)+C$

7. $x^2(2\ln(4x)-1)/4+C$

9. $(3\ln(x)-1)x^3+C$

11. 1 **13.** -2

15. $18-3\sqrt{3}\pi$ **17.** $(e^2+1)/4$

19. $8\ln(2)-4$

21. $(x^2-2x+2)e^x+C$

23. $(x^2-2)\sin(x)+2x\cos(x)+C$

25. $(2x^2-1)\sin(2x)+2x\cos(2x)+C$

27. $(x^3-3x^2+6x-6)e^x+C$

29. $x^4(4\ln^2(x)-2\ln(x)+1/2)+C$

31. $\pi/2-\ln(2)$ **33.** $\pi/\sqrt{3}-1$

35. $\sqrt{3}-\pi/3$ **37.** $10\sqrt{e}-16$

39. $6-2e$

41. $x(\ln(\sqrt{x})-1/2)+C$

43. $x^4(4\ln(1/x)+1)/16+C$

45. $-2^{-x}(x/\ln(2)+1/\ln^2(2))+C$

47. $-\left(\ln^2(x)+2\ln(x)+2\right)/x+C$

49. $x\sec(x)-\ln(|\sec(x)+\tan(x)|)+C$

51. $\cos(x+\pi/4)+x\sin(x+\pi/4)+C$

53. $x\sin(\ln(x))+x\cos(\ln(x))+C$

55. $-4/3$

57. $2\sin(\sqrt{x})-2\sqrt{x}\cos(\sqrt{x})+C$

59. $(x^2-1)\exp(x^2)+C$

61. $e^2/2-13/6$

63. $x^{p+1}(\ln(x)-1/(p+1))/(p+1)+C$ if $p\neq-1$, $\ln^2(x)/2$ if $p=-1$

65. $2\pi/3-\sqrt{3}$

67. $\pi/2+\ln(2)-2$

69. $24\ln(2)-16$

73. $x\ln(x^2-1)-2x+\ln\left(\dfrac{x+1}{x-1}\right)+C$

75. $1-\pi/4$ **77.** $4/\pi$

79. $\ln(2)+4\alpha-4$

87. **a.** $x\,\mathrm{FresnelS}(x)+\dfrac{1}{\pi}\cos\left(\dfrac{\pi}{2}x^2\right)+C$;

b. $x\,\mathrm{FresnelC}(x)-\dfrac{1}{\pi}\sin\left(\dfrac{\pi}{2}x^2\right)+C$;

c. $x^2\,\mathrm{FresnelS}(x)+\dfrac{1}{\pi}x\cos\left(\dfrac{\pi}{2}x^2\right)-\dfrac{1}{\pi}\mathrm{FresnelC}(x)+C$,

d. $x^2\,\mathrm{FresnelC}(x)-\dfrac{1}{\pi}x\sin\left(\dfrac{\pi}{2}x^2\right)+\dfrac{1}{\pi}\mathrm{FresnelS}(x)+C$

89. **a**. $x\,\mathrm{Si}(x)+\cos(x)+C$;
b. $\dfrac{1}{2}(x^2\mathrm{Si}(x)+x\cos(x)-\sin(x))+C$

91. 1.37742... **93.** 0.021772...

95.

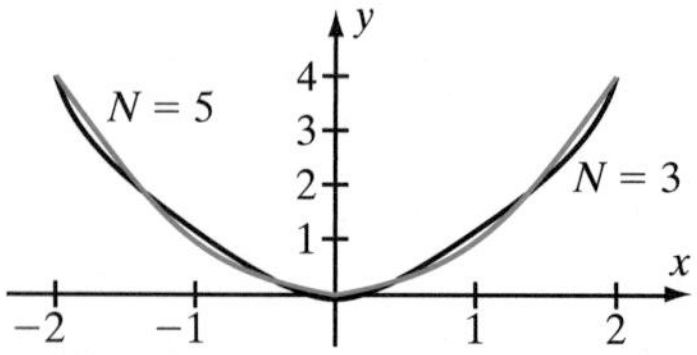

Fourier cosine-polynomial approximations of $f(x)=x^2$.

Section 6.2

1. $x/2 - \cos(x/2)\sin(x/2) + C$ or $\big(x - \sin(x)\big)/2 + C$
3. $x/2 + \cos(2x)\sin(2x)/4 + C$ or $x/2 + \sin(4x)/8 + C$
5. $2\tan(x/2) - x + C$
7. $\cos^3(2x) - 3\cos(2x) + C$
9. $\cos^3(x/3) - 3\cos(x/3) + C$
11. $C - 3\cos^5(x/5) + 10\cos^3(x/5) - 15\cos(x/5)$
13. $C - 3\sin^5(x/3) + 5\sin^3(x/3) + C$
15. $C - \sin^2(x)/2 + \ln(|\sin(x)|)$
17. 4/35 19. 4/63 21. 8/7 25. $3\pi/4$
27. $3\pi/16$ 29. $\pi/32$ 31. $5\pi/16$
33. $\sqrt{3}/64 + \pi/96$
35. $\tan^3(x)/3 - \tan(x) + x + C$
37. $\sec^3(x)/3 + C$
39. 4/3 41. $48\sqrt{3}$ 43. 8/15
45. $2\sqrt{3} - \ln(2 + \sqrt{3})$
47. $2\sec^4(x) + C$
49. $\sec^3(x) - 3\sec(x) + C$
51. $C - \csc^3(x/3)$
53. $C - \csc^3(x/3) + 3\csc(x/3)$
55. $2\sin^3(x) + C$
57. $\tan(x) + \sec(x) + C$
59. $C - \csc(x) - \cot(x) - x$
61. $C - \ln(|\csc(x) + \cot(x)|) + \ln(|\csc(x)|)$
63. $\frac{5}{16}x + \frac{1}{4}\sin(2x) + \frac{3}{64}\sin(4x) - \frac{1}{48}\sin^3(2x) + C$
65. $\frac{1}{16}x - \frac{1}{64}\sin(4x) - \frac{1}{48}\sin^3(2x) + C$
67. $\frac{1}{8}x - \frac{1}{32}\sin(4x) + C$
69. $-\frac{1}{128}\sin^3(2x)\cos(2x) - \frac{3}{256}\cos(2x)\sin(2x) + \frac{3}{128}x + C$
73. 64/35 75. $\pi/4 - 2/3$
81. $k=1: (A^2 + B^2)^{-1/2}\ln(|\sec(x + \phi) + \tan(x+\phi)|) + C$;
$k = 2: (A^2 + B^2)^{-1}\tan(x + \phi) + C$
83. 0.90341... 85. 0.074281... 87. 0.23866...

Section 6.3

1. $C - \sqrt{9 - x^2}$
3. $4\arcsin(x/2) + C$
5. $C - \frac{1}{2}x\sqrt{1 - x^2} + \frac{1}{2}\arcsin(x)$
7. $\frac{x}{\sqrt{1 - x^2}} - \arcsin(x) + C$
9. $\frac{1}{2}x\sqrt{1 - 4x^2} + \frac{1}{4}\arcsin(2x) + C$
11. $2\ln(x^2 + 1) + C$
13. $2\arctan(3x) + C$
15. $2x\sqrt{1 + 4x^2} + \ln(2x + \sqrt{1 + 4x^2}) + C$
17. $\frac{x}{\sqrt{x^2 + 1}} + C$
19. $x\sqrt{x^2 + 1} - \ln\big(x + \sqrt{x^2 + 1}\big) + C$
21. $\pi/6$
23. $\ln\left(\frac{2 + \sqrt{5}}{1 + \sqrt{2}}\right)$
25. $2\sqrt{5} - \sqrt{2} + \ln\left(\frac{2 + \sqrt{5}}{1 + \sqrt{2}}\right)$
27. $3\sqrt{5} - 2\sqrt{10}$
29. $\frac{1}{2}\sqrt{2} - \frac{1}{2}\ln(\sqrt{2} + 1)$
31. $3\arcsin(x) - 2\sqrt{1 - x^2} + C$
33. $\ln(1 + x^2) + 3\arctan(x) + C$
35. $x\sqrt{1 + x^2} + C$
37. $\frac{10x}{9\sqrt{9 - x^2}} - \arcsin\left(\frac{x}{3}\right) + C$
39. $\sqrt{2}\pi/64$
41. $\ln(2)$
43. 4/15
45. $2\ln\left(\frac{5 + \sqrt{21}}{4}\right)$
47. $(8 - 3\sqrt{2})\pi/12 + \ln(\sqrt{2} + 1) - \ln(2 + \sqrt{3})$
49. $\pi/144 + \sqrt{3}/8 - 2/9$
51. $x/(1 + x^2) + \arctan(x) + C$
53. $\sqrt{1 + x^2} - \ln\left(\frac{1 + \sqrt{1 + x^2}}{x}\right) + C$
55. $C - \frac{\sqrt{1 + x^2}}{x} + \ln(x + \sqrt{1 + x^2})$
57. $C - \frac{x}{\sqrt{1 + x^2}} + \ln(x + \sqrt{1 + x^2})$
59. $\arcsin(x - 1) + C$
61. $\sqrt{x^2 + 2x + 2} - \ln(\sqrt{x^2 + 2x + 2} + x + 1) + C$
63. $\frac{2x - 2}{x^2 - 2x + 2} + 2\arctan(x - 1) + C$
65. $3\ln(x^2 - 14x + 58) + 14\arctan\left(\frac{x - 7}{3}\right) + C$
67. $x - 3\ln(x^2 + 4) + 2\arctan(x/2) + C$

69. $3\ln(2x^2+6x+5) - 18\arctan(2x+3) + C$

71. $C - \dfrac{2(x+2)}{(x^2+2x+2)^2} - 3\dfrac{x+1}{x^2+2x+2} - 3\arctan(x+1)$

73. $6\arcsin\left(\dfrac{x}{5}-1\right) - \sqrt{10x-x^2} + C$

75. $x - \dfrac{3}{2}\arctan(x/2) + C$

77. $x - 2\ln(5+4x+x^2) + 4\arctan(x+2) + C$

79. $2\sqrt{x} - 2\ln(1+\sqrt{x}) + C$

81. $C - x + 4\sqrt{x} - 4\ln(1+\sqrt{x})$

83. $2\sqrt{1+x} + \ln\left(\dfrac{\sqrt{1+x}-1}{\sqrt{1+x}+1}\right) + C$

85. $\alpha = \dfrac{1}{2}\ln(3); \beta = \ln(\sqrt{3}) - \ln(1+\sqrt{2})$

87. 0.87670... **89.** 0.98215...

Section 6.4

1. $A/x + B/(x+1)$

3. $A + B/(x-2) + C/(x+2)$

5. $Ax + B + C/(2x-5) + D/(x+4)$

7. $A/x + B/x^2 + C/(x+1) + D/(x-2)$

9. $Ax^2 + Bx + C + D/(x-5) + E/(x-5)^2 + F/(x+3) + G/(x+3)^2$

11. $1/x - 1/(x+1)$

13. $2/(x-2) - 1/(x+2)$

15. $5/(2x-5) + 3/(x+3)$

17. $3/x - 1/x^2 - 2/(x+1) + 1/(x-2)$

19. $-1/(x-2) + 2/(x-2)^2 + 3/(x+3) + 1/(x+3)^2$

21. $(-5/4)/(2x+7) + (7/4)/(2x+9)$

23. $(9/10)/(x-2) + (-37/5)/(x+3) + (23/2)/(x+4)$

25. $-5/x + (45/4)/(2x-1) + (9/4)/(2x+1)$

27. $2\ln(|x-1|) + \ln(|x+1|) + C$

29. $5\ln(|x-3|) + 4\ln(|x+6|) + C$

31. $3x - 3/(x+1) - 6\ln(|x+1|) + C$

33. $5\ln(|x+2|) - 3\ln(|x+3|) - \ln(|x+5|) + C$

35. $-1/x - \ln(|x|) + 2\ln(|x+1|) + C$

37. $-1/(x-2) + \ln(|x-2|) - \ln(|x+2|) + C$

39. $-2\ln(2)$ **41.** $-3\ln(3)$

43. $5\ln(2) - 2\ln(3)$

45. $-1/(2x^2) + 2\ln(|x|) - 2\ln(|x+2|) + C$

47. $1/(x-1) + \ln(|x-1|) + 2/(x-2) + 2\ln(|x-2|) + C$

49. $5/24 + 2\ln(2)$

51. $1/2 + \ln(9/2)$

53. $(1/2)\ln(|\exp(x)-1|) - (1/2)\ln(\exp(x)+1) + C$ or $C - \operatorname{arctanh}(\exp(x))$

55. $\ln(3-\sin(x)) - \ln(2-\sin(x)) + C$

57. $2\ln(|\sqrt{x}-1|) - \ln(x) + C$

59. $\ln(|\sqrt{x+2}-1|) - \ln(\sqrt{x+2}+1) + C$

61. $2x^{1/2} + 6x^{1/6} + 3\ln(|x^{1/6}-1|) - 3\ln(x^{1/6}+1) + C$

67. a. $-2/x + 1/x^2 + 2/(x-1) + 2/(x-1)^2$

69. 0.117435... **71.** 1.681321...

73. $1/(x+1) - 1/(x+2) + 3/(x+3) - 2/(x+4)$

75. $3/(x-1)^2 - 1/(x+1)^4$

Section 6.5

1. $\dfrac{Ax+B}{x^2+1} + \dfrac{Cx+D}{x^2+4}$

3. $\dfrac{Ax+B}{x^2+1} + \dfrac{Cx+D}{x^2+x+1} + \dfrac{Ex+F}{(x^2+x+1)^2}$

5. $\dfrac{Ax+B}{x^2+x+3} + \dfrac{C}{x-4}$

7. $\dfrac{Ax+B}{x^2+4} + \dfrac{Cx+D}{(x^2+4)^2} + \dfrac{Ex+F}{(x^2+4)^3} + \dfrac{G}{x-2}$

9. $\dfrac{A}{x} + \dfrac{B}{x^2} + \dfrac{C}{x-1} + \dfrac{D}{x+1} + \dfrac{Ex+F}{x^2+1}$

11. $\dfrac{1}{x-1} + \dfrac{2x-3}{x^2+1}$

13. $\dfrac{5x}{x^2+2} + \dfrac{2x-3}{x^2+1}$

15. $\dfrac{x}{x^2+1} - \dfrac{2x}{(x^2+1)^2}$

17. $1 + \dfrac{12}{x-3} + \dfrac{3x-4}{x^2+4}$

19. $\dfrac{2x-1}{x^2+4} + \dfrac{x-4}{x^2+3}$

21. $\ln(|x-1|) + \ln(x^2+1) - 3\arctan(x) + C$

23. $\dfrac{5}{2}\ln(x^2+2) + \ln(x^2+1) - 3\arctan(x) + C$

25. $\dfrac{1}{x^2+1} + \dfrac{1}{2}\ln(x^2+1) + C$

27. $x + 12\ln(x-3) + \dfrac{3}{2}\ln(x^2+4) - 2\arctan\left(\dfrac{x}{2}\right) + C$

29. $\ln(x^2+4) - \dfrac{1}{2}\arctan\left(\dfrac{x}{2}\right) + \dfrac{1}{2}\ln(x^2+3) - \dfrac{4}{\sqrt{3}}\arctan\left(\dfrac{x}{\sqrt{3}}\right) + C$

31. $\dfrac{\pi}{2} - \dfrac{1}{\sqrt{2}}\arctan\left(\dfrac{1}{\sqrt{2}}\right)$

33. $\ln(18)$

35. $2\ln(2) - \arctan(2) + \pi/4$

37. $(\pi+1)4$

39. $\ln(|x+1|)+\ln(x^2+2x+2)+C$

41. $5\ln(|x-1|)-\frac{3}{2}\ln(x^2+1+x)-\frac{5}{\sqrt{3}}\arctan\left(\frac{2x+1}{\sqrt{3}}\right)+C$

43. $\ln\left(\frac{x^2+1}{|x^2-1|}\right)-2\arctan(x)-\frac{2}{x-1}+C$

47. $0.077444\ldots$ **49.** $0.013706\ldots$

Section 6.6

1. Divergent **3.** $10\cdot 2^{0.1}$
5. Divergent **7.** $2/3$
9. Divergent **11.** Divergent
13. 6 **15.** Divergent
17. $4\sqrt{3}$ **19.** -1
21. Diverges **23.** $10\cdot 2^{3/5}$
25. $3(5^{1/3}+1)$ **27.** Diverges
29. $3(1-2^{2/3})$ **31.** Diverges
33. Diverges **35.** 0
37. π **39.** $3\cdot 2^{2/3}$
41. $7(\ln(7)-1)$ **43.** Diverges
45. $8(\ln(2)-1)$ **47.** $3-6\ln(2)$
49. $4(\ln(2)-1)$ **51.** $9/8$
53. $\pi^2/8$ **57.** $p<1$
59. Convergent **61.** Convergent
63. Convergent **65.** Divergent
67. Convergent **69.** Divergent

71. **c.** $B(p,1)=\frac{2!}{(p)(p+1)(p+2)}$; $B(p,2)=\frac{3!}{p(p+1)(p+2)(p+3)}$

73. 2.295
75. 2.071
77. 2.597

Section 6.7

1. $2/\sqrt{3}$ **3.** $\sqrt{2}$ **5.** $\pi/2$
7. $1/2$ **9.** $\ln(2)$ **11.** $1/8$
13. $\pi(1+\sqrt{2})$ **15.** $1/(6e^3)$
17. Divergent **19.** $3\pi^2/32$
21. $-1/8$ **23.** 3 **25.** 2
27. 1 **29.** $3e^{4/3}$ **31.** -4
33. $-2/\pi$ **35.** $1/\ln(2)$ **37.** 0
39. $\pi/2$ **41.** 0 **43.** $\pi(1-1/\sqrt{2})$
45. 20000 **47.** 40000 **49.** $\pi+2\ln(2)$
51. $-1/\ln^2(2)$ **53.** $1/2$ **55.** 1
57. Divergent **59.** π **75.** $\sqrt{\pi}$
77. $\mu\sqrt{2\pi}\sigma$ **81.** 0.61 **83.** 0.28

REVIEW EXERCISES FOR CHAPTER 6

1. $3e^4+1$ **3.** $18-3\sqrt{3}\pi$
5. $\ln(3)-4/9$ **7.** $-2/e$
9. π **11.** $1+\ln(3)$
13. $e-2$ **15.** π^2-4
17. $2\pi+21\ln(3)+12\sqrt{3}+4$ **19.** $\pi/8$
21. $5\sqrt{2}$ **23.** $2/15$
25. π **27.** $4/35$
29. $8/45$ **31.** $-16/315$
33. $3\pi/4$ **35.** $5\pi/8$
37. $\pi/48$ **39.** $2\sqrt{3}$
41. $\sqrt{3}$ **43.** $\sqrt{2}/2-(1/2)\ln(1+\sqrt{2})$
45. $3/4$ **47.** $2\ln(1+\sqrt{2})-\sqrt{2}$
49. 3 **51.** $2\ln(2)-1$
53. $\ln(4+2\sqrt{2})$ **55.** $97/15-8\ln(2)$
57. $2\ln(1+\sqrt{2})-\sqrt{2}$ **59.** $-\pi/12$
61. $\ln(2)$ **63.** $\pi/8$
65. $7\sqrt{3}+8\pi$ **67.** $\ln\left(\frac{3+\sqrt{10}}{2}\right)$
69. $4/45$ **71.** $9\sqrt{2}-3\sqrt{3}+\ln\left(\frac{3+\sqrt{8}}{2+\sqrt{3}}\right)$
73. $-1+\frac{1}{\sqrt{2}}\ln(3+2\sqrt{2})$ **75.** $1-\ln(2)$
77. $3\ln(2)-2\arctan(2)+\pi/2$
79. $\ln(2)+\pi/2$
81. $\pi/4+1/2+\ln(2)$
83. $\pi/4+\ln(3/2)$
85. $2\ln(3/2)$
87. 4 **89.** 3 **91.** 9 **93.** $1/4$
95. $\ln(1+\sqrt{2})$
97. $\pi/4-2/3$
99. $2\ln(2)-5/4$

Section 7.1

1. 4π **3.** 2π **5.** $3\pi/2$
7. $\pi^2/4$ **9.** $\pi(e^2-1)/2$ **11.** $32\pi/5$
13. $2\pi/15$ **15.** $108\pi/5$

17. $\pi(14 - 9/\ln(2))/6$
19. $64\pi/15$ **21.** $162\pi/5$ **23.** $20\pi/3$
25. $\pi(e-1)$ **27.** 132π **29.** 4π
31. $40\pi/3$ **33.** $3\pi/2$ **35.** $2\pi(e-1)/3$
37. $\pi/6$ **39.** $92\pi/3$ **41.** $20\pi/3$
43. $101\pi/5$ **45.** $656\pi/15$ **47.** $16\pi/3$
49. $31\pi/30$ **51.** $11\pi/30$ **53.** $8\pi(2-\ln(2))$
55. $14\pi/3$ **57.** $\pi^2/2$ **59.** $\pi(e^2+1)/4$
61. $\pi(e-2)$ **63.** $\pi(1/3 - \ln(4/3))$ **65.** 720π
67. 375π **69.** $375\pi/2$ **71.** $8\pi^2$
73. 2π **75.** $77\pi/30$ **77.** $2\pi R \cdot \pi r^2$
79. $24\pi(\pi-1)$ **81.** $2000/3$ **83.** $500\pi/3$
85. 0.0741 **87.** 9.24g

Section 7.2

1. $3\sqrt{10}$ **3.** $3\sqrt{2}$
5. $(80\sqrt{10} - 13\sqrt{13})/27$
7. $(82\sqrt{82} - 10\sqrt{10})/27$
9. $\ln(2+\sqrt{3})$
11. $5/3$
13. $(13\sqrt{13} - 8)/27$
15. $112/3$ **17.** 10 **19.** $8\pi/3$
21. $4\sqrt{2} - 2$ **23.** $\sqrt{2}(e-1)$ **25.** $39\pi\sqrt{10}$
27. $61\pi/2$
29. $\frac{1}{2}\pi(e^2 + 4 - e^{-2})$
31. $20\pi\sqrt{17}$
33. $\pi(145^{3/2} - 1)/27$
35. $(\sqrt{2} + \ln(\sqrt{2}+1))/2$
37. $\sqrt{2} - \frac{2}{3} + \ln(\sqrt{2}+1) - \frac{1}{2}\ln(3)$
39. $4/3$
41. $\frac{1}{2}(e - 1/e)$
43. $123/64$
45. $\frac{3}{2}\pi\sqrt{9\pi^2+1} + \frac{1}{2}\ln(3\pi + \sqrt{9\pi^2+1})$
47. $1 + \frac{1}{\sqrt{2}}\ln(1+\sqrt{2})$
49. 8 **51.** $\ln(2/\sqrt{3})$ **53.** $21/16$
57. $aE(\varepsilon, \tau)$ **59.** $1179\pi/1024$ **61.** $805\pi/48$
63. $\pi R\sqrt{h^2+R^2}$ **65.** $V = \pi$ **67.** 7.6404
69. 69 cm^2 **71.** 9.385×10^5

Section 7.3

1. $2/\pi$ **3.** $\frac{1}{3}\ln(4)$ **5.** $9/(4\pi)$
7. $14/9$ **9.** 20 **11.** $1/(e-1)$
13. $2/\ln(3)$ **15.** $7/3$ **17.** $7/8$
19. $\sqrt{2}/4$ **21.** $\sqrt{e}/(\sqrt{e}+1)$ **23.** $1\sqrt{3}$
25. $3/80$ **27.** 2 **29.** $3/4$
31. $7/4$ **33.** $5/4$ **35.** 2
37. $375,069$
39. A decrease of about 12,676 people
41. $100.5°$ F **43.** $(e^4 - 1)/4$ **45.** $4(2\sqrt{2} - 1)$
47. $4/(15\pi)$ **49.** $(e^2+1)/(e-1)$ **51.** $\ln(3) + \frac{3}{2}\ln(2)$
53. $\sqrt{3}$ **55.** \$9.21, \$7.69, \$7.91 **57.** $\pi/4$
59. $4/(e^2+1)$ **61.** $3/2$ **63.** $2/(\pi-2)$
65. $\pi/2$ **67.** $\pi/2 - 1$ **69.** $8/5$
71. $1/2$ **73.** $2/\pi$ **75.** $\pi/6$
77. $1 + \ln(2) - \ln(e+1)$
85.

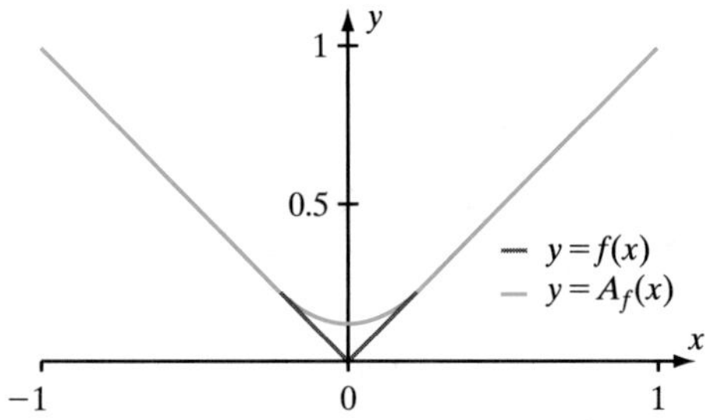

87.

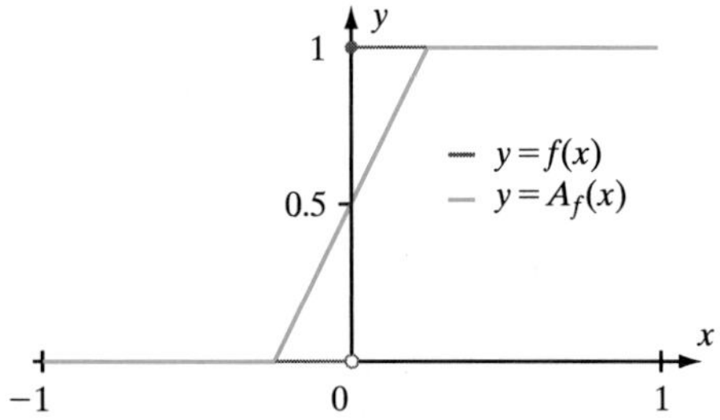

89. $f_{avg} = 0.211$; $c = 1.866$
91. $f_{avg} = 0.069$; $c = 3.041$
93. $0.68\ldots$

Section 7.4

1. -6 **3.** 0 **5.** $-116/15$
7. $1 + 3\ln(2)$ **9.** 4 **11.** $1/4$
13. $16/3$ **15.** $1/20$ **17.** $(2, 2/3)$
19. $(8/(3\pi), 8/(3\pi))$

21. $(-1, 18/5)$

23. $(1/\ln(2), 1/(4\ln(2)))$

25. $(1/(e-1), (e+1)/4)$

27. $(2(2-\sqrt{2})/\pi, (2/\pi)\ln(1+\sqrt{2}))$

29. $(0, 2)$

31. $(8/5, 2)$

33. $(11/10, 104/25)$

35. $(41/20, 27/16)$

37. $(\pi/2 - 1, \pi/8)$

39. $\left(\dfrac{\ln(4)}{\pi}, \dfrac{\pi+2}{4\pi}\right)$

41. $\left(0, \dfrac{128}{105\pi}\right)$

43. $\left(\dfrac{8}{3(3-2\ln(2))}, \dfrac{5}{6(3-2\ln(2))}\right)$

45. $(-2/9, 4/9)$

47. $(-13/30, 46/25)$

49. $(-2/65, 147/130)$

51. $(349/184, 359/92)$

53. $\overline{X} = \overline{x}$

55. $481/30$

57. $5\pi^2 - 4$

59. $3/80$

61. $(b-a)^2/12$

67. $(0.3913\ldots, 1.300\ldots)$

69. $(0.23666\ldots, -0.70397\ldots)$

71. $(0, 263.48)$

Section 7.5

1. 37600 ft-lb

3. 51500148 ft-lb

5. 240 kg

7. 707520000 ft-lb

9. 3150 ft-lb

11. 2.18448×10^5 J

13. 130000 ft-lb

15. 60000 ft-lb

17. 49/3 ft-lb

19. 12 J

21. 140 ft-lb

23. 50 J

25. 8 lb

27. 7.02315×10^5 ft-lb

29. 2.35360×10^6 J

31. 4.7945×10^5 ft-lb

33. 1.4383×10^5 ft-lb

35. 1.7365×10^9 ft-lb

37. 8.0419×10^8 ft-lb

39. 25%

43. 88256 ft-lb

45. 12029 ft-lb

49. An arc of the parabola $W = F^2/(2k)$

51. 3582 ft-lb

53. 5754.9 ft-lb

Section 7.6

9. $y = (2x^{3/2} + C)^2$

11. $y = 1/(C - \ln(|2+x|))$

13. $y = C\exp(x^3)$

15. $y = \dfrac{1}{3}\ln\left(\left|C - \dfrac{3}{2}e^{-2x}\right|\right)$

17. $2\sin(y) = x^2 + C$

19. $y(x) = x^2 + 2$

21. $y(x) = \sin(x) + 2$

23. $y^2(x) = x^2 + 1$

25. $y(x) = 1/(\cos(x) - 1/2)$

27. $y^2(x) = (3x^2+1)^{2/3} - 1$

29. $y^2(x) = 4\exp(2x) - 1$

31. $y(x) = 3\exp(\cos(1/x))$

33. $y(x) = \exp(4x^3/3)$

35. $y^2(x) = 2\arctan(x) + \pi/2$

37. **a.** F_3 **b.** F_6 **c.** F_2
d. F_5 **e.** F_1 **f.** F_4

39. $y = \ln(C - (x+1)\exp(-x))$

41. $y = 1/(1 + C\exp(x^3))$

43. $y = C\exp\left(\dfrac{1}{2}\arctan\left(\dfrac{1}{2}e^x\right)\right)$

45. $y(x) = 8 + f^4(x)$

47. $y(x) = 4 + g(f(x))$

51. $3y^{5/2} + 10y^{3/2} + 15y^{1/2} = 28 - 16 \cdot 0.0001t$, 17500s

53. $\ln(3)/P_0$

59. 144000

61. $P_\infty = 50$ and $k = \dfrac{1}{450}\ln\left(\dfrac{116}{9}\right) \approx 0.00568.$

63. 19274

65. $2.32\exp(-0.000612t)$ moles/l

67. $y(x) = \dfrac{1}{2C}(1 + \sqrt{1+4C^2x^2}),$
$y(x) = \dfrac{1}{2C}(1 - \sqrt{1+4C^2x^2})$

69. $P = 2\sqrt{\dfrac{C}{17}}T^{17/4}$

72. 1/25 s

73. $512/75 \approx 6.83$ ft

75. $4y - 7\ln(y) = 3 - 2x + 6\ln(x)$

79. $L(x) = C\exp\left(Mx + \dfrac{B}{\ln(g)}g^x\right)$

85. $v(t) = \sqrt{\dfrac{mg}{k}}\tan\left(\arctan\left(v_0\sqrt{\dfrac{k}{mg}}\right) - t\sqrt{\dfrac{kg}{m}}\right)$

93. 1.7134

95. 1.76

97. 4.93 weeks

99. **a.** 8.01381×10^{-6}
b. $P(1990) \approx 20788$, $P(1991) \approx 24652$, $P(1992) \approx 28358$, $P(1993) \approx 31707$
c. See Figure 13, Section 7.6
d. 2001

Section 7.7

1. $y = Ce^{3x}$

3. $y = Ce^{-4x}$

5. $y = Ce^{x/2} - 2$

7. $y = x/2 + C/x$

9. $y = 1 + C\exp(-x^2/2)$

11. $y = (x + C)e^{3x}$

13. $y = 2e^{2x} + Ce^{-x}$

15. $y = 4x^2 - 4 + Ce^{-x^2}$

17. $y = 3/2 - e^{-2x}$

19. $y = \left(2(3 + x^2)^{3/2} - 9\right)/x$

21. $y = 2x^{3/2} + 16/x^2$

23. $y = (x - 3)e^x$

25. $y = 2x - 1 + 4e^{-2x}$

27. $y = 3e^x - x^2 - 2x - 2$

29. $m(t) = 20 - 10e^{-t/10}$ kg; $m_\infty = 20$ kg

31. $m(t) = 60e^{-t/40}$ lb; $m_\infty = 0$ lb

33. $70 + 605\sqrt{11}/64 \approx 101.35$ min

35. $m = 10/\ln(10) \approx 4.3429$

37. 8987 days

39. $(193 - 3\sqrt{141})/5\,^\circ\text{F} \approx 31.5\,^\circ\text{F}$

41. $(2t + C)e^{-5t}$

43. $\sin(t) - \dfrac{2}{5}\cos(t) + Ce^{-3t}$

45. 3/2 kg

47. $\alpha = -p\ln(\lambda), \beta = -\ln(\lambda)$

49. α/β

51. $I(t) = \dfrac{\alpha}{\alpha + \lambda} + Ce^{-(\alpha+\lambda)t}$

53. **b.** $m(t) = C\exp(-rt/L)$
c. $m(t) = C\big(L + (\rho - r)t\big)^{-r/(\rho - r)}$

63. 11.4 m/s (downward)

65. **a.** 4.1977 kg/s **b.** 1870 m/s **c.** 238 m/s
d.

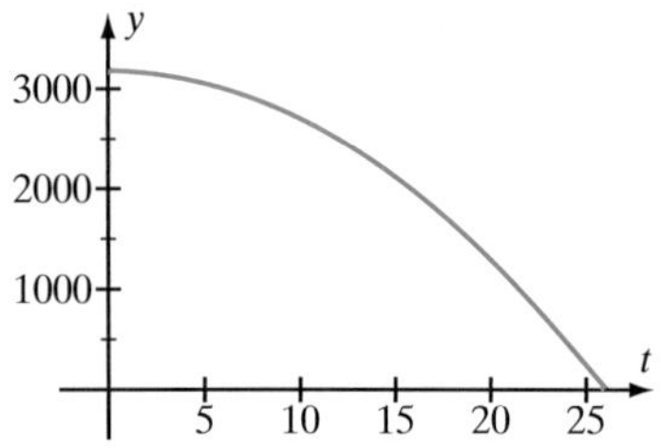

e. 25.422 s, 249.4 m/s

67. $H = 14.6$ m, $v_I = -14.16$ m/s, $T_u = 1.59$ s, $T_d = 1.89$ s

69. **a.** Johnson: $\tau = 1.356$, $P = 8.702$; Lewis: $\tau = 1.457$, $P = 8.099$;
b.

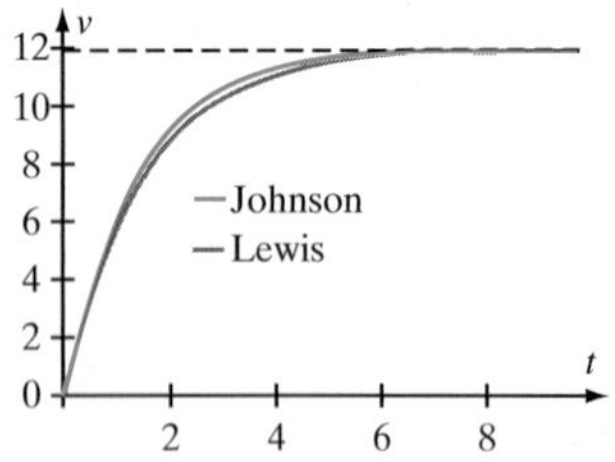

c. 1.18 m

71. 19.1 ft/s

REVIEW EXERCISES FOR CHAPTER 7

1. $\pi/2$

3. $28\pi/15$

5. $7\pi/6$

7. $10\pi/3$

9. $16\pi/3$

11. $\pi/2$

13. $\pi(3 - 4\ln(2))$

15. $333\pi/5$

17. $\dfrac{\pi}{9}(2e^3 + 1)$

19. $\dfrac{1}{2}\pi(\pi + 4)$

21. $(27 - 5\sqrt{5})/24$

23. 5/6

25. 71/80

27. $2\pi\left(\sqrt{2} + \ln(1 + \sqrt{2})\right)$

29. $304\pi/3$

31. $2\sqrt{10}$

33. $5\sqrt{5} - 8$

35. 2/3

37. $\ln(2)$

39. 1/9

41. $\dfrac{5}{3}\ln(2) - 1$

43. 1

45. 2

47. $\dfrac{2}{\pi}(2 - \sqrt{2})$

49. $\dfrac{1}{4}(e^2 + 1)$

51. $\bar{x} = \dfrac{1}{\ln(3/2)} - 1, \bar{y} = \dfrac{1}{12\ln(3/2)}$

53. $\bar{x} = 1, \bar{y} = \pi/4$

55. $\bar{x} = -3/2, \bar{y} = -2$

57. 1299213042 ft-lb

59. 321333 m

61. 71240 J

63. 20/3 ft-lb

65. 108/5 J

67. 324/5 J

69. 80 N

71. 1386.4 J

73. $y = 2(C - 2x - x^2)^{-1}$

75. $y = Ce^x - 2x - 2$

77. $\exp(y) = C + e^{2x}$

79. $y = 1/(C - \arctan(x))$

81. $y = \dfrac{3}{4}e^{2x} - \dfrac{1}{2}x - \dfrac{1}{4}$

83. $y = (e^{2x} - 3)/2$

85. $y = e^x + 4e^{-x}$

87. $m(t) = 40 - 15e^{-2t/25}$ kg, $m_\infty = 40$ kg

89. $m(t) = 100e^{-t/40}$ lb; $m_\infty = 0$ lb

Section 8.1

1. 1/5

3. 2

5. 2/3

7. 2

9. 1

11. 0

13. 0

15. 3/2

17. $\pi/2$

19. e

21. 1, 3/2, 11/6, 25/12, 137/60

23. 3/2, 11/4, 31/8, 79/16, 191/32

25. 2/3, 10/9, 38/27, 130/81, 422/243

27. 1, 3/4, 31/36, 115/144, 3019/3600

29. 3/4 **31.** 3/5 **33.** 10

35. 625/4 **37.** $1 + \sqrt{2}/2$ **39.** 25/4

41. 11/30 **43.** 20/3 **45.** 8/9

47. 17/999 **49.** 403/3300 **51.** 2

53. 3 **55.** 3/2 **63.** 1

65. 0 **67.** 1 **69.** 3/4

71. 1/6 **75.** $1/(r-1)^2$

77. $\pi^2/24$ and $\pi^2/8$

79. The line $y = -x/15$

83. 1 m **85.** $\$1.9 \times 10^{12}$

87. $Q(1 - e^{-kT})$ **93.** 0.4587

95. 1.4581 **97.** 1.1876

Section 8.2

1. No conclusion **3.** Divergence

5. Divergence **7.** No conclusion

9. No conclusion **11.** Divergence

13. No conclusion **15.** Divergence

17. No conclusion **19.** No conclusion

21. may **23.** may

25. may **27.** Converges

29. Converges **31.** Diverges

33. Diverges **35.** Converges

37. Converges **39.** Converges

41. Diverges **43.** Converges

45. Converges **47.** Diverges

49. Converges **51.** Converges

53. Converges **55.** Divergence

57. Divergence **59.** No conclusion

61. Divergence **63.** Divergence

65. Divergence **67.** No conclusion

69. Converges **71.** Diverges

73. Converges **75.** (1/24, 7/96)

77. $\left(\dfrac{1}{2(1+e)^2}, \dfrac{1+3e}{2(1+e)^3}\right)$

83. Converges

95. $M = 13$, $S = 1.202$ **97.** $M = 7$, $S = 0.058$

Section 8.3

1. $a_n = n/(n^3+1) \le b_n = 1/n^2$ and $\sum_{n=1}^{\infty} b_n$ is a convergent p-series $(p = 2 > 1)$.

3. $a_n = 1/(n+\sqrt{2})^2 \le b_n = 1/n^2$ and $\sum_{n=1}^{\infty} b_n$ is a convergent p-series $(p = 2 > 1)$.

5. $a_n = \big(2 + \sin(n)\big)/n^4 \le b_n = 3/n^4$ and $\sum_{n=1}^{\infty} b_n$ is a constant multiple of a convergent p-series $(p = 4 > 1)$.

7. $a_n = (n+2)/(2n^{5/2}+3) \le b_n = 1/n^{3/2}$ and $\sum_{n=1}^{\infty} b_n$ is a convergent p-series $(p = 3/2 > 1)$.

9. $a_n = 2^n/(n3^n) \le b_n = (2/3)^n$ and $\sum_{n=1}^{\infty} b_n$ is a convergent geometric series with (positive) ratio r less than 1.

11. $a_n = 1/(2n\sqrt{n}-1) \le b_n = 1/n^{3/2}$ and $\sum_{n=1}^{\infty} b_n$ is a convergent p-series $(p = 3/2 > 1)$.

13. $a_n = n^3/(n^{4.01}+1) \le b_n = 1/n^{1.01}$ and $\sum_{n=1}^{\infty} b_n$ is a convergent p-series $(p = 1.01 > 1)$.

15. $a_n = 1/n! \le b_n = 2/n^2$ and $\sum_{n=1}^{\infty} b_n$ is a constant multiple of a convergent p-series $(p = 2 > 1)$.

17. $b_n = 1/(2n) \le a_n = 1/(2n-1)$ and $\sum_{n=1}^{\infty} b_n$ is a constant multiple of the harmonic series.

19. $b_n = 1/n \le a_n = 1/\ln(n)$ and $\sum_{n=1}^{\infty} b_n$ is the harmonic series.

21. $b_n = 1/n \le a_n = (2n+5)/(n^2+1)$ and $\sum_{n=1}^{\infty} b_n$ is the harmonic series.

23. $b_n = 1/(4n) \le a_n = 1/\sqrt{10+n^2}$ and $\sum_{n=1}^{\infty} b_n$ is *a* constant multiple of the harmonic series.

25. $b_n = 1/2 \le a_n = (3^n+1)/(3^n+2^n)$ and $\sum_{n=1}^{\infty} b_n$ diverges by the Divergence Test.

27. $b_n = 1/\sqrt{n} \le a_n = (3^n+n)/(\sqrt{n}3^n+1)$ and $\sum_{n=1}^{\infty} b_n$ is a divergent p-series $(p = 1/2 \le 1)$.

29. Diverges **31.** Diverges

33. Converges **35.** Converges

37. Converges **39.** Diverges

41. Converges **43.** Converges

45. Diverges **47.** Converges

49. Diverges **51.** Diverges

53. Converges **55.** Diverges

57. Diverges **59.** Converges

61. Converges **63.** Converges

65. Converges **67.** Converges

69. Converges **71.** Converges

Section 8.4

13. Converges absolutely

15. Converges conditionally

17. Converges conditionally

19. Converges absolutely

21. Converges conditionally

23. Converges absolutely

25. Converges absolutely
27. Converges absolutely
29. Diverges
31. Diverges
33. Converges absolutely
35. Converges conditionally

37. 9999	**39.** 200	**41.** 4
43. 10	**45.** 8	**47.** −0.632
49. −0.424	**57.** 9	**59.** 18

Section 8.5

1. Convergent	**3.** Divergent
5. Convergent	**7.** Divergent
9. Divergent	**11.** Convergent

13. Converges conditionally
15. Converges absolutely
17. Diverges
19. Converges absolutely

21. Diverges	**23.** Diverges
25. Convergent	**27.** Convergent
29. Convergent	**31.** Convergent
33. Convergent	**35.** Converges
37. Diverges	**39.** Diverges
41. Diverges	**43.** Converges

45. Converges conditionally
47. Diverges
49. Converges conditionally
51. Converges conditionally
53. Diverges
55. Converges absolutely
57. Converges conditionally
59. Diverges

Section 8.6

1. $R = 1/2$; $I = (-1/2, 1/2)$
3. $R = 1$; $I = (-1, 1)$
5. $R = 1$; $I = (-1, 1)$
7. $R = 5$; $I = (-5, 5)$
9. $R = 0$; $I = \{0\}$
11. $R = 1$; $I = (-1, 1]$
13. $R = 1$; $I = [-1, 1)$
15. $R = 1/3$; $I = [-1/3, 1/3]$
17. $R = 1$; $I = (-1, 1)$
19. $R = 1/4$; $I = [-1/4, 1/4)$
21. $R = 1$; $\mathrm{I} = [-7, -5)$
23. $R = 1$; $I = [-5, -3]$
25. $R = 1/2$; $I = (-1, 0)$
27. $R = 1$; $I = (-6, -4)$
29. $R = \infty$; $I = (-\infty, \infty)$
31. $(1/2, 3/2)$
33. $(-7/3, -5/3)$
35. $1 + 4x + 9x^2 + 16x^3 + 25x^4$; $I = (-1, 1)$
37. $1 + \frac{1}{\sqrt{2}}(x-1) + \frac{1}{\sqrt{3}}(x-1)^2 + \frac{1}{2}(x-1)^3 + \frac{1}{\sqrt{5}}(x-1)^4$; $I = [0, 2)$
39. $x + \frac{1}{\sqrt{2}}x^2 + \frac{1}{\sqrt{3}}x^3 + \frac{1}{2}x^4$; $I = [-1, 1)$
41. $(x-2) + \frac{1}{6}(x-2)^2 + \frac{1}{27}(x-2)^3 + \frac{1}{108}(x-2)^4$; $I = [-1, 5)$

43. 1/2	**45.** 4	**47.** ∞
49. $(-4, 2)$	**51.** $(1/4, 3/4)$	**53.** $(-1/2, 1/2)$

55. $(-\sqrt{3} - e, \sqrt{3} - e)$
57. $\sum_{n=1}^{\infty} \left(\frac{x-c}{r}\right)^n$ where $c = (\alpha + \beta)/2$ and $r = (\beta - \alpha)/2$
59. $\sum_{n=1}^{\infty} \frac{1}{n}\left(\frac{x-c}{r}\right)^n$ where $c = (\alpha + \beta)/2$ and $r = (\beta - \alpha)/2$
63. 1   **67.** $1/\sqrt{2}$
69.

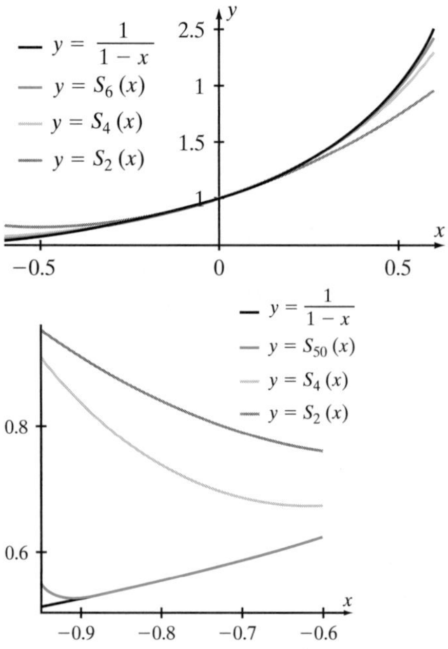

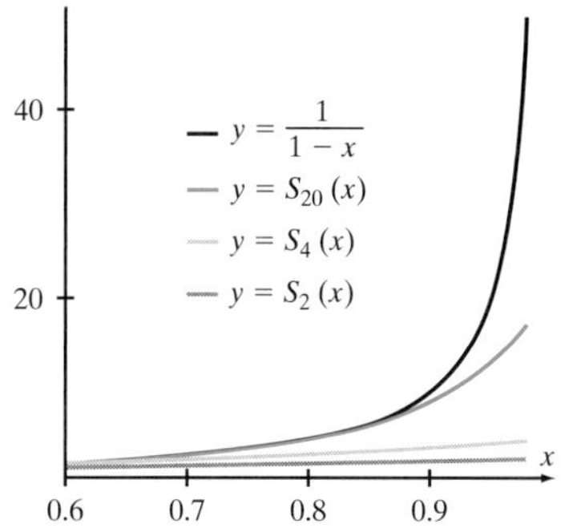

71.

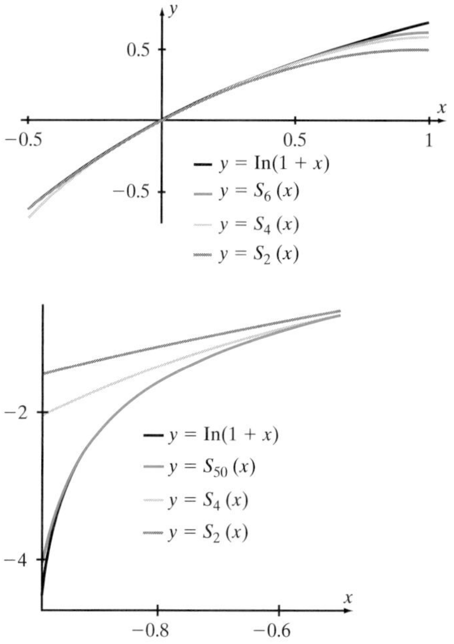

Section 8.7

1. $\sum_{n=0}^{\infty} 2^n x^n$; $R = 1/2$

3. $\sum_{n=0}^{\infty} 2^{-(n+1)} x^n$; $R = 2$

5. $\sum_{n=0}^{\infty} (-1)^n 4^{-(n+1)} x^n$; $R = 4$

7. $1 + \sum_{n=1}^{\infty} 2x^{2n}$; $R = 1$

9. $\sum_{n=0}^{\infty} (-1)^n x^{4n+3}$; $R = 1$

11. $\sum_{m=1}^{\infty} \frac{(-1)^{m-1} 4^m}{m} x^m$; $R = 1/4$

13. $\sum_{m=1}^{\infty} \frac{(-1)^{m-1}}{m} x^{2m}$; $R = 1$

15. $\sum_{m=1}^{\infty} \frac{(-1)^{m-1}}{m(2m+1)} x^{2m+1}$; $R = 1$

17. $\sum_{n=0}^{\infty} \frac{(-1)^n x^{4n+2}}{2n+1}$; $R = 1$

19. $\sum_{n=0}^{\infty} \frac{(-1)^n}{(2n+1)(4n+3)} x^{4n+3}$; $R = 1$

21. $\sum_{n=0}^{\infty} (x-5)^n$; $R = 1$

23. $\sum_{n=0}^{\infty} \frac{(-1)}{2^{n+1}} (x+3)^n$; $R = 2$

25. $\sum_{n=0}^{\infty} \frac{(-1)^n 2^n}{3^{n+1}} (x+1)^n$; $R = 3/2$

27. $\frac{1}{2} - \frac{\sqrt{3}}{2}(x - \pi/3) - \frac{1}{4}(x - \pi/3)^2 + \frac{\sqrt{3}}{12}(x - \pi/3)^3 + \frac{1}{48}(x - \pi/3)^4$

29. $2 + \frac{1}{4}(x-4) - \frac{1}{64}(x-4)^2 + \frac{1}{512}(x-4)^3$

31. $6 + 9(x-1) + 6(x-1)^2 - (x-1)^3 + (x-1)^4$

33. $\frac{1}{8} - \frac{3}{16}(x-2) + \frac{3}{16}(x-2)^2 - \frac{5}{32}(x-2)^3$

35. $3(x+2) - \frac{9}{2}(x+2)^2 + 9(x+2)^3 - \frac{81}{4}(x+2)^4$

37. $x + \frac{1}{3}x^3$

39. $\frac{1}{2} - \frac{1}{4}(x-1)^2 + \frac{1}{4}(x-1)^3$

41. $2 + \frac{1}{12}(x-3) - \frac{1}{288}(x-3)^2 + \frac{5}{20736}(x-3)^3$

43. $\sum_{n=0}^{\infty} (n+1)x^n$

45. $\sum_{n=0}^{\infty} (3n+2)x^n$

47. $\frac{x^2}{1-x}$

49. $\frac{2x}{(1-x)^3}$

51. $\frac{x^3}{(1-x^2)^2}$

53. $a_n = 1/n!$

55. $a_0 = a_1 = 1$, $a_n = 2/n!$ for $n \geq 2$

57. $a_0 = 0$, $a_1 = 1$, $a_n = 2/n!$ for $n \geq 2$

65. **a.** $1 + x + \frac{1}{2}x^2 + \frac{1}{2}x^3 + \frac{1}{8}x^4 + \frac{1}{40}x^5 + \frac{1}{240}x^6 + \frac{1}{1680}x^7$

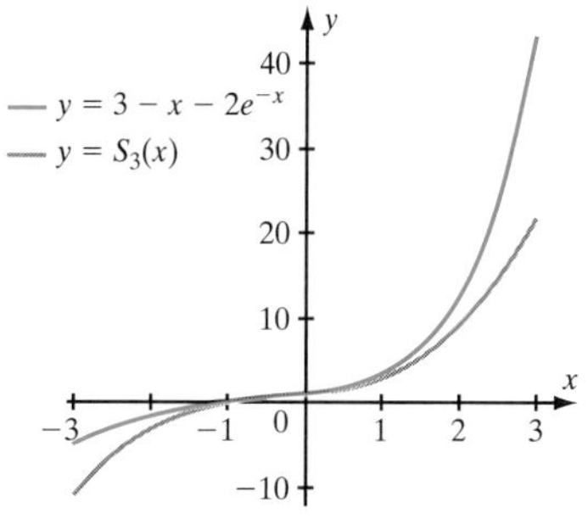

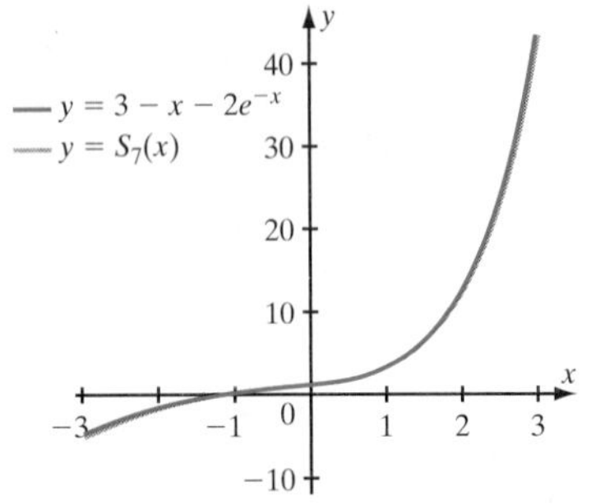

Section 8.8

1. 1.1051708 . . . ; error : $\frac{(0.1)^5}{120}e^s$

3. 0.29552025 ; error : $-\frac{\sin(s)}{720}(0.3)^6$

5. 0.877583 . . . ; error : $\frac{\sin(s)}{6}(0.5-\pi/6)^3$

7. 1.09863 . . .; error : $-\frac{1}{4s^4}(3-e)^4$

9. 0.983 . . .; error : $\frac{3s^2-1}{3(1+s^2)^3}(1.5-\sqrt{3})^3$

11. 2.049375; error : $\frac{1}{16(1+s)^{5/2}}(0.2)^3$

13. **a.** 1.184 **b.** $\frac{15}{16\cdot 4!}(0.4)^4$, or 0.001

15. **a.** −0.19 **b.** $\frac{8}{3!}(0.1)^3$, or 0.0013 . . .

17. **a.** 2.26 **b.** $\frac{1}{3!}(0.2)^3$, or 0.0013 . . .

19. **a.** 0.9164 **b.** $\frac{1}{3e^3}(e-2.5)^3$, or 0.00017 . . .

21. **a.** 0.4 **b.** $\frac{2}{3!}(0.4)^3$, or 0.021 . . .

23. $\sum_{n=0}^{\infty}\frac{(-1)^{n+1}2^{2n-1}}{(2n-1)!}x^{2n-1}$

25. $\sum_{n=0}^{\infty}\frac{(-1)^{n+1}}{2^{2n-1}(2n-1)!}x^{2n+1}$

27. $\sum_{n=0}^{\infty}\frac{(-1)^n}{n!}c_n x^n$ where $c_n = 5\cdot 2^n$ if n is even and $c_n = 4\cdot 3^n$ if n is odd

29. $\sum_{n=0}^{\infty}\frac{(-1)^n}{n!}x^{2n}$

31. $\sum_{n=0}^{\infty}(-1)^n x^{2n+1}$

33. $\sum_{n=0}^{\infty}\binom{1/2}{n}x^{3n}$

35. $1+\sum_{n=1}^{\infty}(-1)^{n+1}\frac{(2n-2)!}{(n-1)!n!2^{2n-1}}(x-1)^n$

37. $\sum_{n=0}^{\infty}\frac{(n+1)e}{n!}(x-1)^n$

39. $\sum_{n=1}^{\infty}\frac{(-1)^{n+1}}{n}(x-1)^n$

41. $\sum_{n=0}^{\infty}(-1)^n\frac{27\ln^n(3)}{n!}(x+3)^n$

43. $1+\frac{3}{4}x-\frac{3}{32}x^2+\frac{5}{128}x^3$

45. $1-\frac{1}{2}x+\frac{3}{8}x^2-\frac{5}{16}x^3$

47. $1-\frac{3}{2}x+\frac{15}{8}x^2-\frac{35}{16}x^3$

49. −1/6 **51.** −1/2 **53.** −2

55. 1/2 **57.** −1/3 **59.** −2

61. $\sum_{n=1}^{\infty}\frac{2^{2n-1}(-1)^{n+1}}{(2n)!}x^{2n}$

63. −1/3 **67.** 1/2 **69.** e **71.** e^2-1

73. Maclaurin series of $\tan(x)$ up to the x^5 term :
$x+\frac{1}{3}x^3+\frac{2}{15}x^5$

75. Maclaurin series of $(1+u)^{-1/2}$: $\sum_{n=0}^{\infty}\frac{(-1)^n(2n)!}{2^{2n}(n!)^2}u^n$;

Maclaurin series of $\arcsin(x)$:
$\sum_{n=0}^{\infty}\left(\frac{(2n)!}{(2n+1)(2^{2n})(n!)^2}\right)x^{2n+1}$

83. 0.2955 **85.** 0.1489

87. $s = 1.07766\ldots$

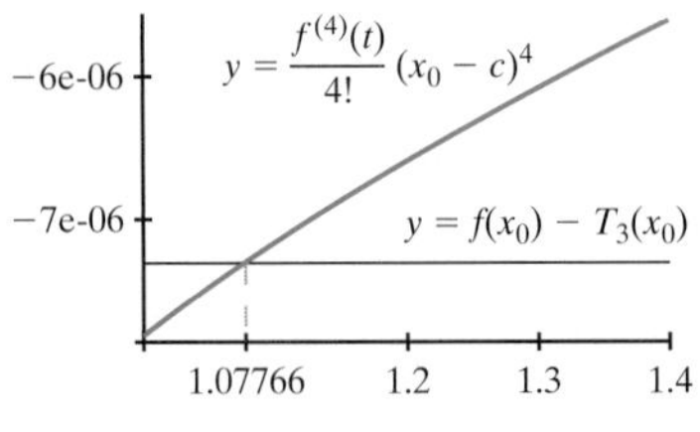

89. $s = 3.23405\ldots$

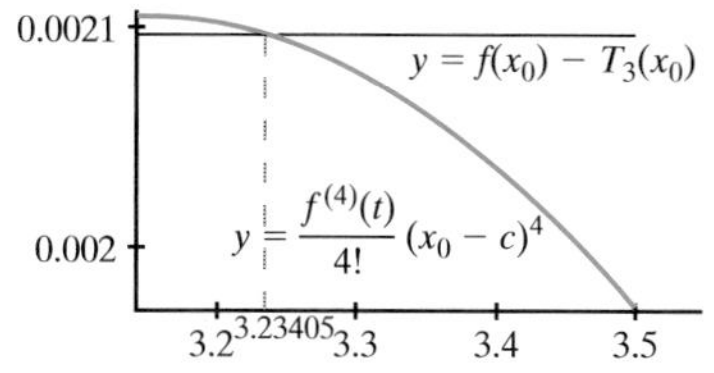

91. 0.01392

93. 0.00188

95. 0.46128

97. 0.33311

99. $y = \dfrac{\sin(\tan(x)) - \tan(\sin(x))}{\arcsin(\arctan(x)) - \arctan(\arcsin(x))}$

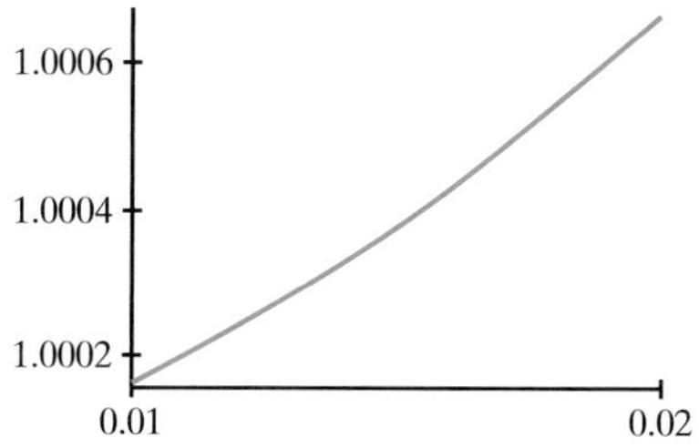

REVIEW EXERCISES FOR CHAPTER 8

1. $S_1 = 3$, $S_2 = 7$, $S_3 = 23/2$, $S_4 = 163/10$, $S_5 = 213/10$

3. 5/2

5. 27/4

7. 1389/2200

9. Converges conditionally

11. Diverges

13. Converges conditionally

15. Diverges

17. Converges absolutely

19. Converges absolutely

21. Converges conditionally

23. Converges absolutely

25. Converges absolutely

27. Converges conditionally

29. Converges absolutely

31. Converges absolutely

33. Diverges

35. Converges absolutely

37. Converges conditionally

39. Converges absolutely

41. Converges absolutely

43. Converges absolutely

45. a) conclusive; b) inconclusive

47. a) inconclusive; b) inconclusive

49. a) conclusive; b) conclusive

51. 0.224...

53. lower estimate: 2; upper estimate: 3

55. $(-5/3, -1)$

57. $[-1, 1]$

59. $(-1, 1]$

61. $(-3/2, 1/2]$

63. $[-3, 0)$

65. $(-2/3, 4/3]$

67. $\sum_{n=0}^{\infty} (-1)^n \dfrac{2}{3^{n+1}} x^n$

69. $\sum_{n=0}^{\infty} \dfrac{27}{81^{n+1}} x^{4n}$

71. $\sum_{n=0}^{\infty} (-1)^n x^{2(n+1)}$

73. $(40 - 10\ln(2) + \ln^2(2))/32$,
$2^{1-s}\ln^2(2)(3 - s\ln(2))/384$

75. $(360 + 12\sqrt{3}\,\pi + \pi^2)/144$,
$-\sin(s)\pi^3/1296$

77. **a)** 2.100008049 **b)** 8.6235×10^{-6}

79. **a)** 3.166666667 **b)** 0.000645287

81. 0.002667

83. 0.1996

85. 1

87. 6

Section 9.1

1. $\langle -1, -4\rangle$ **3.** $\langle -5, 1\rangle$ **5.** $\langle 5, -11\rangle$

7. $\langle 6, -6\rangle$

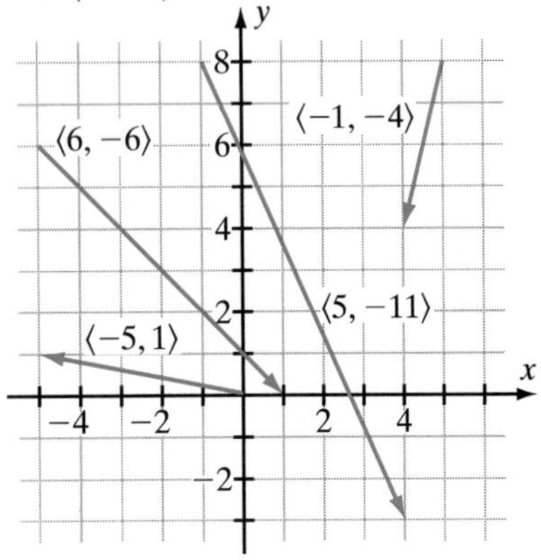

Figure for 1, 3, 5, and 7

9. $\|\overrightarrow{PQ}\| = \sqrt{5}$; $\|\overrightarrow{RS}\| = 3\sqrt{5}$; parallel

11. $\|\overrightarrow{PQ}\| = 3\sqrt{10}$; $\|\overrightarrow{RS}\| = \sqrt{53}$; not parallel

13. $\langle 9, 1\rangle$ See Figure for Exercise 15

15. $\langle 2, 8\rangle$

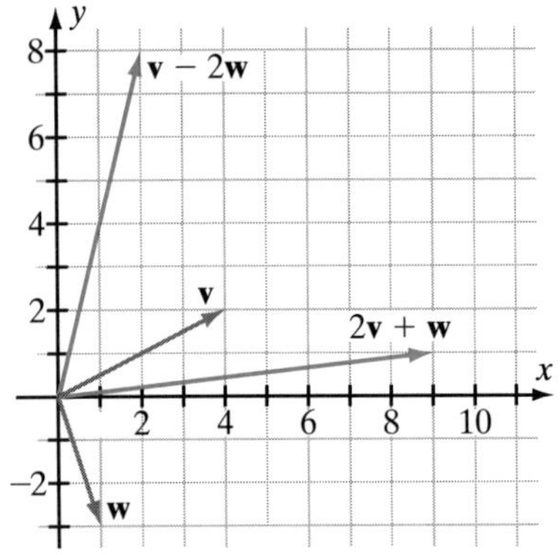

Figure for 13 and 15

17. a. $\langle -5, 12\rangle$ **b.** 13 **c.** $\langle 5, -12\rangle$ **d.** $\langle -5/13, 12/13\rangle$ **e.** $\langle 5/13, -12/13\rangle$

19. a. $\langle 4, -16\rangle$ **b.** $4\sqrt{17}$ **c.** $\langle -4, 16\rangle$ **d.** $\langle 1/\sqrt{17}, -4/\sqrt{17}\rangle$ **e.** $\langle -1/\sqrt{17}, 4/\sqrt{17}\rangle$

21. $\mathbf{v} = -7\mathbf{i} + 2\mathbf{j}$; $\mathbf{w} = -2\mathbf{i} + 9\mathbf{j}$; $-4\mathbf{v} = 28\mathbf{i} - 8\mathbf{j}$; $3\mathbf{v} - 2\mathbf{w} = -17\mathbf{i} - 12\mathbf{j}$; $4\mathbf{v} + 7\mathbf{j} = -28\mathbf{i} + 15\mathbf{j}$

23. $\mathbf{v} = 6\mathbf{i} - 2\mathbf{j}$; $\mathbf{w} = 9\mathbf{i} - 3\mathbf{j}$; $-4\mathbf{v} = -24\mathbf{i} + 8\mathbf{j}$; $3\mathbf{v} - 2\mathbf{w} = 0\mathbf{i} + 0\mathbf{j}$; $4\mathbf{v} + 7\mathbf{j} = 24\mathbf{i} - \mathbf{j}$

25. $\mathbf{F} = \langle 3, 4\rangle$; magnitude: 5

27. $\mathbf{F} = 3\mathbf{i} - \mathbf{j}$; magnitude: $\sqrt{10}$

29. $\langle \sqrt{3}/2, 1/2\rangle$ **31.** $\langle -1, 0\rangle$

33. $\frac{13}{2}\left\langle \frac{12}{13}, -\frac{5}{13}\right\rangle$

35. $\sqrt{13}\left(\frac{3}{\sqrt{13}}\mathbf{i} - \frac{2}{\sqrt{13}}\mathbf{j}\right)$

37. $\langle 7, 9\rangle$ **39.** $\langle 4, 1\rangle$

41. $2\sqrt{10}$ **43.** 5/3

45. −6 **47.** 2

49. 10 **51.** 3/2

53. −6 **55.** $50\sqrt{2}$; $50(1 + \sqrt{3})$

57. 200 m

59. $(-80\sqrt{3} - 200 - 140\sqrt{2})\mathbf{i} + (80 - 140\sqrt{2})\mathbf{j}$ (each component in N)

63. $\lambda\,\langle -b, a\rangle$ for all scalars λ

65. $\xi = (B^2x_0 - ABy_0 - AD)/(A^2 + B^2)$

67. $x = p_1 + t(q_1 - p_1), y = p_2 + t(q_2 - p_2)$

71. 3

73. $\langle 1.85325, 1.44312\rangle$

75. 1.89838

77. Left x-intercept: −0.65421 at $t = -0.98132$; right x-intercept: 1 at $t = 0$; y-intercept: 0.51447 at $t = -0.63598$

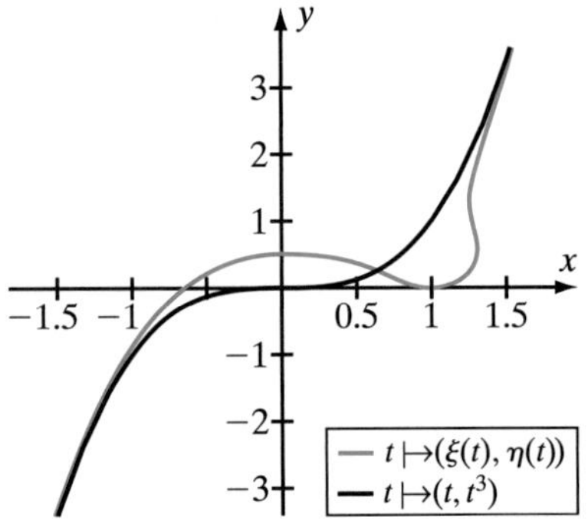

Section 9.2

1.

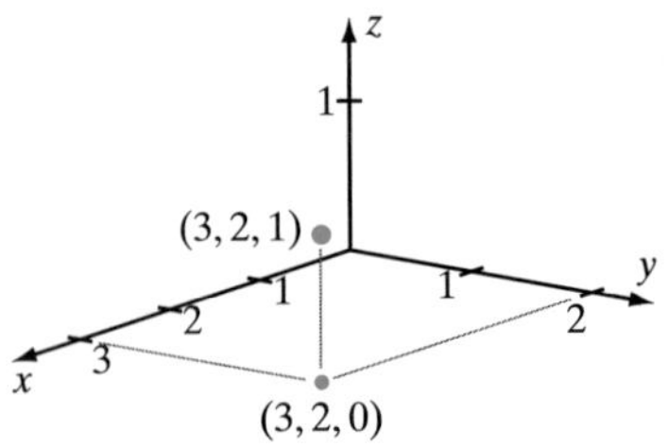

3.

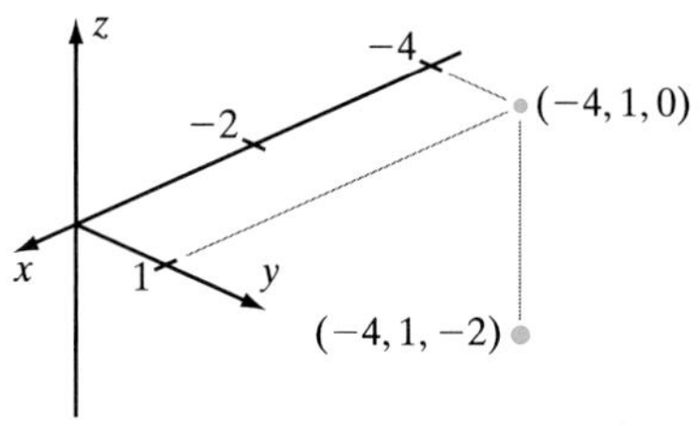

5. $3\sqrt{14}$

7. $3\sqrt{3}$

9. Center: $(-1, -2, -3)$; radius: $\sqrt{22}$

11. Center: $(-1/2, -1/2, -1/2)$; radius: $\sqrt{5}/2$

13. Center: $(1/2, 0, 0)$; radius: $1/2$

15. The coefficients of x^2, y^2, and z^2 are not all equal.

17. The equation has the term $4xy$.

19. The equation has the term $-2xy$.

21. $\{(x, y, z) : (x-3)^2 + (y+2)^2 + (z-6)^2 < 16\}$

23. $\{(x, y, z) : (x-\pi)^2 + (y-\pi)^2 + (z+\pi)^2 = \pi^2\}$

25. $\{(x, y, z) : 4 < (x-2)^2 + (y-1)^2 + z^2\}$

27. $\|\overrightarrow{PQ}\| = \sqrt{41}$; $\|\overrightarrow{RS}\| = 2\sqrt{41}$; $\overrightarrow{PQ}$ and $\overrightarrow{RS}$ *are* parallel.

29. $\|\overrightarrow{PQ}\| = 3$; $\|\overrightarrow{RS}\| = 3\sqrt{5}$; $\overrightarrow{PQ}$ and $\overrightarrow{RS}$ are *not* parallel.

31. $\langle -12, 3, -24\rangle$

33. $\langle 0, -11, 0\rangle$

35. $\langle 6/7, 4/7, 12/7\rangle$

37. $\langle 3, 2, 9\rangle$

39. **a.** $\langle 2, -3, 6\rangle$ **b.** $\langle -2, 3, -6\rangle$ **c.** 7
d. $\langle 2/7, -3/7, 6/7\rangle$ **e.** $\langle 24/7, -36/7, 72/7\rangle$

41. **a.** $\langle 1, -1, -2\rangle$ **b.** $\langle -1, 1, 2\rangle$ **c.** $\sqrt{6}$
d. $\langle 1/\sqrt{6}, -1/\sqrt{6}, -2/\sqrt{6}\rangle$
e. $\langle 2\sqrt{6}, -2\sqrt{6}, -4\sqrt{6}\rangle$

43. $-15\mathbf{i} + 20\mathbf{j} - 5\mathbf{k}$

45. $\mathbf{i} - 6\mathbf{j} + 2\mathbf{k}$

47. $-4\mathbf{j} + \mathbf{k}$

49. $\langle 4, -6, 3\rangle$, $\langle 2, 2, -1\rangle$

51. $\langle 1, 1, -2\rangle$, $\langle 1, -3, 0\rangle$

53. $(x_0, y_0, 0)$, $(0, y_0, z_0)$, and $(x_0, 0, z_0)$, respectively

55. $\{(x, y, z) : y = 0\}$

57. $\{(x, y, z) : 0 < y\}$

59. $\{(x, y, z) : 5 < |z|\}$

61. $\{(x, y, z) : 3 < \sqrt{x^2 + y^2 + z^2}\}$

63. The set of all unit vectors

65. $404/\sqrt{105}$ ft/s

67. The sphere with center $(-1, 2, 3)$ and radius 6

71. The ellipse $x^2 + y^2/9 = 1$, $z = 0$, the ellipse $x^2 + z^2/16 = 1$, $y = 0$, and the ellipse $y^2/9 + z^2/16 = 1$, $x = 0$

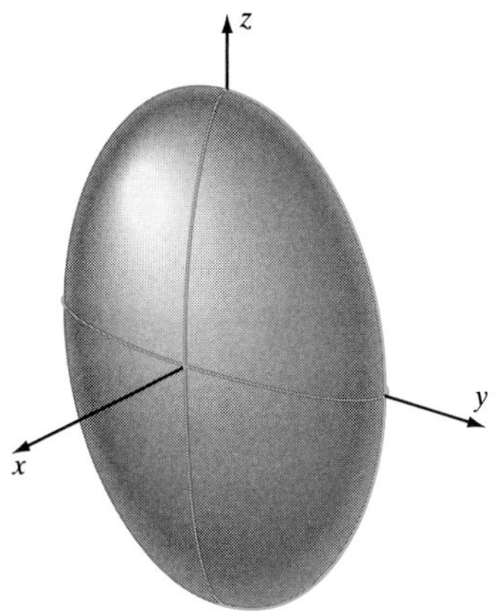

$$x^2 + \frac{y^2}{9} + \frac{z^2}{16} = 1$$

73. Span.

75. Do not span.

77. $-1.8136, -1, 0.4707, 2.3429$

79. 0.6542

Section 9.3

1. 28

3. 4

5. -31

7. $\pi/3$

9. $3\pi/4$

11. $\pi/2$

13. Not perpendicular: Dot product is 1, not 0.

15. Perpendicular: Dot product is 0.

17. $\mathbf{P_v}(\mathbf{w}) = \langle -51/62, 119/62, 17/31\rangle$; $\mathbf{P_w}(\mathbf{v}) = \langle 17/33, 68/33, -68/33\rangle$; $17/\sqrt{62}$; $17/\sqrt{33}$

19. $\mathbf{P_v}(\mathbf{w}) = \langle 4, 8, 12\rangle$; $\mathbf{P_w}(\mathbf{v}) = \langle 2, 4, 6\rangle$; $4\sqrt{14}$; $2\sqrt{14}$

21. $\mathbf{P_v}(\mathbf{w}) = \langle 0, 0, 0\rangle$; $\mathbf{P_w}(\mathbf{v}) = \langle 0, 0, 0\rangle$; 0; 0

23. Direction: $\langle \sqrt{3}/2, -1/2, 0\rangle$; $\alpha = \pi/6$; $\beta = 2\pi/3$; $\gamma = \pi/2$

25. Direction: $\langle 0, 1, 0\rangle$; $\alpha = \pi/2$; $\beta = 0$; $\gamma = \pi/2$

27. Direction: $\langle -1/\sqrt{2}, 0, 1/\sqrt{2}\rangle$; $\alpha = 3\pi/4$; $\beta = \pi/2$; $\gamma = \pi/4$

29. Dot product: -16; lengths 6 and 3; $|-16| \le 6 \cdot 3$ is true.

31. Dot product: 9; lengths $\sqrt{5}$ and $2\sqrt{5}$; $|9| \le \sqrt{5} \cdot 2\sqrt{5}$ is true.

33. $\mathbf{P_u}(\mathbf{v}) = \langle 4, 4, -2\rangle$; $\mathbf{v} - \mathbf{P_u}(\mathbf{v}) = \langle 0, -3, -6\rangle$; $\mathbf{P_u}(\mathbf{v}) \cdot (\mathbf{v} - \mathbf{P_u}(\mathbf{v})) = 0$

35. $\mathbf{P_u}(\mathbf{v}) = \langle -2/\sqrt{3}, -2/\sqrt{3}, 2/\sqrt{3}\rangle$;
$\mathbf{v} - \mathbf{P_u}(\mathbf{v}) = \langle 8/\sqrt{3}, 2/\sqrt{3}, 10/\sqrt{3}\rangle$; $\mathbf{P_u}(\mathbf{v}) \cdot (\mathbf{v} - \mathbf{P_u}(\mathbf{v})) = 0$

37. -2

39. $-4, 4$

45. $100000\sqrt{3}$ ft lb

49. **a.** $6/7$ **b.** $-11/5$ **c.** $-1/5$ **d.** $-9/25$

51. $\overrightarrow{OP} \cdot \mathbf{j} = (\sqrt{3}/2)\|\overrightarrow{OP}\|$

63. $0.158548, 0.700449, 2.282596; 3.141593 \approx \pi$

65. 0.203888

Section 9.4

1. $\langle 9, 11, 12\rangle$

3. $\langle -12, 4, 10\rangle$

5. $\langle 7, 1, 10\rangle$

7. $\langle 32, 0, -4\rangle$

9. $\langle 1/\sqrt{3}, 1/\sqrt{3}, -1/\sqrt{3}\rangle$

11. $\langle -1/\sqrt{6}, -1/\sqrt{6}, 2/\sqrt{6}\rangle$

13. $\frac{\sqrt{2}}{10}\langle 4, 3, -5\rangle$

15. $\sqrt{2}/2$

17. $9/2$

19. $2\sqrt{14}$

21. 7

23. $\mathbf{u} \times (\mathbf{v} \times \mathbf{w}) = \langle -1, -4, 15\rangle$ and $(\mathbf{u} \times \mathbf{v}) \times \mathbf{w} = \langle 40, -3, 42\rangle$

25. $\mathbf{u} \times (\mathbf{v} \times \mathbf{w}) = \langle 11, -33, 11\rangle$ and $(\mathbf{u} \times \mathbf{v}) \times \mathbf{w} = \langle -2, 3, 6\rangle$

27. $\mathbf{u} \times \mathbf{v} = \langle 1, 11, -3\rangle$; $\mathbf{v} \times \mathbf{w} = \langle -11, -1, -7\rangle$; $(\mathbf{u} \times \mathbf{v}) \cdot \mathbf{w} = \mathbf{u} \cdot (\mathbf{v} \times \mathbf{w}) = -40$

29. $\mathbf{u} \times \mathbf{v} = \langle 0, -2, -1\rangle$; $\mathbf{v} \times \mathbf{w} = \langle 7, -12, 1\rangle$; $(\mathbf{u} \times \mathbf{v}) \cdot \mathbf{w} = \mathbf{u} \cdot (\mathbf{v} \times \mathbf{w}) = -7$

31. 5

33. -16

39. 6

41. 36

43. $s = 19, t = -6$

45. $s = 5, t = -1$

47. 65

49. 52

51. **a.** $\langle -3, 3, 3\rangle$ **b.** $\sqrt{3}/2$ **c.** 3 **d.** $1/2$

53. **a.** $\langle -6, 6, 24\rangle$ **b.** $2\sqrt{2}/3$ **c.** 9 **d.** $1/3$

59. $-\mathbf{w}$

61. 30 ft-lb

63. $d(R, \ell) = \| \overrightarrow{PR} \times \overrightarrow{RQ} \| / \| \overrightarrow{PQ} \|$

69. $7x + 22y + 9z = 78$, the graph of which is a plane that passes through P_0 and is perpendicular to $\mathbf{v} \times \mathbf{w}$

71. -7.161655

Section 9.5

1. Normals: $\lambda\langle 1, -3, 4\rangle \lambda \in \mathbb{R}$; $\mathbf{n} = \langle -1, 3, -4\rangle$

3. Normals: $\lambda\ \langle 3, 0, -4\rangle \lambda \in \mathbb{R}$; $\mathbf{n} = \langle -12, 0, 16\rangle$

5. $\lambda\ \langle -5, 7, -4\rangle\ \lambda \in \mathbb{R}$

7. $\lambda\ \langle 1, 4, -2\rangle\ \lambda \in \mathbb{R}$

9. $-3x + 7y + 9z = 79$

11. $2x + 2y + 5z = 18$

13. $-30x - y + 40z = 119$

15. $24x + 20y - 7z = -128$

17. $x = 2 - 3t,\ y = 1 + t,\ z = 9 + 7t$

19. $x = 2t,\ y = 1 - t,\ z = t$

21. $(x - 2)/(-5) = (y - 1)/8 = (z - 5)/2$

23. $x = y - 1 = z/9$

25. $x = 7 + t,\ y = -4 - 3t,\ z = 6 + 7t$

27. $x = -1 + t,\ y = -1 - t,\ z = -8 + t/3$

29. $x - 1 = y = 1 - z$

31. $(x - 1)/2 = 2 - y = 1 - z$

33. $x = 2 - 39t,\ y = 7t,\ z = 12t$

35. $x = 6t,\ y = 1/6 - 8t,\ z = -24t$

37. $(x - 4)/2 = y - 2 = -z$

39. $x = (y + 1)/2 = (z + 3)/4$

41. $(4, -2, -2)$

43. $(2, 2, 7)$

45. $(2, 1, 4)$

47. $(3, 1, 0)$

49. $2/15$

51. $2\sqrt{6}/5$

53. $\sqrt{6}$

55. 3

57. On line

59. Not on line

61. $(-2, -9, 2)$

63. No intersection

65. $\sqrt{2}$

67. $\sqrt{2}$

73. $3x + 4y + z = 14$

75. $(x - 1) + 2(y - 2) + 3(z - 3) = 0$

77. x-intercept: 3; y-intercept: 2; z-intercept: 6

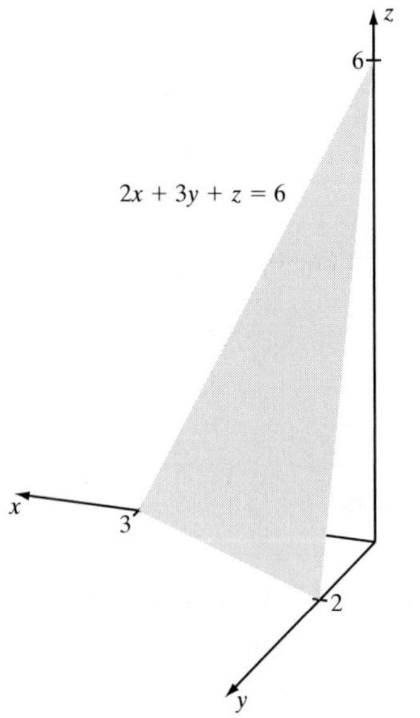

79. x-intercept: 5; y-intercept: 3; z-intercept: 1

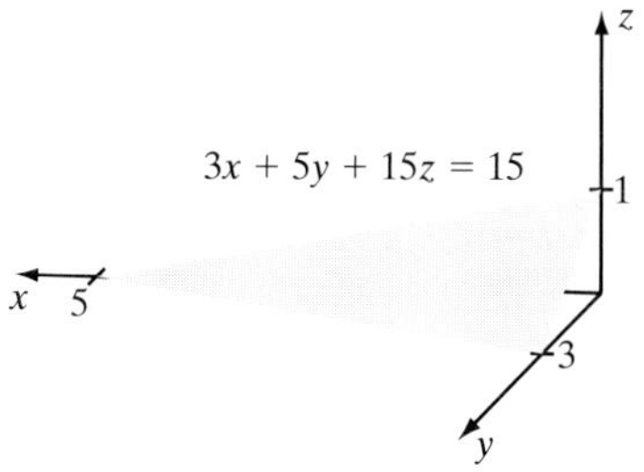

81. **a.** parallel **b.** parallel **c.** not parallel **d.** parallel **e.** parallel

83. **a.** no intersection **b.** the line lies in the plane **c.** $(8, 8, 3)$ is the point of intersection **d.** no intersection

85. $1/\sqrt{2}$

87. $1/\sqrt{2}$

89. $10/\sqrt{3}$

91. $57\sqrt{6}/2$

97.

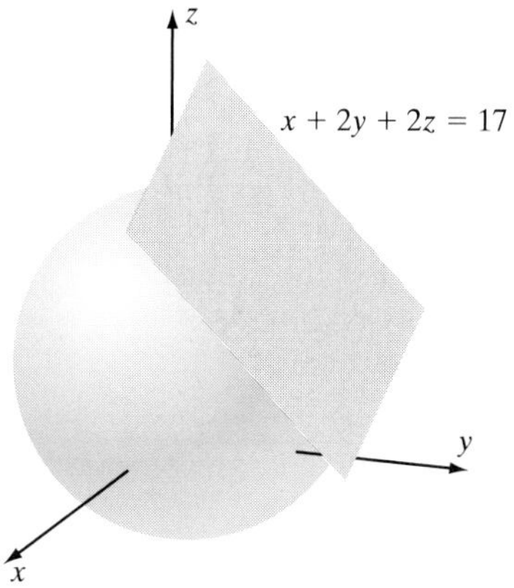

99. $x = (0.808827, 0.617654, 0.573519)$ and $(-0.186758, -1.373516, 3.560274)$

REVIEW EXERCISES FOR CHAPTER 9

1. 9

3. 9

5. Center: $(2, -4, -1)$; radius: 6

7. Center: $(-1/2, -1/2, 1/2)$; radius: 1

9. $9\,\langle 2/3, 2/3, 1/3\rangle$

11. $2\sqrt{3}\,\langle 1/\sqrt{3}, -1/\sqrt{3}, -1/\sqrt{3}\rangle$

13. $\langle 6, 3, 8\rangle$

15. $\langle -60, -66, 48\rangle$

17. $\langle 2, 2, 1\rangle$

19. -7

21. 12

23. 3

25. 2, 3

27. $\arccos(-1/3)$

29. $\arccos(5/6)$

31. $\langle 4, 4, 2\rangle$

33. $\langle 63, 36, 36\rangle$

35. $\mathbf{v}\cdot\mathbf{w} = 62$; $\|\mathbf{v}\| = 7$; $\|\mathbf{w}\| = 9$

37. $\mathbf{v}\cdot\mathbf{w} = 220$; $\|\mathbf{v}\| = 15$; $\|\mathbf{w}\| = 15$

39. $\langle -3, -12, 5\rangle$

41. $\langle -2, -8, 6\rangle$

43. $3\sqrt{2}$

45. $5\sqrt{2}$

47. $5\sqrt{6}$

49. $3\sqrt{10}$

51. 9

53. -2

55. 6

57. 16

59. $-2, 0, 1$

61. $-7/5$

63. -2

65. $1/3$

67. $-3/2$

69. $2/3$

71. $-8/7$

73. $y - z = 1$

75. $2x - 4y - z = 5$

77. $x = 1 + t,\ y = 3,\ z = 5 + 3t$

79. $x = 2t,\ y = -2t,\ z = 1 - 3t$

81. $2(x + 2) = y - 2 = -(z + 1)$

83. $3(x + 2) = 3(y + 3)/2 = -3(z + 5)/2$

85. $(x - 1)/2 = y - 1 = (z + 1)/(-2)$

87. $(x - 1)/2 = (y - 2)/2 = z - 1$

89. $x = 3t,\ y = t,\ z = 2 - 2t$

91. $x = t,\ y = 2 - 2t,\ z = 2 - 3t$

93. $(-1, 8, 5)$

95. $(2, 1, 4)$

97. $(6, 3, -5)$

99. $2\sqrt{5}/3$

Section 10.1

1. The parabola $y = x^2$

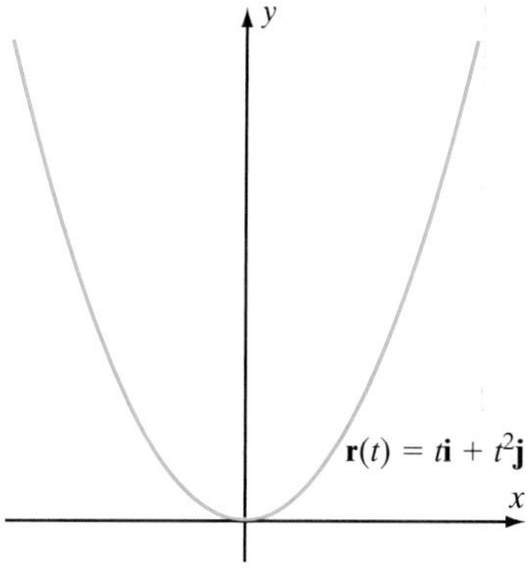

3. The parabola $y = x^2$, $z = 2$

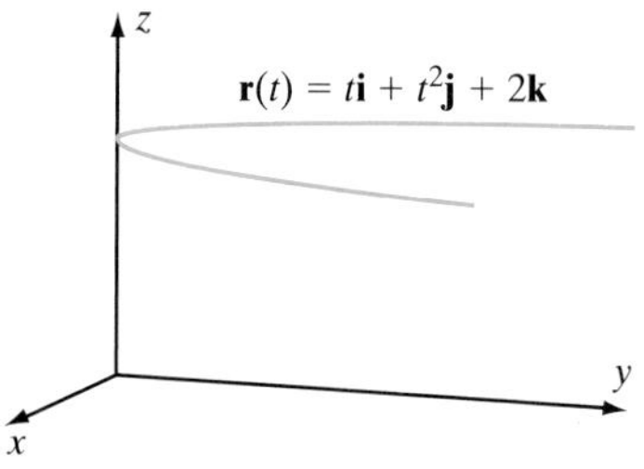

5. The circle $x = 1$, $y^2 + z^2 = 1$

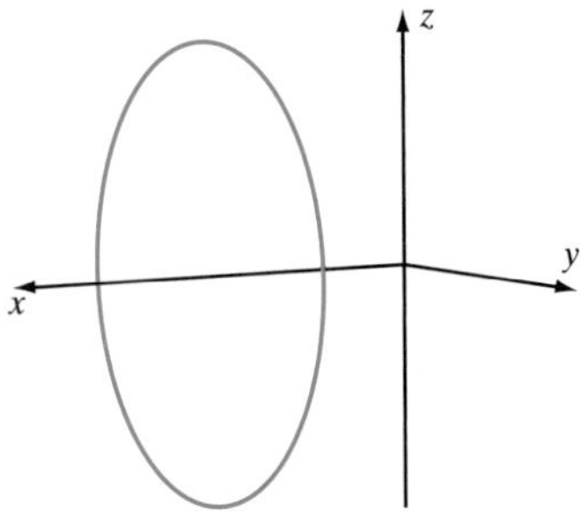

$\mathbf{r}(t) = \mathbf{i} + \cos(t)\mathbf{j} + \sin(t)\mathbf{k}$

7. A straight line segment from $P = (1, 1, 1)$ to $Q = (0, 0, 1)$, followed by the line segment from Q to $R = (4, 4, 1)$

9. One cycle of the graph of $z = \sin(x)$, located in the $x = y$ plane

11. $-\mathbf{i} + (1/2)\mathbf{j}$

13. $\langle 4, 4, 1 \rangle$

15. $(2t^{1/2})^{-1}\mathbf{i} + 2t\mathbf{k}$

17. $\sec^2(t)\mathbf{i} + \csc^2(t)\mathbf{j} + \sec(t)\tan(t)\mathbf{k}$

19. $\mathbf{i} - (t+5)^{-2}\mathbf{j} - 2(t-7)^{-3}\mathbf{k}$

21. $2\mathbf{i} - 3/(4t^{1/2})\mathbf{j} + (15/4)(t+1)^{1/2}\mathbf{k}$

23. $e^{-t}\mathbf{i} - 2(2t^4 + t^2 + 1)/(t(1+t^2))^2\mathbf{j}$

25. $4/(9t^{7/3})\mathbf{i} - ((2t^2 - 1)/(1+t^2)^{5/2})\mathbf{j} - 3/(16t^{7/4})\mathbf{k}$

27. $(t^3/3 + C_1)\mathbf{i} - (2\sqrt{t} + C_2)\mathbf{j} + (\sin(2t)/2 + C_3)\mathbf{k}$

29. $(t\ln(t) - t + C_1)\mathbf{i} - (\cos(t)\sin(t)/2 + t/2 + C_2)\mathbf{j} + (\ln(|\sec(t)|) + C_3)\mathbf{k}$

31. $(t^2+1)\mathbf{i} + (6 - t^3)\mathbf{j} + (t^4 - 15)\mathbf{k}$

33. $(3 + \sin(t))\mathbf{i} + (1 + (1/2)\cos(2t))\mathbf{j} + (\sec(t) - 3)\mathbf{k}$

35. $3e^{3t}\mathbf{i} + 3\sin(3t)\mathbf{j} + (18t/(1+9t^2))\mathbf{k}$

37. $15(1+5t)^2\mathbf{i} - 5\mathbf{k}$

39. $3\,(6+5t)\,e^t$

41. $-9t^{-10}\mathbf{i} + 3t^{-4}\mathbf{k}$

43. $6t^5 + 2t$

45. $5(3t^2e^t + t^3e^t - \ln(1+t^2) - 2t^2/(1+t^2))$

47. $(9t^2 + 10t + 1)\mathbf{j}$

49. $(1+t)e^t\mathbf{j}$

51. 0

53. $\mathbf{i} + \mathbf{j} + \mathbf{k}$

55. $-\mathbf{i} - 2\ln(\pi)\mathbf{j} + \mathbf{k}$

57. Continuous on $\{t \in \mathbb{R} : t \neq 5\}$

59. Continuous on $\mathbb{R}$

65. $\mathbf{r}(t) = \langle t, |t|, 0 \rangle$, $c = (0, 0, 0)$

Section 10.2

1. $\mathbf{v}(t) = \mathbf{i} + 2t\mathbf{j} + 3t^2\mathbf{k}$; $v(t) = \sqrt{1 + 4t^2 + 9t^4}$; $\mathbf{a}(t) = 2\mathbf{j} + 6t\mathbf{k}$

3. $\mathbf{v}(t) = -2t\,\mathbf{i} + 2\mathbf{j} + (-2t/(1+t^2)^2)\mathbf{k}$;
$v(t) = 2\sqrt{1 + t^2 + t^2/(1+t^2)^4}$;
$\mathbf{a}(t) = -2\mathbf{i} + (2(3t^2 - 1)/(1+t^2)^3)\mathbf{k}$

5. $\mathbf{v}(t) = \mathbf{i} + e^t\mathbf{k}$; $v(t) = \sqrt{1 + e^{2t}}$; $\mathbf{a}(t) = e^t\mathbf{k}$

7. $\mathbf{v}(t) = 2t\sin(t^2)\mathbf{i} + 2t\cos(t^2)\mathbf{j} + 3t^2\mathbf{k}$;
$v(t) = |t|\sqrt{4 + 9t^2}$; $\mathbf{a}(t) = (4t^2\cos(t^2) + 2\sin(t^2))\mathbf{i} + (-4t^2\sin(t^2) + 2\cos(t^2))\mathbf{j} + 6t\mathbf{k}$

9. $(2 + 3t)\mathbf{i} - 2t\mathbf{j} + (-5 + t - 16t^2)\mathbf{k}$

11. $(3 + t)\mathbf{i} + (-2 + t)\mathbf{j} + (-1 + t - 16t^2)\mathbf{k}$

13. $x = 1 + s, y = 1 + 2s, z = 1 + 3s$

15. $x = 1 + s, y = 1 - s, z = -1 + 2s$

17. $x = 2 + s, y = -1, z = e + (e/2)s$

19. $x = 5 + s, y = -5, z = 1 - (1/5)s$

21. $x = 1 - (3/5)s, y = 4 + 4s, z = \pi s$

23. $x + 2 = (y + 4)/4 = (z + 8)/12$

25. $x - 5 = 5(1 - z), y = -5$

27. $2(1 - x) = (y - 4)/2, z = 1$

29. $x = 1, y = 2\pi - 2\sqrt{\pi z}$

31. $(x - 1) = 2(y + 4) = 5(z - 1)$

33. Trajectory: $\mathbf{r}(t) = 50\sqrt{3}t\mathbf{i} + (4 + 50t - 16t^2)\mathbf{j}$; time until the arrow hits the ground: $(25 + \sqrt{689})/16$ s; horizontal distance: $25\sqrt{3}(25 + \sqrt{689})/8$ ft; greatest height: $689/16$ ft

35. $-80\mathbf{i} + 44\mathbf{j} + 50\mathbf{k}$

41. Velocity when the arrow hits the ground:
$40\mathbf{i} - 20\sqrt{107}\mathbf{k}$ ft/s; horizontal distance:
$25\sqrt{107} - 75\sqrt{3} \approx 128.7$ ft

43. $2\sqrt{r/g}$

51. $\mathbf{r}'(\pi/3) = -\sqrt{3}\mathbf{i} + (1/2)\mathbf{j}$

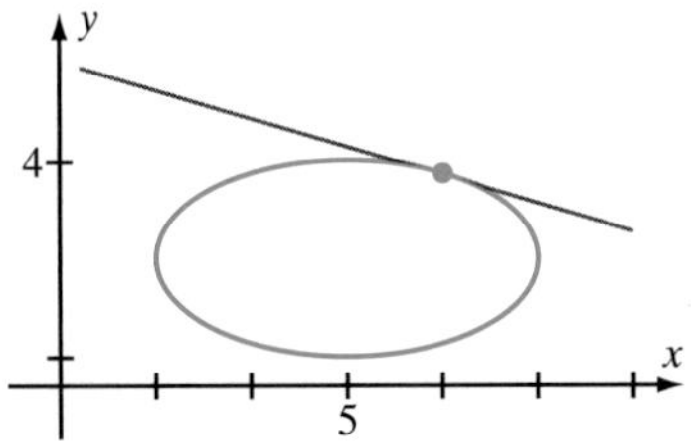

53. $\mathbf{r}'(\pi/4) = 2\mathbf{i} + \sqrt{2}\mathbf{j}$

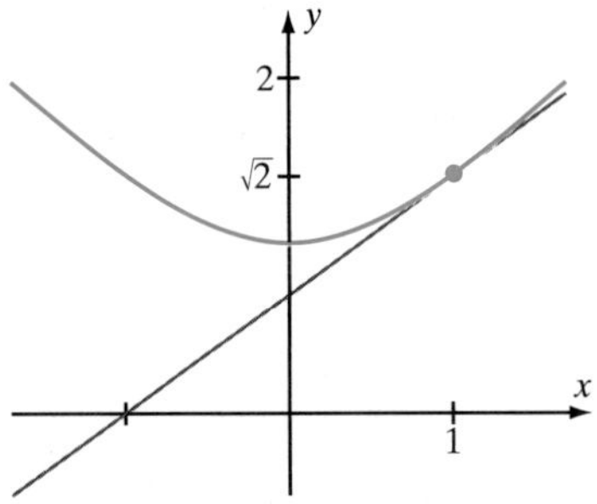

55. $\mathbf{r}'(2) = (e^2 - e^{-2})\mathbf{i} + (e^2 + e^{-2})\mathbf{j}$

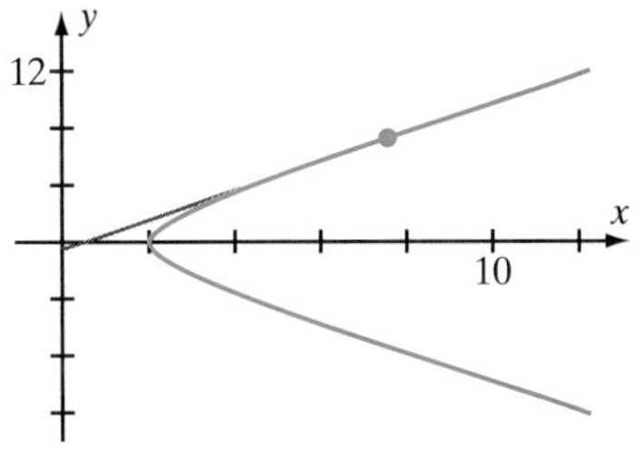

57. $\mathbf{r}'(1/2) = (24/25)\mathbf{i} - (32/25)\mathbf{j}$

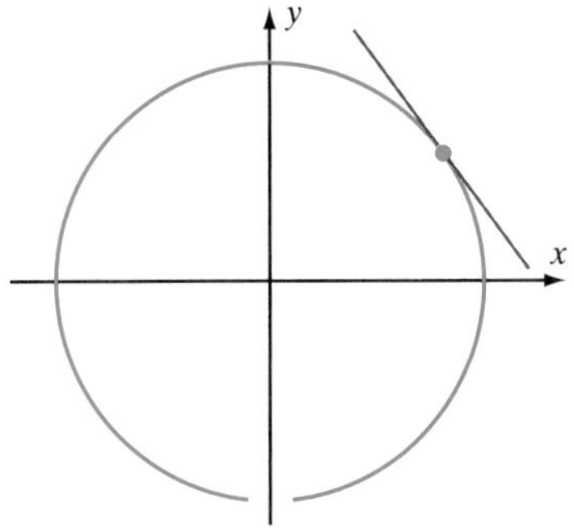

59. Vertical tangent: at $R = (2^{2/3}, 2^{1/3})$; horizontal tangents: at $P = (0, 0)$ and $Q = (2^{1/3}, 2^{2/3})$; tangents parallel to $y = x$: at $S = (1.2067, 0.52540)$ and $T = (0.52540, 1.2067)$

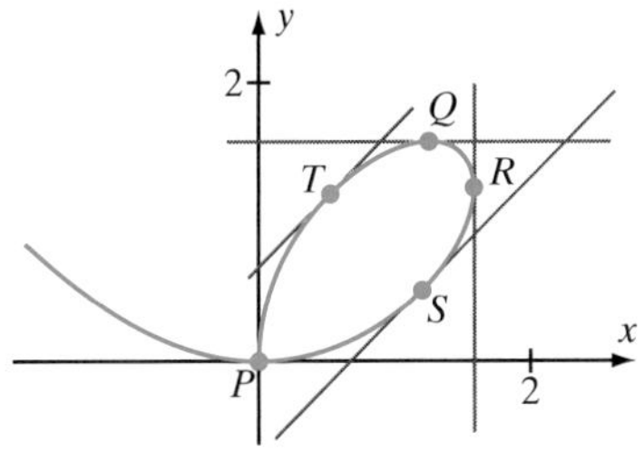

61.

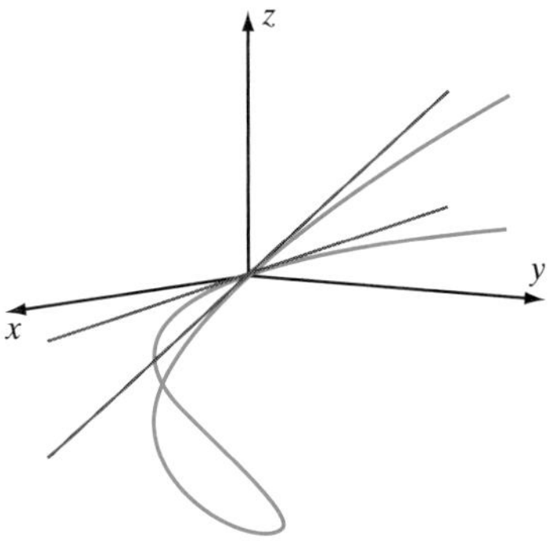

Section 10.3

1. $\mathbf{r}'(t) = \mathbf{i} - 2t\mathbf{j} + 3t^2\mathbf{k}$; $\mathbf{T}(t) = (1 + 4t^2 + 9t^4)^{-1/2}(\mathbf{i} - 2t\mathbf{j} + 3t^2\mathbf{k})$; tangent line: $(x - 1) = (y + 1)/(-2) = (z - 1)/3$

3. $\mathbf{r}'(t) = \mathbf{i} - 1/(2t^{1/2})\mathbf{k}$; $\mathbf{T}(t) = (1 + 1/(4t))^{-1/2}(\mathbf{i} - (t^{-1/2}/2)\mathbf{k})$; tangent line: $1 - x/4 = z + 2, y = 0$

5. $\mathbf{r}'(t) = (e^t + 2t)\mathbf{i}$; $\mathbf{T}(t) = \mathbf{i}$; tangent line: $y = 3, z = 0$

7. $\mathbf{r}'(t) = -\sin(t)\mathbf{i} + \cos(t)\mathbf{j} + \mathbf{k}$; $\mathbf{T}(t) = (1/\sqrt{2})(-\sin(t)\mathbf{i} + \cos(t)\mathbf{j} + \mathbf{k})$; tangent line: $x = \pi/2 - z, y = 1$

9. $\sqrt{2}\pi$

11. 6

13. $3\sqrt{3} - 2\sqrt{2}$

15. Is not

17. Is

19. e^2

21. $e - e^{-1}$

23. $(1 - x)/11 = (y + 1)/8 = (z - 1)/9$

25. $x = 0, z = \pi/2$

27. $x - 1 = y - 1 = -z/2$

29. $(x - 2)/5 = (y - 1)/4 = z - 2$

31. $x - 2y + 3z = 6$

33. $x - y/2 + 4z = 35/2$

35. $x - y = 0$

37. $x - y + 32z = 33$

39.

$$\mathbf{p}(s) = \left(\frac{(27s + 16\sqrt{2})^{2/3} - 8}{9}\right)(\mathbf{i} + \mathbf{j}) + \left(\frac{(27s + 16\sqrt{2})^{2/3} - 8}{9}\right)^{3/2}\mathbf{k}$$

41.

$$\mathbf{p}(s) = \left(\frac{s + \sqrt{s^2 + 4}}{2}\right)\mathbf{i} + \left(\frac{2}{s + \sqrt{s^2 + 4}}\right)\mathbf{j} + \sqrt{2}\ln\left(\frac{s + \sqrt{s^2 + 4}}{2}\right)\mathbf{k}$$

43. Tangent line: $-(x - 6) = \sqrt{3}(2y - 6 - \sqrt{3})$; normal line: $(x - 6) = \sqrt{3}(2y - 6 - \sqrt{3})/12$

45. Tangent line: $y = x/\sqrt{2} + 1/\sqrt{2}$; normal line: $y = -\sqrt{2}(x - 1) + \sqrt{2}$

47. Tangent line: $x = 2$; normal line: $y = 0$

49. Tangent line: $y = -4x/3 + 5/3$; normal line: $y = 3x/4$

51. $\mathbf{T}(\pi/4) = \langle -1/\sqrt{3},\ 0,\ \sqrt{2/3}\ \rangle$;
$\mathbf{N}(\pi/4) = \langle -\sqrt{6}/21, -12\sqrt{3}/21, -\sqrt{3}/21\rangle$;
$\mathbf{B}(\pi/4) = \langle 4\sqrt{2}/7, -1/7, 4/7\rangle$

53. $\mathbf{T}(\pi/4) = \langle 2/\sqrt{7}, \sqrt{2}/\sqrt{7}, 1/\sqrt{7}\rangle$;
$\mathbf{N}(\pi/4) = \langle 0, 1/\sqrt{3}, -2/\sqrt{6}\rangle$;
$\mathbf{B}(\pi/4) = \langle -3/\sqrt{21}, 2\sqrt{2}/\sqrt{21}, 2/\sqrt{21}\rangle$

55. $\mathbf{T}(\ln(2)) = \langle 4/\sqrt{26}, -1/\sqrt{26}, 3/\sqrt{26}\rangle$;
$\mathbf{N}(\ln(2)) = \langle -2/\sqrt{78}, 7/\sqrt{78}, 5/\sqrt{78}\rangle$;
$\mathbf{B}(\ln(2)) = \langle -1/\sqrt{3}, -1/\sqrt{3}, 1/\sqrt{3}\rangle$

59. Simpson's Rule approximation: 255.707

61. Simpson's Rule approximation: 4.69858

Section 10.4

1. a. $\mathbf{T}(s) = \langle \cos(s), \sin(s), 0\rangle$
b. $\mathbf{N}(s) = \langle -\sin(s), \cos(s), 0\rangle$ **c.** $\kappa(s) = 1$

3. a. $\mathbf{T}(s) = (1/\sqrt{2})\langle -\sin(s/\sqrt{2}), -\cos(s/\sqrt{2}), 1\rangle$
b. $\mathbf{N}(s) = \langle -\cos(s/\sqrt{2}), \sin(s/\sqrt{2}), 0\rangle$ **c.** $\kappa(s) = 1/2$

5. a. $\mathbf{T}(s) = \langle (3/5)\cos(s), -\sin(s), (4/5)\cos(s)\rangle$
b. $\mathbf{N}(s) = \langle -(3/5)\sin(s), -\cos(s), -(4/5)\sin(s)\rangle$
c. $\kappa(s) = 1$

7. $\rho(s) = 2$; center: $(-\cos(s/\sqrt{2}), \sin(s/\sqrt{2}), s/\sqrt{2})$

9. $\rho(s) = 1$; center: $(0, 0, 0)$

11. $2/(1 + 4e^{2t})^{3/2}$

13. $t^2/(t^2 + 1)^{3/2}$

15. $|\cos^3(t)|/(2 - \cos^2(t))^{3/2}$

17. $2/(4x + 1)^{3/2}$

19. $|\cos(x)|$

21. $8|x|^3/(x^4 + 1)^2$

23. Radius: $\sqrt{2}(2t^2 + 1)^{3/2}$; center: $(3t^2 + 1, 3t^2 + 1, 1 - 4t^3)$

25. Radius: $(2e^{2t} + 1)^{3/2}/(\sqrt{2}e^t)$;
center: $(2e^t + e^{-t}/2, t - 2e^{2t} - 1, 2e^t + e^{2t}/2)$

27. Radius: $\sqrt{5}(16t^6 + 5)^{3/2}/(60t^2)$;
center: $((28t^6 + 5)/(12t^2), 2t(5 - 8t^6)/15, 4t(5 - 8t^6)/15)$

29. $4x - 2y - z = 3$

31. $x + y - 2z = 0$

33. $(18t^4 + t^2 + 2, 8t^3 + 2t/3)$

35. $(t - 1 - e^{2t}, 2e^t + e^{-t})$

37. $(t + \sin(t), -1 + \cos(t))$

39. $(t - \tan(t), \ln(\sec(t)) + 1)$

41. $(0, 0)$

43. $(-\sqrt{2}, \ln(4))$ and $(\sqrt{2}, \ln(4))$

45. $2\sqrt{2}|ab|/(a^2 + b^2)^{3/2}$

47. $4\sqrt{3}/9$

57.

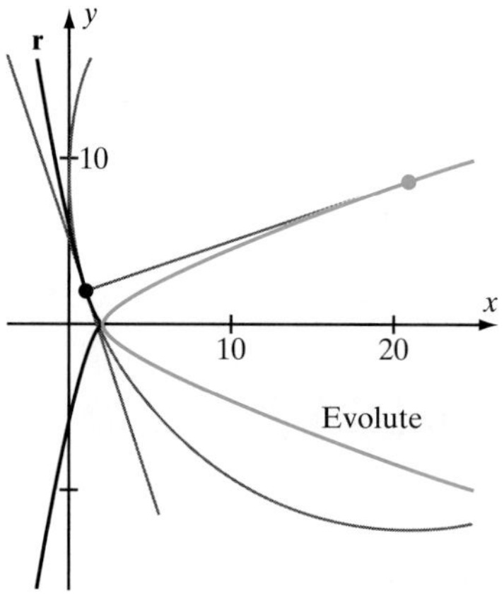

59.

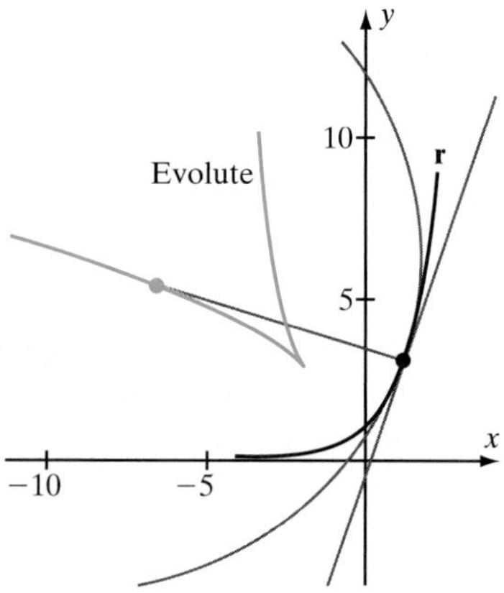

61.

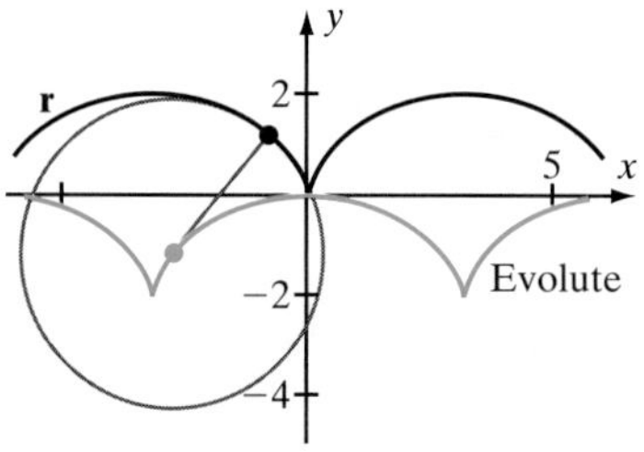

Section 10.5

1. $\mathbf{r}'(t) = \langle 1, -2t\rangle$; $\mathbf{r}''(t) = \langle 0, -2\rangle$;
$\mathbf{T}(t) = (1 + 4t^2)^{-1/2}\langle 1, -2t\rangle$;
$\mathbf{N}(t) = (1 + 4t^2)^{-1/2}\langle -2t, -1\rangle$;
$\kappa_{\mathbf{r}}(t) = 2/(1 + 4t^2)^{3/2}$; $a_T = 4t/\sqrt{1 + 4t^2}$; $a_N = 2/\sqrt{1 + 4t^2}$

3. $\mathbf{r}'(t) = \langle 1-2t, 1+2t\rangle$; $\mathbf{r}''(t) = \langle -2, 2\rangle$;
$\mathbf{T}(t) = (2+8t^2)^{-1/2}\langle 1-2t, 1+2t\rangle$;
$\mathbf{N}(t) = (2+8t^2)^{-1/2}\langle -(1+2t), 1-2t\rangle$;
$\kappa_{\mathbf{r}}(t) = 4/(2+8t^2)^{3/2}$; $a_T = 8t/\sqrt{2+8t^2}$;
$a_N = 4/\sqrt{2+8t^2}$

5. $\mathbf{r}'(t) = \langle 1-\cos(t), \sin(t)\rangle$; $\mathbf{r}''(t) = \langle \sin(t), \cos(t)\rangle$;
$\mathbf{T}(t) = (2-2\cos(t))^{-1/2}\langle 1-\cos(t), \sin(t)\rangle$;
$\mathbf{N}(t) = (2-2\cos(t))^{-1/2}\langle \sin(t), -(1-\cos(t))\rangle$;
$\kappa_{\mathbf{r}}(t) = 1/(2\sqrt{2-2\cos(t)})$; $a_T = \sin(t)/\sqrt{2-2\cos(t)}$;
$a_N = (1-\cos(t))/\sqrt{2-2\cos(t)}$

7. $\mathbf{r}'(t) = \langle 3\sqrt{3}/2, 1/(2\sqrt{t})\rangle$; $\mathbf{r}''(t) = \langle 3/(4\sqrt{t}), -1/(4t^{-3/2})\rangle$;
$\mathbf{T}(t) = (9t^2+1)^{-1/2}\langle 3t, 1\rangle$; $\mathbf{N}(t) = (9t^2+1)^{-1/2}\langle 1, -3t\rangle$;
$\kappa_{\mathbf{r}}(t) = 6\sqrt{t}/(9t^2+1)^{3/2}$;
$a_T = \frac{1}{4}t^{-3/2}(9t^2+1)^{-1/2}(9t^2-1)$;
$a_N = \frac{1}{2}3t^{-1/2}(9t^2+1)^{-1/2}$

9. $\mathbf{r}'(t) = \langle 1, 2t, 3t^2\rangle$; $\mathbf{r}''(t) = \langle 0, 2, 6t\rangle$;
$\mathbf{T}(t) = (1+4t^2+9t^4)^{-1/2}\langle 1, 2t, 3t^2\rangle$;
$\mathbf{N}(t) = (1+4t^2+9t^4)^{-1/2}(1+9t^2+9t^4)^{-1/2}$
$\langle -t(2+9t^2), 1-9t^4, 3t(2t^2+1)\rangle$;
$\kappa_{\mathbf{r}}(t) = 2\sqrt{1+9t^2+9t^4}/(1+4t^2+9t^4)^{3/2}$;
$a_T = 2t(2+9t^2)/\sqrt{1+4t^2+9t^4}$;
$a_N = 2\sqrt{1+9t^2+9t^4}/\sqrt{1+4t^2+9t^4}$

11. $\mathbf{r}'(t) = \langle (3/2)\sqrt{1+t}, -(3/2)\sqrt{1-t}, 1\rangle$;
$\mathbf{r}''(t) = (3/4)\langle (1+t)^{-1/2}, (1-t)^{-1/2}, 0\rangle$;
$\mathbf{T}(t) = 22^{-1/2}\langle 3\sqrt{1+t}, -3\sqrt{1-t}, 2\rangle$;
$\mathbf{N}(t) = \langle \sqrt{(1-t)/2}, \sqrt{(1+t)/2}, 0\rangle$;
$\kappa_{\mathbf{r}}(t) = (3\sqrt{2})/(22\sqrt{1-t^2})$; $a_T = 0$; $a_N = 3\sqrt{2}/(4\sqrt{1-t^2})$

13. $\mathbf{r}'(t) = \langle e^t, -e^{-t}, \sqrt{2}\rangle$; $\mathbf{r}''(t) = \langle e^t, e^{-t}, 0\rangle$;
$\mathbf{T}(t) = (e^{2t}+1)^{-1}\langle e^{2t}, -1, \sqrt{2}e^t\rangle$;
$\mathbf{N}(t) = (e^{2t}+1)^{-1}\langle \sqrt{2}e^t, \sqrt{2}e^t, 1-e^{2t}\rangle$;
$\kappa_{\mathbf{r}}(t) = \sqrt{2}/(e^t+e^{-t})^2$; $a_T = e^t - e^{-t}$; $a_N = \sqrt{2}$

15. $\mathbf{r}'(t) = \langle -\sin(t), \cos(t), 2t\rangle$; $\mathbf{r}''(t) = \langle -\cos(t), -\sin(t), 2\rangle$;
$\mathbf{T}(t) = (1+4t^2)^{-1/2}\langle -\sin(t), \cos(t), 2t\rangle$;
$\mathbf{N}(t) = (1+4t^2)^{-1/2}(5+4t^2)^{-1/2}$
$\langle -\cos(t) + 4t\sin(t) - 4t^2\cos(t), -\sin(t) - 4t\cos(t)$
$- 4t^2\sin(t), 2\rangle$; $\kappa_{\mathbf{r}}(t) = \sqrt{5+4t^2}/(1+4t^2)^{3/2}$;
$a_T = 4t/\sqrt{1+4t^2}$; $a_N = \sqrt{5+4t^2}/\sqrt{1+4t^2}$

17. $a_T = 4t/\sqrt{5+4t^2}$; $a_N = 2\sqrt{5}/\sqrt{5+4t^2}$

19. $a_T = (\exp(2t) - \exp(-2t))/\sqrt{1+\exp(2t)+\exp(-2t)}$;
$a_N = \sqrt{4+\exp(2t)+\exp(-2t)}/\sqrt{1+\exp(2t)+\exp(-2t)}$

21. $(x/2)^2 + (y/3)^2 = 1$

23. $(x/5)^2 + (y/3)^2 = 1$

25. $(x-1)^2/5^2 + (y-2)^2/13^2 = 1$

27. $(x-1)^2/6^2 + (y-2)^2/10^2 = 1$

29. $(x-2/3)^2/(8/3)^2 + y^2/(4/\sqrt{3})^2 = 1$

35. 3.225×10^{-3} days

37. 6.166×10^9 cm

39. $969750\sqrt{3}\,\text{mi}^2/h$

41. 4.600×10^7 km

43. 11.87 Earth yr

53. The graphs of $t \mapsto a_N(t)$ and $t \mapsto \|\mathbf{a}(t)\|$ touch when $a_T(t) = 0$.

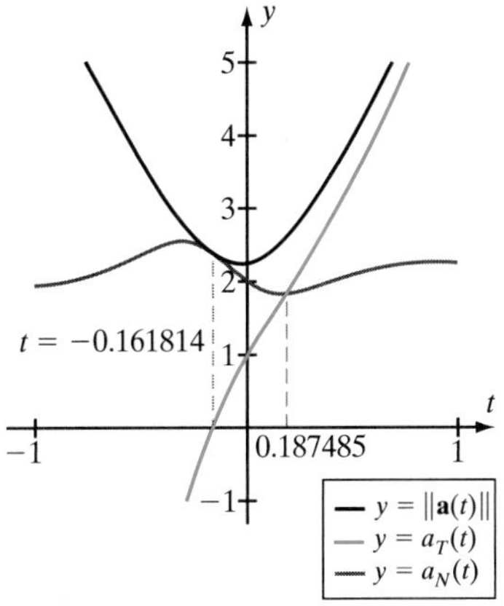

55. The particle slows on the interval (0.608, 0.967). Both components of acceleration decrease on (0.354, 0.6549).

57. Absolute maximum of a_N at 1.63782; absolute maximum of $\|\mathbf{a}\|$ at 0 and 2π

REVIEW EXERCISES FOR CHAPTER 10

1.

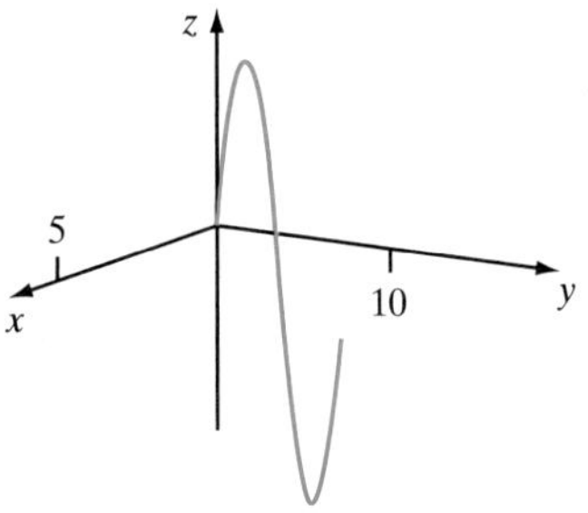

3.

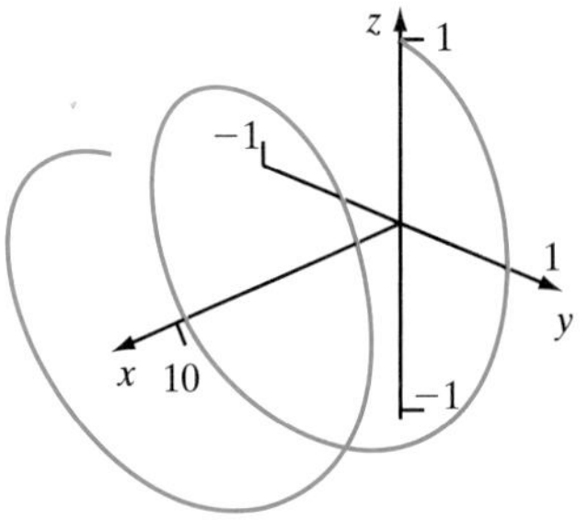

5. $\lim_{t\to 2} \mathbf{r}(t) = \langle e^2, 4, 0\rangle$

7. $\mathbf{r}'(t) = \frac{1}{3}t^{-2/3}\mathbf{i} - 6t\mathbf{j} + 2\mathbf{k}$

9. $\mathbf{r}'(t) = -\frac{1}{t^2}\mathbf{i} - \frac{1}{(t+1)^2}\mathbf{j} + \frac{3}{2}t^{1/2}\mathbf{k}$

11. $e^t\mathbf{i} - \frac{1}{2}e^{2t}\mathbf{j} + \frac{1}{3}e^{3t}\mathbf{k} + \mathbf{C}$

13. $G(t) = \left(\frac{t^3}{3} + \frac{5}{3}\right)\mathbf{i} + \left(-\frac{t^5}{5} - \frac{14}{5}\right)\mathbf{j} + \left(\frac{t^7}{7} + \frac{6}{7}\right)\mathbf{k}$

15. $G(t) = \left(\frac{2}{3}(t+1)^{3/2} + \left[2 - \frac{2}{3}3^{3/2}\right]\right)\mathbf{i} + \left(\frac{t^4}{4} - 5\right)\mathbf{j} + \left(-\frac{3}{2}t^2 + 10\right)\mathbf{k}$

17. $\mathbf{r}'(t) = [(3t^3 + 3t^2 + 3t + 1)e^{3t}]\mathbf{i} + [-(3t^2+1)\cos(t) + (t^3+t)\sin(t)]\mathbf{j} + [5t^4 + 3t^2]\mathbf{k}$

19. $\mathbf{r}''(t) = 4e^{2t}\mathbf{i} - 9e^{3t}\mathbf{j} + 2\sec^2(t)\tan(t)\mathbf{k}$

21. $\mathbf{r}''(t) = -\frac{1}{2}(1-t^2)^{-3/2}(-2t)\mathbf{i} + \left(-\frac{1}{2}\right)(1-t^2)^{-3/2}(-2t)\mathbf{j} + 2\sec^2(t)\tan(t)\mathbf{k}$

23. $\mathbf{v}(t) = 2e^{2t}\mathbf{i} - 4e^{4t}\mathbf{j} - 3e^{-3t}\mathbf{k}$, $v = \sqrt{4e^{4t} + 16e^{8t} + 9e^{-6t}}$, $\mathbf{a}(t) = 4e^{2t}\mathbf{i} - 16e^{4t}\mathbf{j} + 9e^{-3t}\mathbf{k}$

25. $\begin{cases} x = 3 + 6t \\ y = -2 - 6t \\ z = 1 + 4t \end{cases}$

27. $\begin{cases} x = -1 + 0t \\ y = \ln(2) + \frac{1}{2}t \\ z = 0 + \pi t \end{cases}$

29. $\frac{(x-16)}{8} = \frac{(y+8)}{(-3)} = z - 4$

31. $\mathbf{r}(t) = (-t+1)\mathbf{i} + (4t)\mathbf{j} + (-16t^2 - t - 2)\mathbf{k}$

33. $\mathbf{r}(t) = \mathbf{i} + (t-1)\mathbf{j} + (-16t^2 + 2)\mathbf{k}$

35. $\mathbf{r}'(2) = -4\mathbf{i} + 32\mathbf{j} - 12\mathbf{k}$,
$\mathbf{T}(2) = \frac{-4}{\sqrt{1184}}\mathbf{i} + \frac{32}{\sqrt{1184}}\mathbf{j} + \frac{-12}{\sqrt{1184}}\mathbf{k}$

$$\begin{cases} x = -4 + \frac{-4}{\sqrt{1184}}t \\ y = 16 + \frac{32}{\sqrt{1184}}t \\ z = -8 + \frac{-12}{\sqrt{1184}}t \end{cases}$$

37. $\sqrt{8}\pi$

39. 15/4

41. $\begin{cases} x = 0 + 0t \\ y = 1 - 1t \\ z = 3\pi + 0t \end{cases}$

43. tangent line: $\begin{cases} x = t \\ y = 0 \\ z = 3t \end{cases}$
normal line: $\begin{cases} x = 0 \\ y = t \\ z = 0 \end{cases}$

45. tangent line: $\begin{cases} x = 1 + t \\ y = 1 - t \\ z = \sqrt{2}t \end{cases}$
normal line: $\begin{cases} x = 1 + t \\ y = 1 + t \\ z = 0 \end{cases}$

47. $(\langle 0, -1/\sqrt{2}, 1/\sqrt{2}\rangle, \langle -1, 0, 0\rangle, \langle 0, -1/\sqrt{2}, -1/\sqrt{2}\rangle)$

49. $3x - 12y - 20z = 119$

51. $y + z = \pi$

53. $\mathbf{p}'(s) = \frac{3}{5}(-\sin(s))\mathbf{i} + \frac{4}{5}\sin(s)\mathbf{j} + \cos(s)\mathbf{k}$ hence $\|\mathbf{r}'(s)\| = 1$. Then $\mathbf{T}(s) = -\frac{3}{5}\sin(s)\mathbf{i} + \frac{4}{5}\sin(s)\mathbf{j} + \cos(s)\mathbf{k}$, $\mathbf{N}(s) = -\frac{3}{5}\cos(s)\mathbf{i} + \frac{4}{5}\cos(s)\mathbf{j} - \sin(s)\mathbf{k}$, $\kappa(s) = 1$

55. radius of curvature $= 2\sqrt{2}$, center of curvature $= (3, -3, \sqrt{2})$

57. radius of curvature $= 5/2$, center of curvature $= (1, 5/2, 0)$

59. $\frac{|e^x - e^{-x}|}{(3 + e^{2x} + e^{-2x})^{3/2}}$

61. 1

63. $\frac{6}{t(9t + 1/t)^{3/2}}$

65. $-\frac{1}{\sqrt{2}}y + \frac{1}{\sqrt{2}}z = \sqrt{2}\pi$

67. $\mathbf{r}'(t_0) = \mathbf{i}$,
$\mathbf{r}''(t_0) = -2\mathbf{j}$,
$\mathbf{T}(t_0) = \mathbf{i}$,
$\mathbf{N}(t_0) = -\mathbf{j}$,
$\kappa(t_0) = 2$,
$a_T = 0$,
$a_N = 2$

69. $\mathbf{r}'(t_0) = \mathbf{i} - \mathbf{j}$,
$\mathbf{r}''(t_0) = \mathbf{i} + \mathbf{j}$,
$\mathbf{T}(t_0) = \frac{1}{\sqrt{2}}\mathbf{i} - \frac{1}{\sqrt{2}}\mathbf{j}$,
$\mathbf{N}(t_0) = \frac{1}{\sqrt{2}}\mathbf{i} + \frac{1}{\sqrt{2}}\mathbf{j}$,
$\kappa(t_0) = \frac{1}{\sqrt{2}}$,
$a_T = 0$,
$a_N = \sqrt{2}$

71. $\mathbf{r}'(t_0) = \mathbf{i} + \mathbf{j} + \sqrt{2}\mathbf{k}$,
$\mathbf{r}''(t_0) = \mathbf{i} - \mathbf{j}$,
$\mathbf{T}(t_0) = \frac{1}{2}\mathbf{i} + \frac{1}{2}\mathbf{j} + \frac{1}{\sqrt{2}}\mathbf{k}$,
$\mathbf{N}(t_0) = \frac{1}{\sqrt{2}}\mathbf{i} - \frac{1}{\sqrt{2}}\mathbf{j}$,
$\kappa(t_0) = \frac{\sqrt{2}}{4}$,
$a_T = 0$,
$a_N = \sqrt{2}$

73. $a_T = \dfrac{8t}{\sqrt{8t^2+3}}$ $\quad a_N = \dfrac{2\sqrt{6}}{\sqrt{8t^2+3}}$

75. $a_T = \dfrac{4t}{\sqrt{2+4t^2}}$ $\quad a_N = \dfrac{2}{\sqrt{1+2t^2}}$

77. $\dfrac{x^2}{4^2} + \dfrac{y^2}{3^2} = 1$

Section 11.1

1. $V(h, r) = \pi r^2 h$ **3.** $V(h, r) = \pi r^2 h/3$

5. $A(a, b) = (b^3 - a^3)/3$

7. $m(M, \tau, T) = M\exp(-(T/\tau)\ln(2))$

9. $2e^3 + 17$ **11.** $5/2$ **13.** $6/5$

15. 6 **17.** $12\sqrt{17}/5$

19.

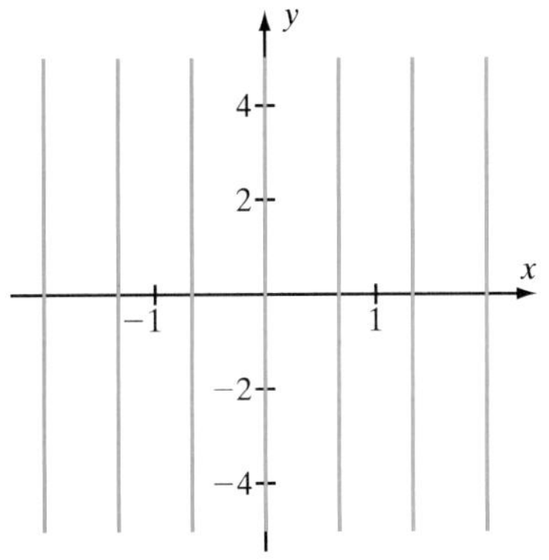

21.

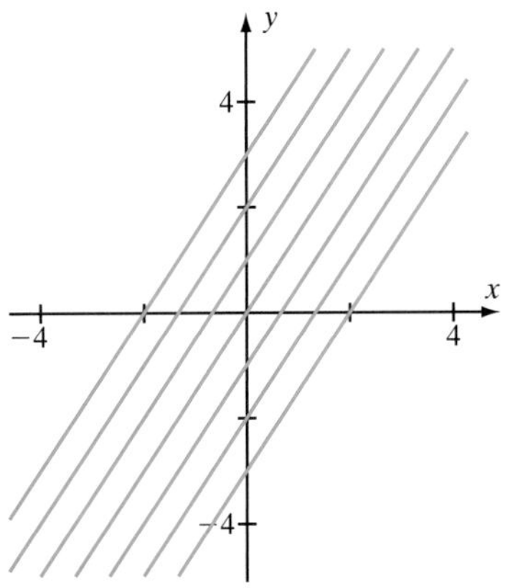

23.

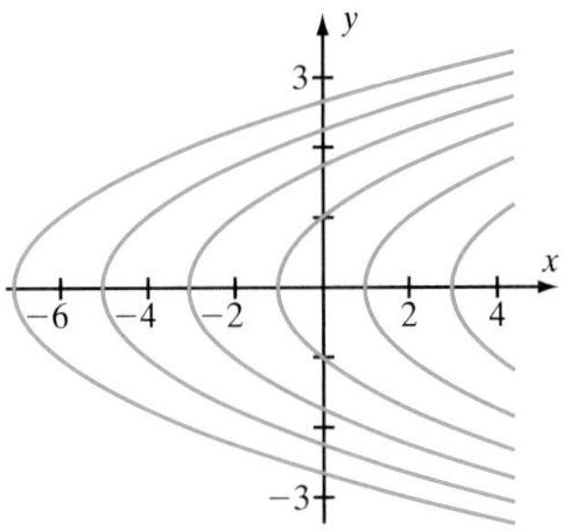

25.

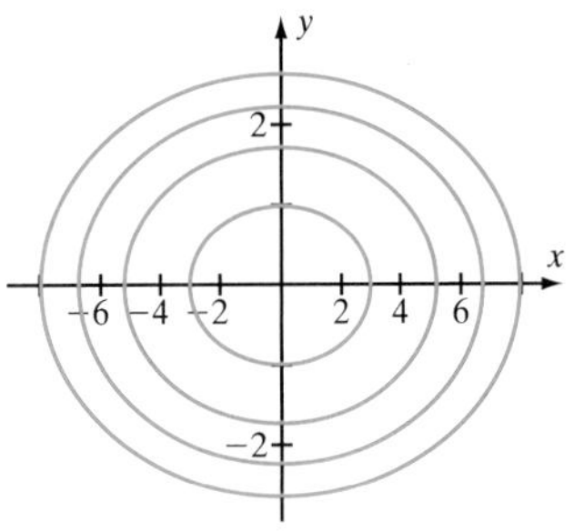

27.

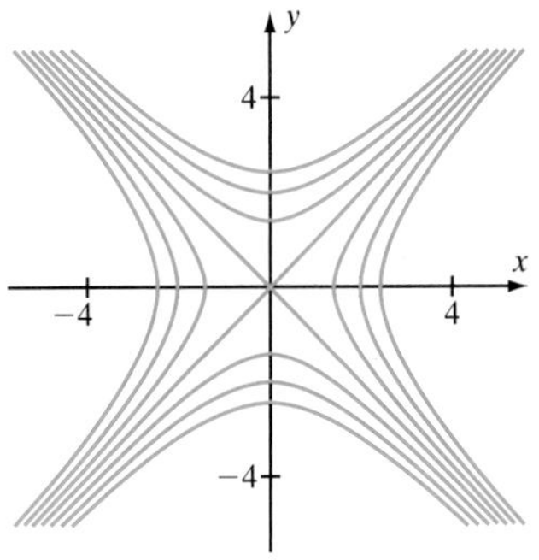

29.

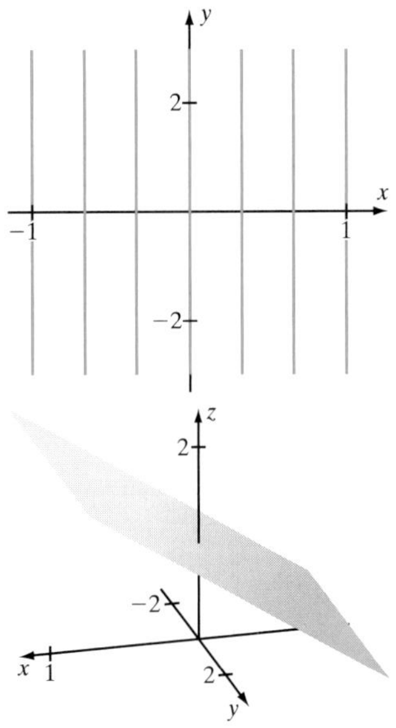

31.

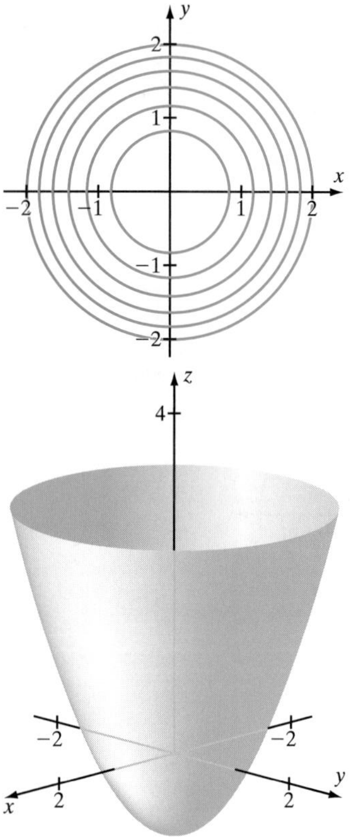
y
2
1
x
−2
−1
1
2
−1
−2
z
4
−2
−2
y
x
2
2

33.

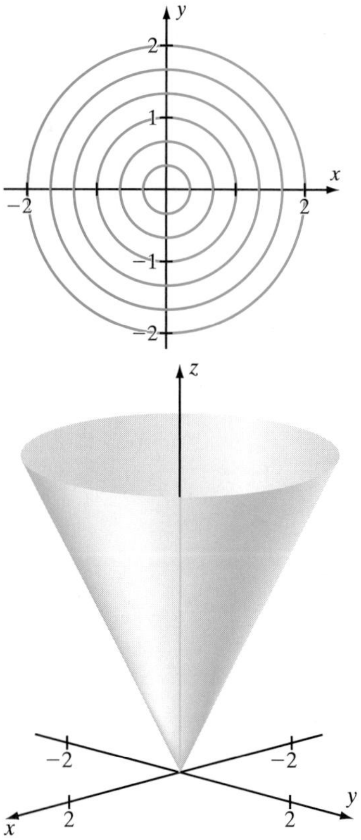
y
2
1
x
−2
2
−1
−2
z
−2
−2
y
x
2
2

35.

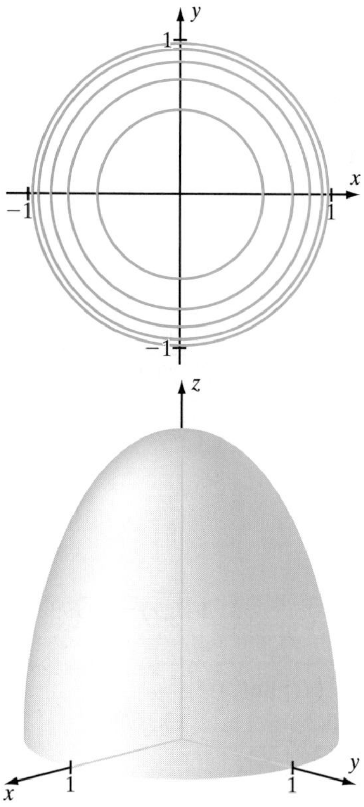
y
1
x
−1
1
−1
z
y
x
1
1

37.

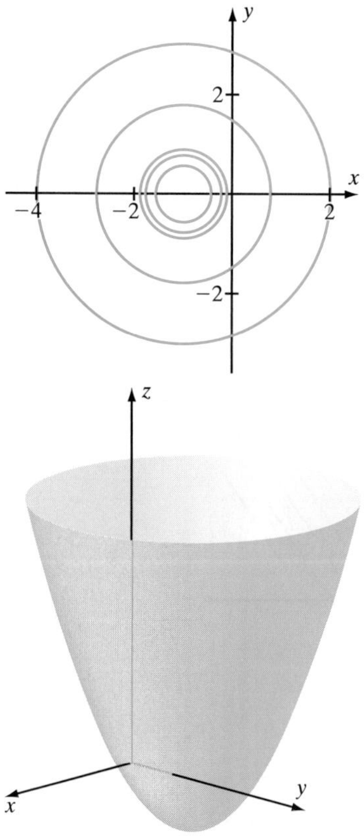
y
2
x
−4
−2
2
−2
z
y
x

39.

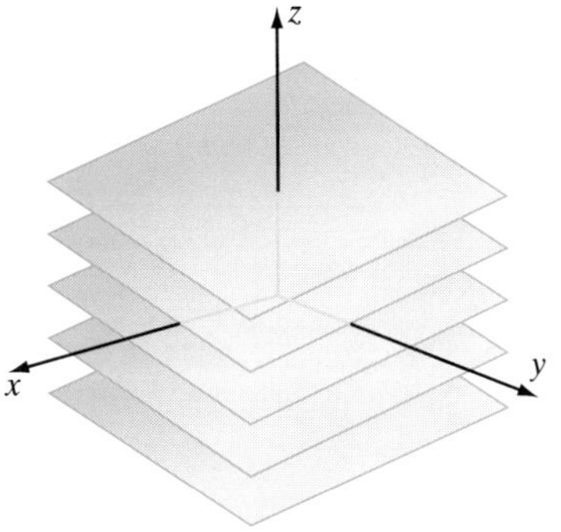

41.

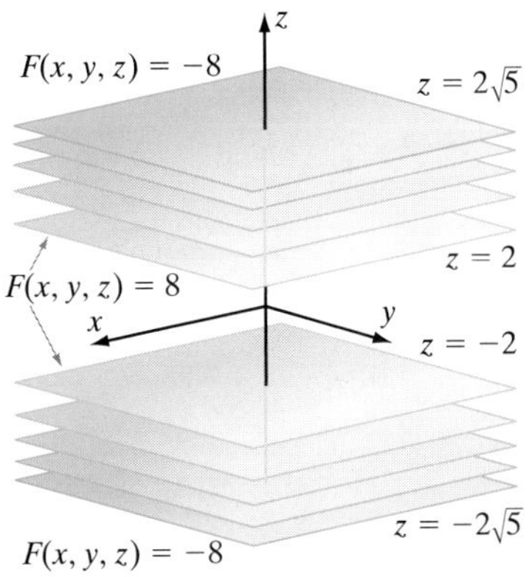

43.

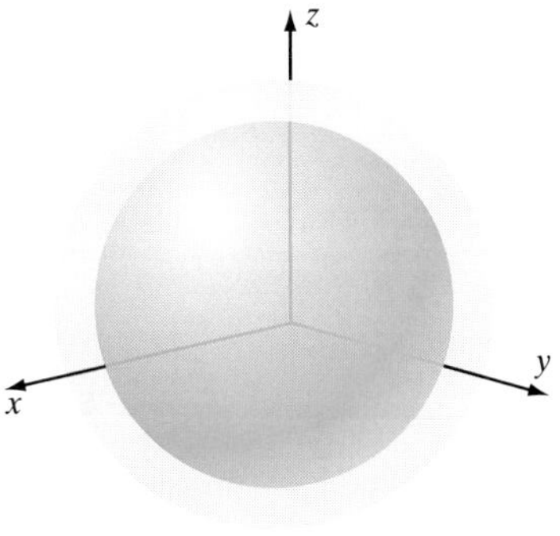

The level sets are concentric spheres. Two are shown.

45. $|xy|/\sqrt{2(x^2 + y^2)}$

47. The level sets of f_1 are the lines $y = x + c$, $c \in \mathbb{R}$. The level sets of f_2 are the pairs of lines $y = x + c$ and $y = x - c$, $c \geq 0$.

49. $a(h, c) = c - 1 - h/\exp(c)$

51.

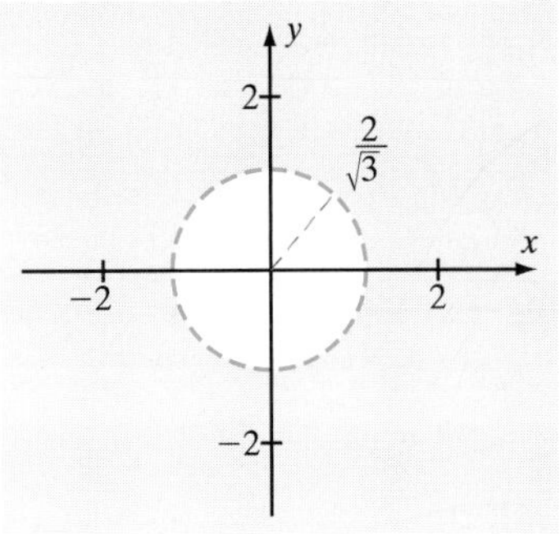

53.

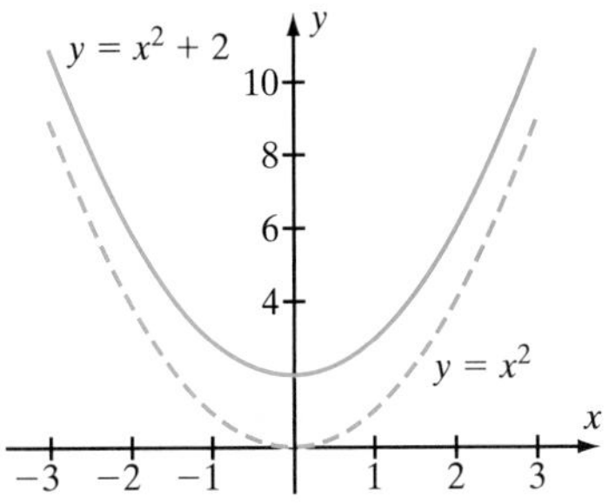

55. The level sets $L_c = \{(x, y) : f(x, y) = c\}$ are nonempty precisely when $-1 \leq c \leq 1$. For c in this interval, the level set L_c is the union of all parabolas of the form $y = x^2 + \arcsin(c) + 2n\pi$ $(n \in \mathbb{Z})$ and $y = x^2 - \arcsin(c) + (2n+1)\pi$ $(n \in \mathbb{Z})$.

57. The level sets $L_c = \{(x, y) : f(x, y) = c\}$ are nonempty precisely when $-1 \leq c \leq 1$. For c in this interval, the level set L_c is the union of all circles of the form $x^2 + y^2 = \arcsin(c) + 2n\pi$ $(n \in \mathbb{Z}, n \geq -\arcsin(c)/2\pi)$ and $x^2 + y^2 = -\arcsin(c) + (2n+1)\pi$ $(n \in \mathbb{Z}, n \geq \arcsin(c)/2\pi) - 1/2)$.

59. The level sets $L_c = \{(x, y) : f(x, y) = c\}$ are nonempty precisely when $c \geq 0$. For c positive, the level set L_c is the hyperbola $y^2/c - x^2 = 1$. For $c = 0$, the level set L_c is the x-axis.

61. $f(x, y) = y/(y - x)$; domain: $\{(x, y) : |x| < |y|\}$

63. $f(x, y) = x^2 + y^3$

65. The level sets $L_c = \{(x,\ y)$: $f(x,\ y) = c\}$ are nonempty precisely when $c \geq 0$. For nonnegative c, we have $L_c = \{(\sqrt{c}(\cos(t) + \sin(t)), \sqrt{c}\sin(t)) : t \in [0, 2\pi)\}$.

67.

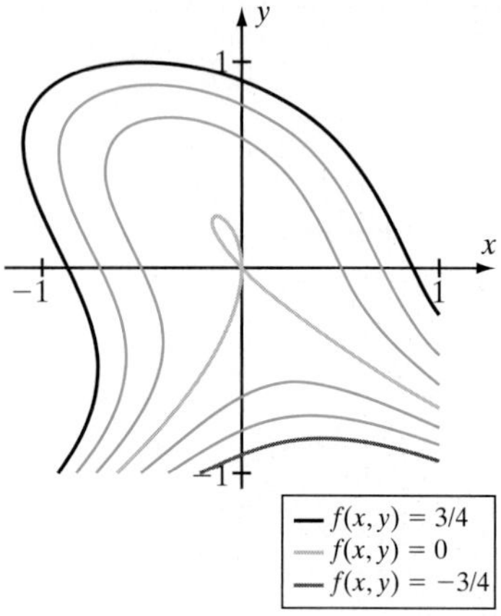

69.

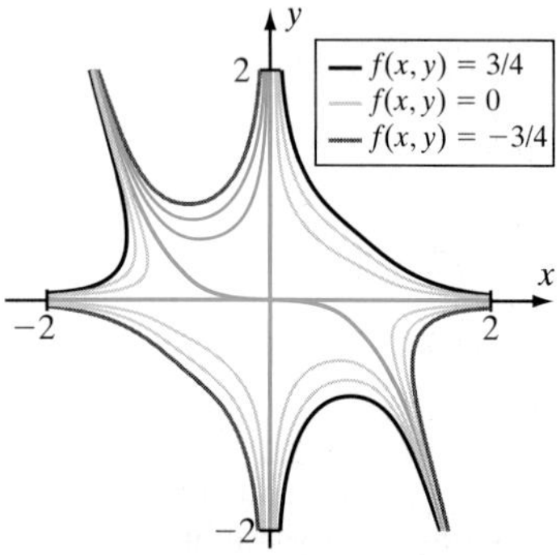

Section 11.2

1.

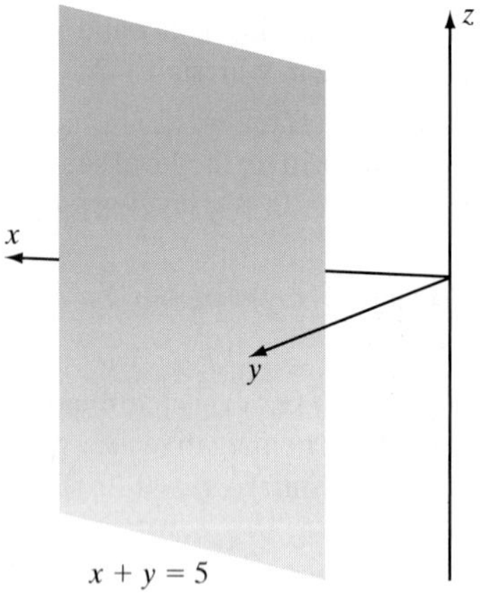

3.

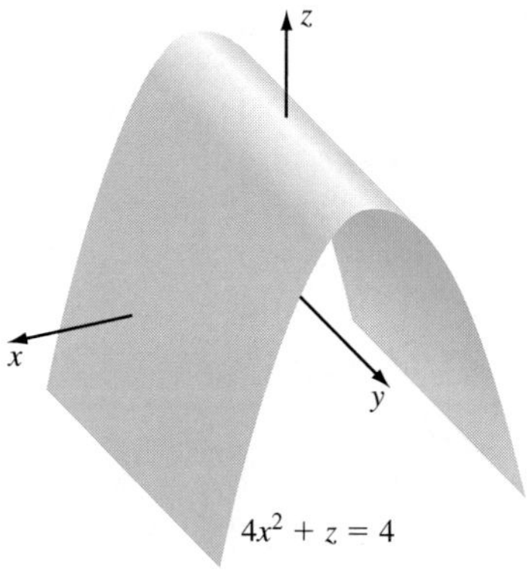

5.

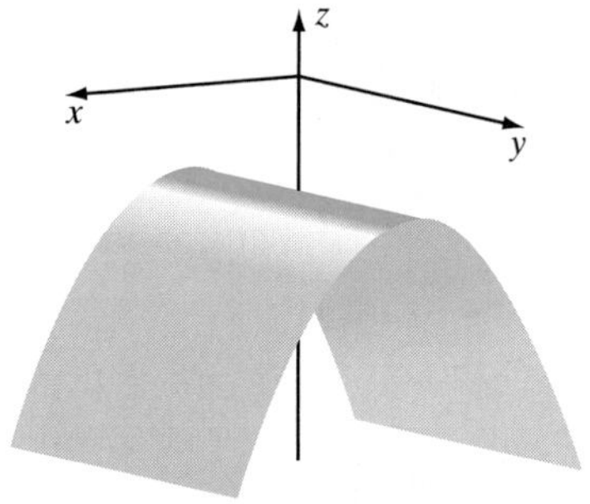

$x^2 + z = -4$

7.

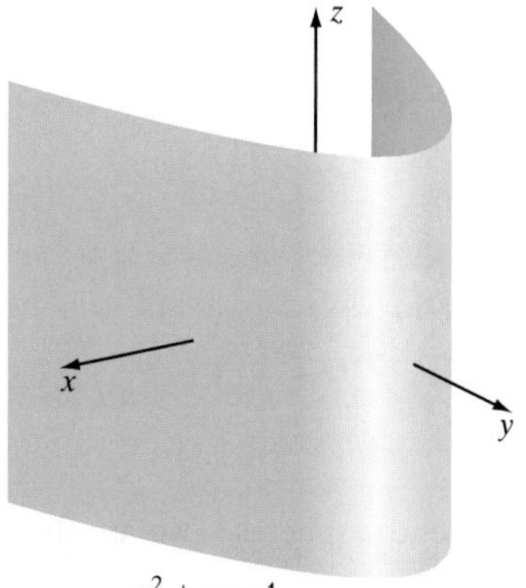

$x^2 + y = 4$

9.

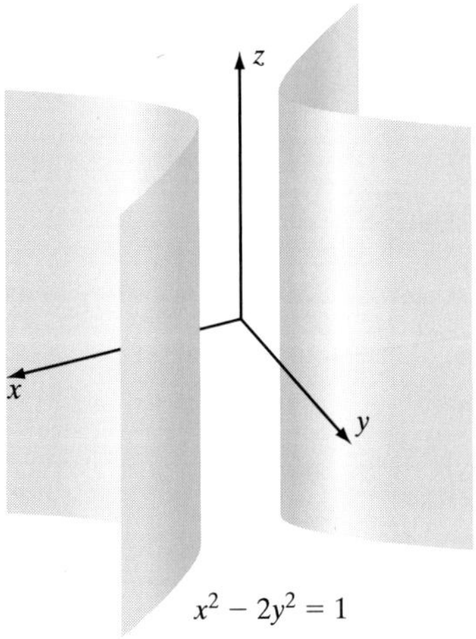

$x^2 - 2y^2 = 1$

11.

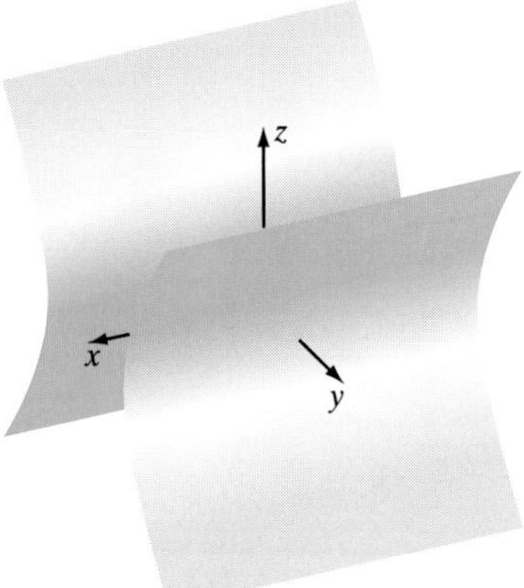

$y^2 - 4z^2 = 1$

13.

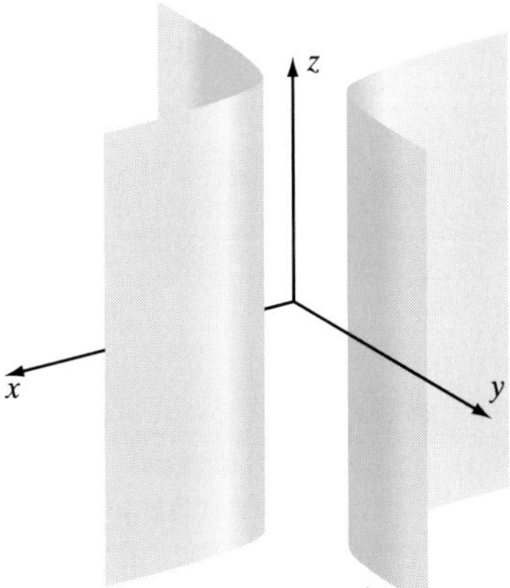

$x - 2xy = 1$

15. Hyperboloid of one sheet, which is not the graph of a function

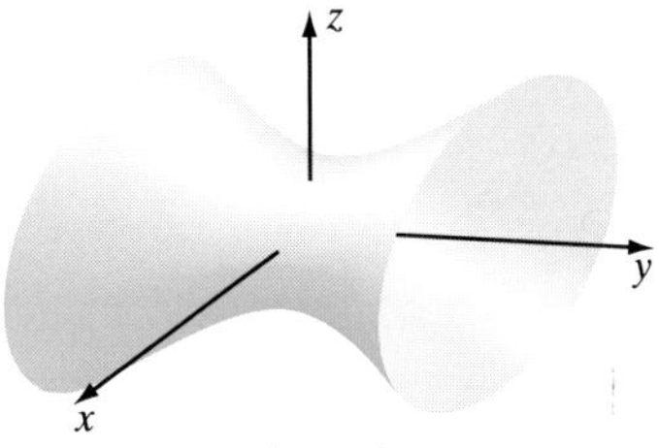

$x^2 - 4y^2 + 9z^2 = 36$

17. Circular paraboloid, which is not the graph of a function

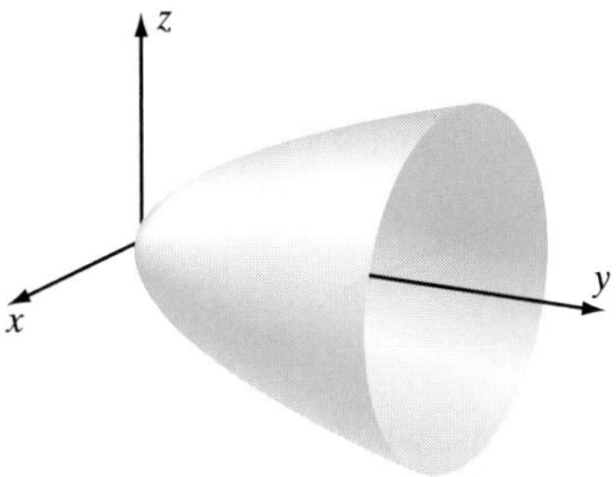

$y - x^2 = z^2$

19. Ellipsoid, which is not the graph of a function

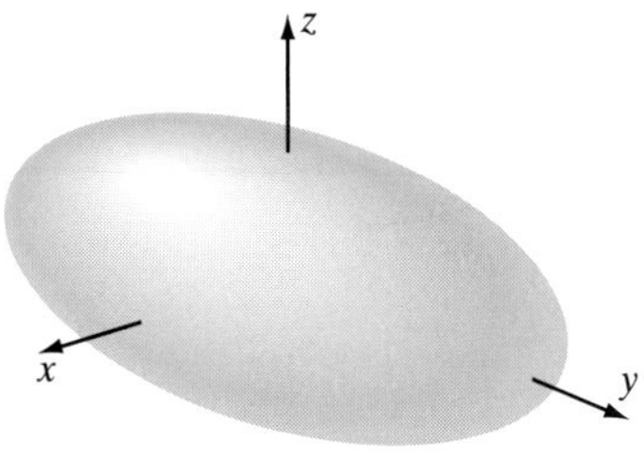

$4x^2 + y^2 + 9z^2 = 25$

21. Circular paraboloid, which is the graph of $f(x, y) = 4 - (x^2 + y^2)$

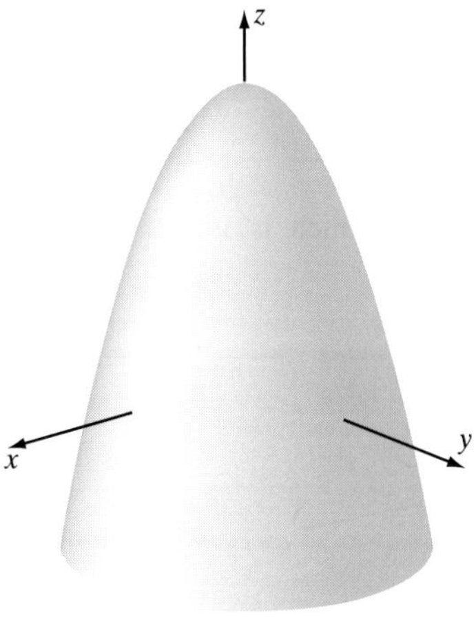

$x^2 + y^2 + z = 4$

23. Elliptic paraboloid, which is the graph of $f(x, y) = 4x^2 + y^2$

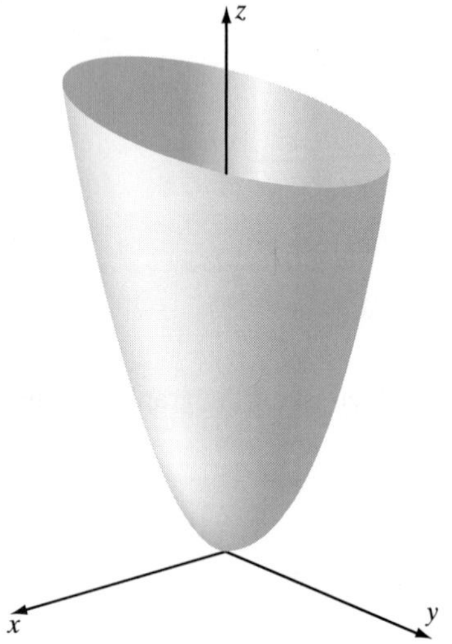

$4x^2 + y^2 = z$

25. Hyperboloid of two sheets, which is not the graph of a function

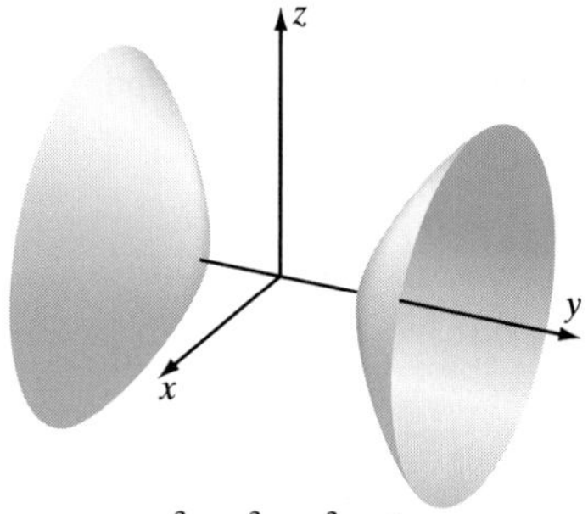

$y^2 - z^2 - x^2 = 1$

27. Elliptic cone, which is not the graph of a function

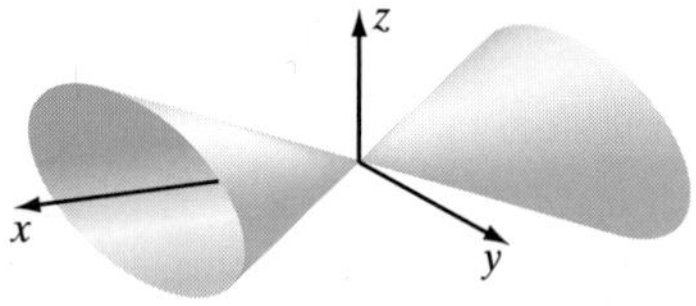

$y^2 + 8z^2 = x^2$

29. Hyperbolic paraboloid, which is not the graph of a function

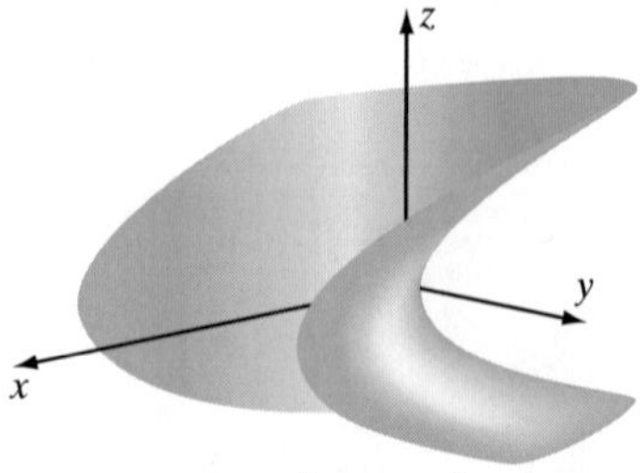

$y^2 - x - z^2 = 0$

31. False

35. The ellipse: $x^2 + y^2/4 = 1$

37.

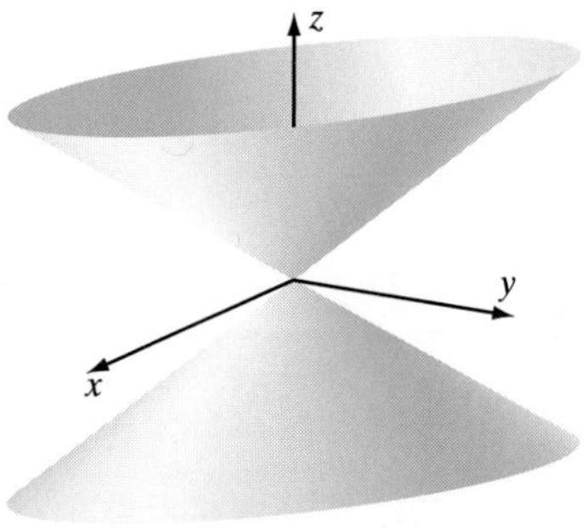

$x^2 + 2xy + 2y^2 - z^2 = 0$

39.

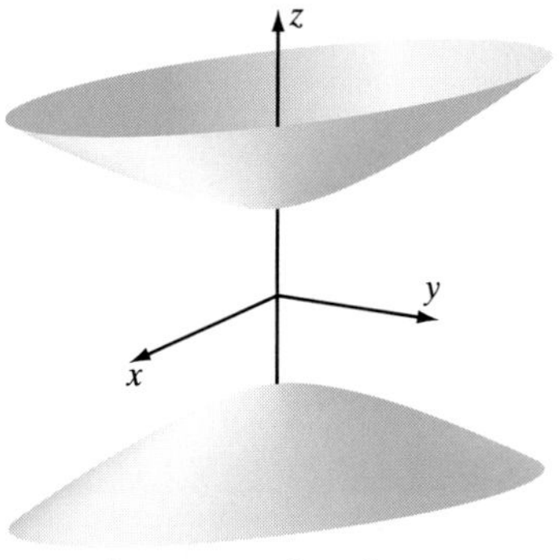

$x^2 + 2xy + 2y^2 - z^2 = -1$

41. $x = -1 + \sqrt{2}\cos(\theta)$, $y = \sin(\theta)$, $z = 2\sqrt{2} - 2\cos(\theta)$, $0 \le \theta \le 2\pi$

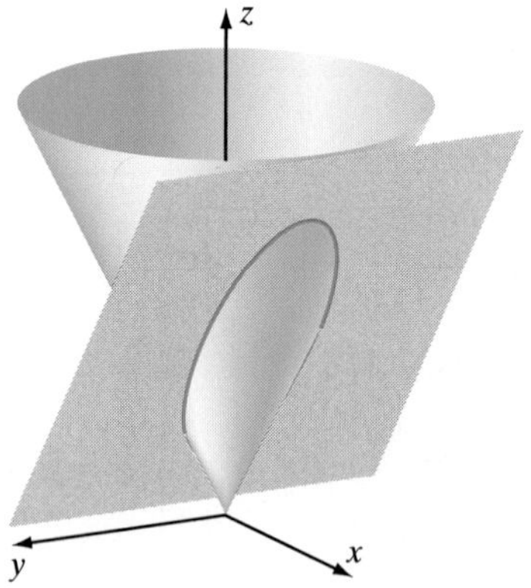

Section 11.3

1. 162

3. $2\sqrt{7}$

5. $\ln(12)$

7. $\pi/2$

9. 2

11. 2

13. 9

21. Continuous

23. Not continuous

29. 0

31. $\lim_{y\to 0} f(y^2, y) = 1/2$

33.

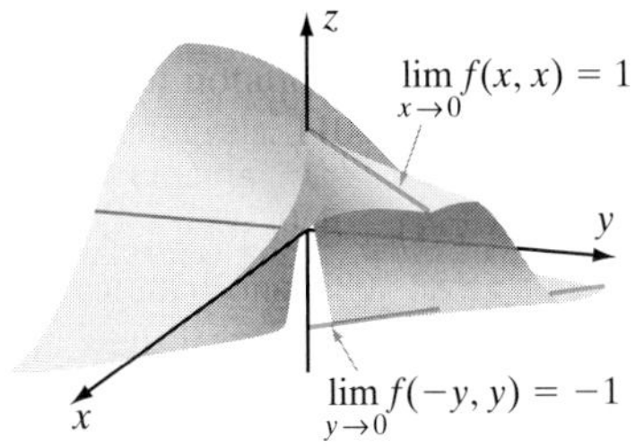

35.

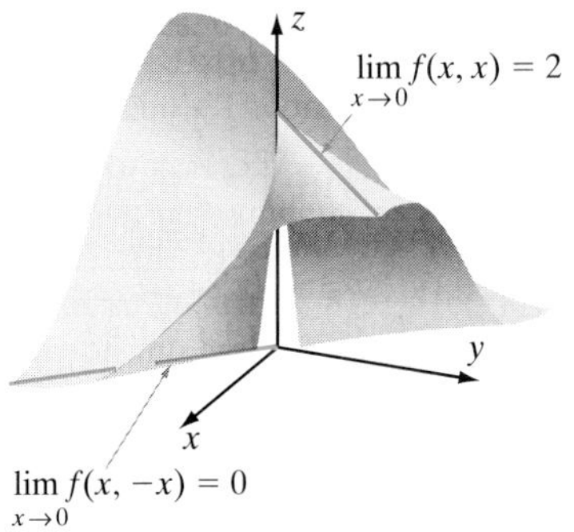

Section 11.4

1. $\varphi(x) = 2x + 1$; $\psi(y) = 10 + y^3$; $\varphi'(5) = 2 = \dfrac{\partial f}{\partial x}(5, 1)$; $\psi'(1) = 3 = \dfrac{\partial f}{\partial y}(5, 1)$

3. $\varphi(x) = 10$; $\psi(y) = 7 - 3y^4$; $\varphi'(2) = 0 = \dfrac{\partial f}{\partial x}(2, -1)$; $\psi'(-1) = 12 = \dfrac{\partial f}{\partial y}(2, -1)$

5. $\varphi(x) = \cos(x/4)$; $\psi(y) = \cos(\pi y^2)$; $\varphi'(\pi) = -\sqrt{2}/8 = \dfrac{\partial f}{\partial x}(\pi, 1/2)$; $\psi'(1/2) = -\pi\sqrt{2}/2 = \dfrac{\partial f}{\partial y}(\pi, 1/2)$

7. $\varphi(x) = x\ln(ex)$; $\psi(y) = \ln(y)$; $\varphi'(1) = 2 = \dfrac{\partial f}{\partial x}(1, e)$; $\psi'(e) = 1/e = \dfrac{\partial f}{\partial y}(1, e)$

9. $\varphi(x) = x^2$; $\psi(y) = 2^y$; $\varphi'(2) = 4 = \dfrac{\partial f}{\partial x}(2, 2)$; $\psi'(2) = 4\ln(2) = \dfrac{\partial f}{\partial y}(2, 2)$

11. $f_{xx}(x, y) = 2y$; $f_{yy}(x, y) = 20y^3$; $f_{xy}(x, y) = 2x - 1 = f_{yx}(x, y)$

13. $f_{xx}(x, y) = 2$; $f_{yy}(x, y) = -2x/y^3$; $f_{xy}(x, y) = f_{yx}(x, y) = 1/y^2$

15. $f_{xx}(x, y) = -\sin(x)/\cos(y)$; $f_{yy}(x, y) = \sin(x)(1 + \sin^2(y))/\cos^3(y)$; $f_{xy}(x, y) = \cos(x)\sin(y)/\cos^2(y) = f_{yx}(x, y)$

17. $f_{xx}(x, y) = -1/(x - y)^2$; $f_{yy}(x, y) = -(x^2 - 2yx + 2y^2)/(xy - y^2)^2$; $f_{xy}(x, y) = 1/(x - y)^2 = f_{yx}(x, y)$

19. $f_{xx}(x, y) = -\cos(x)\sin(y)$; $f_{yy}(x, y) = -\cos(x)\sin(y)$; $f_{xy}(x, y) = -\sin(x)\cos(y) = f_{yx}(x, y)$

21. $D_y f(x, y) = x\sec(y)\tan(y)$; $D_{yy} f(x, y) = x\sec(y)(1 + 2\tan^2(y))$; $D_{yx} f(x, y) = \sec(y)\tan(y) = f_{12}(x, y)$; $f_{11}(x, y) = 0$

23. $D_y f(x, y) = 2\cos(x^3 + 2y)$; $D_{yy} f(x, y) = -4\sin(x^3 + 2y)$; $D_{yx} f(x, y) = f_{12}(x, y) = -6x^2\sin(x^3 + 2y)$; $f_{11}(x, y) = -9x^4\sin(x^3 + 2y) + 6x\cos(x^3 + 2y)$

25. $D_y f(x, y) = 5\sin(y)(\sin(x) - \cos(y))^4$; $D_{yy} f(x, y) = 5(\sin(x) - \cos(y))^3 (4\sin^2(y) + \sin(x)\cos(y) - \cos^2(y))$; $D_{yx} f(x, y) = 20\cos(x)\sin(y)(\sin(x) - \cos(y))^3 = f_{12}(x, y)$; $f_{11}(x, y) = 5(\sin(x) - \cos(y))^3 (4\cos^2(x) - \sin^2(x) + \sin(x)\cos(y))$

27. $D_y f(x, y) = 12(x + 3y)^3$; $D_{yy} f(x, y) = 108(x + 3y)^2$; $D_{yx} f(x, y) = 36(x + 3y)^2 = f_{12}(x, y)$; $f_{11}(x, y) = 12(x + 3y)^2$

29. $D_y f(x, y) = 2y^3/\sqrt{1 + 3x^2 + y^4}$; $D_{yy} f(x, y) = 2y^2(y^4 + 3 + 9x^2)/(1 + 3x^2 + y^4)^{3/2}$; $D_{yx} f(x, y) = -6y^3 x/(1 + 3x^2 + y^4)^{3/2} = f_{12}(x, y)$; $f_{11}(x, y) = 3(1 + y^4)/(1 + 3x^2 + y^4)^{3/2}$

31. $f_x(x, y, z) = 6x^2$; $f_y(x, y, z) = 5y^4 z^8$; $f_z(x, y, z) = 8y^5 z^7$; $f_{xy}(x, y, z) = 0$; $f_{xz}(x, y, z) = 0$; $f_{yz}(x, y, z) = 40y^4 z^7$; $f_{xx}(x, y, z) = 12x$; $f_{yy}(x, y, z) = 20y^3 z^8$; $f_{zz}(x, y, z) = 56y^5 z^6$

33. $f_x(x, y, z) = 5(x + 2y + 3z)^4$; $f_y(x, y, z) = 10(x + 2y + 3z)^4$; $f_z(x, y, z) = 15(x + 2y + 3z)^4$; $f_{xy}(x, y, z) = 40(x + 2y + 3z)^3$; $f_{xz}(x, y, z) = 60(x + 2y + 3z)^3$; $f_{yz}(x, y, z) = 120(x + 2y + 3z)^3$; $f_{xx}(x, y, z) = 20(x + 2y + 3z)^3$; $f_{yy}(x, y, z) = 80(x + 2y + 3z)^3$; $f_{zz}(x, y, z) = 180(x + 2y + 3z)^3$

35. $f_x(x, y, z) = y\exp(xy - yz)$; $f_y(x, y, z) = (x - z)\exp(xy - yz)$;

$f_z(x,y,z) = -y\exp(xy - yz)$;
$f_{xy}(x,y,z) = (1 + xy - yz)\exp(xy - yz)$;
$f_{xz}(x,y,z) = -y^2\exp(xy - yz)$;
$f_{yz}(x,y,z) = -(1 + xy - yz)\exp(xy - yz)$;
$f_{xx}(x,y,z) = y^2\exp(xy - yz)$;
$f_{yy}(x,y,z) = (x - z)^2\exp(xy - yz)$;
$f_{zz}(x,y,z) = y^2\exp(xy - yz)$

37. $f_x(x,y,z) = y/x$; $f_y(x,y,z) = \ln(xz)$; $f_z(x,y,z) = y/z$;
$f_{xy}(x,y,z) = 1/x$; $f_{xz}(x,y,z) = 0$;
$f_{yz}(x,y,z) = 1/z$; $f_{xx}(x,y,z) = -y/x^2$; $f_{yy}(x,y,z) = 0$;
$f_{zz}(x,y,z) = -y/z^2$;

39. $f_x(x,y,z) = 6x(x^2 + y^3 + z^4)^2$;
$f_y(x,y,z) = 9y^2(x^2 + y^3 + z^4)^2$;
$f_z(x,y,z) = 12z^3(x^2 + y^3 + z^4)^2$;
$f_{xy}(x,y,z) = 36xy^2(x^2 + y^3 + z^4)$;
$f_{xz}(x,y,z) = 48xz^3(x^2 + y^3 + z^4)$;
$f_{yz}(x,y,z) = 72y^2z^3(x^2 + y^3 + z^4)$;
$f_{xx}(x,y,z) = 6(x^2 + y^3 + z^4)(5x^2 + y^3 + z^4)$;
$f_{yy}(x,y,z) = 18y(x^2 + y^3 + z^4)(x^2 + 4y^3 + z^4)$;
$f_{zz}(x,y,z) = 12z^2(x^2 + y^3 + z^4)(3x^2 + 3y^3 + 11z^4)$

41. $6yz^2$ **43.** $\sin(z)/x^2$

45. $f_{xx}(x,y) = -4y/(x+y)^3$; $f_{yy}(x,y) = 4x/(x+y)^3$;
$f_{xy}(x,y) = f_{yx}(x,y) = 2(x-y)/(x+y)^3$

47. $f_{xx}(x,y) = 2y\sec^2(x^2y)\left(1 + 4x^2y\tan(x^2y)\right)$;
$f_{yy}(x,y) = 2x^4\tan(x^2y)\sec^2(x^2y)$;
$f_{xy}(x,y) = f_{yx}(x,y) = 2x\sec^2(x^2y)\left(1 + 2x^2y\tan(x^2y)\right)$

49. $f_{xx}(x,y) = -2(x+2y)\left(1 + (x+2y)^2\right)^{-2}$;
$f_{yy}(x,y) = -8(x+2y)\left(1 + (x+2y)^2\right)^{-2}$;
$f_{xy}(x,y) = f_{yx}(x,y) = -4(x+2y)\left(1 + (x+2y)^2\right)^{-2}$

51. y-coordinate: 2; x-coordinate: 8/5

53. y-coordinate: 5/2; x-coordinate: 11/10

61. **a.** Harmonic **b.** Harmonic **c.** Harmonic
d. Not harmonic **e.** Not harmonic **f.** Harmonic

65. $B = -3E$; $C = -3A$

71. $f(x, y) = x^2y + y + C$

73. $f(x, y) = x^2 + y - y^2 + C$

79. $f_x(4,3) = -0.02126$; $f_y(4,3) = -0.09108$

81. $f_{xx}(4,3) = 0.03337$; $f_{yy}(4,3) = 0.07633$

Section 11.5

1. $5s^4 - 4s - s^{-2} + 3s^{-4}$

3. $-\sin(s)$ **5.** $3/(2s)$

7. $-\left(2\sin(s) + 3\cos(s)\right)\exp\left(2\cos(s) - 3\sin(s)\right)$

9. $4s/\left(\cos(s^2) - \sin(s^2)\right)^2$

11. $z_s = 3t\sin(3st)\cos(3st)(2 + 7\cos^5(3st))$;
$z_t = 3s\sin(3st)\cos(3st)(2 + 7\cos^5(3st))$

13. $z_s = 0$; $z_t = 0$

15. $z_s = -2t(3s^4 + t^4)/(s^4 - t^4)^2$;
$z_t = 2s(s^4 + 3t^4)/(s^4 - t^4)^2$

17. $z_s = 2(s-t)(s+4t)/(2s+3t)^2$;
$z_t = -(s-t)(7s+3t)/(2s+3t)^2$

19. $s^2(s-2)^5e^s(s^2 + 7s - 6)$

21. $\frac{1}{2}(6s^5 + 1/s)/\sqrt{s^6 + \ln(s)}$

23. $3s^2(s+1)e^{3s}\cos(s^3e^{3s})$

25. $2^{4s+1}\ln(2)$

27. $2\pi(x-1) + (y-\pi) - \pi(x-1)^2/2 + (x-1)(y-\pi)$

29. $1 - xy$

31. $1 + 4(x - \pi/4) - 2(y - \pi/4) + 8(x - \pi/4)^2$
$- 8(x - \pi/4)(y - \pi/4) + 2(y - \pi/4)^2$

35. Decreasing (at the rate of 0.31 cm^2/s)

37. $g_x(x,y) = 2x\ln(y)\cos\left(x^2\ln(y)\right)$;
$g_y(x,y) = (x^2/y)\cos\left(x^2\ln(y)\right)$

47. $D_1(f)(R)\cdot D_1(f)(P) + D_2(f)(R)\cdot D_1(f)(Q)$
$+ D_2(f)(R)\cdot D_2(f)(Q)\cdot D_1(f)(P)$ where $P = (x,y)$,
$Q = (x, f(P))$, and $R = (f(P), f(Q))$

Section 11.6

1. $\langle \sin(y), x\cos(y)\rangle$

3. $\langle y^2\sec^2(xy^2), 2xy\sec^2(xy^2)\rangle$

5. $\langle -y\sin(xy)\sin(yz), -x\sin(xy)\sin(yz)$
$+ z\cos(xy)\cos(yz), y\cos(xy)\cos(yz)\rangle$

7. $\langle y^2/\cos(y), 2xy/\cos(y) + xy^2\sec(y)\tan(y)\rangle$

9. $\langle yz\cos(xy), xz\cos(xy), \sin(xy)\rangle$

11. -49 **13.** $-(3/2)\sqrt{\pi}$

15. 1/32 **17.** $-1/16$

19. $1/\sqrt{2}$ **21.** $-3\sqrt{3}/8$

23. **a.** $\langle 5/13, 12/13\rangle$ **b.** 13/2
c. $\langle -5/13, -12/13\rangle$ **d.** $-13/2$

25. **a.** $\langle 1/\sqrt{2}, 1/\sqrt{2}\rangle$ **b.** $\sqrt{2}(1 - \ln(2))/4$
c. $\langle -1/\sqrt{2}, -1/\sqrt{2}\rangle$ **d.** $-\sqrt{2}(1 - \ln(2))/4$

27. **a.** $\langle -1/\sqrt{5}, -2/\sqrt{5}\rangle$ **b.** $2\sqrt{5}/3$
c. $\langle 1/\sqrt{5}, 2/\sqrt{5}\rangle$ **d.** $-2\sqrt{5}/3$

29. **a.** $\langle 1, 0\rangle$ **b.** $1/e$ **c.** $\langle -1, 0\rangle$ **d.** $-1/e$

31. **a.** $\langle 2/\sqrt{5}, 1/\sqrt{5}\rangle$ **b.** $80\sqrt{5}$ **c.** $\langle -2/\sqrt{5}, -1/\sqrt{5}\rangle$
d. $-80\sqrt{5}$

33. 36 **35.** $-3/4$

37. $(1 + 3\sqrt{3})/8$

39. **a.** $\langle 3/\sqrt{11}, -1/\sqrt{11}, 1/\sqrt{11}\rangle$ **b.** $4\sqrt{11}$
c. $\langle -3/\sqrt{11}, 1/\sqrt{11}, -1/\sqrt{11}\rangle$ **d.** $-4\sqrt{11}$

41. **a.** $\langle 0, 1/\sqrt{2}, -1/\sqrt{2}\rangle$ **b.** $\sqrt{2}$
c. $\langle 0, -1/\sqrt{2}, 1/\sqrt{2}\rangle$ **c.** $-\sqrt{2}$

45. 47/15

47. $-2/5 + 6\pi/5$

49. $f_x(P) = 3;\ f_y(P) = -1$

51. $f_x(P) = 3\sqrt{3} - 3;\ f_y(P) = -3\sqrt{3} + 3$

55. $1358^{-1/2}\langle -25, -27, -2\rangle$

61.

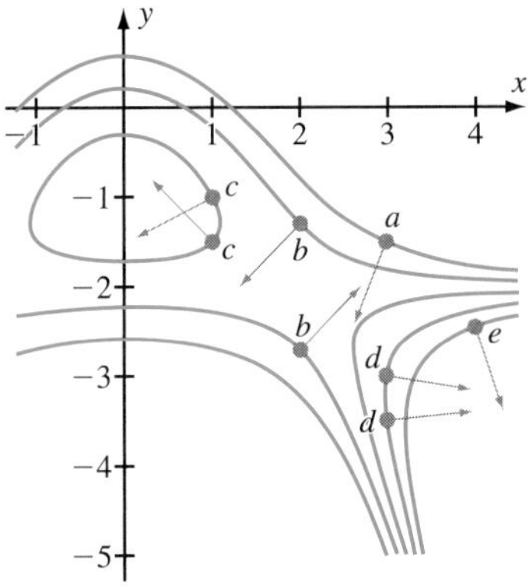

Section 11.7

1. Normal vector: $\langle 6, -47, -1\rangle$; tangent plane: $z = 6x - 47y + 123$

3. Normal vector: $\langle 2/7, 4/7, -1\rangle$; tangent plane: $z = 2x/7 + 4y/7 + \ln(7) - 10/7$

5. Normal vector: $\langle 1/9, -2/9, -1\rangle$; tangent plane: $z = x/9 - 2y/9 + 2/3$

7. Normal vector: $\langle 4e^3, 3e^3, -1\rangle$; tangent plane: $z = e^3(4x + 3y - 8)$

9. Normal vector: $\langle 1/2, 1/2, -1\rangle$; tangent plane: $z = (x + y)/2$

11. Normal vector: $\langle -1/48, 1/48, -1\rangle$; tangent plane: $z = (-x + y + 32)/48$

13. Normal vector: $\langle 2, -16, 2\rangle$; tangent plane: $2x - 16y + 2z = -28$

15. Normal vector: $\langle -1/2, 1/4, -1/2\rangle$; tangent plane: $2x - y + 2z = 0$

17. Normal vector: $\langle 1, 1, 1\rangle$; tangent plane: $x + y + z = e^3 - 1$

19. Normal vector: $\langle 6, 6, 12\rangle$; tangent plane: $x + y + 2z = 2$

21. $(x-2)/12 = (y-1)/(-5) = (z-7)/(-1)$

23. $x - \pi/4 = y = z - \pi/4$

25. $x = 0, y - \pi = 1 - z$

27. 5.26

29. 0.9.

31. 3750 ft^2

33. 1.08 cm^3

35. The summand $f(1, -1)$ has been omitted; the correct equation is $z = 2 + 6(x-1) + 6(y+1)$.

37. The coefficients of x and y should be the evaluations $f_x(P_0)$ and $f_y(P_0)$, not $f_x(x, y)$ and $f_y(x, y)$; the correct equation is $z = -3x + (y+3)$.

41. $\pi/2$

43. $\pi/4$

45. $80p + 9r < 2112/25$ where $100p$ is the percentage of gold and r is the radius

47. $x + 9y + z = -1$

49. $6x - 13y + 12z = -8$

51. $3x - 6y + 4z = 8$

53. $2x + y + z = 3$

55. $2x - y + z = \pi/4$

Section 11.8

1. Local maximum at $(1, 2)$

3. Saddle point at $(0, 0)$; local maximum at $(-2, 2)$

5. Saddle point at $(0, 0)$; local minimum at $(1, 1)$

7. Saddle points at $(0, 0)$, $(0, -3)$, and $(-3, 0)$; local maximum at $(-1, -1)$

9. Local minimum at $(3, 3)$

11. Saddle points at $(0, -1)$ and $(2, -1)$

13. Saddle points at $(0, 0)$, $(0, -8)$, and $(-8, 0)$; local maximum at $(-8/3, -8/3)$

15. Saddle points at $(-2, -1)$ and $(2, -1)$; local maximum at $(0, -\sqrt{5})$; local minimum at $(0, \sqrt{5})$

17. Saddle point at $(0, 0)$; local minima at $(-\sqrt{2}, -\sqrt{2})$ and $(\sqrt{2}, \sqrt{2})$

19. Local minimum at $(0, 0)$

21. Each number is 100/3.

23. Base: a square with side length $4/3^{1/3}$ in.; height: $4 \cdot 3^{2/3}$ in.

25. Equilateral triangle with side length $20/3^{1/4}$

27. Regression line: $y = 0.001956x + 0.2577$; $y(100) \approx 0.4533$ g/mi

29. $y = 6.12x + 21.2$

31. Least squares line: $\ln(y) = -0.14t + 0.0093$; $y(10) \approx 0.25$

39. Saddle point: $(-0.50608337, 0.47573876)$

41. Critical point: $(0, \pi/2)$, which is a global minimum

Section 11.9

1. Maximum value: $6 + \sqrt{34}$; minimum value: $6 - \sqrt{34}$

3. Maximum value: 25; minimum value: 11/3

5. Maximum value: $128/\sqrt{3} - 24$; minimum value: $-128/\sqrt{3} - 24$

7. Maximum value: 37/4; minimum value: -3

9. Maximum value: $16\sqrt{2}$; minimum value: 16

11. Maximum value: $8/(3\sqrt{3})$; minimum value: $-8/(3\sqrt{3})$

13. 223/22

15. -4

17. $4\sqrt{3}/3$

19. $3\sqrt[3]{2\pi V_0^2}$

21. 144

23. $\sqrt{2/3}$

25. $256\sqrt{2}/3$

27. 3

29. Highest: $150 + 180\sqrt{2}$; lowest: $150 - 180\sqrt{2}$

31. $3\sqrt[3]{2\pi k V_0^2}$ times the unit side cost

33. 156/7

35. 11/7

37. 420

39. $x = pT/(A(p+q))$, $y = qT/(B(p+q))$

43. $(x(10000), y(10000)) = (8000/21, 2500/7)$;
$M(10000) \approx 31296.961$;
$\lambda(10000) = (2500/7)^{3/4}/15 \approx 5.477$;
$(x(10001), y(10001)) = (40004/105, 10001/28)$;
$M(10001) \approx 31302.438$; $M(10001) - M(10000) \approx 5.477$ (the same value as $\lambda(10000)$)

51. 3.15694

53. 0.0342767

55. Minimum: -1.01006; maximum: 2.45745

REVIEW EXERCISES FOR CHAPTER 11

1. $A(h,r) = 2\pi rh$, domain of $A: \{(h,r): 0 < h;\ 0 < r\}$

3. $P(\ell_1, \ell_2) = \ell_1 + \ell_2 + \sqrt{\ell_1^2 + \ell_2^2}$, domain of $P: \{(\ell_1, \ell_2): 0 < \ell_1;\ 0 < \ell_2\}$

5. $\dfrac{3}{10}\sin(3)$

7. $8^{3/2}$

9. $2 \cdot 3^{3/2}$

11.

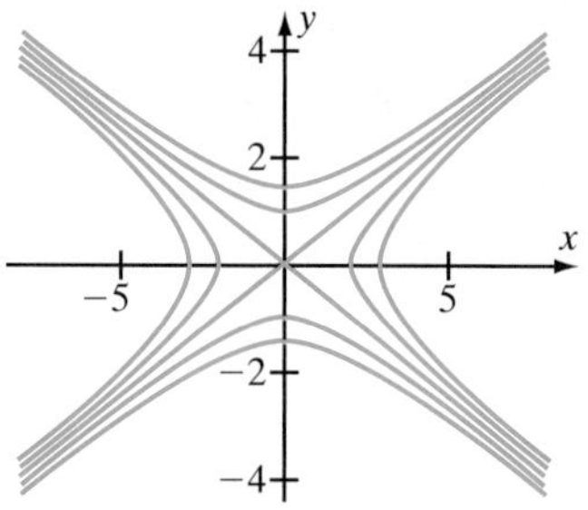

13.

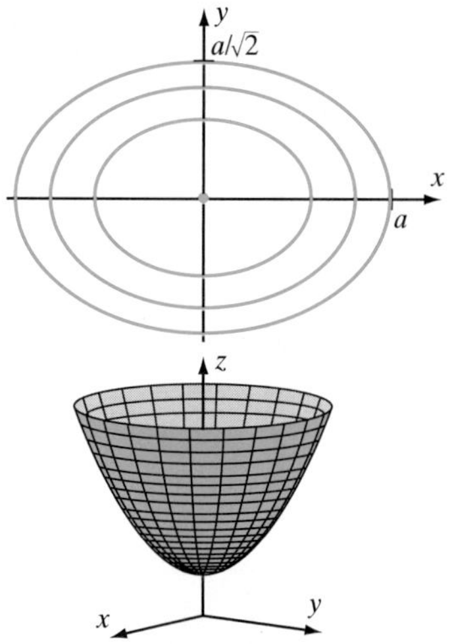

15.

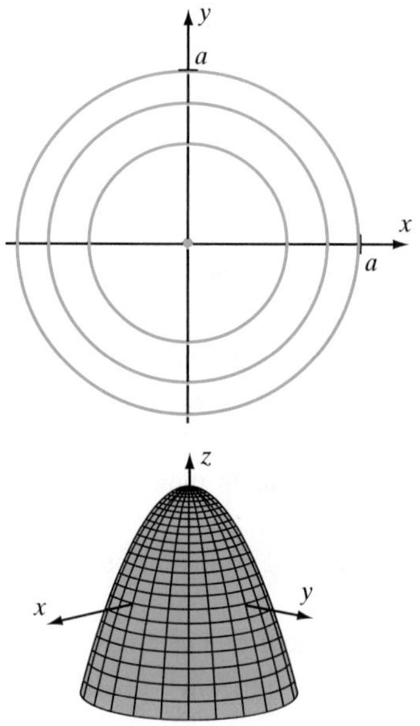

17.

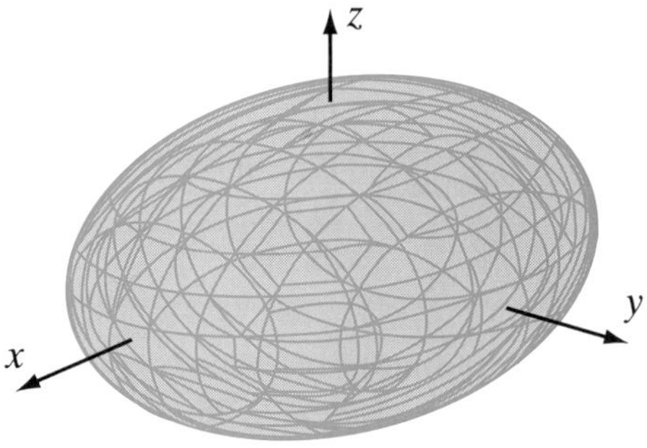

19.

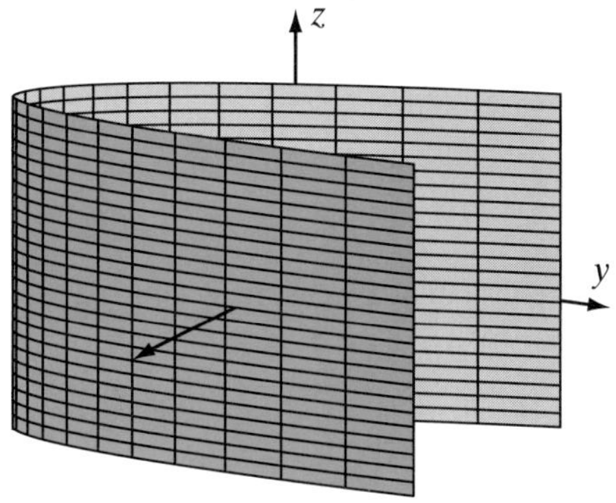

21.

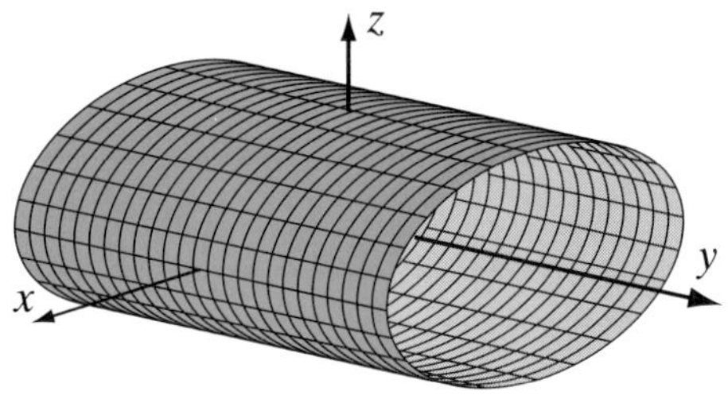

23.

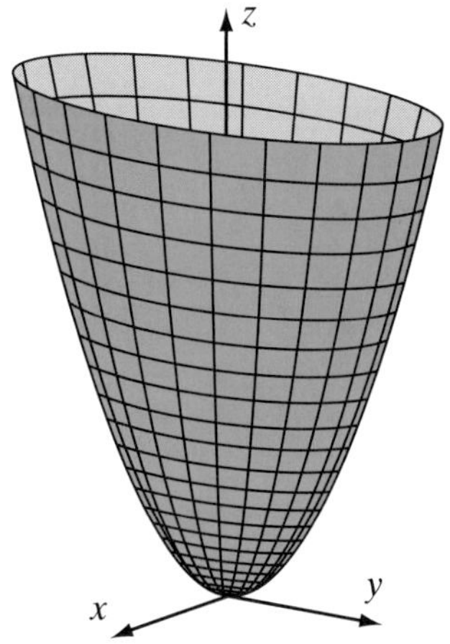

elliptic paraboloid

25. $\lim_{(x,y)\to(1,1)} \dfrac{x^2-2x+1}{xy-2x-y+2} = \lim_{(x,y)\to(1,1)} \dfrac{x-1}{y-2} = 0$

27. $\lim_{(x,y)\to(3,5)} \dfrac{\sin(5x-3y)}{5x-3y} = \lim_{t\to 0} \dfrac{\sin(t)}{t} = 1$

29. $\lim_{(t,t)\to(0,0)} \dfrac{t-t}{t+t} = 0,\ \lim_{(t,2t)\to(0,0)} \dfrac{-t}{3t} = -\dfrac{1}{3}$

31. 0, 0

33. e, 0

35. $f_x = \dfrac{\sin(x)\sin(y)}{\cos^2(x)}, f_{xx} = \dfrac{\sin(y)[\sin^2(x)+1]}{\cos^3(x)},$

$f_y = \dfrac{\cos(y)}{\cos(x)}, f_{yy} = \dfrac{-\sin(y)}{\cos(x)}$

$f_{xy} = f_{yx} = \dfrac{\cos(y)\sin(x)}{\cos^2(x)}$

37. $f_x = 3x^2yz, f_y = x^3z - z\cos(yz), f_z = x^3y - y\cos(yz)$
$f_{xy} = 3x^2,\ \cos(yz), f_{yz} = x^3 - \cos(yz) + yz\sin(yz)$
$f_{xyz} = 3x^2, f_{zz} = y^2\sin(yz)$

39. $f_x = ze^{y^2x} + xzy^2e^{y^2x}, f_y = 2x^2yze^{y^2x}, f_z = xe^{y^2x}$
$f_{xy} = 4xyze^{y^2x} + 2x^2y^3ze^{y^2x}, f_{yz} = 2x^2ye^{y^2x}$
$f_{xyz} = 4xye^{y^2x} + 2x^2y^3e^{y^2x}, f_{zz} = 0$

41. $f_{xyz} = f_{yxz} = f_{zyx} = f_{zxy} = f_{yzx} = f_{yxz} = \cos(xy) - yx\sin(xy)$

43. $\dfrac{dz}{ds} = 4s^3 - 6s^2 + 2s - \dfrac{1}{s^2}$

45. $\dfrac{-2}{(\sin(s)-\cos(s))^2}$

47. $\dfrac{\partial z}{\partial s} = \dfrac{e^t - 2t^2e^{2s}}{se^t - t^2e^{2s}}, \dfrac{\partial z}{\partial t} = \dfrac{se^t - 2te^{2s}}{se^t - t^2e^{2s}}$

49. $\dfrac{dw}{ds} = \dfrac{s^2\ln(s)\cos(s) - s^2e^s\cos(s) + s^2e^s\sin(s) - s\sin(s)}{(s\ln(s) - se^s)^2}$

51. $\dfrac{dw}{ds} = \dfrac{1 - \ln(s) - 2s^3\ln^2(s) - 2s^3\ln(s) + 6s^2e^{6s}}{s\ln(s) - s^4\ln^2(s) + s^2e^{6s}}$

53. $0 + (x-2) - 2(y-1) - \dfrac{1}{2}(x-2)^2 - 3(y-1)^2 +$
$2(x-2)(y-1)$

55. $\nabla f(x,y) = \langle 2x\sin(xy) + x^2y\cos(xy), x^3\cos(xy)\rangle$

57. $\nabla f(x,y) = \langle yze^{xz-y^2} + xyz^2e^{xz-y^2}, xze^{xz-y^2} -$
$2xy^2ze^{xz-y^2}, xye^{xz-y^2} + x^2yze^{xz-y^2}\rangle$

59. $D_{\mathbf{u}}f(P) = \sqrt{2}$

61. direction of greatest increase is
$\mathbf{u} = \left\langle \dfrac{1}{\sqrt{10}}, \dfrac{3}{\sqrt{10}} \right\rangle, D_{\mathbf{u}}f(P) = 4\sqrt{10},$
direction of greatest decrease is
$\mathbf{v} = \left\langle -\dfrac{1}{\sqrt{10}}, -\dfrac{3}{\sqrt{10}} \right\rangle, D_{\mathbf{v}}f(P) = -4\sqrt{10}$

63. direction of greatest increase is
$\mathbf{u} = \left\langle -\frac{1}{\sqrt{1+4\pi}}, \frac{2\sqrt{\pi}}{\sqrt{1+4\pi}} \right\rangle$, $D_{\mathbf{u}}f(P) = \sqrt{1+4\pi}$,
direction of greatest decrease is
$\mathbf{v} = \left\langle \frac{1}{\sqrt{1+4\pi}}, -\frac{2\sqrt{\pi}}{\sqrt{1+4\pi}} \right\rangle$, $D_{\mathbf{v}}f(P) = -\sqrt{1+4\pi}$

65. $D_{\mathbf{v}}f(P) = \frac{4}{3}$

67. direction of greatest increase is
$\mathbf{u} = \left\langle \frac{48}{\sqrt{23312}}, \frac{112}{\sqrt{23312}}, \frac{92}{\sqrt{23312}} \right\rangle$, $D_{\mathbf{u}}f(P) = \sqrt{23312}$,
direction of greatest decrease is
$\mathbf{v} = \left\langle -\frac{48}{\sqrt{23312}}, -\frac{112}{\sqrt{23312}}, -\frac{92}{\sqrt{23312}} \right\rangle$,
$D_{\mathbf{v}}f(P) = -\sqrt{23312}$

69. direction of greatest increase is
$\mathbf{u} = \left\langle \frac{1}{\sqrt{3}}, \frac{1}{\sqrt{3}}, \frac{1}{\sqrt{3}} \right\rangle$, $D_{\mathbf{u}}f(P) = \frac{\sqrt{3}}{2}$,
direction of greatest decrease is
$\mathbf{v} = \left\langle -\frac{1}{\sqrt{3}}, -\frac{1}{\sqrt{3}}, -\frac{1}{\sqrt{3}} \right\rangle$, $D_{\mathbf{v}}f(P) = -\frac{\sqrt{3}}{2}$

71. $2\sqrt{\pi}x + z = 2\pi$

73. $2x + 8y + 16z = 26$

75. $x - y = 1 - \ln(2)$

77. $\begin{cases} x = \frac{\pi}{2} + 0t \\ y = 0 - \frac{\pi}{2}t \\ z = 0 + 1t \end{cases}$

79. $f(3.9, 5.9) \approx 9.42$

81. $f(1.1, 0.1) \approx 0.15$

83. critical point is $(3/4, -2/3)$, positive discriminant with equal, positive f_{xx}, f_{yy}, hence is a local minimum

85. critical points $(4, -1)$ and $(4, 0)$, at $(4, -1)$ the discriminant is negative hence a saddle, at $(4, 0)$ the discriminant is -1 hence a saddle

87. $100^{1/3}$, $100^{1/3}$, $100^{1/3}$

89. critical points are $\left(\sqrt{\frac{126}{25}}, -\frac{12}{5}\right)$, $\left(-\sqrt{\frac{126}{25}}, -\frac{12}{5}\right)$, $(0, 6)$, $(0, -6)$, the first two are minima and the third and fourth are maxima

91. critical point is $(8/9, -16/9)$, and it is the minimum

93. critical points are $(\sqrt{5}, 0)$, $(-\sqrt{5}, 0)$, $(0, 5)$, $(0, -5)$, the last two are minima

Section 12.1

1. 56 **3.** 72

5. $\mathcal{I}(x) = 4x + 24$; $\mathcal{J}(y) = 2 + 4y$

7. $\mathcal{I}(x) = 12x^2$; $\mathcal{J}(y) = 8y + 28$

9. $\mathcal{I}(x) = (e^x - 1)/(2x)$; $\mathcal{J}(y) = (e^{y^2} - 1)/y$

11. 161/2 **13.** 0

15. $e^3 - e^2 - e + 1$ **17.** $-\ln(2)$

19. 0 **21.** 5/3

23. $\pi^2/8$ **25.** 4/3

27. 1/12

29. $(e + 2)\ln(e + 2) - e - 3\ln(3)$

31. $2e - 2$ **33.** $(e^4 - 1)/8$ **35.** $e - 2$

37. $424\ln(2) - 105$

39. $4(\sin(1) - \cos(1))$

41. $(e^4 - 1)/4$ **43.** $\pi^2/32$

45. 2 **47.** 1

49. 0.2248 **51.** 2.231

Section 12.2

1. $\mathcal{R}$ is x-simple only; boundary curves: $\alpha_1(y) = -2y - 1$ and $\alpha_2(y) = -y^2 + 2$ for $-1 \le y \le 3$

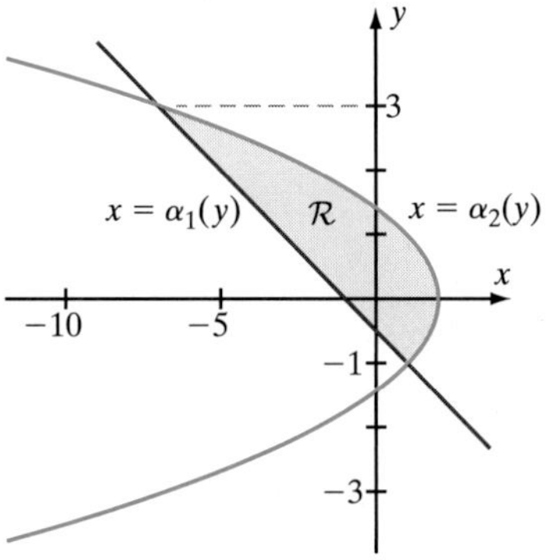

3. $\mathcal{R}$ is x-simple only; boundary curves: $\alpha_1(y) = y^2$ and $\alpha_2(y) = -2y^2 - 3y + 18$ for $-3 \le y \le 2$

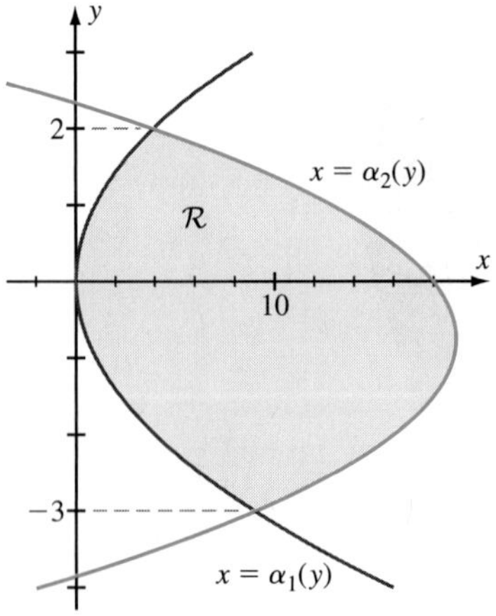

5. $\mathcal{R}$ is y-simple only; boundary curves: $\beta_1(x) = x^{3/2}$ and $\beta_2(x) = 6x - 1$ for $1 \le x \le 3$

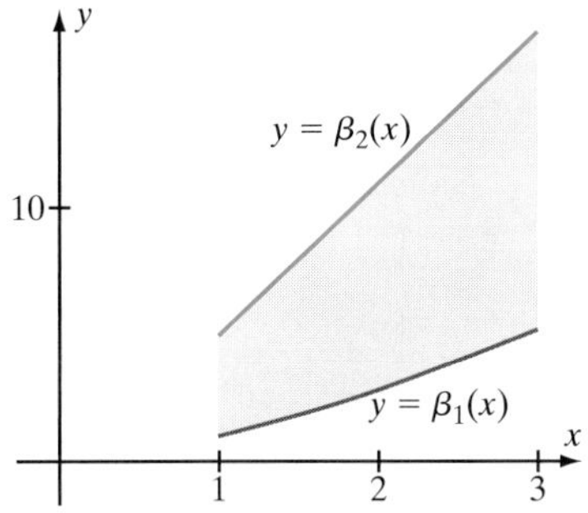

7. -8

9. $78\sqrt{2}/5$

11. $4\pi - 4$

13. $5/\sqrt{2}$

15. 112/15

17. $e^2(13e^6 - 7)/4 - 33/2$

19. $640 - 145\sqrt{3}$

21. 5/6

23. 1

25. $e^4 - 1$

27. 1

29. 1/12

31. $(\sqrt{2}\pi - 5)/4$

33. $(1 - \sin(1))/2$

35. 11/15

37. $-1/24$

39. 9/4

41. 2/3

43. 1/4

45. $(143a + 91b)/3$

47. 68/15

49. 2/3

51. 1/6

53. 1/3

57. 6.6525

59. 0.27694

61. 0.62

63. 0.66

65. Exact: 8/15; approximation: 17/32; absolute error: 0.00208

67. Exact: 4/35; approximation: $(3 + 4\sqrt{2} + 3\sqrt{3})/128$; absolute error: 0.0060591

Section 12.3

1. 1029/4

3. 99/4

5. 48/5

7. 375/4

9. 12

11. 728

13. 813/2

15. 1960/3

17. 108

19. $4e^2 + e^{-3}$

21. 1288/15

23. $2\sin(1) - \sin(2)$

25. 136/15

27. 2744/3

29. 4576/675

31. 64/3

33. 2/3

35. 81π

37. 1/8

39. $\pi/2$

41. $\ln(54/25) - 2\arctan(2) + \pi/2$

43. 1/8

45. 4.7918

47. 3.4931

Section 12.4

1.

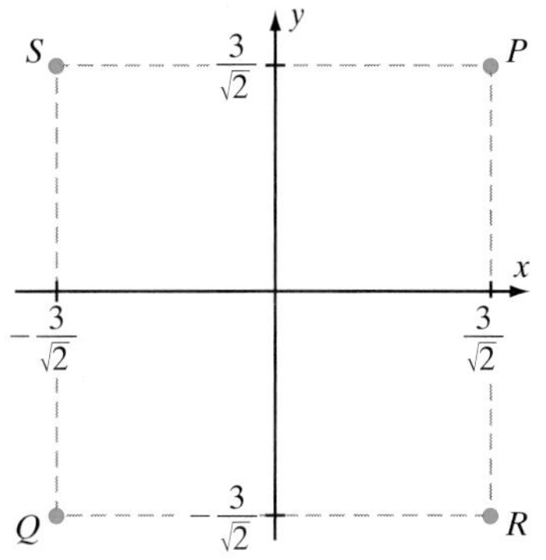

$P = \left(3, \frac{\pi}{4}\right)$ $Q = \left(-3, \frac{\pi}{4}\right)$

$R = \left(3, -\frac{\pi}{4}\right)$ $S = \left(-3, -\frac{\pi}{4}\right)$

3.

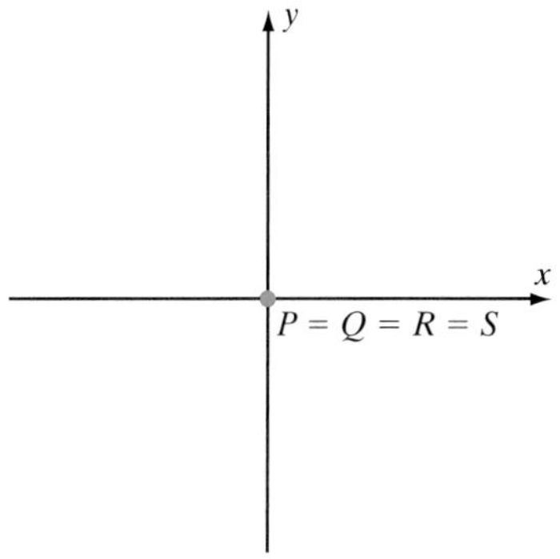

$P = \left(0, \frac{\pi}{2}\right)$ $Q = (0, \pi)$

$R = (0, -\pi)$ $S = \left(0, -\frac{\pi}{2}\right)$

5.

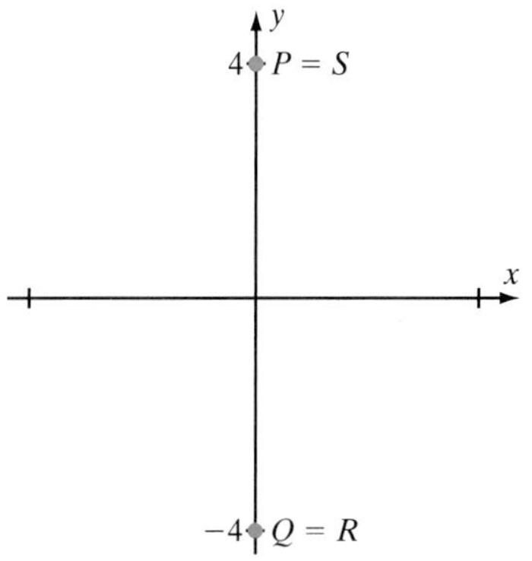

$P = \left(4, -\frac{7\pi}{2}\right)$ $Q = \left(-4, -\frac{7\pi}{2}\right)$

$R = \left(4, \frac{7\pi}{2}\right)$ $S = \left(-4, \frac{7\pi}{2}\right)$

7.

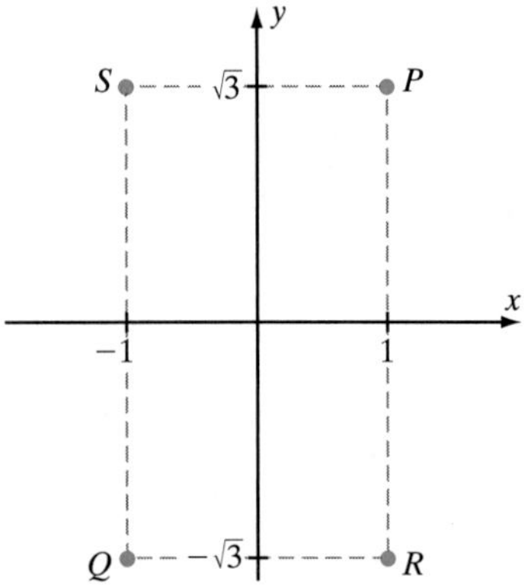

$P = \left(2, \frac{7\pi}{3}\right)$ $Q = \left(-2, \frac{7\pi}{3}\right)$

$R = \left(2, -\frac{7\pi}{3}\right)$ $S = \left(-2, -\frac{7\pi}{3}\right)$

9. $(2\sqrt{2}, 2\sqrt{2})$ **11.** $(0, 0)$

13. $(2\sqrt{3}, 2)$ **15.** $(-3, -3\sqrt{3})$

17. $(4\sqrt{2}, -\pi/4 + 2n\pi), n \in \mathbb{Z}$; $(-4\sqrt{2}, 3\pi/4 + 2n\pi)$, $n \in \mathbb{Z}$

19. $(9, -\pi/2 + 2n\pi), n \in \mathbb{Z}$; $(-9, \pi/2 + 2n\pi), n \in \mathbb{Z}$

21. $(1, \pi + 2n\pi), n \in \mathbb{Z}$; $(-1, 2n\pi), n \in \mathbb{Z}$

23. Symmetric with respect to origin, *x*-axis, *y*-axis

25. Symmetric with respect to origin, *x*-axis, *y*-axis

27. Symmetric with respect to origin, *x*-axis, *y*-axis

29. Symmetric with respect to *x*-axis

31. Not symmetric with respect to origin, *x*-axis, *y*-axis

33.

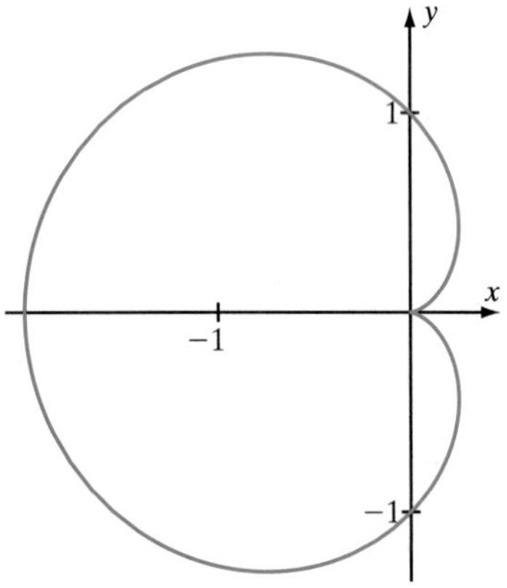

$r = 1 - \cos(\theta)$

35.

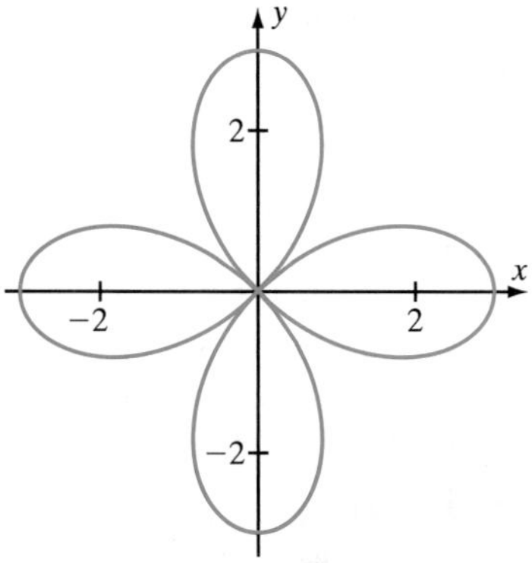

$r = 3\cos(2\theta)$

37.

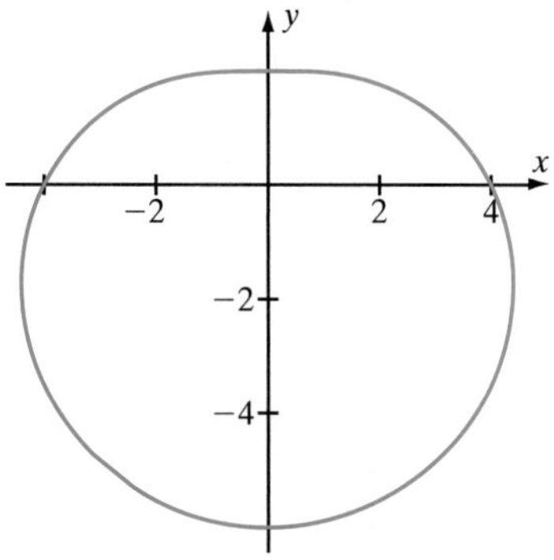

$r = 4 - 2\sin(\theta)$

39.

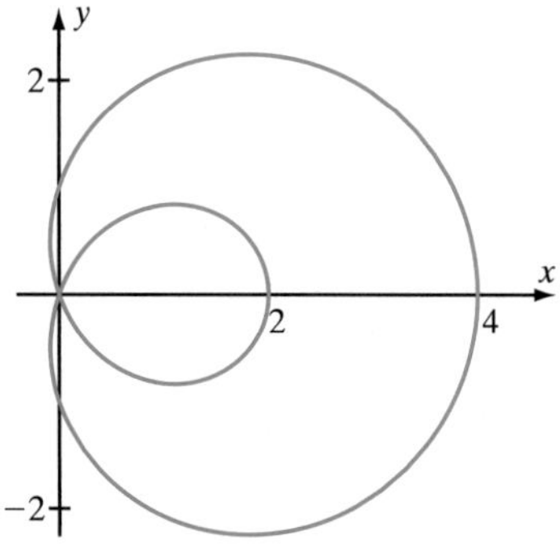

$r = -1 + 3\cos(\theta)$

41.

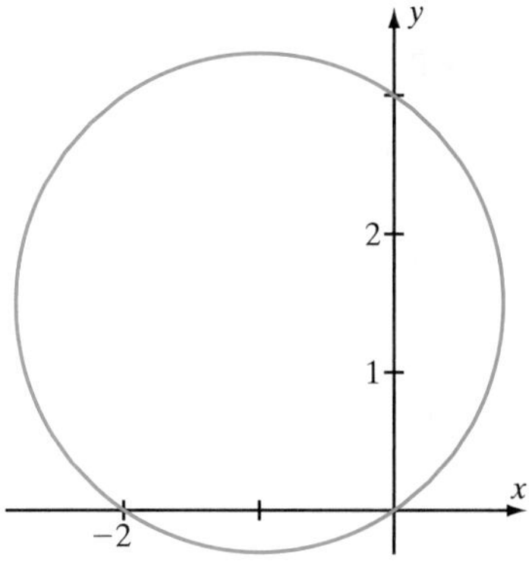

$r = 3\sin(\theta) - 2\cos(\theta)$

43.

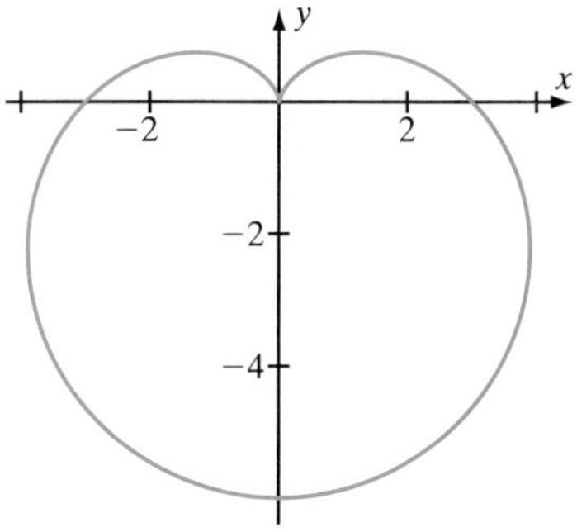

$r = 3 - 3\sin(\theta)$

45. a. $\left(\sqrt{9+\pi^2/4},\ \arctan(\pi/12)+2n\pi\right)$ and $\left(-\sqrt{9+\pi^2/4},\ \arctan(\pi/12)+(2n+1)\pi\right)$ for $n \in \mathbb{Z}$
b. $\left(\sqrt{36+49\pi^2/4},\ -\arctan(7\pi/12)+(2n+1)\pi\right)$ and $\left(-\sqrt{36+49\pi^2/4}, -\arctan(7\pi/12)+2n\pi\right)$ for $n \in \mathbb{Z}$
c. $(\pi\sqrt{1033}/12, -\arctan(32/3)+2n\pi)$ and $(-\pi\sqrt{1033}/12, -\arctan(32/3)+(2n+1)\pi)$ for $n \in \mathbb{Z}$

47. Distance: $\sqrt{r_1^2+r_2^2-2r_1r_2\cos(\theta_1-\theta_2)}$; polar equation of circle: $\sqrt{r^2+9-6r\cos(\theta-\pi/4)}=1$

49. In rectangular coordinates: $(0,0)$, $(1/4, \sqrt{3}/4)$, and $(1/4, -\sqrt{3}/4)$

51. In rectangular coordinates: $(\pm 1/\sqrt{2}, \pm 1/\sqrt{2})$

53. In rectangular coordinates: $(0,0)$ and $(-1/4, \pm\sqrt{3}/4)$

55. The points of the x-axis

57. $\sqrt{2}(\exp(\pi/2) - \exp(-\pi/2))$

59. $\pi/4 - 3\sqrt{3}/8$

61.

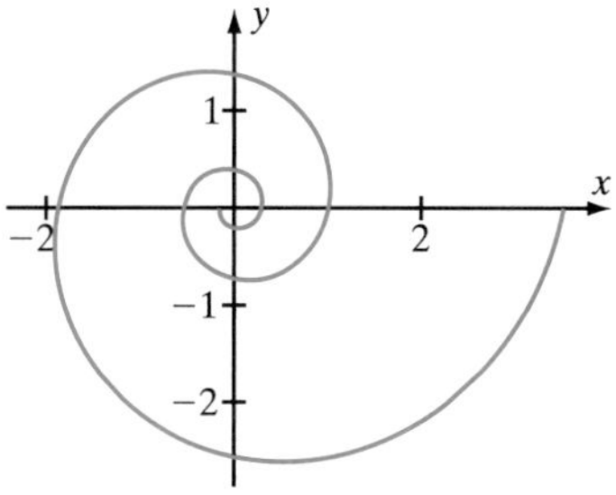

$r = \exp\left(\dfrac{\theta}{5}\right)$

63.

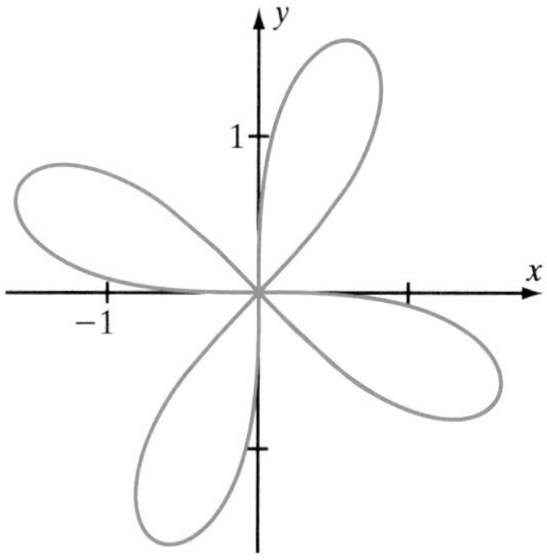

$r^2 = -3\sin(4\theta)$

65.

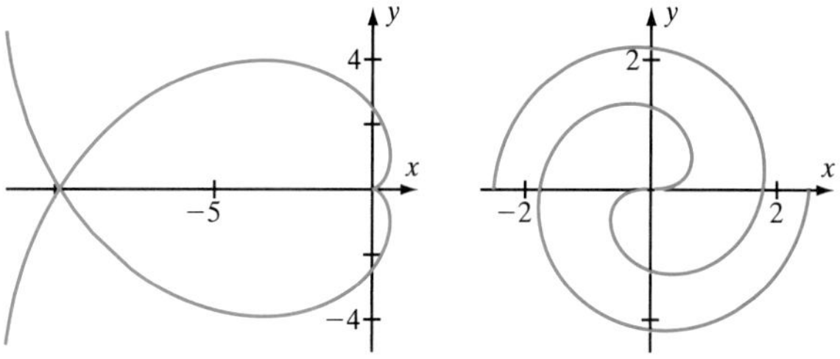

$r = \theta^2;\ -\dfrac{9\pi}{8} \le \theta \le \dfrac{9\pi}{8}$ $\qquad \theta = r^2;\ 0 \le \theta \le 2\pi$

67.

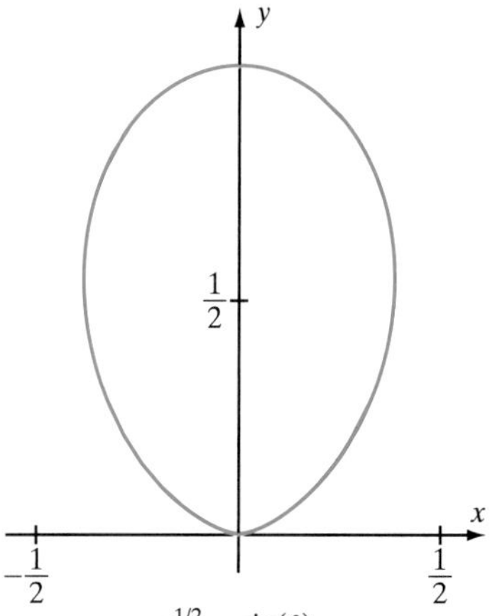

$r^{1/2} = \sin(\theta)$

69.

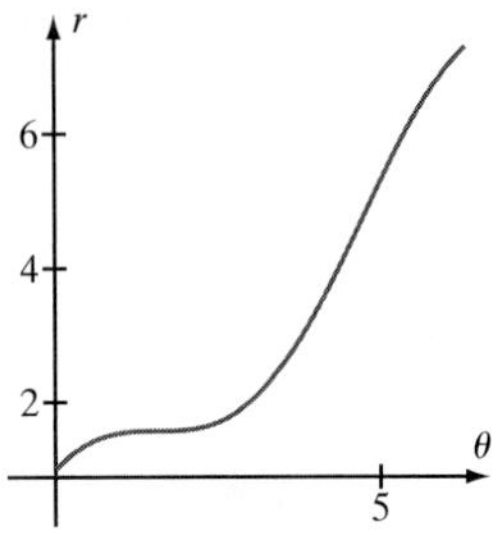

$r = \theta + \cos(\theta)$

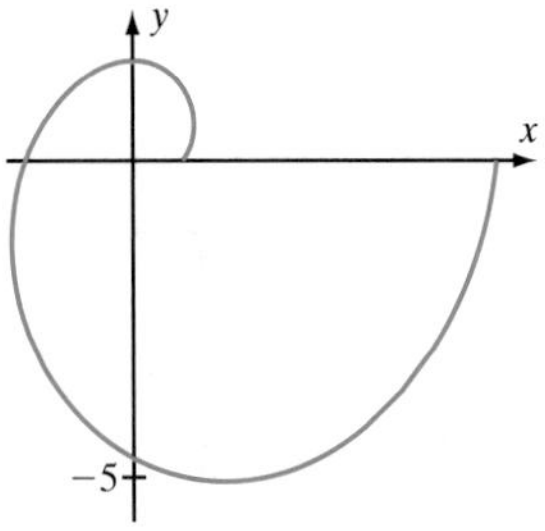

$r = \theta + \cos(\theta)$

Section 12.5

1. $3\pi/2$ **3.** 4π

5. 2 **7.** 35π

9. $\pi/8 - 1/4$ **11.** $\pi/6 + \sqrt{3}/4$

13. $\pi + 4$ **15.** $31\pi/2$

17. 15π **19.** 2/3

21. $-4\sqrt{2}/15$ **23.** 2/3

25. 11π **27.** 8π

29. 32π **31.** $40\pi/3$

33. 18π **35.** $32(2\sqrt{2} - 1)\pi/3$

37. $32\pi/3$ **39.** 29/210

41. 248/27 **43.** $52/9 + (32/3)\ln(2/3)$

45. 12π **47.** $7\pi/12 - \sqrt{3}$

49. $\pi/3 - \sqrt{3}/4$ **51.** $\pi - 3\sqrt{3}/2$

53. $2(\sqrt{5} - 1)\pi$ **55.** $3^{11/3}\pi/16$

57. $\pi^2/32 - 1/8$ **59.** $\ln(2)$

63. $\theta_0 \approx 0.967025$; area: 2.4393

65. 0.24

Section 12.6

1. 27 **3.** 252

5. 3 **7.** $\pi - 4$

9. $\cos(\pi^2) - 1$ **11.** 30π

13. 16/15 **15.** 24π

17. 128π

19. 3; $\mathcal{U}$ is the region in the first octant cut off by the plane $3x + 2y + z = 6$.

21. 1/15; $\mathcal{U}$ is that part of the unit ball $x^2+y^2+z^2 \le 1$ that lies in the first octant.

23. 1/8; $\mathcal{U}$ is that part of the cylinder $x^2+z^2 \le 1$ that lies in the first octant between the planes $y = 0$ and $y = x$.

33. $\pi(a^2 + b^2 + 4c)^2/32$ **35.** $\pi m^6(3m^2 + 4)/60$

37. $(1 - \cos(1))/24$ **39.** 216

41. $\int_0^1 \int_x^1 \int_0^1 f(x,y,z)\,dz\,dy\,dx$;
$\int_0^1 \int_0^y \int_0^1 f(x,y,z)\,dz\,dx\,dy$;
$\int_0^1 \int_0^1 \int_x^1 f(x,y,z)\,dy\,dz\,dx$;
$\int_0^1 \int_0^1 \int_x^1 f(x,y,z)\,dy\,dx\,dz$;
$\int_0^1 \int_0^1 \int_0^y f(x,y,z)\,dx\,dy\,dz$; $\int_0^1 \int_0^1 \int_0^y f(x,y,z)\,dx\,dz\,dy$

43. $\int_{-1}^0 \int_2^{37} \int_{x+1}^{\sqrt{x+1}} y\sqrt{1-x}\,dy\,dz\,dx$;
$\int_2^{37} \int_{-1}^0 \int_{x+1}^{\sqrt{x+1}} y\sqrt{1-x}\,dy\,dx\,dz$;
$\int_{-1}^0 \int_{x+1}^{\sqrt{x+1}} \int_2^{37} y\sqrt{1-x}\,dz\,dy\,dx$;
$\int_0^1 \int_{y^2-1}^{y-1} \int_2^{37} y\sqrt{1-x}\,dz\,dx\,dy$;
$\int_0^1 \int_2^{37} \int_{y^2-1}^{y-1} y\sqrt{1-x}\,dx\,dz\,dy$;
$\int_2^{37} \int_0^1 \int_{y^2-1}^{y-1} y\sqrt{1-x}\,dx\,dy\,dz$

47. 2.0333

Section 12.7

1. 108 **3.** 3

5. 2 **7.** 85

9. 1027/2 **11.** $-(26/3)\ln(2)$

13. (183/448, 65/128) **15.** (156/55, 29/55)

17. $(k/2, 2k/3)$ where $k = (2\ln(2) - 1)^{-1}$

19. 5/3 **21.** 13 **23.** 1/2

25. $M = 72\pi$; $M_{yz} = 0$; $M_{xz} = 0$; $M_{xy} = 576\pi$; center of mass : $(0, 0, 8)$

27. $M = 67/8$; $M_{yz} = -193/40$; $M_{xz} = 25/24$; $M_{xy} = 389/60$; center of mass : $(-193/335, 25/201, 778/1005)$

29. $M = 5/8$; $M_{yz} = = 31/120$; $M_{xz} = 31/40$; $M_{xy} = 31/30$; center of mass : $(31/75, 31/25, 124/75)$

31. $I_{x=0} = \dfrac{48}{5}$; $I_{y=0} = \dfrac{1056}{35}$

33. $I_{x=0} = 1/10$; $I_{y=0} = 13/15$

35. 11 **37.** 4

39. $I_x = 650$; $I_y = 1160$; $I_z = 1530$

41. $I_x = 186$; $I_y = 194$; $I_z = 64$

43. $(\pi/2, 16/(9\pi))$

45. $(e - 2, (e^2 + 1)/8)$

47. $\mathcal{R} = \{(x,y) : 0 < x^2 + y^2 < 1\}$

49. $KE_{\mathcal{U}} = I_x\omega^2/2$; $\rho = \sqrt{I_x/M_{\mathcal{U}}}$

51. $2\pi^2 r^2 R$

53. $(1.16866, 3.52217)$

55. $(0.52368, 0.48447)$

Section 12.8

1. $(4, \pi/6, 5)$

3. $(2\sqrt{2}, 3\pi/4, -1)$

5. $(3, 3\pi/2, -2)$

7. $(2, 5\pi/3, -2)$

9. $(2, \pi/4, \pi/4)$

11. $(4\sqrt{2}, \pi/4, 0)$

13. $(2\sqrt{2}, \pi/6, \pi/4)$

15. $(8\sqrt{2}, 5\pi/6, 3\pi/4)$

17. Rectangular coordinates: $(-3\sqrt{3}/4, 9/4, 3/2)$; cylindrical coordinates: $(3\sqrt{3}/2, 2\pi/3, 3/2)$

19. Cylindrical coordinates: $(2\sqrt{2}, 7/4\pi, -2\sqrt{2})$; spherical coordinates: $(4, 3\pi/4, 7\pi/4)$

21. Rectangular coordinates: $(-\sqrt{2}, -\sqrt{2}, -2)$; spherical coordinates: $(2\sqrt{2}, 3\pi/4, 5\pi/4)$

23. Cylindrical coordinates: $(3, 0, -3)$; spherical coordinates: $(3\sqrt{2}, 3\pi/4, 0)$

25. $32\pi/3$

27. $512\pi/15$

29. $10\pi \ln(2)$

31. 180π

33. 65π

35. -2π

37. 2π

39. $32\pi/3$

41. Rectangular coordinates: $(3\cos(5)\sin(2), 3\sin(5)\sin(2), 3\cos(2))$; cylindrical coordinates: $(3\sin(2), 5, 3\cos(2))$

43. Points in the xz-plane

45.

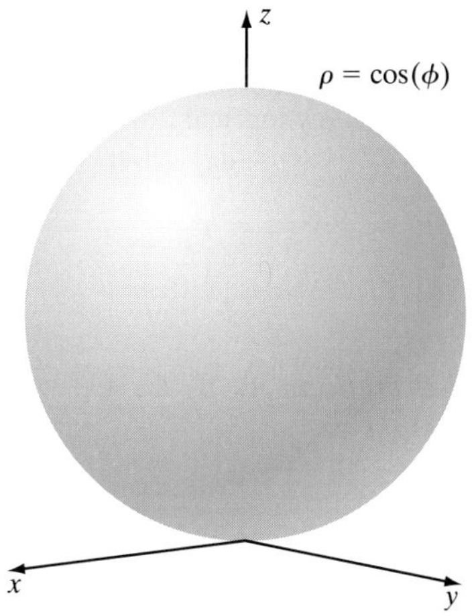

47.

49. 4.797453714

51. 0.6527

REVIEW EXERCISES FOR CHAPTER 12

1. 9

3. $\dfrac{-33}{32}$

5. $\mathbf{I}(x) = \cos(2x - 2\pi) - \cos(2x)$, $\mathbf{J}(y) = -\frac{1}{2}\cos(2\pi - y) + \frac{1}{2}\cos(y)$

7. $-\dfrac{135}{4}$

9. $e^3 - e^2 - e + 1$

11. -2

13. y-simple, $\beta_1(x) = x^2 - 2, \beta_2(x) = x + 1$

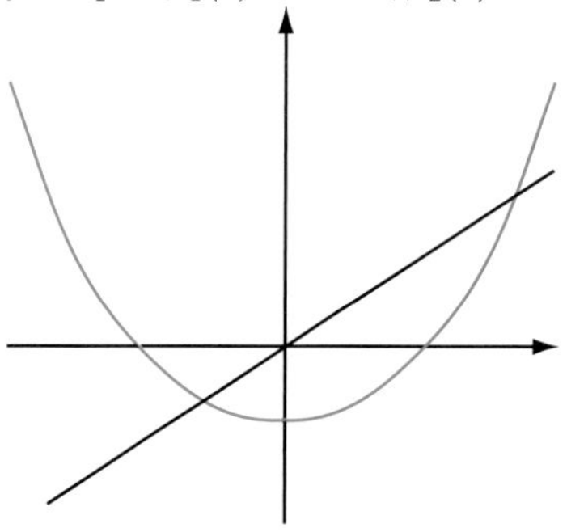

15. either x-simple or y-simple, $\alpha_1(y) = y^2, \alpha_2(y) = 2\sqrt{y}, \beta_1(x) = (x/2)^2, \beta_2(x) = \sqrt{x}$

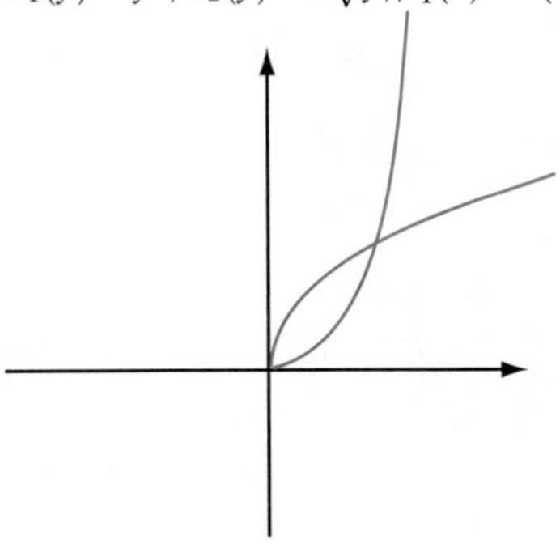

17. $\frac{4}{3} \cdot 4^{10/3}$

19. $\frac{4}{15}$

21. $\frac{\pi\sqrt{2}}{4} - 1$

23. $\frac{1}{2}(1 - e^{-1})$

25. 56

27. $\frac{128}{3}$

29. 12π

31.

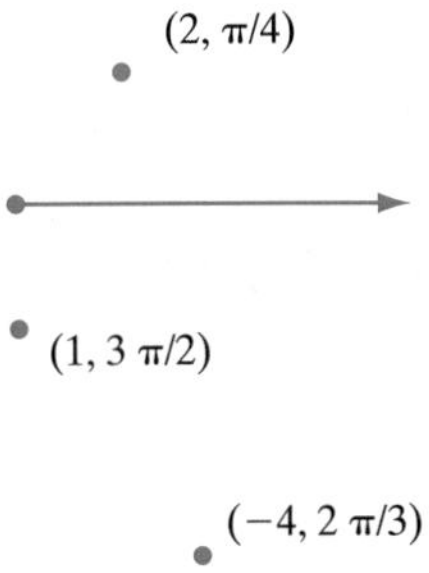

33.

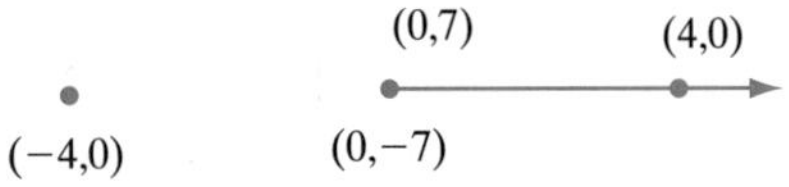

35. (0, 3)

37. $7\pi/4$

39. $\arctan(4/\pi)$

41. symmetric in the origin

43.

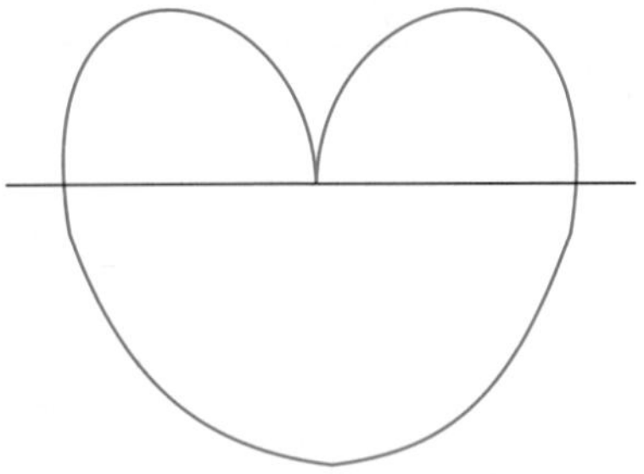

45.

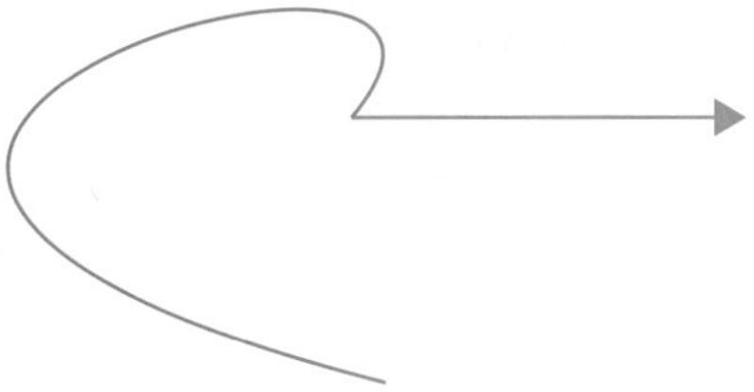

47. 1

49. π

51. 2π

53. $\frac{\pi}{3} + \frac{5\sqrt{3}}{16}$

55. $\frac{8}{3}$

57. $-\frac{\pi}{4} - \frac{29}{6}$

59. $1000\pi \cdot \frac{\sqrt{2}}{3}$

61. $\frac{32}{15}\sqrt{2}$

63. $\frac{7}{2}$

65. $\frac{19}{30}e^5 - \frac{1}{6}e^3 - \frac{7}{15}$

67. $\frac{255}{8}\pi$

69. 32π

71. 2

73. $\frac{441}{2}$

75. $\frac{1126}{105}$

77. $\frac{6 \cdot 2^{13/3}}{13}$

79. $\left(\frac{135}{588 - 352\sqrt{2}}, \frac{1723 - 1632\sqrt{2}}{2058 - 1232\sqrt{2}}\right)$

81. $\left(\frac{63}{576}, \frac{1119}{288}\right)$

83. 42

85. $\left(\frac{1}{2}, \frac{45}{37}, \frac{41}{37}\right)$

87. $\left(\frac{5}{3}, \frac{4}{3}, \frac{38}{15}\right)$

89. $(\sqrt{5}, \arctan(-2), \sqrt{3})$

91. $(5, 0, 0)$

93. $(6\sqrt{2}, \pi/4, 2\pi/3)$

95. spherical : $(4\sqrt{2}, \pi/4, \pi/4)$, cylindrical : $(4, -\pi/4, 4)$

97. $\pi/6$

99. $3\pi/10$

101. $\frac{64\pi}{5}$

Section 13.1

In the answers for this section, A, B, C, C_1, and C_2 represent arbitrary constants.

1.

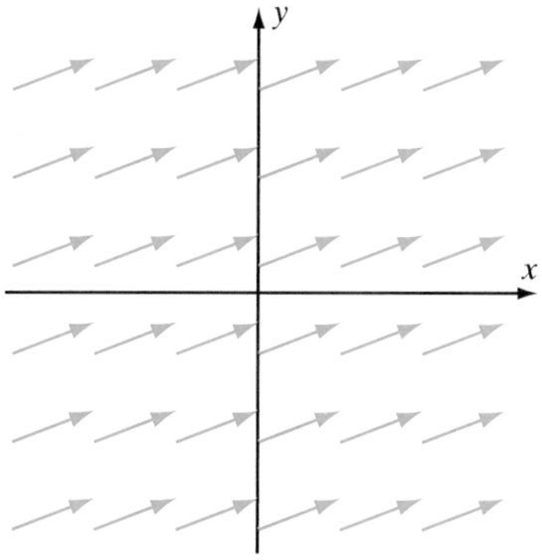

$\mathbf{F}(x, y) = 3\mathbf{i} + \mathbf{j}$

3.

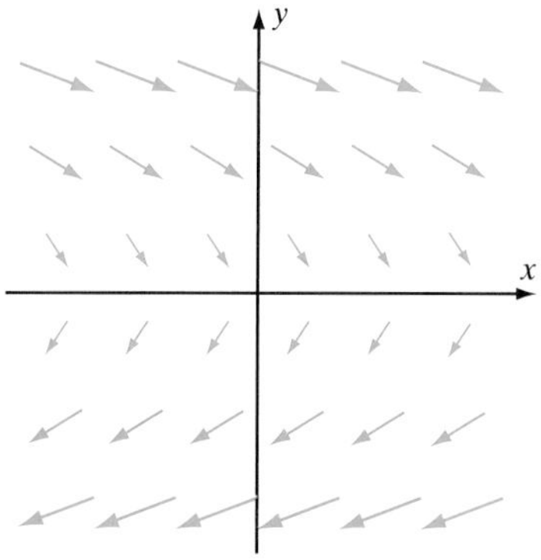

$\mathbf{F}(x, y) = 2y\mathbf{i} - 3\mathbf{j}$

5.

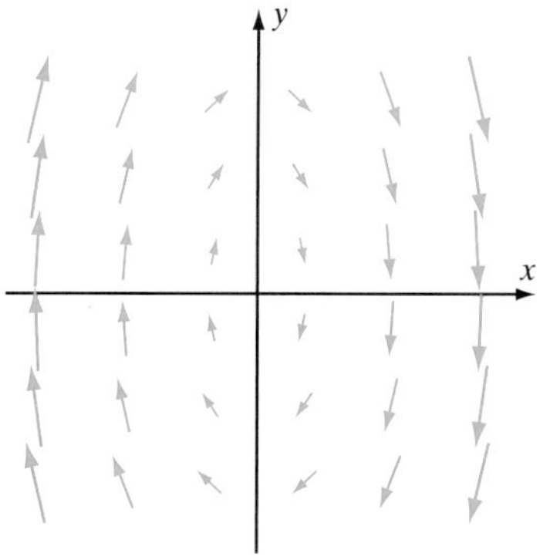

$\mathbf{F}(x, y) = \mathrm{y}\mathbf{i} - 5x\mathbf{j}$

7.

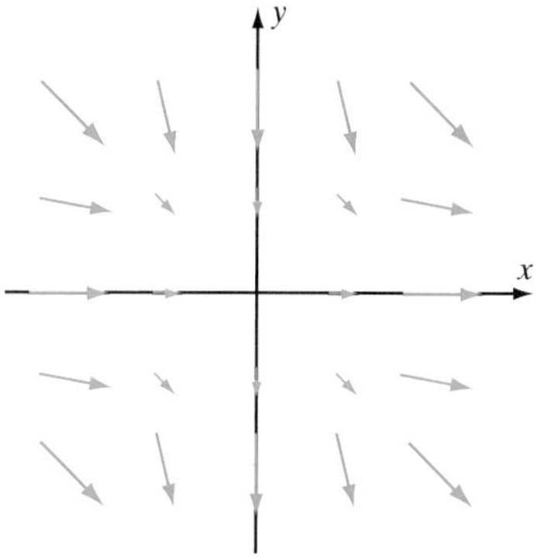

$\mathbf{F}(x, y) = x^2\mathbf{i} - y^2\mathbf{j}$

9. $\alpha\sqrt{x^2 + z^2}\mathbf{j}$ for $-4 < x < 4, 0 < z < 10$

11. $\alpha(y\mathbf{i} - x\mathbf{j})$ for $x^2 + y^2 < 16, 0 < z < 5$

13. $(y + 3)\mathbf{i} + (x - 2y)\mathbf{j}$

15. $(3x^2/(x^3 - y^2))\mathbf{i} + (-2y/(x^3 - y^2))\mathbf{j}$

17. $2xy^3z\mathbf{i} + 3x^2y^2z\mathbf{j} + x^2y^3\mathbf{k}$

19. $\left((x - y)\ln(z)\right)^{-1}\mathbf{i} - \left((x - y)\ln(z)\right)^{-1}\mathbf{j} - \left(z\ln^2(z)\right)^{-1}\ln(x - y)\mathbf{k}$

21. $x + y + C$

23. $xy + C$

25. $xy^3 + x^2y + C$

27. $x^2 + y + C$

29. $x^2yz^3 + 5y + C$

31. $xz^3 - x^2y^2 + yz^2 + C$

33. $(t + C_1)\mathbf{i} + (t + C_2)\mathbf{j}$

35. $C_1e^t\mathbf{i} + (2t + C_2)\mathbf{j}$

37. $(t^2 + 2C_1t + C_2)\mathbf{i} + (t + C_1)\mathbf{j}$

39. $x\exp(2x) + y + C$

41. $y\sin(xy) + C$

43. $xz(y + z)^{-1} + C$

45. $x^y + z\ln(z) - z + C$

49. $-r^{-3}(x\mathbf{i} + y\mathbf{j} + z\mathbf{k})$

51.

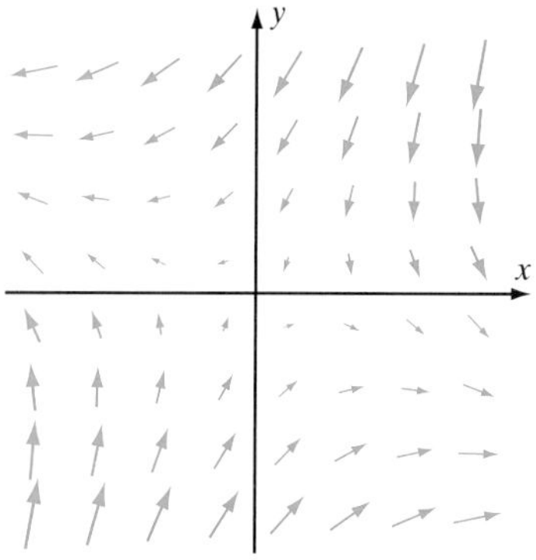

53. $\langle A\exp(2t) + B\exp(3t), -A\exp(2t)\rangle$

55. $\langle A\exp(4t) + B\exp(-t), (-A/4)\exp(4t) + B\exp(-t)\rangle$

57. $-y\mathbf{i} + x\mathbf{j}$

Section 13.2

1. 0
3. $2 - \sin(\pi^{3/2})$
5. 67/896
7. $255/4 + \ln(2/17)$
9. $-2/5$
11. 3/4
13. $\pi/2$
15. 2
17. 1069/35
19. 8/3
21. 2/11
23. 179/180
25. 3
27. -3
29. 2π
31. $(65\sqrt{65} - 5\sqrt{5})/6$
33. 4
35. 0
37. $\int_{C_1} \mathbf{F} \cdot d\mathbf{r}_1 = -4$; $\int_{C_2} \mathbf{F} \cdot d\mathbf{r}_2 = \int_{C_3} \mathbf{F} \cdot d\mathbf{r}_3 = -2$
39. 2π
41. 0
43. $-\pi/4$
45. 3.33904
47. 1.54122
49. 214.420

Section 13.3

In the answers for this section, C represents an arbitrary constant.

1. Potential function: $V(x, y) = C - (x^2/2 + \pi x + y)$
3. Potential function: $V(x, y) = C - xy^3 + y^7$
5. $\mathbf{F}$ is *not* closed on Q.
7. Potential function: $V(x, y) = (x-2)^{-1} - (y - 1/2)^3/3 + C$
9. Potential function: $V(x, y) = \cos(x^2 - y) + C$
11. $\mathbf{F}$ is *not* closed on Q.
13. Potential function: $V(x, y, z) = C - (xyz + x + 2y + 3z)$
15. $\mathbf{F}$ is *not* closed on Q.
17. Potential function: $V(x, y, z) = C - z^2/(x + y - 4)$
19. $\mathbf{F}$ is *not* closed on Q.
21. Potential function: $V(x, y, z) = C - z\tan(x + y)$
23. 3/4
25. $-(2 + 1/e)$
27. $-e^4$
29. $\int_{\overrightarrow{P_0P_1}} \mathbf{F} \cdot d\mathbf{r} = 0$; $\int_{\overrightarrow{P_1P_2}} \mathbf{F} \cdot d\mathbf{r} = 3$; $\int_{\overrightarrow{P_2P_0}} \mathbf{F} \cdot d\mathbf{r} = -3$
31. $\int_{\overrightarrow{P_0P_1}} \mathbf{F} \cdot d\mathbf{r} = 0$; $\int_{\overrightarrow{P_1P_2}} \mathbf{F} \cdot d\mathbf{r} = 1$; $\int_{\overrightarrow{P_2P_0}} \mathbf{F} \cdot d\mathbf{r} = -1$
37. $\{(c, d) : c \le -1, c < d < 1\} \cup \{(c, d) : -1 < c, \max\{c, 1\} < d\}$
39.

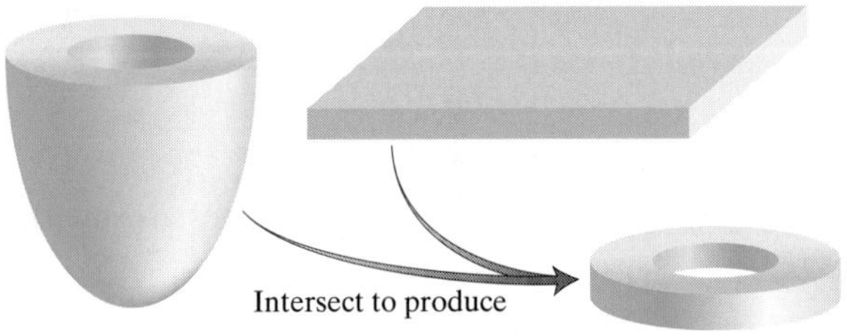

41. 0.6118

Section 13.4

1. $2xy^3$
3. $2x\sin(y) + \cos(y)$
5. $-\sin(x) - 2x\sin(xy)\cos(xy)$
7. $y/(x+y)^2 - 2x/(x-y)^2$
9. $-(x/y^2)\sec^2(x/y) + (y/x^2)\csc^2(y/x)$
11. 7
13. 14
15. $2(x + y + z)$
17. $e^{x-z} - e^{z-y}$
19. $-y\sin(xy) - z\cos(yz)$
21. $xz\mathbf{i} + (x - yz)\mathbf{j}$
23. $(x + ye^{yz})\mathbf{i} + (1/(x+z) - y)\mathbf{j}$
25. $-x\cos(xy)\mathbf{i} + y\cos(xy)\mathbf{j} - (y\sin(xy) + x\sec^2(xy))\mathbf{k}$
27. $0\mathbf{i} + 0\mathbf{j} + 0\mathbf{k}$; closed
29. $0\mathbf{i} + 0\mathbf{j} + 0\mathbf{k}$; closed
31. $(x\sin(z) + \sin(x))\mathbf{i} - y\sin(z)\mathbf{j} + (x\sin(y) - z\cos(x))\mathbf{k}$; *not* closed
33. 0
35. $2x(3y^2 + z + 1)/(z - y^2)^3$
37. 0
39. $-\cos(x)\,\mathbf{i} - \sin(y)\,\mathbf{j}$
41. $\mathbf{i} + z\,\mathbf{j} + (y + 2z)\,\mathbf{k}$
43. $\langle y + z, x + z, x + y \rangle$
45. $-y^{-3}(2x\mathbf{i} + y\mathbf{j} + 2\mathbf{k})$
55. u is harmonic.
61. $(1/2\pi)\int_0^{2\pi} \exp(1 + \cos(t))\cos(3 + \sin(t))\,dt = -2.6910786\ldots = e\cos(3)$

Section 13.5

1. -3π
3. 2π
5. π
7. 0
9. $3\pi/4$
11. $-32/3$
13. $12\ln(3) - 20\ln(2)$
15. 648/5
17. 9
19. 64/3
21. $\pi/3 - \sqrt{3}/4$
23. 1/2
25. 82/693
27. 6π
29. $-1/30$
31. 4/3

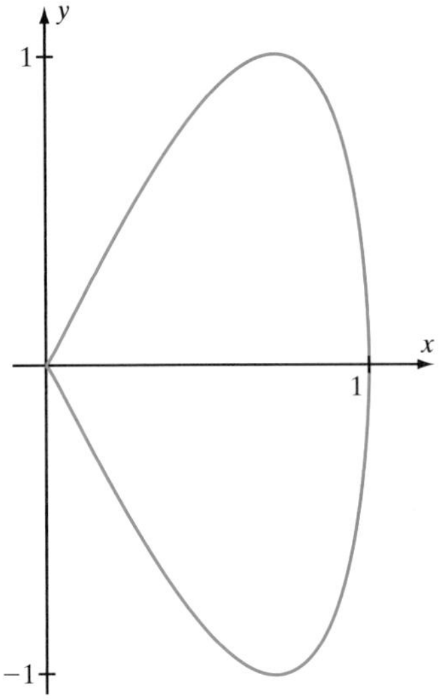

r(t) 5 sin(t)**i** + sin(2t)**j**

33. 3/2

37. −0.070393 **39.** 5.3755

Section 13.6

1. $3\sqrt{26}$ **3.** $8\sqrt{2}\pi$

5. $(17^{3/2} - 1)\pi/6$ **7.** $(37^{3/2} - 5^{3/2})/3$

9. $9\sqrt{38}$ **11.** $(145^{3/2} - 65^{3/2})\pi/24$

13. $5\pi\sqrt{74}$ **15.** $84\pi\sqrt{2}$

17. $4(275 - 6^{5/2} - 7^{5/2})/15$ **19.** $2\pi\sqrt{19}$

21. $3574\pi/5$ **23.** 33π

25. $52/3$ **27.** $135\sqrt{59}/32$

29. $\int_0^1 \int_0^4 \sqrt{1 + 4u^2 + (4uv + 1)^2}\, dv\, du$

31. $\int_0^1 \int_{-1}^1 \sqrt{4\exp(4u) + \exp(2u + 2v) + \exp(2u - 2v)}\, dv\, du$

33. $27\sqrt{26}$ **35.** $8\sqrt{3}$

37. $2\pi \int_a^b f(x)\sqrt{1 + f'(x)^2}\, dx$

41. $2\pi/3$ **43.** $(0, 0, 8/3)$

45. $(0, 0, (289\sqrt{17} - 41)/(170\sqrt{17} - 10))$

47. $2\pi a^2$

49. $2\pi a^2 + 2\pi(ac/e)\arcsin(e)$ where $e = \sqrt{c^2 - a^2}/c$

53. 192

55. 8.510466 **57.** 4.74121

Section 13.7

1. 136 **3.** $2 - 2/e^2 + \ln(2)$

5. $-\pi$ **7.** -42

9. $-22/3$ **11.** -24

13. -9π **15.** 63π

17. -4 **19.** $-3\pi/4$

21. 8π **23.** $\pi/2$

25. $175\pi/4$ **27.** -6π

29. 20π **31.** $-\pi$

35. 18π **37.** -4π

39. -3.550999 **41.** -0.4626517

Section 13.8

1. 0 **3.** π

5. π **7.** 1

9. -8 **11.** 4/15

13. $45\pi/4$ **14.** 65

17. $32\pi/3 - 4\sqrt{3}\pi$ **19.** 256/15

27. $28\pi/3$ **35.** $(\pi/6)\mathbf{i}$

37. $4\pi\mathbf{k}$ **39.** 2.81762

41. 1.68773

REVIEW EXERCISES FOR CHAPTER 13

1.

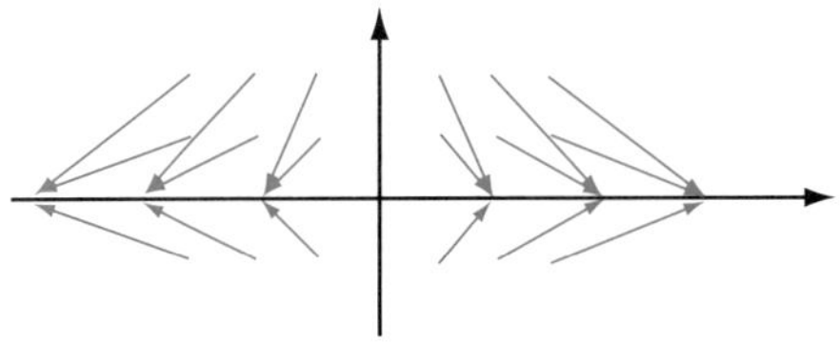

3.

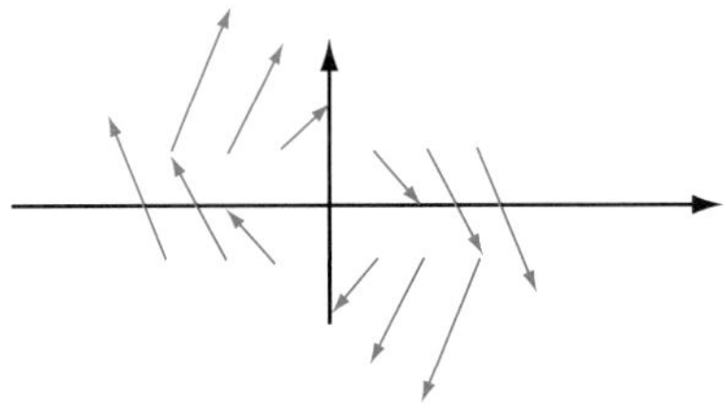

5. $\mathbf{v} = -(y^2 + z^2 + (z + 5))\mathbf{i}$

7. $\nabla u = \langle y^2 - 3x^2y + 1, 2xy - x^3 \rangle$

9. $\nabla u = \langle 3x^2y\cos(x^3y), x^3\cos(x^3y) \rangle$

11. $u(x, y) = x^2y/2$

13. $u(x, y, z) = xz + y$

15. $u(x, y, z) = xy\ln(z)$

17. $\mathbf{r}(t) = e^{2t}\mathbf{i} - 3t\mathbf{j}$

19. 1

21. $\sin(\pi^2) - \sin((\pi/2)^2) + 1$

23. $2-\pi^3/3-\sin(\pi^2)$

25. $-4/3$

27. $-\frac{3}{10}\cdot 3^{2/3}+\frac{1}{10}$

29. 11

31. $\frac{\pi}{8}-\frac{1}{4}-\frac{5\sqrt{2}}{12}$

33. $\frac{17^{3/2}}{4}-\frac{65^{3/2}}{4}$

35. $\frac{\partial M}{\partial y}=2y,\ \frac{\partial N}{\partial x}=-2x,\ 2y\neq -2x$, not closed

37. $\frac{\partial M}{\partial y}=1,\ \frac{\partial N}{\partial x}=0, 1\neq 0$, not closed

39. $\frac{\partial M}{\partial y}=\frac{\partial N}{\partial x}=z,\ \frac{\partial M}{\partial z}=\frac{\partial R}{\partial x}=y,\ \frac{\partial N}{\partial z}=\frac{\partial R}{\partial y}=x$, closed, $u(x,y,z)=xyz-4x+6y+2z$

41. 74

43. $\int_{\overrightarrow{P_0P_1}}\mathbf{F}\cdot d\mathbf{r}=7,\ \int_{\overrightarrow{P_1P_2}}\mathbf{F}\cdot d\mathbf{r}=19,$ $\int_{\overrightarrow{P_2P_0}}\mathbf{F}\cdot d\mathbf{r}=-26$, the sum is 0

45. $\int_{\overrightarrow{P_0P_1}}\mathbf{F}\cdot d\mathbf{r}=4e^6-2e^3,\ \int_{\overrightarrow{P_1P_2}}\mathbf{F}\cdot d\mathbf{r}=e^5-4e^6,$ $\int_{\overrightarrow{P_2P_0}}\mathbf{F}\cdot d\mathbf{r}=-e^5+2e^3$, the sum is 0

47. $\frac{2x}{x^2-y}+\frac{2y}{y^2-x}$

49. $e^{x+2z}-2e^{z-2y}$

51. $y\cos(xy)+z\sin(yz)+xy\sec^2(xyz)$

53. $(-xz-e^{y+z})\mathbf{i}+(yz)\mathbf{j}+\left(\frac{1}{x-y}\right)\mathbf{k}$

55. $\mathbf{curl F}=(xz+y)\mathbf{i}-yz\mathbf{j}-x\mathbf{k}$, not closed

57. $\mathbf{curl F}=(x\cos(xy)+y\sin(yz))\mathbf{i}-(y\cos(xy)-xy\cos(xyz))\mathbf{j}+(-xz\cos(xyz))\mathbf{k}$, not closed

59. $\text{div}(\mathbf{grad}(\mathbf{u}))=[2-4x^2y^2-4y^2z^2]\sin(z^2-x^2)$

61. both sides equal 0

63. both sides equal $-\pi/4$

65. the region is a disc, the integral equals $-\pi$

67. $-\frac{32}{3}$

69. e^4-1

71. 4π

73. 8

75. $2\ln(3/4)$

77. $24\sqrt{46}$

79. $3\sqrt{34}\pi$

81. 12π

83. $-4\sqrt{17}$

85. $\int_1^3\int_0^2\sqrt{8u^4+12v^4+4u^2v^2+8uv^3},\ dudv$

87. $\int_0^{\pi/2}\int_{\pi/3}^{2\pi/3}\sqrt{\sin^2(v)+\cos^2(\mathbf{u}v)\cdot[\mathbf{u}^2+v^2-2\mathbf{u}v+v^2\sin^2(v)]}\,dvd\mathbf{u}$

89. $\frac{45}{2}\sqrt{6}$

91. 0

93. $-\frac{4}{\sqrt{17}}\sqrt{18}[6+e^3-e^{-3}]$

95. $-\sin(1)+\cos(1)$

97. 64π

99. 0

101. $\ln(2)$

INDEX

E

J

M

Q

R

S

W

X

Y